Abbreviations Dictionary

Tenth Edition

Dean Stahl
Karen Kerchelich
Originated by Ralph De Sola

Abbreviations Dictionary

Tenth Edition

Dean Stahl
Karen Kerchelich

Originated by Ralph De Sola

★ Abbreviations ★ Acronyms ★ Airlines and Airports ★ Appellations ★ Astronomical Terminology ★ Bafflegab Divulged (euphemisms explained) ★ Birthstones ★ Chemical Elements ★ Citizen's-Band Call Signs ★ Computer Terms ★ Contractions ★ Corrections Facilities ★ Criminalistic Terms ★ Currencies ★ Diacritical Marks ★ Dysphemistic Place Names ★ Earthquake Data ★ Eponyms ★ Genealogical Terms ★ Geographical Equivalents ★ Government Agencies ★ Greek Alphabet ★ Historical, Musical, and Mythological Data ★ International Conversions ★ International Vehicle License Letters ★ Inventions and Inventors ★ Medical Terms ★ Military Terms ★ National Parks ★ Nations and Capitals ★ Nicknames ★ Numbered Abbreviations ★ Numeration ★ Phobias ★ Ports of the World ★ Principles and Laws ★ Railroads ★ Roman Numerals ★ Russian Alphabet ★ Short Forms ★ Signs & Symbols ★ Slang ★ Steamship Lines ★ Superlatives ★ U.S. Presidents & Vice Presidents ★ Weather Symbols (Beaufort Scale) ★ Wedding Anniversaries ★ Winds and Rains ★ Zip Coding ★ Zodiac

CRC Press
Boca Raton London New York Washington, D.C.

Library of Congress Cataloging-in-Publication Data

Stahl, Dean
    Abbreviations dictionary / Dean Stahl, Karen Kerchelich; originated by Ralph De Sola. -- 10th ed.
        p. cm.
    Rev. ed. of: Abbreviations dictionary / Ralph De Sola. 9th ed. 1995.
    ISBN 0-8493-9003-6 (alk. paper)
    1. Abbreviations, English. 2. Signs and symbols--Dictionaries. 3. Acronyms. I. Kerchelich, Karen. II. De Sola, Ralph, 1908- Abbreviations dictionary. III. Title.

PE1693 .D4 2000
423'.1--dc21

00-058549
CIP

This book contains information obtained from authentic and highly regarded sources. Reprinted material is quoted with permission, and sources are indicated. A wide variety of references are listed. Reasonable efforts have been made to publish reliable data and information, but the author and the publisher cannot assume responsibility for the validity of all materials or for the consequences of their use.

Neither this book nor any part may be reproduced or transmitted in any form or by any means, electronic or mechanical, including photocopying, microfilming, and recording, or by any information storage or retrieval system, without prior permission in writing from the publisher.

The consent of CRC Press LLC does not extend to copying for general distribution, for promotion, for creating new works, or for resale. Specific permission must be obtained in writing from CRC Press LLC for such copying.

Direct all inquiries to CRC Press LLC, 2000 N.W. Corporate Blvd., Boca Raton, Florida 33431.

Trademark Notice: Product or corporate names may be trademarks or registered trademarks, and are used only for identification and explanation, without intent to infringe.

Visit the CRC Press Web site at www.crcpress.com

© 2001 by CRC Press LLC

No claim to original U.S. Government works
International Standard Book Number 0-8493-9003-6
Library of Congress Card Number 00-058549
Printed in the United States of America 2 3 4 5 6 7 8 9 0
Printed on acid-free paper

This is the short and the long of it.

—Shakespeare

Contents

Preface

We live in a new century, a time when speed is king. High-tech devices increase the speed at which we can relay information. Abbreviations, nicknames, jargon, and other short forms save even more time, as well as space and effort. PDAs link to PCs. The Net has grown into data central (as well as shopping mall and grocery store). E-mail is faster than snail mail, cell phones are faster yet, and it's all done 24/7. This is not new, of course. Long-time and widespread use of certain abbreviations, such as *RSVP*, has made them better understood standing alone than spelled out. Certainly we are more comfortable saying *DNA* than *deoxyribonucleic acid*.

Yet the meanings of most abbreviations aren't apparent, a problem made worse by the abundance of abbreviations with a variety of meanings. The letter *a*, for example, stands for about 50 words or phrases; capital *A* stands for about 65. Generally, outside a limited circle of initiates, using abbreviations without first defining them causes confusion. Witness the result when someone who doesn't understand Spanish turns on the tap marked *C* in a shower in Acapulco, Buenos Aires, or Madrid. Hot water steams out instead of cold (no doubt followed by steamy, unabbreviated language).

The book in your hands will help end such confusion. This updated and expanded tenth edition of *Abbreviations Dictionary* contains abbreviations, acronyms, appellations, contractions, numbered abbreviations, and other short forms from computing, high technology, science, and a broad range of other fields, pastimes, groups, and, of course, government and the military—an abbreviation's best friends.

This edition contains many features not found in general reference works, such as Airlines and Airports of the World; Bafflegab Divulged; Emoticons; Eponyms, Nicknames and Geographical Names; Inventions and Inventors; Numbered Abbreviations; Phobias; Principles, Laws, Rules, Theories, and Doctrines; Signs and Symbols Frequently Used; and Superlatives. A turn through these pages reveals the identities of General Day After Tomorrow and the Father of Alligator Wrestling. You'll learn who the canaries of the sea are and the location of the Canadian Galapagos. You'll also find definitions of *chirosphobia, Goldilocks economy,* and *smartsizing.* And if you've wondered what the latter is, here's a clue: It leads to involuntary unemployment.

Extensive reading and listening reveal new short forms daily in newspapers, magazines, and books; on the Internet; and in everyday speech. Rapid developments in technology, science, and medicine yield more. The perpetual creation of agencies, associations, bureaus, commissions, committees, groups, and the like—all with their own abbreviations, acronyms, and other short forms—means only one thing: The end is not in sight. This expanded international tenth edition of *Abbreviations Dictionary* contains more abbreviations and other short forms than ever before.

Dean Stahl and Karen Kerchelich

Acknowledgments

Ralph De Sola began this book in the 1950s and continued to improve it into the ninth edition. Ralph's eclectic career ranged from promoting the new technology of microfilm to teaching English to Vietnamese immigrants, with forays into the fields of zoology and musicology. Books resulted from all these pursuits. As fellow word lovers, we fell into the habit of contributing to *Abbreviations Dictionary* quite naturally several editions ago. Thank you, Ralph, for trusting us to carry on.

Our thanks, too, go to everyone at CRC Press who worked on this edition. Special acknowledgment is due Acquisitions Editor Nora Konopka; we appreciate her unfailing professionalism and good humor.

The authors are most grateful to the following, who contributed entries to this tenth edition of *Abbreviations Dictionary*:

Michael Rose of Aberlady, East Lothian, Scotland, who did the research for *Militarisms: A Compendium of Abbreviations, Synonyms, Acronyms and Slang Words As Used by the Armed Forces of the World* (edited by John W. Mussell), provided a great number of military abbreviations from that book in exchange for material from previous editions of our *Abbreviations Dictionary*.

Margaret Stahl of Sun City, Arizona contributed hundreds of entries on a variety of subjects, including business and finance, agriculture, and medicine.

Laurence McGilvery of La Jolla, California provided numerous entries on books and libraries.

Our thanks for contributions and assistance also go to Debra Brammer, N D, of Seattle, Washington; Rebecca Fowler of Pahoa, Hawaii; David Gilbert of La Grande, Oregon: Louis Isquith, D D S, of Seattle; K E S Kirby of Thimphu, Bhutan; Larie Mallory of Copenhagen, Denmark; John Miller, D C, of Seattle; and Robert Stahl of Center Point, Texas.

Renewed thanks go to those who contributed to previous editions, including:

Arthur E E Ivory of Christchurch, New Zealand, who traded much material found in previous editions of our *Abbreviations Dictionary* for new items in his *Pacific Index of Abbreviations and Acronyms in Common Use in the Pacific Basin*.

Dr Irmaisabel Lovera DeSola of Caracas, Venezuela, who furnished most of the Venezuelan entries and others from around Latin America.

James C Palmer, who compiled and edited the *ERIC Dictionary of Educational Acronyms, Abbreviations and Initialisms*, for a great many entries in the field of education.

Brother John-Charles of the Society of Saint Francis American Province, who provided many liturgical abbreviations. Another contributor was Mary Bucher of Columbus, Ohio.

The *List of Acronyms*, compiled by Linda Blocki of New Mexico, was most welcome.

Other contributors were: Dan Kerlee and Carol Wollenberg in Seattle; Susan Luce in Raleigh, North Carolina; Emmett Murray in Kirkland, Washington; Julie Van Keuren in Olympia, Washington; Hal and Mildred Cary, Dr David Cary and Anna Marie Cary, Julian Gotkewicz, Douglas D McArthur, Chris Mahan, Marie Pauk, John and Fay Silverstein, Simeon Stern, Dr Walter Teutsch and Lucy van Donck in San Diego; Carla De Sola Eaton in Berkeley, California; Dr Ronda De Sola Chervin in Woodland Hills, California; Lorraine Sherkin in Toronto, Canada; Joe Rosner in New York City; and Vanessa Browne near London, England.

We are indebted, finally, to the many reference librarians in Seattle and San Diego who offered valuable ideas and assistance.

How to Use This Book

Definition of Terms

abbreviations abridged contractions such as acdt: accident; AEC: Atomic Energy
Commission; NASA: National Aeronautics and Space Administration.

acronyms words formed from letters in a series of related words such as ABLE:
Activity Balance Line Evaluation; AGREE: Advisory Group on Reliability of
Electronic Equipment; DYNAMO: Dynamic Action Management Operations.

anonyms attempts of authors to enjoy anonymity while maintaining their identity
by such devices as the capitalized diphthong Æ standing for Æon, pen name of
George William Russell.

contractions words shortened by dropping nonpronounced letters or omitted
letter(s) that are indicated by apostrophes as in can't: can not; Lit'l: little; doesn't:
does not; let's: let us.

eponyms designations derived from family names, nicknames or names of places
or persons, e.g., Hapsburg dynasty, *Eroica* symphony, Paris of America
(Montreal), Raynaud's disease.

geographical equivalents entries such as Far East: countries and islands of East
Asia or the Pacific—eastern Siberia, China, Japan, Taiwan, Korea, Indochina, the
Philippines, the Malay Peninsula.

initials FDR: Franklin Delano Roosevelt; HST: Harry S Truman; JFK: John
Fitzgerald Kennedy; LBJ: Lyndon Baines Johnson; initials of all American
presidents are included as well as initials of other noted personalities.

nicknames Al: Alfred; Bea: Beatrice; Hal: Harold; Ike: Dwight David
Eisenhower, Isaac.

place name pseudonyms see *Cannery Row, Main Street, Middletown, Red Gap,
Spoon River, Tortilla Flat, Yoknapatawpha, Zenith* entries.

short forms amps: amperes; Olds: Oldsmobile; pots.: potentiometers.

signs $ & ¢—dollars and cents.

slang shortcuts B-girl: bar girl; C-note: $100 bill; 1-G: $1000.

symbols Al: aluminum; Pt: platinum; Rx: prescription; recipe.

toponyms place names convicts use when stating where they were imprisoned.

STYLE

Arrangement

Everything in this book is arranged in alphabetical and numerical order. For entries containing the same letter, lowercase precedes capital (aa, AA); roman precedes italics (AWA, *A WA);* unpunctuated precedes punctuated (BAE, B.A.E.). An Arabic numeral precedes its Roman equivalent (3, III). The following connectives are ignored in the alphabetical arrangement: & (ampersand), and, by, for, in, of, or, + (plus), the, to. All other articles, particles, prepositions, and the like (between, de, del, di) are treated alphabetically. For example, U of P is alphabetized as UP; *U de ST* appears as if it were UdeSt.

A dollar sign ($) is treated as if it were a lowercase "d," the pound sign (£) as a lowercase "p," and a mu (μ) as a lowercase "m."

In the case of a parenthetical plural ending, the parenthesis will be ignored [e.g., paren(s) is treated as parens].

Capitalization

Capitalization of abbreviations, according to Department of Defense Military Standard 12-B (Mil-Std 12-B), must follow the rules of English grammar. All proper nouns are capitalized. All common nouns are lowercase. Units of weight, measure, and velocity, such as lb, kg, in., cc, mm, rpm, and the like, appear in lowercase to avoid confusion with other letter combinations they resemble.

Many military establishments and officers use full capitals for everything because message machines were provided with only capital letters. That is why many engineering drawings supplied the armed forces contain all abbreviations in capital letters. It is also true that many draftsmen were afraid small letters would fill in, especially a's, b's, e's, g's, o's, and the like. Therefore, they also like to use capital letters. In text, however, 1500 RPM presents a typographical blob, as compared to the more sophisticated 1500 rpm.

At first loran was LORAN. As people became more used to it, it became Loran. Today it is loran. The same is true of other combinations. The trend is to capitalize only those letters standing for proper nouns, running all common nouns in lowercase. Nevertheless, for the sake of readers and researchers, some incorrectly rendered abbreviations appear in this book. Many people have a marked tendency to capitalize everything they think is important. If this tendency is unchecked, confusion follows. All abbreviations and acronyms look alike. So follow the commonsense rules of good grammar and correct usage.

Chemical element symbols, however, have the first letter capitalized: Au (gold), Zn (zinc), etc. The second letter of a chemical symbol always appears in lowercase.

Exceptions

The singular, plural, and tense of the word abbreviated do not alter the abbreviation except in a few instances, such as fig.: figure; figs.: figures; no.: number; nos: numbers; p: page; pp: pages; S. Saint; SS: Saints.

However, readers should be aware that the International *(SI)* System of Measurements calls for the abolition of all pluralized abbreviations. Hence in. stands for inch or inches, lb for pound or pounds, oz for ounce or ounces.

Documentary abbreviations are rendered as follows: FARs (Failure Analysis Reports) or IRs (Inspector's Reports) or RARs (Reliability Action Reports). In the singular they appear as FAR, IR, RAR.

Italics

Items from Latin and other non-English languages, as well as titles of books and periodicals, are usually set in italic type. Many physical symbols are also set in italics to differentiate them from other letter combinations.

Punctuation

Short forms are devised to save time and space and to overcome the necessity of repeating long words and phrases. All punctuation is avoided in modern practice unless the form is taken from Latin or there is some conventional use demanding punctuation, as in the case of academic degrees and a few governmental designations. U.S.A. is the country; USA is the army. D.C. is the District of Columbia; DC is direct current when used as a noun. Cash on delivery is not cod but c.o.d. Similarly, fig., figs., and no. require periods to keep readers from thinking they may be words instead of abbreviations for figure, figures, and number. When in doubt, spell it out.

Capitalization and Punctuation Trends

American as well as British and Canadian publishers appear to be following the trend to capitalize only those letters normally capitalized: proper nouns and important words in titles. They reserve lowercase letters for abbreviations consisting of adjectives and common nouns. This obviates the chaos brought about by those who capitalized all the letters in every abbreviation and then compounded their error by placing unnecessary *full* stops or periods after every letter, as was the custom in bygone times.

Most periods are dropped because it is generally realized that the purpose of all abbreviation is the thoroughgoing promotion of brevity. Years ago, when Rudolf Flesch compiled one of his many useful books, *How To Be Brief—An Index to Simple Writing,* he stated:

> To save even more space, leave out abbreviation periods whenever you can. The British omit them regularly ... Mr, Mrs, Dr, St (Saint), Thos, Chas, jr. Periods are often left out after standard abbreviations like US, UN, FCC, PTA ... following the pattern of most telephone books (e.g., plmbg & heatg supls, atty, flrst, acctnts, svce, rl est).

Signs and Symbols

Frequently used signs and symbols appear in the back of this dictionary. Many are found on typewriters (remember those?) and computer keyboards (&: ampersand— the *and* sign; *: asterisk; ¢: cent; $: dollar, %: percent).

Symbols include the chemical elements (Al: aluminum; Au: gold—from the Latin *aurum;* C: carbon; Sn: tin—from the Latin *stannum).* All are listed in the alphabetical section without special definition to indicate they are not abbreviations but symbols. The chemical elements are also grouped together in the back of this dictionary.

Airlines use two-letter symbols for convenience in baggage handling, ticketing, and scheduling operations. Thus American Airlines is AA, Delta Air Lines is DL, and United Air Lines is UA. These two-letter designations are listed in a separate section at the back of the book.

Railroads and steamship lines are included both in the alphabetical section and in their own sections at the end of the book. Naval craft are designated by many arbitrary symbols. All available are given in the alphabetical section.

A

a abbreviation; about; absent; absolute; abundant; acceleration in feet per second; accidental; account; acre; acronym; acting; adjective; adjutant; adult; aerial; afternoon; age; allele (recessive); altitude; altitude intercept; alto (instruments or voices); amateur; ampere; annealing; annually; anode; ante; anthracite; apprentice; arc; are (unit of metric land); area; argent; arterial blood (symbol); assist; at; atmosphere; attendance; audit(or); aunt; automatic; available; aviation; aviator; axis; azure; distance from leading edge to aerodynamic center; specific absorption coefficient (symbol); specific rotation (symbol)

a′ all (contraction); minute angle; a prime

a″ double prime; second (angle)

a am, an, an der (German—on the, at the); angle of attack; *annus* (Italian—year) (Latin—year); *antes* (Spanish—before); *arteria* (Latin—artery); attenuation constant (symbol); autonomous consumption (macro-economic symbol); (Italian—and)

A absolute; absolute temperature; Absorbance; Absorbancy (symbol); academy; accumulator (computerese); acid; acoustic source; actual weight of an aircraft; address (computer symbol); adenine; adjusted; Admiral; adulterer; adulteress (branded on the foreheads of all convicted of this crime in early New England)—also known as the scarlet letter because branding caused bleeding; Air Branch; aircraft; airman; Alaska Steamship Company; Alcoa Steamship Company; Alfa; allele (dominant); Alpha (code for letter A); alveolar gas (symbol); ambassador; America; American; Americanization; American Stock Exchange (newspaper stock listings); Americanize; Ammunition Examiner (sleeve patch); Amos, The Book of; ampere; amphibian; amplitude; Anchor Line; anode; anterior; Apprentice; April; argon; Army; arterial; Artificer; artillery; Asian; aspect ratio; Assault (military badge); astragal; Assembly Bill; Atlantic; Atomic; atomic weight; attack; August; Australian; Austria (auto plaque); Auxiliary; blood type; bond rating; brassiere cup size; chemical activity; first van der Waals constant; Fraunhofer line due to oxygen; hail (aircraft code); includes extra(s) (in newspaper stock listings); linear acceleration; mean sound absorption coefficient; school grade meaning excellent; shoe size narrower than B; Single-A (baseball); total acidity

Å angstrom unit

A *abajo* (Spanish—down); *abasso* (Italian—down); *absolvo* (Latin—I absolve, I acquit); *alas* (Finnish—down); *albus* (Latin—white); *Alp(en)* [German—Alp(s)]; *Alpe(s)* [French—Alp(s)]; Alpi (Italian—Alps); *alt* (German—old); *Alteza* (Spanish—Highness); *aprobado* (Spanish—approved)— passed an examination; arrival; *arrivare* (Italian—arrival); arrive; *arrivé* (French—arrival); *Atlantic Reporter; auf* (German—up); *Aulus* (Latin—Aulus Gellius) —2nd-century author noted for his *Noctes Atticae* about languages and literature as well as natural history; *aus* (German—out); *avbeta* (Swedish—departure); mountain meadow(s)

Å *aas* (Dano-Norwegian— hills)

a (A) analysis

A+ A-plus; A-positive; Single-A (Advanced–baseball)

A– A-minus; A-negative; atomic (A-bomb)

A-1 air personnel officer; angstrom unit; excellent; first class; first rate; *Lloyd's Register* symbol indicating first rate equipment; personnel section of an air force staff, skyraider single-engine general-purpose attack aircraft flown from aircraft carriers; top quality; tops; very best

A-I (motion pictures) for general patronage

A1c Airman, first class

A/1C Airman First Class

A-1 Skyraider Douglas single-engine attack aircraft (formerly AD)

A-2 air intelligence officer; almost A-1 in quality; intelligence section of an air force staff; just short of the best

A-II (motion pictures) for adults and adolescents only

A₂ aortic second sound; Asian influenza virus

A/2C Airman Second Class

A²C² *see* AACC

A²ᵈ *Atlantic Reporter second series*

A-3 air operations and training officer; operations and training section of an air force

A-111 AAAC Staff; Skywarrior twin-engine turbojet tactical all-weather attack aircraft operating from aircraft carriers; training and operations

A-III (motion pictures) for adults only

A/3C Airman Third Class

A-3 Skywarrior Douglas carrier-based twin-engine jet reconnaissance and light bombing plane (formerly A3D) — USAF B-66 Destroyer

A-4 air material and supply officer; material and supply section of an air force staff; Skyhawk single-engine turbojet attack aircraft operating from aircraft carriers; supply and material

A-IV (motion pictures) for adults with reservations

A-5 planning; supersonic twin-engine turbojet all-weather attack aircraft operating from aircraft carriers

A-6 communications

A-6A Intruder twin-engine turbojet long-range carrier-based low-altitude attack aircraft

A-6 Intruder Grumman carrier-based twin-engine jet low-level attack bomber (formerly A2F)

A-7 Corsair II Ling-Temco-Vought carrier-based single-engine jet light-attack bomber

A-32 Lansen (Swedish— A-32 Lance)—Saab single-seat single-engine jet fighter-interceptor

A-37 radar-homing or television-guided air-to-surface missile

A-60 Saab twin-engine two-place jet trainer-utility aircraft also called the Saab 105

A-68 protein found in the brains of Alzheimer's victims

A-106 Agusta antisubmarine-warfare single-engine single-seat helicopter

A-109 Agusta high-performance eight-seat twin-engine helicopter

aa acetic acid; achievement age; acting appointment; adjectives; alveolar-arterial; always afloat; aminoacetone; approximate absolute; arachidonic acid; armature accelerator; arteries; arctic-alpine; ascending aorta; atomic absorption; author's alteration; average audience; equal parts

aa (AA) achievement age; antiaircraft; ascorbic acid

aA attoampere

a-a air-to-air

a/a antiaircraft

a/A arterial to alveolar oxygen ratio

a & a abbreviations and acronyms; additions and amendments; aid and attendance

aa ana (Greek—of each); *ander andere* (German—among others); *arterias* (Latin—arteries); (Hawaiian—block lava)—pronounced *ah-ah*

AA absolute alcohol; absolute altitude; Academy Award; achievement age; Addicts Anonymous; Administrative Assistant; Adoption Agency; Aerolineas Argentinas (Argentine Airlines); Affirmative Action; Air Armament; Air Attaché; Aircraft Apprentice; Aircraft Artificer; Airman Apprentice; Alcoholics Anonymous; Aluminum Association; Aluminum (Company of) America; American Airlines; American Association; amino acid; aminoacyl;

Ann Arbor (railroad); Ansett Airways; antiaircraft; Appropriate Authority; Argenteum Astrum; arithmetic average; Arlington Annex; Armament Artificer; Army Act; Asian-African; Assistant Adjutant; Athletic Association; author's alteration(s); Automobile Association; Aviation Annex; battery size; bond rating; brassiere cup size; Double-A (baseball); shoe width

A/A Agnostics/Atheists

A.A. Associate in Accounting; Associate in Arts

AA Air Almanac; American Anthropologist; Astronautica Acta (*Journal of the International Astronautical Federation*); *Aviatsionnaya Armiya* (Russian—(Soviet-era) Air Army)

aaa abdominal aortic aneurism; acquired aplastic anemia; acute anxiety attack; amalgam; androgenic anabolic agent; anti-aircraft artillery

aa & a armor, armament, and ammunition

AA&A Armor, Armament and Ammunition

Aaa Alaska (government style is to spell it out)

AAA Abortion Assistance Association; Agriculture Adjustment Act; Agricultural Adjustment Administration; Agricultural Aircraft Association; Alaska; All American Aviation; Allegheny Airlines (3-letter coding); Allied Artists of America; American Academy of Advertising; American Academy of Actuaries; American Academy of Allergy; American Accordionists Association; American Accounting Association; American Airship Association; American Antarctic Association; American Anthropological Association; American Arbitration Association; American Association of Anatomists; American Astronomers Association; American Australian Association; American Automobile Association; antiaircraft artillery; Antique Airplane Association; Appraisers Association of America; Archives of American Art;

Area-Agency on Aging; Argentine Anticommunist Alliance; Armenian Assembly of America; Army Athletic Association; Army Audit Agency; Associated Agents of America; Association of Accounting Administrators; Association of Attenders and Alumni (Hague Academy of International Law); Association of Average Adjusters; Triple-A (baseball)

AAA (AFL-CIO) Actors and Artistes of America

A.A.A. Amateur Athletic Association (British)

A A & A Armor, Armament & Ammunition

AAAA Amateur Athletic Association of America; American Association for the Advancement of Agnosticism; American Association for the Advancement of Atheism; American Association of Advertising Agencies; Antique Appraisal Association of America; Army Aviation Association of America; Associated Actors and Artists of America; Association of Accredited Advertising Agencies (Singapore); Australian Advertising Advisory Authority; Australian Association of Advertising Agencies

AAAAM American Academy of Anti-Aging Medicine

AAAANZ Association of Accredited Advertising Agencies of New Zealand

AAAB American Association of Architectural Bibliographers

AAAC American Association for the Advancement of Criminology; Antiaircraft Australian Army Aviation Corps

AAACE American Association of Agricultural College Editors

AAACE *Alianza Apostolica y Anti-Comunista de España* (Spanish—Apostolic and Anti- Communist Alliance of Spain)—also known as the Triple-A

AAA-CPA American Association of Attorney-Certified Public Accountants

AAAD American Athletic Association for the Deaf, Association of Automotive Aftermarket Distributors

AAAE American Association of Airport Executives, Australian Association of Adult Education

AAAEE American Afro-Asian Educational Exchange

A.A. Ag. Associate of Arts in Agriculture

AAAH American Association for the Advancement of the Humanities

AAAI Affiliated Advertising Agencies International, American Academy of Allergy and Immunology

AAAID Arab Authority for Agricultural Investment and Development

AAAIMH American Association for the Abolition of Involuntary Mental Hospitalization

AAAIP Advanced Army Aircraft Instrumentation Program

AAAIS Antiaircraft Artillery Information Service; Antiaircraft Artillery Intelligence Service

AAAIWA Automobile, Aerospace, and Agricultural Implement Workers of America

aaal abolish all abortion laws

AAAL American Academy of Arts and Letters

AAALAC American Association for Accreditation of Laboratory Animal Care

AAAM American Association of Aircraft Manufacturers; American Association for Automotive Medicine

AAAN American Academy of Applied Nutrition

AAAOC Antiaircraft Artillery Operation Center

AAAR American Association for Aerosol Research; Association for the Advancement of Aging Research

AAARC Antiaircraft Artillery Reception Center

AAARG American Atheist Addiction Recovery Groups

AAAS American Academy of Arts and Sciences; American Academy of Asian Studies;

American Association for the Advancement of Science; Australian Association for the Advancement of Science

A.A.A.S. Associate in Arts and Science

AAASA Association for the Advancement of Agricultural Sciences in Africa; Australian and Allied All Services Association

AAASS American Association for the Advancement of Slavic Studies

AAASUSS Association of Administrative Assistants and Secretaries to United States Senators

AAAUS Amateur Athletic Association of the United States; Association of Average Adjusters of the United States

aab anti-aircraft battery

AAB Aircraft Accident Board; American Air Brake; American Association of Bioanalysts; Army Air Base; Army Artillery Board; Association of Applied Biologists

AABB American Association of Blood Banks

AABC American Amateur Baseball Congress; American Association of Bible Colleges; Associated Air Balance Council; Association for the Advancement of Blind Children; Australian Army Band Corps

AABD Aid to the Aged, Blind, or Disabled

AABEVM Association of American Boards of Examiners in Veterinary Medicine

aabfs amphibious assault bulk fuel system

AABGA American Association of Botanical Gardens and Arboretums

AABH Australian Association for Better Hearing

AABI American Association of Bicycle Importers; Antilles Air Boats Incorporated

AABL Associated Australian Banks of London; Australian Associated Banks in London

aaBm analytical anatomy by the Braille method

AABM Association of American Battery Manufacturers; Australian Association of British Manufacturers

AABNCP Advanced Airborne Command Post

AABP Association of Area Business Publications; Australian Association of Business Publications

AABPDF Allied Association of Bleachers, Printers, Dyers, and Finishers

aabshil aircraft anti-collision-beacon-system high-intensity light(ing)

AABT Association for the Advancement of Behavior Therapy

AABTM American Association of Baggage Traffic Managers

aabp aptitude assessment battery programming

aaby as amended by

aac affirmative action clause; anti-aircraft cannon; automatic aperture control; average annual cost

AAC Aerial Ambulance Company; Aeronautical Advisory Council; Aeronautical Approach Chart; Aircraft Armament Change; Alaskan Air Command; All-American Canal (serving California and Baja California); Aluminum Anodizers Council; Alumnae Advisory Center, American Academy of Criminalistics; American Alpine Club; American Alumni Council; American Archery Council; American Association of Criminology; American Atheist Center; American Cement Corporation (stock exchange symbol); Anglo-American-Corporation; Anti-Aircraft Command; Anti-Aircraft Corps; Area Advisory Committee; Army Apprentices College; Army Air Corps; Association of American Choruses; Association of American Colleges; Australian Agricultural Company; Australian Air Corps; Australian Association of Chiropractors; Automotive

Advertisers Council; Auxiliary Artillery Corps; Auxiliary Army Corps

AAC *Associação Academica de Coimbra* (Portuguese—Coimbra Academic Association); *Auto Avia Contruzione* (Italian automaker: Enzo Ferrari's first company)

A.A.C. *anno ante Christum* (Latin—year before Christ)

AACA American Apparel Contractors Association; American Association of Certified Appraisers; Antique Automobile Club of America; Automotive Air Conditioning Association

AACAP Association of American Colleges Arts Program

AACB Aeronautics and Astronautics Coordination Board; Australian Association of Clinical Biochemists

AACBC American Association of College Baseball Coaches

AACBP American Academy of Crown and Bridge Prosthodontics

aacc all-attitude control capability; automatic approach control complex

AACC American Association of Cereal Chemists; American Association of Clinical Chemists; American Association for Contamination Control; American Association of Credit Counselors; American Automatic Control Council; Army Air Corps Center; Association for the Aid of Crippled Children; Australian—American Chamber of Commerce; Australian Army Catering Corps

A.A.C.C.A Associate of the Association of Certified and Corporate Accountants

AACCLA Association of American Chambers of Commerce in Latin America

AACCP American Association of Colleges for Chiropody-Podiatry

AACD American Association for Counseling and Development

AACDP American Association of Chairmen of Departments of Psychiatry

aace air-to-air combat environment

AACE Airborne Alternate Command Echelon (NATO); American Association of Cost Engineers

AACFO American Association of Correctional Facility Officers

AACFT Army Aircraft

Aach Aachen (German—Aix-la-Chapelle)

AACHS Afro-American Cultural and Historical Society; German geographical place-name equivalent of Aix-la-Chapelle on the Belgian-Dutch borders of Germany

AACI American Association for Conservation Information; Association of Americans and Canadians in Israel

AACJC American Association of Community and Junior Colleges

AACL Association of American Correspondents in London

aacm anti-armor cluster munition

AACM American Academy of Compensation Medicine; Association for the Advancement of Creative Music

AACO Advanced and Applied Concepts Office (USA); American Association of Certified Orthoptists; Assault Airlift Control Office(r)

AACOBS Australian Advisory Council on Bibliographical Services

AACOM Army Area Communications

AACOMS Army Area Communications System

aacp advanced airborne command post

AACP American Academy for Cerebral Palsy; American Academy for Child Psychiatry; American Academy of Clinical Psychiatrists; American Association of Colleges of Podiatry; American Association of Commercial Publications; American Association of Convention Planners; American Association of Correctional Psychologists

AACPA Asian American Certified Public Accountants

AACPP Association of Asbestos Cement Pipe Producers

AACPR American Association for Cleft Palate Rehabilitation

AACR American Association for Cancer Research

AACR Anglo-American Cataloguing Rules

AACR2 Anglo-American Cataloging Rules, Second Edition

AACRAO American Association of Collegiate Registrars and Admissions Officers

AACS Airborne Astrographic Camera System; Airways and Air Communications Service; Army Air Corps Squadron; Army Airways Communications System; Australian Amateur Cine Society

AACSA Anglo-American Corporation of South Africa

AACSB American Association of Collegiate Schools of Business

AACSCEDR Associate and Advisory Committee to the Special Committee on Electronic Data Retrieval

AACSL American Association for the Comparative Study of Law

AACSM Airways and Air Communications Service Manual

AACT American Association of Commodity Traders; Armenian Assembly Charitable Trust

AACTE American Association of Colleges for Teacher Education

AACTP American Association of Correctional Training Personnel

aacu anti-aircraft cooperation unit

AACUBO American Association of College and University Business Officers

AACV Assault Air-Cushion Vehicle

aad active acoustic(al) device; advanced ammunition depot; antibiotics-associated diarrhea

aad (AAD) alloxazine adenine dinucleotide

AAD Aircraft Assignment Directive; American Academy of Dentists; American Academy of Dermatology; Anti-Aircraft Division; Armored Automobile Detachment; Army Air Defense

aada anti-aircraft defended area

AADA Advanced Air Depot Area; American Academy of Dramatic Arts; American Association of Deaf Athletes; Associated Antique Dealers of America; Army Air Defense Area; Australian Automobile Dealers Association

AADAOPA American Association of Dealers in Ancient, Oriental and Primitive Art

AADB Army Air Defense Board

AADC Air Aide-De-Camp; American Association of Dental Consultants; Anti-Aircraft Defense Commander; Army Air Defense Command(er)

AADCCS Army Air Defense Control and Coordination System

AADCP Army Air Defense Command Post

AADE American Association of Dental Editors; American Association of Dental Examiners

AA de L Academia Argentina de Letras (Spanish—Argentine Academy of Letters)

AADGB American Association of District Governing Boards

AA Dip Architectural Association Diploma

AADIS Army Air Defense Information Service

AADLA Art and Antique Dealers League of America

AADM American Academy of Dental Medicine

AADMS Advanced Academic Degree Management System

AADN American Association of Doctors' Nurses

AADOO Army Air Defense Operations Office(r)

AADP American Academy of Denture Prosthetics

AADPA American Academy of Dental Practice Administration

aads advanced air defense system

AADS Advanced Army Defense System; American Association of Dental Schools; American Association of Dermatology and Syphilology; Area Air Defense System; Army Air Defense Site; Army Air Defense System

AADT Anti-Aircraft Dome Trainer

aae (AAE) above airport elevation; acute allergic encephalitis; average annual earnings

AAE American Association of Endodontists; American Association of Engineers; Army Aviation Element; Army Aviation Engineers; Asia Australia Express

AAEA American Agricultural Editors Association

AAEC Association of American Editorial Cartoonists; Australian-American Engineering Corporation; Australian Army Educational Corps; Australian Atomic Energy Commission

AAEDC American Agricultural Economics Documentation Center (USDA)

A.Ae.E. Associate in Aeronautical Engineering

AAEE Aircraft and Armament Experimental Establishment; American Academy of Environmental Engineers; American Association of Economic Entomologists; American Association of Electromyography and Electrodiagnosis

AAEF Australian Air Expeditionary Force; Australian-American Education Foundation

AAEFA Army Aviation Engineering Flight Activity

AAEH Association to Advance Ethical Hypnosis

AAEI American Association of Exporters and Importers

AAEKNE American Association of Elementary-Kindergarten-Nursery Educators

AAELSS Active-Arm External-Load Stabilization System

AAEOCJ American Association of Ex-Offenders in Criminal Justice

AAEP American Association of Equine Practitioners

AAES Advanced Aircraft Electrical System; Aldershot Army Equipment Show; American Association of Engineering Societies; Anti-Aircraft Experimental Section; Australian Agricultural Economics Society; Australian Army Education Service

AAESA Alabama Association of Elementary School Administrators

AAESDA Association of Architects, Engineers, Surveyors, and Draughtsmen of Australia

AAESPH American Association for the Education of the Severely/Profoundly Handicapped

AAEW Atlantic Airborne Early Warning

aaf (AAF) acetylaminofluorine; ascorbic acid factor

a-a-f acetic-alcohol-formalin (fixing fluid)

AAF Allied Air Forces; American Advertising Federation; American Air Filter (company); American Air Force; American Architectural Foundation; American Astronautical Federation; Army Air Field; Army and Air Force; Army Air Forces; Atlantic Amphibious Force; Auxiliary Air Force

AAFA Architectural Aluminium Fabricators Association; Australian Amateur Fencing Association

AA/FA dry-type self-cooled/forced-air-cooled (transformer)

A&AFA Army and Air Force Act

A.A.F.A. Associate in Arts in Fine Arts

AAFAC Army Air Forces Anti-Submarine Command

A.A. Fair Erle Stanley Gardner

AAFB Andrews Air Force Base; Auxiliary Air Force Base

aafc (AAFC) antiaircraft fire control

AAFC Air Accounting and Finance Center; Anti-Aircraft Fire Control; Army Air Forces Center; Army Air Force Classification Center; Association of Advertising Film Companies

AAFCE Allied Air Force, Central Europe

AAFCO Association of American Feed Control Officials; Association of American Fertilizer Control Officials

AAFCWF Army and Air Force Central Welfare Fund; Army and Air Force Civilian Welfare Fund

AAFE Advanced Applications Flight Experiment; American Association of Feed Exporters

AAFEA Australian Airline Flight Engineers Association

AAFEC Army Air Forces Engineering Command

AAFEMPS Army and Air Force Exchange and Motion Picture Service

AAFES Army and Air Force Exchange Service

AAFFA Australian Air Freight Forwarding Association

AAFGL Auxiliary Air Force General List

AAFH Academy of American Franciscan History

AAFI Associated Accounting Firms International

AAFIB Army Air Forces in (Great) Britain

AAFIS Army Air Forces Intelligence School

aafm anti-aircraft field mount

AAFM American Association of Feed Microscopists

AAFMC Army Air Forces Materiel Center

AAFMPS Army and Air Force Motion Picture Service

AAFNDA Army, Air Force, Naval Discipline Act

AAFNE Allied Air Force, Northern Europe

AAFNS Army Air Forces Navigation School

AAFOIC Army Air Forces Officer in Charge

AAFP American Academy of Family Physicians

AAFPF American Academy of Family Physicians Foundation

AAFPS Army and Air Force Pilot School; Army and Air Force Postal Service

AAFRC American Association of Fund-Raising Counsel

AAFRS American Academy of Facial, Plastic, and Reconstructive Surgery

AAFS American Association of Foot Specialists; American Academy of Forensic Sciences

AAFSE Allied Air Force, Southern Europe

AAFSS Advanced Aerial Fire Support System; Advanced Aerial Fire Suppression System

AAFSW Association of American Foreign Service Women

AAFTS Army Air Forces Technical School

AAFU Augmented Assault Fire Unit; All-African Farmers' Union

AAFWB Army and Air Force Wage Board

aag anti-aircraft gun

AAG Aeronautical Assignment Group; Air Adjutant General; Art and Architecture Group (Research Libraries Group); Assistant Adjutant-General; Association of American Geographers; Australian Association of Gerontology

AAGC American Association of Gifted Children

AAGFO American Academy of Gold Foil Operators

aagfs anti-aircraft gunfire simulator

AAGL American Association of Gynecological Laparoscopists

aa&gm anti-aircraft and guided missiles

AAGp Aeromedical Airlift Group (USAF)

AAGP American Academy of General Practice

A. Agr. Associate in Agriculture

AAGR Air-to-Air Gunnery Range

AAGRA Anti-Aircraft Group Royal Artillery

A.Agri. Associate in Agriculture

AAGS All-American Gladiolus Selections; Army Air-Ground System

AAGUS American Association of Genito-Urinary Surgeons

aagw (AAGW) air-to-air guided weapon(s)

aah advanced armed helicopter; advanced attack helicopter

aah (AAH) anti-armor helicopter

AAH Alcoholism Awareness Hour (tv); American Academy of Homiletics; Australian Auxiliary Hospital

aaha awaiting action of higher authority

AAHA American Amateur Hockey Association; American Animal Hospital Association; American Association of Handwriting Analysts; American Association of Homes for the Aging; American Association of Hospital Accountants

AAHC American Academy of Humor Columnists; American Association of Hospital Consultants

AAHCPA American Association of Hispanic Certified Public Accountants

AAHD American Academy of the History of Dentistry

AAHDC American Association of Hospital Dental Chiefs

AAHE American Association for Higher Education; American Association of Housing Educators

A.A.H.E. Associate in Arts in Home Economics

AAHL American Amateur Hockey League (ceased 1917)

AAHM American Association for the History of Medicine; Association of Architectural Hardware Manufacturers

Aahp Army artificial heart pump

AAHP American Association for Hospital Planning; American Association of Hospital Podiatrists; American Association for Humanistic Psychology

AAHPA American Association of Hospital Purchasing Agents

AAHPER American Association for Health, Physical Education, and Recreation

AAHPERD American Alliance for Health, Physical Education, Recreation, and Dance

AAHPhA American Animal Health Pharmaceutical Association

AAHPS Australian Association for the History and Philosophy of Science

AAHQ Allied Air Headquarters

AAHRA Asia and Australia Hotel and Restaurant Association

AAHS American Aviation Historical Society

AAHSLD Association of Academic Health Sciences Library Directors

aai air-to-air identification; angle-of-approach indicator; azimuth angle increment

AAI African-American Institute; Afro-American Institute; Agricultural Ammonia Institute; Akron Art Institute; Alfred Adler Institute; Allied Armies in Italy (World War II); American Association of Immunologists; Arab American Institute

A.A.I. Associate of the Chartered Auctioneers' and Estate Agents' Institute

AAIA Association of American Indian Affairs

A.A.I.A. Associate of the Association of International Accountants

AAIAL American Academy and Institute of Arts and Letters

AAIAN Association for the Advancement of Instruction about Alcohol and Narcotics

AAIB American Association of Instructors of the Blind

AAIC Allied Air Intelligence Center; Australian Advertising Industry Council

AAICD American Association of Imported Car Dealers

AAICJ American Association for the International Commission of Jurists

AAICU Alabama Association of Independent Colleges and Universities

AAID American Academy of Implant Dentistry; American Academy of Implant Dentures; American Association of Industrial Dentists

AAIE American Association of Industrial Editors; American Association of Industrial Engineers

AAII American Association of Individual Investors; Association for the Advancement of Invention and Innovation

AAIM American Association of Industrial Management

AAIMCo American Association of Insurance Management Consultants

AAIMS An Analytical Information Management System

AAIN American Association of Industrial Nurses

AAIND American Association of Independent News Distributors

AAIP Academic Administration Internship Program

AAIPS American Association of Industrial Physicians and Surgeons

AAIS American Association of Insurance Services; Anti-Aircraft Intelligence Service; Associate of the Australian Institute of Secretaries

AAIT American Association of Inhalation Therapists; Associate of the Australian Institute of Travel

AAJ American Association for Justice; Arab Airways, Jerusalem; Axel Axelson Johnson (Johnson Line)

AAJ *Australian Anthropological Journal*

AAJA Afro-Asian Journalists' Association; Asian-American Journalists Association

aajc automatic antijam circuit

AAJC American Association of Junior Colleges

AAJE American Association for Jewish Education

AAJR American Academy for Jewish Research

AAJS American Association for Jesuit Scientists

AAJSA American Association of Journalism School Administrators

AAK Alfred A Knopf

aal above aerodrome level; anterior axillary line; absolute assembly language

AAL Aid Association for Lutherans; Ames Aeronautical Laboratory; Arctic Aeromedical Laboratory; Association of Assistant Librarians; Australian Air League

AALA Afro-American Liberation Army; American Auto Laundry Association; American Automotive Leasing Association; Asian- American Librarians Association

AALAPSO Afro-Asian-Latin American People's Solidarity Organization

AALAS American Association of Laboratory Animal Science

AALASO Afro-Asian Latin American Students' Organization

AALC Advance Airborne Launch Center; African American Labor Center (AFL-CIO); Amphibious Assault Landing Craft (naval symbol)

AALD Australian Army Legal Department

AALE Associate of Arts in Law Enforcement

AALL American Association of Law Libraries

aalmg (AALMG) antiaircraft light machine gun

AALPA Association of Auctioneers and Landed Property Agents

AALPP American Association for Legal and Political Philosophy

AALR American Association for Leisure and Recreation

AALS Active Army Locator System; Air-Cushioned Assault Landing Craft; American Association of Language Specialists; Association of American Law Schools; Association of American Library Schools

AALU Association for Advanced Life Underwriting

aam anti-aircraft missile

aam (AAM) air-to-air missile

AAM Academy of Ancient Music; Acoustics Analysis Memo; American Association of Microbiology; American Association of Museums; Army Aircraft Maintenance; Australian Air Mission

A-AM Afro-American Museum

AAM Arma Aerea de la Marina (Spanish Naval Air Arm)

AAMA American Academy of Medical Administrators; American Amusement Machine Association; American Apparel Manufacturers Association; American Architectural Manufacturers Association; American Association of Medical Assistants; American Automobile Manufacturers Association; Architectural Aluminum Manufacturers Association; Asian American Manufacturers Association

A-A MAN Afro-American Men Against Narcotics

aamb anti-aircraft missile battery

AAMBP Association of American Medical Book Publishers

AAMC American Association of Marriage Counselors; American Association of Medical Clinics; American Association of Medico-Legal Consultants; Army Air Materiel Command; Association of American Medical Colleges; Australian Army Medical Corps

AAMCA Army Advanced Materiel Concepts Agency (USA)

AAMCH American Association for Maternal and Child Health

AAMD American Association on Mental Deficiency; Association of Art Museum Directors

aame (AAME) acetylarginine methyl ester

AAMES American Association for Middle East Studies

AAMF American Association of Music Festivals

AAMFT American Association for Marriage and Family Therapy

aamg (AAMG) antiaircraft machine gun

AAMG American Art Marketing Guild

AAMGA American Association of Managing General Agents

AAMHPC American Association of Mental Health Professionals in Corrections

AAMI American Association of Machinery Importers; American Athletic Motivation Institute; Association for the Advancement of Medical Instrumentation; Association of Allergists for Mycological Investigation

AAMIH American Association for Maternal and Infant Health

AAML Arctic Aeromedical Laboratory

AAMMC American Association of Medical Milk Commissioners

AAMOA Afro-American Music Opportunities Association

AAMP American Academy of Maxillofacial Prosthetics; American Association of Meat Processors

AAMR American Academy on Mental Retardation

AAMRL American Association of Medical Record Librarians

AAMS American Air Mail Society; Australian Army Medical Service

AAMSW American Association of Medical Social Workers

AAMT American Association for Music Therapy

AAMU Army Advanced Marksmanship Unit

A.A. Mus. Associate in Arts in Music

AAMVA American Association of Motor Vehicle Administrators

AAMW Association of Advertising Men and Women

AAMWS Australian Army Medical Women's Service

aan (AAN) aminoacetonitrile; assignment action number(s)

AAN Advance Acquisition Notification; American Academy of Neurology; American Academy of Nutrition; American Association of Neuropathologists; American Association of Nurserymen

A.A.N. Associate in Arts in Nursing

AANA American Association of Nurse Anesthetists; Australian Association of National Advertisers

AANB Alpha-amino-N-butyric acid

AANC American Association of Nutritional Consultants

AANCP Advanced Airborne National Command Post

AANJ Art Administrators of New Jersey

AAN/MA Asthma and Allergy Network/Mothers of Asthmatics

AANNT American Association of Nephrology Nurses and Technicians

AANP American Association of Naturopathic Physicians

AANR American Association of Newspaper Representatives

AANS American Academy of Neurological Surgery; American Association of Neurological Surgery; Australian Army Nursing Service

AANSW Archives Authority of New South Wales

aao amino-acid oxidase; anti-air output

aaO am angeführten Ort (German—in the place cited); an anderen Orten (German—elsewhere, in the place cited)

AAO Academy of Applied Osteopathy; Aircraft Approach Overlay (zone); American Academy of Ophthalmology; American Academy of Optometry; American Association of Orthodontists; Anglo-Australian Observatory

AAO Abastumanskaya Astrofizicheskaya Observatoriya (Russian—Abastumani Astrophysical Observatory)

AAOA Ambulance Association of America

AAOC American Association of Osteopathic Colleges; Antiaircraft Operations Center; Australian Army Ordnance Corps

AAOD Army Aviation Operating Detachment

AAODC American Association of Oilwell Drilling Contractors

AAOG American Association of Obstetricians and Gynecologists

AAOGAS American Association of Obstetricians, Gynecologists, and Abdominal Surgeons

AAOHN American Association of Occupational Health Nurses

AAOM American Academy of Occupational Medicine; American Academy of Oral Medicine

AAOME American Association of Osteopathic Medical Examiners

AAOMS American Association of Oral and Maxillofacial Surgeons

AAONMS Ancient Arabic Order of Nobles of the Mystic Shrine

AAO & O American Academy of Ophthalmology and Otolaryngology

AAOP American Academy of Oral Pathology; Antiaircraft Observation Post

AAOPB American Association of Pathologists and Bacteriologists

AAOPS American Association of Oral and Plastic Surgeons

aaor anti-aircraft operations room

AAOR American Academy of Oral Roentgenology

AAOS American Academy of Orthopaedic Surgery

aap advise if able to proceed; air at atmosphere pressure

AAP Academy of American Poets; Advance Acquisition Plan(ning); Affirmative Action Program; Allied Administrative Publication; American Academy of Pediatrics; American Academy of Periodontology; American Atheist Press; Angina Awareness Program; Apollo Applications Program; Army Ammunition Plant; Association for the Advancement of Psychoanalysis; Association for the Advancement of Psychology; Association for the Advancement of Psychotherapy; Association of American Physicians; Association of American Publishers; Association of Applied Psychoanalysis; Australian Associated Press

A-A P Afro-American Police

AAP Allied Army Procedures (or Publications)

AAPA American Alfalfa Processors Association; American Amateur Press Association; American Association of Physical Anthropologists, American Association of Port Authorities

A-A PA Anglo-American Press Association

AAPAC Arab-American Political Action Committee

AAPB American Association of Pathologists and Bacteriologists

AAPBS Australian Association of Permanent Building Societies

AAPC Advertising Agency Production Club; All-African Peoples' Conference; American Association of Professional Consultants; Assault Amphibious Personnel Carrier; Australian Aluminium Production Commission; Australian Army Psychological Corps

aapcc adjusted average per capita cost

AAPCC American Association of Poison Control Centers; American Association of Psychiatric Clinics for Children

AAPCM Association of American Playing Card Manufacturers

AAPCO Association of American Pesticide Control Officials

AAPD American Academy of Physiologic Dentistry

AAPE American Academy of Physical Education; American Association of Philatelic Exhibitors

AAPF Academy of American Poets Fellowship

AAPG American Association of Petroleum Geologists; Arab-American Press Guild

AAPH American Association for Partial Hospitalization; American Association of Professional Hypnologists

AAPHD American Association of Public Health Dentists

AAPHI Associate of the Association of Public Health Inspectors

AAPHP American Association of Public Health Physicians

AAPHR American Association of Physicians for Human Rights

AAPICU American Association of Presidents of Independent Colleges and Universities

AAPIU Allied Aerial Photographic Interpretation Unit

AAPL Afro-American Policemen's League; American Academy of Psychiatry and the Law; American Artists Professional League; American Association of Professional Landmen; Apple Computer Incorporated (stock exchange symbol)

AAPLE American Academy for Professional Law Enforcement; American Association for Professional Law Enforcement

aapm amphiapomict

AAPM American Academy of Pain Management

AAPMR American Academy of Physical Medicine and Rehabilitation

AAPO All-African Peoples' Organization

AAPOR American Association for Public Opinion Research

AAPP American Association of Police Polygraphists; Association of Amusement Park Proprietors; Australian Association of Psychology and Philosophy

AAPPP American Association of Planned Parenthood Physicians

AAP/PSP Association of American Publishers—Professional and Scholarly Publishing Division

AAPRA All-African People's Revolutionary Army

AAPRCO American Association of Private Railroad Car Owners

AAPRM American Association of Passenger Rate Men

AAPRP All-African People's Revolutionary Party

AAPS American Association of Phonetic Sciences; American Association of Plastic Surgeons; American Association for the Promotion of Science; Association of Alternate Postal Systems; Association for Ambulatory Pediatric Services; Association of American Physicians and Surgeons

AAPSC American Association of Psychiatric Services for Children

AAPSD Alternative Automotive Power System Division (EPA)

AAPSE American Association of Professors in Sanitary Engineering

AAPSS American Academy of Political and Social Sciences

AAPSW Associate of the Association of Psychiatric Social Workers

AAPT American Association of Physics Teachers; Association of Asphalt Paving Technologists

AAPTO American Association of Passenger Traffic Officers

AAPTS Association of America's Public Television Stations

AAPTSR Australian Association for Predetermined Time Standards and Research

aapy average annual percentage yield

AAPY American Association of Professors of Yiddish

AAQM All Arms Quartermaster

AAQMG Acting Assistant Quartermaster-General

AA & QMG Assistant Adjutant and Quartermaster General

aar after action report; against all risks; air-to-air refueling; approved auto repair; average annual rainfall

aar (AAR) antigen-antiglobulin reaction

Aar Aarhus; Australia antigen radioimmunoassay

AAR Abby Aldrich Rockefeller; Aircraft Accident Record; Aircraft Accident Report; American Academy in Rome; American Academy of Religion; Army Area Representative; Association of American Railroads; Association of Authors' Representatives; Australian Associated Resources; Automotive Affiliated Representatives

AARA Australian Association of Reprographic Arts

aa rating average-audience rating (percentage of tv-equipped homes viewing the average minute of a national telecast)

aarb advanced aerial refuelling boom

AARB Australian Road Research Board

AARC Ann Arbor Railroad Company; Association for the Advancement of Released Convicts; Australian Aeronautical Research Committee; Australian Applied Research Centre; Australian Automobile Racing Club

AARCO Afro-Asian Rural Construction Organisation

AARD American Academy of Restorative Dentistry

AARDC Army Aviation Research and Development Command

AARDCO Association of American Railroad Dining Car Officers

AARDS Australian Advertising Rate and Data Service

AARE Australian Association for Research in Education

AARF Australian Accounting Research Foundation

aarg aargang (Dano-Norwegian or Swedish—yearbook)

Aarh Aarhus

AARL Advanced Applications and Research Laboratory; Army Aeromedical Research Laboratory; Australian Academic and Research Libraries

aarp annual advance retainer pay

AARP American Association of Retired Persons; Ancient American Rocket Pioneers

AARPS Air-Augmented Rocket-Propulsion System

AARR Airborne Armored Reconnaissance Regiment; Ann Arbor Railroad

AARRC Army Aircraft Requirements Review Committee

AARRO Afro-Asian Rural Reconstruction Organization

AARS Airborne Armed Reconnaissance Squadron; All-America Rose Selections (award); American Association of Railroad Superintendents; American Association of Railway Surgeons; Army Aircraft Repair Ship; Army Amateur Radio System; Automatic Acquisition Radar Seeker

AART American Association for Rehabilitation Therapy; American Association of Retired Teachers

AARTA American Association of Railroad Ticket Agents

aarv aerial armored reconnaissance vehicle

AARWBA American Auto Racing Writers and Broadcasters Association

aas adjusted air speed; advanced antenna system; amphibious assault ship; aortic arch syndrome; arithmetic assignment statement; automated accounting system

aa's author's alterations

AAs Alcoholics Anonymous members; American Atheists; Asian Americans; author's alterations

AAS Aberdeen Art Society; Academy of Applied Science; Aeromedical Airlift Squadron; Air Armament School; Aircraft Airworthiness Section; All-American Selections; American Amaryllis Society; American Antiquarian Society; American Astronautical Society; American Astronomical Society; American Autobahn Society; Army Air Service; Army Apprentices School; Army Attache System; Arnold Air Society; Ascending Activating System; Association for Asian Studies; Australian Academy of Science; Australian Acoustic Society; Australian Air Services; Australian Art Society; Australian Association of Surgeons; Automatic Adjusted Suspension

A.A.S. *Acadenriae Americanae Socius* (Latin—Fellow of the American Academy of Arts and Sciences)

AASA American Association of School Administrators; Associate of the Australian Society of Accountants

aa-sat advanced anti-satellite

AASB Alabama Association of School Boards; American Association of Small Business

AASC Acupuncture Association of Southern California; Aerospace Applications Studies Committee (NATO); Allied Air Support Command; American Association of Specialized Colleges; Army Area Signal Center; Australian Accounting Standards Committee; Australian Army Service Corps; Australian Army Signal Corps

aascm awaiting action summary court martial

AASCO Association of American Seed Control Officials

AASCU American Association of State Colleges and Universities

aasd anti-aircraft self-destroying

AASD American Association of Social Directories

AASDJ American Association of Schools and Departments of Journalism

AASE American Academy of Sanitary Engineers; American Association of Special Educators; Army Aviation Support Element; Associated Australian Stock Exchanges; Association for Applied Solar Energy

AASEC American Association of Sex Educators and Counselors

AASECT American Association of Sex Educators' Counselors, and Therapists

AASF Advanced Air Striking Force

AASFE American Association of Sunday and Feature Editors

AASG Association of American State Geologists

aasgp amphibious assault ship general purpose

AAS & GP American Association of Soap and Glycerin Producers

AASH American Association for the Study of the Headache; Army Advanced Scout Helicopter

AASHO American Association of State Highway Officials

AASHTO American Association of State Highway and Transportation Officials

AASI Advertising Agency Service Interchange; American Academy for Scientific Interrogation, American Association for Scientific Interrogation

A'asia(n) Australasia(n)

aasir advanced atmospheric sounder and imaging radiometer

aasl anti-aircraft searchlight

AASL American Antiquarian Society Library; American Association of School Librarians; American Association of State Librarians

A & ASL American & Australian Steamship Line

AASLH American Association for State and Local History

AASM Abigail Adams Smith Museum New York City; Associated African States and Madagascar; Association of American Steel Manufacturers

AASMB Australian Association of Stud Merino Breeders

AASND American Association for the Study of Neoplastic Diseases

AASO Association of American Ship Owners

AASP American Antiquarian Society Proceeding; American Association for Social Psychiatry

AASPA American Association of School Personnel Administrators

AASPB American Association of State Psychology Boards

AASPRC American Association of Sheriff's Posses and Riding Clubs

aasr airport and airways surveillance radar

AASR Abhazian Autonomous Soviet Socialist Republic; Adjarian Autonomous Soviet Socialist Republic

AASRC American Association of Small Research Companies

AASRI Arctic and Antarctic Scientific Research Institute

AASRM Ancient and Accepted Scottish Rite Masons

aass advanced airborne surveillance sensor

AASS Afro-American Students Society; American Association for Social Security; *Americanae Antiquarianae Societatis Socius* (Latin—Associate of the American Antiquarian Society)

AASSP Arkansas Association of Secondary School Principals

AAST American Association for the Surgery of Trauma

AASTA Antiaircraft Station; Army Aviation Systems Test Activity

AASTC Associate in Architecture—Sydney Technical College

AASTD Association for the Advancement of the Science and Technology of Documentation

AASU Afro-American Student Union

AASW Australian Association of Scientific Workers; Australian Association of Social Workers

AASWA American Association for the Study of World Affairs

AASWI American Aid Society for the West Indies

aat acute abdominal tympany; advanced authoring tools; after acid treatment; auditory attending task

aat (AAT) alpha-1 antitrypsin

AAT Achievement Anxiety Test; Anglo-Australian Telescope (in NSW); Auditory Apperception Test; Australian Antarctic Territory

A-AT Anglo-Australian Telescope

AAT Art and Architecture Thesaurus (Getty)

AATA American Association of Teachers of Arabic; Anglo-American Tourist Association

AATB Advanced Amphibious Training Base

AATC Advanced Air Training Command; Anti-Aircraft Training Center; Army Aviation Test Center; Army Aviation Test Command; Automatic Air Traffic Control (system)

AATCC American Association of Textile Chemists and Colorists

AATCLC American Association of Teachers of Chinese Language and Culture

AATCO Army Air Traffic Coordinating Office

AATD Australian Association of Teachers of the Deaf

AATDC Army Airborne Transport Development Center

AATE American Association of Teachers of Esperanto; Association of Australian Teachers of English

AATEA American Association of Teacher Educators in Agriculture

AATEFL Australian Association for the Teaching of English as a Foreign Language

AATF American Association of Teachers of French

AATG American Association of Teachers of German

AATH American Association of Teaching Hospitals

AATI American Association of Teachers of Italian

AATM American Academy of Tropical Medicine

AATNU Administration de l'assistance technique des Nations Unies (French—United Nations Technical Assistance Administration)

AATO Army Air Transport Organization

AATOE American Association of Theatre Organ Enthusiasts

AATP American Academy of Tuberculosis Physicians

AATPA American Association of Traveling Passenger Agents

AATPS Australian Association of Temporary Personnel Services

AATRACEN Anti-Aircraft Training Center

AATRI Army Air Traffic Regulation and Identification

AATRIS Army Air Traffic Regulation and Identification System

AATS American Academy of Teachers of Singing; American Association of Theological Schools; American Association for Thoracic Surgery

AATSEEL American Association of Teachers of Slavic and Eastern European Languages

AATT American Association for Textile Technology

AATTA Arab Association of Tourism and Travel Agents

AAT & TC Anti-Aircraft Training and Test Center

AATU Association of Air Transport Unions

AATUF All-African Trade Union Federation

AATV Australian Army Training Team Vietnam

AAU Administrative Area Unit; Air Ambulance Unit; Al-Azhar University; Amateur Athletic Union; Associated Aviation Underwriters; Association of American Universities; Association of Atlantic Universities (Canada); Australian Athletics Union

AAUA Amateur Athletics Union of Australia; American Association of University Administrators

AAUBO Association of Atlantic University Business Officers (Canada)

AAUCG Americans Against Union Control of Government

AAUCS Australian-Asian Universities Cooperation Scheme

AAUN American Association for the United Nations; Australian Association for the United Nations

AAUP American Association of University Presses; American Association of University Professors

AAUQ Associate in Accountancy—University of Queensland

AAUTA Australian Association of University Teachers of Accounting

AAUTI American Association of University Teachers of Insurance

AAUUS Amateur Athletic Union of the U.S.

AAUW American Association of University Women

aav adeno-associated virus; airborne assault vehicle; amphibious assault vehicle; armored anti-aircraft vehicle

AAV Antiaircraft Volunteer

AAVA American Association of Veterinary Anatomists; Army Audio Visual Agency; Australian Automatic Vending Association

AAVB American Association of Veterinary Bacteriologists

AAVC Australian Army Veterinary Corps

AAVCS Automatic Aircraft Vectoring Control System

aavd automatic alternative voice/data

AAVIM American Associations for Vocational Instructional Materials

AAVMC Association of American Veterinary Medical Colleges

Aavn Army aviation

AAVN American Association of Veterinary Nutritionists

AAVP American Association of Veterinary Pathologists

AAVRO American Association of Vital Records and Organizations

AAVS Aerospace Audiovisual Service; American Anti-Vivisection Society

AAVSO American Association of Variable Star Observers

AAVT Association of Audio-Visual Technicians

aaw advanced attack weapon; aeromedical airlift wing; airborne assault weapon; air-to-air warfare; anti-aircraft weapon

AAW Advertising Association of the West; American Atheist Women; Anti-Air Warfare

AAWA American Automatic Weapons Association

AAWB American Association of Workers for the Blind

AAWC Anti-Air Warfare Center; Australian Advisory War Council

AAWD Association of American Women Dentists

AAWEXINPT Antiair Warfare Exercises in Port

AAWg Acromedical Airlift Wing (USAF)

AAWM American Association of Waterbed Manufacturers

AAWO Afro-Asian Workers' Organization

AAWPI Association of American Wood Pulp Importers

AAWR American Association of Women Radiologists

AAWs Anti-Armor Weapons

AAWS American Association of Wardens and Superintendents

AAWSSC Army Atomic Weapons Safety Systems Committee

AAWU Amateur Athletic Western Union

AAXICO American Air Export and Import Company

AAYM American Association of Youth Museums

AAZK American Association of Zoo Keepers

AAZM *Aqueduicto y Alcantarillado de la Zona Metropolitana* (Spanish—Aqueducts and Watercourses of the Metropolitan Area)

AAZN American Association for Zoological Nomenclature

AAZPA American Association of Zoological Parks and Aquariums

ab abbey; abnormal; abortion; about; abscess; abundant; acquisition beacon; adapter booster; aerial burst; afterburner; air blast; airbrake; alcian blue; ambient brine; anchor bolt; antibody; asbestos body; asthmatic bronchitis; at bats; axiobuccal

a/b acid-base (ratio)

a/b (A/B) airborne

a & b applejack and benedictine; assault and battery

ab abril (Spanish—April); (Latin prefix—away from, off); (Persian—water)

aB auf Bestellung (German—on order)

Ab Abbot; Abner; abnormal; Abraham; alabamine

Ab Abade (Portuguese—abbot; fat man)

AB able-bodied seaman; Accessories Bulletin; Admiralty Board; Aid to the Blind; Air Base; Air Board; Air Branch; Airman Basic; Alberta; Anheuser-Busch Incorporated; Arnold Bernstein (steamship line); Aryan Brotherhood; Assembly Bill; Atlantic Beach

A-B Allen-Bradley; Ambrose Bierce; Anton Bruckner

A.B. *artium baccalureus* (Latin—Bachelor of Arts)

A/B Aid to the Blind; Airman Basic

A & B Antigua & Barbuda

AB Analecta Biblica; Army Book; Asia Business

A/B Aktiebolag (Swedish—limited company)

AB-47 Agusta-Bell three-place utility helicopter

AB-204 Agusta-Bell gunship twin-engine helicopter

AB-205 Agusta-Bell ten-place troop-transport helicopter

AB-206 Agusta-Bell five-seat turbine-powered helicopter

aba adrenergic blocking agent; antibacterial activity

ab-a abampere

ABA Aaron Burr Association; Air Brake Association; American Badminton Association; American Bakers Association; American Bamboo Association; American Bandmasters Association; American Bankers Association; American Bar Association; American Beauty Association; American Bell Association; American Berkshire Association; American Billiard Association; American Birding Association; American Bison Association; American Booksellers Association; American Bowhunters Association; American Brazilian Association; American-British-Australian; American Broadcasting Authority; American Buddhist Association; American Bus Association; Annual Budget Authorization; Australian Badminton Association; Australian Bankers Association; Australian Biotechnology Association; Australian Booksellers Association; Australian Boomerang Association; Australian Bowhunters Association; Australian Bridge Association; Ayrshire Breeders Association

ABAA Antiquarian Booksellers Association of America

ab ab. *ab absurdo* (Latin—to the absurd)

abac a basic coursewriter

ABAC Abraham Baldwin Agricultural College

Abaco Great and Little Abaco islands in the Bahamas north of New Providence Island

abact abacterial

ABACUS Air Battle Analysis Center Utility System; Autonetics Business and Control United Systems

-abad (Hindi—city or town)

ABAD Air Battle Analysis Division

ab aet. *ab aeterno* (Latin—until eternity)

ABAFA Association of British Adoption and Fostering Agencies

ABAG Association of Bay Area Governments (San Francisco)

A-bahn *Autobahn* (German—superhighway)

ABAI American Bell Association International; American Boiler and Affiliated Industries

ABAJ Antiquarian Booksellers Association of Japan

ABAJ *American Bar Association Journal*

ABAK *Association di Biblioteka i Archivo di Korsow* (Papiamento—Association of Libraries and Archives of Curaçao)

abamp absolute ampere (10 amperes)

ABANA Artist-Blacksmiths' Association of North America

aband abandoned

abandt abandonment

ABAO *Asociación Bilbaina de Amigos de la Opera* (Spanish—Bilbaoan Association of Friends of the Opera)

abap antibody against panel

ABAR Advanced Battery Acquisition Radar; Alternate Battery Acquisition Radar

ABARE Australian Bureau of Agriculture and Resource Economics

ABAS American Board of Abdominal Surgeons; Australian Buying Advisory Service

abat abattoir

a batt *a battuta* (Italian—by the beat) —musical term

ABATU Advance Base Air Task Unit; Advance Base Aviation Training Unit

ABAUSA Amateur Basketball Association of the United States of America

abb ablating blunt body

abb *abbassamento* (Italian—abatement, decline, diminution, fall of temperature, lowering, subsiding); *abbonamento* (Italian—subscription); *abbuono* (Italian—allowance, bonus, discount)

Abb Abbess; Abbey; Abbot

Abb *Abbildung* (German—illustration)

Abb. *abbas* (Latin—abbot)

Ab of B Archbishop(ric) of Bremen

ABB Akron & Barberton Belt (railroad); Australian Barley Board

ABBA American Blind Bowling Association; American Board of Bio-Analysis; American Brahman Breeders Association; American Bed & Breakfast Association; Australian Brahma Breeders Association

ABBARS American Bread & Breakfast Associated Reservation Services

ABBB Association of Better Business Bureaus

Ab^e *Abbaye* (French—abbey)—monastery

ABBIT Association of Bronze and Brass Founders

Abbild *Abbildungen* (German—illustrations)

ABBIM Association of Brass and Bronze Ingot Manufacturers

ABBMM Association of British Brush Machinery Manufacturers

A.B.B.O. Associate of the British Ballet Organisation

Abbot Vickers self-propelled fortress including 105 mm gun, turret-mounted howitzer, and 7.62 machine gun

Abbotsford British Columbia's Matsqui Institution (for narcotic addicts) at Abbotsford; a town west of Wausau, Wisconsin

Abbot(t) Abbotson

Abbott & Costello Bud Abbott and Lou Costello

abbr abbreviate; abbreviated; abbreviation

ABBRA American Boat Builders and Repairers Association

abbrev *abbreviatura* (Italian—abbreviation)

abbrevia abbreviations

abbreviaz *abbreviazione* (Italian—abbreviation)

abbrevio abbreviomania(c) (al) (ly)

abbrev(s) abbreviation(s)

ABBS Australian Bibliography and Bibliographical Services

abc abecedarium (alphabet primer); acid-balance control; aconite, belladonna, chloroform; advanced base camp; advanced biomedical capsule (ABC); airborne cigar; alphabet; already been chewed; approach by concept; atomic, biological, chemical (ABC); alum, blood, charcoal; automatic bass compensation; automatic brightness control; axiobuccocervical

abc (ABC) advance-booking charter; alarms by carrier (panic-button device alerting fire and police stations)

AbC American-born Chinese

Ab of C Archbishop(ric) of Cologne

ABC Aberrant Behavior Center; A Better Choice (scholarship program for the poor); Abridged Building Classification; Acceptable Biological Catch; Advanced Booking Charters; Aerated Bread Company; Airborne Control; Air Bridge to Canada; Alcohol Beverage Control; Alliance for Better Chicago (public schools); American Ballet

Company; American Baptist Churches; American Bar Center; American Blimp Corporation; American Book (prices) Current; American Bowling Congress; American Brass Company; American, British, Canadian; American Broadcasting Company; American Business Conference; Animal Birth Control; Arts and Business Council; Argentina, Brazil, Chile; Asahi Broadcasting Company; Asian Banking Council; Assisi Bird Campaign; Associated Bottlers Company; Associated Builders and Contractors; Association of Biotechnology Companies; Association of Bituminous Contractors; Association of Bridal Consultants; Atanasoff-Berry Computer; atomic, biological, chemical (warfare); Audit Bureau of Circulation; Australian Band Council; Australian Bankruptcy Cases; Australian Baseball Council; Australian Bowling Council; Australian Bridge Council; Australian Broadcasting Commission; Australian Broadcasting Corporation; automatic bandwidth control; automation of bibliography through computerization; Automotive Boosters Clubs

AB & C Atlanta Birmingham and Coast (railroad)

ABC Abstracts in BioCommerce; Academia Brasileira de Ciencias (Portuguese—Brazilian Academy of Sciences); Spanish newspaper

ABC³ Airborne Battlefield Command and Control Center

ABCA American Building Contractors Association; American Business Communications Association; American-British-Canadian-Australian; Antique Bottle Collectors Association; Army Bureau of Current Affairs; Australian Bushmen's Carnival Association

ABCAL Associated British Cables (NZ)

ABC-ASP American-British-Canadian—Army Standardization Program

abcb air-blast circuit breaker

ABCB American Bottlers of Carbonated Beverages; Australian Broadcasting Control Board

ABCC Association of British Chambers of Commerce; Atomic Bomb Casualty Commission

A-B CC Arab-British Chamber of Commerce

ABCCC Airborne Battlefield Command and Control Center

ABC-Clio American Bibliographical Center—Clio Press (Santa Barbara, California)

ABCCTC Advanced Base Combat Communication Training Center

abcd above and beyond the call of duty; airway (opened), breathing (restored), circulation (restored), definitive (therapy); atomic, biological, chemical, and damage (control); awaiting bad conduct discharge

ABCD Accelerated Business Collection and Delivery (of mail); Action for Boston Community Development; Advanced Base Construction Depot; America, Britain, China, Dutch East Indies (ABCD Powers during World War II); American Society of Bookplate Collectors and Designers

ABCDCAL Alcoholic Beverage Control Department—California

ABCE Adult Basic and Continuing Education

ABC-fm Australian Broadcasting Commission—frequency modulation (system)

ABCI Australian Bureau of Criminal Intelligence; Automotive Booster Clubs International

ABC Islands Admiralty, Baranof and Chicagof, Alaska

ABCL American Birth Control League

ABC Line Antwerp Bulk Carriers (container line)

ABCM Associate of Bandsmen's College of Music; Association for the Bedouin Culture Museum

ABCMR Army Board for Correction of Military Records

abcr atomic-biological-chemical-radiological

abcrw atomic-biological-chemical-radiological warfare

ABC-tv Australian Broadcasting Commission—television

abd abdicated; advanced base depot; advanced base dock; all but dissertation; average body dose (radiation)

ABDA American-British-Dutch-Australian

ABDACOM Advanced Base Depot Area Command; American-British-Dutch-Australian Command

abdom abdomen; abdominal

ABDS Army Bomb Disposal Squad

ABDSP Anza-Borrego Desert State Park

Abduction Abduction from the Seraglio (Mozart's comic opera *Enführung aus dem Serail)*

abe adult basic education; airborne bombing evaluation

ABE American Ballet Ensemble

ABECOR Associated Banks of Europe Corporation

abel air-breathing electric laser

Abel Abelard

abend abnormal end

Abes Abyssinia(n)

ABES Association for Broadcast Engineering Standards

abess abessinisch (German—Abyssinian)

ABET Accreditation Board for Engineering and Technology

ABETS Airborne Beacon Electronic Test Set

ABEU Australian Bank Employees Union

ABEWS Army Battlefield Electronic Warfare Simulator

abf absolute bloody final (beverage); aircraft battle force; annular blast fragmentation

ABF American Bar Foundation; American Beekeeping Federation; Army Benevolent Fund; Australian Bridge Federation

abfab absolutely fabulous
abfc advanced base functional components
ABFP American Board of Family Practice; American Board of Forensic Psychology
abg air base group; air battle group; arterial blood gas
ABGC Australian Banana Growers Council
Abh Abhandlungen (German—transactions)
ABHC Association of Bank Holding Companies
ABHP American Board of Health Physics
abi abstracted business information; agile beam illuminator; application binary interface
ABI American-British Intelligence; Ankle Brachial Index (blood-pressure test); American Butter Institute
ABI Associazíone Bibliotecari Italiani (Italian—Association of Italian Librarians)
ABIC Adaptive Behavior Inventory for Children
ABIH American Board of Industrial Hygiene
abi/inform abstracted business information/information needs
ab init. *ab initio* (Latin—from the beginning)
ABINZ Australian Banking Institute of New Zealand
ABIP Australian Books in Print
ABISA Association of Burglary Insurance Surveyors—Australia
ABJ Association of Broadcasting Journalists
Abk Abkürzung (German—abbreviation)
ABKA American Boarding Kennels Association
Abkürz Abkarzüng (German—abbreviation)
abl ablative
abl abril (Spanish—April)
ABL Atlas Basic Language; Automated Biology Library
ABLA Amateur Bicycle League of America; American Blind Lawyers Association; American Business Law Association

able activity balance line evaluation
ABLE Ability Based on Long Experience; Advocates for Border Law Enforcement
ABLES Airborne Battlefield Light Equipment System
ablon abalone
ABL-SF Asian Business League of San Francisco
abm automated batch mixing
abm (ABM) anti-ballistic missile
ABM Assistant Beachmaster; Adventist Board of Missions; Australian Board of Missions (Anglican); Aviation Boatswain's Mate
ABM Asociación de Banqueros de México (Spanish—Mexican Bankers Association)
ABMA American Brush Manufacturers Association; Army Ballistic Missile Agency
ABMC American Battle Monuments Commission
ABMDA Advanced Ballistic Missile Defense Agency; Army Ballistic Missile Defense Agency
ABMEWS Anti-Ballistic Missile Early Warning System
abmis (ABMIS) airborne ballistic missile system
ABMJ American Board of Missions to the Jews
abmm (ABMM) anti-ballastic missile missile
ABMR Australian Board of Mineral Resources
ABM Treaty Anti-Ballistic Missile Treaty
Abn Airborne
ABN Advance Beneficiary Notice (waiver of liability); Anti-Bolshevik Bloc of Nations
ABNC American Bank Note Company
abndn(d) abandon(ed)
Abn Inf Airborne Infantry
ABNY Association for a Better New York
abo accumulated benefit obligation
Åbo (Finnish—Turku)
ABO Admiralty Berthing Officer (UK)
ABOA Australian Bank Officials Association

ABOC American Board of Optometry Certificate
ABOD Advance Base Ordnance Depot
Abo Eng Aboriginal English
abol abolish(able); abolisher; abolishment, abolition(ary); abolitionism; abolitionist
abortus aborted fetus
abortus pill abortion-inducing pill, developed in France as Mifepristone, generic name: RU-486
abp absolute boiling point
Abp Archbishop
ABP American Business Press
ABPA Advanced Base Personnel Administration; American Book Producers Association; Automotive Body Parts Association
abpg advanced base proving ground
abpo advanced base personnel officer
ABPP American Board of Professional Psychology
ABPS American Board of Plastic Surgery
ABQ Annapolis Brass Quintet
abr abridged; abridgment; amphibian boat reconnaissance; auditory brainstem response
ABR Army Ballistic Research
Abra Abraham
ABRACADABRA Abbreviations and Related Acronyms Associated with Defense, Astronautics, Business, and Radioelectronics (a Raytheon publication)
Abram Abraham
abrb advanced base receiving barracks
ABRC Advisory Board for the Research Councils; Auto Body Representatives Council
abrd advanced base repair depot; advanced base replacement depot; advanced base reshipment depot
ABRES Advanced Ballistic Re-Entry System
abrev abreviatura (Portugimese/Spanish—abbreviation)
abrév abréviation (French—abbreviation)
ABRO Army in Burma Reserve of Officers
ABRS Australian Biological Resources Study

abs able-bodied seaman; absent; absolute value; absorb(ent); abstract; air base squadron; armored boarding steamer

abs (ABS) acrylonitrile-butadiene-styrene

ABs Aryan Brotherhood members

Abs Absatz (German—paragraph)

ABS American Bible Society; American Bureau of Shipping; Anti-lock Braking System; Association of Banks in Singapore; Australian Ballet Society, Australian Book Society; Australian Bureau of Statistics

absap airborne search and attack plotter

abs art abstract art

Abscam Arab scam

Abs cat Abyssinian cat

Abschn Abschnitt (German—chapter or paragraph)

absd advanced base sectional dock; advanced base supply depot

ABSD Army Base Supply Depot; Army Blood Supply Depot

ABSEL Association for Business Stimulation and Experiential Learning

Absen Absender (German—sender)

abse. re. absente reo (Latin—defendant absent)

abs exp abstract expressionism

ABSO Auxiliary Business Service Organization

abso(l) absolute; absolutely

absol absolument (French—absolutely)

ABS-PS Adaptive Behavior Scale—Public School (version)

ABSTI Advisory Board on Scientific and Technical Information (Canadian)

abstr abstract

abs vot absentee voting

absw air-break switch

ABSW American Baptist Seminary of the West

abt about; air-blast transformer

abt (ABT) after beginning of time

Abt Abteilung (German—division or part)

ABT American Ballet Theater; American Board of Trade; Association of Book Travelers; Australian Broadcasting Tribunal

ABTA American Bridge Teachers Association; Australian-British Trade Association

ABTF Airborne Task Force

ABTR Association for Brain Tumor Research

ABTU Advanced Base Torpedo Unit; Air Bomber Training Unit; Army Basic Training Unit

ABU Administrative Base Unit; American Board of Urology; Asian Broadcasting Union; Auto Barrage Unit

abul abulia; abulic

abv absolute value; assault bridging vehicle

abvd adriamycin, bleomycin, vinblastine and dacarbazine

ABWA American Business Women's Association; Associated Business Writers of America

ABWRC Army Biological Warfare Research Center

ABWUA Amateur Boxing and Wresting Union of Australia

Aby Abyssinia (former name of Ethiopia)

ABY AB Byers (stock-exchange symbol)

ABYC American Boat and Yacht Council

ac absolute ceiling; accelerator; account current; acetyl; acetyl-choline; acoustical(ly); acre; activator; acupuncture; adrenal cortex; advisory circular; aerodynamic center; air condition(ed); air conditioning; air conduction; air cool; air cooled; aircraft carrier; also completed; alternating current; analog computer; anodal closure; anthracyclines and cyclophosphamide (chemotherapy); anticorrosive; antiphlogistic corticoid; area controller; arithmetic computation; armed cargo (aircraft); armored car; asbestos cement; asphalt concrete; atriocarotid; attested copy; automatic checkout; automatic

computer; auriculocarotid; auxiliary console; axiocervical; azimuth comparator

ac (AC) average cost

a-c alternating-current

a/c account; account current; air conditioning; aircraft

a & c addenda and corrigenda; arts and crafts

ac (Latin prefix—to, toward)

a.c. ante cibos (Latin—before meals)

a/c ao cuidado de (Portuguese—in care of)

a C avanti Cristo (Italian—before Christ)

Ac acetyl; actinium; altocumulus

AC Adelbert College; Adelphi College; Aden Colony; Administration of Correction (Puerto Rico); Adrian College; Advertising Council; aerodynamic center (symbol); Air Canada; Air Command; Air Commodore; Air Control; Air Corps; Aircraftman; Alabama College; Albion College; Albright College; Allegheny College; Alliance College; Alma College; alternating current; Alverno College; Amarillo College; Ambulance Corps; Amherst College; Anderson College; Andrew College; Annhurst College; anodal contraction or closure; Antioch College; Appeal Cases; Aquinas College; Arcadia College; Area Code; Area Commander; Arithmetic Computation (test); Arkansas College; Armstrong College; Army Circular; Army Cooperation (Squadron); Army Corps; Army Council; Artillery College; Asbury College; Ashland College; Assumption College; Athens College; Athletic Club; Atlantic Council (NATO); Attack Cargo (Vessel); Augusta College; Augustana College; Aurora College; Austin College; Australia Council; Australian Companion (Companion of the Order of Australia); Australian Corps; Australian Cruiser; Averett College; Aviation Cadet; Azusa College

A-C Allis-Chalmers

A/C Air Commodore; aircraft; Aviation Cadet

AC Ação Catolica (Portuguese); *Acción Católica* (Spanish); *Action Catholique* (French); *Azione Cattolica* (Italian—Catholic Action); *Appellation Contrôlée* (French—brand name control of fine wines); *Atlanta Constitution*

A.C. *année courante* (French—current year); *Ante Christum* (Latin—before Christ); *Año Cristo* (Spanish—Year of Our Lord) —A.D.

A & C Antony and Cleopatra

AC3 Audio Codes 3

AC-47 DC-3 Douglas 21-passenger transport also called C-47 Dakota or Skytrain

AC-119 Fairchild-Hiller armed gunship complete with Vulcan cannons and 7.2 mm mini-guns (C-119 Flying Boxcar conversion)

AC-130 Lockheed armed gunship similar to AC-119 but with more guns

aca adenocarcinoma; adjacent channel attenuation; admiral commanding aircraft; anterior cerebral artery; automatic circuit analyzer; awaiting combat assignment; azimuth control amplifier

aca (ACA) advanced combat aircraft

ac a acetic acid

Aca Acapulco (inhabitants—Acapulqueños)

ACA Acapulco, Mexico (airport); Adult Children of Alcoholics; Aero Calif Airlines; Aero Club of America; Aircraft Castings Association; Alberta College of Art; Allied Commission for Austria; Allied Control Austria; Allied Control Authority; American Camping Association; American Canoe Association; American Carnivals; Association; American Casting Association; American Cat Association; American Cemetery Association; American Chiropractic Association; American Civic Association; American Collectors Association; American College of Allergists; American College of Anesthesiologists; American College of Aphothecaries; American Committee on Africa; American Communications Association; American Compensation Association; American Composers Alliance; American Congregational Association; American Correctional Association; American Council of the Arts; American Cryptogram Association; American Crystallographic Association; Americans for Constitutional Action; Anti-Corruption Agency (Singapore); Arms Control Agency; Arms Control Association; Army Caving Association; Art Council Aids; Arts Council of America; Arts Council of Australia; Assembly Constitutional Amendment; Assisting Catering Accountant (military); Associated Chiropodists of America; Association of Correctional Administrators; Atlanta College of Art

A.C.A. Associate of the Institute of Chartered Accountants (of England and Wales)

AC/A accommodative convergence/accommodation ratio

ACAA Agricultural Conservation and Adjustment Administration

ACAAE Australian Council on Awards in Advanced Education

ACAAI Air Cargo Agents Association of India

ACAB Air-Conditioning Advisory Bureau; Army Contract Adjustment Board

acac automated direct analog computer

ACAC Allied Container Advisory Committee; Association of College Admission Counselors; Australian Conciliation and Arbitration Commission; Australian Corporate Affairs Commission

ACACA Army Command and Administration Communication Agency

acad academic; academician; academy

Acad Acadia; Academy

Acad Academia (Latin—academy)

ACAD American Conference of Academic Deans; Autodesk Incorporated

Acad aper Academy of Motion Picture Arts and Sciences aperture (of sound films)

Acad B-A Académie des Beaux-Arts (French—Academy of Fine Arts)

Academic Academic Press

Acad Fran Académie Française (French Academy)

ACADI Association des Cadres Dirigeants de l'Industrie (French—Association of Industrial Executives)

Acadia national park occupying Mount Desert Island, half of Isle au Haut, and Schoodic Point on the Maine coast; old name for French-speaking Canada and Nova Scotia

Acadia(n) Novia Scotia(n); Novia Scotia(n); native Louisianians of French origin

Acad Ins B-L Académie des Inscriptions et Belles-Lettres (French—Academy of Inscriptions and Literature)

Acad mask Academy of Motion Picture Arts and Sciences mask (enclosing the aperture area of sound films)

Acad Med Academy of Medicine

Acad Mgmt Academy of Management

Acad Mus Academy of Music

Acad Pr Academic Press

Acad Pr Ark Academic Press of Arkansas

Acad Sci Académie des Sciences (French—Academy of Science)

Acad Sin Academia Sinica (Chinese Academy of Science)

Acad St Cec Academia di Santa Cecilia, Rome

Acad Therapy Academic Therapy Publications

Acad U Acadia University

ACAE American Council for the Arts in Education; Australian Commission on Advanced Education

AC & AE Association of Chemical and Allied Employees

acaf automatic circuit assurance feature

ACAF Amphibious Corps Atlantic Fleet; Australian Citizen Air Force

ACAG Allied Control Authority for Germany

ACAM Asian Centre of Agricultural Machinery; American College for the Advancement of Medicine; Australian Confederation of Apparel Manufacturers

ACAN Action Committee Against Narcotics; Army Command and Administrative Network

ACAnes American College of Anesthetists

a. cant. after cant frames

Acanth Acanthocephala

acanthite silver sulfide

Acap Acapulco

acap automatic circuit analysis program

ACAP American Council on Alcohol Problems; Annapurna Conservation Area Project; Army Contract Appeals Panel

ACAPA American Concrete Agricultural Pipe Association

Acap gold Acapulco gold (high-grade marijuana of the type grown near the Mexican resort of Acapulco)

a capp *a cappella* (Italian—in chapel style, without musical accompaniment)

Acapulco Acapulco de Juárez (Mexico's leading seaside resort)

ACAR Australian Coal Association Research

ACARD Advisory Council for Applied Research and Development

ACAs Arms Control Associations

ACAS Aboriginal Children's Advancement Society; Advisory, Conciliation, and Arbitration Service; Airborne Collision Avoidance System; Association of Casualty Accounts and Statisticians

AC/AS Assistant Chief of Air Staff

ACAS *Associación Civil Amigos de la Salud* (Spanish—Friends of Health Civil Association)

ACASP Australian Commonwealth Association of Simplified Practice

ACAST Advisory Committee on Applications of Science and Technology (UNESCO)

ACAT Accreditation Council for Accountancy and Taxation

acata acatalectic(al)

ACAUS Association of Chartered Accountants in the United States

ACAV Armored Calvary Assault Vehicle

acb air circuit breaker; asbestos cement board

acb (ACB) aortocoronary saphenous vein bypass; arterialized capillary blood

ACB Advertising Checking Bureau; Advisory Committee on Banking; Airfield Construction Branch; Airman Classification Battery; Amphibious Construction Battalion; Army Classification Battery; Associated Credit Bureaus; Association of Concert Bands; Association of Customers' Brokers; Association of the Customs Bar; Australian Cricket Board

ACB *Association Canadienne des Bibliothèques* (French—Canadian Library Association)

ACBA Academy of Comic Book Artists

ACB of A Associated Credit Bureaus of America

acbad active commission base date

ACBB American Council for Better Broadcasts

ACBCC Australia-China Business Cooperation Committee

ACBFC Academy of Comic-Book Fans and Collectors

ACBL American Commercial Barge Line; American Contract Bridge League; Australian Contract Bridge League

acbm atomic cesium beam maser

ACBM Associated Corset and Brassiere Manufacturers; Aviation Chief Boatswain's Mate

ACBNZ Associated Credit Bureau of New Zealand

ACBO Association of Chief Business Officials

ACBs American Conference of Bishops; Associated Credit Bureaus

ACBS Accrediting Commission for Business Schools

ACBWS Automatic Chemical Biological Warning System

acc accept; accident(al) (ly); accommodate(d); accommodation; accompanied; accompaniment; accompanist; according; account(ed); accounting; accusative; altocumulus castellatus (clouds); alveolar cell carcinoma; amphibious command car; anodal closure contraction; armored car company; artillery control computer; astronomical great circle course (ACC); automatic chroma circuit (tv)

acc (ACC) accumulator

Acc *Lucius Accius* (Latin—name of a tragic poet)

ACC Abilene Christian College; Academy of Canadian Cinema; Accra, Ghana (airport); accumulator; Adirondack Community College; Administrative Committee on Coordination; Air Center Commander; Air Control Center; Air Coordinating Committee; Airport Consultants Council; Allied Control Center; Allied Control Commission; Allied Control Council; Ambulatory Care Center; American Cat Council; American College of Cardiology; American Concert Choir; American Conference of Cantors; American-Chilean Council; American Copper Council; American Craftsmen's Council; Anglican Consultative Council; Army Cadet College; Army Catering Corps; Army Chemical Center; Army Cooperation Command; Assistant Chief Constable; Association of Choral Conductors; Atlantic Coast Conference; Auburn Community College; Australian Chamber of Commerce; Australian Chiropody Council; Australian Computer Conference; Australian Croquet Council

A-C-C Appleton-Century-Crofts

ACCA Aeronautical Chamber of Commerce of America; Air Conditioning Contractors of America; American Clinical and Climatological Association; American College of Clinic Administrators; American Commercial Collectors Association, American Corporate Counsel Association; American Correctional Chaplains Association; American Cotton Cooperative Association; Art Collectors Club of America; Associated Chambers of Commerce of Australia

Accad Accadèmia (Italian—academy)

Accad Ball Accadèmia di Ballo (Italian—dancing academy)

Accad di Mil Accadèmici di Milano (Italian—Academicians of Milan)

Accad Mus Nap Orch Cam Accadèmia Musicale Napoletana Orchestra de Camera (Italian Neapolitan Musical Academy Chamber Orchestra)

Accad Nay Accadèmia Navale (Italian—naval academy)

Accad Sta Cec Accadèmia di Santa Cecilia (Italian—Academy of Santa Cecilia), musical academy and orchestra in Rome

ACCAP Autocoder-to-Cobol Conversion Aid Program

ACCAs American Correctional Chaplains Association members

acc & aud accountant and auditor

ACCC Ad Hoc Committee for Competitive Communications; Advisory Council on College Chemistry; Alternate Command and Control Center; American Council of Christian Churches; Association of Canadian Community Colleges; Association of Community Cancer Centers

ACCCA Association of California Community College Administrators

ACCCE Association of Consulting Chemists and Chemical Engineers

ACCCF American Concert Choir and Choral Foundation

Acc Chem Res Accounts of Chemical Research

ACCCI Associated Chinese Chambers of Commerce and Industry (Singapore); Australia-China Chamber of Commerce and Industry

AC & CCI American Coke and Coal Chemicals Institute

accd accelerated construction completion date

acc dec acceptable deception(s)—half truth(s)

acce acceptance

ACCE American Chamber of Commerce Executives; American Council for Construction Education

accel accelerate; accelerate(d); accelerating; acceleration

accel accelerando (Italian—accelerating)

ACCELS Automated-Circuit Card-Etching Layout System

accepon acceptation (French—acceptance)

access accessory

ACCESS Advanced Customers Connections, an Evolutionary Systems Strategy; Aircraft Communication Electronic Signaling System; American College of Cardiology Extended Study Services; Architects Central Constructional Engineering Surveying Service; Association of Community Colleges for Excellence in Systems and Services; Automated Catalog of Computer Equipment and Software Systems (USA); Automatic Computer Controlled Electronic Scanning System

A.C.C.E.S.S. A Cooperative Community Educational School System

ACCF American Committee for Cultural Freedom; American Council for Capital Formation; Association of Community College Facilities

ACCFA Agricultural Credit Cooperative Finance Administration

AcCH acetylcholine

ACCH Association for the Care of Children in Hospitals

ACCHAN Allied Command Channel (NATO)

acci accidental injury

ACCI American Cottage Cheese Institute; Association of Crafts & Creative Industries

accid accident(al)

ACCION Americans for Community Cooperation in Other Nations

accis accismus

ACCIS Automated Command and Control Information System

accl anodal closure clonus

ACCL American Council of Commercial Laboratories

ACCM American College of Clinic Managers

ACCM P-H Appleton-Century-Crofts Medical (imprint of) Prentice-Hall

ACCME Accreditation Council for Continuing Medical Education

ACCN American Court and Commercial Newspapers

ACCNET Army Command and Control Network

acco account

acco accompagnamento (Italian—accompaniment)

ACCO Associate of the Canadian College of Organists; Association of Child Care Officers

AcCoA acetyl coenzyme A

accom accommodation; accompaniment

accom ad lib accompaniment ad libitum

accom oblto accompaniment obligato

accomp accompaniment; accomplish

ACCOMP Academic Computer Group

ACCOR Australian Coal Corporation

ACCORD Action Coalition to Create Opportunities for Retirement with Dignity

accord accordéon (French—accordion)

ACCORDS Acoustic Correlation and Detection System

Accountemps company providing temporary financial, banking, EDP, credit, bookkeeping and accounting services

ACCIP American College of Chest Physicians

ACCP *Associación de Camaras de Comercio del Perú* (Spanish—Peru Association of Chambers of Commerce)

ACCR American Council on Chiropractic Roentgenography

ACCRA Abortion and Contraception Counselling and Research Association; American Chamber of Commerce Researchers Association; Australian Chart and Code for Rural Accounting; Australian Committee for Coding Rural Accounts

accrd int accrued interest

accred accredited

accres accrescendo or *accresciménto* (Italian—augmented or increasing)

ACCS Acoustic Communication with Submarines; Associate of the Chartered Corporation of Secretaries; Automated Calibration Control System

ACCSA Allied Communications and Computer Security Agency

acct account; accountant; accounting

ACCT Association of Community College Trustees

acctd accented

acctg accounting

acctncy accountancy

accts pay accounts payable

accts rec accounts receivable

accu accurate; automatic combustion-control unmanned

a-c cu alternating-current control unit

accum accumulate

accur accuratissime (Latin—most accurately)

accus accusative

ACCUSE Action Committee for a United States of Europe

accv armored cavalry cannon vehicle

accw alternating current continuous wave

accy accessory

acd absolute cardiac dullness; accord; accordion; acid-citrate-dextrose; active duty commitment; adopted child; advance delivery of correspondence; advice of duration and charges; anodal duration contraction; average daily census; axiodistocervical

acd (ACD) acid citrate dextrose; arms control through defense

ACD Administrative Commitment Document; Allied Chemical Corporation (stock exchange symbol); American Center for Design; American Choral Directors; American College of Dentists; Army Chaplain's Department; Australian College of Dentistry; Australian College of Dermatologists

ACD American College Dictionary

ACDA Advisory Committee on Distinction Awards; American Choral Directors Association; American Component Dealers Association; Arms Control and Disarmament Agency; Asian Centre for Development Administration; Aviation Combat Development Agency

a-c/d-c alternating current/direct current; slang—bisexual

ACDC Australian Counter-Disaster College

ACDCM Archbishop of Canterbury's Diploma in Church Music

ACDE American Council for Drug Education

ACDFA American College Dance Festival Association

ac dis academic dismissal

acdl asynchronous circuit-design language

ACDM Association of Chairmen of Departments of Mechanics

ACDMS Automated Control of Document Management System

ACDPI American Cultured Dairy Products Institute

A Cdre Air Commodore

ACDS Advanced Combat Direction System; Assistant Chief of Defense Staff (Operational Requirements); Australian College of Dental Surgeons

acdt accident

A Cdt Air Commandant

acdu active duty

acdutra active for training

ACDUTRA Active Duty Reserve Army

ace acceptance checkout equipment; acetic; active cornering enhancement; adjusted current earnings; adrenal cortical extract; advanced computing environment; aerospace control environment; air crash equipment; alcohol-chloroformether (anesthetic mixture); animated computer education; assessment of combat effectiveness; attitude control electronics; automatic checkout equipment; automatic circuit exchange; automatic computer evaluation; automatic computing engine

ace (ACE) angiotensin converting enzyme

ACE Action by the Community relating to the Environment; Active Corps of Executives; Advanced Compostition Explorer; African Container Express; Allied Command, Europe; American Cinema Editors; American Coaster Enthusiasts; American Conservatory Theatre; American Council on Education; American Council on Exercise; American Hard Rubber Company (trademark); AMEX Commodities Exchange; Army Certification of Education; Army Corps of Engineers; Assistant City Editor; Association for Community Education; Australian College of Education; Association of Collegiate Entrepreneurs; Auxiliary Corps of Executives; Aviation Construction Engineers

ACEA Air Line Communication Employees Association; American Cotton Exporters Association; Association of Consulting Engineers—Australia

ACEAA Advisory Committee on Electrical Appliances and Accessories

acearts airborne countermeasures environment and radar target simulator

ACEB Army Combat Engineer Battalion

ACEB *Association Canadienne des Écoles Bibliothecaires* (French Canadian Association of Library Schools)

Ace Bks Ace Books

ACEC Ada Compiler Evaluation Criteria; Alcoholism Counseling and Education Center; American Consulting Engineers Council; Army Communications and Electronic Command; Army Communications and Electronic Command (USA)

ACEC *Ateliers de Constructions Électriques de Charleroi* (French—Electrical Construction Workshops of Charleroi)—in Belgium

A.C.Ed. Associate in Commercial Education

ACED Advanced Communications Equipment Depot

ACEEE American Council for an Energy Efficient Economy

ACEF Aboriginal Child Education Fund(ing); Aboriginal Children's Education Fund; Asian Cultural Exchange Foundation; Association of Commodity Exchange Firms; Australian Council of Employers Federations

a-c-e-g (musical mnemonic—all cows eat grass)—bass clef note names of the four spaces

acei angiotensin converting enzyme inhibitor

ACEI Association of Consulting Engineers of Ireland; Association for Childhood Education International

ACEID Asian Centre of Educational Innovation for Development

ACEJ American Council on Education for Journalism

ACEJMC Accrediting Council on Education in Journalism and Mass Communications

ACEL Air Crew Equipment Laboratory

ACELF *Association Canadienne des Éducateurs de Langue Française* (French—Canadian Association of French Language Teachers)

ACEM Aviation Chief Electrician's Mate

a cemb *a cembalo* (Italian—by the harpsichord)

ACE(MF) Army Command Europe (Mobile Force)

ACEN Assembly of Captive European Nations

ACENET Allied Command Europe Communications Network (NATO)

ACENZ Association of Consultant Engineers of New Zealand

ACEORP Automotive and Construction Equipment Overhaul and Repair Plant

ACEP American College of Emergency Physicians; American Council for Emmigrés in the Professions

ACEPD Automotive and Construction Equipment Parts Depot

ACEQ Association of Consulting Engineers of Québec

ACER Australian Council for Educational Research

ACERP Advanced Communications-Electronics Requirements Plan

ACERT Advisory Council for Education of Romanies and other Travellers

aces adjustable convertible-rate equity security (units); aircraft ejection seat; automatic control evaluation simulator

ACES Alternative Consumer Energy Society; Americans for the Competitive Enterprise System; Area Cooperative Educational Service; Association for counselor Education and Supervision; Australian Council for Educational Standards

ACESA Arizona Council of Engineering and Scientific Associations; Australian Commonwealth Engineering Standards Association; Australian Computer Equipment Suppliers Association

ace-s/c acceptance checkout equipment—spacecraft

ACESIA American Council for Elementary School Industrial Arts

acet acetone

ACET Advisory Committee on Electronics and Telecommunications; Australian Council for Education through Technology

ACETA Australian Commercial and Economics Teachers Association

ACE Test American Council on Education Test

acetl acetylene

acetyl-co acetyl-coenzyme A

ACEUR Allied Command, Europe

ACEWR American Committee for European Worker's Relief

acf accessory clinical findings; advanced communications function; aircraft components flight; area confinement facility; automatic claim filing

acf (ACF) air-combat fighter (aircraft)

ACF Active Citizen Force; Advisory Committee on Foodstuffs; African Colonial Forces; Alternate Communications Facility; American Car & Foundry; American Checker Federation; American Chess Foundation; American Choral Foundation; American Culinary Federation; Anglican Charismatic Fellowship; Anti-Crime Foundation; Army Cadet Force; Army College Fund; Army Council Form; Association of Consulting Foresters; Australian Canoe Federation; Australian Chess Federation; Australian Comforts Fund; Australian Conservation Federation

ACF *Académie Canadienne Française* (French—Canadian Academy); *Automobile-Club de France* (French—Automobile Club of France.)

ACFA American Cat Fanciers Association; Association of Commercial Finance Attorneys

ACFAS *Association Canadienne-Française pour l'Avancement des Sciences* (French—French-Canadian Association for the Advancement of Science)

ACFB Australian Canned Fruits Board

ACFC Aviation Chief Fire Controlman

ACFE Air Command Far East

ACFEA Air Carrier Flight Engineers Association

ACFEL Arctic Construction and Frost Effects Laboratory (Greenland)

ACFES Association of Canadian Faculties of Environmental Studies

acfg automatic continuous function generation

ACFHE Association of Colleges for Further and Higher Education

ACFL Atlantic Coast Football League

ACFM Association of Canadian Fire Marshals

ACFN American Committee for Flags of Necessity

ACFO American College of Foot Orthopedics

ACFOD Asian Cultural Forum for Development (Thailand)

ACFP Advisory Commission on Federal Pay

ACFR Advisory Committee of the Federal Register; Advisory Council on Federal Reports; Air Combat Fighter Radar; American College of Foot Roentgenologists

ACFS American College of Foot Surgeons; Air Combat Fleet Support

ACFSA American Correctional Food Service Association

acft aircraft

ac ft acre feet; acre foot

ACFT Aircraft Flying Training

acftc aircraft carrier

ACFTU All China Federation of Trade Unions (PRC)

acg automatic caution guard; automatic control gear

acg (ACG) apex cardiogram

ac-g accelerator globulin

ACG African Cavalry Guard; Airborne Coordinatimig Group; Air Cargo Express (symbol); Airline Carriers of Goods; American College of Gastroenterology; American Council on Germany; Amphibious Control Group; Assistant Chaplain-General; Assistant Commissary-General; Association for Corporate Growth

ACG An Comunn Gaidhealach (The Gaelic, Society)—also called the Highland Society

ACGA American Cranberry Growers' Association; Arizona Cotton Growers Association; Australian Cane Growers Association; Australian Citrus Growers Association

ACGB Arts Council of Great Britain

ACGBI Automobile Club of Great Britain and Ireland

ACGC Australian Cane Growers Council

ACGD Association for Corporate Growth and Diversification

ACGF American Child Guidance Foundation

ACGFC Associate of the City and Guilds Finsbury College

ACGG American Custom Gunmakers Guild

ACGI Associate of the City and Guilds Institute

ACGIH American Conference of Governmental Industrial Hygienists

ACGLA Alcoholism Council of Greater Los Angeles

ACGM Aircraft Carriers General Memorandum

ACGP Army Career Group; Army Carrier Group; Australian College of General Practitioners

ACGPOMS American College of General Practitioners in Osteopathic Medicine and Surgery

ACGS Aerial Cartographic and Geodetic Squadron; American Council on German Studies

ACGSq Aerial Cartographic and Geodetic Squadron (USAF)

ach acetylcholine (Ach); actual obtained achievement; aircraft hand; arm, chest, height; area combined headquarters

ach (Ach) (ACH) acetylcholine; adrenal cortical hormone; automated clearing house

ACh acetylcholine

ACH Association for Computers and the Humanities; Association of Cosmetologists and Hairdressers; Australian Camp Hospital; Australian Commonwealth Horse

ACHA American Catholic Historical Association; American College Health Association; American College of Hospital Administrators; American Council of Highway Advertisers

AC & HBR Algoma Central and Hudson Bay Railway

ACHCA American College of Health Care Administrators

AChD Army Chaplain's Department

ache acetylcholinesterase

ACHE Alabama Commission on Higher Education; American Council for Headache Education

achiev achievement

ach index arm (girth), chest (depth), hip (width) index (of nutrition)

ACHNHP Appomattox Court House National Historical Park

ACHOO Autosomal-dominant Compelling Helio-Ophthalmic Outburst

ACHP Advisory Council for Historic Preservation

ACHPER Australian Council for Health, Physical Education, and Recreation

achr acetylcholine receptor

AChR acetycholine receptor (antibody)

ACHR American Council of Human Rights

achrom achromatism

A Ch S Associate of the Society of Chiropodists

ACHS Association of College Honor Societies

ACHSA American Correctional Health Services Association

ACHTR Advisory Committee for Humid Tropics Research (UNESCO)

ACHU Aircrew Holding Unit

achvit achievement

aci airborne-controlled interception; automatic card identification

aci (ACI) adult correctional institution; anticlonus index

aci *assure contre l'incendie* (French—insured against fire)

ACI Air Cargo Incorporated; Air Combat Information; Air Combat Intelligence; Air

Council Instructions; Alliance Coopérative Internationale (International Cooperative Alliance); Alloy Casting Institute; American Carpet institute; American Concrete Institute; American Cryogenics Incorporated; Army Council Instructions; Associated Colleges of Indiana; Association of Commerce and Industry; Australian Consolidated Industries

ACI Association Cartographique Internationale (French—International Cartographic Association); *Azione Cattolica Italiana* (Italian Catholic Action)

acia asynchronous communications interface adapter

ACIA Associate of the Catering Institute of Australia; Associated Cooperage Industries of America

ACIAA Australian Commercial and Industrial Artists' Association

ACIAS American Council of Industrial Arts Supervisors

ACIASAO American Council of Industrial Arts State Association Officers

ACIATE American Council of Industrial Arts Teacher Education

ACIB Associate of the Corporation of Insurance Brokers

acic acicular

ACIC Aeronautical Chart and Information Center; Allied Captured Intelligence Center; Australian Chemical Industry Council; Auxiliary Combat Information Center

ACICU Arkansas Council of Independent Colleges and Universities

acid acidosis; acidulated drop; automatic classification and interpretation of data; hallucinogenic drug such as LSD-25

acid phos acid phosphatase

acid p'tase acid phosphatase

ACIF All Canada Insurance Federation

ACIGS Assistant Chief of the Imperial General Staff

ACH Associate of the Chartered Insurance Institute

ACIID A Critical Insight Into Israel's Dilemmas

ACIL American Council of Independent Laboratories

acim axis-crossing interval meter

ACIM American Committee on Italian Migration

ACIMS Aircraft Component Management System

ACIO Air Combat Intelligence Office(r); Army Careers and Information Office

acip aviation career incentive pay

ACIP Advisory Committee on Immunization Practices; American College of International Physicians, American Council on International Personnel

ACIPCO American Cast Iron Pipe Company

ACIR Advisory Committee on Intergovernmental Relations; Automotive Crash Injury Research

AC/IREF American Chapter International Real Estate Federation

ACIRL Australian Coal Industries Research Laboratories

ACIS American Committee for Irish Studies; Amphibious Command Information System; Arms Control Impact Statement; Associate of the Chartered Institute of Secretaries; Association for Computing and Information Sciences

acit air-cannon impact tester

ACIU Allied Central Interpretation Unit

ACIV Associate of the Commonwealth Institute of Valuers

ACIWLP American Committee for International Wild Life Protection

ACJ American Council for Judaism

A.C.J. Associate in Criminal Justice

ACJA American Criminal Justice Association

ACJA-LAE American Criminal Justice Association—Lambda Alpha Epsilon

ACJC Assembly Criminal Justice Committee

ACJHSIS Arkansas Criminal Justice Highway Safety Information System

ACJP Airways Corporations Joint Pensions

ACJS Academy of Criminal Justice Sciences

ack acknowledge; acknowledgment; acknowledgment code

ACK accidentally killed; acknowledge; Armstrong Cork (stock exchange symbol)

ack-ack antiaircraft

Ack-Ack Aluminum Company of America (stock exchange nickname)

ackd acknowledged

ackt acknowledgment

acl air-cushion landing; allowable cabin load; application control language

aCl aspiryl chloride

Acl Atlas commercial language

ACL Adjective Check List; Aeronautical Computers Laboratory (USN); American Canadian Line; American Classical League; American Consultants League; American Cruise Line; Association of Cinema Laboratories; Association for Computational Linguistics; Atlantic Coast Line (railroad); Atlantic Container Line; Audit Command Language; Aviation Circular Letter

ACL Automobile Club de Luxembourg (French—Automobile Club of the Grand Duchy of Luxembourg)

ACLA American Citizen's & Lawmen's Association; American Comparative Literature Association; American Cotton Linter Association; Anti-Communist League of America; Australasian Communications Law Association

ACLAM American College of Laboratory Animal Medicine

AClant Allied Command, Atlantic

ACLC Air Cadet League of Canada; Assessment of Children's Language Comprehension

acld aircooled

ACLD Association for Children with Learning Disabilities

aclg air-cushion landing gear

ACLI American Council of Life Insurance

ACLICS Airborne Communications Location, Identification, and Collection System

ACLM American College of Legal Medicine

ACLN American Catholic Lay Network

ACLO Association of Cooperative Library Organizations

aclos (ACLOS) automatic command to line of sight (missile)

ACLP Association of Contact Lens Practitioners

ACLR American Criminal Law Review

acls air-cushion landing system; automatic carrier landing system

ACLS All-weather Carrier Landing System; American Council of Learned Societies; Automatic Carrier Landing System

ACLU American Civil Liberties Union; American College of Life Underwriters; Atlantic Container Line Unit; Australian Civil Liberties Union

aclv accrued leave

acm active countermeasure(s); additional crew members; advanced cluster munition; air combat maneuvering; air cycle machine; anatomy-covering material; anatomy-covering memo; asbestos-covered metal

acm (ACM) advanced cruise missile; anti-armor cluster munition

a-c-m albumin-calcium-magnesium

Ac M Academy of Music, Philadelphia

ACM Air Chief Marshal; Air Commerce Manual; Air Court-Martial; American Campaign Medal; American College of Musicians; American Craft Museum, Associated Colleges of the Midwest; Association for Computing Machinery; Australian Carpet Manufacturers; Australian Consolidated Minerals; auxiliary mine layer (3-letter symbol); Aviation Chief Metalsmith

ACM Automobile Club de Monaco (French—Automobile Club of Monaco)

ACMA Acidproof Cement Manufacturers Association; Air Carrier Mechanics Association; Alumina Ceramic Manufacturers Association; American Certified Morticians Associations; American Circus Memorial Association; American Comedy Museum Association; American Cutlery Manufacturers Association; Associated Chambers of Manufacturers of Australia

ACMC Area Combined Movements Center; Automotive Chemical Manufacturers Council

acme advanced computer for medical (research); attitude control and maneuvering electronics

ACME Adult Community Movement for Equality; Advisory Council on Medical Education; Association of Consulting Management Engineers; Australian Contemporary Music Ensemble

ACMET Advisory Council on Middle East Trade

ACMF Air Corps Medical Forces; Allied Central Mediterranean Forces; American Corn Millers' Federation; Australian Citizen Military Forces; Australian Commonwealth Military Forces

ACM/GAMM Association for Computing Machinery/German Association for Applied Mathematics and Mechanics

ACMI Air Combat Maneuvering Instrumentation system; American Cotton Manufacturers Institute; American Cystoscope Makers, Incorporated; Art and Craft Materials Institute

ACML Association of Canadian Map Libraries

ACMM Aviation Chief Machinist's Mate

A.C.M.M. Associate of the Conservatorium of Music—Melbourne

ACMO Afloat Communications Management Office; Allied Communication Management Organization

acmp accompany

ACMP Amateur Chamber Music Players; Assistant Commissioner of the Metropolitan Police

ACMR Air Combat Maneuvering Range

ACMRR Advisory Committee on Marine Resources Research (FAO)

acmru audio commercial-message repeating unit

ACMS Advanced Configuration Management System; Army Command Management System; Australian Clay Minerals Society; Automated Career Management System

ACMT Advanced Cruise-Missile Technology; American College of Medical Technologists

ACMWA Amon Carter Museum of Western Art (Fort Worth)

acn activities control number; acute conditioned necrosis (ACN); all concerned notified; assignment control number (ACN); automatic celestial navigation (ACN)

ACN Air Commander Norway; American Chain & Cable (stock exchange symbol); American College of Neuropsychiatrists; American Comedy Network; American Council on NATO; Authorized Code Number

A.C.N. Ante Christum Natum (Latin—before the birth of Christ)

ACNA Advisory Council on Naval Affairs; Arctic Institute of North America

ACNAM Associations Council of the National Association of Manufacturers

ACNAS Admiral Commanding North Atlantic Station

ACNB Australian Commonwealth Naval Board

ACNE Action Committee for Narcotics Education; Alaskans Concerned for Neglected Environments

ACNHA American College of Nursing Home Administrators

ACNIL *Azienda Comunale Navigazione Interna* (Italian—City Rapid Transit Shipping Company)

ACNM American College of Nurse Midwifery; American College of Nurse-Midwives

ACNN Air Commander Northern Norway

ACNO Assistant Chief of Naval Operations

ACNOCOMM Assistant Chief of Naval Operations for Communications and Cryptology

ACNOT Assistant Chief of Naval Operations—Transportation

ACNP Assistant Chief of Naval Personnel

ACNS American Council for Nationalities Service; Assistant Chief of Naval Staff; Associated Correspondents News Service

ACNUR *Alto Comisionado de las Naciones Unidas para las Refugiados* (Spanish—High Commission of the United Nations for Refugees)

ACNY Advertising Club of New York

ACNYC Art Commission of New York City

aco anodal closing odor

a co a cargo (Spanish—against)

ACO Administrative Contracting Officer; Admiralty Compass Observatory; Airborne Control Officer; Air Cargo (Leopoldville—Republic of the Congo); Air Controller Officer; American Academy of Optometry; American College of Orgonomy; Army Careers Officer; Arts Center of the Ozarks; Australian College of Ophthalmologists; Australian College of Organists

ACOA Administrative and Clerical Officers Association; Adult Children of Alcoholics

ACOC Air Command Operations Center

ACOCA Army Communication Operations Center Agency

acodac acoustic data capsule

ACODS Army Container Oriented Distribution System (USA)

ACOEM American College of Occupational and Environmental Medicine

ACofN Assistant Controller of the Navy

ACOFO American College of Foot Orthopedists

ACofS Assistant Chief of Staff

ACOFS Australian Council of Film Societies

acog (ACOG) aircraft on ground

ACOG American College of Obstetricians and Gynecologists

ACOHA American College of Osteopathic Hospital Administrators

ACOI American College of Osteopathic Internists

acol(s) acolyte(s)

acom automatic coding machine

ACOM Assistant Chief Ordnanceman; Association for Convention Operations Management; Aviation Chief Ordnanceman

A.Comm. Associate in Commerce

A.Comm.A. Associate of the Society of Commercial Accountants

ACOMPLIS A Computerized London Information Service (GLC)

ACOMR Advisory Committee on Oceanic Meteorological Research (UN)

Acon Aconcagua (South America's highest mountain in the Andes where it rises in Argentina and towers over Valparaiso, Chile)

ACONA Advisory Council on Naval Affairs

ACOOG American College of Osteopathic Obstetricians and Gynecologists

ACOP Airborne Corps Operations Plan; American College of Osteopathic Pediatricians; Association of Chief Officers of Police (England and Wales)

acopp abbreviated Cobol preprocessor

ACOPS Advisory Committee on Oil Pollution of the Sea

ACOR Auditory Cognition of Relationships

ACORD Advisory Council on Research and Development

ACORDE A Consortium on Restorative Dentistry Education

acorn. acronym-oriented nut

ACORN Association of Community Organizations for Reform Now; Associative Content Retrieval Network

ACOS American College of Osteopathic Surgeons; Assistant Chief of Staff

ACOSH Appalachian Center for Occupational Safety and Health; Australian Council on Smoking and Health

ACOT Assistant (Chief of Staff) Organization and Training

acous acoustical; acoustics

acous coup acoustic coupler (device allowing other electronic devices to listen to or make other sounds transmitted by an ordinary telephone)

acousid acoustic seismic intrusion detector

acoust acoustical(ly), acoimstic(s)

acoustint acoustical intelligence

acp acyl-carrier protein (ACP); acid phosphatase; aerodrome control point; aerospace computer program; after conning position; alternate command post; anodal closing picture; aspirin, caffcine, phenacetin; auxiliary control panel; azimuth change pulse

a-c-p aspirin-caffeine-phenacetin

a & cp anchors and chains proved

aCp (ACP) automatic Colt pistol

ACP Academy for Contemporary (Criminal) Problems; African, Caribbean, and Pacific countries; Agricultural Conservation Program; Air Control Point; Airline Carriers of Passengers; Allied Communications Procedure; Allied Communications Publications; American College of Pharmacists; American College of Physicians, American College of Prosthodontists;

Anti-Comintern Pact; Area Characteristic Plan(ning); Area Community Physician; Associated Church Press; Associated Collegiate Press; Association of Clinical Pathologists; Association of Correctional Psychologists; Association of Coupon Processors; Australian Conservative Party; Australian Consolidated Press; Automatic Colt Pistol

ACP *Automóvel Clube de Portugal* (Portuguese—Automobile Club of Portugal)

ACP80(s) Air-Cargo Processing in the '80s

ACPA Affiliated Chiropodists Podiatrists of America; American Capon Producers Association; American Cleft Palate Association; American College Personnel Association; American Concrete Paving Association; American Concrete Pipe Association; American Concrete Pumping Association; American Crop Protection Association; Association of Computer Professionals Australia; Association of Computer Programmers and Analysts; Australasian Corporation of Public Accountants; Australian Clay Products Association

A-CPA Asbestos-Cement Products Association

ACPAE Association of Certified Public Accounts Examiners

a/c pay accounts payable

ACPC American College of Probate Counsel; American Council of Parent Cooperatives

ACPCACP Atlantic, Caribbean, and Pacific Countries Association of Canadian Publishers

ACPD Anti-trust and Consumer Protection Division; Army Control Program Directive

ACPE American Council on Pharmaceutical Education; Association for Continuing Professional Education; Australian College of Physical Education

ACPF Amphibious Corps, Pacific Fleet

ACPFB American Committee for Protection of Foreign Born

ACPI Aviation Crime Prevention Institute

ACPIC American Council for Private International Communications

ACPL American College of Probate Counsel; Assistant Controller, Personnel and Logistics

acpm attitude-control propulsion motor(s)

ACPM American College of Preventive Medicine; American Congress for Preventive Medicine

ACPMR American Congress of Physical Medicine and Rehabilitation

ACPO Association of Chief Police Officers

acpp (ACPP) adrenocorticopolypeptide

ACPP Assistant Chief of Staff, Plans and Policy

acp acoustical paper

ACPRA American College Public Relations Association

ACPs Area Concept Papers

ACPS American Coalition of Patriotic Societies; Arab Company for Petroleum Services (OPEC); Automatic Crash-Program Systems

ACPSAHMWA American Commission for the Protection and Salvage of Artistic and Historical Monuments in War Areas

acpt accept

ACPT Arizona Congress of Parents and Teachers

ACPTA Australian Commonwealth Post and Telegraph Association

acpu auxiliary computer power unit

ACPU Aerial Coast Patrol Unit

acq acquire; acquittal

ACQ *Australian Church Quarterly*

Acq Libr Acquisition(s) Librarian

ACQT Aviation Cadet Qualifying Test

acquis acquisition(s)

acr abandon call and retry; acrylic; adjusted community rate; advanced capabilities radar; aerial combat reconnaissance; air control and reporting; air control radar; airfield-controlled radar; ammunition condition report; antenna coupling regulator; anti-constipation regimen; approach control radar

acr (ACR) acre(age)

ACR Abstracts of Classified Reports; Admiral Commanding Reserves; Advanced Certified Rolfer; Advisory Commission on the Realm; Aircraft Control Room; Algoma Central Railway; Allied Commission on Reparations; American Academy in Rome; American College of Radiology; American College of Rheumatology; Area Characterization Report; Armored Car Regiment; Armored Calvary Regiment

AC & R American Cable and Radio (Corp)

ACR *American Criminal Review*

ACRA American Car Rental Association; American Collegiate Retailing Association; Association of Company Registration Agents

ACRB Aero-Club Royal de Belgique (Royal Belgian Aero Club); Army Council of Review Boards

acrc aircrew reception (receiving) center

ACRC Air Compressor Research Council

acrd accrued

ACRD Australian Council for Rehabilitation of the Disabled

ACRDC Australian Code of Residential Design and Construction

ACRE Acting Commander Royal Engineers; Action Committee for Rural Electrification; Automatic Call-Recording Equipment; Automatic Checkout and Readiness Equipment

a/c rec accounts receivable

ACREC American College of Real Estate Consultants

acre ft acre foot

ACRES Airborne Communication Relay Station

ACRFAET Aircraft Crash Rescue Field Assistance and Evaluation Team (USAF)

acrg acreage·

ACRI Air Conditioning and Refrigeration Institute; American Cocoa Research Institute

ACRiLIS Australian Centre for Research in Library and Information Science (Riverina College of Advanced Education)

acrim (ACRIM) active cavity radiometer irradiance

ACRIM Association for Correctional Research and Information Management

ACRL American Cities Racing League; Association of College and Research Libraries

ACRM American College of Radio Marketing; American Congress of Rehabilitation Medicine; Aviation Chief Radioman

acro acrobat(ic); acrophobe; acrophobia

ACRO Aircraft Control Room Officer; Area Chief Recruiting Officer

acrob acrobatics

acrol acrolect(ic) (al) (ly)

acromeg acromegaly

acron acronym

ACRONYM Allied Citizens Representing Other New York Minorities

acronymiz acronymization; acronymizing; acronymizor(s)

ACRONYMS Acceptable Contractions of Randomly Organized Names Yielding Meritorious Spontaniety

Acropolis Acropolis Hill, Athens, containing the Erectheum, the Parthenon, the Temple of Nike, the Acropolis Museum

across acrostic

ACRP Army Cost Reduction Program

ACRPC Arizona Cotton Research and Protection Council

ACRR American Council on Race Relations

acrs acreage

ACRS Accelerated Cost Recovery System; Advisory Committee on Reactor Safeguards; Association of Commercial Records Centers

ACRT Aviation Chief Radio Technician

acru avionics cooling refrigeration unit

ACRU Advisory Committee on Resource Use

acrv armored command and reconnaissance vehicle

acrw aircrew

ACRW American Council of Railroad Women

ACRY Australian Council of Rural Youth

acs accumulator switch; acoustic communication with submarines; airways communication station; alternating current synchronous; anodal closing sound; asynchronous communications server; autograph card signed; attitude command system; automatic checkout system; automatic control system; auxilliary core system

acs (ACS) antireticular cytotoxic serum; attitude-control system

a-c s (ACS) alternating-current synthesizer

Ac of S Academy of Sciences (USSR); Assistant Chief of Staff

ACs appeal cases

ACS Admiral Commanding Submarines; Admiralty Computing Service; Advanced Communications System; Agriculture Cooperative Service; Aircraft Carrier Squadron; Airline Charter Service; Alaskan Communications System; American Camellia Society, American Cancer Society; American Carnation Society; American Ceramic Society; American Cetacean Society; American Cheese Society; American Chemical Society; American College of Surgeons; American Colonization Society; American Crystal Sugar (company); Armament Control System; Army Community Service; Assistant Chief of Staff; Assistant Chief of Supplies; As-

sociated Counseling Services; Association of Clinical Scientists; Asynchronous Communications Server; Australian Cancer Society; Australian Ceramic Society; Australian Chamber of Shipping; Australian Cinematographers Society; Australian Computer Society; Automatic Control System

AC/S Assistant Chief of Staff

A.C.S. Associate in Commercial Science

ACS *Automobile Club de Suisse* (French—Automobile Club of Switzerland)

ACSA Allied Communications Security Agency (NATO); American Cotton Shippers Association; Association of California School Administrators; Association of Collegiate Schools of Architecture; Australian Corps of Signals Association

ACSB Australian Commonwealth Shipping Board; Aviation Candidates Selection Board

acsc automated contingency support capability

ACSC Air Carrier Service Corporation; Air Command and Staff College; American Council on Schools and Colleges; Association of Casualty and Surety Companies; Australian Coastal Shipping Commission

A-C Scale Anti-Caucasian Scale (measuring negative attitudes towards white persons)

ACSCE Army Chief of Staff for Communications Electronics

ACSCP Association of California State College Professors

ACS/DCI American Chemical Society/Division of Chemical Information

ACSDO Air Carrier Safety District Office(r)

ACSE Association of Consulting Structural Engineers

ACSEA Air Command-Southeast Asia; Allied Command South-East Asia

ACSEF Australian Coal and Shale Employees Federation

acsf artificial cerebrospinal fluid

ACSF Attack Carrier Striking Force

a-c sg alternating-current signal generator

ACSH American Council on Science and Health

ACSI American Consumer Satisfaction Index; Assistant Chief of Staff for Intelligence; Automotive Cooling Systems Institute

ACSIL Admiralty Centre for Scientific Information and Liaison (United Kingdom)

ACSI-MATIC Assistant Chief of Staff-Intelligence (automatic processing system for intelligence information)

ACSL Assistant Cub Scout Leader

acsm acoustic (warfare) support measure(s); advanced conventional stand-off missile

Ac SM Academy Sergeant-Major

ACSM American College of Sports Medicine; American Congress of Surveying and Mapping

ACSMA American Cloak and Suit Manufacturers Association

ACSN Aspartame Consumer Safety Network; Association of Collegiate Schools of Nursing

ACSO Australian Council of School Organisations

ACSOC Acoustical Society of America

ACSOF Air Commando Strategic Offensive Forces

ACSP Advisory Council on Scientific Policy (United Kingdom); Association of Collegiate Schools of Planning

ACSPA Australian Council of Salaried and Professional Associations

A/cs Pay Accounts Payable

acsr aluminum cable, steel reinforced; aluminum conductor, steel-reinforced

A/cs Rec Accounts Receivable

ACSRNAS Armored Car Section Royal Naval Air Service

acss (ACSS) analog computer subsystem; automated color-separation system

ACSS Air Command and Staff School; American Cheviot Sheep Society; Army Chief of Support Services

ACSSAVO Association of Chief State School Audio-Visual Officers

ACSSB American Certified Shake and Shingle Bureau

ACSSN Association of Colleges and Secondary Schools for Negroes

ACSSO Australian Council of State School Organisations

ACSSRB Administrative Center of Social Security for Rhine Boatmen

ACSSS American Council of State Savings Supervisors

acst access time; acoustic; acoustical; acoustics

ACST Army Clerical Speed Test; Australian College of Speech Therapists

acst plas acoustical plaster

acst t acoustical tile

ACSW Academy of Certified Social Workers

acsyn aircraft synthesis

acsys accounting computer system software

act algebraic compiler and translator; apparatus carrier telephone; aspartate carbamoyl transferase

act. acting; action; activated coagulation time; active; actor; actress; actual; actuarial; actuary; actuate; actuating; anti-coagulant therapy; atropine coma therapy; azimuth control torquer

act. (ACT) advanced coronary treatment

ACT Access to Careers in Technology; Action for Children's Television; actual; Adaptive Control of Thought; Advanced Computer Techniques; Airborne Crew Trainer; Air Control Team; Air Council for Training; Alcohol Counseling and Treatment; algebraic compiler and translator; Allied Control Team; American Cell Technology; American College Testing (program); American Conservatory Theatre; Angli-

can Theological College (British Columbia); Armored Cavalry Trainer; Associated Community Theaters; Associated Container Transportation; Association of Christian Therapists; Association of Classroom Teachers; Australian Capital Territory; Australian College of Theology; Australian Commonwealth Territory; Automated Confirmation Transactions; automatic code translation; Aviation Classification Test

acta advanced combat training activity; all combat training area

a cta *a cuenta* (Spanish—on account)

ACTA Aircoach Transport Association; American Community Theatre Association; Associated Container Transportation Australia

Acta Chem Scand *Acta Chemica Scandinavia* (Latin—Scandinavian Chemistry Review)

Acta Crystallog *Acta Crystallographica* (Latin—Crystallography Review)

Acta Math Acad Sci Hung *Acta Mathematica Academiae Scientiarum Hungaricae* (Latin—Journal of the Hungarian Mathematics Academy)

Acta Metall *Acta Metallurgica* (Latin—Metallurgy Review)

Acta Oto-Laryngol *Acta Oto-Laryngologica* (Latin—Otolaryngology Review)

Acta Phys *Acta Physica* (Latin—Physics Review)

Acta Phys Austriaca *Acta Physica Austriaca* (Latin—Austrian Review of Physics)

Acta Phys Pol *Acta Physica Polonica* (Latin—Polish Review of Physics)

Acta Phys Sin *Acta Physica Sinica* (Latin—Chinese Journal of Physics)

ACTAR Asian Centre for Tax Administration and Research

ACTB Aircrew Classification Test Battery

ACTC Air Commerce Type Certificate; Air Corps Training Center; Australian Ceramic Tile Council

A.C.T.C. Art Class Teacher's Certificate

act.ct actual count

acte anodal closure tetanus

ACTF American College Theatre Festival

ACTFL American Council on the Teaching of Foreign Languages

actg acting

ACTG Advance Carrier Training Group

Actg Legal Adv Acting Legal Advisor

acth (ACTH) adrenocorticotrophic hormone

ACTI Advisory Committee on Technology Innovation

act/ic active-in commission

ACTIL Australian Cotton Textile Industries Ltd

ACTION American Council To Improve Our Neighborhoods

act/is active—in service

ACTIS Auckland Commercial and Technical Information Service

activ activity

ACTIV Army Concept Team in Vietnam

Activase see TPA

ACTL American College of Trial Lawyers

ACTM Association of Cotton Textile Merchants of New York

ACTMC Army Clothing, Textile and Material Center

actn (ACTN) adrenocorticotrophin

ACT—NOW AIDS Coalition to Network, Organize, and Win

actnt accountant

ACTNZ Associated Container Transportation New Zealand

acto automatic computing transfer oscillator

ACTO Action Office(r)—USA; Advisory Council on the Treatment of Offenders

act/oc active—out of commission

actol air-cushion takeoff and landing

Acton Lord Acton (1834–1902), remembered for his maxim: *Power tends to corrupt and absolute power corrupts absolutely*

ACTOR Askania cine-theodolite optical-tracking range

act/os active—out of service

actp (ACTP) adrenocorticotrophic polypeptide

ACTR Air Corps Technical Report

ACTRA Alliance of Canadian Television and Radio Artists

ACTRAN Autocoder-to-Cobol translating service

ACTREP Activities Report

actrl acoustic trial(s)

ACTRSWD Asian Centre for Training and Research in Social Welfare and Development

Acts The Acts of the Apostles

ACTS Acoustic Control and Telemetry System; Advanced Communications Tracking Satellite; Air Corps Tactical School; Airline Computer Tracing System; American Catholic Theological Society; American Christian Television System; Association of Career Training Schools; Automatic Cage Transmission System; Automatic Computer Telex System

ACT-SO Afro-Academic Cultural Technological and Scientific Olympics

ACTSU Association of Computer Time-Sharing Users

ACTT Association of Cinematograph and Television Technicians; America's Christmas Train and Trucks

ACTU Association of Catholic Trade Unionists; Australian Council of Trade Unions

ACTUP AIDS Coalition To Unleash Power

actv activate

actv (ACTV) armored cavalry towing vehicle

ACTV American Coalition for Traditional Values

act. vat actual value

act. wt actual weight

ACTWU Amalgamated Clothing and Textile Workers Union

ACTWUA Amalgamated Clothing and Textile Workers Union of America (formerly ACWA and TWUA)

acu acuity; aculeat; acumen; acuminate; acupressure; acupuncture; acute; address con-

trol unit; arithmetic computer; arithmetic and control unit; assault craft unit; automatic calling unit; avionics cooling unit

ACU Abilene Christian University; Abused Child Unit (Los Angeles Police Department); American Church Union; American Congregational Union; American Conservation Union; American Cycling Union; anti-crime unit; Army Canoe Unit; Asian Currency Unit; Association of College Unions; Association of Commonwealth Universities; Australian Church Union; Autocycle Union

ACUA Association of Cambridge University Assistants; Association of College and University Auditors

ACUCAA Association of College, University, and Community Arts Administrators

ACUCM Association of College and University Concert Managers

ACUE American Committee of United Europe

ACUERI American Conservative Union Education and Research Institute

ACUG Association of Computer User Groups

ACUHO Association of College and University Housing Officers

ACUI Association of College Unions International

ACUIIS Association of Colleges and Universities for International Intercultural Studies

ACULE Association of Credit Union League Executives

Acuña Ciudad Acuña (Mexican border town)

ACUNY Associated Colleges of Upper New York

ACUP Association of Canadian University Presses; Association of College and University Printers

acupunc acupuncture; acupuncturist

ACURA Association for the Coordination of University Religious Affairs

ACURL Association of Caribbean University and Research Libraries

ACUs Asian Currency Units

ACUS Administrative Conference of the United States; Atlantic Council of the United States

ACU-SACC American Conservative Union-Save Our Canal Committee

ACUSNY Association of Colleges and Universities of the State of New York

ACUTE Accountants Computer Users for Technical Exchange

acv actual cash value; air-cushion vehicle; alarm check valve; armored command vehicle

ACV Air-cushion vehicle; auxiliary aircraft carrier or tender (3-letter symbol)

ACVAFS American Council of Voluntary Agencies for Foreign Service

ACVC Ada Compiler Validation Criteria; American Council of Venture Clubs; Arms Control Verification Committee; Army Commercial Vehicle Code

acvd acute cardiovascular disease

ACVL Association of Cinema and Video Laboratories

ACVO American College of Veterinary Ophthalmologists

ACVP American College of Veterinary Pathologists

ACVS Australian College of Veterinary Scientists

ACVSF Air-Cooled Vault Storage Facility

acvt armored combat vehicle technology

acvws armored combat vehicle weapon system

acw aircraft control and warning; alternating continuous waves; automatic car wash

ac&w aircraft control and warning

ac/w acetone/water

ACW Air Control and Warning (system); Aircraftwoman; Alcoholism Center for Women; American Chain of Warehouses

AC & W Air Communications and Weather (naval group)

ACWA Amalgamated Clothing Workers of America; Australian Chinese Women's Association

A.C.W.A. Associate of the Institute of Cost and Work Accountants

ACWAI Automatic Car Wash Association International

ACWB Australian Council of Wool Buyers

acwbcn action will be cancelled

ACWC Advisory Committee on Weather Control

ACWF All-China Women's Federation; American Council for World Freedom; Army Central Welfare Fund

ACWL Army Chemical Warfare Laboratory

ACWM Americans for Customary Weights and Measures

ACWO Aircraft Control and Warning Officer

ACWRON Aircraft Control and Warning Squadron

ACWRRE American Cargo War Risk Reinsurance Exchange

ACWS Aircraft Control and Warning System

AC & WS Aircraft Control and Warning Station(s)

ACWW Associated Country Women of the World

acy average crop yield

ACY Akron, Canton & Youngstown (railroad); American Cyanamid Company (stock exchange symbol); Atlantic City, New Jersey (airport)

AC & Y Akron, Canton & Youngstown (railroad)

ACYF Administration for Children, Youth, and Families; Australian Council of Young Farmers

acyl-co A coenzyme A ester (general symbol for an organic compound)

acyro acyrologia; acyrologic(al); acyrology

ad a drink; a drug (addict); active duty; addict; advanced design; advertisement; advertising; advisory directive; aerodynamic decelerator; after drain; airborne doppler; aircraft depot; aircraft direction; air dispatch; air dried; airdrome; ammunition dump; ampere demand (meter); antidumping; area drain; armament depot; armored division; average deviation

ad (AD) address; aggregate demand; athletic director

ad. adopted

a-d advance-decline line

a/d altitude/depth; analog-to-digital

a & d ascending and descending

a & d (A & D) accounting and disbursing

'ad had

ad (Latin prefix—to, toward)—as in adhesion, admixture, adopt, adoring, etc.

a d *a droit* (French—to the right)

a.d. *airis dexter* (Latin—right ear)

a D *ausser Dienst* (German—retired)

Ad Ada; Adah; Adalbert; Adam; Adams; Adán; Addington; Addis; Addison; Adela; Adelaide; Adelard; Adelardo; Adelbert; Adele; Adelina; Adeline; Adelle; Adelsteen; Adeodato; Adlai; Adna; Adolf, Adolfine; Adolfo; Adolph; Adolphe; Adolpho; Adolphus; Adriaan; Adriaen; Adrian; Adriano; Adrianus; Adrien; Adrienne; Aedh; Alzheinmer's disease

AD Accession Document; Aden Airways; Air Defense; Air Department (Admiralty); Air Depot; Air Division; Airdrome; Airframe Design (division); Airworthiness Directive; Alliance Defence; Andorra (Internet code); Appellate Division; Army Depot; Art Deco; Artificer Diver; Assembly District; assistant director; associate director; Astia Document; Atlantic & Danville (railroad); Australian Dame (Dame of the Order of Australia); Australian Democrats; Australian dollar *(telex); Aviatsionnaya Diviziya* (Russian—Aviation Division); destroyer tender (naval symbol)

A-D Albrecht Dürer; Antonin Dvořák

A/D Air Depot; analog-to-digital

A & D Atlantic & Danville (railroad)

AD Acción Democratica (Spanish—Democratic Action Party)—Venezuela's democratic movement begun by Romulo Betancourt

A.D. *Anno Domini* (Latin—in the Year of our Lord); *Agnus Dei* (Latin—Lamb of God)

ad 2 vic. ad duas vices (Latin—for two doses, for two times)

ada action data automation; active duty agreement; actuarial data assembly; air defense artillery; airborne data automation; airtight drywall approach; average daily attendance; average deviation adjustment

ada (ADA) adenosine deaminase

ada adalah (Arabic—equity, justice); *adasi* (Turkish—island); American Dental Association (logotype)

Ada (ADA) computer programming language promoted by the Department of Defense and named for Augusta Ada Byron, Lord Byron's daughter, who designed instructions for a 19th-century mechanical analytical machine

Ad of A Archduchy of Austria; Archduke of Austria

ADA Air Defense Area; American Dairy Association; American Dehydrators Association; American Dental Association; American Dermatological Association; American Diabetes Association; American Dietetic Association; Americans for Democratic Action; Americans with Disabilities Act; Army Discipline Act; Assistant Defense Adviser; Assistant District Attorney; Atomic Development Authority; Australian Dental Association; Australian Department of Agriculture; Australian Design Award; Australian Draughts Association; Automatic Data Acquisition; Automobile Dealers Association

ADA# American Diabetes Association diet number

ADAA Air Defense Action Area; American Dental Assistants Association; Art Dealers Association of America; Australian Development Assistance Agency

ADABAS Adaptable Data Base System (trademark of the German Software AG)-AG standing for *Aktiengesellschaft* (German–jointstock company)

adac automatic direct analog computer

ADAC Acoustic Data Analysis Center; Administratively Directed Access Control; Air Defense Artillery Commander

ADAC Allgemeiner Deutscher Automobilclub (German Automobile Club)

adacc automatic data acquisition and computer complex

adacx automatic data acquisition and computer complex

ADAD Air Defense Artillery Director; Assistant Director, Air Department

ADADs Automatic Dialing and Announcing Devices

Adag adagio (Italian—slowly and expressively)

ADAG Architecture Drawings Advisory Group

ADAGP Association pour la diffusion des arts graphiques et plastiques (French—Association for the Diffusion of Graphic and Plastic Arts)

ADAH Assistant Director of Army Health

ADAIS Aerodynamic Data Analysis and Integration System

adal action data automation language

adaline adaptive linear neuron

adam adamantine; adaptive arithmetical method; advanced data management; air deflection and modulation; area denial artillery munition; associometrics data management (system); automatic distance and angle measurement

A'dam Amsterdam

ADAM Advanced Architecture in Medicine; Agriculture Department Automated Man-

power; Alzheimer's Disease Association of Maryland; Automatic Document Abstracting Method

ADAMHA Alcohol, Drug Abuse, and Mental Health Administration

adamite basic zinc arsenate

adaml advise by airmail

adamm (ADAMW) area defense anti-missile missile

adams airborne data acquisition management system

Adam's Adam's Bridge (30 mile-long island chain linking Ceylon and India); Adam's Peak (7,000-foot-high mountain, Ceylon, also called Samanaliya or Sri Padastanaya)

ADAMS Advanced Design Aluminum Metal Shelter (pre-fabricated ADAMS hut)

Adams Mans Adams Mansion (birthplace and home of John Adams and his son John Quincy Adams in Quincy, Massachusetts)

adandac administrative and accounting purposes

ADAOD Air Defense Artillery Operations Detachment

ADAOO Air Defense Artillery Operations Office(r)

adap adapted

ADAP Airport Development Aid Program (FAA)

ADAPCP Alcohol and Drug Abuse Prevention and Control Program

ADAPS Automatic Display and Plotting System

ADAPSO Association of Data Processing Service Organizations

adapt. adaption of automatically programmed tools

ADAPT Alcohol and Drug Abuse Prevention Team

adapticom adaptive communication

ADAPTS Air-Deliverable Anti-Pollution Transfer System (USCG)

adar advanced design array radar; analog-to-digital-to-analog recording

ADAR Advanced Design Army Radar; Air Defense Area

adare advise date of receipt

ADARF Alcoholism and Drug Addiction Research Foundation (Ontario, Canada)

adars airborne digital acquisition & recording system

adas automatic data acquisition system

ADAS Action Data Automation System; Agricultural Development Advisory Service; Airborne Dynamic Alignment System; Architecture Development and Assessment System

ADASC Auto Dismantlers Association of Southern California (often called ADA)

adash advise date of shipment

ad ast. ad astra (Latin—to the stars)

adat automatic data accumulation and transfer

ADATS Assistant Director, Auxiliary Territorial Service

adaval advise availability

ADAWS Action Data Automation and Weapons System

A-day assault day

adb accidental death benefit; adjusted debit balance; air defense battery

adb automatisk databehandling (Dano-Norwegian—automatic data handling)

adB acceleration decibel(s)

ADB Apollo Data Bank (NASA); Apple Desktop Bus (trademark); Asian Development Bank; Atlantic Development Board (Canada)

A.D.B. Bachelor of Domestic Arts

ADB Australian Dictionary of Biography

ADBA American Dog Breeders Association

ADBC American Defenders of Bataan and Corregidor

adbd active-duty base date

ADBE Adobe Systems Incorporated

ADBM Association of Dry Battery Manufacturers

ADBPA Association pour le Développement des Bibliothèques Publiques en Afrique (French—Association for the Development of Public Libraries in Africa)

adbz active-duty battle zone

adc active-duty commitment; adopted child; advance delivery of correspondence; airborne digital computer; air data computing; albumin, dextrose, catalase; all damn confusion; ammunition dumping craft; animal damage control; anodal duration contraction; automatic data collection; axiodistocervical

ADC Aerodome Defense Corps; Aerophysics Development Corporation; Aerospace Defense Command; Agricultural Development Council; Aid to Dependent Children; Aide-de-Camp; Air Defense Command; Air Development Center; Air Diffusion Council; Alaska Defense Command; American Distilling Company; American Dock Company; analog-to-digital converter; Animal Damage Control; Area Dissemination Coordinator; Army Dental Corps; Army Dress Committee; Asian Development Centre; Assistant Divisional Commander; Association of Diving Contractors; Australian Dairy Corporation; Australian Development Corporation; Australian Diabetic Council; Aviation Development Council

ad&c ammunition, distribution and control

adca advanced-design composite aircraft

ADCA Australian Department of Civil Aviation

adcad airways data collection and distribution

adcap advanced capability

ad cap. ad captandum (Latin—for pleasing, made attractive)

ADCC Air Defense Cadet Corps; Air Defense Command and Control; Air Defense Control Center

adccp advanced data communications control procedure

ADCG Australian Defence Cooperation Group

ADC Gen Aide-de-Camp General

adci adopted child

ADCIS Association for the Development of Computer-based Instruction Systems

ADC/NORAD Air Defense Command/North American Air Defense (Command)

ADCO Alcohol and Drug Control Office(r); American Dredging Company

ADCOC Area Damage Control Center

ADCOM Administrative Command; Aerospace Defense Command

adcon advance concepts; advise or issue instructions to all concerned; analog-to-digital computer

ADCONSEN advice and consent of the Senate (of the United States)

a & d control alcohol and drug control

ADCOP Area Damage Control Party

ADCP Air Defense Command Post

adcs advanced defense communications system

adc's analog-to-digital converters

ADCS Aerospace Defense Command Squadron

ADCSP Advanced Defense Communications Satellite Program

adct assisted-draft crossflow tower

ADCT Art Director's Club—Toronto; Association of District Council Treasurers

ad curtain advertisement curtain (theater)

add adopted daughter; addenda; addendum; address; airborne digital decoder; artificer deep diver; attention deficit disorder; automatic drawing device; average daily dose

ad & d accidental death and dismemberment (insurance)

add addendum (Latin—addition)

ADD Abstracts of Declassified Documents; Addis Ababa, Ethiopia (airport); Administration on Developmental Disabilities; Aerospace Defense Division; Air Defense District; Armaments Design

Department; *Aviastiia Dalnego Deistviia* (Russian—Long-Range Bombing Force)

ADDA Air Defense Defended Area

addar automatic digital data acquisition and recording

ADDAS Automatic Digital Data Assembly System

ADDC Air Defense Direction Center; Association of Deck and Derrick Clubs; Australian Dairy Development Council

Add-Can Addicts-Canada

ADDDS Automatic Direct Distance Dialing System

addee addressee

ad. def. an. *ad defectionem animi* (Latin—to the point of fainting)

ad. deliq. *ad deliquium* (Latin—to fainting)

adder. automatic-digital data-error recorder

ADDF Abu Dhabi Defense Force

ADDH Attention Deficit Disorder with Hyperactivity

ad diag admitting diagnosis

ADDIC Alcoholic and Dependency Intervention Council

addict. addiction

Addis Addis Ababa, Ethiopia

addit additional

add'l additional

ADDL Anti-Digit Dialing League

addm automated drafting and digitizing machine

addn addition

Addo Addo Elephant Park near Port Elizabeth, South Africa

ADDP Air Defense Defended Point

ADDR address

ADDS Alcohol and Drug Dependence Service; American Digestive Diseases Society; American Diversified Dog Society (for mutts); Apollo Document Descriptions Standards (NASA); Assistant Director of Dental Services; Automatic Data Digitizing System; Automatic Direct Distance Dialing System

addsd addressed

addu additional duty

Addy Ada; Adela; Adelaide; Adelina; Adeline

ade air defense emergency; application development environment; automated design engineering; automated drafting equipment; automatic data entry; average daily enrollment

ADE Animal Disease Eradication (Department of Agriculture division); Association of Departments of English; Association for Documentary Editing; Australian Driver Education

ADEA American Driver Education Association

ADEA Age Discrimination in Employment Act

ADEAR Alzheimer's Disease Education and Referral Center

ADEC Accumulated Deductible Employee Contribution

ADECOS Acción Democrática (Spanish—Democratic Action)—Venezuelan political party

ADED Association of Driver Educators for the Disabled

adeda advise effective date

ADEDS Advanced Electronic Display System

adee addressee

Adee Adelaide, Australia

ADEE Air Defense Experimental Establishment

Adeen Aberdeen (shire)

ad effect. *ad effectum* (Latin—until effective)

A de JC Antes de Jesucristo (Spanish—before Jesus Christ)

Adel Adelaide; Adelar(d); Adelbern; Adelbert; Adelbold; Adelfred; Adelfrid; Adelgar; Adelhart; Adelmo; Adelochorda; Adelpho; Adelquist; Adelric; Adelrik; Adelwin

ADELA Atlantic Community Development Group for Latin America

Adélie Adélie Land (French Antarctica)

adem acute disseminated encephalomyelitis

ADEM Air Defense Eastern Mediterranean

ADEMS Advanced Diagnostic Engine Monitoring System

aden augmented deflector exhaust nozzle

Aden part of the Republic of Yemen

ADEN Armaments Development Enfield (aircraft gun)

ADEOS Advanced Earth Observation Satellite

ADEP Air Defense Electronic Partnership; Air Depot

ADEPO Automatic Dynamic Evaluation by Programmed Organizations

adept. a definitely empirical power of theorems; automatic data extractor and plotting table

ADEPT Agricultural and Dairy Educational Political Trust

a des a destra [Italian—at (to) the right]

A de S Acadiémie des Sciences (French—Academy of Sciences)

ADES Automatic Digital Encoding System

ADESA Asamblea Democrática por Sufragio Efectivo (Spanish—Democratic Assembly for Effective Suffrage)

A de T S Alex de Tocqueville Society

ad eund. grad. *ad eundem gradum* (Latin—to the same degree)

ADEW Air Defense Early Warning

adex advanced antisubmarine warfare exercise

ADEX Air Defense Exercise

adf after deducting freight; air direction finder; airborne direction finder; auxiliary detonating fuse

adf (ADF) automatic direction finder

Adf Adolf

ADF Air Defense Force; Air Development Force; American Dance Festival; Antigua Defence Force; Arab Deterrent Forces' Army Distaff Foundation; Asian Development Fund

ADFA Australian Defence Force Academy; Australian Department of Foreign Affairs; Australian Dried Fruits Association

adfc adiabatic film cooling

ADFC Air Defense Filter Center

ADFF Australian Dairy Farmers Federation

ADF & G Alaskan Department of Fish and Game

ADFI American Dog Feed Institute

ADFIAP Association of Development Finance Institutions of Asia and the Pacific

ad fin. *ad finem* (Latin—to the end)

ADFL Association of Departments of Foreign Languages

ADFOR Adriatic Force

ADFS American Dentists for Foreign Service; Australian Documentary Facsimile Society

ADFSC Automatic Data Field Systems Command

ADFW Assistant Director of Fortifications and Works

adg accesory drive gearbox; air defense group; air-driven generator; axiodistogingival

ADG Assistant Director-General; degaussing vessel (3-letter symbol)

ADGA American Dairy Goat Association

adgb accessory drive gearbox

ADGB Air Defence of Great Britain

A & D G C Alliance and Dublin Consumers Gas Company

adge air-defense ground environment

adges air defense ground environment system

ADGMS Assistant Director-General of Medical Services

adgo adagio

ADGPA American Dairy Goat Products Institute

adgo *adàgio* (Italian—gently, peacefully, slowly, softly)

ad grat. acid. *ad gratam acidatem* (Latin—to a pleasing acidity)

ad grat. gust. *ad gratum gustum* (Latin—to an agreeable taste)

ADGRU Advisory Group

adgs air defense ground station

ADGT Assistant Director-General of Transportation

adh (ADH) alcohol dehydrogenase; antidiuretic hormone (vasopressin)

ADH Academy of Dentistry for the Handicapped; Assistant Director of Hygiene; Association of Dental Hospitals

ADHA American Dental Hygienists Association

ADHC Air Defense Hardware Committee (NATO); Australian Department of Housing and Construction

adhca advise this headquarters of complete action

ADHD Attention-Deficit Hyperactivity Disorder

adhib *adhibeatur* (Latin—administer)

ad h.l. *adhunc locum* (Latin—at this place)

ad hoc (Latin—for this special purpose)

ad hom. *ad hominem* (Latin—to the man)

adi acceptable daily intake; adiabat(ic); air defense intercept; air defense interceptor; airborne direction indicator; alien declared intention; antidetonation injection; attitude direction indicator; automatic direction indicator; average daily intake

adi (ADI) area of dominant (radio or tv station) influence

ADI Acoustical Door Institute; Air Defense Interceptor; Air Distribution Institute; American Documentation Institute; Assistant Director of Intelligence; Assistance Doss International

ADI *Agencia para el Desarrollo Internacional* (Spanish—Agency for International Development)—AID

ADIA American Diamond Industry Association

ADIC American Dental Interfraternity Council

ad id. *ad idem* (Latin—both are the same, likewise)

ad ig. *ad ignorantiam* (Latin—to ignorance)

adil (ADIL) air defense identification line

ADIL *Annual Digest of International Law*

adimd advise immediately by dispatch

ad inf *ad infinitum* (Latin—to infinity)

ad init. *ad initium* (Latin—at the beginning)

adinsp administrative inspection

ad int. *ad interim* (Latin—in the interim, meanwhile)

adintelcen advanced intelligence center

Ad Intel Cen Advanced Intelligence Center

ADIOS Automatic Digital Input-Output System

ADIP Advanced Developing Institutions Program

adipu advise whether individual may be properly used in your installation

Adirondacks Adirondack Mountains, New York

ADIS Air Defense Integrated System; Anxiety Disorders Interview Schedule; Association for the Development of Instructional Systems; Automatic Data Interchange System

adit analog digital integrating translator

ADIT Alien Documentation, Identification, and Telecommunications

ADIU Armament and Disarmament Information Unit, Sussex University

ADiv Air Division

ADIZ Air Defense Identification Zone; Air Defense Intercept Zone

adj adjacent; adjectival; adjective; adjoining; adjoint; adjourn(ed); adjust

Adj Adjutant

ADJ adjust

Adj. A. Adjunct in Arts

ADJAG Assistant Deputy Judge Advocate General

Adj Gen Adjutant General

Adjt Adjutant

adkms advanced data and knowledge management systems

adl activities of daily living; armament data line; automatic data link(ing); average decreasing line

ad l administrative law

Adl Adelaide

ADL Adelaide, Australia (airport); Admiral Corporation (stock exchange symbol); Anti-Defamation League (B'nai B'rith); Army Dental Laboratory; Arthur D Little (corporation); Authorized

Data List; Automatic Data Link(ing); Automotive Discount Leasing

ADLA Art Directors' (club) Los Angeles

Adlai Adlai Stevenson II

Ad LB Administrative Law Bulletin

Ad L News Administrative Law News

Adler Adler Planetarium (Chicago)

ad lib. ad libitum (Latin—at one's pleasure)

ADLIPS Automatic Data-Link Plotting System

adlm (ADLM) aerial-delivered land mine

ADLO Air Defense Liaison Office(r)

ad loc. ad locum (Latin—at this passage or place)

ADLOG Advance Command

ADLP Australian Democratic Labour Party

ADLR Assistant Director of Light Railways

ADLS Air Dispatch Letter Service

ADLT Activities of Daily Living Test

ADLTDE Association of Dark Leaf Tobacco Dealers and Exporters

adm activity data method; administration; administrative; administrator; admission; admit; air defense missile; atomic demolition munition; average daily membership

adm (ADM) air decoy missile

adm. admitted

Adm Admiral; Admiralty

ADM Action Description Memorandum; Advanced Diploma in Midwifery; Affiliated Dress Manufacturers; Air Defense Medal; Air Defense Missile; American Defense Medal; American Drug Manufacturers (association); Archer Daniels Midland Company

ADM American Demographics Magazine

adma automatic drafting machine

ADMA Aircraft Distributors and Manufacturers Association; American Drug Manufacturers Association; Australian Direct Marketing

Association; Australian Display Manufacturers Association; Aviation Distributors and Manufacturers Association

admad advise method and date of shipment

Ad Man Advertising Manager

admap advise by mail as soon as possible

admass advertising & mass media effect

ADMC Air Defense Missile Command

Adm Cen Administration Center; Administrative Center; Admiralty Center

Adm Co Admiralty Court

ADME Assistant Director of Mechanical Engineering

ADMI American Dry Milk Institute

ADMIG Australian Drug and Medical Information Group

admin administration; administrative; administrator

Admin administration

AdminInstr Administrative Instructions

AdminO Administrative Order(s)

adminord administrative order

adminplan administrative plan

ADMIRAL Automatic and Dynamic Monitor with Immediate Relocation, Allocation, and Loading (system)

admire. automatic diagnostic maintenance information retrieval

ADMIRES Automatic Diagnostic Maintenance Information Retrieval System

admix administratrix

adml average daily member load

Adml Admiral; Admiralty

ADMMO Atomic Demolition Munitions Mission Officer

admn administration

admon letters of administration

admon administración (Spanish—administration)

Admor Administrador (Spanish—Administrator)

admos automatic device for mechanical order selection

ad mov. ad moveatur (Latin—let it be moved)

admr administrator

adms administrator; advanced depot medical stores

ADMS American Donkey and Mule Society; Assistant Director of Medical Services

ADMSC Automatic Digital Message Switching Centers (DoD)

ADMSLBN Air Defense Missile Battalion (USA)

admsn admission

ADMT Association of Dental Manufacturers and Traders

Adm Ter Administered Territories

Admty Admiralty

admx administratrix

adn (ADN) ácido desoxirribonucleico (Spanish—deoxyribonucleic acid)—dna (DNA)

Adn Aden

ADN Accession Designation Number; *Allgemeiner Deutscher Nachrichtendienst* (General German News Service); Alternative Defense Network; Ashley, Drew & Northern (railroad)

A.D.N. Associate Degree in Nursing

a-DNA alternate deoxyribonucleic acid

ADNA Assistant Director of Naval Accounts

ADNAC Air Defense of the North American Continent

ad naus. ad nauseam (Latin—boring to the point of nausea)

ADNC Air Defense National Center; Air Defense Notification Center, Assistant Director of Naval Construction

ad neut. ad neutralizandum (Latin—until neutral)

ADNI Assistant Director of Naval Intelligence

ADNOC Abu Dhabi National Oil Company

adnok advise if not correct

'ad n't had not

ado advanced development objective; axiodistoclusal

Ado adagio (Italian—slowly and expressively)

ADO Administration Duty Officer; Air Defense Officer; Air Defense Operations; Area Dental Officer

ADOA Air Defense Operations Area

ADOBE Atmospheric Dispersion of Beryllium (program)

ADOC Air Defense Operations Center

ADOD Assistant Director, Operations Division

ADOF Assistant Director of Ordnance Factories

ADOGA American Dehydrated Onion and Garlic Association

adoit automatically directed outgoing intertoll trunk (Bell)

Adolph Adolphus

ADOLT Air Defense Operations Liaison Team

ADONIS Automatic Digital On-Line Instruments System

adop adoption

ADOPT Approach to Distributed Processing Transactions

ADOS Assistant Director of Ordnance Services

adot automatic digital optical tracker; automatically directed outbound trunk (Bell)

ADOT Air Defense Operation Team; Australian Department of Transportation

adoxograph adoxographer; adoxographic(al)(ly); adoxography

adp adenosine diphosphate; advanced data processing; advanced development products; airborne data processor; ammonium dihydrogen phosphate; automatic data processing

ADP Academy of Denture Prosthetics; Agricultural Development Program; Air Defense Position; Airport Development Program; Allied Defense Publication; ammonium dihydrogen phosphate; Animal Disease and Parasite (Research Division—Department of Agriculture); Army Depot Police; Association of Directory Publishers; Automatic Data Processing

ADPA American Defense Preparedness Association

ADPB Australian Dairy Produce Board

ADPC Abu Dhabi Petroleum Company; Automatic Data Processing Center

adpcm adaptive differential pulse code modulation

ADPCM Association for Data Processing and Computer Management

adpe automatic data processing equipment

ADPEA Australian Data Processing Employees Association

ADPESO Automatic Data Processing Equipment Selection Office (USN)

ADPG Air Defense Planning Group

ADPI American Dairy Products Institute

adpl average daily patient load

adplan advancement planning; advertising planning

adpo aircraft depot

ADPO Automatic Data Processing Operations

ad pond. om. *ad pondus omnium* (Latin—to the whole weight)

ADPR Assistant Director of Public Relations

ADPRIN Automatic Data-Processing Intelligence Network (U.S. Bureau of Customs)

ADPs Allied Defense Publications; Artillery Destruction Programs

ADPS Assistant Director, Army Postal Services; Automatic Data Processing System(s)

ADPSD Automatic Data Processing Systems Development

ADPSO Association of Data Processing Service Organizations

ADPSR Architects, Designers, and Planners in Social Responsibility

adpt adapted; adapter

adr address; aircraft direction room; aircraft discrepancy report; asset depreciation range; automatic dividend reinvestment

a-d r analog-to-digital recorder

adr *addresse* (Dano-Norwegian—address); *address* (Swedish—address)

Adr Adrian; Adriatic

Adr *Adresse* (German—address)

ADR Accepted Dental Remedies; adder; Aircraft Direction Room; Air Defense Regiment; Air Defense Region; Alternative Dispute Resolution (program); American Depository Receipt; Australian Defence Representative; Australian Design Rules (automotive); Automatic Dialog Replacement; European Agreement on the International Carriage of Dangerous Goods by Road

ADR *Accepted Dental Remedies; Association pour le Développement de la Recherche* (French—Association for the Development of Research)

ADR 27-a Australian Design Rule 27-a (vehicle emission control)

ADRA Animal Diseases Research Association; Australian Drag Racing Association

adrac automatic digital recording and control

adras aircraft date recovery/analysis system

ADRB Army Disability Review Board; Army Discharge Review Board

ADRC Alzheimer's Disease Research Center

ADRCP Alzheimer's Disease Research Centers Program

ADRDA Alzheimer's Disease and Related Disorders Association

adrde advise reason for delay

ADRDE Air Defense Research and Development Establishment

adren adrenal; adrenalin

adrenals adrenal glands

ADRES Army Data Retrieval System

ADRI Angkatan Darat Republik Indonesia (Indonesian Army)

Adria Adriatic Sea

Adria *die Adria* (German—the Adriatic)—*das Adriatische Meer* (The Adriatic Sea)

Adriatic Adriatic Sea

ADRIS Automatic Dead Reckoning Instrument Systems

adrm airdrome

ADRM Analog-to-Digital Remastering of musical recordings

Adro Alejandro

ADROBN Airdrome Battalion

adrp airdrop

adrs asset depreciation range system

ADRs American Depository Receipts

ADRS Analog-to-Digital Data Recording System; Automatic Document Request Service

adrt analog data recording transcriber

ADRT Assistant Director of Railway Transport

adr tel *adresse telegraphique* (French—telegraphic address)

ads accessory drive system; accurately defined system; activity data sheet; advance data system; advanced dressing station; advertisements; air defense ship; air defense system; alternative delivery system; ammunition delivery system; antibody deficiency syndrome; antidiuretic substance; area, date, subject; autograph document signed; automated data sheet; automatic door seal

AD²ᵈ *Appellate Division-second series*

ADS Aerial Delivery System; Air Defense Sector; Air Dispatch Squadron; Alzheimer's Disease Society; American Daffodil Society; American Dahlia Society; American Dental Service; American Dental Society; American Denture Society; American Dialect Society; Analog & Digital Systems; Association of Diesel Specialists; Association for Dressings and Sauces; Automated Drafting Services

ADS *Academie des Sciences* (French—Academy of Science)

ADSA American Dairy Science Association; American Dental Society of Anesthesiology; Atomic Defense Support Agency

ad. saec. *ad saeculum* (Latin— to the century)

ADSAF Automatic Data System for the Army in the Field

adsap advise as soon as possible

adsarm advanced defense suppression anti-radiation missile

ADSAS Air-Derived Separation Assurance System

ad sat. *ad saturandum* (Latin— to saturation)

ADSATIS Australian Defence Science and Technology Information System

adsc automatic digital switching center; average daily service charge (in hospitals)

ADSC Advanced Section Communication Zone; Aeronautical Division, Signal Corps; Air Defense Software Committee; Army Distinguished Service Cross; Automatic Data Service Center; Association of Drilled Shaft Contractors

ADSCAT Association of Distributors to the Self-service and Coin-operated Laundries and Allied Trades

adscom advanced shipboard communicatiions

adsda advise earliest date

ad sec. *ad sectam* (Latin—at suit of (legal))

AdSec Advanced Section

adshpdat advise shipping data

ADSIA Allied Data Systems Interoperability Agency

adsid air-delivered acoustic-implant seismic-intrusion detector

ADSID Air Defense Systems Integration Division

ADSL Assembly Department Shortage List; Asymmetric Digital Subscriber Line

adsm (ADSM) air defense suppression missile

ADSM American Defense Service Medal; Army Distinguished Service Medal

ADSMO Air Defense Systems Management Office

ADSN Accounting and Disbursing Station Number

ADSO Assistant Division Signal Officer

ADSOC Administrative Support Operations Center

ADSOT Automatic Daily System Operability Test

adss analysis of digitized seismic signals

ADSS Aircraft Damage Sensing System; Australian Defense Scientific Service

ADST Atlantic Daylight Saving Time

ADS&T Assistant Director of Supplies and Transport

adstadis advise status and/or disposition

adst. feb. *adstante febre* (Latin—when fever is present)

adstkoh advise stock on hand

adsu advanced direct support unit

ADSUP Automatic Data Systems Uniform Practice(s)

adsym automobile defog-defrost system model

adt aided tracking; any damn thing (a placebo); automatic damage template; automatic data translator; automatic debit transfer; average daily dose

adT *an demselben Tage* (German—the same day)

ADT American District Telegraph; Applied Drilling Technology; Atlantic Daylight Time

ADT 3 Air Defense Tactical Training Theatre

ADTA American Dental Trade Association

adtac air defense tactical

adtam air-delivered target-activated munitions

ADTC Air Defense Technical Center; Air Defense Test Center; Alzheimer's Disease & Treatment Center; Armament Development Test Center (USAF)

ADTe anodal duration tetanus

adtech advanced decoy technology

ad tert. vic. *ad tertium vicem* (Latin—three times)

ADTF At-Depth Test Facility

ADTI American Dinner Theatre Institute

ADTIC Arctic, Desert, Tropic Information Center

ADTR Australian Department of Tourism and Recreation

ADTS Automatic Data and Telecommunications Service

ADTSEA American Driver Traffic Safety Education Association

adtu automatic digital test unit; auxiliary data translator unit

ADTU Army Dog Training Unit

adu acceleration-deceleration unit; accessory dwelling unit; analog-to-digital (conversion) unit; accumulation-distribution unit; automatic dialing unit

ADU Aircraft Delivery Unit; Army Dog Unit

adult. adulterant; adulterate; adulteration

ad us. *ad usum* (Latin—according to custom)

ad us. ext. *ad usum externum* (Latin—for external use)

adv advance; advanced, advances; advantage; adverb(ial); advertising

a/dv arterio/deep venous (injection)

adv *advokat* (Dano-Norwegian—attorney)

adv. *adversum* (Latin—adversely, against)

Adv. Adventist; Adviser

ADV advance

advac advise acceptance

ad val. *ad valorem* (Latin—according to value)

Adv Appl Mech *Advances in Applied Mechanics*

Adv At Mol Phys *Advances in Atomic and Molecular Physics*

advb adverb; adverbial

Adv Bse Advanced Base

Adv Chem Phys *Advances in Chemical Physics*

adv chgs advance charges

advdisc advance discontinuance of allotment

advec advection

advect advection(al)(ly); advective

adven adventure; adventurer

adversat adversative

advert advertising

advertique advertising antique

adverts advertisements

adv frt advance freight

Adv Intel Cen Advanced Intelligence Center

ad virus adenovirus

ADVISE Area Denial Visual Identification Security Equipment

advl adverbial

advm adaptive delta voice modulation

Adv Magn Reson *Advances in Magnetic Resonance*

adv mtr advertising matter

Advoc Advocate

advof advise this office

advon advanced echelon; advanced operations unit

Adv Phys *Advances in Physics*

adv pmt advance payment

adv poss adverse possession

Adv Quantum Chem *Advances in Quantum Chemistry*

advr advisor

ADVRS Assistant Director of Veterinary and Remount Services

ADVS Assistant Director of Veterinary Services

advst advance stoppage

advt advertise; advertisement; advertiser; advertising

advul air-defense vulnerability

adw assault (with) deadly weapon

ADW Air Defense Warning

ADWA Atlantic Deeper Waterways Association

ADWC Air Defense Warning Condition; Air Defense Weapons Center (USAF)

ADWKP Air Defense Warning Key Point

ADWOC Air Defense Weapons Operation Center

adx automatic data exchange

ADX Adams Express Company (stock exchange symbol); Air Defence Exercise (Australian)

Adyg Adygey

adz advise

ADZ Air Defense Zone

Adzh Adzhar

ae above the elbow; account executive; aeroelectronic; aircraft equipment; air escape; almost everywhere; application engineer, arithmetic element; arithmetic expression; armed experimental; asymmetric epoxidation; auto exposure

a/e absorptivity-emissivity

a & e accident and emergency; aerospace and electronic; armaments and electronics; azimuth and elevation

ae *aetatis* (Latin—aged, at the age of)

AE Agricultural Engineering (Department of Agriculture research division); Air Efficiency (Award); Air Electrical; Air Explorer; Airborne Equipment (naval division); Algoma Eastern Railway; American English; ammunition ship (naval symbol); Ammunition Examiner; Army Education; Assault Engineer; Assistant Engineer; Automatic Electric; Award of Excellence; United Arab Emirates (Internet code)

A-E Adam and Eve; Architect-Engineer; Astro-Eugenics

A.E. Aeronautical Engineer; Agricultural Engineer; Architectural Engineer; Associate in Education; Associate in Engineering

A & E Agricultural and Engineering; Architectural and Engineering; Arts and Entertainment cable television network

AE *Aeon* (pen name of George William Russell); *Aktiebolaget Atomenergi* (Swedish—Atomic Energy Corporation); *American Ephemeris; Atomnaya Energiya* (Russian—Atomic Energy); *Australian Encyclopaedia*

A & E *Adolphuis and Ellis*

aea actual expenses allowable; assignment eligibility and availability

AEA Actors' Equity Association; Adult Education Association; Agriculture Engineers Association; Air Efficiency Award; Aircraft Electronics Association; American Economic Association; American Education Association; American Electrical Association; American Electronics Association; American Enterprise Association; American Entrepreneurs Association; American Export Airlines; Arizona Education Association; Arkansas Education Association; Arts Equity Association; Arts, Education, and Americans; Atomic Energy Authority; Automotive Electric Association

AEAA *Asociación de Escritores y Artistas Americanos* (Spanish—Association of American Writers and Artists)

AEAC Atomic Energy Authority Constabulary

AEAF Allied Expeditionary Air Force

AEAO Airborne Emergency Actions Officer

AEARC Army Equipment Authorizations Review Center

AEAs American Entertainers Abroad

aea sol alcohol-ether-acetone solution

AEAUSA Adult Education Association of the United States of America

AEB Adult Education Board (Singapore); Area Electricity Board; Association of Editorial Businesses; Atomic Energy Bureau; Australian Egg Board

AEBIG Aslib Economics and Business Information Group

aec additional extended coverage; altitude engine control; at earliest convenience; attitude engine control

AEC Adult Education Council; Aeronautical Research Council; Agricultural Economics (division of Department of Agriculture); Airworthiness Examination Committee; Alaska Engineering Commission (Alaska Railroad); Aluminum Extruders Council; American Engineering Council; Anglican Episcopal Church; Army Education Center; Army Educational Certificate; Army Educational Center; Army Educational Corps; Army Electronics Command (formerly Signal Corps); Associated Equipment Company (Army vehicles); Atlantic & East Carolina (railroad); Atlas Educational Center; Atomic Energy Commission; Australian Environment Council

A & EC Atlantic & East Carolina (railroad)

AEC-A Atomic Energy Commission—Albuquerque Operations Office

AEC-AI Atomic Energy Commission—Argonne, Illinois

AEC-ANM Atomic Energy Commission—Albuquerque, New Mexico

AEC-ASC Atomic Energy Commission—Aiken, South Carolina

AECB Arms Export Control Board; Atomic Energy Control Board (Canada)

AEC-BC Atomic Energy Commission–Berkeley, California

AECC Aeromedical Evacuation Control Center; American Evangelical Christian Churches

AEC-CC Atomic Energy Commission-Canoga Park, California

AECE *Asociación Española de Cooperación Europea* (Spanish—Spanish Association for European Cooperation)

AEC-FOA Atomic Energy Commission-Fernal Office Area, Cincinnati, Ohio

AEC-HW Atomic Energy Commission—Hanford, Washington

AECI African Explosives and Chemical Industries

AECIA Australian Electronics Consumer Industry Association

AEC-II Atomic Energy Commission—Idaho Falls, Idaho

AECL Anglo-European Container Line; Atomic Energy of Canada, Limited

AEC-LN Atomic Energy Commission—Las Vegas, Nevada

AEC-LOC Atomic Energy Commission—Lockland Aircraft Reactors Operations, Cincinnati, Ohio

AECM Albert Einstein College of Medicine

AEC-NY Atomic Energy Commission—New York Operations Office

AECO Aeromedical Evacuation Control Officer

AECOM Army Electronic Command

AEC-OR Atomic Energy Commission—Oak Ridge Operations Office

AEC-OT Atomic Energy Commission—Oak Ridge, Tennessee

aecp altitude engine control panel

AECP Airman Education and Commissioning Program

AEC-PP Atomic Energy Commission—Pittsburgh, Pennsylvania

AEC-PR Atomic Energy Commission—Pittsburgh Naval Reactors Operations Office

AECPSUPC Association for the Encouragement of Correct Punctuation, Spelling, and Usage in Public Communications

AEC-RW Atomic Energy Commission—Richland, Washington

AECS Australia-Europe Container Service

AECT Association for Educational Communications and Technology

AEC-UN Atomic Energy Commission—Upton, LI, NY

aed air equipment department; automated external defibrillation; automated external defibrillator; automatic external defibrillator

aed (AED) automatic engineering design

Aed Algol extended for design

A.Ed. Associate in Education

AED Academy for Educational Development; Associated Equipment Distributors; Association of Electronic Distributors

A.E.D. *Artium Elegantium Doctor* (Latin—Doctor of Fine Arts)

AEDB Apollo Engineering Development Board

AEDC Arnold Engineering Development Center

aedcm (AEDCM) advanced electrochemical depolarized concentrator module

AEDD Air Engineering Development Division

AEDE *Association Européenne des Enseignants* (French—European Teachers' Association)

AEDF Australian Executive Development Foundation

AEDO Aircraft Engineering District Office

AEDP Association for Educational Data Processing

AED-RCA Astro-Electronics Division-RCA

AEDS Association of Educational Data Systems; Association of Electronic Data Systems; Atomic Energy Detection System

AEDU Admiralty Experimental Diving Unit

aee absolute essential equipment; absolutely essential equipment; airborne evaluation equipment

Ae.E. Aeronautical Engineer

AEE Alliance for Environmental Education; Army Entrance Examination; Association for Experimental Education; Atomic Energy Establishment

A.E.E. Associate in Electrical Engineering

AEEB Association Européenne de l'Equipement de Bureau (French—European Office Equipment Association)

AEEC Airlines Electronic Engineering Committee

AEEE Army Equipment Engineering Establishment

AEEI Arthur E E Ivory

AEEL Aeronautical Electronic and Electrical Laboratory

AEEN Agence Européenne pour l'Energie Nucléaire (French—European Agency for Atomic Energy)

Ae. Eng. aeronautical engineer

AEEP Association of Environmental Engineering Professors

AEET Atomic Energy Establishment, Trombay (India)

AEEW Action on Employment and Equality for Women; Atomic Energy Establishment—Winfrith

aef air experience flight

AEF Advertising Educational Foundation; Aerospace Education Foundation; Aerospace Expeditionary Forces (Air Force); Aircraft Engine Filter; Aircraft Engineering Foundation; Albert Einstein Foundation; Allied Expeditionary Force; American Economic Foundation; American European Foundation; American Expeditionary Force; Americans for Economic Freedom; Armament Expeditionary Forces; Armenian Educational Foundation; Artists Equity Fund; Australian Expeditionary Force(s); Aviation Engineer(ing) Force

AEFC Australian-European Finance Corporation

A-effect alienation effect

AEFLLC Allied Expeditionary Force Long Lines Control

AEFM Association Européenne des Festivals de Musique (French—European Association of Music Festivals)

AEFORT American-European Friends of ORT (Organization for Rehabilitation through Training)

AEFR Aurora, Elgin & Fox River (railroad)

aeg active element group(ing); air encephalogram(s)

aeg. aeger (Latin—sick)

Aeg Aegean

AEG Association of Engineering Geologists

AEG Allgemeine Elektrizitats Geseläschaft (German—General Electric Company)

Aegean Aegean Sea (arm of the Mediterranean between Greece and Turkey)

Aegeans Aegean Islands (Cyclades, Dodecanese, Sporades, etc.)

AEGIMRDA Army Engineer Geodesy, Intelligence and Mapping Research and Development Agency

Aegis U.S. Navy cruiser class

AEGIS Active Electronic Gimballess Inertial System; Aid for the Elderly in Government Institutions; Airborne Early Warning Ground Integration Segment; Assessment of Effectiveness of Geologic Isolation Systems

AEGp Aeromedical Evacuation Group (USAF)

Aeg S Aegean Sea

AEH A(lfred) E(dward) Housman

AEH Archives of Environmental Health

AEHA Army Environmental Health Agency

AEHA Anuario Español e Hispano-Americano (Spanish and Hispanic-American Annual)

AEHL Army Environmental Health Laboratory

aei automatic error interrogation; average efficiency index; azimuth error indicator

AEI Aerial Exposure Index; Air Express International; American Enterprise Institute; American Express Institute; American Express International; Annual Efficiency Index; Associated Electrical Industries

AEI Association des Ecoles Internationales (French—Association of International Schools); *Associazione Elettrotecnica Italiana* (Italian—Italian Electrical Engineering Association)

AEIB Association for Education in International Business

AEIBC American Express International Banking Corporation

AEIC Advance Earned Income Credit; Association of Edison Illuminating Companies

AEIDC American Express International Development Company

AEIL American Export Isbrandtsen Lines

AEIMS Administrative Engineering Information Management System

A.E.I.O.U. Austria Erit In Orbe Ultima (Latin—Austria will be the world's last survivor)—ancient acrostic of House of Hapsburg

AEIP Allied Electrical Industry Publications

AEIPPR American Enterprise Institute for Public Policy Research

AEIS Association of Electronic Industries in Singapore

AEJ Association for Education in Journalism

AEJI Association of European Jute Industries

aek all-electric kitchen

ael audit error list

AEL Admiralty Engineering Laboratory (UK); Aeronautical Engine Laboratory; Aircraft Engine Laboratory; American Electronic Laboratories; American Emigrants League; American Express Line; Americanism Education League; Animal Education League; Automation Engineering Laboratory

AELC American Evangelical Lutheran Church; Association of Evangelical Lutheran Churches

AELE Americans for Effective Law Enforcement

AELE Association Européenne de Libre-Echange (French—European Free Trade Association)

AELR All England Law Reports

AEL-Rx Appalachian Educational Laboratory—Regional Exchange

AELTC All England Lawn Tennis Club

aem air escort mission; atomic emission monitoring

AEM Advance Engineering Memorandum; Aircraft and Engine Mechanic; American Meter Company (stock exchange symbol); Applications Explorer Mission; Association of Electronic Manufacturers; Association of Evangelical Ministers; Aviation Electrician's Mate

AEMA Asphalt Emulsion Manufacturers Association; Australian Electrical Manufacturers Association

AEMIE Association Européene de Médecine Interne d'Ensemble (French—European Association for Doctors in Internal Medicine)

AEMII Advanced Environment for Medical Image Interpretation

AEMIS Aerospace and Environmental Medicine Information System

AEMP Association of European Management Publishers

AE & MP Ambassador Extraordinary and Minister Plenipotentiary

AEMS American Engineering Model Society

AEMSA Army Electronics Material Support Agency

aen advance evaluation note

aen. aeneus (Latin—made of bronze or copper)

Aen. Aeneid (Virgil's epic poem)

A.En. Associate in English

AEN Asahi Evening News (Japan)

A & EN Arts & Entertainment Network

AENA All-England Netball Association

A.Eng. Associate in Engineering

AEO Air Engineer(ing) Office(r); Appeal Examining Office(r); Assistant Experimental Officer; Australian Electoral Office

AEOB Advanced Engine Overhaul Base; Alternate Escort Operating Base

AEODPs Allied Explosive Ordnance Disposal Publications

AEOE Association for Environmental and Outdoor Education

Aeol Aeolian; Aeolic

Aeol Chamb Play Aeolian Chamber Players

Aeolians Aeolian Islands off Sicily—Lipari, Stromboli, and Vulcano

Aeol Quart Lon Aeolian Quartet of London

AEOO Aeromedical Evacuation Operations Officer

aeop amend existing orders pertaining to

AEOP Allied Explosives Ordnance Publication (NATO)

AEOS Ancient Egyptian Order of Sciots; Astronomical, Earth, and Ocean Sciences

aep accrued expenditure paid; average evoked potential

AEP Addo Elephant Park (South Africa); Adult Education Program; Allied Engineering Publication (NATO); American Electric Power; Army Equipment Policy

AE & P Ambassador Extraordinary and Plenipotentiary

AEP Agence Européenne de Productivite (French—European Production Agency)

AEPB Active Enlisted Plans Branch (U.S. Navy)

AEPC Appalachian Electric Power Company

AEPCO American Elsevier Publishing Company

AEPEM Association of Electronic Parts and Equipment Manufacturers

AEPG Army Electronic Proving Ground

AEPI American Educational Publishers Institute

aep(s) auditory-evoked potential(s)

AEPs Allied Engineering Publications; Allied Equipment Publications

AEPS Aircrew Escape Propulsion System; American Electroplaters Society

aeq age equivalent

aeq. aequales [Latin—equal(s)]

AEQA Alabama Environmental Quality Association

AEqPs Allied Equipment Publications

aer aldosterone excretion rate; alteration equivalent to a repair; auditory-evoked response; average evoked response

AER Abbreviated Effectiveness Report; Aeronautical Engineering Report; After Engine Room; Airman Effectiveness Report; Army Emergency Relief; Army Emergency Reserve; Assault Engineer Regiment; Association for Education by Radio; *Association Européenne pour l'Etude du Probleme des Rèfugies* (French—European Association for the Study of the Refugee Problem)

aera aeration

AERA American Educational Research Association; American Engine Rebuilders Association; Australian Endurance Riders Association; Automotive Engine Rebuilders Association

AERB Army Education Requirements Board

AERC Association of Executive Recruiting Consultants

aercab advanced escape/rescue capability; advanced aircrew escape/rescue capability

AERDL Army Electronics Research and Development Laboratory

Aer.E. Aeronautical Engineer

AERE Atomic Energy Research Establishment

AERI Agricultural Economics Research Institution; Automotive Exhaust Research Institute

aerl aerial

AERL Aero-Elastic Research Laboratory (M.I.T.)

Aer Méx Aero México (Spanish—Air Mexico formerly Aeronaves de México)

AERNO Aeronautical Equipment Reference Number

aero aerographer; aeronautical; aeronautics

AERO Association of Electronic Reserve Officers

aerobatics aeronautical acrobatics

aerobee aerojet/bumblebee (naval missile)

aerob(s) aerobic exercise(s)

aerocade aerial parade; aviation parade (massed formations, stunts aloft)

Aero Commander U-4 transport aircraft

aerodyn aerodynamics

Aero E Aeronautical Engineer

Aer Of Aerological Officer

AEROFLOT Aero Flotilla (Soviet Air Lines)

aerol aerological

aeromed aeromedical

aeromod aerodynamic modelling

aeromus aeronautical museum

aeron aeronautical

aéron *aéronautique* (French— aeronautical)

Aeron Aeronaut; Aeronautics

aeronaut aeronautical(ly), aeronautics

AERONAVES *Aeronaves de México* (Spanish—Airships of Mexico)

AERONORTE *Empresa de Transportes Aereos Norte do Brasil* (Spanish—North Brazil Airways)

Aero O/Y Finnair (Finnish Airlines)

Aerop *Aeropuerto* (Spanish— airport)

aeropost aerodynamic postprocessing

AEROS Aerometric and Emissions Reporting System; Artificial Earth Research and Orbiting Satellite

AEROSAT Aeronautical Communications Satellite System

aerosp aerospace

aerospace aeronautics + space

aerospacecom aerospace communication(s)

AEROTAL *Aerolineas Territoriales de Colombia* (Spanish—Territorial Airlines of Colombia)

aerotel airplane hotel (hangar)

Aerovias "Q" Aerovias Cubana (Cuban Airlines)

aers aircraft equipment requirement schedule

AERS Atlantic Estuarine Research Society

AERT Association for Education by Radio-Television

AERU Agricultural Economics Research Unit (NZ)

aes aeromedical evacuation system; annual expectation of sales; auger electron spectroscopy

Aes *Aesop* (Greek fabulist); (Latin—bronze or copper) used by numismatists to denote bronze or copper coins or coins of such colors

AES Abrasive Engineering Society; Aerospace Electrical Society; Aerospace and Electronic Systems; Agricultural Estimates (division of Department of Agriculture); Agricultural Experiment Station; Aircraft Electrical Society; Airways Engineering Society; American Electrochemical Society; American Electroencephalographic Society; American Electroplaters Society; American Entomological Society; American Epidemiological Society; American Epilepsy Society; American Equilibration Society; American Ethnological Society; American Eugenics Society; Apollo Extension System; Armored Engineer Squadron; Army Education Scheme; Army Exchange Service (PX); Atlantic Estuarine Society; Audio Engineering Society; Australian Educational Secretariat; Australian Entomological Society

AE & S Air Equipment and Support

A & ES Arson and Explosion Squad

AESBOW Association of Engineers and Scientists of the Bureau of Weapons (USN)

AESC American Engineering Standards Committee; Association of Executive Search Consultants

Aescul Aesculapius (Greek god of medicine)

AESD Acoustic Environment Support Detachment (USN Office of Naval Research)

AESE Association of Earth-Science Editors

AESHS Alfred E Smith High School

AESM Albert Einstein School of Medicine

AESMC Automotive Exhaust Systems Manufacturers Council

AESO Aircraft Environmental Support Office (USN)

AESOP An Evolutionary System for On-line Processing; Artificial Earth Satellite Observation Program; Automated Engineering and Scientific Optimization Program (NASA)

AESq Aeromedical Evacuation Squadron (USAF)

AESQ Air Explorer Squadron

AESRS Army Equipment Status Reporting System

AESS American Ethnic Science Society

AEST Aeromedical Evacuation Support Team

AESTE Association for the Exchange of Students for Technical Experience

aesth aesthete; aesthetic; aesthetician; aesthetics

AESU Aerospace Environmental Support Unit

aet absorption-equivalent thickness

aet. *aetatis* (Latin—at or of the age of)

AET Australian Eastern Time

A.E.T. Associate in Electrical Technology; Associate in Electronic Technology

AET *Aerlinte Eireann Teoranta* (Irish Airlines)

AETA American Educational Theatre Association

AETD Aero-Electronic Technology Department (USN)

AETE Aerospace Engineering Test Establishment (Canada)

AETFAT *Association pour l'Etude Taxonomique de la Fiore d'Afrique Tropicale* (French—Association for the Taxonomic Study of African Tropical Flora)

A et M *Arts et Métiers* (French—arts and crafts)

AETM Aviation Electronic Technician's Mate

AETN American Educational Television Network

AEtPs Allied Electronic Publications

AETR Advanced Engineering Test Reactor

AETS Association for the Education of Teachers in Science

AETT Australian Elizabethan Theatre Trust

aeu accrued expenditure unpaid

AEU Amalgamated Engineering Union; American Ethical Union; Asia Electronics Union

aeuia alleluia (Italian—hallelujuh)

aev armored engineer vehicle; articles of extraordinary value; avian erythroblastosis virus

aev (AEV) aerothermodynamic elastic vehicle

AEV Asociación de Escritores Venezolanos (Spanish—Association of Venezuelan Writers)

AEVA Australian Equine Veterinary Association

aevac air evacuation

aew armored electronic warfare (vehicle)

AEW Admiralty Experimental Works; Airborne Early Warning

AEWB Army Electronic Warfare Board (USA)

AEW & C Airborne Early Warning and Control

AEWCAP Airborne Early Warning Combat Air Patrol

AEWES Army Engineers Waterways Experiment Station

AEWHA All-England Women's Hockey Association

AEWIS Army Electronic Warfare Information System (USA)

AEWL Association of Employers of Waterside Labour (Australian)

AEWLA All-England Women's Lacrosse Association

AEWRON Airborne Early Warning Squadron

AEWS Advanced Earth Satellite Weapon System (USAF); Aircraft Early Warning System (DoD)

AEWSPS Aircraft Electronic Warfare Self-Protection System

AEWTF Aircrew Electronic Warfare Training Facility

AEWTU Airborne Early Warning Training Unit

aex automatic electronic exchange (facilitating telephony)

AExO Assistant Experimental Officer

af air filter; army form; audio-frequency; automatic following; auxiliary force

af (AF) ale firkin; audio fidelity; autofocus

a-f anti-foam; audio-frequency

a/f *a favor* (Spanish—in favor)

af afgang (Danish—departure); *anno futuro* (Italian—next year); (Latin prefix—movement toward a central point)

af. affidavit

a.f. *ad finem* (Latin—to the end)

aF attofarad(s)

Af Africa; Afrikaans; African(s)

Af Académie française (French Academy); *Armée française* (French Army)

AF Admiral of the Fleet; Advance Freight; Afghanistan (Internet code); Africa(n); Air Force; air freight; Allied Forces; Anglo-French; Armored Force; Arthritis Foundation; Aviation Photographer's Mate; provision stores ship (2-letter symbol)

A-F Anglo-French

A.F. Admiral of the Fleet

A/F Air Field

A & F Agriculture and Forestry (Senate Committee)

A of F Admiral of the Fleet

AF35m auto-focus 35mm (camera)

afa advanced fleet anchorage; automatic fire alarm; azimuth follow-up amplifier; dry-type forced-air cooled (transformer)

AFA Actors Fund of America; Advertising Federation of America; Advertising Federation of Australia; Aerophilatelic Federation of the Americas; Air Force Academy; Air Force Association; Alien Firearms Act; American Family Association; American Farriers Association; American Fashion Association; American Federation of the Arts; American Finance Association; American Floorcovering Association; American Florists Association; Athletic Footwear Association; American Forensic Association; American Forestry Association; American Foundrymens Association; American Freedom Association; Armed Forces Act; Army Finances Association;

Army Flight Activity; Association of Federal Appraisers; Association of Flight Attendants; Australian Field Artillery; Australian Foundation for Alcoholism

AF of A. Advertising Federation of America

A.F.A. Associate in Fine Arts

AFAA Adult Film Association of America; Aerobics and Fitness Association of America; Air Force Audit Agency; Allied Forces Airborne Association; Automatic Fire Alarm Association

AFAAEC Air Force Academy and Aircrew Examining Center

AFAB Army Families Advice Bureau

afac (AFAQ) airborne forward air controller

AFAC Air Force Armament Center; Airborne Forward Air Controller; American Fisheries Advisory Committee; Arkansas Foundation of Associated Colleges

afactplan affirmative action plan(ning)

AFADD Australian Foundation for Alcoholism and Drug Dependence

AFADI Air Force Air Defense Interceptors

AFADL Air Force Avionics Development Laboratory

AFADO Association of Food and Drug Officials

AFAFC Air Force Accounting and Finance Center

AFAG Airforce Advisory Group

AFAIDSR American Foundation for AIDS Research

AFAIM Associate Fellow of the Australian Institute of Management

AFAITC Armed Forces Air Intelligence Training Center

AFAK Armed Forces Assistance to Korea

AFAL Air Force Avionics Laboratory

afam airfield attack ammunition; automatic frequency-assignment model

AF & AM Ancient Free and Accepted Masons

AFAMRL Air Force Aero Medical Research Lab

Af-Am(s) African-American(s); Afro-American(s)

afap artillery-fired atomic projectiles

AFAP Armed Forces Arabian Peninsular; Australian Federation of Airline Pilots

AFAPL Air Force Aero-Propulsion Laboratory

afar airborne fixed-array radar

AFAR Advanced Field Army Radar; Azores Fixed Acoustic Range (NATO)

AFAR Australian Foreign Affairs Record

Afars and Issas formerly French Somaliland, now the Republic of Djibouti

AFAS Air Force Aid Society; Air Force and Society; Armed Forces Art Society; Automated Frequency-Assignment System

AFAS Association française pour l'avancement des sciences (French—Association for the Advancement of Science)

AFASE Association for Applied Solar Energy

AFA-SEF Air Force Association—Space Education Foundation

AFASIC Association For All Speech-Impaired Children

AFATL Armed Forces Armament and Testing Laboratory

AFAUD Air Force Auditor General

afb acid-fast bacillus; anti-friction bearing

afb afbeelding (Dutch—illustration)

AFB Air Force Base; American Farm Bureau; American Foundation for the Blind; Army Fire Brigade

AFBA Armed Forces Broadcasters Association

AFBF American Farm Bureau Federation

AFBI Australian Fibre Box Industry

AFBMA Antifriction Bearing Manufacturers Association

AFBMD Air Force Ballistic Missile Division

AFBNM Agate Fossil Beds National Monument (Nebraska)

AFBS American and Foreign Bible Society; Ansett Flying Boat Services

AFBSD Air Force Ballistic Systems Division

afc antibody forming cells; automatic frequency control

afc (AFC) average fixed cost

AFC Air Force Cross; American Football Conference; American Forest Council; Apollo Flight Control (NASA); Area Forecast Center; Army Field Code; Australian Film Commission; Australian Flying Corps

AFCAI Associate Fellow of the Canadian Aeronautical Institute

AFCAL Association Française de Calcul (French—French Calculus Association)

AFCBS Australian Federation of Commercial Broadcasting Stations

afcc assault fire command console

AFCC Air Force Combat Command; Air Force Communications Center; Air Force Cost Center; Australian Federal Cycling Council; Australian Federation of Construction Contractors

AFCCB Air Force Configuration Control Board

AFCCDD Air Force Command and Control Development Division

AFCCP Air Force Component Command Post

AFCD Air Force Cryptologic Depot

afce automatic flight-control equipment

AFCE Associate in Fuel Technology and Chemical Engineering; Allied Forces Central Europe

AFCEA Armed Forces Communications and Electronics Association

AFCEC Australian Federation of Civil Engineering Contractors

AFCent Allied Forces, Central Europe

afcfs advanced fighter-control flight simulator

AFCI American Foot Care Institute

AFCL Africa Container Lines

AFCM Air Force Commendation Medal

AFCMA Aluminum Foil Container Manufacturers Association; Australian Fibreboard Container Manufacturers Association

AFCMC Air Force Contract Maintenance Center

AFCMD Air Force Contract Management Division

AFCMO Air Force Contract Management Office

AFCMR Armed Forces Central Medical Registry

AFCN American Friends of the Captive Nations

afco automatic fuel cutoff

AFCO Admiralty Fleet Confidential Order; Air Force Contracting Office(r); Australian Federation of Consumer Organisations

AF Compt Air Force Comptroller

AFCOMS Air Force Commissary Service

AFCOMSECCEN Air Force Communications Security Center

AFCON Air Force Controlled (units)

AFCOS Air Force Combat Operations Staff; Armed Forces Courier Service

AFCP Association of Free Community Papers

AFCPMC Air Force Civilian Personnel Management Center

AFCR American Federation for Clinical Research

AFCRC Air Force Cambridge Research Center

AFCRL Air Force Cambridge Research Laboratories

AFCRS Army Full Crew Research Simulator

afcs advanced facer canceler system; avionic flight control system

AFCs Air Force Circulars

AFCS Active Federal Commissioned Service; Adaptive Flight Control System; Air Force Communications Service; Armed Forces Courier Service; Army Facilities Components System; Artillery Fire Control Simulator; Automatic Flight Control System

AFCSC Air Force Command and Staff College

AFCS E & I Air Force Communications Service–Engineering and Installation

AFCSA Air Force Center for Studies and Analyses

AFCSL Air Force Communications Security Letter

AFCSM Air Force Communications Security Manual

AFCSP Air Force Communications Security Pamphlet

AFCUL Australian Federation of Credit Union Leagues

AFCW Association of Family Case Workers

AFCWF Air Force Civilian Welfare Fund

afd accelerated freeze drying; alternate full day; auxiliary floating dock

afd afdeling (Dano-Norwegian, Dutch—department, division, section)

AFD Admiralty Floating Dock; Air Force Depot; Armed Forces Day; Association of Food Distributors; Association of Footwear Distributors; mobile floating drydock (naval symbol)

AFDA Air Force Discipline Act; American Flag Day Association; Australian Defence Force Academy

AFDAA Air Force Data Automation Agency

AFDAP Air Force Directorate of Advanced Technology

AFDATACOM Air Force Data Communications System

AFDATASTA Air Force Data Station

AFDB African Development Bank; Air Force Decorations Board; large auxiliary floating drydock (naval symbol)

afdc artillery fire date computer

AFDC Aid for Families with Dependent Children; Air Force Department Constabulary; Airborne Forces Development Center; Australian Film Development Corporation

AFDCB Armed Forces Disciplinary Control Board

AFDCMI Air Force Policy on Disclosure of Classified Military Information

AFDCS American First Day Cover Society

AFDCUF Aid to Families with Dependent Children of Unemployed Fathers

AFDC-UP Aid to Families of Dependent Children—for Unemployed Parents

AFDDA Air Force Director of Data Automation

AFDE American Fund for Dental Education

AFDEA American Funeral Directors and Embalmers Association

AFDFS Air Force Department Fire Service

AFDH American Fund for Dental Health

AFDL small auxiliary floating drydock (naval symbol)

AFDM medium auxiliary floating drydock (naval symbol)

AFDO Air Force Duty Officer; Association of Food and Drug Officials; Assistant Fighter Director Officer

AFDOA Armed Forces Dental Officers Association

AFDOUS Association of Food and Drug Officials of the United States

AFDP Air Force Development Plan

AFDRB Air Force Disability Review Board; Air Force Discharge Review Board; Air Force Disciplinary Review Board

AFDRD Air Force Director of Research and Development

AFDRQ Air Force Director of Requirements

AFDS Air Fighting Development Squadron; Auxiliary Fighter Director Ship

AFDSC Air Force Data Services Center

AFDU Air Fighting Development Unit

AFDW Air Force District of Washington

afe authorization for expenditure

AFE Administración de Ferrocarriles del Estado (Spanish—State Railway Administration)

AFEA American Farm Economic Association; American Film Export Association

AFEB Armed Forces Epidemiological Board

AFEE Airborne Forces Experimental Establishment

AFEES Armed Forces Examining and Entrance Stations

AFELIS Air Force Engineering and Logistics Information System

AFEM Armed Forces Expeditionary Medal

AFEMS Air Force Equipment Management System

AFEOC Air Force Emergency Operations Center

AFEOS Air Force Electro-Optical Site

AFER Air Force Engineering Responsibility

AFERB Air Force Educational Requirements Board

AFERO Asia and the Far East Regional Office (FAO)

AFES Admiralty Fuel Experimental Station; Air Force Exchange Service; American Far Eastern Society; Armed Forces Examining Stations

AFESA Air Force Engineering and Services Agency

AFESC Air Force Engineering and Services Center

AFETR Air Force Eastern Test Range

AFEX Air Forces Europe Exchange (PX)

aff above finished floor; affairs

AFF Aden Field Force; affinity; Army Field Forces

AFFA Air Freight Forwarders Association; Angels Forever, Forever Angels (slogan of Hell's Angels motorcycle gang)

affaire affaire de coeur (French—affair of the heart)—love affair

AFFC Air Force Finance Center

affd affixed; afford(able); affordability

aff'd affirmed

affd per cur affirmed by the court

AFFDL Air Force Flight Dynamics Laboratory

AFFE Air Force Far East; Allied Forces Far East; Army Forces Far East

affec affectation; affection; affective

affet affettuoso (Italian—tenderly, with pathos)

afff aqueous film-forming foam

AFFFA American Forged Fitting and Flange Association

AFFI American Frozen Food Institute

Affie Alfred

affil affiliated

affirm affirmative

AFFIRM Association for Federal Information Resources Management

AFFJ American Fund for Free Jurists

affl affluent

AFFL Agricultural Finance Federation, Limited

afflat afflatus

AFFLC Air Force Film Library Center

AFFOR Air Force Forces (joint task force element)

affores afforestation

affret affrettando (Italian—speeding the tempo)

affrs artillery free flight rocket system

AFFS American Federation of Film Societies

AFFSC Army Force Flight Safety Committee

afft affidavit

AFFTC Air Force Flight Test Center; Air Force Flying Training Command

afg above finished grade; analog function generator

afg afgang (Dano-Norwegian—departure)

Afg Afghan; afghani (currency); Afghanistan; Afghans

AFG Allied Freighter Guard; Army Freighter Guard (NATO)

AFGC American Forage and Grassland Council

AFGCM Air Force Good Conduct Medal

aft Dem Jam Afghanistan Democrateek Jamhuriat (Pashto—Afghan Democratic Republic)

AFGE American Federation of Government Employees

Afghan Afghanistan(i)

afghani monetary unit of Afghanistan

Afghanistan Republic of Afghanistan, *Doulat i Jamhouri ye Afghánistán*

AFGIS Aerial Free Gunnery Instruction School

AFGL Air Force Geophysics Laboratory

AFGM American Federation of Grain Millers

AFGU Aerial Free Gunnery Unit

AFGW American Flint Glass Workers

AFGWC Air Force Global Weather Central

afh anterior facial height

AFH Air Force Hospital; American Foundation for Homeopathy; Associated Federated Hotels; Australian Field Hospital

AFHC Air Force Headquarters Command

AFHF Air Force Historical Foundation; American Foot Health Foundation

AFHQ Air Force Headquarters; Allied Forces Headquarters; Armed Forces Headquarters

AFHRL Air Force Human Resources Laboratory

AFHW American Federation of Hosiery Workers

afi amaurotic familial idiocy

AFI Air Filter Institute; Air Force Intelligence; American Film Institute; American Filter Institute; American Forest Institute; American Friends of Israel; American Fur Industry; Armed Forces Institute; Association of Federal Investigators; Association of Food Industries; Atlantic Refining Company (stock exchange symbol); Australian Film Institute; Australian Foundry Institute; Australian Frontier Incorporated; Auxiliary Force India

AFIA Air Force Intelligence Agency; American Feed Industry Association; American Footwear Industries Association; American Foreign Insurance Association

AFIAS Associate Fellow of the Institute of the Aerospace Sciences

afib atrial fibrillation

afic aficionado (Spanish—admirer, devotee, fan)

AFIC Air Force Intelligence Center; Australian Fishing Industry Council

AFICCS Air Force Interim Command and Control System

AFICE Air Forces—Iceland

AFIED Armed Forces Information and Education Division

AFII American Federation of International Institutes

AFIIM Associate Fellow of the Institute of Industrial Managers

AFINE Association Française pour l'Industrie Nucleaire d'Equipement (French—French Association for the Nuclear Equipment Industry)

AFINS Airways Flight Inspector

AFIO Association of Former Intelligence Officers

AFIP Air Force Intelligence Publication; Armed Forces Information Program; Armed Forces Institute of Pathology

AFIPS American Federation of Information Processing Societies

AFIR Air Force Installation Representative

AFIRAN Africa-Indian Ocean Region Air Navigation

AFIRO Air Force Installations Representative Officer

AFIS Air Force Intelligence Services; American Forces Information Service; Armed Forces Information School; Armed Forces Information System; Automated Field Interview System; Automated Fingerprint Identification System

AFISC Air Force Inspection and Safety Center

afism aluminum-free inorganic suspended material

AFISR Air Force Industrial Security Regulations

afit airblast fuel-injection tube

AFIT Air Force Institute of Technology

AFITAE Association Française d'Ingénieurs et Techniciens de l'Aéronautique et de l'Espace (French—French Association of Aeronautical and Aerospace Engineers and Technicians)

AFJA Armed Forces Judo Association

AFJAG Air Force Judge Advocate General

AF JINTACCS Air Force Joint Interoperability of Tactical Command and Control Systems

AFJKT Air Force Job-Knowledge Test

afk *afkorten, afkorting* (Dutch—abbreviation)

Afk Afrikaans

afl abstract family of languages; anti-fatty liver; atrial flutter

afl aflevering (Dutch—part)

AFL Aeroflot (Soviet Air Lines); Air Force Letter; Air Force List; American Federation of Labor; American Football League; Applied Fisheries Laboratory (University of Washington); Arena Football League; Association for Family Living; Australasian Federation League; Australian Fertilizers Limited

AFLA Adolescent Family Life Act; Amateur Fencers League of America; American Foreign Law Association; Asian Federation of Library Associations; Automotive Fleet and Leasing Association

AFLAC American Family Life Assurance Company

AFLANT Air Forces Atlantic

AFLAT Air Force Language Aptitude Test

aflatox aflatoxin

AFLC Air Force Logistics Command

AFL-CIO American Federation of Labor and Congress of Industrial Organizations

AFLCPs Air Force Logistics Command Pamphlets

afld airfield

affir advanced forward-looking infrared

AFLM Accredited Farm and Life Member

AFLO Airborne Force Liaison Officer

AFLP American Farmer Labor Party; Armed Forces Language Program

AFLRL Army Fuels and Lubricants Research Laboratory

AFLS Air Force Library Service

AFLSA Air Force Longevity Service Award

AFLSC Air Force Legal Services Center

aflt afloat

afm all flaps meet; antifriction metal; atomic-force microscope; automatic flight management

AFM Air Force Manual; Air Force Medal; Air Force Museum; American Federation of Musicians; Associated Fur Manufacturers

AFMA American Film Marketing Association; American Footwear Manufacturers Association; American Fur Merchants Association; American Furniture Manufacturers Association; Armed Forces Management Association

AFMA Air Force Manual of Abbreviations

AFMAB Armed Forces Medical Advisory Board

AFMBE Australian Federation for Medical and Biological Engineering

AFMBT Artificial Flower Manufacturers Board of Trade

AFMC American Floral Marketing Council; Armed Forces Medical Corps

AFMDC Air Force Missile Development Center

AFME American Friends of the Middle East

AFMEA Air Force Management Engineering Agency

AFMEC African Methodist Episcopal Church

AFMed Allied Forces, Mediterranean

AFMF Air Fleet Marine Force

ARMFP Aircraft Fleet Marine Force Pacific

AFMH American Foundation for Mental Hygiene

AFMIC Air Force Materials Information Center

AFML Air Force Materials Laboratory; Armed Forces Medical Library

AFMMFO Air Force Medical Materiel Field Office

afmo afectísimo (Spanish—most affectionate)

AFMPA Armed Forces Medical Publication Agency

AFMPC Air Force Military Personnel Center

afmr antiferromagnetic resonance

AFMR American Foundation for Management Research; Armed Forces Master Records

AFMS Air Force Medical Service; American Federation of Minerological Societies; Australian Farm Management Society

AFMSC Air Force Medical Specialist Corps

AFMTC Air Force Missile Test Center

AFMVOP Air Force Motor Vehicle Operator Test

AFMW Australian Federation of Medical Women

afn active filter network

AFN Air Force Finance Center; Alaska Federation of Natives; American Forces Network; Armed Forces Network

AF of N Alaska Federation of Natives

AFN Afrique du Nord (French—North Africa)

AFNA Accordion Federation of North America; Air Force with Navy; American Foundation for Negro Affairs

AFNB Armed Forces News Bureau

AFNC Air Force Nurse Corps

AFNE Allied Forces, Northern Europe; Americans For Nuclear Energy

AFNEI Allied Forces Netherlands East Indies

AFNIL Agence Francophone pour la Numérotation Internationale du Livre (French—French Agency for the International Numbering of Books)

AFNOR Association Française de Normalisation (French—French Standards Association)

AFNorth Allied Forces, Northern Europe

afo animal-feeding operation

AFO Accounting and Finance Office(r); Admiralty Fleet Order; Airports Field Office; Army Forwarding Officer; Atlantic Fleet Organization

AFOAR Air Force Office for Aerospace Research

AFOAS Air Force Office of Aerospace Sciences

AFOAT Air Force Office for Atomic (energy)

AFOB American Foundation for Overseas Blind

AFOC Air Force Operations Center

AFOECP Air Force Officer Education and Commissioning Program

AFOG Asian Federation of Obstetrics and Gynaecology

AFOIC Air Force Officer in Charge

AFOMS Air Force Office of Medical Support

AFOQT Air Force Officer Qualifying Test

AFOR Air Force Operations Room

AFORG Air Force Overseas Replacement Group

a fort *a fortiori* (Italian—with greater force)

AFOS Advanced Field Operations System; Automation of Field Operations and Services

AFOSI Air Force Office of Special Investigations

AFOSP Air Force Office of Security Police

AFOSR Air Force Office of Scientific Research

AFOTEC Air Force Operation Test and Evaluation Center

AFOUA Air Force Outstanding Unit Award

afp annual firing practice; anterior faucial pillar; automated filter photometer

afp (AFP) alphafetoprotein

AFP Agence France-Presse (successor to Havas); Air Force Pamphlet; Air Force Police; Alternate Flight Plan; American Family Publishers; American Federation of Police; Anglican Fellowship of Prayer; Annual Funding Program; Armed Forces of the Philippines; Armed Forces Police; Authority for Purchase

AF of P American Federation of Police

afpa automatic flow process analysis

AFPA American Fighter Pilots Association; Aquarama and Fairmount Park Aquarium; Australian Fire Protection Association

AFPAO Air Force Property Accountable Office(r)

AFPAV Air Force Pavement

AFPB Air Force Personnel Board

af/pc automatic frequency/phase-controlled (loop)

AFPC Air Force Personnel Council; Air Force Procurement Circular; American Food for Peace Council; Armed Forces Policy Council

AFPCB Armed Forces Pest Control Board

AFPD Armed Forces Police Detachment

AFPE American Foundation for Pharmaceutical Education; American Foundation for Political Education

AFPH American Federation of the Physically Handicapped

AFPI Air Force Procurement Instructions; American Forest Products Industries

AFPO Abortion and Family Planning Organization

AFPP Air Force Procurement Procedures

AFPPA American Federation of Poultry Producers Associations

AFPR Air Force Plant Representative

AFPRB Armed Forces Pay Review Board

AFPRO Action for Food Production; Air Force Plant Representative's Office

AFPs American Freeway Patrol cars

AFPS Armed Forces Press Service; Army Field Park Squadron

afpt artillery fire plan table

AFPT Air Force Personnel Test

AFPTRC Air Force Personnel and Training Research Center

AFPU Air Force Postal Unit; Australian Federation of Police Unions; Army Field Punishment Unit; Army Film and Photographic Unit

AFQ *Association Forestière Québeçoise* (French—Quebec Forestry Service)

AFQA Air Force Quality Assurance

AFQC Air Force Quality Control

AFQT Armed Forces Qualification Test

afr acceptable failure rate; airframe; air-fuel ratio; alternating frequency rejection; applicable federal rate; artillery flash ranging; automatic field/format recognition; away from reactor

afr *afrikansk* (Dano-Norwegian—African)

Afr Africa; African; Africans; Afrikaans (South African Dutch)

A Fr Algerian franc

A-Fr Anglo-French

AFR Air Force Regulation(s); Air Force Reserve

AFR *Australian Financial Review*

Afr Eng African English

afra average freight rate assessment

AFRA American Farm Research Association

AFRAeS Associate Fellow of the Royal Aeronautical Society

AFRAM Afro-American

A-frame capital-A-shaped support frame

Aframerican African + American

AFRASEC Afro-Asian Organization for Economic Cooperation

Afrasia Africa + Asia

AFRB Air Force Retiring Board

AFRBA Armed Forces Relief and Benefit Association

AFRBO Air Force Review Boards Office

AFRBSG Air Force Reserve Base Support Group

AFRC Air Force Records Center; Air Force Regional Civil (engineer); Air Force Research Center; Armed Forces Recreation Center

AFRCC Air Force Rescue Coordination Center; Air Force Reserve Coordination Center

AFRCSTC Air Force Reserve Combat Support Training Center

afrd acute febrile respiratory disease (AFRD)

AFRD Air Force Research Division; Air Force Reserve Division; Association of Fund-Raising Directors

AFRDS Association of Fund Raisers and Direct Sellers

A-freak acid (LSD) user

AFRes Air Force Reserve

AFRESM Armed Forces Reserve Medal

AFRESNAVSQ Air Force Reserve Navigation Squadron

AFRFI American Friends of Religious Freedom in Israel

afri acute febrile respiratory illness (AFRI)

AFRI Applied Forest Research Institute

Afric Africa; African

African languages Hausa in Central and West Africa; Swahili in parts of East Africa; Yoruba and Ibo in West Africa; Rwanda in southern Central Africa; Somali in East Africa; Xhosa and Zulu in South Africa

Africs Africans

afrik *afrikansk* (Dano-Norwegian—African)

Afrik Afrikaans; Afrikaner State (northern Cape Province of South Africa)

Afrik Afrikaans (Dutch dialect spoken in South Africa, the language of the Boers)

Afrik Eng Afrikaans English

afrm airframe

AFRMA American Fancy Rat and Mouse Association

Afr Nat Cnl African National Council

AFRO African Regional Office (FAO)

Afro-Am Afro-American(ese)

Afro-Amer Afro-American

Afro-American African-American

Afro-Bras Afro-Brasiliero (Afro-Brazilian)

Afro-Carib Afro-Caribbean; Afro-Caribeño

Afro(s) Afro-American(s)—Black(s), Negro(es)

AFROTC Air Force Reserve Officers Training Corps

AFRPL Air Force Rocket Propulsion Laboratory

AFRR Air Force Reserve Region

AFRRG Air Force Reserve Recovery Group

AFRRI Armed Forces Radiobiology Research Institute

afr's auditor freight receipts

AFRS Air Force Reserve Sector; Armed Forces Radio Service

AFRTC Air Force Reserve Training Center

AFRTS Armed Forces Radio-Television Service

AFRTVS Armed Forces Radio and Television Services

AFRVN Air Force of the Republic of Viet Nam

afs aerial fire support; aforesaid; atomic fluorescence spectroscopy; audio frequency shift

afs *afsender* (Danish—sender)

AFS Active Fusing System; Advanced Flying School; Air Force Specialty; Air Force Station; Air Force Supply; Airline Feed System; Airways Facilities Shop; Alaska Ferry Service; American Federation of Scientists; American Feline Society; American Fern Society; American Field Service; American Fisheries Society; American Folklore Society; American Foundrymen's Society; American Fuchsia Society; Andrew File System; Army Fire Service; Atlantic Ferry Service; Aviation Facilities Service; Azimuth Follow-up System; combat store ship (naval symbol)

AFSA Air Force Sergeants Association; American Federation of School Administrators; American Fire Sprinkler Association; American Flight Strips Association; American Foreign Service Association; Armed Forces Security Agency

AFSAB Air Force Science Advisory Board

AFSARC Air Force System Acquisition Review Council

AFSAS American Federation of School Administrators and Supervisors

AFSATCOMS Air Force Satellite Communications System

AFSAW Air Force Special Activities Wing

Af-Sax Afro-Saxon (black person of part Anglo-Saxon parentage)

AFSB American Federation of Small Business

AFSBO American Federation of Small Business Organizations

AFSC Air Force Service Command; Air Force Specialty Code; Air Force Staff College; Air Force Supply Catalog; Air Force Systems Command; American Federation of Soroptimist Clubs; American Friends Service Committee; Armed Forces Staff College

AFSCC Air Force Special Communications Center; Armed Forces Supply Control Center

AFSCF Air Force Satellite Control Facility

AFSCM Air Force Systems Command Manual

AFSCME American Federation of State, County, and Municipal Employees

AFSCN Air Force Satellite Control Network

afsd aforesaid

AF/SD Air Force/Space Division

AFSE Allied Forces Southern Europe (NATO)

AFSec Air Force Section

AFSF Air Force Stock Fund

AFSHC Association of Financial Services Holding Companies

AFSIL Accommodations for Students in London

AFSINC Air Force Service Information and News Center

AFSJT Air Force Strategic Jet Tankers

afsk audio-frequency shift keying

AFSM Association for Food Service Management

AFSMAG Air Force Section—Military Advisory Group

AFSN Air Force Serial Number; Air Force Service Number; Air Force Stock Number

AFSNCOA Air Force Senior Noncommissioned Officers' Academy

AFSOG Air Force Special Operations Group

AFSouth Allied Forces, Southern Europe

AFSPACECOM Air Force Space Command

AFSRS Australian Flying Saucer Research Society

afss automatic fire suppression system

AFSS Air Force Security Service; Air Force Service Statement

AFSSC Armed Forces Supply Support Center

AFSSD Air Force Space Systems Division

AFSSO Air Force Special Security Office

AFSTC Air Force Space Test Center

AFSU Auxiliary Ferry Service Unit

AFSUB Army Air Forces Anti-Submarine Command

AFSWA Armed Forces Special Weapons Agency

AFSWC Air Force Special Weapons Center

AFSWP Armed Forces Special Weapons Project

aft after; afternoon; at, near, or toward the rear; automatic fine tuning

aft. (AFT) automatic fund transfer

Aft Aftenposten (Danish—Evening Post, Oslo)

AFT Air Freight Terminal; American Federation of Teachers; Annual Field Training (USA)

AFT (AFL-CIO) American Federation of Teachers

AFTA Air Force Tactical Airlift; Atlantic Free Trade Area; Australian Federation of Travel Agents

AFTAC Air Force Technical Applications Center

aftads advanced field artillery tactical data system

AFTAU American Friends of Tel Aviv University

aftb afterburner

AFTB Air Force Test Base

AFTC Airborne Flight Training Command; American Fair Trade Council; American Fox Terrier Club; American Free Trade Clubs

AFTCC Air Force Troop Carrier Command

AFTDU Airborne Forces Tactical Development Unit

AFTE American Federation of Technical Engineers

AFTEC Air Force Test and Evaluation Center

AFTF Air Force Task Force

AFTFA Air Force Tactical Fighter Aircraft

AFTI Advanced Fighter Technology Integration (USAF); Allied Forces Technology Integration

AFTIA Armed Forces Technical Information Agency

AFTLI Association of Feeling Truth and Living It

AFTM American Foundation for Tropical Medicine

AFTMA American Fishing Tackle Manufacturers Association

aftn afternoon

AFTN Aeronautical Fixed Telecommunications Network

afto afecto (Spanish affectionate, fond)

AFTO Air Force Technical Order

AFTOSB Air Force Technical Order Standardization Board

aftp additional flight-training period

AFTR American Federal Tax Reports

AFTRA American Federation of Television and Radio Artists

AFTRC Air Force Technical Training Command

AFTRCC Aerospace Flight Test Radio Coordinating Council

afts automatic frequency tone shift

AFTS Advanced Flying Training School; Aeronautical Fixed Telecommunications Service; Aseptic Fluid Transfer System; Australian Film and Television School

AFTTH Air Force Technical Training Headquarters

AFTU Arizona Federation of Teacher Unions

afu all fucked up

AFU Advanced Flying Units; American Fraternal Union; Assault Fire Unit (U.S. Army)

AFU Association Fonciere Urbaine (French—Urban Land Association)

AFUA Annual Fuel Utilization Agency

AFUD American Foundation for Urologic Disease

afue annual fuel utilization efficiency

AFULE Australian Federated Union of Locomotive Enginemen

AFUS Air Force of the United States; Armed Forces of the United States

AFUW Australian Federation of University Women

afv armored fighting vehicle; armored force vehicle

AFVA Air Force Visual Aid

AFVBM American Federation of Violin and Bow Makers

AFVC Auxiliary Force Veterinary Corps

AFvg Anglo-French variable geometry

AFVN Armed Forces Vietnam Network

AFVOA Aberdeen Fishing Vessel Owners Association

afw army field workshop

AFW Association for Family Welfare

AFWA Air Force with Army

AFWAL Air Force Wright Aeronautical Laboratories

AFWE Air Forces Western Europe (NATO)

AFWETS Air Force Weapons Effectiveness-Testing System

AFWL Air Force Weapons Laboratory; Armed Forces Writers League

AFWN Air Force with Navy

AFWOFS Air Force Weather Observing and Forecasting System

AFWR Atlantic Fleet Weapons Range

AFWST Armed Forces Women's Selection Test

AFWTR Air Force Western Test Range (see WTR)

Afyon Afyonkarahisar (Turkish—Black Castle of Opium)—town in western central Turkey where much of the world's opium has been grown

afz afzender (Dutch—from)

AFZ Australian Fishing Zone

ag advanced guard; against; agar-agar; agency; agent; aggie; aggressive; agribusiness; agricultural; agriculture; agrobiology; agroindustrial; agrology; agronomy; albumen gland; alternate geologies; armed guard; armor grating; atrial gallop; automatic gun; axiogingival

a-g air-to-ground; anti-gas

a/g air-to-ground; albumin globulin ratio

a g à gauche (French—to the left)

Ag Agostino

Ag argentum (Latin—silver)

AG Adjutant General; Aeronautical Standards Group; Air Group; Air Gunner; Apparel Guild; Army Group; Artists Guild; Attorney General; Auditor General; escort research vessel (naval symbol); miscellaneous auxiliary vessels (naval symbol); sonar research ship (naval symbol); technical research ship (naval symbol)

AG Aktien Gesellschaft (German—company, joint stock company); Alberghi per la Gioventu (Italian—Youth Hostels); Arkansas Gazette; Asamblea General (Spanish—General Assembly); Astronomische Gesellschaft (German—Astronomical Society)

aga accelerated growth area; appropriate for gestational age

a/g/a air-to-ground-to-air

AGA Abrasive Grain Association; Accredited Gemologists Association; Adjutants General Association; Alabama Gas (symbol); American Galvanizers Association; American Gas Association; American Gastroenterological Association American Gastroscopic Association; American Genetic Association; American Glassware Association; American Goiter Association; American Gold Association; Australian Garrison Artillery, Australian Gas Association

AGAA Art Galleries Association of Australia

AGAAC Acuerdo General sobre Aranceles Aduaneros y Comercio (Spanish—General Accord Custom's Duties and Commerce)

AGAC American Guild of Authors and Composers

Aga cooker Aktiebolaget gas-accumulative cooker

agacs air-ground-air communications system

AGAFBO Atlantic and Gulf American Flag Berth Operators

AGAI Army General and Administrative Instructions

AGAL Australian Government Analytical Laboratory

AGARD Advisory Group for Aeronautical Research and Development (NATO)

AGAS Air Ground Aid Service; Australian Government Advertising Service

agate chalcedony

Agathon Agathon Press

agave. automatic gimballed antenna vectoring equipment

agb accessory gear box; any good brand

AGB Audits of Great Britain (television survey); icebreaker (3-letter symbol)

AGBAD Alexander Graham Bell Association for the Deaf

AGBI Artists' General Benevolent Institution

agbio agrobiology

agbm antiglomerular basement membrane antibody

AGBUC Association of Governing Boards of Universities and Colleges

agbus agribusiness; analog ground bus

agc air-ground communications; automatic gain control

AGC Adjutant General's Corps; Aerojet-General Corporation; American Grassland Council; amphibious force flagship (naval symbol); Amphibious Group Command; Armed Guard Center; Associated General Contractors; astronomical great circle (course); Australian Government Centre; Australian Guarantee Corporation

AGC Amgueddfa Cenedlaethol Cymru (Welsh—National Museum of Wales)

agca automatic ground-controlled approach

AGCA Associated General Contractors of America

AGCan Auditor General of Canada

agcl automatic ground-controlled landing

AGCM Army Good Conduct Medal

AGCMWA Amon G. Carter Museum of Western Art

agcol agricultural college

AGCRSP Army Gas-Cooled Reactor Systems Program

AGCSB Atlantic-Gulf Coastwise Steamship Freight Bureau

AGCSD Attorney General's Consumer Services Department

AGCT Army General Classification Test

AGCTS Armed Guard Center Training School

Ag-Cu al silver-copper alloy (U.S. coin facing)

agcy agency

agd agreed; axial gear differential

AGD Academy of General Dentistry; Adjutant General's Department; American Gage Design; Auditor General's Department

AGD Australian Government Digest

AGDA American Gasoline Dealers Association; American Gun Dealers Association

AGDC Assistant Grand Director of Ceremonies

AGDE escort research ship (naval symbol)

Ag. Dei Agnus Dei (Latin—Lamb of God)

Ag Dept Agriculture Department

AGDS American Gage Design Standard; deep submergence support ship (naval symbol)

age (AGE) aerospace ground environment; aerospace ground equipment; automatic guidance electronics

Age The Age (Melbourne)

Ag.E. Agricultural Engineer

AGE Admiralty Gunnery Establishment; AG Edwards; Amarillo Grain Exchange; Asian Geotechnical Engineering (Thailand)

A.G.E. Associate in General Education

AGEC Army General Equipment Command

A.G.Ed. Associate in General Education

AGED Advisory Group on Electronic Devices

ageft agricultural electronic fund transfer

AGEH hydrofoil research ship (naval symbol)

AGEHR American Guild of English Handbell Ringers

agents provocs *agents provocateurs* (French—secret agents) persons hired to provoke others to commit crimes so arrest and conviction can follow

AGEO Apple-of-God's-Eye Obsession

ageocp aerospace ground equipment out of commission for parts

AGEP Advisory Group on Electronic Parts

AGER environmental research ship (naval symbol)

agerd aerospace ground equipment recommendation data; aerospace ground equipment requirements data

AGERS Auxiliary General Electronics Research Ship(s)

AG & ES American Gas & Electric System

AGET Advisory Group on Electronic Tubes

AGF Adjutant General to the Forces; Army Ground Forces; miscellaneous command ship (naval symbol)

AGFA *Aktiengesellschaft für Anilinfabrikation* (German—Corporation for Aniline Manufacture)

ag. feb. *aggrediente febre* (Latin—when fever increases)

AGFF frigate research ship (naval symbol)

Ag and Fish Ministry of Agriculture and Fisheries

AGFRTS Air and Ground Forces Resources and Technical Staff (U.S. Army)

AGFSRS Aircraft Ground Fire Suppression and Rescue System (DoD)

agg agammaglobulinaemia(c); agglutination(ed); aggravate(ed); aggregate(d); aggregation

aggie agriculture

aggies agate playing marbles; students of agricultural colleges or schools

agglut agglutination (ed)

aggr aggregate

AGGR Air-to-Ground Gunnery Range

aggred. feb. *aggrediente febre* (Latin—while fever is developing)

aggro aggression; aggressiveness

aggs anti-gas gangrene serum

AGGS American Good Government Society

Aggy Agatha; Agnes

AGH Addison Gilbert Hospital, Gloucester, Mass.; Auditor General's Hotline; Australian General Hospital

AGHE Association for Gerontology in Higher Education

agi adjusted gross income; airborne gunlaying (radar); auxiliary (vessel) general intelligence

AGI Adjusted Gross Income; Air Gunnery Instructor; American Geographical Institute; American Geological Institute; Annual General Inspection; Anti-Gas Instructor

AGI *Agenzia Giornalistica Italiana* (Italian News Agency); *Associazione Guide Italiane* (Italian Girl Guides' Association)

AGIC Air-Ground Information Center

AGIFORS Airlines Group of International Federation of Operations Research Societies

agil airborne general illumination light

agile. airborne general illumination light; analytic geometry interpretative language

AGILE Autonetics General Information Learning Equipment

ag imps hnd agricultural implements hand

ag imps ot hand agricultural implements other than hand

ag'in' against

agind agroindustrial

AGIO Australian Government Insurance Office

AGIP *Azìenda Generale Italiana Petroli* (Italian—National Italian Oil Company)

agipa adaptive ground-implemented-phased array

agit agitate(d); agitation; agitato; agitator

agit. *agitatum* (Latin—shaken)

agit. ante sum. *agita ante sumendum* (Latin—shake before using)

agit. a. us. *agita ante usum* (Latin—shake before using)

agit. bene *agita bene* (Latin—shake well)

agit-prop agitation and propaganda

agit. vas. *agitato vase* (Latin—shaking the vessel)

agl above ground level; acute granulocytic leukemia; airborne gunlaying; aminoglutethimide; automatic gunlaying

Agl *Argelia* (Spanish—Algeria)

AGL Australian Gas Light (company); lighthouse tender (3-letter symbol)

agland(s) agricultural land(s)

AGLC Air-to-Ground Liaison Code

AGLINET Agricultural Libraries Information Network (UN)

aglm agglomerate

AGLO Air-Ground Liaison Officer

AGLS Association of General and Liberal Studies

AGLSP Association of Graduate Liberal Studies Programs

A-glue airplane glue

agm air-launched guided missile

agm (AGM) air-to-ground missile

AGM Admiralty General Message; American Guild of Music; Annual General Meeting (of shareholders); Australian Glass Manufacturers; Award of Garden Merit; missile range instrumentation ship (naval symbol)

AGM-53A North American Rockwell Condor air-to-surface missile

AGMA American Gear Manufacturers Association; American Guild of Musical Artists; Athletic Goods Manufacturers Association

AGMA (AFL-CIO) American Guild of Musical Artists

AGMIS Adjutant General Management Information System

Agmk African green monkey kidney

agmr advanced gas-cooled maritime reactor

AGMR major communications relay ship (naval symbol)

agn active galactic nuclei; acute glomerulonephritis; again; agnomen

Agn Augustín

Agñ Agaña, Guam

AGN Aerojet-General Nucleonics

Agncy Agency (postal place name abbreviation)

agnos agnostic; agnosticism

agnos agnostos (Greek—unknowable), origin of agnostic invented in 1869 by Thomas Henry Huxley

AGNS Allied General Nuclear Services

ago. atmospheric gas oil

ago agitato (Italian—agitated); *agosto* (Spanish—August)

Ago agosto (Italian/Spanish—August)

AGO Adjutant General's Office; Air Gunnery Officer; American Guild of Organists; Attorney General's Office; Attorney General's Opinion; Auditor General's Office

AGOR Auxiliary General Oceanographic Research (vessel)

AGOS Air Gunnery Officers School; ocean surveillance ship (naval symbol)

agp above-ground pool; accelerated graphics port; automatic guidance programming

AGP Academy of General Practice; Achievement Goals Program (to improve test scores in minority schools);

Achievement Guidance Program; Adjutant General's Pool; Army Ground Pool; motor torpedo boat tender (naval symbol)

AGP Archives of General Psychiatry

AGPA American Group Psychotherapy Association

AGPC Adjutant General Publications Center

agpe angle plate

agpi automatic ground position indicator

AGPL Administração-Geral do Porto de Lisboa (Portuguese—Port of Lisbon Authority)

ag prov agent provocateur

AGPS Australian Government Publishing Service

agr agree(ment); agricultural; agriculture

agr (AGR) advanced gas-cooled graphite-moderated reactor

AGRA Army Group Royal Artillery; Australian Garrison Royal Artillery

agrar agrarian; agrarianism; agrarians

a/g ratio albumin-globulin ratio

Agra U Agra University

agrbl agreeable

agrd agreed

AGRE Army Group Royal Engineers; Atlantic Gas Research Exchange

AGREE Advisory Group on Reliability of Electronic Equipment

agrep agricultural research project(s)

AGREX Agricultural Guarantee Fund Expenditure

AGRF American Geriatric Research Foundation

agri agricultural; agriculturalist; agriculture; agriculturist

agribusiness agricultural business (large-scale farming)

agric agriculture

Agric E Agricultural Engineer

agricrime agricultural crime (theft of crops and/or equipment)

agridollars agricultural (market) dollars

agri-indus agricultural-industrial (complex)

agrimech agriculture mechanized

agripower agricultural power

AGRIS Agricultural Information System

AGRM Adjutant General—Royal Marines

agrmt agreement

agro aggravation; agrobiological; agrobiologist; agrobiology; agrologic; agrological; agronomical; agronomics; agronomist; agronomy

agro agronomiae (Dano-Norwegian—agronomy)

agrobio agrobiologic(al)(ly); agrobiologist; agrobiology

agrogeol agrogeology

agroind agroindustrial(ly); agroindustrialization; agroindustrialize(r); agroindustry

agron agronomy

agronome (Russian—agricultural expert)

agros agrostology

AGRS American Graves Registration Service

ags adrenogenital syndrome; agencies

ags angelsächsisch (German—Anglo-Saxon)

ags (Ags) antigens

ags (AGS) alternation gradient synchrotron

Ags Aguascalientes (inhabitants—Hidrocalidos)

AGs Attorneys General

AGS Abort Guidance System; Academic Guidance Service; Africa General Service medal; Aircraft General Standards; Alabama Great Southern (railroad); Allied Geographic Section; American Gem Society; American Geographical Society; American Geriatrics Society; American Goat Society; American Gynecological Society; Army General Staff; Army Ground Staff; Army Guard School; Army Gymnastic Staff; Association of Graduate Schools; Australian Geographic Society; Australian Geomechanics Society; surveying ship (naval symbol)

A.G.S. Associate in General Studies

AGSA Art Gallery of South Australia

AGSAN Astronomical Guidance System for Air Navigation

AGSI Automatic Government Source Inspection

AGSM American Gold Star Mothers; Associate of the Guildhall School of Music; Australian Graduate School of Management

AGSRO Association of Government Supervisors and Radio Officers

AGSS American Geographical and Statistical Society; auxiliary research submarine (naval symbol)

agst against

agt agent; agreement

agt (AGT) antiglobulin test

AGT Art Gallery of Toronto; Association of Geology Teachers; transport oiler (naval symbol)

AGTA American Gem Trade Association; Australian Geography Teachers Association

AGTC Airport Ground Traffic Control

AGTE Association of Group Travel Executives

AGTELIS Automated Ground Transportable Emitter Location and Identification System

AGTELS Automated Ground Tactical Emitter Location System

AGTF Alternate Geology Test Facility

agto agosto (Portuguese and Spanish—August)

AGTOA American Greyhound Track Operators Association

agtt (AGTT) abnormal glucose tolerance test

agtv advanced ground transportation vehicle

AGU American Geophysical Union; Army Gymnastics Union

Aguacates Aguacate Mountains, Costa Rica

Aguas Aguascalientes

Agu Cur Agulhas Current

Agunalaksh (Aleut—Unalaska)—shores where the sea breaks on the Aleutian Islands

agv aniline gentian violet

AGV Automatic Guided Vehicle

AGVA American Guild of Variety Artists

AGvga Anglo-German variable-geometry aircraft

agw actual gross weight; air-to-ground warfare; allowable gross takeoff weight

AGWA Australian Government Workers Association

AGWAC Australian Guided Weapons and Analog Computer

AGWD Australian Government Weekly Digest

AGWI American Gulf and West Indies (steamship line)

agy agency

agz actual ground zero

ah abdominal hysterectomy; acetohexamide; after hatch; alter heading; amenorrhea and hirsutism; aminohippurate; antihalation; antihyaluronidase; anti-submarine helicopter; armed helicopter; arterial hypertension; assault helicopter; astigmatism hypermetropic; attack helicopter

a-h ampere-hour

a/h at home

a & h accident and health; alive and healthy

aH attoHenry

Ah ampere-hour; hyperopic astigmatism

AH Adult Humor; Aircraft Handler; Airfield Heliport; Alfred Holt's Blue Funnel Line (house flag and funnel mark); Animal Husbandry (division of Department of Agriculture); Army Helicopter; Army Hospital; hospital ship (naval symbol)

A-H Alaska-Hawaii Time; American-Hawaiian Line; Arrow-Hart & Hegeman Electric Company

A & H Arm and Hammer (trademark)

AH Akademiya Nauk (Russian—Academy of Sciences)

A.H. Anno Hebraico (Latin—in the Hebrew Year); *Anna Hegirae* (Latin Year of Hegira)—Moslem

AH-I Huey Cobra gunship military aircraft carrying machine-gun pods on its stub wings, a 7.62mm minigun in its nose plus a grenade launcher

AH-64 attack helicopter

aha acquired hemolytic anemia; all have automobiles; alpha-hydroxy acid; autoimmune hemolytic anemia

AHA Adirondack Historical Association; American Hardboard Association; American Heart Association; American Hereford Association; American Historical Association; American Hospital Association; American Hotel Association; American Humane Association; American Humanist Association; American Hypnotherapy Association; Area Health Authority; Association of Handicapped Artists; Association for Humane Abortion; Australian Hospital Association; Australian Housewives Association

ahab attacking hardened air bases

aham advanced heavy anti-tank missile

AHAM Association of Home Appliance Manufacturers

ahas (AHAS) acetohydroxy acid synthase

AHAUS Amateur Hockey Association of the U.S.

ahaws advanced heavy anti-tank weapons system; advanced heavy assault weapons system

ahb active hostile battery

AHB Air Historical Branch

ahc acute haemorrhagic conjunctivitis

AHC Academy of Hospital Counselors, American Hardware Corporation; American Hockey Coaches; American Horticultural Council; American Hospital Corps; Army Hospital Corps; Assault Helicopter Company; Australian Heritage Commission

ahca (AHCA) American Health Care Association

AHCEI American Histadrut Cultural Exchange Institute

AHCo Assault Helicopter Company (USAF)

AHCPR Agency for Health Care Policy and Research

ahd acetaldehyde dehydrogenase; ahead; airhead; aired head; arteriosclerotic heart disease; atherosclerotic heart disease; audio high density; auto-immune haemolytic disease

AHD American Heritage Dictionary

A-H DT Alaska-Hawaii Daylight Time

ahe acute hemorrhagic encephalomyelitis

AHE Association for Higher Education

A.H.E. Associate in Home Economics

AHEA American Home Economics Association; American Hungarian Educators Association

AHEAD Army Help for Education and Development

AHEC American Hardwood Export Council; Army Higher Education Center

AHEL Army Human Engineering Laboratory (USA)

AHEM Association of Hydraulic Equipment Manufacturers

AHEPA American Hellenic Educational Progressive Association

AHES American Humane Education Society

ahf anti-hemophilic factor

Ahf Argentinian hemorrhagic fever

AHF American Health Foundation; American Heritage Foundation; American Hobby Federation; American Homeowners Foundation; American Hungarian Foundation; Associated Health Foundation

AHF American Hospital Formulary; Azod Hind Fouj (Indian National Army)

AHFCR Anderson Hospital for Cancer Research

AHFS American Hospital Formulary Service

ahg active homing guidance; antihemolytic globulin; antihuman globulin

ahg (AHG) antihemophilic globulin

AHG American Housing Guild

ahh alpha-hydrazine analog of histidine; arylhydrocarbon hydroxylase

AHHA Allied Home Health Association

AHHS Alexander Hamilton High School

AHI American Health Institute; American Honey Institute; American Hospital Institute; Animal Health Insurance

AHIL Association of Hospital and Institution Libraries

AHIP Army Helicopter Improvement Program; Art History Information Program; Australian Health Insurance Program

AHIRS Australian Health Information and Research Service

AHIS American Hull Insurance Syndicate

AHIWG Ad Hoc Intelligence Working Group (NATO)

ahl alcohol-induced hyperlipidemia

a.h.l. ad hunc locum (Latin—at this place)

AHL Alaska Historical Library; American Hockey League; Associated Humber Lines; Association for Holistic Living

ahle acute hemorrhagic leukoencephalitis

AHLMA American Home Laundry Manufacturers Association

ahls antihuman-lymphocyte serum

ahm ampere-hour meter

Ahm Ahmadabad; Arnhem

AHM American Health Magazine

ahma advanced hypersonic manned aircraft

AHMA American Hardware Manufacturers Association; American Hemisphere Marine Agencies; American Holistic Medical Association; American Hotel and Motel Association

AHMC Association of Hospital Management Committees

Ahmed Mohammed Ahmed ibn-Seyyid Abdullah—the Mahdi

AHMI Appalachian Hardwood Manufacturers Incorporated

AHMPS Association of Headmistresses of Preparatory Schools

ahms(s) Asiatic homosexual male(s)

AHMS American Home Missionary Society

AHMSA Altos Hornos de México (Spanish—Great Ovens of Mexico)—steel mills

AHN Assistant Head Nurse

AHNA Accredited Home Newspapers of America; American Holistic Nurses Association

a-hole ass hole, anus

AHOP Assisted Home Ownership Plan

A-horiz A-horizon (soil rich in organic matter just below the surface)

ahp acute hemorrhagic pancreatitis; air at high pressure; air horsepower; aviation horsepower

AHP Accountable Health Plan; Allied Health Professional; American Home Products; American Horse Publications; Assistant House Physician; Association for Humanistic Psychology

AHPA American Herbal Products Association; American Horse Protection Association

AHPC American Heritage Publishing Company

AHPR Academy of Hospital Public Relations

ahps auxiliary hydraulic power supply

ahq alternate headquarters; area headquarters

AHQ Air Headquarters; Allied Headquarters; Army Headquarters

AHQE Air Headquarters Egypt

AHQWD Air Headquarters Western Desert

ahr acceptable hazard rate

AHR Academy of Human Rights; Association for Health Records

AHRC Association for the Help of Retarded Children; Australian Housing Research Council; Australian Humanities Research Council

AHRGB Association of Hotels and Restaurants of Great Britain

AHRQ Agency for Health-care Research and Quality

ahrs attitude heading reference system

ahs ablative heat shield; airborne hardware simulator

AHS Aerospace High School; American Harp Society; American Hearing Society; American Helicopter Society; American Hibiscus Society; American Home Security; American Horticultural Soci-

ety; American Hospital Supply (stock exchange symbol); American Humane Society; American Hypnodontic Society; Assistant House Surgeon; Association for Humanistic Studies; Aviation High School; Aviation Historical Society

AHS *Anno Humanae Salutis* (Latin—the Year of Human Salvation)

AHSA American Hampshire Sheep Association; American Horse Shows Association; Art, Historical, and Scientific Association; Aviation Historical Society of Australia

AHSB Authority Health and Safety Branch

AHSC American Hospital Supply Company

A-H Scale Anti-Hispanic Scale (measuring negative attitudes toward persons of Latin American origin)

AHSCo Assault Helicopter Support Company (USAF)

ahse assembly, handling, and shipping equipment

AHSM Academy for Health Services Marketing

AHSNZ Aviation Historical Society of New Zealand

ahsr air height-surveillance radar

AHSRC American High-Speed Railway Corporation

AHSS Association of Home Study Schools

AHSSPPE Association of Handicapped Student Service Programs in Postsecondary Education

A-H ST Alaska-Hawaii Standard Time

aht acoustic homing torpedo; antihyaluronidase titer

AHT Animal Health Technician; Animal Health Trust; Augmented Histamine Test

AHTD Army Headquarters Transport Division

ahtm(s) Asiatic heterosexual male(s)

AHTN Association of Hospital Television Networks

ahu after hangup

ahv armored heavy vehicle

ahv (AHV) alternative fuel vehicle

a.h.v. *ad hunc vocem* (Latin— at this word)

AHV *Altos Hornos de Vizcaya* (Spanish—Great Ovens of Vizcaya)—steel mills

Ahvenanmaa *(Finnish—Ahvenanmaa Islands)—called Åland by the Swedes*

AHWA Association of Hospital and Welfare Administrators

AHWG Ad Hoc Working Group (USA)

ahwt advanced heavyweight torpedo

ai accidentally incurred; achievement via independence; action item; airborne intercept; aircraft identification; aircraft instruments; anti-icing; aortic incompetence; aortic insufficiency; apical impulse; articulation index; artificial insemination; artificial intelligence; axioincisal; azimuth indicator

a & i abstracting and indexing; accident and indemnity

a. i. *ad interim* (Latin—in the interim)

AI Aaland Islands; Admiralty Instructions; Admiralty Islands; Adult Institutions (New Hampshire); Airborne Infantry; Air India; Air Inspector; Air Installation(s); Airways Inspector, Alianza Interamericana (Inter-American Alliance); Allegheny International; American Institute; Amnesty International; Andaman Islands; Appraisal Institute; Arctic Institute; Army Intelligence; Aspen Institute (Aspen, Colorado); Astrologers International; Avery Island

AI *Altesse impériale* (French—imperial highness)

A/I Aptitude Index

A & I Afars and Issas, Republic of Djibouti (formerly French Somaliland); agricultural and industrial (college or school or subjects); Arts and Industries

aia advise if able; anti-icing additive

AiA Accuracy in Academia

AIA Aerospace Industries Association; Aestheticians International Association; Allergy Information Association; Allred Interaction Analysis; American Institute of Accountants; American Institute of Aeronautics; American Institute of Architects; American Insurance Association; Anthracite Industry Association; Antivenin Institute of America; Archeological Institute of America; Arctic Institute of America; Assistant Inspector of Armaments; Association of Insolvency Accountants; Association of Insurance Attorneys; Australian Insurance Association; Australian Italian Association; Auto International Association

A.I.A. Associate of the Institute of Actuaries

AIAA Aerospace Industries Association of America; American Industrial Arts Association; American Institute of Aeronautics and Astronautics

aiaam advanced intercept air-to-air missile

AIAC Air Industries Association of Canada

AIAD Acronyms, Initialisms & Abbreviations Dictionary

AIADA American Imported Automobile Dealers Association; American International Automobile Dealers Association

AIAE Association of Institutes of Automobile Engineers

AIAESD American International Association for Economic and Social Development (AIA)

AIAG Automotive Industry Action Group

AIAL Associate of the Institute of Arts and Letters

AIAM Association of International Automobile Manufacturers

AIAOS Academic Instructor and Allied Officer School

AIAP Ardmore Industrial Air Park

AIArb Associate of the Institute of Arbitrators

AIAS Australian Institute of Aboriginal Studies; Australian Institute of Agriculture and Science

AIAT Attitude-Interest Analysis Test

AIAW Association for Intercollegiate Athletics for Women

aib aminoisobutyric acid

AIB Academy for International Business; Accident Investigative Branch; Accident Investigative Bureau; Admiralty Interview Board; American Institute of Baking; American Institute of Banking; Anti-Inflation Board; Assassination Information Bureau; Australian Infantry Battalion; Australian Institute of Building

A.I.B. Associate of the Institute of Bankers

AIB Association des Industries de Belgique (Association of Belgian Industries); *Associazione Italiana Biblioteche* (Italian Library Association)

aiba amino-isobutyric acid

AIBA American Industrial Bankers Association

AIBA Association Internationale de Boxe Amateur (French—International Amateur Boxing Association)

AIBC Architectural Institute of British Columbia

AIBCS American Intersociety Board of Certification of Sanitarians

AIBD Associate of the Institute of British Decorators and Interior Designers; Association of International Bond Dealers

aibf advanced internally blown jet flap

aibm (AIBM) anti-intercontinental ballistic missile

AIBM Association Internationale des Bibliothèques Musicales (French—International Association of Music Libraries)

AIBP Associate of the Institute of British Photographers

AIBS American Institute of Biological Sciences

aic air information center; air information codification; air intercept control; aminoimidazole carboxamide

aic (AIC) aircraft in commission

AIC Accelerator Information Center; Advanced Intelligence Center; Agricultural Institute of Canada; Aircraft Industries Center; Allied Intelligence Center; Allied Intelligence Committee; American Immigration Control; American Indian Center; American Institute of Chemists; American Institution of Cooperation; Ammunition Identification Code; Arab Information Center; Arab Investment Company; Arkansas Intercollegiate Conference; Army Industrial College; Army Intelligence Center; Art Information Center; Art Institute of Chicago; Artificial Illumination Centre; Australian Institute of Cartographers; Australian Institute of Criminology; Automotive Information Council

AICA American Institute of Commemorative Art; Australasian Institute of Cost Accountants

AICA Association Internationale des Critiques d'Art (French—International Association of Art Critics); *Associazione Italiana per il Calco*

Automatico (Italian—Italian Association for Automatic Data Processing)

aicar amino-imidazolecarboxamide ribonucleotide

AICB Association Internationale Contre le Bruit (French—International Association Against Noise)

aicbm (AICBM) anti-intercontinental ballistic missile

aicc antibody-induced cell-mediated cytoxicity

AICC All-India Congress Committee; Association of Independent Corrugated Converters

AICCC American Institute of Child Care Centers

AICCU Association of Independent California Colleges and Universities

AICE Agency for Information and Cultural Exchange (formerly the USIA); American Institute of Chemical Engineers; American Institute of Consulting Engineers

AI-CE Atomic International Combustion Engineering

aicf auto-immune complement fixation

AICF America-Israel Cultural Foundation; American Immigration Control Foundation; American Indian College Fund

aich automatic integrated container handling

AIChE American Institute of Chemical Engineers

AICI Association of Image Consultants International

AICK Associated Independent Colleges of Kansas

AICM Association of Independent Conservators of Music; Australian Institute of Credit Management

AICMA Association Internationale des Constructeurs de Matériel Aéronautique (French—International Association of Builders of Aeronautical Material)

AICMDM Association of Independent Copy Machine Dealers and Manufacturers

AICO Action Information Control Office(r)—USN; American Insulator Corporation

AICP Association of Independent Commercial Producers

AICPA American Institute of Certified Public Accountants

AICPOA Advanced Intelligence Center Pacific Ocean Area

AICQ Associazione Italiana per il Controllo della Qualità (Italian —Italian Association for Quality Control)

AICR American Institute for Cancer Research

AICRO Association of Independent Contract Research Organizations

aics air inlet control system

AICS Air Induction Control System; Air Intercept Control School; American Institute of Ceylonese Studies; Association of Independent Colleges and Schools

A.I.C.S. Associate of the Institute of Chartered Shipbrokers

AICS Association Internationale du Cinéma Scientifique (French—International Scientific Film Association)

AICTA Associate of the Imperial College of Tropical Agriculture

AICU Association of International Colleges and Universities

AICUM Association of Independent Colleges and Universities in Massachusetts; Association of Independent Colleges and Universities of Michigan

AICUN Association of Independent Colleges and Universities of Nebraska

AICUO Association of Independent Colleges and Universities of Ohio

AICUZ Air Installation Compatible Use Zone

aicv armored infantry combat vehicle (AICV)

aid acute infectious disease; applications interface device; artificial insemination donor; avalanche injection diode

Aid Aideen

AID Agency for International Development; Airline Interline Development; Aircraft Intelligence Department; American Institute of Decorators; American Instructors of the Deaf, Arkansas Information Dissemination; Army Information Digest; Army Intelligence Department; Artificial Insemination Donor; Association for International Development; Automatic Interaction Detector (program)

A & ID Acquisition and Improvement District

AID Acronyms and Initialisms Dictionary; Association Internationale des Documentalists et Techniciens de l'Information (French—International Association of Documentalists and Information Technicians)

aida attention-interest-desire-action (marketing formula); automatic instrumented diving assembly; automobile information data advertising

aida (AIDA) automatic intruder-detector alarm

AIDA Advanced Integrated Circuit Design Aids; American Indian Development Association; Associated Independent Dairies of America; Australian Industries Development Association

AIDATS Army In-flight Data Transmission System

AIDC American Industrial Development Council; Arkansas Industrial Development Commission; Asian Industrial Development Council; Association of Information and Dissemination Centers; Australian Industries Development Corporation

AIDD American Institute of Design and Drafting

aide airborne insertion display equipment; aircraft installation diagnostic equipment

AIDE Adapted Identification Decision Equipment; American Institute of Driver Education; Arizona Information Dissemination for Educators; Automated Image Device Evaluation

AIDE Association Internationale des Distributions d'Eau (French—International Water Supply Association)

aidecs automatic inspection device for explosive charge shell

AIDI Associazione Italiana per la Documentazione e l'Informazione (Italian—Italian Association for Documentation and Information)

AIDIA Associate of the Industrial Design Institute of Australia

AIDIS Asociación Interamericana de Ingeneria Sanitaria (Spanish—Inter-American Association of Sanitary Engineering)

AIDL Auckland Industrial Development Laboratory; Australian Industrial Development Laboratories

AIDMED Assistant for Interacting with Multimedia Medical Databases

AIDP Advanced Institutional Development Program; Associate of the Institute of Data Processing

AIDP Association Internationale de Droit Pénal (French—International Association of Penal Law)

AIDRB Army Investigational Drug Review Board

aids acute infectious diseases; aircraft integrated data system

aids (AIDS) acquired immune deficiency syndrome

AIDS Abstracts Information Dissemination System; Account Identification and Description Services; acquired immune deficiency syndrome; Action Information Display System; Activity Information Data System; Administrative Information Data System; Advanced Integrated Data System; Aerospace Intelligence Data System; Aircraft Intrusion Detection System; Air Force Intelligence Data System; Alabama Information Development System; American Institute for Decision Services; Army Information and Data Systems; Atlantic Institute for Defense Studies; Automated Identification Division System; Automatic Inventory Dispatching System

AI & DSC Army Information and Data System Command

AIDSCOM Army Information Data Systems Command

AIDT All-Inclusive Deed of Trust

AIDUS Automated Information Directory Update System; Automated Input and Document Updating System

AIE American Institute of Esthetics; Australian Institute of Export

AIEA Agence Internationale de l'Energie Atomique (French—International Atomic Energy Agency); *Agencia Internacional de Energia Atómica* (Spanish—International Atomic Energy Agency)

AIEAL Australian Institute of Engineering Associates Ltd

AIEC Association of Iron Exporting Countries

AIECF American Indian and Eskimo Cultural Foundation

A.I.Ed. Associate in Industrial Education

AIEE American Institute of Electrical Engineers

AIEF Association Internationale des Etudes Françaises (French—International Association for French Studies)

AIEI Association of Indian Engineering Industry

aiep amount of insulin extracted from the pancreas

AIER American Institute for Economic Research

AIERI Association Internationale des Etudes et Recherches sur l'Information (French—International Association for Mass Communication Research)

AIES Accreditation and Institutional Eligibility Staff

AIESEC Association Internationale des Etudiants en Sciences Economiques et Commerciales (French—International Association of Students in Economics and Commerce)

AIEST Association Internationale d'Experts Scientifiques du Tourisme (French—International Association of Scientific Experts in Tourism)

AIETA Airborne Infra-red Equipment for Target Analysis

aif automated intelligence file

AIF Air Intelligence Force; American Industry Foundation; American Institute of France; Amphibian Imperial Forces; Army Industrial Fund; Assembly Inactive File; Atomic Industrial Forum; Atomic International Forum; Australian Imperial Forces; Australian Institute of Fuel

AIF Agencia Internacional de Fomento (Spanish—International Development Agency); *Agenzia Internazionale Fides* (Italian—International Faith Agency—Vatican State News Service); *Alliance Internationale des Femmes* (French—Women's International Alliance); *Asociación Internacional de Fomento* (Spanish—International Development Association)—IDA

AIFA Associate of the International Faculty of Arts

AIFCS Airborne Interception Fire-Control System

AIFD Alaska Institute for Fisheries Development; American Institute of Floral Designers; American Institute of Food Distribution

AIFE Associate of the Institution of Fire Engineers

aiff audio interchange file format

AIFLD American Institute for Free Labor Development

AIFM Association Internationale des Femmes Médecins (French—International Association of Women Doctors)

AIFO Army Instructions and Force Orders

AIFR American Institute of Family Relations

aifs advanced indirect fire system; advanced infantry fire system

AIFS Advanced Instruction Flying School; American Institute for Foreign Study

AIFSA American Institutions Food Service Association

AIFST Australian Institute of Food Science and Technology

AIFT American Institute for Foreign Trade; Americans for Indian Future and Traditions

AIFTA Anglo-Irish Free Trade Area

AIFUCTO All-India Federation of University and College Teachers Organization

aifv (AIFV) armored infantry fighting vehicle

aig all inertial guidance; angle of inner gimbal

Aig Aiguille (French—needle, peak)

AIG Address Indicating Group; Adjutant Inspector General; American International Group (Incorporated); Assistant Inspector-General; Assistant Instructor in Gunnery

AIG Association Internationale de Géodesia (French—International Geodesy Association)

AIGA American Institute of Graphic Arts

AIGAM Australian Institute of Graphic Art Management

AIGCM Associate of the Incorporated Guild of Church Musicians

AIGS Agricultural Investment Grant Scheme

AIGT Association for the Improvement of Geometrical Teaching

aih artificial insemination by husband

AIH American Institute of Homeopathy; Aspen Institute of the Humanities; Australian Institute of Horticulture

AIH Association Internationale de l'Hotellerie (French—International Hotel Association)

aiha autoimmune hemolytic anemias

AIHA American Industrial Hygiene Association

AIHC American Industrial Health Conference

AIHE Association for Innovation in Higher Education

AIHEC American Indian Higher Education Consortium

AIHED American Institute for Human Engineering and Development

AIHR Australian Institute of Human Relations

AIHS American Indian Historical Society; American Irish Historical Society; Aspen Institute of Human Studies; *Association Internationale d'Hydrologie Scientifique* (French—International Association of Scientific Hydrology); Australian Institute of Health Surveyors

AIHSC Auto Industry Highway Safety Committee

AII Air India International; Army Intelligence Interpreter; Australian Insurance Institute

AIIA Association of International Insurance Agents; Australian Institute of Incorporated Accountants; Australian Institute of International Affairs

AIIC Army Imagery Intelligence Corps; Associate of the Insurance Institute of Canada

AIID American Institute of Interior Designers

AIIDC Authorized Item Identification Data Collaborator

AIIDR Authorized Item Identification Data Receiver

AIIDS Authorized Item Identification Data Submitter

AIIE American Institute of Industrial Engineers

AIIM Association for Information and Image Management

AIIMS All-India Institute of Medical Sciences

AIInfSc Associate of the Institute of Information Scientists

AIIP Australian Institute of Industrial Psychology

aiis advanced infra-red image seeker

AIIS American Institute for International Steel; American Institute of Indian Studies; American Intraocular Implant Society

AIK Assistance-in-Kind (funds)

AIKD American Institute of Kitchen Dealers

ail aileron

Ail Aileen

AIL Aeronautical Instruments Laboratory; Air Intelligence Liaison; Airborne Instruments Laboratory; American Institute of Laundering; American Israeli Lighthouse; Art Institute of Light; Association of International Libraries; Australian Institute of Librarians; Aviation Instrument Laboratory

A.I.L. Associate of the Institute of Linguistics

AILA American Institute of Landscape Architects; Australian Institute of Landscape Architects

A.I.L.A. Associate of the Institute of Landscape Architects

AILA *Association Internationale de Linguistique Applique* (French—International Association of Applied Linguistics)

AILAS Automatic Instrument Landing Approach System

AILC American Indian Law Center

AILC *American International Law Cases*

Ailie Aileen; Alice; Alicia; Alison; Helen; Helena

AILO Air Intelligence Liaison Office(r)

AILS Advanced Integrated Landing System

AILSA American Indian Law Students Association

ailuroph ailurophile (fancier or lover of cats); ailurophobe (one who fears or hates cats)

aim active inert missile; acrotriangulation (by observation of) independent models; advanced intercept missile; air intercept missile; air-isolated monolithic (circuit); armored infantry mechanized

AiM Anglicans in Mission

A i M Accuracy in Media; Adventures in Movement

AIM Abstracts of Instructional Materials; Academy Introduction Mission (USCG); Accuracy In Media; Advanced Informatics in Medicine; Aide Inter-Monasteries; Amcrican Indian Movement; Almond Industry Market; American Institute of Management; American Institute of Musicology; Army Installation Management; Association of Incentive Marketing; Association for Information Management; Association for the Integration of Management; Automatic Identification Manufacturers; Australian Institute of Management; Australian Institute of Metals

AIM *Abstracts of Instructional Material; Airman's Information Manual*

AIM-9 Sidewinder air-to-air missile

AIM47A Hughes air-to-air missile

aima as interest may appear

AIMA All-India Management Association

AIMACC Air Material Command Compiling (system)

AIMACO Air Materiel Command Compiler (language)

AIM Bankers Asian International Merchant Bankers (Singapore)

AIMBW American Institute of Men's and Boy's Wear

AIMC Accredited Investment Management Consultant; American Indian Medical Clinic; American Institute of Medical Climatology; Association of Internal Management Consultants; Association of Interstate Motor Carriers

AIMCAL Association of Industrial Metallizers, Coaters and Laminators

aime (AIME) average indexed monthly earnings

AIME American Institute of Mechanical Engineers; American Institute for Mining Engineers; Association for Informational Media and Equipment

AIMES Association of Interns and Medical Students

AIMF American International Music Fund; Australian Institute of Metal Finishing

AIMH Academy of International Military History

AIMILO Army/Industrial Material Information Liaison Office(s)

AIMIT Associate of the Institute of Musical Instrument Technicians

AIML All-India Muslim League

AIMLT Australian Institute of Medical Laboratory Technology

AIMM Association of Importers–Manufacturers for Muzzleloading; Australasian Institute of Mining and Metallurgy

AIMME American Institute of Mining and Metallurgical Engineers

AIMMPE American Institute of Mining, Metallurgical, and Petroleum Engineers

aimo air mold; audibly-instructed manufacturing operations (AIMO)

AIMP Association of Independent Music Publishers

aimp air intercept missile package

AIMPA *Association Internationale de Météorologie et de Physique de lAtmosphère* (French—International Association of Meteorology and Atmospheric Physics)

AIMPE Australian Institute of Marine and Power Engineers

AIMR Association for Investment Management and Research

AIMRA Agricultural and Industrial Manufacturers Representatives Association

AIMS Advanced Intercontinental Missile System; Air Traffic Control Radar Beacon/Identification Friend or Foe/Mark XII/System; American Institute for Marxist Studies; American Institute for Mathematical Statistics; American Institute for Mental Studies; American Institute of Merchant Shipping; American Institute of Musical Studies; Association of Independent Maryland Schools; Association for International Medical Study; Australian Institute of Marine Science; Australian Institute of Marine Studies; Automated Instructional Materials Services; Automatic Industrial Management System

AIMT Australian Institute of Medical Technologists

A.I.M.T.A. Associate of the Institute of Municipal Treasurers and Accountants

AIMU American Institute of Marine Underwriters

ain acute interstitial nephritis; approved item name

AIN Aircraft Identification Number; American Institute of Nutrition; Association of Interpretive Naturalists; Australian Institute of Navigation

aina automated immunoephelometric assay

AINA American Indian Nurses Association; American Institute of Nautical Archeology; American Israel Numismatic Association; Arctic Institute of North America

A-Ind Anglo-Indian

AINDT Australian Institute for Non-Destructive Testing

AINEC All-India Newspaper Editors' Conference

AINL Association of Immigration and Naturalization Lawyers

AINO Assistant Inspector of Naval Ordnance

AINS Assateague Island National Seashore (Maryland and Virginia)

AINSE Australian Institute of Nuclear Science and Engineering

ainsuf aortic insufficiency

ain't am not, are not, has not, have not, is not

AINWR Aleutian Islands National Wildlife Refuge (Alaska)

AINZ Advertising Institute of New Zealand

aio activity, interest, opinion (marketing factors); activity-interest-option (marketing factor scores)

Aio Aioi

AIO Action Information Organization; Air Installation Office; Air Intelligence Organization; Allied Interrogation Organization; American Institute of Organbuilders; Amer-

icans for Indian Opportunity; Anglican Information Office; Arecibo Ionospheric Observatory; Artillery Intelligence Officer

AIOB American Institute of Oral Biology

AIOEC Association of Iron Ore Exporting Countries

AIOPI Association of Information Officers in the Pharmaceutical Industry

aip ablative insulative plastic; accident insurance policy; acute intermittent porphyria; average intravascular pressure

aip (AIP) aldosterone-induced protein; automated imagery processing

AIP Aeronautical Information Publication; Aerovias Panama (Panamanian airline); Allied Intelligence Publication; American Independent Party; American Institute of Physics; American Institute of Planners; American Institute for Psychoanalysis; Attention and Information Processing; Australian Institute of Petroleum; Australian Institute of Physics; Automatic Investment Plan

A-I-P Afghanistan-Iran-Pakistan

AIP *Association Internationale de Papyrologues* (French—International Association of Papyrologists); *Association Internationale de Pediatrie* (French—International Pediatric Association)

AIPA American Indian Press Association; American Institute for the Prevention of Addiction

AIPA *Association Internationale de la Psychologie Adlerienne* (French—International Association of Adlerian Psychology)

AIPAC American-Israeli Public Affairs Committee

AIPB Arnerican Institute of Professional Bookkeepers

AIPC All Indian Pueblo Council; Australian Institute of Pest Control

AIPC *Association Internationale de Prophylaxie de la Cécité* (French—International Association for the Prevention of Blindness); *Association Internationale des Ponts et Charpentes* (French—International Association of Bridges and Scaffolds)

AIPCEE *Association des Industries du Poisson de la Communauté Economique Européenne* (French—Association of Fishing Industries of the European Economic Community)

AIPCN *Association Internationale Permanente des Congrés Navigation* (French—Permanent Sailing International Association of Congresses)

AIPCR *Association Internationale Permanente des Congrés de la Route* (French—Permanent Road International Association of Congresses)

AIPE American Institute of Park Executives; American Institute of Plant Engineers; American Institute for Professional Education

AIPG American Institute of Professional Geologists

AIPHE Associate of the Institution of Public Health Engineers

AIPLU American Institute for Property and Liability Underwriters

AIPO American Institute of Public Opinion; Asian Inter-Parliamentary Organisation

AIPR American Institute of Pacific Relations; Australian Institute of Parks and Recreation

AIPs Allied Intelligence Publications; Association of Irish Priests

AIPS AIDS-induced panic syndrome; Australian Institute of Political Science; Automatic Indexing and Proofreading System

AIPS *Association Internationale pour la Prevention du Suicide* (French—International Association for the Prevention of Suicide)

AIPT Assistant Instructor Physical Training

AIQ Associate of the Institute of Quarrying

AI & Q Animal Inspection and Quarantine

AIQS Australian Institute of Quantity Surveyors

A.I.Q.S. Associate of the Institute of Quantity Surveyors

air airborne intercept radar; average injection rate

a-i-r artist-in-residence

AIR Action for Industrial Recycling; Air Control Products; All-India Radio; American Institute of Refrigeration; American Institute of Research; Arkansas Intermediate Reformatory; Army Institute of Research; Army Intelligence Reserve; Association for Immigration Reform; Association for Institutional Research; Australian Industrial Refractories; Australian Institute of Radiography

AIR *Amnesty International Report; Asociación Interamericana de Radiodifusión* (Spanish—Interamerican Broadcasters Association)

AIR-2A Douglas air-to-air rocket fitted with a nuclear warhead and called Genie

AIRA Air Attaché; American Independent Refiners Association

airac (AIRAC) aeronautical information regulation and control

AIRAH Australian Institute of Refrigeration, Air Conditioning, and Heating

AIRB Alabama Inspection and Rating Bureau; Arkansas Inspection and Rating Bureau; Australian Industrial Relations Bureau

AIRBALTAP Allied Air Forces Baltic Approach (NATO)

airbm (AIRBM) anti-intermediate-range ballistic missile

AIRBR *Association Internationale du Registre des Bateaux du Rhin* (French—International Association of Rhine Ship Registry)

AIRCAL Air California

Air Can Air Canada (formerly Trans-Canada Air Lines)

aircat automated integrated radar control for air traffic

Air Cav Airmobile Cavalry

Air Cdr Air Commander

AIRCENT Allied Air Forces, Central Europe

AIRCEY Air Ceylon

Air Cmdre Air Commodore

AIRCO Air Reduction Chemical Company

AIRCOM AirForce Communication Complex

AIRCOMNET Air Communications Network

AIRCOMS Airways Communications System

aircond air condition(ed); air conditioning

AIRDEF Air Defense (NATO division)

AIRDEP Air Deputy (NATO)

AIREA American Institute of Real Estate Appraisers

AIREASTLANT (Naval) Air Forces Eastern Atlantic (NATO)

AiRepDn Aircraft Repair Division

airew airborne infrared early warning

AIRFA American Indian Religious Freedom Act

airfil air filter(s)

Air Force I Air Force One (aircraft reserved for or used by the president of the U.S.)

AIRII *Association Internationale des Recherches Hydrauliques* (French—International Association of Hydraulic Research)

air hp air horsepower

AIRI Atomic Industry Research Institute

AIRIMP ATC/IATA (*q.q.v.*) reservations interline message procedures

Air Jam Air Jamaica

AIRL Aeronautical Icing Research Laboratory

AIRLANT Air Forces Atlantic Fleet

AIRLEX Air Landing Exercise

Air LO Air Liaison Officer

AIRLORDS Airlines Load Optimization Recording and Display System

Air Mad Air Madagascar

airmada airplane aramada

airmap air monitoring, analysis, and prediction

AIRMIC Association of Insurance and Risk Managers in Industry and Commerce

airmiss aircraft-in-flight collision barely missed

air/mmh acoustic intercept receiver/multiniode hydrophone

AIRMOVEX Air Movement Exercise

Air NG Air National Guard

Air Niu Air Niugini (national airline of Papua New Guinea)

AIRNON Allied Air Forces in Northern Norway (NATO)

AirNorth Allied Air Forces, Northern Europe

Air NZ Air New Zealand

AIRO Army in India Reserve Officers

AIROPNET Air Operational Network

AIRPAC Air Pacific

AIRPASS Aircraft Interception Radar and Pilots Attack Sight System

airpl airplane(s)

Air Port Air Portugal

Air Prog Air Progress

airpt art airport art

airr (AIRR) adjusted internal rate of return

Air Res Squad Air Reserve Squadron

airs. (AIRS) advanced inertial reference sphere

AIRS Aircraft Inventory Reporting System; Airline Interline Reservations System; Automatic Image Retrieval System

AIRSONOR Allied Air Forces in Southern Norway (NATO)

AIRSouth Allied Air Forces, Southern Europe

Air-Std Air Force International Standard

airsurance air insurance

Air Svc Air Service

airtel air + hotel, airport hotel

airvan airmobile van

AIRWORK Airwork Atlantic Limited

AIRX American Industrial Radium and X-ray Society

ais ablating inner surface; active isolated stretching; agreed industry standard; answer in sentence; average insurance set

ais (AIS or Lunik III) automatic interplanetary station

AIS Abbreviated Injury Scale; Academic Instructors School (USAF); Administrative and Information Services; Advanced Information System; Aeronautical Information Service; Air Intelligence Service; Alexander I Solzhenitsyn; American Indian Scholarships; American Indian Studies; American Interplanetary Society; American Iris Society; American Israeli Shipping (Zim Lines); Army Intelligence School; Association of Iron and Steel; Australian Information Service; Australian Institute of Sport; Australian—Iron and Steel

AI & S Army Intelligence and Security

AIS *Association Internationale de la Savonnerie et de la Detergence* (French—International Association of Soaps and Detergents); *Association Internationale de Sociologie* (French—International Sociology Association)

aisa analytical isoelectrofocusing scanning apparatus

AISA Associate of the Incorporated Secretaries Association; Australian Institute of Systems Analysis

AISA *Association Internationale pour la Sécurité Aérienne* (French—International Air Security Association)

AISB Artificial Intelligence and Simulation of Behavior

AISC American Institute of Steel Construction; Association of Independent Software Companies; Australian Institute of Steel Construction

AISE American Intercultural Student Exchange; Association of Iron and Steel Engineers

AISE *Association Internationale des Sciences Economiques* (French—International Association of Economic Sciences)

AISI American Iron and Steel Institute

AISJ *Association Internationale des Sciences Juridiques* (French—International Association of Juridical Sciences)

AISM *Association Internationale des Sociétiés de Microbiologie* (French—International Association of Microbiology Societies)

AISS *Association Internationale de la Science du Sol* (French—International Solar Science Association)

AIST Australian Institute of Science Technology

aisv advanced infantry support vehicle

ait advanced individual training; advanced infantry training; auto-ignition temperature

AiT *Anjuman-i-Tarikh* (Historical Society of Afghanistan)

AIT Agency for Instructional Technology; Agency for Instructional Television; American Institute in Taiwan; American Institute of Technology; Army Intelligence

Translator; Australian Institute of Travel; Automatic Information Test

AIT *Académie Internationale du Tourisme* (French—International Academy of Tourism)

AITA Act Inside the Army (antiwar society); Air Industries and Transport Association; Australian Industrial Truck Association

AITC Action Information Training Center; American Institute of Timber Construction

AITC *Association Internationale des Traducteurs de Conference* (French—International Association of Conference Translators)

aitd all-inclusive trust deed

AITD Australian Institute of Training and Development

AITF Art Information Task Force

AITI Aero Industries Technical Institute; American Institute of Technical Illustrators

AITO Association of Independent Tour Operators

AITVS Association of Independent Television Stations

aiu abort interface unit; absolute iodine uptake; advanced instrumentation unit

AIU Action for Interracial Understanding; Aero Insurance Underwriters; American International Underwriters

AIU *Alliance Israelite Universelle* (French—Universal Israelite Alliance)

AIUM American Institute of Ultrasound in Medicine

AIUS Australian Institute of Urban Studies

AIUSA Amnesty International in the United States of America

aiv accelerated inverse voltage

AIV *Association Internationale de Volcanologie* (French—International Association of Vulcanology)

AIVAF American-Israeli Vocal Arts Foundation

AIVF Association of Independent Video and Filmmakers

aiw auroral intrasonic wave

AIW Atlantic Intracoastal Waterway (Cape Cod to Florida Bay)

AIWC Air Intelligence Watch Condition

AIWM American Institute of Weights and Measures

Aix Aix-cn-Provence

AIX Advanced Interactive Executive—a version of UNIX (trademark IBM)

Aix-la-Chap Aix-la-Chapelle (French—Aachen)

Aix-les Aix-les Bains (French)

AIY Ayrshire Imperial Yeomanry

AIYW Association for International Youth Work

aiz air intercept zone

aj ankle jerk; antijamming; apple juice; attack jet

aj a jini (Czech—and others)

AJ Air Jordan; Alma & Jonquieres (railroad); Americans for Justice; Andrew Jackson (7th U.S. President); Andrew Johnson (17th U.S. President); Associate Justice

A.J. Associate in Journalism

AJ American Jurisprudence; Architects Journal; l'Arméé Juife (French—Jewish Army)–anti-Nazi resistance group

AJ-37 Swedish Thunderbolt multimission combat aircraft also called Viggen

AJA American Jail Association; American Jewish Archives; American Judges Association; Australian Journalists Association

A-JA Anglo-Jewish Association

AJA American Journal of Anatomy

AJAC Automatic Jamming Avoidance Circuitry

AJAC American Journal of Alzheimer's Care

AJADD Australian Journal of Alcoholism and Drug Dependence

AJAG Assistant Judge Advocate General

ajai antijamming anti-interference

AJAs Americans of Japanese Ancestry

AJAS Australian Journal of Applied Science

AJASS African Jazz Art Society Studios

Ajax Douglas Nike-Ajax surface-to-air missile; mytholog ical Greek hero of the Trojan Wars and title of a play by Sophocles

AJAZ American Jewish Alternatives to Zionism

AJB Arthur J(ames) Balfour

AJB Association des Juifs de Belgique (French—Association of the Jews of Belgium); *Australian Journal of Botany*

AJBC American Junior Bowling Congress

AJBP Association of Jewish Book Publishers

aJc antes de Jesucristo (Spanish—before Jesus Christ)

AJC Altus Junior College; American Jewish Committee; American Jewish Congress; Anderson Junior College; Australian Jockey Club

AJCA Association of Juvenile Compact Administrators

AJCC Alternate Joint Communications Center

AJCL Australia-Japan Container Line

AJC-RC American Jewish Committee-Records Center

AJCSA All-Japan Cotton Spinners Association

AJCU Association of Jesuit Colleges and Universities

AJCW Association of Jewish Center Workers

AJDC American Joint Distribution Committee; Asian-Japan Development Corporation

AJE Adult Jewish Education

AJEI Australia-Japan Economic Institute

AJF Asian-Japan Forum; Australia-Japan Foundation

AJHS American Jewish Historical Society; Andrew Jackson High School

AJI American Justice Institute

AJIF Australia-Japan International Finance

AJIL American Journal of International Law

AJIS Automated Jail Information System

AJJUST Automated Juvenile Justice System Technique

AJL Association of Jewish Libraries; Association of Junior Leagues

AJLAC American Jewish League Against Communism

AJNHS Andrew Johnson National Historic Site, Greeneville, Tennessee

ajo antijam operator

ajp alarm and jettison panel

AJP Additional-Jobs Programme (New Zealand)

AJPA American Jewish Press Association

AJPO Ada Joint Program Office

AJPRS American Jewish Public Relations Society

AJR Association of Jewish Refugees

AJR American Journalism Review; Australian Jurist Reports

AJRC American Junior Red Cross

AJRJ Association of Japanese Residing in Japan

AJS American Judicature Society; Australia–Japan Society

AJS American Journal of Sociology

AJSJ American Justinian Society of Jurists

AJSR Australian Journal of Scientifuc Research

AJSS Australian Joint Staff Service

ajvd abrupt junction varactor doubler

AJVR American Journal of Veterinary Research

AJWS American Jewish World Service

AJY Association for Jewish Youth

AJYB American Jewish Year Book

ak above the knee; ass kisser

a k *alter kocker* (Yiddish colloquialism—old man)

Ak Auckland, New Zealand

AK Alaska; Alaska Coastal—Ellis Airlines; Arthur Koestler (Hungarian journalist); Australian Knight (Knight of the Order of Australia); cargo ship (naval symbol)

AK *Armia Krajowa* (Polish—Home Army); *Avtomat Kalasnikov* (Russian—submachine gun)

AK 47 automatic rifle

aka above-the-knee amputation; also known as

Aka *Akasaka* (Tokyo district)

AKA American Kiteflyers Association; Associated Klans of America; Australian Karate Association; Australian Kidney Association; cargo vessel, attack (3-letter coding)

Akad *Akademie* (German—Academy); *Akademi* (Dano-Norwegian—Academy)

Akad Nauk Akademiya Nauk (USSR Academy of Sciences)

AKAG Albright-Knox Art Gallery

ak amp above-the-knee amputation

Akan Akan National Park on Hokkaido Island, Japan

AKBS Advanced Kinematic Bombing System

AKC American Kennel Club; Associate King's College

AKCF Arthur Kill Correctional Facility

AKERR Allied Kinetic Energy Recovery Rope

AKES Aga Khan Educational Services

AKF American-Korean Foundation; Australian Koala Foundation

AKG *Astronomischen Gesellschaft Katalog* (German—Astronomical Society Catalog)

AKHS Aga Khan Health Services

AKI Alfred Kinsey Institute; American Kynol Incorporated

akk *akkusativ* (Dano-Norwegian—accusative)

Akk Akkadian

Akord *Akkordeon* (German—accordion)

AKL *Algemene Kunstzijde Unie* (Artist's Union); Auckland, New Zealand (airport)

AKLD Auckland

AKM Soviet military weapon capable of firing 600 rounds per minute

AKN King Salmon, Alaska (airport)

Akr Akron

Akr *Akra* (modern Greek—cape); *Akrotirion* (Modern Greek—Cape)

AKR vehicle cargo ship (naval symbol)

Akropolis (Greek—Upper City)—Acropolis Hill section of Athens

AKRSP Aga Khan Rural Support Programme

AKS general stores issue ship (3-letter symbol)

AKSEA Alaska Society of Enrolled Agents

Akt *Aktiebolag* (Swedish—limited company)

AKT Akita Television (Japan)

Akt Ges *Aktiengesellschaft* (German—corporation or joint stock company)

Aktieb *Aktiebolag* (Swedish—limited company)

Akties *Aktieselskab* (Swedish—joint stock company)

akubra (Australian—hat; not a short form but a trademark)

Akust Zh *Akusticheskii Zhurnal* (Russian—Acoustics Journal)

AKV cargo ships and aircraft ferries (3-letter symbol)

akwic author and key word in context

al absolute limit; accommodation ladder; action limit; adaption level; air landing; air lock; albumin; alcohol; alias; all lengths; almond; annual leave; assisted living; autograph letter; axiolingual

al *alumnos* (Spanish—pupils, students)

a l *aprés livraison* (French—after delivery)

a/l airlines

al-63 anti-louse powder

aL assumed latitude

Al accommodation ladder; air lock; Alan; Albert; Albin; alcohol; Alden; *Alemanha* (Portuguese—Germany); Alex(ander); Alf, Alfred; alias; Allan; Allen; Alley; Allied; all lengths; Alton; aluminum; Alva; Alvah; Alvin; Alvina; Alym; annual leave; autograph letter

Al. Book of Alma

AL Abraham Lincoln; Accession List(s); Acoustics Laboratory; Admiralty Letter; Aeronautical Laboratory; Air Liaison; Aircraft Laboratory; Aircraft Logistics; Alabama; Albania (Internet code); Aleutian low; Allegheny Airlines; Aluminum Limited; aluminum (machine shop symbol); American League; American Legion; Amyloidosis; *Angkatan Laut* (Indonesian—Naval Forces); Anglo-Latin; Annual Lease; Annual Leave; Arab League; Architectural League; Army List; Assumed Latitude; Astronomical League; Aviation Electronicsman

A-L Allegheny-Ludlum; Anglo-Latin

A/L airlift

AL *América Latina* (Portuguese or Spanish—Latin America)

A.L. *Anno Lucis* (Latin—in the Year of Light)

AL-60 Lockheed Associates Conestoga—six-seat piston-powered transport aircraft

ala alanine; alpha-linolenic acid; alternate living arrangement; aminolevulinic acid; axiolabial

ala (ALA) alighting area

Ala Alabama; Alabamian; Alameda(n)

ALA Alternate Living for the Aged; Amalgamated Lithographers of America, American Laminators Association; American Landrace Association; American Landscape Architects; American Laryngological Association; American Latvian Association; American Legion Auxiliary; American Liberal Association; American Library Association; American Lighting Association; American Livestock Association; American Lung Association; Arkansas Library Association; Army Launching Area; Assembly of the Lbrarians of the Americas; Association of Legal Administration; Australian Lebanese Association; Australian Legal Aid; Authors League of America

ALA (I) Amalgamated Lithographers of America (Independent)

A.L.A. Associate in Liberal Arts; Associate of the (British) Library Association

ALAA Associate of the Library Association of Australia

alaar air-launched air-recoverable rocket

A-lab acid (LSD) laboratory

alabaster calcite (onyx marble); variety of gypsum

ALABEL American Library Association Board of Education for Librarianship

alabol algorithmic and business-oriented language

ALAC Alcoholic Liquor Advisory Council (New Zealand); Artificial Limb and Appliance Centre; Australian Labor Advisory Council

ALACP American League to Abolish Capital Punishment

alad abnormal left axis deviation; aminolevulinic acid dehydrase; automatic liquid agent detector

ALAD Army Legal Aid Department

aladdin atmospheric layer and density distribution of ions and neutrons

ALADI *Asociación Latinoamericano de Integración* (Spanish—Latin American Integration Association)

ALAEA Australian Licensed Aircraft Engineers Association

alag axiolabiogingival

Al Ahr Al Ahram (Arabic—The Pyramids)—Cairo's daily paper

alairs advance low-altitude infrared-reconnaissance sensor

ALA-ISAD American Library Association–Information Science and Automation Division

Alaj Alajuela, Costa Rica

alal axiolabiolingual

ALAL Association of Legal Aid Lawyers

ALALC Asociación Latinoamericana de Libre Comercio (Spanish—Latin American Free Trade Association)

ALAM Associate of the London Academy of Music

Alan Alain; Allan; Allen

Åland (Swedish—Aland Islands)–between the Gulf of Bothnia and the Baltic Sea separating Sweden from Finland

Alanders Aland islanders

Alands Aland Islands

alanon alcoholics' anonymous

ALAO Australian Legal Aid Office

alara as low as reasonably achievable

alarm air-launched anti-radiation missile; automatic light aircraft readiness monitor

ALARM Assessment of Language and Reading Maturity

alarr air-launched air recoverable rocket

Alas. Alaska; Alaskan

ALAS Army Library Automated Systems; Army Light Aircraft Squadron; Automated Literature Alerting System

A.L.A.S. Associate in Letters, Arts, and Sciences

alasca all aspect capability

Alas Cur Alaska Current

Alas DST Alaskan Daylight Saving Time

Alasia Australasia

Al-Ass Al-Assifa (Syrian terrorist group)

Alas ST Alaskan Standard Time (150th meridian west of Greenwich)

Alastair Alexander

Al-At Alma-Ata (former name for Almaty, capital of Kazakhstan)

ALAT Army Language Aptitude Test

ALATB Air-Landing Anti-Tank Battalion

alateen program for teenagers affected by someone else's drinking *(see* prealateen)

a la v a la vista (Spanish—at sight, payable upon presentation)

alaw advanced light anti-tank weapon

alb albanisch (German—Albanian); albumin; aluminum-bronze

alb (ALB) air-land battle

ALB Army Language Bureau

Alb Albania; Albanian; Albany; Albert; Alberta; Albertan; Albion; Albalasserdarn

ALB American League Baseball

ALBA Aluminium Bahrain; American Lawn Bowling Association

Albac Albacete

Alban Albania; Albanian

Alban Albanensis (Latin—of St. Albans)

Albania Republic of Albania (smallest of the Balkan nations, once called Illyria by the Romans)

Albanian Ports (north to south) Shengjin, Durres, San Nicolo, Vlore

albany adjustment of large blocks with *any* number of photos, points, or images, using *any* photogrammetric-measuring instrument on *any* computer

Albatross Grumman amphibian transport aircraft; name of a series of oceanographic survey ships flying the American flag; Piaggo P-166M coastal patrol aircraft built in Italy

ALBE Air League of the British Empire

Alben Isaac Albeniz

Alber Alberic; Albern; Albert(ino)

Albert Albert Canal connecting Antwerp and Liege; Albert Einstein; Albert National Park in Zaire; Albert Nyasa (Albert Lake, Africa's third largest); Halbert; Halbertus

Alberta Girls Alberta Institution for Girls

albi air-launched ballistic interceptor; air-launched booster intercept

Albion Correctional Albion State Institution and Western Correctional Facility, Albion, NY

ALBIS Australian Library-Based Information System

albm (ALBM) air-launched ballistic missile

Alb Mus Albany Museum, Grahamstown, South Africa

Albn Albanian

Albq Albuquerque

Albr Albrecht

ALBS Anti-Lock Braking System

Albt Albert

Albturist Albanian Tourism

Albuq Albuquerque

atbus all bureaus (naval coding)

alc alcohol; alternate level of care; approximate lethal concentration; assault landing craft; assembly language coding; avian leukosis complex; axiolinguocervical

a l c à la carte (French—on the menu)

ALC Accredited Land Consultant; Air Logistics Center (USAF); Alabama Central (railroad); American Licorice Company; American Life Convention, American Locomotive Company; American Lutheran Church; Area Logistics Center; Area Logistics Command; Armament Logis-

tics Center; Armament Logistics Command; Army Legal Corps; Asian Law Collective; Associated Lutheran Charities

ALCA American Leather Chemists Association; Associated Landscape Contractors of America

ALCAC Airlines Communications Administrative Council

AlCan Alaska-Canada

ALCAN Aluminum Company of Canada

Alcan Highway road between Dawson Creek, British Columbia, and Fairbanks, Alaska

Alcanzamos Alcanzamos por f'n la Vctoria (Spanish—Finally we attained victory), Panama's national anthem

alcapp automatic list classification and profile production

alcc airlift coordination center

ALCC Airborne Launch-Control Center

ALCC Asociación de Libre Comercio del Caribe (Spanish—Caribbean Free Trade Association)

Alc^{de} Alcalde (Spanish—justice of the peace, mayor)

alch approach-light contact height

alchem alchemy

alcid alcohol + acid

alcism alcoholism (addiction to alcohol)

ALCL Association of London Chief Librarians

alcm (ALCM) air-launched cruise missile

ALCM Associate of the London College of Music

ALCMs Air-Launched Cruise Missiles

ALCO Air-Lift Liaison Coordination Officer

ALCOA Aluminum Company of America

alcoh alcohol

alcohol ethyl alcohol (C_2H_5OH)

alcolic alcoholic

alcom algebraic compiler; algebraic computer

ALCOM Alaska Command

alcon all concerned

ALCOP Alternate Command Post

alcotrician alcohol + nutrition (adverse effects of alcohol on good nutrition)

Alcott Amos Bronson Alcott or his daughter Louisa May

alcr aluminum crown (dental)

alcs (ALCS) airborne launch control system

ALCS American League Championship Series; Authors' Lending and Copyright Society

ALCTP Academic Library Consultant Training Program

ald a later date; acceptable limit for dispersion; aldolase

Ald Aldabra; Alderman; Aldermanic

ALD Acral Lick Dermatitis; Assistive Listening Device

ALDA Air Line Dispatchers Association; American Land Development Association; Australian Land Development Association

ALDCC Active Lift Distribution Control Center

aldehyde al(cohol) dehy(drogenated)—dehydrogenated (oxidized) alcohol

aldep automated layout design program

Alder Alderic; Alderley

Alderson minimum-security Federal Reformatory for Women at Alderson, West Virginia

ALDEV African Land Development

aldh aldehyde dehydrogenase

ALDHA Appalachian Long-Distance Hikers Association

Aldm Alderman

aldo aldosterone

Aldo Teobaldo; Teobaldo Manuzio, 16th-century Venetian printer and typographer

aldp automatic language-data processing

ALDS Apollo Launch-Data System

ale annual loss expectancy

Ale Alemania (Spanish—Germany)

ALE Association for Liberal Education

ALEA Airline Employees Association

alec algebraic components and coefficients

ALEC American Legislative Exchange Council; American Lutheran Evangelical Churches

Alec(k) Alexander

ALECS Automated Law-Enforcement Communications System

ALECSO Arab League Educational, Cultural, and Scientific Organization

alegar ale + vinegar (vinegar derived from ale)

Ale^jo Alejandro

Aleksandr Aleksandr Solzhenitsyn (Russian novelist)

ALEOA American Law Enforcement Officers Association

Alep (Turkish—Aleppo)—Syrian city

Ale RD República Democrática Alemana (Spanish—German Democratic Republic)—East Germany

Ale RF República Federal Alemana (Spanish—Federal German Republic)—Germany

alerfa alert phase

ALERT Affiliated League of Emergency Radio Teams; Automatic Linguistic Extraction and Retrieval Technique

ALERT II Automatic Law Enforcement Response Time

ale(s) additional living expense(s)

Ales Alessandro

ALESCO American Library and Educational Service Company

Aleut Aleutian; Aleutian Islands

Aleut Cur Aleutian Current

Aleutians Aleutian Mountains; Aleutian islanders; Aleutian Islands

Aleut Is Aleutian Islands

A-levels advanced levels (of educational tests)

alex alert exercise; alexandrine (verse)

alex (ALEX) alert exercise

Alex Alexander; Alexandra; Alexandria

Alexa Alexandra

Alexanders Alexander Archipelago; Alexander cocktails; Alexander islanders; Alexander Islands of southeastern Alaska

Alex City Alexander City, Alabama

alf advanced landing field; auxiliary landing field; automatic letter facer

Alf Alfa; Alfhild; Alfonso; Alford; Alfred; Alfred(o)

ALF American Life Federatiorm; American Life Foundation; Animal Liberation Front; Arab Liberation Front; Association of Libertarian Feminists; Australasian Labour Federation; Australian Labor Federation

Alfa letter A radio code

ALFA Anomma Lombarda Fabbrica Automobili

ALFC Aboriginal Land Fund Commission (Australia)

ALFCE Allied Land Forces in Central Europe (NATO)

ALFI American League of Financial Institutions

Alfie Alfred

Alfo Alfonso

ALFORD Appalachian Laboratory for Occupational Respiratory Diseases

ALFS U.S. Navy's Airborne Low-Frequency Sonar program

ALFSEA Allied Land Forces, South-East Asia

ALFSH Allied Land Forces in Schleswig-Holstein (NATO)

alft airlift

alg advanced landing ground; algae; algal; algebra; algebraic; allergic; allergical; allergy; along; alongside; antilymphocyte globulin (ALG)

alg algemeen (Dutch—generally or universally); *algerisch* (German—Algerian)

Alg Algeria; Algiers

ALG Air Algérie; Algiers, Algeria (airport); Automotive Lease Guide

ALGC Association of Local Government Clerks

Alge Algeciras

Algeb algebra

Alger Algernon

Algeria Democratic and Popular Republic of Algeria (North African Arab nation), *El Djemhouria El Djazairia Demokratia Echaabia* (Arabic name); *République Algérienne Démocratique et Populaire* (French name)

algett allegretto (Italian—brisk or jolly)

Algie Algernon

algins algae derivatives

alglyn aluminum glycinate

algol algorithmic language

Algonquin Algonquin Peak in the Adirondacks; Algonquin Provincial Park, Ontario

algtto allegretto (Italian—brisk or jolly)

ALGU Association of Land Grant Colleges and Universities

ALGWA Australian Local Government Women's Association

Algy Algernon

alh advanced light helicopter; anterior lobe horrnone; anterior lobe of the hypophysis

Alh Alhambra

ALH Aldous Leonard Huxley; Assam Valley Light Horse; Australian Light Horse

Alham Alhambra

ALHFAM Association of Living Historical Farms and Agricultural Museum (Washington, D.C.)

ALHS Abraham Lincoln High School

Alht Apollo lunar hand tool

Alhtc Apollo lunar hand tool carrier

ali air-launched interceptor

ali alibi (Latin—elsewhere)

'*ali* (Arabic—high)

Ali Alicante

ALI American Law Institute; American Library Institute

ALIA Royal Jordanian Airlines

Alianza La Alianza Federal de las Mercedes (Spanish—Federal Alliance of Mercedes)— New Mexican organization founded by Reies Lopez Tijerina to reclaim Mexican land acquired by the United States

Alic Alicante

ALIC Association of Life Insurance Counsel

alice (ALICE) automatic laundering instrument control equipment

Alice The Alice–Alice Springs, Northern Territory, Australia; Allis-Chalmers Manufacturing Company (stock exchange slang)

ALICE Artillery Line Communications Equipment

Alick Alexander

ALICS Advanced Logistics Information and Control System (USAF)

alien alienist

'**Alifax** (Cockney—Halifax)

align. alignment

alim (ALIM) air-launched interceptor missile

ALIMD Association of Life Insurance Medical Directors

ALIMDA Association of Life Insurance Medical Directors of America

Al Imp Reps Alert Implementation Reports

Aline Adeline

alirt (ALIRT) adaptive long-range infrared tracker

ALIS Advanced Life Information System; Automated Library Information Service; Automated Library Information System

alit automatic line insulation tester

Alitalia Italian national airlines (AZ)

ALITALIA Italian International Airline

A.Litt. Associate in Letters

ALJ Administrative Law Judge

ALJ Australian Law Journal, Australian Library Journal

aljak aluminum-jacketed coaxial cable

ALJC Alice Lloyd Junior College

ALJH Association of Libraries of Judaica and Hebraica (in Europe)

ALJR Australian Law Journal Reports

al-Jum al-Iraq al-Jumhuriyah al Iraqiyah (Arabic—Republic of Iraq)

al-Jum al-Jaz ad-Dim ash Shab al-Jumhuriya al-Jazairiya ad-Dimuqratiya ash-Shabiya (Arabic—Democratic and Popular Republic of Algeria)

alk alkali

Alkali Soviet air-to-air radar-guided homing missile (NATO)

Alkan Valentin Alkan, Charles-Valentin Morhang

alki alcohol; homeless alcoholic

alkie(s) alcoholic(s)

alk phos alkaline phosphatase

alkums air-launched cruise missiles (ALCMs)

all above lower limit; acute lymphoblastic leukemia; acute lymphocytic leukemia; airborne laser laboratory (ALL); allergy

al.l. alia lectio (Latin—a different reading)

All Alley; Alloa; Aloha

Al-L Alsace-Lorraine

ALL Admiralty Lines Limited; Airborne Laser Laboratory; American League of Lobbyists; American Life Lobby against abortion

ALL Admiralty List of Lights

All 8va all'ottava (Italian—in the octave)

all'8va all'ottava (Italian—an octave higher)

ALLA Allied Long Lines Agency (NATO)

Allagash Allagash River and Allagash Wilderness Waterway, Maine

allcat all critical atmospheric turbulence (programs)

alld airborne laser locator designator; allowed

Alld Allahabad

alleg allegation; allegiance; allegoric; allegorical; allegory

Alleghenies Allegheny Mountains of Pennsylvania, Maryland, Virginia, and West Virginia

Allem Allemagne (French—Germany)

allergol allergologic(al)

alig allgemein (German—general)

allgem allgemein (German—general)

All H All Hallows (Halloween)

all hands all hands on deck (everyone)

Allie Alice; Alison

Alligators Alligator Rivers of Australia's Northern Territory (East, South, and West Alligator)

all'ingr all'ingrosso (Italian—wholesale)

Allison Allison Division, General Motors

allit alliteration; alliterative

ALLNAVSTAS All Naval Stations

allo allonym

allo (Greek prefix—other)—allele, allopathic, allopatric, allophone, alloplasm, allotetrapoid

Allo allegro (Italian—lively, quickly)

alloc allocate; allocation

allop allophone

all'ott all'ottava (Italian—an octave higher)

allow. allowance

allp audiolingual language programming

All Quiet All Quiet on the Western Front (Lewis Milestone film, 1930)

All S All Souls College, Oxford

ALLS Apollo Lunar Logistic Support

All Saint's All Saint's Day (November 1)

allstat all-purpose statistical (package)

Alltto allegretto (Italian—lively but less so than *allegro)*

allu allude; allusion; allusively

allus allusion

alluv alluvial; alluvium

Ally Pally Alexandra Palace in North London

alm. alarm

aim almindelig (Dano-Norwegian—common, frequent, plain, simple)

Alm Almeria

Alm Almirante (Spanish—Admiral; German—mountain pasture)

ALM American Legion of Merit; American Leprosy Missions

A & LM Arkansas & Louisiana-Missouri (railroad)

ALM Antilliaanse Luchtvaart Maratschappij (Dutch—Antillean Airline Company)

A.L.M. Artium Liberalium Magister (Latin–Master of Liberal Arts)

ALMA Aircraft Locknut Manufacturers Association; American Loudspeaker Manufacturers Association; Association of Literary Magazines of America

ALMAC Association of Labor-Management Administrators and Consultants

ALMACA Association of Labor-Management-Administrators and Consultants on Alcoholism

ALMAJCOM All Major Commands

al-Mam al-Urd al-Hash al-Urd al-Mamlakah al-Urdunniyah al Urdun (Arabic–Hashemite Kingdom of Jordan)

ALMC Army Logistic Management Center

alme acetyl-lysine methyl ester

Almer Almeric

almi anterior lateral myocardial infarct

ALMIDS Army Logistics Management Integrated Data System

ALMs Amindivi, Laccadive, and Minicoy Islands off India's Malabar Coast

ALMS Analytic Language Manipulation System

ALMT Association of London Tailors

Almte Almirante (Spanish—Admiral)

aln ammunition lot number; anterior lymph node

aln (ALN) accounting line number

alnico aluminum, nickel, copper (magnet alloy containing iron and cobalt)

alnmt alignment

alnot alert notice

ALNP Abraham Lincoln National Park

ALNZ Air League of New Zealand

alo axiolinguoclusal

alo' alow

Alo Alonso

ALO Accreditation Liaison Officer; Agency Liaison; Air Liaison Office(r); Allied Liaison Office(r); Aloha Airlines; Amalgamated Lace Operatives; American Liaison Office(r); Army Liaison Office(r)

ALOA Amalgamated Lace Operatives of America; Amalgamated Lithographers of America; Assembly of Librarians of the Americas; Associated Locksmiths of America

aloc air lines of communication; allocation

ALOC Air Line of Communication

ALOE A Lady Of England—Charlotte Maria Tucker

alof' aloft

aloft. airborne light optical fiber technology

ALOHA Aboriginal Lands of Hawaiian Ancestry; Aloha Airlines

ALON Air Liaison Officer Net

alo'-'n'-alof' alow and aloft (everywhere aboard ship—in the lower rigging and in the upper rigging)

ALOO Albuquerque Operations Office

alor advanced lunar orbital rendezvous

alos average length of stay

alot allotment

aloteen alcoholic teenagers (rehabilitation program)

alotm allotment

ALOTS Airborne Lightweight Optical Tracking System

Alouette Aerospatiale armed helicopter made in 4-passenget and 6-passenger versions

ALOW Air Electrical Officer's Writer

Aloys Aloysius

alp anterior lobe (of) pituitary; assembly language program (data processing); autocode list processing; automated language processing

Alp Alphen; Alpine

ALP Air Liaison Party; Allied Liaison and Protocol; Ambulance Loading Post; American Labor Party; Antigua Labour Party; Australian Labor Party; Australian Liberal Party; Automated Learning Process; Automated Library Program

ALP Agence Lao Press (French—Lao Press Agency)

ALPA Air Line Pilot's Association

ALPAC Automatic Language Processing Advisory Committee (National Research Council)

alpak algebra package

ALPB American Lutheran Publicity Bureau

ALPC Army Logistics Policy Council; Australian Library Promotion Council

ALPCA Auto License Plate Collectors Association

Alpen (German—Alps)—Richard Strauss's *Alpine Sympho*ny

ALPGA Australian Liquefied Petroleum Gas Association

Alph Alphonse
alpha alphabetical
Alpha letter A radio code
ALPHA Action League for Physically Handicapped Advancement
alphameric alphanumeric and alphabetic-numeric
alphametic alphabet arithmetic
alphanumeric alphabetical-numerical
alpha order alphabetical order
ALPHAS Automatic Literature Processing, Handling, and Analysis System
ALPL Advanced Lunar Projects Laboratory
alpo (ALPO) Apollo lunar polar orbiter
ALPO Allen Products; Amalgamation of Left Political Organizations; Association of Lunar and Planetary Observers
ALPOWAD Alaska Power Administration
Alps Alpine Mountains of south-central Europe
ALPS Accidental Launch Protection System; Advanced Linear Programming System; Automated Library Processing Services; Automatic Landing Positioning System
ALPSP Association of Learned and Professional Society Publishers
ALPT Army Language Proficiency Test
Alpujarras Alpujarras Mountains of Almería and Granada in Spain
ALPURCOMS All-Purpose Communications System
al-Qahira (Arabic—Cairo)— means victorious
ALQAS Aircraft-Landing Quality-Assessment Scheme
alr. aliter (Latin—otherwise)
ALR Australian League of Rights
ALR *American Law Reports*
ALRA Abortion Law Reform Association; Agricultural Labor Relations Act
alraam advanced long-range air-to-air missile

ALRANZ Abortion Law Reform Association of New Zealand
ALRB Agriculture Labor Relations Board; Agriculture Labor Relations Bureau
ALRC Anti-Locust Research Center; Australian Law Reform Commission
alri airborne long-range input
ALRI Angkatan Laut Republik Indonesia (Indonesian Navy)
ALRM Aboriginal Legal Rights Movement (Australia)
ALROS American Laryngological, Rhinological, and Otological Society
a l r p de V M *a los reales pies de Vuestra Majestad* (Spanish—at the royal feet of Your Majesty)
ALRS *Admiralty List of Radio Signals*
ALRTF Army Long-Range Technological Forecast
als active laser seeker; advanced life support; advanced light source; air logistics service; aldolase; antilymphocytic serum; autographed letter signed
als. alias
als (ALS) amyotrophic lateral sclerosis (Lou Gehrig's disease); automatic line supervision
ALS Aboriginal Legal Service; Advanced Laser System; Advanced Launch System; Alton & Southern (railroad); American Littoral Society; anti-lymphocyte stimulator; Approach Light System; Area Licensing Scheme (Singapore); Australian Literary Society; Australian Literary Studies; auxiliary lighter (naval symbol)
A.L.S. Associate of the Linnean Society
ALSA American Land Sailing Association; American Lap Swimmers Association (Incorporated); American Law Student Association

ALSAA Americans of Lebanese–Syrian Ancestry for America
alsam (ALSAM) air-launched surface-attack missile
Alsat Alsatian
ALSC American Lumber Standards Committee; Association for Library Service to Children
ALSCP Appalachian Land Stabilization and Conservation Program
alse aviation life-support equipment
AlSEA Alabama Society of Enrolled Agents
Al seg al segno (Italian—return to the sign)
alsep (ALSEP) apollo lunar surface experiments package
Alsk Alaska(n)
Also Also Sprach Zarathustra (German—*Thus Spake Zarathustra*)—symphonic poem by Richard Strauss
ALSO Alex Lindsay String Orchestra (New Zealand)
alsor air-launch sounding rocket
alss airline system simulator
ALSS Advanced Location Strike System; Airborne Location and Strike System; Apollo Logistics Support System
ALST Alaska Standard Time
alt academic learning time; airborne laser tracker; alanine aminotransferase; alter(ation); altered; altering; alternative; alternator; altimeter; altitude
alt altitud; altura (Spanish—altitude, height)
Alt alternating (light); Altoona
Alt Altesse (French–Highness)
ALT Aboriginal Lands Trust; Aer Lingus (Irish Air Lines); alteration
ALT *Australian Law Times*
Alta Alberta
ALTA Agricultural Landlords and Tenants Act; American Land Title Association; American Library Trustee Associa-

tion; Army Lawn Tennis Association; Association of Local Transport Airlines

altac algebraic translator and compiler

altair (ALTAIR) ARPA (q.v.) long-range tracking and instrumentation radar

Altais Altai Mountains

altan alternate alerting network

altare automatic logic testing and recording equipment

Altay high mountains rising above northern edge of Gobi Desert in Central Asian portion of Russia

alt-ch alternate-channel

ALTCOMIND Alternate Commander, Indian (USN)

ALTCOMLANT Alternate Commander, Atlantic (USN)

ALTCOMPAC Alternate Commander, Pacific (USN)

ALTCS Arizona Long-Term Care System (Medicaid)

altd altered

alt. dieb. alternis diebus (Latin—alternate days)

ALTDS Army Laser Target Designating System

altener alternative energy

alter. alteration; alternate

Alt F Fl alternating fixed and flashing (light)

Alt F Gp Fl alternating fixed and group flashing (light)

Alt Fl alternating flashing (light)

Alt Got Alternate Gothic

Alt Gp Occ alternating group occulting (light)

Alt Gr Fl alternating group flashing (light)

altho although

alt. hor. alternis horis (Latin— at alternate hours)

altim altimeter

altm altimeter

alt. noc. alternis noctibus (Latin—on alternate nights)

altnr alternator

Alt Oce alternating occulting (light)

ALTPR Association of London Theatre Press Representatives

altran algebraic translator

altrec automatic life testing and recording of electronic components

altru altruism; altruist; altruistic

ALTS Advanced Lunar Transportation System; Airborne Laser Tracker System

alt set. altimeter setting

ALTUC All-India Trade Union Congress

alt udk alt udkonme (Dano-Norwegian—all published)

alt xyl alto xylophone

alu (ALU) arithmetic and logic unit

Alucon Aluminium Conductors (New Zealand)

alue admissible linear unbiased estimator

alum. alumna; alumnae; alumni; alumnus; hydrated potassium aluminum sulfate

alv alveolar; avian leukosis virus

alv (ALV) avian leukemia virus(es)

älv (Swedish—river)

ALV Annual Lease Value

alv. adstrict. alvo adstricto (Latin—bowels being constipated)

ALVAO *Association des Longues Vivantes pour l'Afrique Occidentale* (French— West African Modern Languages Association)

alv. deject. alvi dejectiones (Latin—intestinal discharges)

Alver Alvern(on)

Alvº Alvaro

alvx alveolectomy

alw allowance; arch-loop whorl

Alweg Axel Lennert Wenner-Gren (Swedish industrialist's name applied to monorailroad systems)

alwin algorithmic wiswesser notation

ALWL Army Limited War Laboratory

alwt advanced lightweight torpedo

Alx Alexandria

aly alloy

Aly Alley

Alyce Girls Alyce D. McPherson School for (delinquent) Girls at Ocala, Florida

ALYESKA Alaska Pipeline Service

Alz. Alzheimer's disease

alz alzamento (Italian—heaving, lifting, raising)

am active militia; air-cooled motor; ammeter; amplitude modulation; asymmetric multiplier; attack missile; auditory (sequential) memory

am (AM) air-locked module

a.m. *ante meridiem* (Latin—before noon)

a/m auto/manual

a & m agricultural and mechanical; ancient and modern; architectural and mechanical; archy and mehitabel

am amerikansk (Dano-Norwegian—American)

Am Amazonas; America; American; americium; myopic astigmatism (symbol)

Am. American; *Amós* (Spanish—The Book of Amos)

AM Academy of Management; Aeromexico; Aeronaves de México (Mexican Airlines); Air Marshal; Air Mechanic; Air Medal; Air Ministry; Albert Medal; Alexander Mackenzie (Canada's second Prime Minister); amplitude modulation; angular momentum; Armenia (Internet code); Army Manual; Arthur Meighen (Canada's tenth and twelfth Prime Minister); Australian Member (Member of the Order of Australia); Auxiliary Minesweeper; Aviation Medicine; Aviation Structural Mechanic; Award of Merit; large minesweeper (naval symbol); metric angle (symbol)

A-M Addressograph Multigraph; Alpes-Maritimes

A.M. Air Mail

A/M Aviation Medicine

A & M Agricultural and Mechanical; Agricultural and Mechanical College of Texas; Ancient and Modern (hymns)

A of M Academy of Music

AM Almacenes Maritimos; Alpes Maritimes (Maritime Alps); *Almanaque Mundial* (Spanish—World Almanac)

A.M. Anno Mundi (Latin—in the year of the world); *artium magister* (Latin—Master of Arts); *Ave Maria* (Latin—Hail Mary)

a/m¹ amperes per square meter

AM-3C Aeritalia-Aermacchi single-engine three-place armed-trainer aircraft

ama actual mechanical advantage; against medical advice; antimitochondrial antibody; asset management account

amᵃ amiga (Spanish—female friend)

AMA Academy of Model Aeronautics; Acoustical Materials Association; Aerospace Medical Association; Agricultural Marketing Administration; Air Materiel Area; Aircraft Manufacturers Association; Amarillo, Texas (airport); Amateur Trapshooting Association; Ambulance Manufacturers Association; American Machinery Association; American Management Association; American Maritime Association; American Marketing Association; American Matthay Association; American Medical Association; American Ministerial Association; American Monument Association; American Motel Association; American Motorcycle Association; American Municipal Association; Apparel Manufacturers Associatiion; Arena Managers Association; Army Mounteering Association; Association of Metropolitan Authorities; Australian Medical Association; Australian

Meteorological Association; Automobile Manufacturers Association

A & MA Advertising and Marketing Association

AMAA Adhesives Manufacturers Association of America; Army Mutual Aid Association; Association of Medical Advertising Agencies

AMAB Army Medical Advisory Board

AMACAB Allied Military Administration Civil Affairs Branch

Am Acad Pol Soc Sci American Academy of Political and Social Science

Am Acad Rel American Academy of Religion

AMA/CMA American Medical Association/California Medical Association

AMACO American Medical Assurance Company

AMACUS Automated Microfilm Aperture Card Updating System

amad aircraft-mounted accessory drive

Amad Amadeus

AMA-DE American Medical Association–Drug Evaluation(s)

Amad Quart Amadeus Quartet

amads airframe mounted accessory drive system

AMAE American Museum of Atomic Energy; Association of Mexican-American Educators

AMAERF American Medical Association Education and Research Foundation

amag aircraft mounted accessory gearbox

amal amalgam; amalgamate; amalgamation

AMAL Aero-Medical Acceleration Laboratory; American Medical Acceleration Laboratory

amalg amalgamated

amalgam mercury and silver mixture

a M (a/M) am Main (German—on the Main River)

amams advanced medium-range anti-tank missile system

Aman Agaf Modiin (Hebrew—Military Information Bureau)–Israel

amap advanced multiprogramming-analysis

AMAR Annual Major Additions Rate

AMARC Army Materiel Acquisition Review Committee

AMARS Air Mobile Aircraft Refueling System; Automatic Message Address Routing System

amas advanced multi-purpose armor system

AMAS American Military Assistance Staff; Automatic Message Accounting System

Am Assn Blood American Association of Blood Banks

Am Assn Coll Pharm American Association of Colleges of Pharmacy

Am Assn Comm Jr Coll American Association of Community and Junior Colleges

amat amateur

AMATC Air Materiel Armament Test Center

amatol ammonia & toluene (explosive)

A-matter advance matter (written in advance of a newspaper story)

AMATYC AmericanMathematical Association of Two-Year Colleges

AMAUS Aero Medical Association of the United States

AMAWA American Medical Association Women's Auxiliary

amaws advanced medium assault weapon system

AMAX American Metal Climax

Amaz Amazonas (Brazilian state)

amb amber; ambient; ambulance

amb. ambulatory

Amb Ambassador; Ambrose Bierce (American author)

Amb Amberes (Spanish—Antwerp)

AMB Admiralty Medical Board; Airways Modernization Board; A M Best (insurance rating); Australian Meat Board

AMB Asociación Mundial de Boxeo (Spanish—World Boxing Association); *Associação Médica Brasileira* (Portuguese—Brazilian Medical Association); *Association Maritime Belge* (French—Belgian Maritime Association)

amba andelsselskab med begraenset ansvar (Dano-Norwegian—limited liability cooperative)

AMBA American Mold Builders Association; Association of Military Banks of America

AMBAC American Bosch Arma Corporation; American Municipal Bond Assurance Corporation

AMBAE Association of Master of Business Administration Executives

Am Bankr Reps American Bankruptcy Reports

Am Baptist American Baptist Historical Society

AMBBA Associated Master Barbers and Beauticians of America

Amb Brdg Ambassador Bridge (Detroit)

Amb Col Ambassador College

ambel ambiguity eliminator

Amber Amberes (Spanish—Antwerp)

Amb Ex Ambassador Extraordinary

Amb Ex/Plen Ambassador Extraordinary and Plenipotentiary

ambi (Latin prefix—both)—ambidextrous

Am Bibl American Bibliographic Center—Clio Press

ambidex ambidextrous

ambig ambiguity; ambiguous

ambisex ambisextrous (bisexual)

ambish ambition

ambit algebraic manipulation by identity translation

ambiv ambivalence; ambivalent

Am Bk American Book Company

Am Bk Prices American Book Prices Current

ambl ambulatory

amblads advise method, bill of lading, and date shipped

Amb Lib Ambrosian Library (Milan)

Ambo Ambrose

Ambon Amboina, Indonesia

Am Booksellers American Booksellers Association

Ambos Amboseli National Park (East Africa)

Amboys New Jersey's Perth Amboy and South Amboy

Am Brass Quin American Brass Quintet

Ambridge American Bridge (company)

AMBRL Army Medical Biomechanical Research Laboratory

arnbros ambrosia

Ambrosian Ambrosian Library (Milan)

ambt ambulant

ambu ambulance

ambul ambulacral; ambulacrum; ambulance; ambulant; ambulate(d); ambulating; ambulation; ambulatorily; ambulatory

Amburgo (Italian—Hamburg)

amc arthrogryposis multiplex congentia (AMC); automatic mixture control; axiomesiodistal

amc (AMC) armed merchant cruiser; armed missile cruiser; armored mortar carrier; auxiliary minesweeper, coastal

AMc coastal minesweeper (3-letter naval symbol)

AMC Aerospace Manufacturers Council; Air Mail Center; Air Materiel Command; Aircraft Manufacturers Council; Air Ministry Constabulary; Air Mobility Command; Albany Medical Center; Albany Medical College; Alternate Media Center; American Maritime Cases; American Mining Congress; American Mission to the Chinese; American Motors Corporation; American Movie Classics; American Music Center; American Music Conference; Animal Medical Center; Appalachian Mountain Club; Army Materiel Command; Army Medical Center; Army Medical Corps; Army Missile Command; Army Mobility Command; Army Munitions Command; Association of Management Consultants; automatic message counting; Avionics Maintenance Conference

AMCA Air Movement and Control Association; American Medical College Association; American Mosquito Control Association; Australian Management Consultants Association

AMCALMSA Army Materiel Command Automated Logistics Management Systems Agency

Am Camping American Camping Association

Am Can American Can

AMCAS American Medical College Application Service

AMC-ASC Air Materiel Command-Aeronautical Systems Center

AMCAWS Advanced Medium-Caliber Aircraft Weapon System

amcbh auxiliary machine casing bulkhead

AMC & BW Amalgamated Meat Cutters and Butcher Workmen

AMCD American Medical Center at Denver

AMCEA Advertising Media Credit Executives Association

AMCFSA Army Materiel Command Field Safety Agency

Am Chem American Chemical Society

AMCI & SA Army Materiel Command Installations and Service Agency

amcl amended clearance

AMCL African Metals Corporation Limited; Association of Metropolitan Chief Librarians

AMCLDC Army Materiel Command Logistic Data Center

AMCLSSA Army Materiel Command Logistics Systems Support Agency

amcm airborne mine countermeasures; aldershot mine countermeasures

AMCM Air Materiel Command Manual; Army Materiel Command Memorandum

AMCMFO Air Materiel Command Missile Field Office

AMCO American Manufacturing Company

AMCOA AiResearch Manufacturing Company of Arizona

AMCOM American Stock Exchange Communications

Am Con American Consul(ate)

AMCOS Aldermaston Mechanized Cataloguing and Ordering System; Australian Mechanical Copyright Owners Society

AMCP Allied Military Communications Panel

AMCPI Army Materiel Command Procurement Instruction(s)

AMCR Air Materiel Command Regulation(s); Army Materiel Command Regulation(s)

AMCRD Air Materiel Command Research and Development; Army Materiel Command Research and Development

AMCS Airborne Missile Control System; Association of Military Colleges and Schools

AMCSA Army Materiel Command Support Activity

AMCSOF Army Combat Surveillance Office

AMCST Associate of the Manchester College of Science and Technology

AMCTB Associated Motor Carriers Tariff Bureau

AMCU Australian Malaria Control Unit

am. cur. amicus curiae (Latin—a friend at court)

amd age-related macular degeneration; air movement designator; alpha-methyldopa; axiomesiodistal

AMD Accident Model Document; Admiralty Machinery Department; Advanced MicroDevices; Aero-Mechanics Department (USN); Aerospace Medical Division; Air Movement Data; Air Movement Directive; Army Medical Department; Atomic and Molecular Data

AMD Aerospace Material Document

AMDA Advanced for Mutual Defense Assistance; Advances for Mutual Defense Assistance; Airlines Medical Directors Association

AMDB Agricultural Machinery Development Board

AMDC Army Mechanized Demonstration Column

AMDEA Associated Manufacturers of Domestic Electrical Appliances

Am Dec American Decisions

AMDEC Associated Manufacturers of Domestic Electric Cookers

Amdel Australian Mineral Development Laboratories

Am Dent American Dental Association

AMDF Army Master Data File

a.m. Dg. ad majorem Dei gloriam (Latin—to the greater glory of God)—also A.M.D.G.

AMDI Associazione Medici Dentisti Italiani (Italian—Association of Italian Medical Dentists)

Am Dig American Digest of Public International Law Cases

AmdlEvac aeromedical evacuation

Amdoc American Doctors (organization)

Am Doc Inst American Documentation Institute

amdp ammunition distribution point

AMDS Advanced Missions Docking Subsystem (NASA); Association of Military Dental Surgeons; Automatic Message Distribution System

amdsbsc amplitude-modulation double-sideband suppressed carrier

amdt amendment

ame angle-measuring equipment; automatic microfiche editor

AmE American English

AME Admiralty Mining Establishment (UK); Aero-Medical Evacuation; African Methodist Episcopal; Air Ministry Examination; Aviation Medical Examiners

A.M.E. Advanced Master of Education

AMEA Association of Machinery and Equipment Appraisers

AMEB Australian Music Examinations Board

amec aft master-events controller

AMEC Airframe Manufacturing Equipment Committee; Australian Minerals and Energy Council

amecd antimechanized

amech account mechanical (failure or malfunction)

ameda automatic-microscope electronic-data accumulator

AmedD Army Medical Department

AMEDDPAS Army Medical Department Property Accounting System

AmedP Army Medical Publication(s)

AmedS Army Medical Service

AMEE Admiralty Marine Engineering Establishment

AMEG Association for Measurement and Evaluation in Guidance

AMEIC Associate Member of the Engineering Institute of Canada

AMEL Acro Medical Equupment Laboratory

amelior amelioration

Am Elsevier American-Elsevier Publishing Company

AMEM African Methodist Episcopal Mission; Association of Marine Engine Manufacturers

Am Emb American Ambassador; American Embassy

AMEME Association of Mining, Electrical, and Mechanical Engineers

AMEMIC Association of Mill and Elevator Mutual Insurance Companies

amend. amendment(s)

Am Engr American Engineer

Amenia Girls Amenia Center for (delinquent) Girls at Amenia, New York

amens amenities

amér américain (French—American)

amer amerikansk (Dano-Norwegian—American)

Amer America; American

AMERADC Army Mobility Equipment Research and Development Center

Amerasians American-Asians (offspring of Americans and Asians)

Amer Blk American Black

Amer-Eng American-English

Amer F American French

AMERICAL Americans in New Caledonia (Army division)

American Grove New Grove Dictionary of American Music

American Virgins U.S. Virgin Islands

Americans United Americans United for Separation of Church and State

Americas Western Hemisphere; including North, Central, and South America

America's Tropical Islands Florida Keys; Guam; Hawaiian Islands; Padre Island, Texas; Puerto Rico; U.S. Virgin Islands

Americo-Libs Americo-Liberians (Liberian descendants of American Blacks)

Ameridish American Yiddish

Amerind American & Indian (American Indian or Eskimo)

Amer Ind American Indian

Amerindians American Indians

Ameringlish American English

Ameritech American Information Technologies

Amer Men Sci American Men of Science

AmerSp American Spanish

Amer Spec American Spectator

Amer Std American Standard

Amer St Papers American State Papers

Amer Trauma Soc American Trauma Society

AmeS American Meteorological Society

AMES Air Ministry Experimental Station; Association of Marine Engineering Schools

Ameslan American Sign Language

AMETA Army Management Engineering Training Agency

AMETS Army Meteorological System; Artillery Meteorological System

AMEWA Associated Manufacturers of Electric Wiring Accessories

Amex American Stock Exchange

AMEX American Express

AMEX Agencia Mexicana de Noticias (Mexican News Agency)

Amexco American Express Company

AMEZ African Methodist Episcopal Zion

AMEZC African Methodist Episcopal Zionist Church

amf (AMF) airmail facility

AMF ACE (Allied Command Europe) Mobile Force; AIDS Medical Foundation; Air Material Force; Air Mobile Force; American Machine and Foundry; Arab Monetary Fund; Arctic Marine Freighters; Australian Marine Force; Australian Military Forces

AMFA Aircraft Mechanics Fraternal Association

AMF(A) Allied Mobile Force (Air)—NATO

AmFAR American Foundation for AIDS Research

Am Feed American Feed Manufacturers Association

AMFGC Association of Midwest Fish and Game Commissioners

AMFI Aviation Maintenance Foundation International

AMFIC Automatic Microfilm Information System

AMFIS American Microfilm Information Society; Automatic Microfilm Information System

AMF(L) Allied Mobile Force (Land)—NATO

am/fm amplitude modulation/frequency modulation

AmFr American French

Am Friends American Friends Service Committee

amg automatic magnetic guidance; axiomesiogingival

AMG Aircraft Machine Gunner; Albertus Magnus Guild; Allied Military Government; Australian Map Grid

amgb airframe mounted gearbox

Am Geol American Geological Institute

Am Geophysical American Geophysical Union

AMGN Amgen Incorporated

AMGNY Associated Musicians of Greater New York

AMGO Assistant Master-General of Ordnance

AMGOLD Anglo-American Gold Investment Trust
AMGOT Allied Military Government
AMGS Acceleration Monitoring Guidance System
Am Guidance American Guidance Service
amh amharisch (German—Amharic); astigmatism with myopia predominating; automated medical history
Amh Amharic
AMH Australian Military Hospital
AMHA American Motor Hotel Association; Anchorage Museum of History and Art
AMHCI Associate Member of the Hotel and Catering Institute
Am Heart American Heart Association
Am Heritage American Heritage Publishing Company
AMHIS American Marine Hull Insurance Syndicate
Am Hist Res American History Research Associates
Am Home Prod American Home Products
AMHS Alaska Marine Highway System; American Material Handling Society; Australian Maritime Historical Society
AMHT Automated Multiphasic Health Testing
ami acute myocardial infarction; advanced manned interceptor; air mileage indicator; alternative mortgage instrument; amitriptyline; automatic monthly investment; auxiliary minesweeper inshore; axiomesioincisal
AMI Advanced Manned Interceptor; American Management Institute; American Marine Institutes; American Meat Institute; American Medical International, Inc.; American Military Institute; American Museum of Immigration; American Mushroom Institute; American Music Institute; Association of Medi-

cal Illustrators; Association for Multi-Image; Australian Marketing Institute; Australian Motor Industries
AMI Aeronautica Militare Italiana (Italian Air Force); *Association Montessori Internationale* (French—International Montessori Association)
AMIA American Metal Importers Association; American Mutual Insurance Alliance
AMIADB Army Member—Inter-American Defense Board
AMIAE Associate Member of the Institute of Automobile Engineers
AMIAMA Associate Member of the Incorporated Advertising Managers Association
AMIAP Associate Member of the Institution of Analysts & Programmers
AMIC Aerospace Materials Information Center; Air Movement Information Center (NATO); Army Methods of Instruction Center; Australian Mining Industry Council
AMICA Automobile Mutual Insurance Company of America
AMICE Associate Member of the Institution of Civil Engineers
AMICH American Marine Insurance Clearinghouse
AMICI Association Mondiale des Interprétes de Conférences International (French—World Association of International Conference Interpreters)
AMICO American Measuring Instrument Company
AMICOM Army Missile Command
AMIDS Advanced Multispectral Image Descriptor System; Area Manpower Instructional Development System
amigo ants, mice, gophers (electromagnetic device affecting the neurological system of such pests)
AMIGOS Americans Interested In Giving Others a Start

AMII Association of Musical Instrument Industries
AMILO Army-Industuy Materiel Information Liaison Offiice
amilpri amilprilose hydrochloride (rheumatoid arthritis drug)
AMIN Advertising and Marketing International Network
AMINA Association Mondiale des Inventeurs (French—World Association of Inventors)
Am Ind American Indian
Am Indus Arts American Industrial Arts Association
AMINOIL American Independent Oil (company)
aminos amino acids (protein building blocks)
Am Inst American Institute
Am Inst Disc American Institute of Discussion
AMINTAPHIL American Section, International Association for Philosophy of Law and Social Philosophy
AM International Addressograph-Multigraph International
AMIO Arab Military Industrialization Organization
AMIOP Associate Member of the Institute of Printing
AMIPA Associate Member of the Institute of Practitioners in Advertising
Amirantes Amirante Islands
AMIR Air Mission Intelligence Report
AMIRS Alternative Mortgage Instruments Study
AMIS Agricultural Markets Intelligence System; Aircraft Movement Information Section; American Musical Instrument Society; Army Management Information System; Automated Mask Inspection System
Amistad Amistad National Recreation Area near Del Rio, Texas
AMJ American Muslims for Jerusalem

AMJ Assemblée Mondiale de la Jeunesse (French)–World Assembly of Youth)

Am Jour Sci American Journal of Science

Am J Phys American Journal of Physics

Am J Psy American Journal of Psychiatry

aml acute monocytic leukemia; acute myeloblastic leukemia; acute myelocytic leukemia; acute myelogenous leukemia; acute myoblastic leukemia; armored missile launcher

aml (AML) adjustable-mortgage loan; amplitude-modulated link

Am L American Lawyer

AML Aberdeen Marine Laboratory; Admiralty Materials Laboratory; Aeromedical Laboratory; Allied Military Liaison; American Mail Line; Applied Mathematics Laboratory

AML-60 French four-wheeled armored car with 7.5mm machine guns and a 60mm mortar

AML-90 French four-wheeled armored car with 7.5mm machine guns and a 90mm mortar

AMLC Aerospace Medical Laboratory (USAF)

Am Lib Dir American Library Directory

Am Librarians American Librarians' Agency

Am-Lib(s) Americo-Liberian(s)

AMLO Assistant Military Landing Officer

amls antimouse lymphocyte serum

AMLS American Medico-Legal Society; Master of Arts in Library Science

amm agnogenic myeloid metaplasia; ammonia; ammunition; anti-missile missile (AMM)

AMM Air Mining Mission; American Military Mission; Amman, Jordan (airport); Anti-Missile Missile; Associated Millinery Men; Aviation Machinist's Mate

AM & M Applied Mathematics and Mechanics

AMM Association Medicale Mondiale (French—World Medical Association)

AMMA Adult Movies and Magazines Association; American Museum of Marine Archeology; Assistant Masters and Mistresses Association

Am Mach American Machinist

Am Malacologists American Malacologists

Am Management American Management Association

Am Map American Map Company

Am Math Soc American Mathematical Society

AMMC Aviation Materiel Management Center

Am Media American Media

Am Metal Mkt American Metal Market/Metalworking News

Am Meteorite American Meteorite Laboratory (Denver)

ammeter amperemeter (current-measuring instrument)

AMMI American Merchant Marine Institute; American Museum of the Moving Image

AMMINET Automated Mortgage Management Information Network

AMMIS Aircraft Maintenance Manpower Information System (USAF)

AMMISCA American Military Mission in China

amml acute myelomonocytic leukemia

AMMLA American Merchant Marine Library Association

ammo ammunition

Ammo American Motors

AMMO Air Ministry Meteorological Office

ammobr ammunition bearer

ammon ammonia

Ammon Ammonite

ammonia water ammonium hydroxide (NH_4OH)

Am Motors American Motors

AMMPE American Mining, Metallurgical, and Petroleum Engineers

ammrpv advanced multimission remotely piloted vehicle

Am Mus Mag American Museum of Magic

amn airman; ammunition

amnes amnesia(c)(al)(1y)

AMNH American Museum of Natural History

amnip adaptive man-machine nonarithmetic information processing

amnl amplitude-modulation noise level

AmnM Airman's Medal

AMN & PA Australian Monthly Newspapers and Periodicals Association

amnswp acoustic minesweeping

AMNZIE Associate Member of the New Zealand Institution of Engineers

AMNW Air Ministry Navigational Warning

amo *amigo* (Spanish—male friend); axiomesio-occlusal

AMO Accredited Management Organization; Advance Material Order; Admiralty Monthly Order; Aircraft Material Officer; Air Ministry Order; Air Movements Officer; American Medical Optics; Army Medal Office; Army Medical Officer; Area Medical Officer; Aviation Medical Officer

AMOA Amusement and Music Operators Association

amo (AMO) air mail only; alternant molecular orbit

amob automatic meterological oceanographic buoy

AMOCO American Oil Company

amol acute monocytic leukemia

Amon Carter Amon Carter Museum of Western Art

AMOP Association of Mail Order Publishers

amor amorphous

AMORC Ancient Mystic Order Rosae Crusis (Rosicrucian Order)

amorph amorphous

amort. amortizable, amortization, amortize(d), amortizement, amortizing

amos antireflection-coated metal-oxide semiconductor

AMOS Acoustic, Meteorological, and Oceanographic Survey; American Monument and Outdoor Sculpture Database; Associated Migrant Opportunity Services; Automatic Metrrological Observation Station

AMOSC Authorized Military Occupational Specialty Code

amp acid mucopolysaccharide; adenosine monophosphate (hormonal chemical); amperage; ampere; amphetamine; ampicillin; amplification; amplifier; amplitude; ampule; amputation; average mean pressure

amp (AMP) automatic multipattern (camera meter)

AMP Aerospace Medical Panel; Air Mail Pioneers; Air Member for Personnel; American Museum of Photography; Army Mine Planter; Association of Media Producers; Aurora Memorial Park (Philippines); Automated Mathematics Program; Aviation Modernization Program

AMPA American Medical Publishers Association; Associate of the Master Photographers Association; Australian Magazine Publishers Association

AMPAC American Medical Political Action Committee

AMPAS Academy of Motion Picture Arts and Sciences

AMPC Automatic Message Processing Center; Auxiliary Military Pioneer Corps

AMPCO Associated Missile Products Corporation; Association of Major Power Consumers of Ontario

AMPD Air Ministry Plans Department

Am Peace American Peace Society

ampdr advanced multi-mode pulse Doppler radar

Ampersand Ampersand Press (Princeton)

Ampersand NYC Ampersand Press (New York City)

AMPFIPA American Military Precision Flying Teams Association

amph amphibian; amphibious; amphimict; amphoric;

Amph Amphibia

AMPH Association of Management in Public Health

amphet amphetamine (stimulant)

amphetamine alpha-methylphenethylamine

amphets amphetamines

amphi (Greek prefix—both, both sides of, two)—amphibian, amphibolite, amphibological

amphib amphibia(n); amphibious

amphibex amphibious exercise

amphig amphigoric; amphigorical; amphigorist; amphigory

Am Philatelic American Philatelic Society

Am Philos Soc American Philosophical Society

Amphoto American Photographic Book Publishing Co.

amp hr ampere hour

ampi annual military personnel inspection

AMPI Associated Milk Producers, Incorporated; Associated Music Publishers, Incorporated

Ampico American Piano Company

ampl a macroprogramming language; amplifier; amplitude

ampt ampliata (Italian—enlarged); amplus (Latin—large)

AMPOL American Petroleum

ampp advanced microprogrammable processor

ampr advanced multipurpose radar; automatic manifold pressure regulator

AMPR Aeronautical Manufactures Planning Report; Airframe Manufacturers Planning Report; Area Manpower Planning Report; Area Manpower Planning Review

amps amperes; ampules; atmospheric, magnetospheric, and plasmas in space

AMPS Accrued Military Pay System; Advanced Mobile Phone Service; American Metered Postage Society; Army Mine Planter Service; Army Motion Picture Service; Associated Music Publishers; Association for Media Psychology; Automatic Message Processing System

AMPSS Advanced Manned Precision Strike System

AMPT alpha-methyl paratyrosine

AMPTC Arab Maritime Petroleum Transport Company

AMPTEs Active Magnetosphere Particle Explorers

AMPTP Association of Motion Picture and Television Producers

amp-turns ampere-turns

Am Public Health American Public Health Association

ampul. ampulla (Latin—ampule)

ampus (s) amputee (s)

AMQ American Medical Qualification; Army Married Quarters

AMQUA American Quaternary Association

amr (AMR) automatic message routing

AMR Abnormal Mission Routine; Advanced Material Request; Airman Military Record; Atlantic Missile Range; Auckland Mounted Rifles; Auxiliary Machinery Room (USN)

AMR Applied Mechanics Reviews

A.M.R. Master of Arts in Research

AMRA American Medical Records Association; Army Materials Research Agency; Australian Model Railways Association

AMRAAM Advanced Medium-Range Air-to-Air Missile

AMRAC Anti-Missile Research Advisory Council

Am Radio American Radio Relay League

AMRC Advanced Metals Research Corporation; Army Mathematics Research Center; Automotive Market Research Council

AMRCA American Miniature Racing Car Association

amrcr anti-mine reconnaissance castor roller

AMRCUS Alternative Marriage and Relationship Council of the United States

AMRD Army Missile and Rockets Directorate

AMR & DL Air Mobility Research and Development Laboratory (USA)

AMRE Air Ministry Research Establishment

Am Record American Record Collectors Exchange

Am Red American Red Cross

Am Res American Research Council

AMREX American Real Estate Exchange

AM & RF African Medical and Research Foundation

AMRINA Associate Member of the Royal Institution of Naval Architects

AMRIP Avionics Module Repair Improvement Program

Amrit Amritsar

AMRL Aerospace Medical Research Laboratories; Army Medical Research Laboratory

AMRNL Army Medical Research and Nutrition Laboratory

AMRO Amsterdam-Rotterdam (bank); Association of Medical Record Officers

amrpd applied manufacturing research and process development

AMRS Air Ministry Radio Station; American Moral Reform Society; Army Medical Reserve Store; Australian Media Research Services

amrss advanced medium-range sonar system

ams accelerator mass spectronometry; advanced mapping spectrometer; aggravated in military service; auditory memory span; automated multiphasic screening; automatic music scan; auxiliary minesweeper

Ams Amsterdam

Ams auxiliary motor minesweeper

AMS Acute Mountain Sickness; Administration and Management Services; Administrative Management Society; Advanced Marketing Services; Aeronautical Material Specification; Agricultural Marketing Service; Airship Management Services; American Management Systems; American Marketing Service; American Mathematical Society; American Meteor Society; American Meteorological Society; American Microscopical Society; American Mineral Spirits; American Montessori Society; American Museum of Safety; American Musicological Society; Amsterdam, Netherlands international airport; Anglican Men's Society; Army Map Service; Army Medical Service; Army Medical Staff; Association of Messenger Services; Association of Museum Stores; Australian Museum, Sydney, Australian Medical Services; Aviation Marketing Services (Australia); Avionics Maintenance Squadron

AM & S Australian Mining and Smelting

AMS *Acta Medica Scandinavica*

amsa (AMSA) advanced manned strategic aircraft

AMSA American Metal Stamping Association; American Museum of Social Anthropology; Association of Metropolitan Sewerage Agencies; Australian Marine Sciences Association; Australian Medical Students Association

AMSACP Advanced Multistage Axialflow Compressor Program (NASA)

amsam anti-missile surface-to-air-missile

Am Sam American Samoa

AMSANZ Aviation Medicine Society of Australia and New Zealand

amsat amateur satellite

AmSAT American Society for the Alexander Technique

AMSAT Amateur Satellite

AMSC Army Medical Specialist Corps

Am Sch Athens American School of Classical Studies at Athens

Am School American Scholar

Am Sci & Eng American Science and Engineering, Inc.

AMSCO American Mineral Spirits Company; American Sterilizer Company

amsd anti-missile ship defense

AMSE Associate Member of the Society of Engineers

amsef anti-mine-sweeping explosive float

AMSGA Association of Manufacturers and Suppliers for the Graphic Arts

AMSH Association for Moral and Social Hygiene

AMSI Admiralty Merchant Shipping Instructions

AMSIR Agriculture, Marine, Scientific, and Industrial Research Ministry

AMSIS Air Ministry Secret Intelligence Summary

amsl above mean sea level

AMSMH Association of Medical Superintendents of Mental Hospitals

AMSO Air Member for Supply and Organisation (RAF)

AMSOC American Miscellaneous Society

Am Soc Afr Cult American Society of African Culture

Am Soc HRAC Eng American Society of Heating, Refrigerating, and Air-Conditioning Engineers

Am Society Pr American Society Press

Am Soc Indxrs American Society of Indexers

Am Soc Metals American Society for Metals

Am Soc Not American Society of Notaries

Am Soc Soc American Sociological Society

Am Soc Tool and Mfg Eng American Society of Tool and Manufacturing Engineers

Am Sp American Spanish (Latin American); American Speech

AMSP Army Master Study Program

AMSPAR Association of Medical Secretaries, Practice Administrators, and Receptionists

AMSq Avionics Maintenance Squadron (USAF)

amsr automatic missile site radar

amss advanced minehunting sonar system

ams s autographed manuscript signed

AMSS Advanced Meterological Sounding System); Automatic Music Select System

AMSSEE Area Museum Service for South-Eastern England

AMSSFG Association of Manufacturers of Small Switch and Fuse Gear

a mss s autographed manuscripts signed

amst advanced medium STOL (short takeoff and landing) transport

Amst Amsterdam

AMST Association of Maximum Service Telecasters

AMSTAT News *American Statistical Association News*

Amst Concert Amsterdam Concerto

Amster Amsterdam(mer)

AMSU Amphibious Maintenance Support Unit

AMSUS Association of Military Surgeons of the United States

Am Sym Orch American Symphony Orchestra

amt air member for training; alpha-methyltyrosine; amethopterin; amount; amphetamine

amt (AMT) alternative minimum tax

AMT Academy of Medicine, Toronto, Canada; Advanced Manufacturing Technology; Aerial Mail Terminal; Air Mail Transfer; American Medical Technologists; Area Management Team; Astrograph Mean Time

A.M.T. Associate in Mechanical Technology; Associate in Medical Technology; Master of Arts—Teaching

amta airborne moving target attack

AMTA American Massage and Therapy Association

amtank amphibious tank

AMTC Airframe Manufacturing Tool Comunittee; Army Mountain Training Center

A.M.T.C. Art Master's Teaching Certificate

AMTCL Association for Machine Translation and Computational Linguistics

AMTD Automatic Magnetic Tape Disseminatiom (Service)

AMTDA Agricultural Machinery and Tractor Dealers Association; American Machine Tool Distributors Association

Am Technical American Technical Society

Am Tech Soc American Technical Society

AMTEG Australian Metal Trades Export Group

Am Tel & Tel American Telephone and Telegraph

amtex air mass-transportation experiment

AMTF Air Mobile Task Force

Am Theatre Assoc American Theatre Association

amti airborne moving target indicator

Amtorg *Amerikanskaya Torgovlya* (Russian—American Trading Company)

AMTPI Associate Member of the Town Planning Institute (UK)

amtrac amphibious tractor

Amtrak American railroad tracks, the National Railway Passenger Corporation

amtran automatic mathematical translator

amt(s) amphetamine(s)

AMTS Army Mechanical Transport School; Associate Member of the Television Society

AMTSD Air Ministry Technical and Stores Department

amtt active moving-target tracking

amu air mileage unit; air mission unit; astronaut maneuvering unit; atomic mass unit

AMU Advanced Marksmanship Unit; Alaska Methodist University; American Malacological Union; American Marksmanship Unit; Arab Maghreb Union; Army Marksmanship Unit; Associated Midwestern Universities; Association of Marine Underwriters

AMUA Associate in Music—University of Adelaide

AMUBC Association of Marine Underwriters of British Columbia

Am U Field American Universities Field Staff

Amund Amundsen (Canadian–Arctic gulf named for Norwegian explorer Roald Amundsen)

Am Univ Artforms American Universal Artforms

AMURT Anada Marga Universal Relief Team (India)

A.Mus. Associate in Music

A.Mus.A. Associate in Music-Australia

A.Mus.C. Associate in Music-Canada

A.Mus.L.C.M. Associate in Music-London College of Music

A.Mus.N.Z. Associate in Music-New Zealand

A.Mus.S.A. Associate in Music-South Africa

A.Mus.T.C.L. Associate in Music-Trinity College of Music-London

amv alfalfa-mosaic virus; avian myeloblastitis virus

AMV *Association Mondiale Vétérimaire* (French—World Veterinary Association)

AMVAP Associate Manufacturers of Veterinary and Agricultural Products

AMVER Atlantic Merchant Vessel Report; Automated Mutual Assistance Vessel Rescue (USCG)

AMVERS Automated Merchant Vessel Reporting System

AMVETS American Veterans

AMVM Administrative Motor Vehicle Management

AMVOP Army Motor Vehicle Operators Permit

amw actual measurement weight

AMW Air Ministry Warden; Antimissile Warfare; Association of Married Women

AMWA American Medical Women's Association; American Medical Writers' Association

AMWC Association of Workers for Maladjusted Children

AMWDD Admiralty Miscellaneous Weapons Development Depot

Am West American West Publishing Company

AMWG Academy of Master Wine Growers

AMWM Association of Manufacturers of Woodworking Machinery

AMWSU Amalgamated Metalworkers and Shipwrights Union (Australia)

AMX-13 French light tank carrying SS-11 antitank guided missiles and a 75mm gun

AMX-30 French medium tank carrying a 105mm gun plus machine guns (antiaircraft and ground)

AMX-105 French self-propelled 105mm howitzer

AMX-155 French self-propelled 155mm howitzer

AMX-VTT French armored personnel carrier (crew of 2 plus 12 troops)

amy amytal (barbituate depressant and sedative)

Amy Amelia; Amoy, China

amyl amyl nitrate

amys amyl nitrate

an acrylonitrile

an. above named; airman; annual; anode

an' and

a/n acidic and neutral

an (Greek—lacking, not, without)—anaerobic, anemic, anonymous, anorexia

an. anno (Latin—year); *ante* (Latin—before)

An Annam; Annamese

A$_n$ normal atmosphere

AN Acid Number; Advance Note; Aerodynamic Note; Air Force-Navy; Airmail Notice; Air Navigation; Air Navigator; Air Reduction (stock exchange symbol); alphanumeric; Anglo-Norman; Apalachicola Northern (railroad); Army-Navy; net laying vessel (naval symbol); Aryan Nation; Netherlands Antilles (Internet code)

A.N. Associate in Nursing

A & N Army and Navy

AN-2 Soviet Antonov 14-passenger biplane nicknamed Colt by NATO forces

AN-12 Soviet Antonov 100-passenger cargo plane nicknamed Cub by NATO

AN-14 Soviet Antonov 6-seat transport aircraft nicknamed Clod by NATO

AN-14M Soviet Antonov 15-passenger turboprop plane

AN-22 Soviet Antonov 22 (super transport plane)

AN-26 Soviet Antonov 50-passenger transport plane nicknamed Coke by NATO

ana anesthesia; anesthesiac

ana (ANA) antinuclear antibodies

ana (Greek prefix—up or up against)—analogy, analysis, anatomy

Ana Anaheim; Anita; Anna; Annabel(la)

'Aña Agafia, Guam

ANA Adminstration for Native Americans; Agricultural Numerical Annexes; Air Force-Navy Aeronautical; All Nippon Airways; American Nature Association; American Neurological Association; American Newspaper Association; American Numismatic Association; American Nurses' Association; Anti-Nicaragua Association; Arab Network of America; Armenian National Army; Army-Navy Aeronautical; Assistant Naval Attaché; Association of National Advertisers; Association of Nurse Administrators; Australian National Airways; Australian Natives Association

ANA Asociación Nacional Automovilista (Spanish—National Automobile Association); *Automotive News Almanac*

ANAAS Australian and New Zealand Association for the Advancement of Science

anab anabasis

anac anachronism; anachronistic

ANACHEM Association of Analytical Chemists

anacol anacoluthon

anacom analog computer

anacreon anacreontic(s); anacreontist

anacru anacrusis

ANADIS Australian National Animal Disease Information System

anaesth anaesthesia; anaesthetic(s); anaesthesiologist; anaesthesiology

ANAF Army, Navy, Air Force

anag anagram; anagrammatic(al)(ly); anagramist; anagrams

Anaheim Anaheim Stadium, Anaheim California

ANA/HEW Administration for Native Americans—HEW

ANAHL Australian National Animal Health Laboratory

ANAI Asociación Nacional de Administradores de Inmuebles (Spanish—National Association of Real Estate Administrators)

anal analogy; analysis; analytical

Anal Chem Analytical Chemistry

analg analgesic

anal psychol analytical psychology

analyst psychoanalyst

analyt analytical

Anambas Anambas Islands in the South China Sea where Indonesians permitted Vietnamese boat-people to land while awaiting international aid

anap agglutinatiom negative, absorption positive

ANAP Asociación Nacional de los Agricultores Pequeños (Spanish—National Association of Small Farmers)

ANAPO Alianza Nacional Popular (Spanish-Popular National Alliance)—Colombia

ANARC Association of North American Radio Clubs

anarch anarchist; anarchism; anarchy

ANARE Australian National Antarctic Research Expeditions

ANAS Australian National Airlines Commission

anat anatomical; anatomist; anatomy

Anat Anatomy

anath anathema; anathematize

anatran analog translator

anav area navigation

anb annual number belonging

ANB Alaska Native Brotherhood; Army-Navy-British Standard

ANB Australian National Bibliography

anbs (ANBS) armed nuclear bombardment satellite

ANBS Air Navigation and Bombing School

ANB & TC American National Bank and Trust Company

anc all numbers calling; ancient

anc. ancestor; ancestry

Anc Ancona

ANC African National Congress; African National Council; Air Force-Navy-Civil; Alaska Native Coalition; American News Company; Anchorage, Alaska (airport); Area Naval Commander; Arlington National Cemetery; Army and Navy Civil Committee on Aircraft; Army Nurse Corps; Australian Newspaper Council

ANCA Allied Naval Communications Agency; American National Cattlemen's Association

ANCAM Association of Newspaper Classified Advertising

ANCAP Administratión Nacional de Combustibles Alcohol y Portland (Spanish—National Administration of Alcohol and Portland Fuel)

ANCAR Australian National Committee for Antarctic Research

ancc anodal closure contraction

ANCC Australian National Cattlemen's Council

ANCCAC Australian National Committee on Computation and Automatic Control

anch anchorage

Anch Anchorage, Alaska

anchor. alphanumeric character generator

Anchorage Youth McLaughlin Youth Center, Anchorage, Alaska

Anchors Anchorage, Alaska

Anchor Your Age founded in 1915 in Anchorage

ancienn anciennement (French—formerly)

ANCIRS Automated News Clipping, Indexing, and Retrieval System

ANCLD Australian National Committee on Large Dams

Anco Ancohuma

ANCO Anderson-Collingwood (tanker service)

ANCOA Aerial Nurse Corps of America

ANCOM Andean Common Market

ANCON Asociación Nacional para la Conservación de la Naturaleza (Spanish—National Association for the Conservation of Nature)

ancova analysis of covariance

ancr aircraft not combat ready

ANCs African National Congress members

ANCS American Numerical Control Society

ANCSA Alaska Native Claims Settlement Act

ANCUN Australian National Committee for the United Nations

ANCW Australian National Council of Women

ANCXF Allied Naval Commander Expeditionary Force

and almordisch (German—Old North German); *andante* (Italian—of moderate speed)

And Andalucía; Andaman Islands; Andorra (whose capital is Andorra la Vella); Andromeda

And Andromeda (Latin—mythological Ethiopian princess whose name adorns a constellation)

AND Army-Navy Design

AND Australian News Digest

Anda Andaman Sea between India and Malaysia in the Indian Ocean

ANDA Australian National Dance Association

andalusite aluminum silicate

Andamans Andaman islanders; Andaman Islands

Andamans and Nicobars Andaman and Nicobar Islands off Burma in the Bay of Bengal

ANDB Air Navigation Development Board

ANDC Australian National Dairy Committee

Andes Cordillera de los Andes; Los Andes; mountain chain of South America

ANET Ancient Near-Eastern Texts

And I Andaman Islands

Andie Andrew

Andno andantino (Italian—slower than andante)

Ando Andorra; Andorran

Andorra Valleys of Andorra (Pyrenees principality)

Andover Hawker-Siddeley troop transport designated HS-748 and holding 40 paratroopers or 58 regular soldiers; Phillips Andover Academy

Andr Andromeda

ANDRA Agence National pour la Gestion des Dechets Radioactifis (French—National Radioactive Management Agency)

Andre Andrea

Andreaofs Andreaof Islands

Andrews twenty-dollar bills bearing the portrait of President Andrew Jackson

andro androstenedione; androsterone

andro (Greek prefix—old)—androgen(ic)(al)(1y)

androg androgyn (girlish male)

Andryusha (Russian nickname—Andrei)—Andrew; Andy

Ands Andreas

ANDSA *Almacenes Nacionales de Deposito SA* (Spanish—National Depository Warehouse Company)

andte anodal duration tetanus

Andte andante (Italian—of moderate speed)

Andy Andrew

andz anodize

ane acoustic noise environment

ANEAC Australian National Energy Advisory Committee

anec anecdotal; anecdote(s)

Aneda (Latin—Edinburgh)

ANEDA Association Nationale d'Etudes pour la Documentation Automatique (French—National Association for Automatic Documentation Studies)

ANEF American Nepal Education Foundation

ANEQ Association Nationale des Etudiants du Québec (French—National Association of Students of Québec)

ANERA American Near East Refugee Aid

anes anesthesia; anesthesiologist; anesthesiology; anesthetician; anesthetic(s)

anesth anesthetic

anesthesiol anesthesiology

an. ex anode excitation

anf anchored filament; antinuclear factor(s); antinuclear force

ANF Advance Notification Form; Allied Nuclear Force; American Nurses Foundation; Atlantic Nuclear Force (NATO); Atomic Nuclear Forum; Australian Nursing Federation

anfe (ANFE) aircraft not fully equipped

anfi automatic noise-figure indicator

ANFIA *Associazione Nazionale fra le Industrie Automobilistiche* (Italian—National Association of Automobile Industries)

anfo ammonium nitrate fuel oil (explosive)

ang angiogram; angle; angular

ang angaende (Danish, Norwegian, Swedish—concerning)

Ang Anchorage; Angel (phonograph records); Anglo; Angola

ANG Air Force-Navy-Army Guided Missiles; Air National Guard; American Newspaper Guild; Australian National Gallery; Australian New Guinea

ANGAU Australian New Guinea Administrative Unit

Ang Chamb Orch Angelicum Chamber Orchestra

Angela Angelica

Angeles Port Angeles, Washington, opposite Victoria, British Columbia

angew angewandte (German—applied, put into practice)

Ångfart Ångfartyaas (Swedish—steamship company)

Angie Angela; Angelina; Angeline

angiol angiology

Angkor Angkor Thom (Walled City) or Angkor Vat (Temple City)—Cambodia

Angi Anglican

Angl Angleterre (French—England)

Anglic Anglican, Anglicism

ANGLICO Air and Naval Gunfire Liaison Company

Anglo- combining form meaning England

Anglo (Spanish—an English-speaking person)

Anglo-Afg Anglo-Afghan

Anglo-Afr Anglo-African

Anglo-Amer Anglo-American

Anglo-Ant Anglo-Antarctic(an); Anglo Antillean

Anglo-Arab Anglo-Arabian

Anglo-Arg Anglo-Argentine

Anglo-Art Anglo-Arctic

Anglo-Aus Anglo-Australian; Anglo-Austrian

Anglo-Bah Anglo-Bahaman

Anglo-Barb Anglo-Barbadian

Anglo-Bas Anglo-Basque

Anglo-Bel Anglo-Belizian

Anglo-Belg Anglo-Belgian

Anglo-Bhu Anglo-Bhutanese

Anglo-Bol Anglo-Bolivian

Anglo-Bots Anglo-Botswana

Anglo-Braz Anglo-Brazilian

Anglo-Bul Anglo-Bulgarian

Anglo-Bur Anglo-Burman; Anglo-Burundian

Anglo-CA Anglo-Central American
Anglo-Cam Anglo-Cameroonian
Anglo-Can(ad) Anglo-Canadian
Anglo-Cat Anglo-Catalan
Anglo-Cath Anglo-Catholic
Anglo-Cey Anglo-Ceylonese
Anglo-Chi Anglo-Chinese
Anglo-Chil Anglo-Chilean
Anglo-Col Anglo-Colombian
Anglo-Cub Anglo-Cuban
Anglo-Cyp Anglo-Cypriot
Anglo-Czech Anglo-Czechoslovak(ian)
Anglo-Dah Anglo-Dahomean
Anglo-Dan Anglo-Danish
Anglo-Du Anglo-Dutch
Anglo-Ecu Anglo-Ecuadorean
Anglo-Egypt Anglo-Egyptian
Anglo-Eng Anglo-English
Anglo-Epis Anglo-Episcopal(ian)
Anglo-Ethio Anglo-Ethiopian
Anglo-Fin Anglo-Finnish
Anglo-Fr Anglo-French
Anglo-Gael Anglo-Gaelic
Anglo-Gam Anglo-Gambian
Anglo-Ger Anglo-German
Anglo-Gr Anglo-Greek
Anglo-Guy Anglo-Guyancse
Anglo-Hond Anglo-Honduran
Anglo-Hung Anglo-Hungarian
Anglo-Ice Anglo-Icelandic
Anglo-Ind Anglo-Indian
Anglo-Indo Anglo-Indonesian
Anglo-Ir Anglo-Iranian; Anglo-Iraqi; Anglo-Irish
Anglo-Isr Anglo-Israeli
Anglo-Itat Anglo-Italian
Anglo-Jam Anglo-Jamaican
Anglo-Jap Anglo-Japanese
Anglo-Jew Anglo-Jewish
Angle-Jor Anglo-Jordani
Anglo-Ken Anglo-Kenyan
Anglo-Kuw Anglo-Kuwaiti
Anglo-Lat Anglo-Latin
Anglo-Mal Anglo-Malawian; Anglo-Malaysian; Anglo-Maltese
Anglo-Mald Anglo-Maldivian
Anglo-Mex Anglo-Mexican
Anglo-N Anglo-Norse
Anglo-Nep Anglo-Nepalese
Anglo-Nig Anglo-Nigerian

Anglo-Nor Anglo Norwegian
Anglo-Norm Anglo Norman
Anglo-NZ Anglo-New Zealand
Anglo-Pak Anglo-Pakistani
Anglo-Para Anglo-Paraguayan
Anglo-Per Anglo-Persian; Anglo-Peruvian
Anglo-Pol Anglo-Polish
Anglo-Port Anglo-Portuguese
Anglo-Rho Anglo-Rhodesian
Anglo-Rom Anglo-Romanian
Anglo-Rus(s) Anglo-Russian
Anglo(s) Anglo-Saxon(s)
Anglo-SA Anglo-South African; Anglo-South American
Anglo-Sam Anglo-Samoan
Anglo-Sax Anglo-Saxon
Anglo-Scot Anglo-Scottish
Anglo-SL Anglo-Sierra Leonean
Anglo-Som Anglo-Somali
Anglo-Sov Anglo-Soviet
Anglo-Span Anglo-Spanish
Anglo-Sud Anglo-Sudanese
Anglo-Swi Anglo-Swiss
Anglo-Tanz Anglo-Tanzanian
Anglo-Tob Anglo-Tobagan
Anglo-Togo Anglo-Togolese
Anglo-Ton Anglo-Tongan
Anglo-Trin Anglo-Trinidadian
Anglo-Turk Anglo-Turkish
Anglo-Ugan Anglo-Ugandan
Anglo-Uru Anglo-Uruguayan
Anglo-Ven Anglo-Venezuelan
Anglo-W Anglo-Welsh
Anglo-Yem Anglo-Yemini
Anglo-yugo Anglo-Yugoslav(ian)
Anglo-Zamb Anglo-Zambian
Anglo-Swe Anglo-Swedish
Angl-Yingl *Anglish-Yinglish* (Yiddish in American life and literature
Angola Republic of Angola
Ang Pam Akla *Ang Pambansang Aklatan* (Filipino—The National Library)—in Manila
ang pec *angina pectoris* (Latin—strangling of the chest)—heart attack
Ang-Sax Anglo-Saxon

Ang-Sax Amer Anglo-Saxon America (Canada, United States, Bahamas, Belize, Guyana, Jamaica, Trinidad, West Indian islands, Bermuda)
ANGTS Alaska Natural Gas Transportation System
Angus Aeneas
ANGUS Acoustically Navigated Geophysical Underwater Survey; Air National Guard of the United States
anh anhydrite; anhydrous
Anh *Anhang* (German—appendix)
ANH Australian National Highway; Australian National Hotels
ANHA American Nursing Home Association
ANHAL Australian National Humanities and Arts Library
anhed anhedral
ANHC Army Native Hospital Corps
ANHF Australian National Historic Site
anhyd anhydrous
ani automatic number identification
ani *atmosphère normale internationale* (French—international normal atmosphere)
ANI Army-Navy-Industry; Australian National Industries
A & NI Andaman and Nicobar Islands
ANI *Agéncia Nacional de Informaçao* (Portuguese—National Information Agency); *Agencia Nacional de Informaciones* (Spanish—National Information Agency)
ANIB Australian News and Information Bureau
ANICA *Associazione Nazionale Industrie Cinematografiche e Affini* (Italian—National Association of Cinematographic and Related Industries)
ANICO American National Insurance Company
anil aniline
ANILCA Alaska National Interest Lands Conservation Act

aniline phenyl amine

anim animal; animate; animism

anim animato (Italian—animated)

ANIM Association of Nuclear Instrument Manufacturers

animad animadversion

ANIP Army-Navy Instrumentation Program

aniv anniversary

aniv anniversario (Spanish—anniversary)

ank ankle

ank ankomen (Dutch—arrival); ankomst (Danish—arrival); ankunft (German—arrival)

Ank Ankara

ANK Ankara, Turkey (airport)

Ankerplatz Ankerplatz der Freude (German—Anchorage of Joy)–St Pauli's Reeperbahn section of Hamburg

anl anneal; annoyance level (aircraft noise); automatic noise limiter

ANL Argonne National Laboratory; Australian National Library; Australian National Line, net-laying ship (naval symbol)

A-N L Anti-Nazi League

ANLCA Alaska Native Land Claims Act

anld annealed

ANLINE Australian National Line

an. lt anchor light

anlys analysis

anm Amnaerkning (Danish, Norwegian, Swedish—footnote, note, remark, observation)

Anm Anmerkung (German—footnote, note)

ANM Admiralty Notice to Mariners; Alliance of Non-Profit Mailers; Anacostia Neighborhood Museum; Australian Newsprint Mills

ANM Académie Nationale de Musique (French—National Academy of Music); *Admiralty Notices to Mariners*

ANMA Advertising Mail Marketing Association

ANMC American National Metric Council

ANMCC Alternate National Military Command Center

ANMEF Australian Naval and Military Expeditionary Force

anmi air navigation multiple indicator

ANMI Allied Naval Maneuvering Instructions (NATO)

ANMPE Associación Nacional de Municipalidades del Perú (Spanish—National Association of Municipalities of Perú)

ANMRC Australian Numerical Meteorology Research Centre

ANMS Automated Notice to Mariners System

ann annotated; announce(ment); announcer; annual(ly); annuity; annunciator

ann annonce (Dano-Norwegian—advertisement, announcement)

ann. anni (Latin—years); *anno* (Latin—year)

Ann Anastacia; Angela; Angelina; Angeline; Anita; Anna; Annabelle; Annelida; Annetta; Annette; Annie; Antoinette

Ann Annalen (German—annals); *Annales* (French—annals); *Annali* (Italian—annals)

ANN All Nippon News (network)

Anna Annabella; Annapolis, Maryland; Annette

ANNA Army, Navy, NASA, Air Force

ANNAF Army, Navy, NASA, Air Force

Ann Arbor Pub Ann Arbor Publishers

Ann Chim Phys Annales de Chimie et de Physique (French—Annals of Chemistry and Physics)

ANNECS Automated Nikkei News Editing

Ann Fluid Dyn Annals of Fluid Dynamics

Anng Annapolis graduate

Ann Geophys Annales de Geophysique (French—Annals of Geophysics)

ANNIE Applications of Neutral Networks for Industry in Europe

Ann Inst Henri Poincaré *Annales de l'Institut Henri Poincarj* (French—Annals of the Henri Poincaré Institute)

anniv. anniversarium (Latin—anniversary)

Ann Math Annals of Mathematics

annot. annotated; annotation

Ann Oto Rhino Laryngol Annals of Otology, Rhinology, and Laryngology

Ann Phys Annalen der Physik (German—Annals of Physics); *Annales de Physique* (French—Annals of Physics); *Annals of Physics* (New York)

Ann Rept Annual Report

Ann Rev Nucl Sci Annual Review of Nuclear Science

annu annual; annuale; annuario

annuit annuitant

Annuit Coeptis Annuit coeptis novus ordo seculorum (Latin—He has favored our undertakings—a new secular order of the ages is born) Motto on the reverse of the great seal of the United States

annul. annulment

Annunc Annunciation

ano above-named officer

ano. another

ANO Air Navigation Office; Air Navigation Officer; Anti-Narcotics Office; Area Nursing Officer

anoc anodal opening contraction

ANOC Authorized Notice of Change

anod. anodize

anom anomia; anomiac; anomiacal

Anon. anonymous (Latin—nameless)

anop assembly no operation

ANOP Australian Nationwide Opinion Polls

ANOPP Aircraft Noise Prediction Program

anorex anorexia nervosa

anorm aircraft not operationally ready—maintenance

anor nerv *anorexia nervosa*
(Latin) Psychological and endocrine disorder characterized by a pathological fear of
weight gain leading to faulty
eating habits, malnutrition,
and weight loss

anors aircraft not operationally
ready—supplies

ANOS Australian Native Orchid Society

anot annotate

A.N. Other another (person)

anov analysis of variance

anova analysis of variance

anp aircraft nuclear propulsion

A-np A-norprogesterone

ANP Aberdare National Park
(Kenya); Acadia National Park
(Maine); Adult Nurse Practitio-
ner; Aircraft Nuclear-propul-
sion Program; Akan National
Park (Japan); Albert National
Park (Zaire); American Nazi
Party; Angkor National Park
(Cambodia); Arusha National
Park (Tanzania); Associated
Negro Press; Australian Na-
tionalist Party; Awash National
Park (Ethiopia)

ANP *Administración Nacional
de Puertos* (Spanish—Co-
lombia's National Administration of Ports); *Algemeen
Nederlandsch Persbureau*
(Dutch—Netherlands Press
Bureau)

ANPA American Newspaper
Publishers Association; Appalachian National Park Association; Australian National
Publicity Association; Australian Newspaper Proprietors
Association

ANPAT American Newspaper
Publishers Abstracting
Technique

ANPI *Associazione Nazionale
Partigiani d'Italia* (Italian—
National Association of Ital-
ian Partisans)

ANPO Aircraft Nuclear Propulsion Office

anpod antenna-positioning
device

ANPP Aircraft Nuclear Propulsion Program

ANPPF Aircraft Nuclear
Power Plant Facility

ANPPIA *Associazione Nazionale Perseguitani Politici Italiani Antifascisti* (Italian—National Association of Italian
Antifascist Political Victims)

ANPRM Advanced Notices of
Proposed Rule Making
(FAA)

ANPs Allied Navigation Publications

ANPS American Nail Producers Society

anpsi animal psi (animal extrasensory perception and psychokinesis)

anpt aeronautical national taper
pipe threads

ANPWS Australian National
Parks and Wildlife Service

ANQUE *Asociación Nacional
de Quimicos de España*
(Spanish—National Chemi-
cal Association of Spain)

anr another

ANR Agriculture and Natural
Resources; American Natural
Resources (formerly Ameri-
can Natural Gas); American
Newspaper Representatives;
Antwerp, Belgium (airport);
Australian National Railways

ANR *Asociación Nacional Republicana* (Spanish—Na-
tional Republican Association)–Paraguay's Colorado
Party

ANRA Amistad National Recreation Area (Texas); Ar-
buckle National Recreation
Area (Oklahoma)

anrac aids navigation radio
control

ANRAO Australian National
Radio Astronomy Observatory

ANRC American National
Red Cross; Animal Nutrition
Research Council; Australian
National Railways Commis-
sion; Australian National Re-
search Council

ANRF Australian Nomads Research Foundation

ANRPC Association of Natural
Rubber Producing Countries

ANRT *Association Nationale
de la Recherche Technique*
(French—National Associa-
tion of Technical Research)

ans answer; answered; answering; autograph note signed;
autonomic nervous system

Ans Anselm; Anselmo

ANS Advanced Navigation
School; Advanced Network-
ing Services (commercial Internet provider); Air Navigation School; American Name
Society; American Nuclear
Society; American Numis-
matic Society; American Nutrition Society; Army Newspaper Service; Army News
Service; Army Nursing Service; Astronomical Netherlands Satellite (first joint
United States–Netherlands
satellite)

ANS *Agencia Noticiosa Sapor-
iti* (Argentine press service)

ansa aminonapthosulfonic
acid; automatic new structure
alert

ANSA Australian National
Sportfishing Association

A(N)SA American (National)
Standards Association

ANSA *Agenzia Nazionale
Stampa Associata* (Italian—
National Press Association
Agency)

ansam (ANSAM) antimissile
surface-to-air missile

ANSC American National
Standards Committee

ANSCA Alaska Native Claims
Settlement Act

A-N Scale Anti-Negro Scale
(measuring negative attitudes)

ANSCO Anthony and Scovill
(New York camera and film
mnanufacturer merged with
AGFA to become Agfa-
Ansco and more recently
GAF—General Aniline and
Film Corporation)

Ansel Anselm

Anseri Anseriformes (ducks,
geese, swans)

ANSETT Ansett Airways

ANSETT-ANA Ansett Australian National Airways

ANSI American National Standards Institute; Australian National Standards Institute

ANSIC Aerospace Nuclear Safety Information Center

ANSL Australian National Standards Laboratory

ANSO Assistant Naval Stores Officer

ANSP Academy of Natural Sciences of Philadelphia; Australian National Socialist Party

ANSS American Nature Study Society

ANSSL Australian National Social Sciences Library

ANSSMFE Australian National Society of Soil Mechanics and Foundation Engineering

ANST Appalachian Nature Scenic Trail (Maine to Georgia)

ANSTEL Australian National Scientific and Technological Library

answ (ANSW) antinuclear submarine warfare

ant. antenna(s); anterior; anticipated; antilog; antilogarithm; antiquarian; antique; antiquities; antiquity; antonym

ant anterior (Spanish); *antología* (Spanish—anthology)

ant. antico (Italian–antique); *antiporta* (Italian—half-title)

Ant Antigua; Antillean; Antillea—West Indian Federation; Antilles; Antlia (constellation); Antwerp

Ant Antilia Pneumatica (Latin—air pump, whose name adorns a constellation)

ANT American National Theater; AN Tupolev (Soviet aircraft designer's initials and designation of planes he designed); Australian Northern Territory

ANT Australian National Times

ANTA American National Theater and Academy; Australian National Travel Association

antag antagonistic

Antarc Antarctic; Antarctica

Antarc O Antarctic Ocean

ant. ax line anterior axillary line

Ant & Bar Antigua and Barbuda

ANTC Australian National Television Council

Ant-C antennapedia complex

Ant Chil Antarctica Chilena (Chilean Antarctic)

Ant & Cl Antony and Cleopatra

Ant Cur Antilles Current

ant. d anterior diameter

ante (Latin prefix—before, in front of); (Latin—before)

antec annual technical conference

ANTELCO Administración Nacional de Telecomunicaciones (Spanish—Paraguayan National Telecommunication Administration)

antennafier antenna + radiofrequency amplifier

antennamitter antenna + transmitter

antennaverter antenna + converter

Antf Antofagasta

Ant f Antillean florin (guilder)

anthol anthological(ly); anthologist, anthologize, anthology

anthro anthropogeography; anthropological; anthropologist; anthropology; anthropometry; anthropomorphism; anthropophagy

anthroco anthrocosis (sickness due to coal-dust inhalation)

anthrop anthropology

anthropol anthropologist; anthropology

anthropom anthropometry

Anthroposophic Anthroposophic Press

Anthy Anthony

anti (Latin prefix—against)—anticommunist

antibio(s) antibiotic(s)

antichlor anti + chlorine; antichloristic

anticli anticlimactic(al)(ly); anticlimax; anticlinal; anticline; anticlinorium

antidis antidisestablishmentarianism

antid(s) antidote(s)

antifreeze grain or methyl alcohol (CH_3OH) mixture

Antig Antigua

Antigua West Indian island nation, officially Antigua and Barbuda including nearby Redonda

antikv antikvarisk (Dano-Norwegian—antiques)

Antillas (Spanish—Antilles) West Indies

Antillas Mayores (Spanish—Greater Antilles) Cuba, Hispaniola, Jamaica, Puerto Rico

Antillas Menores (Spanish—Lesser Antilles) Leeward and Winward Islands extending from the Virgin Islands to Aruba

Antillas Neerlandesas (Spanish—Netherlands Antilles) Aruba, Bonaire, Curacao, Saba, Sint Eustatius, and half of Sint Maarten

Antilles West Indian Islands excluding Bahamas

antilog antilogarithm

antimag antimagnetic

antimat antimatter

antinuke anti nuclear (energy, power, or war)

Antioch Antioch Press

antip antiparasitic; antiparticle; antipasti; antipasto; antipathetic; antipathy; antiperiodic; antipersonnel; antiperspirant; antipodal; antipode; anti-poetic; antipollution; anti-poverty; antiproton; antipsychotic; antipyretic; antipyrine

antiphon antiphonal(ly)

Antipodes Australia and New Zealand; rocky islands off Dunedin, New Zealand

antipol antipollutant; antipollution

antiporn antipornographic; antipornography

antiq antiquarian, antique; antiquities; antiquity

antiq. antiquary

antiq égypt antiquité égyptienne (French—Egyptian antiquity)

antiq gr antiquité grecque (French—Grecian antiquity)

antiq hébr antiquité hébraique (French—Hebraic antiquity)

antiq rom antiquité romaine (French—Roman antiquity)

antiquar antiquarian

Antiques Antiques Publications

anti-Sem anti-Semitic

antisex antisexual

antivox antivoice-operated transmission

ant. jentac. ante jentaculum (Latin—before breakfast)

Antl Antlia

Ant Lat Antique Latin

ant. ld antique laid

Anto Antofagasta (Chilean province)

Antº Antonio

anton antonym

Anton Antonio; Antony

Ant Ops Antarctic Operations

Ant Pen Antarctic Peninsula (once called Graham Land)

ant. pit. anterior pituitary

ant. prand. ante prandiulmatin (Latin—before dinner)

antr apparent net transfer rate

Antr Antrim

Antrims Antrim Mountains of Northern Ireland

ant(s). antonym(s)

ANTS Advanced Naval Training School

ant. sup. spine anterior superior spine

antu alpha-naphthyl-thiourea (rat poison)

ANTU Advanced Naval Training Unit; Atlantic (Line container) Unit

Antuer Antuérpia (Portuguese—Antwerp)

Antw Antwerpen (Dutch, Flemish, German—Antwerp)

ant. wo antique wove

ANU Australian National University (Canberra); St. John's, Antigua (3-letter code)

anug acute necrotizing ulcerative gingivitis

ANUP Australian National University Press

an anverdelse (Dano-Norwegian—application, use)

anvo accept no verbal orders

ANWA Abstracts of New World Archeology

ANWC American News Women's Club

ANWG Apollo Navigation Working Group (NASA)

ANWR Agassiz National Wildlife Refuge (Minnesota); Aransas NWR (Texas); Arctic Natural Wildlife Refuge in Alaska; Arrowhead NWR (North Dakota); Audubon NWR (North Dakota)

anx. annex

Anx Annex (postal place-name abbreviation)

ANX Anixter Brothers (stock exchange symbol)

ANXF Allied Naval Expeditionary Force

anytimeteller anytime bank teller

Anz Anzania (African name for South Africa)

ANZ Air New Zealand; Australia and New Zealand (bank)

ANZA Association of New Zealand Advertisers

ANZAAS Australian and New Zealand Association for the Advancement of Science

ANZAC Australia and New Zealand Army Corps

ANZAM Australia, New Zealand, and Malaysia (defense pact)

ANZAMRS Australia and New Zealand Association for Medieval and Renaissance Studies

ANZAR Australian and New Zealand Association of Radiology

ANZASA Australian and New Zealand American Studies Association

ANZ Bank Australia and New Zealand Bank

ANZCAN Australia-New Zealand-Canada (submarine cable system)

ANZCP Australian and New Zealand College of Psychiatrists

ANZDEC Asian-New Zealand Development Consultants

ANZECO Australia and New Zealand Exploration Company

ANZECS Australian and New Zealand-Europe Container Service

ANZESC Australian and New Zealand Eastern Shipping Conference

ANZGAM Australian and New Zealand Graduates Association of Malaysia (Singapore)

ANZHES Australian and New Zealand History of Education Society

ANZIA Associate of the New Zealand Institute of Architects

ANZIC Associate of the New Zealand Institute of Chemists

ANZJS Australian and New Zealand Journal of Sociology

ANZLA Associate of the New Zealand Library Association

ANZPC Australia and New Zealand Passenger Conference

ANZPPA Australia and New Zealand Professional Photographers Association

ANZQR Australia and New Zealand Bank Quarterly Review

ANZS Africa New Zealand Service

ANZSNM Australian and New Zealand Society of Nuclear Medicine

ANZSOM Australian and New Zealand Society of Occupational Medicine

ANZSOS Australian and New Zealand Society of Oral Surgeons

ANZSTS Australia and New Zealand Society for Theological Study

ANZUK Australia, New Zealand, United Kingdom (cultural, military, and trading alliance)

ANZUS Australia, New Zealand, United States (mutual security pact)

ao access opening; account of; acoustic-optic; air ordnance; anodal opening; anterior oblique; anti-oxidant; aorta; aortic opening; area of operation(s); auxiliary oiler; axioocclusal

ao (AO) accuracy only; account of; arresting officer

a/o (A/O) account of

A° anno (Latin—year)

AO Administration Office; Admiralty Order; Airdrome Office(r); American Optical (company); Angola (Internet code); Arkansas & Ozarks (railroad); Army Order; Australian officer (Officer of the Order of Australia); Autonomous Oblast; Aviation Ordnanceman; fleet tanker (2-letter naval designation); Order of Australia

A/O Administrator's Office (EPA)

AO *Ahonim Ortalik* (Turkish–Anonymous Company)–joint stock company; Avtonómnaya Oblast (Russian—Autonomous Region)–province

A-O *Auslands-Organisation* (German—Overseas Organization)–Hitler's overseas public relations agency

aoa abort once around; amphibious objective area; angle of attack; at or above

AoA Administration on Aging (HEW)

AOA Air Officer Administration; American Oceanology Association; American Optometric Association; American Ordnance Association; American Orthopedic Association; American Orthopsychiatric Association; American Osteopathic Association; Amer-

ican Overseas Airlines; American Overseas Association; Australian Optometrical Association; Australian Orthopaedic Association

AOAC Association of Official Agricultural Chemists; Association of Official Analytical Chemists

AOAD Army Ordnance Arsenal District

a O (a/O) *an der Oder* (German—on the Oder River)

aoas angle-of-attack sensor

AOATC Atlantic Ocean Air Traffic Control

aob alcohol on breath; angle on the bow; any other business; at or below

aob (AOB) annual operating budget

AOB Advanced Operational Base

AO-BIRMDis Army Ordnance-Birmingham District

AOBMO Army Ordnance Ballistic Missile Office (USA)

AO-BOSTDis Army Ordnance-Boston District

AOBs Antediluvian Order of Buffaloes

AOBS Army Outward Bound School

AOBSR Air Observer

AOBTA American Oriental Bodywork Therapy Association

aoc anodal opening contraction; any other color; attention operation characteristic

AoC Architect of the Capitol (D.C.)

A o C Academy on Computers

AOC Air Officer Commanding; Air Operations Center; Airport Operators Council; American Optical Company (stock exchange symbol); American Orthoptic Council; Arabian Oil Company; Army Ordnance Corps; automatic output control; Aviation Office Candidate

AOC *Anno Orbis Conditi* (Latin—year the world was

created); *Appellation d'Origine Contrôllée* (French—controlled designation of origin)

AOCA American Osteopathic College of Anesthesiologists

AOCCOS Australian Organisations Coordinating Committee for Overseas Students

AOCF Association of Outplacement Consulting Firms

AO-CHIDis Army Ordnance-Chicago District

AOCI Airport Operators Council International

AO C-in-C Air Officer Commander-in-Chief

aocl anodal opening clonus

AO-CLEVDis Army Ordnance-Cleveland District

aocm advanced optical counter measures

aocm (AOCM) aircraft out of commission for maintenance

AOCMU Air Officer Commanding Maintenance Units

AOCO Atomic Ordnance Cataloging Office

aocp (AOCP) aircraft out of commission for parts

AOCs American Olympic Committee members; Association of Old Crows

AOCS Airline Operational Control Society; American Oil Chemists' Society; Atlantic Outer Continental Shelf

aod acknowledgment of delivery; advanced ordnance depot; arterial occlusive disease; as of date; automatic opening device (parachute)

A o D Airport of Departure

AOD Academy of Oral Dynamics; Air Officer of the Day; Army Ordnance Depot

ao diag acridine-orange diagnosis (cancer)

AODs All Ordnance Destruct System; Ancient Order of Druids

aoe airborne operational equipment; auditing order error

AoE Aerodrome of Entry

A o E Airport of Entry

AOE Association of Overseas Educators; Auxiliary Oiler, Explosives; fast combat support ship (naval symbol)

AOEHI American Organization for the Education of the Hearing Impaired

AOEM Automotive Original Equipment Manufacturers

AOER Army Officer's Emergency Reserve

AOF Afrique Occidentale Francaise (French—French West Africa); Ancient Order of Foresters; Australian Olympic Federation

AOFS Active Optical Fuzzing System

AOFU Army Operational Film Unit

aog (AOG) aircraft on ground

AOG Assembly of God church; Atlantic Oceanographic Group; Australian Oil and Gas; gasoline tanker (naval symbol)

AOG Appellation d'Origine Garantie (Algeria—Guaranteed Designation of Origin)

AOGM Army, of Occupation of Germany Medal

AOH Ancient Order of Hibernians

aoi accent on information; angle of incidence; area of interest

A o I Aims of Industry

AOIB Anglo-Oriental International Bank

AOIPS Atmospheric and Oceanic Information Processing System

aoiv automatically operated inlet valve

ao. J.C. aprés Jésus-Christ (French—after Christ) A.D.

AOJC Association of Orthodox Jews in Communications

AOJDD Anglican Orthodox Joint Doctrinal Discussions

aok all okay; everything in good order

AOK Armee-Oberkonunando (German—Army High Command)

aol absent over leave

AOL America On Line; American-Oriental Lines; Atlantic Oceanography Laboratories

AO-LADis Army Ordnance-Los Angeles District

aolo advanced orbit laboratory operations

AOLP Action Organization for the Liberation of Palestine

AOM Army of Occupation Medal; Association of Operative Millers

A.O.M. Master of Obstetric Art

AOMA American Occupational Medical Association; Apartment Owners and Managers Association

AOMAA Apartment Owners and Managers Association of America

AOMC Army Ordnance Missile Command

AOME Assistant Ordnance Mechanical Engineer

AOMSA Army Ordnance Missile Support Agency

AOMSC Army Ordnance Missile Support Center

aon all or none

aonb area of outstanding natural beauty

AONB Area of Outstanding Natural Beauty

AONS Air Observers Navigation School

AO-NYDis Army Ordnance-New York District

aoo anodal opening odor

AOO American Oceanic Organization; Amphibious Operations Officer; Aviation Ordnance Officer

AOOA Aircraft Owners and Operators Association

AOOC Administrative Office Overview Committee

aop air observation post; anodal opening picture; aortic-pressure pulse; artillery observation post

AOP American Order of Pioneers; Apprenticeship Outreach Program; Army Observation Post; Association of

Optical Practitioners; Association of Osteopathic Publications

AOPA Aircraft Owners and Pilots Association; American Orthotic and Prosthetic Association

AOPC Australian Overseas Projects Corporation

AOPE Associated Organization for Professionals in Education

AOPEC Arab Organization of Petroleum Exporting Countries

aopf active optical proximity fuse

AO-PHILDis Army Ordnance-Philadelphia District

AOPL Association of Oil Pipe Lines

aoProf auszerordentlicher Professor (German—associate professor or special lecturer)

A Ops Air Operations

AOPs Allied Ordnance Publications

AOPU Asian Oceanic Postal Union (China, Korea, Philippines, Thailand)

aor air operations room; amphibious operations room; auxiliary oiler, replenishment

AOR Adventure Outdoor Resorts

AORC Automotive Occupant Restraints Council

AORS Abnormal Occurrences Reporting System

aoq average outgoing quality

AOQC Australian Organisation for Quality Control

aoql average outgoing quality limit

aor album-oriented rock music (FM radio performance); angle of reflection; aorist; area of responsibility

a/or and/or

AOR Army Operational Research; Australian Oil Refining; auxiliary oil replenishment (naval symbol)

Aorangi (Maori—Cloud Piercer)–Mount Cook, New Zealand's tallest towering to 3764 meters or 12,349 feet

AORB Aviation Operational Research Branch

AORG Army Operational Research Group (United Kingdom)

AORL Apollo Orbital Research Laboratory

AORN Association of Operating Room Nurses

AOrPA American Orthopsychiatric Association

AORS Abnormal Occurrences Reporting System; Army Operations Research Symposium

AORT Association of Operating Room Techniques

AORTF American Organization for Rehabilitation through Training Federation

aort regurg aortic regurgitation

aort sten aortic stenosis

aos acquisition of signal; add or subtract; angle or sight; anodal opening sound

AOS Air Observer School; American Opera Society; American Opthalmological Society; American Orchid Society; American Oriental Society; American Otological Society; Australian Overseas Smelting; Azimuth Orientation System

AOSA American Orff-Schulwerk Association

AOSC Association of Oilwell Servicing Contractors

A-O Scale Anti-Oriental Scale (measuring negative attitudes)

AOSE American Order of Stationary Engineers

AOSO Advanced Orbiting Solar Observatory

aosp advanced on-board signal processor; automatic operating and scheduling program

AOSPS American Otorhinologic Society for Plastic Surgery

AOSs Ancient Order of Shepherds

AO-STLDis Army Ordnance-St Louis District

AOSTRA Alberta Oil Sands Technology and Research Authority

AOSW Association of Official Shorthand Writers

aot active on target; anodal opening tetanus

Aot Askania optical tracker

AOT Alameda-Oakland Tunnel; Association of Occupational Therapists

AOTA American Occupational Therapy Association; Australian Overseas Transport Association

AOTC Aspen Opera Theater Center; Australian Overseas Trading Corporation; Aviation Officers Training Corps

aotd active optical target detector

Aotearoa (Maori—Long White Cloud)—New Zealand

AOTE Amphibious Operational Training Element

ao technique acridine-orange technique (two-color fluorescent test for cancer)

AOTOI American Organization of Tour Operators to Israel

AotOS Admiral of the Ocean Sea (U.S. Merchant Marine award)

AOtS American Otological Society

AOTU Amphibious Operational Training Unit

aou apparent oxygen utilization; azimuth orientation unit

AOU American Ornithologists' Union; Australian Ornithologists Union

AOUSC Administrative Office of the United States Courts

AOUW Ancient Order of United Workmen

aov analysis of variance; any other variety; artillery observation vehicle

AOVP Australian Ordnance Vehicle Park

AOW Army Ordnance Workshop; Articles of War

AOWF Army Officers Widows' Fund

ap above proof; access panel; acid phosphatase; acid proof, action potential; acute proliferation; aerial port; aiming point; aircraft park; airplane; alkaline phosphatase; alum precipitated; aminopeptidase; ammonium perchlorate; ammunition point; angina pectoris; antepartum; anterior pituitary; anteroposterior; anti-personnel; aortic pressure; appendectomy; appendices; appendix; apple; apprentice; arithmetic progression; armor piercing; arterial pressure; artificial pneumothorax; as prescribed; association period; attached processor; attaching part(s); author's proof, auxiliary patrol; axiopulpal; (Welsh prefix—son of)

ap (AP) advanced placement; automatic payment; average product

a/p accounts payable; after perpendicular; air port (porthole); angle point; authority to pay; authority to purchase; autopilot

a & p agricultural and pastoral; anterior and posterior; apogee and perigee (apex and antapex); auscultation and percussion

a$_p$ geomagnetic index

ap anno passato (Italian—last year)

ap. apud (Latin—according to)

a.p. ante prandium (Latin—before a meal)

Ap Apothecary

Ap Apus (Latin—Bird of Paradise)

Ap. Apocalypse

Ap. Apostolus (Latin—Apostle)

AP Acquisition Plan; Additional Premium; Advanced Placement; Air Police; Air Publication; Airport; Allied Publication; American Party; American President Lines; Andra Pradesh; Artillery of Pakistan; Assistant Paymaster; Associated Press; Australia Party; Aviation Pilot; (electronic) Annual Position (form); personnel transport (naval symbol)

A-P American Plan (includes meals)

A/P allied papers; authority to pay

A & P Agricultural and Pastoral Society of New Zealand; Atlantic & Pacific Travel International; Great Atlantic & Pacific Tea Company

AP *Acción Popular* (Spanish—Popular Action); *Administration Pénitentiare (French—*Penitentiary Administration); Almonester y Pontalba (New Orleans: Don Andres Almonester and his daughter the Baroness de Pontalba); *Arbeidøpartiet (*Norwegian—*Det Norske Arbeiderpartiet—*The Norwegian Labor Party); *Atlanska Plovidba* (Russian—Atlantic Press); *Australia Post*; *Aviapolk* (Russian—Air Regiment)

A.P. *a protester* (French—to be protested later)

apa aidosterone-producing adenoma; aminopenicillanic acid; antipernicious anemia factor; axial pressure angle

APA Administrative Procedure Act; Adult Parole Authority; Advertising Photographers of America; Aerovias Panamá Airways; Agricultural Publishers Association; Airline Passenger Association; Alaska Power Authority; American Patients Association; American Payroll Association; American Pharmaceutical Association; American Philological Association; American Philosophical Association; American Photoengravers Association; American Physiotherapy Association; American Pilots Association; American Planning Association; American Plywood Association; American Podiatry Association; American Polygraph Association; American Poultry Association; American Press Association; American Protective Association; American Psychiatric Association; American Psycho-analytical Association; American Psychological Association; American Psychosomatic Association; American Psychotherapy Association; American Pulpwood Association; American Pyrotechnics Association; Animation Producers Association; Anthracite Producers Association; Anti-Papal Association; Apache Railway; Architectural Photographers Association; Architectural Precast Association; Army Parachute Association; Associated Press of Australia; Association of Paroling Authorities; Association of Physicians of Australasia; Association for the Prevention of Addiction (London); Australian Provincial Insurance; Automobile Protection Agency; transport attack vessel (naval symbol)

APA *Austria Presse Agentur* (German—Austrian Press Agency)

APAA Australian Port Authorities Association; Automotive Parts and Accessories Association

APAAC Asian-Pacific-American Advocates of California

APAC Aerial Photographic Analysis Center

apache analog programming and checking

APACHE Accelerator for Physics and Chemistry of Heavy Metals; Application Package for Chemical Engineers

APACI *Associación Peruana para el Avance de la Ciencia* (Spanish—Peruvian Association for the Advancement of Science)

APACL Asian Pacific AntiCommunist League; Asian People's Anti-Communist League

apacs adaptive planning and control sequence (marketing)

APADA Australian Petroleum Agents and Distributors Association

APADS Automatic Programmer and Data System

APAE Association of Public Address Engineers

apaf antipernicious anemia factor

apaf-1 apoptotic protease activating factor-1

APAFA ASEAN Professional Association on Food and Agriculture

APAG American Photographic Artisans Guild; Atlantic Policy Advisory Group; Atlantic Political Advisory Group (NATO)

APAIS Australian Public Affairs Information Service

APAL American Puerto-Rican Action League

apam anti-personnel/anti-material (bomb)

AP & AM Adler Planetarium and Astronomical Museum

APANZ Associate of the Public Accountants of New Zealand

APAP American People for American Prisoners

APAP *Asociación Peruana de Agencias de Publicidad* (Spanish—Peruvian Association of Advertising Agencies)

apar apparatus

APAR Automatic Programming and Recording

apart. apartment

a-part alpha particle(s)

apart apartheid (Afrikaans—apartness, racial segregation)—South African government policy

APAs American Polygraph Association members; Anti-Papal Association members and sympathizers

APAS Automatic Performance Analysis System

APASCO Australian Planning and Systems Company

apat auxiliary propelled anti-tank

APATS Antenna Pattern Test System (USA)

APAVIT *Associación Peruana de Agencias de Viajes y Turismo* (Spanish—Peruvian Association of Travel and Tourism Agencies)

apb air-portable bridge; atrial premature beat; auricular premature beat; 4-aminophosphonobutyric acid

apb (APB) all-points bulletin

APB Accounting Principles Board; Associated Press Broadcasters; barracks ship, self-propelled (naval symbol)

APB All-Points Bulletin

APBA American Pet Boarding Association; American Power Boat Association; Association Press Broadcasters Association

apbc armor piercing ballistic cap

APBPA Association of Professional Ball Players of America

APBS Accredited Poultry Breeding Scheme; Australian Permanent Building Society

APBSD Advanced Post Boost System Development

a/p c autopilot capsule

apc alien property custodian; ammonium perchlorate; approach power compensator; armor-piercing capped; aspirin, phenacetin and caffeine; automatic picture control

apc (APC) average propensity to consume

APc coastal transport (3-letter symbol)

APC Aeronautical Planning Chart; Aerospace Primus Club; Agricultural Productivity Commission; Air-Pollution Control; American Parents Committee; American Peace Corps; American Philatelic Congress; American Plastics Council; American Pop(ular) Culture; Area Positive Control; Arkansas Polytechnic College; Armament Practice Camp; armored personnel carrier; Army Pay Corps; Army Petroleum Center; Army Policy Council;

Aseptic Packaging Council; Association of Private Camps; Association of Pulp Consumers; Australian Police College; Australian Postal Commission; Automotive Presidents Council; Avian Propagation Center

apca antiparietal cell antibody

APCA Air Pollution Control Association; American Petroleum Credit Association; American Planning and Civic Association; Anglo-Polish Catholic Association; Australian Physical Culture Association

APCB Air Pollution Control Board

apcbc armor-piercing carbide ballistic cap

apc-c aspirin, phenacetin, caffeine—with codeine

APCC Asian and Pacific Coconut Community

APCD Air Pollution Control District

APCG Association of Pacific Coast Geographers

apche armor-piercing capped high explosive; automatic programmed checkout equipment

apci armor-piercing capped incendiary

apcit armor-piercing capped incendiary with tracer

APCK Association for Promoting Christian Knowledge

APCL Association of Professional Color Laboratories

apcm authorized protective connecting module

APCM Asiatic-Pacific Campaign Medal

APCN Anno Post Christian Natum (Latin—year after Christ's birth)

APCO Air Pollution Control Office; Alabama Power Company; Associated Public Safety Communications Officers

apcr armor-piercing-composite rigid

apCr activated protein C resistance

APCS Air Photographic and Charting Service; Army Postal and Courier Service; Associative Processor Computer System

apct armor-piercing capped with tracer

a/p ctl autopilot control

apc virus adenoidal, pharyngeal, conjuctival virus

APCYA A Presidential Classroom for Young Americans

apd action potential duration; aiming point determination; anteroposterior diameter

APD Admiralty Press Division; Air Pollution Division (U.S. Dept. Agriculture); Air Procurement District; Army Pay Department; high-speed troop transport (3-letter naval symbol)

APDA American Parkinson's Disease Association; Appliance Parts Distributors Association

APDC Agricultural Pest Destruction Council; Albany Port District Commission

Ap Del Apostolic Delegate

APDF Asian-Pacific Dental Federation

apdl algorithmic processor description language

Apdo Apartado (Spanish—post-office box)

apds armor-piercing discarding sabot

ApdS American Pediatric Society

APDSMS Advanced Point Defense Surface Missile System

APDU Association of Public Data Users

APDUSA African People's Democratic Union of Southern Africa

apdy appropriate duty

ape adapted physical educator; aerial port of embarkation (APE); aminophylline, phenobarbital, ephedrine; anterior pituitary extract; apparent effect; automatic photomapping equipment

APE aerial port of embarkation; Amalgamated Power Engineering

A.P.E. Air Pollution Engineer

APEA Association of Professional Engineers (Australia); Australian Petroleum Exploration Association

APEC Asia-Pacific Economic Cooperation; Automotive Products Emissions Committee

Apeco American Photocopy Equipment Company

APELR Association for the Preservation of English Language Rights

Apennines Apennine Mountains of the Italian Peninsula

aper aperture

APER Air Pollutant Emissions Report

apers antipersonnel

APES Advanced Planning Enemy Section; American Philatelic Expertizing Service

APETS Australia-Papua New Guinea Education and Training Scheme

apex advance-purchase excursion (airline fare); air-pollution exercise; assembler and process executive

APEX Advance-Purchase Excursion (Plan) Association of Professional Executive and Management Staff

apf acidproof floor; animal protein factor

APF Alaska Permanent Fund; All People's Front; American Progress Foundation; Anglican Pacifist Fellowship; Asia Pacific Forum; Asian Packaging Federation; Association of Pacific Fisheries; Association of Protestant Faiths; Australian Parachute Federation

APF Australian Pharmaceutical Formulary

APFA American Pipe Fittings Association; American Professional Football Association (name changed to National Football League in 1922); Associated Poultry Farmers of Australia; Association of Professional Flight Attendants

APFA Association des Professeurs Franco-Americains (French—Association of Franco-American Professors)

APFC Asia-Pacific Forestry Commission; Association of Physical Fitness Centers

APFF American Police and Fire Foundation

APFM Australian Postgraduate Federation in Medicine

APFO Association on Programs for Female Offenders

APFRI American Physical Fitness Research Institute

apfsds armor-piercing fin-stabilized discarding sabot

apg anti-personnel grenade

Apg Appingedam

APG Aberdeen Proving Ground; Air Proving Ground; American Pewter Guild; Army Planning Group; Army Proving Ground; Association of Professional Genealogists; Australian Proving Ground

APG Autoridad Portuaria de Guayaquil (Spanish—Port Authority of Guayaquil)

APGA Alabama Personnel and Guidance Association; American Personnel and Guidance Association; Apple and Pear Growers Association; Arizona Personnel and Guidance Association; Arkansas Personnel and Guidance Association

APG-BRL Aberdeen Proving Ground—Ballistics Research Laboratory

APGC Air Proving Ground Center

apgcu autopilot ground control unit

APG/HEL Aberdeen Proving Ground—Human Engineering Laboratory

APG/OBDC Aberdeen Proving Ground—Ordinance Bomb Disposal Center

APGOEF Air Proving Ground—Eglin, Florida

APGp Aerial Port Group

aph antepartum hemorrhage; anterior pituitary hormone (APH); aphelion; apethic

Aph Aphasia

APH Access Permit Holder; transport fitted for evacuation of wounded (3-letter symbol)

A.P.H. A(lan) P(atrick) Herbert

APhA American Pharmaceutical Association

APHA American Printing History Association; American Protestant Hospital Association; American Public Health Association; Australian Pneumatic and Hydraulic Association

APHB American Printing House for the Blind; Army Pearl Harbor Board

aphc armor-piercing hard-core

APHCA Animal Production and Health Commission for Asia, Far East and South West Pacific

aphe armor-piercing high-explosive

aphet aphetic

aphhw all-purpose hand-held weapon

APHI Association of Public Health Inspectors

APHIA Association for the Promotion of Humor in International Affairs

APHIS Animal and Plant Health Inspection Service

APhO Area Pharmaceutical Officer

Aph of Lath Aphorism of Lathem ("There is nothing edible or potable failing to find someone to take it as a sovereign remedy for some disease and upon the earnest recommendation of some eminent physician.")

aphor aphorism; aphorist(ic) (ally); aphorize

aphp antipseudomas human plasma

aphros aphrodisiac(s)

APHS Arizona Pioneer Historical Society; Australian Psychology and Hypnotherapy Association api acceptable periodic inspection; air position indicator; armor-piercing incendiary tracer

API Academic Press, Inc; Alabama Polytechnic Institute; American Paper Institute; American Petroleum Institute; American Pipe Institute; American Potash Institute; American Press Institute; Animal Protection Institute; Application Program Interface; armor-piercing incendiary; Associated Photographers International; Australian Pharmaceutical Industries; Australian Planning Institute

API *Association Phonetique Internationale* (French—International Phonetic Association); *Associazione Pionieri Italiani* (Italian—Italian Boy Scouts Association)

APIA Animal Protection Institute of America

APIC Allied Press Information Center; Apollo Parts Information Center; Army Photo Interpretation Center; Association for Practitioners in Infection Control

APICORP Arab Petroleum Investments Corporation

APICP Association for the Promotion of the International Circulation of the Press

APICS American Production and Inventory Control Society

APICSC Atlantic-Pacific Interoceanic Canal Study Commission

apicult apiculture

APID Army Photo Interpretation Detachment

APIDC Andhra Pradesh Industrial Development Corporation

APIF Automated Process Information File

APIJ *Australian Planning Institute Journal*

A-pill abortion pill

APIM *Association Professionelle Internationale des Midecins* (French-International Professional Association of Physicians)

APIN Atlas Propulsion Information Notice

apipocc appropriating property in possession of (a) common carrier

APIS Army Photographic Intelligence Service; Australian Professional Interpreting Society

APITCA American Producers of Italian Type Cheese Association

APIU Army Photo Intepretation Unit

apivr artificial pacemaker ventricular rhythm

APJ American Power Jet (company)

APJE Association of Philosophical Journals Editors

APJI Army Parachute Jump Instructor; Association for the Protection of Jewish Immigrants

apl a programming language; adult performance level; aluminum-polythene laminate; automatic premium loan

apl (APL) anterior pituitary-like hormone

a/pl armorplate

Apl Appledore

APL Aden Protectorate Levies; Advanced Programming Language; Approved Products List; A Programming Language (mathematically oriented computer language developed by Kenneth E. Iverson of IBM); Air Pacific Limited; Air Provost Marshal; Akron Public Library; Albany Public Library; Albuquerque Public Library; American Pioneer Line; American President Lines; Applied Physics Laboratory; Army Promotion List; Assembly Parts List; Augusta Public Library; Australian Pensioners League; barracks ship (naval symbol)

AP-LS American Psychology-Law Society

A-PL All-Purpose Linotype

APLA American Patent Law Association; Armenian Progressive League of America; Atlantic Provinces Library Association

apilat anteroposterior and lateral

APLC Automated Parking Lot Control

Aplcrs Applecross

apld aspiration percutaneous lumbar diskectomy

apld (APLD) automatic program locate device

APLD Association of Professional Landscape Designers

APLE Association of Public Lighting Engineers

APLIC Association of Parliamentary Librarians in Canada

apll analog phased-locked loop(s)

apins applications

APLQ *Agence de Presse Libre du Québec* (French—Free Press Agency of Québec)

APLS American Plant Life Society

AP-LS American Psychology-Law Society

APLU American President Line (container) Unit

apm apomict; associative principle for multiplication; average portfolio maturity

apm (APM) antipersonnel missile

APM Academy of Physical Medicine; Academy of Psychosomatic Medicine; Air Power Museum; Assistant Paymaster; Assistant Provost Marshal; Association for Psychoanalytic Medicine; Australian Paper Manufacturers; Australian Paper Mills

apma advance payment of mileage authorized

APMA Absorbent Paper Manufacturers Association; Automatic Phonograph Manufacturers Association

APMAC A.P. Moller Associated Concerns

APMC Academy of Psychologists in Marital Counseling; Andhra Pradesh Mining Corporation; Asia and Pacific Maritime Cooperation Centre

a/p mcu autopilot monitor and control unit

APME Associated Press Managing Editors (Association)

APMF Australian Paper Manufacturers Federation

APMG Assistant Postmaster General

APMHC Association of Professional Material Handling Consultants

apmi area precipitation measurement indicator

APMIS Automated Project Management Information System

APMM Allied Political and Military Mission

APMR Association for Physical and Mental Rehabilitation

APMT Antenna Pattern Measuring Test (USA)

apn artificial pneumothorax; average peak noise

Apn. Apocryphon

APN American Practical Navigator

APNA Asia-Pacific News Agencies

APNEC All-Pakistan Newspaper Employees Confederation

APNP Arthur's Pass National Park (South Island, New Zealand)

APNR American Professional Needlework Retailers

apo airport; apochromatic; apogee; apolipoprotein

APO Accountable Property Office(r); Acting Pilot Officer; Advanced Post Office; Air Force (Army) Post Office; American Potash & Chemical (stock exchange symbol); Animal Procurement Office(r); Area Patrol Office(r); Area Petroleum Office(r); Army Pay Office; Army Post Office; Association of Physical Oceanographers; Australian Post Office

APOAM Acting Petty Officer Air Mechanic

apob airplane observation

apoc apocopated; apocopation

Apoc Apocalypse; Apocrypha; Apocryphal

APOC Army Point of Contact; Army Post Office Corps

APOD Aerial Port of Debarkation

APOE Aerial Port of Embarkation

apol apologete; apologetic(al); apologetics; apologia; apologise; apoligist(s); apologize; apology

APOLLO Article Procurement and On-Line Local Ordering

APON Association of Pediatric Oncology Nurses

APOP Australian Public Opinion Polls

APOR Advisory Panel on Operational Research

apos apostrophe

APOS Advanced Polar Orbiting Satellite

apost apostacy; apostate

a post. a posteriori (Latin—derived from or relating to reasoning from observed facts)

Apostle Islands national lakeshore, Wisconsin

Apostles Apostle Islands, Lake Superior; New Testament disciples

apota automatic positioning of telemetering antenna

apotek apoteket (Danish–apothecary)—drugstore

apoth apothecaries'(weight); apothecary

A-powered atomic-powered

app accelerated pacification program; apparatus; apparel; apparent; appeal; appeared; appelate; appendage; appended; appendix; apperception; appetite; appetizer; applause; applied; appointed; apprehended; apprentice; approach; appropriate; appropriation; approval; approve; approximate

app. application; apprentice; approximately

App Appellate; Lucius Apuleius

App Apparat (German—apparatus); Lucius Appuleius (Roman philosopher)

App. Apostoli (Latin—Apostles)

APP Advanced Placement Program; Air Parcel Post; Algonquin Provincial Park (Ontario); Army Procurement Procedure; Association of Professional Photogrammetrists; automatic priority processing

APP Alianza Para Progreso (Spanish—Alliance for Progress)

appa advise present position and altitude

APPA American Penal Press Association; American Probation and Parole Association; American Public Power Association; American Pulp and Paper Association

APPAG All Party Penal Affairs Group (Britain)

Appalachia eastern Kentucky, southeastern Ohio, eastern Tennessee, and western West Virginia; region in the Appalachian Mountains extending from Québec to northern Georgia

Appalachians Appalachian Mountains

appar apparatus; apparently

APPAUC Association of Physical Plant Administrators of Universities and Colleges

APPC Advance Procurement Planning Council; Advanced Program-to-Program Communications

app. code application code

appd approved

APPDI American Professional Pet Distributors Incorporated

App Div Appellate Division (NY Supreme Court Reports)

APPECS Adaptive Pattern-Perceiving Electronic Computer System

appellat appellative

appi advanced planning procurement information; Application Program Interface

APPITA Australian Pulp and Paper Industries Technical Association

appl applause; appliance; applicable; application; applied

APPL Advance Procurement Planning List(s)

applan. applanatus (Latin—flattened)

applc appliance

APPLE Advanced Propulsion Payload Effects (NASA); Association of Public and Private Labor Employees

Appleton Appleton-Century-Crofts

applican. applicandus (Latin—applied, to be applied)

appln application

Appl Opt Applied Optics

Appl Phys Lett Applied Physics Letters

Appi Spectrosc Applied Spectroscopy

APPM Association of Publication Production Managers; Associated Pulp and Paper Mills

APPMA American Pet Products Manufacturers Association

appmt appointment

appn appropriation

App^{no} Appennino (Italian—Appenines)

a/p poi autopilot positioning indicator

APPP Association of Planned Parenthood Physicians

appr approval; approve; approved

appr. appraisement

APPR Army power package reactor

apprd. appeared; apprised

appren apprentice

appro approval

approp appropriation

approx approximate(d); approximately; approximation; approximative

apps appendices; appendixes

APPS Australian Physiological and Pharmacological Society; Australian Plant Pathology Society

appt appoint; appointment

apptd appointed

App Thorn Appleton Thorn prison, Lancashire, England

APPU Australian Primary Producers Union

APPWP Association of Private Pension and Welfare Plans

appx appendix

appy appendectomy

APQC American Productivity and Quality Center

apr amoebic prevalence rate; annual percentage rate; anterior-pituitary reaction; apprentice

apr aprile (Italian—April)

apr (APR) aerial photographic reconnaissance

Apr April

Apr Aprel (Russian—April)

APR Air Pictorial Service; Air Priority Raging; Airman Performance Report; American Public Radio; Annual Percentage Rate; Annual Progress Reports; Army Public Relations; Association of Petroleum Re-Refiners; Association for Promoting Retreats; Association of Publishers' Representatives; Automatic Place and Route

Apra San Luis de Apra (Guam's principal port)

APRA Aircraft Resources Production Agency; American Popular Revolutionary Alliance (Peru); Australian Performing Rights Association; Australian Plastics Research Association; Automotive Parts Rebuilders Association

APRA Alianza Popular Revolucionaria Americana (Spanish—Popular American Revolutionauy Alliance)—Peru's Aprista Party of Haya de la Torre

aprax apraxia(l)

APRC Army Physical Review Council; Automotive Public Relations Council

APRC Anno Post Roman Conditam (Latin—year after the foundation of Rome)

APRDC Army Polar Research and Development Center

APRE Air Procurement Region–Europe; Army Personnel Research Establishment

APREF Asia Pacific Real Estate Federation

après-40 after 40 years of age

après JC après Jesus Christ (French—after the birth of Jesus Christ)

APRF Army Pulse Radiation Facility

APRFE Air Procurement Region–Far East

a pri. a priori (Latin—derived from or relating to self-evident propositions)

APri air priority

april automatically programmed remote indication logged

A/Prin Assistant Principal

Apr J-C Après Jésus-Christ (French—after Christ) A.D.

APRL American Philatelic Research Library; American Prosthetic Research Laboratory

Aprmay April and May

aprmd appointment recommended

APRO Aerial Phenomena Research organization; Army Pay Records Office; Army Personnel Research Office; Association of Progressive Rental Organizations

apróx apróximadamente (Spanish—approximately)

AprS American Proctologic Society

APRS Association of Professional Recording Studios

aprt airport

APRTA Associated Press Radio and Television Association

aprthd Apartheid (Afrikaans—apartness)

APRU Asia Pacific Research Unit

aprx approximately

aps acceleration propagation system accessory power supply; adenosine phosphosulfate; advanced photo system; attending physician statement; autograph postcard signed; auxiliary power supply; auxiliary propulsion system

aps (APS) average propensity to save

Aps Apus

ApS AnpartsSelskab (Dano-Norwegian—limited company)

APS Aboriginal Protection Society; Academy of Political Science; adenosine phosphosulfate; Alternative Press Syndicate; American Metal Products (stock exchange symbol); American Pediatric Society; American Pheasant Society; American Philatelic Society; American Philosophical Society; American Physical Society; American Physiological Society; American Phytopathological Society; American Plant Selections; American Poinsettia Society; American Polar Society; American Proctologic Society; American Prosthodontic Society; American Psychosomatic Society; American Purchasing Society; Army Pigeon Service; Army Pilot School; Army Postal Service; Association of Photo Sensitizers; Association of Productivity Specialists; Atmospheric Protection System; Australian Pig Society; Australian Psychological Society; Australian Public Service; submarine transport (naval symbol)

APS Algerie Presse Service (French–Algerian Press Service); *Associated Press Stylebook and Libel Manual*

APSA Aerolíneas Peruanas, South America; American Political Science Association;

Associate of the Photographic Society of America; Association of Professional Scientists of Australia; Australian Pharmaceutical Sciences Association; Australian Political Studies Association; Australian Pre-School Association

A & PSA Aden and Protectorate of South Arabia

APSA Aerolíneas Peruanas SA (Spanish—Peruvian Airlines Corporation); *Azufrera Panamericana SA* (Spanish—Panamerican Sulfur Corporation)

APsaA American Psychoanalytic Association

APSB Aid to the Potentially Self-supporting Blind; Australian Public Service Board

APSC Army Personnel Selection

APSE Abstracts of Photographic Science and Engineering

APSF Alfred P. Sloan Foundation; Australian Public Service Federation

APS/HEW Administration for Public Services–HEW

APSL Acting Paymaster Sub-Lieutenant

APSO Assistant Polaris Systems Officer

APSq Aerial Port Squadron

APSS Army Printing and Stationery Service; Association for the Psychophysiological Study of Sleep; Association of Professional Sleep Societies

A.P.S.T. Associate in Public Service Techumology

APsychoA American Psychoanalytic Association

apsvds armor-piercing super-velocity discarding sabot

APsychosomS American Psychosomatic Society

APsychpthA American Psychopathological Association

apt afterpeak tank (APT) airborne pointer-tracker; alum-precipitated toxoid; apartment; armor-piercing

with tracer; automatic photometric telescope; automatic picture transmission; automatically programmed tool(s)

apt. airportable

apt apartadero (Spanish—platform)

APT Advanced Passenger Train; Airman Proficiency Test; Applied Performance Test(s); Automatic Photoelectric Telescope; Automatic Picture Transmission; Automotive Professional Training

APTA American Physical Therapy Association; American Pioneer Trails Association; American Platform Tennis Association; American Public Transit Association

APTC Allied Printing Trades Council; Army Physical Training Corps

APTD Aid to the Permanently and Totally Disabled

Aptdo apartado (Spanish—post office box)

apte advance passenger train express (149 mph British turbine-powered train)—APTE

apth apthong (a silent letter such as the *p* in pneumatic)

apti actions per time interval

APTI Association of Principals of Technical Institutions

APTIC Air Pollution Technical Information Center

apto aluminum plastic tearoff (container cover)

apto apartaniento (Spanish—apartment)

Apto apartaniento (Spanish—apartment)

APTP Association of Part-Time Professionals

apt(s) apartment(s)

APTs Advanced Passenger Trains

APTS Army Physical Training Staff; Australian Parachute Training School; Automatic Picture Transmission System

aptt activated partial thromboplastin time

APTU African Postal and Tele-communications Union; Australian Postal and Telecommunications Union

apu aircraft power unit; assessment of performance unit; auxillary power unit

APU Alaska Pacific University; American Pacific University; Army Postal Unit; Asian Parliamentary Union

APUA Alliance for the Prudent Use of Antibiotics

apud amine precursor uptake, decarboxylase

apu-hs automatic program unit–high speed

apu-ls automatic program unit–low speed

apv amino-phosphonovaleric acid; automatic-patching verification

APV All-Purpose Vehicles; Avenida Presidente Vargas, Rio de Janeiro, Brazil

APVE Association of Professional Vocal Ensembles

apw anti-personnel weapon; architectural projected window

APW Accelerated Public Works; American Prisoner of War; Apia, Western Samoa (airport)

APWA American Public Welfare Association; American Public Works Association

APWC Association of Professional Writing Consultants

APWD Army Psychological Warfare Division

APWP Accelerated Public Works Program

APWSS Asian-Pacific Weed Science Society

APWU American Postal Workers Union

apx appendix; approximate(ly)

apy annual percentage yield

APZ Assiniboine Park Zoo

aq accomplishment quotient; achievement quotient; any quantity; aqueous

a-q aircraft quality

aq. aqua (Latin—water)

AQ achievement quotient; Antarctica (Internet code); aviation fire-control technician (USAF symbol); Schreiner Aerocontractors (Hague)

AQ Australian Quarterly

AQA Australian Quadriplegic Association

AQAB Air Quality Advisory Board

AQAPs Allied Quality Assurance Publications

Aqar Aquarius

aq. astr. aqua adstricta (Latin—ice)

aq. bull. aqua bulliens (Latin—boiling water)

aq. cal. aqua calida (Latin—warm water)

AQCL Analytical Quality Control Laboratory

aq. com. aqua communis (Latin—ordinary water)

AQCR Air Quality Control Region (EPA)

a.q. dest. aqua destillata (Latin—distilled water)

aqdm air quality display model

AQE Airman Qualifying Examination

aq. ferv. aqua fervens (Latin—hot water)

aq. fluv. aqua fluvii (Latin—river water)

aq. font. aquafontana (Latin—spring water)

aq. fort. aqua fortis (Latin—nitric acid)

aq. frig. aqua frigida (Latin—cold water)

aqgv azimuth-quantized gated video

AQHA Australian Quarter Horse Association

Aqil Aquila

aql acceptable qualifying levels; acceptable quality level; approved quality level

Aql Aquila

aq. mar. aqua marina (Latin—sea water)

AQMD Air-Quality Management District (Southern California)

aq. ment. pip aqua menthae piperitae (Latin—peppermint water)

AQMG Assistant Quartermaster General

AQMP Air Quality Master Plan

aqn azimuthal quantum number

aq. niv. aqua nivalis (Latin—snow water)

AQP Association for Quality and Participation

aq. pluv. aqua pluvialis (Latin—rain water)

aq. pur. aqua pura (Latin—pure water)

Aqr Aquarius

AQREC Army Quartermaster Research and Engineering Command

aq. regia aqua regia (Latin—royal water)—hydrochloric and nitric acid

aqs additional qualifying symptoms

AQT Applicant Qualification Test; Aviation Qualification Test

aq. tep. aquia tepida (Latin—tepid water)

aqu aqueous

Aqu Aquila (Latin—Eagle constellation)

Aqua Aquarius (Latin—Water Carrier constellation)

aqua. aquaria; aquarium; aquatic

aquacade aquatic parade (or exhibition of diving, swimming, water sports)

aquacult aquaculture

aquar aquarium

Aquar Aquarius

aque aqueduct

Aqued Aqueduct

Aqueduct Aqueduct Books

aqui aquifer

ar absentee ratio; achievement ratio; acid resisting; active resistance; aerial reconnaissance; alarm reaction; all rail; all risks; allocated reserve; amphibious reconnaissance; analytical reagent; antireflection; aromatic; arrival; artificial respiration; artillery radar; as required; aspect ratio; assault rifle; auditory reception; automatic rifle

ar (Ar) address register; armed robbery; average revenue

ar arabisch (German—Arabian); *avis de reception* (French—return receipt)

a/r accounts receivable; all risks; armed robbery; at the rate of

a & r approved and removed; artists and repertory; assault and robbery; assembly and repair

a/R am Rhein (German—on the Rhine River)

Ar Arab; Arabia; Arabian; Arabic; Aragon; argon; Aries; aryl

Ar Arabic; *Arroyo* (Spanish—brook, creek, or rivulet)

AR Aberdeen & Rockfish (railroad); Adelaide Rifles; Administrative Ruling; Aerodynamic Report; Aerolineas Argentinas (Argentine Airlines); Aeronautical Radio-navigation; Air Reserve; Airman Recruit; Airship Rigger; Algonquin Regiment; Amendment Request; Annual Report; Argentina (Internet code); Arkansas (ZIP code); Army Regulation(s); Army Reserve; repair ship (naval symbol); Assam Regiment; Atlantic Records

A/R Antwerp-Rotterdam (range of ports)

A & R Angus and Robertson; assembly and repair

AR Agencja Robotnicza (Polish—Workers' Press Agency); *Aller et Retour* (French—roundtrip); *American Rationalist* (freethought magazine); *Andata-Ritomo* (Italian—roundtrip)

A.R. Anno Regni (Latin—In the Year of the Reign of)

A/R Aksjerederi (Norwegian—shipping company)

ara aerial rocket artillery; airborne rocket artillery; assigned responsible agency (DoD)

Ara Ara (a three-letter constellation without an abbreviation); Arabic; arabinose (symbol); Argentina

ARA Academy of Rehabilitative Audiology; Aerospace Research Association; American Radio Association; American Railway Association; American Rationalist Association; American Recovery Association; American Relief Association; American Rental Association; American Republics Area; American Revenue Association; American Rheumatism Association; Applied Research of Australia; Arcade & Attica (railroad); Area Redevelopment Administration; Army Rifle Association; Artists' Representatives Association; Association of Radiologists of Australasia; Association of River Authorities; Auckland Regional Authority; Australian Regular Army; Australian Retailers Association; Automatic Retailers of America

A.R.A. Associate of the Royal Academy

A/R/A Antwerp-Rotterdam-Amsterdam (range of ports)

ARA Acçao Revolucionaria Armada (Portuguese—Armed Revolutionary Action); *Armada República Argentina* (Argentine Navy)

ARA (AFL-CIO) American Radio Association

ara (ARA) aerial rocket artillery

Arab. Arabia; Arabian; Arabic

ARAB Australian Radio Advertising Bureau

Arab Emirates United Arab Emirates including the seven Trucial Sheikdoms along the southern shore of the Persian Gulf with Abu Dhabi its capital

Arabian Pen Arabian Peninsula, including southern Arabia

Arab League League of Arab States (Algeria, Bahrain, Egypt, Iraq, Jordan, Kuwait, Lebanon, Libya, Morocco, Oman, Qatar, Somalia, Sudan, Syria, Tunisia, United Arab Emirates, Yemen)

ARABS Australian Racing and Breeding Stables

Arabsat Arab Communications Satellite

ARAC Accredited Review Appraisers Council; Aerospace Research Applications Center; Army Radar Approach Control; Associate of the Royal Agricultural College; Association of Rain Apparel Contractors; Australian Refugee Advisory Council

arach arachnology

Arach Arachnida

ARACI Associate of the Royal Australian Chemical Institute

arad airborne radar and doppler

ARAD Associate of the Royal Academy of Dancing

ARADCOM Army Air Defense Command

ARAeS Associate of the Royal Aeronautical Society

ARAF Associated Regional Accounting Firms

Arafat Yasir Arafat (Palestinian leader)

Arafura Arafura Sea between Australia and New Guinea

Arag Aragón

ARAgS Associate of the Royal Agricultural Society

ARAIA Associate of the Royal Australian Institute of Architects

aral automatic record analysis language

Aral Aral Sea

Aram Aramaic

ARAM Association of Railroad Advertising and Marketing

A.R.A.M. Associate of the Royal Academy of Music

Aramco Arabian-American Oil Company

Arans Aran Islands off County Clare, Ireland

Aransas Aransas National Wildlife Refuge near Rockport, Texas

ARANY Abortion Rights Association of New York

ARAPCS Association for Research, Administration, Professional Councils, and Societies

Ararat Mount Ararat (Turkey's highest mountain and place where Noah's Ark is believed to have landed; Armenian name *of* the mountain the Turks call Agri Dagi on the border of Soviet Armenia)

aras ascending reticular activating system

ARAS Accept (each person you encounter), Respect (each person whatever their position), Affect (each person with the warmth of your heart), Support (each person in the place they are now); Ascending Reticular Activating System; Associate of the Royal Astronomical Society

Arava Israeli IAI-201 light transport plane

arb arbitrageur; arbitrary; arbitration; arbitrator; auxiliary repair battle damage; labor arbitrator

arb arbeid(er) [Dano-Norwegian—work(s)]; *arbejde* (Dano-Norwegian—job, labor, work)

Arb Arbroath

ARB Accident Records Bureau (NYC Police Dept.); Air Registration Board; Air Research Bureau; Air Reserve Base; Air Resources Board; Armored Rifle Battalion; Army Rearming Base; Army Retiring Board; Asian Reserve Bank; ASTIA Report Bibliography; battle damage repair ship (naval symbol)

ARB Accounting Research Bulletin

ARBA American Railway Bridge and Building Association; American Road Builders Association; Association Retail Bakers of America

ARBA American Reference Books Annual

arb & aw arbitration and award

ARBBA American Railway Bridge and Building Association

arbit arbitrageur

ARBM Association of Radio Battery Manufacturers

arbo arthropod-borne (viral diseases)

arbor. arboriculture

arbor. virus arthropod-borne virus

ARBP Associated Reinforcing Bar Producers

ARBs Air Resources Boards (pollution-control agencies)

ARBS Angular-Rate Bombing System; Associate of the Royal Society of British Sculptors

Arb Trib Arbitration Tribunal

arbtrn arbitration

arbtror arbitrator

Arbuckle Arbuckle National Recreation Area near Sulphur, Oklahoma

arb. vit. arbor vitae (Latin—tree of life) any of various evergreens with overlapping or compressed scale leaves often grown for ornament in hedges

arc acoustic research center; advanced reading copy; aircrew reception center; anti-riot cartridge; Association for Retarded Citizens; auxiliary repair cable

arc. arcade; auto-refrigerated cascade

arc arco (Italian—bow, indicating end of *pizzicato* passages)

Arc Arachon; Arcade; Archaic; Arctic

ARC Aboriginal Research Club; Addicts Rehabilitation Center; Agricultural Relations Council; Agricultural Relations Council; Agricultural Research Council; AIDS-Related Complex(es); Air Rescue Center; Air Reserve Center; Aircraft Radio Corporation; Airlines Recording Corporation; Airworthiness Requirements Committee; Alcoholic Rehabilitation Center; American Red Cross;

Ames Research Center (NASA); Amistad Research Center (New Orleans); Annual Review Committee (NATO); Appalachian Regional Commission; Arthritis and Rheumatism Council; Asian Research Center (Harvard); Association of Rehabilitation Centers; Association of Retail Confectioners; Association for Retarded Citizens; Association of Railway Communicators; Atlantic Research Corporation; Atomedic Research Center; automatic radio control; cable laying or repair ship (naval symbol)

ARC Armada República de Colombia (Spanish—Colombian Navy); *Association des Restauratrice-Cuisinières* (French—Association of Women Restaurant-Cook Owners)

ARCA American Rainwater Catchment Association; Associate of the Royal College of Art

ARCAA Associate of the Royal Canadian Academy of Arts

ARCAD Association of Renault and Chrysler for Automotive Development

arcade automatic radar-control-and-data equipment

ARCADE Ampere Remote Control Access Data Entry

ARCAM Army Reserve Components Achievement Medal

Arc Arch Arctic Archipelago (Canadian Arctic)

ARCAS Automatic Radar Chain Acquisition System

ARCB Air Resources Control Board; Army Regular Commissions Board; Association of Reserve City Bankers

arccos arccosine

arccot arccotangent

arccsc arccosecant

Arc Cur Arctic Current

arce amphibious river-crossing equipment

ARCE American Record Collectors Exchange

ARCen Air Reserve Center

ARCFCP Alliance for Responsible CFC Policy (chlorofluorocarbons)

arch. archaic; archipelago; architect(s); architectural; architecture

Arch archive; archivum; archiwum; archway

Arch archipiélago (Spanish—archipelago)

ARCH Articulated Computing Hierarchy

archaeol archaeologist; archaeology

Archam Archambault; prison in Archambault, Quebéc

Arch-Bish Archbishop

archcrit arch critic; architectural critic(ism)

Archd Archdeacon; Archduke

Arch de Cln Archipelago de Colón

Arch Dig Architectural Digest

Archdn Archdeacon

Arch E Architectural Engineer

archeo archeological; archeologist; archeology

archeo (Latin prefix—beginning)—archaic, archeologist, archeology

archeol archeology

Archeol Archeology

Archeoz Archeozoic

Arches Arches National Monument, Utah

archi archival; archive; archivist

ARCHI Asociación de Radiodifusoras de Chile (Spanish—Association of Chilean Broadcasters)

Archie Archibald

archip archipelago

archipel (Dutch or French—archipelago)

archit architecture

Archit Architecture

Archive Archive Press

Archives Soc Hist Archives of Social History

ARCHON Architecture for Cooperative Heterogeneous Online Systems

Arch Neurol Archives of Neurology

Arch Rec Bks Architectural Record Books

archv archive

Archy Archibald

ARCI Addiction Research Center Inventory; American Railway Car Institute

ARCIC Anglican-Roman Catholic International Commission

Arclos Army Close support

ARCM Associate of the Royal College of Music

ARCNet Attached Resource Computer Network

ARCNS American Red Cross Nursing Services

arco arcata (Italian—bow) stroke of a bow (music)

ar. co. artillery company

Arco Arco Publishing Company

Arc O Arctic Ocean Command

ARCO Associate of the Royal College of Organists; Atlantic Richfield Company

ARCOM Army Reserve Command

ARCOMET Area Commander's Meeting (NATO)

ARCON Advanced Research Consultants

ARCOPS Arctic Operations

arcos arc cosine

ARCOS Anglo-Russian Cooperative Society; Automation of controlled drug Reports and Consolidated Orders System

ARCOV Army Combat Operations Vietnam

arcp air refueling control

ARCR Arthritis and Rheumatism Council for Research

ARCRL Agricultural Research Council Radiobiological Laboratory

ARCs Alcoholic Rehabilitation Centers

ARC(S) AIDS–Related Complex(es)

ARCS Achievement Rewards for College Scientists; Air Resupply and Communication

Service; Alternative Remedial Contracting Strategy; Australian Red Cross Society

A.R.C.S. Associate of the Royal College of Science; Associate of the Royal College of Surgeons

ARCSA Aviation Requirements for the Combat Structure of the Army

arcsec arcsecant

arcsin arcsine

ARCST Associate of the Royal College of Science and Technology

arct air refueling control time

Arct Arctic

arctan arctangent

Arctic Unicorn one-tusked narwhal of the Canadian arctic

ARCUK Architects' Registration Council of the United Kingdom

ARCUP Atlantic Region Canadian University Press (news cooperative)

ARCVS Associate of the Royal College of Veterinary Surgeons

arc/w arcweld

ard acute reference doses; acute respiratory disease; anti-recovery device (mines; auxiliary repair dock

ar & d aeronautical research and development; air research and development

Ard Ardrossain

ARD Accelerated Rehabilitated Disposition; Accelerated Rural Development; Air Reserve District; American Research and Development (corporation); Appalachian Regional Development; Armaments Research Department; Army Renegotiation Division; Association of Research Directors; auxiliary floating dock (naval)

ARD Arbeitsgemeinschaft Rundfunkanstalten Deutschland (German—German National Broadcasting)

AR & D air research and development

arda analog recording dynamic analyzers

ARDA Advanced Reactor Development Associates; American Railway Development Association; American Resort Development Association

ARDB Australian Resources Development Ban

ARDC Aberdeen Research and Development Center; Action Resource Development Center; Agricultural Refinance and Development Corporation; Air Research and Development Command; American Racing Drivers Club

ARDCM Air Research and Development Command Manual

ARDCO Applied Research and Development Company

arddie analysis, requirements determination, design, development, implementation, and evaluation

ARDE Armament Research and Development Establishment (Ministry of Supply)

Ardennes Ardennes Forest of Belgium, France, and Luxembourg

ARDG Army Research and Development Group (USA)

ARDG(E) Army Research and Development Group (Europe)

ARDG(FE) Army Research and Development Group (Far East)

ARDIS Army Research and Development Information System

ardm auxiliary repair dock, medium

Ardnamurchan Point Ardnamurchan—Scottish headland

ardo ardito (Italian—bold, daring, fearless)

ard's analog recording dynamic analyzers

ARDS Aviation Research and Development Service

ARDU Analytical Research and Development Unit

ARDXC Australian Radio DX Club

are (ARE) air reactor experiment

Are Arecibo (Puerto Rican municipality)

ARE Admiralty Research Establishment; Arab Republic of Egypt; Association for Research and Enlightenment

A.R.E. Associate in Religious Education

AREA Acrovias Ecuatorianas (Ecuadorian Airways); American Railway Engineering Association; American Recreational Equipment Association; Army Reactor Experimental Area; Australian Remedial Education Association

a'ready all ready

AREC Amateur Radio Emergency Corps

AREE Admiralty Regional Electrical Engineer

AREFS Air Refueling Squadron

AREI Associate of the Real Estate and Stock Institute (Australia)

ARENA Adoption Resource Exchange of North America

ARENA Alianza Republicana Nacionalista (Spanish—National Republican Alliance)—rightist party in El Salvador; *Alianca Renovadora Nacional* (Portuguese—National Renovating Alliance)—political party in Brazil

aren't are not

ARENTS ARPA Environmental Test Satellite

Areop Areopagite; *Areopagitica* (Milton's pamphlet advocating freedom of the press); Areopagus

arestem a recording stray energy monitor

ARETO Arab Republic of Egypt Telecommunications Organization

AREUEA American Real Estate and Urban Economics Association

arf acute renal failure; acute respiratory failure; all-round fire; automatic return fire; dog's bark

ARF Addiction Research Foundation; Advertising Research Foundation; African Research Foundation; Air Reserve Foundation; Air Reserve Force(s); American Radio Forum; American Rationalist Federation; American Rehabilitation Foundation; American Retail Foundation; American Rose Foundation; Armour Research Foundation; Arthritis and Rheumatism Foundation; Australian Road Federation

ARFA Allied Radio Frequency Agency

ARFC Air Reserve Flying Center

ARFCOS Armed Forces Courier Service

ARFDC Atomic Reactor and Fuel Development Corporation

ARFL Australian Rugby Footbal League; Australian Rules Football League

arfo after receipt firm order

arfor area forecast

ARFPC Air Reserve Forces Policy Committee

ARFU Australian Rugby Football Union

arfx automatic riveting fixture

arg anti-riot gun; argent; argot; argument; argumentation; argumentative; argumentator (a controversialist); argus; arresting; arresting gear

arg (Arg) arginine

arg argang (Dano-Norwegian—yearbook); *argol* (Mongolian—dried camel or cattle dung fuel); *arguendo* (Latin—in arguing)

Arg Argentina; Argentinian; Argyll

Arg Argelia (Spanish—Algieria)

ARG Aerolineas Argentinas; repair ship; Amphibious Ready Group; internal combustion engine

ARG American Record Guide

arga appliance, range, adjust (data processing)

arg. ad bac. *argumentum ad baculum* (Latin—argument of the club) appeal to force

arg. ad hom. *argumentum ad hominem* (Latin—personal attack)

arg. ad ignor. *argumentum ad ignorantum*—(Latin—argument from ignorance)

arg. ad miser. *argumentum ad misericordiam* (Latin—appeal to pity)

argads advanced radar-directed gun air defense system

arg. ad veri. *argumentum ad vericundiam* (Latin—argument based on misuse of authority)

ARGCA American Rice Growers Cooperative Association

Argel *Argelia* (Spanish—Algeria); *Argeli* (Portuguese—Algeria)

Argen Argentine; Argentinian

Argentina Argentine Republic, *República Argentina*

Argie(s) Argentino(s)

ARGMA Army Rocket and Guided Missile Agency

argmt arrangement

Argosy Argosy-Antiquarian Limited; Hawker–Siddeley four-engine turboprop transporting 54 paratroopers or 69 regular troops

args anti-radar guidance sensor

ARGS American Rock Garden Society

argus advanced research on groups under stress

Argus Canadair long-range reconnaissance plane designated CL-38

ARGUS Automatic Routine Generating and Updating System

Argyll Argyllshire

arh ammunition railhead

arh (ARH) advanced reconnaissance helicopter

a Rh am Rhein (German—on the Rhine)

ARH heavy-hull repair ship (3-letter symbol)

ARHA Associate of the Royal Hibernian Academy

arh/ir anti-radiation homing/infrared

arhm asphalt rubber hot-mix

ARHS Associate of the Royal Horticultural Society; Australian Railway Historical Society

ari airborne radio installation

Ari *Aries* (Latin—Ram constellation); Aristotle

ARI Acne Research Institute (Newport Beach, California); Air-Conditioning and Refrigeration Institute; Alcoholic Rehabilitation, Inc.; Aluminum Research Institute; American Reciprocal Insurers; American Refractories Institute; American Russian Institute; Army Research Institute; Atherosclerosis Research Institute; Automatic Radio Information

ARIA Administration, Ryukyus Island, Army; Accounting Research International Association; Adult Reading Improvement Association; American Risk and Insurance Association

ARIANA Ariana Afghan Airlines

ARIB Asphalt Roofing Industry Bureau

ARIBA Associate of the Royal Institute of British Architects

ARIC Associate of the Royal Institute of Chemistry

ARICRSU American Russian Institute for Cultural Relations with the Soviet Union

ARICS Associate of the Royal Institution of Chartered Surveyors

ARIEL Automated Real-Time Investments Exchange Limited

ARIEM Army Research Institute of Environmental Medicine

aries astronomical radio interferometric earth surveying

ARIES Advanced Radar Information Evaluation System

arima autoregressive integrated moving average

ARINA Associate of the Royal Institution of Naval Architects

ARINC Aeronautical Radio Incorporated

ARIO Association of Retired Intelligence Officers

arip automatic rocket impact predictor

aris (ARIS) advanced range instrumentation ships

ARIS Activity-Reporting Information System; Advanced Research Instrument System; Aerial Radio Instrument System; Aircraft Research Instrumentation System

ARISE Adult Recovery in a Supportive Environment; A Reusable Infrastructure for Software Engineering

Arist Aristotle

ARIST *Annual Review of Information Science and Technology*

Arista high school honor society

Aristide Jean-Bertrand Aristide (Haitian)

Aristo Aristocles; Aristol; Aristotle

aristocat(s) aristocratic cat(s)

Aristoph Aristophanes

aristo(s) aristocrat(s)

ARISTOTLE Annual Review and Information Symposium on the Technology of Training, Learning, and Education (DoD)

arit *aritmética* (Portuguese or Spanish—arithmetic)

A.R.I.T American Registered Inhalation Therapist

arith arithmetic(al)(ly); arithmetician

Arith Arithmetic

Ariz Arizona; Arizonian

Ariz Hist Found Arizona Historical Foundation

Arizona Girls Arizona Girls School at Phoenix (correctional facility)

ARJIS Automated Regional Justice Information System

Ark Arkansas; Arkansan

Ark City Arkansas City, Arkansas

Arkf Fys *Arkivfor Fysik* (Danish—Physics Archives)

ARKIA Israel Inland Airlines

arl acceptable reliability level; air run landing; average remaining lifetime

ARL Abortion Rights League; Admiralty Research Laboratory (UK); Aeromedical Research Laboratory; Aeronautical Research Laboratory; Aerospace Research Laboratory; American Reefer Line; American Republics Line; American Roque League; Americans for Religious Liberty; Anesthesia Research Laboratories; Applied Research Laboratory (Johns Hopkins University); Association of Research Libraries; Australian Rugby League; landing craft repair ship (3-letter naval symbol)

ARLA Arab Latin American Bank

Arlanda Stockholm, Sweden's airport

ARLD Army Logistics Research and Development

ARLHS Australian Railway and Locomotive Historical Society

Arlington Arlington Books (Louisville, KY); Arlington House (New Rochelle, NY); Arlington National Cemetery in Arlington, Virginia; Arlington Stadium, Texas

Arlington House Robert E. Lee's home overlooking the Potomac

Arlington Hse Arlington House

ARLIS Arctic Research Laboratory Island (USN); Art Libraries Society

ARLIS/ANZ Art Libraries Society/Australia and New Zealand

ARLIS/NA Art Libraries Society/North America

ArLO Army Liaison Officer

ARLO Air Reconnaissance Liaison Officer; Art Reference Libraries of Ohio

Arlon Luxembourg, Belgium's provincial capital

ARLSEA Active-Retired Lighthouse Service Employees Association

arm (ARM) adjusted-rate mortgage

arm. anti-radar missile (ARM); anti-radiation missile; area radiation monitor(s); armature; arming; armor(ed)

arm armenisch (German—Armenian)

Arm Armagh; Armand(o); Armenia(n)

Ar.M. Architecturae Magister (Master of Architecture)

ARM Abortion Rights Mobilization; Accredited Resident Manager; Animal Rights Movement; Aryan Resurgence Movement (neo-Nazi organization); Association of Rotational Molders; Auditory Rehabilitation Mobile; Australian Reform Movement

A.R.M. Allergy Relief Medicine

arma armature

ARMA American Bosch Arma Corporation; Anglican Renewal Ministries Australia; Association of Records Managers and Administrators; Australian Rubber Manufacturers Association

a & r man artist and repertory man (supervising phonograph record production)

Armand Armando

ARMCM Associate of the Royal Manchester College of Music

ARMCOM Armament Command (USA)

ARMCOMSAT Arab Communications Satellite System

armd armored

armd (ARMD) age-related macular degeneration

ARMD Army Rockets and Missile Division

Armen Armenia(n)

armet area forecast (given in metric system)

armgrd armed guard

ARMH Academy of Religion and Mental Health

ARMI American Rack Merchandisers Institute; American Research Merchandising Institute; Army Resources Management Institute; Associated Risk Managers International

Armin Armindo

ARMIS Agricultural Research Management Information System

ARMIT Associate of the Royal Melbourne Institute of Technology

arml airmail

armm analysis and research methods for management

ARMM Association of Reproduction Materials Manufacturers

ARNMA American Railway Master Mechanics' Association

ARMMS Automated Reliability and Maintainability Measurement System

armon armonica (Italian—accordion)

A-R/MONP Ayers Rock/Mount Olga National Park (Northern Territory, Australia)

ARMOP Army Mortar Program

armor armorikanisch (German—Armorican)—Celtic language of Brittany

Armor Armoric

armpl armorplate

armr armorer

ARMS Advanced Receiver Model System; Aerial Radiological Measuring Survey; Amateur Radio Mobile Society; Automated Records Management System

arm-saf arm-safe (switch)

armt armament

ARMU Associated Rocky Mountain Universities

army disease drug addiction

Army NG Army National Guard

Army ROTC Army Reserve Officers' Training Cops

arn (ARN) *ácido ribonucleico* (Spanish—ribonucleic acid)—rna (RNA)

a Rn am Rhein (German—on the Rhine)

Arn Arnold

ARN Stockholm, Sweden (Arlanda Airport)

ArNa Army with Navy

ARNA Arab Revolution News Agency

ARNAT Asia Research News Analysis Team

arng arrange

ARNG Army National Guard

A Rn I Association of Rhodesian Industries

Arnie Arnold

ARNM Aztec Ruins National Monument

ARNMD Association for Research in Nervous and Mental Disease

Arno Arnold; Arnon; Arnot

ARNO Association of Retired Naval Officers

aro after receipt of order; airborne range only

ARO Air Radio Office(r); Applied Research Objective; Army Research Office; Army Routine Order; Asian Regional Organization; Association for Research in Ophthalmology; Association of Roentgenological Organizations

AROCC Association for Research of Childhood Cancer

arod airborne range and orbit determination

ARO-FE Army Research Office—Far East

arom aromatic; artificial rupture of membranes

aromath aromatherapist; aromatherapy

AR-ONP Ayers Rock-Olgas National Park (Australia)

arp airborne radar platform; airport reference point; alternator research package; anti-radiation projectile; (cartoonist's symbol—dog's bark)

ARP Acreage Reduction Program; Advanced Research Project(s); Aeronautical Rec-

ommended Practice(s); Air Raid Precautions; American Registry of Pathologists; Ammunition Refilling Point; Area Redevelopment Program; Association for Realistic Philosophy; Australian Reptile Park (New South Wales); Australian Republican Party

ARP Anti-Revolutionaire Partij (Dutch—Anti-Revolutionary Party)

ARPA Advanced Research Projects Agency; Association of Representatives of Professional Athletes

ARPAC Army Personnel and Administration Center

ARPANET Advanced Research Projects Agency (computer) Network

ARPAS Air Reserve Pay and Allowance System

ARPAT Advanced Research Projects Agency Terminal (defense system)

ARPC Air Reserve Personnel Center

arpd (ARPD) applied research planning document

arpege air-pollution episode game

ARPEL Asistencia Reciproca Petrolera Estatal Latinoamericana (Spanish—Latin American State Petroleum Reciprocal Assistance)—international agency

ARPI Automotive Refrigeration Products Institute

arpl a retrieval-process language

Arpo arpeggio (Italian—producing the tones in a chord successively rather than simultaneously)

ARPO Association of Resort Publicity Offmcers

ARPOA Association of Railway Professional Officers of Australia

Arprt Airport (postal place-name abbreviation)

arps agricultural report production system

ARPS Adjustable Rate Preferred Stock; Advanced Radar Processing System; Arab Physical Society; Associate of the Royal Photographic Society; Association of Railway Preservation Societies; Australian Radiation Protection Society; Australian Royal Photographic Society

ARPSA Army Postal Service Agency

Arpt Airport

ARPT American Registry of Physical Therapists

ARP Tests Aptitude Research Project Tests

a-r pulse apical-radial pulse

arpv advanced remotely piloted vehicle

arq automatic repeat request

arq arquitecto (Spanish—architect); *arquitectura* (Spanish—architecture); *arquiteto* (Portuguese—architect); *arquitetura* (Portuguese—architecture)

ARQ automatic error correction; automatic request for repetition

Arqto Arquitecto (Portuguese or Spanish—architect)

arquo arquipélago (Portuguese—archipelago)

arr airborne radio receiver; arrange(ment); arrestor; arrival; arrive; arriving

arr (ARR) arrest(ed)

arr. arrived

arr arrecife (Spanish—reef)

Arr arrondissement (French—district)

ARR Air Regional Representative; Air Reserve Record(s); American Right to Read; Armored Reconnaissance Regiment; Army Retail Requirements; Australian Rifle Regiment

ARR Anno Regni Regis (or *Reginae*) (Latin—in the year of the king's or queen's reign)

ARRA Amateur Radio Retailers Association; Australian Rough Riders Association

ARRB Australian Road Research Board

ARRC Air Reserve Records Center; Associate of the Royal Red Cross

ARRCS Air-Raid Reporting Control Ship

ARRDE Army Radar Research and Development Establishment

ARRDO Australian Railway Research and Development Organisation

arre arrecife (Spanish—reef, roadbed, rocky road)

ARRE Assault Regiment Royal Engineers

ARRES Automatic Radar-Reconnaissance Exploitation System

arrex arriving ex ...

ARRF Australian Reading Research Federation; Automatic Recording and Reduction Facility

ARRGp Aerospace Rescue and Recovery Group (USAF)

Arri Arnold and Richter (reflex motion-picture camera)

ARRL American Radio Relay League

arr n arrival notice

arro arroyo (Spanish—creek, brook, stream)

arro seco arroyo seco (Spanish—dry creek, brook bed, streambed, or riverbed)

arrowhead symbol used to indicate direction

Arrowhead Arrowhead Stadium, Kansas City, Missouri

arrs airborne radar ranging system

ARRS Aerospace Rescue and Recovery Service; Aircraft Refueling and Rearming System; American-Russian Research Society

ARRSq Aerospace Rescue and Recovery Squadron (USAF)

ARRT American Registry of Radiologic Technologists

ARRTC Aerospace Rescue and Recovery Training Center (USAF)

arrv armored repair and recovery vehicle

ARRWg Aerospace Rescue and Recovery Wing (USAF)

arry arrythmia

ars active repeater satellite; aerospace research satellite; air reconnaissance support; arsenal; amateur radio service; arysufatase-A; asbestos roof shingles

Ars Arsenal

ARs Action Requests

ARS Advanced Record System; Aerospace Research Satellite; Agricultural Research Service; Airborne Reconnaissance Squadron; Airail Service (monorail); Air Rescue Service; American Recorder Society; American Records Society; American Recreation Society; American Repair Society; American Rescue Service; American Rhododendron Society; American Rocket Society; American Rose Society; Arizona Revised Statutes; Army Radio School; Army Relief Society; salvage ship (naval symbol)

ARS Annual Report to Shareholders

ARSA Aeronautical Repair Station Association; Associate of the Royal School of Art

A.R.S.A. Associate of the Royal Scottish Academy

arsab arsonist sabotage; arsonist saboteur

arsabs arsonist saboteurs

ARSC Association of Recorded Sound Collections

ARSD Advanced Reentry System Deployment; salvage lifting ship (naval symbol)

ARSDEP Asian Regional Skills Development Programme

ArSEA Arkansas Society of Enrolled Agents

arsen arsenal

arsg arising

ARSH Associate of the Royal Society for the Promotion of Health

ARSI Associate of the Royal Sanitary Institute

arsin arc sine

ARsl Arsenal (postal place-name abbreviation)

ARSL Associate of the Royal Society of Literature

ARSM Associate of the Royal School of Mines

ARSP Aerospace Research Support Program

arspa aerial reconnaissance and surveillance penetration analysis

ARSPH Associate of the Royal Society for the Promotion of Health

arsr air route surveillance radar

ARST Aerial Reconnaissance and Security Troop; salvage craft tender (naval symbol)

ARSTAF Army Staff

ARSTRAC Army Strike Command

ARSU Alcohol Rehabilitation Services Unit (Navy Regional Medical Center in Long Beach, California)

ARSV armored reconnaissance scout vehicle (USA)

art advanced research and technology; airborne radiation thermometer; annual renewal term; art assembly; arterial; artery; article; articulate; articulation; artifact; artificial; artillery; artisan; artist; artistic; artistry; automatic reporting telephone

'art heart

art. artillery

art artikel (Dano-Norwegian—article)

Art Arthur; Arturo

Art Artikel (German—article)

ART Accredited Record Technician; Air Reserve Technician; American Repertory Theater; Arithmetic Reading Test; Arithmetic Reasoning Test; Artificer; Art Resources in Teaching; Assisted Reproductive Technologies; Aviation Radio Technician

ARTA American River Touring Association; Association of Retail Travel Agents

artac advanced reconnaissance and target acquisition capabilities

ARTADS Army Tactical Data Systems

artb (ARTB) automatic return to base

ARTC Addiction Research and Treatment Center; Addiction Research and Treatment Corporation; Air Route Traffic Control

ARTCC Air Route Traffic Control Center

Art C-Part articles of co-partnership

artcrit art critic(ism)

art deco *arts décoratifs* (French —decorative arts)—decorative style of the 1920s and 1930s emphasizing bold outlines, geometric forms, streamlining, and strong colors

art. dlr artistic director

Art Div Artillery Division

ARTDO Asian Regional Trade Development Organisation

Art E Artificer Engineer

ARTE Admiralty Reactor Test Establishment

ARTEMIS Automatic Retrieval of Text through European Multipurpose Information Systems

ARTEP Army Training and Evaluation Program; Asian Regional Team for Employment Promotion

artesian artesian well (of the type originating in Artois, France)

ARTF Advanced Radiation Technology Facility; Asian Rice Trade Fund; Australian Road Transport Federation

Arth Arthropoda; Arthur; Arthurian

arthro *arthron* (Greek— joint)—arthritis, arthropod

artic articulate(d); articulation

Articles Articles of Agreement between a ship's crew and its master

Artie Artemas; Artemisia; Artemus; Arthur; Artur; Arturo; Artus

artif artificer(s); artificial(ly)

Artigas José Gervasio Artigas—defender of Uruguayan independence

ARTINS Army Terrain Information System

art. insem artificial insemination

Art Kill Arthur Kill waterway between New Jersey and Staten Island, New York

art nou art nouveau

art nr *artikelnummer* (Dano-Norwegian—article number)

arto air run takeoff

art⁰ *articulo* (Italian—article); *artículo* (Spanish—article); *artigo* (Portuguese—article)

ARTO Advanced Radiation Technology Office; Area Railway Transport Office

ARTOC Army Tactical Operational Control; Army Tactical Operations Central

ARTP Army Rocket Transportation System

art. pf artist's proof

artrac advanced range testing, reporting, and control

artron(s) artificial neuron(s)

arts. articles

ARTS Advanced Radar Traffic Control System; Automatic Radar Traffic Control System

ArtSci Arts and Sciences

artsem artificial insemination

arts graph *arts graphiques* (French—graphic arts)

ARTSM Association of Road Sign Makers

ARTS & P Australian Radio Technical Sevices and Patents

artt automatic rubber tensile tester

ARTT Annual Review Travelling Team (NATO)

artu automatic range tracking unit

Arturo Arturo Toscanini

arts automated radar terminal system

arty artillery

aru analog remote unit; audio response unit

Aru Aruba

ARU Air Reserve Unit; American Railway Union; Army Rugby Union; Australian Railway Union

arv (ARV) advanced reentry vehicle; aeroballistic reentry vehicle

Arv AIDS-related virus

Arv *Arvoisa* (Finnish—esteemed)

arv armed reconnaissance vehicle

ARV aircraft engine overhaul and structural repair ship; American Revised Version; armored recovery vehicle

ARVA aircraft repair ship for aircraft (4-letter designation)

ARVE aircraft repair ship for engines (4-letter designation)

ARVH aircraft repair ship for helicopter (naval symbol)

ARVIA Associate of the Royal Victorian Institute of Architects

ARVN Army of the Republic of Vietnam

ARVO Association for Research in Vision and Ophthalmology

ARVSG Air Reserve Volunteer Support Group

arw attitude reaction wheel

ARW Air Raid Warden; Air Raid Warning; Air Reserve Wing (Canada)

ARWC Army War College

ARWEN Anti-Riot Weapon (shoots rubber bullets)

ARWH Air Reserve Wing Headquarters

ARWS Antiradiation Weapon System; Associate of the Royal Society of Painters in Water Colours

Aryabhata Indian spacecraft named for the fifth-centuty astronomer

Arz *Arzobispo* (Spanish— Archbishop)

ARZ Active Reconnaissance Zone

Arzbpo *Arzobispo* (Spanish— Archbishop)

as air support; anti-ship (missile); application system; assault squadron

as airscoop; air-to-surface missile; alloy steel; alto sax; antiseptic; aortic stenosis; asymmetric

a-s ascendance-submission

a/s account sales; after sight; airspeed; alongside; antisubmarine

a & s accident and sickness (insurance)

as. aanstaande (Dutch—next)

a.s. auris sinistra (Latin—left ear)

a/ls. aux soins de (French—in care of)

As altostratus; arsenic; Asia; Asian; Asiatic; astigmatism; aunicles; Australia(n)

A$_s$ atmosphere standard

AS Abilene & Southern (railroad); Academy of Science(s); Admiral Superintendent; Aeronautical Standard(s); Air Service; Air Speed; Air Staff, Air Station; Air Surveillance; Airports Service; air-to-surface missile; Alaska Airlines; Alaska Statutes; Ambulatory SurgiCenter; American Samoa; American Samoa (Internet code); Anglo-Saxon; antisubmarine; Apprentice Seaman; Army Security; Army Staff; Asperger Syndrome; Associated Steamships; Australian Standard(s); Auto Squad (police); submarine tender (naval symbol)

A.S. Antonius Stradivarius (initials usually accompanied by a Maltese cross, both enclosed in a double circle); Associate in Science

A-S Allied-Signal Corporation

A/S alongside (barge, cargo carrier, lighter)

A & S Alton & Southern (railroad); Arts and Sciences

A of S Academy of Science

AS Anonim Sirket (Turkish—joint stock company); *Arabia Saudita* (Spanish—Saudi Arabia); *Aviaeskadra* (Russian—air squadron)

A/S Aksjeselskap (Norwegian—limited company); *Aktieselkab* (Danish—joint stock company)

AS-11 Nord air-launched antitank missile

AS-12 Nord automatic-telecommand antitank missile

AS-20 Nord air-to-surface radio-controlled missile

AS-30 improved version of AS-20 with longer range and heavier warhead

AS-33 Nord air-launched inertial-guidance missile

asa acetylsalicylic acid (aspirin); antistatic additive; antisubmarine aircraft; appropriate superior authority; azimuth servo assembly

aSa anti-Soviet agitation

A-S a Adams-Stokes attack

ASA Acoustical Society of America; Actuarial Society of America; Adaptive Simulated Annealing; African Studies Association; Alaska Airlines; Aluminum Siding Association; Amateur Softball Association; Amateur Swimming Association; American Scientific Affiliation; American Shorthorn Association; American Sightseeing Association; American Society for Aesthetics; American Society on Aging; American Society of Agronomy; American Society of Anesthesiologists; American Society of Appraisers; American Society of Artists; American Society of Auctioneers; American Sociological Association; American Sociometric Association; American Softball Association; American South African Line; American Soybean Association; American Sportscasters Association; Ameri-

can Standards Association; American Statistical Association; American Stockyards Association; American Studies Association; American Subcontractors Association; American Sugar Alliance; American Sunbathing Association; American Supply Association; American Surgical Association; Anthroposophical Society of America; Army Sailing Association; Army Seal of Approval; Army Security Agency; Army Ski Association; Assistant Secretary of the Army; Associated Stenotypists of America; Association for the Study of Abortion; Association of Secretaries in Asia; Association of Southeast Asia; Atomic Security Agency; Australian Shareholders Association; Australian Sheep-breeders Association; Australian Society of Anaesthetists; Australian Society of Authors; Automotive Service Association; Aviation Supply Annex

ASA Aerovias Suid Americana (Spanish—South American Airways)

A of SA (ASA) Association of Southeast Asia

ASAA Amateur Softball Association of America; American Society for the Abandonment of Acronyms; Armenian Students Association of America; Asian Studies Association of Australia

ASAALH Association for the Study of Afro-American Life and History

ASAB Association for the Study of Animal Behavior

ASAC American Society for African Culture; American Society of Agricultural Consultants; Anti-Submarine Air Controller; Army Study Advisory Committee; Assistant Special Agent in Charge

asacs airborne surveillance and control system; anti-submarine air close support

ASAE American Society of Agricultural Engineers; American Society of Association Executives; American Society of Automotive Engineers

ASAES Army Small-Arms Experimental Station

ASAF Australian Small-Arm Factory

ASAFS Army Security Agency Field Station

AS of AF Assistant Secretary of the Air Force

ASAH American Society of Association Historians

Asahi Asahi Shimbun (Japanese—Rising Sun Newspaper)

ASAIHL Association of Southeast Asian Institutions of Higher Learning

ASAIO American Society for Artificial Internal Organs

ASALA Armenian Secret Army for the Liberation of Armenia; Associate of the South African Library Association

asalm (ASALM) advanced strategic air-launched multimission missile

ASAM American Society for Abrasive Methods; American Society of Asset Managers

ASAN Adriatica Società per Azioni di Navigazióne (Italian—Adriatic Society for Navigation)

ASAnes American Society of Anesthesiologists

ASAO Association for Social Anthropology in Oceania

asap analog system assembly pack; as soon as possible

asap (ASAP) antisubmarine attack plotter

ASAP Aircraft Synthesis Analysis Program; Airlines of South Australia Proprietary; American Society for Association Publishing; American Society for Automation in Pharmacy; Antenna-Scatterer Analysis Program; antisub-

marine attack plotter; Australian Society of Animal Production

ASAPs Alcohol Safety Action Projects

ASAPS American Society for Aesthetic Plastic Surgery; Anti-Slavery and Aborigines Protection Society

asar advanced synthetic aperature radar

asar (ASAR) advanced surface-to-air ramjet

ASARCO American Smelting and Refining Company

arars advanced synthetic array radar system

asas all-source analysis system

ASAS Advertising Standards Authority of Singapore; American Society of Abdominal Surgery; American Society of Animal Science; Army Security Agency School; Australian Staffing Assistance Scheme

ASASAU Anti-Submarine Air Search and Attack Unit

asat air search and attack team

a-sat (A-Sat) anti-satellite

ASAT anti-satellite

ASATs antisatellite weapons

asatt advanced small-axial-turbine technology

ASATTU Anti-Submarine Attack Teacher Training Unit

ASAU Air Search-and-Attack Unit

ASAWS Advanced Surface-to-Air Weapon System

asb aircraft safety beacon; antisubmarine barrier; asbestos

asb (ASB) anxiety scale for the blind

as & b aloin, strychnine, and belladona (pills)

ASB Administration and Storage Building; Air Safety Beacon; Air Safety Board; Air Staff Board; Aircraft Safety Beacon, American Society of Bacteriologists; Associated Student Body; Australian Savings Bonds; Australian Shipping Board

A.S.B. Associate in Science in Business; Associate in Specialized Business

ASBA Association of Ship Brokers and Agents; Australian Sheep Breeders Association

ASBAH Association for Spina Bifida and Hydrocephalus

asb c asbestos covered

ASBC American Society of Biological Chemists

ASBC American Standard Building Code

ASBCA Armed Services Board of Contract Appeals

ASBCO American Ship Building Company

ASBD Advanced Sea-Based Deterrent Program; American Society of Bank Directors

ASBDA American School Band Directors Association

ASBE American Society of Bakery Engineers

A.S.B.E. Associate in Science in Basic Engineering

asb & i aloin, strychnine, belladona, and ipecac

asbl assemble

ASBO Association of School Business Officials

ASBPA American Shore and Beach Preservation Association

ASBPE American Society of Business Press Editors

asc air support carrier; altered state of consciousness; arteriosclerosis; arteriosclerosistic; ascarid; ascaridian; ascend; ascender; ascending; ascension; ascent; ascertain; ascertainable; automatic sequence control; automatic switching center; auxiliary switch closed

as & c aerospace surveillance and control

Asc Ascidian

A.Sc. Associate in Science

ASC Adelaide Steamship Company; Adhesive and Sealant Council; Aeronautical Systems Center; Aiport Security Council; Air Service Command; Air Support Command; Air Support Control; Air Systems Command; Alabama State College; Alaska Steamship Company; Albany State College; All Souls College; American Samoa Commission; American Security Council; American Silk Council; American Society of Cinematographers; American Society of Criminology; American Society for Cybernetics; American Society of Cytologgy; Area/Site Characterizations; Arizona State College; Arkansas State College; Armed Services Committee; Army School of Cookery; Army Selection Center; Army Service Corps; Army Subsistence Center; Asian Socialist Conference; Assam Sebundy Corps; Associated Sandblasting Contractors; Associated Schools of Construction; Associated Specialty Contractors; Australian Schools Commission; Australian Shippers Council; Automotive Sales Council

A & SC Adhesive and Sealant Council

asca automatic science citation alerting; automatic subject citation alert

ASCA Advanced Satellite for Cosmology and Astrophysics; Alaska State Council on the Arts; American School Counselor Association; American Senior Citizens Association; American Speech Correction Association; Associate of the Society of Company and Commercial Accountants; Association for Science Cooperation in Asia; Association of State Correctional Administrators

ASCAA Automobile Seat Cover Association of America

ASCAC Antisubmarine Contact Analysis Center

ASC/AIA Association of Student Chapters/American Institute of Architects

ascap at-sea calibration procedure

ASCAP American Society of Composers, Authors, and Publishers

ASCAT Antisubmarine Contact Analysis Team

ASCATS Apollo Simulation Checkout and Training System

ASCB American Society of Cell Biology; Army Sports Control Board

ASCC Adams State College of Colorado; Air Standardization Coordinating Committee; American Society for Concrete Construction; American Society for the Control of Cancer; Army Strategic Communications Command; Association of Senior Citizens Clubs; Australian Society of Cosmetic Chemists

ASCD Association for Supervision and Curriculum Development; American Society of Computer Dealers

ASCE Allied Supreme Commander, Europe; American Society of Civil Engineers

ASCEA American Society of Civil Engineers and Architects

ASCEF American Security Council Education Foundation

ASCEP Australian Society of Clinical and Experimental Pharmacologists

ascert ascertain(ed)

ASCET American Society of Certified Engineering Technicians

ascg anti-submarine carrier group

ASCHAL American Society of Corporate Historians, Archivists, and Librarians

ASCHE American Society of Chemical Engineers

ASCI American Society for Clinical Investigation; Army Service Corps of India

ASCIE American Standard Code for Information Exchange

ASCII American Standard Code for Information Interchange (pronounced *asky*)

ascl airborne sonobuoy communication link

ASCL Australia Straits Container Line

ASCLA Association of Specialized and Cooperative Library Agencies

ASCLD American Society of Crime Laboratory Directors

ASCLS American Society of Contact Lens Specialists

ASCLU American Society of Chartered Life Underwriters

ascm (ASCM) antiship capable missile

ASCM Association of Steel Conduit Manufacturers

ASCMA American Sprocket Chain Manufacturers Association

ASCN American Society of Clinical Nutrition

asco automatic sustainer cutoff

Asco Automatic Switch Company

ASCO American Society of Contemporary Ophthalmology; Arab Satellite Communications Organization; Australian Services Canteen Organization

ASCom Army Service Command

ASCOPE ASEAN Council on Petroleum

ASCOC Assessment of Systems and Components for Optical Communications

ascorbic acid antiscorbutic (anti-scurvy)

ascore automatic shipboard checkout and readiness equipment

A Scot type-A Scottish influenza virus

ASCP American Society of Clinical Pathologists; American Society of Consulting Pharmacists; American Society of Consulting Planners; Army Service Corps of Pakistan

ASCPC American Society of Clinical Pharmacology and Chemotherapy

ASCPT American Society for Clinical Pharmacology and Therapeutics

aser. ascriptum (Latin—ascribed to)

ASCR Association of Specialists in Cleaning and Restoration

ASCRO Active Service Career for Reserve Officers

asc's (ASCs) altered states of consciousness

ASCs All Savers Certificates

ASCS Agricultural Stabilization and Conservation Service, American School of Classical Studies (Athens); American Society of Corporate Secretaries; Automatic Stabilization and Control System

ASCU Air Support Control Unit; Association of State Colleges and Universities

ascus atypical squamous cells of undetermined significance

ascvd arteriosclerotic cardiovascular disease; atherosclerotic cardiovascular disease

A Sc W Association of Scientific Workers

asd air surveillance drone; aldosterone secretion defect; armament supply depot; atrial septal defect; avionic system division

ASD Admiralty Salvage Department; Aeronautical Systems Division; Army Schools Department; Army Shipping Document; Artillery Spotting Division; Assistant Secretary of Defense; Association of Steel Distributors; Aviation Supply Depot

ASD (PA & E) Assistant Secretary of Defense (Program Analysis and Evaluation)

ASD Association Suisse de Documentation (Swiss Association of Documentation); *Avionic System Design*

ASDA American Safe Deposit Association; American Seafood Distributors Association; American Sleep Disorders Association; American Stamp Dealers Association; Asbestos and Danville (railroad); Association of Structural Draftsmen of America; Atomic and Space Development Authority; Australian Stamp Dealers Association

ASDAE Association of Seventh-Day Adventists Educators

ASD-ALA Adult Services Division American Library Association

AsDB Asian Development Bank

ASDC Aeronomy and Space Data Center (NOAA); Association of Sleep Disorders Centers; Automobile Safe Driving Center

asde aircraft surface detection equipment

a/s de aux soins de (French—in care of)

asder airfield surface-detection radar

ASDF Air Self-Defense Force (Japanese Air Force)

ASDG Aircraft Storage and Disposition Group

asdi automatic selective dissemination of information

ASDIC Anti-Submarine Detection Investigation Committee (British sonar); Armed Services Documents Intelligence Center

ASDIRS Army Study Documentation and Information Retrieval System

ASDM Apollo-Soyuz Docking Module

asdng ascending (flow chart)

asdr airport surface-detection radar

ASDR American Society of Dental Radiographers; American Society of Dermatological Retailers

ASDS American Society of Dental Surgeons

A/S D/S Akties Dampskibsselskab (Danish—steamship company, limited)

ASDT Australian Society of Dairy Technology

ASD(T) Assistant Secretary of Defense (Telecommunications)

ase airborne search equipment; automotive service excellence

ASE Admiralty Signals Establishment; Amalgamated Society of Engineers; American Society of Enologists; American Steel Equipment; American Stock Exchange; Army School of Education; Association of Science Education; Australian Society of Engineers; Australian Stock Exchange(s)

AS & E American Science and Engineering

ASEA American Society of Engineers and Architects; Association of South-East Asia

ASEAN Association of Southeast Asian Nations (Brunei, Indonesia, Malaysia, the Philippines, Singapore, Thailand)

ASEANTA ASEAN Travel Association

ASEB Aeronautics and Space Engineering Board; Assam State Electricity Board

ASEC All Saints' Episcopal College; American Standard Elevator Code

ASECA Association for Education and Cultural Advancement (South Africa)

ASECS American Society for Eighteenth-Century Studies

ASED Aviation and Surface Effects Department

ASEE American Society for Ecological Education; American Society of Electrical Engineers; American Society for Engineering Education; American Society for Environmental Education; Anti-Submarine Experimental Establishment

ASEET Associate in Science in Electronic Engineering Technology

ASEI American Sports Education Institute

ASEM Anti-Ship Euro-Missile

ASEP American Society for Experimental Pathology; American Society of Electroplated Plastics; ASEAN Environment Program(s)

aseptics aseptically packaged liquids

ASERL Admiralty Services Electronic Research Laboratory

ASESA Armed Services Electro-Standards Agency

ASESB Armed Services Explosive Safety Board

ASESS Aerospace Environment Simulation System

aset aeronautical satellite earth terminal

ASET Assistant Secretary for Energy Technology; Author System for Education and Training

ASETC Armed Services Electron Tube Committee

asew airborne and surface early warning

asex asexualization

ASEWS Airborne and Surface Early Warning System

asf additional selection factor; amperes per square foot; amphibious strike forces

a-s-f aniline-formaldehyde-sulfur

AsF America's Future

ASF Advisory Support Force; African Swine Fever; Aircraft Services Facility; Alaskan Sea Frontier; American Scandinavian Foundation; American Schizophrenia Foundation; Americans for Safe Food; Ammunition Storage Facility; Army Service Forces; Army Stock Fund; Association of State Foresters; Australian Soccer Federation; Australian Speleological Federation; Automotive Safety Foundation

ASFA American Steel Foundrymen's Association; Association of Superannuation Funds of Australia

ASFC All Sports Federation of China; Atlantic Salt Fish Commission (Canada)

ASFCO American Soda Fountain Company

ASFD American Society of Furniture Designers

ASFDO Anti-Submarine Fixed Defences Officer

asfe accelerometer scale factor error

ASFE American Society For Aesthetics; Association of Specialized Film Exhibitors

ASFEC Arab States Fundamental Education Center

ASFFHF Association of Science Fiction, Fantasy, and Horror Films

ASFH Albert Schweitzer Friendship House

asfip accelerometer scale factor input panel

asfir active swept-frequency interferometer radar

ASFIS Aquatic Science and Fisheries Information Service (FAO)

ASFM American Sexual Freedom Movement

ASFMRA American Society of Farm Managers and Rural Appraisers

ASFP Association of Specialized Film Producers

ASFS Australia-Soviet Friendship Society

ASFSA American School Food Service Association

asfts airborne systems functional test stand

asfx assembly fixture

asg assignment

ASG Aeronautical Standards Group (Air Force and Navy); American Saint Gobain (glass); American Society of Genealogists; American Society of Genetics; Army Service Group

ASGA Advertising Specialty Guild of America

ASGB Aeronautical Society of Great Britain

ASGBI Association of Surgeons of Great Britain and Ireland

asgd assigned

ASGE American Society for Gastrointestinal Endoscopy

asgmt assignment

asgn assign; assignment

ASGp Aeronautical Standards Group (USAF)

ASGP Australian Society of General Practitioners

ASGRO Armed Services Graves Registration Officer

ASGS American Scientifiic Glassblowers Society

asgw Air-to-surface guided weapon

ASGW Association for Specialists in Group Work

ash advanced scout helicopter; anti-submarine helicopter

ash. airship; armature shunt; (ASH) aerial scout helicopter

Ash American short hair (cat); Ashbel; Ashburton; Ashbury; Ashdown; Asher; Asheto; Ashkenazi (East European Yiddish); Ashkenazim (Jews, generally of East European origin, who write in Hebrew but speak a Yiddish distinctly German, Hungarian, Polish, Russian, Ukrainian); Ashley; Ashman; Ashton; Ashur; Ashville; Ashvillian; Soviet infrared and radar-homing missile (NATO)

Ash Asahi Shimbun (leading Japanese newspaper)

AsH hyperopic astigmatism

ASH Action on Smoking and Health; American Society of Hematology; Army School of Hygiene; Australian Society of Herpetologists; Australian Stationary Hospital

A-S-H Allen-Sherman-Hoff

A & SH Argyll and Southerland Highlanders

ASHA Aerospace Human Factors Association; American School Health Association; American Social Health Association; American Social Hygiene Association; American Speech and Hearing Association

ASHACE American Society of Heating and Air-Conditioning Engineers

ASHBM Associate Scottish Hospital Bureau of Management

ASHC All-States Hobby Club

ash can Ashcan School of 20th century painters and photographers including The Eight

ashd arteriosclerotic heart disease

ASHE American Society of Hospital Engineers; Association for the Study of Higher Education

Ashfo'd Ashford Remand Prison Center, London area

ASHG American Society of Human Genetics

ASHH American Society for the Hard of Hearing

ASHI American Society of Home Inspectors; Association for the Study of Human Infertility

ASHM Association of Scottish Hospital Matrons

Ashken Ashkenazim (Hebrew—Jews of central and northern Europe)

Ashland Youth Federal Youth Center (for delinquents) at Ashland, Kentucky

Ash Mus Ashmolean Museum

ashp airship

ASHP American Society of Hospital Pharmacists

ASHPA American Society for Hospital Personnel Administration

ASHRA American Spa and Health Resort Association

ASHRAE American Society of Heating, Refrigerating, and Air-Conditioning Engineers

ASHS Advanced Study of Human Sexuality; American Society for Horticultural Science

ASHU Airline Stewards and Hostesses Union (New Zealand)

asi airspeed indicator; azimuthal speed indicator

ASI Adrenal Stress Index; Advanced Scientific Instruments; Aero-Space Institute; Aerospace Studies Institute; Africa Service Institute; Air Society International; Amended Shipping Instruction(s); American Society of Indexers; American Specifications Institute; American Statistics Index; American Statistics Institute; American Swedish Institute; Asian Statistical Institute (Japan); Audience Studies, Incorporated; Australian Shipbuilding Industries; Aviation Simulations International

ASIA Airlines Staff International Association; Army Signal Intelligence Agency; Australian Scientific Industry Association; Australian Stevedoring Industry Authority; Automotive Service Industry Association

ASIAC Aerospace Structures Information and Analysis Center; Asian International Acceptances and Capital

ASIAL Australian Security Industry Association Ltd.

Asian-Am(s) Asian-American(s)

ASIC Air Service Information Circular; Application Specific Integrated Circuit; Australian Standard Industry Classification

ASID American Society of Interior Designers; Australian Society of Implant Dentistry

ASIDIC Association of Scientific Irmformation Dissemination Centers

ASIE American Society of International Executives

ASIF Airlift Service Industrial Fund

ASI & H American Society of Ichthyologists and Herpetologists

ASII American Science Information Institute

ASIL American Society of International Law

ASILS Association of Student International Law Societies

ASIM American Society of Insurance Management; American Society of Internal Medicine

ASIMET Asociacion de Industrias Metalurgicas (Spanish—Association of Metallurgical Industries)—Chile

a sin a sinistra [Italian—at (to) the left]

ASI/NATO Advanced Study Institute/NATO

ASIO Australian Security Intelligence Organization

asip aircraft structural integrity program

ASIP Army Stationing and Installation Plan

ASIRC Aquatic Sciences Information Retrieval Center (U of RI)

ASIRT Association for Safe International Road Travel

asis anterior superior iliac spine

ASIs American Society of Indexers

ASIS Abort-Sensing Implementation System; American Society for Industrial Security; American Society for Information Science; ammunition stores issue ship (naval designator); Association of Small Island States; Australian Security Intelligence Service

asist advanced scientific instruments symbolic translator

ASIST Alzheimer Support, Information, and Service Team

ASIWPCA Association of State and Interstate Water Pollution Control Administrators

ASJ Asiatic Society of Japan

ASJA American Society of Journalists and Authors

ASJJA Association of State Juvenile Justice Administrators

ASJMC Association of Schools of journalism and Mass Communications

ASJR Army Summary Jurisdiction Regulations

ASJSA American Society of Journalism School Administrators

ask. amplitude shift keying

Ask American standard keyboard (typewriter)

ASK Associated Students of Kansas; Association for Social Knowledge

ASKA Automatic System for Kinematic Analysis

askg asking

Askham G Askham Grange (female offender's prison in Yorkshire, England)

ASKS Automatic Station-Keeping System

ASKT American Society of Knitting Technologists

asl abandon ship ladder; above sea level

asl altslavisch (German—Old Slavic)

Asl (ASL) American sign language

ASL Acting Sub-Lieutenant; American Association of State Libraries; American Scantic Line; American Shuffleboard Leagues; American Sign Language; Anti-Saloon League; Australian Society for Limnology

A-SL Abelard-Schuiman Limited

ASLA American Savings and Loan Association; American Society of Landscape Architects; Arizona State Library Association; Australian School Library Association

ASLAB Atomic Safety and Licensing Appeal Board

ASLB Atomic Safety and Licensing Board (AEC)

AS & LB American Savings and Loan Bank

ASLBP Atomic Safety and Licensing Board Panel (NRC)

ASLE American Society of Law Enforcement; American Society of Lubrication Engineers

ASLEC Association of Street Lighting Erection Contractors

ASLEF Associated Society of Locomotive Engineers and Firemen

ASLEP Apollo Surface Lunar Experiments Package

ASLH American Society for Legal History

ASLHA American Speech-Language-Hearing Association

ASLIB Association of Special Libraries and Information Bureaus

ASLIS Association of Special Libraries and Information Services

ASLNY Art Students League of New York

aslo assembly layout

ASLO American Society of Limnology and Oceanography; Australian Scientific Liaison Office (London)

ASLP Association of Special Libraries in the Philippines

ASLR American Short Line Railroads

ASLRA American Short Line Railroad Association

aslt advanced solid logic technology; assault(ing)

aslv assurance sur la vie (French—life insurance)

ASLW Amalgamated Society of Leather Workers

asm advanced surface-to-air missile; air-to-surface missile; anti-ship missile; anti-submarine missile; assembly; available-seat mile

AsM myopic astigmatism

ASM Acting Sergeant-Major; Air-to-Surface Missile; Alaska State Museum; American Society of Mamalogists; American Society for Metals; American Society for Microbiology; Antarctic Service Medal; Artificer Sergeant-Major; Association for Systems Management

asma anti-smooth muscle antibody

ASMA Aerospace Medical Association; American Society of Music Arrangers; American Student Media Association

ASMAR Astilleros Maritimos (Spanish—Maritime Shipyards)—Chile

asmbl assemble (flow chart)

asmblr assembler

ASMC American Society of Music Copyists; Army Supply and Maintenance Command (formerly Quartermaster Corps)

asmd (ASMD) anti-ship missile defense

asmd/ew antiship-missile defense/electronic warfare

ASME American Society of Magazine Editors; American Society of Mechanical Engineers; Association for the Study of Medical Education; Australian Society for Music Education

As Mem Associate Member

ASMFC Atlantic States Marine Fisheries Commission

ASMFS American Society of Maxillo-Facial Surgeons

ASMH Association for Social and Moral Hygiene

asmi airfield surface movement indication

ASMI Alaska Seafood Marketing Institute

ASMM American Supply and Machinery Manufacturers

ASMMA American Supply and Machinery Manufacturers Association

ASMP American Society of Magazine Photographers; Army Survival Measures Plan

ASMPA Armed Services Medical Procurement Agency

ASMPE American Society of Motion Picture Engineers

ASMR Australian Society for Medical Research

ASMRO Armed Services Medical Regulating Office

ASMS Advanced Surface Missile System

ASMSA Army Signal Material Support Agency

asmt assortment

ASMT American Society of Medical Technologists; Army School of Mechanical Transport

ASMWS Army Snow and Mountain Warfare School

asn average sample number

asn (ASN) asparagine (amino acid)

Asn Association

As of N Assistant Secretary of the Navy

ASN Allotment Serial Number; American Society of Naturalists; Army Serial Number; Army Service Number; Asiatic Steam Navigation; Assistant Secretary of the Navy

ASN (R & D) Assistant Secretary of the Navy (Research and Development)

ASNA Advertising Specialty National Association

asnap automatic-steerable null-antenna processor

ASNC Atlantic Steam Navigation Company

ASNDE Associate of the Society of Non-Destructive Examination

ASNE American Society of Naval Engineers; American Society of Newspaper Editors; Assistant Secretary for Nuclear Energy (DoE)

ASNLH Association for the Study of Negro Life and History

ASNP Army Student Nurse Program

As & Ns Andamans and Nicobars (Andaman and Nicobar Islands)

ASNSW Anthropological Society of New South Wales; Art Society of New South Wales; Astronomical Society of New South Wales

ASNT American Society for Nondestructive Testing

aso allele-specifiic oligonucleotide; arteriosclerosis obliterans; auxiliary switch open

ASO Adelaide Symphony Orchestra; Advanced Solar Observatory (in space); Aeronautica Supply Office(r); Air Signal Officer; Air Staff Officer; Air Staff Orientation; Air Surveillance Officer; Akron Symphony Orchestra; Albany Symphony Orchestra; Albuquerque Symphony Orchestra; American School of Orthodontists; American Sokol Organization; American Symphony Orchestra; antistreptolysin O; Area Safety Officer; Area Supply Office(r); Area Supplies Officer; Armament Supply Officer; Assistant Secretary's Office; Assistant Section Officer; Athens Symphony Orchestra; Atlanta Symphony Orchestra; Australian Security Organisation; Australian Society of Orthodontists; Aviation Supply Office(r); Aviation Safety Officer

ASOA American Society on Aging; Australian Shipping Officers Association

ASOC Air Support Operations Center

asoj anti-standoff jammer (missile)

ASOK Angfartygas Svenska Ostasiatiska Kompaniet (Swedish—Swedish East Asiatic Steamship Line)

ASOL American Symphony Orchestra League

ASOP Atomic Standing Operation Procedures

ASOPA Australian School of Pacific Administration

ASOR American Schools of Oriental Research

ASOS American Society of Oral Surgeons; Automated Surface Observing Systems

aso titer antistreptolysin titer

asp acoustic signal processor; active server page; advanced signal processor; affirmative self protection; air stores park; ammunition supply point; anti-social personality; anti-submarine patrol; application service provider; aspartic acid; aspen; automatic servo plotter; automatic switching panel; automatic system procedure

a s p *accepté sous protêt* (French—accepted under protest)

Asp American selling price

ASP Alexander Sergeyevich Pushkin; Allied Standing Procedures; Amalgamated Society of Printers; American Schulzhund Products; American Society of Parasitologists; American Society of Pharmacognosy; American Society of Photogrammetry; Ammunition Supply Point; Antisubmarine Patrol; Apollo Space Program; Archival Security Program; Arizona State Prison; Assistant Superintendent of Police; Association of Seattle Prostitutes; Association of Shareware Professionals; Astronomical Society of the Pacific; atmosphere-sounding projectile; Atomic Strike Plan; Audio Signal Processor; Australian Socialist Party; Australian Society for Parasitology; Australian Society of Prosthodontists; Automatic Schedule Procedure

A-S P Anglo-Saxon Protestant

A/S/P Aleksandr Sergeyvich Pushkin, apostle of freedom and father of Russian literature

A.S.P. accepté sans protêt (French—accepted without protest)

ASPA Alloy Steel Producers Association; American Salvage Pool Association; American Society of Pension Actuaries; American Society of Professional Appraisers; American Society for Public Administration; Australian Sugar Producers Association

ASPAC Asia and South Pacific Area Council; Asian and Pacific Council

A-span anticipation span (eye–voice span); capital-A-shaped span

ASPAP Australian South Pacific Aid Programme

aspb assault support patrol boat

ASPB Armed Services Petroleum Board

aspc accepté sous protêt pour acompte (French—accepted under protest for account)

ASPC American Sheep Producers Council

ASPCA American Society for the Prevention of Cruelty to Animals

ASPCC American Society for the Prevention of Cruelty to Children

ASPE American Society of Plumbing Engineers

ASPERS American Society of Professional Draftsmen

aspect acoustic short-pulse echo-classification techniques

ASPER Assembly System for Peripheral Processors; Assistant Secretary (of Labor) for Police Evaluation and Research

ASPERS Armed Services Procurement Regulations

ASPET American Society -for Pharmacology and Experimental Therapeutics

aspf annual service practice firings

ASPF Association of Specialized Film Producers; Association of Superannuation and Pension Funds

ASPFA Association of Superannuation and Provident Funds of Australia

asph asphalt; asphaltic

ASPH Australian Society of Professional Hypnotherapists

asphalt solid bitumen pitch

asphaltum mineral pitch

asphic asphaltic

asph mac asphalt macadam

asphy asphyxia

ASPI American Society for Performance Improvement

ASPIRE Associated Students Promoting Individual Rights for Everyone; Art Shared by People Investing in Relevant Education

ASPIRIN Automatic System for Passenger Reservation by Notation

aspis application software prototype implementation system

aspj advanced self-protection jammer; airborne self-protection jammer

ASPL American Society for Law Pharmacy; Associated Steamships Proprietary Limited

ASPLP American Society for Political and Legal Philosophy

ASPM American Society of Paramedics

ASPM Armed Services Procurement Manual

aspn asparagme

ASPO American Society of Planning Officials; Avionics System Project Officer

ASPOD Autonomous Space Processor for Orbital Debris

aspp alloy-steel protective plating

ASPP American Society for the Perfection of Punctuation; American Society of Picture Professionals; American Society of Plant Physiologists; American Society of Polar Philatelists; Australian Society of Plant Physiologists

ASPPA Armed Service Petroleum Purchasing Agency

ASPPO Armed Services Procurement Planning Office

ASPPT Association for Suppliers of Printing and Publishing Technologies

ASPQ Association Suisse pour la Promotion de la Qualite (French—Swiss Association for Quality Improvement)

ASPR American Society of Psychical Research; Armed Services Procurement Regulations; Association of South Polar Research

ASPRL Armament Systems Personnel Research Laboratory (USAF)

aspro ass prostitute (male homosexual)

ASPRS American Society of Plastic and Reconstructive Surgeons

asps advanced sleep-phase syndrome; annular suspension and pointing system

ASPs Anglo-Saxon Protestants; Assistant Superintendents of Police

ASPS Acoustic Ship-Positioning System

ASPSPOM American Society for the Preservation of Sacred, Patriotic, and Operatic Music

ASPT American Society of Plant Taxonomists; Army School of Physical Training

ASPTC Army Support Center

ASQ Anthropological Society of Queensland; Anxiety Scale Questionnaire

ASQ Administrative Science Quarterly

ASQC American Society for Quality Control

ASQDE American Society of Questioned Document Examiners

asr air staff requirement; airport surveillance radar; air-sea rescue; answer and receive; apical sensory region; automatic send-receive; armed strike reconnaissance; assessor; available supply rate

ASR Accounting Series Release; American Society of Rocketry; Army Scripture Reader; Army Status Report; Army Support Regiment; Association of Southeastern Railroads; Aviation Safety Regulation(s); submarine rescue vessel (naval symbol)

asra athwartships reference axis; automatic stereophonic recording amplifiier

asraam advanced short range air-to-air missile

asradi adaptive surface-signal recognition-and-direction indicator

ASRAPS Acoustic Sonar Range Prediction System

ASRB Australian Sales Research Bureau

asrc (ASRC) air-sea rescue craft

ASRC Alabama Space and Rocket Center; Atmospheric Sciences Research Center .

asrd aircraft shipment readiness date

ASRE Admiralty Signal and Radar Establishment (UK); American Society of Refrigeration Engineers

ASREC American Society of Real Estate Counselors

asrf air-sea rescue flight

ASRI Aluminum Smelters Research Institute

ASRL Aero-elastic and Structures Research Laboratory (M.I.T)

ASRM American Society of Range Movement

asro (ASRO) astronomical roentgen observatory (satellite)

asroc (ASROC) antisubmarine rocket

ASRP American Society for the Republic of Panama

ASRPP American Society for Research in Psychosomatic Problems

asrs air-surveillance radar system

a-s rs air-sea rescue service

ASRS Aviation Safety Reporting System

ASRT Air Support Radar Team; American Society of Radiologic Technologists

asrv angle-stop radiator valve; armored scout reconnaissance vehicle

ASRY Arab Shipbuilding and Repair Yard (Bahrain)

ass anterior superior spine; assault supply ship; assurance

Ass Assyrian

ASS Accordion Symphony Society; Anglo-Swedish Society; Army Special Staff; Associated Scholastic Society; Associated Sociologists Society; Australian Security Service; Aviation Security System

A-SS Anti-Slavery Society

A.S.S. Associate in Secretarial Science; Associate in Secretarial Studies

ASSA American Society for the Study of Allergy; Army Signal Supply Agency; Astronomical Society of South Australia; Australian Society of Security Analysts

ass. & rob. assault and robbery

ASSArthr American Society for the Study of Arthritis

ASSASSIN Agricultural System for Storage and Subsequent Selection of Information

assassrep assassination report

A-S Scale Anti-Semitism Scale (measuring negative attitudes)

Assateague Assateague Island National Seashore linking Maryland and Virginia

ASSC Airborne Systems Support Center

assce assurance

ASSCO American Steam Ship Company

Ass Com Gen Assistant Commissary General

assd assigned

ASSE American Society of Safety Engineers; American Society of Sanitary Engineers

assem assemble

Assem God Assemblies of God

assess. analytical studies of surface effects of submerged submarines

ASSESS Airborne Science-Spacelab Experiments-Simulation System

ASSET Aerothermodynamic Elastic Structural System Environmental Tests; Auto-

mated Supported for Software Engineering Technology

ASSGB Association of Ski Schools in Great Britain

ASSH American Society for Surgery of the Hand

ASSIFONTE Association de l'Industrie de la Fonte de Fromage (French—Association of the Processed Cheese Industry)

assigt assignment

assim assimilated

ASSIST Assessment of Information Systems and Technologies (in medicine)

assist. assistant

assm anti-ship supersonic missile; anti-surface ship missile

assmt assessment

assn association

Assn Brain Injured New York Association for Brain Injured Children

Assn Brit Zool Association of British Zoologists

assnce assurance

Assn Clin Biochem Association of Clinical Biochemists

Assn Consumer Res Association for Consumer Research

assnd assigned

Assn Ed Comm Tech Association for Educational Communications and Technology

Assn Pr Association Press

Assn Sch Busn Association of School Business Officials of the United States and Canada

Assn Study Anim Behav Association for the Study of Animal Behaviour

Assn Supervision Association for Supervision and Curriculum Development

Assn Tchr Ed Association of Teacher Educators

Assn Under Man Association for the Understanding of Man

ASSO Anti-Submarine Support Operations

ASSOBANCA Associazione Bancaria Italiana (Italian—Italian Bankers' Association)

assoc associate; associated; association

Assoc Bk Associated Booksellers

Assoc Coun Arts Associated Councils of the Arts

Assoc Eng Associate in Engineering

ASSOCHAM Associated Chambers of Commerce

Assoc IEE Associate of the Institution of Electrical (Electronic) Engineers

Assoc I Min E Associate of the Institute of Mining Engineers

Assoc INA Associate of the Institute of Naval Architects

Assoc ISI Associate of the Iron and Steel Institute

Assoc Met Associate of Metallurgy

Assoc Pr Associated Press

Assoc Sci Associate in Science

assoc w associated with

asson assonance

assp application specific standard part

ASSP All Saints Sisters of the Poor

ASSPHR Anti-Slavery Society for the Protection of Human Rights

ASSR American Society for the Study of Religion

ASSS American Society for the Study of Sterility; Australian Society of Soil Science

asst anti-ship surveillance and targeting; assist; assistance; assistant

ASST American Society for Steel Treating

ASST *Aziendo de Stato per i Servizi Telefonici* (Italian—State Telephone Service)

Asst Att Gen Assistant Attorney General

Asst Chf Engr Assistant Chief Engineer

asstd assented; assorted

Asst Engr Assistant Engineer

Asst Pur Assistant Purser

Asst Stwd Assistant Steward

A-S Study Ascendance-Submission (reaction)

assu (ASSU) air support signal unit

ASSU American Sunday School Union

assw antistrategic submarine warfare

assy assembly

Assyr. Assyrian

Assyr-Babyl Assyro-Babylonian

ast absolute space-time; air staff target; aspartate aminotransferase

ast (AST) advanced supersonic transport; average spring tides

Ast astigmatism; Astoria(n); Asturian; Asturias

AST Academic Salaries Tribunal; Adult Sexual Theme; Aerial Survey Team; Air Service Training; Air Surveillance Technician; Alaska State Troopers; Alaskan Standard Time; American Radiator and Standard Sanitary (stock exchange symbol); Army School of Transport; Army Satellite Tracking; Army Specialized Training; Arts Society of Tasmania; Association for Student Training; Astronomical Society of Tasmania; Atlantic Standard Time

asta aerial surveillance and target acquisition

ASTA Aerial Surveillance and Target Acquisition; American Seed Trade Association; American Society of Travel Agents; American Spice Trade Association; American String Teachers Association; Army Strategy and Tactics Analysis; Australian Science Teachers Association

ASTA *Allgemeiner Student-enausschuss* (German—General Students Committee)

ASTAC Australian Shipping, Trading, and Chartering

ASTANO *Astilleros y Talleres del Noroeste* (Spanish—Dockyards and Workshops of the Northwest)

ASTAP Advanced Statistical Analysis Program

ASTAS Antiradar Surveillance and Target Acquisition System

astc (ASTC) airport surface traffic control

ASTC American Society of Theater Consultants; Appalachian State Teachers College; Arkansas State Teachers College; Aroostook State Teachers College

A.S.T.C. Associate of the Sydney Technical College

astd anti-ship torpedo defense

ASTD American Society of Teachers of Dancing; American Society for Training and Development; American Society of Training Directors

ASTE American Society of Tool Engineers

astec advanced solar turboelectric concept; advanced solar turboelectric conversion

ASTEC Antisubmarine Technical Evaluation Center; Australian Science and Technology Council

ASTECNAVAIR Assistant Secretary of the Navy for Air

a sten aortic stenosis

ASTEO *Association Scientifique et Technique pour l'Exploration des Oceans* (French—Scientific and Technical Association for the Exploration of the Oceans)

ASTER Advanced Spaceborne Thermal

ASTF Aeropropulsion System Test Facility; Aerospace Structures Test Facility

asth asthenopia

astheno asthenosphere

asti antispasticity index (ASTI); antisubmarine training indicator

ASTI American School of Technical Intelligence

ASTI *Applied Science and Technology Index*

ASTIA Armed Services Technical Information Agency

astig astigmatic; astigmatism; astigmatizer; astigmatoscope; astigmatoscopy; astigmia; astigmometer; astigmoscope

astinator procrastinator

ASTIP Army Scientific and Technical Information Program

ASTM American Society for Testing and Materials; American Society of Tropical Medicine

ASTME American Society of Tool and Manufacturing Engineers

ASTMH American Society of Tropical Medicine and Hygiene

ASTMS Association of Scientific, Technical, and Managerial Staffs

asto antistreptolysin

as tol as tolerated (by the patient)

astor (ASTOR) antisubmarine torpedo

A-story top story of the tallest tree

ASTP American Society of Tax Professionals; Apollo-Soyuz Test Project; Army Specialized Training Program

ASTPPTP American Society for the Preservation of Professional Tax Professionals

astr astronomy

ASTR American Society of Therapeutic Radiologists

astra advanced structure analyzer; advanced system for radiological assessment; automatic scheduling with time-integrated resource allocation

ASTRAC Arizona Statistical Repetitive Analog Computer

astrag astragalus (ankle bone connecting with heel bone)

ASTREA Air Support to Regional Enforcement Agencies (helicopter surveillance)

astrion astrionic(al)(ly); astrionics

Astrl Australia

astro astrograph(ic); astrolabe; astrology; astrometry; astronautics; astronomer; astronomical; astronomy; astrophysics

Astro Astronautics

ASTRO Air-Space Travel Research Organization

astro-ad-anon astrological adventures anonymous

astrobio astrobiological; astrobiologist; astrobiology

astrochem astrochemical(ly); astrochemist(ry)

astrochronics astrochronological relatives

Astrodome Astrodome Stadium, Houston, Texas

astrodyn astrodynamic(al)(ly); astrodynamic(ist)

astrog astrogeological; astrogeologist; astrogeology

astrogen astrogenealogy

astrogeo astrogeology of celestial bodies

astrol astrology

Astrol Astrology

astromonk astronautical monkey (specimen used in biological tests)

astron astronomer; astronomnic(al)(ly); astronomy

Astron Astronomy

astronaut astronautical(ly); astronautics

Astron J *Astronomical Journal*

Astron Nachr *Astronomische Nachrichten* (German—Astronomical News)

Astron Zh *Astronomicheskii Zhurnal* (Russian—Astronomical Journal)

Astro Obsv Astrophysical Observatory

astrophys astrophysics

Astrophys Astronomy and Astrophysics

Astrophys J *Astrophysical Journal*

Astrophys Lett *Astrophysical Letters*

ASTS Alabama State Training School (for female delinquents at East Lake near Birmungham)

as'ts assists

ASTSECNAV Assistant Secretary of the Navy

astt (ASTT) action-speed tactical trainer

ast t astronomical time

ASTT American Society of Traffic and Transportation

A.S.T.T. Associate in Science Teacher Training

ASTU Air Support Training Unit

asu all screwed up

asu (ASU) administrative systems unit; acromedical staging unit

ASU Aircraft Storage Units; American Secular Union; American Student Union; Arab Socialist Union; Arizona State University; Asuncion, Paraguay (airport); Atheist Student Union; Australian Swimming Union; (Lat Am St) Arizona State University Center for Latin American Studies

ASU-57 Soviet self-propelled 57mm gun on tracked chassis

ASUA Amateur Swimming Union of the Americas

ASUC American Society of University Composers; Associated Students of the University of California

asupt advanced simulator for undergraduate pilot training

ASUSSR Academy of Sciences of the USSR

ASUTS American Society of Ultrasound Technical Specialists

ASUUS Amateur Skating Union of the U.S.

asv acceleration switching valve; airborne radar for detecting surface vessels; aircraft-to-surface vessel; angle stop valve; armored support vehicle; avian sarcoma virus

asv (ASV) automatic self-verification

a-s v anti-snake venom; arteriosuperficial venous

a/sv arterio/superficial venous

ASV American Society of Viticulture; American Standard Version; Anthropological Society of Victoria; Astronomical Society of Victoria

ASVA Associate of the Society of Valuers and Auctioneers

ASVAB Armed Services Vocational Aptitude Battery

asveo advance space vehicle engineering operation

ASVT Applications Systems Verification Test

ASVU Army Security Vetting Unit

asw air/sea warfare; antisubmarine warfare

ASW Association of Scientific Writers; Association of Social Workers; Australian Standard White (wheat)

ASW (LR) Antisubmarine Warning (long-range)

ASW (SR) Antisubmarine Warning (short-range)

AS & W American Steel and Wire (gage)

ASWA American Society of Women Accountants; Anthropological Society of Western Australia

A/S WA Aviation/Space Writers Association

aswacs anti-submarine warfare air close transport

asw/aaw antisubmarine warfare/anti-air warfare

ASWBPL Armed Services Whole Blood Processing Laboratory

ASWC Antisubmarine Warfare Center (NATO)

aswcr airborne surrveillance warning and control radar

aswcsi anti-submarine warfare combat system integration

ASWD Army Special Weapons Depot

ASWDU Air-Sea Warfare Development Unit

ASWE Admiralty Surface Weapons Establishment

ASWEPS Anti-Submarine Warfare Environmental Prediction System

ASWEX Anti-submarine Warfare Exercise

aswf arithmetic-series weight function(s)

ASWG American Standard Wire Gage; American Steel and Wire Gage

aswgw anti-submarine wire-guided weapons

ASWI Antisubmarine Warfare Installations (NATO)

ASWIPT Antisubmarine Warfare In-Port Training (NATO)

aswlr anti-submarine warning long-range

ASWPL Association of Southern Women for the Prevention of Lynching

ASWRC Antisubmarine Warfare Research Center (NATO)

ASWS Audubon Shrine and Wildlife Sanctuary

ASWSOW Anti-Submarine Warfare Standoff Weapon

ASWSS Anti-submarine Warfare Schoolship (USN)

ASIVTDS Anti-submarine Warfare Tactical Data System

asy asylum

asym asymmetrical

async asynchronous

ASYTS Armed Services Youth Training Scheme

ASZ American Society of Zoologists

ASZD American Society for Zero Defects

at accounting tabulating (card); advanced technology (computer); aircraft technician; airtight; air transport; alternative technology; anti-tank; anti-torpedo; appropriate technology; artillery tractor; asphalt; asphaltic; asphalt tile; atmosphere (technical); atomic (AT); automatic transmission; auxiliary tug

at. (AT) approximate technology

a t a tempo (Italian—at the speed written)

at. attendant

at.% atomic percent

a/t action/time; antitank; antitorpedo

a & t acceptance and transfer; assemble and test

At ampere-turn; astatine

AT Adirondack Trail; Adult Theme; Advanced Technology; Advanced Trainer; Air Travel; Ammunition Technician; Appalachian Trail; Atherton Tablelands (Queensland parks); Army Transport; Atlantic Time; Austria; Autogenic Training

A T Absolute Time (compact disc)

A/T American terms

AT Antico Testamento (Italian—Old Testament)

a-t anti-tank

A-T Alef-Tav (Hebrew—from the first to the last letter of the alphabet)—similar to the English expression from A to Z

at 3 anti-thrombin III

AT₇ hexachlorophene (disinfectant)

AT₁₉ dihydrotachysterol

AT-26 Aermacchi jet-trainer ground-attack aircraft also known as Xavante

ata academic travel abroad; actual time of arrival; air-to-air; air training area; azimuthal torque amplifier

ata admission temporair (French—temporary admission)

ata (ATA) advanced tactical aircraft

ATA Advertising Typographers Association; Air Transport Association; Air Transport Auxiliary; Albanian Telegraph Agency; Amateur Trapshooting Association; American Taxicab Association; American Taxpayers Association; American Telemarketing Association; American Theatre Association; American Thyroid Association; American Tinnitus Association; American Title Association; American Topical Association; American Transit Association; American Translators Association; American Tree

Association; American Trucking Association; American Tunaboat Association; Anatomical Transplant Association; Applied Technology Associates; Area Transportation Authority; Army Temperance Association; Army Transportation Association; Asia Teachers Association; Association of Talent Agents; Atlantic Treaty Association; Australian Taxpayers Association; Australian Toolmakers Association; Australian Translators Association; auxiliary ocean tug (naval symbol)

A.T.A. Associate Technical Aide

ATA Agence Telegraphique Albanaise (French–Albanian News Service); *Art Through the Ages* by Helen Gardner

ATAA Advertising Typographers Association of America; Air Transport Association of America; Amateur Trapshooting Association of America

atac air-to-air combat

ATAC Air Transport Association of Canada; Anatomical Transplant Association of California; Army Tank Automotive Center; Army Tank and Automotive Command

ATACS Army Tactical Communications System

atad absent on temporary additional duty

ATAD Air Transport and Delivery (service)

ATAE Association of Tutors in Adult Education; Automotive Trade Association Executives

ATAF Allied Tactical Air Force; American Trucking Associations Foundation

ATAFCS Airborne Target-Acquisition and Fire-Control System

ATAG Air Training Advisory Group

ATAI Air Transport Association International

ATAIU Allied Tactical Air Intelligence Unit

ATAJ Association of Transport Advisers of Japan

ATALA Association pour l'Etude et de la Linguistique Appliquee (French—Association for the Study of Applied Linguistics)

atam air-to-air missile

ATAO Application for Taxpayer Assistance Order

ATAM Association for Teaching Aids in Mathematics

atan arc tangent

atar above transmitted and received; (ATAR) antitank aircraft rocket

ATAR Automated Travel Agents Reservation

atars advanced tactical airborne reconnaissance system

ATARS Anti-Terrain-Avoidance Radar System

ATAs American Tinnitus Association members

ATAS Academy of Television Arts and Sciences; Air Transport Auxiliary Service

atasm advanced tactical air-to-surface missile

atav atavism; atavist; atavistic(al)(ly)

atb amphibious training base; asphalt tile base; at the time of bombing

atb (ATB) advanced technology bomber; stealth bomber

ATB Air Transportation Board; Army Tank Battalion

A & TBCB Architectural and Transportation Barriers Compliance Board

ATBI Allied Trades of the Baking Industry

atbm average time between maintenance

atbm (ATBM) advanced technology ballistic missile; anti-tactical ballistic missile

atbt acoustic telemetry bathythermometer

atbyropt at buyer's option

atc ablative thrust chamber; acoustic test(ing) chamber; acoustical tile ceiling; aerial

tuning condenser; allergic to combat; anti-torpedo craft; approved type certificate; armored troop carrier; automatic temperature control; automatic tint control (tv)

atc (ATC) all-terrain cycle; automatic train control; average variable cost

ATC Advertising Training Center; Air Traffic Conference; Air Traffic Control; Air Training Command; Air Training Corps; Air Transport Command; Air Transportation Corps; Aircraft Technical Committee; Airport Traffic Control; Airway Traffic Control; Alcohol Treatment Center; All Things Considered (radio program); Alpine Tourist Commission; Appalachian Trail Conference; Armament Test Center; armored troop carrier; Army Training Center; Army Transportation Corps; Associated Traffic Clubs; Associated Travel Clubs; Athletic Training Council; Australian Tariff Council; Australian Tourist Commission; mini-armored patrol vehicle (naval symbol)

atca (ATCA) advanced tanker-cargo aircraft

ATCA Air Traffic Conference of America; Air Traffic Control Association; Allied Tactical Communications Agency; American Theater Critics Association

ATCAS Air-Traffic-Control Automated System

atcase aspartate transcarbamylase

ATCB Air Traffic Control Board

ATCC Air Traffic Control Center; American Type Culture Collections; Automatic Train Control Center

ATCDE Association of Teachers in Colleges and Departments of Education

atce ablative thrust chamber engine

atcen air traffic control evaluation unit

ATCF Automobile and Touring Club of Finland

ATCGP Air Training Command Ground Personnel

atch armored troop-carrier helicopter; attach; attaching; attachment

atchd attached

ATCL Associate of Trinity College of Music–London

ATCMD Atlanta Contract Management District

ATCMS Advanced Technology Cruise Missile Study

ATCO Air Traffic Coordinating Office(r)

ATCOM Atoll Commander

ATCOS Atmospheric Composition Satellite

ATCP Antarctic Treaty Consultative Parties

ATCRBS Air Traffic Control Radar Beacon System

atcru air-traffic-control radar unit

atc's airtight containers; any terrain motorcycles

ATCU Association of Texas Colleges and Universities

atd absent (on) temporary duty; actual time of departure; anthropomorphic test dummy; armored tank destroyer

atd *a tak dale* (Czech—et cetera)

ATD Actual Time of Departure; Aid to the Totally Disabled; Armament Test Division; Art Teachers Diploma

atda augmented target docking adapter

ATDA American Train Dispatchers Association; Army Training Device Agency; Australian Telecommunications Development Association

atdc after top dead center (valve setting)

atdp attitudes toward disabled persons

ATDS Airborne Tactical Data System; Association of Teachers of Domestic Science; Automated Data and Telecommunicatitons Service

ATDU Air Transport Development Unit; Armored Trials and Development Unit

ate altitude transmitting equipment; automatic test equipment

Ate Almirante (Spanish—admiral)

ATE Associated Telephone Exchanges; Association of Teacher Educators; Automatic Telephone and Electric (New Zealand company)

ATEA American Technical Education Association; American Toy Export Association; Australian Telecommunications Employees Association

ATEC Air Transport Electronics Council; Aviation Technician Education Council

A.Tech. Associate in Technology

ATEM Aircraft Test Equipment Modification; Anti-Tank Euro Missile

ATEMIS Automated Traffic Engineering and Management Information System

A temp *a tempo* (Italian—in the speed written)

Aten Atenas (Portuguese or Spanish—Athens); *Atene* (Italian—Athens), *Athenes* (French—Athens)

ATEN Association Technique pour la production et l'utilisation de l'Energie Nucleaire (French—Technical Association for the Production and Use of Nuclear Energy)

At Energ Atomnaya Energiya (Russian—Atomic Energy)—journal

ATEP Aboriginal Teacher Education Programme

A Term Air Terminal

ATES Advanced Techniques integration into Efficient Scientific Software

ATESL Association of Teachers of English as a Second Language

ATEWS Advanced Tactical Electronic Warfare System

Atex Atlantic tradewind experiment

atf accounting tabulating form; actual time of fall

ATF Acceptance Test(ing) Facility; Advanced Technical Fighter; Advanced Technology Fighter; Air Task Force; Alcohol, Tobacco, and Firearms (bureau); Airborne Task Force; Allied Task Force; Alternative Test Facility; American Type Founders; Amphibious Task Force; Australian Teachers Federation; ocean tug (3-letter symbol)

ATFAC American Turpentine Farmers Association Cooperative

ATFCNN Allied Task Force Commander–Northern Norway (NATO)

atfi attitudes toward feminist issues scales

atfr automatic terrain-following radar

ATFS Association of Track and Field Statisticians

atg air-to-ground

ATG Accordion Teachers Guild; Alaskan Territorial Guard; Army Technical Group

atgar (ATGAR) anti-tank guided air rocket

at gemss anti-tank ground-emplaced mine-scattering system

ATGM Anti-Tank Guided Missile

ATGSB Admission Test for Graduate Study in Business

atgw (ATGW) antitank guided weapon(s)

ath above the horizon; atheism; atheist(ic); athletic

äth äthiopisch (German—Ethiopian)

Ath Athens

ATH Athens, Greece (airport); Philadelphia Athletics

AT-H August Thyssen-Hutte

Athab Athabasca(n)

athe allotetrahydrocortisol

ath dfld atheism defiled

Athel Athel Line

Athen Athenian

atheol atheological; atheologist; atheology

athodyd aerothermodynamic duct (ramjet engine)

athsc atherosclerosis

athw athwartship

ati actual time of interception; aerial tuning inductance; aptitude-treatment interaction; average total inspection

ATI Air Technical Intelligence; American Technology Institute; American Television Institute; Ansett Transport Industries; Army Training Instruction; Asbestos Technical Institute; Asbestos Textile Institute; Australian Textile Institute

A & TI Agricultural and Technical Institute

ATI Aero Transporti Italiani (Italian—Italian Air Freight Line); Air Technical Index; *Azienda Tabacchi Italiani* (Italian—Italian State Tobacco Board)

ATIC Aerospace Technical Intelligence Center; Air Technical Intelligence Center; Antigua Tourist Information Center; Australian Tin Information Centre

ATIGS Advanced Tactical Inertial Guidance System

ATII Associate of the Taxation Institute Incorporated

ATIL Air Target Intelligence Liaison program (USAF)

atiob as this is our best

ATIP Alaskan Talent, Information, and Practices

atis automatic terminal information service

ATIS Adirondack Trail Improvement Society; Air Technical Intelligence Study

ATISC Air Technical Intelligence Services Command (USAF)

ATIU Air Technical Intelligence Unit

ATJ Association of Teachers of Japanese

ATJS Advanced Tactical Jamming System

atk attack

a-tk anti-tank

atl analog threshold logic

Atl Atlanta; Atlantic; Australia

Atl Atlantico (Italian; Portuguese, or Spanish—Atlantic); *Atlantique* (French—Atlantic); *Australia* (Spanish—Australia)

ATL Acoustic Test(ing) Laboratory; Adult T-cell leukemia; Adult T-cell lymphoma; Alexander Turnbull Library (Wellington, NZ); Associated Truck Lines; Atlanta, Georgia (airport); Atlantic Tankers Limited; Automatic Totalisators Limited; Brooklyn Atlantics

ATLA Air Transport Licensing Authority; American Theological Library Association; American Trial Lawyers Association

ATLAJ American Trial Lawyers Association Journal

Atlanta capital of Georgia; U.S. Penitentiary at Atlanta, Georgia

Atlanta Youth Atlanta Youth Development Center (for female juvenile delinquents) in Atlanta, Georgia

Atlantic Ridge submerged mountain range extending from Antarctica to Iceland in the mid-Atlantic

ATLANTIC Atlantic Refining Company

Atlantol Atlantologic(al)(ly); Atlantolgist(ic)(al)(ly); Atlantology

atlas anti-tank laser-assisted system

Atlas Atlas Mountains of Algeria and Morocco

ATLAS Abbreviated Test Language for Avionic Systems; Automated Tape Label Assignment System; Automatic, Tabulating, Listing, and Sorting System

Atlas-Agena two-stage launch vehicle

Atlas-Centaur first American high-energy launch vehicle for space exploration—D-Series Atlas boosts Centaur space vehicle

Atlas-E intercontinental ballistic missile designed to place a thermonuclear warhead on a 9000-mile-distant target

atlas fol atlas folio—a book about 25 inches high

Atlas icbm first American intercontinental ballistic missile

ATLB Air Transport Licensing Board (UK)

Atl C Atlantic City

ATLD Air-Transportable Loading Dock

ATLIS Army Technical Library Improvement Studies; Automatic-Tracking Laser-Illumination System

Atl O Atlantic Ocean

Atl Pil Aut Atlantic Pilotage Authority

ATLS Air Transport Liaison Section

Atl Sym Atlanta Symphony

atm air-turbine motor; anti-tank mine; anti-tank missile; atmosphere (normal)

at. m atomic mass

a&tm ammunition and toxic materials

at/m ampere turns per meter

At/m ampere turns per meter

ATM Apollo Telescope Mount; Association of Teaching Aids in Mathematics; Associated Tobacco Manufacturers; Automated Teller Machine

ATM Amateur Telescope Making; Azienda Tranviaria Municipale (Italian—Municipal Rapid Transit Board)

ATMA Adhesive Tape Manufacturers' Association; American Textile Machinery Association

atm ab atmospheres absolute

ATMAC Air Traffic Management Automated Center

ATMC Army Transportation Materiel Command; Automotive Training Managers Council

atm/d (ATM/D) automatic teller machine deposit

atmds antitank mine-dispensing system

ATMI American Textile Manufacturers Institute

atmos atmosphere; atmospheric(al)(1y)

ATMOSPHERE Atmospheric Monitoring System for Protection of Health and Environment from Risky Emissions

atm press atmospheric pressure

atmr advance transport, medium-range

ATMS Air Traffic Management System; Automatic Teller Machine System(s); Automatic Transmission Measuring System

ATMU Aircraft Torpedo Maintenance Unit

atm/w (ATM/W) automatic teller machine cash withdrawal

ATMX railcar used by the Department of Energy for shipping defense waste

atn actual test number; acute tubular necrosis; augmented transition network

ATN Alabama, Tennessee, and Northern (railroad)

ATNA Australian Trained Nurses' Association

atnav acoustic-transponder navigation

atndt attendant

at. no. atomic number

ATNP Atherton Tablelands National Parks (Queensland)

ato according to others; assisted takeoff, arsenic trioxide; automatic train operation

ATO Academy of Teachers of Occupations; Alternative Trade Organization; Ammunition Technical Officer; Australian Taxation Office; ocean tug, old (3-letter symbol)

ATOA Australian Transport Officers Association

ATOC Allied Tactical Operations Center

atoll acceptance, test, or launch language

atomdef atomic defense

atomdev atomic device

atoms. automated technical order maintenance sequence(s)

ATOM August Twenty-One Movement (Philippines)

ATOMSTATREP Atomic Status Report

aton at once

atorp antitorpedo; atomic torpedo

ATOS American Theatre Organ Society; Association of Temporary Office Services

atot actual time over target

atp actual time of penetration; array transform processor; authority to proceed

atp (ATP) adenosine triphosphate, material found in almost all terrestrial life

atp a tout prix (French—at any price)

ATP Acceptance Test(ing) Procedure; adenosine triphosphate; Admissions Testing Program; Advanced Technology Program; Allied Technical Publication; Army Training Program; Associated Tax Processors; Association of Tennis Professionals

ATP Accorde Transports Perissable (French) European agreement on international shipment of perishable foodstuffs

atpa auxiliary turbopump assembly

ATPAS Association of Teachers of Printing and Allied Subjects

ATPase adenosine triphosphate

atpcc attitudes toward parental control of children

atpd ambient ternperature and pressure–dry

ATPE Association of Teachers in Penal Establishments

ATPI American Textbook Publishers Institute

ATPM Association of Toilet Paper Manufacturers

at pres at present

atps ambient temperature and pressure–saturated with water vapor

atpu air transport pressurizing unit

atr advanced test reactor; advanced tactical radar; air transport rating; antitransmit-receive; transmitter-receiver

atr (ATR) after tax rate; audiotape recorder

Atr Achilles tendon reflex

ATR Admiralty Test Rating; Advanced Test Reactor; Air Transport Rack; Association of Teachers of Russian; ocean tug, rescue (3-letter naval symbol)

ATR Admission Temporaire Roulette (French—Temporary Admission on Wheels); *Anglican Theological Review*

ATRA American Television and Radio Artists; American Tort Reform Association; Automatic Transmission Rebuilders Association

atran automatic terrain recognition and navigation

atrax air-transportable communications complex

atrc anti-tracking control

ATRC Air Traffic Regulation Center

atrd automatic target recognition device

atr fib atrial fibrillation

atrial atrial fibrillation

atrid automatic target recognition, identification, and detection

atrima as their respective interests may appear

ATRIS Air Traffic Regulation Identification System (USA)

atrl anti-tank rocket launcher; atrial

atrls actual time of release

atrm after torpedo room

ATRM American Tax Reduction Movement

atro actual time of return to operation

atrop atrophy

A Tr Ps *Allied Training Publications* (NATO)

atrr advanced threat-reactive receiver; allocated transfer risk reserve

ATRS Australian Tape Recording Society

atrso accepts transfer as offered

atrt anti-transmit-receive tube

ats absolute temperature scale; advanced technological satellite; air-to-ship; air-turbine starter; anxiety-tension state; astronomical time switch

ATs Achievement Tests

ATS Acoustic Transmission System; Acquisition and Tracking System; Administrative Terminal System; Advanced Technological Satellite; Aeronautical Training Society; Air Tactical School; Air Traffic Services; Air Transport Service; American Theological Society; American Therapeutic Society; American Travel Service; American Trudeau Society; Anglican Truth Society; Applications Technology Satellite; Army Technical School; Army Transport Service; Association of Theological Schools; Automatic Transfer Service (bank accounts); Auxiliary Territorial Service; salvage tug (naval symbol)

ATSBNZ Associated Trustee Savings Banks of New Zealand

ATSC Air Technical Service Command; American Traffic Safety Council

ATSDR Agency for Toxic Substances and Disease Registy

ATSE Alliance of Theatrical Stage Employees

ATSF Advanced Television Systems Committee

ATSF *Avion de Transport Supersonique Futur* (French—Supersonic Airplane of the Future—French-British Concorde)

AT & SF Atchison, Topeka, and Santa Fe (railway)

ATSFSD Air Traffic Service Flight Services Division (FAA)

atsit automatic techniques for the selection and identification of targets

ats/jea automated test system/jet engine accessories

atsm advanced tactical stand-off missile

ats/m air-turbine starter/motor

ATSOA American Truck Stop Owners Association

ATSOCC Applications Technology Satellite Operations Control Center (NASA)

ATS's Advanced Technological Satellites

ATSSA American Traffic Safety Services Association

AtST Atlantic Standard Time

ATSU Association of Time-Sharing Users

att attach; attempt; attorney

a t & t all tacos and tamales (American Southwestern roadside-stand short form); always talking and talking

Att Attic(a)

ATT Army Training Test

AT & T American Telephone & Telegraph

A & TT Alcohol and Tobacco Tax

atta *atenta* (Spanish—attentively)

ATTA Association of Travel and Tourist Agents (Singapore)

ATTAIN Applicability in Transport and Traffic of Artificial Intelligence

ATTC American Towing Tank Conference

ATTDU Air Transport Tactical Development Unit

atten attenuation, attenuator

ATT & F Alcohol, Tobacco Tax, and Firearms (Division of U.S. Treasury Dept.)

AttGen Attorney General

Att Gen Attorney General

ATTI Association of Teachers in Technical Institutions

Attica Facility Attica Correctional Facility (for males) at Attica, New York

attn attention

atto attorney

ato *atento* (Spanish—attentively); 10^{-18}

attr attractive

attrd attributed

attrest(s) attitude arrest(s)—made by law-enforcement officers who dislike the attitude(s) of the person(s) arrested

attrib attributive

attrit attrition

ATTS Automatic Telemetry Tracking System

ATTW Aircrew Training and Test Wing

atty attorney

atty & c attorney and client

Atty Gen Attorney General

AT type adenine and thymine type

atu alien tax unit; atomic time unit

Atu *Atinosphärenüberdruck* (German—atmospheric excess pressure)

ATU Alliance of Telephone Unions; Amalgamated Transit Union; Anchorage Telephone Utility; Anglo-Turkish Union; Anti-Terrorist Union; Anti-Terrorist Unit; Arab Telecommunications Union

atum antitank nonmetallic

ATURM Amphibious Training Unit–Royal Marines

atv (ATV) all-terrain vehicle; armored transport vehicle

ATV Amateur Television; Associated Tele Vision

ATV *Akademiet for de Tekniske Videnskaber* (Danish—Academy of Technical Sciences)

atvm attenuator thermo-element voltmeter

atvr armored transport vehicle reconnaissance

at. vol atomic volume

atw (ATW) antitank weapon

at/w atomic hydrogen weld

at w/od automatic transmission with overdrive

ATW Advanced Training Wing; American Theater Wing; Atlantic & Western (railroad)

at/wb ampere turns per weber

ATWE Association of Technical Writers and Editors

ATWg Air Transport Wing (USAF)

atws adjustable thermal wire stripper; automatic track while scanning

at. wt. atomic weight

ATWU Australian Textile Workers Union

atx air taxi

at. xpl atomic explosion

ATYP Australian Theatre for Young People

A Typ I Association Typographique Internationale (French—International Typographic Association)

atyropt at your option

ATZ Air Traffic Zone

au activity unit; angstrom unit; antitoxin unit; arbitrary unit(s); author; azauridine

au aurum (Latin—gold)

a.u. aures unitas (Latin—both ears); *au usum* (Latin—according to custom)

Au angstrom unit; astronomical unit; gold (symbol)

AU Aarhus Universitet (University of Aarhus); Air University; Alfred University; Allen University; American University; Andrews University; Army Unit; Assumption University; astronomical unit; Atheists United; Atlanta University; Auburn University; Australia (Internet code)

AÜ Ankara Üniversitesi (University of Ankara)

A/U advanced undersea weapons

A & U Allen & Unwin Publishers

Au[198] radioactive gold (symbol)

AU-23A Fairchild piston-powered STOL aircraft

AUA American Unitarian Association; American Urological Association; Aruba, Nether-

lands West Indies (airport); Associated Unions of America; Austrian Airlines

A.U.A. Associate of the University of Adelaide

AUAF Association of University Affiliated Facilities

AUAS Academy of Underwater Arts and Sciences

aub alstublieft (Dutch—please)

Aub Aubrey

AUB American University of Beirut

AUBC Association of Universities of the British Commonwealth

AUBER Association for University Business and Economic Research

AUBTW Amalgamated Union of Building Trade Workers

Auburn Facility Auburn Correctional Facility (for males) at Auburn, New York

auc average unit cost

a.u.c ab urbe condita (Latin—from the founding of the city, usually refers to Rome)

AUC Aberystwyth University College; American University of the Caribbean; American University Club; Australian United Corporation; Australian Universities Commission

AU of C American University of Cairo

AUCA American Unitarian Christian Association

AUCANNZUKUS Australia, Canada, New Zealand, United Kingdom, United States

AUCANUKUS Australia, Canada, United Kingdom, United States

AUCAS Association of University Clinical Academic Staff

AUCC Association of Universities and Colleges of Canada

AUCECB All Union Council of Evangelical Christian Baptists

Auck Auckland

Aucklands Auckland Islands

AUCOA Association of United Contractors of America

AUCSRLFRVWAM All-Union Central Scientific Research Laboratory for the Restoration of Valuable Works of Art in Museums

auct auction(eer)

auct auctorum (Latin—of authors)

AUCTU All-Union Council of Trade Unions

aud audible; audit; audition; auditor; auditorium

Aud (AUD) Australian dollar(s)

Aud[a] audiencia (Spanish—court of justice, hearing)

AUDACIOUS Automatic Direct Access to Information with On-line UDC System

audar autodyne detection and ranging

aud disb auditor disbursements

AUDDITS Automated Dynamic Digital Test System

Audel Theodore Audel

Aud Gen Auditor General

Aud Gen Nav Auditor General of the Navy

Audie Audry

auding auditory hearing, listening, and understanding

audio audiofrequency; audiogenic; audiogram; audiology; audiometer; audiometry; audiophile; audiovisual; audiovisual aids; et cetera

audiol audiologist; audiology

audiom audiometer; audiometric(al)(ly); audiometrist

audiovis audiovisual; audiovisual aids

audre audio response; automatic digit recognizer

AUEC Association of University Evening Colleges

AUEW Amalgamated Union of Engineering Workers

AUF Australian Underwriters Federation

Aufdr Aufdrucke (German—imprint)

Aufl Auflage (German—edition)

AUFL Americans United For Life

AUFS American Universities Field Staff

AUFUSAF Army Unit for United States Air Force

aug augment; augmentation; augmentative

Aug Augsburg; August; Augusta; Augustan

Augember August and September

Augie August; Augusta, Georgia; Augustine; Augustus

augm augmentation

augm augmente (French—augmented)

augra authority granted

August Augustine; Augustus

aui adaptive user interface

AUI Associated Universities Incorporated

auj aujourd'hui (French—today)

Auk Auckland

aul above upper limit

AUL Aberdeen University Library; Air University Library; Americans United for Life; American United Life (insurance)

AULC American University Language Center

auld auld lang syne (Scottish—old times fondly remembered)

AULI As You Like It

AULLA Australasian Universities Language and Literature Association

aum (AUM) air-to-underwater missile

aum aumentado (Spanish—augmented)

AUM Animal Unit Month; Association of University Managers

AUMLA Australian Universities Modern Language Association

a. u. n. abesque ulla nota (Latin—without annotation)

AUNT Alliance for Undesirable but Necessary Tasks

auntie. automatic unit for national taxation and insurance (UK)

AUO African Unity Organization; Atlantic Union Oil

AUP Acceptable Use Policy; Andrews University Press; Australian United Press

AUPE Amalgamated Union of Public Employees (Singapore)

AUPF Australian Uranium Producers Forum

AUPG American University Publishers Group

AUPHA Association of University Programs in Hospital Administration

AUPO Association of University Professors of Ophthalmology

AUPOSTCOM Australian Postal Commission

aur auricle; auricular; auricularis; aurum

Aur Auriga

AUR Adelaide University Regiment; Association of University Radiologists

AURA Association of Universities for Research in Astronomy; Automated Reasoning Assistant

aureq authority is requested

aur fib auricular fibrillation

Auri Auriga

AURI Angkatan Udara Republik Indonesia (Indonesian Air Force)

auric auricular

AURISA Australian Urban and Regional Information Systems Association

aurist. auristillae (Latin—ear drops)

aurora australis (Latin—southern lights)

aurora borealis (Latin—northern lights)

Aus Austin; Austria; Austrian

AUs Area Units (New Zealand)

AUS Army of the United States; Austin, Texas (airport); Australian Union of Students

AUSA Assistant United States Attorney; Association of the United States Army; Australian Universities Sports Association

ausc auscultation

AUSCS Americans United for Separation of Church and State

AusE Australian English

Ausg Ausgabe (German—edition)

Au sh Australian serum hepatitis

AUSHC Australian High Commission

AUSIMM Australian Institute of Mining and Metallurgy

Aus Slang Australian Slang

AUSLFL All-Union State Library of Foreign Literature (Moscow)

AUSS Association of University Summer Sessions

AUSSAT Australian Satellite

Aussie(s) Australian(s)

Aust Australia; Australian

Aust Alps Australian Alps of New South Wales and Victoria

Aust Cur Australian Current

Aust$ Australian dollar

austen austenitic

Austen Australian sten gun

Aust Engl Australian English (Cockney with Australian accents)

Auster Auster-Beagle light liaison aircraft

Austin Augustina; Augustine; capital of Texas

AUSTIRAN Australian-Iran Shipping Company

Aust J Phys Australian Journal of Physics

AUSTRAFORD Ford Motors of Australia

Austrail Railways of Australia

austral unit of Argentina currency

Austral Australian

Australas Australasian

Australasian Australia, Tasmania, New Zealand, and islands of Melanesia

Australs Austral Islands of Polynesia, also called the Tubuais

Austria Republic of Austria, *Republik Österreich*

AUSTRIATOM Austrian Atomic Energy Group

Austro-Hungarian Empire Austria, Bohemia, Bosnia, Croatia, Moravia, Bukovina, Transylvania, Galicia, Hungary, and part of Yugoslavia and Italy (1867–1918)

Austronesia islands of South Pacific from Madagascar in Indian Ocean to Hawaiian Islands in the Pacific

AUSUDIAP Association of U.S. University Directors of International Agricultural Programs

aut autore (Italian—author)

Aut Autriche (French—Austria)

AUT American Union Transport; Association of University Teachers

AUTA Association of University Teachers of Accounting

AUTE Association of University Teachers of Economics

AUTEC Atlantic Underwater Test Evaluation Center

AUTELCOM Australian Telecommunications Commission

auth authentic; authenticate; authenticity; author; authority; authorization; authorize(d)

Auth Authority

authab authorized abbreviation (USAF)

Auth Ver Authorized Version

AUT(I) Association of University Teachers (Ireland)

autiobio autobiograph; autobiographer; autobiographic(al); autobiography

autmwtr ck automatic water check

auto. automobile; automatic; automotive

auto (Latin prefix—self)—automobile (self-moving vehicle)

autobird automatic bird feeder

autocade automobile parade

AUTOCAP Automobile Consumer Action Programs(s)

autocat automatic cat feeder

autocolor automatic color (tv)

autocom automated combustor (design code)

autodidac autodidact(ic)(al) (ly)

autodin automatic digital network

autodoc automatic documentation

autodog automatic dog feeder

autog autograph

auto. lean automatic lean

autom automatic, automation; automobile; automotive

autom automobile (Italian—automobile); *automóvel* (Portuguese—automobile); *automóvil* (Spanish—automobile)

automag automatic-loading magnum (handgun)

Auto Mag Automobile Magazine

automap automatic machining program

automast automatic mathematical analysis and symbolic translation

automát automática; automático (Spanish—automatic)

automatic automatic revolver

automation automatic operation

automtn automation

auton autonomous; autonomy

autonet automatic network

autop automatic pistol; autopsy

autopet automatic pet feeder

AUTOPIC Automatic Personal Identification Code

autopilot automatic pilot

autopistol automatic pistol

AUTOPOLIS Automatic Policing Information System

autoprom autoprompter (tape)

autoprompt automatic programming of machine tools

AUTOPROS Automated Process Planning System

AUTOPSY Automatic Operating System (IBM)

auto pts automobile parts

autoqest automatic generation of requests

auto. recl automatic reclosing

auto. rich automatic rich

autorotic(s) automobile neurotic(s)

autos automobiles; automatics

autosate automatic data systems analysis technique

autoscript automated system for composing, revising, illustrating, and typesetting

auto s & cv automatic stop-and-check valve

AUTOSERVCEN Automated Service Center

autosevocom automatic secure voice communication(s)

autospec automated specification(s)

autospot automatic system for positioning tolls

AUTOSTATIS Automatic Statewide Theft Inquiry System (California)

autostrad automated system for transportation data

autosyn automatically synchronous

autotr autotransformer

autotran automatic translation

auto trans automatic transmission

autovon automatic voice network

auto w/od automatic (transmission) with overdrive

au tr aural training

autran automatic target-recognition analysis

AUTRANAVS Automated Transponder Navigation System

AUT(S) Association of University Teachers (Scotland)

AUT(W) Association of University Teachers (Wales)

AUU Association of Urban Universities

auv administrative use vehicle; armored utility vehicle

Au virus Australian antigen

AUVMIS Administrative Use Vehicle Management Information System (USA)

AUVS Association of Unmanned Vehicle Systems

auw airframe unit weight

auw (AUW) advanced underwater weapon(s)

auwc advanced undersea weapon circuitry

AUWE Admiralty Underwater Weapons Establishment

aux auxiliary, auximones, auxocardia

aux (Greek—grow, increase)— *auxiliar*; (Spanish—auxiliary)

aux m auxiliary machinery

AUXOPS Auxiliary Operational Members (USCG)

auxrc auxiliary recording control

av acid value; anteversion; aortic valve; armored vehicle; arteriovenous; assessed valuation; atrioventricular; auriculoventricular; average; average; aviator; avoirdupois

av (AV) audiovisual

a-v atriventricular; audio-visual

av avril (French—April)

a v a vista (Italian—at sight)

a/v (A/V) ad valorem (Latin—as valued)

aV attovolt

Av Avenue; Aves; Avestan; Avian; Avila(n)

Av avenida (Portuguese or Spanish—avenue); *Avrum* (Yiddish—Abraham)

AV American viewpoint; Antonio Viivaldi; arteriovenous; Artillery Volunteers; audiovisual; Authorized Version; aventurine (green); large seaplane tender (naval symbol)

AV alta voltagem (Portuguese—high voltage); *alto voltaggio* (Italian—high voltage); *alto voltaje* (Spanish—high voltage); *Avtomat Kalashnikov* (Russian—Kalashnikov automatic)—Soviet assault rifle

A. V. Anno Vixit [Latin—he (she) lived (a given number of) years]

AV-8B U.S. Marine Corps fighter-bomber jump-jet (capable of taking off and landing vertically)

ava arteriovenous anastomosis; azimuth versus amplitude

ava (AVA) automatic voice alarm

AVA Academy of Vocal Arts; *Aerodynamische Versuchsanstalt*; Alaska Visitors Association; American Video Association; American Vocational Association; Asbestos Victims of America; AudioVisual Aids; Australian Veterinary Association; Australian Volunteers Abroad; Award(s) in the Visual Arts

A-V A All-Volunteer Army

AVAC Asociación Venezolana para la Avance de la Ciencia (Spanish—Venezuelan Association for the Advancement of Science)

AVADS Autotrack Vulcan Air Defense System

av/af anteverted/anteflexed

avail. available; availability

aval availability; available

Aval Pen Avalon Peninsula in Newfoundland

'Avana Havana

avasi abbreviated visual-approach slope indicator

AVASIS Abbreviated Visual Approach Slope Indicator System

avb avbeta (Swedish—departure)

AVB advanced aviation base ship (naval symbol)

avbl armored vehicle bridge launcher

avc (AVC) audio-visual connection; average variable cost

av C avanti Cristo (Italian—before Christ)

AVC American Veterans Committee; Antelope Valley College; Army Veterinary Corps; Association of Virginia Colleges; Association of Visual Communications; Association of Vitamin Chemists; Audio-Visual Center; automatic volume control; average variable cost

AVCA acceleration vector control; allantoid vaginal cream; Australian Volunteer Coastguard Association

AvCad Aviation Cadet

avcat aviation high-flash turbine fuel

Av Cert Aviation Certificate

avcs atrioventricular conduction system

AVCS Advanced Videcon Camera Systems; Assistant Vice Chief of Staff

avd air vehicle detection; automatic voice data; automatic voltage digitizer

avd avdeling (Dano-Norwegian—part, section)

AvD Automobil Club von Deutschland (German—German Automobile Club)

AVD Army Veterinary Department; high-speed seaplane tender (3-letter naval symbol)

Avda avenida (Spanish—avenue)

AVDA American Venereal Disease Association; American Veterinary Distributors Association

a-v difference arteriovenous concentration difference

AVDO Aerospace Vehicle Distribution Office(r)

avdp avoirdupois

avdth average depth

ave automatic volume expansion

'ave have

Ave Avenue

AVE Asociación Venezolana de Ejecutivos (Spanish—Venezuelan Association of Executives)

AVEA American Veterinary Exhibitors Association

avec amplitude vibration exciter control

AVEM Association of Vacuum Equipment Manufacturers

AVENSA Aerovias Venezolanas (Spanish—Venezuelan Airlines)

AVERA American Vocational Education Research Association

Averroes Abul-ibn-Roshd

Aves Los Aves—Bird Islands off Venezuela, west of Curaçao in the Caribbean

avf arteriovenous filstula; azimuthally varying field

AVF All-Volunteer Force; American Vineyard Foundation

reassignment

avfuel aviation fuel

avg average

Avg Avgust (Russian—August)

AVG Army Volunteer Group

Av Gar Avant Garde

avgas aviation gasoline

avge average

avgp armored vehicle general-purpose

avh acute viral hepatitis

Avh Avhandlinger (Swedish—transactions)

AvH Alexander von Humboldt

avhrr advanced very-high-resolution radiometer

avi airborne vehicle identification; air velocity index; automatic vehicle identification; aviation

AVI American Virgin Islands; Audio-Visual Institute; Automatic Vehicle Identification

AVI Association Universelle d'Aviculture Scientifique (French—Universal Association of Scientific Aviculture)

Aviaco Aviación y Comercio (Spanish airline)

AVIADOR Avery Videodisc Index of Architectural Drawings on RLIN (Research Libraries Information Network)

AVIANCA Aerovias Nacionales de Colombia (National Airlines of Colombia)

AVIATECA Empresa Guatemalteca de Aviacion (Guatemalan Aviation Enterprise)

avica advanced video endoscopy image communication and analysis

AVID Advancement Via Individual Determination (educational program); American Veterinary Identification Device; Audio-Visual Instruction Department

avigation aircraft navigation

aviob aviation observation

avionics aviation and astronautics electronics

AVIP Association of View-data Information Providers

AVIS Active Vibration Isolation System

AVISCO American Viscose Corporation

AVISPA Aerovias Interamericanas de Panamá (Spanish—Interamerican Airways of Panama)

avit (AVIT) audiovisual instruction(al) technology

av JC avant Jésus Christ (French—before Jesus Christ)—B.C.

avl average versus length

avl. available

av l average length

AVL Asheville, North Carolina (airport)

avla amphibious vehicle landing area; amphibious vehicle launching area

AVLA Audio-Vimsual Language Association

Av Labs Aviation Laboratories (USA)

avlb armored vehicle-launched bridge

AVLH Assam Valley Light Horse

AVLINE Audiovisuals On-Line (computer retrieval system)

avlm anti-vehicle land mine

avloc airborne visible-laser optical communication

AVLS Automatic Vehicle Location System

avlub aviation lubricant

avm automatic voting machine

avm (AVM) arteriovenous malformation

AVM Air Vice Marshal; Allied Victory Medal; guided-missile ship (naval symbol)

AVMA American Veterinary Medical Association; Audio-Visual Management Association

AVMF Aviatsiya Voenno Morskikh Flota (Russian—Soviet Naval Aviation)

avn atrioventricular node; aviation

Avn Avonmouth

AVN Air Vietnam

AVNA Australian Visiting Nurses Association

AVNMED Aviation Medicine (DoD)

av node arterioventricular node

AVNOJ Antifasisticko Vijece Narodnog Oslobidjenja Jugoslavije (Serbo-Croat—Anti-Fascist Council of National Liberation of Yugoslavia)

avnrt atnoventricular nodal re-entrant tachycardia

avo ampere-volt-ohm; avocado

AVO Allam Védelmi-Osztály (Hungarian—Hungarian Secret Soviet Police); avoid verbal orders

Avog Avogadro

avoid airfield vehicle obstacle indication device

AVOIDS Avionic Observation of Intruder Danger System

avoil aviation oil

avoir avoirdupois

avolo automatic voice link observation

avos avocados

avozvots average Australian voters

avp arginine vasopressin

AVP Assistant Vice President; seaplane tender, small (3-letter symbol); Wilkes-Barre/Scranton airport at Avoca, Pennsylvania

AVP Aruba Volkspartie (Dutch—Aruba People's Party)

avpv anteroventral periventricular nucleus

avr aircraft rescue vessel; armored vehicle reconnaissance; aortic valve replacement

AVR Army Volunteer Reserve

AVRA Audio-Visual Research Association

Avram (Hebrew—Abraham)

AVRE Armored Vehicle Royal Engineers

AVRI Animal Virus Research Institute

AVRO A.V. Roe (Ltd.)

AVRO *Algemeene Vereniging Radio Omroep* (Dutch—General Broadcasting Association)

AVROS *Algemeene Vereniging van Rubberplanters ter Oostkust van Sumatra* (Dutch—General Association of Rubber Plantations of the East Coast of Sumatra)

avrp atrioventricular refractory period; audiovisual recording and presentation

AVRS Army Veterinary Remount Service; Audiovisual Recording System

avrt atrioventricular reentrant tachycardia

avs adaptive variable suspension; aerospace vehicle simulation; area vocation school(s)

AVS American Vacuum Society; Association for Voluntary Sterilization; aviation supply ship (naval symbol)

A-V S Anti-Vivisection Society

AVSA African Violet Society of America

AVSC Audio-Visual Support Center (USA)

AVSECOM Aviation Security Command (Philippines)

AVSL Assistant Venture Scout Leader; Association of Visual Science Librarians

avst automated visual-sensitivity test(er)

AVSYCOM Aviation Systems Command (USA)

avt audiovisual tutorial; arginine vasotocin

Avt Allen vision test

AVT Adult Vocational Training; auxiliary aircraft transport (naval symbol); Aviation Medicine Technician

avta automatic vocal transaction analyzer

AVTC Army Vocational Training Center

av tmp average temperature

AVTP Adult Vocational Training Program

AVTRW Association of Veterinary Teachers and Research Workers

avtur aviation turbine fuel

AVUS *Automobile Versuchs and Untersuchungs Strecke* (German—Automobile Test Track)

avv *avvocato* (Italian—advocate)—lawyer

av vales atrioventricular (heart) valves

av w average width

AVWV *Antilliaans Verbond van Werknemers Verenigingen* (Dutch—Antillean Confederation of Workers' Unions)

AVX Avalon Bay, Catalina Island, California (airport)

aw abandoned woman; above water; acid waste; actual weight; air-to-water; air warning; air weapon; amphibious warfare; anterior wall; antiwear; atomic warfare; atomic weight; automatic weapons(s)

aW attowatt

Aw *Antwerpen* (Dutch—Antwerp)

a/w actual weight; all-water; all-weather

a & w alive and well

AW Air Work, Ltd.; American Welding; Articles of War; Aruba (Internet code); distilling ship (naval symbol)

A-W Addison-Wesley Publishers

A & W Atlantic & Western (railroad)

awa absent without authority; advise when able

AWA Air Warfare Analysis; All-Weather Attack; Aluminum Wares Association; America West Airlines, American Warehousemen's Association; American Watch Association; American Waterfowl Association; American Wine Association; American Woman's Association; American Wrestling Alliance; Army Wives Association; Association of Women in Architecture; Aviation/Space Writers Association

AWA *All the World's Aircraft*

awac airborne warning and control

AWAC American Women's Army Corps

AWACS Airborne Warning and Control Systems

AWADS All-Weather Aerial Delivery System

AWAE Automotive Wholesalers Association Executives

AWAIK Abused Women's Aid in Crisis

AWAL American-West African Line

AWAM Association of West African Merchants

AWANS Aviation Weather and Notice to Airmen System

awar area-weighted average resolution

aware advance warning equipment

AWARE Actively Working Against Racism and Ethnocentrism; Addiction Workers Alerted to Rehabilitation and Education (NYC); Association for Women's Active Return to Education

AWARS Airborne Weather and Reconnaissance System

AWAS Acoustic Wave Analysis System; Australian Women's Army Service

AWASA Australian Women's Army Service Association

AWASM Associate of the Western Australia School of Mines

awasp advanced weapons ammunition supply point

awb air waybill

AWB Agricultural Wages Board (UK); Arizona White Battalion (neo-Nazi group); Australian Wheat Board; Australian Wine Board; Australian Wool Board; Australian Wool Bureau

AWBA American Wholesale Booksellers Association; American World Boxing Association

AWB/CN Air Waybill or Consignment Note

AWC Air War College; American Watershed Council; Air Warfare Coordination; American Wool Council; Anaconda Wire & Cable (stock exchange symbol); Area Wage & Classification (office); Arizona Western College; Army War College; Army Weapons Command; Army Work Corps; Australian Whaling Commission; Australian Wool Corporation

AWC Amgueddfa Werin Cymru (Welsh Folk Museum)

AWCC Australian Wine Consumer's Cooperative

awcls all-weather carrier landing system

awcm acoustic warfare countermeasures

AWCO Area Wage and Classification Office

awcs agency-wide coding structure

AWCS Air Weapons Control System

AWCSSI Army War College Strategic Studies Institute

AWCU Association of World Colleges and Universities

awcw airborne warning and control system

awd all-wheel drive; award

AWD Action for World Development; Air Warfare Division; Air Worthiness Division

AWDA Automotive Warehouse Distributors Association

awdats artillery weapons data transmission system

AWDCS Alternate Waste Disposal Concepts Study

awdr advanced weapon-delivery radar

awdrey atomic weapons detection, recognition and yield

awe accepted weight estimate; advise when established; average weekly earnings

AWEA American Wind Energy Association; Australian Wind Energy Association

AWEASVC Air Weather Service

AWED American Woman's Economic Development

A Weld I Associate of the Welding Institute

AWES Army Waterways Experiment Station; Association of Western European Shipbuilders

awf all-weather fighter; awful(ly)

a wf acceptable work-load factor; adrenal weight factor

AWF African Wildlife Foundation; American Wildlife Foundation; Australian Wheat-growers Federation

AWFS All-Weather Fighter Squadron

AWG American Wire Gage; Art Workers Guild; Australian Writers Guild

AWH Association of Western Hospitals; Australian Women's Hospital

AWHA Australian Women's Home Army

AWhitman Albert Whitman Company

AWHPS Association of White House Press Secretaries

awi anterior wall infarction

AWI Air Warfare Instructor; All-Weather Interceptor; American Watchmakers Institute; Animal Welfare Institute; Architectural Woodwork Institute; Australian Welding Institute; Australian Wire Industry; Australian Wool Industries

AWIA American Wood Inspection Agency

awips advanced weather information processing system

AWIPS Advanced Weather Interactive Processing System

AWIRA American Wax Importers and Refiners Association

AWIS Association of Women in Science

AWIU Agricultural Workers Industrial Union; Allied Workers International Union; Aluminum Workers International Union

awiy as we informed you

awk awkward

AWK Wake Island (airport)

awkm a wonderfully knowledgeable man

awl absent with leave; administrative weight limitation; allweather landing; artesian well lease

Awl Soviet infrared or radarguidance system (NATO nickname)

AWL Animal Welfare League

AWLC Association of Women Launderers and Cleaners

AWLF African Wildlife Leadership Foundation

AWLNET Area Wide Library Network

AWLOGS Army Wholesale Logistic System

AWLS All-Weather Landing System

awm automatic washing machine; awaiting maintenance

AWM American War Mothers; Association of Women Mathematicians; Australian Wallcovering Manufacturers; Australian War Memorial

AWMA American Wholesale Marketers Association

AWMF Andrew W. Mellon Foundation

awmi anterior wall myocardial infarction

AWMPF Australian Wool and Meat Producers Federation

awn awning

AWN Activation Work Notice; Automated Weather Network

AWngSvc Air Warning Service

AWNL Australian Women's National League

AWNP American White Nationalist Party

AWNY Advertising Women of New York

AWO Accounting Work Order; Admiralty Weekly Order; American Waterways Operators

awol a wolf on the loose; absent without leave; absent without official leave

AWOP All-Weather Operations Panel

awp amusements with prizes; automatic withdrawal plan

A & WP Atlanta and West Point (railroad)

AWPA Academy of Wind and Percussion Arts; American Wire Producers Association; American Wood Preservers Association; American Wood Products Association; Australian Women Pilots Association

AWPB American Wood Preservers Bureau

A-WPC Addison-Wesley Publishing Company

AWPI American Wood Preservers Institute

AWPL Australia-West Pacific Line

AWPPW Association of Western Pulp-Paper Workers

AWPs Allied Weather Publications

awr adaptive waveform recognition

AWR Arctic Wildlife Refuge (Alaska); Australian Wire Rope

AWRA American Water Resources Association; Australian Welding Research Association; Australian Wool Realization Agency

AWRC Australian Water Resources Council

AWRE Atomic Weapons Research Establishment

AWRF American Wire Rope Fabricators

AWRIS Army War Room Information System

AWRNCO Aircraft Warning Company (Marines)

AWRO Atomic Weapon Retrofit Order

AWRT American Women in Radio and Television

aws adjustable wire stripper; advanced warning system

AWS Air Warning Service; Air Warning Squadron; Air Warning System(s); Air Weapon Systems; Air Weather Service; Aircraft Warning Service; Aircraft Warning System; Alston Wilkes Society; American War Standards; American Watercolor Society; American Weather Service; American Welding Society; Amphibious Warfare School; Army Welfare Service; Association of Winery Suppliers; Atlas Weapon System; Attack Warning System; Automatic Warning System; Automatic Weather Station; Aviation Weather Service

AWSA American Water Ski Association; American Woman Suffrage Association; Association of Wisconsin School Administrators

AWSCPA American Woman's Society of Certified Public Accountants

AWSF Australian Wholesale Softgoods Federation

AWSG Army Work Study Group

AWSJ Asian Wall Street Journal

AWSM Association for Women in Sports Media

AWSOS Army Women's Services Officers School

AWSP Association of Washington School Principals

AW & ST Aviation Week & Space Technology

awt advanced waste treatment

AWT Aero-elastic Wind Tunnel; Arctic Warfare Training; Associate in Wildlife Technology

AWTA American Working Terrier Association; Australian Wool Testing Authority

AWTAO Association of Water Transportation Accounting Officers

AWTC Army Watership Training Center

AWTE Association for World Travel Exchange

AWTEW All's Well That Ends Well

AWTI Air Weapons Training Installation

awtss all-weather tactical strike system

awu atomic weight unit

AWU Aluminum Workers Union; Australian Workers Union

AWVS American Women's Volunteer Service

aww average weekly wage

AWWA American Water Works Association; Asian Women's Welfare Association; Australian Water and Wastewater Association

awwf all-weather wood foundation(s)

AWWU American Watch Workers Union

awx (AWX) all-weather aircraft

awy airway

ax attack experimental (aircraft)

ax. axiom(atic); axes; axis

AX A-12 fighter plane; American Air Export & Import Company (stock exchange symbol)

A/X Armament Exchange

AXAF Advanced X-ray Astrophysics Facility

axbt aircraft-expendable bathythermograph

axd auxiliary drum

axe (AXE) automatic electronic exchange

AXF Advanced X-ray Facility

axfl axial flow

axgrad axial gradient

axidnt accident

axio axiological(ly); axiologist; axiology; axiom; axiomatic(al)(ly)

axmin(s) axminster(s)

AXO Assistant Experimental Officer

Axon Axelson (Swedish—son of Axel)

AXP Allied Exercise Publication

Ay Ayala

Ay Ayios (Modern Greek—Holy)

AY Allied Youth; Ayrshire Yeomanry

AYA American Yachtsmen's Association

ayat ayatollah (Arabic—interpreter of Muslim law)—a religious leader among Shiite Muslims

AYC Albany Yacht Club; American Yacht Club; American Youth Congress; Arthur Young & Company; Atlantic Yacht Club; Audubon Yacht Club

AYD American Youth for Democracy

ayer (Malay—water); (Spanish—yesterday)

Ayer NW Ayer and Son

Ayers Ayers Rock National Park, in Australia's Northern Territory, features a colossal red sandstone rock—the world's largest monolith

ayf anti-yeast factor

AYF Australian Yachting Federation

AYH American Youth Hostels

AYHA American Youth Hostel Association; Australian Youth Hostels Association

AYI Academic Year Institute (NSF)

Ayla Aylett; Aylmar; Aylmer; Aylsworth; Aylward; Aylwin

AYLC Association of Young Launderers and Cleaners

Aym Aymara

AYM Ancient York Mason; Ancient York Masonry

AYM-YWHAs Association of Young Men-Young Women's Hebrew Associations of Greater New York

AYP Alaska-Yukon Pacific; Alaska-Yukon Pioneers

AYPE Alaska-Yukon Pacific Exposition

ayr all-year 'round

Ayr Ayrshire

AYSA American Yarn Spinners Association

AYSC Army Youth Selection Center

AYSO American Youth Soccer Organization

AYT Army Youth Team

aytng anything

az azure

az (AZ) azimuth

a Z aan Zee (Dutch—on sea); *auf Zeit* (German—on account, on credit)

a/z aan zee (Dutch—on sea)

Az azimuth; Azores; Aztec; Aztecan; azure

AZ Azote (Greek—nitrogen)

AZ Active Zone; Alitalia (Italian International Airlines); Alzheimer's disease; Arizona; Azerbaijan (Internet code)

A-Z Ascheim-Zondek (pregnancy test)

A to Z from the beginning to the end; thoroughly

AZ Akademisch Ziekenhuis (Dutch—Academic Hospital)

AZA American Zionist Assocmation

Azania South Africa's name according to African nationalists

AZAPO Azania People's Organization (militant South African Blacks)

azas adjustable-zero adjustable span

Azb Azerbaijan; Azerbaijani; Azerbaijanian

AZC American Zionist Council

azel azimuth elevation

Azer Azerbaijan

AZF American Zionist Federation

azg azaguanine

AZGS Azusa Ground Station

azi azimuth

Az I Azores Islands

AZI American Zinc Institute

Azië (Dutch—Asia)

az ld azure laid (paper)

azm azimuth

azon azimuth only

Azores Azores Islands in the North Atlantic far to the west of Portugal

Azorín José Martínez Ruiz

Azov Sea of Azov (landlocked body of water between Ukraine and Russia)

Azr Azores

azran azimuth and range

AZRI Arid Zone Research Institute

azrock asbestos rock

Azru Aztec Ruins National Monument

azs automatic zero set

AzSEA Arizona Society of Enrolled Agents

azt azusa transponder

azt (AZT) azidothymidine, one of several drugs used in the treatment of AIDS

Azt Aztec; Aztecan

AZT Ascheim-Zondek Test

aztc azusa transponder coherent

Aztec Ruins Aztec Ruins National Monument, New Mexico

A-Z Test Ascheim-Zondek Test (for pregnancy)

aztran azimuth from transit

Azuero Pen Azuero Peninsula in northwestern Panama

azur azauridine

azusa azimuth, speed, altitude

az wo azure wove (paper)

azy azyme (matzos, unleavened bread)

B

b bachelor; baby; banns; base; bass (instruments or voices); bats; bicuspid; birth; bituminous; black; blinkers; blue; boatswaine; bondsman; book; born; brass; breadth; bridge; brother; bulb (camera exposure device); bus; wing span (symbol)

b. *bis* (Latin—twice)

b *bas* (French—down); span

B annual rate plus stock dividend (symbol in newspaper stock listings); Bacillus; bad; balboa (Panamanian currency); Baltic; bandwidth; Barber Lines; bass; bassoon; bastard; Baume; Baume scale; bay; Beatrice (Beatrice Foods); Beech; Belgium (auto plaque); belted; Bendix; Benoist scale; unit of marijuana measurement consisting of just enough to fill a small matchbox; benzene; body; Boeing; boils at; bolivar (Venezuelan currency); boliviano (Bolivian currency); bomber; bonded; borderline; boron; Boston; bowels; Bravo—code for letter B; British; brightness (symbol); Brother; Bruning; Buddhist; Bulb (camera setting); Bull Lines; buoyancy; Burroughs; contrabass; flux density (symbol); Fraunhofer line caused by terrestrial oxygen

B/ balboa (Panamanian currency unit)

°B degrees Baumé

B *Baai* (Afrikaans or Dutch—bay); *Bad* (German—bay); *Bahia* (Spanish—bay); *Baia* (Portuguese—bay); *Baie* (French—bay); *Baja* (Spanish—lower); *bajar* (Spanish—to descend) *Ban* (Indo-Chinese—bay); *Bay* (Turkish—Mister) *Bir* (Arabic—cistern, well); *Bucht* (German—bay); *bueno* (Spanish—good); *Bukhta* (Russian—bay)

B' *Ben* (Hebrew—son, son of)

b 1 booster 1

B+ B-plus; B-positive

B–1 B-minus; B-negative (battery terminal); North American-Rockwell strategic supersonic bomber

B₁ thiamine vitamin

B-1B Rockwell International long-range supersonic swept-wing bomber

b 1 p booster 1 pitch

b 1 y booster 1 yaw

b 2 booster 2

b2b (B2B) business to business

b2c (B2C) business to consumer

B₂ riboflavin vitamin

B-2 Advanced Technology Fighter aircraft

B2F Boeing 320 fanjet airplane; Boeing 720 fan jet airplane

b 2 p booster 2 pitch

b 2 y booster 2 yaw

B3F Boeing 320 aircraft

b4 before

b7d buyer (has) seven days (to pay)

B7D buyer has seven days to pay

B7F Boeing 707 fan jet airplane

B8H Boosey and Hawkes

B-25 World War II light bomber called the Mitchell

B-26 modernized Douglas B-26 Invader renamed Counter Invader

B-47 Stratojet all-weather strategic medium bomber

B-52 Stratofortress all-weather intercontinental strategic heavy bomber

B-57 Canberra two-place twin-engine turbojet all-weather tactical bomber

B-58 Hustler strategic all-weather supersonic bomber

B-66 Destroyer twin-engine turbo-jet tactical all-weather light-bombardment aircraft

B 77 Bratislava 77 (viral) strain

B-707 one of a Boeing aircraft series

B-747 Boeing jumbo jet aircraft

ba baby addict; bachelor; bad attitude; baptized; base line; bath; bath(room); batting average; blind approach; bomb aimer; broken appointment

b-a bare ass(ed); naked; unclothed

b/a backache; billed at; boric acid; budget authority; budget authorization; budget authorized

b.a. *balneum arenae* (Latin—sand bath)

Ba Baia (Portuguese—Bahia); barium (symbol)

BA Bank of America; Basic Airman; Bellas Artes (Fine Arts); Belt Association; Bengal Army; Berkshire Athenium; Bettmann Archive; Bombay Army; Boreal-Asiatic; Boston & Albany (railroad); British Academy; British Admiralty; British Airways (airline code); British Army; British Association (for the Advancement of Science); Budget Authority; Buenos Aires; Bureau of Accounts; Bureau of Apprenticeship; Busted Aristocrat (an officer reduced to the ranks)

B-A Basses-Alpes

B.A. *Baccalaureus Artium* (Latin—Bachelor of Arts)

B/A Bank of America; British American (oil company)

B & A Bangor & Aroostook (railroad); Boston & Albany (railroad)

BA Bayerische Landesbank (German—Bavarian National Bank); *Biological Abstracts; Bonne Action* (French-Good Deed); *Bowker Annual; Business Automation; Britannica Atlas*

baa benzoyl arginine amide; bleat of a sheep; brigade administrative area

Baa Baal; Baalam

BAA Brewers Association of America; British Acetylene Association; British Airports Authority; British Archeological Association; British Astronomical Association; Bureau of African Affairs

B.A.A. Bachelor of Applied Arts

BAAA British American Arts Association; British Association of Accountants and Auditors

BAAB British Amateur Athletic Board

BAAC Bank of Agriculture and Agricultural Cooperatives (Thailand)

BAADS Bangor Air Defense Sector

BAAF Brigade Airborne Alert Force

BAAG British Army Aid Group

Baal Baalbek

BAAL Black Academy of Arts and Letters; British Association for Applied Linguistics

BAAP Bilateral Aid to Asia and Pacific (program)

BAAR Board of Aviation Accident Research

BAAS British Association for the Advancement of Science

bab (Arabic—gate, strait)

Bab Barbara; Babbitt; Baboo (Anglo-Inthan dialect); Babylon; Babylonia; Babylonian; WS Gilbert

BAB British Airways Board; BT Babbitt (Babo cleanser)

BABA British Antiquarian Bookseller's Association

Babars Babar Islands of Indonesia

babb babbit metal

Babbie Barbara

babbitt babbitt metal (named for its inventor, Isaac Babbitt of Taunton, Massachusetts)

Babb's computer Babbage's computer (the first computer)

BABEL Broadcasting Across the Barriers of European Languages

Babette Elizabeth

BABI Brooke Army Burn Institute (San Antonio, Texas)

Babines Babine Mountains of British Columbia

bab met babbitt metal

Babo Boolean approach for bivalent optimization; BT Babbitt detergent scouring powder

babord (Norwegian—larboard) port side

Babs blind approach beacon system

Bab(s) Barbara

BABS Babbage Society; Blind-Approach Beacon System

BABT Brotherhood of Associated Book Travelers

Baby Babylon(ia); Babylonian

bac bacilli; bacillus; bacteria; bacterial; bacterial antigen complex; bacteriologist; bacteriology; blood-alcohol concentration; blood-alcohol content; buccoaxiocervical

bac **(BAC)** binary asymmetric channel

bac bachot (French abbreviations for baccalaureat)—*bachot* also means ferryboat

Bac. Baccalaureus (Latin—Bachelor)

BAC Bendix Aviation Corporation; Black Action Committee; Blair Athol Coal; blood alcohol concentration, Boeing Airplane Company; born-again Christian; British Aircraft Corporation; British Association of Chemists; Bureau of Air Commerce; Business Advisory Council (U.S. Department of Commerce); Business Archives Council of Australia

BAC Baile Atha Cliath (Gaelic—Dublin)

BAC-145 British Jet Provost trainer aircraft

BACA British Aerospace Commercial Aircraft; Business and Consumer Affairs

BACAH British Association of Consultants in Agriculture and Horticulture

BACAIC Boeing Airplane Company Algebraic Interpretive Computing

BACAICS Boeing Airplane Company Algebraic Interpreter Coding System

BACAL Butter and Cheese Association Limited

BACAN British Association for the Control of Aircraft Noise

bacat barge aboard catamaran

BACAT Barge Canal Traffic

bac bag bactine bag

B.Acc. Bachelor of Accountancy

BACC British-American Collector's Club

BA & CC Billiards Association and Control Council

BACCHUS Boost Alcohol Consciousness Concerning Health of University Students; British Aircraft Corporation Commercial Habitat Under the Sea

BACD Boeing Airplane Company Design

bace basic automatic checkout equipment

BACE British Association of Consulting Engineers; Bureau of Agricultural Chemistry and Engineering

BACGA British-Australian Cotton Growing Association

bach bachelor

Bach (German—brook, stream)

B.A. Chem. Bachelor of Arts in Chemistry

bach girl(s) bachelor girl(s)

bachot (French—baccalaureat, also means "ferryboat")

Bach Soc Bach Society

BACI Brazilian-American Cultural Institute

BACIE British Association for Commercial and Industrial Education

back. backwardation

'backs wetbacks (illegal immigrants from Mexico)

BACM British Association of Colliery Management

BACMA British Aromatic and Compound Manufacturers Association

BACNATO British Atlantic Committee of NATO

BACO British Aluminum Company

BACP Blair Athol Coal Pty

BACS Ben Asia Container Service

bact bacteria; bacteriological; bacteriologist; bacteriology; bacterium

BACT Best Available Control Technology

bacter bacteriologist

bacteriol bacteriologic(al)(ly); bacteriologist; bacteriology

bactrian bactrian camel (two-hump camel of Asia)

BACU Battle Area Control Unit

Bad Badajoz

Bad (German—bath)—short form for more than a hundred Austro-German hydrotherapeutic resorts ranging from *Bad Abbach* to *Bad Zwischenahn*

BAD Bantu Administration and Development; Base Air Depot; Benefit Assessment District; Berlin Airlift Device; Black, Active, and Determined; British Army Delegation; British Association of Dermatology

BADA Base Air Depot Area; British Antique Dealers' Association

BADAS Binary Automatic Data Annotation System

badc binary asymmetric dependent channel

BADD Bartenders Against Drunk Driving; Blacks Against Drunk Driving; Bothered About Dungeons and Dragons

baddies bad ones

baddie(s) bad guy(s)—incorrigible criminal(s)

Baden Baden-Baden

Baden (German—Baths)— *Baden bei Wien* (Baden near Vienna) and Baden-Baden near Karlsruhe, Germany

BADGE Basic Air Defense Ground Environment

BADGES Base Air Defense Ground Environment System

badhouse bawdyhouse

Badian(s) Barbadian(s)

B.Admin. Bachelor of Administration

Bad Reich Bad Reichenhall (German bath close to Salzburg, Austria)

BADS British Association of Dermatology and Syphilology

Bad-Württemberg (German—Baden-Württemberg)

bae Beacon antenna equipment

Ba e barium enema

Bae British Aerospace

BAE Board of Architectural Examiners; Bureau of Agricultural Economics; Bureau of American Ethnology

BA of E Badminton Association of England

B.A.E. Bachelor of Aeronautical Engineering; Bachelor of Agricultural Engineering; Bachelor of Architectural Engineering; Bachelor of Art Education; Bachelor of Arts in Education

BAE Buque Armada Ecuatoriana (Spanish—Ecuadorian Naval Ship)

BAEA British Actors' Equity Association

BAEC British Agricultural Export Council

B.A.Econ. Bachelor of Arts in Economics

B.A.Ed. Bachelor of Arts in Education

BAED British Airways European Division

B.Ae.E. Bachelor of Aeronautical Engineering

BAEF Belgian-American Educational Foundation; British-American Educational Foundation

Ba enem barium enema

BAEng. Bureau of Agricultural Engineering

baer (BAER) brainstem auditory-evoked response

baf baffle; bunker adjustment factor

ba & f budget, accounting, and finance

BAF Admiral Bradley Allen Fiske (inventor of many naval instruments); British Air Force; Burma Air Force; Burundi Air Force

BAFCom Basic Armed Forces Communication Plan

BAFF British Air Forces in France

bafflegab ambiguous expressions *(see* Bafflegab addendum)

BAFG British Air Forces in Greece

bafgab bafflegab—*see* Bafflegab addendum

BAFL Baltic-American Freedom League

BAFM British Association of Forensic Medicine; British Association of Friends of Museums

BAFMA British and Foreign Maritime Agencies

BAFO British Air Forces of Occupation; British Army Forces Overseas

BAFS British Academy of Forensic Science

BAFSEA British Air Forces, South-East Asia

BAFSC British Association of Field and Sports Contractors

BAFSV British Armed Forces Special Vouchers

BAFT Bankers Association for Foreign Trade

BAFTA British Academy of Film and Television Arts

BAFTM British Association of Fishing Tackle Makers

bag. bagasse; baggage; ballistic attack game; buccoaxiogingival

Bag Baghdad

B.Ag. Bachelor of Agriculture

BAG Beaverbrook Art Gallery

BAGA British Amateur Gymnastics Association

BAGBI Booksellers Association of Great Britain and Ireland

BAGDA British Advertising Gift Distributors Association

B.Ag.E. Bachelor of Agricultural Engineering

bagg buffered azide glucose glycerol

B.Agr. Bachelor of Agriculture

BAGR Bureau of Aeronautics General Representative

B.Agr.Eco. Bachelor of Agricultural Economics

B.Agric. Bachelor of Agriculture

BAGS Bombing and Gunnery School

B.Ag.Sc. Bachelor of Agricultural Science

Bah Bahamas; Bahia; Bahrain

BAH Bahrain Island, Persian Gulf (airport); British Airways Helicopters

B-A H British-American Hospital

Baha'l (Abdul) Baha Bahai

Bahm Bahamas (whose capital is Nassau)

Bahamas Commonwealth of the Bahamas off coast of Cuba (discovered by Columbus on October 12, 1492)

Bahams Bahamas

Bahia São Salvador de Bahia

Bah Ind Bahasa Indonesian (national language)

BAHOH British Association of the Hard of Hearing

BAHPA British Agricultural and Horticultural Plastics Association

Bahr Bahrain (whose capital is Manama)

Bahrain Bahrain Island in the Persian Gulf

Bahrains Bahrain Islands in the Persian Gulf between Qatar and Saudi Arabia

BAHS British Agricultural History Association

baht unit of currency in Thailand

Ba I Bahama Islands

BAI Bank Administration Institute; Bank of America International; Barrier Industrial Council; Book Association of Ireland; British Airports International; Bureau of Animal Industry

B.A.I. *Baccalaureus in Arte Ingeniaria* (Latin—Bachelor of Engineering)

baib beta-amino-isobutyric (acid)

BAIC Bureau of Agricultural and Industrial Chemistry

baid boolean array identifier

BAIE British Association of Industrial Editors

BAINS Basic Advanced Integrated Navigation System

BAIO British Air Intelligence Officer

bait. bacterial automated identification technique

B.A.J. Bachelor of Arts in Journalism

Baja Baja California (Spanish—Lower California)

Bajan Barbarian (inhabitant of Barbados); Barbadian English; Barbadian people; pertaining to Barbados

Baja Norte Baja California Norte (Spanish—Northern Baja California—Mexican state including Ensenada, Mexicali, Tecate, and Tjuana

Baja Sur Baja California Sur (Spanish—Southern Baja California)—Mexican territory including Cabo San Lucas, La Paz, and Loreto

B.A. Jour. Bachelor of Arts in Journalism

Bajuns Barbadans

bak bakery

Bak Baku, Azerbaijan

baka African pygmy

Bakery Workers Bakery and Confectionery Workers International Union of America

baktr baktrisch (German—Bactrian)

bal balance; balcony; baloney; blood-alcohol level; broad absorption line

bal (BAL) basic assembly language (computer programming)

Bal Baleares; Ballarat; Balthasar; Baltimore; British anti-lewisite

BAL Baltimore Lord Baltimores; Baltimore, Maryland (Friendship Airport); Barclays Australia Ltd; Belgian African Line; Bonanza Airlines (3-letter coding); Borneo Airways Ltd.

bala balalaika

balance. basic and logically applied norms-civil engineering

Balanchine Georgy Melitonovich Balanchivadze

bal. arenae balneum arenae (Latin—sandbath)

balast balloon astronomy

Balaton Lake Balaton, central Europe's largest lake, nicknamed the Hungarian Ocean

Balb Balboa

balboa unit of currency in Panama

balc balconette; balconied; balcony

Bald Baldwin

Baldie Archibald; Baldassare; Baldomero; Balduin; Baldur; Baldwin; Baldwina

baldie(s) bald person(s)

Baldy Baldemar; Baldram; Baldred; Baldrey; Baldric; Baldur; Baldwin

Balearics Balearic islands of Majorca, Minorca, Ibiza, Formentera, Aire, Aucanada, Botafoch, Cabrera, Dragonera, Pinto, and El Rey

Balgol Burroughs algebraic compiler

BALH British Association for Local History

balid ballistics identification

Balkans Balkan mountain peoples, and states in south eastern Europe (Albania, ... garia, Greece, Romania, ... key, Serbia, and Monten...

ball. ballast

Ball Ballerup; Balliol Oxford

Ball Coll Balliol Coll... ford

Ballenys Balleny Isl...

ballots. bibliograph... tion of large libra...

ballute balloon pa...

bally ballyhoo

Ballyhouras Ba... of southern I...

bat. mar. b a l ... (Latin—sal... water bath...

Bal-Mol B a... (.45-ca... semi-au...

BALO B... ficer

balop b...

B Alp b...

balpa

bal...

BAI...

B-alpes Basses-Alpes
bals balsam
bals. balsamum (Latin—balsam)
Bals Balearic Islands
B.A.L.S. Bachelor of Arts in Library Science
BALSA Black American Law Students Association
Balt Balthasar; Baltic; Baltimore
BALTAPs Baltic Approaches (NATO)
balth balthazar (16 bottle capacity)
balthum balloon temperature and humidity
Balti Baltimore (slang)
Baltic Baltic and Mercantile Shipping Exchange (in London); Baltic Sea
Baltic(s) Estonia, Latvia, Lithuania
Baltis Baltistan (Karakoram Range between N Kashmir and NW Tibet)
Balto Baltimore
Balts Baltic countries and peoples on the Baltic Sea (Denmark, Estonia, Finland, Germany, Latvia, Lithuania, Poland, Russia, Sweden); Baltic peoples; Balto-Slavs (East Prussians, Estonians, Latvians, Lithuanians); Balto-Slavic-speaking peoples
Balt Sym Baltimore Symphony
Baluch Baluchistan
baluchi tribal rug (often woven in Afghanistan)
balun balance-to-balance (network)
balute balloon parachute
bal. vap. balneum vapour (Latin—steambath, vapor bath)
bam bio-activated mineral lotion; broadcasting am
Bam Bamberger
BAM Baikal-Amur-Magistral (railroad); BankAmerica Corporation (stock-exchange symbol); Basic Access Method; broadcasting AM; Brooklyn Academy of Music
B-A-M Baikal-Amur-Magistral (railway of eastern Siberia)
'Bama Alabama
BAMA British Air Motoring Association; British Amsterdam Maritime Agencies
bamb bamboo

bambi (BAMBI) ballistic missile bombardment interceptor
bame benzoylarginine methyl ester
BAMIRAC Ballistic Missile Radiation Analysis Center
BAMM Black Afro Militant Movement
BAMO BuAer Material Officer
bamp basic analysis and mapping program
BAMR BuAer Maintenance Representative
BAMS British-American Minesweepers
B.A.M.S. Bachelor of Ayurvedic Medicine and Surgery
BAMTM British Association of Machine Tool Merchants
B.A.Mus. Bachelor of Arts in Music
BAMW British Association of Meat Wholesalers
ban. best asymptotically normal; bond anticipation note
Ban Bancor (Venezuela); Bantu; Byron Bancroft Johnson
BAN Base Activation Notice; British Association of Neurologists
BAN Biblioteka Akademii Nauk (Russian—Library of the Academy of Sciences)—St Petersburg
Banaca Banco Nacional de Credito Agrícola (Spanish National Agricultural Credit Bank)—Mexico
Banace Banco Nacional de Credito Ejidal (Spanish—Public Land Credit National Bank)—Mexico
Banacoex Banco Nacional de Comercio Exterior (Spanish—National Bank of Foreign Commerce)
Banafoco Banco Nacional de Fomento Cooperativo (Spanish—National Bank for Cooperative Promotion)
Banamex Banco Nacional de México (Spanish—National Bank of Mexico)
banana build almost nothing anywhere near anyone
Bananagate Honduran scandal involving high officials bribed to lower export taxes on bananas

Banana Republics countries of Central and northern South America; Jamaica often included
Banape Banco Nacional de la Pequeña Empresa (Spanislm—National Small Business Bank)—Mexico
BANC British Association of National Coaches
Banco *El Banco* (Spanish—The Bank)—World Bank for Reconstruction and Development
Bancomer Banco de Comercio (Spanish—Bank of Commerce)
Bancoop Banco Nacional de las Cooperativas del Perú (Spanish—National Bank of Peruvian Cooperatives)
Banc. Sup. Bancus Superor (Latin—Upper Bench)—King's or Queen's Bench
Band Bandung
Bandas Banda Islands of Indonesia
Bandeirante Brazilian 12-passenger transport honoring frontier pioneers, Bandeirantes
Bandelier Bandelier National Monument and cliff-dweller Indian reservation in New Mexico west of Santa Fe
Banffs Banffshire
BANEWS British Army News Service
Bang Bangalore; Bangladesh (whose capital is Dhaka)
Bangla Bangladesh (formerly East Pakistan)
Bangladesh People's Republic of Bangladesh—formerly East Pakistan
bang(s) bombing(s); explosion(s)
bang torp bangalore torpedo
BANHC Bengal Army Native Hospital Corps
banir bombing and navigation inertial reference
Banjul formerly Bathurst, The Gambia
bank. banking
Bank Bangkok; The Bank of England
BANK International Bank for Reconstruction and Development
BankCal Bank of California
bankcy bankruptcy

Bankers Bankers Publishing (Boston); Bankers Trust (New York)

Bankhead Bankhead National Forest in northwest Alabama

BANKPAC Bankers Political Action Committee

banks. bank holidays (West Indian English)

Banks the Banks (shallow fishing banks offshore Canada—the Grand or Newfoundland Banks, or Georges Banks off New England)

banks clgs bank clearings

bankster(s) banker gangster(s) who deprive(s) you of your deposits under the cloak of bank management

Bann Bannockburn

Banobras Banco Nacional de Obras y Servicios Publicos (Spanish—National Bank of Public Works and Services)—Mexico

Banpeco Banco Peruano de los Constructores (Spanish—Peruvian Constructor's Bank)

Banrural Banco Nacional de Credito Rural (Spanish—National Rural Credit Bank)—Mexico

ban's bond anticipation notes

BANS Basic Air Navigational School; Bright Alphanumeric Subsystem; British Association of Numismatic Societies

BANSA Banco del Ahorro Nacional (Spanish—National Savings Bank)—Mexico

BANSW Band Association of New South Wales

Bantam Bantam Books; Swedish antitank guided missile

BANTSA Bank of American National Trust and Savings Association

BANU British Army News Unit

B.A. Nurs. Bachelor of Arts in Nursing

BANWR Bosque Apache National Wildlife Refuge (New Mexico)

BANZ Bahrain-New Zealand Trading and Storage Company

BANZARE British, Australian, New Zealand Antarctic Research Expedition

bao basal acid output

BAO Bankruptcy Annulment Order; British Army of Occupation; British Association of Otolaryngologists; British-American Oil

B.A.O. Bachelor of the Art of Obstetrics; Bachelor of Arts in Oratory

BAOD British Airways Overseas Division

bao-mao basal acid output to maximal acid output (ratio)

BAOP British Atlantic Ocean Possessions (Ascension, St Helena, and Tristan da Cunha islands)

BAOR British Army on Rhine

bap baptism; baptized; base auxiliary power; beginning at a point; biotechnology action program; blood-agar plate; brachial artery pressure

bap billets a payer (French—bills payable)

Bap Baptist; Baptista; Baptiste

BAP Bankers Association of the Philippines; Booksellers Association of Philadelphia

B A & P Butte, Anaconda & Pacific (railroad)

BAP Barco de la Armada Peruana (Spanish—Ship of the Peruvian Navy)

BAPA British Airline Pilots' Association

BAP & C British Association of Print & Copyshops

BAPCO Bahrain, Petroleum Company

bapd base ammunition and petroleum depot

bape baseplate

B.A.P.E. Bachelor of Arts in Physical Education

BA Phys Med British Association of Physical Medicine

BAPL Bettis Atomic Power Laboratory (AEC)

BAPM British Association of Physical Medicine

BAPO British Army Post Office

B.App.Arts Bachelor of Applied Arts

B.App.Sci. Bachelor of Applied Science

BAPS Beacon-Automated Processing System; British Association of Pediatric Surgeons; British Association of Plastic Surgeons; Bureau of Air Pollution Sciences

BAPSA Broadcast Advertising Producers Society of America

bapt. baptism; baptized

Bapt Baptist

BAPT British Association of Physical Training

baq basic allowance for quarters

BAQ Barranquilla, Colombia (airport)

BAQC Bureau of Air Quality Control

bar. baritone; barometer; barometric

bar (BAR) battery acquisition radar; buffer address register

bar billets à recevoir (French—bills receivable)

Bar Baroque; Baruch, Book of

Bar Barone (Italian—Baron)

B.Ar. Bachelor of Architecture

BAR Base Address Register (computer); Board of Airline Representatives (Singapore); Board of Anthropological Research; Broadcast Advertisers' Reports; Browning automatic rifle; Bureau of Aeronautics Representative; Bureau of Automotive Repair

BARA Bureau d'Analyse et de Recherche Appliquées (French—Bureau of Analysis and Applied Research)

Barak Israeli version of the French Mirage military aircraft

barb. barbarian; barbary ape; barbary horse; barbecue; barber; barbiturate

Barb Barbados Islands; Barbara; Barbary

barbi baseband radar bag initiator

Barbie Barbara

barbies barbiturates

bar-b-q barbecue

Barbra Barbara

barbs. barbiturates

barbus barbudos (Spanish—bearded ones)

barc barge amphibious resupply cargo; barometric rate computed

Barc Barcelona; Barclay

BARC Bay Area Reference Center; Bay Area Research Collective; British Aeronautical Research Committee

Barca Barcelona

Bare Con Cam Barcelona Conjunto Cameristico (Spanish—Barcelona Chamber Ensemble)

B.Arch. Bachelor of Architecture

B.Arch.E. Bachelor of Architectural Engineering

B.Arch. & T.P. Bachelor of Architecture and Town Planning

BARCS Battlefield Area Reconnaissance System

Barents Barents Sea in the Arctic between Norway and Russia

barg(s) bargain(s)

bari baritone; baritone saxophone

Bari Bari delle Puglie, Italy

Bariloche San Carlos de Bariloche, Argentina

Barisans Barisan Mountains of Sumatra

barite barium sulfate

Bark Barker; Barkhain

Barlinnie Glasgow, Scotland's great prison

Barme Bartolome

bar mitz bar mitzvah (Hebrew—one who is responsible for the Commandments)—religious ritual marking a boy's 13th birthday; similar ceremony for girls called *bas mitzvah* or *bat mitzvah*

barn. bombing and reconnaissance navigation

Barn Barnard

BARN Body Awareness Resource Network

Barna Barcelona

Barney Barnabas; Barnett; Bernard; Bernardino; silver cigarette box engraved with drawings of Barney Google and Snuffy Smith (awarded to the year's best cartoonist)

BARNS Bombing and Reconnaissance Navigation System

Barnum P(hineas) T(aylor) Barnum; 19th-century American impresario and showman who brought his circus to all parts of the United States; presented the Swedish soprano Jenny Lind, the midget General Tom Thumb, Jumbo the enormous elephant, the first hippopotamus ever shown in America; also established model industrial and workers community in Bridgeport, Connecticut

BARONS Business-Accounts Reporting Operating Network System

b & arp bare and acid resisting paint

BARP British Association of Retired Persons

barq barquentine

Barq Barranquilla

barr barrister

BARR British Association of Rheumatology and Rehabilitation

Barrens Barren Grounds of northern Canada west of Hudson Bay in treeless tundra; Pine Barrens of New Jersey

Barrow Point Barrow (Alaska's most northerly point of land and settlement)

Barry Barrymore

BARS Backup Attitude-Reference System; Ballistic Analysis Research System; British Association of Residential Settlements

BARSR Biblioteca Academiei Republicii Socialiste Romania (Academic Library of the Socialist Republic of Romania)—in Bucharest

BARSTUR Barding Sands Underwater Test Range

bart bartender; barter

Bart Baronet; Bartel; Barth(el); Barthold; Bartholomew; Bartimeus; Bartlet(t); Bartley; Bartolo; Bartolomeo; Barton; Bartram

BART Bay Area Rapid Transit (San Francisco); Brooklyn Army Terminal (New York)

BARTD Bay Area Rapid Transit District

Bartenders Union Hotel and Restaurant Employees and Bartenders International Union

BARTOC Bus Advanced Real Time Operational Control

Bartolo Bartolomeo (Italian—Bartholomeo)

Barts Saint-Barthelemy in the French West Indies; Saint Bartholomew's Hospital in London

Barty Bartholomew

BARU Bay Area Revolutionary Union

barv beach armored recovery vehicle

bas basenji; basic airspeed; basic allowance for subsistence; basilica; basophil(s); basset; beam-alignment sensor; benzyl analog of serotonin

bas (BAS) baths

Bas Basel; Basil; Basilica; Basilicata; Bass Strait; Bastogne; Bastrop; Basuto; Basutoland

BAs British Airways; Business Agents (of unions)

B.As Buenos Aires

BAS Basic Allowance for Subsistence; Behavioral Approach Scale; Brazilian-American Society; British Acoustical Society; British Antarctic Survey; British Arachnological Society; British Army Staff

B.A.S. Bachelor of Agricultural Science; Bachelor of Applied Science

BAS Bulletin of the Atomic Scientists

Basa Baronessa (Italian—Baroness)

BASA British Architectural Students' Association; Buckeye Association of School Administrators

BASAF British and South Africa Forum

BASAM British Association of Grain, Seed Feed, and Agricultural Merchants

B. A. Sc. Bachelor of Applied Science

BASC Berlin Air Safety Center; Booth American Shipping Corporation

base b bascule bridge

basd basic active service date

base. (BASE) basic semantic element

BASE Bank-Americard Service Exchange; buildings, antennas, spans, earth formations (term used by parachutists); Business Assessment Study and Evaluation

Base 2 binary

Base 8 octal

Base 10 decimal

Base 16 hexadecimal

BASEC British Approvals Service for Electric Cables

BASEEFA British Approvals Service for Electrical Equipment in Flammable Atmospheres

BASF Badische Anilin und Soda Fabrik

bash. body acceleration given synchronously with the heartbeat

BASH Bird Aircraft Strike Hazard; Bulimia Anorexia Self Help

BASI British Association of Ski Instructors

basic. (BASIC) battle-area surveillance and integrated communications; beginner's all purpose symbolic instruction code (computer language)

BASIC British-American Scientific International Commercial (English)

BASIC Biological Abstracts Subjects in Context

BASICO Behavior Science Corporation

basicpac basic processor and computer

basictng basic training

BASIE British Association for Commercial and Industrial Education

basil. basilect(ic)(al)(ly)

BASIL Barclays Advanced Staff Information Language

basis. (BASIS) bibliographic author of subject interactive researches

BASIS-H BASIS history

BASIS-P BASIS political science, public administration, urban studies, amd international relations

BASIS-S BASIS sociology

bask baskisch (German—Basque)

Bask Baskir(ia)

Bask(er) Baskerville

BASMA Boot and Shoe Manufacturers' Association

bas mitz see bar mitz

BASO Base Accountable Supply Officer; Brigade Air Support Officer; Bureau of Aeronautics Shipping Order

BASOC Brigade Air Support Operations Center

basops base operations

baso(s) basophile(s)

baspm basic planning memorandum

BASR Bureau of Applied Social Research (Columbia University)

BASRA British Amateur Scientific Research Association

bass. bassoon

BASS Basic Analog Simulation System; Bass Anglers Sportsman Society

B.A.S.S. Bachelor of Arts in Social Science

bass con basso continuo (Italian—continuous bass)—figured bass background

Bassie Sebastian

BASSR Bashkirian Autonomous Soviet Socialist Republic; Buriat Autonomous Soviet Socialist Republic

bast. bastard; bastardization; bastardize; bastardly; bastard title; bastardy

Bast Sebastian

bas tit. bastard title (half title)

Basto Sebastiano

ba sw bell-alarm switch

BASW British Association of/for Social Workers

basys basic system

bat battalion anti-tank; body adjustment test; brown adipose tissue

bat. battery; battle; best available technology

Bat Bartholomew; Battista

BAT Beaux Arts Trio; Blind Approach Training; Body Adjustment Test; Boeing Air Transport; Bolshoi All-azimuth Telescope; British American Tobacco; Bureau of Apprenticeship and Training

BA & T Bureau of Apprenticeship and Training

B-AT British-American Tobacco

BATAB Baker and Taylor Automated Bookordering

BATC British Amateur Television Club

batch. bachelor

bat. chg battery charger; battery charging

BATDIV battleship division

bate base activation test equipment

batea best available technology economically available

bates battlefield artillery target engagement system

b-a test blood-alcohol test

BATF Bureau of Alcohol, Tobacco, and Firearms (U.S. Treasury)

BATFOR battle force

bath. bathroom; best available true heading

batho bathometer

bathy bathymeter; bathysphere; bathyscaphe

BATLSK British Army Training Liaison Staff, Kenya

BATM British Admiralty Technical Mission

bat mitz see *bar mitz*

bato balloon-assisted takeoff

B.A.T.P Bachelor of Arts in Town Planning

batreadcompi battle readiness and competition instructions

BATRON battleship squadron

batrop baratropic

Bat Rou Baton Rouge

bats ballistic aerial target system

Bats British-American Tobacco (stock-exchange sobriquet)

BATS Business Air Transport Service

BATSU British Army Training and Support Unit

batt batter; batteries; battery

batt. battalion

BATT British Army Training Team

BATUS British Army Training Unit Suffield

bau basic assembly unit; British absolute unit (BTU, Btu)

Bau Bauer; Bauhaus

BAU Bangladesh Agricultural University; British Association Unit

BAUA Business Aircraft Users' Association

Baubie (Scottish—Barbara)

baud telecommunication unit measuring speed of signalling and equal to one code element or pulse per second—named for its French inventor, JME Baudot (1845–1903)

BAUS British Association of Urological Surgeons

bav bon h vue (French—good at sight)—sight draft

Bav Bavaria; Bavarian

BAV Biblioteca Apostolica Vaticana (Latin—Apostolic Vatican Library), in Rome's Vatican City

BAVA Bureau of Audio-Visual Aids (NY)

BAVE Bureau of Audio-Visual Educational (Calif.)

BAVIP British Association of Viewdata Information Providers

BAVTE Bureau of Adult, Vocational, and Technical Education (HEW)

baw bare aluminum wire

BAWA British Amateur Wrestling Association

BAWHA Bide-A-Wee Home Association

BAWR Bertrand Arthur William Russell

BAWRA British Australian Wool Realization Association

BAX Burlington Air Express

bay. bayonet

bay cand dc bayonet candelabra double contact

Bay Bayern (German—Bavarian)

Bbados Barbados

Bayer Bayerisch (German—Bavarian)

bayr bayrisch (German—Bavarian)

BAYS British Association of Young Scientists

baz. bazooka

bazuco Colombian cocaine product

bb bail bond; ball bearing; ball beaters; ball-beating contests; balloon barrage; baseball, bases on balls; basketball, billiards, cricket, football, hockey, ping-pong, soccer, tennis, volleyball; bank burglar(y); baseline-to-baseline measure in typography; bases on balls (walks); bayonet base (lamp or socket); below bridges; bill book; blocking back; blood bank; blood buffer (base); blue bloaters; border bearded; both bones; both to blame; breakthrough bleeding; breast biopsy; buffer base; bungling bureaucrat; double black; pellet fired from or made for a bb gun

bb (BB) billboard (television script)

b-b black bordered; bogie-bogie (single-unit locomotive)

b/b bail; bottled in bond; breadboard

b & b bed and board; bed and breakfast; benedictine and brandy

b or b brass or bronze (cargo)

b to b back to back

b2b (B2B) business to business

bb babord (Swedish—port side)

BB Bailey Bridge; Barbados (Internet code); battleship; Before Bach; B'nai B'rith; Brigitte Bardot; Bureau of the Budget

BB (DCO) Barclays Bank (Dominion, Colonial and Overseas)

B B former Columbia president Belisario Betarhcur Cuartas

B-B Bora-Bora

B.B. Bernard Berensen; Bjørnstjernè Børnson; Boys' Brigade

B & B Brown and Bigelow

B of B Bureau of the Budget

BB Banco do Brasil (Portuguese—Bank of Brazil)

BB 62 USS *New Jersey*

bba born before arrival

BBA Balanced Budget Act; Big Brothers of America; Black Business Alliance; born before arrival (of the midwife or doctor); British Bankers' Association

B.B.A. Bachelor of Business Administration

bbac bus-to-bus access circuit(ry)

BBAC British Balloon and Airship Club

Bbados Barbados

b-ball basketball

b/bar bull bar (fender)

BBAR Broad-Band Anti-Reflection

bbb banker's blanket bond; basic boxed base; bed, breakfast, and bath; beleaguered by bullets and burglars; blood brain barrier; triple black

BBB Bach, Beethoven, Berlioz; Bach, Beethoven, Brahms; Bach, Beethoven, Bruckner; Best Berlin Broadcast; Best British Briar (pipes); Better Business Bureau

BBBC British Boxing Board of Control

BBBS Boat Builders' Benefit Society

BB & BU Bagel Boilers and Bakers Union

bbc barrels, boxes, and crates (cargo); broad-band chaff (anti-radar foil); bromobenzylcyanide (gas)

BBC Bank of British Columbia; Beautiful British Columbia; Best British Cinema; Better Breathers Club (for those with pulmonary disease); Billionaire Boys Club; Biwako Broadcasting Corporation (Japan); British Broadcasting Corporation; Brown Boveri Couporation; Buchanan Borehole Colleries

BBC-1 British Broadcasting Corporation's first television network featuring cinema and sports

BBC-2 British Broadcasting Corporation's second television network featuring classical music

BBCC Big Bend Community College

BBCCS B'nai B'rith Career and Counseling Services

BBC dissociation Braid-Berheim-Charcot dissociation

BBC Eng BBC English

BBC English cultured way of speaking English

BBCF British Bacon Curers' Federation

BBCL Bermuda Broadcasting Company Limited

BBCM Bandmaster—Bandsmen's College of Music

BBCMA British Baby Carriage Manufacturers' Association

BBC Northern British Broadcasting Corporation's Northern Symphony

BBCS Browne Book-Charging System

BBC Scot British Broadcasting Corporation's Scottish Symphony

BBCSO British Broadcasting Corporation Symphony Orchestra

BBC Sym British Broadcasting Corporation Symphony

bbcw bare beryllium copper wire

BBCWS BBC World Service

bbd baby born dead; beta-binomial distribution; big bass drum; bucket-brigade device; bulletin board

bbdc before bottom dead center

BB(DCO) Barclays Bank (Dominion, Colonial, Overseas)

BBDO Batton, Barton, Durstine, and Osborne

BBE Board of Barber Examiners

BBEA Brewery and Bottling Engineers Association

bb & em bed, breakfast, and evening meal

bbf boron-based fuel

BBF Biblioteca Benjamin Franklin (Mexico City); Boilermakers, Blacksmiths, Forgers (union)

BBFC British Board of Film Censors

bbg bundle branch block

BBG Bermuda Botanical Gardens; Brooklyn Botanic Garden

BBGA British Broiler Growers' Association

bb-gun airgun shooting bb's (ball bearings)

BBHC Buffalo Bill Historical Center (Cody, Wyoming)

BBHF B'nai B'rith Hillel Foundations

BBI Barbecue Briquet Institute; Brandeis-Bardin Institute

B Bibl *Bachelier en Bibliothalconomie* (French—Bachelor in Library Science)

BBINA Bed and Breakfast Inns of North America

bbing bureaucratic bumbling

BBiP *British Books in Print*

BBIRA British Baking Industries Research Association

B Bisc Bay of Biscay

bbj ball-bearing joint

bbk breadboard kit

bbl barrel

BBL Bahia Blanca; Bangkok Bank Ltd; Barclay's Bank Limited; Big Brothers League

bbl roll barrel roller

bbls/day barrels per day

bbm big bowel movement; break-before-make

b & b magazine periodical featuring breasts and buttocks

BBMRA British Brush Manufacturers Research Association

BBNNR Braunton Burrows National Nature Reserve (England)

BBNP Big Bend National Park (Texas)

BBNR Back Bay National Refuge (Virginia)

Bbo Bilbao

B-Bomb benzedrine bomb (slang—benzedrine inhaler)

bboy bound boy

B-boy busboy; mess sergeant

bbp boxes, barrels, packages, (cargo); break bulk point; building block principle

BBP Beech Bottom Power Company

BBP *Boletim de Bibliografa Portuguesa* (Portuguese—*Bulletin of Portuguese Bibliography*)

b & b pericarditis bread-and butter pericarditis

BBPR Bianfi, Barbiano, Peresutti, and Rogers (avant-garde Italian architects)

bbq barbecue

BBQ Brooklyn, Bronx, Queens

bbr balloon-borne radio

BBRM British Bombing Research Mission

BBRR Brookhaven Beam Research Reactor (AEC)

BBRS Balloon-Borne Radio System

bbs baby-bottle syndrome; ball bearings; barrels of basic sediment; basic battle skills; box bark strips

Bbs British biscuits

BBS Barber Blue Sea (steamship line); Bermuda Biological Station; Bhutan Broadcasting Service; Brunei Broadcasting Service; Bulletin Board System

B.B.S. Bachelor of Business Science

BBSATRA British Boot, Shoe and Allied Traders Research Asociation

BBSE Board of Behavioral Science Examiners

BBSI Beauty and Barber Supply Institute; British Boot and Shoe Institution

bbsj ball-bearing swivel joint

B & B SNC British and Burmese Steam Navigation Company

BBSR Bermuda Biological Station for Research

bbsu bid bond service undertaking

bbs & w barrels of basic sediment and water

bbt basal body temperature; bombardment

BBTA British Bureau of Television Advertising

BB & TC Bahamas Broadcasting and Television Commission

BBU Bagel Boilers Union; Blue-Brick University (one of the highest academic traditions such as Cambridge, Harvard, Oxford, or Yale)

bbw bare brass wire

BBW B'nai B'rith Women

BBWAA Baseball Writers' Association of America

BBX Bluebird (stock-exchange symbol)

BBYO B'nai B'rith Youth Organization

bbz bearing bronze

bc back country; back course; back-cross; bad check; ballistic camera; base (shield) connection; basic computer; between centers; bills for collection; binary code; binary counter; birth control; bogus check; bolometric correction; bolt circle; bone connection; book club; born in colony; bottom (dead) center; broadcast control; budgeted cost; building center; burden center; bursa copulatrix

bc (BC) bio-conversion; bulk carrier

b/c bales of cotton; bills for collection; birth control; broadcast

b & c building and contents

b2c (B2C) business to consumer

Bᶜ *Banc* (French—bank, reef or sandbank)

BC Bacone College; Bad Conduct; Baja California, Mexico; Bakersfield College; Balliol College (Oxford); Bank Clearing; Bankruptcy Court; Bard College; Barnard College; Barrington College; Barry College; Baruch College; Bates College; Battery Commander; Battle Cruiser; Beaver College; Beckley College; Before Christ; Belgian Congo; Belhaven College; Bellarmine College; Belmont College; Beloit College; Benedict College; Bengal Cavalry; Bennett College; Bennington College; Berea College; Berry College; Bethany College; Bethel College; Bill Clinton; Bishop College; Blackburn College; Blinn College; Bliss Classification; Bliss College; Bloomfield College; Bluefield College; Bluffton College; Bombay Cavalry; Bomber Command; Borough Council; Boston College; Bourget College; Bowdoin College; Boys Club; Brandon College; Brasenose

College (Oxford); Bren Carrier; Brenau College; Brentwood College; Brescia College; Brevard College; Briarcliff College; Bridgewater College; British Columbia; British Commonwealth; British Council; Brooklyn College; Bruyere College; Bryant College; Burdett College; Bus Controller; Butler College

B-C Barber-Colman

B.C. Bachelor of Chemistry; Bachelor of Commerce; Baja California; Before Christ; British Columbia

B & C Banking and Currency (Senate Committee)

B of C Bank of Canada; Bureau of the Census

BC Baja California (Spanish—Lower California), northern section is a state whose capital is Mexicali while the southern part is a territory whose capital is La Paz; *Banco Central* (Spanish—Central Bank), Spain's largest; *Biological Conservation*

B C basso continuo (Italian—continuous bass background)

bca best cruise altitude; blood color analyzer

bca barrica (Spanish—cask, keg); *biblioteca* (Portuguese or Spanish—library)

*B*ca *Boca* (Portuguese or Spanish—mouth, river mouth)

BCA Bank Credit Analyst; Battery Control Area; Billiard Congress of America; Blue Cross Association; Boys' Clubs of America; British Caledonian Airways; British Canoe Association; British Colonial Airlines; Bureau of Consular Affairs; Bureau of Consumer Affairs; Business Committee for the Arts

B/C of A British College of Aeronautics

bcaa branched-chain amino acids

BCAA British Columbia Automobile Association

BCAB Birth Control Advisory Bureau; British Computer Association for the Blind

bcac biology classroom activity checklist

BCAC British Conference on Automation and Computation

BCA/DoS Bureau of Consular Affairs—Department of State

BCAir British Commonwealth Air Force

BCAL British Caledonian Airways

BCALA Black Caucus of the American Library Association

BCAM Breast Cancer Awareness Month (annual)

BCAP Beacon Collision-Avoidance Program

BCAPT Braverman-Chevigny Auditory Projective Test

BCAR British Civil Airworthiness Requirements; British Council for Aid to Refugees

BC-ARS Birth Control-Abortion Referral Service

BCAS Beacon-compatible Collision-Avoidance System; British Compressed Air Society

b'cast(ing) broadcast(ing)

BCAT Birmingham College of Advanced Technology

BCATP British Commonwealth Air Training Program

bcb binary code box; broadcast band; button-cell battery

BCBC British Cattle Breeders' Club

bcbh boiler casing bulkhead

BC/BS Blue Cross/Blue Shield

bcc beam-coupling coefficient; blind carbon copy; body-centered cubic

BCC Battery Control Center; Battery Control Central; Berkshire Community College; Board of Crime Control (Montana); British Communications Corporation; British Council of Churches; British Crown Colony; Broadcasting Complaints Commission; Broadcasting Corporation of China; Bronx Community College; Bureau Central de Compensation; Bureau of Charities and Corrections (South Dakota); Burlington Community College; Bus & Coach Council; Business Co-operation Center

BCCA Beer Can Collectors of America; British Cyclo-Cross Association; Broadcast Cable Credit Association

BCCBP Biblioteca de Cataluña y Central de Biblioteca Populares (Spanish—Library of Catalonia and Central Public Library)—Calle Carmen, Barcelona

BCCCUS British Commonwealth Chamber of Commerce in the United States

bccd buried-channel charge-coupled device

BCCF British Cast Concrete Federation

BCCG British Cooperative Clinical Group

BCCI Bank of Credit and Commerce International

BCCLA British Columbia Civil Liberties Association

BCCO Base Consolidation Control Office(r)

BCCR Banco Central de Costa Rica (Spanish—Central Bank of Costa Rica)

BCCS British Columbia Coastal Service

bcd barrels per calendar day; binary-coded data

BCD Bad Conduct Discharge; Batelle's Columbus Division; British Columbia Dragoons

BCDA Biscuit and Cracker Distributors Association

bcdc binary coded decimal counter

BCDDP Breast Cancer Detection Demonstration Project

B-cdf B-cell differentiation factor

BCDIC Binary-Coded Decimal Interchange Code

bce (BCE) book club edition

Bce Belice (Spanish—Belize) formerly British Honduras

BCE before the Christian era

BCE (B.C.E.) Before the Common Era

BCEL British Commonwealth Ex-Services League

bcer. birth certificate

BCESL British Commonwealth Ex-Services League

bcf bromo-chloro-fluoromethane

BCF Battle Cruiser Fleet; Battle Cruiser Flotilla; Battle Cruiser Force; British Commonwealth Forces; British Cruiser Force

BCF *Bataillon Canadien Français*

BCFK British Commonwealth Forces, Korea

BCFMA Broadcast Cable Financial Management Association

bcfsk binary code frequency shift keying

BCG Board for Certification of Genealogists

B-cgf B-cell growth factor

Bch Beach

BCH Bachelor of Clinical Hypnotherapy; Bomber Command Headquarters; Boot Command High; British Columbia Horse; British Columbia Hussars

BCHA British Columbia Health Association

BCHCM Board of Certified Hazard Control Management

bchcmbr(s) beachcomber(s)

BCHP *Banco Central Hipotecario del Perú* (Spanish—Central Mortgage Bank of Peru)

bci broadcast interference

BCI Bat Conservation International; Battery Council International; Birth Control Institute; Bureau of Correctional Institutions (Iowa); Bureau of Criminal Identification

BCIS Bureau of Collection and Investigation Services

BCIU Border of Crime Intervention Unit

BCKB British Commonwealth Korean Base

bcl broadcast listener

BCL Battelle's Columbus Laboratories; Bermuda Container Line; Bougainville Copper Limited

B.C.L. Bachelor of Civil Law

b clar basse clarinette (French—bass clarinet)

BCLA Birth Control League of America

BCLO Bomber Command Liaison Officer

bcm basic control monitor; brightest cluster member

BCM Berklee College of Music; Bougainville Copper Mine (Papua, New Guinea)

BCM *Banco Comercial Mexicano* (Spanish—Mexican Commercial Bank)

BCMA Bank Capital Markets Association

B & CMA Biscuit and Cracker Manufacturers Association

bcme bichloromethylether

bcmg becoming

bcn beacon

BCN British Commonwealth of Nations

BC-NET Business Cooperation Network

BCNY Bond Club of New York

BCNZ Broadcasting Corporation of New Zealand

bco binary-coded octal

BCOA Bituminous Coal Operators Association

BCOF British Commonwealth Occupation Force

B.Com. Bachelor of Commerce

BCOO Bomber Command Operation Order

BCOS British Chief of Staff

BCP Bonus Community Plan (telephone service)

bcp boom clearance party

BCP *Banco de Credito del Perú* (Spanish—Credit Bank of Peru); *Book of Common Prayer*

BCPA British Copyright Protection Association

BCPE Board of Certification in Professional Ergonomics

BCPM British Columbia Provincial Museum (Victoria, British Columbia)

BCPs Black College Presidents

BCPSM Board of Certified Product Safety Management

BCPT Breast Cancer Prevention Trial

BCPU Border Crime Prevention Unit

BCR British Columbia Railway; British Columbia Regiment

BCRLRA Beverage Container Recycling and Litter Reduction Act

BCRP *Banco Central de Reserva del Perú* (Spanish—Central Reserve Bank of Peru)

bcs because

BCS Battle Cruiser Squadron; Black Christian Students; Bowl Championship Series; British Conchological Society; Budget Control System

B-C-S Bardeen-Cooper-Schrieffer

B.C.S. Bachelor of Criminal Science

BCSC British Columbia Steamship Company

BCSP Board of Certified Safety Professionals

BCSR Board of Certified Shorthand Reporters

BCSS Bard Cardiopulmonary Support System

bcst broadcast

bct basic combat training

BCT Bell Canyon Test

B & CT Bank and Corporation Tax

bctmp bleached chemi-thermal mechanical pulping

BCTWU Bakery, Confectionery and Tobacco Workers Union

bcu big closeup; bus-control unit

bcwp budgeted cost of work performed

bcz because

bd bank draft; barrack department; base depot; battle dress; behavior(ally) disorder(ed); bills discontinued; birth date; blessed; board; boat deck; bomb disposal; bond; boom defense; bound; bridge deck

b & d bondage and discipline (sado-masochism)

bd. buried

b/d broker/dealer

bd bind (Dano-Norwegian—volume)

Bd Bahrain dinar; board

BD Banking Department; Bangladesh (Internet code); blowing dust (aircraft code)

B.D. Bachelor of Divinity

B.D. *Bona Dea* (Latin—good goddess) Roman goddess of chastity and fertility

B/D Banker's Draft

BDA bomb damage assessment; Broadcast Designers Association

BDAC Bureau of Drug Abuse Control

bdc binary decimal counter; bottom dead center

BDC Berlin Documentary Centre (National Archives)

BDCC British Defence Coordinating Committee

bdd body dysmorphic disorder

BDE Board of Dental Examiners

bde. brigade

BDF Bangladesh Forces; Belize Defence Force; Birth Defects Foundation

BDHI Bearing, Distance, Heading Indicator

bdi both days included

BDI Beck Depression Inventory

BDI Bundesverband der Deutschen Industrie (Federation of German Industries)

bdl baseline demonstration laser

bdle bundle

bdm bomber defense missile

BDM Bund Deutsche Mädchen (Hitler Youth, Females)

bdm's births, deaths, marriages

BDMS Bulk Direct Mail Service

BDO Boom Defence Officer

bdp biparietal diameter

BDPA Black Data Processing Associates

BDPC Breakthrough Drug Pricing Committee

b dr bass drum

bdr. bombardier

bdra basic daily ration allowance

bdrk bomb damage repair kit (for runways)

bdrm bedroom

BDRU Baor Drops Trials Unit

bds brands; bomb-disposal squad

bds bis in die summendus (Latin—take twice daily)

BDS British Defence Staff

B.D.S Bachelor of Dental Surgery

B.D.Sc. Bachelor of Dental Science

BDSO British Defence Sales Organisation

B-D Squad Bomb-Disposal Squad

BDSS Bayesian Decision-Support System

BDSW British Defence Staff, Washington

bdt. birth date

bdt bundt (Dano-Norwegian—bundle)

bdt's back-door trots (diarrhea)

BDU Bomb Disposal Unit; Bombing Development Unit

bdv boom defense vessel

be base excess; beam expander; beige; below elbow; bent; bilingual education; binding cdge; black enterprise

b/e bill of entry; bill of exchange

b & e beginning and ending; breaking and entering

Be Baume; Berkeley (University of California Library at Berkeley)

BE Beginning Experience (retreat program for separated, divorced and widowed); Belgium (Internet code); Black English; Bleriot Experimental (aircraft); Borough Engineer; British Embassy; British English

B/E bill of exchange

BE Buque Escuela (Spanish—Schoolship)

be4 before

Bea Beatrice

BEA Broadcast Education Association; Bureau of Economic Analysis

BEAM Brain Electrical-Activity Mapping

BEAR Bureau of Electronic and Appliance Repair

BEAR HUG Basic Extended Acronym Human Users Guide

Beatrix Beatrix Wilhelmina Armgard, Queen of the Netherlands, 1980–

Beau Beauford; Beaufort; Beaumont; Beauregard

beautility beauty + utility

BEBA Bureau of Economic and Business Affairs (U.S. State Department)

bec because

bec. became

BEC Bose-Einstein Condensation

BECA Bureau of European and Canadian Affairs

BECC Boeing Engineering & Construction Company

BECL Bangkok Expressway Company Limited

Bedloe's Bedloe's Island in New York's Upper Bay, also called Liberty Island, site of the Statue of Liberty

bedoc beds occupied

bed(s). bedroom(s)

Beds Bedfordshire

Bed Smet Bedrich Smetana

Bed-Stuy Bedford-Stuyvesant

BEE Basic Earthquake Education; Basic Economic Education

B.E.E. Bachelor of Electrical Engineering

BEE Bulletin of Environmental Education

Beeb BBC (British Broadcasting Corporation)

beec birmary error-erasure channel

Beech 99 Beechcraft seventeen-seat aircraft

Beecham Beauchamp

Beech F-33C five-place Bonanza training airplane

Beech Queen Air Beechcraft Seminole transport aircraft

beef. business-and-engineering-enriched fortran

beefalo beef cattle + buffalo (hybrid)

BEEP Bureau European del' Education Populaire (French—European Bureau of Adult Education)

Beethoven Nine 9 Beethoven symphonies (1st, 2nd, 3rd *Eroica*, 4th, 5th, 6th *Pastorale*, 7th, 8th, 9th *Choral*)

bef before; blunt end first; buffered emitter follower

Bef Befehl (German—command, order)

BEF Bonus Expeditionary Force; British Equestrian Federation; British Expeditionary Force

BEFA British Emigrant Families Association

befm bending form

befo' before

beg. begin; beginning

beg begynde(lse) [Dano-Norwegian—begin(ning)]

BEG Belgrade, Yugoslavia (airport); Bureau of Economic Geology

BEG Bank Europaeischer Genossenschafsbanken (German—European Cooperative Bank)

Begl Begleitung (German—accompaniment)

begr begrundet (German—established)

BEH Bureau of Education for the Handicapped

BEHA British Export Houses Association

behav behavior; behavioral; behaviorist(ic)

Behavioral Res Behavioral Research Laboratories

BEHC Bio-Environmental Health Center

behp bisethylhexyl phthalate

bei butanol-extractable iodine

BEI Bridgeport Engineering Institute

BEI British Education Index

BEIA Board of Education Inspectors' Association

Beibl Beiblatt (German—supplement)

beif beifolgend (German—sent
herewith)

Beih Beiheft (German—supplement)

beil beiliegend (German—enclosed)

Beil Beilage (German—appendix, supplement)

beir biological effects of ioniz-
ing radiation

BEIS British Egg Information
Service

Beisp Beispiel (German—example or illustration)

Beitr Beitrag (Gerrnan—contribution)

bek bekendgørelse (Dano-Norwegian—announcement)

bel below; 10 decibels

bel (Turkish—pass)

Bel Bela; Belcher; Belden;
Belem; Belen; Belfast; Bel-
ford; Belgian; Belgium; Belgrade (capital of Yugoslavia);
Belham; Belize (whose capital is Belmopan); Bellamy;
Bellanca; Belmont; Belorus-
sia; Belos; Beltram

Bel Bacharel (Portuguese—
Bachelor)—academic degree

BEL Belém do Pára, Brazil
(airport)

Béla Béla Bartok; Béla Kun;
Béla Lugosi

BELAIR Belgian Air Staff
(NATO)

Belar Belarus (whose capital is
Minsk)

Belau Republic of Belau (formerly Palau)

Belchers Belcher Islands in
Hudson Bay just north of
James Bay

belcrk bellcrank

Bel & Dr Bel and the Dragon

B.Elec. & Tel. Eng. Bachelor
of Electronics and Telecommunication Engineering

Belém (Portuguese—Bethle-
hem)—also the Amazon
River port of Belém do Pará;
Lisbon suburb

bel ex bel example (French—
fine example)—fine copy of a
book, engraving, map, etc.

Belf Belfast; Belfastian(s)

belg belgisk (Dano-Norwegian—Belgian)

Belg Belgian; Belgium

Belg Belgica (Portuguese or
Spanish—Belgium); *Belgio*
(Italian—Belgium)

Belgium Kingdom of Belgium
(lowland nation on the North
Sea); *Royaume de Belgique*
(French); *Koninkrijk België*

Belglais Belgian-English

Belgolux Belgium and Luxembourg

Bell Bell Aircraft; Bell System
(American Telephone and
Telegraph and associated
companies collectively called
Ma Bell or Mother Bell)

Bell 47 Sioux utility helicopter
built by Bell Aircraft

Bell 204 gunship helicopter
nicknamed Huey as it is designated UH-1

Bell 206 five-seat turbo-pow-
ered helicopter also called Jet
Ranger or Sea Ranger

bella belladonna (drug stimulant)

Bella Arabella; Isabella

*Bellas Artes Instituto Nacional
de Bellas Artes* (Span-
ish—National Institute of
Fine Arts)—Mexico City

Belle Bella; Arabella; Isabella

Bellevue New York City hospital
noted for its psychiatric ward

BELLMATIC Bell Laboratories Machine-Aided Technical Information Center

bells bell-bottom pants

Bell's Cr C Bell's Crown
Cases (British)

*Bell Sys Tech J Bell System
Technical Journal*

Belmo Belmopan

BELNAV Belgian Naval Staff
(NATO)

Belsen Bergen-Belsen (World
War II concentration camp
town in northwest Germany)

*Belvac Société Belge de Vacu-
ologie et de Vacuotechnique*
(French—Belgian Society for
Vacuum Science and Tech-
nology)

bem (BEM) behavior engineering model

Bem Bemerkung (German—
comment, note, observation)

BEM British Empire Medal

B.E.M. Bachelor of Engineering of Mines

BEMA Bakery Equipment
Manufacturers Association;
Business Equipment Manu-
facturers Association

BEMB British Egg Marketing
Board

BEME Brigade Electrical and
Mechanical Engineer

bemews basic missile early
warning system

BEMO Base Equipment Management Office

bems bug-eyed monsters (science-fiction jargon)

BEMS Bakery Equipment
Manufacturers Society; Bio-
electromagnetics Society;
Building Energy Manage-
ment Systems

BEMSA British Eastern Merchant Shippers Association

ben. bene (Latin—good, well);
benedictio (Latin—blessing)

Ben Benard; Benedict; Bengali;
Beni (Bolivian province); Be-
niah; Benito; Benjamin; Ben-
net(t); Benno; Beno; Be-noni;
Bentley; Benvenuto

Ben Ben Hur

*BEN Bureau d' Etudes nucle-
aires* (French—Belgian Bureau of Nuclear Studies)

Benavides originally Ben
David

Bend Bendigo

benday benday (photoengraving) process named for a
19th-century American
printer, Benjamin Day

BENDEX Beneficial Data Exchange (linking Social Security Administration with state
welfare agencies)

B en Dr Bachelier en Droit
(French—Bachelor of Law)

bends caisson disease

bene benzine

BENECHAN Benelux Subarea
Channel (NATO)

Bened Benedict; Benedictine

benef beneficiary

Benef Benefice

*Ben Eil Benedenwindse Eilan-
den* (Dutch—Leeward Islands)

Benelux economic union of
Belgium, Netherlands, and
Luxembourg

Bene't Benedict

benev benevolent

Beng Bengal; Bengali

B.Eng. Bachelor of Engineering

bengals bengal tigers; thick cigars

Bengis Bengalis

Bengs Bengalis

B.Eng.Sci. Bachelor of Engineering Science

B.Eng.Tech. Bachelor of Engineering Technology

Ben-Gur David Ben-Gurion (Israeli statesman)

BENHA Board of Examiners of Nursing Home Administrators

Benin People's Republic of Benin (West African nation formerly called Dahomey)

Benj Benjamin

Benja Benjamin

Benjn Benjamin

Benjy Benjamin

Bennie Benjamin

bennies benzedrine stimulants

benny benzedrine

Benny Benjamin

ben sug beneficial suggestion

Bento Baruch; Benito

b & ent & pl breaking and entering and petty larceny

benz benzedrine; benzine

Benziger Benziger, Bruce, and Glencoe

BeO beryllium oxide

BEO Borough Education Office(r)

beoc battery echelon operating control

BEOG Basic Educational Opportunity Grant

bep biomolecular engineering program; break-even point

BEP Bureau of Engraving and Printing

BE & P Bureau of Engraving and Printing

B.E.P. Bachelor of Engineering Physics

BEP Brevet d'Études Professionelles (French—Professional Studies Diploma)

B EpA British Epilepsy Association

BEPC Beijins Electron Positron Collider

BEPI Budget Estimates Presentation Instructions

BEPO British Experimental Pile Operation

bepoc Burrough's electrographic printer-plotter for ordnance computing

BEPQ Bureau of Entomology and Plant Quarantine

bepti bionomics, environment, plasmodium, treatment, immunity (factors in malaria epidemiology)

BEPZ Bataan Export Processing Zone (Philippines)

beq bequeath

BEQ Bachelor Enlisted Quarters; Background and Experience Questionnaire

beqd bequeathed

beqt bequest

ber basic encoding rules; beyond economic repair; buffer (flow chart)

ber berechnet (German—computed)

Ber Berber; Berlin; Berwickshire

Ber Bericht (German—report)

BER Berlin, West Germany (Tempelhof airport); Bureau of Economic Regulation

BERA Business Education Research Associated

Berb Berber

Berb Berberia (Spanish—Barbary Coast)

Ber Bunsenge Phys Chem Berichte der Bunsengesellschaft für Physikalische Chemie (German—*Report of the Bunsen Society for Physical Chemistry*)

BERC Biomedical Engineering Research Corporation; Black Economic Research Center

BERCO British Electric Resistance Company

BERCON Berlin Contingency (NATO)

Berdoo San Bernardino, California

berg iceberg (sometimes written 'berg)

Berg Berger; Bergin

Berg (Dutch or German—mountain)

Bergen Bergen-Belsen (site of Nazi German concentration camp); Norway seaport city; Bergen op Zoom (Dutch port); county in New Jersey; former name of Mons, Belgium

Bergman Ingmar Bergman (Swedish film-writer director); Ingrid Bergman (Swedish actress); Torbern Bergman (Swedish chemist and physicist)

BERH Board of Engineers for Rivers and Harbors

Beria Lavrenty Pavlovich Beria (director of Stalin's secret service)

Bering sea linking the Arctic Ocean with the North Pacific

Beringia Bering and Chukchi Seas area

Berk Berkeley

Berks Berkshire

Berl Berlin

Berl Tid Berlingske Tidende (Danish—Berling's Times)— a leading daily newspaper

Berm Bermuda (whose capital is Hamilton); Bermuda Islands

Bermudas Bermuda Islands

Bern Berna; Bernal(do); Bernan; Bernard(o); Bernarr; Berner; Bernhard; Bernold

Bernh Bernhard

Bernie Bernard

Berno Bernardo

Ber Phil Berlin Philharmonic

Bert Albert; Alberta; Albertina; Bertel; Bertha; Berthel; Bertillon (system); Bertin; Bertol; Berton; Bertram; Bertrand; Bertwin; Cuthbert; Delbert; Elbert; Elberta; Filbert; Gilbert; Herbert; Hilbert; Ibert; Lambert; Norbert; Philbert; Roberta; Wilbert; Zilbert

bertm berth term

Ber Tri Bermuda Triangle (North Atlantic shipwreck area extending from Bermuda to Cape Hatteras to Key West and back to Bermuda)

Berts Bertillon Measurements

BERU Building Economics Research Unit

Berw Berwick

beryl beryllium aluminum silicate

bes balanced electrolyte solution

bes besonders (German—especially)

Bes Bessel's functions

BES Biological Engineering Society; Bureau of Employment Security; Bureau of Environmental Statistics

B.E.S. Bachelor of Engineering Science

B es A Bachelier des Arts (French—Bachelor of Arts)

BESA Birmingham, England, Small Arms; British Engineering Standards Association

BESE Bureau of Elementary and Secondary Education

BeShT Baal Shem-Tov (Israel Ben Eliezer)

besi bus electronic-scanning indicator

B es L Bachelier des Lettres (French—Bachelor of Letters)

BESL British Empire Service League

BESN British Empire Steam Navigation (company)

BESRL Behavioral Science Research Laboratory (USA)

bess binary electromagnetic signature

Bess Bessemer; Mrs Harry S Truman

B es S Bachelier es Sciences (French—Bachelor of Science)

BESS Bank of England Statistical Summary

Bessie Bethlehem Steel (Wall Street slang); Elizabeth

Bessy Elizabeth

BESSY Bestell System (German teleordering)

Best Bestellung (German—order); *bestyrelse* (Dano-Norwegian—board, direction)

BEST Basic Essential Skills Training; Believing, Encouraging and Studying Together; Black Efforts for Soul in Television; British Expertise in Science & Technology

best ofr best offer

bet. best estimate of trajectory; between

Bet Beirut; Betsy; Elizabeth

BET Biker Enforcement Team (of law-enforcement officers); Black Entertainment Television; British Electric Traction

beta battlefield exploitation and target acquisition

Beta Betamax (video-cassette system)

BETA Billion-Channel Extra-Terrestrial Assay; Business Equipment Trade Association

BETFOR British Element Trieste Forces

Beth Bethlehem; Elizabeth

Bethlehem Bethlehem Steel Corporation

Bethnel Bethnel Green, London

Beth Steel Bethlehem Steel

betr better

betr betrefend (German—concerning)

Bets Betsy

BETS Business Entities Tax System

Betsy Elizabeth

BETTS Bolt Extrusion Thrust Termination System

Betty Elizabeth

betw between

BEU Black Economic Union; British Empire Union

BEUC Bureau Européen des Unions Consommateurs (French—Bureau of European Consumer Unions)

bev bevel; beverage; billion electron volts

bev (BeV) billion electron volts

Bev Beva; Bevan; Beveridge; Beverley; Beverly; Bevis

BEV Black English Vernacular; Blake E Vance

Bevans Charles I. Bevans' compilation of *Treaties and Other International Agreements of the United States of America*

BEW Board of Economic Warfare

BEWT Bureau of East-West Trade

bex broadband exchange

bexec budget execution

BEY Beirut, Lebanon (airport)

bez bezahlt (German—paid); *bezüglich* (German—referring to)

Bez Bezirk (German—district)

bezw beziehungsweise, beziehungsweise (German—respectively)

bf back feed; batsmen faced; beer firkin; before; blackface (minstrel makeup); blackface (sheep); bloody fool; board feet, board foot; body flap; boiler feed; boldface (type); bold face; both faces; boy friend; broadleaf forest; buffered; butter fat

b-f beat-frequency

b/f black female; brought forward; brown female

b & f bell and flange

bf bassa frequenza (Italian—bass frequency); *bestemtform* (Dano-Norwegian—definite shape); *bonum Jactum* (Latin—good deed; approved); *bouillon filtrate* (French—filtered bouillon)

b.f. bonafide (Latin—genuine, sincere)—in good faith; without deception; without fraud

Bf black female; Bowser factor

BF Battle Fleet; Battle Force; Bengal Fusiliers; black female; Bombay Fusiliers; Bowser factor; Breadalbane Fencibles; Bristol Fighter; British Forces; Burkina Faso; Burrhus Frederic Skinner (behaviorist)

BF Banque de France (French—Bank of France);

B.F. Bachelor of Forestry

BF Beogradska Filharmonica (Serbo-Croat—Belgrade Philharmonic)

bfa basal forebrain area

BFA Black Faculty Association; British Fellmongers' Association; British Film Academy; British Forces, Antwerp; Broadcasting Foundation of America; Bureau of Financial Assistance

B.F.A. Bachelor of Fine Arts

bfaln buffer boundary alignment

BFAP British Forces—Arabian Peninsula

BFAR Bureau of Fisheries and Aquatic Resources (Philippines)

bfb breakage-fusion-bridge

BFB Bureau of Forensic Ballistics

BFBC British Forces Broadcasting Service

BFBPW British Federation of Business and Professional Women

BFBS British Forces Broadcasting Service; British and Foreign Bible Society

bfc benign febrile convulsion

BFC British Forces in Cyprus; Broadcasting and Film Commission

BFCA British Federation of Commodity Associations

BFCC Brothers for Christian Community

BFCF Bremerton Freight Car Ferry

BFCS British Friesian Cattle Society

BFCSD Brewery, Flour, Cereal, Soft Drink and Distillery (Workers of America)

bfct boiler feed compound tank

bfcy beneficiary

bfd big fucking deal

BFD Blind Fire Director
BFDC Bureau of Foreign and Domestic Commerce
bfe beam-forming electrode; buyer furnished equipment
BFEA Bureau of Far Eastern Affairs (U.S. Department of State)
BFEBS British Far Eastern Broadcasting Society
BFF Berlin Field Force
BFFA British Film Fund Agency
BFFC British Federation of Folk Clubs
BFFI British Forces, Falkland Islands
bfg brute-force gyro
Bfg Bank für Gemeinwirtschaft (German—Bank for Municipal Management)
BFG BF Goodrich; British Forces, Germany
BFH British Field Hospital
BFHMF British Felt Hat Manufacturers' Federation
BFHS Benjamin Franklin High School
bfi battlefield interdiction: beam-forming interface
BFI British Film Institute; Business Forms Institute; Seattle, Washington (Boeing Field)
BFIA British Flour Industry Association
BFICC British Facsimile Industry Compatability Committee
BFIG British Forces in Greece
bfl back focal length
BFL Barber Fern Line; Belgian Fruit Line; Blue Funnel Line (Holt's); Books For Libraries
BFLF Biblioteca della Facolta di Lettere e Filosofia (Italian—Library of the Faculty of Letters and Philosophy)—in Florence
BFM British Forces, Malta
BFM Ballet Folklórico de México (Spanish—Folklore Ballet of Mexico)
BFMA Business Forms Management Association
BFMB British Field Message Book
BFMF British Federation of Music Festivals; British Footwear Manufacturers' Federation

BFNHRA British Food Manufacturing Industries Research Association
BFMO Base Fuels Management Officer (USAF)
BFMP British Federation of Master Printers
Bfn Bloemfontein
BFN British Forces Network
BFNE British Forces, Near East
bfo beat-frequency oscillator; blood-forming organs
Bfo Buffalo
BFO Bath Festival Orchestra; Bureau of Field Operations
bfoq bona-fide occupational qualification
B.For. Bachelor of Forestry
bform budget formulation
B.For.Sci. Bachelor of Forestry Science
bfozp best-fit optic Z-plane
bfp batters facing pitcher; biological false-positive (reactions); boiler feedpump
BFP British Fishing Port (registration symbols appearing on the bows of British fishing vessels and indicating their home ports)—(see British Fishing Port appendix)
BFPA British Film Producers Association
BFPC British Farm Produce Council
bfpdda binary floating-point digital-differential analyzer
bfpgp batter facing pitcher per game for the player
bfpgt batters facing pitcher per game for the team
BFPO British Field Post Office
BFPPS Bureau of Foods, Pesticides, and Product Safety (FDA)
bfpv bona fide purchaser for value
bfr biologic false reactor; blood flow rate; bone formation rate; buffer
Bfr Belgische frank (Dutch—Belgian franc)
B Fr Belgian franc
bfr(s) belgisk(e) franc(s) [Dano-Norwegian—Belgian franc(s)]
BFRS Bio-Feedback Research Society
bfR sol buffered Ringer's solution

BFS Belfast, Northern Ireland (airport); Board of Foreign Scholarships; British Frontier Service; Bureau of Family Services; Bureau of Federal Supply
B.F.S. Bachelor of Foreign Service
BFS Bundesanstalt für Flugsicherung (German—Air-Traffic Control Authority)
BFSA Black Faculty and Staff Association; British Fire Services Association
BF set British Field (radio)
BFSS British and Foreign Sailors' Society
bft basic fitness test; binary file transmission; bio-feedback training
BFT Bentonite Flocculation Test
B.F.T. Bachelor of Foreign Trade
BFTA British Fur Trade Alliance
BFTC Boeing Flight Test Center
BFTS British and Foreign Temperance Society
BFUP Board of Fire Underwriters of the Pacific
BFUSA Basketball Federation of the United States of America
BFUW British Federation of University Women
BfV Bundesamt für Verfassungsschutz (German—Federal Office for the Protection of the Constitution)—German FBI roughly equivalent to the Special Branch in Britain)
bfw boiler feedwater
BFYSG British Forces Youth Service, Germany
bg back gear; bag; battle group; beach group; before girls (entered the armed forces); big grin; bluish-green; bombardment group; buccal ganglion; buccogingival; business girl
bg (BG) background (behind tv performers); block grant
b/g bonded goods
bG bluish green
Bg Bengal; Bengalese; Bengali
Bg Berg (German—mountain); *Rogen* (German—arch or bow)
BG Benny Goodman; Birmingham Gage; Birmingham (wire) gauge; Bombay Grenadiers; Bren Gunner; Brigade of Gurkhas; Brigadier-Gen-

eral; British Guiana; Bulgaria (Internet code); Butchart Gardens (Victoria, British Columbia)

B-G Bach Gesellschaft; David Ben-Gurion

B & G Barton and Guestier; Bing and Grøndahl; buildings and grounds

BG Bibliothèque publique et universitaire de Genève (French—Public and University Library of Geneva)

bga blue-green algae (virus)

BGA Better Government Association; British Gliding Association

BGAS Boys and Girls Aid Society

BGAU Berlin Garrison Administrative Unit

B-G b Bordet-Gengou bacillus

BGB Booksellers of Great Britain

BGB Bürgerliches Gesetzbuch (German—Complete Civil Law)

bgc blood group class

BGC Bren Gun Carrier; British Gas Corporation

BGCC Bowling Green College of Commerce

BGCCC Board of Governors of the California Community Colleges

Bge barrage (artificial dam)

BG & E Baltimore Gas and Electric

B.G.E. Bachelor of Geological Engineering

BGEN Biogen Incorporated

B.Gen.Ed. Bachelor of General Education

BGF Banana Growers' Federation; Black Guerrilla Family

BGFE Boston Grain and Flour Exchange

BGFO Bureau of Government Financial Operations

bgg booster gas generator; bovine gamma globulin

BGGRA British Gelatine and Glue Research Association

BGGS Brigadier-General General Staff

bgh bovine growth hormone

bght bought

BGI Bechtel Group Inc; Bridgetown, Barbados (airport)

BGIRA British Glass Industry Research Association

bgirl bound girl

B-girl bar girl

Bgk Bangkok

bgl below ground level

B.G.L. Bachelor of General Laws

BGLA Business Group for Latin America

bglb brilliant-green lactose broth

bglr burglar

bgl(s) bagel(s); beagle(s); bugle(s)

bgrn background music

BGM Bethnal Green Museum; Binghamton, New York (airport); British Guiana Militia; Burma Gallantry Medal

BGMA British Gear Manufacturers' Association; British Guiana Militia Artillery

bgmn baggageman

Bgn Bergen

BGN Board on Geographic Names

BGNR Barren Grounds Nature Reserve (New South Wales)

BGNY Bookbinders' Guild of New York

Bgo Bugo

bgp below-ground pool

bgr bombing and gunnery range

bgrv (BGRV) boost-glide reentry vehicle

bgs bags

bg(s) back gear(s); bag(s)

Bgs Brightlingsea

BGS Brigadier, General Staff; British Geriatrics Society; Brotherhood of the Good Shepherd

B&GS Bombing and Gunnery School

BGS Bundevgrenz Schutz (German—Frontier Troops)—West German NATO forces

bgsa blood-granulocyte specific activity

BGSC Belfer Graduate School of Science (Yeshiva University); Boise-Griffin Steamship Company

BGSM Bowman Gray School of Medicine

BGSU Bowling Green State University

bgt bought

Bgt Bight

Bgt Bugt (Danish—Bay)

BGT Battle Group Trainer; Bender Gestalt Test; British Guiana Time

BGTA Birmingham Group Training Association

BGTT Borderline Glucose Tolerance Test

B Gu British Guiana

BGU Ben Gurion University (Beersheba); Bowling Green University

BGU Biblioteca General da Universidade (Portuguese—University General Library)—in Coimbra

BgUL Bibliothèque générale de l'Université de Liège (French—General Library of the University of Liege)

BGVF British Guiana Volunteer Force

bgw (BGW) battlefield guided weapon

BGW Baghdad, Iraq (airport)

bh bloody hell (British expletive); boiler house; breast height; Brinell hardness

bh bougie-heure (French—candlehour); *brysholder* or *busteholder* (Dano-Norwegian—bra or brassiere)

Bh Brinell hardness

BH Bahrain (Internet code); Base Hospital; Bath & Hammondsport (railroad); Benjamin Harrison (23rd President U.S.); Bill of Health; Brigade Headquarters; Brinell hardness; British Honduras; Broken Hill; magnetization curve (symbol)

B/H Bill of Health; Bordeaux-Hamburg (range of ports)

B & H Bell and Howell; Breitkopf and Härtel

B of H Board of Health

BH Bonne Humeur (French—Good Humor); *Boston Herald*

B H Ben Hur

bha base helix angle

bha (BHA) butylated hydroxyanisole

BHA Bank Holders of America; Bengal Horse Artillery; Bombay Horse Artillery; Boston Housing Authority; British Homeopathic Association; British Honduras Airways

bh ad broach adapter

B.H.Adm. Bachelor of Hospital Administration

BHAFRA British Hat and Allied Feltmakers' Research Association

B'ham Birmingham

BHB British Hockey Board

BHBNM Big Hole Battlefield National Monument

B Hbr boat harbor

BHBS British Honduras Broadcasting Service

bhc beaching cradle; benzene hexachloride (BHC)

BHC Barbers' Hairdressers' Cosmetologists (and Proprietors' Union); Black Hawk College; British High Commissioner; British Hovercraft Corporation; Brotherhood of the Holy Cross

BHCIUS Barbers, Hairdressers, and Cosmetologists International Union of America

BHCSA British Hospitals Contributory Scheme Association

bhd beachhead; bulkhead

BH$ British Honduras dollar

BHD British Forces in Greece; Bronx House of Detention

BHDF British Honduras Defence Force; British Hospital Doctors Federation

B.H.E. Bachelor of Home Economics

B of HE Board of Higher Education

B'head Birkenhead

BHEW Benton Harbor Engineering Works

Bhf Bahnhof (German—station)

BHF Berliner Handels und Frankfurter (bank); British Heart Foundation

bhf(s) black homosexual female(s)

bhfx broach fixture

B H& G Better Homes and Gardens

BHGMF British Hang Glider Manufacturers Federation

bhi brain-heart infusion

BHI Better Hearing Institute; British Horological Institute

BHI British Humanities Index; Bureau Hydrographique Internationale (French—International Hydrographic Bureau)

bhib beef-heart infusion broth

BHIS Burroughs Hospital Information System

BHISSA Bureau of Health Insurance, Social Security Administration

BHK type-B Hong Kong influenza virus

bhl biological half-life

BHL Borax Holdings Limited

Bhm Birmingham, England; Buddhism

BHM Birmingham, Alabama (airport); Bureau of Health Manpower

BHMIA Bald-Headed Men of America; British Hard Metal Association

BHMC Bell & Howell/Mamiya Company

BHMH Benjamin Harrison Memorial Home (Indianapolis, Indiana)

BHMRA British Hydromechanics Research Association

bhm(s) black homosexual male(s)

BHMS Bishop's Home Mission Society

bhn bephenium hydroynaphthoate

Bhn Bremerhaven; Brinell hardness number

BHNWR Bombay Hook National Wildlife Refuge (Delaware)

B Hond British Honduras

B-hor B-horizon (soil zone below A-horizon)

B.Hort. Bachelor of Horticulture

B.Hort.Sci. Bachelor of Horticultural Science

bhp biological hazard potential; boiler horsepower; brake horsepower

BHP Borehole Plugging Program; Broken Hill Proprietary

bhp hr brake horsepower hour

BHPRD Bureau of Health Planning and Resource Development (HEW)

Bhpric Bishopric

bhq battalion headquarters

BHQ Brigade Headquarters

bhr basal heart rate; biotechnology and human research

Bhr Bahrain

BHRA British Hotels and Restaurants Association; British Hydromechanics Research Association

B & HRO Biotechnology and Human Research Office (NASA)

bhs betahemolytic streptococcus

Bhs Bohus

BHS Balboa High School; Boys High School; British Home Stores; British Horse Society; Bureau of Health Services; Burlesque Historical Society; Bushwick High School

B & HS Bonhomie and Hattiesburg Southern (railroad)

B.H.Sci. Bachelor of Household Science

BHSS Bronx High School of Science

bhst bottom-hole static temperature

bht baht tical (Thai monetary unit)

bht (BHT) butylated hydroxytoluene

BHTA British Herring Trade Association

bhtm(s) black heterosexual male(s)

Bhu Bhutan

Bhutan Kingdom of Bhutan (Asian Himalayan nation); *Druk-yul* (its name in the official language called Dzonkha Bhután)

Bhv Bhavnagar

BHVG British Honduras Volunteer Guard

bh/vh body hematocrit/venous hemocrat (ratio)

Bhvn Bremerhaven

bhw boiling heavy water

BHW Boston Hospital for Women

BHYC Boothbay Harbor Yacht Club

B.Hyg. Bachelor of Hygiene

bi background investigation; bacteriological index; base ignition; base of prism in; biopsy; brother in law; burn index; bodily injury; buffer index

bi (BI) binary

bi. bisexual

b/i battery inverter

b & i bankruptcy and insolvency; base and increment

b or i brass or iron (cargo)

Bi biot; bismuth (symbol)

B^i Beni or *Beni* (Arabic—sons of)

BI Babson Institute; background investigation; Bahama Islands; Beer Institute; Bermuda Islands; Braniff International; British India; British Isles; Brookings Institution; Bureau of Investigation; Burundi (Internet code); Business International; National Biscuits (stock exchange symbol)

B of I Bureau of Investigation

BI *Banca d'Italia* (Bank of Italy)

BIA Barbecue Industry Association; Bicycle Institute of America; Binding Industries of America; Board of Immigration Appeals; Braille Institute of America; Brazilian International Airlines; Brick Institute of America; Building Industry Association; Bureau Issues Association; Bureau of Indian Affairs; Bureau of Insular Affairs; Bureau of Inter-American Affairs; Business Improvement Area

BI & A Bureau of Intelligence and Research (US Department of State)

BIAA Bureau of Inter-American Affairs (US Department of State)

BIAC Business and Industry Advisory Committee (NATO)

BIAE British Institute of Adult Education

BIALL British and Irish Association of Law Librarians

bialy bialystok roll (holeless onion-flaked bagel)

bias battlefield illuminator airborne system

BIAS Brooklyn Institute of Arts and Sciences

BIATA British Independent Air Transport Association

bib. bibliography; biographical inventory blank; bottled in bond

bib. (BIB) baby incendiary bomb

bib *biblioteca* (Italian, Latin, Portuguese, Romanian, Spanish—library); *bibliotecario* (Spanish—librarian); *biblioteka* (Albanian, Bulgarian, Macedonian, Polish, Russian, Serbo-Croatian, Slovene, Ukrainian); *bibliotek* (Dano-Norwegian or Swedish); *bib-*

lioteket (Dano-Norwegian or Swedish); *bibliotheek* (Dutch or Flemish); *bibliotheka* (Latin); *bibliotheke* (Greek); *Bibliothek* (German); *Bibliothèque* (French)

bib. *bibe* (Latin—drink)

Bi Bible; Biblical

Bib *Biblica* (Latin—Bible)

BIB Biennale of Illustrations Bratislava (international exhibition of children's book illustrations); Board for International Broadcasting; Broadcast Information Bureau; Bureau of International Broadcasting

BIB *Berliner Institut für Betriebsführung* (German—Berlin Business Management Institute)

BIBA Babson Institute of Business Administration; British Insurance Brokers' Association

Bib Amb *Biblioteca Ambrosiana* (Italian—Ambrosian Library)—in Milan

Bib Apo Vat *Biblioteca Apostolica Vaticana* (Italian—Vatican Library)

bib b *biblioteksbind* (Dano-Norwegian—library binding)

bibber winebibber

Bib Bod *Bibliotheca Bodmeriana* (Latin—Bodmer Library)-in Cologny/Geneva contains first editions of Cervantes, Dante, Goethe, Homer, Shakespeare, a Gutenberg Bible and one of the three recorded copies of Luther's *Disputio pro Declaratione Indulgentiarum* from 1517

BIBC British Isles Bowling Council

Bib Cen *Biblioteca Central* (Spanish—Central Library)—in Mexico City's Ciudad Universitaria

Bib Ecu *Biblioteca Ecuatoriana* (Spanish-Ecuadorian Library)—in Quito, also known as Padre Aurelio Espinosa Pólit

Bib Esc *Biblioteca de San Lorenzo el Real de El Escorial* (Spanish—Library of Royal San Lorenzo of the Es-

corial Palace)—monastic library within the Escorial Palace in the Guadarramas near Madrid

BIBF British and Irish Basketball Federation

bi or bin (Latin *bini*—two by two; Latin *bis*—twice)—binary, binoculars, binomial, bipolar

bibl bibliotec-; bibliotek-; bibliothec-; bibliothek; bibliothèque

biblio bibliographical imprint or note; biblioclasm (book destruction); biblioclast (book destroyer); bibliogenesis (book production); bibliognost (bibliographic expert or book expert); bibliogony (book production); bibliograph (bibliographer); bibliographer; bibliographic(al); bibliography

biblioc biblioclasm; biblioclast

bibliog bibliographer; bibliographic(al); bibliography

bibliograph bibliographer; bibliographee; bibliography

biblioklept bibliokleptomania(c)

bibliol bibliolater (person with excessive admiration or reverence for books); bibliolatrous (characterized by bibliolatry); bibliolatry (book worship); bibliological; bibliologist; bibliology (scientific description and study of books)

bibliom bibliomancy; bibliomane (collector of books); bibliomania; bibliomaniac; bibliomanist

bibliop bibliopegic (relating to book binding); bibliopegist; bibliopegy; bibliophagist; bibliophile; bibliophilia; bibliophobe; bibliophobia; bibliopole

bibliopsy bibliopsychology (study of authors, books, and readers as well as their interrelationships)

bibliothec bibliotheca (bibliographer's catalog or a library); bibliothecal (belonging to the library); bibliothecar (librarian); bibliothecary

bibliother bibliotherapeutic; bibliotherapist; bibliotherapy

bibliotrain railroad car—converted into a mobile library

*bibl mun bibliothèque munici-
pale* (French—city library,
public library)
Bib Mus Biblioteca Musicale
(Italian—Musical Li-
brary)—in Rome's Via dei
Greci
Bib Nac Biblioteca Nacional
(Spanish—National Library)
Bib Nur Biblioteka Narodowa
(Polish—National Li-
brary)—in Warsaw
Bib Nat Bibliothèque Nationale
(French—National Li-
brary—Paris)
*Bib Naz Bra Biblioteca Nazio-
nale Braidense* (Ital-
ian—Braidense National Li-
brary) in Milan
*Bib Naz Cen Biblioteca Nazio-
nale Centrale* (Italian—Na-
tional Central Library)
Bib Pal Biblioteca de Palacio
(Spanish—Palace Li-
brary)—Madrid
BIBRA British Industrial Bio-
logical Research Association
bibs. bibliographies
Bib Soc Ain Bibliographical
Society of America
Bib Soc Can Bibliographical
Society of Canada
*Bib Sor Bibliotèque de la Sor-
bonne* (French—Sorbonne
Library)
*Bib Uni Biblioteca Universi-
taria* (Spanish—University
Library)
bic bank instrument contract;
bank investment contract
Bic Société Bic (ballpoint pen
factory founded by Baron
Marcel Bich)
BIC Barrier Industrial Council;
Bengal Irregular Cavalry;
Bronx Irish Catholic; Bureau
of International Commerce
*BIC Bureau International des
Containers* (French—Inter-
national Container Bureau)
bicarb sodium bicarbonate
BICC Battlefield Information
Control Center; British Insu-
lated Callenders Cables
BICEMA British Internal
Combustion Engine Manu-
facturers Association
bicept book indexing with con-
text and entry points from text
BICERA British Internal Com-
bustion Engine Research Association

BICERI British Internal Combustion Engine Research Institute
bichrome sodium bichromate
BICs Business and Innovation
Centers
BICS British Institute of Cleaning Science
BICTA British Investment
Casters' Technical Association
bicv biconcave
bicx biconvex
Bicycle Bicycle Thieves
bicyea best ice cream you ever
ate
bicyplane bicycle-powered airplane (first cross-Channel
flight achieved June 12, 1979
by the *Gossamer Albatross*
designed by Paul MacCready
of Pasadena, California
pedalled and piloted by Bryan
Allen)
bid. (BID) brought in dead
b.i.d bis in die (Latin—twice
daily)
Bid Bideford
BId Bureau of Identification
BID Business Improvement
District
*BID Banco Interamericano de
Desarrollo* (Spanish—Inter-
american Development Bank)
B.I.D. Bachelor of Industrial
Design
B. of I.D. Bachelor of Interior
Design
bidap bibliographic data pro-
cessing program
Biddy Bridget; Briged; Brigid
bidec binary-to-decimal con-
verter
BIDP Basic Institutional De-
velopment Program
BIDS Brittle hair, Impaired in-
telligence, Decreased fertility
and Short stature
BIE Bureau of Industrial Eco-
nomics (DoC); *Bureau Inter-
national d'Education*
(French—International Bu-
reau of Education); *Bureau
International des Expositions*
(French—International Bu-
reau of Expositions)
B.I.E. Bachelor of Industrial
Engineering
Bieder Biedermeier
BIEE British Institute of Electrical Engineers
bien biennial

BIEN Basic Income European
Network
BIEPR Bureau of International
Economic Policy and Re-
search
BIET British Institute of Engineering Technology
*BIETA Biblioteca Interameri-
cana de Estadistica Teórica y
Aplicada* (Spanish—Inter-
american Library of Theoretical and Applied Statistics)
bif buyer-induced failure
BIF Bombardier's Information
File; British Industries Federation
biff binary interchange file format
BIFMA Business and Institutional Furniture Manufacturers Association
*BIFN Banque Internationale
pour le Financement de
l'Énergie Nucléaire*
(French—International Bank
for the Financing of Nuclear
Energy)
bif O opium
bifs biologically integrated
farm system
BIFUS Britain, Italy, France,
United States
big. best in group; bigamist; bigamy; biological isolation garment
BIG Bartoni International Gallery (Melbourne); Basic Industries Group; Beneficial Insurance Group; Better
Independent Grocers; Blacks
In Government
BIG Bazak Israel Guide
Big Belts Big Belt Mountains
of Montana
Big Cats lions, tigers, jaguars,
leopards, and pumas
big demo big demonstration
biggies big ones
big H big house (slang—penitentiary such as San Quentin
or Sing Sing); hernia; heroin
Big Horns Big Horn Mountains, Wyoming
BIGI Bond Investors Guaranty
Insurance
Big Mac Municipal Assistance
Corporation
big mo big momentum
big O's obsolescence, overcrowding, overburdening,
omigod!
bigr bigger

bigs biological isolation garments

bigst biggest

Big Ten also known as the Western Conference; college teams formed around football in 1896

big. unlwfl—trig awf bigamy is unlawful—trigamy is awful, explained Ogden Nash

bih benign intracranial hypertension

BIB Beth Israel Hospital

BIHA British Ice Hockey Association

bihor. bihorium (Latin—two hours)

BII Beckman Instruments Incorporated; Biosophical Institute Incorporated

BIIA British Institute of Industrial Art

BIICL British Institute of International and Comparative Law

Bij Benjamin

bijb bijbelse term (Dutch—biblical term)

bijv bijvoorbeeld (Dutch—for example)

bi k bilge keel

Bik. Bikkurim

bike bicycle

BIKE Biotechnology Information Knot for Europe

bikers motorcyclists

biki bikini

bil bilateral; billet; billion; block input length

b-i-l brother-in-law

Bil Bilbao

BIL Billings, Montana (airport); Braille Institute Library; Brierly Investments Limited; British India Line; Bulk Items List

BILA Bureau of International Labor Affairs

bilat bilateral

Bilders Bilderbergers (now called Tri-Laterals)

bildg bill of lading

bildl bildeich (German—figuratively)

bile. balanced-inductor logical element

BILG Building Industry Libraries Group

bili bilirubin

biling bilingual(ism); bilingualist(ic)(al)(ly)

bilj biljarttern (Dutch—billiards)

bil k bilge keel

bill billede (Dano-Norwegian —illustrations)

Bill Billie; Billy; William

BILL Bofors Infantry Light and Lethal (missile)

BIO Brigade Intelligence Officer

bill. acad billiard academy

Billie William

billion (American—a thousand million, 10^9); (British—a million million , 10^{12})

Bill of Rights first ten amendments to the *Constitution of the United States*

Billy William

bil(s) billion(s)

BILS British International Law Society; Butterworth Industrial Laws Service

bilt built

bim ballistic intercept missile; beginning of information marker

bi-m bi-monthly

bim bimensile (Italian—semimonthly); *bimestrale* (Italian—bimonthly); *bimestre* (Italian—two-month period)

Bim Barbadan

BIM British Institute of Management

B.I.M. Bachelor of Indian Medicine

BIM Bord Iascaigh Mhara (Gaelic—Sea Fisheries Board)

bimac bi-stable magnetic core

BIMCAM British Industrial Measuring and Control Apparatus Manufacturers Association

BIMS Business Information Management System

BIMT Bahama Islands Ministry of Tourism

bin. bank identification number; binary

BINA Broadcast Institute of North America

BINA Bureau International des Normes de l'Automobile (French—International Bureau of Automobile Standards)

binac high-speed electronic digital computer

b-in-B banned in Boston (and therefore a best-seller)

BINCOS Binder Control System

BIND Berkeley Internet Name Domain

bind. binding

B.Ind. Bachelor of Industry

B.Ind.Ed. Bachelor of Industrial Education

BINDT British Institute of Non-Destructive Testing

Binet Binet-Simon (psychological test)

Bing Binghampton

Binj Benjamin

BINL Basic Inventory of Natural Language

BINM Buck Island National Monument, St Croix, Virgin Islands

binocam binocular and camera (combination instrument)

binocs binoculars

bins (Cockney contraction—binoculars)

BINS Barclays Integrated Network System

binsum brief intelligence summary

BINW Blackbeard Island National Wildlife Refuge (Georgia)

BINZ Bankers Institute of New Zealand

bio biographical; biography; biological; biology

bio (Latin prefix—life)— biology, the study of life and living things

Bio Biology

BIO Base Installation Officer; Bedford Institute of Oceanography; Biological Information-Processing Organization; Brigade Intelligence Officer; Broadcasting Information Office

BIOA Bureau of International Organization Affairs (U.S. Department of State)

bioact bioactive; bioactivity

bioastro bioastronaut(ic)(al) (ly)

bioauto bioautograph(ic)(al) (ly)

biocam binocular camera

biochem biochemical; biochemist; biochemistry

Biochem Biochemistry

biochron biochronometry

biocid biocidal; biocide

bioclean biologically clean

biocon biocontamination

biocyb biocybernetics

biodef biological defense

biodeg biodegradability; biodegradable; biodegradation; biodegrade; biodegraders; biodegrading

biodeg(s) biodegradable(s)

biodes biodestructible

biodet biodeterioration

bioelectrog bioelectrogenesis

bioelectron bioelectron(ic)(al)oy); bioelectronics

bioeng bioengineer(ing); biological engineer(ing)

bioenv bioenvironment(al)(1y); bioenvironmentalist

bioex bioexperiment(ation)

biog biographer; biographical; biography

biogeo biogeology

biogeog biogeographer; biogeographic(al); biogeography

bioinstru bioinstrument(al)(ly); bioinstrumentation

biol biological; biologist; biology

biol *biologi* or *biologisk* (Dano-Norwegian—biology or biologist)

Biol Biology

Biol Abstr *Biological Abstracts*

BIOLWPNSYS Biological Weapon System (USA)

biomass mass of biological material

bio-mass biological mass source of ethanol and methanol from crops and trees

BIOMASS Biological Investigations of Marine Antarctic Systems and Stocks

biomath biomathematician; biomathematics

biomed biomedical; biomedicine

bionics biology + electronics

bio-org bio-organic(al)(ly)

biophys biophysical; biophysicist; biophysics

biopol(s) biopolymer(s)

bior business input-output rerun

biore bioresearch(er)

BIOREP Biological Attack Report

bios basic input/output system; biologically integrated orchard system

bios (BIOS) biological satellite

BIOS Basic Input/Output System; Biological Investigations of Space

biosat biosatellite

biosci bioscience; bioscientific; bioscientist

biosen(s) biosensor(s)

BIOSIS Biosciences Information Service of *Biological Abstracts*

biospel biospeleologist(ic)(al)(ly); biospeliology

biostat biostatistic(s)

biot biotron(ic)(al)(ly)

BIOT British Indian Ocean Territory (Aldabra, Chagos Archipelago, Des Roches, Diego Garcia, Farquhar, Mauritius, Seychelles)

BIOTA Biological Institute of Tropical America

biotec biotechnical(ly); biotechnological(ly); biotechnologist; biotechnology

biotech biotechnologist; biotechnology

biotel biotelemetric; biotelemetry

biotrans biotransformation; bio-transformer

biowar biological warfare

bip background interference procedure(s); bacterial intravenous protein; balanced in plane; binary image processor; bismuth iodoform paraffin; books in print; borough-interborough problem(s)

bip (BIP) body improvement plan

BiP *Books in Print*

BIP Board for International Broadcasting; Border Industrial Program; British Industrial Plastics; British Institute of Physics

BIP *Banco Industrial del Perú* (Spanish—Industrial Bank of Peru)

BIPAD Bureau of Independent Publishers and Distributors

bipco built-in-place components

bipd biparting doors

biphet biphetamine (drug stimulant)

bipkwele former unit of currency in Equatorial Guinea

BIPL Burma Industrial Products Limited

BIPM *Bureau international des Poids et Mesures* (French—International Bureau of Weights and Measures)

BIPO British Institute of Public Opinion

bipp bismuth, iodoform, paraffin paste

BIPP British Institute of Practical Psychology

BIPS British Integrated Programme Suite

BIPS *Bibliographic Information Publication System*

bipyr bipyramidal

biquin biquinary

bir basic incidence rate; built-in robes (closets)

Bir *Birmania* (Italian or Spanish—Burma); *Birmânia* (Portuguese—Burma)

BIR Board of Inland Revenue; Board of Internal Revenue; British Institute of Radiology; Bureau of Intelligence and Research; Bureau of Internal Revenue

BIRC Bio-Integral Resource Center

BIRD *Banque Internationale pour la Reconstruction et le Développement* (French—International Bank for Reconstruction and Development)

BIRDDOG Basic Investigation of Remotely Detectable Deposits of Oil and Gas

birdie battery integration and radar display equipment

BIRE British Institution of Radio Engineers

B.Ir.Eng. Bachelor of Irrigation Engineering

BIRF Brewing Industry Research Foundation

BIRF *Banco Internacional de Reconstrucción y Fomento* (Spanish—International Bank for Reconstruction and Development)

BIRG Basic Income Research Group (England)

birg(ing) bask(ing) in reflected glory

birl girlish boy (transvestite)

Birm Birmingham

Birmingham notation (*see* GKD-notation)

BIRMO British Infra-Red Manufacturers' Association

BIRMPDis Birmingham Procurement District (U.S. Army)

BIRMS Battelle Interactive Resources Management System (computerized)

BIRP Beverage Industry Recycling Program

birr monetary unit of Ethiopia

BIRS Basic Indexing and Retrieval System; British Institute of Recorded Sound

birt bolt installation and removal tool

birthquake population explosion

bis best in show; bissextile

Bis Bismarck; Bissau

BIS Bank for International Settlements; Bismarck, North Dakota (airport); Board of Inspection Survey (USN); British Imperial System; British Information Service; British Interplanetary Society; Business Information System

bis in 7d. bis in septem diebus (Latin—twice in seven days, twice weekly)

BISA British International Studies Association

BISAC Book Industry System Advisory Committee

bisad business information systems analysis and design

BISAKTA British Iron, Steel, and Kindred Trades Association

BISAM Basic-Indexed Sequential-Access Method

BisArch Bismarck Archipelago

bisc biscuit

Bisc Biscayan

BISCA Building Industry Subcontractors Association

Biscayne national park south of Miami, Florida, consisting of islands leading to Key West

BISCO British Iron and Steel Corporation

bis in d. bis in dies (Latin—twice daily)

BISDN Broadband Integrated Services Digital Network

bisett bisettimanale (Italian—bi-weekly)

bisex bisexual(ism); bisexualist(ic)(al)(ly); bisexually

BISF British Iron and Steel Federation

BISFA British Industrial and Scientific Film Association

BISG Book Industry Study Group

Bish Bishop

bishaw bicycle rickshaw

Bish Mus Bishop Museum

Bishop Bishop Museum; Bishop Museum Press

BISITS British Iron and Steel Industry Translation Service (BISRA)

BISL British Information Service Library

Bismarck North Dakota's capital; Prince Otto Eduard Leopold von Bismarck-Schönhausen—the Iron Chancellor

Bismarck Pen North Dakota Penitentiary at Bismarck

Bismarcks Bismarck Islands

BISN British India Steam Navigation (company)

bisp between ischial spines; bispinous (interspinous diameter)

BISPA British Independent Steel Producers Association

BISRA British Iron and Steel Research Associates

BISS Battlefield Identification System Study (NATO)

BISTA Bureau of International Scientific and Technological Affairs (U.S. Department of State)

Bister Bicester

bisw bisweilen (German—sometimes)

bisync binary synchronous computer

bit. binary digit; bituminous

BIT Bradford Institute of Technology; British Independent Television; Built-In Test; *Bureau International du Travail* (French—International Labor Organization)

BITA British Industrial Truck Association

BITB Building Industry Training Board

bitblt binary digit block transfer

BITC Bahamas International Trust Company

BITCH Black Intelligence Test of Cultural Homogeneity

bite. base installation test equipment; built-in test equipment

BITE Base Installation Test Equipment

bitm bituminous

BITM Birla Industrial and Technological Museum

bitn bilateral iterative network

BITNET Because It's Time Network

bito burnishing tool

BITO British Institution of Training Officers

bit(s) binary digit(s)

BITTC Building Industry Technicians Training Council

bitu benzyl-thiourea

BITU Bustamante Industrial Trade Union

bitum bituminous

bitumd bituminized

bituminous soft coal

biu basic income unit; bus interface unit

B-I U Bar-Ilan University

BIU Bureau International des Universités (French—International University Bureau)

biv bivouac

BIV Banco Industrial de Venezuela (Spanish—Industrial Bank of Venezuela)

bivar bivariant (function generator)

bivs biologically integrated vineyard system

bi-w bi-weekly

BIW Bath Iron Works; Boston Insulated Wire (and Cable Company)

BIWF British Israel World Federation

BIWS Bureau of International Whaling Statistics

bix binary information exchange

BIX McGraw-Hill trademark

BIY Bedfordshire Imperial Yeomanry

biz bizarre; business

BIZ Bank für Internationalen Zahlungsausgleich (Bank for International Settlements)

bizad business administration

bizjet business-type jet airplane

bizmac business machine computer

bizman business man

BIZNET American Business Network (database of Chamber of Commerce)

B^j Burj (Arabic—bluff, cliff, fort, tower)

bj back judge (football); biceps jerk; blow job (fellatio)

b & j bone and joint

BJ Benin (Internet code); Benito Suarez; Byron Jackson (Borg-Warner)

B.J. Bachelor of Journalism

B & J Burke & James

B of J Bank of Japan

BJA Burlap and Jute Association; Bureau of Justice Assistance,

b/Jan binding expected in January (for example)

bjc book jacket cover

BJC Baltimore Junior College; Bismarck Junior College; Boise Junior College; Brevard Junior College; Bureau of Juvenile Correction (Delaware)

BJCEB British Joint Communications-Electronics Board

BJCO British Joint Communications Office

BJCPA Baptist Joint Committee on Public Affairs

bjf batch-job format

BJ1P Better Jobs and Incomes Program

B Jon Ben Jonson

BJOS British Journal of Occupational Safety

BJp Bence Jones protein

BJP Bharatiya Janata Party (India)

BJS Bureau of Justice Statistics

BJSM British Joint Services Mission

bjt bipolar junction transistor

BJTRA British Jute Trade Research Association

BJU Bob Jones University

B.Juris. Bachelor of Jurisprudence

bk balk; bank; below the knee; black; book; brake; break

Bk berkelium; Brook

B^k Bank

Bk Buku (Indonesian or Malay—hill, mountain)

B-K Blaw-Knox

B/K oscillator Barkhausen/Kurz oscillator

BK Biblioteka Kombëtare (Albanian—National Library)—Tirana

bka below-knee amputation

BKA Bundeskriminalamt (German—Federal Criminal Ministry)

bkble bookable; bookmobile

bkbndg bookbinding

bkbndr bookbinder

bkbrd bakboord (Dutch—port) port side

bkc benzalkonium chloride (BKC)

bkcy bankruptcy

bkd blackboard

bk di brake die

bkfst breakfast

bkg banking; bookkeeping; breakage

bkgd background

bkge brokerage

bkhs blockhouse

BKII Vsesoyuenaya Kommunisticheskaya Partiya (Russian—All-Union Communist Party)

BKK Bangkok, Thailand (airport)

bkkpr bookkeeper

bklr black letter

bklt booklet

Bklyn Brooklyn

Bklyn Brdg Brooklyn Bridge

Bklyn HTF Brooklyn Homicide Task Force

Bklyn Mus Brooklyn Museum

Bklyn Phil Brooklyn Philharmonic Symphony

bkm buckram

BKM Moscow, former USSR (Bykovo Airport)

bkn broken

Bkn Birkenhead

BKN broken (aircraft code)

bkpg bookkeeping

bkpr bookkeeper

bkpt bankrupt

bkr baker; beaker; breaker

bks bunks; barracks; books; brakes

BKS British Kinematograph Society

BKS Biotech Knowledge Sources

bk sh bookshelves

Bks for Libs Books for Libraries

bksp backspace (flow chart)

bkt basket; bracket

Bkt Bukit (Malay—hill, hilly street)

bkt(s) basket(s)

bktt below knee to toe

bkw breakwater

bkwp below-knee walking plaster (cast)

bl balls; bank larceny; baseline; bats left (baseball); billet; bleed(ing); black; blood; blood loss; blue; body length; bomb line; breech-loading; buccolin-gual; butt line; buttock line

bl. bibliography

b/l basic letter; bill of lading (B/L); blueline; blueprint

b & l ball and lever; business and loan

bl blad, blank (Dano-Norwegiarm—leaf, sheet, blank)

Bl Burkitt's lymphoma (BL)

Bl Blasinstrument (German—wind instrument; woodwind); Blatt(er) [German—leaf; leaves, page(s)]; Böluk (Turkish—company)

BL Barrister-at-Law; Basutoland; Bengal Lancers; Blessed Lady; Boatswain Lieutenant; Bonanza Airlines; British Legion; British Leyland; British Library (formerly British Museum Reading Room); British Library (London); Bundelkund Legion

B-L Belgium-Luxembourg

B.L. Bachelor of Letters

B/L Bill of Lading

B & L Bausch & Lomb; Building and Loan (association or bank)

bl a blandt andet (Dano-Norwegian—among other things); blandt andre (Danish—among other things)

Bla Belawan; Brasilia

BLA Bangladesh Library Association; Black Liberation Army; Board of Landscape Architects; Bombay Library Association; British Legal Association; British Liberation Army; British Library Association

B.L.A. Bachelor of Landscape Architecture; Bachelor of Liberal Arts

BL-AA Biblioteca Luis-Angel Arango (Spanish—Luis-Angel Arango Library)—Bogotá, Colombia's library named for a former bank president

BLAC British Light Aviation Center

BLACC British and Latin American Chamber of Commerce

black. blackmail

Black Black's United States Supreme Court Reports

Blackfeet Blackfeet Indians

Black Islands see Melanesia

Blackstairs Blackstairs Mountains of Ireland

Blackstone Sir William Blackstone's Commentaries on the Laws of England

blad blotting pad

blade. basic level automation of data through electronics

BLADES Bell Laboratories Automatic Design System

BLAISE British Library Automated Information Service

Blake Blakely; Blakeman; Blakeslee

Blan Blanca; Blanchard; Blanco; Bland(on)

Blanca Blanche

B.Land.Arch. Bachelor of Landscape Architecture

Blaskets Blasket Islands on Ireland's Atlantic coast

blast blastos (Greek—sprout) —blastoderm, blastodisc, blastopore, osteoblast

Blast Blastoidea

BLAST Black Legal Action for Soul in Television

Bla Sta Blackfriars Station

BLAT British Life Assurance Trust

Blatch Blatchford's United States Circuit Court Reports

BLAV British Latin American Volunteers

Blaxican(s) Black Mexican(s)

BLB Boothby-Lovelace-Bulbulian (oxygen mask)

blc balance; boundary-layer control

BLC Bengal Light Cavalry; British Lighting Council; Bombay Light Cavalry

blchd bleached

blchg bleaching

bl cult. blood culture

bld blood; blood and lymphatic system; bloody; bold; boldface

BLD Burglary Larceny Division (NYPD)

bldg building

Bldg Engr Building Engineer

bldi blank die

bldr builder

BLE Brotherhood of Locomotive Engineers

B & LE Bessemer and Lake Erie (railroad)

bleap bought ledger and expenditure analysis package

BLEDCO Brooklyn Local Economic Development Corporation

Blemish Belgian & Flemish

blems blemishes (acne, blackheads, pimples)

blenno blennorrhea

BLESMA British Limbless Ex-Service Men's Association

bless. bath, laxative, enema, shampoo, and shower

bleu blind landing experimental unit

BLEU Belgium-Luxembourg Economic Union

bleve boiling-liquid expanding vapor explosion

Blf Bluff

BLF & E Brotherhood of Locomotive Firemen and Enginemen

blg betalactoglobulin

Blg Bulgarian

BLG Burke's Landed Gentry

BLH Baldwin-Lima-Hamilton; Bihar Light Horse

BLHA British Linen Hire Association

BLHS Ballistic Laser Holographic System

BLI Bliss & Laughlin Industries; Buyers Laboratory Incorporated

B.L.I. Bachelor of Literary Interpretation

BaLI Bank Leumi le-Israel (Bank Association of Israel)

BLIA Buddha's Light International Association

B.Lib.S. Bachelor of Library Science

B.Lib.Sci. Bachelor of Library Science

Blick Blickensderfer portable typewriter

Blinder NATO code name for Soviet Tu-22 bomber

blip background-limited infrared photoconductor

blip. background-limited infrared photography

BLIP Big Look Improvement Program

BLIS Bell Laboratories Interpretive System

BLISS Baby Life-Support Systems

B-lite baton-flashlight combination

B.Litt. Baccalaureus Literarum (Latin—Bachelor of Literature)

Blitz Blitzkreig (German—lightning war)

bliz blizzard; blizzardly; blizzardous

blk black; block; blocking; bulk

Blk Block

blkcnt block count (flow chart)

blkd bulkhead

Blk Eng Black English

blk lt black light

blk rt bulk rate

Blk(s) Black(s)

blksh blackish

blksmith blacksmith

blkstp blackstrap (molasses)

bll below lower limit

BLL Butyrka, Lefortovo, and Lubyanka (Moscow's most dreaded prisons)

BLLD British Library Lending Division (Boston Spa)

BLLRCS Bureau of Library and Learning Resources and Community Services (Office of Education)

blm bilayer lipid membrane

blm besa la mano (Spanish—a kiss to your hand)

BLM British Leather Manufacturers; British Leland Motor (corporation at one time merging Austin, British Motor Moldings, Jaguar, Morris, Riley, Rover, Triumph, Wolseley); Bureau of Land Management (General Land Office and Grazing Service)

B.L.M. Bachelor of Land Management

BLM Bonniers Literary Magasin (Bonnier's Literary Magazine)

BLMA British Lead Manufacturers' Association

BLMC British Leyland Motor Corporation

BLMRA British Leather Manufacturers' Research Association

BLMRCP Bureau of Labor-Management Relations and Cooperative Programs

BLMS Book-Library-Management System

bln balloon; bronchial lymph nodes

Bln Berlin

blnk blank (flow chart)

blnkt blanket

BLNR Benton Lake National Refuge (Montana)

BLNWR Big Lake National Wildlife Refuge (Arkansas); Bitter Lake NWR (New Mexico); Buffalo Lake NWR (Texas)

BLNY Book League of New York; Booksellers League of New York

blo blower
BLO Bombardment Liaison Officer; British Liaison Officer
BLOB Ban Large Office Buildings
Bloch Pub Bloch Publishing Company
block. blockade
blodi block diagram (compiler)
blokops blockade operations
bloody (Early English—By Our Lady)
Bloomies Bloomingdale's
blooper blunder and error
Blos Blossom
BLOT British Library of Tape Recordings
blou blouse
B-love being love (unselfish accepting love of another person, according to Maslow)
BLOWS British Library Of Wildlife Sounds
blp back-loading point
blp *besa los pies* (Spanish—a kiss to your feet)
BLP British Labor Party
BLPES British Library of Political and Economic Science (London)
bl pr blood pressure
blr beyond local repair; boiler; breech-loading rifle; broadline system
BLR Ballistic Research Laboratories (USA); Business & Legal Reports Incorporated
BLRA British Launderers' Research Association
BLRD British Library Reference Division (British Museum Library)
blrmkr boilermaker
BLROA British Laryngological, Rhinological, and Otological Association
blrp boilerplate
bls bales; barrels; binary light switch; blood sugar
bls *bawang. luya, siling labuyo* (Philippines— tea)
BLS Boston Latin School; Brooklyn Law School; Buccal-Lingual Syndrome; Bureau of Labor Standards; Bureau of Labor Statistics
B.L.S. Bachelor of Library Science; Bachelor of Library Service

B.L.S. *Benevolenti Lectori Salutern* (Latin—Salutations to the Kind Reader)
BLSA Black Law Students Association
BL & SA Bank of London and South America
BLSC Basic Logistics Support Company
BLSGMA British Lampblown Scientific Glassware Manufacturers' Association
blsh bluish
blsn blowing snow
blstg pwd blasting powder
blstl billet steel
blsw barrels of load salt water
blswd barrels of load salt water per day
blt blood type; built
blt (BLT) bottom-loading transporter
b-l-t bacon, lettuce, and tomato (sandwich)
BLT Battalion Landing Team; Before Large Telescopes
Bltc Baltic
bltg belting
bltn(s) built-in(s)
blu blue
B-L u Bessey-Lowry units
Blubo *Blut und Boden* (German—blood and soil)
BLUCB Bancroft Library of the University of California at Berkeley
blue bullet blue-tipped bullet color-coded to indicate its incendiary purpose
Blue Ridge Blue Ridge Mountains of the Appalachian range; national parkway in North Carolina and Virginia
Blues Blue Mountains
blunt marijuana joint rolled in a cigar
BLV British Legion Village
Blvd Boulevard
BLW Baldwin-Lima-Hamilton; Bount Land Warrant
BLWA British Laboratory Ware Association
Bly Blyth
BLYMSA *Banco de Londres y México SA* (Spanish—Bank of London and Mexico Corporation)
Blz Belize (formerly British Honduras); Belizian
blz *bladzijde* (Dutch—page)

bm basal metabolism; basement membrane; beam; board measure; body mass; bone marrow; book of the month; bowel movement; boy mechanic; buccal mass; buccomesial
bm (BM) bench mark; buffer mark (flow chart); buffer modules
b/m (B/M) bill of material; black male; brown male
bm *bez mista* (Czech—no place of publication)
b.m. *balneum maris* (Latin— bath in sea water); *bene merenti* (Latin—to the well-deserving)
Bm beam; birthmark; black male; board measure; bowel movement; Burma; Burmese
BM Beachmaster; Bermuda (Internet code); Boatswain's Mate; Boston & Maine (railroad); Bravery Medal (Australia); Brian Mulroney— Canada's 23rd prime minister Brigade Major; British Museum; Bronze Medal (USA); Brooklyn Museum; Bureau of Medicine; Bureau of Mines; Bureau of the Mint; Business Manager
B-M Bolinder-Munktell; Bristol-Myers
B.M. Bachelor of Medicine; Bachelor of Music
B & M Beaufort & Morehead (railroad); Boston & Maine (railroad)
B of M Bank of Montreal; Bishop(ric) of Münster; Bureau of Mines
BM *Banca Mondiale* (Italian—World Bank); *Banco de México* (Spanish—Bank of Mexico); *Banco Mundial* (Portuguese or Spanish—World Bank); *Banque du Monde* (French—World Bank); *Beata Maria* (Latin—Blessed Mary)
bma brigade maintenance area
BMA Bald Men of America; Baltimore Museum of Art; Bank Marketing Association; Bible Memory Association; Bicycle Manufacturers' Association; British Medical Association; British Military Administration; British Mil-

itary Authority; Stockholm, Sweden, airport (3-letter code)

BMAD Black Mothers Against Drugs

B.Mar.E. Bachelor of Marine Engineering

B.Mar.Eng. Bachelor of Marine Engineering

BMASR Bureau of Military Application of Scientific Research

bmat beginning of morning astronomical twilight

BMAT British Military Advisory Team

B.Math. Bachelor of Mathematics

BMATT British Military Advisory Training Team

BMB Ballistic Missile Branch (USA); British Medical Board; British Metrication Board

BMB British Medical Bulletin

B-M B Baader-Meinhof Bande (German—Baader-Meinhof Gang)—terrorist Red Army Group

BMBW Bundesministerium für Bildung und Wissenschaft (German—Ministry for Education and Science)

bmc blockhouse monitor console

BMC Ballistic Missile(s) Center; Ballistic Missiles Committee; Book-of-the-Month Club; Box Maker's Certificate; British Mountaineering Council; Bryn Mawr College; Business Microcomputer (Technology Group)

BMCC Blue Mountain Community College

BMCS Ballistic Missile Cost Study; Bureau of Motor Carrier Safety

bmd births, marriages, deaths; bone marrow depression

BMD Ballistic Missile Defense; Becker Muscular Dystrophy; Bengal Medical Department; Bombay Medical Department; Bureau of Medical Devices

B-M-D Blow-Me-Down, Nova Scotia

BMDATC Ballistic Missile Defense Advanced Technology Center (USA)

BMDCA Ballistic Missile Defense Communications Agency

BMDEAR Ballistic Missile Defense Emergency Action Report

BMDITP Ballistic Missile Defense Integrated Training Plan

BMDM British Museum Department of Manuscripts

BMDMB Ballistic Missile Defense Missile Battalion (USA)

BMDMP Ballistic Missile Defense Master Plan

bmdns basic mission, design number, and series (aircraft)

BMDO Ballistic Missile Defense Operations; Bomb and Mine Disposal Officer

BMDOA Ballistic Missile Defense Operations Activity

BMDPM Ballistic Missile Defense Program Manager

BMDPO Ballistic Missile Defense Program Office(r)

bmdr bombardier

bmds ballistic missile defense system

BMDSCOM Ballistic Missile Defense System Command (USA)

BMD System Ballistic Missile Defense System

bme biomedical engineering; bulk memory element

BME Blue Mountains Expeditions; Brotherhood of Marine Engineers

B.M.E. Bachelor of Mechanical Engineering; Bachelor of Mining Engineering; Bachelor of Music Education

BMEC British Marine Equipment Council

B. Med. Bachelor of Medicine

B.M.Ed. Bachelor of Music Education

B.Med.Biol. Bachelor of Medical Biology

B.Med.Sc. Bachelor of Medical Science

BMEF British Mechanical Engineering Federation

BMEG Building Materials Export Group

BMEL Barber Middle East Line

bmep brake mean effective pressure

BM & ESA Building Materials and Equipment Southeast Asia (Singapore)

BMET Biomet Incorporated

B.Met. Bachelor of Metallurgy

B.Met.E. Bachelor of Metallurgical Engineering

BMEWS Ballistic Missile Early Warning System

bmf broken mixed fannings (tea)

BMF Business Master File

BMFA Boston Museum of Fine Arts

bmg business management game

BMG Browning Machine Gun

B.Mgt.Eng. Bachelor of Management Engineering

BMH British Military Hospital

bmi ballistic missile interceptor (BMI); body mass index

BMI Barley and Malt Institute; Battelle Memorial Institute; Book Manufacturers Institute; Boulton's Mounted Infantry; Broadcast Music Incorporated; Broadway Memorial Institute

B.Mic. Bachelor of Microbiology

BMIC British Music Information Centre (London); Broadcast Music Incorporated (Canada)

BMIC Bureau of Mines Information Circular

B.Min.E. Bachelor of Mining Engineering

BMIP Basic Medical Insurance Plan

BMIS Business Management Information System

BMJ British Medical Journal

bmk birthmark; bookmark(er)

bmkr boilermaker

BML Belfast & Moosehead Lake (railroad); Bodega Marine Laboratory (University of California); Bougainville Mining Limited; British Maritime League; British Museum Library (London)

B.M.L. Bachelor of Modern Languages

B & M L Belfast & Moosehead Lake (railroad)

BMLA British Maritime Law Association

BMLG Branch and Mobile Libraries Group

BMM Belfast, Mersey and Manchester Steamships; British Military Mission

BMMA British Mantle Manufacturers' Association

BMMA Biblioteca Municipal Mário de Andrade (Portuguese—Mario de Andrade Municipal Library)—named in honor of Brazil's musician-poet promoter of modernism

BMMFF British Man-Made Fibres Federation

BMMO Birmingham and Midland Motor Omnibus

bmn bone marrow necrosis

Bmn Bremen

BMN British Merchant Navy

BMNH British Museum (Natural History)

BMNP Bale Mountains National Park (Ethiopia); Blue Mountains National Park (New South Wales)

BMNT beginning morning nautical twilight

bmo business machine operator

BMO Ballistic Missile Office

bmoc big man on campus

bmom base maintenance and operations model

B'mouth Bournemouth

bmp best management practices; brake mean power; buttermilk powder

BMP Beefy-Meaty-Peptide; Bricklayers, Masons and Plasterers' (Union); Burma Military Police; Business Migration Program (Australian)

BMP Banco Minero del Perú (Spanish—Mining Bank of Peru)

BMP-76PB Soviet amphibious armored-infantry combat vehicle; also designated BTRM

BMPA British Metalworking Plantmakers' Association

BMPIUA Bricklayers, Masons, and Plasterers International Union of America

bmpp benign mucous-membrane pemphigus

BMPS British Medical Protection Society; British Musicians Pension Society

BMQ Best Memory Quiz

BMQA Board of Medical Quality Assurance

bmr basal metabolic rate; bomber

BMR Basal Metabolism Rate; Bihar Mounted Rifles; Border Mounted Rifles; Bureau of Mineral Resources

BMRA Brigade Major Royal Artillery; British Manufacturers' Representatives' Association

BMRB British Market Research Bureau

BMRL Building Materials Research Laboratories

BMRR British Museum Reading Room

BMRS Ballistic Missile Recovery System

bms balanced magnetic switch; ballistic missile ship

BMs Black Muslims; Boatswain's Mates

BMS Boston Museum of Science; British Malachological Society; British Ministry of Supply; Buffalo Museum of Science; Bureau of Medical Services; Bureau of Medicine and Surgery

B.M.S. Bachelor of Marine Science; Bachelor of Medical Science

BMSA British Medical Students' Association

BMSE Baltic Mercantile and Shipping Exchange

BMSG British Merchant Service Guild

BMSS British and Midlands Scientific Society

bmt bone-marrow transplantation

BMT Basic Military Training; Boston & Maine Transportation (railroad); Brooklyn-Manhattan Transit (subway system)

B.M.T. Bachelor of Medical Technology

BMTA Boston Metropolitan Transit Authority

BMTFA British Malleable Tube Fittings Association

BMTP Bureau of Mines Technical Paper

BMTS Ballistic Missile Target System

BMTV Ballistic Missile Test Vessel

BMU British Medical Union

B.Mus. Bachelor of Music

bmv bromegrass-mosaic virus

BMVM British Military Volunteer Service

BMW Bayerische Motoren Werke (German—Bavarian Motor Works)

BMWE Brotherhood of Maintenance of Way Employees

BMWS Ballistic Missile Weapon System

bmx (BMX) bicycle motorcross

BMYC Baltimore Motor Yacht Club

bmz basement membrane zone

bn bassoon; battalion; between; billion; bloody nuisance; branchial neuritis

Bn (BN) binary number (system)

bn bijvoeglijk naamwoord (Dutch—adjective)

Bn beacon (daybeacon); bearing; Benjamin

Bn Bayan (Turkish—Miss, Mrs)

B^n Bassin (French—basin, pond)

BN Beverage Network; blowing sand (aircraft code); Braniff, Brief Nudity; Brunei Darussalam (Internet code); Bureau of Narcotics; Burlington Northern (merger of Chicago, Burlington, and Quincy, Frisco—St Louis and San Francisco, Great Northern, Northern Pacific, Spokane, Portland, and Seattle railroads)

B-N Bloomington-Normal, Illinois

B.N. Bachelor of Nursing

B & N Barnes & Noble; Bauxite & Northern

B of N Bureau of Narcotics

BN Biblioteca Nacional (Portuguese or Spanish—National Library); *Biblioteca Nazionale* (Italian—National Library); *Bibliothèque National* (French—National Library)

bna (BNA) beta-naphthylamine

BNA Black Network Alliance; Blackwell North America; Brazil Nut Association; British Naturalists' Association; British Naval Attaché; British North America; British North America Philatelic Society; British North Atlantic; British Nursing Association; Bureau of National Affairs; Nashville, Tennessee (airport)

BNA Basle Nomina Anatomica (Basel Anatomical Nomenclature)

BNAF British North Africa Force

B'nai Brith Benai Berith (Hebrew—Sons of the Covenant)

BNAPS British North American Philatelic Society

BNAs British Naval Attachés

BNAU Bulgarian National Agrarian Union

B.Nav. Bachelor of Navigation

BNB British National Bibliography; British North Borneo (Sabah)

BNB British National Bibliography

BNBC British National Book Centre; British North Borneo Company

BNC Bayonet Neill-Concelman (antenna feedline connection)

BNC Biblioteca Nacional de Chile (Spanish—National Library of Chile); *Biblioteca Nacional de Colombia* (Spanish—National Library of Colombia)

B.N.C. Brasenose College (Oxford)

BNCC Bay de Noc Community College

BNCF Biblioteca Nazionale Centrale Firenze (Italian—National Central Library—Florence)

bnchbd benchboard

BNCM Bibliothèque Nationale du Conservatorie de Musique (French—National Library of the Conservatory of Music—Paris)

BNCO British Non-Commissioned Officer

BNCOR British National Committee for Oceanographic Research

BNCS British Numerical Control Society

BNCSR British National Committee for Space Research (Royal Society)

BNCVE Biblioteca Nazionale Centrale Vittorio Emanuele II (Italian—Victor Emanuel IInd Central Library)—in Rome

b/nd binding—no date available

Bnd Bend

BND Bundesnachrichtendienst (German—Federal Intelligence Service)

BNDD Bureau of Narcotics and Dangerous Drugs

bndl bundle

Bndr Bandmaster

Bndr S-L Bandmaster—Sub-Lieutenant

bndsmn. bondsman

bndy bindery; boundary

bne but not exceeding

BNE Bank of New England; Board of National Estimates (CIA); Brisbane, Australia (airport); Buffalo Niagara Electric Corporation

BNEC British National Export Council; British Nuclear Energy Conference

BNES British Nuclear Energy Society

BNE & SAA Bureau of Near Eastern and South Asian Affairs (U.S. Department of State)

bnf bomb nose fuse

Bnf Banff

BNF Brand Name Foundation; Braniff International Airways

BNF British National Formulary

BNFC British National Film Catalogue

BNFEX Battalion Field Exercise

BNFL British Nuclear Fuels Limited

BNFMF British Non-Ferrous Metals Federation

BNFMRA British Non-Ferrous Metals Research Association

BNFMTC British Non-Ferrous Metals Technology Centre

BNFP Barnwell Nuclear Fuel Plant

BNFSA British Non-Ferrous Smelters' Association

Bng Bangor

BNGA British Nursery Goods Association

BNG-DL Bibliothèque Nationale du Grand-Duche de Luxembourg (French—National Library of the Grand Duchy of Luxembourg)

BNGM British Naval Gunnery Mission

BNGS Bomb Navigation Guidance System

bnh burnish

BNHA Badlands Natural History Society

BNHQ Battalion Headquarters

BNHS British National Health Service

BNI Bechtel National Inc; Bengal Native Infantry; Black Nation of Islam; Bombay Native Infantry

BNIB British National Insurance Board

BNJ Bonn, Germany (Cologne-Bonn airport)

BNJM Biblioteca Nacional José Martí (Spanish—José Martí National Library)—Havana's library named for Cuba's apostle of independence in the late nineteenth century

Bnk bank

bnkg banking

BNL Brookhaven National Laboratory

BNL Banco Nazionale del Lavoro (Italian—National Bank of Labor); *Biblioteca Nacional de Lisboa* (Portuguese—National Library of Lisbon); *Bibliothèque Nationale du Liban* (French—National Library of Lebanon)—in Beirut

BNLI Bengal Native Light Infantry

BNLO British Naval Liaison Officer

bnm bajo el nivel del mar (Spanish—below sea level)

BNM Badlands National Monument (South Dakota); *Biblioteca Nacional de México* (Spanish—National Library of Mexico—Mexico City)

BNM Banco Nacional de México (Spanish—National Bank of Mexico); *Biblioteca Nacional de México* (Spanish—National Library of Mexico)—in Mexico City; *Biblioteca Nazionale Marciana* (Italian—Marcian National Library)—in Venice

bno barrels of new oil; bladder neck obstruction; but not over

BNO Bank of New Orleans

BNOC British National Oil Corporation; British National Opera Company

bnp (BNP) bruttonationalprodukt (Dano-Norwegian—gross national product)

BNP Bahamas NP (West Indies); Bako National Park (Sarawak); (Spanish—National Bank of Panama) Banco Nacional de Panamá; Banff NP (Alberta); Belair NP (South Australia); Bontebok NP (South Africa)

BNP *Banque Nationale de Paris* (French—National Bank of Paris)

bnpa binasal pharyngeal airway

bnr burner

BNR Blue Note Records

BNRDC British National Research Development Corporation

BNRVR Bengal-Nagpore Railway Volunteer Rifles

BNS Bank of Nova Scotia; Bathymetric Navigation System; British Naval Staff; British Nylon Spinners

B.N.S. Bachelor of Natural Science; Bachelor of Naval Science

B of NS Bank of Nova Scotia

B.N.Sc. Bachelor of Nursing Science

BNSM British National Socialist Movement

bnst bassoonist

Bnt Burntisland

BNTL British National Temperance League

BNU *Banco Nacional Ultramarino* (Portuguese—Overseas National Bank)

B Nurs Bachelor of Nursing

B-nut B-shaped nut

BNV *Biblioteca Nacional de Venezuela* (Spanish—National Library of Venezuela)—in Caracas

BNVE *Biblioteca Nazionale Vittorio Emanuele III* (Italian—Victor Emanuel III Library)—in Naples

BNW Battelle-Northwest; Bureau of Naval Weapons

BNWR Blackwater National Wildlife Refuge (Maryland); Bowdoin National Wildlife Refuge (Montana); Brigantine National Wildlife Refuge (New Jersey)

Bnx Bronx

BNX British Nuclear Export Executive

BNY Bank of New York

BNYI Brooklyn Navy Yard Incinerator

BNZ Bank of New Zealand

bnzn benzoin

bo base (of prism) out; battalion orders; beat oscillator; blackout; body odor; bowel obstruction; bowels open; brigade ordnance; buccoocclusal; buyers' option

bo. boarder; bottom; bought

bo' bore; brother

b/o back order; boiloff; brought over; budget outlay

b & o belladonna and opium

'bo hobo (vagrant)

Bo Bolivia; Bolivian

BO Baltimore & Ohio (stock exchange symbol); Base Order; black oil (bunker oil fuel); Board of Ordnance; body odor; Bolivia (Internet code); box office; British Officer; branch office; broker's order; Bureau of Ordnance

B.O. Bachelor of Oratory

B & O Baltimore & Ohio Railroad; Bang & Olufsen; business and occupations (tax)

BO *Boletín Oficial* (Spanish—Official Bulletin)

BO-5 Messerschmidt-Bolkow-Blohm five-seat helicopter

boa born on arrival; breakoff altitude

Boa Balboa, Panama

BoA Bookkeepers of America

B o A Board of Accountancy

BOA Basic Ordering Agreement; Boat Owners Association; British Optical Association; British Orthopedic Association; British Osteopathic Association; British Overseas Airways (BOAC)

BOA (Disp) British Optical Association (Dispensing Certificate)

BOAC Better on a Camel; British Overseas Airways Corporation

BOAdicea British Overseas Airways digital information computer for electronic automation

BOADS Boston Air Defense Sector

BOAE Bureau of Occupational and Adult Education (Office of Education)

BOAFG British Order of Ancient Free Gardeners

'board aboard; all aboard; on board; starboard

Boardwalk Atlantic City, New Jersey's famous promenade

BOAS British Orphans Adoption Society

BOAT Business Operational and Administrative Training (program)

boat dk boat deck (lifeboat-boarding deck)

boatel boat + hotel (waterside hotel or motel)

boats. boatswain (bo'sun)

BOAT/US Boat Owners Association of the United States

bob back of the book; best of breed

Bob Robert

BoB Bay of Bengal

B o B Bookbuilders of Boston

BOB Bank One Ballpark; Bureau of the Budget

BOBA British Overseas Banks Association

Bobbie Robert

Bobbs Bobbs-Merrill

Bobby Robert(a); London policeman, named after Sir Robert Peel, who organized the London police

BOBC British Outward Bound Centre (Norway)

b-o-b cult ban-on-bathing cult

Bob La Fol Robert La Follette

BOBMA British Oil Burner Manufacturers Association

bobo bourgeois bohemian

bo-bo bo-bo-type locomotive

bobr boring bar

boc back outlet central; blowout coil; body on chassis; butoxycarbonyl

Boc Boccaccio

B o C Board of Cosmetology; Bureau of Correction (Pennsylvania); Bureau of Corrections (Virgin Islands)

BOC Bank of China; British Ornithologists Club; Brooklyn Opera Company; Burmah Oil Company

BOCA Building Officials and Code Administrators; Building Officials Conference of America

boca(s) (Spanish—gulf(s); inlet(s); mouth(s))

BOCCI Bureau of Organized Crime and Criminal Intelligence (California)

B.Occu.Ther. Bachelor of Occupational Therapy

bocd barrels of oil per calendar day

BOCE Board of Customs and Excise

BOCES Board of Cooperative Educational Services

BoCHS Bureau of Community Health Services

BOCI *Bloque de Obreros, Campesinos e Intelectuales* (Spanish—Bloc of Workers, Farmers, and Intellectuals)

BoCI Bureau of Criminal Investigation

BOCM British Oil and Cake Mills

B & O—C & O Baltimore and Ohio—Chesapeake and Ohio (merged railroad)

Boc Rat Boca Raton, Florida

BOCS Board of Cooperative Services

bod base ordnance depot; beneficial occupancy date; biochemical oxygen demand; biological oxygen demand; blackout door

bod *bodega* (Spanish—wineshop); *bodoniana* (Italian—Bodoni-style type)

Bod Bodaway; Bodel; Boden; Bodil; Bodleian; Bodnar; Bodo; Bodoni

BoD Board of Directors; Bureau of Drugs

Bodl Bodleian Library

bod lang body language (communication via body movements or postures)

Bodley Bodley Head

b-o d(s) box-office disaster(s)

Bod units Bodansky units

boe back outlet eccentric

B E Bank of England

B o E Bank of England; Board of Equalization

BOE Board of Osteopathic Examiners

BOE *Boletín Oficial del Estado* (Spanish—Official State Bulletin)

Boeing 707 four-engine long-range jet transport

Boerst Boerestaat (Boer State in South Africa)

BOES Branch Ordinary Enquiry System

bof basic oxygen furnace; beginning of file; binary oxide film

bof *beurre, oeifs, fromages* (French—butter, eggs, cheeses)—big butter-and-egg man

BoF Bureau of Foods

B-o-F Books-on-File

B of N *Birth of a Nation* (David Wark Griffith's three-hour classic)

bofu *bordfunker* (German—radio-radar operator)

Bog Bogotá

BoG Board of Governors

B o G Board of Governors

BOG Bogotá, Colombia (airport); Boston Opera Group

bogaz boghaz (Turkish—strait)

boggan toboggan

bogh *boghandel* (Dano-Norwegian—bookstore, booktrade)

bogie unidentified aircraft

boh breakoff height

Boh Bohemia(n)

BoH Bank of Hawaii

B O'H Bernardo O'Higgins

Bohem Bohemia; Bohemian

B o HF Bureau of Home Furnishings

bohica bend over—here it comes again

BoHM Bureau of Health Manpower

BoHP & RD Bureau of Health Planning and Resources Development

BOHS British Occupational Hygiene Society

boi basis of issue; break of inspection

boi (BOI) branch output interrupt

Boi Boise

BoI Board of Investment

BOI Boise, Idaho (airport)

B & OI Bank and Office Interiors

BoIA Board of Immigration Appeals

BOIBIA *Business One Irwin BIA*

BOIC Boarding Officer in Charge

BOIESA Bureau of Oceans and International Environmental and Scientific Affairs (U.S. Department of State)

boil. boiling

boil.pt. boiling point

Boi Phil Boise Philharmonic

Bois (French—woods)—Bois de Boulogne park, racetrack, and recreation area of Paris

Boise St Univ Boise State University

boj booster jettison

BoJ Bank of Japan

bo juice body-odor deodorant

BOK Book-of-the-Month Club

bokhara bokhara or bokhara-style rug

Boko Bohner & Kohle

'boks springboks

bol beginning of life; bill(s) of lading; bollard(s)

bol (BOL) block output length

bol. *bolus* (Latin—large pill)

Bol Bolivar (Venezuelan province and hero); Bolivia; Bolivian; boliviano

Bol *Bol'shaya* or *Bol'shoy(e)* (Russian—big)

bol-148 (also BOL-148) d-2-bromolysergic acid tartrate (hallucinogen)

Bol cols Bolivarian colors (yellow, blue, and red as in the flags of Colombia, Ecuador, and Venezuela)

BOLD Bibliographic On-Line Display (document retrieval system); Blind Outdoor Leisure Development

BOLDS Burroughs Optical Lens Docking System

bolf barge off loading facility

Bolivar monetary unit of Venezuela; Simón Bolívar (Guayaquil, Ecuador's airport)

Bolivar Pen Bolivar Peninsula north of Galveston

Bolivia Republic of Bolivia (landlocked Andean nation named for Simón Bolívar who liberated it from Spain) *República Boliviana*

boliviano monetary unit of Bolivia

Bolivianos *Bolivianos—el hado es propicio* (Spanish—Bolivians, destiny is propitious) national anthem

bolo be on the lookout (for a criminal at large)

bolo(s) bolshevik(s)

Bolo(s) Bolshevik(s)

bolovac bolometric voltage and current (voltage measurement)

hols bolster

BOLSA Bank of London and South America

bolshie(s) bolshevik(s)

bolt. beam-of-light transistor

BOLT Basic Occupational Language Training; Basic Occupational Literacy Test

boltop better on lips than on paper (a kiss)

Bolv Bolivia; Bolivian

bom beginning of the month; bomb(ardier); business office must

Bom Bombay

BoM Bureau of Mines; Bureau of the Mint

BOM Boiler Operations & Management; Bombay, India (airport)

BOMA Building Owners and Managers Association

BOMAP Barbados Oceanographic and Meteorological Analysis Project

Bomarc Boeing long-range surface-to-air missile bearing nuclear warhead

BOMARC Boeing-Michigan Research Center

BOMAS Business Opportunity and Management Advisory Service

bomb. bombardment

Bomb Bombardier

Bombay Hook Bombay Hook National Wildlife Refuge near Dover, Delaware

bombex bombing exercise

BOMC Book of the Month Club

Bom Com Bomber Command

BoMD Bureau of Medical Devices

BOMEX Barbados Oceanographic and Meteorological Experiment; Board of Medical Examiners

bomfog brotherhood of man under the fatherhood of god

BOMI Box Office Management International

Bompo Bompensiero (Frank Bompensiero—San Diego, California's mob boss slain in 1977)

bomrep bombing report

bomroc bombardment rocket

BoMS Bureau of Medical Services

bomst bombsight

Bon Bonin Islands

BON Bonaire, Netherlands West Indies (airport); British Organization for Non-parents

Bon Air Girls Bon Air School (for delinquent) Girls at Bon Air, Virginia

bond. bonding

bone(s) trombone(s)

Bo'ness Borrowstounness

Boni Boniface

BONI Bombay Native Infantry

Boniato Santiago de Cuba's prison noted for inhuman treatment of prisoners

Bonins Bonin Islands (Ogasawaras)

boo brigade ordnance officer

Boo Bootes

B o O Board of Optometry

B o OE Board of Osteopathic Examiners

Book Bookman

bookie(s) bookmaker(s)

bookmobile book + automobile (mobile branch library within a truck)

Bookstax Bookstax of Britain

Booklist Booklist and Subscription Books Bulletin

boom boomerang

boomers baby boomer generation

boonies boondocks

BOOST Broadened Opportunities for Officer Selection and Training (USN)

Boot Bootes

Boöt Boötes (Latin—Herdsman constellation)

Boothia Boothia Peninsula in the Canadian Arctic, northernmost extension of North America

bop balance of payments; basic oxygen process(ing); bebop; best operating procedure; broken orange pekoe; buy our product(s)

b-o-p balance of payments

Bop Buffalo orphan prototype (virus)

BoP Bay of Pigs (invasion); Bay of Plenty

B o P Board of Pharmacy; Bureau of Prisons (U.S. Department of Justice)

BOP Behavior Observation Program

BoPa Borgelige Partisaner (Danish—Middleclass Partisans)—underground resistance against occupying German forces during World War II

BoPat Border Patrol

bopd barrels of oil per day

bopf broken orange pekoe fannings

B o PM Board of Podiatric Medicine

bops blowout preventer stack(s)

B o PS Board of Personnel Services

bopt broken orange pekoe tea

B.Opt. Bachelor of Optometry

Boq Boquerón (Paraguayan—department)

BOQ Bachelor Officers' Quarters; Base Officers' Quarters

bor boring; bowels open regularly

Bor Boris; Borough

BOR Board of Review; Borg-Warner (stock exchange symbol); Bureau of Outdoor Recreation

BORAD British Oxygen Research and Development

Borains people of Belgium's Borinage mining district

boram block-oriented random access memory

borax sodium tetraborate

borazon boron nitrogen compound harder than diamond; boron nitride heated and pressed with a catalyst

Bore Ro-Ro Bore Roll-on Roll-off Line

boricua(s) *borinqueño(s)* (Spanish-American slang—Puerto Rican(s))

borino(s) *borinqueño(s)* [Spanish-American slang—Puerto Rican(s)]

BORL Borland International Incorporated

BORN FREE Build Options, Reassess Norms, Free Roles through Educational Equity

boro borough

Boro Borough

Boro' Borough

Borromeans Borromean Islands in Lake Maggiore

bos basic oxygen steel

bos' bosun (boatswain)

Bos Bosphorus; Boston

Bos Bosanski (Serbo-Croatian—Bosnian)

BoS Bureau of Ships

BOS Boston, Massachusetts (airport); Boston Red Stockings; British Oil Shipping

Boschaps Boston Symphony Chamber Players

Bösend Bösendorfer

Bos-Her Bosnia-Herzegovina (formerly part of Yugoslavia, formerly part of Austria-Hungary)

bo's'n boatswain

Bo'sn Boatswain

Bosna (Yugoslav—Bosnia-Herzegovina)

Bosnia Bosnia-Herzegovina (part of the Austro-Hungarian Empire, now a federated republic in Yugoslavia)

Bosnywash Boston-New York-Washington, D.C. corridor

Bosox Boston Red Socks (baseball team)

BosPops Boston Pops Orchestra

boss basic orthographic syllabic structure

BOSS Bioastronautic Orbital Space System; Boeing Operational Supervisory System; Bureau of State Security (South Africa's Secret Service)

Bost Boston

Boston Spa British Library Lending Division in Boston Spa, Wetlmerby, West Yorkshire

Boston Tech Boston Technical Publishers

BOSTPDis Boston Procurement District (U.S. Army)

bo'sun boatswain

Boswash Boston-to-Washington

bot balance of time (to be served by a convict); balance of trade; botanic; botanical; botanist; botany; bottle; bottled; bottom; bottomed; bottoming; bought

bot (BOT) beginning of tape

b-o-t build-operate-transfer (toll roads)

bot botanik or *botanisk* (Dano-Norwegian—botany or botanist)

Bot Botany; Botswana (whose capital is Gaborone)

BoT Bank of Tokyo

B o T Board of Trade (British); Board of Transport (NATO)

BOT Board of Trade; Board of Trade unit; Board of Trustees; Books On Tape

B.O.T. Bachelor of Occupational Therapy

BOTAC Board of Trade Advisory Committee

botan botanic(al)(ly); botanist; botany

BOTB British Overseas Trade Board

bot & can bottle and can

botel boat hotel

BOTEX British Office for Training Exchange

both. bombing over the horizon

botmg bottoming

BOT-ohm Board of Trade ohm

botox botulinium toxin

botp both of this parish

botr battalion orderly room

BOTR British Other Rank

bot(s) bottle(s)

Bots Botswana

Botswana Republic of Botswana (landlocked South African country)—formerly Bechuanaland

botu botulism

BOTU Board of Trade Unit

boty bike of the year

Bou Boulogne-sur-Mer

BOU Bank Officers Union; Boat Operating Unit; British Ornithologists' Union

boul boulevard

Boul' Mich' Boulevard St Michel in the student quarter of Paris

Boulogne Boulogne-sur-Mer (French—Boulogne by the Sea)—English Channel port

bound. boundaries; boundary

'bout about

bov best of variety; bovine; bovril; brown oil of vitriol

Bov Eil Bovenwindse Eilanden (Dutch—Windward Islands, Aruba, Bonaire, Curaçao)

Bovid Bovidde (Latin— bovines: cows, goats, oxen, other ruminants, sheep)

bovinol bovinologic(al)(ly); bovinologist; bovinology

bow bag of water (amniotic sac); base ordnance workshop; blackout window; born out of wedlock

bo & w barrels of oil and water

bowdler bowdlerize

Bowker RR Bowker Company

bowla bowlathon

BOWO Brigade Ordnance Warrant Officer

Box Post Office Box

boyc boycott (named for C C Boycott), not having any dealings—commercial or so-

cial—with a company, country, person, or their products or services

bozo brawny intellectual lightweight

bp back pressure; bandpass; baptized; base pairs; basic pay; bathroom privileges; beautiful people; bedpan; before present; behavior pattern; below proof, benzypyrene; between perpendiculars; bills payable; biotic potential; biparietal; bipolar; birthplace; bishop; black pimp; blood pressure; blueprint; boiling point; broken pekoe; broncho-plural; buccopulpal

bp (BP) back projection (tv slide-or-film background projection)

b/p baking powder; bills payable; blood pressure; blueprint

b & p bare and painted; beer and pretzels; budgetary and planning

b of p balance of payments

bp Bergstrom Paper Company; *buono per* (Italian—good for)

b.p. bonum publicum (Latin— the public good)

Bp Babinski's phenomenon; Bishop

Bp Boerenpartij (Dutch—Farmers' Party)

BP Basic Protocol; Beach Party (amphibious military operation); Beacon Press; Be Prepared (Boy Scout and Girl Guide motto); *Beschleunigter Personenzug* (German—express train); Board of Parole; Bombay Pioneers; Border Patrol; British Petroleum; British Pharmacopoeia; British Public; Broadcast Pioneers; Bureau of Power; Bureau of Prisons (U.S. Department of Justice); Burns Philp Lines

B & P Ben and Penn (Benjamin Franklin and William Penn); Bombay Pioneers

B-P Basses-Pyrénées; Bermuda Plan (breakfast only); Lord Robert S Baden-Powell, founder of the Boy Scouts

B.P. Bachelor of Pharmacy; Bachelor of Philosophy

B/P Bills Payable

B of P Bishop(ric) of Passau; Bureau of Prisons

BP Bassposaune (German—bass trombone); *Battleship Potemkin; Berliner Philharmoniker* (German—Berlin Philharmonic); *Biblioteca Pubblica* (Italian—Public Library); *Biblioteca Pública* (Portuguese or Spanish—Public Library); *Boerenpartij* (Dutch—Farmers' Party); *Boite Pôstale* (French—post office box); *Box Poste* (French—mail box); *British Pharmacopoeia*

bp 120/80 lar blood pressure 120 (systolic)/80 (diastolic) left arm reclining

bpa boronated phenylalanine; broadband power amplifier

Bpa Bahnpostamt (German—Railway Post Office)

BPA Baltimore Publishers Association; Beach Protection Authority; Bedding Plants Australia; Biological Photographers Association; Blanket Purchasing Agreement; Bonneville Power Administration; Book Publishers Association; British Pediatric Association; British Ports Association; Broadcasters Promotion Association; Brunswick Port Authority; Bureau of Public Affairs (U.S. State Department); Bureau of Public Assistance; Bush Pilots Association (Australia); Business Press Association; Business Publications Audit (of circulation)

B.P.A. Bachelor of Professional Arts

BPA Banco Portugués do Atlántico (Portuguese—Portuguese Bank of the Atlantic)

BPAA Bowling Proprietors' Association of America

B/PAA Business/Professional Advertising Association

BPAC Budget Program Activity Code; Business Publications Audit of Circulation

BP-ACT Blueport Associated Container Transportation

BPA/DoS Bureau of Public Affairs—Department of State

B.Paed. Bachelor of Paediatrics

BPAGB Bicycle Polo Association of Great Britain

bpam basic partitioned access method

BPANZ Book Publishers Association of New Zealand

BPAO Branch Public Affairs Office(r)

BPAS British Pregnancy Advisory Service

BPASC Book Publishers Association of Southern California

bpay bill(s) payable

bpb bank post bills; blanket position bond; boom patrol boat; bromophenol blue

BPBD Bill Posters, Billers and Distributors (Union)

BPBF British Paper Box Federation

BPBI British Plaster Board Industries

BPBIF British Paper and Board Industry Federation

BPBIRA British Paper and Board Industry Research Association

BPBM Bernice P Bishop Museum (Honolulu)

BPBMA British Paper and Board Makers Association

bpc back-pressure control; beach patrol craft; book prices current; book and periodical circulation

BPC British Pharmaceutical Codex; British Phosphate Commission; British Printing Corporation; British Purchasing Commission; Business and Professional Code

BPC Bataillon De Parachutistes De Choc (French—Red Berets)

b-p cartridge barricade-penetrating cartridge

BPCC British Printing and Communication Corporation

bpcd barrels per calendar day

BPCF British Precast Concrete Federation

BPCI Bulk Packaging and Containerization Institute

BPCR Brakes on Pedal Cycle Regulations

BPCRA British Professional Cycle Racing Association

bpctca best practicable control technology currently available

bpcu bus power-control unit

bpd barrels per day; boxes per day; bronchopulmonary dysplasia

B.Pd. Bachelor of Pedagogy

BPD Bureau of the Public Debt

bpd & a basic planning data and assumption

BPDC Berkeley Particle Data Center; Books and Periodical Development Council (Canadian)

BPDMS Basic Point-Defense Missile System

BPDP Brotherhood of Painters, Decorators, and Paperhangers

bpe bit-plane encoding; black powder express (cartridge)

BPE Bureau of Postsecondary Education (Office of Education)

B.P.E. Bachelor of Physical Education

BPE-LCA Board of Parish Education—Lutheran Church in America

B.Pet.E. Bachelor of Petroleum Engineering

bpf batter's park factor; bottom pressure fluctuation; broken pekoe fannings

bpf bon pour francs (French—good for francs)

BPF British Pacific Fleet; British Polio Fellowship

bpg break pulse generator

BPGC Building Performance Guarantee Corporation

Bpge bearing per gyro compass

bph barrels per hour; benign prostatic hyperplasia; benign prostatic hypertrophy

B.Ph. Bachelor of Philosophy

BPh British Pharmacopoeia

B.P.H. Bachelor of Public Health

B.Pharm. Bachelor of Pharmacy

B.P.H.E. Bachelor of Physical and Health Education

B.Phil. Bachelor of Philosophy

BP & HL Brown Picton and Hornby Libraries (Liverpool)

B.Phys. Bachelor of Physics

B.Phys.Ed. Bachelor of Physical Education

B.Phys.Thy. Bachelor of Physical Therapy

bpi bits per inch; bytes per inch

BPI Bernreuter Personality Inventory; British Pacific Islands; Brooklyn Polytechnic Institute; Bureau of Public Information; Business Periodicals Index

BPICA *Bureau Permanent Internationale des Constructears d'Automobiles* (French—Permanent International Bureau of Automobile Manufacturers)

B picture moving picture designed as a second or supporting feature in a cinema program

b-pid book-physical inventory difference

BPIF British Printing Industries Federation

BPISAE Bureau of Plant Industry, Soils, and Agricultural Engineering

BP & JC FL Birmingham Public and Jefferson County Free Library

BPKT Basic Programming Knowledge Test

bpl birthplace

Bpl Barnstaple

BPL Belfast Public Library; Binghamton Public Library; Birmingham Public Library; Boston Public Library; Brass Pounders League; Bridgeport Public Library; Brooklyn Public Library; Buffalo Public Library

BP Lib Broadcast Pioneers Library

BPLS Boston Public Latin School

bpm barrels per minute; beats per minute; best practical means; blows per minute; breaths per minute

BPM British Prime Minister

BPM *Bulletin of Paleomalacology*

BPMA British Photographic Manufacturers Association; British Printing Machinery Association; British Pump Manufacturers Association

BPMA/DoS Bureau of Politico-Military Affairs—Department of State

BPMD Battelle Project Management Division

BPME Broadcast Promotion and Marketing Executives

BPMF British Postgraduate Medical Federation; British Pottery Manufacturers' Federation

BPMS Blood Pressure Measuring System

bpn bloody public nuisance

Bpn Balikpapan

BPNHM Banff Park Natural History Museum

BPNMA British Plain Net Manufacturers' Association

BPO Base Post Office; Base Procurement Office; Berlin Philharmonic Orchestra; Boston Pops Orchestra; British Post Office; British Postal Order; Brooklyn Philharmonia Orchestra, Brooklyn Post Office; Buffalo Philharmonic Orchestra

BPO *Berliner Philharmonisches Orchester* (German—Berlin Philharmonic Orchestra)

BPOE Benevolent and Protective Order of Elks

BPOEW Benevolent and Protective Order of Elks of the World

bpp bus-protection panel

BPP Black Panther Party; Botswana People's Party

BPP *Banco Popular del Perú* (Spanish—Popular Bank of Peru), *British Parliamentary Papers*

BPPMA British Power Press Manufacturers Association

BPR Bureau of Public Roads

BPR *Bloque Popular Revolucionario* (Spanish—Popular Revolutionary Block)—El Salvador; *Book Publishing Record* (periodical)

BPRA Book Publishers' Representatives' Association

bprf bulletproof

bprs brief psychiatric rating scale

bps bits per second; bytes per second

bp(s) black pimp(s)

bp's beautiful people

bpS broken pekoe Souchong

BPs Book Publishers (sales reports); Burns Philp steamships

B.Ps Bachelor in Psychology

BPS Balanced-Pressure System; Basic Programming System; Bayou Preservation Society; Benchmark Portability System; Border Patrol Sector; Border Patrol Station; British Police Service; British Psychological Society; Bureau of Product Safety

B$_{psc}$ bearing per standard compass

bpsd barrels per steam day

bpsm bulk pre-sorted mail

BPsS British Psychological Society

B$_{p\ stg\ c}$ bearing per steering compass

bpsk burst pulse shift key

BPSO Base Personnel Staff Officer

B.Psych. Bachelor of Psychology

bpt battle practice target; boiling point

bpt. baptized

bpt (BPT) bound plasma tryptophan

BPT Board of Prison Terms; British Petroleum Tanker; Bureau of Prison Terms

B.P.T. Bachelor of Physiotherapy

bpti bovine pancreatic trypsin inhibitor

bptv battleship propulsion test vehicle

bpu base production unit; bus protection unit

BPU British Powerboating Union

BPUNP *Biblioteca Pública de la Úniversidad de la Plata* (Spanish—Public Library of the University of La Plata)

bpv bovine papilloma virus; bullet-proof vest

BPWA Business and Professional Women's Association

bpwr burnable poison water reactor

B-Pyr Basses-Pyrénées

bq beauty quotient; boiler quality

Bq Becquerel

BQ Bachelor's Quarters; Basic Qualification; Basically Qualified (member of USCG Aux)

B Q *Bibliotèque nationale du Québec* (French—National Library of Quebec)— Montreal

Bqa Barranquilla

BQL Bank of Queensland

BQLI Brooklyn, Queens' Long Island

BQMS Battery Quartermaster Sergeant

BQSF British Quarrying and Slag Federation

b quark bottom quark

bque barque
Bquilla Barranquilla
br bank rate; bank robber(y); bats right; batting runs; berth; bill of rights; bills receivable; branch; bread (slang— money); breath; breeder reactor; broken; bromine; brown; builder's risk; butadiene rubber
br. brother
br (BR) bedroom; bedroom steward; branch (flow chart); break (request signal)
b/r bills receivable
b & r budget and reporting
b or r bales or rolls (freight)
br *bez roku* (Czech—no date, no year)
Br Branch; Bridge; Britain; British
Br Bachiller (Spanish—Bachelor)—academic degree; *Bratsche* (German—viola); *Bredning* (Danish—bay); *Brücke* (German—bridge); *Burun* (Turkish—nose, Point)
BR Baluch Regiment; Barbados Regiment; Baton Rouge; bearing; Bermuda Regiment; Boca Raton; branch; Brazil (auto plaque); Breeder Reactor; bridge; British; British Railways; British Resident (commissioner); British United Airways; Brockville Rifles; bromine; brown (buoy); Bureau of Reclamation
B-R Bas-Rhin; Business Route
B/R Bills Receivable; Bordeaux or Rouen
B of R Bureau of Reclamation; Bureau of Rehabilitation
BR Banco di Roma (Italian—Bank of Rome)
B.R. Bancus Reginae (Latin— Queen's Bench); *Bancus Rex* (Latin—King's Bench)
BR-1150 Breguet maritime-patrol aircraft, also called Atlantique
bra bombing restricted area; brain; brassiere
Bra Beira
Bra Brasil (Portuguese or Spanish—Brazil)
BrA Bibliothèque royale Albert I (French—Albert lst Royal Library)—Brussels library called *Koninklijke Bibliotheek Albert I* in Flemish

BRA Bankruptcy Reform Act; Bee Research Association; Boston Redevelopment Authority; Bougainville Revolutionary Army (Papua, New Guinea); Brigadier Royal Artillery; British Records Association; British Robot Association; Building Renovating Association
Brab La Brabanfconne (French—the Brabant) national anthem of Belgium
brac base realignment and closure; basic rest activity cycle
BRAC Brotherhood of Railway and Airline Clerks
brachi brachion (Greek— arm)—brachiation, brachiopod, brachium
brachycephs brachycephalics (short-skulled people)
Bra Cur Brazil Current
Brad Bradburn; Bradbury; Braden; Bradfield; Bradford; Bradley; Bradner; Bradshaw; Bradstreet; Brady
Bradshaw's Bradshaw's Railway Guide
brady bradycardia
Brady Bradenton, Florida; Brady Glacier, Alaska; Brady Lake, Ohio; Brady Mountains, Texas; Dr Brady C Hartman
Bráh. Bráhmana
braid. bidirectional reference array internally deprived
BRAIN Basic Research in Adaptive Intelligence and Neurocomputing
BRAINS Behavior Replication by Analog Instruction of Nervous System
Bram Abraham
Br.Am. British America safety lock invented by Joseph Bramah
Brambach Radiumbad Brambach in Saxony
Brampton Women Vanier Centre for (criminal) Women at Brampton, Ontario
BRANCHHYDRO Branch Hydrographic Office
Brand Brandenburg
brane bombing radar navigation equipment
Brangus 3/8 Brahman + 5/8 Angus cattle

BRANZ Basketball Referees Association of New Zealand; Building Research Association of New Zealand
bras ballistic rocket air suppression
bra(s) brassiere(s)
Bras Brasenose College, Oxford; Brasil; Brasileiro
Bras Brasil (Portuguese or Spanish—Brazil); *Brasile* (Italian—Brazil)
BRAs Bosom-Rehabilitation Associates
BRASCFHESE Brotherhood of Railway, Airline, and Steamship Clerks, Freight Handlers, Express, and Station Employees
BRASCO Brigade Royal Army Service Corps Officer
Bras Coll Brasenose College—Oxford
Brasilsat Brazil's satellite
brass breathe-relax-aim-squeeze-shoot
b-r-a-s-s breathe, relax, aim, squeeze, shoot (marksman's acronym)
BRASS Bottom Reflecting Active Sonar System
brasses brass instruments: bugles, cornets, French horns, horns, mellophones, trombones, trumpets, tubas, wagner horns
brass knucks brass knuckles
BRASTACS Bradford Scientific, Technical, and Commercial Service
brat bananas, rice, applesauce, toast (diet)
Brat bi-drive recreational all-terrain transporter; Bratislava (capital of Slovakia)
BRAVE Boeing Robotic Air Vehicle
Bravo letter B radio code
braz Brazil; Brazilian
Braz Brazil(ian)
Brazil Federative Republic of Brazil (South America's largest country), *Republica Federativa do Brasil*
Brazza Brazzaville
Brb Borba (Yugoslavia— Struggle)—leading newspaper in former Communist-controlled Yugoslavia
BRB Benefits Review Board; British Railways Board; Builders' Registration Board

brbc bovine red blood cells

Brbds Barbados

BRBMA Ball and Roller Bearing Manufacturers Association

brbzc brass, bronze, or copper (cargo)

brc business reply card

Br.C. British Columbia

BRC Balcones Research Center (University of Texas); Base Residence Course; Bengal Royal Cavalry; Bolivia Railway Company; British Research Council; Broadcast Rating Council; Broadcast Research Council; Brotherhood of Railway Carmen

BRCA Brotherhood of Railway Carmen of America

BRCCP British Royal Commission on Capital Punishment

Brch Branch

BRCMA British Radio Cabinet Manufacturers' Association

Br Col British Columbia

BRCS British Rail Catering Service; British Red Cross Society

BRCUSC Brotherhood of Railway Carmen of the United States and Canada

Br Cwlth British Commonwealth

brd base remount depot; basic retirement date; board; bomb-release distance; broad

BRD *Bundesrepublik Deutschland* (Federal Republic of Germany)

BRDC British Racing Drivers' Club

brdcst broadcast

brdf bidirectional reflectance distribution function

BRDM Soviet amphibious reconnaissance vehicle carrying three men and antitank missiles

Brdrck *Bäredreck* (German—bear dung)—licorice

Br Du bromodeoxyuridine

Brdw Broadwood

Brc Bremen; Bremerhaven

BrE British English

BRE Bureau of Readjustment Education

B.R.E. Bachelor of Religious Education

brec bills receivable

B/Rec Bills Receivable

breccia pyroclastic volcanic rock

Breck Breckinridge; Brecknockshire

Brecon Breconshire (Brecknockshire)

BRECSU Building Research Energy Conservation Unit

Breguet 765 Sahara flying transport for 145 troops

Breguet 1150 Atlantique maritime-patrol aircraft

brek breakfast

BREL British Rail Engineering Limited

'brella umbrella

Brem Bremen; Bremerhafen; Bremerhaven; Bremerton

BREMA British Radio Equipment Manufacturers Association

BrEn Brno-Enfield

Brenner Brenner Pass in the Alps connecting Bolzano, Italy with Innsbruck, Austria

Brennero Brenner Pass

Brent Brentford and Chiswick

Br'er Brother

Bres Breslau

bret *bretonisch* (German—Breton)

Bret Brittany; Breton

brev brevet; breviary; breviate; brevier

brev *breveté* (French—patent); *brevetto* (Italian—patent)

brev. *breviarium* (Latin—abridgement or breviary)

brew. brewer; brewery; brewing

Brew Brewer; Brewster

brew'd brewed

Brewer's *Brewer's Dictionary of Phrase and Fable*

brf brief, briefing

BRF Bass Research Foundation; Bible Reading Fellowship; British Road Federation

BRFC British Record Fish Committee (of rod anglers)

brg bearing; brewing; bridge; brigantine

Brg Bridge

BrG British Guiana

BRG *Bibliotheek van de Rijksuniversiteit te Gent* (Flemish—Royal University Library of Ghent)

brghd bridgehead

brghm brougham (pronounced *broom)*

Brgo Spgs Borrego Springs

Br Gu British Guiana

BrH British Honduras

BRH Brussels, Belgium (airport); Bureau of Radiological Health

BRHL British Rail Hovercraft Limited

BrHon British Honduras

BRHS Bay Ridge High School; Betsy Ross High School

Bri Brian; Bridge; British(er)(s); Briton(s)

Br I British Isles

BRI Babson's Reports Incorporated; *Banque des Reéglements internationaux* (French—Bank of International Settlements); Biological Research Institute; Brain Research Institute; Brand Rating Index; Bristol Royal Infirmary; Building Research Institute; Bureau of Rehabilitation Inc; Burlington-Rock Island (railroad)

BRI *Banque des Règlements Internationaux* (French—Bank for International Settlements); *Brand Rating Index*

BRIA Biological Research Institute of America; Bread Research Institute of Australia

BRIC British Columbian Resource Investment Corporation

Bricklayers Union Bricklayers, Masons, and Plasterers International Union of America

BRICS British Rail Inter-City Service

BRICSHST British Rail Inter-City Service High-Speed Train

BRIDGE Biotechnology Research for Innovation, Development and Growth in Europe

BRIDGEX Bridge Construction Exercise

brig brigantine; ship's prison

Brig Brigade; Brigadier

Brig Gen Brigadier General

Brigitte Bridget

BRIGLEX Brigade Landing Exercise

Brigton Glasgow's Bridgetown

Brilab bribery-labor (FBI investigation's code name)

brill *brillante* (Italian—brilliant)

BRIMEC British Mechanical Engineering Federation

BRINCO British Newfoundland Corporation

BRINDEX British Independent Oil Exploration (Companies Association)

BRINDIV British-Indian Division

Bris Brisbane

Brisb Brisbane

Brissie Brisbane

Brist Bristol

brit britisk (Dano-Norwegian—British)

Brit Britain; Britannia; British

Brit Britannica (*Encyclopaedia Britannica*)

Britain Great Britain (England, Scotland, and Wales)

Brit Book Centr British Book Centre

Brit and For British and Foreign State Papers

Britcoms British comedians; British comedies

BRITCON British Contingent

BRITE Basic Research in Industrial Technologies in Europe

BRITFORLEB British Forces in Lebanon

Britic Briticism

Brit Info British Information Services

British pertaining to the British Commonwealth, the British Empire, or the British people (English, Scottish, and Welsh)

British Am Bks British American Books

British Bk Ctr British Book Center

British Commonwealth British Commonwealth of Nations: Great Britain and Northern Ireland; British dominions, republics, and dependencies

British Dependencies Bermuda, Cayman Islands, Falklands, Gibraltar, South Georgia, South Sandwich, South Shetlands, Turks, and Caicos Islands

British India colonial India (1757–1948)

British Virgins British Virgin Islands

Brit J Psychiat British Journal of Psychiatry

Brit J Surg British Journal of Surgery

brit. met britannia. metal (tin, copper, antimony alloy—sometimes bismuth, lead, and zinc)

Brit Mus British Museum

Brit Pat British Patent

Brit Phos Comm British Phosphate Commmssmon

BritRail British Railways

Brit-Rail Hover British Railways Hovercraft

Brits British; Britons

Brit(s) British(ers); Briton(s)

Brits (Dutch—British)

Brit Sam British Samoa

BRITSHIPS British Shipbuilding Integrated Production System

Britt *Britannorum* (Latin—of the Britons)

Brit Telecom British Telecommunications

Brix Brixham; Brixton

Br J App Phys British Journal of Applied Physics

brk brick

Brk Brook; Brooklyn

BRK The Bridge on the River Kwai (movie)

brkf breakfast

brklyr bricklayer

brkmn breakman

brks breakers

brkt bracket

brkwtr breakwater

brl bomb-release line

br/l brown line positive

BRL Babe Ruth League; Ballistic Research Laboratories; Beecham Research Laboratories; Bible Research Library; *Bibliotheek der Rijksuniversiteit te Leiden* (Dutch—Library of the Royal University in Leyden); British Research Library

BRL 1241 Beecham Research Laboratories formula 1241 (methicillan)

BRL 1341 Beecham Research Laboratories formula 1341 (penbritin)

brlg bomb radio longitudinal generator-powered

brlp burlap

hrl sys barrier ready light system

brm bedroom; biological response modifier

BRM British Racing Motors

BRMA Board of Registration of Medical Auxiliaries; British Rubber Manufacturers' Association

BRMBR Bear River Migratory Bird Refuge (Utah)

BRMC Business Research Management Center (USAF)

BRMCA British Ready-Mixed Concrete Association

BRMF British Rainwear Manufacturers' Federation

brn brown

Brn Bahrain; Brunei; Sultanate of Brunei

BRN Board of Registered Nurses

BRNC Britannia Royal Naval College (Dartmouth)

bng bearing; browning; burning

BRNP Blue Ridge National Parkway

brnsh brownish; burnish

Brnx Bronx

brnz bronze; bronzing

bro broach; bronchoscopy; brother

brO brownish orange

Bro Brother

BRO Brigade Routine Order(s)

Broads The Broads (Norfolk Broads)—England's east coast holiday resort area

Broadway main north-south avenue of New York City; New York's theater district

broast(ed) broil(ed) + roast(ed)

broc brocaded

Brod Broderick; Brodie; Brody

broficon broadcast fighter control

bro-i-l brother in law

BROILER Biopedagogical Research Organization on Intensive Learning Environment Reactions

bro-in-law brother-in-law

brok broker; brokerage

brom bromide; bromidic; bromo; bromo-seltzer

bromat bromatology (treatise on foods)

brome bromeliad

bromidrosis bromohydrosis

bromo bromidrosis; bromoform; bromo-seltzer

bromo-seltzer (bromide + seltzer)

bromot bromotology [study of smell(s)]

bronc bronco (Spanish—small half-wild horse)

bronch bronchial; bronchitis; bronchoscopic; bronchoscopist; bronchoscopy

broncho bronchus (Greek—windpipe)—bronchi(tis)

Bronx HTF Bronx Homicide Task Force (NYPD)

Bronx Zoo New York Zoological Gardens (Bronx Park)

bronze 92% copper, 6% tin, 2% zinc

Brookings Brookings Institution

Brooklyn HTF Brooklyn Homicide Task Force (NYPD)

Brookwood Girls Brookwood Center for (delinquent) Girls at Claverack, New York

Bros brothers

brosch broschiert (German—stitched)

Brose Ambrose

brot brought

brotel brothel + hotel

BROU Banco de la República Oriental del Uruguay (Spanish—Bank of the Oriental Republic of Uruguay)

browners brown nosers

brownulated granulated brown sugar

Brown U Pr Brown University Press

Brownwood Girls State Home, Reception Center, and School for Delinquent Girls at Brownwood, Texas

brp bathroom privileges

BRP Breeder Reactor Program

BRPF Bertrand Russell Peace Foundation

brph bronchophony

brPk brownish pink

BRPL Baton Rouge Public Library

brpp basic radio propagation prediction(s)

BRR Brockville Regiment of Rifles

Br Rys British Railways

brs brass

brs (BRS) break request signal

br's bedrooms

Brs Bristol

Br S Bedroom Steward

BRS Bertrand Russell Society; Bibliographic Retrieval Services (database); Bomber Recorder System; British Road Services; British Roentgen Society; Brotherhood of Rail-

way Signalmen; Bureau of Railroad Safety; Business Radio Service; Buyers' Research Syndicate

BRSA British Railway Staff Association

BR & SC Brotherhood of Railway and Steamship Clerks

BRSCC British Racing and Sports Car Club

brskn brass knuckles

br snds breath sounds

br sounds breath sounds

brst burst

Br std British standard

brstr burster

brt bright

brt (BRT) bruttoregisterton (Dano-Norwegian—registered gross tonnage)

Brt Brest

BRT Brotherhood of Railroad Trainmen

B.R.T. Before Recorded Time

BRT Belgische Radio en Televisie (Belgian Radio and Television); *Brutto-Register-Tonnen* (German—gross register tons)

BRTA British Regional Television Association; British Road Tar Association

BRTC British Rail Travel Centre

BR & TC Bermuda Radio and Television Company

brt fwd brought forward

brtg bartering

Bru Bruce; Brunei; Bruno; Brutus

BRU Brussels, Belgium (National Airport)

BRU Bibliotheek der Rijksuniversiteit te Utrecht (Dutch—Library of the Royal University in Utrecht)

B.Ru.Eng. Bachelor of Rural Engineering

BRUFMA British Rigid Urethane Foam Manufacturers Association

brum brougham

Brum Brummagen (Birmingham, England's nickname)

Brum Brumaire (French—Foggy Month)—beginning October 22 second month of the French Revolutionary Calendar

Brun Brunei

brunch(eon) breakfast-lunch(eon)

Bruns Brunswick

Brunsw Brunswick

Brun U Brunel University

B.Rur.Sci. Bachelor of Rural Science

Brus Bruselas (Spanish—Brussels); *Bruselle* (Italian—Brussels); *Brussel* (Dutch or Flemish—Brussels); *Brüssel* (German—Brussels)

BRUTE British Universal Trolley Equipment

brux bruxism; bruxitic

Brux Brussels

Brux Bruxelas (Portuguese—Brussels); *Bruxelles* (French—Brussels)

brv (BRV) ballistic reentry vehicle

BRVMA (BVA) British Radio Valve Manufacturers' Association

Brw Barrow

BRW British Relay Wireless

Brx Bronx

bry bryology

Bry Barry; Bryant

Bryce Bryce Canyon National Park, Utah; Mount Bryce, British Columbia

bryol bryology

Bryth Brythonic

brz bronze

brzg brazing

bs balance sheet; beam splitter; bill of sale; blood sugar; bluestone; bomb service; bomber support; bonded single-silk (insulation); border scouts; both sides; bowel sound; breath sound; bullshit

bs bancus superior (Latin—upper bench)

bs (BS) backspace (data-processing character); binary subtract(ion)

b's boomerangs

b/s back stamp

b/s (B/S) bill of sale

b & s beams and stringers; bell and spigot; boosters and sustainers; brandy and soda

b-S by-Sea

Bs bolivares (Venezuelan currency); bolivianos (Bolivian currency)

BS Bahamas (Internet code); Battle Squadron; Battle Star; Bethlehem Steel; Berlin Sector; Birmingham Southern (railroad); British Standard;

Broadcast Satellite; Bureau of Ships; Bureau of Standards; Burma Star

B-S Bedford-Stuyvesant

B.S. Bachelor of Science

B/S Bill of Sale

B & S Bank and Savill (steamship line); Brown and Sharpe; Butterfield and Swire

BS Bayerische Staatsbibliothek (German—Bavarian State Library)

bsa bismuth-sulphite agar; body surface area; bovine serum albumin; brown strain apparent

BSA Bank Stationers Association; Bibliographical Society of America; Biofeedback Society of America, Birmingham Small Arms; Black Stuntmen's Association; Blind Service Association; Blinded Soldiers Association (Australia); Botanical Society of America; Boy Scouts Association; British School of Athens; British South Africa; Brotherhood of St Andrew; Bruckner Society of America; Bureau of Supplies and Accounts

B.S.A. Bachelor of Agricultural Science

BSAA British South American Airways

BSA(A) British School of Archeology (Athens)

B.S.A.A. Bachelor of Science in Applied Arts

BSAC Big Sky Athletic Conference; British South Africa Company; Brotherhood of Shoe and Allied Craftsmen

B.S.Adv. Bachelor of Science in Advertising

B.S.A.E. Bachelor of Science in Aeronautical Engineering; Bachelor of Science in Architectural Engineering

BSAF British Sulphate of Ammonia Federation

BSAG Bristol Social Adjustment Guides

B.S.Agr. Bachelor of Science in Agriculture

BSAM Basic Sequential Access Method

B.S.A.M. Bachelor of Suddha Ayurvedic Medicine

BSAOT Bell System American Orchestras on Tour

BSAP British South Africa Police

B.S.Arch. Bachelor of Science in Architecture

B.S.Arch. Eng. Bachelor of Science in Architectural Engineering

B.S.Art Ed. Bachelor of Science in Art Education

Bs As Buenos Aires

BSAS British Ship Adoption Society

BSAVA British Small Animals Veterinary Association

bsb body surface burned

bsb (BSB) backspace block

Bsb Brisbane

BSB Backer Spielvogel Bates (Worldwide); Brasilia, Brazil (airport); British Satellite Broadcasting; Brotherhood of St Barnabas

BSBA British Starter Battery Association

B.S.B.A. Bachelor of Science in Business Administration

BSBC British Social Biology Council

bsbg burst and synchronous bit generator

BSBI Botanical Society of the British Isles

BSBSPA British Sugar Beet Seed Producers' Association

B.S.Bus. Bachelor of Science in Business

bsc basic; basic-message switching center; binary synchronous communication

Bsc British standard channel (steel)

B.Sc. Bachelor of Science

BSC Bank of Southern California; Bank Street College; Beltsville Space Center; Bemidji State College; Bengal Staff Corps; Bethlehem Steel Corporation; Bibliographical Society of Canada; Biological Stain Commission; Biomedical Sciences Corporation; Bloomsburg State College; Bluefield State College; Booth Steamship Company; British Security Coordination; British Society of Cinematographers; British Steel Corporation; British Supply Council; Business Service Center

B.S.C. Bachelor of Science in Commerce

BSCA Bureau of Security and Consular Affairs (U.S. Department of State)

B.Sc.Acc. Bachelor of Science in Accounting

B.Sc.Ag. & A.H. Bachelor of Science in Agriculture and Animal Husbandry

B.Sc.Agr.Bio. Bachelor of Science in Agricultural Biology

B.Sc.Agr.Eco. Bachelor of Science in Agricultural Economics

B.Sc.Agr.Eng. Bachelor of Science in Agricultural Engineering

B.Sc.Ag(ri)(c) Bachelor of Science in Agriculture

BSCAI Building Service Contractors Association International

B.Sc.Arch. Bachelor of Science in Architecture

B.Sc.B.A. Bachelor of Science in Business Administration

BSCC British Society for Clinical Cytology

B.Sc.C.E. Bachelor of Science in Civil Engineering

B.Sc.Chem.E. Bachelor of Science in Chemical Engineenng

B.Sc.Dent. Bachelor of Science in Dentistry

B.Sc.Dom.Sc. Bachelor of Science in Domestic Science

BSCE Bank Street College of Education

B.S.C.E. Bachelor of Science in Civil Engineering

B.S.Ch. Bachelor of Science in Chemistry

B.S.Chm. Bachelor of Science in Chemistry

B-school(s) business school(s)

bscn bit scan

B.Sc.Nurs. Bachelor of Science in Nursing

BSCO British Security Coordination Office

B.S.Comm. Bachelor of Science in Commerce

BSCorp British Steel Corporation

BSCP Brotherhood of Sleeping Car Porters; Business Service Centers Program

BSCP British Standard Code of Practice

BSCRA British Steel Castings Research Association

BSCS Biological Sciences Curriculum Study

B.Sc.S.S. Bachelor of Science in Secretarial Studies

BSCU Blue Star (Line) Container Unit

B.Sc.Vet.Sc. Bachelor of Science in Veterinary Science

bsd barrels per steaming day; base supply depot; beam-steering device; bit storage density; blast-suppression device; burst-slug detection

BSD Ballistic Systems Division (USAF); Bank of San Diego; Berkeley Software Distribution; British Space Development

B.S.D. Bachelor of Science in Design

BSDA British Spinners and Doublers Association

BSD Bancorp Bank of San Diego Bank Corporation

bsdc binary symmetric dependent channel

BSDC British Space Development Company; British Standard Data Code

B.S. Dent. Bachelor of Science in Dentistry

bsdg breveté sans garantie du gouvernement (French—patented without government guarantee)

B.S.D.H. Bachelor of Science in Dental Hygiene

bsdl boresight datum line

BSDU Bomber Support Development Unit

bse base support equipment; bovine spongiform encephalopathy; breast self-examination (cancer control)

BSE Base Support Equipment; Birmingham & Southeastern (railroad); Broadcast Satellite Experiment; Broadcasting Satellite for Experimental Purposes; Building Service Employees (Union); Bureau of Steam Engineering

B.S.E. Bachelor of Sanitary Engineering; Bachelor of Science Education; Bachelor of Science Engineering

B & SE Birmingham & Southeastern (railroad)

B.S.Ec. Bachelor of Science in Economics

B.S.Ed. Bachelor of Science in Education

B.S.E.E. Bachelor of Science in Electrical Engineering

B.S.El.E. Bachelor of Science in Electronic Engineering

B.S.Eng. Bachelor of Science in Engineering

b's'er bullshitter

BSES British Schools Exploring Society

bsf back scatter factor; bulk shielding facilities

bsf (BSF) beta-s-fetoprotein

BSF Basic Skill Films; Black Student Fund; British Salonika Force; British Shipping Federation

B.S.F. Bachelor of Science in Forestry

BSFA British Sanitary Fireclay Association; British Steel Founders' Association; Brotherhood of St Francis of Assisi; Building Science Forum of Australia

bsfe brake specific fuel consumption

BSFC Baltic States Freedom Council

BSFF Buffer Stock Financing Facility

B.S.Fin. Bachelor of Science in Finance

BSFL British Shipping Federation Limited

B.S.For. Bachelor of Science in Forestry

B.S.F.S. Bachelor of Science in Foreign Service

BSFT Basalt Spent Fuel Test(ing)

BSF & W Bureau of Sport Fisheries and Wildlife

BSG British standard gage

B.S.G.E. Bachelor of Science in General Engineering; Bachelor of Science in Geological Engineering

B.S.Gen.Nur. Bachelor of Science in General Nursing

B.S.Geog. Bachelor of Science in Geography

B.S.Geol. Bachelor of Science in Geology

B.S.Geol.Eng. Bachelor of Science in Geological Engineering

B & S glands Bartholin and Skene's glands

bsh bushel

BSH British Society of Hypnotherapists; British Standard of Hardness

B.S.H.A. Bachelor of Science in Hospital Administration

B.S.H.E. Bachelor of Science in Home Economics

B.S.H.Eco. Bachelor of Science in Home Economics

B.S.H.Ed. Bachelor of Science in Healtlm Education

BSHS British Society for the History of Science

bsi basic shipping instructions; bound serum iron

BSI Baker Street Irregulars; British Sailors' Institute; British Standards Institute

BSI Business Survey Index

BSIA Better Speech Institute of America

BSIB Boy Scouts International Bureau; British Society for International Bibliography

bsic binary-symmetric independent channel

BSIC British Ski Instruction Council

B.S.I.E. Bachelor of Science in Industrial Engineering

BSIHE British Society for International Health Education

B.S.Ind.Art Bachelor of Science in Industrial Art

B.S.Ind.Chem. Bachelor of Science in Industrial Chemistry

B.S.Ind.Ed. Bachelor of Science in Industrial Education

B.S.Ind.Eng. Bachelor of Science in Industrial Engineering

BSIP British Solomon Islands Protectorate

BSIPDF British Solomon Islands Protectorate Defence Force

B.S.I.R. Bachelor of Science in Industrial Relations

BSIRA British Scientific Instrument Research Association

BSIs Baker Street Irregulars

BSIS BioScience Information Services

BSIU British Society for International Understanding

bsj balanced swivel joint; ball-and-socket joint

B.S.J. Bachelor of Science in Journalism

BSJA British Show Jumping Association

B.S.Jr. Bachelor of Science in Journalism

bsk basket(s)
Bskrvile Baskerville
bskt basket
bsl billet split lens
bs/l bills of lading
Bsl Bislig Bay
BSL Bankers Security Life; Barber Steamship Lines; Behavioral Sciences Laboratory; Black Star Line; Blue Sea Line; Blue Star Line; Boatswain Sub-Lieutenant; British Sign Language; Building Service League; Bull Steamship Lines
BSLA Bus Services Licensing Authority (Singapore)
B.S.Lab.Rel. Bachelor of Science in Labor Relations
bslb ball-and-socket lower bearing
bsln ball-and-socket lower bearing
bsl(s) bushel(s)
B.S.L.S. Bachelor of Science in Library Science; Bachelor of Science in Library Service
bsm bi-stable multivibrator; bottom sonar marker
BSM Battery Sergeant Major; Birmingham School of Music; Branch Sales Manager; Bronze Star Medal
BSM beso sus manos (Spanish—I kiss your hands)— respectfully yours
BSMA British Skate Makers' Association
B.S.Mar.Eng. Bachelor of Science in Marine Engineering
BSMD Business System Marketing Division
B.S.M.E. Bachelor of Science in Mechanical Engineering; Bachelor of Science in Mining Engineering; Bachelor of Science in Music Education
B.S.Med. Bachelor of Science in Medicine
B.S.Med.Rec. Bachelor of Science in Medical Records
B.S.Med.Rec.Lib. Bachelor of Science in Medical Records Librarianship
B.S.Med.Tech. Bachelor of Science in Medical Technology
B.S.Mct. Bachelor of Science in Metallurgy
B.S.Met.Eng. Bachelor of Science in Metallurgical Engineering

B.S.Mgt.Sci. Bachelor of Science in Management Science
B.S.Min. Bachelor of Science in Mincrology; Bachelor of Science in Mining
B.S.Min.Eng. Bachelor of Science in Mining Engineering
BSMMA British Sugar Machinery Manufacturers Association
bsmt basement
B.S.Mus.Ed. Bachelor of Science in Music Education
bsmv barley-stripe-mosaic virus
bsn bassoon; bowel sounds normal
BSN Baker School of Navigation; Broadcasting System of Niigata (Japanese)
B.S.N. Bachelor of Science in Nursing
BSN Bayerische Staatsoper— Nationaltheater (German— National Theater—in Munich)
bsna bowel sounds normal and active
BSNA Bureau of Salesmen's National Associations
B.S.N.A. Bachelor of Science in Nursing Administration
B.S.Nat.Hist. Bachelor of Science in Natural History
BSNDT British Society for Non-Destructive Testing
BSNH Boston Society of Natural History; Buffalo Society of Natural History
B.S.N.I.T. Bachelor of Science in Nautical Industrial Technology
B.S.Nurs. Bachelor of Science in Nursing
B.S.Nurs.Ed. Bachelor of Science in Nursing Education
bso battalion standing orders; blue stellar objects
BSO Baltimore Symphony Orchestra; Bamberg Symphony Orchestra; Birmingham Symphony Orchestra; Bombay Symphony Orchestra; Boston Symphony Orchestra; Bournemouth Symphony Orchestra; Budapest Symphony Orchestra
BSOA British Sexual Offenses Act
B.S.Occ.Ther. Bachelor of Science in Occupational Therapy

B.Soc.Sci. Bachelor of Social Science
B.Soc.St. Bachelor of Social Studies
B.Soc.Wk. Bachelor of Social Work
BSOIW Bridge, Structural and Ornamental Iron Workers
B.S.Opt. Bachelor of Science in Optometry
B.S.O.T. Bachelor of Science in Occupational Therapy
bsp bromosulphalein
Bsp British Standard pipe
BSP Bering Sea Patrol; Border Security Police (NATO); Boy Scouts of the Philippines; British Society for Parasitology; Brotherhood of St Paul; Brunei Shell Petroleum; Bulgarian Socialist Party
B-S-P Bartlett-Snow-Pacific (foundry division)
B.S.P. Bachelor of Science in Pharmacy
BSP Bureau de Sécurité Publique (French—Bureau of Public Security)
BSPA Basic Slag Producers' Association; Black Students Psychological Association
B.S.P.A. Bachelor of Science in Public Administration
B.S.P.E. Bachelor of Science in Physical Education
B-Specials Belfast's special soldiers (attached to the Ulster Special Constabulary)
B.S.Per. & Pub.Rel. Bachelor of Science in Personnel and Public Relations
B.S.Pet. Bachelor of Science in Petroleum
B.S.Pet.Eng. Bachelor of Science in Petroleum Engineering
B.S.P.H. Bachelor of Science in Public Health
B.S.Phar. Bachelor of Science in Pharmacy
B.S.Pharm. Bachelor of Science in Pharmacy
B.S.P.H.N. Bachelor of Science in Public Health Nursing
B.S.Phys.Ed. Bachelor of Science in Physical Education
B.S.Phys.Edu. Bachelor of Science in Physical Education
B.S.Phys.Ther. Bachelor of Science in Physical Therapy
bspl behavioral science programming language

BSPL Blue Star Port Lines
BSPM Battlefield Systems Project Management
BSPMA British Sewage Plant Manufacturers Association
BSPP Burmese Socialist Program Party
B.S.P.T. Bachelor of Science in Physical Therapy
bsp test bromsulphalein test
B.Sp.Thy. Bachelor of Speech Therapy
bspw bare silver-plated wire
Bsq Basque
BSQ Bachelor Sergeant Quarters
bsr backspace recorder; balloon-supported rockets (rockoons); basal skin resistance; basic service rate; battle short relay; blood sedimentation rate; blue-streak request; bore sight restricted
Bsr Basra (Busreh)
BSR British Society of Rheology
B.S.R. Bachelor of Science in Rehabilitation
BSRA British Ship Research Association
BSRC Battelle Seattle Research Center; Biological Serial Record Center
BSRD Behavioral Sciences Research Division
B.S.Rec. Bachelor of Science in Recreation
B.S.Ret. Bachelor of Science in Retailing
bsrf brain stem reticular formation
BSRI Bem Sex Role Inventory
BSRIA Building Services Research and Information Association
BSRL Boeing Scientific Research Laboratories
B.S.R.T. Bachelor of Science in Radiological Technology
bss balanced salt solution; basic shaft system; beam-steering system; black-silk suture; buffered saline solution
BSS Bibliothèque Saint-Sulpice (Montreal); Biological and Social Sciences (NSF); British Sailors Society; British Security Service; British Standard Specification; Bronze Service Star

(USA); Bureau of School Systems; Bureau of State Services
B.S.S. Bachelor of Sanitary Science; Bachelor of Science in Science; Bachelor of Secretarial Science; Bachelor of Social Science(s)
Bssa Baronessa (Italian—Baroness)
B.S.S.A. Bachelor of Science in Secretarial Administration
B.S.Sc. Bachelor of Sanitary Science
B.S.Sc.Eng. Bachelor of Science in Science Engineering
B.S.See.Ed. Bachelor of Science in Secondary Education
B.S.Sec.Sci. Bachelor of Science in Secretarial Science
BSSG Biomedical Sciences Support Grant
BSSM Blue Star Ship Management
BSSML Blue Star Ship Management Limited
BSSO British Society for the Study of Orthodontics
B.S.Soc.Serv. Bachelor of Science in Social Service
B.S.Soc.St. Bachelor of Science in Social Studies
B.S.Soc.Wk. Bachelor of Science in Social Work
bssp broadband solid-state preamplifier
BSSP Battelle Seminars and Studies Program
BSSR Bureau of Social Science Research; Byelorussia Soviet Socialist Republic
BSSS British Society of Soil Science
B.S.S.S. Bachelor of Science in Secretarial Studies; Bachelor of Science in Social Science
B.S.S.Sc. Bachelor of Science in Social Science
B.S.Struc.Eng. Bachelor of Science in Structural Engineering
bssw bare stainless-steel wire
bst beam-steering transducer; blood serological test(ing); breast stimulation test (fetal test); brief stimulus therapy
bst. (BST) bovine somatotropin
b s & t blood, sweat, and tears
b/st bill of sight

BST Bering Standard Time; Blood Serological Test; British Summer Time
BSTA British Surgical Trades Association
BSTAS Battlefield Surveillance and Target Acquisition System
BSTC Ball State Teachers College
bstd bastard
B.S.Text. Bachelor of Science in Textiles
bst lt blue stern light
bstm biaxial shock-test machine
B-story below the A-story (30 to 110 feet up)
bstr booster
B.S.Trans. Bachelor of Science in Transportation
bstrk bomb service truck
bstr rkt booster rocket
BSU Black Students Union; Boat Support Unit; British Standard Unit(s)
bsub ball-and-socket upper bearing
B.Sur. Bachelor of Surgery
B.Surv. Bachelor of Surveying
bsut beam-steering ultrasonic transducer
bsv blown saves (baseball); Boolean simple variable
BSV Batten-Spielmyer-Vogt (syndrome)
B.S.Voc.Ag. Bachelor of Science in Vocational Education
bsw barrels of salt water
bs & w basic sediment and water
BSW Biological Society of Washington; Boot and Shoe Workers (union); Botanical Society of Washington; British Standard Whitworth
B.S.W. Bachelor of Social Work
BSWB Boy Scouts World Bureau
bswd barrels of salt water per day
BSWE Boy Scouts in Western Europe
BSWG British Standard Wire Gage
BSWIA British Steel Wire Industries Association
bt baby talk; *Bacillus thuringiensis* (biological pesticide); bathtub; bathythermograph; bedtime; bent; bishop's transcripts; bitemporal; blue tet-

razolium (stain); boat; boat-tail; body temperature; bombing table; bottom time; bought; brain tumor; broader term; brought; bulk transport; byssal thread

bt (BT) basic typing

b & t bacon and tomato sandwich; (cabin or room with) bath and toilet; bridges and tunnels

b of t balance of trade

Bt baronet

Bt Bukit (Indonesian or Malay—height, hill)

B-t *Bacillus thuringiensis* (biological pesticide)

BT Babylonian Talmud; basic trainer, Bering Time (Samoa); Bhutan (Internet code); British Telecon/Telecommunications; Broader Topic; Burgtheater (Vienna)

B & T Baker & Taylor

B of T Bank of Tokyo; Board of Trade

BT Berlingske Tidende (Danish—Berling's Times—Copenhagen); *Bolshoi Teatr* (Russian—Bolshoi theater) Moscow concert hall; *Brevet Technique* (French—Technical Diploma); *Burgtheater* (German—Court Theater) Vienna opera house

BT-13 Vultee two-place basic-trainer aircraft used during World War II

bta better than average; biological terrain assessment

bta (BTA) best time available (for tv broadcast)

BTA Blood Transfusion Association; Board of Tax Appeals; Border Trade Alliance; Boston Transportation Authority; Brazilian Travel Agency; Brith Trumpeldor of America; British Tourist Authority; British Travel Association; British Troops in Austria; British Tuberculosis Association

BTAM Basic Telecommunications Access Method; Basic Terminal Access Method

BTANZ British Trade Association of New Zealand

BTAO Bureau of Technical Assistance Operations (UN)

BTAP Bond Trade Analysis Program

BTASA Book Trade Association of South Africa

btb braided tube bundle; bus tie breaker

BTB Barbados Tourist Board; Belgian Tourist Bureau; British Troops in Berlin

BTBA Blood Transfusion Betterment Association

BTBL Braille and Talking Book Library

BTBS Book Trade Benevolent Society

btc below threshold change; beryllium thrust chamber; bus tie control

BTC Bankers Trust Company, Basic Training Center; Bethlehem Transportation Company; Board of Transport Commissioners; British Textile Confederation; Building Trades Council

B.T.C. Bachelor of Textile Chemistry

btca biblioteca (Spanish—library)

BTCC Bloom Township Community College; Board of Transportation Commissioners for Canada; Broome Technical Community College

B.T.C.P. Bachelor of Town and Country Planning

BTCV British Trust for Conservation Volunteers

btd bomb testing device

BTDB Bermuda Trade Development Board

btdc before top dead center

btdl basic-transient diode logic

bte battery terminal equipment; better than expected; blunt trailing edge; Boltzmann transport equation; bourdon tube element; Brayton turboelectric engine; bulk tape eraser

bte breveté (French—patent)

BTE Board of Teacher Education; Board of Transport Economics; British Troops in Egypt

B.T.E. Bachelor of Textile Engineering

BTEA British Textile Employers Association

B.Tech. Bachelor in Technology

Btee Brayton turboelectric engine

BTEF Book Trade Employers' Federation

B.Tel.E. Bachelor in Telecommunications Engineering

BTEMA British Tanning Extract Manufacturers' Association

BTES Beginning-Teacher Evaluation Study

B.Text. Bachelor of Textiles

btf balance to follow; barrels of total fluid; bomb tail fuse

b/tf balance transferred

BTF British Trawlers Federation

btg ball-tooth gear; battery timing group; beacon trigger generator; burst transmission group

BTG British Technology Group; British Troops in Germany; Building Trades Group

btgi ball-tooth gear joint

bth bath; bathroom; beat the hell; berth; beyond the horizon

B.Th. Bachelor of Theology

BT-H British Thompson-Houston

BTHS Brooklyn Technical High School

B th u British thermal unit (btu, Btu, BTU)

bti bank-and-turn indicator; bridgetape isolator

Bti Bacillus thuringiensis israelensis (mosquito-control substance)

BTI Bandung Technical Institute; Benefit Tours International; British Troops in Iraq

BTI British Technology Index

BTIA British Tar Industries Association

BTIPR Boyce Thompson Institute for Plant Research

btj ball-tooth joint

BTJ Board of Trade Journals

btk buttock

BtK Bacillus thuringiensis kurstaki

btk l buttock line

btl beginning tape label; behind the lens (camera); bottle

BTL Bell Telephone Laboratories; Bus Transceiver Logic

BTLC British Troops in Low Countries

BTLS Bell Telephone Laboratories System

btlv biological threshold limit value

BTLZ British Telecom Lempel Ziv

btm bottom

btm (BTM) bromotrifluoromethane (fire extinguisher)

Btm Bottom (postal abbreviation)

BTMA Bow Tie Manufacturers Association; British Typewriter Manufacturers Association

BTME Babcock Test of Mental Efficiency

BTMSQ Billy Tipton Memorial Saxophone Quartet

btn button

Btn Batangas

BTN Brussels Tariff Nomenclature

BTNA British Troops in North Africa

BTNI British Troops in Northern Ireland

bto big-time operator; bombing through overcast

bto bruto (Spanish—gross weight); *brutto* (Dano-Norwegian—bulk or gross weight); *bulto* (Spanish—bulk)

BTO Battalion Transport Officer; Branch Transportation Office(r); Brigade Transport Officer; British Trust for Ornithology

bto(s) big time operator(s)

B-to-B business to business

BTOW Boiler Technician of the Watch (USN)

B-town Bean Town (Boston—sailor's sobriquet)

btp body temperature and pressure

BTP Bailment Test Program; British Transport Police; Bush Terminal Piers

B.T.P. Bachelor of Town Planning

btps body temperature and pressure-saturated

btr better; bus tie relay; bus transfer

BTR Baton Rouge, Louisiana (airport); Bureau of Trade Regulation

BTR British Tax Review

BTR-40 Soviet armored personnel carrier and scout car for 10 troops including the driver

BTR-50 Soviet amphibious personnel carrier for 15 troops including the driver

BTR-60P Soviet amphibious armored personnel carrier including 12.7mm machinegun

B.T.R.A. Bachelor of Town and Regional Planning

BTRM Soviet armored-infantry combat vehicle armed with 76.2 gun and antitank missile

btrmlk buttermilk

btry battery

bts back to school; base of terminal service (USAF); Boolean time sequence

BTS Blood Transfusion Service; British Tanzania Society; British Textile Society

BTSA British Tensional Strapping Association

BTSB Bound-to-Stay-Bound Books

BTSC British Transport Staff College

BTSS Basic Time-Sharing System

BTTA British Thoracic and Tuberculosis Association

bttns battens

btu (BTU, Btu) British thermal unit

BTU Board of Trade Unit

btv basic transportation vehicle

btw between; by the way

BTW Belasting Toegevoegde Waarde (Dutch—value added tax)

BTWHS Booker T. Washington High School

btwn between

btx benzene, toluene, xylene

BTX Bildschirmtext (German—viewdata interactive videotext system)

bty battery

B-type Basedow type

bu base (of prism) up; base unit; base up; biological urge; brick unprotected; brilliant uncirculated; bromouracil; builder; burglary; bushel

bu. burgundy; buried

Bu Bulgaria; Bulgarian; Bureau (United States Navy); butyl

Bü Büyük (Turkish—big)

BU Baker University; Baylor University; Bishop's University; Bloomsburg University; Board of Underwriters; Boston University; Bradley University; Brandeis University; Brown University; Brunel University; Bucknell University; Burma (symbol); Butler University

B & U Beechey and Underwood

BU Bolletmino Uffciale (Italian—Official Gazette)

BUA Belfast Urban Area; British United Airways

BuAer Bureau of Aeronautics (USN)

BUAF British United Air Ferries

BUAV British Union for the Abolition of Vivisection

Bubs Bubbles

buc buccal; buccaneer; buccinator

BUC Bangor University College

bucc buccal

BUCCS Bath University Comparative Catalogue Study

Buchar Bucharest

buck buckram

Buck House Buckingham House (Buckingham Palace—London residence of British royalty)

Buck Island Buck Island Reef National Park off the north shore of St Croix, American Virgin Islands

Bucknell U Pr Bucknell University Press

Buck Pal Buckingham Palace

Bucks Buckinghamshire

Bucks Co Hist Bucks County Historical Society

BUCOP British Union Catalogue of Periodicals

bucu burring cutter

bud beautiful, useful, durable

bud. budget

Bud Buddha; Buddhism; Buddhist; Buddy; Budweiser

BUD Basic Underwater Demolition Team; Budapest, Hungary (airport)

Buda across the Danube from Pest, together known as Budapest

Bud(dy) Brother

BUDFIN Budget and Finance Division (NATO)

budgie(s) budgerigar(s)

BuDocks Bureau of Yards and Docks (USN)

Budpst Budapest

budr bromodeoxyuridine

bue built-up edge
BUE Buenos Aires, Argentina (Ezeiza airport)
Buen Buenaventura
BUET Bangladesh University of Engineering and Technology
buf buffer(ed)
Buf Buffalo (city and port)
BUF British Union of Fascists; Buffalo, New York (airport)
BUFF Big Ugly Fat Fellow (Air Force nickname for the eight-engine B-52 bomber)
Buffalo Acad Buffalo Fine Arts Academy
bufg buffing
bufno buffers, number of
Buf Phil Buffalo Philharmnonic
bug. black urban ghetto
Bug Bugatti; standard-model Volkswagen (also called the Beetle)
BUG Brooklyn Union Gas (company)
BUGA-UP Billboard Utilizing Graffitists Against Unhealthy Promotions
Büg Nay Mong Bügd Nayramdakh Mongol (Khalka—Mongolian People's Republic, now Mongolia)
BUH Bucharest, Rumania (airport)
BUIA British United Island Airways
buic (BUIC) backup interceptor control
BUIC Bureau (of Naval Personnel) Unit Identification Code (USN)
build. building
builds. builders
buisys barrier-up indicating system
Buk Bukit (Malay—hill, hilly street)
Buk Bukhta (Russian—bay)
bul below upper limit; bulletin
Bul Bulgaria(n)
BUL Bombay University Library
BUL Bibliotèque de l'Université Laval (French—Laval University Library)—Québec
Bulg Bulgaria; Bulgarian
bulg bulgarisch (German—Bulgarian)
Bulgaria Republic of Bulgaria, Republika Bulgaria

bull. bulla (Latin—leaden seal, a papal pronouncement bearing such a seal)
Bull bulletin
BULL Bank Users Legislative Lookout
Bull Acad Sci Bulletin of the Academy of Sciences
Bull Am Astron Soc Bulletin of the American Astronomical Society
Bull Am Phys Soc Bulletin of the American Physical Society
Bull Astron Inst Neth Bulletin of the Astronomical Institutes of the Netherlands
Bull Chem Soc Jp Bulletin of the Chemical Society of Japan
bullet(s) bullet train(s)
bulli. bulliat (Latin—let it boil)
Bull NYZS Bulletin of the New York Zoological Society
bull(s) bulletin(s)
Bull Seismol Soc Am Bulletin of the Seismological Society of America
bullsh Australian contraction of bullshit
buloga business logistics game
BULVA Belfast and Ulster Licensed Vintner's Association
BUMA Bureau voor Muziek-Auteursrecht (Dutch—Author's Rights Bureau)
BuMed Bureau of Medicine and Surgery
bump-and-run bump-and-run mugger-team technique
Bu M & S Bureau of Medicine and Surgery (USN)
B.U.M.S. Bachelor of Urani Medicine and Surgery
Bun Bunbury, Western Australia
BUN blood urea nitrogen
buna butadiene + natrium (synthethic rubber)
BUNAC British Universities North America Club
Bund German-American Volks-bund (pre-World War II alliance of Hitler's supporters in the U.S. now supplanted by the American Nazi Party); secret cells of Jewish Social Democrats in Russian Lithuania and Poland in 1897 who organized the General Union of Jewish Workers known as the Bund

Bund Deut Bundesrepublik Deutschland (German—Federal Republic of Germany)
bunsenite nickel oxide
Bunty Barbara
bunwich bun + sandwich (sandwich made in a bun)
BuOrd Bureau of Ordnance (USN)
bup backup plate; bull pup
BUP Boston University Press; British United Press
BUPA British United Provident Association
Bupers Bureau of Personnel (USN)
BuPers Bureau of Personnel (USN)
bupp backup plate perforated
Buppies black urban professionals
BuPubAff Bureau of Public Affairs
bur built-up roof(ing); bureau
bur. buried
Bur Burma; Burmese; Burundi (whose capital is Bujumbura)
BUR Bottom-Up Review; Burbank, California (Lockheed Airport)
Buran unmanned flight of the first Soviet space shuttle
'burbs suburbs
BURCEN Bureau of the Census
burd biplane ultralight research device
Burd suc Burdick suction
BuRec Bureau of Reclamation
Bur Eco Aff Bureau of Economic Affairs (U.S. Department of State)
Bur Eur Aff Bureau of European Affairs (U.S. Department of State)
burg burgess;. burgomaster
Burg Burgess; Burgo; Burgos; Burgwald
Burg Burgtheater (German—Castle Theater) Vienna opera house
burger(s) hamburger(s)
burgle(d) burglarize(d)
burgrep burglary report
Bur Intl Aff Bureau of International Affairs
Burk Burke; Burkhardt
Burke's Burke's Peerage
Burk Fas Burkina Faso
burl. burlesque
Burl Burleigh; Burley; Burlingame

Burl N Burlington Northern and St Louis San Francisco (railway merger)

Burm Burmese

Burma Socialist Republic of the Union of Burma (mountainous Asian country between India and Malaysia)

Burn amount of fuel needed for a flight (aircraft code); Burnell; Burnett; Burney

Burnaby Lower Mainland Regional Correctional Centre in British Columbia's Burnaby

burner afterburner of jet plane

Burns & Allen George Burns (Nathan Birnbaum) and Gracie Allen

buro bureau

burocrap bureaucratic excess

burp. backup rate of pitch

Bur Pub Aff Bureau of Public Affairs (U.S. Department of State)

Burs Bursar

Bursting Bursting Day (February 18, when the sea ice bursts apart and crumbles in Iceland's icy waters)

Burun Burundi; Burundian

Burundi Republic of Burundi (Central African land)

bus. business; omnibus

Bus autobus; Busan; business

BuSanda Bureau of Supplies and Accounts (USN)

BUSARB British-United States Amateur Rocket Bureau

busbar omnibus bar

Busch Busch Stadium, St. Louis

buscrit business critic(ism)

BUSF British Universities' Sports Federation

BuS glands Bartholin's, urethral, Skene's glands

bush. bushing(s)

Bush George Herbert Walker Bush, 41st President and 43rd Vice President of the United States

BuShips Bureau of Ships (USN)

bus hrs business hours

busk(s) busker(s)

BUSM Boston University School of Medicine

Bus Mgr Business Manager

busn business

Busn Intl Business International

BuSpar buspirone

bust(ed) arrest(ed)—slang

Bus W Business Week

buswrec ban unsafe school-buses which regularly endanger children

but. butter; button

but. butyrum (Latin—butter)

BUT British United Traction

bute butazolidin (phenylbuta-zone)

Buten Mus Buten Museum of Wedgewood

Butterick Butterick Publishing

Bu-Tyur Butyrskaya Tyurma (Russian—Butyrki Prison)—one of Moscow's major prisons

buv backscatter ultraviolet

buvs backscatter ultraviolet spectrometer

BUW Biblioteka Uniwersytecka w Warszawie (Polish—Warsaw University Library)

buwc basic underseas weapon circuit

BuWeps Bureau of Weapons (USN)

buy. buyer; buying

buz buzzer

bv balanced voltage; bellows valve; biologic(al) value; blow valve; blood vessel; blood volume; boiling vessel; bonnet valve; breakdown voltage; breviary; bronchovesicular

bv (BV) breakdown voltage

b/v brick veneer

bv bijvoorbeeld (Dutch—for example)

b.v. balneum vaporis (Latin—steambath, vapor bath)

Bv Benvenuto

B/v book value

BV Bouvet Island (Internet code); Bureau Veritas (French ship-classification bureau)

B + V Blohm und Voss (ship-builders)

BV Bayerische Vereinsbank (German—Bavarian Union Bank); *Besloten Vennootschap* (Dutch—closed corporation, private partnership)

BV. Beata Virgo (Latin—Blessed Virgin); *bene vale* (Latin—a good farewell); *bene vixit* (Latin—he lived a good life)

BV-202 Norwegian Army armored personnel carrier

BVA British Veterinary Association

B.V.A. Bachelor of Vocational Adjustment; Bachelor of Vocational Agriculture

BVAL Blackman's Volunteer Army of Liberation

bvbrf blood vessel of bronchial filament

BVC Buena Vista College; Bushveldt Carabiniers

bvd beacon video digitizer; boys vear dem (New York City slang)

BVD Bradley, Vorhees & Day; trademark brand of men's underwear (also: BVDs)

BVD Binnenlandse Veiligheidsdienst (Dutch—Internal Security Service)—FBI-type organization in the Netherlands

BVDs trademark brand of men's underwear

BVDT Brief Vestibular Disorientation Test

Bve Buenaventura

B.V.E. Bachelor of Vocational Education

Bventura Buenaventura

B/ventura Buenaventura, Colombia

b ver back verandah

BVES British Voluntary Euthanasia Society

B.Vet.Med. Bachelor of Veterinary Medicine

B.Vet.Sci. Bachelor of Veterinary Science

B.Vet.Sur. Bachelor of Veterinary Surgery

BVF Barbados Volunteer Force

BVG Berliner Verkehrs-Betriebe (German—Berlin Traffic Carrier)—Berlin's transit system

bvh biventricular hypertrophy

BVH British Van Heusen

bvi blood vessel invasion

BVI Better Vision Institute; British Virgin Islands

Bville Bougainville (Papua New Guinea)

BVJ British Veterinary Journal

BVL Bowler Victory Legion

BVLA British Volunteers, Latin America

bvm broncho-vascular markings

B.V.M. Bachelor of Veterinary Medicine

B.V.M. Beata Virgo Maria (Latin—Blessed Virgin Mary)

BVMA British Valve Manufacturers Association

BVMGT Bender Visual-Motor Gestalt Test

B.V.M.S. Bachelor of Veterinary Medicine and Surgery

BVN Bund der Verfolgten des Nazi Regimes (German—League of Persons Persecuted by the Nazi Regime)

BVNP Bolusan Volcano National Park (Luzon, Philippines)

bvo brominated vegetable oil

bvp beacon video processor; booster vacuum pump; boundery value problem

BVP British Visitors Program; British Volunteer Programme

BVPS Beacon Video Processing System; Booster Vacuum Pump System

bvr balanced valve regulator; beyond visible range; black void reactor

BVR Bangalore Volunteer Rifles ; British Vehicle Registration (symbols appearing on automotive vehicle license plates)—*see* British Vehicle Registration Symbols *in appendix*; Bureau of Vocational Rehabilitation

BVRC Bermuda Volunteer Rifle Corps

BVRO Base Vehicle Reporting Officer

BVRR Bureau of Veterans Reemployment Rights

BVRS Breadboard Visual Reference System

BVS Best Vested Socialists; Bevier & Southern (railroad)

B-V S Brisch-Vistem System (Visican punched-cards)

B.V.S. Bachelor of Veterinary Science; Bachelor of Veterinary Surgery

B.V.Sc. Bachelor of Veterinary Science

B.V.Sc. & A.H. Bachelor of Veterinary Science and Animal Husbandry

bvt brevet; brevetted

bvv bovine vaginitis virus

bvw binary voltage weigher

bw ballistic warning; bandwidth; batting wins; best of winners; bid wanted; biological warfare; birth weight;

body water; body weight; both ways; braided wire (armor)

b/w backed with; black-and-white

b & w black and white; bread and water

bw bijwoord (Dutch—adverb); *hitte wenden* (German—please turn over)

bW blood Wassermann

BW Bank of the West; Barrack Warden; Bendix Westinghouse; Biological Warfare; biological weapon; Black Watch; Borg-Warner; Botswana (Internet code); Business Week

B W Bendix Westinghouse Automotive Air Brake; Borg-Warner

B & W Babcock and Wilcox; Barker and Williamson; Burmeister and Wain

B of W Bishop(ric) of Würzburg

BW Bitte Wenden (German—please turn over); *Business Week*

bwa backward-wave amplifier; bent-wire antenna

BWA Baptist World Alliance; Baseball Writers Association; Boxing Writers Association; British West Africa; Building Waterproofers Association

BWAA Bowling Writers Association of America

BWAF Beijing Workers' Autonomous Federation

BWAL Barber West African Line

BWAM Brain-Wave Activity Measurement

Bway Broadway

BWB British Waterways Board

BWB Bundestampt für Wehrtechnik und Beschaffung (German—Federal Office for Military Technology and Procurement)

bwc basic weight calculator; broadband waveguide oscillator

BWC Battered Women's Coalition; Biological Weapons Convention; British War Cabinet

BWCA Boundary Waters Canoe Area

BWCC British Weed Control Conference

bwcdi best we can do is

BWCI Beauty Without Cruelty, Incorporated

bwcp bench welder control panel

bw-cw biological warfare–chemical warface

bwd bacillary white diarrhea; backward; barrels of water per day

BWD Baldwin Wallace College; British War Cabinet

B & WE Bristol and West of England

BWF Baha'i World Faith; Beyond War Foundation

Bwg Bowling

BWG Birmingham Wire Gage

bwh barrels of water per hour

BWH Book Week Headquarters

BWI Baltimore Washington International (airport); Boating Writers International; British West Indies

bwia better walk if able

BWIA British West Indian Airways

BWIS British West Indian dollar

BWIP Basalt Waste Isolation Project; Black Women In Publishing

BWIPO Basalt Waste Isolation Project Office

BWIR British West India Regiment

BWISA British West Indies Sugar Association

BWIU Building Workers Industrial Union

bwk brickwork; bulwark

bwl belt work line

BWL Biological War Laboratory

bwlt bow light

bwm barrels of water per minute

BWM British War Medal; Broom and Whisk Makers (union)

BWMA British Woodwork Manufacturers Association

BWMB British Wool Marketing Board

BWN Basic Weather Network; Brown Company (stock-exchange synrbol)

bwo backward-wave oscillator

bwoc big woman on campus

B'worth Butterworth

bwos backward-wave oscillator synchronizer

bwot backward-wave oscillator tube

bwp ballistic wind plotter

BWP Basic War Plan

bwpa backward-wave parametric amplifier

BWPA British Wood Preserving Association; British Wood Pulp Association

bwpd barrels of water per day

bwph barrels of water per hour

bwr (BWR) boiling-water reactor

BWRA British Water Research Association; British Welding Research Association

BWRC Biological Warfare Research Center

BWRS British War Relief Society

BWRWS Biological Warfare Rapid Warning System (USA)

bws beveled wood siding

BWS Bandipur Wildlife Sanctuary (India); Bank of Western Samoa; Batch Weighing System; Battered Wife Syndrome; Battered Women's Service; Battlefield Weapons System; Beaufort Wind Scale; Better World Society; Biological Weapons System; British Watercolour Society

BW & S Boyd, Weir & Sewell

BWSF British Water Ski Federation

BWSL Battlefield Weapons Systems Laboratory

bwso backward wave sweep oscillator

BWSR Bruno Walter Society Recording(s)

bwt both-way trunk

BWT Boeing Wind Tunnel

BWTA British Women's Temperance Association

BWTP Bureau of Work-Training Programs

bw-tv black-and-white television

bwu blue whale unit

bwv back-water valve

BWV *Bach-Werke-Verzeihnis* (German—Bach's Works Catalog)

BWVA British War Veterans of America

bwvs black-and-white vertical stripes

BWW Bad Weather Watch (Coast Guard)

BWWA British Water Works Association

bx biopsy; box; electrical cable contained in flexible tubing (bx cable)

Bx Beatrix; Bos (post-office box); Brix; Bronx

BX Base Exchange (USAF); Bellingham-Seattle Airways (2-letter code)

BXA Bureau of Export Administration

Bx-arts *Beaux-arts* (French—fine arts)

BX-C bithorax complex

bx cable insulated wires within flexible tubing

bxd boxed

bxk broadband X-band klystron

bx k box keel

BXL Bakelite Xylonite Limited

Bxm Brixham

Bxng D Boxing Day (first weekday after Christmas in England, Wales, Northern Ireland, Canada, South Africa, New Zealand, and Australia when boxes of gifts are given lettercarriers, newspaper carriers, and all others who provide services)

Bx Pk Bronx Park

bxs boxes

by buy stop limit

by. billion years; brilliant yellow (litmus paper for testing alkalinity)

b-y bloody

By Buryat(ic); Byron(ic)

BY Bedfordshire Yeomanry; Belarus (Internet code); blowing spray

BYC Baltimore Yacht Club; Bayside Yacht Club; Bengal Yeomanry Cavalry; Bensonhurst Yacht Club; Berkshire Yeomanry Cavalry; Beverley Yacht Club; Boston Yacht Club; Brewers Yeast Council; Bridgeport Yacht Club; Bronx YachtClub; Buffalo Yacht Club

BYCH Buckinghamshire Yeomanry Cavalry Hussars

bydv barley yellow dwarf virus

Bye Byelorussia; Byelorussian

Byelorussia White Russia bordering on Latvia, Lithuania, and Poland

byfml by first mail

byg buying

BYMS British Yard Minesweepers

bynam byname (nickname)

byo bring your own

Byo Bulawayo

byob bring your own beer; bring your own booze; bring your own bottle

byod bring your own drinks

byog bring your own girl

byp bypass

Byp Bypass

bypro(s) by-product(s)

byr(s) billion year(s)

byssin byssinosis

byt bright young things (British younger set)

byte eight adjacent binary digits that a computer processes as a unit

Byu Bayou (postal place-name abbreviation)

BYU Brigham Young University

Byz Byzantine

Byzantine Empire eastern segment of the Roman Empire

bz blank when zero; buffer zone; buzzer; (cartoonist's symbol— buzzing, sawing, snoring)

Bz benzene; benzodiazepine; benzoyl; Brazil; Brazilian

Bz Beobachtungszinimer (German—examining room)—hospital observation room

BZ Air Congo (Brazzaville, Congo Republic); B'nai Zion; Belize (Internet code); Belize Zoo

B/Z British Zone

BZ *Bild Zeitung* (German— Picture Newspaper)

Bza Bizerta

BZA Board of Zoning Adjustment

bz brigade (Danish—*besaet brigade*)—occupiers claiming squatter's rights on vacant property

bzbx brazing box

Bze Belize

bzfm brazing form

bzfx brazing fixture

Bzi Benghazi

bzw *beziehungsweise* (German—respectively)

bzz cartoonist's symbol -buzzing; sawing; snoring

Bzzrds Buzzards Bay

bzzz same as bzz

C

c calorie (small); cancel; candle; canine; capacity; capillary blood (symbol); carbon; cash; catcher; cathode; caudal; cent; cent-tavo; center; centi (prefix); centime; centimeter; central; certified; cervical; cervix; chapter; charm; chest; child; chord length (symbol); circa; cirrus; clearance; clonus; closed; closure; clouds; cloudy; coarse; coast; cocaine; codicil; coefficient; cold; color; colored; colorless; common; compensation; competitor; complement; conductor; contact; continuously; contraction; control; cook (military arm badge); cortex; cousin; cranial; crystal(line); cube; cubic; cubical; cycle(s); cylinder(s); cytidine; cytochrome; cytosine; heat capacity per mole (symbol); see; speed of light (symbol)

c (C) convict

c cibus (Latin—meal); circa (Latin—about); colón; colones (currency in Costa Rica and El Salvador); congius (Latin—gallon); cum (Latin—with)

c/ cargo (Spanish—total, weight); contra (Spanish—against, versus)

C calculated weight (symbol); calorie (large); candle; capacitance; capacitor; Cape; Captain; carat; carbon; Cardinal; cargo or transport airplane;

cargo vessel; carton; case; cathode; cavalry; celestial; Celsius; Celtic; Centigrade; century; cervical; chairman; Chaptered (Law); Charlie—code for letter C; Chief; Christ(ian); coast; cccaine; cold; college; colored; combat aircraft; commander; compliance; computer programming language; concentration; concentration of a gas in blood (symbol); Conservative; consul; control; Convair; copyright; Cosmopolitan Shipping; coulomb; council; course; Curie's constant; Fraunhofer line characteristic of hydrogen (symbol); hundredweight (symbol); liquidating dividend (in stock listings in newspapers); molecular heat (symbol); see (popular phonetic spelling)

C. carbohydrates; cocaine; Conservative (political party)

"C" Costa Line

°C degree Celsius; degree centigrade

C Cabo (Spanish—cape); Cap (French—cape); centum (Latin—one hundred); Col (French or Italian—highpass, pass); (Latin—Gaius)

C+ C-plus

C++ computer programming language

C– C-minus

Cº Cabeço (Portuguese—hillock, knoll, mound); Comisario (Spanish—Commisariat)

C_1 first class

C^1 bacteriologic complement

c 1º canto primo (Italian—first soprano part or voice)

C l, C 2, C 3, etc. cervical nerves or vertebrae 1, 2, 3, etc.

C I, C II, C III, etc. cranial nerves, I, II, III etc.

C_1, C_2, C_3, etc. cytochromes 1, 2, 3, etc.

C_2 command and control; second class

C2MOS cloned complimentary metal-oxide semiconductor

$C^2 D^2$ (ARDC) Command and Control Development Division

C-3 mentally or physically defective (British equivalent of American 4-F)

C_3 command, control, communications; third class

C^3 I Command, Control, Communications, and Intelligence

C.3.3. cell 3, 3rd landing, gallery C (occupied by Oscar Wilde while in Reading Gaol and the nom de plume he used there)

c3cm command control communication counter-measures

C3S College Chemistry Consultants Service

C4 Convair 440 airplane; crown quarto ($7-1/_2 \times 10$ inches)

C-4 military explosive

C5 Convair 580 turboprop airplane

C-5A Lockheed military cargo transport airplane

C-6 hexamethonium

C₆H₆ benzene

C8 crown octavo (5×7–$1/_2$ inches)

c8ᵛᵃ coll'ottava (Italian—in octaves)

C-9 McDonnell-Douglas twin-engine jetliner designed for medical evacuation, named Nightingale to honor Florence Nightingale

C-10 decamethonium

C¹1, C¹2, C¹3, etc. complements of complements

C₁₂H₂₂O₁₁ cane sugar

C¹⁴ radioactive carbon (used in determining age of objects)

C₁₇H₂₁NO₄ cocaine (also known as blow, coke, flake, freeze, happy dust, nose candy, lady, Peruvian, white girl)— derived from the leaves of the coca plant *(Erythroxylon coca)*

C 19 ster steroids containing 19 carbon atoms

C 21 ster steroids containing 21 carbon atoms

C 33 Oscar Wilde's identification number in Reading Gaol

C⁴I Command, Control, Communication, Computers and Intelligence

C-42 Brazilian Neiva four-seat utility aircraft called Regente

C-45 Beechcraft four-passenger transport plane

C-46 Curtiss-Wright World War II Commando 36-passenger transport

C-47 Douglas DC-3 Dakota or Skytrain 21-passenger air transport

C-54 Douglas DC-4 44-passenger transport called Skymaster

C-95 Brazilian 12-passenger transport aircraft named Bandeirante honoring frontier pioneers

C-118 Douglas DC-6 92-passenger transport also called Liftmaster

C-119 Fairchild-Hiller Flying Boxcar carrying 62 paratroopers or an equal weight of cargo

C-121 Lockheed Constellation or Super-Constellation transport carrying 63 or 99 passengers, respectively

C-123 Provider twin-engine assault transport

C-124 Globemaster heavy cargo four-engine transport airplane

C-130 Hercules medium-range cargo and troop transport airplane powered by four turboprop engines; Lockheed four-engine transport aircraft for military use

C-131 Convair 48-passenger military transport adapted from 24/440 commercial airliners

C-133 Cargomaster heavy four-engine turboprop cargo transport airplane

C-135 Boeing Stratofreighter military transport carrying 126 troops or equivalent cargo

C-140 Jet Star support-type transport aircraft powered by four turbojet engines

C-141 Starlifter large cargo transport airplane powered by four turbojet engines

C200 Committee of 200

C-212 Casa 15-seat aero medical or paratrooper transport plane made in Spain and called Aviocar

¢ cedi—monetary unit of Ghana; centmonetary unit of the United States; colon monetary unit of Costa Rica and El Salvador

ca cable; calibrated altitude; cancer; capital account; capital asset; carbonic anhydrase; carcinoma; cardiac arrest; cash; catecholamine (neurotransmitter); cathode; caudal; centare; cervoaxial; chances accepted, errorless (baseball); chronological age; circa; civil affairs; civil authorities; clerical aptitude; cold agglutinin; common antigen; compounded annually; controlled atmosphere; convening authority; coronary artery; council accepted; covert action; credit account; croup associated; current account; current assets

ca (CA) cancer; carcinoma

ca' calf; call (Scottish contraction)

c-a computer aided

c/a capital account; center angle; coated abrasive; current account

c&a (C&A) command and administration

c & a classification and audit

ca125 cancer antigen 125

ca2729 cancer antigen 27, 29

ca *circa* (Latin—about); *corrente alternada* (Portuguese—alternating current); *corriente alterna* (Spanish—aternating current)

cᵃ *compañia* (Spanish—company)

c a *coll'arco* (Italian—to be bowed)

Ca calcium; Canada; Canadian

Ca *Compagnia* (Italian— company)

Ca' *Casa* (Venetian—house)

Cᵃ *Cabeça* (Portuguese—head, headland); *Companhia* (Portuguese—company); *Compañia* (Spanish—company)

CA California; Canada (Internet code); Canadian Army; Capital Airlines; Central America; Certificate of Airworthiness; Certification Authority; Certified Acupuncturist; Chargé d'Affaires; Chartered Accountant; Chemical Abstracts; Chief Accountant; Church Army; Civil Affairs; Coast Artillery; Cocaine Anonymous; Combat Aircrew; Combat Aircrewman; Commercial Agent; Compensation Act; Comptroller of the Army; Conditional Award; Confederate Army; Construction Authority; Construction Authorization; Consular Agent; Consumers Association; Controlled Atmosphere; Convening Authority; Coordinating Agency; County Attorney; Court Auditor; Court of Appeals; Cranial Academy; heavy cruiser (naval symbol)

CA (Aust) Institute of Chartered Accountants in Australia

C.A. Chartered Accountant

C & A Clemens and August Breeninkmeyer's international house of fashion

C of A College of Aeronautics; Commonwealth of Australia

CA *Centre Agricole* (French—Agricultural Center)—prison farm; *Chemical Abstracts; corriente alterna* (Spanish—alternating current); *Companhia de Navegação Carregadores Açoreanos* (Azore Line)

caa caging amplifier assembly; circular aperture antenna; combat available aircraft; computer amplifier alarm; crime aboard aircraft

CAA Canadian Authors' Association; Canadian Automobile Association; Cantors Assembly of America; Caribbean Atlantic Airlines; Central African Airways; Chester Allan Arthur (21st President U.S.); Chief Aircraft Artificer; Chief of Army Aviation; Civil Aeronautics Administration; Civil Aeronautics Authority; Civil Aviation Authority; Clean Air Act; Collectors of American Art; College Art Association; Colonial Athletic Association; Commercial Apiarists Association; Community Action Agencies; Community Aid Abroad; Congressional Assassination Act; Cooperative Alumni Association; Correctional Administrators Association; Council on American Affairs; Creative Artists Agency; Cremation Association of America; Custom Agents Association

C.A.A. Civil Aviation Authority (United Kingdom)

CAAA Canadian Association of Advertising Agencies; College Art Association of America; Composers, Authors, and Artists of America

CAAB California Avocado Advisory Board

CAABU Council for the Advancement of Arab-British Understanding

CAAC China Aeronautics Airline Corporation (nicknamed China Airlines Always Cancels); Civil Aviation Administration of China; Customs and Allied Affairs Committee

CAADAC California Association of Alcoholism and Drug Abuse Counselors

CAADRP Civil Aircraft Airworthiness Data Recording Program (UK)

CAAE Canadian Association for Adult Education

CAAF Combined Allied Air Forces

CAAFS Chinese Academy of Agricultural and Forestry Sciences

CAAIS Computer-Assisted Action Information System(s)

caar compressed-air-accumulator rocket

CAAR Committee Against Academic Repression

CAARC Commonwealth Advisory Aeronautical Research Council

CAAs Community Action Agencies

CAAS Ceylon Association for the Advancement of Science; Civil Aviation Authority of Singapore; Computer-Assisted Acquisition System; Connecticut Academy of Arts and Sciences

CAASE Computer-Assisted Area Source Emissions

CAAT Campaign Against Arms Trade; Canadian Academic Aptitude Test; Colleges of Applied Arts and Technology

CAATO Combined Army Air Transport Organization

CA Att Civil Air Attaché

CAAV Central Association of Agricultural Valuers

cab cabal; cabbage; cabin; cabinet; cable; cabochon; cabriolet; calibration; captured air bubble; cellulose acetate butyrate; controlled amortization bond; taxicab

cab (CAB) cellulose acetate butyrate; coronary artery bypass

Cab Cabell; Cabot; NATO nickname for Soviet Lisunov transport plane designated Li2

CAB Cabletelevision Advertising Bureau; Canadian Association of Broadcasters; Charles A(ustin) Beard; Circulation Audit Board; Citizens Advice Bureau; Civil Aeronautics Board; Civil Aeronautics Bulletin; Clark Air Base; Commonwealth Agricultural Bureau; Consumer Affairs Bureau; Contract Appeals Board (Veterans Administration)

CABA Charge Account Bankers Association

cabaf currency and bunker adjustment factor

cabal *cabbala* (Hebrew—something secret)

CABAS City and Borough Architects Society

CABB Captured Air-Bubble Boat (naval)

Cabe Cabedelo (easternmost point of Brazil and all South America close to João Pessoa)

CABE California Association of Bilingual Education

CABEI Central American Bank for Economic Integration

CABEZA California and Baja California Enterprise Zone Authority

cabg (CABG) coronary artery bypass grafting

CABI Commonwealth Agricultural Bureau International

CABIC Copper and Brass Information Centre (Australia)

CABIN Campaign Against Building Industry Nationalization

CABLE Computer-Assisted Bay Area Law Enforcement (San Francisco)

cablecast broadcast by cable tv; cablecaster; cablecasting

cablecast(ing) cable television telecast(ing)

cablese cablegram language (abbreviated, telegraphic, truncated style)

cable tv community-antenna television

cablevision cable television

CABM Commonwealth of Australia Bureau of Meteorology

CABMA Canadian Association of British Manufacturers and Agencies

CABMS Chinese-oriented Antiballistic Missile System

Cabo Cabo San Lucas, Baja California; San José del Cabo, Baja California

CABO Council of American Building Officials

Cabo de H. *Cabo de Hornos* (Spanish—Cape Horn)southernmost point of Chile and all South America

cabot cabotage (coastal navigation)

Ca bp calcium-binding protein

CABRA Copper and Brass Research Association

Cabral Pedro Alvarez Cabral, discoverer of Brazil

cabs. cabbages

cab(s) cabochon(s)

CABS Children's Adaptive Behavior Scale; Computer-Augumented Block System; Computerized Annotated Bibliographic System; Current Awareness in Biological Sciences

cabtmkr cabinetmaker

CABWA Copper and Brass Warehouse Association

cac cardiac-accelerator center

Cac Caceres

CAC California Administration Code; California Advisory Council (on Vocational Education); California Aeronautics Commission; California Arts Commission; California Arts Council; Canadian Armmoured Corps; Central Arbitration Committee; Chief of Air Corps; City Administration Center; Civic Administration Center; Civil Administration Commission; Coast Artillery Corps; College Admissions Center; Colonial Ammunition Company; Combat Air Crew; Commander Air Center; Commission of Accreditation for Corrections; Community Administration Council; Commuter Aircraft Corporation; Consumer Advisory Council; Consumer Affairs Council; Consumers' Association of Canada; Continental Air Command; Corrective Action Commission; Corrective Action Committee; Cosmetology Accrediting Commission; County Administration Center

CAC *Comité de Acción Cultural* (Spanish—Cultural Action Committee)

CACA Canadian Agricultural Chemicals Association; Central After-Care Association

cacb compressed-air circuit breaker

CACB Council Against Cigarette Bootlegging

cacc cathodal closure contraction

CACC Central Atlantic College Conference; Civil Aviation Communications Center; Corrective Action Control Section; Council for the Accreditation of Correspondence Colleges

CA-CC Christian Anti-Communist Crusade

CACCE Council of American Chambers of Commerce in Europe

CACCI Confederation of Asian Chambers of Commerce and Industry

CACDA Combined Arms Combat Development Activity

CACE California Association for Childhood Education; Chicago Association of Consulting Engineers

CACEX *Carteira do Comercio Exterior* (Portuguese—Foreign Commerce Department)—Bank of Brazil

CACF Colombian-American Culture Foundation

Cach *Cachoeira* (Portuguese—rapids, waterfall)

cache. computer-controlled automated cargo-handling envelope

CACHE Computer Aids for Chemical Engineering Education

cachi *cachivache* (Spanish—broken crockery, foolish or worthless person, poor quality, pots and pans, utensils)—dialect heard around Buenos Aires also called *porteño*

CACJ California Attorneys for Criminal Justice

CACL Canadian Association of Children's Librarians

CACM Central American Common Market

Caco *Cacoliche* (pidgin Argentine-Spanish including many Italian words)

CACO Casualty Assistance Call Office(r)

CaCO₃ calcium carbonate (limestone)

CACOHIS Computer Aided Community Oral Health Information System

cacoph cacaphonic; cacophony

cacp cartridge-actuated compaction press

CACPAF Continental Association of Certified Public Accounting Firms

CACS California Aqueduct Control System

CACSW Citizen's Advisory Council on the Status of Women

CACTI Common Agricultural Customs Transmission of Information

CACTUS *Capteur Accelerometrique Capacitif Triaxial Ultra-Sensible* (French—Ultra-Sensitive Triaxial Capacitive Accelerometric Detector)

CACUL Canadian Association of College and University Libraries

CACVE California Advisory Council on Vocational Education

CAC & W Continental Aircraft Control and Warning

cad cadastral; cadaver; caddie; cadenza; cadet; cadmium; caaridge-activated device; cartridge-actuated device; cash against disbursements; cash against documents; cash available for distribution; central ammunition depot; computer-aided design; contract award date; coronary artery disease

cad (CAD) computer-aided design

c.a.d. cash against disbursements

cad *cadenza* (Italian—solo passage near end of a concerto movement)

c-a-d *c'est-à-dire* (French—that is to say)

Cad Cadell; Cadiz; Cadmar; Cadmus; Cadogan; Cadwalader or Cadwallader (often pronounced *Calder*)

CAD California Association of the Deaf; Civil Affairs Division; Civil Air Defense; Claude Archille Debussy; Combat Air Division; Commission Against Discrimination; Computer-Aided Dispatch (police); Computer-Assisted Design; Consumer Affairs Department; Crown Agents Department

cada clean air dot angle

CADA Centre d'Analyse Documentaire pour l'Archélogie (French—Document Analysis Center—Archaeology)

CADAFE Compañia Anónima de Administración y Fomento Electrico (Spanish—Corporation for Electrical Administration and Development)

CADAM Computer-graphic Augmented Design and Manufacturing (registered trademark of Cadam, Inc.)

CADAN Centre d'Analyse Documentaire pour Afrique Noir (French—Document Analysis Center—Africa)

cadav cadaver(ous)

cadc central air data computer

CADC Canadian Army Dental Corps; Continental Air Defense Command; Corrective Action Data Center

cad/cam computer-aided design/computer-aided manufacturing

CADCEP California Alcohol and Drug Counselors Education Program

cadco core and drum corrector

cadd computer-aided design drafting

CADDIA Cooperation in Automation of Data and Documentation for Imports/Exports and Agriculture

Caddie Charlotte

Cad(dy) Cadillac

cade computer-aided design engineering; computer-aided design evaluation; computer assisted data engineering; computer-assisted data evaluation

cadet. computer-aided design experimental translator

Cadet old Russian acronym for Constitutional Democratic Party or one of its members

Cadets Constitutional Democrats (in czarist Russia)

cadf commutated antenna direction finder

CADF Central Air Defense Force; Contract Administrative Data File

cadfiss computation and data flow integrated subsystems

'Cadian(s) Acadian(s)

CADIG Coventry and District Information Group

CADIN Continental Air Defense Integration North

cadis coronary artery disease

CADIZ Air Defense Identification Zone

CADL Christian Anti-Defamation League

CADM CONUS (Continental United States) Air Defense Modernization

CADO Central Air Documents Office (USAF); Central American Development Organization; Current Actions Duty Office(r)

Ca'd'Oro Casa de Oro (Italian—House of Gold)

'cado(s) avocado(s)

CADPE Comprehensive Alcohol-Drug Prevention Education

CADPIN Customs Automatic Data Processing Intelligence Network (U.S. Bureau of Customs)

CADPOS Communications and Data Processing Operation System

cadr clean-air delivery rate

cadre. current awareness and document retrieval for engineers

cads. cellular-absorbed-dose spectrometer

CADS Central Air Data System (USAF); Containerized Ammunition Distribution System (USA)

cadss combined analog-digital systems simulator

CADSYS Computer-Aided Design System

cadte cathodal duration tetanus

Cadwal Cadwallader

cae carrier aircraft equipment; computer-aided engineering; computer-assisted electrocardiography; computer-assisted enrollment

Cae Caelum

CAE Canadian Aviation Electronics; Columbia, South Carolina (airport); Common Applications Environment; Council on Anthropology and Education

CA & E Council on Anthropology and Education

CAE Cóbrese al Entregar (Spanish—cash on delivery)

CAEA California Aviation Education Association; Chartered Auctioneers and Estate Agents

CAEAI Chartered Auctioneers and Estate Agents Institute

CAED Canadian Association of Equipment Dealers

Ca edta calcium disodium ethylene diamine tetra-acetate

caef casualty air evacuation flight

Cael Caelum

CAEL Council for the Advancement of Experimental Learning

CAEM Conseil d'Assistance Economique Mutuelle (French—Council for Mutual Economic Assistance)

Caer (Cornish or Welsh—fortress)—Caermarthen, Caernarvon, Caerphilly, Caerwent, and Caerwys—the Caers

CAER Centre for Applied Economic Research (New South Wales)

Caern Caernarvonshire

caerul. caeruleus (Latin—cerulian)—sky blue

caes compressed-air energy storage

Caes Caius Julius Caesar

CAES Canadian Agricultural Economics Society; Connecticut Agricultural Experiment Station

CA&ES College of Agriculture & Environmental Sciences

caesar computerized automation by electronic system with automated reservations

CAET Corrective Action Evaluation Team

CAEU Casualty Air Evacuation Unit; Council of Arab Economic Unity

CAEWW Carrier Airborne Early Warning Wing (USN)

caf cafeteria; caffeine; centralized authorization file; clerical, administrative, and fiscal; cost and freight; cost, assurance, and freight; currency adjustment factor; cytoxan + adriamycin + 5-fluorouracil

caf coût, assurance, fret (French—cost, assurance, freight)

CAF Cambodian Armed Forces; Canadian Armed Forces; Central African Federation; Centralized Authorization File; Ceylon Air Force; Chief Air Fitter; Citizen Air Force; Conventional Armed Forces

CAF Corporación Andina de Fomento (Spanish—Andean Promotion Corporation)

CAFA Chicago Academy of Fine Arts

CAFAF Commander, Amphibious Forces Atlantic Fleet

CAFB Canadian Air Force Base; Clark Air Force Base

CAFCINZ Campaign Against Foreign Control in New Zealand

cafd contact analog flight display

cafe (CAFE) corporate average fuel economy

cafe. computer-aided film editor; corporate average fuel economy

CAFE Conventional Armed Forces in Europe; Corporate Average Fuel Economy

CAFEA-ICC Commission on Asian and Far Eastern Affairs—International Chamber of Commerce

CAFEI Central American Fund for Economic Integration

cafetorium cafeteria-auditorium

caff caffeine

CAFG Commander, Air Forces Gulf

cafga computer applications for the graphic arts

CAFGU Citizens Armed Forces Geographical Unit (Philippines)

CAFI Commercial Advisory Foundation in Indonesia

CAFIC Combined Allied Forces Information Center

CAFIT Computer-Assisted Fault Isolation Test(ing)

cafm commercial air freight movement

CAFMS Continental Association of Funeral and Memorial Societies

cafo concentrated animal feeding operation

CAFO Command Accounting and Finance Office; Confidential Admiralty Fleet Order

CAFPF Commander, Amphibious Forces Pacific Fleet

CAFR Comparative Annual Financial Report

C Afr Fed Central African Federation

CAFS Cartridge-Actuated Flame System

CAFSC Control Air Force Specialty Code

CAFTA Council of Australian Food Technology Associations

CAFU Civil Aviation Flying Unit

CAFVTC Canadian Army Fighting Vehicles Training Centre

cag chronic atrophic gastritis; constant aerial glide; constant altitude glide; cytosine, adenine, and guanine (chemical building blocks of DNA)

Cag Cagliari; Cagliostro

CAG Cancer Assessment Group; Carrier Air Group; Civil Air Guard; Combat Arms Group; Commander Air Group; Commissioner's Advisoy Group; Composers-Authors Guild; Computer Analysis Group; Concert Artist Guild; Corrective Action Group; heavy guided-missile cruiser (naval symbol)

CAGA California Asparagus Growers Association; Commercial and General Acceptance

CAGE Convicts' Association for a Good Environment; Cut down on drinking? Annoyed by criticism of drinking? Guilty feelings about drinking? Eye-opener? (questionnaire to detect drinking problem)

cagel consolidated aerospace ground equipment list

CAGEO Council of Australian Government Employee Organisations

CAGI Compressed Air and Gas Institute

CAGRA Commander Army Group Royal Artillery

CAGRE Commander Army Group Royal Engineers

CAGS Canadian Arctic Gas Study

CAGW Citizens Against Government Waste

cah congenital adrenal hyperplasia

CAH Community of All Hallows; Conzinc Asia Holdings

cahd coronary atherosclerotic heart disease

CAHOF Canadian Aviation Hall of Fame

CAHS Comprehensive Automation of the Hydrometeorological Service

CAHT Canadian Association for Humane Trapping

cai computer-aided instruction; computer-assisted instruction; confused artificial insemination

Cai Cairo; Gonville and Caius College, Cambridge

C-a I Computer-assisted Instruction

CAI Canadian Aeronautical Institute; Canadian Airlines International; Career Apparel Institute; Career Assessment Inventouy; Computer Applications Incorporated; Confederation of Australian Industry; Configuration Audit Inspection (USA); Container Aid International; Cruelty to Animals Inspectorate; Culinary Arts Institute

CAI Club Alpino Italiano (Italian—Italian Alpine Club)

CAIA California Association of Independent Accountants; Customs Agents Institute of Australia

CAIB Certified Associate of the Institute of Bankers

caic computer-assisted indexing and classification

CAIC Civil Aviation Information Circular; Coalition for the Apparel Industry of California

Cai Col Gonville and Caius College—Cambridge

Caicos Caicos and Turks Islands (in British West Indies southeast of the Bahamas and north of Hispaniola)

CAIG Canadian Aircraft Insurance Group

CAIL Coal and Allied Industries Limited

CAIMAW Canadian Association of Industrial, Mechanical, and Allied Workers

CAIN CAtaloging-INdexing (National Agricultural Library data base)

CAINS Carrier Aircraft Inertial System

caint counter-air and interdiction

caiop computer analog input-output

CAIP Computer-Assisted Indexing Program (UN)

CAIQ Chamber of Automotive Industries of Queensland

CAIR Council on American-Islamic Relations

CAIRA Central Automated Inventory and Referral Activity (USAF)

CAirC Caribbean Air Command

CAIRS Central Automated Inventory and Referral System (USAF); Computer-Assisted Information Retrieval System; Computer-Assisted Interactive Resources Scheduling System

CAIS Canadian Association for Information Science; Center for Advanced International Studies (Univ of Miami); Central Abstracting and Indexing Service; Common Agreement on Investment and Society; Connecticut Association of Independent Schools

Caith Caithness

CAITS Chemical Agent Identification Training Set

caj calked joint

CAJ Center for Administrative Justice; Confederation of ASEAN Journalists

caje consolidated antijam equipment

'cajun Acadian (native of Louisiana)

cak conical alignment kit; cube alignment kit

CAK Akron, Ohio (airport)

CAK-C Concept Assessment Kit-Conservation

cal caliber; calorie (small); computer-aided learning; computer-assisted learning; conversional algebraic language

cal calando (Italian—calming); carbine automatique légère (French—light automatic carbine)—CAL

Cal Calabar; Calabozo; Calabria; Calafat; Calahan; Calais; Calamar; Calbert; Calcutta; Caldecott; Calder; Caldwell; Cale; Caleb; Caledonia(n); Calgary; Calhoun; Caliente; California; Calixto; Calkins; Call; Callaghan; Callahan; Callao; Callcott; Callyhan; Calorie (large); Calpurnius; Calumet; Calvagh; Calvary; Calven; Calvert; Calvin; Calvus

CAL Catholic Aviation League; Center for Applied Linguistics; China Airlines; Citizens Action League; Commonwealth Acoustic Laboratories; Computer Accounting Limited; Computer-Assisted Learning; Conference-Approved Literature (Australia); Continental Airlines; Conversational Algebraic Language; Cornell Aeronautical Laboratory; Cyprus Airways; Point Arguello (California) tracking station

CAL Comandos Armados por Liberación (Spanish—Armed Commandos for Liberation)—Puerto Rican underground militants fighting for decolonialization

cala calabozo (Spanish—cell, dungeon, jail)

CALA Chinese-American Librarians Association; Civil Aviation Licensing Act

calaham California ham (picnic ham)

calamine smithsonite (zinc carbonate)

CALANS Caribbean and Latin American News Service

CalArts California Institute of the Arts

CALAS Computer-Assisted Language Analysis System

calb computer-assisted line balancing

C_{alb} albumin clearance

calbr calibration

calc calculate(d); calculation; calculator; calculus

calc (CALC) calculate; calculator (flow chart)

Calc Calcutta

Calc Calçada (Portuguese—Street)

Calcasieu Calcasieu Lake or Calcasieu Pass in southwestern Louisiana where the lake waters flow into the Gulf of Mexico

calcd calculated

CALCOFI California Cooperative Oceanic Fishery Investigation

Calcomp California Computer Products

Calc Univ Calcutta University

cald calculated; caldera

CALDA Canadian Air Line Dispatchers Association

CALDAC California Debt Advisory Commission

CALDEA California Driver Education Association

Calder Cadwalader; Cadwallader

CALE Canadian Army Liaison Executive

CALEA Canadian Air Line Employees Association; Commission on Accreditation for Law Enforcement Agencies

Caled Caledonia

Caled Can Caledonian Canal

Caledonia (Latin—Scotland)

Caledonian Caledonian Canal bisecting northern Scotland and connecting the Atlantic Ocean with the North Sea; pertaining to Scotland and things Scottish

calef. calefactus (Latin—warmed)

calen calendar; calender

Ca/EPA California Environmental Protection Agency

CALEV Compañía Anómina Luz Electrica de Venezuela (Spanish—Electric Light of Venezuela Corporation)

Calex Calexico (California border city)

Cal Expo California Exposition (permanent show at Sacramento)

CALFAA Canadian Air Line Flight Attendants Association

calfin calendered finish

Calg Calgary

Calhan Calahan

Calhoun Catquahoun; originally Colquhoun

calib calibrate; calibration

calibn calibration

caliche calcium carbonate crust (or) dust—CaC0₃

Caliente Agua Catiente, Mexico; Nevada (delinquent) Girls Training Center at Caliente, Nevada; racetrack town adjacent to Tjuana, in Baja California

Cal-ID California Identification System

Calif California; Californian

CALIF California

Calif Cur California Current

Calif Hist California Historical Society

Calif Rev Pr California Review Press

Calipuerto Cali Aeropuerto (Cali, Colombia)

cal_lt calorie (International Table calorie)

CALIT California Institute of Technology (also Caltech or CIT)

CALL Canadian Association of Law Libraries; Community Access Library Line; Community Action for Limited Learners; Composite Aeronautical Load List(ing); Counselling at the Local Level (SBA)

CALLA Common Acute Lymphoblastic Leukemia Antigen, J5

callas calla lillies

calli calliope

Calli Callimachus of Alexandria (bibliographer-poet-scholar)

callig calligrapher; calligraphic; calligraphy

call-in call-in radio or television program soliciting audience participation; call-in telephone call advising of an anticipated absence due to illness, etc.

calm. collected algorithms for teaming machines

calm *calmato* (Italian—calmly; tranquilly)

CALM Child Abuse Listening Mediation; Citizens Against Legalized Murder; Computer-Assisted Library Mechanization

CALMA California Marine Associates

Cal Maritime California Maritime Academy

Calmex California-Mexico

CALMS Computer Automatic Line Monitoring System

caln calculation

calo *calando* (Italian—softer and slower, bit by bit)

calogsim computer-assisted logistics simulation

calomel mercurous chloride (Hg₂C12)

CAL/OSHA California Occupational Safety and Health Administration

CALPA Canadian Air Line Pilots Association

CalPAW California Parks and Wildlife Initiative

Cal Pen California Peninsula, containing Baja or Lower California

CALPERS California Public Employees Retirement System

CALPIRG California Public Interest Research Group

Cal Poly California Polytechnic

CalREN-2 California Research and Education Network

CALRI Central Artificial Leather Research Institute

Cal Rptr *California Reporter*

CALS Canadian Association of Library Schools

CALSO California Transport

CalSpace California Space Institute

CalTec California Institute of Technology

CALTEX California-Texas Petroleum; Overseas Tankship Corporation

cal_th calorie (thermochemical calorie)

CALTIP Californians Turn In Poachers (who fish with two poles or hunt out of season)

CALTRAC California Track

Caltrans California Department of Transportation

CALURA Corporation and Labour Unions Returns Act

calv calving

Calv Calvin; Calvin Coolidge (American president); Calvinism; Calvinist

Calvary Calvary Hill outside Jerusalem where its Aramaic name is Golgotha—Place of the Skull

Cal-VDAC California Venereal Disease Advisory Council

Calz *Calzada* (Spanish—boulevard, highway)

cam camber; camouflage; catapult aircraft armed merchantman; catapult aircraft merchant (ship); chemical agent monitor; circular area method; cockpit area microphone; commercial air movement; comprehensive achievement monitoring; computer-addressed memory; computer-assisted manufacturing; content-addressed memory; continuous air monitor; contract air mail

cam (CAM) central address memory; checkout and automatic monitoring; computer aided manufacturing

ca'm calm

Cam Camagüey (Cuban Province) Cambodia; Cambodian; Camden; Camelopardalis (Latin-Giraffe constellation); Cameron; Cameroons; Campbell; Campechanos; Campeche

C_am amylase clearance

CAM Certified Administration Manager; Civil Aeronautics Manual; Civil Aviation Medicine; Composite Army-Marine; Computer-Aided Management; Computer-Aided Manufacture; Consumers Action Movement; Contract Air Mail; Contract Audit Manual; Cooperative Autoworks Malaysia; Cost-Account Manager; Course-A-Month

cama centralized automatic message accounting

CAMA Children's Aid Movement of Australia; Civil Aerospace Medical Association

camal (CAMAL) continuous air borne missile alert

C'Amalie Charlotte Amalie

CAMALS Cambridge Algebra System

CAMAR Competition of Agriculture and Management of Agricultural Resources

CAMARC Computer Aided Movement Analysis in a Rehabilitation Context

camb camber; cambium; cambric linen or tea

Camb Cambrian; Cambridge

Cambod Cambodia; Cambodian

Cambodia former French Indo-Chinese colony, Peoples Republic of Kampuchea

Cambrian Cambrian Airways

Cambrians Cambrian Mountains, Wales

Cambridge UP Cambridge University Press

Cambs Cambridgeshire and Isle of Ely

CAMC Canadian Army Medical Corps; Cartoon Art Museum of California

CAMCE Computer Aided Multimedia Courseware Engineering

camcorder camera and tape recorder device

CAMDA Car and Motorcycle Drivers Association

camel. common automatic manifest language; computer-assisted machine loading; credit, assets, management, earnings, liquidity

CAMEO Capitol Area Motion Pictures Education Organization (D.C.)

Camer Cameroon (whose capital is Yaoundé)

camera cooperating agency method for event reporting and analysis

CAMERA Committee for Accuracy in Middle East Reporting in America

Cameroon United Republic of Cameroon (equatorial African nation); *République Unie du Cameroun*

CAMESA Canadian Military Electronics Standards Agency

Camford Cambridge and Oxford

Cam High Camden High School; Cameron Highlanders

CAMI Civil Aeromedical Institute; Columbia Artists Management, Incorporated

camiknick camisole-knickerbocker combination

Camillo Escamillo

CAMIS Computer-Assisted Makeup and Imaging System

caml cargo aircraft mine laying

Caml Camelopardus

CAML Canadian Association of Music Libraries

cam luc camera lucida

CAMM Canadian Association of Medical Microbiologists; Council of American Master Mariners

CAMMIS Command Aircraft Maintenance Manpower Information System

cam obs camera obscura

camof camouflage

camol computer-assisted management of learning

camp. computer-assisted menu planning; cosmopolitan art—modern and personalized; cyclic adenosine monophosphate

cAMP cyclic adenosine 3′,5′-monophosphate

Camp Campeche (inhabitants—Campechanos); Campion Hall, Oxford

CAMP Campaign Against Marijuana Planting; Campaign Against Moral Persecution; College-Assistance Migrant Program; Computer Applications of Military Problems; Continuous Air Monitoring Program; Course-A-Month Program; Craft Attitude Monitoring Package

campan campanological; campanologist; campanology

Camp Hall Campion Hall—Oxford

Campoformido Campo Formio

CAMPS Centralized Automated Military Pay System; Cooperative Area Manpower Planning Systems

CAMPSA *Compañia Arrendataria del Monopolio de Petroleos* (Spanish—Petroleum Company)

CAMP Test Christie-Atkins-Munch-Peterson Test

CAMPUS Comprehensive Analytical Methods for Planning in University Systems

CAMQAB Consumer Affairs Medical Quality Assurance Board (California)

CAMRA Campaign for Real Ale

CAMRC Child Abuse and Maltreatment Reporting Center

cams. cybernetic anthropomorphous machines

CAMS Chinese Academy of Medical Sciences; Coastal Anti-Missile System; Com-

munication, Advertising, and Marketing Studies (System); Confederation of Australian Motor Sports; Continuous Air-Monitoring System

camship catapult aircraft merchant ship

CAMSI Canadian Association of Medical Students and Interns

Cam Soc Camden Society

CaMV cauliflower mosaic virus

can customer account number

can. canal; canalization; canalize; cancel; canceled; cancellation; canister; cannon; canon; canopy; canto; canvasback (duck)

can. (CAN) cancel character (data processing)

can canto (Italian—melody, song)

Can Caen; Canada; Canadian; Canal; Canberra; *Cancer* (Latin-Crab constellation); Canfield; Cano; Canyon

Can Canal (French, Portuguese, Spanish—canal); *Canale* (Italian—canal); Cañon (Spanish—canyon)

Can. Canaanite

Can. *Cantoris* (Latin—cantor's or preceptor's side of the choir)

CAN Canberra, Australia (airport); Citizens Against Noise: Committee of an Advanced Nature; *Compagne Auxiliare de Navigation*; Corporate Angel Network; Customs Assignment Number

CANA Canadian Army; Clergy Against Nuclear Arms; Cremation Association of North America

CANABRIT Canadian Naval Joint Staff in Britain; Canadian Navy Joint Staff in Great Britain

CANACIN *Camara Nacional de la Industria* (Spanish—National Chamber of Industries)—Mexico

CANACO *Camara Nacional de Comercio* (Spanish—National Chamber of Commerce)

CANACOR Canadian Agro-Industrial Corporation

Canad Canadian

Canada formerly the Dominion of Canada (largest nation in the western hemisphere and second largest in the world)

Canad Fr Canadian French (French Canadian)

Canadian Galápagos southern tip of Queen Charlotte Islands, British Columbia

Ca Na F Campaña Nacional Fronterizo (Spanish—National Frontier Campaign)

CANAIRDEF Canadian Air Force Defense Command

CANAIRDIV Canadian Air Force Division

CANAIRHED Canadian Air Force Headquarters

CANAIRLIFT Canadian Air Force Transport

CANAIRLON Canadian Air Force Joint Staff—London, England

CANAIRMAT Canadian Air Force Material Command

CANAIRNEW Canadian Air Force—Newfoundland

CANAIRNORWEST Canadian Air Force—Northwest, Edmonton

CANAIRPEG Canadian Air Force—Winnipeg

CANAIRTAC Canadian Air Force Tactical Command

CANAIRTRAIN Canadian Air Force Training Command

CANAIRVAN Canadian Air Force—Vancouver

CANAIRWASH Canadian Air Force Joint Staff—Washington, D.C.

Canajan Canadian speech

Canal Zone Panama Canal Zone

Can-Am Canadian-American

Can-Am Cup Canadian-American Challenge Cup (automobile racing)

Canar Cur Canaries Current

Canaries Canary Islands in the Atlantic off southern Morocco

CANAS Canadian Naval Air Station

CANAVAT Canadian Naval Attaché

CANAVCHARGE Canadian Naval Officer in Charize

Canaveral Cape Canaveral, Florida also called Cape Kennedy; national seashore on Florida east coast

CANAVHED Canadian Naval Headquarters

CANAVSTORES Canadian Naval Stores

CANAVUS Canadian Naval Joint Staff in United States

Canb Canberra

canc cancel; canceled; cancellation; cancelling

Canc Cancer (constellation)

Canc. Cancellarius (Latin—Chancellor)

CANCARAIRGRP Canadian Carrier Air Group

CANCEE Canadian National Committee for Earthquake Engineering

CANCIRCO Cancer International Research Cooperative

CANCOMARLANT Canadian Maritime Commander, Atlantic

CANCOMARPAC Canadian Maritime Commander, Pacific

Can Cus Canadian Customs

cand candelabra; candidate

Can$ Canadian dollar

cande command and edit (computer program)

CANDEP Canadian Naval Depot

candf cost and freight

candi cost and insurance

CANDLES Children of Auschwitz Nazis' Deadly Lab Experiment Survivors

Candn Canadian

CANDOC Canadian Electronic Document Ordering Service

cand sc candelabra screw

Candu Canadian deuterium uranium

CANDU Chelsea Against Nuclear Destruction United

CANDUR Canadian Deuterium-Uranium Reactor

Candy Candice

CANDY Cigarette Advertising Normally Directed to Youth

CANE Computer-Assisted Navigation Equipment

Canea Khania

CanE Canadian English

CANEL Connecticut Advanced Nuclear Engineering Laboratory

Can-End Canton and Enderbury Islands

cane sugar saccharose or sucrose

canew competitively aggressive novelly engineered watercraft

Can F Canadian French

CANF Combined Allied Naval Forces; Cuban-American National Foundation

CANFARMS Canadian Farm Management Data System

CANFORCEHED Canadian Forces Headquarters

Can Fr Canadian French

CANHR California Advocates for Nursing Home Reform

Can I Canary Islands

CANI Committee on Non-discrimination and Integrity

Canid Canidae (Latin—dog family)

Can Imm Cen Canada Immigration Centre

canis canister

Can J Chem Canadian Journal of Chemistry

Can J Phys Canadian Journal of Physics

Can J Res Canadian Journal of Research

Can-Jud Canadian-Judeo (Canadian Jewish)

CANLANT Canadian Atlantic

Can Ltd Canadair Limited (operating unit of General Dynamics Corporation)

Can. maj. Canis Major (Latin—Greater Dog)—astronomical constellation

Can Man Cen Canadian Manpower Centre(s)

Can Met Ser Canadian Meteorological Service (EAES)

CANMET Canada's Center for Mineral and Energy Technology

Can. min. Canis minor (Latin—Lesser Dog)—astronomical constellation

canna. cannabinoid; cannabinol; cannabis

cannib cannibal(ism); cannibalistic; cannibalization; cannibalize

CANO Chief Area Nursing Officer

canola Canada oil

Canola Canadian oil, rapeseed

Can/ole Canadian on-line enquiry

Canon City Colorado State Penitentiary at Canon City

Can Op Canadian Opera Company (Toronto)

CANP Campaign Against Nuclear Power; Civil Air Notification Procedure

Can Pac Canadian Pacific

Can Pen Ser Canadian Penitentiary Service; Canadian Pension Commission

cans. canvasbacks (ducks); custom-assigned numbers

CANSA California Association of Nurses in Substance Abuse

CANSAV Canadian Save the Children Fund

Can/sdi (CAN/SDI) Canadian selective dissemination of information

CANSG Civil Aviation Navigational Services Group

Can St Cannon Street (rail terminal)

Can Syrm Canadian Symphony

cant canto (Italian—song; singing; voice part; instrument string tuned to the highest pitch)

cant. cannot; cantaloupe

can't can not; cannot

cant. canticum (Latin—canticle or hymn of praise)

Cant Canterbury; Canton; Cantonese

cantab cantabile (Italian—singable or songlike)

Cantab Cantabria (Spanish province)

Cantab. Cantabrigiensis (Latin—of Cambridge)

Cantabrians Cantabrian Mountains of northern Spain

CANTAP Canadian Technical Awareness Programme

CANTAT Canadian Transatlantic Telephones

can.tb canine tuberculosis

cant b cantilever bridge

Cant Chin Cantonese Chinese

Can Telsat Canadian Telecommunications Satellite System

cantran cancel(led) in transmission

Can Tram Comm Canadian Transport Commission

cants cantaloupes

CANTT Cantonment Telegraph

CANTU Compañía Anónima Nacional Telefonos de Venezuela (Spanish—National Telephone Corporation of Venezuela)

Cantuar. Cantuaria or *Cantuariensis* (Latin—of Canterbury)

can't win mnemonic abbreviation for community property states—California, Arizona, Nevada, Texas, Wyoming, Idaho, New Mexico

canu can you

CANU Constabulary Anti-Narcotics Unit (Philippines)

CANUKUS Canada—United Kingdom—United States

CANUS Canada—United States

CANUSE Canadian-United States Eastern (electric power interconnection)

CANUSPA Canada, Australia, New Zealand, and United States Parents Association

canv canvas

Can Ven Canes Venatici (Latin—Hunting Dogs constellation)

CANY Correctional Association of New York

Canyon de Chelly Canyon de Chelly National Monument (cliff-dweller ruins in northern Arizona)

Canyonlands Canyonlands National Park surrounding the junction of the Colorado and Green Rivers in southeastern Utah

CANYPS Canadian Yellow Pages Service

canz canzone; canzonetta

CANZ Composers Association of New Zealand

cao chronic airway obstruction

CAO Canadian Association of Optometrists; Central Accounting Office(r); Chief Accounting Office(r); Chief Administrative Officer; Civil Affairs Office(r); County Administration Office(r); Crimean Astrophysical Observatory (USSR); Cultural Affairs Office(r)

caoc cathodal opening contraction

CAOC Consumers' Association of Canada

CAOCS Carrier Aircraft Operational Compatibility System

Ca OCl Cathodal Opening Clonus

CAOF Canadian Army Occupation Force

CAOGA Crown Agents for Oversea Governments and Administrations

CAOOAA Civil Air Operations Officers Association of Australia

CAORB Civil Aviation Operational Research Branch

CAORE Canadian Army Operational Research Establishment

CAOS Completely Automatic Operational System

CAOSOP Coordination of Atomic Operations— Standard Operating Procedures

CAOT Canadian Association of Occupational Therapy

cap capacitance; capacity; capital(ize); capital letter; capsule; caput; catabolite activator protein; cellulose acetate proprionate; chlor-aceto-phenone (gas); client assessment package; combat aircraft prototype

cap (CAP) computer-aided production; consumer account protection

'cap handicap

cap capìtolo (Italian—chapter); *capítulo* (Portuguese or Spanish—chapter); (French—cape)

cap. captivity; captured

cap. capiat (Latin—take); *capsula* (Latin—capsule)

c/a/p codice di avviamento postale (Italian—mailing code)— zip coding

Cap capitol, captain; Captain; Caspar; Charles A; Pearce

Cap. Captain; Chapter—Number of Act of Parliament

Cap Capitán (Spanish—captain); *Capricornus* (Latin—Horned Goat Constellation)

CAP California Assessment Program (educational); Canadian Association of Pathologists; Central Arizona Project; Citizens Against Pornography; Civil Addict Program; Civil Air Patrol; College of American Pathologists; Combat Air Patrol; Common Agricultural Policy; Commonwealth Association of Planners; Community Action Program; Community Artists Project; Community Advancement

Program; Comprehensive Assessment Program; Computer Address Panel; Consumer Action Panel; Cooperating Accountability Project

CAP Certificat d'Aptitude Professionelle (French—Certificate of Professional Aptitude)

CAPA California Association of Port Authorities; Canadian Association of Purchasing Agents; Commission on Asian and Pacific Affairs; Confederation of Asian and Pacific Accountants; Council of Australian Postgraduate Associations

capac capacity; cathodic protection

CAPAC Canadian Association of Primary Air Carriers; Composers, Authors, and Publishers Association of Canada

capal computer-and-photographic-assisted learning

CAPAR Cost-Account Problem Analysis Report

CAPC Civil Aviation Planning Committee; Canadian Army Pay Corps; Canadian Army Provost Corps

capche component automatic programmed checkout equipment

cap com capsule communicator

capcon(s) captured conversation(s)—electronically recorded tape of speech between two or more persons; recorded conversation

CAPCR Canadian Armoured Personnel Carrier Regiment

capd (CAPD) continuous ambulatory peritoneal dialysis

Cape The Cape—short form for the Cape of Good Hope, Cape Hatteras, Cape Horn, Cape Verde, or other well-known headlands, such as: Cape Aguilhas, Cape Ann, Cape Blanc, Cape Blanco, Cape Breton, Cape Camorin, Cape Canaveral (Kennedy), Cape Charles, Cape Clear, Cape Cod, Cape Columbia, Cape Cornwall, Cape Cruz, Cape Disappointment, Cape Fear, Cape Finisterre, Cape Flattery, Cape Guardafue, Cape Law, Cape Leeuwin, Cape Maisi, Cape May, Cape Mendocino, Cape Muharmmad, Cape Palliser, Cape Race, Cape Sable, Cape San Antonio, Cape San Lucas, Cape Spear, Cape St Cape Vincent, Cape Wrangell, Cape Wrath, Cape York

CAPE California Association of Polygraph Examiners; Classification and Placement Examination; Confederation of American Public Employees; Council for American Private Education; Course and Professor Evaluation

Cape Breton Highlands Breton Highlands National Park, Cape Breton Island

Cape-Cairo Cape Town-to-Cairo Highway; Cape Town-to-Cairo Railway

Cape Cod National Seashore on coast of Massachusetts

Cape Cod National Cape Cod National Seashore conservation and recreation reservation on Cape Cod, Massachusetts

Cape Colony Cape of Good Hope Colony (South Africa)

CAPED California Association of Postsecondary Educators of the Disabled

Cape Hatteras national seashore in North Carolina coast

Cape Lookout national seashore in North Carolina

Cape Province Cape of Good Hope Province

Cape Roca Cabo da Roca, Portugal (Europe's westernmost point)

capertsim computer-assisted program evaluation review technique simulation

CAPES College Association for Public Events and Services

Cape Verde Islands Republic of Cape Verde (small island nation off Africa's westernmost tip) *República de Cabo Verde*

Cape Verdes Cape Verde Islands off the Senegal coast of West Africa, called Ilhas do Cabo Verde by the Portuguese

CAPEXIL Chemicals and Allied Products Export Promotion Council

CAPF Canadian Army Pacific Force

CAPH California Association of Physically Handicapped

Capitol Reef Capitol Reef National Park, Utah

CAPL Canadian Association of Public Libraries; Controlled Assembly Parts List

CAPLOT Canadians Against PLO Terrorism

capm capital asset pricing model

CAPM Computer-Aided Patient Management

cap. moll. *capsula mollis* (Latin—soft capsule)

CAPMS Central Agency for Public Mobilization and Statistics

Cap'n Captain

Cap^n Capitán (Spanish—captain)

capo (Italian—*capobanda*—bandmaster; *capo cameriere*—chief steward; *capo fabbrica*—factory foreman or overseer; *caporione*—ringleader; *capo stazione*—station master)

CAPO Canadian Army Post Office; Chief Administrative Pharmaceutical Office(r)

CAPOSS Capacity Planning and Operation Sequencing System

C App Chartered Appraiser

CAPPA Crusher and Portable Plant Association

capp^n capellán (Spanish—chaplain)

CAPPS Council for the Advancement of the Psychological Professions and Sciences

cap & puncless capitalization and punctuationless American author e e cummings

cap. quant. vult capiat quantum vult (Latin—allow the patient to take as much as he will)

Capr Capricornus

capri computerized advance personnel requirements and inventory

Capric Capricorn (constellation)

capris capri pants

Cap-Rouge Maison Notre Dame de la Garde facility for juvenile delinquents at Cape-Rouge, Québec

caps collection account processing system; computer-assisted passenger screening

caps. capital letters

caps. (CAPS) computer-assisted problem solving

caps. capsule (Latin—capsule)

CAPs Community Action Programs; Consumer Action Panels

CAPS Casette Programming System; Cashiers Automatic Processing System; Children of Aging Parents; Clearinghouse on Counselling and Personnel Services; Coastal Aerial Photolaser Survey; Collins Adaptive Processing System; Combat Air Patrol Support; Computer-Aided Pipe Sketching System; Computer-Aided Project Study; Computer-based Aid-to-Aircraft Project Studies; Conventional Armaments Planning System; Creative Artists Public Service

capsep capsule separation

caps and lower case capital letters and lower case letters

CAPSS Computer-Assisted Public Safety System

caps and small caps upper case capital letters and small capital letters

capt caption

capt. captivity; captured

Capt(.) Captain

CAPT Clearinghouse for Applied Performance Testing

CAPTAINS Character and Pattern Telephone Access Information Network System

CAPTIVE Collaborative Authoring Production and Transmission of Interactive Video for Education

Captn Captain

captor capsulated torpedo

Capulin Capulin Peak, Capulin Mountain National Monument, Mount Capulin—all in New Mexico

capun capital punishment

capy capacity; capybara

Caq Caquetá (Colombian department, not be be confused with Cúcuta on the border of Venezuela)

CAQ Craft Association of Queensland

caqa (CAQA) computer-aided quality assurance

car baggage car, boxcar, buffet car, cattle car, club car, coal car, dining car, electric car, elevated car, freight car, mail car, motor car, observation car, parlor car, passenger car, prison car, pullman car, railroad car, refrigerator car, sleeping car, steam car, street car, subway car, surface car, tourist car, trolley car; computer-assisted retrieval

car. carat; carton; cloudtop altitude radiometer

car (CAR) channel address register

Car Caradoc; Carberry; Carbury; Carel; Carew(e); Carey; Carina (constellation); Carleton; Carlow; Caroline Islands; Carter; Carvel; Carver

Car. Carpenter

Car. Carolus (Latin—Charles)

CAR California Association of Realtors; Canadian-American Rights; Canadian Association of Radiologists; Central African Regiment; Central African Republic; Chief Airship Rigger; Children of the American Revolution; Civil Air Regulation(s); Civil Air Reserve; Commonwealth Arbitration Reports; Contract Authorization Request; Corrective Action Request; US Army, Caribbean (area)

CAR Cadena Azul de Radiodifusión (Spanish—Blue Broadcast Chain); *Comité Agricole Régional* (French—Regional Agricultural Committee)

cara combat air rescue aircraft

CARA California Association for Research in Astronomy; Center for Applied Research in the Apostolate; Chinese-American Restaurant Association; Combat Air Rescue Aircraft

CARAC Civil Aviation Radio Advisory Committee

CARAL California Abortion Rights Action League

Carav Caravelle

carb carbon; carbonacious; carbonate; carburetor; carburize

CARB California Air Resources Board

carbage car-floor garbage

carbecue car + barbecue (device for melting waste out of junked automobiles)

carb 14 carbon 14

carbo carbohydrate

carboloy carbon-cobalt-tungsten alloy

carbonado black or grayish-black industrial diamond; meat scored before grilling over charcoal

carbon dioxide carbonic acid gas

Carbonif Carboniferous

carbontet carbon tetrachloride

carbopol carboxypolymethylene

carborundum silicon carbide (SiC)

carb(s) carburetor(s)

carbs. carbohydrates

CARBS Computer-Assisted Rationalized Building System

CARC Canadian Arctic Resources Committee

carcin (Latin prefix—cancer)— carcinogenic

Carcross Caribou Crossing

card cardamon; cardinal; certificate for amortizing revolving debt

card (CARD) compact automatic retrieval device; compact automatic retrieval display

Card Cardiganshire; Cardinal

CARD Campaign Against Racial Discrimination; Civil Aeronautics Research and Development; Compact Automatic Retrieval Device (or Display)

CARDA Continental Airborne Reconnaissance for Damage Assessment (USAF)

cardamap cardiovascular data analysis by machine processing

CARDE Canadian Armament Research and Development Establishment

cardi (Latin prefix—heart)— cardiac arrest (heart failure)

CARDI Colorado Allergy and Respiratory Disease Institute

cardioac cardioacceleration; cardioaccelerator(y); cardioactive; cardioactivity

cardiog cardiogenesis; cardiogenetic; cardiograph(ic); cardiography

cardiol cardiolith(ic); cardiologist(ic)(al)(ly); cardiology; cardiolysin

cardiomeg cardiomegaly

cardiomyo cardiomyopathy

cardiopul cardiopulmonary

cardiov cardiovascular

cardiover cardioversion (electric-shock therapy)

cardiv carrier division

CARDIV Carrier Division (naval)

Cardl Cardenal (Spanish—Cardinal)

CARDPACS Card Packet System

card(s). (CARDs) computer-aided research device(s)

Cards Cardinals

CARDS Combat Aircraft Recording and Data System; Computer-Assisted Recording of Distribution Systems

care communicated authenticity, regard, empathy

care. continuous aircraft reliability evaluation

Care Caretaker

CarE Caribbean English

CARE Citizens Association for Racial Equality; Comprehensive Assessment and Referral Evaluation; Consumer Awareness Retailer Effort; Continental Association of Resolute Employers; Cooperative for American Relief Everywhere; Cooperative for American Remittances to Everywhere; Cottage and Rural Enterprises

CAREIRS Conservation and Renewal Energy Inquiry and Referral Service

CAREL Central Atlantic Regional Educational Laboratory

CARES Computer-Assisted Regional Evaluation System

CARF Canadian Amateur Radio Federation; Canadian Arthritis and Rheumatism Society; Central Altitude Reservation Facility

CARG Caribbean Ready Group (USN); Corporate Accountability Research Group (Nader's)

Cargomaster Douglas 200-passenger military transport designated C-133

cargotainer cargo container

Cari Carina

CARI Civil Aeromedical Research Institute

C.A.R.I. Cuban Agricultural Reform Institute

Carib. Caribbean

CARIBAIR Caribbean Atlantic Airlines

CARIBANK Caribbean Development Bank

Caribbean Caribbean Area (islands in or washed by the Caribbean Sea); Caribbean Sea

Caribbean countries Antigua and Barbuda, Barbados, Dominica, Dominican Republic, Grenada, Haiti, Jamaica, St Kitts and Nevis, St Lucia, St Vincent and the Grenadines, Trinidad and Tobaggo

CARIBCOM Caribbean Command

Carib Cur Caribbean Current

Caribe (Spanish—Carib language, Caribbean Sea)

Caribous Caribou Mountains of British Columbia

CARIBSEAFRON Caribbean Sea Frontier

caric caricature; caricaturist

CARIC Contractor All-Risk Incentive Contract (USAF)

Caricom Caribbean Community Anguilla, Antigua, Barbados, Belize, Dominica, Grenada, Guyana, Jamaica, Montserrat, St Kitts-Nevis, St Lucia, St Vincent, Trinidad and Tobago

CARIFESTA Caribbean Festival of the Arts

CARIFTA Caribbean Free Trade Association

CARIH Children's Asthma Research Institute and Hospital (Denver)

CARIPLO Cassa di Risparmio delle Provincie Lombarde (Italian—Savings Bank of the Province of Lombardy)

CARIS Computerized Agricultural Research Information System; Current Agricultural Research Information System (FAO)

carl computer-assisted reference locator

Car L Carpenter Lieutenant

Carl Carla; Carle; Carleton; Carlisle; Cario(s); Carlton; Carlyle

CARL Canadian Academic Research Libraries; Chatfield Applied Research Laboratories; Colorado Association of Research Libraries; Computer Audio Research Laboratory

Carla Carlotta; Caroline

Carl Gustaf 9mm Swedish submachine gun firing parabellum bullets; recoilless 84mm antitank weapon

Carm Carmathenshire; Carmichael; Carmo; Carmody

CARML County and Regional Municipality Librarians (Ontario)

carmrand civilian application of the results of military research and development

Carn Caernarvonshire

Carnegie Inst Carnegie Institution of Washington

Carnegie Tech Carnegie Institute of Technology

carni carnival

Carnics Carnic Alps

carnie(s) carnival(s); carnival workers

Caro Carolina; Caroline

Carol Carola; Carole; Carolina; Caroline; Carolyn

CAROL Caribbean Overseas Lines

Carolina Art Carolina Art Association

Carolina Pop Ctr Carolina Population Center

Carolinas North and South Carolina

Carolines Caroline Islands (Kusaie, Palau, Ponape, Truk, Yap) in the Western Pacific

carot centralized automatic recording on trunks (Bell)

carp carpenter; carpentry; carpet(ing); computed air-release point; construction of aircraft and related procurement

Carp Carpathian; Carpentaria

CARP Campaign Against Restrictive Parking; Campaigns Against Rising Prices; computed air-release point

CARPAS Comisión Asesora Regional de Pesca el Atlantico Sud-Occidental (Spanish—Regional Fisheries Advisory Commission for the Southwest Atlantic)

Carpathians Carpathian Mountains

Carpaths Carpathian Mountains

carpilf cargo pilferage

carp(s) stage carpenter(s)

Carps Carpathian Mountains

carr carriage (flow chart); carrier

Carrasco Montevideo, Uruguay's airport

Carrie Carolina; Caroline

Carrie Mac California Association of Realtors Mortgage Assistance Corporation

CARRIS Companhia Carris de Ferro de Lisboa (Portuguese—Lisbon Street Railway)

CARR-L Computer-Assisted Reporting and Research—Louisville (Kentucky)

Carroll of Carrollton Charles Carroll of Carrollton, Maryland—self-identified signer of the *Declaration of Independence*

cars cable relay stations

cars. collateralized auto receivable securities; community antenna relay service

CARS Canadian Arthritis and Rheumatism Society; Central American Research Station (for disease control); Combat Arms Regimental System; Community Antenna Relay Station; Community on Alcohol and Road Safety; Computer-Aided Routing System; Computer-Assisted Reliability Statistics

Carson Carson City; Nevada State Penitentiary in Carson City

Carson City Women's Women's Prison at Carson City, Nevada

CARSTRIKFOR Carrier Striking Force

cart. carrot; cartage; carton; collision-avoidance radar trainer

cart carta (Italian, Portuguese, Spanish—card, chart, document, page of music)

Cart Carter; Cartwright

CART Cargo Automation Research Team; Central Automated Replenishment Techniques; Championship Auto Racing Teams; Coalition Against Regressive Taxation; Complete Automatic Rating Technique; Complete Automatic Reliable Testing

CARTB Canadian Association of Radio and Television Broadcasters

Carter James Earl Carter, 39th President of the United States

Carth Carthage; Carthaginian; Carthusian

cartobib cartobibliographer; cartobibliography

cartog cartographer; cartographic; cartography

cartoonitorial cartoon editorial

CARTS Computer-Automated Reserved Track System

Car Z Caribische Zee (Dutch—Caribbean Sea)

cas calibrated airspeed; casual; casualty; close air support; commence average speed; communications application specification

cas (CAS) cooperative applications satellite

ca's catecholamines; combat actions; covert actions

Cas Caracas; Casimir; Caslon; Cassiopeia (constellation); Castle

Cas ciencias (Spanish— sciences)

CAS Consumers Associations; Cooperative Associations

CAS California Academy of Sciences; Cambrian Airways (symbol); Capital Assistance Scheme; Casualty Actuarial Society; Center for Administrative Studies; Center for Auto(mobile) Safety; Change Analysis Section; Chemical Abstracts Service; Chicago Academy of Sciences; Chief of Air Staff; Children's Aid Society; Chinese Academy of Sciences; Civil Affairs Section; Civil Air Surgeon; Clean(er) Air System(s); Coast Artillery School; Collision Avoidance System (aircraft); Combat Arms School; Commercial Air Service; Computer Acquisition System; Contemporary Art Soci-

ety; Contract Administration Services; Controlled American Sources; Cost Accounting Standards; Courier Air Services; Current Australian Serials; Current Awareness Service; Customs Agency Service

C.A.S. Certificate of Advanced Studies

ca.sa. capias ad satisfaciendum (Latin—writ of execution)

Casa Casablanca

CASA Campaign Against Psychiatric Abuse (in the former USSR); Canadian Association of School Administrators; Canadian Automatic Sprinkler Association; Catgut Acoustical Society of America; Citizens Against Sneakin' Aroun'; Contemporary Art Society of Australia; Court-Appointed Special Advocates; Crafts Association of South Australia

CASA Construcciones Aeronauticas, SA (Spain)

Casa Grande Casa Grande National Monument near Phoenix, Arizona

CASANZ Clean Air Society of Australia and New Zealand

CASAO Chartered Accountants Students Association of Ontario

CASB Cost-Accounting Standards Board

CASBS Center for Advanced Study in the Behavioral Sciences (Stanford)

casc computer-assisted cartography

CASC Canadian Army Service Corps; Certified Alfalfa Seed Council; Ceylon Army Service Corps; Council for the Advancement of Small Colleges

CASCADE Citizens and Scientists Concerned About Dangers to the Environment

Cascades Cascade Mountains extending from British Columbia to California

Cascadia region between Vancouver, British Columbia and Eugene, Oregon

cascan casualty cancelled

CASCOMP Comprehensive Airship Sizing and Performance Computer Program cascor casualty corrected

CASCU Cooperative Association of Suez Canal Users

CASD Center for Applied Studies in Development

casdac computer-aided ship design and construction

CASDO Computer Applications Support and Development Office (USN)

casdos computer-assisted detailing of ships

case computer-aided software engineering

case. common-access switching equipment, computer-automated support equipment

Case Casey

CASE Coalition of Agencies Serving the Elderly; Coalition for Safe Energy; Colorado Association of School Executives; Commission on the Accreditation of Service Experiences; Committee on Academic Science and Engineciing; Committee on the Atlantic Salmon Emergency; Conference Association of Society Executives; Coordinated Aerospace Supplier Evaluation; Council of Administrators of Special Education; Council for the Advancement of Secondary Education; Council for Advancement and Support of Education; Counselling Assistance to Small Enterprises

CaSEA California Society of Enrolled Agents

CASEA Center for the Advanced Study of Educational Administration

CASETT Cases of Settlements and Removals

casevac casualty evacuation

CASEX Close Air Support Exercise; Combined Aircraft Submarine Exercise

CASF Canadian Active Service Force; Combat Alert Strike Force; Composite Air Strike Forces

CASGTC California Academy of Sciences Geology Type Collection

cash. cashew; cashier

Cash Cashlin; Cassius

CASH Catalog of Available and Standard Hardware; Citizen Action for Safer Harlems; Commission for Administrative Services in Hospitals

cash ad cash advance

Casi Casimir

CASI Canadian Aeronautics and Space Institute; Convenient Automotive Services Institute

CASIA Chemical Abstracts Subject Index Alert

CASIG Careers Advisory Service in Industry for Girls

Casl Caslon

Ca S-L Catering Sub-Lieutenant

CASLE Commonwealth Association of Surveying and Land Economy

CASLIS Canadian Association of Special Libraries and Information Services

casm close air support missile; cycling air sampling monitor

CASMT Central Association of Science and Mathematics Teachers

CASNP Canadian Association in Support of Native Peoples

CASO Civil Aviation Safety Order

CASOE Computer Accounting System for Office Expenditures

casoff control and surveillance of friendly forces

CASOS Center for Advanced Study in Organization Science (U of Wisconsin)

Casp Caspar(d)

CASP Capability Support Plan; Cape Arago State Park (Oregon); Country Analysis and Strategy Paper (U.S. State Department)

Caspar Cambridge analog simulator for predicting atomic reactions

CASPER Contact Area Summary Position Estimate Report

CASPERS Computer-Automated Speech-Perception-System

Caspian Caspian Sea (landlocked body of water between Iran and Turkmenistan, Kazakhstan, Russia, and Azerbaijan)

Cas Reps Casualty Reports

CASRO Council of American Survey Research Organizations

cass cassette; cassowary; combat air surveillance system; command active sonobuoy system

Cass Casimir; Cassander; Cassandra; Casseus; Cassidy; Cassius; Casso

Cass Cassiopeia (Latin—Lady in the Chair constellation)

CASS California Agricultural Statistics Service; Center for Astrophysics and Space Scientists

CASSA Continental Army Command Automated System Support Agency (USA)

Cassi Cassiopeia

CASSI Chemical Abstracts Service Source Index

CASSIOPE Computer Aided System for Scheduling, Information and Operation of Public Transport in Europe

CASSIS Communication and Social Science Information Service (Canada)

CASSO Command Active Sonobuoy System; Connecticut Association of Secondary Schools

CASSR Chuvash Autonomous Soviet Socialist Republic

cast castanets; computer applications and systems technic; computer-augmented scanning technics

Cast Castel; Castile; Castilian; Castillon; Castimir; Castislav; Castle; Castor

CAST Center for Application of Sciences and Technology; Citizens Adversity Support Team; Contemporary Artists Serving the Theater; Council for Agricultural Science and Technology

CAST Clearinghouse Announcements in Science and Technology

CASTE Collision-Avoidance System Technical Evaluation

CASTLE Computer-Assisted System for Theater-Level Engineering

Cast la Man Castilla la Mancha (Spanish province)

castor cask storage (for transporting radioactive material)

CASTOR College Applicant Status Report

Castro San Francisco neighborhood

CASTS Canal Safe Transit System

Cast y León Castilla y León (Spanish province)

CASW Council for the Advancement of Science Writing

casws close air support weapons system

cat carburetor air temperature; catalog; catamaran; catapult; catboat; category; caterpillar tractor; caudé auxiliary telescope; clear air turbulence; combined arms teamwork; compressed air tunnel; computer-assisted transcription; computer of averaged transients; computerized axial tomography; customer-activated terminal

cat (CAT) choline acetyl-transferase; computer-aided typesetting; computer-assisted test(ing); computerized adaptive test(ing); computerized axial test(ing); computerized axial tomography

Cat Catalán; Catalina; Catalonia; Catalonian; Cataluña; Catalunya; Catamarca; Catania; Cataño; Catarina; Caterino; Catasauqua; Catawba; Caterpillar Tractor; Catesby; Catlett

Cat Cat Duet (Rossini's comic work for two voices); Catalan (language spoken in the Spanish province of Catalonia, France, and Andorra)

CAT California Achievement Test; California Advocacy for Trollops; Child's Apperception Test; Children's Apperception Test; Civil Air Transport; Civilian Actress Technician; Clerical Aptitude Test; Cognitive Abilities Test; College Ability Test; Colleges of Advanced Technology; Commercial Airlift Contract; Computer-Aided Training; Computer-Aided Translation; Computer-Aided Typesetting; Computer-Assisted Teaching; Computerized Axial Tomography; Consolidated Atomic Time; Container Associates

Transport; Control and Assessment Team; Corrective Action Team

CAT Comisaría de Abastecimientos y Transportes (Spanish—Commisariat of Supply and Transport)

Cata Catumarca (Argentine province)

CATA Canadian Air Transport Association; Canadian Air Transportation Administration

catal catalog; catalogue

Catal Catalan; Catalonia; Cataluña

Catalina Catalina de Güines southeast of Havana, Cuba; Santa Catalina Island off Long Beach, California

catawump catawumpus (catamount, mountain lion)

catc computer-assisted test construction

CATC Coast Artillery Training Center; Commonwealth Air Transport Council; Continental (Oil), Atlantic (Refining), Tidewater (Oil), and Cities (Service) (combined in mutual drilling)

CATCALL Completely Automated Technique for Cataloging and Acquisition of Literature for Libraries

CATCC Canadian Association of Textile Colorists and Chemists; Carrier Air Traffic Control Center

CATCH Citizens Against The Concorde Here

CATCO Catalytic Construction Company

CATCS Central Air Traffic Control School

CATD Cooperative Association of Tractor Dealers

cate comprehensive automatic test equipment

CATE Currermt ARDC (Air Research and Development Command) Technical Efforts (program)

catec catechism; catechist(ic)(al)(ly)

Cater Trac Caterpillar Tractor

Ca Test Calcium Test (dental)

CATF California Tax-Free; Canadian Achievement Test in French; Central American Task Force

cath cathartic; cathedral; catheter; catheterize

Cath Cathedral; Catherine; Catholic; St Catherine's College, Cambridge; St Catherine's College, Oxford

CATHAY Cathay Pacific Airways

Cathedrals Cathedral Caverns near Grant, Alabama

cath fol cathode follower

Cathie Catherine

Cath Lib Assn Catholic Library Association

Cathlibs Catholic liberals

cathol catholic; catholically; catholicly; catholicalness; catholicness; catholicate; catholice; catholicity

CAT houses Central American Tropical (barracks)

Cath U Pr Catholic University of America Press

Cathy Catherine

CATIB Civil Air Transport Industry Training Board

CATIE Centro Agronómico Tropical de Investigación y Enseñanza (Spanish—Tropical Agriculture Center of Investigation and Teaching)—Turrialba, Costa Rica

catk counterattack

catlg catalog(ue)

CATM Canadian Achievement Test in Mathematics

CATNIP Computer-Assisted Technique for Numerical Index Preparation

Catoctins Catoctin Mountains of Maryland and Virginia

CATOR Combined Air Transport and Operations Room

CATP Computer-Assisted Typesetting Process

CATPAC Catering, Planning, Accounting and Control

catproc catalog(ue) procedure

CATRA Cutlery and Allied Trades Research Association

Catracho(s) Honduran(s)

CATRALA Car and Truck Renting and Leasing Association

CATs Civic Action Teams

CATS Canadian Auxiliary Territorial Service; Canon Auto Tuning System; Certificates of Accrual on Treasury Certificates; Civil Affairs Training School (USN); Comprehensive Analytical Test

System; Compute Air-Trans Systems; Computer-Assisted Test Shop; Computer-Automated Test System (AT & T); Concerned About Teen Sexuality

CATSA Cooperative Air Transport System for Antarctica

CAT-scan computerized axial tomography scan that depicts body structures not shown by conventional X-ray procedures

Catskills Catskill Mountains, southeastern New York

CATSS Cataloguing Support System

catt conveyorized automatic tube tester

cattalo cattle + buffalo—hybrid

CATTCM Canadian Achievement Test in Technical and Commercial Mathematics

CAT test Computerized Axial Tomography Test

CATTS California and Texas Telecommunications System

Catty Catherine

CATU Combined Aircrew Training Unit

catv cabin air temperature valve; cable television; community antenna television

catva computer-augmented total-value assessment

cau command arithmetic units

Can Caucasian

CAU Child Abuse Unit (police department); Congress of American Unions; Consumer Affairs Union; Consumer Affairs Unit

Caucasus Caucasus Mountains between the Black Sea and the Caspian Sea

Cauc(s) Caucasian(s)

caud caudal; caudate

caud (Latin prefix—tail)—caudal appendage

CAUL Committee of Australian University Librarians

cauli cauliflower

caus causation; causative

CAUS Color Association of the United States

CAUSA Confederation of Associations for the Unity of the Societies of the Americas

CAUSA Compania Aeronautica Uruguay SA

causat causative

'cause because

CAUSE College and University Systems Exchange; Comprehensive Assistance to Undergraduate Science Education (National Science Foundation); Counselor Advisor University Summer Education

caust caustic

caustic potash potassium hydroxide (KOH)

caustic soda sodium hydroxide (NaOH)

caut caution

CAUT Canadian Association of University Teachers

CAUTION Citizens Against Unnecessary Tax Increases and Other Nonsense (St Louis citizens)

cav cavalier; cavalry; cavitation; cavity; congenital absence of vagina; congenital adrenal virilism; continuous airworthiness visit; constant angular velocity

cav (CAV) class attendance verification; construction assistance vehicle

cav. caveat (Latin—warning, writ of suspension)

c.a.v. curia advisare vult (Latin—the court cares to consider)

Cav Cavanagh; Cavanaugh; Cavell; Cavendish

Cav Cavaliere (Italian—Knight)

cav brk cavity brick

CAVC Canadian Army Veterinary Corps

cavd completion, arithmetic, vocabulary, directions (test)

Cav Div Cavalry Division

CAVE California Association of Vocational Educators; Catholic Audio-Visual Educators Association; Consolidated Aquanauts Vital Equipment

CAVEA Connecticut Audio-Visual Education Association

caveat code and visual entry authorization technic

cav. emp. caveat emptor (Latin—let the buyer beware)—also appears as *c.e.*

Cavendish Cavendish Laboratory (Cambridge University)

CAVI Centre Audio-Visuel International (French—International Audio-Visual Center)

caviol caviology

ca virus croup-associated virus

CAVN Compañía Anonima Venezolana de Navegación (Spanish—Venezuelan Steamship Line)

cavu ceiling and visibility unlimited

caw cam-action wheel; channel address word

CAW Carrier Air Wing; Church of All Worlds

c-a w conflict-alert warning

CAW Cables and Wireless (Company); Californians Against Waste

CAWA Canadian-American Women's Association

CAWC Committee on Air and Water Conservation (American Petroleum Institute)

CAWD Canadian-American Wolf Defenders

CAWE California Association of Work Experience Educators

cawg coaxial adapter waveguide

CAWG Clean Air Working Group

cawgs covert all-weather gun system

CAWM College of African Wildlife Management

CAWP Center for the American Woman and Politics

CAWS Central Aural Warning System; Conflict Alert Warning System; Conservation and Wildlife Studies

CAWSPS Computer-Aided Weapon Stowage Planning System (USN)

CAWU Clerical and Administrative Workers' Union

cax community automatic exchange (telephone)

Caxton Caxton Printers, Ltd

Cay Cayenne; Cayman

Cayen Cayenne (coastal French Guiana) — *see* Inni

Cayes Haitian seaport also called Aux Cayes or Les Cayes

Caymans Cayman Islands (Grand Cayman, Little Cayman, Cayman Brac)

cb cast brass; catch basin; cement base; center of buoyancy; chemical and biological; circuit breaker; collective bargaining; common battery; continuous breakdown

cb (CB) container base

c-b circuit breaker

c/b caught and bowled; circuit breaker

c & b collating and binding

c of b confirmation of balance

Cb base Capacitance; columbium (symbol); cumulo-nimbus

Cb *contre-basse* (French—double bass)

CB Cadet Battalion; Cape Breton (island); Caribair (airline); Caribbean-Atlantic Airlines; Carte Blanche; Cavalry Brigade; Cemetery Board; Census Bureau; Chairman of the Board (of directors); Chief Boilermaker, Children's Bureau; Church of the Brethren; citizen's band (radiofrequency); Companion of the Bath; compass bearing; confidential book; confidential bulletin; confinement to barracks; Construction Battalions (hence the nickname "seabees"); Consultants Bureau; contrabass(o) (double-bass viol); Control Branch; Counter Battery; County Borough; Cumulative Bulletin; Currency Bond; large cruiser (naval symbol); William Cullen Bryant

C-B (Sir Henry) Campbell-Bannerman

C.B. *Chirurgiae Baccalaureus* (Latin—Bachelor of Surgery)

C & B Clemens and Brenninkmeyer; Cleveland and Buffalo (steamship line)—*See and bee*

C of B Commonwealth of the Bahamas

CB *Carte Blanche* (French—white card) full discretionary power; *col basso* (Italian—with the bass); *contrabasso* (Italian—double bass)

C-B *Creditanstalt-Bankverein* (German—Credit Institution and Bank Association)—Austria's largest banking institution

cba chemical bond approach; chronic bronchitis with asthma; cost-benefit analysis

cba (CBA) colliding beam accelerator

CBA California Benefit Association; California Broadcaster's Association; Canadian Bankers Association; Canadian Booksellers Association; Caribbean Atlantic Airlines; Christian Booksellers Association; Clydesdale Breeders Association; College of Business Administration; Community Broadcasters Association; Consumer Bankers Association; Continental Basketball Association

CBA *Chemical-Biological Activities*

CBAA Canadian Business Aircraft Association

cbaf cobalt-base alloy foil

CBAICP Chemical and Biological Accident and Incident Control Plan (USA)

c/bale cents per bale

cbam concerns-based adoption model

cbar counterbore arbor

CBARC California Border Area Resource Center

CBAT Central Bureau of Astronomical Telegrams

cbb commercial blanket bond

CBB Chesapeake Bay Bridge (Maryland)

CBB *Centre Belge du Bois* (French—Belgian Forestry Research Center)

CBBA Christian Brothers Boys Association

CBBB Council of Better Business Bureaus

CBBI Cast Bronze Bearing Institute

CBBII Council of the Brass and Bronze Ingot Industry

CBBT Chesapeake Bay Bridge-Tunnel (Maryland to Virginia)

cbc cipher-block chain; complete blood count

cbc (CBC) combined blood count; complete blood count

cb&c corned beef and cabbage

CBC Canadian Broadcasting Corporation; Caribbean Broadcasting Company; Central Bank of China; Central Bible College; Ceylon Broadcasting Corporation; Children's Book Council; Christmas Bird Count (Audubon Society); Columbia Basin Council; Commendation for Brave Conduct; Commercial Banking Company; Commonwealth Banking Corporation; Congressional Black Caucus; Congressional Budget Committee; Contraband Control; Corset and Brassiere Council; Cyprus Broadcasting Corporation; large tactical-command ship (naval symbol)

CBC *Cadena Raja California* (Spanish—Baja California Radio Chain) Mexican border network

CBCA California Black Commission on Alcoholism

cbcc common bias—common control

CBCII California Bureau of Criminal Identification and Investigation

CB circuit common-base amplifier for junction transistors

CB Club Citizen's-Band (radio) Club

cbcm cheque book-charging method

CBCMA Carbonated Beverage Container Manufacturers Association

CBCS Commonwealth Bureau of Census and Statistics

cbet circuit board card tester

CBCT Customer-Bank Communication Terminal

cbcu counterbore cutter

cbd cash before delivery; closed bladder drainage; common bile duct

Cbd Cambodian

CBD Central Business District; Construction Battalion Detachment

CBD *Commerce Business Daily*

cbdn can be done

CBDNA College Band Directors National Association

CBDS Carcinogenesis Bioassay Data System

cbe cesium bombardment engine; chemical binding effect; circuit board extractor; competency-based education; compression bonding encapsulation

CBE Central Bomber Establishment; Cheese Bureau of England; Community-Based Education; Competency-Based Education; Computer-Based Education; Conference of Biological Editors; Council for Basic Education; Council of Basic Education; Council of Biology Editors

C.B.E. Commander of the Order of the British Empire; Companion of the Order of the British Empire

CBEA California Business Educators Association

CBEL Cambridge Bibliography of English Literature

CBEMA Canadian Business Equipment Manufacturers Association; Computer and Business Equipment Manufacturers Association

CBer operator or owner of a Citizen's-Band radio

CB-er(s) citizen's radio-frequency band (short-wave two-way) broadcaster(s)

CBEST California Basic Skills Test

cbf cerebral blood flow; coronary blood flow

CBF Children's Blood Foundation; Commander British Forces

CBFAP Commander British Forces, Arabian Peninsula

CBFBS Cyprus British Forces Broadcasting Service

CBFC Commander British Forces in Cyprus; Commonwealth Bank Finance Company

CBFCA Commander, British Forces, Caribbean

CBFG Commander British Forces, Gulf

CBFHK Commander British Forces, Hong Kong

cbfm constant bandwidth frequency modulation

cbft cubic feet

cbg (CBG) corticosteroid-binding globulin; transcortin

CBG Committee for Better Government

CBG Compagnie des Bauxites de Guinée (French—Bauxite Company of Guinea)

CBH Cooperative Bulk Holding

C B & H Continent between Bordeaux and Hamburg

CBHE Connecticut Board of Higher Education .

cbi complete background investigation; compound batch identification; computerbased information

CBI Cape Breton Island; Carbonated Beverage Institute; Caribbean Basin Initiative; Central Bureau of Investigation; Chesapeake Bay Institute; Chicago Bridge and Iron (company); China-Burma-India (theater of war); Coffee Brewing Institute; Confederation of British Industries; Confederation of British Industry; Council of the Building Industry; Council of Burma Industries

CB&I Chicago Bridge and Iron (company)

CBI Cumulative Book Index

CBIA California Building Industry Association

cbid counter bid

CBIS Campus-Based Information System (NSF); Communist Bloc Intelligence Service; Computer-Based Instruction(al) System

cbit (CBIT) contract bulk inclusive tour (travel plan)

cbj common bulkhead joint

CBJO Coordinating Board of Jewish Organizations

cbk checkbook

cbl cable; cement-bond log; commercial bill of lading

cb/l commercial bill of lading

c bl carte blanche (French—white card)—full power to act

CBL Clare Boothe Luce; Configuration Breakdown List; Chesapeake Biological Laboratories .

CB of L Chartered Bank of London

cbls carrier bomb light store

cbm chemical biological munitions; confidence-building measure; conventional buoy mooring; cubic meter(s)

cbm Kubikmeter (German—cubic meter)

CBM Christian Blind Mission

CBMA Carbonated Beverage Manufacturers Association

CBMC Corregidor-Bataan Memorial Commission

CBME Combined Bureau Middle East

CBM-I Common Bahasa Malay-Indonesian

CBMIS Computer-Based Management Information System

CBMM Council of Building Materials Manufacturers

CBMPE Council of British Manufacturers of Petroleum Equipment

CBMQA California Board of Medical Quality Assurance

CBMS Conference Board of Mathematical Sciences

cbmu current bit monitor unit

CBMU Canadian Board of Marine Underwriters

CBMUA Canadian Boiler and Machinery Underwriters Association

cbn chemical, bacteriological, nuclear; courses by newspaper

C bn contrabassoon

CBN Christian Broadcasting Network; Columbia Carbon Company (stock-exchange symbol); Commonwealth Broadcasting Network

CBNE California Bureau of Narcotics Enforcement

CBNM Custer Battlefield National Monument

CBNS Commander, British Naval Staff

CBNY Chemical Bank, New York

cbo collateralized bond obligation; combined bomber offensive; compensation by objectives

Cbo Colombo

CBO Community-Based Organizations; Conference of Baltic Oceanographers; Congressional Budget Office (U.S.A.); Counter-Battery Officer

CBOA Commonwealth Bank Officers Association

cboc completion bed occupancy care

CBOE Chicago Board Options Exchange

C-bomb cobalt bomb

cbore counterbore

CBOT Chicago Board of Trade

cbp ceramic beam pentode; constant boiling point

CBP Centro de Biologia Piscatória (Portuguese—Piscatorial Biological Center—Lisbon)

CBPA Catholic Book Publishers Association; Connecticut Book Publishers Association

CBPBG Commonwealth Bureau of Plant Breeding and Genetics

CBPC Canadian Book Publishers' Council

CBPDC Canadian Book and Periodical Development Council

CB & PGNCS Circuit Breaker and Primary Guidance Navigation Control System

CBPI Canadian Business Periodicals Index

CBPM Certified Bonnie Prudden Myotherapist

CBPO Consolidated Base Personnel Office

cbpp contagious bovine pleuropneumonia

CBQ Children's Behavior Questionnaire; Civilian Bachelor Quarters

C B & Q Chicago, Burlington & Quincy (railroad)

CBQ Catholic Biblical Quarterly

cbr change board register; chemical, biological, radiological; computer-based reference; crude birth rate

Cbr Calabar

CBR Canberra, Australia (airport); Center for Brain Research (University of Rochester)

CBRA Chemical, Biological, Radiological Agency

CBRA Consolidated Budget Reconciliation Act

CB radio citizen's band radio (26.965 to 27.405 MHz)

C-branding chromosome branding

CBRC Crichton Behavioral Rating Scale

CBRE Chemical, Biological, and Radiological Element

CBRI Central Building Research Institute

CBRL Chemical, Biological, and Radiation Laboratories (Ottawa)

cbrn chemical, biological, radiological, and nuclear

CBRN Caribbean Basin Radar Network

CBRO Chemical, Biological, Radiological Officer

CBRS Canadian Bond Rating Service (Montreal); Child Behavior Rating Scale; Citizen's Band Radio Service

CBRTGW Canadian Brotherhood of Railway, Transport, and General Workers

cbrw chemical, biological, radiological warfare

cbs chronic brain syndrome; concrete-block stucco

cb's citizen's-band transceivers

cBs concerned Black students

CBs cost-of-living benefits

CBS Canadian Biochemical Society; Central Bureau of Statistics (Jerusalem); Columbia Broadcasting System; Common Beliefs Survey; Commonwealth Bureau of Soils; Confraternity of the Blessed Sacrament; Currumbin Bird Sanctuary (Queensland); Custom Bucket Service

CBS Centraal Bureau de Statistiek (Dutch Central Statistical Bureau)

cbse caboose

CBSO City of Birmingham Symphony Orchestra; City of Bournemouth Symphony Orchestra; Czechoslovak Broadcasting Symphony Orchestra

cbt cesium beam tube; computer-based training; criminal breach of trust

cbt (CBT) cognitive-behavioral therapy

CBT Chicago Board of Trade; Computer-Based Testing; Connecticut Bank and Trust (company)

CBT Centre Belge de Traductions (French—Belgian Translations Center)

CB & TC Connecticut Bank & Trust Company

CBTE Competency-Based Teacher Education

cbts cesium beam time standard

cbu cluster bomb unit

CBU Chicago Board of Underwriters

c/bush cents per bushel

cbv central blood volume; circulating blood volume; corrected blood volume

CB-VD citizen's-band radio-contracted venereal disease (resulting from sexual pickups made along byways and highways)

CBVHS Clara Barton Vocational High School

cbw chemical-biological warfare

CBW Canadian Black Watch

CBX Computerized Branch Exchange

cbx's (CBXs) computerized business exchanges (telephone service)

cby carboy

Cbz carbobenzoxy

cc camp chair; canonical correlation; carbon copy (or copies); centuries; chapters; chief complaint (medical history chart); civil commotions; close-captioned; close control; closed cup; closing coil; coastal command; cognitive complexity; collector capacitance; color code; combat command; combustion chamber; command and control; command center; complex conjugate; compounded continuously; condemned cell; continuation clause; contrasting color; county court; credit card; cubic centimeter(s); current cost; customs clearance

cc (CC) check charge; chief complaint

c/c center to center; corner/card

c & c capets and curtains; caviar and champagne; command and control; consultation and concurrence

c of c certificate of occupancy; cost of construction

c-to-c center-to-center

c.c. corpora cardiaca (Latin—cardiac body)—heart

c/c compte courant (French); *conta corrente* (Portuguese); *conto corrente* (Italian); *cuenta corriente* (Spanish)—current account

Cc cirrocumulus

Cc. Confessores (Latin— Confessors)

CC Cadet Corps; Caius College; Calvin Coolidge (30th President U.S.); Cape Corps; Caterpillar Club (people who have parachuted to save their lives); Chamber of Commerce; Chief Constable; Circuit Court; City Center; Clare College (Cambridge); Coastal Commission; Cocos (Keeling Islands—Internet code); Colby College; Common Cause; Community College; Company Commander; Corps Commander; County Clerk; County Commissioner;

County of City (Aberdeen, Dundee, Edinburgh, Glasgow); Critical Care

C-C Coca-Cola

C & C Columbia & Cowlitz (railroad); Command and Control; Computer and Communications

C-by-C Come-by-Chance, Newfoundland

C of C Conclave of Cardinals; Count(y) of Cleves

CC corriente continua (Spanish—direct current)

CC-106 Canadair version of the Britannia called Yukon

CC-109 Canadian-built medium-range transport designed by General Dynamics and known as the Cosmopolitan

CC-115 Canadian twin-engine turboprop transport called the Buffalo

cca carrier-controlled approach; cellular cellulose acetate (plastic); chimpanzee coryza agent; chromated copper arsenate; cold cranking amperage; crop-condition assessment; current-controlled amplifier

CCA Cacchetti Council of America; California Central Airlines; California Chiropractic Association; California Commission on Aging; California Confederation of the Arts; California Correctional Association; Camp and Cabin Association; Canadian Cat Association; Canadian Construction Association; Central City Association; Chief of Civil Affairs; Chemical Communications Association; Circuit Court of Appeals; Citizens for Clean Air; Citizens' Councils of America; Cleaning Contractors Association; Coca-Cola Amatil (Hungary); Colorado Correctional Association; Comics Code Authority; Committee for Conventional Armaments; Commonwealth Correspondents Association; Community College Association; Community Concerts Association; Community Corrections Act; Conquest of Cancer Act; Conservative Clubs of America; Consumers Cooperative Association; Container Corporation of America; Continental Control Area; Coordinating Committee on Alcoholism; Copywriter's Council of America; Corduroy Council of America; Corrections Corporation of America (private operator of jails in the U.S.); Crafts Council of Australia; Cruising Club of America; Current Cost Accounting

C & CA Consumer and Corporate Affairs (Canada)

CCAA California Collegiate Athletic Association; Cement and Concrete Association of Australia; Chefs de Cuisine Association of America; Collector Car Appraisers Association

CCAB Canadian Circulation Audit Board

CCAC California College of Arts and Crafts; Canadian Casualty Assembly Centre

CCAD Commerce and Consumer Affairs Department

CCAE California Council for Adult Education

CCAF Commander-in-Chief—Atlantic Fleet; Community College of the Air Force

CCAGW Council for Citizens Against Government Waste

CCAHC Central Council for Agricultural and Horticultural Cooperation; Central Council for Agricultural and Horticultural Cooperatives

CCAI Chamber of Commerce of the Apparel Industry; Chemical Coaters Association International; Chinese Children Adoption International

CCAIA California Council of the American Institute of Architects

CCAIR Citizens for Corporate Accountability and Individual Rights

CCAIT Community College Association for Instruction and Technology

CCAM Colby College Art Museum

ccap communication capability application program

CCAP Citizens Crusade Against Poverty

CCAPS Circuit-Card Assembly and Processing System

CCAQ Consultative Committee on Administrative Questions (UN)

CCAR Central Conference of American Rabbis

CCAs California Correctional Association members; Cruising Club of America members

CCAS Council of Colleges of Arts and Sciences

ccat conglutinating complement absorption test

CCATF Commander Combined Amphibious Task Force

CCATS Chief Controller Auxiliary Territorial Service; Communications, Command, and Telemetry Systems

cca unit chicken-cell agglutination unit

ccb command control block; convertible circuit breaker; cubic capacity of bunkers

CCB California Canadian Bank; Civil Cooperation Bureau (South African Secret Military); command-and-control boat (naval symbol); Command Communications Boat; Configuration Control Board; Criminal Courts Building

CCBD Council for Children with Behavioral Disorders

cc black conductive channel black

CCBI Central City Business Institute

CCBM Copper Cylinder and Boiler Manufacturers

CCBO Cape Clear Bird Observatory (Ireland)

CCBS California Canadian Banks; Clear Channel Broadcasting Service

ccbv central circulating blood volume

CCBW Commission on Chemical and Biological Warfare

ccc cathodal closure contraction; central computer complex; command control console; computer-command control

CCC California Cheese Company; California Conservation Corps; California Correctional Center; Calorie

Control Council; Camp Coast to Coast; Canadian Chamber of Commerce; Cape Cod Canal; Car Care Council; Caribbean Conservation Corporation; Caribbean Cultural Center; Central Collegiate Conference; Central Community Center; Central Control Commission; Central Criminal Court; Changcheng (Great Wall) Computer Corporation; Chimpanzee Crisis Concern; Chopin Cultural Center; Christ Church College; Christian Chamber of Commerce; Christian Compassion Center; Citizens Crime Commission; Civil Construction Corps; Civilian Conservation Corps; Columbian Carbon Company; Combat Cargo Command; Commercial Credit Corporation; Commissioner's Coordination Council; Commodity Credit Corporation; Commonwealth Credit Corporation; Community Care Center; Consumer Credit Counselors; Consumers' Consultative Committee; Copyright Clearance Center; Corning Community College; Coronary Care Certificate; Corpus Christi College; Council of Conservative Citizens; County Cadet Commandant; Cox Cable Communications, Crime and Correction Commission; Crime and Correction Committee; Customs Cooperation Council; Cuyahoga Community College

CC & C Command Control and Communications (USAF)

CCC Consejo de Cooperación Cultural (Spanish—Council of Cultural Cooperation)—of the Council of Europe; *Cwmni Cyfyngedic Cyhoeddus* (Welsh—Public Limited Company)

C.C.C. Constitutio Criminalis Carolina (Latin—Carolinian Criminal Code)

CCCA Classic Car Clubs of America; Conservative Christian Churches of America

CCCB Component Change Control Board (DoD)

CCCC Cape Cod Community College; Cape Colony Cyclists Corps; Committee for the Conservation and Care of Chimpanzees; Conference on College Composition and Communication; Council of Car Care Centers

CCCCO Chicago Coordinating Council of Community Organizations

CCCCSA California Community College Community Service Association

CCCD California Commission of College Districts

CCC-FID Central Classification Committee—Fédération Internationale de Documentation

CCC Highway Cleveland-Columbus-Cincinnati Highway

CCCI Computer Control Company, Inc

CC circuit common-collector amplifier for junction transistors

CCCJ California Council on Criminal Justice

cccl cathodal closure clonus

CCCM Central Committee for Community Medicine

CCCN Customs Cooperation Council Nomenclature

CC Co Commercial Cables Company

CCCO Coordinating Council of Committee Organizations

CCCP California Coalition for Capital Punishment; Citizens Crime Commission of Philadelphia; Council on Cooperative College Projects

CCCP (Russian transliteration—USSR)—*Soyuz Sovetchikh Sotsialisticheckikh Respublik* (Union of Soviet Socialist Republics)

CCCPS Chicago College of Chiropody and Pedic Surgery

CCCR Center for Crime Control Research; Communications and Command Control Requirements

CCCS Concerned Citizens for Community Standards; Consumer Credit Counseling Services

CCCT California Community College Trustees

ccd cash concentration and disbursement; charge-coupled device; computer-controlled display

ccd (CCD) charge-coupled device (photographic)

CCD Center for Curriculum Development; Central Council for the Disabled; charge coupled device; Circuit Court Decision; Commander Coastal Defense; Confraternity of Christian Doctrine; Cost Center Determination; Criminal Conspiracy Division

CCDA Commercial Chemical Development Association

CC-DA Committee on Cataloging Description and Access (American Library Association)

CCDB Canadian Car Demurrage Bureau

CCDC Center City Development Corporation; Central Citizens' Defence Committee; Centre City Development Corporation; Commission on Crime, Delinquency, and Corrections (Nevada)

CCDN Central Council for District Nursing

CCDU Coastal Command Development Unit

cce carbon-chloroform extract; control and computation equipment

CCE California Cooperative Extension; Council for a Competitive Economy

CCE Casa de la Cultura Ecuatoriana (Spanish—House of Ecuadorian Culture)

CCEA Cabinet Council on Economic Affairs; Chief Control Electrical Artificer

CCEBS Committee for the Collegiate Education of Black Students

CCED Center for Compatible Economic Development; County Council Electoral Division

ccei composite cost-effectiveness index

CCEL Chief Control Electrician

CCEM Comprehensive Career Education Model

CCEMN Chief Control Electrical Mechanician

CCES Catholic Church Extension Society

CCET Carnegie Commission on Educational Television

CCETSW Central Council for Education and Training of Social Workers

CCETT Centre Commun d' Etudes de Télévision et de Télécommunication (French—Public Center for the Study of Television and Telecommunication)

ccf cephalin-cholesterol flocculation; compound comminuted fracture; congestive cardiac failure; chronic cardiac failure; concentrated complete fertilizer

CCF Canadian Commonwealth Federation; Captain Coastal Forces; Chinese Communist Forces; Citizens Council Forum; Combined Cadet Force; Common Cold Foundation; Congressional Clearinghouse on the Future; Cooperative Commonwealth Federation; Credit Commercial de France

CCFA Combined Cadet Force Association

CCFATU Coastal Command Fighter Affiliation Training Unit

CCFC California Cut Flower Commission; Citizens Committee for a Free Cuba

ccfe commercial customer-furnished equipment

CCFG California Contemporary Fashion Guild

CCFIS Coastal Command Flying Instructors School

ccfm cryogenic continuous-film memory

CCFPE Center for Commercial-Free Public Education

ccfr constant current flux reset

CCFR Chicago Council on Foreign Relations

CCG Canadian Coast Guard; Choral Conductors Guild; Control Commission of Germany

CCGB Cycling Council of Great Britain

CCGBE Control Commission for Germany British Element

CCGE California Council for Geographic Education

CCGEA Community College General Education Association

CCGNY Community Council of Greater New York

CCGS Canadian Coast Guard Service; Canadian Coast Guard Ship

ccgt closed-cycle gas turbine

cch cubic capacity of holds

Cch Christchurch, New Zealand

CCH Chaminade College of Honolulu; Commerce Clearing House; Computerized Criminal Histories

C of CH Chief of Chaplains

CCHE California Coordinating Council for Higher Education; Central Council for Health Education; Coordinating Council for Higher Education

CC-HEW Clinical Center—Health, Education, and Welfare

CCHF Children's Country Holidays Fund

CCHK Crown Colony of Hong Kong

CCHP Consumer Choice Health Plan

CCHPP California Council for the Humanities in Public Policy

cc/hr cubic centimeters per hour

CCHS Christopher Columbus High School; Computerized Criminal Histories System (FBI)

CCHW Citizen's Clearinghouse for Hazardous Wastes

cci chronic coronary insufficiency; circuit condition indicator; concentric coordinate incident; corrugated, cupped, or indented (cargo)

CCI California Correctional Institution; Certified Cellelectrology Instructor; Citizens Committee of Investigation (into President Kennedy's assassination); Community Concerts, Incorporated; Computer Consoles Inc; Connecticut Correctional Institution; Conservative Caucus, Inc; Cotton Council International

CCI Central Campesina Independiente (Spanish—Independent Peasant Central)—political party in Mexico

CCIA Computer and Communications Industry Association; Consumer Credit Insurance Association

CCIAP Cooperative Committee on Interstate Air Pollution (New Jersey-New York)

ccib computerized central information bank

CCIB Chinese Commodities Inspection Bureau; Computerized Central Information Bank; Cook County Inspection Bureau

ccig cold cathode ion gage

CCIL Commander's Critical Item List (USA)

CCIM Certified Commercial Investment Member

CCINC Cabinet Committee for International Narcotic Control

ccip continuously computed impact point (USAF)

CCIPT Canadian Center for Investigation and Prevention of Torture

CCIR Comité Consultatif International de la Radiodiffusion (French—International Consultative Committee on Broadcasting)

ccirid charge-coupled infrared imaging device

CCIRN Coordinating Committee for Intercontinental Research Networks

ccis command and control information system; common channel Interoffice signaling

CCIS Command Control Information System

CCIT California Council for International Trade

CCIT Comité Consultatif International Telegraphique (French—International Telegraph Consultative Committee)

CCITT Consultative Committee for International Telephony and Telegraphy

CCIW Canada Centre for Inland Waters; College Conference of Illinois and Wisconsin

CCJ Center for Correctional Justice (Harvard Law School); Center for Criminal

Justice (Washington, D.C.); Circuit Court Judge; Cook County Jail (Chicago); County Court Judge

CCJC Canadian Centre for Justice Statistics; Chicago City Junior College; Cook County Junior College; Custer County Junior College

CCJCA California Community and Junior College Association

CCJO Consultative Council of Jewish Organizations

cck (CCK) cholecystokinin

CCK Centre College of Kentucky

cck-pz (CCK-PZ) cholecystokinin-pancreozymin

cckw counterclockwise

ccl cancel(ed); cancellation

CCL Canadian Congress of Labour; Caribbean Cruise Lines; Center for Creative Leadership; Commissioner for Crown Lands; Commodity Control List; Convective Condensation Level; Council for Civil Liberties

CC$_{14}$ carbon tetrachloride

C-clamp C-shaped clamp

C-class Soviet nuclear-powered submarines nicknamed Charlies by NATO; capable of underwater missile launchings

CCLC Cooperative College Library Center

c clef alto clef (on the third line); soprano clef (on the first line); tenor clef (on the fourth line)

CC List Critical Condition List

cclkws counterclockwise

CCLM Coordinating Council of Literary Magazines

ccln consignment note control label number

CCLN Council for Computerized Library Networks

CCLP Cabinet Council on Legal Policy

ccl's criminal criminal lawyers

CCLs Court of Claims

CCLS Canadian Council of Library Schools

ccm cubic centimeter(s); counter-countermeasure(s)

ccm *Kubikzentimeter* (German—cubic centimeter)

CCM California College of Medicine; Canadian Cycle Manufacturers; Caribbean

Common Market; Certified Club Manager; Council of Communication Management

CCMA Canadian Council of Management Association; Commander Corps Medium Artillery; Community College Media Association

ccmc coincident-current magnetic core

CCMC College-Conservatory of Music of Cincinnati; Cross-Country Motor Club

ccmd continuous-current monitoring device

CCMD Chicago Contract Management District

cc/min cubic centimeters per minute

CCMP Ceylon Corps of Military Police; Coalition of Concerned Medical Professionals

CCMR Central Contract Management Region

CCMS California College of Mortuary Science; Chicago Chamber Music Society; Committee on the Challenges of Modern Society (NATO)

ccmt catechol-O-methyl transferase

CCMTC Crown Cork Manufacturers' Technical Council

ccmv (CCMV) cowpea chlorotic-mottle virus

ccn coronary care nurse; coronary care nursing

CCN Cloud Condensation Nuclei; Command Control Number; Community of the Cross of Nails; Contract Change Notice; Contract Change Notification

CCN Companhia Colonial de Navegação (Portuguese—Colonial Navigation Company)

CCNA Canadian Community Newspapers Association

CC-NASW California Chapter, National Association of Social Workers Incorporated

CCNDT Canadian Council for Non-Destructive Testing

CCNI Cía Chilena de Navegación Interoceanica (Spanish—Chilean Interoceanic Navigation Company)

CCNM Chaco Canyon National Monument

CCNMA California Chicano News Media Association

CCNP Callao Cave National Park (Luzon, Philippines); Carlsbad Caverns National Park (New Mexico)

CCNR Citizens Committee on Natural Resources; Consultative Committee for Nuclear Research

CCNS Cape Cod National Seashore (Massachusetts)

CCNSC Cancer Chemotherapy National Service Center

CCNV Community for Creative Non-Violence

CCNWR Cross Creeks National Wildlife Refuge (Tennessee)

CCNY Carnegie Corporation of New York; City College of the City University of New York

cco current-controlled oscillator

cc/o certificate of consignment/origin

Cco Curaçao

CCO Casualty Clearing Officer; Chicago College of Osteopathy; Chief of Combined Operations; Clandestine Communist Organization; Commonwealth Communications Organization; Comprehensive Certificate of Origin; Convoy Control Officer; Council of Consulting Organizations; Crossroads Chamber Orchestra

CCOA California Correctional Officers Association; County Court Officers' Association; Crafts Council of Australia

CCOC Command Control Operations Center (USA)

CCOD Coalition to Cease Ocean Dumping

CCOF California Certified Organic Farmers

CCOFI California Cooperative Oceanic Fisheries Investigations

c conc cast concrete

CCOHS Canadian Centre for Occupational Health and Safety

CCOU Construction Central Operations Unit

ccp control change proposal; credit card purchase

ccp conto corrente postale (Italian—current postal account)

CCP Caribbean Conservation Program; Central Cataloging Project; Certificate in Computer Programming; Chinese Communist Party; Code of Civil Procedure; Code of Criminal Procedure; Commonwealth Centre Party; Consolidated Cryptologic Program; Court of Common Pleas; Cultural Center of the Philippines

ccpa cloud chamber photographic analysis

CCPA Council of Canadian Personnel Associations; Court of Customs and Patent Appeals; Credit Card Protection Agency

CCPC Community Crime-Prevention Centers

CCPE Canadian Council of Professional Engineers

CCPF Commander-in-Chief—Pacific Fleet

CCPG Chemical Corps Proving Ground

CCPI California Consumer Price Index

cc-pill compound-cathartic pill

CCPIT China Commission for the Promotion of International Trade

CCPL Corpus Christi Public Library

CCPO Central Civilian Personnel Office

CCPO Comité Central Permanent de l'Opium (French—Permanent Central Opium Committee)

CCPOA California Correctional Peace Officers Association

CCPOST California Commission on Peace Officer Standards and Training

CCPP Citizen Commission on Pension Policy

ccpr coherent cloud physics radar

CCPR Central Council of Physical Recreation

CCPS Center for Chemical Process Safety; Certified Crime Prevention Specialist; Consultative Committee for Postal Studies; Council of Commonwealth Public Service

CCPSHE Carnegie Council on Policy Studies in Higher Education

CCPSO Council of Commonwealth Public Service Organisations

ccr closed-cycle refrigerator; combat crew; command control receiver; complex chemical reaction; computer character recognition; consumable case rocket; control circuit resistance; credit card reader; cross-channel rejection; crystal can relay; cube corner reflector

C_{cr} creatinine clearance

CCR Center for Constitutional Rights; Central Commission for the Navigation of the Rhine; Commission on Civil Rights; Contract Change Request

CCRA Commander Corps of Royal Artillery

CCRB Civilian Complaint Review Board

CCRB Cooperatief Centraal Raiffeisen-Boerenleenbank (Dutch—Raiffeisen's Central Cooperative Farmer's Loan Bank)—the Netherlands

ccrc continuing care retirement community

CCRC Combat Crew Replacement Center

CCRDC Chemical Corps Research and Development Command

CCRE Canadian Council for Research in Education; Commander Corps of Royal Engineers

CCRESPAC Current Cancer Research Project Analysis Center

CCRF City College Research Foundation

CCRKBA Citizens Committee for the Right to Keep and Bear Arms

CCRMA Center for Computer Research in Music and Acoustics (Stanford University)

CCRN Critical Care Registered Nurse

ccros card-capacitor read-only storage

C Cr P Code of Criminal Procedure

CCR & R covenants, conditions, restrictions, and reservations

CCRs covenants, conditions and restrictions

CCRS Canadian Centre for Remote Sensing

ccrt cathodochromic cathoderay tube

CCRT Check Collectors Round Table

ccru complete crew

CCRU Common Cold Research Unit

ccs central computer and sequencer; collective call sign; column code suppression; command, control, support (military function); computer control station(s); continuous commercial service; custom contract service(s)

cc's carcasses

cc & s central computer and sequencer

Ccs Caracas (inhabitants called Caraqueños)

CCs Community Centers; Community Colleges

CCS Calculus of Communicating Systems; Canadian Cancer Society; Canadian Chaplains' Service; Cape Cod System; Caracas, Venezuela (Maiquetia Airport); Casualty Clearing Station; Catholic Community Service; Center for Chinese Studies (University of California); Center for Cuban Studies; Chemical Corporation of Singapore; Chief Commissary Steward; Church of Christ, Scientist; Civil Communications Section; Classification and Compensation Society; Clergy Consultation Service; Combined Chiefs of Staff; Controller of Communication Services; Council of Communication Societies; Customer Conversion Statistics

CCSA Canadian Committee on Sugar Analysis; Community College Service Association; Community College Student Association

CCSB Credit Card Service Bureau

CCSC Central Connecticut State College; Central Coordinating Staff, Canada

CCSD Counter Counter-Strategic Defense

CCSEA Council of the Church of Southeast Asia

cc/sec cubic centimeters per second

ccsem computer-controlled-scanning electron microscope

ccsep cement-coated single epoxy

CCSF City College of San Francisco

CCSL Communications and Control Systems Laboratory

CCSN Center City Shelter Network

CCSO Corpus Christi Symphony Orchestra

ccsr cash-to-common-stock ratio

CCSS Charles Camille Saint Saëns; Cleveland-Cliffs Steamship (company)

CCSSO Council of Chief State School Officers

CCST Chelsea College of Science and Technology

cct cathodal closing tetanus; chocolate-coated tablet; controlled cord traction

CCT Clarkson College of Technology; Combat Control Team; Common Customs Tariff; Cumberland College of Tennessee

C & CT Chemistry and Chemical Technology

CCTA Central Computer and Telecommunications Agency; Community College Trustees Association

CCTC Chinese Cultural and Trade Center; Columbia County Teachers College

ccte cathodal closure tetanus

CCTE Canadian Council of Teachers of English

cctep cement-coated triple epoxy

cctf combat cargo task force

CCTF California Correctional Training Facility

CCTI Composite Can and Tube Institute

CC & TI Community College and Technical Institute

cctks cubic capacity of tanks

CCTP Center City Transportation Program; Coronary Care Training Project

ccts closed cycle thermal system

CCTS Canaveral Council of Technical Societies; Combat Crew Training School

CCTT Cornell Critical Thinking Test

cctv closed-circuit television

CCTWg Combat Crew Training Wing (USAF)

ccu chart comparison unit; color-control unit; coronary care unit

ccu (CCU) camera-control unit (television); cardiac care unit; coronary care unit; correctional custody unit (U.S. naval vessel's brig or jail)

Ccu Calcutta

C-C u Cherry-Crandall units

CCU Calcutta, India (airport); California Conservative Union; Community College Unit; Confederation of Canadian Unions; Cooperative Care Unit; Council for Canadian Unity

CCUA Credit Card Users of America

CCUDA Community College Urban District Association

CCUL California Credit Union League

CCUN Collegiate Council for the United Nations

CCURR Canadian Council on Urban and Regional Research

CCUS Chamber of Commerce of the United States

c/cut crosscut

ccv closed-circuit voltage; coolant control valve

ccv (CCV) control-configured vehicle

ccw carrying a concealed weapon; channel command word; counterclockwise

CCW Caldwell College for Women; Circulation Control Wing; Citizens Crime Watch; Combat Crew Wing

CCWA Consultative Commission on Workers Affairs

cc wr hdr canvas-covered wire rope handrail

ccws counterclockwise

ccxd computer-controlled X-ray diffractometer

ccy currency

CCY Chief Communications Yeoman

cd Caesarian delivery; candela; canine distemper; cash discount; center door; certificate of deposit; change directory; civil defense; coastal defense; coin dimpler; cold drawn; commercial dock; communicable disease; companion dog; confidential document; conjugate diameter (pelvic inlet); contagious disease; convulsive disorder; convulsive dose; cord; corona discharge; countdown; curative dose

cd (CD) compact digital; compact disc; compact-disc player; compact-disc record(ing); companion dog; compounded daily

c-d countdown

c/d cash against documents; cigarettes per day; cigars per day

c/d (C/D) carried down (bookkeeping); certificate of deposit

c & d carpets and drapes; censorship and documents; collection and delivery

cd *cadde* (Turkish—street); *colla destra* (Italian—with the right hand); *corriente directa* (Spanish—direct current)

c.d. *conjugata diagonalis* (Latin—diagonal conjugate)—pelvic inlet diameter

Cd cadmium; candela; caudal; coefficient of drag

C $ cordoba (Nicaraguan monetary unit)

Cd *ciudad* (Spanish—city)

CD Canadair turboprop airplane; Canadian (Forces) Decoration; Certificate of Deposit; Certified Doula; Civil Defense; coastal defense radar (for surface-vessel detection); Collision Detect; communicable disease; Community Development; confidential document; Congressional District; Continental Divide; Control Data; Corrections Department (New Mexico); Corrections Division (Hawaii, Oregon); countdown; Customs Declaration

C.D. Chancery Division

C/D Commercial Dock; consular declaration; Customs Declaration

C & D Chemist and Druggist; collection and delivery

CD Centre de Détention (French—Detention Center); *Centre Démocrate* (French—Democratic Center); *Commission du Danube* (French—Danube Commission); *Computer Design*; Corps Diplomatique (French—Diplomatic Corps)

C & D Crime and Delinquency

cd_{50} median curative dose (abolishing symptoms in 50 percent of all test cases)

Cd_{115} radioactive cadmium

cda chain data address; command and data acquisition

cda (CDA) chenodeoxycholic acid

CDA California Dental Association; California Dietetic Association; Canadian Dental Association; Canadian Dietetic Association; Catholic Daughters of America; Child Development Association; Colonial Dames of America; Control Data Australia; Copper Development Association

CDA Christen Democratisch Appèl (Dutch—Christian Democratic Appeal)—political party; *Compañía Dominicana de Aviación* (Spanish—Dominican Aviation Company)

CD Act(s) Contagious Diseases Act(s)

CDAE Civil Defense Adult Education

Cd A Eng Commissioned Air Engineer

CDAI Crohn's Disease Activity Index

CD Aim Commissioned Airman

Cd Airn Commissioned Airman

Cdale Carriedale

CDAP Civil Defense Auxiliary Police

CDARC Chelsea Drug Addiction and Research Centre

CDAS Civil Defense Ambulance Service

cdb caliper disk brake; capacitance decode box; cast double base; central data bank; could be; current data bit

Cd B Commissioned Boatswain

CDB Caribbean Development Bank; Combat Development Board; Combat Development Branch

cdba clearance divers breathing apparatus

CDBA California Dining And Beverage Association

cdbd cardboard

CDBG Community Development Block Grant

CD/BMI Columbus Division/Battelle Memorial Institute

Cd Bndr Commissioned Bandmaster

cdc calculated date of confinement; call direction code; career development course; casualty decontamination center; cell division cycle; command and data-handling console; conservation data center

CDC Cadaver Disposal Center; California Debris Commission; California Democratic Council; California Department of Corrections; Canada Development Corporation; Canadian Dental Corps; Caribbean Defense Command(er); Centers for Disease Control; Certificate of Disposition of Classified Documents; Cesspool Detergent Chemistry; Citizens' Defense Corps; Citizens Democracy Corps; Civil Defense Coordinator; Civil Defense Council; Combat Development Command; Command Destruct Control; Commissioners of the District of Columbia; Commonwealth Development Corporation; Communicable Disease Center; Community Development Corporation; Configuration Data Control; Conservation Data Center; Control Data Corporation; Control Distribution Center; Criminal Diagnostics and Counseling

C.D.C. Commonwealth Development Corporation (formerly Colonial Development Corporation)

CDC Centro de Documentação Científica (Portuguese—Scientific Documentation Center)

CDCA chenodeoxycholic acid

cdce central data-conversion equipment

CD circuit common-drain circuit for field-effect transistors

cdcm carbon-dioxide concentration module

Cd Cmy O Commissioned Commissary Officer

Cd C O Commissioned Communications Officer

Cd Con Commissioned Constructor

CDCP Construction and Development Corporation of the Philippines

CDCR Center for Documentation and Communication Research; Control Drawing Change Request

CDCs Community Development Corporations

CDCS Civil Defense Countermeasures System; Construction Dollar Control System

CDCT Centro de Documentación Científica y Téchnica (Spanish—Center of Scientific and Technical Documentation)—Mexico City

cdcu community-development credit union

cdd central data display; chart distribution data; coded decimal digit; color data display; command-destruct decoder; computer-directed drawing; cosmic dust detector; cratering demolition device

CDD Certificate of Disability for Discharge

cddd comprehensive dishonesty, disappearance, and destruction (insurance policy)

cddi computer-directed drawing instrument

CDDP Canadian Department of Defense Production

cde carbon dioxide economizer; contamination-decontamination experiment

cde (CDE) canine distemper encephalitis

CDE Central Document Exchange; Chemical Defense Establishment; Conference on Disarmament in Europe; Cornell-Dubilier Electronics

C.D.E. Certificate in Data Education

C de C (CDC) Canyon de Chelly

CDEE Chemical Defense Experimental Establishment

C de F Collège de France (French—College of France)

C de G Ciudad de Guatemala (Spanish—Guatemala City); *Croix de Guerre* (French—War Cross)

CDEG Chicago District Electric Generating Corporation

CDEI Control Data Education Institutes

C de J Compañía de Jesus (Spanish—Company of Jesus)—Society of Jesuits

cdek computer data entry keyboard

Cd El O Commissioned Electrical (Electronic) Officer

C del S Corriere della Sera (Evening Courier—Milan)

C de M Ciudad de México (Spanish—Mexico City)

C-de-N Côtes-de-Nord

Cd Eng Commissioned Engineer

CDEOS Civil Defense Emergency Operations System

CDER Center for Death Education and Research

CDES Chemical Defense Experimental Station

cdev control panel device

cdf command decoder film; command decoder filter; confined detonating fuse; constant current fringes; continuous desk file; cumulative distribution function

CDF California Department of Forestry; Canadian Department of Forestry; Channel Definition Format; Children's Defense Fund; Civil Defence Force; Coastal Defense Force; Collider Detector at Fermilab; Colorado Department of Forestry; Community Development Foundation; Congregation for the Doctrine of the Faith (conservative Catholics); Connecticut Department of Forestry; Council for the Defense of Freedom

CDFA California Department of Food and Agriculture (statistics service); California Dried Fruit Association

CDFC Commonwealth Development Finance Company

CDFGI Charles Darwin Foundation for the Galápagos Islands

CDFI Community Development Financial Institutions

CD film camouflage detection film

cdfr (CDFR) commercial demonstration fast reactor

CDFRS Charles Darwin Foundation Research Station, Academy Bay, Santa Cruz, Galápagos

CDFS Chief of Defence Force Staff

CDFSB Canadian Dairy Foods Service Bureau

cd/f² candela per square foot

cd fwd carried forward

Cdg Cardigan; Cardiganshire

CDG Coder-Decoder Group (USA); Costume Designers Guild

CDGA California Date Growers Association

CD & GB TC Chicago, Duluth and Georgian Bay Transit Company

Cd Gr Commissioned Gunner

Cd Gr O Commissioned Gunnery Officer

cdh constant differential height

CDH College Diploma in Horticulture

CDHP California Dental Health Plan

CDHS Comprehensive Data-Handling System

cdhv could have

cdi course deviation indicator

cdi (CDI) capacitor discharge ignition

cd-i compact disk-interactive

CDI Center for Defense Information; Children's Depression Inventory; Classified Defense Information; Comprehensive Dissertation Index; Contractor Demonstration Inspection; Conventional Defense Improvement (NATO)

CDIAC Carbon Dioxide Information Analysis Center

CDIB Certified Degree of Indian Blood

CDIC Canada Deposit Insurance Corporation

Cd In O Commissioned Instructor Officer

C Dip F & A Certified Diploma in Finance and Accounting

c div cum dividend

Cd J Ciudad Juárez (Spanish—Juarez City) (inhabitants—*Juaristas*)

CDJ California Department of Justice

CDJ Comité de Défense des Juifs (French—Committee of the Defense of Jews)

cdk containers (carried on) deck

c$k consumer's survival kit

cdl canal defense light; commercial driver's license; common display logic; construction, demolition and land-clearing

Cdl Cardinal

CDL Central Dockyard Laboratory (UK); Christian Defense League; Citizens for Decency through Law; Citizens for Decent Literature; Country and Democratic League

CDLA Computer Dealers and Lessors Association

CDLC Canadian Dental Laboratory Conference

CDLS Canadian Defense Liaison Staff

cdm contributing to the delinquency of a minor

CDM Coalition for a Democratic Majority; Consolidated Diamond Mines (South Africa)

cd/m² candela per square meter

cdma code division multiple access

CDMA Chain Drug Marketing Associates

Cd/m² candelas per square meter

Cd M-a-A Commissioned Master-at-Arms

CDMB Civil Defense Mobilization Board

CDMMA Canadian Direct Mail Marketing Association

CDMSWA Consolidated Diamond Mines of South-West Africa

cdn (CDN) condition

cDNA complementary deoxyribonucleic acid

CDN Chicago Daily News

CDNAC Canadian Daily Newspaper Advisory Council

CDNPA Canadian Daily Newspaper Publishers Association

CDNRA Coulee Dam National Recreation Area (Washington)

CDNS Chicago Daily News Service

cdnt could not

cdo (CDO) chronic drunkenness offender

Cdo Commando

Cd O Commissioned Officer

C-d'O Côte-d'Or

CDO California Disaster Office; Civil Defense Organization; Community Development Office(r)

CDOA Car Department Officers Association

Cd Ob Commissioned Observer

Cd O E Commissioned Ordnance Engineer

CDOGS Council for the Defence of Government Schools

Cd O O Commissioned Ordnance Officer

cdos controlled date of separation

cdo(s) commando(s)

cdp checkout data processor; communications data processor; complete decongestive physiotherapy; contract definition phase; cytidine 5-diphosphate

CDP Centralized Data Processing; Certified Data Plan; Critical Decision Point

C.D.P. Certificate in Data Processing

CD Pack Compact Disc Packaging

CDPA Civil Defense Preparedness Agency

cdpc central data-processing computer

CDPC California Delinquency Prevention Commission

CDPD Cellular Digital Packet Data

CDPE Continental Daily Parcels Express

cd pl cadmium plate

cd player compact-disc player

cdp's comprehensive dwelling policies

CDQ Choice Dilemma Questionnaire

CDQAP California Dairy Quality Assurance Program

cdr command-destruct receiver; composite damage risk (audiometry); critical design review

cdr (CDR) crude death rate

Cdr Commander

C d R *Casa di Risparmio* (Italian—Savings Bank)

CDR Change Design Request; Conceptual Design Report; Countdown Deviation Request; Critical Design Review(s)

CDR *Comité Defensa Revolucionario* (Spanish—Revolutionary Defense Committee)—Cuba

CDRA Canadian Drilling Research Association; Committee of Directors of Research Associations; Corps of Drivers Royal Artillery

CDRB Canadian Defense Research Board

CDRBTE Canadian Defense Research Board Telecommunication Establishment

CDRC Civil Defense Regional Commission(er)

Cdre Commodore

CDRE Chemical Defense Research Establishment

CDRF Canadian Dental Research Foundation

CDRI Central Drug Research Institute

cdrl contract data requirements list

cdrill center drill

Cdrngtn C Codrington College

Cd R O Commissioned Radio Officer

cdrom (CD Rom) compact-disc read-only memory

cdromxa compact disk read only memory extended architecture

CDRS Charles Darwin Research Station

CDRSC Children's Depression Rating Scale for Classrooms

cds cards; circular date stamp; cold-drawn steel; combat direction system; container delivery service; single cotton double silk (insulation)

cds (Cd S) cadmium sulfide

cd's (CDs) certificates of deposit; compact-disc players; compact-disc record(ing)s— played by a laser beam

C d S *Circolo della Stampa* (Italian—Press Club); *Codice della Strada* (Italian—Highway Traffic Code); *Consiglio di Sicurezza* (Italian—Security Council)

CDs Catholic Documents; compact discs

CDS California Dental Service; Cataloging Distribution Service; Center for Degree Studies; Central Defense Staff; Civil Defence Services; Civil Defense Staff; Climatological Data Sheet; Commander, Destroyer Squadron; Community Dispute Services; Comprehensive Display System (radar)

cdsa comprehensive digestive stool analysis

Cd S B Commissioned Signals Boatswain

cdse computer-driven simulation environment

CdSh Commissioned Shipwright

CDSHA Country Day School Headmasters Association

Cd S O Commissioned Supply Officer

CDSO Commonwealth Defense Service Organization

CDSP Church Divinity School of the Pacific

CDSP *Current Digest of the Soviet Press*

CDSs Civil Disobedience Squads

CDSS British Post Office trade mark covering telecommunications and telephonic apparatus, instruments, and installations; Compressed Data Storage System; Customers' Digital Switching System

CDST Central Daylight Saving Time

cdt command-destruct transmitter; conduct; conductor

Cdt Cadet; Commandant

CDT Canadian Department of Transport; Center for Democracy and Technology; Central Daylight Time

CDT *Catholic Dictionary of Theology*

CDT (ADA) Council on Dental Therapeutics (American Dental Association)

C.D.T. Certified Dental Technician

Cdte Comandante (Spanish—Commander)

CDTE Council for Distributive Teacher Education

cd tec compact-disc technology

Cdt Mid Cadet Midshipman

cdts constant-depth temperature sensor

cdu cable distribution unit; central display unit; chemical dependency unit; control display unit

CDU Cable Distribution Unit; Christian Democratic Union (Germany); Civil Disobedience Unit; coastal defense (radar) unit; Computer Display Unit

CDU Christlich-Demokratische Union (German—Christian Democratic Union)—political party

CDUEP Civil Defense University Extension Program

cdv cadaver; computed dollar value; current domestic value

Cdv Commonwealth dollar value

CDV Civil Defense Volunteer(s)

CD-V compact disc—video

cdv carte de visite (Spanish—visiting card, sometimes with photograph)

cdvr cadaver

cdw catcher's day's work; charge density wave; chilled drinking water; collision damage waiver

CDW Civil Defense Warning; Collision Damage Waiver

CD & W Colonial Development and Welfare

Cd Wdr Commissioned Wardmaster

cdwe could we

Cd W O Commissioned Writer Officer

c dwr chest of drawers; chilled drinking water return

CDWS Civil Defense Wardens Service

cdwt cordwelt

cdx companion dog excellent; control differential transmitter

cdz concordant zone

Cdz Cádiz

ce capillary electrophoresis; carbon equivalent; career education; center of effort (na-

val architecture); center entrance; circular error; combat exhaustion; compression engine; constant error; consumption entry; converting enzyme; counterespionage; critical examination; cum entitlement

ce (CE) counterespionage

c-e communications-electronics

c/e custom entry

c & e commission and exchange; customs and excise

c of e coefficient of elasticity

c.e. caveat emptor (Latin—let the buyer beware); *curvée extra* (French—special sort) special quality

Ce Ceará; cerium; Ceylon

CE Canada East; Canadian Engineers; Certified Exchangors; Chemical Engineering; Chief Engineer; Chief Executive; Christian Endeavor; Church of England; circular error; City Editor; Common Era; compass error; Contributing Editor; Corps of Engineers; cost effectiveness; Counselor of Embassy; Customer Engineer

C-E communications electronics

C.E. Chemical Engineer; Civil Engineer

C/E Chancellor of the Exchequer; Chief Engineer

C of E Church of England (Protestant Episcopal); Corps of Engineers

CE Catholic Encyclopedia

C E Chemical Engineering; *Comedy of Errors*

C.E. Christian Era; Civil Engineer

cea circular error average

cea (CEA) carcinoembryonic antigen

CEA California Earthquake Authority; Canadian Education Association; Canadian Electrical Association; Canadian Export Association; Captain's Endowment Association (police); Cement Employers Association; Certified Environmental Auditor; Chief Electrical Artificer; Childbirth Education Association; Classified Employees Association; College English Asso-

ciation; Combined Educational Associations; Combustion Equipment Associates; Commodity Exchange Authority; Congressional Education Associates; Connecticut Educational Association; Conservation Education Association; Cooperative Education Association; Correctional Education Association; Cost-Effective Analysis; Council of Economic Advisers; Council for Energy Awareness; Council of Engineering Associations; County Employees Association

CEA Comisión Económica de Africa (Spanish—Economic Commission of Africa); *Commissariat à l'Energie Atomique* (French—Atomic Energy Commission)

CEAA Center for Editions of American Authors; Council of European-American Associations

CEAC Commission for European Airspace Coordination; Consulting Engineers Association of California

CEAC Commission Européenne de l'Aviation Civile (French—European Civil Aviation Commission)

CEAFU Concerned Educators Against Forced Unionism

CEAN Community Energy Action Network

CEANAR Commission on Education in Agriculture and National Resources

CEAPD Central Air Procurement District

CEARC Computer Education and Applied Research Center

CEARD Commission for Attention to Refugees and Displaced Persons (Guatemala)

CEAT Canadian English Achievement Test

ceb cryogenic expulsive bladder

Ceb Cebu

CEB Central Electricity Board; Continuing Education Books; Council on Employee Benefits

CEB Comité Electrotechnique Belge (Belgian Electrotechnical Committee)

cebar chemical, biological, radiological warfare

CEBS Certified Employee Benefit Specialist; Church of England Boys' Society; Commonwealth Experimental Building Station

Ceb-Vis Cebu-Visayan

CEC California Energy Commission; Canadian Electrical Code; Catholic Education Council; Central Economic Committee; Ceramic Educational Council; Chief Executive Commissioner; Civil Engineer Corps; Coal Experts Committee; Commission of the European Community; Commodity Exchange Commission; Commonwealth Economic Committee; Commonwealth Edison Company; Communications and Electronics Command; Consolidated Edison Company; Consolidated Electrodynamics Corporation; Consultung Engineers Council; Continental Entry Chart(s); Correctional Economics Center; Council for Exceptional Children

CECA Communauté Européenne du Charbon et de l'Acier (French—European Coal and Steel Community); *Comunidad Europea del Carbon y del Acero* (Spanish—European Coal and Steel Community)

CECC California Educational Computer Consortium

Cece Cecil

CECEW Catholic Education Council for England and Wales (often truncated to CEC—Catholic Education Council)

CECH Citizenship Education Clearinghouse

Cechy (Czechoslovakian—Bohemia)

CECIL Compact Electronic Components Inspection Laboratory

CE circuit common-emitter amplifier for junction transistors

CECLA Comisión Especial de Coordinación Latinoamericana (Spanish—Special Commission for Latin American Coordination)

CECR Center for Environmental Conflict Resolution; Central European Communication Region (USAF)

CECs California Ecology Corpsmen

CECS Church of England Children's Society; Communications Electronics Coordinating Section

CECS Comisión Especial de Consulta sobre Seguridad (Spanish—Special Commission for Security Consultation)

ced capacitance electronic disc; communications-electronics doctrine; computer entry device; cooling effects detector

c-e-d carbon-equivalent-difference

c & ed clothing and equipment development

Ced Ceda; Cedomil; Cedric; Cedron

CED Committee for Economic Development; *Communauté Européene de Defense* (French—European Defense Community); Communications-Electronics Doctrine (USAF manuals); Council for Economic Development

CED Centro Elletronnico di Documentazione (Italian—Electronic Documentation Center)—in Rome

CEDA California Economic Development Agency; Canadian Electrical Distributors Association

CEDA Comisión Económica de Oeste Asiatico (Spanish—Economic Commission of West Asia); *Confederación Española de Derechas Autonomas* (Spanish—Spanish Confederation of Autonomous Rights)—Catholic party

cedac central differential analyzer control; cooling effect detection and control

CEDAL Centro de Estudios Democráticos de America Latina (Spanish—Latin American Center of Democratic Studies)

CEDAM Conservation, Exploration, Diving, Archeology, Museums (organization)

CEDAR Council for Educational Development and Research

Cedar Breaks Cedar Breaks National Monument in Utah's Wasatch Mountains

CEDB Component Event Data Bank

CEDDA Center for Experimental Design and Data Analysis

cedi unit of currency in Ghana

CEDI Centre Européen de Documentation et d'Information (French—European Documentation and Information Center); *Centro Europeo de Documentación e Información* (Spanish—European Documentation and Information Center)

CEDIC Church Estates Development and Improvement Company

CEDO Centre for Educational Development Overseas (UK)

CEDPA California Educational Data Processing Association

ced's captured enemy documents

cee computer-enhanced education

CEE Center for Environmental Education; Central Engineering Establishment; Certificate of Extended Education; Citizens for Excellence in Education; College of Extended Education; Common Entrance Examination; Cultural Environment Emergency

CEE Comunidad Económica Europea (Spanish—European Economic Community)

CEEA Communauté Européenne de l'Energi Atomique (French—European Atomic Energy Community)

CEEB College Entrance Examination Board

CEEC Council for European Economic Cooperation

CEECC Consolidated-Edison Energy Control Center

Ceece Cecil

CEEED Council on Environment, Employment, Economy, and Development

ceefax see the facsimile; see the facts

CEEP *Centre Européen d'Etudes de Population* (French—European Center for Population Studies)

CEev Central European encephalitis virus

cef cellular-expansion factor; chicken-embryo fibroblasts

CEF Canadian Expeditionary Force; Child Evangelism Fellowship; Children's Emergency Fund(ing); Citizens for Educational Freedom; Citizens for Energy and Freedom

CEF *Corps Expéditionnaire Français* (French—French Expeditionary Corps)

C of EF Count(y) of East Friesland Country

CEFA Council for Educational Freedom in America

ceff controlled energy flow forming

cefo complete equipment flying order

CEFP Council of Educational Facility Planners

CEFT Children's Embedded Fissures Test(ing)

CEFTRI Central Food Technological Research Institute

CEG Coalition for Economic Growth

CEGB Central Electricity Generating Board

CEGEP *Collège d'Enseignement Géneral et Professionnel* (French—General and Professional College Teaching)

CEGGS Church of England Girls' Grammar School

CEGJA Coalition to End Grand Jury Abuse

CEGS Church of England Grammar School

CEHHS Charles Evans Hughes High School

CEHS Civilian Employee Health Service

cei contract end item

CEI Certified Environmental Inspector; Charles Edward Ives (innovative American composer); Claremont Economics Institute; Cleveland Electric Illuminating Company; Commission Electrotechnique Internationale (International Electrotechnical Commission); Communications-Electronics Instruction;

Cost Effectiveness Index; Council of Engineering Institutions

C & EI Chicago & Eastern Illinois (railroad)

CEI *Chemical Engineering Index; Commission Electrotechnique Internationale* (French—International Electrotechnical Commission)

CEIE Commission on Excellence in Education(U.S.)

CEIF Community Employment Initiatives Fund; Council of European Industrial Federations

ceil ceiling

cein contract end-item number

CEIP Carnegie Endowment for International Peace; Communications-Electronics Implementation Plan

C-E-I-R Corporation for Economic and Industrial Research

CEIS California Education Information System; Cost and Economic Information System

ceisd customer engineering instruction system diagram

CEIWT Central Europe Inland Waterways Transport (NATO)

cej cement-enamel junction

CEJEDP Central Europe Joint Emergency Defense Plan (NATO)

CEJNSA Council of European and Japanese National Shipowners Associations

cel celesta; celluloid; cellulose

c-e-l carbon-equivalent-liquid

Cel Celeban; Celebes; Celsius

CEL Constitutional Educational League; Cryogenics Engineering Laboratory

cela celadon

CELAC *Comisión Económica de Latinoamérica y el Caribe* (Spanish—Economic Commission of Latin America and the Caribbean)

cel acet cellulose acetate

CELADE *Centro Latinoamericano de Demografía* (Latin American Demographic Center)

CELDS Computerized Environmental Legislative Data System

celeb celebrate; celebration; celebrity

celebs celebrities

celest celestial

Celia Cecilia

celintrep accelerated intelligence report

ccll celluloid

CELL Case Existological Laboratories Limited; Continuing Education Learning Laboratory

celli cellos (violoncellos)

'cellist(s) violoncellist(s)

cell violoncello

cellokon *cellokonzert* (German —cello concerto)

cellphone cellular telephone

cell phone cellular telephone

celnav celestial navigation

cel nitr cellulose nitrate

celo chicken embryo lethal orphan (virus)

CELOS *Centrum voor Landbouwkundig Onderzoek in Suriname* (Dutch—Center for Agricultural Research in Surinam)

cels (Cels) celsius

CELS Continuing Education for Library Staffs

cel sheet cellulose (plastic) sheet

CELSS Closed Ecological Life-Support System

celt classified entries in lateral transposition

Celt Celtic

CELT Comprehensive English Language Test(ing)

celtuce celery-lettuce (lettuce-derived vegetable whose stalks taste like celery)

cem captured enemy material; cement; cement asbestos; cemetery; channel electron multiplier; communication-electronics and meteorological; control electrical mechanic

CEM Center for Entrepreneurial Management; Council of European Municipalities

CEM *Confederación Evangelical Mundial* (Spanish—World Evangelical Confederation)

CEMA Canadian Electrical Manufacturers Association; Connecticut Educational Media Association; Consumer Electronics Manufacturers

Association; Converting Equipment Manufacturers Association; Conveyor Equipment Manufacturers Association; Council for Economic Mutual Assistance; Council for the Encouragement of Music and the Arts

CEMAA Council for Egg Marketing Authorities of Australia

cem ab cement asbestos board

cemad coherent echo modulation and detection

cemb cembalo (Italian— harpsichord)

CEMB Communications-Electronic-Meteorological Board (USAF)

CEMCO Continental Electronics Manufacturing Company

cemf counter-electromotive force

cem fl cement floor

CEMI Computer Exchange of Museum Information

c-e mix chloroform-ether mixture

CEMLA Centro de Estudios Monetarios Latinoamericanos (Spanish—Center of Latin American Monetary Studies)

CEMO Command Equipment Management Office

cemon customer engineering monitor(ing)

cem p cement paint

CEMPIMS Communications Electronics Meteorological Program Implementation Management System (USAF)

cem plas cement plaster

CEMR Canadian Energy, Mines, and Resources

CEMREL Central Midwestern Regional Educational Laboratory

CEMS Church of England Mens' Society

CEMT Conferencia Europea de Ministros de Transporte (Spanish—European Conference of Ministers of Transport)

cen center; census; central; centralization; centralize

Cen Cenozoic (last major era of geologic time); Centaurus (constellation)

CEN Captive European Nations; Central Airlines; Certified Emergency Nurse; Philadelphia Centennials (National Association)

CEN Comité Européen de Coordination des Normes (French—European Committee of the Coordination of Standards)

CENA Coalition of Eastern Native Americans

CENACO Centro Nacional de Computación (Spanish—National Computation Center)

Cenacolo Il Cenacolo (Italian—Refectory, Supper Room)—Leonardo da Vinci masterpiece—*L'Ultima Cena*—The Last Supper

CENAMEC Centro Nacional para el Mejoramiento de la Enseñanza de la Ciencia (Spanish—National Center for the Betterment of the Teaching of Science)—Venezuelan society

CENCOMMURGN Central Communications Region

CENCOMS Center for Communication Sciences (USA)

CENDES Centro de Enseñanza para el Desarollo (Spanish—Center of Learning for Development)

CENDHRRA Center for the Development of Human Resources in Rural Asia

CENDIT Centre for Development of Instructional Technology

Cen Eccl Censura Ecclesiastica (Latin—Ecclesiastical Censure)

CENEUR Compañia Española de Navegación Marítima (Spanish—Spanish Maritime Navigation Company)

CENFAM Centro Nazionale di Fisica dell'Atmospera e Meteorologia (Italian—National Center of Physics of the Atmosphere and Meteorology)

C Eng Chartered Engineer; Chief Engineer

CENIM Centro Nacional de Investigaciones Metalúrgicas (Spanish—National Center for Metallurgical Research)

cenio census input output

cenog computerized electro-neuro-ophthalmograph

cens censor; censorship; census

CENS China Economic News Service

censor. centrifugal solids recovery

cent central

cent. centrifugal; century

cent. centum (Latin—hundred)

Cent Centaurus; Century

Cent Centaur (Latin—Centaur constellation)

Cent. Af. Central African Republic (whose capital is Bangui)

CENTA Committee for Establishing a National Testing Authority

centac(s) central tactical report(s)

CENTACS Center for Tactical Computer Sciences (USA)

CENTAG Central European Army Group

centen centennial

Centennial(s) Coloradan(s)

centi $^{10-2}$

CENTO Central Treaty Organization (Great Britain, Iran, Pakistan, Turkey)

Central Archives records of the NSDAP on microfilm in the Hoover Institution on War, Revolution, and Peace, Stanford University, California

centrex central exchange

cents. centuries

cent(s) céntimo(s)—one-hundredth of a peseta

CENTURY Humanist Century (tabloid)

ceo chick embryo origin; comprehensive electronic office

ceo (CEO) chief executive officer

CeO Chairman ex-Officio

CEO Chief Education Office(r); Chief Engineer's Office; Chief Executive Officer; Chief Executives Organization; County Employees Organization

CEOA Central European Operating Agency

CEOAS Corps of Engineers Office of Appalachian Studies (USA)

CEOED Compact Edition of the Oxford English Dictionary

CEOs Chief Executive Officers (conglomerate and multinational corporations)

CEOSL Confederación Centroamericana de Organizaciones Sindicales Libres (Spanish—Central American Confederation of Free Trade Unions)

cep circle of equal probability; circle of error probability

'cep' except

Cep Cepheus

CEP Capability Evaluation Plan; Chicano Education Project; Civil Emergency Planning (NATO); Color Evaluation Program; Concentrated Employment Program; Continuing Education Program; Coordinated Examination Program; Council on Economic Priorities; Council on Educational Policy

CEPA Chicago Educational Publishers Association; Civil Engineering Program Applications; Consumers Education and Protective Association

CEPACS Customs Entry Processing and Cargo System

CEPAL Comisión Económica Para América Latina (Spanish—Economic Commission for Latin America)—UNs ECLA

CEPB Civil Emergency Planning Bureau (NATO)

CEPC City of Erie Port Commission

CEPC Comité Européen pour les Problémes Criminels (French—European Committee on Crime Problems)

CEPD Career Education Planning District; Council for Economic Planning and Development

CEPDs Communications Electronics Policy Directives (NATO)

CEPE Central Experimental and Proving Establishment; *Corporación Estatal Petrolera Ecuadorian* (Spanish—Ecuadorian State Petroleum Corporation)

CEPEC Center for Professional Executive Career Development and Counselling

CEPEX Controlled Ecosystem Pollution Experiment

CEPG Cambridge Economic Policy Group

Ceph Cepheus

CEPH Centre d'Etude du Polymorphisme Humain (French—Center for the Study of Human Polymorphism)

cephal (Latin prefix—head)—cephalic

ceph floc cephalin flocculation (test)

CEPM Center for Educational Policy and Management

CEPO Central Engineering Projects Office (NATO); Corps of Engineers—Portland, Oregon

CEPR Center for Educational Policy Research

ceps civil engineering problems

CEPS Central Europe Pipeline System (NATO); Commonwealth-Edison Public Service; Cornish Engines Preservation Society

Cepsa Compañía Española de Petróleos (Spanish—Spanish Petroleum Company)

cept chemically enhanced primary treatment

'cept accept; except

CEPT Conférence Européenne des Administrations des Postes et des Télécommunications (French—European Conference of Posts and Communications)

CEPTA Committee to End Pay Toilets in America

'cepted accepted; excepted

'cepting accepting; excepting

'ception deception; exception; perception; reception

cept(s) concept(s); precept(s)

CEQ Council on Environmental Quality (appointed by the President of the United States)

CEQA California Environmental Quality Act

ceqom combined electron quench and optical masker

cer ceramic; conditioned emotional response

cer. certificate

c & er combustion and explosives research

CER Center for Educational Reform; Certification Evaluation Review; Combat Effectiveness Report; Community Educational Resources

CERA Chief Engine-Room Artificer

CERA/ACCE Canadian Educational Researchers Association/Association Canadienne des Chercheurs en Education

CERACS Comparative Evaluation of the Different Radiating Cables and Systems (technologies)

ceram ceramic; ceramicist, ceramics

ceramal ceramic + alloy

CERB Coastal Engineering Research Board (USA)

cerc centralized engine-room control

CERC Coastal Engineering Research Center; Coastal Engineering Research Council; Concurrent Engineering Research Center

CERCA Commonwealth and Empire Radio for Civil Aviation

CERCLA Comprehensive Environmental Response, Compensation and Liability Act

CERDS Charter on the Economic Rights and Duties of States

Cer.E. Ceramic Engineer

cerebro (Latin prefix—brain)—cerebral, cerebrospinal fluid

CERES Center for Research and Education in Sexuality; Clouds and the Earth's Radiant Energy System; Coalition for Environmentally Responsible Economics

CERF Canine Eye Registration Foundation; Coastal Education and Research Foundation

CERI Center for Educational Research and Innovation; Clean Energy Research Institute (University of Miami)

CERIC Central ERIC

CERIF Common European Research Project Information Format

CERL Central Electricity Research Laboratories; Coastal Engineering Research Laboratory; Cooperative Educational Research Laboratory

CERLAL *Centro Regional para el fomento del Libro en America Latina* (Spanish— Regional Center for the Development of Books in Latin America)

cermet ceramic-metallic (powders fused to form solid nuclear fuel elements)

CERN Center for Nuclear Research

CERN Commission Européenne pour la Recherche Nucléaire (French—European Commission for Nuclear Research)

CEROILFOOD China National Cereals, Oils, and Foodstuffs Import and Export Corporation

CERP Current Economic Reporting Program

CERP Centre Européen des Relations Publiques (French— European Center of Public Relations)

CERR Commonwealth Employees Redeployment and Retrenchment; Conference on Education and Race Relations

CE/RRT Central Europe Railroad Transport (NATO)

cert certificate; certified; certify

cert. certiorari (Latin—to certify; to be informed)—a writ from a superior court to a lower court to produce papers needed for a review of a case

CERT Christian Emergency Relief Team; Communications Effectiveness Response Test; Computer Emergency Response Team; Cost-Effective Rapid Transit; Council of Energy Resources Tribes

CE/RT Central Europe Road Transport (NATO)

certif certificate(d)

cert inv certified invoice

certs certificates

cerv cervical

ces central excitatory state; compressor end seal; constant elasticity of substitution; constructive error source

CEs Council of Europe members

CES California Employment Security; Certified Environmental Specialist; Closed Ecological System; College of Extended Studies; Commercial Earth Station; Committee on Earth Sciences; Commonwealth Education Scheme; Commonwealth Employment Service; Comprehensive Export Schedule; Conference on European Security; Consolidated Electronic Services; Consumer Electronics Show; Cost Effectiveness Study; Council for European Studies; Crew Escape System

CES Certifcat d'Etudes Supérieures (French—Advanced Studies Certificate); *Consejo Económico y Social* (Spanish—Economic and Social Council)—UN

CESA Canadian Engineering Standards Association; Commercial Education Society of Australia; Cooperative Educational Service Agency

CESAALA Charles E Stevens American Atheist Library and Archives (Austin, Texas)

CESAME Commonwealth Employment Service Animated Memory

CESAP Comisión Económica y Social de Asia y el Pacifico (Spanish—Economic and Social Commission of Asia and the Pacific)

CESAR Capsule Escape and Survival Applied Research

CESAR Compagnie d'Etudes des Stations Air-Route (French—Company for the Study of Airfields)

CESC Calcutta Electric Supply Corporation

cesemi computer evaluation of scanning electron microscopic image

CESG Communications-Electronics Security Group

cesi closed-entry socket insulator

CESI Council for Elementary Science International

cesk cable end-sealing kit

Cesko Soc Ceskoslovenská Socialistická (Czech—Czechoslovak Socialist Republic)

CESMM Civil Engineering Standard Method of Measurement

CESO Canadian Executive Service Overseas; Civil Engineer Support Office (USN)

CESO-W Council of Engineers and Scientists Organizations—West

c esp con espressione (Italian— with expression)

CESP Centrais Eléctricas de São Paulo

cesr conduction electron spin resonance

CESR Canadian Electronic Sales Representatives

cess assess; assessment; cessation; cession(aitre); cessionary; cessment; cesspipe; cesspit; cesspool; success

cess. cesspit; cesspool; excrement

Cess Cecil

CESS Council of Engineering Society Secretaries; Crew Escape Subsystem

CESSAC Church of England Soldiers, Sailors, and Airmens Clubs

Cessna 180 6-passenger utility aircraft

Cessna 185 6-passenger utility aircraft called the Cessna 185 E Skywagon

Cessna 310 6-passenger aircraft designated U-3

Cessna FR-172 French four-place rocket launcher aircraft

CEST Career Education Study Trip(s)

Cestr Chester

Cestr. Cestrensis (Latin—of Chester)

cet capsule-elapsed time; combat engineering tractor; controlled environmental test(ing); corrected effective temperature; cumulative elapsed time

Cet Centus; Cetus (constellation)

CET Center for Employment Training; Central European Time; Cerebral Electrotherapy; Certified Electrical Technician; Certified Electronics Technician; Common External Tariff (European Communities); Control Equipment Technician; Council for Educational Technology

CET *Collèges d'Enseignement Technique* (French—Technical Education Colleges)

CETA Chinese-English Translation Assistance; Cleaning Equipment Trade Association; Comprehensive Employment and Training Act

CETA *Centre d'études pour la Traduction* (French-Center for the Study of Translation)

CETAG *Centre d'études pour la Traduction, Grenoble* (French—Center for the Study of Translation, Grenoble)

CETAP *Centre d'études pour la Traduction, Paris* (French—Center for the Study of Translation, Paris)

CETC Combat Engineering Training Center; Community and Education Center

CETC *Centro de Estudios Tecnicos* (Spanish—Center for Technical Studies)

CETDC China External Trade Development Council

CETEC Consolidated. Engineering Technology Corporation

CETEDOC *Centre de Traitement Electronique des Ducments* (French—Center of Electronic Treatment of Documents)

CETEKA *Ceskoslavenská Tisková Kancelár* (Czechoslovakian Press Bureau)

CETEM Comprehensive Elementary Teacher Education Models

CETEX Committee on Contamination of Extra-Terrestrial Exploration (NASA)

CETF Clothing and Equipment Test Facility (USA)

CETHV Council for the Education and Training of Health Visitors

ceti communications with extraterrestrial intelligence

CETIS *Centre de Traitement de l'Information Scientifique* (French—Center for Processing Scientific Information)

CETME *Centro de Estudios Tecnicos de Materiales Especiales* (Spanish—Center of Technical Studies of Special Materials)

CETO Center for Educational Television Overseas

cetol cetologic(al); cetologist; cetology

cet. par. *ceteris paribus* (Latin—other things being equal)

CETRA *Centre des Etudes Transatlantigues* (French—Center for Transatlantic Studies)

CETS Church of England Temperance Society; Commission on the Education of Teachers of Science; Contractor Engineering and Technical Services

ceu continuing education unit

CEU Christian Endeavor Union; Constructional Engineering Union

CEUCA Customs and Economic Union of Central Africa

CEUs Continuing Education Units

CEUSA Committee for Exports to the U.S.A.

cev combat engineer vehicle; convoy escort vessel; cryogenic explosive valve

cevat combined environmental, vibration, acceleration, temperature

cevi contract exhibit vendor item

cew circular electric wire

CEW Church-Employed Women; Continuing Education for Women; Cosmetic Executive Women

cewi combat electronic warfare and intelligence

cewrm communications-electronics war-readiness material

cex charge exchange; civil effects exercise

CEX Corn Exchange Bank (stock-exchange symbol)

Cey Ceylon; Singhalese

CEY Century Electric (stock-exchange symbol)

CEYC Church of England Youth Council

Ceyl Ceylon

Cey Rs Ceylon rupees

cf calf binding; carried forward; carrier frequency; carry forward; cement floor; center field; center of flotation; center forward; central files; central filing; centrifugal force;

certificates (in newspaper stock listings); communication factor; compact fluorescent; complement fixation; conception formulation; conditional freedom; confer; continuous flow; continuous focusing; contract formulation; corrugated furnaces; cost and freight; counterfire; counting fingers; cubic feet, cubic foot; cystic fibrosis

c/f carried forward

c & f clearing and forwarding; cost and freight

c-to-f center-to-face

cf. *confer* (Latin—compare)

c.f. *cantus firmus* (Latin—fiixed song)

Cf californium

Cf. *Confessor* (Latin—Confessor)

CF Canadian Forces; Canadian Fusiliers; Cape Fear (railroad); Central African Republic (Internet code); Chaplain to the Forces; Chief of Finance; Coastal Frontier; Colorado Fuel & Iron (stock-exchange symbol); Commonwealth Fund(ing); Conservation Foundation; Consolidated Freightways; Corresponding Fellow

C/F Contract Formulation

C of F Chief of Finance

CF *Carlo Felice* (Italian) theater in Genoa; *Chemin de Fer* (French—Railroad); *Club de Fútbol* (Spanish—Football Club)

CF-5 Canadian version of the F-5 jet fighter

CF-86 Australian-built Canadian version of the F-86 jet fighter called Sabre

CF-100 Avro two-seat jet interceptor called the Canuck

CF-101 Canadian-built version of the F-101 jet interceptor named Voodoo

CF-104 Canadian version of the F-104 interceptor called Starfighter

cfa cavalry field ambulance; complement-fixing antibody; configural frequency analysis; cowl flap angle; crossed-field amplifier

cFa complete Freund's adjunct

CfA Coalitions for America

CFA California Farmer Advocate; California Forestry Association; Canadian Federation of Agriculture; Canadian Field Artillery; Canadian Football Association; Canadian Forestry Association; Canadian Freight Association; Cape Field Artillery; Cat Fanciers' Association; Center for Astrophysics; Chartered Financial Analyst; Chilled Foods Association; Clearing and Forwarding Agents; Commercial Finance Association; Commission on Fine Arts; Commonwealth Firemen's Association; Community Facilities Administration; Consumer Federation of America; Correctional Facilities Association; Council for Foreign Affairs; Country Fire Authority

CF & A Chief of Finance and Accounting (USA)

C & FA Cookery and Foods Association

CFA *Colonies Française d'Afrique* (French—French Colonies of Africa); *Communauté Financière Africaine* (currency)

CFAA Circus Fans Association of America

CF&AC California Food and Agricultural Code

c factor cleverness factor

CFAD Commander Fleet Air Defense

CFADC Canadian Forces Air Defence Command; Controlled Fusion Atomic Data Center

cfae contractor-furnished aerospace equipment

CFAE Council for Financial Aid to Education

CFAE *Centre de Formation en Aérodynamique Expérimentale* (French—Training Center for Experimental Aerodynamics)

CFAF California Financial Aid Form

CFA franc unit of currency in Benin, Burkina Faso, Cameroon, Central African Republic, Chad, Comoros, Republic of Congo, Gabon, Ivory Coast, Niger, Senegal, Togo

CFAITC California Foundation for Agriculture in the Classroom

CFAL Current Food Additives Legislation

CFANS Canadian Forces Air Navigation School

CFAP Canadian Foundation for the Advancement of Pharmacy

cfar constant false alarm rate

CFAR Chicago for AIDS Rights

CFAS Curriculum Frameworks Assessment System

CFAT Carnegie Foundation for the Advancement of Teaching

CFAW Canadian Food and Allied Workers

cfb cipher feedback

cfb (CFB) circulating fluidized bed (combustion)

CFB California Farm Bureau; Canadian Forces Base; Center for Family Business; Commonwealth Forestry Bureau; Consumer Fraud Bureau; Council of Foreign Bondholders

cf black conductive furnace black

CFBS Canadian Federation of Biological Societies

CFBT Canadian Forces Base Toronto

cfc campus-free college; capillary filtration coefficient; colony-forming cells; complex facility console; contained financial contraction

cfc(s)(CFC(s)) chlorofluorocarbon

cf & c cost, freight & commission

CFC Canadian Forestry Corps; Catholics for a Free Choice; Chartered Financial Counselor; Christian Fellowship Church; Chrysler Financial Corporation; Citizens for a Free Cuba; Combined Federal Campaign (USA); Committee for a Free China; Consolidated Freight Classification

CFC-113 chlorofluorocarbon

CFCA Canterbury Farmers Cooperative Association; Communications Fraud Control Association

cfcb card feed circuit breaker

CFCC Canadian Forces Communications Command

CFCF Central Flow Control Facility

cfd control functional diagram; cubic feet per day

CFD Consumer Fraud Division

CFDA Catalog of Federal Domestic Assistance; Council of Fashion Designers of America

CFDC Canadian Film Development Corporation

CFDD Contract Furniture Dealer Division

CFDM Cognizant Functional Division Manager

CFDT *Confederation Française et Democratique du Travail* (French—French Democratic Confederation of Labor)

CFDTS Cold-Flow Development Test System (AEC)

cfe complex features (realty); contractor-furnished equipment

CFE Canadian Forces Europe; Central Fighter Establishment; College of Further Education; Conventional Forces in Europe; Corps of Fiji Engineers

CFE *Comisión Federal de Electricidad* (Spanish—Federal Electricity Commission)

CFEME Canadian Forces Environmental Medicine Establishment

CFESA Commercial Food Equipment Service Association

cff computer forms feeder; counter flip-flop; critical flicker frequency; critical fusion frequency

Cff Cardiff

CFF Cat Fanciers Federation; Commission for the Future; Compensatory Financing Facility

CFF *Chemin de Fer Fédéraux* (French—Swiss Federal Railroad)

CFFA Chemical Fabrics and Film Association

cffc counterflow film cooling

CFFC Catholics For a Free Choice

cfg constant frequency generator; cubic feet of gas

CFG Camp Fire Girls

cfgd cubic feet of gas per day

cfgh cubic feet of gas per hour

cfgm cubic feet of gas per minute

cfh cubic feet per hour

CFH Council on Family Health

CFHO Canada-France-Hawaii Observatory

CFHQ Canadian Forces Headquarters

CFHS Canadian Federation of Humane Societies

CFIIT Canada-France-Hawaii Telescope

cfi cost, freight, and insurance

CFI Canadian Film Institute; Canadian Forest Inventory; Chief Flying Instructor; Committee on Foreign Intelligence (CIA); Corporate Financial Instruction; Counselor Function Inventory; Court of First Instance

CF & I Colorado Fuel and Iron

CFI Corporación Financiera Internacional (Spanish—International Finance Corporation)—IFC

CFIA Cavity Foam Insulation Association; Center for Independent Action; Component Failure Impact Analysis

CFIAB Canadian Federation of Insurance Agents and Brokers

CFIC Canned Food Information Council

CFIEM California Forces Institute of Environmental Medicine

CFIL Connections for Independent Living

C-FIIE Coalition for Immigration Law Enforcement

CFIP Chamber of Furniture Industries of the Philippines

CFIT Culture Fair Intelligence Test(ing)

CFJS Center For Judicial Studies

cfl cease-fire line; compact fluorescent lamp; context-free language

CFL Canadian Football League; Carnegie Free Library; Consolidated Fertilizers Limited; Container Fleets Limited

CFL Chemins de Fer Luxembourgeois (French –Luxembourg State Railways)

cflg counter flashing

cfm chlorofluoromethane; confirm; confirmation; confirmed; cubic feet per minute; cubic feet per month

CFM Cadet Forces Medal; Canterbury Frozen Meat (New Zealand); Council of Foreign Ministers

CFMA Central Financial Management Activities; Construction Financial Management Association

CFMC Consumer-Farmer Milk Cooperative

CFMS Canadian Forces Medical Service

CFMUA Cotton Fire and Marine Underwriters Association

Cfn. Craftsman

CFN Compagnie France-Navigation

CFNI Caribbean Food and Nutrition Institute

CFNP Community Food and Nutrition Programs

cfo calling for orders; chlorinated flour oil; coast for orders

Cfo Channel for orders; Coast for orders

CFO Chief Financial Officer; Chief Fire Officer; Commonwealth Fisheries Offices; Complex Facility Operator

CFOA Chief Fire Officers Association

CFOL Corporate Files on Line

cfp certified financial planner; chartered financial planner; cold frontal passage; computer forms printer; contractor-furnished property; cystic fibrosis of the pancreas

CFP Certified Feldenkrais Practitioner; Common Fisheries Policy; Consumer Fraud Protection

CFP Colonies Française du Pacifique (French—French Colonies of the Pacific); *Communauté Française de Pacifique* (French—French Community of the Pacific); *Compagnie Française des Petroles* (French—French Petroleum Company); *Cours du Franc Pacifique* (French—French Pacific francs)

CFPAE Council of Food Processors Association Executives

CFPC College of Family Physicians of Canada

CFPF Central Food Preparation Facility (USA)

CFPO Center for Populations Options

CFPO Compagnie Française des Phosphates de l'Océanie (French—French Oceana Phosphates Company)

CFPS Captain Fisheries Protection Squadron; Central Food Preparation System (USA)

CFPTS Coalition For Peace Through Strength

cfr catastrophic failure rate; chauffeur; crash fire rescue

cfr confronta (Italian—compare)

CFR Center for Future Research; Code of Federal Regulations; Contact Flight Rules; Coorong Fauna Reserve (South Australia); Council on Foreign Relations

CFRC Canadian Forces Recruiting Centre

CFR engine Cooperative Fuel Research (Council) engine (for measuring quality of fuels)

cfrg carbon-fiber-reinforced glass

cfrgc carbon-fiber-reinforced glass ceramic

cfrp carbon fiber reinforced plastic

CFRPA California Fire Rescue and Paramedic Association

CFRS Central Fisherics Research Station

cfs chronic fatigue syndrome;, completely-finished sets; cubic feet per second

cf's confessions of fornication (colonial-style abbreviation originating in Massachusetts and used before the American Revolution)

CFS Canadian Forestry Service; Central Federal Savings; Central Flying School; Container Freight Station; Contract Field Service; Council of Fleet Specialists

CFS Chemins de Fer Fédéraux Suisses (French—Swiss Federal Railways)

CFSA College Food Service Association

CFSAN Center for Food Safety and Applied Nutrition (FDA)

CFSC Canadian Forces Staff College

CFSP Common Foreign and Security Policy

CFSPL Canadian Forces Special Projects Laboratory

CFSR Commission on Financial Structure and Regulation (White House); Contract Funds Status Report

CFSTI Clearinghouse for Federal Scientific and Technical Information

cft clinical full time; combat fitness test; complement fixation test; craft; craftsman; cubic feet; cubic foot

CFT California Federation of Teachers; Colorado Federation of Teachers; Concept Formation Test; Cooperative Field Test(ing)

CFT Compagnie Française de Télévision (French—French Television Company)

CFTA Cattle Food Trade Association

CFTAU Canadian Friends of Tel Aviv University

cftb controlled-flight test bed

CFTB Commonwealth Forestry and Timber Bureau

CFTC Commodity Failures Trading Commission; Commodity Futures Trading Commission; Commonwealth Fund for Technical Cooperation

c-f tests complement-fixation tests

CFTH Compagnie Française Thomson-Houston

cftmn craftsman

CFTR Citizens For The Republic

cfts captive firing test set(s)

CFTSD Canadian Forces Technical Services Detachment

cfu colony-forming units

CFU Central Functional Unit; Commonwealth Film Unit; Consumer Fraud Unit

CFUA Croatian Fraternal Union of America

cfv cavalry fighting vehicle; conventional friend virus

cfvd constant-frequency variable dot

CFWA Canadian Fruit Wholesalers Association

CFWI County Federation of Women's Institutes

CFWIS Central Fighter Weapons Instructor School

cfy clarify

CFZ Contiguous Fisheries Zone

cg cardiogreen; center of gravity; centigram; cerebral ganglion; choking gas (phosgene); chorionic gonadotropin; chronic glomerulonephritis; colloidal gold; complete games; comer guard

c/g coincidence guidance

c of g center of gravity

cg Zentigram (German—centigram)

CG Captain General; cargo glider aircraft (DoD symbol); Central of Georgia (railroad); Certified Genealogist; Chaplain General; Coast Guard; Coldstream Guards; Commandant-General; Commanding General; Commissary General; Congo (Internet code); Connecticut General (Life Insurance Company); Consul General; Covent Garden; guided-missile cruiser (naval symbol)

CG (ROH) Covent Garden (Royal Opera House)

C of G Central of Georgia (railway); College of Guam (Agaña)

CG Consumer Guide; Croix de Guerre (French—War Cross)

C G cassa grande (Italian—bass drum)

cga cargo (proportion of) general average; catabolite gene activator; color graphics adaptor

CGA Canadian Garrison Artillery; Canadian Gas Association; Cape Garrison Artillery; Coast Guard Auxiliary; Coast Guard Academy; Commonwealth General Assurance; Compressed Gas Association; Corcoran Gallery of Art

CGAC Coast Guard Air Corps

CGADC Commanding General, Air Defense Command

CGAIRFMLANT Commanding General, Air Fleet Marine Force, Atlantic

CGAS Coast Guard Air Station; Cornell Guggenheim Aviation Safety Center

CGB Canadian Geographic Board

CGBR Central Government Borrowing Requirement

cgc ceramic gold coating; critical grid current

CGC Certified Good Citizen; Chinese Grandmother Carryon (luggage); Coast Guard Cutter; Continental Grain Company

CGCARC Commanding General, Continental Army Command

CG circuit common-gate amplifier for field-effect transistors

cgd chronic granulomatous disease

cgd (CGD) cow grazing day (13.5 kg pasture matter or feed in average bale of hay)

cge carriage

CGE Central Government Expenditures

CG & E Cincinnati Gas and Electric Company

CGE Compagnie Générale d'Electricité (French—General Electric Company)

CGEC Committee on Global Ecology Concern (U.S.—USSR)

cge fwd carriage forward

CGEL & PB Consolidated Gas, Electric Light and Power Company of Baltimore

C Gen Consul General

cge pd carriage paid; charge paid

cgf center of gravity factor; chemotaxis-generating factor; coarse-glass frit

CGF College of Great Falls

CGFA Columbus Gallery of Fine Arts; Consolidated Gold Fields of Australia

CGFMFLANT Commanding General, Fleet Marine Force, Atlantic

cgfp calcined gross fission product

CGFSA Consolidated Gold Fields of South Africa

cgg continuous grinding gage

CGG Canadian Grenadier Guards

cgh computer-generated hologram

cgh (CGH) chorionic gonadotrophic hormone

CGH Cape of Good Hope (medal); São Paulo, Brazil (Congonhas Airport)

C of GH Cape of Good Hope

CGHB Cape of Good Hope Bank

CGHSB Cape of Good Hope Savings Bank

cgi computer-generated imagery; computer graphics interface; corrugated galvanized iron; cruise guide indicator

CGI Capital Guaranty Insurance; Chief Ground Instructor; Chief Gunnery Instructor; City and Guilds of London Institute

CGIAR Consultative Group on International Agricultural Research

CGIC Comisaria General de Investigación Criminal (Spanish—Commissariat General of Criminal Investigation)—Spain's Interpol office

CGIL Confederazione Generale Italiana del Lavoro (Italian—Italian General Confederation of Labor)

C-girl call girl (prostitute); hundred-dollar girl

cgit compressed-gas-insulated tube

C G Jung Foun C G Jung Foundation for Analytical Psychology

cgk grid cathode capacitance

cgl center-of-gravity locator; continuous-gas laser; controlled ground landing; corrected geomagnetic latitude (CGL)

cgl (CGL) chronic granulocytic leukemia

CGL Canadian Gulf Line; Central Gulf Lines

CGL Confederazione Generale del Lavoro (Italian—General Confederation of Labor)

CGLAT Cassel Group Level of Aspiration Test

CGLI City and Guilds of London Institute

cg lkr cleaning gear locker

cgm centigram(s); ciliated groove to mouth; computer graphics metafile

cgm (CGM) central gray matter

CGM Conspicuous Galantry Medal

CGMA Covent Garden Market Authority

CGMIS Commanding General's Management Information System

cGMP cyclic GMP

CGMW Commission for the Geological Map of the World

cgn chronic glomerulonephritis

Cgn Cartagena, Colombia (British maritime abbreviation) (see Ctg)

CGN Cologne, Germany (airport); nuclear-powered guided-missile cruiser (naval symbol)

CGNM Casa Grande National Monument

cgo cargo

Cgo Chicago

CGO Committee on Government Operations; Connecticut Grand Opera

Cgo Chkr Cargo Checker

cg/oq cerebral glucose oxygen quotient

CGOT Canadian Government Office of Tourism

CGOU Coast Guard Oceanographic Unit

cgp choline glycerophosphatide; chorionic growth hormone prolactin; circulating granulocyte pool; grid plate capacitance

CGP College of General Practitioners; Comparative Guidance and Placement Program

CGP Current Geographical Publications

CGPM Conférence Générale des Poids et Mesures (French—General Conference of Weights and Measures)

CGPN Catalog of Galactic Planetary Nebulae

CGPP Comparative Guidance Placement Program

CGPS Canadian Government Purchasing System

CGPSq Cartographic and Geodetic Processing Squadron (USAF)

cgr captured gamma ray; crime on government reservation

CGR Canadian Garrison Regiment

CGRA Canadian Good Roads Association; Chinese Government Radio Administration (Taiwan)

CGRDO Coast Guard Radio

CGRLS Coast Guard Radio Liaison Station

CGRM Commandant General—Royal Marines; Commando Group Royal Marines

cgs centimeter gram second

CGS Canadian Geographical Society; Central Glider School; Central Gulf Steamship (corporation); Central Gunnery School; Chief of General Staff; Coast and Geodetic Survey; Council of Graduate Schools (in the U.S.)

C & GS Coast and Geodetic Survey

CGSA Carriage of Goods by Sea Act; Computer Graphics Structural Analysis

CGSAC Commanding General, Strategic Air Command

CGSB Canadian Government Specifications Board

C & GSC Command and General Staff College

cgse centimeter-gram-second electrostatic

cgsfu ceramic glazed structural facing units

cgsm centimeter-gram-second-electromagnetic

CGSS Cryogenic Gas Storage System

CGSSC Columbia Gas Service Corporation

CGST Caribbean Graduate School of Theology

CGSTC Centro Giovanile Scambi Turistici e Culturali (Italian—Youth Center for Tourism and Culture)

cgsub ceramic glazed structural unit base

CGSUS Council of Graduate Schools in the United States

cgt capital gains tax(ation); chorionic gonadotropin; combustible gas tracer; gains tax(ation)

cgt (CGT) corrected geomagnetic time

CGT Compagnie Générale Transatlantique (French Line); Confederación General del Trabajo (Spanish—General Confederation of Labor); Confederation du Travail (French—General Confederation of Labor)

CGTA *Companie Générale de Transports Aériens* (French— Air Algeria)

CGTAC Commanding General, Tactical Air Command

CGTB Canadian Government Travel Bureau

CGTEL Coast Guard Teletype

CGTSF *Compagnie de Télégraphie San Fils* (French wireless company)

cgtt (CGTT) cortisone glucose tolerance test

cgtv (CGTV) command guidance test vehicle

cgu ceramic glazed units

CGU Canadian Geophysical Union

CGUSACE Commanding General, United States Army Corps of Engineers

CGUSACOMZEUR Commanding General, United States Army, Communications Zone, Europe

CGUSARMC Commanding General, United States Army Materiel Command

CGUSCONARC Commanding General, United States Continental Army Command

CGUSFET Commanding General, United States Forces— European Theater

cgv critical grid voltage

cgvs ciliated groove to ventral sac

CGW Chicago Great Western Railway; Coast Guard Women

Cgy Cagayan de Oro

ch case harden; center halfback (field hockey); chain; chain home (radar); champagne (color); champion; chancellor; chancery; change; chapter; chest; chief; child; chipped; choke; choline; church; coat hook

ch (CH) critical hours (when broadcast signals can cause interference)

ch. children

c/h cards per hour

c & h cocaine and heroin; cold and hot

ch *chambre* (French—room); *château* (French—wine estate); *cheque* (French, Portuguese, or Spanish—check)

ch. *chori* (Latin—choruses)

Ch Chad (whose capital is N'Djamena); Chancery; Channel; Chile; Chilean; China; Chinese; choreographer; Christchurch (New Zealand or Oxford or other); church; Clearinghouse

Ch. Chaplain

Ch. *Chirurgiae* (Latin—surgery)

CH Camp Hospital; Carnegie Hall; Chaucer (Geoffrey Chaucer); Chicago Helicopter (airways); Chiracahua Mountains; College Heights; Companion of Honor; compass heading; concentration of hydrogen ions in moles per liter (symbol); Court House; Custom House; Switzerland (autoplaque and Internet code)

C-H Crouse-Hinds;Cutler-Hammer

C.H. Companion of Honour

C and H California and Hawaiian Sugar Company

CH *Church History; Confederatio Helvetico* (Latin—Swiss Confederation)

CH₂O formaldehyde

CH₃COOH acetic acid

CH-46 Boeing-Vertol twin-rotor helicopter called Sea Knight

CH-47 Boeing-Vertol helicopter called Chinook

CH-53 Sikorsky heavy-assault helicopter called Sea Stallion

CH-54 Sikorsky crane helicopter called Sky Crane or S-64

CH-113 Canadian version of Boeing-Vertol helicopter designated CH-46 and called Labrador

cha cable-harness analyzer; congenital hypoplastic anemia; cyclohexylamine

cha (CHA) cyclohexylamine

Cha Chamaeleon (constellation); Chamber maid; Charles

CHA California Hospital Association; Canadian Horse Artillery; Catholic Health Association; Catholic Hospital Association; Chattanooga, Tennessee (airport); Chicago Helicopter Airways; Child Health Association; Community Health Association; County Health Authorities

CHABA Committee on Hearing and Bio-Acoustics (US Army)

chabak *chabakano* (Philippine Spanish dialect)

C-habit cocaine habit

Chaco Canyon Chaco Canyon National Monument near Bloomfield, New Mexico

chacom chain of command

chad code to handle angular data

Chad Chadburn; Republic of Chad (landlocked North African country) *République du Tchad*

CHAD Combined Health Agency Drive

CHADS Chicago Air Defense Sector

CH AE Chief Artificer Engineer

Chafarinas Chafarinas or Zafarinas Islands (in the Spanish Mediterranean off Morocco and southeast of Melilla)

CHAFB Chanute Air Force Base

chaffroc (CHAFFROC) chaff rocket

chag compact high-performance aerial gun

Chagos Chagos Archipelago

CHAIN California Housing, Action, and Information Network

Chair Chairman

Chairp Chairperson

chal challenge; chalumeatu (ancient clarinet)

chal *chaleur* (French—heat, warmth)

Chald Chaldean

CHALFA Charlottesville/Albemarle Foundation for the Encouragement of Artists

chalicos chalicosis (sickness caused by metallic-dust inhalation)

chalk calcium carbonate ($CaCO_3$)

cham chamfer; champagne; champion; combustion, heat, mass

Cham *Chamaeleon* (Latin— Chamaeleon constellation)

chamb chamber

Chamb Chamberlain

Chamb Ency *Chambers's Encyclopaedia*

Chamber Chamber of Deputies

Chambers Chambers Dictionary of Science and Technology

Chambly Girls Girls' Cottage School (for delinquents) at Chambly, Québec

chammy (English slang—champagne)

champ champion(ship)

Champ Beauchamp

CHAMP Character Manipulation Procedure(s); Civilian Health and Medical Program; Community Health Air Monitoring Program

Champ Intl Champion International

champion. compatible hardware and milestone program for integrating organizational needs

Champs Champs Elysées (French—Elysian Fields)—main boulevard of Paris

CHAMPUS Civilian Health and Medical Program for the Uniformed Services

CHAMPVA Civilian Health and Medical Program of the Veterans Administration

chan channel

chan. chancery

Chan Channel

Chanc Chancellor; Chancery

CHANCE Complete Help and Assistance Necessary for College Education

CHANCOM Channel Command (NATO)

CHANCOMTEE Channel Committee (NATO)

Chandeleurs Chandeleur Islands of Louisiana

'change exchange; produce exchange; stock exchange

'Change Royal Stock Exchange in London

Chan Isl Channel Islands (Alderney, Guernsey, Jersey, Sark)

Channel The Channel (Beagle, Bristol, English, Old Bahama, Saint George's, Santa Barbara, Ten Degree)

Channel Islands Park islands off California

Channels Channel Islanders; Channel Islands

chans chanson (French—song)

CHANSEC Channel Committee Secretary (NATO)

Chao Phraya Krung Thep (Bangkok) river

CHAOS Committee for Halting Acronymic Obliteration of Sense; Consortium for the Hastening of the Annihilation of Organized Society

CHAOTIC Computer-and-Human-Assisted Organization of a Technical Information Center (NBS)

chap. chapter; cychophosphamide, hexamethylmelamine, doxorubicin and cisplatin (chemotherapy for ovarian cancer)

Chap Chaplain

CHAP Certified Hospital Admissions Program; Charring Ablation Program (NASA); Child Health Assistance Program

chapa chaparral

Chap(pie) Chapin; Chapman; Chappell

chaps. chaparajos (Spanish—open-backed leather overalls)

CHAPS Children Have A Potential Society; Clearing Houses Automated Payments System (trademark)—London; contractor-held Air Force property

Chapter 7 liquidation

Chapter 11, etc. bankruptcy—Chapter 11, etc., of the Bankruptcy Act of the U.S.

char character; characteristic; charcoal; charwoman

char (CHAR) character (data processing)

Char Charter

Char Amal Charlotte Amalie

Charbray Charolais-Brahman cattle

charc charcoal

char-flip charitable family limited partnership

Chard. Chardonnay

Charl Charlottenburg

Charles University University of Prague

Charley Charles

Charlotte Corday Marie-Anne-Charlotte Corday d'Armont

Charm Charmian

char reac character reaction (sometimes simply cr)

chars characters

char(s) charwoman; charwomen

chart. charta (Latin—paper)

chart. bib. charta bibula (Latin—blotting paper)

chart. cerat. charta cerata (Latin—waxed paper)

chartul. chartula (Latin—small paper)

Char X Charing Cross (rail terminal)

chas chassis

Chas Charles

CHAS Catholic Housing Aid Society

chase. cut holes and sink 'em (navalese acronym for sinking old ammunition cases or obsolescent barges or boats)

Chase Chase Manhattan Bank

Chasn Charlestown

Chat château (French—castle)

Chat Choo-Choo Chattanooga Choo-Choo (restaurant)

chat mtg chattel mortgage

Chat(ty) Charlotte

Ch^{au} Chateau (French—castle, country mansion)

Chauc Geoffrey Chaucer

chaud chemical audit

chauf chauffeur

Chávez Jorge Chávez International Airport of Lima, Peru

chb complete heart block

Chb Cherbourg; Chiba

Ch B Chief of Bureau

CH B Chief Boatswain

Ch.B. Chirurgiaé Baccalaureus (Latin—Bachelor of Surgery)

chbd chalkboard

CHBI Commission on Health Benefits and Integration

ChBuAer Chief of the Bureau of Aeronautics

ChBuDocks Chief of the Bureau of Yards and Docks

ChBuMed Chief of the Bureau of Medicine and Surgery

ChBuOrd Chief of the Bureau of Ordnance

ChBuPers Chief of the Bureau of Naval Personnel

ChBuSanda Chief of the Bureau of Supplies and Accounts

ChBuShips Chief of the Bureau of Ships

ChBuWeps Chief of the Bureau of Weapons

chc choke coil

Ch of C Chamber of Commerce

CHC Chicago House of Correction; Christchurch, New Zealand (airport); Community Health Council; Comprehensive Health Center; Congressional Hispanic Caucus

ch cab china cabinet

CHCF Component Handling and Cleaning Facility

Chch Christchurch

Ch Ch Christ Church College, Oxford

CHC$_{13}$ Chloroform

CHCMD Chicago Contract Management District

CH CR Chief Carpenter

chd chaldron; childhood disease(s); congestive heart disease; coronary heart disease

Ch D Charles Darwin

Ch.D. Chirurgiae Doctor (Latin—Doctor of Surgery)

C-H d Chediak-Higashi disease

CHD Charles Halliwell Duell

Ch d'A Chargé d'Affaires

ch de f chemin de fer (French—railroad)

CHDF Civilian Home Defense Force

chdl computer hardware description language

chdm cyclohexanedimethanol

CHDP Child Health and Disability Prevention

CHDR Center for Health-care Dispute Resolution

che cholinesterase

che (CHE) channel end(ing)

c & he consumer and home-making education

Che Chetverg (Russian—Thursday); Ernesto (Che) Guevara

Ch^e Chapelle (French—Chapel)

C^{he} Chaine (French—chain)

Ch E Chief Engineer

Ch.E. Chemical Engineer

CHE Chete Game Reserve; Chewore Game Reserve; Chizarira Game Reserve

C-head coke head (slang—cocaine addict)

Cheaha Cheaha Mountain or Cheaha State Park south of Anniston, Alabama

CHEAP Computerized Health Education Assessment Program

CHEAR Council on Higher Education in the American Republics

Cheb. Chebeague Island (in Casco Bay off coast of Maine)

chec checked; checkered

CHEC Citizens Helping Eliminate Crime; Commonwealth Human Ecology Council

Chech Checheno-Ingush Republic (whose capital is Groznyy)

Checo Checoslovaquia (Spanish—Czechoslovakia)

Chee Chee-chee (Anglo-Indian dialect)

cheesesan cheese sandwich

cheesewich cheese sandwich

CHEF Contour-clamped Homogeneous Electric Field

Cheka Chrezvychainaya Kommissiya po Borbe s Kontrrevolutisiei i Sabotzhem (Russian—Extraordinary Commission for Combating Counterrevolution and Sabotage)—original Soviet Secret Police founded December 20, 1917, at Lubianka Prison in Moscow (q.v.—VOT)

CHEL Cambridge History of English Literature

Chelm (ancient Jewish town in Poland known in folklore as the Town of Fools); short form for Cheltenham

Chelon Chelonia

cheloniol cheloniologic(al) (ly); cheloniologist; cheloniology

chelons chelonians (tortoises, terrapins, turtles)

Chelt Cheltenham

chem chemical; chemist; chemistry

Chem Chemistry

chemanal chemical analysis

Chem E Chemical Engineer(ing)

Chem.E. Chemical Engineer

Chem Econ Chemical Economic Services

Chem Ed Chemical Education Publishing Company

Chem Educ Chemical Education Publishing Co

Chem Elements Pub Chemical Elements Publishing Co

chem etch chemically etched; chemical etching

CHEMI Chemical Engineering Modular Instruction

chemly chemically

Chem & Met Eng Chemical and Metallurgical Engineering

chem mill chemically milled; chemical milling

chemo chemotherapist; chemotherapy

chemonuc chemonuclear

chemos chemosphere; chemospheric(al)(1y)

chemosens chemosensory

chemoster chemosterilant; chemosterilization; chemosterilize(d)

chemosurg chemosurgical(ly); chetnosurgery

chemotax chemotaxonomic(al)(1y); chemotaxonomist; chemotaxonomy

chemoth chemotherapy

Chem Phys Chemical Physics

Chem Phys Lett Chemical Physics Letters

Chem Pub Chemical Publishing Company

Chem Rev Chemical Reviews

Chem Rubber Chemical Rubber Company

CHEMS Chemical Education Materials Study

chemsearch chemicals selected for equal, analogous, or related characters

chemsol chemical solution (for decontamination)

CHEMTREC Chemical Transportation Emergency Center

chem war. chemical warfare

CHEN Chail Nashim (Hebrew—Women's Force of the Israeli Army); chen is the Hebrew word for grace

CHEOPS Chemical Operations System

CHEP Commonwealth Handling and Equipment Pool

Cher Cherilyn; Cherilyn Sarkisian

chert ironstone sedimentary rock

Cherv Cherville; Chervin

Ches Cheshire

Cheskey(s) Czechoslovakian(s)

chesky cherry-flavored whiskey

CHESS Center for Health Systems Research and Analysis; Comprehensive Health Enhancement Support System

chester(s) (Early English— city, old fortification, town)— short form for such places as Manchester, Winchester, and even Tadcaster and Worcester

Chet Chester

chev chevron

Chev Chevalier (French— Knight)

Cheviots Cheviot Hills between England and Scotland

Chevron Standard Oil of California

Chev(y) Chevrolet

Chewko Chewing Tobacco Company

Chey Cheyenne

chf congestive heart failure; critical heart flux

Chf Chief; Crimean hemorrhagic fever

Ch F Chaplain of the Fleet

CHF Carnegie Hero Fund; Coalition for Health Funding

CHFA California Housing Finance Agency

ch-factor chutzpah factor (degree of guts or nerve)

Chf Bkr Chief Baker; Chief Bookkeeper

ChFC Chartered Financial Consultant

CHFC Carnegie Hero Fund Commission

Chf Engr Chief Engineer

Chf Libr Chief Librarian

Chf M Sgt Chief Master Sergeant

Chf Off Chief Officer

Chf Pur Chief Purser

Chf Stwd Chief Steward

Chf Surg Chief Surgeon

ch fwd charges forward

Chf Wt Ofcr Chief Warrant Officer

chg change; charge

Chg Chittagong

CHGC Committee for Hand Gun Control

chgd charged

Chgo Chicago

chgph choreographer; choreographic; choreography

chg pl change plane

CH GR Chief Gunner

chgs charges

chh cartilage-hair hypoplasia

CH & H Continent between Havre and Hamburg

chi clutch hitting index; specific magnetic susceptibility

Chi Chicago; Chichester; China; Chinese

Chi. China (whose capital is Beijing)

CHI Catastrophic Health Insurance; Chicago; Chicago White Stockings (National Association); Crouse-Hinds (stock-exchange symbol)

Chia Chiapas

CHIA Canadian Health Insurance Association

CHIAA Crop-Hail Insurance Actuarial Association

chic cermet hybrid integrated circuit; compact high intensity cooler

Chic Chicago

Chicagorican Chicago Puerto Rican

Chicano (diminutive nickname for *Mexicano*)

CHICC Community Health Information Classification and Coding

Chich Chichester

chick. chicken

Chick Chickering

chickensand chicken sandwich

chickenwich chicken sandwich

chick(s) chicken(s)

Chico Francisco

Chicom Chinese communist

Chicos Chinese communists

Chi$ Chilean peso

Chidic Chinese dictionary

Chief Chief Engineer

CHIEF Controlled Handling of Internal Executive Functions; Customs Handling of Import and Export Freight

Chih Chihuahua (inhabitants— Chihuahuenses, chihuahua dogs characteristic of this area—*chihuahueños*)

Chih. Chihuahua (Mexican state or small dog)

chil children('s)

Chil Chile (whose capital is Santiago)

CHI-LAX Chicago—Los Angeles

Chil Cur Chilean Current

child (CHILD) children having individual learning difficulties

child. computer having intelligent learning and development

Children's Children's Crusade (in 1212 when 90,000 children from France and Germany set out to free the Holy Land); Children's Opera [Hansel and Gretel by Engelbert Humperdinck (1854– 1921)]; Children's Hospital

Chile Republic of Chile, *República de Chile*

chilidog chile-con-carne sauced hotdog (frankfurter)

chili(es) chili pepper(s)

Chillicothe Institute Chillicothe Correctional Institute at Chillicothe, Ohio

Chillicothe School Training School for Girls at Chillicothe, Missouri

Chilterns Chiltern Hills of England

Chilton Chilton Book Company

chim chimica (Italian—chemistry); *chimie* (French—chemistry)

Chimbo Chimborazo, Ecuador; Chimbote, Peru

Chi Met Chicago Metropolitan Correctional Center

CHI-MIA Chicago-Miami

chimponaut chimpanzee astronaut (primate used in space travel experiments)

chimp(s) chimpanzee(s)

chin. chinchilla

chin chinesisch (German—Chinese)

Chin China; Chinese

Chin Chinese (world's leading language in terms of numbers)

CHIN Canadian Heritage Information Network

China People's Republic of China (communist-controlled mainland); Republic of China (nationalist island of Taiwan once known as Formosa plus Matsu, Quemoy, and the Penghus or Pescadores

CHINALIGHT China National Light-Industry Products Import and Export Corporation (mainland China)

China Nac China Nacionalista (Spanish—Nationalist China)—offshore China also known as Formosa or Taiwan

China Sea(s) East China Sea and South China Sea

Chinat Chinese nationalist

CHINATEX China National Textiles Import and Export Corporation (mainland China)

CHINATIVE China National Native Produce (mainland China)

Chi Nats Chinese Nationalists

CHINATUHSU China National Trading Corporation (mainland produce and animal by-products)

Chinese restaurant syndrome monosodium glutamate (msg) symptoms first associated with eating Chinese food containing msg, resulting in chest pain, dizziness, headache, and numbness

Chin J Phys Chinese Journal of Physics (*Acta Physica Sinica—Wuli Xuebao*)

CHINKUNG China National Trading Corporation for Light Industrial Products (mainland China)

chins. children in need of supervision

Ch Insp Chief Inspector

Chinsyn Chinese-English synthesis-oriented machine translation system

chinu chinook (warm, dry wind on the lee slope of a mountain or mountain range)

CHI-NY Chicago—New York

Chios English equivalent of Khios island in the Aegean

Chip Chipre (Portuguese or Spanish—Cyprus)

CHIP Chieftain and Challenger Improvement Program; Children's Health Insurance Plan; Community Housing Improvement Programme (New Zealand)

CHIPDis Chicago Procurement District (US Army)

Chipitt Chicago-to-Pittsburgh (complex of cities)

chippy fish-and-chip shop

Chippy Chipping Norton, England

chips children in need of protection and services

chips. coherent high-intensity photon source

Chips ship's carpenter

ChiPs California Highway Patrol cops

CHIPS Chemical Engineering Information Processing System; Clearing House Interbank Payment System; Clearing Houses Interbank Payments System (trademark)—New York

chir chiropody

chir chirurgia (Italian—surgery)

Chir. Doc. Chirurgiae Doctor (Latin—Doctor of Surgery)

Chiricahua Chiricahua National Monument in southeastern Arizona

Chiricahuas Chiricahua Mountains of Arizona

Chiricano(s) Panamanian(s)

chiro chirography; chiropractic; chiropractor

CHIRP Center to Help Instill Respect and Preservation; Community Housing Improvement and Revitalization Program

Chis Chiapas (inhabitants—Chiapanecos)

CHI-SAN Chicago—San Diego

CHI-SEA Chicago—Seattle

CHI-SFO Chicago—San Francisco

Chish Chisholm

Chisox Chicago White Sox (baseball team)

Chi Sym Chicago Symphony

chit chitty (Hindustani—voucher signed to cover small debts for drinks, food, tobacco, etc.)

Chita Conchita

Chi-Trib Chicago Tribune

chiv chivalry

chix chickens

Ch J Chief Justice

CHJM Carnegie Hall— Jeunesses Musicales

CHJMKHK Chung-Hua Jen-Min Kung-Ho Kuo (People's Republic of China)

chk check

chkpt checkpoint

chkr checker

chl chain home low (radar); chloroform; confinement at hard labor

Chl Chalna

CHL Central Hockey League

chlamy chlamydia (sexually transmitted bacterial infection)

Chla Vsta Chula Vista

chlb chlorobutanol

Ch Lbr Chief Librarian

chldn. children

ch-lkr chiffonier-locker

Ch^{lle} Chapelle (French—Chapel)

chlor chloride; chlorination; chlorine

chloro chloroform; chlorophyll; chloroprene

chloro chlorus (Greek—green)—chlorine, chlorophyll

chloroform trichloromethane (CHC13)

chloroprene synthetic rubber (C_4H_5Cl)

chlw commercial high-level waste

chm chamber; checkmate

Chm Chairman; Chairwoman; Choirmaster; Choirmistress

Ch.M. Chirurgiae Magister (Latin—Master of Surgery)

CHM Cleveland Health Museum; Community of the Holy Myrrhbearers

C-H M Cooper-Hewitt Museum

CH MAA Chief Master At Arms

chmbl center high-mounted brake light

CHMC Children's Hospital Medical Center (Boston)

CHMDDA Cooper-Hewitt Museum of Design and Decorative Arts

ch-mir chiffonier-mirror

CHMK Chung-Hua Min-Kuo (Republic of China)

chmn chairman

ChMNH Chicago Museum of Natural History

chmp chairperson

Chn Cochin

chn. children

Chn China (Spanish abbreviation)

CHN College of the Holy Name; Community of the Holy Name

C-H-N carbon, hydrogen, nitrogen, oxygen, phosphorus, sulfur (compounds)

CHNAVPERS chief, Naval Personnel

CHNAVSECMAAG Chief, Navy Section, Military Assistance Advisory Group

Chne Chaîne (French—Chain)—mountain range

chngd changed

CHNOPS mnemonic for carbon, hydrogen, nitrogen, oxygen, phosphorus, sulfur

chns chains

CHNS Cape Hatteras National Seashore (Buxton, North Carolina)

CHNSRA Cape Hatteras National Seashore Recreational Area

CHNSY Charleston Naval Shipyard (South Carolina)

CHNT Community Health Nurse Teacher/Tutor

ChNZAgCo China New Zealand Agricultural Consultants

cho (Cho) containers carried in hold

ch/o child of

Cho Chosen (Korea)

CHO Cameron Highlanders of Ottawa; carbohydrate (generalized formula); Christian Haitian Outreach; Community Health Organization

CHOBS Chief Observer (USN)

choc chocolate

chocbar(s) chocolate bar(s)

chocmalt chocolate malted milk

choco chocolate

chocs chocolate candies; chocolate drops; chocolates

CHOD Chief of Defense

choirm choirmaster

choke choke hold (bar hold or carotid hold)

CHOKE Care How Others Keep the Environment

chol cholesterol

chol (Latin prefix—bile)—cholecyst, cholera, cholesterol(ic), cholic

chol est cholesterol esters

Cho Min Inm Kon Chosun Minchu-chui Inmin Konghwaguk (Democratic People's Republic of Korea) North Korea

CHOMPS Canine Home Protection System

Chonos Chonos Islands

CHOP Change of Operational Control

CHOPS Chief of Operations

chor choral; choreographer; choreographist; choreography; chorus; choruses

Chord Chordata

chorégr chorégraphie (French—choreography)

C Horn Cur Cape Horn Current

chortle chuckle and snort

Chou Chou En-lai

cho/vac cholera vaccine

chovr changeover

chovy anchovy

chp child psychiatry; community health plan; comprehensive health plan(ning)

Chp Chepstow

ChP Choiseul Province (Solomon Islands)

Chp Chipre (Spanish—Cyprus)

CHP California Highway Patrol; Chihuahua Pacific (railroad—Ferrocarril de Chihuahua al Pacifico)

CH P Chief Paymaster

CHPA California Highway Patrol Academy

chpae critical human performance and evaluation

ch pd charges paid

Chpn Chairperson

CHPP Cypress Hills Provincial Park (Saskatchewan)

ch ppd charges prepaid

chpx chickenpox

chq cheque

CHq Corps Headquarters

CHQ Company Headquarters

chr character; chrome; chromium; chromobacterium; chronic

c hr candle-hour

chr. christened

Chr Choir; Christ; Christ College, Cambridge; Christian; Church

Chr Chronicles

CHR Commission on Human Rights; Connecticut Hard Rubber (company)

CHRB California Horse Racing Board

Chr Coll Christ College—Cambridge

chrg charge

CHRG Citizens Health Research Group

CHRIE Council on Hotel, Restaurant, and Institutional Education

chris. christened

Chris Christian(a); Christiana; Christina; Christopher

CHRIS Cancer Hazards Ranking and Information System

Chrissie Christina; Christine

Christ. Christian; Christianity; Christmas

Christ Her Christian Herald

christie Christiania turn

Chrlstn Charleston

Chrlstn SC Charleston, South Carolina

chrm chrome

Chrm Chairman; Chairwoman

chrom (Latin prefix—color)—chromatic, chromoplast, chromosome, chromosphere

chromite iron chromate

chromolith chromolithograph(y)

chromo(s) chromolithograph(s); chromosome(s)

chron chronogram; chronograph; chronology; chronometer; chronometry

Chron Chronicle(s)—First Book of Chronicles; Second Book of Chronicles

chrono chronologic(al)(1y); chronology; chronometer; chronometric(al)(1y)

chrono order chronological order

chro pltd chrome plated

Chrp Chairperson

Chrs Chambers; Christians; Churches

CHRS Charming Shoppes Incorporated

Chrys Chrysler

chrysanthemum nationalist symbol of China and Japan; symbol of the Orient Overseas Line

chrysoberyl beryllium aluminate

chrysocolla hydrous copper silicate

chrysoprase chalcedony gemstone

chs chapters; crime on the high seas

Chs Chambers; Charles; Chester

Ch of S Chamber of Shipping

C-H s Chediak-Higashi syndrome

CHS Canadian Hydrographic Service; Charleston, South Carolina (airport); Chicago Historical Society; Childrens Home Society; Citizens for

Highway Safety; Community Health Service (HEW); Community of the Holy Spirit; Confederate Hammer Skins (neo-Nazi group, Dallas, Texas); Connecticut Herpetological Society; Cristobal High School; Curtis High School

CHSA Chest, Heart, and Stroke Association

CH SB Chief Signal Boatswain

CH SCH Chief Schoolmaster

ch'ship championship

Ch Skr Chief Skipper

CHSL Cleveland Health Sciences Library

Ch. Slav. Church Slavic

CHSM China Service Medal

CHSS Children's Hypnotic Susceptibility Scale; Cooperative Health Statistics System

Ch Supt Chief Superintendent

cht cylinder head temperature

Cht Chittagong

chtg charting

CHTNP Chittagong Hill Tracts National Park (Bangladesh)

cht tanks collect, hold, transfer (raw sewage) tanks (used by naval vessels to overcome harbor pollution when in port)

chu (CHU) central heating unit

Chu Centrigrade heat unit

CHU Christelijk-Historische Unie (Dutch-Christian Historical Union)—political party

CHUA Canadian Hail Underwriters Association

chub chubasco (violent storm on west coast of Guatemala and Mexico)

Chub Chubut (Argentine province)

Chuck Charles

Chugach Chugach National Forest in Alaska

Chugaches Chugach Mountains of Alaska

Chukchi Chukchi Peninsula and the Chukchi Sea in the Arctic between Alaska and Siberia where the peninsula is located

Chulajuana Chula Vista-Tijuana area of southwestern-most California and north-westernmost Tijuana

CHUM Computing and the Humanities

Chumley (British contraction—Chalmondeley)

CHUMS Cancer Hopefuls United for Mutual Support; Care and Help for Unmarried Mothers

Chung Chungking

Chung Min Chung-hua Minkuo (Mandarin Chinese—Republic of China)—offshore China

Chunnel Channel Tunnel (under the English Channel)

Chuqui Chuquicamata

Chur Churchill College, Cambridge

Churchill Sir Winston Churchill

chut cable households using tv (audience survey)

'chute parachute

Chuuk Truk (Micronesian)

ch v check valve

chw chilled water; cladding hull waste; cold-and-hot water; constant hot water

CHW Charleston, West Virginia (airport)

CH & W Canadian Health and Welfare

CHWA California Health and Welfare Agency

Chwdn Churchwarden(ess)

chx chiro-xylographic; chlorhexidine

Chx Châteaux (French—wine estates)

chy chimney

C Hy Commission for Hydrology

chyd churchyard

Chy Div Chancery Division

ci cardiac index; cardiac insufficiency; cast iron; cerebral infarction; chemotherapeutic index; clonus index; coefficient of intelligence; colloidal iron; color index; compression ignition; confidential informant; contamination index; contrast index; coronary insufficiency; cost and insurance; counterintelligence; criminal informant; crystalline insulin; cubic inch

ci (CI) clinical investigator; consular invoice; cooperative individual (informant)

c-i criminal-investigation

ci. child

c.i. (C.I.) consular invoice

c/i carriage-to-interference (ratio); configuration interface

c/i (C/I) certificate of insurance

c & i commercial & industrial; cost and insurance; cowboys and indians

Ci cirrus; curie (unit of activity in radiation dosimetry); input capacitance

Ci cerveau isolé (French— isolated intellect, intellectual)

CI Carnegie Institute; Cayman Islands; Channel Islands; Chlorine Institute; Color Index; Combustion Institute; Communist International; Confidential Informant; Conservation International; Consumers Institute; Cotton Incorporated; Cranberry Institute; Crown of India, Imperial Order; Curtis Institute; Ivory Coast (Internet code)

C.I. Lady of the Imperial Order of the Crown of India

C/I Certificate of Insurance; Consular Invoice

C & I Currier and Ives

C of I Church of Ireland (Roman Catholic)

CI Colour Index; Comédie-Italienne (Paris)

cia captured in action; cash in advance; chief inspector of armaments; child(ren) in arms; computer interface adaptor

Cia Compagnia (Italian—Company); *Companhia* (Portuguese—Company); *Compañía* (Spanish—Company)

CIA Caribbean International Airways; Catering Institute of Australia; Central Institute of Australia; Central Intelligence Agency; Commerce and Industry Association; Cook Island Airways; Correctional Industries Association; Cotton Insurance Association; Cowboy and Indian Alliance; Culinary Institute of America

CIA Comité International d'Auschwitz (French—International Auschwitz Committee), *Conseil International des Archives* (French—International Council on Archives)

CIAA Cheese Importers Association of America; College Inventory of Academic Adjustment; Coordinator Inter-American Affairs

CIAB Canadian Immigration Appeal Board

C[iac] *Campania* (Spanish—company)

CIAC Canadian Independent Adjusters Conference; Career Information and Counseling (USAF)

CIAL Communauté Internationale des Associations de la Librairie (French—International Community of Booksellers' Associations)

CIAM Computerized, Integrated, Automated, Manufacturing

CIAM Congreso Internacional de Arquitectura Moderna (Spanish—International Congress of Modern Architecture)

CIANY Commerce and Industry Association of New York

CIAO Congress of Italian-American Organizations

CIAP Comité Interamericano de la Alianza para el Progreso (Spanish—Inter-American Committee of the Alliance for Progress)—ICAP

CIAPS Customer-Integrated Automated Procurement System

CIAS California Institute of Asian Studies; Council for Inter-American Security

CIASSR Cecheno-Ingush Autonomous Soviet Socialist Republic

CIAT Communications Installation Advisory Team

CIAT Centro Interamericano de Administradores Tributarios (Inter-American Center of Revenue Administrators); *Centro Internacional de Agricultura Tropical* (Spanish—International Center of Tropical Agriculture)

CIAU Canadian Interuniversity Athletic Union

CIAW Commission on Intercollegiate Athletics for Women

cib. cibus (Latin—food)

CIB California Industries for the Blind; Canadian Infantry Brigade; Canadian International Bank; Central Intelligence Board; Cognac Information Bureau; Combat Infantryman Badge; Commonwealth Investment Bank; Criminal Intelligence Bureau; Criminal Investigation Bureau

CIB COBOL Information Bulletin (USAF)

CIBA Corporation of Insurance Brokers of Australia

CIBC Canadian Imperial Bank of Commerce; Council on Interracial Books for Children

CIBG Canadian Infantry Brigade Group

cibha congenital inclusion body hemolytic anemia

cibhp closed-in-bottom hole pressure

CIBNC Cook Islands Broadcasting and Newspaper Corporation

CIBNZ Corporation of Insurance Brokers of New Zealand

CIBS Chartered Institution of Building Services; Cosmetic Industry Buyers and Suppliers

cic cardio-inhibitor center; cloud in cell; command input coupler; critical item code

Cic Marcus Tullius Cicero

CIC Calcium Information Center; Canadian Intelligence Corps; Career Initiatives Center; Caribbean Investment Corporation; Cedar Rapids & Iowa City (railroad); Center for Instructional Communications (Syracuse University); Central Inspection Commission; Central Intelligence Center; Change Identification Control; Chemical Institute of Canada; Combat Information Center; Combat Intelligence Center; Combined Intelligence Committee; Command Information Center; Commander-in-Chief; Commission on Interracial Cooperation; Committee on Institutional Cooperation; Commonwealth Industrial Court; Commonwealth Infor-

mation Centre; Community Information Centre(s); Consumer Information Center (Pueblo, Colorado 81009); Continental Insurance Companies; Cooperative Insurance Corporation; Council of Independent Colleges; Counter-Intelligence Corps; Coupon Information Center; Crime Intelligence Center; Criminal Investigation Command; Critical Issues Council; Curacao Information Center; Customer Identification Code

CIC Comité International de la Conserve (French—International Canning Committee); *Conseil International des Compositeurs* (French—International Council of Composers); *Consejo Interamericano Cultural* (Spanish—Interamerican Cultural Council); *Cymdeithas yr Iaith Cymraeg* (Welsh Language Society)

CICA Canadian Institute of Chartered Accountants; Council of International Civil Aviation

CICA Centro de Investigaciones Ciencias Agronómicas (Spanish—Agronomic Sciences Investigation Center)

CICAR Cooperative Investigations of the Caribbean and Adjacent Regions (UNESCO)

CICAS Computer-Integrated Command-and-Attack Systems

CICB Criminal Injuries Compensation Board

CICC California Institute of Color Consulting; Criminal Injuries Compensation Commission (Hawaii)

CICCU Cambridge Inter-Collegiate Christian Union

CICESE Centro Investigación y Educación Superior de Ensenada (Spanish—Center of Investigation and Higher Education of Ensenada)

Cicestr. Cicestrensis (Latin—of Chichester)

CICI Composite Index of Coincident Indicators

CICJ *Comité International pour la Coopération des Journalistes* (French—International Committee for the Cooperation of Journalists)

CICMA Canadian Insurance Claims Managers Association

CICO Combat Information Center Office(r)

CICOM *Centro de Comercialización Nacional e Internacional* (Spanish—Center of National and International Marketing)

CICP Capital Investment Computer Program; Center Program(ming); Committee to Investigate Copyright Problems

CICRIS Cooperative Industrial and Commercial Reference and Information Service

CICs Community Inprovement Corpsmen; Community Improvement Corpswomen

CIC's Change Information Control (numbers)

CICS Committee for Index Cards for Standards; Customer Information and Control System; trademark of International Business Machines

CICSB Coalition of Indian-Controlled School Boards

CICS/VS Customer Information Control System/Virtual Storage

CICT Commission on International Commodity Trade

CICT *Conseil International du Cinéma et de la Télévision* (French—International Council of Cinema and Television)

cicu cardiology intensive care unit (CICU); computer-integrated converter unit; coronary intensive care unit (CICU)

CICU Commission for Independent Colleges and Universities

CICUNM Council of Independent Colleges and Universities of New Mexico

CICUP Commission for Independent Colleges and Universities of Pennsylvania

CICV Council of Independent Colleges in Virginia

CICYP *Consejo Interamericano de Comercio y Producción* (Spanish—Interamerican Council of Commerce and Production)

cid cash in drawer; charge-injection device; chick infective dose; commercial item description; cubic-inch displacement

cid (CID) cytomegalic inclusion disease

CID Center for Industrial Development; Central Institute for the Deaf; Change in Design; Classification of Instructional Disciplines; Commission for International Development; Committee for Imperial Defence; Council for Independent Distribution; Criminal Investigation Department (Scotland Yard); Ciminal Investigation Division; Counterintelligence Division

CID *Centre d'Information et de Documentation* (French—Center for Information and Documentation—Belgium); *Colegio Interamericano de Defensa* (Spanish—Inter-American Defense College)

cida cometary and interstellar dust analyzer

CIDA Canadian International Development Agency

CIDA *Comité Interamericano de Desarollo Agricola* (Spanish—Inter-American Committee of Agricultural Development)

CIDALC *Comité International du Cinéma d'Enseignement et de la Culture* (French—International Committee of Film Education and Culture)

CIDC Cyogenic Information and Data Section; Curriculum and Instructional Development Center

cide (Latin suffix—destroy)—germicide, insecticide

CIDEM *Consejo Interamericano de Música* (Spanish—Inter-American Music Council)

CIDG Civilian Irregular Defense Groups; Civil Indigenous Defense Group (Vietnam)

CIDH *Comisión Interamericana de Derechos Humanos* (Spanish—Inter-American Commission of Human Rights)

cidi crimping die

cidnp chemically induced dynamic nuclear polarization

CIDOC *Centro Intercultural de Documentación* (Spanish—Intercultural Documentation Center)

cids cellular immunity deficiency syndrome

CIDS Chemical Information and Data System

cidstat civil disturbance status (USA reporting activity)

cie coherent infrared energy; counter immunoelectrophoresis; customer-initiated entry; customs input entry

Cie *Compagnie* (French—company)

CIE Center for Independent Education; Chrysler Institute of Engineering; Cleveland Institute of Electronics; Commonwealth Institute of Entomology; Companion of the Order of the Indian Empire

C.I.E. Companion of the Order of the Indian Empire

CIE *Centro de Informções do Exército* (Portuguese—Military Intelligence Center)—Brazil; *Comité Interamericano de Educación* (Spanish—Inter-American Committee of Education); *Commission Internationale D'Eclairage* (French—International Commission on Light)

CIEA California Indian Education Association

CIEA *Centro Internacional de Estudios Agricolas* (Spanish—International Center of Agricultural Studies)

CIEBM Committee on the Interplay of Engineering with Biology and Medicine

CIEC *Centre International d'études Criminologiques* (French—International Center of Criminological Studies)

CIECC *Consejo Interamericano para la Educación, la Ciencia, y la Cultura* (Spanish—Inter-American Council for Education, Science, and Culture)

CIEE Companion of the Institution of Electrical Engineers; Council on International Educational Exchanges

Cie Gle Transadantique *Compagnie Générale Transatlantique* (French Line)

CIEM Conseil International pour l'Exploration de la Mer (French—International Commission for the Exploration of the Sea)

CIEN Comisión Interamericana de Energía Nuclear (Spanish—Inter-American Commission for Nuclear Energy)

CIENES Centro Interamericano de Enseñaza de Estadística (Spanush—Inter-American Center for the Study of Statistics)

CIENT Cambridge and Isle of Ely Naturalist Trust (England)

CIEO Catholic International Education Office

ciep counterimmunoelectrophoresis

CIEP Council on International Economic Policy

CIER Centro Interamericano de Educación Rural (Spanish—Inter-American Center of Rural Education)

CIES Comparative and International Education Society; Council for International Exchange of Scholars

CIES Consejo Interamericano Economico y Social (Spanish—Inter-American Economic and Social Council)

CIESMM Commission Internationale pour l'Exploration Scientifique de la Mer Méditerranée (French-International Commission for the Scientific Exploration of the Mediterranean Sea)

CIESPAL Centro Internacional de Estudios Superiores de Periodismo para America Latina (Spanish—International Center for Advanced Studies of Journalism in Latin America)

CIESS Chief Inspector of Engineer and Signal Stores

CIET Centro Interamericano de Estudios Tributarios (Spanish—Inter-American Center of Revenue Studies)

CIETA Calcutta Import and Export Trade Association

CIETA Centre International d'Etude des Textiles Anciens (French—International Center for the Study of Ancient Textiles)

cif cash in fist; central index(ing) file; central integration facility; corporate income fund; cost, insurance, and freight; customer information file

CIF California Interscholastic Federation; Canadian Institute of Forestry; Construction Industry Foundation

CIF Commission Interaméricaine des Femmes (French—Interamerican Commission of Women); *Conseil International des Femmes* (French—International Council of Women)

CIFA Courtauld Institute of Fine Arts

cifane cost, insurance, freight, and slight difference in exchange

CIFAR Central Institute of Foreign Affairs Research

CIFAS Consortium Industriel Franco-Allemand pour Symphonie (French—Franco-German Industrial Consortium for Symphonie)—communication satellite linking systems between points in Africa, the Americas, Europe, and the Middle East

cif & c cost, insurance, freight, and commission

CIFC Council for the Investigation of Fertility Control

cif & e cost, insurance, freight, and exchange

cifci (CIF and C & I) cost, insurance freight (plus) commission and interest

CIFE Central Index File—Europe; Combined Intelligence, Far East

CIFEJ Centre International du Film pour l'Enfance et la Jeunesse (French—International Center of Films for Children and Young People)

CIFF Cannes International Film Festival; Comprehensive International Freight Forwarders

cif & i cost, insurance, freight, and interest

CIFIUS Current Interagency Committee on Foreign Investment in the United States

cifLt cost, insurance, and freight, London terms

CIFO Catalogue of Interface Features and Options

cig cigarette

CIG Comité International de Géophysique (French—International Geophysics Committee); Commonwealth Industrial Gases

CIGA Compagnia Italiana dei Grandi Alberghi (Italian—Italian Great Hotels Company)

CIGAR Common Interactive Graphics Application Routine (USA)

CIGGT Canadian Institute of Guided Ground Transportation

CIGNA Connecticut General Corporation combined with the Insurance Company of North America

CI Gov Cook Islands Government

CIGS Chief of the Imperial General Staff (Great Britain)

cigsmug cigarette smuggler

CIGTF Central Inertial Guidance Test Facility

cih carbohydrate-induced hyperglyceridemia

CIH Central India Horse; Certified Industrial Hygienist

CIHR Clinical Institute for Human Relations

CII Chartered Insurance Institute; Coffee Information Institute; Combat Information Intelligence; Council of International Investigators

CIIA Canadian Institute of International Affairs

CIIB Consumers Insurance Information Bureau

CIIC Chemical Industry Institute of Toxicology; Counter Intelligence Interrogation Center

СшА *Soedinennye Shtaty Ameriki* (Russian—United States of America)—U.S.A.

c-i info criminal-investigation information

CIIR Central Institute for Industrial Research

CIIS California Institute of International Studies

CIIT Chemical Industry Institute of Toxicology

cij control joint

CIJ *Consejo Interamericano de Jurisconsultos* (Spanish—Inter-American Council of Legal Consultants)

CIJE *Current Index to Journals in Education*

CIKCU Council of Independent Kentucky Colleges and Universities

cil control interpreter language; core-image libray; currentinhibit logic

Cil Cilicap

CIL Canadian Industries Limited; Center for Independent Living; Commissioner of Irish Lights; Council for Interinstitutional Leadership

C/I/L *Computer/Informationl Library Sciences*

cila casualty insurance logistics automated

CILA *Centro Interamericano de Libros Académicos* (Spanish—Inter-American Center of Academic Books)

CILES Central Information Library and Editorial Section

Cilla Priscilla

cilop conversion in lieu of procurement

CILSA Chief Inspector of Land Service Ammunition

CILT Center for Information on Language and Teaching

cim capital investment model; communication-interface module(s); computer-input microfilm(ing); computer-integrated manufacturing; conductance-increase mechanism; continuous-image microfilm(ing)

CIM California Institution for Men; Canadian Institute of Management; Canadian Institute of Mining; Canadian Institute of Music; Channel Island Militia; Chief Inspector of Machinery; Commission

for Industry and Manpower; Computer Integrated Manufacture; Computer-Integrated Manufacturing; Curtis Institute of Music; Council of Independent Managers

C & IM Chicago & Illinois Midland (railroad)

CIM *Centro Italiano della Moda* (Italian—Italian Fashion Center); *Conseil International de la Musique* (French—International Music Council); *Consejo Internacional de Mujeres* (Spanish—International Council of Women)

CIMA Certified Investment Management Analyst; Construction Industry Manufacturers Association; Coordinated Investigation of Micronesian Anthropology

Cimabue Cenni di Pepo

cimb *cimbalom* (Hungarian—dulcimer)

CIMB Construction Industry Management Board

CIMBA Contractor Installation Make-or-Buy Authorization

CIMC California Institution for Men in Chino (state prison); Certified Investment Management Consultant; Commander's Internal Management Conference

CIMC *Colegio Interamericano Médicos y Cirujanos* (Spanish—Interamerican College of Physicians and Surgeons)

cimco card image correction

CIMCO Congo International Management Corporation

CIME Council of Industry for Management Education

CI Mech E Companion of the Institution of Mechanical Engineers

CIMIC Civilian Military Co-operation

CIMIS California Irrigation Management Information System

cimm constant-impedance mechanical modulation

CIMM Canadian Institute of Mining and Metallurgy

CIMMS Civilian Information Manpower Management System (USN)

CIMMYT *Centro Internacional de Mejoramiento de Maíz y Trigo* (Spanish—International Center for the Improvement of Corn and Wheat)

CIMP *Conseil International de la Musique Populaire* (French—International Folk Music Council)

CIMR Commander's Internal Management Review

cims chemical ionization mass spectrometry

CIMS Computer-Integrated Manufacturing System; Convair Integrated Management System

CIMTP *Congrés International de Médecine Tropicale et de Paludisme* (French— International Congress of Tropical Medicine and Malaria)

cimu compatibility-integration mockup

cin cervical intra-epithelial neoplasia; code identification number; component identification number; cost item number

cin (CIN) communication identification navigation

c$_{in}$ insulin clearance

cin *cinéma* (French—motion picture theater)

Cin Cincinnati

CIN Change Incorporation Notice; Change Instrumentation Notice; Cooperative Information Network

CIN *Chemical Industry Notes*

CINAT Cook Islands National Art Theatre

CINB & T Continental Illinois National Bank and Trust

Cinc Cincinnati

C-in-C Commander-in-Chief

CINC Commander-in-Chief

CINCAF Commander-in-Chief, Allied Forces

CINCAFE Commander-in Chief, Air Forces Europe

CINCAFLANT Commander-in-Chief, Air Force Atlantic Command

CINCAFMED Commander-in-Chief, Allied Forces Mediterranean

CINCAFSTRIKE Commander-in-Chief, Air Force Strike Command

CINCAL Commander-in-Chief, Alaskan Command

CINC ATL FLT Commander-in-Chief, Atlantic Fleet

CINCEASTLANT Commander-in-Chief, Eastern Atlantic

CINCENT Commander-in-Chief, Central Europe

CINCEUR Commander-in-Chief, Europe

CINCHAN Commander-in-Chief—Channel (NATO)

CINCHF Commander-in-Chief, Home Fleet (British)

CINCHOMEFLT Commander-in-Chief, United Kingdom Home Fleet

Cinci Cincinnati

CINCIBERLANT Commander-in-Chief, Iberian Atlantic

Cincin Cincinnati

CINCLANT Commander-in-Chief, Atlantic

CINCLANTFLT Commander-in-Chief, Atlantic Fleet

CINCMEAFSA Commander-in-Chief, Middle East, Southeast Asia, Africa South of the Sahara

CINCMED Commander-in-Chief, Mediterranean

CINCMELF Commander-in-Chief, Middle-East Land Forces

CINCNELM Commander-in-Chief, U.S. Naval Forces in Europe, the Eastern Atlantic, and the Mediterranean

CINCNORAD Commander-in-Chief, North American Defense Command

CINCNORTH Commander-in-Chief, Northern Europe

CINCONAD Commander-in-Chief, Continental Air Defense Command

CINCPAC Commander-in-Chief, Pacific

CINCPACFLT Commander-in-Chief, Pacific Fleet

CINCRDAF Commander-in-Chief, Royal Danish Air Force

CINCRDN Commander-in-Chief, Royal Danish Navy

CINCRNAF Commander-in-Chief, Royal Norwegian Air Force

CINCRNORN Commander-in-Chief, Royal Norwegian Navy

CINCSOUTH Commander-in-Chief, Southern Europe

CINCSTRIKE Commander-in-Chief, United States Strike Command

CINCSWPA Commander-in-Chief, South-West Pacific Area

CINCUKAIR Commander-in-Chief, United Kingdom Air Forces

CINCUNC Commander-in-Chief, United Nations Command

CINCUSAFE Commander-in-Chief, United States Air Forces in Europe

CINCUSAFLANT Commander-in-Chief United States Air Force, Atlantic

CINCUSAFSTRIKE Commander-in-Chief–United States Air Force Strike

CINCWA Commander-in-Chief Western Approaches

CINCWESTLANT Commander-in-Chief, Western Atlantic

Cincy Cincinnati

Cindy Cinderella; Cynthia

cine cinema; cinematography

CINE Council on International Nontheatrical Events

CINECA Cooperatiive Investigation of the Eastern Central Atlantic

cinemactor cinema actor

cinemactress cinema actress

cinerama cinematic panorama (three-dimensional film)

CINFAC Counterinsurgency Information Analysis Center

CINFO Chief of Information

CINM Channel Islands National Monument (Southern California)

cinn cinnabar

Cinn Cincinnati

cinna cinnamon

cinnabar mercuric sulfide (HgS)

Cinn Sym Orch Cincinnati Symphony Orchestra

CINO Chief Inspector of Naval Ordnance

CINOA Confédération Internationale des Négociants en Oeuvres d'Art (French—International Confederation of Art Dealers)

CINPDis Cincinnati Procurement District (US Army)

cins child(ren) in need of supervision

CINS CENTO Institute of Nuclear Science

Cin Sym Cincinnati Symphony

CINTA Compañía Nacional del Turismo (Spanish—National Travel Company)—Chilean Airline

Cinty Cincinnati

CINVA Centro Interamericano de Vuvienda y Planteamiento (Spanish—Inter-American Center of Housing and Planning)

cio central input/output (multiplexer)

CIO Careers Information Officer (recruiter); Chief Information Officer; Church Information Office(r); Congress of Industrial Organizations

CIO Commission Internationale d'Optique (French—International Optical Commission)

CIOCS Communications Input-Output Control System

CIOMS Council for the International Organization of Medical Sciences

ciopw charcoal, ink, oil, pencil, and watercolor (title of a book illustrated and written by e e cummings in 1931)

CIOSL Confederación Internacional de Organizaciones Sindicales Libres (Spanish—International Confederation of Free Trade Union Organizations)

cip cast-iron pipe; cataloging in publication; certified inhalation protection; cipher

cip (CIP) capital investment program; commercially important passenger

C i P Cataloging in Publication (Library of Congress program)

CIP Canadian International Paper; Career Internship Program; Carriage/Freight and Insurance Paid (to); Cataloging-in-Publication; Chartered Investment Planner; Citizens Involvement Project; Civilian Institution Program; Communications and Information Policy; Composite Interface Program; Consolidated Intelligence Program; Cook Islands Party; Coordinated Instrument Package; Cost Improvement Proposal

CIP Comisión Interamericana de Paz (Spanish—Inter-American Peace Commission)

CIPA Canadian Industrial Preparedness Association; Chartered Institute of Patent Agents; Committee for Independent Political Action

CIPAC Collaborative International Pesticides Analytical Council (UK)

CIPASH Committee for an International Program in the Atmospheric Sciences and Hydrology

cipc cast-in-place concrete

CIPC Christmas Island Phosphate Commission

CIPCE Centré d'Information et de Publicité des Chemins de Fer Européens (French—Information and Publicity Center of the European Pailways)

CIPE Center for International Private Enterprise

CIPE Centro Interamericano para la Promoción de las Exportaciones (Spanish—Inter-American Center for the Promotion of Exports); *Consejo Internacional de la Pelicula de Ensena* (Spanish—International Council for Educational Films)

CIPEC Conseil Intergouvernmental des Pays Exportateurs de Cuivre (French—Intergovernmental Council of Copper-Exporting Nations)

CIPFA Chartered Institute of Public Finance and Accountancy

ciph cipher

CIPHER Calculations of Patient and Hospital Education Resources

ciphony enciphered telephony

CIPL Canada India Pakistan Line; Commission of Inquiry into Public Libraries (Australian)

CIPL Comité International Permanent de Linguistes (French—Permanent International Committee of Linguists)

CIPM Council for International Progress in Management

Cipo Cipriano

CIPO Conseil International pour la Préservation des Oiseaux (French—International Council for the Preservation of Birds)

CIPP Cataloging-in-Publication Program (Library of Congress); Comprehensive Income and Price Policy

CIPP Conseil Indo-Pacifique des Pêches (French—IndoPacific Fisheries Council)

CIPR Commission Internationale de Protection Contre les Radiations (French—International Commission on Radiological Protection)

CIPRA Cast-Iron Pipe Research Association

CIPRA Commission International pour la Protection des Régions Alpines (French—International Commission for the Protection of Alpine Regions)

CIPREC Canadian Institute of Public Real Estate Companies

CIPs Commercially Important Persons

CIPS Canadian Information Processing Society

CIPT Canadian Institute for the Prevention of Torture

CIQ Customs, Immigration, Quarantine

cir circle; circuit; circular

cir. circa (Latin—about)

cIR crime on Indian Reservation

Cir Circimus; Circinis (constellation); Circle; Circus

CIR Commission on Intergovernmental Relations; Commissioner of Internal Revenue; Consumer Information Report; Cost Information Report; Court of industrial Relations; Current Industrial Reports

CIRA Committee on International Reference Atmosphere; Conference of Industrial Research Associations

CIRA Centro Interamericano de Reforma Agraria (Spanish—Inter-American Center of Agrarian Reform)

CIRADS Counter-Insurgency Research and Development System

cir ant. circular antenna

cir bkr circuit breaker

circ circle; circular; circulate; circulation; circumcision; circumference; circumferential(ly); circumstance; circus

Circ Circimus; Circle; Circus

CIRC Central Information Reference and Control; Cross Interleave Reed-Solomon Code

CIRC Centre International de Recherché sur le Cancer (French—International Center for Cancer Research)

CIRCA Computerized Information Retrieval and Current Awareness

circad circadic; circadian; circadianly

circal circuit analysis

CIRCALS Circle Analysis System

CIRCAS Contract Information Retrieval and Cost Accounting System

Circi Circinus (Latin—Pair-of-Compasses constellation)

circltr circular letter

circs circumstances

circum circumcision; circumference

circum (Latin prefix—around)—circumnavigation

circum haema circumorbital haematoma (a black eye)

Circumv Stz Circumvesuviana Stazione (Neapolitan railroad station serving Herculaneum, Mt Vesuvius, and Pompeii)

Circus circular intersection (Oxford Circus, Piccadilly Circus, St Giles Circus, etc.)

circuscade circus parade

CIREI Commercial Investment Real Estate Institute

Ciren Cirencester (Sisister)

CI Rep Communist International Representative

CIRES Chief Inspector of Royal Engineers Stores; Cooperative Institute for Research in Environmental Studies

CIRF Corn Industries Research Foundation

CIRF Centre International d'Information et de Recherche sur la Formation Professionelle (French—Vocational Training and Research Center)

CIRIA Construction Industry Research and Information Association

CIRIEC Canadian International Centre of Research and Information on Public and Cooperative Economy

CIRIS Completely Integrated Range-Instrumentation System (NASA)

CIRJP Commission on International Rules of Judicial Procedure

CIRM Centro Internazionale Radio-Médico (Spanish— International Medical Radio Center); *Comissão Interministerial de Recursos do Mar* (Portuguese—Interministerial Commission for the Resources of the Sea)

CIRO Consolidated Industrial Relations Office

CIRP City Improvement and Restoration Program; Cooperative Institutional Research Program

CIRS Certified Insurance Rehabilitation Specialist

CIRVIS Communication Instructions for Reporting Vital Intelligence Sightings (of ufo's from aircraft)

cis carcinoma in situ; cataloging in source; central inhibitory state; contact image sensor

cis (CIS) cataloging in source

ci's conflict indicators

Cis Cecilia

CIs Current Investigations

CIS Cancer Information Service; Career Information System; Catholic Information Society; Center for International Studies (MIT); Central Industrial Secretariat; Central Instructor School; Chartered Industries of Singapore; Chartered Institute of Secretaries; Commonwealth of Independent States; Communications and Information Systems; Congressional Indexing Service; Congressional Information Service;

Cost Inspection Service; Council for Inter-American Security; Cranbrook Institute of Science

CIS-12 Commonwealth of 12 Independent States (Armenia, Azerbaijan, Belarus, Georgia, Kazakhstan, Kyrgyzstan, Moldova, Russia, Tajikistan, Turkmenistan, Ukraine, Uzbekistan)

CISA Canadian Industrial Safety Association; Cook Islands Sports Association; Council for Independent School Aid

CISA Commission Internationale pour le Sauvetage Alpin (French—International Commission for Alpine Rescue)

CISAC Confédération Internationale des Auteurs et Compositeurs (French—International Federation of Authors and Composers)

cisam compressed index sequential access method

CISC Commission for Investigation of Special Crimes; Complex Instruction Set Computer; Complex Instruction Set Computing

CISCA Ceilings and Interior Systems Construction Association

Cisco San Francisco

CISCO Civil Service Catering Organization; Commercial and Industrial Security Corporation (Singapore)

cisd critical incident stress debriefing

CISE Colleges, Institutes, and Schools of Education (Library Association)

CISF Confédération Internationale des Sages-Femmes (French—International Confederation of Midwives)

CISHEC Chemical Industry Safety and Health Council

CISI Command Inspection System Inspection (USAF)

CISI Compagnie Internationale de Service et Information (French—International Service and Information Company)

CISIR Ceylon Institute of Scientific and Industrial Research

Cisister Cirencester, England

CISL Confederazione Italiana Sindacati Lavoratori (Italian—Confederation of Labor Syndicates)

CISLE Centre International des Syndicalistes Libres en Exil (French—International Center of Free Trade Unionists in Exile)

cislun cislunar; cislunarian; cislunarite

CISPES Committee in Solidarity with the People of El Salvador

CISR Center for International Systems Research; Commonwealth Inscribed Stock Registry

CISS Computerized Information Storage System

Cissie Cecilia

Cissy Cecilia

Cist Cistercian

CISTI Canada Institute of Scientific and Technical Education

CISV Children's International Summer Village

cit citation; cited; citizen(ship); citrate; a citric acid; compression in transit; computer interface terminition; configuration identification table(s); counterintelligence team

cit (CIT) call-in time

cit citat (Dano-Norwegian—quotation)

Cit Citadel

CIT Calcutta Improvement Trust; California Institute of Technology (Cal Tech); Carnegie Institute of Technology; Case Institute of Technology; Center for Information Technology; Central Institute of Technology; Chartered Institute of Transport; Coal Industry Tribunal (Australian); Conference of Interpreter Trainers; Continental Inclusive Tour; Counterintelligence Team; Court of International Trade; Cranfield Institute of Technology

CIT Compagnia Italiana di Turismo (Italian—Italian Travel Bureau)

CITA Commercial-Industrial-Type Activity; Cook Islands Tourist Authority

CITAB Computer Instruction and Training Assistance for the Blind

CIT (ARIA) Commission on Insurance Terminology (American Risk and Insurance Association)

CITARS Crop Identification Technology Assessment for Remote Sensing (NATO)

CITB Construction Industry Training Board

CITC Canadian Institute of Timber Construction; Cook Islands Trading Corporation

cite. compression ignition and turbine engine

CITE Consolidated Index of Translations into English; Co-ordinating Information for Texas Educators; Council of the Institute of telecommunication Engineers; Current Information on Tapes for Engineers

CITEL Comisión Interamericana de Telecomunicaciones (Spanish—Inter-American Telecommunication Commission)

CITEP Canadian Indian Teacher Education Projects

CITES Convention on International Trade in Endangered Species

CITGO Cities Service Gulf Oil

cithp closed-in tubing head pressure

Citi Citibank

Citibank First National City Bank

CITIC China International Trust and Investment Corporation

Citicorp First National City Bank Corporation

CITIS Centralized Integrated Technical Information System

CITL Canadian Industrial Traffic League

CITO Charter of the International Trade Organization

cito disp. cito dispensetur (Latin—dispense rapidly)

CITP Civilian Industrial Technology Program

CITRAIL Centre Inter-Regional de Transit Rail-Route (French Canadian)

citricult citriculture

citrine false topaz (quartz with ferric iron)

cits central integrated test system

CITS China International Travel Service

CITT Canadian Institute of Traffic and Transportation

citu (CITU) coronary intensive-care unit

Cit U City University

City The City—business and financial section of London

City Ed City Editor

CIU Central Interpretation Unit; Coopers' International Union; Criminal Intelligence Unit (police)

CIUL Council for International Urban Liaison

CIUS Conseil International des Unions Scientifiques (French—International Council of Scientific Unions)

civ civil; civilian; civilization; civilize

CIV Central Inspectorate of Vehicles; City Imperial Volunteers (London); Commonwealth Institute of Valuers

CIV Commission Internationale du Verre (French—International Glass Commission)

CIVA Cook Islands Visitors Association

Civ Air NM Civil Aircraft National Marking(s)

civd cold-induced vasodilation

Civ E Civil Engineering

civ eng civil engineering

Civ Eng Civil Engineer

civies civilian clothes; civilians

civiling civilingenier (Dano-Norwegian—civil engineer)

CIVIS Centro Italiano per i Viaggi d'Instruzione per Studenti (Italian—Italian Center for Students' Educational Travel)

civ svc civil service

civ svt civil servant

civvies civilian clothes; civilians

civvy civilian

ciw current instruction word

CIW California Institution for Women; Catering Instruction Wing

CIWF Compassion in World Farming

CIWMB California Integrated Waste Management Board

CIWS Close-in Weapons System

CIX Commercial Internet Exchange

Ci(x) cosine integral

cixa constant infusion excretory urogram

cj clip joint; conjectural; construction joint

CJ Chief Justice; Civil Jail; County Judge; Court of Judiciary

C of J Collector of Junk

CJ Computer Journal

CJA Carpenters and Joiners of America; Criminal Justice Act

CJA Criminal Justice Abstracts

C-jam cocaine

CJB Constructors John Brown (British shipbuilders)

CJC Colby Junior College; Community Junior College

CJC Corpus Juris Canonici (Latin—Code of Canon Law)

CJCA California Junior College Association

CJCiv Corpus Juris Civilis (Latin—Code of Civil Law)

CJCs Criminal Justice Councils

CJD Creutzfeldt-Jakob Disease

C-J disease Creutzfeldt-Jakob disease

cje corretaje (Spanish—brokerage)

CJE Citizens for Jobs and Energy

CJF Carlos J. Finley

CJFWF Council of Jewish Federations and Welfare Funds

CJI Concrete Joint Institute

CJI Comite Juridico Interamericano (Spanish—Inter-American Juridical Committee)

CJIS Criminal Justice Information System (Rhode Island)

CJLF Criminal Justice Legal Foundation

CJM Congregation of Jesus and Mary

C-joint cocaine joint

CJP Criminal Justice Publications

CJR Cecil John Rhodes

CJR Columbia Journalism Review

CJRL Criminal Justice Reference Library (Austin)

cjs cotton, jute, or sisal (cargo)

CJS Canadian Joint Staff; College of Jewish Studies
CJS Corpus Juris Secundum
CJTF Commander Joint Task Force
ck cask; certified kosher; check; coke; cork; creatine kinase
ck ceekay (Spanish-American slang—cocaine)
Ck chalk; Creek
CK Cook Islands (Internet code); cyanogen chloride (poison gas)
C K Cape Kennedy
ckb cork base
ckbd cork board
CKC Canadian Kennel Club
CKCJP Center for Knowledge in Criminal Justice Planning
CKCL Chicago-Kent College of Law
ckd completely knocked down
CKD Certified Kitchen Designer
CKE Central Kingdom Express (*see* Ori Exp)
ckf cork floor
ckfm checking form
ckg. checking
c/Kg coulombs per kilogram
ckga checking gage
CKIC Chemical Kinetics Information Center (NBS)
ckm cents per kilometer
ck-MB creatinine kinase, MB isoenzyme
C/K mol coulombs per kilmol
CKMTA Cape Kennedy Missile Test Area
cko checking operator
ck os countersink other side
ckout checkout
CKPG Containerboard and Kraft Paper Group
ckpt cockpit
cks casks; checks
CKS Chiang Kai-Chek (international airport)
ckt circuit
CKT Chung-Kuo Kung-ch'an Tang (Chinese—Chinese Communist Party)
ckt bd circuit board
ckt bkr circuit breaker
ckt cl circuit closing
ck tp check template
ck ts countersink this side
ck viv check valve
ckw clockwise
cl carload; ceiling level; ceiling value; center line; centiliter; chest and left arm (cardiol-

ogy); chloride; city limits; clarinet(s); class; classification (baseball); clause; clavicle; clear; clearance; climb; clinic; close; closure; comparison level; conceptual level; confidence limits; corpus luteum; critical list
cl (CL) control leader (data processing)
c/l combat loss
c/l (C/L) carload lot; cash letter
cl. clip
cl. classis (Latin—class or collection)
Cl chlorine; chlorine gas; Cloister; Close
Cl Calle (Spanish—street)
CL Cambridge Library; Canadian Legion; Capital Airlines; Certified Lymphologist; Chile (Internet code); chlorine; chlorine gas; Communications Lieutenant; Cooperative League; Critical List; light cruiser (2-letter naval symbol)
C-L Canadair Limited (Division of General Dynamics); Consultation-Liaison
C/L craft loss (insurance)
C & L Canal and Lake; Coopers and Lybrand
C of L Count(y) of Lippe
CL. Clericus (Latin—cleric or clergyman)
CL-1 Computer Language One
CL-13 Canadair-built F-86 Sabre aircraft
CL-28 Canadair-built long-range reconnaissance version of the Britannia
CL-41 Canadair-built jet-trainer aircraft nicknamed Tutor
cla center line average; communication link analyzer; conjugated linoleic acid
Cla Clare College, Cambridge
CLA California Library Association; Canadian Library Association; Canadian Lumbermen's Association; Catholic Library Association; Chinese Librarians Association; Coin Laundry Association; College Language Association; Commercial Law Association; Computer Law Association; Connecticut Library Association; Conservative Library

Association; Copyright Licensing Agency; Culinary Institute of America
C.L.A. Certified Laboratory Assistant
CLAA anti-aircraft light cruiser (4-letter naval symbol)
CLA-ACB Canadian Library Association—*l'Association Canadienne des Bibliothéques*
Clack Clackmannan(shire)
cl ad collet adapter
CLAE Council of Library Associations Executives
CLAH Conference of Latin American History
CLAIMS Class Codes Assigned Index Method Search
CLAIRA Chalk Lime and Allied Industries Research Association
clalt comparison level for alternatives
clam (CLAM) chemical low-altitude missile
clam. chemical low-altitude missile
clamato clam-and-tomato juice
clamsan clam sandwich
clamwich clam sandwich
cland lit clandestine literature (underground)
cland press clandestine press
CLAO Contact Lens Association of Ophthalmologists
CLAP Citizens Lobbying Against Prostitution
clar clarification; clarify; clarinet
Clar Clarence
Clara Clarabelle; Clarissa; Clarita
Clare Clara; Clarita
Clar(en) Clarendon
clark combat launch and recovery kit
Clark William Andrews Clark Memorial Library of the University of California at Los Angeles
CLARNICO Clark, Nichols, and Coombes (confectioners)
Clarrie Clarice; Clarissa
clas classification; classify; congenital localized absence of skin
c-l-a-s crowd-lift-actuate-swing (tractor backhoe control)
CLAS Chartered Land Agents Society; Computer Library Applications Service

CLASB Citizens League Against the Sonic Boom

CLASC Confederación Latinoamericana de Sindicalistas Cristianos (Spanish—Latin American Confederation of Christian Trade Unionists)

clasn classification

clasp. (CLASP) computer liftoff and staging program

CLASP Center for Law and Social Policy; Citizens Local Alliance for a Safer Philadelphia; Client's Lifetime Advisory Service Program; Computer Language for Aeronautics and Space Programming; Computer Launch and Separation Problem

CLASPS Coded Label Additional Security and Protection System

class. classification

Class Classical

CLASS California Library Authority for Systems and Services; Carrier Landing Air Stabilisation System; Christian Leaders And Speakers Seminars; Class Action Study and Survey; Close Air-Support System; Closed-Loop Accounting for Store Sales; Computer-based Laboratory for Automated School Systems; Cooperative Library Agency for Systems and Services; Current Literature Alerting Search Service

class A's class-A narcotics (addictive drugs such as opium and its derivatives)

class B's class-B narcotics (almost non-addictive drugs such as codeine and nalline)

CLASSIC Classroom Interactive Computer

classif classification

CLASSMATE Computer Language to Aid and Stimulate Scientific, Mathematical, and Technical Education

class M's class-M narcotics (non-addictive drugs)

classn classification

class X's class-X narcotics (drugs containing small amounts of narcotics such as cough syrups with non-narcotic and almost non-addictive codeine)

clat communication line adapters for teletype

CLAT Confederation of Latin American Teachers

clav clavecin; clavichord; clavicle

clave autoclave; steamclave (sterilizer)

clavicemb clavicembalo (Italian—clavichord)

claw. clustered atomic warhead

CLAW Community Law Workshop (New Zealand)

clax claxon

Clay Claypool (northern Indiana farming center)

Claymont Girls Woods Haven-Kruse School for (delinquent) Girls at Claymont, Delaware

Clb Caleb

CLB Church Lads' Brigade; Configuration Liaison Board

clbbb complete left bundle branch block

clbr calibration

CLBs Combat Lessons Bulletins

c & lc capital and lower case letters

CLC Canadian Labour Congress; Canners League of California; Chiriqui Land Company; Cost of Living Council; task-fleet command cruiser (naval symbol)

CLCB City of Liverpool College of Building; Committee of London Clearing Banks

CLCCS Cammel-Laird Cable-Control System

cl/cll counseling learning/community language learning

CLCMD Cleveland Contract Management District

CL & Co Cammell Laird and Company (shipbuilders)

CL Cr Communications Lieutenant-Commander

clcs current-logic-current switching

clct collector

cl ct claims court

CLCT City of Liverpool College of Technology

cld cancelled; chronic liver disease; chronic lung disease; cleared; colored; cooled; cost laid down

cld (CLD) called (line)

CLD Central Library and Documentation

CLDAS Clinical Laboratory Data Acquisition System

cldwn cooldown

cldy cloudy

CLE City of London Engineers; Cleveland Forest City (National Association); Cleveland, Ohio (Hopkins Airport)

Clea Cleopatra

CLEA Canadian Library Exhibitor's Association

clean. comprehensive lake-eco-system analyzer

CLEAN Committee for Leaving the Environment of America Natural; Commonwealth Law Enforcement Assistance Network (Pennsylvania)

CLEAPSE Consortium of Local Education Authorities for the Provision of Science Equipment

CLEAR Center for Lake Erie Area Research; Civic Leaders for Ecological Action and Responsibility; Closed-Loop Evaluation and Reporting (system); Committee to Leach the Environment of Acid Rain; County Law Enforcement Applied Regionally

Clearwaters Clearwater Mountains of Idaho

clec closed-loop ecological cycle; competitive local-exchange carrier

CLEC Citizen/Labor Energy Coalition

Clem Clemens; Clement; Clementina; Clementine

CLEMARS California Law Enforcement Mutual-Aid Radio System

Clemte Clemente

CLENE Continuing Library Education Network and Exchange

cleo clear language for expressing orders

Cleo Cleopatra

CLEO Council on Legal Education Opportunity

cleopatra comprehensive language for elegant operating system and translator design

CLEP College-Level Education Program; College-Level Examination Program

CLEPR Council on Legal Education for Professional Responsibility

cler clerical; controlled letter contract reduction

CLERC Citizen's Law Enforcement Research Committee

cleric. clerical(s); clerical error; clericalism; clericality; clerically

CLES Customs Law-Enforcement Service

CLETS California Law Enforcement Telecommunications System

CLEU Coordinated Law Enforcement Unit

Clev Cleveland

Cleve Cleveland

Cleveland Cleveland Stadium

Cleve Orch Cleveland Orchestra

CLEVPDis Cleveland procurement District (US Army)

CLEW Chicago Law Enforcement Week

clf capacitive loss factor; central liquidity facility

CLF Chicano Liberation Front; Church of the Larger Fellowship (Unitarian Universalist); Commander Land Forces; Commonwealth Literary Fund(ing)

CLFC Commander Land Forces, Cyprus

CLFFI Commander Land Forces, Falkland Islands

CLFNE Conservational Law Foundation of New England

Clfs Cliffs

clg calling; ceiling; clearing

Clg College

CLG Civil Labor Group; light guided-missile cruiser (3-letter symbol)

CLGA Composers and Lyricists Guild of America

CLGES California Life Goals Evaluation Schedules

clgp (CLGP) cannon-launched guided projectile

clgsfu clear glazed structural facing units

clgsub clear glazed structural unit base

cl gt cloth gilt

CLGW Cement, Lime and Gypsum Workers (union)

CLGW Columbia Lippincott Gazetteer of the World

Cl H Clare-Hall, Cambridge

CLH Calcutta Light Horse; Canadian Light Horse

CLH Croix de la Légion d'Honneur (French—Cross of the Legion of Honor)

CLHU Computation Laboratory of Harvard University

cli central life interests; coin-level indicator; cost-of-living index

CLI Cornwall Light Infantry; Cost-of-Living Index

CLIA Clinical Laboratory Improvement Act; Cruise Lines International Association

C-library circulating library

CLIC Corporate Library Information Centre (Canadian)

clics computer-linked information for container shipping

Cliff Clifford; Clifton

clim climatic

CLIMAPS Climate Long-range Investigation, Mapping, and Prediction Study

climat climatological; climatologist; climatology

Climatol Climatology

CLIMPO Contract Liaison and Master Planning Office

clin clinic; clinical; clinicial; clinometer

CLIN Contract Line Item Number

clin/d clinical death

clink (generic nickname—prison)—also the nickname for brothels and in London, where it originated in Clink prison, also stands for the Southwark Fair depicted by Hogarth

clin path clinical pathology

clin proc clinical procedures

Clint Clinton

Clinton William Jefferson Clinton, 42nd President of the United States

Clinton Men Clinton Correctional Facility at Dannemora, New York

Clinton Women Correctional Institution for Women at Clinton, New Jersey

clip. compiler language for information processing; confused, lacerated, incised, and punctured (wounds)

CLIP Cancel Launch in Progress (USAF); Country Logistics Improvement Program (USAF)

clips. clippings; computer launch interference problems

CLIS Central Library and Information Services; Clearinghouse for Library Information Sciences

clit clitoral; clitoridectomy; clitoris

C. Litt. Companion of Literature

CLIX Compression Labs Incorporated

clj control joint

Clj Callejon (Spanish—alley, blind alley, cul-de-sac, lane)

CLJ California Law Journal

CLJC Copiah-Lincoln Junior College

clk clerk; clock

CLK hunter-killer cruiser (naval symbol)

clkg caulking

clks clockwise

clkws clockwise

cll cholesterol lowering lipid; chronic lymphoblastic cytic leukemia; chronic lymphocytic leukemia; circuit load logic; community language learning

CLL Chief of Legislative Liaison

cllo cuartillo (Spanish—fourth of a real, pint)

Cllr Councillor

clm column; culumnar

c-lm common-law marriage

Clm Culham

CLM Canadian Liberation Movement

CLMA Cigarette Lighter Manufacturers Association; Contact Lens Manufacturers Association

CLML Current List of Medical Literature

CLMS Clinical Laboratory Monitoring System; Company Lightweight Mortar System

cln colon; corrective lens

Cln Colón

clnc clearance

CLNG City of London National Guard

CLNP Crater Lake National Park (Oregon)

clnr cleaner

CLNS Cape Lookout National Seashore (North Carolina)

clnt coolant

CLNTS China Lake Naval Test Station

CLNWR Crescent Lake National Wildlife Refuge (Nebraska)

clo closet; cloth; clothing; cod liver oil; computerized loan origination

Clo Callao

CLO Cali, Colombia (Calipuerto airport); Chief Liaison Officer; Citizens for Law and Order; Civic Light Opera (Los Angeles); Cornell Laboratory of Ornithology

CLOB Composite Limit Order Book

CLOC Computerized Logging and Outage Control

CLOCE Contingency Lines of Communication Europe

CLODS Computerized Logic-Oriented Design System

clog. change log (for software); computer-logic graphics

clor container loaded at owner's risk

clora closed-form ray analysis

clos closure; command to line of sight

clousy cloudy—lousy (weather)

clp common law procedure; control line platform; criminal law and procedure

clp (CLP) command language processor

Clp Cornell list processor (language)

CLP Cargo Loss Prevention; Carnegie Library of Pittsburgh; Country Liberal Party

CLPA Common Law Procedure Acts

cl pal cleft pallet

clpr caliper

clq cognitive laterality quotient

clr center of lateral resistance; clear; clearing; cooler

clr (CLR) computer language recorder

CLR California Law Review; Central London Railway; City of London Rifles; Council on Library Research; Council on Library Resources; County of London Regiment

CL & R Canal, Lake, and Rail

CLR Common Law Reports; Commonwealth Law Reports

CLRA Consumer Law Reform Association

CLRB Canada Labour Relations Board

CLRI Council on Library Resources Incorporated

clrm classroom

clr test chloride test

CLRU Cambridge Language Research Unit

CLRV Canadian Light Rail Vehicle

cls coils

cls (CLS) close (flow chart)

CLS Certificate in Library Science; Congressional Liaison System; Country Library Service

CLSA Conservation Law Society of America; Contact Lens Society of America; Contact Lens Society of Australia

CLSB California Library Services Board

CLSC Chautauqua Literary and Scientific Circle

CLSCS Cain-Levine Social Competency Scale

clsd closed

clsg closing

CLSG Contact Lens Study Group

CLSI Computer Library Services, Inc

clsl chronic lymphosarcoma leukemia

CLSP Cape Lookout State Park (Oregon)

clsr closure

clst clarinettist

clsx close-loop support extended

clt communications line terminals

CLT Certified Landscape Technician; Charlotte, North Carolina (airport)

CLT Canadian Law Times

CLTA Canadian Library Trustees Association; Canterbury Lawn Tennis Association; Chinese Language Teachers Association

CLTC Continental Lawyers Title Company

C Lt-Cdr Communication Lieutenant-Commander

cltgl climatological

cltgr climatographer

cltv closed-loop television

clu central logic unit; circuit lineup

CLU Chartered Life Underwriter

CLUB Central Library of the University of Baghdad

Club Med Club Méditerranée

CLUC Central Library Union Catalog

CLUE Comprehensive Loss Underwriting Exchange

CLUM Civil Liberties Union of Massachusetts

CLUMIS Cadastral and Land-Use Mapping Information System

clurt come let us reason together (mediator's motto)

CLUS continental limits United States

CLUSA Cooperative League of the USA

clut color look up table

clv clevis; constant linear velocity

Clv Cleveland

clvd calved; clavichord

clvs calves

Clw Collingwood

CLW Council for a Livable World

clwg clear wire glass

clws clockwise

Cly Clyde; Clydebank

CLY County of London Yeomanry

Clydebank Scotland's shipyard city on the River Clyde northwest of Glasgow

clz copper, lead, or zinc (cargo)

cm center of mass; centimeter(s); circular mil; circular muscle; cochlear microphonic; compounded monthly; contrast media; costal margin; countermortar; coupling multiplier; cruise missile; mechanic (symbol)

cm2 square centimeter

cm3 cubic centimeter

cm (CM) command module; communications multiplexor

c'm' come

c/m color modulation (tv); communications multiplexer; control and monitoring; corrected manifest current month

c & m care and maintenance; cocaine and morphine

cm carat métrique (French—metric carat); *Zentimeter* (German—centimeter)

c.m. *causa mortis* (Latin—cause of death); *cras mane* (Latin—tomorrow morning)

cM centimorgan (measure of genetic length)

Cm curium

Cm *Camino* (Spanish—highway)

CM absolute coefficient of pitching moments (symbol); Cameroon (Internet code); Canada Medal; Canadian Militia; Certificate of Merit; Certificated Master (or Mistress); Circulation Manager; Clyde-Mallory (steamship line); Command Module; Commercial Message, Common Market; Configuration Management; Corporate Member(ship); Corresponding Member; Court Martial; Cow Month (New Zealand); mine layer (naval symbol)

C-M Charente-Maritime

C.M. central meridian; *Chirurgiae Magister* (Latin—Master of Surgery)

C/M Curtis/Mathes

C of M Certificate of Merit; Count(y) of Mark

CM *Collegiate Microcomputer; Correo Maritimo* (Spanish—sea mail)

cm² square centimeter

cm³ cubic centimeter

CM4 Comet 4 jet airplane

cma civil-military affairs

Cma Camilla

C^ma *Cima* (French or Itatian—summit)

C Ma Canis major

CMA Cable Makers Australia; California Maritime Academy; California Medical Association; Canadian Manufacturers Association; Canadian Medical Association; Candle Manufacturers Association; Canterbury Manufacturers Association; Cash Management Account(ing); Casket Manufacturers Association; Central Monetary Authority; Certified Medical Assistant; Chemical Manufacturers Association; Childrenswear Manufacturers Association; Chinese Manufacturers Association; Chocolate Manufacturers Association; Cigar Manufacturers Association; Cleveland Metal Abrasive (company); Clothespin Manufacturers Association; Colleges of Mid-America; Colorado Mining Association; Commonwealth Medical Association; Confederate Memorial Association; Continental Marketing Association; Contract Manufacturers Association; Controller Military Accounts; Corps Maintenance Area; Corps of Military Accountants; Country Music Association; Court of Military Appeals; Crucible Manufacturers Association

CMA (C.M.A.) Certified Management Accountant

CMA *Compañia Mexicana de Aviación* (Spanish—Mexican Aviation Company)

CMAA Chief Master At Arms; Cleveland Musical Arts Association; Club Managers Association of America; Cocoa Merchants' Association of America; Comics Magazine Association of America; Construction Management Association of America; Courts-Martial Appeal Act; Crane Manufacturers Association of America

CMAAC Certified Medical Assistant Administrative and Clinical

CMAAO Confederation of Medical Associations in Asia and Oceania

cmab clothing maintenance allowance, basic

CMAB California Milk Advisory Board

CMAC Capital Military Assistance Command; Catholic Marriage Advisory Council; Computer Management and Control; Courts-Martial Appeal Court

CMAD Computer Manufacture and Design

cmai clothing maintenance allowance, initial

C Maj Canis Major

CMAL Clothing Monetary Allowance List; Coal Mines Authority Limited

CMAR Can't Manage A Rifle

C/marca Cundinamarca, Colombia

CMAS Council for Military Aircraft Standards

CMAS *Confédération Mondiale des Activités Subaquatiques* (French—World Confederation of Sub-aquatic Activities)

CMAT Canadian Mathematics Achievement Test; Commonwealth Military Advisory Team (Ghana)

CMAV *Coalition Mondiale pour l'Abolition de la Vivisection* (French—World Coalition for the Abolition of Vivisection)

Cmax maximum capacitance

cmb carbolic methylene blue; chloromercuribenzoate; coastal motor boat; core-mantle boundary

Cmb Colombo

CMB Chase Manhattan Bank; Cheese Marketing Board; Cina Motor Bus; coastal motor boat; Colombo, Ceylon (airport); Combat Maneuver Battalion(s)

CMB *Compagnie Maritime Belge* (Royal Belgian Lloyd Line); *Consejo Mundial de Boxeo* (Spanish—World Boxing Commission); *cuyas manos beso* (Spanish—whose hands I kiss)—very respectfully yours

CMBARMTNG Combined Arms Training

CMBG Canadian Mechanized Brigade Group

CMBI Caribbean Marine Biological Institute

CMBO Cape May Bird Observatory

cmbs commercial mortgage-backed securities

CMBs Certified Mortgage Loan Brokers

cmbt combat

cmc code for magnetic characters; computer-mediated communications; contact-making clock; coordinated manual control; cruise-missile countermeasures

cmc (CMC) carboxymethyl cellulose

CMc coastal mine layer (naval symbol)

CMC California Men's Colony; Canadian Marconi Company; Canadian Museum of Civilization (Ottawa); Canadian Music Council; China Machinery Company; Commandant of the Marine Corps; Commercial Metals Company; Computer-Mediated Communication; Computer Musician Coalition

cmca cruise-missle carrier-aircraft

CMCC Canadian Memorial Chiropractic College; Classified Matter Control Center

cm-cellulose carboxymethyl cellulose

CMCHS Civilian-Military Contingency Hospital System

cmcr continuous melting, casting, and rolling

CMCR Compagnie Maritime des Chargeurs Réunis

CMCSK Comcast Corporation

cmct communicate; communication

cmd command; common meter double

CMD California Moderate Democrats; Center for Management Development; Center for Moral Democracy (NYC); Central Marine Depot; Chief Medical Director; Contract Management District

cmdg commanding

Cmdr Commander

CMDR Council for Microphotography and Documentary Reproduction

Cmdre Commodore

Cmdt Commandant

cmdty commodity

cme continuing medical education; coronal mass ejection

CME California Motor Express; Center for Musical Experience; Central Medical Establishment; Chicago Mercantile Exchange (fornerly Chicago Butter and Egg Board); Chicago Merchandise Exchange; Courtesy Motorboat Examination (U.S. Coast Guard)

CME Conférence Mondiale de l'Energie (French—World Power Conference)

CMEA Chief Marine Engineering Artificer; Council for Mutual Economic Assistance (also called CEMA or COMECON)

CMEC Council of Ministers of Education (Canadian)

CMERD Centre for Medical Education, Research, and Development (New South Wales)

c'mere come here

CMERI Central Mechanical Engineering Research Institute (India)

cmet coated metal

cmf calcium-and-magnesium-free; countermortar fire; cylindrical magnetic film; cytoxan + methotextrate + 5-fluorouracil

cmf (CMF) cyclophosphamide methotrexate 5-fluorouracil (anticarcinogen)

CMF California Medical Facility; Canadian Military Forces; Central Mediterranean Fleet; Central Mediterranean Forces; Ceylon Military Forces; Citizen Military Forces; Commonwealth Military Forces; Composite Medical Facility

CMFNZ Chamber Music Federation of New Zealand

CMFRI Central Marine Fisheries Research Institute

cmfsw calcium-and-magnesium-free seawater

cmg call me gov(ernor); control-moment gyroscope; cystometrogram

CMG Color Marketing Group; Computer Management Group; Congress Medal for Gallantry; Corning Museum of Glass

C.M.G. Companion of the Order of St Michael and St George

CMGC Canadian Machine-Gun Corps

CMGH Cleveland Metropolitan General Hospital

cmh countermeasures homing

CMH Columbus, Ohio (airport); Combined Military Hospital; Congressional Medal of Honor; Council for the Mentally Handicapped; County Mental Health

cmha confidential, modified handling authorized

CMHA Canadian Mental Health Association

CMHC Central Mortgage and Housing Corporation; Community Mental Health Center(s)

CMHCA Community Mental Health Centers Act

CMHQ Canadian Military Headquarters

CMHPA Cloves Memorial Hall for the Performing Arts (Indianapolis)

cmi cable microcell integrator; carbohydrate metabolism index; cellular-mediated immune (response); container master information; cumulative monthly issue

cmi (CMI) computer-managed instruction

C Mi Canis Minor

CMI Can Manufacturers Institute; Career Maturity Inventory; Case-Mix Index; Christa McAuliffe Institute; Christian Michelson Institute (for Science and Free Thought—Bergen, Norway); Church Management Institute (Episcopal); Cleaning Management Institute; Command Maintenance Inspection (US Army); Commonwealth Mining Investments; Commonwealth Mycological Institute; Communications Management Inc; Consolidated Metal Industries; Cornell Medical Index; Cultured Marble Institute

CMI Comité Météorologique Internationale (French—International Meterological Committee); *Commission Mixte Internationale* (Spanish—International Joint Commission); *Cornell Medical Index*

CMIA Coal Mining Institute of America; Cultivated Mushroom Institute of America

cmid cytomegalic inclusion disease

cmif career-management individual file

CMIK Choson Minjujuui In'min Konghwaguk (North Korea)

cmil circular mil

c/min cycles per minute

Cmin minimum capacitance

C Min Canis Minor

CMIP Common Management Information Protocol

CMIU Cigar Makers' International Union

CMJ Church Mission to the Jews; Church's Ministry among the Jews

CMJ Computer Music Journal

cml chemical; circuit micrologic; commercial; current mode logic

cml chronic myeloblastic leukemia

cml (CML) chronic myelocytic leukemia

CML Central Music Library; Colonial Mutual Life; Container Marine Lines

CML Camara Municipal de Lisboa (Portuguese—Lisbon Town Council)

CMLA Canadian Music Library Association; Central Medical Library Association; Chief Martial Law Administrator (Bangladesh); Civilian Maimed and Limbless Association

CmlC Chemical Corps

cml def chemical defense

CMLEA California Media and Library Educators Association

cmlops chemical operations

CMLS Cleveland-Marshall Law School

CM/LSCNP Cradle Mountain/ Lake Saint Clair National Park (Tasmania)

CMLU Container Marine Lines (container) Unit

cmm commemorative; cubic millimeter(s); cutaneous malignant melanoma

CMM Capability Maturity Model; Center for Male Medicine; Central Methodist Mission; Chief Machinist's Mate (USN); Commander of the Order of Military Merit; Commission for Maritime Meteorology (WMO)

cmma clothing monetary maintenance allowance

CMMA Concrete Mixer Manufacturers Association

CMMBE Comissão Militar Mista Brasil-Estados Unidos (Portuguese—Joint Brazilian-American Military Commission)

cmmch combat Mach change

cmme chloromethyl-methyl ether

CMMGB Canadian Motor Machine-Gun Brigade

CMMM Chase Manhattan Money Museum (New York City)

CMMNA Catholic Major Markets Newspaper Association

cmmnd command(ing)

CMMP Commodity Management Master Plan; Corps of Military Mounted Police

CMMS Columbia Mental Maturity Scale

cmn commission; cystic medial necrosis

CMN Children's Miracle Network; Common Market Nationals; Common Market Nations

CMN Common Market News; *Conselho Monetario Nacional* (Portuguese—National Monetary Council)—Brazil

cmn-aa cystic medial necrosis of the ascending aorta

cmnce commence

CMNH Cleveland Museum of Natural History

CMNM Capulin Mountain National Monument; Craters of the Moon National Monument

cmnr commissioner

cmo calculated mean organism; cardiac minute output; computer microfilm output

CMO Chief Medical Officer; Collateralized Mortgage Obligation; Commonwealth Medical Officer; Configuration Management Office; Contract Management Office(r)

c'mon come on

cmos (CMOS) complementary metal-oxide semiconductor

cmp chronic mental patient; corrugated metal pipe; cost of maintaining product

cmp (CMP) compare (flow chart); computation(al)

CMP Canadian Military Police; Catoctin Mountain Park (Maryland); Christian Movement for Peace; Church Music Publishers; Commissioner of the Metropolitan Police; Company of Mission Priests; Competitive Medical Plan; Controlled Materials Plan; Cornell Maritime Press; Corps of Madras Pioneers; Corps of Military Police

CMPC California Manpower Planning Council; Canadian Military Police Corps

CMPC Compañia Manufacturera de Papeles y Cartones (Spanish—Paper and Carton Manufacturing Company)

CMP(I) Corps of Military Police (India)

cmpd compound; compounded; compounding

cm pf cumulative preference; cumulative preferred (shares)

cmpl complement (flow chart)

cmpld compiled

cmplt complete

cmpltn completion

cmplx complex

Cmpn Companion

Cmpn IAP Companion of the Institution of Analysts & Programmers

cmpnt component

CMPO Calcutta Metropolitan Planning Organisation

cmps centimeters per second

cmpt component

cmptr computer

cmr cerebral metabolic rate; common-mode rejection

CMR Canadian Mounted Rifles; Cape Mounted Rifles; Ceylon Mounted Rifles; Communications Monitoring Report; Consolidated Mail Room; Contract Management Region; Convention Merchandises Routiers; Court-Martial Reports; Court of Military Review

cmrO$_2$ cerebral metabolic rate for oxygen

CMRA Chemical Management and Resources Association

CMRB Chemicals and Minerals Requirements Board

Cmrd comrade

CMRE California Marriage-Readiness Evaluation

CMRF Childrens Medical Research Foundation

cmrg cerebral metabolic rate of glucose

CMRL Chamber of Mines and Research Laboratories

CMRN Cooperative Meteorological Rocket Network

CMRNWR Charles M. Russell National Wildlife Range (Montana)

cmro cerebral metabolic rate of oxygen

CMRO County Milk Regulations Office(r)

cmrr common mode rejection ratio

cmrs computer-managed rotation schedules

CMRs Classified Material Receipts

CMRS Countermeasures Receiver System

CMRW Coalition for the Medical Rights of Women

cms coastal minesweeper; complete matched set; conversational monitor system

cm/s centimeters per second

c.m.s. cras mane sumendus (Latin—to be taken tomorrow morning)

CMS California Malacozoological Society; California Museum of Science; Center for Measurement Science (George Washington University); Central Monitoring Stations (Africa); Chicago Medical School; Chief Master Sergeant; Christian Medical Society; Church Missionary Society; College Music Society; Computer Management System; Configuration Management System; Consumers and Marketing Service; Contemporary Music Society; Contractor's Management System; Conversational Monitor System; Correctional Medical Systems; County Medical Services

CMS Compagnie Maritime de la Seine (French—Seine Maritime Company)

CMSA Consolidated Metropolitan Statistical Areas

CM & SA Canning Machinery and Supplies Association

CMSC Cape Medical Staff Corps; Central Missouri State College; Corps of Military Staff Clerks

CMSEP Contractor Management System Evaluation Program

CMSER Commission on Marine Science, Engineering, and Resources

CMSG Canadian Merchant Service Guild

CMSgt Chief Master Sergeant

CMSI California Museum of Science and Industry; Council of Mutual Savings Institutions

CMS & I California Museum of Science and Industry

cm/sm command module/service module

CMSN China Merchants Steam Navigation (company)

CMSTP & P Chicago, Milwaukee, St Paul and Pacifiic (railroad)

cmt comment; confluence of major and tributary (rivers); cut, make, and trim (clothing)

CMT California Mastitis Test; California Motor Transport; Camden Marine Terminals; Certified Medical Transcriptionist; Charcot-Marie-Tooth; Chief Medical Technician; Compulsory Military Training; Concert Memory Test; Country Music Television; Current Medical Terminology; Current Mortuary Tables

CMT Confédération Mandiale du Travail (French—World Confederation of Labor)

CMTA Chinese Musical and Theatrical Association

CMTC Citizens Military Training Camp

cmt/conc cement or concrete

CMTCS Computer Management Transaction Control System

CMTCU Communications Message Traffic Control Unit

cmte committee

Cmte committee

Cmto Caminito

Cmto Caminito (Spanish—small street)

cmu central markup unit; chlorophenyldimethylurea; concrete masonry unit

CMU Central Michigan University; Church Management Institute

C-M U Carnegie-Mellon University

CMUA Commercial Motor Users Association

cmv current market value

cmv (CMV) cytomegalovirus

CMV Controlled Mechanical Ventilation

cmvm contact-making (or breaking) voltmeter

CM von W Carl María von Weber

CMVPB California Motor Vehicles Pollution Board

CMVS Cavalry Mobile Veterinary Section

cmw critical minimum weight

CMW Citizens for Migrant Workers

CMWA County Meat Works Association

cmy civilian man-years; cyan, magenta, yellow

CMY Cape Mounted Yeomanry

cmz concordant memory zone

CMZ Compagnie Maritime du Zaire (French—Zaire Maritime Company)

CMZS Corresponding Member of the Zoological Society

cn canal; cannon; cellulose nitrate; condensation nuclei; combined nomenclature; coordination number; credit note

cn (CN) chloroacetophenone

c/n carbon-to-nitrogen ratio; carrier-to-noise ratio

c/n (C/N) credit note

c.n. cras nocte (Latin—tomorrow night)

cN centinewton

Cn canon; contract number; cumulonimbus

CN absolute coefficient of yawing moments (aerodynamic symbol); Carl Nielsen; Central Airlines; China (Internet code); Chinese Nationalists; Code Napoléon; Commercial Node (zone); Cornmonwealth Nations; compass north; Confederate Navy; cosine of the amplitude (mathematical symbol)

C/N Consignment Note; Cover Note

C & N communication and navigation

CN Canadian National-Grand Trunk Railways

cna code not allocated

CNA California Nurses Association; Canadian Nuclear Association; Canadian Numismatic Association; Canadian Nurses Association; Center for Naval Analyses (Franklin Institute); Central News Agency (Nationalist China); Central Northern Airways; Certified Nurses Aide; Certified Nurses Assistant; Chemical Notation Association; Chief of Naval Air; Chief of Naval Aviation; Community Newspaper Association; Confederate Navy of America

CNA Comisión Nacional del Azúcar (Spanish—National Sugar Commission)—Mexico

CNAA Council for National. Academic Awards

CNAC China National Aviation Corporation

CNACO Canada National Arts Centre Orchestra

CNAD Center for a New American Dream

CNADS Conference of National Armaments Directors

CNAEA California Narcotic Addict Evaluation Authority

CNAF Chinese Nationalist Air Force

CNAN Compagnie Navale Afrique du Nord (French—North Africa Naval Company)

CNAS Chief of Naval Air Services; Civil Navigation Aids System

CNASA Council of North Atlantic Shipping Associations

CNATra Chief of Naval Air Training

CNAV Canadian Naval Auxiliary Vessel

CNAVSTA Charleston Naval Station (South Carolina)

CNB Central Narcotic Bureau (Singapore); Crocker National Bank

CNBC Consumer News and Business Channel

Cnbr Canberra

Cnbry Canterbury

CNBS Comisión Nacional Bancaria y Seguros (Spanish—National Banking and Security Commission)—Mexico

cnc central navigation computer; computer numerical control; consecutive number control

Cnc Cancer

CNC Christopher Newport College; Counter Narcotics Center

Cncl(r) Council(or)

CNCMH Canadian National Committee for Mental Hygiene

CNCO China Navigation Company

CNCP Canadian National/Canadian Pacific (telecommunications)

cncr concurrent

cnct connect(ion)

cnd condition(ed); conduit

CND Campaign for Nuclear Disarmament; Commission on Narcotic Drugs (UN)

CND Code Names Dictionary

CNDC Canadian National Defence College

cndi commercial nondevelopment items

cn di combination die

CNDP Communications Network Design Procedure(s)

cnds condensate

cne chronic nervous exhaustion

CNE Canadian National Exhibition; Commission on National Elections

cnee consignee

Cnel Coronel (Spanish—Colonel)

CNEL community noise equivalent level

C'nelia Cornelia

CNEN Comisión Nacional de Energia Nuclear (Spanish—National Nuclear Energy Commission)

CNEngO Chief Naval Engineering Officer

CNEO Chief Naval Engineer Officer

CNEP Cable Network Engineering Program (Bell)

CNES Centre National d'Etudes Spatiales (French—National Center for Space Studies)

CNET Chief of Naval Education and Training

CNET Centre National d'Etude des Télécommunications (French—Telecommunication National Study Center)

CNEXO Centre pour d'Exploitation des Océans (French—Center for the Exploitation of the Oceans)

cnf confine

CNF Caribbean National Forest (Puerto Rico); Cleveland National Forest (near San Diego, California); Commonwealth Nurses' Federation

cng compressed natural gas

cng (CNG) compressed natural gas

CNG Connecticut Natural Gas

CNGA California Natural Gas Association

CN-gas cyanide gas

CNGB Chief, National Guard Bureau; Chief of the National Guard Bureau

CNGO Committee on Non-Governmental Organizations (UN)

CNGT Commonwealth New Guinea Timbers

CN-GT Canadian National Railways-Grand Trunk Western

CNH Community Nursing Home

cnhd congenital nonspherocytic hemolytic disease

CNHI Committee for National Health Insurance

CNHM Chicago Natural History Museum (Field Museum of Natural History)

CNI Chief of Naval Information; Chief of Naval Intelligence; Coalition for Networked Information; Communications—Navigation and Identification; Community Nutrition Institute

CNI Centro de Información Nacional (Spanish—National Information Center)—Chilean security police; *Centro Nacional de Informaciones* (Spanish—National Information Center)—Chile's intelligence service

CNIB Canadian National Institute for the Blind; Champagne News and Information Bureau

CNIE Comisión Nacional de Inversión Extranjiera (Spanish—National Commission for Foreign Investment)—Mexico

CNIF Conseil National des Ingénieurs Français (French—National Council of French Engineers)

CNIN California Narcotic Information Network

CNIPA Committee of National Institutes of Patent Agents

CNJ Central of New Jersey (railroad)

CNJA Chief Naval Judge Advocate

cnl cancel(lation); cardiolipin natural lecithin

cnl (CNL) circuit net loss

CNL Canadian National Library (Ottawa); Commonwealth National Library (Canberra)

CNLA Canadian National Library Association; Council of National Library Associations

CNLIA Council of National Library and Information Associations

CNM Cabrillo National Monument; Certified Nurse Midwife; Chief of Naval Materiel; Chiridahua National Monument; Colombo National Museum; Colorado National Monument

CN-M Certified Nurse-Midwife

CNMB Central Nuclear Measurements Bureau

cnmbp chronic nonmalignant back pain

CNMI Commonwealth of the Northern Mariana Islands (in the western Pacific)

CNMP Comprehensive Nutrient Management Plan

CNN Cable News Network; Certificated Nursery Nurse

CNN Compagnie de Navigation Nationale (French—National Navigation Company)

CNNI Cable News Network International

CNNR Caerlaverock National Nature Reserve (Scotland); Cairngorms National Nature Reserve (Scotland)

cno computer non-operational

C^{no} Como (Italian—peak, summit)

CNO Chief of Naval Operations; Chief Nursing Officer

CNOA California Narcotics Officers Association

CNOAE Council of National Organizations of Adult Education

CNOBO Chief of Naval Operations Budget Office

cnop conditional no operation

cnor consignor

C-note $100 bill

CNP Canyonlands National Park (Utah); Caramoan NP (Philippines); Center for National Policy; Chief of Naval Personnel; Cleveland NP (South Australia); Colonial NP (Virginia); Corbett NP (India); Council for National Policy; Cyril Northcote Parkinson

CNP Compañia Navegación Peruana (Peruvian Navigation Company)

CNPA California Newspaper Publishers Association

CNPB Canadian National Parole Board

CNPC China National Petroleum Corporation

cn/pnl contractor's panel

CNPP Centre National de Prévention et de Protection (French—National Prevention and Protection Center)

CNPS California Native Plant Society; Central National Philharmonic Society (Beijing)

cnr carrier-to-noise ratio; composite noise rating; corner

Cnr Corner

CNR Canadian National Railway; Civil Nursing Reserve; Coleford Nature Reserve (South Africa); Council of National Representatives

CNR Consiglio Nazionale delle Ricerche (Italian—National Research Council)

CNRA Curecanti National Recreation Area (Colorado)

CNRS Centre National de la Recherche Scientifique (French—National Center for Scientific Research)

cnrt concrete

cns central nervous system; construction; continuous net settlement

cns (CNS) central nervous system

c.n.s. cras nocte sumendus (Latin—to be taken tomorrow night)

Cns Cairns

CNS Catholic News Service; Center for New Schools; Chief of the Naval Staff; Community Nursing Services; Congress of Neurological Surgeons; Copley News Service

CNS Chubu Nippon Shinibun (Central Japan Newspaper)

CNSA Carl Nielsen Society of America

CNSC China National Software Corporation

cnsg consolidated nuclear steam generator

cnsl console (flow chart)

CNSS Center for National Security Studies

CNSSO Chief Naval Supply and Secretariat Officer

Cnst Pty Constitution Party

cnstr canister

cnstrn construction

CNSWTG Commander, Naval Special Warfare Task Group

cnt (CNT) celestial navigation trainer

CNT Canadian National Telegraphs; Composite Negotiating Text(s)

CNT Compañia Nacional de Teléfonos (Spanish—National Telephone Company); *Confederación Nacional de Trabajo* (Spanish—National Confederation of Labor); *Conselho Nacional de Telecommunicação* (Portuguese—National Telecommunications Council)—government controlled radio and television for all Brazil

CNTB Colombia National Tourist Board

CNTCA Canadian National Railway—Transcanada Airlines

cntn contain

CNTO Centocor Incorporated

Cn Tower Canadian National Tower, Toronto, Ontario

cntr container; contribute; contribution; counter

Cntr Centaur (space vehicle)

cntrfugl centrifugal

cntrl central; control(ler)

cntrs containers

CNTU Canadian National Trade Unions; Confederation of National Trade Unions

CNUCE *Centro Nazionale Universitàrio di Culcol Elettronico* (Italian—National University Center of Electronic Calculation)

Cnut King Canute II of Denmark and England

cnv contingent negative variation

CNV Cape Canaveral, Florida (tracking station)

CNVA Committee for Non-Violent Action

cnvc conveyance

cnvr conveyor

cnvt convict

C & NW Chicago and North Western (railway)

CNWDI Critical Nuclear Weapons Design Information

CNWR Camas National Wildlife Refuge (Idaho); Chassahowitzka NWR (Florida); Chatauqua NWR (Illinois); Chincoteague NWR (Virginia); Columbia NWR (Washington)

CNX Canadian National Exposition

CNYP Central New York Power (corporation)

co carbon monoxide; cardiac output; castor oil; central office; cervicoaxial; choir organ; cleanout; coenzyme; combined operations; command order; company; conscientious objector; consolidated obligation; consultation; convenience outlet; corneal opacity; criminal offense; crossover(s); cutoff, cutout; output capacitance

Co. county

co (CO) close/open (to official correspondence)

c-o cutoff

c/o care of; carried over; cash order; complains of

co *compagno* (Italian—company); (Latin prefix—together)—copulation

co. *compositus* [Latin—compound(ed)]

Co cobalt; Colombia; Colombian; Colombiano; Columbia; Columbian; company; County

C/o complained of

Co *cerro* (Spanish—hill or peak)

CO carbon monoxide; Certificate of Origin; Chief Officer; Cleveland Orchestra; Colombia (Internet code); Colonial Office; Colorado; Commanding Officer; Commissioners Office; Communications Officer; conscientious objector; Continental Airlines (2-letter code); Controller's Office; Copyright Office; Correctional Officer; Criminal Office; Crown Office(r)

C/O cash order; Chief Officer

C & O Chesapeake & Ohio (railroad)

C of O Count(y) of Oldenburg

co 1mo *canto primo* (Italian—first treble)

CO₂ carbon dioxide

Co⁶⁰ radioactive cobalt

coa certificate of authority; condition on admission; corps of ordnance artificers

coA coenzyme A

CoA College of the Atlantic; Committee on Accreditation (ALA); Council of the Americas

C o A Committee on Accreditation (ALA); Court of Appeal

COA Canadian Orthopedic Association; Change Order Account; Chattanooga Opera Association; Conchologists of America; Connecticut Opera Association; Control Operating Authority (NATO); Cordova Airlines; Correctional Officers Association

CO(A) Change Order (Aircraft)

COA *Comunidad Oriental Africana* (Spanish—East African Community)

coac clutter-operated anti-clutter receiver

COAC Commanding Officer, Atlantic Coast

Coad Coadjutor

COADS Command and Administration System (USA)

coag coagulant; coagulate; coagulation

coag time coagulation time

Coah Coahuila (inhabitants—Coahuileños or Coahuilenses)

Coal. Coalition

coalit govt coalition government

co/alr cooper/aluminum

coam coaming; customer-owned-and-maintained equipment

coam equip customer-owned-and-maintained equipment (data processing)

CO-AMP Cost Optimization Analysis of Maintenance Policy

coas crewman optical alignment sight

COAS Council of the Organization of American States

Coast The Coast (Pacific coast of Canada and the United States)

Coastal Eastern East-Coast-of-the-United-States English reflecting cultural influences

Coastal Express Norwegian steamship line linking ports from Bergen to Kirkenes

Coast Line Atlantic Coast Line Railroad

coax coaxial

c-o-b close of business

CoB Chief of Base (CIA)

C o B Chief of Boat (submarine)

COB Change Order Board; COBOL (Common Business-Oriented Language); Command Operating Budget

COBA Correction Officers Benevolent Association

COBAS Council of Black Architectural Schools

cobble(s) cobblestone(s)

COBE Cosmic Background Explorer Satellite

COBF Cobol-F (program)

cobh carboxyhemoglobin

cobility cobol utility (program)

coblib cobol library

C & O-B & O Chesapeake and Ohio-Baltimore & Ohio (merged railroads)

COBOL common business-oriented language

cobra cabinet officer briefing action

cobra. (COBRA) coolant boiling in rod arrays

COBRA Consolidated Omnibus Budget Reconciliation Act (of 1985); Counter Battery Radar

COBRA Computadores Brasileiros (Portuguese—Brazilian Computers)

COBRAY Cobra + Moray (anti-terrorist academy)

COBSI Committee on Biological Sciences Information

COBTU Combined Over-the-Beach Terminal Unit

coc cathodal opening clonus; cathodal opening contraction; cocaine; coccygeal; combination-type oral contraceptive; combined operations center

Coc Cleveland open cup; Cocos Islands

CoC Chamber of Commerce

C o C Council on Competitiveness

COC Canadian Opera Company; Combat Operations Center

coca cocaina (Spanish—cocaine)

coca-colon coca-colonization; coca-colonize; coca-colonizer

C & O Canal Chesapeake and Ohio Canal

COCAST Council for Overseas Colleges of Art, Science, and Technology

cocb crossed olivochochlear bundles

cocc coccyx

cocci coccidioidomycosis

coccy coccidioidomycosis

COCESS Contractor-Operated Civil Engineer Supply Store

coch coach(es)

coch. cochleare (Latin—spoonful)

Coch Cochin

coch. ampl. cochleare amplum (Latin—tablespoonful)

COCHASE Code for Coupled-Channel Schrödinger Equations

coch. infant. cochleare infantis (Latin—teaspoonful)

coch. mag. cochleare magnum (Latin—tablespoonful)

coch. med. cochleare medium (Latin—dessertspoonful)

coch. parv. cochleare parvum (Latin—teaspoonful)

COCI Council on Consumer Information

cock. cockney (dialect of London's East End and waterfront residents born within sound of the bells of the Church of Saint Mary-leBow—Bow bells)

Cock Cockburn; Soviet Antonov 350-passenger plane (NATO)

cockapoo cocker spaniel-poodle mix-breed dog

cockaterr cocker-terrier mixed breed dog

Cock Engl Cockney English (spoken by people in London's East End and traditionally born within sound of Bow bells)

coci cathodal opening clonus

C & OC NM Chesapeake and Ohio Canal National Monument

co-co carried-on-carried-off (break-bulk cargo)

cocohol coconut-oil-extended ethanol (diesel fuel)

COCOM Coordinating Committee for Export to Communist Area(s); Coordinating Committee on Multilateral Export Controls

COCOMO Constructive Cost Model

COCOSEER Coordinating Committee on Slavic and East European Library Services

cocp closed olivocochlear potential

cocr cylinder overflow control record

COCS Container-Operating Control System

coct. coctio (Latin—boiling)

Co Cts County Courts

COCU Churches of Christ Uniting; Consultation on Church Union (of Episcopalians, Methodists, Presbyterians, and others)

cod cancellation of debt; carrier on-board delivery; cause of death; central ordnance depot; change over delay; chemical oxygen demand; cleanout door; codeine; collect on delivery

c-o-d cargo-on-deck

cod. codicil

c.o.d. cash-on-delivery

CoD Chief of Detectives

Co D Costume Designer

COD coding

COD Concise Oxford Dictionary

coda cash or deferred arrangement

CODA Came Out Decades Ago; Committee on Drugs and Alcohol

codac coordination of operating data by automatic computer

CODAC Community Organization for Drug Abuse Control

CODAF Commission on Border Development and Friendship (U.S.-Mexican)

codag combined diesel and gas (turbine machinery)

codan carrier-operated device anti-noise

CODAP Client-Oriented Data-Acquisition Process; Control Data Assembly Program

CODAS Customer-Oriented Data System

CODASYL Conference on Data Systems Languages

CODATA Committee on Data for Science and Technology

CODC Canadian Oceanographic Data Center

codd codices

CODE Collaborating on Drug Education; Commission on Declining Enrollments; Committee on Donor Enlistment

Code Blue (medical jargon) frequently used to mean cardiac arrest

codec coder decorder

coded. computer-oriented design of electronic devices

CODEF Chairman of Defense Committee

CODELCO Corporación del Cobre (Spanish—Copper Corporation)—Chile

codel(s) congressional delegation(s)

CODELS Computer Development System

Code N Code Napoléon

CODES Computer-Oriented Data Entry System

codic computer-directed communication(s)

codiphase coherent digital-phased array system

CODIS Combined DNA Index System

codit computer direct to telegraph

codl cash on the dotted line

cod. memb. *codex membranacius* (Latin—book printed or written on skin or vellum)

CoDoC Cooperation in Documentation and Communication

CODOFIL Council for the Development of French in Louisiana

codog combined diesel or gas

CODOT Classification of Occupations and Directory of Occupational Titles (UK)

CODSIA Council of Defense Space Industries Association

coe cab over engine (truck); close of escrow (realty); crude oil equivalent

coe (COE) crossover electrophoresis

CoE Corps of Engineers

COE Commonwealth Office of Education; Corps of Engineers; Council for Occupational Education; Council on Optometric Education

CO(E) Change Order (Electronic)

COE Conseil Aécuménique des Eglises (French—World Council of Churches)

coea cost and operational effectiveness analysis

COEA Chief Ordnance Electrical Artificer

COEA Consejo de la Organisación de los Estados Americanos (Spanish—Council of the Organization of American States)

coed coeducation(al); female student

coed (COED) computer-operated electronic display

co-ed co-editor

co-edn(s) co-edition(s)

COEDS Char Oil Energy Development Systems

COEES Central Office Equipment Engineering System (Bell)

coef coefficient

coel (Greek *koilos*—cavity, hollow)—coelenterate, coelomate, coelostat(ic)

Coel Coelenterata

COENCO Committee for Environmental Conservation .

COEP Community Outreach Educational Program

COEPS Cortically-Originating Extra-Pyramidal System

COESA Committee on Extension of the Standard Atmosphere (United States)

coet (COET) crude oil equalization tax(ation)

COEV Canadian Ocean Escort Vessel

coew combined operations experimental wing

cof cause of failure; coefficient of friction; cost of funds

CoF Chaplain of the Fleet; Chief of Finance

COF Captain of the Fleet

c of a coat of arms

COFACE Confederation of Family Organizations in the European Community

cofad computerized facilities design

cofc container on flat car

coff cofferdam

COFFEE Community Organization for Full-Employment Economy

c/offer counter offer

Coffs Coffs Harbour, Australia

COFHE Consortium on Financing Higher Education

COFI Committee on Fisheries (FAO); Cost of Funds Index

COFIDE Corporación Financiera de Desarrollo (Spanish—Financial Development Corporation)

COFIPS Central Ohio Federation of Informatioim Processing Societies

COFIS Canadian On-Line Financial Information Service

COFO Council of Federated Organizations (CORE, NAACP, SCLC, SNC)

COFPHE Capital Outlay Fund for Public Higher Education

co/fr counter offer

COFRC Chevron Oil Field Research Company

cofron copper iron (patent medicine mixture)

COFSAF Chief of Staff, U.S. Air Force

cofx component fixture

cog center of gravity

cog. cognate

c-o-g coal to oil to gas

CoG Council of Governments

COG Change Our Gender; Change Our Goal; Council of Governments

cogag combined gas and gas

CoGARD Coast Guard

cogas coal-oil-gas

cogb certified official government business

cogent. compiler and generalized translator

cogita computerized general I.Q. test(ing)

cogn cognomen

cognit cognition(al)(ly); cognitive(ly)

cogn w cognate with

cogo coordinate geometry

COGP Commission on Government Procurement

cog/prsl cognizant personnel

cogs cost of goods sold

cogs. combat-oriented general support

COGS Computer Oriented Geological Society; Continuous Orbital Guidance System

COGSA Carriage of Goods by Sea Act

cogtt cortisone-primed oral glucose tolerance test

coh cash-on-hand; coefficient of haze; coheir

COH carbohydrate (generalized formula); Corporal of Horse

COHA Council on Hemispheric Affairs

COHATA Compagnie Haitienne des Transports Aériens (French—Haitian Air Transport Company)

cohb carboxyhemoglobin

COHb Carboxyhemoglobin

Co Hd coral head

COHD Copyright Office History Document

coher cohere(d); coherence; coherency; coherer; cohering; coherent(ly)

COHM Copyright Office History Monograph

coho coherent oscillator

COHO Council of Health Organization

Cohoun Colquhoun

cohq combined operations headquarters

COHSE Confederation of Health Service Employees

COHST Certified Occupational Health Safety Technician

coi classroom observation instrument; crack-opening interferometry

COI Central Office of Information; Certificate of Origin and Interest; Coordinator of Information

COI Comité Olimpico Internacional (Spanish—International Olympic Committee); *Commission Oceanagraphique Intergouvernementale* (French—Intergovernmental Oceanographic Commission)

COIC Canadian Oceanographic Identification Center; Combined Operations Intelligence Center

CoID Council of Industrial Design

coif coiffure

COIMS Council for International Organizations of Medical Sciences

coin. coinage; counterinsurgency—anti-guerrilla warfare

coin. (COIN) complete operating information

COIN Counterinsurgency

Coin & Curr Coin and Currency Institute

coin gold 90% gold, 10% copper

coin-op coin-operated

COINS Computer and Information Sciences; Computerized Information System(s); Control in Information Systems; Cooperative Intelligence Network System

coin silver 50 to 92.5% silver with balance of copper or other metals

co-intel counterintelligence

COINTELPRO Counterintelligence Program (FBI)

COIR Commission on Intergroup Relations (NYC)

Cois François

COIT Central Office of the Industrial Tribunal (UK)

COIU Congress of Independent Unions

COJ Court of Justice

COJO Conference of Jewish Organizations

Cok Cochin

coke coca drink; cocaine

Coke Coca Cola; NATO nickname for the Soviet Antonov AN-26 350-passenger transport plane

Coke Rep Coke's English King's Bench Reports

cokesmoke(s) cocaine smoker; cocaine smoking

col chain overseas low (radar); colon; colonial; colonic; colonist; colonization; colonize; colony; color; coloring; colorist; colors; column

col (COL) computer-oriented language

c-o-l (COL) cost of living

col. colatus (Latin—strained, as through a filter); *collum* (Latin—collar); *colon* (Latin—large intestine)

co-L co-latitude

Col Colchester; Colima; Coliseum (London); College; Cologne; Colombia(no); Colón (Panamanian port city facing Christóbal); Colonel; Colossians, Epistle to the; Columba (constellation); Columbia(n); Columbus; Coronel

COL Computer Oriented Language

cola (COLA) cost-of-living adjustment; cost-of-living allowance

cola. cost-of-living allowance

cola colonia (Spanish—colony)

COLA Committee on Latin America; Committee on Library Automation (ALA); Cost of Living Adjustment

COLAC Central Organization of Liaison for Application of Circuit

Col Alb College of the Albermarle

COLAs Cost-of-Living Adjustments

colat. colatus (Latin—strained)

col bh collision bulkhead

col C col canto (Italian—follow the voice)

COLC Cost of Living Council

cold. chronic obstructive lung disease; colored

Col$ Colombian peso

COLDEMAR Compañia Colombiana de Navegación Maritima (Spanish—Colombian Maritime Navigation Company)

colen. colentur (Latin—let them be strained, strain them)

Col Ency Columbia Encyclopedia

Col Ent Exam College Entrance Examination

coleop coleoptera; coleopterist

colet. coleatur (Latin—let it be strained, strain it)

COLI Corporate-Owned Life Insurance

colidar coherent light detection and ranging

Colin Nicholas

colingo compile online and go (data processing)

coll collected; collection; collective; collect(or); colloid(al); colloquial(ism)

Coll College; Collegiate

collab collaboration; collaborator

collabo(s) collaborator(s)

coll agc collection agency

collat collateral

collect. collection; collective; collectively

Coll Ency Colliers' Encyclopedia

Collier-Macmillan Collier-Macmillan Library Service

Collins Wm Collins Sons & Co

Collins Bay Canadian penitentiary on Collins Bay near Kingston, Ontario

Coll L Collection Letter

colloq colloquial(ism); colloquium

coll'ott coll'ottava (Italian—play in octaves, with the octave)

collr collector

collun. collunarium (Latin—nose wash)

collut. collutorium (Latin—mouthwash)

coll vol collective volume

Coll Wooster College of Wooster

colly colliery

collyr. collyrium (Latin—eyewash)

colm column

Colm Columba

COLMIS Collection Management Information System

colo colophon (printer's or publisher's device, symbol, or trademark)

Colo Colorado; Coloradan

Colo Colossians

Colo Assoc Colorado Associated University Press

colog cologarithm

colograph color lithograph

Colom Colombia; Colombian

Colom Christovão Colom (Portuguese—Christopher Columbus)

Colombia Republic of Colombia (South American two-ocean nation), *República de Colombia*

Colombian West Indies Saint Andrew and Providence islets, seven groups of coral cays on Nicaragua's Mosquito Coast

Colombo Christoforo Colombo (Italian—Christopher Columbus)

colón unit of currency in Costa Rica and El Salvador

coloph colophon

Colorado Springs U.S. Air Force Academy at Colorado Springs, Colorado

colorectal colon rectal (area or cancer)

coloreds colored persons (South Africans of mixed blood)

Colo St U Comm Colorado State University Institute in Technical and Industrial Communications

col p color page

COLPA Commission on Law and Public Affairs

colrad collegiate research and development

COLREG Regulations Governing Collisions

COLS Communications for On-Line Systems

Col-Sgt Colour-Sergeant

colspd collapsed

Col Sym Columbia Symphony

colt computerized on-line test(ing)

COLT Council on Library Technology; Council on Library-Media Technical-Assistants

Colu Columba

Columbia School Columbia Training School at Columbia, Mississippi

Columbis Columbidae (doves and pigeons)

Columbus House Columbus Workhouse and Women's Correctional Institution at Columbus, Ohio

com comedy; comma; command; commercial; commission; committee; common; communication(s); complement; compliment; componentized architecture

com. commentary; commoner; communicate; companion

com (COM) computer output on microfilm; computer-output microfilm(ing)

com comunicaciones (Spanish—communications)

com or *con* or *cor* (Latin *cum*—together, with)—compound, confine, correct

com. commemoratio (Latin—commemoration)

Com Coma Berenices (constellation); Commander; Comoro Islands; Republic of Comoros (whose capital is Moroni)

Com. Commissioner

COM Chief Operations Manager; Church of Melanesia; Council of Ministers

COMA Coke Oven Managers' Association; Council On Minority Aging

comac continuous multiple-access comparator

COMACH Confederación Maritima de Chile (Spanish—Maritime Confederation of Chile)

COMAEGEAN Commander, Aegean

comaf commodore amphibious forces

COMAINT Command Maintenance

COMAIR Commercial Airways

COMAIRBALTAP Commander Baltic Air Forces (NATO)

COMAIRCENT Commander, Allied Air Forces, Central Europe

COMAIRCENTLANT Air Commander, Central Atlantic

COMAIRCHAN Maritime Air Commander, Channel

COMAIREASTLANT Air Commander East Atlantic Area

COMAIRESTLANT Air Commander, Eastern Atlantic

COMAIRLANT Commander, Air Force, Atlantic

COMAIRNON Commander, Allied Air Forces, Northern Norway

COMAIRNORLANT Air Commander, Northern Atlantic

COMAIRNORTH Commander, Allied Air Forces, Northern Europe

COMAIRSONOR Commander, Allied Air Forces, Southern Norway

COMAIRSOUTH Commander, Allied Air Forces, Southern Europe

Comalco Commonwealth Aluminum Company (Australia)

COMAMF (1) Commander Ace Mobile Force (land)

COMANDOS Construction and Management of Distributed Office Systems

COMANSEC Computation and Analysis Section (Canadian Defense Research Board)

COMANTDEFCOM Commander, United States Antilles Defense Command

comar computer aerial reconnaissance

COMARAIRMED Commander, Maritime Air Forces Mediterranean

COMARC Cooperative Machine Readable Cataloging

COMARRHIN Commander, Maritime Rhine

COMART Commander, Marine Air Reserve Training

comat computer-assisted training

COMAT Committee on Materials

COMATS Commander Military Air Transport Service

Com Aus Commonwealth of Australia

comb. combat; combination; combine; combing; combustion

Com Bah Commonwealth of the Bahamas

COMBALTAP Allied Command Baltic Approaches (NATO)

COMBARFORCLANT Commander, Barrier Forces, Atlantic

COMBATCRULANT Commander, Battleship-Cruiser, Atlantic Fleet

combi combination

combine. combined harvester

COMBISLANT Commander, Bay of Biscay, Atlantic

COMBLACKBASE Commander, Black Sea Defense Sector

combo combination (of musicians, or of a safe)

COMBO Combined Arts of San Diego

combo lock(s) combination lock(s)

COMBOSFORT Commander, Bosphorus Fortifications

COMBQUARFOR Combined Quarantine Force

Com Brit Comunidad Británica (Spanish—British Commonwealth of Nations)—Great Britain and former colonies

COMBRITELBE Commander, British Naval Elbe Squadron

COMBRITRHIN Commander, British Naval Rhine Squadron

combs. combinations

combu combustion

combust combustion

COMCANLANT Commander, Canadian Atlantic

COMCARIBSEAFRON Commander, Caribbean Sea Frontier

COMCAT Computer Output Microfilm Catalog

COMCEN Communications Center

COMCENTLANT Commander, Central Atlantic

ComCm Communications counter-measures and deception

COMCRUDESFLOT Commander Cruiser-Destroyer Flotilla

COMCRUDESPAC Commander Cruisers and Destroyers in the Pacific (USN)

COMCRULANT Commander, Cruisers, Atlantic

comd command

COMDARFORT Commander, Dardanelles Fortifications

COMDESFLOT Commander, Destroyer Flotilla

COMDEV Commonwealth Development

comdg commanding

Com Dom Commonwealth of Dominica

Comdr Commander

Comdt Commandant

COME Chief Ordnance Mechanical Engineer

comeas countermeasures

COMEASTSEAFRON Commander, Eastern Sea Frontier

COMECON Council of Mutual Economic Assistance (of communist nations)

COMED Communications Editing Unit

COMEDBASE Commander, Mediterranean Defense Sector

COMEDI Commerce Electronic Data Interchange

COMEDS Continental Meteorological Data System

COMEINDORS Composite Mechanized and Document Retrieval System

COMERMEX Commercial Mexicano (Banco) (Spanish—Commercial Bank of Mexico)

Com Err Comedy of Errors

comet. computer operated management evaluation technique

COMET Committee for Middle East Trade; Computer-Operated Management Evaluation Technique; Construction Organization Membership Education Training; Controllability, Observability, and Maintenance Engineering Technic

COMETS Computer-Operated Multifunction Electronic Test Station

COMETT Community Program for Education and Training in Technology

COMEX Commodity Exchange (NY)

COMEXCO Committee for Exploitation of the Oceans

COMFAIRELM Commander, Air Fleet, Eastern Atlantic and Mediterranean

COMFAIRWINGLANT Commander, Fleet Air Wing, Atlantic

Com Fran Cumunidad Francesa (Spanish—French Community of Nations)—France and former colonies

comfy comfortable

Com-Gen Commissary-General

COMGENEUCOM Commanding General, European Command

COMGENTHIRDAIR Commanding General, Third Air Division

COMGENUSAFE Commanding General, U.S. Air Forces, Europe

COMGENUSAREUR Commanding General, U.S. Army, Europe

COMGERNORSEA Commander, German North Sea Subarea

COMGIBLANT Commander, Atlantic Approaches Gibraltar

Com/I Commercial Invoice

COMIBERLANT Commander, Iberian Atlantic Area

COMIBOL Corporacíon Minera de Bolivia (Spanish—Bolivian Mining Corporation)

COMICEDEFOR Commander, United States Iceland Defense Force

COMICS Computer-Oriented Managed-Inventory Control System

COMIL Chairman of Military Committee

comin' coming

COMINCH Commander-in-Chief, United States Fleet

COMIND Commander, Indian (USN)

Cominform Communist Information Bureau

comint communications intelligence

Comintern Communist International; Cominform

Com Int Sec Committee on Internal Security (formerly House Committee on Un-American Activities—HUAC)

Comiskey Comiskey Park, Chicago

Com Isl Comoro Islands

comis° comisario (Spanish—commissary, delegate, deputy, manager, police inspector)

comit computer operations management information training

COMITEXTIL Coordination Committee for the Textile Industry in the EEC

COMJUWATF Commander, Joint Unconventional Warfare Task Force

comkd completely knocked down

coml commercial

COMLA Commonwealth Library Association

COMLAND Commander Land Forces

COMMANDCENT Commander, Allied Land Forces, Central Europe

COMLANDEAST Commander, Allied Land Forces, Southeastern Europe

COMLANDMARK Commander, Allied Land Forces, Denmark

COMLANDNON Commander, Allied Land Forces, Northern Norway

COMLANDNORWAY Commander, Allied Land Forces, Norway

COMLANDSOUTH Commander, Allied Land Forces, Southern Europe

COMLANT Commander, Atlantic (USN)

COMLOGNET Combat Logistics Network

comm commerce; commercial; commission; committee; commonwealth; commune; communication; commutator

comm. commissioners

comm. commune (Latin—all the people, the community)

Comm community

Comm. Commodore

Comm Commendatore (Italian—Commander, knight)—equivalent to the British Sir

COMMAIRGIBLANT Commander, Maritime Air, Gibraltar, Atlantic

Commanders Commander Islands in the Bering Sea

Com Mat Cen Communication Materials Center (Columbia University)

Comm Bio Pest Committee for Biological Pest Control (San Ysidro, California)

COMMCEN Communications Center

commd command(ing); commissioned

commdg commanding

Commdr Commander

Commdt Commandant

commem commemoration; commemorative

Comments Nucl Part Phys Comments on Nuclear and Particle Physics

Commerce Department of Commerce; High School of Commerce

COMMFEX Communications Field Exercise

commfu complete and utterly monumental foulup

commi communism; communist

commie commissary; communist

COMMIR Commissioner of Inland Revenue

Commiss Commissary

comm'l commercial

commn commission

commo communications

Commo Office of Communications (CIA)

commod commodity

Commons House of Commons

common salt sodium chloride (NaCl)

Commonwealth British Commonwealth of Nations: the United Kingdom, Australia, Bahamas, Bangladesh, Barbados, Botswana, Canada, Cyprus, Ghana, Grenada, Guyana, Fiji, India, Jamaica, Kenya, Lesotho, Malawi, Malaysia, Malta, Mauritius, Nauru, New Zealand, Nigeria, Sierra Leone, Singapore, Sri Lanka, Swaziland, Tanzania, The Gambia, Tonga, Trinidad and Tobago, Uganda, Western Samoa, Zambia, and their dependent territories

Commr Commissioner

commstitch communications failure detecting and switching (equipment)

commun communication

commun dis communicable disease

Commun Math Phys Communications in Mathematical Physics

Commun Pure Appl Math Communications on Pure and Applied Mathematics

commuterport(s) commuter-type airport(s)

commy commissariat; commissary; communist

commz communications zone

comn common

ComNAB Commander, Naval Air Bases

COMNAV Commander Naval Forces

COMNAVBALTAP Commander Baltic Naval Forces (NATO)

COMNAVBASE Commander, Naval Base

COMNAVBREM Commander, Bremerhaven Naval Group

CONAVCAG Commander, Naval Forces, Central Army Group Area and Bremerhaven

COMNAVCENT Commander, Allied Naval Forces, Central Europe

COMNAVCRUITCOMINST Commander, Naval Recruiting Command Instructions

COMNAVFORCESMARIANAS Commander, Naval Forces, Marianas Islands

COMNAVFORJAPAN Commander, Naval Forces, Japan

COMNAVGERBALT Commander, German Naval Forces, Baltic

COMNAVNORCENT Commander, Northern Air Forces, Central Europe

COMNAVNORTH Commander, Allied Naval Forces, Northern Europe

COMNAVSONOR Commander, Allied Naval Forces, Southern Norway

COMNAVSOUTH Commander, Naval Forces, Southern Europe

COMNAVSUPPACT Commander, Naval Support Activity

comnd commissioned

COMNEATLANT Commander, Northeast Atlantic

COMNON Commander, Allied Forces, Northern Norway

COMNORASDEFLANT Commander, North American Anti-Submarine Defense Force, Atlantic

COMNORLANT Commander, Northern Atlantic

COMNORSEACENT Commander, North Sea Subarea, Central Europe

comnr commissioner

Como Commodore; Como-doro, Rivadavia (Argentine naval hero and seaport name); Comoro

comp accompaniment; accompany; comparative; comparator; compare; comparison; compass; compensate; compensation; compensatory; compilation; compile(d); compiler; complimentary; compose(d); composition; compositor; compound(ed); comprehension; comprehensive; comptroller; computer; rhetorician's mark meaning false comparison

comp (COMP) complainant

comp. company

comp. compositus (Latin—compounded of)

COMP College of Osteopathic Medicine of the Pacific; Conceptually-Oriented Mathematics Program

comp a compressed air

compac computer-output microfilm package; computer program for automatic control

COMPAC Commander, Pacific (USN); Commonwealth Pacific Telephone Cable (linking Australia, New Zealand, and Pacific Ocean islands with the rest of the world)

COMPACS Computer-Output Microforms Program and Concept Study (USA)

compact. compatible algebraic compiler and translator; computer planning and control technique

COMPACT Computator Planning and Control Technique

compand compress to expand (radio communication term)

compar comparative

compare. computerized performance and analysis response evaluator; console for optical measurement and precise analysis of radiation from electronics

COMPASS Comprehensive Assembly System: Computerized Movement Planning and Status System

COMPATFOR Commander, Patrol Forces

COMPATFORNORLANT Commander, Patrol Forces, Northern Subarea, Atlantic

comp case compensation case

Comp Curr Comptroller of the Currency

compd compound

compdes compensator design; competitive design

COMPELS Computerized Evaluation and Logistics System (USA)

compen compensate; compensation; compensatory

compend compendious; compendium

Compendex Computerized Engineering Index

compf composition floor

Comp Gen Comptroller General

compl complaint; complete; compilation; compiled

Compl A Lovers Complaint

COMPLEX Committee on Planetary and Lunar Exploration

complic complications

complt complainant; complaint

comp mar companionate marriage

COMPMR Commander, Pacific Missile Range

compn composition

compo compensation; component; composer; composite; composition; compositor

compool common pool; communications pool(ing)

compos components; composers; composites; compositions; compositors

compound A 11-dehydrocorticosterone

compound B corticosterone

compound E cortisone

compound F cortisol

compound S 11-deoxycortisol

Compr compressor

compreg compressed-impregnated (wood)

comprosl compound procedural scientific language

comp(s) complimentary ticket(s)

compt catecholomethyltransferase; compartment; comptroller

Compt Comptroller

Comptes Rend. Comptes rendus de l'Académie des Sciences (Proceedings of the Academy of Science)

comp time compensatory time off in lieu of overtime wages

COMPTUEX Composite Training Unit Exercise

compu computable; computability; computation(al); computer; computerization; computerize

COMPUSEC Computer Security

CompuServ Computer Service (network)

Compu Sex Computer Sex (erotic messages conveyed to home-computer owners)

comput computer

computa computational

computes. computers

computime computer-computed time

Comr Commissioner

COMRAC Combat Radius Capability (DoD)

com rcm command reconnaissance

comirel community relations

COMRNDN Commander, Riverine Division (USN)

COMRNFLOT Commander, Riverine Flotilla (USN)

COMRNRON Commander, Riverine Squadron (USN)

coms communications support

COMS College of Osteopathic Medicine and Surgery (Des Moines)

comsab communist sabotage; communist saboteur

comsabs communist saboteurs

COMSAMAR Commander, Straits and Marmara Defense Sector

Comsat Communications Satellite (corporation)

comsat(s) communications satellite(s)

ComSeaFron Commander Sea Frontier (USN)

comsec communications security

COMSEC Communications Security

COMSECONDFLT Commander, Second Fleet (USN)

COMSENEX Combined Sensor Tracking Exercise

COMSER Commission on Marine Science and Engineering Research

Com Ser Cen Community Service Center

COMSEVFLT Commander, Seventh Fleet (USN)

COMSIXFLT Commander, Sixth Fleet (USN)

comsn commission

comsoal computer method of sequencing operations for assembly lines

comstar communications satellite network

Comsteel Commonwealth Steel

comstock comstockery

COMSTRATRESCENT Commander, Strategic Reserve, Allied Land Forces, Central Europe

COMSTRIKFLTLANT Commander, Striking Fleet Atlantic (USN)

COMSTRIKFORSOUTH Commander, Naval Striking and Forces Support, Southern Europe

COMSTS Commander Military Sea Transport Service

COMSUBEASTLANT Commander, Submarine Force, Eastern Atlantic

COMSUBLEDNOREAST Commander, Submarines, Northeast Mediterranean

COMSUBPAC Commander, Submarines, Pacific

comsy commissary

comsymp communist sympathizer

comt comptroller

comt (COMT) catechol-O-methyltransferase

COMTAC Command Tactical (USN)

COMTAFDEN Commander, Tactical Air Force, Denmark

comte committee

COMTEC Computer Micrographics Technology (group)

com tech communications technician

COMTRAC Computer-aided Traffic Control

comtran commercial translation; computer translation

COMUSAF Commander U.S. Air Forces

COMUSAFSO Commander, U.S. Air Force, Southern Command

COMUSFORAZ Commander, U.S. Forces, Azores

COMUSJAPAN Commander, U.S. Forces, Japan

COMUSKOREA Commander, U.S. Forces, Korea

COMUSMACV Commander, U.S. Military Assistance Command Vietnam

COMUSRHIN Commander, U.S. Rhine River Patrol

COMUSTDC Commander, U.S. Taiwan Defense Command

Com Ver Common Version (of the Bible)

com wc command weapon carrier

Comy-Gen Commissary-General

Com Z Communications Zone

con confidence (game, man); conned; conning; consolidated; control; conversation; convict

con. conservation

con (CON) constant (flow chart)

con (Latin prefix—together or with)—confab, conference

con. contra (Latin—against)

Con Concord(e); Congo (Republic of the; whose capital is Brazzaville); Connie; Conservative; Constance; Consuela; consul; consultation

CON Certificate of Need; Conservative; Conservative Party

con8va. con ottava (Italian—with octaves)

CONAC Continental Air Command

CONACS Contractor's Accounting System

CONACYT Consejo Nacional de Ciencia y Tecnologia (Spanish—National Council for Science and Technology)

CONAD Continental Air Defense Command

CONADE Consejo Nacional de Desarrollo (Spanish—National Development Council)

conaloc continuity and logic

ConArC Continental Army Command

CONASA Council of North Atlantic Shipping Associations

conbat converted battalion antitank

conc concentrate; concentration; concentric; concrete

CONCACAF Confederación Centro Americano y Caríbean Futbal (Spanish—Central American and Caribbean Soccer Association)

concb concrete block

conc c concrete ceiling

conc clg concrete ceiling

concd concentrated; concerned

concentr concentrate(d)

CONCEPT Computation Online of Networks of Chemical Engineering Processes

Concertg Concertgebouworkest (Dutch—Concert Hall Orchestra)—Amsterdam's celebrated symphony orchestra

conc f concrete floor

conc fl concrete floor

concg concentrating

conch. conchology

concis. concisus (Latin—cut)

concn concentration

concomp conversational computation project

Con Con Constitutional Convention

Con Cpt Constructor Captain

concr concrete

cond condenser; condition; conductivity; conductor

condeep(s) concrete deepwater structure(s)

condit conditional

condiv continental divide

condo(s) condominium(s)

condr conductor

cond ref conditioned reflex

cond resp conditioned response

condrill concrete drill(ing)

conds condoms

conductimetric conductance + metric

CONE Chamber Orchestra of New England; Collectors of Numismatic Errors

CONEA Confederation of National Educational Associations

Con Ed Consolidated Edison

CONEFO Conference of New Emerging Forces

CONEG Conference of Northeastern Governors

conelrad control of electromagnetic radiation

Con Eng Convent English

CONESCAL *Centro Regional de Construcciones Escolares para America Latina* (Spanish—Regional Center of School Construction for Latin America)

con esp con espressione (Italian—with expression)

co-netic high-permeability nonshock-sensitive (alloy developed for maximum attenuation at low flux density)

conex connection(s); container export

conex (CONEX) connection(s)

Coney Coney Island (amusement center on southwest side of Brooklyn, New York)

conf confer; conference; confidential

conf. confer (Latin—compare)

Conf Confucian; Confucius

CONF Conference Papers Index

confab confabulation; confabulate

CONFAD Concept of a Family of Army Divisions

confec. confectio (Latin—confection)

Conf Econ Prog Conference on Economic Progress

confed confederation

Confed Confederate

Confederacy Confederate States of America (Virginia, North and South Carolina, Georgia, Florida, Alabama, Mississippi, Louisiana, Texas, Arkansas, Tennessee—and temporarily in Kentucky and Missouri)

confer. conference; conferred

confi confidant(e); confidence; confidential

CONFICS Cobra Night Fire Control System (helicopter)

confid confidential

confid(l)(ly) confidence; confidential(ly)

confit(s) confiture(s)

confr confectioner

cong congress(ional)

cong. congius (Latin—gallon)

Cong Congress; congregation; Vietcong

congal (cuarto) con gal [Mexican-American—(room) with girl]—house of prostitution

con game confidence game; confidence trick(ery)

Cong Christ Congregational Christians

Cong Digest Congressional Digest

congen common specification statements generator; congenial; congenital(ly)

Con Gen Consul General

Cong Fr Congolese franc

Congl Congregational

conglom(s) conglomerate(s); conglomerateur(s); conglomerator(s)

Congo Republic of the Congo (formerly the French Congo in west central Africa), *République Populaire du Congo*

Cong Orat Congregation of the Oratory

Congrats congratulations

Cong Rec Congressional Record

Congreg Congregationalist

Cong Staff Congressional Staff Directory

Cong U Congregational Union (England and Wales)

CONGU Council of National Golf Unions

conics conic sections

conif coniferous

conj conjugal; conjugate; conjunction; conjunctivitis

conject. conjecture

CONLIS Committee on National Library and Information Systems

Con Lt Constructor Lieutenant

Con Lt-Cdr Constructor Lieutenant-Commander

con man confidence man; swindler

CONMAROPS Concept of Maritime Operations

conn connection; connective; connector

Conn Connecticut; Connecticuter

CONN Connellan Airways

CONNECT Connecticut On-Line Enforcement Communication and Teleprocessing (computerized criminal file)

Connie Conrad; Constance; Consuela; Cornelia; Cornelius

Conn Turn Connecticut Turnpike

Conny Constance

co/no current operator/next operator

conobjtr conscientious objector

CONOCO Continental Oil Company

con of consisting of

conopt constrained optimization

conq conquer(ed); conquering; conqueror; conquest

Conr Conrad

conrad contour radar data

CONRAD Contraceptive Research and Development

ConRail Consolidated Rail Corporation (government-sponsored railroads including the Ann Arbor, Central Railroad of New Jersey, Erie-Lackawanna, Lehigh and Hudson River, Lehigh Valley, Penn Central, Reading)

con rod connecting rod

cons conservatory; consider; consist; consonant

con(s) confidence (games); consultant(s); conviction(s); convict(s)

cons. construction

cons. conserva (Latin—a preserve)

Cons Conservative

CONSAL Congress of Southeast Asian Librarians

CONSCIENCE Committee on National Student Citizenship in Every National Case of Emergency

Consc° Consejo (Spanish—Council)

con sect conic section

Cons Eng Consulting Engineer

CONSER CONversion of SErials (Council on Library Resources project)

conserv conservation; conservationist; conservatoire; conservatory

Conserv Conservatoire; Conservatory

cons. et prud. consilio et prudentia (Latin—by counsel and prudence)

Cons Gen Consul General

consgt consignment

conshelf continental shelf

conship control by ship

conshore control from shore

consid consideration

consig consignee

Con S-Lt Constructor Sub-Lieutenant

consltnt consultant

consol consolidated
Consol Consolidated Coal
consolex consolidation exercise
consols consolidated annuities
CONSORT Conversation System with On-Line Remote Terminals
consperg. *consperge* (Latin—dust, sprinkle)
conspic conspicuous
const constant; constitution; constitutional; construction; constructor
Const Constable; Constitution; Constructor
Const Constitution (of the United States)
constab constabulary
Constan Constantine; Constantinople (Istanbul)
Constance Lake Constance called Bodensee by the Austrians and the Germans
constit constituent(s); constitution(al)
constn constitution; construction
constocs contingency support stocks
constr construction; constructor
construct. construction(al)(ly)
Const US Constitution of the United States
consub continental-shelf submersible
consult. consultant
consumcrit consumer critic(ism)
consv conservation; conserve
cont contact; content(s); continent(al); continue(d); continuous(ly); contract(or); control(ler)
cont. contra (Latin—against); *contusus* (Latin—bruised, confused)
Cont Continent; Continental
contac continuous action
contag contagious
containerport container-ship seaport (equipped for handling and storing containers)
contam contaminant; contaminate; contamination
CONTAM Committee on Nationwide Television Audience Measurement
contax consumers and taxpayers
contb contraband
contbg contributing

cont. bon. mor. contra bonos mores (Latin—contrary to good manners)
Cont Cong Continental Congress
contd contained; continued
cont drift continental drift
coutemp contemporary
contempo contemporary
conter. contere (Latin—rub together)
conter US conterminous United States (48 states having common boundaries)
Cont Eur Continental Europe
Cont Eur & Br I Continental Europe and British Isles
contg containing
Cont HH continental range of ports from Havre to Hamburg
cont hp continental horsepower
Conti Constantine; Constantinople
contigs contiguous states (48 abutting states that form the continental U.S.)
contig US contiguous United States (48 states that border one another)
contin continental; continuous
contin continuo (Italian—continuous); *continuetur* (Latin—let it be continued)
contin US continental United States (Alaska plus the 48 states occupying much of the North American continent)
contl continental
contr contract; contracted; contraction; contractor
contra against; contra-indicated
contra (Latin prefix—against or opposite)—contradict, contraception
CONTRA Coalition of Non-Theist Religious Alternatives
contrail condensation trail
contralat contralateral
contran control translator
contraprop contra + propeller; contrarotating propeller
contrapun contrapuntal(ist); contrapuntistic)
contra(s) contraceptive(s)
Contras Contra Sandinistas (Spanish—anti-Sandinistas)—Nicaraguan rebels
contr. bon. mor. contra bonos mores (Latin—contrary to good manners)

contr. contract
cont. rem. continuetur remedia (Latin—let the remedy be continued)
contrib contribution; contributor
contrit. contritus (Latin—broken, ground, macerated)
cont shelf continental shelf of a continent extending into the sea before it descends sharply
CONTU Commission on New Technological Uses of Copyrighted Works (Library of Congress)
contus. contusus (Latin—bruised, confused)
cont w continuous window
CONU Contrans (container) Unit
conurb(s) conurbation(s)
Con US (CONUS) Continental United States
CONUS Intel Continental United States Intelligence (USA)
conv convalescent; convention; conventional; conversation(al); convertible
Conv convention, conversion
Conv Convento (Spanish—convent)
convair conveyed by air
convce conveyance
conv encl convector enclosure
convert convertible
convex convoy exercise
convg convergence
ConVis Convention and Visitors Bureau
Convis Bur Convention and Visitor's Bureau
convl conventional
convn convenient
convt convert(ible)
conv^te conveniente (Spanish—convenient)
CONWR Crab Orchard National Wildlife Refuge (Illinois)
coo controller of ordnance
Coo Coos Bay
Coo Coo blimey (Cockney—God blind me)
CoO Chief of Operations; Committee on Organization
CoO (COO) Chief of Outpost (CIA)
COO Chief Operating Officer; Chief Ordnance Offficer
cooc contact with oil or other cargo

COOC Commission on Organized Crime

COOH (carboxyl group found in all organic acids)

cook. cookery

Cooks Cook Islanders; Cook Islands; Cook's Tours (Thomas Cook and Son, Ltd)

cool. coolant

Cool-Kal Coolgardie-Kalgoorlie

coon(s) coonhound(s)-contraction of racoon hounds

'coon(s) racoon(s)

coop. cooperation

co-op cooperative; customer on-line order processing

Coop Cooper

COOP ED Cooperative Education

coopg cooperage

Co-op L Cooperative League

COOPLAN Continuity of Operations Plan (USN)

Coop Rep Guy Cooperative Republic of Guyana

Co-op U Co-operative Union

coorauth coordinating authority

coord coordinate; coordination; coordinator

COORS Communications Outrage Restoration Section

COOS Chemical Orbit-to-Orbit Shuttle (NASA)

cop capillary osmotic pressure; casing operating pressure; constable on patrol (origin of cop); copper; copyright; customer owned property; policeman (slang)

cop (COP) computer optimization package

cop. coefficient of performance; commencement of passage; custom of port

c-o-p change of plea

Cop Copernican; Coptic

Cop *Copenhague* (French, Portuguese, Spanish—Copenhagen)

COP Career Opportunity Program; Certificate of Participation; Certificate of Proficiency; City of Prineville (railroad); Coalition on Police; Combat Outpost; Commissary Operating Program; Community-Oriented Policing; Continuity of Operations Plan; Cox's Orange Pippin; Custom of the Port

Copa Copacabana

COPA Council on Postsecondary Accreditation

COPA *Compañía Panameña de Aviación* (Spanish—Panamanian Aviation Company)

copac continuous operation production allocation and control

COPAFS Council of Professional Associations on Federal Statistics

COPAL Cocoa Producers' Alliance

COPANT *Comisión Panamericana de Normas Tecnicas* (Spanish—Panamerican Commission for Technical Standards)

COPAO Council of Philippine-American Organizations

COPAR Corrective or Preventive Action Report

COPARS Contractor-Operated Automotive Parts Store (DoD)

COPAS Council Of Petroleum Accountants Societies

copay. copayment

copc combined operations planning committee

copd chronic obstructive pulmonary disease; coppered

COPD chronic obstructive pulmonary disease

COPDAF Continuity of Operations Plan—Department of the Air Force

cope chronic obstructive pulmonary emphysema

COPE Champions of Private Enterprise; Committee on Political Education (AFL-CIO); Committee for Original People's Entitlement (Canadian Eskimo's claim to Canadian land); Concerned Organization of Parents to Educate; Congress on Optimum Population and Environment; Council on Population and Environment

COPEC *Compañía Petrolera Chilena* (Spanish—Chilean Petroleum Company)

COPEI *Comité Organizador del Partido Electoral Independiente* (Spanish—Organization Committee of the Independent Electoral Party)—Venezuela's Social Christian Party

Copen Copenhagen

COPES College Occupational Programs Educational System; Committee on Program Evaluation and Support; Conceptually-Oriented Program in Elementary Science

COPH Congress of Organizations of the Physically Handicapped

COPIAT Coalition to Preserve the Integrity of American Trademarks

COPICS Copyright Office Publication and Interactive Cataloging System (Library of Congress)

COPL Council of Planning Librarians

copo copolymer

COPO Council of Philatelica Organizations

copp cobaltiprotoporphyrin; combined operations pilotage party

Copp Copperplate

COPP Conservation Organization Protesting Pollution

copperas ferrous sulfate; green vitriol

COPPS Committee on Power Plant Siting (Nat Acad Engineering)

COPR Center for Overseas Pest Research; Critical Officer Personnel Requirement (USAF)

COPRED Consortium on Peace Research, Education and Development

cops coppers; policemen (slang)

COPs Coalition on Police members

COPS California Organization of Police and Sheriffs; Chief of Operations (CIA); Committees Organized for Public Service; Community-Oriented Policing Services; Concerns of Police Survivors

Copsa Mica (Romanian—Black Town) ink-soaked town near chemical plants

Copt Coptic

coptec controller overload prediction technic

copter(s) helicopter(s)

co-ptr co-partner

copu copulate; copulation; copulatory

COPUL Council of Prairie University Libraries

copy. copyright

coq cost of quality

coq. coque (Latin—boil)

co Q coenzyme Q

coq. s.a. coque secundum artem (Latin—boil correctly)

coq. in s.a. coque in suffuciente aqua (Latin—boil in sufficient water)

coq. sim. coque simul (Latin—boil together)

cor cold-regulated (genes); contactor, running; corner; cornet; correction

cor corno (Italian—horn)

cor. corpus (Latin—body)

Cor Corinthians; Corona; Coronado; Coroner; Corsica; Coruña

Cor Corea (Portuguese or Spanish—Korea); *Corvus* (Latin—Crow constellation)

COR Commonwealth Oil Refineries

COR Comisión(es) de Orientación Revolucionaria [Spanish—Revolutionary Orientation Committee(s)]—Cuba

cora conditioned orientation reflex audiometry

Cor A Corona Australis

CORA Corporación de la Reforma Agraria (Spanish—Agrarian Reform Corporation)—Chile

coral. class-oriented ring-associated language

Coral Coral Sea; Coral Sea Island Territory beyond Australia's Barrier Reef

CORAL Coherent Optical Radar Laboratory (USAF)

Cor Aus Cornona Australis (Latin—Northern Cross constellation)

Cor B Corona Borealis

cor bd corner bead

corbfus copy of reply to be furnished us

Corbin Corbin on Contracts by Arthur L Corbin

Cor Bor Corona Borealis (Latin—Northern Cross constellation)

Corbu Le Corbusier (nickname of Edouard Jeanneret-Gris)

Corc Cornell computing (language)

Cor Chr Col Corpus Christi College—Cambridge

CORCO Commonwealth Oil Refining Company (Puerto Rico)

cord computer on-line devices

cord. cordillera (Spanish—mountain range)

Cord Cordelia; Córdoba

C of Ord Chief of Ordnance

CORD Center for Occupational Research and Development; Commissioned Officer(s) Residency Deferment; Congress on Research in Dance

cordat coordinate data set

cordic coordinate rotation digital computer

Cordilleras Cordillera Mountains of the Americas

CORDIPLAN Oficina Central de Coordinación y Planificación (Spanish—Central Office of Coordination and Planning)

CORDIS Community Research and Development Information Service

cordoba monetary unit of Nicaragua

cordpo correlated radar data printout

cords. corduroy pants; corduroy trousers

CORDS Civil Operations and Revolutionary Development Support

core computed oriented reporting efficiency

CORE Competitive Operational Readiness Evaluation (Air Force); Congress of Racial Equality

corex coordinated electronic countermeasures exercise

corf classroom observational rating form

corf(ing) cut(ting) off reflected failure

CORF Comprehensive Outpatient Rehabilitation Facility

corfam microporous artificial leather

corflu correction fluid

CORFO Corporación de Fomento (Spanish—Development Corporation); *Corporación de Fomento de la Producción* (Spanish—Production Development Corporation)—Chile

CORG Combat Operations Research Group

CORGI Confederation for Registration of Gas Installers

corin corinthian

Coriol Coriolanus

CORL Canadian Operations Research Society

CORLS Central Ontario Regional Library System

CORM Council for Optical Radiation Measurements

CORMA Corporación de la Madera (Spanish—Wood Corporation)

cormant cormorant

CORMAR Coral Reef Management and Research

Cor Mem Corresponding Member

Corn Cornelius; Cornish; Cornwall

corned-beefsan corned-beef sandwich

corned-beefwich corned-beef sandwich

Cornell Maritime Cornell Maritime Press

Cornell U Pr Cornell University Press

Corner House Central Mining and Finance Corporation (South Africa)

Corney Cornelia; Cornelius

Cornie Cornelia; Cornelio; Cornelis; Cornelisz; Corneliu; Cornelius; Cornewall; Cornwall; Cornwallis

Corning Mus Corning Museum of Glass

Corns Corn Islands in the Caribbean

coroll corollary

coron coronary

Coron Convair 990 Coronado (aircraft)

Coronados Coronado Islands (*Los Coronados*) south-southwest of San Diego

COROT Committee on Regular Officer Training

corp (Latin prefix—body)—corporation, *corpus delicti* (the body of the crime)

Corp corporal; Corporation; Corpus Christi College, Cambridge or Oxford

Corp Coll Corpus Christi College—Oxford

Corpl Corporal

Corpn Corporation

CORPOANDES Corporación de los Andes (Spanish—Andes Corporation)

CORPORIENTE *Corpo-ración de Oriente* (Spanish—Corporation of the East)—Venezuela

corppin corporeal pin (tuberculin testing)

Corpus Corpus Christi, Texas

CORPUS Corps of Reserve Priests United for Service

corr correction; correspondence; corrosion; corrugate

corr *corregido* (Spanish—corrected); *corriage* (French—corrected)

Corr *Corriere della Sera* (Italian—Daily Courier)—Milan's leading newspaper

CORRA Combined Overseas Rehabilitation Relief Appeal

corr case corrugated case

corregate correctable gate

correl correlated; correlation; correlative

corres correspondence; correspondent; corresponding

corresp corresponding

corrig corrigenda

Corr Memb Corresponding Member

corros corrosive

corrosive sublimate mercuric chloride

corrᵗᵉ *corriente* (Spanish—current month)

corrupt. corruption

Cors Corners; Corsica; Corsican

CORS Canadian Operational Research Society

corsa (CORSA) cosmic-ray satellite

Cor Sec Corresponding Secretary

cort cortex; cortical

cort. cortisol

cort. *cortex* (Latin—bark)

CORT Council On Radio and Television

CORTEX Computer-based Optimization Routines and Techniques for Effective X

Corfissoz *Aeropuerto Ernesto Cortissoz* (Barranquilla, Colombia's airport)

CORU *Coordinación de Organizaciones Revolucionarias Unidas* (Spanish—Coordination of United Revolutionary Organizations)—Cuban exiles

corundolite emery

corundum aluminum oxide

Corv Corvette; Corvus

CORVA California Off-Road Vehicle Association

Cory Cornelia

cos cash-on-shipment; contactor, starting; cosine; cosmic; cosmogany; cosmography; cosmology; cosmopolitan

co's career officers

Cos Consul; Counties

Cos *Kosinus* (German—cosine)

CoS Chief of Staff; Chief of Station

C-o-S Clacton-on-Sea

COS Canadian Ophthalmological Society; Central Opera Service; Chamber of Shipping; Chief of Section; Colorado Springs, Colorado (airport); Conservative Opportunity Society; Corporation for Open Systems; Czechoslovak Ocean Shipping

COS *College Outline Series*

cosa combat operational support aircraft

co sa *come sopra* (Italian—as above)

COSA Chief of Staff, Army

COSAD Classroom Observation System for Analyzing Depression

cosag combined steam and gas (turbine machinery)

COSAL Consolidated Shipboard Allowance List

COSAMREG Consolidation of Supply and Maintenance Regulations

COSA NOSTRA Computer-Oriented System And Newly Organized Storage-To-Retrieval Apparatus

cosar compression scanning-array radar

COSAS Congress of South African Students

COSATI Committee on Scientific and Technical Information (Federal Council for Science and Technology)

COSATU Congress of South African Trade Unions

COSBA Computer Services and Bureaus Association

COSBAL Coordinated Shore-based Allowance List

COSBO Council on Small Business Organizations

COSCO China Ocean Shipping Company

COSCOE Congress of Seniors and Coalition of Elders

cosd command supply depot

COSD Council of Organizations Serving the Deaf

Cos de Mar *Costa de Marfil* (Spanish—Ivory Coast)

COSE Council of Smaller Enterprises

CoSEA Colorado Society of Enrolled Agents

COSEBI *Corporación de Servicios Bibliotecarios* (Spanish—Librarian Services Corporation)—Puerto Rico

cosec cosecant

COSEC Coordinating Secretariat of National Unions of Students

cosecy company secretary

cosfad computerized safety and facility design

COSFPS Commons, Open Spaces, Footpaths Preservation Society

cosh hyperbolic cosine (symbol)

COSHTI Council for Science and Technological Information

COSI Center of Science and Industry (Columbus, Ohio); Committee on Scientific Information

Cosie Kathleen

COSINE Committee on Computer Science in Electrical Engineering Education

COSIP College Science Improvement Program

COSIRA Council for Small Industries in Rural Areas

cosis care of supplies in storage

COSLA Chief Officers of State Library Agencies

cosm cosmetic; cosmetics; cosmetologist; cosmetology

cosma computerized service for motor freight activities

COSMD Combined Operations Signals Maintenance Department (Division)

COSMEP Committee of Small Magazine Editors and Publishers

cosmetol cosmetologist(ic); cosmetology

COSMIC Chief of Staff, Military Intelligence Committee; Computer Programmes Information Center (Univ of Georgia); Computer Software Management and Information Center

COSMIS Computer System for Medical Information Services

cosmo cosmoline; cosmopolitan

cosmog cosmogony; cosmographical; cosmography

cosmograph(s) composite photograph(s)

cosmonaut. cosmonautic(al)(ly); cosmonautics

cosmor component open/short monitor

COSMOS Coast Survey Marine Observation Station; Cost Management with Metrics of Specification

co so come sopra (Italian—as above)

COSPAR Committee on Space Research (International Council of Scientific Unions)

COSPEC Christian Organizations for Social, Political, and Economic Change

COSPUP Committee on Science and Public Policy (National Academy of Sciences)

cosr cutoff shear

COSR Committee on Space Research

coss. consules (Latin—consuls)

COSSA Consortium of Social Science Associations

cossac cooled spectral shared-aperture concept

COSSAC Chief of Staff to the Supreme Allied Commander

cost. contaminated oil settling tank; costume

COST Congressional Office of Science and Technology; Costco Wholesale Corporation; Cost-Oriented Systems Technique; Council Opposing Signal Theft; (European) Cooperation in the Field of Scientific and Technical research

Costa Costa Rica (whose capital is San José)

costar conversational on-line storage and retrieval

Costa Rica Republic of Costa Rica (Central American nation), *República de Costa Rica*

COSTEP Commissioned Officer Student Training and Extern Program

coster costermonger

COSTPRO Canadian Organization for the Simplification of Trade Procedures

COSTS Committee on Sane Telephone Service

COSVN Central Office for South Vietnam

COSW Citizen's Organization for a Sane World

coswap coaxial switch and alternator panel

COSWL Committee on Status of Women in Librarianship

COSY Checkout Operating System

COSYWOG Communications System Working Group

cot. card or tape reader; cathodal opening tetanus; cotangent; cotter; cotton

Cot Côte d'Ivoire (whose capital is Abidjan)

COT Consecutive Overseas Tour

COTA Certified Occupational Therapy Assistant; confirming telephone or message authority

COTAL Confederación de Organizaciones Turísticas de la América Latina (Spanish—Confederation of Tourist Organizations of Latin America)

COTAM Commandement du Transport Aerien Militaire (French—Military Air Transport Command)—Air Force

cotan cotangent

cotar correction tracking and ranging

CotB Commonwealth of the Bahamas

COTC Canadian Officers' Training Corps; Canadian Overseas Telecommunications Corporation

cote cathodal opening tetanus

COT & E Contractor Operation Test and Evaluation

cotfin cotton finish(ed)

cotg component tooling gage

coth hyperbolic cotangent (symbol)

COTH Council on Teaching Hospitals

cotics narcotics

cotnsd cottonseed

Coto Cotopaxi

CotP Captain of the Port

COTPAL Comité Tecnico Permanente sobre Asuntos Laborales (Spanish—Permanent Technical Committee for Labor Matters)

COTR Contracting Officers' Technical Representative

COTRANS Coordinated Transfer Applications System

cots. checkout test set; cottages

cot's classical organizational theories

'cot(s) apricot(s)

C o t S College of the Sea

COTS Commercial, Off-The-Shelf

Cotswolds Cotswold Hills of south-central England

Cott Cottesloe

COTT Central Organization for Technical Training

Cottians Cottian Alps between France and Italy

cott(s) cottage(s)

coty car of the year

cou civilian observer unit; clip-on unit; coupon

COU Coalition Unionist

couch couchant

couldn't could not

Coun Council; Councillor; Counsellor; County

Coun Biology Eds Council of Biology Editors

Coun Exc Child Council for Exceptional Children

couns. counsellor

Counterpart trademark of Fifth Generation Systems—Baton Rouge, Louisiana

COUP Congress of Unrepresented People

cour courant (French—current)

Court Courtenay; Courtland; Courtney

cous cousin

COUSA Confederation of Ontario University Staff Associations

cov concentrated oil of vitriol; covenant; cutout valve; cover

c-o v cross-over value

Cov Covell; Covenant; Coventry

COVA Combined Organization for the Visual Arts

COVE Citizens Opposed to the Violation of the Environment

coven. covenant

covers. coversed sine

COVET Cooperative Venture in the Education of Teachers

covff coverings, facing, or floor (cargo)

COVINCA *Corporación Venezolana de la Industria Naval* (Spanish—Venezuelan Corporation of the Naval Industry)

COVIRA Computer Vision in Radiology

cov pl coverplate

cow chlorinated organics in wastewater; computer on wheels; crude oil washing

COW Coventry Ordnance Works

COWAR Committee on Water Research

Coward Coward, McCann and Geohegan

COWEAEX Cold-Weather Exercise (military)

cowl. cowling

Cowles Cowles Education Corporation

COWLEX Cold-Weather Landing Exercise (military)

COWPS Council on Wage and Price Stability

COWRR Committee on Water Resources Research

cox cyclo-oxygenase (enzyme)

Cox Coxwain

cox'n coxswain (pronounced as contracted)

Cox's *Cox's Criminal Cases*

coxsec coexsecant

Coy Company

coydog(s) coyote(s) + dog(s)— mixed-breed canine(s)

COYOTE Call Off Your Old Tired Ethics

Coyte Coyte Lines

coz cousin (colloquial contraction)

Coz Cozumel Island, Mexico

cozi communication zone [indicator(s)]

cp camp; candlepower; capillary pressure; carrier packed; cellulose propionate; center of pressure; centipoise; central processor; cerebral palsy; cesspool; chemically pure; chloropurine; chloroquinine and primaquine; chronic pyelonephritis; circumpolar; claw plate; closing pressure; cochlear potential; code of practice; coldpunch(ed); combination product; com-

bining power; command post (CP); commercial paper; compare; compound; compressed; concrete-piercing; constant pressure; copy-protected; copyright page; cor pulmonale; cover point (lacrosse); creatine phosphate

cp (CP) carotid pulse; cerebral palsy; construction permit

c/p carport; change package; composition/printing; control panel

c/p (C/P) charter party

c & p carriage and packing; collated and perfect

cP centipoise; polar continental air

Cp Caucasian pimp; Chicano pimp; Chinese pimp; Compline

CP Canadian Pacific; Canadian Press (news agency); Central Province (Solomon Islands); cerebral palsy; Characterization Plan; charter party; chemically pure; Civil Parish; Combined Protocol; Command Paymaster; Common Pleas; Communist Party; Community Physician; Conservative Party; Constitution Party; copilot; Country Party; Crescendo Publishers; cytoxan/prednisone

C-P Colgate-Palmolive

C & P Compensation and Pension

C of P Captain of the Port

CP *Caminhos de ferro Portuguese* (Portuguese Railways); *Centre Pénitentiaire* (French—Penitentiary Center)

cpa claims payable abroad; closest point of approach; cost planning and appraisal; critical path analysis

c-p a cattle-prod approach (electric-shock stimulation)

CPA California Protection Agency; Canadian Pacific Airlines; Canadian Petroleum Association; Canadian Psychological Association; Canaveral Port Authority; Cathay Pacific Airways, Catholic Press Association; Certified Puiblic Accountant; Chartered Patent Agent; Chicago Publishers Association; Civilian Production Adminis-

tration; Classroom Publishers Association; Cocoa Producers Alliance; Combat Pilots Association; Combined Pensioners Association; Commonwealth Parliamentary Association; Commonwealth Preference Area; Communist Party of Australia; Computer Press Association; Connecticut Prison Association; Consumer Protection Agency; Control of Pollution Act; Council of Professional Associations; Country Press Association; Creditors Protection Association

CPA *Community Planning Act*

CPAA *Current Physics Advance Abstracts*

CPAAI Certified Public Accountant Associates International

CPAB California Prune Advisory Board

CPAC Center for Protection Against Corrosion; Conservative Political Action Conference; Corrosion Prevention Advisory Center

CPACS Coded Pulse Anti-clutter System

CPAE Certifiied Public Accountant Examination

cpaf cost plus award fee

cpafsk continuous phase amplitude frequency shift keying

CPAG Collision Prevention Advisory Group

C$_{pah}$ para-aminohippurate clearance

CPAI Canvas Products Association International

CP Air Canadian Pacific Air

C Pal Crystal Palace

cpam continental polar air mass

CPAM Committee of Purchasers of Aircraft Material

CPAO Country Public Affairs Office(r)

cpap continuous positive airway pressure

CPAP Committee on Pan-American Policy

CPAR Cooperative Pollution Abatement Research (Canadian)

CPARS Compact Programmed Airline Reservation System

CPAS Catholic Prisoners' Aid Society

CPAUS & C Catholic Press Association of the United States and Canada

cpaws computer-planning and aircraft-weighting scales

cpb cardiopulmonary bypass; casual payments book; cetyl pyridinium bromide; competitive protein-binding (clearance); corporation for public broadcasting

cpb (CPB) charged-particle beam(s)

cpb cuyos pies beso (Spanish—whose feet I kiss)

Cpb Campbelltown

CPB California Prune Board, Casual Payments Book; Central Planning Bureau; Consumer Protection Bureau; Corporation for Public Broadcasting

CPB Centraal Plan Bureau (Dutch—Central Planning Bureau)

cpba competitive protein-binding analysus

cpbl capability; capable

CPBMP Committee on Purchases of Blind-Made Products

cpc card-programmed calculator; chemical protective clothing; chronic passive congestion; clinicopathological conference (CPC); coated-paper copier; commercial property coverage; common projectile capped; compound parabolic concentrator; computer-production control

CPC California Polytechnic College; Canadian Parachute Corps; Canadian Postal Corps; Canadian Provost Corps; Canterbury Promotion Council; Cessna Pilots Center; China Petroleum Company; China Productivity Council; Church Periodical Club; City Planning Commission; City Police Commissioner; City Projects Council; Cogswell Polytechnical College; College Placement Council; Communist Party of China; Consumers Power Company; Creole Petroleum Corporation

CPC Congreso Panamericano de Carreteras (Spanish—Pan-American Highway Congress)

cpca coolant pumping and conditioning assembly

CPCC Central Piedmont Community College

c-p-c cycle circumspection-preemption-control (or choice) cycle

CPCG Comité Panamericano de Ciencias Geofícicas (Spanish—Panamerican Committee of Geophysical Sciences)

CPCGN Canadian Permanent Committee on Geographical Names (Ottawa)

CPC(M-L) Communist Party of Canada (Marxist-Leninist)

CPCN Canadian Pacific—Canadian National (telecommunications)

cpcs cabin pressure control system

CPCS Canadian Pacific Consulting Services

CPCU Chartered Property and Casualty Underwriter

c-p cycle constant-pressure cycle

cpd charter pays dues; commissioner of public debt; compound; contact potential difference; contagious pustular dermatitis; container-padded delivery

cpd (CPD) charter (party) pays (port) dues

CPD Committee on the Present Danger; Community Planning and Development; Consumer Protection Division; County Probation Department

CPD Catalog of the Public Documents

CPDA Council for Periodical Distributors Associations

cpdd command-post digital display

CP Div Cinque Ports Division

CPDL Canadian Patents and Developments Limited

C-P D L Christian-Patriots Defense League

CPDM Cameroon People's Democratic Movement

cpDNA cytoplasm deoxyribonucleic acid

cpds compounds

CPDS Computerized Preliminary Design System

cpe chlorinated polyethylene; chronic pulmonary emphysema; circular probable error; compensation, pension, and education; coupe, customer premises equipment; customer-provided equipment; cytopathic effect; cytopathogenic effect

cpe (CPE) central programmer and evaluator

CPE Center for Packaging Education; Central Park East; Certificate for Proficiency in English; Certified Professional Ergonomist; Certified Property Exchanger; Chief Polaris Executive (missiles); Clinical Pastoral Education; College of Physical Education; Continuing Professional Education; Contractor Performance Evaluation

CPEA Confederation of Professional and Executive Associations; Cooperative Program for Educational Administration

CPEC California Post-secondary Education Commission

cpe d vle coupe de ville

CPEG Contractor Performance Evaluation Group

CPEHS Consumer Protection and Environmental Health Service

c pen code pénal (French—penal code)

CPEP Contractor Performance Evaluation Plan

CPEQ Corporation of Professional Engineers of Quebec

CPERB California Public Employees Relation Board

cpf conditional peak flow; control program facility; cost per flight

CPF Central Processing Facility; Central Provident Fund; Church Pension Fund; Commission on Federal Paperwork; Committee to Protect the Family; Commonwealth Police Force

cpfa (CPFA) cyclopropenoid fatty acid

cpff (CPFF) cost plus fixed fee

CPFS Council for the Promotion of Field Studies

cpg controlled-pore glass; cotton piece goods

CPG Collectibles and Platemakers Guild; College Publishers Group

CPGA California Personnel and Guidance Association; Colorado Personnel and Guidance Association; Connecticut Personnel and Guidance Association

CPGB Communist Party of Great Britain

Cpgc course per gyro compass

cph cards per hour; cycles per hour

CPH Certificate of Public Health; Copenhagen, Denmark (airport); Corps of Public Health

C-PH Columbia-Presbyterian Hospital

CPHA Canadian Public Health Association

CP & HA Canadian Port and Harbour Association

CPHC Central Pacific Hurricane Center (Honolulu)

cpi characters per inch; chronic public inebriates; clutch pitching index; commercial performance index; constitutional psychopathic inferior; consumer price index; corps of permanent instructors; crash position indicator

CPI California Personality Inventory; California Psychological Inventory; Canadian Pacific Investments; Chemical Processing Industries; Chief Pilot Instructor; Committee on Public Information; Communist Party of India; Conference Papers Index; Consolidated Plastic Industries; Consumer Price Index; Credit Professionals International

cpia close-pair interstitial atom

CPIA Chemical Propulsion Information Agency

cpiaf (CPIAF) cost-plus-incentive-award fee

cpib chlorophenoxyisobutyrate

CPIB Corrupt Practices Investigation Bureau

CPIC Canadian Police Information Centre

cpif character position in frame

cpif (CPIF) cost plus incentive fee

CPILS Correlation-Protected Integrated Landing System

CPIM Curaçaosche Petroleum Industrie Maatitschappij (Dutch—Curaçao Petroleum Industry Society)

cpin crankpin

CPI-U Consumer Price Index—Urban

CPI-U-NSA Consumer Price Index—Urban—Not Seasonally Adjusted

CPI-W Consumer Price Index—revised

CPJ Communist Party of Japan (also called JCP)

CPJI Cour Permanente de Justice Internationale (French—Permanent Court of International Justice)

cpk (CPK) creatinine phosphokinase

CpK plate-cathode capacitance

cpkg cents per kilogram

cpl cement plaster; characters per line; common program language; complete; completion

Cpl Corporal

CPL Calgary Public Library; Canadian Pacific Limited; Cats' Protection League; Central Public Library; Certified Parts List; Certified Products List; Charleston Public Library; Charlotte Public Library; Chattanooga Public Library; Chicago Public Library; Cincinnati Public Library; Civilian Personnel Letter; Cleveland Public Library; Clio Press Limited (Oxford); Colonial Products Laboratory; Columbus Public Library; Commercial Pilot's License; Commonwealth Parliamentary Library; Commonwealth Public Library; Coronado Public Library; Council of Planning Librarians; Council of Prison Locals; Crew Procedures Laboratory

CPLA California Palace of the Legion of Honor

cpld coupled (flow chart)

cplg coupling

cplmt complement

cplr center of pillar; coupler

CPLS Canberra Public Library Service; Certified Professional Legal Secretary

Cplt copilot

cpm captured personnel and material; cards per minute; commutative principle of multiplication; condensed particulate matter; counts per minute; critical path method; cycles per minute

cpm (CPM) cost per thousand

cp/m control program/microcomputers

CPM Center for Preventive Medicine; Central Pacific Minerals; Certified Property Manager; Certified Purchasing Manager; Chief Postmaster; Colonial Police Medal (British); Communist Party of Malaya; Critical Path Method

CP/M Control Program for Microprocessors (trademark of Digital Research)

CPMA Computer Peripheral Manufacturers Association

CPMC Columbia-Presbyterian Medical Center

CPMEEW Council for Postgraduate Medical Education in England and Wales

CPML Communist Party, Marxist-Leninist

CPMS Civilian Personnel Management System; Computer Performance Monitoring System

cpn chronic pyelonephritis; coupon

Cpn Copenhagen

CPN Communistische Partij van Nederland (Dutch—Netherlands Communist Party)

Cpnhgn Copenhagen

CPNP Cape Perth National Park (Western Australia)

CPNZ Communist Party of New Zealand

cpo command pay office; cost proposal outline

CPO Calgary Philharmonic Orchestra; Center for Population Options; Chief Petty Officer; Chief Post Office; Civil Post Office; Civilian Personnel Office(r); Community Post Office; Community Producers Organization; Comprehensive Planning Organization; Constitutional Protection Office (Germany); County Planning Office(r); Czech Philharmonic Orchestra

CPOA California Peace Officers Association; Chief Petty Officers Association

CPOG Canadian Pacific Oil and Gas

cpp critical path plan

CPP Caltech Population Program; Canada Pension Plan; Center for Policy Process; Chemical Processing Plant; Communist Party of the Philippines; Critical Path Planning

CPP Civilian Personnel Pamphlet

CPPA Canadian Pulp and Paper Association

cppb continuous positive-pressure breathing

CPPB Canada Pension Plan Benefits; Commonwealth Prickly Pear Board

cppc cost plus a percentage of cost

CPPCA California Probation, Parole, and Correctional Association

cppd calcium pyrophosphate dihydrate

CPPD Collaborative Program for Professional Development

CPPL Canadian Pacific Princess Lines (Vancouver-Nanaimo run)

CPPP Center for Public Policy Priorities (Austin, Texas)

CPPR Cassel Psychotherapy Progress Record

cpps critical path planning and scheduling

CPPS Comisión Permanente para la Explotación y Conservación de las Riquezas Maritimas del Pacifico Sur (Spanish—Permanent Commission for the Exploitation and Conservation of the Maritime Riches of the South Pacific)

CPQ Children's Personality Questionnaire

cpr cardiopulmonary resuscitation; conditional prepayment rate; copper

cpr (CPR) chemical propulsion rocket

CPR Canadian Pacific Railway; Cape Peninsula Rifles; Carlos Peña Romulo; Center for Public Resources; Central Premonitions Registry; Cobourg Peninsula Reserve

(Australian Northern Territory); Committee on Polar Research; Cost Performance Report; Council for Public Responsibility

CPRA Council for the Preservation of Rural America

CP Rail Canadian Pacific Rail

CPRC Ceylon Planters Rifle Corps

CPRE Council for the Preservation of Rural England

CPRF Cancer and Polio Research Fund

CPRG Computer Personnel Research Group

CPRI Council for the Protection of Rural Ireland

CPR-nummer Centrale Person Register nummer (Dano-Norwegian—Central Person Register number)

CPRS Council for the Protection of Rural Scotland

CPRSA Cape Peninsula Road Safety Association

CPRV Cinque Ports Rifle Volunteers

CPRW Council for the Protection of Rural Wales

CPR-WBS Cost Performance Report—Work Breakdown Structure

cps cartridge pneumatic starter; characters per second; constitutional psychopathic state; coupons; creative problem solving; critical path scheduling; cycles per second

Cp(s) Caucasian pimp(s); Chicano pimp(s); Chinese pimp(s)

CP's Command Posts

CPS Cable Programming Services; California Physician's Service; California Production Service; Canadian Pacific Steamships; Canadian Penitentiary Service; Catholic Pamphlet Society; Center for Population Studies (Harvard); Central Park South; Central Philharmonic Society (Beijing); Certified Professional Secretary; College Placement Service; College Press Service; Commission on Presidential Scholars; Commonwealth Public Service; Computer Processing Service(s); Condensate Polishing System; Congrega-

tional Publishing Society; Consumer Price Survey; Consumer Purchasing Service; Conversational Programming System; Current Population Survey

C.P.S. *Custos Privati Sigilli* (Latin—Keeper of the Privy Seal—Great Britain)

CPS Compendium of Pharmaceuticals and Specialities; Conseil Permanent de Sécurité (French—Permanent Security Council)

CPSA Canadian Political Science Association; Civil and Public Services Association (UK); Clay Pigeon Shooting Association; Commonwealth Public Service Association

CPSAA Commonwealth Public Service Artisans Associations

cpsac cycles-per-second alternating current

C$_{psc}$ course per standard compass

CPSC Consumer Product Safety Commission

CPSCU College of Physicians and Surgeons—Columbia University

cpsd cross-power spectral density

cpse counterpoise

cpsi causing pressure shut in

CPSI Council of Profit-Sharing Industries

CPSL Canadian Pacific Steamship Line

CPSLCS Complete Power Signalling Local Control System

CPSM Colonial Prison Service Medal (British)

CPSP Cove Palisade State Park (Oregon)

CPSR Calibration Procedure Status Report (Polaris); Computer Professionals for Social Responsibility; Contractor Procurement Systems Review

CPSS Certificate in Public Service Studies; Common Program Support System

C$_{p\,stg\,c}$ course per steering compass

CPSU California Polytechnic State University; Combined Public Service Unions; Communist Party of the Soviet Union

cpt carpet(ed); casement-projected transom; change-parity time; chest physiotherapy; cockpit procedure trainer; coding physician treatment; continuous performance task; counterpoint; critical path technique; current procedural terminology

cpt (CPT) California Public Television; critical path technic

Cpt Capitaine (French—Captain)

CPT Canadian Pacific Telegraphs; Cape Town, South Africa (Malan Airport); Civilian Pilot Training; Communist Party of Thailand; Continuing Performance Test(ing)

CPT Current Physics Titles

C.P.T. Contador Público Titulado (Spanish—Certified Public Accountant)

CPTA Cinque Ports Training Area

CPTB Clay Products Technical Bureau

CPTL Canadian Pacific Transport Limited

Cptn Captain

cptng mats rgs carpeting, mats, or rugs

cptr capture; carpenter; carpentry

CPTS California Public Television Stations; Coalition for Peace Through Strength; Council of Professional Technological Societies

CPTV Connecticut Public Television

cpu (CPU) central processing unit

CPU California Pacific University; Canadian Paperworkers Union; Central Processing Unit; Commonwealth Press Union; Crime Prevention Unit

CPUBINFO Chief of Public Information Division (NATO)

CPUC California Public Utilities Commission

cpue catch per unit effort

CPUSA Communist Party USA

cpv command post vehicle; common pressure vessel

cpv (CPV) cytoplasmic polyhedrosis virus

CPV Combination Pump Valve; Communist Party of Vietnam

CPV Compañía Peruana de Vapores (Peruvian Steamship Line)

cpvc critical pigment volume concentration

cpvc (CPVC) chlorinated polyvinyl chloride

CPVPL Charles Patterson Van Pelt Library (University of Pennsylvania)

cpw commercial projected window

cPw polar continental air warmer than underlying surface

CPW California Press Women; Central Park West

CPWH Committee for the Preservation of the White House

cpx complex

CPX Command Post Exercise

cpy copy

CPY Communist Party of Yugoslavia

cpz chlorpromazine

CPZ Central Park Zoo

cq chloroquine quinine; circadian quotient; come quick; compounded quarterly; conceptual quotient; copy correct; copy (spelled) correctly

cq (CQ) class quotient (lowerclass, upperclass, etc.)

CQ call to quarters (radio signal meaning message following is intended for all receivers); Charge of Quarters; Conditionally Qualified

CQ Caribbean Quarterly; Congressional Quarterly

CQC Citizens for a Quiet City

CQCA Central Queensland Coal Associates

cqcm cryogenic quartz-crystal microbalance

CQD wireless distress signal

cqi continuous quality improvement

cqm chloroquine mustard

CQM Camp Quartermaster; Chief Quartermaster; Company Quartermaster

CQMS Company Quartermaster Sergeant

cqr secure anchor (British short form for a plowshare-shaped single-fluke anchor)

CQR Customer Quality Representative

CQR Church Quarterly Review

CQs Citizens for Quieter Cities

CQS California Q-Set

CQSW Certificate of Qualification in Social Work

cqt circuit; correct

CQT College Qualifiication Test

CQU College Qualification (test)

CQUCC Commission on Quantities and Units in Clinical Chemistry

cr calculus removal; calculus removed; cardiorespiratory; carriage return; cathode ray; center; center of resistance; chest and right arm; chlorine resistant; church report; class rate; clinical research; clot reaction; coefficient (of fat) retention; cold roll; cold-rolled; colon resection; combat ready; complete remission; complete round; compression ratio; conditioned reflex; conditioned response; continuing resolution; control rating; copy recorder; cranial; created; creatinine; credit; creek; cresyl red; crew; critical; critical ratio; crown; crownrump; cruise

cr (CR) carriage return (data processing); conditional release (parole); conditioned reflex; conditioned response; critical ratio

cr curia regis (Latin—the king's court)

c-r cognitive restructuring

c/r carrier's risk; company risk; correction requirement(s)

c & r cops and robbers

cr. crux (Latin—cross)

c/r cuenta y riesgo (Spanish—for account and risk of)

Cr chromium; Commander; creatinine; creditor

Cr Contador (Spanish—Bookkeeper, Cashier, Purser)

Cr. Credo (Latin—I believe, the creed); *Ceskoslovensky rozhlas* (Czechoslovak—Radio)

CR Camping Reserve; Carriers Risk; Central Registry; Change Recommendation; Characterization Report; Chief Ranger; Classified Register; Columbia Records; Columbia River; Combat Ready; Commendation Ribbon; Commonwealth Railways (Australia); Compound Risk; Connaught Rangers; Contract Requisition; cost reimbursement; Costa Rica; Costa Rican; Current Rate

C-R Crouse-Hinds; Cutler-Hammer

C/R Chicago Rawhide (manufacturing company)

C & R convoy and routing

C of R Count(y) of Ravensberg

CR Centre de Réadtation (French—Rehabilitation Center); *Ceskoslovenska Republika* (Czechoslovakian Republic); *Corpus Reformatorum; Computing Reviews; Consumer Reports*

C R comptes rendus (French—proceedings, report)

C.R. Carolina Regina (Latin—Queen Caroline); *Carolus Rex* (Latin—King Charles); *Civis Romanus* (Latin—Citizen of Rome); *Custos Rotulorum* (Latin—Roll Keeper)

cra central retinal artery

Cra Carretera (Spanish—highway); *Contadora* (Spanish—Bookkeeper, Cashier, Purser)

Cr A Commander at Arms; Corona Australis

CRA California Redwood Association; California Republican Assembly; Canadian Rheumatism Association; Cave Research Associates; Church Records Archives; Coal Research Association; College of Radiologists of Australia; Colorado River Aqueduct; Colorado River Authority; Commander Royal Artillery; Community Redevelopment Agency; Community Reinvestment Act; Concentrated Rehabilitation Area; Continuing Resolution Authority; Convair Recreation Association; Conzinc Riotinto of Australia

CRA Centres de la Recherche Appliqué (French—Applied Research Centers)

C.R.A. Conzinc Riotinto of Australia

CRAB Central Registry at Bethesda; coastal-research amphibious buggy

CRABS Close-Range Analytical-Bundle System

crabsan crab sandwich

crabwich crab sandwich

CRAC Careers Research and Advisory Center; Commander Royal Armoured Corps; Community Research Action Center; high-potency cocaine

CR Acad Sci Comptes Rendus Hebdomadaires des Seances de l'Academie des Sciences (French—Weekly Reports of Meetings of the Academy of Sciences)

crac-coc crack-cocaine

CRACK Children Requiring a Caring Kommunity

CRAD Committee for Research into Apparatus for the Disabled; Contracted Research Development

craf comet rendezvous asteroid flyby

CRAF Civil Reserve Air Fleet

CRAFT Commonwealth Rebate for Apprentice Full-time Training; Computerized Relative Allocation of Facilities Technic; Cooperative Research Action for Technology; Cycle Reporting and Fatigue Tracking

CRAG Combat Readiness Air Group

CRAGS Chemical Records and Grading System

cram. card random access memory

CRAM Contractual Requirements Recording, Analysis, and Management

CRAMM trademark of CCTA—London

cran cranial; craniology; cranium

cranapple cranbery-and-apple juice

Cranch's Cranch's United States Supreme Court Reports

craniol craniologic(al)(ly); craniologist; craniology

craniom craniometry

cranple cranberry-apple cider or juice

cran(s) cranberries; cranberry

Cranston Juvenile (delinquent) Diagnostic Center at Cranston, Rhode Island

CRAOC Commander Royal Army Ordnance Corp

crap crapola; completely ridiculous anthropic principle

CRAR Critical Reliability Action Request

cras coder and random access switch

CRASC Commander—Royal Army Service Corps

CRASH Center for Reproductive and Sexual Health; Citizens Rally and Appeal to Save Our Homes; Citizens to Reduce Airline Smoking Hazards; Community Resource and Self Help; Community Resources Against Street Hoodlums (Los Angeles Police Department detail)

crast. crastinus (Latin—of tomorrow)

'crastinator(s) procrastinator(s)

Crat Crater

crat craton (immobile center of a continent)

Crater Lake national park in Oregon

Craters of the Moon Craters of the Moon National Monument, Idaho

C-rat(s) C-ration(s)

CRAV Compañía Refineria de Azucar Viña del Mar (Spanish—Viña del Mar Sugar Refining Co)

CRAW Combat Readiness Air Wing (USN)

cray(s) crayfish(es)

CRAY Cray Computer Corporation

crb central radio bureau; curb; curbing

crb (CRB) chemical, radiological, biological (warfare)

Cr B Corona Borealis

CRB Central Reproduction Bureau; Certified Residential Broker; Change Review Board; Civilian Review Board; Commission for Relief in Belgium; Commodities Research Bureau; Cooper River Bridge (Charleston, South Carolina); Country Radio Broadcasters; County Roads Board

crbbb complete right bundle branch block

CRBC Chinese Road and Bridge Company

cr bl credit balance

cr & br crown and bridge (dental)

CRBRP Clinch River Breeder Reactor Plant

CRBS Customer Records and Billing System

crc candida-related complex; cavity rim cap (contraceptive device); combined recruiting center; complete round chart; crop revenue coverage; cyclic redundancy check

CrC control and reporting center; Crew Chief

CRC California Rehabilitation Center; Certified Recreation Counselor; Chemical Rubber Company; Civil Rights Commission; Commonwealth Reply Coupon; Computer Recycling Center; Consolidated Rail Corporation; Consolidated Railroads of Cuba; Control and Reporting Center; Coordinating Research Council; Corrosion Reaction Consultant; CRC Press

CRCA Canadian Rodeo Cowboys Association

CRCC Consolidated Record Communications Center (USA)

CRCE Centaur Reliability Control Engineering; Chicago Rice and Cotton Exchange

crchf crew chief

CRCNJ Central Railroad Company of New Jersey

crcp continuously reinforced concrete paving

CRCP Certificate of the Royal College of Physicians

CRCR Center for Rate-Controlled Recordings

CRCRS Civil Rights Community Relations Service

CRCS Canadian Red Cross Society; Certificate of the Royal College of Surgeons

Crct Circuit

CRCT Commander Royal Corps of Transport

crd central recruiting depot; chronic renal disease; chronic respiratory disease; complete reaction of degeneration

cr&d contractural research & development

Cr$ cruzeiro (Brazilian—monetary unit)

CRD Central Registration Depository; Community Relations Department; Crop Research Division (USDA)

CRDHE Center for Research and Development in Higher Education

crdl cradle

CRDL Chemical Research and Development Laboratories; Contractor Data Requirements List

crdm control-rod device mechanism

CR & DP Cooperative Research and Development Program

CRDS Clarence Ralph De Sola; Colgate-Rochester Divinity School

CRDSD *Current Research and Development in Scientifc Documentation*

cre combat-readiness evaluation; corrosion resistant

Cre Crescent

CRE Center for Radical Education; Commander Royal Engineers; Commission for Racial Equality; Congress of Racial Equality; Counselor of Real Estate

CREA California Real Estate Association; Chief Radio Electrical Artificer; Clearinghouse of Resources for Educators of Adults

C Real Ciudad Real

cream of tartar potassium acid tartrate ($KHC_4H_6O_6$)

creat creatine

CREAT Combined Resources for Editing Automated Teaching

CREATE Computational Requirements for Engineering, Simulation, Training, and Education (USAF time-sharing computer complex)

Creation Sci Creation Science Research Center

Creative Ed Creative Educational Society

crectte *creciente* (Spanish—crescent, growing)

cred credit; creditor

credd customer requested earlier due date

CREDO Chaplain's Religious Education Development Organization

Creek The Creek—oilfields scattered along the creeks of western Pennsylvania

CREEP Committee to Re-elect the President (Nixon)

CREF College Retirement Equities Fund

CREFAL *Centro Regional de Educación Fundamental para la America Latina* (Spanish—Regional Center of Fundamental Education for Latin America—United Nations organization)

CREG Cancer Research Emphasis Grants

CREI Capitol Radio Engineering Institute

crem cremation

cremains cremation remains

CREME Commander Royal Electrical and Mechanical Engineers

CREMI *Credito Minero y Mercantil* (Spanish—Mining and Mercantile Credit)

crem mus crematorium music (Beethoven's Marcia funebre from his *Eroica* Symphony, Berlioz's Death March from *Les Troyens,* Chopin's Funeral March, Handel's Death March from *Saul,* Mozart's Masonic Funeral Music, Rachmaninoff's *Isle of the Dead,* Richard Strauss's *Tod und Verklaerung,* Wagner's funeral music from *Siegfried*)

cremo crematorium

CREN Corporation for Research and Educational Networking

CREO Central Real Estate Office; Crystalline Regions Exploration Office (ONWI)

crep. *crepitus* (Latin—crepitation)

crepe(s) crepe(s) suzette

cres corrosion-resistant stainless steel; crescent; crescentic

cres *crescendo* (Italian—expanding, swelling)

Cres Crescent

CRES Center for Reproduction of Endangered Species (San Diego Zoological Society); Center for Research in Engineering Science (University of Kansas); Corrosion Resistant Stainless Steel

CREST Calcinosis, Committee on Reactor Safety Technology; Reynaud's phenomenon, Esophageal motility disorders, Sclerodactyly and Telangiectasia

cresc crescendo (Italian—increasing, swelling)

Crescendo Crescendo Publishing Company

creso crescendo (Italian—increasing, swelling)

cress garden cress; watercress

CRESS Combined Reentry Effort in Small Systems; Computer Reader Enquiry Service System

crest. crew-escape and rescue techniques (USAF)

Cret Cretaceous

CrewTAF Crew Training Air Force

crf capital recovery factor; carrier frequency; chemical releasing factor; continuous reinforcements; control-reference file; corticotrophic-releasing factor; corticotropin releasing factor

crf (CRF) corticotropin-releasing factor

CRF Cadet Royal Fusiliers; Cancer Research Foundation; Citizens Research Foundation; Constitutional Rights Foundation; Credit Research Foundation

CRFA Czechoslovak Rationalist Federation of America

CRFG California Rare Fruit Growers

crfs copper reverbatory furnace slag

crf's change request forms

crg carriage

CRG Cave Research Group; Cooperative Republic of Guyana (formerly British Guiana)

crh corticotropin releasing hormone

cri chemical rust inhibitor; cold running intelligibility; color rendering index; criminal; criterion-referenced instruction

CRI Caribbean Research Institute; Carpet and Rug Institute; Chinese Regiment of Infantry; Coconut Research Institute; Committee for Reciprocity Information; Communications Research Institute; Composers Recordings Incorporated

CR & I Chicago River and Indiana (railroad)

CRI Croce Rossa Italiana (Italian Red Cross)

CRIB Computerized Resources Information Bank

CRIC Canon Regular of the Immaculate Conception

CRICAP Carpet and Rug Industry Consumer Action Panel

CRIEPI Central Research Institute of the Electrical Power Industry

CRIF Comité Représentatif des Israélites de France (French—Representative Committee of the Jews of France)

CRILC Canadian Research Institute of Launderers and Cleaners

CRILI Center for Research in Learning and Instruction

crim criminal; criminalism; criminalist; criminologist; criminology

crim con criminal conversation (British euphemism—adultery)

Crimea Crimean Peninsula between the Sea of Azov and the Black Sea

criminol criminologist; criminology

criminotechnol criminological technology (using electronic and photographic devices and techniques to apprehend criminals and secure evidence needed for their conviction)

criminotic criminal neurotic

crip cripple

CRI & P Chicago, Rock Island and Pacific (railroad)

CRIPA Civil Rights of Institutionalized Persons Act

crips cripples

crip(s) crippler(s)—gangster(s) noted for crippling victims

CR & IR Chicago River and Indiana (railroad)

Cris Cristina; Cristine; Cristóbal; Cristopher

CRIS Command Retrieval Information System; Conference Resource and Information Services; Current Research Information System

crisco cream received in separating cottonseed oil

CRISP Computer Resources Integrated Support Plan; Cosmic Radiation Ionization Spectrographic Program (NASA)

crit critic; critical; criticality; criticism

CRITICOMM Critical Intelligence Communications System

crits critical reactor experiments

Crk Creek; Cork

crkc crankcase

crl cross register line; crown rump length

CRL California Republican League; Cambridge Research Laboratory; Cardiac Research Laboratory; Center for Research Libraries; Chemical Research Laboratory; Civil Rights Law(s); County Rugby League; Crown Renewable Lease

C.R.L. Certified Record Librarian; Certified Reference Librarian

C & RL College and Research Libraries

CRLA California Rural Legal Assistance; Canadian Railway Labor Association

CRLLB Center for Research on Language and Language Behavior (Univ Mich)

crls carelessness

crm confidence rulemaking; count rate meter; counter-radar missile; cross-reacting material; crucial reaction measure(ment); customer-relationship management

cr/m crew member

CRM Certified Records Manager; Cockpit Research Management; Combat Readiness Medal; Communications/Research/Machines (publisher); Counter-Radar Missile

CRM Consumer Research Magazine

CRMA City and Regional Magazine Association; Cotton and Rayon Merchants Association

crmch cruise Mach change

CRMD Children with Retarded Mental Development

CRME Council for Research in Music Education

Crml Carmel

crmn crewman

crmnls criminalism; criminalist; criminalistics; criminals

crmoly chrome molybdenum

CRMP Corps of Royal Military Police

crmr continuous-reading meter relay

crms close-range missile system

CRMT Community Resources Management Team (parole and probation)

CRMWD Colorado River Municipal Water District

crn crane; crown

Crn (The) Crown (The Monarchy)

CRN Council for Responsible Nutrition; Course Reference Number; Customs Registered Number

CRNA Certified Registered Nurse Anesthetist

CRNL Chalk River Nuclear Laboratories (Canada)

CRNLE Center for Research in the New Literatures in English (Australian)

CRNM Capitol Reef National Monument

cr note(s) credit note(s)

CRNP Cape Range National Park (Western Australia)

CRNPTG Commission on the Review of the National Policy Toward Gambling

CRNSS Chief of the Royal Naval Scientific Service

CRNWR Cape Romain National Wildlife Refuge (South Carolina); Clarence Rhode National Wildlife Range (Alaska)

cro cathode-ray oscilloscope; contract research organization

Cro Croatia (whose capital is Zagreb)

Cr O chrome oxide (recording tape)

CRO Carnarvon, Australia (tracking station); Chief Recruiting Officer; Commonwealth Relations Office; Community Relations Office; Contractor's Resident Office; County Recorder's Office; Criminal Records Office

CrO₂ chromium dioxide (recording tape coating)

Croat. Croatia; Croatian

C Rob Christopher Robinson's Reports of Cases Argued and Determined in the High Court of Admiralty

CROC Committee for the Rejection of Obnoxious (tv) Commercials

CROC Confederación de Revolucionarios Obreros y Campesinos (Spanish—Confederation of Revolutionary Workers and Peasants)

crock. crockery; crocks (English slang—broken-down animals or athletes)

Crockett Girls Crockett State School for (delinquent) Girls at Crockett, Texas

Croco Crocodilia

crocodiliol crocodiliologic(al)(ly); crocodiliologist; crocodiliology

croc(s) crocodilian(s)—alligator(s), caiman(s) or cayman(s), crocodile(s), gavial(s)

Croix St Croix, American Virgin Islands

cro'jack crossjack

crom control read-only memory

Crom Cromwell

CROM Confederación Regional de Obreros Mexicanos (Spanish—Regional Confederation of Mexican Workers)

CROP Christian Rural Overseas Program; Community Response in Opposition to Poverty

cross. crossing

Cross King's Cross (Sydney, Australia's nightlife section also called The Cross)

CROSS Committee to Retain Our Segregated Schools (Arkansas); Computerized Rearrangement of Special Subjects

CROSSBOW Computerized Retrieval of Organic Structures Based on Wiswesser

'crosse lacrosse; lacrosse stick

CROW Conditions of Roads and Weather

CROWCASS Combined Registry of War Criminals and Security Suspects

Crowell Crowell Collier; Thomas Y Crowell

Crown Crown Publishers

Crozets Crozet Islands in the South Indian Ocean

crp cathode-ray tube/keyboard printer

Crp C-reactive protein

CrP creatinine phosphate

CRP Committee to Re-elect the President (Nixon's fund-raising organization also known as CREEP); Conservation Reserve Program; Control and Reporting Post; Corpus Christi, Texas (airport); Cost Reduction Program; Crime Restitution Program; Crisis Relocation Plan

CRP Cruz Roja Peruana (Spanish—Peruvian Red Cross)

CRPD Chicago Regional Port District

cr pl chromium plate

CRPL Central Radio Propagation Laboratoy

Cr Pr Criminal Procedure

CRPR Child-Rearing Practices Report

CRPS Center for Research on Population and Security

crr constant ratio roll

CrR Croix-Rouge (French—Red Cross)

CRR Cost Reduction Representative

CRRA Component Release Reliability Analysis

CRRB Centaur Reliability Review Board

CRRC Costa Rica Railway Company

crrd conceptual reference repository description

CRREL Cold Regions Research and Engineering Laboratory (USA)

CRRERIS Commonwealth Regional Renewable Energy Resources Information System

CRRES Combined Release and Radiation Effects Satellite

crrl contour roller

CRRS Combat-Readiness Rating System (USAF)

crs coast radio station(s); cold-rolled steel; colon-rectal surgery; creditors; credits; crew reserve status

cr's character reactions

Crs Cristóbal, Panama

CRs counter-revolutionaries (sometimes appears as KRs)

CRS Calibration Requirements Summaries; Career Service Status (USAF); Certified Residential Specialist; Childcare Resource Service; Child Rearing Study; Coast Radio Service, Commonwealth Rehabilitation Service; Community Relations Service; Computing Research Station; Congressional Research Service; Consumer Reservation System; Corrective and Rehabilitation Squadron (USAF)

CRS Counseil de la Recherche Scientifique (French—Scientific Research Council)—Quebec; Corps Républicain de la Sécurité (French—Republican Security Corps)—anti-riot squads

CRSA Canadian Retail Shipment Association; Cold Rolled Sections Association; Concrete Reinforcement Steel Association; Connecticut River Salmon Association

CRSC Center for Research in Scientific Communications (Johns Hopkins)

CRSG Classification Research Study Group

CRSI Concrete Reinforcing Steel Institute

CRSM Certified Real Estate Marketer

crsp criminally receiving stolen property

CRSP Colorado River Storage Program

crspd. correspond; correspondence

CRSR Center for Radiophysics and Space Research (Cornell University)

CRSS Certifed Real Estate Security Sponsor; Collectors of Religion on Stamps Society; Community Refugee Settlement Scheme (Australia)

crst syndrome calcification and clinical signs of Raynaud's phenomenon, scleroderma, and telangiectasis

crt cargo-restraint transporter; cathode-ray tube; charitable trust; cold-rolled and tempered; complex reaction time

Crt Court; Crater

CRT Canadian Railway Troops; Certified Radiologic Technician; Combat Readiness Training; Criterion-Referenced Tests

cr tan lthr chrome-tanned leather

CRTC Canadian Radio-Television Commission; Cavalry Replacement Training Center

CRT/DRE C-repeat/drought-responsive element

crtgc cartographer

crtkr caretaker

crtl criticality

crtn correction

crtog cartographer; cartographic; cartography

cr tp contour template

CRTPB Canadian Radio Technical Planning Board

crt's cathode-ray tubes

CRTS Commonwealth Reconstruction Training Scheme

crtu combined receiving and transmitting unit

cru clinical research unit; combined rotating unit; crucible; cruise

Cru Crux

CRU Cecil Rhodes University; Civil Resettlement Unit; Collective Reserve Union; Crime Reduction Unit

Cru Base Cruiser Base

CRUBATFOR cruisers, battle force

CRUD Chalk River Unidentified Deposit

CRUDESLANT Cruiser-Destroyer Forces, Atlantic

CRUDESPAC Cruiser-Destroyer Forces, Pacific

CRUDIV cruiser division

CRUEL Commission on Reform of Undergraduate Education and Living (Univ Ill)

crug corrugated

cruis cruiser; cruising

cruise cruise control

CRULANT Cruiser Forces, Atlantic

CRUPAC Cruiser Forces, Pacific

cru's collective reserve units (international banking currency)

CRUSK Center for Research on Utilization of Scientific Knowledge (Univ Mich)

Crust Crustacea

crustas ice-encrusted cocktails

Crux (Latin—Cross constellation)

cruz cruzeiro (Brazilian currency)—also appears as *C, Cr, Cruz, Crz*

Cruz(an) St Croix Island (or person from there)—American Virgin Islands

CRUZEIRO Serviços Aéreos Cruzeiro do Sul (Southern Cross Air Service—Brazil)

crv central retinal vein; combat reconnaissance vehicle

Crv Corvus

CRV California Redemption Value; Corvette aircraft

crvan chrome vanadium

CRVC Cambridgeshire Rifle Volunteer Corps

cr. vesp. cras vespere (Latin—tomorrow evening)

crvf congestive right ventricular failure

crw counter-revolutionary warfare

CRW Clean Radwaste (System); Commission on Rural Water

CRWA Charles River Watershed Association

CRWG Computer Resources Working Group

CRWM Committee on Radioactive Waste Management (NAS-NRC)

CRWPC Canadian Radio Wave Propagation Committee

cry. crystal(s)

CRY Citizen's Redirecting Youth

cryng carrying

cryobio cryobiological(ly); cryobiologist; cryobiology

cryochem cryochemical(ly); cryochemist(ry)

cryoelectro cryoelectronic(al)(ly); cryoelectronicist; cryoelectronics

cryogen cryogenic(al)(ly)

cryolite sodium aluminum fluoride

cryon cryonic(s)

cryosurg cryosurgeon; cryosurgic(al)(ly); cryosurgery

crypt. cryptography

crypt (Latin prefix—hidden)— cryptogram; cryptographer; cryptographic

crypta cryptanalysis; cryptanalyst

crypto cryptograph; cryptographer; cryptographic; cryptography

cryptocom cryptocommunism; cryptocommunist

cryptofasc cryptofascism; cryptofascist

cryptonet crypto-communication network

crypton(s) cryptonym(s)

cryptos cryptocommunists; cryptofascists; cryptograms

cryptozool cryptozoological(ly); cryptozoologist(s); cryptozoology

crys crystal; crystalline; crystallization; crystallize; crystallography; crystalloids

crysnet crystallographic computing network

cryst crystal; crystalline; crystallography

crystal methamphetamine

Crystal Meth Capital Crystal Methamphetamine Capital (San Diego, California)

crystd crystallized

crystn crystallization

cs caesarean section; capital stock; carbon steel; cast steel; cast stone; casual; caught stealing (baseball); center section; central supply; cerebrospinal; cirrostratus; close support; cognitive style; color stabilizer; common steel (projectile); complementary symmetry; concentrated strength; conditioned stimulus; copy signed; coronary sinus; corticosteroid; crucible steel; cryptographic system; current series; current strength; cutting specification(s); cycloserine

cs (CS) central service; close-up shot (waist-up tv picture); conditioned stimulus

c/s case; cycles per second

c & s clean and sober

cs céntimos (Spanish—centimes, hundredths)—coins worth a hundredth part of any unit; *come sopra* (Italian—as

above); *con safos* (Spanish—American slang—impervious to attack, the same to you, you're stuck with it); *cours* (French—course, currency, current price); *cuartos* (Spanish—apartments, fourths)—coins worth a fourth part of any unit

c s calla sinistra (Italian—left hand); *con sordino* (Italian—with the mute)

cS centistoke(s)

Cs cesium; Chinese visitors; cirrostratus; standard capacitance; source capacitance

CS Cadbury Schweppes; Call Sign; Carrier Sense; Casualty Station; Chemical Society; Chief Secretary; Chief of Staff; Cincinnati Southern Railway; Civil Service; Colonial Secretary; Commonwealth Secretariat; Communications Station; Communications System; contract surgeon; Cooperative Society; Correspondence School; Credit Suisse (bank); Cryptographic System; Cultural Survival; current series; current strength; cutting specifications

C/S call signal; certificate of service; Currency Surcharge

C & S Chicago and Southern (Delta Airlines); Citizens and Southern (bank); Colorado and Southern (railroad)

C of S Chief of Staff; Chief of Service; Church of Scotland (Presbyterian)

CS Centraal Station (Dutch— Central Station); *Consejo de Seguridad* (Spanish—Security Council)—UN; Customs Service

C.S. Custos Sigilli (Latin— Keeper of the Seal)

Cs¹³⁷ radioactive cesium

CS-3A Sakura-3A (Japanese communications satellite)

CS-3B Sakura-3B (Japanese communications satellite)

csa compounded semiannually

CSA Cambridge Scientific Abstracts; Canadian Standards Association; Canterbury Society of Arts; Casualty Surgeons Association; Central South Australia; Central Surgical Association; Chief of

Staff, Army; Chloro-Sulphonic Smoke Apparatus; College of Surgeons of Australasia; Commercial Service Authorization; Common Services Agency; Commonwealth Sugar Agreement; Communication Service Authorization; Community Services Administration; Community of St Andrew; Computer Sciences of Australia; Confederate States of America; Confederate States Army; Consolidated Steamship Agency; Contractor Service Action; Controlled Substances Act

C & SA Counterinsurgency and Special Activities (Joint Chiefs of Staff)

CSA Ceskoslovenske Aerolinie (Czechoslovakian Airline)

CSAA California State Automobile Association; Central Station Alarm Association; Child Study Association of America; Committee on Space Astronomy and Astrophysics; Council of Specialized Accrediting Agencies

CSAC Cameron State Agricultural College; Citizens' Stamp Advisory Committee; Conners State Agricultural College

CSADC Canadian—South African Diamond Coporation

CSAE Canadian Society of Agricultural Engineering

CSAF Chief of Staff, United States Air Force

CSAL Central Scientific Agricultural Library (Moscow)

CSANZ Cardiac Society of Australia and New Zealand

CSAO Civil Service Association of Ontario

CSAP Canadian Society of Animal Production; Career Skills Assessment Program

csar communication satellite advanced research

CSAR Comité Secret de l'Action Révolutionnaire (French—Secret Committee of Revolutionary Action), the Cagoule active during World War II in aiding the Nazis

CSATU Congress of South African Trade Unions

CSAV Compañía Sud America de Vapores (Chilean Line)

csb calcium silicate brick; chemical stimulation (of the brain); concrete splash block

Csb Casablanca

CSB Canterbury Savings Bank; Central Statistical Board; Christian Service Brigade; Committee for Safe Bicycling; Commonwealth Savings Bank; Copra Stabilization Board

C.S.B. Bachelor of Christian Science

CSB Centro Simón Bolívar (Spanish—Simón Bolívar Center), metropolitan management investment in Caracas, Venezuela

CSBA California School Board Association; Chief Sick Berth Attendant

CSBE California State Board of Education

CSBG Community Services Block Grant; Concerned Seniors for Better Government

CSBM Confidence and Security-Building Measures

CSBPA California Shore and Beach Preservation Association

csbs course-setting bomb sight

CSBs Canada Savings Bonds

CSBS Conference of State Bank Supervisors

csc cartridge storage case; change schedule chart; combined selection center; concrete and steel combination; cosecant

c & sc capital and small capital letters

CSC Canadian Shippers Council; Canadian Space Centre; Central Security Control; Central Security Council; Charles Schwab Corporation; Child Safety Council; Citizens Service Corps; Civil Service Commission; Civilian Screening Center; Colorado State College; Combat Support Company; Command and Staff College (USAF); Commissariat Staff Corps; Commonwealth Scientific Committee; Commonwealth Steel Company; Communications Satellite Corporation; Community Service Center;

Computer Science Corporation; Consolidated Coal Company (stock exchange symbol); Conspicuous Service Cross; Continuous Service Certificate

CSCA Central States Corrections Association; Civil Service Clerical Association

CSCAR Citizens for Sensible Control of Acid Rain

CSCAW Catholic Study Circle for Animal Welfare

CSCB Civil Service Cadet Battalion

CSCC Civil Service Commission of Canada

CSCCL Center for Studies in Criminology and Criminal Law (University of Pennsylvania)

CSCD Center for Studies of Crime and Delinquency; Community Service Center for the Disabled

CSCE Coffee, Sugar and Cocoa Exchange; Conference on Security and Cooperation in Europe

CSCFE Civil Service Council for Further Education (UK)

csch hyperbolic constant; hyperbolic cosecant

CS Ch E Canadian Society for Chemical Engineering

CSCI Computer Software Configuration Item

CS circuit common-source amplifier for field-effect transistors

CSCJ Center for Studies in Criminal Justice

CsCl cesium chloride

cscn character scan(ning)

CScO Chief Scientific Officer

CSCO Cisco Systems Incorporated

CSCP Christian Science Committee on Publications

CSCS Cost, Schedule, and Control System; Crusader S-wire Container Service

C/SCSC Cost-Schedule Control Systems Criteria

cscu countersink cutter

csd closed shelter deck(ing); constant-speed drive; controlled-slip differentials; convection suppression device(s); cortical spreading depression

CSD Civil Service(s) Department; Commonwealth Society for the Deaf; Consumer Correctional Services Department; Consumer Service(s) Division; Convair San Diego (Division of General Dynamics Corporation); Correctional Services Department; Corrective Services Department

CSDA Canadian Stamp Dealers Association; Concrete Sawing and Drilling Association

CSD-ALA Children's Services Division—American Library Association

csdc computer signal data converter

csdcu constant speed drive control unit

CSDE California State Department of Education; Central Servicing Development Establishment

CSDI Center for the Study of Democratic Institutions

CSDIC Combined Services Detailed Interrogation Centre

CSDP Coordinated Ship Development Plan (USN)

CSDPH California State Department of Public Health

CSDS Chicago Sewage Disposal System

csdv closed shelter-deck vessel

cse combat support element; combined services entertainments; course

cs & e crew station and escape

Cse Causse (French—limestone plateau)

CSE Calcutta Stock Exchange; Canadian Standard English; Central Signals Establishment; Child Support Enforcement; Cincinnati Stock Exchange; Citizens for a Sound Economy; Certificate of Secondary Education; Committee on Special Education

CSEA California Society of Enrolled Agents; California State Electronics Association; California State Employees Association; Combat System Engineering Authorization

CSEAA Civil Service Employees Association of America

csect control section; cross section

c-sect cesarian section

csed coordinated ship electronics design

CSEE Canadian Society for Electrical Engineering

CSEF Canadian Siberian Expeditionary Force

csei concentrated solar-energy imitator

CSEIP Center for the Study of the Evaluation of Instructional Programs

CSEL Consolidated Support Equipment List(ing)

CSEPA Central Station Electrical Protection Association

cseq/cseqt consequences/consequent

CSERB Computers, Systems, and Electronic Research Board

CSEU Confederation of Shipbuilding and Engineering Unions

csf cerebrospinal fluid; colony-stimulating factors

Csf one hundred cubic feet

CSF California Scholarship Federation; Center for Southern Folklore; colony-stimulating factor; Community of St Francis; Correctional Service Federation

CSF Campagnie Générale de Télégraphie Sans Fil (French—Wireless Telegraph Company)

CSFA California State Firefighters' Association; Canadian Scientific Film Association; Citizens Scholarship Foundation of America

CSFAC Colorado Springs Fine Arts Center

CSFB Credit Suisse First Boston (bank)

CSFE Canadian Society of Forest Engineers

CSFPA Central Station Fire Protection Association

CSFS Commonwealth Scholarship and Fellowship Scheme

CSFT Climax/Granite-Spent Fuel Test(ing)

csf-Wr cerebrospinal fluid-Wassermann reaction

csg casing; combat support group

CSG Capital Systems Group; Configuration Steering Group; Council of State Governments

CSG Centre Spatial Guyanais (French—Guiana Space Center)

CSGA Canadian Seed Growers Association; Central States Gas Corporation

CS-gas civil(ian)-security or cyanide-simulating gas also called Mace or tear gas, causes temporary blindness, burning, tearing, choking, coughing, stinging, and vomiting; used to control unruly mobs

CSGBI Cardiac Society of Great Britain and Ireland

csgn consign

csgnd consigned

c/sgnd countersigned

csgnee consignee

csgng consigning

csgnmt consignment

C/Sgt Colour Sergeant

CSGUS Clinical Society of Genito-Urinary Surgeons

csh calcium silicate hydrate; cash

CSH Combat Support Hospital

cshaft crankshaft

CSHP Canadian Society of Hospital Pharmacists

csi contractor standard item

CSI Calculus Surface Index; Campus Studies Institute; Child Study Institute; Common Security Interface; Companion of the Order of the Star of India; Construction Specification Institute; Container Status Information; Customer Satisfaction Index

C.S.I. Companion of the Order of the Star of India

CSI Cinematique Scientifique Internationale (French— International Scientific Film Library)

CSIA Chimney Safety Institute of America

csic customer specific integrated circuit

CSIC Consejo Superior de Investigaciones Cientificas (Spanish—Superior Council of Scientific Investigations)

CSICC Canadian Steel Industries Construction Council

CSICOP Committee for the Scientific Investigation of Claims of the Paranormal

csid consider

csidd considered

csidl considerable

csidn consideration

CSIE Center for the Study of Information and Education

CSIES Community Sex Information and Education Service

CSigO Chief Signal Officer

csink countersink

CSIP Committee for the Scientific Investigation of the Paranormal

CSIR Council of Scientific and Industrial Research; Council for Scientific and Industrial Research (South Africa); Council of Scientific and Industrial Research (India)

CSIRA Council for Small Industries in Rural Areas

CSIRO Commonwealth Scientific and Industrial Research Organization (Australia)

CSIROLCA Commonwealth Scientific and Industrial Research Organization Laboratory Craftsmens Association

CSIRONET Commonwealth Scientific and Industrial Research Organization Computing Network

CSIROTA Commonwealth Scientific and Industrial Research Organization Technical Association

CSIS Canadian Security Intelligence Service; Center for Strategic and International Studies (Georgetown University)

CSISRS Cross-Section Information Storage and Retrieval System (AEC)

CSIT Chapin Social Insight Test

CSIVP California State Influenza Vaccine Program

CSJ Council for Soviet Jews

CSJ Christian Science Journal

CSJB Community of St John Baptist

csk cask; countersink; countersunk

CSK Cooperative Study of the Kuroshio (UNESCO)

CSK Consumer Survival Kit (public tv program)

csk hd countersunk head

csko countersink other side

csl common strategic launcher; computer simulation language; computer-sensitive language; console

csl (CSL) crane stores lighter

CSL Canada Steamship Lines; Cedar Springs Library; Chicago Short Line (railroad); Cinderella Softball League; Circle of State Librarians; Colorado State Library; Communications Sub-Lieutenant; Computer Science Laboratory; Consumer Service Litigants

CSL Centre de Semi-Liberté (French—Semi-Liberty Center)—halfway house for criminals; Conseil Supérieur du Livre (French—Better Book Council)

CSLA Canadian School Library Association; Church and Synagogue Library Association

CSLATP Canadian Society of Landscape Architects and Town Planners

CSLB Contractor's State License Board

CSLC California State Lands Commission

CSLEA Center for the Study of Liberal Education for Adults

CSLICC Counseling Service of the Long Island Council of Churches

CSLO Canadian Scientific Liaison Office; Combined Services Liaison Office(r)

CSLP Center for Short-Lived Phenomena (Smithsonian)

cslr consular

CSLS Civil Service Legal Society (UK)

CSLT Canadian Society of Laboratory Technologists

csm cerebrospinal meningitis; chronic symptomatic maladjustment; close support missile; combustion space monitor; command service module (CSM); constant speed motor; controlled speed motor; corn-soya-milk (mixture)

CSM Campaign Service Medal; Central States Motor Freight Bureau; Chief Stipendiary Magistrate; China Service Medal; Christian Socialist Movement; Colorado School of Mines; Command and Service Module; Command Sergeant-Major; Commission for Synoptic Meteorology; Company Sergeant-Major; Correctional Service of Minnesota; Cosmopolitan School of Music

CSM Christian Science Monitor

csma (CSMA) carrier-sensed multiple access

CSMA Chemical Specialities Manufacturers Association

CSMA/CD Carrier Sense Multiple Access with Collision Detection

CSMC Catholic Students' Mission Crusade; Council for the Single Mother and her Child

CSMFTA Central and Southern Motor Freight Tariff Association

CSMI Company Sergeant-Major Instructor

csmith coppersmith

CSM-LM Command Service Module-Lunar Module (Apollo spacecraft)

CSMMG Chartered Society of Massage and Medical Gymnastics

CSMP Comprehensive School Mathematics Program; Continuous System Modeling Program

c/smp(s) counter sample(s)

CSMPS Computerized Scientific Management Planning System

CS/Ms Commander, Submarines

CSMSW Carver School of Missions and Social Work

csn colloidal suspension

csn. cousin(s)

CSN Canadian Switched Network; Community of the Sacred Name; Confederate States Navy; Contract Serial Number; Control Symbol Number

CSN Companhia Siderurgica Nacional (Portuguese—National Steel Company)

CSNAR Charles Sheldon National Antelope Refuge (Nevada)

CSNDA Center for the Studies of Narcotic and Drug Abuse (National Institute of Mental Health)

CSNH Cincinnati Society of Natural History

CSNI Committee for the Safety of Nuclear Installations

CSNMDU Center for the Study of Non-Medical Drug Use

CSNO California School Nurses Organization

CSNWR Carolina Sandhills National Wildlife Refuge (South Carolina)

CSN&Y Crosby, Stills, Nash and Young

cso central signoff, chained sequential operation; combined sewer overflow

C^{so} Corso (Italian—street)

CSO Cairo Symphony Orchestra; Canadian Shipwatchers Organisation; Canberra Symphony Orchestra; Cargo Security Office; Central Selling Organisation (diamonds sold in London); Central Statistical Office; Charlotte Symphony Orchestra; Chattanooga Symphony Orchestra; Chicago Symphony Orchestra; Chief Signal Officer; Chief Staff Officer; Chief Surgical Officer; Cincinnati Summer Opera; Cincinnati Symphony Orchestra; Clothing Supply Office(r); Columbia Symphony Orchestra; Columbus Symphony Orchestra; Command Signal Office(r); Commissioners Standard Ordinary (table); Commonwealth Scientific Office; Community Service Office(r); Community Service Organization; Community Standards Organization; Consumer Services Organization; Montevideo, Uruguay (Carrasco airport)

csocr code-sort optical-character recognition

CSOP Commission to Study the Organization of Peace (UN)

csoro conical span on receive only

CSOs Community Service Officers; Community Service Organizations

csp central switching point; channel substrate planar; concurrent spare parts; constant-speed drive; continued service plan; cross system product

Csp Caspar; Caspean

CSP Center for the Study of the Presidency; Certified Safety Professional; Chartered Society of Physiotherapy; Communicating Sequential Processes; Connecticut State Police; Continuous Sampling Plan; Contractor Support Program; Corporation Standard Practice

C.S.P. Congregation of St Paul

CSPA California State Psychological Association; Civil Service Pensioners' Alliance; Columbia Scholastic Press Association; Council of State Planning Agencies

C-SPAN Cable-Satellite Public Affairs Network

CSPAA Columbia Scholastic Press Advisers Association

CSPB California State Personnel Board

CSPC California State Polytechnic College

CSPCA Canadian Society for the Prevention of Cruelty to Animals

CSPCo Caledonian Steam Packet Company

C/SPCS Cost-Schedule Planning Control Specification

CSPE Columbia Storage Power Exchange

cspep carry straps parachutists' equipment

CSPI Center for Science in the Public Interest

CSPM Communications Security Publications Memorandum

cspp corrugated structural plate pipe

CSPP Community Shelter Planning Program

cspr chlorosulphonated polyethylene rubber

CSPR(s) Christian Science Practitioner(s)

CSPS Cable-Suspended Pumping Station; Christian Science Publishing Society

csqm climax stock quartz monsonite

CSQs College Student Questionnaires

csr circumsolar radiation; command signal regulator; compulsive security ritual; corrected sedimentation rate; corrugated steel reinforce-

ment; craniosacral release; customer service representative

C-S r Cheyne-Stokes respiration

CSR Canadian Scottish Regiment; Certified Shorthand Reporter; Chartered Stenographic Reporter; Civil Service Requirement; Colonial Sugar Refining; Commonwealth Strategic Reserve

CSRA Central Savannah River Area (Planning and Development Commission); Civil Service Reform Act (of 1978)

CSRC Communication Science Research Center (Batelle Memorial Institute—Columbus, Ohio)

CSRF Childrens Surgical Research Fund

CSRG Commonwealth Special Research Grant

csrl common strategic rotary launcher

CSRL Center for the Study of Responsive Law

CSRO Consolidated Standing Route Order (USA)

CSRP Canadian Sprinkler Risk Pool; Cognitive Systems Research Program

CSRS Cooperative State Research Service

CSRUIDR Chemical Society Research Unit in Information Dissemination and Retrieval

CSRV Civil Service Rifle Volunteers

css center spar station; computer systems simulator; content scrambling system; controlstick steering

CSS Calcutta School Society; Center for Strategic Studies; Central Security Service (DoD); Chinese Surface to Surface (missile); City Shuttle Service; Civil Service System; Clandestine Services Staff (CIA); Coded Switch System (to arm nuclear weapons); College Scholarship Service; Combat Service Support (USA); Commit Sequence Summary; Community Service Society; Comparative Shopping Service; Computerized Shipping Ser-

vice; Confederate States Ship (C.S.S.); Contractor Storage Site

C.S.S. Charles Stuart Calverley (nineteenth-century satirist whose works appear under the initials shown)

CSS *Custom Statistical Service*

C^{ssa} *Contessa* (Italian—Countess)

CSSA Cactus and Succulent Society of America; Central States Speech Association; Central Supply Support Activity

cssb compatible single sideband

CSSB Cedar Shake and Shingle Bureau; Civil Service Supply Board

CSSC California Seismic Safety Commission

CSSCG Container Systems Standardization-Coordination Group

C S-S Co Cunard Steam-Ship Company

CSSD Central Sterile Supply Department

CSSDA Council of Social Science Data Archives

CSSDC Canadian Society for the Study of Diseases in Children

CSSE Canadian Society for the Study of Education

cssl continuous system simulation language

CSSL Central Sierra Snow Laboratory (Norden, California)

CSSLRP Commonwealth Secondary School Libraries Research Project

cssm compatible single-side-band modulation

CSSM Council of State Supervisors of Music

CSSO Combined State Services Organization; Consolidated Surplus Sales Office

CSSP Center for Studies of Suicide Prevention; Committee on Solar and Space Physics; Customer Standard Settlement Program

CSSR Cost Schedule Status Report

CSSRC Canadian Social Science Research Council

CSSS Canadian Soil Science Society; Council of State Science Supervisors

csst computer system science training

CSSU Crime Scene Search Unit

cst cargo ships and tankers; centistokes; channel status indicator; combined station power; convulsive shock therapy

c's/t certificates of title

CST Cat Survival Trust; Celeban Standard Time; Central Standard Time; Conventional Stability Talks; Council for Science and Technology

CSta consolidating station

CSTA Canadian Society of Technical Agriculturists; Canterbury Science Teachers Association; Coal and Slurry Technology Association; Correspondence School Teachers Association

cs & tae combat surveillance and target acquisition equipment (DoD)

CSTAL Confederación Sindical de Trabajadores de America Latina (Spanish—Trade Union Confederation of the Workers of Latin America)

CSTC Charles Schwab Trust Company; Charleston Submarine Training Center (South Carolina); Coordinating Scientific and Technical Council (UN); Coppin State Teachers College

C'sted Christiansted, St Croix

cstg casting

CSTI California Specialized Training Institute (for coping with terrorism); Chattanooga State Technical Institute

CStJ Commander, Order of St John of Jerusalem

cstmr customer

cstol combined short takeoff and landing; controlled short takeoff and landing

cstr canister

CSTS Combined Systems Test Stand; Computer Science Time Sharing

cstv community-supported television

C/Stwd Chief Steward

csu catheter specimen of urine; central statistical unit; channel service unit; circuit switching unit(s); constant-speed unit

CSU California State University; Casualty Staging Unit; Civil Service Union; Colorado State University; Combined State Unions; Connecticut State University; Crime-Suppression Unit

CSU Christlich-Soziale Union (German—Christian Social Union)—political party

CSUB Center for the Study of Urban Poverty (UCLA)

CSUC California State Universities and Colleges; California State University at Chico

CSUCA Consejo Superior Universitaria Centroamericano (Superior Council of Central American Universities)

CSUF California State University at Fresno

CSUH California State University at Humboldt

csul consult

CSULA California State University at Los Angeles

CSULB California State University at Long Beach

csuld consulted

csulg consulting

CSUN California State University at Northridge

CSUS California State University at Sacramento

CSUSA Copyright Society of the U.S.A.

CSUSB California State University at San Bernardino

CSUSD California State University at San Diego

CSUSF California State University at San Francisco

CSUSJ California State University at San Jose

CSUSM California State University at San Marcos

csv combat support vehicle; comma separated value

csvli cash surrender value of life insurance

CSV Community Service Volunteer

csw channel status word(ing); continuous seismic wave; conventional standoff weapon

CSW Certified Social Worker; Commission on the Status of Women

CSWA Chinese Seamens Welfare Association

CSWAE Commission on the Status of Women in Adult Education

CSWE Council on Social Work Education

CSWI Commission for Synoptic Weather Information

csws crew-served weapon sight; corps support weapon system

Cswy Causeway

USX Chessie and Seaboard (railroads consolidated)

csz copper, steel, or zinc (freight)

ct cable transfer; carat; caught; cellular therapy; cent; center; center tap; central timing; ceramic tile; circuit; coagulation time (blood test); coated tablet; coffee table; combat team; compressed tablet; compute topography; conning tower; contrast threshold; control transformer; corrective the~apist; corrective therapy; count; countertenor; court; credit; current; current transformer

ct (CT) computed tomograph(y); corrective therapist; corrective therapy

ct. citation; county

c/t conference terms

c & t classification and testing

ct. centum (Latin—hundred)

Ct celtium; Court

CT Canadian Terms; Capital Territory (airline code); Central Time; Certificate of Title; Chrysler Technologies; Cognitive Therapy; Community Transit; Communist Terrorists; Connecticut; Copy Typist; Credit Transfer; Sir Charles Tupper (Canada's seventh Prime Minister)

C/T California Terms

C of T Certificate of Title; Count(y) of Tyrol

CT Corrections Today

CT-4 New Zealand-built Airtrainer aircraft

cta call time adjustor; cash to assume; catamenia (menstruation); companion trainer aircraft; current transformer assembly; cystine trypticase agar

cta (CTA) cyano-trimethyl-androsterone

cta communiquer à toutes adresses (French—circulate to all addresses); *cuenta* (Spanish—account)

c.t.a. cum testamento annexo (Latin—with the will annexed)

Ct A Control Area

CTA California Taxpayers Association; California Teachers Association; Canadian Tuberculosis Association; Caribbean Tourist Association; Catterick Training Area; Chaplain to the Territorial Army; Chemical Toilet Association; Chicago Transit Authority; Colorado Teachers Association; Commercial Travellers Association; Committee on Thrombolytic Agents (blood test); Consolidated Tape Association; Container Truckers Association; Council for Technical Advancement; Covered Threads Association

cta corr^{te} cuenta corriente (Spanish—current account)

cta cte cuenta corriente (Spanish—current account)

CTAF Crew Training Air Force

cta/ir control area/instrument restricted

cta/iv control area/instrument visual

CTAL Container Terminals Australia Ltd

CTAL Confederación de Trabajadores de America Latina (Spanish—Confederation of Latin American Workers)

ctam continental tropical air mass

Ct Ap Court of Appeals

CTAU Catholic Total Abstinence Union

cta/ve control area/visual excepted

ctb cement-treated base; ceramic-tile base; coastal torpedo boat

CTB Cable Television Bureau; California Test Bureau; Canadian Tourist Board; Commercial Traffic Bulletin; Commonwealth Telecommuni-

cations Board; Commonwealth Trading Bank; Comprehensive Test Ban; Corporation for Television Broadcasts

CTB Centre Technique de Bois (French—Wood Research Center)

CTBA California Toll Bridge Authority; Commonwealth Trading Bank of Australia

ctbid(s) counterbid(s)

ctbm cetyl-trimethyl-ammonium bromide

ctbore counterbore

CTBRD Commonwealth Taxation Board of Review Decisions

CTBS Comprehensive Tests of Basic Skills

CTBT Comprehensive Test Ban Treaty

ctc cadet training center; carbon tetrachloride; combat training center; combined training center; commando training center; contact; cut tear curl

ctc (CTC) centralized traffic control; central train control; chlortetracycline

CTC California Tankers Company; California Transportation Commission; Canadian Tire Corporation; Canadian Transport Commission; Canberra Technical College; Catholic Teachers College; Central Test Control; Certified Travel Consultant; Charter Travel Company; Chicago Teachers College; Chicago Technical College; Citizens Tax Council; Citizens Training Camp; Citizens Training Corps; Commissariat and Transport Corps; Concordia Teachers College; Corn Trade Clauses; Curaçao Trading Company; Cyclists Touring Club

C&TC Commissariat and Transport Corps

CTC Congrés du Travail du Canada (French—Canadian Congress of Labour)

CTCA Canadian Telecommunications Carriers Association; Channel and Traffic Control Agency

CTCB Contract Technical Compliance Board

ctcd contacted

ctcg contacting

Ct Cl Court of Claims

CTCL Community and Technical College Libraries

CTCOSBA Cape Town Computer Services and Bureaux Association

CTCP Contract Task Change Proposal

CTCRM Commando Training Centre Royal Marines

Ct Crim Aps Court of Criminal Appeals

CTCs Community Treatment Centers (US Bureau of Prisons)

CTCSS Continuous-Tone Coded-Squelch System

ctd charge transfer device; coated; crated; cumulative trauma disorder; current, temperature, depth

c-t-d conductivity-temperature-depth

CTD Central Training Depot; Classified Telephone Directory; Convention Travel Document; Corrective Therapy Department

CTDA Ceramic Tile Distributors Association

CTDAS Canadian Trade Document Alignment System

ctdc control track direction computer

CTDC Chemical Thermodynamics Data Center (NBS)

ctdh command and telemetry data handling

CT & DM Canadian Transportation and Distribution Magazine

CTDO Central Technical Documents Office (USN)

cte coefficient of thermal expansion

cte corriente (Spanish— current)

Cte Comte (French—Count)

Cte Conte (Italian—Count)— Earl

CTE Car Tours in Europe; Citizens for Tax Equity

CTE Compañía Transatlántica Espanola (Spanish Line)

CTEB Council of Technical Examining Bodies

CTEC Chemical Transportation Emergency Center

Ctee Committee

Cten Ctenophora

Cteno Ctenocephalides (fleas)

CTEP Community Transportation Enhancement Program

CTES Computer Telex Exchange System (RCA)

Ctesse Comtesse (French—Countess)

CTETOC Council for Technical Education and Training for Overseas Countries

CT Exam Computed Tomography Examination

ctf certificate; combined task force; correction to follow; cytotoxic factor

Ctf Colorado tick fever

CTF Canadian Teachers Federation; Cayman Turtle Farm; Chaplain to the Forces; Commander Task Force; Correctional Training Facility

CTFA California Tree Fruit Agreement; Cosmetics, Toiletry, and Fragrance Association

CTFC Commodity Futures Trading Commission

ctfe chlorotrifluorethylene

CTFE Colleges of Technology and Further Education (subsection of the University and Research Section of the Library Association)

ctfet counterfeit

ctfm continuous-transmission frequency-modulated (sonar)

CTFT Centre Technique Forestier Tropical (French—Tropical Forest Technical Center)

ctfy certify

ctg cartage; cartridge; cutting

Ctg Cartagena, Spain (*see* Cgn)

CTG Center Theatre Group; Commander Task Group; Commercial Travellers Guild; Components Technology Group

ctge cartage; cartridge; cottage

ctgf clean tanks, gas free

CTGI Canadian Test of General Information

Cth Commonwealth

CTH Cape Town Hussars; Chalmers Tekniska Högskola (Swedish—Chalmers Institute of Technology); Corporation of Trinity House

CTHA Chinese Tourist Hotel Association

Cthse Courthouse

cti Container Transport International (trademark)

CTI Central Technical Institute; Ceramic Tile Institute; Container Transport International; Contract Technical Instructor; Cooling Tower Institute

CTI Communication Technology Impact

CTIA Caravan Trades and Industries Association; Cellular Telecommunications Industry Association; Committee to Investigate Assassinations

CTIAC Concrete Technology Information Analysis Center (USA)

CTIC Cable Television Information Center

CTIU Container Transport International (container) Unit

CTJ Citizens for Tax Justice

ctk capacity-ton kilometer

cTK tropical continental air colder than underlying surface

ctl castellate; cental; central; complementary transistor logic; constructive total loss; control

ctl (CTL) checkout test(ing) language

Ctl central

CTL Certified Tool List; Cincinnati Testing Laboratories; Container Terminals Limited

CTLA California Trial Lawyers Association; Council of Tree and Landscape Appraisers

ctlg catalog

ctlo constructive total loss only

ctm capacity ton mile; centrifugal turning moment; communications terminal modules

CTM Contract Termination Manual; Contractor Technical Meeting

CTM Confederación de Trabajadores de México (Spanish—Confederation of Workers of Mexico)

CTMA Collapsible Tube Manufacturers Association; Commercial Truck Maintenance Association; Country Timber Merchants Association (Australia)

ctmc communications controller; communications terminal modules

ctmdr clamptop metal drum

CTMM California Test of Mental Maturity

ctn carton; cotangent

C Tn Cape Town (British maritime contraction)

CTN Canton Island (tracking station)

ctnd contained

CTNE Compañía Telefonica Nacional de España (Spanish—National Telephone Company of Spain)

ctng containing

ctnrs containers

ctns cartons

ctn's confectioners, tobacconists, newsagents

CTNS Chicago Tribune News Service

cto cancelled to order; concerto

c^{to} conto (Italian—account); *cuarto* (Spanish—fourth)

CTO Central Telegraph Office; Central Treaty Organization; Chief Technical Officer; City Ticket Office; Cognizant Transportation Office; Combined Transport Operator; Container Transport Operator; Courier Transfer Officer

CTOA Commonwealth Telephone Officers Association; Creative Tour Operators Association

ctocu central technical order control unit

ctofr counteroffer

ctol conventional takeoff and landing

ct ord court order

ctp central transfer point; close to profit; corporate trade payment

ctp (CTP) cytidine triphosphate

CTP Canterbury Timber Products; Certified Trager Practitioner; Columbia Television Pictures; Computer Technology Program

CTP Centre de Tutelle Pénale (French—Penal Surveillance Center)—for recidivists; *Confederación de Trabajadores del Peru* (Spanish—Confederation of Workers of Peru)

CTPL Commission for Teacher Preparation and Licensing

CTPOA Commonwealth Telephone and Phonogram Officers Association

ctprt counterpart

ctpt counterpoint

CTPTA Centro Tropical de Pesquisas y Tecnologías de Alimentos (Tropical Center of Food Research and Technology)

ctptal contrapuntal

ctptst contrapuntist

ctr center; close target reconnaissance; commuter trip reduction; contour; controlled thermonuclear reactor; counter; cutter

Ctr Center

CTR Cambridge Territorial Regiment; Cash Transaction Report; Controlled Thermonuclear Reactor; Currency Transaction Report

CTRA Coal Tar Research Association

Ctr Appl Ling Center for Applied Linguistics

Ctr Appl Res Center for Applied Research in Education (New York)

Ctr Byz Center for Byzantine Studies

CTRC Caribbean Tourism Research Center

Ctr Calif Pub Center for California Public Affairs (Claremont)

Ctr Chin Stud Center for Chinese Studies (Berkeley, California)

Ctr Cont Celeb Center for Contemporary Celebration (Chicago)

Ctr Cont Poetry Center for Contemporary Poetry (La Crosse, Wisconsin)

Ctr Info Am Center for Information on America

ctr/iv control zone/instrument visual

ctrl control (flow chart)

Ctr Land Arch Center for Landscape Architecture

Ctr Marital Sexual Center for Marital and Sexual Studies (Long Beach, Calif)

Ctr Mig Center for Migration Studies (New York)

ctrofr counteroffer

ctrofrdcl counteroffer declined

CTRP Controlled Thermonuclear Research Program

CTRP Confederación de Trabajadores de la République de Panamá (Spanish—Confederation of Workers of the Republic of Panama)

Ctr Pol Process Center for Policy Process (DC)

Ctr Pre-Col Center for Pre-Columbian Studies (DC)

Ctr Sci Pub Center for Science in the Public Interest (DC)

Ctr Sci Study Rel Center for the Scientific Study of Religion (Chicago)

Ctr S & SE Asian Center for South and Southeast Asian Studies (Ann Arbor, Mich)

CTRU Colonial Termite Research Unit

Ctr Urb Pol Res Center for Urban Policy Research (New Brunswick, NJ)

ctr/ve control zone/visual exempted

ctry country

cts carats; carpal tunnel syndrome; cents; clear to send; computer typesetting; contralateral threshold shift (audiometry); crates

cts (CTS) communications technology satellite(s)

cts centavos (Spanish—cents); *centimes* (French—cents); *centimos* (Spanish—cents)

Cts courts

CTS Canadian Thoracic Society; Captive Trajectory System; Card-to-Magnetic Conversion System; Catholic Truth Society; Central Transportation System; Centralized Title Service; China Travel Service; Commercial Teachers Society; Commonwealth Teaching Service; Commonwealth Time Service; Component Test(ing) Set; Computer Test(ing) Site; Computerized Type System; Consolidated Tin Smelters; Consolidated Translation Survey; Container Terminal Station; Contract Technical Services; Contractor Technical Service; Conversational Terminal System; Cosmic Top Secret; Courier Transfer Station; Custom Track Service

CTSA Catholic Theological Society of America; Coordinated Transportation Service Agency; Crucible and Tool Steel Association

CT Scan computerized tomography scan

CtSEA Connecticut Society of Enrolled Agents

Ct Sess Court of Sessions

ctsp contract technical services personnel

CTSS Compatible Time-Shared System

ctt capital transfer tax(ation); compressed tablet triturate; computerized transaxial tomography

CTT Columbia Technical Translations

CTT Correios e Telecommuniações de Portugal (Portuguese—Postal and Telegraph Services of Portugal)

ct ta control tape

CTTB Central Trade Test Board

CTTC Canadian Technical Training Corps; Central Telegraph Test Center

Cttee Committee

CTTF California Turtle and Tortoise Club

CTTO Central Trials and Tactics Organization

CTTP Complementary Trade Training Program

ctu centigrade thermal unit; central terminal unit; components test unit

CTU Combat Training Unit; Commander Task Unit

CTU (AFL-CIO) Commercial Telegraphers' Union

C-tube C-shaped tube

CTUS Carnegie Trust for the Universities of Scotland

CTUY Confederation of Trade Unions of Yugoslavia

ctv (CTV) cable television; color television

CTV Canadian Television; Children's Television

CTV Confederación de Trabajadores de Venezuela (Spanish—Confederation of Venezuelan Workers)

ctvo centavo (Spanish—cent)

CTVT Childrens Television Theater

CTVW Children's Television Workshop

ctw counterweight

cTw tropical continental air warmer than underlying surface

CTW Children's Television Workshop

ctwt counterweight

ctx computer telex exchange (RCA system); corporate trade exchange

Ct X Court Exhibit

Cty City; County

ctz chlorothiazide

CTZ Corps Tactical Zone

ct zone chemoreceptor trigger zone

cu cleanup; clinical unit; close-up; condition unknown; contents unknown; control unit; conversion unit; cube; cubic; cumulus

cu (CU) container unit; credit union

c-u see you

c/u cada uno (Spanish—each one)

Cu Cuba; Cuban; cumulus; cuprum (Latin—copper)

C_u urea clearance

CU Cambridge University; Capital University; Carleton University; Church Union; City University; Clafkin University; Clark University; Colgate University; Colorado University; Columbia University; Commercial Union; Concordia University; Consumers Union; Cooper Union; Cooperative Union; Cornell University; Creighton University; Cuba (Internet code); Cumberland University; Customs Union

Cu_2SO_4 copper sulfate

Cu-7 copper-constructed 7-shaped intrauterine device

cua central unit assembly; computer unit assembly

CUA Canadian Underwriters Association; Catholic University of America; Center for Urban Analysis; Council on Urban Affairs

CUAC Cambridge University Athletic Club

cuad cuadrado (Spanish—square)

CUAFC Cambridge University Association Football Club

CUAG Computer Users Associations Group

CUAP Catholic University of America Press

CUAS Cambridge University Agricultural Society; Cambridge University Air Squadron

CUAV Citizens United Against Violence

cub. control unit busy; cubic

cúb cúbico (Spanish—cubic)

Cu b copper band

CUB advanced unit base; Carlton and United Breweries; Citizens Utility Board; Consumers Utility Board; Council for UHF Broadcasting

Cuba Republic of Cuba (largest West Indian island, República de Cuba)

Cubana (Spanish—Cuban)—feminine

CUBANA Compañía Cubana de Aviación (Spanish—Cuba Air Company)

Cubano (Spanish—Cuban)—masculine

CUBC Cambridge University Boat Club; Cambridge University Boxing Club

CUBE Concertation Unit for Biotechnology in Europe

cubo conduct unbecoming an officer

CUBS Congress for the Unity of Black Students

cuc chronic ulcerative colitis

CUC Canadian Unitarian Council; Canberra University College; Canterbury United Council

CUCA Carpet and Upholstery Cleaning Association

cu cap. cubic capacity

CUCC Cambridge University Cricket Club

Cuch Cuchillas (Spanish—mountain chain, range)

cu cm cubic centimeter

CUCNY Citizens Union of the City of New York

CUCS Columbia University Community Services

cud congenital urinary (tract) deformities

'cuda(s) barracuda(s)

Cuddy Cuthbert

CUDN Common User Data Network

cuds barracuda

CUDS Cambridge University Dramatic Society

cue coastal upwelling experiment; computer update equipment; configuration utilization efficiency; control unit end; correction update extension

CUE Center for Urban Education; Coastal Upwelling Experiment; Concentrated Urban Enforcement (of gun control)

CUEA Coastal Upwelling Ecosystem Analysis

CUEBS Commission on Undergraduate Education in the Biological Sciences

CUED Council for Urban Economic Development

Cuen Cuenca

CUEPACS Congress of Unions of Employees in the Public and Civil Services

CUERAC Computer Controlled Random-Access Cartridge Libraries

CUERD Committee for Upgrading Environmental Radiation Data

CUES College and University Environment Scales

CUEW Congregational Union of England and Wales

CUF California University Farm; Canadian Universities Foundation

CUF Companhia Uniao Fabril (Portuguese—United Manufacturing Company)—Iberian conglomerate whose company street in Barreiro is named Rua do Acido Sulfúrico (Sulfuric Acid Street)

CUFC Consortium of University Film Centers

CUFF Catholics United for the Faith

'cuffed handcuffed

'cuffs handcuffs

cu ft cubic feet; cubic foot

cu ft min cubic feet per minute

cu ft sec cubic feet per second

cug closed-user group; cystourethrogram

CUGC Cambridge University Golf Club

CUHC Cambridge University Hockey Club

CUHK Chinese University of Hong Kong

cui character-based user interface

CUIC California Unemployment Insurance Code; Canadian Unemployment Insurance Commission

CUIHC California Urban Indian Health Council

cu in cubic inch

cuis cuisine (French—cookery, kitchen)

cuj. cujus (Latin—of which)

cuj. lib. cujus libet (Latin—of any you wish)

CUK São Paolo, Brazil (Combica Airport)

cukes cucumbers

CUKT Carnegie United Kingdom Trust

cul culinary

c-u-l see you later

CUL Cambridge University Libraries; Catholics United for Life; China Union Lines; Columbia University Library; Cooper Union Library; Cornell University Library

c-u later see you later

cull. cullage; cullboard; culling; cullion

CULP California Union List of Periodicals

cult. cultural; culture

CULT Chinese University Language Translation (system)

cult. anthro(s) cultural anthropologist(s); cultural anthropology

CULTC Cambridge University Lawn Tennis Club

culv culvert

cul vul(s) culture vulture(s)

cum central unit memory; cumulative

cu m cubic meter

CUM Cambridge University Mission

CUM Centro Universitario México (Spanish—Mexico University Center)

CUMA Canadian Urethane Manufacturers Association

Cumb Cumberland

Cumberland Island national seashore on coast of Georgia

Cumberlands Cumberland Caverns, Tennessee; Cumberland Islands off Queensland,

Australia; Cumberland Mountains extending from Alabama to Virginia

cum/d cubic meters per day

cum d(iv) cum dividend (with dividend)

cume. cumulative (audience)

cumes cumulative audience survey (radio/tv)

cu mm cubic millimeter

Cummins Farm Cummins Prison Farm in Arkansas

CUMMM Council of Underground Mining Machinery Manufacturers

Cum Nursing Lit Cumulative Index to Nursing Literature

Cump Tecumseh

cum pref cumulative preference

CUMS Cambridge University Musical Society

cum/sec cubic meters per second

cumshaw (Chinese—grateful thanks)—quasi-legal barter system; under-the-table graft

cu mu cubic micron

cumu cumulus

CUMWA Consortium of Universities in the Metropolitan Washington Area

cun cuneiform

Cun Cunard (steamship line including *Aquitania*, *Lusitania*, and *Mauretania*)

CUN Convent van Universiteits-bibliothecarissen in Nederland (Dutch—Association of University Librarians in the Netherlands)

CUNA Credit Union National Association

cuni cupro-nickel (coin alloy)

cu-nim cumulo-nimbus (clouds)

CUNSA Canadian University Nursing Students Association

CUNY City University of New York

CUOG Cambridge University Opera Group

CUOTC Cambridge University Officers Training Corps

cup. cupboard

CUP Cambridge University Press; Canadian University Press; Columbia University Press; Conditional Use Permit; Cornell University Press

CUPA College and University Personnel Association

CUPBEQ Canadian University Press Québec Region

CUPE Canadian Union of Public Employees

cupper cup-tie-er (athletic matches played for a trophy cup)

CUPR Catholic University of Puerto Rico

cuprite cuprous oxide

cupronic copper-nickel alloy

CUPS Consolidated Unit Personnel Section

CUPW Canadian Union of Postal Workers

cur. curiosa; curiosity; currency; current

Cur Curaçao (maritime abbreviation)

CUR Cambridge University Rifles; Council on Undergraduate Research; Curaçao, Netherlands West Indies (Plesman Airport)

CURA Center for Urban Research and Action; Comprehensive Urban Renewal Area

CURAC Coal Utilization Research Advisory Committee

curat curative

curat. curatio (Latin—dressing, wound dressing)

CURB Campaign on the Use and Restriction of Barbiturates

CURE Citizens United for Racial Equality

C-U-R-E Care, Understanding, Research (organization for the welfare of drug addicts)

CURES Coalition for Urban/Rural Environmental Stewardship; Computer Utilization Reporting System

CURF Citizens Union Research Foundation

CURFC Cambridge University Rugby Football Club

curio curiosa; curiosity

CURLS College, University, and Research Libraries Section (California Library Association)

CURMCO City Urban Renewal Management Corporation (NYC)

curr currency; current

curric curriculum

curt. current (Scottish—instant); curtain

Curt Curtis; Quintus Curtius Rufus (Roman historian)

CURTS Common-User Radio Transmission System

curv cable-operated unmanned recovery vehicle

CURV Cable-controlled Underwater Research Vehicle

cury currently

Curzon Lord Curzon (George Nathaniel Curzon)—Viceroy and Governor General of India

cus customer

Cus Customs

CUS Cambridge Union Society; Continental United States

CUSA Conservative United Synagogue of America; Council of Unions of South Africa

cusecs cubic feet per second

Cush Cushing's Massachusetts Reports

Cus Ho Custom House

CUSIP Committee on Uniform Security Identification Procedures (for computer user protection); Committee for Uniform Securities Information

CUSIP No. CUSIP Number (assigned to every issue of a bond or stock)

cusm customs

CUSM Columbia University School of Medicine

cusmr customer

CUSO Canadian University Service Overseas

CUSP Central Unit for Scientific Photography

CUSR Canada/United States Region

CUSRPC Canada-United States Regional Planning Committee

CUSRPG Canada-United States Regional Planning Group

CUSS Continental, Union, Shell, Superior (oil compnies' deep-sea oil-drilling ship)

cust custard; custodian; custody; custom(s)

Cust Ct Customs Court

custod custodian

custs custards; customers

CUSW/NAS Committee on Undersea Warfare—National Academy of Sciences

cut control unit terminal

CUT California United Terminal

CUTC Cambridge University Training Corps

CUTF Commonwealth Unit Trust Fund

cutg cutting

Cuth Cuthbert; Cuthwald; Cuthwold

cuti cutis (Latin—skin)—cutaneous, cuticle, cutin

cutl cutlass

'cutor prosecutor

CUTS Computer-Utilized Turning System

Cuu Chihuahua

CUUI Center for US—USSR Initiatives

CUUS Consumers Union of the United States

cuv current use value

CUW Committee on Undersea Warfare (DoD)

CUWARFA California Urban Waterfront Area Restoration Financing Authority

Cux Cuxhaven

cu yd cubic yard

cuz cousin; because

cv caloric value; capital value; cardiovascular; carrier vehicle; cataclysmic variable; catalog value; check valve; chief valve; closing volume; coefficient of variation; collection voucher; command vehicle; concave; constant velocity; contributing valve; convertible; convertible security; culture vulture

cv caballo de vapor (Spanish), *cavallo vapore* (Italian), *cavalo vapor* (Portuguese), *chevalvapeur* (French)—horsepower (also appears as *CV*); *curriculum vitae* (Latin—biographical résumé)

c.v. conjugata vera (Latin—true conjugate)—pelvic inlet diameter; *cras vespere* (Latin—tomorrow evening); *cursus vitae* (Latin—course of life)

Cv Cove; molecular heat (symbol); specific heat at constant volume (symbol)

CV Cape Verde (Internet code); aircraft carrier (2-letter naval symbol); Central Vermont (railroad); Chula Vista; City Volunteers; collection voucher; combat vehicle; Commercial Village (zone); Community of the Visitation;

Community Volunteers; Convair; Cross of Valour (Australia)

C-V Convair (Division of General Dynamics)

C/V Certificate of Value

CV Cabo Verde (Spanish—Cape Verde Islands); *cheval-vapeur* (French—horsepower)

CV4 Convair 440 airliner

cva cerebrovascular accident stroke); costovertebral angle

CVA attack aircraft carrier (naval symbol); Civilian Voluntary Agency; Columbia Valley Authority

CVA Centro Venezolano América (Spanish—Venezuelan-American Center), cultural display promoted by the U.S. Embassy

CVAA Centre de Vulgarisation Aéro-Astronautique (French—Center for the Popularization of Astronautics)

CVAC Consolidated Vultee Aircraft (now Convair)

cvae coordinated vocational academic education

CVALI Crime Victims Legal Advocacy Institute

CVAN nuclear-powered aircraft carrier (naval symbol)

CVAS Configuration Verification and Accounting System

cvb combined very-high-frequency band

CVB large aircraft carrier (naval symbol)

C & VB Convention and Visitors Bureau

CVBC Cape Volunteer Bearer Corps

c-v-c consonant-vowel-consonant

CVC Cauca Valley Corporation; Clinch Valley College; Consolidated Vacuum Corporation

CVCB Crime Victims Compensation Board (New York)

cvcc compound vortex-controlled combustion (Japanese automotive engine designed to reduce air pollution)

cvcm collected volatile condensable material

CVCP Committee of Vice-Chancellors and Principals

cvcr control van connecting room

cvd cardiovascular disease; cash versus documents; central vehicle depot; chemical vapor deposition; coordination of valve development; countervailing duty; coupled vibration dissociation; current-voltage diagram

CVDE Columbia-Viking Desk Encyclopedia

CVDLS California Veterinary Diagnostic Laboratory System

cve (CVE) customer-vended equipment

CVE aircraft carrier, escort (naval symbol); Council for Visual Education

C Ven Canis Venatici

CVF Caravelle fan jet airplane; Cyprus Volunteer Force

CVF Corporación Venezolano de Fomento (Spanish—Venezuelan Promotion Corporation)

CVG Cincinnati, Ohio, airport; Corporación Venezolana de Guayana (Spanish—Venezuelan Corporation of Guyana)

cvh combined ventricular hypertrophy

cvh (CVH) compound-valve hemispherical head (cam-in-head automotive powerplant)

CVH Center for Visual History

CVHC coastal helicopter aircraft carrier (naval symbol)

CVHS Center on Violence and Human Survival (John Jay College); Chelsea Vocational High School

cvi center of visual impact; cerebrovascular insufficiency; common variable immunodeficiency

CVI Cape Verde Islands; College of the Virgin Islands

CVIA Commercial Vehicle Industry Association

C viruses Coxsackie viruses

CVIS Computerized Vocational Information System

cvk centerline vertical keel

CVL Caravelle jet airplane; small aircraft carrier (naval symbol)

CVLAI Crime Victims Legal Advocacy Institute

cvli cash value life insurance

CVM Company of Veteran Motorists

CVMA Canadian Veterinary Medical Association

cvn carbohydrate vitamin nitrogen; convene

C Vn Canis Venatici

CVN nuclear-powered aircraft carrier (naval symbol)

c.v.o. conjugata vera obstetrica (Latin—conjugate obstetric diameter)

CVO Chief Veterinary Office(r)

C.V.O. Commander of the Royal Victorian Order

C/VO Certificate of Value and Origin

c voc colla voce (Italian—with the voice)

cvou (CVOU) cardiovascular observation unit

cvp cars, vans, pickups; central venous pressure; climate, vegetation, productivity

CVP Central Valley Project (California)

CVP Corporación Venezolana de Petroleo (Spanish—Venezuelan Petroleum Corporation)

CVPIA Central Valley Project Improvement Act

cvr cardiovascular renal; cardiovascular-respiratory; cerebrovascular resistance; combat vehicle reconnaissance; continuous video recorder

cvr (CVR) cockpit voice recorder; crystal video receiver

cvrd cardiovascular renal disease

cvrd hpr covered hopper (freight car)

cvs cardiovascular surgery; cardiovascular system; computer vision syndrome

cvs (CVS) chorionic villus sampling

CVS antisubmarine warfare support aircraft carrier (3-letter symbol); Change Verification System

CVSA Commercial Vehicle Safety Alliance

cvsd continuously variable slope-delta (modulation)

CVSM Canadian Volunteer Service Medal

cvt chemical vapor transport; constant-voltage transformer; continuously variable transmission; controlled variable time (fuze); convertible

CVT training aircraft carrier (naval symbol)

c/vta cuenta de venta (Spanish—bill of sale)

Cvt Gdn Covent Garden (Royal Opera House)

cvtr charcoal viral transport medium

CVV conventional oil-powered aircraft carrier (naval symbol)

CVW attack carrier air wing (naval symbol)

CVWS Combat Vehicle Weapon System

cw call(s) waiting; canistered waste(s); cardiac work; carrier wave (Morse); casework(er); chemical warfare (CW); chest-wall(s); child welfare; children's ward; cladding waste(s); clockwise; commission and warrant; continuous wave; conventional wisdom; copperweld (copper-covered steel); cubic weight

c-w chronometer time minus watch time

c/w chainwheel; counterweight

c & w country and western (music)

CW Canada West; Channel Airways; chemical warfare; Church Warden; Civil War; continuous wave

C-W Curtiss-Wright

C & W Cable and Wireless

C of W College of Wooster (Ohio)

CW Computer World

CWA Central Wholesalers Association; Civil Works Administration; Clean Water Act; Communication Workers of America; Concerned Women for America; Concerned Women of America; Construction Writers Association; Country Womens Association; County Water Authority; Crime Writers Association; Customer Work Authorization

CWA Clean Water Act

CWAA Cotton Warehouse Association of America

CWAC California Wildlife Advisory Committee; Canadian Women's Army Corps; City-Wide Anti-Crime Unit (sometimes called Quacks)

CWAM Cliffs Western Australian Mining Company

cwar continuous-wave acquisition radar

cwas contractor-weighted average share

CWAS Central Welfare Advisory Service

cwb clay-water-base (oil well) mud

CWB Canadian Wheat Board; Central Wages Board; Child Welfare Board

CWBS Cost Work Breakdown Structure

cwbts capillary whole blood true sugar

cw-bw chemical warfare—biological warfare

CWC Canadian Welfare Council; Central Wesleyan College; Chemical Weapons Convention

CWCA California Women's Commission on Alcoholism

CWCC Civil War Centennial Commission

c & w ck caution and warning check

CWCO China Wire and Cable Company

CWCP Combat Wing Command Post

cwd chronic wasting disease; civilian war dead; clerical work data

cwe current working estimate

CWE Commonwealth Edison

C'wealth Commonwealth

CWEP Community Work Experience Program

CWET Community Work Experience Training

CWETA California Worksite Education and Training Act

CWF California Wildlife Federation; Central Wool Facility; Christian Women's Fellowship; Cornell Word Form(ation)

cwfm continuous-wave frequency modulated

cwfp clean wool fibers present

CWFT Cornell Word Form Test

cwg corrugated wire glass

CWG California Writers Guild

CWGC Commonwealth War Graves Commission

cwgt counterweight

CWHSSA Contract Work Hours and Safety Standards Act

cwi cardiac work index; clear word identifier

CWI California Wine Institute; Colonial Williamsburg Incorporated; Country Women's Institute

cwik cutting with intent to kill

CWINC Central Waterways, Irrigation, and Navigation Commission

CWIS Campus-Wide Information System; Chaim Weizmann Institute of Science

CWISS Child Welfare Information Services System

cwit concordance words in title

cwl calm waterline

CWL Catholic Women's League

CWLA Child Welfare League of America

C & W Ltd Cables and Wireless Limited

cwm (CWM) commercial waste management

CWMH Colonial War Memorial Hospital

CWMTU Cold Weather Materiel Test Unit

CWNA Canadian Weekly Newspapers Association

cwo cash with order; continuous-wave oscillator

CWO Canadian War Office; Chief Warrant Officer

CWOD California's War On Drugs

cwp childbirth with(out) pain; circulating water pump; communicating word processor; community work plan

CWP Communist Workers Party

CWPEA Childbirth Without Pain Education Association

CWPLs Childbirth Without Pain Leagues

CWPP Commercial Waste Packaging Program

CWPS Center for Women Policy Studies; Council on Wage and Price Stability

CWPU Central Water Planning Unit

cwr continuous welded rail

CWR California Western Railroad; Crusade for World Revival

CWRA California Water Resources Association

cwrb (CWRB) canistered waste-receiving building

CWRC Chemical Warfare Review Commission

CWRE Chemical Warfare Research Establishment

CWRSM Case-Western Reserve School of Medicine

cws clockwise; cold-water soluble; countersunk woodscrew; curve warning signal

cw & s crushed, washed, and screened

Cws chemical weapons; Cowes

CWS California Water Service; Canadian Welding Society; Canadian Wildlife Service; Chandraprabhia Wildlife Sanctuary (India); Chemical Warfare Service; Child Welfare Services; Church World Service; Clearinghouse on Women's Studies; College Work Study; College World Series; Cooperative Wholesale Society; Cunard-White Star (steamship line)

C-WS Crop-Weather Service

CWSC Canterbury Winter Sports Club; Central Washington State College

cwsfp commercial waste spent-fuel packaging

cw sig gen continuous wave signal generator

CWSP College Work-Study Program

CWSS Center for Women's Studies and Services

Cwsy Causeway

cwt centurm weight; counterweight; hundredweight

CWT Central War Time; Command Word Trap; Community Work Training; Consumers for World Trade; Container Warehousing and Transportation; Cooperative Wind Tunnel

CWTC California World Trade Center

cwtd continuous-wave target detector

cwtdc continuous-wave target detection console

cwu composite weighted work unit

CWU California Western University; Chemical Workers Union; Church Women United; Culinary Workers Union

cwv continuous-wave video

CWV Catholic War Veterans

cww cruciform wing weapon

CWWC Concerned Women in the War on Crime

cwy clearway

cx cervix; chest x-ray; complex; connection; control transmitter; convex; correct copy (instruction to the printer)

cx (CX) central exchange

Cx Caxton; Caxton Printers; unknown

Cx Caixa (Portuguese—Box)—post office box; also written *cx*

CX Cathay Pacific Airways; Christmas Island (Internet code); outsize cargo-transport aircraft

C-X Cargo X (new airlift aircraft for Rapid Deployment Forces)

Cxo Calexico

cxr carrier

cXr chest X-ray

cxs consort parallax servo

CXT Common External Tariff

cy calendar year; capacity; copy; cubic yard; currency; current year; cyanide; cyanogen; cycle; cylinder(s)

Cy City; cyanogen; Cyprus; Cyrus

CY Chief Yeoman; Communications Yeoman; Container Yard; Cyprus (Internet code)

cya cover your ass (protect yourself)

CYA California Youth Authority; Canadian Yachting Association; Carded Yarn Association; Catholic Youth Adoration (Society); Chicano Youth Association; Covenant Youth of America

cyan cyanamid; cyanic; cyanide; cyanogen; cyanotype

cyan (Latin prefix—blue)—cyanometer (device for measuring the blue in the sky)

cyanide cyanide of potassium

cyath. cyathus (Latin—cup, ladle, glass)

cyath. vin. cyathus vinarius (Latin—wineglassful)

cyb cybernetic; cyberneticist; cybernetics

CYB Canada Year Book

cyber cybernetics

cybercult cybercultural(ly); cyberculture

cyberlog cybernetic logistics

cybernat cybernated; cybernation(al)(ly)

cyborg(s) cybernetic organism(s)

cyc curb your curiosity; cyclazogine (narcotic antagonist used in curing victims of addiction); cycle; cyclorama

CYC Capital Yacht Club; Chicago Yacht Club; Cleveland Yacht Club; Colorado Youth Center; Columbia Yacht Club; Company of Young Canadians; Corinthian Yacht Club

CYCA Clyde Yacht Clubs Association; Cruising Yacht Club of Australia

Cycl Cyclostomata

Cyclades Cyclades Islands

cyclams cyclamates

cyclaz cyclazocine

cycle bicycle; motorcycle

cyclecade bicycle parade; motorcycle parade; tricycle parade

cyclo cyclopedia; cyclopedic; cyclophosphamide; cyclopropane; cyclorama

cyclon cyclonometer

cyclon gas cyanide, chlorine, and nitrogen

cyclos (Greek—circle, ring)—bicycle, circle, cyclorama

CY/CY Container Yard to Container Yard

c yd cubic yard(s)

Cyd cytidine

CYDEF Cyrenaica Defence Force

CYEE Central Youth Employment Executive (UK)

CYFA Club for Young Friends of Animals

cyflo cylinder overflow

Cyg Cygnus (Latin—Swan constellation)

CYHA Canadian Youth Hotels Association

cyk consider yourself kissed

cyke cyclorama

cyl cylinder; cylindrical; cylindroid

CYL Chinese Youth League

cyl l cylinder lock

cyls cylinders

cym cymbal(s)

Cym Cymraeg (Welsh); Cymric

CYMA Catholic Young Men's Association

Cymb Cymbeline

Cymr Cymric (Welsh—Wales)

CYMRU Welsh television

CYMS Catholic Young Men's Society

cyn cyanide

Cyn Canyon; Cynthia

cyni cynical; cynicism

cynol cynologic(al)(ly); cynologist; cynology

CYO Catholic Youth Organization; Civic Youth Orchestra; Community Service Corps

Cyp Cyprian; Cypriote; Cyprus

CYP Commonwealth Youth Programme; Couple Years of Protection; Cyprus Airways

Cyprus Republic of Cyprus (Mediterranean island country split between Greek and Turkish settlers *Kypriaki Dimokratia* (Greek name for Cyprus); *Kibris Cumhuriyeti* (Turkish name)

CYR Carleton York Regiment

CYRA Commission Yellowfin Regulatory Area

Cyrano Cyrano de Bergerac

cys cysteine; cystoscopy

cys (CYS) cystine (amino acid)

CYS Cheyenne, Wyoming airport; Chief Yeoman of Signals

CYSA Combed Yarn Spinners Association

CYSS Community Youth Support Scheme (Australia)

cysti (Latin prefix—bladder or sac)—cystoscope (device for examining the bladder)

cysto cystoscope; cystoscopic examination

CYSYS Center for Cybernetics System Synergism

cyt cytology

Cyt cytosine

cytac control of tactical aircraft

cytd calendar year to date

cyto (Latin prefix—cell)—cytology

cytoeco cytoecologic(al)(ly); cytoecologist; cytoecology

cytol cytological; cytologist; cytology

Cytol Cytology

cytomorph cytomorpho-logic(al)(ly); cytomorphologist; cytomorphology

cytopatho cytopatho-genic(al)(ly); cytopathogenicity

cytophoto cytophotometer; cytophotometric(al)(ly); cytophotometry

cytostat cytostatic(al)(1y)

cyto syst cytochrome system

cytotech cytotechnic(al)(ly); cytotechnician; cytotechnologist; cytotechnology

cz coryza; cubic zirconia

Cz Czech; Czechoslovakia; Czechoslovakian

CZ Canal Zone; combat zone; communications zone; Consolidated Zinc

C-Z Crown-Zellerbach

C.Z. Canal Zone

CZ *Ceska Zbrojovka* (Czechoslovak Arms Factory)

Cza Constanza

CZA Canal Zone Authority; Coastal Zone Authority

CZAG Committee for Zero Automobile Growth

CZBA Canal Zone Biological Area

CZC Canal Zone College

CZC *Canal Zone Code* (legal)

CZCA Coastal Zone Conservation Act

czcs coastal-zone color scanner

czd calculated zenith distance; combat zone distance

Czech Czechoslovakia; Czechoslovakian

Czech J Phys *Czechoslovak Journal of Physics*

Czech Phil Czech Philharmonic

Czech Rep Czechoslovak Republic inaugurated January 1, 1993

Czecho Czechoslovakia (during the Austro-Hungarian Empire divided into Bohemia, Moravia, and Slovakia)

CZF Canadian Zionist Federation; Canal Zone Forces

CZG Canal Zone Government

czi (CZI) crystalline zinc insulin

CZI Canal Zone Institute

CZJC Canal Zone Junior College

Cz kr Czechoslovakian kronen (monetary unit)

CZL-M Canal Zone Library-Museum (Balboa Heights)

CZm compass azimuth

CZMA Coastal Zone Management Act

Czml Cozumel

C-Zone commercial zone

CZP Chicago Zoological Park (Brookfield Park); Consolidated Zinc Proprietary

CZ Pen Canal Zone Penitentiary

czr combat zone radius

CZRs Canal Zone Regulations

CZS Chicago Zoological Society

CZSG Canal Zone Study Group

C-Z strain Carr-Zilber (viral) strain

c-Z-t chirp-Z-transform

czy crazy

D

d angular deformation (symbol); daily; date; daughter; day; death; deciduous; declination; defense; degree; delete; denied; density (symbol); dented; depth; desert scrub; destroyer; detectivity; dextrorotatory; died; differentiation; dime; dinar; diopter; dissipation; disturbed site; divorced; dorsal; double (occupancy); dragoons; drain (symbol); drive; driver; drizzling; duration; dyne; grating space in calcite (symbol); liter (symbol); pence (symbol); penny (symbol)

d (D) dead; died; demand; deposit; district

d' surname prefixes such as da, de, di, etc

'd (contraction—could, did, had, would)

d decimus (Latin—tenth); der (German—the); demissione (Latin—on the demise); denarii (Latin—pennies); denarius (Latin—penny); dexter (Latin—right); dictum (Latin—remark or observation)

D dead space gas (symbol); December; degree of curve (symbol); Delta—code for letter D; democracy; Democrat(ic); density; Denver; department; derivation; determinant (symbol); Detroit; deuterium; Deuteronomic; diameter; dielectric flux density (symbol); Dietzgen; diffusing

capacity (symbol); diffusion constant (symbol); dioptric power (symbol); director aircraft; disaster; disaster broadcasting; dissipation factor (symbol); dollar; dose; Douglas; down; drag (symbol); drain (symbol); drone control version (symbol); Dublin; Dutch; electrostatic flux density (symbol); Fraunhofer lines caused by sodium (symbol); propeller diameter (symbol)

D' surname prefixes such as Da, De, Di, Do, Du, etc.

D Dagh (Persian—Daglar (Turkish—mountain range); Dagi (Turkish–mountain range); Dag (Turkish—mountain); Damas (Portuguese or Spanish—ladies); Damen (German—ladies); darin (German—in); Darreh (Persian—valley); Daryaceh (Persian—lake); Dauer (German—bulb type camera shutter stop); dehors (French—out); départ (French—departure); derecha (Spanish—right); Deus (Latin—God); dexter (Latin—right); Don (Spanish—Sir)—Mr in its most formal meaning; dun (Danish—down)

D. Don (Spanish—Sir) Mr

d_1 diffusing capacity–lung

d 1/2 d dispatch money payable at one-half demurrage rate

$D_1, D_2, D_3,$ etc. 1st dorsal vertebra, 2nd dorsal vertebra, 3rd dorsal vertebra, etc.

D_2O deuterium oxide (heavy water)

d2s & cm dressed two sides and center matched (lumber)

d2s & m dressed two sides and matched (lumber)

d2s & sm dressed two sides and standard matched (lumber)

D3 Douglas DC-3 airplane

D4 Douglas DC-4 airplane

D-5-HS dextrose 5 percent in Hartman's Solution

D-5-S dextrose 5 percent in saline (solution)

d_5w 5 percent dextrose in water

D6 Douglas DC-6 airplane

D7 Douglas DC-7 airplane

D8F Douglas D8F fan jet airplane

D8S Douglas super DC-8 fan jet airplane

D9S Douglas super DC-9 fan jet airplane

D-18 Beechcraft four-passenger transport also designated C45

D 40 iopax (uroselectan)

D'66 Democrats 1966 (Dutch political party)

D of '98 Daughters of '98

D-150 Dimension 150 (150-degree field of vision achieved by deeply curved motion-picture screen)

da daughter; day; days after acceptance; deca (symbol); defined adult; delayed action; delayed arming; density altitude; deposit account; desk accessory; developmental age; direct action; discharge afloat; discriminant analysis; district attorney; do not answer; documents against acceptance; documents attached; doesn't answer (the telephone); dopamine; double acting; double aged; drift angle

da (DA) diphenylchlorasine (deadly gas); directional antenna; dopamine

d-a direct-action (adjective)

d/a digital-to-analog

d/a (D/A) deposit account

d in a (found) dead in automobile (or) airplane

d-to-a digital-to-analog

da dansk (Dano-Norwegian—Danish); *dette år* (Dano-Norwegian—this year)

dA deciampere; deoxyadenosine; differential of amplification (symbol): differential of area (symbol)

dA der Altere (German—senior)

Da Danish; Danmark

Dᵃ Doña (Spanish—lady, woman of rank)

DA Data Administrator; Daughters of America; Dealers Alliance; Defense Act; Defense Aid; Dental Apprentice; Department of Agriculture; Department of the Army; Dermatology Associates; direct action (DA as a noun, d-a as an adjective); District Attorney; Division Artillery; does not affect; Dominion Atlantic (railroad); Dragon Airways; drift angle (symbol)

D-A Devin-Adair

D.A. Diploma in Aesthetics; Diploma in Anesthetics; Doctor of Arts

D/A digital-to-analog

D of A Defenders of Animals; Department of Agriculture

DA Dalniya Aviatsiya (Russian—Long-Range Aviation); *Dissertation Abstracts*

daa data access arrangement; direct access arrangement

DAA Danish Atlantic Association; Dental Assistants Association; Department of Ab-

original Affairs (Australian); Deputy Assistant Adjutant; Diploma of the Advertising Association; Direct Action Associates

DAA Défense Anti-Aérienne (French—Anti-Aircraft Defense)

D/AA Days After Acceptance; Documents Against Acceptance

DAAC Director Army Air Corps

DAACA Department of the Army Allocation Committee–Ammunition

DAAG Deputy Assistant Adjutant General

DAA & QMG Deputy Assistant Adjutant and Quartermaster General

Da Afg Da Afghnestan (Pashto—Afghanistan)

dab daily audience barometer; delayed-action bomb; diaminobenzidine; dimetylaminoazobenzene

DAB Daytona Beach, Florida (airport); Digital Audio Broadcasting

DAB Deutsches Apothekerbuch (German—Pharmacopoeia); *Dictionary of American Biography*

dabco diazabicyclooctane

DABPN Diplomate American Board of Psychiatry and Neurology

DABS Discrete Address Beacon System

DABSIPCS Discrete Address Beacon System with Intermittent Positive Control System

dac dangerous air cargo; data acquisition and control; data assistance and control; deductible average clause; delivery against cost; digital-to-analog converter; digital arithmetic center; direct air cycle; dynamic amplitude control

Dac Dacca

DAC Daughters of the American Colonists; Debt Advisory Commission; Defenders of the American Constitution; Department of Adult Corrections (Alaska); Department of the Army, Civilian; Disaster Assistance Center; Discre-

tionary Access Control; Douglas Aircraft Company; Durex Abrasives Corporation

daca (DACA) diphenylaminochloroarsine

DACA Drug Abuse Control Amendments

DACAN Douglas Aircraft Company of Canada

DACAR Data Acquisition and Communication Techniques and Their Assessment for Road Transport

dacbu data acquisition and control buffer unit

D.Acc. Doctor of Accountancy

DACC Dangerous Air Cargoes Committee

DACCC Defense Area Communications Control Center

DACCEUR Defense Area Communications Control Center Europe (NATO)

DACG Deputy Assistant Chaplain-General; Deputy Assistant Commissary-General

dachs dachsbracke (Swedish basset); dachshund (German-badger dog)

dachsaterr dachshund-terrier mixed-breed dog

dacks slacks (sport pants) made of dacron

DACL Depression Adjective Check List(s)

DACO Douglas Aircraft Corporation Overseas

DACOM Datascope Computer Output Microfilmer

dacon digital to analog converter

d/a converter device converting digital input to analog input

dacor data correction

DACOS Deputy Assistant Chief of Staff

DACOWITS Defense Advisory Committee on Women in the Services

dacr dacron (synthetic fiber)

DACRP Department of the Army Communication Resources Plan

DACs Department of the Army Civilians; Desegregation Assistance Centers

DACS Data Acquisition and Correction System; Director of Aerospace Combat Systems

dact dactyl(ic); dactylology; dactylus; dissimilar aircraft combat training

dacty dactylography; dactyloscopy

dactygram dactylogram (fingerprint)

dactyl dactylogic(al)(ly); dactylologist(ic); dactylology

dacum designing a curriculum

dad daddy (father); database action diagram; design-approval data; digital audio definition; dispense as directed; double-acting door

Dad Daddy; Dadiangas

DAD Director, Air Division; Directorate of Armament Development; Double Atmospheric Density (rocket); Drop A Dime (anti-drug program)

DADATS Deputy Assistant Director, Auxiliary Territorial Service

DADEE (Dynamic-Analog-Differential-Equation Equalizer)

DADGMS Deputy Assistant Director, General of Medical Services

DADIT Daystrom Analog-to-Digital Integrating Translator

D.Adm. Doctor of Administration

DADME Deputy Assistant Director, Mechanical Engineering

DADMS Deputy Assistant Director, Medical Services

DADOS Deputy Assistant Director, Ordnance Services

DADPR Deputy Assistant Director, Public Relations

DADQ Deputy Assistant Director of Quartering

dads (DADS) dual air density satellite

DADS Deputy Assistant Director of Supplies; Director Army Dental Service

dadsm direct-access device (for) space management

DADVS Deputy Assistant Director of Veterinary Services

D.Ae. Doctor of Aeronautics

DAE Diploma in Advanced Engineering; Director of Aircraft Equipment; Director of Army Education; Division of Adult Education

DAE Dictionary of American English

daea dimethyl aminoethyl acetate

DAEC Danish Atomic Energy Commission

DAEDARC Department of the Army Equipment Data Review Committee

D.Ae.Eng. Doctor of Aeronautical Engineering

daemon data-adaptive evaluator and monitor

DAEP Division of Atomic Energy Production

DAER Department of Aeronautical Engineering Research

D.Ae.Sc. Doctor of Aeronautical Science

daf delayed auditory feedback; described as follows; discharge afloat

DaF Delivered at Frontier

DAF Danish Air Force; Department of the Air Force; Desert Air Force; Dutch Air Force

DAF Dansk Arbejdsgiverforening (Danish—Employers, Confederation); *van Doorne Auto Fabriek* (Dutch—van Doorne's Auto Factory), autos and trucks

dafa data accounting flow assessment

DAFA Data Accounting Flow Assessment

dafc digital automatic frequency control

DAFCCS Department of the Air Force Command and Control System

DAFFO Dunsk Forening til Fremme af Opfindelser (Danish Society for Encouraging Inventions)

daffs daffodils

daff(y) daffodil

DAFIE Directorate for Armed Forces Information and Education

dafm discard-at-failure maintenance

DAFM Department of the Army Field Manuals

DAFO Division Accounting and Finance Office

DAFS Department of Agriculture and Fisheries (Scotland); Duty Air Force Specialty

DAFSC Duty Air Force Specialty Code

DAFSO Department of the Air Force Special Order

daft digital/analog function table

dag aquadag; decagram; dysprosium aluminum garnet

dag dagh (Turkish—mountain)

Dag Dagestan(i); Dag Hammarskjöld; Dagmar; Dagna

Dag Dagbladet (Oslo's Daily Blade)

D.Ag. Doctor of Agriculture

DAG Deputy Adjutant General; Deputy Attorney General; diacylglycerol

DAG Deutsche Angestellten-Gewerkschaft (German Employees Union)

dagc delayed automatic gain control

dagl daglig (Dano-Norwegian—daily); *dagligdags* (Dano-Norwegian—ordinary)

dagmar defining advertising goals for measured advertising results; drift-and-ground speed-measuring radar

Dagmar Dagmar Godowsky

Dag Nyh Dagens Nyheter (Sweden's Daily News)

D.Agr. Doctor of Agriculture

DAGRA Deputy Adjutant-General Royal Artillery

D.Agr.Eng. Doctor of Agricultural Engineering

D. Agr.Sc. Doctor of Agricultural Science

DAGS Department of Accounting and General Services

dah disordered action of the heart

Dah Dahomey

DAH disordered action of the heart

Dahlaks Dahlak Islands in the Red Sea off Eritrea, Ethiopia

DAHST Dial A Hearing Screening Test

dai (DAI) death from accidental injuries

Dai David

DAI Dayton Art Institute; Drug Abuse Information

DAI Dissertation Abstracts International

DAIA Department of Aboriginal and Island Affairs (Australian)

DAICS Data-Acquisition and Instrument-Control System

daigc direct-aqueous-injection gas chromatography

DAIM Data Analysis Information Memo

DAIMC Defense Advanced Inventory Management Course (USA)

DAIMS Department of the Army Integrated Material Support (USA)

DAIR Driver Aid Information and Routing (System)

DAIRE Delaware Application of Information and Research in Education

DAIS Defense Automatic Integrated Switching System; Digital Avionics Integration Systems

DAISY Data Acquisition and Interpretation System; Decision-Aiding Information System; Displacement-Automated Integrated System

DAISY-201 Double-Precision Automatic Interpretive System

DAJAG Deputy Assistant Judge Advocate General

Dak Dakota; Dakotan

DAK Deutsche Afrika Korps (German—German Africa Corps)

Dakoming Dakota + Wyoming

Dakotas North and South Dakota

Dak Ter Dakota Territory

Dak Zoo Dakota Zoo (Bismarck, North Dakota)

dal decaliter; delta-aminolevulinic acid

d'AL d'Amico Line

Dal Dallas; Dalmatia; Dalmatian

DAL Dallas, Texas (Love Field); Delta Air Lines; Department of Agriculture Library

DAL Deutsche Afrika Linien (German—Africa Line)

dala delta-amino-levulinic acid

dalapon dialphapropionic acid (herbicide)

Dalarna Swedish truncation of Dalecarlia, the lake district of folklore

Dalarna (Swedish—Dalecarlia), derived from *Dalkarl* — Man from Dalarna—Sweden's agricultural valley area

dalasi monetary unit of Gambia

DALATS Data Logging and Transmission System

DALE Drug Abuse Law Enforcement

D'Alembert Jean Le Rond d'Alembert

dalgt daylight

Dalh Dalhousie

dalite bsmt daylight basement

Dall Dallas' Reports—U.S. Supreme Court

dalpo do all possible

dalr dry adiabatic lapse rate

DALRLV Department of the Army Logistics Readiness Liaison Visit(s)

DALRTF Department of the Army Long-Range Technological Forecast(ing)

dal s dal segno (Italian—from the sign)

DALS Director of Army Legal Services; Distress Alerting and Locating System

dal seg dal segno (Italian—from the sign)

Dal Sym Orch Dallas Symphony Orchestra

dalvp delay enroute authorized chargeable as ordinary leave provided it does not interfere with reporting on date specified and provided individual has sufficient accrued leave

dam damage; data-addressed memory; decameter; degraded amyloid; diacetyl monoxime; divided and mashed

dam (DAM) direct-access method; down-range antimissile (program)

Dam Damascus; Damman

DAM Damascus, Syria (airport); Dayton Art Museum; Denver Art Museum

DAM Dictionary of Abbreviations in Medicine

DAMCO Dampier Mining Company

dame data acquisition and monitoring equipment; digital automatic-measuring equipment

Dame Kiri Dame Kiri Te Kanawa

DAMIS Department of the Army Management Information System

DAMOS Data Moving System

dAMP deoxyadenylic acid

DAMP Down-Range Anti-Missile Measurement Project

Dampiers Dampier Islands in the Indian Ocean off north-Western Australia

DAMPS Data Acquisition Multiprogramming System

DAMRIP Department of the Army Management Review and Improvement Program

dams dental amalgam mercury syndrome

DAMS Defense Against Missiles System; Deputy Assistant Military Secretary

DAMT Draw-A-Man Test

DAMWO Department of the Army Modification Work Order

dan dekanewton

dän dänisch (German—Danish)

Dan Daniel (name); Daniel, Book of; Danish; Danmark (Denmark)

Dan Daniel (Book of)

DAN Dan-Air Service

DANA Diffraction Analysis System

DANBIF Danske Boghandleres Importrfrening (Danish Booksellers Importation Association)

dancin' dancing

DANCOM Danube Commission (Austria, Bulgaria, Czechoslovakia, Hungary, Romania, the former USSR, Yugoslavia)

Dand(ie) Andrew

Dandy Andrew

DANFS Dictionary of American Naval Fighting Ships

Danglish Danish-English

dang mod dangling modifier

Dani Daniel

Danl Daniel

Danl W W Daniel Webster

Danm Danmark (Denmark)

Dannebrog (Danish-Danish cloth)–Denmark's flag reputed to be the oldest national symbol in western Europe

Dannemora Clinton Correctional Facility (for males) at Dannemora, New York

Dan-Nor Dano-Norwegian

Danny Daniel

DANR Department of Agriculture and Natural Resources

dans dansyl chloride (fluorescent dye)

DANS Director of Army Nursing Services

Dansker Dane; Danish sailor

DANTES Defense Activity for Non-Traditional Education Support (USN)

DANZ Dyslexia Association of New Zealand

D and D Decontamination and Decommissioning

dao double action only; duly authorized officer, paldao (Philippine wood)

DAO Dairy Advisory Office(r); Defense Attachés Office; District Accounting Office(r); District Aviation Office(r); Division Air Office(r); Division Ammunition Office(r); Dominion Astrophysical Observatory (Victoria, British Columbia)

DAOC Deputy Air Officer Commanding

DAOT Director of Air Organization and Training

dap data analysis package; data automation proposal; digital audio processor; direct-agglutination pregnancy (test); do anything possible; documents against payment; draw a person

dap (DAP) diaminopimelic acid; diammonium phosphate; dihydroxyacetone phosphate

d-a-p draw-a-person (psychological test)

DAP Democratic Action Party; Design Analysis Phase; Development Academy of the Philippines; Director of Administrative Planning; Division of Air Pollution (US Public Health)

DAPA Drug Abuse Programs of America

DAPD Directorate of Aircraft Production Development

DAP & E Diploma in Applied Parasitology and Entomology

dapi (DAPI) diamidinophenylindole

DA Plan Deposit Administration Plan

DAPM Deputy Assistant Provost Marshal

DAPMC Defense Advanced Procurement Management Course (USA)

dapon diallyl phthalate resin

Da Ponte Lorenzo Da Ponte (Mozart's librettist, born Emanule Conegliano)

DAPP Data Acquisition and Processing Program

D.App.Sci. Doctor of Applied Science

dapr digital automatic pattern recognition

daps downed airman power source (USN)

DAPS Direct-Access Programming System; Director of Army Postal Services

dapsone diaminodiphenyl sulfone

dapt daptazole; direct-agglutination pregnancy test (DAPT)

DAPT Direct Latex Agglutination Pregnancy Test; Draw-a-Person Test

DAQ Director of Army Quartering

DAQMG Deputy Assistant Quartermaster General

dar damned average raiser (good student whose high marks raise the grading scale); defence acquisition regulation; digital aids recorder; dressed all 'round

Dar Dar-es-Salaam

Dar Dar-es-Salaam (Arabic— There is the Peace)—capital and seaport of Tanzania; *Darié* (Panamanian province of Colombia to 1903)

DAR Daily Activity Report; Data Automation Request; Data Automation Requirements; Daughters of the American Revolution; Dead Animal Removal; Department of Animal Regulation; Director of Army Recruiting; Directorate of Atomic Research; Dominion Atlantic Railway

DARAS Direction and Range Acquisition System

Darb Darby(shire)

D & ARC Drug and Alcohol Resource Center

DARCEE Demonstration and Research Center for Early Education (Peabody College)

D.Arch. Doctor of Architecture

D.Arch.E. Doctor of Architectural Engineering

DARCOM Development and Readiness Command (USA)

dard data acquisition requirements document

DARD Directorate of Aircraft Research and Development

Dardan Dardanelles

Dardanelles Dardanelle Straits called Hellespont by the Greeks and Canakkale Bogazi by the Turks; link the Aegean Sea with the Sea of Marmara, the Bosporus, and the Black Sea

dare data automatic reduction equipment; data automation research and experimentation; destination arrival research engineering; documentation automated retrieval equipment; doppler automatic reduction equipment

DARE Detroit Association for Rational Enquiry; Drug Abuse Research and Education (UCLA's neuropsychiatric institute); Drug Abuse Resistance Education; Drug Assistance, Rehabilitation, and Education; Drug Awareness Resistance Education

DARE Dictionary of American Regional English [Frederic (Gomes) Cassidy–Editor-in-Chief]

daren't dare not

DARES Data Analysis and Reduction System

DARF Defense Atomic Research Facility

Darlings short form for the Darling Ranges of Western Australia

DARME Directorate of Armament Engineering (Canada)

darms digital alternate representation of music scores

DARPA Defense Advanced Research Projects Agency

DARR Department of the Army Regional Representative; Drawing and Assembly Release Record

Darren Darwen, England

dars differential absorption remote sensing (laser)

DARs Design Assist Reports; Development Appraisal Reports; Digital Adaptive Recording System

darss diode-array rapid-scan spectrometer

dart datagraphic automated retrieval technique(s); deployable automatic relay terminal; detection, action, and response technique(s); development advanced rate techniques; disappearing automatic retaliation target

DART Dallas Area Rapid Transit; Data-Analysis Recording Tape; Direct Access to Rapid Transit; Direct Access to Regional Transit; Disaster Assisted Radio Teams; Dublin Area Rapid Transit

Dartmouth Dartmouth College at Hanover, New Hampshire; Massachusetts fishing port near New Bedford; Nova Scotia port and rail terminus across Halifax harbor from Halifax; Royal Naval College at Dartmouth near Plymouth on the English Channel

DARTS DoE Audit Report Tracking System; Dynamically-Actuated Road Transit System

Darw Darwin College, Cambridge

Darwin formerly Port Darwin, Australia

Darwin Pr Darwin Press

das data analysis station; dekastere; delivered alongside ship; dial-assistance switchboard

das (DAS) dextroamphetamine sulfate (stimulant)

da's domestic afflictions (menses)

DAs Design Assist Reports

DAS Data Acquisition System; Data Analysis System; Defense Atomic Support Agency; Defense Audit Service; Digital Analog Simulator; Digital Attenuator System; Director(ate) of Administrative Services; Director(ate) of Aerodrome Standards; Director of Air Services (Admiralty); Distributed Authorization System; Division of Apprenticeship Standards

DAS *Departamento Administrativo de Seguridad* (Spanish—Security Administration Department); *Dictionary of American Slang*

dasa dual aerospace servoamplifier

DASA Defense Atomic Support Agency

DASA-TP Defense Atomic Support Agency–Technical Publication(s)

DASC Defense Automotive Supply Center; Direct Air Support Center

D.A.Sci. Doctor of Agricultural Science

dasd direct access storage device

DASD Direct-Access Storage Device; Director of Army Staff Duties

DASDL Data and Structure Definition Language

dash dashboard; drone antisubmarine helicopter

Dash Dashiell Hammett

DASH Delta Airlines Special Handling (of small packages); Drug Abuse Services of Hawaii

dasi diffusion of arsenic in silicon

dasm (DASNI) delayed-action space missile

DASNET Data Switching Network

daso (DASO) development and shakedown operations

dasp discrimination among sound patterns; double-arm magnetic spectrometer

DASP Director(ate) of Advanced Systems Planning

dass defined antigen substrate sphere

DASS Direct Air Support Squadron (USAF)

DASSR Dagestan Autonomous Soviet Socialist Republic

DAST Division for Advanced Systems Technology

DAST *Detective-Agents-Science Fiction-Thriller*

dastard destroyer anti-submarine transportable array detector

DASTL Defense Atomic Support Agency Technical Letters

DASU Defence Animal Support Unit

DASWE Director of Admiralty Surface Weapons Establishment

dat date after tomorrow; dative; datum; delayed-action tablet; differential agglutination titer; diffused-alloy transistor

dAt (DAT) dementia of the Alzheimer type

dat (DAT) differential aptitude test(ing); digital audio tape

d a t diet as tolerated

DAT Dental Aptitude Test; Development Acceptance Test (USA); Differential Aptitude Test; Docking Alignment Test (NASA)

DATA Data Acquisition and Technical Analysis; Defense Air Transportation Administration; Development and Technical Assistance (UN); Dial-a-Teacher Assistance (telephone-service program); Draughtsmen's and Allied Technicians' Association

datac digital automatic tester and classifier

DATAC Development Areas Treasury Advisory Committee

datacom data communications

datacor data correction; data correlator

datan data analysis

datanet data network

datap data transmission and processing

datar digital automatic tracking and ranging

datastor data storage

DATC Developmental and Training Center

datda (DATDA) diallytartardiamide

DATDC Data Analysis and Technical Development Center

datel data + telecommunication

datico digital automatic tape intelligence checkout

datin data inserter

DATM *Department of the Army Technical Manual*

DATO Disbursing and Transportation Office; Discover America Travel Organizations

datom data aids for training, operations, and maintenance

dator digital (data), auxiliary (storage), track (display), outputs (and) radar (display)

DATOR Data Operational Requirements Board (NATO)

datran data transmission

datrec data recording

datrix direct access to reference information

dats data accumulation/transmittal sheet

DATS Dynamic Accuracy Test System

DATSC Department of the Army Training and Support Committee

dau data adapter unit; digital acquisition unit

daugr. daughter

dau-i-l daughter-in-law

D.Au.Eng. Doctor of Automobile Engineering

dau(s) daughter(s)

DAUS Dispatch Agency of the United States

dav data above voice

dav *davaerende* (Dano-Norwegian—then)

Dav David

DAV Dayanana Anglo Veolic (Fiji school); Disabled American Veterans

DAVA Defense Audiovisual Agency

davc delayed automatic-volume control

Dave David

Davey David; General David C. Jones

DAVI Department of Audio-Visual Instruction (National Education Association)

davidite uranium ferric ferrous iron titanate

Davie David

da Vinci Leonardo da Vinci (Rome—Italy's airport)

Davis Mountains West Texas range of Rocky Mountain system running south through Big Bend National Park into Mexico; named after Jefferson Davis

D.Av.Med. Diploma in Aviation Medicine

DAVNO Division Aviation Office(r)

DAVRS Director of Army Veterinary and Remount Services

Davy David

DAW Director of Air Warfare; Directorate of Atomic Warfare

Daw al-Bah *Dawait al-Bahrayn* (Arabic—State of Bahrain)

DAWE Daughters Already Well-Endowed

dawid device for automatic word identification and discrimination

DAWN Drug Abuse Warning Network

Daw Qat *Dawlet al-Qatar* (Arabic—State of Qatar)

DAWS Director of Army Welfare Services

DAWT Director of Air Warfare and Training

dax dachsund

Day Birthday; Dayton; Daytona; Natal Day; President's Birthday; President's Day (holiday)

DAY Dayton, Ohio (airport)

Day Phil Dayton Philharmonic

daysoap(s) daytime (tv) soap opera(s)

db daily bulletin; damned bad; database; day book; dead body; decibel(s); defensive back; dextran blue; diameter baudelocque (external pelvic conjugate diameter); diffused base; diode block; disability; distobuccal; distribution board; distribution box; domestic boiler; double bass(es); double bayonet-base (lamp); double bed; double braid(ed); double breasted; double-biased (relay); down(ward) bound; draw bar; drop by for a few minutes; dry bulb; dynamic brake

db (DB) delayed broadcast

d/b documentary bill

d & b dead and buried

d in b (found) dead in bed

dB decibel; differential of susceptance (symbol)

Db dubhium (ytterbium symbol)

DB Data Bank; Date of Birth; David Brown (tractors); Daytona Beach; Declining Balance; Defined Benefit (retirement plan); Detective Bureau; Disciplinary Barracks; Dispersal Base; Dodge Brothers; Dominion Breweries

D-B Daimler-Benz

D & B Dun & Bradstreet

D of B Daughters of Bilitis

DB *Danmarks Biblioteksforening* (Danish Library Association); *Danske Bank* (Danish Bank); *Deutsche Bank* (German Bank); *Deutsche Bundesbahn* (German State Railways); *Domesday Book; Dresdner Bank* (German—Dresden Bank)

D.B. *Divinitatis Baccalaureus* (Latin—Bachelor of Divinity)

dba database administrator; design-basis accident; doing business as/at

dba (DBA) Dibenzanthracene; dihydro-dimethyl-benzopyran butyric acid

d b a doing business as

dBa adjusted decibels

Dba Dubai

DBA Database Administrator; Duke Bar Association

D.B.A. Doctor of Business Administration

dbacc debit account(ing)

DBAE Discipline-Based Art Education

dbam data-base-access method(ology)

DBAP Darien Book Aid Plan

d/bar draw bar

DBAS Development Bank of American Samoa

DBAT Dating Behavior Assessment Test

dbb deals, battens, boards; detector back bias; dinner, bed, breakfast; distance between bends

db & b deals, boards, and battens

dbbal debit balance

dbbd (DBBD) dibromopolybutadiene

dbc diameter bolt circle; dry breast care; dye-binding capacity

dBc decibels referred to the carrier

DBC Demerara Bauxite Company; Detective Book Club; Drums and Bugle Corps

D.B.C. Doctor of Beauty Culture

DBCA Du Bois Clubs of America

dbcl dilute blood clot lysis

DBCM De Beers Consolidated Mines

DBCO Dairy Board Consulting Office(r)

dbcp (DBCP) dibromochloropropane

DBCSO DeBeers Central Selling Organization

db/dc database/data communications

dbcu data bus control unit

dbd death by drugs (execution by lethal injection); double-base diode

D Bd Distribution Board; Drug Board

dbe design-basis earthquake; design-basis event; double-bell euphonium (marching band tuba)

dbe (DBE) dibasic ester

d.b.e. *de bene esse* (Latin—conditionally)

'dbe *adobe* (Spanish—clay-and-straw building material)

D.B.E. Dame Commander of the Order of the British Empire

dbeats despatch payable at both ends (for) all time saved

dbed (DBED) dibenzyl-ethylene-diamine (penicillin)

D.B.Ed. Doctor of Business Education

dbelts despatch payable both ends (for) all laytime saved

dbf design-basis fire; design basis flood

dBf decibel femtowatt; divorced Black female

DBG Division of Basic Grants

dbh diameter breast high

DBHNT Detective Bureau Hostage Negotiating Team (NYPD)

dbhp drawbar horsepower

dbi database index; development-at-birth index (DBI)

Dbi Dubai

DBib Douay Bible

D.Bi.Chem. Doctor of Biological Chemistry

D.Bi.Eng. Doctor of Biological Engineering

D.Bi.Phy. Doctor of Biological Physics

dbir double built-in (ward)robes

D.Bi.Sc. Doctor of Biological Sciences

DBIU Dominion Board of Insurance Underwriters

DBJC Daytona Beach Junior College

dbk debark; drawback

dBK decibels referred to 1 kilowatt

DBK Daiichi Bussan Kaisha (Japanese steamship line); Dobeckmun (company)

dbkn debarkation

dbl double; doubler

DBL Disability Benefit Law; Displaced Business Loan (SBA)

dbl act. double acting

dblb double room with bath

dbl bsn double bassoon

dbl eleph fol. double elephant folio–books about 50 inches high

dbir doubler

dbls double room with shower

dbm decibels per milliwatt; diabetic management

dBm divorced Black male; decibel referred to one milliwatt

dBmV decibels referred to I millivolt

DBM Data Base Management; Division of Biology and Medicine (Atomic Energy Commission)

D.B.M. Diploma in Business Management

DBM *Deutches Bundes Marine* (German Federal Navy)

db meter decibel meter

DBMS Data Base Management System; Director of Base Medical Services

Dbn Durban

dbo dead blackout; disassembled by owner; distobuccooc-clusal; dreadful body odor

dbomp database organization and maintenance processor

DBOS Data-Based Operating System

D-box distribution box

dbp design-basis probability; diastolic blood pressure; distobuccopulpal; drawbar pull

DBP Development Bank of the Philippines; Division of Beaches and Parks

DBP *Dicionário Bibliografico Portugues* (Portuguese Bibliographic Dictionary)

db part double-beaded partition

DBPO Data Buoy Project Office

dbr double book rack

D Br Defendant's Brief

DBR Division of Building Research

dbrap decibels above reference acoustic power

dbre *diciembre* (Spanish—December)

DBRL DeBeers Research Laboratory

dbrn data bank release notice; decibels above reference noise

DBRS Dominion Bond Rating Service (Toronto)

db rts debenture rights

dbs damn bloody soon; deep-brain stimulation; despeciated bovine serum; double bottoms

dbs (DBS) direct broadcast satellite

db's dirty books; dune buggies

DBS Development Bank of Singapore; Direct Broadcast Satellite; Distressed British Seaman (provided free passage home); Division of Biological Standards

DBSI Development Bank of the Solomon Islands

dbsm decibels per square meter

dbsr double bed sitting room

dbst double bituminous surface treatment

DBST Double British Summer Time

dbt debit; design-basis tornado; dry-bulb temperature

DBT David Brown Tractors

dbtd debited

dbtfl doubtful

dbtg debiting

dbtl doubtful

dbt ductile-brittle transmission temperature

dbtu (DBTU) dibutylthiourea

dbtw design-basis tornado and windstorm

dbuf dry-buffed (leather)

d-bug debug; debugged; debugging

dbur data bank update request

dbv decibel referred to 1 volt

DBV *Deutscher Bibliotheksverband* (German—Library Association); *Deutscher Bund für Vogelschutz* (German Bird-shooters Bund)

dbw differential ballistic wind

dBw decibels referred to 1 watt

dbx design-basis explosion

dBx decibels above reference coupling

dc data collection; dead center; death cell; deck cargo; decontamination; deep discount issue (in bond listings); defined contribution; deposited carbon; deviation clause; device control; diagonal conjugate; diesel car; differential calculus; digital computer; direct credit; direct cycle; directional coupler; disorderly conduct; door closer; double cap; double certificated; double column; double contact; double crochet; double crown; down center; draft card; drawing change; drift correction; drill collar (oil well); dry coniferous (forest)

dc (DC) cancrizans of the duration series; debit card(s); diagonal conjugate

d-c direct-chill (casting); direct-current (adjective)

d/c deviation clause; double column (bookkeeping)

d & c dilation and curettage

dc *da capo* (Italian—again); Dick Cavett

d/c *dinero contante* (Spanish—cash)

dC *depois de Cristo* (Portuguese—after Christ); *dopo Cristo* (Italian—after the birth of Christ)

DC Dana College; Dartmouth College; Davidson College; Death Certificate; decimal classification; Defiance College; Dental Corporation; Dental Corps; Department of Commerce; Deputy Chief; Deputy Clerk; Deputy Commander; Deputy Commissioner; Deputy Consul; Deputy County Clerk; Deviation Clause; Diagnostic Center; Dickinson College; Diners Club; Dining Car; direct current (when used as a noun); District of Columbia (D.C.); District Commissioner; District Court(house); Divisional Commander; Doane College; Doctor of Chiropractic; Dominican College; Donnelly College; Dordt College; Douglas Commercial (aircraft); Downing College (Cambridge); Drury College; Duchesne College; Dumbarton College; Dyke College; D'Youville College

D-C Denver-Chicago (truck line); Dow-Corning (chemical products)

D/C drift correction

D&C Dean and Chapter; Detroit and Cleveland (steamship line); Doctrine and Covenants

D of C Daughters of the Confederacy; Department of Commerce; Department of Communications (DoC); Department of Correction(s); District of Columbia (D.C.); Duchy (Duke) of Carinthia; Duchy (Duke) of Carniola

DC *Democrazia Christiana* (Italian—Christian Democracy)—political party; *Distrito Capital* (Spanish—Capital District)

D C *da capo* (Italian—again)

DC-X Delta Clipper-Experiment

DC-1 Defense Condition-1 (war)

DC1, DC2, DC3, etc. device-control characters (data processing)

DC-2 through DC-5 Defense Condition-2 through Defense Condition-5 (stages of military alert short of war)

DC-3 Douglas 21-passenger twin-engine transport aircraft also known as the C47, Dakota, or Skytrain

DC-4 Douglas 44-passenger four-engine transport aircraft also called C-54 or Skymaster

DC-6 Douglas 64 to 92-passenger transport also known as C-118 Liftmaster because of its cargo-carrying capacity

DC-8 Douglas DC-8 jet airplane

DC-9 Douglas twin-jet short-range airplane

DC-9 Super 80 McDonnell-Douglas commercial jetliner

DC-10 McDonnell-Douglas jumbo jetliner

dca deoxycholate citrate sugar; dollar cost averaging

Dca Dacca

DCA Dachshund Club of America; Dalmatian Club of America; Damage Control Assistant; Defence Costs Agreement (Hong Kong); Defense Communications Agency; Department of Civil Aviation; Department of Consumer Affairs; desoxycorticosterone acetate; Diamond Council of America; Diapulse Corporation of America; Digital Computers Association; Director of Civil Aviation; Disassembly Compliance and Analysis; Disc Company of America; Distribution Contractors Association; District Court of Appeals; Division of Consumer Affairs; Document Content Architecture; Dominion Curling Association; Drafting Contractors Association; Drawing Change Authority; Drug Control Agency; Dynamics Corporation of America; Washington, D.C. (Ronald Regan National Airport)

DCA *Défense Contre Aéronefs* (French—anti-aircraft defense)

DCAA Defense Contract Audit Agency

DCA/A Disassembly Compliance and Analysis/Abbreviated

dcac define and control airplane configuration

dcac-mrm define and control airplane configuration/manufacturing resource management

DCADA District of Columbia Alley Dwelling Authority

D.C.Ae. Diploma of the College of Aeronautics

DCAEUR Defense Communications Agency, Europe

DCAF Design Corrective Action Form

DCAO Deputy County Advisory Officer

DCAOC Defense Communications Agency Operations Center

d cap double foolscap (paper)

DCAP Disadvantaged Country Areas Program

DCAR Defense Contract Administration Services Region; Design Corrective Action Report; Disassembly Compliance and Analysis Report

DCAS Data Collection and Analysis System; Defense Contract Administration Services; Defense Control Administration Services; Deputy Chief of Air Staff; Deputy Controller Anti-Submarine (warfare)

DCASO Defense Contract Administration Services Office

DCASR Defense Contract Administrative Service Region

DC-AST McDonnell Douglas Advanced Supersonic Transport

DCAT Drug, Chemical and Allied Trades

DCATA Drug, Chemical, and Allied Trades Association

D Cath Documentation Catholique (French—Catholic Documentation)

dcb data control block

DCB Decimal Currency Board (British)

D.C.B. Dame Commander of the Most Honourable Order of the Bath

DCBD Division for Children with Behavioral Disorders (Council for Exceptional Children)

DCBJ District of Columbia Bar Journal

DCBRE Defense Chemical, Biological, and Radiation Establishment

dcc dark curtain closed; double concave; double cotton covered

dcc (DCC) decade counter code; deleted from colorectal carcinoma

DCC Damage Control Center; Day Care Center; Defense Concessions Committee; Deputy Chief Constable; Design Change Control; Disease Control Center; Dispatch Crew Commander; Dutchess Community College

DCCA Design Change Cost Analysis

DCCAH District of Columbia Commission on the Arts and Humanities

DCCB Defense Center Control Building (USA)

DCCC Democratic Congressional Campaign Committee; Domestic Coal Consumers Council

DCCDCA Day Care and Child Development Council of America

d & c color drug and cosmetic color (synthetic dye)

dccl digital charge-coupled logic

DCCP Design Change Control Program; Directorate of Communication Components Production

DCCS Digital Command Communications System

DCCT Diabetes Control and Complications Trial

dccu data communications control unit

dcd differential current density

DCD Daitch Crystal Dairies; Damage Control Diagram(s) (USN); Design Change Document; Director of Combat Development; Directorate of Civil Disturbance

D.C.D. Diploma in Chest Diseases

DCD Dansk Central för Dukumentation (Danish Center for Documentation)

DCDMA Diamond Core Drill Manufacturers Association

DCDPO Directorate for Civil Disturbance Planning and Operations (USA)

dcdr decoder

D Cdr Deputy Commander

dcds double cotton double silk

DCDS Deputy Chief of the Defense Staff; Digital-Control Design System

dcdt direct-current differential transformer

dce dairy cow equivalent; data conversion equipment; differential compound engine; domestic credit expansion

dce (DCE) data-communications equipment

DCE Distributed Computing Environment; Division of Career Education; Division of Compensatory Education

DCE Dictionary of Christian Ethics

D.C.E. Doctor of Civil Engineering

DCEA Dictionary of Civil Engineering Abbreviations

dcel direct-current electroluminescence

D.C.E.P. Diploma of Child and Educational Psychology

dcf deal-cased frame; direct centrifugal flotation; discounted cash flow

DCF Deputy Chief; Donner Canadian Foundation

dcfem dynamic crossed-field electron multiplication

dcfm discounted cash flow method

dcfp dynamic cross-field photomultiplier

dcg dancing; decigram; displacement cardiograph; dynamic cardiogram

dcg (DCG) deoxycorticosterone glucoside

DCG Deputy Chaplain-General; Descendants of Colonial Governors

dcgm decorticated groundnut meal

DCGS Deputy Chief of the General Staff

dch dicyclohexyl

D. Ch. Doctor Chirugiae (Latin—Doctor of Surgery)

DCH Diploma in Child Health; Doctor of Clinical Hypnotherapy

dcha derecha (Spanish—right)

DCHCL Dropsie College for Hebrew and Cognate Learning

D.Ch.E. Doctor of Chemical Engineering

dchn dicyclohexylamine nitrate

D.Ch.O Diploma in Ophthalmic Surgery

DCHV Domiciliary Care for Homeless Veterans

dci dichloroisoprenaline; dischloroisoproterenol; double-column inch; driving car intoxicated

DCI Defense Council Instruction; Department of Citizenship and Immigration; Des Moines and Central Iowa (railway); Director of Central Intelligence

DCIC Defense Ceramic Information Center

dcid decide

DCID Department of Commercial and Industrial Development

DCIEM Defence and Civil Institute of Environmental Medicine (Canadian)

DCIGS Deputy Chief of the Imperial General Staff

DCII Defense Central Index of Information

dcis ductal carcinoma in situ

dcisn decision

D.Civ.L. Doctor of Civil Law

DCJ Dade County Jail (Miami, Florida); Department of Criminal Justice; District Court Judge

DCJC Dawson County Junior College

dckng docking

dcl decaliter; declaration; declarative; decline

DCL Dartmouth College Library; Detroit College of Law; Deuterium of Canada, Limited; Distillers Company Limited

D.C.L. Doctor of Canon Law; Doctor of Civil Law

DCLA Deputy Chief of Staff, Logistics and Administration (NATO)

dcld declined

DCLE Department of Criminal Law Enforcement (Florida)

dclg declining

DCLI Duke of Cornwall's Light Infantry

dclrt decelerate

dcls deoxycholate citrate lactose saccharose (agar); disclose

D.Cl.Sci. Doctor of Clinical Science

dclsd disclosed

dclsg disclosing

dclsr disclosure

DCLTC Dry Cargo Loading Technical Committee (NATO)

dcltr declines transfer (offered)

DCLU Developing Countries Liaison Unit

dcm decameter; defense combat maneuvers; digital capacitance meter

DCM Deputy Chief of Mission; Director of Civilian Marksmanship; Directorate of Classified Management; Distinguished Conduct Medal; District Court Martial; Dominican Campaign Medal; Don't Come Monday (you are dismissed)

D.C.M. Doctor of Comparative Medicine

DCMA Defense Contract Management Association; District of Columbia Manpower Administration; Dye Color Manufacturers Association

DCMBA Dairy, Confectionery, and Mixed Business Association

D.C.M.G. Dame Commander of the Order of St. Michael and St. George

demi disclosure of classified military information

dcmps degaussing compass

dcmptr degaussing computer

DCMs Deputy Chiefs of Missions

DCMS Deputy Commissioner of Medical Services

dcmsn decommission

dcmt document

dcmu dichlorophenyldimethylurea

dcn delayed conditioned necrosis; double crown

DCN Data Change Notice; Defense Communication Network; Design Change Notice; Drawing Change Notice

DCNEO Deputy Chief Naval Engineering Officer

DCNI Department of the Chief of Naval Information

D.Cn.L. Doctor of Canon Law

DCNO Deputy Chief of Naval Operations

DCNS Deputy Chief of Naval Staff

dco doppler cutoff, double crossover (genes); draft collection only

Dco diffusing capacity-carbon monoxide

DCO Dallas Civic Opera; Data Change Order; Deputy Chief of Staff, Operations (NATO); Director of Combat Operations; Director of Combined Operations; District Control(ling) Office(r); Dominion, Colonial, and Overseas (Department of Barclays Bank); Duke of Cambridge's Own; Duke of Connaught's Own

DCOBE Dame Commander — Order of the British Empire

DCOG Diploma of the College of Obstetricians and Gynecologists

d. & coh. daughter and co-heiress

d col double column

D. Com. Doctor of Commerce

D.Com.L. Doctor of Commercial Law

D. Comp. L. Doctor of Comparative Law

dcop displays, controls, and operation procedures

DCOR Defense Committee on Research (USAF)

DCOS Deputy Chief of Staff

dcp dental continuation pay; depot condemnation percent; development cost plan; discrete component parts

dcp (DCP) dicalcium phosphate

DCP Department of Consumer Protection; Diploma in Clinical Pathology; Disaster Control Plan; Division of Consumer Protection

dcpa (DCPA) dicylcopentenyl acrylate

DCPA Defense (Department's) Civil Preparedness Agency

DCPANDP Deputy Chief of Staff, Plans and Policy (NATO)

DC Path Diploma of the College of Pathologists

dcpd (DCPD) dicyclopentadiene

DCPG Defense Communications Planning Group

DCPL District of Columbia Public Library

DCPO Deputy Chief of Staff, Personnel and Organization (NATO)

DCPR Defense Contractors Planning Report; Dry Cleaning Plant Registration

dcp's development concept papers

dcpta (DCPTA) dichlorophenoxy triethylamine

dcr data conversion receiver; decrease; decreasing; direct cortical response; division credit rebate

DCR Design Change Request; Design Characteristic Review; District Chief Ranger; District Court Report(s); Division of Computing Research; Drawing Change Request

DCRB Design Change Review Board

DCRCH Duke of Connaught's Royal Canadian Hussars

DCRE Deputy Commandant–Royal Engineers; Deputy Commander Royal Engineers

DCRI Duke Clinical Research Institute

DCRLA District of Columbia Redevelopment Land Agency

DCR (MU) Diploma—College of Radiographers (Medical Ultrasound)

DCR (NM) Diploma—College of Radiographers (Nuclear Medicine)

DCRO Dyers and Cleaners Research Organization

DCRR Drawing Change Recorder Request

DCR (R) Diploma—College of Radiography (Radiography)

DCR (T) Diploma—College of Radiographers (Radiotherapy)

dcs dorsal column stimulator; double cotton silk

DCs Douglas Commercial-type airplanes

DCS Damage Control School (USN); Data Control System; Deaf Community Services; Defense Communications System; Department of Children's Services; Department of Correctional Services (Nebraska, New York); Deputy Chief of Staff, Digital Command System; Direct Coupler System; Director of Community Services; Distillers Corporation—Seagrams; Domestic Contact Service (CIA)

DC of S Deputy Chief of Staff

D.C.S. Doctor of Christian Science; Doctor of Commercial Science

DCSA Department and Chain Store Association

DCSAB Distinguished Civilian Service Awards Board

DCSADN Defense Communication System Automatic Digital Network

DCSC Defense Construction Supply Center

DCSCD Deputy Chief of Staff for Combat Developments (NATO)

DCSCOMPT Deputy Chief of Staff, Comptroller (NATO)

DCSF Dry Caisson Storage Facility

DCSFOR Deputy Chief of Staff, Force Development (NATO)

DCSL Deputy Chief of Staff, Logistics (NATO); District Cub Scout Leader

DCSM Deputy Chief of Staff, Materiel (NATO)

DCSMIS Deputy Chief of Staff, Management Information System (NATO)

DCSO Deputy Chief Scientific Officer; Deputy Chief Signals Officer

DCSOI Deputy Chief of Staff for Operations and Intelligence (NATO)

DCSOPS Deputy Chief of Staff for Operations

dcsp digital control signal processor

DCS/P Deputy Chief of Staff for Personnel

DCSPA Deputy Chief of Staff, Personnel and Administration (NATO)

DCS/P & O Deputy Chief of Staff for Plans and Operations

DCS/P & R Deputy Chief of Staff for Programs and Resources

dc sr *da capo senza replica* (Italian—from the beginning without repeat); *da capo senza repetizione* (Italian—from the beginning play repeated parts once)

DCS/R & D Deputy Chief of Staff for Research and Development

DCSRDA Deputy Chief of Staff–Research, Development, and Acquisition (USA)

DCSRM Deputy Chief of Staff for Resource Management (NATO)

DCSROTC Deputy Chief of Staff for Reserve Officers' Training Corps

DCS/S & L Deputy Chief of Staff for Systems and Logistics

DCSS Deputy Chief of the Secret Service

DCST Deputy Chief of Supplies and Transport

DCSTS Deputy Chief of Staff for Training and Schools

dct depth-charge thrower; depth-control tank; distal convoluted (kidney) tubule; document(ary); documentation

DCT Department of Commerce and Trade

DCT *Division de Contre-Topilleurs* (French—Division of Destroyers)

DCTC District of Columbia Teachers College; Dodge County Teachers College

DCTD Diploma in Chest and Tuberculous Diseases

dcti direct-coupled transistor logic

DCTSC Defense Clothing and Textile Supply Center

dctv digital color television

dcu digital computer unit; display and control unit; dynamic checkout unit

dcu (DCU) dichloral urea (herbicide)

dcutl direct-coupled unipolar transistor logic

DCUC Defense Credit Union Council

dcv double cotton varnish

DCVO Dame Commander of the Royal Victorian Order; Deputy Chief Veterinary Officer

dcw dead carcass weight

DCW Detroit Chemical Works

dcwv direct-current working volts

dcx double convex

DCZ District of the Canal Zone

dd days after date; day's date; deadline date; decreased (sexual) desire; deep-drawn; deferred delivery; delayed delivery; delivered; dental discomfort; development directive; developmentally disabled; differential diagnosis; digital display; direct dial; discharged dead; double draft; drunk driver(s); drydock; dual damping; due date; due diligence; duplex drive (tank); dutch door

dD differential of electric displacement (symbol)

d-d dumb-dumb; dum-dum

d'd deceased

d/d dated; delivered at dock(s); demand draft; detergent dispersant; domicile to domicile; due date

d...d damned

d & d deaf and dumb; death and decay; death and dying; defiled and deflowered; diarrhea and dehydration; drinking and drugging; drunk and disorderly; dungeons and dragons (game)

d & d (D & D) decontamination and decommissioning; development and demonstration

d-to-d dawn-to-dusk (daylight patrol); dusk-to-dawn (night patrol)

dd *dags dato* (Dano-Norwegian—days to date); *direttissimo* (Italian—express train)

dd (DD) direct deposit; double density

d.d. *dono dedit* (Latin—he gave as a gift)

Dd David; Drydock

DD Deputy Director; destroyer (naval symbol); Detective District; Detective Division; Development Directive; dichloropropane dicholoropropylene (insecticide); Dishonorable Discharge; Disk Defender trademark

D.D. Doctor of Divinity

D & D Devonshire and Dorset (regiment); Dungeons and Dragons

DD *Divinitatis Doctor* (Latin—Doctor of Divinity); *Doctores* (Spanish—Doctors); *Dottores* (Italian—Doctors); *Doutores* (Portuguese—Doctors)

DD-2 Second Development Decade (1971- 1980)

DD-214 Department of Defense Honorable Discharge (form DD-214)

dda duty deposit account

dda (DDA) digital differential analyzer

DDA Dangerous Drug Act; Deputy Director of Administration (CIA); Digital Dealers Association; Disabled Drivers Association; Display and Decision Area; Duty Deferment Account

ddalv days delay enroute authorized chargeable as leave

DDANS Deputy Director Army Nursing Service

DDAS Digital Data Acquisition System

DDATS Deputy Director Auxiliary Territorial Service

DDAU *Doctoral Dissertations Accepted by American Universities*

ddavp (DDAVP) decamino-D arginine vasopressin

ddavp desmopressin acetate

D-Day day of attack; Decimalisation Day (Feb 15, 1971 when British money was decimalized)

ddb defined dollar benefit; dodecylbenzene

ddbms distributed database management system

ddbsa dodecylbenzene-sulfonic acid

DDB Double Declining Balance

ddc data documentation costs; decision-difficulty checklist; direct digital control; double-deck car(s)

DDC corvette (naval symbol); Data Device Corporation; Defense Documentation Center (now DTIC–Defense Technical Information Center); Defensive Driving Course; Dewey Decimal Classification; Diamond Dealers Club; Diecasting Development Council; Digital Development Corporation

DDC *Docteur en droit canonique* (French—Doctor of Canon Law)

ddce digital data-conversion equipment

DDCI Deputy Director of Central Intelligence (CIA)

ddcmp digital-data communication—message protocol

ddc's deck decompression chambers

DDCs Desk and Derrick Club members (petroleum professionals)

ddcv deep-draft caisson vessel

ddd darling discipline of the decade (computer science courses in the 1975–1985 era); deadline delivery date; detail data display; dichlorodiphenyldichloroethane;

digital data distributor; digital display driver; drink, drank, drunk (alcoholic's progress); dynamic dummy director

d.d. in d. *de die in diem* (Latin—from day to day)

DDD Department of Decentralization and Development; direct distance dialing

DDD *dat, dicat, dedicat* (Latin—he gives, devotes, dedicates)

ddda decimal digital differential analyzer; dodecanedioic acid

dd/dc diamond differential direct current

DDDIC Department of Defense Disease and Injury Code

dddm dihydroxydichlorodiphenylmethane

DDDS Deputy Director of Dental Services

dde direct data entry; dual-displacement engine; dynamic data exchange

dde (DDE) dichlorodiphenyl-dichloroethylene

DDE Designated Destroyer Escort (Canadian); dichlorodiphenyldichloroethylene (insecticide less toxic than DDT); Dwight David Eisenhower (34th President U.S.)

D De L Daniel De Leon

D de l'U *Docteur de l'Université* (French—Doctor of the University of Paris)—the Sorbonne

DDEM Dwight D. Eisenhower Museum

DDEP Defense Development Exchange Program

ddf design disclosure format; double defruit

DDF Dental Documentary Foundation; Dominica Defence Force

DDF *Departamento del Distrito Federal* (Spanish—Federal District Department)

ddg (DDG) digital display generator

DDG Deputy Director General; guided missile destroyer (naval symbol)

DDGSE Deputy Director General—Signals Equipment

DDGSR Deputy Director General of Signals Equipment

ddh diamond drill hole; dichlorodimethylhydantoin

DDH Digital Data Handling (system); Diploma in Dental Health

DDHA Detective Division Homicide Assault Squad

ddi depth deviation indicator; digital data indicator; discrete digital input; document disposal indicator

DDI Deputy Director, Intelligence (CIA); Drug Didanosine (anti-viral)

dd-ing double dipping

ddis data display

DD & J Dragons for Defense and Justice

ddl data description language; digital data link

ddl (DDL) data definition language

DDL Data Disclosure List; Det Danske Luftfartsselskab (The Danish Airways)

ddm data demand module; dialkyl dihexadecymalonate; diaminodiphenylmethane; difference in depth of modulation; digital drawing monitoring; distributed data management; n-dodecyl mercaptan

DDM Diploma in Dermatological Medicine; Distributed Data Management

DDME Deputy Director of Mechanical Engineering

DDMI Deputy Director of Military Intelligence

DDMOI Deputy Director of Military Operations and Intelligence

DDMS Deputy Director of Medical Services

DDMT Deputy Director of Military Training

Ddn Dunedin, NZ

DDN Data Network; nuclear-powered destroyer (naval symbol)

ddnc direct digital numerical controller

DDNI Deputy Director of Naval Intelligence

DDN NIC Defense Data Network Network Information Center

ddnp diazodinitrophenol

ddo despatch money payable (for) discharging only

DDO David Dunlap Observatory (Ontario); Deputy Director of Operations (CIA)

D.D.O. Diploma in Dental Orthopedics

DDOD Deputy Director Operations Division

D-dog detector dog (U.S. Customs)

DDOS Deputy Director of Ordnance Services

ddp digital data processor; directors' deferral plan; dodecyl phthalte

ddp (DDP) distributed data processing

DDP Data Distribution Point (NATO); Declaration of Design Performance; Delivered Duty Paid; Department of Defense Production (Canadian); Deputy Director, Plans (CIA); Design Development Plan; Devalued Dollar Planning

DDPH Diploma in Dental Public Health

DDPR Deputy Director of Public Relations

DDPS Discrimination Data Processing System

ddr direct debit

DDr Doktor, Doktor (Austrian-German—person with two doctor's degrees)

DDR Darling Downs Regiment; Devon and Dorset Regiment; radar picket destroyer (3-letter naval symbol)

D.D.R. Diploma in Diagnostic Radiology

DDR Deutsche Demokratische Republik (German Democratic Republic)

DDRA Deputy Director—Royal Artillery

DDRD Deputy Directorate of Research and Development

DD R & D Department of Defense Research and Development

ddr&e design, develop, research and evaluate

DDRE Danish Defense Research Establishment

DDR & E Defense Development Research and Engineering

D+D Regt Devonshire & Dorset Regiment

DDRM Deputy Director of Repair and Maintenance

ddrr directional discontinuity ring radiator

DDRS Declassified Documents Reference System

dds diaminodiphenysulfone; digital data service; digital display scope, digital dynamics simulator; distributed data system

d/d's developer/demonstrators

DDS Deep-Diving System; Demos D-Scale; Department of Developmental Services; Deployable Defense System; Deputy Director of Support (CIA); Dewey Decimal System; Disability Determination Services; Documentation Distribution System; Domestic Disclosure Spreadsheet; Drug Delivery System

D.D.S. Doctor of Dental Science; Doctor of Dental Surgery

D.D.Sc. Doctor of Dental Science

DDSD Deputy Director of Staff Duties

DDSC Dewey Decimal System of Classification

DDSG Donau-Dampfschiffahrts-Gesellschaft (German—Danube Steamship Travel Service)

dd & shpg dock dues and shipping

ddso diamino-diphenyl sulphoxide

DDSR Deputy Director of Scientific Research

DDSSA Digital Distributed System Security Architecture trademark

DDST Denver Developmental Screening Test; Deputy Director of Supplies and Transport; Double Daylight Saving Time (two hours ahead)

DDS & T Deputy Director of Science and Technology (CIA)

DDSTs Denver Developmental Screening Tests

ddt deduct; digital data transceiver; digital data transmitter; digital debugging tape(s); drop dead twice (epithet); ductus deferens tumor; dynamic debugging technique

ddt (DDT) direct-decision therapy

DDT dichlorodiphenyl-trichloro-ethane (insecticide)

ddt & e design, development, test, and evaluation

DDTE Deputy Director, Test and Evaluation (NASA)

DDTF Dynamic Docking Test Facility (NASA)

ddtl dreary desk-top lunch

DDTS Dynamic Docking Test System (NASA)

DDTV Dry Diver Transport Vehicle (naval)

ddu data diagnostic unit; data display unit; design diagnostic unit; display driver unit; distribution data unit

ddv deck drain valve

ddvp (DDVP) dimethyldichlorovinylphosphate

DDVS Deputy Director of Veterinary Services

ddw displaying a deadly weapon

DDWE & M Deputy Director of Works, Electrical and Mechanical

DDWP Deputy Director Weapons Polaris

DDx differential diagnosis

DDY *Devlet Demiryollari* (Turkish Railways)

de data entry; decision element; deckle edge; deckle edging; defensive end; deflection error; development engineering; diatomaceous earth; diesel-electric; digestive energy; direct elimination; direct entry; double end; double entry; dream elements; drive end; duration of ejection

dE differential of voltage (symbol)

d/e date of establishment

d & e dilation and evacuation

de *det er* (Norwegian—that is); (Latin prefix—down, from)—descent, description

DE Deere (stock exchange symbol); Delaware; delayed early (genes); Department of Education; Department of Employment; Department of the Environment; destroyer escort (naval symbol); Dextrose Equivalent; District Engineer; Dynamite Engineer(ing); Germany (Internet code)

DE *Dáil Éireann* (Irish Gaelic—Assembly of Ireland)—Ireland's House of Representatives

D.E. Doctor of Economics

D of E Department of Energy; Department of the Environment; Department of the Environment (UK)

D of E *Dictionary of Electronics*

dea data encryption algorithm; dehydroepiandrosterone

dea (DEA) dehydroepiandrosterone

Dea Deacon

DEA Dance Educators of America; Davis Escape Apparatus; Department of External Affairs; Detectives Endowment Association; Digital Equipment Australia; Drug Enforcement Administration

deac deacon; diethylaluminum chloride

DEACONS Direct English Access and Control System

dead (DEAD) destruction-entrusted automatic devices

Dead Horses Dead Horse Mountains between Mexico and Texas in the Big Bend Area where it is also called Sierra del Caballo Muerto

DEADS Detroit Air Defense Sector

deae diethylaminoethyl

DEAE Division of Eligibility and Agency Evaluation (Office of Education)

DEAE-cellulose diethylaminoethyl cellulose deal; decision evaluation and logic

deal decision, evaluation and logic

dealer prep dealer preparation

DEAN Deputy Educators Against Narcotics

dear diamonds, emeralds, amethysts, rubies

dearg. pil. *deargentur pilulae* (Latin—let the pills be silvered)

DEAS Delaware Educational Accountability System

Death Valley Death Valley National Monument on the border of California and Nevada

DEAUA Diesel Engineers and Users Association

deaur. pil. *deaurentur pilulae* (Latin—let the pills be gilded)

deb debenture; debit; debut(ante); diethylbutanediol

DEB Dental Examining Board

Debbie Deborah

Deb(by) Deborah

de Bc Honoré de Balzac

debil debilitate(d); debilitating; debilitation; debility

debk debark; debarkation

De Brücke (German—The Bridge)—pessimistic expressionism popular in Germany in the 1920s and 1930s

deb(s) debenture(s); debutante(s)

deb. spis. *debita spissitudine* (Latin—of the correct consistency)

deb stk debenture stock

dec decant; decanter; deceased; deciduous; decimal; decimeter; decision; declination; decompose(d); decorate; decoration; decorator; decrease(d); won by judge's decision

dec. *decani* (Latin—sung by the choir on the deacon's side of the church); *décembre* (French—December); *décor* (French—decoration, stage scenery); *decubitus* (Latin—lying down)

Dec Decca; December

DEC Deductible Employee Contribution; Dental Examination Centre; Department of Environmental Conservation; Detroit Edison Company; Developmental Education Center; Digital Electric Corporation; Digital Equipment Corporation; Disaster Emergency Committee; District Export Council; Dominion Executive Council

deca- 10

DeCA Defense Commissary Agency

DECA Distributive Education Club of America

decad decadence; decadency; decadent(ly)

decaf decaffeinated

decal decalcomania

DECAL detection and classification of an acoustic lens

decap(ped) decapitation(ed), behead(ed)

decasyl decasyllable; decasyllabic

decb data event control block

Decca Decca Navigation System

Deccan Deccan Plain of southern India

DECCO Defense Commercial Communications Office

decd deceased

decdest deceased estate

decel deceleration

DECHEMA Dechema Chemical Engineering and Biotechnplogy Abstracts Database

deci 1^{0-1}

decid deciduous

DECIDE Define the problem precisely, Enumerate two groups of decision factors, Collect relevant information, Identify the best alternatives, Develop and implement a detailed plan, Evaluate the decision

decim decimeter

decis decision

decit decimal digit

decl declension

DECL Direct Energy Conversion Laboratory (NASA)

declon declaration

decm defensive electronic counter-measure

DECMD Detroit Contract Management District

decn decision; decontamination

DECnet Digital Equipment Corporation trademark

deco direct energy conversion operation

deco (DECO) decreasing consumption of oxygen

decoct decoction

decomm decommissioning (date)

decomp decomposition

DECOMPS Decomposition Mathematical Programming System

decon decontaminate; decontamination

D. Econ. Doctor of Economics

decor decorate; decoration; decorative

decr decrease; decrement(al)(ly)

Decr *Decreto* (Italian, Portuguese, Spanish—Decree)

decres *decrescendo* (Italian—contracting, subsiding)

decrim decriminalization (al)(1y); decriminalize(r)

DECS Direct Evacuation Control System (air filtration)

decsn decision

DEC Station Digital Equipment Corporation Station

dec stories detective stories

DECTRA Decca Track and Range

decu data-exchange control unit

Decuary December and January

decub. *decubitus* (Latin—lying down)

DECUS Digital Equipment Computer Users Society

ded date expected delivery; dedendum; dedicate; dedicated; deduct; deducted; deduction; diesel engine driven

Ded Dedan; Dedham; Dedric(k)

D. Ed. Doctor of Education

DED Department of Economic Development

DED *Data Element Dictionary* (USA)

de d. in d. *de die in diem* (Latin—from day to day)

dedic dedicate(d)(1y); dedicating; dedication; dedicative; dedicator(y)

dedl data element description list

dedn deduction

DEDP Director of Executive Development Programs

deduct. deduction

dee digital events recorder; discrete event evaluator

dee (DEE) diethoxyethylene

DEE Diploma in Electrical (Electronic) Engineering

dee-dee deaf and dumb

Deedee Dorothy

Dee High Doctor of Hygiene

dee jay disc jockey

deeks duck decoys

DEEP Development Economic Education Program; Diffusion of Exemplary Educational Practices; Directly Elected European Parliament

deep 6 burial at sea; disposing of anything unwanted in at least six fathoms of water

Dee Pee Doctor of Pharmacy

Dee R doctor

dees dynamic electromagnetic environment simulator

Deeside River Dee valley around Aberdeen

deet diethyl toluamide (insecticide)

def decayed, extracted and filled; defeated; defecate; defecation; defect; defection; defective; defector; defendant; defense; defensive; defer; deferred; deficiency; deficient; define; definite; definition; definitive; deflagrate; deflagrating; deflagration; deflect; deflecting; deflection; defoliate; defoliating; defoliation; defrost; defroster; defrosting; defunct, defunction; defunctive; delayed electronic feed

déf *déficit* (Spanish—deficit)

def. *defunctus* (Latin—deceased)

def art. definite article

defcon defense condition; defensive concentration

Def Con-1 Defense Condition 1 (war)

Def Con-2 through Def Con-5 stages of military alert short of war

Def Con(s) Defense Condition(s): Def Con I–war, Def Con II–attack imminent, Def Con III–highest state of readiness for war, Def Con IV-readiness alert, etc.

defec defective

Defense Department of Defense

defi deficiency

defib defibrillate

deric deficiency; deficit

defin definition; definitive

defl deflate; deflation; deflect; deflection

deflor defloration

deform. deformity

DEFREPNAMA Defense Representative North Atlantic and Mediterranean

defs definitions

DEFSIP Defense Scientists Immigration Program

deft. defendant; dynamic error free transmission (DEFT)

DEFY Drug Education For Youth

deg degenerate; degeneration; degree(s); diethanolglycine

deg (DEG) diethylene glycol

DEG guided-missile escort ship (naval symbol)

D & EG Development and Engineering Group

DEG Derechos Especiales de Giro (Spanish—Special Drawing Rights)

de ga depth gage

de Gaulle Charles de Gaulle (Paris, France's airport)

degen degeneration

deglut. deglutiatur (Latin—let it be swallowed)

degn diethylene glycol dinitrate

degrad(s) degradable(s)

degsvc degaussing service(s)

degust degustation (savoring, tasting)

de gustibus de gustibus non est disputandum (Latin—there is no disputing about tastes)

DE-H destroyer escort—hydrofoil

deha (DEHA) diethylhydroxylamine

DEHCD Department of Environment, Housing, and Community Development

DeHoCo Detroit House of Correction

DEHS Division of Emergency Health Services

dehyd dehydrate(d)

dei design engineering identification; development engineering inspection; double electrically isolated

DEI Digital Electronics Incorporated; Dutch East Indies

DEIC Diver Equipment Information Center; Dutch East India Company

Deich Bib *Deichmanske Bibliotek* (Norwegian—Deichman's Library)—Oslo

DEIR Department of Employment and Industrial Relations

deis design engineering inspection simulation; design engineering inspection simulator

DEIS Defense Energy Information System; Draft Environmental Impact Statement

dej dento-enamel junction

dek data encryption key

Dek Dekábr (Russian—December)

deka 10

dekag dekagram

dekal decaliter

dekam decameter

Deke Deacon; Donald

dekon economic declaration (Indonesian)

del delegate; delegation; delete; deletion; deliberate; deliberation; delineate; delineated; delineation; deliver(y)

del (DEL) delete character (data processing)

del. delineavit (Latin—he or she drew it)

Del Delaware; Delawarean; Delhi

Del Delphinus (Latin—Dolphin Constellation)

del acct delinquent account

delcap delay capacity

DELCO Dayton Engineering Laboratory Company

deld delivered

dele delete

deleat. deleatur (Latin—delete)

deleg delegation

del ent delete entirely

delg delivering

deli delicatessen

delib deliberate; deliberation

delic delicatamente (Italian—delicately)

deli-market delicatessen and market

DELIMCO German-Liberian Mining Company

delin delineate(d); delineating; delineation; delineative; delineator; delineatrix; delinquencies; delinquency; delinquent; delinquently; delinquents

delinq delinquent

deliq deliquescent

De L Isls De Long Islands

Dell Dell Publishing Company

DELL Dell Computer Corporation

Dells The Dells (the Dells of Wisconsin), scenic gorge of the Wisconsin River

Del-Mar-Va Delaware-Maryland-Virginia (Eastern Shore peninsula)

Delmarvia the Delaware-Maryland-Virginia peninsula

delmes delay message

Del Mus Nat Hist Delaware Museum of Natural History

D. Elo. Doctor of Elocution

De Longs De Long Islands in the Arctic

delphi declaiming eclectic liberalism possessively, hotly, instantaneously

delpho deliver(y) by telephone

Del Rio San Felipe del Rio

dels deliveries

DELS Direct Electrical Linkage System

delt delete; deletion

de lt deck edge light(ing)

delt. delineavit (Latin—he or she drew it)

delta. detailed labor and time analysis

Delta letter D radio code

DELTA Daily Electronic Lane Toll Audit; Developing European Learning through Technological Advance

deltic delay line time compression

delu delusion

delv deliver

Delv Delvalle

delvd delivered

dely delivery

dem demand; dementia; democracy; democrat; democratic; demodulate; demodulator; demolish; demolition; demonstrate; demonstration; demonstrative; demote; demotion; demur; demurrage; demy; differential element movement

dem (DEM) demerol

Dem Demerera (British Guiana); democracy; Democrat; democratic; Democratic Party

Dem. Dema'i

DEM Department of Environmental Management

DEM Development-Études-Marketing (French—Marketing Studies Development)

DEMA Diesel Engine Manufacturers Association

demac deck and engine mechanic

dem adj demonstrative adjective

Demba Demarara bauxite

dem cap democratic capitalism

de/me decoding memory

DEME Director of Electrical (Electronic) and Mechanical Engineering

DEMETER Digital Electronic Mapping of European Territory

demij demijohn

demimond demimondaine; demimonde

demirep demireputation; a woman of loose morals

DEMKO Dansk Elektrische Materialkontrol (Danish Board for Approving Electrical Equipment)

demo demolition; demonstration (model)

Demo Democrat(ic)

demob demobilization; demobilize

demobed demobilized

democ democracy; democrat; democratic; democratization; democratize; democratizer

demod demodulator

demogr demographer; demographic(al); demography

demon. demonology; demonstrate; demonstration; demonstrator

demonol demonologic(al)-(ly); demonologist(ic)-(al)(ly); demonology; [*see* maj dem(s)]

demonstr demonstrative

demos demonstrators

demo(s) demolition(s); demonstration(s); demonstrator(s)

Demos Democrats

DEMOS Director(ate) of Estate Management Overseas

dem pro demonstrative pronoun

dem pug dementia pugilistica (Latin—punch drunk)

Dem(s) Democrat(s)

DEMS Defensively Equipped Merchant Ship; Development Engineering Management System

demur demurrage

den denotation; dental; dentist; dentistry

den Denier (German—denier)

Den Denbighshire; Deniz; Denmark; Denver

Den Denizi (Turkish—lake, sea)

D. En. Doctor of English

DEN Denver, Colorado airport

Denali National Park in Alaska—contains Mt McKinley, North America's highest peak (20,320 ft)

denat denatured

Denb Denbighshire

dend dendrology

D en D Docteur en Droit (French—Doctor of Law)

dendro dendrometer

dendrol dendrology

D. Eng. Doctor of Engineering

D.Eng.Sc. Doctor of Engineering Science

DENIS Deep Near-Infrared Survey

D en M Docteur en Médecine (French—Doctor of Medicine)

Denmark Kingdom of Denmark (Scandinavian nation) *Kongeriget Danmark*

Denny Denis; Dennis

Dennys Dennys Lascelles

denom denomination

denot denotation; denotative(ly); denote(ment)

dens density

dent dental; dentist; dentistry; denture

dent. dentur (Latin—give, let it be given)

Dent JM Dent & Sons Ltd

D. Ent. Doctor of Entomology

dentac dental accounting

Dent Corps Dental Corps

Dent Hyg Dental Hygienist

Denticare Dental Care

dent. tal. dos. dentur tales doses (Latin—give of such doses)

DEO District Engineering Office; District Engineers Office; Divisional Education Office(r); Divisional Entertainment Office(r)—British Army; Duke of Edinburgh's Own

DEOR Duke of Edinburgh's Own Rifles

DEOS Director of Equipment and Ordnance Stores

DEOVR Duke of Edinburgh's Own Volunteer Rifles

dep depart; department; departure; dependency; dependent; depilate; depilatory; deponent; depose; deposit; depositor; depot; depotize; deputy; diethylphthalate; do everything possible

dep. depuratus (Latin—purify)

Dep Deputy

Dep Département (French—Department); *Député* (French—Deputy)

DEP Defense Electronic Products (RCA); Department of Employment and Production; Department of Environmental Protection; Department of Export Promotion

depa diethylene phosphoramide

DEPA Defense Electric Power Administration

depart. department; departure

Dep CFO Deputy Chief Fire Officer

dep con departmental control

depcru dependent's (daylight) cruise (USN)

dep ctf deposit certificate

Dep Dir Deputy Director

depend. dependent; dependency

depen-undepen dependably undependable

depi differential equations pseudocode interpreter

depil depilate, depilation, depilator(y)

Dep Insp Deputy Inspector

dep inst depot installed

depi depilate(d) depilation; depilator(y); deplete; depletion(ary); deploy(ed); deployment

deplab depilatory laboratory

DEPMIS Depot Management Information System (USA)

depn dependency; dependent

DEPNAV Naval Deputy (NATO)

depo deposit

depod deposited

depog depositing

depon deponent

depor depositor

depos depositary

deposn deposition

depr depreciation; depreciative; depression

DEPRA Defense European and Pacific Redistribution Activity

DEPS Departmental Entry Processing System; Diploma in Economics and Political Science

DEPSACLANT Deputy Supreme Allied Commander, Atlantic (NATO)

DepSO Departmental Standardization Office

dept depart; department; departure; deponent; depot; deputy

dep't (contraction—department)

Dept Ag Department of Agriculture

deptr departure

Dept State Bull Department of State Bulletin

DePU De Paul University; De Pauw University

deputn deputation

Depy Deputy

DEQ Department of Environmental Quality

der derivation; derivative; derived; dermatine; derrick(s); designated engineering representative

der derecha (Spanish—right); *dernier* (French—last)

Der Derringer

DeR reaction of degeneration

DER Department of Environmental Resources; Development Engineering Review; Draft Environmental Report; radar picket escort ship (naval symbol)

DERAP Development Economics Research and Advisory Service

Derb(s) Derby; Derbyshire

Derby. Derbyshire

DERBY Derby Aviation

Derbys Derbyshire

DERE Dounreay Experimental Reactor Establishment

dereg deregulation

Derek Theodoric

deric de ea re ita censuere (Latin—concerning that matter have so decreed)

deriv derivation; derivative

derm dermatitis; dermatology; dermatophyte

derm (Latin prefix—skin) — dermatology, epidermis

dermat dermatology

dermatol dermatologic(al)(ly); dermatologist; dermnatology

Derniers Dernieres Islands

derog derogatory

deros date eligible for return from overseas; date of estimated return from overseas service

DERR Duke of Edinburgh's Royal Regiment

Derrick Theodoric

Derry Londonderry

DERT Division Electronique, Radioélectricité et Télécommunications (French—Electronic, Radioelectric, and Telecommunications Division)

derv diesel-engine road vehicle

des data encryption standard; descend(ed); descending; desert; design; designate; designation; designator; designer; desire; dessert; diesel electric submarine; diethylstilbesterol; direct entry scheme

des (Des) diethylstilbestrol (morning-after contraceptive)

des descubrimiento (Spanish—discovery)

Des Des Moines; Desmond

Des Desfiladero (Spanish—canyon); *Desierto* (Spanish—desert); (German—D-flat)

DES Department of Economic Security; Department of Education and Science; destroyer (naval symbol); Director of Educational Services; Director of Engineering Stores; Dispersed Emergency Station; Doctor of Esoteric Studies; Draft Environmental Statement; Drug Education Specialist; IBM trademark

DESAC Destroyer Sonar Analysis Center (USN)

desal desalinization; desalinize(r)

desat desaturated

DESAT Defense (Department) Small (Business) Advanced Technology (Program)

DESATO Defensive Antisatellite Operations

DESB Devereaux Elementary School Behavior Rating Scale

Des Base Destroyer Base

desc descend(ant)

DESC Defense Electronics Supply Center

descr description

descron description

desco descuento (Spanish—discount)

desdg descending (flow chart)

DESDIV Destroyer Division (naval)

DeSEA Delaware Society of Enrolled Agents

desfex desert field exercise

desfirex desert firing exercise

desg designate; designation

desi designated hitter

DESI Division for Economic and Social Information (UN)

desid desiderata; desideratum

desider desiderative

desig designate; designer

D ès L Docteur ès Lettres (French—Doctor of Literature)

DESLANT Destroyer Forces—Atlantic

desp despatch

DESP Department of Elementary School Principals

DESPAC Destroyer Forces Pacific

despd despatched

despg despatching

despot design performance optimization

Desq View trademark of Quarterdeck International of Santa Monica, California

DesRCA Designer of the Royal College of Art

DESRON destroyer squadron

dess dessiatine

d è S Docteur ès Science (French—Doctor of Science)

D ès S Dar ès Salaam

DESS destroyer schoolship (naval symbol)

dest destination; destroy; destroyer; destruction

dest. destilla (Latin—distilled); *destra* (Italian—right)

DEST Diplôme de l'École Supérieure Technique (French—Diploma of the Technical Institute)

destdist destructive distillation

destil. destilla (Latin—distill)

destination SPPK destination Singapore, Penang, and Port Klang

destn destination

destr desires to transfer; destructor

destr fir destructive firing

desubex destroyer/submarine anti-submarine warfare exercise

DESY Deutsches Elektronen Synchrotron (German Electronic Synchrotron)

det detach; detachment; detail; detective; detector; determinant; determine; determiner; detonator; double end trimmed

det (DET) diethyltryptamine (quick-acting hallucinogen drug)

det. detur (Latin—let it be given)

Det. Detective; Detroit

DET Design Evaluation Testing; Detroit, Michigan (Detroit City Airport)

DETA Direção de Exploração dos Transportes Aéreos (Mozambique airline)

detab decision table

detab/X decision table(s)/experimental

DETAPS Decision Table Processing System

detcom(s) detected communist(s)

Det Con Detective Constable

detd determined

det gar detached garage

det. in dup. detur in duplo (Latin—give twice as much; let twice as much be given)

detectionary dictionary of detectives (mystery-fiction type)

determin determination

DETEST Demystify Established Standardized Tests

DETG Defense Energy Task Group

Det Insp Detective Inspector

detl detail

detm determine

DETMAHOG Deliver-the-Mail/Holy-Grail (dichotomous theory of problem protection practiced by adept bureaucracies worldwide)

detn detention

detox detoxification; detoxification center (for alcoholic and narcotic addicts)

detoxcen detoxification center

detr detector

detrins detailed routing instructions

Detroit Inst Detroit Institute of Arts

d. et s. detur et signatur (Latin—let it be given and labelled)

Det Sgt Detective Sergeant

Det Sup Detective Superintendent

Det Sym Orch Detroit Symphony Orchestra

deu data exchange unit; digital evaluation unit; display electronics unit

DEU Data Exchange Union; Drug Epidemiology Unit

DEUA Diesel Engines and Users Association

deuce digital electronic universal computing engine

Deuce The Deuce (New York City's 42nd Street)

Deut Deuteronomy

Deut Deuteronomy

deutan deuteranomolous; deuteranope

Deutschland Deutschland, Deutschland über alles (German—Germany, Germany above all)—national anthem

dev develop; developer; development; deviate; deviation; deviator

dev (DEV) duck embryo vaccine

Dev Devon; Devonian; Devonshire; Eamon De Valera

De V De Vilbiss

DEV Development Well

deva development acceptance

DEVCO Development Committee

devd device data set residence

devel developer; development

Dev-Genc Devrimci-Gencler (Turkish—Revolutionary Youth)

devil development of integrated logistics

Devils The Devils of Loudon by Aldous Huxley

Devils Postpile Devils Postpile National Monument

Devils Tower Devils Tower National Monument on Wyoming's Belle Fourche River

Devin Devin-Adair

devis devised

devlp develop

devlpd developed

devlpg developing

devlpmt development

dev^mo^ *devotissimo* (Italian—devotedly yours)—yours truly

Devon Devonshire

devp develop

devpt development

devs developers; devotions

DEVSIS Development of Science Information Systems

Dev-Yol Devrimoi-Yol (Turkish—Revolutionary Path)

dew. dewpoint

DEW Demineralization Water (decontamination subsystem); Distant Early Warning

dewat deactivated war trophy

dewd detailed elementary wiring diagram(s)

De Witt De Witt Clinton High School

DEWIZ Distant Early Warning Identification Zone

DEW Line Distant Early Warning Line

DEWS Diagnostic Evaluation of Writing Skills

dex dexamethasone; dextroamphetamine tablet

dext. dexter (Latin—right)

Dex Dexter

D. Ex. Doctor of Expression

DEXA Dual Energy X-Ray Absorptiometry

dexan digital experimental airborne navigator; distant experimental airborne navigator

dexe dexedrine

dexies dexedrine tablets (stimulant drugs)

d. ex m. deus ex machina (Latin—god from a machine)—introduction of a god-like device to resolve a play or problem

dext. *dexter* (Latin—right)

dextrose glucose ($C_6H_{12}O_6H_2O$)

dez diethyl zinc

dez dezembro (Portuguese — December)

Dez Dezember (German—December)

Dezhda Nadezhda

df damage free; dead freight; decayed and filled; decontamination factor; defensive fire; defogging; degree(s) of freedom; dense film; derrick floor; diamond flap; direction finder; direction finding; disposition form; double feeder; double foolscap; double fronted; draft; drinking fountain; drive fit; drop forge; drop forging; dummy fuze; dummy fuze; dunnage free

dF differential of field intensity (symbol); differential of force (symbol)

d/f defogging; direct flow; double fleece; double fronted

d & f determination and finding

d/f dias fecha (Spanish—days from date)

Df Douglas fir

DF Dean of the Faculty; Defender of the Faith; Destroyer Flotilla; deuterium fluoride; Dongfeng (Chinese missile); Dublin Fusiliers

D-F Dansk-Franske; deflection factor (symbol)

D of F Department of Fisheries

DF Distrito Federal (Spanish—Federal District)

D.F. *Defensor Fidei* (Latin — Defender of the Faith)

dfa digital fault analysis

DFA Dairy Farmers Association; Department of Foreign Affairs; Dimensional Fund Advisor; Division Freight Agent; Drop Forging Association

D.F.A. Doctor of Fine Arts

DFAC Dried Fruit Association of California

DFAR Daily Field Activity Report

dfb damage-free bulkheads; distributed feedback; distribution fuse board; dunnage-free bulkheads

dfc data format converter; discriminant function coefficient; dry-filled capsules

DFC Department Frequency Coordinator; Development Finance Corporation; Distinguished Flying Cross

dfclt difficult

dfcs digital flight-control software

DFCT Deputy Federal Commissioner of Taxation

dfcty difficulty

dfd data function diagram; defend(ed); deferred

DFU Dogs For Defense

dfdr digital flight-data record(er)

DFDS Det Forende Dampskibsselskab (Danish United Steamship Company, Limited, Denmark)

DFDT difluoro-diphenyl trichloroethane (insecticide)

dfe derivative fighter engine; derrick floor elevation; double fish eye (buttons)

DFE Department of Further Education

dff dilutent-free formulation

dfg digital function generator; diode function generator

DFG Department of Fish and Game

dfga distributed floating-agate amplifier

DFGJPC Daniel and Florence Guggenheim Jet Propulsion Center

DFH Danmarks Fiskeri og Havundersogelser

dfi definite; direct-flame impingement

DFI Deep Foundations Institute; Director(ate) of Food Investigation

DFIB Data Function Information Book

DFIC Dairy Foods Information Center

d/fing direction finding

DFISA Dairy and Food Industries Supply Association

dfitw dead flat in the water (becalmed ship)

dfiy definitely

D fl Dutch florins

DFL Daily Flight Log; Democrat Farmer Labor

DFL Deutsche Forschungsanstalt für Luft und Raumfahrt (German— German Air and Space Research Institute)

dfld defiled; deflated

DFLP Democratic Front for the Liberation of Palestine

DFLS Day Fighter Leaders' School

DFM Distinguished Flying Medal

dfml dust flux monitor instrument

DFMR Dazian Foundation for Medical Research

DFMS Domestic and Foreign Missionary Society

DFMSR Directorate of Flight and Missile Safety Research

dfn distance from nose

dfndt defendant

DFNWR Deer Flat National Wildlife Refuge (Idaho)

DFO Department of Fisheries and Oceans (Canadian); District Field Office(r)

d forg drop forging

dfp (DFP) diisopropyl phosphofluoridate

DFP Detroit Free Press

DFPA Douglas Fir Plywood Association; Daughters of Founders and Patriots of America

dfq day frequency

dfr decreasing failure rate; defrost(ing); digital flight recorder; dropped from rolls

Dfr Dounreay fast reactor

D fr Djibouti franc

DFRA Drop Forging Research Association

DFRC Dryden Flight Research Center (NASA)

dfr(d) defer(red)

DFRDBA Defence Forces Retirement and Death Benefits Authority (Australian)

dfrn differential

dfrs differs

dfs distance finding station

DFs Duty Frees (tobacco products)

DFS Dirección Federal de Seguridad (Spanish—Federal Security Directorate)—Mexico's famed *Federales,* the Feds or Federals

D.F.Sc. Doctor of Financial Science

DFSC Defense Fuel Supply Center

DFSM Defence Force Service Medal; Distinguished Fire Service Medal

dfsr diffuser

dft deaerating feed tank; defendant; discrete fourier transform; distributed function terminal; draft

DFT Diagnostic Function Test; Director of Flying Training

dfti distance from touchdown indicator

dftmn draftsman

dft(s) draft(s)

d-f tube double-flare tube

dfu data file utility; dead fetus in uterus; dummy flying unit

dfus diffuse

DFW Dallas-Fort Worth, Texas (airport); Director of Fortifications and Works

dfwes direct fire weapons effect simulator

dg dark ground; decigram(s); defensive guard; degenerate(d); deoxyglucose; diagnosis; diastolic gallup; digestive gland; diglyceride; disk grind; distogirgival; double glass; double groove; dry grassland; durable gum

dG differential of conductance (symbol)

d/g dangerous goods; decomposed granite; directional gyroscope; displacement gyroscope

DG Diego Garcia; Director General; Dragoon Guards

DG *Déclaration de Guerre* (French—Declaration of War); *Deutsche Grammophon*

D.G. *Dei Gratia* (Latin—By the Grace of God)

dga (DGA) diglycolamine

DGA Democratic Governors Association; Directors Guild of America

DGAA Distressed Gentlefolk's Aid Association

DGAMS Director General of Army Medical Services

DGAR Director-General Army Requirements

DGAS Double-Glazing Advisory Service

DGAVS Director-General Army Veterinary Services

DGB *Deutscher Gewerkschaftsbund* (German Federation of Trade Unions)

DG Bank *Deutsche Genossenschaftsbank* (German Cooperative Bank)

DGBAS Director General of Budget, Accounting and Statistics

dgbus digital ground bus

DGC Dangerous Goods Classification; Data General Corporation; Duty Group Captain

D.G.C. Diploma in Guidance and Counseling

DGCA Director General of Civil Aviation

DGCE Director General of Communications Equipment

dgd double glass doors

DGD Director Gunnery Division

DGDC Deputy Grand Director of Ceremonies

DGD & M Director General Dockyards and Maintenance

DGE Director-General of Equipment; Directorate General of Equipment; Division of Geothermal Energy (DoE)

DGEIS Draft Generic Environmental Impact Statement

DGFV Director-General of Fighting Vehicles

DGG *Deutsche Grammophon Gesellschaft* (German— German Gramophone Record Company)

dgi disseminated gonococcal infection

DGI Date Growers Institute; Director General of Information; Director General of Inspection; Directorate of General Intelligence

DGI *Directorio General de Inteligencia* (Spanish— Directorate General of Intelligence)—Cuban branch of the former Soviet KGB

DGIP Division of Global and Interregional Projects (UN)

DGL deglycyrrhizinated licorice

DGLA dihomo-gamma linolenic acid

Dgls Douglas

Dglsh Daglish

dgm decigram; destroyer guided missile

DGM Diploma in General Medicine; Director General of Manpower; Director(ate) of General Mobilization

DGMS Director General of Medical Services

DGMT Director General of Military Training

dgmw double-gimbal momentum wheel

DGMW Director General of Military Works

Dgn Dragoon(s)

dgnast (DGNAST) design assist

dgnl diagonal

Dgo Durango

DGO Diploma in Gynecology and Obstetrics; Director-General of Organization

DGP Director-General of Personnel; Director General of Production

DGPH Data/Graphics Processor Hybrid

DGPS Director General of Personnel Services

dgpwc digital ground proximity warning computer

dgr danger(ous)(ly); door gunner

d Gr *der Grosse* (German— the Great)

DGR Director of Graves Registration

DGR *Dirección General de Radiocomunicaciones* (Spanish—General Administration of Radio Communications)— Bolivian broadcasting control

DGRR *Deutsche Gesellschaft für Raketentechnik und Raumfahrt* (German—Society for Rocket Technique and Space Flight)

dgs double green silk

DGs Directorates General

DGS Degaussing System; Diploma in General Surgery; Director General of Ships; Director-General of Signals

DGSC Defense General Supply Center

DGSE *Directorat Général Sécurité External* (French— General Directorate of External Security); *Directorio General de Seguridad del Stado* (Spanish—Directorate General of State Security) Nicaragua's secret police

DGSRD Director(ate) General of Scientific Research and Development

DGSS Director General Secret Service

DGST Director-General of Supplies and Transport

Dgt Dumaguette

DGT Director General of Training

DGT *Dirección General de Turismo* (Spanish—Administration of Tourism)

DGTA Director-General Territorial Army; Dry Goods Trade Association

DGTM Director-General of Transport and Movements

DGTTT *Dirección General de Transporte y Transito Terrestre* (Spanish—Ministry of Communications)

DGW Director-General of Works; Director General of Weapons

dgz designated ground zero

dh dead heat; deadhead; decision height; dehydrogenase (DH); delayed hypersensitivity; designated hitter; differential of height (symbol); double header (baseball); double hung; drill hole

dH differential of magnetic field intensity (symbol)

dh (DH) designated hitter

d & h daughter and heiress; dressed and headed

dh *das heisst* (German—that is to say)

Dh Moroccan dirham(s)

DH Deccan Horse; Declaration of Homestead; De Havilland (aircraft); Department of Health

D.H. Doctor of Humanities

D & H Delaware & Hudson (railroad)

D of H Degree of Honor; Degree of Honour

dha dihydroxyacetone; docosahexaenoic acid

dha *dicha* (Spanish—good luck, happiness)

DHA Dhahran, Saudi Arabia (airport); Drug Houses of Australia

D & HAA Dock and Harbour Authorities Association

DHAC De Havilland Aircraft of Canada Limited

D-handle D-shaped handle

dha (DHAP) dihydroxyacetone phosphate

dhard dehaired (skins)

dhas (DHAS) dehydroepiandrosterone sulfate

dhc (DHC) dihydrochatcone

DHC De Havilland Canada; Department of Housing and Construction; Detroit House of Correction

DHC-3 Canadian version of De Havilland Otter utility aircraft

DHC-6 Canadian De Havilland Twin Otter transport aircraft

DH Canada De Havilland Aircraft of Canada Limited

dhcv down-hole control valve

dhd distillate hydrosulfurization

dhdd digital high-definition display

dh di drophammer die

dhe data-handling equipment

dhea (DHEA) dehydroepiandrosterone

dheas (DHEAS) dehydroepiandrosterone sulfate

DHEW Department of Health, Education, and Welfare

DHF Dag Hammarskjöld Foundation

dhfr dihydrofolate reductase

D. Hg. Doctor of Hygiene

DHHS Department of Health and Human Services

DHI Dental Health International; Door and Hardware Institute

DHI *Deutsches Hydrographisches Institut* (German— German Hydrographic Institute)

dhia dehydro-isoandrosterol (DHIA)

DHIA Dairy Herd Improvement Association

Dhic dihydro-isocodeine (DHIC)

DHIY Denbighshire Hussars Imperial Yeomanry

dhl distemper, hepatitis, leptospirosis (vaccine)

DHL Dag Hammarskjöld Library (UN in NYC)

D.H.L. Doctor of Hebrew Letters; Doctor of Hebrew Literature

Dhllp direct high-level language processor

dhl-p distemper, hepatitis, leptospirosis-parainfluenza (vaccine)

dhlpp distemper, hepatitis, leptospirosis, parainfluenza, parvovirus (vaccine)

dhlw defense high-level waste(s)

DHM Detroit Historical Museum

dhma dehydroxymandelic acid (DHMA)

DHMPGTS Department of Her (His) Majesty's Procurator General and Treasury Solicitor

dhn dynamic hardness number

DHN Department of Hospital Nursing

dho *dicho* (Spanish—said)

DHO deuterium hydrogen oxide; District Health Office(r); Downhill Only (ski club)

d'Holbach Paul Henry Thiry d'Holbach

dhon dishonor(able)

D.Hor. Doctor of Horticulture

dhp developed horsepower

DHP *Diplôme en Hygiène Publique* (French—Diploma in Public Health)

dhpg dehydroxyphenylglycol (DHPG)

dhq mean diurnal high water inequality

DHQ District Headquarters; Division Headquarters; Divisional Headquarters

dhr delayed hypersensitivity reaction(s)

DHR Division of Housing Research

Dhr *de heer* (Dutch—Mr)

DHRC Dufferin Haldimand Rifles of Cananda

dhrr *de herrer* (Dano-Norwegian—the gentlemen)

dhs destroyer helicopter system; dihydrostreptomycin; dry heat sterilization.

dh's deadheads (freeloaders who never buy a ticket or pay their own way)

DHS Department of Health Services; Department of Hypertension and Stress; Detroit High School; Diploma in Horticultural Science; District High School; Dublin High School

D.H.S. Doctor of Health Science(s)

DHSA Diploma–Health Service Administration

DHSC Department of Health and Social Security

Dhsm dihydrostreptomycin (DHSM)

DHSS Department of Health and Social Security

dht distillate hydrotreating

dht (DHT) dihydrotestosterone

dhtv downhole television

DHUD Department of Housing and Urban Development

D.Hum.L. Doctor of Humane Letters

dhw domestic hot water; double-hung windows

DHX Dependable Hawaiian Express

D. Hy. Doctor of Hygiene

di daily inspection; de-ice; diameter; diametral; dielectric isolation; diplomatic immunity; direction indicator; display indicators; document identifier; double imperial

dI differential of current (symbol)

di (DI) diabetes insipídus; double indemnity; inversion of the duration series

di (Latin prefix—two)—dipole antenna

d i *das ist* (German—that is)

Di Diana; Diane; didymium; Dinorah

DI Defense Intelligence; Denizyollari Isletmesi (Turkish Maritime Lines); Department of the Interior; Departmental Instruction(s); Detective Inspector; Deterioration Index (annual rate for the deterioration of a mailing list to the point it ceases to be deliverable); Diffusion Index; direct ignition; Director of Infantry; Director of Intelligence; Distinctive Insignia; District Inspector; Divergence Indicator; Division Instruction; Divisional Inspector; Drill Instructor (USMC); Dvorak International

D-I Dai-Ichi

D of I Daughters of Isabella; Declaration of Independence; Department of Insurance; Department of the Interior; Division of Intelligence

DI-5 Defense Intelligence (British agency)

D.I. *Dies Irae* (Latin—Day of Wrath)

dia date of initial appointment; diagram; diameter; diaphone; diathermy; document interchange architecture; due in assets

dia (Greek prefix—passing through or through)—diabetes, diagnosis, dialysis, diaphragm

DIA Defense Intelligence Agency; Defense Investigative Agency; Department of Institutions and Agencies (NJ); Design and Industries Association; Designated International Accounts; Detroit Institute of the Arts; Document Interchange Architecture; Dulles International Airport (Washington, D.C.); Dying in Action

diab diabetic

DIAB Defense Internal Audit Board

diac di-iodothyroacetic acid (DIAC)

DIAC Defense Industry Advisory Council; Dixie Intercollegiate Athletic Conference

diacrit diacritic(al)(ly)

di ad die adapter

DIAFS Distributed Internet Applications for Financial Services

diag diagnose; diagnosis; diagnostic; diagnostician; diagonal; diagram

dial. dialect; dialectical; dialectician; dialectics

DIAL Disc Interrogation and Loading (system)

dial-a-mation dial-a-cremation

dial-a-porn dial-pornography (erotic telephone service)

dialec dialectic(al)(1y); dialectician(s); dialectics; dialectologist(s); dialectological(ly); dialectology

dialgol dialect of algol (q.v.)

diam diameter

DIAMANG *Companhia de Diamantes de Angola* (Portuguese—Angolan Diamond Company)

diamat dialectical materialism

diamond carbon

dian digital analog

DIAND Department of Indian Affairs and Northern Development (Canada)

diane digital-integrated attack and navigation equipment (DIANE)

DIANE Direct Information Access Network for Europe

diap. *diapason* (Greek—consonant harmony, octave)

diaph diaphragm

diaphor diaphoresis

DIAR Defense Intelligence Agency Regulation

dias defense-integrated automatic switch

DIAS Drug Information and Assistance Service; Dublin Institute for Advanced Studies; Dynamic Inventory Analysis System

diast diastolic

diat diathermy

DIAT Dundee Institute of Art and Technology

diath diathermy

dib dead in bed (not physically but sexually); diameter inside bark

DIB Department of Information and Broadcasting; Department Information Bulletin; Diagnostic Interview for Borderlines

DIB *Dictionary of International Biography*

DIBA Domestic and Internal Business Administration; Dominion Investment and Banking Association

dibah (DIBAH) disobutylaluminum hydride

dibas dibasic

dibb double-income baby boomers

DIBR Dartnell Institute of Business Research

dic data insertion converter; data item category; defense identification code; dependency and indemnity compensation; dictionary; digital integrated circuit; digital integrating computer; disseminated intravascular coagulopathy; disseminated intravascular coagulation; drunk in charge; inverted cancrizans of the duration series

d & ic dependency and indemnity compensation

dic *dicembre* (Italian—December); *diciembre* (Spanish—December)

DiC diesel cargo vessel

DIC Dai Nippon Ink and Chemicals; Diplomate of the Imperial College (London); Direct Importing Company (New Zealand); Diving Information Center (USN)

DICAP Direct-Current Circuit Analysis Program

DICASS Directional Command Active Sonobuoy System

dicautom automatic dictionary look-up

DICB Demolition Industry Conciliation Board

dice digital intercontinental-conversion equipment; digital-interface countermeasure equipment; direct-installation coaxial equipment

DICEF Digital Communications Experimental Facility (USAF)

DIChem. Diploma of Industrial Chemistry

dichlorvos dimethyldichlorovinyl phosphate (insecticide)

dicht *dichterlijk* (Dutchpoetic)

dick detective

Dick Richard

dickel dime and nickel

Dickie Dickman; Richard

Dickon Richard(son)

dick(s) detective(s)

Dicky Richard; Tricky Dicky

DICNAVAB *Dictionary of Naval Abbreviations*

dicorap directionally controlled rocket assisted projectile

dicot(s) dicotyledon(s)

dict dictated; dictation; diction; dictionary

dicta dictaphone

DICTA Diploma of the Imperial College of Tropical Agriculture

Dict Amer Slang Dictionary of American Slang

dictsort dictionary sorter

did data item description; dead of intercurrent disease; didactic; direct in dialing

did (DID) drum information display

Did Didot

DID Department of Industrial Development; Detail Issue Depot; Director of Intelligence Division; Division of Institutional Development; Drainage and Irrigation Department

DID Daily Intelligence Digest

dida differential in-depth analysis

didac didactic(al)(ly); didacticism; didactics

didad digital data display

DIDAMES Distributed Industrial Design and Manufacturing of Electronic Subassemblies

DIDAS Dynamic Instrumentation Data Automobile (Automotive) System

DIDC Depository Institutions Deregulation Committee

didd diddle

dident distortion identity

DIDMCA Depository Institutions Deregulation and Monetary Control Act

didn't did not

di/do data input/data output

DIDS Digital Information Display System

die died in emergency room (DIE); diesel (engine devised by Dr Rudolf Diesel)

DIE Diploma in Industrial Engineering; Diploma of the Institute of Engineering; Division of International Education

DIEA Dictionary of Industrial Engineering Abbreviations

dieb. alt. diebus alternus (Latin—on alternate days)

dieb. secund. diebus secundis (Latin—every second day)

dieb. tert. diebus tertius (Latin—every third day)

Diego (Mexican-American truncation–San Diego) San Diego, California

die Kö Königsallee (main street of Dusseldorf)

diel dielectrics; diesel electric

di el diesel electric

DIEME Director(ate) of Inspection of Electrical (Electronic) and Mechanical Equipment

DIEPO Dieterich-Post

diet. dietary; dietetic(s); dietician

di. et fi. dilacto et fideli (Latin—to his beloved and faithful)

DIEX Dirección de Identifcación y Extranjería (Spanish—Directorate of Identification and Immigration)

dif data interchange format; differ; difference; differential; differentiation-inducing factor; diffuse(er)(s)

dif (DIF) discriminant function

DIF Defense Industrial Fund; Descriptive Item File; Discriminant Function (auditing system for income-tax returns); District Inspector(ate) of Fisheries

DIF Desarrollo Integral de la Familia (Spanish—Integral Family Development)

difamps differential amplifiers

difar direction finding and ranging; directional frequency analysis and recording

difce difference

diff difference; differential; differential white blood count

diff calc differential calculus

diff diag differential diagnosis

diffr diffraction

diffu diffusion

DiFr diesel fruit vessel

dift different

diftl differential

dig. digamist; digamy; digest; digestion; digestive

dig. digeratur (Latin—let it be digested)

DIG Deputy Inspector General; Discussion Interest Groups

DIGA Dynamics International Gardening Association

digas digastrie

DIGEPOL Dirección General de Policías (Spanish—General Directorate of Police)–Venezuela

digger(s) gold digger(s); gold miner(s)

digi digital

Digi Digiform (business forms typesetter)

DIGI DSC Communications Corporation

digicom digital communications (system)

DIGIRALT digital radar altimetry

digital IC digital integrated circuit

digres digression(al)(ly); digressionary; digressive(ly); digressiveness

digrm digit/record mark(ing)

dig r-o digital readout

digs. archeological excavation; diggings

DIGs Development Import Grants

DIGS Delta Inertial Guidance System

di-H hydrogen

DIH Diploma of Industrial Health; Division of Indian Health

diic dielectric-isolated integrated circuit

DIHJHU Department of International Health–Johns Hopkins University

Dij Dijon

di ji drill rig

dik drug-identification kit

DIKB Dai-Ichi Kangyo Bank

diks double income kids

dil dilute; dissolve

dil. dilue (Latin—dilute); *dilutus* (Latin—diluted)

DIL Deliverable Items List; Director of International Logistics; Division of Insured Loans

d-i-l daughter-in-law

dilat dilatation; dilate; dilation (ed)

dild diluted

dil'd diluted

dilet dilettante

dilligaf do I look like I give a fuck?

diligas do I look like I give a shit?

diln dilution

diluc. diluculo (Latin—at daybreak)

dilut.   *dilutus* (Latin—dilute)

dim   defense information memo; description, installation, and maintenance; digital dimmer memory; dimension; dimensional; dimension(al)(ly); diminutive

dim   *dimanche* (French—Sunday); *dimidius* (Latin—one half); *diminuendo* (Italian—diminishing gradually)

DIM   Dialogue Inter-Monasteries; Diploma in Industrial Management; Dmitry Ivanovich Mendeleyev

DIMA   Detroit Institute of Musical Art

DIMAN   Distributed International Manufacturing

Dimashq   (Arabic—Damascus)—capital of Syria

dimate   depot-installed maintenance automatic test equipment

DIMD   *Dorland's Illustrated Medical Dictionary*

DIMDI   *Deutsches Institut für Medizinische Dokumentation und Information* (German—German Institute for Medical Documentation and Information)

dime   dual independent map encoding

DIME   Development of Integrated Monetary Electronics; Division of International Medical Education (Assn Amer Med Colleges)

DIMES   Defense Integrated Management Engineering Systems

dimin   diminish; diminution; diminutive

DIMIS   Depot Installation Management Information System (USA)

dimn   dimension

di'mon'   diamond

dimorph   dimorphous

DIMPE   Distributed Integrated Multimedia Publishing Environment

dimple   deuterium-moderated pile low energy

DIMS   Data Information and Manufacturing System; Director International Military Staff Memo (NATO); Disorders of Initiating and Maintaining Sleep

din   dinar(s); dining room; dinner; do it now

din   *dinar* (Yugoslavian monetary unit)

Din   *Dinsdag* (Dutch—Tuesday)

DIN   Data Identification Number

DIN   *Deutsche Industrie Norm* (German Industry Standard)—film rating sometimes written *din* and said to mean *das ist nonn* (this is standard); *Deutsches Institut für Normung* (German Standards Institute)

Dina   *Dinamarca* (Portuguese or Spanish—Denmark)

DINA   *Dirección de Inteligencia Nacional* (Spanish—Directorate of National Intelligence)—Chilean secret police

dinar   monetary unit of Algeria, Bahrain, Iraq, Jordan, Kuwait, Libya, South Yemen, Tunisia, Yugoslavia

diner   dining car

Ding   Jay Norwood Darling

D.Ing.   *Doctor Ingeniatiae* (Latin—Doctor of Engineering)

dinin'   dining

dinks   double incomes, no kids

dino(s)   dinosaur(s)

Dinosaur   Dinosaur National Monument, Colorado and Utah

DINP   Dunk Island National Park (Queensland)

din rm   dining room

DINS   Dormant Inertial Navigation System

dio   diocese; diode; direct input-output

DIO   Director of Industrial Operations; Director(ate) of Intelligence Operations; District Intelligence Office(r); Duty Intelligence Officer

diob   digital input-output buffer

dioc   dioceasan; diocese

diode   digital input-output display equipment

Dion   Dionisio

diop   di-iso-octyl phthalate (plasticizer); diopter; dioptrics

dior   diorama

dios   diver lockout submersible

DIOS   Distributed Input-Output System

diox   dioxygen

dip   desquarnative interstitial pneumonia; digital inline pins; dipeptide; diphtheria; diphthong; diplex; diplococcus; diploma; diplomacy; diplomat; dipsomania(c); dissemination and improvement of practice; dual inline package; (slang for pickpocket); stupid person

DIP   Document Improvement Program (DoD)

DIPA   Diploma of the Institute of Park Administration

DipAD   Diploma in Art and Design

Dip Agr   Diploma in Agriculture

Dip A Ling   Diploma in Applied Linguistics

Dip AM   Diploma in Applied Mechanics

Dip Amer Bd P & N   Diplomate of the American Board of Psychiatry and Neurology

Dip AMS   Diploma in Ayurvedic Medicine and Surgery

Dip Anth   Diploma in Anthropology

Dip App Sci   Diploma in Applied Science

Dip Arch   Diploma in Architecture

Dip Ars   Diploma in Arts

Dip Bac   Diploma in Bacteriology

Dip BMS   Diploma in Basic Medical Sciences

Dip CAM   Diploma in Communications, Advertising, and Marketing

Dip Card   Diploma in Cardiology

Dip Com   Diploma in Commerce

dipcrit   diplomatic critic(ism)

Dip DP   Diploma in Drawing and Painting

Dip DS   Diploma in Dental Surgery

DIPEC   Defense Industrial Plant Equipment Center

Dip Eco   Diploma in Economics

Dip Ed   Diploma in Education

Dip Eng   Diploma in Engineering

Dip FA   Diploma in Fine Arts

Dip For   Diploma in Forestry

Dip G & O   Diploma in Gynaecology and Obstetrics

Dip GT Diploma in Glass Technology

diph diphtheria

Dip HA Diploma in Hospital Administration

Dip HE Diploma in Higher Education; Diploma in Highway Engineering

diph tet diphtheria tetanus

diphth diphthong

diph tox diphtheria toxin

diph tox ap diphtheria toxin alum precipitated

Dip Hus Diploma in Husbandry

dipj distal interphalangeal joint

Dip J Diploma in Journalism

dipl diplomacy; diplomat; diplomatic

Dipl Diplom (German—Diploma)

Dip L Diploma in Languages

Dip Lib Diploma in Librarianship

Dip Lib Sci Diploma in Library Science

diplo diploma; diplomacy; diplomat; diplomatic; diplomatics; diplomatism; diplomatist

diplo diplomatico (Spanish—diplomat); *diplotienda* (Spanish—diplomat store)—special store catering only to diplomats; (Greek *diploos* — two-fold)—diploid, diplomacy, diplomat(ic)

Dip ME Diploma in Mechanical Engineering

Dip MFOS Diploma in Maxial, Facial, and Oral Surgery

Dip Mgmnt Diploma of Management

Dip Micro Diploma in Microbiology

Dip Mus Edu Diploma in Musical Education

Dip NA & AC Diploma in Numerical Analysis and Automatic Computing

Dip NS Edu Diploma in Nursery School Education

Dip NZLS Diploma of the New Zealand Library Service

Dip OL Diploma in Oriental Learning

Dip Phar Diploma in Pharmacology

Dip Phys Edu Diploma in Physical Education

Dip P & OT Diploma in Physical and Occupational Therapy

Dip Pub Adm Diploma in Public Administration

Dip RADA Diploma of the Royal Academy of Dramatic Art

Dip RSAM Diploma of the Royal Scottish Academy of Music

diprt discharge printed

dips dipeptides; diphtheria patients; diphthongs; diplexes; diplomas; diplomats; dipsomaniacs; dynamic isotope power system

DIPS Dendenkosha's Information Processing System; Development Information Processing System

dipsey deep-sea lead (line for measuring depths)

Dip SMS Diploma in School Management Studies

dipso dipsomania(c); drunkard

Dip Sp Ed Diploma in Special Education

Dip SS Diploma in Social Studies

Dip SW Diploma in Social Work

Dip T Teachers Diploma

Dip T & CP Diploma in Town and Country Planning

Dip Tec Diploma in Technology

Dip TEFL Diploma in Teaching English as a Foreign Language

dipth diphthong

Dip The The Diploma in Theology

Dip TP Diploma in Town Planning

dipu diputado (Spanish—deputy)

Dip VFM Diploma in Valuation and Farm Management

dIQ deviation IQ

dir direct; direction; director; directory

dir (DIR) digital instrumentation radar

dir. directione (Latin—directions); *direxit* (Latin—directed by)

Dir Director(ate); Dirham(s)—Moroccan money

DIR Detailed Inspection Report; Diesel Inspector's Report

dir conn direct-connect

dir coup directional coupler

dircty directly

direct. directory

D.Ir.Eng. Doctor of Irrigation Engineering

Dir Gen Director General

Dir Gen Direttore Generate (Italian—General Manager)

dirham monetary unit of Morocco and United Arab Emirates

Dirk Derek; Everett McKinley Dirksen

dir max directional maximum

dir min directional minimum

diron direction

dir. prop. directione propria (Latin—with proper directions)

DIRT Department of Industrial Relations and Technology (Australian)

dis daily issue store; delivered into store; disability; disable(d); discharge; disciple; discipline; disclosure; disconnect(ed); discontinue(d); discount(ed); disease(d); disrespect; distal; distance; distant; distribute(d); distribution

dis (Latin prefix—apart, away from)—disable, disarticulate

Dis Disney (Walt Disney); Disneyland; Disraeli (Benjamin Disraeli); Pluto

DIs Department(al) Instructions

DIS Dairy Industry Society; Defense Intelligence School; Defense Intelligence Service; Defense Intellgence Staff; Defense Investigative Service; Department of Industrial Services; Diagnostic Interview Schedule; Disney Productions (stock exchange symbol); Distribution Advisory Service; Distribution Information System; Dow Industrial Service; Drug Instruction Service; Ductile Iron Society

disab disable; disabled

disabl disability

disac digital simulator and computer

DISAM Defense Institute of Security Assistance Management

disap disapprove

Disap Disappointment (cape in Washington state, lake in western Australia, islands in French Polynesia)

disas disaster

disassy disassembly

dish disburse; disbursement

disbmt disbursement

disbt disbursement

disc discus throw

disc (DISC) direct-injection stratified charge (automobile engine)

disc dimension of schooling questionnaire; discography; disconnect; discontinue; discophile; discount; discover(ed)

DISC Defense Industrial Supply Center; Distribution Stock Control System; Domestic International Sales Corporation

discd discounted

discg discounting

disch discharge; discharging

dischd discharged

dischg discharging

discip disclipline

disc jock(s) disc jockey(s)

disco disc jockey; discotheque; discotheque music

DISCO Defense Industrial Security Clearance Office

discol discolored

discom digital selective communication(s)

discomb discombobulate(d); discombobulation

discon disconnect; disorderly conduct

DISCON Defense-Integrated Secure Communications Network (Australian)

discontd discontinued

discos discotheques

discr discriminator

discron discretion

discrtn discretion

DISCs Domestic International Sales Corporations

disct discount

discum discumbobulate(d); discumgalligumfricate(d)

DISCUS Distilled Spirits Council of the United States

discus (see **DSSCS**)

Discuss Faraday Soc Discussions of the Faraday Society

DISD Data and Information System Division

DISE Digital Systems Education

disemb disembark

disg disagreeable

dishon dishonest; dishonesty; dishonorable; dishonorably

disi door insulating system index

DISI Dairy Industries Society International

disid disposable seismic intrusion detector

disin disinfectant; disinfection

dis int discrete integrator

DISIP *Dirección de Seguridad e Inteligencia Policiales* (Spanish—Directorate of Police Security and Intelligence)

disk *diskonto* (Norwegian—discount)

Disk Watcher trademark of RG Software Systems

disloc dislocation

dism dismiss; dismissal

Dismals Dismal Gardens near Phil Campbell, Alabama

dismd dismissed

dis/min disintegrations per minute

DISNET Domain-Independent Information Service Network

diso die shoe

disod disodium

disord disorder

DISOSS Distributed Office Support System

disp dispatch; dispensary; dispensatory; dispenser; displacement; display; disposal; disposition

disp. *dispensa* (Latin—dispense)

dispen dispensatories; dispensatory

displ displacement

dispr dispatcher

disq disqualified

disr disrated

diss disassembly; dissent; dissenter; dissertation

DISS Director(ate) of Information Systems and Settlement (stock exchange)

dissd dissolved

dissec dissection

dis/sec disintegrations per second

dissed disrespectful

dissern disseminate; disseminated

dissert dissertation(s)

dissim dissimilated; dissimilation

disson dissonance; dissonant(ly)

disspla display integrated software system and plotting language

dissyl dissyllable

dist distance; distant; distribute; distribution; distributor; district

dist. *distancia* (Spanish—distance); *distilla* (Latin—distill)

Dist District

distab disestablish(ment)(tarian)(ism)

Dist Ad District Administrator

distads administrative districts

distar direct instruction

DISTAR Direct Instruction System for Teaching Arithmetic and Reading

Dist Atty District Attorney

distb distillable

distbtr distributor

Dist Ct District Court

distil distillation; distilled; distilling

Dist J District Judge

distn distillation

distng distinguish; distinguishing

Dis TP Distinction in Town Planning

distr distribute; distribution

DISTRAMS Digital Space Trajectory Measurement System

distran diagnostic fortran

distrib distribution; distributive; distributor

DISTRIPRESS *Fédération International des Distributeurs de Presse* (French—International Federation of Wholesale Book, Newspaper, and Periodical Distributors)

Dists Districts

DISUM Daily Intelligence Summary (USAF)

disy disyllabic

dit domestic independent tour; dual input transponder

dit (DIT) diodotyrosine

DIT Defining Issues Test; Detroit Institute of Technology; Drexel Institute of Technology; Durham Institute of Technology

DIT *Deutscher Investment Trust*

DiTa diesel tanker vessel

ditar digital telemetry analog recording

DITC Disability Insurance Training Council

Ditch The Ditch, 3100-mile long (4989-kilometer-long) Intracoastal Waterway along the Atlantic coast from Boston to Key West, as well as from the St Marks River in Florida to Brownsville, Texas

Dithy dithyramb(ic)(al)(1y); dithyrambs

ditmco data information test material checkout

DITN Diabetes in the News

Dito Ernesto

diu data interface unit; digital interface unit

DIU Diversion Investigation Unit

div data in voice; digits in voice; divergence; diverse; divide; divided; dividend; divinity; divisibility; division; divisor; divorce; divorced

div divisi (Italian—divided)

Div Divide (postal abbreviation); Divine; Divinity; Division

DIV Derived Investment Value

divab digital input/voice answer back

DIVAD Divisional Air Defense

Div Arty Division Artillery

divd dividend

divde dividende (French—dividend)

Div E Division Engineer

divear diving instrumentation vehicle for environmental and acoustic research

div. en p. aeq. divide in partes aequales (Latin—divide into equal parts)

divi(s) dividend(s)

divn division

divnl divisional

div. inparaeq. dividatur in partes aequales (Latin—divide into equal parts)

Div Jum Divehi Jumhuriyya (Divehi—Republic of the Maldives)

divs digital video express; dividends

divvy divide, dividend

diw dead in the water

DIW Deutsches Institut für Wirtschaftforschung (German—German Institute for Economic Research)

Dix Dixie; Fort Dix, New Jersey

diy do it yourself

DIY Derbyshire Imperial Yeomanry

diz dizionario (Italian—dictionary)

dizz dizziness

dj diffused junction; disc jockey; dust jacket

d J der Jüngere (German—junior); dieses Jahres (German—of this year)

Dj Djawa (Indonesian—Java); Djebel (Arabic—mount, mountain)

DJ David Jones (Australian department store chain); Department of Justice; District Judge; Divorce Judge; Djibouti (Internet code); Don Jail (Toronto); Dow Jones Industrial Average

D of J Department of Justice; Dominion of Jamaica

DJ Deutsches Jungvolk (German—East German Youth); Divehi Jumhuriyya (Divehi Arabic—Republic of Maldives)—Maldive Islands

D.J. Doctor Juris (Latin—Doctor of Law)

Dja Djakarta

DJA Disabled Journalists of America

DJAD Department of Justice Antitrust Division

DJAG Deputy Judge Advocate General

DJCD Department of Justice Civil Division; Department of Justice Criminal Division

DJCP Division of Justice and Crime Prevention (Virginia)

DJCRD Department of Justice Civil Rights Division

djd degenerative joint disease

dJf divorced Jewish female

Dji Djibouti (whose capital is Djibouti)

DJI Dow Jones Industrials

DJIA Dow Jones Industrial Average

Djib Djibouti (formerly Afars and Issas Territory also known as French Somaliland)

Djibouti Republic of Djibouti (formerly French Somaliland)

Djinn Sud-Aviation two-seat helicopter built in France

DJJ Department of Juvenile Justice

Djkta Djakarta (Batavia), Java

Djl Djalan (Malay—road or street)

DJLNRD Department of Justice Land and Natural Resources Division

dJm divorced Jewish male

DJNR Dow Jones News Retrieval

DJO Den Jyske Opera (Danish National Opera)

Djokja Djokjakarta, Java, Indonesia

D. Journ. Doctor of Journalism

dj's (DJs) disc jockeys

djs desperate job seeker

DJs Department of Justice investigators

D.J.S. Doctor of Juridical Science

DJTA Dow Jones Transportation Average

DJTD Department of Justice Tax Division

DJUA Dow Jones Utility Average

D.Jur. Doctor of Jurisprudence

DJWWB Dowodztwo Jednostki w Wielkiej Brytanii (Polish—Units Command in Great Britain)

dk dark; decay; deck; didn't know; diseased kidney(s); dock; dog kidney; don't know; drop kick; duck; duct keel; dusky

d & k dining and kitchen

DK Danny Kaye; Denmark (Internet code)

D & K Dalhoff and King

DK Danmark (Danish—Denmark)

dka diabetic ketoacidosis

DKAA Donna Karan for Administrative Assistants (fashion trying too hard)

DKB Dai-ichi Kangyo Bank

DKB Det Kongelige Bibliotek (Danish—The Royal Library)—Copenhagen

DKC De Kalb College

dk di dinking die

dkftcol dark fast color

dkg decking; dekagram(s)

dkga (DKGA) diketogulonic acid

dkgreol dark-ground color

dk hse deck house

DKI *Det Kriminalistiriske Institute* (Danish—The Criminalistic Institute)—Copenhagen

DKK *Danmark Kroner* (Danish crowns)

dkl dekaliter

dkm dekameter

dkm² square dekameter

dkm³ cubic dekameter

dkp deck passenger(s); diketopiperazene

DKP Democratic Korea Party

DKP *Danmarks Kommunistiske Parti* (Danish Communist Party); *Deutsche Kommunistische Partei* (German Communist Party)

Dkr Dakar

DKr Danish krone(r)

DKR Dakar, Senegal (aiport)

DKs dekastere

DKS Deputy Keeper of the Signet; Direct Keying System

dkt docket

DKT dipotassium tartrate

DKTC Door-Kewaunee Teachers College

DKV *Deutscher Kanu Verband* (German—German Canoe Association)

DKW *Dampf Kraft Wagen* (German—steam powered vehicle); *Das Kleine Wunder* (German—The Little Wonder—automobile); *Deutsche Kraftfahrt Werks* (German—German Power-drive Works)

dkyd dockyard

dl data link; day letter; dead load; deadlight; deciliter; delay line; demand loan; designer links; difference limen (threshold); disabled list; dog license; dollar; double acetate; drawing list; driver's license

dL differential of inductance (symbol)

d-l dextro-levo

d/l data link; demand loan

Dl Daniel

DL Danger List; Delta Air Lines (2-letter symbol); Department of Labor; difference of latitude; Djakarta Lloyd; Drawing List; Drill Leader; Drury Lane (London theater); frigate (naval symbol)

D-L Deputy-Lieutenant

D/L De Luxe

D of L Department of Labor; Department of Labour; Department of Law; Drill Leader; Duchy (Duke) of Lancaster; Duchy (Duke) of Lorraine; Duchy (Duke) of Luneburg

DL *Danske Lov* (Danish Law)

dla distolabial

Dla Douala

DLA Decorative Lighting Association; Defense Land Agency; Defense Logistics Agency; District Licensing Authority; Divisional Land Agent (UK); Documentation, Libraries, and Archives Director(ate)

dlab disc label

dlai distolabioincisal

D.Lang. Doctor of Languages

D-L antibody Donath-Landsteiner antibody

DLAS Defence of Literature and the Arts Society

d lat difference in latitude

DLAT Defense Language Aptitude Test (USA)

dlb's dead-letter boxes

dlc data link control; digital loop carrier; direct lift control; down left center

DLC Democratic Leadership Council; Disaster Loan Corporation; Duquesne Light Company

DLCO Deck Landing Control Officer; Desert Locust Control Office

DLCO-EA Desert Locust Control Organization-East Africa

dld deadline date; delivered

dle data link escape; disseminated lupus erythematosus (DLE)

DLE Department of Law Enforcement

dlea double leg elbow amplifier

D.L.E.S. Doctor of Letters in Economic Studies

dlet delete

dletd deleted

dletg deleting

D Lett *Docteur en Lettres* (French—Doctor of Letters)

DLF Development Loan Fund(ing); Disabled Living Foundation

DLG David Lloyd George; guided-missile frigate (naval symbol)

DLG *Deutsche Landwirtschafts Gesellschaft* (German—German Agricultural Society)

DLGA Decorative Lighting Guild of America

DLGCD Department of Local Government and Community Development

DLGN nuclear-powered guided missile frigate (naval symbol)

DLH *Deutsche Lufthansa* (German airline)

DLI Defense Language Institute; Department of Labour and Industry (Australian); Durham Light Infantry

DLIA Dental Laboratories Institute of America

D-library duplicating library

dlimp descriptive language for implementing macroprocssors

dlir depot-level inspection and repair

DLIS Desert Locust Information Service

D. Litt. *Doctor Litterarum* (Latin—Doctor of Letters, Doctor of Literature)

DLJ Donaldson, Lufkin & Jenrette Securities Corporation

dll dial long line; dynamic link library

DLL *Deutsche Levante-Linie* (Levant Line); Donaldson Line Limited

dllf design limit load factor

dlli dulcitol lysine lactose iron (DLLI)

dlM *des laufenden Monats* (German—this month)

DLM Daily List of Mails; Depot-Level Maintenance

DLMA Decorative Lighting Manufacturers Association; Downtown Lower Manhattan Association

dlmmjvs *domingo, lunes, martes, miércoles, jueves, viernes, sábado* (Spanish—Sunday, Monday, Tuesday, Wednesday, Thursday, Friday, Saturday)

dln document locator number

DLNS Department of Labour and National Service (Australian)

DLNWR Des Lacs National Wildlife Refuge (North Dakota)

dlo difference in longitude; dispatch loading only; distolinguo-occlusal

D'Lo The Lord (town in Mississippi)

DLO Dead Letter Office; Difference of Longitude; District Legal Office(r); Divisional Legal Officer

D.L.O. Diploma in Laryngology and Otology

DLOC Division Logistical Operation Center

DLOCA Department of Law Office Consumer Affairs

d lock dial-lock

d long difference in longitude

D-love deficiency love (exploitative and possessive love of another person)

DLOY Duke of Lancaster's Own Yeomanry

dlp date of last payment; defense of life and property; digital light processor; distolinguopulpal; double-large post; mean diurnal low- water inequality

DLP Democratic Labour Party (Barbados); Director of Laboratory Programs (USN)

DLPA Decorative Laminate Products Association DL-phenylalanine

DLPS Department of Law and Public Safety (New Jersey)

dlq deliquescent; mean diurnal low water inequality

dir dealers; discharge, land, and reload; discharged, landed, and reshipped; dollar; double lift restow; double-lens reflex (camera)

d-l-r discharge-load-reposition (containers)

DLR District Land Registrar; Driving Licenses Regulations

DLR *Distrito de la Luz Roja* (Spanish—Red Light District)

dlra door lock rotary actuator

DLRA Divorce Law Reform Association

DLRO District Labor Relations Office(r)

dlrs dollars

DLRs *Dominion Law Reports* (Canadian)

dis debt liquidation schedule; dollars

dls *dólares* (Spanish—dollars)

DLs Defence Lists

DLS Debt Liquidation Schedule; Defense Legal Services (agency); Director of Legal Services; District Law Society

D.L.S. Doctor of Library Science; Doctor of Library Service

DLSA Directorate of Land Services Ammunition

D.L.Se. Doctor of Library Science

DLSC Defense Logistics Service Center

DLSEF Division of Library Services and Educational Facilities (U.S. Office of education)

dls/shr dollars per share

dlt deck landing training; dry long tons

dlt (DLT) data-loop transceiver (data processing)

dlt *dans le texte* (French—in the text)

DLT Development Land Tax(ation); Discrimination Learning Test

D-L T Donath-Landsteiner Test

dlts deep-level transient spectroscopy

DLTS Deck Landing Training School

dlu digitizer logic unit

dlvd delivered

dlvr deliver; delivery

dlvry delivery

DLW Diesel Locomotive Works

DL & W Delaware, Lackawanna and Western (railroad)

dlwg daily weight gain

dix deluxe

dly daily; delay; dolly

dlyd delayed

dm data management; dead meat; decimeter(s); delta modulation; demand meter; development milestone; diabetes mellitus (DM); diabetic mother; diagnostic monitor; diastolic murmur; diesel-mechanical; diphenylamine-arsine chloride (Adamsite war gas); direct mail; direct mon-

itoring; distance multiplier; double medium; draftsman; dry matter

dM decimorgan (unit of distance on a genetic map); differential of mutual inductance (symbol)

d/m date and month; day and month; density/moisture

d & m dressed and matched

d m *destra mano* (Italian—right hand)

d M *dieses Monats* (German—this month)

DM Deputy Master; Des Moines; Design Manual; Deutsche Mark (German mark—currency unit); Devon Militia; Dominica (Internet code); Drafting Manual; Driver Mechanic; Du Mont (television network); Dungeon Master; Dungeon Module; light minelayer, high-speed (naval symbol)

D.M. Doctor of Mathematics; Doctor of Medicine; Doctor of Music; Doctor of Musicology

D & M Detroit and Mackinac (railroad)

D of M Duchy (Duke) of Milan

DM *Daily Mail; Deutsche Mark* (German mark)

din² square decimeter

din³ cubic decimeter

dma direct memory access; direct memory access; dynamic mechanical analysis

DMA Dance Masters of America; Defence Manufacturers Association (Australian); Defense Mapping Agency; Delicatessen Managers Association; Dental Manufacturers of America; Direct Mail Association; Direct Marketing Association; Division of Military Application; Dog Museum of America; Dominion Marine Association

DMAA Direct Mail Advertising Association

DMAAC Defense Mapping Agency Aerospace Center

dmac dimethylacetamide (DMAC)

DMAC Des Moines Art Center

dmae dimethylaminoethanol

D.Ma.Eng. Doctor of Marine Engineering

D Mag   *D Magazine*

DMAHC Defense Mapping Agency Hydrographic Center

DMAIAGS Defense Mapping Agency Inter-American Geodetic Survey

D-man drug-enforcement officer

DMAODS Defense Mapping Agency Office of Distribution Services

dmards disease-modifying anti-rheumatic drugs

D Mark Deutsche Mark (German mark)—currency unit

DMATC Defense Mapping Agency Topographic Center

D.Math. Doctor of Mathematics

d-max density maximum

dmb dual-mode bus

Dmb Dumbarton

dmba dimethylbenzanthracene (DMBA)

dmbc direct material balance control

DMBC Detroit Motor Boat Club

dmbl demobilization; demobilize; demobilized

dmc dichlorodiphenyl methyl carbinol; digital microcircuit(ry); dimetimylcarbinol (DMC)–insecticide; direct manufacturing cost(s); dough moulding compound

DMC Del Mar College; Democratic Movement for Change; Developing Member Country; District Materials Center

DMCA Direct Marketing Credit Association

DMCG Direct Marketing Creative Guild

dmcl (DMCL) device media control language

dmctc dimethylchlortetracycline (DMCTC)

DM & CW Diploma in Maternity and Child Welfare

dmd demand; diamond; disc memory drive

Dmd Duchenne's muscular dystrophy

D.M.D. Dentariae Medicinae Doctor (Latin—Doctor of Dental Medicine)

DMDC Defense Manpower Data Center

dmdd demanded

dmdg demanding

dme distance measuring equipment; durable medical equipment

DME Designated Medical Examiner; Direct Marketing Enterprises Incorporated; Director of Mechanical Engineering; Director of Medical Education

DMEA Defense Minerals Exploration Administration

DMEA Dictionary of Mechanical Engineering Abbreviations

D.Mec.E. Doctor of Mechanical Engineering

D.Mech. Doctor of Mechanics

dmed digital message entry device

D.Med. Doctor of Medicine

D.M.Ed. Doctor of Musical Education

D-men drug-enforcement officers; narcotics officers

Dmet distance-measuring equipment and tacan

D.Met. Doctor of Metallurgy

DMET Director(ate) of Marine Engineering Training

D.Met.Eng. Doctor of Metallurgical Engineering -

D. Meteor. Doctor of Meteorology

dmetu (DMETU) dimethylethylthiourea

dmf decayed, missing, or filled (teeth)

DMF Decorative Marble Federation

DMFA Direct Mail Fundraisers Association

DMFOS Diploma in Maxillo-Facial and Oral Surgery

dmfs decayed, missing and filled surfaces

dmft decayed, missing or filled teeth

dmg damage; damaged; damaging; dimethylglycine

DMG Defense Marketing Group; Division of Mines and Geology

D of M-G Duchy (Duke) of Mecklenburg-Güstrow

DMGO Division(al) Machine Gun Officer

dmh drop manhole

DMH Director of Mental Hygiene; Division of Mental Hygiene

dm/ha dry matter per hectare

DMHS Director of Medical and Health Services; Dolley Madison High School

dmi defense mechanisms inventory; deferred maintenance item

DMI Data Machines Incorporated; Data Management Inquiry; Department of Manufacturing Industry; Design Management Institute; Director(ate) of Military Intelligence

DMIAAI Diamond Manufacturers and Importers Association of America, Incorporated

DMIC Defense Metals Information Center (Batelle Memorial Institute)

D.Mi.Eng. Doctor of Mining Engineering

D.Mil.S. Doctor of Military Science

d-min density minimum

DMIR Duluth Mesabi and Iron Range (railroad)

DMJ Diploma in Medical Jurisprudence

dml demolish; demolition

dml (DML) data manipulation language; dimyristoyl lecithin

D.M.L. Doctor of Modern Languages

d mld depth moulded

DMLS Doppler Microwave Landing System

DMLT Diploma in Medical Laboratory Technology

dmm digital multimeter; depot maintenance manual; domestic mail manual

DMM Defense Market Measures; Directorate of Materiel Management

dmma (DMMA) Direct Mail/Marketing Association

dmmf dry mineral matter free

dmmp (DMMP) dimethylmethyl phosphonate

dmn dimension; dimensional

Dmn Drammen

dmna (DMNA) dimethylnitrosamine

Dmn Fst Damnation of Faust

DMNH Delaware Museum of Natural History; Denver Museum of Natural History

dmnstr demonstrator

dmo demetallized oil

DMO Deputy Medical Officer; Director of Military Operations; District Medical Officer

dmod displacement-measuring optical device

DMO & I Director of Military Operations and Intelligence

dmp difference of meridional parts; dimethylphthalate (insect repellent also abbreviated DMP)

DMP Developing Mathematical Processes; Diagnostic and Maintenance Processor; Director of Manpower Plannmng; Dublin Metropolitan Police

dmpa depomedroxyprogesterone (DMPA)

DMPA Dublin Master Printer's Association

DMPB Diploma in Medical Pathology and Bacteriology

dmpea (DMPEA) dimethoxyphenylethylamine

dmpi desired mean point of impact

dmpl digital microprocessor language

DMPL Des Moines Public Library

dmpp dimethylphenylpiperazinium

DMPP Duck Mountain Provincial Park (Manitoba and Saskatchewan)

dmpr damper

DMPS Deepwater Motion Picture System

dmpt dual mode power train

DMR Data Management Routines; Defective Material Report; Department of Main Roads; Diploma in Medical Radiology; Director of Materials Research; Director of Medical Research

DMRC Deering Milliken Research Corporation

DMRD Diploma in Medical Radio-Diagnosis; Directorate of Materials Research and Development

DMRE Diploma in Medical Radiology and Electrology

DMRT Diploma in Medical Radio-Therapy

dms dermatomyositis; diacritical marking system (DMS); digital multiplex switching;

dimethylsulfate; direct molded sole (Army boots); drums

DMS Data Management System; Decision Making System; Defense Mapping School; Denominational Ministry Strategy; Director of Medical Services; Director of Military Survey; Disk Monitoring System; Display Management System; Division of Medical Standards; Dominion Mutual Securities

D.M.S. Doctor of Medical Science

D of M-S Duchy (Duke) of Mecklenburg-Schwerin

dmsa dimercaptosuccinic acid

D.M.Sc. Doctor of Medical Science

DMSC Defense Medical Supply Center

DMSDS Direct Mail Shelter Development System

DMSE Developing Models for Special Education

DMSGR Dowd's Morass State Game Reserve (Victoria, Australia)

dmsh diminish

DMSI Directorate of Management and Support of Intelligence

dmso (DMSO) dimethyl sulfoxide

DMSP Defense Meteorological Satellite Program

DMSS Data Multiplex Subsystem; Director of Medical and Sanitary Services

dmst demonstrate; demonstration

dmstn demonstration

dmstr demonstrator

dmt demountable; dimethyltryptamine–DMT (dangerous hallucinogen); dual mode transmission

DMT Department of Motor Transport(ation); Director(ate) of Military Training; District Mounted Troops

dmti digital moving target indicator

DM & TS Department of Mines and Technical Surveys

dmu dual maneuvering unit

DMU Des Moines Union (railway)

dmwr depot maintenance work requirement

D.Mus. Doctor of Music

D.Mus.A. Doctor of Musical Arts

D.Mus.Ed. Doctor of Musical Education

DMV Department of Motor Vehicles

D.M.V. Doctor of Veterinary Medicine

dmwr depot maintenance work requirement

dmy dummy

DmZ (DMZ) Demilitarized Zone

dn debit note; decinem; dekanem; delta amplitude (symbol); dibucaine number; dicrotic notch; died near; down; downward

d'n damn

d/n (D/N) debit note

d & n dumb and numb (insensitivity factor)

d ... n damn

d/N dextrose/nitrogen (ratio)

Dn Dale; Daniel; Dragoon(s)

Dn Don (Spanish—title equivalent to "Sir")

DN Department of the Navy; Division Notice

D.N. Diploma in Nursing; Diploma in Nutrition

D of N Daughters of the Nile

D.N. Dominus Noster (Latin—Our Lord)

dna did not attend; does not answer

Dna Doña (Spanish— Lady)—Mrs

DNA Defense Nuclear Agency; deoxyribonucleic acid (chromosome and gene cornponent); Digital Network Architecture; Director of Naval Accounts

DNA Deutscher Normenausschusz (German Committee on Standards)

DNAD Director of Naval Air Division

DNAN Department Number Assignment Notice

DNANR Department of Northern Affairs and National Resources

D.N.Arch. Doctor of Naval Architecture

dna(s) docena(s) (Spanish—dozen(s))

dNase (also DNAse, DNAase) deoxyribonuclease

DNAW Director, Naval Air Warfare

dnb dead, non-battle; departure from nucleate boiling; dinitrobenzene

DNB Dance Notation Bureau; Distribution Number Bank

D.N.B. Diplomate of the National Board of Medical Examiners

DNB Dictionary of National Biography

dnc did not compete; direct numerical control

DnC Der Norske Creditbank (The Norwegian Credit Bank)—also shown as *DNC*

DNC Democratic National Committee; Domestic National Committee; Director of Naval Construction

dncb dinitrochlorobenzene (DNCB)

DNCCC Defense National Communications Control Center

DNCMD Dayton Contract Management Office

dn ctl down control

dnd died a natural death

Dnd Dunedin

DND Department of National Defense; Department of National Development; Director of Navigation and Direction; Division of Narcotic Drugs (UN)

DN & D Director of Navigation and Direction

DNDS Director of Naval Dental Services

dne douane (French—customs)

DNE Director of Naval Equipment; Director of Nursing Education

D.N.Ed. Doctor of Nursing Education

D.N.Eng. Doctor of Naval Engineering

DNES Director of Naval Education Service

DNET Democracy Network; Director of Naval Engineering Training

dnf did not finish

dnfb dinitrofluorobenzene

DNFCT Director of Naval Foreign and Commonwealth Training

dnft directional non-force technique

DNGW Director of Naval Guided Weapons

DNHW Department of National Health and Welfare (United Kingdom)

dni distributable net income

DNI Director of Naval Intelligence

DNI Dana Normalisasi Indonesia (Indonesian Institute of Standards)

DNIC Data Network Identification Code

DNII Dirección Nacional de Información e Inteligencia (Spanish—National Direction of Information and Intelligence)—Uruguay

dnj drone noise jammer

DNJ Det Norske Justervesen (Norwegian Bureau of Weights and Measures)

D.N.J.C. Dominus Noster Jesus Christus (Latin—Our Lord Jesus Christ)

Dnk Dunkirk

dnka did not keep appointment

dnl do not load; dynamic noise limiter

DNL Det Norske Luftfartselkap (Norwegian Airlines)

dnm data name

DNM Dinosaur National Monument; Director of Naval Manning

DNMR Director of Naval Manpower Requiremens

DNMS Director(ate) of Naval Medical Services; Division of Nuclear Materials Safeguards

DNO Director of Naval Operations; Director of Naval Ordnance; District Naval Office(r); District Nursing Officer

DNO Den Norske Opera (The Norwegian Opera)—Oslo

DNOA Director of Naval Officer Appointments

dnoc dinitro-ortihocresol (DNOC)

DNOR Director of Naval Operations Requirements

d/note debit note

D-Note $500 bill

D-Notices Defense Notices

D-notice system British defense-notice system for protecting state secrets with the cooperation of the press

dnp deoxyribonucleoprotein; do not publish

DNP 2,4-dinitrophenol; Dinder National Park (Sudan)

dnpm dinitrophenyl morphine (DNPM)

D.N.P.P. Dominus Noster Papa Pontifex (Latin—Our Lord the Pope)

dnpt (DNPT) dinitrosopentamethylene tetramine

dnr do not reduce; does not run; do not renew; dynamic noise reduction

dnr (DNR) do not resuscitate

D/N r dextrose-to-nitrogen ratio

DNR Department of National Revenue; Department of Natural Resources; Dialed Number Recorder; Director(ate) of Naval Recruiting

DNRC Department of Natural Resources and Conservation

d/n ratio ratio of dextrose (glucose) to nitrogen in the urine

dns determination of non-significance; did not start; dinoyl sebacate (DNS)

Dns Downs

DNS Decimal Number System; Department of National Savings (British); Director of Naval Signals; Domain Name System; Domain Naming Service; Doppler Navigation System

DNSA Diploma in Nursing Administration; Director of National Security Affairs

D.N.Sc. Doctor of Nursing Science

dnslp downslope

DNSS Defense Navigation Satellite System

DNSY Director of Naval Security

dnt dinitrotoluene

DNT Director(ate) of Naval Training

Dntn downtown

DNTO Danish National Travel Office

dntp diethyl-nitrophenyl thiophosphate (DNTP)—insecticide

Dnus. Dominus (Latin—Lord)

DNV Det Norske Veritas (Norwegian ship classifier)

dnwind downwind

DNWR Darling National Wildlife Refuge (Florida); Delta NWR (Louisiana); Desert NWR (Nevada)

DNWS Director(ate) of Naval Weather Service(s)

do daily only; day(s) off, defense optics; delivery order; diamine oxidase (DO); diesel oil; direct order; dissolved oxygen; ditto; dropout; dual ownership

do' door

d-o dropout

d/o daughter of; delivery order

do. *dictum* (Latin—as before, the same); *ditto* (Italian—the same)

d:o: dito (Swedish—ditto)

d O der (die, das) Obige (German—the aforementioned)

Do Dominican; Dominican Republic; Dominican or Santo Domingan; Dornier

DO Defense Order; Department of Oceanography; Design Office; Director of Operatitons; Disbursing Office(r); District Office(r); Division(al) Office(r); Dominican Republic; Dominion Observatory; Dominion Office(r); Driver Operator; Duty Officer

D.O. Doctor of Optometry; Doctor of Osteopathy

D/O Disbursing Officer

D & O Directors & Officers

DO-27 Dornier 6-passenger utility aircraft built in West Germany and also called Skyservant

DO Denominacão de Origem (Portuguese—Denomination of Origin); *Denominacionés de Origen* (Spanish—Denomination of Origin)

doa date of arrival; date of availability; dead on arrival; direction of approach; disposal of assets; dissolved oxygen analysis

DoA Department of Agriculture; Department of the Army (DOA)

DOA Dead on Arrival; Draft on Arrival

Doac Dubois oleic albumin complex

DOAE Defence Operational Analysis Establishment (UK)

DOAL Deutsche Ost Afrika Linie (German East Africa Line)

DOARS Donnelley Official Airline Reservations System

dob date of birth; degree of bend(ing); diameter overbark; disbursed operating base; doctor's order book

dob (DOB) 2.5-dimethoxy- 4-bromoamphetamine (hallucinogen causing blood-vessel constriction leading to possible loss of limbs)

DoB Daughters of Bilitis

DOB Date of Birth; doctor's order book; Do Our Best (Boy Scout and Girl Guide slogan)

DOB Deutsche Oper Berlin (German Opera of Berlin)

Dob(bin) Robert

Dobbs School for Girls State Training School for (delinquent) Girls at Kinston, North Carolina

'dobe adobe

dobe(s) doberman dog(s)

dobra monetary unit of São Tome and Principe

D Obst RCOG Diplomate of the Royal College of Obstetricians and Gynaecologists

doc data optimizing computer; deoxycholate; desoxycorticosterone (DOC); died of other causes; diesel oil cement; direct operating cost; doctor; doctoral; document; documentary; documentation; drive(s) other cars; dyadic orthotic cranioplasty

Doc doctor

DoC Department of Commerce

D o C Department of Correction (Arkansas, Connecticut, Delaware, Indiana, Massachusetts, North Carolina, Tennessee); Department of Corrections (Arizona, California, District of Columbia, Florida, Guam, Idaho, Illinois, Kansas, Kentucky, Louisiana, Maine, Michigan, Minnesota, Mississippi, Missouri, New Jersey, Rhode Island, South Carolina, Texas, Vermont, Washington, West Virginia); Division of Corrections (Utah, Wisconsin)

D-o-C Doctors-on-Call

DOC Department of Commerce; Department of Communications; District Officer in Command; District Officer Commanding; Doctors Ought to Care

DOC Denominazione di Origine Controllata (Italian—Place of Origin Controlled)

doca data of current appointment; deoxycorticosterone acetate (DOCA)

doce date of current enlistment

DOCED Documentation Edition

Doc.Eng. Doctor of Engineering

docg desoxycorticosterone glucoside (DOCG)

DOCG Denominazione di Origine Controllata Garantita (Italian—Controlled Denomination of Origin Guaranteed)

DOCIT Directors of Central Institutes of Technology (Australian)

DOCLINE Document Delivery On-Line (computer service)

doen documentation

Doc.Pol.Sci. Doctor of Political Science

DOCS Department of Correctional Services (NY)

Doct. Doctor (Latin—Doctor)

Doctᵃ Doctora (Spanish—Doctor)—feminine

docu document(ary)

docubio documentary biographee; documentary biographer; documentary biography

docudrama documentary drama

docum document; documentary; documentation; documented

documto documento (Spanish—document)

DOCUS Display-Oriented Computer Usage System

dod date of death; died of disease; dust of desuetude

Dod Dodecanese

DoD Department of Defense

DOD date of death; Department of Defense; died of disease; Director Operations Division; Domestic Operations Division (CIA)

DODA Door and Operator Dealers Association

DODAS Digital Oceanographic Data Acquisition System

DoDCI Department of Defense Computer Institute

Dodd Dodd, Mead

DODD Department of Defense Directive

DODDAC Department of Defense Damage Assessment Center

DoDDS Department of Defense Dependent Schools

Dod(dy) Dorothy; George

Dodec Dodecanese

Dodecanese Dodecanese Islanders; Dodecanese Islands

DODI Department of Defense Instruction

dodpit date of departure

Dodson's Dodson's Reports

doe date of enlistment; depends on experience; dyspnea on exercise; dyspnea on exertion

DoE Department of Education; Department of Energy; Department of the Environment; Director(ate) of Education

DOE Department of Education; Department of Energy

DoEd Department of Education

DOEOIS Design and Operational Evaluation of Disributed Offices Information Servers

DoEn Department of Energy

DOES Disk-Oriented Engineering System; Disorders Of Excessive Somnolence

doesn't does not

dof degrees of freedom; delivery on field

DoF Department of Finance

DOF Defense Optics Facility; Developmental Optics Facility; Director of Ordnance Factories

dofab damned old fool about books

dofic domain-originated functional integrated circuit

DOFL Diamond Ordnance Fuze Laboratories

dog disgruntled old graduate; hot dog; frankfurter; sausage(s)

doG difference of Gaussians

Dogger Dogger Bank in the North Sea off England's east coast

dogm dogmatic; dogmatism; dogmatist

DOGMAD Dissatisfied Owners of General Motors Automotive Diesels

doh direct operating hours

Doh Doha

dohc double overhead cam; dual overhead cam

DOHSA Death On High Seas Act

doi date of inquiry; dead of injuries; descent orbit insertion; division operating income

DoI Department of Industry; Department of the Interior (DoInt is better); Director(ate) of Information

D o I Department of Institutions (Montana); Department of the Interior; Director of Institutions (North Dakota); Division of Institutions (Oklahoma)

doin' doing

DoInt Department of the Interior

do/it digital output/input translator

DoJ Department of Justice

D ø K Det østasiaatiske Kompagni (Royal Danish East Asiatic Company)

Dok Akad Nauk Doklady Akademii Nauk (Russian—Proceedings of the Academy of Science)

dol dear old lady; display-oriented language; dollar

dol (DOL) dioleoyl lecithin

dol dolce (Italian—sweet); *dolor* (Latin or Spanish—pain) —the *dol* is the unit of pain; *dolore* (Italian—pain)

Dol Dolph (Adolf); dolphin; Dorothea; Dorothy

DoL Department of Labor (Do Lab is better)

D o L Department of Labor; Department of Labour

D.o.L. Doctor of Oriental Learning

D & O L Director's and Officer's Liability

Do Lab Department of Labor

dolciss dolcissimo (Italian— very sweetly)

Dolf Adolph; Adolphus; Rudolph

dolichocephs dolichocephalics (long-skulled people)

Doll Dorothy

dollar monetary unit of the Bahamas, Canada, Fiji, Guyana, Jamaica, Liberia, New Zealand, Singapore, Solomon Islands, Trinidad and Tobago, the United States, and Zimbabwe

dollies dolophine pills

dolo dolophine (methadone hydrochloride used as a morphine substitute in withdrawing addicts from heroin)

Dolomites Dolomite Alps of northeastern Italy

Dolores Dolores Hidalgo, Guanajuato, Mexico

Dolph Adolph, Bardolph, Rudolph, Zardolph

DOLPHIN Dump Obsolete Laws–Prove Hypocrisy Isn't Necessary

dols dollars

dom date of marriage; digestible organic matter; dirty old man; dissolved organic matter; division-owned material(s); domestic; domicile; dominant; dominion; drawn over mandrel

dom domenica (Italian—Sunday); *domingo* (Portuguese or Spanish—Sunday)

Dom Domain; Domenico; Dominic; Dominican; Dominican Republic; Dominion

Dom. Dominicus (Latin—of the Lord, as in *Dies Dominica*—the Lord's Day)

DOM Date of Marriage; dimethoxyalpha methyl phenethylmine (psychedelic drug also called STP); Diocese Of Melanesia

D.O.M. Deo Optimo Maximo (Latin—to God the Best and the Greatest)

DOMAINS Deep-Ocean Manned Instrument Station(s)

Dom Bk *Domesday Book*

Dom Can Dominion of Canada

Dom Day Dominion Day (celebrated in Canada July l)

dom econ domestic economy (home economics)

DOMEI-KAIGI Zen Nihon Rodo Sodomei Kumiai Kaigi (Japanese—Confederation of Labor)

DOMES Deep-Ocean Mining Experimental Study

dom ex domestic exchange
Dom Fiji Dominion of Fiji
domi domicile
domina distribution-oriented management information analyzer
Dominica Commonwealth of Dominica (formerly a British Windward Island and the most northern of the Windwards in the Caribbean)
Dominican Republic eastern half of Hispaniola in the West Indies, *República Dominicana*
Dominion Dominion Day or Canada Day (July 1)
DOMIS Directory of Materials Data Information Services
DOMMDA Drawing Office Material Manufacturers and Dealers Association
dom° domingo (Spanish—Sunday)
Dom° Domingo (man's name)
DOMO Dispensing Opticians Manufacturing Organization
Dom Pedro II Dom Pedro de Alcantara, emperor and president of Brazil
Dom.Proc. Domus Procerum (Latin—House of Lords)
Dom Rep Dominican Republic
DOMS Diploma in Ophthalmic Medicine and Surgery
domsat domestic communication satellite; domestic satellite carrier
dom sci domestic science
don' don't (do not)
don. donec (Latin—until)
Don Donald; Donaldo; Donegal
Don Donderdag (Dutch—Thursday); *Don Giovanni* (Mozart opera); *Donnerstag* (German—Thursday); (Spanish—Lord and Master, from the Latin—*dominus*); *Don Quixote* (fantastic variations for cello and orchestra by Richard Strauss); *The Don*—Mozart's two-act comic opera—*Don Giovanni*
DoN Department of the Navy
DON Diploma in Orthopaedic Nursing
DONA Doulas of North America
Donalbane Donald Bane

Donbas Donets Basin in the Ukraine
donec alv. sol. fuerit donec alvus soluta fuerit (Latin—until the bowels move)
Doneg Donegal (sometimes Don)
Donets Donets Basin or Donbas of the Ukraine
dong monetary unit of Vietnam
Dong Phan Van Dong
donk donkey; donkeyback; donkeyboiler; donkey boy; donkey breakfast (sailor's straw-stuffed mattress); donkeycart; donkey crosshead; donkey engine(man); donkey house; donkeyman; donkey pump; donkey puncher; donkey sled; donkey stack; donkeywork(man)
Donnie Donald
Don Q Don Quixote
DONS Department of National Security (South Africa)
don't do not
do-nut doughnut
doo diesel oil odor
DOO Directing Ordnance Officer; Director—Office of Oceanography
doom deep ocean optical measurement
dop dermo-optical perception; designated overhaul point; developing-out paper; dressing-out percentage
dop (DOP) diocytl phthalate
D o P Department of Prisons (Nevada)
dopa dynamic output printer analyzer
dopa (DOPA) dihydroxyphenylalanine
dopadic dope addict
dopase dopa oxidase
D. Oph. Doctor of Ophthalmology
D.Ophth. Doctor of Ophthalmology
dopl doplene (Czech—enlarged)
dopp ped doppio pedale (Italian—double pedal)—musical term
d-o psychiatrists directive-organic psychiatrists
D.Opt. Doctor of Optometry
dor date of rank; dead on road; dental operating room; digital optical recording; doric; dormitory

Dor Dorado; Doric; Dorothy
DoR Department of Rehabilitation
D. Or. Doctor of Oratory
DOR Department of Offender Rehabilitation (Georgia); Director(ate) of Operational Research
dora dynamic operators research apparatus; dynamic operators response apparatus
Dora Deborah; Dorothea; Dorothy; Eudora; Theodora
DORA Defence of the Realm Act
doran Doppler range and navigation
Dord Dordogne
DORDEC Domestic Refrigerator Development Council
Doric(k) Theodoric(k)
Dorie Doris; Theodora; Theodore
Doris Doreen; Dorothea; Dorothy; Eudora; Theodora
DORIS Direct Order Recording and Invoicing System
DORL Developmental Orbital Research Laboratory
dorm(s) dormitory; dormitories
dorna deoxyribose nucleic acid
Dors Dorset; Dorsetshire
DORS Defense Operational Requirements Staff
Dorset Dorsetshire
Dort Dordrecht
DORT Detroit Objective Reference Test
D Orth Diploma in Orthodontics; Diploma in Orthoptics
dos date of sale; date of separation; denial of service; dosage; dose; dosimetric; dosimetry; dosiology
dos. dosis (Latin—dose)
Dos John Dos Passos
DoS Department of State
DOS Date of Separation; Defence Operations Staff; Department of State; Digital Operation System; Disk Operating System
D.O.S. Doctor of Ocular Science; Doctor of Optical Science; Doctor of Optometric Science
Dosc Dubois oleic serum complex
DOSCO Dominion Steel and Coal Corporation

DOSES Development of Statistical Expert Systems

Dosh Univ Doshira University

dosim dosimetry (measurement of radiation doses)

do'sn't does not

DOSS Deep-Ocean Search System

DOST Dictionary of the Older Scottish Tongue

dosv deep ocean survey vehicle

DOS/VSE Disk Operation System/Virtual Storage Extended

dot. deep ocean transponder; deep-ocean technology; deep-oceanic turbulence; directly observed treatment; dotation(al); draft(s) on treasury

Dot Dorothy; Dotty

DoT Defense of the Territory; Department of Telecommunications; Department of Tourism; Department of Trade (United Kingdom); Department of Transport (Canada); Department of Transport(ation); Department of Transportation (US); Department of the Treasury

D o T Defense of the Territory; Department of Trade; Department of Transport; Department of Transportation

DOT Deep Oil Technology (company); Department of Overseas Trade; Diploma in Occupational Therapy; Direct Order Turnaround system; Director of Operational Training

DOT Dictionary of Occupational Titles

DOTC Department of Transportation Classification

DOTE Department of the Environment

DOTIPOS Deep Ocean Test-in-Place and Observation System

DOTM Department of Ordnance, Torpedoes, and Mines

dots directly observed treatment, short course

Dott Dottore (Italian—Doctor)

D o T & T Dominion of Trinidad and Tobago

Dotty Doreen; Dorothea; Dorothy; Eudora

dou (DOU) definitive observation unit

Douay Douay Version of the Bible (published at Douai, France in 1609)

double-B double-backed; double-banked; double-barreled; double-bass; double-bedded; double-benched; double-bonded; double-bottomed; double-breasted; double-brooded

Double D Doubleday

double-X doublecross; double quality; double quantity; double thickness; doubleweight; two-X; XX

doubt. doubtful

Doug Douglas(s)

Doug fir Douglas fir

dov data over voice; double oil of vitriol (sulphuric acid); dried on vine (grape varieties)

Dov Dover; Dovid

Dov Dovid (Yiddish—David)

DOV Defence of Village (educational program)

dovap Doppler velocity and position

Dover Delaware's capital named after an English Channel port; Dover Publications

dow died of wounds; dowager; dowel; dowelled

Dow Dow Jones Industrial Average; Dowager

D o W Defenders of Wildlife

DOW Died of Wounds; Dow Chemical Company; Dow Chemicals

DoWaPO Dictionary of Word and Phrase Origins

dowb deep ocean work boat

Dow Kuw Dowlat al-Kuwait (Arabic—State of Kuwait)

Down Downing College, Cambridge

dows dowsing; dowsers

doy day of year

doz dozen

dozer bulldozer

dp damp proof(ing); dash pot (relay); data processing; data processor; deck piercing; deep penetration; deep pulse; deflection plate; departure point; dewpoint; diametral pitch; diastolic pressure; differential of pressure (symbol); differential protection; diffusion pressure; digestible protein; diphosgene (deadly gas); diproprionate; disability pension; disabled person; dis-

criminatory power; diphosphate; displaced person; distopulpal; distribution point; donar's plasma; double paper; double plays; double pole; drip-proof; drop point; dual purpose; dump; durable press; potential difference (symbol)

dp (DP) data processing; dementia praecox

d/p delivery papers; documents against payment; door-to-port/port-to-door (delivery)

d & p developing and printing; development and printing; drain and purge

d.p. directione propria (Latin—with proper direction)

d/p días plazo (Spanish—pay days)

d. in p. divide in partes (Latin—divide)

DP by direction of the President; Dan Pezze; Democratic Party; Department of the Pacific; Detrucking Point; Dinosaur Park (Drumheller, Alberta); Director of Photography; Director of the Port; Displaced Person

D-P Data-Phone

D.P. dementia praecox; Doctor of Pharmacy; Doctor of Podiatry

D & P Deberny and Peignot

D of P Daughters of Pennsylvania; Daughters of Pocahontas; Director of Planning; Director of Plans; Duchy (Duke) of Prussia

DP Denver Post

D.P. Domus Procerum (Latin—House of Lords)

dpa deferred payment account; diagnostic prescriptive arithmetic

dpa (DPA) diphenylamine; dipicolinic acid

dPA di Pietro Aretino

Dpa Diputada (Spanish—Deputy)—feminine

DPA Director of Postal and Courier Services; Data Processing Agency; Data Protection Authority; Diabetes Press of America; Directory Publishers Alliance; Discharged

Prisoners Association; Division of Performing Arts; Division of Public Affairs

D.P.A. Doctor of Public Administration

DPA Deutsche Presse Agentur (German news agency); *Doulat i Padshahi ye Afghanistan* (Kingdom of Afghanistan)

d. in p. aeq. divide in partes aequales (Latin—divide into equal parts)

dpars data processing automatic record standardization

DPAS Discharged Prisoners' Aid Society

D Path Diploma in Pathology

dpb deposit passbook; disinfection byproduct

DPB Department of Printed Books (British Museum Library); Domestic Purposes Benefit

dpbc double pole both connected

dpc damp-proofing course; data processing computer; data processing control; double paper single cotton

DPC Daniel Payne College; Data Processing Center; Defence Planning Committee; Defense Plant Corporation; Defense Procurement Center; Defense Procurement Circular; Defense Production Chief; Deputy Police Commissioner; Desert Protective Council; Displaced Persons Commission; Dissemination Policy Council; District Police Commissioner; Division Planning Corporation; Domestic Policy Council; Duke Power Company

DPCE Data Processing Customer Engineering

dpcm differential pulse-code modulation

DPCP Department of Prices and Consumer Protection (British)

DPCS Director of Postal and Courier Services

dpct differential protection current transformer

dpd data project directive; diffuse pulmonary disease

DPD Data Products Division (Stromberg-Carlson); Department of Public Dispensary; Diploma in Public Dentistry

DPD Data Processing Digest

dpdc double paper double cotton

dp di dimple die

DPDS Defense Property Disposal Service

dpdt double-pole double-throw

dpe data processing equipment; digital processing effects; digital production effects; direct plate exposure

d-p-e development-printing-enlargement

Dpe Dieppe

DPE Diploma in Physical Education; Director of Primary Education; Director of Public Education

D.P.E. Doctor of Physical Education

D.Ped. Doctor of Pedagogy

DPED Department of Planning and Economic Development

DP/ED Data Processing for Education

dpe service developing-printing-enlarging service

DPEWS Designed-to-Price Electronic Warfare System

dpf deferred pay fund

DPf Deutsche Pfennig (German—pfennig)

DPF Drug Policy Foundation

dpfc double pole front connected

dpft double-pedestal flat-top (desk)

dpg data processing group; deck plate girder; digital pattern generator

dpg (DPG) diphosphoglyceric acid

DPG Dugway Proving Ground

DPGA Delaware Personnel and Guidance Association

DPGs Development Planning Groups

dph diamond pyramid hardness; diphenylhydantoin (DPH)

D. Ph. *Doctor Philosophiae* (Latin—Doctor of Philosophy)

DPH Department of Public Health; Department of Public Highways; Diploma in Public Health; Domestic Packing House

D.P.H. Doctor of Public Health

D.Pharm. Doctor of Pharmacy

DPHD Diploma in Public Health Dentistry

dphf differential protection high frequency

D.Phil. Doctor of Philosophy

DPHN Diploma in Public Health Nursing

d'phone dictaphone

D.Ph.Sc. Doctor of Physical Science

D Phys Med Diploma in Physical Medicine

dpi data processing installation; disposable personal income; dots per inch

DPI Department of Primary Industries; Department of Public Information; Department of Public Instruction; Disorderly Persons Investigation; Distillation Products Industries; Division of Plant Industry

DPIF Drug Product Information File

DPII Dairy Products Improvement Institute

dp-ing data processing; durable pressing

dpir data processing and information retrieval

dpl death place; deferred pastoral lease; deferred payment license; diploma; diplomat; dual propellant loading; duplex

dpl (DPL) dipalmitoyl lecithin

DPL Dallas Public Library; Dayton Power and Light; Dayton Public Library; Delhi Public Library; Denver Public Library; Detroit Public Library; diplomatic corps (license plate)

DP & L Dallas Power and Light

DPL Den Polytekniske Laeranstalt (Danish—The Polytechnic Institute)—Copenhagen

DP & LC Dundee, Perth & London (shipping) Company

dplf differential protection low frequency

DPLU Department of Planning and Land Use

dplx duplex

dpm data processing machine; digital power meter; disintegrations per minute; disruptive pattern material (military clothing); disruptive pattern material (military clothing); documents per minute

DPM Data Processing Manager; Deputy Prime Minister; Deputy Provost Marshal; Development Program Manual; Diploma in Psychological Medicine

D.P.M. Doctor of Podiatric Medicine

DPMA Data Processing Management Association

dpn diamond pyramid number

dpn (DPN) diphosphopyridine nucleotide

dpn+ oxidized diphosphopyridine nucleotide

dpng deepening

dpnh (DPNH) reduced diphosphopyridine (same as nadh or NADH)

d pnl distribution panel

DPNM Devil's Postpile National Monument

dpo development planning objective

Dpo Depot (postal abbreviation)

Dpo Diputado (Spanish— Deputy)—masculine

DPO Dayton Philharmonic Orchestra; Discontinued Post Office; Distributing Post Office; District Pay Office(r)

dpob date and place of birth

D.Pol.Eco. Doctor of Political Economy

D.Pol.Sci. Doctor of Political Science

dpp deferred payment plan

DPP Democratic Progressive Party; Director of Public Prosecutions; Disease Prevention Program

DPPS Department of Public Printing and Stationery

DPQCA Dairy Products Quality Checked Association

dpr day press rates; double lapping of pure rubber

DPR Democratic Peoples' Republic; Department of Pesticide Regulation; Differential Police Response; Director(ate) of Public Relations

DPRGR Dewan Perwakilan Ratjat-Gotong Rojong (Indonesian—Mutual Cooperation House of Representatives)

DPRI Disaster Prevention Research Institute

DPRK Democratic People's Republic of Korea (North Korea)

DPRORM Drafting, Pay and Records Office, Royal Marines

DPRS Dynamic Preferential Runway System

dps disintegrations per second; double-pole snap switch

dp's (DPs) displaced persons

dp & s data processing and software

dPs displaced Palestinians

DPs Detention Pens

DP's displaced persons

DPS Data Processing Service; Data Processing Station; Data Processing System; Defence Policy Staff; Defense Printing Service; Department of Public Safety; Director of Personal Services; Division of Primary Standards; Domestic Policy Staff

DPS 6 Trademark of Honeywell Corporation

DPSA Data Processing Supplies Association

DPSB Defense Production Supply Board (NATO)

DPSC Defense Personnel Support Center; Defense Petroleum Supply Center

DPSCS Department of Public Safety and Correctional Services (Maryland)

dpsk differential phase shift keying

DPSS Data Processing Subsystem; Department of Public Social Services

dpst deposit

dp st double pole, single throw

DPsy Diploma in Psychiatry; Diploma in Psychology

D. Psych. Doctor of Psychology

D.Psy.Sci. Doctor of Psychological Science

dpt deeper-pool (pay) test (oil well); department; deponent; deposition; depth

dpt (DPT) dipropylphytamine

DPT Design Proof Test(ing); Director of Physical Therapy

dpta diethylene triamine pentaacetic acid

Dpto Departamento (Spanish—Department)

dpt vaccines diphtheria, pertussis, tetanus vaccines

dptw double-pedestal typewriter (desk)

dpty deputy

dpu data processing unit

D.Pub.Adm. Doctor of Public Administration

dpv dry pipe valve; drive propulsion vehicle; duty-paying value

dp/w drawbar pull/weight (ratio)

dpw dried poultry waste

DPW Department of Public Works

DPWA Data Processing Work Assignment

DPWG Defense Planning Working Group (NATO)

DPWO District of Public Works Office

dpx duplex

dpy dipyridamole

dq definite quantity; deterioration quotient; developmental quotient; direct question(s)

dQ differential of figure of merit (symbol); differential of quantity (symbol)

DQ Dairy Queen

dqd digital quadrature detection

dqm data quality monitors

DQMG Deputy Quartermaster General

DQMS Deputy Quartermaster Sergeant

DQU Deganawidah Quetzalcoatl University (University of California at Davis)

dqy don't quit yet

dr debit; defense regulations; differential rate; dining room; door; double-reduction; drachma; dram; draw; drawn; dress regulations; drill; drive; drum

dR differential of resistance (symbol)

dr (DR) data register; dead reckoning; delivery room

d/r deposit receipt

Dr debtor; doctor; Drenthe (Dutch province); Drive; drachma (Greek monetary unit)

DR Data Report; Date of Rank; Daughter of the Revolution; Dead Reckoning; Deficiency Report; Dental Record; Dental Recruit; Design Requirements; Despatch Rider; Detailed Report; Development Report; Diocesan Registry; Document Report; National Distillers and Chemical Corporation (stock exchange symbol); reaction of degeneration (symbol)

D/R date of rank; dead reckoning

Dr Drenthe (Dutch province)

DR Deutsche Reichsbahn (German State Railway)

dra dead-reckoning analyzer

dr & a data reporting and accounting

dra derecha (Spanish—right)

Dra Doctora (Spanish—woman doctor); *Draco* (Latin—Dragon constellation)

Dr^a Doctora (Spanish—doctor)—feminine form; *Doutora* (Portuguese—doctor)—feminine form

DRA Defence Rifle Association (S. Africa); Democratic Republic of Afghanistan; Director, Royal Artillery; Division of Ratepayer Advocates; Dude Ranchers Association

Drac Draco

DRAC Director of the Royal Armoured Corps

drachma monetary unit of Greece

DRACO Driver and Accident Coordinated Observer

dr ad drill adaptor

Drª D^na Doctora Doña (Spanish—Madam Doctor)

Dr Ae.Sc. Doctor of Aeronautical Science

Dr Agr. Doctor of Agriculture

drai dead-reckoning analog indicator

drain. drainage

dram dynamic random-access memory

dram. drama; dramatic; dramatist

dram (DRAM) detection radar automatic monitoring

dram. pers. *dramatis personae* (Latin—cast of characters)

DRAMs Dynamic Random Access Memories

dr ap dram, apothecaries'

Draper Utah State Prison at Draper

drapes draperies

Dr Arne Thomas Arne

dras derechas (Spanish—duties, fees, tariffs)

Dr Atl Gerardo Murillo

dr av dram avoirdupois

Drav Dravidian

draw direct read after write; drawing

drb design requirements baseline

Drb Durban

DRB Defense Research Board (Canada); Defence Resources Board; Discharge Review Board; Druggists' Research Bureau

DRBC Delaware River Basin Commission

dr bg drill bushing

Dr Bi.Chem. Doctor of Biological Chemistry

DRBU Dharma Realm Buddhist University

Dr Bus.Adm. Doctor of Business Administration

drc damage-risk criteria (noise exposure limits); down right center (driving, lighting, or seating)

DRC Denver Railway Car Company; Department of Rehabilitation and Correction (Ohio); District Recruiting Command(er); Driver Re-education Course; Drug Referral Center; Drug Rehabilitation Center; Dutch Reformed Church; Dynamics Research Corporation

Dr C Cabinet of Dr Caligari

dr canon droit canon (French—canon law)

drch drachma

Dr Chem. Doctor of Chemistry

dr ck drill chuck

DRCOG Diploma of the Royal College of Obstetricians and Gynaecologists

Dr Com. Doctor of Commerce

dr com droit commun (French—common law)

dr comm droit commercial (French—commercial law)

dr cout droit coutumier (French—common law)

DrD Doctor Don (Spanish—Sir Doctor)

DR & D Defense Research and Development

DRDBMS Distributed Relational Database Management System

DRDO Defense Research and Development Organization

drdp detection radar data processing

DRDT Daily Record of Dysfunctional Thoughts; Division of Reactor Development and Technology (AEC)

drdto detection-radar data takeoff

dre dead reckoning equipment; digital rectal exam

DRE Defense Research Establishment (Canada); Department of Real Estate (California); Director of Religious Education

DR & E Defense Research and Engineering

D.R.E. Doctor of Religious Education

DREA Defense Research Establishment, Atlantic

drec detection-radar electronic component

Dr Ec. Doctor of Economics

dred. dredging

DREE Department of Regional Economic Expansion (Canada)

drek dead reckoning

Dren Drenthe (Dutch province)

Dr Eng. Doctor of Engineering

Dr Ent. Doctor of Entomology

DREO Defense Research Establishment, Ottawa

DREP Defense Research Establishment, Pacific

dres dietary risk evaluation system

Dres Doctores (Spanish—Doctors)

DRES Defense Research Establishment, Suffield

Dr ès L. Docteur ès Lettres (French—Doctor of Letters)

Dr ès S. Docteur ès Sciences (French—Doctor of Sciences)

DRET Defense Research Establishment, Toronto

DREV Defense Research Establishment, Valcartier

Drew Andrew; Charles E Drew Postgraduate Medical School

drews (DREWS) direct readout equatorial satellite

drf depression range-finder; differential reinforcement; dose reduction factor

DRF Deafness Relief Foundation; Direct Relief Foundation

dr féod *droit féodal* (French—feudal law)

drftmn draftsman

dr fx drill fixture

drg diagnosis—related group; dorsal root ganglion; drawing(s); drogue; during

DRG Detroit Rubber Group; Dickinson Robinson Group

DRGM *Deutsches Reichgebrauchsmuster* (German—registered design)

DRGs Diagnosis-Related Groups

D & RGW Denver and Rio Grande Western (railroad)

drh differential reinforcement of high (response rates)

DRH Division of Radiological Health

Dr h.c. *Doctor honoris causa* (Latin—honorary doctor)

dr hd drill head

Dr Hor. Doctor of Horticulture

Dr Hy. Doctor of Hygiene

dri data rate indicator; data reduction interpreter; direct reduced iron; drive

DRI Dairy Research Institute; Data Resources Inc; Defense Research Institute; Denver Research Institute; dietary reference intakes; Direct Relief International; Document Retrieval Index

drib deoxyribose

DRIC Dental Research Information Center; Dispute Resolution Information Center

drid direct-readout image disector

drift. driftwood

DRIFT Diagnostic Retrieval Information For Teachers

Drigo Rodrigo

drill. drilling

DRINC Dairy Research Incorporated

D-ring capital-D-shaped ring

Dr Ing. *Doktor-Ingenieur* (German—Doctor of Engineering)

drinkin' drinking

drip digital ray and intensity projector

DRIP Dividend Reinvestment Plan

drir direct readout infrared

DRIS Department of Defense Retail Interservice Support Program

DRIVE Dedicated Road Infrastructure for Vehicle Safety in Europe; Developing Resources for Instructors of Vocational Education; Digital Raster Imaging, Viewing, and Editing system

Dr J *Dr Jekyll* (and Mr Hyde)

dr jg drill jig

Dr J.Sc. Doctor of Judicial Science

Dr Jur. *Doctor Juris* (Latin—Doctor of Law)

drk dark; display request keyboard

DRK *Deutsches Rotes Kreuz* (German Red Cross)

Drk-Yul *Druk-Yul* (Dzongka—Kingdom of Bhutan)

drl data retrieval language; differential reinforcement of low (response rates)

DRL Design Report Letter; Diamond Research Laboratory; Differential Reinforcement of Low Rate

DRLG Danish Royal Life Guards

Dr Lit. Doctor of Literature

DRLS Dispatch Rider Letter Service

drm direction of relative movement

DRM Drafting Room Manual

dr mar *droit maritime* (French—maritime law)

Dr Med *Doktor der Medizin* (German—Doctor of Medicine)

Dr Med. *Doctor Medicinae* (Latin—Doctor of Medicine)

Dr Mus. Doctor of Music

drn drawn

Drn Dairen; Darien

DRN Daily Reports Notice; Detroit River Navigation

drna (DRNA) deoxyribose nucleic acid

Dr Nat.Sci. Doctor of Natural Science

drnt diagnostic roentgenology

dro daily routine orders; destructive readout; digital readout

dro (DRO) differential reinforcement of other behavior; double-room occupancy

dro *derecho* (Spanish—custom duty, right)

DRO Disablement Resettlement Office(r)

drod delayed readout detector

DRO-LA Defense Research Office-Latin America (USA)

DROLS Defense RDT&E (Research, Development, Test and Evaluation) On-Line System

dromdi direct readout miss-distance indicator

'drome aerodrome; airdrome

'Drome Hippodrome

dron data reduction

DROPS Demountable Offloading and Pick-up System

dros date returned from overseas

dros *derechos* (Spanish—duties, fees, tariffs)

drp dead reckoning position; differential reinforcement of paced (response rates)

DRP Democratic Republican Party (Korean); Diebold Research Program; Dividend Reinvestment Plan (stock report)

DRP *Deutsche Reichspartei* (German Reich Party); *Deutsches Reichspatent* (German—patent)

DRPA Delaware River Port Authority

DRPC Defense Research Policy Committee

dr pén *droit pénal* (French—penal law)

Dr P.H. Doctor of Public Health

Dr Phil. *Doktor der Philosophie* (German—Doctor of Philosophy)

DRPL Del Rio Public Library

Dr Pol.Sc. Doctor of Political Science(s)

DRPP Director(ate) of Research Programs and Planning

drps digital random program selector; drapes

drq discomfort relief quotient

DRR Drawing Release Record

Dr Ra.Eng. Doctor of Radio Engineering

DRRB Data Requirements Review Board (DoD)

Dr Rec. Doctor of Recreation

Dr Re.Eng. Doctor of Refrigeration Engineering

DRRI Defense Race Relations Institute (DoD)

dr rom   *droit romain* (French—Roman law)

d-r-r-r-r-r-um   snaredrum roll

drs   data-reduction situation; data reduction system; detecting ranging set; digital range safety; drawers; drowsiness

DRs   Development Rights; Discrepancy Reports

DRS   Dairy Research Station; Data Reduction System; Data Relay Station; Debtor Reporting System; Development Reference Service; Diagnostic Research System; Diagnostic Rework Sheet(s); Document Retrieval System

DRSAM   Diploma of the Royal Scottish Academy of Music

drsc   direct radarscope camera

Dr Sc.   Doctor of Science

Dr Sci.   Doctor of Science

DRSCS   Digital Range-Safety Command System

dr sh   drill shell

drsmkr   dressmaker

drsn   drifting snow

DRSO   Danish Radio Symphony Orchestra

drsr   dresser

DRSS   Discrepancy Report Squawk Sheet

drt   data review technique; dead reckoning tracer; dead reckoning trainer

dr t   dram troy

Drt   Dartmouth

DRT   Diagnostic Rhyme Test; Dog Rescue Team

DRTC   Documentation Research and Training Center

DRTE   Defense Research Telecommunications Establishment (Canada)

Dr Tech.   Doctor of Technology

Dr Theol.   Doctor of Theology

Dr Theol.   *Doktor der Theologie* (German—Doctor of Theology)

dr tp   drill template

dr trav   *droit du travail* (French—labor law)

dru   digital register unit; digital remote unit

Dru   Drusila

drub   digital remote unit buffer

D. Ru.Eng.   Doctor of Rural Engineering

Drug Abuse   Drug Abuse Council

Drug Rehab   Drug Rehabilitation Program

drum.   drumlin (hill formed by gravel and rocks deposited by glacier)

drums   kettledrums, bass drums, field or tenor drums, side or snare drums, tomtoms or bongo drums

Dr und Vrl   *Druck und Verlag* (German—printed and published by)

Dr Uni.Par.   Doctor of the University of Paris

D. Rur.Sci.   Doctor of Rural Science

DRUs   Directing Reporting Units (Air Force)

drv   data-recovery vehicle; daily reference value

Drv   Drive

DRV   Democratic Republic of Vietnam (North Vietnam); Dumbarton Rifle Volunteers

DRVN   Democratic Republic of Vietnam

drvr   driver

dr vs   drill vise

drw   defensive radio (logical) warfare; drawing

DRW   Darwin, Australia (airport)

drwg   drawing

DRWS   Dirty Radwaste System

DRWW   Distillery, Rectifying, Wine Workers (union)

drx   drachma (Greek monetary unit)

dry.   drying

dry alum   aluminum and potassium sulfate

dry disco   alcohol-free, drug-free, tobacco-free discotheque

Dr Z   *Doctor Zhivago*

drzl   drizzle

ds   days after sight; day's sight; dead-air space; debenture stock; decanning scuttle; decimal selector; density standard; detached service; dietary supplement; dilute strength; dioptric strength; direct support; discarding sabot; dissociation; document signed; domestic service; donar's serum; double silk; double stout; double strength;

double-screened, double-stitch(ed); downspout; draft stop; dressing station

ds (DS)   data set (data processing); duration series

d-s   dead slow (ship's engine signal)

d.s.   died single; document signed

d/s   dextrose in saline

d & s   demand and supply; dermatology and syphilology; distribution and supply

ds   *destro* (Italian—right)

Ds   Down syndrome; dysprosium (symbol)

Ds.   *Deus* (Latin—God); *Durchführumgssatz* (German—development of a sonata)

DS   Date of Service; Delphian Society; Delta Society; Dental Surgeon; Department of Sanitation; Department of State; Design Standard(s); Detached Service; Direct Support; Directing Staff; Director of Services; Discarding Sabot; Distributed Services; Drill Ship; Drug Store; Durham & Southern (railroad)

D-S   Deux-Sèvres; Ditlev-Simonsen Lines

D & S   Durham & Southern Railway

D of S   Daughters of Scotia; Department of State; Duchy (Duke) of Savoy; Duchy (Duke) of Silesia; Duchy (Duke) of Styria

DS   *Danske Standariseringsraad* (Danish Standards Institute); *Dictionary of Spirituality*

D S   *dal segno* (Italian—return to the sign)

D/S   *Dampskip* (Norwegian—steamer, steamship)

DS1   Deep Space 1

dsa   data set adapter; dial service assistance; digital storage architecture; dimensionally-stabilized anode; discrete sample analyzer

dsa (DSA)   digital subtraction angiography

DSA   Danish Sisterhood of America; Dante Society of America; Defense Shipping Authority; Defense Supply Agency; Defense Supply Association; Democratic Socialists of America; Dental Surgery Assistant; Department of Substance Abuse; Deputy Sheriff's Association; Design Schedule Analysis; District Senior Adviser; Division Service Area; Drillsite Supervisors Association; Drum Seiners Association; Duluth, South Shore and Atlantic (railroad); Duodecimal Society of America

DSAA Defense Security Assistance Agency

DSAB Dictionary of South African Biography

dsabl disable; disability

DSACEUR Deputy Supreme Allied Command, Europe

DSAHBK Defense Supply Agency Handbook

DSAM Defense Supply Agency Manual

DSANZ Direct Selling Association of New Zealand

DSAO Diplomatic Service Administration Office(r)

DSAP Data Systems Automatic Program; Data Systems Automation Program; Defense Systems Application Program

DSARC Defense Systems Acquisition Review Council

dsas dial-service-assistance switchboard

D/S A/S Dampskipaksjeselskap (Norwegian—joint stock steamship company, limited)

dsasbl disassemble

DSASO Deputy Senior Air Staff Officer

DSAT Defensive Antisatellite Operations

dsb double sideband; double-strength B (quality glass)

DSB Danske Stats Baner (Danish State Railways); Defence Signal Bond; Defense Science Board; Drug Supervisory Body (UN)

DSBA Delaware School Boards Association

dsbg disbursing

dsbn disband

dsbse double-sideband suppressed-carrier

dsc double silk-covered (wire); downstage center; dynamic standby computer

D.Sc. Doctor of Science

DSC Defense Supply Corporation; Delaware State College; Depot Supply Center; Die Casters' Conference; Distinguished Service Cross; Document Service Center

D.S.C. Doctor of Christian Science; Doctor of Commercial Science; Doctor of Surgical Chiropody

D & SC Defense and Space Center (Westinghouse)

DSCA Douglas Social Credit Association

dscb data set control block

DSCC Deep Space Communications Complex

D.Sc.Com. Doctor of Science in Commerce

DSCDP Delaware State Central Data Processing

D.Sc.Eco. Doctor of Science in Economics

D.Sc.Eng. Doctor of Science in Engineering

D Sch Dmitri Shostakovich (in his *Tenth Symphony* uses his initials to form a four-note theme, applying German letters D, S for Es—E-flat, C, and H—German for B natural)

D.Sch.Mus. Doctor of School Music

D.Sc.Hyg. Doctor of Science in Hygiene

DSC (I) Die Sinkers' Conference (International)

D.Sc.I. Doctor of Science in Industry

D.Sc.Jur. Doctor of the Science of Jurisprudence

D.Sc.L. Doctor of the Science of Law

DSCM Diploma of the Sydney Conservatorium of Music

DSCMD Dallas Contract Management District

D.Scn. Doctor of Scientology

D. Sc. Os. Doctor of the Science of Osteopathy

DSCP Detailed Site Characterization Plan

D.Sc.Pol. Doctor of Political Science(s)

DSCR Detailed Site Characterization Report

dscs direct-set cheese starter

DSCS Defense-Satellite Communication System(s)

dsct descendant

dsd dry surgical dressing

DSD Daily Staff Digest; Defence Signals Directorate (Australian); Director of Staff Duties; Depressive Spectrum Disease; Depressive Spectrum Disorder; Director of Signals Division

ds/dd double-sided, double-density

DSDP Deep Sea Diving Project; Deep Sea Drilling Program; Deep Sea Drilling Project

DSDS Deep Sea Diving School (USN)

d'o & d's disbelievers and doubters

dsdu data signal display unit

dse data-storage equipment; depot support equipment; development support equipment

D.S.E. Doctor of Science in Economics

DSE Departamento de Seguridad del Estado (Spanish—Department of State Security)—Cuba

DSEA Davis Submerged Escape Apparatus; Delaware State Education Association

dsf day-second-feet (or foot)

Dsf Dusseldorf

DSF Dainippon Silk Foundation; Daughters of St Francis of Assisi; Division of Sea Fisheries

dsfc direct side force control

dsg designate; designation; disodium 5-guanylate

DSG Deputy Surgeon General

DSG Deutsche Schlaf-und Speisewagen Gesellschaft (German—Sleeping-and-Dining-Car Company)

dsgl desgleichen (German—ditto)

dsgn design; designed; designer

dsgnd designated

D/Sgt Drill Sergeant

dsh deliberate self-harm; domestic short hair (cat)

DSHC Defence Service Homes Corporation

dshe downstream heat exchanger

DSHEA Dietary Supplement Health and Education Act

d s'horn dairy shorthorn

dsi data systems inquiry; digital speech interpolation

DSI Dairy Society International; Dalcroze Society Incorporated; Distilled Spirits Institute; Drinking Straw Institute

DSIATP Defense Sensor Interpretation and Application Training Program

DSIF Deep-Space Instrumentation Facility

dsipt dissipate

DSIR Department of Scientific and Industrial Research

DSIs Directorate of Service Intelligence members or operatives

DSIS Directorate of Scientific Information Services

D-site decoy site

dsj differential space justifier

Dsk Dvorak simplified keyboard

D Sk *Daily Sketch*

dsl deep scattering layer; diesel; doppler speed log

DSL Dampier Salt Ltd; Deep Scattering Layer; Defence Standards Laboratory; Delta Steamship Lines; Dickinson School of Law; Digital Subscriber Line; Dominican Steamship Line

D & SL Denver and Salt Lake (railroad)

DSL *Directory of Special Libraries and Information Centers*

DSLC Defense Logistics Services Center

D-sleep desynchronized sleep; rem sleep

dsl elec diesel electric

ds lt deck surface light

dsltd dry-salted (hides)

dsm demand-side management; dense-staining material; dried skim milk

d & sm dressed and standard matched (lumber)

DSM Des Moines, Iowa (airport); Development Shop Memorandum; Distinguished Service Medal; District Sales Manager; Divisional Sergeant Major

DSM *Diagnostic and Statistical Manual* (of mental disorders)

DSM-III *Diagnostic and Statistical Manual* (of mental disorders–third edition)

dsmb data safety monitoring board

DSMC Defense System Management College

dsmd dismissed

D.S.Met.Eng. Doctor of Science in Metallurgical Engineering

DSMG Designated Systems Management Group

DSM Project Development of Substitute Materials (Manhattan Engineer District secret project from 1942 to 1947, responsible for development of A-bomb)

dsn design

DSN Deep Space Network; Department of School Nurses (NEA)

DSNA Dictionary Society of North America

dsnd descend

dsndi descend immediately

dsnrv double-swivel-nose reentry vehicle

DSNWR De Soto National Wildlife Refuge (Iowa)

dso data set optimizer; deck stowage only; direct shipment order; direct shipping ore

D.So. Doctor of Sociology

DSO Dallas Symphony Orchestra; Defence Systems Officer; Defense Security Officer; Defense Systems Operator; Denver Symphony Orchestra; Detroit Symphony Orchestra; Distinguished Service Order; District Security Office(r); District Service Office(r); District Staff Office(r); District Supply Office(r); Division Signal Officer; Duluth Symphony Orchestra

D.S.O. Doctor of the Science of Oratory

DSOAG Deputy Senior Officer Assault Group

DSOC Democratic Socialist Organizing Committee

D.Soc.Sci. Doctor of Social Science

dsorg data set organization

D.So.Se Doctor of Social Service

dsp digital signal processor; double silver-plated

dsp (DSP) digital signal processor

d.s.p. *decessit sine prole* (Latin—died without issue)

DSP Defense Standardization Program; Defense Support Program; Democratic Socialist Party; Detroit Steel Products; Director of Selection and Personnel; Division Standard Practice

DS & P Duell, Sloan & Pearce

DSPA Deep-Submersible Pilots Association

dspch dispatch; dispatcher

d spec(s) design specification(s)

dsph diopter spherical

dspl disposal

d.s.p.l. *decessit sine prole legitima* (Latin—died without legitimate issue)

dspln disciplinary; discipline

d.s.p.m. *decessit sine prole mascula* (Latin—died without male issue)

d.s.p.m.s. *decessit sine prole mascula superstite* (Latin—died without surviving male issue)

dspn disposition

dspo disposal; dispose; disposition

dsprsl dispersal

dsps delayed sleep-phase syndrome

d.s.p.s. *decessit sine prole superstite* (Latin—died without surviving issue)

DSPS Dynamic Ship-Positioning System

DSPT Dominican School of Philosophy and Theology

d.s.p.v. *decessit sine prole virile* (Latin—died without male issue)

dsq discharged to sick quarters

DSQ Director of Supplies and Quartering

D-squad death squad

dsr data set ready; depolymerized scrap rubber; digit storage relay; digital stepping recorder

dsr (DSR) dynamic spatial reconstructor

ds & r data storage and retrieval; document search and retrieval

DSR Danmarks Radio (Danish radio and tv); desired fuel load (aircraft code); Detroit

Street Railways; Director of Scientific Research; District Sales Representative

DSRC David Sarnoff Research Center (RCA)

DSRD Director(ate) of Signals Research and Development

DSRK Deutsche Schiffs Revision und Klassifikation (German Ship Revision and Classification)

dsRNA double-stranded ribonucleic acid

d's & r's dailies and rushes (motion-picture film editing)

DSRS Data Storage and Retrieval System

dsrv (DSRV) deep-submergence rescue vehicle

dss decision support system; developmental sentence scoring; dioctyl sodium sulfosuccinate; direct station selection; documents signed; dry surface storage; dyregulation spectrum syndrome

Dss Deaconess

DSS Data Systems Services; David (Solomon) Schwab; Deaf Supportive Services; Decision Support System; Defence Signal Staff; Defense Supply Service; Department of Social Services; Department of Supply and Services; Developmentally Stable Strategy; Digital Spread Spectrum; Director of Sea Transport; Director of Social Services; Director(ate) of Statistical Services; Distributed Secure System

DS & S Data Systems and Statistics

D.S.S. Doctor of Sacred Scripture; Doctor of Social Science

D S S & A Duluth, South Shore & Atlantic (railroad)

dssc double-sideband suppressed-carrier

DSSc Diploma in Sanitary Science

DSSC Defense Subsistence Supply Center

DSSCS Defense Special Security Communications System

DSSD Dry Surface Storage Demonstration

DSSH Department of Social Services and Housing

DSSL Delta Steamship Lines

DSSN Disbursing Station Symbol Number

DSSO Defense Surplus Sales Office; Duty Space Surveillance Officer

dssp deep-sea submergence project

DSSRG Deep Submergence System Review Group

DSSS Division of Special Schools and Services

DSSSP Division of Student Support and Special Programs (Office of Education)

DSSV Deep Submergence Search Vehicle

dst doorstop; drop survival time

dst (DST) dexamethasone-suppression test

DST Daylight Saving Time; Dermatology and Syphilology Technician; Desensitization Test (for allergies); Director of Supplies and Transport; Double Summer Time

DS(T) Discarding Sabot (Training)

DST Défense et Sécurité du Territoire (French equivalent of FBI)

DS & T Directorate of Science and Technology (CIA)

D.S.T. Doctor of Sacred Theology

d-std vehicle driver-seated vehicle

D.St.Eng. Doctor of Structural Engineering

d-stg vehicle driver-standing vehicle

dstl distill

dstn destination

DSTO Defense Sciences and Technology Organization (Australian); District Supply and Transport Officer; Divisional Sea Transport Office(r)

DSTP Director, Strategic Target Planning

dstpn dessert spoon

dstr distribution; distributor

dsu dissemination services unit; drum storage unit

dsu/csu data service unit/channel service unit

DSUE Dictionary of Slang and Unconventional English

dsuh direct suggestion under hypnosis

dsuphtr desuperheater

D.Sur. Doctor of Surgery

D.Surg. Dental Surgeon

dsv double silk varnish

dsw door switch

DSW Department of Social Welfare

D.S.W. Doctor of Social Welfare

DSWP Director of Surface Weapons Project

dsws division support weapon system

dsz decrement and skip on zero (calculator)

D Sz Diego Suarez

dt data transmission; date, dead time; defensive tackle; delerium tremens; detective; differential of time (symbol); dinette; diphtheria tetanus; double throw; double time; drain tile; dual tires

dt (DT) deep tank(s)

d-t deuterium-tritium; double-throw

d/t deaths (total ratio); dictaphone typist

d of t deed of trust

dt doit (French—debit)

Dt duration tetanus

Dt. Deuteronomy (Book of the Bible)

DT Daylight Time; Dental Technician; Department of Tourism; Department of Transportation; Department of the Treasury; Detroit Terminal (railroad); Director of Transport(ation); Directorate of Tests; Distance Test; Distance Test(ing); Dylan Thomas

D.T. Dental Technician; Doctor of Theology

DT Daily Telegraph (London); *Danmarks Turistrad* (Danish Tourist Board)

dta daily travel allowance; development test article; differential thermal analysis; distributing terminal assembly; double tape armored cable

DTA Defense Transportation Administration; Democratic Turnhalle Alliance (multiracial); Development Test Article; Differential Thermal Analysis; Diploma in Tropical Agriculture

D.T.A. Democratic-Turnhalle Alliance (of South-African oriented Namibians)

DTA *Divisão de Exploração dos Transportes Aéreos* (Portuguese—Air Transport Exploration Division)

dtas diffuse thalamic activating system

DTASW Department of Torpedo and Anti-Submarine (DTR) Warfare; Director, Torpedo, Anti-Submarine and Mine Warfare

dtbc disturbance

dtc depository transfer check; depository trust company; deposit-taking company; design to cost; direct-to-consumer

DTC Day Training Center; Department of Trade and Commerce

DTC *Deutscher Touring Club* (German Touring Club); *Dictionnaire de Théologie Catholique* (French—*Dictionary of Catholic Theology*)

DTCD Diploma in Tuberculosis and Chest Diseases

D.T.Chem. Doctor of Technical Chemistry

DTCN *Direction Technique des Constructions Navales* (French—Technical Direction of Naval Construction)

DTCS Digital Test Command System

dt c sk don't countersink

dtcw data transfer command word

dtd dated; direct to disc (recording system)

d.t.d. *detur talis dosis* (Latin—let such a dose be given)

DTD Diploma in Tuberculosis; Director(ate) of Technical Development

DTD *Decoratie voor Trouwe Dienst* (Dutch—Decoration for Loyal Service)

DTDRS Direct-to-Disc Recording System

dte data terminal equipment; development; test(ing) and evaluation; diagnostic test equipment; digital television equipment; diploma test of empathy (DTE)

dte (DTE) data terminal equipment

D.Tech. Doctor of Technology

DTEE Division of Technology and Environmental Education

D.T.Eng. Doctor of Textile Engineering

D Ter Dakota Territory (before 1889)

dtf daily transaction file

DTF Deep Test(ing) Facilities; Dental Traders' Federation; Division of Training and Facilities; Domestic Traffic Federation; Domestic Textiles Federation

dtfc differential temperature-flow controller

dtfcd define the file for card

dtfcn define the file for console

dtfda define the file for direct access

dtfdi define the file for device independence

dtfdr define the file data recorder

DT & FE Department of Technical and Further Education (Australian)

dtfis define the file for indexed sequential (files)

dtfmr define the file for magnetic reader

dtfmt define the file for magnetic tape

dtfor define the file for optical reader

dtfph define the file for physical input-output multiplexer

dtfpr define the file for printer

dtfpt define the file for paper tape

dtfsd define the file for sequential direct-access storage device

dtfsr define the file for serial device file

dtg data time group; date time group(ing); display transmission generator

Dtg *Dienstag* (German—Tuesday)

DTGW Director of Guided Weapons Trials

dth delayed-type hypersensitivity

D.Th. Doctor of Theology

DTH Dance Theatre of Harlem; Diploma in Tropical Hygiene

D.Theol. Doctor of Theology

D ThPT Diploma in Theory and Practice of Teaching

dti dial test indicator

DTI Department of Trade and Industry (UK); Direct Trader Input

DT & I Detroit, Toledo and Ironton (railroad)

DTIC Defense Technical Information Center

d-time dream time

dt-sit drunkard

dtl detail; detailed; diode transistor logic

dd (DTL) diode-transistor logic

DTL Detroit Testing Laboratory

dtm designed to meet; duration time modulation

Dtm Dortmund

DTM Diocesan Travelling Mission; Diploma in Tropical Medicine; Director of Transport and Movements

D.T.M. Doctor of Tropical Medicine

DTMB David Taylor Model Basin

DTMBAL David Taylor Model Basin Aerodynamics Laboratory

dtmf dual-tone multifrequency (telephone)

DTMH Diplomate of Tropical Medicine and Hygiene

DTMI Dairy Training and Merchandising Institute

dt mld draft moulded

DTMO Design Test and Mission Operations; Development Test and Mission Operations

dtmp deoxythymidylic acid

DTMS Defense Traffic Management Service

dtn detain; double-tinned

dtn (DTN) diphtheria toxin, normal

DTN Defense Telecommunications Network; Defense Teleprinter Network

DTN *Drug Trade News*

DTNM Devil's Tower National Monument

DTNSRDC David Taylor Naval Ship Research and Development Center (USN)

dto detailed test objective; ditto; dollar tradeoff; due to

dto *descuento* (Spanish—discount)

dt° *direito* (Portuguese—right)

DTO Dental Therapist of Ontario; Director(ate) of Trade and Operations; Disbursing and Transport(ation) Office(r)

DTO *Dansk Teknisk Oplysningstjeneste* (Danish—Technical Information Service)

dtol digital test-oriented language

dtp data type punch; diphtheria, tetanus, pertussis (whooping cough)—combined vaccination

dtp (DTP) directory tape processor

DTP Desk Top Publishing; distal tingling on pressure

dtpa diethylenetriamine pentaacetic acid

dtpb divider time pulse distributor

DTPEWS Design-to-Price Electronic Warfare System

DTPH Diploma in Tropical Public Health

dtps diffuse thalamic projection system

dtr data tape recorder; data terminal ready; daughter; deep tendon reflexes; demand totalizing relay; double tax(ation) relief direct terminal repeat

dtr (DTR) distribution tape reel (data processing)

d/tr documents against trust receipt

DTR Diploma in Therapeutic Radiology

DTRA Defense Technical Review Agency (USA)

DTRC David Taylor Research Center

dtrm determine

DTRP Diploma in Town and Regional Planning

dtrt deteriorate; do the right thing

Dtrt Detroit

dts data transmission service; dense tar surfacing

dt's deep tanks; delerium tremens; dementia tremors

DTS Data Transmission System; Defense Telephone Service; Defense Transportation System; Dynamic Test Station

D & TS Detroit and Toledo Short Line (railroad)

Dtsch Deutsch (German—German)

DTSD Director of Training and Staff Duties

DTSG Data Transmission Study Group

DTSS Dartmouth Time-Sharing System

dtt diphtheria tetanus toxin; duplicate title transferred

DTT Dictionary of Technical Terms

D of TT Dominion of Trinidad and Tobago

dt/tm delayed-time/telemetry

dttp thymidine diphosphate

dtu data transfer unit; data transformation unit

DTU Delft Technical University

dtur departure

dtv digital television; diver transport vehicle

DTV Deutsche Taschenbuch Verlag (German—German Pockethook Publisher)

DTVM Diploma in Tropical Veterinary Medicine

DTVP Developmental Test of Visual Perception

DTW Dance Theater Workshop; Detroit, Michigan (Detroit Metropolitan Airport)

DTWPD Director of Tactical Weapons Policy Division

dtx detoxification

Dtz Dutzend (German—dozen)

DTZ Division Tactical Zone (USA)

Dtzd Dutzend (German—dozen)

du density unknown; depleted uranium; diagnosis undetermined; died unmarried; digital unit; distribution unit; dog unit; drug use; duodenal ulcer; duty cycle

Du Ducal; Duchy; Duke; Dutch

Du Dutch

DU Dalhousie University; Deakin University; Denison University; diagnosis undetermined; Dillard University; Drake University; Drew University; Drexel University, Ducks Unlimited; Duke University; Duquesne University

du 26 ct du 26 mois courant (French—the 26th of this month)

dua digital uplink assembly; directory user agent

DUA Digitronics Users Association

DUADS Duluth Air Defense Sector

DUAH Department of Urban Affairs and Housing

dual dynamic universal assembly language

DUAL Data Use and Access Laboratories

DUAP Dante University of America Press

dub diameter underback; double; dubber; dubbing; dubious

dub. dubius (Latin—dubious)

Dub Dublin

DUB Dublin, Eire (airport)

DUBC Durham University Boat Club

Dubl Dublin; Dubliner

DUBS Durham University Business School

duc demonstration unity capsule

DUC Datatron Users Organization; Distinguished Unit Citation; Durban University College

D.U.C. Doctor of the University of Calgary

Duck Mountain Duck Mountain Provincial Park in western Manitoba and adjacent Saskatchewan

DUCS Deep Underground Communications System

duct. ductile

duct (Latin prefix—conduct or lead)—conductor, ductless

dud (DUD) dependably undependable

Dud Dudley

dudat due date

due distal upper extremity

DUF Democratic Unification Party (Korean); Drug Use Forecasting (drug-control program)

Duff Duffield; Duffle; McDuff

DUFFEL Dutch Far East Lines

Du Fl Dutch Flemish

dufus blundering person, a dip

DUH Duke University Hospital

dui driving under the influence (of alcohol and/or drugs)

DUI Driving Under the Influence (of alcohol, drugs, or narcotics)

duit2me (DUIT2ME) do it to me

Duke Marmaduke; The Duke, actor John Wayne

dukw (DUKW) code letters, pronounced *duck,* for an amphibious automotive vehicle

DUKW amphibious truck

Dul Duluth

DUL Duke University Library; Durham University Library

Dulag Durchgangslager (German—prisoner-of-war transit camp)

dulc. dulcis (Latin—sweet)

Dulles John Foster Dulles International Airport named for a former secretary of state and serving Washington, D.C.

du'log duolog (conservation wherein the conversants talk without listening to one another)

dum died unmarried

d. um. died unmarried

DUM Dublin University Mission(aries)

DUMAND Deep Underwater Muon And Neutrino Detection

Dumb Dumbarton

DUMB Deep Underground Missile Basing; Defensive Umbrella

DUMBO Down Under the Manhattan Bridge Underpass; seaplane used for rescue work (naval symbol)

DUMC Duke University Medical Center

Dumf Dumfries

Dumf & Gall Dumfries and Galloway

dums deep unmanned submersibles

dun. dunnage

Dun Dun Laoghaire (Dunleary); Dunbar; Duncan; Dundalk; Dundas; Dundee; Dundrennan; Dunedin; Dunellen; Dunelm; Dunfermline; Dungarvan; Dungeness; Dunglas; Dunglison; Dunlap; Dunlop; Dunmore; Dunn; Dunnachie; Dunning; Dunnsville; Dunoon; Dunscore; Dunsmuir; Dunstable; Dunstan; Dunvegan; Dunwood; Dunwoody

Dunb Dunbarton

dunc deep underwater nuclear counter

Dunc Duncan

D.Univ. Doctor of the University

Dunk Dunkerque (Dunkirk)

dunna don't know

DUNS Data Universal Numbering System

duo. duodecimo

duod duodenum

duodec duodecimo

duol duologue

dup duplicate; duplicating; duplication

DUP Democratic Unionists Party (Northern Ireland); Diplomate of the University of Paris; Duquesne University Press

D.U.P. Docteur de l'Université de Paris (French—Doctor of the University of Paris)—the Sorbonne

du pa duplicating pattern

DUPA Drug Users Parent Aid

dup^do duplicado (Spanish—duplicate)

dupe. duplicate; duplicate copy

dupe. neg duplicate negative

dupes. duplicates; duplicate copies

dupl duplicate; duplication

dupli duplicate; duplicated; duplication

DUPONT EI du Pont de Nemours & Company

dur duration

dur (Latin prefix—hard)—durable

dur. duris (Latin—hard)

Dur Durango; Durban; Durham

Duraks Durak Ranges of northernmost Western Australia

duralumin durable aluminum-copper-magnesium-manganese alloy

Durant Will and Ariel Durant

DURD Department of Urban and Regional Development

dur. dolor. durante dolore (Latin—as long as the pain lasts)

Durf Durfee's Reports

durg during

durgc during climb

durgd during descent

Durh Durham

Dur Mus Durban Museum

DUS Düsseldorf, Germany (airport)

DUSA Defense Union of South Africa

DUSA Dispensatory of the United States of America

dusam dummy surface-to-air missile

DUSC Deep Underground Support Center (USAF)

dus/testing distinctness, uniformity, and stability testing

DUSW Director(ate) of the Undersurface Warfare Division

dut device under test(ing); dunnage untreated

Dut Dutch; Dutch Harbor

Dutch Leewards Aruba, Bonaire, Curaçao

Dutch provinces Drenthe, Friesland, Gelderland, Groningen, Limburg, North Brabant, North Holland, Overijssel, South Holland, Utrecht, Zeeland, plus the Netherlands Antilles

Dutch Windwards Saba, Sint Eustatius, Sint Maarten

Dutton EP Dutton & Co

Dutz Dutzend (German—dozen)

duv data under voice

duvd direct ultrasonic visualization of defects

DUW Director of Underwater Weapons

dv daily value; dependent variable; device; differential of velocity (symbol); digital video; dilute volume; direct vision; disabled veteran; distemper virus; distinguished visitor; dive; double vibrations; double vision

dV differential of voltage (symbol)

d.v. dorsiventral

d/v declared value

d & v diarrhea and vomiting

d/v demonstration/validation

d/v dias vista (Spanish—days at sight)

DV Diploma in Venereology; Douay Version

D/V Discovery Vessel

D.V. Deo volente (Latin—God willing)

dva dynamic visual acuity

DVA Department of Veterans Affairs

DVA Distribuidora Venezolana de Azucareros (Spanish—Venezuelan Sugar Growers Distributing Organization)

D.V.A. Doctor of Visual Aids

D-value death value (minutes needed for a lethal agent or environment to kill the population)

dvars doppler velocity altimeter radar set

dva test duration of voluntary apnoea test

DVC Delaware Valley Conference; Diablo Valley College; Deputy Vice-Chancellor

DVCSA Delaware Valley College of Science and Agriculture

dvd delayed, very delayed (nickname for digital video disc after many false starts); direct-view device

DV & D Diploma in Venereology and Dermatology

dvd (DVD) digital versatile disc, digital video disc

DVDP Dry Valley Drilling Project

dvd-r (DVD-R) digital versatile disc–record; digital video disc–record

dvd-rw digital versatile disc–rewrite; digital video disc–rewrite

dve device end; digital video effect

Dve Drive

DVECC Disease Vector Ecology and Control Center

d Verf der Verfasser (German—the author)

DVES Defense Value Engineering Services

dvfr defense visual flight rules

dvg digital video generator

DvH Dietrich von Hildebrand

DVH Diploma in Veterinary Hygiene; Division for the Visually Handicapped

d-vhs (D-VHS) digital–very high speed

dvi digital video interactive; digital video interface

DVI Deuel Vocational Institution (California)

dvin deviation

dvl direct voice line

DVLC Driver and Vehicle Licensing Centre

DVTF Domestic Violence Task Force

dvlp development

dvm digital voltmeter

d.v.m. decessit vita matris (Latin—he died during his mother's lifetime)

D.V.M. Doctor of Veterinary Medicine

D.V.M.S. Doctor of Veterinary Medicine and Surgery

DvN D. Van Nostrand

DVNM Death Valley National Monument

Dvnport Devonport

Dvo Davao

DVO Divisional Veterinary Office(r)

dvom digital volt ohmeter

dvp delivery versus payment; differential value profile; direct vision panel

d.v.p. decessit vita patris (Latin—he died during his father's lifetime)

DVPH Diploma in Veterinary Public Health

dvppi daylight-view plan-position indicator

dvr driver

DVR Division of Vocational Rehabilitation

D.V.R. Doctor of Veterinary Radiology

dvrg diverge

DVRP Domestic Violence Recovery Program

dvrsn diversion

dvs descriptive video services

dvs det vill säga (Swedish—that is); *det vil si* (Norwegian—that is); *det vil sige* (Danish—that is)

DVS Director of Veterinary Services; Division of Vital Statistics

D.V.S. Doctor of Veterinary Surgery

D.V.Sc. Doctor of Veterinary Science

DVSL District Venture Scout Leader

DVSM Diploma of Veterinary State Medicine

dvst direct-view storage tube

dvt deep vein thrombophlebitis; deep venous thrombosis

DVTE Division of Vocational and Technical Education

DVTI De Vry Technical Institute

dvtl dovetail

Dvwp *Deo volente,* weather permitting (God willing, weather permitting)

dw data word(ing); deadweight; delivered weight; developed width; diameter width; dishwasher; distilled water; double weight; dry wine; dumbwaiter; dust wrapper

d/w dextrose in water; dock warrant

DW Defenders of Wildlife; Department of Waters; Duke of Wellington's (troops)

DW Deutsches Wörterbuch (German Word Book)

dwa double wire armor(ed)

DWA Deadly Weapons Act; Distributive Workers of America

DWAA Dog Writers' Association of America

DWAAF Director of Women's Auxiliary Air Forces

dwb double with bath

DWB Doctors Without Borders

dwba direct-wire burglar alarm

dwc deadweight capacity

dwcc deadweight cargo capacity

DWCHS De Witt Clinton High School

DWCP Detroit-Wayne County Port

dwd died while drinking; driving while drunk; dumbwaiter door

dw di draw die

DWDL Donald W Douglas Laboratory

dwdm dense wave division multiplexing

dweg died while eating gumbo

dwel dwelling

DWES Director of Weapons Equipment Surface

DWEU Director of Weapons Equipment Underwater

dwf divorced white female

dw fm draw form

dwg drawing; dwelling

DWG Diamond Walnut Growers

dwg-ho dwelling house

DWGNRA Delaware Water Gap National Recreation Area (New Jersey and Pennsylvania)

dwi driving while intoxicated

DWI Descriptive Word Index; Durable Woods Institute; Durham Wheat Institute; Dutch West Indies (Netherlands Antilles)

DWIC Disaster Welfare Inquiry Center

Dwig Dwiggins

dwim do what I mean

dwk death-wish kids (gangsters)

DWK *Deutsche Gesellschaft für Wiederraufarbeitung von Kernbrennstoffen* (German—Society for Reprocessing Nuclear Waste); *Die Wit Kommando* (Afrikaans—The White Commando)

dwl derived working level; designed waterline; displacement waterline; dowel

dwm dangerous waste material; deadweight machine(ry); divorced white male

DWM *Deutsche Waffen und Munitionsfabriken* (German—German Arms and Ordnance Factory)

dwn down

dwndfts downdrafts

dwntn downtown

dwo delta-wing orbiter

DWOP Denver War On Poverty

dwp deepwater port; dyna whirlpool

DWP Department of Water and Power

D W & P Duluth, Winnipeg & Pacific (railroad)

DWPF Defense Waste Processing Facility (Savannah River)

dwpnt dewpoint

dwr drawer

DWR Department of Water Resources; Duke of Wellington's Regiment

DWRAF Director of Women's Royal Air Force

dws drinking-water standards; drop wood siding; double white silk

DWS Department of Water Supply

DWSAP Drinking Water Source Assessment Program

DWSC Director of Welfare and Service Conditions

DWSG & E Department of Water Supply, Gas, and Electricity

DWSO Drainage and Water Supply Office(r)

dwt deadweight ton(nage)(s); deck watch time; denarius weight; double weight; pennyweight

DWT *Deutsche Gesellschaft für Wehrtechnik* (German—German Society for Defense Technology)

dw tk drinking water tank

dwuld dewooled (skins)

dwv drain, waste, and vent

dww downward

dwz *dat wil zeggen* (Dutch—that is)

dx de luxe; dextran; diagnosis; differential of reactance (symbol); distance (ratio); double cash ruled; duplex; static (symbol)

dx (DX) defense exhibit

Dx diagnosis (medical)

DX Aerotaxi (Colombia); distance radio reception or transmission; Sun Ray Mid-Continent Oil (stock exchange symbol)

dxc data exchange control

DXC Penn-Dixie Cement (stock exchange symbol)

dxd discontinued

dxda-mc ductile metals experimental diamond abrasive-metal clad

dxer (DX-er) long-distance radio receptionist

dx-ing long-distance (radio) communicating

dxm dexamethasone

dxr deep X-ray

dxrt deep X-ray therapy

dXt deep X-ray therapy

dy delivery; demy (paper); died young; dock yard; duty; dysplasia; penny (nails)

dY differential of admittance (symbol)

Dy Dylan; dysprosium

Dy *Douay Bible* (Roman Catholic English translation of the Latin Vulgate made at Douay and Rheims in 1610)

D-y *Druk-yul* (Bhutanese—Bhutan)

DY Derbyshire Yeomanry; De Young Memorial Museum; Druk-Yul (Kingdom of Bhutan)

DYA Department of Youth Authority (California)

dyana dynamics analyzer

dyb do your best; dynamic braking

DYB Do Your Best

dy bf hl day before holiday

DYC Detroit Yacht Club; Dominion Yeast Company

dyd dockyard

dydff dyed and fully finished (leather)

dye. dyeing

dyf damned young fool

DYF Democratic Youth Front

dy fi hl day following holiday

DYFS Division of Youth and Family Services

dykes diagonal wire cutters

dymaxion dynamic maximum

DYMM MH De Young Memorial Museum

DYMM (Malay—His Highness the Ruler or Her Highness the Ruler)

dyn dynamic; dynamics; dynamo; dynamometer; Dynasty; dyne

dyna dynamite

dynam dynamic; dynamics; dynamite; dynamo

dynamit dynamic allocation of manufacturing inventory and time

dynamo. dynamic model

DYNAMO Dynamic Action Management Operation

dynasoar dynamic soaring (space flight)

dynatac dynamic adaptive total area coverage

dyncm dyne centimeter

dynmt dynamite

dyno dynamite, undiluted drugs; dynamometer

DYO Duke of York's Own

dypso dypsomania(c)

dy r dynamic response

DYRMS Duke of York's Royal Military School

dys (Latin prefix—bad, difficult; painful)—dysentery, dyspepsia

DYS Department of Youth Services; Department of Youth Services (Alabama); Division of Youth Services; Division of Youth Services (Arkansas)

dysac digitally simulated analog computer

dysen dysentery

dyslex dyslexia; dyslexic

dysm dysmenorrhea

dysp dyspepsia

dysphem dysphemistic(al)(ly); dysphernism(s)—antonym(s) for euphemism(s)

dystac dynamic storage analog computer

dystal dynamic storage allocation language

dysto dystopia(n)

dystope(s) dystopian(s)—slum dweller(s) leading a fear-filled and wretched existence

dyu do your utmost
DYW Dynamic Youth Workers
dz dizygotic; dizziness; dizzy; dozen; drizzle; drop zone
dz *deppelzentner* (German—100 kilograms); *distance zénthale* (French—zenith distance); *dozzina* (Italian—dozen)
dZ differential of impedance (symbol)
d Z *der* Zeit (German—of the time)

Dz *Deniz* (Turkish—sea)
DZ Department of Zoology; Drop Zone
D.Z. Doctor of Zoology
DZA dizygotic twins raised apart; Drop Zone Area
DZF *Deutsche Zentrale für Fremdenverkehr* (German—German National Tourist Association)
dzg dizygotic
Dzl Delfzijl (Dutch port)

dzne *douzaine* (French—dozen)
D.Zool. Doctor of Zoology
dzt digit zero trigger
DZT *Deutsche Zentrale für Tourismus* (German—Directorate for Tourism)
dz twins dizyotic (fraternal) twins
D-Zug *Durchgangszug* (German—express train, through train)
Dzun Dzungaria

E

e base for natural logarithms 2.7182818; coefficient of impact (symbol); ear; early; east(ern); eating; electromagnetic; electron; emitter; emotion; empty; emulsifier; emulsion; end (football); engines; equity; error; errors; estimated; evening; exa(E)—^{1018} (one quintillion); excellence; exertion; experimenter; experimental group; longitudinal strain per unit length (symbol); numerical value of electron charge in an electron or proton (symbol)

'e he

e angle of downwash (symbol); natural logarithmic (Napierian) base; (Portuguese—and); (Spanish—and)—used when the following word begins with *i* or *hi* as in *Juana e Ignacio* or *padre e hijo*

el *envío* (Spanish—sent)

E American Export-Isbrandtsen Lines; declared or paid in preceding 12 months (stock listings); Eagle Airways; Earth; east; eastern; eccentricity of a curve (symbol); Ecclesiastical; Echo—code for letter E; Edinburgh; efficiency; einsteinium; electric field strength (symbol); Elohist; emmetropia; energy (symbol); engineer; engineering; England; English; Equator; equatorial; erbium; estimated weight (symbol); excellent; Exchequer; exempt; expired gas (symbol); eye; Fraunhofer line caused by iron (symbol); instantaneous value alternating current (symbol); modulus of elasticity (symbol).

E east; Einstein unit of energy; (symbol); electromotive force (symbol); (Latin—Egregius); *en* (Dutch, Portuguese, Spanish—in); Envoy Extraordinary and Minister Plenipotentiary; *est* (French or Italian—east); *este* (Portuguese or Spanish—east); *etelä (Finnish—south); experiment (symbol); voltage (symbol)*

E1 estrone (estrogen)

E2 estradiol (estrogen); voltage drop across an impedance

E^1 Lhotse I (27,890-ft adjoining peak of Mount Everest)

EI, E2, etc. East One, East Two, etc. (London postal zones)

E-2 Hawkeye airborne early-warning and fighter-control aircraft

E^2 Lhotse II(27,560-ft adjoining peak of Mount Everest)

E3 estriol (estrogen)

E-14 Hispano Saeta twin-engine jet trainer, also designated HA-200

E 107 tribromoethanol (anesthetic)

E 605 parathion (deadly insecticide)

ea each; early antigen; editor's alteration; ends annealed; enemy aircraft; enlistment allowance

ea (EA) educational age; enrolled agent

e/a (E/A) experimental aircraft

EA East Africa(n); Eastern Air Lines; Economic Adviser; educational age; Egyptian Army; Electrical Artificer; Electronic Artificer; Electronic Associates; Engineer Admiral; Environmental Action; Environmental Agency; Environmental Assessment; expectancy age; experimental aircraft

E/A Ecology Action; enemy aircraft

EA *Ente Autonomo* (Italian—Autonomous Corporation)

EA-6B Grumman electronic-intelligence-gathering aircraft named Intruder

eaa essential amino acid (EAA); ethylene acrylic acid (EAA)

EAA Education Amendment Act; East African Artillery; Electrical Appliance Association; Electrical Artificer (Air); Electronics Association or Australia; Employment Agents Association; Engineer in Aeronautics and Astronautics; Engineers and Architects Association; Equipment Approval Authority; Everglades Agricultural Area; Experimental Aircraft Association; Export Advertising Association

E.A.A. Engineer in Aeronautics and Astronautics

EAA Encyclopedia of American Associations

EAAA European Association of Advertising Agencies

EAAAP Electrical Artificer Air Apprentice

EAAC East African Airways Corporation; East African Armored Corps

EAAEC East African Army Educational Corps

EAAFRO East African Agriculture and Forestry Research Organization

EAAM European Association for Aquatic Mammals

EAAMC East African Army Medical Corps

EAAOC East African Army Ordnance Corps

EAAP European Association for Animal Production

EAASC East African Army Service Corps

EAB Enemy Activities Branch; Ethnic Affairs Bureau; European American Bank; European Asian Bank

EABn Engineer Aviation Battalion

eabrd electrically-actuated band-release device

eac end around carry; erythrocyte antibody complement; estimate at completion

EAC East African Community (Kenya, Tanzania, Uganda); East Asiatic (Line) Container; East Asiatic Company; Eastern Air Command; Estate Agents Cooperative; European Accident Code; European Atomic Commission

eaca (EACA) epsilon-aminocaproic acid

EACC-PDH Employee Assistance Certification Commission-Professional Development Hours

eacd eczematous allergic contact dermatitis

E & A Co Eastern and Australian Steamship Company

ea content effective-agent content

EACR Environmental Area Characterization Report

EACSO East African Common Services Organization

EACU East Asiatic (Line) Container Unit

ead encoded archival description; equipment allowance document; error adjusted; estimated availability date; extended active duty

ead. eadem (Latin—the same)

EAD Employer Association of Detroit

EADB East African Development Bank

EADF Eastern Air Defense Force

eadi electronic attitude and direction indicator

eae experimental allergic encephalitis; experimental allergic encephalomyelitis

EAE East African Engineers; East African English

EAEBP European Association of Editors of Biological Periodicals

EAEC East African Economic Community; European Atomic Energy Community

EAEG European Association of Exploration Geophysicists

EAEI Ecology Action Educational Institute

EAEMC European Airline Electronic Maintenance Committee

EAEME East African Electrical and Mechanical Engineers

EAES Environment-Atmospheric Environment Service (Canada); European Atomic Energy Society

eaf emergency action file

EAFB Eglin Air Force Base (near Pensacola, Florida)

EAFC Eastern Area Frequency Coordinator; Eastern Association of Fire Chiefs

EAFE Europe, Australia and Far East (index)

EAFFRO East African Freshwater Fishery Research Organization

EAG Edmonton Art Gallery; experimental assistant, gunnery

EAGGF European Agricultural Guidance and Guarantee Fund

Eagle Springs Girls Samarkand Manor for females at Eagle Springs, North Carolina

EAHC East African Harbours Corporation; East African High Commission

eahf eczema, asthma, and hay fever

EAI East Asian Institute (Columbia University); Education Audit Institute

EAIC East African Industrial Council

EAID Equipment Authorization Inventory Data

EAJC Eastern Arizona Junior College

eal electromagnetic amplifying lens; estimated average life

Eal English as an additional language

EAL East Asiatic Line; Eastern Air Lines; Ethiopian Airlines

EALA East African Library Association

eam electronic accounting methods; emergency action message

eam (EAM) electrical accounting machine; electronic accounting machine; electronic automatic machine(ry)

EAM Eastern Atlantic and Mediterranean; Emergency Action Message

EAM Ethniko Apelevtherotiko Metopo (Greek—National Liberation Front)

EAMC East African Medical Corps

EAME European, African, Middle Eastern

EAMECM European-African Middle Eastern Campaign Medal

eamedpm (EAMEDPM) electric accounting machine and electronic data processing machine

EAMF European Association of Music Festivals

EAMFRO East African Marine Fisheries Research Organization

EAMLS East African Military Labor Service

EAMP Envoy Extraordinaire and Minister Plenipotentiary

EAMPA East Anglian Master Printers' Alliance

EAMS Empire Air Mail Scheme

EAmst Elsevier Amsterdam

EAMTC European Association of Management Training Centers

EAN Emergency Action Notification

EANA Esperanto Association of North America

EANDC Edgewood Arsenal Nuclear Defense Center; European American Nuclear Data Center

e and e escape and evasion

EANS Emergency Action Notification System (radio broadcasting)

EANSW Electricity Authority of New South Wales

eaon except as otherwise noted

eaos expiration of active obligated service

eap engines, armament, and pyrotechnics; equivalent air pressure; eye artifact potential

eap (EAP) erythrocyte acid phosphatase

EAP Edgar Allan Poe; Emergency Action Procedure; Employee Assistance Programs; English for Academic Purposes; Environmental Analysis and Planning

EAP École d'Administration Penitentiaire (French—Penitentiary Administration School)

EAPA Employment Aptitude Placement Association

EAPC East African Pioneer Corps

EAPD Eastern Air Procurement District

EAPG Eastern Atlantic Planning Guidance (NATO)

EAPR European Association for Potato Research

eaps engine air particle separator

'eap(s) heap(s)

EAPTC East African Posts and Telecommunications Corporation

ear electronic array radar; electronic analog resolver; electronically agile radar; estimate after release

Ea-R Entartungs-Reaktion (German—degeneration reaction)

EAR East African Railways; Edwin Arlington Robinson

EARB Engineering Associates Registration Board

EARC East African Railways Corporation; East African Reconnaissance Corps; Eastern Air Rescue Center;

EARDHE European Association for Research and Development in Higher Education

EAR & H East African Railways and Harbours

EARI Equipment Acceptance Requirements and Inspection

Earnie Ernest; Ernestine; Ernesto

earom electrically alterable read-only memory

earp equipment anti-riot projector

EARS East African Reconnaissance Squadron; Electronic Airborne Reaction System; Electronically Agile Radar System; Emergency Airborne Reaction System;

earssn early season

Earth Planet Sci Lett Earth and Planetary Science Letters

Earth Station trademark of Earth Computer Technologies

eas electronic article surveillance; emergency alert system; equivalent airspeed; estimated air speed

EAs East African shilling

EAS Early American Society; Enterprise Allowance Scheme; Executive Assignment Service; Extended Area Service

EASA Electrical Apparatus Service Association; Engineers Association of South Africa

EASC East African Service Corps

EASE Emigrant's Assured Savings Estate; European Association of Science Editors

easemt easement

EASEP Early Apollo Scientific Experiments Payload

EA sh East African shilling

easi easily

easl engineering analysis and simulation language

EASS Engine Automatic Stop-and-Start System

east. easterly; eastern

East east of the Mississippi, eastern states of the U.S.; Far East

EAST Eastern Australian Standard Time; European Assistance for Science and Technology

EASTAF Eastern Transport Air Force

East African Community Kenya, Tanzania, Uganda

East Bloc Cold War alliance of Albania, Bulgaria, Czechoslovakia, East Germany, Hungary, Poland, Romania, Yugoslavia

Eastcommrgn Eastern Communications Region

East L East Lothian

East Lake Girls Alabama State Training School (for female delinquents) at East Lake near Birmingham

EASTLANT Eastern Atlantic Area

East Los East Los Angeles, California

East Phil Eastern Philharmonic (North Carolina); Eastman Philharmonia

EASTROLANT Eastern Tropical Atlantic

EASTROPAC Eastern Tropical Pacific

East Sutton Park borstal for delinquent girls in Kent, England

easy. efficient assembly system; expense-account spending money

EASY Early Acquisition System (USA); Engine Analyzer System

eat. earliest arrival time (EAT); earnings after taxes; estimated arrival time (EAT); expected approach time (EAT)

e/at. electrons per atom

EAT earliest arriving time; Economic and Technical Committee(s); Experiments in Art and Technology

EATA East Asia Travel Association

EATC Ecology and Analysis of Trace Contaminants

EATCS European Association for Theoretical Computer Science

EATRO East African Trypanosomiasis Research Organization

EATS Empire Air Training Scheme; Equipment Accuracy Test Station; Extended Area Tracking System (USN)

EATTA East Africa Tourist Travel Association

eau extended arithmetic unit

EAU Emergency Assistance Unit

EAU Emiratos Arabes Unidos (Spanish—United Arab Emirates)

EAVE European Audiovisual Entrepreneurs

Eavg average voltage (symbol)

EAVRO East African Veterinary Research Organization

EAVS Edinburgh Artillery Volunteer Corps

eaw Electrical Association for Women; equivalent average words

EAW Electrical Association for Women

EAWOP European Association for Work and Organizational Psychology

EAWP Eastern Atlantic War Plan (NATO)

EAWS East African Wildlife Society

eax electronic automatic exchange

eb electron beam; electronics box; elementary body; emergency brake; engine burn; environmental buoy

e-b estate-bottled

e/b eastbound

eb point d'ebullition (French—boiling point)

Eb battery voltage (symbol); Ebba; Ebed; Eben; Ebenezer; erbium (symbol)

EB Avitour Airlines; Easter Bunny; Eesti Vabariik (Estonian Republic); Electricity Board; Extension Bulletin

E-B Electric Boat (Division of General Dynamics)

E & B Ellerman and Bucknall (Ellerman Lines)

EB Encyclopaedia Britannica; Engineering Bulletin

eb 1 s edge bead one side (lumber)

eb 2 s edge bead two sides (lumber)

EBA Education Boards Association; Energy Business Association; English Bowling Association

EBAA Eye-Bank Association of America

ebacs engine bleed air control system

EBAILL European Bureau for the Allocation of International Long Lines

ebam electron-beam-addressed memory

EBAM Electron-Beam Addressed Memory

ebar edited beyond all recognition

EBAR EB Aabys Rederi (Norwegian freight line)

EBB Elias Baseball Bureau; Elizabeth Barrett Browning; plate-supply voltage (symbol)

ebc enamel bonded single cotton

EBC Educational Broadcasting Corporation; European Bibliographical Center (Oxford, England)

EBCB European Bank of Computer programs in Biotechnology

ebcdic extended binary-coded decimal interchange code

EBCDIC Extended Binary Coded Decimal Interchange Code (pronounced *ebsidick*)

ebce experience-based career education

ebct electron beam computer tomography

ebd education by discussion; effective biological dose

ebd ebenda (German—in the same place)

EBDCs ethylenebisdithiocarbamates

ebds enamel bonded double silk

EBEC Encyclopedia Britannica Educational Corporation

Eben Ebenezer

Eber Eberard; Eberhard; Eberhart; Ebert

EBES Electron-Beam Exposure System; Electronic Banking Economics Society

ebf electronically blown flap; erythroblastosis foetalis; externally blown flap

EBF Encyclopedia Britannica Films

ebfa electron-beam fusion accelerator

ebi earnings before interest; equivalent background input

ebi (EBI) emetine bismuth iodide

EBI Emerson Books, Incorporated

EBIC European Banks International Corporation (lowercase logotype appears as *ebic*)

ebicon electron-bombardment induced conductivity

ebis electron-beam imaging system

ebit earnings before interest and taxes; electron-beam ion trap

ebitda earnings before interest, taxes, depreciation and amortization

ebiv electron-beam-induced voltage

ebk embryonic bovine kidney

EBL Eastern Basketball League

ebm electronic bearing marker; expressed breast milk

EBM Empresa Bacaladera Mexicana (Mexican Codfishing Enterprise)

EBMC English Butter Marketing Company

EBMUD East Bay Municipal Utility District

EBNA Epstein Barr Nuclear Antigen

EBNI Electricity Board for Northern Ireland

Ebnr Ebenezer

EBNY Edition Bookbinders of New York

E-boat enemy boat

Ebor. Eboracensis (Latin—of York); *Eboracum* (Latin—York)

ebp enamel single paper bonded

ebpa electron-beam parametric amplifier

EBQ Empire Brass Quintet

ebr electron-beam recorder; experimental breeder reactor

EBR Emu Bay Railway; Engineering Business Report

EBR-75 Panhard armored car carrying a 75mm gun

EBR-90 Panhard armored car carrying a 90mm gun

EBRA Engineer Buyers' and Representatives' Association

EBRD European Bank for Reconstruction and Development; Export Business Division (U.S. Department of Commerce)

EBRI Employee Benefit Research Institute

ebs electrical brain stimulation; electron-bombarded semiconductor; electron-bombarded silicon; enamel single cotton

eb(s) eager beaver(s)

EBS Emergency Bed Service; Emergency Broadcast System; English Bookplate Society; Ethiopian Broadcasting Service

EBSA Estuarine and Brackish Water Sciences Association

EBSR Eye-Bank for Sight Restoration

EBSRVR East Bengal State Railways Volunteer Rifles

ebt earth-based tug (NASA); electron-beam technique; electronic benefits transfer; examination before trial

EBTI European Binding Tariff Information

EBU European Broadcasting Union

ebul ebullition

EBv Epstein-Barr virus

EBV Epstein-Barr Virus

E-B V Epstein-Barr Virus

ebw electron-beam welder; exploding bridge wire

EBW Elwyn Brooks White

ebwr (EBWR) experimental boiling-water reactor

EBYC European Bureau for Youth and Childhood

ec earth closet (toilet using earth instead of water); economics; electric(al) coding; electrolytic corrosion; electronic calculator; electronic combat; electronic computer; emergency capability; emulsifiable concentrate; enamel coated; engineering change; enteric coated; entering complaint; error correcting; ethyl cellulose; expansive classification; expiratory center; extended coverage; extension and conversion; extension course; exterior closet; extra choice (wool)

e-c ether-chloroform (mixture)

e/c estrogen-to-creatinine (ratio)

ec en cuento (Spanish—on account)

e.c. exempli causa (Latin—for example)

Ec Ecclesiastic; Ecuador; Ecuadorian

EC Earlham College; East African Airways; East Carolina (railroad); East Caribbean; East Central; East Coast; Eastern College; Eastern Command; Ecuador (Internet code); Edgewood College; Electricity Council; Elizabethtown College; Elmhurst College; Elmira College; Elon College; embryonal carcinoma; Emergency Commission(er); Emergency Coordinator; Emerson College; Emmanuel College; Energy Commission; Engineer Captain; Engineering Change; Engineering Construction; Environmental Control; Episcopal Church; Erskine College; Essex College; Established Church; Eureka College; European Communities; European Community; Evangel College; Evansville College; Executive Committee; Executive Council; Exeter College; Explorers Club

EC (followed by numbers) Enzyme Commission (numbers indicate enzyme classification)

E-C Erckmann-Chatrian (collaborators: Emile Erckmann and Alexandre Chatrian)

E & C Engineering and Construction

EC Encyclopedia Canadiana; Era Cristiana (Spanish—Christian Era); *Etoile du Courage* (French—Star of Courage)

EC1, EC2, etc. East Central One, East Central 2, etc. (London postal zones)

eca electronic control amplifier; electronics control assembly

ECA Earthmovers and Contractors Association; Economic Commission for Africa (UN); Economic Control Agency; Economic Cooperation Administration; Educational Communication Association; Educational and Cultural Affairs; Electrical Contractors Association; Engineering Change Analysis; Engineering Contractors Association; Epidemiologic Catchment Area; European Confederation of Agriculture; Exchange Control Act

ECA Empresa de Comercio Agricola (Spanish—Agricultural Commerce Enterprise)

ECAB Early Case Assessment Bureau; Employees' Compensation Appeals Board

ECAC Eastern College Athletic Conference; Electromagnetic Compatibility Analysis Center; Extra-Curricular Activities Center

ecad error check analysis diagram

ECAFE Economic Commission for Asia and the Far East (UN)

ecal equipment calibration

ecam extended communications access method

ecan excitation, calibration, and normalization

ECAP Electronic Circuit Analysis Program; Environmental Compatability Assurance Program (USN)

ECARS Electronic Coordinatograph Readout System

ECAS Electrical Contractors Association of Scotland

ecat emission computerized axial tomography

ECB E(benezer) Cobham Brewer; Electronic Codebook; Energy Conservation Board (US); European Central Bank

e & cb 1 s edge and center bead one side (lumber)

e & cb 2 s edge and center bead two sides (lumber)

ecbo enteric cytopathogenic bovine orphan (virus)

ecc eccentric; electrically-continuous cloth; electrodeposited composite coat(ing); emergency cardiac care; emergency combat capability; equipment classification control; equipment configuration control; equipment control classification; error correction code; execute control cycle

ecc (ECC) electrocorticogram; exchange control copy

ecc eccetera (Italian—et cetera)

Ecc Eccellenze (Italian—Excellency)

Ecc grid-supply voltage (symbol)

ECC Economic Council of Canada; Educational Cultural Complex; Elderly Citizens Club(s); Electronics Capital Corporation; Emergency Conservation Committee; Employees Compensation Commission; End Conscription Campaign (coalition of South African civil-rights organizations); Energy Control Center; English Conservation Center; Ethnic Communities Council; European Coordinating Committee; European Coordinating Council; European Cultural Center; European Cultural Commission

ECCA East Caribbean Currency Authority

ECCA Empresa Consolidada Cubana de Aviación (Spanish—Consolidated Cuban Aviation Enterprise)

ECCAA Executive Chefs de Cuisine Association of America

ECCC English Country Cheese Council

ECCCM Electronic Counter-Counter-Counter Measure

ECCCS Emergency Command Control Communications System

ECCDA Eastern Connecticut Clam Diggers Association

ECCE Exchange and Cooperation between Culture and Enterprise

eccen eccentric; eccentrics

ECCET Engineering Casualty Control Evaluation Team (USN)

Ecc. Hom Ecce Homo (Latin—Behold the Man)

ECCI Executive Committee Communist International

eccl ecclesiastic(al)

Eccl Ecclesiastes

eccles ecclesiastic; ecclesiastical

Ecclus. Ecclesiasticus

eccm electronic counter-counter-measures

eccmo electronic counter-countermeasures operation(s); electronic counter-countermeasures operator(s)

ec^{co} eclesiástico (Spanish—clergyman, ecclesiastic, ecclesiastical, priest)

ECCP East Coast Coal Port; Engineering Concepts Curriculum Project; European Commission on Crime Problems

ECCR Engineering Calibration Cycle Request

ECCs Emitter-Coupled Circuits

ECCS Emergency Core Cooling Systems (AEC)

eccsl emitter-coupled-current steered logic

ECCT Eddy Current Conductivity Test(ing)

ECCTYC English Council of the California Two-Year Colleges

ECCU English Cross-Country Union

ecd early closing day; endocardial cushion defect; estimated completion date

ec&d electronic cover and deception

ec & d electronic components and devices

ECd East Caribbean dollar; monetary unit of Antigua and Barbuda, Grenada, St Kitts and Nevis, St Lucia, St Vincent and the Grenadines

ECD Energy Conversion Devices

EC & D Electronic Components and Devices

ecdc electrochemical diffused-collector transistor

ECDC Economic Cooperation among Developing Countries

ecdi electronic course deviation indicator

ECDIN European Chemical Data and Information Network

ecdn electrical cables down

ECDP Estimating Controlled Data Package

ECDU European Christian Democratic Union

ece eligible capital expenditure; extended coverage endorsement

ECE Early Childhood Education; Economic Commission for Europe (UN)

ECEO Economic Crime Enforcement Office (U.S. Dept. of Justice) to combat white-collar crimes

ECES Educational and Career Exploration System

ecf extracellular fluid

ECF Edgar Cayce Foundation; Electrical Contractors Federation; Enhanced Connectivity Facilities; Episcopal Charismatic Fellowship; European Cultural Foundation; Ex-Communist Forces

ecf-a eosinophil chemotactic factor of anaphylaxis

ECFA Evangelical Council for Financial Accountability

ECFC Employers Council on Flexible Compensation

ECFI Eastern Caribbean Farm Institute

ECFMG Educational Council for Foreign Medical Graduates

ECFMS Educational Council for Foreign Medical Students

ecg export credit guarantee(s)

ecg (ECG) electrocardiogram; electrocardiograph(y)

ECG electrocardiogram

ECGAI Education Council of the Graphic Arts Industry

ECB European Central Bank

ECCE Exchange and Cooperation between Culture and Enterprise

ECGB East Coast of Great Britain

ECGC Empire Cotton Growing Corporation

ECGD Export Credit Guarantee Department

ech echelon; engine compartment heater

ECH European Country Hotels

echo. enteric cytopathogenic human orphan (virus)

echo. (ECHO) electronic computing, hospital oriented

Echo letter E radio code; NATO nickname for Soviet missile-carrying nuclear-powered submarine designated E-class

ECHO Efficient Car-Handling Operations (railroad); Electronic Case-Handling in Offices; European Commission Host Organization; Evidence for Community Health Organization; Experimental Contract Highlight Operation

ECHR European Commission on Human Rights; European Court of Human Rights

ECHS Evander Childs High School

eci employment-cost index; extracorporeal irradiation

ECI Electronic Communications Incorporated; Extension Course Institute (Air University)

ECIAL Enseñanza de las Ciencias y de la Ingeniería en la América Latina (Spanish Teaching of Science and Engineering in Latin America)

ECIC Export Credits Insurance Corporation (Canada)

ECICS Export Credit Insurance Corporation of Singapore

EC-IIP European Communities International Investment Partners

ECIS Error-Correction Information System (NASA)

ECISS European Committee for Iron and Steel Standards

ECITO European Central Inland Transport Organization

ECIUSAF Extension Course Institute, USAF

ECIY Earl of Chester's Imperial Yeomanry

ECJ Erie County Jail (Buffalo); European Court of Justice

ECJA Eastern Collegiate Judo Association

eck embryonic chicken kidney

ECK Brooklyn Eckfords (National Association)

Eck(ie) Alexander; Alexandra; Alexis; Hector; Hecuba

ecl eclipse; electrocardiograph log; electronic crash locator (aircraft); extended center line

ecl (ECL) emitter-coupled logic

ecl éclairage (French—lighting)

ECL Electronic Components Laboratory; Engineering Computing Laboratory; Entertainment Corporation Limited; Equipment Component List; Europe-Canada Line

ECL Encyclopedia of Comparative Letterforms

ECLA Economic Commission for Latin America (UN)

ECLAIR European Collaborative Linkage of Agriculture and Industry through Research

ECLAS European Commissioners' Library Automated System

E-class NATO designation for a Soviet class of missile-carrying nuclear-powered submarines also known as Echo

eclec eclectic; eclecticism

ecli eclipse; ecliptic

eclo emitter-coupled logic operator

ecm electric coding machine; electrochemical machining; electronic countermeasure(s); ends matched, center (lumber); extended core memory

ECM electronic countermeasures; Engineering Change Management; European Common Market

EC & M Electric Controller and Manufacturing (company)

e/cm³ electrons per cubic centimeter

ECMA Engineering College Magazines Associated; European Computer Manufacturers Association

Ecmalgol European Computer Manufacturers Association Algorithmic Language

ECMCA Eastern-Central Motor Carriers Association

ecmd electronic countermeasures display

ecme electronic countermeasures equipment

ECME Economic Commission for the Middle East (UN)

ecmex electronic countermeasures exercise

ECMF Electric Cable Makers' Federation

ECM & FS East Coast Marine and Ferry Service

ECMHP East Coast Migrant Health Project

e-c mix. ether-chloroform mixture

ecmo (ECMO) enteric cytopathogenic monkey organ (virus); extracorporeal membrane oxygenation (pulmonary therapy)

ECMR Eastern Contract Management Region

ECMRA European Chemical Market Research Association

ECMS Engine Condition Monitoring System

ECMSA Electronics Command Meteorological Support Agency (USA)

ecmtng electronic countermeasures training

ECMWF European Centre for Medium-range Weather Forecast

ecn electronic communications network

ECN Engineering Change Notice

ECNA East Coast of North America

ECNOS Eastern Atlantic, Channel, and North Sea (orders for ships given by NATO)

eco ecological; ecologist; ecology; economic; economist; economics; electron-coupled oscillator; exempted by commanding officer

eco (Latin prefix—home or house)—ecology, economy, ecosphere, ecosystem

ECO East Coast Overseas; Economic Corporation Organization; Effective Citizens Organization; Emergency Commissioned Officer; Engineering Change Order; Environment and Conservation Organizations; Environmental Control Organization; European Coal Organization

ECOA Equal Credit Opportunity Act; Equipment Company of America

ECOCEN Economic Cooperation Center

ecocrit economic critic(ism)

ecodoom large-scale ecological destruction

ecodoomster(s) predictor(s) of large-scale ecological destruction

ECOFIN Economic and Finance

ecofuel ecology fuel (made from garbage and other wastes generated by man and his domestic animals)

ecog electrocorticogram

ecogeo ecogeographer; ecogeographic(al)(ly); ecogeography

ECOIN European Core Inventory (of chemicals)

ecol ecology
Ecol Ecology
ecolcrit ecological criticism; ecology critic
E coli Escherichia coli (intestinal bacillus)
Ecol Soc Am Ecological Society of America
ecom (ECOM) electronic computer-originated mail
ECOM Electronic Computer-Originated Mail (postal service); Electronics Command (USA)
ECOMINAS Empresa Colombiana de Minas (Spanish—Colombian Mining Enterprise)
econ economic; economics; economist; economy
e con. e contrario (Latin—on the contrary)
Econ Economics
Econ Jrnl Economic Journal
economan effective control of manpower
economet econometric(al)(ly); econometrician; econometrics; econometrist
Econ Rev Economic Review
ECOP Extension Committee on Organization and Policy
ECOPETROL Empresa Colombiana de Petróleos (Spanish—Colombian Petroleum Enterprise)
ecophys ecophysiologic(al)(1y); ecophysiologist; ecophysiology
ecopow(s) economic superpower(s)—U.S.A., Japan, Germany
Eco Pty Ecology Party
ECOR Engineering Committee on Ocean Resources
ECORS Eastern Counties Operational Research Society
EcoSoc Economic and Social (Council)
ecosupow(s) economic superpower(s)—U.S.A., Japan, Germany
eco system ecological system economic system
Eco-Tech Economic-Technology
ecotopia(n) ecologically ideal utopia(n)
ecotour. ecotourism; ecotourist
ecou electric clip-on unit; electronic clip-on unit

ECOWAS Economic Community of West African States (Benin, Burkina Faso, Cape Verde Islands, Ivory Coast, Gambia, Ghana, Guinea, Guinea-Bissau, Liberia, Mali, Mauretania, Niger, Nigeria, Senegal, Sierra Leone, Togo)
ECP Engineering Change Proposal; Estonian Communist Party; Examiner of Commercial Practices, Executive Control Program
ECPA Electronic Communications Privacy Act (1986; prohibits monitoring cellular telephones); Energy Consumers and Producers Association; Evangelical Christian Publishers Association
ECPAC East County Performing Arts Center
ECPD Engineers Council for Professional Development
ecpiu electronic circuit plug-in unit
ecpnl equivalent continuous perceived noise level
ecpo enteric cytopathogenic porcine orphan (virus)
ECPO Environmental Characterization Projects Office
ecpog electrochemical potential gradient
ECPR European Consortium for Political Research
ecp(s) external casing packer(s)
ECPS European Center for Population Studies
ecpt egress cockpit procedure trainer
ECPTA European Conference of Postal and Telecommunication Administrations
ECQAC Electronic Components Quality Assurance Committee
ecr electronic cash register; enemy contact report; energy consumption rate; error cause removal; extended chromosome region; external channels ratio
ECR Engineering Change Request; Environmental Characterization Report; European Commercial Register
E Cr Engineer Commander
ECRB Export Control Review Board

ECRC Earth Colonization Research Center; Electronic Component Reliability Center; Engineering College Research Council
ECRL Eastern Caribbean Regional Library
ecro erection counter readout
ECRO European Chemoreception Research Organization
ECRs Enemy Contact Reports
ECRS Empty Car Routing System
ecs electrocerebral silence; electroconvulsive shock; emergency care services; emperor's clothes syndrome; ends cut square; error correction servomechanism; error correction signals; extended core storage
ECS Education Commission of the States; Electrochemical Society; Electronic Composing System; Employment Counseling Service; Engineering Change Schedule; Engineering Change Sheet; Environmental Control Systems; Episcopal Community Services; Equipment Concentration Sites; Equipment Configuration Study; Etched Circuit Society; European Communications Satellite; Experimental Communications Satellite (NASA)
ECS Échantillons Commerciaux—(French—commercial samples)
ECSA East Coast of South America; European Communication Security Agency; Expanded Clay and Shale Association
ECSC European Coal and Steel Community
ECSCF Eastern Connecticut State College Foundation
ECS/HCS Education Career Services/Health Career Service
ECSIL Experimental Cross-Section Information Library (University of California—Livermore)
ecss extendable computer system simulator
ECST European Convention on the Suppression of Terrorism

ECSTC Elizabeth City State Teachers College

ECSU Educational Cooperative Service Unit

ect electro-convulsive therapy; engine cutoff time; enteric coated tablet; estimated completion time

ect (ECT) electroconvulsive treatment

ECT Edge Crush Test; Environmental Control Technology (DoE division); Equivalence Conversion Training; European Container Terminus

ECTA Economics and Commercial Teachers Association; Electrical Contractors' Trading Association; Electronic Components Test(ing) Area

ectl emitter-coupled transistor logic

ecto (Latin prefix—external, outer, outside)—ectoderm

ectohorm ectohormonal; ectohormone

ectomy (Latin suffix—surgical removal)—tonsillectomy

ectr endoscopic carpal-tunnel release

ECTS Economic Communtiy Transfer System

ecu ecumania(c); ecumenism; electronic computing unit; engine compatability unit; environmental control unit; extra closeup; extreme closeup

ecu (ECU) extra closeup; extreme closeup

Ecu Ecuador; Ecuadorean; European currency unit

Ecu (ECU) European currency unit

ECU East Carolina University; Economic Crime Unit; English Church Union; European Customs Union

Ecua Ecuador; Ecuadorean

Ecuador Republic of Ecuador, *Republica del Ecuador*

ecube energy conservation using better engineering

ECUC Education Credit Union Council

Ecu Con Ecumenical Conference; Ecumenical Council

ecufuel eucalyptus-tree fuel

ECUK East Coast of United Kingdom

Ecum Ecumenical

ecumen ecumenical(ism)(ist); ecumenicist; ecumenicity; ecumenics; ecumenism

ECUSA Episcopal Church of the U.S.A.

ecusat ecumenical satellite

ECUSATCOM Ecumenical Satellite Commission

ecux. executrix

ecv estimated cash value; extracellular virus

ecv (ECV) energy conservation vehicle

ECVP European Community Visitors' Program

e & cV 1 s edge and center-V one side (lumber)

e & cV 2 s edge and center-V two sides (lumber)

ecw extracellular water

ECW Episcopal Church Women

ECWA Economic Commission for Western Asia (UN)

ECY European Conservation Year

ECYC Earl Of Chester's Yeomanry Cavalry

ECYO European Community Youth Orchestra

ecz eczema(tic)

ed edge distance; edit; edited; edition; editor; editorial; educate; educated; education; educational; educator; effective dose; electronic displays; emergency department; emotionally disturbed; enemy dead; enumeration district; error detecting; erythema dose; excused from duty; ex-dividend; existence doubtful; extra dividend; extra duty

ed (ED) erectile dysfunction

e & d (E & D) exploration and development

ed edición (Spanish—edition); *édition* (French—edition); *edizione* (Italian—edition)

e_d price elasticity of demand

Ed Edgar; Editor; Edmond; Edmund; Edson; Edvard; Edvardson; Edward; Edwin

Ed. Editor

Ed edifucio (Spanish-building)

E$ Eurodollar (American dollar deposited in Europe)

ED Consolidated Edison Company (stock exchange symbol); Eastern District; Economics Division; Education

Department; Efficiency Decoration; Elder Dempster Line; Electric Dynamic; Engineering Data; Engineering Depot; Engineering Design; Engineering Draftsman; Export Declaration

E-D Electro-Dynamics (division of General Dynamics); Elsevier-Dutton

E.D. Doctor of Engineering

ed₅O median effective dose

eda early departure authorized; equipment design agent; equivalent design axles; erection digital assembly

E d A *Ejercito del Aire* Spanish-Air Force)

EDA Economic Development Administration (of United States Department of Commerce); Electrical Development Association; Employment Development Act; Environmental Development Administration; Environmental Development Agency

edac error detection and correction

EDAC Engineering Decision Analysis Company; Evaluation, Dissemination, Assessment Centers

E da M Escuatrão da Morte (Portuguese—Death-Squad)—Brazilian right-wing terrorists

EDANA European Disposables and Non-Wovens Association

EDAQ Electrical Development Association of Queensland

EDARR Engineering Drawing and Assembly Release Record

edb emergency dispersal base(s); end of data block(s); ethene dibromide (EDB); extended double base

edb (EDB) ethylene dibromide (cancer—productive)

edb elektronisk databehandling (Dano—Norwegian—electronic data handling)

Edb Edinburgh

Ed.B. Bachelor of Education

EDB Economic Development Board; Energy Development Board

edbiz educational business

EDBP Epidemiology, Demography, and Biometry Program

Edc electronic digital computer; emergency digital computer; energy distribution curve; engine-driven compressor; error detection and correction; estimated date of completion; estimated date of confinement; extra dark color

EDC Eastern Defense Command; Economic Development Corporation; Educational Development Centers; Educational Development Corporation; Educational Development Council; Engineering Data Consultants; European Defense Community; European Documentation Centres; Export Development Corporation

EDCC Environmental Dispute Coordination Commission

edcl electric discharge coaxial laser

edcn education

edcom editor-compiler

edcp engineering design change proposal

E/DCP Equipment/Document Change Proposal

EDCPF Environmental Data Collection and Processing Facility (USA)

EDCs Economic Development Committees

edcsa effective date of change of strength accountability

edcv enamel double cotton varnish

edcw external-device control word(ing)

edd electronic data display; envelope-delay distortion; estimated delivery date; expected date of delivery

edd *ediderunt* (Latin—published by)

edd. *editiones* (Latin-editions)

Ed. D. Doctor of Education

EDD Eastman Dental Dispensary; Economic Development Division (Singapore); Employment Development Department; Engineering Data Depository; Engineering Development Department; Engineering Development and Design; Engineering and Development Directorate (NASA)

EDD *English Dialect Dictionary*

eddf error detection and decision feedback

Eddie Edgar; Edmund; Edoardo; Edouard; Edsel; Eduard; Eduardo; Edvard; Edward; Edwin; Edwina

EDDS Electronic Devices Data Service

E-D DS Elsevier-Dutton Distribution Services

Eddy Edgar; Edmund; Edward; Edwin; Edwina

ede electronic defense evaluator; emitter dip effect

ede (Latin prefix—swelling)—edema

EDE Electrical Design Engineering; Electronic Design Engineering; Elevator Design Engineering; Engineering Design Establishment

edent edentate; edentulous

EDEO Episcopal Division Ecumenical Officers

edexs education of exceptional students

EDF Environmental Defense Fund; European Development Fund; Everyman Defense Fund

Edg Edgar

EDG Economic Development Group; Export Development Group(ing)

Edgar The Edgar (bust of Edgar Allan Poe given for the best mystery novel)

EDGAR Electronic Data Gathering, Analysis, and Retrieval

EDGB Export Development Grants Board

edge. electronic data-gathering equipment

edh efficient deck hand

EDH Ego-Distonic Homosexuality

edhe experimental data-handling equipment

edi electron-diffraction instrument; electronic data interchange

EDI Economic Development Institute; Edinburgh, Scotland (airport); Electronic Data Interchange; Engineering Department Instruction

EDIA Engineering Department Instruction Amendment

edic electric diesel injection control

edict engineering document information collection technique

Edie Edith

EDIL Electronic Document Interchange between Libraries

Edim *Edimburgo* (Portuguese or Spanish—Edinburgh)

Edin Edinburgh

Ed-in-Ch Editor-in-Chief

edinet education instruction network

EDIP European Defence Improvement Program (NATO)

edis electronic delayed impact sensing (fuse)

EDIS Engineering Data Information Service; Engineering Data Information System

E-disk emulated-disk

edit editing; edition; editor; editorial

EDIT Estate Duties Investment Trust

editar electronic digital tracking and ranging unit

EDITH Exit Drills in the Home (for fire prevention)

EDITS Electronic Data Information Technical Service; Experimental Digital Television System

edl edition de luxe; electric(al) discharge laser

EDL Elder Dempster Lines; Every-Day Life (psychological test); Executive Data Link

EDLD Employee Daily Labor Distribution

EDLNA Exotique Dancers League of North America

edlp everyday low prices

edm early diastolic murmur; electrical-discharge machining; electromagnetic discharge measuring; electrostatic discharge machining

Edm Edmund; St Edmund's House, Cambridge

Ed. M. Master of Education

EDM Engineering Development Mode

EDM *Engineering Drafting Manual* (USAF)

EDMICS Engineering Data Management Information Control System

Edm & Ips St Edmundsbury and Ipswich

Edmn Edmonton

Edmo Edmonton (inhabitants—Edmontonians)

EDMS Engineering Data Microreproduction System

edn edition; electrodesiccation

Edn Edwin

EDN Engineering Department Notice

EDNA Emergency Department Nurses Association

Ednbgh Edinburgh

EDNY Eastern District, New York

edo effective diameter of objective, error demodulator output; error detector output

EDO Employee Development Officer; Engineering Duty Officer; Engineering Duty Only

edoc effective date of change

Ed Op Edmonton Opera Association

EDOPAC Enlisted Personnel Distribution Office Pacific Fleet

edp emotionally disturbed person; engineering data processing

edp (EDP) electronic data processing

EDP Educational Data Processing; Engineering Design Proposal

EDPA Exhibit Designers and Producers Association

EDPAA Electronic Data-Processing Auditors Association

edpac electronic data processing air conditioning

EDPC Electronic Data-Processing Center

edp crimes electronic data-processing crimes

edpe electronic data processing equipment

ed-ped-psych-soc education-pedagogy-psychology-sociology

edpm electronic data processing machine(s)

EDPRESS Educational Press Association of America

EDPS Electronic Data Processing System

EDPT Electronic Data Processing Test

Ed & Pub *Editor and Publisher*

edr electrical distance recorder; electrodermal response; electronic decoy rocket; equivalent direct radiation

EDR Engineering Data Requirements; Engineering Division Regulation(s); Equipment Damage Report; European Depository Receipt

EDRA Environmental Design Research Association

EDRI Electronic Distributors Research Institute

edrl effective damage risk level

EDRs European Depository Receipts

EDRS Education Document Reproductive Service; Engineering Data Retrieval System

edrt effective date of release from training

eds editors; enamel double silk; estimated date of separation

ed's endangered species

EDs Explosive Disposal specialists

E-Ds Ehlers-Danlos syndrome

EDS Electronic Data Systems; Electronic Devices Society; Engineering Data Sheet; English Dialect Society; Environmental Data Service; Episcopal Divinity School; European Demonstration Scheme

edsac electronic delayed-storage automatic computer

edsat educational television satellite (EDSAT)

EDSC Engineering Data Support Center (USAF)

Ed. Spec. Educational Specialist

edst elastic diaphragm switch technology

EDST Eastern Daylight Saving Time

edsv enamel double silk varnish

edt effective date of training; electronic data transfer

EDT Eastern Daylight Time

edta ethylene diamine tetraacetic (acid)

EDTA European Dialysis and Transplant Association

edtr experimental, developmental, test, and research

edts electronic data transfer system

edtsr electronic dial tone speed register

edu electronic display unit; experimental diving unit

EDU European Democratic Union; European Drugs Unit

educ education; educational; educator

Educ Education

Educational Film Educational Film Library Association

Educ Digest *Educational Digest*

educom education communication(s)

Educ Pr Educational Press; Educational Press Association of America

Educ Pub Educational Publications Services; Educational Publishers

educrat educational bureaucrat

educrit educational critic(ism)

educ(s) eductor(s)

edui extreme driving under the influence

EDUPLAN *Oficina de Planeamiento Integral de Educación* (Spanish—Office of Integral Planning in Education)

edutainment educational entertainment (via tv)

edutele educational television

edutherap educational therapist; educational therapy

edv end-diastolic volume

edv *eau de vie* (French—water of life)—cognac

edvac electronic discrete variable automatic computer

edw energy dump window

Edw Edward

Edw VII King Edward VII

Edw VIII King Edward VIII

Edward G Edward G Robinson (Emmanuel Goldberg)

eDx electrodiagnosis

'Eduy. '*Eduyyot*

ee eased edges (lumber); electric eye (camera); embryo extract; employee; entry establishment; environmental education; equine encephalitis; exoelectron(ic)(al) emission; expiration of enlistment; expressed emotion; extended edition; extra efficiency; eye and ear

ee (EE) errors excepted

e/e electrical/electronic

e & e evacuation and evasion; evasion and escape; eye and ear

e-to-e end-to-end

'ee thee

EE Early English; Electrical Engineer(ing); Electronics Engineer(ing); Envoy Extraordinary; Estonia (Internet code)

EE Estado Espaniol (The Spanish State); *Euer Ehrwürden* (German—Your Reverence)

E.E. Electrical Engineer

eea engineering evaluation article

EEA Electronic Engineering Association; Emergency Employment Act; Environmental Education Act; Ethical Education Association; European Economic Area; European Environment Agency

EEA Electrical and Electronic Abstracts

EEAC Energy and Education Action Center

EEAIE Electrical, Electronic, and Allied Industries of Europe

eeat end-of-evening astronomical twilight

EEB Eastern Electricity Board; Educational Employees Board

EEB Enosis Ellenon Bibliotekarion (Modern Greek— Greek Library Association)

EEBA Energy Efficient Building Association

eec electronic engine control

EEC East Erie Commercial (railroad); Education Exploration Center; English Electronic Computers; European Economic Community (Belgium, Britain, Denmark, France, Germany, Greece, Ireland, Italy, Luxembourg, Netherlands, Portugal, Spain); European Energy Charter

EECA Engineering Economic Cost Analysis

eecom electrical, environmental, and communications

EECS Electrical Engineering and Computer Service

e.e. cummings Edward Estlin Cummings

eed elastic energy density; electrical explosive device

eee eastern equine encephalitis; eastern equine encephalomyelitis

EEE Environmental-Ecological Education (program)

EEEP Entry Employment Experience Program

EEF Egyptian Expeditionary Force

eefi essential elements of friendly information

eeg(EEG) electroencephalogram; electroencephalograph

EEG Environmental Education Group

ee/ha ewe equivalents per hectare

EEI Edison Electric Institute; Environmental Equipment Institute; Essential Elements of Information

EEIA Electrical and Electronic Insulation Association

EEIBA Electrical and Electronic Industries Benevolent Association

EEIC Electrical/Electronics Insulation Conference

EEIG European Economic Interest Grouping

EEIS Evanston Early Identification Scale

EEL Ecology and Epidemiology Laboratory; Engineering Electronic Laboratory; English Electric Limited; Evans Electroselenium Limited

eelv evolved expendable launch vehicle

eem Electronic Engineers Master (catalog)

EEMP Envoy Extraordinary and Minister Plenipotentiary

eems enhanced expanded memory specification

EE & MP Envoy Extraordinary and Minister Plenipotentiary

een exceptional educational needs

e'en even; evening

E Eng Early English

E-engine(s) electric engine(s)

EENT end, evening nautical twilight; eye, ear, nose, and throat

EENWR Exe Estuary National Wildlife Refuge (England)

eeo equal employment opportunity

ee & o excuses, errors, and omissions

EEO Energy Efficiency Office

EEOC Economic Employment Opportunity Committee; Equal Employment Opportunity Commission

eep early entrance program; electronic evaluation and procurement; electronic event programmer(s); emergency essential personnel

EEP Energy Emergency Plan(ning); Export Enhancement Program

eepnl estimated effective-perceived noise level

eeprom electrically erasable programmable read-only memory

eer energy efficiency rating; energy efficiency ratio; explosive echo ranging

e'er ever

EER Engineering (Equipment) Evaluation Report; Experimental Ecological Reserves

EERA Electrical Equipment Representatives Association

EERC Earthquake Engineering Research Center (NSF)

EERI Earthquake Engineering Research Institute; Experience, Education, and Research Institute

EERL Electrical Engineering Research Laboratory (University of Texas)

EEROM: Electronically Erasable Read-Only Memory

ees electronic environment simulator

EES Electrical Engineering Squadron; Engineering Experiment Station; Enlisted Evaluation System; European Economic Space; European Exchange System.

E E Somerville and V Martin Edith Somerville and Vivian Martin

EESS Encyclopedia of Engineering Signs and Symbols

eet estimated elapsed time

EET Eames Eye Test; Eastern European Time; Education Equivalency Test; Engineering Evaluation Test(ing)

E-et-L Eure-et-Loire

EETS Early English Text Society

EEU European Economic Unit

EETU Electrical Electronic Telecommunication Union

EEUA Engineering Equipment Users Association

EEUU Estados Unidos (Spanish—United States)

EEV English Electric Valve (company)

EEVC English Electric Valve Company

eex electronic egg exchange (computer program)

EEXCEL Educational Excellence for Children with Environmental Limitations

eez (EEZ) eastern economic zone; exclusive economic zone

ef each face; electroflotation; elevation finder; equivalent focal length; expectant father; experimental flight; exposure factor; extra fine; extremely fine

ef (Ef) electrofocus

e&f ebb and flood (of seabeach tide)

EF Eastern Fleet; Educational Foundation; Emergency Fleet; Engineering Foundation; Expeditionary Force

E & F Elders and Fyffes (steamship line)

EF Europaiske Faellesskaber (Danish—European Common Market)

efa essential fatty acids

EFA Eastern Finance Association (publishers); Editorial Freelancers Association; Empire Forestry Association; Environmental Financing Authority; Epilepsy Foundation of America; European Free Associations; Evangelical Friends Alliance

EFA Empresa Ferrocarriles Argentinos (Spanish—Argentine Railway Enterprise)

EFAS Enroute Flight Advisory Service

efc earth fixed coordinate; electronic feedback carburetor; engineered for color (tv); Evergreen Fir Corporation (initials)

e & fc examined and found correct

EFC Eastern Football Conference; Educational Facilities Center; Electronic Fabrication Center; Escort Force Commander; European Forestry Commission; European Freedom Council

EFCB Emergency Financial Control Board

EFCX Evergreen Freight Car Express

efd electro fluid dynamics; excused from duty

EFDO European Film Development Office; European Film Distribution Office

EFDSS English Folk Dance and Song Society

efe early fuel evaporation; endoctrinal fibro-elastosic; expected field emergence

EFEA European Free Exchange Area

EFEA Empresa Ferrocarriles del Estado Argentino (Spanish—Argentine State Railways)

EFEC Efforts From Ex-Convicts (Washington, D.C.'s parole project)

eff effect; effective; efficiency

eff effeto (Italian—bill, promisory note)

EFF Educational Freedom Foundation; Electronic Frontier Foundation; European Furniture Federation

effcy efficiency

effect. effective; effectivity

effer efferent

EFFI Educational Foundation for the Fashion Industries

Effie award for effective advertising; Euphemia

Effigy Mounds Effigy Mounds National Monument on the Mississippi in northeastern Iowa

effl efflorescent

eff wd effective wind

EFG Educational Fee Grant(s); Edward FitzGerald

EFH Eileen F Hodges

efi electronic flight instruments; electronic fuel injection

ef & i engineer, furnish, and install

EFI Educational Forces Inventory; Electronic Fuel Injection (system); Expeditionary Forces Institute

EFIB European Freight Inspection Bureau

EFIC Export Finance and Insurance Corporation

EFICS European Forestry Information and Communication System

eficon electronic financial control

EFILWC European Foundation for the Improvement of Living and Working Conditions

EFIN Electronic Filing Number

EFINS Enrico Fermi Institute for Nuclear Studies (Univ of Chicago)

efl effective focal length; emitter-follower logic

Efl (EFL) English as a foreign language

EFL Educational Facilities Laboratories

EFLA Educational Film Library Association

EFLC Engineers Foreign Language Circle

efm eight-to-fourteen modulation; electronic fuel management; electronic fuel metering

efm (EFM) electronic fetal monitor(ing)

EFM European Federalist Movement

efmamjjasond enero, febrero, marzo, abril, mayo, junio, julio, agosto, septiembre, octubre, noviembre, diciembre (Spanish—January, February, March, April, May, June, July, August, September, October, November, December)

EFMCNTA Elastic Fabric Manufacturers Council of the Northern Textile Association

EFMG Electric Fuse Manufacturers Guild

EFMM Education for Mission and Ministry

EFNEP Expanded Food and Nutrition Education Program

EFNIR Exhibition/Festival for New Instrumental Resources

EFNS Educational Foundation for Nuclear Science

EFOs errors, freaks, and oddities

efp effective filtration pressure; electric(al) fuel propulsion; end of flight plan

efp (EFP) electronic field production (on-location videotape production); emergency firing panel

EFPA Educational Film Producers Association

EFPW European Federation for the Protection of Waters

efr effective filtration rate; engine firing rate

EFR Electronic Failure Report(ing)

EFRC Edwards Flight Research Center

E Fris East Frisian

efs economic farm surplus; equivalent standard fillet

EFS Edinburgh Festival Society; Emergency Feeding Service; Emergency Fire Service(s)

EFSA European Federation of Sea Anglers

EFSC European Federation of Soroptimist Clubs

EFSS Emergency Food Supply Scheme

eft earliest finish time

eft (EFT) electronic funds transfer

EFT Electronic Funds Transfer; Embedded Figures Test; Engineering Flight Test

EFTA Electronic Funds Transfer Association; European Free Trade Association (Austria, Finland, Iceland, Luxembourg, Norway, Sweden, Switzerland)

EFTC Electrical Fair Trading Council

eftf efterfölger (Dano—Norwegian—successor)

Eftf(lg) Efterfölgere (Dano—Norwegian—successor)

EFTI Engineering Flight Test Instrumentation

eftm eftermiddag (Norwegian—after noon)—p.m.

efto encrypt for transmission only

EFT-POS Electronic Funds Transfer initiated at Point of Sale

EFTS Electronic Funds-Transfer System; Elementary Flying Training School

efu energetic feed unit; environmental force unit

EFU European Football Union

EFU Europäische Frauenunion (German—European Women's Union)

efv equilibrium flash vaporization

EFVA Education Foundation for Visual Aids

EFWA Eastern Farmworkers Association

eg electrogalvanized; electronic guidance; erythrocyte ghost

e.g. exempli gratia (Latin—for example)

Eg Egypt; Egyptian

EG Egypt (Internet code); Electrographic (copier); Engineers Guild; Equatorial Guinea (formerly Spanish Guinea); Evaluation Group(ing); grid voltage (symbol)

ega enhanced graphics adaptor

EGA Elizabeth Garrett Anderson (hospital); Ethics in Government Act; European Golf Association

egad oh god

egad. electronegative gas detector

egads electronic ground automatic destruct sequencer (system for destroying malfunctioning missiles)

egal egalitarian(ism)

EGAT Electricity Generating Authority of Thailand

Egb Egbert

e-g-b-d-f (musical mnemonic—every good boy does fine)—treble clef note names of the five lines

egc electrogalvanized coated

egcr experimental gas-cooled reactor

EGCRNR Eilat Gulf Coral Reef Nature Reserve (Israel)

EGCS English Guernsey Cattle Society

egd electrogasdynamics; esophagogastroduodeno—scopy

egdg electrogasdynamic generator

EGDS Equipment Group Design Specifications

ege expected grade equivalent

ege eau, gaz, electricite (French—water, gas, electricity)

Egeo Mar Egeo (Spanish—Aegean Sea)

E Ger East Germany

egf epidermal growth factors

egg. electrogastrogram

EG & G Edgerton, Germeshausen & Grier

eggler egg + dealer (an egg dealer)

eggsan egg sandwich

eggwich egg sandwich

EGH Elbert Green Hubbard *(Little Journeys)*

ECI Export Consignment Identifier

EGIFO Edward Grey Institute of Field Ornithology

Egip Egipto (Portuguese or Spanish—Egypt)

Egit Egitto (Italian—Egypt)

egl egentlig (Dano-Norwegian—actual, proper, real)

EGL Eglin, Florida (tracking station); European Guarantee Loan(s)

egm extraordinary general meeting

EGM Empire Gallantry Medal; Extraordinary General Meeting (of shareholders)

EGmc East Germanic

EGMEX Eastern Gulf of Mexico

Egmonts Egmont Islands in the Chagos Archipelago northwest of Diego Garcia

EGMRSA Edible Gelatin Manufacturers Research Society of America

EGNR Ein Gedi Nature Reserve (Israel's Dead Sea oasis)

ego earth guide only

EGO Ankara Elektrik, Hava gazi ve Otobüs Isletme Milessesesi (Ankara Electricity, City-Gas, and Bus Traffic Department); Eccentric Orbiting Geophysical Observatory; Educational Growth Opportunities

egomac effect of gravity on methane-air combustion

egot erythrocyte glutamic oxaloacetic transminase (enzyme)

egp embezzlement of government property; exhaust gas pressure

EGP Experimental Geodetic Payload

EGPA Egyptian General Petroleum Authority

EGPC Egyptian General Petroleum Corporation

EGPS Electric Ground Power System

egr egress; exhaust gas recirculation

egr (EGR) erythrocyte glutathione reductase

EGRET Explorer Gamma-Ray-Experiment Telescope (NASA)

egrs extragalactic radio source

EGS Electronic Guidance Section; Employment Guarantee Scheme

EGSA Electrical Generating Systems Association

EGSP Electronics Glossary and Symbol Panel

egt exhaust gas temperature

Egy Egypt (whose capital is Cairo)

Egyp Egypt; Egyptian; egyptology

Egypt Arab Republic of Egypt (North African nation)

Egypt. Egyptian

egyptol egyptology

eh educationally handicapped

e/h exercise-head

e & h environment and heredity

eH oxidation-reduction potential (symbol)

Eh heater voltage (symbol)

EH Electra House; Ernest Hemingway; extra hazardous; Western Sahara (Internet code)

EH Enciclopedia Hoepli (Italian—Hoepli's Encyclopedia)

eha enroute high altitude

EHA Economic History Association; Education of the Handicapped Act; Environmental Health Assistant; Environmental Health Association; Equipment Handover Agreement

EHB Environmental Hearing Board

ehbf extrahepatic blood flow

EHBQ Eerste Hulp bij Ongelukken (Dutch—First-Aid Organization)

ehc enterohepatic circulation; enterohepatic clearance

ehc (EHC) external heart compression

E & HC Emory and Henry College

EHCD Environment, Housing, and Community Development (Australian Department of)

ehd electrohydrodynamics

ehd (EHD) epizootic hemorrhagic disease

EHDM Enhanced Hierarchical Development Methodology

ehec (EHEC) ethylhydroxyethylcellulose

EHES Environmental Health Engineering Services (USA)

ehf extreme high-frequency— 30,000–300,000 mc

ehf (EHF) epidemic hemorrhagic fever

EHF Experimental Husbandry Farm

ehg extra high grade

EHG Edvard Hagerup Grieg

EHH Ernst Heinrich Haeckel

EHHS Erasmus Hall High School

EHI Emergency Homes, Incorporated

EHIS Emission History Information System

EHJ Emanuel Haldeman-Julius

ehl effective half life

Ehl English as a home language

EHL Eastern Hockey League; Environmental Health Laboratory (USAF)

EHLASS European Home and Leisure Accident Surveillance System

E/h/m eggs per hen per month

EHMA Electric Hoist Manufacturers Association

EHMI Environmental Hazards Management Institutes

EHMS Engine Health Monitoring System

EHN Exploring Human Nature

EHOG European Host Operators Group

ehp effective horsepower; electric horsepower; extra-high potency

EHP Eric Honeywood Partridge (British lexicographer)

EHPT Eddy Hot-Plate Test

ehr enhanced reflector

ehs environmental health and safety

EHS Emergency Health Service; Environmental Health Services; Experimental Horticultural Station (UK)

ehsi electronic horizontal-situation indicator

EHSP Environment Health Safety Program

eht extra-high tension

EHTRC Emergency Highway Traffic Regulation Center

ehv extra-high voltage

EHV Empresa Hondureña de Vapores (Honduran Steamship Line)

EHVIST Ethical and Human Value Implications of Science and Technology

ehw extreme high water

EHW Environmental Health Watch

ehws electric hot water service; extreme—high-water-level spring tides

e/h/yr eggs per hen per year

ei electrical insulation; end item; enemy intelligence; engineering installation; estrogen index; exposure index

ei (EI) environment(al) illness

e-i electromagnetic interference; electronic interface; electronic interference; extraversion-introversion

e/i endorsement irregular

e by i execution by injection (of air or poison)

e^i elasticity of demand

Ei Eire (Irish Free State); input voltage (symbol)

Ei encéphale isolé (French— isolated intellectual)

EI East Indies; Electro Institute; Embrittlement Index; Essex Institute; Eugène Ionesco; Eunice Institute

EI Engineering Index

eia enzyme immunoassay; equine infectious anemia

EIA East Indian Association; Electronic Industries Association; Empire Industries Association; Energy Information Administration; Engineering Institute of America; Environmental Impact Assessment; Enzyme-Immune Assay

EIA Environmental Information Abstracts

EIAC Environmental Information Analysis Center

EIAJ Electronics Industry Association of Japan

EIAP Employer Information Access Project

EIAR Environmental Impact Analysis Report

EIB Ernst Ingmar Bergman; European Investments Bank; Export-Import Bank; Expert Infantryman Badge

EIB Economisch Instituut voor de Bouwuijverheid (Dutch— Economics Institute of the Building Industry); Elsevier International Bulletins

EIBA Electrical Industries Benevolent Association

EIBs Elsevier International Bulletins

EIBUS Export-Import Bank of the United States

EIBW Export-Import Bank of Washington

eic electronic engine control emotional inertia concept; equipment installation and checkout

EIC Earned Income Credit; East India Company; Ecology International Corporation; Education Information Center; Energy Information Center; Engineering Institute of Canada; Environmental Industry Council; European Investment Center; Exhibitors in Cable

EICBL Eastern Independent Collegiate Basketball League

EICF European Investment Casters' Federation

e-i children emotionally-impaired children

eicm employer's inventory of critical manpower

EICR Eppley Institute for Cancer Research (Omaha)

EICS East India Company's Service

eid electron impact desorption; end item description

EID East India Dock; End Item Delivery; End Item Description; Engineering Item Description

EIDA Engineering Industries Development Agency

EIDC East Indian Defense Commission; East Indian Defense Committee (Canad)

EIDE Enhanced Integrated Device Electronics

EIDEBOEWABEW Economic Intelligence Division of the Enemy Branch of the Office of Economic Warfare Analysis of the Board of Economic Warfare

Eidg Eidgenössisch (Swiss—federal)

eid it emergency identification light

EIDs East India Docks (London)

eie end-item equipment

EIES Electronic Information Exchange System

EIF Elderly Invalids Fund; Executive Inventory File

EIFA Eastern Intercollegiate Football Association

EIFAC European Inland Fisheries Advisory Committee (FAO)

eiff enemy identification—friend or foe

eifs exterior insulation and finish system

eig eigenlijk (Dutch—proper)

Eig Eigenvalue

EIG Exchange Information Group

Eight Great Eight Great Islands of Japan (largest islands of the Japanese archipelago)

EII East Intercourse Island (Australia)

EIII Electrical Industry Information Institute

eil electron injection laser

Eil Eileen; English as an international languagge

Eil eiland(en) [Afrikaans or Dutch—island(s)]

EIL Eletronic Instruments Limited; Experiment in International Living; extra-intestinal lesions

EIMA Exterior Insulation Manufacturers Association

Eimac Eitel-McCullough

EIMO Electronic Interference Management Office

EIMS Engineering Installation Management System; European Innovation Monitoring System

EIN Employer Identification Number; Equipment Installation Notice

EIN Empresa Insulana de Navegacão (Island Navigation Line)

E-in-C Engineer-in-Chief

E Ind East Indian; East Indies

EINECS European Inventory of Existing Chemical Substances

EINP Elk Island National Park (Alberta)

einschl einschliesslich (German—including)

eint electronic intelligence

Einw Einwohner (German—inhabitants' population)

EINZ Export Institute of New Zealand

EIO Emergency Information Office(r)

EIOL European Infrastructure for Open Learning

EIP Environmental Improvement Program; Experiment Implementation Plan

EIPC European Institute of Printed Circuits

eiph exercise-induced pulmonary hemorrhage

eir earned income relief (tax)

EIR East Indian Railway; Emergency Information Readiness; Engineering Information Request; Environmental Impact Report; Equipment Interchange Receipt; Executive Intelligence Review

EIRMA European Industrial Research Management Association

EIRNS Executive Intelligence Review News Service

eirnv extra incidence rate in non-vaccinated groups

eiro evaluation of infra-red-optics

EIRs Environmental Impact Reports

EIRS Education Information Resources Service

eirv extra incidence rate in vaccinated groups

eis electrical intersection splice; electron impact spectroscopy; end interuption sequence; executive information system; extended instruction set

EIS Early Implementation System; Economic Information Systems; Education Information Services; Electronic Ignition System; Electronic Inquiry System; End Item Specification; Environmental Impact Statement; Epidemic Intelligence Service (HEW); Export Intelligence Service; Exxon Information System

EISA Extended Industry Standard Architecture

Eisted (Welsh—Eisteddfod—session, meeting for competition)

e-i student(s) emotionally impaired student(s)

eit engineer in training

ei & t emplacement, installation, and test(ing)

EIT Electrical Information Test

EITA Electric Industrial Truck Association

EITB Engineering Industry Training Board

EITC Earned Income Tax Credit

eitp environmental interaction theory of personality

EITS Educational and Industrial Testing Service

EIU Economist Intelligence Unit

EIVT European Institute for Vocational Training

EIWS Engineering Installation Workload Schedule

ej elbow-jerk; expansion joint

ej ejemplo (Spanish—example)

E-J Endicott-Johnson

EJ ERIC Journal

EJA Executive Jet Aviation

EJC Edison Junior College; Engineers Joint Council; Engineers Junior College; Everett Junior College

EJCC Eastern Joint Computer Conference

eject. ejector

EJ & E R Y Elgin, Joliet & Eastern Railway

EJMA Educational Jewelry Manufacturers Association; Expansion Joint Manufacturers Association

EJN Edicott Johnson (stock exchange symbol)

ejp excitatory junction potential

EJS Engineering Job Sheet

EJT Engineering Job Ticket

EJTA Emergency Jobs Training Act

ejusd. ejusdem (Latin—of the same)

ek even keel; single enamel single cellophane (insulation symbol)

eK etter Kristi (Norwegian—after Christ)

Ek cathode voltage (symbol)

EK Eastman Kodak

EK Eisernes Kreuz (German—Iron Cross)—military decoration; *Esperanto Klubo* (Esperanto Club)

ekc epidemic keratoconjunctivitis

EKCO EK Cole (Limited)

EKD Evangelische Kirche in Deutschland (German—Protestant Church in Germany)

Eken (Swedish slang—Stockholm)

ekg electrokardiogram (electrocardiogram); electrocardiography

EKG electrokardiogram

eKr efter Kristus (Dano-Norwegian-after Christ) A.D.

eks eksempel (Dano-Norwegian—example)

EKSC Eastern Kentucky State College

EKSD Esperanto Klubo San Diego

ekskl eksklusive (Dano—Norwegian—exclusive)

eksp expederet or *ekspedition* (Dano—Norwegian expedite or expedition)

ekspl eksemplar (Dano-Norwegian—example or sample)

eku earliest known usage

EKU Eastern Kentucky University

ekv electron kilovolt

eKv exoatmospheric kill vehicle

EKVF East Kent Volunteer Fencibles

ekw electrical kilowatt(s)

eKy electrokymogram

el each layer; educational level; elastic level; elect(ed); electric(ity); electroluminescence; element(ary); elevated; elevation; elongation; extra line

el eller (Dano-Norwegian—or)

El Elbert; Elevated Railroad; Elias; Elvie; Elvira

EL Eastern League; Electrical Laboratory; Electronics Laboratory; Engineer Lieutenant; English Leicester (sheep); Epworth League; Erie-Lackawanna (railroad); Everyman's Library, Export Licence; Explicit Language

EL Empresa do Limpopo (Limpopo Line)

E-L Erie-Lackawanna (railroad)

e12 elongation in 2 inches

ELA English Language Amendment; Equipment Leasing Association

elab elaborate(d); elaborately; elaborating; elaboration; elaborative

ELAC East Los Angeles College

e lacte. e lact (Latin—with milk)

EL AL El Al Israel Airlines

ELAIN Export License Application Information Network

ELAM Escuela Latinoamericana de Matemáticas (Spanish—Latin American School of Mathematics)

ELAP Emergency Legal Assistance Project

ELAPR Estado Libre Asociado de Puerto Rico (Spanish—Associated Free State of Puerto Rico)—the Commonwealth of Puerto Rico's official name

elas elastic; elasticity; emergency logistical air support

ELAS Ethnikos Laikos Apelephterotikos Stratos (Greek—Hellenic People's Army of Liberation)

Elasm Elasmobranchia

elasmobranchs elasmobranch fishes (cartilaginous fishes such as chimaeras, dogfishes, rays, and sharks)

EL-ay Los Angeles, California

elb electronic lean burn; emergency locator beacon

Elb Egyptian pound

El of B Elector(ate) of Bavaria; Elector(ate) of Brandenburg

ELB English Language Battery

E.L.B. Bachelor of English Literature

elba emergency location beacon aircraft

El Banco (Spanish—The Bank)—World Bank for Reconstruction and Development

El'brus Mount El'brus (Europe's highest mountain in the Caucasus range)

ELBS English Language Book Society

elc extra-low carbon (electrodes)

El C El Centro

ELC Electronic Location Center

ELCA Evangelical Lutheran Church in America

El Cap El Capitan Dam; El Capitan Reservoir

elcar electric car

El Cen El Centro

El Sal El Salvador (whose capital is San Salvador)

ELCID Enforcement of Law Through Court Intervention and Diversion

elct electronics

elct rm electronics room

eld edge-lighted display; elder; eldest; electric load dispatcher; equal letter distance (hidden words in Bible); extra-long distance

Eldercare plan providing medical care for the elderly

el-dl electric locomotive-diesel locomotive

ELDO European Launcher Development Organization

ELDS Editorial Layout Display System

Elean Eleanor

Eleanor Mrs Anna Eleanor Roosevelt—wife of President Franklin Delano Roosevelt

elec electric; electrical; electrician; electricity; electroelectuary

Elec Elector, Electorate; Electra; Electricity

ELEC Election Law Enforcement Commission; European League for Economic Cooperation

ELEC/DR Electric Residential Service

Elec Engr Electrical Engineer; Electronic Engineer

elec pt electric(al) point

elecpub electronic publishing (dissemination of information via any electronic distribution means)

elect electrical

elect. election; elector; electoral; electrolyte; electrolytic

elect. electuarium (Latin—electuary)—confectioned drug; lollipop

elec tech electrical technician; electronic technician

elect. in. electricity installed

electn electrician

elect. pd electricity planned

ELECTRA Electrical, Electronics, and Communications Trades Association; trademark of the London Electricity Board

electraac electronic auto analysis clinic

electro electrocute; electrocution; electrotype

electrochem electrochemistry

electrocortico electrocorticograph(ic)(al)(ly), electrocorticography

electroderm electrodermal(ly)

electroenceph electroencephalography

electrogas electrogasdynamic(s)

electrogen electrogenic(al)(ly); electrogenesis

electrohyd electrohydraulic(s)

electrohydraul electrohydraulic(al)(ty)

electrol electrolysis

electrolev electronic levitation; electronically levitated

electromusic electronic music

electron. electronic(s)

Electron Lett Electronics Letters

electro-ocu electro-oculogram

electrophonics electro-phonic instruments

electrophys electrophysics

electroret electroretinograph(ic)(al)(ly); electroretinography

electro(s) electrotype(s)

electrosen electrosensitive; electrosensitivity

electrostat electrostatic copy; electrostatic printing

Electrovette electric-battery-powered Chevette

electrum 50% gold, 50% silver

electy electricity

elek electric(al); electronic

Elekt Elektrizität (German—electricity)

elektr elektriciteit (Dutch—electricity)

elem element; elementary

elephantocade elephant parade

eleph fol elephant folio-books about 23 inches high

e-less e-less novel written by British author Ernest Vincent Wright in 1933 with more than 50,000 words without the letter e; George Perec's *La Disparition,* published in French in 1969, is also e-less

elev elevated; elevation; elevator

elex electronics; electronics exercise

elf early lunar flare; electric light fitting; extra low frequency; extremely low frequency

El F El Ferrol

ELF Early Lunar Flare; Earth Liberation Front; Electronic Filing; Eritrean Liberation Front

ELFA Electric Light Fittings Association

elfc electroluminescent ferroelectric cell

ELF-HELP Educators, Librarians, and Families—Helping Loving Prescholars

elg electrolytic grinder; emergency landing ground

Elg Elgar; Elgin

El G El Paso Natural Gas Company

ELG European Liaison Group (USA)

elgas electricity and gas

ELGB Emergency Loan Guarantee Board

elhi elementary and high school (textbooks)

eli electricity installed; electronic line indicator

Eli Elias; Elihu; Elijah; nickname for a student or alumnus of Yale University

ELI English Language Institute; Environmental Law Institute

ELIA English Language Institute of America

ELIC Electric Lamp Industry Council

Elien. Eliensis (Latin—of Ely)

elig eligible

Elij Elijah

elim eliminate; eliminated; elimination

ELIM Evangelical Lutherans in Mission

elin exhibit line item number

elint electronic intelligence

elints electronic intelligence gathering vessels

elip electrostatic latent image photography

Elis Elisabeth

elisa enzyme-linked immunosorbent assay

Elise Elizabeth

elix elixir

Eliz Elizabeth(an)

Eliza Elizabeth

Elizabeth II Elizabeth Alexandra Mary of Windsor (Queen of United Kingdom of Great Britain and Northern Ireland and Her Other Realms and Territories)

Elizabeths Elizabeth Islands; queens named Elizabeth

Elk Hills U.S. Navy's petroleum reserve at Elk Hills, California

Elk Island Elk Island National Park east of Edmonton, Alberta

ell. elbow; ellipsoid(al); elliptic(al)

ell eller (Spanish—or)

Ell English language learning

Ella Eleanor; Eleanora; Eleanore; Isabella

ELLA European Long Lines Agency

Ell Dim Elliniki Dimokratia (Greek—Hellenic Republic) Greece

Ellen Eleanor(a)(e)

Ellerman Ellerman Lines Ltd

Ellices Ellice Islands now called Tuvalu

Ellie Alice

el lign eller lignende (Dano-Norwegian—or similar)

ellip elliptic; elliptical; elliptically

Ellis Ellis Island Immigration Examination Station, New York

ELLIS Ellis Air Lines

el lt electric light; electric lighting

Elly Eleanor

elm. element; energy-loss meter

ELM Eastern Atlantic and Mediterranean; Edgar Lee Masters

elma electromechanical aid

Elma Elizabeth Mary; Wilhemina

ELMA Empresa Lineas Maritimas Argentinas (Argentine Lines)

Elmer R Rice Elmer Reizenstein

ELMG Engine Life Management Group (USN)

elmint electromagnetic intelligence

Elmira Men's Reception Center (for male prisoners) at Elmira, New York

ELMO Engineering and Logistics Management Office (USA)

elmobile electric automobile

ELMS Earth Limb Measurement Satellite; Experimental Library Management System

El Mus East London Museum

E Ln East London

ELN Ejército de Liberación Nacional (Spanish—Army of National Liberation)—Bolivian and Colombian group

ELNA Esperanto League of North America

El Ng Ela Nguema (formerly San Fernando)

elngn elongation

El Niño El Niño Current (warm ocean current sweeping northward from the west coast of South America to the west coast of North America)

ELNM Edison Laboratory National Monument, West Orange, New Jersey

elo elocution; eloquence

Elo Eloheimo

ELO Electric Light Orchestra; English Language Office (r)

eloc elocution(ary); elocutionist(ic)(al)(ly)

ELOI Emergency Letter of Instruction

Eloise European large-orbiting instrumentation for solar experimentation

elong elongate; elongation

E long east longitude

eloq eloquence; eloquent(ly)

elos estimated length of stay

E Loth East Lothian

elox electrical spark erosion

elp electricity planned

ELP El Paso, Texas (airport)

El Pais (Spanish—The Country)—daily morning newspaper published in Madrid

eipc electroluminescence photo conductor

ELR Engineering Laboratory Report

elra electronic radar

ELRACS Electronic Reconnaissance Accessory System

elrat electrical ram air turbine

El Reno Federal Reformatory, El Reno, Oklahoma

ELRO Electronics Logistics Research Office (USA)

els extreme long shot

Els Elsinore (Helsingör); Elspeth

El of S Elector(ate) of Saxony

ELs elevated railways

ELS Escabana and Lake Superior (railroad)

Elsa Elizabeth

El Salv El Salvador (Spanish Republic of El Salvador)

El Salvador Republic of El Salvador (Central America's smallest nation), *República de El Salvador*

elsbm exposed location single buoy mooring

ELSE European Life Sciences Editors

elsec electronic security

ELSEGIS Elementary and Secondary General Information System

elsets elements sets

Elsev Elsevier (family of Dutch printers and publishers dating from the 16th century)—also spelled Elzevir

Elsev App Sci Elsevier Applied Science

Elsevier Sci Elsevier Scientific Publishing Co

Elsev NH Elsevier North Holland

Elsev Sci Elsevier Science Publishing Company

Elsev Seq Elsevier Sequoia

El Sgndo El Segundo

elsie electronic letter-sorting and indicator equipment; electronic location of status indicating equipment; electronic signalling and indicating equipment; emergency life-saving instant exit

Elsie Elizabeth

Elspet(h) (Scottish—Elizabeth)

ELSS Electronic Legislative Search System; Emplaced Lunar Scientific Station

elsse electronic sky screen equipment

elsur electronic surveillance

elt electrometer; emergency locator transmitter

Elt European letter telegram

E Lt Engineer Lieutenant

ELT English Language Teaching

E Lt-Cdr Engineer Lieutenant Commander

ELTDA English Language Teaching Development Aid

eltec electrical technician; electronic technician

ELTI English Language Teaching Institute

ELTJ English Language Teaching Journal

ELU English Lacrosse Union

elv expendable-launch vehicle; extra-low voltage

ELVIS Export Licensing Voice Information System

elw extreme low water

elwh elsewhere

El WLD Electrical World

elws extreme-low-water-level spring tides

elxr elixir

Ely easterly

ELYC East Lothian Yeomanry Cavalry

Elz *(see* Elsev)

ELZ Environmental Living Zone

em efficiency modulation; electromagnetic(al)(ly); emanation; embargo; emergency maintenance; emergency mobilization; enlisted man; expanded metal; width of the capital letter M

em (EM) electron microscope; electron microscopy; end of medium character (data processing)

e/m specific electronic mass

e & m endocrine and metabolism; erection and maintenance

e of m error of measurement

'em them

em *eftermiddag* (Dano-Norwegian—after midday)

Em Emily; Emma; Emmanuel; Emy; maximum voltage

EM Earl Marshal; Education Manual; Edward Medal; Efficiency Medal; Electrical Mechanic; Electrician's Mate; electromagnetic (symbol); Energy Mastery; Engineer Manager; Engineering Memorandum; Enlisted Man (Men); Etna & Montrose (railroad); European Movement; External Memorandum

E-M Electric Machinery (company); Electro-Motive (corporation); Embden—Meyerhof (glycolitic path)

E.M. Engineering of Mines;, Engineer of Mining

EM *Estado-Maior* (Portuguese—general staff, headquarters); *Estado Mayor* (Spanish—general staff, headquarters); *Excerpta Medica* (Elsevier logo-type)

E-M *Etat-Major* (French— Headquarters)

E.M. *Equitum Magister* (Latin—Master of Horse)

EM 1 C Electrician's Mate First Class (USN)

EM 2 C Electrician's Mate Second Class (USN)

EM 3 C Electrician's Mate Third Class (USN)

EMA Electronics Manufacturers Association; Employment Management Association; Engine Manufacturers Association; Envelope Manufacturers Association; Environmental Management Association; Environmental Media Association; European Marketing Association; European Monetary Agreement; Evaporated Milk Association; Expediting Management Association; Exposition Management Association; Extended Mission Apollo

E MacD Edward MacDowell

EMAD Engine Maintenance Assembly and Disassembly

EMAIA Electrical Meter and Allied Industries Association

E-mail electronic mail

Em Ar Un *Emiratos Arabes Unidos* (Spanish—United Arab Emirates)

EMAS Eco-Management and Audit Scheme; Emergency Medical Advisory Service; Emergency Message Authentication System; Employment Medical Advisory Service (UK)

emat expendable mobile acoustic target

EMATS Emergency Message Automatic Transmission System

Emax maximum voltage (symbol)

emb embankment; embargo; embark; embarkation; embassy; embroidered; embroidery; embryo; embryology

Emb *Embalse* (Spanish—dam or reservoir)

Emb Embankment; Embassy

EMB Egg Marketing Board; Energy Mobilization Board

EMBA Executive Master of Business Administration

emball *emballasje* (Norwegian—packing)

EMBERS Emergency Bed Request System

embgo embargo

embk embark

Embkmt Embankment

embkn embarkation

EMBL Eniwetok Marine Biological Laboratory

embo emboss(ed); embossing

EMBO European Molecular Biology Organization

embr embroider(y)

embry embryology

embryol embryology

EMBS Energy Management Bumper System

embsy embassy

emc electromagnetic capability; electromagnetic compatibility; engineered military circuit; equilibrium moisture content

emc (EMC) electromagnetic control; encephalomyocarditis

EMC Education Media Council; Einstein Medical Center; Electronic Material Change; End Mollycoddling in America; Engineering Maintenance Center; Engineer(ing) Maintenance Control; Engineering Manpower Commission; European Monitoring Center; Evergreen Marine Corporation

EMCA Ethnic Minority Council of America

EMCC European Municipal Credit Community

EMCCC European Military Communications Coordinating Committee

EMCE Eastern Montana College of Education

emcee(s) master(s) of ceremonies

EMCF European Monetary Cooperation Fund

EMCMF Embarked Mine Countermeasures Force

emcon emission control

EMCS Energy Management and Control Systems

EMCU Evergreen Maritime Container Unit

emcv encephalomyocarditis virus

emd electric-motor-driven

Emd Emden

EMD Energy Management Display; Engineering, Manufacturing and Development

E-MD Electro-Motive Division (General Motors)

EMID Export Market Development Incentive

emdp electromotive difference of potential

EMDP Export Market Development Programme

EME Electrical and Mechanical Engineering; Electrical and Mechanical Engineer(s); Extraordinary Minister of the Eucharist

EMEA Electrical Manufacturers Export Association

EMEB East Midlands Electricity Board

EMEC Electronics Maintenance Engineering Center

EMELEC trademark of East Midlands Electricity Board

EMEM Electrical and Mechanical Engineer

emend. emendate(d); emendating; emendation(s); emendator(s); emendatory; emender(s)

emend. emendatis (Latin—corrected, edited, emended)

emer emergency

Emer Emeritus

Emerald Triangle Humboldt, Mendocino, and Trinity counties, California, where potent marijuana is grown

emerg emergency

emergcons emergency conditons

emerit. emeritus (Latin—retired with honor)

E.Met. Engineer of Metallurgy

E-Meter electrical-resistance galvanometer

EMETF Electromagnetic Environment Test Facility (USA)

EMEU East Midlands Educational Union

emf electromagnetic field; electromagnetic force; electromotive force; erythrocyte maturing factor, every morning fix

EMF European Management Forum; European Motel Federation; Exccerpta Medica Foundation

E.M.F. E(dward) M(organ) Foster

emg electromyogram; electromyography

EMG Economic Monitoring Group

EMG Estado-Maior General (Portuguese—Staff General); *Estado Mayor General* (Spanish—Staff General)

emgb engine mounted gear box

emgcy emergency

emh electrical, mechanical, and hydraulic

EMH Efficient Market Hypothesis

EMHA Electronically-Monitored Home Arrest

emi electromagnetic interference

EMI Electrical and Musical Industries; Equipment Manufacturers Institute; Equipment Manufacturing Incorporated; Experiences in Mathematical Ideas

EMI Edizione Musicali Italiane (Italian Musical Publications)

emia (Latin suffix—a condition of the blood)—anemia

emic electromagnetic interference cell; emergency maternity and infant care

emid electromagnetic intrusion detector

E Midl East Midland

EMIETF Ethnic Materials Information Exchange Task Force

emig emigrant; emigration

EMILY Early Money Is Like Yeast

Emin Eminence

emip equivalent means investment period

Emirates United Arab Emirates (on the Trucial Coast of the Persian Gulf)

emis emission

EMIS Electromagnetic Intelligence System; Electronic Materials Information Services; Engineering Maintenance Information System

emit enzyme-multiplied immunoassay

EMIT Engineering Management Information Technique

EMJC East Mississippi Junior College

Emjo Emmanuel Jobe

emK elektromotorische Kraft (German—electromotive force)

eml electromagnetic levitation; equal matrix languages

Eml Emily

EML Earthquake Mechanisms Laboratory; Equipment Modification List

EML Everyman's Library

em log electromagnetic log

EMLTS Electromagnetic Levitation Transportation System (wheelless railway)

emm electromagnetic measurement; expanded memory manager

Emm Emmanuel; Emmanuel College, Cambridge

EMM Extended Memory Module

emma electron microscopy and microanalysis

Emma Emma Adelheid Welheimine Therese, Queen of The Netherlands 1879—1890; Queen Emma Bridge, Willemstad Harbor, Curaçao

EMMA Exceptional Merit Media Award

Emm Coll Emmanuel College Cambridge

Emmie Emma; Emy; Emmy

Emml Emmanuel

EMᵐᵒ Eminetísimo (Spanish—Most Eminent)—masculine ecclesiastical title applied to cardinals

EMMS Electronic Mail and Message System; Electronic Music Management System

EMMSA Envelope Makers and Manufacturing Stationers Association

E Mn E Early Modern English

EMNM El Morro National Monument

emo educational-management organization

EMO Emergency Measures Organization; Emergency Services Organization; Engineering Maintainability Organization; Equipment Move Order

EModE Early Modern English

emol emolumentos (Portuguese or Spanish—emoluments, official fees)

EMOL Excerpta Medica On-Line

Emos Earth's mean orbital speed

emot emotion(al)

emoticon emotion + icon

E-motor(s) electric motor(s)—submarine

EMOW Electrician's Mate of the Watch (USN)

emp electromagnetic pulses; empennage; end-of-month payment

emp. employee

emp. *emplastrum* (Latin—adhesive, a plaster)

e.m.p. *ex modo prescripto* (Latin—in the manner prescribed)

Emp Emperor; Empire; Empress

EMP Experience Music Project (rock 'n' roll museum in Seattle, Washington)

EMPAC Engineering Management Planning and Control

emp agcy employment agency

empath empathetic; empathy

EMPC Educational Media Producers Council

empd employed

Emperor Range mountains traversing Bougainville in the Solomon Islands

emph emphasis

emphy emphysema; emphysematous; emphyteusis; emphyteuta; emphyteutic

EMPI European Motor Products Incorporated

EMPIRE Early Manned Planetary Interplanetary Round Trip Experiment

empl emplace; emplacement; employ; employee; employer; employment

empld employed

EMPOCOL *Empresa Puertos de Colombia* (Columbian Port Works)

EMPORCHI *Empresa Portuaria de Chile* (Spanish—Chilean Port Enterprise)

EMPPO European and Mediterranean Plant Protection Organization

Empress of India Britain's Queen Victoria of Hanover (1837–1901)

empro emergency proposal

empsked employment schedule

empsz emphasize

emp. vesic. *emplastrum vesicatorium* (Latin—a blistering plaster)

emq electromagnetic quiet

emr educable rnentally retarded; electromagnetic radiation; electromagnetic resonance; electromagnetic riveting

EMR Eastern Mediterranean Region; Emerson Electric (stock exchange symbol); Engineering Master Report; Engineering Model Report; Enlisted Manning Report

EM & R Equipment Maintenance and Readiness

em-related emission-related (smog)

EMRIC Educational Media Research Information Center

EMRODA Electronic Maintenance Repair Operation Distributors Association

EMRS East Malling Research Station

ems emergency medical services; eosinophilia myalgia syndrome; ethylmethane sulfonate; expanded memory specification; expected mean squares

ems (EMS) electronic muscle stimulation

Ems Bad Ems

EMS Econometric Society; Electronic Message System; Emergency Medical Service; Engineering Material Specification; European Monetary System; Expanded Memory Specification; Export Marketing Service; Express Mail Service

EMSA Electron Microscope Society of America

EMSC Educational Media Selection Center; Electrical Manufacturers Standards Council

EMSO European Mobility Service Office (USA)

EMSP Enhanced Modular Signal Processor

EMSS Electromechanical Subsystem

EMSU Environmental Meteorological Support Unit

emt electrical metallic tubing; emergency medical technique; equivalent megatonnage

EMT Emergency Medical Technician; Evaluation Modality Test

EMTA Electro-Mechanical Trade Association

EMT-A Emergency Medical Technician—Ambulance

EMTALA Emergency Medical Treatment & Active Labor Act

EMTN European Meteorological Telecommunications Network

EMT-P Emergency Medical Technician—Paramedic

emtr emitter

emu. electromagnetic unit

Emu European monetary unit

EMU Eastern Michigan University; Economic and Monetary Union

EMU *Europese Monetaire et Economische Unie* (Dutch—European Monetary and Economic Union)

emul emulsion

emuls. *emulsio* (Latin—emulsion)

emut electric multiple-unit train

emux electronic multiplexer; electrical multiplux subsystem

emv electromagnetic vulnerability; electron megavolt

Emy Emilia; Emily

en endemic; enema; enemy; exceptions noted; width of one half the captial letter M

en (Greek—in, into)—encephalitis, energy, entropy, environment

En Engineer, English; voltage remaining at null (symbol)

EN Esquimalt and Nanaimo (railway)

EN *Emissora Nacional* (Portuguese—National Broadcast); *Estrada Nacional* (Portuguese or Spanish—National Highway); *Evening News*

En 1 c Engineman first class

ena experimental negotiating agreement; extractable nuclear antigen

ENA Eastern North America; English Newspaper Association; Evening News Association

ENA *Escuela Nacional de Agricultura* (Spanish—National School of Agriculture); *L'Ecole Nationale d'Administration* (French-National Administration School) France's civil-service academy

ENAB Evening Newspaper Advertising Bureau

ENAF *Empresa Nacional de Fundiciones* (Spanish—National Smelters Enterprise)

ENAFRI *Empresa Nacional de Frigorificos* (Spanish—National Freezer Enterprise)

enam enamel; enameled; enamels

ENAMI *Empresa Nacional de Mineria* (Spanish—National Mining Enterprise)—Chile

ENAP *Empresa Nacional del Petróleo* (Spanish—National Petroleum Enterprise)

ENAPUPERU *Empresa Nacional de Puertos del Peru* (Spanish—National Enterprise of the Ports of Peru)

ENASA *Empresa Nacional de Autocamiones* (Spanish—National Trucking Enterprise)

ENATA *Empresa Nacional de Tabaco* (Spanish—National Tobacco Enterprise)

ENATEL *Empresa Nacional de Telecomunicaciones* (Spanish—National Telecommunication Enterprise)

ENB East New Britain; English National Board (of health visiting, midwifery, and nursing)

ENBPS *Ente Nazionale per le Biblioteche Populari e Scolastiche* (Italian—National Organization of Popular and Scholastic Libraries)

enc enclosed

ENC *Empresa Nacional de Carbon* (Spanish—National Coal Enterprise)

ENCA European Naval Communications Agency

encap encapsulate(d); encapsulation

Enc Can *Encyclopedia Canadiana*

ENCI *Empresa Nocional de Comercializacion de Insumos* (Spanish—National Enterprise for the Commercialization of Raw Materials)

Enc Jud *Encyclopedia Judaica*

encl enclose; enclosed; enclosure

encld enclosed

enclg enclosing

enclit enclitic

enclo enclosure

ENCO Energy Company (Humble Oil & Refining)

encom encomiast(ic); encomium(s)

ENCORE Encouragement, Normalcy, Counseling, Opportunity, Reaching out, En-

ergies revived (YWCA program for women who have undergone breast surgery); European Network of Catchments Organized for Research on Eurosystems

ENCOTEL *Empresa Nacional de Correos y Telegrafos* (Spanish—Post and Telegraph National Enterprise)

ENCP European Naval Communications Plan (NATO)

enct encounter

ency encyclopedia

Ency Assn *Encyclopedia of Associations*

Ency Brit *Encyclopaedia Britannica*

end. endorsement

end (Latin prefix—within)—endoderm

ENDE *Empresa Nacional de Electricidad* (Spanish—National Electricity Enterprise)

ENDESA *Empresa Nacional de Electricidad SA* (Spanish—National Electricity Enterprise Corporation)

ENDEX Environmental Data Index

endis endispiece (antonym for frontispiece, correct term is tailpiece)

end mth end of month

endo endocrine; endocrinology

endo *endon* (Greek—within)—endocrine, endodermic, endoskeletal

endocrin endocrinological; endocrinologist; endocrinology

endocrino endocrinologic(al)(1y); endocminologist; endocrinology; endocrinopath(ic)(al)(1y); endocrinopathy; endocrinosis; endocrinotherapy; endocrinous

EndocSoc Endocrine Society

endor electron nuclear double resonance

endor(s) endorsement(s)

endow. endowment

endp endpaper(s)

ends. endpapers

ENDS Euratom Nuclear Documentation System

ENDS *Empresa Nacional de Semillas* (Spanish—National Seed Enterprise); *Environmental Data Services Limited*

end tel *endereço telegráfico* (Portuguese—cable address)

endv *endvidere* (Dano—Norwegian—furthermore)

endvr endeavor

endvrg endeavoring

end wk end of week

end yr end of year

ene ethylnorepinephrine

ene *enero* (Spanish—January)

ENE east northeast

ENE *Escuela Nacional de Economica* (Spanish—National School of Economics)

ENEA European Nuclear Energy Association

ENEL *Ente Nazionale per l'Energia Elettrica* (National Electric-Power Company of Italy)

enem. *enema* (Greek—injection)

ener energize

ENERGAS *Empresa Nacional de Gas* (Spanish—National Gas Enterprise)

energe *energicamente* (Italian—energetically)

ENEWS Effectiveness of Navy Electronic Warfare System

ENEX Engineering Export Association

ENF European Nuclear Force

en fav de *en faveur de* (French—in favor of)

enf(d) enforce(d)

enft enforcement

eng electronic news gathering, electronystagmogram; engine; engineer(ing)

eng (ENG) electronic news gathering (tv news reports)

eng *engelsk* (Dano-Norwegian—English)

Eng Engineer; England; English

Eng *Engineering* (British periodical)

Engañol English-Spanish

ENGBCA Engineers Board of Contract Appeals (USA)

Eng. D. Doctor of Engineering

Eng Div Engineering Division

eng dvr engine driver

EngE English in England

Eng/efp camera electronic news-gathering/electronic field-production camera

eng err engineering error

eng fnd engine foundation

enggmt engagement

Eng hrn English horn (a low oboe and neither English nor horn)

engin engineering
Eng Index *Engineering Index*
engitist engineer + scientist
engl englisch (German—English)
Engl England; English
Englewood Federal (delinquent) Youth Center at Englewood, Colorado
English Lit English literature
Eng Lit English Literature
Eng News-Rec *Engineering News-Record*
eng° *engenheiro* (Portuguese-engineer)
engr engineer
eng rm engine room
engrv engravei, engraving
Eng. Sc. D. Doctor of Engineering Science
ENGSS Engineering Schoolship (USN)
EN-H Elsevier North-Holland
ENI Ente Nazionale Idrocarburi (Italian—National Fuel Agency); *Escuela Nacional de Ingenieria* (Spanish—National School of Engineering)
eniac electronic numerical integrator and computer
ENIAC Electronic Numerical Integrator and Calculator
ENIC Ente Nazionale Industrie Cinematografiche (Italian—National Association of Film Producers); *Ente Nazionale della Cinofilia Italiana* (Italian—National Organization of Italian Dog Lovers)
ENICO Exxon Nuclear Idaho Company
ENIDS Ethnic Name Identification System
Enigma Elgar's *Enigma* Variations for Orchestra with an enigmatic program wherein the composer dedicates its movements to his friends described by their initials
ENIM Ente Nazionale dell' Istruzióne Media (Italian—National Organization for Intermediate Instruction)
ENIT Ente Nazionale Industrie Turistiche (Italian—National Tourist Industry)
enk enkelvoud (Dutch—singular)
enl enlarged; enlist
E n l English as a national language

enlgd enlarged
en ml end mill
Enmod Environmental Modification (Convention)
ENMOD Entity Module
ENMU Eastern New Mexico University
Enn Quintus Ennius (Roman poet)
ENNWR Eastern Neck National Wildlife Refuge (Maryland)
eno. enough
eno enero (Spanish—January)
en° enero (Spanish—January)
Elno estacionamiento no (Spanish—no parking)
ENO English National Opera
Enoch Pratt Enoch Pratt Free Library
enol Enology
Enos Book of Enos
ENOS European Network of Ocean Stations
Eno's Eno's Fruit Salts
ENP Egmont National Park (North Island, New Zealand); Etosha NP (South-West Africa); Everglades NP (Florida)
ENPA Ente Nazionale Protezione Animali (Italian—National Society for the Protection of Animals)—Italy
ENPI Ente Nazionale Prevenzione Infortuni (Italian National Institution for the Prevention of Accidents)
ENPMA Eastern National Park and Monument Association
enq enquire; enquiry
enr en route; equivalent noise resistance
enr (ENR) extrathyroidal neck radioactivity
ENR Emissora Nacional de Radiodifusão (Radio Portugal)
E & NR Esquinmalt and Nanaimo Railway
enrep environmental research programs
ENRI Electronic Navigation Research Institute
enrpae enroute (to/from) public affairs event
enrt enroute
Ens Ensign
Ens Ensenadas (Spanish—inlets, small bays)
ENS *Empresa Naviera Santa;* European Nuclear Society; experimental navigation ship

ENSA Entertainments National Service Association
ensen ensenada (Spanish—cove)
Ensen Ensenada
ensi equivalent—noise sideband input
ENSIDESA Empresa Nacional Siderurgica SA (Spanish—National Steel Works)
ENSIP Engine Structural Integrity Program (USAF)
en-skids National Security Intelligence Directives
ENSO El Niño/South Oscillation
ensu. ensuing
ent ear, nose, and throat; enter; entrance
ent ental (Dano-Norwegian—singular)
ENT Aerolineas Argentinas (Argentine Airlines); Ear, Nose, and Throat; EuroNATO Training
ENTA Environmental Test Area
entd entered
ENTE Ente Nazionale per l'Energia Elettrica (Italian—National Electric Energy Enterprise)
ENTEL Empresa Nacional de Telecomunicaciones (Spanish—National Telecommunications Enterprise)
ENTELEC Energy Telecommunications and Electrical Association
entero (Latin prefix—intestine)—enteritis
enterobact enterobacterial(ly); enterobacteriologist; enterobacterium
enteropath enteropathogenic(al)(ly)
enterov enterovioform
Entführung Entführung aus dem Serail (German—Abduction from the Seraglio)-Mozart opera
ent hall entrance hall
entl entitle
entn entertain
entom entomology
entomol entomologic(al)(ly); entomologist; entomology
entr entrance
entspr entsprechend (German—corresponding)
Ent Sta Hall Entered at Stationers' Hall

ENTURPERU Empresa Nacional de Tourismo del Perú (Spanish—National Tourist Enterprise of Peru)

ent-vio entero-vioform (anti-diarrhetic)

ENUF Everybody Now Undo Foul-ups

E-numbers Pan-European code numbers identifying a range of different food additives

enur enuresis

enus end user

enutech enuresis technology (controlling bedwetting)

env envelop; envelope; environ; environment(al)(ly); envoy

Env Envoy

Env Ext Envoy Extraordinary

ENVI Envirosphere Company

environ. environment; environmental; environmentalism; environmentalist

ENWR Erie National Wildlife Refuge (Pennsylvania); Eufaula National Wildlife Refuge (Alabama)

ENY Elsevier New York

enz enzovoort (Dutch—et cetera)

enza influenza

En Zed(er)(s) Zealand(er)(s)

eo end of operation; engine oil

e-o electro-optical; even-odd

e & o errors and omissions

e.o. ex officio (Latin—by virtue of office)

Eo Ecuadorian escudo(s); escudo(s) (Portuguese currency); output voltage (symbol)

E₀ electric affinity (symbol)

EO Eastern Orthodox; Education Officer; Employers Organization; Engineer Officer; Engineering Order; Entertainments Office(r); Examnining Office(r); Executive Office(r); Executive Order

E & 0 Eastern and Oriental

eoa effective on or about; end of address; examination, opinion, advice (medical)

EOA Economic Oil Association; Education Officers Association; Essential Oil Association

EOARDC European Office of the Air Research and Development Command (USAF)

eob end of block (character); expense operating budget; explanation of benefits

EOB Executive Office Building

EOBs Explanation of Benefits

eoc electric overhead crane; emergency operations center; emotional-organization; end of card; eocene

EOC Economic Opportunity-Commission; Educational Opportunity Center; Electronic Operations Center; Enemy Oil Committee; Equal Opportunities Commission; Executive Officers Council

EOCC Engineering Operational Casualty Control (USN)

EOCI Electric Overhead Crane Institute

eocm electro-optical countermeasures

eocp engine out of commission for parts

eod entry on duty; every other day; explosive ordnance demolition; explosive ordnance disposal

eod (EOD) explosive ordnance device

eodad end-of-data-set address

EODAP Earth and Ocean Dynamic Applications Program (NASA)

EODG Explosive Ordnance Disposal Group

EODP Engineering Order Delayed for Parts

eodta explosives ordnance disposal training area

EODU Explosive Ordnance Disposal Unit

eoe earth orbit ejection; equal opportunity employer

e & oe errors and omissions excepted

e and oe errors and omissions excepted

EOE Enemy-Occupied Europe

eof (EOF) end of file

EOF Earth Orbital Flight

eog effect on guarantees; electro-oculogram; electro-oculograph; electro-oculography; electro-olfactogram

EOG English Opera Group

eogb (EOGB) electro-optical glide bomb(ing); electro-optical guided bomb(ing)

EOGs Educational Opportunity Grants

eoh end of overhaul; equipment on hand

EOH Emergency Operation Headquarters

eohp except otherwise herein provided

EOIC Ethylene Oxide Industry Council

eoj (EOJ) end of job

eol effective operational length; end of life; expression-oriented language

Eol Eolic

EOL Ex Oriente Lux (The Light of the Orient—The Oriental Society)

eolb end-of-line block

eolm electro-optical light modulator

eom end of month; every other month; extra-ocular movements

eom (EOM) end of message (data processing)

EOM Employment Office Manager(s)

EOMB Explanation of Medicare Benefits

eomi end-of-message incomplete

eoms end-of-message sequence

EONR European Organization for Nuclear Research

EOO Equal Opportunity Office

eooe error or omission excepted

EOOW Engineering Officer of the Watch (USN)

eop earth orbit plane; end of part; end of passage

EOP Educational Opportunity Programs; Engineering Operational Procedure; Equal Opportunity Program; Equipment Operations Procedure; Executive Office of the President; Experiment of Opportunity Program

EOPs Extended Opportunity Programs

EOPS Equal Opportunity Programs and Services; Extended Opportunity Program and Services

eoq economical ordering quantity; end of quarter

EOQC European Organization for Quality Control

eor end of reel; explosive ordnance reconnaissance

eor (EOR) end of record; end of run

EOR Earth Orbit Rendezvous; Equaled Olympic Record

EORSA Episcopalians and Others for Responsible Social Action

EORTC European Organization for Research on the Treatment of Cancer

eos eligible for overseas service; end operation suppress; end of segment

EO's Engineering Orders

EOS Earth Observing System (satellite); Earth Observation Satellite; Earth Orbiting Satellite; Earth Orbiting Shuttle (NASA); Electro-Optical System; Engine Overhaul Shop; Equal Opportunity Specialist; European Orthodontic Society

EOSDIS Earth Observing System Data and Information System

eosins eosinophils

eosp economic order and stocking procedure

EOSS Earth Orbital Space Station; Engineering Operational Sequencing System

eot end of tape; end of transmission; enemy-occupied territory; engine order telegraph

EOT Eagle Ocean Transport

eot/bot end of tape/beginning of tape

EOTP European Organization for Trade Promotion

eots electro-optical tracking system

eou electro-optical unit; enemy objectives unit

EOU Epidemic Observation Unit

eov economic order van; end of volume

eow electro-optical warfare; engine(s) over wing(s); every other week

EOW Empty Operational Weight (aircraft code)

EOx Elsevier Oxford

ep easy projection; electric primer; electrically polarized; electrophysiology procedure; electroplate; electroplated; electroplating; electropneumatic; end page; endpaper(s); entrucking point; epoxy; estimated position; evoked potential; exit pupil; experienced playgoer; explosion-

proof; extended play (records); external publication; extreme pressure

ep (EP) extended play (45 rpm phonograph disc)

e/p endpaper

e&p earnings and profits

e & p exploration and production (area)

Ep peak voltage (symbol); plate voltage (symbol)

e p en passant (French—in passing*)*

e.p. editio princeps (Latin—first edition)

Ep. Episcopus (Latin—Bishop or overseer)

EP Eagle-Picher; Employee Plan; Engineering Personnel; Engineering Publications; entrucking point; estimated position; exceptions passed; European Parliament

E-P European Plan (no meals)

E & P Extraordinary and Plenipotentiary

EP Ecole Polytechnique (French—Polytechnic School); *Environmental Pollution*

E & P Editor & Publisher

EP1 Egyptian Pattern Mark 1 (mine)

epa economic price adjustment; eicosapentanoic acid; electron probe analyzer; estimated profile analysis

EPA Eastern Psychological Association; Economic Planning Agency; Educational Paperback Association; Emergency Powers Act; Empire Parliamentary Association; Empire Press Agency; Employment Protection Act; Energy Policy Act; Engineering Practice Amendment; Environmental Planning Authority; Environmental Protection Agency; Equal Pay Act; European Productivity Agency; Evangelical Press Association; Executive Protective Agency

EPAA Educational Press Association of America; Emergency Petroleum Allocation Act; Employing Printers Association of America

EPAC Electronic Parts Advisory Committee

EPACCI Economic Planning and Advisory Council for the Construction Industries

epam (EPAM) elementary perceiver and memorizer

epaq electronic parts of assessed quality

EPAT Every Pupil Achievement Test

epb equivalent pension benefit

EPB Effective Physiological Base; Electronic Planning Board; Environmental Periodicals Bibliography

epbm electroplated base metal

EPBX Electronic Private Branch Exchange

epc easy processing channel; edge-punched card; electronic program control; electroplate on copper; engine performance computer; every poor cluck

EPC Economic and Planning Council; Economic Policy Committee; Economic Policy Council; Education Products Center; Education Promotion Certificate; Educational Publishers' Council; Environmental Policy Center; Esso Petroleum Company; European Planning Council; European Political Community; European Political Cooperation

epca external-pressure circulatory assist

EPCA Energy Policy and Conservation Act; European Petro-Chemical Association

EPCAF El Paso Coalition Against the Fence (*see* Tortilla Curtain)

epc black easy-processing channel black

ep cells epithelial cells

epcg endoscopic pancreaticocholangiography (EPCG)

EPCOT Experimental Prototype Community of Tomorrow

epcp electric plant control panel

EPCRA Emergency Planning and Community Right-to-know Act

epcrbs emergency-position communication radio beacons

EPCS Equitable Pioneers Cooperative Society

epd earliest practicable date; enzyme potentiated desensitization; excess profits duty

ep & d electric power and distribution

epd en paz descanse (Spanish—may he rest in peace)

EPD Excellent Policy Duty (citation)

EPDA Education Professions Development Administration; Exhibit Producers and Designers Association

epdc economic power dispatch computer

EPDC Electric Power Development Corporation

ep disc extended-play (45 rpm) disc

epdm epidemiological; epidemiologist; epidemiology

epdm (EPDM) ethylene propylene diene monomer

epe electrical pans and equipment; electronic parts and equipment

EPE Editorial Projects for Education

EPEA Electrical Power Engineers Association; Environmental Protection Encouragement Agency (Hamburg, Germany)

epedemiol epedemiology

EPEM Electric Parts and Equipment Manufacturers

epenth epenthesis; epenthetic

epf exopthalmos-producing factor

EPF Employees Provident Fund; European Packaging Federation

EPF Empresa Petrolera Fiscal (Spanish—State Petroleum Enterprise)—Peru

EPFL Enoch Pratt Free Library (Baltimore)

ÉPFL École Polytechnique Fédérale de Lausanne (French—Federal Polytechnic School of Lausanne)

epg eggs per gram (parasitology); electropneumogram

EPG Economic Policy Group; Electronic Proving Ground (US Army)

EPGA Emergency Petroleum and Gas Administration

epgs electrical power generating system

EPGS Export Programme Grants Scheme (New Zealand)

Eph Ephraim

Eph Ephesians

EPHC Eastern Pacific Hurricane Center

ephmer ephemeral; ephemerides; ephemeris

epi electronic position indicator; emotional-physiologic illness

epi (Latin prefix—after, in addition, upon)—epicardial, epidemic, epidermis, epilogue, epithelium

EPI Edwards Personality Inventory; Emergency Public Information; Environmental Policy Institute; Expanded Program on Immunization; Eysenck Personality Inventory

epi. epinephrine

epic. electronic printer image construction; electron-positron intersecting complex

EPIC Early Purchase Individual Contract; Education Professional for Indian Children; El Paso Intelligence Center; Electronic Properties Information Center; Elyria Project for Innovative Curriculum; End Poverty in California; Exchange Price Indicators; Exports Payments Insurance Corporation

epicen epicenter; epicentral(ly)

Epict Epictetus

epid epidemic

EPIE Educational Products Information Exchange

EPIEI Educational Products Information Exchange Institute

epig epigastric; epigeal; epigeous; epigenesis; epigenetic; epigenic; epiglottal; epiglottic; epiglottis; epigone; epigonic; epigonism(s); epigonus; epigram; epmgrammatic(al)(ly); epigrammatism; epigrammatist(s); epigrammatize; epigrammatized; epigrammatizing; epigraph(er); epigraphic(al)(ly); epigraphist(s); epigraphy; epigynous; epigyny

epil epilogue

epineph epinephrine

epingrad equal participation in the great American dream

EPIP European Pattern, Indian Produced (tent)

Epiph Epiphania; Epiphany

epirb emergency position-indicating radio beacon

epirb (EPIRB) emergency position-indicating radio beacon

epis episiotomy

Epis Episcopal(ian)

Epist. Epistola (Latin—epistle or letter)

epistem epistemic(al)(ly); epistemological(ly); epistemologist(s)

epistom. epistomium (Latin—stopper)

epit epitaph; epitome

EPIT Equipment Procurement and Installation Team

epith epithelial; epithelium

epithal epithalamic; epithalamion

epivag epivaginitis

epl early programming language; extreme pressure lubricant

EPL Engineering Parts List; Erie Public Library; Evaluated Products List; Evansville Public Library

EPL Ecole Polytechnique de Lausanne (French—Polytechnic School of Lausanne)

EPLF Eritrean People's Liberation Front

EPLOT Enhanced Performance Lasers for Optical Transmission

epm electric pedestrian mover; explosions per minute

epm en propia mano (Spanish—in good hands, the right way)

EPM Easy Pickin's Mine (Imperial County, California); Environmental Program Manager

epma electron-probe micro analysis

EPMS Engine Performance Monitoring System; Engineering Project Management System

epn effective-perceived noise

epn (EPN) ethyl paranitrophenyl

epnd effective-perceived noise decibels

epndb equivalent perceived noise decibels

epndbl effective-perceived noise decibel level

EPNG El Paso Natural Gas

epnl effective perceived noise level(s); equivalent perceived noise level

epns electroplated nickel silver

EPNS English Place-Name Society

epo exclusive provider organization; experimental processing operation

EPO Emergency Planning Office(r); Energy Policy Office; European Patent Office

EPOC Earthquake Prediction Observation Center; Eastern Pacific Ocean Conference

EPOCA Environmental Project on Central America

EPOCH European Program on Climatology and Natural Hazards

EPOCS Effectual Planning for Operation of Container Systems

epon eponym(s)—designation(s) derived from proper names of families, places, or persons

EPOQUE European Parliament On-Line Query System

epos electronic point of sale

EPOS European PTT Open Learning System

EPOSS Environmental Protection Oil Sands System

epp end plate potential; erythropoietic protoporphyria; excess personal property

epp edellä puolenpäiven (Finnish—before noon)

Epp peak-to-peak voltage (symbol)

Epp. Episcopi (Latin—Bishops or overseers)

EPP Earth Physics Program; European Pallet Pool

EPPL El Paso Public Library

EPPO European and Mediterranean Plant Protection Organization

EPPR Engineering Procurement Proposal Request

EPPS Edwards Personal Preference Schedule; Engineering Procurement Planning Sheet

EPQ Eysenck Personality Questionnaire

epr earnings-to-price ratio; electronic paramagnetic resonance; engine pressure ratio;

ethylene propylene rubber; evaporator pressure regulator; external power relay

epr (EPR) electric propulsion rocket

EPR Einstein-Podolsky-Rosen effect; Engineering Power Reactor; Engineering Purchase Request; Essential Performance Requirements; External Planning Regent(s)

EPRA Early Planning for Retirement (Australia); Eastern Psychiatric Research Association

EPRC Educational Policy Research Center (Syracuse University)

EPRDF Ethiopian People's Revolutionary Democratic Front

EPRI Electric Power Research Institute

EPRL Electric Power Research Laboratory

eprom erasable programmable read-only memory

EPRS Engineering Proposal Requirement Specification

eps earnings per share; electric power supply; electron proton spectrometer; emergency power supply; emergency power system; encapsulated postscript system; exophthalmos producing substance

eps (EPS) energetic particle(s) satellite(s); expanded polystyrene (insulation); extrapyramidal side effect(s)

ep's epithelial cells

EPS El Paso Southern (railroad); Emergency Power System; Emergency Pressurizing System; Emergency Procurement Service; Engineering Purchase Specification; Entry Processing Station; Escape Propulsion System

EPSA Energy Products and Services Administration

EPSA Empresa Publica de Servicios Agropecuarios (Spanish—Public Enterprise Agricultural Services)

epsdt early and periodic screening, diagnosis, and treatment

EPSEP Empresa Publica de Servicios Pesqueros (Spanish—Public Enterprise Fishing Services)

EPSIS Education Program and Studies Information Services

epsp excitatory postsynaptic potential

Eps Vle Epsom Vale

ept egress procedures trainer; ethylene-propylene terpolymer; excess profits tax; external pipe thread

EPT Early Pregnancy Test(ing); Excess Profits Tax(ing)

EPTA Expanded Program of Technical Assistance (UN)

epte existed prior to entry

EPTG Electronic Publication Technology Group

EPTI Export Performance Taxation Incentive

epts existed prior to service

epu electrical power unit; electronic power unit; emergency power unit; entry processing unit

EPU Empire Press Union; European Payment Union

EPUL Ecole Polytechnique de l'Université de Lausanne (French—Polytechnic School of the University of Lausanne)

Epus Episcopus (Latin—Bishop)

cput events-per-unit-time

EPUY Education Program for Unemployed Youth

EPVT English Picture Vocabulary Test

epw enemy prisoner of war

epwm electroplated white metal

EPZ Ecole Polytechnique de Zürich (French—Polytechnic School of Zurich); Export Processing Zone

eq encephalization quotient; equal; equalization quotient; equation; equivalent; *(also see* EQ)

Eq Equator

EQ educational quotient; enthusiasm quotient; ethnic quotient

EQA Environmental Quality Act (California)

EQAA Environmental Quality Advisory Agency

EQAD Electrical (Electronic) Quality-Assurance Directorate

EQB Environmental Quality Board

EQC Environmental Quality Council
eqcc entry-query-control console
Eq Guin Equatorial Guinea
eqi environmental quality index
eqiv equivalent
eqm equal-flow manifold
eqn equation; equine
eqn prdx equine paradox (the fact that there are more horses' asses than horses)
eqp equip; equipment
eqpmt equipment
eqpt equipment
eqq electric quadripole-quadripole
E.Q.R. Eques Romanus Eques (Latin—Roman cavalry or knights)
eqs equations
eqs (EQS) equivalents (20-foot containers)
EQSC Environmental Quality Study Council
eqt equivalent training
Eq T equation of time
eq tr equipment trust
equ equate; equation
Equ Equerry; Equuleus
Equ Equuleus: (Latin—colt constellation)
Equa Equator; Equatorial
Equa C Cur Equatorial Counter-current
EQUAP Engineering Qualification Approval Program
Equa Pac Equatorial Pacific
equat equator; equatorial
Equatorial Guinea Republic of Equatorial Guinea (West African island and mainland country), *República de Guinea Ecuatorial*
Equatorial islands Borneo, Galápagos, Halmahera, Kiribati, Maldives, Moluccas, Nauru, São Tomé and Principe, Sulawesi (Celebes), Sumatra
Equatorial nations Brazil, Colombia, Congo, Ecuador, Gabon, Indonesia, Kenya, Peru, Somalia, Uganda, Zaire
Equatorials Equatorial Islands in the central and South Pacific Ocean, also called the Line Islands
Equi Equidae (Latin—horse family)—domestic breeds, wild horses, zebras
equil equilibrium

equin equinox
equinol equinologic(al)(1y); equinologist; equinology
equip. equipment
EQUIP Environmental Quality Incentives Program
equipt equipment
Equity Actors' Equity Association
equiv equivalent
eq & wd earthquake and war damage (insurance)
er earned runs (baseball); echo ranging; electronic reconnaissance; emergency rescue; employee relations; employer; endoplasmic reticulum; enhanced radiation; enhanced recovery (oil well); error; established reliability; estrogen receptor; external resistance; extremely rough
e/r editing/reviewing; en route
'er her
Er erbium; Eritrea; Eritrean; voltage drop across a resistance (symbol)
ER East Riding; East River; Edwardus Rex (King Edward); Effectiveness Report; Elgin Regiment; Elizabeth Regina (Queen Elizabeth); Emergency Request; Emergency Rescue; Emergency Reserve; Emergency Room; Engine Room; Engineering Release; Engineering Report; Environmental Report; Equipment Requirement; Evaluation Report; Expense Report; Expert Rifleman; Explosives Report; Express Route; Extended Range; External Report
ER Ecumenical Review
E by R English by Radio
E.R. Elizabeth Regina (Queen Elizabeth)
ER-200 high-speed train between Leningrad and Moscow
era electrically reconfigurable array; evoked response audiometry
era. electronic reading automation; electronic ring accelerator
era (ERA) earned run average
ERA Earthquake Risk Analysis; Economic Regulatory Administration; Economic Regulatory Agency; Echophysical Research Association; Electrical Research Association;

Electronic Realty Associates; Electronic Representatives Association; Energy Resources of Australia; Engine Room Artificer; Engineering Research Association; Equal Rights Amendment; Equitable Reserve Association; Eritrean Relief Organization; European Ramblers Association; Evaporative Rate Analysis
ERA Equal Rights Amendment
ERAA Equipment Review and Authorization Activity
eraam extended range anti-armor mine
ERAI Embry-Riddle Aeronautical Institute
eram extended range air munition
ERAP Economic Research and Action Project
ERAP Enterprise de Recherches et d'Activités Petrolienes (French—Petroleum Research Development Enterprise)
eraps expendable reliable acoustic path sonobuoy
Eras Erasmus
erase. electromagnetic radiation source elimination
eraser. (ERASER) elevated radiation seeker rocket
ERASMUS European Community Action Scheme for the Mobility of University Students
erb economic requirement batching; electron beam recording; emergency radio beacon; enlisted record brief; epigram record bureau; equivalent rectangular bandwidth
'Erb Herbert
Erb Erbitten (German—ask for, beg for, request)
ERB Economic Research Bureau; Educational Records Bureau; Electricians Registration Board; Engineering Review Board; Engineers Registration Board; Environmental Review Board; Equipment Review Board
ERBE Earth Radiation Budget Experiment
cr bh engine room bulkhead
erbf effective renal blood flow
erbm (ERBM) extended-range ballistic missile

ERBS Earth Radiation Budget Satellite

erc earnings-related compensation; en-route chart; equatorial ring current; equipment record card; expendability repair classification

ERC Economic Research Council; Economic Resources Corporation; Educational Resources Center; Electronics Research Center (NASA); Elmira Reception Center (for male prisoners in Elmira, NY); Employee Relocation Council; Employment Rehabilitation Center; Enlisted Reserve Corps

ERCA Educational Research Council of America

erdc equine respiratory disease complex

ERC & I Economic Reform Club and Institute

ERCO Electric Reduction Company

ercp endoscopic retrograde cholangio-pancreatography

ercr electronic retina-computing reader

ERCS Emergency Rocket Communications System

erd emergency return device; equivalent residual dose

ERD Earth Resources Data; Emergency Reserve Decoration; Equipment Requirements Data

ERDA Electrical and Radio Development Association; Electronics Research and Development Agency; Energy Research and Development Administration

ERDC Earth Resources Data Center; Electronic Research and Development Command (USA)

ERDE Explosives Research and Development Establishment

ERDF European Regional Development Fund

ERDIP Experimental Research and Development Incentives Program

ERDL Engineering Research and Development Laboratory

ERDS European Reliability Data System

ere expected repository environment(s); extra-regimentally employed

'ere here

ERE Edison Responsive Environment

ERE *Encyclopaedia of Religion and Ethics*

EREC Energy Efficiency and Renewable Energy Clearinghouse

erect. erection

'Ereford(shire) [Cockney contraction—Hereford(shire)]

EREM Earth Resources Experiment Program

EREP Earth Resources Experiment Package (NASA)

erf error function

ERF Education and Research Foundation; Eye Research Foundation

ERFA European Radio-Frequency Agency

ERFAA European Radio-Frequency Allocation Agency

erfc error function complement

ERFPI Extended-Range Floating-Point Interpretive System

erg electrical resistance gage; unit of mechanical energy or work (derived from the word *energy)*

erg (ERG) erase gap

erg. electroretinogram

ERG electro-magnetic rail gun; Energy Research for the Governors; Energy Research Group

ergo ergonomics

ERGOM European Research Group on Management

ergon ergonomic; ergonomical; ergonomics

ergp emergency removal gate pass

ergs (ERGS) earth geodetic satellite (USAF)

ERGS Electronic Route Guidance System

Erh Erhard

E & R: Hist Soc Evangelical and Reformed Historical Society

Eri Eridamus; Eridanus (constellation)

ERI Earthquake Research Institute (Tokyo University); Economic Research Institute; Environmental Research Institute; Erie, Pennsylvania (airport)

E.R.I *Edwardus Rex et Imperator* (Latin—Edward, King and Emperor)

eric electronic remote and independent control; energy rate input controller

ERIC Educational Resources Information Center (US Office of Education)

ERICA Experiment on Rapidly Intensifying Cyclones over the Atlantic

ERIC/AE Educational Resources Information Center/Adult Education

ERIC/CAPS Educational Resources Information Center/Clearinghouse on Counseling and Personnel Services

ERIC/CE Educational Resources Information Center/Clearinghouse in Career Education

ERIC/CEA Educational Resources Information Center/Clearinghouse on Educational Administration

ERIC/CEC Educational Resources Information Center/Clearinghouse for Educational Change

ERIC/CEM Educational Resources Information Center/Clearinghouse on Educational Management

ERIC/CHE Educational Resources Information Center/Clearinghouse on Higher Education

ERIC/CHESS Educational Resources Information Center/Clearinghouse for Social Studies and Social Science

ERIC/CIR Educational Resources Information Center/Clearinghouse on Information Resources

ERIC/CLL Educational Resources Information Center/Clearinghouse on Languages and Linguistics

ERIC/CLS Educational Resources Information Center/Clearinghouse for Library and Information Sciences

ERIC/CRESS Educational Resources Information Center/Clearinghouse on Rural Education and Small Schools

ERIC/CRIER Educational Resources Information Center/ Clearinghouse on Retrieval Information and Evaluation on Reading

ERIC/CUE Educational Resources Information Center/ Clearinghouse on Urban Education

ERIC/ECE Educational Resources Information Center/ Clearinghouse on Early Childhood Education

ERIC/IR Educational Resources Information Center/ Clearinghouse for Information Resources

ERIC/IRCD Educational Resources Information Center/ Information Retrieval Center on the Disadvantaged

Ericofon Ericsson telephone

ERIC/RCS Educational Resources Information Center/ Clearinghouse on Reading and Communication Skills

ERIC/SMEAC Educational Resources Information Center/Clearinghouse for Science, Mathematics, and Environmental Education

ERIC/TME Educational Resources Information Center/ Clearinghouse on Tests, Measurement, and Evaluation

Erid Eridamus

Erie Erie-Lackawanna (railroad)

ERIE Eastern Regional Institute for Education

Erie Phil Erie Philharmonic

erild earth rotation in lunar distances

ERIM Environmental Research Institute of Michigan

ERIN Environmental Resources Information Network

ERISA Employee Retirement Income Security Act

Erit Eritrea

ERJA E R Johnson Association

erk enroute kit; extracellular signal-related kinase

erl emergency reference level

Erl Erläuterung (German—explanatory note)

ERL Environmental Research Laboratories

erm elastic reservoir moulding; ermine

Erm European red mite

ERM European Rate Mechanism; exchange rate mechanism

erma electronic recording machine accounting

ERMES European Radio Messaging System

Erms root-mean-square voltage (symbol)

ERMS Educational Resource Management System

Ern Ernest; Ernst

ernic earnings-related national insurance contribution

ernie electronic random-numbering-and-indicating equipment

Ernie Ernest

ERNIE Electronic Random Number Indicator Equipment

ERO Eastman-Rochester Orchestra

eroa economic rehabilitation in occupied area(s)

eroduction(s) erotic production(s)

ero-gro erotic and grotesque (writing)

erom erasable read-only memory

E-room engine room

EROPA Eastern Regional Organization for Public Administration

eropt error option(s)

EROS Earth Resources Observation Satellite; Eliminate Range-Zero System; European River Ocean System; Experimental Reflector Orbital Shot (space probe)

EROSP Earth Resources Observation Systems Program

erot erotic; erotica; erotical(ly); eroticism; erocicist; eroticization; eroticize; eroticizing; erotism(s); erotogenic(s); erotologic(al)(1y); erotologist; erotology

erotol erotologist; erotology

erp early receptor potential; effective radiated power; electro rustproofing; enterprise resource planning; event-related potential; eye reference position

ERP Easy Revolving Plan; Emerson Radio & Phonograph (stock exchange symbol); European Recovery Program

ERP Ejército Revolucionario Popular (Spanish—Popular Revolutionary Armed Force)—Argentina; *Ejército Revolucionario del Pueblo* (Spanish—People's Revolutionary Army)—Argentina

ERPC Eastern Railroads Presidents Conference

erpf effective renal plasma flow

ERPM East Rand Proprietary Mines

ERPSL Essential Repair Stock List

err. error; erroneous

ERR Engineering Release Record; Engineering Research Report

err & app error and appeals (legal)

errc expandability, recoverability, repairability cost

ERRDF Earth Resources Research Data Facility (NASA)

erron erroneous(ly)

ERRS Environmental Response and Referral Service

ers economic research service; emergency road service

ers (ERS) environmental research satellite

ERS Earth Regeneration Society; Economic Research Service; Educational Research Service; Edwards Rocket Site; Emergency Relocation Site; Ergonomics Research Society; Experimental Research Society

ER & S Electrolytic Refinery and Smelting (company)

ersa extended range strike aircraft

ERSC Engineer and Railway Staff Corps

ersir earth-resources shuttle imaging radar

E-R S O Eastman-Rochester Symphony Orchestra

ersos (ERSOS) earth-resources-survey operational satellite

ERSP Earth Resources Survey Program (NASA)

ERSR Equipment Reliability Status Report

ert electrical resistance temperature; electrical resistance thermometer; extended research telescope

ert (ERI) estrogen replacement therapy

ERT Emergency Room Technician; Exhibits Round Table

ERTA Economic Recovery Tax Act; Energy Research and Technology Administration

ERTC European Regional Test Center (NATO)

ERTS Earth Resources Technology Satellite; European Rapid Train System

eru emergency reaction unit

ERU English Rugby Union

erv electronic repair vehicle; expiratory reserve volume

ERV English Revised Version

ERVAD Engineering Release for Vendor Article Data

erw electro-resistance welding

erw (ERW) enhanced radiation weapon (neutron bomb)

erw *erweiterte* (German—enlarged, extended)

ERWS Engineering Release Work Sheet

erx electronic remote switching

ery erysipelothrixia

ER Yorks East Riding, Yorkshire

ERY East Riding Yeomanry

erythro (Latin prefix—red)—erythrocyte

es echo sounding; effect size; eldest son; electric starting; electrical sounding; electrostatic; enamel single silk (insulation); engine speed; engine-sized (paper); equal section; estimate; exploratory shaft

es (ES) ejection sound

e/s early shorn (sheep); en suite

es esempio (Italian—example)

e_s price elasticity of supply

Es einsteinium; Essen; screen voltage (symbol)

ES Eagle Squadron; East Sussex; Eastern States; Econometric Society; Educational Specialist; El Salvador; Electrochemical Society; Ellis Air Lines; Employee Suggestion; Endocrine Society; Engineering Study; Entomological Society; Environmental Studies; Espirito Santo; Ethnological Society; Etymological Society; Experiment(al) Station; Extension Service; Spain (Internet code)

ES El Salvador (Spanish abbreviation)

E.S. Esoteric Section (Theosophical Society)

esa engine starter assembly

ESA Ecological Society of America; Economic Stabilization Agency; Economic and Statistical Analysis; Electric(al) Supplies Authority; Electrical Supply Authorities; Electrolysis Society of America; Employment Standards Administration; Engineers and Scientists of America; Engine Service Association; Entomological Society of America; Epiphyllum Society of America; European Space Agency; European System of (integrated) Accounts; Endangered Species Act; Euthanasia Society of America; Exceptional Service Award; Export Screw Association

ES & A English, Scottish, and Australian (Bank)

ESAA Electrical Supply Authorities Association; Electricity Supply Association of Australia; Emergency School Aid Act

ESAAB Energy Systems Acquisition Advisory Board

ESAB Energy Supplies Allocation Board (Canada)

ESAC Environmental Systems Applications Center

esaddi estimated safe and adequate daily dietary intake

ESAE1 Electric Supply Authority Engineers Institute

ESA-IRS European Space Agency—Information Retrieval Service

E Sam Eastern Samoa (American Samoa)

ESANZ Economic Society of Australia and New Zealand; Electrical Supply Authorities of New Zealand; Ergonomics Society of Australia and New Zealand

ESAP Emergency School Assistance Program

ESAPP Energy System Acquisition Project Plan

esar electronically-steered array radar

ESARS Employment Service Automated Reporting System

ESAs Eastern Socially Attractives (Ivy League graduates)

ESA System Easy, Speedy Accounting System

eSat except Saturday

ESAWC Evaluation Staff, Air War College

esb electrical stimulation (of the) brain; electric storage battery

ESB Economic Stabilization Board; Electricity Supply Board; Electric Storage Battery (company); Empire State Building

ESBA Eastern Sovereign Base Area; English Schools' Badminton Association

ESBBA English Schools' Basket Ball Association

esc electronic service change; electronic spark control; elongation-sensitive cell; escadrille; escalator; escape; escape character; escort; escrow; escutcheon; evanescent; space charge; extended core storage; extra sex combs

esc (ESC) escape character (data processing)

esc escompte (French—discount)

Esc escudo (Portuguese currency)

ESC Economic and Social Council (UN); Education Service Center; Education Systems Incorporated; Electronic Security Command; Electronic Systems Command (USN); Electronics Systems Center; Energy Security Corporation; European Shippers Council; Executive Service Corps; Extended-Service Commission

esca electron spectroscopy for chemical analysis

ESCA English Schools' Cricket Association; English Schools' Cycling Association; Exposition Service Contractors Association

Escales (French—ports of call)—Jacques Ibert symphony

escap escapologist; escapology

ESCAP Economic and Social Commission for Asia and the Pacific (UN)

Escarp Escarpment

ESCAT Emergency Security Control of Air Traffic

eschat eschatology

ES/CIP Employee Suggestion/Cost Improvement Proposal

escl esclamazione (Italian—exclamation); *exclamativo* (Italian—exclamative); *esclusivo* (Italian—exclusive)

ESCL Elias Sourasky Central Library (Tel Aviv); Evans Signal Corps Laboratory

ESCMA Electric Steel Conduit Manufacturers' Association

escn electrolyte-and-steroid-produced cardiopathy characterized by necrosis

esc₀ escudo (Portuguese or Spanish—coat of arms, Portuguese monetary unit, shield)

Esco Escocia (Spanish—Scotland); *Escócia* (Portuguese—Scotland)

ESCO Educational, Scientific, and Cultural Organization (UN)

Escom Electrical Supply Commission

ESCORTDIV escort division

ESCOW Engineering and Scientific Committee on Water (New Zealand)

escp expendable surface-current probe

ESCP Earth Science Curriculum Project

ÉSCP École Supérieure de Commerce de Paris (French—Paris College of Commerce)

escr escrow

escrit^a escritura (Portuguese or Spanish—assignment, contract, deed, writ)

escrnía escrabania (Spanish—notary's office)

escrno escribano (Spanish—notary)

escr^no escribano (Spanish—court clerk, notary, scribe)

escs escudos (Portuguese or Spanish—coats of arms, Portuguese monetary units, shields)

ESCS Economics, Statistics, and Cooperatives Service

escudo monetary unit of Cape Verde and Portugal

esd echo-sounding device; electronic smoke detector; electrostatic discharge; energys-

torage device; equipment supply depot; estimated shipping date; estimated standard deviation; extended school day

esd (ESD) echo-sounding device; external symbol dictionary (data processing)

Esd (ESD) English as a second dialect

ESD East San Diego; Education Service District; Electronic Systems Division (USAF); Emergency Service Division (NYPD)

ESDA Earth-Science Data Acquisition

ESDAC European Space Data Analysis Center (Darmstdat)

ESDAG Earth-Science Data Acquisition Guidelines

esdi enhanced small device interface

esdp external stores data package

Esdr Esdras (The Book of Esdras)

esE electrostatische Einheit (German—electrostatic unit)

Ese Ensenada

ESE east southeast

ESEA Electrical Supply Engineers Association; Elementary and Secondary Education Act

ESECA Energy Supply and Environmental Coordination Act

ESEF Electrotyping and Stereotyping Employers Federation

esep extreme somatosensory evoked potential

esf electrostatic focusing; erythropoietic stimulating factor; extended superframe

ESF Eastern Sea Frontier; Economic Support Fund; Engineering Specification Files; European Social Fund; Extended Spooling Facility

esfc extended specific fuel consumption

esfp environment-sensitive fracture process(es)

esfswr extra-special flexible steel wire rope

esg electrically suspended gyro(scope); electronic-sweep generator; extended-sweep generator

e sg e seguente (Italian—and the following one)

Esg English standard gage

ESG Environmental Support Group

esgm electrostatically supported gyro monitor

esh equivalent solar hour(s)

ESH European Society of Haematology

ESHL Eastern Seaboard Herpetological League

eshp equivalent shaft horsepower; established standard horsepower

esi emergency stop indicator; equivalent spherical illumination; externally specified indexing

ESIL European Standard Inventory List (NATO)

ESIS Executive Selection Inventory System

Esk Eskimo

Eskie(s) Eskimo(s)

esl expected significance level

Esl (ESL) English as a second language

ESL Eagle Shipping Ltd; Earth Sciences Laboratory; Eastern Steamship Lines; Engineering Societies Library

E S-L Engineer Sub-Lieutenant

ESL Endangered Species List

ESLAB European Space Laboratory (Delft)

esle engineering special laboratory equipment

ES/LES Equipment Section/Loaded Equipment Section

ESLO European Satellite Launching Organization

esm electronic support measures; electrostatic memory; ends standard matched (lumber)

ESM East Surrey Militia; Eastman School of Music; Engineering Shop Memo

ESMA Electric Sign Manufacturers Association; Electronic Sales-Marketing Association; Engraved Stationery Manufacturers Association; Episcopal Society for Ministry on Aging

ESMC Eastern Space and Missile Center, Cape Canaveral, Florida

ESMI Energy Studies Measurement Instrument

ESMRI Engraved Stationery Manufacturers Research Institute

esm's electronic-support measures

esn essential

esn (ESN) educationally subnormal

ESN Elastic Stop Nut (corporation); Electronic Serial Number; Engineering Shipping Notice; English-Speaking Nations (NATO); Executive Suite Network

esna electrical survey net adjuster

ESNA Elastic Stop Nut Corporation of America; Empire State Numismatic Association

ESNE Engineering Societies of New England

ESN-H Elsevier North-Holland

esntl essential

ESNZ Entomological Society of New Zealand

ESO Educational Services Office(r); Electronic Supply Office(r); Embarkation Staff Office(r); Emergency Services Organization, Engineering Service Order; Engineering Stop Order

ESOA Employee Stock Ownership Association

ESOAA Eight Sheet Outdoor Advertising Association

ESOC European Space Operations Center

Esol English for speakers of other languages

ESOMAR European Society for Opinion Surveys and Market Research

ESOP Employees Stock Ownership Plan

esoph esophageal; esophagus

esor electronically scanned optical receiver

esot esoteric; esoterica; esoterical(ly); esotericism(s)

ESOT Employee Stock Ownership Trust

esp easy solution possible; echeloned series processor; electro-magnetic surface profiler; electronic stability program; electronic still photography; electro-sensitive paper; electro-sensory panel; engine sequence panel; enhanced-ser-vice provider; equal sector probability; especially; extra-sensory perception

esp (ESP) electro-selective pattern (light meter system for cameras); electrosensitive programming

e & sp equipment and spare parts

esp espressivo (Italian—expressive)

Esp Esperanto; Esplanade

Esp (ESP) English for special purposes

Esp Espagne (French—Spain); *España* (Spanish—Spain); *Español* (Spanish—Spanish)

ESP East Sepik Province; Eastern State Penitentiary (Philadelphia, Pennsylvania); Elsevier Science Publishing; Emerson Select Protection; English for Special Purpose(s); Equipment Status Panel; Extrasensory Perception

ESP Ecole des Sciences Politiques (French—School of Political Science)

espa electronically steered phased array

ESPA Elementary School Principal's Association; Evening Student Personnel Association

ESPAC Elementary School Principal's Association of Connecticut

Espantuguese Spanish-Portuguese

ESPAW Elementary School Principal's Association of Washington

ESPC Elsevier Scientific Publishing Company (Amsterdam); Essential Services Protection Corps

espec especial(ly)

Esper Esperanto

espg espionage

espi electronic speckle-pattern interferometer

Esplish Spanish-English

ESPN Entertainment and Sports Programming Network

ESPOA Electricity Supply Professional Officers Association

ESPOIR European Shoe Program On Instant Response

ESPQ Early School Personality Questionnaire

espr espressivo (Italian—expressive)

ESPR English Society for Psychical Research

espress espressivo (Italian—expressive)

ESPRI Education Service of the Plastics and Rubber Institute

ESPRIT European Strategic Program for Research and Development in Information Technology

esq extra-special quality

esq esquerdo (Portuguese—left)

Esq Esquire

ESQ Entomological Society of Queensland

ESQA English State Quarries Association

esq. esquerdo (Portuguese—left)

Esqrr Esquire

ESQST Ego-Strength Q-Sort Test

esr effective signal radiated; electrical skin resistance; electron spin resonance; electronic slide rule; electronically scanned radar; electron spin resonance; electro-slag resmelting; equivalent series resistance; erythrocyte sedimentation rate

ESR Engineering Societies Library; Engineering Stop Release; Engineering Summary Report

ESRANGE European Space Research (northern rocket range)—Kiruna

esrc electronics recovery control; engine surge recovery control

ESRC European Science Research Council

ESRD End-Stage Renal Disease

ESRG Earth Sciences Review Group

ESRI Economic and Social Research Institute (Dublin)

ESRIN. European Space Research Institute

ESRO European Space Research Organization

ESRP Environmental Standard Review Plans

ESRU Environmental Sciences Research Unit

ess electronic switching system; empty solution set; engine start system; essences; essential expendable sound source; evolutionary stable strategy

ess. essence

ess. essentia (Latin—escence)

eSS except Saturday and Sunday

Ess Essex

ESS Eastern Searoad Service; Educational Services Section; Electrical Standards System; Electronic Switching System; Elementary School Science; Elementary Science Study; Emplaced Scientific Station; Employment Security System; English Speaking Society; Evaluation SAGE Sector; Experimental SAGE Sector

ESS Encyclopedia of the Social Sciences

essa environmental survey satellite (weather satellite)

ESSA Environmental Science Services Administration—Central Radio Propagation Laboratory, Coast and Geodetic Survey, Weather Bureau (Department of Commerce); environmental survey satellite

Essandess Simon and Schuster

essb electrical self-stimulation of the brain

ESSENTIAL European Systems Strategy for the Evolution of New Technology in Advanced Learning

Essequibos Essequibo Islands (in the Essequibo River estuary off Guyana)

ESSEX Effects of Sub-Surface Explosions (USA)

ESSI European Software and Systems Initiative

Essie Esther

ess neg essentially negative

ESSO Esso Shipping; Standard Oil

ESSPO Electronic Support System Project Office

ess pos essentially positive

ESSR Estonian Soviet Socialist Republic

ESSS Electronic Security Surveillance System

essu electronic selective switching unit

ESSWACS Electronic Solid State Wide-Angle Camera System

est earliest start time; elastic surface transformation; electrolytic sewage treatment; erhard seminar training; establish; established; establishment; estate; estimate; estimated; estimation; estimator; estuary; eternal static pressure; extended standard theory

est (EST) electrical stimulating treatment; electroshock therapy

est estación (Spanish—station); *estimado* (Spanish—estimated)

Est The Book of Esther; Estates (postal abbreviation); Estonia (whose capital is Tallinn); Estuary

Est Estado (Spanish—State); (French—east)

EST Eastern Standard Time; Eastern Summer Time; Electrical Stimulating Treatment; English in Science and Technology; Enlistment Screening Test; Enroute Support Team; Epidemiology and Sanitation Technician; Erhard Seminars Training

esta electrical system test area

ESTA Energy Systems Trade Association

estab established

Estab Establishment

estab est established estimate

establ. establishment

Established Church Established Church of England

estab tip establecimiento tipográfico (Spanish—publishing company)

estar estimated arrival

estb establish

estbl establishment

est'd. estimated

ESTEC European Space Technology Center

ESTF Exploratory Shaft Task Force; Exploratory Shaft Test Facility

estg estimating

esth esthetics

Esth Esthonia; Esthonian

Esth Esther

Esther Hester

Esthr Book of Esther

ESTI European Solar Test Installation; European Space Technology Institute

estm estimate

estmd estimated

estmg estimating

estmn estimation

estn estimation

estn estnisch (German—Estonian)

Estoc Estocolmo (Portuguese or Spanish—Stockholm)

ESTP Earth Science Technical Plan

ESTPP Earth Science Teacher Preparation Project

ESTRACK European Space Satellite Tracking and Telemetry Network

Estr B Estero Bay

estriff encryptic-secure tracking-radar identification friend or foe

Ests Estates

est wt estimated weight

esu educational service unit(s); electrostatic unit; evolutionary significant unit

ESU Emporia State University; English-Speaking Union

E-SU English-Speaking Union

e sub excitor substance

E Suffolk East Suffolk

eSun except Sunday

ESUNA Ethiopian Students Union of North America

Esup suppressor voltage (symbol)

E Sussex East Sussex

E-SUUS English-Speaking Union of the United States

esv earth satellite vehicle; enamel single varnish (insulation code)

ESV Earth Satellite Vehicle; Experimental Safety Vehicle

ESW Ethical Society of Washington

eswl equivalent single-wheel loading

eswl (ESWL) extracorporeal shock-wave lithotripsy

Esx Essex

esy extended school year

et edge thickness; educational therapy; educational training; effective temperature; electric telegraph; electric telegraphy; electric typewriter; electrical time; electrical transcription; electronic tests; electronic transaction; engineering test; engineering testing; equation of time; evapotranspiration

et (ET) elapsed time; electronic timing; ephemeris time; external tank; extra terrestrial

e/t (E/T) ergotamine tartrate; ergotin tartrate

e t en titre (French—in the tide)

Et Ethyl; Etienne

ET East Texas (Pulp & Paper Company); Eastern Time; Educational Therapy; Electronics Technician; Employment and Training choices; English Text; English translation; Entertainment Tax; ephemeris time; Ethiopia (Internet code); Ethiopian Airlines; European Theater (of war); Exchange Telegraph

ET Extra Terrestrial (symphonic suite by John Williams)

eta electronic transfer account; estimated time of arrival; expect to arrive

ETA Educational Telecommunications for Alaska; Electronic Traders Association; Employment Training Administration; English Teachers Association; European Teachers Association; Exception Time Accounting; Express Transport Association

ETA Euzkadi to Azkatasuna (Basque—Nation and Liberty)

Etab Etablissement (French—business establishment or factory)

ETAB Environmental Testing Advisory Board (Dow)

ETAC Environmental Technical Applications Center

et al. et alibi (Latin—and elsewhere); *et alia* (Latin—and others)

ETAI Electronics Technicians Association International

e-tailers electronic retailers

e-tailing electronic retailing

ETAN East Timor Action Network

ETAP Expanded Technical Assistance Program

ETAQ English Teachers Association of Queensland

etas escort towed array sonar

ETAS Escort-Towed Array System

ETASS Escort-Towed-Array Sonar System

etb early to bed; end of transmission block; estimated time of berthing

etb (ETB) end of transmission block character (data processing)

ETB Engineering Test Basis

ETBO Engineering Test Base Office

etc earth terrain camera; effluent treatment cell; electronic temperature control; electronic travel computer; electronic typing calculator; employee time card; estimated time of completion; export trading company; extraterrestrial civilization

etc. *et cetera* (Latin—and so forth)

e t c en tout cas (French—in any case)

ETC Electrical Technician Certificate; Electro Tech Corporation; Electronic Technician Certificate; Emergency Training Center; Engine Technical Committee; Environmental Testing Corporation; Episcopal Travel Club; European Translations Center; European Travel Commission

ETC. A Review of General Semantics (Official Organ of the International Society for General Semantics)

ETCC Eastern Tank Carrier Conference

ETCE Empresa Transportes Colectivos del Estado (Spanish—State Collective Enterprise Transport)

etcg elapsed-time code generator

etcrrm electronic teleprinter cryptographic regenerative repeater mixer

etd estimated time of departure; extension trunk dialing

ETD End of Train Device; English Teaching Division

ETDS Electronic Theft Detection System

ete estimated time enroute

ete este (Spanish—east)

ETE Experimental Tunnelling Establishment

ETE Escuela Technica del Ejército (Spanish—Technical School of the Army)

ETEE Education Technologies for European Enterprises

ETEMA Engineering Teaching Equipment Manufacturers Association

eter estimated time enroute

etf electron-transferring flavor-protein; enhanced tactical fighter; enhanced technology fighter; exchange-traded fund

Étg Étang (French—lagoon, pond)

etgm estimate to get money

eth ether; ethical; ethics; ethmoid; ethmoidal; ethnic; extraterrestial hypotheses (explaining close encounters of the third kind, such as ufos); extraterrestrial hypothesis

Eth Ethiopia; Ethiopian; Ethiopic

ETH Eidgenössiche Technische Hochschule (German—Swiss Federal Institute of Technology)

ethanol ethyl alcohol or grain alcohol (C_2H_5OH)

Eth$ Ethiopian dollar

eth dat ethic dative

Ethel Ethelberg; Ethelberta; Ethelburg; Ethelda; Etheldrid; Ethelind; Etheljean; Ethelrede; Ethelsa; Ethelwyn

ether ethyl ether ($CH_2H_5)_2O$

Ethernet Xerox trademark

Ethiop Ethiopia; Ethiopian

ethno ethnology

ethnoc ethnocide

ethnog ethnography

ethnograph ethnograph(er); ethnographic(al)(1y); ethnography

ethnol ethnologist(ic)(al)(1y); ethnology

ethnomus ethnomusicologist; ethnomusicology

ethnomusi ethnomusic(al)(ly); ethnomusicologist; ethnomusicology

ethnophaul ethnophaulism (study of international slurs); ethnophaulist(ic)(al)(ly)

ethnosci ethnoscience; ethnoscientific(al)(1y); ethnoscientist(s)

etho ethylene oxide

ethog ethogram; ethographer; ethographic; ethography

ethol ethologic(al)(1y); ethologist(ic)(al)(ly); ethology

eti elapsed-time indicator; estimated time of interception

Eti Etiopia (Italian, Spanish—Ethiopia); *Etiopia* (Portuguese—Ethiopia)

ETI Electric Tool Institute; Electronic Technical Institute; Emerging Technologies Initiative; Equipment and Tool Institute

ETIA European Tape Industry Association

ETIC English Training Information Centre (London)

e-time execution time

etio etiocholandone

etiol etiology

ETIS-MARFO European and Technical Information Service in Machine-Readable Form

etk (ETK) erythrocyte transketolase

etkm every test known to man

etl emergency tolerance limit; ending tape label; etching by transmitted light

ETL Electrical Testing Laboratory; Electro-Technical Laboratory; Endorsed Tools List; Engineering Test Laboratory; Essex Terminal (railroad); European Test Laboratory

etlt equal to or less than

ETM Electronic Technician's Mate; Element Test and Maintenance

ETMA English Timber Merchants Association

ETMA-A Engineering Tooling and Manufacturing Aide

ETMWG Electronic Trajectory Measurements Working Group

etn equipment table nomenclature

ETN Eastern Technical Net (USAF)

EtNu ethylnitrosourea

eto electric truck operator; estimated takeoff; estimated time off; ethylene oxide

ETO Earth to Orbit; Energy Technology Office; European Theater of Operations; European Transport Organization; Executive Training Office(r)

etoc expected total operating cost

Et OH estimated turnaround point; estimated turning point; ethyl alcohol; extratemporal perception

ETOPS Extended-range Twin-Engine Operations

ETOUSA European Theater of Operations, United States Army

etp extratemporal perception

etp (ETP) electron transfer particle

ETP Eastern Tropical Pacific; Education and Training Program; Effluent Treatment Plant; Engineering Test Plan; Evaluation Test Plan; Executive Training Program

ETPI Eastern Telecommunications Philippines Inc

et-pnl engine test panel

ETPO Eastern Tropical Pacific Ocean

ETPRO Employment Tax Problem Resolution Office

ETPS Empire Test Pilots School

etr effective thyroid ratio; estimated time of return; export traffic release

etr (ETR) engineering test reactor

Etr Etruscan

Etr entrada (Spanish—entrance)

ETR Eastern Test Range; Easy-growth Treasury Receipts; Electric Target Range; Engineering Test Reactor; Export Traffic Release; External Technical Report

etra estimated time to reach altitude

ETRC Educational Television and Radio Center; Engineering Test Reactor Critical Facility

etro estimated time of return to operation

ETRs Encrypted Traffic Reports

etry entirely

ets electronic telegraph system; estimated time of sailing; expiration of term of service; expiration term of service; expiration of time of service

Ets Etablissements (French—establishments)

ETS Educational Television Stations; Educational Testing Service; Electronic Telegraphic System; Engine Test Stand; Engineering Task Summary; Engineering Test Satellite

ETSA Electricity Trust of South Australia

ETSC East Tennessee State College; East Texas State College

et seq. et sequens (Latin—and following)

ETSI European Telecommunications Standards Institute

etsp entitled to severance pay

etsq electrical time superquick

ETSS Engineering Time-Sharing System; Entry Time-Sharing System; Experimental Time-Sharing System

ETSU Energy Technology Support Unit

ett early thrust termination; electromagnetic thickness tool; exercise tolerance test(ing)

ett (ETT) evasive target tank

ETT Elizabethan Theatre Trust; Explosion Tear Test(ing)

etta electronic temperature trip and alarm

Etta Henrietta

ETTA English Table Tennis Association

ETTDC Electronics Trade and Technology Development Corporation

et to extractor tool

et tp etch template

ETTU English Table Tennis Union

etu electron tube; ethylene thiourea

ETU Electrical Trades Union; Emergency Treatment Unit

ETUC European Trade Union Confederation

et ux. et uxor (Latin—and wife)

etv educational television; engine test vehicle

etv (ETV) educational television

ETV Educational Television; *Electrotechnischer Verein* (German—Electrotechnical Society); Engine Test Vehicle

etvm electrostatic transistorized voltmeter

etw empty tank weight; end-of-tape warning

etw etwas (German—something)

ETW Equipment Trials Wing

ETWN East Tennessee & Western North Carolina (railroad)

etx (ETX) end of text character (data processing)

etym etymologic(al)(ly); etymologist(ic)(al)(ly); etymology

E-type Jungian extrovert type

eu electron unit; emergency unit; expected utility; external upset (oil well)

eu (Greek—good or well)—eubacterium, eucalyptus, euphoria

Eu entropy unit (symbol); Euler unit; Europe; European; europium; Eustace; Eustatia

EU Emory University; Evacuation Unit; European Union; Everyman's University (Tel Aviv); Experimental Unit

EU Estados Unidos (Spanish—United States); *Europa Unie* (French—United Europe)

E-U Etats-Unis (French—United States)

eua examination under anesthetic

Eua European unit of account

EUA Eastern Underwriters Association

EUA Estados Unidos da América (Portuguese—United States of America); *Estados Unidos de América* (Spanish—United States of America); *Etats Unis Amérique* (French—United States of America)

EUB Evangelical United Brethren

EUB Estados Unidos do Brasil (Spanish—United States of Brazil)

euc end-use check(ing)

EUC Euclid (railroad)

eucd emotionally unstable character disorder

Eucl Euclid

EUCLEX European Cloud and Radiation Experiment

EUCLID Experimental Use Computer—London Integrated Display

EUCOM European Command

euc(s) eucalyptus tree(s)

EUDISED European Documentation and Information System for Education

euf eufemismo (Italian, Portuguese, Spanish—euphemism)

EUF European Union of Federalists

eufe eufemismo (Italian, Portuguese, Spanish—euphemism)

EUFTT European Union of Film and Television Technicians

Eug Eugene; Eugenia

eugen eugenics

Eugn eugenics

Eugº Eugenio

EUI Energy Utilization Index; European University Institute

EUI Enciclopedia Universal Ilustrada (Spanish—Universal Illustrated Encyclopedia)

EUL Edinburgh University Library

EUL Everyman's University Library

EUM European Mediterranean

EUM Entr'aide Universitaire Mondiale (French—World University Service); *Estados Unidos Mexicanos* (Spanish—United States of Mexico)

EUM-AFTN European-Mediterranean Aeronautical Fixed Telecommunications Network

EUMOTIV European Association for the Study of Economic, Commercial, and Industrial Motivation

EUMR Emergency Unsatisfactory Material Report

Euni Eunice

EUP Edinburgh University Press; English Universities Press

euph euphemism(s); euphemistic(al)(ly); euphemize(d); euphemizer(s); euphemizing

euphé euphémisme (French—euphemism)

Euphe der Euphemismus (German—euphemism)

euphem euphemism; euphemistic(al)

euphem euphémique (French—euphemistic); *euphémisme* (French—euphemism)

Euphie Euphemia

euphon euphonic; euphonically;euphony

eur europaeisk (Dano-Norwegian—European)

Eur Europe; European

EUR Erasmus Universiteit Rotterdam (Dutch—Erasmus University of Rotterdam)

Eurafrica Europe and Africa

Eurail European Railways

Eurailpass European tourist railroad pass

EURAM European Research in Advanced Materials

EURAS European Anodisers Association

Eurasafrica Europe, Asia, and Africa

EURASBANK European Asian Bank

Eurasia Europe and Asia; from the Caspian Sea and the Caucasus Mountains to the Ural Mountains

Eurasian(s) person(s) of European and Asian parents

Euratom European Atomic Energy Community—a six-nation atomic energy pool consisting of France, Germany, Italy, Belgium, Netherlands, and Luxembourg

EURATOM European Atomic Energy Community

Eur Ct H R European Court of Human Rights

EUREKA European Research Coordination Agency

eurex enriched uranium extraction

EURIMA European Insulation Manufacturers Association

Eurip Euripides

EURIPA European Information Providers Association

EURO European Regional Office (FAO)

EURO-AIM European Organization for an Audiovisual Independent Market

Eurobonds European bonds

EUROCAE European Organization of Civil Aviation Electronics

Euro-Can(s) European-Canadian(s)

EUROCARE European Conservation and Restoration

EUROCEAN European Oceanographic Association

Eurochemic European chemical processing of irradiated fuels

Euro Com European (NATO) Communications

EUROCOM European Coal Merchants Union

Eurocom(s) European communism; European communist(s)

EUROCOOP European Community of Cooperative Societies

EUROCORD European Cord, Rope, and Twine Industries

eurocrat European bureaucrat

EURODICAUTOM European Automated Dictionary

EURODIDAC European Association of Manufacturers and Distributors of Educational Materials

Euro$(s) European dollar(s)

Eurodol(s) European dollar(s)

EUROFAR European Future Advanced Rotorcraft

Eurofima European Company for the Financing of Rolling Stock

Eurofinance *Union International d'Analyse Economique et Financière* (French—International Union of Economic Analysis and Finance)

EUROFINAS European Financial Houses

Eurogroup Belgium, Denmark, Germany, Greece, Italy, Luxembourg, Netherlands, Norway, Portugal, Spain, Turkey, United Kingdom

Eurolex full-text electronic legal-research network

Euro Log European (NATO) Logistics

Euro Long-Term European (NATO) Long-Term Operations

Euromart European Common Market

Euro Med European (NATO) Military Medicine

Euromissiles European-deployed medium-range nuclear missiles

EURONET European Network (data-transmission)

EURONETT Evaluating User Reaction on New European Transport Technologies

Europ European railway car pool

Europhot European professional photographers

EUROS European Register of Ships

Eurosac European paper sack manufacturers

Eurosat European application satellite systems

EUROSPACE European Space Study Group

Eurostat European Communities Statistical Office

eurotainer European-owned container

EUROTECNET European Technical Network

Euroterro European terrorism; European terrorist

EUROTEST European Association of Testing Institutions

EUROTOPP European Transport Planning Process

Eurotories European Tories (conservative parties such as Britain's Conservatives and Germany's Christian Democratic Union)

Eurotox European Committee on Toxicity Hazards

EUROTRA European Translation System

EUROTRAP European Transport Planning System

Eurotunnel railway tunnel linking England and France

Eurovision European Television

EUS Eastern United States; Engineering Undergraduates Society

EUSA Eighth United States Army

EUSAFEC Eastern United States Agricultural and Food Export Council

Euseb Eusebius Pamphili

EUSIDIC European Association of Scientific Information Dissemination Centers

Eus Sta Euston Station

eutec eutectic; eutectoid

EUTELSAT European Telecommunications Satellite organization

euth euthanasia(n), euthanasic(al)(ly), euthanatize

euthan euthanasia

euv energetic ultraviolet; equivalent ultraviolet; expected utility value; extreme ultraviolet

EUVE Extreme Ultraviolet Explorer satellite

euvsh equivalent ultraviolet solar hour

euw engine(s) under wing(s)

EUW European Union of Woman

Eux Euxine

Euxine Sea Black Sea

ev earned value; efficient vulcanization; electric vehicle; electron volt; enclosed and

ventilated; escort vessel; evangelical; exposure value; extremely violent

ev *electrón-voltio* (Spanish—*electron v*olt)—also appears as *eV; en ville* (French—local); *evangelisch* (German—Protestant)

eV electronvolt

eV *eingetragener Verein* (German—registered society)

Ev Evenkian; Everett

Ev *Eingang vorbehalten* (German—rights reserved)

Ev. *Evangelium* (Latin—the Gospel)

EV Elivie (Italian Heliways); English Version; Erne Valley; Everett (railroad)

eV 1 s edge-V one side (lumber)

eV 2s edge-V two sides

eva electronic velocity analyzer; ethyl-vinyl acetate; external vehicular activity; extra-vehicular activity; extravehicular ambulation

EVA Educational Voucher Authority; Electrical Vehicle Association; Engineer Vice Admiral; Essex Volunteer Artillery

evac evacuate; evacuation

evacship evacuation ship

eval evaluate; evaluation

Evan Evangelical; Evangelist

Evans Phil Evansville (Indiana) Philharmonic

evap evaporate; evaporation; evaporator; evaporize

evapd evaporated

evaptr evaporator

evata electronic visual auditory training aid

EVC Educational Video Corporation; Electric Vehicle Council; Engineer Volunteer Corps

EVCA European Venture Capital Association

evce evidence

evco electron vibration cutoff

EVCP Engineering Value Control Proposal

EVCS Extravehicular Communications System

evd extended voluntary departure (immigration)

EVDF Eugene V. Debs Foundation

eve evening

Eve Eveleen; Evelina; Eveline; Evelyn; Everarda; Everett; Everette; Everina; Everline

evea extravehicular engineering activities

evenin' evening

event. eventuell (German—possibly)

Ever Everest—world's highest mountain towering over the Himalayas of Nepal and Tibet

Everglades Everglades National Park in Florida

evf electronic viewfinder

evg evening

EVG Europäische Verteidigungsgemeinschaft (German—European Defense Community)

evi evidence

EVI Extreme Value Index

evict. evaluation of intelligence collection tasks

evid evidence

evir evirato (Italian—emasculate)—eunich

e viv. disc. e vivis discessit (Latin—departed from life)

EVL E(dward) V(errall) Lucas

evln evolution

ev-luth evangelisch-luterisch (German—Evangelical Lutheran)

evm extraneous vegetable matter

evm (EVM) earth-viewing module

evminfin everglazed minicare finish

EVMS Eastern Virginia Medical School

evmu extra-vehicular material unit

evng evening

evol evolution; evolutionary; evolutionist

evop evolutionary operation

EVP Executive Vice President

evr electronic video recording; electronic voice recording

EVR Edinburgh Volunteer Rifles

EVRC Eton Volunteer Rifle Corps

evrep event recording potential

EVRS Electronic Video Recording System

evs expected value saved

evs (EVS) extravehicular system

EvS Environmental Science

EVs electric vehicles

EVS Electronic Voice Switching (system); Electronic-optical Viewing System

evsd energy-variant sequential detection

evss extravehicular space suit

evstc (EVSTC) extravehicular suit telemetry and communications

evt educational and vocational training; effective visual transmission; equiviscous temperature; eventually; extra-value trimmed (meat)

evt eventuel (Dano-Norwegian—possible)

E v T E van Tongeren

EVT Engineering Verification-Test(ing)

EVT Europäische Vereinigung für Tierzucht (German—European Association for Animal Production)

evtl eventuell (German—eventually, perhaps, possibly)

EVV Evansville, Indiana (airport)

EVW European Voluntary Workers

EVX Electric Vehicle Experimental

evy every

evythg everything

ew each way; earthenware; effective warmth; electrically welded; electronic warfare; equivalent widths; extensive wound

ew (EW) earth watch

e/w equipped with

Ew Ewart; Ewbanke; Ewell; Ewen; Ewing

Ew Euere or *Eure* or *Eurer* (German—your)—abbreviation used in titles

EW early warning; electronic warfare; Emergency Ward; Engineer's Writer; enlisted woman; enlisted women

E & W England and Wales

EWA East and West Association; East-West Airlines; Education Writers Association; Electrical Wholesalers Association

ewac electronic warfare anechoic chamber

EWACS Electronic Wide-Angle Camera System

EWAD Early Warning Air Defense

EWAS Economic Warfare Analysis Section

ewb estrogen withdrawal bleeding

ewc electric water cooler

ewc (EWC) electronic warfare coordinator

EWC East-West Center (University of Hawaii)

EWCB Electrical Workers and Contractors Board

EWCRP Early Warning Control and Reporting Post

ewd elementary wiring diagram

EWD Economic Warfare Division

ewdt early warning data transmission

ewe. electronic warfare element

ewec electromagnetic wave energy conversion

ewes electronic warfare evaluation simulator

EWES Engineering Waterways Experiment Station

ewex electronic warfare exercise

ewexipt electronic warfare exercise in port

ewf equivalent weight factor

EWF Earth, Wind, and Fire (music group); Electrical Wholesalers Federation

EWG Executive Working Group (NATO)

EWG Europäische Wirtschaftsgemeinschaft (German—European Common Market)

ewgcir early-warning ground-control-intercept radar

EWHS Eli Whitney School

ewi education with industry; entered without inspection

Ewi English winter index

EWI Earl Warren Institute; Executive Women International

ewicb electronic-warfare interface-connection box

ewl evaporative water loss

EWL Ellerman's Wilson Line

EWLD Engineering Weekly Labor Distribution

ewma exponentially weighted moving average

EWMC Eli Whitney Metrology Center

EWO Electrical and Wireless Operators; Electronic Warfare Officer; Electronic Weapons Officer; Emergency War Order; Engineering Work Order; Essential Work Order

EWOC Eligible Worker Owned Cooperative

EWO-DS Engineering Work Order-Drawing Summary

ewops electronic warfare operations

EWOS Electronic Warfare Operational System (USAF)

EWP Emergency War Plan

EWPI Eysenck-Withers Personality Inventory

EWPs Electronic Warfare Plans

ewr early-warning radar

EWR Electrical Wiring Regulations; Engineering Work Request; Equaled World Record; Newark, New Jersey (airport)

EWRC European Weed Research Council

EWRT Electrical Women's Round Table

ews early warning system; experienced workers standard

ew's edge weapons (sharp bladed daggers, cutlasses, knives, machetes, swords, etc.)

EWS Emergency Water Supply; Emergency Welfare Service; European Wars Survey

EWSC Eastern Washington State College; Electric Water Systems Council

EWSF European Work Study Federation

ewsl equivalent single-wheel load(ing)

ewsm electronic-warfare support measures

EWT Eastern War Time (advanced time)

EWTN Eternal Word Television Network

EWU Eastern Washington University

eww extended work week

EWWS Electronic Warfare Warning System

ex etc.; exact(ed); exacting; ex-actitude; exactly; examination; examine(d); examiner; examining; example; ex-cess(ive); exclusive; exclusively; exclusivity; exe-cute(ed); executing; exercise; exercising; experiment(al's); extra(neous)

ex (Latin prefix—out of)—excision; (Latin—from)

Ex Excelsior; Exchange; Exchequer; Exeter; Exmoor; Exmouth; Extremadura; Exuma; voltage drop across a reactance (symbol)

Ex Exodo (Spanish—The Book of Exodus); *Exodus*

EX experimental broadcasting

exacct expense account

ex af. ex ajfnis (Latin—of affinity)

exafs extended X-ray-absorption final structure

exag exaggerate; exaggerated; exaggeration

Ex Agt Executive Agent

exam examination; examine; examiner

examd examined

exametnet experimental meteorological sounding rocket network

examg examining

examn examination

examr examiner

exams examinations

ex aq. ex aqua (Latin—out of water)

exbedcap expanded bed capacity

Ex B/L exchange bill of lading

EXBT Exabyte Corporation

exc excavate; excellent; ex-cept(ion)(al)(ly); exciter

exc. excellency; excepted; exchange

exc. excudit (Latin—he engraved it)

Exc Excélencia (Spanish—Excellency); Excellency

Exc Excélsior (Mexico City); *Excelencia* (Spanish—Excellency)

exca excavate; excavation

Exc^a Excelencia (Spanish—Excellency)

ex cath. ex cathedra (Latin—from the seat of authority)

Excel Excelsior

EXCEL Corporation for Excellence in Public Education; Ex-offender Coordinated Employment Lifeline (Indiana's parole project)

exch exchange

ex champ ex-champion; former champion

Excheq exchequer

exch oper exchange operator

exchq exchequer

exchr extra charge

excl exclude; exclusion; exclusive; exclusivity

excl exclusief (Dutch—not included)

exclam exclamation; exclamatory

exclt excellent

exclu exclusive(ly); exclusivity

Exc^ma Excelentísima (Spanish—Most Excellent)—feminine

Excmo Excelentísimo (Spanish—Most Excellent)

Exc^mo Excelentísimo (Spanish—Most Excellent)—masculine

Ex Cncl Executive Council

Ex Co Executive Council

Ex Com Executive Committee

ex-con(s) ex convict(s); former convict(s)

ex cont from contract

excp except(ion)(al)(ly); execute channel program

ex cp ex coupon

excpt except(ion)(al)(ly)

excr excrescent

Excrpt Med Excerpta Medica

excs excess

exct execution

excv exclusive

exd examined

EXDAMS Extendable Debugging and Monitoring System

ex det explosives detector

ex div ex dividend

Ex Div Experimental Division

Ex Doc Executive Document

Exe Exeter

exec execute(d); execution; executive; executive officer; executor

exec (EXEC) execute statement (data processing)

Exec Dir Executive Director

Exec Off Executive Officer

execs executives

Exec Sec Executive Secretary

exeod expects to enter on duty

exer exercise

exes expenses

Exet Coll Exeter College-Oxford

Exeter Exeter College (Oxford), Phillips Exeter Academy

exf external function

ex f extremely fine

ex fac ex factory

ex fy extra fancy

exg existing

ex ga external gage

EXGO Export Guarantee Office(r)

ex gr. *exempli gratia* (Latin—for example)

exh exhaust

exhib exhibit; exhibition; exhibitor

exhib. *exhibeatur* (Latin—let it be shown)

exhn exhibition

exh t exhaust turbine

exh v exhaust vent

ex hy extra heavy

EXIAC Explosives Information and Analysis Center (USA)

Ex-Im Export-Import Bank

EXIMBANK Export-Import Bank

ex int ex interest

exis existential; existentialism; existentialist

exist. existing

EXIT Ex-offenders In Transit (Maine's parole project)

exkl *exklusiv* (German—excepted, not included)

ex lib. *ex libris* (Latin—from the library of)

Ex^{ma}SR^{a}D *Excelentíssima Senhora Dona* [Portuguese—Mrs (precedes full name in formal style)]

EXMAN Experimental Manipulation (of Forest Ecosystems in Europe)

ex-mer ex-meridian

Ex^{mo}SR *Excelentíssimo Senhor* [Portuguese—Mr (precedes full name in formal style of address)]

exmr examiner

ex n(ew) excluding new shares

ex-nupt(s) ex-nuptial(s)-person(s) born out of wedlock

ExO executive officer; executive order

Ex O Experimental Office(r)

EXO European X-ray Observatory

exobio exobiologic(al)(ly); exobiologist; exobiology

exocrin exocrinologic(al)(ly); exocrinologist; exocrinology

Exod *Exodus*

ex off. *ex officio* (Latin—by authority of his office)

Exon. *Exonia* (Latin—Exeter)

exonum exonumia(l)(ly); exonumic(al)(ly); exonumist(s)

exonym foreign-language place name such as Londres (Spanish for London)

Ex O P Executive Office of the President

exopac excatmospheric jettisonable control wafer

exor executor

exord exercise order

exos (EXOS) exospheric satellite

exosat (EXOSAT) European X-ray observatory satellite

exot exotic

exotheo exotheologic(al); exotheologist(s); exotheology

exox. executrix

exp expansion; expenditure; expense; experience; experiment(al); exponential; export; Exposition; exposure; express; expulsion

exp *expreso* (Spanish—express)

ex p. *ex parte* (Latin—on one side only)

EXP Exchange of Persons (UNESCO office)

expate(s) expatriate(s)

exp'd experienced

expdivun experimental diving unit

expdn expedition

expdt expiration date

exped expedite; expedition

exper experiment; experimental

Expert Expanded Pert (program evaluation and review technique)

expi export performance taxation incentive

exp-imp export-import

expir expiratory; expiration

exp jt expansion joint

expl explain; explanation; explanatory; explosimeter; explosimetric; explosion; explosive(s)

expl *exemple* (French—example)

explan exercise plan

EXPLIC Export License

explo explosion; explosive

exploit. exploitation

explor exploration

Explora Exploratorium

explos explosion; explosive

expn exposition

expnd expenditure

expo expose; exposition

exp o expermmental order(s)

expo, expreso (Spanish—express)

Expo 67 1967 exposition in Montreal

Expo 70 1970 exposition at Tokyo

expol expanded polysterene (light-weight packing moulding)

export *exportaciones* (Spanish—exports)

expr expiration; expire; expression

expr *expressif* (French—expressive)

ex-Pres ex-President

EXPRESO *Expreso Aéreo Interamericano,* (Spanish—Interamerican Air Express)

expressway express highway

exps expenses

expsl expected runs batted in by slot number (baseball)

expt experiment

exptl experimental

expto expedite travel order

exptr exporter

expul expulsion

expur expurgate(d)

Expwy Expressway

Expy Expressway

ex-quay free on quay

exr executor

ex r ex rights

exray expendible relay

exrx executrix

exs expenses; expropriations

exs. executors

ex's expenses

exsec exsecant

exshi expedite shipment

ex ship delivered out of the ship

exspec exercise specification(s)

exst exempt sales tax

Ex Sta Experimental Station

ext extend; extension; exterior; external; extinguish; extinguisher; extra

ext (EXT) extraction (dental)

ext. *extend* (Latin—spread); *extractum* (Latin—extract)

Ext Extended; Extension

ExT *Expository Times*

extal extra time allowance

extd extracted

ext d & cc external drug and cosmetic color

Extel Exchange Telegraph (press agency)

EXTEL Exchange Telegraph (British news agency)

extemp extemporaneous(ly)

exten extension

extend. extensus (Latin—spread)

extern external; externally

EXTERRA Extraterrestrial Research Agency (USA)

ext fl extract fluid (fluid extract)

extg extinguish(er)

extgh extinguish

exting extinguished

ext. liq. extractum liquidum (Latin—liquid extract)

extm extended telecommunications module

ex tm. ex testamento (Latin—in accord with the testament)

extn extraction

extr extract; extrude; extruded; extrusion

Extr Extremadura

extra extraordinary

extrad extradition

extradop extended range doppler

extradovap extended-range doppler velocity and position

extrap extrapolate; extrapolated; extrapolating; extrapolation; extrapolative; extrapolator

extra sess extra session (legislature)

extrd extruded

extrem extremity

Extrem Extremadura

extro extroversion; extrovert

extrx executrix

extsn extension(al)

exurb exurban; exurbanite; exurbia; exurbian

ex works out of the factory (factory price exclusive of delivery charge)

exx examples; executrix

e_{xy} cross-elasticity of demand

Exy Expressway

Exz Exzellenz (German—Excellency)

EY Essex Yeomanry

eyawtkas everything you always wanted to know about sex

EYC Eastern Yacht Club; Encinal Yacht Club; European Youth Campaign

eyco estimated yearly cost of operation

EYD Ejaan Yang Disempurnakan (Indonesian—Improved Spelling System)

EYE European Year of the Environment

Eyedentify trademark of Eyedentify Incorporated

EYOA Economic and Youth Opportunities Agency

EYR East Yorkshire Regiment

EYS Ecumenical Youth Service

EYW Key West, Florida (airport)

ez easy; eczema; electrical zero

e-z easy

e/z equal zero

Ez Ezekiel; Ezra; The Book of Ezra

EZ Eastern Zone; Emile Zola; Extraction Zone

EZ Einelige Zwillinge (German—monozygotic twins)

EZ Duzit Easy Does It

Ezeiza Ezeiza International Airport (Buenos Aires, Argentina)

Ezek The Book of Ezekiel

Ezek Ezekiel

Ezi Ezias; Eziel; Eziongaber

EZI Electrolytic Zinc Industries

EZPERT Easy Programme Evaluation and Review Technic

Ezr Ezra

EZ terms easy terms

EZU Emiliano Zapata Unit (Chicano terrorists); *Europäische Zahlungsunion* (German—European Payment Union)

F

f faded; fair; family; farm; farthing; fast; father (capitalized in religious orders); fathom; feast; feet; female; feminine; femto; ferry; filment; final target; finance charge; financial responsibility; fine; first class (travel); flat; flying; focal length; fog; folio; following; following page; force; forecastle; founded; franc(s); frequency; freshwater; frost; fuel; fugacity; fullback (football), full function; fumble; latitude factor (symbol); relative humidity (symbol)

f/ relative aperture of a lens (also shown as f:)

f fecit (Latin—he did); filius (Latin—son); forte (Italian—loud); fundada; fundado (Spanish—founded); für (German—for)

f/ fardo(s) [Spanish—bale(s); bundle(s); package(s)]

F dealt in flat (bond listings); Fahrenheit; Fairchild; farad; Faraday; Faraday constant (symbol); Farrell Lines; fathom(s); February; Federal League; Fellow; fermi; field of vision (symbol); fighter; filament (symbol); fire; fishing mortality rate; fixed; fixed broadcast; fixed broadcasting; flagship; florin; fluorine; Foreign (newspaper stock listings); formal(ity); formula; Foxtrot—code for letter F; fractional concentration in gas (symbol); France;

franc(s); Fraunhofer line (caused by hydrogen); freedom; freedom, degree of (symbol); free energy (symbol); French; Friday; fuel; furlong(s); Fumess Lines; fuse (symbol); Grumman; function; longitude factor; Society of Friends

F3 Form, Fit, Function; subsequent generations

fa fielding average (baseball); first attack (lacrosse); fluorescent angiography; free alongside

F. fats (dietary symbol)

°F degree Fahrenheit

F Federal Reporter; feria (Latin, Portuguese, Spanish—fair or market); fine (Italian—the end); fora (Portuguese—out); framkomst (Swedish—arrival); Frauen (German—women); freddo (Italian—cold); frio (Portuguese, Spanish—cold); froid (French—cold); fuera (Spanish—out); fuori (Italian—out); (Latin—Filius)

F- fluoride ion

F 1 Formula One (automobile sport)

F-1 Fury single-engine jet fighter-bomber flown from aircraft carriers

F_1F_1 layer [lower of two atmospheric layers wherein the F region of the ionosphere splits during the day at heights varying from 90 to 150 miles

(145 to 241 kilometers) above the earth's surface]; first filial generation

F_1O decimetric solar flux (symbol)

F 1C Fireman 1st Class (USN)

F1S finish one side

F 2 Formula Two (automobile sport)

F_2 F_2 layer [upper of two atmospheric layers wherein the F region of the ionosphere splits during the day at heights varying from 150 to 250 miles (241 to 402 kilometers) above the Earth's surface; second filial generation

F^2 prostaglandin alpha (abortion-producing hormone)

F^{2d} Federal Reporter, second series

F2S finish two sides

F-3 Demon single-engine supersonic all-weather jet fighter

F4 Phantom II twin-engine all-weather supersonic jet fighter-bomber

F4p fortran 4 plus

F-4U Chance-Vought single-engine fighter popular during World War II and called the Corsair

F-5 Northrup Freedom Fighter twin-jet aircraft

F-6 Skyray single-engine supersonic all-weather jet fighter

F6F Grumman single-seat piston-powered fighter aircraft named Hellcat

f8 factor 8

F-8 Crusader single-engine all-weather supersonic jet fighter

f/8 @ 1/50th camera lens aperture ratio 8 at an exposure of 1/50th of a second

F-9 single-engine single-seat jet fighter aircraft made by the Shen Yang Aircraft Production Complex of the People's Republic of China

F9F-2 Grumman Panther single-engine single-seat naval fighting aircraft

F9F-6 Grumman carrier-based transonic fighter aircraft called Cougar

F 11 fluorocarbon (concentrations and emissions)

F-11 Tiger single-engine supersonic jet fighter

f-12 freon (refrigerant)

F-13 dope; drugs; narcotics

F-14 swing-wing jet fighter aircraft nicknamed Tomcat and carried on some U.S. naval vessels

F-15 Eagle supersonic-jet fighter aircraft

F-16 high-performance low-cost air-combat fighter aircraft produced by Convair's Fort Worth Division

F-16B two-place fighter/trainer aircraft

F-18 all-weather fighter and attack airplane

F-18A McDonnell-Douglas Hornet strike fighter aircraft

F-18L McDonnell-Douglas Northrup multirole fighter aircraft

F-27 Fokker Friendship (aircraft)

F-27M Fokker Troopship built in the Netherlands

F-28 Fokker turbojet aircraft

F-47 Republic fighting aircraft developed during World War II and called Thunderbolt

F-51 North American fighter aircraft developed during World War If and called Mustang

f/64 Group f/64 (photographers Ansel Adams, Imogen Cunningham, Edward Weston, Willard Van Dyke, and their followers)

F-80 Lockheed Shooting Star jet fighter-bomber

F-84 Republic Thunderjet fighter-bomber

F-86 North American Sabre single-engine jet fighter aircraft

F-89 Scorpion all-weather interceptor with twin turbojet engines

F-100 Super Sabre supersonic turbojet fighter

F-101 Voodoo supersonic twin-engine turbojet aircraft

F-102 Delta Dagger single-engine supersonic turbojet interceptor

F-104 Starfighter supersonic single-engine turbojet fighter

F-105 Thunderchief supersonic single-engine turbojet tactical fighter

F-106 Delta Dart supersonic single-engine turbojet interceptor aircraft

F-111 twin-engine turbojet tactical fighter-bomber all-weather interceptor aircraft (TFX)

F-111A variable-geometry supersonic fighter-bomber (TFX)

F-404 General Electric turbofan jet engine

fa family allowance; fatty acid; field activities; filterable agent; fire alarm; first aid; first attack; flight attendant; fluorescent antibody; folic acid; fortified aqueous; free acid; free aperture; frequency agility; friendly aircraft; fuel-air (ratio)

fa (FA) fatty acid; fluvic acid

f/a friendly aircraft; fuel-air ratio; further advances

f & a finance & administration; fire and allied (insurance); fore and aft

fa. father

fa firma (Dano-Norwegian—company or firm); *foregående (forrige) år* (Dano-Norwegian—previous year); (Italian—fourth tone, *D* in diatonic scale, *F* in fixed-do system)

få forrige år (Dano-Norwegian—last year)

fᵃ factura (Spanish—invoice)

fA femtoampere

fA forrige Aar (Danish—last year)

Fa Faeroes

Fa Firma (German—firm, business)

FA Factory Act; Factory Automation; Failure Analysis; Farm Advisor; Field Allowance; Field Ambulance; Field Artillery; Final Acceptance; Financial Adviser; Fine Art(s); Fireman Apprentice; Football Association; Forecast Area; Foreign Affairs; Frankford Arsenal

F-A fighter-attack (aircraft)

F/A friendly aircraft

F & A Finance and Accounting; Financing and Accounting

F of A Foresters of America; Freethinkers of America

FA Flota Argentina (de Navegación Fluvial)—Argentine River Navigation Line; *Forze Armate* (Italian—Armed Forces); *Frontoviya Aviatsiya* (Russian—Frontal Aviation)—Soviet air force

F-A-18 McDonnell-Douglas fighter-attack aircraft named Hornet

faa field artillery airborne; formalin, acetic acid, alcohol (mixture); free of all average

FAA Federal Aviation Administration; Federal Aviation Agency; Fifth Avenue Association; Film Artists' Association; Fleet Air Arm; Foreman's Association of America; Foundation for Aboriginal Affairs; Foundation for American Agriculture; Fraternal Actuarial Association; Free Afghanistan Alliance

FAA Finska Angparrygys (Finnish Steamnship Line)

Faaa Papeete, Tahiti's airport

FAAA Fellow of the American Academy of Allergy

FAAAS Fellow of the American Academy of Arts and Sciences; Fellow of the American Association for the Advancement of Science

FAABMS Forward Army AntiBallistic Missile System

FAADS Federal Assistance Award Data System; Forward Air-Defense Area System

FAAG First Advertising Agency Group

FAAI Filipinos for Affirmative Action, Inc.

FAAN Fellow of the American Academy of Nursing; First Advertising Agency Network

FAAO Federation of American Arab Orianizations; Finance and Accounts Office (US Army)

FAAOS Fellow of the American Academy of Orthopaedic Surgeons

FAAP Federal Aid to Airports Program

FAAPS Fine Art, Antique and Philatelic Squad (Scotland Yard)

faar forward area alerting radar

FAAR Feminist Alliance Against Rape

fab fable; fabric; fabricate; fabrication; fabulist; fabulous; first-aid box; file access block

fab fabrique (French—factory); *fanco à bord* (French—free on board); *frei an bord* (German—free on board)

Fab Fabio; Fabius; Fabre; Fabrian; Fabrice; Fabrizio

FAB Facilities Advisory Board; Fijian Affairs Board; Fleet Air Base; Fourth Avenue Booksellers (NYC); Frederic Auguste Bartholdi; French-American-British

FAB Força Aérea Brasileira (Portuguese-Brazilian Air Force)

FABAS Farm Amalgamations and Boundary Adjustment Schemes

fabbr fabbrica (Italian—factory)

FABI Fédération Royale des Associations Belges d'Ingénieurs (French—Royal Federation of Belgian Engineering Associations)

fabl fire alarm bell

FABMDS Field Army Ballistic Missile Defense System

FABMIDS Field Army Ballistic Missile Defense System

fabr fabricate; fabrication

Fab Soc Fabian Society

FABSS Fellow of the Architectural and Building Surveyors' Society

FABU Fleet Air Base Unit

fabx fire alarm box

fac facade; facial; facility; facsimile; factor; factory; faculty; fast as can; field accelerator; forward air cargo; forwarding agents commmssion

fac. factum similie (Latin—facsimile)

Fac Faculty

FAC Factor (Max, stock exchange symbol); Federal Advisory Council; Federal Aviation Commission; Financial Administrative Control; Financial Affairs Commission; Fleet Air Control; Forward Air Controller; Frequency Allocation Committee; Friday Afternoon Club (collegiate drinking group)

FACA Federal Advisory Committee Act; Fellow of the American College of Anaesthetists; Fellow of the American College of Angiology; Fellow of the Association of Chartered Accountants

FAC(A) Forward Air Controller (Airborne)

FACA Federación Argentina Cooperativa Agrarias (Spanish—Argentine Agrarian Cooperative Federation)

FACAI Fellow of the American College of Allergists

FACAn Fellow of the American College of Anesthesiologists

FACB Federation of Australian Commercial Broadcasters

facc fear, anxiety, cynicism, confusion

FACC Fellow of the American College of Cardiology; Florida Association of Community Colleges; Ford Aerospace and Communications Corporation

FACCA Fellow of the Association of Certified and Corporate Accountants

FACCC Faculty Association of the California Community Colleges

faccm fast-access charge-coupled memory

FACCP Fellow of the American College of Chest Physicians

facd foreign area consumer dialing

FACD Fellow of the American College of Dentistry

FACDS Fellow of the Australian College of Dental Surgeons

face field artillery computer equipment; forced-air-cooled electronics

FACE Facilities and Communications Evaluation (USA); mnemonic for remembering the space notes of the treble clef—F, A, C, E

FACEM Federation of Associations of Colliery Equipment Manufacturers

FACES Fortran Automatic-Code- Evaluation System

facet facetious(ly)

FACFI Federal Advisory Committee on False Identification

FACFO Fellow of the American College of Foot Orthopedics

FACFP Fellow of the American College of Family Physicians

facg fast attack-craft gun

FACG Fellow of the American College of Gastroenterology

FACHA Fellow of the American College of Health Administrators; Fellow of the American College of Hospital Administrators

FACI First Article Configuration Inspection

facil facility

facile fire and casualty insurance library edition

FACIT Fast Automatic Conversion with Integrated Tools

facl facilitate

facm fast attack-craft missile

FACMTA Federal Advisory Council on Medical Training Aids

FACO Fellow of the American College of Otolaryngology

FACOG Fellow of the American College of Obstetricians and Gynecologists

facp fast attack-craft patrol; forward air control point

FACP Fellow of the American College of Physicians; Fellow of the American College of Prosthodontists

FACPM Fellow of the American College of Preventive Medicine

fac pwr ctl facility power control

fac pwr mon facility power monitor

fac pwr pnl facility power panel

FACR Fellow of the American College of Radiology

facs facsimile(s)

fac's facilities

FACS Family and Community Services; Faxon's Automatic Claim System; Federation of American-Controlled Shipping; Fellow of the American College of Surgeons; Financial Accounting and Control System; Floating-Decimal Abstract Coding System

FACSAF Fleet Air Control and Surveillance Facility (USN)

FACSFAC Fleet Air Control and Surveillance Facility

facsim facsimile(s)

facsim(s) facsimile(s)

FACSM Fellow of the American College of Sports Medicine

fact factory; fast attack-craft torpedo; flexible automatic circuit tester; fully-automatic compiler translator

fact factura (Spanish—bill of lading, invoice)

FACT Facilitation and Coordination Therapy; Factor Analysis Chart Technique; Family Action Council of Texas; Fast Access Current Text Bank; Federal Air Conditioning Technologies; Federation of Automated Coding Technologies; Financial Accounting Control Technique; Flanagan Aptitude Classification Test; Flight Acceptance Composite Test(ing); Fully-Automatic Compiler Translator; FullyAutomatic Compiling Technique

facta factura (Spanish—invoice)

FACTS Facilities Administration Control and Time Schedule; Federation of Australian Commercial Television Stations; Financial Accounting and Control Techniques for Supply

facw facultative wetlands

facty fact filled; factory

fad force activity designator; fracture analysis diagram; free air delivered; free air delivery; funds available for distribution

fad (FAD) flavine adenine dinucleotide; funding authorization document

FAD Families Against Drunks; Fleet Air Defense; Food and Agricultural Department

FADA Federal Asset Disposition Association

fadac field artillery digital automatic computer

FADC Federal Alien Detention Center

FADD Fight Against Dictating Designers

F Adm Fleet Admiral

FADM Functional Area Documentation Manager (USAF)

FADN Farm Accounting Data Network

FADO Fellow of the Association of Dispensing Opticians

FADOSS Fly Away Deep Ocean Salvage System

fadsid fighter-aircraft-delivered seismic intrusion detector

fadsorog false and dangerous systems of religion or government

FADT First Article Demonstration Test(ing)

fae fetal alcohol effect; figural after-effect; fine alignment equipment; forward air express; fuel air explosive

Fae Faeroese

FAE Federation of Arab Engineers; Foundation for Accounting Education; Fund for the Advancement of Education

FAE Federación de Amigos de Enseñanza (Spanish—Federation of the Friends of Teaching); *Fuerza Aérea Ecuatoriana* (Spanish—Ecuadorian Air Force)

FAEA Federation of ASEAN Economic Associations

FAECC Fellow of the Accountants and Executives Corporation of Canada

FAECT Federation of Architects, Engineers, Chemists, and Technicians

faer faerøsk (Dano-Norwegian—Faeroese)

Faer Faeroe Islands

Faeroes Faeroe Islands in the North Atlantic

fae's fuel air explosives

faeshed fuel-air-explosivesystem helicopter delivered

FAETUA Fleet Airborne Electronic Training Unit, Atlantic

FAETUP Fleet Airborne Electronic Training Unit, Pacific

faf final approach fix; financial-aid form; first article flow; flyaway factory; forage acre factor; forward air freight; free at field; fuzing, arming, and firing

FAF Fafnir Bearings (stock-exchange symbol); Financial Accounting Foundation; Financial Analysts Federation; Fine Arts Foundation; French American Foundation

FAFSA Free Application for Federal Students Aid

FAFT First Article Factory Test(s)

FAG Failure Analysis Group; Finance and Accounting Group (USAF); Fine Arts Gallery; Finished Americans Group; Fiscal Activities Guide

Faga Fagatoa (American Samoa's seat of government facing Pago Pago harbor)

Fagatogo American Samoa's capital on Tutuila Island adjacent to Pago Pago

fagms (FAGMS) field artillery guided missiles

FAGO Fellow of the American Guild of Organists

fag(s) faggot(s)

fag(s) fagotto(s) [Italian—bassoon(s)]

FAGS Federation of Astronomical and Geophysical Permanent Services; Fellow of the American Geographical Society

fagt first available government transportation

fagtrans first available government transportation

FAGU Fleet Air Gunnery Unit

fah failed to attend hearing

FAHA Finnish-American Historical Archives

fahqmt fully automatic high-quality machine translation

Fahr Fahrenheit

fai final acceptance inspection; first article inspection; frequency-azimuth intensity; fresh air intake

FAI Fairbanks, Alaska (airport)

FAI Fédération Abolitionniste Internationale (French—International Abolitionist Federation); *Fédération Aéronautique Internationale* (French—International Aeronautics Federation); *Federacion Anarquista Iberica* (Spanish—Iberian Anarchist Federation)

FAIA Fellow of the American Institute of Architects

FAIAS Fellow of the Australian Institute of Agricultural Science

FAIB Fédération des Associations Internationales Establies en Belgique (French—Federation of International Associations Established in Belgium)

FAIC Federation of Australian Investment Clubs; Fellow of the American Institute of Chemists

FAIDS Feline-Acquired Immune- Deficiency Syndrome

FAIEx Fellow of the Australian Institute of Export

FAIHA Fellow of the Australian Institute of Hospital Administration

FAII Fellow of the Australian Insurance Institute

FAIM Fellow of the Australian Institute of Management

FAIME Foreign Affairs Information Management Effort (Dept State)

fain. functional air index number

FAIO Field Army Issuing Office(r)

FAIP Fellow of the Australian Institute of Physics

FAIPM Fellow of the Australian Institute of Personnel Management

fair. fairing; fast-access information retrieval

Fair. Fairbanks

FAir fleet air

FAIR Fair Access to Insurance Requirements; Fairness and Accuracy In Reporting; Federation for American Immigration Reform; Financial Assistance for Independent Rehabilitation; Fleet Air (Wing); Friends in America for Independence of Rhodesia

FAIRA Foundation for Aboriginal and Island Reserve Action (Australia)

Fairbanks Institute Northern Region Correction Institute at Fairbanks, Alaska

FAIRELM Fleet Air Eastern Atlantic and Mediterranean

FAIRS Fair and Impartial Random Selection System (military draft); Federal Aviation Information Retrieval System

fairships fleet airships

FAIS Fellow of the Amalgamated Institute of Secretaries

FAITH Fending Alone In The Home (Girl Scout program)

fak fly-away kit; freights all kinds

Fak Faktura (German—invoice)

FAK Federasie van Afrikaanse Kultuurvereniginge (Afrikaans–Federation of Afrikaans Cultural Societies)

fak-pak freight all kinds (in a box on wheels)

faks faksimile (Dano-Norwegian—facsimile)

Fakt Faktura (German-–invoice)

fal fusil automatique légère (French—light automatic rifle)—FAL

Fal Falmouth

F a L Fathers-at-Large

FAL Frequency Allocation List; Frontier Airlines

FAL Frente Argentino de Liberación (Spanish—Argentine Liberation Front)—pro-Cuban

FALA Federation of Asian Library Associations; Federation of Australian Literature and Art

Falashas (Amharic—strangers)—ones without a place

Falconi Falconiformes (eagles, falcons, hawks, vultures)

'falfa alfalfa

Falk Cur Falkland Current

Falk Isl Falklammd Islands (Islas Maldivas)

Falklands Falkland Islands and Dependencies (South Georgia, South Sandwich Islands, South Shetlands)

fallex fall exercises

fall(s) waterfall(s)

Falls The Falls (any waterfall place-name such as Angel Falls, Niagara Falls, Victoria Falls, Yosemite Falls, etc.)

fallwarn fallout warning

FALN Fuerzas Armadas de Liberación Nacionul (Spanish—Armed Forces of National Liberation)

FALNP Fuerzas Armadas de Liberación Nacional Puertoriqueñia (Spanish—Armed Forces for Puerto Rican National Liberation)

FALS Ford Authorized Leasing System

falset falsetto

fam familiar; familiarization (of flights); family; foreign air mail; free at mill

Fam Famagusta; Family

FAM Federal Air Marshal; Federation of Apparel Manufacturers; Football Association of Malaysia; foreign airmail; forward air mail; Free and Accepted Masons

F & AM Free and Accepted Masons

FAMA Federal Agriculture Marketing Authority; Fellow of the American Medical Association; Fire Apparatus Manufacturers Association

FAMA Fábrica Argentina de Materiales Aerospaciales (Spanish—Argentine Factory of Aerospace Materials)

FAMAS Flutter and Matrix Algebra System

FAMC Fitzsimons Army Medical Center

fame fatty-acid methyl ester(s); financial accounting made easy

FAME Farmers Allied Meat Enterprises Cooperative; Fine Arts Magnet Program; Future American Magical Entertainers

FAMEM Federation of Associations of Mining Equipment Manufacturers

FAMEME Fellow of the Association of Mining, Electrical, and Mechanical Engineers

famex familiarization exercise

Fam Cir Family Circle

FAMHEM Federation of Associations of Materials Handling Equipment Manufacturers

famil familiar(izing)

FAMIS Financial and Management Information System

F-am-M Frankfurt-am-Main (Frankfurt-on-Main)

FAMM Families Against Mandatory Minimums (prison sentences)

Fam-Med Family Medicine

famos floating-gate avalance-injection metal oxide semiconductor

FAMOS Fleet Applications of Meteorological Observations from Satellites

FAMOUS French-American Mid-Ocean Undersea Study (of an Atlantic reef off the Azores)

fam per para familial periodic paralysis

fam phys family physician

fam rm family room

FAMS Fellow of the Ancient Monuments Society

FAMSF Fine Arts Museum of San Francisco

FAMU Florida A & M University

fan. fanatic (usually in sense of enthusiast); fantasia; fantasy

Fan Fannie; Fanny; Frances

FANA Federation of Australian Nurserymens Associations

FANAC Fabrica Nacional de Aceite (Spanish—National Oil Factory)

FANALOZA Fabrica Nacional de Loza (Spanish—National Porcelain Factory)

Faneuil Faneuil Hall meeting house in Boston's Dock Square where colonial Americans met to plot the revolution

FANK Forces Armées Nationalés Khmères (French—Khmer National Armed Forces)—Cambodian armed forces

Fannie Mae Federal National Mortgage Association

Fan(ny) Frances; Francisca; Frasquita

FANPT Freeman Anxiety and Psychosomatic Test

FANS Food and Nutritional System; Fresh Air for Non-Smokers

Fanshaw Featherstonehaugh

fant fantasia; fantasy

Fant Fantasia

fantabulous fantastic + fabulous

fantac fighter analysis tactical air combat

FANU Flota Argentina de Navegación de Ultramar (Spanish—Argentine High Seas Navigation Line)

FANY First-Aid Nursing Yeomanry

FANYS First Aid Nursing Yeomanry Service

FANZAAS Fellow of the Australian and New Zealand Association for the Advancement of Science

fanzines fan + magazines

fao finish all over

FAO Farm Advisory Office(r); Field Audit Office(r); Finance and Accounts Office(r); Fleet Accountant Officer; Fleet Administration Office(r); Food and Agriculture Organization (UN); Free Albania Organization

F & AO Finance and Accounts Office (US Army)

faop full away on passage

FAOR Functional Analysis of Office Requirements

fap final approach; fixed action pattern; floating arithmetic package

fap (FAP) familial adenomatouns polyposis; final anthropic principle; fixed action pattern

fAp full American plan

FAP Failure Analysis Program; Family Assistance Plan; Family Assistance Program(ming); First Aid Post; Foreign Assistance Program; Frequency Allocation Panel

FAP Força Aérea Portuguesa (Portuguese Air Force); *Fuerza Aerea del Perú* (Spanish-Peruvian Air Force); *Fuerzas Armadas Peronistas* (Spanish—Peronist Armed Forces)—Argentine guerrilla group

FAPA Filipino-American Political Association

FAPC Food and Agriculture Planning Committee (NATO)

FAPHA Fellow of the American Public Health Association

FAPHI Fellow of the Association of Public Health Inspectors

FAPI First Article Production Inspection

FAPIG First Atomic Power Industry Group

FAPP Federation of Associations of Periodical Publishers

FAPR Federal Aviation Procurement Regulations

FAPREC Federación de Asociaciones de Padres, Representantes, y Educadores Católicós (Spanish—Federation of Associations of Fathers, Representatives, and Catholic Educators)

FAPRS Federal Assistance Programs Retrieval System

FAPS Fellow of the American Physical Society

FAPT Fellow of the Association of Photographic Technicians

faq fair average quality; free at quay; frequently asked question

FAQ Free at Quay

faqs fair average quality of season

far. false alarm rate; farad; Faraday; faradic; farriery; farthing; finned air rocket; floor/area ratio; forward-acquisition radar

Far Faraday; Farsi (Iranian language)

FAR Failure Analysis Report; Federal Acquisition Regulations; Federal Agent's Report; Federal Aviation Regulations; Financial Accounts Receivable; Fisheries and Aquaculture Research; Florida Association of Realtors; Foundation for Australian Resources; finned air rocket; flight aptitude rating

FAR Fuerzas Armadas Rebeldes (Spanish—Rebel Armed Forces)—Guatemala; *Fuerzas Armadas Revolucionarias* (Spanish—Revolutionary Armed Forces)—Cuba

FARA Foreign Agents Registration Act

FARACS Faculty of Anaesthetists of the Royal Australian College of Surgeons

FARADA Failure Rate Data (BuWeps Program)

Farallones Farollon Islands off San Francisco

Farasans short form for the Farasan Islands of the Red Sea off Saudi Arabia

FARB Federation of Australian Radio Broadcasters

FARC Federal Addiction Research Center; Federal Archives and Records Center

FARC Fuerzas Armadas Revolucionarias de Colombia (Spanish—Armed Revolutionary Forces of Colombia)

Far East countries and islands of East Asia or the Pacific– eastern Siberia, China, Japan, Taiwan, Korea, Indochina, the Philippines, the Malay Peninsula

FARELF Far East Land Forces

faret fast reactor test

FARI Foreign Affairs Research Institute

Farl Farley

FARL Frick Art Reference Library; Lebanese Armed Revolutionary Factions

farm farmacia (Spanish—pharmacy)—drugstore

farmobile farm automobile

FARN Fuerzas Armadas de Resistencia Nacional (Spanish—Armed Forces of Nadonal Resistance)—El Salvador guerrillas

Farnes Farne Islands off England's Northumberland coast

FARNET Federation of American Research Networks

faro. flow(ed, ing) at rate of

FARO Flare-Activated Radiobiological Observatory; Fuel Melting and Release Oven

Faroes Faerøerne (Faroe Islands)

FARP Fronte Antifascista e di Rinascita Populare (Italian–Antifascist Front and Popular Revival)—left-wing group

Far Pom Farther Pomerania (coastal Poland)

Farrar Farrar, Straus and Giroux

Fars Faristan

fas fetal alcohol syndrome; first and seconds; forward acquisition sensor; free alongside ship; frozen at sea

FAS Facility Activity Schedule; Farm Advisory Service; Federal Agricultural Service; Federal Air Surgeon; Federation of American Scientists; Fellow of the Antiquarian Society; Fellow of the Society of Arts; Financial Advisors Service; Food Advice Service; Foreign Agricultural Service; Free Alongside Ship; Frequency Assignment Subcommittee

FASA Federation of ASEAN Shipowners Associations; Fellow of the Acoustical Society of America; First Audit(or) of Sheriff's Accounts; Florida Association of School Administrators

FASAP Fellow of the Australian Society of Animal Production

FASB Financial Accounting Standards Board

FASBA Florida Association of School Business Administrators

fasc fascicule (French—part); *fasciculus* (Latin—little bundle)

FASC Federation of ASEAN Shippers Councils; Free Standing Ambulatory Surgical Center

fascam family of scatterable mines

FASCE Fellow of the American Society of Civil Engineers

fasci (Latin prefix—band)—fascia board

fascic fascicle

FASCO Forward Area Support Coordination Office(r)

fase fundamentally-analyzable simplified English

FASE Fellow of the Antiquariarm Society–Edinburgh

FASEB Federation of American Societies of Experimental Biology

FASG Fellow American Society of Genealogists

fasgrolia fast-growing language of initialisms and acronyms

fash (FASH) forward area support helicopter

FASH Fraternal Association of Steel Haulers

FASII Federation of Associations of Small Industries in India

FASL Florida Association of School Librarians

FASOC Forward Air Support Operations Center

FASPAC Ford Asia Pacific

faspl fair average sample

FASPM Flotte Administrative des Ties Saint Pierre et Miquelon

FASS Fine Alignment Sub-System

fa-ssn fast attack sub-surface nuclear

fast. fuel and sensor tanks; fully automatic switching teletype; functional assessment stages

fast (FAST) facial affect scoring technique; facility for automatic sorting and testing; failure analysis by statistical technics; field data applications, systems, and technics; file analysis and selection technics; fleet-sizing analysis and sensitivity technic; flexible algebraic scientific translator; forecasting and scheduling technic; formula and statement translator; free and single tourist

FAST Factor Analysis System; Fast Answers about State Taxes; First Atomic Ship Transport; flexible algebraic scientific translator; Forecasting and Assessment in the Field of Science and Technology; freight accounting system tracing; French Advances in Science and Technology (newsletter)

FASTM Freight Automated System for Traffic Management

fastnr fastener

FASWAC Food and Service Workers Association of Canada

fat fire and theft; fatigue; final assembly test(ing); fixed asset transfer; free alongside terminal; free at terminal; full annual toll

FAT Family Adjustment Test; File Allocation Table; Flight Test Station; Folk Arts Theater; Fresno, California (airport)

fata fatigue test(ing) article

Fatah Harakat-Tahrir Falastin (Arabic—Palestinian terrorist underground organization)—Arabic acronyms such as this have inverted initials

fatdog fatty hotdog

fa technique fluorescent antibody technique

fatfurters fat-filled frankfurters
fath father; fathom
fath-in-law father-in-law
FATIS Food and Agriculture Technical Information Service
FATS Factory Acceptance Test Specification; Fast Analysis of Tape Surfaces; Fiji Air Travel Service; Firearms Training Systems
fatt fattura (Italian—invoice)
fau faucet; field action units; fixed asset utilization; forced air unit
fau (FAU) fine-alignment unit
FAU Flight Attendants Union; Florida Atlantic University; Friends' Ambulance Unit; Fugitive Apprehension Unit
F & AUA Fire and Accident Underwriters Association
FAUI Federation of Australian Underwater Instructors
FAUL Five Associated University Libraries (Binghamton, Buffalo, Cornell, Rochester, Syracuse)
Faunty Fauntleroy
FAUSA Federation of Australian University Staff Associations
FAUSST French-Anglo-U.S. Supersonic Transport
faustite basic hydrated zinc aluminum phosphate (zinc-rich form of turquoise)
fav favor; favorable; favorite
FAVA Fixed Asset Valuation Assignment
FAVO Fleet Aviation Officer
fav's far-away visitors
FAW Fellowship of Australian Writers
FAWA Factory Assist Work Authorization; Federation of Asian Women's Associations
Fawcett Fawcett World Library
FAWCO Federation of American Women's Clubs Overseas
fawg free at wharf gate
FAWS Flight Advisory Weather Service
FAWU Fishermen and Allied Workers Union
fax facilities (tv technical equipment such as cameras, lights, microphones); facsimile transmission; facsimile(s); facts; fuel air explosion; photo facsimile transmission
Fax Faxon

FAX fixed aeronautical station
Faxon Fetherstoneaugh
fax sheet facilities sheet (tv production)
Fay Fagele; Faith; Fanny
Fayette La Fayette
FAZ Frankfurter Allgemeine Zeitung (German—Frankfurt's Universal Newspaper)
fb film bulletin; flat bar; flat bottom (rails); fog bell; foreign body; freight bill; fringe benefits; full American breakfast; fullback; full board; fullback; fully bleached
f-b full-bore (greater than .22 caliber)
f/b feedback; flock book; front to back (ratio)
f & b fire and bilge; fumigation and bath
fB family Bible; female Black
f/B female Black
FB Fenian Brotherhood; Fernandina Beach; fighter bomber; Film Bulletin; Fire Brigade; Fisheries Board; Flying Boat; Forth Bridge; Free Baptist
FB-111 Convair strategic-bomber version of the F- 111 with variable-geometry wings
fba fighter-bomber aircraft; fighter-bomber attack; fluorescent brightening agent
FBA Farm Buildings Association; Federal Bar Association; Fellow of the British Academy; Fibre Box Association; Freshwater Biological Association; Fur Brokers Association
FBAA Fellow of the British Association of Accountants and Auditors
f'ball football
FBBDO Fibre Building Board Development Organization
FBBO Fellow of the British Ballet Organisation
fbc fallen building clause; fluidized bed combustion; fully-buffered channel; fully-buxomed charmer
FBC Family Benefit Capitalization; Federal Broadcasting Corporation; Fiji Broadcasting Commission; First Boston Corporation; Fukui Broadcasting Company (Japan)

FBCM Federation of British Carpet Manufacturers; Federation of British Cutlery Manufacturers
FBCP Fellow of the British College of Physiotherapists
FBCS Fellow of the British Computer Society; Foreground-Background Operating System
fbcw fallen building clause waiver
fbd freeboard
FB & D Ford, Bacon and Davis
FBEA Fellow of the British Esperanto Association
FBF Federal Buildings Fund; Frankfurt Book Fair
fbfm frequency feedback frequency modulation
FBFM Federation of British Film Makers
FBG Federation of British Growers
fbh fire-brigade hydrant
FBHI Fellow of the British Horological Institute
fbhp flowing bottom hole pressure (oil well)
FBHTM Federation of British Hand Tool Manufacturers
FbI foreign-born Irish
FBI Fast Boats Incorporated; Federal Bureau of Investigation; Federation of British Industries; Food Business Institute; full-blooded Irishman, Icelander, Indian, Indonesian, Iranian, Iraqi, Israelite, Italian, or Ivory Coaster
FBIA Fellow of the Bankers' Institute of Australasia
FBIM Fellow of the British Institute of Management
FBIRA Federal Bureau of Investigation Recreation Association
FBIRE Fellow of the British Institution of Radio Engineers
FBIs Forgotten Boys of Iceland (American armed forces personnel stationed in Iceland)
FBIS Fellow of the British Interplanetary Society; Foreign Broadcast Information Service (CIA)
fbk flat back (lumber); fast buck
FBKS Fellow of the British Kinematograph Society

fbl forged billet; form block line

FBL Federal Barge Lines; Furness Bermuda Line

FBLA Future Business Leaders of America

fbm feet board measure; fetal breathing movements; fleet ballistic missile; forward branch mail

FBM Fleet Ballistic Missile

FBMP Fleet Ballistic Missile Project (Polaris-Poseidon)

FBMWS Fleet Ballistic Missile Weapon System

FBN Federal Bureau of Narcotics

fbnrv fixed bent-nose reentry vehicle

fbo fixed-base operation; for benefit of; foreign building office

FBOA Fellow of the British Optical Association

fboe frequency band of emission

fbop flowery broken orange pekoe

FBOU Fellow of the British Ornithologists' Union

fbp fetal biophysical profile; final boiling point

FBP Federal Bureau of Prisons; Federation of Podiatry Boards

FBPI Franklin Book Programs, Incorporated

FBPMC Federation of British Police Motor Clubs

FBPN Feminist Business and Professional Network

FBPS Fellow of the British Psychological Society; Forest and Bird Protection Society

fbr fast burst reactor; fiber

fbr (FBR) fast breeder reactor; fast burst reactor

FBR Full Bibliographic Record(ing)

FBRAM Federation of British Rubber and Allied Manufacturers

fbrk firebrick

fbrl final bomb release line

fbro *febrero* (Spanish—February)

FBRS Farm Business Recording Scheme

fbs fasting blood sugar; fighterbomber strike

fbs (FBS) frontal bovine serum

fb's fullbacks

FBS Fellow of the Botanic(al) Society; Fighter Bomber Squadron; Forward-Base System(s); Franco-Belgian Services; Fukuoka Broadcasting System

FBSC Fellow of the British Society of Commerce

FBSE Fellow of the Botanical Society–Edinburgh

FBSM Fellow of the Birmingham School of Music

FBTT Federal Board of Tea Tasters

f/bu flowing/buildup (oil well)

FBu Burundi Franc(s)

FBU Fire Brigades Union, Oslo, Norway (Fornebu Airport)

FBUI Federation of British Umbrella Industries

fbw full bandwidth

FBW Fighter-Bomber Wing

FBW System Fly-by-Wire System

fby future budget year

fbyracc for buyer's account

fc carrier frequency (symbol); facilities control; fairly common; file cabinet; filter center; financial consultant; fire clay; fire-control; first cross(ing); flat cable; follow copy; football club; foot-candle; foreign consultant; franc; front- connected; functional code; fund code

fc (FC) field champion; fixed cost

f/c for cash; fill and check; fixed contract; flight control; foolscap; free and clear; frequency converter

f & c fire and casualty (insurance); fish and chips; full and change (tides)

fc *ferrocarril* (Spanish—railroad, railway)

f.c. *fidei commissum* (Latin—bequeathed in trust); *fieri curavit* (Latin—donor directed this be done)

Fc fractocumulus

FC Facilitated Communication; Fairbury College; Farm Credit; Federal Cabinet; Federal Conference; Federal Convention; Fencing Club; Fenn College; Fighter Command; Finch College; Findlay College; fre control; Fisheries Convention; Fitzwilliam Col-

lege; Fontbonne College;_ Foothill College; Franconia College; Frederic Chopin; Frederick College; Free Church (Scotland); French Canada; French Canadian

F-C Franche-Comté

FC *Ferrocarril(es)* [Spanish railroad(s)]

fca frequency control and analysis; functional configuration audit

FCA Facility Change Authorization; Farm Credit Administration; Federal Credit Administration; Federated Confectioners' Association; Federation of Canadian Artists; Federation of College Academics; Fellow (of the Institute) of Chartered Accountants; Fiji College of Agriculture; Finance Corporation of Australia; Financial Corporation of America; Fishermen's Cooperative Association; Foster Care Association; Freight Claim Agent; Freight Claim Association

FCA *Fédération Canadienne de l'Agriculture* (French—Canadian Federation of Agriculture)

FCAA Federal Clean Air Act; Florence Crittenton Association of America

FCAAA Federal Council of Australian Apiarists Association

FCAATSI Federal Council for the Advancement of Aborigines and Torres Strait Islanders

FCACS Federal Civil Agencies Communicatimons System

FCAI Federal Chamber of Automotive Industries

f cant. forward cant frames

fcap foolscap

f/cap foolscap

FCAP Fellow of the College of American Pathologists

F-car(s) French car(s)

F Cas *Federal Cases*

FCAS Federal Council of Agricultural Societies; Fellow of the Casualty Actuarial Society

FCASA Foreign Correspondent's Association of South Africa

FCASI Fellow of the Canadian Aeronautics and Space Institute

fcb free-cutting brass

FCB Facility Clearance Board; Flight Certification Board; Foote, Cone and Belding communications; Foundation for Commercial Banks; Freight Container Bureau; Frequency-Coordinating Body

FCBA Federal Communications Bar Association

fcbd fibrocystic breast disease

fcbu foreign currency banking unit

fcc face-centered cubic; facilities control console; fire-control computer; fire-control console; flat conductor cable; flight-control console; fluid catalytic cracking; fluid convection cathode; freight control computer

fcc (FCC) first-class certificate

FCC Fairbanks Correctional Center (Alaska); Farm Credit Corporation (Canada); Farm Credit Council; Federal Communications Commission; Federal Council of Churches; Federal Court of Canada; First-Class Certificate; Flight Coordination Center; Florida Citrus Commission; Foreign Correspondents Club

FC of C Foundation Company of Canada

FCC Food Chemicals Codex

FCCA Federal Court Clerks Association; Fellow of the Association of Certifiued Accountants; Four Cylinder Club of America

fccc fire-control control console

FCCCA Federal Council of Churches of Christ in America

FCCD Florida Council on Crime and Delinquency

fcci fuel-cladding chemical interaction

fcck fire-control check

FCCO Fellow of the Canadian College of Organists

fccp (FCCP) carbonylcyanide p-trifluoromethoxyphenyl-hydrazone

FCCP Fellow of the College of Chest Physicians

FCCS Federal Cost-Control Survey; Fellow of the Corporation of Certifued Secretaries

FCCSS Fire-Control Control Subsystem

fccu fluid catalytic cracking unit

fcd failure-correction coding; function circuit diagram

FCDA Federal Civil Defense Administration

F & CD—IR Failure and Consumption Data—Inspector's Report

FCDNA Field-Command Defense Nuclear Agency (DoD)

FCDR Failure Cause Data Report

FCDU Foreign Currency Deposit Unit

fce food conversion efficiency

FCE Florida Citrus Exchange; Foreign Currency Exchange; French-Canadian Enterprises

FCE Fondo de Cultura Economica (Spanish—Foundation of Economic Culture)

FCECA Fishery Committee for the Eastern Central Atlantic

fcepc fire-control electrical package container; flight-control electrical package container

FCEX Fruit Growers Express

fcf front-end communications facility; functional check flight

fcg facing

FCG Foreign Clearance Guide

fcga facility gage

FCGB Forestry Committee of Great Britain

FCGI Fellow of the City and Guilds of London Institute

FCGP Fellow of the College of General Practitioners

fcgpc flight-control gyro-package container

fcgr fatigue crack growth rate

FCGS Freight Classification Guide System

fchl flight control hydraulics lab

FChS Fellow of the Society of Chiropodists

fci fuel-coolant interaction

FCI Federal Correctional Institution; Fellow of the Clothing Institute; Fluid Controls Institute; Franklin College of Indiana

FCI Federación Cynological Internacional (Spanish—International Cynological Federation); *Federazione Calcistica Italiana* (Italian Football Association); *Federazione Ciclista Italiana* (Ital-

ian Cycling Association); *Federazione Colombotila Italiana* (Italian Carrier-pigeon Fanciers' Association);

FCIA Fellow of the Canadian Institute of Actuaries; Fellow of the Coporation of Insurance Agents; Foreign Credit Insurance Association; Franchise Consultants International Association; Friends of Cast-Iron Architecture

FCIB Fellow of the Corporation of Insurance Brokers

FCIC Fairchild Camera and Instrument Corporation; Farm Crop Insurance Corporation; Fellow of the Chemical Institute of Canada

FCICA Floor Covering Installation Contractors Association

FCIF Flight Crew Information File

FCII Fellow of the Chartered Insurance Institute

fcim farm, construction, and industrial machinery

FCIP Federal Crime Insurance Program; Fire Company Inspection Program

FCIPA Fellow of the Chartered Institute of Patent Agents

FCIs Fast Coastal Interceptors (US Coast Guard)—fast boats; Federal Correctional Institutions

FCIS Fellow of the Chartered Institute of Secretaries; Florida Council of Independent Schools; Foreign Counterintelligence System (FBI)

FCIT Fellow of the Chartered Institute of Transport

FCIV Fellow of the Commonwealth Institute of Valuers

FCJ Foreign Criminal Jurisdiction

FCJC Flit Community Junior College

FCJS Federal Criminal Justice System

fcl freon coolant loop; front connecting loop; full container load

FCL Foundation for Christian Living

F-class NATO designation for a Soviet class of attack submarines also known as Foxtrot

f clef bass clef

f-c los fire-control line of sight

FCLS Family Colonization Loan Society

fclty facility

fcly face lying

fcm fat-corrected milk; futures commission merchant

FCM Ferrocarril Mexicano (Mexican Railway); Firestone Conservatory of Music (Akron)

FCMA Finch College Museum of Art; Fishery Conservation and Management Act

FCMI Federation of Coated Macadam Industries

FCMIE Fellow of the Colleges of Management and Industrial Engineeririg

FCMS Fellow of the College of Medicine and Surgery

FCMSBR Federal Coal Mine Safety Board of Review

FCN Federal Catalog Number; Friendship, Commerce, and Navigation

FCNA Fellow of the College of Nursing—Australia

FCNM *Ferrocarriles Nacionales de México* (Spanish—National Railroads of Mexico)

fco cleanout flush with finished floor; cutoff frequency; fair copy; franking privilege; free postage

fco *franco* (Italian—free)

Fco Francisco; Franco

F^co Francisco (Spanish—Francis)

FCO Facility Change Order; Fire Control Officer; Fleet Construction Offfcer; Fleet Constructor Officer; Rome, ItalY (Leonardo da Vinci airport, formerly Fiumicino)

F & CO Foreign and Commonwealth Office

fcos *francos* (Spanish—francs)

fcp final common pathway; foolscap

FCP Family Circle Publications; Fellow of the College of Preceptors

FCP *Ferrocarril de Chihuahua al Pacifico* (Spanish—Chihuahua Pacific Railroad)

FCPA Fellow of the Canadian Psychological Association; Foreign Contempt Practices Act

FC Path Fellow of the College of Pathology

FCPC Federal Committee on Pest Control

fc pl face plate

FCPO Fleet Chief Petty Officer

FCPS Fellow of the College of Physicians and Surgeons; France and Colonies Philatelic Society

fcr forward contactor; full cold rolled (steel sheeting)

FCR Facility Capability Report; Facility Change Request; Field Contact Report; Fire Control Room; First City Regiment; Flight Configuration Release; Flinders Chase Reserve (South Australia); Forwarders Certificate of Release

FCR *Free China Review*

FCRA Fellow of the College of Radiologists of Australia; Fellow of the Corporation of Registered Accountants; Fair Credit Reporting Act

FCRLS Flight-Control Ready Light System

fcs forged carbon steel; fraction charge states; francs

fc & s free of capture and seizure (insurance)

FCS Facsimile Control System; Farm Credit System; Farmer Cooperative Service; Fellow of the Chemical Society; Financial Control System; Fire Control School; Fire Control Station; Fire Control System

F/CS Flight-Control System

fcsad free of capture, seizure, arrest or detainment (shipping insurance)

fcsb fire-control switchboard

FCSBC *Ferrocarril Sonora-Baja California* (Spanish—Sonora-Baja California Railroad)

FCSC Foreign Claims Settlement Commission

FCSCUS Federal Claims Settlement Commission of the United States

FCSI Foodservice Consultants Society International

fcsle forecastle

fcsm fire-control system module

fc sm functional simulation

FCSP Fellow of the Chartered Society of Physiotherapy

fcsrcc free of capture, seizure, riots and civil commotion (shipping insurance)

fc & s and r & cc free of capture, seizure, riots, and civil commotion

fcst forecast

FCST Federal Council for Science and Technology (Executive Office of the President)

fcsu fire-control simulator unit

fcswbd fire-control switchboard

fct factory; filament center tap; fraction thereof; function

FCT Federal Capital Territory; Federal Commission(er) of Taxation

FCTB Fellow of the College of Teachers of the Blind

FCTC Fleet Combat Training Center (USN)

fcte fire-control test equipment

fctry factory

fcts fire-control test set; firing-circuit test set; flight-control test stand

FCTU Fiji Council of Trades Unions

fcty factory

fcu fare calculation unit; fare construction unit; fire-control unit; fuel-control unit

FCU Federal Credit Union(s); Federated Clerks Union

FCUA Federal Credit Union Administration

FCUS Federal Credit Union System

FCUSAA Federated Council of University Staff Associations of Australia

fcv fill-and-check valve

FCV Feline Calcivirus

FCW Fire-Control Workshop

FCWA Fellow of the Chartered Institute of Cost and Works Accountants

fcy fancy

fcy pks fancy packs

FCZ Fishery Conservation Zone; Forward Combat Zone

FCZ *Ferrocarril de Coahuila y Zacatecas* (Spanish—Coahuila and Zacatecas Railroad)

fd face of drawing; faculty development; fan douche; fatal dose; field; field dependence; field discharge; finite difference; fine dust (tea); fjord; first defense (la crosse); flame detector; flight deck; floor drain; focal dispatch; focal distance; forced draft; framed; free delivery; free discharge; free dispatch; freeze-dried; front of dash; full dress (formal attire or uniform); full duplex; functional description; fund

fd (Fd) ferredoxin

f/d father and daughter; free dock

f & d faced and drilled; fill and drain; findings and determination; fire and flushing; freight and demurrage

Fd Ferdinand; Fiord (Fjord)

F$ Fiji dollar

FD Federal Defender; field drum; Finance Department; Fire Department; Fleet Duties; Flight Director; Flying Dutchman (yacht); Forestry Department; Foundation for the Disabled; Free Democrat

F.D. Fidei Defensor (Latin—Defender of the Faith)

fd₅₀ median fatal dose

fda forensic distortion analysis; flight-direction attitude; fronto-dextra anterior; fully drawn account; functional demonstration and acceptance

FDA Fisheries Development Authority; Food and Drug Administration

FDA Fraternidade Descendencia Americana (Portuguese—Fraternity of American Descendants)—offspring of Confederate refugees in Brazil

FDAA Federal Disaster Assistance Administration

FDAMA Food and Drug Administration Modernization Act

FDATC Flying Division, Air Training Command

fdau flight-data acquisition unit

FDAWU Food, Drinks, and Allied Workers Union

fdb field dynamic braking; forced-draft blower

FDB Fiji Development Bank

fdc fire-direction center (FDC); first-day cover (postage stamp); flight-director computer; formation density content (oil well log)

fdc fleur de coin (French—mint condition)

FDC Facility Design Criteria; Fire- Detection Center; Flight Data Center; Forsyth Dental Center (Harvard); Friends of Democratic Cuba; Furniture Development Council

FD & C Food, Drug, and Cosmetic (Act)

FDCC Fort Dodge Community College

F D & C-color Food, Drug, and Cosmetic (Act) color

FDCD Facility Design Criteria Document

FDCs Federal Detention Centers (Florence, Arizona and El Paso, Texas)

FDCT Franck Drawing Completion Test

fdd franc de droits (French—free of charge)

FDD Foundation Documentaire Dentaire (Dental Documentation Foundation)

fddc (FDDC) ferric dimethyl dithiocarbonate

FDDI Fiber-optic Data Distribution Interconnect; Fiber-optic Distributed Data Interface

fddl frequency division data link

fddlp. frequency division data link printout

fde field decelerator

FDE Faber Dictionary of Euphemisms

FDEA Federal Drug Enforcement Administration

f/deck flat deck (truck)

F del P Ferrocarril del Pacífico (Spanish—Pacific Railroad)

F del S Ferrocarril del Sureste (Spanish—Southeast Railway–Tabasco, Campeche, Veracruz, Yucatan)

F de P Ferrocarril de Panama (Spanish—Panama Railroad)

F de PS General Francisco de Paula Santander–South American liberator assisting Bolívar

F de S Ferrovie dello Stato (Italian State Railways)

F de T Fulano de Tal (Spanish—So-and-So)

Fdez Fernández

FDF Footwear Distributors' Federation

FDFU Federation of Documentary Film Units

fdg funding

FDH Federal Detention Headquarters

FDHO Factory Department–Home Office

fdi fat depth indicator; field discharge

FDI Farm Dairy Instructor; Federal Department of Information (Malaysia); *Fédération Dentaire Internationale* (French—International Dental Federation); Fir Door Institute; Frederick Douglass Institute (Museum of African Art)

FDIC Federal Deposit Insurance Corporation; Fire Department Instructor's Conference

FDIF Fédération-Démocratique Internationale des Fémmes (French—International Democratic Federation of Women)

FDIM Federación Democrática Internacional de Mujeres (Spanish—International Democratic Federation of Women)

FDIT Federal Daily Income Trust

FDJ Freie Deutsche Jugend (German—Free German Youth)

FDL Fast Deployment Logistic(s)—naval logistic(s)—naval cargo carrier(s); fleet deployment logistic ship (naval symbol); Flight Dynamics Laboratory; Foremost Defended Localities

F & DL Food and Drug Laboratory

fd ldg forced landing

FDLE Florida Department of Law Enforcement

FDLI Food and Drug Law Institute

FDLP Federal Depository Library Program

FDLS Fast Deploymermt Logistics Ship

fdm frequency division multiplexing

FDM Formal Development Methodology; Unisys trademark

FDM Forenede Dansk Motorejere (Danish—Federation of Danish Motorists)

fdma frequency division multiple access

FDMA Fibre Drum Manufacturers Association

F-DMA Farm-Direct Market Association

FDMBB Ferruccio Dante Michelangelo Benvenuto Busoni (Ferruccio Busoni—for short)

FDMHA Frederick Douglass Memorial and Historical Association

FDMS Flight Data Management System (USAF)

fdn foundation

FDN Field Designator Number

FDN Frente Democrática Nacional (Spanish—National Democratic Front); Fuerza Democrática de Nicaragua (Spanish—Nicaraguan Democratic Force)

fdnb (FDNB) fluorodinitrobenzene

Fdo Ferdinando

FDO Fighter Duty Officer; Fleet Dental Officer

F do I Foz do Iguaçu (Portuguese—Mouth of the Iguazu)—three miles above the gigantic Iguazu Waterfalls shared by Argentina, Brazil, and Paraguay

fdor four door (vehicle)

FDOS Floppy Disc Operating system

fdp fibrin degradation products; fibrinogen degradation products; field-developed program; foreign duty pay; forward defense post; funded delivery period

fdp (FDP) fructose 1, 6-diphosphate

f/dp field despatch

FDP foreign duty pay; frontodextra posterior

FDP Freie Demokratische Partei (German—Free Democratic Party)

FDPA Fogg Dam Protected Area (Australian Northern Territory)

FDPC Federal Data Processing Center(s)

Fd PO Field Post Office

fdr feeder; field data recorder; functional demonstration requirement

fdr (FDR) flight data recorder

f dr fire door

Fdr Founder

FDR Franklin Delano Roosevelt—thirty-second President of the United States

FDRA Footwear Distributors and Retailers of America

FDRHS Franklin Delano Roosevelt High School

FDRL Franklin D Roosevelt Library (Hyde Park, New York)

FDRMC Franklin Delano Roosevelt Memorial Commission

FDRS Fire Department Rescue Squad; Flight Data Recording System; Flight Display Research System

fdry foundry

fds fixed disc store

FDS Faraday dark space; Fellow in Dental Surgery; fighter-director ship

FDSRCPS Glas Fellow in Dental Surgery of the Royal College of Physicians and Surgeons of Glasgow

FDSRCS Fellow in Dental Surgery of the Royal College of Surgeons

FDSRCS Edin Fellow in Dental Surgery of the Royal College of Surgeons of Edinburgh

FDSRCS Eng Fellow in Dental Surgery of the Royal College of Surgeons of England

fdt first destination transportation; frequency doubling technology; fronto-dextra transverse

FDT Failure Diagnostic Team

fdte force development testing and experimentation

FDTL'O François Dominique Toussaint L'Ouverture (Haitian patriot who freed his country from Napoleon's control)

fdtn foundation

FDU Fairleigh Dickinson University

f/d vlv fill-and-drain valve

fdw feed water

fdx (FDX) full duplex (data processing)

FD-Zug Fernschnellzug (German—long-distance express train)

fe feather-edged (lumber); female employee; fighter escort; finite element; fire extinguisher; first edition; fisheye (buttons); flanged ends; for example; format effector; front end; further education

fe (FE) format effective character (data processing)

f/e fortnight ending

f & e facilities and equipment

Fe ferrum (Latin—iron)

FE Far East; ferroelectric; Fighter Escort; Flight Engineer; Foreign Editor; Friends of the Earth

f & e flood and ebb

F & E Feamley & Eger [Fern Ville (steamship) Lines]

F of E Friends of the Earth

FE Fonetic English (for spelling words as they sound)

Fe$_2$O$_3$·H$_2$0 rust

fe3dgw finite-element three-dimensional ground water (model)

Fe^{5}2/^{3} radioactive iron

fea finite element analysis; front-end analysis

FEA Failure Modes and Effects Analysis; Federal Energy Administration; Federal Executive Association; Federation of Enployment Agencies; Fiji Electricity Authority; Freelance Editorial Association; French Equatorial Africa

FEAA Federal Employees Appeal Authority

FEAD Fondo Especial de Asistencia para el Desarrollo (Spanish—Special Assistance Fund for Development)

FEAF Far East Air Force

Fe-Ag iron-silver blend

FEA(I) Federal Employees Association (Independent)

FEAL Fast data-Encryption ALgorithm

FE al P Ferrocarril Eléctrico al Pacifico (Spanish—Costa Rican electric railway)

FEAMIS Foreign Exchange Accounting and Management Information System

FEANI Fédération Européenne d'Associations Nationales d'Ingénieurs (French—Federation of European National Associations of Engineers)

Fearkar Farquhar

feat frequency of every allowable term

feath feather(ed)(ing)

Feathers Featherstone

Featherstone Featherstone Prison near Wolverhampton northwest of Birmingham, England

FEAU Florida Education Association United

feb functional electronic block

feb febrero (Spanish—February)

feb. febris (Latin—fever)

Feb February

FEB Federal Executive Boards; Field Engineering Bulletin; Financial and Economic Board; Flying Evaluation Board

feba (FEBA) forward edge of battle area

Febarch February and March

febb febbraio (Italian—February)

FEBC Far Eastern Broadcasting Company

feb.dur. febre durante (Latin—as long as fever lasts)

FEBF Far East Bridge Federation

feb⁰ febrero (Spanish—February)

febr (Latin prefix—fever)—febrile

FEBs Federal Executive Boards

FEBS Federation of European Biochemical Societies

FEBTC Far East Bank and Trust Company

fec feckless; forward error correction; forward exchange control

fec foi, espérance, charité (French—faith, hope, charity)

fec. fecit (Latin—he made)

FEC Facilities Engineering Command; Faculty Exchange Center; Far East Command; Federal Election Commission; Federal Election Council; Federal Electoral Council; Federal Electric Corporation; First Edition Club; Florida East Coast (railway); Foreign Exchange Certificate; Free Europe Committee; Free-standing Emergency Center

FECA Facilities Engineering and Construction Agency; Fiji Employers Consultative Association

FECB Foreign Exchange Control Board

FECEP Fédération Européenne des Constructeurs d'Equipement Petrolier
(French—European Federation of Petroleum Equipment Constructors)

FECIT Federación Española de Centros de Iniciativas y Turismo (Spanish Federation of Centers of Initiative and Tourism)

feck (Scottish abbreviation—effect, efficacy, value)

FECL Fleet Electronics Calibration Laboratory

fecm ferret electronic countermeasures

FECM Fellowship of the Elder Conservation of Music

feco fringes of equal chromatic order

FECONS Field Engineer Control System

fe cr ferrichrome (recording tape)

FECS Fédération Européenne des Fabricants de Céramiques Sanitaires
(French—European Federation of Manufacturers of Sanitary Ceramics)

FECU Far Eastern Container Unit

FECUA Farmers Educational and Cooperative Union of America

fed. federal; federal law-enforcement officer; federal narcotics agent; federated; federation

Fed Federal; Federalist (Party); Federation; The Fed-The Federal Reserve Board

Fed Federación (Spanish—Federation)

FED Facilities Engineering Department; Field Experience Data; Fuel Element Design

FEDA Foodservice Equipment Distributors Association

FEDAPT Foundation for the Extension and Development of the American Professional Theatre

FEDC Federation of Engineering Design Consultants

FEDECAM Federación de Cameras de Comercio (Spanish—Federation of Chambers of Commerce)

FEDECAMARAS Federación Venezolana de Cameras y Asociaciones de Comercio y Producción (Spanish—Venezuelan Federation of Chambers of Commerce and Manufacturers)

FEDECAME Féderación Cafetalera de America (Spanish—Coffee-Growers' Federation of America)

FEDEX Federal Express

FEDFU Federated Engine Drivers and Firemens Union

FEDLINK Federal Library and Information Network

Fed Mal Federation of Malaya; Federation of Malay States; Malaysia

Fed Mal Sta Federated Malay States

fedn federation

fed narc federal narcotics agent

Fed Ref Federal Reformatory

Fed Reg Federal Register

Fed Rep Federal Reporter

Fed Rep Nig Federal Republic of Nigeria

FEDRIP Federal Research in Progress

Feds federal excise tax collectors; federal law-enforcement officers

Feds Federales (Spanish—federal police, federal troops)

FEDS Fixed Exchangeable Disc Store; Foreign Economic Development Service

FEDSIM Federal Computer Performance Evaluation and Simulation Center (GSA)

Fed-Spec Federal Specification(s)

Fed-Std Federal Standard

FEDWIRE US Federal Reserve Service trademark

Fedya Fyodor

FEE Foundation for Economic Education; Foundation for Environmental Education

feeb feeble; feebleminded

FEEB Fleet Electronic Effectiveness Branch (USN)

FEEL ferroelectric-electroluminescent

Feeney Leonard Feeney

FEER Far Eastern Economic Review

fef fast-extrusion furnace

FEF Foundation for Educational Futures; Foundry Educational Foundation

FEFC Far Eastern Freight Conference

FE & FO Francis E and Freeland O Stanley of Stanley Steamer fame

FEGA Film Editors Guild of Australia

FEGLI Federal Employees Group Life Insurance

FEHB Federal Employees Health Benefit

FEHD Fair Employment and Housing Department; Far Eastern Hotel Development

fei for engineering information; for engineer's informnation; fluidic explosive initiator

FEI Farm Equipment Institute; Financial Executives Institute; Flight Engineers International; Free Enterprise Institute

FEI Fédration Équestre Internationale (French—International Equestrian Federation)

FEIA Flight Engineers International Association

FEICRO Federation of European Industrial Cooperative Research Organizations

FEIS Fellow of the Educational Institution of Scotland; Final Environmental Impact Statement; Florida Educators Information Service

FEIP Front-End for Echographic Image Processing

fekg fetal electrocardiogram

feks for eksempel (Dano-Norwegian—for example)

f eks for eksempel (Dano-Norwegian—for example)

fel fellow; front-end loader; front-end loading

fel (FEL) free-electron laser

Fel Felicita; Felix

FEL Financial Enterprises Limited; Food Engineering Laboratory (USA)

FELA Federal Employers Liability Act

FELABAN *Federación Latino-americana del Caribe de Asociaciones de Exportadores* (Spanish—Latin American Federation of Caribbean Exporters Associations)

FELCRA Federal Land Consolidation and Rehabilitation Authority (Malaysian)

FELDA Federal Land Development Authority

feldspar barium, calcium, potassium, or sodium silicates (mineral mixtures such as orthoclase)

FELF Far East Land Forces

Felid Felidae (Latin—cat family)

felinol felinotogic(al)(1y); felinologist; felinology

Felix Felixstowe

Fell Fellow

FELL Finland, Estonia, Latvia, and Lithuania

fella fellaheen (Arabic—tillers)—peasant farmers of Egypt, Syria, and nearby lands

FELO Far Eastern Liaison Office

f/e loader front-end loader; front-end loading

Felsto Felixstowe

felv feline complex leukemia virus(es)

FeLV Feline Leukemia Virus

fem female; feminine; femur; femoral; field-effect mode; fuel efficiency monitor

fem femininum (Danish-Norwegian—feminine)

fem. feminea (Latin—female); *femoris* (Latin—femur, thigh)

f.e.m. (fem or FEM) fuerza electromotriz (Spanish—electromotive force)

FEM Finite Element Method

FEM Fédération Européenne des Motels (French—European Motel Federation)

FEMA Farm Equipment Manufacturers Association; Federal Emergency Management Administration; Federal Emergency Management Agency; Fire Equipment Manufacturers Association; Flavor and Extract Manufacturers Association; Food Equipment Manufacturers Association; Foundry Equipment Manufacturers Association

femboy(s) feminine boy(s)

fem. ext. femur externum (Latin—external thigh)

fern gen circum female genital circumcision

FEMIC Fire Equipment Manufacturers Institute of Canada

fem. int. femur internum (Latin—inner thigh)

fernlib feminine liberation (women's liberation); feminine liberationist

femm femminile (Italian—feminine)

femo femoral

FEMSA Fire Equipment Manufacturers and Suppliers Association

FEMSACO Fabrica Electromecanica SA (Spanish—Electromechanical Factory)

fem-sem feminine seminary (women's college)

femto 10^{-15}

FEMUSI Federación Mundial de Sindicatos de Industrias (Spanish—World Federation of Industrial Unions)

Fen Fenner; Fenwick; Fenwood

fenc fencing

fender bender fender-bending automotive vehicle accident

F/Eng Flight Engineer

Feng Yun I Chinese weather satellite

Fenno-Scandinavia Finland, Greenland, Iceland, Norway, Sweden, Denmark

FENSA Film Entertainment National Service Association

FENSA Fabrica Nacional de Electrodomesticos SA (Spanish—National Factory for Domestic Electric Appliances Incorporated)

Fen-Scan Fenno-Scandia; Fenno-Scandinavian

Fen St Fenchurch Street (rail terminal)

Fenway Fenway Park Stadium, Boston

FEO Federal Energy Office; Federal Executive Office; Federation of Economic Organizations; Fleet Engineer(ing) Office(r)

FeO₂ ferric oxide (recording tape coating)

feov force end of volume

fep final evaluation phase; fore edges painted; front-end processing; formal evaluation phase; front-end processor

FEP Fair Employment Practice; Federal Employees Program; Financial Evaluation Program

FEP *Federación de Estudiantes del Perú* (Spanish—Pernvian Students Federation)

FEPC Fair Employment Practices Commission; Federation of Electric Power Companies

FEPE *Fédération Européenne pour la Protection des Eaux* (French—European Federation for the Protection of Waters)

FEpow Far East prisoner of war

fer forward engine room

fer ferre (Latin—to bear)—fertile, fertilization, fertilize

fer. ferrum (Latin—iron)

Fer Ferdinand; Fermanagh; Ferris

FERA Federal Emergency Relief Administration; Foreign Exchange Regulation Act

Fer. Aet. Ferrea Aetas (Latin—Iron Age)—last of the four ages of the human race—the Plutonian period marked by avarice, crime, and cunning in the absence of honor, justice, or truth

FERC Federal Energy Regulatory Commission, Franco-Ethiopian Railway Company

fer con ferrule-contact

Ferd Ferdinand

Ferde Grofé Ferdinand Rudolph von Grofé

Ferdie Ferdinand

FERF Financial Executives Research Foundation

Ferg Fergus(on)

Fergie Fergus

FERIC Florida Educational Resources Information Center

FERIT Far East Regional Investigation Team

Ferm Fermanagh

fermentol fermentology

Fernando Rey Fernando Casado d'Aranbillet Velga

Fern^do Fernando (Spanish—Ferdinand)

FERNS Federal Reserve Notes

Fernspr Fernsprecher (German—telephone)

ferp family educational rights and privacy

FERP Far Eastern Refugee Program

FERPA Family Education Rights and Privacy Act

FERPC Far Eastern Research and Publications Center

Ferr *ferrovia* (Italian—railroad)

FERRUPS Galatrek trademark

fert fertility; fertilization; fertilizer

fertd fertilized

FERTIPERU *Fertilizantes del Perú* (Spanish—Fertilizers of Peru)

fertz fertilizer

ferv. fervens (Latin—boiling)

Ferv *Fervidor* (French—Glowing Month)—synonym sometimes used for *Messidor* (see *Mess*)

fes festival(s); foil, épée, saber (fencing); fundamental electrical standards

FES Factor Evaluation System; Farm Employment Scheme; Federation of Engineering Societies; Fellow of the Entomological Society; Fellow of the Ethnological Society; Final Environmental Statement; Fisheries Experiment Station; Flat Earth Society; Florida Engineering Society

FESA Fonefic English Spelling Association

FESCO Far Eastern Shipping Company

FESE Far East Stock Exchange (Hong Kong)

FESIIP Fifth Estate Security Information Project

FESO Federal Employment Stabilization Office

FESPAC Far East and South Pacific

fess functional endoscopic sinus surgery

'fess confession

FESS Flywheel-Energy Storage System

'fession confession

'fessor professor

fest festival; festive; festivities; festivity

fest. festivus (Latin—festive or gay)

FEST Federation of Engineering and Shipbuilding Trades (British)

fesv feline sarcoma virus

fet field-effect transistor; forced expiratory time

FET Federal Estate Tax; Federal Excise Tax

FET *Falange Española Tradicionalista* (Spanish—Spanish Traditional Falange)—fascist organization

FETF Flight Engine Test Facility (National Reactor Test Station, Idaho)

fetol fetological; fetologist; fetology

fets field-effect transistors

FETS Far East Trade Service

FETU Federation of Entertainment Trade Unions

feu forty-foot equivalent container unit

FEU Federated Engineering Union

FEU *Federación de Estudiantes Universitarios* (Spanish—Federation of University Students)

feud. feudal; feudalism; feudalistic

fev fever(ish); forced expiratory volume

fev fevereiro (Portuguese—February); *février* (French—February)

FEV Future Electric Vehicle

fev 1 forced expiatory volume in 1 second

fev2 forced expiratoy volume in 2 seconds

fev3 forced expiratory volume in 3 seconds

FEVA Federal Employees Veterans Association

Fevr Fevral' (Russian—February)

FEW Federally-Employed Women

FEWA Farm Equipment Wholesalers Association

fex fleet exercise

FEX Foreign Exchange Service (telephonic)

fext far-end crosstalk

fey forever yours

Fez el Bali (Arabic—Fez the Old)—capital of Morocco

ff far field; fat-free; file finish; filtration factor; fixed focus; flip-flop; folded flat; following folios; form factor; form feed; fortissimo; foster father; french fried; front focal (length); front focus; fuel flow; full fashioned; full field

ff (FF) form feed character (data processmng); folios

f/f face to face; fat and forward (sheep); flip-flop; full force

f&f fat and forward (stock); fire and flushing; fire-and-forget missile; fitttings and fixtures; furniture and fixtures

f to f face to face; foe to foe; friend to friend

ff følgende (Dano-Norwegian—following or next); *folgende Seiten* (German—following pagges); *fortissimo* (Italian—very loud)

fF femtofarad

Ff fragmenta–the digest of Justinian's *Corpus Juris Civilis;* French franc(s)

Ff Fortsetzung folgt (German—to be continued)

FF Federated Farmers; Field Force(s); Field Foundation; fleet flagship (naval symbol); Ford Foundation; Foreign Friend (tourist to the People's Republic of China); Formula Ford; Fragrance Foundation; Freight Forwarder; Frontier Force(s)

F & F Faber & Faber

F of F field of fire; Firth of Forth

FF Faith and Freedom; Fianna Fail (Irish–Republican Party); *fratres* (Latin—brothers); *frères* (French—brothers)

ffa flexible factory automation; foreign freight agent; free for all; free of fatty acid; free foremgn agency; free from alongside; for further assignment; full freight allowed

FfA Fund for Animals

FFA Fellow of the Faculty of Actuaries; Fire Fighters Association; Foreign Freight Agent; Forum Fisheries Agency; Foundation for Foreign Affairs; Future Farmers of America; Future Fuels of America (hydrogen, methanol, etc.)

FFAC Freshwater Fisheries Advisory Council

FFACT Fiber, Fabric and Apparel Coalition for Trade

FFAP Full Face Air Purifying (respirator)

ffar folding-fin aircraft rocket; forward-fighting aircraft rocket

FFARACS Fellow of the Faculty of Anaesthetists of the Royal Australasian College of Surgeons

FFARCS Fellow of the Faculty of Anaesthetists of the Royal College of Surgeons

FFAS Fellow of the Faculty of Architects and Surveyors; Full Face Air Supplied (respirator)

ffb fat-free body

FFB Federal Financing Bank; Fellow of the Faculty of Building

ff black fine furnace black

ffbp free-fall bomb pod

ffc first flight cover; free from chlorine

ffC foreign friend of China

FFC Farmers Federation Cooperative; Federal Facilities Corporation; Federal Fire Council

FFC Folklore Fellows Communications

FFCB Federal Farm Credit Board

ff cc ferrocarriles (Spanish—railroads)

FFCC Nales Ferrocarriles Nacionales (Spanish—Colombian National Railways)

FFCDPA Federal Field Committee for Development Planning in Alaska

FFCM Fellow of the Faculty of Community Medicine

FFCS Federal Farm Credit System

FFCSA Florida Fresh Citrus Shippers Association

ffd focus film distance; forward floating depot; fuel failure detection; functional flow diagram

ffda fiber fineness distribution analyzer

FFDA Flying Funeral Directors of America

FFDCA Federal Food, Drug and Cosmetic Act

FFDRCS Fellow of the Faculty of Dental Surgery of the Royal College of Surgeons

FFE Fight for Free Enterprise; Fire Fighting Equipment (company)

ffex field firing exercise

fff fat, forty, and female

f,ff following folios (following pages)

fff forte fortissimo (Italian—very, very loud)

FFF Fellowship of First Fleeters; Flight Freedoms Foundation; Four Freedoms Foundation; Frozen Food Foundation

ffff forte forte fortissimo (Italian—very, very, very loud)

ffft symbol for the sound of a pump spray

ffg friendly foreign government

FFG-7 guided-missile frigate

ffgt firefighter; firefighting

ffh fast freqency hopping; formerly-fat housewife; formerly-fat husband

FFHC Federation of Feminist Health Centers; Freedom from Hunger Campaign

FFHMA Full-Fashioned Hosiery Manufacturers of America

FFHom Fellow of the Faculty of Homeopathy

ffi free from infection; fuel-flow indicator

FFI Fiji Forest Industry; Finance for Industry (Bank of England); Flanders Filters Incorporated; Freight Forwarders Institute; Frozen Food Institute; Frozen Foods Industries

FFI Forces Françaises de l'Intérieur (French—French Forces of the Interior) underground soldiers fighting against the Germans in occupied France during World War II

ffim far-field image maximizer

ff ind fact-finding index

ffl federal firearms license; field failure; fixed and flashing

F Fl fixed and flashing (light)

FFL Feminists for Life

FFL Forces Françaises Libres (French—Free French Forces)

FFLA Federal Farm Loan Association

FFLI Frozen Food Locker Institute

ffly faithfully

ffm floating fecal material

FFM Fellowship for Freedom in Medicine; Fulton Fish Market

FFMC Federal Farm Mortgage Corporation

FFML Franklin Ferguson Memorial Library

ffn free-floating nozzle

FFN nuclear-powered frigate (naval symbol)

FFNM Fort Frederica National Monument (Georgia)

ffo funds from operations; furnace fuel oil

FFOPA Federal Firearms Owners Protection Act

ffp ferromagnetic fine particles; firm fixed price; fuel fabrication plant

f & fp fraud and false pretenses

FFP Forest Fires Prevention; Free Flight Plan

F & FP Force and Financial Program

ffpa free from prussic acid

FFPB Flora and Fauna Protection Board

FFPS Fellow of the Faculty of Physicians and Surgeons

F & FPS Fauna and Flora Preservation Society

FFPSG Fellow of the Faculty of Physicians and Surgeons

ffr foreign force reduction; freeflight rocket; frequency following response

FFR Fellow of the Faculty of Radiologists; Fleay's Fauna Reserve (Queensland)

FFRF Freedom From Religion Foundation

ffrr full frequency range recording

ffr(s) *fransk(e) franc(s)* [Dano-Norwegian—French franc(s)]

ffs fat-free solids

FFs first families

FFS Family Finamcial Statement; Fee for Service (provider plan); *Ferrovia Federali Svizzere* (Italian—Swiss Federal Railways); Financial Forecasting System; Free-Flying System; Fruit Frost Service

FFSA Fund Family Shareholder Association Incorporated

ffss full-frequency stereophonic sound

FFSS *Ferrovia dello Stato* (Italian—State Railways)

fft fast fourier transform; for further transfer

FFT Formation Flight Trainer (USAF); Functional Field Test(er)

FFTA Foundation of the Flexographic Technical Association

FFTB Freight Forwarders Tariff Bureau

FFTC Food and Fertilizer Technology Center

FFTF Fast Flux Test Facility

ff/tot fuel-flow totalizer

fftr firefighter

FFU Feminist Free University; Fire Fighters Union

ffv flexible-fuel vehicle; foreign fishing vessel

FFV First Families of Virginia

FFVA Florida Fruit and Vegetable Association

FFVMA Fire-Fighting Vehicle Manufacturers Association

ffw fast flood watch

F f W Foundation for Wellness

FFW Failure-Free Warranty

ffwd fast forward; full-speed forward

ffwm free-floating wave meter

ffy federal fiscal year

Ffy Faithfully

FFY Fife and Forfar Yeomanry

FFZ Free Fire Zone (USA)

fg fencing; filter gate; fine grain(ed); fire glaze(d); fiscal guidance; flashgun; flat grain(ed); fog; friction glaze(d); frog(ged); fuel gas; full gilt; fully good

fg *faubourg* (French—suburb)

FG Federal Government; Field Goal; Fire Guard(s); Fitzroy Gardens

F & G Farmers and Graziers

FG *Fine Gael* (Irish—United Ireland Party)

fga foreign general average; free of gencral average

FGA Fellow of the Gemological Association; Fighter, Ground Attack; Freer Gallery

FGAA Federal Government Accountants Association

FGAJ Fellow of the Guild of Agricultural Journalists

FGBMFI Full Gospel Business Men's Fellowship International

fgc facility group control; fine-grained concrete

f & gc failure and guilt complex

FGC Fish and Game Code; Friends General Conference

FGCM Field General Court Martial

fgcr (FGCR) fast gas-cooled reactor

FGCSO Florida Gulf Coast Symphony Orchestra

FGCSSWA Federation of Glass, Ceramic, and Silica Sand Workers of America

fgd flue-gas desulfurization

FGD Fish and Game Department

FGDS *Fédération de la Gauche Démocrate et Socialiste* (French—Federation of the Democratic and Socialist Left)

FGEX Fruit Growers Express

fgf fibroblast growth factor; fully good, fair

fgfop tippy golden orange pekoe

FGG-1 First-Generation Fuel-cell System

FGGE First GARP Global Experiment

FGI Federation of German Industries

FGIC Financial Guaranty Insurance Corporation

fgim figures or images

FGIS Federal Grain Inspection Service

FGJS Farm Groups Joint Secretariat

FGL Federico García Lorca

f/glass fiberglass

fgm (FGM) field guided missile

FGMC Federal Government Micrographics Council

FGMD Fairchild Guided Missile Division

fgn foreign; foreigner

FGN Family Group Number(s)

FGNRA Flaming Gorge National Recreation Area (Utah and Wyoming)

FGO Fellow of the Guild of Organists; Fleet Gunnery Officer

Fg Off Flying Officer

FGP Foster Grandparent Program

FGR Franklin Game Reserve

FGRA Family Group Record Archives

f & g's folded-and-gathered signatures

FGS Fatstock Guarantee Scheme; Federation of Geological Societies; Fellow of the Geological Society

FGSA Fellow of the Geological Society of America

FGSM Fellow of the Guildhall School of Music

fgt freight

FGT Federal Gift Tax

FGTO French Government Tourist Office

FGU Fuel Geoscience Unit

fh firehose; flathead; foghorn; forehatch; full hole (oil well)

fh (FH) familial hypercholesterolemia

f/h freehold(er)

f.h. fiat haustus (Latin—make a draft)

FH Fair Haven; Family History; Far Hills; Fashion Hills; Field Hospital

FH² dihydrofolic acid

FH⁴ tetrahydrofolic acid

FH⁵ Firehouse Five

FH-1100 Fairchild-Hiller observation helicopter

fha filterable hemolytic anemia

fha fecha (Spanish——date)

FHA Farmers Home Administration; Federal Highway Administration; Federal Housing Adrninistration; Fellow of the Institute of Health Service Administrators; Finance Houses Association; Fine Hardwoods Association; Friends Historical Association; Future Homemakers of America

FHAA Field Hockey Association of America

FHAI Federal Housing Authority Insurance

FHAS Fellow of the Highland and Agricultural Society (Scotland)

FHASA Forces Hydroelectriques de l'Andorre (French—Andorra Hydroelectric Power)

fhb family hold back

FHBC Federation of Historical Bottle Clubs

FH/B USA Freedom House/Books USA

fhc family history center; firehose cabinet

FHC Freed-Hardeman College

FHCI Fellow of the Hotel and Catering Institute

FHCIMA Fellow of the Hotel and Catering Institutional Management Association

ffcm familial hypertrophic cardiomyopathy

fhd first-hand distribution; fixed-head disc

fhdo fechado (Spanish—dated)

F & HE Fridays and Holidays Excepted (Moslem)

fhf (FHF) fulminant hepatic failure

FHF Federation of Hardware Factors

FHFB Federal Housing Finance Board

fhh fetal heart heard

FHI Family Health International; Fraser-Hickson Institute; Fuji-Hakone-Izu (national park on Honshu, Japan)

FHI Fédération Halterophile Internationale (French—International Weightlifting Federation)

FHIP Family Health Insurance Plan

FHKSC Fort Hays Kansas State College

fhl family history library

FHL Friends Historical Library (Swarthmore)

FHLB Federal Home Loan Bank

FHLBB Federal Home Loan Bank Board

FHLBs Federal Home Loan Banks

FHLBS Federal Home Loan Bank System

FHLC Forest Hill Learning Centre (Toronto)

fhld freehold

FHLKH VIII The Famous History of the Life of King Henry VIII

FHLMC Federal Home Loan Mortgage Corporation

FHNWR Flint Hills National Wildlife Refuge (Kansas)

f-holes f-shaped sound holes in tops of stringed instruments such as violins, violas, cellos, double basses

fhp fractional horsepower; friction horsepower

FHP Family Health Plan

FHPC FHP International Corporation

FHPRP Farrily Housing Program Review Panel

fhr fetal heart rate; firehose rack

FHR Federal House of Representatives

fhs fetal heart sounds

FHS Fellow of the Heraldry Society; Forest History Society

fhsg family housing

fht fetal heart tone

FHT Fellowship Houses Trust

FHTA Federated Home Timber Association

fhtl first-class hotel

FHTNC Fleet Home Town News Center

FHU Foundation for Human Understanding

fhv forhenvaerende (Dano-Norwegian—former)

FHVMA Flowers and Hughes Values for Marriage Analysis

FHWA Federal Highway Administration

fhws flat-headed wood screw

FHWU Federated Hotel Workers Union

fhy fire-hydrant

fi fade in; failed item; family interview; female impersonator; field independence; field ionization; fire insurance; fixed interval; free in; fuel injection; fuel inspection; for instance

f&i finance and insurance

fi (FI) foreign intelligence

fi finsk (Dano-Norwegian—Finnish)

Fi Fidel; Finland; Finnie; Finnish

Fi. Fireman

FI Faeroe Islands; Falkland Islands; Field Interview; Fiji Islands; Finland (Internet code); Fiscal Interinediary; Fog Index; Formaldehyde Institute; Fourth International; Franco-Iberian; Franklin Institute

Fof I Fruit of Islam (Black Muslim disciplinary corps)

fia financial inventory accounting; full interest admitted

FIA Factory Insurance Association; Faculty Insurance Association; Fashion Industries of Australia; Federal Insurance Administration; Federal Intelligence Agency; Federated Ironworkers Association; Feline Infectious Anemia; Fellow of the Institute of Actuaries; Flatware Importers Association; Flight Information Area; Footwear Industries of America; Forging Industry Association; Futures Industry Association

FIA Fédération Internationale de l'Automobile (French—International Automobile Federation); *Federazión Internazionale Automobilística* (Italian—International Automobile Association); *Freedom of Information Act*

FIAB Foreign Intelligence Advisory Board

FIAB Fédération Internationale des Associations de Bibliothécaires (French—International Federation of Librarian Associations)

FIABCI Federation Internationale des Administrateurs de Biens Conseils Immobiliers (French—International Federation of Real Estate Administrators)

FIAC Federation of International Amateur Cycling; Flanders Interaction Analysis Categories

FIAC Fédération Interaméricaine des Automobile Clubs (French—Interamerican Federation of Automobile Clubs)

FIAE Food Industry Association Executives

FIAF Federación Interamericana de Filatelia (Spanish—International Federation of Philately); *Fédération Internationale des Archives du Film* (French—International Federation of Film Archives)

FIAI Fédération Internationale des Associations d'Instituteurs (French—International Federation of Teachers' Association)

FIAJ Fédération Internationale des Auberges de la Jeunesse (French—International Federation of Youth Hostels)

FIAJY Fellowship in Israel for Arab-Jewish Youth

FIAL Fellow of the Institute of Arts and Letters

FIAM Fellow of the International Academy of Management

FIAMA Fellow of the Incorporated Advertising Managers' Association

FIAMS Fellow of the Indian Academy of Medical Sciences

FIANEI Fédération Internationale d'Associations Nationales d'Éléves Ingénieurs (French—International Federation of National Associations of Engineering Students)

FIANZ Fellow of the Institute of Actuaries of New Zealand

FIAP Fédération Internationale del' Art Photographique (French—International Federation of the Photographic Art); Fellow of the Institution of Analysts & Programmers

FIAPF Fédération Internationale des Associations de Producteurs de Films (French—International Federation of Film Producers)

FIAR Fabbrica Italiana Apparecchi Radio (Italian—Italian Radio Apparatus Factory)

FIArb Fellow of the Institute of Arbitrators

fias free in and stowed

FIAS Flanders Interaction Analysis System

fiaT fix it again Tony (angry Flat-owners' acronym)

FIAT Fabrica Italiana Automobili, Torino (Italian—Italian Automobile Factory-Turin)

FIAV Fédération Internationale des Agences de Voyage (French—International Federation of Travel Agencies)

fiawol fandom is a way of life

fib. fibro cement; fibula; focused ion beam; free into barge; free into bond; free into bunkers; fibrinogen

FIB Fellow of the Institute of Bankers; First Interstate Bank (formerly UCB); Fishing Industry Board; Franklin Institute of Boston

FIB Fédération des Industries Belge (French—Federation of Belgian Industries); *Félag Islenzkra Bifreidaeigenda* (Icelandic Automobile Association)

FIBA Fédération Internationale de Basketball Amateur (French—International Federation of Amateur Basketball)

FIBCA Flexible Intermediate Bulk Container Association

FIBEX First International Biomass Experiment

FI Bio Fellow of the Institute of Biology

FIBM Fellow of the British Institute of Management

FIBOT Fair Isle Bird Observatory Trust

FIBP Fellow of the Institute of British Photographers

fibrd fiberboard

fibril fibrillation

FIBST Fellow of the Institute of British Surgical Technicians

fic fiction; freight, insurance, carriage; frequency interference control

FIC Farm Improvement Club(s); Federal Information Center(s); Federal Insurance Corporation; Federal Insurance Counsel; Fellow of the Institute of Chemistry; Flight Information Center; Forest Industries Council; Foundation for International Cooperation; Freedom of Information Committee

FIC Federación Internacional de Carreteras (Spanish—International Highway Federation)

fica food immune complex assay

FICA Federal Insurance Contributions Act; Food Industries Credit Association

FICA Ferrocarriles Internacionales de Centro America (Spanish International Railways of Central America); *Foreign Intelligence Surveillance Act*

FICAP Furniture Industry Consumer Advisory Panel

FICB Federation of International Commercial Broadcasters

FICBs Federal Intermediate Credit Banks

FICCI Federation of Indian Chambers of Commerce and Industry

FICD Fellow of the International College of Dentists

FICE Fellow of the Institute of Civil Engineers

FICeram Fellow of the Institute of Ceramics

FIC/HEW Fogarty International Center–HEW

fi/ci foreign intelligence/counterintelligence

FICO Ford Instrument Company

FICOA Film Instruction Company of America

FICP Federal Information Centers Program

fic(s) afcionado(s)[Spanish—devotee(s)]

FICS Fellow of the International College of Surgeons; Fellow of the Institute of Chartered Shipbrokers

FICSA Federation of International Civil Servants Associations

fict fiction; fictitious

fict. fictilis (Latin—made of pottery)

FICWA Fellow of the Institute of Cost and Works Accountants

fid fiduciary; force identification; free induction decay

Fid Fidji (Spanish—Fiji)

FID Falkland Island Dependencies; Federation of International Documentation; Fellow of the Institute of Directors; Field Intelligence Department

FIDA Federal Industrial Development Authority

FIDA Fondo Internacional para el Desarrollo de la Agricultura (Spanish—International Fund for the Development of Agriculture)

fidac film input to digital automatic computer

fidal fixed-wing insecticide-dispersal apparatus, liquid (USNs defoliant spraying system)

FIDCR Federal, Interagency Day Care Requirements

FIDE Fédération Internationale des Échecs (French—International Chess Federation)

Fidel Fidel Castro

fidel. fidelity

FIDEL Frente Izquierda de Liberación (Spanish—Leftist Liberation Front)

FIDER Foundation for Interior Design Education Research

fidivan fiber-diameter video analyzer

Fidnet Federal Intrusion Detection Network

fido fog investigation dispersal operation; freaks, irregulars, defects, oddities (created by minting errors); fugitive information data organizer

FIDO Facility for Integrated Data Organization; Federal Island Development Organization; Fire Incident Data Organization; Flight Dynamics Officer

FIDOR Fibre Building Board Development Organisation

FIDP Fellow of the Institute of Data Processing

FIDS Falkland Islands Dependencies Survey; Foolproof Identification System

FIE Feline Infectious Enteritis

FIE Fédération Internationale d'Escrime (French—International Fencing Federation)

FIED Fellow of the Institution of Engineering Designers

FIEE Fellow of the Institution of Electrical Engineers

FIEJ Fédération Internationale des Éditeurs de Journaux et Publications (French—International Federation of Editors of Journals and Publications)

FIEL Fundación de Investigaciones Económicas Latinoamericanas (Spanish—Foundation for Latin American Economic Investigations)

FIEN Forum Italiano dell' Energia Nucleare (Italian–Italian Nuclear Energy Forum)

FIER Foundation for Instrumentation Education and Research

FIERE Fellow of the Institute of Electronic and Radio Engineers

FIES Fellow of the Illuminating Engineering Society

fif ferric ion free

FIF First Investment Fund; Friends of Irish Freedom

fi. fa. fieri facias (Latin—see it done)

FIFA Fédération Internationale de Football Associations (French—International Federation of Football Associations)

FIFCLC Fédération Internationale des Femmes de Carrières Libérales et Commer-

ciales (French—International Federation of Liberal and Commercial Career Women)

Fife Fifeshire

FIFE Fellow of the Institution of Fire Engineers

FIFE Fédération Internationale Féline (French—International Feline Federation)

fifo first in, first out (inventory)

FIFO Flight Inspection Field Office(r)

FIFRA Federal Insecticide, Fungicide, and Rodenticide Act

FIFSP Fédération Internationale des Fonctionnaires Supérieurs de Police (French—International Federation of Senior Police Officers)

FIFTU Fédération Internationale des Femmes Diplômées des Universites (French—International Federation of University Women)

fig figuratively; figure

Fig. Figur(en) [German—figure(s)]; *Le Figaro* (Paris' old est daily newspaper)

FIG Farmers Insurance Group

FIG Federazione Italiana Golf (Italian—Italian Golf Association)

FIGA Fretted Instrument Guild of America

FIGB Federazione Italiana Gioco Bocce (Italian—Italian Bocce Ball Association)

FIGC Federazione Italiana Gioco Calcio (Italian—Italian Football Federation)

FIGCM Fellow of the Incorporated Guild of Church Musicians

figlu formiminoglutamate

FIGM Fellow of the Institute of General Managers

FIGO Fédération Internationale de Gynécologie et d'Obstétrique (French—International Federation of Gynecology and Obstetrics)

figs (FIGS) figures shift (data processing)

fig(s). figure(s); finger-sized banana(s)

Fig(s) figure(s)

FIGS Future Income and Growth Securities

figt fully inclusive group tour

fih fat-induced hyperglycemia; free in harbor

FIH Fédération Internationale des Hôpitaux (French—International Federation of Hospitals)

FIHVE Fellow of the Institution of Heating and Ventilating Engineers

FII Fellow of the Imperial Institute; Foreign Investment Institute

FIIA Fellow of the Institute of Industrial Administration

FIIAL Fellow of the International Institute of Arts and Letters

FIIC Fellow of the Insurance Institute of Canada

FIICU Federation of Independent Illinois Colleges and Universities

fiigmo forget it, I've got my orders

FIIGS Federal Item Identification Guide System

FIIM Fellow of the Institute of Industrial Management

FIIN Federal Item Identification Number

FI Inf Sc Fellow of the Institute of Information Scientists

FIIP Fellow of the Institute of Incorporated Photographers

FIIT Federal Individual Income Tax

FIJ Fellow of the Institute of Journalists

FIJ Fédération Internationale de Judo (French—International Judo Federation); *Fédération Internationale des Journalistes* (French—International Federation of Journalists)

FIJET Fédération Internationale des Journalistes et Écrivains du Tourisme (French—International Federation of Travel Journalists and Writers)

Fiji Dominion of Fiji (island nation in the western South Pacific)

Fijis Fiji islanders; Fiji Islands

FIJL Fédération Internationale des Journalistes Libres (French—International Federation of Free Journalists)

fil filament; fillet; fillister; filter; filtrate

f-i-l father-in-law

Fil Filbert; Filemón; Filiberto, Filinto; Filipevna; Filipp; Filippino; Filippo; Filley; Fillmore; Filmore; Filomena; Filpot; Filpotts

Fil Filipinas (Spanish—Philippines)

FIL Fellow of the Institute of Linguists

FILA Fellow of the Institute of Landscape Architects

Fil-Am Filipino-American

fild federal item logistics data

fildr federal item logistics data record

File Filemón (Spanish—Philemon)—The Epistle of St Paul to Philemon

FILE Fellow of the Institute of Legal Executives

file 13 trashcan; wastebasket

fil h fillister head

Fili Filipinas (Portuguese or Spanish—Philippines)

FILIM Fédération Internationale des Langues et Litteratures Moderne (French—International Federation of Modern Languages and Literature)

fill. filling

Film Q Film Quarterly

f. inl. father-in-law

filo first in, last out

filos filosofia (Italian or Portuguese—philosophy); *filosofía* (Spanish—philosophy)

filt filter; filtrate; filtration

filt. filtra (Latin—filter)

FILT Fédération Internationale de Lawn Tennis (French—Lawn Tennis International Federation)

fim facing identification mark; fault isolation manual; field ion microscope

Fim Finnish mark(s)

FIM Fellow of the Institute of Metallurgists; Flight Information Manual; Foundation for International Meetings

FIM Fédération Internationale des Musiciens (French—International Federation of Musicians); *Fédération Internationale Motocycliste* (French—International Motorcycle Federation)

FIMA Fellow of the Institute of Municipal Administration; Financial Institutions Market-

ing Association; Food Industries of Malaysia; Forging Ingot Makers' Association; Friendly International Males' Association

FIMC Fellow of the Institute of Management Consultants

FIME Fellow of the Institute of Mechanical Engineers

FIMF Fédération Internationale de Médecine Fisica (French—International Federation of Physical Medicine)

FIMI Fellow of the Institute of the Motor Industry

FIMIT Fellow of the Institute of Musical Instrument Technology

FIMLT Fellow of the Institute of Medical Laboratory Technology

FIMS Fédération Internationale de Médecine Sportive (French—International Federation of Sporting Medicine)

FIMT Fellow of the Institute of the Motor Trade

FIMTA Fellow of the Institute of Municipal Treasurers and Accountants

fin. finance; financial; financier; finish

fin. finis (Latin—the end)

Fin Finistère; Finland; Finnic; Finnish

Fin Finnish

FIN Fellow of the Institute of Navigation

fina following items not available

FINAC Fast Interline Non-Active Automatic Control (automatic teletype service)

FINAST First National Stores

FINCANTIERI Società Finanziaria Cantleri Navali (Italian—Dockyards Finance Company)

fincl financial

FIND Friendless, Isolated, Needy, Disabled (older people)

FIND Federal Item Name Directory

fin dec final decree

fine b fine boomerang

FINEBEL France, Italy, Netherlands, Belgium, and Luxembourg (economic agreement)

fined finished

FINELETTRICA *Società Finanziaria Elettrica* (Italian—Electric Power Finance Company)

fines fine particulates

fin fl finished floor

FINFO Flight Inspection National Field Office (FAA)

fing finishing

F-ing fucking (slang—copulating)

Fingal Finn Mac Cumhail (semimythical Irish fighter whose Hebrides hideaway is described in Mendelssohn's *Fingal's Cave* overture)

finif field-induced negative ion formation

Finisterre Cape Finisterre northern Spain's westernmost cape

Finlan *Finlândia* (Italian or Spanish—Finland); *Finlandia* (Portuguese—Finland)

Finland Republic of Finland (north European land) *Suomen Tasavalta* (Finnish name); *Republiken Finland* (Swedish name)

FINMARE *Società Finanziaria Marittima* (Italian—Maritime Shipping Finance Company)

finn finnisch (German—Finnish)

Finn Finnish

FINNAIR Finish Airlines

Finnglish Finnish + English

Fi-No-Tro Finmark-Nord Troms (fish processing)

FINS Fire Island National Seashore

Fin Sec Financial Secretary

FINSIN Finlandia Sinfonietta

FINSINDER *Società Finanziaria Siderùrgica* (Italian—Iron and Steel Financing Society)

F Inst Fellow of the Institute; Fellow of the Institution

F Inst F Fellow of the Institute of Fuel

F Inst P Fellow of the Institute of Physics

F Inst Pet Fellow of the Institute of Petroleum

F Inst SP Fellow of the Institute of Sewage Purification

f insulin fibrous insulin

Fin-Syn Federal Communications Commission–Financial and Syndication (rules)

FINTEL Financial Times Publishing Group

Fin-Ug Finno-Ugric

fio for information only; free in and out

FIO Fleet Information Office; Flight Information Office(r)

FIO Fédération Internationale l'Oléiculture (French—International Olive Growers Federation)

fio and stowed (trimmed) free in and out and stowed (trimming paid for)

Fiordland Fiordland National Park (southwest corner of New Zealand's South Island); Norway's nickname

fios free into owner's store; free in and out stowage

flot free in and out trimmed

fip fair in place; fi'pence (fivepence); fi'penny (fivepenny); fire insurance policy

FIP Federal Implementation Plan; Feline Infectious Peritonitis; Fleet Introduction Program; Flight Instruction Program; Forestry Incentives Program

FIP Fédération Internationale de Philatélie (French—International Philatelic Federation); *Fédération Internationale des Phonothèque* (International Federation of Record Libraries); *Fuerze Interamericana de Paz* (Spanish—Interamerican Peace Force)

FIPA Fellow of the Institute of Practitioners in Advertising

FIPAGO Fédération Internationale des Fabricants de Papiers Gommes (French—International Federation of Manufacturers of Gummed Paper)

FIPD Fellow of the Institute of Professional Designers

FIPJF Fédération Internationale des Producteurs de Jus de Fruits (French—International Federation of Fruit Juice Producers)

FIPLV Fédération Internationale de Professeurs des Langues Vivante (French—International Federation of Professors of Living Languages)

FIPM Fédération Internationale de Psychothérapie Médicale (French—International Federation of Medical Psychotherapy)

FiPo Fire and Police (Research Association)

FIPP Fondation Internationale Pénale et Pénitentiare (French—International Penal and Penitentiary Foundation); *Fédération Internationale de la Presse Périodique* (French—International Federation of the Periodical Press)

FIPR Fédération Internationale de Patinage à Roulettes (French—International Roller-Skating Federation)

fips female iron-pipe size

FIPS Federal Information Processing Standards

FIPSE Fund for the Improvement of Postsecondary Education

FIPTP Fédération Internationale de la Presse Technique et Periodique (French—International Federation of the Technical and Periodical Press)

FIQ Fédération Internationale des Quilleurs (French—International Bowling Federation)

FIQS Fellow of the Institute of Quantity Surveyors

fir financial inventory report; firkin; flight information region; flight information requirement; floating-in rate(s); fuel indicator reading; full indicator reading; future issue requirement(s)

FIR Fabrication Information Report; Field Interrogation Record; Flight Information Report

FIRA Federal Investment Review Agency; Foreign Investments Review Agency; Furniture Industry Research Association

FIRAA Fire Insurance Research and Actuarial Association

FIRB Fire Insurance Rating Bureau; Florida Inspection and Rating Bureau; Foreign Investment Review Board

FIRCE *Fiscalizaçãdo e Registro de Capitais Estrangeiros Banco Central do Brasil* (Portuguese—Central Bank of Brazil's Foreign Capital Regulations)

FIRE Fellow of the Institution of Radio Engineers

FIREBRICK Federal Inter-Agency River Basin Committee

fireclay sedimentary rock containing chlorite-kaolinite with illite

fire damp methane

Fire Island National Seashore, Long Island, New York

fires. firearms

FIRES Fire Inspection Reporting and Evaluation System

fir ex fire extinguisher

FIRFLT First Fleet

FIRI Fellow of the Institute of the Rubber Industry; Fishing Industry Research Institute

FIRME *Fondo de Inversiones Rentables Mexicanas* (Spanish—Mexican Rental Investments Fund)

FIRPA Foreign Investment in Real Property Tax Act

FIRREA Financial Institutions Reform, Recovery, and Enforcement Act

FIRST Fair and Immediate Resources for Students Today; Fast Interactive Retrieval System Technology; Financial Information Reporting System; Foster Initial Reading Skills in Time

firta far infrared technical area

FIRT Fertilizer Industry Round Table

FIRTO Fire Insurers Research and Testing Organization

Firton Girton College, Cambridge

fis family income supplement; foam in salvage; foreign instrumentation signals; free in store; freight, insurance, and shipping (charges)

fis *fisica* (Italian—physics); *fisica* (Portuguese or Spanish—physics)

FIS Facial Identification Systems; Field Instruction System; Fighter Interceptor Squadron; Financial Information System (DoE); Flight Information Service

FIS *Fédération Internationale de Sauvetage* (French—International Life-Saving Federation)

FISA Fellow of the Incorporated Secretaries Association; Food Industries Suppliers Association; Foreign Intelligence Surveillance Act (United States; signed 1978)

FISAR Federal Institute for Snow and Avalanche Research

FISARS Fleet Information Storage and Retrieval System (USN)

FISC Federation of Infant School Clubs; Financial Industries Service Corporation

fisc irre fiscal irresponsibility

FIS countries France, Ivory Coast, Senegal

FISD *Fédération Internationale de Sténographic et de Dactylographie* (French—International Federation of Stenography and Typewriting)

FISE *Fédération Internationale Syndicale de l'Enseignement* (French—International Federation of Teachers' Unions); *Federazione Italiana Sports Equestri* (Italian Federation of Equestrian Sports)

fish fishery; fishes; fishing

fish fluorescent in-situ hybridization

FISH Friends in Service Here

fishwich fish sandwich

FISIPE *Fibras Sintéticas de Portugal* (Portuguese Synthetic Fibers of Portugal)

fisk *fiskeri* (Dano-Norwegian—fishery)

FISL Federally Insured Student Loan(s)

fiss (Latin prefix—split)—fissure

FIST Federation of Interstate Truckers; Field Intelligence Simulation Test; Fugitive Investigative Strike Team (for catching criminals at large)

fisteg fiscal integrity

FI Struct E Fellow of the Institute of Structural Engineers

fit fabrication in transit; foreign inclusive tour; foreign independent traveler; foreign independent trip; formation interval tester (oil well); free

of income tax; free in truck; freely independent traveller; fully inclusive tour

fit (FIT) foreign independent travel (tour)

FIT Families In Touch; Fashion Institute of Technology; Federal Income Tax; Fellow of the Institute of Transport; Fiji Institute of Technology; Footscray Institute of Technology; Foreign Independent Tours; Fixed Individual Tariff; Forest Industries Telecommunications

FIT *Fédération Internationale des Traducteurs* (French—International Federation of Translators)

FITA *Fédération Internationale de Tir l'Arc* (French—International Archery Federation)

fits flexible image transport system; foreign individual travellers; fuel injection timing system

fitw federal income tax withholding

fitwh federal income tax with holding

Fitz Fitzedward; Fitzgerald; Fitzgreen(e); Fitzhugh; Fitzjames; Fitzjohn; Fitzmaurice; Fitzrandolph; Fitzroy; Fitzsim(m)ons; Fitzwilliam(s)

Fitzbill Fitzwilliam

Fitzw Fitzwilliam College, Cambridge; Fitzwilliam Library (Cambridge)

Fitzw Coll Fitzwilliam College—Cambridge

FIU Federation of Information Users; Florida International University; Forward Interpretation Unit (US Army)

FIV Feline Immunodeficiency Virus; Fellow of the Institute of Valuers

FIV *Fondo de Inversiones de Venezuela* (Spanish—Investments Fund of Venezuela)

fiva fluid inject valve actuator

FIVA *Fédération Internationale des Vehicules Anciens* (French—International Federation of Antique Vehicles)

five forced inspiratory vital capacity

fiw frec in wagon

FIWC Fiji Industrial Workers Congress

fix. fixture
fixed fixed-rate (mortgage)
Fiz Fizika (Russian–physics)
Fiz Elem Chastits At Yadra
   Fizika Elementarnykh Chas-
   tits i Atomnogo Yadra (Rus-
   sian—Journal of Particles and
   Nuclei)
Fiz Met Fizika Metallov i Met-
   allovedenie (Russian–Physics
   of Metals and Metallography)
Fiz Nizk Temp Fizika Nizkikh
   Temperatur (Russian—Jour-
   nal of Low-Temperature
   Physics)
Fiz Plazmy Fizika Plazmy
   (Russian—Plasma Physics)
Fiz Tekh Poluprovodn Fizika i
   Tekhnika Poluprovodnikov
   (Russian—Semiconductor
   Physics and Technology)
Fiz Tverd Tela Fizika Tver-
   dogo Tela (Russian—
   Solid-State Physics)
fj flush joint
Fj Fjord
FJ Fiji Airways; Fiji (Internet
   code); Flying Junior
F-J Fisher-John
FJA Future Journalists of
   America
FJC Federal Judicial Center;
   Fullerton Junior College
Fjd Fjord
Fjd(s) Fiji dollar(s)
FJH Franz Josef Haydn
Fji Fiji
FJI Fellow of the Institute of
   Journalists
FJIC Federal Job Information
   Center
FJNM Fort Jefferson National
   Monument
FJOL Federal Job Opportu-
   nity List
fjp first job program
FJS First Jersey Securities;
   Fulton J Sheen
fk flat keel; fork
Fk Frank
FK Falkland Islands (Internet
   code); Fluid Kinetics; Franz
   Kafka; Fujita Airways
FK Frankfurt Kassenverein
   (German—Frankfurt Clear-
   inghouse)
fka formerly known as
FKBD Fort Knox Bullion De-
   pository
FKBI Fourdrinier Kraft Board
   Institute

FKC Fellow of King's College
Fkd Frankford
FKI Federation of Korean In-
   dustries
FKJC Florida Keys Junior
   College
*FKL Frauen Konzentrations-
   Lager* (German—Women's
   Concentration Camp)
Fkn Franklin; Frederikshavn
fkr før Kristus (Dano-Norwe-
   gian—before Christ)
Fks Fredrikstad
FKSII Fresh Kills, Staten Island, Incinerator
FKSNS Fort Kent State Normal School
FKTU Federation of Korean
   Trade Unions
FKWR Florida Keys Wildlife
   Refuges
fl farmland; flanker (football);
   flash(ing); flash(ing) light;
   flight level; flood(ing);
   floor(ing); fluorescent level;
   flourish; flow(ing); flow line;
   fluid loss; fluid(s); flush(ing);
   flute; focal length; fol-
   low(ing); footlambert; foreign langage; forklift
f/l freight liner
f & l fuel and lubricants
fl flaske (Dano-Norwegian—
   bottle, flask); *flauti* (Italian—
   flute, flutes); *flauto; flores*
   (Latin—flowers; *florin*
   (Dano-Norwegian or
   Dutch—florin); *floruit*
   (Latin—he flourished)
f.l. falsa lectio (Latin—false
   reading)
fL foot-lambert
Fl Fall (postal abbreviation);
   Flemish; fluorine
Fl Fleuve (French—large river)
FL Federal League; Ferdinand
   Laeisz; First Lady; Flag Lieu-
   tenant; Flight Lieutenant;
   Florida; focal length; Football
   League; foreign language;
   French Line (ships included
   are *France, Normandie,
   Paris*); Frontier Airlines (2-
   letter code)
F.L. Franz Liszt
F for L Feminists for Life
FL Fürstentum Liechtenstein
   (Principality of Liechtenstein)
fla fronto-laeva anterior
f.l.a. fuat lege artis (Latin—ac-
   cording to the rules of art)

Fla Florida; Floridian
FLA Federal Loan Adminis-
   tration; Federal Loan Agency;
   Fellow of the Library Associ-
   ation; Florida; Florida East
   Coast Railway (symbol);
   Foam Laminators Association
*FLA Frente de Libertação
   Açoriana* (Portuguese—Azo-
   rian Liberation Front)
FLAA Fellow of the Library
   Association of Australia
flab flabby
fl abwth flush armor balanced
   watertight hatch
FLAC Florida Automatic
   Computer (USAF)
FLACCS Florida Climate and
   Control System
Fla Cur Florida Current
flag. flageolet
FLAI Fellow of the Library
   Association of Ireland
FLAIR Floating Airport;
   Food-Linked Agricultural Industrial Research
FLAIRS Fleet Locating and
   Information Reporting System (for police-patrol vehicles)
flak fondest love and kisses
flak Fliegerabwehrkanone
   (German—anti-aircraft cannon, anti-aircraft shrapnel)
flake colorful and eccentric
   person
flam flamlandisch (German—
   Flemish)
flame inflammatory message
   (via computer)
FLAME Facility Laboratory
   for Ablative Materials Evaluation; Facts and Logic About
   the Middle East
flam(s) flamenco (songs); flam-
   ing(s); flammable(s)
flang flowchart language
FLAP Fatal Light Awareness
   Program; flap setting (aircraft
   code); Flores Assembly Program
FLAPS Flexibility Analysis of
   Piping Systems
flar florward-looking airborne
   radar
FLARE Family, Life, America
   and Responsible Education
   under God
FLAS Fellow of the Land
   Agents Society

FLASH Foreign Fishing Vessels Licensing and Surveillance Hierarchical fnformation System (Canadian)

flats fellow astronaut trainees

Flats Durango, Colorado's slums

flav flavor(ing)

flav. *flavus* (Latin—yellow)

flb flight-line bunker; foot-lambert

FLB Federal Land Bank

FLBAs Federal Land Bank Associations

flbin floating-point binary

flbm (FLBM) fleet-launched ballistic missile

fl bp filter, band-pass

fl bs filter, band-suppression

f&lc frequency and load controller

FLC Federal Library Committee (See FLICC); Foundation Library Center

flcc flight-control computer

FLCM Fellow of the London College of Music

FLCO Fellow of the London College of Osteopathy

fl crs flat cars

fld failed; field; flowered; fluid

Fld Field (postal abbreviation)

FLD Friends of the Lake District

Fld Com DNA Field Command, Defense Nuclear Agency

fld dr field drum

fldec floating-point decimal

fldg folding

fldg chr folding chair(s)

fl di flare die

FL & DI Food Law and Drug Institute

fldl field length (flow chart)

fldo final limit, down

fldop field operations

fl dr fluid drain

Flds Fields

fldxt fluid extract

Fl e Flemish ell (unit of measure)

flea. flux logic element array

fleact fleet activities

flee fast-linkage editor

FLEEC Federal Libraries' Experiment in Cooperative Cataloging

fleetex fleet exercise

Flem Flemish

fleming(s) fleming-gear hand-propelled lifeboat(s)

Flem(s) British slang for Belgian(s)

fles foreign language in elementary school

FLES Foreign Languages in Elementary Schools (linguistic teaching program)

FLETC Federal Law Enforcement Training Center

FLETRABASE Fleet Training Base (USN)

Flev Flevoland (Dutch province)

FLEWEACEN Fleet Weather Center

FLEWEAFAC Fleet Weather Facility

flex. flexible

FLEX Federal Licensing Examination

flexo flexographic

flex sig flexible sigmoidoscopy

flf final limit, forward; flip flop

FLF Freedom Leadership Foundation

flg failing; flagging; flange; flashing; flooring; flying

FLG Flagship (USN)

FLGA Fellow of the Local Government Association

FLGB Federal Loan Guarantee Board

flgd flanged

flgstn flagstone

flh familial lefthandedness; final limit, hoist

fl hd flathead

flhls flashless

fl hp filter, high-pass

flia *familia* (Spanish—family)

flib friggin little itinerant bastard(s)

FLIC Film Library Information Council

FLICC Federal Library and Information Center Committee (formerly Federal Library Committee)

flick(s) flicker(s), [motion picture(s)]

flicon flight control

flier fluid-logic industrial control relay

fliden flight data entry

FLIM Flight Mechanics Internal Memorandum

Flinders Flinders Ranges of South Australia

flint variety of chalcedony

flint (FLINT) floating-interpretive language

Flint Flintshire

Flints Flintshire

flip. film library instantaneous presentation

FLIP Flexible Loan Insurance Program; Flight Information Publication; Floated Lightweight Inertial Platform; Floating Instrument Platform; Free-form Language for Image Processing

Flip(s) Filipino(s)

flir forward-look infrared

FLIRT Federal Librarians Round Table

flit functional literacy

fliv flivver

FLIWR Functional Listing and Interconnection Wiring Record

flkprt flock printed

Flks Falkland Islands

Fll final limit, lower

FLL Finanglia Line Ltd; Fort Lauderdale, Florida (airport); Friends Library, London

fllar forward-looking light attack radar

fl ld floor load

Flli *fratelli* (Italian—brothers)

fl lp filter, low-pass

flm functional-level manager

Flm Flemish

FLM Free Library Movement

FLM *Fédération Luthérienne Mondiale* (French—Lutheran World Federation)

FLMI Fellow of the Life Management Institute

flmthwr flame thrower

fl/mtr flow meter

fln fallen; following landing numbers

Fln Flensburg

FLN *Frente de Liberación Nacional* (Spanish—National Liberation Front); *Front de Liberation Nationale;* (French—National Liberation Front)—official Algerian party

FLNC *Frente de Liberación Nacional de Cuba* (Spanish—National Liberation Front of Cuba)

flng falling

FLNM Fort Laramie National Monument

flo floodlight(s)

Flo Flobert; Florence; Florentz; Florian; Floris

Fl O Flight Officer

FLO Foreign Liaison Office(r)

float. floating offshore attended terminal
floatel floating motel
floc floccule; flocculent; floccus
FLOC For Love of Children
flodac fluid-operated digital automatic computer
Fl Offr Flying Officer
FLOG Fleet Logistics Air Wing
FLOOD Fleet Observation of Oceanographic Data (USN)
flop. floating octal point
flops floating-point operations per second
flor floriculture
flor flores (Latin—flowers); floruit (Latin—he flourished)
Flor Floréal (French—Flowery Month)—beginning April 20—eighth month of the French Revolutionary Calendar
Flor(a) Florence
Florence Federal Detention Headquarters at Florence, Arizona
Floribbean Floridian-Caribbean (resort area)
Florrie Flora; Florence
florsent fluorescent
floss. flossing (dental care)
Floss(ie) Florence
flot flotation; flotilla; flotsam
Flota Flota Oceanica Brasileira (Portuguese—Brazilian Oceanic Fleet)
flotel floating hotel
fl ovth flush oiltight ventilation hole
flox flourine + liquid oxygen
Floy Florence
fl oz fluid ounce
flp family limited partnership; fault location panel; frontolaeva posterior
FLP Free Library of Philadelphia
flpl fortran-compiled list-processing language
fl pl. flore pleno (Latin—in full bloom)
FLPMA Federal Land Policy and Management Act
fl prf flameproof
FLPS First Log Procurement Status
fl pt flashpoint; fluid pint

FLQ Front de Libération Quebecois (French—Front for the Liberation of the people of Quebec)
flr failure; final limit, reverse; flame resistant; flare(s); floor; florin; flow rate; flower; forward-looking radar
FLR Florence, Italy Airport
FLR Federal Law Reports
FLRA Federal Labor Relations Authority
flrg flooring
flrng flash ranging
FLRR Federal Labor Relations Reporter
flrs flares; flowers; forward looking radar set
fl/rt flow rate
FLRT Federal Librarians Round Table
fls floors; forward-looking sonar
Fls Falls (postal abbreviation); Flushing
FLS Fellow of the Linnaean Society; Flashing Light System
FLSA Fair Labor Standards Act
flsc flexible linear-shaped charge; flight shape charge
FISEA Florida Society of Enrolled Agents
FLSEP Family Life and Sex Education Program
flsh flesh (side) leather
flshd fleshed (skins)
FLSO Fort Lauderdale Symphony Orchestra
FLSP Fort Lincoln State Park (North Dakota)
flst flautist; flutist
flt filter; fleet; flight; float; flotation; fork-lift truck; frontolaeva transverse
flt flertall (Dano-Norwegian—majority or plural)
Flt Flats (postal abbreviation); Fleetwood
FLT Foreign-Language Teaching
F/Lt Flight Lieutenant
Flt Adm Fleet Admiral
fltbest fleet broadcast (USN)
Fltcher C Fletcher College
fltck flight check
Flt Cmdr Flight Commander
Fltg floating
flt ld sim flight-load simulator
Flt Lt Flight Lieutenant
Flt No. Flight Number
fltp flight template
flt/pg flight programmer

flt pln flight plan
fltr floater
flts flights
Flt Sat Com Fleet Satellite Communications System (USN)
FLTSATCOM Fleet Satellite Communications (DoD)
Flt Sgt Flight Sergeant
Flt Sgt Nav Flight Sergeant Navigator
fltstrikex full general-emergency striking force (USN)
flu fault location unit; final limit, up; first line unit; influenza
fluc fluctuant; fluctuate; fluctuating; fluctuation
FLUG Flugfelag Islands (Iceland Airways)
flummery foolish humbuggery (named after British custard made of flour or oatmeal boiled with water until almost too thick to swallow)
fluor fluor-apatite; fluorescence; fluorescent; fluorite; fluorspar; fluotaramite; synonym of flourite
fluorspar calcium fluoride (CaF2)
flur fluorescent
fluss flüssig (German—fluid)
flv foreign leave
flW follow(s)
FLW Frank Lloyd Wright
flwd followed
flwg following
flwop forced landing without power
flx flexible
fly flinty; flying; flyweight
FLY Flying Tiger Line
flying wing B-2 advanced technology bomber; stealth bomber
FlyTAF Flying Training Air Force
FLZO Farband-Labor Zionist Order
fm face measurement; facial measurement; facing matter; fan marker; farm; farmer; fathom; fathometer; female white; femtometer(s); fermi; fine measurement; foster mother; form; free mulatto; frequency modulation; frequency multiplier; from; front matter; fumigation
f-m frequency modulation
f/m feet per minute
f & m foot-and-mouth disease

fm *femmess mariées* (French—married women); *formiddag* (Dano-Norwegian—before noon)—a.m; *formiddagen* (Swedish—before noon) a.m.

f.m. *fiat mistura* (Latin—make a mixture)

f/M female Mexican

Fm fermium

F/m unit of permittivity

FM Fed Mart; Federated States of Micronesia; Ferrocarril Mexicano (Mexican Railroad); Field Manual; Field Marshal; Fire Marshal; Flight Mechanic; Foreign Minister; frequency modulation

F & M Franklin and Marshall College

F.M. *Fraternitas Medicorum* (Latin—Physicians Fraternity)

fma forward maintenance area

FMA Federal Maritime Administration; Felt Manufacturers Association; File Manufacturers Association; Financial Management Association; Financial Marketing Association; Fish Marketing Authority, Flour Mills of America; Food Machinery Association; Ford Motor Argentina; Forging Manufacturers Association; Fragrance Materials Association; Fulfillment Managememt Association

FMACC Foreign Military Assistance Coordinating Committee

FMAI Financial Management for Administrators Institute

fman foreman

FMANA Fire Marshals Association of North America

FMAO Farm Machinery Advisory Office(r)

FMAS Foreign Marriage Advisory Service

fmb (FMB) fast missile boat

FMB Federal Maritime Board; Felix Mendelssohn Bartholdi

fmbid firm bid

FMBRA Flour Milling and Baking Research Association

FMBSA Farmers and Manufacturers Beet Sugar Association

FMC Failure Mode Center (Reliability Laboratory); Federal Maritime Commission; Federated Motor(ing) Clubs; Federated Mountain Clubs; Federation of Mothers Clubs; Felt Manufacturers Council; Food Machinery Corporation; Ford Motor Company

FMC *Federación de Mujeres Cubanas* (Spanish—Federation of Cuban Women)

fmca forming cam

FMCA Family Motor Coach Association; Fire Mark Circle of the Americas

FM Can Ford Motor of Canada

FMCC Fulton-Montgomery Community College

F McH NM Fort McHenry National Monument

FMCI Forms Manufacturers Credit Interchange

FMCL Fleet Mechanical Calibration Laboratory

FMCS Federal Mediation and Conciliation Service

fm cu form cutter

fmcw frequency-modulated continuous wave

fmd foot-and-mouth disease

FMD Federated Metals Division- American Smelting and Refining; Fisheries Management Division; Fixtures Manufacturers and Dealers; *Flota Mercante Dominicana* (Spanish—Dominican Steamship Line); Forward Metro Denver

fmdf fixed mirror-distributed focus

fm di form die

FMDP Financial Management for Data Processing

fme frequency-measuring equipment

FME Foreign Materials Exclusion

FMEA (FEA) Failure Modes and Effects Analysis

FMEC Fur Merchants Employers Council

FMECA Failure Mode, Effects, and Criticality Analysis

fmer factory mutual engineering and research

fmet-N formylmethionine

fmeva floating-point means and variance

fmf fetal movement felt; field maintenance factor

fMf (FMF) familial Mediterranean fever

FMF Fleet Marine Force; Food Manufacturers' Federation; Foreign Military Financing

FMF-A Fleet Marine Force—Atlantic

fmfb frequency-modulation feedback

F'MFIC Federation of Mutual Fire Insurance Companies

FMFLANT Fleet Marine Force—Atlantic

FMF-P Fleet Marine Force—Pacific

FMFPAC Fleet Marine Forces–Pacific

fmfs fat in the moisture-free substance

fmg foreign medical graduate

FMG *Flota Mercante Grancolombiana* (Spanish—Colombian national steam ship lines); franc(s) Malagasy

FMGJ Federation of Master Goldsmiths and Jewelers

fmh (FMH) fat-mobilizing hormone

FMH Friends Meeting House

FmHA Farmers Home Administration

FMHCSS Federal Mobile Home Construction and Safety Standard

FMHHS Fort McHenry Historic Shrine (Baltimore)

fmi field maintenance instructions

FMI Farmers Mutual Insurance; Fiber Materials Inc; FM Intercity (relay broadcasting); *Fonds Monétaries Internationals* (French—International Monetary Fund); Food Marketing Institute; Freight Management International

FMI *Fondo Monetario Internacional* (Spanish—International Monetary Fund)

fmicw frequency-modulated intermittent-continuous-wave (radar)

FMIG Farmers Mutual Insurance Group; Food Manufacturers' Industrial Group

FMIS Functional Management Inspection System

fmj full metal jacket

fmk full-mouth radiograph

Fmk Finnmark; Finnish markka (currency unit)

fml formal

FML Factory Mutual Laboratories; Fermi National Laboratory (Batavia, Illinois)

FMLA Family Medical Leave Act

FMLN *Farabundo Martí Liberación Nacional* (Spanish— Augustín Farabundo Martí National Liberation Front)— leftist rebels in El Salvador

FMLNF *Farabundo Marti National Liberation Front* (Spanish—Salvadoran Marxist guerrillas)

fmly formerly

fmly k a formerly known as

FMM Federation of Malay Manufacturers; French Military Mission

FMMA Floor Machinery Manufacturers Association

fmmd form mandrel

FMME Fund for Multinational Management Education

fmn formation

fmn (FMN) flavin mononucleotide

FMN *Fédération Motorcycliste Nationale* (French— National Motorcycling Federation); *Ferrocarril Mexicano del Norte* (Spanish—Northern Mexican Railroad)

FMNH Field Museum of Natural History

FMNM Fort Matanzas National Monument

FMO Fleet Mail Office; Fleet Medical Officer; Flight Medical Officer

FMOF First Manned Orbital Flight (NASA)

fmofr firm offer

fmp first menstrual period; functional maintenance procedure; funny-man prop

FMP Fairbanks Morse Pump; Family Medicine Program; Final Management Plan; Fourth Malaysia Plan; Frontier Mounted Police; Fuel Management Panel

FMPA Fellow of the Master Photographers' Association

FMPA *Fédération Mondiale pour la Protection des Animaux* (French—World Federation for the Protection of Animals)

FMPC Feed Materials Production Center

FMPE Federation of Master Process Engravers

FMPEC Financial Management Plan for Emergency Conditions (USA)

fm/pm phase-modulated telemetering system

fm prot fine-mesh (cover) protected

FMPS Fairbanks Morse Power Systems

fmr fair market rent; farmer; fast metabolic rate; ferromagnetic resonance; former; former(ly)

FMR Field Maintenance Reliability; Field Materials Request; Franco Maria Ricci (or his magazine *FMR)*

F-M-R Friend-Moloney Rauscher (virus)

FMRA Fertilizer Manufacturers Research Association

FMRC Financial Management Research Center

FMR Corp. Fidelity Management & Research Corporation

fmri (FMRI) functional magnetic resonance imaging

fm rl form roll

fmrly formerly

fmrr (FMRR) financial management rate of return

FMRS Federal Mediation and Reconciliation Service; Foreign Member of the Royal Society

fms false-memory syndrome; fathoms; fat-mobilizing substance; fibromyalgia syndrome; flush metal saddle; foreign military sales; free-machining steel; frequency-multiplexed subcarrier

fm's formerly-married persons

FMS Federal Mining and Smelting (company); Federated Malay States; Field Music School; Financial Management Service; Financial Management System; Financial Managers Society; Flexible Manufacturing System; Floating Machine Shop; Fort Myers Southern (railroad); Frequency Monitoring System; Friends Mission Society

FMS *Fédération Mondiale des Sourds* (French—World Federation of the Deaf); Foreign Military Sales

fmsa frequency measuring spectrum analyzer

FMSA Federal Managers Support Agency; Fellow of the Mineralogical Society of America

FMSF False Memory Syndrome Foundation; Foreign Military Sales Financing

FMSI Friction Materials Standards Institute

FMSL Fort Monmouth Signal Laboratory

FMSM *Fédération Mondiale pour la Santé Mentale* (French—World Mental Health Federation)

fmswr flexible mild-steel wire rope

fmt flush metal threshold

fmt (FMT) format (flow chart)

FMT Factory Marriage Test; Fligght Management Team (NASA)

fm to. form tool

FMTS Field Maintenance Test Station

F & MTVHS Food and Maritime Trades Vocational High School

fmu force measurement unit; freight multiple unit

fmv fair market value

FMVCP Federal Motor Vehicle Control Program

FMVSS Federal Motor Vehicle Safety Standard

FMWC Federation of Medical Women of Canada

FMWS Fairbanks Morse Weighing Systems

fmx full-mouth radiography

fn fence; flatnose (projectile); footnote; formerly nested; free negro; fusion

f/n freight note

fn *fête nationale* (French—national holiday)

Fn Factonimbus

F$_n$ Fibonacci number(s)

FN Flight Nurse; Fridtjof Nansen

FN *Fabrique Nationale* (French—National Factory)—Belgian arms firm's initials appearing on all its products; *Forenede Nationer* (Danish—United Nations)

FN4RM/62FAB Belgian four-wheeled armored vehicle armed with a 60mm mortar and two machineguns or a 90mm cannon

fna fine-needle aspiration; for necessary action

FNA following named airmen; French North Africa

FNAA Fellow of the National Association of Auctioneers

FNAF Federal Nigerian Air Force

FNAL Fermi National Accelerator Laboratory

FNB First National Bank; Food and Nutrition Board

FNBC First National Bank of Chicago

FNBP Far North Bicentennial Park (Anchorage)

fnc finance

FNC Federal Networking Council

FNC *Federación Nacional de Cafeteros* (Spanish—National Federation of Coffee Growers—Colombia); *Ferrocarriles Nacionales de Colombia* (Spanish—National Railroads of Colombia)

FNCB First National City Bank

fncg financing

fncl financial

FNCR *Ferrocarril del Norte de Costa Rica* (Spanish—Northern Railway of Costa Rica)

fnd found; foundation; foundered

FND Flinders Naval Depot (Australia)

fndd founded

fndg founding

fndn foundation

fndr founder

fndrs fenders

fndry foundry

FNDTS Fellow of the Non-Destructive Testing Society

fne fine

fnf flying needle frame

fng fuckin' new guy

fnh flashless nonhygroscopic (gunpowder)

FNH *Ferrocarril Nacional de Honduras* (Spanish—National Railway of Honduras)

FNIC Food and Nutrition Information and Educational Materials Center

FNIE *Fédération Nationale des Industries Électriques* (French—National Federation of Electrical Industries)

FNIF Florence Nightingale International Foundation

FNIMC Florida Normal and Industrial Memorial College

fnl final

FNL Friends of the National Libraries

FNLA *Frente Nacional de Libertacão de Angola* (Portuguese—Angolan National Liberation Front)

FNLO French Naval Liaison Office(r)

fnl qtr final quarter

fnly finally

fnlz finalize

FNM *Ferrocarriles Nacionales de México* (Spanish—National Railroads of Mexico); Financial Network Manager

FNMA Federal national Mortgage Association

FNN Fiji News Network; Financial News Network

FNNWR Fort Niobrara National Wldlife Refuge (Nebraska)

FNO Fleet Navigation Officer; following-named officers

FNOA following-named officers and airmen

fnp fusion point

fnp (FNP) floating nuclear-power plant

FNP Family Nurse Practitioner; Fiordland National Park (South Island, New Zealand); Fundy National Park (New Brunswick, Canada)

FNPF Fiji National Provident Fund

FNRC Federal Nuclear Regulatory Commission

FNRJ Federation Narodna Republik Jugoslavija (name for former Yugoslavia)

fns flask-nitrogen supply

FNS Food and Nutrition Service; Frontier Nursing Service

FNSAE Fellow of the National Society of Art Education

fnsh finish

fnshd finished

fnshg finishing

fnshr finisher

FNTO Finnish National Travel Office

fnu first name unknown

FNU *Forces des Nations Unies* (French—United Nations Forces)

f number focal length of a lens

f-number diameter of a lens aperture in relation to its focal length

FNV *Financiera Nacional de la Vivienda* (Spanish—National Housing Finance)

FNWA Foreign National Weather Agency

FNWC Fleet Numerical Weather Center (USN)

FNWF Fleet Numerical Weather Facility

FNZDT Federation of New Zealand Dancing Teachers

FNZLA Fellow of the New Zealand Library Association

FNZSA Fellow of the New Zealand Society of Accountants

FNZSID Fellow of the New Zealand Society of Industrial Designers

fo faced only; fade out; fast operating; filter output; firm offer; firm order; flat oval; folio; formal offer; free out; free overside; freight on; fuel oil; full out terms; for orders

fø damping factor (symbol)

fo (FO) full organ

fo' for; four

f/o for credit of; firm offer; free overside; for orders

f^b folio

f^o *firmato* (Italian—signed)

f/O female Oriental

Fo Fornax

F pure parental type

FO Faroe Islands (Internet code); Federal Office(r); Federal Official(dom); Field Office(r); Field Operations; Field Order; Finance Office(r); Finance Officer; Fisheries Office(r); Flag Officer; Flying Officer; Foreign Office; Forward Observer

F.O. Foreign Office

f/o follow on

F/O Flight Officer; Flying Officer

FOA Farmers Organization Authority; Financial Operations Association; Football Officials Association; Foreign Operations Administration; Foresters of America; Friends of Animals

FOAC Flag Officer, Aircraft Carrier(s)

fob feet out of bed; freight on board

fob. feet out of bed; fresh off the boat; front of body (hoist); fuel on board; full of baloney

fo & b fuel oil and ballast
f.o.b. free on board; fuel on board
FoB Friends of Bill; Friends of the Bureau (FBI)
F o B Faculty of Building
FOB Federal Office Building; Forward Operating Base; Free on Board
Fobcnlf free on board cars, named point, lighterage free
fobcnp free on board cars, named point
FOBFO Federation of British Fire Organisations
fobot free on board, owners trim
FOBS Fractional-Orbit Bonbardment System
fobse free on board, sacks extra
fobsi free on board, sacks included
fobt fecal occult-blood test
foc final operation capability; flag of convenience; focal; focus(ing); free of charge; free on car(s); free on container(s); full operational capability; functional operational capability
f.o.c. free of charge; free on car(s); free on container(s)
FoC Father of the Chapel (printer's union)
FOC *Ferrocarriles Occidentales de Cuba* (Spanish—Western Railroads of Cuba); Flight Operations Center
FOCA Federation of Citizens Associations; Formula-One Constructors Association
FOCAS Ford Operating Cost Analysis System
fochr free of charge
FOCI Farrand Optical Company, Incorporated
FOCIS Financial On-Line Central Information System
FOCLA Federation of Country Local Associations
focmg forthcoming
FOCOL Federation of Coin Operated Launderettes
FOCS Freight Operation Control System
FOCK Fleet-Oriented Consolidated Stock List
Fo'c's'le forecastle
FOCT Flag Officer Carrier Training

FOCUS Federation of Community United Services; Financial and Operations Combined Uniform Single Report
fod fodder; foreign object damage; foreign object debris (airplanes); free of damage
fod (FOD) foreign object damage
f.o.d. free of damage
FOD Flag Officer, Denmark; follow-on destroyer
fo/do fuel oil/diesel oil (consumed daily)
FOD FREE Free of Dirt (dirt-free aircraft hangar)
foe fuel oil equivalent
FOE Fraternal Order of Eagles; Friends of the Earth
fof free on field (airmail)
FOF *Facts-On-File*
FOFA Follow-On Forces Attack
F of F Firth of Forth
fofr firm offer
fog flow of gold
FoG Friends of Gill
FOG Field Operations Group (US Army); Flag Officer, Germany; Florida Orange Growers
FOGA Fashion Originators Guild of America
FOGAIN *Fondo de Garantia y Fomento a la Industria Mediana y Pequeña* (Spanish—Fund for the Guarantee and Promotion of Medium and Small Industry)—Mexico
Fog Sig fog signal (station)
foh front of house
foi freedom of information
foi (FOI) fighter officer interceptor; follow-on interceptor
F o I Freedom of Information
FOI (station) Operations Intelligence; Fighter Officer Interceptors; Fruit of Islam (Black Nationalists)
FoIA Freedom of Information Act
FOIC Flag Officer in Charge
foil. file-oriented interpretive language
FOIR Field-of-Interest Register
fok fill or kill; free of knots (lacing, ropestring, twine)
f.o.k. free of knots
fol folio; folios; follow; following; follows; free-on-lorry oil and lubricants

fol. *folium* (Latin—leaf); *folia* (Latin—leaves)
FOL Federation of Labor; Federation of Labour (New Zealand); Folsom State Prison (California); Foreign Office Library; Friends of the Land
fold. folding
folg *folgend* (German—following)
Folkes Folkestone
folkl folklore
foll followed; followed by; following
folnoaval following (items) not available
fols folios; follows
Folsom California State Prison at Folsom
fom fat off mothers (sheep); fault of management; figure of merit
fomaj force maleure
FOMC Federal Open Market Committee
FOMCA Federation of Malaysian Consumers Association
FOMEX *Fondo Nacional de las Exportaciones* (Spanish—National Fund for the Promotion of Exports); *Fondo para el Fomento de las Exportaciones de Productos Manufacturados* (Spanish-Fund for the Promotion of Export of Manufactured Products)
FOMIN *Fondo Nacional de Fomento Industrial* (Spanish—National Fund for Industrial Promotion)
fomm functionally-oriented maintenance manual(s)
FoMoCo Ford Motor Company
fomth for one month
FON Fiber Optic Network
FONADE *Fondo Nacional de Desarrollo* (Spanish—National Development Fund)
FONASBA Federation of National Associations of Shipbrokers and Agents
FONATUR *Fondo Nacional de Fomento al Turismo* (Spanish—National Fund for Tourist Promotion)
fonc *fonctionnaire* (French—bureaucrat or office holder)
Fondo *El Fondo* (Spanish—The Fund)—International Monetary Fund—IMF
fonecon telephone conversation

FONEI Fondo Nacional de Equipamiento Industrial (Spanish—National Fund for Industrial Equipment)

fonet fonetica (Portuguese or Spanish—phonetics)

fonét fonética (Italian—phonetics)

F on F Facts-on-File

fono photograph

fonoff foreign office

Fons Alphonse; Fonseca

fonsi finding of no significant impact

Fontanka Fontanka Canal linking the port with the main section of St Petersburg and the Neva River

FONZ Friends of the National Zoo

fo° folio (Spanish—folio)

FOO Forward Observation Officer

foob (FOOB) firing out of the battery (artilley project)

fool's gold pyrites (copper, iron, tin, etc.)

foot(s) footnote(s)

fop flowery orange pekoe

fop forward observation post

f/op furing/observation port

FOP feminization of poverty; Fraternal Order of Police

fopt fiber-optics photon transfer

f.o.q. free on quay

for foreign; foreigner; forensic; forest; forester; forestry; forint (Hungarian monetary unit); free on rail; free on road

f.o.r. free on rail

for (Latin prefix— opening)— foramen

For Formosa(n); Fornax

Fødr Føroyar (Danish—Faroe Islands)

FOR Fellowship of Reconciliation; Final Outturn Report; Flying Objects Research; Foundation for Ocean Research

forac for action

FORACS Fleet Operational Readiness Accuracy Check Site

forast formula assembler translator

FORATOM Forum Atomique Européen (French—European Atomic Forum)

for. bal forensic ballistics

FORBID Federatie van Organisaties op het gebied van Bibliotheek-Informatieen Dokumentatiewezen (Dutch—Federation of Organizations on Libraries, Information, and Documentation Services)

forbloc fortran-compiled blockoriented (simulation programme)

for. bod foreign body

forcap forward combat air patrol

for & cc free of riots and civil commotion

for'd forward

Ford Gerald R Ford—38th President of the United States, 40th Vice President; automobile manufacturer Henry Ford; English playwright John Ford; movie director John Ford; historian Paul Leicester Ford

FORD Families Opposed to Revolutionary Destruction; Fix Or Repair Daily; Found On the Road Dead

FORDS Floating Ocean Research and Development Station

'fore before

FORE Foundation of Record Education; Fraternity of Recording Executives.

fore 1/4s fore-quarters (meat cuts)

foren forensic(ally); forensic medicine

fores'l foresail

FOREST Freedom Organisation for the Right, to Enjoy Smoking Tobacco

FOREWAS Force and Weapon on Analysis System

forf forfeit; forfeiture

forf forfattare, författarinna (Swedish—author, authoress)

förf forfatter (Dano-Norwegian—author)

forg forger; forgery; forging

f org (F Org) full organ

forint monetary unit of Hungary

fork forkortelse (Dano-Norwegian—abbreviation) *forkortning* (Swedish—abbreviation)

fork. forkortelse (Danish—abbreviation)

for. lang foreign language(s)

form format; formation; former(ly)

form formiddag (Dano-Norwegian—morning, before midday)

forma fortran matrix analysis

formac formula manipulation compiler

formal. formaldehyde; formalin

formalin HCHO

format. fortran matrix abstraction technique(s)

for med forensic medicine

For Min Foreign Minister; Minister of Foreign Affairs

formn foreman

FORMS Federation of Rocky Mountain States

for'm'st foremast

formul formulary

forpac forecasting passengers and cargo

For Pol Foreign Policy

forr forretning (Dano-Norwegian—business or store)

for'rd forward

for.rts foreign rights

Forsch Forschung (German—research)

FORSIC Forces Intelligence Center

forsk forskellig (Dano-Norwegian—different, distinct, unlike)

forls'l foresail

FORSTAT Force Status and Identity Report (USAF)

fort. fortification; fortify; fortnight(ly); fortress; full-out rye terms (grain trade)

fort. fortis (Latin—strong)

fortel formatted teletypewriter

Fort Frederica Fort Frederica National Monument on Saint Simon's Island off Brunswick, Georgia

Fort Jeff Fort Jefferson National Monument on the Dry Tortugas in the Gulf of Mexico west-northwest of Key West

Fort Laramie Fort Laramie National Monument on the Oregon Trail in southeastern Wyoming

Fort Leavenworth U.S. Disciplinary Barracks at Fort Leavenworth, Kansas

fortly fortnightly

Fort Matanzas Fort Matanzas National Monument near St Augustine, Florida, built by the Spaniards in 1736

Fort McHenry Fort McHenry National Monument in Baltimore Harbor where the *Star Spangled Banner* was written

Fort Meade Maryland operations center of the National Security Agency

for. tox forensic toxicology

Fort Pulaski Fort Pulaski National Monument at the mouth of the Savannah River

FORTRAN formula translation computer language

For-Trans Ford Foundation Transfer Student Project

FORTRANS Formula Translating System

fortransit formula translator internal translator

Fort Riley U.S. Army Correctional Training Facility at Fort Riley, Kansas

forts fortsaettelse (Dano-Norwegian—continuation or sequel)

Forts Fortsetzung (German—continuation)

Fort Savage New York City's East Harlem police precinct

Fortschr Phys Fortschritte der Physik (German—Advances in Physics)

fortsim fortran simulation

Fort Sumter national monument in Charleston Harbor (South Carolina)—first shot of Civil War fell on this fort

fort. twn fortified town

Fortune Five Hundred *Fortune Magazine's* annual listing of the 500 leading corporations

Fort Union Fort Union National Monument near Santa Fe, New Mexico

Forth Worth Federal Correctional Institution at Fort Worth, Texas

forum formula for optimizing through realtime utilization of multiprogramming

FORUM Federation of Retired Union Members

For Whom For Whom The Bell Tolls

forwn forewoman

'forz sforzando (Italian—emphasized forcefully)

fos fossil; free on station; free on steamer; fuel-oxygen scrap; full of shit

fos (FOS) faint object spectography; full operational status

f.o.s. free on station; free on steamer

F-o-S Frinton-on-Sea

FOS File Organization System; Fisheries Organization Society; Fuel Oil Supply (company)

FOSATU Federation of South African Trade Unions

fosdic film optical sensing device for input to computers

FOSFA Federation of Oil Seed and Fats Association

fos fls fossil fuels (coal, natural gas, oil, etc.)

FOSG Factory Outlet Shopping Guide

FOSH Foshing (airlines)

FOSI Florida Ocean Sciences Institute

fosplan formal space-planning language

foss fear-of-success syndrome

Fos sur Mer (French—Fos by the Sea)—Marseilles port

fot free of tax(ation); free on truck; frequency optimum traffic; fuel-oil transfer

f.o.t. free on truck

fot fotographie (Dutch—photography)—plus all derivatives

FOT Fraternal Order of Police

fot & e follow-on test(ing) and evaluation

F o t L Friends of the Library

FOTL Follow-on-to-Lance (launcher)

FOTM Friends of Old-Time Music

foto photograph(ic)

foto fotografia (Italian or Portuguese—photography) *fotografía* (Spanish—photography)

Foto (Jewtongo—Fort)—native nickname for Paramaribo, Surinam

fotog fotografua (Italian or Portuguese-photography) *fotografia* (Spanish—photography)

fotsu forward observer target survey unit

fo'ty forty

found foundation; foundling; foundry

Found Econ Educ Foundation for Economic Education

Found Phys Foundations of Physics

Foun Mot Dent Foundation for Motivation in Dentistry

fount fountain

Foun Than Foundation of Thanatology

FOUO For Official Use Only

Four Cs Community-Coordinated Child Care

Four H Four H (hand, head, heart, and health) Club

Four Horsemen Four Horsemeim of the Apocalypse (each mounted, respectively, on a white horse symbolizing pestilence, a red horse—war, a black horse–famine, a pale horse—death); title of a novel by Vicente Blasco Ibañez– *Los cuatro jinetes del Apocalipsis*

Four Mountains Islands of time Four Mountains

FOUSA Finance Office(r), United States Army

fov field of view

fov (FOV) flyable orbital vehicle

fow first open water; free on wagon; free on warehouse; free on wharf

f.o.w. first open water; free on wagon

Fowm Fibre Optic Well Monitoring System

Foxes Fox Islands off southwestern tip of Alaska

Foxtrot letter F radio code

Foy Fowey

fp factory pass; family plan(ning); fast peak; field punishment; film pack; fine paper; fine print; fire plug; fire policy; fireplace; first performance; first performed; first proof; fix point; fixed price; flame-proof(ed); flash point; flat pad(ded); flat pattern; flat point(ed); flight pay; flight plan; floating (open) policy; flower people; flowery pekoe (tea); focal plane; food poisoning; foot path; foot pound(s); forte piano; forward perpendicular; free piston; freezing point; fresh paragraph; frontispiece; full page; full point; full price; fully paid

fp (FP) family practitioner; flavoprotein

f/p flat pattern

f.p. fiat potio (Latin—make a potion)

Fp power-loss factor (symbol)

FP Family Physician; Federal Parliament; former pupil; Franklin Pierce (14th President U.S;); Free Press

FP Ferrocarril del Pacifico (Spanish—Pacifc Railroad); *Fuerza Publica* (Spanish—Police Force)—Panamanian

F/P Fire Policy (insurance)

FP Freiheitliche Partei (German—Freedom Party)—Austrian party

fp4c full page four colors

fpa fibrinopeptide-A; fluorescent pen aerosol; focal plane array; free of particular average

FPA Family Planning Association; Federal Preparedness Agency; Federation of Motion Picture Producers in Asia; Flexible Packaging Association; Flying Physicians Association; Foreign Policy Association; Foreign Press Association; Forest Products Association; Franklin Pierce Adams; Free Pacific Association; Freemantle Port Authority; Freethought Press Association

fpaa free from particular average, absolutely

FPAA Family Planning Association of Australia

fpaAc free of particular average, American conditions

FPAD Fund for Peaceful Atomic Development

fpaEc free of particular average, English conditions

fpaf fixed-price award fee

FPAS Federal Property and Administrative Services; Fellow of the Pakistan Academy of Sciences

FPASA Federal Property and Administrative Services Act

FPAT Family Planning Association of Tasmania

fpaueb free from particular average unless caused by (stranding, etc.)

FPB Fast Patrol Boat

FPBA Folding Paper Box Association

FPBAI Fellow of the Publishers' and Booksellers' Associations in India

FPBG fast patrol boat, guided-missile (USN)

FPBRS Fels Parent Behavior Rating Scale(s)

fpc fish protein concentrate; fixed price contract; fixed price call; flat plate (solar) collector(s); flight progress chart; full-page color (ad); for private circulation

fp-C flash point–Celsius

FPC Facility Power Control; Family Planning Center; fast patrol craft; Federal Pacific Electric (stock exchange symbol); Federal Power Commission; Federal Prison Camp; Fiji Pine Commision; Food Packaging Council; Friends Peace Committee; Frozen Pea Council

FPCA Federal Pay Comparability Act (of 1970); Federal Post Card Application (for absentee ballot)

fpcc flight propulsion-control coupling

FPCC Fair Play for Cuba Committee

FPCE Fission Products Conversion and Encapsulation (AEC plant)

FPCI Federal Penal and Correctional Institutions

FPCL Fronte Paisanu Corsu di Liberazione (Italian—Corsican Peoples Liberation Front)

FPCS Fire Power Control Subsystem; Full-Page Composition System

FPD Federal Public Defender

FPD Fundación Panamericana de Desarrollo (Spanish—Pan-American Development Foundation)

FPDA Finnish Plywood Development Association; Five-Power Defence Arrangement (Malaysian); Fluid Power Distributors Association

FPDC Federal Procurement Data Center

fpdi flight path deviation indicator

FPDO Federal Public Defender Organization(s)

fpe fixed price with escalation

FPE Foundation for Personality Expression; Full Personality Expression

FPEA Fellow of the Physical Education Association

FPEB Family Planning Evaluation Branch (USPHS)

FPEBT Fire Prevention and Engineering Bureau of Texas

fpec four-pile-extended cantilever (platform)

FPED Farm Production Economics Division (USDA)

FPF French Protestant Federation

fpga field programmable gate array

fph feet per hour (oil well drilling)

FPH Federal Pacific Hotels (Australian)

FPHA Federal Public Housing Authority

F Pharm S Fellow of the Pharmaceutical Society

fphs fallout protection in homes

F Ph S Fellow of the Philosophical Society

F Phy S Fellow of the Physical Society

fpi faded prior to interception; family pitch in; fixed price incentive; fuel-pressure indicator

FPI Federal Prison Industries; Fellow of the Plastics Institute; Food Processors Institute; Foodservice and Packaging Institute

FPI Fédération Prohibitioniste Internationale (French—International Prohibitionist Federation)

FPIA Family Planning International Assistance

fpif fixed-price-incentive firm

fpil full premiurn if lost

f. pil. fiat pilulae (Latin—make pills)

fpis fixed-price incentive successive; forward propagation by ionospheric scatter

FPJMC Four-Power Joint Military Commission

fpl final protective line; fire plug; fireplace

FPL Family Protection Law; feline panleukopaenia; Florida Power and Light; Forest Products Laboratory

FPL *Fuerzas Populares de Liberación* (Spanish—Popular Forces of Liberation)—El Salvador

FPLA Fair Packaging and Labelling Act

fplce fireplace

fpm facility power monitor; feet per minute; fissions per minute; frequency pulse modulation

FPM *Federal Personnel Manual*

FPML Forest Products Marketing Laboratouy

FPMR Federal Property Management Regulation(s)

FPMSA Food Processing Machinery and Supplies Association

FPMT Filter Paper Microscopic Test

FPNM Fort Pulaski National Monument

fpn fine print note

fpo fixed price open

FPO Field Post Office; Field Project Office; Fleet Post Office; Fleet Postal Organization; Florida Philharmonic Orchestra

FPOA Federal Probation Officers Association

fpoe first port of entry

fpoh food prepared outside the home

FPOP Family Planning Organization of the Philippines

fpp facility power panel; fixed-pitch propeller; floating-point processor; forward(ing) parcel(s) post

FPP Family Planning Program; Foster Parents Plan; Foster Parents Program; Friendly Peoples Proviso

FPPB Family Planning and Population Board (Singapore)

FPPC Fair Political Practices Commission

fppe fluorescent pen-post emulsified

FPPS Flight Plan Processing System; Full-Page Phototypesetting System

fpr feet per revolution; fixed price redeterminable; flat-plate radiometer; forward parcels rail

FPR Factory Problem Report; Farm Publications Reports; Field Personnel Record; Fishing Ports Registration

FPRC Fair Play for Rhodesia Committee

fprf fireproof

FPRF Fats and Proteins Research Foundation

FPRI Foreign Policy Research Institute (University of Pennsylvania)

FPRL Forest Products Research Laboratory

FPRS Federal Property Resources Service; Forest Products Research Society

fps feet per second; focus projection and scanning, foot per second; foot-pound-second; frames per second

f'ps former priests

FPs Flying Physicians; Flying Psychologists

FPS Farm Placement Service; Fauna Preservation Society; Federal Protection Service; Fellow of the Pharmaceutical Society; Fellow of the Philharmonic Society; Fellow of the Philological Society; Fellow of the Philosophical Society; Fence Protection System; Financial Planning System; Fire Protection System; Fluid Power Society

FPSA Fellow of the Photographic Society of America

FPSAA Federated Public Service Assistants Association

FPSE Federation of Public Service Employees

FPSL Fellow of the Physical Society of London

FPSO Fleet Publication Supply Office

fpsps feet per second per second

fpt female pipe thread; fill, puddle, and tamp; fixed price tenders; forepeak tank; full power trial

FPT Flight Proof Test(ing); Four Picture Test

fptm fluorescent pen-tank method

fpts forward propagation by tropospheric scatter

FPTU Federation of Progressive Trade Unions

fpu field pickup unit; floating point unit

FPU Food Preservers Union

fpv fixed-price vendor

FPWA Federation of Professional Writers of America; Federation of Protestant Welfare Agencies

FPWT Flight Plan Weight (aircraft code)

fq fiscal quarter

FQ French Quarter (New Orleans)

FQ *Faerie Queene* (Edmund Spenser allegory)

fqawt flush quick-acting watertight

fqcy frequency

FQDN Fully Qualified Domain Name

FQL Food Quality Laboratory

FQO Federation of Quarry Owners

FQPA Food Quality Protection Act

FQS Federal Quarantine Service

fque *fabrique* (French—factory)

fr family room; fast release (relay); father; fielding runs (baseball); field relay; fire; fire retardant; fixed ratio; fixed response; flaring and retracting; flight request; flow rate; frame; franc; franline; frayed; frequency response; frequent; from; front; fruit

f/r fixed response; flat rack; freight release; front to rear

f & r feed and return (plumbing); force and rhythm (pulse)

fr *franco* (Spanish—franc); *franc(s)* or *fransk* [Dano-Norwegian—franc(s) or French]

fr. *folio recto* (Latin—front of the sheet)

f.r. *folio recto* (Latin—front of the sheet)—righthand page

Fr Father; fragmentum; France; francium; Franco; Franklin; Frau (German—Missus); French; Friar; Friday; Friesian(s); Frisian(s); Froude number

F/r restricted first-class (travel)

Fr *Franca* (Portuguese—France); *Frankrijk* (Dutch—France); *Frau* (German—Missus); *Fray* (Spanish—Friar); *Fredag* (Danish—Friday)

FR Facilities Request; Family Registry; Feather River (railroad); Federal Reformatory;

Federal Register; Federal Reserve; Field Report; fighter reconnaissance (aircraft); Final Report; Fireman Recruit; flash red– enemy aircraft nearby; Fleet Reserve; Folkways Records; France (Internet code); Freight Release; Friden (stock exchange symbol)

F of R Fellowship of Reconciliation

FR *Federal Register*

F.R. *Forum Romanum* (Latin— Roman Forum)

FR-172 French-built four-place rocket-launching counter-insurgency aircraft

fra forward refueling area; functional residual air

fra factura (Spanish—in voice)

Fra Francis

Fra Francia (Spanish— France)

Fra. frater (Latin—brother; monk)

FRA Federal Railroad Administration; Federal Reserve Act; Fleet Reserve Association; Food Retailers Association; Footwear Research Association; Frankfurt-am-Main (airport)

frac frationator reflux analog computer

FRAC Food Research and Action Center

FRACA Failure Reporting. Analysis, and Corrective Action

FRAC Arts Foundation for Research in the Afro-American Creative Arts

FRACHE Federation of Regional Accrediting Commissions of Higher Education

FRACI Fellow of the Royal Australian Chemical Institute

FRACP Fellow of the Royal Australian College of Physicians

FRACS Fellow of the Royal Australian College of Surgeons

fract fraction; fracture

fract. dos. fracta dosi (Latin— in divided doses)

FRAD Fellow of the Royal Academy of Dancing

FRAeS Fellow of the Royal Aeronautical Society

frag fragile; fragment; fragmentary; fragmentation; fragmented

frago fragmentary order; fragmented order

FRAgS Fellow of the Royal Agricultural Societies

FRAHS Fellow of the Royal Australian Historical Society

FRAI Fellow of the Royal Anthropological Institute

FRAIA Fellow of the Royal Australian Institute of Architects

FRAIC Fellow of the Royal Architectural Institute of Canada

'fraid afraid

fram ferromagnetic random-access memory

FRAM Fellow of the Royal Academy of Music; Fleet Rehabilitation and Maintenance (USN)

FRAME Fund for the Replacement of Animals in Medical Research

Framingham Massachusetts Correctional Institution (for female) at Framingham, Massachusetts

fran framed-structure analysis; franchise

Fran France; Frances; Francis; Franciscan

Fran Francia (Spanish— France)

franc monetary unit of Andorra, Belgium, France, Franciscan; Luxembourg, Madagascar, Monaco, Rwanda, and Switzerland

France French Republic

Francine Frances

Francisco I Francisco Indalécio Madero

Franco Francisco Paulino Hermenegildo Teodulo Franco-Bahamonde—Spanish dictator

Franco Francisco (Spanish— Francis)

frangi(s) frangipani(s)

Franglais francais + anglais (French + English)

Fran-Jud Franco-Judeo (French-Jewish)

Frank Frank; Frankford; Frankfort; Frankfurt; Frankish; Franklin

Frank Frankrike (Norwegian—France)

Franklinite ferric iron and zinc crystalline compound

frank(s) frankfurter(s)

Frans Francis

FRAP Federal Rules of Appellate Procedure; Fellow of the Royal Academy of Physicians

FRAP Frente Revolucionario de Acción Popular (Spanish—Revolutionary Popular Action Front)—Chile

FRAPS Farm Record Analysis Pilot Scheme

Fras Francis

FRAS Fellow of the Royal Asiatic Society; Fellow of the Royal Astronomical Society

Frasca Francesca

Frasco Francisco

frat fraternity

frat fratello (Italian—brother)

FRAT Free Radical Assay Technique (heroin-morphine test)

FRATADD Foundation for Research and Treatment of Alcoholism and Drug Dependence (Australian)

frate formula for routes and technical equipment

frater fraternity brother

frats fraternities

fratting fraternizing

Frau Die Frau Ohne Schatten (German—The Woman without a Shadow)—Richard Strauss opera

fraud. fraudulent

frav first available

Fraxi Pisanus Fraxi (Herbert Specer Ashbee)

FRB Federal Reserve Bank; Federal Reserve Board; Fisheries Research Board

frbb fracture of both bones; free room, board, and beverages

FRBC Fisheries Research Board of Canada

fr bel from below

FRBk Federal Reserve Bank

FRBNY Federal Reserve Bank of New York

FRBs Federal Reserve Banks

FRBS Fellow of the Royal Botanic Society; Fellow of the Royal Society of British Sculptors

frc fiber-reinforced concrete; functional residual capacity

f.r.c. free carrier

FRC Facility Review Committee; Fasteners Research Council; Federal Radiation Council; Federal Radio Commission; Federal Records Center; Federal Republic of Cameroon; Filipino Rehabilitation Commission; Flag Research Center; Flight Research Center; Foreign Relations Committee; Foreign Relations Council; Forwarder's Receipt Certificate; Fuels Research Council

FRCA Fellow of the Royal College of Art; Fire Retardant Chemicals Association

FRC-AAP Freedom-to-Read Committee–Association of American Publishers

Fr-Can French-Canadian

FRCAT Fellow of the Royal College of Advanced Technology

fr & cc free of riots and civil commotion

FRCD Fellow of the Royal College of Dentists

frcd's floating-rate certificates of deposit

FRCGP Fellow of the Royal College of General Practitioners

FRCI Fellow of the Royal Colonial Institute

FRCM Fellow of the Royal College of Music

FRCO Fellow of the Royal College of Organists

FRCOG Fellow of the Royal College of Obstetricians and Gynaecologists

FRCP Federal Rules of Civil Procedure; Fellow of the Royal College of Physicians

FRCPath Fellow of the Royal College of Pathologists

FRCP(C) Fellow of the Royal College of Physicians of Canada

FRCPE Fellow of the Royal College of Physicians of Edinburgh

FRCPGlas Fellow of the Royal College of Physicians of Glasgow

FRCPI Fellow of the Royal College of Physicians of Ireland

FRCP Lond Fellow of the Royal College of Physicians of London

FRCPSG Fellow of the Royal College of Physicians and Surgeons of Glasgow

FRC Psych Fellow of the Royal College of Psychiatrists

FRCR Fellow of the Royal College of Radiologists

FRCrP Federal Rules of Criminal Procedure

FRCs Federal Regional Councils

FRCS Fellow of the Royal College of Surgeons

FRCSC Fellow of the Royal College of Science

FRCS(C) Fellow of the Royal College of Surgeons of Canada

FRCSE Fellow of the Royal College of Surgeons of Edinbrugh

FRCSGlas Fellow of the Royal College of Surgeons of Glasgow

FRCSI Fellow of the Royal College of Surgeons of Ireland

FRCSL Fellow of the Royal College of Surgeons of London

FRCTS Fast Reactor Core Test Facility

frcu frequency reference control unit

FRCVS Fellow of the Royal College of Veterinary Surgeons

frd formerly restricted data; friend; friendly

Frd Ford (postal abbreviation)

FRD Federal Reserve District; Federal Rules Decisions

FR Dist Federal Reserve District

Frdn Friedenau

fre free energy region

fre *fracture* (French—invoice)

Fre Freemantle; French

Fre *Freitag* (German—Friday)

FRE Federal Rules of Evidence

FREB Federal Real Estate Board

FR Econ S Fellow of the Royal Economic Society

FR Econ Soc Fellow of the Royal Economic Society

fred figure-reader electronic device

Fred Alfred; Alfredo; Freddie; Frederic; Frederick; Fredric; Fredrick; Wilfred

Freda Winifred

Fred(die) Frederica; Fredrica

Freddie Mac Federal Home Loan Mortgage Corporation

Fred(dy) Alfred; Frederick; Wilfred

Fredk Frederick

Fr-edk D Frederick Douglass

Fredo Alfredo

Free Freeway

FREE Florida Resources in Education Exchange

freebd freeboard

freebies free services; free things; free tickets

FREED Foundation for Research and Education in Eugenics and Dysgenics

Freedman's Bureau Bureau of Refugees, Freedmen, and Abandoned Lands (set up after the Civil War in the United States)

Free Lib Phila Free Library of Philadelphia

freem freeman; freemen; freemason(ry)

Free-0 Freemantle, Western Australia

Freep *Free Press* (Los Angeles underground newspaper)

Free Soc Freethinkers Society

freeture freedom, the wave of the future

freeway toll-free express highway

FREI Fellow of the Real Estate Institute

Freib Freiburg (Germany)

freid *freidenker(ei)* (German—freethinker; freethinking)—latitudinarian(ism)

FREIT Finite-Life Real Estate Investment Trust

FRELIMO *Frente de Libertação de Moçambique* (Portuguese—Mozambique Liberation Front)

FRELP Flexible Real Estate Loan Plan

Frem Fremantle

frem. voc. *fremitus vocalis* (Latin—vocal fremitus)

French Antilles Désirade, Guadeloupe, Les Saintes, Marie Galant, Martinique, Saint Barthélmy, and part of Saint Martin

French Can French Canadian

Frenglish frenchified English

FREntS Fellow of the Royal Entomological Society

freon tf trifluorotrichloroethane (solvent)

FREP Fleet Return Evaluation Program

freq frequency; frequent; frequentative; frequently

FrEqAfr French Equatorial Africa

freq m frequency meter

fres fire-resistant

fres frères (French—brothers)

FRES Fellow of the Royal Entomological Society

frescanar frequency scan radar

fresh. freshman; freshmen

Freud. Freudian

frev fast reverse

frf flight-readiness firing; follicle-stimulating-hormone releasing factor; frequency response function

fr-f french-fried (potatoes)

FRF Freedom-to-Read Foundation; Fringe Reduction Facility

FRFPS Fellow of the Royal Faculty of Physicians and Surgeons

Frf(s) French franc(s)

FRFS Fast Reaction Fighting System

Frg Forge (postal abbreviation)

FrG Federal Republic of Germany (the former West Germany)

Fr G Frans Guyana (Dutch—French Guiana)

FRG Facility Review Group; Federal Republic of Germany (the former West Germany)

FRGS Fellow of the Royal Geographical Society

frgt freight

FRHB Federation of Registered House Builders

frhgt free height

ER Hist S Fellow of the Royal Historical Society

FR Hort S Fellow of the Royal Horticultural Society

Fr hr French horn

Frhr Freiherr (German—Baron)

Fr hrn French horn

FRHS Fellow of the Royal Horticultural Society

fri feeling rough inside

Fri Friday; Friesland (Dutch province)

FRI Fellow of the Royal Institution; Fels Research Institute; Forest Research Institute; Friends of Rhodesian Independence

FRIA Fellow of the Royal Irish Academy

FRIAI Fellow of the Royal Institution of Architects of Ireland

FRIAS Fellow of the Royal Incorporation or Architects of Scotland

Frib Fribourgh (Switzerland)

FRIBA Fellow of the Royal Institute of British Architects

fric frication; fricative; fritcatruce; fricatrix; friction; frictional

FRIC Fellow of the Royal Institute of Chemistry

Frick Frick Collection (New York City)

FRICS Fellow of the Royal Institution of Chartered Surveyors

frict friction

FRIDA Framework for Integrated Dynamic Analysis

fridg frigidaire (refrigerator)

fridge(s) refrigerator(s)

Friedrh Friedrichshafen

Fried Test Friedman Test (for pregnancy)

Friends Society of Friends (Quakers)

Friends Meet Friends Meeting

fries friesisch (German—Frisian)

Fries Friesland (Dutch province); Friesic

Friesn Friesian (cattle, language, or people)

frig refrigerator

frig. frigidus (Latin—cold)

FRIGS Fellow of the Royal Imperial Geographical Society

FRIIA Fellow of the Royal Institution of International Affairs

Frim Frimaire (French—Sleety Month)—beginning November 21st—third month of the French Revolutionary Calendar

FRINA Fellow of the Royal Institution of Naval Architects

fringe. file-and-report information-processing generator

Fringlish French + English (English interladen with French exressions and words)

fring(s) french onion ring(s)

f'r instance for instance

FRIPA Fellow of the Royal Institution of Public Administration

FRIPHH Fellow of the Royal Institute of Public Health and Hygiene

Fris Friesland; Frisia; Frisian

frisco fast-reaction integrated submarine control

Frisco (navalese—San Francisco)—no San Franciscan will use this nickname

FRISCO St Louis-San Francisco Railway

Frisco Bay (sailor's slang—San Francisco Bay)

Frisia (Latin—Friesland)—in the Netherlands

Frisians Frisian islanders or the Frisian Islands in the North Sea

Fritalux France, Italy, and Benelux nations

frits fritters

Fritz Friedrich

frjm full-range joint movement

frk fröken (Swedish—Miss)

Frk Fork (postal abbreviation); Frankfort

Frk Froken (Dano-Norwegian—Miss)

Frks Forks (postal abbreviation)

frl fractional; fuselage reference line

Frl El Ferrol

Frl Fräulein (German—Miss)

FRL Fuel Research Laboratory

FRLL Farrell Lines (container unit)

frm fault reporting manual; fiberglass-reinforced metal; fireroom; former(ly); frame; framing; frequency meter

FRM Federal Reformatory for Men

FRMA Floor Rug Manufacturers Association

FRMCM Fellow of the Royal Manchester College of Music

FR Met Soc Fellow of the Royal Meteorological Society

FRMIT Fellow of the Royal Melbourne Institute of Technology

frmn formation

frmr former

Frms Farms (postal abbreviation)

FRMS Federation of Rocky Mountain States; Fellow of the Royal Microscopical Society

frn floating rate note

FRN Federal Republic of Nigeria; Federal Reserve Note

frna foreign rations not available

FRNHS Fort Raleigh National Historic Site

FRNM Foundation for Research on the Nature of Man

FRNS Fellow of the Royal Numismatic Society

FRNSA Fellow of the Royal Navy School of Architects

Frnz Fernandez

FRNZIH Fellow of the Royal New Zealand Institute of Horticulture

'fro Afro

FRO Fellow of the Register of Osteopaths; Fire Research Organization; Friends Religious Order

FROC Federated Russian Orthodox Clubs

frof fire risk on freight

frog free rocket over ground

FROGIE Fellowship to Resist Organized Groups Involved in Exploitation

from full range of movement

from. full range of movement

fron frontal; frontalis

FRONAPE Frota Naccional de Petroleiros (Portuguese— National Petroleum Fleet— Brazil)

front. frontispiece

FRONT BC Frontera (Fronteriza) Baja California (Spanish—Baja California Frontier)

Frontera Girls California Institution for Women at Frontera

frosh freshman; freshmen

frp fiberglass reinforced plastic; fiberglass-reinforced polyester; forward refueling area

frp (FRP) follicular regulatory protein (hormone)

FRP Fuel Reprocessing Plant; Fundamental Research Press

frpf fireproof

frpl fireplace

frplc fireplace

frplce fireplace

frpng fireproofing

Fr Pol French Polynesia (whose capital is Papeete)

FRPS Fellow of the Royal Photographic Society

FRPSL Fellow of the Royal Philatelic Society of London

frq frequent(ly)

FRR Facilities and Rearrangement Request

FRRA Facilities and Rearrangement Request and Authorization

frs first rank symptom(s); flight reference selector; francs

frs (FRS) first readiness state

Frs Fresno; Frisian

Fr S French Somaliland (French Territory of the Afars and the Issas)

FRS Federal Reserve System; Fellow of the Royal Society; Financial Relations Society; First-Rank Symptoms (Schneiderian); Fisheries Research Society; Foundation Research Service; Freethought Reprint Series; Frequency Response Survey; Fuel Research Station

FRSA Fellow of the Royal Society of Arts

FRSAI Fellow of the Royal Society of Antiquaries of Ireland

frsc full range source code

FRSC Fellow of the Royal Society of Canada

FRSCM Fellow of the Royal School of Church Music

FRSE Fellow of the Royal Society of Edinburgh

FRSGS Fellow of the Royal Scottish Geographical Society

FRSH Fellow of the Royal Society of Health

FRSI Fellow of the Royal Sanitary Institute

FRSL Fellow of the Royal Society of Literature; Fellow of the Royal Society—London

FRSM Fellow of the Royal Society of Medicine

FRSNA Fellow of the Royal School of Naval Architecture

FRSNZ Fellow of the Royal Society of New Zealand

Fr Som French Somaliland

FRSPS Fellow of the Royal Society of Physicians and Surgeons

FRSS Fellow of the Royal Statistical Society

FRSSA Fellow of the Royal Scottish Society of Arts

FRS(SA) Fellow of the Royal Society of South Africa

FRSSI Fellow of the Royal Statistical Society of Ireland

FRSSS Fellow of the Royal Statistical Society of Scotland

Frst Forest (postal abbreviation)

FRSTAT Fringe Software System

FRSTM & H Fellow of the Royal Society of Tropical Medicine and Hygiene

frt free return trajectory; freight; fruit

frt før vor tidregning (Dano-Norwegian—before time was reckoned)

FRT Family Relations Test

FRTC Fast-Reactor Training Center

frt/fwd freight forward

frtiso floating-point root isolation

frto flight radio telephone operator

Fr To French Togoland

FRTO Federated Road Transport Organization(s)

frt ppd freight prepaid

frtr freighter

fru frequency reference unit; fructose; fruit sugar

FRU Federal Reserve Unit; Fiji Rugby Union

fruat. frustrillatum (Latin—in small bits)

fruc. fructus (Latin—fruit)— sometimes abbreviated *fr.*

Fruc Fructidor (French— Fruitful Month)—August 18 through September 16— twelfth month of the French Revolutionary Calendar whose remaining five days were called Sansculottides and named respectively for the Virtues, Genius, Labor, Reason, and Rewards

fruct fructification, fructify, fructose, fructuous

frug fmgal(ity), frugally

frugal fortran rules used as a general applications language

frumpie formerly-radical upwardly-mobile professional in elections

FRUS *Foreign Relations of the United States*

FRUSA Flexible Rolled-Up Solar Array

frust. *frustillatim* (Latin—in small portions)

fru veg fruits and/or vegetables

frv (FRV) flight-readiness vehicle

FRVIA Fellow of the Royal Victorian Institute of Architects

FRW Federal Reformatory for Women (Alderson, West Virginia)

frwd foreword, forward

FRWI Framingham Relative Weight Index

frwis frost warnings issued

frwk framework

Frwy Freeway

frx firex

Fry Ferry (postal abbreviation); Freeway (highway abbreviation)

FRYC Fall River Yacht Club

fr yr gdnce for your guidance

frz *französisch* (German—French)

FRZS Fellow of the Royal Zoological Society

FRZS (NSW) Fellow of the Royal Zoological Society of New South Wales

FRZS(Scot) Fellow of the Royal Zoological Society of Scotland

fs facsimile; factor of safety; far side; fee simple; film strip; final settlement; fin stabilized; fire station; flight service; flying status; foot second; foreign service; foresight; freight supply; front scalloped; front spar; sulfur trioxide chlorsulfonic acid (commercial short form or symbol)

fs (FS) file separator character (data processing)

f/s feet per second; first-stage

fˢ *francos* (Spanish—francs)

fs *faites suivre* (French—please forward)

Fs fractostratus

FS Faraday Society; Feasibility Study; Federal Specification(s); Field Security; Field Service; Fighter Squadron; Financial Statement; Fire Station; Flight Sergeant; Fog Signal (Station); Foreign Ser-

vice; Forest Service; Franciscan Studies; Franz Shubert; Free State; Freedom School; freight supply (vessel); Friendly Society; Friends Society; Friendship Store(s); small freighter (naval symbol)

F-S Fenno-Shipping

F.S. Father of Sion

F/S Financial Statement

FS *Filharmonisk Selskap* (Norwegian—Philharmonic Orchestra); *Forente Staterna* (Swedish—United States)

fsa family separation allowance; flexible spending account; fuel storage area

fsa (FSA) fetal sulfoglyco-protein

f.s.a. *fiat secundum artem* (Latin—let it be done skillfully)

FSA Farm Security Administration; Federal Security Administration; Federal Security Agency; Federal Supply Classification; Federation of South Arabia; Fellow of the Society of Antiquaries; Fellow of the Society of Arts; Field Survey Association; Finance Service—Army; Financial Security Assurance; Fire Support Area; Flax Spinners Association; Florida Student Association; Fluid Sealing Association; Fraternal Scholastic Association.; Free Selectors Association; Free Society Association; Freethinkers Society of America; Friendly Societies—Act; Future Scientists of America

F & SA Farmers and Settlers Association (Australian)

FSAA Family Service Agency of America; Family Service Association of America; Flight Stewards Association of Australia

FSAC Freight Station Accounting Code

FSAG Fellow of the Society of Australian Genealogists

fsaga first sortie after ground alert

FSAICU Federation of State Associations of Independent Colleges and Universities

FSAL Fellow of the Society of Antiquaries of London

FSALA Fellow of the South African Library Association

f.s.a.r *fiat secundum artem regulas* (Latin—let it be prepared according to the rules of the art)

FSAM Flying Swiss Ambulance Maldives; Free South Africa Movement

FSAR Final Safety Analysis Report

FSAS Fellow of the Society of Antiquaries of Scotland

FSAScot Fellow of the Society of Arts of Scotland

FSASM Fellow of the South African School of Mines

fsb forward space block

f.s.b. (FSb) federal savings bank

FSB Federal Specifications Board; Federal Supplemental Benefits (program); Field Selection Board; Final Staging Base; Floating Supply Base

FSBA Florida School Boards Association

FSBC *Ferrocarril Sonora Baja California* (Spanish—Sonora—Baja California Railway)

fsbl feasible

fsbly feasibility

fsbo for sale by owner

fse foreign service credit

fse (FSC) fast strike craft

FSC Family Services Bureau; Federal Safety Council; Federal Salary Commission; Federal Stock Catalog; Federal Stock Code; Federal Supplemental Compensation; Federal Supply Classification; Federal Supply Code; Federal Supreme Court; Fiji Sugar Corporation; Five Star Corporation; Flight Service Center; Flying Status Code; Food Standards Committee; Foreign Sales Corporation; Foreign Service Credits; Foundation for Student Communication; Foundation for the Study of Cycles; Friends Service Council; Frostburg State College

FSC *Federal Supply Catalog*

FSCA Fellow of the Society of Company and Commercial Accountants

f/scap foolscap

FSCC Federal Surplus Commodities Corporation; Ferrous Scrap Consumers Coalition; Fire Support Coordination Center; Food Surplus Commodities Corporation

fsce fire-support coordination element

FS Cen Flight Service Center

fscl fire-support coordination line

FSCM Federal Supply Code for Manufacturers

fscp foolscap

FSCS Fire Support Coordination Section; Flight Service Communications System

fsd flying spot digitizer; foreign sea duty; full-scale deflection; full-scale development; functional sequence diagram

fsd (FSD) focus skin distance

FSD Federal Systems Division; Flight Service Director; Fuel Supply Depot; Sioux Falls, South Dakota (airport)

FSDC Fellow of the Society of Dyers and Colourists

fse field-support equipment, forward support element

FSE Federation of Stock Exchanges; Fellow of the Society of Engineers; Field Service Engineer

FSEA Food Service Executives Association

FSEC Florida Solar Energy Center

fsed full scale engineering development

FSER Field Service Engineering Report

FSERI Federal Solar Energy Research Institute

FSES Federal-State Employment Service

fsf forward space file

FSF Fleet Servicing Facility; Flight Safety Foundation; Forensic Sciences Foundation; Free Software Foundation

FSFA Federation of Specialized Film Associations

fsg first-stage graphitization

FSG Federal Supply Group; Fellow of the Society of Genealogists; Friends School Group

FS & G Farrar, Straus & Giroux

FSGB Foreign Service Grievance Board

FSgt Flight Sergeant

FSGT Fellow of the Society of Glass Technology

fsh (FSH) follicle-stimulating hormone

FSHM Fellow of the Society of Housing Managers

fshrf (FSHRF) follicle-stimulating hormone releasing factor

fshrh (FSHRH) follicle-stimulating hormone releasing hormone

FSHS Friendly Societies Health Services

fsh stk fish steak

fsi free-standing insert; foam stability index

FSI Fastbreak Syndicate Incorporated; Federal Stock Item; Fellow of the Sanitary Institute; Fellow of the Surveyors' Institution; Foreign Service Institute; Foundation Sciences Inc; Free Sons of Israel

FSIA Fellow of the Society of Industrial Artists

FSIC Federal Savings Insurance Corporation; Foreign Service Inspection Corps (US Department of State)

FSIO Foreign Service Information Office(r)

FSIP Federal Services Impasses Panel

FSIS Food Safety and Inspection Service (USDA)

FSJC Fort Smith Junior College

fsk frequency shift keying

FSK Fatigue Scales Kit

fsklf frequency shift keying low frequency

fskof for the sake of

fsl fire services levy; formal semantic language; frequency-selective limiter

FSL First Sea Lord; Folger Shakespeare Library; Food Science Laboratory (USA)

FSLA Federal Savings and Loan Association

FSLAC Federal Savings and Loan Advisory Council

FSLAs Federal Savings and Loan Associations

FSLIC Federal Savings and Loan Insurance Corporation

FSLN *Frente Sandinista de Liberación Nacional* (Spanish—Sandinista National Liberation Front)—Nicaragua

fslracc for seller's account

fsm field-strength meter; flying-spot microscope

FSM Federated States of Micronesia; Fiji School of Medicine; Fort Smith, Arkansas (airport); Free Speech Movement

FSM *Fédération Syndicale Mondiale* (French World Federation of Trade Unions)

FSMA Friendly Societies Medical Association; Full Service Maintenance Agreement

FSMB Federation of State Medical Boards

FSMC Flora Stone Mather College

FS Method Federal Standard Method

FSMI Food Service Marketing Institute

fsmtc full-size moving target carrier

FSMWO Field Service Modification Work Order

FSN Federal Stock Number

FSNA Fellow of the Society of Naval Architects

FSNC Federal Steam Navigation Company

FSNM Fort Sumter National Monument

FSNP Fuyot Spring National Park (Philippines)

FSNWR Fish Springs National Wildlife Refuge (Utah)

FSNY Free Synagogue of New York

fso field service operation(s)

FSO Field Security Office(r); Fleet Signals Officer; Flint Symphony Orchestra; Florida Symphony Orchestra; Flying Safety Officer; Foreign Safety Officer; Foreign Service Office(r); Fuel Supply Office(r)

FSOs Foreign Service Officers

FSOTS Foreign Service Officers Training School

fsp fiber saturation point; fibrin split products; flat salary payroll; foreign service pay

FSP Family Survival Project; Field Security Police; Food Stamp Program; Freedom Socialist Party; Freeway Service Patrol

FSPB Field Service Pocket Book; Forward Support Patrol Base
FSPT Federation of Societies for Paint Technology
fs & q functions, standards, and qualifications
F Sq Flying Squadron
FSQS Food Safety and Quality Service
fsr flight safety research; free of strikes and riots; full-scale repository
FSR Fellow of the Society of Radiographers; Field Service Report; Field Service Representative; Foreign Service Reserve
FSRA Federal Sewage Research Association
FSRS Frequency Selective Receiver System
fss finite solution set (mathematics)
FSS Federal Supply Schedule; Federal Supply Service; Fellow of the Statistical Society; Field Support System; Fire Support Station; Flight Service Station; Flight Standard Service; Forward Scatter System; Friday(s), Saturday(s), Sunday(s)
FSSA Fire Suppression Systems Association
FSSC Federal Standard Stock Catalog; Foreign Student Service Council
FSSCT Forer Structured Sentence Completion Test
fssd foreign service selection date
fssp fuel system supply point
FSSP Friendly Sons of Saint Patrick
FSSS Fuel Set Subsystem
fsst flying spot-scanner tube
FSSU Federated Superannuation Scheme of Universities
fsswt full-scale subsonic wind tunnel
fst forged steel; full-scale tunnel
Fst *Funkstation* (German—radio station)
FST Follow-on Soviet Tank; Franciscan School of Theology; Free Southern Theater
FSTA Food Science and Technology Abstracts
fstacoe fleet special test and checkout equipment

FSTC Farmington State Teachers College; Fayatteville State Teachers College
FS & TC Foreign Science and Technology Center (US Army)
FSTD Fellow of the Society of Typographic Designers
F'sted Frederiksted, St Croix
FSTL Future Strategic Target List
FSTMB *Federación Sindical de Trabajadores Mineros de Bolivia* (Spanish—Syndicalist Federation of Working Miners of Bolivia)
f-stop camera diaphragm setting of an f-number stop
FSTPP Foreign Service Team Preceptorship Program
fsts fuze set test set
FSTS Federal Secure Telephone Service
FSTWP Fellow of the Society of Technical Writers and Publishers
Fsy firstly
fsu freak student union
FSU Family Service Unit; Florida State University; Friends of the Soviet Union
FSuH Fridays, Sundays, Holidays
F Supp *Federal Supplement*
fsv final-stage vehicle
fsv *for så vidt* (Dano-Norwegian—as far as)
FSVA Fellow of the Society of Valuers and Auctioneers
fsw feet of sea water; final status word(ing); flexible steel wire (cable); forward swept wing
FSWA Federation of Sewage Works Associations
F & SWMA Fine and Specialty Wire Manufacturers Association
fswr flexible steel wire rope
fswt free-surface water tunnel
FSX Fighter-Support Experimental (advanced F-16 fighter airplane)
ft feet; firing table; fixed tannin; flat; flat top; flush threshold; foot; formal training; free of tax(ation); free throw; free trade; frequent traveller; full-terms; fumetight; full time
f-t follow through
f/t freight ton; full time
f & t fire and theft

ft. *fiat* (Latin—let it be made)
ft. fort
Ft Fort; forint (Hungarian currency unit)
Ft *Folyoirat* (Hungarian—journal, review)
FT Field Test; Flying Test; Flying Tiger Lines (2-letter coding); Functional Test(ing)
FT *Financial Times* (London); *Freethought Today*
ft² square feet; square foot
ft³ cubic feet; cubic foot
ft³/min cubic feet per minute
ft³/s cubic feet per second
FT4 Free Thyroxine T4
fta failure to appear (in court); fatigue test(ing) article; film training aid; fluorescent treponemal antibody; fullthrottle altitude
ftA fuck the Army (to hell with the rules)
FTA Federal Transit Administration; Finnish Travel Association; Flexographic Technical Association; Free Trade Agreement (Canada-U.S;); Free Trade Area; Free Trade Association; Free Transport Association; Future Teachers of America
fta-abs fluorescent treponemal antibody absorption (test for syphilis)
FTAC Foreign Trade Arbitration Commissmon
FTAF Flying Training Air Force
ftam file transfer access and management
FTAT Fluorescent Treponemal Antibody Test
ftb fails to break
FTB fleet torpedo bomber; Forestry and Timber Bureau; Franchise Tax Board; Fukui Television Broadcasting (Japan)
ftbd fit to be detained; full-term born dead
ft black fine thermal black
ftbm foot board measure
ftbrg footbridge
FTBS Free Throwers Boomerang Society
ftc fast time constant; final turn collision
ft c foot-candle
FTC Fair Trade Commission; Farmers Trading Company; Federal Telecommunications Laboratories; Federal Trade

Commission; Fleet Training Center; Flight Test Center; Flying Training Command

FTCA Federal Tort Claims Act

ft. cata. *fiat cataplasma* (Latin—make a poultice)

FTCC Flight Test Coordinating Committee; French Telegraph Cable Company

ft cd foot candela

FTCD Fellow of Trinity College—Dublin

ft. cerat. *fiat ceratum* (Latin—make a cerate)

ft. chart. *fiat chartulae* (Latin—let powders be made)

FTCL Fellow of Trinity College of Music–London

ft col fast color

ft. colly. *fiat collyrium* (Latin—make an eyewash)

ftcolovprt fast color overprint

ftd fails to drain; flight test(ing) direction

FTD Field Training Detachment; Florists' Telegraph Delivery; Foreign Technology Division; Fuel Testing Department

FTDA Fellow of the Theatrical Designers and Craftsmens Association; Florists' Transworld Delivery Association (formerly Florists' Telegraph Delivery)

FTDC Fellow of the Society of Typographic Designers of Canada

ft di flattening die

ftdr friction-top drum

fte fracture transition elastic; full-time equivalence; full-time equivalent

FtE Free the Eagle

ftee full-time equivalency enrollment

ft. emuls. *fiat emulsio* (Latin—make an emulsion)

ft. enem. *fiat enema* (Latin—make an enema)

FTESA Foundry Trades Equipment and Supplies Association

F test Fisher Test (forestry)

ftet fulf-time equivalent terminals

ftf face to face

FTF Failure To File

FTF *Flygtekniska Forsoksantalten* (Swedish—Aeronautical Research Institute of Sweden)

ftfet four-terminal field-effect transistor

f-t fibers fast-switch muscle-cell fibers

ftg fitting; footing

FTG Fleet Training Group (USN); Fuji Texaco Gas

ft. garg. *fiat gargarisma* (Latin—make a gargle)

ftgfop finest tippy golden flowery orange pekoe

FTGSVC Fleet Training Group Services

ft hd flathead

fth(m) fathom

fthp flowing tubing head pressure (oil well)

ft/hr feet per hour

fti federal tax included; fixed time indicator; fixed time interval; free thyroxine index; frequency time indicator; frequency time intensity

FTI Facing Tile Institute; Federal Tax Included; Fellow of the Textile Institute; Functional Test(ing) Instruction(s)

FTIG Fort Indiantown Gap (USA)

FTII Fellow of the Taxation Institute Incoporated

FTIMA Federal Tobacco Inspectors Mutual Association

FT Index *Financial Times Index*

ft. infus. *fiat infusum* (Latin—make an infusion)

ft. injec. *fiat injectio* (Latin—make an injection)

ftir functional terminal innervation ratio

Ft-ir Fourier transform-infra-red spectroscopy

ftit fan turbine inlet temperature

FTIT Fellow of the Institute of Taxation

ftk forward track kill

ftka failed to keep appointment

ftl faster than light

ft l foot-lambert

FTL Federal Telecommunications Laboratory; Flight Test Letter; Flying Tiger Line

ft lb foot pound

ft-Lb foot-lambert

ft-lbf foot-pound force

Ftle Fremantle

ft. linim. *fiat linimentum* (Latin—make a liniment)

FTLS Formal Top-Level Specification

ftm flat tension mask; fractional test meal; functional testing machine(ry)

FTM Flight Test(ing) Manual; Flight Test(ing) Model; Flying Training Manual

FTM *Federación de Trabajadores de México* (Spanish—Federation of Mexican Workers)

FTMA Federation of Textile Manufacturers Associations

ft. mas. *fiat massa* (Latin—make a mass)

ft. mus. dlv. in pil. *fiat massa dividenda in pilulas* (Latin—make a mass and divide into pills)

ft md flattening mandrel

ft/min feet (foot) per minute

ft. mist. *fiat mistura* (Latin—make a mixture)

ftn fortification

Ftn Fountain (postal abbreviation); Freetown (maritime abbreviation)

FTN Facsimile Transmission Network; Floculus Target Neuron (a single nerve cell)

ftnd full-term normal delivery

fto flexible time off

f{}^{to} *firmato* (Italian—signed)

FTO Field Test(ing) Operations; Field Training Officer (police); Fleet Torpedo Officer; Fleet Training Officer; Fleet Training Officer (naval); Franciscan Third Order

FTOs Field Training Officers

F-town Frenchtown; Funtown

ftp field terminal platform (oil well); final-turn pursuit (aircraft); folded, trimmed, and packed (books); full-time personnel (civil service)

FTP Failure To Pay; Field Test Plan; File Transfer Protocol; Fleet Training Publication; Flight Test Program; Functional Test(ing) Procedure

FTP *Francs Tireurs Partisans* (French—Partisan Sharpshooters)—communists in the anti-Nazi underground of France during World War II

FTPAA Film and Television Production Association of Australia

ft-pdl foot poundal

ft/pf foot-pound force

ft. pil. *fiat pilulae* (Latin—make pills)

ft/pnl fuel-tanking panel

ftpo for testing purposes only

FTPR *Federación del Trabajo de Puerto Rico* (Spanish—Federation of Labor of Puerto Rico)

FTPS Fellow of the Technical Publishing Society

ft. pulv. *fiat pulvis* (Latin—make a powder)

ftr fighter; fixed-transom; flat-tile roof; fusion test reactor

F Tr flag tower

FTR Final Technical Report; flag tower (chart and map designation); Flight Test Report; Fruehauf (stock exchange symbol); Functional Test Report; Functional Test Request

FTRA Freight Train Riders of America

ftrac full-tracked (vehicle)

FTRF Freedom-to-Read Foundation

ftro fighter operations

FTRO Flight Test Release Order

ftrp fighter plans

fts favorite-track selection

ft/s feet (foot) per second

FTS Federal Telecommunications System; Federal Telephone System; Field Test Support; Flying Traffic Specialist; Flying Training School; Forged Tool Society; Funeral Telegraph Service; Furnishing Trades Society

ft/s² foot per second squared

FTSC Federal Telecommunications Standards Committee

ft. sec foot second

ft. so. *fiat solutio* (Latin—make a solution)

ft. suppos. *fiat suppositorium* (Latin—make a suppository)

ftt field test telescope; formation tester tool (oil well); framed timber trestle; full-time temporary (civil-service employee); functional test tool

FTT Fever Therapy Technician; Five Task Test

fttp full-time temporary personnel

fttr filter

ft & tw combination flat top and typewriter (desk)

ftu field transfer unit; fuel tanking unit

Ftu Freeman time unit

FTU Federation of Trade Unions; Field Torpedo Unit; First Training Unit

FTUC Fiji Trades Union Congress;

ft. ung. *fiat unguentum* (Latin—make an ointment)

FTUR Flight Test Unsatisfactory Report

FTV Flight Test Vehicle; Fukushima Television; Functional Test Verification

ftw free-trade wharf

Ft W Fort Worth

FTX *Federal Tax Code*

Fty Factory

F-type Jungian feeling type

FTZ Foreign Trade Zone; Free Trade Zone

FTZB Foreign Trade Zones Board

fu Farmers Union; feed(ing) unit; frame unprotected (insurance classification)

f/u fine used (postage stamps)

Fu Finsen unit

F-u fuck you (underground slang—very insulting epithet)

FU Fairfield University; Farmers Union; Fisk University; Fordham University; Franklin University; *Freie Universitat* (Berlin Free University); Friends University; Furman University

FUA Farm Underwriters Association; Future Urban Area

FUB *Freie Universität Berlin* (German—Free University, Berlin)

fubar fouled up beyond all recognition

fubb fouled up beyond belief

fue full usable capacity

FUC *Ferrocarriles Unidos de Yucatan* (Spanish—United Railroads of Yucatan)

FUCA Federal Unemployment Compensation Act

fuchsite chrome mica

fucm (FUCM) full-utility cruise missile

fudr fluorodeoxyuridine

FUDR Failure and Usage Data Report

FUE Federated Union of Employers

FUEL Fuel-Users Emergency Line

FUEN Federal Union of European Nationalities

Fuente (Spanish—fountain, source, spring)—short form for such places as Fuente Alamo, Fuente de Cantos, Fuente-Palmera, Fuente Vaqueros. etc.

Fuentes Carlos Fuentes (Mexican novelist)

fuetap (concrete) formed under elevated temperature and pressure

fufo fly under, fly out

FUG-1966 Hungarian-built armored vehicle based on Soviet model

Führer (German—leader)—Hitler's title

FUIB Fire Underwriters Inspection Bureau

Fuji Alberto Fujimori, former president of Peru; Fujinoyama, Fujisan, or Mount Fuji (Japan's highest peak)

Fujita Leonardo Fujita

Fujiyama Europeanized form for Fujisan or Mount Fuji–Japan's highest peak–3775 meters or 12,388 feet above sea level

Ful Fulcran; Fulgence; Fulgencio; Fulke; Fuller; Fullerton; Fulton; Fulvia; Fulvius

FULICO Fidelity Union Life Insurance Company

fulnm full name

fum fuming

FUM Friends United Meeting

fumi fumigant; fumigate; fumigation

fumtu fouled up more than usual

fun funeral; funerary

FUNAI *Fundáçáo Nacional do Indio* (Portuguese—National foundation of the Indian)

funamb funambulation; funambulist (tightrope or tight-wire walker)

func function(al)

funct function; functional; functionally

fund. fundamental; fundamentalism; fundamentalist

fund. *fundador* (Spanish—founder)

FUND International Monetary Fund

fundies fundamentalists

Fundy Bay of Fundy; Fundy National Park on the north shore of the Bay of Fundy in New Brunswick, Canada

fungi. fungicide

Funk Funk & Wagnalls

FUNK Front Uni National du Kampuchea (French—Khmer National United Front)— Cambodia and Khmer forces

Funk & W Funk & Wagnalls

FUNM Fort Union National Monument

FUNNs For Your Nieces and Nephews

F-U-N trio flourine, uranium, nitrogen tests for relative dating

FUNU Force d'Urgence de Nations Unies (French— United Nations Emergency Force)

fuo fever (of) unknown origin

fup fusion point

f/up follow up

FUP Friends United Press; Furman University Press

FUP Frente Unido del Pueblo (Spanish—People's United Front)—Colombia

fuposat follow-up on supply action taken

fur. furlong; further

fur furiant (Czech—Bohemian folk dance)

FUR Follow-up Report

FURG Fundacão Universidade de Rio Grande (Portuguese—Rio Grande University Foundation)

furl. furlough

Für Liech Fürstentum Liechtenstein (German Principality of Liechtenstein)

furlong furrow long (one eighth mile or 220 yards— 201.17 meters), originally the average length of a plowman's furrow

furmr furthermore

furn furnace; furnish(es)(ed) (ing)(ings); furniture

Furn Furnace (postal abbreviation)

furngs furnishings

furnit furniture

furn pts furniture parts

Fur Seals Fur Seal Islands (Alaska's Pribilofs)

Furtwängler Gustav Heinrich Ernst Martin Wilhelm von Furtwängler

fus far ultraviolet spectrometer; firing unit simulator; fuselage; fusing

FUS Feline Urological Syndrome

FUSA Flinders University of South Australia

FUSE Far-Ultraviolet Spectroscopic Explorer; Federation for United Science Education

fusf federal universal service fee

FuSf Fortsetzung und Schluss folgen (German—to be concluded in the next issue)

FUSRAP Formerly Utilized Sites Remedial Action Program

FuSV Fujinami sarcoma virus

fut future

Fut Futura

FUTA Federal Unemployment Tax Act

FUTC Fidelity Union Trust Company

futs firing unit test set

fuv far ultraviolet

FUW Farmers' Union of Wales; Federation of University Women

fv femtovolt; fire vent; flots vacant; flush valve; forward visibility; fuel valve; future value

f-v flow volume

f.v. folio verso (Latin—back of the sheet)—lefthand page

fv. folio verso (Latin—back of the sheet)

FV Falck's Flyvetjeneste (Copenhagen); fishing vessel; Fruit and Vegetable (US Department of Agriculture)

FV-432 British armored personnel carrier called Trojan

FV-1609 advanced model of the FV-432

FVA Fellow of the Valuers Association

F/VA Film/Video Arts

FVB Fiji Visitors Bureau

fvc forced vital capacity

FVCQFRA Fruit and Vegetable Canning and Quick Freezing Research Association

FVDE Fighting Vehicles Design Establishment

f vd & w firearms, venereal disease, and whiskey

FVl Fellow of the Valuers' Institution

FVMMA Floor and Vaccum Machinery Manufacturers Association

FVNM Fort Vancouver National Monument

fvo for valuation only

FVPA Flat Veneer Products Association

FVPRA Fruit and Vegetable Preservation Research Association

fvq full variable quality

FVR Feline Viral Rhinotracheitis

fvrbl favorable

FVRDE Fighting Vehicles Research and Development Establishment

fvs fighting vehicle system

fv's fashion victims

f.vs. fiat venaesectio (Latin— perform a venesection)

FVS Forer Vocational Survey

FVSC Fort Valley State College

fvt family vewing time

fw fielding wins (baseball); fieldworker; fire wall; fixed wing; flash welding; fore-wing; formula weight; fresh water; fresh weight; front wiring

f & w feed and water; feeding and watering

fw Funk & Wagnalls

f/W female White

FW Fairbanks Whitney (stock exchange symbol); Focke-Wulf; Fog Whistle; Fort Worth; Foster Wheeler

F & W Funk and Wagnalls

fwa financial working arrangement; first word address; fluorescent whitening agent

FWA Family Welfare Association; Farm Workers Association; Federal Works Agency; Forest Workers Association; French West Africa; Future Weapons Agency

FWAA Football-Writers Association of America

FWAS Fort Wayne Art School

FWAT Fish and Wildlife Advisory Team (Alaskan)

fwb four-wheel brake; four-wheel braking; free-wheel bicycle; front-wheel bicycle; furnished with bed

FWB Fort Worth Belt (railroad); Free-Will Baptists

fw ball. freshwater ballast

fwc free woman of color (United States—pre-Civil War designation); full weight contents

FWC Farmers Wholesale Cooperative; Federal Warning Center; Foster Wheeler Corporation

FW & C Furners, Withy & Company

FWCC Friends' World Committee for Consultation

FWCI Foundation of the Wall and Ceiling Industry

fwd forward(ing); four-wheel drive; freshwater damage; freshwater draught; front-wheel drive

F W & D Fort Worth & Denver (railroad)

FwdBL forward bomb line

fwdct fresh water drain collecting tank

fwdg forwarding

fwdr forwarder

fwe finished with engines

f-w-e finished with engine(s)

FWeldI Fellow of the Welding Institute

FWFM Federation of Wholesale Fish Merchants

fwg following

FWGE Fort Worth Grain Exchange

fwh flexible working hours; free-wheeling hubs

FWHC Feminist Women's Health Center

FWHF Federation of World Health Foundations

FWI Federation of West Indies; French West Indies; Financial Women International

FWID Federation of Wholesale and Industrial Distributors

fwiw for what it's worth

fwl foilborne waterline

FWL Foreign Workers Levy; Foundation for World Literacy

FWLERD Far West Laboratory for Educational Research and Development

FWO Facilities Work Order; Fleet Wireless Officer

FWOA Fort Worth Opera Association

FWONA Free World Outside North America

fwop furloughed without pay

FWOTSC First Woman on the Supreme Court

fwp filament-wound plastic(s)

FWP Federal Writers' Project

FWP Fish, Wildlife and Parks

FWPCA Federal Water Pollution Control Administration

FWPO Federal Wildlife Permit Office; Fort Wayne Philharmonic Orchestra

FWQA Federal Water Quality Administration; Federal Water Quality Association

fwr full-wave rectifier; full-wave reflector

F-W r Felix-Weil reaction

FWRC Federal Water Resources Council

FWRM Federation of Wire Rope Manufacturers

fws filter wedge spectrometer; flight warning system

FWS Fighter Weapons School; Fish and Wildlife Service

F & WS Fish and Wildlife Service

FWSG Farm Water Supply Grant

FWSO Fort Worth Symphony Orchestra

FWSSUSA Federation of Worker's Singing Societies of the U.S.A.

fwt fair wear and tear; featherweight

FWT Free World Trade

FWTBT Hemingway's *For Whom The Bell Tolls*

fwth flush watertight hatch

FWU Food Workers Union

FWWS Fire-Weather Warning Service

Fwy Freeway

fx extraneous (television) effects; fixed; foreign exchange; foxed; fractured; fractures; frozen section; special effects

fx *for eksempel* (Dano-Norwegian—for example)

Fx fracture (bone)

FX Foreign Exchange

F.X. Francis Xavier

fxd fixed; foxed

fxg fixing

fxle forecastle

fxr fixer(-upper)

fy (FY) fiscal year

Fy Ferry

FY fiscal year; Ferdinand(e) Ysabella

FY2 fuck you too (graffitic defacement)

fya first-year algebra; for your attention

FYC Federal Youth Center; Florida Yacht Club

FYDP First-Year Development Program; Five-Year Defense Plan

fyg for your guidance

fyi for your information

fyig for your information and guidance

fym farmyard manure

Fyod Fyodor Dostoyevski (Russian writer)

fyou fuck you

FYP Five-Year Plan; Four-Year Plan; etc.

FYPB Five-Year Planning Base (USA)

fypi for your personal information

FYPP Five-Year Procurement Program (USA)

fyr for your reference

fyrace for your account

fys *fysik or fysisk* (Dano-Norwegian—physics, physical)

fytd fiscal year to date

FYTP Five-Year Test Program

Fyz Fyzabad

fz freeze; freezing; fuze (ordnance explosive device)

fz *forzando* (Italian—accented strongly)

Fz Fernández; Franz

FZ Franc Zone; Free Zone; French Zone; Frugal Zealot (Amy Dacyczyn, tightwad)

FZA Fellow of the Zoological Academy; Fellow of the Zoological Association

fzdz freezing drizzle

fzfg freezing fog

FZGB Federation of Zoological Gardens of Great Britain and Ireland

FZGBI Fellow of the Zoological Gardens of Great Britain and Ireland

FZIA First Zen Institute of America

fzr freezer

fzra freezing rain

FZS Fellow of the Zoological Society of Scotland

FZSL Fellow of the Zoological Society, London

FZSScot Fellow of the Zoological Society of Scotland

G

g gage; game(s); garage; gateway; gelding; gender; general ability; general intelligence; generator; gilbert; glucose; goal; goalkeeper (field hockey); gold; good; grain; gram; grand; grasslands; gravitational acceleration (symbol); gravity; grease; great; green; grey; gross; ground; grunting; guide; gun; gyromagnetic ratio (symbol); Lande factor (symbol); offensive and defensive guard (football)

g (G) giga-(prefix meaning billion); glucose

g acceleration of gravity (symbol); gloom (gloomy weather symbol)

g/ giro (Spanish—bank check)

G conductance (symbol); control grid (symbol); deflection factor (symbol); dividends and earnings in Canadian dollars (in newspaper stock listings); Fraunhofer line caused by iron (symbol); gap; gate (symbol); gauss; gear; general audiences (all ages admitted); German(ic); Germany; Gibbs function (free energy symbol); giga; gigabyte; glider; go; God (on Masonic emblems); Golf—code for letter G; good; Goodyear; gourde (Haitian unit of currency); government (broadcasting); Grace (steamship line); gravitational constant (symbol); Green Line; Greene Line; Greenwich; Greyhound (bus line); guanine; guineas; gulden (Netherlands guilder); gulf; Gulf oil (stock exchange symbol); Newtonian gravitational constant (symbol); specific gravity (symbol); thousand-dollar bill

G (G) government spending

G *Gade* (Danish—street); *Galica* (Latin—Gaul or Germania); *Gasse* (German—street); *Gata* (Swedish—street); *Gate* (Norwegian—street); *gawa* (Japanese—river, stream)—also shown as *kawa; Gebel* (Arabic—mountain); *Göl* (Turkish—lake); *Golfe* (French—gulf); *Golfo* (Italian or Spanish—gulf); *Gôlfo* (Portuguese—gulf); *Gora* (Russian—hill, mountain); *Góra* (Polish—hill, mountain); *Guba* (Russian-bay); *Gunung* (Indonesian or Malay—mountain)

G-1 Army or Marine Corps personnel section; conductance (symbol); personnel officer

G-2 military intelligence section of Army or Marine Corps; military intelligence officer

G-3 operations and training section of Army or Marine Corps; operations and training officer

G3P glyceraldehyde 3-phosphate

G-4 logistics officer or section of U.S. Army or Marine Corps; undercover anti-terrorist group within the Royal Canadian Mounted Police

G$_4$ dichlorophen (bactericide and fungicide)

G-5 civil affairs section of Army; civil affairs officer

G6P glucose 6-phosphate

G6PD glucose-6-phosphate dehydrogenase

G-6-pdd glucose-6-phosphate dehydrogenase deficiency

G-7 Group of Seven (Britain, Canada, France, Germany, Italy, Japan, United States)—banks; Group of Seven Industrial Nations (Britain, France, Germany, Italy, Japan, Sweden, United States)

G-8 Group of Eight (United States, Japan, Great Britain, Germany, France, Italy, Canada and Russia)

G$_{11}$ hexachlorophene (antibacterial agent)

G-15 Group of Fifteen (Algerian, Argentina, Brazil, Egypt, India, Indonesia, Jamaica, Malaysia, Mexico, Nigeria, Peru, Senegal, Venezuela, Yugoslavia, Zimbabwe)

G-91 Fiat-built single-engine jet fighter-bomber

ga gage; gas amplification; gastric analysis; gauge; general average; general aviation; glide angle; go ahead; goals against; grand aunt; greenhouse annual; ground to air; ground alert; ground attack

g/a general average; ground-to-air

g & a general and administrative

Ga gallium; Georgia; Georgian; Ghana (tribe)

G^a García

GA Gabon (Internet code); Gage Man; Gamblers Anonymous; Garrison Adjutant; Garrison Artillery; Gemmological Association; General Accounting; General Agent; General Assembly (UN); General Assignment; Geographical Association; Geologists Association; Georgia; Georgia (railroad); Glen Alden (stock exchange symbol); Government Actuary; Government Agency; Grant Award; Graphic Arts; Gypsum Association

G-A General Atomic (Division of General Dynamics)

gaa ground-aided acquisition

GAA Gaelic Athletic Association; Gay Activists Alliance Gemmological Association of Australia; General Aviation Association; Gift Association of America; Gravure Association of America

GAA *Glossary of Aeronautical Abbreviations*

GAAC Graphics Arts Advisers Council

gaafr governmental accounting, auditing, and financial reporting

GAAP Generally Accepted Accounting Principles

Ga As gallium arsenide

GAAS Generally Accepted Auditing Standards

GAATV Gemini-Atlas-Agena Target Vehicle

gab gabardine; gabbing; gabble; gable; girth above buttress

Gab Gabon Republic (République Gabonaise); Gabriel

GAB Games and Amusement Board; General Adjustment Bureau; General Arrangements to Borrow

GABA gamma-aminobutyric acid

Gabba Wollongabba, Brisbane

Gabby Gabriel; Gabriella; Gabrielle

Gabe Gabriel

Gabl Gabriel

Gabon Gabonese Republic (West African country)

Gabr Gabriel; Gabriella; Gabrielle

Gabriel Gabriel Garcia Márquez (Latin American novelist)

gac granular-activated carbon; grilled American cheese (sandwich)

GAC General Acceptance Corporation; General Advisory Committee; General Apprenticeship Committee; General Atomic Company; General Average Certificate; Geological Association of Canada; Georgia Association of Colleges; Goodyear Aircraft Corporation; Gulfstream Aerospace Corporation; Gustavus Adolphus College

GACHAL *Gush Herut Liberalim* (Hebrew—Herut-Liberal. Bloc)—right-wing party

g/a con general average contribution

gad glutamic acid decarboxylase

GAD Gases Applications Development; Generalized Anxiety Disorder; Great American Desert

g/a dep general average deposit

Gadis Gaditanas (dancing girls of Cadiz who perfected abdominal dancing to a fine art representing fertility rites and child bearing)

GADNA *Gdud Noar* (Hebrew—Youth Corps)

GADO General Aviation District Office

gadpet graphic data presentation and edit(ing)

Gads Gadsden

GADS Goose Air Defense Sector

Gae Gaelic

GAE General American English; Georgia Association of Educators

GAEC Goodyear Aircraft and Engineering Corporation; Grumman Aircraft Engineering Corporation

gael *gaelisch* (German—Gaelic)

Gael Gaelic

GAER Gay Alliance for Equal Rights

Gaet Gaetano

gaf General Aniline & Film Corporation (trademark); guaranteed airfare

GAF General Aniline & Film; Government Aircraft Factories

GAFB Goodfellow Air Force Base

GAFD Guild of American Funeral Directors

gaffer (motion-picture and tv slang—chief electrician)

gaffer and gammer grandfather and grandmother

GAFLAC General Accident Fire and Life Assurance Corporation

GAFTA Grain and Feed Trade Association

gag. gaging

g/a/g ground-air-ground

GAG Graphic Artists Guild

gag mm gage thickness (in millimeters)

gags glycosaminoglycans

GAHH Good American Helping Hands

gai guaranteed annual income; guided affected imagery

GAI Government Affairs Institute

gaia go and inspect aircraft

GAIA Graphic Arts Information Association

GAIF General Assembly of International Federations

Gail Abigail

GAIN Greater Avenues for Independence (from welfare)

GAINS Growth And Income Securities

GAIS Georgia Association of Independent Schools

GAIU Graphic Arts International Union

GAJ Guild of Agricultural Journalists

GAJC Georgia Association of Junior Colleges

GAK Garlock (stock-exchange symbol)

Gaki Akutagawa Ryunosuke

Gaku Univ Gakushuin University

gal galileo (unit of acceleration); gallon (unit of capacity); generic array logic

Gal Epistle to the Galatians; Galacia(ns); Galatians; Galveston; Galway

Gal Galatians; Général (French—General)

GAL Gdynia America Line; General Assembly Library (Wellington, NZ); Guggenheim Aeronautical Laboratory; Guild of American Luthiers; Guinea Airways

G A & L General Aircraft and Leasing (Division of General Dynamics Corporation)

GALA Gay Atheists League of America

Galap Galápagos Islands

GALAP Graphic Arts Literacy Action Program

Galápagos Galápagos Islands or Galápagos tortoise(s)

galaxy. general automatic luminosity and x y (measuring machine)

gal cap gallon capacity

GALCIT Guggenheim Aeronautical Laboratory, California Institute of Technology

Gal Col Gallaudet College

Gale Gale Research Company

galena lead suflide

Galileo Galileo Galilei

gall. gallery

gall gallego (German—Galician)—Spanish provincial language and people

Gall Galleria (Italian—gallery or tunnel)

Galliforms Galliformes (grouse, quail, turkeys)

Gallifs Galliformes (grouse, quail, turkeys)

Gall C Gallaudet College

Gallo-Rom Gallo-Romance

Gall U Gall det University

Gallup Dr George Horace Gallup of Gallup Poll fame

GALOP Gay London Police Monitoring Group

gal per min gallons per minute

gals gallons

gals (GALS) generalized assembly-line simulator, geographic adjustment by least squares

gal(s) girl(s)

galt gut-associated lymphoid tissue

Galtees southern Ireland's Galty Mountains

galumphing galloping and triumphing

galv galvanic; galvanism; galvanize(d); galvanometer

Galv Galveston

Galv i galvanized iron

galvnd galvannealed

galvo(s) galvanometer(s)

Galvy Galveston

Galw Galway

gam gammon (sailor's gossip, seamen's talkfest); gamut; guided-aircraft missile

gam (GAM) ground-to-air missile

gam gamos (Greek—marriage)—gamete(s)

Gam Gamaliel; Gambia (whose capital is Banjul)

GAM Guest Aerovías Mexico; Guided-Aircraft Missile

GAMA Game Manufacturers Association; General Agents and Managers Association; Gas Appliance Manufacturers Association; General Aviation Manufacturers Association; Graphic Arts Marketing Associates

GAMAA Guitar and Accessories Manufacturers Association of America

gamba viola da gamba

Gambia Republic of The Gambia (West African coastal country)

Gambiers Gambier Islands in the South Pacific

Gamblers Anon Gamblers Anonymous

gamblin' gambling

Gamboa Penitentiary (close to the midsection of the Panama Canal)

GAMC General Agents and Managers Conference

Gamerceo Gallup American Coal Company—coal-mining town near Gallup, New Mexico

GAMES General Architecture for Medical Expert Systems

GAMET Gyro Accelerometer Misalignment Erection Test

GAMIS Graphic Arts Marketing Information Service

Gamla Stan (Swedish—Old Town)—old Stockholm

gamm gimbal angle matching monitor

GAMM Gesellschaft für Angewante Mathematik und Mechanik (German—Association for Applied Mathematics and Mechanics)

GAMMA Guns and Magnetic Material Alarm (anti-hijacking device)

Gam Pra Bang Gama Praj tantrï Bangladesh (Bengali—People's Republic of Bangladesh)

GAMTA General Aviation Manufacturers and Traders Association

gan generating and analyzing networks

gan (GAN) gyrocompass automatic navigation

gan ganado (Spanish—cattle; livestock)

GAN Generalized Activity Network

Gandhi *Mahatma* (Hindustani Great Souled)—Mohandas Karamchand Gandhi

Gandhi's Mahatma Gandhi's Birthday (October 2)

GANE General Agreement on a New Economy

ganefo games of new emerging forces

gang. ganglia; ganglion

Ganges Ganges—Brahmaputra Delta, sacred waters of the Ganges, India

GANZ Gas Association of New Zealand

ganzl gänzlich (German—complete, entire)

gao general alert order

GAO General Accounting Office; General Administrative Order; General American Oil (company); General American Overseas (corporation); General Auditing Office; Government Accounting Office

GAO Glavnaya Astronomicheskaya Observatoriya (Russian—Main Astronomical Observatory)

gaof gummed all over flap

GAOR General Assembly Official Records (UN)

gap. gallium arsenic phosphorus; guidance autopilot

gap. (GAP) gross agricultural product

Gap The Gap-Delaware Water Gap between New Jersey and Pennsylvania on the upper Delaware River or Semangko Gap north of Kuala Lumpur in Malaysia; Pennington Gap

GAP General Assembly Program; Global Action Plan; Government Aircraft Plant; Grandparents As Parents; Great American Public; Great

Atlantic & Pacific (Tea Company); Group for the Advancement of Psychiatry

GAP *Grupo de Amigos Personales* (Spanish—Group of Personal Friends)—former Chilean President Allende's bodyguard; *Gruppo d'Azione Partagiana* (Italian—Partisan Action Group)

gapa ground-to-air pilotless aircraft

Ga-Pac Georgia-Pacific

GAPAN Guild of Air Pilots and Air Navigators

GAPCE General Assembly of the Presbyterian Church of England

GAPEFA Graphic Arts Platemaking Employers Federation of Australia

GAPL Group Assembly Parts List

gapo giant armpit odor; gorilla armpit odor

gapp growth and profit plan(ing)

GAPR Grant Application Request

GAPs Geographic Applications Programs

gapsfas graduate and professional students financial statement

gapt graphical automatically programmed tools

gaq general average quality; good average quality

gar garage; garfish; garpike; garrison; gavial; gharial; ground avoidance radar; guided aircraft rocket

gar (GAR) gross annual return; growth analysis and review; guided air(craft) rocket

GAR Gioacchino Antonio Rossini; Grand Army of the Republic; Guided Aircraft Rocket; Gustavus Adolphus Rex (King Gustav II of Sweden)

Gara Garamond

garade gathers, alarms, reports, displays, and evaluates

garb. garbage; green, amber, red, blue (airway priority color code)

GARB Garment and Allied Industries Requirements Board

gar-barge garbage barge

garbd garboard

garbol garbologic(al)(1y); garbologist(ic)(al)(1y); garbology (archeological study of man's discards)

garbologist garbage collector

garbz *garbanzos* (Spanish-chickpeas)

GARC Graphic Arts Research Center

G. Arch. Graduate in Architecture

Garcia Diego Garcia (Anglo-American naval base in the Chagos Archipelago or Oil Islands of the Indian Ocean)

gard gamma atomic radiation detector; garden; gardener; gardening; general address reading device; guard

GARD Gamma Atomic Radiation Detector

gards gardenias

gard(s) garden(s)

GA REs *General Assembly Resolution* (UN)

garg. *gargarisma* (Latin—gargle)

garg(s) gargoyle(s)

Garifuna Black Carib (spoken in Belize)

garioa government and relief in occupied areas

Garmo Garmo Peak (formerly Stalin Peak and the highest in the USSR)—also called Communism Peak

garp growth at a reasonable rate

GARP Global Atmospheric Research Program

garr garrote

G.A.R.S. Gustavus Adolphus Rex Sueciae (Gustavus Adolphus King of Sweden)

gar str garboard strake

Gart Garrett

GARUDA Garuda Indonesia Airways

Gary Gareth; Garvey

gas gasoline; general adaptation syndrome

ga & s general average and salvage

g-a s general-adaptation syndrome

GAs Gamblers Anonymous

GAS Georgia Academy of Science; Ghana Academy of Sciences; Government of American Samoa; Grant's Acronymical Shorthand *(see* JERK); Great American

Smokeout (campaign to get smokers to kick the habit); Guild of All Souls

GASAA Graphic Arts Services Association of Australia

gasahol nine parts gasoline and one part alcohol (fuel extender mixture)

GASB Governmental Accounting Standards Board

GASBO Georgia Association of School Business Officials

GASC German-American Securities Corporation

GASCO General Aviation Safety Commission; General Aviation Safety Committee

GASDA Gasoline and Automotive Service Dealers Association

gasdyn gas dynamic; gas dynamicist; gas dynamics

GaSEA Georgia Society of Enrolled Agents

gaser gamma-ra laser

GASF Graphic Arts Sales Foundation

GAS/GR Gas Service

GASH guanidine aluminum sulfate hexahydrate

gasid gas-acid (indigestion)

GASL General Applied Science Laboratories

GA S & L Great American Savings & Loan

gaso gasoline

gasoff gasoline ripoff

gasohol Brazilian gasoline made from alcohol; gasoline + alcohol (fuel-extender fluid)

gasp. gravity-assisted space probe

Gasp Gaspar(o)

GASP General Aviation Safety Panel; Global Air Sampling Program; Greater (Washington, D.C.) Alliance to Stop Pollution (air and water); Group Against Smog and Pollution; Group Against Smokers' Pollution; Group Against Smoking Pollution

Gaspar Jasper

gasphyxiation gas + asphyxiation (death by gas)

GASS Gimbal Assembly Storage System

GASSAR General Atomic Standard Safety Analysis Report

gast gastric

Gast Gaston

gastro gastronomy

gastro (Latin prefix—relating to the stomach)—gastrointestinal

gastroc gastrocnemius

gastroenterol gastroenterology

gastrol gastrolithiasis; gastrolith(ic)(al); gastrolithogroph(y); gastrologist; gastrology

gat gatling gun; ground air transmitter; gun; revolver

gat (GAT) generalized algebraic translator

g-at. gram-atom

gat gata (Swedish—Street); (Dano-Norwegian—channel)

GAT Georgetown Automatic Translation; Greenwich Apparent Time

GATA Graphic Arts Technical Association

gatac general assessment tridimensional analog computer

GATAP Generally Accepted Tax-Accounting Principles

GATB General Aptitude Test Battery

GATCO Guild of Air Traffic Control Officers

Gate The Gate (harbor entrance such as the Golden Gate at San Francisco, the Lion's Gate at Vancouver, the Silver Gate at San Diego)

GATE Gifted and Talented Education program; Group to Advance Total Energy (American Gas Association)

GATF Graphic Arts Technical Foundation

'gator(s) alligator(s)

gatri gamma technology research irradiator

GATS Guidance Acceptance Test Set

GATT General Agreement on Tariffs and Trade

Gatti Guilio Gatti-Casazza

Gatun Girls Gatun Prison for Women and Juveniles at Gatun, Panama

g at. wt gram atomic weight

GATX General American Transportation Corporation (tank car marking)

GAU Gay Academic Union

GAUFCC General Assembly of Unitarian and Free Christian churches

Gaul. Gaulish

Gautama Siddhartha Gautama, or Buddha (founder of Buddhism)

gav gavage(r); gavel; gavelkind; gavial; gavotte; gross annual value

g/ay general average

GAVRS gyrocompass attitude vertical reference system

gav(s) gavial(s); gavotte(s)

gaw guaranteed annual wage

gawam great American wife and mother

gawr gross axle weight rating

Gay Gaylord

gaylib gay liberation(ist)

Gay Lib Gay Liberation Movement

gayola homosexual payola (forced payments made by homosexual establishments to crime syndicates offering them protection)

gaz gazette; gazetteer

GAZ (Russian—*Gorki Avtomobilnii Zavod*)—Gorki Automobile Factory producing the Volga sedan-type auto

gb gall bladder; games behind; generator breaker; glide bomb; goodbye; grid bearing; gun bed

gb (GB) code name for sarin (high-toxicity warfare chemical)—$C_4H_{10}FO_2P$

g-b goof-ball (barbiturate pill)

g/b ground based

gB greenish blue

Gb gigabyte; gilbert

GB Gas Board; General Board; General Bronze (corporation); Georges Bizet; Girls' Brigade; Great Basin; Great Books; Great Britain; gunboat (naval symbol)

GB *Gran Bretaña* (Spanish—Great Britain)

gba give better address

GBAD Great Britain Allied and Dominion

gbb glossopharyngeal breathing

GBBA Glass Bottle Blowers Association

GBBCS Ground Base Beam Control System

GBC General Binding Corporation; Gibraltar Broadcasting Company; Green Belt Council of Greater London; Greenland Base Command

GB & C General Battery and Ceramic (corporation)

GB COLL George Brown College

gbd grain boundary dislocation

gb'd goofballed (underground slang—drugged)

GBDC Grand Bahama Development Company

g-b-d-f-a (musical mnemonic—good boys do fine always)—bass clef note names of the five lines (g-b-d-f-a)

GBDO Guild of British Dispensing Opticians

gbe gilt bevelled edge; ginkgo biloba extract

G.B.E. Dame or Knight of the Grand Cross of the British Empire

gbef gallbladder ejection fraction

GBF *Gakujitsu Bunken Fukyukai* (Japanese Society of Scientific Documentation and Information); Great Books Foundation

GBG General Baking (Stock exchange symbol); Golden Bay Group

gb gas US Army symbol for a colorless and odorless nerve gas of extreme lethality also referred to as general biological gas, goodbye gas, or gruesome business gas

GBGB Gaming Board for Great Britain

gbh girth breast height; grievous bodily harm

gbh (GBH) gamma benzene hydrochloride

GBHC Governor Bacon Health Center

gbi great bodily injury

GBI Georgia Bureau of Investigation; Grand Bahama Island (tracking station)

GB & I Great Britain and Ireland

gbiu geoballistic input unit

g/bl government bill of lading

GBL Georgian Bay Line; government bill of lading

gbm gliblastoma multiforme; glomerular basement membrane; good bowel movement

GBMA Golf Ball Manufacturers Association

GBMC Greater Baltimore Medical Center

GBMD Global Ballistic Missile Defense

GBNE Guild of British Newspaper Editors

GBNM Glacier Bay National Monument

gbo goods in bad order

G-bomb gravitational bomb

gbop golden broken orange pekoe

GBOTA Greyhound Breeders, Owners, and Trainers Association

g/box gearbox

GBp Great Britain pound

GBPA Grand Bahama Port Authority

gbr give better reference; gun, bomb, rocket

gbr gebräuchlich (German—usual)

GBR Great Barrier Reef

GBR Guinness Book of Records

GBRMP Great Barrier Reef Marine Park

GBRMPA Great Barrier Reef Marine Park Authority

gbs gall-bladder series

gb's goofballs (barbiturates)

G-B s Guillain–Barré syndrome

GBS General Business Systems; George Bernard Shaw; Gifu Broadcasting System (Japanese); Guillain-Barré Syndrome; Guyana Broadcasting Service

GBSM Guild of Better Shoe Manufacturers

GBST Grass Block Substitution Test

GBSTC General Beadle State Teachers College

GBSUA Greater Blouse, Skirt and Undergarment Association

gbu glide bomb unit

gbu (GBU) guided bomb unit

GBV Gustahlwerk Bochumer Verein (Krupp Steel)

GBW Guild of Bool Workers

GB & W Green Bay & Western (railroad)

g'bye goodbye

g by pos games by position

gc gas cheek; gas controller; general cargo; general contract(or); geographical coordinates; gigacyle; glucocorticoid; going concern; gonorrhea case; good condition; good conduct; great circle; grid course; ground clear-

ance; ground control(led); guidance control; gun carriage; gun control

g&c guidance and control

Gc great tropic range; gyro-compass

GC Gallaudet College; Gannon College; Garden Council; Gas Council; Gaston College; General Command; Geneva College; Georgetown College; Gettysburg College; Girton College; Glendale College; Gliding Club; glucorticold; Goodard College; Gold Club; Gordon College; Goshen College; Goucher College; Government Chemist; Graceland College; Grambling College; Greensboro College; Greenville College; grid course (symbol); Grinnell College; Grover Cleveland (22nd and 24th President U.S.); Guilford College; Gustave Charpentier

G.C. George Cross; gonorrhea case

G & C Gonville and Caius (Cambridge)

GC Guardian Civil (Spanish—Civil Guard); *Guardia Costa* (Spanish—Coast Guard)

gca group capacity assessment

gca (GCA) ground-controlled approach

GCA Girls' Clubs of America; Government Contract Committee; Graphic Communications Association; Green Coffee Association; Greeting Card Association

GCAA Government Corporations Athletic Association

GCAHS Guggenheim Center for Aviation Health and Safety

g cal gram calorie

G Capt Group Captain

G-car(s) German car(s)

G-Cass Gomes-Cáceres; Gomez-Gáceres

gcb ground circuit breaker

GCB Glen Canyon Bridge

G.C.B. Knight of the Grand Cross, Order of the Bath

GCBA Gold Course Builders of America

GCBS General Council of British Shipping

GCC Grand Canyon College; Ground Control Center; Gulf Coast College; Gulf Cooperation Council (six Arab nations bordering the Persian Gulf)

G & CC Gonville and Caius College-Cambridge

GCCA Graduate Careers Council of Australia; Greater Clothing Contractors Association

GCCC Goshen County Community College

GCCF Governing Council of the Cat Fancy (Great Britain)

GCCS Government Code and Cypher School (nicknamed Government Golf, Cheese, and Chess Society)

gcd general and complete disarmament; great circle distance; greatest common divisor

GCD Grand Coulee Dam

gce ground cooperational equipment

GCE Gas City Empire; General Certificate of Education; General College Entrance (diploma or examination)

GCEC Greater Colombo Economic Community

gcf greatest common factor

GCFI Gulf and Caribbean Fisheries Institute

gcfbr gas-cooled fast breeder reactor

GCFT Gonorrhea Complement Fixation Test

gcg gas-chamber green (institutional paint color)

GCGR Giant's Castle Game Reserve (South Africa)

gch. grandchildren

G.C.H. (Knight) Grand Cross of Hanover

GCHQ Government Communications Headquarters

gci gray cast iron; ground-controlled interception

gci (GCI) gas chromatograph intoximeter (test for drunk drivers)

GCI General Cognitive Index; Grand Canary Island (tracking station); ground controlled interception

GCIA Granite Cutters' International Association

G.C.I.E. Knight Grand Commander of the Order of the Indian Empire

gcip guidance correction input panel

GCIS Ground Control Interception Squadron

gcitng ground-control intercept training

GCIU Graphic Communication International Union

GCJC Gulf Coast Junior College

gcl general control language

gcl. grandchildren

gcl (GCL) ground-controlled landing

GCL Guild of Cleaners and Launderers; Gulf Caribbean Lines

GCL *Guide to Catholic Literature*

G-class Soviet diesel-powered submarines fitted for launching missiles and nicknamed Golf by NATO

G clef treble clef

GCLH Grand Cross of the Legion of Honour

gcm gas-cut mud (oil well); greatest common measure; greatest common multiple; gyrocompass module

GCM General Court-Martial; Gian Carlo Menotti; Good Conduct Medal; Grand Cayman, Cayman Islands (airport)

GCMA Government Contract Management Association

GcmG God calls me God

G.C.M.G. Knight Grand Cross of the Order of Saint Michael and Saint George

GCMI Glass Container Manufacturers Institute

GCMO General Court-Martial Order

gcmps gyro(scope) compass

GCMRU General Control of Mosquitoes Research Unit (India)

gcms gas chromatograph mass spectrometer

GC/MS Gas Chromatography/Mass Spectometry

GCN Greenwich Civil Noon

GCNA Guild of Carillonneurs in North America

GCNM Grand Canyon National Monument

GCNP Grand Canyon National Park (Arizona)

GCNRA Glen Canyon National Recreation Area (Arizona and Utah)

GCO Greater Coin Operators; Guidance Control Officer; Gun Control Officer

gcos general comprehensive operating supervisor

GCOS Great Canadian Oil Sands

GCPL Glasgow Corporation Public Libraries

gcr generator control relay; great circle route

gcr (GCR) gas-cooled graphitemoderated reactor; ground-controlled radar

GCR Geological Characterization Report; Great Central Railway

g crg gun carriage

GCRI Gilette Company Research Institute; Glasshouse Crops Research Institute

GCRO Grand Council and Register of Osteopaths

gcs gate-controlled switch; generator control switch; gram-centimeter-second

gc's genetic girls (real girls)

gc/s gigacycles per second

Gc/s gigacycler per second

GCS Game Conservation Society; General Computer Systems; Georgia Consumer Services; Grant and Contract Service; Group Computer Service(s)

GCSCO Göta Canal Steamship Company

GCSE General Certificate of Secondary Education

Gc/sec gigacycles per second

g-csf granulocyte macrophage colony-stimulating factor

G.C.S.G. Knight Grand Commander of the Order of Saint Gregory the Great

G.C.S.I. Dame or Knight Grand Commander of the Star of India

gct ground-control unit

GCT General Classification Test; Glamorgan College of Technology; Greenwich Civil Time

GCTC Green County Teachers College

gcte guidance computer test equipment

GCTS Ground Communication Tracking System

gcu generator control unit; ground control unit

GCU Glasgow Choral Union

G.C.V.O. Dame or Knight of the Grand Cross of the Victorian Order

gcw gross combination weight (of tractor and loaded trailer)

GCW Gridiron Club of Washington, D.C.

gd general duties; golden dust; good; good delivery; grade; grading, granddaughter; gravimetric density; ground; ground detector; guard; guardian; gum disturbance

g-d god-damned; granddaughter

g/d gallons per day

g & d galvanized and dipped

gd *gade* (Danish—street)

Gd gadolinium

G-d God (Hebraic contraction)

G^d *Grand* (French—big, large, principal)

GD Goal Delivery; General Discharge; General Dispensary; General Dynamics (corporation); Geologic(al) Depository (for spent fuel); George Dewey; Grand Duchy; Grenada (Internet code)

G-D General Dynamics Corporation; Gudermannian or hyperbolic amplitude (symbol); Gunnery Division

G & D Garcia & Diaz (steamship line); Grosset & Dunlap (publisher)

GD *Globe-Democrat*

gda gun-damage assessment; gun-defended area; gunned accelerator

g'day good day (Australia)

GDA General Dynamics Ardmore

GD/A General Dynamics/Astronautics

GDBA Guide Dogs for the Blind Association

GDBMS Generalized Data Base Management System

gdc geneva double curtain (grape trellis); geocentric dust cloud

GDC General Dynamics Convair; *Gesellschaft Deutscher Chemiker* (German—Society of German Chemists)

Gd Ch *Grand Chaeur* (French—full choir or full organ)

GDCL General Dynamics Candair Limited

GD/Convair General Dynamics/Convair

GD/D General Dynamics/ Daingerfield

gddm graphical data display manager

GDDQ Group Dimensions Descriptions Questionnaire

gde gilt deckle edging; gross domestic expenditure

Gde gourde (Haitian monetary unit)

G^{de} Grande (Italian, Portuguese, Spanish—big, large, principal)

GDE General Dynamics Electronics; Graduate Diploma in Extension

GD/EB General Dynamics/ Electric Boat

GDED General Dynamics Electro Dynamic

G de F *Gaz de France* (French—Gas of France)

g del's golden-delicious apples

GDEUT Guidance Digital Evaluation Test

GDFB Guide Dog Foundation for the Blind

GD Fort Worth General Dynamics Fort Worth

GDFW General Dynamics Fort Worth

GDGA General Dynamics General Atomic

gdh growth and development hormone

gdh (GDH) glutamate dehydrogenase

GDHS Ground Data Handling System

gdi god-damned independent (college student failing to join a fraternity or a sorority)

GDI General Dynamics International; Graphics Device Interface (language in Microsoft Windows)

GDIFS Gray and Ductile Iron Founders' Society

GDIS General Dynamics International Service

Gdk Gdansk (Danzig)

gdl ground dynamic laser

Gdl Guadalajara; Guadalajareños (inhabitants)

GDL *Grand-Duche de Luxembourg* (French—Grand Duchy of Luxemburg); Guadalajara, Mexico (airport)

GDLC General Dynamics Liquid Carbonic

gdling good looking

GDLST General Dynamics Low-Speed Tunnel

gdm gestational diabetes mellitus

GDM General Dynamics Manufacturing

gdml gas dynamic mixing laser

GDMM *Grove Dictionary of Music and Musicians*

GDMO General Duty Medical Officer

GDMS General Dynamics Material Service

gdn garden

gdn. guardian

Gdn Gardener; Godown; Guardian

GDNA *Gesellschaft Deutscher Naturforscher und Arzt* (German—Society of German Naturalists and Physicians)

gdnce guidance

gdnf glial cell-line derived neurotrophic factor

Gdnk Gdansk (Danzig)

gdnr gardener

Gdns Gardens

gdo gate-dip oscillator; grid-dip oscillator; gun direction officer

GDO Guild of Dispensing Opticians

gdop geometric dilution of precision

gdp general development plans; graphic display processor; grounded into double play (baseball); guanosine diphosphate; gun director pointer

gdp (GDP) gross domestic product; guanosine diphosphate

GDP General Defense Plan; General Dynamics Pomona; Guanosine diphosphate

GDP(D) General Dynamics Pomona (Daingerfield)

GDPS Global Data Processing System (WMO); Government Document Publishing Service

gdr guard rail

GDR German Democratic Republic

gds. goods; ground demonstration system

Gds Guards

GDS Global Deterioration Scale; Gradual Dosage Schedule; Graphic Data System; Graphic(al) Display System; Greater Danube Society

Gdsk Gdansk (Danzig)

Gdsm Guardsman; Guardsmen

gdsob god-damned son of a bitch

GDSS General Dynamics Space Systems

gdt graphic display terminal

gd & t guidance dimensioning and tolerancing

GDTP Geologic(al) Disposal Technology Program

gdu graphic display unit

GDU Guide Dog Users

gdwnd gradient wind

Gdy Gdynia

ge gas ejection; gastroenterology; greater than or equal to; gilt edge(s); good evening; gross energy; gyroscope error

Ge German; Germanic; germanium; Germany

GE Garrison Engineer; General Election; General Electric; Great Exuma; Group Engineer

GEA Garage Equipment Association; Gravure Engravers Association; Greater East Asia

GEACS Great East Asia Coprosperity Sphere

GE-ANPD General Electric Aircraft Nuclear Propulsion Development

gear. gearing

geb *geboren* (German—born); *gebunden* (German—bound)

Geb *Gebergte* (Afrikaans or Dutch—mountain range); *Gebirge* (German—mountains)

GEB General Education Board; Gerber Products (stock exchange symbol); Grain Elevators Board; Guiding Eyes for the Blind

gebco general bathymetric chart of the oceans

GEBECOMA *Groupement Belge des Constructeurs de Matériel Aérospatial* French—Belgian Group of Aerospace Builders

Gebr *Gebroeders* (Dutch—brothers); *Gebirüder* (German—brothers)

gec *gecartonneerd* (Dutch—bound in boards)

GEC General Electric Company

GECAM *Geréncia de Operações de Cambio* (Portuguese—Exchange Operations Agency)

GECC General Electric Credit Corporation

gecom general(ized) compiler

GECOMIN General Congolese Ore Company

gecref geographic reference (worldwide geographic reference system, also appears as GECREF)

ged gedampft (German—muted)

GED General Education Diploma; General Educational Development (testing service); General Equivalency Diploma; Geologic-(al) Exploration Department; Graduate Equivalency Degree; Ground Environmental Development

Geda Goodyear electronic differential analyzer

GEDCOM Genealogical Data Communication

GEDP General Educational Development Program (USA)

gedr gedrukt (Dutch—printed)

GEDT General Educational Development Test

GEE Generic Environmental Evaluation

GEEC General Egyptian Electricity Corporation

Geech Geechee (dialect spoken along the Ogeechee River, Georgia)

GEEIA Ground Electronics Engineering Installation Agency

geek geomagnetic electrokinetograph

Ge. Eng. Geological Engineer

geep goat + sheep (hybrid)

GEEP General Electric Electronic Evaluator

gef gonadotrophin enhancing factor

GEG Spokane, Washington (airport)

GEGAS General Electric Gas (process)

gegr gegründet (German—founded)

GEHP George Eastman House of Photography (Rochester)

Geh Rat Geheimrat (German—Privy Councillor)

GEI Giovani Esploratori Italiani (Italian—Italian Boy Scouts)

GEIA Generic Environmental Impact Assessment; Ground Equipment Electronics Installations Agency

GEIC Gilbert and Ellice Islands Colony

GEICO Government Employees Insurance Company

GEIS Generic Environmental Impact Statement

Geist Kon Geistliches Konzert (German—sacred concerto)

geistl geistlich (German—spiritual)

gek geomagnetic electrokinetograph

gek gekürzt (German—abbreviated)

GEKTUSA Grand Encampment of the Knights Templar of the United States of America

gel gelatine; gelatinous; gelding

GEL General Electric Laboratory; Great Eastern Line

gelat gelatinous

Geld Gelderland (Dutch province)

GELISH Ground Emitter Location Identification System—High

Gell Aulus Gellius (Roman—grammarian)

gel. quav. gelatina quavis (Latin—in some jelly)

gem. ground-effect machine; guidance evaluation missile

gem. (GEM) growing equity mortgage

Gem Gemini

Gem Gemini (Latin—Twins constellation)

GEM Gas Equipment Manufacturers; General Education Model; Growing Equity Mortgage

GEMA Gymnastic Equipment Manufacturers Association

GEMAC General Electric Measurement and Control

GEMCO Grazing Export Meat Company; Groot Eylandt Mining Company

GEMCOS Generalized Message Control System

gemi grating efficiency measurement instrument

Gemini two-man spacecraft

gemms geophysical exploration manned mobile submersible

GEMMWU General, Electrical, Mechanical, and Municipal Workers' Union

gems growth, economy, management, and customer satisfaction

gem's ground-effect machines

GEMS Geostationary European Meteorological Satellite; Global Environmental Monitoring System; Goodyear Electronic Mapping System; Group Export Marketing Scheme

Gemy General Motors Corporation

gen gender; genealogy; genera; general; generally; generating; generator; generic; genetic(s); genital; genitive; gentian; genus

Gen General; Gennadi; Genoa; Genoese

Gen Genesis

GEN Oslo, Norway (Gardermoen Airport)

gen av general average

Gen Cls Comm General Claims Commission (U.S. and Mexico or U.S. and Panama)

Gend Gendarme (French—Policeman)

genda general data analysis and simulation

Gen Dyn General Dynamics Corporation

Gene Eugene; Eugenia

geneal genealogy

gen eng genetic engineering

General (Portuguese or Spanish—General)—short form for many Latin American places ranging from General Acha in Argentina to General Zuazua in Mexico

genet genetic; geneticist; genetics

Genet Janet Flanner's pen name

gen. et sp. nov. genus et species nova (Latin—new genus and species)

Geneva Girls Illinois State Training School for (delinquent) girls at Geneva; a similar girls training school at Geneva, Nebraska

Gen Hosp General Hospital

genic (Latin suffix—produced from, producing), *carcinogenic* (cancer producing)

GENIP Geographic Education National Implementation Project

GENIRAS Generalized Information Retrieval System

genit genitive

genl general

gen led general ledger

Gen¹ General (Spanish—General)

Gen Mgr General Manager

genn gennaio (Italian—January)

gen. nov. genus novum (Latin—new genus)

genoc genocide

gen prac general practice

gen proc general procedure

gen pub general public

genr generate; generation; generator

Gen Rel Gravit General Relativity and Gravitation

genrl general

Gensek Generalnyi Sekretar (Russian—Secretary General)—leader of the secretariat of the Central Committee of the Communist Party

Gen Supt General Superintendent

gent gentleman

Gent East Flanders' provincial capital in Belgium

Gen Tel & El General Telephone and Electric

gents gentlemen; gentlemen's

Gen-X Generation X

Gen-Xer member of Generation X

genvst general visiting (aboard ships of the USN inviting the public)

GENZL Geothermal Energy New Zealand Limited

geo genetically engineered organism; geocentric; geochemistry; geodesy; geodetic; geodynamics; geognosy; geography; geology; geometry; geophysics; geopolitics; geostatic; geosynchronous equatorial orbit; geothermal (and all their derivatives)

Geo George

Geo I King George I (England, 1714–1727)

Geo II King George II (England, 1727–1760)

Geo III King George III (England, 1760–1820)

Geo IV King George IV (England, 1820–1830)

Geo V King George V (England, 1910–1936)

Geo VI King George VI (England, 1936–1952)

GEO Georgetown, Guyana (Atkinson Field)

GEOC General Estate and Orphan Chamber (trust company)

Geochim Cosmochim Acta Geochimica et Cosmochimica Acta (Latin—Geochemical and Cosmochemical Review)

geod geodesic; geodesist; geodesy; geodetic; geodynamic(s)

Geo Dat Pt Geodetic Datum Point (North America's geodetic datum point is the National Ocean Survey's triangulation station at Meades Ranch in Osborne County, Kansas)

Geod. E. Geodetic Engineer

geodss ground electro-optical deep-space surveillance

Geof Geoffrey; Geoffroy

Geoffrey Jeffrey

geog geographer; geographical; geography

geogr geografi or geografisk (Dano-Norwegian—geography or geographer)

geograph geographical

geohy geohygiene

GEOIS Geographic Information System

geol geologic; geological; geologist

geol geologi or geologisk (Dano-Norwegian—geology or geologist)

Geol Geology

Geol.E. Geological Engineer

Geol Surv Geological Survey

geom geometric; geometry

Geom Geometry

geomed geometric editor

geomorph geomorphologic(al); geomorphologist; geomorphology (land form study)

geon (GEON) gyro-erected optical navigation

geoph geophysics

geophy geophysical; geophysics

geopol geopolitical; geopolitics

geor Georgian

Geordie George; Newcastleon-Tyne, England

Georef World Geographic Reference System

Georg Georgia(n)

George George Herbert Walker Bush—41st President of the United States

Georges one-dollar bills bearing the portrait of President George Washington

Georgie George

Georgies one-dollar bills bearing the portrait of President George Washington

Georgy George

geos generator, earth orbital scene; geodetic earth-orbiting satellite; geodetic orbiting satellite

GEOS Geodetic Orbiting Satellite; Geodynamics Experimental Ocean Satellite

GEOSECS Geochemical Ocean Sections Study

gep gross energy product

GEPAC General Electric Programmable Automatic Comparator

Geph Gephyra

GEPI Gestioni e Partecipazioni Industriali (Italian—Industrial Management and Participation)

GEPURS General Electric General Purpose (Computer)

ger gerund; gerundial; gerundival; gerundive

Ger German; Germanic; Germany (whose capital is Berlin)

GER Great Eastern Railway

gera geratic(al)(1y); geratologic(al)(1y); geratologist; geratology

ger grndng gerund grinding (pedagogic pedantry)

gerd gastroesophageal reflux disease

geriat geriatrics

germ. ground-effect research machine

Germ German(y)

Germ Germinal (French—Seedy Month)—beginning March 21st—seventh month of the French Revolutionary Calendar and also title of a novel by Zola

germi germicide

GERNORSEA German Naval Forces in the North Sea

Geron Geronime

gerontol gerontology

Gerry Gerald; Gerard; Gerhard

Gersis General Electric range safety instrumentation system gert graphical evaluation and review technique

Gert Gertie; Gertrude

Geru Gerusalemme (Italian—Jerusalem)

Ger Yid German Yiddish (Ashkenazi)

ges gesetzlich (German—registered)

Ges (German—G-flat); *Gesellschaft* (German—association, company, society)

GES General Engineering Service; Government Economic Service; Great Eastern Shipping

GE-S Gold Exchange—Singapore

GESAMP Group of Experts on the Scientific Aspects of Marine Pollution

gesch geschützt (German—registered)

Gesch Geschichte (German—history)

GESCO General Electric Supply Corporation

gesp general extrasensory perception

gespeg (Micmac Indian—end of the earth)—Quebec's Gaspé Peninsula

gest gas-explosive simulation technique

gest gestorben (German—dead, deceased)

Gestapo Geheime Staatspolizei (German—State Secret Police)

get. ground-elapsed time; ground-engaging tool(s)

GET Getty Oil (stock exchange symbol); Great Eastern Television (Australian); Gross Error Test(ing) Guaranteed Education Tuition

get 1/2 gastric emptying half-time

GETIS Ground Environment Technical Information System

getlo get locally

getma get from local manufacturer; purchase for local manufacturer

getol ground-effect takeoff and landing

Getty The Getty (J Paul Getty Museum in Santa Monica, California)

GEU Genetic Evaluation and Utilization

gev giga electron volt (10^9 electron volts)

gev (GEV) ground-effect vehicle

GeV billion electron volts; giga electron volt

GEVIC General Electric Variable Increment Computer

gew gram equivalent weight

gew gewoonlijk (Dutch—as a rule, generally, usually)

Gew Gewehr (German—rifle)

gewürz gewürztraminer (wine)

gez gezeichnet (German—signed)

Gez Gezira (Arabic—island)

GEZ Gosudarstvennoe knigoisdatelstvo (Russian—State Publishing House)

gf gap filler; generator field; girl friend; globular fibrous; glomerular filtrate; goldfield; government form; green feed; ground fog; growth factor; growth fraction; guiltfree

g-f globular-fibrous

gf. grandfather

gf games finished; goals for

Gf Gottfried

GF Galois Field; General Fireproofing; General Foods; Georgia & Florida (railroad); Guggenheim Foundation; Guiana (Internet code)

G F Guyane Française (French Guiana)

G & F Georgia & Florida (railroad)

gfa gas-filled ass; good fair average; good freight agent; gunfire area

GFA Game Fishing Association; Gardens For All; Gasket Fabricators Association; Glider Flying Area; Gliding Federation of Australia

GFA Générale Française (de Construction) Automobile

GFAA Game Fishing Association of Australia

g factor general factor

gfae government-furnished aerospace equipment; government furnished aircraft equipment

gfam graphics flutter analysis methods

gfbop golden flowery broken orange pekoe

GFC Gorilla Foundation of California

gfci ground-fault circuit interrupter

GFCM General Fisheries Council for the Mediterranean (FAO)

gfcs gun fire control system

gfd general functional description; government funded development

GFD General Freight Department

GFDL Geophysical Fluid Dynamics Laboratory

gfe government-furnished equipment

gff granolithic finish floor

GFG Good Food Guide

GFH George Frideric Handel

gfi gas-flow indicator; ground-fault interrupter

GFI General Felt Industries

GFIC Georgia Foundation for Independent Colleges

Gfk Gustafsvik

Gfl Genfle (Gävle)

GFL Glossary Function List

gfm government-furnished materiel; government-furnished missile

gfme government-furnished missile equipment

GFMVT General Foods Moisture Vapor Test

GFO General Freight Office

gfop golden orange pekoe

g-force(s) gravity force(s)

G forces acceleration forces

gfp government-furnished parts; government-furnished property

gfr gap-filled radar; glomerular filtration rate

GFR German Federal Republic; Government Facilities Request

gfrc glass-fiber reinforced cement; glass-fiber reinforced concrete

gfrp glass-fiber reinforced plastic

gfs forward transconductance (symbol)

GFS Girls Friendly Society

gfst ground fuel start tank

gft graphical firing table

GFTU General Federation of Trade Unions

gfu glazed facing units

gfut ground fuel ullage tank

GFWC General Federation of Women's Clubs

gfwh gas-fired water heat

gg gamma globulin; gas generator; gender gap; go girls; great gross

g-g ground-to-ground

gg *gange* (Dano-Norwegian—multiply)

Gg Georgian

GG George Gershwin (Jacob Gershvin); Georgia (Internet code); Government Gazette; Governor General

G-G Goodrich-Gulf (chemicals)

G & G Gordon and Gotch

GGA Girl Guides Association; Gulf General Atomic

GGAC Gulf General Atomic Company (formerly General Atomic division of General Dynamics)

Ggb *Gorilla gorilla beringei* (the mountain gorilla)

GGB Golden Gate Bridge

GGB & HD Golden Gate Bridge and Highway District

G & GBR Geologists and Geophysicists Board of Registration

ggc ground guidance computer

GGC Golden Gate College

GGCST Gleb-Goldstein Color Sorting Test

ggd great granddaughter

gge garage; generalized glandular enlargement

ggf great-grandfather; ground gained forward

ggg gadolinium gallium garnet

g.g.g. *gummi guttae gambiae* (Latin—gamboge)—cathartic

Ggg *Gorilla gorilla gorilla* (the lowland gorilla)

gggf. great-great-grandfather

gggm. great-great-grandmother

GGHNP Golden Gate Highlands National Park (South Africa)

GGI Guided Group Interaction

g gl ground-glass

ggm. great-grandmother

ggm (GGM) ground-to-ground missile

GGNRA Golden Gate National Recreation Area (San Francisco)

Ggo Gallego

GGOC Goldovsky Grand Opera Company

g gr great gross

GGR Gambill Goose Refuge (Texas); Ground Gunnery Range

ggrs great grandson

ggs great grandson; gained sideways

g-g's go-go girls

Gg's Ganges gavials; Ganges gharials

GGS Ground Guidance System

GGSM Graduate of the Guildhall School of Music

ggt gamma glutamyl transferase

ggts gravity-gradient test satellite

gh genealogical helper; grid heading; growth hormone; guardhouse

gh (GR) growth hormone

Gh Ghana, Commonwealth of

GH General Hospital; Grosvenor House

GH *Good Housekeeping*

Gha Ghana (whose capital is Accra)

GHA Greenwich Hour Angle

GHAA Group Health Association of America

GhAF Ghanian Air Force

GHAMS Greenwich Hour Angle of the Mean Sun

Ghan Afghan Express

Ghana Republic of Ghana (West African nation), formerly the Gold Coast and British Togoland

GHANA Ghana Airways

ghar *ghariyal* (Hindu—gavial)—long-snouted fisheating crocodilian

Gharb *al-Gharb* (Arabic—the west)—southwest Portugal, Algarve

GHATS Greenwich Hour Angle of the True Sun

G-Hb alyco-hemoglobin

GHB gamma hydroxybutyrate; Good Housekeeping Bureau

ghc guidance heater control

GHC Gray Harbor College; Group Health Cooperative

GHDVHS Grace H Dodge Vocational High School

ghe ground handling equipment

G H & H Galveston, Houston & Henderson (railroad)

GHI Good Housekeeping Institute

GHMC Good Harvest Marine Company

GHMS Graduate in Homeopathic Medicine and Surgery

GhN Ghana Navy

ghost. global horizontal sounding technique

GHP Group Health Program

GHP *German Homeopathic Pharmacopoeia*

g/hphr gallons per horsepower hour

GHQ General Headquarters

g/hr gallons per hour

ghrf growth hormone-releasing factor

ghrh growth hormone releasing hormone

GHS Galileo High School; Girls High School

Ght Ghent

ghu gas hydraulic unit

Ghub *Ghubba* (Arabic—bay, cove)

GHWB George Herbert Walker Bush, 41st President of the United States

ghx ground heat exchanger

GHz gigahertz (gigacycle per second)

gi galvanized iron; gastrointestinal; general issue; generation interval; gill; globulin insulin; government issue; gross income; gross inventory

g-i granuloma inguinale

Gi Giles; Guy; input conductance (symbol)

GI Air Guinée; American Soldier (from *gi*—general issue or government issue); Garden Island; General Infantry; General Intelligence; Gibraltar (Internet code); Gideons International; Gimbel Brothers (stock exchange symbol); Gingival Index; Gold Institute; Government of India

GI *Gesellschaft für Informatik* (German—Society for Data Processing)

gia grant-in-aid

GIA Garuda Indonesian Airways; Gemological Institute of America; Goodwill Industries of America; Gregorian Institute of America; Gummed Industries Association

GIAC Gateway Intercollegiate Athletic Conference

GIAHA Gilcrease Institute of American History and Art (Tulsa)

Gian Gian Carlo Menotti (American composer)

Giants Giants Stadium, East Rutherford, New Jersey

giardiasis diarrhea due to contaminated food or water

gib guy in the back

Gib Gibraltar; Gibraltarian

GIB Gibraltar, British Crown Colony (airport); Gulf International Bank

GIBAIR Gibraltar Airways

Gib(bie) Gilbert

gibb(s) gibbon(s)

Gibfo Gibraltar for orders

GIBMED Gibraltar Mediterranean Command (NATO)

gibs guy in the back seat

gib(s) gibbon(s)—smallest of the anthropoid apes found in the forests of southeast Asia

Gibs Gibraltarians

Gibson Gibson Desert of east-central Western Australia

Gib-tv Gibraltar television

gic ground intercept control

GIC General Investment Corporation; Government Information Center; Guaranteed Income Contract; Guaranteed Investment Certificate (Canada); Guaranteed Investment Contract

GICA Green Island Coral Atoll (Queensland)

gi'd prepared for military-type inspection

Gid Gideon

GID General Intelligence Division; Group Identification

GIDAP Guidance Inertial Data Analysis Program

GIDC Georgia Information Dissemination Center

GIDEP Government-Industry Data Exchange Program

Gide André Gide

gi distress gastro-intestinal distress

gidp grounded into double play

GIDS Generic Intelligent Driver Support System

GIEE Graduate of the Institution of Electrical Engineers

gif (GIF) growth hormone-inhibiting factor

GIF Graphical Interchange Format; Rio de Janeiro, Brazil (Galeo Airport)

GIFAS Groupement des Industries Françaises Aéronautiques et Spatiales

(French—French Aeronautical and Aerospace Industry Association)

gift (GIFT) gamete intra-fallopian transfer

GIFT Glasgow International Freight Terminal

giga 10^9

Gig Harbor Purdy Treatment Center for Women at Gig Harbor, Washington

gigo garbage in, garbage out (computer term)

GII Global Information Infrastructure

GIIS Graduate Institute of International Studies (Geneva)

GIJ Guild of Irish Journalists

Gil Gilbert; Gilchrist; Gilder; Gildo; Gilead; Giles; Gilford; Gilland; Gillespie; Gillian; Gilman; Gilmore; Gilroy

Gila mons Gila monster (*Heloderma suspectum*), found in arid Arizona and New Mexico, and the Mexican species (*Heloderma horridum*), found in northern Mexico and called the beaded lizard

Gilberts Gilbert and Ellice Islands in the Pacific including the Line Islands and Phoenix Islands

Gill(y) Gillian

Gilo Gilberto

gim general information management; gimmick

GI Mech Eng Graduate of the Institution of Mechanical Engineers

gimic guard-ring-implanted monolithic integrated circuit

GIMLCS Generalized Information Management Language and Computer System

gimp general image manipulation program

gimp. gimbal position(ing)

GIMPEX Guyana Import-Export

GIMPS Great Internet Mersenne Prime Search

GIMRADA Geodesy, Intelligence and Mapping Research and Development Agency (US Army)

gin giugno (Italian—June)

Gin Ginebra (Spanish—Geneva)

Gina Genevieve; Virginia

ging gingival; gingivitis

ging. gingiva (Latin—gum)

Ginnie Mae nickname of the Government National Mortgage Association

Ginny Virginia

gins aborigine girls

G Inst T Graduate of the Institute of Transport

GINTRAP (European) Guide to Industrial Trading Regulations and Practice

GI Nue Eng Graduate of the Institution of Nuclear Engineers

Ginza center of downtown Tokyo

gio giovedi (Italian—Thursday)

GIO Government Information Organization; Government Insurance Office; Guild of Insurance Officials

g ion gram ion

Giorgione Giorgio, Barbarelli

giorn giornaliero (Italian—daily); *giornalist* (Italian—journalist)

Giov Giovanna; Giovanni

gip gas in place (oil well); gastric inhibitory polypeptide; get(ting) into publication(s); get(ting) into publishing

GIP Great Indian Peninsular (railway)

GIPE Generation of Interactive Programming Environments

Gipps Gippsland, Victoria, Australia

GIPR Great Indian Peninsula Railway

GIPSY General Information Processing System

giq giant imperial quart (of beer)

gir girder

GIR Gulf Interior Region

giraffe graphic interface for finite elements

GIRB Georgia Inspection and Rating Bureau

gird ground-installed recording data

GIRDHS Ground Installation Reconnaissance Data Handling System

girl generalized information retrieval language

GIRLS Generalized Information Retrieval and Listing System

Girls' Cottage Girls' Cottage School (for delinquents) at Chambly, Québec

Girls' Town correctional facility at Tecumseh, Oklahoma

giro autogiro

GIRU General Intelligence and Reconnaissance Unit (Israel's anti-terrorist commando force is GIRU 269)

gis galvanized iron sheet(ing); gastrointestinal series; geographical information systems

gi's gastrointestinal troubles (diarrhea)

GI'S enlisted men; enlisted soldiers in the US Army

GIS General Mills (stock exchange symbol); Generalized Information System; Geographic Information Systems; Geoscience Information Society; Global Information System; Government Information Service(s)

Gisep Giuseppe

GISP Greenland Ice Sheet Program; Guided Independent Study Program

gi spasm gastrointestinal spasm

GISS Goddard Institute of Space Studies (NASA)

git guitar

git (GIT) group inclusive tour; group insurance tour (travel plan)

GIT General Information Test; Georgia Institute of Technology

Gita Bhagavad-Gita

Gitmo Guantánamo Naval Base (Guantánamo Bay, Cuba)

giu geoballistic input unit

giu giugno (Italian—June)

GIUK Greenland, Iceland, United Kingdom

Gius Giuseppe

giv. given; giving

Givhans Ferry Givhans Ferry State Park (near Charleston, South Carolina)

GIW Gulf Intracoastal Waterway

gj gigajoule (1000-million joules); grapefruit juice

GJB George Jackson Brigade (underground black extremists)

GJC Galdhöppigen Jotunheimen Climbers; Gibbs Junior College; Grand Junction Canal

GJD Grand Junior Deacon; germanium junction diode

GJM Gregor Johann Mendel

Gjn Gijon

GJO Grand Junction Office

Gk Greek

GK Gaol Keeper

GKC Gilbert Keith Chesterton

GKD-notation Gordon-Kendall-Davison notation for chemical formulas (sometimes called Birmingham notation)

Gk I Greek Isles

GKIAE Gossurdarstveinny Komitet po Ispolzovaniyu Atomnoi Energi (Russian—State Committee for the Use of Atomic Energy)

GKN Guest, Keen, and Nettlefold

GK & N Guest, Keen, & Nettleworth

GKS Graphical Kernel System

gkw god knows what

gl gas lifting (oil well); general ledger; glass; glazed; gloss; gold lease; gothic letter; ground level; gun layer; gun license

gl (GL) general liability

g/l grams per liter; guideline

Gl Glagolitic; glucinium

Gl. Gloria in excelsis Deo (Latin—Glory be to God in the highest)

GL Germanischer Lloyd's (German ship classifier); Goldstar Lines; Government Laboratory; Great Lakes (load line mark); Greek line; Greenland (Internet code)

G.L. Graduate in Law

GL Gamle (Swedish—old); *Glacier* (French—glacier, ice field)

gla gingiovolinguo-axial

GLA gamma-linolenic acid; General Laboratory Associates; Georgia Library Association; Glasgow, Scotland (airport)

GLAA Greater London Arts Association

GLAAD Gay and Lesbian Alliance Against Defamation

glab glabrous

glac glacial

GLAC Greek Library Association of Cyprus

Glacier place-name in British Columbia or Montana; Glacier Bay National Monument, Glacier Highway, or Glacier Island in Alaska, Glacier Mountain in Colorado, Glacier National Park in British Columbia and Montana, Glacier Peak in Washington

glaciol glaciologist(ic)(al) (1y); glaciolographic(al)(1y); glaciolography; glaciology

Glad Gladstone; Gladwin; Gladys

GLAD Gay and Lesbian Advocates and Defenders

'Glades Florida's Everglades

glads gladiolas

Glam Glamorganshire

GLAMO Great Lakes Association of Marine Operators

Glamorgan Glamorganshire

gland. glandular

gland. glandula (Latin—gland)

glas glasnost (Russian—openness)

Glas Glasgow; Glaswegian

GLASLA Great Lakes—St Lawrence Association

glasphalt glass + asphalt (paving)

glass silicon dioxide—SiO_2

glass. glassware

glassie glass playing marble

glassteel glass + steel (skyscrapers)

Glaswegian(s) Glasgow person(s)

glau glaucous

glauberite calcium sodium sulfate

glauber's salt sodium sulfate

glauc glaucoma

Glav Red Glavnyi Redaktor (Russian—Editor-in-Chief)

glb glass block

GLB Girls' Life Brigade; Greater London Borough (City of London)

GLBA Great Lakes Booksellers' Association

glbs globes

GLBSA Greater London Building Surveyors Association

glc gas-liquid-chromatographic; generator line contractor; global loran (navigation) chart(s); glucose

glc (GLC) G-force-induced loss of consciousness

GLC Genealogical Library Catalog; Greater London Council; Great Lakes Carbon; Great Lakes Colleges; Great Lakes Commission

glcA gluconic acid

GLCA Great Lakes College Association

glcm (GLCM) ground-launched surface-to-surface cruise missile

GLCM Graduate of the London College of Music

glcN glucosamine

GLCSC Gay and Lesbian Community Service Center

glcUa glucuronic acid

gld gilded; glider; gold; guilder

Gld Gelderland (Dutch province)

Gld Cst Gold Coast

GLDP Greater London Development Plan

gld pltd gold plated

gldr guilder

GLE *Grand Larousse Encyclopedie* (French—Great Larousse Enclyclopedia)

gleep graphite low-energy experimental pile

GLEF Gay and Lesbian Emergency Fund

GLERL Great Lakes Environmental Research Laboratory

Glesca Glasgow

GLF Gay Liberation Front

GLFB Greater London Fund for the Blind

GLFC Great Lakes Fisheries Commission

Glf Mex Gulf of Mexico

Glf Str Gulf Stream

GLHA Great Lakes Harbor Association

gli glider

gli (GLI) glucagon-like immunoreactive factor from gastrointestinal gastro-intestinal mucosa

GLI General Time (stock exchange symbol); Great Lakes Institute (University of Toronto)

glickums ground-launched cruise missiles (GLCMs)

Gliéré Reinhold Moritsevich Gliéré

GLIS Greater London Information Service

glit glittering

glitch unexpected transient

glitz *glitzy* (Yiddish—glitter)—golden glitter

Glitz *Galitzianer* (Yiddish—Galician)—person of Judaic origin from Austrian or Polish Galicia

glld ground laser locator designator

GLLO Great Lakes Licensed Officer's Organization

glm graduated learning method; graduated length method

glm *grand livre du mois* (French—great book of the month)—best-seller

GLM Gay Liberation Movement

GLMI Great Lakes Maritime Institute

gln (GLN) glutamine (amino acid)

Gln Glen (postal abbreviation)

GLNTC Great Lakes Naval Training Center

GLO General Land Office; Goddard Launch Operations (NASA); Ground Liaison Office(r); Gunnery Liaison Office(r)

g LO₂t ground liquid-oxygen tank

glob globular; globule

GLOBE Gay, Lesbian Or Bisexual Employees (union)

globecomm global communications

GLOBECOMS Global Communications System

glock glockenspiel

GLOE Gay and Lesbian Outreach to Elders

glomb glide bomb

glomex global oceanographic and meteorological experiment (GLOMEX)—1975–1980

GLOMR Global Low-Orbiting Relay satellite

GLONASS Global Navigation Satellite System

Gloria Gloria Swanson

Gloria *Gloria al bravo pueblo* (Spanish—Glory to the brave people)—Venezuelan anthem

Glos Gloucestershire

gloss. glossary

gloss (Latin prefix—tongue)—glossary, glossopharyngeal

glossies slick-paper magazines

Gloster Gloucester

Glostr Glostrup

glotrac global tracking

Glou Gloucester(shire)

Gloucestr Gloucester

glow. gross liftoff weight

GLOW Gay and Lesbian Older Way; Gorgeous Ladies of Wrestling

glp general layout plan(ning); general letter packet

GLP Good Laboratory Practice(s); Greater London Plan

GLP *Great Lakes Pilot*

GLPA Gay and Lesbian Press Association; Great Lakes Pilotage Administration

glq greater than lot quantities

glr gas liquid ratio (oil well)

Glr Gloucester

g'ls goals

Gls Glasgow

GLS Georgetown Law School; Graduate Library School; Greene Line Steamers (Mississippi); Gypsy Lore Society

giso group legal services organization

GLSOA Great Lakes Ship Owners Association

glt gilt; greetings letter telegram; guide light

gin glutamic acid

Gluck Alma Gluck (Mrs Efrem Zimbalist)

gluco or glyco (Greek *glykys*—sweet)—glucose, glycerol, glycogen, glycoprotein

glulam(s) glue-laminated wooden beam(s)

glv globe valve

GLV Gemini Launch Vehicle

GLW Corning Glass Works (stock exchange symbol)

glwb glazed wallboard

GLWQB Great Lakes Water Quality Board (Canada-U.S.)

gly glycerine; glycerol glycogen

gly (GLY) glycine (amino acid)

Gly Gulley; Gully

glycerol glycerine—C₃H₅(OH)₃

glyc-pos glycerine suppositories

glykr *glykrrhiza* (Greek—licorice)—herbal root

glyp glyphography; glyptics; glyptography

glyph hieroglyph

glyph(s) hieroglyph(s)

GLZ General Bronze Corporation (stock exchange symbol)

gm general medicine; general mortgage; genetically modified; gold medal; good morning; gram; gross margin; guard mail; guided missile; mutual conductance (symbol); transconductance (symbol)

gm (GM) group mark (data processing); group mark(ing)

g/m gallons per minute

gm. grandmother

GM Gambia (Internet code); General Manager; General Medicine; General Motors; Grand Master; Guided Missile; Gunner's Mate; Gustav Mahler

G-M Geiger-Müller (detector)

G.M. George Medal; Gold Medal

GM metacentric height (symbol)

G & M *Globe and Mail* (Toronto)

gm² grams per square meter (paper weight)

GMA Gallery of Modern Art; Glass Manufacturers Association; Government Modification Authorization; Grocery Manufacturers of America; Grocery Manufacturers of Australia; Growth Management Act

GMAA Gold Mining Association of America

gmac gaining major air-command

GMAC General Motors Acceptance Corporation

GMAD General Motors Assembly Division

GMAIC Guided Missile and Aerospace Intelligence Committee

G-man FBI law-enforcement officer also known as a special agent

GMAS Ground Munitions Analysis Study

GMAT Graduate Management Admissions Test; Greenwich Mean Astronomical Time

GMATS General Motors Air Transport Section

gm-aw gram atomic weight

gmb good merchandise brand

GMB Georg Morris Brandes (originally Cohen)

GMBE Grand Master (of the Order of the) British Empire

GmbH *Gesellschaft mit beschrankter Haftung* (German—incorporated, limited liability company)

gmbl gimbal

gmc guaranteed mortgage certificate; gun motor carriage

Gmc Germanic

GMC General Medical Council; General Motors Corporation; George Mason College; Global Marine Corporation; Guggenheim Memorial Concerts; Guided Missile Committee; Gulf Maritime Company

gm-cal gram-calorie; gram-centimeter

GMCC Geophysical Monitoring for Climatic Change

GMCO General Mathematics Computing Option

g-m counter Geiger-Müller counter for measuring cosmic rays arid radioactivity

GMCS Group Medicare Cooperative Society

gm-csf (GM-CSF) granulocyte-macrophage colony-stimulating factor

gmd (GMD) green-monkey disease

GMDA Golf Manufacturers and Distributors Association

GMDRL General Motors Defense Research Laboratories

GMDSS Global Maritime Distress and Safety System

G-men FBI law-enforcement officers

g met gun-metal

gmf general mail facility

GMF Glass Manufacturers Federation

GMFC General Mining and Finance Corporation

gmfp guided-missile firing panel

Gmh Grangemouth

GM-H General Motors-Holden (Australia)

GMHC Gay Men's Health Crisis

GMI General Motors Institute

GMIA Gelatin Manufacturers Institute of America

gmidg garnish moulding

gmk grand master keyed; green monkey kidney

GMK Gold Mines of Kalgoorlie

GMK *Gomei Kaisha* (Japanese—Mercantile Partnership)

gm/l grams per liter

GML Gold Mining Lease

gmldg garnish molding

gm-m gram-meter

GMMC Godden Memorial Medical Centre (Fiji)

GMNNR Glasson Moss National Nature Reserve (England)

GMNP Guadalupe Mountains National Park (Texas)

GMNZ General Motors New Zealand

gmo genetically modified organism

Gmo *Guillermo* (Spanish—William)

GMO Government Medical Office(r); Guided Missile Office(r)

GM & O Gulf, Mobile & Ohio (railroad)

g mol g molecule

GMOO Guided Missile Operations Office(r)

gmp glycomacropeptide; good manufacturing practices; guaranteed minimum price

gmp (GMP) guanosine monophosphate

GMP General Medical Practice; Greater Manchester Police; Green Mansion Properties

gmpa gas-metal-plasma arc

gmpg ground nautical miles per gallon

GMPI Guilford-Martin Personnel Inventory

gmq good merchantable quality

gmr ground mapping radar

GMRD Guided Missiles Range Division (Pan American World Airways)

gm rm games room

gms guidance monitor set

gms (GMS) geostationary meteorological satellite

gm & s general, medical, and surgical

Gms Gomori's methenamine silver (stain); Grimsby

GMS General Maintenance System; General Medical Services

GMSB Guided Missile System Branch

GMSC Guangdong Manpower Service Corporation

GMSL Group Management Service Limited

GMST General Military Subjects Test

GMT General American Transportation (stock exchange symbol); Greenwich Mean Time; Greenwich Meridian Time

GMT *Geo Marine Technology; Geriatric Medicine Today*

GMTC General Motors Technical Center; Glutamate Manufacturers Technical Committee

GMTL Goudy Memorial Typographic Laboratory (Newhouse Communications Center—Syracuse University)

gmts guided missile test set

g-m tube Geiger-Müller tube

GMU George Mason University

gmv gram molecular volume

GMV Government Motor Vehicle; Government Motor Vessel

gmw gram molecular weight

GMWU General and Municipal Workers Union

gn general; glen; golden number; good night; green; guide number; guinea (21 shillings); gun

g:n glucose-nitrogen (ratio)

Gn. *Genesis (Book of)*

GN Great Northern (railroad), great novel; Guinea (Internet code)

G.N. Graduate Nurse

G & N Gippsland and Northern

GN *Gas Natural* (Spanish—natural gas); *Guardia Nacional* (Spanish—National Guard)

GN₂ gaseous nitrogen

GN₂ sla gaseous nitrogen storage area

g N₂ stor gaseous nitrogen storage

GNA Ghana News Agency; Great Northern Insured Annuity Corporation

GNAL Georgia Nuclear Aircraft Laboratory

GNAS Grand National Archery Society

gnat. global network of automatic telescopes

GNB *Good News Bible*

gnc general nuclear war

gn & c guidance, navigation, and control

GNC General Nursing Council; General Nutrition Center; Good Neighbor Council(s)

gnd gross national demand; ground

gndck ground check

gnd ht xgr ground heat exchanger

gne gross national effluent

gne (GNE) gross national expenditure

gni (GNI) gross national income

gnl general

GNL Georgia Nuclear Laboratory

GNM Ghana National Museum

GNMA Government National Mortgage Association

GNN Great Northern Nekoosa

g noz grease nozzle

gnp (GNP) gross national product

g np gas, nonpersistent

GNP Geriatric Nurse Practitioner; Glacier National Park (in British Columbia and in Montana); Gombe National Park (Tanzania); Gorongoza National Park (Mozambique); gross national product

GNP & BL Great Northern Pacific & Burlington Lines

GNPC Great Northern Paper Company

gnr goods not received; gunner; gunnery

gnr (GNR) gaseous nuclear rocket

GNR Great Northern Railway

GNRA Gateway National Recreation Area (New York City's designation by the Department of the Interior)

g/n ratio glucose-nitrogen ratio

gnrh (GnRH) gonadotropin-releasing hormone

gnrl general

gnry gunnery

gns guineas

Gns Guernsey

GNS General Naval Staff

GNSRA Great North of Scotland Railway Association

GNT Grand National Teams

GNT *Gesellschaft für Nukleartransporte* (German—Nuclear Transport Society)

GNTC Girls' Nautical Training Corps

gnte *gerente* (Spanish—manager)

GNTO Greek National Tourist Organization

GNTP Georgia Narcotics Treatment Project

gnw (GNW) gross national welfare

Gny Sgt Gunnery Sergeant

go gas operated; gear oil; general obligation (bond); growth opportunities; output conductance (symbol)

go' gore

g/o gearbox oil

go. governess

Go gadolinium; Gothic

Gº Gonzalo (Spanish)

GO General Office; general order(s); George Orwell (Eric Blair); Governor's Office; Group Officer; Gulf Oil (stock exchange symbol)

GO₂ gaseous oxygen

goa gone on arrival; gyro(scope) output amplifier

GOA Gun Owners of America

GOAL Gay Organized Alliance for Liberation; Gun Owners Action League

goar ground-observer aircraft recognition

GOAT Give Our Animals Time (organization devoted to saving the endangered goats on California's offshore islands)

gob. gobbledygook; good ordinary brand

gob *gobierno* (Spanish—government)

Gob *Gobernador* (Spanish—Governor)

G o B Government of Belize (formerly British Honduras)

GObC Ground Observers Corps (Canada)

Gobi Gobi Desert of China and Mongolia; great desert of Central Asia in Mongolia

gobº *gobierno* (Spanish—government)

Gobr *Gobernador* (Spanish—Governor)

GOBS Guardians of Better Speech

goc gas-oil content (oil well); generator oil cooler

GOC General Occupational Classification; General Officer Commanding; General Optical Council; Greek Orthodox Church; Ground Observer Corps; Gulf Oil Company

GOC in C General Officer Commanding in Chief

GO City Greater Omaha, Nebraska

goco government-owned contractor-operated

god (GOD) government observing device

g.o.d. good old days

GODAS Graphically Oriented Design and Analysis System

GODCO Gulf Oman Oilfields Development Company

GODE Gulf Organization for the Development of Egypt (funded by Kuwait and Saudi Arabia)

godf. godfather

godm. godmother

GODORT Government Documents Round Table

God Save God Save the King (*Queen*) British national anthem

godsd godsdienst (Dutch—religion)

Godzone God's own country

goe gas, oxygen, ether (mixture); ground operating equipment

GOE General Ordination Examination

GOES Geostationary Operational Environmental Satellite

gof good old Friday; government-owned facilities

G o F Gang of Five

GOFAR Global Ocean Floor Analysis and Research

gogo government-operated government-owned

Gogol Nikolai Vasilyevich Gogol-Yanovsky

gogs goggles

goi gross operating income

Goi Goidelic

GoI Government of Indonesia

GOI Gallup Organization Incorporated

goin' going

GOIN Gosudarstvienny Okeanograficheskiy Institut (Russian—State Oceanography Institute)

gol general operating language

GOLB Gosudarstvennaya Ordena Lenina Biblioteka (Russian—Lenin State Library)

gold. geometric on-line definition

GOLD Grandparents Offering Love and Direction

Golda Golda Meir (Israel's first woman prime minister)

Gol de Cá d Golfo de Cádiz (Spanish—Gulf of Cadiz)

Gol de Cal Golfo de California (Spanish—Gulf of California)—also called Sea of Cortés or Vermillion Sea

Gol de Coro Golfete de Coro (Spanish—Small Gulf of Coro)

Gol de Méx Golfo de México (Spanish—Gulf of Mexico)

Gol de Uru Golfo de Uruba (Spanish—Gulf of Uruba)

Gol de Val Golfo de Valencia (Spanish—Gulf of Valencia)

Gol del Dar Golfo del Darien (Spanish—Gulf of Darien)

Golden Rule *What is hateful to thee do not do unto thy neighbor*—Hillel (30 B.C.-10 A.D.) stated in explaining essence of the Torah; *Do unto others as you would have them do unto you*

Goldie Gold; Golden; Goldilocks; Goldsborough; Goldsmith; Goldsworthy; Goldwin; Goldwyn

Gol di Gen Golfo di Genova (Italian—Gulf of Genoa)

Gol di Ven Golfo di Venezia (Italian—Gulf of Venice)

gold(s) gold bond(s); gold coin(s); gold medal(s)

Gol du Lion Golfe du Lion (French—Gulf of Lyons)

Golf letter G radio code; Soviet G-class submarines with missle-launching capability (NATO)

Gollancz Victor Gollancz Ltd

Gol Ven Golfo de Venezuela (Spanish—Gulf of Venezuela)

gom (GOM) government-owned material

Gom God's own medicine (opiates)

G.O.M. Grand Old Man (sobriquet for William Ewart Gladstone)

goma general officer money allowance

GOMA Good Outdoor Manners Association

gomer get out of my emergency room

go'n' going

gon goniff (Yiddish—thief)

gond(s) gondola(s); railroad car(s); car(s)

Gone Gone With The Wind

gonna (American slang—going to)

GONP Gal Oya National Park (Ceylon)

GONS Gun Orientation Navigation System

Gonz Gonzàles

Goo Goole

GOO Get Oil Out (of Santa Barbara, California)

goobs going out of business sale(s)

Goochland State Industrial Farm for Women (convicts) at Goochland, Virginia

goodbye god be with you (contracted)

Good H Good Housekeeping

goodies good ones

Goodwins Goodwin Sands off Kent near the North Sea entrance to the English Channel

goof. general on-line oriented function

goof(er) stupid person

googol 10 raised to the 100th power (10^{100})

googoo good government political plank

goon goonda (Hindi—hired killer)

GOP Grand Old Party (Republican)

GO & P Griffith Observatory and Planetarium

gor gas/oil ratio; general operational requirement(s); gorilla

Gor Gordon; Gorham; Gorki; Gorman

GOR General Operating Room; General Operational Requirements

GORA Government Oil Refineries Administration

Gordie Gordon

Goree Unit Women's Prison at Huntsville, Texas

GORF Goddard's Optical Research Facility

g org great organ

goric paregoric (an opiate narcotic)

gorill(s) gorilla(s)

Gorki Maxim Gorki (Russian novelist Alex Peslikoff); Russian city named for Maxim Gorki from 1932 to 1991, now called Nizhni Novgorod

gorm gormandize(r)

gos gosudarstvo (Russian—state)

Gos *Gosudarstvo* (Russian—State)

GOS General Operating Specification(s); George Orwell Society; Global (weather) Observing Systems

Gos Alb Gossamer Albatross (*see* bicyplane)

GOSIP Government Open Systems Interconnection Profile

GOSS Ground Operational Support System

gost government-owned special tooling

GOST Goddard Satellite Tracking

Gösta Björling Karl Gustaf Björling

got. (GOT) glutamic oxaloacetic transaminase

Got Gothenburg (Göteborg)

gotcha got you

Göteborg Gothenburg

Goten (German naval contraction—Gotenhafen)—Gdynia's name during World War-II Nazi occupation

goth. gothic type

Goth. Gothic

Göt(h) Göteborg (Gothenburg)

gothic gothic script or gothic type

gotran go fortran

got-roy gotong-royong (Indonesian—cooperation, mutual aid)

gott glucose tolerance test

Gott Gottingen

gotv get out the vote

gou gourde (Haitian currency)

Gou Goudy

Goudy Frederic William Goudy (American printer and designer of 90 typefaces)

goulard water lead lotion

gourde monetary unit of Haiti

Gouv Gouverneur

gov government

Gov Governor

GOVA Guide to Opportunities in Volunteer Archaeology

goveclop government closest to the people

govg governing

Gov Gen Governor General

Gov Is Governor's Island

Gov Pres President Bill Clinton

govt government

govtalk government talk

Govt Print Government Printer

GOW Grand Old Woman (Queen Victoria)

GOWA Guild of Watchmen of Australia

gox gaseous oxygen

goya & kod get off your ass and knock on doors

Goyo Gregorio

gp galley proofs; gas; persistent; general paralysis; general practice; general practitioner; general public; general purpose; geographic position; glide path; government property; grateful patient; gratitude patient, greenhouse perennial; ground pneumatic; group; guinea pig; gun pointer; plate conductance (symbol)

g-p general purpose; graduated-payment

gp. grandparents

gp grand prix (French—grand prize)

g/p giro postal (Spanish—money order)

Gp Group

GP Gallup Poll; Gaspesian Park (Quebec); general public; Georgia-Pacific (stock exchange symbol); Giacomo Puccini; glutathione peroxidase; Guadalcanal Province (Solomon Islands); Guadeloupe (Internet code); Gulf Province

G-P Georgia-Pacific (forest products); Gunier-Preston zone

GP Generalpause (German—general pause)—musical term

gpa grade-point average

GPA Gas Processors Association; Gay Press Association; Genealogical Publications of Australia; General Practitioners' Association; General Public Accounting; Governmental and Public Affairs; Guinnesse Peat Aviation

GPAALS Global Protection Against Accidental Launch System

gpac grade point average category

GPAC Great Plains Athletic Conference

gpad gallons per acre per day

gpae general-purpose aerospace equipment

GPAEVD Greater Philadelphia Alliance for the Eradication of Venereal Disease

GPAI Genealogical Periodical Annual Index

GPAL Gold Producers Association Limited

GPALS Global Protection Against Limited Strikes

gparm graduated-payment adjustable-rate mortgage

gpate general-purpose automatic test equipment

GPATS General-Purpose Automatic Test System

gpb glossopharyngeal breathing

GPB Gosudarstvennaya Publichnaya Biblioteka (Russian—State Public Library)

gpc gallons per capita; game play counselors; general purpose computer; germanium point-contact; giant papillary conjunctivitis; gypsum-plaster ceiling

gpc (GPC) general physical condition

GPC Genuine Parts Company; Georgia Power Company; Gulf Park College

Gp Capt Group Captain

gpcd gallons per capita per day

Gp Cmdr Group Commander

Gp Comdr Group Commander

GPCR Great Proletarian Cultural Revolution (China)

GPCT George Peabody College for Teachers

gpcu ground power-control unit

gpd gallons per day

GPDA Gypsum Plasterboard Development Association

gpdc general-purpose digital computer

GPDS General-Purpose Display System

GPDST Girls' Public Day School Trust

gpdw gypsum-plaster dry wall(ing)

gpe good phonetic equivalents

GPE General Precision Equipment, Global Perspectives in Education; Guided Projectile Establishment

Gp. Eng. Geophysical Engineer

gperf ground passive electronic reconnaissance facility

GPES Ground Proximity Extraction System

gpete general-purpose electronic test equipment

gpf gasproof; general purpose forces

GPF Generic Packaging Facility

Gp Fl group flashing (light)

GPFS General-Purpose Financial Statement

gpg grains per gallon

gph gallons per hour; graphite

G.Ph. Graduate in Pharmacy

GPH Game Packing House; Grand Pacific Hotel (Suva)

GPHI Guild of Public Health Inspectors

gphs general purpose heat source

gpi general paralysis of the insane (symptom of tertiary syphilis); ground-position indicator (aviation)

gpi (GPI) general price index; glucosephosphate isomerase

GPI General Printing Ink; Gingival Periodontal Index; Glass Packaging Institute; Gordon Personal Inventory; Government Property Inventory; Graphical Programming Interface

GPIA Generic Pharmaceutical Industry Association

GPIB General Purpose Interface Bus

gpid guidance package installation dolly

GPII Geist Picture Interest Inventory

gp int qk fl group interrupted quick flashing

gpl generalized programming language; geographic position locator; grams per liter; gypsum lath

GPL General Precision Laboratory

GPLC Guild of Professional Launderers and Cleaners

gply gingivoplasty

gpm gallons per minute; gross profit margin

gpm (GPM) graduated payment mortgage

GPM General Preventive Medicine; Grand Past Master

gpmg general-purpose machinegun

g-p mortgage graduated-payment mortgage

GPMS Gross Performance Measuring System (USAF)

GPN Graduate Practical Nurse

GPNITL Great Plains National Instructional Television Library

gpo group-purchasing organization; gun position officer

GPO General Post Office; Government Printing Office; Great Plains Organization

Gp Occ group occulting (light)

gpp galley page proofs; graphic part programmer; ground protection panel

gpp (GPP) graduated property payment(s)

GPP Gordon Personal Profile; Gross Provincial Product (Canadian)

gppl gypsum plaster

GPPT Group Personality Projective Test

gpr ground power relay

GPR Glider Pilot Regiment

GPR *Grand-Positif-Récit* (French—great choir and swell coupled)—organ music

G.P.R. *Genio Populi Romani* (Latin—Genius of the Roman People)

GPRA General Practice Reform Association

gps gage pressure switch; gallons per second; general problem solver; general-purpose solver; ground plane simulator; guidance power supply

gp's galley proofs; guinea pigs

g-p's general practitioners (GPs)

Gps general-parents motion pictures (for youngsters only with parent's consent)

GPs Great Performances

GPS General Practitioners Society; Gibbs-Poole-Stockmeyer (algorithm); Global Positioning System; Graduated Pension Scheme; Great Persons Society; Greater Public Schools

GPSA Gas Processors Suppliers Association; General Practitioners Society of Australia

gpsdw general-purpose scientific document writer

gpse general-purpose simulation environment

gpss general-purpose systems simulator

GPSS General Purpose Simulation System

gpt gas power transfer; glutamic-pyruvate transaminase; guidance position tracking; gypsum tile

gpt (GPT) glutamic pyruvic transaminase

GPT Grayson Perceptualization Test; Guild of Professional Toastmasters; Guild of Professional Translators

gpte general-purpose test(ing) equipment

gp th group therapy

gptr guidance power temperature regulator

gpu general processor unit; ground power unit

GPU General Postal Union

GPU *Gosudarstvennoe Politicheskoe Upravlenie* (Russian—State Political Administration)—secret police

GPV Gun Powder Van

GPV *Gereformeerd Politiek Verbond* (Dutch—Reformed Political Union)—Calvinist party

gpw gross plated weight; gypsum-plaster wall

GPW Geneva Convention Relative to Treatment of Prisoners of War

gpwc ground proximity warning computer

GPWS Ground Proximity Warning System

gpx generalized programming extended

GPX Greyhound Package Express

GPY Government Property Yard

GQ Equatorial Guinea (Internet code); general quarters

gqa give quick answer; government quality assurance; grain quality analyzer

GQG *Grand Quartier Général* (French—General Headquarters)

GQNM Gran Quivira National Monument

GQR Gauss Quadrature Rule

GQS General Quarter Sessions

Gquil Guayaquil

gr games in relief (baseball); gear; generator relay; glutathione reductase; grab rod; grade; grain; gram; grammar; grand; granite (color); grass runway; grave record; gravity; gray; great(er); grind(er); gross; ground; group; gunner

g-r gamma ray

gr. graduate; grant

gr gravida (Latin—gravid)—pregnant

Gr Grashof number; Great (postal abbreviation); Grecian; Greece; Greek; Grove

Gr Graben (German—ditch, trench); Greek; *Groot* (Afrikaans—big, great); *Gross(e)* (German—big, great, vast)

GR B.F. Goodrich (stock exchange symbol); General Radio; General Reconnaissance; General Reserve; Government Report; Grand Recorder; Grasse River (railroad); Graves Registration; Greece (Internet code); Group Report; Gunnery Range

GR Georgius Rex (King George); Guardia Republicana (Spanish—Republican Guard)

gra geared rotary actuator

GRA Girls Rodeo Association; Government Reports Announcements; Governmental Research Association; Grass Roots Association; Guaranteed Return Annuity; WR Grace & Company (stock exchange symbol)

gr ab grade ability

GRAB Group Rooms Availability Bank (hotel-motel convention service)

GRACE Grace Agencies; Grace Chemicals; Grace Line; WR Grace and Company (stock exchange symbol); graphic arts composing equipment; group routing and exchange equipment (telephone)

grad gradient; grading; graduate

grad (GRAD) graduate résumé accumulation and distribution

grad. graditim (Latin—by degrees)

Grad IAE Graduate of the Institution of Automobile Engineers

Grad IM Graduate of the Institution of Metallurgists

Grad Inst BE Graduate of the Institution of British Engineers

Grad Inst P(hys) Graduate of the Institute of Physics

Grad Inst R(frg) Graduate of the Institute of Refrigeration

Grad IRI Graduate of the Institution of the Rubber Industry

GPA & I Government Reports Announcement and Index Journal

Grad NDTS Graduate (member) of the Non-Destructive Testing Society

Grad RIC Graduate (member) of the Royal Institute of Chemistry

grad(s) gradient(s); graduate(s)

GRADS Great Falls Air Defense Sector

Grad SE Graduate of the Society of Engineers

Grad Soc Eng Graduate of the Society of Engineers

gradu gradual(ly); graduate(d); graduating

graf graphic additions to fortran

graf(s) graphic addition(s); paragraph(s)

Grahams Grahamstad or Grahamstown in South Africa

Grail Soviet shoulder-fired surface to air missile called SA-7 (NATO)

gral general (Spanish—general)

Gral General (Spanish—General)

gram. grammar; grammatical; gramophone

'gram cablegram; radiogram; telegram

gram (Latin suffix—record)—radiogram

Gram Grammar; Grandfather; Grandpa(pa)

Gram Gramaphone

gramm grammatical

Grammy National Academy of Recording Arts and Sciences award; replica of a gramophone

Grampians Grampian Hills of Scotland or the Grampian Mountains of Australia

gramp(s) grandfather

GRAMS Ground Recording and Monitoring System

gran granite; granular; granulated sugar

gran. granulatus (Latin—granulated)

Gran Granada; Granjon

GRAN Global Rescue Alarm Net

Grand Canyon the Grand Canyon of the Colorado in the Grand Canyon National Park in Arizona, the Grand Canyon of the Arkansas in Colorado where it is also called the Royal Gorge, the Grand Canyon of Santa Elena in the Big Bend National Park in Texas, the Grand Canyon of the Snake River in Idaho, the Grand Canyon of the Tuolumne in California, the Grand Canyon of the Yellowstone in the Yellowstone National Park in Wyoming

Grand Coulee Grand Coulee Dam; Grand Coulee Valley in eastern Washington

grando grandioso (Italian—grandiose)

Grand Teton Grand Teton Mountain; Grand Teton National Monument in northwestern Wyoming

Gran-Duc Lux Grand-Duché de Luxembourg (French—Grand Duchy of Luxembourg)

Granny Grandmother

Gran Quivira Gran Quivira National Monument in central New Mexico

grapden graphic data entry

Grapes Grapes of Wrath

graph. graphology

graph (Latin suffix—record, recording)—radiograph

grapheme written language symbol representing an oral language code

Graphics 'O' Level commercial art with some drawing or fine art

graphite carbon

graph rec graphic record(ing)

GRAPO Grupos Antifascista Para Octubre (Spanish—October 1st Antifascist Groups)—Spain

gr ar grinding arbor

GRAR Government(al) Report Authorization and Record

gras generally recognized as safe

GRAS generally regarded as safe

graser gamma-ray laser

grasp. graphics-augmented structural-post processing

grat graticule

grats congratulations
GRATS Gang-Related Active Traffickers Suppression
grav gravimetric; gravitation; gravity
grazo grazioso (Italian—gracious)
grb gamma-ray burst; granolithic base
GRB Gerakan Rakjat Baru (Indonesian—New People's Movement); *Guide to Reference Books*
GRBI Gardeners' Royal Benevolent Institution
grbm (GRBM) global-range ballistic missile
Gr Br Grande Bretagne (French—Great Britain); Great Britain
Gr Brit Great Britain
GRBS Gardeners' Royal Benevolent Society
grc generator relay control; glass-reinforced cement
GRC Gale Research Company; Gerontology Research Center; Government Research Corporation; Gulf Research Corporation
GRC Gendarmarie Royale du Canada (French—Royal Gendarmarie of Canada)—Royal Canadian Mounted Police
GRCB Greyhound Racing Control Board
GRCM Graduate of the Royal College of Music
grcol ground color
Gr Cpt Group Captain
grd granddaughter; grind; ground; ground detector; guard
Grd Ground (postal abbreviation)
Gr d Greek drachmae
Gr D Grand Duchy
GRD Geophysics Research Directorate
GRDC Gulf Research and Development Company
grdl gradual(ly)
Grdn The Guardian
grd tot grand total
gre ground reconstruction equipment
Gre Greece (whose capital is Athens)
Gre Grecia (Spanish—Greece)
GRE glucocorticoid response element; Graduate Record Examination; Guardian Royal Exchange

Great Amer Great American Federal Savings Bank; Great American First Savings Bank
Great Communicator Ronald Reagan
Great Creole General Pierre Custave Toutant de Beauregard
Great Dic Great Dictator
Great Dividing Great Dividing Range of Australia's New South Wales and Queensland
Great Lon Greater London
Great Sandy Great Sandy Desert of South and Western Australia
Great Smokies Great Smoky Mountains of North Carolina and Tennessee
Great Smoky Great Smoky Mountains; Great Smoky Mountains National Park in North Carolina and Tennessee
Great Vic Great Victoria Desert of Western Australia
Great-West Great-West Life and Annuity Insurance
GREB Graduate Records Examination Boards
Grec Grécia (Italian or Spanish—Greece); *Grecia* (Portuguese—Greece)
GRECC Geriatric Research, Education, and Clinical Center (Seattle)
Greece Hellenic Republic (Balkan nation whose history antedates classical antiquity), *Elliniki Dintokratia*
Greece's Principal Port Piraeus
Greek Fabulist Aesop
Greeks Greek Islands; Greek people
Green Green College (Oxford or elsewhere); Greenland (whose capital is Nuuk)
GREEN General Research in the Environment for Eastern European Nations
green flag all-clear signal; express; go-ahead
Greenhouse Greenhouse Effect
Greenland Arctic Greenland north of the Arctic Circle
green light all-clear signal; go-ahead signal; safety signal; starboard side of aircraft, ships, or other vessels
Green Mts Green Mountains of Vermont

Greenock Girls prison for female offenders in Greenock, Scotland
green onions scallions
greeny conservationist or environmentalist
Grefco General Refractories
Greg Gregorian; Gregory
Grg° Gregorio (Spanish—Gregory)
gr el greatest elongation
gren grenade
Gren Grenada; Grenada (whose capital is St George's)
Grenada State of Grenada (West Indian island nation)
Grenadines Bequia, Cannouan, Carriacou, and Mustique islands in British Windward Islands
Grendr Grenadier
grep gets repeating patterns
Grepo Grenzpolizei (German—border-control police)
Greta Greta Garbo; Margaret
Gretchen Marguerite
Greyf Greyfriars College, Oxford
grf. grandfather
grf (GRF) growth hormone-releasing factor
gr f grass firm (on runway)
GRF Gerald Rudolph Ford—thirty-eighth President of the United States; Graphic Reproduction Federation; Grassland Research Foundation; Gravity Research Foundation
gr Fl grosse Flöte (German—full-size flute)
GRFMA Grand Rapids Furniture Market Association
grfrp graphite fiberglass-reinforced plastic
gr fx grinding fixture
grg generalized reduced gradient; gravimetric rain gage
grh (GRH) gonadotrophin-releasing hormone
GRI Gas Research Institute; General Religions International (publishing division of Humanists of South Jersey); Geothermal Resources International; Government Reports Index; Government of the Ryukyu Islands
G.R.I. Georgius Rex et Imperator (Latin—George, King and Emperor)
GRID gay-related immune deficiency

grif (GRIF) growth hormone-inhibiting factor

Grif Griffin; Griffith; Griffiths

griff griffin

GRIN Germplasm Resources Information Network

GRIP Grass Roots Improvement Program; Greenland Icecore Project

griphos general retrieval and information processor humanities-oriented studies

Gris Griswold

GRIST Grazing Incidence Solar Telescope

grit. gradual reduction in temperature; gradual reduction in tensions

grits boiled grits; hominy grits; rockahominie in Algonquian Indian

GRITS Goddard Range Instrumentation Tracking System (NASA)

griz grizzly

grizz grizzly bear

GRJC Grand Rapids Junior College

Grk Greenock

Gr-L Graeco-Latin

Gr L Gunner Lieutenant

gr lp ground lamp

Gr Lt Gunner Lieutenant

GRL Gross Regional Loss

grm gaseous radiation monitor; gram; gross rent multiplier; guidance rate measuring

grmo. grandmother

grmp generalized report module program

grn green

g/r/n goods received note

Gr.N. Graduate Nurse

grnd ground

Grnd Grand

grndr grinder(s)

grnl giornalista (Italian—newspaperman)

GRNL Gay Rights National Lobby

Grnld Greenland

grns green skins

grnsh greenish

grnt guarantee

gro gross

Gro Grocer(y); Groningen; Grove; Guerrero

Gro. grocer

GRO Gamma Ray Observatory; Greenwich Royal Observatory

GROBDM General Register Office of Births, Deaths, and Marriages

groc grocer(y)

Groen Groenlandia (Italian or Spanish—Greenland); *Groenlandia* (Portuguese—Greenland)

GROIN Garbage Removal Or Income Now

Grolier Grolier Society

grom grommet

Grom Andrei Gromyko

Gron Groningen (Dutch province)

gros. grossus (Latin—coarse, gross)

Grose Francis Grose's *Classical Dictionary of the Vulgar Tongue*

Grosset Grosset & Dunlap

Groucho Groucho Marx (Julius Marx)

Group of Seven *see* G-7

Group W Westinghouse Broadcasting

Grove's Sir George Grove's *Dictionary of Music and Musicians*

GROW Gay Rights for Older Women; Group Recovery Organizations of the World

Growlers Growler Mountains of southwestern Arizona

growsy grumpy and drowsy

grp glass-reinforced plastic (fiberglass); graphite-reinforced plastic; ground relay panel; group repetition panel

Grp Group

GRP Gross Regional Profit

grp(s) group(s)

grp's gross rating points

grr growler

GRR Grand Rapids, Michigan (airport); Graphic Reproduction Request

grreg graves registration

GrReg graves registration

grs grains; grandson; grass; greens

gr s grass soft (on runway)

gr-s government rubber plus styrene (buna-S synthetic rubber)

GRs Government Regulations; Granitic Regions

GRS General Railway Signal; Graves Registration Service; Great Red Spot

GRSE Guild of Radio Service Engineers

Gr S-Lt Gunner Sub-Lieutenant

GRSM Graduate of the Royal Schools of Music (Royal Academy of Music and the Royal College of Music)

GRSP General Revenue Sharing Program

grst gross ton(s)

grsy greasy

grt gross register(ed) tonnage (tons); grout

grtee guarantee

grtg grating

grtm gross-ton mile

gr tons gross tons

grtr greater

gr tr graphite treatment

gr Tr grosse Trommel (German—bass drum)

Gru Grus; Gruyère

GRU Glavnoye Razvedyvatelnoye Upravlenie (Russian—Intelligence Directorate of the Red Army)—(*q.v.* VOT)

grub. grubby; grubstreet

grub. (GRUB) grocery update and billing

GRULA Grupo Latino, Americano (Spanish—Latin American Group)—UN power bloc

gr'ups grownups

Grux (Latin—Crane constellation)

Grv Grove

gryl gravel(ly)

GRW Greco-Roman Wrestling

Grwd Grunewald

gr wt gross weight

gry grocery; gross redemption yield

gs galvanized steel; games started; gastric shield; gauss; german silver; glandular segment; glide slope; grand slalom; grandson; grave stone; ground speed; guardship; guineas

gs (G8) gold standard; group separator character (data processing)

g-s grandson

g/s gallons per second; games scheduled, season

Gs force of gravity; general motion pictures (for the general public); German silver; Gomes

GS General Schedule (civil service classification system); General Secretary; General Service; General Sessions;

General Staff, General Studies; General Support; Geochemical Society; Geological Survey; Gerontological Society; Gillette (stock exchange symbol); Girl Scouts; Glow Start (tractor); Government Servant; Government Service; Grand Secretary; Ground Staff; Ground Station; Guidance Station; Gunnery School; Gunnery Sergeant

GS1 Gustav Siegfried Eins (British radio station broadcasting propaganda to Germany in World War II)

G-S Gallard-Schlesinger

G & S Gilbert and Sullivan (Sir William Schwenck Gilbert, librettist, and Sir Arthur Seymour Sullivan, composer)

GS Garda Siochana (Gaelic— Police Force)—Ireland's police force

gsa gross soluble antigen

gsa (GSA) general (travel) sales agency; general (travel) sales agent

GSA Garden Seed Association; Gas Service Agents; General Services Adininistration; Genetics Society of America; Geological Society of America; Geriatric Services of America; Girl Scouts of America; Gourd Society of America; Greenhouse Suppliers Association

G & SA Gulf and South American (steamship line)

GSABCA General Services Administration Board of Contract Appeals

GSAI General Services Administration Institute

gsap gunsight aiming point

GSAPBS General Services Administration Public Building(s) Service(s)

gsb gypsum sheathing board

gsb (GSB) go subroutine

GSB Government Savings Bank

GSBA Georgia School Boards Association

GSBAA General Service Board of Alcoholics Anonymous

gs bot glass-stoppered bottle

gsbr gravel-surface built-up roof

gsc gas-solid chromatography; geodetic spacecraft; ground speed continue; guidance systems console

GSC General Staff Corps; Genetics Society of Canada; Geological Survey of Canada; Gold Star Cable (Korean); Group Study Course; Group Switching Center; Gulf State Conference

GSCBA Georgia State College of Business Administration

GSCP Generic Site Characterization Plan

GSCT Goldstein-Scheerer Cube Test

GSCW General Society of Colonial Wars; Georgia State College for Women

gsd general system description; genetically significant dosage; grid sphere drag

GSD General Services Department; General Supervisor's Directive; General Supply Depot

GSDFJ Ground Self-Defense Force Japan

GSDNM Great Sand Dunes National Monument

gse government-sponsored enterprise; ground support equipment

gse (GSE) government-supplied equipment; ground-service equipment; ground-support equipment

GSE Graduate School of Education (Harvard University)

GSED Ground Support Equipment Division (USN)

GSEL Ground-Support Equipment List

GSERD Ground-Support Equipment Recommendation Data

GSES Geocentric Solar Ecliptic System (NASA)

GSE/TD General Systems Engineering/Technical Direction

gsf general scientific framework

GSF General Support Force (USAF); Government Superannuation Fund

GSFC Goddard Space Flight Center

GSFG Group of Soviet Forces in Germany

GSFLT Graduate School Foreign Language Test

GSFSR Ground Safety and Flight Safety Requirements

gsfu glazed structural facing units

GSG Grenzschut zgruppe (German—Border Protection Group)—anti-terrorist commando force

GSGB Geological Survey of Great Britain

GSGS Geographical Section-General Staff

GSGS maps General Staff, Geographical Section (British War Office) maps covering Africa, Asia, the East Indies, and Europe

gsh good study habits

GSH glutathione

gshr grand-slam home run(s)

gshv globe stop hose valve

gsi gas installed; glide scope indicator; graphic structure input; gross scheduled income; ground speed indicator

GSI General Safety Inspection; General Safety Inspector; General Service Infantry; General Steel Industries; Geological Survey of Israel; Geophysical Services International; Government Source Inspection

G & SI Gulf and Ship Island (railroad)

gsid ground-emplaced seismic intrusion detector

g sil german silver

GSIP Global Share Investment Program

GSIS Government Service Insurance System; Group for the Standardization of Information Services

gskt gasket

gsl guaranteed student loan

GSL Geological Society of London; Graphics Subroutine Library; Great Slave Lake Railway; Guaranteed Student League

GS & LA Guam Savings and Loan Association

gslcv globe stop lift check valve

GSLP Guaranteed Student Loan Program

gsm global system for mobile (communications); good sound merchantable; grams per square meter; gross sales monthly; ground-supplied material

GSM General Sales Manager; Gibson Spiral Maze; Guildhall School of Music

GSMD General Society of Mayflower Descendants; Guildhall School of Music and Drama

GSMFC Gulf States Marine Fisheries Commission

GSML General Stores Material List

GSMNP Great Smoky Mountains National Park (Tennessee and North Carolina)

GSMOL Golden State Mobilehome Owners League

GSMS Geocentric Solar Magnetospheric System (NASA); Graduate Student of the Management Society

GSNC General Steam Navigation Company

GSNWR Great Swamp National Wildlife Refuge

gso ground speed outbound

GSO General Staff Officer; Girls Service Organisation, Greensboro, North Carolina (airport); Ground Safety Officer

GSOST Goldstein-Scheerer Object Sorting Test

gsp gas paid for; gas planned

GSP Generalized System of Preferences; Good Service Pension

GSPA Gulfport State Port Authority

G-spot Grafenberg spot — erotic vaginal area present in some women

GSPOT Geometric Spot Analysis System

gsps guidance spare power supply

gsr galvanic skin reflex; galvanic skin resistance; galvanic skin response; general service reinforcement(s); ground speed return; ground surveillance radar

GSRI Gulf South Research Institute

GSRS General Support Rocket System

gsrv globe stop radiator valve

gss guidance system simulator

GSS General Service School; General Social Services; General Supply Schedule; Geo-Stationary Satellite; Gilbert and Sullivan Society; Global Surveillance System; Grumman Standard Specification

GSSF General Supply Stock Fund

GSSH Grand Street Settlement House

GSSL Genoa, Savona, Spezia, and Leghorn (ports)

GSSLNCV Genoa, Savona, Spezia, Leghorn, Naples, Civetta, and Vecchia (ports)

GSSR Georgian Soviet Socialist Republic; Ground Support System Review

GSSS Ground Support System Specification(s)

GSST Goldstein-Scheerer Stick Test

gst garter stitch (knitting); generation-skipping transfer; graphic stress telethermometry; ground special tools

GST General Service Test; General Staff Target; Goods and Services Tax (Canadian), Greenwich Sidereal Time; Guamanian Standard Time

GSTC Gorham State Teachers College

gste guidance system test equipment

GSTP Generalized System of Tariff Preferences

G-string capital-G-shaped string-like genital covering worn by exotic entertainers; violin's lowest string

gsts guidance system test set

gstu guidance system test unit

gsu glazed structural units; ground-support unit

GSU General Service Unit; Georgia State University; Gulf States Utilities

gsub glazed structural unit base

gsuc ground stub-up connection

g-suit antigravity suit worn during supersonic flight

GSUSA Girl Scouts of the USA

GSUSDA Graduate School, United States Department of Agriculture

gsv globe stop valve

GSV Grumman Submersible Vehicle; Guided Space Vehicle

gsvr ground-to-surface vessel radar

gsw gunshot wound

GSW Fort Worth, Texas (Greater Southwest International Airport)

GSW 1812 General Society of the War of 1812

GS & WR Great Southern and Western Railway

gt gas turbine; gastight; gilt; gilt top; glass tube; grand total; grease trap; great; greater than; greetings telegram; gross tonnage; gross ton(s); ground transmit(ter); group technology; gun target; gut tripe

g/t gooseneck tunnel; grams per ton; granulation time; granulation tissue

gt gate (Norwegian—street)

gt. gutta (Latin—drop)

Gt Great; Greenwich time

Gt Groot (Afrikaans—big, large, vast)

GT General Tariff; Global Telecommunications; Good Templar; Goodyear Tire & Rubber (stock exchange symbol); Grand Trunk (Pakistan road); Gran Turismo; Grand Tiler; Grupo de Transportes (Transport Group); Guatemala (Internet code)

GIT Gas Turbine (vessel)

GT Gamle Testamente (Dano-Norwegian—Old Testament); *Gran Turismo,* (automobile)

gta gas-tungsten arc; graphic training aid

GTA Gatt Textiles Arrangement; Geography Teachers Association; Gospel Truth Association; Government Telecommunications Agency; Graduate Teaching Assistant; Gun Trade Association

GTAP General Technical Assistance Program

gtaw gas turbine arc welding

Gtb Godthaab, Greenland's Capital

GTB Government Tourist Bureau

GTBC Guild of Teachers of Backward Children

Gt Br Great Britain

Gt Brit Great Britain

gtc gain time control; gas turbine compressor; good till cancelled

GTC Girls Training Corps; Government Training Center; Government Travel Center; Guam Territorial College; Guild of Television Cameramen; Gulf Transport Company (railroad)

GTCs Government Training Centres (UK)

gtd geometrical theory of diffraction; guaranteed

GTDS Goddard Trajectory Determination System (NASA)

gte general total energy; gilt top edge; ground test(ing) equipment; guidance test(ing) equipment; gunner tracking evaluator

gt-e gas turbo-electric

gte *gerente* (Spanish—manager)

GT & E General Telephone and Electronics (Corporation)

GT & EA Georgia Teachers and Education Association

GTEC Georgia Institute of Technology

gtee goatee; guarantee

gtee od guaranteed overdraft

GTEIS General Telephone and Electronics Information System

GT & EL General Telephone and Electronics Laboratories

GTEP Guaranteed Training Equipment Program

gtev (GTEV) gas-turbine electric vessel

gtf glucose tolerance factor

GTF Gang Task Force (police function); General Trust Fund; Global Telecommunications Fund; Government Test Facilities; Government-Test(ing) Facility; Granite Test Facility; Great Falls, Montana (airport)

gtg gas to gasoline

GTG Sappho's daddy

gt-gr. great-grand

gth go to hell

gth (GTH) gonadotrophic hormone

gthtgr (GTHTGR) gas-turbine high-temperature gas-cooled reactor

gti general transportation importance

GTI Grand Turk Island (tracking station)

GTIL Government Technical Institute Library

GTIO German Tourist Information Office

GTL Glass Technology Laboratories

Gt Lakes Great Lakes

Gt Ldn Greater London

gtm gas turbine module; gas turbine motor; good this month; gross ton(nage) mile(s)

GTM General Traffic Manager

GTMA Gauge and Tool Makers Association

Gt Man Greater Manchester

Gtmo Guantanamo Bay

GTMS Graphic Text Management System

gtn glomerulo-tubulo nephritis

GTN *Government Training News*

GTNP Grand Teton National Park (Wyoming)

gto gate turnoff

gto (GTO) geostationary transfer orbit

g to go to (calculator)

Gto Gunajuato

GTO Government Team of Officials

GTO *Gran Turismo, Omologato* [hard-top type of high-performance auto certified *(omologato)* to enter Gran Turismo automobile race]

gtol graphic takeoff language; ground takeoff and landing

gtow gross takeoff weight

gtp ground test(ing) plotter

gtp (GTP) guanosine triphosphate

GTP General Test Plan

gtr gantry test rack; greater; ground test(ing) reactor

GTR Grand Trunk Railway; Gurkha Transport Regiment(als)

GT-R Grand Touring-Racing (version)

Gtr Ant Greater Antilles

gtrp general transpose

gts guidance test(ing) set

gts (GTS) geostationary technology satellite

gt's grand touring cars

g/t/s gas-turbine ship

Gts Gateshead

GTS gas turbine vessel (3-letter code); General Telephone System; General Theological Seminary; Global Telecommunications System (WMO); Government Transport Ser-

vice (Pakistan buses); Greenwich Time Signal; Ground Transport System; Guadalcanal Travel Service; Guinean Trawling Survey

GTSC German Territorial Southern Command (NATO)

GTSI Government Technology Services, Inc

gtss gas turbine self-contained starter

GTSTD Grid Test of Schizophrenic Thought Disorder

gtt glass transition temperature

gtt (GTT) gelatin-tellurite-taurocholate

git. *guttae* (Latin—drops)

gtT gone to Texas (one jump ahead of the sheriff)

GTT Glucose Tolerance Test

gtu guidance test unit

GTU Graduate Theological Union

gtv gate valve; gravity vacuum transit

gtv (GTV) gas turbine vessel

GTV Gumma Television (Japan)

gtw good this week; gross ton(nage) weight

GTW Grand Trunk Western (railroad)

Gtwy Gateway

gty gritty

Gtz Galatz

gu gastric ulcer; genitourinary; geographically unsuitable; glycogenic unit; great uncle

Gu Guinea; Gujarat; Gujarati

Gu *Göteborgs Universitetsbiblioteket* (Swedish—Gothenburg's University Library)

GU genito-urinary; Georgetown University; Glasgow University; Gonzaga University; Griffith University; Guam

gua. guardian

Gua Guatemala (whose capital is Guatemala City)

GUA Guatemala City; Guatemala (airport)

Guad Guadalcanal, Solomon Islands; Guadeloupe

Guadal Guadalajara

Guadalupe Mountains Guadalupe Mountains National Park, east of El Paso, Texas

Guadalupes Guadalupe Mountains of New Mexico and Texas

Guadarramas Guadarrama Mountains of central Spain (*Sierra de Guadarrania*)

Guajira Pen Guajira Peninsula in northeastern Colombia

Guam Pacific Island possession of United States; inhabited by Guamanians

Guam ST Guamanian Standard Time

Guana Guanacaste (Costa Rican province); Guarajuato

'Guana Iguana Island, British Virgin Islands

Guanacastes Guanacaste Mountains of northwestern Costa Rica (*Cordillera de Guanacaste*)

Guanajay Cuban prison in Guanajay southwest of Havana in Pinar del Rio province

'guana(s) iguana(s)

GUANOMEX *Guanos y Fertilizantes de México* (Spanish—Guanos and Fertilizers of Mexico)

guar guarantee

Guar Guarani (Brazil)

guarani monetary unit of Paraguay

guard. guaranteed

GUARD Government Employees United Against Discrimination

Guardian *The Guardian* (leading British newspaper published in London and Manchester)

guarg guaranteeing

Guat Guatemala(n)

GUATEL *Empresa Guatemalteca de Telecomunicaciones* (Spanish—Guatemalan Telecommunications Enterprise)

Guatemala Republic of Guatemala (Central American country), *República de Guatemala*

Guay Guayaquil

Guaya Guayaquil

gub generalized upper boundary

GUBC Guyana United Broadcasting Company (Radio Demerara)

gubernalection gubernatorial election

Guer Guerrero

GUGK *Glavnoje Upravlenije Geodesii i Kartografii* (Russian—Administrative Agency for Geodesy and Cartography)

GUGMS *Glavnoje Upravlenije Gidrometeorologicheskoi Sluzhby* (Russian—Administrative Agency of the Hydrometeorological Service)

Gug Mus Guggenheim Museum

Gui Guinea (whose capital is Conakry)

GUI Golfing Union of Ireland; Graphical User Interface

Gui-Bis Guinea-Bissau (whose capital is Bissau)

Gui Cur Guinea Current

guid guidance

guide guidance for users of integrated data equipment

guidn guidance

Gui E *Guinea Ecuatorial* (Spanish—Ecuatorial Guinea)

guil guilder(s)

Guil Guillaume

Guild The Newspaper Guild (American union)

guilder monetary unit of Netherlands, Netherlands Antilles, and Surinam

Guildhall Lib Guildhall Library (London)

Guild Prof Trans Guild of Professional Translators

Guillaume Appolinaire Guillaume Appolinaire de Kostrowitsky

Guill° Guillermo (Spanish—William)

guillo guillotine (named in France to honor its perfector Dr Joseph-Ignance Guillotine)—last used in 1977 and banned in 1981 with the outlawing of capital punishment

guin guinea(s)

Guinea Republic of Guinea (West African nation), *République de Guinée*

Guinea-Bissau Republic of Guinea-Bissau (former West African colony of Portugal)

Guip Guipuzcoa

guit guitar

guit *guitarra* (Spanish—guitar); *guitarrazo* (Spanish—blow struck with a guitar); *guitarrear* (Spanish—strumming the guitar); *guitarrista* (Spanish—guitar player); *guiterrera or guiterrero* (Spanish—guitar maker or guitar player)

Guj Gujarat; Gujarati

Gul Ad Gulf of Aden

GULAG Chief Administration of Corrective Labor Camps, Prisons, Labor, and Special Settlements of the Soviet Secret Police (*q.v. VOT*)

Gul Alas Gulf of Alaska

Gul Both Gulf of Bothnia

GULC Georgetown University Law Center

Gul Cal Gulf of California

Gulf Gulf of Adalia, Aden, Alaska, Alexandretta, Aqaba, Boothia, Bothnia, Cadiz, California, Cambay, Campeche, Canada, Carpentaria, Cattaro, Chihli, Chiriqui, Cutch, Darien, Eilat, Finland, Fonseca, Gabes, Genoa, Guayaquil, Guinea, Honduras, Izmir, Kotor, Kutch, Lepanto, Lions, Maine, Manaar, Maracaibo, Martaban, Mexico, Nicoya, Oman, Panama, Paria, Quarnero, Santa Catalina, Siam, Sidra, Smyrna, St Lawrence, Suez, Taranto, Tehuantepec, Tonkin, Venice; Gulf Oil, Spencer Gulf

GULF Gays United for Liberty and Freedom

Gulf Coast coastline of Texas, Louisiana, Mississippi, Alabama, and Florida

Gul Fin Gulf of Finland

Gul Hon Gulf of Honduras

Gulf Islands islands off Vancouver, B.C.; national seashore along Florida and Mississippi

Gulfs Gulf Islands off Pascagoula, Mississippi

Gulf States Florida, Alabama, Mississippi, Louisiana, and Texas along the Gulf of Mexico; Iran, Iraq, Kuwait, Saudi Arabia, Bahrain, Qatar, United Arab Emirates, and Oman along the Persian Gulf

Guli Gulielma

Gull Gullah (dialect of coastal South Carolina, Georgia, and Florida)

Gul Om Gulf of Oman

gulp (data-processing slang—a succession of bytes)

gulp. (GULP) general utility language processor

GULP General Utility Library Program

Gul Par Gulf of Paria

Gul St L Gulf of St Lawrence

Gul Teh *Gulfo de Tehuantepec* (Spanish—Gulf of Tehuantepec)

Gul Ton Gulf of Tonkin

gum. (GUM) genito-urinary malignancy

Gum Guam (container port)

GUM Guam (airport)

GUM *Gosurdarstvennoe Universalny Magasin* (Russian—State Universal Store)

gumbo mucilaginous alluvial mud; slang for okra or a person of Black and Cajun ancestry

Gumbo patois spoken in Louisiana

Gumbo French dialect prevailing in parts of Louisiana

gums gum trees (eucalyptus)

gun. guncotton; guncrete; gunnery; gunpowder

gun *gunung* (Malay—mountain)

gun dip gunboat diplomacy

gun'l gunwale

Gun Sgt Gunnery Sergeant

GUNSS Gunnery Schoolship (USN)

guo government use only

GUO Greater Union Organization

GUOOF Grand United Order of Odd Fellows

gup guppy

GUPCO Gulf Petroleum Corporation; Gulf of Suez Petroleum Company

guppie(s) gay urban professional(s)

guppy. greater underwater propulsive-powered (guppy-shaped) submarine

gups guppies

gup(s) gay urban professional(s)

GURC Gulf Universities Research Corporation

Gurneyites Gurney Quakers

Gus August; Augustus; Gustafi, Gustave; Gustavus

GUs Guns Unlimited

GUS Globe Universal Services; Great Universal Stores; Grocers United Stores

Gus Rom Gustavo Romero

Gussies Great Universal Stores

gust gustation; gustatorily; gustatory; gustily; gustiness; gusto; gusty

Gustus Augustus

gut. gutter

Gut Gutenberg; The Gut (Valetta's redlight street on Malta)

GUT Grand Unified Theory

Gutenberg Johannes Gensfleisch (German—John Gooseflesh)—the inventor of movable type

GUTS Georgians Unwilling to Surrender

gutt. *gutta* (Latin—drop)

guttat. *guttatim* (Latin—drop by drop)

gutt. quibus *guttis quibusdam* (Latin—a few drops)

guv governor

GuV *Gerecht und Volkommen* (German—correct and complete)

guv'nor governor

Guy Guayaquil; Guido; Guyana; Guyana (whose capital is Georgetown); Guyon

Guy *Guyana* (Spanish—Guiana)

Guyana Cooperative Republic of Guyana (formerly called British Guiana or Demerara, South America's only English-speaking nation)

Guyane *Guyane français* (French Guiana)

Guybau Guyana Bauxite

Guy Esse Guyana Essequibo (between Guyana's Essequibo River and Venezuela's eastern border)—area claimed by Venezuela for more than a century

Guy's Guy's Hospital (London)

gv gate valve; gentian violet; government valuation; gravimetric volume; grid variation; ground visibility

gv (GV) granulosis virus

gv *grande vitesse* (French—fast-freight train); *gran velocidad* (Spanish—high velocity)

Gv Gustav

GV gigavolt; Giuseppe Verdi; Göita Verken (steel company); Graphic Violence; grid variation

gva general visceral afferent

GVA Geneva, Switzerland (airport); Grants by Voluntary Agencies

gvb gelatine veronal buffer

GVB Guam Visitors Bureau

GVC Grand View College

gve general visceral efferent

Gve Grove; Gustave

GVF *Grazhodanskii Vozdushnyi Flot* (Russian—Civil Air Fleet)

gvh graft versus host (reaction in bone marrow transplants)

gvhd (GVHD) graft-versus-host disease

gvhr graft versus host reaction(s)

gvhrr geosynchronous very high-resolution radiometer

GVI Gas Vent Institute

gvl gravel

GVL Global Van Lines

gvm generating voltmeter; gross vehicle mass

GVMDS Ground-Vehicle Mine-Dispensing System

GVN George V Novotny; Government of Vietnam

gvo gross value of output

GVP General Vice President

GVP *Gereformeerd Politiek Verbond* (Dutch—Reformed Political Union)

GVRD Greater Vancouver Regional District

GVS Government Vehicle Service

gvt government; gravity vacuum transit; gravity vacuum tube

GVT Ground Vibration Test(ing)

gvty gingivectomy

gvw gross vehicle weight

gvwr gross vehicle weight rating

gw gigawatt(s); green weight; ground wave(s); guerilla warfare

g/w gross weight(s)

GW George Washington—first President of the United States; Guinea-Bissau (Internet code); Gulf War

G-W Globe-Wernicke

G & W Gulf and Western

G + W Gulf and Western

GWA Girl Watchers of America; Golden West Airlines

GWA *Goodes World Atlas*

GWAA Golf Writers Association of America

GWB George Washington Bridge

gwc gas-water contact (oil well)

GWC George Washington Carver

GWCHS George Washington Carver High School

GWCM George Washington Carver Museum

gwcswbd gunnery weapon-control switchboard

gwe gigawatts electrical

gwen ground-wave emergency network

Gwen Gwendolyn

Gwenda Gwendolen

Gwennie Gwendolen

GWG George Washington Geist

gwh gigawatt hour

gwh/day gigawatt hours per day

GWHNWR Great White Heron National Wildlife Refuge (Florida)

GWHS George Washington High School; George Westinghouse High School

GWI Grinding Wheel Institute; Ground Water Institute

G'wich Village Greenwich Village

Gwin Gwinett

GWK Grenswisselk-Kantoren

GWMNP George Washington Memorial National Parkway

GWO General Wage Order

GWOA Guerrilla Warfare Operational Area

gwp (GWP) gross world product

GWP *Government White Paper*

GWPA *Grote Winkler Prins Atlas* (Dutch—Great Winkler Prins Atlas)

GWR General War Reserves; Great Western Railway

gwrbi game-winning run(s) batted in

GWRI Ground Water Resources Institute

gws grid-wire sensor

GWS Geneva (Convention for the Amelioration of the) Wounded and Sick (in Armed Forces in the Field); George Washington School; Gir Wildlife Sanctuary (India)

gwt glazed wall tile

GWTA Gift Wrappings and Tyings Association

G W T W *Gone With The Wind*

GWU George Washington University

GWVA Great War Veterans Association

GWWD Greater Winnipeg Water District (Railway)

Gwyn Gwynedd; Gwynne

gx (GX) government exhibit

gxmtr guidance transmitter

gxt graded exercise test(ing)

gy gray; gunnery; gyro; gyro-car; gyrocompass; gyrodyne; gyroscope

gY greenish yellow

Gy gray

GY Guyana (Internet code)

gya got yuh again (slang for caught you again)

GYE Guayaquil, Ecuador (airport)

gym gymnasium; gymnastics

Gym Gymnastics

GYM General Yard Master; Guyamas, Mexico (tracking station)

gymstic gymnastic(s)

gyn gynecology

gyn *gyne* (Greek—woman)— gynecologist; gynecology; misogyny; polygyny

G.Y.N. gynecologist

gynae(col) gynaecological; gynaecologist; gynaecology

Gyn Dyn General Dynamics

gynecol gynecologic(al)(1y); gynecologist; gynecology

gyp gypsum; gypsy; cheat or swindle (slang)

GYP Guild of Young Printers

gyp bd gypsum board

gypsiol gypsiologic(al)(1y); gypsiologist(s); gypsiology

gypsum calcium sulfate ($CaSO_4$ $2H_2O$)

'gyptian(s) Egyptian(s)

gyro gyrocompass; gyroplane; gyroscope

gyrocop gyrocopter

gyrocopter autogyro helicopter (rotary-wing aircraft driven forward by a conventional propeller)

gyrodyn gyrodynamic(al)(1y); gyrodynamicist; gyrodynamics

GYS Co Great Yarmouth Shipping Company

gz ground zero

Gz Gomez

GZ Girozentrale Vienna

G-Z Guilford-Zimmerman test(ing)

GZ *Girozentrale Vienna* (German—Vienna Central Exchange)—Austrian international bank

GZG *Gutegemeinschaft Zinngerat* (German—Pewter Quality Society)

GZn grid azimuth

GZT Greenwich Zone Time

H

h hail; halfback (football); hard; hardening; hardness; hazy; heath; hectare; hecto; height; heir; heiress; heritability; hexadecimal; high(er); hind; hit(s); home; horse; hour(s); house; hundred(s); hurdles; husband; hydrant; hydraulic(s); hydrodynamic head (symbol); hydrolysis; Planck's constant (symbol); Planck's element of action (symbol)

(h) per hypodermic

h altitude (symbol); atmospheric head (symbol)

H amateur broadcasting (symbol); ceiling (symbol); declared or paid after stock dividend or split-up (in newspaper stock listings); Fraunhofer line produced by calcium (symbol); Hamiltonian function (symbol); Hangarage; Harbor; hard; hardness; harmonic (symbol); hatch; Hauch; headlines; heart disease potential; heat; heater; helicopter; henry; heroin; Hill; Hindu; Hinduism; Hispanic; Holtzkriecht unit (symbol); horizontal component of the Earth's magnetism (symbol); hospital; hot; Hotel-code for letter H; humidity; hydrogen; hyperopia; intensity of magnetic field (symbol); latent hypermetropia (symbol); magnetizing force unit function (symbol); maximum altitude (symbol);

McDonnel Aviation; Minneapolis–Honeywell (trademark); very hazy (symbol)

H Hacienda (Spanish—customs service, treasury); *haut* (French—high); *heet* (Dutch—hot); *Herren* (German or Swedish—gentlemen); *herrer* (Norwegian—gentlemen); *het* (Norwegian—hot); *hinaus*(German—out); *hombres*(Spanish—men); *Hoyre* (Norwegian—Right)—Conservative Party

H- *Hauptstimme* (German—principal voice)—12-tone term

H¹ protium

H¹+ proton

H-1 symbol for radar air navigation system

H-2 Australian macadamia

H² deuterium (heavy hydrogen symbol)

H₂O water

H₂O₂ hydrogen peroxide

H₂S hydrogen sulfide

H₂SO₄ sulfuric acid

H₃ procaine hydrochloride (symbol)

H³ tritium

H₃BO₃ boric acid

H-13 Bell three-place helicopter named Sioux and made in Britain, Italy, and Japan

H-19 Sikorsky transport helicopter called Chickasaw or U-19

H-23 Hiller utility helicopter used by USA and called Raven

H24 hard rolled and partially annealed (half hard)

H-34 Sikorsky troop-transport helicopter called Choctaw

H-37 Sikorsky heavy helicopter called Mojave

H-43 Kaman utility helicopter called Huskie

H-53 Sikorsky CH-53 Stallion assault helicopter

ha hahnium, unnilpentium; hardy annual; hatch(way); hectare; heir apparent; high altitude; high angle; home address; hostile aircraft; hour angle; hour aspect

ha (HA) humic acid

ha' half

h.a. hoc anno (Latin—in this year)

Ha hahnium (element 105); Haiti(an); Hawaii(an)

Ha (German pronunciation for B sharp)

HA Hatch Act; Hawaiian Airlines; Headquarters Administration; Health Alliances; Heavy Artillery; Historical Association; Horse Artillery; Hospital Apprentice; House of Assembly; Housing Authority

H-A Hautes-Alpes

H/A Havre-Antwerp (range of ports)

HA Hardware Age

HA-200 Hispano Saeta jet trainer

HA-220 Hispano Saeta ground-attack jet fighter

haa heavy anti-aircraft; heavy antiaircraft artillery; height above airport

haa (HAA) hepatitis-associated antigen

HAA Helicopter Association of America; Home Automation Association; Hospital Activity Analysis; Hotel Accountants Association; Humanist Association of America

HAAC Harper Adams Agricultural College

HAAFE Hawaiian Army and and Air Force Exchange

haandb haandbog (Dano-Norwegian—handbook)

Ha'uretz (Hebrew—The Land)—Israeli newspaper

Haar Haarlem (Dutch—Harlem) provincial capital of North Holland in the Netherlands

haa's heterocyclic aromatic amimes

haat height (of tv transmission antenna) above average terrain

haatc high altitude air traffic control

haaw heavy anti-tank assault weapon

hab high-altitude bombing; habitat; habitation

hab habitantes (Spanish-inhabitants)

Hab Habakkuk, Habana (Spanish-Havana); The Book of Habakkuk

HAB Hazards Analysis Board (USAF)

HAB Handels Aktie Bolag (Swedish-Limited Trading Company)

HABA Hardwood Agents and Brokers Association; Health And Beauty Aids

hab. corp. habeas corpus (Latin—may you have the body)—prisoner's right to be brought before the court so its judge may decide on the legality of the detention

habe habeas corpus

habit. habitat (Latin—it inhabits)

HABITAT Homeless Americans and Individuals Taking Action Together

habs high-altitude bombsight

HABS Historic American Buildings Survey

habt habeat (Latin-let him have)

hac high acceleration cockpit; high alumina cement

HAc acetic acid

HAC Hawkesbury Agricultural College; Heart of America Conference; Helicopter Aircraft Command(er); Hines Administrative Center; Honourable Artillery Company; Hughes Aircraft Company

hacc high alumina cement concrete

HACC Harrisburg Area Community College

Haccp (HACCP) hazard analysis critical control points

hace (HACE) high-altitude cerebral edema

hack hackney coach; hackney horse; taxicab

hack hacking (illegally breaking into computerized systems)

Hack Hackbrett (German—cymbalom or dulcimer)

Hackworth Hackworth's Digest of International Law

hacls (HACLS) harpoon-type aircraft command and launch subsystem missile

HACR Heating, Air Conditioning & Refrigeration (circuit breaker)

HACS Hyperactive Child Syndrome

HACTL Hong Kong Air Cargo Terminal Limited

HACU Hansa (Line) Container Unit

had head acceleration device; heat-actuated device (thermostat); hereinafter described

Had Hadley

H/A or D Havre-Antwerp or Dieppe (grain trade)

H^{ada} Hacienda (Spanish—estate, farm, ranch)

HADA Hawaiian Defense Area

hadbn had been

HADC Holloman Air Development Center

HADES Hypersonic Air Data Entry System

HADIS Huddersfield and District Information Service

HADIZ Hawaiian Air Defense Identification Zone

hadn't had not

HADR Hughes Air Defense Radar

hads hypersonic air data sensor

hae hereditary angioedema

ha'e (Gaelic contraction—have)

Haeck Ernst Heinrich Haeckel; Haeckelian; Haeckelism

HAECO Hong Kong Aircraft Engineering Company

HAER Historic American Engineering Record

haes high-altitude-effects simulation

haf high-abrasion furnace; high-altitude fluorescence

HAF Hebrew Arts Foundation; Helicopter Assault Force; Hellenic Armed Forces; Helms Athletic Foundation; Helvetia-America Federation

HAFB Homestead Air Force Base (Florida)

haf black high-abrasive furnace black

HAFMED Headquarters-Allied Forces Mediterranean

HAFO Home Accounting and Finance Office (USAF)

HAFRA Hat and Allied Feltmakers Research Association

HAFSE Headquarters, Armed Forces, Southern Europe

HAFTB Holloman Air Force Test Base

Hag The Hague

Hag Haggai; Hagigah (Book Of)

HAG Hardware Analysis Group

HAGB Helicopter Association of Great Britain

hagio hagiogracies, hagiogracy, hagiographer, hagriographic(al)(1y), hagiographist, hagiography, hagiolater, hagiolatrous, hagiologic(al)(1y), hagiologies, hagiologist(ic)(al)(1y), hagiology, hagiolotry, hagioscope, hagioscopic

hagiol hagiology

Hague The Hague

Hai Haiti (whose capital is Port-au-Prince)

HAI Helicopter Association International; Holocaust Awareness Institute (University of Denver); Hospital Audiences Incorporated

HAIA Hearing Aid Industry Association

haic hetero-atom in context

haid hand-emplaced acoustic intrusion detector

HAIL Hague Academy of International Law

H & A Ins Health and Accident Insurance

hair high-accuracy instrumentation radar

hairdrsr hairdresser

HAISS High-Altitude Infrared Sensor System

hait *haitisch* (German—Haitian)

Hait Haitian

Haiti Republic of Haiti (French-speaking West Indian nation occupying western half of Hispaniola), *République d'Haiti*

HAJ Hanover, Germany (airport)

Hak Hakka; Hakodate

HAKASH *Hayl Kashish* (Hebrew—Army of Elders)—Israel's senior-citizen corps

Haken *Hakenkreuz* (German-hooked cross); swastika

Hak Soc Hakluyt Society

hal halogen(ic); handicapped assistance loan; helmet audio link(age)

Hal Halawa; Halbert; Halcott; Haiden; Halensee; Halex; Halford; Halfrid; Haliburton; Hallam; Halleck; Hallett; Halogen; Halsey; Halsom; Halstead; Halvar; Halworth; Harold; Harry

Hal *Hallah*

HAL Hamburg-Amerika Line (Hamburg-America Line); Hamburg-Atlantic Line; Hardboards Australia Limited; Hawaiian Airlines; Holland America Line

Halawa Halawa Jail at Aiea on Oahu, Hawaii

HALDIS Halifax and District Information Service

Haleakala Haleakala National Park and Haleakala Volcano on the Hawaiian island of Maui

half-g half-gallon(s)

Hali Halifax

halite rock salt (sodium chloride)

Hallé Halle Orchestra in Manchester, England

Halle a/S *Halle an der Saale* (German—Halle on the Salle River)

Halle Bach Wilhelm Friedemann Bach

Hallowell Girls Stevens School for female juvenile delinquents at Hallowell, Louisiana

hallu hallucinant; hallucinate; hallucination; hallucinogen; hallucinogenic

halluc hallucination

hallus hallucinations; hallucinogens

halo high-altitude large optics; high-altitude low opening

Hal Orch Hallé Orchestra (Manchester)

HALS Harwell Automated Loans System

HALT Help Abolish Legal Tyranny; High-Altitude Laser Targeting; Houston Anti-Litter Team

haltata high-and-low-temperature-accuracy testing apparatus

Halterm Halifax Container Terminal

halv hamster leukemia virus

Halv *Halvøy* (Dano-Norwegian—peninsula)

ham hammer throw; hardware-associated memory

Ham Hamal; Haman; Hamblin; Hamburg; Hamed; Hamilton; Hamitic; Hamlet; Hamlin; Hamlyn; Hammerfest; Hamnet; Hamon

HAM Hamburg, Germany (airport)

HA & M *Hymns Ancient and Modern*

ham and ham and eggs

Hamb Hamburg

Haml *Hamlet, Prince of Denmark*

hamlet ham omelet

hamletom ham, lettuce and tomato (sandwich)

Hamm Hammerfest, Norway

hamma' hammer

hammer and sickle communist symbol; the crossing of the proletarian hammer and the agrarian sickle

Hammersleys short form for the Hammersley Mountains of Western Australia

ham 'n' eggsan ham-and-egg sandwich

ham 'n' eggwich ham-and-egg sandwich

Hamp Hampton Roads; Lionel Hampton

Hamptons Bridgehampton, Easthampton, Southhampton, and Westhampton, Long Island, New York; collective short form for all Hamptons such as Bridgehampton, East Hampton, Hampton Bays, Southampton, West Hampton, and West Hampton Beach—all at the eastern end of Long Island, New York, plus the original English estates and homestead place-names such as Hampton, Hampton Bishop, Hampton Court Palace, Hampton Heath, Hampton in Arden, Hampton Lovett, Hampton Lucy, Hampton Poyle, Hampton Wick, and Northampton as well as the great port of Southampton, and adjacent Southampton Water plus all other Hamptons

hams. hour angle of the mean sun

hamsan ham sandwich

hamt human-aided machine translation

HAMTC Hanford Atomic Metal Trades Council

hamwich ham sandwich

han' hand

Han Handel; Handel Society; Hanna; Hannibal; Hanover(ian)

hand. Handling

HAND Hawaii Association for National Defence

Handb Phys *Handbuch der Physik* (German—Handbook of Physics)

hande hydrofoil analysis and design

Handl *Handlingar* (Swedish-transactions)

HANDS High-Altitude Nuclear-Detection Studies

hane hereditary angioneurotic edema; high-altitude nuclear effects; high-altitude nuclear explosion

HANES Health and Nutrition Examination Survey

HANG Hawaii Air National Guard

H-A N-G Hamburg-American North-German (United steamship lines)

Hank Henry

hanki Handkerchief

Hanover Girls Jane Porter Barrett School for (delinquent) Girls at Hanover, Virginia

Hans Johann(es)

HANS High-Altitude Navigation System

han't has not; have not

Hants Hampshire

Hanuk Hanukkah (Hebrew—Feast of Candle Lights)

HANZ Hotel Association of New Zealand

hao hardware action officer; high-altitude observation

HAO High Altitude Observatory; Horticultural Advisory Office(r)

haoa height angle of attack

hap happening; heading axis perturbation

hap (RAP) high-altitude platform

Hap Henry Harley Arnold (WW-II commander U.S. Army Air Corps)

HAP Health Access Project; Home Attendant Program

HAPAG Hamburg-American Line

HAPAG Hamburg-Amerikanische Packetfahrt Aktien Gesellschaft (German—Hamburg-American Packet Company)

hapd happened

hapdar hardpoint demonstration array radar

hapdec hard point decoy

HAPE High-Altitude Pulmonary Edema

ha'penny halfpenny

ha'p'orth half-pennyworth

happ high air pollution potential

haps happenings

hap's housing assistance payments

har harbor; harmonic

Har Harbin; Harbor; Harbour; Harold; Harwich

HAR Harrisburg, Pennsylvania (airport)

HARAO Hartford Aircraft Reactor Area Office

Harare, Zimbabwe formerly Salisbury, Rhodesia

harb harbor

Harbison Girls Harbison Correctional Institution for Women at Irmo, South Carolina

Harbrace Harcourt Brace Jovanovich

HarBraceJ Harcourt Brace Jovanovich

Har Bus R Harvard Business Review

HARB Historic Architectural Review Board

HARC Human Affairs Research Center

harcft harbor craft

Harcourt Harcourt Brace Jovanovich

hard hardwater

hard. hardware

Hard Hardangerfelen (Norwegian—Hardanger fiddle)

HARD Hardware Resources for Development

Harden (British contraction Harwarden)

hard porn hard-core pornography

hardtack ship's biscuits

Hardwick Girls Georgia Rehabilitation Center for Women at Hardwick

hare. high-altitude ramjet engine

Hare Soviet Mi-1 utility helicopter

HAREP Harbour Repairs

HARES High-Altitude Radiation Environment Study

Har Hakarmel (Hebrew—Mount Carmel)

HARIS High-Altitude Radiation Instrument System

harma harmattan (warm dry wind on the west coast of Africa)

harm. harmonic; harmony

harm (HARM) high-speed anti-radiation missile; hyper-velocity anti-radiation missile

Har Mag Harvard Magazine

HARM high-speed anti-radiation missile; Humans Against Rape and Molestation

harn harness

HAR n Hartford Dark Blues (National Association)

harn lthr harness leather

harp harpoon; harpsichord; harpsichordist; heater above reheat point; heating, air conditioning, refrigeration, plumbing; high-altitude relay point; high-altitude research probe

Harp Halpern's anti-radar point

HARP Helmlich-Armstrong—Rieveschi-Patrick (aerospace heart pump); Honeywell Acoustic Research Program

Harp Baz Harper's Bazaar

Harper Harper & Row

harps. harpsichord

harpsi harpsichord

Har-Row Harper and Row

Harry Harold; Henry

hart (HART) hyper-velocity anti-aircraft rocket tactical

Hart Hartford

HART Halt All Racist Tours; Highway Advisory Radio Tactical; Honolulu Area Rapid Transit

Hartran Hartwell Atlas fortran

Hart Sym Orch Hartford Symphony Orchestra

HARU Harrison Line (container) Unit

harv (HARV) high-altitude research vehicle

Harv Harvard; Harvey

Harw Harwarden *(Harden)*

HARYOU Harlem Youth Opportunities Unlimited

has high-altitude sample

Has Haselhorst

HAs Hells Angels; Hispanic Americans; Housing Assistants

HAS Health Advisory Service; Helicopter Air Service; Hellenic Affiliation Scale; Hospital Adjustment Scale; Hospital Administrative Services; Housing Alternatives for Seniors

HASAWA Health and Safety at Work Act

HASC House (of Representatives) Armed Services Committee

HASCO Haitian-American Sugar Company

HASD Humanist Association of San Diego

hash. hashish

Hashbury Haight-Ashbury (district of San Francisco)

Hasid Hasidim (Hebrew—godly pious people)

HASL Health and Safety Laboratory (Atomic Energy Commission)

hasn't has not

hasp. hardware-assisted software polling; high-altitude sampling program; high-altitude space platform

HASP Hawaiian Armed Services Police; Houston Automatic: Simulator of Peripherals

haspa high-altitude superpressure-powered aerostat

hasr high-altitude sounding rocket

hast high-altitude supersonic target

Hastings Hastings House; Hastings-on-Hudson

Ha strain Harris (viral) strain

hasvr high-altitude space-velocity radar

hat height above touchdown; high-altitude temperature

HAT hypoxanthine, aminopterin, thymidine

HATA Hong Kong Association of Travel Agents

hato handling tool

hatoff highest astronomical tide of the foreseeable future

hatom highest astronomical tide of the month

hatoy highest astronomical tide of the year

HATRA Hosiery and Allied Trades Research Association

hatrack. hurricane and typhoon tracking

HATREMS Hazardous and Trace Emissions System

HATRICS Hampshire Technical Research Industrial and Commercial Service

hats. hour angle of the true sun

HATS Helicopter Advanced Tactical System

Hatteras Cape Hatteras, the Cape Hatteras National Seashore Recreational Area, Hatteras Inlet, Hatteras Island, the village of Hatteras—all part of North Carolina's Outer Banks

Hau Hausa

Hauptw Hauptiverk (German—great or chief work)

haust. haustus (Latin—a draught)

haut hautboy (oboe)

hav haversine

hAv hepatitis A virus

Hav Havre

HAV Havana, Cuba (airport); Hepatitis A Virus

HAVEN Help Addicts Voluntarily End Narcotics

HAV/IgM Hepatitis A Virus immunoglobulins—specific

haven't have not

havoc histogram average ogive calculator

Havre Havre de Grace, Maryland; Le Havre (de Grace), France

haw highly active waste (radioactive); hour angle west

haw (HAW) heavy anti-tank assault weapon

Haw Hawaii; Hawaiian

HAW Kauai, Hawaii (tracking station)

HAWA Hammond Ambassador World Atlas

Hawaii Volcanoes Hawaiian national park

hawb house air waybill

HAWC Help for Abused Women and Children, Salem, Mass.

HAWE Honorary Association for Women in Education

HAWEIT Harnburg-Wechsler Intelligence Test

hawk (HAWK) homing-all-the-way kill (missile)

Hawks Nest Hawks Nest State Park, West Virginia

Haw'n Hawaiian

Hawna Hawaiiana

Hawn Isl Hawaiian Islands

HawTel Tel Hawaii Telephone (company)

Hawthorn Hawthorn Books; Missouri state flower

hax hrir/apt interface (high-resolution infrared radiometer/automatic picture transmission)

hay hay-de-guy, hay-de-guise (French—a dance)

Hay Hayle

HAYP Hire-A-Youth Program

haystaq have you stored answers to questions?

haz hazard; hazardous

haz con hazard control

HAZMAT Hazrdous Materials Response and Assessments Division

hb halfback; halfbound; half bow; half breadth; handbook; hard black; hardback (book); hardy biennial; hatchback; heavy barrel; heavy bombardment; heavy bombing; hemoglobin; herringbone; high band; hit batsmen (baseball); hollow bar; homing beacon; horizontal bands; horizontal bombing; hose bib; human being

h/b handbook

Hb hemoglobin; Herbarium

H^b deuterium (heavy hydrogen symbol)

Hb Habakkuk; Hoboe (German—oboe)

HB Hawke's Bay; Hawthorn Books; Hector Berlioz; High Bridge; Household Bank; House (of Representatives) Bill

H-B Huebner-Bleistein (process)

H & B Humboldt and Bonpland

HB Hindi Bhiarat (Hindustani—Republic of India)

Hba Habana (Spanish—Havana)

HBA Hispanic Bankers Association; Hoist Builders Association; Hollywood Bowl Association; Home Baking Association; Home Builders Account; Honest Ballot Association; Hospital Benefit Association; Housing Builders Association

h'back hatchback

h/back hardback

h B ag hepatitis B antigen

H-bar capital-H-shaped bar

HbA1c glycosolated hemoglobin

HBAVS Human Betterment Association for Voluntary Sterilization

hbb hollow-bored bar

hbc high breaking capacity; historically black college; historic black colleges

hbcu historically black colleges and universities

HBcAb Hepatitis B core antigen antibody

HBcAg Hepatitis B core antigen

HBC Hit By Car; Hokkaido Broadcasting Company; House Budget Committee; Hudson's Bay Company

HbCO carbon monoxide hemoglobin

hbd hardboard; has been drinking; headboard; herein-before described

hbd (HBD) hydroxybutyrate dehydrogenase

HBD Harbor Board; Harbour Board

hbe hard-boiled egg(s); his bundle electrogram

H-beam capital H-shaped beam

HBeAb Hepatitis B e antigen antibody

hbf hepatic blood flow

Hbf fetal hemoglobin

Hbf Hauptbahnhof (German—depot, main station)

HBF Hospital Benefit Fund

Hbg Hamburg; Harrisburg; Helsingborg (Hälsingborg)

HBG Henry B(arbosa) Gonzalez; Hongkong Bank Group; Huntington Botanical Gardens

HBIG Hepatitis B Immune Globulin

HBJ Harcourt Brace Jovanovich

hbk halfback; hardback (book); hatchback; hollow back (lumber); hollowback

Hbk Hoboken

HB & K Humboldt, Bonpland, and Kunth (botanists)

hblv (HBLV) human B-cell lymphotropic virus

hbm hard bowel movement; health belief model

HBM His (Her) Britannic Majesty

hbn hazard beacon

HBNNR Hickling Broad National Nature Reserve (England)

HBNWR Holla Bend National Wildlife Refuge (Arkansas)

hbo hyperbaric oxygen

Hbo Hoboken

HbO₂ Oxyhemoglobin

HBO Home Box Office

HBOs Health Benefit Organizations

H/board hardboard; headboard

HBOG Hudson's Bay Oil and Gas

HBOI Harbor Branch Oceanographic Institution, Fort Pierce, Florida

H-bomb hydrogen bomb

HBO S Oxyhemoglobin

hbp high blood pressure; hit by pitch (baseball)

HBPS Home Building Plan Service

hbr has been reviewed

Hbr Harbor

HBR Hudson Bay Railway

HBR Harvard Business Review

hbr acw has been reviewed and concurred with

hbs (HBS) hulking building syndrome

hb's halfbacks

Hbs sickle-cell hemoglobin

HBsAb Hepatitis B surface antigen antibody

HBsAg Hepatitis B surface antigen

HBS Harvard Business School; Hawaiian Botanical Society; Hope Botanic Gardens

Hbt Hobart

HbT Homeobotanical Therapist

HB & T Houston Belt and Terminal (railroad)

Hbt's human-breast tumors

hBv hepatitis B virus

HBV Hepatitis B Virus

H & BV Houston and Brazos Valley (railroad)

hbw highspeed black-and-white (photography)

Hbwr Halden boiling heavy water reactor

hby hereby

hc habitual criminal; hand control; hard copy; heating cabinet; hexachlorethane; high carbon; high-capacity; highly commended; house call; hydrocarbons

hc (HC) hard copy

h/c hard cover(ed); held covered

h & c heroin + cocaine; hot and cold (running water)

h.c. hac nocte (Latin—tonight); *honoris causa* (Latin—out of respect for); *hors commerce* (French—not for sale, privately printed)

Hc computed altitude; Hermitian conjugate; Huntington's chore

HC Hagerstown College; Hague Convention; Hamilton College; Hamline College; Hanover College; Harding College; Harpur College; Hartford College; Hartnell College; Hartwick College; Hastings College; Haverford College; Health Certificate; Heidelberg College; Helicopter Council; Hendrix College; Hershey College; Hertford College; Hesston College; High Commission(er); High Councillor; Higher Certifi-

cate; Highest Commendation; Highway Code; Hillsdale College; Hiram College; Hood College; Hope College; Hospital Consult(ation); Hospital Corps; House of Commons; House of Correction(s); Housing Commission; Housing Corporation; Howard College; Humphreys College; Hunter College; Huntingdon College; Huntington College; Huron College; Hussan College; Hutchinson College; hydrocarbon(s)

HC Holy Communion

H-C Harbison-Carborundum

H.C. High Commission

H of C House of Commons; House of Correction

HC Hartford Courant; haute-contre (French—high tenor)

HC-54 Douglas C-54 modified for search-and-rescue missions

hca held by civil authorities; hydroxcitric acid

HCA Habitat Conservation Area; High Conductivity Association; Hobby Clubs of America; Hospital Corporation of America; Hotel Corporation of America; Hunting-Clan Air Transport

HC(A) Helicopter Coordinator (Airborne)

HCAAS Homeless Children's Aid and Adoption Society

hcap handicap

H-caps heroin capsules

HCAs heterocyclic amines (potent carcinogens); heterocyclic aromatic amines

hcb hard-core base; hard-covered book; heating and cooling of buildings; hollow concrete block(s)

HCB House of Commons Bill

HCBS Home and Community-Based Services

hcc hydraulic cement concrete

hcc (HCC) 25-hydroxycholecalciferol (vitamin D³ metabolite)

HCC Hebrew Culture Council; Holyoke Community College

HCCA Health Care Consumers Association

HCCJ Harvard Center for Criminal Justice

hcd hard-copy device; high current density

hcd (HCD) human chorionic gonadotropin

HCD Housing and Community Development

HCDCS Harmonized Commodity Description and Coding System

HC Deb House of Commons Debates

hce human-caused error

HCEA Healthcare Convention and Exhibitors Association

HCEEP Handicapped Children's Early Education Project

hcef henceforth

HC & ES Hull Chemical and Engineering Society

hcex high-speed color exterior

hcf haemolytic complement fixation; hardened compacted fibers; height-correction factor; high carbohydrate fiber (diet); high cycle fatigue; highest common factor; Hundred cubic feet

hcf (HCF) high-carbon ferrochrome

HCF Health Care Financing; Honorary Chaplain to the Forces; Hospital Contribution Fund(ing); Hungarian Cultural Foundation

HCFA Health Care Financing Administration

hcfc (HCFC) halogenated chloro fluorocarbon

hcg horizontal location of center of gravity; human chorionic gonadotropin pregnancy test

hch (HCH) hexachlorocyclohexane (insecticide)

HCH Herbert Clark Hoover (31st President U.S.)

HCHI Hand Chain Hoist Institute

HCHP Harvard Community Health Plan

HCI Handgun Control Inc; Home Center Institute; Hotel and Catering Institute; Hughes Communications Incorporated

HCIL Hague Conference on International Law

HCIMA Hotel and Catering Institutional Management Association

HCIS House Committee on Internal Security

HCITB Hotel and Catering Industry Training Board

HCJ High Court of Justice

HCJC Howard County Junior College

hcl high cost of living; horizontal center line

h cl hanging closet

HCl hydrochloric acid (muriatic acid)

HCL Hod Carriers, Building and Common Laborers

HCLIP Home Care Live-In Plan

H-class Soviet missile-launching nuclear-powered submarines called Hotel by NATO

hcm half-cycle magnetizer

HCM Her (His) Catholic Majesty

HCM Ho Chi Minh (Chinese— He Who Shines)

HCMC Ho Chi Minh City

hcmm heat-capacity map mission (NASA)

HCMPA Home Counties Master Printers' Alliance

hcmr heat-capacity mapping radiometer

HCMT Ho Chi Minh Trail

HCMW Hatters, Cap and Millinery Workers (union)

hcn hydrocyanic acid

HCn hydrocyanic acid

HCN House (of Representatives) Committee on Narcotics

HCNZ Housing Corporation of New Zealand

hco hydrogenated coconut oil; health-care options

HCO Harvard College Observatory; Headquarters Catalog Office

HCO₃ bicarbonate ion

hcp habitat conservation plan; handicap; hexachlorophene; hexagonal close-packed; high card point; humidity control(ler's) panel

HCP Honors Cooperative Program

HCP House of Commons Proceedings

HCPNI Hardware Cloth and Poultry Netting Institute

hcpp health-care prepayment plan

HCPT Historic Churches Preservation Trust

hcptr helicopter

HCR High Chief Ranger

HCRAO Hat Creek Radio Astronomy Observatory (University of California)

HCRC Honeywell Corporation Research Center

hcrit hematocrit

HCRS Heritage Conservation and Recreation Service

hcrw hot and cold running water

hcs high-carbon steel

hcs (HCS) human chorionic somatomammotropin; hydrological communications satellite

hc's hard cover books

HCs Hebrews converted to Roman Catholicism; hydrocarbons

HCS Hallé Concerts Society; Harvey Cushing Society; Home Civil Service; Hydromechanical Control System

HC & S Hawaiian Commercial and Sugar (company)

HCSA Hospital Consultants and Specialists Association; House (of Representatives) Committee of Space and Astronautics

hcsht high-carbon steel heat treated

HCSI Health Correspondence Schools International

hct heater center tap; hematocrit

hct. hematocrit

H Ct High Court

HCT High Commission Territories; Huddersfield College of Technology

HCTBA Hotel and Catering Trades Benevolent Association

hcu homing comparator unit; hydraulic cycling unit

HCUA Honeywell Computer Users Association

HCV Hepatitis C Virus; Housing Commission of Victoria

HCVA Historic Commercial Vehicle Association

HCVC Historic Commercial Vehicle Club

hcvd hypertensive cardiovascular disease

hd half day; hand; hard-drawn; head; hearing distance; heavy duty; high density; hogshead; homeodomain; horse-drawn; hourly difference; hundred; hurricane deck

hd (HD) half duplex (data processing); harmonic distortion (audio); harmonic definition (video)

h-d heavy-duty; high-density

h/d holddown

h.d. *hora decubitus* (Latin—at bedtime)

Hd Head

Hd (HD) Huntington's disease

Hd *Hochdruck* (German—high pressure)

HD Hansen's Disease (leprosy); Harbor Defense; Harbor Drive; Historic District; Historical Division; Home Defense; Honorable Discharge; Hoover Dam

H.D. Hilda Doolittle

H/D Havre-Dunkirk (range of ports)

H & D Hurter & Driffield (photo emulsion speed)

hda high-duty alloy(s); horizontal danger angle; hydroxydopamine

HDA High Duty Alloys

hdatz high-density air traffic zone

HDB Housing Development Board

hdbk handbook

hdbn had been

hdc half double crochet; high-density cotton; high-dose chemotherapy; holder in due course

HDC Helicopter Direction Center; Housing Development Corporation

HD Clinic Hansen's Disease Clinic (for lepers)

hd cr hard chromium

hdd head-down display; heavy-duty detergent

HDD Higher Dental Diploma

hddr high-density digital recording

HDDS High-Density Data System

HDE Higher Diploma in Education

h de c *hidratos de carbono* (Spanish—carbohydrates)

hded heavy-duty enzyme detergent

hdf high-frequency direction finder

hdg heading

HDGA Hot Dip Galvanizers Association

HDH Hawker de Haviland

H-D-H Holland-Dozier-Holland (Brian Holland, Lamont Dozier, Eddie Holland)—successful songwriters for Motown music groups

hdhc high-density hydrocarbon(s)

HDHD Hawaiian District Harbors Division

hdhl high-density helicopter landing (USA)

HDHQ Hostility and Direction of Hostility Questionnaire

HDI Human Development Index; Humane Development Institute

hdip hazardous-duty incentive pay

H Dip E Higher Diploma in Education

H disease Hart's disease

hdk husbands don't know

h dk hurricane deck

hdkf handkerchief

hdl handle; hardware description language

hdl (HDL) high-density cholesterol; high-density lipoproteins

HDL Harry Diamond Laboratory (US Army Diamond Ordnance Fuze Laboratory); Hydrologic Data Laboratory (USDA)

HDLC High-level Data Link Control

hdlg handling

hdlr handler

hdls headless

hdlw hearing distance, watch at left ear

hdm high-duty metal

HDM Hierarchical Development Methodology

hdmi high-density multichip interconnect

HDML Harbor Defense Motor Launch

hdmr high-density moderated reactor

hdn harden

hdn (HDN) hemolytic disease of the newborn

HDO Home Dish Only (Satellite Networks Incorporated)

H Doc House Document

hdp (HDP) hexose diphosphate

hdpe high-density polyethylene

hdpg half deck plate girder

hdqrs headquarters

hdr handrail; header

hd & r human development and relationships

HDR Humanitarian Daily Ration

HDRA Heavy-Duty Representatives Association

HDRI Hannah Dairy Research Institute

HDRSS High-Data-Rate Storage System(s)

hdrv human diploid-cell rabies vaccine

hdrw hearing distance, watch at right ear

hds heat detectors; holidays; hundreds; hydrodesulfurization

Hds Holidays (of Obligation)

HDS Hospital Discharge Survey; Human Development Services

Hd Schm Head Schoolmaster

hdsp hardship

hdst high-density shock tube

HDST Hawaiian Daylight Saving Time

HDT Henry David Thoreau

hdta high-density-traffic airport

HDTI Human Development Training Institute

hdtm heavy-duty target mechanism

HDTMA Heavy-Duty Truck Manufacturers Association

hdtp hardtop

HDTS Harbor Drive Test Site (Convair Ramp)

hdtv (HDTV) high-definition television

hdu hemodialysis unit

h/duty heavy duty

hdv heavy-duty vehicle; high dollar value; human delta virus

hdw hardware

Hdwbch *Handwörterbuch* (German—pocket dictionary)

hdw c hardware cloth (wire screen)

hdwd hardwood

hdwe hardware

hd whl hand wheel

hdwre hardware

hdx (HDX) half duplex (data processing)

he hammerless ejector; heat engine; heavy enamel; height of eye; high explosive, horizontal equivalent; hub end; human enteric; hydrogen embrittlement

h & e hemotoxylin and eosin; heredity and environment

h.e. hic est (Latin—this is)
He Hebraic; Hebrew; helium; Hertz
He. Book of Helaman
HE Her Eminence; Her Excellency; high explosive; His Eminence; His Excellency; Hollis & Eastern (railroad); Human Engineering; Hydraulics Engineer(ing)
H.E. His Eminence; His Excellency
HE Human Events
HEA Higher Education Act; Home Economics Association; Horticultural Education Association
heaa high-explosive anti-aircraft (shell)
HEAA Home Economics Association of Australia
head(s). headache(s)
HEADS-UP Health Care Delivery Simulator for Urban Populations
heaf heavy end aviation fuel
HEAF Higher Education Assistance Foundation
heafs high-explosive antitank fin-stabilized
HEAL Health Education Assistance Loans; Home Environment Aid for Living; Human Ecology Action League
HEALT Helicopter Employment and Assault Landing Table
HEAO High-Energy Astronomical Observatory
heap high-explosive armor-piercing (shell)
HEAR Hospital Emergency Administrative Radio
HEARS Higher Education Administration Referral Services
HEARU Higher Education Advisory and Research Unit
heat heat escape lessening posture; heating; high-explosive anti-tank (projectile)
HEATH Higher Education and the Handicapped
Heathrow airport near London
heavy water deuterium oxide
Heb Epistle of Paul the Apostle to the Hebrews: Hebraic; Hebrew
Heb Hebrew (classical language)
HEBA Home Extension Building Association

hebc heavy enamel bonded single cotton
hebd hebdomadal (weekly)
hebdo hebdomad(al)(1y); hebdomadaries; hebdomadary
hebdom. hebdomas (Latin—week)
hebdomag hebdomadal magazine—weekly magazine
hebdp heavy enamel bonded double paper
hebds heavy enamel bonded double silk
Heb Ety Hebretasebawit Etyopia (Amharic-Socialist Ethiopia)—formerly Abyssinia
hebr hebräisch (German—Hebraic)
Hebr Hebraic; Hebrew; Hebrides
Hebrides Hebrides Islands off Scotland
hec heavy-enamel single-cotton (insulation)
Hec Hasselblad electric camera; Hector; Hector Berlioz (French composer-conductor); Hecuba; Hollerith electronic computer
HEC Hawaii Electric Company; Health Education Council; Hydro-Electric Commission
HEC Hautes Étuder Commerciales (French—High Commercial Studies) France's leading business school; *Heure de l'Europe Central* (French—Central European Time)
HECC Higher Education Coordinating Council (St Louis library network)
he cls b heating coils in bunkers
he cls ct heating coils in cargo tanks
hecm home-equity conversion mortgage
HECO Hawaiian Electric Company; Hydro-Electric Commission of Ontario
he comp helium compressor
hect hectare; hectoliter
Hect Hector
HECT Hydro-Electricity Commission of Tasmania
hecticity hectic activity
hecto 10^2
hectog hectograin
hectol hectoliter
hectom hectometer

hector. heated experimental carbon thermal oscillator reactor (HECTOR)
hed horizontal electric dipole
hed (HED) high-energy detector
he'd he had; he would
HED Haupt-Einheits Dosis (German—unit skin dose)—X-rays
HEDCO Hawaii Economic Development Corporation
HEDCOM Headquarters Command
hedis health-care employer data information sets
HEDL Hanford Engineering Development Laboratory
hedrb high-energy deep rechargeable battery
hed('s) hearing-ear dog(s)
HEDS Hall-Effect Distribution System
hed sked headline schedule
hedsv heavy-enamel doublesilk varnish (insulation)
Hedy Hedvig; Hedwig
HEE High Express Emotion
HEEA Home Economics Education Association
HEED Health and Education (department or ministry)
heei high-energy electronic ignition
heent head, ears, eyes, nose, throat
HEEP Highway Engineering Exchange Program
HEERA Higher Education Employer-Employee Relations Act
hef heifer; high-energy fuel
HEF High-Energy Fuel; Hospital Employees Federation
HEFA Higher Education Facilities Act
HEFC Higher Education Facilities Commission
heg heavy-enamel single-glass (insulation)
HEGIS Higher Education General Information Survey; Higher Education General Information System
HEH Her (His) Exalted Highness
HEHF Hanford Environmental Health Foundation (AEC)
HEHL Henry E Huntington Library
hei high-explosive incendiary; holographic exposure index

HEI Heat Exchange Institute; Hotel Enterprises Incorporated

HEI H/F Eimiskipafelag Islands (Icelandic Steamship Company)

HEIAC Hydraulic Engineering Information Analysis Center (USA)

HEIAS Human Engineering Information and Analysis Service (Tufts U)

HEIB Home Economists In Business

HEIC Honourable East India Company

HEICN Honourable East India Company Navy

HEICS Honourable East India Company Service

Heidel Heidelberg

Heidelberg Ruprecht-Karl-Universität popularly called the University of Heidelberg

Hein Heinersdorf

Heinesen William Heinesen (Danish novelist)

heip high-explosive incendiary plug

heir app heir apparent

heir pres heir presumptive

heisd high-explosive incendiary self-destroying

heit high-explosive incendiary with tracer

heitdisd high-explosive incendiary tracer dark ignition self-destroying

heitsd high-explosive incendiary tracer self-destroying

hek heavy-enamel single-cellophane (insulation)

hek (HEK) human embryo kidney

hel helicopter; home-equity line

hel (HEL) hen's egg-white lysozyme; high-energy laser (beams); human embryonic lung

Hel Helen; Helena; Helsinki (Heisingfors); Helvetia (Switzerland)

Hel Hellas (Norwegian—Greece)

HEL Hartford Electric Light; Helsinki, Finland (airport); Human Engineering Laboratories (USA); Hydraulic Engineering Laboratory

HeLa Helen Lake (tumor cells)

HELCIS Helicopter Command Instrumentation System

HELCO Hilo Electric Company

heli helicopter; heliport

helio heliochrome; heliodon; heliodor; helioelectric; helioengraving; heliogram; heliograph; heliogravure; heliology; heliostat; heliothcrapy; heliotrope; heliotype

HELIOS Handicapped people in the European community Living Independently in an Open Society

helipad helicopter landing pad

he'll he will

Hell Hellerup

HELL Higher Education Learning Laboratory

Hellen Hellenic; Hellenism; Hellenistic

hellfire (HELLFIRE) helicopter-launched fire-and-forget missile

helminthol helminthology

helo helicopter; heliport

Heloise Heloise Fulbert (abbess-scholar best remembered for her love of Abelard—a monk living in a nearby monastery he founded in 1112)

helosid helicopter-delivered seismic intrusion detector

help high-energy-level pneumatic automobile bumpers

HELP Helicopter Electronic Landing Path; Help Elderly Locate Positions; Help Establish Lasting Peace; High School Education Law Project; Highway Emergency Locating Plan; Home Environment and Living Program; Home Equity Living Plan; Homophile Effort for Legal Protection

HELPR Handbook of Electronic Parts Reliability

hel rec health record

Hels Helsingborg, Sweden; Helsingør, Denmark; Helsinki, Finland

Hel San Helsingin Sanomat (Helsinki's News)

Helv Helvetia; Helvetica

Helv Chim Acta Helvetica Chimica Acta (Latin—Swiss Review of Chemistry)

Helv Phys Acta Helvetica Physica Acta (Latin—Swiss Review of Physics)

hem. hemerocallis; hemoglobin; hemorrhage; hemorrhoid

hem (HEM) hybrid electromagnetic wave

HEM Ernest Hemingway

hema (Latin prefix—blood)—hematology

HEMA Heavy Engineering Manufacturers Association

hematol hematolith(ic)(al)(1y); hematologist; hematology; hematolymphangioma; hematolysis; hematolytic

HEMF Handling Equipment Maintenance Facility (USN)

hemi (Latin prefix—half)—hemisphere

hemi engine hemispherical combustion chamber engine

he. missile high-energy missile

hemlaw (HEMLAW) helicopter-mounted laser weapon

hemloc heliborne-emitter-location countermeasures

hemo (Latin—blood)—hemoglobin, hemophilia, hemorrhage, hemostat

hemolysis hemocytolysis

hempos hereditary erythroblastic multinuclearity (with) positive acidified serum

Hem Soc Hemlock Society

Hen Henrietta; Henry; Soviet Ka-15 light-utility helicopter

Hen V King Henry V

Hen VIII King Henry VIII

HENA Home Economics and Needlework Association

Hence Henderson

H'english Limey English

HENILAS Helicopter Night Landing System

Henk Hendrik

Henri (French—Henry)—Henri Vieuxtemps

Henriqz Henriquez

Henry Henry Ford Commercial College; Patrick Henry Commercial College; Patrick Henry High School; Patrick Henry State Junior College

Henry B Henry B Gonzalez of San Antonio, Texas

Hen wlad Hen Wlad fy Nhadau (Welsh—Land of my Fathers) national anthem of Wales

HEOC Honeywell Electro-Optics Center

HEOP Higher Education Opportunity Program

*HEOr Heure de l'Europe Ori-
ental* (French—Eastern European Time)
heos (HEOS) high eccentric
orbiting satellite
hep hepatic; high-energy phosphate; high-explosive plastic
Hep Hepburn; Hepple; Hep-
worth
HEP Have Error-free Product;
High School Equivalency
Program(s); Higher Educa-
tion Panel
hepa high-efficiency particulate
air
HEPA High Efficiency Particulate Arresting
HEPAC High-Energy Physics
Advisory Council
hepaf high-efficiency particulate air filter
HEPALIS Higher Education
Policy and Administration Library and Information Service
hepat high-explosive plastic
antitank
hepat (Latin prefix—liver)—
hepatitis
HEPC Hydro-Electric Power
Commission
HEPCAT Helicopter Pilot
Control and Training
HEPCC Heavy Electrical Plant
Consultative Council
hepdnp high-explosive point-
detonating nose plug
hepl high-energy pulse laser
HEPL High Energy Physics
Laboratory
HEPP Hoffman Evaluation
Program and Procedure
her. Heraldry
her (HER) high-energy rotor
(helicopter)
her. *heres* (Latin—heir)
her-2 human egf-receptor-re-
lated receptor
Her Hercules (constellation);
Hereford; Hereford(shire)
*HER Harvard Educational Re-
view*
hera high explosive rocket assisted
Hera (Greek—Juno)—god-
dess of the heavens
HERA Heavy Engineering Research Association; Housewives for ERA (Equal Rights
Amendment)
HERALD Highly-Enriched
Reactor—Aldermaston

herb. herbaceous; herbarium
Herb Herbert
Herc Hercules (constellation)
HERC Humber Estuarial Research Committee
Herdez (Spanish contraction—
Hernandez)
herd° herdeiro (Portuguese—
heir)
herdr herdruk(ken) [Dutch—
reprint(s)]
HERE Hotel Employees and
Restaurant Employees; Hu-
man Endurance Range Ex-
tender
hered heredity
hereds herederos (Spanish—
heirs)
Heref Herefordshire
Herefs Herefordshire
here's here is
Here & Worcs Hereford and
Worcester
herf high-energy rate, forging
herfs high-energy-rate forging
systems
HERI Higher Education Research Institute
herj high explosive ramjet
herm hermetically
hermes heavy element and radioactive material electromagnetic separator (HER-
MES)
Hermes (Greek—Mercury) the
messenger
HERMES Helicopter Energy
and Rotor Management Sys-
tem
hero heroin; hazards of electromagnetic radiation to ordnance; hot experimental reactor of 0 (zero power)—also
appears as HERO
hero heroína (Spanish—heroin)
HERO Hazard of Electromagnetic Radiation of Ordnance;
Hazardous Emanations of Radiation to Ordnance, Health
Educational Robot; Historical
Evaluation and Research Organization; Home Economics
Research Organization
Herod. Herodotus
Herois Herois do mar (Portuguese—Heroes of the Sea)
Portugal's anthem
herp herpetologist; herpetology
HERP Human Exposure Ro-
dent Potency
*HERP Bulletin of the New York
Herpetological Society*

herpetol herpetologic(al)(1y);
herpetologist; herpetology
HERPOCO Hercules Powder
Company
herps herpetological books, papers, specimens; herpetologists
*Herr Kaleun Herr Kapitän-
leutnant* (German—Mr Cap-
tain Lieutenant)—U-boat
commander
hers. herself
HERS Heart and Estrogen
Study; Higher Education Resource Services; Home Economics Reading Service
Hersch Herschel
herst herstellung (German—
manufacture)
Hertf Hertford College, Oxford
HERTIS Hertfordshire County
Council Technical Informa-
tion Service
Herts Hertfordshire
HERU Higher Education Research Unit
Herv Hervest; Hervey
Hervey Allen William Hervey
Allen
hes heavy enamel single silk
(insulation)
he's he has; he is
Hes Hesba; Hesione, Hes-
per(ian)(s); Hesketh; Hesperides; Hesperus; Hessels;
Hessian(s); Hessin; Hester;
Hesther
HES Hawaiian Entomological
Society; Health Economic
Service; Health Examination
Survey; History of Economics Society
HESCA Health Sciences Communication Association
hesd high-explosive self-destroying
hesh high-explosive squash
head
HESIS Hazard Evaluation System and Information Service
hess human-engineering systems simulator
hest heavy-end aviation fuel
emergency service tanks
h'est highest
Hest Hester
HEST High-Explosive Simulation Test
he. stor helium storage
hesv heavy-enamel single-silk
varnish (insulation)

het heavy equipment transporter

Hetch Hetchy Hetch Hetchy Dam; Hetch Hetchy Lake (both in Yosemite National Park)

hetdi high-explosive tracer dark ignition

hetero (Latin prefix—different, other)—heterosexual

heterocl heteroclite

heterog heterogeneous

heterosex heterosexual(ity); heterosexuals

HETS High-Energy Telescope System; Hyper-Environmental Test System

Hetty Hester

heu highly enriched uranium; hydroelectric units

heur heuristic (problem solution by trial and error)

Heure L'Heure Espagnole (French—The Spanish Hour)—Ravel opera

HEVAC Heating, Ventilating, and Air Conditioning Manufacturers Association

hevr heavier

Hew Heward; Hewett; Hewitt; Hewlett; Hewson; Hugh; Hugo

HEW Health, Education, and Welfare (US department); Housing, Education, and Welfare (Philippines)

HEWPR Health, Education, and Welfare (department) Procurement Regulations

hex hexachord; hexagon(al); uranium hexafluoride

hex (HEX) hexadecimal

hex. hexadecimal

hexa hexamethylene tetramine

hexag hexagon(al)

hex hd hexagonal head

hey health and environment; heavy

Hez Hezekiah

hf hageman factor; half; hard finish(ed); hard firm; height finding; high frequency (3000 to 30,000 kc); hind foot; hold fire; home freezer; hook fast; horse and foot (cavalry and infantry); hot finished; hyper filtration; hyperfocal

h/f held for

Hf Hafnium

HF Handwriting Foundation; Home Fleet; Home Forces; hydrofluoric acid; hydrogen fluoride

H of F Hall of Fame

H/F Hlutatfjelagid (Icelandic—limited company)

hfa hard factory automation

Hfa Haifa

IIFA Headquarters Field Army; Holiday Fun Association; Hollywood Film, Archive; Hourly Faculty Association

HFAA Hoistein-Friesian Association of America

HFAC Human Factors Association of Canada

HFARA Honorary Foreign Associate of the Royal Academy

hf bd half-bound

hf bd cf half bound in calfskin (calf leather back and corners)

hf bd cl half bound in cloth (cloth back and corners or cloth sides)

hf bd mor half bound in morocco (morocco leather back and corners)

HFLBLB Hokkaido Farmland Bride Liaison Bureau

hfbr high flux beam reactor

hfc hard-filled capsules; high-frequency current

HFC Household Finance Corporation; Human Freedom Center

HFCC Henry Ford Community College

hf cf half-calf

hf cl half-cloth (binding)

hfcs high-fructose corn sweetener; high-fructose corn syrup

HFCs hydrofluorocarbons

HFCT Hawaii Federation of College Teachers

Hfd Hereford

hf-df high-frequency direction finder

hfe human factors (in) electronics; human factors engineering

hff horizontal falling film

HFFF Hungarian Freedom Fighters Federation

hfg heavy free gas

HFGA Hall of Fame for Great Americans

hfh half-hard (steel)

HFH Habitats for Humanity

hfi hydraulic fluid index

HFI Helicopter Foundation International

HFIA Heat and Frost Insulators and Asbestos Workers Union; Home Furnishings International Association

hfim high-frequency instruments and measurements

hfir high flux isotope reactor

HFL Human Factors Laboratory (NBS)

hfm hold for money

HFM Henry Ford Museum

hfmd hand-foot-and-mouth disease

hfmf home-furnish monolithic floor

hf mor half-morocco

hfmr high-fat milk replacer

hfo heavy fuel oil; high-frequency oscillator; hole full of oil

Hford Hereford

HFORL Human Factors Operations Research Laboratory

Hfors Helsingfors

hfp hostile fire pay

h & f pool heated and filtered (swimming) pool

HFPA Hollywood Foreign Press Association

HFPS Home Fallout Protection Survey

hfr heifer; held for release; high-frequency range; high frequency recombination; hold for release; high flux reactor

hfr high-frequency recombinants

HFR (Sir Edward) Hallstrom Faunal Reserve (New South Wales)

HFRA Honorary Fellow of the Royal Academy

h-f radar height-finder radar

HFRB Hawaii Fire Rating Bureau

Hfrz Halbfranzband (German—halfbound in calf)

hfs high-fructose syrup(s); hot-finished seamless; hyperfine structure

Hfs Helsinki (Helsingfors)

HFS Human Factors Society

HFSP Human Frontier Science Program

HFSSA Historical Firearms Society of South Africa

Hfssb high-frequency single sideband

hft high-frequency transducing; hot flow test(ing)

hft hefte (Dano-Norwegian—part, issue)

Hft Heft (German—part)

HFT Hawaii Federation of Teachers; Heavy Fire Team; Human Factors Team

HFTS Human Factors Trade Studies (USU)

hfupr hourly fetal urine production rate

hfw hole full of water

Hfx Halifax

hg hand generator; hectogram; heliogram; high grade; hunter green; hydrostatic gage; hypobranchial gland

h & g harden and grind

Hg *hydrargyrum* (Latin—mercury)

Hg Haggai

H^g Hegység (Hungarian—mountain, mountainous)

HG Haute-Garonne; Her (His) Grace; H(erbert) G(eorge) (Wells); High German; Home Guard; Horse Guards

H-G Haute-Garonne

HG House & Garden

hga high gain antenna

HGA Heptagonal Games Association, Hobby Guild of America; Holological Guild of Australia; Hop Growers of America; Hotel Greeters of America; Hungarian Gypsy Association

h-galv hot-galvanize

hgb hemoglobin

HGB Handelsgesetzbuch (German—Commercial Law Code)

Hgb hemoglobin

Hgb M Methemoglobin, methemoglobinemia (blood test)

HGCA Home-Grown Cereals Authority

HgCI$_2$ bichloride of mercury; mercuric chloride

HGD Hourglass Device

HGDH Her (His) Grand Ducal Highness

hge heavy gold electroplate; hogshead

HGEA Hawaii Government Employees Association

hgf (HGF) hyperglycemic-glucogenolytic factor

HGF Human Growth Foundation

HGFA Hang Gliding Federation of Australia

hg ga height gage

HGH human growth hormone

HGHCA Hotels, Guest Houses, and Caterers' Association

HGI Henry George Institute

HGJP Henry George Justice Party

Hglds Highlands

HGM Human Gene Mapping workshop

HGMM Hereditary Grand Master Mason

HGMN Hair Growing Marketing Network

hgo hepatic glucose output

Hgo Hidalgo

HGOA Houston Grand Opera Association

HGOAA Hobby Greenhouse Owners Association of America

hgor high gas-oil ratio

HGP Humbug Gulch Press; Human Genetics Program

HGPRT Hypoxanthine-guanine phosphoribosyltransferase

hgps high-grade plow steel

hg pt hard-gloss paint

hgr hangar; hanger

HGR Hluhluwe *(shloosh-loo-way)* Game Reserve (northern Zululand, now part of Kwa-Zulu-Natal Province, South Aftica)

hgs hangars; hangers

Hgs Haugesund

HGS Hydrological Growing Season

hgsw horn gap switch

hgt height; hogget

HGTAC Home Grown Timber Advisory Committee

HGTB Haiti Government Tourist Bureau

hgts heights; hoggets

H^{gts} Heights

hgv heavy goods vehicle

Hgw Hanford (basalt) ground water

HGW Herbert George Wells

Hgy Highway

Hgz Hoogezand

hh half-hard; handhole; hands high (horse's height); heavily hinged; heavy hydrogen; hired hand

h/h half height; hard of hearing; house-to-house (search or transport)

h to h heel-to-heel

hh hojas (Spanish—leaves)

hH heavy hydrogen

HH double-hard (pencils); Harry Hansen; Helen Hunt Jackson; Her (His) Highness; His Holiness; Howard Hanson; Hugo Hurndinger; Huntington Hartford

H/H Havre-Hamburg (range of ports)

H & H Handy & Harman; Holland & Holland

HH Herren (German—Gentlemen)

HH-52 Sikorsky 12-passenger helicopter

HH-53 Sikorsky Sea Stallion CH-53 assault helicopter

hha half-hardy annual

hhb half-hardy biennial

HHBS Hereford Herd Book Society

hhcc higher-harmonic circulation control

hhd hogshead

HH. D. *Humanitatis Doctor* (Latin—Doctor of Humanities)

hhdws heavy handy deadweight scrap

hhe health hazard evaluation

hhf household furniture

HHFA Housing and Home Finance Agency

HHFTH Happy Horsemanship for the Handicapped (foundation)

hhg household goods

hhh triple hard

HHH Hubert Horatio Humphrey; triple-hard (pencils)

HHHC Hunt the Hunters Hunt Club (Amory Foundation funded)

HHHIPA Hubert H Humphrey Institute of Public Affairs

Hhhs Hincherton hayfever helmets

HHI Habitat for Humanity International; Hellenic Hydrobiological Institute; Highland Home Industries; Hyundai Heavy Industries

H-hinge capital-H-shaped hinge

HHJ Helen Hunt Jackson (author of *Ramona*)

HHK Honor Hong Kong

HHKA Husband's Housemaid's Knee Association

hhld household

HHMS His Hellenic Majesty's Ship

hhmu hand-held maneuvering unit

HHNSR Hudson Highlands National Scenic Riverway

H-hour hostile operations commencement hour

hhp half-hardy perennial; hydraulic horsepower

HHP Holistic Health Practitioner

HHPL Herbert Hoover Presidential Library

HHR Health and Human Resources

HHRA Heartland Human Relations Association

HHRC Health and Human Resource Center

H & HRR Harlem and Hudson River Railroad

HHS Haaren High School; Hawaiian Humane Society; Health and Human Services; Hunter High School

hhsd holographic horizontal situation display

HHSP Highland Hammock State Park (Florida)

hht (IJHT) high-temperature helium turbine

HHT Hom-Hellersberg Test

hhtg house(hold) heating

hhtv hand-held thermal viewer

HHUMC Hadassah-Hebrew University Medical Center

HHV-8 human herpes virus 8

hhw (HHW) household hazardous waste

HHW higher high water

HHWI higher high water interval

hi contracted form of "hail"; hemagglutination inhibition; high; high intensity; horizontal interval; hospital insurance; humidity index

hi (HI) hyperglycemic index

h & i harassing and interdictory (artillery fire)

h.i. hic iacet (Latin—here lies)—also appears on tombstones as H.I.

Hi Hering illusion; High (postal abbreviation); Hindi; Hiram

H¹ Hasi (Arabic—waterhole)—also appears as *Hasy*

HI Hammersley Iron; Hampton Institute; Handwriting Institute; Harris Intertype; Hat Institute; Hawaii; Hawaiian Islands; Heat Index; Henrik lbsen; Holiday Inns; Hoover Institution; Household International; Hudson Institute; Humidity Index; Hydraulic Institute; Hydronics Institute

hia hold in abeyance

HIA Handkerchief Industry Association; Headwear Institute of America; Hobby Industry Association; Home Improvement Association; Homeopathic Institute of Australia; Horological Institute of America; Hospital Industries Association; Housing Improvement Association; Housing Industry Association; Hungarian Imperial Association

HIAA Health Insurance Association of America; Health Insurance Association of Australia

HIAB Hydrauliska Industri AB (Swedish—Hydraulic Industry Company)

hiac high acuity

hi-ac high accuracy

HIAD Handbook of Instructions for Airplane Designers

HIAG Hifsorganisation auf Gengenseitigkeit (German—Mutual Aid Organization)

HIAGSED Handbook of Instruction for Aircraft Ground Support Equipment Designers

HI/AMBBA Hair International/Associated Master Barbers and Beauticians of America

HIAP Health Insurance Advocacy Program

HIAS Hebrew Sheltering and Immigrant Aid Society

HIAVED Handbook of Instructions for Aerospace Vehicle Equipment Design

Hib *Haemophilus influenzae* type B; Hibemia (Ireland); Hibernian (Irish)

HIB Herring Industry Board

Hibbd Halbband (German—half binding)

hibex high-acceleration booster experiment

HIBR Huxley Institute for Biosocial Research

HIBT Howard Ink-Blot Test

hic head injury criterion; hearing-impaired children; hot isostatic compaction; hybrid integrated circuit; hydrologist in charge

HIC Heart Information Center; Herring Industries Council

HICAP Health Insurance Counseling and Advocacy Program

hicapcom high-capacity communications

hicat high-altitude clear-air turbulence

hic jac hic jacet (Latin—here lies)

Hicks Hicksite Quakers

hiclass hierarchical classification

Hi Com High Command; High Commission; High Commissioner

HICS Hardened Intersite Cable System

hid hallucinations, illusions, and delusions; headache, insomnia, depression (syndrome); high-intensity discharge (lamps)

Hid Hidalgo

HIDA Health Industry Distributors Association

hidal helicopter insecticide-dispersal apparatus, liquid

hidalgo hijo de algo (Spanish—son of someone)

Hidalgo Miguel Hidalgo y Costilla (Padre Hidalgo)

HIDB Highlands and Islands Development Board (Scotland)

HIDTA High Intensity Drug Trafficking Areas

hidvl high-intensity-discharge vapor lamp

hi-E high efficiency

HIE Hibernation Information Exchange; Histrionic Instruction Education

hier hieroglyphics

Hier. Hierosolma (Latin—Jerusalem)

HIES Hadassah Israel Education Services

hif higher intellectual function

HIF Health Information Foundation

hifar high-flux Australian reactor (HIFAR)

hifc hog intrinsic factor concentrate

hi-fi high-fidelity

Hi Fi High Fidelity and Musical America

hiflex high flexibility

HIFNY Hospitality Industry Foundation of New York

hifo highest in, first out

hifor high-level forecast

hig hermetically sealed integrating gyroscope; higgler

Hig Higgins; Higginson

HIG Hartford Insurance Group; Hawaii Institute of Geophysics

HIGED Handbook of Instruction for Ground Equipment Designers

high-def high-definition

higher 3-Rs remedial reading, remedial writing, remedial arithmetic

Highlands Highlands of the Hudson; Highlands of the Navesink close to where Henry Hudson first landed in 1609; Highlands of Scotland

highpro high protein (diet)

high-Q high quality

high-Z high-impedance

High Sierras higher Sierra Nevada Mountains of California

High Tatras high Tatra Mountains of Czech Republic's Carpathians

high tech high technology

HIH Her (His) Imperial Highness

HII Health Industries Institute; Health Insurance Institute

HIID Harvard Institute for International Development

hijack hijacked; hijacker; hijacking

hik hiking

hil high intensity lighting; high lift

Hil Hilary; Hilda

hila health insurance logistics automated

hilac heavy-ion linear accelerator

Hilary Rules orders and forms modifying pleading and practice in English superior courts of common law

hilat high latitude satellite

HILC Hampshire Inter-Library Center (Amherst, Mount Holyoke, and Smith colleges)

Hilda Hildegarde

Hill The Hill (Capitol Hill in Washington, D.C.)

hilla (HILLA) high-input low-labor agriculture

hi-lo high-low

HILT Hardware In The Loop

Hil-Vis Hiligaynon-Visayan

him high impact; horizontal impulse

HIM Her (His) Imperial Majesty

HIMA Health Industry Manufacturers Association

Himal Himalaya (Sanskrit—Abode of Snow)

Himalayas Himalaya Mountains between India and Tibet

himat high-maneuverable advanced-fighter technology

hi mi high mileage

HIMM Hallberg Index of Male Menopause

HIMS Heavy Interdiction Missile System

Hin Hindi

HIN Health Insurance Network

Hinck Hinckley

hind hindustanisch (German—Hindustani)

Hind Hindi; Hindu; Hindustani

Hindenburg General Paul Ludwig Hans Anton von Benackendorf und von Hindenburg

hinga anhinga (fish-eating wading bird)

Hinglish Hindi + English (English interlarded with Hindi expressions and words)

hinil high noise-immunity logic

HINP Hundred Islands National Park (Philippines)

Hint Hinton; Hinton Test (for syphilis)

HINWR Hawaiian Islands National Wildlife Refuge

hio hypoiodite

hiomt (HIOMT) hydroxyindole-O-methyltransferase

H-ion hydrogen ion

HI-OVIS Highly Interactive Optical Video Information System

hip hierarchical information processor; high-impact pressure; hot isostatic pressure; humanizing, individualizing, and personalizing

hip (HIP) historically-informed performance

hips high-impact polystyrene

Hip Hippolyte; Soviet Mi-8 transport helicopter

HIP Health Insurance Plan; Help for Incontinent People; Hoover Institution Press; Houston's Informed Parents

HIPAA Health Insurance Portability Accountability Act

HIPACS Hospital Picture Archiving and Communication System

hipar high-power acquisition radar

HIPCO Hunt International Petroleum Company

HIPERNAS High Performance Navigation System

hipi high-performance intercept(ion)

hipo hierarchy plus input process output

hi-po high-potential (manager)

hipoe high-pressure oceanographic equipment

hipot high potential

Hipp Hippocrates

hippo(s) hippopotamus(es)

hi pres high pressure

hiptoc high-power testing of optical components

hi-q high iq (IQ)

hir hydrostatic impact rocket

HIR Harbour Improvement Rate; Heron Island Resort (Queensland); Honiara (Solomon Islands airport)

HIRA Health Industry Representatives Association

hiran high-precision shoran

HIRB Health Insurance Registration Board

HIRC Housing Industry Research Committee

HIRE Help Through Industry Retraining and Employment; Hooking Is Real Employment

hirel high reliability

HIRI Hawaiian Independent Refinery Incorporated; Home Improvement Research Institute

hirl high-intensity runway lights

Hirohito Emperor Hirohito Showa (Japan's 124th emperor in direct lineage)

Hiroshige Ando Hiroshige (19th-century Japanese landscape painter)

HIRS High-Impulse Retrorocket System; Holographic Information Retrieval System

HIRS/smrd High-Impulse Retrorocket System/spin-motor rotation detector
Hirt Aulus Hirtius (Roman historian)
his. (HIS) histidine (amino acid); history
h.i.s. *hic iacet sepultus* (Latin—here lies buried)—also appears as h.i.s.
Hi-S Hi-Standard (firearms)
HIS Health Interview Survey; Horticultural Improvement Scheme; Hospital Information System; Human Intrusion Studies
HISA Hawaii International Services Agency; Headquarters and Installation Support Activity (USA)
HISAM Hierarchical Indexed Sequential Access Method
His-Am(s) Hispanic American(s)
HISC House Internal Security Committee (formerly House Un-American Activities Committee-HUAC)
HiSEA Hawaii Society of Enrolled Agents
his'n his own
Hisp Hispaniola
HISPA History of Sport and Physical Education (association)
Hispan Hispanic
Hispano Hispanoamericano (Spanish American); Hispano-Suiza (automobile)
HISSG Hospital Information Systems Sharing Group
hist historical; history
hist. historian
hist *historie* or *historisk* (Dano-Norwegian—history or historian)
Hist historic(al); History
Hist Abs *Historical Abstracts*
Hi Stan 22 plus High Standard .22-caliber automatic plus silencer (assassin's special)
Hist Dist Historic District
histn historian
histo histoplasmosis
histo (Latin prefix—tissue or web)—histology
histocrit historical critic(ism)
histol histologic(al)(1y); histologist; histology
HISU Hoover Institution on War, Revolution, and Peace at Stanford University

hit high-intensity tutoring; homing intercept(ion) technology
hi-T high torque
Hit Holtzman inkblot technique
HIT Health Inca Tea (with coca leaves); Health Indication Test; Hitachi Innovative Technology; Hong Kong International Terminals
Hitac Hitachi computer
Hitch Hitchborn(e); Hitchcock
hi-tec(h) high technology
hi-temp high temperature
Hitler Adolf Schicklgruber
Hit Pom Hither Pomerania (coastal Germany)
HITS Homicide Investigation Tracking System; Hyperlink-Induced Topic Search
Hitt Hittite
HIU Hypnosis Investigation Unit
HIUS Hispanic Institute of the United States
HIUS *Historisches Institut der Universitdt Salzburg* (German—Historical Institute of the University of Salzburg)
HIUV *Historisches Institute der Universitdt Vienna* (German—Historical Institute of the University of Vienna)
hiv (HIV) human immunodeficiency virus
hiv *hiver* (French—winter)
hi-vision high-definition television
hivos high-vacuum orbital simulator
hi wat high water
Hiwi *Hilfsfreiwilliger* (German—auxiliary volunteer); *Hilfswillige* (German—volunteers)
HIWRP The Hoover Institution on War, Revolution and Peace
hj high jump
h & j hyphenation and justification
HJ Honest John (short-range unguided missile); Howard Johnson
H. J. *hic jacet* (Latin—here lies); *Hitler Jugend* (German—Hitler Youth)
HJBS Hashemite Jordan Broadcasting Service
HJC Hershey Junior College
HJD Heliocentric Julian Date; Hospital for Joint Diseases

hjed heliocentric Julian ephemeris date
HJPA Holmes Junge Protected Area (Australian Northern Territory)
H J Res House Joint Resolution
H.J.S. *hic jacet sepultus* (Latin—here lies buried)
hk. Housekeeper
h-k hand to knee
Hk Hakenkreuz (German—hooked cross) swastika
HK Heckler and Kock (firearms); Hong Kong
HK Handelskammer (German—Chamber of Commerce); *Helsingin Kaupunginorkesteri* (Finnish—Helsinki City Symphony Orchestra)
HKA Hong Kong Airways
HKCEC Hong Kong Catholic Education Council
hk cells human kidney cells
HKCL Hong Kong Container Line
Hkd Hakodate
HK$ Hong Kong dollar
HKDR Hong Kong Depository Receipt
HKECIC Hong Kong Export Credit Insurance Corporation
HKEL Hong Kong Export Lines
hkf handkerchief
HKFE Hong Kong Futures Exchange
H Kg Hong Kong
HKG Hong Kong, British Crown Colony (airport)
HKGMA Hosiery and Knit Goods Manufacturers Association
HKH Hans (Hendes) Kongelige Højhed [Dano-Norwegian—His (Her) Royal Highness]
HKI Helen Keller International
HKIL Hong Kong Islands Line
HKJ Hashemite Kingdom of Jordan
HKL Halldor Kilyan Laxness
HKLA Hong Kong Library Association
hkm high-velocity kill mechanisin
h-k m (H-K M) hunter-killer missile
HKMA Hong Kong Management Association
H'Kong Hong Kong
HKP Hong Kong Polytechnic

HKPO Hong Kong Philharmonic Orchestra

HK & S Hong Kong and Shanghai Bank

HKSE Hong Kong Stock Exchange

HKTA Hong Kong Tourist Association

HKTC Hong Kong Training Council

HKTDC Hong Kong Trade Development Council

HKU Hong Kong University

hkups hookups

HK virus Hong-Kong type of influenza virus

HKX Hong Kong Express (container service)

Hky hand knitting yarn(s)

hl hand lantern; hands lost (baseball); hard labor; heavy lift; hectoliter; high level; high lift, hinge line; holiday

h-l highest and lowest (quotations)

h/l heading line; high or low

h & l door hinge resembling ligature of capital H and capital L

hl heilig (German—holy)

h.l. hoc loco (Latin—in this place)

HI latent hyperopia

HL Haute-Loire; Herpetologists League; Home Lines; Homestead Lease; Honours List; House of Lords; Hygienic Laboratories; Hygienic Laboratory

H-L Haute-Loire

H & L Harbour and Light Department

H of L House of Lords

HL House of Lords Cases

hla human lymphocyte antigen(s); horizontal line array; human leukocyte antigen

hla (HLA) histocompatibility antigens; homologous leucocytic antibodies; human leucocytic antigen

HLA Hawaii Library Association; Human Life Amendment

HL & AG Henry E Huntington Library and Art Gallery

HLAHWG High-Level Ad-Hoc Working Group (NATO)

H-land Headland

hlb hydrophile-lipophile balance

HLB Hotel Licensing Board

HLBB Home Loan Bank Board

hlc health locus of control

HLC Hapag-Lloyd Container (steamship line); Hospital Library Council (Dublin)

HLCAS House of Lords Cases

HLCU Hapag-Lloyd Container Unit

hld held; hold; holder

HLD Harold Handley Page (aircraft)

hldg holding

hl di hole die

HLDI Highway Loss Data Institute

hlds holdings

hldw high-level defense waste(s)

hlem horizontal-loop electromagnetic method

HLF Human Life Foundation

hlg halogen

HLG High-Level Group

hlge handling equipment

hlgp heavy-lift general purpose

HLH Haroldson Lafayette Hunt

HLHS Heavy-Lift Helicopter System

HLI Highland Light Infantry

HLIC Housing Loans Insurance Corporation

hll high-level language

HLL Hellenic Lines Limited

HLLAPI High Level Language Application Program Interface

hllw high-level liquid waste(s)

hl IX high-level language X

HLM Henry Louis Mericken

HLM *Habitations à loyer modéré* (French—moderately priced housing)

HLMR Hunter-Leggitt Military Reservation

HLNP Hattah Lakes National Park (Victoria, Australia)

h/l number hydrophile/lipophile number

hlnw high-level nuclear waste(s)

HLNWR Havasu Lake National Wildlife Refuge (California); Hutton Lake National Wildlife Refuge (Wyoming)

hlo horizontal lockout

H Lords House of Lords

hip (HLP) hyperlipidemia

hlpr helper

HLPR Howard League for Penal Reform

hlr heart-lung resuscitation

HLRS Homosexual Law Reform Society

hls heavy liquid separation; heavy logistics support; hills; holes

hl S heifige Schrift (German—holy scripture)

Hls Hills (postal abbreviation)

HLS Harvard Law School; Heavy Logistics Support; High-Level Scheduler

hl sa hole saw

hlse high-level single-ended

HLSS Harry Lundeberg School of Seamanship

HLSUA Honeywell Large Systems Users Association

hlsw high-level solidified waste(s)

hlt halt; halter

HLT Holborn Law Tutors

HLTF High-Level Task Force

hlth prof health professions

HLTs highly leveraged transactions

hlttl high-level transistor-translator logic

hltw high-level transuranic waste(s)

Hlu Honolulu

hlv herpes-like virus; heavy lift vehicle

HLVA Hospital Lady Visitors Association

hlw higher low water; high-level waste

Hlw *Halbleinwand* (German—half-bound cloth)

HLW higher low water

HLWI higher low water interval

HLW-ICB High-Level Waste-Interface Control Board

HLWIP High-Level Waste Immobilization Program

hlwn highest low-water neap tides

HLWRP Hoover Library on War, Revolution, and Peace (Stanford University)

Hlzbl *Holzbläser* (German—woodwinds)

hm hallmark; harmonic mean; heavy metal; hectometer; hired man; hollow metal; horizontal (position) multiplier

hm? how many?; how much?

h & m hit and miss; hull and machinery

h.m. hoc mense (Latin—in this month)

Hm Haymarket (London theater); manifest hypermetropia; manifest hyperopia

HM Harbour Master; Haute-Marne; Head Master; Head Mistress; Her (His) Majesty; Herman Melville; Home Missions; Houghton Mifflin

H-M Haute-Marne

HM Heure de Moscou (French—Moscow Time)

hm² square hectometer

hm³ cubic hectometer

hma high memory area

Hma Hiroshima

HMA Hardwood Manufacturers Association; Head Masters Association; Her (His) Majesty's Airship; Hoist Manufacturers Association; Home Manufacturers Association

H & MA Hotel and Motel Association

HMAA Horse and Mule Association of America

HMAC Her (His) Majesty's Aircraft Carrier

HMAI Handbook of Middle-American Indians

HMANA Hawk Migration Association of North America

HMARC Houston Metropolitan Archives and Research Center

HMAS Her (His) Majesty's Australian Ship

HMAV Her (His) Majesty's Armed Vessel

hmb homatropine methyl bromide (HMB)

HMB Home Mission Board; Hops Marketing Board

HMBA Hotel and Motel Brokers of America

HMBDV Her (His) Majesty's Boom Defence Vessel

HMBI Her (His) Majesty's Borstal Institution

HMBP Heavy Machine Building Plant

hmc heavy media cyclone; heroin-morphine-cocaine (mixture); howitzer motor carriage

hmc (HMC) hydroxymethyl cystosine

HMC Harvey Mudd College; Her (His) Majesty's Customs; Hospital Management Committee

hmcc housewife/mother career concept

HMCG Her (His) Majesty's Coastguard

HMC & H Hahnemann Medical College and Hospital

HMCIF Her (His) Majesty's Chief Inspector of Factories

HMCN Her (His) Majesty's Canadian Navy

HM Comm Historical Manuscripts Commission

HMCS Her (His) Majesty's Canadian Ship

HMCSC Her (His) Majesty's Civil Service Commissioners

HMCyS Her (His) Majesty's Ceylonese Ship

hmd hollow metal door; humid; hydraulic mean depth

hmd (HMD) hyaline membrane disease

HMD Her (His) Majesty's Destroyer

HMDBA Hollow Metal Door and Buck Association

hmde hanging-mercury-drop electrode

hmdf hollow metal door and frame

hmdi (HMDI) hexamethylene diisocyanate

HMDS Hazardous Material Data System

hmf hollow metal frame

HMF Her (His) Majesty's Forces

hmf black high-modulus furnace black

HMFI Her (His) Majesty's Factory Inspectorate

hmg heavy machine gun; high modulous graphite

hmg (HMG) human menopausal gonadotrophin

HMG heavy machine gun; Her (His) Majesty's Government; High Mobility Group (proteins)

HMGI Hotel-Motel Greeters International

HMHS Her (His) Majesty's Hospital Ship; Horace Mann High School

hmi heavy maintenance interval

HMI Hahn-Meitner Institut; Her (His) Majesty's Inspector; Hughes Medical Institute

HMI Himpunan Mahasiswa Islam (Indonesian—Islamic Students Society)

HMIA Haitian Migrant Interdiction Operation

HMIC Her (His) Majesty's Inspectorate of Constabulary

HMIS Hazardous Materials Identification System; Her (His) Majesty's Indian Ship; Her (His) Majesty's Inspector of Schools

HMIT Her (His) Majesty's Inspector of Taxes

HMK His Majesty the King

HML Harper Memorial Library (University of Chicago); Horace Mann—Lincoln Institute

HMLR Her (His) Majesty's Land Registry

hmlt hamlet

HMM Her (His) Majesty's Minister; Hyundai Merchant Marine

hmma (HMMA) 4-hydroxy-3-methodxy-mandelic acid

HMML Her (His) Majesty's Motor Launch

HMMS Her (His) Majesty's Motor Mine Sweeper

HMMWV High-Mobility Multi-Purpose Wheeled Vehicle (successor to the Jeep)

HMNAO Her (His) Majesty's Nautical Almanac Office

HMNAR Hart Mountain National Antelope Refuge (Oregon)

hmnts humanities-the arts, language, literature, music, philosophy

HMNZS Her (His) Majesty's New Zealand Ship

hmo heart minute output; hypothethical mean organism

HMO Health Maintenance Organization; Hospital Medical Office(r)

HMOA Health Maintenance Organization Act (of 1973)

HMOCS Her (His) Majesty's Overseas Civil Service

hmo's (HMOs) health maintenance organizations

HMOW Her (His) Majesty's Office of Works

hmp handmade paper

hmp (HMP) hexose monophosphate

HMP Habitat Management Plan; Her (His) Majesty's Penitentiary; Her (His) Majesty's Prison

H.M.P hoc monumentum posuit (Latin—he erected this monument)

HMPMA Historical Motion Picture Milestones Association

HMQ Her Majesty the Queen

HMRC Heineman Medical Research Center

HMRCS Her (His) Majesty's Royal Canadian Ship

H & M RR Hudson & Manhattan Railroad (Hudson Tubes)

HMRT Her (His) Majesty's Rescue Tug

hms hours, minutes, seconds; hypothetical mean strain

hms. himself

HMs Her (His) Majesty's

HMS Harvard Medical School; Her (His) Majesty's Service, Ship, or Steamer; Home Marketing Services; Home Mission Society

H & MS Headquarters and Maintenance Squadron

HMS *Hotel and Motel Systems*

HMSA Hawaii Medical Service Association

HMSAS Her Majesty's Special Air Service

HMSG Hirshhorn Museum and Sculpture Garden

HMSO Her (His) Majesty's Stationery Office

HMSS Hospital Management Systems Society

hmstd homestead

hmt hydromechanical transmission

HMT Her (His) Majesty's Transport; Her (His) Majesty's Trawler; Her (His) Majesty's Treasury; Her (His) Majesty's Tug

HMTA Hotel-Motel Association

hmu (HMU) hydroxymethyl uracil

HMV His Master's Voice (phonograph records)

hmw high molecular weight

hmwp high molecular weight polyethylene

hmwpe high-modular-weight polyethylene

hmy too little

h.n. *hac nocte* (Latin—tonight)

Hn Herman(n); Horn

H^n Horn

HN Head Nurse; Hoff und Nationaltheater (Munich); Honduras (Internet code)

Hna Habana

HNA Hawaii Newspaper Agency

HNBI Hellenic National Broadcasting Institute

hnc hypothalamic-neurohypophysical complex

HNC Harbors and Navigation Code; High National Council; Higher National Certificate; Human Nature Cooperative; Human Nature Council; Human Nutrition Center; Human Nutrition Council

Hnd *The Hindu* (Madras)

HND Haneda-Tokyo (airport); Higher National Diploma

hndbk handbook

HNDCPD handicapped

hnddiprt hand-discharge printed

hndflshd hand-fleshed (skins)

hndlg handling

hndlg/shpng handling & shipping charges

hndlr handler

hndovprt hand overprint(ed)

hndprt hand print(ed)

HNEI Hawaii Natural Energy Institute

HNF Home Nursing Foundation

hn fm hand form

HNG Hawaii National Guard; Houston Natural Gas

HNIS Human Nutrition Information Service

Hnl Honolulu

HNL Honolulu, Hawaii (airport)

hnml hindumeal

HNMS High NATO Military Structure

HNNNR Herma Ness National Nature Reserve (Scotland)

Hno Hanover

HNO^2 nitrous acid

HNO^3 nitric acid

Hnos Hermanos (Spanish—brothers)

hnp high needle position

HNP Haleakala National Park (Maui, Hawaii)

HNP *Herstigte Nasionale Partij* (Afrikaans—Reformed National Party)—South African segregationists

hnr handwritten numerical recognition; hiss noise reduction

HNRC Human Nutrition Research Center

hnrna (HNRNA) heterogeneous nuclear ribonucleic acid

hnrs honors

hn(s) horn(s)

HNT *Hof und Nationaltheater* (German—Court and National Theater)—Munich

HNTLA Hiskey-Nebraska Test of Learning Aptitude

HNYS Hughes Night Vision System

HNWR Hagerman National Wildlife Refuge (Texas); Horicon National Wildlife Refuge (Wisconsin)

ho hands out (baseball); held over; heterothallic; hoist; hold over; holds; homothallism; hostilities only; house

h & o hook and oil (damage to cargo)

'ho' whore

Ho Ho Chi Minh; holmium; Honduran; Honduras; Hondureño; House (of Representatives)

HO Head Office; Health Office(r), Home Office (British); Hydrographic: Office (USN)

HO *Handelsorganisation* (German—trade organization)

hoa hands off—automatic

HOA Homeowners A (insurance policy); Home Owners Association

hoax (Contraction—hocus pocus)

hob height of burst; horizontal oscillating barrel; human observation(al) blunder

Hob Anthony van Hoboken (Dutch chronologist-enumerator of Haydn's music); Hobart, Tasmania; Hoboken (Belgian seaport near Antwerp, place near Waycross, Georgia, port city in New Jersey)

HOB Homeowners B (insurance policy); House Office Building

Hoban Holborn

hobe honeycomb before expansion

hobgob(s) Hobgoblin(s)

Hob-Job Hobson-Jobson (similar-sounding words to those of other languages with some or complete loss of meaning, e.g., white rhino really the Dutch *weid rhino*—a wide-mouthed rhinoceros and really not white)

hobo (HOBO) homing bomb

Hobo Hoboken

Hobohemia Hobo bohemia (skid-row areas such as Manhattan's Bowery)

HOBOS Homing Bombing System

HOBS High Orbital Bombardment System

Hobt Hobart

hoc heavy organic chemical(s); held on charge; hospital outcall

hoc (HOC) hydrofoil ocean combatant

HoC House of Commons

HOC Homeowners C (insurance policy); Hydrology Overview Committee

hock. Hockheimer (Rhine wine)

HOCPC Home Office Crime Prevention Center

H.O.C.S. Hostem Occidit, Civem Servavit (Latin—A foe he slew, a citizen he saved)—inscription found on Roman civic crowns

hoc vesp. hoc vespere (Latin—this evening)

hod. hyperbaric oxygen drenching

HoD Head of Department

HOD Hoffer-Osmond Diagnostic Test

HOD Test Hoffer, Osmond, and Desmond Test (for schizophrenia)

hoe holographic optical element; homing overlay experiment

hof home owner's fee

HoF Hall of Fame

Hoff Hoffinan; Hoffinann; Hofman reflex

Hoffmann Ernst Theodor Amadeus Hoffmann

hofin hostile fire indicator

HOFSL Home Office Forensic Science Laboratory (London)

HOG heavy ordnance gunship

HO-gage 5/8-inch track gauge (model railroads)

HOGD Hermetic Order of the Golden Dawn

ho & gem heavy oil and gascut mud

hoh hard of hearing

HoH Head of Household

HOHI Home Ownership and Home Improvement (loan program)

HOI Headquarters Operating Instruction

HOI *Handbook of Operating Instructions*

hoj home on jamming

HoJo Howard Johnson

H o K House of Keys

hoke hokum

Hokusai Katsushika Hokusai (19th-century Japanese engraver-illustrator-teacher)

hol. holiday; hollow; holly

Hol Holland; Hollander

Hol Holanda (Portuguese or Spanish—Holland)

HoL House of Lords

HOL Higher Order Language

HOLC Home Owners Loan Corporation

hold. Holding

holidaze alcohol-or-drug-induced daze characterized by incidence of over-the-holidays accidents and fatalities

holl hollandais (French—Dutch); *hollandsk* (Dano-Norwegian—Dutch)

Holl Holland; Hollander

holland hope our love lives and never dies

Holländer Der Fliegende Holländer (German—The Flying Dutchman)—Wagner opera

hollands hollands gin (juniper flavored)—also called Dutch gin as it originated in the Netherlands

Hollands Holland & Holland (British gunmakers)

Hollyw'd Hollywood

Holmes Sherlock Holmes

holo holograph

holo (Latin prefix—entire or whole)—holocaust

holog hologram; holograph(ic) (al)(1y); holography

hol-ry whole rye

hols holidays

holsum wholesome

Holt Holt, Rinehart & Winston

HOLUA Home Office Life Underwriters Association

holupk holiday upkeep

Hol Via Holbom Viaduct (rail terminal)

Holw Hollow

hom. homing; hominy; hornonym

Hom. Homer; Homerton College, Cambridge

Hom. Homilia (Latin-homily, sermon)

HOMA Houston Oil and Minerals

Home Home Office (England and Wales)

HOME Home Opportunities Made Equal; Home Ownership Made Easy; Homemakers Organized for More Employment; Housing Opportunities & Maintenance for the Elderly; Highly Optimized Microscope Environment; Holy Order of Mother Earth

home ec home economics

homeo homeopath; homeopathic; homeopathy

homeo (Latin—same or similar)—homeostasis, homogeneous, homogenized, homologous

Homer. Homeric

HOMES mnemonic for remembering the five Great Lakes—Huron, Ontario, Michigan, Erie, Superior

Hominoid Hominoidea [primate superfamily including the great apes (chimpanzees, gorillas, orangutans), the gibbons, and humans]

homo homeopath; homeopathic; homeopathy; homosexual; homosexuality

homo (Latin prefix—man)—homosexual

homoeo homoeopath(ic); homoeopathy

homolat homolateral

Homo ludens creative, festive, leisure-loving, playful person

homomilk homogenized milk

homop homophobia (anti-homosexual hysteria)

homosex homosexual; homosexuality

homrep homicide report

hon honey; honor; honorable; honorarium; honorary; honored

Hon Honduran; Honduras; Hondureño; Honolulu; Honorable

Hon'ble Honourable

Hon Consul Honorary Consul

hond honored

Hond Honduran; Honduras

hon. dis. honorable discharged

Hong Kongese person(s) of British and Chinese descent

Hono Honolulu
honor. honorary
honora. honorably
hons honors
Hon See Honorary Secretary
HOO House Officer Observer
hood. Hoodlum
'hood neighborhood
Hook Hook Point, Ireland; Hooker; The Hook—Hook of Holland *(Hoek van Holland);* Hooky Nail; Sandy Hook, New Jersey
Hoosier Dome Hoosier Dome Stadium, Indianapolis
Hoover Hoover Dam southeast of Las Vegas, Nevada; Hoover Institution (Stanford University)
hop. high oxygen pressure; holding procedures
Hop Hopkin; Hopkins; Hopkinson; Hopwood
HOP Hong Kong Outline Plan; Hydrographic Office Publication(s)
HOPE Harbingers of Productive English; Health Opportunity for People Everywhere; Helping Other Parents Educate; Help Organize Peace Everywhere; Hispanic Organization of Professional and Executives; Homeownership and Opportunity for People Everywhere
HOPEG Hotel and Public Building Equipment Group
HOPES High-Oxygen-Pulping Enclosed System
Hopkins Mt Hopkins Observatory, Amado, Arizona
hoppers grasshoppers
HOQ Hysteroid-Obsessoid Questionnaire
hor home of record; Horace; Horatio; horizon; horizontal; Horologium (constellation); horology
Hor Horayot
HoR House of Representatives
H-O-R Hoover-Owens Rentschler (engines)
Horace Quintus Horatius Flaccus
hora decub. hora decubitus (Latin—at bedtime)
hora intem. hora intermedius (Latin—at the intermediate hours)
hora som. hora somni (Latin—at bedtime)

HO & RC Humble Oil and Refining Company
HoReCa Hotel, Restaurant, and Cafe Keepers
horen horizontal enlarger
horiz horizontal
horn french horn (in the British Isles, Canada, or the United States)
Horn The Horn (Cape Horn—southernmost South America); Hornblower
horo horoscope
Horo Horologium (constellation)
horol horology
Hor Q Horatius Quintus Flaccus (Roman poet)
HORSA Hut Operation Raising School-leaving Age
horse. (HORSE) hydrofoil-operated rocket submarine
Horsemonger Horsemonger Lane Gaol (notorious London prison in the early 1800's)
hort horticulture
Hort Horticulture
horti horticultural; horticulturalist; horticulture
hortic horticultural; horticulture; horticulturist
HORU Home Office Research Unit
hor. un. spatio horae unius spatio (Latin—at the end of an hour)
hos human operator simulator; higher order software
'ho's whores
Hos The Book of Hosea; Hosea
Hos. Hosea (Book of)
H-o-S Holland-on-Sea
HOS Hawaiian Orchid Society
HOSA Home Owner Services Administration
HOSC History of Science Cases
HOSCOM Hospital Comparisons
hose. Hosiery
Hosp Hospital
hosp ins hospital insurance
hot. human old tuberculin
Hot high-subsonic optically guided tube-launched Franco-German antitank missile
HOT Hamilton-Oshawa-Toronto (industrial complex); Hot Springs, Arkansas (airport)
ho/ta hold tank(s)

HOTAC Hotel Accommodation (London hotel service)
hot arts hot art dealers (trafficking in stolen works of art)
HOT Car Hands Off This Car (antitheft program)
HOTLIPS Honorary Order of Trumpeters Living in Possible Sin
HOTOL Horizontal Takeoff and Landing (aircraft)
HOTS Higher-Order Thinking Skills
Hotz Hotzenplatz (also known as Osoblaha by its Czechoslovakian citizens)
Hou Houston
HOU Houston, Texas (airport)
Houghton Houghton Mifflin
Hounds Houndsditch
Hous Houston
House House of Representatives in the United States; House of Commons in England; London's Stock Exchange; Oxford University's Christ College
House of D (Women's) House of Detention (NYC)
Hou Sym Orch Houston Symphony Orchestra
houv houvere (Finnish—charity)
hov high-occupancy vehicle
Hovensweep Hovensweep National Monument in southwestern Colorado and southeastern Utah
how. Howitzer
How *Howard (U.S. Supreme Court Reports)*
HoW Happiness of Womanhood
HOW Home-Owners' Warranty
Howard U Pr Howard University Press
howtar howitzer-mortar
HOW-TO Housing Operation with Training Opportunity (OEO)
Hox Hoxie
Hoxford(shire) (Cockney—Oxfordshire)
hp half pay; hardy perennial; health physics; high-efficiency particulate; high pass; high performance; high potency; high potential; high power; high pressure; hire purchase; hit by pitch (baseball); holiday pay; hollowpoint; horizontal

parallax; horizontally polarized; horsepower; hot press(ed); hybrid perpetual; hyperbolic; hypoid(al)

h/p house to pier

h & p history and physical (examination); hydraulic and pneumatic

HP Haute-Pyrénées; Highway Patrol; historic park; House Physician; Houses of Parliament

H-P Handley-Page; Haute-Pyrénées; Hewlett-Packard

HP Homeopathic Pharmacopoeia

hpa high power amplifier; horn-parabola antenna; human-powered aircraft; hydraulic pneumatic area; hypothalamus, pituitary, adrenals

HPA History, Physical, Admit; Hospital Physicists Association

HPAAS High-Performance Aerial Attack System

hpac hydropress accessor

HPAC Hawaii Performing Arts Company

HPAL Holland Pan-American Line

hpb hinged plotting board

HPB Helena Petrovna Blavatsky, cofounder of the Theosophical Society

HPBA Housing Patrolmen's Benevolent Association

hpc history of the present complaint; hunt, peck, and cuss (United States Army—typing)

HPC Hercules Powder Company; Highland Park College

hpc black hard-processing channel black

HPCC High-Performance Control Center

hpchd harpsichord

hpchdst harpsichordist

HPCI Home Price Comparison Index

HPCL Hindustan Petroleum Corporation Limited

HP Club Homing Pigeon Club

hp cyl high-pressure cylinder

hpd high-performance diesel; hydraulic pump discharge

H-P d Hough-Powell digitizer

HPD Hawaii Police Department; Housing Preservation and Development

HPDC High-Pressure Data Center (NBS)

HPDF High-Performance Demonstration Facility

hpdo high-performance diesel oil

hpe high-power effect(s)

hper health, physical education, and recreation

HPERB Hawaii Public Employment Relations Board

hpew high-power(ed) early warning

hpf highest possible frequency; high-powered field; hydropress form

HPF Horace Plunkett Foundation

hpfm hydropress form

hpfs high-performance file system

hpftp high-pressure fuel turbopump

hpfts high-pressure fuel turbopump

hpg (HPG) human pituitary gonadotrophin

HPGA Hawaii Personnel and Guidance Association

Hp Ge high-purity Germanium

H$_{pgc}$ heading per gyro compass

hp hd high-pressure high-density

hp hr horsepower hour

hpi height position indicator; high-power illuminator; history of present illness; homing-position indicator

HPI Handicap Problems Inventory; Heifer Project International; Hydrocarbon Processing Industry

hpl high(est) point level; human parotid lysozyme; human placental lactogen

Hpl Hartlepool

HPL Halifax Public Library; Hamilton Public Library; Hartford Public Library; Houston Pipe Line; Houston Public Library

hplc high-pressure-liquid chromatography

hpll hybrid phase-locked loop

hplr hinge pillar

hpls hopeless

hpm high polymer molecular; human potential movement; high-power microwave (weapon)

H-P m Harding-Passey melanoma

HPMA Hardwood Plywood Manufacturers Association

hpmv high-pressure mercury vapor

hpn home parental nutrition; horsepower nominal

hpns high-pressure nervous syndrome

hpo high-pressure oxygenation

HPO Hamilton Philharmonic Orchestra; Helsinki Philharmonic Orchestra; Highway Post Office

HPOL High-Power Optical Laboratory

H-pole H-shaped telegraph or telephone pole

hpotp high-pressure oxygen turbopump

hpox high-pressure oxygen

hpp hydraulic pneumatic panel

HPP Harvard Project Physics; Hawker Pacific Proprietary

HPPA Horses and Ponies Protection Association

HPPB Historic Pensacola Preservation Board

h-p plan hire-purchase plan (British equivalent of American installment-plan purchasing)

HPPP High-Priority Production Program

hpr high penetration resistant; high-power(ed) radar; hopper; hot particle rolling

HPR House of Pacific Relations

HPRF Hypersonic Propulsion Research Facility

HPRO High Public Relations Officer

HPRP High-Performance Reporting Post; High-Power(ed) Radar Post

hps high primary sequence; high protein supplement; high-pressure sodium; high-pressure steam; hot-pressed sheet

HPS Harlem Preparatory School; Health Physics Society; High Protestant Society

Hpsc heading per standard compass

hpsn hot-pressed silicon nitride

hpst harpist

Hpstgc heading per steering compass

hpt high point; high-pressure test

HPTA High-Power Test Area; Hire Purchase Trade Association

hptn hypertension

Hptw Hauptwerk (German—great work)

hpu hot-plate unit; hydraulic pumping unit; hydraulic power unit

hpv high-passage virus; human-powered vehicle

HPV Human papillomavirus

HPVA Hardwood Plywood and Veneer Association

HPVA Human-Powered Vehicle Association

hpv-de high-passage virus (grown-in) duck embryo

hpv-dk high-passage virus (grown in) dog kidney

hpw high-power window

H-pylori Helicobacter pylori

hq headquarters

hq (HQ) high quality

h.q. hoc quaere (Latin—see this)

Hq Headquarters

H-Q Hydro-Quebec

HQASC Headquarters Air-Support Command (NATO)

HQBA Headquarters Base Area

hqc hydroxyquinoline citrate

HQCC Headquarters Coastal Command (UK)

HQ Comdt Headquarters Commandant

HQ COMD USAF Headquarters Command, USAF

HQD Harold Q(uinten) Driscoll

HQDA Headquarters, Department of the Army

hqdt handling qualities during tracking

HQDTMS Headquarters, Defense Traffic Management Service

HQEARC Headquarters, Equipment Authorization Review Center (USA)

HQFC Headquarters, Fighter Command (NATO)

HQFT Humanist Quest For Truth

HQMC Headquarters-Marine Corps

Hq's (Hq's) headquarters

HQSC Headquarters, Signal Command (UK)

HQSTC Headquarters, Strike Command (UK)

HQT Humanist Quest for Truth

HQTC Headquarters, Transport Command (UK)

Hqtrs Headquarters

HQ USAF Headquarters, USA-F

hr hairspace; handling room; handrail; heat resisting; height range; high resilient; high resonance-proof; high risk; hinge remnant; home run; homing relay; hook rail; hoserack; hospital recruit; hot rolled; hour; human relations; relative humidity (symbol)

h(r) hail and rain (meteorological symbol)

h/r heart rate; home-road ratio

hr% home-run percentage

hr herr (Swedish—Sir)—Mr

Hr Herr (Danish or German—Mr, Sir); *Horner* (German—brass instruments; horns)

HR Highland Regiment; Hillside Review (overlay zone); Holmes Ribgrass; Hospital Recruit; House of Representatives; House Resolution; Human Resources; International Harvester (stock exchange symbol)

H-R Haut-Rhin; Hertzsprung-Russell

H & R Harper & Row; Harrington & Richardson; Herweg & Romine

HR Hauptrhythmus (German—outstanding rhythm)—12-tone term; *Hellenic Register* (Greek ship-classification book); *House* (of Representatives) *Resolution*

hra housing review account; high right atrium

hra (HRA) hypersonic research airplane

hRA hardness Rockwell A (scale)

Hra Herra (Finnish—Mister)

HRA Hardware Retailers Association; Health Resources Administration; Historical Records of Australia; Human Resources Administration; Hunters' Rights Association; Hypnotic Research Association

HRA Historical Records of Australia

H & RA Homeowners and Renters Assistance

HRAA Hypnotic Research Association of Australia

HRAF Human Relations Area File

HRAG Helena Rubinstein Art Gallery

hRB hardness Rockwell B (scale)

HRB Highway Research Board; Highway Research Bureau; Housing and Redevelopment Board

hrbr harbor

hrbr vu harbor view

hrc high rupturing capacity

hRC hardness Rockwell C (scale)

HRC Hardwood Research Council; Herpes Resource Center; Holy Roman Church; Horse Racing Commission; Humacao Regional College; Human Relations Commission; Human Resources Center; Human Rights Campaign; Human Rights Commission (OAS); Humanities Research Council

HRCC Humanities Research Council of Canada

HRCF Human Rights Campaign Fund(ing)

HRCI Human Resource Certification Institute

hrd hard; high roughage diet

HRD Hertzprung-Russell Diagram; Human Resources Development

HRDA Human Resources Development Agency

HRDF Human Resource Development Foundation

HRDI Hospital Reserve Disaster Inventory

HRDL Hudson River Day Line

hrdly hardly

hrdwd hardwood

hrdwr hardware

hre hormone-responsive element; hypersonic research engine

hre (HRE) high-resolution electrocardiography

HRE Holy Roman Empire

HREBU Hotel and Restaurant Employees and Bartenders Union

H reflex Hoffnian reflex (of the tibial nerve)

H Rept House Report

HRes House Resolution (US House of Representatives)

HRET Hospital Research and Educational Trust

HREU Hotel and Restaurant Employees Union

hrf high rate of fire; home-run factor

Hrf Harfe (German—harp)

HRF Hat Research Foundation

HRFA Hudson River Fishermen's Association

hrf's health-related facilities

hrg hearing

HRG Halford, Robins, and Codfrey

HRGs Health Research Groups

hrh high resistance hold

HRH His (Her) Royal Highness

hri height-range indicator

hri (HRI) high-resolution images

HRI Horticultural Research International; Hotel Reservations International; Human Relations Inventory

HRIP Highway Research in Progress

H.R.I.P. *hic requiescit in pace* (Latin—here rests in peace)

hrir high-resolution infrared radiometer

hrirs high-resolution infrared-radiation sounder

HRIS Highway Research Information Service; Human Resource Information System

hrl horizontal reference line

Hrl Harlingen

HRL Hughes Research Laboratories; Human Resources Laboratory

Hrm Herman

HRM Human Resources Manager

HRMA Hampton Roads Maritime Association

Hr Ms Haar Majesteits Schip (Dutch—Her Majesty's Ship)

Hrn Herren (German—gentlemen); *Horner* (German—brass instruments; horns)

HRNTWT High-Reynolds-Number Transonic Wind Tunnel

HRO Housing Referral Office (USAF)

hrp horizontal radiation pattern

HRP Hampton Roads Ports; Human Reliability Program; Huntsville Research Park

HRPA Hudson River Pilots Association

HRPP Human Rights Protection Party (Samoa)

HRPS Human Resource Planning Society

hrr higher reduced rate (taxation)

HRRA Human Resources Research Organization

HRRC Human Resources Research Center

HRRL Human Resources Research Laboratory

HRRO Human Resources Research Office

hrs high-resolution spectrograph; high-resolution spectrometer; hot-rolled steel; hours

HRS Hamilton Rating Scale; Health and Rehabilitation Services; Health Resources Statistics; Human Resource System; Hydraulics Research Station; Hydrostatic Research System

HRSA Health Resources and Services Administration; Honorary Member of the Royal Scottish Academy

hrsg herausgegeben (German—edited or published)

Hrsg Herausgeber (German—editor)

hrsi high-temperature reusable surface insulation

HRSRS Hartbeestehoek Radio Space Research Station

hrt high-resolution track(er)

HRT Honolulu Rapid Transit; Hormone Replacement Therapy

hrts high risk test site

HRTS Hollywood Radio and Television Society

hrtwd heartwood

Hrtz Ha'aretz (Hebrew—The Land)—Israeli newspaper

HRU Hydrological Research Unit

Hrv hypersonic research vehicle

HRVC Hudson River Valley Commission

HR & W Holt, Rinehart & Winston

HRWMC House of Representatives Ways and Means Committee

Hry Henry

HRYC Halifax River Yacht Club; Hampton Roads Yacht Club

HRZ Hertz Corporation (stock exchange symbol)

hs half strength; hardstand; head suppression; heating surface; hide substance; high-speed; hinged seat; horizontal shear; horizontal stripe(s); hot stuff; hypersonic

hs hora somni (Latin—at bedtime)

hs (HS) hardened site

h/s hand stamp(ing)

h.s. *hic situs* (Latin—laid here); *hoc sensu* (Latin—in this sense)

Hs Henriques

Hs Handschrift (German—manuscript)

HS Hakluyt Society; Haute-Saône; Haute-Savoie; Hawker Siddeley; Head Start; High Secretary; High School; historic site; Home Secretary; Home Service; Hospital Ship; House Surgeon; Hunterian Society; hydrofoil ship (naval symbol)

H-S Haute-Saône; Haute-Savoie

H & S Health and Safety (Code); Home & School

HS Hauptsatz (German—principal theme in a sonata)

H.S. *hic sepultus* or *hic situs* (Latin—here lies buried)

HS-30 German armored-personnel carrier

HS-125 Hawker-Siddeley Dominie jet transport

HS-748 Hawker Siddeley support aircraft; Hawker-Siddeley troop transport carrying 40 paratroopers or 50 regular soldiers

hsa heat source assembly; human serum albumin; hypersonic aircraft (HSA)

HSA Hawker Siddeley Aviation; Health Service Area; Health Services Administration; Health Systems Agency; Herb Society of America; Hispanic Society of America; Holly Society of America; Home Servicemens Association; Hospital Savings Association; Huguenot Society of America; Hunt Saboteurs Association

HSAA Health Sciences Advancement Award

HSAC Helicopter Safety Advisory Conference; House (of Representatives) Science and Astronautics Committee

HSA & D High School of Art and Design

HSAL Hispanic Society of America Library (NYC)

hsb human sexual behavior

HSBC Hong Kong and Shanghai Banking Corporation

hsbr high-speed bombing radar

hsc (HSC) engine high-swirl combustion engine

HSC Health and Safety Code

H-S-C Hand-Schfiller-Christian (disease)

HSCA Health Sciences Communication Association

HSCC Historical Society of Southern California

H Sch High School

Hschonhsn Hohenschonhausen

HS-Co A reduced coenzyme, A

hscp high-speed card punch

hscr high-speed card reader

hsct high-speed compound terminal

HSCT High Speed Civil Transport (aircraft); High-Speed Commercial Transport

HSCTB Heavy and Specialized Carriers Tariff Bureau

hsctt high-speed-card teletypewriter terminal

hsd hard-site defense; high-speed diesel (oil)

HSD Hawker Siddeley Dynamics; Health Services Department

hsda high-speed data acquisition

HSDB High Speed Data Bus

HSDE High-School Driver Education

HSDG Hamburg-Sudamerika Dampfschiffahrts - Gesellschaft (Columbus Line)

HSDM Harvard School of Dental Medicine

hse herpes simplex encephalitis; house

hse (HSE) syndrome hemorrhagic shock and encephalopathy syndrome

Hse House (postal abbreviation)

HSE Health and Safety Executive

H.S.E. hic sepultus est or *hic situs est* (Latin—here lies buried)

HSERF High-Score Educational Research Foundation

H & SF Heart and Stroke Foundation

HSFI High School of Fashion Industries

hsg housing

Hsg Helsingör (Elsinore)

HSG Hawker Siddeley Group

hsgt high-speed ground transport

HSGTC High-Speed Ground Test Center

HSGTP High-Speed Ground Transportation Program

HSH Her (His) Serene Highness

h & s hole hellhole and smell-hole

Hshld Household Bank

hsi heat-stress index; horizontal situation indicator

HSI Health Services Ine; Home and School Institute; Hotel Systems International; Humane Society International; Humanist Science Incorporated; Human Suffering Index

HSIA Halogenated Solvents Industry Alliance

HSIS Highway Safety Information Service

hsk housekeeper; housekeeping (flow chart); housekept

HSK Honorary Surgeon to the King

hskpg housekeeping

hskpr housekeeper

hsl herpes simplex labialis (HSL); hytran simulation language

HSL Hawaii State Library; Huguenot Society of London

hsla high-strength low-alloy (steel)

HSLA Home and School Library Association

HS Lab Health Service Laboratory (USA)

hslkfin high silky finish

hsltd hand salted

HSLWI Helical Spring Lock Washer Institute

hsm high-speed memory; historic structure museum; hydraulic start motor; hypersonic motor

hsm (HSM) holosystolic murmur

HSM Her (His) Serene Majesty; Historical Society of Montana; Host Security Module

HSMA Hotel Sales Management Association

HSMAI Hospital Sales and Marketing Association International

HSMB Hydronautics Ship Model Basin

HSMHA Health Services and Mental Health Association

HSN Home Shopping Network

HSNP Hot Springs National Park

HSNR Huleh Swamp Nature Reserve (Israel)

HSNY Handel Society of New York

hso high specific outlet

HSO Haifa Symphony Orchestra; Hamburg Symphony Orchestra; Hartford Symphony Orchestra; Hitachi Symphony Orchestra; Honolulu Symphony Orchestra; Houston Symphony Orchestra

HSORS High-Seas Oil-Recovery System

hsp high-speed printer

h of sp hybrid of species

H-S p Henoch-Schönlein purpura

HSP Historic State Park; Historical Society of Pennsylvania; Hospital Surgical Plan

HSP Haute Société Protestant (French—High Protestant Society)

HSPA Hawaiian Sugar Planters' Association; High School of the Performing Arts

h-span hydrolized starch-polyacrylonitrile copolymer graft(ing)

hspf heating seasonal performance factor

HSPG Hansard Society of Parliamentary Government

HSPH Harvard School of Public Health

HSPQ High-School Personality Questionnaire

hsptp high-speed paper-tape punch

hsptr high-speed paper-tape reader

HSQ Historical Society of Queensland; Honorary Surgeon to the Queen

hsr high-speed reader; high-speed rewind

hsr (HSR) high-speed rail(road); high-speed railway

HSR Health Service Region (USA)

HSRA High-Speed Rail Association

hsrc high-speed rail concept

HSRC Health Sciences Resource Center (Canadian)

HSRI Health Systems Research Institute; Highway Safety Research Institute

hsro high-speed repetitive operation

hsr(s) homogeneously staining region(s)

hss high-speed steel

hss (HSS) hydrological-sensing satellite

H-S s Hallervorden-Spatz syndrome

HSS History of Science Society; Hungarian State Symphony

HSSA History of Science Society of America

HSSO Hungarian State Symphony Orchestra

hsss high-speed stainless steel; high-strength stainless steel

hsst (HSST) high-speed surface transport (vehicle floats above its track on a magnetic cushion)

HSSTS Hit-Speed Surface Transport system

hst highest spring tide; high-speed train; hoist; hydrostatic transmission; hypersonic transport

Hst (HST) Hubble space telescope

H St Hugo Stinnes (steamship line)

HST Harry S Truman—thirty-third President of the United States; Hawaiian Standard Time; hypersonic transport

HSTA Hawaii State Teachers Association

HSTC Henderson State Teachers College

h/stead homestead(er)(s)

HSTI Hartford State Technical Institute

HSTL Harry S Truman Library

Hstn Houston

HSTRU Hydraulic System Test and Repair Unit

hsts horizontal stabilizer trim setting(s)

HSTS House Subcommittee on Traffic Safety

hstvl high survivability test vehicle— lightweight

HSU Hardin-Simmons University

H substance histamine-like capillary vasodilator

H-substance histamine-like substance

HSUL Haile Selassie University Libraries (Addis Ababa, Ethiopia)

HSUNA Humanist Student Union of North America

HSUS Humane Society of the United States

hsv heat-suppression valve

hsv (HSV) herpes simplex virus

HSV Huntsville, Alabama (airport)

HSWA Hazardous and Solid Waste Amendment

hswf housewife

HSWP Hungarian Socialist Workers Party

HSWT High-Speed Wind Tunnel

hszd hermetically-sealed zener diode

ht half title; halftime; halftone; hard top; heat; heat treat; heat treatment; heat-treated; heavy formex; heavy tank; heavy traffic; height; height telling; high temperature; high tensile; high tension; high tide; high treason; hired transport; hollow tile; hybrid tea (rose); hydrotherapy; hypertropia; hypodermic tablet

ht (HT) horizontal tabulation (data processing)

h & t harden(ed) and temper(ed); hospitalization and treatment; hospitalize and treat

h.t. *hoc tempore* (Latin—at this time); *hoc titulo* (Latin—under this title)

Ht total hypermetropia; total hyperopia

H^t Haut (French—high, upper)

HT Haiti (Internet code); Hawaiian Telephone; Hawaiian Territory; Hawaiian Theater; Hawaiian Time; Height Technician; High Treasurer; Horse Transport; Hospital Train

HT-2 trichothecense toxin used in biochemical warfare

hta heavier than air; high temperature accelerant

HTA Hardcourt Tennis Association; Harris Tweed Association; Horticultural Trades Association

Htal Hospital (Spanish—hospital)

htb hautboy (oboe); high-tension battery

HTB Highway Tariff Bureau; Horserace Totalisator Board

htc head(ing) to come; headline to come; heat transfer coefficient; hydraulic temperature control

htd heated

HTD Hospital for Tropical Diseases

htd pl heated pool

htd rm heated room

HTDS Host Target Development System

H^{te} Haute (French—high, upper)

Hte-Gar Haute-Garonne

Hte-L Haute-Loire

Hte-M Haute-Marne

H^{ter} Hinter (German —behind, rear)

Hte-Sao Haute-Saône

Hte-Sav Haute-Savoie

Htes-Pyr Hautes-Pyrénées

htfc high-temperature fuel cell

ht fx heat treat fixture

htg heating

htgcr (HTGCR) high-temperature gas-cooled reactor

htgr high-temperature gas reader

Htg & Vent Heating & Ventilating

hth (HTH) holiday travel hostility

hthp high temperature high pressure

hti high-temperature isotope

HTI Hand Tools Institute; High Twelve International

htk headline to come

htl hearing threshold level; heavy-traffic licence; high threshold logic

Htl Hotel

htld high-technology light division

htls high torque low speed

htlv human T-cell leukernia virus

HTLV-I human T-cell lymphotropic virus

htm heat transfer medium; high-temperature metallography

HTMC High Temperature Materials Corporation

html hypertext markup language

Htn Hamilton, Bermuda

hto high-temperature oxidation; horizontal takeoff

htofore heretofore

htohl horizontal take-off and horizontal landing

htol (HTOL) horizontal-take-off-and-landing

h top hardtop

HTOT High-Temperature Operating Test

H-town Hagerstown, Maryland

H-Town Hartford, Connecticut

htp high-test peroxide

h-t p half-title page

h-t-p house-tree-person (psychological drawing test)

HTP House-Tree-Person (test); Humor Test of Personality

htr heater

htr (HTR) high-temperature reactor

HTR Highway Traffic Regulation(s)

HTR Harvard Theological Review

htrac half-track

htrb high-temperature reverse bias

HTRDA High-Temperature Reactor Development Associates

htres heat resistant

htrf hypothalmic-releasing factor

H Trin Holy Trinity

hts half-time survey; heights; high-tensile steel

Hts Heights

HTS Huntington, West Virginia (airport)

HTSA Highway Traffic Safety Administration

htst high-temperature short-time (pasteurization)

htt (HTT) heavy tactical transport

http hypertext transfer protocol

ht tr heat treat

htu heat transfer unit

htv (HTV) hypersonic test vehicle

HTV Hiroshima Television

htvt heating and ventilating

htw high-temperature water

HT & W Hoosac Tunnel & Wilmington (railroad)

ht wkt hit wicket

htxgr heat exchanger

hu hunter (color); hyperemia unit

Hu Hungarian; Hungary

HU Haifa University; Harvard University; Hebrew University; Howard University; Hungary (Internet code)

HUA Highway Users Association; Housing and Urban Affairs

HUAC House Un-American Activities Committee

Huascán Huascarán (Peru's highest mountain)

HUB Humboldt Universität zu Berlin (German—Humboldt University of Berlin)—library on Berlin's Clara-Zetkin Strasse

hubba hubbas (crack-type cocaine)

Hubble Hubble Space Telescope

Hubert H Humphrey Hubert H Humphrey Stadium, Minneapolis-St Paul

HUC Hebrew Union College

HuCal Human combinatorial antibody library

HUCIA Harvard University Center for International Affairs

HUJIR Hebrew Union College Jewish Institute of Religion

hucks huckleberries

Huck(y) Huckleberry Finn

hucr highest useful compression ratio

hud head-up display

Hud Huddleston; Hudson

HUD Hong Kong United Dockyard; Housing and Urban Development

HUDC Housing and Urban Development Corporation

hud-eu head-up display electronic unit

Hud Inst Hudson Institute

HUDG History at the Universities Defence Group

HUDPR Housing and Urban Development Procurement Regulations

hudson hudson seal (imitation seal made of dyed muskrat fur)

Hudson Angus Hudson, the butler in *Upstairs, Downstairs;* river rising in the Adirondacks and flowing to New York Bay

Hudson Girls New York School for (delinquent) Girls at Hudson

Hudson Internat Legis Hudson's International Legislation

Hudson World Court Hudson's World Court Reports

Hud Val Phil Hudson Valley Philharmonic

hudwac head-up-display weapon-aiming computer

Huel Huelva

Hueneme Port Hueneme, California

Hues Huesca

huff-duff high-frequency direction finder

HUFIT Human Factor Laboratories in Information Technologies

HUFSM Highway Users Federation for Safety and Mobility

Huggin Hugh; Hugo

HUGHES Hughes Aircraft Company

hugo highly unusual geophysical operations

HUGO Human Genome Organization

HU H Hughes Hall, Cambridge

HUJ Hebrew University of Jerusalem

huk (HUK) hunter-killer

HUKFORPAC Hunter-Killer Forces—Atlantic (USN)

HUKFORPAC Hunter-Killer Forces—Pacific (USN)

huks (HUKS) hunter-killer submarine(s)—USN

Huks Hukbong Mapagpalqyang Bayan (Philippine Communist Armed Forces)

Hul Hulbert; Huldreich; Hulton

Hul Hullin

HUL Harvard University Library; Helsinki University Library; Hokkaido University Library

Hull in England, Kingston-upon-Hull

Hully Hulbert

HULTIS Hull Technical Interloan Scheme

hum human; humane; humanism; humanities

hum. *humaniora* (Latin—humanities)—also appears as *H. U. M.*

Hum Humbert; Hummel; Humphrey; Humphreys; Humphry

Huma L'Humanité (French-communist daily paper)

human eng human engineering

HUMARIS Human Materials Resources Information System

Humb humberside

HUMBLE Humble Oil (Company)

Hum Con Humanist Conference

humer humerus

humi humidity

humint human intelligence

hummer. high-mobility multipurpose wheeled vehicle

hummer(s) humming bird(s)

Hummon Herman

Humph Humphrey

HUMRRO Human Resources Research Office

hums humanitarian reasons

Hum Soc Humane Society

hun hundred

Hun Hungarian; Hungary

Hun Hungria (Portuguese—Hungary); *Hungría* (Spanish—Hungary); *Hun's NY Supreme Court Reports*

hund hundred

Hung Hungaria; Hungarian; Hungarica; Hungary

Hungary Hungarian Republic (central European nation), *Magyar Népköztársaság*

Hunnen Die Hunnenschlacht (German—The Battle of the Huns)—Liszt's Symphonic Poem No. 11

Hunt Hunter; Huntington; Huntley; Huntly

hunth hundred thousand

Huntington Huntington Library, Art Gallery, and Botanical Gardens at San Marino, California

Hunts Huntingdonshire; Huntsville

Huntsville Girls Goree Unit Women's Prison at Huntsville, Texas

HUP Harvard University Press; Hospital of the University of Pennsylvania

HUPAS Hofstra University Pro Arte Symphony

hur hurricane

hur (HUR) homes using radio

Hurb Hurban (Hebrew—Holocaust)

hurcn hurricane

hurevac hurricane evacuation

HURRAH Help Us Reach and Rehabilitate America's Handicapped (HEW program)

hus hemolytic uremic syndrome (HUS)

hus. husband

Hus Johannes Hus von Husinetz

HUSAT Human Sciences and Advanced Technology

husb husbandry

Huss Jan Huss known in German as Johannes von Husinetz—his death by being burned at the stake for heresy led to the Hussite War from 1419 to 1434

hustle helium-underwater speech-translating equipment

hut (HUT) homes using television

hutch humidity-temperature charts

Hutch Hutcheson; Hutchings; Hutchins; Hutchinson; Hutchison

hutv home(s) using television

Hux Huxley

hv heavy; high velocity; high voltage

h-v high-voltage

h & v heating and ventilating

h.v. hoc verbum (Latin—this word)

HV Hardness Vickers (symbol); Health Visitor; Hospital Visit

H-V Haute-Volta (French—Upper Volta)

hva homovanillic

Hva Huelva

HVA Health Visitors' Association

hvac heating, ventilating, and air conditioning; high-voltage alternating current

hvap hyper-velocity armor-piercing

hvar (HVAR) high-velocity aircraft rocket

HVB Hawaii Visitors Bureau

hvbn have been

hvc hardened voice circuit

hv & c heating, ventilating, and cooling

HVCA Heating and Ventilating Contractors' Association

HVCC Hudson Valley Community College

HVCR High Vice Chief Ranger

hvd high-velocity detonation; hypertensive vascular disease

hvdc high-voltage direct current

hvdn have done

hve home video entertainment

HVEC High Voltage Engineering Corporation

hvem high-voltage transmission electron microscopy

hvf high viscosity fuel

hvg having

hvgo heavy-vacuum gas oil

hvh herpesvirus hominis

hvh (HVH) herpesvirus horninis (type 1 transmitted by mouth and marked by cold sores and fever blisters; type 2 transmitted venereally and characterized by genital lesions)

Hvh Herpesvirus hominus (Latin—herpes simplex virus)

H V H *Hoek van Holland* (Dutch—Hook of Holland)

hvhmd holographic visor-helmet mounted display

hvi high viscosity index

HVI Hartman Value Inventory; Home Ventilating Institute

H'ville Huntsville, Alabama

hvJ hemagluttinating virus of Japan

H v K Herbert von Karajan

hvl half-value layer

HVL Hanseatic Vaasa Line; Heitor Villa-Lobos

hvm hydraulic valve module

HVMC Harbor View Medical Center

Hvn Haven

HVNP Hawaii Volcanoes National Park

HVO Hawaiian Volcano Observatory

HVOT Hooper Visual Organization Test

hvp high-value package; high-value product; hydrolyzed vegetable protein

HVP Hudson Vitamin Products

HVPO Hudson Valley Philharmonic Orchestra

hvps high-voltage power supply

hvpve high-voltage photovoltaic effect

hvr high-vacuum rectifier

HVRA Hawaiian Volcano Research Association

hvrap (HVRAP) hyper-velocity rocket-assisted projectile

hvsa high-voltage slow activity

hvss horizontal volute spring suspension

hvtl high-voltage transmission line

hvtp high-velocity target-practice

HVWS Hebrew Veterans of the War with Spain

hvy heavy

hw hardwood (floors); headwaiter; headwind; heartworm; herewith; high water; hindwing; hot water

h/w hardware; husband and wife

hW hectowatt

Hw Hauptwerk (German— great work)

HW Helsinki Watch; high water

H-W Harbison-Walker (refractories)

H & W Harland and Wolff (Belfast shipbuilders); Hereford and Worcester

Hway Highway

Hwb Handwörterbuch (German—pocket dictionary)

hwe hot-water circulating

HWC Heriot-Watt College

hwctr heavy-water components test reactor

hwd hardwood

H'w'd Hollywood

HWDYKY How Well Do You Know Yourself? (psychological test)

hwf & c high water full and change

hwgcr (HWGCR) heavy-water-moderated gas-cooled reactor

hwi high water interval

HWI Helical Washer Institute

hwl high-water line

HWL Hardy-Weinberg Law—fundamental concept in population genetics formulated by Godfrey H Hardy and Wilhelm Weinberg; Henry Wadsworth Longfellow; Hutchison Whampoa Limited

hwLB high water London Bridge

hwlwr (HWLWR) heavy-water-moderated boiling light-water-cooled reactor

hwm high-water mark; high-wet modulus

HWM Hiram Walker Museum (Windsor, Ontario)

HWMC House Ways and Means Committee

HWMD Hazardous Waste Management Division (EPA)

hwmnt high-water mark neap tide

hwmont high-water mark ordinary neap tide

hwmost high-water mark ordinary spring tide

hwmst high-water mark spring tide

hwnt high-water neap tide

HWO Homosexual World Organization

hwocr (HWOCR) heavy-water (moderated) organic-cooled reactor

hwont high-water ordinary neap tide

Hwood Hollywood

hwos high-water ordinary springs

hwost high-water ordinary spring tides

hwp height and weight proportional

hwq high-water quadrature; tropic high-water inequality

hwr (HWR) heavy water reactor (AEC)

hws hot-water soluble; hot-water system

HWS Hazardous Waste Service; Hurricane Warning Service

H & WSC Hobart and William Smith Colleges

HWSS Hot Water Service System

hwst high-water spring tide

HWT Herald and Weekly Times

H-W U Heriot-Watt University

hwvr however

HWW Hochschule für Welthandel, Wien (School for World Trade—Vienna)

hwwc hand wash with care

Hwy Highway

hx heat exchanger; hexode; history

hx (HX) headroom extension

Hx history (medical case)

Hxd Hardinxveld

hXr head X-ray

hy heavy; henry; high yield; hundred yards; hydrant

Hy Henry; Highway; Hiram; Hyman

HY Highway

Hy Hasy (Arabic—waterhole)—also appears as *Hasi*

HY *Helsingin Yliopisto* (University of Helsinki)

Hya Hydra (constellation)

hyb hybrid

hyball hydraulic ball

HYC Harlem Yacht Club; Hartford Yacht Club; Haverhill Yacht Club

hycol hybrid computer link

hycon hydraulic control

hycotran hybrid computer translator

hyd hydrate; hydraulic(s); hydrostatics

Hyd Hyderabad; Hydrus (sometimes abbreviated Hyi for the genitive Hydri)

hydac hybrid digital-analog computer

hydapt hybrid digital-analog pulse time

h-y dash hundred-yard dash

hydel hydroelectric(al)

hyd/pnu hydraulic/pneumatic

hydr hydrographer

hydrarg. hydrargyrum (Latin—mercury)

HYDRAS Hydrographic Diffital Positioning and Depth Recording System

hydraul hydraulic(s)

hydraweld hydraulic-drawn welded (steel tubing)

hydro hydrodynamic group of hydrodynamics (slang); hydroelectric; hydroelectrical; hydrographic; hydrology; hydrostatic

hydro (Latin prefix—water)— hydrodynamics

HYDRO Hydrographic Office

hydrodyn hydrodynamics

hydroelec hydroelectric

hydrog hydrography

HYDROIND Hydrography of the Indian Ocean

hydrol hydrology

HYDROLANT Hydrography of the Atlantic Ocean

hydrom hydromechanics

hydromag hydromagnetic(s)

hydromagnetics magnetohydrodynamics

HYDROPAC Hydrography of the Pacific Ocean

hydros hydrostatics

hydrot hydrotherapy

hydrox hydroxyline

HYDRSS High Data Rate Storage System (NASA)

hydt hydrant

hydx hydroxide(s)

hyf (HYF) hydrofoil

HYF Hong Kong and Yaumati Ferry

hyfes hypersonic flight environmental simulator

hyg hygiene; hygienic; hygroscopic

hygas hydrogen gasification

hygrom hygrometer; hygrometist

hygst hygienist

Hyk Helsingin yliopiston kirjasto (Finnish—Helsinki University Library)

hyl (Hyl) hydroxylysine

hyla hybrid language assembler

HYMA Hebrew Young Men's Association

hymnol hymnologist; hymnology

hynmm have you not made a mistake

hyp hyperbola; hyperbolic; hyphen; hyphenate; hyphenation; hypochondria(c); hypothesis; hypothetical

hyp (Hyp) 4-hydroxyproline

Hyp Hypolite

HYP Harvard, Yale, and Princeton

hype high performance (bullet); hyperbole; hypertension; hypodermic (underground slang—person who injects drugs with a hypodermic syringe)

hyper hypercritical

hyper (Latin prefix—above, beyond, excessive)—hypercritical

hyperb hyperbole; hyperbolic

hyperdip hyperdiploid(al); hyper diploidy

hyperdop hyperbolic doppler

hypersex hypersexual(ity)

hypert hypertape (flow chart); hypertension

hypn hypertension

hypno hypnotism

hypnot hypnotic; hypnotist

hypo hypochondria; hypochondriac; hypochondriacal; hypodermic (injection or needle); hyposulfite of soda (sodium thiosulfate—NaS_2O_3 $5H_2O$)

hypo (Latin prefix—below or under)—hypodermic

hypodip hypodiploid(al); hypodiploidy

hypoth hypothesis

hypro hydroxyproline

hys hysteria; hysteric; hysterical; hysterics

HYSAS Hydrofluidic Stability Augmentation System

hyst hysteresis; hysteria

hystad hydrofoil stabilizing device

hyster hysterectomy

hysterec hysterectomic (sterilization); hysterectomy (removal of the uterus)

HYSTU Hydrofoil Special Trials Unit (USN)

HYSURCH Hydrographic Survey and Charting System

hytemco high-temperature coefficient nickel-iron alloy

hy tr heat treat

hyv's high-yielding varieties (of grain)

hz haze; heritability zone; herpes zoster

hz (Hz) hertz (one cycle per second); hertzian

Hz Henriquez; hertz (cycles per second)

Hzbl Holzbläser (German—woodwind instruments or players)

Hzk Hezekiah

hzy hazy

I

i angle of incidence (symbol); ice; icy; incident ray (symbol); incisor; indigo; infant(icide); inmate; instantaneous current (symbol); instantaneous value (symbol); interceptor; interest; interlaced; intransitive; invoice; irregular; isotopic fine structure (symbol); moment of photographic plate (symbol); optically inactive (symbol); rate of interest (symbol); Van't Hoff factor (symbol); vapor pressure constant (symbol)

i (I) inversion (12-tone matrix)

i' in

i Illinois Central symbol; Imperial Savings

i. id (Latin—that)

I acoustic intensity (symbol); candlepower or intensity of luminosity (symbol); conduction current (symbol); convection current (symbol); Ido (artificial language); in; inclination; Independent; India—code for letter I; Indian; industrial broadcasting; inertia; infantry; Inspector; inspired gas (symbol); Institute; Institution; Instructor; Intelligence; interstate; intrinsic semiconductor (symbol); iodine; ionic strength (symbol); Ireland; Irish; Island; Isthmian Line; Italian Line; Italy—auto plaque; izzard; luminous intensity (symbol); paid this year, dividend omit-

ted, deferred, or no action taken at last dividend meeting (in newspaper stock listings)

I (I) investment (macroeconomics symbol)

I Ile (French—Island, Isle); Imperator (Latin—Emperor); Imperatrix (Latin—Empress); Imperium (Latin—Empire); in (German or Italian—in); inde (Danish—in); Infdelis (Latin—infidel or unbeliever); Isle (French—island); itd (Finnish—east); izquierda (Spanish—left)

I-l luminous intensity (symbol)

I¹2⁸ radioactive iodine

I¹3⁰ radioactive iodine

I¹3¹ radioactive iodine

I²L integrated-injection logic

I³ Illinois Innovators and Inventors

ia immediately available, impedance angle; indicated altitude; infra-audible; initial allowance; initial appearance; international angstrom; intra-arterial; intra-articular

i & a indexing and abstracting; integration and assembly

i.a. in absentia (Latin—in the absence of)

i A im Auftrage (German—by order, for, under instruction)

Ia anode current (symbol); Ingegerda

IA Import Administration; Incorporated Accountant; Indian Army; Industrial Arts; Infected Area; Inspection Ad-

ministration; Installment Agreement; Institute of Actuaries; Instructional Aide; Intercoiffure America; Internal Affairs; International Angstrom; Iowa; Iraqi Airways; Irrigation Area; Irrigation Association; Isaac Asimov

I/A Insurance Auditor; Isle of Anglesey

I of A Inspector of Anatomy; Institute of Accountants; Institute of Acoustics; Instructor of Artillery

IA International Atlas (Rand McNally)

IAA Independent Airlines Association; Indian Association of America; Inspector Army Aircraft; Institute for Alternative Agriculture; Insurance Accountants Association; Interamerican Accounting Association; Interment Association of America; International Academy of Astronautics; International Acetylene Association; International Advertising Association; International Apple Association; International Association of Allergology; Intimate Apparel Associates; Inventors Association of Australia

IAA Instituto do Acuar e do Alcool (Portuguese—Sugar and Alcohol Institute); International Aerospace Abstracts

IAAA Institute of Air Age Activities; Intermarket Association of Advertising Agencies; International Airforwarders and Agents Association

IAAAA Intercollegiate Association of Amateur Athletes of America

IAAABBP International Association of African and American Black Business People

IAAB Inter-American Association of Broadcasters

IAABA International Association of Aircraft Brokers and Agents

IAAC International Agriculture Aviation Center; International Antarctic Analysis Center

IAACC Inter-Allied Aeronautical Control Commission

IAADFS International Association of Airport Duty-Free Stores

IAAE Institution of Automotive and Aeronautical Engineers

IAAER International Association for the Advancement of Educational Research

IAAF International Amateur Athletic Federation

IAAFA Inter-American Air Force Academy

IAAHU International Association of Accident and Health Underwriters

IAAI International Airports Authority of India; International Association of Arson Investigators

IAALD International Association of Agricultural Librarians and Documentalists

IAAM International Association of Auditorium Managers; International Association of Automotive Modelers

IAANZ Institute of Actuaries of Australia and New Zealand

IAAO Interlochen Arts Academy Orchestra; International Association of Assessing Officers

IAAOPA International Association of Aircraft Owners and Pilots Associations

IAAP International Association of Applied Psychology

IAAPA International Association of Amusement Parks and Attractions

IAAPEA International Associations Against Painful Experiments on Animals

IAAS Incorporated Association of Architects and Surveyors; Institute of Advanced Arab Studies; International Association of Agricultural Students

IAASE Inter-American Association of Sanitary Engineering

IAASS International Association of Applied Social Science

iab increasing assurance benefits

IAB Industrial Advisers to the Blind; Industrial Arbitration Board; Industry Advisory Board; Inter-American Bank; International Air Bahama; International Association of Bureaucrats; Internet Activities Board; Internet Architecture Board

IABA Inter-American Bar Association; International Association of Aircraft Brokers and Agents

IABBE International Association for Better Basic Education

IABC International Association of Business Communicators

IABF International Association of Business Forecasting

IABG International Association of Botanic Gardens

IAB-ICSU International Abstracting Board—International Council of Scientific Unions

IABLA Inter-American Bank for Latin America; Inter-American Bibliographical and Library Association

IABM International Association of Broadcast Monitors

IABO International Association of Biological Oceanography

IABPAI International Association of Blue Print & Allied Industries

IABPC International Association of Book Publishing Consultants

IABSE International Association for Bridge and Structural Engineering

IABSIW International Association of Bridge and Structural Iron Workers

IABTI International Association of Bomb Technicians and Investigators

iac innovative academic courses; integrating assembly contractor; integration, assembly, checkout; interview after combat

IAC Ibrahim Ali Commission (Malaysia); Indian Airlines Corporation; Industrial Applications Centers (of National Aeronautics and Space Administration); Industrial Arbitration Court; Industry Advisory Commission; Industry Assistance Commission; Information Analysis Center; Institute of Amateur Cinematographers; Intelligence Advisory Committee (CIA); Intermediate Air Command; International Anti-counterfeiting Coalition; Interview After Combat; Irish Air Corps

IACA Independent Air Carriers Association; Indian Arts and Crafts Association; Inter-American College Association

IACAC Inter-American Commercial Arbitration Commission

IACB Indian Arts and Crafts Board; International Advisory Committee on Bibliography (UNESCO); International Association of Convention Bureaus

IACC International America's Cup Class; International Anti-Counterfeiting Coalition; International Association of Conference Centers; Italy-America Chamber of Commerce

IACC Instituto Argentino de Control de la Calidad (Spanish—Argentine Institute for Quality Control)

IACCD Inter-American Confederation of Continental Defense

IACCI International Association of Credit Card Investigators

IACCP Inter-American Council of Commerce and Production

IACD International Association of Clothing Designers

IACDLA International Advisory Committee on Documentation, Libraries, and Archives (UNESCO)

IACE International Air Cadet Exchange

IACES International Air Cushion Engineering Society

IACHR Inter-American Commission on Human Rights

IACI Inter-American Childrens Institute; Irish-American Cultural Institute

IACID Inter-American Center for Integral Development

IACM International Association of Circulation Managers; International Association of Concert Managers

IACOMS International Advisory Committee on Marine Sciences (FAO)

IACP International Association of Chiefs of Police

IACPR International Association of Corporate and Professional Recruiters

IACP & AP International Association for Child Psychiatry and Allied Professions

IACRL Italian-American Civil Rights League

IACRS International Association of Concrete Repair Specialists

IACS International Annealed Copper Standard; International Association of Cooking Schools; International Association of Counseling Services; Irish-Australian Cultural Society

IACSS International Association for Computer Systems Security

IACT Illinois Association of Classroom Teachers; Indiana Association of Cities and Towns

IACTE Iowa Association of Colleges for Teacher Education

IACUSD International Association of College and University Security Directors

IACVB International Association of Convention and Visitor Bureaus

IACW Inter-American Commission for Women

iad initiation area discriminator; installation, assembly, or detail; integrated automatic documentation

IAD Dulles International Airport (Washington, DC; Integrated Area Development; Internal Affairs Department; Internal Affairs Division (LAPD and NYPD); International Agricultural Distribution; International Astrophysical Decade—1965–1975

I-A D Inter-American Dialogue

IADB Inter-American Defense Board; Inter-American Development Bank

IADC Image Assisted Data Capture; Inter-American Defense College; Inter-American Drug Commission; International Association of Dredging Companies; International Association of Drilling Contractors

IADD International Association of Diecutting and Die-making

IADF Inter-American Association for Democracy and Freedom

IADIS Irish Association for Documentation and Information Services

iadl (IADL) instrumental activities of daily living

IADL International Association of Democratic Lawyers; Italian-American Defense League

IADO Iranian Agriculture Development Organization

IADPC Inter-Agency Data Processing Committee

IADR International Association for Dental Research

IADS Integrated Air Defense System; International Association of Dental Students; International Association of Department Stores

iadt initial active duty training

iae in any event; integral absolute error

IAE Institute of Army Education; Institute of Automobile Engineers; Institution of Automobile Engineers; International Animal Exchange

IAE *Institut Atomnoi Energii* (Russian—Atomic Energy Institute)

IAeA Institution of Aeronautical Engineers

IAEA Inter-American Education Association; International Association for Educational Assessment; International Atomic Energy Agency

IAEC Israel Atomic Energy Commission

IAECOSOC Inter-American Economic and Social Council

IAEE International Association of Earthquake Engineers

IAEI International Association of Electrical Inspectors

IAEL International Association of Electrical Leagues

IAES International Association of Electrotypers and Stereotypers

IAESP Indiana Association of Elementary School Principals; Iowa Association of Elementary School Principals

IAESTE International Association for the Exchange of Students for Technical Experience

IAET In-Flight Acromedical Evacuation Team

IAEVG International Association for Educational and Vocational Guidance

IAEWP International Association of Educators for World Peace

iaf immobilizing accelerating factor; interview after flight

Iaf audio-frequency current (symbol)

IAF Industrial Areas Foundation; Inter-American Foundation; International Abolitionist Federation (for abolition of prostitution); International Association of Firefighters; International Astronautical Federation; Israeli Air Force

I-AF Inter-American Foundation

IAFAE Inter-American Federation for Adult Education

IAFC International Association of Fire Chiefs

iafd intentionally administered fatal dose(s)

IAFD International Association of Food Distribution

IAFE International Association of Fairs and Expositions

IAFF Institute of Australian Flora and Fauna; International Association of Fire Fighters

iafi infantile amaurotic family idiocy

IAFMM International Association of Fish Meal Manufacturers

IAFP International Association for Financial Planning

IAFV Infantry Armed Fighting Vehicle

IAFWNO Inter-American Federation of Working Newspapermen's Organizations

IAG Institute of Australian Geographers; Interagency Advisory Group; International Association of Geodesy; International Association of Gerontology

IAGA International Association of Geomagnetism and Aeronomy; International Association of Golf Administrators

IAGB & I Ilcostomy Association of Great Britain and Ireland

iagc instantaneous automatic gain control

IAGC International Association for Geochemistry and Cosmochemistry; International Association of Geophysical Contractors

IAGFCC International Association of Game, Fish, and Conservation Commissioners

IAGLP International Association of Great Lakes Ports

IAGM International Association of Garment Manufacturers

IAGMA International Assembly of Grocery Manufacturers Associations

I Agr E. Institution of Agricultural Engineers

IAGS Inter-American Geodetic Survey

IAH Houston International Airport (Texas); Inter-American Highway; International Asian Highways; International Association of Hydrology

IAHA Inter-American Hotel Association; International Association of Hospitality Accountants

IAHF International Aerospace Hall of Fame

IAHIC International Association of Home Improvement Councils

IAHM International Association of Head Masters

IAHP Institutes for the Achievement of Human Potential; International Association of Horticultural Producers

IAHR International Association for Hydraulic Research

IAHS International Association of Hydrological Sciences

IAI Icelandic Airlines Incorporated; Independent Accountants International; International African Institute; International Association for Identification

IAI-201 Israeli Arava light transport plane

IAIA Institute of American Indian Arts

IAIAS Inter-American Institute of Agricultural Sciences

IAICM International Association of Ice Cream Manufacturers

IAIE Inter-American Institute of Ecology

IAII Inter-American Indian Institute

IAIP International Association of Independent Producers

IAIs Israeli Aircraft Industries

IAIS Industrial Aerodynamics Information Service (UK)

iaiyh it's all in your head

IAJAM Industrial Association of Juvenile Apparel Manufacturers

ial initial; initial appearance; initialism; instrument approach and landing; interlaminar adhesive layer; international algebraic language

IAL Icelandic Airlines; Imperial Airways Limited; International Aeradio Limited; International Algebraic Language; International Arbitration League; International Association of Limnology; Irish Academy of Letters

IAL Icelandic Airlines-Loftleider

IALA International Association of Lighthouse Authorities

IALC International Association of Lions Clubs; International Association of Lyceum Clubs

IALD International Association of Lighting Designers

IALL International Association of Law Libraries

i allg im allgemeinen (German—generally; in general)

IALP International Association of Logopedics and Phoniatrics

IALS International Association of Legal Science

iam interactive algebraic manipulation

IAM Indian-Artifact Magazine; Institute of Appliance Manufacturers; Institute of Aviation Medicine; International Academy of Management; International Academy of Medicine; International Association of Machinists; International Association of Meteorology

IAMA International Abstaining Motorists Association; Intimate Apparel Manufacturers Association

IAMAM International Association of Museums of Arms and Military History

IAMAP International Association of Meteorology and Atmospheric Physics

IAMAT International Association for Medical Assistance to Travelers

IAMAW International Association of Machinists and Aerospace Workers

IAMB International Association of Microbiologists

IAMC Indian Army Medical Corps; Institute for Advancement of Medical Communication; Institute of Association Management Companies; Inter-American Music Council

IAMCA International Association of Milk Control Agencies

IAMCL International Association of Metropolitan City Libraries

IAMCR International Association for Mass Communication Research

IAME International Association for Modular Exhibitry

IAMFE International Association on Mechanization of Field Experiments

IAMFS International Association of Milk and Food Sanitarians

IAML International Association of Music Libraries

IAMLT International Association of Medical Laboratory Technologists

IAMM International Association of Master Mariners; International Association of Medical Museums

IAMO Inter-American Municipal Organization

IAMP Inter-Agency Motor Pool

IAMPO International Association of Mechanical and Plumbing Officials

IAMPTH International Association of Master Penmen and Teachers of Handwriting

IAMR Institute of Arctic Mineral Resources

IAMS International Association of Microbiological Societies; International Association of Municipal Statisticians

IAMSO Inter-African and Malagasy States Organization

IAMTCT Institute of Advanced Machine Tool and Control Technology

IAMTF Inter-Agency Maritime Task Force

IAMWF Inter-American Mine Workers Federation

Ian (Gaelic—John)

IAN Instituto Agrario Nacional (Spanish—National Agrarian Institute)

IANA Institute of Alaska Native Arts; Inter-African News Agency

IANAP Interagency Noise Abatement Program

IANC International Airline Navigators Council; International Anatomical Nomenclature Committee

IA & ND Indian Affairs and Northern Development (Canada)

IANE Institute of Advanced Nursing Education

IANEC Inter-American Nuclear Energy Commission

Ian F Ian Fleming

Ian Xen Iannis Xenakis

IANSA Industria Azucarera Nacional SA (Spanish—National Sugar Industry Corporation)

IANSW Institute of Architects of New South Wales

iao intermittent aortic occlusion

IAO Incorporated Association of Organists

IAOC Indian Army Ordnance Corps

IAOL International Association of Orientalist Libraries

IAOR International Abstracts in Operations Research

IAOS International Association of Oral Surgeons; Irish Agricultural Organization Society

IAOT International Association of Organ Teachers

iap interceptor aim point(s); intermittent acute porphyria; intra-abdominal pressure

IAP Institute of Agricultural Parasitology; Institute of Australian Photographers; Institute of Australian Photography; Institution of Analysts & Programmers; International Academy of Pathology; International Academy of Proctology

IAPA Industrial Accident Prevention Association; Inter-American Parliamentary Association; Inter-American Parliamentary Organization; Inter-American Police Academy; Inter-American Press Association; International Airline Passengers Association; International Association of Police Artists

IAPB International Association for the Prevention of Blindness

IAPC Institute for the Advancement of Philosophy for Children; International Association of Pet Cemeteries; International Association of Political Consultants; International Association for Public Cleansing

IAPCO International Association of Professional Congress Organizers

IAPCU Iowa Association of Private Colleges and Universities

IAPD International Association of Plastics Distributors

IAPES International Association of Personnel in Employment Security

IAPESGW International Association of Physical Education and Sports for Girls and Women

IAPG Interagency Advanced Power Group; International Association of Physical Geography

IAPH International Association of Paper Historians; International Association of Ports and Harbors

IAPHA International Association of Port and Harbor Authorities

IAPHC International Association of Printing House Craftsmen

IAPI Institute of American Poultry Industries

IAPI Instituto Argentino de Producción Industrial (Spanish—Argentine Industrial Production Institute)

IAPIP International Association for the Protection of Industrial Property

IAPL International Association of Penal Law

IAPM International Academy of Preventive Medicine; International Association of Progressive Montessorians

IAPMO International Association of Plumbing & Mechanical Officials

IAPN International Association of Professional Numismatists

IAPO International Association of Physical Oceanography

IAPP International Association of Police Professors

IAPPW International Association of Pupil Personnel Workers

IAPR Indian Air Patrol Reserve

iaps inductosyn angle position simulator

IAPs Industry Application Programs

IAPS Incorporated Association of Preparatory Schools; International Affiliation of Planning Societies; International Association for the Properties of Steam

IAPSC Inter-African Phytosanitary Commission

IAPSO International Association of Physical Sciences of the Oceans

IAPT International Association for Plant Taxonomy

IAPTA International Allied Printing Trades Association

IAPW International Association of Personnel Women

iaq indoor air quality

IAQ Independent Activities Questionnaire

IAQC International Association of Quality Circles

IAQR Indian Association for Quality and Reliability

iar intersection of air routes

IAR Institute for Air Research; Integrated Avionics Racks

IARA Inter-Allied Reparations Agency

I Arb Institute of Arbitrators

IARC Indian Agricultural Research Council; International Agency for Research on Cancer

IARD Information Analysis and Retrieval Division (American Institute of Physics)

IARF International Association for Liberal Christianity and Religious Freedom

IARFP International Association of Registered Financial Planners

IARI Indian Agricultural Research Institute; Industrial Advertising Research Institute

IARIGAI International Association of Research Institutes for the Graphic Arts Industry

IARIW International Association for Research into Income and Wealth

IAROO International Association of Railway Operating Officers

IARP Indian Association for Radiation Protection; Inflation Accounting Research Project

IARQ Intellectual Achievement Responsibility Questionnaire

IARS International Anesthesia Research Society

IARSS Illinois Association of Regional Superintendents of Schools

IARU International Amateur Radio Union

ias immediate access storage; indicated airspeed; instrument approach system

ia's infant addicts

IAS Indian Administrative Service; Industrial Arbitration Service; Infrared Astronomy Satellite; Institute for Advanced Study; Institute of the Aeronautical Sciences; Institute of Aerospace Sciences; Institute of American Strategy; Institute of Andean Studies, Instrument Approach System; Intelligent Authoring Systems; International Accountants Society; International Association of Siderographers; International Audiovisual Society; International Aviation Service; Investors Arbitration Service

IASA Idaho Association of School Administrators; Illinois Association of School Administrators; Insurance Accounting and Systems Association; International Air Safety Association; International Association of Sound Archives; Iowa Association of School Administrators

IASB Iowa Association of School Boards

IASBO Indiana Association of School Business Officials; Iowa Association of School Business Officials

IASC Indian Army Service Corps; Inter-American Safety Council; International Association of Seed Crushers

IASCA International Auto Sound Challenge Association

IASCH Institute for Advanced Studies in Contemporary History (formerly Wiener Library)

iasd interatrial septal defect

IASDI Inter-American Social Development Institute

IaSEA Iowa Society of Enrolled Agents

IASG Inflation Accounting Steering Group

IASH International Association of Scientific Hydrology

IASI Inter-American Statistical Institute

IASL Illinois Association of School Librarians; International Association for the Study of the Liver; International Association of School Librarians; Irish Association of School Librarians

IASLIC Indian Association of Special Libraries and Information Centers

iasor ice and snow on runway

IASP International Association for Social Progress; International Association for Suicide Prevention; International Association of Scholarly Publishers

IASPEI International Association of Seismology and Physics of the Earth's Interior

IASPO International Association of Senior Police Officers

IASPS International Association for Statistics in Physical Sciences

IASS Insurance Accounting and Statistical Society; International Association for Shell Structures; International Association of Soil Science

IASSS International Association for Shell and Spatial Structures

IASSW International Association of Schools of Social Work

IASTE International Association for the Study of Traditional Environments

iasy international active sun years

iat inside air temperature

IAT Individual Acceptance Test(ing); Institute for Applied Technology; Institute of Atomic Physics (Peking); International Academy of Tourism

IATA International Air Transport Association; International Amateur Theatre Association

iatc inlet air temperature control

IATC International Association of Tool Craftsmen

iatd is amended to delete

IATE Illinois Association of Teachers of English; International Association for Television Editors; International

Association for Temperance Education; International Association of Travel Exhibitors

IATJ International Association of Travel Journalists

IATL International Association of Theological Libraries

IATM International Association for Testing Materials

IATME International Association of Terrestrial Magnetism and Electricity

IATP Individual Aircraft Tracking Program

iatr is amended to read

IATSE International Alliance of Theatrical Stage Employees

IATTC Inter-American Tropical Tuna Commission

IATUL International Association of Technical University Libraries

iau intrusion alarm unit

IAU International Association of Universities; International Astronomical Union

IAUPE International Association of University Professors of English

IAUPL International Association of University Professors and Lecturers

IAUPPR Inter-American University Press of Puerto Rico

IAUPR Inter-American University of Puerto Rico

IAUR Institute of Art and Urban Resources

IAV International Association of Volcanology

IAVA Industrial Audio-Visual Association

iavc instantaneous automatic volume control

IAVCB International Association of Visitors and Convention Bureaus

IAVCEI International Association of Volcanology and Chemistry of the Earth's Interior

IAVE Interaction Analysis for Vocational Educators

IAVFH International Association of Veterinary Food Hygienists

IAVG International Association for Vocational Guidance

IAVI International AIDS Vaccine Initiative

IAVMS International Academy of Vibrational Medical Science

IAVRS International Audio-visual Resource Service (UNESCO)

IAVTC International Audio-Visual Technical Center

iaw in accordance with

IAW International Alliance of Women

IAWA International Association of Wood Anatomists

IAWCM International Association of Wiping Cloth Manufacturers

IAWL International Association for Water Law

IAWMC International Association of Workers for Maladjusted Children

IAWP International Association of Women Police

IAWPC International Association on Water Pollution Research and Control

IAWPR International Association on Water Pollution Research

IAWS Intercollegiate Association for Women Students; Irish Agricultural Wholesale Society

IAWWW *International Authors and Writers Who's Who*

IAZ Inner Artillery Zone

ib illegal behavior; in bond; inbound; incendiary bomb; inclusion body; index of body build; infectious bronchitis; inner bottom; instruction book; instructional brochure; intermediate bearded (iris); invoice book(let); inward bound

i & b improvements and betterments

ib. *ibidem* (Latin—in the same place)

i b *im besonderen* (German—in particular)

Ib Ibadan; plate current (symbol)

IB Imperial Bank; Imperial Beach; incendiary bomb; Infantry Battalion; Information Bulletin; Information Bureau; Intelligence Branch; International Baccalaureate; International Bank(ing); international broadcast(ing); plate power-supply current (symbol)

I of B Institute of Bankers; Institute of Biology

IB *Iberia Lineas Agreas de Espafia* (Spanish—Iberian Airlines of Spain); *Istanbul Bankasi* (Turkish—Istanbul Bank)

iba important bird area

IBA Independent Bakers Association; Independent Bankers Association; Independent Bar Association; Independent Broadcasting Association; Independent Broadcasting Authority (United Kingdom); Institute for Bioenergetic Analysis; Institute for Briquetting and Agglomeration; Institute of British Architects; Institute of Business Appraisers; International Banana Association; International Bar Association; International Bauxite Association; International Biometric Association; International Briqueting Association; Investing Builders Association; Investment Bankers Association

IBAA Independent Bankers Association of America; Investment Bankers Association of America; Italian Baptist Association of America

IBAC Identity Based Access Control

IBAE Institution of British Agricultural Engineers

IBAHP Inter-African Bureau for Animal Health and Protection

IBAM Institute of Business Administration and Management

IBAP Intervention Board of Agricultural Produce

I-bar capital-I-shaped metal bar

IBAR Inter-African Bureau of Animal Resources

IBAS Indonesian Business Association of Singapore

IBASS Intelligent Business Applications Support System

IBAU Institute of British–American Understanding

ibb intentional bases on balls (baseball)

IBB Illinois Inspection Bureau; Institute of British Bakers; International Bowling Board; International Brotherhood of Bookbinders

IBBD *Instituto Brasileiro de Bibliografía e Documentação* (Portuguese—Brazilian Institute of Bibliography and Documentation)

IBBISBBFH International Brotherhood of Boilermakers, Iron Ship Builders, Blacksmiths, Forgers, and Helpers

ibbm iron body bronze (or brass) mounted

IBBY International Board on Books for Young People

ibc iron binding capacity

ibc (IBC) integrated broadband communication; intermediate bulk carrier; intermediate bulk container

IBC Insurance Bureau of Canada; International Biographical Centre; International Broadcasting Corporation; Iwate Broadcasting Company (Japan)

IBC *Instituto Brasileiro do Café* (Portuguese—Brazilian Coffee Institute)

IBCA Industry Bar Code Alliance; Institute of Burial and Cremation Administration

IBCE International Book Club Edition

IBCs Institutional Biosafety Committees

IBCS Integrated Battlefield Control System (USA)

IBCUSCAN International Boundary Commission, United States and Canada

ibd inflammatory bowel disease; interest-bearing debentures; interest-bearing deposit

IBD Inflammatory Bowel Disease; Institute of British Decorators; Institute of Business Designers; International Bank of Detroit

ibda indirect bomb damage assessment

ibe integrity basis earthquake; inventory by exception

IBE Institute of British Engineers; International Bureau of Education

I-beam capital-I-shaped metal beam

IBEC International Bank for Economic Cooperation; International Basic Economy Corporation

IBECC *Instituto Brasileiro de Educação Ciencia e Cultura* (Portuguese—Brazilian Institute of Science Education and Culture)

IBEE International Builders Exchange Executives

IBEG International Book Export Group

iben incendiary bomb with explosive nose

Iber Iberia Airlines of Spain; Iberia(n); Iberic(a)(n); Iberville

IBERIA *Lineas Aéreas de España* (Spanish—Iberia Airlines of Spain)

Iberian Pen Iberian Peninsula containing Portugal and Spain

IBERLANT Iberian Atlantic

ibes integrated building and equipment scheduling

IBES Illinois Bureau of Employment Security; Institutional Broker's Estimate System

IBEW International Brotherhood of Electrical Workers

ibf internally blown flap; international banking facility; international bond fund

IBF Institute of Banking and Finance; Institute of British Foundrymen; International Boxing Federation

IBFD International Bureau of Fiscal Documentation

IBFI International Business Forms Industries

IBFMP International Bureau of the Federations of Master Printers

IBFO International Brotherhood of Firemen and Oilers

IBFs International Banking Facilities

ibg inter-block gap

IBG Institute of British Geographers

IBHA Insulation, Building, and Hardwood Association

IBhd initial beachhead

IBHE Illinois Board of Higher Education

ibi information-based indicia; invoice book, inward

IBI Illinois Bureau of Investigation; Indiana Bureau of Investigation; Insulation Board Institute

IBI *Instituto Bancario Italiano* (Italian—Italian Banking Institute)

ibid international bibliographical description

ibid. *ibidem* (Latin—in the same place)

IBiol Institute of Biology

IBIS International Book Information Service

IBJ Industrial Bank of Japan

IBK Institute of Bookkeepers

IBK *Institut für Bauen mit Kunststoffen* (German—Institute for Building with Plastics)

ibkr icebreaker

IBL Institute of British Launderers; Irish Biscuits Limited

ibm induced bowel movement

ibm (IDM) intercontinental ballistic missile

IBM Institute for Burn Medicine; International Business Machines

IBM *Industrias Biologicas Mexicana* (Spanish—Mexican Biological Industries)

IBMA Independent Battery Manufacturers Association

IBM JRD *IBM Journal of Research and Development*

IBMR International Bureau for Mechanical Reproduction

ibn identification beacon

IBN *Institut Belge de Normalisation* (French—Belgian Standards Institute)

ibnr incurred but not reported

ibo invoice book, outward

IBO International Baccalaureate Office

IBOB International Brotherhood of Old Bastards

ibol integrated business-oriented language

ibop (IBOP) international balance of payments

IBOP International Brotherhood of Operative Potters; International Business Opportunity Program

ibp initial boiling point

IBP Institute of British Photographers; International Biological Program; Iowa Beef Processors

IBPAT International Brotherhood of Painters and Allied Trades (U.S. and Canada)

IBPGR International Board for Plant Genetic Resources

IBPI International Bureau for Protection and Investigation

IBPOEW Improved Benevolent and Protective Order of Elks of the World

ibr information-bearing radiation; integral boiling reactor

ibr (IBR) infectious bovine rhinotracheitis

IBR Institute of Behavioral Research; Institute of Biosocial Research

IBRA Indonesian Bank Restructuring Agency; International Bible Reading Association

IBRD International Bank for Reconstruction and Development (World Bank)

ibrl initial bomb-release line

IBRM Institute of Boiler and Radiator Manufacturers

IBRMR Institute for Basic Research on Mental Retardation

IBRO International Bank Research Organization; International Brain Research Organization; International Brewers' Research Organization

ibs inflatable boat, small; international belt skimmer (for cleaning oil slicks)

ibs (IBS) ionospheric beacon satellite; irritable bowel syndrome

IBS Ibaraki Broadcasting System; Indian Boy Scouts; Institute of Basic Standards; Institute of Biblical Studies; Institute of Buddhist Studies; Intercollegiate Broadcasting System; International Bach Society; International Bank of Singapore; Irritable Bowel Syndrome; Israel Broadcasting Service

IBSA Inanimate Bird-Shooting Association; International Barber Schools Association; International Bible Student Association

IBSC Iranian-British Shipping Company

IB Scot Institute of Bankers in Scotland

IBSGR Isiolo Buffalo Spring Game Reserve (Kenya)

ibsr individual base stock requirements; individual battle shooting range

EBSS Imperial Bureau of Soil Science

IBST Institute of British Surgical Technicians

IBSTP International Bureau for the Suppression of Traffic in Persons

ibt in-barrel time; initial boiling-point temperature

IBT Industrial Bio-Test (laboratories); International Brotherhood of Teamsters

IBTA Individualized Behavior Therapy for Alcoholics; International Baton Twirlers Association

IBTCWH International Brotherhood of Teamsters, Chauffeurs, Warehousemen, and Helpers

IBTCWHA International Brotherhood of Teamsters, Chauffeurs, Warehousemen, and Helpers of America

ib test inkblot test (Rorschach test)

IBTS International Bicycle Touring Society

IBTTA International Bridge, Tunnel, and Turnpike Association

ibu imperial bushel

IBU Inland Boatmen's Union; International Broadcasting Union

ibv infectious bronchitis vaccine; infectious bronchitis virus

IBVL *Instituut voor Bewaring van Landbowprodukten* (Dutch—Institute for Storing and Processing Agricultural Products)

ibw information bandwidth

IBW International Boiler Works

IBWA International Bottled Water Association

IBWCUSMEX International Boundary and Water Commission, United States and Mexico

IBWM International Bureau of Weights and Measures

IBWS International Bureau of Whaling Statistics

ibx (IBX) intermediate branch exchange

IBY International Book Year (1972)

Ibz Ibiza

ic immediate constituent; in calf, in charge of, in command; in conference; ice crystals; index correction; informal communication;

inspected and condemned; inspiratory capacity; inspiratory center; instruction counter; instrument correction; insulation contract, integrated circuit; interference control; interior communication; intermediate language; internal classification; internal combustion; internal communication; internal connection; international control; interrupting capacity; interstitial cells; interstitial cystitis; intracerebral; intracutaneous

ic. icon (Latin—figure, woodcut)

i-c integrated circuit

i/c in charge; in command; ice cream; intercom (intercommunication via interphone)

i&c inspected and condemned

i & c installation and checkout; installation and construction; instrumentation and control

i.c. inter cibos (Latin—between meals)

Ic Iceland; Icelander; Icelandic; transistor collector current (symbol)

IC Idaho College; Identity Card; Ignatius College; Illinois Central (railroad); Illinois College; Immaculata College; Industrial Court; Information Center; Information Circular; integrated circuit; Integrating Contractor; Intelligence Corps; Interchemical Corporation; Intercity; International Control; Iola College; Iona College; Iron Curtain; Itaska College; Itawamba College; Item Code; Ithaca College

I-C Indo-China; Indo-Chine; Indo-Chinese; Iran-Contra scandal

I.C. Iesus Christus (Latin—Jesus Christ); Institute of Charity (Rosminian)

I & C Ictinus and Callicrates (designers of the Parthenon)

IC 4-A Intercollegiate Amateur Athletic Association of America

ica ignition control additive; information, counseling and assistance; instrument compressed air

ICA Illinois Correctional Association; Imperial Corporation of America; Imperial Savings Association; Inclinator Company of America; Indiana Correctional Association; Industrial Communication Association; Industrial Coordination Act; Institute of Chartered Accountants; Institute of Contemporary Arts; Institute of Criminal Anthropology; Insurance Council of Australia; Intermuseum Conservation Association; International Cartographic Association; International Carwash Association; International Ceramic Association; International Chefs' Association; International Chiropractors Association; International Christian Aid; International Claims Association; International Coffee Agreement; International Commercial Arbitration; International Commodity Agreement; International Communication(s) Agency; International Communication Association; International Cooperative Administration; International Cooperative Alliance; International Cooperative Association; International Council on Archives; International Credit Association; Iowa Corrections Association

I of CA Institute of Chartered Accountants

ICA *Ingenieros Civiles Asociados* (Spanish—Associated Civil Engineers)

icaa integrated cost-accounting application(s)

ICAA Institute of Chartered Accountants of Australia; Insulation Contractors Association of America; International Council on Alcohol and Addictions; Invalid Children's Aid Association; Investment Counsel Association of America

ICAAAA Intercollegiate Association of Amateur Athletes of America

ICAB International Council Against Bullfighting

ICAC Independent Commission Against Corruption (Hong Kong); Institute of Clean Air Companies

icad integrated control and display

icade interactive computer-aided design evaluation

icadg interactive computer-aided design and graphics

ICADS Integrated Control and Display System

ICAE International Commission on Agricultural Engineering; International Council for Adult Education

ICAESD International Center for African Economic and Social Documentation

ICAEW Institute of Chartered Accountants in England and Wales

ICAF Industrial College of the Armed Forces; International Committee on Aeronautical Fatigue

ICAFI International Commission on Agriculture and Food Industries

ICAI Institute of Chartered Accountants in Ireland; International Commission of Agricultural Industries

ICAITI *Instituto Centroamericano de Investigacion y Tecnologia Industrial* (Spanish—Central American Institute of Industrial Investigation and Technology)

icam integrated computer-aided manufacturing

ICAM Institute of Corn and Agricultural Merchants; International Civil Aircraft Marking(s)

I Can Information Canada

ICAN Insurance Consumer Action Network; International Commission for Air Navigation

ICANN Internet Corporation for Assigned Names and Numbers

ICAO International Civil Airlines Operation; International Civil Aviation Organization

ICAP Institute of Certified Ambulance Personnel; Integrated Criminal Apprehension Program (computerized system); Inter-American

Committee of the Alliance for Progress; International Civil Aviation Policy

ICAP *Instituto Cubano de Amistad con los Pueblos* (Spanish—Cuban Institute for Friendship with Peoples)

ICAPR Interdepartmental Committee on Air Pollution Research (UK)

ICAIPS Integrated Carrier Acoustic Prediction System (for aircraft carriers)

ICAR Indian Council of Agricultural Research; International Committee Against Racism

ICARMO International Council of the Architects of Historical Monuments

I-car(s) Italian car(s)

Icarus *International Journal of the Solar System*

ICARUS Inter-Urban Control and Roads Utilization Simulation

icas intermittent commercial and amateur service

ICAS Institute for Chartered Accountants in Scotland; Interdepartmental Committee for Atmospheric Sciences; Intermittent Commercial and Amateur Service; International Council of the Aeronautical Sciences; International Council of Aerospace Sciences

ICASALS International Center for Arid and Semi-Arid Land Studies

ICASB Iowa Council of Area School Boards

ICATS Intermediate-Capacity Automated Telecommunications System (USAF)

icav intracavity

icb interlocking concrete block; international competitive bid(ding)

ICB Indian Coffee Board; Industrial and Commercial Bank(ing); Institute of Collective Bargaining; Institute of Comparative Biology; Interface Control Board; International City Bank; International Container Bureau

ICBA International Community of Booksellers Associations

ICBC Institute of Certified Business Counselors; Insurance Corporation of British Columbia; International Commercial Bank of China

ICBIF Inner-City Business Improvement Forum

icbm (ICBM) intercontinental ballistic missile

ICBO International Conference of Building Officials; Interracial Council for Business Opportunities

icbp intracellular binding proteins

ICBP International Council for Bird Preservation

ICBR Institute for Child Behavior Research

ICBS Interconnected Business System; International Call-Boy Service (for women)

icbt intercontinental ballistic transport

icc improved contemporary comparison; income capital certificate; institute cargo clauses; integrated circuit computer; international catalog card (3 × 5 inches or 7.5 × 12.5 centimeters); intracompany correspondence

ic & c invoice cost and charges

ICC Indian Claims Commission; Indiana Collegiate Conference; Instrumentation Control Center; International Chamber of Commerce; International Control Commission; International Correspondence Course(s); Interstate Commerce Commission; Inuit Circumpolar Conference

I.C.C. Isthmian Canal Commission

icca initial cash clothing allowance

ICCA Independent Computer Consultants Association; Infants' and Children's Coat Association; International Congress and Convention Association; International Consumer Credit Association; International Corrugated Case Association

ICCAD International Center for Computer-Aided Design

ICCAIA International Coordinating Council of Aerospace Industries Associations

ICCAT International Commission for the Conservation of Atlantic Tunas

ICCB Illinois Community College Board

ICCC International Center for Comparative Criminology

iccd intensified charge-coupled device; internal coordination control drawing

ICCD Information Center on Crime and Delinquency

ICCF International Correspondence Chess Federation

ICCI Inter-Continental Computing Incorporated

ICCL International Council of Cruise Lines

ICCO International Carpet Classification Organization

ICCP International Conference on Cataloguing Principles

ICCR Indian Council for Cultural Relations; Interfaith Center on Corporate Responsibility; International Charge Card Registry

ICCS Institute in Culture and Creation Spirituality; International Center of Criminological Studies

ICCSL International Commission of the Cape Spartel Light

ICCTA Illinois Community College Trustees Association

IC & CY Inns of Court and City Yeomanry

icd immune complex disease; impulse control disorder; interface control document; interface control drawing; investment certificate of deposit (ICD)

ic & d installation, checkout, and demonstration

ICD Industrial Cooperation Division; Industry Cooperation Division; Inland Clearance Depot; Inland Container Depot; Institute for the Crippled and Disabled; International Classification of Disease; International College of Dentists; International Cooperative Distributors

ICD *International Classification of Diseases*

ICDA *International Classification of Diseases, Adapted for Use in the United States*

icdh (ICDH) isocitric dehydrogenase

ICDO International Civil Defense Organization

ICDRG International Contact Dermatitis Research Group

icd's investment certificates of deposit

ICDS Information Collection Dissemination System

ice ice, compression, and elevation; in-car entertainment; increased combat effectiveness; input-checking equipment; instruction-curriculum environment; internal combustion engine; inventory control effectiveness

ice. (ICE) crystallized smokable derivative of methamphine

Ice Iceland; Icelander; Icelandic

ICE Indigenous Council for the Environment (Alaska); Institute for Consumer Ergonomics; Institution of Chemical Engineers; Institution of Civil Engineers; Instruction-Course Evaluation; Instructor-Course Evaluation; Inter-City Express (train); International Cometary Explorer; International Cultural Exchange

ICE *Instituto Costarricense de Electricidad* (Spanish—Costa Rican Electric Institute)

ICEA Instrument Contracting and Engineering Association

ICEBAC International Council of Employers of Bricklayers and Allied Craftsmen

ICEC Illinois Citizens Education Council

ICECAP Infrared Chemistry Experiments-Coordinated Auroral Program (DoD)

ice(d)-T iced tea

ICED International Council for Educational Development

ICEED International Center for Energy and Economic Development

ICEF International Children's Emergency Fund; International Council for Educational Films

ICEG Insulated Conductors' Export Group

ICEI International Combustion Engine Institute

Icel Icelandic

ICEL International Committee on English in the Liturgy; International Council of Environmental Law

Iceland Republic of Iceland (island nation in the North Atlantic) *Lydveldio Island*

ICEM International Commission for European Migration

icepack. individual career exploration pack(age)

ICEPS International Center for Economic Policy Studies

ICER Information Centre of the European Railways

I Ceram Institute of Ceramics

ICEs International Customs Examinations

ICES Instructor and Course Evaluation System; Integrated Civil Engineering Systems; International Council for the Exploration of the Sea

ICESC Industry Crew Escape Systems Committee

ICET Institute for the Certification of Engineering Technicians; International Center of Economy and Technology; International Council on Education for Teaching

ICETEX *Instituto Colombiano de Especialización Tecnica en el Exterior* (Spanish— Colombian Institute of Technical Specialization Abroad)

ICETT Industrial Council for Educational and Training Technology

ICEWATER Inter-Agency Committee on Water Resources

icf inertial confinement fusion; intermediate care facilities; intracellular fluid

ICF Ice Cream Federation; Independent Cat Federation; Inter-bureau Citation of Funds; International Canoe Federation; International Cynological Federation; Iowa College Foundation

ICF *Ingénieur Civil de France* (French—Civil Engineer of France)

ICFA Independent College Funds of America; International Chicken Flying Association; International Cystic Fibrosis Association

ICFC Industrial and Commercial Finance Corporation

icff intercommunication flip-flop

ICFNC Independent College Fund of North Carolina

ICFNJ Independent College Fund of New Jersey

icfp immune complex by C1q binding (fluid phase)

ICFP Institute of Certified Financial Planners

ICFPW International Confederation of Former Prisoners of War

ICFR Intercollegiate Conference of Faculty Representatives (Big Ten)

ICFTU International Confederation of Free Trade Unions

ICFU International Council on the Future of the University

icg icing

ICG Industries Consultative Group; Interface Coordination Group; International Commission on Glass; International Congress of Genetics; Interviewers Classification Guide; Iowa Corn Growers

ICGB International Cargo Gear Bureau

ICGE International Center of Genetic Epistemology

ICGEBnet The International Centre for Genetic Engineering and Biotechnology network (computer resource for molecular biology)

ICGS Icelandic Coast Guard Service; International Call-Girl Service (Amsterdam)

ICGSCA Infants', Children's and Girls' Sportswear and Coat Association

ich ichthyology; in-calf heifer

ich (ICH) infectious canine hepatitis

Ich Ichabod

ICHA International Collegiate Hockey Association

ICHAM Institute of Cooking and Heating Appliance Manufacturers

ICHCA International Cargo Handling Coordination Association

IChemE Institute of Chemical Engineers

ICHEO Inter-University Council for Higher Education Overseas

ichnol ichnolite; ichnologist; ichnology

ICHPER International Council for Health, Physical Education, and Recreation

ichs ichthyologists

I/C Hs Iran-Contra Hearings

ICHS International Committee of Historical Sciences

ichth ichthyology

ichthyol ichthyologic(al)(ly); ichthyologist; ichthyology

ICI Imperial Chemical Industries; Institution of Chemistry in Ireland; International Commission on Illumination; Interpersonal Communication Inventory; Investment Casting Institute; Investment Company Institute; Investment Costing Institute

ICIA Interagency Committee on International Athletics; International Communications Industries Association; International Credit Insurance Association; International Crop Improvement Association

ICIANZ Imperial Chemical Industries of Australia and New Zealand

ICIAP Interagency Committee on International Aviation Policy

ICIC International Copyrights Information Center

ICID International Commission on Irrigation and Drainage

ICIE International Council of Industrial Editors

ICIECA Interagency Council on International Educational and Cultural Affairs

ICIMP Interagency Committee for International Meteorological Programs

ICIP International Conference on Information Processing

ICIPE International Center for Insect Physiology and Ecology

ICIREPAT International Cooperation in Information Retrieval among Examining Patent Offices

ICIS International Cargo Information System

ICITA International Chain of Industrial and Technical Advertising Agencies; International Cooperative Investigation of the Tropical Atlantic

ICITAP International Criminal Investigative Training Assistance Program

ICITO Interim Commission for the International Trade Organization

ICJ Institute of Creative Judaism; Institute of Criminal Justice; International Commission of Jurists; International Court of Justice

ICJC Institute of Criminal Justice and Criminology

ICJP Irish Commission for Justice and Peace

ICJW International Council of Jewish Women

icky sticky

icl incoming correspondence log(ging); input capacitorless (circuitry or hi-fidelity)

ICL Institute for Continued Learning; International Computers Limited; International Confederation of Labor; International Containers Limited; Israel Chemicals Limited; International Collectors Library

ICL Institut de Chimie de Lyon (French—Lyon Industry of Chemistry)

ICLA International Committee on Laboratory Animals

ICLD International Center for Law in Development

Iclnd Iceland

ICLP Institute of Criminal Law and Procedure (Georgetown University)

I & CLQ International and Comparative Law Quarterly

ICLs Income Contingent Loan

ICLS Irish Central Library for Students

icm increased capability missile; intercostal margin; interference control monitor

icm (ICM) increased capability missile

ICM Increased Capability Missile; Indian Campaign Medal; Industrial and Construction Machines; Institute of Computer Management; International Confederation of Midwives

ICMA Independent Cable Makers Association; Institute of Cost and Management Accountants; International Circulation Managers Association; International City Manager's Association

ICMAD Independent Cosmetic Manufacturers and Distributors

ICME International Conference on Medical Electronics

ICMF Indian Cotton Mills Federation

ICMPH International Center of Medical and Psychological Hypnosis

icmps induction compass

ICMR Indian Council of Medical Research

ICMREF Interagency Committee on Marine Science, Research, Engineering, and Facilities

ICMs Instructional Curriculum Maps

ICMS International Commission on Mushroom Science

ICMUA International Commission on the Meteorology of the Upper Atmosphere

ICN Instrumentation and Calibration Network; International Chemical and Nuclear (corporation); International Council of Nurses

I.C.N. in Christi Nomine (Latin—in Christ's name)

ICNA Infection Control Nurses Association

ICNAF International Commission for the Northwest Atlantic Fisheries

ICND International Commission on Narcotic Drugs

ICNI Integrated Communication/Navigation/Identification

ICNT Informal Composite Negotiated Text

ICNV International Committee on Nomenclature of Viruses

ico in case of; iconology

Ico collector cutoff current (symbol)

ICO Immediate Commanding Officer; Immediate Controlling Office(r); Instrumentation Control Office(r); Interagency Committee on Oceanography; International Coffee Organization; International Commission for Optics; International Computer Orphanage; International Congress of Ophthal-

mology; Islamic Conference Association; Israel Chamber Orchestra

ICOA International Castor Oil Association

ICOD International Centre for Ocean Development

ICOGRADA International Council of Graphic Design Associations

ICOM International Council of Museums

ICOMIA International Council of Marine Industries Associations

ICOMOS International Council on Monuments and Sites

icon. iconic; iconoclasm; iconoclast; iconography

iconoc iconoclast (breaker of political or religious images)

ICONS Information Center on Nuclear Standards; Isotopes of Carbon, Oxygen, Nitrogen, and Sulfur (AEC)

ICOO Iraqi Company for Oil Operations

icop imported crude oil processing

ICOPA International Conference of Police Associations

icor incremental capital-output ratio

ICOR Intergovernmental Conference on Oceanic Research (UNESCO)

I Corr Tech Institution of Corrosion Technology

ICOS International Committee of Onomastic Sciences

ICOT Institute of Coastal Oceanography and Tides; Institute for New Generation Computer Technology

icp intracranial pressure; inventory control point

ICP Industry Cooperative Program; Infection-Control Procedure; Insurance Conference Planners; International Center of Photography; International Commerce Promoters; International Council of Psychologists, Inter-University Cooperation Programs

ICP Institut de Chimie de Paris (Chemical Institute of Paris)

ICPA International Commission for the Prevention of Alcoholism; International Conference of Police

Associations; International Cooperative Petroleum Association

ICPC International Criminal Police Commission (Interpol)

ICPCCP Illinois Council of Public Community College Presidents

ICPHS International Council for Philosophical and Humanistic Studies

ICPI Insurance Crime Prevention Institute

ICPM Institute of Certified Professional Managers

i/c/pm/m incisors, canines, premolars, molars (dentition formula, e.g., i 4/4 means 4 upper and 4 lower incisors, c 2/2 means 2 upper and 2 lower canines, etc.)

icpo intercompany payment orders

ICPO International Criminal Police Organization (Interpol)

ICPP Idaho Chemical Processing Plant (AEC)

ICPRB Interstate Commission on the Potomac River Basin

ICPS International Congress of Photographic Science; International Credit Protection Services

ICPSR Inter-university Consortium for Political and Social Research

ICQ International Career Quotient; Invested Capital Questionnaire

icr increase; increment; instrumentation control rack; intelligent character recognition; ion cyclotron resonance

icr (ICR) iron-core reactor

ICR Independent Congo Republic; Institute of Cancer Research; Institute for Cooperative Research; International Council for Reprography

ICRA Industrial Chemical Research Association; International Copper Research Association

ICRC International Committee of the Red Cross

ICRDA Independent Cash Register Dealers Association

ICRDB International Cancer Research Data Bank

ICRF Imperial Cancer Research Fund

ICRH Institute for Computer Research in the Humanities (NYU)

ICRI Illinois Committee for Responsible Investment; International Crop Research Institute

icrm intercontinental reconnaissance missile (ICRM)

ICRM Institute of Certified Records Managers

ICRO International Cell Research Organization

ICRP International Commission on Radiation Protection; International Commission on Radiological Protection

ICRSC International Council for Research in the Sociology of Cooperation

ICRT Individualized Criterion-Referenced Test(ing)

ICRU International Commission on Radiological Units and Measurements

ICRUM International Commission on Radiological Units and Measurements

ics installment credit selling; intercostal space; interim contractor support; intracranial stimulation

ic's immediate constituents; integrated circuits

ic's (ICs) integrated circuits

ICS Imperial College of Science and Technology; Indian Chemical Society; Indian Civil Service; Information Centers Service; Inner Continental Shelf, Institute of Chartered Shipbrokers; Institute of Chartered Shipbuilders; Institute for Contemporary Studies; Institution of Computer Sciences; Integrated Command System; Integrated Container Service; Intelligence Community Staff, Interagency Communications System; Inter-Communications System; International Cardiovascular Society; International Chamber of Shipping; International Clarinet Society; International College of Surgeons; International Container Service; International Contract Specialists; International Correspondence Schools; International Crime Statistics;

International Telephone and Telegraph Communications System; Interway Container Service

ICSA Institute of Chartered Secretaries and Administrators; International Chain Salon Association

ICSAC International Confederation of Societies of Authors and Composers

ICSB International Center of School Building

ICSC Independent Colleges of Southern California; International Council of Shopping Centers; Inter-oceanic Canal Study Commission

ICSDW International Council of Social Democratic Women

icse intermediate current stability experiment

ICSE International Committee for Sexual Equality

ICSEAF International Commission for the Southeast Atlantic Fisheries

ICSEM International Center of Studies on Early Music

ICSEMS International Commission for the Scientific Exploration of the Mediterranean Sea

icsh (ICSH) interstitial cell-stimulating hormone

ICSH International Committee for Standardization in Haematology

icsi intracytoplasmic sperm injection

ICSI International Conference on Scientific Information

Icsia I can't stand it anymore

ICSIC Integrated Communications System for Intensive Care

ICSID International Center for the Settlement of Investment Disputes; International Council of Societies of Industrial Design

ICSLS International Convention for Safety of Life at Sea

icsm instant corn-soya milk

ICSOM International Conference of Symphony and Opera Musicians

icsp immune complex by C1q binding (solid phase)

ICSP International Council of Societies of Pathology

ICSPE International Council of Sport and Physical Recreation

ICSPRO International Calcium Silicate Products Research Organization

icss intracranial self stimulation

ICSS International Council for the Social Studies

ICSSD International Committee for Social Sciences Documentation

ICSST Institute of Child Study Security Test

ICST Imperial College of Science and Technology; Institute for Computer Sciences and Technology

ICS & T Imperial College of Science and Technology

ICSTA International Cooperative Study of the Tropical Atlantic

ICSTS Intermediate Combined System Test Stand

ICSU Integrated Container Services Unit; International Council of Scientific Unions; Interway Container Service (container) Unit

ICSW Interdepartmental Committee on the Status of Women

icswbd interior communications switchboard

ict icterus; identity conversion training; inflammation of connective tissue-I insulin coma therapy (ICT); interface control tool

ic/t integrated computer/telemetry

ICT Inter-City Train; International Cablecasting Technologies Incoporated; International Computers and Tabulators; Wichita, Kansas (airport)

ICT International Critical Tables

ICTA Imperial College of Tropical Agriculture; International Center for the Typographic Arts; International Council of Travel Agents

ICTB International Customs Tariffs Bureau

icte integration and calibration test(ing) equipment

ICTF International Cocoa Trade Federation

I.C.TH.U.S. *Iesous Christos, Theou Huios, Soter* (Latin—Jesus Christ, the Son of God, the Saviour)

ICTMM International Congresses of Tropical Medicine and Malaria

ICTN Industry Center for Trade Negotiations

ICTP International Center for Theoretical Physics

ICTR International Center of Theatre Research

ICTS Intermediate-Capacity Transit System

Ictus. Iurisconsultus (Latin—attorney, counsellor-at-law)

ic tv integrated-circuit television

icu indicator console unit; intensive care unit (medical)

icu (ICU) intensive care unit

Icu I see you

ICU International Code Use; International Cultural Understanding; International Cultural University

ICUA Institute for College and University Administrators

ICUF Independent Colleges and Universities of Florida

ICUI Independent Colleges and Universities of Indiana

ICUM Independent Colleges and Universities of Missouri

ICUMSA International Commission for Uniform Methods of Sugar Analysis

ICUS inside continental United States

ICUT Independent Colleges and Universities of Texas

icv improved capital value; intracellular virus

ICVA International Council of Voluntary Agencies

ICVS International Crime Victim Survey

icw in connection with; interrupted continuous wave; intracellular water

ICW India-China Wing (World War II); Institute of Child Welfare; Inter-American Commission of Women; International Chemical Workers; International Commission on Whaling; International Council of Women

ICWA Indian Council of World Affairs; Institute of Cost and Works Accoun-

tants; Institute of Current World Affairs; International Coil Winding Association

ICWG International Cooperative Women's Guild

ICWL International Creative Writers League

ICWM International Committee on Weights and Measures

ICWP International Council of Women Psychologists

ICWU International Chemical Workers Union

ICX International Cultural Exchange

ICY International Cooperation Year (1965)

ICZ Intertropical Convergence Zone

ICZN International Code of Zoological Nomenclature; International Commission on Zoological Nomenclature

id idea; identification; immediate delivery; import duty; independent development; independent distributor; individual difference; induced draft; industrial design; infectious disease; infective dose, inside diameter; instructional development; intercept director; interest deductible; intradermal; island; islander; item description

id (ID) import duty; instruction decoder

i & d incision and drainage

id. idem (Latin—the same)

Id Iraqi dinar (monetary unit of Iraq)

I'd I could; I had; I should; I would

ID Idaho; Identifier; Import Declaration; Indonesia (Internet code); Institute of Distribution; Insurance Department; Intelligence Department; Interior (US department); Interior Department; Iraqi dinar (currency unit)

Id$_{50}$ median infective dose

ida integrated digital avionics; integrodifferential analyzer

ida (IDA) iminodiacetic acid

Ida Idah; Idaho; Idalah; Idalia; Idalina; Idaline

IDA Individual Development Account; Industrial Development Agency; Industrial Development Authority; Industrial Diamond Association;

Industrial Distribution Association; Institute for Defense Analyses; Institute for Design Analysis; Institute of Directors in Australia; Interchange of Data between Administrations; Intercollegiate Dramatic Association; International Development Assistance; International Development Association; International Discotheque Association; International Documentary Association; International Downtown Association; International Drapery Association; International Dredging Association; Irish Dental Association

IDA Import Duty Act

IDAA Industrial Diamond Association of America; International Doctors in Alcoholics Anonymous

idac interim digital-analog converter

id. ac idem ac (Latin—the same as)

IDAC Import Duties Advisory Committee

IDAI Industrial Development Authority of Ireland

IDA Ireland Industrial Development Authority of Ireland

IDAS Information Display Automatic System

IDASF Institute for a Democratic Alternative for South Africa

idast interpolated data and speech transmission

idb illicit diamond buyer; illicit diamond buying; integrated data base; intercept during burning

IDB Industrial Development Board; Industrial Development Bond; Inter-American Development Bank; Internal Drainage Board; International Data Base; Islamic Development Bank(ing); Israel Diamond Building (Ramat Gan)

IDB Interpreter's Dictionary of the Bible

IDBMA International Data Base Management Association

IDBP Industrial Development Bank of Pakistan

IDBT Industrial Development Bank of Turkey

idc interest during construction

IDC Imperial Defense College; Industrial Development Commission; Industrial Development Corporation; Industry Development Commission; Intelligence Documentation Center; Intercontinental Dynamics Corporation; Interdepartmental Committee; Interdepartmental Communication; Inter-Departmental Correspondence; International Danube Commission; International Drycleaners Congress; Iowa Development Commission

IDC Internationale Dokumentationgesellschaft für Chemie (German—International Chemical Documentation Society)

IDCA Industrial Design Council of Australia; International Development Corporation Agency

id card identification card

ID-card identification card

IDCAS Industrial Development Center for Arab States

IDCDA Independent Dealer Committee Dedicated to Action

idcf indirect command file

IDCF Industrial Development Completion Form

IDC(orp) International Disposal Corporation

IDCSP Initial Defense Communications Satellite Program

IDCSS Initial Defense Communications Satellite System

i-d curve intensity-duration curve

idd identified; industrial diamond drill; interface designation drawing

idd (IDD) insulin-dependent diabetes; international direct dialing

i/d/d illicit diamond dealer

IDD International Direct Dialing; Island Development Department

IDD Industrielle Designere Danmark (Danish—Denmark Industrial Design)

IDDBA International Dairy-Deli-Bakery Association

IDDD International Direct Distance Dialing

IDDD International Demographic Data Directory

iddm insulin-dependent diabetes mellitus

IDDS International Dairy Development Scheme; International Digital Data Service

ide industry-developed equipment; integrated drive electronics

IDE International Development Enterprises; Industrial Development Executive; Institute of Diesel Engineers; Israel Desalination Engineering

idea intelligent data equipment adaptor

IDEA Illinois Drug Education Alliance; Institute for the Development of Educational Activities; Institute of Diesel Engineers of Australia; Instructional Development and Effectiveness Assessment; International Dance-Exercise Association; International Downtown Executives Association; International Drug Enforcement Association

IDEAS Intregrated Design and Analysis System

I de C Islas del Cisne (Spanish—Swan Islands)

IDEC Industrial and Domestic Equipment Corporation; Interior Design Educators Council; International Drug Enforcement Conference

IDECC Interstate Distributive Education Curriculum Consortium

ideea information and data exchange experimental activities

IDEEA International Design for Extreme Environments Association

idef intercept during exo-atmospheric fall

I de F Institut de France (French—Institute of France)

iden identification; identify; identity

ident identification; identify; identity

IDEP Interagency Data Exchange Program; Interservice Data Exchange Program

IDES Interactive Data Entry System

IDET Institute for Development, Employment, and Training

idex initial defense experiment

idf intermediate distribution frame; international distress frequency

idf (IDF) integrated data file; interceptor day fighter

IDF International Dairy Federation; International Democratic Fellowship; International Dental Federation; International Diabetes Federation; International Drilling Federation; Israeli Defense Force (Israeli Secret Service or Zahal)

IDFA International Dairy Foods Association

IDFC Indo-Pacific Fisheries Council

IDFF Internationale Demokratische Frauenfederation (German—Women's International Democratic Federation)

idfm induced directional fm

idg integrated-drive generator

idgprt indigo print

ID grinding internal grinding

idgs integrated drive generator system

idh (IDR) isocitrate dehydrogenase

IDHCA International District Heating and Cooling Association

id he. index head

IDHEC Institut des Hautes Etudes Cinimatographiques, Paris (French—Paris Institute of Higher Cinematographic Studies)

IDHS Intelligence Data Handling System

idi improved data interchange; inter-division invoice

IDI Industrial Designers' Institute

IDI Institut de Droit International (French—International Law Institute); *Instituto Venezolano de Derecho Imobiliario* (Spanish—Venezuelan Institute of Withholding Law)

IDIA: Industrial Design Institute of Australia

IDIB Industrial Diamond Information Bureau

IDIL Institute for the Development of Indiana Law

IDIMS Interactive Digital Image Manipulation System

idio (Latin prefix—distinct, self, separate)—idiograph, idiomatic, idiot

idiot. instrumentation digital online transcriber

IDIU Interdivisional Information Unit; Interdivisional Intelligence Unit

idl (IDL) intermediate-density lipoproteins

IDL Instrumentation Development Laboratories; International Date Line; Ira D Levine; New York, New York (Kennedy International Airport—Idlewild)

IDLE Idaho Department of Law Enforcement

idlh immediately dangerous to life or health

IDLIS International Desert Locust Information Service

id lt identification light

idm illicit diamond mining; integrated direct metering

IDM Institute of Defense Management (Indian)

IDMA Indian Drug Manufacturers Association; International Dancing Masters Association; Isaac Delgado Museum of Art

IDMS Integrated Database Management System

I.D.N. In Dei Nomine (Latin—in God's name)

idne inertial-doppler navigation equipment

IDNL Indiana Dunes National Lakeshore (Indiana)

ido industrial diesel oil

IDO Intelligence Division Office; Interim Development Order; Interim Development Ordnance; International Disarmament Organization

idoc inner diameter of outer conductor

IDOC Illinois Department of Corrections

IDOE International Decade of Ocean Exploration—1970–1980

idon. vehic. idoneo vehiculo (Latin—in a suitable vehicle)

idot instrumentation on-line transcriber

IDOT Illinois Department of Transportation

idp industrial data processing; information data processing; input data processing; input data processor; integrated data processing; intermodulation distortion percentage; interplanetary dust particle

idp (IDP) inosine diphosphate

IDP Immediate Decision Plan; Independent Development Project; Industrial Development Bank; Institute of Data Processing; Integrated Data Processing; International Driving Pernit

IDPE Incorporated Data Processing Executives

IDPS Incremental Differential Pressure System

idr industrial development revenue; intercept during reentry; international depositary receipt

idr idraulica (Italian—hydraulics)

IDR Incremental Design Review; Infantry Drill Regulations; Institute for Desert Research; Institute for Dream Research

IDRC International Development Research Centre; International Drycleaning Research Committee

IDRDS International Directory of Research and Development Scientists

IDRS International Double Reed Society

ids illicit diamond smuggling; inadvertent destruct; input data strobe; integrated data store; intelligent driver support; interdiction and strike; intermediate drum storage

IDs Instructural Designs

IDS Inertial Doppler System; Information Dissemination System; Institute of Dental Surgery; Instructional Dimensions Study; Interior Design Society; Interior Designers Society; International Development Services; International Development Strategy; International Documents Service; Investigative Dermatological Society; Investors Diversified Services

IDSA Indian Dairy Science Association; Industrial Designers Society of Arnerica

IDSC Information Dissemination Service Center

IDSCS Initial Defense Satellite Communication System

IdSEA Idaho Society of Enrolled Agents

IDSO International Diamond Security Organization

Idss drain current at zero gate voltage (symbol)

idt identification disposition tag(ging); inter-division transfer

idt *in de text* (Dutch—in the text)

IDT Industrial Design Technology; Industrial Detergents Trade; Instrument Definition Team

IDTA International Dance Teachers Association

IDTS Instrumentation Data Transmission System

idu interactive data-base utilities; intermittent drive unit; iododeoxyuridine

IDU idoxuridine; International Democratic Union; International Dendrology Union

idur intercept during unpowered rise

i Durchshn *im Durchschnitt* (German—on an average)

IDV International Distillers and Vintners

IDX *Index to Dental Literature*

IDZ *Internationales Design Zentrum* (German—International Design Center)

ie immediate early (genes); index error; initial equipment; inside edge; ion exchange

i-e internal-external; introversion-extroversion

i/e ingress/egress

i & e identification and exposition (lines)

i.e. *id est* (Latin—that is); *inside english* (journal of the English Council of California Two-Year Colleges)

Ie emitter current (symbol)

IE Indo-European; Industrial Engineering; Industrial Espionage; Information and Education; Institute of Education; Institute of Electronics; Institute of Engineers; Institution of Electronics; Institution of Engineers; Ireland (Internet code)

I-E Indo-European; Internal-External Scale

I.E. Industrial Engineer

I & E Information and Education; Inspection and Enforcement

I of E Institute of Export

IE *Immunitäts Einheit* (German—immunizing unit)

iea intravascular erythrocyte aggregation

IEA Idaho Education Association; Illinois Education Association; Index of Economic Activity; Industrial Environmental Association; Institute of Economic Affairs; Institute of Engineers—Australia; International Association for the Evaluation of Educational Achievement; International Economic Association; International Energy Agency; International Entrepreneurs Association; International Epidemiological Association; International Ergonomics Association; International Evaluation of Education; International Exhibitors Association; International Study of Educational Achievement

IEA Indian Education Act (1972)

IEAF Imperial Ethiopian Air Force

I EARN International Educational Resource Network

IEAS Institute of East Asian Studies

IEB Institute of Economic Botany; International Energy Bank; International Environmental Bureau (of the Non-Ferrous Metals Industry); International Executive Board

iec injection electrode catheter; integrated electronic component; integrated electronic control; intra-epithelial carcinoma

iec (IEC) inherent explosion clause

IEC Independent Electrical Contractors; Industry-Education Council; International Education Center; International Electrochemical Commission; International Electrotechnical Commission; Interstate Economic Committee

IEC *Institut d'Etudes Centrafricaines* (French—Institute of Central African Studies)

IECDF International Economic Cooperation Development Fund

IECE Institute of Electronic and Communication Engineers

IECEJ Institute of Electronic and Communication Engineers of Japan

IECI Institute for Esperanto in Commerce and Industry

iecm internal electronic counter-measures

IECO International Engineering Company

IECOK International Economic Consultative Organization for Korea

IECP Intermediate Engineering Change Proposal

iec's integrated electronic components

ied improvised explosive device; individual effective dose

IED Institute for Educational Development; Institution of Engineering Designers; Integrated Electronics Division (USA Electronics Command)

IEDC International Energy Development Corporation

IEDR Integrated Electric-Drive Propulsion

ied's income equalization deposits

IEDs Intermediate Educational Districts

iee inner enamel epithelium

IEE Institute of Electrical Engineers; Institute of Electronic Engineering; Institute for Environmental Education; Institute of Environmental Engineers; Institution of Electrical Engineers

Ieee I expect everything eventually

IEEE Institute of Electrical and Electronics Engineers

IEEE J Quantum Electron

IEEE Journal of Quantum Electronics

IEEE Trans Antennas Propag

IEEE Transactions on Antennas and Propagation

IEEE Trans Electron Devices

IEEE Transactions on Electronic Devices

IEEE Trans Inf Theory *IEEE Transactions on Information Theory*

IEEE Trans Instrum Meas
IEEE Transactions on Instrumentation and Measurement
IEEE Trans Magn *IEEE Transactions on Magnetics*
IEEE Trans Microwave Theory Tech *IEEE Transactions on Microwave Theory and Techniques*
IEEE Trans Nucl Sci *IEEE Transactions on Nuclear Science*
IEEE Trans Sonics Ultrason
IEEE Transactions on Sonics and Ultrasonics
ieef ion-integrated evaporation filter
IEEIE Institution of Electrical & Electronics Incorporated Engineers
IEEJ Institute of Electrical Engineers of Japan
IEES International Educational Exchange System
IEETE Institution of Electrical and Electronics Technician Engineers
ief integrated electronic flash(ing); isoelectric focusing hb electrophoresis
IEF Indian Expeditionary Force; International Ecumenical Fellowship; International Exhibitions Foundation; International Eye Foundation
IEG Information Exchange Group
IEHA International Economic History Association
iei indeterminate engineering items
IEI Industrial Education Institute; Industrial Engineering Institute; Institute of Electrical Inspectors; Institution of Engineering Inspection; Institution of Engineers of Ireland; Iran Electronics Industries
IEIC Iowa Educational Information Center
IEKV *Internationale Eisenbahn-Kongress-Vereiningung* (German—International Railway Congress Association)
IEL Industrial Engineermng Limited; Industrial Equity Limited; Institute for Educational Leadership
iem *iemand* (Dutch—a man, somebody, someone)

IEMA Iowa Educational Media Association
IEMCAP Intrasystem Electromagnetic Compatibility Analysis Program (USAF)
IEME Inspectorate of Electrical and Mechanical Engineering
IEMS Institute of Experimental Medicine and Surgery; Integrated Emergency Management System
IEN Imperial Ethiopian Navy
IEO *Instituto Español de Oceanografía* (Spanish—Spanish Oceanographic Institute)
IEOM Institute of Environment and Offshore Medicine
ieop immunoelectro-osmophoresis
iep immunoelectrophoresis; iso-electric point
IEP Individual Education Plan; Individual Educational Program; Institute of Earth Physics; Institute of Experimental Psychology
IEP *Institut d'Etudes Politiques* (French—Institute of Political Studies)
IEPA International Economic Policy Association
IEPG Independent European Program Group
IEPs Individualized Education Programs
ieq index of environmental quality
ier installation enhancement release
Ier (Dutch—Irishman)
IER Industrial Equipment Reserve; Institute for Econometric Research; Institute of Educational Research; Institute of Engineering Research; Institute of Environmental Research; Interim Engineering Report
IERC Indian Education Resources Center (Bureau of Indian Affairs); International Electronic Research Corporation
IERE Institution of Electronic and Radio Engineers
Ierl *Ierland* (Dutch—Ireland)
IERT Institute for Education by Radio-Television
IER Test Institute of Educational Research Test (intelligence)

ie & s institutional, environmental, and safety
IES Illuminating Engineering Society; Independent Educational Services; Indian Educational Service; Industrial Electronic Systems; Information Exchange Service; Institute of Environmental Sciences; Institute of European Studies; Institution of Engineers and Shipbuilders; Intensive Employment Services
IE-S Institution of Engineers—Singapore
IESA Illuminating Engineering Societies of Australia
IESC International Executive Service Corps
IES-DC Information Exchange System - Data Collections
IESG Internet Engineering Steering Group
IESS Institution of Engineers and Shipbuilders in Scotland
iet interest equalization tax
IET Initial Engine Test; Institute of Educational Technology
IETF Internet Engineering Task Force
IETG International Energy Technology Group
I-et-L Indre-et-Loire
IETRA Interhuman Embryonic Transfer and Restablilization Agency
IETS Institute for Ecumenical Theological Studies
I-et-V Ille-et-Vilaine
IEWS Integrated Electronic Warfare System
if ice fog; immunofluorescence; information feedback; infrastructure; initiation factor; instrument flight; intermediate frequency; internal flush(ing); interstitial fluid; intrinsic factor
if. (IF) interferon; internal focusing
i-f in-flight; intermediate frequency
i & f inverter and cathode follower
if *iflge* (Danish—according to)
i.f. *ipse fecit* (Latin—he did it himself)
If filament current (symbol); Ifni; Sidi Ifni (Spanish West Africa)
IF grid current (symbol)
I-F Isotta-Fraschini

ifa indirect fluorescent antibody; instrumental fuel assembly; integrated file adapter

I f A Institutt for Atomenergi (Norwegian-Atomic Energy Institute)

IFA Industrial Forestry Association; Industry Film Association; Institute of Foresters of Australia; Intercollegiate Fencing Association; International Federation of Actors; International Federation on Aging; International Fertility Association; International Festivals Association; International Fiscal Association; International Football Association; International Footprints Association; International Formalwear Association; International Franchise Association

IFA Institut Fiziki Atmosfery (Russian—Atmospheric Physics Institute)

IFAA International Federation of Advertising Agencies; International Flight Attendants Association; International Flow Aids Association

IFABC International Federation of Audit Bureaus of Circulations

IFAC International Family Association of Canada; International Federation of Automatic Control

IFAD International Fund for Agricultural Development

IFALPA International Federation of Air-Line Pilots' Associations

IFAN Institut Français d'Afrique Noire (French— French Institute of North Africa, Dakar, Ivory Coast)

IFAP Industrial Foundation for Accident Prevention; International Federation of Agricultural Producers; International Foundation for Airline Passengers

IFAPA International Federation of Airline Pilots Association

IFAR International Foundation for Art Research

IFAS Institute for American Strategy; International Federation of Aquarium Societies

IFATCA International Federation of Air Traffic Controllers Associations

IFATCC International Federation of Associations of Textile Chemists and Colourists

IFATE International Federation of Airworthiness Technology and Engineering

IFAW International Fund for Animal Welfare

IFAWPCA International Federation of Asian and Western Pacific Contractors Associations

ifb invitation for bid(s)

IFB International Federation of the Blind; Invitation for Bid(s)

IFBA International Fire Buff Association

IFBB International Federation of Bodybuilders

IFBD International Foundation for Bowel Dysfunction

IFBPW International Federation of Business and Professional Women

IFBS Interrupted Feedback System

IFBWW International Federation of Building and Woodworkers

ifc independent fire control; inflight collision; inner front cover; institute freight clause(s); integrated fire control

IFC Inland Fisheries Commission; Intellectual Freedom Committee; International Finance Corporation; International Fisheries Commission; International Freighting Corporation

IFC-ALA Intellectual Freedom Committee—American Library Association

IFCATI International Federation of Cotton and Allied Textile Industries

IFCC International Federation of Camping and Caravanning; International Federation of Clinical Chemistry

IFCCTE International Federation of Commercial, Clerical, and Technical Employees

IFCJ International Federation of Catholic Journalists

IFCL International Fixed Calendar League

IFCO Interreligious Foundation for Community Organization

ifcr interface control register

IFCS Improved Fire-Control System; International Federation of Computer Sciences

IFCU International Federation of Catholic Universities

i.f.d. in flagrante delicto (Latin—in the heat of the evil deed)—caught in the act

IFD International Federation of Documentatiuon

IFDA International Foodservice Distributors Association; International Furnishings and Design Association

IFDP Institute for Food and Development Programs

IFE Industrial Foundation on Education; Institution of Fire Engineers; Institute of Financial Education

IFEBP International Foundation of Employee Benefit Plans

IFEBS Integrated Foreign Exchange and Banking System

IFEC International Foodservice Editorial Council

IFEE Institute for Free Enterprise Education

IF'EMS International Federation of Electron Microscope Societies

IFEP Instituttet for Elektronikmateriels Palideliged (Danish—Electronic Materials Reliability Institute)

IFEW Inter-American Federation of Entertainment Workers

iff integrated fish farming; invert fluid fill(ing)

iff (IFF) identification friend or foe

if & f intermediate flush and fill

IFF Institute of Freight Forwarders; Institute for the Future; Intermediate Financing Facility; International Flavors and Fragrances (corporation); International Freedom Foundation

IFFA Independent Federation of Flight Attendants; International Federation of Film Archives; International Frozen Food Association

IFFC Integrated Fire and Flight Control

IFFCO Indian Farmers Fertilizer Cooperative

IFFF Internationale Frauenlige far Frieden und Freiheit (German—International Women's League for Peace and Freedom)

IFFGD International Foundation for Functional Gastrointestinal Disorders

IFFJ Independent Federation of Free Journalists

IFFJP International Federation of Fruit Juice Producers

IFFM Independent Feature Film Market

iffn identification friend, foe, or neutral

IFFNM Internationale Ferienkurse für Neue Musik (German—International Vacation Courses for New Music)— held in Darmstadt

IFFPA International Federation of Film Producers Associations

IFFS International Federation of Film Societies

IFFTU International Federation of Free Teachers' Unions

ifg instrument flight guide

IFG International Fashion Group

IFGA International Federations of Grocers' Associations

IFGO International Federation of Gynecology and Obstetrics

ifh in-flight helium

IFHE International Federation of Home Economics

IFHP International Federation for Housing And Planning

IFHTM International Federation for the Heat Treatment of Materials

IFI Industrial Fasteners Institute; Institutional Functioning Inventory; International Fabricare Institute

IFI Instituto de Fomento Industrial (Spanish—Institute for Industrial Promotion)

IFIA International Federation of Ironmongers' and Iron Merchants' Associations; International Fence Industry Association

IFIAS International Federation of Institutes for Advanced Studies

IFIC International Food Information Council

IFIESR International Foundation for Industrial Ergonomics and Safety Research

IFIF International Foundation for Internal Freedom (hallucinogenic experimenter's society founded by Richard Alpert and Timothy Leary)

IFINDU I find you

IFIP Iguazu Falls International Park (shared by Argentina, Brazil, and Paraguay)— Argentinians spell it Iguazu, Brazilians—Igauçu, Paraguayans—Iguassu; International Federation of Information Processing

IFIPS International Federation of Information Processing Societies

I Fire E Institution of Fire Engineers

IFIS Integrated Flight Instrument System

IFJ International Federation of Journalists

IfK Institut für Kriminologie (German—Institute of Criminology); *Institut for Kriminologi* (Norwegian—Institute of Criminology)

IFKM Internationale Föderation für Kurzschrift 'ind Maschinenschreiben (German—International Federation of Shorthand and Typewriting)

IfL Institut für Landeskunde (German—Geographical Institute)—at Bad Godesberg

IFL Imperial Fascist League; Institute of Fluorescent Lighting; International Federation of Labor; International Friendship League

IFLA International Federation of Landscape Architects; International Federation of Library Associations

iflet interim focal-length optical tracker

IFLFF Internationale Frauenliga - für Frieden und Freiheit (German—International Women's League for Peace and Freedom)

IFLWU International Fur and Leather Workers Union

ifm intermediate frame memory

IFM Industrial Facility Manager; Institute for Forensic Medicine

IFMA International Facility Management Association; International Federation of Margarine Associations; International Foodservice Manufacturers Association

IFMBE International Federation for Medical and Biological Engineering

IFMC International Folk Music Council

IFME International Federation of Medical Electronics; International Federation of Municipal Engineers

IFMEO International Fish Meal Exporters Organization

if/mf intermediate frequency/ medium frequency

IFMI Irish Federation of Marine Industries

IFMP International Federation of Medical Psychotherapy

IFMS Integrated Financial Management System

IFMSA International Federation of Medical Students Associations

ifn information; interferon

IFN Institut Français de Navigation (French—French Institute of Navigation)

IFNAFSS International Feminist Network Against Female Sexual Slavery

IFNB Idaho First National Bank

IFNE International Federation for Narcotic Education

if nec if necessary

ifo in favor of; in front of; identified flying object; intermediate fuel oil

IFOFSAG International Fellowship of Former Scouts and Guides

IFOP Institut Français d'Opinion Publique (French— French Institute of Public Opinion)

IFOR International Fellowship of Reconciliation

IFORS International Federation of Operational Research Societies

IFOSA International Federation of Stationers' Associations

ifov instantaneous field of view; instrument field of view

ifp in-flight performance; in-flight printer; international fixed public broadcast band

IFP Imperial and Foreign Post; Independent Feature Project; Institute of Fluid Power; International Federation of Purchasing

IFP Institut Français du Pitrole (French—French Petroleum Institute)

IFPA Independent Free Papers of America; Industrial Film Producers Association; Institute for Foreign Policy Analysis; International Fire Photographers Association

IFPAAW International Federation of Plantation, Agricultural, and Allied Workers

IFPCW International Federation of Petroleum and Chemical Workers

IFPI International Federation of the Phonographic Industry

ifpm in-flight performance monitor

IFPM International Federation of Physical Medicine

IFPMA International Federation of Pharmaceutical Manufacturers Associations

IFPMM International Federation of Purchasing and Materials Management

IFPO International Freelance Photographers Organization

IFPP Imperial and Foreign Parcel Post

IFPRA International Federation of Park and Recreation Administrators

IFPRI International Food Policy Research Institute

IFPS Institute for Foreign Policy Studies

IFPTO International Federation of Popular Travel Organizations

IFPTS Intertype Fototronic Photographic Typesetting System

IFPW International Federation of Petroleum Workers

ifq individual fishing quota

ifr infrared; inflight refueling

ifr (IFR) internal function register

i-f-r image-to-frame ratio

IFr Internationaler Frauenrat (German—International Council of Women)

IFR Institute for Food Research; Instrument Flight Rules

IFRA International Foundation for Research in the Field of Advertising; International Fund-Raising Association

IFRB International Frequency Registration Board

IFRC International Fusion Research Council

IFREMER Institut Français de Recherches pour l'Exploitation des Mers (French Research Institute for the Exploitation of the Seas)

IFRF International Flame Research Foundation

IFRT Institute for Fitness Research and Training; Intellectual Freedom Round Table

ifru interference rejection unit

IFRU Institut Français du Royaume-Uni (French Institute of the United Kingdom)

ifs installable file system; integral flange and skirt

IFS Indian Forest Service; Instrument Flight System; International Federation of Surveyors; International Foundation for Science; International Freight Services; Irish Free State

IFS International Financial Statistics

IFSA Inflight Food Service Association; International Federation of Sound Archives

IFSDA International Federation of Stamp Dealers' Associations

IFSDP International Federation of the Socialist and Democratic Press

IFSEA International Federation of Scientific Editors Associations; International Food Service Executives Association

IFSEM International Federation of Societies for Electron Microscopy

IFSF Independent Fuel Storage Facility; Irradiated-Fuels Storage Facility

IFSIT In-Flight Safety Inhibit Test

IFSMA International Federation of Ship Master Associations

IFSO In-Flight Safety Officer

IFSP International Federation of Societies of Philosophy

IFSPO International Federation of Senior Police Officers

IFSPS International Federation of Students in Political Sciences

ifss infinite solution set

IFSS Instrumentation Flight Safety System; International Fertilizer Supply Scheme; International Flight Service Station(s)

IFSSO Irish Free State Stationery Office

IFST Institute of Food Science and Technology; International Federation of Shorthand and Typing

IFSTA International Fire Service Training Association

IFSW International Federation of Social Workers

ift inflight text

IFT Indiana Federation of Teachers; Institute of Food Technologists; International Federation of Translators; International Foundation for Telemetering; International Foundation for Timesharing; International Frequency Tables

IFT Institut für Tieflagerung (German—Institute for Geologic Disposal)

IFTA International Federation of Travel Agencies

IFTC International Film and Television Council

iftcd in-flight thrust-calculation deck

IFTF Inter-Faith Task Force

IFTPP International Federation of the Technical and Periodical Press

iftr in-flight thrust reverser

IFTR International Federation for Theatre Research

IFUW International Federation of University Women

ifv (IFV) infantry fighting vehicle

IFVME Inspectorate of Fighting Vehicles and Mechanical Equipment

IFWHA International Federation of Women's Hockey Associations

IFWL International Federation of Women Lawyers

IFWTWA International Food, Wine and Travel Writers Association

IFZ Industrial Free Zone

ig ignition; immunoglobulin; inertial guidance

ig (Ig) (IG) immunoglobulin

i/g in ground

Ig grid current (symbol)

IG gate current (symbol); Illustrators Guild; Indo-Germanic; Inspector General

IG *Interessengemeinschaft* (German—pool, trust)

iga integrating gyro(scope) accelerometer

IgA Immunoglobulin A

IGA Independent Grocers' Alliance; Integrated Grant Administration; International Geneva Association; International Geographical Association; International Glaucoma Association; International Golf Association; International Graduate Achievement

IGAEA International Graphic Arts Education Association

i gal imperial gallon

IGAM *Internationale Gesellschaft für Allgemeinmedizin* (German—International Society of General Medicine)

IGAS International General Aviation Society; International Graphic Arts Society

I Gas Eng Institution of Gas Engineers

IGB International Gravimetric Bureau

IGB *International Geophysics Bulletin*

igc intellectually gifted children

IGC Institute for Graphic Communication; Intergovernmental Copyright Committee; International Geophysical Cooperation

IGCA Industrial Gas Cleaning Association; International Garden Centre Association; Israel Government Corporation Authority

IGCC Insulating Glass Certification Council; Inter-Governmental Copyright Committee

igce independent government cost estimate

IGCI Industrial Gas Cleaning Institute

IGCM Incorporated Guild of Church Musicians

i/g/d illicit gold dealer

IGD Inspector General's Department

ig det ignition detector

IGDS Iodine Generating and Dispensing System

ige in ground effect; individually guided education; instrumentation ground equipnment

IgE Immunoglobin E

IGE International General Electric

iges initial graphics exchange specification

IGF International Grieg Festival

igf-1 (IGF-1) insulin-like growth factor l

IGFA International Game Fish Association

I.G. Farben *Interessengemeinschaft der Farbenindustrie* (German—German Dye Trust)

ig. fat. *ignis fatuus* (Latin—foolish fire)—will-o'-the-wisp; marsh gas

igfet insulated gate field-effect transistor

igg ill-gotten gains

IgG Immunoglobulin G

IG&GA International Grooving and Grinding Association

IGH Incorporated Guild of Hairdressers

IGI Institutional Goals Inventory

IGI *I Grandi Interpreti* (Italian—The Great Interpreters)—of classical music; *International Genealogical Index*

IGIA Interagency Group on International Aviation

IGIS International Guild for Infant Survival

igl information grouping logic

igl *iglesia* (Spanish—church)

iglª *iglesia* (Spanish—church)

igla *iglesia* (Spanish—church)

IgM Immunoglobulin M

IGM *Internationale Gesellschaft für Moorforschung* (German—Society for Research on Moors)

ign ignite; ignition; ignorant

ign. *ignotus* (Latin—unknown)

Ign Ignacio; Ignatius; Ignatz; Ignazio

IGN International Great Northern (railroad)

Ign⁰ *Ignacio* (Spanish—Ignatius)

IGO Independent Garage Owners; Intergovernmental Organization

igor injection gas-oil ratio; intercept ground optical recorder

igortt intercept ground optical recorder tracking telescope

IGOSS Integrated Global Ocean Station System

igp inside gravel pack (oil well)

IGP Industrial Government Party; Inspector General of Police

igpm imperial gallons per mile; imperial gallons per minute

igpp (IGPP) interactive graphics packaging program

IGPP Institute of Geophysics and Planetary Physics (UCLA)

igr. *igitur* (Latin—therefore)

IGRA Indian Gaming Regulatory Act

igrf international geomagnetic reference field

IGROF *Internationale Rorschach Gesellschaft* (German—International Rorschach Society)

IGRS Irish Geneological Research Society

igs integrated gas spacer

igs (IGS) interactive graphics system

IGS Imperial General Staff; Inert Gas System; Inertial Guidance System; Infogram Service; Institute of General Semantics; Institute of Geological Sciences; International Geranium Society; International Graphoanalysis Society

igse interim ground-support equipment

IGSEAP Inertial Guidance System Error Analysis Program

IGSESS International Graduate School for English-Speaking Students

IGSHPA International Ground Source Heat Pump Association

IGSS gate reverse current (symbol)

IGSS *Instituto Guatemalteco de Seguridad* (Spanish—Guatemalan Social Security Institute)

IGST Intergovernmental Committee on Science and Technology

igt (IGT) interactive graphics terminal

IGT Institute of Gas Technology; International Game Technology

IGTO India Government Tourist Office; Israel Government Tourist Office; Italian Government Tourist Office

IGU International Gas Union; International Geographical Union

igv inlet guide vane

IGWF International Garment Workers Federation

IGWT In God We Trust

IGWU International Garment Workers Union

IGWUA International Glove Workers Union of America

IGY International Geophysical Year (July 1957 through December 1958)

ih in home (lacrosse); inpatient hospital; inside height

ih (IH) infectious hepatitis

i.h. *iacet hic* (Latin—here lies)

Ih heater current (symbol); holding current (symbol)

IH International Harvester

I of H Institute of Hydrology

IH *International Humanism*

iha (IHA) idiopathic hyperaldosteronism

IHA Institute of Hospital Administrators; International Hahnemannian Association; International Hotel Association; International House Association

IHAB International Horticultural Advisory Bureau

IHAR Institute for Human-Animal Relationships

IHAS Integrated Helicopter Avionics System

IHATIS International Hide and Allied Trades Improvement Society

IHB International Hydrographic Bureau (Monaco)

IHBR Indiana Harbor Belt Railroad

ihc interstate highway capability

IHC Intercontinental Hotels Corporation; International Help for Children

IHCA International Hebrew Christian Alliance

IHCC International Harvester Credit Company

IHCD International Holocaust Commemoration Day

ihd (IHD) ischemic heart disease

IHD Institute of High Fidelity; Institute of Human Development; International Health Division (Rockefeller Institute for Medical Research); International Hydrological Decade (1965–1974)

IHDS Interstate Highway and Defense System

IHE Institute of Highway Engineers; Institute of Home Economics; Institute of Human Ecology; Institutions of Higher Education

I-head capital-I-shaped head (gasoline engine)

IHEU International Humanist and Ethical Union

ihf interesting historic figure

IHF Independent Health Food; Industrial Health Foundation; Industrial Hygiene Foundation; Institute of High Fidelity; International Health Foundation; International Hockey Federation; International Hospital Federation

IHFA Industrial Hygiene Foundation of America

IHFAS Integrated High-Frequency Antenna System

ihff inhibit halt flip-flop

IHFM Institute of High-Fidelity Manufacturers

IHFRA International Home Furnishings Representatives Association

IHHA International Halfway House Association

ihhs idiopathic hypertrophic subaortic stenosis

IHI Ishikawajima-Harima Heavy Industries

IHK *Internationale Handel skammer* (German—International Chamber of Commerce)

IHL International Homeopathic League

IHM Institute of Hotel Marketing; Institute of Housing Managers

I.H.M. Immaculate Heart of Mary

IHMA Industrialized Housing Manufacturers Association

iho in-house operation

IHOP International House of Pancakes

IHOU Institute of Home Office Underwriters

ihp indicated horsepower; ischemic heart disease

IHP Integrated Humanities Program

IHPA International Hardwood Products Association

ihph indicated horsepower hour

ihp/hr indicated horsepower hour

IHQ International Headquarters

IHR Institute of Historical Research; Institute for Historical Review; Institute of Human Relations

IHRA Independent Human Rights Association

IHRB International Hockey Rules Board

ihrd international rubber hardness degree(s)

IHRLA International Human Rights Law Group

ihs independent hemopathic syndrome; integrated healthcare system; integrated heat sink; integrated home systems; intellectually handicapped society

i.h.s. a variant of I.H.S., Jesus

IHS Christian symbol and monogram for Jesus; Immigration Historical Society; Indian Health Service; Indian Health Services; Infrared Homing System; Institute for Humane Studies; Institute of Hypertension Studies; International Horn Society; Interstate Highway System; Irish Hospitals Sweepstakes; Ivory Hunters Society

I.H.S. *Iesus Hominum Salvator* (Latin—Jesus Savior of Men); *In Hoc Signo* (Latin—In This Sign), Jesus

ihsa iodinated human serum albumin

IHSA Institute of Health Service Administrators; Italian Historical Society of America

ihsbr improved high-speed bombing radar

ihss idiopathic hypertrophic subaortic stenosis (IHSS)

IHSS In-Home Supportive Services; Integrated Hydrographic Survey System

IHT Institute of Handicraft Teachers

IHT *International Herald Tribune*

Ihu I hate you

IHU Irish Hockey Union; Interservice Hovercraft Unit

ihv intravenous hyperalimentation

IHVE Institute of Heating and Ventilating Engineers

ihx intermediate heat exchanger

IHY International Historical Year

IHYC Indian Harbor Yacht Club; Indian Harbour Yacht Club

ii illegal immigrant; individualized instruction; ingot iron; initial issue; injectivity index (oil well); interest included; inventory and inspection

i & i intercourse and intoxication; introduce and interview

Ii input current (symbol)

II Ikebana International; Instituto Interamericano (Interamerican); Irish Institute; Islamic Institute

I-I Mikhail Mikhaylovich Ippolitov-Ivanov (Russian composer)

I/I Inventory and Inspection (Report)

I & I instruction and inspection

II *Instituto Interamericano* (Spanish—Interamerican Institute)

iia if incorrect advise; inertial instrument assembly; innerinch adjustment; integrated irradiance analyzer

IIA Aerlinte Eireann (3-letter symbol for Irish Airlines); Incinerator Institute of America; Independent Innkeepers Association; Information Industry Association; Institute of Industrial Arts; Institute of Internal Auditors; Insurance Institute of America; International Information Administration; Invention Industry Association

IIAA Independent Insurance Agents Association

IIAC Industrial Injuries Advisory Council

IIAF Imperial Iranian Air Force

IIAG Interbureau Insurance Advisory Group

IIAL International Institute of Arts and Letters

IIAPCO Independent Indonesian-American Petroleum Company

IIAR International Institute of Ammonia Refrigeration

IIAS International Institute of Administrative Services

IIASA International Institute of Applied Systems Analysis

IIASR Israel Institute of Applied Social Research

IIB International Investment Bank; International Institute of Bankers; Internordic Investment Bank

IIB *Institut International de Bibliographie* (French—International Institute of Bibliography); *Institut Internationale des Brevets* (French—International Patent Institute)

IIBA International Intelligent-Buildings Association

IIC Independent Insurance Conference; Insurance Institute of Canada; International Institute of Communications; International Institute for the Conservation of Historic and Artistic Works; International Inter-City (train)

IICA Indians Into Communications Association; Institute of Instrumentation and Control—Australia; International Ice Cream Association

IICA *Instituto Interamericano de Ciencias Agrícolas* (Spanish—Inter-American Institute of Agricultural Sciences)

I-ICB Isolation-Interface Control Board

IICLRR International Institute for Children's Literature and Reading Research

IICM International Institute of Convention Management

IICS International Interactive Communications Society

iid impact ionization diode; infrared intrusion detection; interior intrusion device

IID Internal Investigation Division

IID *Institut International de Documentation* (French—International Documentation Institute)

IIDA Irish Industrial Development Authority

IIDS Interior Intruder Detection System

IIDS *Instituto Interamericano de Desarrolo Social* (Spanish—Interamerican Social Development Institute)

IIE Institute of Industrial Engineers; Institute for International Economics; Institute for International Education; International Institute of Embryology

IIE *Instituto Interamericano de Estadistica* (Spanish—Inter-American Institute of Statistics)

IIEA International Institute for Environmental Affairs

IIEC Inter-Industry Emission Control (program)

IIEG Interest Inventory for Elementary Grades

IIEL *Institut Internationale d'Études Ligures* (French—International Institute for Ligurian Studies)

IIEP International Institute of Educational Planning

IIET Inspection Instructions for Electron Tubes

IIF Institute of International Finance; *Institut International du Froid* (French—International Institute of Refrigeration)

IIFA International Institute of Films on Art

IIfA International Institute for Inhalant Abuse

IIFT Indian Institute of Foreign Trade

IIGF Imperial Iranian Ground Forces

IIGH *Instituto Interamericano de Geografía e Historia* (Spanish—Interamerican Institute of Geography and History)

IIHCEHV International Institute of Health Care, Ethics, and Human Values

IIHF International Ice Hockey Federation

IIHR Institute for International Human Resources

IIHS Insurance Institute for Highway Safety

III Insurance Information Institute; Inter-American Indian Institute; International Institute of Interpreters (UN); International Isostatic Institute

III Instituto Indigenista Interamericano (Spanish—Inter-American Indigenist Institute); *International Intertrade Index*

IIIC International Irrigation Information Center (Israeli)

IIJR Illinois Institute of Juvenile Research

iil integrated injection logic

IIL Institute of Industrial Launderers; Intelligence International Limited

IILC *International Instituut voor Landaanwinning en Cultuurtechniek* (Dutch— International Institute of Land Reclamation and Cultivation)

IILRI International Institute for Land Reclamation and Improvement

IILS International Institute for Labour Studies

IIM Indian Institutes of Management

IIME Institute of International Medical Education

IIMS Intensive Item Management System

IIMSD International Institute for Music Studies and Documentation

IIMT International Institute for the Management of Technology

IIN Item Identification Number

IIN Instituto Interamericano del Niño (Spanish—Inter-American Children's Institute); *Instituto Italiano di Navigazione* (Italian—Italian Institute of Navigation)

IInfSc Institute of Information Scientists

IINZ Insurance Institute of New Zealand

IIOE International Indian Ocean Expedition

IIOOF International Independent Order of Odd Fellows

IIOS International Indian Ocean Survey

iip index of industrial production; individualized instructional planning

IIP Indian Imperial Police; Institute International de la Presse (International Institute of the Press); International Ice Patrol; International Institute of Peace; International Institute of Philosophy

IIP Institute International de la Presse (French—International Institute of the Press)

IIPER International Institution of Production Engineering Research

IIPL Independent Investor Protective League

IIPs Individualized Instruction(al) Program(s)

iir imaging infra-red; isobutylene isoprene rubber

IIR International Institute of Reflexology; International Institute of Refrigeration

IIRA International Industrial Relations Association

IIRE International Institute for Resource Economics

IIRM Irish Immigration Reform Movement

iirs instrumentation inertial reference set

IIRS Institute for Industrial Research and Standards (Erie)

ii's illegal immigrants

IIs Iberial Inquisitions in Spain and Portugal; Immigration Inspectors

IIS Indian Institute of Science; Industrial Inquiry Service; Institute of Information Science; Institute of Information Scientists; Insurance Institute of Singapore; Integrated Instrument System; Interactive Instructional System

IIS Institut International de la Soudre (French—International Institute of Welding); *Institut International de la Statistique* (French—International Institute of Statistics); *Internationales Institut der Sparkassen* (German—International Institute of Savings Banks)

IIS & EE International Institute of Seismology and Earthquake Engineering

IISG Internationaal Instituut voor Sociale Geschiedenis (Dutch—International Institute of Social History)—in Amsterdam

IISL International Institute of Space Law

IISL Istituto Internazionale di Studi Liguri (Italian—International Institute for Ligurian Studies)

IISO Institution of Industrial Safety Officers

IISR International Institute for Submarine Research

IISRP International Institute of Synthetic Rubber Producers

IISS International Institute of Strategic Studies

IIST International Institute for Safety in Transportation

IISWM International Institute of Iron and Steel Wire Manufacturers

iit independent inclusive tour; individual inclusive tour

IIT Illinois Institute of Technology; Indian Institutes of Technology; Israel Institute of Technology

IIT Institut International du Théâtre (French—International Institute of the Theater)

IITB Indian Institute of Technology—Bombay

IITM Indian Institute of Technology—Madras

IITRAN Illinois Institute of Technology Translators

IITRI Illinois Institute of Technology Research Institute

IITYWYBAD? If I tell you will you buy a drink?

IIW International Institute of Welding

iiwfm if it weren't for me

iiwfy if it weren't for you

iJ im Jahre (German—in the year)

IJ IJssel; Institute of Journalists; Irish Jurist

I of J Institute of Jamaica

IJ Internationale Jugendbibliothek (German—International Youth Library)—Munich

IJA Institute of Jewish Affairs; Institute of Judicial Administration; International Judiciary Association

IJA International Journal of the Addictions

IJAF International Journal of American Folklore

IJC International Joint Commission (Canada–U.S.); Itawamba Junior College

IJE Institute for Journalism Education

IJF International Judo Federation

I-J FC Iselin-Jefferson Financial Company

IJIAP International Juridical Institute for Animal Protection

IJISID Imperial Japanese Institute for the Study of Infectious Diseases

IJK *Internationale Juristen-Kommission* (German— International Jurist Commission)

IJLP Islamic Jihad for the Liberation of Palestine

IJMA Infant and Juvenile Manufacturers Association

IJMS *Israel Journal of Medical Sciences*

IJO Independent Jewelers Organization; International Juridical Organization (for developing countries of the third world)

IJOA International Juvenile Officers Association

ijp inhibitory junction potential

IJPA International Jelly and Preserve Association

IJPPR Institute for Jewish Policy Planning and Research

IJR Institute for Juvenile Research

IJS Institute of Jazz Studies; Institute of Jesuit Sources; Institute of Jewish Studies

IJszee (Dutch—Ice Sea)— Arctic Ocean, Polar Sea

IJVA *International Journal of Verbal Aggression* (*Maledicta*)

ik inner keel

ik *ikke* (Danish—not)

Ik Ichabod

IK cathode current (symbol)

IK *Immune Korper* (German— immune bodies)

IKAR *Internationale Kommission für Alpines Rettungswesen* (German—International Commission for Alpine Rescue)

IKAROS Intelligence and Knowledge-Aided Recognition of Speech

IKB Isambard Kingdom Brunel

IKCs International Keratorefractive Centers

ike iconoscope; ikebana; ikebanism; ikebanist(ic)

Ike Dwight David Eisenhower, 34th president of the United States; Isaac

Ike *USS Eisenhower* (nuclear-powered supercarrier)

IKG *Internationale Kommission für Glas* (German—International Commission for Glass)

IKI *Internationale Kali-Institut* (German—International Potash Institute)

IKN *Internationale Kommission für Numismatik* (German—International Numismatic Commission)

I-K-P In-Ko-Pah (Park or Mountains in California)

IKPK *International Kriminal-Polizei-Kommission* (German—International Criminal Police Commission)

I kr Icelandic krona (monetary unit)

ikrd inverse kinetics rod drop

IKRK *Internationale Komitee vom Roten Kreuz* (German—International Committee of the Red Cross)

ik unit infusoria killing unit

IKV *Internationaler Krankenhausverband* (German— International Hospital Federation)

IKV-91 Hagglund and Soner tank destroyer made in Sweden

il illustrate; illustrated; illustration; illustrator; including loading; incoming letter; independent laboratory; inside layer; inside left; inside length; instrument landing; interline; interlinear; interlinearly; interpretive language

il (IL) interleukin

Il illinium; Italian Line (ships included *Conte di Savoia* and *Rex)*

Il *Illiad*

IL current in an inductor (symbol); Identification List(ing); Illinois; Import License; Incres Line; Independent Laboratory; Instruction Leaflet; International Logistics; Interocean Line; Israel (auto plaque and Internet code)

I/L Import License

I & L Installations and Logistics

I of L Institute of Linguists

IL *Institut Littéraire*

Il-12 Soviet Ilyushin transport called Coach by NATO

Il-14 Soviet Ilyushin transport called Crate by NATO

Il-18 Soviet Ilyushin transport called Coot by NATO

Il-28 Soviet Ilyushin jet bomber called Beagle by NATO

Il-38 Soviet Ilyushin transport called May by NATO

Il-62 Ilyushin 62 aircraft

ila insulin-like activity; insurance logistics automated

ila (ILA) instrument landing approach

ILA Illinois Library Association; Independent Literary Agent; Indian(a) Library Association; Indonesian Library Association; Institute of Landscape Architects; International Laundry Association; International Law Association; International Leprosy Association; International Linguistic Association; International Longshoremen's Association; Iowa Library Association; Iranian Library Association; Iraq Library Association; Israel Library Association

ILAA Independent Literary Agents Association; International Legal Aid Association

ILAAS Integrated Light Aircraft Avionics System; Integrated Light Attack Avionics System; International League Against Anti-Semitism

ILAB International League of Antiquarian Booksellers

ILAFA *Instituto Latinamericano del Fierro y del Acero* (Spanish—Latin American Institute of Iron and Steel)

ILAMA International Lifesaving Appliance Manufacturers Association

ILAP Individualized Language Arts Program

ILAR Institute of Laboratory Animal Resources

ilas interrelated logic accumulating scanner

ILAS Institute of Latin American Studies; Instrument Low-Approach System

ilb inshore lifeboat

ilc irrevocable letter of credit

ilc (ILC) instruction length code

ILC Independent Learning Center; Individualized Learning Center; International Law Commission (UN)

ILCA Insurance Loss Control Association; International Labor Communications Association; International Livestock Centre for Africa

ILCNY I Love a Clean New York

ILCOP International Liaison Committee of Organizations for Peace (UN)

ILCW Inter-Lutheran Commission on Worship

ild indentation load deflection; instructional logic diagram

ILD International Labor Defense

ILDA International Laser Display Association

Ildef⁰ Ildefonso (Spanish)

ildf integrated logistic data file

ildt item logistics data transmittal

ile individual learning expectations; isoleucine

ile (ILE) isoleucine (amino acid)

Il^e *Illustre* (Spanish—illustrious)

ILE Institution of Locomotive Engineers

ILEA Inner London Education Authority; International League of Electrical Associations

ILEI Index of Leading Economic Indicators

ILEI *Internacia Ligo de Esperantistaj Instruistoj* (International League of Esperanto Instructors)

ILERA International League of Esperantist Radio Amateurs

ILESA International Law Enforcement Stress Association

ileu isoleucine

ilf inductive loss factor

Ilf Ilya Arnoldovich Feisliber

ILF International Landworkers Federation

ILFI International Labor Film Institute

ILFO International Logistics Field Office (USA)

ILGA Institute of Local Government Administration; International Lesbian Group Association

ILGPNWU International Leather Goods, Plastics, and Novelty Workers Union

ILGWU International Ladies' Garment Workers' Union

ILH Imperial Light Horse; International League of Honolulu; Interscholastic League of Honolulu

ILHR International League of Human Rights

ILI Indiana Limestone Institute; Institute of Life Insurance; Inter-African Labor Institute; International Language Institute

ILIA Indiana Limestone Institute of America

ILIAS Inforonics Library Automation Services

ILIC International Library Information Center

ill. illusion; illusionary; illusionist; illustrate; illustrated; illustration; illustrator

ill. illustrissimus (Latin—most illustrious)

Ill I lack love; Illinois; Illinoisan

I'll I shall; I will

ILL Institute of Languages and Linguistics; Institute of Lifetime Learning; Inter-Library Loan; Interstate Loan Library

ILLA Irish Ladies Lacrosse Association

ILLC Inner London Library Committee

illegals illegal aliens

illegit illegitimate

ILLIAC Illinois Automatic Computer

ILLINET Illinois Library Information Network

illit illiterate

ILLRI Industrial Lift and Loading Ramp Institute

Ill St Hist Lib Illinois State Historical Library

illu illustrate

illud illustrated

illum illuminant; illuminate; illumination

illun illustration

illus illustrated; illustration; illustrator

illw intermediate-level liquid waste

ilm insulin-like material

ILM International Literary Management

ILMA Incandescent Lamp Manufacturers Association; Independent Lubricant Manufacturers Association

Ilmo Illustrissimo (Italian—Most Illustrious)

Il^mo Illustrísimo (Spanish—Most Illustrious)

ILMP International Literary Market Place

ILN Illustrated London News

ilo in lieu of

Ilo Iloilo

I lo iodine lotion

ILO International Labour Office (UN); International Labor Organization

ILOA Industrial Life Officers Association

I Loco E Institution of Locomotive Engineers

I Loco Eng Institution of Locomotive Engineers

ILOOSEU I loose you

Ilopango San Salvador, El Salvador's airport

iloue in lieu of until exhausted

ilp instant linear programming

ILP Independent Labour Party; Israel Labor Party

ILPA Independent Labor Press Association

ILPES Instituto Latinoamericano de Planificación Económica y Social (Spanish—Latin American Institute for Economic and Social Planning)

ILPH International League for the Protection of Horses

ILQ International Law Quarterly

ilr (ILR) independent local radio

ILR Industrial and Labor Relations; Institute of Library Research; International Luggage Registry

ILR Instituut voor Landbowtechniek en Rationalisatie (Dutch—Institute for Agricultural and Planning Technics); *International Law Reports*

ILRA International Log Rolling Association

ILRAD International Laboratory for Research into Animal Diseases

ILRC Indian Law Resources Center

ILREC International League for the Rational Education of Children

ILRI Indian Lac Research Institute

ILRM International League for the Rights of Man

ILS Incorporated Law Society; Industrial Locomotive Society; Instrument Landing System; Integrated Logistic Support; International Latitude Service; International Lunar Society

ILSA Insured Locksmiths and Safemen of America

ilsam international language for servicing and maintenance

ILSC International Learning Systems Corporation

IISEA Illinois Society of Enrolled Agents

ILSI International Life Sciences Institute

ILSMT Integrated Logistic Support Management Team

ILSP Integrated Logistic Support Plan(ning)

ILSR Institute for Law and Social Research; Institute for Local Self-Reliance

ilsw interrupt-level status word

ilt interferometric landmark tracker; in lieu thereof

ilt (ILT) infectious laryngotracheitis

ILT Illinois Terminal (railroad)

ILTF International Lawn Tennis Federation

ILTS Institute of Low Temperature Science; Integration Level Test Series

iltw intermediate-level transuranic waste

ILU Institute of Life Insurance; Institute of London Underwriters

ilv induced leukemia virus(es)

Ilvu I love you

ilw intermediate-level wastes

ILWC International League of Women Composers

ILWU International Longshoremen's and Warehousemen's Union

ILZ Illinois Zinc (company)

ILZRO International Lead Zinc Research Organization

im immature; imperial measure; impulse modulation; infectious mononucleosis; inner marker; installation maintenance; installment mortgage; instant message; intensity modulation; interactive multimedia; intermodulation; intramuscular

im (IM) inland marine (insurance); interceptor missile

i&m improvement and modernization

'im (American slang—him)

im in dem (German—in the); **imeni** (Russian—in the name of); (Latin—in)—imprint(ing)

Im Imperial; maximum current (symbol); meter current (symbol)

I'm I am

IM impulse modulation; Industrial Management; Institute of Metallurgists; Institute of Metals; intermediate modulation; Inventory Manager

I of M Institute of Medicine

IM *Index Medicus*

ima ideal mechanical advantage

Iᵐᵃ *prima* (Italian—first)

IMA Ignition Manufacturers Institute; Independent Music Association; Indian Military Academy; Indianapolis Museum of Art; Individual Medical Account; Indonesian Mining Association; Industrial Marketing Association; Industrial Medical Association; Institute of Management Accountants; Institute for Mediterranean Affairs; Institute of Municipal Administration; Integrated Modular Avionics; Interbank Merchants Association; International Management Association; International MIDI (Musical Instrument Digital Interface) Association; International Mineralogical Association; International Minilab Association; Islamic Mission of America; Issues Management Association

IMA *Instituto Mobiliare Italiano* (Italian—Italian Security Institute)

IMAA Indochinese Mutual Assistance Association

IMAAWS Infantry Man Portable Anti-Armor Weapon System

imac integrated microwave amplifier converter

IMAC International Metals and Commodities

IMACA International Mobile Air Conditioning Association

imag imaginary

IMAGE Instruction in Motivation Achievement and General Education

IMAGES Improving Morale And Giving Excellent Service

IMAGI Index Measuring Accurate Growth of Inflation

IMAJ International Management Association of Japan

IMAR Inner Mongolia Autonomous Region (of the People's Republic of China)

IMarE Institute of Marine Engineers

IMARPE *Instituto de Mar del Perú* (Spanish—Sea Institute of Peru)

IMARS Institutional Management for Accountability and Renewal System

IMAS Integrated Management Accounting System; International Marine and Shipping Conference

IMAU International Movement for Atlantic Union

IMAU *Instituto Municipal de Aseo Urbano* (Spanish—Municipal Institute of Urban Sanitation)

IMAURO Integrated Model for the Analysis of Urban Route Optimization

IMAWU International Molders and Allied Workers Union

imb interaction of man and the biosphere

IMB Institute of Marine Biochemistry; Institute of Marine Biology; International Maritime Bureau

IMB *Internationaler Metallarbeiterbund* (German—International Metalworkers Federation)

ImbarsbIdbib I may be a rotten sod but I don't believe in bullshit (early 19th century British slang)

IMBC Independent and Multi-Cultural Broadcasting Corporation

IMBD International Migratory Bird Day (in May)

IMBE Institute for Minority Business Education

IMBLMS Integrated Medical and Behavioral Laboratory Measurement System (NASA)

IMBO *Institutt for Marin Bi-ologi* (Norwegian—Institute for Marine Biology)

imc image motion compensation; instrument meteorological condition; intermediate metal(lic) conduit

IMC Industrial Management Center; Infant Mortality Commission; Informational Media Center; Institute of Management Consultants; Institute of Measurement Control; Instructional Materials Center; Instructional Media Center; International Management Council; International Maritime Committee; International Medical Corps; International Meteorological Committee; International Minerals & Chemical; International Mining Corporation; International Missionary Council; International Monetary Conference; International Music Council; Iran Meat Corporation

IMC *Instituto Mexicano del Café* (Spanish—Mexican Coffee Institute)

IMCA Institute of Management Consultants in Australia; Insurance Marketing Communications Association; Investment Management Consultants Association

imcc item management control code

IMCC Integrated Mission Control Center

IMCE *Instituto Mexicano de Comercio Exterior* (Spanish—Mexican Institute of Foreign Commerce)

IMCEA International Military Community Executives Association

IMCI Interracial Music Council, Incorporated

imco improved combustion

IMCO Inter-Governmental Maritime Consultative Organization

IMCOS International Meteorological Consultant Service

IMCOV Iron Mines Company of Venezuela

IMCS Interactive Manufacturing Control System

IMD Indian Medical Department; Inventory Management Division

IMDA Independent Medical Distributors Association; Indirect Missile Damage Assessment; International Magic Dealers Association

IMDC Integrated Mission/Display Computer; Internal Message Distribution Center

IMDGC International Maritime Dangerous Goods Code

IMDS International Microform Distribution Service

imdt immediate(ly)

imdtty it's my duty to tell you

ime independent medical examination

ime (IME) international magnetospheric explorer

IME Indo-Malaysian Engineering; Institute of Makers of Explosives; Institute of Marine Engineers; Institute of Mechanical Engineers

I&ME Indiana and Michigan Electric Company

I of ME Institution of Mining Engineers

I Mech E Institution of Mechanical Engineers

IMEG International Management and Engineering Group

IMEO Interim Maintenance Engineering Order

imep indicated mean effective pressure

IMER Institute for Marine Environmental Research

I Met Institute of Metals

IMET International Military Education and Training

imf intermediate fuel

imf (IMF) integrated maintenance facility

IMF International Metalworkers Federation; International Monetary Fund; International Motorcycle Federation; Interstate Motor Freight (stock exchange symbol); Israel Music Foundation

IMFC Investment and Merchant Finance Corporation

IM FI International Mineral Fiber Institute

IMFJC International Metalworkers Federation Japan Council

im/fm intensity modulated/frequency modulated

imfrad integrated multiple-frequency radar

IMF/SDR International Monetary Fund—Special Drawing Rights

imfu immense military fuckup

img informational media guarantee

IMG International Marxist Group

IMGP Internal Medicine Group Practice

IMH Institute of Materials Handling

imhe international management in higher education

imho in my humble opinion

IMHT Institute for Material Handling Teachers

imi improved manned interceptor; intermediate maintenance instructions

IMI Ignition Manufacturers Institute; Imperial Metal Industries; Imperial Mycological Institute; International Masonry Institute; Irish Management Institute; Israel Military Industries; Israeli Military Intelligence

IMI *Instituto Mobiliare Italiano* (Italian—Italian Assets Institution)—credit bank

IMIA International Marketing Institute of Australia

IMIB Inland Marine Insurance Bureau

imid inadvertent missile ignition detection

imieo initial mass in earth orbit

IMIMI Industrial Mineral Insulation Manufacturers Institute

IMINCO Iran Marine International Oil Company

IMinE Institute of Mining Engineers

IMINOCO Iranian Marine International Oil Company

imint imagery intelligence

imit imitate; imitation

IMIT Institute of Musical Instrument Technicians

imitac image input to automatic computers

imit lea imitation leather

iml inside mold layer; inside mold line

Iml Imanuel

IML International Music League; Irradiated Materials Laboratory; Island Merchants Limited (Cook Islands)

IMLS Institute of Medical Laboratory Sciences

IMLT Institute of Medical Laboratory Technology

imm immediate; immune; immunization; immunologist; immunology; impairing a minor's morals; impairing the morals of a minor; intermediate maintenance manual

Imm Immingham

IMM Institute of Mining and Metallurgy; Integrated Maintenance Management; International Mercantile Marine; International Monetary Market (Chicago)

immac inventory management and material control

Immarsat International Maritime Satellite

immat immature; immaturity

immed immediate

immedly immediately

immie imitation marble; lowgrade playing marble

immig immigrant; immigration

immob immobilization; immobilize

IMMS International Material Management Society

IMMT Integrated Maintenance Management Team

IMMTS Indian Mercantile Marine Training Ship

immun immunity; immunization

immunol immunologist; immunology; imunologic(al)(ly)

immy immediately

imn indicated mach number

IMNS Imperial Military Nursing Service

IMNX Immunex Corporation

imo imitation (slang short form); immobilized

Imo Imogen(e)

IMO Integrated Marketing Organization; Inter-American Municipal Organization; International Maritime Organization; International Meteorological Organization (World Meteorological Organization)

imos inadvertent modification of the stratosphere

imp inflight motion pictures

imp. impedance; imperative; imperfect; imperial; implement; implementation; import; importation; imprint; improve; improvement

imp (IMP) indeterminate mass particle; inertial measuring platform; inflatable micrometeoroid paraglide;

imp. impotentia (Latin—impotent); *imprenta* (Spanish—printing office, printing press); *imprimatur* (Latin—let it be printed); *imprimé* (French—printed); *imprimis* (Latin—especially, particularly)

Imp Imperator (Latin—supreme ruler)

Imp. Imperator (Latin—Emperor); *Imperatrix* (Latin—Empress)

IMP Instrumented Mobile Platform (oceanographic drone boat); Integrated Mediterranean Program; International Match Point; International Monitoring Probe (space instrument); Interplanetary Monitoring Platform (space vehicle)

IMP Instituto Mexicano del Petroleo (Spanish—Mexican Petroleum Institute)

imp. 8 imperial octavo ($7^1/_2 \times$ 11-inch or 19×28-cm book size)

IMPA International Master Printers Association; International Motor Press Association; International Museum Photographers Association; International Myopia Prevention Association

impact. implementation planning and control technique

IMPACT Improving Public Awareness of Concepts of Telecommunications; Information Market Policy Actions; Interdisciplinary Model Programs in the Arts for Children and Teachers

impatt impact avalanche and transit time

Imp B Imperial Beach

IMPC Institutional and Municipal Parking Congress

impce importance

imper imperative

imperat imperative

imperf imperfect; imperforate

Imperial Imperial Savings Association

impers impersonal

imp-exp import-export

impf imperfect

impg importing; impregnate

imp. gal imperial gallon

IMPI International Microwave Power Institute

impig impignorate; impignorated; impignorating; impignoration

impi imperial; implement

implic import license

import important

import importaciones (Spanish—imports)

imposs impossible

impr improvement

impr impresión; imprenta (Spanish—edition, printing office)

impracl impracticable

impreg impregnate(d); impregnation

IMPRESS Inter-disciplinary Machine Processing for Research and Education in the Social Sciences

imprim. imprimatur (Latin—let it be printed)

Impr Nat Imprimerie Nationale (French—National Printing Office of France)

improp improper(ly)

improv improvement

imps interplanetary measurement probes

Imps Imperial Tobacco Company

IMPS Inpatient Multidimensional Psychiatric Scale; Institute of Management Public Speaking

Imp Sav Imperial Savings

impt important

imptd imported

imptr importer

Imptypco Imperial Typewriter Company (also appears as ITC)

impv imperative

impvt improvement

impx impaction

imqc imported merchandise quantity control

imr internal mold release

imr (IMR) infant mortality rate

IMR Individual Medical Report; Institute of Marine Resources; Institute of Masonry

Research; Institute for Materials Research; Institute for Medical Research; Institute for Mortuary Research; Institute for Motivational Research; Institute for Muscle Research; International Medical Research

IMRA Incentive Manufacturers Representatives Association; Industrial Marketing Research Association; International Manufacturers Representatives Association; International Mass Retail Association

IMRADS Information Management, Retrieval, and Dissemination System

imran international marine radio aids to navigation

IMRC International Marine Radio Company

IMRF Independent Manufacturers Representatives Forum

IMRL Individual Material Readiness List(ing)

IMRO Inspection Minor Rework Order; Interior Macedonian Revolutionary Organization

IMRS Inpatient Multidimensional Rating Scale

imrt intensity-modulated radiation therapy

ims industrial methylated spirit(s); inertial measurement set

im's intramuscular injections

IMS Index Management System; Indian Medical Service; Individualized Mathematics System; Industrial Management Society; Industrial Mathematics Society; Information Management System; Institute of Management Sciences; Institute of Marine Science; Institute of Mathematical Statistics; Institute of Museum Services; Institute on Man and Science; International Magnetic System; International Magnetosphere Study; International Military Staff; International Musicological Society; International Mythological Society

IMSA International Management Systems Association; International Motor Sports Association; International Municipal Signal Association

IMSC International Maritime Satellite Corporation; International Military Sports Council

IMSCOM International Military Staff Communication (NATO)

IMS/HEW Institute of Museum Services—HEW

IMSI International Maple Syrup Institute

IMSL Independent Measurement Standards Laboratory; International Mathematical and Statistical Library

IMSM Institute of Marketing and Sales Management

IMSO Institute of Municipal Safety Officers

IMSR Isle of Man Steam Railway

imss integrated manned-system simulator

IMSS Integrated Manned Systems Simulator; International Museum of Surgical Science

IMSS *Instituto Mexicano del Seguro Social* (Spanish—Mexican Social Security Institute)

imt independent model triangulation

IMT International Military Tribunal (Nuremberg)

IMTA Imported Meat Trade Association; Institute of Municipal Treasures and Accountants; International Marine Transit Association

IMTC *Instituto Municipal de Transporte Colectiva* (Spanish—Municipal Institute of Collective Transport)—metropolitan bus system

IMTD Inspectors of the Military Training Directorate

IM Tech Institute of Metallurgists Technician

IMTFE International Military Tribunal for the Far East

IMTP Industrial Mobilization Training Program

IMTS International Message Telecommunications Service

imu (IMU) inertial measurement unit

IMU International Mailers Union; International Maritime Union; International Mathematical Union

IMUA Inland Marine Underwriters Association

I Mun E Institution of Municipal Engineers

imusc intramuscular

imv imperative; improve; intermittent mandatory ventilation

IMVS Institute of Medical and Veterinary Science

imw international map of the world

IMW Institute of Masters of Wine

imwxprt imitation wax prints

IMX Inquiry Message Exchange

Im Yem Imamate of Yemen

IMZ *Internationales Musikzentrum* (German—International Music Center)

in indemnity (time loss or other); input

in. inch(es); interest

in. (In) inulin

i/n item number

In India; Indian; indium; Indus; Instructor

In *Indre* (Norwegian—inner, interior, inside)

IN India (Internet code); Indiana; Indian Navy; Institute of Neurobiology (Göteborg); Interested Negroes

I & N Immigration and Naturalization

I of N Institute of Navigation

in.² square inch(es)

in.³ cubic inch(es)

ina international normal atmosphere

INA Indian National Army; Inspector Naval Aircraft; Institute of National Affairs; Institute of Nautical Archaeology; Institution of Naval Architects; Insurance Company of North America, Iraqi News Agency; Israeli News Agency

inabi inability

inacdutra inactive duty training

INACESA *Industria Nacional de Cemento SA* (Spanish—National Cement Industry Company)

INACH *Instituto Antártico Chileno* (Spanish—Chilean Antarctic Institute)

inactv inactivate; inactivation; inactive

INAEA International Newspaper Advertising Executives Association

InAF Indian Air Force

inah (INAH) isonicotinic acid hydrazide

INAH Instituto Nacional de Antropología e Historia (Spanish—National Institute of Anthropology and History)—Mexico

INALPRE Instituto Nacional de Preinversión (Spanish—National Investment Control Institute)—Bolivia

INAM Istituto Nazionale Assicurazione Malattie (Italian—National Health Insurance Board)

INAME International Newspaper Advertising and Marketing Executives

inanim inanimate; inanimative

INANTIC Instituto Nacional de Normas Tecnicas Industriales y Certificatión (Spanish—National Institute of Technical Standards)

INAO Institut Nationale des Appellations d'Origine (French—National Institute of Authentic Names)

inappbl inapplicable

INAR Institute of Northern Agricultural Research

in'ards innards

INAS Inertial Navigation and Attack System(s); Interbank National Authorization System

INATAPROBU International Association of Professional Bureaucrats

inaud inaudible

inaug inaugurate; inaguration

inaug diss inaugural dissertation (thesis for doctor's degree)

in bal. in ballast

inbd inboard

inbu internal navigation battery unit

INBUCON International Business Consultants

inc in cloud; inclosure; include; income; increase; incumbent

inc. incomplete

Inc Inchon; Incorporated

In C Instructor Captain

INC Indian National Congress; Industrial National Corporation; International Narcotics Control; International Nickel Company; International Numismatic Commission; Island Navigation Company (tankers)

INC Instituto Nacional de Cultura (Spanish—National Institute of Culture)—Lima, Peru

inca inventory control and analysis

Inca Incahuasi

INCA Information Council of the Americas; International Newspaper Color Association; Inventory Control and Analysis

incair including air

incalz incalzando (Italian—increasing dynamics and tone)

incan incandescent

incap incapacitant; incapacitating

INCAP Instituto de Nutrición de Centroamerica y Panamá (Spanish—Institute of Nutrition of Central America and Panama)

incaps incapacitating agents

InCAR International Committee Against Racism

incb inclusion body

INCB International Narcotics Control Board

incct incorrect

inccty incorrectly

incd incendiary; incident

incdt incident

ince insurance

INCE Institute of Noise Control Engineering

incfmy inconformity

inch. inchoative; integrated chopper

In-Ch Indo-China

incho inchoate

inchoat inchoative

incid incidence; incident; incidental

incid mus incidental music

INCING International Copyright Information Center

INCIRS International Communication Information Retrieval System

incl incline; inclose; inclosure; include; including; inclusive

incl inclusivement (French—inclusively)

incl. included

incld included

incln inclusion

inclntr inclinator

inclr intercooler

inclsv inclusive

INCMD Indianapolis Contract Management District

INCO International Nickel Company

incog incognito

INCOLSA Indiana Cooperative Library Services Authority

INCOMAG International Communication Agency

INCOMEX International Computer Exhibition

INCOMEX Instituto Colombiano de Comercio Exterior (Spanish—Colombian Institute of Overseas Commerce)

incomp incomplete

incompat incompatible; incompatibility

incompl incomplete

incor incorrect

INCORA Instituto Colombiano de Reforma Agraria (Spanish—Colombian Institute of Agrarian Reform)

incorp incorporated

Incorp Incorporated; Incorporation

incorr incorrect

inco(s) incorrigible(s)

incpt intercept

incr increase; increased; increasing; increasingly; increment; incremental

INCRA International Copper Research Association

INCREF International Children's Rescue Fund

incrim incriminate; incrimination; incriminatory

INCSR International Narcotics Control Strategy Report

incumb incumbent

incun incunabula

incur. incurable

ind independent; index; indicate; indicative; indicator; indirect; indigo; indorse; indorsement; industrial; industry; investigational new drug

ind (IND) investigational new drug

in d. in diem (Latin—daily)

Ind India; Indian; Indiana; Indianapolis; Indianian; Indo-; Indonesian; Indus (constellation); Industries; Industry

Ind Indiano (Italian—Indian, Indian Ocean); *Indico* (Portuguese or Spanish—Indian, Indian Ocean); *Indien* (French—Indian, Indian Ocean)

IND India (auto plaque); Indianapolis, Indiana (airport)

INDA International Non-wovens and Disposables Association

indac industrial data acquisition and control

INDALUM Industria del Aluminio (Spanish—Aluminum Industry)

INDASAT Indian Scientific Satellite

INDAX Interactive Data Exchange

Ind Day Independence Day

Ind Dem Independent Democrat

Ind. E. Industrial Engineer

indecl indeclinable

INDECO Industrial Development Corporation; International Development and Construction Corporation

indef indefinite

indef art. indefinite article

indefops indefinite operations

indem indemnify; indemnity

inden indenture; indentured; indenturing

Ind Eng Industrial Engineer(ing)

Ind Eng Chem Industrial and Engineering Chemistry

Ind & Eng Chem Industrial and Engineering Chemistry

indep independent

INDEP Industria Nacional del Plomo (Spanish—National Lead Institute)

Ind-et-L Indre-et-Loire

Index Index Librorum Prohibitorum (Latin—Index of Forbidden Books); *Index on Censorship*

indi indicate; indication

India letter I radio code; Republic of India (Asian nation); *Bharat* (India's name in Hindi)

Indiana Girls Indiana Girls School (for juvenile delinquents, at Indianapolis)

Indianap Indianapolis

indic indicative; indicator

indic indicateur (French—informer)

indicolite blue tourmaline

indies independents

Indies East Indies; West Indies

Ind. Imp. Indiae Imperator (Latin—Emperor of India)

Indira Indira Ghandi (India's first woman prime minister)

INDITECNOR Instituto Nacional de Investigaciones Tecnologicas y Normalización (Spanish—National Institute for Technical and Standardization Investigation)

indiv individual

indivl individual

indiv psychol individual psychology

Ind L Independent Liberal

Ind Lab Independent Labor (party)

indm indemnity

Ind Med Index Medicus

Ind Mgr Industrial Manager

indn indication (flow chart)

Indo Indonesia; Indonesian

Ind O Indian Ocean

INDO Instituto Nacional de Denominaciónes de Origen (Spanish—National Institute of Denomination of Origin)

Indo-Afr Indo-African

Indo-Amer Indo-American

Indo-Austral Indo-Australasian

indoc indoctrinate; indoctrination

Indoc Indochina; Indochinese

Indo-Chi Indo-China; Indo-Chinese

indocin indomethacine

indocum indocumentado (Spanish—undocumented) person without identification papers

Indo-Eur Indo-European

Indo-Ger Indo-German(ic)

Indo-Mal Indo-Malayan

Indon Indonesia(n)

Indon Indonesian (modified Malay language)

Indonesia Republic of Indonesia (Asian island nation), *Republik Indonesia*

Indonesias Indonesian Islands

Indo-Pak India-Pakistan; Indo-Pakistan(i)

INDOSUEZ Banque de l'Indochine et de Suez (French—Bank of Indochina and Suez)

Indpls Indianapolis

indpol industrial pollutant; industrial pollution

ind quest indirect question

indr indicator (flow chart)

indre indenture

ind reg induction regulator

Ind Rep Independent Republican

IND.S.C. Indian Survivors' Certificates

Ind Sym Indianapolis Symphony

Indt indent

Ind. T. Indian Territory

Ind Ter Indian Territory (now Oklahoma)

indtr indentor

induc inductance; induction

INDUGAS Industra de Gas (Spanish—Gas Industry)

Ind U Pr Indiana University Press

indus industrial; industry

Indus (Latin—Indian constellation)

indust industrial; industrialization; industrialize; industrialized; industry

indvl individual

Ind. wc Indian widow's certificate

Indy Indianapolis; Indianapolis Speedway

Indy-style Indianapolis-style

inec inverted emulsifiable concentrate

ined. ineditus (Latin—unpublished)

INED Institute for New Enterprise Development

ineffv ineffective

ineffy ineffectively

inel inelastic

INEL Idaho National Engineering Laboratory (ERDA); Intelligent Electronics Incorporated

INEOA International Narcotic Enforcement Officers Association

Iness Inverness-shire

INETr Interbank Network for Electronic Transfer

in ex. in extenso (Latin—at length)

inf infant(ile); infantry; infect(ious); inferior; infinitive; infinity; influence; information; interferon; intervertebral foramina

inf. informed

inf (INF) interceptor night fighter; intermediate-range nuclear force

inf. infra (Latin—below, beneath); *infunde* (Latin—pour into)

Inf Infirmary

Inf Inférieur (French—lower, nether)

INF Intermediate Nuclear Force; Intermediate-Range Nuclear Forces; International Naturist Federation; International Nudist Federation

INFA Institut pour l'Etude du Fascisme (French—Institute for the Study of Fascism)

INFAMA International Fair Promotion and Marketing

INFANTS Iroquois Night Fighter And Night Tracker System

infarc infarction

Inf Bat Infantry Battalion

infce international nuclear-fuel-cycle evaluation

Inf Div Infantry Division

infe inferior

INFE International Newspaper Financial Executives

infect. infection; infectious

infib infibulate; infibulation

infin infinitive

infirm. infirmary

infl inflammable; inflorescence; influence(d)

in-fl in-flight

inflor inflorescence

influ influence; influential

infm information

infmy infirmary

INFN Istituto Nazionale di Fisica Nucleare (Italian—National Institute of Nuclear Physics)—Italy

info inform; information

INFO International Fortean Organization

Info Can Information Canada

infol (INFOL) information-oriented language

infoline information line (telephone)

INFONAC Instituto de Fomento Nacional (Spanish—Institute for National Production)

inforem inventory forecasting and replenishment module(s)

INFORFILM International Information Film Service

Informbureau Communist Information Bureau (Cominform)

Informburo (Soviet) Information Bureau

Informex Informaciones Mexicanas (Mexican Information Service)

INFORS International Federation of Engineers

INFORSA Industrias Forestal SA (Spanish—Forest Industries Corporation)

INFOSEC Information Security

INFOTERM International Information Center for Terminology (UNESCO)

info theory information theory

infr inferior

infra below

infra (Latin prefix—beneath)—infraspinal; infrastructure

infra dig. infra dignitatem (Latin—beneath one's dignity, undignified)

infral information retrieval automatic language

infraptum. *infrascriptum* (Latin—written below)

Infrared Phys Infrared Physics

infric. infricetur (Latin—let it be rubbed in)

infross information requirements of the social sciences

inft infant

infus infusible

infx inspection fixture

ing inguinal

ing ingégnere (Italian—engineer); *ingegneria* (Italian—engineering); *i ngeniør* (Dano-Norwegian—engineer)

Ing Ingmar

Ing Ingénieur (French—engineer); *Ingenieur* (German—engineer)

ING International Newspaper Group

inga inspection gage

INGA Interstate Natural Gas Association

INGAA Interstate Natural Gas Association of America

INGEOMINAS Instituto de Investigaciones Geologico Mineras (Spanish—Institute of Geological Sources Research)

Ingg Inggeris (Malay—English)

Ingl Inghilterra (Italian—England); *Inglaterra* (Portuguese or Spanish—England)

Ingm Berg Ingmar Bergman

INGO International Non-Governmental Organization

INGR Intergraph Corporation

ingred(s) ingredient(s)

Ingria Ingermanland

Ingrid Ingrid Bergman

inh (INH) isonicotinic hydrazide

inh. inherited

Inh Inhaber (German—proprietor)

INH Instituto Nacional de Hipódromos (Spanish—National Institute of Racetracks)

inhab inhabitant(s)

inhal inhalation

in. Hg inch of mercury

inhib inhibition; inhibitory

INHP Independence National Historical Park

INHS Indian Naval Hospital Ship

INI Indianapolis Newspapers Incorporated; Industrial Nurses Institute

INI Institut National De l'Industrie (French—National Institute of Industry); *International Nursing Index*

INIA Instituto Nacional de Investigaciones Agricolas (Spanish—National Institute of Agricultural Research)

INIBP Instituto Nacional de Investigaciones Biológico-Pesqueras (Spanish—National Institute of Fish Biology Research)

INIC Instituto Nacional de Investigaciones Cientifica (Spanish—National Institute for Scientific Investigation)

INIF Instituto Nacional de Investigaciones Forestales (Spanish—National Institute for Forestry Research)

in./in. inch per inch

in init. in initio (Latin—in the beginning)

INIP Instituto Nacional de Investigaciones Pecuarias (Spanish—National Institute for Cattle Research)

INIS International Nuclear Information System

init initial

init. initiate

initv initiative

inj inject; injection; injections; injure; injury

inj. injectio (Latin—inject, injection)

inj. enema inji ciatur enema (Latin—inject an enema)

inj. hyp. injectio hypodermica (Latin—hypodermic injection)

inj mldg injection moulding

inkl inklusiv (German—inclusive)

inl initial

inl inlichtingen (Dutch—information)

In L Instructor Lieutenant

INL Independent Newspapers Limited

INLA International Nuclear Law Association; Irish National Liberation Army

inlaw (INLAW) infantry laser weapon

in.-lb inch-pound

In L-Cdr Instructor Lieutenant Commander

in lim. in limine (Latin—at the outset)

in litt. in litteris (Latin—in correspondence)

in loc. in loco (Latin—in the place)

in. loc. cit. in loco citato (Latin—in the place cited)

Inlt Inlet (postal abbreviation)

inly initially

INM Institute of Naval Medicine; International Narcotics Matter; Irish National Museum

INMA International Newspaper Marketing Association

inmarsat international marine satellite

Inmarsat International Maritime Satellite

INMARSATORG International Maritime Satellite Organization

INMAS Intensifikasi Massnal (Indonesian—Mass Intensification)

INMED Indians into Medicine

in mem. in memoriam (Latin—in memory of)

In Mem In Memoriam—Sir Arthur Sullivan's Overture in C

i n mi international nautical mile(s)

INMM Institute of Nuclear Materials Management

inn. inning

Inn Inni (hinterland French Guiana) — *see* Cayen

Inn. Innoshima

INN Instituto Nacional de Normalización (Spanish—National Institute of Standards); *Instituto Nacional de Nutrición* (Spanish—National Institute of Nutrition)

innerv innervated; innervation

Innis Inniskilling

INNOTECH Institute for Educational Innovation and Technology

inns. Innings

INO Inspectorate of Naval Ordnance

inoc inoculation; inoculate

INOC Iraq National Oil Company

INOCO Indonesian Nippon Oil Corporation

inop inoperative

inorg inorganic

Inorg Chem Inorganic Chemistry

INOS Instituto Nacional de Obras Sanitarias (Spanish—National Institute of Sanitation—Venezuela)

in-out input-output

inp inert nitrogen protection

INP Inyanga National Park (Rhodesia)

INPA International Newspaper Promotion Association

INPA Instituto Nacional por Pesquisar au Amazonas (Portuguese—National Institute for Research on Amazons)

in partibus in partibus infidelium (Latin—in the region of the unbelievers)

INPFC International North Pacific Fisheries Commission

inph interphone

in p. inf. in partibus infidelium (Latin—in the region of the unbelievers)

INPO Institute of Nuclear Power Operations

INPOLSE International Police Services (CIA)

inpr in progress

in pr. in principio (Latin—in the first place)

Inprecorr International Press Correspondence

in prep in preparation

in pro in proportion

inprons information processing in the central nervous system

inps if not previously sold

INPS Istituto Nazionale di Previdenza Sociale (Italian—National Institute of Social Security)

in pulm. in pulmento (Latin—in gruel)

inq index of nutritional quality; inquiry

Inq Inquisidor (Spanish—inquisitor, investigator)

INQUA International Association on Quaternary Research

inr impact noise rating; impact noise ratio; intelligence and research

in'r inner

i-n r interference-to-noise ratio

INR Institut National de la Radio (French—National Radio Institute); Institute of Natural Resources; Intelligence and Research

INRA Instituto Nacional de la Reforma Agraria (Spanish—National Institute of Agrarian Reform)

in ref in reference (to)

in req information requested

In Res Indian Reservation

I.N.R.I. Iesus Nazarenus Rex Iudaeorum (Latin—Jesus of Nazareth, King of the Jews)

INRIA Institut National de Recherche Informatique et Automatique (French—National Institute of Informative and Automatic Research)

INRO International Natural Rubber Organization

ins inches; inscribe(d); inscription; inspector; insular; insulate(d); insulation; insurance; insure(d)

ins (INS) inertial navigation system

ins. insert

in's in his

in./s inch(es) per second

ins in das (German—into the)

in s in situ (Latin—in the original place)

Ins Insecta; Inverness

INS Immigration and Naturalization Service; Indian Naval Ship; Inertial Navigation System; Institute of Naval Studies; Institute of Nuclear Sciences; Institute of Nutritional Sciences; Integrated Navigation System; International News Service

I & NS Immigration and Naturalization Service

INSA International Shipowners Association

INSA *Industria Nacional de Neumaticos SA* (Spanish—National Tire Industry Corporation)

InsACS Interstate Airway Communication Station

Ins Agt Insurance Agent

INSAIR Inspector of Naval Aircraft

INSAT Indian National Satellite

insav interim shipyard availability

INSCAIRS Instrumentation Calibration Incident Repair Service

insce insurance

INSCO Intercontinental Shipping Corporation

inscr inscribed; inscription

INSDC Indian National Scientific Documentation Center

INSDOC Indian National Scientific Documentation Center

insd val insured value

INS & E Institute of Nuclear Science and Engineering

InSEA Indiana Society of Enrolled Agents

INSEA International Society for Education Through Art

in./sec inches per second

insecti insecticide(s)

insectories insect nurseries

INSEL International Nickel Southern Exploration Limited

INSENG Inspector of Naval Engineering Material

insep inseparable

Ins Gen Inspector General

insh inspection shell

insig insignificant

insinuendo insinuate + innuendo

INSIS Inter-Institutional Integrated Services Information System

INSJ Institute for Nuclear Study—Japan

insl insulate; insulation

INSMACH Inspector of Naval Machinery

INSMAT Inspector of Naval Material

INSNAVMAT Inspector of Navigational Material (USN)

insol insoluble

insolv insolvent

insoly insolubility

INSORA Instituto de Organización y Administración de Empresas (Spanish—Institute for the Organization and Administration of Enterprises)

INSORD Inspector of Naval Ordnance

insp inspect; inspected; inspection; inspector; inspiration; inspire; inspired

Insp Inspector

in-spec within specifications

INSPECC Information Services in Physics, Electrotechnology, Computers, and Control

INSPECT Infrared System for Printed-Circuit Testing; Inquiry into Pollution and Environmental Conservation; Integrated Nationwide System for Processing Entries from Customs Terminals

INSPEL International Journal of Special Libraries

INSPETRES Inspector of Petroleum Resources

Insp Gen Inspector General

inspir. inspiretur (Latin—let it be inspired)

INSPIRE Institute for Public Interest Representation

Inspr Inspector

INSRADMET Inspector of Radio Materials

insrnc insurance

INSRP Interagency Nuclear Safety Review Panel

inst. install; installation; installment; instant; instantaneous; institute; institution; instruct; instruction; instructor; instrument; instrumentation; instrumented

Inst Institute; Institution

INSTAAR Institute of Arctic and Alpine Research

INSTAB Information Service on Toxicity and Biodegradability

insta-cam instant camera (tv)

instar inertialess scanning, tracking, and ranging

INSTARS Information Storage and Retrieval System

Inst CE Institute of Civil Engineers

Inst Ceram Institution of Ceramics

inst ctl instrumentation control

instd instead

Inst Dirs Institute of Directors

Inst EE Institute of Electrical Engineers

Inst F Institute of Fuel

Inst Gas Eng Institute of Gas Engineers

Inst Gen Sem Institute of General Semantics

Inst HE Institute of Highway Engineers

Inst Int Educ Institute of International Education

instl institutional

instln installation

instm instrument; instrumentation; instrumented

Inst ME Institute of Mechanical Engineers

Inst Mediaeval Mus Institute of Mediaeval Music

Inst Met Institute of Metals

Inst Mod Lang Institute of Modern Languages

instn institution(al)

INSTN Institut National des Sciences et Techniques Nucléaires (French—National Institute of Science and Nuclear Techniques)

instns instructions

Inst P Institute of Physics

Inst Pat Institute of Patentees

Inst Pckg Institute of Packing

Inst Pet Institute of Petroleum

Inst Plan & Res Institute for Planning and Research

instpn instrument panel

Inst P S Institute of Purchasing and Supply

instr instruct; instruction; instructor; instrument(s)

instru instrumentation

instruct. instruction; instructor

Instru Soc Am Instrument Society of America

Inst W Institute of Welding

Inst WE Institute of Water Engineers

insuf insufficient

insul insulation

insur insurance

insurd insured
INSURV Board of Inspection and Survey
in sync in synchronization; perfectly synchronized
int (INT) initial (flow chart)
int intérêt (French—interest)
int. intake; integer; integral, intention; interest; interested; interior; interjection; internal; international; interred; intersection
Int International
INt Isaac Newton telescope
INT Air Inter *(Lignes Aériennes Intérieures);* Interpool
INTA International New Thought Alliance
INTA Instituto Nacional de Tecnologia Agropecuaria (Spanish—National Institute of Agricultural Technology)
INTACS Integrated Tactical Communications Systems
INTACT Infants Need to Avoid Circumcision Trauma
INTAF Internal Affairs (ministry)
int. al inter alia (Latin—among other things)
INTAL Industria Nacional de Tejidos de Alambre (Spanish—National Wire Netting Industry)
INTAMEL International Association of Metropolitan City Libraries
INTASGRO Interallied Tactical Study Group (NATO)
INTC Intel Corporation
int. cib. inter cibos (Latin—between meals)
intcl intercoastal
intcol intelligence collecting; intelligence collection
int comb. internal combustion
Int Com Illum International Commission on Illumination
intcp intercept; interception; interceptor
int dec interior decorator
Int Doc Serv International Documents Service (Columbia University)
INTECOM International Council for Technical Communication
Integ Ed Assoc Integrated Education Associates
intel intelligence

intelpost (INTELPOST) international post (computerized postal service)
intelsat international telecommunications satellite
Intelsat International Telecommunications Satellite Organization
INTELSAT International Telecommunications Satellite Corporation; International Telecommunications Satellite Organization
intelting intelligence training
INTEM Instituto Interamericano de Educacion Musical (Spanish—Inter-American Institute of Musical Education)
Intend Intendente (Spanish—manager, police commissioner, provincial governor, superintendent, supervisor)
intens intensive
inter intercalation; interest; intermediate; interrogation
inter (Latin prefix—among or between)—interborough, international
Interarmco International Armament Corporation
INTERASMA Association Internationale d'Asthmologie (French—International Association for the Study of Asthma)
Interavia World Review of Aviation and Astronautics
Interchem Interchemical Corporation
intercom intercommunication system
interd interested
INTERDATA Interdata Computers
interdict. intelligence detection and interdiction countermeasures
interdisc interdisciplinary
INTER-EXPERT International Association of Experts
INTEREXPO International Expositions
interf interference
INTERFILM International Church Film Center
INTERFLORA International Florists (telegraphic service)
interg interesting
Interior US Department of the Interior
interj interjection

INTERMAC International Association of Merger and Acquisition Consultants
INTERMARC International Machine-Readable Catalog
Intermex International Mexican Bank
InterMilPol International Military Police (NATO)
intern. internal
internat international; internationalism; internationalist
International Date Line 180° longitude
internet internetwork
INTERNOISE International Conference on Noise Control Engineering
interp interpolation; intepretation; interpreter
Interpace International Pipe and Ceramics
Interpen/IAB Intercontinental Penetration Force/International Anti-communist Brigade
INTERPHOTO Fédération Internationale des Négociants en Photo et Cinéma (French—International Federation of Photograph and Cinema Merchants)
Interpol International Criminal Police Commission
INTERPOL International Criminal Police Organization (Saint-Cloud, France)
interr interrogative
interrog interrogation; interrogative
INTERSTENO International Federation of Short Hand and Typewriting Stenographers
INTERTANKO International Association of Independent Tanker Owners
Intertel International Television
INTERTELL International Intelligence Legion
intertwangled intertwined + wangled
inter/w intersection with
intest intestinal; intestine
INTEXT International Textbook Company
intfc interference
intg interrogate; interrogator
inth intrathecal
Int Harv International Harvester
inti monetary unit of Peru

intif *intifadah* (Arabic—uprising)—Palestinian uprising against Israel

intip integrated information processing

INTIPS Integrated Information Processing System

intj interjection

Int J Mag *International Journal of Magnetism*

Int J Quantum Chem *International Journal of Quantum Chemistry*

Int J Theor Phys *International Journal of Theoretical Physics*

intl international

intl comb. internal combustion

Intl Ctr Envir International Center for Environmental Research

Intl Film Bur International Film Bureau

Intl Law Reps International Law Reports

Intl Legal Mats *International Legal Materials*

Intl Review International Review Service

Intl Univs Pr International Universities Press

intmed intermediate

int med (Int Med) internal medicine

intmt intermittent

int. noct. *inter noctem* (Latin—during the night)

intns intransit

INTO Irish National Teachers' Organization

intops interdiction operations

Intourist Soviet Tourist Office

intox intoxicant; intoxicate; intoxicated; intoxication

Int Pap International Paper

intpr interpret; interpretation; interpreter

int qk fl interrupted quick-flashing light

intr intransitive; intruder; intrasion

intra (Latin prefix—inside or within)—intracranial, intrastate

INTRACO International Trading Company

intran input translator

intrans intransitive

in trans in transit

in trans. *in transitu* (Latin—in transit)

intransit intransitive

Int Rep Intelligence Report

Int Rev Internal Revenue

intrex information transfer complex

intrmt interment

intro introduce; introduced; introducing; introduction; introductory; introversion; introvert

introd introduction

introd *introduzione* (Italian—introduction)

intropta. *introscripta* (Latin—written within)

intro(s) introduction(s)

intrp interrupt(ion) -

intrpt interpret(ation); interrupt(ion)

intrvlmtr intervalometer

intsf intensification; intensify

int sig interval signal

int std d international standard depth

Int Sum Intelligence Summary

INTU Interpool (container unit)

INTUC Indian National Trades Union Congress

intv independent television

intvlmtr intervalometer

intvw interview

Int Wildlife *International Wildlife*

inu internal navigation unit

INU Nauru Island (airport)

I Nuc E Institute of Nuclear Engineering

INUIDS *Instituto de las Naciones Unidas para la Investigación del Desarrol Social* (Spanish—United Nations Institute for the Investigation of Social Development)

inurn inurnment

InUS inside the United States

in ut. *in utero* (Latin—within the uterus)

inv invent; inventor; inventory; inverse; inversion; invert; inverter; investment; invoice

inv. *invenit* (Latin—he devised it)

Inv Inverness

INV *Instituto Nacional de la Vivienda* (Spanish—National Housing Institute)

INVAID Integration of Computer Vision Techniques for Automatic Incident Detection

inval invalid(ate)

invc invoice

invcd invoiced

invert. invertebrate

inves investigate; investigation; investigator

investig investigate; investigation; investigator

invest(s) investigation(s)

inv. et del. *invenit et delineavit* (Latin—devised and drawn)

invic. *invictus* (Latin—unconquerable)—title of a poem by William Ernest Henley—*Invictus*

in vit. *in vitro* (Latin—within glass, within a test tube or other laboratory glass vessel)

in viv. *in vivo* (Latin—within a living body)

inv. obj. investment objective

invol involuntary

invos *in vivo* optical spectroscopy

invt inventory

invtn invitation

invtrx inventrix

INWATS Inward Wide Area Telephone Service

INWR Imperial National Wildlife Refuge (Arizona); Iroquois National Wildlife Refuge (New York)

INX Inexco Oil (stock-exchange symbol)

INZP *Index to New Zealand Periodicals*

io integrated optics; interest only; intraocular; ion engine

io (IO) inverted original (12-tone)

i/o image/orthicon; in and/or over; inboard-out-board (motorboat engine); input/output; instead of

i & o input and output

Io ionium; output current (symbol)

I/o input/output

IO Chagos Islands (Internet code); India Office; Information Officer; Inspecting Officer; Inspection Office(r); Intelligence Office(r); Intercept Office(r); Irish Office; Issuing Office(r)

I/O Inspection Order; Investigating Officer

ioa instrument-operating assembly; instrumentation-operating area

IOA Institutional Overlay zone; Intelligence Oversight Act; International Omega Association

IOAM Institute of Appliance Manufacturers

IOAT International Organization Against Trachoma

ioau input/output access unit

iob input/output buffer; internal operating budget

I o B Institute of Bakers; Institute of Bankers; Institute of Bookkeepers; Institute of Brewers; Institute of Builders

IOB Institute of Brewing; Intelligence Oversight Board (CIA)

IOBB Independent Order of B'nai B'rith

IOBC Indian Ocean Biological Center; International Organization for Biological Control of Noxious Animals and Plants

IOBI Institute of Bankers in Ireland

IOBS Institute of Bankers in Scotland

iobyte input/output byte

ioc immediate or cancel; initial operational capability; in our culture

i-o c input-output channel(s)

I o C *Index on Censorship*

IOC Indian Ocean Commission (Madagascar, Mauritius, and the Seychelles); Institute of Chemistry; Intergovernmental Oceanographic Commission; International Olympic Committee; Interstate Oil Compact

IOCA Interstate Oil Compounders Association

IOCC Interstate Oil Compact Commission

IOCI Interstate Organized Crime Index

ioco industry-owned contractor operator

iocs interoffice comment sheet

IOCS Input-Output Control System

IOCU International Office of Consumers Unions; International Organization of Consumer Unions

IOCV International Organization of Citrus Virologists

IOD Imperial Order of the Dragon; International Operations Division

IODE Imperial Order of Daughters of the Empire

I o E International Office of Education; Isle of Ely

IOE Institute of Education, International Office of Epizootics; International Organization of Employers

IOEC International Order for Ethics and Culture

ioem invert oil emulsion mud (oil well)

I o F Institute of Fuel

IOF Independent Order of Foresters; International Oceanographic Foundation; International Olympic Federation; International Olympic Foundation

iofb I only fire blanks (sterile male); intraocular foreign body

IOFC Indian Ocean Fishery Commission

IOFI International Organization of the Flavor Industry

IOFSI Independent Order of the Free Sons of Israel

ioga industry-organized goverment-approved

IOGP Independent Oil and Gas Producers

IOGT International Order of Good Templars

ioh item(s) on hand

IOH Institute of Heraldry

ioho in our humble opinion

ioi internal operating instruction

IOI Industrial Oxygen Incorporated; International Ocean Institute; Israel Office of Information

I o J Institute of Journalists

IOJ International Organization of Journalists

IOJD International Order of Job's Daughters

iol intraocular lens

I o L Institute of Librarians

IOL India Office Library (London); Interoffice Letter

iol('s) interocular lens(es)

IOLTA Interest on Lawyers' Trust Accounts

IO Ltd Imperial Oil Limited

iom input/output multiplexer

IoM Isle of Man

IOM Institute for Organization Management; Institute of Medicine; Institute of Metallurgists; Institute of Metals

I.O.M. *Iovi Optimo Maximo* (Latin—Jove the Greatest Superlative)

IOMA International Oxygen Manufacturers Association

IOMC International Organization for Medical Cooperation

IOME Institute of Marine Engineers

i/o media input-output media

IOMM & P International Organization of Masters, Mates and Pilots

IOMR International Offshore Multihull Rule

IOM SPC Isle of Man Steam Packet Company

IOMTR International Office for Motor Trades and Repairs

iomux input/output multiplexer

ion integrated on-line network

Ion Ionic

ION (pseudonymic initials— George Jacob Holyoake); Institute of Navigation

IONDS Integrated Operational Nuclear Detection System

Ioanians Ionian Islands

iont in order not to

IOO Inspecting Ordnance Officer

IOOC International Olive Oil Council; Iranian Oil Operating Companies; Irish Organization of Celts

IOOF Independent Order of Odd Fellows

IOOTS International Organization of Old Testament Scholars

iop input-output processor; intraocular power; irrespective of percentage

i & op in-and-out processing

IoP Institute of Poverty; Isle of Palms; Isle of Pines

I o P Institute of Packaging; Institute of Petroleum; Institute of Physics; Institute of Plumbing; Institute of Printing

IOP Institute of Petroleum; Integrated Obstacle Plan; International Organization of Paleobotany; Iranian Oil Participants; Irish Organization of Papists

IOPAB International Organization for Pure and Applied Biophysics

IOPC Interagency Oil Policy Committee

IOPK Independent Order of Panamanian Kangaroos

IOP & LOA Independent Oil Producers and Land Owners Association

IOPS Input/Output Processing System

IOQ Institute of Quarrying

ior input/output register; item on request

I o R Institute of Roofing

IOR Independent Order of Rechabites (Quaker abstainers); International Ocean Rule; International Offshore Rules

IOR Istituto per le Opere de Religione (Italian—Institute for the Operation of Religion)—the Vatican Bank

IORD International Organization for Rural Development

IORM Improved Order of Red Men

IORS International Orders' Research Society

ios integrated office system

Ios Isles of Scilly; Isles of Shoals; Isles of the Sea

IOS Inspection Operation Sheet; Institute of Oceanographic Science; Institute of Oceanographic Services; International Organization for Standardization; Investors Overseas Services

IOSA Incorporated Oil Seed Association; International Oil Scouts Association; Irish Offshore Services Association

IOSHD International Organization for the Study of Human Development

IOSM Independent Order of the Sons of Malta

IOSOT International Organization for the Study of the Old Testament

IoT Institute of Transport; Isle of Thanet

iota inbound-outbound traffic analysis; information overload testing aid

IOTA Institute of the Americas; Institute of Traffic Administration

IOTC International Originating Toll Center

iot&e initial operational test and evaluation

IOT & E Interim Operational Test and Evaluation

Iotthy I'm only trying to help you

IOTTSG International Oil Tanker and Terminal Safety Guide

iou immediate operation use; industrial operations unit; inevitability of the unpredictable

I.O.U. I owe you

IO UBC Institute of Oceanography—University of British Columbia

I.O.U.s (plural of I.O.U.)

IOUSP Instituto Oceanográfico da Universidade de São Paulo (Portuguese— Oceanographic Institute of the University of São Paulo)

IOV Instituto Oceanográfico de Valparaíso (Spanish— Oceanographic Institute of Valparaíso)

IOVPT Internationale Organisation für Vakuum-Physik und Technik (German—International Organization for Vacuum Science and Technology)

IOVST International Organization for Vacuum Science and Technology

iow in other words

IoW Isle of Wight

I o W Isle of Wight

IOW Institute of Welding

iox instructional objectives exchange

IOY Iron Ore Year

IOZV Internationale Organisation für Zivilverteidigung (German—International Organization for Civil Defense)

ip identification point; illegal possession; impact predictor; improvement purchase; incentive pay; india paper; induced polarization; industrial photographer; industrial photography; information provider; initial phrase; initial point; injured person; innings pitched; input primary; installment plan(ning); integer

programming; intermediate pressure; iron pipe; plate current (symbol)

ip (IP) imposter phenomenon; insurance payment

i/p input

i & p indexed and paged

iP in Preussen (German—in Prussia)

Ip Ipanema

I£ Israeli pound

IP ice pellets (aircraft code); Imperial Preference; Institute of Petroleum; Instructor Pilot; Insular Police; International Pictures; Internet Protocol; Interpool (container unit); Isabel Province (Solomon Islands); plate current (symbol)

IP Institut Pasteur (French—Pasteur Institute); *Isla de Pinos* (Spanish—Isle of Pines)

I-P Indian-Pacific (transcontinental train linking Perth, Australia with Sydney);

I & P Island and Peninsular (development bank)

I & P Izvestia and *Pravda* (Russian—*News* and *Truth*)

ipa including particular average; initial perceptual alphabet; intermediate power amplifier; internal power amplifier; international phonetic alphabet (IPA)

ipa (IPA) isopropanol; isopropyl alcohol

IPA Illinois Principals Association; Independent Petroleum Association; Independent Physicians Association; Independent Practice Association; Independent Publishers Association; Independent Publishers of Australia; Individual Practice Association; Institute for Physics of the Atmosphere; Institute of Propaganda Analysis; Institute of Public Administration; Institute of Public Affairs; International Police Association; International Peace Academy; International Pediatric Association; International Phonetic Alphabet; International Phonetic Association; International Platform Association; International Police Academy; International Police Archives (Manchester Central Library); International Police Association; International Press Association; International Psychoanalytical Association; International Publishers Association *(see* TALA); Investment Partnership Association

IPA *Information Please Almanac; International Pharmaceutical Abstracts*

IPAA Independent Petroleum Association of America; Institute of Patent Attorneys of Australia; International Plan of Action on Aging (UN); International Prisoners Aid Association; Interstate Professional Applicators Association

ipac isopropyl acetate

IPAC Independent Petroleum Association of Canada; Iranian Pan-American Oil Company

IPACK International Packaging Material Suppliers Association

IPACS Integrated-Power/Attitude-Control System

IPAD Integrated Program for Aerospace Vehicle Design

IPAI Information Processing Association of Israel; International Primary Aluminum Institute

IPAR Initial Product Assessment Report; Institute of Personality Assessment and Research

IPARA International Publishers Advertising Representatives Association

IPARS International Programmed Airline Reservation System

IPAT Institute for Personality and Ability Testing

IPATA Independent Pet and Animal Transportation Association

ipb illustrated parts breakdown

IPB International Peace Bureau; Islamic Party of Britain

ip & be initial program and budget estimate

ipbm (IPBM) interplanetary ballistic missile

IPBMM International Permanent Bureau of Motor Manufacturers

ipc industrial process control; illustrated parts catalog; innings pitched corrector; in-

ter-personal communication (between jailed inmates); inter-process communication; isopropyl carbanilate

IPC Illinois Power Company; Industrial Process Control; Industrial Property Committee; Institute of Paper Chemistry; Institute of Pastoral Care; Institute of Printed Circuits; Integrated Programme for Commodities; Intelligence Priorities Committee (CIA); Inter-African Phytosanitary Commission; International Pacific Corporation; International Packings Corporation; International Paper Chemists; International Petroleum Company; International Polar Commission; International Poplar Commission; Iraq Petroleum Company; Isopropyl Carbanilate

IPCA Industrial Pest Control Association; International Petroleum Credit Association

IPCAIL International Pacific Corporation Australian Investments Limited

IPCC Intergovernmental Panel on Climate Change (UN)

ipce independent parametric cost estimate

IPCEA Insulated Power Cable Engineers Association

IPCI International Potato Chip Institute

IPCL *Institut du Pétrole des Carburants et Lubrifiants* (French—Petroleum Institute for Motor Fuel and Lubricants)

IPCO International Paper Company

IPCPA Institute of Private Clinical Psychologists of Australia

IPCR Immediate Past Chief Ranger; Institute of Physical and Chemical Research

IPCS Institution of Professional Civil Servants; Integrated Propulsion Control System; International Peace Corps Secretariat

IPCU Intensive Psychiatric Care Unit

ip cyl intermediate-pressure cylinder

ipd individual package delivery; insertion phase delay

IPD Institute for Professional Development; Institute of Professional Designers

IPD *In Praesentia Dominorum* (Latin—In the Presence of the Lords)

IPDA International Periodical Distributor's Association

IPDC International Program for the Development of Communication

I pd cash I paid cash

ipe industrial plant equipment; interpret parity error

IPE Institution of Plant Engineers; Institute of Production Engineers

IPE *Instituto de Providéncia do Estado* (Portuguese—State Loan Institute); *International Petroleum Encyclopedia*

IPEC International Petroleum Exploration Company; Interstate Parcel Express Company

ipecac ipecacuanha

IPEU International Photo Engravers' Union

ipf idiopathic pulmonary fibrosis; initial production facilities

IPF Irish Printing Federation

IPFA Institute of Public Finance Accountants

IPFC Indo-Pacific Fisheries Council

IPFF International Planned Parenthood Federation

ipfm impact form; integral pulse frequency modulation

ipg immediate participation guarantee (insurance plan)

IPG Income Property Group; Independent Publishers' Group; Information Policy Group

IPGA Illinois Personnel and Guidance Association; Iowa Personnel and Guidance Association

IPGCU International Printing and Graphic Communications Union

IPGH *Instituto Panamericano de Geografía e Historia* (Spanish—Pan-American Institute of Geography and History)

iph impressions per hour; inches per hour; interphalangeal

IPHC International Pacific Halibut Commission

IPHE Institute of Public Health Engineers

ipi individually planned instruction; intelligent peripheral interface; interior point intermodal

i.p.i. in partibus infidelium (Latin—in the region of unbelievers)

IPI Industrial Production Index; Institute of Poultry Industries; International Patent Institute; International Press Institute; Interpositional Implant

IPI Intelligence Publications Index (DIA)

IPICS Initial Production and Information Control System

IPIECA International Petroleum Industry Environmental Conservation Association

IPIP Information Processing Improvement Program

IPIR Initial Photographic Interpretation Report (USAF); Institute for Public Interest Representation; Integrated Personnel Information Report(s)

IPISD Interservice Procedures for Instructional Systems Development

IPKF Indian Peace-Keeping Force

IPKI *Ikatan Pendukung Kemerdekaan Indonesia* (Indonesian Malay—Upholders of Indonesia's Independence)

ipl information program loading; initial program load(er)

ipl (IPL) information processing language

IPL Illustrated Parts List; Integrated Parts List; Italian Pacific Line

I Plant Eng Institution of Plant Engineers

IPLE Institute for Political/Legal Education

IPLGY Institute for the Protection of Lesbian and Gay Youth

ip log(ging) induced polarization log(ging)

ipm impulses per minute; inches per minute; inches per month; incidental phase modulation; interruptions per minute; inventory policy model

ipm (IPM) integrated pest management

IPM Institute of Personnel Management; Institute of Police Management; Institute for Police Management; Integrated Post Management

IPMA In-Plant Management Association; International Personnel Management Association; International Planned Music Association

ipmin inches per minute

IPMP Industrial Plant Modernization Program

ipms impact predictor monitor set

IPMS International Polar Motion Service; Inter-Personal Messaging Service

ipn inspection progress notification; integrated provider network

ipo initial public offer(ing)(s); instant profit opportunity

ipo (IPO) input processing output

IPO Installation Production Order; International Projects Office (NATO); Israel Philharmonic Orchestra

IPO Instituut voor Perceptie Onderzoek (Dutch—Institute for Perception Research)

ipod initial phase of ocean drilling

IPOD Interstate Project on Dissemination

IPOEE Institution of Post Office Electrical Engineers

IPOH Instituto Panamericano de Geografía e Historia (Spanish—Panamerican Institute of Geography and History)

ipos intellectual property owners

IPOT Imperial Philharmonic Orchestra of Tokyo

ipp imaging photopolarimeter; impact prediction point; india paper proof(s); intrapleural pressure

Ipp Ippolito

IPP Immediate Past President; Islamic People's Party; Ivan Petrovich Pavlov

ippa inspection, palpitation, percussion, auscultation

IPPA International Pentecostal Press Association; International Planned Parenthood Association

IPPAU International Printing Pressmen and Assistants' Union

ippb intermittent positive-pressure breathing

ippb/i intermittent positive pressure breathing/inspiratory

IPPF International Penal and Penitentiary Foundation

IPPJ Institute of Plasma Physics—Japan

IPPL International Primate Protection League

IPPNW International Physicians for the Prevention of Nuclear War

IPPP Industrial Property Policy Program

IPPPE Institute on Public Policy and Private Enterprise

ippr intermittent positive pressure respiration

IPPTA Indian Pulp and Paper Technical Association

IPPTT *Internationale du Personnel des Postes, Télégraphes et Téléphones* (French—International Postal, Telegraph, and Telephone Personnel)

ippv intermittent positive pressure ventilation

ipq intimacy potential quotient

IPQ International Petroleum Quarterly

ipr inches per revolution; inflow performance relationship (oil well); interpersonal process recall

IPR Individual Pay Record; Industrial Public Relations; Institute of Pacific Relations; Institute for Philosophical Research; Institute for Public Representation

IPR International Public Relations

IPRA International Professional Rodeo Association; International Public Relations Association

IPRC Institute of Puerto Rican Culture

IPRE Incorporated Practitioners in Radio and Electronics

IPRO International Patent Research Office

I-PRO Independent Professional Representatives Organization

I Prod Eng Institute of Production Engineers

IPRR Integrated Personnel Requirement Report

ips inches per second; interceptor pilot simulator; interruptions per second; iron pipe size

Ips Ipswich

IPS Incremental Purchasing System; Independent Preparer Services Incorporated; Indian Police Service; Industrial Planning Specification; Industrial Promotion Service; Inertial Positioning System; Information Processing Society; Institute of Pacific Studies; Institute for Policy Studies; Institute of Population Studies (Japan); Institute of Private Secretaries; Institute of Public Safety; Institute of Public Supplies; Integrated Publishing System; Intensive Probation Supervision; International Phenomenological Society; International Pipe Standard; International Preview Society; Interpretive Programming System; Introductory Physical Science; Ionospheric Prediction Service; Israel Prison Service

IPS *Instituto Poligrafico dello Stato* (Italian—State Printing and Stationery office)

IPSA Independent Passenger Steamship Association; Independent Pool Service Association; Independent Postal System of America; International Political Science Association

IPSB Institute of Psycho-Structural Balancing

IPSC International Pacific Salmon Committee

IPSCE Inventory of Psychic and Somatic Complaints in the Elderly

IPSF International Pharmacy Students Federation; International Piano Symphony Foundation

IPSFC International Pacific Salmon Fisheries Commission

IPSJ Information Processing Society of Japan

IPSM Institute of Purchasing and Supply Management

ipsp inhibitory postsynaptic potential

IPSP Industrial Personnel Security Program

IPSS International Packet Switching Service; International Pilot Study of Schizophrenia

IPSSB International Processing Systems Standards Board

IPSSG International Printers Supply Salesmen's Guild

ipt indexed, paged, titled; internal pipe thread

IPT Initial Production Test (USA); Institute of Photographic Technology; International Planning Team (NATO); Interpersonal Psychotherapy

IPT *Instituto Panameño de Turismo* (Spanish—Panamanian Institute of Tourism)

IPTA International Patent and Trademark Association

IPTCCS Integrated Pipeline Transportation and Coal-Cleaning System

iptg isopropylthiogalactoside

ipth (IPTH) immunoreactive parathyroid hormone

ipto independent power takeoff

IPTPA International Professional Tennis Players' Association

ipts international practical temperature scale

IPTS Improved Programmer Test Section; International Practical Temperature Scale

IPTT *Internationale du Personnel des Postes, Télégraphes et Téléphones* (French—International Postal, Telegraph, and Telephone Personnel)

iptv intelligent personal television; interactive personal television

Iptv Internet protocol television

ipu input preparation unit

IPU Institute for Public Understanding; International Paleontological Union; Inter-Parliamentary Union

ipv inactivated poliomyelitis vaccine; infectious pustular vaginitis; infectious pustular vulvovaginitis; integrated pressure vessel

ipv *in plaats van* (Dutch—in place of)

IPVRA International Professional Vinyl Repair Association

IPW interrogation prisoner of war

ipy inches penetration per year; inches per year

IPY International Polar Year

IPZ Investment Promotion Zone

IPZE *Instituto Poligrafico dello Stato* (Italian—State Poligraphic Institute)—issues paper money and stamps

i.q. *idem quod* (Latin—the same as)

Iq Iraq

IQ Import Quota; intelligence quotient; Intelligent Quattro (European project for intelligent hydraulically operated vehicles); Iraq (Internet code)

I.Q. I Quit (smoking)

I of Q Institute of Quarrying

IQA Institute of Quality Assurance

IQCA Irish Quality Control Association

IQCT Institute for Quality Control Training

i.q.e.d. *id quad erat demonstrandum* (Latin—that which was to be proved)

iqf individually quick frozen; instant quick frozen

IQHL Institute for Quality in Human Life

IQI Instructional Quality Inventory

I Qk interrupted quick flashing (light); interrupted quick (light)

iqmf image-quality merit function

IQP Institute for Quality and Productivity

iqr interquartile range

iqrp interactive query and report processor

iq & s iron, quinine, and strychnine

IQS Institute of Quality Surveyors; Institute of Quantity Surveyors

IQSA Institute of Quantity Surveyors of Australia

IQSY International Quiet Sun Year (1964–1965)

Iqu Iquique

ir ice on runway; incidence rate; industry remarketer; information retrieval; infrared; inherited runners (baseball); inland revenue; inside radius; inside right; instantaneous relay; instrument rating; instrument reading; insulation resistance; intelligence resource(s); internal resistance; interrogator responder; investor relations

ir (IR) inflation rate

i-r infra-red

i/r interchangeability and replaceability

i & r information and retrieval; intelligence and reconnaissance; interchangeability and replaceability

ir irisch (German—Irish)

i R im Ruhestand (German—in retirement)

Ir Iran (whose capital is Teheran); Irania; Ireland; iridium; Irish

Ir Iers [Dutch—Ireland, Irish(man) (woman)]

IR Index Register; Indian Reservation; Indian Reserve; Industrial Relations; Informal Report; Information Request; Inspection Rejection; Inspector's Report; Institutional Research; Intelligence Report; Internal Revenue; International Relations; International Representative; Invention Report; Investigation Record; Iran (Internet code)

I-R Ingersoll-Rand

I & R Initiative and Referendum; Intelligence and Reconnaissance

ira independent retirement account (IRA)

ira (IRA) immunoregulatory alpha globulin

Ira Dr Ira D Levine; Iraq

IRA Indian Reorganization Act; Indian Rights Association; Individual Retirement Account; Individual Retirement Arrangement; Institute of Registered Architects; Intercollegiate Rowing Association; International Racquetball Association; International Reading Association; International Recreation Association; International Reprographics Association; Iranian Airways; Irish Republican Army; Israel Railway Administration

IRAB Institute for Research in Animal Behavior

irac mnemonic abbreviation helpful in making legal or logical presentations; letters stand for issue, rule, application, and conclusion

IRAC Indochina Refugee Assistance Program; Industrial Relations Advisory Committee; Intelligence Resources Advisory Committee; Interdepartmental Radio Advisory Committee; Interfraternity Research and Administrative Council

iracq instrumentation radar acquisition

irad independent research and development

IRAD Institute for Research on Animal Diseases

iraf image reduction and analysis facility; interferogram requirements and analysis funnel

iran inspect and repair as necessary

Iran formerly Persia; Imperial Government of Iran (ancient Asian nation), *Keshваré Shahanshahiyé Iran*

IRAN Individual Retirement Annuity

IRANAIR Iran National Airlines

Iran(ian) Persia(n)

IRANOR Instituto Nacional de Racionalización y Normalización (Spanish—National Institute of Rationalization and Standards)

IRAP Indochinese Refugees Assistance Program; Industrial Research Assistance Program

Iraq Republic of Iraq (Persian Gulf country formerly called Mesopotamia), *al Jumhouriya al 'Iraqia*

Iraquia Iraq (Mesopotamia)

IR/AR Inspector's Report/Action Request

iras (IRAS) infrared astronomical satellite; infrared-measuring astronomical satellite

IRAs Individual Retirement Accounts; Irish Republican Army members or supporters and sympathizers

IRASA International Radio Air Safety Association

IRASE Institute of Refrigeration and Air Conditioning Service Engineers

iraser infrared amplification by stimulated emission of radiation

irb (IRB) in-shore rescue boat

IRB Indiana Rating Bureau; Individual Retirement Bond; Industrial Relations Bureau; Industrial Revenue Bond; Industrial Review Board; Institutional Review Board; Insurance Rating Board; International Resources Bank; Irish Republican Brotherhood

IRBDC Insurance Rating Bureau of the District of Columbia

IRBEL Indexed References to Biomedical Engineering Literature

irbm (IRBM) intermediate range ballistic missile

irc infrared countermeasures; international reply coupon; item responsibility code

IRC Immigration Restriction Council; Indonesian Red Cross; Industrial Recreation Council; Industrial Relations Center; Industrial Relations Committee; Industrial Relations Counselors; Industrial Relations Council; Industrial Reorganization Corporation; Inebriate Reception Center; Information Resources Center; Institutional Research Council; Instructional Resources Center; Internal Revenue Code; International Radiation Commission; International Railways of Central America (stock exchange symbol); International Rainwear Council; International Red Cross; International Relations Committee; International Relief Committee; International Rescue Committee; International Research Council; International Resistance Company; International Rice Commission; Internet Relay Chat

IRCA Immigration Reform and Control Act; International Railway Congress Association; International Railways of Central America; International Remodeling Contractors Association

IRCAM *Institut de Recherche et de Coordination Acoustique-Musique* (French—Institute of Research and Acoustic-Music Coordination)

IRCAR International Reference Center for Abroation Research

irccd infrared charge-coupled device

IRCD Information Retrieval Center on the Disadvantaged

i-r charts infrared correlation charts

ircm infrared countermeasures

IRCO Industrial Rustproof Company

IRCP International Commission on Radiological Protection

IRCs Inebriate Reception Centers (where nonviolent abusers of alcohol and other drugs accept coffee and counseling in lieu of being jailed)—pilot facility in San Diego, California

IRCS International Research Communications System

ird income in respect of a decedent

ird (IRD) internal research and development; international resource development

ir & d independent research and development; international research and development

IRD Institute on Religion and Democracy (anti-communist); Institute of Reading Development; International Resource Development

IRD *Instituto Rubin Dario*

IR & D International Research and Development

IRDA Industrial Research and Development Authority

IRDC International Research Development Centre (Canadian)

ir & dg industrial research and development grants

IRDI Indian Resources Development and Internship

irdm illuminated runway distance marker

IRDN Illinois Resource and Dissemination Network

IRDOE Institute for Research and Development in Occupational Education

irdome infrared dome

irds idiopathic respiratory-distress syndrome

irdu infrared detection unit

Ire Ireland

IrE Irish English

IRE Institute of Radio Engineers; Institute for Responsive Education; Investigative Reporters and Editors

IREC International Real Estate Corporation (Singapore); Irrigation Research and Extension Commission

IREE Institute of Radio and Electronic Engineers

IREF International Real Estate Federation

IREI International Real Estate Institute

Ireland Irish Republic (North Atlantic island nation), *Eire*

IREM Institute of Real Estate Management

irene indicating random electronic numbering equipment

Ir Eng Irish English

IREQ Institute of Research—Quebec

irer infrared extra rapid

IREX International Research and Exchanges (New York-based board arranging scholarly interchanges between the United States and other countries

irf instrument reliability factor; interrogation repetition frequency

Irf radio-frequency current (symbol)

IRF International Road Federation; International Rowing Federation

IRFAA International Rescue and First Aid Association

IRFB *Internationaler Regenmantelfabrikantenverband* (German—International Rainwear Fabric Council)

IRFC Ingersoll-Rand Finance Corporation

IRFM *Industrias Reunidas Francisco Matarazzo* (Spanish—Francisco Matarazzo's Reunited Industries)

IRFU Iriah Rugby Football Union

irg interrecord gap

IRG Interdepartmental Regional Group

IRG *Internationale des Résistans á la Guerre* (French—War Resisters' International)

Ir Gael Irish Gaelic

IRGDLP International Research Group on Drug Legislation and Programs (Geneva, Switzerland)

irgl immunoreactive glucagon

IRGRD International Research Group on Refuse Disposal

IRH *internationalen de Roten Hilfe* (German—International Red Aid)—Red Fighting Fund

irha injured as a result of hostile action

irhd international rubber hardness degrees

IRHIS Intelligent Adaptive Information Retrieval System on Hospital Information System

iri immunoreactive insulin

Iri Irina

IRI Industrial Reconstruction Institute; Industrial Research Institute; Information Resources Inc.; Institute of the Rubber Industry; Islamic Republic of Iran

IRIA Infrared Information and Analysis

IRIC Inter-Regional Insurance Conference

IRICA Industrial Research Institute for Central America

iricbm (IRICBM) intermediate-range intercontinental ballistic missile

irid iridescent

IRIG Inter-Range Instrumentation Group

IRIR Interchangeability Replaceability Information Report

iris infrared interferometer spectrometer

IRIS Industrial Research and Information Service; Infrared Information Symposia; Integrated Reconaissance Intelligence System

Irish Irish Gaelic

Irish FP Irish Fishing Port (registration symbols displayed on the bows of fishing vessels)

Irish VR Irish Vehicle Registration (symbols on automotive vehicle licenses)

IRJC Indian River Junior College

irl information retrieval language

Irl Ireland; *Irlanda* (Italian, Portuguese, Spanish—Ireland)

Irl *Irlande* (French—Ireland)

IRL Illustrations Requirements List

IRLA Independent Research Library Association

Irland Irlandais (French—Irish)

IRLC Illinois Regional Library Council

IRLCS International Red Locust Control Service

IRLI Immigration Reform Law Institute

IRLS Interrogation Recording Location System

irm information resource management; infrared measurement; innate release mechanism; intermediate range monitor

IRM Improved Risk Mutuals; Information Resource Management; Islamic Republic of Mauritania

IRM (NYU) Institute of Rehabilitation Medicine (New York University)

irma information revision and manuscript assembly; integrated revenue and marketing applications

IRMA Individual Reverse Mortgage Account

IRMMH Institute of Research into Mental and Multiple Handicaps

IRMP Intermountain Regional Medical Program

IRMPC Industrial Raw Materials Planning Committee (NATO)

IRMRA Indian Rubber Manufacturers Research Association

IRMS Imperial Russian Musical Society; Information Retrieval and Management System

IRN Independent Radio News

IRNA Islamic Republic News Agency

IRNP Isle Royale National Park (Michigan)

iro in rear of

IRO Industrial Recycling Organization; Industrial Relations Office(r); Information Resource Office(r); Inland Revenue Office(r); Internal Revenue

Officer(r); International Refugee Organization; International Relief Organization

IRO-ALA International Relations Office-American Library Association

IROC International Race of Champions (autos)

irod instantaneous readout detector

IRODP International Registry of Organization Development Professionals

iron. ironic(al)

Iron Curtain barrier between eastern and western Europe (1946–1989)

iron pyrites sulfide of iron

iron rust iron oxide

IROPCO Iranian Offshore Petroleum Company

iros ipsilateral routing of signal

IROS Increase Reliability of Operational Systems

irp inherited runners scoring percentage (baseball); initial receiving point

ir£ Irish pound

IRP Individualized Reading Program; Information Reporting Program; Information Resources Press

IRPA International Radiation Protection Association

irpm individual risk premium modification

IRPS Institute of Reconstructive Plastic Surgery (NYU); International Religious Press Service (Vatican City)

IRPT Inland Rivers, Ports and Terminals

irq interrupt request

Irq Iraq (whose capital is Baghdad)

irr infrared rays; infrared reflectance; interest rate return; internal rate of return; irredeemable; irregular(ity)

irr (IRR) integral rocket ramjet; internal rate of return

IRR Individual Ready Reserves; Institute of Race Relations

IRRA Industrial Relations Research Association

IRRC Investor Responsibility Research Center

irrd international road research documentation

IRRD International Road Research Documentation

IRRDB International Rubber Research and Development Board

irreg irregular

irres irrespective

irrev irrevocable

irrg irregular

irrgty irregularity

irrgy irregularly

IRRI International Rice Research Institute

irrig irrigation

IRRN Illinois Research and Reference Center (libraries)

IRRP Icefield Ranges Research Project

irr/ssm (IRR/SSM) integral rocket ramjet/surface-to-surface missile

IRRT International Relations Round Table

irs incremental range summary; independent rear suspension; inherited runners scoring (baseball)

IRs Inspector's Reports

IRs *Irish Reports*

IRS Indian Register of Shipping; Industrial Rubber Sales; Ineligible Reserve Section; Information Retrieval System; Interface Requirements Specification; Internal Revenue Service; International Recruiting Service; International Referral System; International Rorschach Society; Irrigation Research Station

I & RS Information and Research Services

IRS-1A Indian-built remote-sensing satellite

IRSA Industrial Relations Society of Australia

IRSE Institution of Railway Signal Engineers; International Reactor Safety Evaluation

IRSF Inland Revenue Staff Federation

IRSG International Rubber Study Group

IRSID *Institut des Recherches de la Sidérurgie Française* (French—French Steel Research Institute)

IRSNB *Institut Royal des Sciences Naturelles de Belgique* (French—Royal Belgian Institute of Natural Sciences)

IRSP Irish Republican Socialist Party

IRSS Instrumentation Range Safety System

IRST Infrared Search and Track

irt infrared temperature; infrared tracker; integrated retail terminal; intermediate range technology; interresponse time; interrogator responder transponder

IRT Institute for Rapid Transit; Institute of Reprographic Technology; Interborough Rapid Transit (subway system)

IRT Industria Radio y Television (Spanish—Radio and Television Industry)

IRTA Illinois Retired Teachers Association; International Reciprocal Trade Association

IRTAC International Round Table for the Advancement of Counseling

IRT Corp Instrumentation/ Research/Technology Corporation

IRTE Institute of Road Transport Engineers

IRTP Integrated Reliability Test Program

irts infrared target seeker

IRTS International Radio and Television Society

IRTU International Railway Temperance Union

IRTWG Inter-Range Telemetering Working Group

iru industrial rehabilitation unit; inertial reference unit; infrared unit; international radium unit; international rat unit

IRU International Road Transport Union

irupt interrupt

iruptd interrupted

iruptn interruption

iruptng interrupting

irv inspiratory reserve volume

Irv Irvin; Irvine; Irving; Irwin

Irve Irving

Irving Irving Trust Company; Sir Henry Irving; Washington Irving

Irw Irwin

IRW Iowa Reformatory for Women

IRWC International Registry of World Citizens

irwl interchangeability/replaceability working list

is his; ingot sheet; integrally stiffened; intercoastal space; interim storage; internal shield; international status; interval signal; island; isle

i&s installation and service; investigation and suspension; iron and steel

i & s inspection and security; inspection and survey; interchangeability and substitutability

Is Islam; Islamic; Island; Isle; Israel; Israeli; Israeli pound; screen current (symbol); source current (symbol)

Is *Isaías* (Spanish—Isaiah)

I(s) *isla(s)* Spanish—island(s)

Iˢ *Îles* (French—islands); *Ilhas* (Portuguese—islands); *Islas* (Spanish—islands)

IS Iceland (internet code); Identification Section (NY-PD); Igor Stravinsky; Independent Seminar; Independent Study; Index of Sexuality; Indian Summer (freeboard marking); Information Services; Information Systems; Instruction(s) to Ship; insertion sequences; Interim Study; Irish Society

I of S Institute of Sound; Isle of Skye

IS 201 Intermediate School 201 (for example)

isa international standard atmosphere

Isa Isaiah, The Book of the Prophet

Isa Indonesia (Spanish—Indonesia); *Isaiah*

ISA Incest Survivors' Anonymous; Independent Showmen of America; Industry Standard Architecture; Institute of Strategic Affairs; Institution of Surveyors—Australia; Instruction Set Architecture; Instrument Society of America; Insulating Siding Association; Intelligence Support Activity (US Army); Intermediate Service Agency; Internal Security Act; International Schools Association; International Scientific Affairs; International Seabed Authority; International Security Affairs; International Security Assis-

tance; International Service Agencies; International Shipmasters Association; International Sign Association; International Silk Association International Society of Appraisers; International Silo Association; International Society of Aboriculture; International Sociological Association; International Standards Association; International Student Association; International Sugar Agreement; Israel Space Agency

ISA Information Science Ab-. stracts; Irregular Serials and Annuals

ISAA Institute of Shops Acts Administration

Isab Isabella

ISAB Institute for the Study of Animal Behavior

ISAB Internationaler Studenbund (German—International Student Society)

ISABPS Integrated Submarine Automated Broadcast Processing System

ISAC International Security Affairs Committtee

ISACP Italian Society of Authors, Composers, and Publishers

ISAD Information Science and Automation Division (ALA)

ISADPM International Society for the Abolition of Data-Processing Machines

ISAE Internacia Scienca Asocio Esperantista (International Esperantist Scientific Association)

IsAF Israeli Air Force

isaf black intermediate super-abrasive furnace black

ISAFP Intelligence Service of the Philippine Armed Forces

ISAGA International Simulation and Gaming Association

ISAHM International Society for Animal and Human Mycology

ISALPA Incorporated Society of Auctioneers and Landed Property Agents

ISAM Independent School Association of Massachusetts; Indexed Sequential Access Method; Institute for Studies in American Music; Integrated Switching and Multiplexing

ISAP Institute for the Study of Animal Problems

ISAPC Incorporated Society of Authors, Playwrights, and Composers

ISAPM International Shipowners Association of Peninsular Malaysia

isar information storage and retrieval

ISAR International Society for Astrological Research

isarc installation shipping and receiving capability

ISAS Industrial Sales and Service; Institute of Social and Administrative Studies; Institute of Space and Aeronautical Science; Isotopic Source Assay System

ISAT International Society of Analytical Trilogy

ISAW International Society of Aviation Writers

isb independent sideband; intermediate sideband

ISB International Society of Bassists; International Society of Biomechanics; International Society of Biometeorology

ISBA Incorporated Society of British Advertisers; Indiana School Boards Association

ISBB International Society for Bioclimatology and Biometereology

ISBD International Standard Book Description

ISBD(M) International Standard Bibliographic Description for Monographic Publications

ISBD(NBM) International Standard Bibliographic Description for Non-Book Materials

ISBD(S) International Standard Bibliographic Description for Serial Publications

ISBE International Society for Business Education

ISBN International Standard Book Number

ISBNA International Standard Book Numbering Agency

ISBOA Idaho School Business Officials Association

ISBP International Society for Biochemical Pharmacology

isbr interior salt basin region

ISBS Icelandic State Broadcasting Service; International Scholarly Book Services

isc intermediate slack compensation; interstate commerce; intrasite cabling; item status code; item status coding; in such case

ISC Icelandic Steamship Company; Idaho State College; Imperial Service College; Imperial Staff College; Indian Staff Corps; Indiana State College; Indoor Sports Club; Industrial Security Commission; Institute for the Study of Conflict; Inter-American Society of Cardiology; International Salt Company; International Science Center; International Sericultural Commission; International Society of Cardiology; International Softball Congress; International Statistical Classification; International Sugar Council; International Supreme Council (World Masons); Interseas Shipping Corporation; Interservice Sports Council; Interstate Sanitation Commission

IS&C International Systems and Controls

ISCA Industrial Specialty Chemical Association; International Sailing Craft Association; International Senior Citizens Association

iscan inertialess steerable communication antenna

ISCB International Society for Cell Biology

ISCC Inter-Society Color Council

ISCDD International Scheme for the Coordination of Dairy Development

ISCE International Society for Christian Endeavor

ISCEBS International Society of Certified Employee Benefit Specialists

ISCED International Standard Classification of Education

ISCEH International Society for Clinical and Experimental Hypnosis

ISCERG International Society for Clinical Electroretinography

ISCET International Society of Certified Electronics Technicians

ISCII International Standard Code for Information Interchange

ISCM International Society for Contemporary Music

ISCO Independent Schools Careers Organization; International Standard Classification of Occupations

ISCOR Iron and Steel Industrial Corporation (South Africa)

ISCOS Institute for Security and Cooperation in Outer Space

ISCP International Society of Clinical Pathology

ISCPET Illinois Statewide Curriculum Study Center in the Preparation of Secondary School English Teachers

ISCRP International Society of City and Regional Planners

ISCS Information Service Computer System; Intermediate Science Curriculum Study; International Society of Communications Specialists

ISCTP International Study Commission for Traffic Police

isd induction system deposit; inhibited sexual desire; installation start date; instructional systems development; integrated symbolic debugger

isd (ISD) international standard depth

ISD Independent School District; Information Systems Design; Institute for the Study of Diplomacy, Georgetown University; Instructional Systems Design; Instructional Systems Development; Intermediate School District; Internal Security Department; Internal Security Division (U.S. Dept of Justice); International Subscriber Dialing

ISDA International Security and Detective Alliance

ISDAIC International Staff Disaster Assistance Information Coordinator (NATO)

ISDD Institute for the Study of Drug Dependence (London, England)

ISDI International Social Development Institute

ISDN Integrated Services Digital Network

ISDO International Staff Duty Officer (NATO)

ISDRA International Sled Dog Racing Association

ISDS Inadvertent Separation Destruct System; International Serials Data System; International Sheep Dog Society

ISDSI Insulated Steel Door Systems Institute

ISDS/IC International Center of the International Series Data System (UNESCO)

ise integral square error

ISE Indian Service of Engineers; Institute for Services to Education; Institute for Sex Education; Institute of Social Ethics; Institute of Space Engineering; Institution of Sanitary Engineers; Institution of Structural Engineers; International Stock Exchange; Irish School of Ecumenics

ISE Instituto de Seguros del Estado (Spanish—Institute of State Insurance)

ISEA Industrial Safety Equipment Association; Iowa State Education Association

ISEAS Institute of Southeast Asian Studies

ISEE International Sun-Earth Explorer (NASA/ESRO)

ISEEP Infrared-Sensitive Element Evaluation Program

ISEF International Science and Engineering Fair

ISEK International Society of Electrophysiological Kinesiology

ISELS Institute of Society, Ethics, and Life Sciences

ISEP Instructional Scientific Equipment Program; International Summer Education Project; International Society for Educational Planners; Interservice Experiments Program

isepc installation specification; insulation specification

iseq input sequence check(ing)

ISES International Ship Electric Service Association; International Society of Explosives Specialists; International Solar Energy Society; International Special Events Society

ISETU International Secretariat of Entertainment Trade Unions

ISEU International Stereotypers' and Electrotypers' Union

isf intermittent storage flood(ing); interstitial fluid

ISF Industrial Space Facility; Intermediate-Scale Facilities; International Science Foundation; International Shipping Federation; International Society for Fat Research; International Society of Financiers; International Softball Federation; International Solidarity Fund; International Spiritualist Federation

ISFA Institute of Shipping and Forwarding Agents; Intercoastal Steamship Freight Association; International Scientific Film Association

ISFAA International Society of Fine Arts Appraisers

ISFL International Scientific Film Library

ISFMS Indexed Sequential File Management System

ISFR Institute for the Study of Fatigue and Reliability

ISFSC International Society of Free Space Colonizers

isfsi independent spent-fuel storage installation

ISFSI International Society of Fire Service Instructors

isg imperial standard gallon

ISGE International Society of Gastroenterology

ISGM Isabella Stewart Gardner Museum

ISGS International Society for General Semantics

ISGW International Society of Girl Watchers

Ish Isham; Ishbel; Ishmael

ISH International Society of Hematology

ISHAM International Society for Human and Animal Mycology

ISHI Institute for the Study of Human Issues

ISHL Illinois Social Hygiene League

ISHR International Society for Human Rights

ISHS International Society for Horticultural Science

isi indeterminate short internode; industrial standard item; internally specified index(ing); interstimulus interval

ISI Indian Standards Institution; Information Sciences Institute (USC); Institute for Scientific Information; Intercollegiate Society of Individualists; Intercollegiate Studies Institute; International Statistical Institute; Iron and Steel Institute

ISIB Inter-Services Ionospheric Bureau

isic (ISIC) immediate superior in command

ISIC International Standard Industrial Classification; International Student Identity Card

ISID International Society of Interior Designers

ISIM International Society of Internal Medicine

isip inertial system indication position

ISIP Iron and Steel Industry Profile Service

ISIR International Society for Invertebrate Reproduction

isirta I'm sorry, I'll read that again (BBC comedy)

isis (ISIS) ionospheric studies

ISIS Independent Schools Information Service; Individualized Science Instructional System; Instant Sales Indicator System; Institute of Scrap Iron and Steel; Integral Spar Inspection System; Integrated Scientific Information Service; Integrated Set of Information Systems; Integrated Ship Instrumentation System; Integrated Statistical Information Service; International Satellites for Ionospheric Studies; International Science Information Service; International Shipping Information Services; International Society of Introduction Services; International Species Identification System; Investigative Support Information System (FBI)

ISIT Institute for Studies in International Terrorism (SUNY)

ISIYM International Society of Industrial Yarn Manufacturers

ISJP International Society for Japanese Philately

isk insert storage key

ISK Isambard Kingdom Brunel

ISK *Internationale Seidenbau Kommission* (German—International Sericulture Commission)

ISKC International Society for Krishna Consciousness

ISKON International Society of Krishna Consciousness

iskr identification, station keeping, and rendezvous

isl intermediate-stage letter; island

isl (ISL) intersatellite link

isl *islandsk* (Dano-Norwegian—Icelandic); *isländisch* (German—Icelandic)

Isl *island* (Norwegian—Iceland); *Islanda* (Italian—Iceland); *Islandia* (Spanish—Iceland); *Islândia* (Portuguese—Iceland)

ISL Iceland Steamship Company; Interseas Shipping Lines; Iranian Shipping Lines; Irish Shipping Limited

I S-L Instructor Sub-Lieutenant

ISLA International Survey Library Association

Isle Royale Isle Royale National Park, Michigan

ISLFD Incorporated Society of London Fashion Designers

ISLIC Israel Society of Special Libraries and Information Centers

isl of Lan islands of Langerhans

ISLLSS International Society for Labor Law and Social Security

Islm Rep Pak Islamic Republic of Pakistan

isln isolation

ISLRS Inactive Status List Reserve Section

isl's initial stock lists; islands

ISLSCP International Satellite Land Surface Climatology Program

isls L islands of Langerhans

Islw Indian spring low water

ISLWF International Shoe and Leather Workers Federation

ism industrial, scientific, medical wave length; interpretive structural modeling; interstellar matter

ism (Latin suffix—condition or state)—capitalism, communism, rheumatism

ISM Imperial Service Medal; Incorporated Society of Musicians; Industrial Sugar Mills; Institute of Sales and Marketing; Institute of Sports Medicine; Institute of Supervisory Management; International Society for Musicology

ISMA International Superphosphate Manufacturers Association

ISMAP Integrated System for the Management of Agricultural Production

ISMDA Independent Sewing Machine Dealers Association

ISME International Society for Musical Education

ISMEC Information Service in Mechanical Engineering

ISMH International Society of Medical Hydrology

ISMI Institute for the Study of Mental Images

ISMLS Interim Standard Microwave Landing System

ISMR Independent Snowmobile Medical Research (organization)

ISMRC Inter-Services Metallurgical Research Council

ISMS Inherently-Safe Mining System(s)

ISMUN International Student Movement for the United Nations

Is N (Sir) Isaac Newton

ISN International Society for Neurochemistry

ISNAC Inactive Ships Navy Custody

ISNP International Society of Naturopathic Physicians

isn't is not

iso incentive stock option; isolate; isolation; isolator; isotope; isotopic; in spite of

iso (ISO) infrared space observatory

iso (Latin suffix—equal or like)—isometric, isotonic

iso. isolated (camera)

ISO Imperial Service Order; Indianapolis Symphony Orchestra; Individual System Operation; Information Service(s) Office(r); Information Systems Office (Library of Congress); Insurance Ser-

vice(s) Office(r); International Science Organization; International Standardization Organization; International Standards Organization; International Sugar Organization; Irish Symphony Orchestra

ISO-4 international code for the abbreviation of periodical titles

ISO-833 international list of periodical title word abbreviations

isobu isobutyl

is/oc individual system/organization cost

ISOC Internal Security Operation Command; Internet Society

isochr isochronal

ISODOC International Center for Standards in Information and Documentation

isogone isogonal line

isogons isogonic lines

isol isolate; isolation

Isol Isolation; Isolator

isoln isolate; isolated; isolation

isolr isolationer

isom isometric(s)

ISOMATA Idlewild School of Music and the Arts; Idyllwild School of Music and Art

Isomorph isomorphic(al)(ly); isomorphism

ison isolation network

ISONET International Standards Organization Network

ISOO Information Security Oversight Office

isordil isorbide dinitrate

ISORID International Information System on Research in Documentation

ISORT Interdisciplinary Student-Originated Research Training (NSF)

ISOS International Ship Operating Services

isot isotropic

iso wd isolation ward

isp intraspinal

isp (ISP) internet service provider

Isp specific impulse (symbol)

ISP Idaho State Penitentiary; Imperial Smelting Process; Index of Social Position; Indiana State Police; Industrial Security Program; Institute of Social Psychiatry; Institute of Store

Planners; Integrated Support Plan(ning); Interamerican Society of Psychology

ISP Internationale des Services Publics (French—Public Services International)

ISPA International Screen Publicity Association; International Sleep Products Association; International Small Printers' Association; International Society for the Protection of Animals; International Sporting Press Association; International Squash Players Association

ISPC International Statistical Program Center (AID)

ISPE Institute and Society of Practitioners in Electrolysis

ISPEMA Industrial Safety Personal Equipment Manufacturers' Association

ISPHS International Society of Phonetic Sciences

ISPM International Solar Polar Mission; International Staff Planners Message (NATO)

ISPMEMO International Staff Planners Memo (NATO)

ISPO Instrumentation Ships' Project Office; International Society for Prosthetics and Orthotics; International Statistical Programs Office

ISPP Inter-Services Plastic Panel

isps international standard paper sizes

ISPS International Society of Phonetic Sciences

ISPT Initial Satisfactory Performance Test(ing)

isq in status quo

ISQA Israeli Society for Quality Assurance

isr information storage and retrieval; interrupt service routine

Isr Israel; Israeli

Isr Israel (Portuguese or Spanish—Israel); *Israele* (Italian—Israel)

ISR Indian State Railways; Institute for Sex Research; Institute for Social Research; Institute of Surgical Research; International Sanitary Regulations; International Society of Radiology

IS & R Information Storage and Retrieval (system)

I & SR Institutional and Staff Relations

ISRAD Institute for Social Research and Development

Israel State of Israel (Middle Eastern country), *Medinat Israel* known as Judea Palestine before the Christian era

ISRB Idaho Surveying and Rating Bureau; Inter-Services Research Bureau

ISRC International Synthetic Rubber Company

ISRD International Society for the Rehabilitation of the Disabled

ISRF International Squash Rackets Federation

ISRI Israeli Shipping Research Institute

ISRR Institute of Social and Religious Research

ISRRT International Society of Radiographers and Radiological Technicians

ISRSA International Synthetic Rubber Safety Association

ISRT In-School Resources Teacher

isru information search and retrieval unit

ISRU International Scientific Radio Union

iss ideal solidus structures; ion-scattering spectroscopy; issue

iss (ISS) immune system suspected; ionospheric sounding satellite

ISS Industry Standard Specifications; Inertial Sensor System; Information Service Specialist; Inspection Surveillance Sheet; Institute for Socioeconomic Studies; Institute of Salesian Studies; Institute of Space Sciences; Institute of Space Studies; Institutional Staff Services; Integrated Start System; International School Service; International Seismological Summary; International Shoe Company (Stock Exchange Symbol); International Social Service; International Space Station; International Students Society; International Sunshine Society; Israeli Secret Service

ISSA Information Systems Security Association, International Sanitary Supply Association; International Slurry Surfacing Association; International Social Security Association

ISSAS Interactive Structural Sizing and Analysis System

ISSB Inter-Services Security Board

ISSC Institute for the Study of Social Conflict; International Social Science Council

ISSCA Institute of Steel Service Centres of Australia

ISSCAAP International Standard Statistical Classification of Aquatic Animals and Plants

ISSCB International Society for Sandwich Construction and Bonding

ISSCO Integrated Software Systems Corporation

ISSCT International Society of Sugar Cane Technologists

ISSL Initial Spares Support List

ISSLIC Israel Societies of Special Libraries and Information Centers

ISSMFE International Society of Soil Mechanics and Foundation Engineering

ISSMIS Integrated Support Services Management Information System

ISSMP&D International Society for the Study of Multiple Personality and Dissociation

ISSMS Integrated Support Services Management System (USA)

ISSN International Standard Serial Number

ISSOE Instructional Support System for Occupational Education

ISSOL International Society for the Study of the Origin of Life

ISSPA International Sport Show Producers Association

issr information storage, selection, and retrieval

ISSRA Individual Social Security Retirement Account

ISSS International Society for the Study of Symbols; International Society of Soil Science

ISSSC International Society for the Suppression of Savage Customs

ISSSTE *Instituto de Seguridad y Servicios Sociales de los Trabajadores* (Spanish—Workers' Institute of Social Security and Social Services)

ISST International Society of Skilled Trades

IS Standards International Safety Standards

ist insulin shock therapy; interstellar travel

is't is it

ist *istituto* (Italian—institute)

Ist Istanbul

IST Indian Standard Time; Industrial Steel and Tube (consolidated); In-Service Training; Institute of Science and Technology (University of Michigan); Institute on Strategic Trade (Washington, D.C.); Institute for the Study of Terrorism; International Society of Toxicology; Istanbul, Turkey (airport)

IST *International Steam Table*

IS & T *International Science and Technology*

ista intelligence, surveillance and target acquisition

ISTA Indiana State Teachers Association; Industrial Science and Technology Agency; International Seed Testing Association

Istan Istanbul

istar information storage translation and reproduction

ISTAT *Istituto Centrale di Statistica* (Italian—Central Institute of Statistics)

ISTB Interstate Tariff Bureau

ISTC Inland Society of Tax Consultants; Interdepartmental Screw Thread Committee; International Shade Tree Conference; Iron and Steel Trades Confederation

ISTD Imperial Society of Teachers of Dancing; Institute for the Study and Treatment of Delinquency; International Society of Tropical Dermatology

ISTEA Intermodal Surface Transportation Efficiency Act; Iron and Steel Trades Employers' Association

ISTEM Inter-Seminary Theological Education for Ministry

istes integrated spacecraft total energy system

isth isthmian; isthmus

Isth Isthmian; Isthmus

ISTI Iowa State Technical Institute

istim interchange of scientific and technical information in machine language

ISTM International Society for Testing Materials

ISTO Italian State Tourist Office

istom interstate transportation of obscene matter

I Struct E Institute of Structural Engineers

istse integral square time square error

ISU Idaho State University; Indiana State University; International Seamen's Union; International Shooting Union; International Skating Union; International Society of Urology; International Space University; International System of Units; Iowa Southern Utilities; Iowa State University; Italian Service Unit; Southern Iowa Railway (railroad coding)

I-sub inhibitor substance

ISUE4U I sue for you (lawyer's license plate)

ISUM Intelligence Summary

Isup suppressor current (symbol)

ISUP Iowa State University Press

ISUST Iowa State University of Science and Technology

isv independent software vendor

ISV Institute for the Study of Violence (Brandeis U); International Scientific Vocabulary

ISVA Incorporated Society of Valuers and Auctioneers

ISVR Institute of Sound and Vibration Research

ISVS International Secretariat for Volunteer Service

isw interstitial water

ISW Institute for Solid Wastes

ISWA International Science Writers Association; International Solid Wastes and Public Cleansing Association

ISWG Imperial Standard Wire Gauge

ISWM Institute of Solid Wastes Management

ISWNE International Society of Weekly Newspaper Editors

isy intrasynovial

ISY International Space Year

isz increment and skip on zero

it inferotemporal (cortex); Intermediate Technology; slang term for sex appeal

it. information theory; inspection tag; internal thread; international tolerance; inventory transfer; item; itemization(s); itemize(d); in transit

it. (IT) information technology; internal tank; intertuberous; transposed inversion (12-tone)

1/t intensity duration

it *italienisch* (German— Italian); *italiensk* (Dano-Norwegian—Italian); *item* (Spanish—item)

i.t. *in transitu* (Latin—in transit)

i/t *in transitu* (Latin—in transit)

It Italian; Italy

It *Italia* (Italian, Portuguese, Spanish—Italy); Italian

I(t) indicial response (symbol)

IT Idaho Territory; Immunity Test; Imperial Territory; Imperial Typewriter; Income Tax; Indian Territory; Information Technology; Inner Temple; Institute of Technology; International Telephone and Telegraph (Wall Street slang); Italy (Internet code)

ita initial teaching alphabet; inner transport area; interface test adapter(s); international teaching alphabet; international telegraph alphabet

Ita *Italia* (Italian, Portuguese, Spanish—Italy)

ITA Income Tax Act; Independent Teachers Association; Independent Television Authority; Industrial Training Act; Industrial Truck Association; Industry and Trade Administration; International Tea Agreement; International Temperance Association; International Thermographers Association; International Tin Agreement; International

Touring Alliance; International Trade Administration; International Twins Association

ITA Institut du Transport Agrien (French—Air Transport Institute)

ITAA Information Technology Association of America; International Transactional Analysis Association

ITAB Industry Technical Advisory Board

ITAC Interconnect Association of Canada

ITACS Integrated Tactical Air Control System

itae integrated time and absolute error

ITAI Institute of Technical Authors and Illustrators

ital italic; italicize; italics

ital italiensk (Dano-Norwegian—Italian)

Ital Italia; Italian; Italy

Ital Italia (Spanish—Italy)

ITAL Information Technologies and Libraries

Italia Italia Società di Navigazione (Italian Line); (Italian, Latin, Spanish—Italy)

Italian Penin Italian Peninsula, often called the Italian Boot

Italsat Italy International Communications Satellite

italy I trust and love you

Italy Italian Republic (European nation whose civilization antedates the Roman Empire), *Repubblica Italiana*

ITAM Instituto Tecnologico Autonomo de México (Spanish—Technological Institute of Mexico)

ITAP Interim Track Analysis Program

itar interstate and foreign travel (or transportation) in aid of racketeering enterprises

ITAR International Traffic in Arms Regulations; Interstate and Foreign Travel (or Transportation) in Aid of Racketeering Enterprises

ITASS Interim Towed-Array Surveillance System

Itavia Italian Aviation (domestic airline)

itax italics

itb (ITB) integrated tug barge

ITB Industrial Training Board; Integrated Tug Barge; International Theft Bureau; International Time Bureau; Invitation to Bid; Irish Tourist Board; Irish Tourist Bureau

ITB Internationaler Turnerbund (German—International Gymnastic Federation)

itbh internal broach

ITBL Integrated Transportation Bill of Lading

IT & BL Island Tug & Barge, Ltd

ITBS Iowa Test of Basic Skills

itc installation time and cost; investment tax credit

itc (ITC) Intertropical Confluence

ITC Illinois Terminal Company (railroad); Imperial Tobacco Company; Inclusive Tour Charter; Independent Television Corporation; Industrial Training Council; Infantry Training Center; International Tea Council; International Tin Council; International Toastmistress Clubs; International Trade Commission; International Traders Clubs; International Training Center; Island Trading Company

IT & C Industry, Trade, and Commerce (Canada)

ITCA Independent Television Companies' Association; Interamerican Technical Council of Archives; Intercollegiate Tennis Coaches Association; International Typographic Composition Association; Invest to Compete Alliance

ITCA Instituto Tecnologico Centroamericano (Spanish—Central American Technical Institute)

itcan inspect, test, and correct as necessary

ITCC International Technical Cooperation Center

itcm integrated tactical countermeasures

ITCP Integrated Test and Check-out Procedures

ITCRM Infantry Training Center—Royal Marines

ITCUA International Telephone Credit Union Association

ITCV Inter-Tropical Convergence Zone

it'd it had; it would

ITD International Telephone Directory

itda indirect target damage assessment

ITDC Indian Tourist Development Corporation

ITDNS Integrated Tow Operating Digital Network Service

ITDP Institute for Transportation and Development Policy

ITE Institute of Telecommunication Engineers; Institute of Terrestrial Ecology; Institute of Traffic Engineers

ITEC International Thoroughbred Exposition and Conference

ITED Iowa Tests of Educational Development

item illustrated tools and equipment manual

i-time instruction time

ITEP Indian Teacher Education Project; Integrated Test-Evaluation Program

iter iteration; iterative

ITER International Thermonuclear Experimental Reactor

itf inland transit floater (insurance); in trust for

ITF Integration Task Force; International Television Federation; International Tennis Federation; International Trade Federation

ITF Institut Textile de France (French—Textile Institute of France); *Internationale Transportarbeiter Föderation* (German-International Transportworkers Federation)

ITFCA International Track and Field Coaches Association

ITFCS Institute for Twenty-First Century Studies

itfs instructional television fixed service

ITFS Instructional Television Fixed Service; International Television Fixed Service

ITG International Trumpet Guild

itga internal gage

ITGWF International Textile and Garment Workers Federation

ithp increased take-home pay

iti intertial interval

ITI Inagua Transports Incorporated; Integrated Task Indices; International Technical Institute; International Technical Institute of Flight Engineers; International Theatre Institute; International Thrift Institute

ITIB Iceland Tourist Information Bureau

ITIC International Tsunami Information Center

ITIM Interchurch Trade and Industry Mission

i-time instruction time

itin itinerary

itis (Latin suffix—inflammation)—bronchitis, meningitis

Iti(s) Italian(s)

ITITA Instituto Veterinario de Investigaciones Tropicales y de la Altura (Spanish—Veterinarian Institute of tropical and High-Altitude Research)

ITI/US International Theatre Institute of the United States

itk intetkøn (Dano-Norwegian—neuter)

itl integrate-transfer-launch

Itl Italian

it'll it will

ITLMCF Instrument Technicians Labor-Management Cooperation Fund

itlx italics

itm inch trim moment; information transfer module

ITM Institute of Travel Managers; Institute of Tropical Medicine

ITMA Institute of Trade Mark Agents; International Twelve-Meter Association

ITMA It's That Man Again (Tommy Handley's most popular World-War-II BBC series)

ITMF International Textile Manufacturers Federation

I.T.M.I. International Talent Management Incorporated

ITMRC International Travel Market Research Council

ITN Independent Transportation Network; International Television Network

ITN Independent Television News

ITNA Independent Television News Association

ITNS Integrated Tactical Navigation System (USN)

ITO India Tourist Office; Interim Technical Order; International Terminal Operators;

International Trade Organization (UN); Invitational Travel Orders

ITOA Independent Taxi Owners Association

ITOFCA Industrial Trailer-on-Flatcar Associates

ITOL International Thomson Organisation, Ltd

itom interstate transportation of obscene matter

itp (ITP) idiopathic thrombocytopenic purpura; immune thrombocytopenic purpura; inosine triphosphate

ITPA Illinois Test of Psycholinguistic Abilities; International Truck Parts Association

ITPP Institute of Technical Publicity and Publications

ITPS Income Tax Payers' Society

itq invitation to quote

itq's in-text questions

itr incremental tape recorder; integrated test requirement(s); inverted terminal repeat

ITR Indiana Toll Road

ITRA International Truck Restorers Association

ITRC International Terrorist Research Center (El Paso, Texas); International Tin Research Council

ITRP Institute of Transportation and Regional Planning

ITRS Income Tax Refund Service

ITRU Industrial Training Research Unit

its intelligent transportation system

its (ITS) invitation to send (data processing)

it's it has; it is

Its Italians

ITS Idaho Test Station; Industrial TEMPEST Scheme; Institute for Transportation Studies; Integrated Trajectory System; Interactive Training System; Intermarket Trading System; International Technogeographical Society; International Teleproduction Society; International Thespian Society; International Tracing Service; International Trade Secretariat; International Transportation Service

itsa interstate transportation of stolen aircraft

ITSA Institute for Telecommunication Sciences and Aeronomy

itsb interstate transportation of strikebreakers

itsc interstate transportation of stolen cattle

ITSC International Telecommunications Satellite Consortium

itse integral time square error

ITSEC Information Technical Security Evaluation Criteria

itsmv interstate transportation of a stolen motor vehicle

itsp interstate transportation of stolen property

ITSS Integrated Tactical Surveillance System

itt instant-touch tuning

ITT Institute of Textile Technology; Insulin Tolerance Test

IT & T International Telephone and Telegraph

ITTA International Table Tennis Association

ITTC International Television Trading Corporation

ITTCS International Telephone and Telegraph Communications System

ITTE Institute of Transportation and Traffic Engineering

ITTF International Table Tennis Federation

Itti Imar Arab Ittih d al-Imarat al-Arabiyah (Arabic—United Arab Emirates)

ITTP Indian Teacher Training Program

ITTTA International Technical Tropical Timber Association

itu (ITU) intensive therapy unit; intrauterine transfusion

ITU Income Tax Unit; International Telecommunications Union; International Typographical Union

ITUA Industrial Trades Union of America

ITURM International Typographical Union Ruling Machine

itv industrial television; instructional television; internal television

ITV Independent Television

ITVA Instructional Television Authority; International Television Association

ITVN Independent Television News

ITVs Independent Television programs broadcast from off-shore Great Britain

ITVS International Television Service

itw initial training wing

ITW Illinois Tool Works

ITWF International Transport Workers Federation

itx inclusive tour excursion(s)

I-type Jungian introvert type

Itz Itzik

Itz Itzik (Yiddish-Isaac)

ITZC Intertropical Convergence Zone

ITZEL Irgun Tzvai Le'umi (Hebrew—National Military Organization)

ITZN International Trust for Zoological Nomenclature

iu immunizing unit(s); indicator unit; interface unit; internal upset (oil well); international unit(s)

i of u inevitability of the unpredictable

IU Indiana University; Indianapolis Union (railroad); international unit; International Utilities

IÜ Istanbul Üniversitesi (Universityo) (University of Istanbul)

IUA International Union of Architects

IUAA International Union Against Alcoholism; International Union of Advertisers' Association; International Union of Alpine Associations

IUAES International Union of Anthropological and Ethnological Sciences

IUAI International Union of Aviation Insurers

IUAIWA International Union of Allied Industrial Workers of America

IUAJ International Union of Agricultural Journalists

IUAO International Union for Applied Ornithology

IUAPPA International Union of Air Pollution Prevention Associations

IUAT International Union Against Tuberculosis

IUB International Union of Biochemistry; Interstate Underwriters Board

IUBS International Union of Biological Sciences

IUC International Union of Chemistry; International Union for Conservation

IUCc International Union of Crystallography

iucd intrauterine contraceptive device

IUCL Istanbul University Central Library

IUCNNR International Union for Conservation of Nature and Natural Resources

IUCr International Union of Crystallography

IUCSTP Inter-Union Commission on Solar-Terrestrial Physics

IUCSTR Inter-Union Commission on Solar and Terrestrial Relationships

IUCW International Union for Child Welfare

iud intrauterine device; intrauterine diaphragm

Iud Iudicum (Spanish—Epistle of St Paul to the Hebrews)—Book of the Jews

IUD Institute for Urban Development

iudr idoxuridine

IUDTPNAPUSCAN International Union of Dolls, Toys, Playthings, Novelties, and Allied Products of the United States and Canada

IUDZG International Union of Directors of Zoological Gardens

IUE International Ultraviolet Explorer (space vehicle); International Union of Electrical Workers; International Union for Electroheat

IUEC International Union of Elevator Constructors

IUEF International University Exchange Fund

IUER & MW International Union of Electrical, Radio & Machine Workers

IUES International Ultraviolet Explorer Satellite

IUF International Union of Food and Allied Workers' Associations

IUFA International Union of Family Organizations

IUFDT International Union of Food, Drink, and Tobacco Workers' Association

IUFOST International Union of Food Science and Technology

IUFRO International Union of Forest Research Organizations

IUGG International Union of Geodesy and Geophysics

iugr intrauterine growth retardation

IUGS International Union of Geological Sciences; International Union of Geological Services

IUHE International Union for Health Education

IUHFI International Union of Housing Finance Institutions

IUHPS International Union of the History and Philosophy of Science

IUHS International Union of the History of Science

IUIS International Union of Immunological Societies

IUL Ibadan University Library (Nigeria); Indiana University Library (Bloomington)

IULA International Union of Local Authorities

IULIA International Union of Life Insurance Agents

iumbb if unworkable, make best bid

IUMC Indiana University Medical Center

IUMI International Union of Marine Insurance

IUMK Istanbul Üniversitesi Merkez Kütüphanesi (Turkish—Istanbul University Central Library)

IUMM & SW International Union of Mine, Mill and Smelter Workers

IUMP International Union of the Medical Press

IUMSA International Union for Moral and Social Action

IUMSWA Industrial Union of Marine and Shipbuilding Workers of America

IUNS International Union of Nutritional Sciences

IUOE International Union of Operating Engineers

IUOPA International Union of Practitioners in Advertising

IUOPAB International Union of Pure and Applied Biophysics

IUOT Indiana University Opera Theater

IUOTO International Union of Official Travel Organizations

IUOW Industrial Union of Workers

iup intrauterine pregnancy

IUP Indiana University Press; International University Press; Irish Universities Press; Israel Universities Press

IUPA International Union of Police Associations; International Union of Practitioners in Advertising

IUPAC International Union of Pure and Applied Chemistry

IUPAP International Union of Pure and Applied Physics

IUPLAW International Union for the Protection of Literary and Artistic Works

IUPM International Union for Protecting Public Morality

IUPN International Union for the Protection of Nature

iups installed user programs

IUPS International Union of Physiological Sciences

IUPW International Union of Petroleum Workers

IUR International Union of Railways

I U Res Ctr Indiana University Research Center (for the Language Sciences)

IURN Institut Unifé de Recherche Nucléaires (French—Unified Institute of Nuclear Research)

ius inertial upperstage; interim upperstage

IUs international units

IUS Institute of Urban Studies, International Union of Students; International Urban Society; International Urban Studies

IUSA Institute of the U.S.A. (Soviet office that was charged with analyzing American news and political statements concerning the USSR)—Russian counterpart of American kremlinologists

IUSM Indiana University School of Music

IUSP International Union of Scientific Psychology

IUSS Institute of United States Studies

IUSSI International Union for the Study of Social Insects

IUSSP International Union for the Scientific Study of Population

IUT Instituts Universitaires de Technologie (French—University Institutes of Technology)

IUTAM International Union of Theoretical and Applied Mechanics

iutip if unworkable; telegraph idea of price

IUUAAAIWA International Union of United Automobile, Aerospace, and Agricultural Implement Workers of America

IUUCLG International Union, United Cement, Lime & Gypsum Workers

IUVD International Union Against Venereal Diseases

IUVDT International Union against the Venereal Diseases and the Treponematoses

IUVSTA International Union for Vacuum Science Techniques and Applications

IUW Industrial Union of Workers

IUWCC Inshore Undersea Warfare Control Center (USN)

IUWWML International Union of Wood, Wire, and Metal Lathers

iv increased value; initial velocity; initial visit; interactive video; intravenous(ly); intravertebral; inverted vertical (engine); invoice value; ivory

i/v increased value; instrument/visual

i.v. in verbo (Latin—under the word)

i V in Vertretung (German—as a substitute, by proxy)

IV Identity Verification; Imperial Valley; initial visit; Ivan; Ivy

IV Islas Virgenes (Spanish—Virgin Islands)

iva inspection visual aid

Iva Godiva

IVA Independent Voters Association; International Volleyball Association

IVAAP International Veterinary Association for Animal Production

ivala integrated visual approach and landing aids

Ivaran Ivar Anton Christensen's steamship company

IVBF International Volley-Ball Federation

IVBH Internationale Vereinigung für Brückenbau und Hochbau (German—International Association for Bridge and Structural Engineering)

ivc inferior vena cava; intermittent vertical chambers

IVC Imperial Valley College

ivcd intraventricular conduction defect

ivc's inner van connectors

Iv Cst Ivory Coast

ivd interactive video disk; interpolated voice data; intervertebral disk

IVDP Identification, Validation, and Dissemination Process

ivds independent variable depth sonar

Ive Ivan; Iven

I've I have

IVE Institute of Vitreous Enamellers

IVECO Industrial Vehicles Corporation

I've had it (popular American contraction—I have had enough of it)

I-vets Iceland veterans

ivf idiopathic ventricular fibrillation; intravenous fluid; in-vitro fertilization

IVF Innocent Victims Fund

IVFZ International Veterinary Federation of Zootechnics

IVGMMA International Violin, Guitar Makers, and Musicians Association

ivhm in vessel handling machine

IVHS Intelligent Vehicle Highway System

IVHSA Intelligent Vehicle Highway Society of America

IVI Independent Voters of Illinois

IVIC Instituto Venezolano de Investigaciones Científicas (Spanish—Venezuelan Institute of Scientific Investigations)

IVICO Integrated Video Code

IVIG Intravenous Immunoglobulin Therapy

IVIS Integrated Vacuum Instrumentation System; International Visitors Information Service

ivjc intervertebral joint complex

IVJH *Internationale Vereinigung für Jugendhilfe* (German—International Union for Child Welfare)

IVK *Institute för Vaxtforskning och Kyllagring* (Swedish—Institute for Foodstuff Research and Refrigeration—Sweden)

IVL Ivaran Lines

IVL *Internationale Vereinigung für Theoretische und angewandte Linmologie* (German—International Association of Theoretical and Applied Limnology)

IVLA International Visual Literacy Association

IVLU Ivaran Lines (container) Unit

IVMB *Internationale Vereinigung der Musikbibliotheken* (German—International Association of Music Libraries)

ivmu inertial velocity measurement unit

ivo in view of

Ivory Coast Republic of the Ivory Coast (West African nation)

ivox (IVOX) intravascular oxygenator

ivp initial vapor pressure; inspected variety purity (certified seeds); intravenous pyelogram; intravenous pyelography

IVP *Instituto Venezolano de la Petroquimica* (Spanish—Venezuelan Petrochemical Institute)

ivr instrumented visual range; interactive voice response

IVR International Vehicle Registration (symbols displayed on automotive license plates)

IVR *Internationale Vereinigung des Rheinschiffsregisters* (German—International Association of Rhine Ships Registers)

ivs intraventricular septum

iv's intravenous feedings; intravenous injections

IVS Integrated Versaplot Software; International Voluntary Service

IVS *Instituto Venezolano de los Seguros Sociales* (Spanish—Venezuelan Institute of Social Security)

ivsd interventricular septal defect

ivsi instantaneous vertical speed indicator

IVSS *Internationale Vereinigung für Soziale Sicherheit* (German—International Association for Social Security)

IVSU International Veterinary Students Union

ivt intravehicular transfer; intravenous transfusion

ivu intravenous urography

ivu (IVU) *imposta valore aggiunto* (Italian—value-added tax)

IVU International Vegetarian Union

IVU *Instituto de Vivienda Urbana* (Spanish—Institute of Urban Housing)

ivv independent verification and validation

ivvs instantaneous vertical-velocity sensor

iw index word; indirect waste; individually wrapped; inside width; instruction word(ing); isotopic weight; ivory woodpecker

iw (IW) index word

i/w interchangeable with; in work

iW *innere Weite* (German—inside diameter)

IW Aero Trasporti Italiani (2-letter coding, Italian Air Transport)

IWA Inland Waterways Association; Institute of World Affairs; Insurance Workers of America; International Wheat Agreement; International Woodworkers of America

IWA *International Wheat Agreement*

IWAHMA Industrial Warm Air Heater Manufacturers Association

IWAIU Industrial Workers of America International Union

IWBP Integration with British Party (Gibraltar)

iwc in which case

IWC Inland Waterways Corporation; International Watch Company; International Whaling Commission; International Wheat Council

IWCA International World Calender Association

IWCC International Wrought Copper Council

IWCCA Inland Waterways Common Carriers Association

IWCI Industrial Wire Cloth Institute

IWCS Integrated Wideband Communications System

IWCT International War Crimes Tribunal

IWD International Waterways and Docks; International Women's Day (March 12)

iwdm intermediate water depth mine

IWE Institution of Water Engineers

IWFI Italian Wine and Food Institute

IWG Imperial Wire Gauge; Interface Working Group; International Working Group (NATO)

IWGA International Wheat Gluten Association

IWGC Imperial War Graves Commission

IWGM Intergovernmental Working Group on Monitoring (or Surveillance)—UN

IWGMP Intergovernmental Working Group on Marine Pollution (UN)

IWHS Institute of Works and Highway Superintendents

IWI Inventors' Workshop International

iwistk issue while in stock

IWIU Insurance Workers International Union

iwl insensible water loss

IWL Institute of World Leadership

IWLA Izaak Walton League of America

IWM Imperial War Museum; Institute of Works Managers

IWMA International Working Men's Association

iwmi inferior wall myocardial infarction

IWML Imperial War Museum Library

Iwo Iwo Jima (Sulfur Island)

IWO Institute of World Order; International Wine Office; International Workers Order

IWP Indicative World Plan (FAO)

IWPA International Word Processing Association

IWPC Institute of Water Pollution Control

IWPO International Word Processing Organizations

IWPPA Independent Waste Paper Processers Association

IWPS Institute of War and Peace Studies (Columbia)

IWRI Informal Word Recognition Inventory; International Wildfowl Research Institute

IWRMA Independent Wire Rope Manufacturers Association

IWS Inland Waterway Service; Institute of Wood Science; Intelligent Warning System; International Wool Secretariat

IWSA International Water Supply Association

IWSB Insect Wire Screening Bureau

IWSc Institute of Wood Science

IWSC Irrigation and Water Supply Commission

IWSG International Wool Study Group

IWSP Institute of Work Study Practitioners

IWST Integrated Weapon System Training

IWT Indus Water Treaty; Inland Water Transport; Institute of Women Today; International Working Team (NATO)

IWT *Industriewerke Transportsystem* (cargo container system)

IWTA Inland Water Transport Authority

IWTD Inland Water Transport Department

IWTO International Wool Textile Organization

iwu illegal wearing of uniform

IWU Illinois Wesleyan University; Insurance Workers Union

IWV *Internationale Warenhaus Vereinigung* (German—International Department Store Association)

IWVA International War Veterans Alliance

iwvmts interim water velocity meter test set

iww inland waterway

IWW Industrial Workers of the World; Intracoastal Waterway

IWWP *International Who's Who in Poetry*

IWY International Women's Year (1975–1984)

Ix current in a reactance (symbol)

IX unclassified vessel (2-letter naval code)

I.X. *Iesous Christos* (Greek—Jesus Christ)

ixc interexchange

IXSS unclassified miscellaneous submarine (letter symbol)

Ixta Ixtaccihuatl

iy ionized yeast

Iy current in an admittance (symbol)

IY Imperial Yeomanry; International Petroleum (stock exchange symbol)

IYB *International Year Book*

IYC Inland Yacht Club; International Year of the Child

IYDP International Year of Disabled Persons (1981)

IYEO Institute of Youth Employment Officers

IYF International Youth Federation

IYHF International Youth Hostel Federation

IYL International Youth Library

IYRU International Yacht Racing Union

iyswim if you see what I mean

i y v *ida y vuelta* (Spanish—round trip)

I y v *ida y vuelta* (Spanish—round trip)

iz izzard; zed

Iz current in an impedance (symbol); Izar; Izar: Izmir (Smyrna)

IZ Institute of Zoology; Israel Zangwill

IZ I Zingari (Italian—The Gypsies)

izd *izdanie* (Russian—edition)

Izd *izdatl'* (Russian—publisher)

IZD *Internationaler Zivildienst* (German—International Voluntary Service)

izdat *izdatel* (Russian—publisher)

IZL *Irgun Z'vai Leumi* (Hebrew—National Army Organization)

izqº *izquierda* (Spanish—left)

izq *izquierdo* (Spanish—left)

izs insulin zinc suspension

IZTO Interzonal Trade Office (NATO)

Izv *Izvestia* (Russian—News)—official newspaper of the Presidium of the Supreme Soviet—published in Moscow

Iz: Wa: Izaak Walton, *The Compleat Angler*, used colons after his initials

Izzie Isador; Isadora; Isadore; Isidro; Isodoro; Ysidra

Izzy (*see* Izzie)

J

j inner quantum number (symbol), jack; joint (marijuana); joist(ing); junior; junk(ie); square of minus 1 (symbol); unit vector in y direction (symbol)

j journal; jour(nal) (French—day, newspaper)

j. juris (Latin—of law); *jus* (Latin—law)

J action variable (symbol); advance ratio (symbol); electric current density (symbol); emissive power (symbol), flux (symbol); gram-equivalent weight (symbol); heat transfer factor (symbol); Jacob; Jacobean; Jacobian; Jaeger; Jaen; Jamaica; Jamaican; January; Japan; Japanese; Jesuit; jet; Jew; Jewish; joint; joule; Judaic; Judaism; Juliet—code for letter J; Julliard; July; Junction; junction devices; June; North American Aviation (symbol); polar movement of inertia (symbol); radiant intensity (symbol)

J Jabal (Arabic or Persia—mountain, mountain range); *Jebel* (Arabic—mountain, mountains); *Jejunium* (Latin—fast, hunger); *Jibal* (Arabic—mountain range); *Jogi* (Estonian—river); *Jøkel* (Norwegian—glacier); Joki (Finnish—river); *Jökull* (Icelandic—glacier); *Journal* (French—journal)

J-1 personnel section of joint military staff

J-2 intelligence section of joint military staff

J-3 operations and training section of joint military staff

J-4 logistics section of joint military staff

J4 fuel jet-engine fuel derived from coal oil or kerosene)

J-5 Plans and Policy (Joint Chiefs of Staff)

J-6 Communications, Electronics (Joint Chiefs of Staff)

J-32 Saab Lansen jet interceptor aircraft

J-35 Saab double-delta-wing supersonic fighter or fighter-bomber built in Sweden and named Draken (Dragon)

J-37 Swedish Thunderbolt or Viggen jet fighter aircraft

ja jack$_a$ck adapter; jetavator assembly; job analysis; joint account; joke awful

ja (JA) jump address

j/a (J/A) joint account

j & a junk and abandon; junked and abandoned

Ja Jacob; Jacque(s); James; Japan; Japanese

JA Jamaica(n); Japan Association; Jewelers of America; Jewish Agency; John Adams (2nd President of U.S.); Judge Advocate; Junior Achievement; Justice of Appeal

JAA Japan Aeronautic Association; Japan Asia Airways; Joint Airways Association; Joint Aviation Authorities

JAAA Japan Amateur Athletic Association

JAAB Joint Airlift Allocations Board

JAAF Joint Army-Air Force

JAAFU Joint Anglo-American Foulup; Joint Anglo-American Fuckup

JAALD Japanese Association of Agricultural Librarians and Documentalists

JAAOC Joint Anticraft Operation Center (NATO)

JAAR Journal of the American Academy of Religion

jaarg jaargang (Dutch—annual volume)

JAARS Jungle Aviation and Radio Service

JAAS Jewish Academy of Arts and Sciences

Jab Jabal; Jabalpur; Jabez; Jabneel

JAB J Ashleigh Burke; Joint Amphibious Board; Junior Advisory Board

JABA Jefferson Area Board for Aging

JABC Japan Audit Bureau of Circulations

jac jet aircraft coating

Jac Jacobean; Jacobite; Jacobus; Jacumba

Jac. Book of Jacob

Jac. Jacobus (Latin—James)

JAC Japan Advisory Committee; Joint Advisory Committee; Joint Apprenticeship Council

JAC Journal of Applied Chemistry

JACA Japan Air Cleaning Association

JACC Joint Admissions Centre for Colleges (Australia); Journalism Association of Community Colleges

JACCC Japanese-American Cultural and Community Center; Joint Air Control and Coordination Center (USAF)

Jace Jason

jack jackass

Jack Jackson; Jacob; John

Jack Murphy Jack Murphy Stadium in San Diego, California

JACKPOT Joint Airborne Communications Center and Command Post

jack(s) jackass(es)–male donkey(s)—*see* hinny

JACL Japanese-American Citizens League

JACM *Journal of the Association for Computing Machinery*

JACOB Junior Achievement Corporation of Business

Jacq Jacques; Jacquin

Jacq's loom Jacquard's loom

JACS *Journal of the American Chemical Society*

JACT Joint Association of Classical Teachers

Jad Jadavpur, India

JAD Julian astronomical day

JADA Japan Automobile Dealers Association

JADA *Journal of the American Dental Association*

JADB Joint Air Defense Board

JADE Japanese Air Defense Environment

jadeite sodium aluminum silicate

JADF Japan Air Defense Force

jaditbhkycc just a drop in the basket helps keep your city clean (anti-litter-civic-responsibility campaign)

JADPU Joint Automatic Data Processing Unit (shared by the Home Office and the Metropolitan Police of London)

Jadroplov *Jadranska Slobodna Plovida* (Yugoslav Great Lakes Line)

J Adv Judge Advocate

J Adv Gen Judge Advocate General

JAE Joint Atomic Exercise (NATO)

JAEC Japan Atomic Energy Commission; Joint Atomic Energy Committee (U.S. Congress)

JAEIA Japan Atomic Energy Industrial Association

JAEIC Joint Atomic Energy Intelligence Committee

JAEIP Japan Atomic Energy Insurance Pool

JAERI Japan Atomic Energy Research Institute

JAES Japan Atomic Energy Society

JAF Japan Automobile Federation; Jordanian Air Force; Judge Advocate of the Fleet

jafa just another frigging academic

JAFC Japan Atomic Fuel Corporation

jaff chaff and electronic jamming

Jaffna Jaffnapatam

JAFPUB Joint Armed Forces Publication

jag. jaguar; jaguarundi

Jag Jaguar

JAG James Abram Garfield (20th President U.S.); Judge Advocate General

JAG-A Judge Advocate General–Army

JAGC Judge Advocate General's Corps

JAGD Judge Advocate General's Department

JAG-N Judge Advocate General–Navy

jag(s) jaguarundi(s)

JAGS Judge Advocate General's School

JAH John Adams House

Jahrb *Jahrbuch* (German—yearbook)

Jahrg *Jahrgang* (German—annual publication, vintage of the year); year's growth

jai juvenile amaurotic idiocy

JAICI Japan Association for International Chemical Information

JAIEG Joint Atomic Information Exchange Group

JAIF Japan Atomic Industrial Forum

JAIL Justice Against Identification Laws

JAIMS Japan-America Institute of Management Science

JAIS Japan Aircraft Industry Society

jak jackfruit (durian)

Jak Jakarta (Batavia)

Jake Jacob

Jal Jalisco (inhabitants nicknamed tapatios as they excel in dancing the tapatio, jarabe)

Jal *Jalan* (Malay—Lane, Road, Street)

JAL Japan Air Lines; Jet Approach and Landing Chart

JAL *Journal of Academic Librarianship*

JALMA Japan Leprosy Mission for Asia

JALTOS Japan Air Lines Computerized Air Cargo Terminal System

jam job analysis memo

jam. jamming

Jam Jamaica

Jam Cre Jamaican Creole

Jam Eng Jamaican English

JAM James A. Michener; Joint Action for Mission; Joslyn Art Museum; Sir John Alexander Macdonald (Canada's first and third Prime Minister)

JAMA Japan Automobile Manufacturers Association

JAMA *Journal of the American Medical Association*

JAMAG Joint American Military Advisory Group

Jamaica English-speaking West Indian island nation

Jam Arab Lib Sha Ish al-Jamahiriyah al-Arabiya al-Libya al-Shabiya al-Ishtirakiya (Arabic—Socialist People's Libyan Arab Republic)

Jam Arab Sour *al-Jamhouriya al Arabia as-Souria* (Arabic—Syrian Arab Republic)

JAMC Japan Aircraft Manufacturers Corporation

J Am Ceram Soc *Journal of the American Ceramic Society*

J Am Chem Soc *Journal of the American Chemical Society*

Jam Dim Som *Jamhuriyadda Dimugradiga Somaliya* (Arabic—Somali Democratic Republic)

jamex jamming exercise

J Am Geriatric Soc *Journal of the American Geriatrics Society*

JaMi Jacksonville–Miami (metropolitan area including Fort Lauderdale, Hollywood,

Tampa, and St. Petersburg)—also called Metro or Metro Area

Jamie James

J Am Inst Electr Eng *Journal of the American Institute of Electrical Engineers*

Jam Ken *Jamhuri ya Kenya* (Swahili—Republic of Kenya)

JAMMAT Joint Military Mission for Aid to Turkey

jammies pyjamas

jamocha java & mocha (prison argot-coffee)

jams pyjamas

jamsan jam sandwich

Jamsat Japan radio amateur satellite

JAMS Judicial Arbitration and Mediation Services

JAMSTEC Japan Marine Science and Technology Center

jamtrac jammers tracked by azimuth crossings

JAMTS Japan Association of Motor Trade and Service

jamwich jam sandwich

Jam Mwu Tan *Jamhuri ya Mwungano wa Tanzania* (Swahili—United Republic of Tanzania)

Jam Sud *Jamhuryat as-Sudan* (Arabic—Republic of the Sudan)

jan janitor; janitorial

Jan Janice; Jansen; Janson; January; John

JAN Jackson, Mississippi (airport); Job Action Network; Joint Army-Navy

JANAF Joint Army-Navy Air Force

JANAIR Joint Army-Navy Aircraft Instrument Research

JANAP Joint Army-Navy-Air Force Publication

JANAST Joint Army-Navy-Air Force Sea Transport

JANBMC Joint Army-Navy Ballistic Missile Committee

JANET Joint Academic NETwork

JanFeb January and February

JANFU Joint Army-Navy Foulup; Joint Army-Navy Fuckup

JANIS Joint Army-Navy Intelligence Surveys

jan mer *jangan merokok* (Malay—no smoking)

Janna Johanna

Jans Janson

JANS Jet Aircraft Noise Survey; Joint Army-Navy Specification

JANSPEC Joint Army-Navy Specification

JANSRP Jet Aircraft Noise Survey Research Program

JANSTD Joint Army-Navy Standard

JANUS Joint Academic Network Using Satellite

janv *janvier* (French—January)

JAOS *Journal of American Oriental Society*

jap japanned

jap *japanisch* (German—Japanese); *japansk* (Dano-Norwegian—Japanese)

Jap Japan; Japanese; Jasper

Jap *Japón* (Spanish—Japan); Japanese language

JA£ Jamaican pound

JAP Joint Acceptance Plan(ning); Joint Apprenticeship Program

J-AP Jewish-American Prince(ss)

JAPAC Japan Atomic Power Company, Joint Air Photo Center

Japan Asia's most productive country—*Nippon* or *Nihon*

JAPATIC Japan Patent Information Center

JAPC Joint Air Photo Center

JAPCO Japan Atomic Power Company

Jap Cur Japan Current

Japdic Japanese dictionary

Japex Japan Petroleum Exploitation Company

JAPEX Japan Express

JAPIA Japan Auto Parts Industries Association

Japlish Japanese & English

J App Crystallogr *Journal of Applied Crystallography*

J Appl Phys *Journal of Applied Physics*

Jap Soc Japan Society

jar. jargon(ish)

Jar. Book of Jarom

JARC Joint Air Reconnaissance Center (NATO)

Jardines' Jardine, Matheson & Company

JARE Japanese Antarctic Research Expedition

jarg jargon; jargonese; jargonist; jargonistic; jargonize

Jarg Soc Jargon Society

JARI Japan Automotive Research Institute

JARL Japan Amateur Radio League

JARO Johore Area Rehabilitation Orgarnization

JARS Journalization and Recovery System

JARTS Japan Railway Technical Service

Jas James; Jason

JAS Jamaica Agricultural Society; Japan Agricultural Standards; Japan Association of Shipbuilders; Japan Astronautical Society; Japanese Amateur (radio) Satellite; Jewish Agricultural Society; Jordanian Agricultural Society

J-A S Japan-Australia Society

JASA *Journal of the Acoustical Society of America*

JASC Japan-Asia Sea Cable

JASDF Japan Air Self-Defense Force

JASG Joint Advanced Study Group

JASI Joint Asian Surgical Industries

JASIN Joint Air-Sea Interaction (oceanographic experiment)

JASIS *Journal of the American Society for Information Science*

jasp jasper; jasperoid

Jasp Jasper

Jasper Jasper National Park

Jaspr Jasper

JASRAC Japanese Society of Rights of Authors and Composers

jastop jet-assisted stop

jasu jet aircraft starting unit

JAT Jugoslovenski Aero-Transport (Yugoslav Airlines)

JATC Joint Apprenticeship Training Committee

JATCA Joinery and Timber Construction Association

JATCC Joint Aviation Telecommunications Coordination Committee

JATCRU Joint Air Traffic Control Radar Unit

JATMA Japan Automobile Tire Manufacturers Association

J Atmos Sci *Journal of Atmospheric Sciences*

J Atmos Terr Phys Journal of Atmospheric and Terrestrial Physics

jato jet-assisted takeoff

JATP Jazz at The Philharmonic

JATS Joint Air Transportation Service

J Audio Eng Soc Journal of the Audio Engineering Society

jaund jaundice

JAUNT Jefferson Area United Transportation

jav javelin throw

Jav Java; Javanese

Java Djawa

JAVA Jamaica Association of Villas and Apartments

JAVHS Jane Addams Vocational High School

Ja, vi elsker Ja, vi elsker dette landet (Norwegian—Yes, we love this land)—national anthem

JAWA Jane's All the World Aircraft

JAWS Japan Animal Welfare Society; Jet Advance Warning System

Jax Jacksonville, Florida

JAX Jacksonville, Florida (airport)

jaycee (JC) Junior Chamber of Commerce

jazzercise jazz-induced exercise

jb jet black; jet bomb (JB); job blank (form); joint board; junction box

Jb Jacob

Jb Jahrbuch (German—annual, yearbook)

Jb. Job (Book of)

JB Jacksonville Beach; James Buchanan (15th President U.S.); Jodrell Bank; John Bull (British Empire personified); Joint Board; Stetson hat (after its original maker—JB Stetson)

J-B Jacques Barzun; Jean-Baptiste; Johannes Brahms

J.B. *Jurum Baccalaureus* (Latin—Bachelor of Laws)

JB Jerusalem Bible

JBa Jacques Barzun

JBA Japan Binoculars Association; Junior Bluejackets of America

JBAA Journal of the British Archeological Association

JBAKC John Brown Anti-Klan Committee

J-bar capital-J-shaped bar (as used in ski tow lifts)

JBC Jamaica Broadcasting Corporation; Japan Broadcasting Corporation; (*q.v.* NHK); Jewish Book Council

JB & C John Brown and Company (shipbuilders)

JBCA Jewish Book Council of America

JBCNS Joint Board of Clinical Nursing Studies (England and Wales)

JB & Co John Brown and Company (shipbuilders)

JBCSA Joint British Committee for Stress Analysis

jbd jet blast deflector

JBe Japanese B encephalitis

Jber Jahresbericht (German—annual report)

JBES Jodrell Bank Experimental Station (Cheshire, England)

JBG Jewish Board of Guardians

JBHS John Bartram High School

JBIA Jewish Braille Institute of America

J-bird jailbird, convict

JBL James B Lansing (sound equipment)

JBL Journal of Biblical Literature; Journal of Business Law

JBMA John Burroughs Memorial Association

JBMMA Japanese Business Machine Makers Association

J-boat large yacht, often 76 feet or longer; small racing boat sailed by youngsters

J-bolt capital-J-shaped bolt

J-box J-shaped bleaching box; junction box

JBP Jewel Bearing Program; John Boynton Priestley

JBPA Japan Book Publishers Association

JBPI Japanese Bicycle Promotion Institute

JBPS Jamaica Banana Producers Steamship

J'burg Johannesburg

JBRJ Jardim Botánico do Rio de Janeiro (Portuguese—Botanic Gardens of Rio de Janeiro)

JBS Japan British Society; John Birch Society

JBSW Joseph Bulova School of Watchmaking

jbt jail(ing) before trial

JBT Jewelers Board of Trade

JBUC *Jardín Botánico de la Universidad Central* (Spanish—Botanical Gardens of the Central University)—in Caracas, Venezuela

JBUSDC Joint Brazil-United States Defense Commission

JBUSMC Joint Brazil-U.S. Military Commission

JBYC Jamaica Bay Yacht Club

jc joint compound

Jc Junction

JC Brother John Charles SSF; Jackson College; Jacksonville College; Jamestown College; Jeanswear Communication; Jefferson City; Jefferson College; Jersey City; Jesus College (Cambridge); Jet Club; Job Corps; Jockey Club; Johannesburg Consolidated; Johnstown College; Joliet College; Judson College; Juniata College; Junior Chamber (of Commerce, members called *Jaycees);* Juvenile Corps; Juvenile Court

J.C. Jesus Christ; Julius Caesar

JC Jewish Chronicle

J-C Jésus-Christ (French—Jesus Christ)

J.C. *Juris Consultus* (Latin—Juris Consult)

JCA Jewelry Crafts Association; Johore Consumers Association; Joint Commission on Accreditation (of colleges and universities); Joint Communication Activity; Joint Communications Agency; Joint Construction Agency; Junior College of Albany

JCAA Japanese Civil Aviation Authority

JCAB Japan Civil Aviation Bureau

JCAE Joint Committee on Atomic Energy

JCAH Joint Committee on Accreditation of Hospitals

JCAHO Joint Commission on Accreditation of Health-care Organizations

JCAM Joint Commission on Atomic Masses

JCAP Joint Conventional Ammunition Panel (DoD)

JCAR Joint Commission of Applied Radioactivity

J-car(s) Japanese car(s)

JCB Japan California Bank; Japan Credit Bank; Japan Credit Bureau; Joint Consultative Board (NATO)

J.C.B. *Juris Canoni Baccalaureus* (Latin—Bachelor of Canon Law); *Juris Civilis Baccalaureus* (Latin—Bachelor of Civil Law)

JCBC Junior College of Broward County

JCBL John Carter Brown Library (of Americana) —Brown University, Providence, Rhode Island

JCBSF Joint Commission for Black Sea Fisheries

JCC Jamestown Community College; Japan Cotton Center; Jefferson Community College; Jesus Christ Christian Aryan Nations Church (neo-Nazi organization); Jewish Community Center; Job Corps Center; John C Calhoun; John Caldwell Calhoun; Joint Communications Center; Joint Communications' Committee; Junior Chamber of Commerce

JC of C Junior Chamber of Commerce

JCCA Joint Conex Control Agency

JCCI Japan Chamber of Commerce and Industry

JCCRG Joint Command Control Requirements Group

JCCSO Jewish Community Center Symphony Orchestra

JCD *Journal of Crime and Delinquency*

J.C.D. *Juris Canonici Doctor* (Latin—Doctor of Canon Law); *Juris Civilis Doctor* (Latin—Doctor of Civil Law)

JCE Johannesburg College of Education; Junior Certificate Examination

JCE *Journal of Chemical Education*

JCEC Joint Communication Electronics Committee

JCED Japan Committee for Economic Development

JCEE Joint Council on Economic Education

JCEG Joint Communications Electronics Group

JCENS Joint Communication Electronic Nomenclature System

JCER Japan Center for Economic Research

JCET Joint Combined Exchange and Training; Joint Council on Educational Television

JCFA Japan Chemical Fibres Association

J Chem Phys *Journal of Chemical Physics*

J Chem Soc *Journal of the Chemical Society*

J Chim Phys *Journal de Chimie Physique* (French— *Journal of Chemical Physics*)

JCHO Joint Commission on Healthcare Organizations

JCI Joint Communications Instruction; Junior Chamber International

JCIA Japan Camera Industry Association; Japan Chemical Industry Association

JCIC Johannesburg Consolidated Investment Company

JCIEABJ Joint Commission for the Investigation of the Effects of the Atomic Bomb in Japan

JCII Japan Camera Inspection Institute

JCJC Jasper County Junior College; Jefferson County Junior College

Jck Jacksonville

jcl job-control language

Jcl Johnny come lately

JCL Job Control Language; John Crerar Library

J.C.L. *Juris Canonici Licentiatus* (Latin—Licentiate in Canon Law)

JCLA Joint Council of Language Associations

J-class Soviet diesel-powered missile-launching submarines nicknamed Juliet

JCLCPS *Journal of Criminal Law, Criminology, and Police Science*

JCLS Junior College Libraries Section

jcm jettison control module

JCM Joint Committee on Microcards

JCMC Joint Conference on Medical Conventions

JCN Job Change Notice

JCNM Jewel Cave National Monument

JCO José Clemente Orozco

JCOA Jazz Composers Orchestra Association

JCOC Joint Combat Operations Center; Joint Command Operations Center; Joint Committee on the Organization of Congress

J Comput Phys *Journal of Computational Physics*

JCOSA Joint Chiefs of Staff in Australia

jcp jettison control panel; jungle canopy penetration

JCP Japan Communist Party; J.C. Penney; Joint Committee on Printing (Congress); Junior Collegiate Players; Justice of the Common Pleas

JCPCI Junior College of Packer Collegiate Institute

JCPI Japan Cotton Promotion Institute

JCPS Joint Center for Political Studies (Washington, D.C.)

JCR Junior Common Room

JCR *Journal of Coastal Research*

JC3RCP Joint Committee of the three Royal Colleges of Physicians (Edinburgh, Glasgow, London)

JCRFD Joint Commission for Regulation of Fishing on the Danube

JCRR Joint Commission on Rural Reconstruction

J Cryst Growth *Journal of Crystal Growth*

JCs Job Corpsmen

JCS Jewish Community Center(s); Joint Chiefs of Staff, Joint Commonwealth Societies

JCS *Jesus Christ Superstar*

JCSat Japan Communications Satellite

JCS-ACA Joint Chiefs of Staff-Automatic Conference Arranger

JCS-IDTN Joint Chiefs of Staff-Interim Data Transmission Network

JCSMR John Curtin School of Medical Research

JCS-PUBS Joint Chiefs of Staff—Publications

JCSRE Joint Chiefs of Staff Representative, Europe (NATO)

JCSUK Jersey Cattle Society of the United Kingdom

jct junction

Jct Junction (postal abbreviation)

JCT Joint Committee on Taxation

JCTC Japanese Cultural and Trade Center; Juneau County Teachers College

jctn junction

jct pt junction point

JCU James Cook University; John Carroll University

JCUDI Japan Computer Usage Development Institute

JCUNQ James Cook University of North Queensland

J-curve shape formed on a graph when a country's currency drops in value and the trade balance worsens before it gets better

JCUS Joint Center for Urban Studies (MIT and Harvard); Judicial Conference of the United States

JCW JC Williamson

JCWA Japan Child Welfare Association

JCWI Joint Council for the Welfare of Immigrants

jd joined, joint dictionary; junior debutante; jury duty; juvenile delinquency; juvenile deliquent

jd jemand (German—someone, somebody)

Jd Jordanian dinar (monetary unit of Jordan)

JD Julian day; Junior Deacon; Junior Dean; Justice Department

J.D. Doctor of Jurisprudence; *Jurus* or *Jurum Doctor* (Latin—Doctor of Law or Laws)

JDA Japan Defense Agency; Japan Domestic Airline; Jefferson Davis Association; Jewish Direct Action

J-day Judas Day (Wednesday before Good Friday when Judas is believed to have betrayed Jesus)

JDB Japan Development Bank

JDC Joint Distribution Committee; Juvenile Delinquency Control; Juvenile Detention Center

JDCC Juneau-Douglas Community College

J/deg joule per degree

JDHS Jefferson Davis High School

JDI Juvenile Delinquency Index

JDIC Justice Data Interface Controller

jdl job description language

JDL Jewish Defense League

jdm joint drafting manual

JDO Jewish Defense Organization

JDP John Dos Passos

JDPA Japan Dairy Products Association

JDR Japanese Depository Receipts

jds job data sheet

jd's juvenile delinquents

JDS Job Diagnosis Survey; John Dewey Society; Joint Defense Staff (NATO); Justice Data System

J.D.S. Doctor of Juridical Science

JDSCS Joint Defense Space Communications Station (Nurrangar, Australia)

JDSFA Japan Self-Defense Forces Academy

JDSRF Jim Dandy's Still and Refreshment Factory (Australian definition for the Joint Defense Space Research Facility near Alice Springs)

Jdt. Judith

je job estimate

jé jésus (French—paper of super-royal size)

jea joint export agent

JEA Jesuit Educational Association; Joint Engineering Agency; Journalism Education Association

jebm jet engine base maintenance

JEC Jardine Engineering Corporation; Joint Economic Committee (Congress)

JECC Japan Electronic Computer Company

JECMA Japan Export Clothing Makers Association

JECMOS Joint Electronic Countermeasures Operation Section (NATO)

Jed Jedediah

jed joint engineering design

J Ed Journal of Education

JED Japan Engineering Development

JEDEC Joint Electron Device Engineering Council

JEDPE Joint Emergency Defense Plan, Europe (NATO)

JEDS Japanese Expeditions to the Deep Sea

JEE Japan Electronics Engineering

jeep (from GP meaning general purpose) 4-wheel-drive quarter-ton utility vehicle

Jeep graduated payment mortgage

JEEP Joint Emergency Evacuation Plan

jeepney Filipino-built jitney bus

Jef(f) Geoffrey; Geoffroy; Jefferson; Jeffery; Jeffrey; Jeffry

Jeff City Jefferson City, Missouri

Jeff D Jefferson Davis

Jefferson's Thomas Jefferson's Birthday (April 13)

jefrn (JEFW jet engine field maintenance

JEFR Japan Experimental Fast Reactor

JEG John Edward Gray; Joint Exploratory Group (NATO)

Jeho Jehosaphat

JEH Journal of Ecclesiastical History

JEI Japan Electronics Industry

JEIA Japanese Electronic Industries Association

JEIDA Japan Electronic Industry Development Association

jeim jet engine intermediate maintenance

JEIPAC Japan Electronic Information Processing Automatic Computer

JEJ Japan Economic Journal

jejun jejunectomy; jejunitis; jejunostomy

J Electrochem Soc Journal of the Electrochemical Society

jem jet engine modulation

Jem Jemima

JEMC Joint Engineering Management Conference

Jen Jennifer

JEN Junta de la Energia Nuclear (Atomic Energy Board)

JEN Journal of Emergency Nursing

Jen Jih Jen-min Jih-pao (People's Daily)—published in Peking by Communist Party of China

Jennie Jennifer

Jenny Jennette; Jennifer

jentac. jentaculum (Latin—breakfast)

JEOCN Joint European Operations Communications Network

JEOL Japan Electron Optics Laboratory

JEPI Junior Eysenck Personality Inventory

JEPIA Japan Electronic Parts Industry Association

JEPO Jet Engine Project Office

JEPOSS Javelin Experimental Protection Oil Sands System

JEPS Joint Exercise Planning Staff (NATO)

Jer Jersey

Jer. Jeremiah, The Book of the Prophet

Jer Jeremiah; Jeroesjalaim (Dutch—Jerusalem)

Jer Bro Jerry Brown (Edmund G Brown Jr)

JER Japan Economic Review

JERC Joint Electronic Research Committee

Jere Jeremiah; Jerry

JERI Japan Economic Research Institute; Joint Economic Research Institute

jerk ineffectual fool

JERK Journalist's Easy Road to Knowledge (routed through an impasse of acronyms such as GAS—Grant's Acronymical Shorthand)

jerky beef jerky; buccan; charqui; jerked beef

jerob jeroboam (4-bottle capacity)

Jeronº Jerónimo (Spanish—Jerome)

JERS Japanese Ergonomics Research Society

Jes Jessica; Jesus; Jesus College, Cambridge

JES James Ewing Society; Japan Electroplating Society; Japan Engineering Standards; Job Entry Subsystem; John Ericsson Society

JES Journal of Ecumenical Studies

JESA Japanese Engineering Standards Association

Jes Coll Jesus College—Cambridge

jes min just a minute

Jesp Jespersen

Jess Jessica

jes sec just a second

JESSI Joint European Summicron Silicon

Jesus Jesus College, Oxford

jet black lignite; jet-engine aircraft

jet. jetsam

JET Joint European Torus (program to develop nuclear fusion as an efficient energy source); Joint European Tours; Joint European Transport

JETCO Jamaican Export Trading Company

JETDLAG Joint European Development of Tunable Diode Laser Absorption Spectometry for the Measurement of Atmospheric Gases

JETDS Joint Electronics Type Designation System

JETEC Joint Electron Tube Engineering Council

jet fag jet flight fatigue

Jeth Jethro

jetma jet mechanic

jet-p jet-propelled; jet propulsion

JETP Journal of Experimental and Theoretical Physics

JETRO Japan Exterior Trade Research Organization; Japan External Trade Organization

JETS Joint Enroute Terminal Systems (air traffic); Junior Engineering Technical Society

jett jettison

jeu jeudi (French—Thursday)

Jev Japanese encephalitis virus

JEVA Japan Electric Vehicle Association

Jew. Jewish

Jewel The Jewel (La Jolla, California); Jewel Cave National Monument near Custer in southwestern South Dakota

Jewtong Jewtongo (Surinam dialect spoken by former slaves who picked it up from their masters who spoke Dutch, English, Portuguese, or Spanish within their hearing in the belief they would not understand)

JEZ Johannes Enschede en Zonen

JEZ Journal of Experimental Zoology

jf distant fog (meterological symbol)

j/f jigs and fixtures; journal folio

JF Japan Fund; Jewish Federation; Joint Force

JFACT Joint Flight-Acceptance Composite Test

JFAI Joint Formal Acceptance Inspection (NATO)

jfb jet flying belt

jfc Japan Food Company

JFC Japan Film Center

JFCC Japanese Federation of Culture Collections of Micro organisms

JFCS Jewish Family and Child Service

JFEA Japan Federation of Employers Association

jfet junction field-effect transistor

JFK John F Kennedy international airport, New York City; John Fitzgerald Kennedy—35th President of the United States

JFKCAS John F Kennedy College of Arts and Sciences (Trinidad)

JFKCPA John F Kennedy Center for the Performing Arts, Washington, D.C.

JFKMF John F Kennedy Memorial Forest (near Jerusalem, Israel)

JIFKMH John F Kennedy Memorial Highway (Baltimore, Maryland to Wilmington, Delaware)

JFKML John F Kennedy Memorial Library

JFKSC John F Kennedy Space Center

JFKYCC John F Kennedy Youth Correctional Center

jfl joint frequency list

J Fluid Mech Journal of Fluid Mechanics

JFM Jeunesses Fédéralistes Mondiales (French—Young World Federalists)

JFMAMJJASOND January, February, March, April, May, June, July, August, September, October, November, December

JFMIP Joint Financial Management Improvement Program

JFNP John Forrest National Park (Western Australia)

JFO San Francisco, California (heliport)

jfp joint frequency panel

JFP Jobs For Progress

JFPS Japan Fire Prevention Society

jfr jevnfr (Dano-Norwegian—compare)

JFR Joint Fiction Reserve

JFRC James Forrestal Research Center

JFRCA Japanese Fisheries Resources Conservation Association

JFRO Joint Fire Research Organisation (UK)

jfs jet fuel starter

JFS Japan Fishery Society; Jewish Family Service

JFS Jane's Fighting Ships

JFSOC Junior Foreign Service Officers Club

JFTC Joint Fur Trade Committee

JFU Jersey Farmers' Union

JFV Jobs for Veterans

JG junior grade

jga juxtaglomerular apparatus

JGA Japan Golf Association

JGC Japan Gas Chemical; Japan Gasoline Company

JGD John George Diefenbaker (Canada's 17th Prime Minister)—also known as Dief the Chief

jg di joggle die

JGE Journal of General Education

J Geophys Res Journal of Geophysical Research

J-girl joy girl (prostitute)

jgn junction gate number

JGNP Japanese Gross National Product

JGR Jaldapara Game Reserve (India); Jamaica Government Railway

Jgs. Judges (Book of)

JGS Joint General Staff (NATO)

JGSA John G Shedd Aquarium

JGSDF Japan Ground Self-Defense Force

jg sm joggle shims

JGTC Junior Girls Training Corps

JGW Junior Grand Warden

JGWTC Jungle and Guerrilla Warfare Training Center (USA)

jh juvenile hormone

Jh Jahresheft (German—yearly publication)

J & H Jack & Heintz

JH Jugendherberge (German—youth hostel)

jha job hazard analysis

JHA John Howard Association; Justice and Home Affairs

JHAH John Howard Association of Hawaii

JHAI John Herron Art Institute

Jhb Johannesburg

JHC John Hammcock Center

Jhd Japanese haakon dahl (log measure of 11.7 cubic feet being 100 Jbd)

JHDA Junior Hospital Doctors Association

Jhdf Japanese haakon-dahl feet

JHE Journal of Higher Education

jhg joule heat gradient

JHH Johns Hopkins Hospital

JHI Jacob Hiatt Institute; Jesuit Historical Society; Jewish Historical Society

jhj jail-house juvenile (delinquent)

JHL John Harvard Library

JHMI Johns Hopkins Medical Institutions

JHMO Junior Hospital Medical Officer

JHO Jam Handy Organization; Japan Hydrographic Office

JHOS Johns Hopkins Oceanographic Studies

jhp jacketed hollow point

JHP Jackson Hole Preserve; Johns Hopkins Press

JHS John Howard Society; Judaic Heritage Society; Junior High School

J.H.S. Jesus Hominum Salvator (Latin—Jesus Savior of Men)

JHU Johns Hopkins University

JHUL Johns Hopkins University Library

JHUP Johns Hopkins University Press

JHUSHPH Johns Hopkins University School of Hygiene and Public Health

JHUSM Johns Hopkins University School of Medicine

JHVH Jehovah [transliteration of Hebrew tetragrammaton Yhwh, Yahwah, or Jahvah, used by Hebrew tribes in 3rd century BCE because they thought "Jehovah" was too sacred to pronounce]

JHWH (*see* JHVH)

ji jet interaction; junction isolation; junction isolator

JI Aerovias Sudamericanos (symbol)

JI Japan Interpreter

JIA Japanese Interchange Association

JIAS Jewish Immigration Aid Society

JIAWG Joint Integrated Avionics Working Group

jib. job-information block

Jib Jibouti

JIB Jack-in-the-Box; Japan International Bank; Joint Intelligence Bureau

JIBA Japan Institute of Business Administration

JIBC Japan International Biological Program

JIBICO Japan International Bank and Investment Company

jíb(s) jíbaro(s) [Spanish—peasant farmer(s)]—Puerto Rican(s)

Jibuti Djibouti

jic jet-induced circulation; jet-induced combustion

JIC Jewelry Industry Council; Joint Industrial Council; Joint Industry Council; Joint Intelligence Center; Joint Intelligence Committee

JICA Japan International Cooperation Agency; Joint Intelligence Collecting Agency

JICS Joint Interpreting and Conference Service

JICST Japan Information Center of Science and Technology

JICTAR Joint Industry Committee for Television Advertising Research

JID Journal of Infectious Diseases; Junta Interamericana de Defensa (Spanish—Inter-American Defense Board)

JIDA Japan Industrial Designers Association; Jewelry Industry Distributors Association

JIDC Jamaica Industrial Development Corporation

JIDPO Japan Industrial Design Promotion Organization

JIE Junior Institution of Engineers

JIEA Japan Industrial Explosives Association

JIFA Japanese Institute of Foreign Affairs

JIFE Junta Internacional de Fiscalización de Estupefacientes (Spanish—International Council for the Investigation of Narcotics)

JIG Joint Intelligence Group

JIIST Japan Institute for International Studies and Training

JILA Japanese Institute of Landscape Architects; Joint Institute for Laboratory Astrophysics

Jill Jillian

Jim James

JIM Jakarta Informal Meetings; Japan Institute of Metals; Junior Index of Motivation

JIMA Japan Industrial Management Association

JIMA Journal of the Israel Medical Association

Jimmu Jimmu Tenno—first emperor of Japan who began his reign in 660 BCE

Jimtown Jamestown (California, New York, or North Dakota)

JIMS Job Information Matrix System

jimson weed Datura stramonium

J Inorg Nucl Chem Journal of Inorganic and Nuclear Chemistry

JINR Joint Institute for Nuclear Research

jins juveniles in need of supervision

JIO Joint Intelligence Organization

JIOA Joint Intelligence Objectives Agency

JIP Joint Installation Plan(ning)

JIPS Japanese Information Processing Service

JIR Jewish Institute of Religion; Job Improvement Request

JIRA Japan Industrial Robot Association

JIRP Juneau Icefield Research Project

jirv jet-interaction reentry vehicle

jis job instruction sheet

JIS Jail Inspection Service; Jamaica Information Service; Japan Industrial Standard; Jewish Information Society; Joint Intelligence Staff

JISA Japan Industrial Safety Association

JISAO Joint Institute for the Study of the Atmosphere and Oceans

JISC Japanese Industrial Standards Committee

JISEA Japan Iron and Steel Exporters Association

JISF Japan Iron and Steel Federation

JISP Jack Island State Park (Florida)

jit jitney bus, just in time

JIT Job Instruction Training

jizz jizzim (slang—semen)

jj jaw jerk

JJ Jean Juarès (French Socialist assassinated on eve of World War I); Jesse Jackson; Judges, Justices

J-J Jean-Jacques

J & J Johnson & Johnson

J-J Jen-min Jih-pao (Chinese—people's daily communist-controlled Peking newspaper)

JJA John James Audubon

JJC Juvenile Justice Center (Los Angeles); Juvenile Justice Clearinghouse

JJCA Sir John Joseph Caldwell Abbott (Canada's fourth Prime Minister)

JJCCJ John Jay College of Criminal Justice

JJHL John Jay Hopkins Laboratory for Pure and Applied Science (General Atomic Division of General Dynamics Corporation)

JJHS John Jay High School

jj's joyless jobs (pimping and prostituting)

Jj's Jewish jokes

JJS James Joyce Society

JJS Journal of Jewish Studies

JJSC Juvenile Justice Standards Committee

JJSS Jean-Jacques Servan-Schreiber

J-J S-S Jean-Jacques Servan-Schreiber

jk just kidding

JK Jack Kerouac; Sun World (airline code)

J/°K joule(s) per degree Kelvin (unit of entropy)

J & K Jammu and Kashmir (University)

Jka Jakarta

JKC Jane Kathryn Conrad; Japan Kennel Club

jkg joules per kilogram

JKG John Kenneth Galbraith (economist and diplomat)

J/kg°K joule(s) per kilogram degree Kelvin

JKP James Knox Polk (11th President U.S.)

JKS Julius Kayser (stock-exchange symbol)

jkt jacket

Jkt Jakarta

JKT Jakarta, Indonesia (airport); Job Knowledge Test

jl just looking (pseudo customer)

Jl Joel

JL J Lauritzen (steamship line); Japan Airlines (2-letter code); Johnson Line; Jones and Laughlin; Joseph Lewis

Jla Julia

JLA Jamaica Library Association; Japan Library Association; Jewish Librarians Association; Jordan Library Association

JLB Jewish Lads' Brigade; John Logie Baird (tv's inventor)

JLC Japan Logistical Command; Jewish Labor Committee; Joint Logistics Command(ers)

JLCU Johnson Line Container Unit

Jlem Jerusalem

JLG Joint Liaison Group (Hong Kong)

JLMIC Japan Light Machinery Information Center

Jln Jalan (Malay—Lane, Road, Street)

JLOIC Joint Logistics, Operations, Intelligence Center (NATO)

J Low Temp Phys Journal of Low Temperature Physics

JLP Jamaica Labour Party

JLPPG Joint Logistics and Personnel Policy Guidance

jlr(s) jeweler(s)

JLRSS Joint Long-Range Strategic Study

jls jewels

JLS Jail Library Service (California State Library); Junior Literary Society

Jlt Juliet

J Lumin *Journal of Luminescence*

JM Jamaica (Internet code); James Madison (4th President U.S.); James Monroe (5th President U.S.); Japan Mail; Jardine Matheson; Jewish Museum; José Martí

J-M Johns-Manville

J-M Jiyu-Minshuto (Japanese-Liberal Democratic Party)

JMA Japan Management Association; Japan Medical Association; Japan Meterological Agency; Jewish Music Alliance; joule per meter (impact strength) squared

J Macromol Sci *Journal of Macromolecular Science*

J Math Phys *Journal of Mathematical Physics* (published in New York City); *Journal of Mathematics and Physics* (published in Cambridge, MA)

JMB J(ames) M(atthew) Barrie

JMBA *Journal of the Marine Biological Association*

JMC Japan Metals and Chemicals; Japan Monopoly Corporation; Jefferson Medical College; Jerusalem Music Centre; Joint Maritime Commission; Joint Maritime Congress

JMCC Joint Mobile Communications Center (NATO)

JMD M(alaby) Dent

JMDC Japan Machinery Design Center

JMD & S Joseph Malaby Dent and Sons

J Mech Phys Solids *Journal of the Mechanics and Physics of Solids*

jmed jungle message encoder decoder

JMF Jewish Music Forum; Juilliard Musical Foundation

JMG Jewelry Manufacturers Guild

JMHS James Madison High School; James Monroe High School; John Muir High School

JMI Japan Machinery and Metal Inspection; John Muir Institute

JMIA Japan Mining Industry Association

JMIF Japan Motor Industrial Federation

JMJ Jesus, Mary, and Joseph

JMMA Japan Materials Management Association

JMMC James Madison Memorial Commission

JMMF James Monroe Memorial Foundation

JMMII Japan Machinery and Metal Inspection Institute

J Mol Spectro *Journal of Molecular Spectroscopy*

JMP Jen Men Piao (Chinese—People's Bank Dollar)

jmpr jumper

JMPTC Joint Military Packaging Training Center

JMRMA John and Mable Ringling Museum of Art

JMRP Joint Meteorological Radio Propagation

JMRT Junior Members Round Table

JMS James Madison Society; Japan Medical Society; Johannesburg Musical Society

JMSA Japanese Maritime Safety Association

JMSDF Japanese Maritime Self-Defense Force

JMSO Joint Meetings of Seafarers' Organization

jmt (JMT) jointly managed trust

JMT Job Methods Training

JMTBA Japan Machine Tool Builders Association

JMTR Japan Material Testing Reactor

JMUSDC Joint Mexico-United States Defense Commission

jn join; junction

j-n jet navigation

Jn John

Jn Juan (Spanish—John)

JNA Jordan News Agency

JNA *Jena Nomina Anatomica; Jugoslovenska Narodna Armija* (Serbo-Croat—Yugoslav People's Army)

JNB Johannesburg, South Africa (airport)

Jnc Junction

JNC Joint Negotiating Committee

JNCA Junior Naval Cadets of America

jnd joined; just noticeable difference

JND Juvenile Narcotics Division

JNDC Jamaica National Dance Company

JNDNWR J.N. (Ding) Darling National Wildlife Refuge (Florida)

jne ja niin edespäin (Finnish—and so on)

JNEC Jamaican National Export Corporation

JNF Japan Nuclear Fuel (company); Jewish National Fund

JNH *Journal of Negro History*

JNI *Journal of the Nautical Institute*

JNIP Jamaican National Investment Promotion

Jnl Journal

JNL Japanese National Laboratory

jnls journals

jntst journalist

JNM *Journal of Nuclear Medicine*

JNN Japan News Network

jnnd just not noticeable difference

Jno John

JNODC Japanese National Oceanographic Data Center

J Non-Cryst Solids *Journal of Non-Crystalline Solids*

JNOV *Judgment Non Obstante Veredicto* (Latin—judgment notwithstanding a verdict)

JNP Jasper National Park (Alberta)

JNPGC Japan Nuclear Power Generation Corporation

jnr joiner; junior

Jnr Jesurun

JNR Japanese National Railways

jns just noticeable shift

Jns Johannes

JNS Japan Nuclear Society; Jet Noise Survey

JNSDA Japan Nuclear Ship Development Agency

jnt joint; junction; juncture

JNT John Napier Turner—Canada's 17th Prime Minister if counted by name and 22nd if counted by terms in office

JNTA Japan National Tourist Association

JNTO Japan National Tourist Office; Japan National Tourist Organization

jnt stk   joint stock

JNU   Juneau, Alaska (airport)

J Nucl Energy   *Journal of Nuclear Energy*

J Nucl Mater   *Journal of Nuclear Materials*

JNUL   Jewish National and University Library (Jerusalem)

JNV   *Junta Nacional do Vinho* (Portuguese—National Wine Board)

jnwpu   joint numerical weather-prediction unit

JNZ   Jewelers of New Zealand

jo   journalist

Jo   Joel; Joseph; Josephine

JO   Job Order; Jordan (Internet code); Juilliard Orchestra; Jupiter Orbiter

JO   *Journal Officiel* (French—*Official Journal*); *Justie Ombudsman* (Swedish—representative of justice)

Joa   Joachim

JOA   Joint Operating Agreement

joalh   *joalharia* (Portuguese—jewelry store); *joalheiro* (Portuguese—jeweler); *joalheria* (Portuguese—jewelry store)

Jo Bapt   John the Baptist

joblib   job library

jo block(s)   johannson block(s)

jobman   job management

JOBS   Job Opportunities and Basic Skills; Job Opportunities in the Business Sector

Jo'burg   Johannesburg

joc   jocuse; jocular

JOC   Japan Olympic Committee; Joint Operations Center; Joint Opposition Council

jock   jockey; jockstrap

jock(s)   jockstrap(s)—nickname for physical education student(s)

Jock(s)   Scot(s)

joco   jocose

JOCV   Japan Overseas Cooperation Volunteers (Peace Corps)

jod   joint occupancy date

JODC   Japanese Oceanographic Data Center

Jo Div   John the Divine

Joe   Joel; Joseph; Josephine

JOE   Juvenile Opportunities Extension

Joe C   Joe Clark (Canada's 16th Prime Minister)

JOERA   Japan Optical Engineering Research Association

Joe Robbie   Joe Robbie Stadium, Miami

Jo Evang   John the Evangelist

joey   baby kangaroo

J-off   jack off (underground slang—masturbate)

jog.   joggle

JOG   Joint Operations Group; Junior Ocean Group (*jay-oh-gees*—smallest sailing cruisers); Junior Offshore Group

Jogja   Jogjakarta

Jogjakarta   Djokjakarta

Joh   St John's College, Cambridge

Joh   *Johann(es)* (German—Hans, John)

Johan   Johannesburg

J'burg   Johannesburg

John   The Gospel According to John; St John, American Virgin Islands; St John, New Brunswick, Canada

John B   John B Stetson (hat)

John D.   John D Rockefeller, Sr

JOHNNIAC   John von Newman's integrator and Automatic Compiler

Johns H   Johns Hopkins University

Joh Seb Bach   Johann Sebastian Bach

JOI   Joint Oceanographic Institution

JOIDES   Joint Oceanographic Institutions for Deep Earth Sampling

JOIDESP   Joint Oceanographic Institutions Deep Earth Sampling Program

join.   joinery

JOIN   Job Orientation in Neighborhoods; Jobs or Income Now

Joint   American Jewish Joint Distribution Committee

Joio   Norman Dello Joio

JOIS   Japan On-Line Information System

JOK   Oakland, California (heliport)

jol   job organization language

JOLA   *Journal of Library Automation*

JoLoPo   José Lopez Portillo

Jolyon   Joseph Lyons

JOM   Job-Oriented Manual

JOM   Johnson O'Malley Act

jom-Isl-Irân   *Jomhori-e-Islami-e-Irân* (Farsi—Islamic Republic of Iran)

JOMO   Junta of Militant Organizations (Black Nationalists)

Jon.   The Book of Jonah

Jona   Jonathan

Jonathan   Jonathan David

JONS   *Juntas de Ofensiva Nacional Sindicalista* (Spanish—United National Syndicalist Offensive)–fascist anti-syndicalists

JONSDAP   Joint North Sea Data Acquisition Program

JONSIS   Joint North Sea Information Systems

JONSWAP   Joint North Sea Wave Project

JOOD   Junior Officer of the Deck

JOOM   Junior Observers of Meteorology

JOP   Joint Operating Plan(ning); Joint Operations Procedure

JOPCN   Job Order Program Control Number (USA)

JOPM   Joint Occupancy Plan Memo; Joint Operation Procedure Memo

JOPR   Joint Operation Procedure Report

JOPS   Joint Operating Study

JOPS   *Journal of the Patent Office Society*

J Opt Soc Am   *Journal of the Optical Society of America*

Jor   Jordan (whose capital is Amman)

JOR   Jet Operations Requirements

Jord   Jordan

Jord   *Jordânia* (Portuguese—Jordan); *Jordania* (Spanish—Jordan)

Jordan   Hashemite Kingdom of Jordan (Middle East country formerly called Transjordania), *Al Mamlaka al Urduniya al Hashemiyah*

JORG   Joint Oceanographic Research Group

Jos   Joseph; Joshua; Josiah; Jossie

JOS   Junior Ordinary Seaman

Josa   Josepha; Josephine

José Ferrer   José Vicente Ferrer de Otero y Cintrón

Josh.   The Book of Joshua

Josh   Joshua

Joshua Tree Joshua Tree National Monument north of the Salton Sea in southern California

JOSS JOHNNIAC Open-Shop System

Josy Joseph

jot. jump-oriented terminal; junction optimization technique

JOT Joint Observer Team

JOTS Job-Oriented Training Standards

JOUAM Junior Order of United American Mechanics

JOULE Joint Opportunities for Unconventional or Long-Term Energy

jour journal; journalese; journalism; journalist; journalistic; journey

journ journal; journalese; journalism; journalist; journalistic; journey

Journal-Net International Journalist Network

JOV Japanese Overseas Volunteers

Jove Jupiter or Zeus

JOVE Job Placement on the Job Training Vocational Education Educational Assistance; Jupiter Orbiting Vehicle for Exploration

JOVIAL Jules' Own Version of IAL (International Algebraic Language)

jp jet penetration; jet pilot; jet power; jet propulsion; junior partner; precipitation in sight but not at weather station reporting (symbol)

j & p joists and planks

Jp Japan(ese)

JP Japan (Internet code); Japan Press (news agency); Jaya Prakash Narayan, Prime Minister of India; Jet Pilot; Justice of the Peace

J.P. Jayaprakash Narayan; J Pierpont Morgan

JP Jerusalem Post

JP-4 jet propellant 4

jpa jack panel assembly

JPA Jamaica Press Association; Japan Petroleum Association; Japan Procurement Agency; Joint Passover Association; Joint Powers Agreement

JPB Joint Planning Board; Joint Production Board; Joint Purchasing Board; Judah Philip Benjamin

JPBHS Judah P Benjamin High School

jpbs jettison pushbutton switch

JPC Jan Pieterszoon Coen; Japan Productivity Center; Jet Propulsion Center; Joint Planning Center; Joint Planning Council; Joint Production Council; Joint Publishers Committee; Judge of the Prize Court

JPCAC Joint Production, Consultative, and Advisory Committee

JPCC Joint Petroleum Coordination Center (NATO)

JPCRSP John Pennekamp Coral Reef State Park (Florida)

JPDC Japan Petroleum Development Corporation

JPDR Japan Power Demonstration Reactor

JPF Jewish Peace Fellowship

j-p fuel jet-propulsion fuel

JPG J Peter Grace; Job Proficiency Guide; Joint Planning Group (NATO)

JPGA Japan Professional Golf Association

JPGM Paul Getty Museum

J Phys Journal de Physique (French—*Journal of Physics*)

J Phys Chem Journal of Physical Chemistry

J Phys Chem Ref Data Journal of Physical and Chemical Reference Data

J Phys Radium Journal de Physique et le Radium (French—*Journal of Physics and Radium*)

J Phys Soc Jpn Journal of the Physical Society of Japan

JPI Joint Packaging Instruction

JPIA Japan Plastics Industry Association

JPJ John Paul Jones

JPL Jacksonville Public Library; Java Pacific Line; Jet Propulsion Laboratory (California Institute of Technology); Job Parts List

J Plasma Phys Journal of Plasma Physics

JPM J(ohn) P(ierpont) Morgan

JPMA Juvenile Products Manufacturers Association

Jpn Japan(ese)

Jpn J Appl Phys Japanese Journal of Applied Physics

Jpn J Phys Japanese Journal of Physics

Jpns Japanese; Japan's

JPO Japan Patent Office; Joint Petroleum Office; Joint Program Office; Joint Project Offices; Junior Police Officer

J Pol Eco Journal of Political Economy

J Polymer Sci Journal of Polymer Science

jpp jälkeen puolenpäiven (Finnish—afternoon, P.M.)

JPPS Japan Pearl Promoting Society

JPPSOWA Joint Personal Property Shipping Office (Washington, D.C.)

JPR Joint Procurement Regulation(s)

J Prob Judge of Probate

JPRS Joint Publications Research Service

JPRST (guo) Joint Publications Research Service Translations (government use only)

JPs Jesuit priests

JPS Jet Propulsion Systems; Jewish Publication Society; Johannesburg Philharmonic Society; Joint Planning Staff; Juvenile Probation Services

J-P S Jean-Paul Sartre

JPSA Jewish Publication Society of America

JPSA Journal of Police Science and Administration

JPSO Jamaica Philharmonic Symphony Orchestra

jpt jet pipe temperature

JPT Journal of Petroleum Technology

JPTDS Joint Photographic Type Designation System

jpto jet-propelled take-off

JPV Japan Peace Volunteers

jpw job processing word

JP-X jet-propellant rocket fuel

JPz4-5 West German tank-destroyer tracked vehicle

jq job questionnaire

JQ Japan Quarterly; Journalism Quarterly

JQA John Quincy Adams (6th President U.S.)

JQAH John Quincy Adams House

J Quant Spectros Radiat Transfer Journal of Quantitative Spectroscopy and Radiative Transfer

jr jinx ratio

Jr Journal; Junior

JR Joint Resolution

JR Journal of Religion; *Jugoslav Register* (of shipping)

J.R. Jacobus Rex (Latin—King James)

jra junior rheumatoid arthritis

JRA Japan Racing Association; Japan Ryokan Association; Japanese Red Army (terrorist group)

JRAI Journal of the Royal Anthropological Institute

Jr Asst Pur Junior Assistant Purser

JRATA Joint Research and Test Activity

JRB New York, New York (Wall Street Heliport)

JRC Jamaica Railway Corporation; Japan Red Cross; Joint Research Centre (nine institutions in four European countries working on nuclear research); Joint Rivers Commission; Junior Red Cross

JRCA Junior Ruritan Clubs of America

JRCD Journal of Research in Crime and Delinquency (published semi-annually by NCCD)

jrci jamming radar coverage indicator

JRCS Jet Reaction Control System

JRD Riverside, California (heliport, 3-letter code)

JRDB Joint Research and Development Board

JRDC Japan Research and Development Corporation

JREA Japanese Railway Engineering Association

JREDS Jordan Royal Ecological Diving Society

J Res Nat Bur Stand Journal of Research of the National Bureau of Standards

JRF Job Request Form; Judicial Research Foundation

jrg jaargang (Dutch—year)

jr gr junior grade

Jr HS Junior High School

JRHS Julia Richman High School

jri jail release information

JRI Japan Research Institute

JRIA Japan Radioisotope Association; Japan Rocket Industry Association; Japan Rubber Industry Association

jrl journal

JRMO Junior Resident Medical Officer

JRN Japan Radio Network

Jro Jerome

JROTC Junior Reserve Officers' Training Corps

JRPG Joint Radar Planning Group

JRR Japan Research Reactor

JRRC Joint Regional Reconnaissance Center (NATO)

J.R.R. Tolkien John Ronald Reuel Tolkien

JRS Jerusalem, Jordan (airport)

JRSMA Japan Rolling Stock Manufacturers Association

JRSWG Joint Reentry System Working Group

JRT Jaguar Rover Triumph Inc; Job Relations Training

JRTUR Jugoslovenska Radio-Televisija Udruzenja Radiostancia (Yugoslav Association of Radio and Television Stations)

Jrw Jarrow-on-Tyne

j's joints (of marijuana)

j/s jamming-to-signal ration; joules per second

Js Jesuits

JS Al-Jainhourya as-Souriya (Syria); Jan Sibelius; Japan Society; Jet Study; Johnson Society; Judeo-Spanish; Judgement Summons; Judicial Separation; Junior Sailor

J-S Judeo-Spanish

JS-2 Soviet heavy tank of World War II vintage

JS-3 Soviet post-WWII heavy tank

JSA Jewelers Security Alliance; Journeymen Stone Cutters Association; Junior Statesmen of America

jsact jetstream anti-countermeasure trainer

JSACT Joint Strategic Air Control Team

JSAE Japan Society of Automotive Engineers

JSAP Japan Society of Applied Physics

JSB Jewish Society for the Blind; Jewish Statistical Bureau; Johann Sebastian Bach

JSBs Joint Stock Banks

JSC Jackson State College; Japan Science Council; Johnson Space Center (NASA); Joint Staff Council; Joint Standing Committee; Joint Stock Company; Justice Statistics Clearinghouse

JS-C Jesus College–Cambridge (also appears as JCC, J.C.C., and Jes Coll or Jes. Coll.)

JSCA Journeyman Stone Cutters Association

JSCC Japan Securities Clearing Corporation

J.Sc.D. Doctor of Juristic Science

J-school journalism school

J Sci Instrum Journal of Scientific Instruments

JSCM Joint Service Commendation Medal

JSCP Joint Strategic Capabilities Plan

JSCR Job Schedule Change Request

JSCS Joint Strategic Connectivity Staff

JS & CS Jewish Family and Child Services

JSD Jackson System Development

J.S.D. Jurum Scientiae Doctor (Latin—Doctor of the Science of Laws)

JSDA Japan Self-Defense Agency

JSDFA Japan Self-Defense Forces Academy

JSDFs Japan Self-Defense Forces

JSDT Sir John Sparrow David Thompson (Canada's 5th Prime Minister)

JSDTI John S Donaldson Technical Institute (Trinidad)

JSE Johannesburg Stock Exchange

JSEA Japan Ship Exporters Association

JSEE Japanese Society for Engineering Education

JSEM Japan Society of Electrical Discharge Machining; Japan Society for Electron Microscopy

JSESPO Joint Surface Effect Ships Program

Jsey Jersey

JSF Japan Scholarship Foundation; Jewish Student Federation; Junior Statesman Foundation

JSFC Japanese-Soviet Fisheries Commission

JSGMF John Simon Guggenheim Memorial Foundation

JSGMRAM Joint Study Group for Material Resource Allocation Methodology

jsi job satisfaction inventory

JSI Japanese Studies Institute

JSIA Japan Software Industry Association; Justice System Improvement Act

JSIF Japan Spinners Inspecting Foundation

JSIIDS Joint-Services Interior-Intruder Detection System

JSL Jurong Shipyard Limited

JSLB Joint Stock Land Bank(s)

JSLE Japan Society of Lubrication Engineers

JSLS Joint Services Liaison Staff

JSM Joint Staff Mission; Juilliard School of Music

JSMA Joint Sealers Manufacturers Association

JSMB Joint Sealift Movements Board

JSMDA Japan Ship Machinery Development Association

JSME Japan Society of Mechanical Engineers

JSMEA Japan Ship Machinery Export Association

J-smoke (underground slang —marijuana cigarette)

JSNP Japan Satellite News Pool

JSO Jackson Symphony Orchestra; Jacksonville Symphony Orchestra; Joint Services Organization; Judgement Summons Order

J Soc Arts *Journal of the Society of Arts*

JSOP Joint Strategic Objectives Plan

J Sound Vab *Journal of Sound and Vibration*

jsp jacketed soft point

JSP Japan Socialist Party

JSP *Jadranska Slobodena Plovida* (Yugoslavian Shipping Line)

JSPA Japan Screen Printing Association

JSPB Joint Staff Pension Board (UN)

JSPC Joint Strategic Plans Committee

J Speech Hear Disorders Journal of Speech and Hearing Disorders

J Speech Hear Res *Journal of Speech and Hearing Research*

jspf jet shots per foot

JSPF Joint Staff Pension Fund (UN)

JSPG Joint Strategic Plans Group

JSPS Japan Society for the Promotion of Science; Japan Sword Preservation Society

JSQC Japan Society for Quality Control

JSQS Japan Shipbuilding Quality Standard

jsr jump to subroutine

jsrt joint short range technology

JSS Johnson Scan Star (Johnson, East Asiatic, and Blue Star lines); Joint Services Standard

JSSA Japan Science Student Awards

JSSC Joint Services Staff College; Joint Strategic Service Committee

JST Japan Standard Time; Javanese Standard Time; Job Safety Training

J Stat Phys *Journal of Statistical Physics*

JSTB Jesuit School of Theology at Berkeley (California)

JSTC Japan-Singapore Training Center

J-stick joystick (underground slang—marijuana cigarette)

JSTPB Joint Strategic Target Planning Board

JSTPS Joint Strategic Target Planning Staff

JSU Jewish Student Union

JSU-122 Soviet 122mm assault-gun howitzer (SU-122)

JSU-152 Soviet assault-gun howitzer (SU-152)

J-S unit Junkerman-Schoeller unit (of thyrotrophin)

JSW Japan Steel Works

JSWPB Joint Special Weapons Publications Board

JSY Jersey Airlines

jt joint; joint tenancy; junction

JT Air Oregon (2-letter code); Jamaica Air Service (symbol); John Thomas (British slang—penis); John Tyler (10th President U.S.); joint tenancy; Juvenile Templar

JT *Japan Times* (Japan's oldest English newspaper); *Jerusalem Talmud*

JTA Japan Toilet Association; Jewish Telegraphic Agency (news service)

JTAC Joint Technical Advisory Committee

JTAD Joint Tactical Aids Detachment

JTAG Joint Test Action Group

jt agt joint agent

jt auth joint author

jtb joint bar

JTB Jamaica Tourist Board; Japan Travel Bureau; Jute Trade Board

JTBI Japan Travel Bureau International

JTC Japan Tobacco Corporation; Joint Technical Committee; Joint Telecommunications Committee; Junior Training Corps

JTCGALNNO Joint Technical Coordinating Group for Air-Launched Non-Nuclear Ordnance (DoD)

JTCGAS Joint Technical Coordinating Group for Aircraft Survivability (DoD)

jt comp joint compiler

jtda joint track data storage

jtde joint technology demonstration engine

J-teacher journalism teacher

Jt Ed Joint Editor

JTES Japan Techno-Economics Society

JTF Joint Task Forces

JTF-4 Joint Task Force 4

JTFOA Joint Task Force Operating Area

Jth. Apocryphal Book of Judith

Jthh justice the helping hand

JTI *Journal of the Textile Institute; Jydsk Teknologisk Institut* (Danish—Jutland Technological Institute)

JTIDS Joint Tactical Information Distribution System (USAF and USN)

JTII Japan Telescopes Inspection Institute

jtl just too late

jtly jointly

JTM & H *Journal of Tropical Medicine and Hygiene*

jtrns jamb-template machine screws

JTNM Joshua Tree National Monument

jto jump takeoff

JTO Jordan Tourist Office

J-town Juarez

JTPA Job Training Partnership Act

JTPT Job Task Performance Test

jt r joint rate

JTR Joint Termination Regulation; Joint Travel Regulation; Jordan Travel Research

JTRC Joint Theater Reconnaissance Committee (NATO)

JTRE Joint Tsunami Research Effort

JTRU Joint-Services Tropical Research Unit

JTS Jamaica Theological Seminary; Jewish Theological Seminary; Job Training Standards

JTS *Journal of Theological Studies*

JTSA Jewish Theological Seminary of America

JTSG Joint Trials Subgroup (NATO)

jtst jet stream

jt stk joint stock

JTTA Japan Table Tennis Association

jt ten joint tenancy; joint tenant

jt ten. joint tenant(s)

JTWC Joint Typhoon Warning Center

jtwros joint tenant with right of survivorship

J-type Jungian judging type

ju jackup (oil well); joint use

Ju June; Junkers

JU Jacksonville University; Jadavpore University

JU *Jeunesse Universelle* (French—World Youth)

JU-52 German Junkers transport developed before World War II and used by many airlines

juana marijuana

Juana *Juana la Loca* (Spanish—Crazy Jane)—nickname of the demented and lisping daughter of Ferdinand and Isabella; when queen of Castile in 1504 her courtiers flattered her by lisping in the manner still called Castilian; title of an opera by Gian Carlo Menotti—*Juana la Loca*

Juan Carlos Juan Carlos de Bourbon—chief of state and king of Spain

Juan Fernández Islas Juan Fernández (South Pacific islands Robinson Crusoe, Santa Clara and Alejandro Selkirk)

Juárez Ciudad Juárez (formerly El Paso del Norte)

jub jubilate

Jub. *Jubilees*

juco junior college

jucund. *jucunde* (Latin—pleasantly)

jud judgment; judicial; judo

jud. judicious

judic. judicial; judicious

Jud Judah; Judaic; Judaism; Judean; Judson

J.U.D. *Juris Utriusque Doctor* (Latin—Doctor of Civil and Canon Law)

Jud-Alg Judeo-Algerian (Algerian Jewish)

Jud-Amer Judeo-American (American Jewish)

Jud-Arg Judeo-Argentinian (Argentine Jewish)

Jud-Ash Judeo-Ashkenazic (Ashkenazic Jewish) or the oriental branch of Yiddish-speaking Jews of eastern Europe (*see* Jud-Sep)

Jud-Aus Judeo-Austrian (Austrian Jewish)

Jud-Aust Judeo-Australian (Australian Jewish)

Jud-Bel Judeo-Belgian (Belgian Jewish)

Jud-Bol Judeo-Bolivian (Bolivian Jewish)

Jud-Bra Judeo-Brazilian (Brazilian Jewish)

Jud-Bul Judeo-Bulgarian (Bulgarian Jewish)

Jud-Can Judeo-Canadian (Canadian Jewish)

Jud-Chi Judeo-Chilean (Chilean Jewish)

Jud-Chr Judeo-Christian (biblical and historic connection between Jews and Christians)

JUDCLA *Juventud Demócrata Cristiana Latino-Americana* (Spanish—Latin American Christian Democratic Youth)

Jud-Col Judeo-Colombian (Colombian Jewish)

Jud-CR Judeo-Costa Rican (Costa Rican Jewish)

judcrit judicial critic(ism)

Jud-Cub Judeo-Cuban (Cuban Jewish)

Jud-Cur Judeo-Curaçoan (Curaçoan Jewish)

Jud-Czech Judeo-Czechoslovakian (Czechoslovakian Jewish)

Jud-Dan Judeo-Danish (Danish Jewish)

Jud-Dut Judeo-Dutch (Dutch Jewish)

Jude The General Epistle of Jude

Jud-Ecu Judeo-Ecuadorean (Ecuadorean Jewish)

Jud-Egy Judeo-Egyptian (Egyptian Jewish)

Jud-Eng Judeo-English (English Jewish)

Judes Judesmo (Ladino)

Jud-Eth Judeo-Ethiopian (Ethiopian Jewish)

Jud-Fin Judeo-Finnish (Finnish Jewish)

Jud-Fre Judeo-French (French Jewish)

Judg. The Book of Judges

Judg *Judges*

Judge Adv Gen Judge Advocate General

Jud-Ger Judeo-German (German Jewish)—Yiddish dialect retaining many German words although written in Hebrew

Jud-Gib Judeo-Gibraltarian (Gibraltarian Jewish)

Jud-Gre Judeo-Grecian (Grecian Jewish)

judgt judgment

Jud-Guat Judeo-Guatemalan (Guatemalan Jewish)

Jud-His Judeo-Hispanic (Hispanic Jewish)—Portuguese-Spanish Jewish

Jud-HK Judeo-Hong Kongese (Hong Kongese Jewish)

Jud-Hung Judeo-Hungarian (Hungarian Jewish)

Jud-Ind Judeo-Indian (Indian Jewish)

Jud-Ire Judeo-Irish (Irish Jewish)

Jud-Irn Judeo-Iranian (Iranian Jewish)

Jud-Isr Judeo-Israeli (Israeli Jewish)

Jud-Itl Judeo-Italian (Italian Jewish)

Jud-Jam Judeo-Jamaican (Jamaican Jewish)

Jud-Jap Judeo-Japanese (Japanese Jewish)

Jud-Jor Judeo-Jordanian (Jordanian Jewish)

Jud-Lad Judeo-Ladino (Ladino Jewish)–Ladino-speaking Jews who settled in Muslim lands around the Mediterranean following the expulsion of the Jews from Portugal and Spain; Ladino combines medieval Castilian with Arabic, Hebrew, Turkish, and other elements local to places where they settled

Jud-Leb Judeo-Lebanese (Lebanese Jewish)

Jud-Mex Judeo-Mexican (Mexican Jewish)

Jud-Mor Judeo-Moresque (Moorish Jewish) also called Judeo-Moroccan (Moroccan Jewish)–Ladino-speaking Jews whose ancestors came to Morocco and other parts of northwest Africa when expelled from Spain by the Inquisition

Jud-Nor Judeo-Norwegian (Norwegian Jewish)

Jud-NZ Judeo-New Zealand (New Zealand Jewish)

JUDO Jews United to Defend Ourselves

Jud-Pan Judeo-Panamanian (Panamanian Jewish)

Jud-Par Judeo-Paraguayan (Paraguayan Jewish)

Jud-Per Judeo-Peruvian (Peruvian Jewish)

Jud-Pol Judeo-Polish (Polish Jewish)

Jud-Port Judeo-Portuguese (Portuguese Jewish)

Jud-Rho Judeo-Rhodesian (Rhodesian Jewish)

Jud-Rom Judeo-Romanian (Romanian Jewish)

Jud-Rus Judeo-Russian (Russian Jewish)

Jud-SAf Judeo-South African (South African Jewish)

Jud-Scot Judeo-Scottish (Scottish Jewish)

Jud-Sep Judeo-Sephardic (Sephardic Jewish)—Portuguese-Spanish Jewish or the occidental branch of European Jews who settled in Portugal and Spain before expulsion by the Inquisition (*see* Jud-Ash)

Jud-Sin Judeo-Singaporan (Singaporan Jewish)

Jud-Slav Judeo-Slavic (Slavic Jewish)

Jud-Span Judeo-Spanish (Spanish Jewish)—also called Ladino or Spanish Yiddish, a dialect composed of medieval Spanish plus Arabic, Hebrew, and Turkish terms; Ladino is heard around the Mediterranean from Morocco to the Balkans, Greece, and Turkey; Ladino is written in Hebrew characters

Jud-Sur Judeo-Surinamer (Surinamer Jewish)

Jud-Swe Judeo-Swedish (Swedish Jewish)

Jud-Swiss Judeo-Swiss (Swiss Jewish)

Jud-Syr Judeo-Syrian (Syrian Jewish)

Jud-Tun Judeo-Tunisian (Tunisian Jewish)

Jud-Tur Judeo-Turkish (Turkish Jewish)—Ladino-speaking Jews who settled in the Turkish Empire after their expulsion from Spain and Portugal by the Holy Inquisition

Jud-Uru Judeo-Uruguayan (Uruguayan Jewish)

Jud-Ven Judeo-Venezuelan (Venezuelan Jewish)

Judy Judith

Jud-Yem Judeo-Yemenite (Yemenite Jewish)

Jud-Yug Judeo-Yugoslavian (Yugoslavian Jewish)

jue jueves (Spanish—Thursday)

Juec Jueces (Spanish—Judges)

juev jueves (Spanish—Thursday)

Jug Jugoslavia (Yugoslavia)

JUG Joint Users Group

JUGC Jugolinija Container

Jugolinija Yugoslav Line

Jugoslav(ia)(n) Yugoslav(ia)(n)

Jugoslavija Yugoslavia

Jug(s) Jugoslavia(n)(s)

J.U.G.S. Justice, Unity, Generosity and Service

juil juillet (French—July)

jul julho (Portuguese—July); *julio* (Spanish—July)

Jul July

Jules Julian

Jul Caes Julius Caesar

Juliana Juliana Louise Emma Marie, Queen of the Netherlands, 1948–1980

Julians Julian Alps (northwestern Yugoslavia)

Juliet J-class Soviet-era submarines (diesel-powered and missile launching); letter J radio code

Jul^n Julián (Spanish—Julius)

Julust July and August

Jum Arab Yam al-Jumhuriyat al-Arabiyah al-Yanianiyah (Arabic—Yemen Arab Republic)—North Yemen, merged with Yemen on May 23, 1990

Jum Dji Jumhourīyya Djibouti (Arabic—Republic of Djibouti)

JUMIP Juror Utilization and Management Incentive Program

Jum 'Iraqia al Jumhouriya al' Iraqia (Arabic—Republic of Iraq)

Jum Lub al-Jumhouriya al-Lubnaniya (Arabic—Republic of Lebanon)

Jum Misr Jumhourīyya Misr al Arabiya (Arabic—Arab Republic of Egypt)

JUMPS Joint Uniform Military Pay System

Jum Qum Itt Islam Jumhurīyat al-Qumur al-itthadiyah al-Islamiyah (Swahili Arabic—Federal Islamic Republic of the Comoros)

Jum Tunis al-Jumhuriyah at-Tunisiyah (Arabic–Republic of Tunisia)

Juin Yam Dim Jumhuriyat al-Yaman ad-Dimuqratiyah (Arabic—People's Democratic Republic of Yemen)—South Yemen merged with North Yemen on May 23, 1990

jun juniore (Italian–junior); *junio* (Spanish—June)

Jun June; Juneau; Junior

Jun Julián (Spanish—Julius)

Junc Junction

Junct Junction

jun part. junior partner

junr. junior

Junuly June and July

Jup Jupiter

jur juridical

jur juridisch (Dutch—juridical); *juridisk* or *jurist* (Dano-Norwegian—legal or lawyer)

Jur Jurassic

Jur Juridisch (German—juridical)

jura la jura (Mexican gangster Spanish—the law, the police)

Juras Jura Mountains between France and Switzerland

Jur.D. Juris Doctor Latin–Doctor of Law

jurimet(s) jurimetrician(s)~ jurimetric(s)

juris jurisdiction

JURIS Justice Retrieval and Inquiry System (U.S; Department of Justice); Juvenile Referral Information System

jurisd jurisdiction

jurisp jurisprudence

Jus justice(s)

jus' just

Jus Justin

jusc. jusculum (Latin—broth)

JUSCIMPC Joint United States—Canada Industrial Mobilization Planning Committee (NATO)

JUSE Japanese Union of Scientists and Engineers

J-U.S. FC Japan-United States Friendship Commission

JUSMAG Joint United States Military Advisory Group; Joint United States Military Aid Group to Greece

JUSMAP Joint United States Military Advisory and Planning Group

JUSMG Joint United States Military Group

JUSMMAT Joint United States Military Mission for Aid to Turkey

JUSPAO Joint United States Public Affairs Office

juss jussive

Juss Jussieu

just. justification

Just Justinian

Justice Department of Justice; Hall of Justice; United States Department of Justice

Justin Justin cowboy boots (made by Joe Justin in Fort Worth, Texas)

JUSTIS Japan-United States Textile Information Service

JUT Joint European Torus

juv juvenile

Juv Juvenal

juve juvenile

juve delinq juvenile delinquent

juve gang juvenile gang

juven juvenile; juvenilization; juvenilized; juvenilizing

juvie juvenile delinquent; juvenile hall; juvenile law-enforcement officer

JUWTFA Joint Unconventional Warfare Task Force-Atlantic

JUWTF—P Joint Unconventional Warfare Task Force—Pacific

jux juxtapose; juxtaposition

jv japanese vellum; joint venture; jugular vein; jugular venom

Jv Java; Javanese

JV Jules Verne; Junior Varsity

JVA Jordan Valley Authority

J Vac Sci Technol Journal of Vacuum Science and Technology

JVC Japan Victor Company; Jewelers Vigilance Committee

jvp japanese vellum proofs; jugular venous pulse

jvp (JVP) jugular venous pulse

JVS Jewish Vocational Service; Joint Vocational School

jw jacket water; jugwell (hydrocarbon storage well); junior wolf (a young philanderer)

J & W Jackass & Western (8 miles of rail track; Jackass Flat, Nevada)

JW Jehovah's Witnesses

JWA Japan Whaling Association

jwac jacket water aftercooled

J-walk jaywalk (cross streets against traffic lights at any part of the street except the pedestrian crossing)

J-walker jaywalker

JWB Jewish Welfare Board Joint Wages Board; Joint Welfare Board

jwc junction wire connector

JWC Joint Working Committee

JWCA Japan Watch and Clock Association

JWDS Japan Work Design Society

JWEF Joinery and Woodwork Employers Federation

JWGA Joint War Games Agency

JWI Jack Winter (stock-exchange symbol)

JWJ James Weldon Johnson

JWJL JW Jagger Library (Cape Town)

jwl jewel; jeweler

JWL Johnston Warren Lines

jwlr jeweler

jwlry jewelry

j & wo jettison and washing overboard

JWO Jardine Waugh Organisation

JWPAC Joint Waste Paper Advisory Council

JWPT Jersey Wildlife Preservation Trust

JWR Joint War Room

JWR Jane's World Railways

JWs Jehovah's Witnesses

JWS Japan Welding Society

JWT J Walter Thompson (advertising agency)

JWTC Jungle Warfare Training Center

JWU Jewelry Workers' Union

JWV Jewish War Veterans (of the United States)

JW von G Johann Wolfgang von Goethe

JX Bougainville Air Service (2-letter code)

J.X. Jesus Christ

Jy Jenny; July; Jury

JY British United Channel Islands Airways (2-letter coding)

JYC Judicial Youth Corps

JYL Jugolinja-Yugoslav Line

Jyll Jylland (Danish—Jutland)

Jylland (Danish–Jutland)

JZP Jersey Zoological Park

JZS Jersey Zoological Society

JZS Jugoslovenski Zavod za Standardizacijd (Jugoslavian Standards Institution)

K

k Boltzmann constant; carat (karat); cathode or vacuum tube; coefficient of alienation; compressibility factor; cumulus (symbol); force constant; keel; kill; killed; kilo; kilo(gram)(s); king; knot(s); kweer (homosexual); place kicker or punter (football); reaction velocity constant; reproduction factor; thermal conductivity; torsion constant; unit vector in Z-direction

k (K) unit of computer memory capacity = 1000 (or 1024 in binary system of bytes, characters, or words)

k kontra (German—against, an octave lower); units of capital (microeconomics)

K 1024 storage bytes; capacity (symbol); centuple calorie (symbol); curvature (symbol); declared or paid this year and cumulative issue with dividends in arrears (in newspaper stock listings); dielectric constant (symbol); equilibrium constant (symbol); Fraunhofer line produced in part by calcium (symbol); go ahead (radiotelegraph symbol); hip; kaiser; Karman constant (symbol); Kawasaki Line; kelvin; Kelvin; Kerr constant; Kidde Fire Protection; kilobyte (symbol denoting 1024 units of stored matter); Kilo—code word for letter K; kilohm(s);

kilometer(s); Kindergarten; King(dom); Kiwanis International; Knabe; Köchel, cataloger of Mozart's music; kopec(s); kosher; krone; kroner; luminous efficiency (symbol); modulus of cubic compressibility (symbol); pilotless aircraft (symbol); potassium (kalium); proportionality constant (symbol); radius of gyration (symbol); smoke (aircraft code); strike-out (baseball); tanker (naval symbol); thousand

°K degree(s) Kelvin

K kade (Dutch—embankment, quay); *Kadenz* (German—cadenza); *kald* (Norwegian—cold); *kall* (Swedish—cold); *kalt* (German—cold); *koel* (Dutch—cold); *köld* (Danish—cold); *Köln* (German—Cologne); *kvinde* (Danish—women); *kvinne* (Norwegian—women); *kvinde* (Swedish—women); *kylmä* (Finnish—cold)

K² Mount Godwin Austen, Kashmir (28,250-ft mountain, second highest in the world)

K²K KEK to Kamioka

K⁵ Kunlun Mountain known on the Chinese-Kashmir border as Muztagh

K-9 Corps Canine Corps (staffed by police dogs)

k9p dog piss; urine produced by coyotes, dogs, foxes, hyenas, jackals, wolves, and other canines

K–12 kindergarten through 12th grade

K-61 Soviet amphibious-assault vehicle

K98k German carbine (World War II)

ka cathode(s); kiloampere(s)

k/a ketogenic to antiketogenic (diet ratio)

Ka auroral absorption index (symbol); kathode

Ka Komppania (Finnish—company)

KA Kapok Association; Karhumaki Airlines (Finland)

K-A King-Armstrong (units)

K of A King(dom) of Aragon

Ka-15 Soviet light-utility helicopter nicknamed Hen

Ka-18 Soviet utility-transport helicopter nicknamed Hog

Ka-20/Ka-25k Soviet helicopters built for military or commercial use with Ka-20 nicknamed Harp and Ka-25k nicknamed Hormone

Ka-25 Soviet armed helicopter called Hormone by NATO

kaa keep-alive anode

KAA Kwikasair (Australasia)

kaad kerosene, alcohol, acetic acid, dioxane (insect larva killer)

Kaap Kaapstad (Afrikaans—Cape Town)

Kab Kabel; Kabul

KAB Keep America Beautiful

KAC Kuwait Airways Corporation

KACC Kaiser Aluminum Chemical Corporation; Kansas Association of Community Colleges

KACF Korean American Cultural Foundation

KACIA Korean-American Commerce and Industry Association

KADA Kemubu Agricultural Development Authority; Kemuta Agricultural Development Authority

Kadet(s) (see *KD*)

Kae Katherine

KAESP Kansas Association of Elementary School Principals

Kaf kaffir

KAF Kenya Air Force

KAFB Kirtland Air Force Base

kaffir kaffir bean (African cowpea); kaffir beer; kaffir bread (*Encephalartos* fruit); kaffir (cat reputedly the ancestor of the common domestic cat); kaffir crane (black-plume gray crane of southern Africa); kaffir piano (southern African marimba); kaffir plum (edible fruit from southern Africa also called kaffir date or kaffir date plum)

Kafka Franz Kafka

Kagan Kaganovich

Kaganovich Lazar Moisevich Kaganovich

KAH Kahului Railroad

KAIIN Third word of Sen Nihon Kaiin Kumiai—the All Japan Seamen's Union

Kaimanawas Kaimanawa Mountains of New Zealand's North Island

Kajiwara Takuma Kajiwara

KAK *Kungliga Automobil Klubben* (Swedish—Royal Automobile Club)

kakis kakistocracy (government by the worst men)

kal kalamein

kal. *kalendae* (Latin—calends, the first day of the month)

Kal Kalana; Kalmar; Kalgoorlie

Kal *Kalendae* (Latin—first day of the Roman month)—day when debts were paid; *Kalium* (Latin—potassium)

KAL Korean Air Lines

Ka Lae Hawaii's southernmost point also called South Cape or South Point

Kalahari Kalahari Desert or Kalahari National Park in South Africa

kald kalamein door

kaleido kaleidoscope

Kaleun *Kapitänleutnant* (German—Commander)

Kali Kalimantan (Borneo)

Kal Nun *Kalaallit Nunaat* (Greenlandic—Greenland)

Kam *Kampong* (Malay—Village); *Kampuchea* (Khmer—Kampuchean People's Republic)—formerly Cambodia

KAM Kimball Art Museum (Fort Worth)

Kamarans Kamaran Islands in the Red Sea

kamb *kamboganisch* (German—Cambodian)

Kamchatka Pen Kamchatka Peninsula on the Bering Sea

Kamk keyed alike and master keyed

kamp known as male prostitute

Kamp Kampuchea (Cambodia)

Kampuchea Democratic Kampuchea (formerly Cambodia)

Kan Kansas; Kanpur

Kan *Kanal* (German—canal); *Kanaal* (Afrikaans or Dutch—canal)

Kanal Der Kanal (German—The Channel)—The English Channel

Kanchen Kanchenjunga; (28,146-foot-high mountain in the Himalayas, third highest in the world)

Kangeans Kangean islanders or Kangean Islands in the Java Sea north of Bali

kang(s) kangaroo(s)

Kano Eitoku Kano (late 16th-century Japanese painter)

Kans Kansas; Kansan

k antigen capsular antigen

KANU Kenya African National Union (party)

KANUPP Karachi Nuclear Power Plant

Kao kaolin

Kao Kaohsiung

KAO Kuiper Airborne Observatory

kaocon kaopectate concentrate

Kaoh Kaohsiung

kaolin aluminum silicate (Al_2O_3 $2SiO_2$ · $2H_2O$); kaolinite

kaos killing as an organized sport

kap knowledge, attitude, practice

kap *kapitel* (Dano-Norwegian—capital); *kapitel* (Swedish—chapter)

Kap (German— cape); *Kapital* [German—capital (money)]; *Kapitel* (Danish and German—chapter)

KAP Chinese Ministry of Public Security—external counterintelligence and internal secret police force of the People's Republic of China

KAPG Kluwer Academic Publishers Group

KAPL Knolls Atomic Power Laboratory

Kapo *Kamaradschafts Polizei* (German—police fellowship)—organization of non-political prisoners in prison camps

Kar Karachi; Karafuto

Kar *Karabiner* (German—carbine)—short rifle

K/Ar potassium-argon dating

KAR King's African Rifles

KARAI Karhumaki Airways (Finland)

Karakorums Karakorum Mountains of Kashmir

Kara Kum and Kyzyl Kum deserts of southern Russia

Karawankens Karawanken Alps between Austria and Yugoslavia

Karel Karelia; Karelian

Kar Fin Karelian Finland

Karimunjawas Karimunjawa Islands off the north coast of Djawa or Java

Karimuns Karimun Islands between Singapore and Sumatra

Karmen Karlsruhe-Rutherford

Kas Kansas

KAS Kentucky Academy of Science; Kroeber Anthropological Society

KASA Kentucky Association of School Administrators

KASC Knowledge Availability Systems Center

Kash Kashmir

KASSR Kalmyk Autonomous Soviet Socialist Republic; Karelian Autonomous Soviet Socialist Republic; Komi Autonomous Soviet Socialist Republic *kast kastilianisch* (German—Castilian)

kat katalog or *katolsk* (Dano-Norwegian—catalog or Catholic); *katalonisch* (German—Catalonian)

Kat Katmandu; Katowice

Kat Katar (Spanish—Quatar)

KAT Kenosha Auto Transport

kath katholisch (German—Catholic)

Kate Katharine; Katherine

Kath Katherine

Kath Katholik (German—Catholic)

Kathy Katharine; Kathleen; Kathryn

Katie Catherine; Katherine

Katmai Alaska's Katmai National Monument or its Valley of Ten Thousand Smokes or its Katmai Volcano creating the foregoing smoky valley in the Aleutian Peninsula

Kats Katangese

Katteg Kattegat (North Sea between Jutland peninsula of Denmark and west coast of Sweden)

KATUSA Korean (soldier) attached to (the) United States Army

Katy Missouri-Kansas-Texas Railroad

Katyn Katyn Forest massacre of 15,000 Polish officers and other prisoners of the Soviet Union, 1940

Katzen (German—Cats)—highest mountain in the Odenwald—Katzenbuckel or the village of Katzenellenbogen (Cats' Elbows) with its ancestral castle once inhabited by the counts and countesses Katzenellenbogen

Katzet Konzentrationslager (German—concentration camp)

Kauf Kaufman

kauk kaukasisch (German—Caucasian)

K-A units King-Armstrong units

KAVAS Knowledge Acquisition Visualization and Assessment Study

Kawa Kawasaki

Kawarthas Kawartha Lakes of southeastern Ontario

kay knockout (*kayo*—spelled abbreviation of ko); okay (truncated slang)

Kay Catherine

Kay-Cee Kansas City

Kaz Kazak(stan)

Kazak Kazakhstan(i)

Kazoo Kalamazoo

kb keyboard; kilobase; kilobit(s); kilobyte (1024 characters); kitchen and bathroom; kite balloon; knee brace

k & b kitchen and bathroom

Kb Kontrabass (German—double bass)

KB Knight Bachelor; Koninkrijk Belgie (Flemish—Kingdom of Belgium)

K.B. King's Bench; Knight of the Order of the Bath

K of B King(dom) of Bavaria

KB Kongelige Bibliotek (Danish—Royal Library)—in Copenhagen; *Koninklijke Bibliotheek* (Dutch—Royal Library)—in The Hague; *Koninkrijk Belgie* (Flemish—Kingdom of Belgium); *Kungliga Biblioteket* (Swedish—Royal Library)—Stockholm

kba killed by air

KBAI Koninklijke Bibliotheek Albert I (Flemish—Albert Ist Royal Library)—see *BrA*

K-band 10,900–36,000 mc

kbar kilobar(s); 1 kbar equals approx 14,500 lbs per square inch

KBART Kings Bay Army Terminal

KBASSR Kabardino-Balkar Autonomous Soviet Socialist Republic

KBC King's Bench Court; Kyushu Asahi Broadcasting

kbd keyboard

KBD King's Bench Division

kbe keyboard encoder; keyboard entry; knotted both ends

K.B.E. Knight Commander of the Order of the British Empire

kbes knowledge-based expert system

kbh killed by helicopter

Kbh København (Dano-Norwegian—Copenhagen)

Kbhvn Köbenhavn (Copenhagen)

KBI Keyboard Immortals (record label); Klan Bureau of Investigation (Ku Klux Klan)

KBIM Kongres Buruh Islamic Merdeka (Indonesian—Islamic Trade Union Congress)

K-bit unit of computer storage capacity equal to 1024 bytes

KBL Kabul, Afghanistan (airport)

KBL Kilusang Bagong Lipunan (Filipino—Philippines New Society Movement)

kbm keyboard monitor

KBNWR Klamath Basin National Wildlife Refuges (California and Oregon)

K Bon Klein Bonaire (Netherlands Antilles)

kbp kilobase pair

KBP Koala Bear Park (Adelaide)

kbps kilo bits per second

kbs kilobits per second

KBS Kinki Broadcasting System (Japan); Korean Broadcasting System

KBS Kämbränslesakerhet (Swedish—Nuclear Fuel Safety Project)

KBSI Kongres Buruh Seluruh Indonesia (Indonesian—Indonesian Trade Union Congress)

KB & TS Kuwait Broadcasting and Television Service

kbtu kilo British thermal unit (1,000 btu's)

kbv kauri-butanol value

KBW Klan Border Watch (along the Mexican border)

kc kilocycle(s); koruna (Czechoslovakian monetary unit)

Kc Kyle classification (social sciences)

KC Kalamazoo College; Kansas City; Keble College, Oxford; Kendall College; Kennedy Center; Kennel Club; Kenyon College; Keuka College; Keystone College; Keystone Shipping Company (flag code); Kilgore College; King College; King's College; Kirksville College (of osteopathy and surgery); Knox College; Knoxville College

K.C. King's Counsel; Knight Commander

K of C Knights of Columbus

KC-10A advanced tanker-cargo aircraft

KC-50 tactical aerial tanker for refueling aircraft in flight

KC-97 Stratofreighter strategic tanker-freighter equipped for inflight refueling

KC-130 Lockheed Hercules tanker aircraft

KC-135 Stratotanker multipurpose aerial tanker-transport

KCA Kitchen Cabinet Association

KCA Keesings Contemporary Archives

kcal kilocalorie(s)

kcas knots calibrated air speed

k/cb keel combined with centerboard

KCB Kenya Commercial Bank

K.C.B. Knight Commander of the Order of the Bath

KCBT Kansas City Board of Trade

kcc kathodic closure contraction; keyboard common contact

KCC Kellogg Community College; Kenai Community College; Kennedy Cultural Center; Kern County College; Ketchikan Community College; Kingsborough Community College; King's College, Cambridge

KC & C Kembla Coal and Coke

KCCD Kentucky Council on Crime and Delinquency

KCCE Keystone Center for Continuing Education

KCCI Korean Chamber of Commerce and Industry

kcd kilocandelas

KCDMA Kiln, Cooler, and Dryer Manufacturers Association

kcf thousand cubic feet

KCFF Korean Cultural and Freedom Foundation

Kch Kuching

KCH King's College Hospital; Knight Commander of the Order of Hanover

K.C.H.S. Knight Commander of the Order of the Holy Sepulchre

kCi kiloCurie(s)

KCI Key Club International

KCIA Korean Central Intelligence Agency

K.C.I.E. Knight Commander of the Indian Empire

KCl potassium chloride

KCL Kai Curry-Lindahl; King's College, London; Kirchoff's Current Law

KCLA Known Coal-Leasing Area(s)

KCLY Kent and County of London Yeomanry

KCM Kansas City Museum

KCMA Kitchen Cabinet Manufacturers Association

KC Mag Kansas City Magazine

kcmG kindly call me God

K.C.M.G. Knight Commander of the Order of Saint Michael and Saint George

KCM & O Kansas City, Mexico & Orient (railroad)

kcmx keyset central multiplexer

KCNA Korean Central News Agency

KCNP Kings Canyon National Park (California); Ku-ring-gai Chase National Park (New South Wales)

KCNS King's College, Nova Scotia

KCOBE Knight Commander—Order of the British Empire

KCP Key Curriculum Project

KCPA Kaolin Clay Producers Association; Kennedy Center for the Performing Arts

KCPL Kansas City Public Library

KCPO Kansas City Philharmonic Orchestra

kcps kilocycles per second

KCR Kowloon-Canton Railway

Kcs Czechoslovakian koruna(s); kilocycles per second; thousand characters per second (symbol)

kc/s kilocycles per second

KCS Kansas City Southern (railroad)

KCS Kansas City Star

KCSC Kansas Cosmophere and Space Center

kc/sec kilocycles per second

K.C.S.G. Knight Commander of Saint Gregory the Great

KCSI Knight Commander of the Star of India

KCSO Kansas City Symphony Orchestra

KCT Kansas City Terminal (railroad)

kcte kathodic closure tetanus

K Cur Klein Curaçao (Netherlands Antilles)

KCVO Knight Commander of the Victorian Order

kd key depression(s); killed; kiln dried; knocked down; known distance; pilotless aerial target (code)

Kd Konrad; Kuwait dinar(s)

KD Kidderpore Docks (Calcutta)

K of D King(dom) of Denmark

KD Kampuchea Democratique (French—Democratic Kampuchea)—formerly Cambodia; *Kongeriget Danmark* (Kingdom of Denmark); *Konstitutsionnodemokraticheskaya partiya* (Russian—Constitutional Democratic Party)—party of the KDs or Kadets liquidated by the Bolsheviks under Lenin

KDA Kongelige Danske Aeroklub (Danish—Royal Danish Flying Club)

KDAK Kongelig Dansk Automobile Klub (Danish—Royal Danish Automobile Club)

K Dan Vidensk Selsk, MatFys Medd Kongelige Danske Videnskabernes Selskab, Matematisk-Fysiske Meddelelser (Danish—Royal Scientific Society, Mathematical and Physics Announcements)

KDAR Klein-Drohne Anti-Radar device

K-day basic date for introduction of convoy system or lane; carrier aircraft assault day

KDB Korea Development Bank

KDC Key Distribution Center

kdcl knocked down in carload lots

KDD Kokusai Denshin Denwa (Japan's Overseas Radio and Cable System)

kdf knocked-down flat

K d F Kraft durch Freude (German—Strength through Joy)—Nazi holiday association

KDG King's Dragoon Guards

KDHNM Kill Devil Hill National Memorial

KDI Kwaliteitsdienst voor de Industrie Stichting (Dutch—Industrial Quality Control Society)

kdlcl knocked down in less than carload lots

kdly kindly

kdm kingdom

KDM *Kongelige Danske Marine* (Danish—Royal Danish Navy)

K d N *Koninkrijk der Nederlanden* (Dutch—Kingdom of the Netherlands)

Kdo Kasado

K-do *Kamarado* (Esperanto—comrade)

KDOM Kosovo Diplomatic Observer Mission

KDP potassium dihydrogen phosphate

KDs Kadets

kdv kiln-dried veneer

ke kinetic energy

K$_e$ exchangeable body potassium

KE Kaiser Engineers; Kenya (Internet code)

K-E Krafft-Ebing

K + E Keuffel & Esser

K of E King(dom) of England; Knights of Equity

KEA Kentucky Education Association; Kiwifruit Exporters Association; Knitwear Employers Association

Keams Keams Canyon, Arizona, Hopi Indian Reservation headquarters

keas knots equivalent airspeed; knots estimated airspeed

Keb Coll Keble College—Oxford

KEBK Korea Exchange Bank

Keble Keble College, Oxford

Kech Kechua (Quechua)

KECO Korea Electric Company

KEDDS Kansas Education Dissemination/Diffusion System

Kee Keelung

Keel Keeling

KEEP Kentucky Environmental Education Program

KEF Keflavik Airport, Iceland

KEGG *Kyoto Encyclopedia of Genes and Genomes*

KEHF King Edward's Hospital Fund

KEK Fort Wayne Kekiongas (National Association)

Kel Kiel (British maritime abbreviation)

Kel. Kelim

KELP Kindergarten Evaluation of Learning Potential

KELTS Key English Language Teaching Scheme

kem *kemisk* (Dano-Norwegian—chemical)

KEMA Kitchen Equipment Manufacturers Association

KEMA *Keuring van Electrotechnische Materialen* (Dutch—Testing Institute for Electrochemical Materials)

Kemp Richard M Kemp, American naturalist; Kemp's sea turtle, sometimes known as Kemp's Bastard, thought to be a cross between Green and Loggerhead sea turtles

Ken Kendal(l); Kendrick(k); Kenelm; Kenilworth; Kenley; Kenna(rd); Kennedy; Kennet; Kenneth; Kennit; Kenny; Kenric(k); Kensell; Kensington; Kent(on); Kentuckian; Kentucky; Kenward; Kenwood; Kenya; Kenyan; Kenyon

Ken *Kenia* (Spanish—Kenya)

Kenai Kenai Fjords National Park, Alaska

Kennedy John F Kennedy (his brothers and others); John F Kennedy International Airport (New York)

Kens Kensington

Kent Kentucky

Kentuck Kentucky

Kenyatta Jomo Kenyatta (Kamau Ngengi)

Keogh Keogh Retirement Plan

kep key-entry processing

kep' kept

Kep *Kepulauan* (Indonesian or Malay—archipelago)

KEPCO Kyushu Electric Power Company

KEPZ Kaohsiung Export Processing Zone

Ker. Keritot

Kerala southwestern India including most of the Malabar and Travancore coasts

kerat keratometric(al)(ly); keratometry

Kerguelens Kerguelen Islands in the subantarctic South Indian Ocean

kerk *kerkelijke term* (Dutch—ecclesiastical term)

Kermadecs Kermadec Islands

kern kernan

kero kerosene

KERO Kuwait Emergency Reconstruction Office

KESCO Kowloon Electricity Supply Company

Ket. Ketubbot

keto ketonaemia; ketogenic; ketone; ketonuria; ketoses; ketosis

ketol ketone alcohol (compound)

kev kilo electron volt; 1,000 electron volts

keV kiloelectronvolt(s)

Kev Kelvin; Kevin

KEVs King's Empire Veterans

kew (KEW) kinetic energy weapon

Kew Kew Gardens (Royal Botanic Gardens outside London)

Kew Gar Kew Gardens

Kew Obs Kew Observatory

key keep extending yourself

keyboard computer input device, to input computer data; keyboard instruments: celestas, clavichords, concertinas, harpsichords, organs, piano accordions, pianolas, pianos, virginals

keyper key personnel; keywords permuted

Keys the Keys (the Florida Keys)

kf kitchen facilities; koff

KF Kaiser-Frazer; Kellogg Foundation; Kent Foundation; Kidney Foundation; *Kooperativa Forbunded* (Federation of Cooperatives—Sweden); Kresge Foundation

K of F King(dom) of France

KF *kleine Flöte* (German—small flute or piccolo); *Konservative Folkeparti* (Danish—Conservative Party); *Kontrafagott* (German—double bassoon); *Kooperativa Forbunded* (Swedish—Federation of Cooperatives)

KFA Kenya Farmers Association; Krishnamurti Foundation of America

KFAS Kuwait Foundation for the Advancement of Sciences

KFASSR Karelo-Finnish Autonomous Soviet Socialist Republic (formerly the Karelia of Finland)

Kfc Kentucky fried chicken

KFC Kentucky Fried Chicken; Kropp Forge Company

KFEA Korean Federation of Education Associations

kff keep from freezing

KFH Kaiser Foundation Hospitals

KFL Kenya Federation of Labour

kfm kaufmännisch (German—commercial)

Kfm Kaufmann (German—merchant)

KFNP Kaieteur Falls National Park (Guyana)

kfo killing federal officer

KFP Kristelig Folkeparti (Norwegian—Christian People's Party)

KFPC Kansas Foundation for Private Colleges

K-F s Klippel-Feil syndrome

KFSR Karakul Fur Sheep Registry

KFT Kansai Fishing Tackle

KFUK Kristelig Forening for Unge Kvinder (Danish—Young Women's Christian Association)

KFUM Kristelig Forening for Unge Maend (Danish—Young Men's Christian Association)

Kfz Kraftfahrzeug (German—motor vehicle)

kg keg; kilogram; known gambler

/kg per kilogram

kG kilogauss

Kg Kirghiz(ian)

Kg Kampong (Malay—village); *Kompong* (Indo-Chinese—landing place, riverside)

KG Kelly Girl; Kyrgyzstan (Internet code)

K-G Kanematsu-Gosho Ltd.

K.G. Knight of the Order of the Garter

K of G King(dom) of Granada

KG Kommanditgesellschaft (German—limited partnership)

KG5 King George V School

KGA Kitchen Guild of America

KGB Komitét Gosudársvennoi Bezopásnosti (Russian—Committee of State Security, Soviet Secret Police)

KGBW Kewaunee, Green Bay, and Western (railroad)

kgC kilogram-calorie

KGC Knights of the Golden Circle

K.G.C. Knight of the Grand Cross

kg cal kilogram calorie

kg-cal kilogram calorie

K.G.C.B. Knight of the Grand Cross of the Bath

kg/cm kilograms per centimeter

kg cum kilograms per cubic meter

kgf kilogram-force

Kgf Kriegsgefangener (German—prisoner of war)

KGFS King George's Fund for Sailors

kg/hl kilograms per hectoliter

kg/hr kilograms per hour

KGIS Kuder General Interest Survey

KGJT King George's Jubilee Trust

KGK Kabushiki Goshi Kaisha (Japanese—joint stock limited partnership of members with unlimited liability and shareholders with limited liability)

kgl kongelig (Dano-Norwegian—royal)

Kgl Königlich (German—royal)

kgm kilogram meter

kg/m² kilograms per square meter

kg/m³ kilograms per cubic meter

Kg/M³ kilograms per cubic meter

kg/ms kilograms per meter second

Kgn Kingston, Jamaica

KGNP Kalahari Gemsbok National Park (South Africa); Katherine Gorge National Park (Australian Northern Territory)

kgps kilograms per second

kgra known geothermal resource area

kgs kegs; kilograms

kg/s kilograms per second

KGS Kate Greenaway Society; Kigezi Gorilla Sanctuary (Uganda); Korean Geological Survey

kgs/ha kilograms per hectare

KG St J Knight of Grace of the Order of Saint John of Jerusalem

kg U kilogram of uranium

KGV Knight of Gustavus Vasa

KGVDs King George V Docks (London)

KGWS Keoladeo Ghana Wildlife Sanctuary (India)

kh keyhole

kH kilohertz

Kh Khmer (Cambodia)

Kh Khawr (Arabic—creek, inlet, ravine, water-course)

KH Cambodia (Internet code); King's Hussars; Knut Hamsun; Knight of Hanover

K-H Kelsey-Hayes

K of H King(dom) of Hungary

KH Karen Hayesod (Hebrew—United Israel Appeal); *Københavns Handelsbank* (Danish—Copenhagen's Commercial Bank); *Kuput Holim* (Hebrew—Health Insurance Fund)

KH-4 Kawasaki all-purpose helicopter similar to the Bell 47

KH-11 American-made intelligence-gathering satellite

KH IV King Henry IV

KH VI King Henry V1

kha killed by hostile action

Khar Kharkov

Khazar Bahr-ul-Khazar (Arabic—Caspian Sea)

KHC Karen Horney Clinic

KHDS King's Honorary Dental Surgeon

Khi Karachi

KHI Karachi, Pakistan (airport)

Khingans Kinghan Mountains of northeast China

K-H layer Kennelly-Heaviside layer

KHL Koninklijke Hollandsche Lloyd (Dutch—Royal Holland Lloyd)

KHM King's Harbour Master

Khn Knoop hardness number

KHNS King's Honorary Nursing Sister

Khois. Khoisan

khp kilohorsepower (hour)

KHP King's Honorary Physician

KHPC Karen Horney Psychoanalytic Clinic

Khr Khrebet (Russian—mountain range)

KHRI Kresge Hearing Research Institute

KHS Kennedy High School; King's Honorary Surgeon

khz (kHz) kilocycle(s)/second; kilohertz, formerly kilo cycle(s) per second

ki karyopyknotic index; kilo; kitchen

KI Kathleen Investments; Kiribati (Internet code); Kiwanis International; potassium iodide

K-I Kaiser-Illin

K of I King(dom) of Ireland; King(dom) of Italy

KI Kol Israel (Hebrew—Voice of Israel)—broadcasting service; *Kommunisticheskii Internatsional* (Russian—Communist International); *Komunisticna Internacijonala* (Yugoslav—Communist International)

kia (KIA) killed in action

Kia Kligler iron agar

kias knots indicated airspeed

KIB Kansas Inspection Bureau; Kentucky Inspection Bureau

Kib Cum Kibris Cumhuriyeti (Turkish—Republic of Cyprus)

Kibo Mount Kibo (Africa's highest peak also called Kilimanjaro)

KIC Kaktovik Inupiat Corporation

KICF Kentucky Independent College Foundation

kid. kidney

KID Key Industry Duty

kidult kid adult (older person who enjoys juvenile entertainment)

kidvid children's television program; children's tv or video programs

kidzines magazines for children

Kierkegaard Søren Kierkegaard (Danish philosopher)

Kiev Kiev-class 40,000-ton Soviet aircraft carrier

Kifis Kollsman integrated flight instrument system

K-i-H Kaiser-i-Hind (Emperor of India medal)

KIICC *Kommunisticheskaya Partiya Sovetskogo Soyuza* (Russian—Communist Party of the Soviet Union)

Kikdl Krokodil

kikú (Japanese—chrysanthemum)—symbol of Japan's highest order for men

kil (Dutch—channel, estuary, strait)—as in Arthur Kill, French Kill, and Kill van Kull along the shores of New York's Staten Island

kild kilderkin(s)

Kild Kildare

Kili Kilimanjaro

Kilk Kilkenny

Kill van Kill van Kull (waterway between Bayonne, New Jersey and Port Richmond, Staten Island, New York)

kilo Kilogram; 10^3

Kilo letter K radio code

kilobrick(s) kilo-weight brick(s) of marijuana measuring about $1/2 \times 5 \times 12$ inches ($64 \times 127 \times 300$ millimeters)

kilohm kilo-ohm

kilovar kilovolt-ampere (reactive)

Kim Kimball; Kimballton; Kimberley; Kimberly; Kimble; Kimbolton; Kimborough; Kimbrough; Kimiwan; Kimmell; Kimmins; Kimmswick; Kimsquit

K i M Knudsen i Marken

Ki-mi-ga yo wa (Japanese—May thy peaceful reign last long)—national anthem

kin kinesisk (Dano-Norwegian—Chinese)

Kin Frank McKinney Hubbard; Kingston, Ontario (maritime contraction)

KIN Kingston, Jamaica (airport); Kinross

kina monetary unit of Papua, New Guinea

Kinc Kincardinel

kind. kindergarten

kine kinema (variation of cinema)

Kines J M Keynes (pronounced as italicized)

kinesi kinesics; kinesiologist; kinesiology

King Kingston

King of the Congo Leopold II of Belgium (1835–1909)

kingd kingdom

Kingdome Kingdome Stadium, Seattle

King James King James Version of the Bible (authorized by King James I of England in 1611)

King Karls King Karl Islands in the Norwegian sector of the Arctic

King Leopolds King Leopold Ranges of northern Western Australia

Kings X Sta King's Cross Station (rail terminal)

King's King's College (Cambridge, Columbia)

Kings Canyon Kings Canyon National Park in central California

Kings Point United States Merchant Marine Academy at Kings Point, New York

Kinr Kinross-shire

kinsym kinematic synthesis

KINTEL K Laboratories (instruments and television)

Kintetsu Kinki Nippon Railway Company, Ltd

kip thousand pounds (from contraction of kilo and pound)

Kip Kipling (Joseph Rudyard Kipling)

KIP Kennedy Institute of Politics (Harvard)

kip ft thousand foot pounds

KIPP Knowledge Is Power Program

KIPS Knowledge Information Processing System(s)

kiq (KIQ) key intelligence questions

Kir Kirghiz; Kirghizia; Kirghizian; Kiribati

Kircud Kircudbrightshire (*Kircoobrisheer*)

Kiri Kiribati (whose capital is Tarawa)

Kiribati Republic of Kiribati (Gilbert and Ellice islands colony in the equatorial Pacific, includes Tarawa)

Kirk Kirkham; Kirkland; Kirkudbright (*Kircoobri*); Kirkwood

Kirov Sergei Mironovich Kostrikov; Soviet name for Viatka

Kirsten Kirsten Flagstad (Wagner soprano)

KISA Karaoke International Sing-Along Association; Korean International Steel Associates

kisc knowledge industry system concept

kismif keep it simple—make it fun

KI smog potassium-iodide smog (automobile induced)

KISO Kol Israel Symphony Orchestra

KISR Kuwait Institute for Science Research

kiss keep it simple, stupid

kiss. keep it simple, sir; keep it simple, stupid

KISS Kids In Safety Seats (automobile safety program); Knowledge-Based Interactive Signal Monitoring System

KIST Korean Institute for Science and Technology

kit. key issue tracking; kitchen(ette); kitten; kitty

Kit Kitty

KIT Kentucky and Indiana Terminal (railroad); Korean International Telecommunications

KIT Koninklijk Insturit voor de Tropen (Dutch—Royal Institute for the Tropics)

kitch kitchen

Kitch Kitchener

KITCO Kwajalein Import and Exporting Company

kiteoon kite + balloon

kitin' kiting (money)

kits. kittens

kitsch kitschen (German—thrown together)—commercial art or art objects cheapened by vulgarity; e.g., miniature reproduction of the Venus de Milo with an alarm clock set in her belly

Kitsch Kitschmensch (German—kitschman)—anyone creating, dealing in, or displaying artistic rubbish—junk art

KIVI Koninklijk Instituut van Ingenieurs (Dutch—Royal Institution of Engineers)

KIWA Keurings Instituut voor Waterleiding Artikelen (Dutch—Inspection Institute for Waterworks Equipment)

kj killer judo; kilojoule; kimberly joint (lumbing); knee jerk; kraut joint; krystal joint (pcp)

k-j knee-jerk(s)

kJ kilojoule

KJ Kahlil Jibran (Gibran)

KJ King James (version of the Bible)

KJB Korea-Japan Board

KJC Kaiser Jeep Corporation; Keystone Junior College

K John Life and Death of King John

KJ.St.J. Knight of Justice, Order of Saint John of Jerusalem

KJV King James Version

kk killer karate

k-k knee-kicks (knee-jerks)

kK kilokelvin

KK Karl Kautsky (German Socialist); Kazakhstan (Internet code)

K-K Krupp-Koppers

K of K Kitchener of Khartoum

KK Kabushiki Kaisha (Japanese-joint stock company of shareholders with limited liability); *Kaiserlich Königlich* (German—Imperial Royal)

K.K. Kahal Kadosh (Hebrew—Holy Congregation)

KKASSR Kara-Kalpak Autonomous Soviet Socialist Republic

KKH Karakoram Highway

KKI Keren Kayemeth le Israel (Hebrew—National Fund of Israel)

kkk kikes (Jews), *koons* (Afro-Americans), *katholics* (Roman Catholics) according to the Ku Klux Klan

KKK Ku Klux Klan (secret organization antagonistic to certain racial & religious groups)

KKK Kataas-taasan Kagalan-ggalan-gang Katipunan (Filipino—Mightiest Warriors Fighting for Freedom); *Kinder, Kirche, Küche* (German—Children, Church, Kitchen)—traditional three Ks of Teutonic womanhood

KKKK Kansai Kisen Kabushiki Kaisha; Kawasaki Kisen Kabushiki Kaisha (steamship lines)

KKKK København s Kul og Koks Kompagnie (Danish—Copenhagen Coal and Coke Company)

KKKKs Knights of the Ku Klux Klan

KKKUK Ku Klux Klan in the United Kingdom

KKL Karlander Kangaroo Line

kklmg koom, koom, lauz, mere gain (Yiddish—come, come, let us go)

KKMKI Kungliga Karolinska Mediko-Kirurgiska Institutet (Sweden—Caroline Medico-Surgical Institute-Stockholm)

KKO Korps Kommando (Bahasa-Indonesian-Malay—Commando Corps)—marine corps

Kkr Karlskrona

kkv (KKV) kinetic kill vehicle

kl key length; kiloliter

kl klasse or *klokken* (Dano-Norwegian—class or o'clock); *klockan* (Swedish—o'clock)

Kl Klarinette (German—clarinet); *Klasse* (German—class); *Klein(e)* (German—little, small)

KL Key Largo; Klebs-Loeffler; Knutsen Line; Kuala Lumpur; Kwik Lok

KL King Lear

kla Klavier; klystron amplifier

KLA Kansas Library Association; Karachi Library Association; Kentucky Library Association; Korean Library Association; Kosovo Liberation Army

Klamaths Klamath Mountains bordering California and Oregon

Klan Ku Klux Klan (*q.v.* KKK)

Klar Klarinette (German—clarinet)

Klaus Nikolaus

Klav Klavier (German—piano); *Klavierkonzert* (German—piano concerto)

klax klaxon

K-L bacillus Klebs-Loeffler bacillus (diphtheria)

KLC Kaingaroa Logging Company

kld kaelder (Dano-Norwegian—basement or cellar)

Klebs Klebsiella

Klem Klemens; Klement; Klementi; Kliment

klepto kleptomania(c, al)

kl Fl kleine Flöte (German—piccolo)

Klg Keelung

klh keyhole limpet hemocyanin (KLH)

KLIAU Korean Land Improvement Association Union

klic key letter in context

klicks kilometers

klieg klieg light (named for German-American inventor-brothers Anton and J H Kliegl)

klim (milk spelled backwards) dried milk

K Line Kawasaki Kisen Kaisha

KLM Koninklijke Luchtvaart Maatschappij (Dutch—Royal Dutch Airlines)

Klmpb Klampenborg

Kln Köln (Cologne)

klo klystron oscillator

k-lo *kello* (Finnish—hour, o'clock)

KLPA Knuckeys Lagoon Protected Area (Australian Northern Territory)

KLr Kuala Lumpur

kls key lock switch

k-l-s kidney-liver-spleen

KLSE Kuala Lumpur Stock Exchange

klt kiloton (nuclear equivalent, 1,000 tons of high explosives)

klto knurling tool

Kluxer member of the Ku Klux Klan (*q.v.* KKK)

km kilometer

kM kilomega (^{109} giga)

Km Kingdom

KM Comoros (Internet code); Kaffrarian Museum; Kearny Mesa; Khedivial Mail (steamship line)

K-M Krauss-Maffei

K.M. Knight of Malta

K & M King and Martyr (Charles Ist's sobriquet)

km² square kilometer

km³ cubic kilometer

KMA Kalgoorlie Mining Associates; Kinematograph Manufacturers Association

KMAG United States Military Advisory Group to the Republic of Korea

KMB Kowloon Motor Bus

kmc kilomegacycle

KMD Kentucky Manpower Development

kmef keratin, myosin, epidermin, fibrin (proteins)

KMF Koussevitzky Music Foundation

km/h kilometers per hour

KMH Kleinhans Music Hall (Buffalo)

KMI Kentucky Military Institute

KMIDC Korean Marine Industry Development Corporation

KMIT four-letter name bestowed by Trotsky on the Bolshevik ministry of foreign affairs; when foreign journalists demanded to know what KMIT stood for his aides confided it was Yiddish for *küss mir im tuchus* (kiss my ass)

km/l kilometers per liter

KMMA Korean Merchant Marine Academy

KMO Kobe Marine Observatory

KMP Kaiser Metal Products; Kearny Mesa Plant (Convair)

KMPA Korean Maritime and Port Administration

kmph kilometers per hour

kmps kilometers per second

Kmr Khorramshahr

KMR Kwajalein Missile Range

kms kilometers

KMS Kansas Medical Society; Keeve M Siegel

KMT Key Master Terminal; Kuomintang

KMTC Korea Marine Transport Company

KMUB *Karl-Marx-Universitäts Bibliothek* (German—Karl-Marx University Library)—on Beethovenstrasse in Leipzig

KMUL *Karl Marx Universität Leipzig* (German—University of Leipzig)

kmv killed measles-virus vaccine

kmw kilomegawatt

KMW *Karlstads Mekaniska Werkstad* (Swedish iron foundry)

kmwhr kilomegawatt-hour

kn kilonewton; knife (philatelic stationery); knot; krone; kronen

kn. known

Kn Knight

KN Saint Kitts and Nevis (Internet code)

KN *Koninkrijk der Nederlanden* (Kingdom of the Netherlands); *Kongeriket Norge* (Kingdom of Norway)

K-N Know-Nothing (political party)

K of N King(dom) of Naples; King(dom) of Navarre; King(dom) of Norway

KNA Kenya News Agency; Korean National Airlines

KNA *Kongelig Norsk Automobilklub* (Norwegian—Royal Norwegian Automobile Club)

KNAC *Koninklijke Nederlandse Automobiel Club* (Dutch—Royal Dutch Automobile Club)

KNAN *Koninklijke Nederlandse Akademie voor Naturwetenschappen* (Dutch—Royal Netherlands Academy of Sciences)

K'naw Kanawha River

KNB Kita-Nihon Broadcasting

KNC Kalamazoo Nature Center

Knd Kandla

K-NEA Kansas-National Education Association

KNF Key Notarization Facility

KNGR Kruger National Game Reserve

Kng X King's Cross (rail terminal)

kni knife

Knick Knickerbocker Club

knickers knickerbockers

KNIP *Komite Nasional Indonesia Pusat* (Indonesian Central National Committee)

knit. knitted; knitting

KNK *Kita Nippon Koku* (Northern Japan Airlines)

Knls Knolls

KNM Katmai National Monument; Kenya National Museum; *Kongelige Norske Marine* (Norwegian—Royal Norwegian Navy)

KNMI *Koninkliji Nederlands Meteorologisch Instituut* (Dutch—Royal Netherlands Meteorological Institute)

KNMR Kenai National Moose Range (Alaska)

KNO Kano, Nigeria (tracking station)

KNOC Kuwait National Oil Company

Knockmealdowns Knockmealdown Mountains of southern Ireland

KNOJ *Korpus Narodne Odbrane Jugoslavje* (Serbo-Croat—National Defense Corps of Yugoslavia)

Knopf Alfred A Knopf

knork knife + fork (combination utensil)

Knott's Knott's Berry Farm

Know-Nothings American political party members (1853-56 attempted to keep control of the government in hands of native-born citizens; members professed ignorance of the party's activities)

KNP Kafue National Park (Zambia); Kalahari NP (South Africa); Kalbarri NP (Western Australia); Kanha NP (India); Kejimkujik NP (Nova Scotia); Kinabalu NP (Sabah); Kinchega NP (New South Wales); Kootenay NP (British Columbia); Korean

National Party; Kosciusko NP (New South Wales); Kruger NP (South Africa)

KNPC Kuwait National Petroleum Company

KNPI Kundu's Neurotic Personality Inventory

KNR Kinki Nippon Railway; Korean National Railroad

KNSM Koninklijke Nederlandsche Stoomboot Maatschappij (Dutch—Royal Netherlands Steamship Company)

kn sw knife switch

Knt Knight

KNT Knight-Knott Hotels (stock-exchange symbol)

KNT Koninklijke Nederlandsche Toeristenbond (Dutch—Royal Netherlands Touring Club)

KNTC Korean National Tourism Corporation

knu knuckle

Knud(sen) (pronounced *Nud* or *Nudsen*)

KNUFNS Kampuchean National United Front for National Salvation

KNUST Kwame Nkrumah University of Science and Technology

Knut Knut Faldbakken (Norwegian novelist)

KNVD Koninklijk Nederlands Verbond van Drutkkerijen (Dutch—Royal Netherlands Printing Association)

KNVL Koninklijke Nederlandse Vereniging voor Luchtvaart (Dutch—Royal Netherlands Flying Club)

KNWR Kirwin National Wildlife Refuge (Kansas)

KNX Kinney Company (stock-exchange symbol)

Knxv Knoxville

ko keep off; keep out; kilohm; kit order; knockout (KO)

k-o knockout

Ko Korea; Korean

KO kickoff (football); knockout (boxing); Kodiak Airways (2-letter coding)

KO Komische Oper (German—Comic Opera)—Berlin opera company and opera house

K o A Kampgrounds of America

KOA Kentucky Opera Association

Kob Kobe (British Maritime contraction)

Køb København (Dano-Norwegian—Copenhagen)

Koblenz Coblenz

kobol keystation on-line business-oriented language

Kobuk Kobuk Valley National Park, Alaska

KOC Kollmorgen Optical Corporation; Kuwait Oil Company

kod kickoff drift

ko'd knocked out

KODAK trade name for Eastman Kodak photographic products

k-o drops knockout drops (chloral hydrate sedative)

koe knotted one end

kOe kiloOersted(s)

KOEX Korean Exhibition Center

kog kindly old gentleman

KOG Kansas, Oklahoma & Gulf (railroad)

KOH potassium hydroxide

KOHEMA Korean Heavy Machinery Industries

kohm kilohm

KOJ Kagoshima (airport)

Kok Cochrane

KOKS Kul og Koks Selskab (Danish—Coal and Coke Company)

Kol Kolonia, Ponape (Trust Territory of the Pacific)

KOLA Keep Old Los Angeles

Kolloid Z Z Polm Kolloid Zeitschrift und Zeitschrift Polymere (German—Colloidal and Polymer Periodicals)

Kol Nid Kol Nidre (Aramaic—all promises and vows be nullified and forgiven)—prayer recited and sung in synagogues on eve of Yom Kippur or played by cello and orchestra as arranged by Max Bruch

KOM Knight of the Order of Malta

Komei (Japanese—Komeito) —Buddhist party

komm kommunal or *kommune* or *kommunistik* (Dano-Norwegian—communal or commune or communist)

Komp Kompanie (German—company)

Kon Konstant; Konstantin

Kon Bel Koninkrijk België (Flemish—Kingdom of Belgium)

Kon Dan Kongeriget Danmark (Danish—Kingdom of Denmark)

Kon Den Kongeriget Danmark (Danish—Kingdom of Denmark)

Kong Kristian Kong Kristian stod ved Højen Mast (Danish—King Christian stood by the lofty mast)—national anthem of Denmark

Kon Ned Koninkrijk der Nederlanden (Dutch—Kingdom of the Netherlands)

Kon Nor Kongerik Norge (Norwegian—Kingdom of Norway)

Konr Konrad

KONR Komitet Osvobozhdeniya Narodov Rossii (Russian—Committee for the Liberation of the Peoples of Russia)

kons konservativ (Dano-Norwegian—conservative)

Konst Konstantin

Kon Sver Konungariket Sverige (Swedish—Kingdom of Sweden)

konz konzentriert (German—concentrated)

KOON Knight of the Order of Orange Nassau

Kootenay Kootenay Lake or Kootenay National Park in British Columbia; Kootenay River flowing from British Columbia to Idaho and Montana

kop kopeck(s)

Kop Kopenhagen (Dutch, Flemish, German—Copenhagen)

KOP Koppers (company)

KOP Kansallis-Osake-Pankii (Finnish-National Bank)

kops keep off pounds sensibly

KOPS Keep Our Precinct Safe

kor knowledge of results

kor koreanisch (German—Korean)

Kor Korea; Korean; Korean Air; The Koran

Kör Körfez(i) (Turkish—bay, gulf)

KORDI Korean Ocean Research and Development Institute

Korea Democratic People's Republic of Korea (North Korea) and Republic of Korea (South Korea)—*Chosun Minchuchui Inmin Konghwa-Guk* (North Korea) and *Daehan-Minkuk* (South Korea)

Korean Pen Korean Peninsula containing North and South Korea

korfez korfezi (Turkish—bay or gulf)

Korin Ogata Korin (early 18th-century Japanese decorative artist)

Kor N North Korea (whose capital is Pyongyang)

KORR King's Own Royal Regiment

Kor S South Korea (whose capital is Seoul)

KORSTIC Korean Scientific and Technological Information Center

koruna monetary unit of Czechoslovakia

kos kilograms; kilos

KOSB King's Own Scottish Borderers

Kosci Mount Kosciusko (Australia's highest mountain)

kosmo kosmonavt (Russian—cosmonaut)

Koste André Kostelanetz

Kot Kotakinablalu

KOTRA Korea Trade Promotion Corporation

Koussi Serge Koussevitzky

kov key-operated valve

Kow Koweit (Spanish—Kuwait)

KOW Knight of the Order of William

KOWACO Korean Water Resources Development Corporation

KOYLI King's Own Yorkshire Light Infantry

kp key personnel; kick plate; kill probability; kilopond; king post; kitchen police (KP); knotty pine; (KP) keypunch

kp Kochpunkt (German—boiling point)

KP Korea (Internet code); Korean People's (Republic)

K.P. Knight of St Patrick

K of P King(dom) of Poland; King(dom) of Portugal; King(dom) of Prussia; Knights of Pythias

KP Kommunistische Partei (German–Communist Party); *Komsomolskaya Pravda* (Russian—Young Communist League Truth) Moscow newspaper; *Kuvendi Popullore* (Albanian People's Assembly)

kPa kilopascal (pressure unit)

KPA Korean People's Army; Kraft Paper Association

kpc keypunch cabinet; kiloparsec

KPC Kangaroo Protection Committee; Koblenz Procurement Center; Korean Productivity Council

kp & d kick plate and drip

KPD Kommunistische Partei Deutschland (German—Communist Party of Germany)

KPDR Korean People's Democratic Republic

KPFC Kuwait Pacific Finance Company

KPFSM King's Police and Fire Service Medal

KPGA Kansas Personnel and Guidance Association; Kentucky Personnel and Guidance Association

kph kilometers per hour; knots per hour

kpi kips per inch

kpic key phrase in context

KPJ Kommunisticka Partija Jugoslavije (Serbo-Croat—Communist Party of Yugoslavia)

kpl kilometers per liter

KPL Knoxville Public Library

kpm kathode pulse modulation

KPM King's Police Medal

KPM Koninklijke Paketvaart Maatschappij (Dutch—Royal Packet Company)—inter-island shipping line

KPMB Kunming Phosphorus Mining Bureau (China)

Kpmtr Kapellmeister (German—conductor)

KPNO Kitt Peak National Observatory

KPNWR Kern-Pixley National Wildlife Refuge (California)

kpo keypunch operator

KPO Korean Post Office

kpos keep pounds off sensibly (weight-reduction program)

KPP Keeper of the Privy Purse

kpps kilopulses per second

kpr keeper; knots per revolution

Kpr Kodak photo resist

KPR Korean Presidential Ribbon

KPRA Korean People's Revolutionary Army (North Korea)

kps kips (thousand pounds) per square foot

kpsi kips (thousand pounds) per square inch

KPSS Kommunisticheskaya Partiya Sovetskovo Soyuza (Russian—Communist Party of the Soviet Union)—CPSU

Kpt Kaptajn (Danish—captain)

Kpt Kapitein (Dutch—captain)

KPU Kenya People's Union (party)

kq line squall

K & Q king and queen

kr keel rider; kiloroentgen

k & r kidnapping and ransom—(insurance)

Kr krypton

KR Khmer Rouge (Red Cambodian)—Communist insurgents in Cambodia; King's Regulations; Korean Registry (of ships); krona (Icelandic or Swedish monetary unit); krone (Danish or Norwegian monetary unit)

KR III King Richard III

krad kilorad

Krag Krag-Jörgensen rifle

Krak Krakatoa, Indonesia

Krakow Cracow

K-ration Calorie ration (lightweight emergency meal)

Kraut(s) Anglo-American slang for German(s)

K-R bb Krebs-Ringer bicarbonate buffer

KRC Knight of the Red Cross

krd kilorutherford

KREEP K (potassium) REE (rare-earth elements) P (phosphate)—yellow brown glassy lunar material

krem kremlin (Russian—citadel)

Krema Krematorium (German—crematory)

kreml (Russian—fortress)—the Moscow Kremlin, a citadel and seat of Soviet government

Krete Crete

Kreuzb Kreuzberg

KRF Kentucky Research Foundation

Krh Karachi

KRI Kyle Railway Inc

KRIC Korean Reinsurance Company

Kriegies *Kriegsgefangenen* (German—war prisoners)

KRIM Danish association for penal reform

Kripo *Kriminalpolizei* (German—Criminal Investigation Department)

Kris Kristian; Kristopher

Krist Kristian; Kristijonas; Kristmann; Kristofer

Krist *Kristallnacht* (German—night of broken glass)—night when Nazi gangs broke glass shop windows and killed or captured Jewish merchants in Germany and Austria

KRN Knight-Ridder Newspapers

kroat *kroatisch* (German—Croatian)

KROM the Norwegian association for penal reform

krona monetary unit of Sweden

krone monetary unit of Denmark

kroner monetary unit of Norway

kronur monetary unit of Iceland

Krop Kropotkin (Peter Kropotkin)

krp key resource people

KRP Keogh Retirement Plan

KRR King's Royal Rifles

krs Korus (Turkish—piastre)

Krs Kristiansand

KRs (*see* CRs)

KRS Kerato-Refractive Society; Kinematograph Renters Society

KRSB Kindergarten Reading Screening Battery

krt cathode-ray tube

KRT Khartoum, Sudan (airport)

KRU Krueger Brewing (stock-exchange symbol)

Kruger Kruger National Park (South Africa's big game and wildlife reservation named for Oom Paul Kruger, President of the Transvaal)

KRUM the Swedish association for penal reform

Krung Thep Bangkok

Krungthep *Krungthep Mahanakhon Bovorn Ratanakosin Mahintharayutthaya Ma-*

hadilokpop Noparatratchanthani Burirom Udomratchanivetmahasathan Amornpiman Avatarnsathit Sakkathattiyavisnukarmprasit (Siamese—full name of the capital city of Bangkok, Thailand)

Krung Thep (Siamese—Bangkok)

Krupp Krupp von Bohlen (German armament and steel firm)

Krupskaya Nadezhda Konstantinovna Krupskaya Lenin

ks drifting snowstorm (symbol); keep (type) standing; knowledge structure(s)

k's kilobytes; kilograms; kilometers; kilos; kilowatts; New Zealand slang for kilometers per hour (pronounced *kays*)

Ks kyats (Burmese money)

K-s King-size

KS Kansas; Kaposi's Sarcoma; Key Session; King's Scholar; King's School; Kipling Society; *Konungariket Sverige* (Swedish—Kingdom of Sweden); Korea Shipping

K of S King(dom) of Scotland; King(dom) of Siam (Thailand); King(dom) of Spain; King(dom) of Sweden

ksa kite-supported antenna

KSA Kingdom of Saudi Arabia; Knitwear and Sportswear Association

KSA *Kommission für die Sicherheit von Atomlagen* (German—Commission for the Safety of Nuclear Power Plants)

KSAA Keats-Shelley Association of America

KSB *Kypriakos Synthesmos Bibliothicarion* (Modern Greek—Library Association of Cyprus)

KSBA Kentucky School Boards Association

KSC Kansas State College; Kennedy Space Center; Kentucky State College; Korean Shipping Coporation; Kutztown State College

KSC *Komunistická Strana Ceskoslovenska* (Czechoslovak Communist Party)

KSEAFA Korea and South-East Asia Forces Association

KSEC Korea Shipbuilding and Engineering Corporation

ksf kips (thousand pounds) per square foot

KSF Kulkyne State Forest (Victoria, Australia)

KSFUS Korean Student Federation of the United States

KSG Kennedy School of Government

K.S.G. Knight of Saint Gregory the Great

K sh Kenya shilling(s)

KSHU Kaposi's sarcoma virus

ksi kips (1000 pounds) per square inch

KSI *Keshvare Shuhanshahiye Iran* (Iran—Persia); Kingdom of Saudi Arabia

ksia thousand square inches absolute

K-size King-size

k/sk (truncated fixed) keel (and) swing keel (combined)

KSK ethyl iodoacetate (tear gas)

ksl kidney, spleen, liver

KSL Kifikl Drug (stock-exchange symbol)

KSLI King's Shropshire Light Infantry

KSM King's Service Medal; Korean Service Medal

KSM *Kommunisticheskii Soyuz Molodozhi* (Russian—All-Union League of Communist Youth)—Komsomol or Young Communist League—YCL; *Kungliga Svenska Marinen* (Royal Swedish Navy)

ksml kosher meal

KSN Kit Shortage Notice

KSNP Khao Salob National Park (Thailand)

KSO Kalamazoo Symphony Orchestra; Knoxville Symphony Orchestra

ksoc key symbol out of context

ksr keyboard send-receive (set); keyboard send-receive (unit)

K.S.S. Knight of Saint Sylvester

KSS *Kommission zur Stahlenschutz* (German—Radiation Protection Commission); *Komunisticka Strana Slovenska* (Communist Party of Slovakia)

KsSEA Kansas Society of Enrolled Agents

KSSR Kazak Soviet Socialist Republic; Kirghizian Soviet Socialist Republic

KSSU Kiev IG Shevchenko State University (University of Kiev)

kst keyseat

kst kom schwitzen tod (German—come sweet death) —philosophical saying attributed to Kant and others

KST King-Seeley Thermos (company)

KSTC Kansas State Teachers College

K-S Test Kveim-Siltzbach Test

K.St.J. Knight of the Order of Saint John of Jerusalem

ksu key service unit

KSU Kansas State University; Kent State University; Kentucky State University

KSUAAS Kansas State University of Agriculture and Applied Science

ksv kinetic safety vehicle

KSY King Seeley (stock-exchange symbol)

kt karet (caret); key telephone; kiloton (nuclear equivalent, 1000 tons of high explosives); knot

kt. knight

Kt Knight

K$_t$ stress concentration factor

KT Käimtnerthor-Theater (Vienna); Kentucky & Tennessee (railway); Knight of the Order of the Thistle; Knight Templar; Missouri-Kansas-Texas (Katy Route Railroad)

K-T Kazin-Turkic

K.T. Knight of the Thistle

K/T Kretaceous/Tertiary (boundary between Cretaceous and Tertiary geologic periods)

K of T Kingdom of Tonga

KTA Kindergarten Teachers Association; Knitted Textile Association; Korean Traders Association

KTAAK Korea Trading Agents Association

ktas knots true airspeed

Ktb Kriegstagebuch (German—war diary)

KTB Kluwer Technical Books

Kt.Bach. Knight Bachelor

KTC Key Telephone System; Keystone Tankship Corporation; Key Translation Center; Kindergarten Teachers College; Kodiak Tracking Station

KTH Kungliga Tekniska Hegskolan (Royal Institute of Technology, Stockholm)

K through 12 kindergarten through high school

ktl kai ta loipa (Greek—et cetera)

KTM Keretapi Tanah Melayu (Malayan Railway)

KTN Ketchikan, Alaska (Annette Island airport)

Kto Konto (German—account)

K-Town Knoxville, Tennessee

ktr keyboard typing reperforator

Ktr Katorzhane (Russian compound)—prison compound reserved for people sentenced to hard labor

K-truss K-shaped truss

kts knots

KTS Kagoshima Television Station; Key Telephone Systems; Kwajalein Test Site

KTSA Kahn Test of Symbol Arrangement

KTTC Kingston-upon-Thames Technical College

ktu kill the umpire

KTX Keith Railway Equipment (railway code)

Ku Karmen unit(s)

Kü Küçk (Turkish—little, small)

KU Kalmar Union; Kansas University; Keio University; Kutztown University; Kuwait Airways (2-letter symbol)

KU Københavns Universitet (Danish—Copenhagen University)

kub kidney(s)-ureter(s)-bladder

ku'd knocked up (made pregnant)

KUD Koperasi Unit Desa (Indonesian—Village Cooperative Unit)

Ku'dam Kurfürstendamm (main street of Berlin)

KUED Kodak's Unitized Engineering Drawing (system)

K u H Kingston upon Hull (official name for Hull)

Kuk Kukrail Crocodile Breeding Center in Uttar Pradesh, India (protects gavials)

KUK Kollege of Universal Knowledge

KUL Kabul University Library (Kabul, Afghanistan); Karachi University Library (Pakistan); Kyoto University Library (Japan)

Kun Kunsan

K unit Kimball unit

Kunluns Kunlun Mountains of Tibet

Kur (British maritime contraction of Kure); Kurdish; Kurile Islands

Kuria Murias Kuria Muria Islands in the Arabian Sea

Kuril Cur Kurile Current (Oyashio)

Kuriles Kurile Islands in the northwest Pacific

Kurland Courland

Kuro Kuroshio (Japanese—Black Salt)—warm ocean current of the North Pacific Ocean

KURRI Kyoto University Research Reactor Institute

Ku's Karmen units

kutd keep up to date

KUU Kungliga Universitet i Uppsala (Swedish—Royal University of Uppsala)

Kuw Kuwait

Kuwait State of Kuwait (Persian Gulf gas-and-oil producer), *Dowlat al Kuwait*

Kuyb Kuybyshev

kv kilovolt

kv kvinde (Dano-Norwegian—woman)

kV kilovolt; kV/s

KV Köchel-Verzeichnis (German—Kochel Catalog)—catalog of Mozart's compositions

KV-107 Japanese-made Sea-Knight-type helicopter built by Kawasaki

kva kilovolt ampere

kVA kilovolt(s)/ampere

KVA Korean Veterans Association; *Kungliga Vetenskaps Akademien* (Swedish—Royal Swedish Academy of Sciences)

kvah kilovolt-ampere-hour

kvam kilovolt ampere meter

Kvan Electron Kvantovaya Electronika (Russian—Journal of Quantum Electronics)

kvar kilovar; kilovolt ampere reactive

kvarh kilovar hour

kvcp kilovolt constant potential

K-Vets Korean War Veterans of the United States

kvg keyed video generator

KVHS Kanawha Valley Historical Society

KvK Kill van Kull

KVL Kirchoff's Voltage Law

kvm kilovolt meter

KVNP Kidepo Valley National Park (Uganda)

kvp kilovolt peak

KVP Katholieke Vokspartij (Dutch—Catholic People's Party)

KVW Kansas City Kaw Valley (railroad)

kw killer weed (pcp sprinkled on leaves and smoked); kilowatt

kw kwacha(s) (Zambian monetary unit)

kW kilowatt(s)

KW Kellogg West; Key West; King-Wilkinson; Korean War; Kuwait (Internet code)

K-W Keith-Wagener

kwac key word and context

K-W AG Kitchener-Waterloo Art Gallery

Kwaj Kwajalein

Kwan Kwantung

kwat key well allowable transfer

KWB Keith, Wagener, Barker (classification)

KWC Kentucky Wesleyan College

kwe kilowatts electrical

KWest Key West

K-W findings Keith-Wagener (ophthalmoscopic findings)

kwh kilowatt hour

kWh kilowatt hour

kwhr kilowatt hour

Kwi Kuwait

KWI Kosher Wine Institute

kwic key word in context

kwip keyword in permutation

KWIPS Thousands of Whetstone Instruction mixes Per Second

kwit key word in text; key word in title

kwm kilowatt meter

KWMA Kirtland's Warbler Management Areas (Michigan)

KWNWR Key West National Wildlife Refuge (Florida)

kwoc key word out of context

kwot key word out of title

KWP Korean Workers Party

KWPL Kitchener-Waterloo Public Library

kwr kilowatts reactive

KWS Kaziranga Wildlife Sanctuary (India)

KWSM Korean War Service Medal

kwt key word in title; kilowatts thermal

Kwt Kuwait

KWT King William's Town

KWU Kansas Wesleyan University

kwuc keyword and universal decimal classification

KWVZAB Ko-operative Wijnhoutwers Vereeninging van ZuidAfrika Beperkt (Dutch—Cooperative Wine Farmers Association of South Africa, Limited)

kwy keyway

kxu kilo-x-unit

ky cocoa; key; keyer; keying (device); kyat

Ky Kentuckian; Kentucky

KY Cayman Islands (Internet code); Kentucky (zip code); Kol Yisrael (Israel Broadcasting Service); (underground slang-federal hospital in Lexington, Kentucky where drug addicts are treated)

kyhd keyboard

kybo keep your bowels open (keep healthy)

kyc know your customer

KYC Klan Youth Corps; Knickerbocker Yacht Club

kyd Kilo yard

kyeri know your endorsers—require identification (advice to all who cash checks)

Kyle Kyle Railway

kymo kymograph; kymography

KYNP Khao Yai National Park—(Thailand)

Kyo Kyoto

Kyocera Kyoto Ceramics

Kyot Univ Kyoto University

Kypros Cyprus

Kyr. Kyrie eleison (Greek—Lord, have mercy upon us)

Kyrg Kyrgyzstan

kyrie kyrie eleison (Latin—Lord have mercy on us)

KySEA Kentucky Society of Enrolled Agents

kytoon kite balloon

kz duststorm or sandstorm

kz konzentrations (Dano-Norwegian—concentration camp)

Kz Kazakh(stan)

KZ King Zulu

KZ Konzentrationslager (German—concentration camp)

Kz's unlicensed citizen's-band radio-interference jammers

K z S Kapitan zur See (German—Sea Captain)—naval rating

L

l azimuthal or orbital quantum number (symbol); elbow (plumbing); land; large; late; latent heat per unit mass (symbol); lateral; latitude; law; leaf, league; left-handed pitcher (baseball); left or port (L or P); length; levorotatory; liaison; license; light(ning); lignite; limen; line; link; lire; liter; load; locus; lodger; loose; losses; losing pitcher; lost; low; lumen

l (L) luminosity

l/ *letra* (Spanish—letter)

l* lumen

l *lectio* (Latin—reading); *links* (German—left); units of labor (microeconomics)

L Bell Aircraft (symbol); center line (symbol); coil or inductance (symbol); drizzle (aircraft code); elevated railroad (EL); inductance (symbol); kinetic potential (symbol); Labor; Labour; lactobacillus; lago; Lagramge function; lake; lake vessel; Lamar State College of Technology; lambert; lameness; langmuir; Laplace transform; Latin; launching; law (the police task force, vice squad, etc.); left (port side); lempira (Honduran currency unit); Leo; Leon; Liberal; lift (symbol); lift force; light; Lima—code for L; Linnaeus; Lions International; listed securities; loch; London; longitude; loran; Lorentz unit; lottery;

lough; Luckenbach Lines; Luxembourgh (auto plaque); Lykes Lines; mean life (symbol); rolling moment (symbol)

L (L) demand for money (macroeconomics symbol)

L *Lago* (Spanish—lake); *lähteä* (Finnish—departure); *lämmin* (Finnish—warm); *länsi* (Finnish—wheat); (Latin—Lucius); *laudes* (Latin— praises); *levato* (Italian—raised); *libra* (Latin—pound); *Life Magazine*; *links* (German—left); *liejada* (Spanish—arrival); *Loteria* (Spanish—lottery)

L-1 first language

l1 first lumbar vertebra

l2 second lumbar vertebra

L-2 second language

l/3 lower third

l3 third lumbar vertebra

L-3 lousy trio

L-4 military version of the Piper Cub

14, 15, etc. fourth lumbar vertebra; fifth lumbar vertebra

lØ no evidence of lymphatic invasion

l1 superficial lymphatics invaded

l2 deep lymphatics invaded

L7 Hollywood slang for old-fashioned person or *square* as capital-letter L and figure 7 may be combined to form a square

L-19 Cessna Bird Dog liaison aircraft

L-100 Lockheed four-engine transport aircraft for civilian use

L-188 Lockheed Electra turbo-prop transport plane

L-1011 Lockheed's jumbo jet-liner

la landing account; large aperture(s); latex agglutination; lava; lead antimony; leave allowance; left angle; left atrium; left auricle; light alloy; light amphibian; lighter than air; lightning arrestor; long-acting; low alcohol; low altitude

la (LA) linoleic acid; longitudinal acoustic

l/a landing account; letter of advice; letter of authority; lighter than air

l & a left and above; light and accommodation

la A in fixed-do system; (Italian—the); sixth tone in diatonic scale

la. laborer

l.a. *lege artis* (Latin—according to the art)—as directed

l/a *lettre d'avis* (French—letter of advice)

La Lane; lanthanum; Lao; Laos; Laotian; Louisiana; Louisianian

La *Lebensalter* (German—chronological age); *Luisiana* (Spanish—Louisiana)

LA Latin America(n); Laos (Internet code) Leasehold Area; Legal Advisor; Legislative Assembly; Leschetizky Association; Letter of Activation; Library Association; License Application; Licensing Act; Licensing Authority; Lieutenant-at-Arms; light-alcohol beer brewed by Anheuser-Busch; Local Authority; Los Angeles; Louisiana; Louisiana & Arkansas (railroad); Louvain Association; Lower Alabama Oocular placename)

L-A Loire-Atlantique (formerly Loire-Inférieure)

L/A Launch Area; Lloyd's Agent

L & A Louisiana & Arkansas (railroad)

LA Linea Aéria de Chile (Spanish—Air Line of Chile)

LA 400 400 women of Los Angeles who raised 4 million dollars for its music center

laa light anti-aircraft

LAA League of Advertising Agencies; Library Association of Australia; Life Assurance Advertisers; Los Angeles Airways

LAABF Ladies Auxiliary of the American Beekeeping Federation

LAACC Light Antiaircraft Control Center

LAAD Latin America Agribusiness Development

LAADS Los Angeles Air Defense Sector

LAAF Libyan Arab Air Force

LAAG Latin American Anthropology Group

laam (LAAM) levo-alpha acetylmethadol (alternative to methadone for treatment of drug addiction)

laar liquid-air accumulator rocket

LAAS Los Angeles Air Service

lab label; labeling; labor; laboratory; load/lorry aboard barge

Lab Laboratory; Labour(ite); Labrador

LAB Labor; Labour; The Legal Advisory Boards; Labour Party; Licquor [sic.] Administration Board; Liquor Administration Board; Lloyd Aereo Boliviano (Bolivian airline); low-altitude bombing

LAB Lloyd Aéreo Boliviano (Spanish—Bolivian Air Lines)

LABA Laboratory Animal Breeders Association

LABAN Lakasang Bayan (Fili-pino-Peoples Power)

Lab Cur Labrador Current—cold Arctic current flowing southward along Atlantic coast of Canada and northern New England

LABEN Laboratori Elettronici e Nucleari (Italian—Electronic and Nuclear Laboratories-Milan)

labe(s) label(s)

labi (Latin prefix—lip)—labial

Labor US Department of Labor

Laborers' Union Laborers' International Union of North America

lab proc laboratory procedure(s)

labr. laborer

labrador label-address routine

labrv (LABRV) large ballistic reentry vehicle

labs laboratories

Lab(s) Labrador retriever(s)

LABS Low-Altitude Bombing System

labv (LABV) large ballistic (reentry) vehicle

lac lacquer; lacrimal; lactation; large-aperture component(s); limits of acceptable change; linear amplitude continuous; lunar aeronautical chart; shellac

lac (LAC) load accumulator

Lac Lacerta; Lacertilia

LAC Laboratory Animals Center; Leading Aircraftsman; League of Arab Countries; Liberty Amendment Committee; Library Association of China; Lockheed Aircraft Corporation

LAC Lineas Aéreas Chaqueñas (Spanish—Aero Chaco)

LACA Ladies Apparel Contractors Association; Latin American Coffee Agreement

LACAC Latin American Civil Aviation Commission

LACAP Latin American Cooperative Acquisitions Project

LACATA Laundry and Cleaners Allied Trades Association

LACBW La Crosse Boiling-Water (reactor)

lacc lathe chuck

LACC Los Angeles City College

Laccadives the Laccadive Islands in the Arabian Sea

LACE liquid-air cycle engine

lacertiliol lacertiliologic(al)(ly); lacertiliologist; lacertiliology

LACES London Airport Cargo Electronic Scheme; Los Angeles Council of Engineering Societies

Lachie Lachlan

LACIE Large Area Crop Inventory Experiment

LACIRS Latin American Communication Information Retrieval System

LACJ Los Angeles County Jail

LACL Latin American Citizens League

LACM Latin American Common Market; Los Angeles County Museum

LACMA Los Angeles Conservatory of Music and Arts; Los Angeles County Museum of Art

LACMedA Los Angeles County Medical Association

LACO Los Angeles Chamber Orchestra

LA Co Art Mus Los Angeles County Art Museum

laconiq laboratory computer on-line inquiry

Lacp Lloyd's anchors and chains proved

LACP London Association of Correctors of the Press

lacr low-altitude coverage radar

lacri (Latin prefix—tears)—lacrimose

LACSA Líneas Aéreas Costarricenses (Spanish—Costa Rican Airlines)

lact lease automatic custody transfer

LACTC Los Angeles County Transportation Commission

lactns lactations

lacto-ovo(s) lacto-ovo vegetarian(s) (confining diet to milk, milk products, eggs, and vegetables)

lactovegan vegetarian whose diet includes milk products

lacv (LACV) light-amphibious air-cushion vehicle

LACW Leading Aircraftswoman

lad ladder; liquid agent detector; logarithmic analog-to-digital; logistic approval data; low-altitude discharge; lunar atmosphere detector

lad (LAD) language-acquisition device

Lad Ladino (Spanish-Jewish language from Morocco and the Balkans to Turkey; mestizos of pure Spanish descent from Chile to Mexico, including Central America)

Lad (Spanish dialect spoken by many persons of Judaic origin who were forced to flee Spain during the Inquisition)

LAD Language Acquisition Device; Library Administration Division (American Library Association); Light Air Detachment

ladar laser detection and ranging

ladd low-altitude drogue delivery

ladder life-assurance direct entry and retrieval

LADE Líneas Aereas del Estado (Spanish—State Airlines, Argentina)

LADECO Línea Aéreo del Cobre (Spanish—Copper Air Line)

LADIES Life After Divorce Is Eventually Sane (Hollywood ex-wife group); Los Alamos Digital-Image-Enhancement Software

Ladies' Garment Workers International Ladies' Garment Workers Union

Ladies Home J Ladies Home Journal

ladir low-cost arrays for detection of infrared

Lad lang Ladino language (Spanish-Yiddish spoken by people of Spanish-Jewish origin in the Balkans, Greece, the Near East, North Africa, and Turkey)

LADO Latin American Defense Organization; Latin American Development Organization

Ladoga Lake Ladoga east of St Petersburg and called Ladoshskoye Ozero by the Russians

ladp ladyship

Ladrones Marianas Islands

lads: low-altitude defense system

Lad(s) Ladino(s)—Europeanized Central American mestizo(s) of Spanish descent; their Judeo-Spanish dialect developed by refugees from the Holy Inquisition (1487 to 1834), also called Judeo-Spanish, Spagnuolo, or Spanish Yiddish

LADSIRLAC Liverpool and District Scientific, Industrial, and Research Library Advisory Council

L Adv Lord Advocate

LADWP Los Angeles Department of Water and Power

LAE Leadership Ability Evaluation

laetrile laevo-mandelonitrile-beta-glucuronic acid

laev. laevus (Latin—left)

laf laminar air flow

Laf Lafayette

LaF Louisiana French

LAF Legislative Action Fund; Living Arts Foundation

LAF L'Académie Française (French—The French Academy)

Lafayette Marie Joseph Paul Yves Roch Gilbert du Motier (Marquis de Lafayette)

LaFr Louisiana French

LAFB Lincoln Air Force Base

LAFC Latin-American Forestry Commission

LAFCO Local Agency Formation Commission

LAFD London Association of Funeral Directors

Lafe Lafayette

LAFE Laboratorio de Fisica Espacial (Portuguese—Space Physics Laboratory)

La Fen La Fenice (Italian—The Phoenix)—Venice opera house

La Font LaFontaine

LaFr Louisiana French

LAFS Los Angeles Funeral Society

LAFTA Latin American Free Trade Area; Latin American Free Trade Association

lafts laser and flir test set

lafv (LAFV) light-armored fighting vehicle

lag. lagan

lag. lagena (Latin—bottle, flask)

Lag Lagoon; Laguna

Lag Laguna (Spanish—lagoon or lake)

La G La Guaira

LAG Layton Art Gallery; Librarians Automation Group

LAGB Linguistics Association of Great Britain

LAGE Los Angeles Grain Exchange

LAGEOS Laser Geodetic Satellite

LAGIC Life and General Insurance Committee

lags (LAGS) laser-activated geodetic satellite

Lags Lagunas

LAGS Los Angeles Geographic Society

Lagunas Laguna Mountains of California

Lah Lahore

LAH Licentiate Apothecaries Hall

La Habana Vieja (Spanish—old Havana)—capital of Cuba

LAHAWS Laser Homing and Warning System

LAHC Los Angeles Harbor College; Los Angeles Harbor Commission

LAHD Los Angeles Harbor Department

lahs low-altitude high speed

LAHS Local Authority Health Services

lai leaf area index

LAI Library Association of Ireland

LAI Linea Aèree Italiane (Italian—Italian Air Lines)

LAIA Latin American Integration Association

LAIC Lithuanian-American Information Center

LAICA Los Angeles Institute of Contemporary Art

LAIICS Latin American Institute for Information and Computer Sciences

LAINS Low-Altitude Inertial Navigation System

LAIRS Labor Agreement Information Retrieval System

LAIS Loan Accounting Information System (AID)

LAIT Logistics Assistance and Instruction Team

LAIV Loss Adjusters Institute of Victoria

la jura (Mexican-Spanish—law enforcement; the law)

La J La Jolla

LAJ Los Angeles Junction (railroad)

Lake Clark national park in Alaska

laks lakrids (Danish—licorice)

LAL Langley Aeronautical Laboratory (Langley Research Center)

LAL Laboratoire de l'Accelerateur Linéaire (French—Linear Accelerator Laboratory)

LA-LB Los Angeles-Long Beach (ports)

La Leche La Leche League International

lali lonely aged of low income

lalsd language for automated logic and system design

LALUCS Local Authority Land Use Classification System

lam laminate

lam (LAM) load accumulator with magnitude

Lam The Book of Lamentations; Lamarck; Lambretta

Lam Lamentations

La M *La Monnaie* (French—the mint) Brussels theater

LAM Lamarck; Lambert; Latin American Mission; London Academy of Music

L.A.M. Liberalium Artium Magister (Latin—Master of Liberal Arts)

LAMA Latin American Management Association; Latin American Manufacturers Association; Lead Air Materiel Area; Library Administration and Management Association; Light Aircraft Manufacturers Association; Locomotive and Allied Manufacturers' Association; London Annual Monthly Average

Lamb Lambert; Lamberto; Lambertus; Lambeth

LAMB lentigines, atrial myxoma, mucocutaneous myromas and blue nevi

LAMBC Los Angeles Motor Boat Club

Lambeau Lambeau Field, Green Bay, Wisconsin

lambsan lamb sandwich

lambwich lamb sandwich

LAMC Letterman Army Medical Center; Los Angeles Metropolitan College; Los Angeles Music Center

LAMCO Liberian-American-Swedish Mineral Corporation

LAMDA London Academy of Music and Dramatic Art

LAME Licensed Aircraft Maintenance Engineer

LAMIDA Lancashire and Merseyside Industrial Development Associamioum

LAMM Los Angeles Master Morticians; Lutheran-American Melancthon Movement

lamma laser microprobe mass analyser

lammr large-antenna multifrequency microwave radiometer

LA MOCA Los Angeles Museum of Contemporary Art

Lamp Lampeter

LAMP Library Additions and Maintenance Program; Low-Altitude Manned Penetration; Lunar Analysis and Mapping Program

LAMPP Los Alamos Molten Plutonium Program (AEC)

Lamps ship's lamp trimmer

LAMPS Center for the Study of Legal Authority and Mental Patient Status; Light Airborne Multi-purpose System

LAMS Launch Acoustic Measuring System

LAMSACC Local Authorities Management Services and Computer Committee

lamsim launcher-and-missile simulator

lan listing agent's name

lan (LAN) local area network

Lan Lancaster; Lansing

L An Los Angeles

LAN Local Apparent Noon

LAN Latin American Newspapers (bibliographic reference); *Línea Aérea Nacional de Chile* (Chilean National Airlines)

lanac laminair air navigation and anti-collision

Lanarks Lanarkshire

Lan Bag Lansing Bagnall

lanby large automatic-navigation buoy

Lane Lancaster

Lance Lancelot; Ling-Temco-Vought MGM-52A surface-to-surface tactical missile

Lance Cpl Lance Corporal

Lanes Lancashire

land. landscaping

LandCraB landing craft and bases

LANDJUT Land Forces (Schleswig-Holstein and) Jutland (NATO)

LANDNONOR Land Forces Northern Norway (NATO)

LANDNORTH Land Forces Northern Europe (NATO)

'Lando Orlando, Florida

Land of Peace Thailand

Land of Religion India

Land of Shining Mountains Montana

landprop landlord proprietor

Lands Landsmaal (Norwegian national language)

landscp landscape

Lands Down Under Australia and New Zealand

LANSCE Los Alamos Neutron Scattering Center

LANDSONOR Land Forces Southern Norway (NATO)

Land Time Forgot Australia

Landw Landwirtschaft (German—agriculture)

LANDZEALAND Land Forces Zealand (NATO)

Lan Fus Lancashire Fusiliers

lang language

Lang Langbridge; Langdon; Lange; Langer; Langford; Langhorne; Langlois; Langson; Langston; Languedoc

Langley Langley, Virginia headquarters of the CIA

LANICA Líneas Aéreas de Nicaragua (Spanish—Air Lines of Nicaragua)

LANL Los Alamos National Laboratory

Lan Reg Lancashire Regiment

LANS Land Area Navigation System

LANSA Líneas Aéreas Nacionales SA (Spanish—National Airlines Corporation)—Peru

Lant Atlantic (naval short form)

'Lanta Atlanta

lantirn low altitude navigation and targeting infrared night

LANTIRNS Low-Altitude Navigation, Targeting Infra-Red Navigation System

LANWR Laguna Atascosa National Wildlife Refuge (Texas); Lake Andes National Wildlife Refuge (South Dakota)

LANY Linseed Association of New York

lao legislative analyst's office

Lao Laos (whose capital is Vientiane); Laotian

LAO Legal Assistance Office(r); Licentiate of the Art of Obstetrics

LAOAR Latin American Office of Aerospace Research

LAOD Los Angeles Ordnance District (USA)

Laos Lao People's Democratic Republic (Indo-Chinese country)

LAOT Los Angeles Opera Theater

lap leucine aminopeptidase; leukocyte alkaline phosphatase (stain); link access procedure

lap. laparotomy; launch analyst's panel; learning activity package; left atrial pressure

Lap Lapland; Lappish

LaP Las Palmas (British maritime abbreviation)

La P La Paz (Bolivian capital along with Sucre)

La Pam *La Pampa* (Argentine province)

LAP La Paz, Mexico (airport); Laboratory of Aviation Psychology (Ohio State University); Library Awareness Program; *Líneas Aéreas Paraguayas* (Spanish—Paraguayan Air Lines)

laparo laparoscope; laparoscopic (sterilization); laparotomy

lap chole laparoscopic cholecystectomy

LAPC Los Angeles Pacific College; Los Angeles Pierce College

LAPCO Lavan Petroleum Company

lapd lollipop aggression psychological disorder

LAPD Los Angeles Police Department

LAPDis Los Angeles Procurement District (US Army)

LAPES Low-Altitude Parachute-Extraction System

L A Phil Los Angeles Philharmonic

LAPI Los Angeles Philharmonic Institute

lapid lapidary

lapid *lapideum* (Latin—stony)

LAPL Los Angeles Public Library

lap lin *lapsus linguae* (Latin—slip of the tongue)

LAPO Los Angeles Philharmonic Orchestra

Lapp Lappish; member of a people of northern Scandinavia, Finland, and Russia

LAPS List Assembly Programming System

LAPT London Association for the Protection of Trade

LAPTA Local Authorities Passenger Transport Association

La Pucelle *La Pucelle d'Orléans* (French—The Maid of Orleans)—Joan of Arc

laput light-activated programmable unijunction transistor

laq lacquer

lar left arm reclining; liquid argon; local-acquisition radar

lar (LAR) light-artillery rocket; long-range radar

Lar Lara; Larina; Larry

LAR Lease Application Request; Library Association of Rhodesia; Life Assurance Relief; Limit Address Register

lara (LARA) light armed reconnaissance aircraft

Lara Agustin Lara (Mexican composer of *Granada, Marina Bonita*, and *Solamente una vez*)

LARA League of Americans Residing Abroad; Licensed Agency for Relief of Asia

laram line-addressable random-access memory

larat low-altitude radar altimeter

La Raza (Hispanic-American Spanish—The Race) *La Raza Unida* (Spanish—The United Race)—Mexican-American political organization

larbord (Norwegian—larboard)—originally a vessel's loading side opposite starboard hence the portside

larc lighter, amphibious, resupply, cargo (vehicle); logic alarm radio clock

Larc Livermore automatic research computer

LARC Langley Research Center; League Against Religious Coercion; Library Automation and Consulting; Libyan-American Reconstruction

Commission; Lindheimer Astronomical Research Center; Local Alcoholism Reception Center(s)

larct last radio contact

larf low-altitude radar fuzing

larg *largamente* (Italian—broadly); *largeur* (French—width); *largo* (Italian—slow)

Large Print Large Print Publications

largo. *larghetto* (Italian—moderately slow)

LARIAT Laser-Radar Intelligence-Acquisition Technology

LARIC Later Reading In-Service Course

LARO Latin American Regional Office (FAO)

Larousse Pierre-Athanase Larousse

larp line automatic reperforator; local and remote printing

La R-P La Rochelle-Pallice

lar rep larceny report

lars laminar angular-rate sensor

Lars Lawrence

LARS Laboratory for Applications of Remote Sensing (Purdue); Light Artillery Rocket System; Low-Altitude Radar System

LARSP Language Assessment, Remediation and Screening Procedure

LART Los Angeles Rapid Transit

LARZ Los Angeles Revitalization Zone

larv (LARV) low-angle reentry vehicle

larva (LARVA) low-altitude research vehicle

larvi larvicidal; larvicide

laryng laryngological; laryngologist; laryngology

laryngol laryngology

las large astronomical satellite; liberal arts and sciences; lookout aiming sight; low-alloy steel; lower airspace

las *lassú* (Hungarian—slow introductory passages leading to fast section, *friss*, of a csárdas or rhapsody)

LAs Latin Americans

LAS Land Agents Society; large astronomical satellite; Las Vegas, Nevada (airport); League of Arab States; Lebanese-American Society; Legal Aid Society; Library Association of Singapore; Lord Advocate of Scotland

LA & S Liberal Arts and Sciences

lasa large-aperture seismic array

LASA Latin American Shipowners Association

LASAIL Land-Sea Interaction Laboratory

LASBO Louisiana Association of School Business Officials

LASC Los Angeles State College

LASCO Latin American Unesco Science Cooperation Office

lascot large-screen color television

lascr light-activated silicon-controlled rectifier

lascs light-activated silicon-controlled switch

LASEA Louisiana Society of Enrolled Agents

LASEORS London and South Eastern Operational Research Society

laser light amplification by stimulated emission of radiation; lucrative approach to support expensive research

LASER London and South Eastern Library Region

LASER Los Angeles Skeptics Evaluative Report

LASERS London and South Eastern Regional Library System

LASH Legislative Action on Smoking and Health; Lighter Aboard Ship (cargo system)

lasi landing-site indicator

LASIE Library Automated System Information Exchange

lasik laser-assisted in situ keratomileusis

LASIK Laser in Situ Keratomileusis

LASL Los Alamos Scientific Laboratory

LASMCO Liberian American-Swedish Minerals Company

LASO London and Scottish Marine Oil

LASO Latin American Solidarity Organization; Los Angeles Sheriff's Office; Los Angeles Society of Ophthalmology

lasp low-altitude space platform

LASRA Leather and Shoe Research Association

lasrm **(LASRM)** low-altitude short-range missile

lass. lighter-than-air submarine simulator (LASS)

LASS launch-area support ship; Local Authority Social Services; Los Angeles Special Services (telephone)

LASSCO Los Angeles Steamship Company

Lassen Volcanic California national park containing Lassen Peak

'lasses molasses

lassie low-airspeed sensing and indicating equipment

lasso laser search-and-secure observer

LASSO Latin American Student Studies Organization

last last sale price, closing price

La State U Pr Louisiana State University Press

lasv **(LASV)** low-altitude surface vehicle; low-altitude supersonic vehicle

lat lateral; latitude

lat **(LAT)** lowest astronomical tide

lat lateinsich (German—Latin); *latin* (Dano-Norwegian—Latin); *latitud* (Spanish—latitude)

lat. latus (Latin—wide)

Lat Latin; Latvia; Latvian

Lat Latin (classical language of Roman antiquity and the base of Romance languages)

LAT Latex Agglutination Test(ing); Learning by Advanced Telecommunications; Linseed Association Terms; Local Apparent Time; Taxader Airport (Bogotá)

LAT Los Angeles Tmes

LATA Local Access and Transport Areas; Los Alamos Technical Associates

lat. admov. lateri admoveatum (Latin—apply to the side)

LATARS Laser-Augmented-Target Acquisition and Recognition System

LATCC London Air Traffic Control Center

LATCRS London Air Traffic Control Radar Station

lat. dol. lateri dolenti (Latin—to the painful side)

later (Latin prefix—side)—lateral

LATH Laos and Thailand Military Assistance

lat ht latent heat

Latin America Portuguese-speaking Brazil and Spanish-speaking countries such as the Central American republics, Cuba, the Dominican Republic, Mexico, Puerto Rico, the South American republics; understood variously as: Spanish-speaking America plus Brazil; all of the Americas south of the United States; Mexico, Central and South America and islands of the Caribbean

latm living as though married

latoff lowest astronomical tide of the foreseeable future (assuming anyone can foresee the future)

latom lowest astronomical tide of the month

latoy lowest astronomical tide of the year

lats long-acting thyroid stimulator

LATS Los Angeles Times Syndicate

latssn late season

LATTC Los Angeles Trade-Technical College

LATUF Latin American Trade Union Federation

Latv Latvia; Latvian

LATWPNS Los Angeles Times Washington Post News Service

lau. launderer; laundry

Lau Laura; Lauretta

LAUA Lloyd's Aviation Underwriters' Association

Lauch Lauchlin

laughing gas nitrous oxide (N_2O)

LAUK Library Association of the United Kingdom

Lau Lib Laurentian Library (Florence)

laun launched

Laun Launceton, Tasmania

Launce Lancelot

laup **(LAUP)** laser-assisted uvulopalatoplasty program (treats snoring)

laund launder; laundry

laundromat automatic coin-operated laundry

Laur Laurence

Laurentians Laurentian Mountains of southern Québec where they are also called the Laurentides as is Laurentides Park

Laurie Laurence

LAUSC *Linguistic Atlas of the United States and Canada*

LAUSD Los Angeles United School District

lav lavatory; light armored vehicle; lymphadenopathy virus

LAV Lalomalava (Western Samoan airport); light-armored vehicle; *Línea Aeropostal Venezolana* (Spanish—Venezulean Airmail Line); lymphadenopathy-associated virus

Lava Beds Lava Beds National Monument in northern California

LAVC Los Angeles Valley College

lavm loran automatic vehicle monitoring

law lawyer; left attack wing (lacrosse); light anti-tank weapon; light assault weapon; low-altitude weapon

Law Lawrence

LAW Lawyers Against Wigs; League of American Wheelmen; League of American Writers; Legal Aid Warranty; light anti-tank weapon; Local Air Warning

Law Lat Law Latin

Lawr Lawrence; Lawrencian

Law Rept Law Report(s)

Lawrie Lawrence

LAWRS Limited Airport Weather Reporting System

LAWS Leadership and World Society

lax. laxative

LAX Los Angeles, California (International Airport)

Lax-Chi Los Angeles—Chicago

Lax-NO Los Angeles—New Orleans

Lax-NY Los Angeles—New York

Lax-San Los Angeles—San Diego

Lax-Sea Los Angeles—Seattle

Lax-Sfo Los Angeles—San Francisco

Lax-Tor Los Angeles—Toronto

Laz Lazarus

LAZ Los Angeles Zoo

laze lava haze

lb landing barge; lavatory basin; left back (field hockey); letter box; lifeboat; linebacker; link belt(ing); linoleum base; local battery; low band; lumen band; pound

lb (LB) line buffer

l-b lemon-and-butter (sauce)

l & b land and building(s); left and below

lb *libra* (Latin—pound)

l.b. *lectori benevolo* (Latin—to the kind reader)

Lb lambert (unit of brightness)

LB Labrador; landing barge; large burgh; Lebanon (Internet code); Leonard Bernstein; light bomber; Lloyd Brasileiro (Brazilian Steamship Line); Local Board; Long Beach; Longview Bridge (Columbia River, Washington); Luther Burbank

L-B Link-Belt

L,B Little, Brown Publishing Company

L.B. *Baccalaureus Litterarum* (Latin—Bachelor of Letters)

lba lifting-body airship

Lba Luba (formerly San Carlos)

LB & AL Lever Brothers and Associates Limited

L-band 390–1550 mc

lb ap apothecaries' pound

L-bar capital-L-shaped bar

lb av avoirdupois pound

LBB Lubbock, Texas (airport)

lbbb left bundle branch block

lb/bhp-hr pounds per brake horsepower hour

lbbsb left bundle branch system block

Lbc Lübeck

LBC Letter Book Copy; Liberian Broadcasting Corporation; London Broadcasting Company; Lutheran Baptist Convention

lb cal pound calorie

LBCC Long Beach City College

lbcd left border of cardiac dullness

LBCH London Bankers' Clearing House

lb chu pound centigrade heat unit

LBCL Liberty Baptist College

LBCM Licentiate of Bandsmen's College of Music

lb/cu ft pounds per cubic foot

lbd learning and/or behavior disordered; left border of dullness; lifeboat deck; little black dress; lower bovine distemper; lower-back disorder

LBD League of British Dramatists

L/Bdr Lance Bombardier

L-beam capital-L-shaped beam

LBEB Laboratory of Brain Evolution and Behavior

lbf lactobacillus bulgaricus factor; pound-force

LBF Louis Braille Foundation (for blind musicians)

lbf-ft pound-force foot

lbf/in.² pound-force per square inch

lb ft pound foot

lb ft² pound per square foot

lb ft³ pound per cubic foot

lbg liquefied butane gas

LBG Paris, France (Le Bourget Airport)

lbh length, breadth, height

LBHD Long Beach Harbor Department

lb/hr pounds per hour

LBHS Luther Burbank High School

LBI Library Binding Institute; Licensed Beverage Industries; Lloyds Bank International

LBI *Lands Bókasafn Islands* (Icelandic—National Library of Iceland)—in Reykjavik

LBIMS Leban-Bartenieff Institute of Movement Studies

lb in. pound inch

lb in.² pound per square inch

lb in.³ pound per cubic inch

lbir laser-beam-image reproducer

LBJ Lyndon Baines Johnson—thirty-sixth President of the United States

LBJL Lyndon Baines Johnson Library (Austin)

LBJSHP Lyndon B Johnson State Historic Park (Texas)

LBJTMC Lyndon B Johnson Tropical Medical Center (American Samoa)

LBK landing barge, kitchen

lbl label (flow chart)

LBL Lawrence Berkeley Laboratories; Lloyds Bank Limited; London British Library

lb/lb pound per pound

lbld labelled

lblg labelling

lbm lean body mass; liquid (loose) bowel movement

lb m pound mass

lb/m pounds per minute

Lbm Lifeboatman

lb-mol pound-mole (mass)

LBMS London Boroughs Management Services

Lbn *Libano* (Spanish—Lebanon)

lbnpd lower-body negative-pressure device

lbo (LBO) leveraged buyout

LBO Lima, Peru (Limatambo Airport)

lboe lime-base oil emulsified

lbp length between perpendiculars; low back pain; low blood pressure

LBP Lanier Business Products; Lester Bowles Pearson (Canada's eighteenth Prime Minister); London Borough Polytechnic

LBPL Long Beach Public Library

lbr labor; laser-beam recorder; lumber

Lbr Labrador; Librarian

Lbr *Liberia* (Spanish abbreviation)

LBR Library Bill of Rights

lbs pounds (from the Latin—*Librae*)

lb/s pounds per second

l.b.s. *lectori benevolo salutem* (Latin—to the kind reader, greetings)

LBS landing barge support; Lean Burn System; Libyan Broadcasting Service; Lifeboat Station; London Boroughs Association; London Botanical Society

LBSC Long Beach State College

LB & SCR London, Brighton and South Coast Railway

lbs H₂O mm pounds of water per million standard cubic feet (natural gas)

LBSM Licentiate of Birmingham and Midland Institute School of Music

lb/sq in. pounds per square inch

lbs sq ft pounds per square foot

lbt laser-beam transmissiometer; lupus band test

lb t pound(s) thrust; pound(s) troy

LBTF Long Beach Test Facility

LBTS Land-Based Test Site

Lbu Labuan

LBV landing barge, vehicle

lbw leg before wicket; low body weight; low-speed black-and-white (photography)

lc label clause; laundry chute; lead-covered; leading case(s); left center; legal currency; letter card; level crossing; light case; light change; line-carrying; liquid crystal; load carrier; load center; load constant; load controller; localized corrosion; locked-closed; locus caeruleus; low calorie; low carbon; lower case; single acetate single cotton

l-c launch control; low calorie; low carbohydrate

l/c letter of credit; lower center

l.c. *loco citato* (Latin—in the place cited)

Lc corrected middle latitude

LC inductance-capacitance (symbol); Lackawanna College; Ladycliff College; Lafayette College; Lake Central Airlines; Lakehead College; Lakeland College; Lambuth College; Lance Corporal; Lander College; landing craft; Lane College; Laredo College; large-capitalization core (stocks); Lassen College; L'Assumption College; launch(ing) control; Law Court(s); Lawrence College; Lee College; Legal Committee; Legislative Council; Lesley College; Letter Contract; Lewis College; Library of Congress; Lieutenant Commander; Limestone College; Lincoln Center; Lincoln College; Lindenwood College; line of communication; Linfield College; Livingstone College; London Clause; London Club (of criminologists, freelance investigators, and members of the media concerned with controversial trials and unsolved crimes); Longwood College; Loras College; Louisburg College; Louisiana College; Lower Canada; Loyola College; Luther College; Luzon College(s); Lycoming College; Lynchburg College; Saint Lucia (Internet code)

L-C Liquid-Carbonic (Division of General Dynamics)

L/C Letter of Credit

L of C Library of Congress

LC *A Lover's Complaint*

lca low-cost automation

LCA Lake Carriers Association; Lake Central Airlines; landing craft—assault; Landscape Contractors Association; Launcher Control Area; Learning Corporation of America; Library Club of America; Licensed Company Auditor; Life Communicators Association; Lutheran Church in America

LCAC Landing-Craft Air Cushion

LCACS Landing-Craft Air Cushion Ships

LC-ADD *Library of Congress—American Doctoral Dissertations*

lcal lowercase alphabet length

lcao linear combination of atomic orbitals

LCAO Leadership Council of Aging Organizations

LCARA Love Canal Area Revitalization Agency

lcat (LCAT) lecithin-cholesterol acyltransferase

L & C ATA Laundry and Cleaners Allied Trades Association

L Cav Lucy Cavendish Collegiate Society, Cambridge

lcb longitudinal position of center of buoyancy

LCB Liquor Control Board; London and Continental Bankers

LCBBC Liquor Control Board of British Columbia

LCBM Liquor Control Board of Manitoba

LCBO Liquor Control Board of Ontario

LCBS Liquor Control Board of Saskatchewan

lcc lateral center of gravity; ledger card computer; life cycle cost; light curtain closed

LCC landing craft, control (3-letter symbol); Lands Conservation Council; Lansing Community College; Launch Control Center; League of California Cities; Life Cycle

Center; London County
Council; Lower Columbia
College

L & C C Lewis and Clark
College

LCcc Library of Congress catalog card

LCCC Library of Congress
Computer Catalog; Lorain
County Community College;
Lucas County Corrections
Center (Ohio)

lcce life-cycle cost estimate

LCCI London Chamber of
Commerce and Industry

LCCJ Louisiana Council on
Criminal Justice

LCCMARC Library of Congress Current MARC (file)

lccs low cervical caesarian
section

LCCS Launcher Captain Control System; Lucy Cavend-
ish Collegiate Society
(Cainbridge)

lccv (LCCV) large crude-car-
rying vessel (oil tanker)

lcd lead cadmium diode; liquid
crystal diode; lowest common denominator

lcd (LCD) liquid-crystal display

LCD Lord Chamberlain's Department; Lord Chancellor's
Department

l/c derv lowercase derivative
(angora, axminster, bakelite,
bunsen burner, canada balsam, castile soap, china clay,
congo red, cordovan leather,
delftware, etc.)

LCDHWIU Laundry, Cleaning, and Dye House Workers
International Union

lcdo licenciado (Spanish—licensed)

Lcdo Licenciado (Spanish—
lawyer)

LCDs Lower Court Decisions

lcdtl load-compensated diode-
transistor logic

lce lance; left center entrance

LCE Licentiate in Civil Engineering

lces least-cost estimating and
scheduling

lcf least common factor; liquid
complex fertilizer; longitudinal position of center of flotation; low cycle fatigue; lowest
common factor

LCF landing craft, flak; launch
control facility

LCFA Lower California Fisheries Association

l-c f-s pr last-come first-served
preemptive résumé

LCFTA London Cattle Food
Trade Association

lcg liquid-cooled (under) garment; longitudinal position of
center of gravity

LCG British armored landing
craft; Load Classification
Group

LCGB Locomotive Club of
Great Britain

LCGC London Community
Gospel Choir

lch launch

L Ch Licentiate in Surgery

LCH landing craft—heavy

LCHQ Local Command Headquarters (NATO)

lchr launcher

lci legally-correct interpretation;
locus of control interview

LCI landing craft, infantry;
Learner-Centered Instruction;
Liquid Crystal Institute (Kent
State University); Livestock
Conservation Incorporated

LCIP London Centre for International Peacebuilding

lcircon letter of credit irrevocable and confirmed

LCIS Library of Computer and
Information Sciences

LCJ Lord Chief Justice

LCJ *Louisville Courier-Journal*

LCJC Lake City Junior College

Lcks Locks (postal abbreviation)

lcl less than carload lot; less
than container lot; lifting
condensation level; lo-
cal(izer); loose container
load; lower card lever; lower
control limit

lcl (LCL) lowest charge level

LCL Liberal and Country
League (Australia); Licenti-
ate in Canon Law; Licentiate
in Canonic Law; Lincoln
Center Library; Licentiate in
Common Law

L-C-L Levinthal-Coles-Lillie
(bodies)

LCL *La Casa del Libro* (Spanish—House of the Book)—
Puerto Rico's typographic
arts museum on San Juan's
Calle del Cristo

LCLA Lutheran Church Library Association

L-C-L Levinthal-Coles-Lillie
(bodies)

lci/ci limited calendar life/con-
trolled item

LCLs Liverpool Central
Libraries

LCLS Livestock Commission-
Levy Scheme

lcm lead-coated metal; least
common multiple; left coastal
margin; limit-cycle monitor;
liquid curing media; logistics
composite model; lost circulation materials (oil well);
lowest common multiple

lcm (LCM) large-core memory; lymphocytic choriomen-
ingitis

LCM landing craft, mechanized;
landing craft—medium; London College of Music

LCMARC Library of Congress MARC (files beginning
in 1968)

LCMC Loosely Coupled, Mul-
tiComputer

lcmm life-cycle management
model

lcmp launcher control and
monitoring panel

LCMS Launch Control and
Monitoring System; Lutheran
Church Missouri Synod

lcn local civil noon

Lcn Lincoln

LCN La Cosa Nostra (Italian—
Our Thing)—The Mafia;
Load Classification Num-
ber(ing)

LCNM Lehman Caves Na-
tional Monument (Nevada)

LCNN Land Commander
Northern Norway (NATO)

LCNY Linguistic Circle of
New York

LCNYC Lincoln Center (New
York City)

LCO Landing Craft Officer;
Launch Control Officer; London College of Osteopathy

lcoc launch control officer's
console

L Col Lieutenant Colonel

l-commerce location commerce

lcos lead computing optical sight

lcp language conversion program; last complete program;
local coastal plan; low-cost
production

LCP landing craft, personnel; Latvian Communist Party; Liberal Country Party (Australia); Library Company of Philadelphia; Licentiate of the College of Preceptors; Livable Cities Program; Local Coastal Program; London College of Printing

LCPA Lincoln Center for the Performing Arts

L Cpl Lance Corporal

LCPL landing craft, personnel, large (naval symbol)

LCPR landing craft, personnel, ramped (naval symbol)

lcps last card program start

LCPS Licentiate of the College of Physicians and Surgeons

LCP & SA Licentiate of the College of Physicians and Surgeons of America

LCP & SO Licentiate of the College of Physicians and Surgeons of Ontario

lcr limited carrier(s) risk; locus control region

l/cr letter of credit

l/cr lettre de crédit (French— letter of credit)

L Cr Lieutenant Commander

LCR inductance-capacitance-resistance (symbol); landing craft, rubber; Line Condition Report

LCRA Lower Colorado River Authority

l/crt lamb crutchings

LCRT Lincoln Center Repertory Theater

lcs launch-control simulator

lcs (LCS) large-core storage

lc's (LCs) liquid crystals

LCs Langerhans cells

LCS Laboratory of Computer Sciences (M.I.T.); landing craft, support (naval vessel); League Championship Series (baseball); Library Computer System

lcsa letter of credit (negotiable by drafts at) sight airmail

LCSA Lewis and Clark Society of America

LCSH Library of Congress Subject Headings

lcss land combat support set

LCSS London Council of Social Service

LCST Licentiate of the College of Speech Therapists

LCSW Licensed Counseling Social Worker

lct last card total; less than truckload lot

LCT landing craft—tank; latest closing time; less than truckload lot; Local Civil Time; Loughsborough College of Technology

LCT Laboratoire Central de Télécommunications (French—Central Telecommunications Laboratory)

LCTC Langlade County Teachers College; Leicester College of Technology and Commerce; Lewis and Clark Trail Commission

lctp launcher control test panel

lcty locality

lcu launch-control unit; lower control unit

lcu (LCU) large closeup

LCU landing craft, utility

LCUSA Lutheran Council in the United States of America

lcv low calorific value

LCV landing craft, vehicle; League of Conservation Voters

LCVP landing craft, vehicle, personnel

lcv's light commercial vehicles

LCWS Landing Configuration Warning System

lcx launch complex

lcxt large cosmic-X-ray telescope

LCY League of Communists of Yugoslavia

LCYC Lemon Creek Yacht Club

LC zone land conservation zone

ld lactate dehydrogenase; ladies day; land; language disability; lead; learning disability; learning disabled; lethal dose; library development; lid; lifeboat deck; low density; light difference; light drinker; line of departure; line of duty; list(ing) date; load; load draft; Lord; low door; lower deck

ld (LD) laser disc; learning-disabled; lethal dose

l-d low-density

l/d length to diameter (ratio); life to drag (ratio)

l & d labor and delivery; loans and discounts; loss(es) and damage(s)

l.d. lepide dictum (Latin— wittily related)

Ld Leopold; Limited

LD Labor (US department); Lake District; lighting director; Lime Disease; line of departure; line of duty; London Docks; Low Dutch; lower berth (double occupancy)

L-D Leishman-Donovan (bodies)

L/D Letter of Deposit

L.D. Litterarum Doctor (Latin—Doctor of Letters)

LD-3 lower-deck container

ld₅₀ median lethal dose

LD-10 lethal dose for 10 percent of the animals tested

LD-50 lethal dose for 50 percent of the animals tested

lda land development aircraft; landing distance available; learning disabilities average; left dorso-anterior; line drawing amplifier; localiser direction(al) aid

lda (LDA) light defense aircraft

L^{da} Limitada (Portuguese or Spanish—Limited); *Licenciada* (Spanish—lawyer)— feminine form of *L^{do}* — *Licenciado*

LDA Ladies Darts Association; Laser Disc Association; Lead Development Association

ldac lunar-surface data-acquisition camera

L da P Lorenzo da Ponte (orig. Emanuele Conegliano)

L da. V Leonardo da Vinci

ldb light distribution box

LDBHS Louis D Brandeis High School

ldc large document copier; leased directive circuits; less developed country; long-distance call; lower dead center

ldc (LDC) latitude data computer

LDC Late Developing Country; Laundry and Dry Cleaning (union); Less Developed Countries; Light Direction Center; List of Design Changes; Local Defense Center; Local Development Corporation

LD & C Louis Dreyfus & Compagnie

LDCMMA Laundry and Dry Cleaners Machinery Manufacturers Association

ldc's (LDCs) least-developed countries; less-developed countries

ldd laser development device

lddo long-distance diesel oil

LDDS Low-Density Data System

LDEF Long-Duration Exposure Facility

ldel land development encouragement loans

L Dent Sci Licentiate in Dental Science

Lderry Londonderry

L de V Lope de Vega

LDF Legal Defense Fund (NAACP); Lime Disease Foundation; Local Defense Force(s); Low-frequency Direction Finder

ldg landing; loading; lodging

Ldg Lodge (postal abbreviation)

LDG *Los Desastres de la Guerra* (Spanish—*The Disasters of War* by Francisco Goya)

ldg & dly landing and delivery

Ldge Lodge

ldg gr landing gear

ldglts leading lights

ldgs lodgings

ldh lactate dehydrogenase

Ld'H *Légion d'Honneur* (French—Legion of Honor)

LDH *Ligue des Droits de l'Homme* (French—League for the Rights of Man)

ldhc locker-door hydraulic cylinder

ldi lecture-driven instruction

ldk lower deck

LDKG *Life and Death of King John*

ldl learning disabilities limited; loudness discomfort level; low-density lipoprotein

ld lmt load limit

LDMA London Discount Market Association

Ld May Lord Mayor

ld mk landmark

ldmwr limited depot maintenance work requirements

Ldn London; Londoner

LDN Long-Distance Network

ldo light diesel oil; long-distance oil

L^{do} *Licenciado* (Spanish—lawyer, licentiate holding master's degree)

LDO Licensed Deck Officer; Limited-Duty Officer

L-dopa levodihydroxyphenyla-lanine (Parkinson's disease treatment drug)

LDOS Lord's Day Observance Society

ldp left dorso-posterior; loan deficiency payment; logistics data package

Ldp Ladyship; London daily price; Lordship

LDP landed duty paid; Liberal Democratic Party (Japanese)

ldpe low-density polyethylene

ldr launder; laundry; leader; lecture-discussion-recitation; ledger; light-dependent resistor; list(ing) date received; lodger

l-d-r labor-delivery-recovery

LDR Large Deployable Reflector; London Digital Recording

l/d ratio length to diameter ratio; lift-to-drag ratio (of aircraft)

LDRC Lumber Dealers Research Council

ldri low data rate input

L-drivers learner-drivers

Ldr's London depository receipts

LDRTA Long-Distance Road Transport Association

ldry laundry

lds loads

lds (LDS) large disc store; large disk storage

Lds Leeds

LDs Learning Disabilities

LDS Latter Day Saints (Church of Jesus Christ of); Licentiate in Dental Surgery; Line Drawing System

LDSc Licentiate in Dental Science

ld/sd look down/shoot down

LDSR League of Distilled Spirits Rectifiers

LDSRA Logistics—Doctrine Systems and Readiness Agency (USA)

LDSRCPS Licentiate in Dental Surgery of the Royal College of Physicians and Surgeons

LDSRCS Licentiate in Dental Surgery of the Royal College of Surgeons

l&d store liquor and delicatessen store

ldt logic design translator

LDTF Large Dynamic Test Facility

LDTO Long-Distance Telegraph Office

ldtr long-dwell-time radar

LDV Local Defense Volunteer

ldw left defense wing (lacrosse); loss damage waiver

ldx long-distance xerography

ldy laundry

Ldy Londonderry

LDY Lancashire and Derbyshire Yeomanry; Leicestershire and Derbyshire Yeomanry

le launch(ing) equipment; leading edge; left eye; less than or equal to; library edition; limit of error; limited edition; local elder; low explosive

l/e lifetime earnings

l.e. *lupus erythematosus* (skin disease)

Le Lebanese; Lebanon

LE Labor Exchange; Labour Exchange; light equipment; Limited Edition; low explosive

Lea Leacock (Stephen Butler Leacock—Canadian humorist)

lea leather

LEA Limited English Ability; Local Education Authority; Local Education(al) Agency; Locomotive Engineers Association; Loss Executives Association; Lutheran Education Association

LEAA Lace and Embroidery Association of America; Law Enforcement Assistance Administration

LEAA *Law Enforcement Assistance Act*

LEAD Law Students Exposing Advertising Deception; License Education on Alcohol & Drugs

LEADER Lehigh Automatic Device for Efficient Retrieval

LEADS Law Enforcement Agencies Data System (Illinois)

Lea & F Lea & Febiger

LEAF Law, Equality, and Freedom (association)

leaf(s) leaflet(s)

LEAJ Law Enforcement and Administration of Justice (President's Commission on)

Leamington Royal Leamington Spa in central England

Leander British class of all-purpose frigates

Leanu Nikolaus Lenau (Nikolaus Niembsch von Strehlenau)

leap liftoff elevation and azimuth programmer

LEAP Lambda Efficiency Analysis Program; Language for the Expression of Associative Procedures; Law, Education, and Participation; Legislative Electoral Action Program; Loan and Educational Aid Program

LEAPS Laser Engineering and Applications for Prototype Systems; Law Enforcement Agencies Processing System (Massachusetts); London Electronic Agency for Pay and Statistics; Long-Term Equity Anticipation Securities

Lear The Tragedy of King Lear

Lear 23 Lear jet transport

LEAR Low Energy Antiproton Ring

leas leading-edge actuation system; lower-echelon automatic switchboard

Leasat Leased Satellite (communications service)

LEAST Learning Systems Standardization

Leavenworth U.S. Penitentiary at Leavenworth, Kansas

leaverats leave rations

Leb Lebanese; Lebanon

LEB London Electricity Board

Lebanese Lebanon-grown brownish-red hashish; people of Lebanon

Lebanon Republic of Lebanon (Middle East country), *al-Jumhoūrīya al-Lubnaniya*

le bodies lupus erythematosus bodies (LE bodies)

LEBS London Emergency Bed Service

lec local exchange carrier; lunar equipment conveyor

L Ec Ecclesiastic Latin

LEC Lake Erie College; Law and Economics Center (University of Miami); Law Enforcement Center; Livestock Equipment Council

LECE Ligue Européenne de Coopération Economique (French—European League for Economic Cooperation)

le cells lupus erythematosus cells (LE cells)

LECG Law and Economics Consulting Group

lech lecher; lecherous; lechery

LECLU Law Enforcement Civil Liberties Unit

Le Corbu Le Corbusier

Le Corbusier Charles Edouard Jeanneret-Gris

lect lecture

lect. *lectio* (Latin—lesson)

Lect Lecturer

lectr lecturer

'lectric electric

led (LED) light-emitting diode

led light-emitting dial; light-emitting diode(s)

Led Ledbetter; Ledyard

L Ed Lawyer's Edition (U.S. Supreme Court Reports)

LED Library Education Division (American Library Association)

LEDA Local Employment Development Action

LEDC League for Emotionally Disturbed Children

LEDETS Law-Enforcement Detachments (Coast Guard)

led's light-emitting diodes

lee. laser energy evaluator

Lee Leroy

LEE Low Expressed Emotion

LEEA Law Enforcement Education Agency

LEEGS Law Enforcement Explorer Girls

Lee I Leeward Islands

leep loop electrosurgical excision procedure

LEEP Law Enforcement Education Program

Leeward Islands Anguilla, Antigua, Barbuda, British Virgin Islands, Montserrat, Nevis, Redonda, Saint Christopher (St Kitts)

Leewards Leeward Islands

lef leading-edge flap; light-emitting film

LEF Life Extension Foundation; Lincoln Educational Foundation

LEF Liberté, Egalité, Fraternité (Liberty, Equality, Fraternity—slogan of the French Revolution)

LEFTA Labour (Party) Economic, Finance, and Taxation Association

leg. legal; legate(e); legation; legislation; legislative; legislature

leg (LEG) liquefied energy gas

leg. *legato* (Italian—smoothly flowing)

Leg Leghorn

Leg Legierung (German—alloy)

LEG Law Enforcement Group

LEG (UN) Legal Affairs (department of United Nations)

legat FBI agent or office working in an overseas legation of the United States; legation

LEGCO Legislative Council

leg com legally committed

legcrit legal critic(ism)

legg leggiero (Italian—lightly and rapidly)

legis legislative; legislature

legit legitimate

LEGT Lycée d'Enseignement Général et Technologique (French—High School of General Education and Technology)

legumes legumbres (Spanish-American truncation—beans, greenstuff, vegetables)

leg. wt legal weight

LEH Licentiate in Ecclesiastical History

LEHI Lehigh University

Lehman Caves Lehman Caves National Monument in eastern Nevada

lei land exclusive of improvements

LEI Leading Economic Indicator; Life-Expectancy Inventory; Life Extension Institute

Leic Leicester

leichtl leichtlöslich (German—readily soluble)

Leics Leicestershire

Leip Lepzig

lei's (LEIs) leading economic indicators

LE Is Leeward Islands

LEIS Law Enforcement Information System

Leit Leitrim

LEIU Law Enforcement Intelligence Unit

lej longitudinal expansion joint

lek monetary unit of Albania

lel low-energy laser; lower explosive limit

LEL Labor Electoral League; Laureate in English Literature; Letitia Elizabeth Landon

LELDC Law Enforcement Legal Defense Center

lem lateral eye movements; layered-earth model; lemon (ade); logical end of media

lem (LEM) lunar excursion module

Lem Lemuel

LeM *Le Monde* (The World)—Paris

LEM Lunar Excursion Module

LEMA Lifting Equipment Manufacturers' Association

lemac leading edge mean aerodynamic chord

LEMDA Lighting-Electrical Materials Distributors Association

lemo lemonade

lempira monetary unit of Honduras

LEMSIP Laboratory for Experimental Medicine and Surgery in Primates

Lemurio Lemuriologic(al)(ly); Lemuriologist(ic)(al)(ly); Lemuriology

len length; low-entry networking

Len Leningrad, formerly Petrograd, formerly St Petersburg; Lensky; Leonard

LEN Law Enforcement News

LENA Lower Eastside Neighborhoods Association

LENA *Labatorio Energia Nucleare Applicata* (Italian—Applied Nuclear Energy Laboratory)

LENDS Library Extends Catalog Access and New Delivery System

Lenin Vladimir Ilich Ulyanov

Leninpor Lenin Port (Leningrad Harbor, now St Petersburg Harbor)

lenit. *leniter* (Latin—gently)

Len Lib Lenin Library (Moscow)

Lenny Leonard; Leonard Bernstein

Lenslok J Frost trademark

Lenson Levensohn; Levenson; Levinson; Levinsky

lento *lentando* (Italian—increasingly slow)

Leo Leo N Tolstoy (Russian author); Leonard; Leonese; Leonidas; Leonine; Leopold; Leopoldville

Leo (Latin—Lion constellation)

LEO Low Earth Orbit

leone monetary unit of Sierra Leone

LEOPARD Law Enforcement Operations and Activities to Reduce Drugs

leopon leopard + lioness (hybrid offspring of male leopard and lioness)

LEOT Laser/Electro Optics Technology

lep lepton (collective term embracing anti-neutrino, electron, neutrino, photon, positron); lowest effective power

lep (LEP) large-electron positron collider

Lep Lepus (constellation)

Lep *Lepus* (Latin—Hare constellation)

LEP Labor Education Project; Library of Exact Philosophy; Limited English Proficiency

LEP *Lycée d'Enseignement Professionnel* (French—High School of Professional Education)

LEPA Law Enforcement Planning Agency

lep. dict. *lepide dictum* (Latin—well said)

lepid(s) lepidopterist(s)

LEPMA Lithographic Engravers and Plate Makers Association

Lepmus *Lepramuseet* (Norwegian—Leprosy Museum)—Bergen museum reflecting Dr Armanser Hansen's struggle against leprosy, also known as Hansen's disease

Lepontines Lepontine Alps along the Italo-Swiss border

LEPORE Long-Term and Expanded Program of Oceanic Research and Exploration

LEPRA Leprosy Relief Association (British)

le prep lupus erythematosis preparation

lepro leprology; leprosarium; leprose; leprosy

lep(s) lepidopterist(s)

Leps Lepus (constellation)

lept (LEPT) long-endurance patrolling torpedo

Lepto *Leptospira*

ler life expectancy reduction

Ler Lerida

LeRC Lewis Research Center (NASA)

LERC Laramie Energy Research Center; Law Enforcement Resource Center

les lesbian; local excitatory state

les (LES) lower esophageal sphincter

lEs limited English speaking

Les Lescombe; Lesley; Leslie; Lesotho (whose capital is Maseru); Lester

LES Launch Escape System; Lincoln Experimental Satellite

lEsa limited English-speaking ability

LESA Licensing Executives Society of Australia; Lunar Exploration System—Apollo

lesb lesbian(ism)

lesbie lesbian

lesbo lesbian; lesbianism

Leschy Theodor Leschetizky (Teodor Leszetycki)—Polish composer-pianist

Les L *Licensie es Lettres* (French—Licentiate in Letters)

Les Lip Leslie Lipson

les Mis *les Misérables* (French—the wretched)—Victor Hugo's novel about street people

Leso Lesotho (formerly Basutoland)

LESOP Leveraged Employee Stock Ownership Plan

Lesotho Kingdom of Lesotho (formerly called Basutoland, landlocked South African country)

les Ricains *les Americans* (French slang—The Americans)

LESS Least-cost Estimating and Scheduling Survey

Les Sc *Licensie es Sciences* (French—Licentiate in Science)

Lesser Antilles Leeward and Windward Islands extending from the Netherlands Antilles (Aruba, Bonaire, Curaçao) to the Virgin Islands

Lesser Sundas Lesser Sunda Islands east of Bali in Indonesia

lessie(s) lesbian(s)

Lester Leicester

let. letter; linear energy transfer

Let. Jer. *Letter of Jeremiah*

Let Lettish (Latvian)

LET Leader Effectiveness Training; Logical Equipment Table

LETAC Law Enforcement Training Advisory Council

L-et-C Loir-et-Cher

letch slang shortcut—lecher; lecheress; lecherous; lecherous feeling for; lechery

letfo letter follows

L-et-G Lot-et-Garonne

let's let us

LETS Low-Energy Telescope System

lett letter(s)

lett letteratura—(Italian—literature); *letterlijk* (Dutch—literally); *lettisch* (German—Latvian)

Lett Lettish

Letter Carriers Union National Association of Letter Carriers

letterk letterkunde (Dutch—literature)

Lettie Leticia

Lett Nuovo Cimento Lettere al Nuovo Cimento (Italian—Communications Regarding New Findings)

Letts Lettish peoples (Latvians)

Letty Leticia; Letitia

Letz Letzeburgesch (Flemish dialect of Luxembourg)

leu monetary unit of Romania

leu (LEU) leucine (amino acid)

leuk (Latin prefix—white)—leukocytes

leuko leukos (Greek—white)—leukemia, leukocyte(s), leukorrhea—the whites

lev lever; low emission vehicle; monetary unit of Bulgaria

lev (LEV) lunar excursion vehicle

lev levert (Norwegian—delivered)

lev. levis (Latin—light)

Lev The Book of Leviticus; Leo; Leon

Lev Leviticus

levant levant or morocco leather (also called levant morocco and characterized by its prominent grain and high quality prized by bookbinders and book lovers alike)

Levant eastern Mediterranean lands such as Israel, Lebanon, and Syria

levis Levi Strauss' reinforced denim workclothes, particularly dungaree trousers with heavily stitched-and-riveted pockets

levit. leviter (Latin—lightly)

Lew Lewis; Llewellyn

le'ward leeward

Lewisburg U.S. Penitentiary at Lewisburg, Pennsylvania

lex lexical; lexicographer; lexicography; lexicon

Lex Lexington

LEX Lexington, Kentucky (airport)

Lex-Fay Lexington-Fayette, Kentucky

lexi lexical; lexicographer; lexicographic(al)(ly); lexicography; lexicolater; lexicological(ly); lexicon(s)

lexic lexical(ic)(ally)

lexico lexicographer

lexicog lexicographer, lexicography

lexicon. lexiconist(ic)(al)(ly)

lexig(s) lexigram(s)—word symbol(s)

Lexington Lexington, Kentucky's U.S. Public Health Service Hospital for narcotic addicts

Lexington Bks Lexington Books—Division of DC Heath

lexiphan lexiphanic(al)(ly); lexiphanicism; lexiphanist

Lex Phil Lexington Philharmonic (Kentucky)

lexis (LEXIS) legal data base on-line to head

LEXIS Lexicography Information Service

l/ext lower extremity

Ley Leyden

LEY Liberal European Youth

Leyd Leyden

lez(es) lesbian(s)

lezz lesbian

lf lawn faucet; leaf; ledger folio; left field; left front; life float; light face type; line feed; linear feet; linear foot; linoleum floor; load factor; low frequency (30–300 kc)

lf (LF) line feed (data-processing character); line feed character (data processing)

l/f left front; light fittings

Lf Loaf (postal abbreviation)

LF Launch(ing) Facility; Law French; Lindbergh Field; Local Force (Red China)

lfa last field address; left fronto-anterior

LFA Land Force, Airmobility (NATO); Light Freight Agent; Low Flying Area

LFAS Loose-Fitting Air Supplied

LFB Licensed Fishing Boat; London Fire Brigade

LFBC London Federation of Boys' Clubs

lfc laminar flow control; level of free convection; low-frequency current

l-fc low-frequency current

LFC Lutheran Free Church

lfd least fatal dose; low fat diet

lfd laufend (German—current, consecutive)

Lfd Laufend (German—current, rent)

LFE Laboratory For Electronics

lffp laser fusion feasibility project

lfg live foal guaranteed

Lfg (Lfrg) Lieferung (German—installment, party delivery)

LFI Lethal Force Institute

LFICS Landing Force Integrated Communications System (USMC)

lfl lower flammable limit

LFL Lesbian Feminist Liberation (society); Lithuanian Freedom League

lfm low-power fan marker

lf/mf low-frequency medium-frequency

lfo light fuel oil; low-frequency oscillator

LFO Licentiate of the Faculty of Osteopathy

lfp left fronto-posterior

LFP Lindbergh Field Plant (Convair)

LFPC Louisiana Foundation for Private Colleges

LFPP Louisiana Family Planning Program

LFPS Licentiate of the Faculty of Physicians and Surgeons

lfq light-foot quantizier

L fr Luxembourg franc(s)

LFR inshore fire-support ship (naval symbol)

LFRC League for Fighting Religious Coercion

lfrd low-friction reliability deviation

lfre liquid-fuel ramjet engine

LFS amphibious fire-support ship (naval symbol)

LFSIR Linear Feedback Shift Register

lft leaflet; left fronto-transverse; linear feet; linear foot; low-flush toilet; low-frequency transducing

l/ft² lumens per square foot

LFTB Liquid Fuels Trust Board

LFTU Landing Force Training Unit

LFU Light Fighting Unit; Lunar Flying Unit

lg lagoon; landing; landing gear; languages(s); large; large grain; leg bye; length; long; long grain; low grade

l/g locked gate

Lg Landgrave; Landgraviate

LG Landing Ground; large-capitalization growth; Leathercraft Guild; Leipzig Gewandhaus; Lloyd George; Low German

L/G Letter of Guarantee

LGA La Guardia, New York City airport; Local Government Association

LGAA Local Government Auditors Association

LGAT Local Government Appeals Tribunal

lgav league average

lgb laser-guided bomb

Lgb Long Beach, California

LGB Long Beach, California (airport)

L-G-B Landry-Guillain-Barré (syndrome)

LGC Laboratory of the Government Chemist; Local Government Commission

L-G C Lockheed-Georgia Company

LGCC Letchworth Garden City Corporation

LGCR Location Geological Characterization Report

lgd leaderless group discussion

LGD London Gaol Delivery

lge large

LGEA Local Government Electricity Association

LGEB Local Government Examination Board

LG&E Corp Louisville Gas & Electric Corporation

L-Gen Lieutenant-General

L Ger Low German

Lg of H-D Landgrav(iate) of Hesse-Darmstadt

Lg of H-K Landgrav(iate) of Hesse-Kassel

LGIO Local Government Information Office

LGk Late Greek

lgm little green men (supposedly inhabiting extraterrestrial planets)

LGM Lloyd's Gold Medal

LGM *Laboratorium voor Grondmechanica* (Dutch—Soil Mechanics Laboratory)

LGMB Lady Godiva Marching Band

lgn lateral geniculate nuclei

Lgn Lagoon; Leghorn

LGO Lamont Geological Observatory (Columbia University); Local Government Office(r); Local Government Ordinances

LGOC London General Omnibus Company

lgp laser guided projectile; liquefied petroleum gas; low ground pressure

lgp (LGP) lasergraphic plotter

Lgp Legaspi Albay

LGPA Livestock and Grain Producers Association

lgr leasehold ground rent; ligroin

L Gr Late Greek

LGRA Local Government Reports of Australia

L Gr Ec Ecclesiastic Late Greek

LGRs Local Government Reports

lgs (LGS) laser geodynamics satellite

Lgs Lagos

LGS Landing Guidance System

LGSM Licentiate of the Guildhall School of Music

Lgt Light (postal abbreviation) LGT Liggett Group (stock-exchange symbol)

LGTB Local Government Training Board

lgth length

lg tn long ton

lg tpr long taper

lg-type ed large-type edition

LGU Ladies Golf Union

lgv lymphogranuloma venereum (venereal disease)

LGW London, England (Gatwick Airport); Longines-Witnauer (watches)

lg wh & br landing gears, wheels, and brakes

lh last hope; learning handicapped; left halfback (field hockey); left hand; left hind; lighthouse; lightly hinged; linked hearts; liquid hydrogen; lower half, lower hold

lh (LH) lactogenic hormone; lateral hypothalamus; left hand; luteinizing hormone

l/h labor hour; lamp holder; liters per hectare; low to high

lH *linke Hand* (German—left hand)

LH Liberty House; lighthouse; Lufthansa (airline)

L + H Lamport & Holt (Line)

L.H. left hand

LH₂ liquid hydrogen

lha lateral hypothalamic area; lower-half assembly

LHA landing ship, helicopter, assault; local hour angle

lhams lower hour angle of the mean sun

LHAR London-Hamburg-Antwerp-Rotterdam (range of ports)

LHAs multipurpose amphibious-warfare ships (naval symbol)

lhats lower hour angle of the true sun

lhb left halfback; lost heartbeat (attractive woman)

LHBANA Log House Builder's Association of North America

lhc large hadron collider

LHC Large Hedron Collider; Lease Housing Coordinator; Lord High Chancellor

LHCJEA London and Home Counties Joint Electric Authority

lhd left-hand drive; load-haul-dump(ing) machinery

L.H.D. *Litterarum Humanorum Doctor* (Latin—Doctor of Human Letters); *In Litteris Humanioribus Doctor*, (Latin—Doctor in Humane Letters)

lhdc lateral homing depth charge

lh dr lefthand drive

lHe liquid helium

Lhe liquid helium

L Heb Late Hebrew

L'heed Lockheed

LHG Library History Group

LHI Library of the Hoover Institution (on War, Revolution, and Peace)—Stanford, California

LHI Ligue Homeopathique Internationale (French—International Homeopathic League)

L-hinge capital-L-shaped hinge

lhm letterhead memo(randum)

LHMC London Hospital Medical College

LHNCBC Lister-Hill National Center for Biomedical Communications

LHO Local Health Office(r); Livestock Husbandry Office(r)

l hold(er) lease hold(er)

LHON Leber's hereditary optic neuropathy

LHOOQ Elle á Chaud au Col (French slang—She of the hot crotch)—Marcel Duchamp's crude nickname for da Vinci's Mona Lisa

lhp left-handed pitcher

lhr lumen hour(s)

l/hr liters per hour

LHR London, England (Heathrow Airport)

L & HR Lehigh and Hudson River (railroad)

lhrf (LHRF) luteinizing hormone releasing factor

lhrh (LHRH) luteinizing hormone-releasing hormone

LHRT Library History Round Table

lhs laser heterodyne spectrometer; lefthand side

LHS Lafayette High School

LHSC Lock Haven State College

lhsv liquid hourly space velocity

LHT Lord High Treasurer

lh th lefthand thread

lhtl luxury class hotel

LHW League of Hispanic Women; lower high water

LHWCA Longshore and Harbor Workers' Compensation Act

LHWI lower high water interval

lhwnt lowest high water neap tide

li left inner (field hockey); lifting index; line item; link; lira; lithograph; lithographer; lithography; lived; living; longitudinal interval

l/t letter of intent

li (Li) liability

Li *Limburg* (Dutch province); lithium

LI Lakshadweep Islands; Leeward Islands; Letter of Introduction; Liberia; Liberian; Liechtenstein (Internet code); Lions International; Locksmithing Institute; Long Island (L.I.)

L-I Loire-Inférieure

LI Lingvo Internacia (Esperanto—International Language); *Lydveldid Island* (Icelandic—Republic of Iceland)

Li-2 Soviet Lisunov transport plane called Cab

lia liaison

LIA Laser Institute of America; Lead Industries Association; Leather Industries of America; Lebanese International Airways; Livestock Improvement Association; Long Island Association

LIA Ligue Internationale d'Arbitrage (French—International Arbitration League)

LIAA Life Insurance Association of America

liab liability

LIALS Long Island Airport Limousine Service

LIAMA Life Insurance Agency Management Association

LIAT Leeward Islands Air Transport

lib liberal; liberalism; liberation(ist); libertarian(ism); liberty; librarian; library

lib libretto (Italian—operatic text)

lib. liber (Latin—book); *libra* (Latin—pound)

Lib Liberal; Liberal Party; Liberation(ist); Liberty Party; Libra (constellation); Libya; Libyan

Lib Libano (Italian—Lebanon); *Libano* (Portuguese or Spanish—Lebanon)

LIB Let's Ignite Bras

Lib Auto Res Con Library Automation Research Consulting Associates

LIBBA Long Island Beach Buggy Association

Libby Elizabeth

lib cat. library catalog

libcon libertarian conservative (term invented by John Chamberlain to describe the liberal-conservative views of Pablo Casals, Milovan Djilas, John Dos Passos, Max Eastman, James T Farrell, Sidney Hook, Alberto Moravia, Allen Tate, Edmund Wilson, etc.)

LIBCON/E Library of Congress/English

Lib Cong Library of Congress

Lib Dem Liberal Democratic Party (formerly Social Democratic Party of the United Kingdom)

libe librarian; library

LIBE Liga Internacia de Blindaj Esperantisol (International League of Blind Esperantists)

libec light behind camera

lib ed library edition

Liberace Wladziu Valentino Liberace

Liberia West African coastal country adjacent to Sierra Leone—founded by United States in 1822 and settled by freed American Negroes

Libert Libertarian

LIBGIS Library General Information Survey

LIBID London Interbank Bid Rate

Lib-Lab Liberal-Labour (Australian coalition)

Lib J Library Journal

Libn Librarian

Libor London interbank-offered rate

LIBOR London Interbank Offered Rate

Lib Parl Library of Parliament

libr librarian; library

libr libretto (Italian—opera or oratorio text)

Libra (Latin—Balance constellation)

LIBRA Living In the Buff Recreational Associates

LIBRE Living In the Buff Residential Enterprises (nudist apartments and beaches)

Lib Res Library Research Associates

LIBRIS Library Information System (Swedish on-line retrieval system)

Lib(s) Liberal(s)

LIBS Library Information and Bibliographic System

Lib Sig Library Signs
Lib Soc Sci Library of Social Science
libst librettist (Italian—libretto autimor)
Libs Unl Libraries Unlimited
Lib UN Library of the United Nations (New York headquarters)
Libya People's Socialist Libyan Arab Republic (North African country populated by Arab Berbers), *Al-Jumhuria al-Arabia allibya*
lic license; linear integrated circuit; low-intensity conflict
Lic Licentiate
Lic Licenciado (Spanish—lawyer, licentiate holding master's degree)
LIC Lands Improvement Company; Liquor Industry Council; Local Import Control
LICA Ligue Internationale Contre le Racisme et l'Antisemitisme (French—International League Against Racism and Anitsemetism)
LICC London Institute for Contemporary Christianity; Long Island Council of Churches
licd licensed
Lic D Licenciado Don (Spanish—Sir Lawyer)
lic dlr licensed dealer
LICeram Licentiate of the Institute of Ceramics
licm left intercostal margin
Lic Med Licentiate in Medicine
Lic Phil Licentiate in Philosophy
LICTBOSS Life-Cycle Theory of Bureaucratic Ossification
Lic Theol Licentiate in Theology
lid local improvement district
LID League for Industrial Democracy
L & ID London and India Docks
lidar laser radar device (for measuring wind direction and speed); laser-impulsed radar; light detection and ranging (laser-beam air pollution or smog measuring device)
LIDAR Light Detection and Range (research program)
LIDB Logistics Intelligence Data Base

LIDC Lead Industries Development Council; Livestock Industry Development Council
Liddie Lydia
LIDH Ligue Internationale des Droits de l'Homme (French—International League for the Rights of Man)
LIDO Logistics Inventory Disposition Order
lidoc lidocaine (xylocain)
lids lift improvement devices
Lie Liechtenstein (whose capital is Vaduz); Liepaya
LIE Liberal Intellectual Establishment (Philip Wylie's description of the befuddled and often nonsensical liberals of his time)
LIEAP Low-Income Energy-Assistance Program
Liech Liechtenstein
Liechtenstein Principality of Liechtenstein (Alpine country), *Fürstentum Liechtenstein*
Lief Lieferung (German—issue)
LIEJA Long Island Equal Justice Association
LIEMA Long Island Electronics Manufacturers Association
LIESA Long Island Episcopal Schools Association
Lieut Lieutenant
Lieut Col Lieutenant Colonel
Lieut Comdr Lieutenant Commander
Lieut Gen Lieutenant General
Lieut Gov Lieutenant Governor
lif left iliac fossa
LIF Lone Indian Fellowship
LIFA Life Insurance Federation of Australia
life laser-induced fluorescence of the environment
life see LIFFE
LIFE Ladies Involved For Education; League for International Food Education; Love Is Feeding Everyone
lifes laser-induced fluorescence and environmental sensing
Life Sta Lifeboat Station (US Coast Guard)
liff laser-induced fluorescence fluorimetry
LIFFE London International Financial Futures Exchange
LI Fire Eng Licentiate of the Institution of Fire Engineers
lifmop linearly frequency-modulated pulse

lifo last in, first out
LIFPL Ligue Internationale de Femmes pour la Paix et la Liberté (French—International League of Women for Peace and Freedom)
LIFR Low Instrument Flight Rules
lift logically-integrated fortran translator
LIFT Leveraged Investments For Tomorrow; Loans and Investments for a Future Together; London International Festival of Theatre; London International Freight Terminal
lig ligament; ligature
Lig Liguria(n); Limoges
Lige Elijah
liger offspring of lion and tigress
light. lighting; lightning
lightex searchlight illumination exercise
lign lignende (Dano-Norwegian—similar)
lignite brown coal
liguid. liguidation
Liguori Liguori Publications
Ligurians Ligurian Alps or Ligurian Apennines of northwestern Italy or the people of the region around the Gulf of Genoa
lih left inguinal hernia; light intensity high
LIHEAP Low-Income Heating and Energy Assistance Program
LIHDC Low Income Housing Development Corporation
LII Librarians' Index to the Internet
Lik Obs Bull Lick Observatory Bulletin(s)
lil light intensity low; lilliputian; little
li'l little
Lil Lilian; Lillian; Lily
LIL Lunar International Laboratory (proposed in 1961 by Dr Theodore von Karman)
lila life insurance logistics automated
LILA Ligue Internationale de la Librairie Ancienne (French—International League of the Old Library)
lilangeni monetary unit of Swaziland
LILCO Long Island Lighting Company

Lilly Lilian; Lillian

lilo last in, last out

LILS Lead-in-Light System (airport term)

lim light intensity marker; limber; limit(er); line induction motor; line insulation monitor; line interface module; linear induction motor; linear-induction motor(s); liquid injection moulding; locator inner marker

Lim Limburg (Dutch province); Limerick; Limón (Costa Rican province)

LIM Lima, Peru (Callao International Airport); limit for weight restriction (aircraft code)

Lima letter L radio code; (pronounced *leema*)

LIMA Long Island Museum Association

LIMAC Linden Industrial Mutual Aid Council

lim dat limiting date

lime calcium oxide (CaO)

LIMEAN London Interbank Median Averages Rate

limestone calcium carbonate ($CaCO_3$)

limewater calcium-hydroxide solution—$Ca(OH)_2$; limejuice and water mixture

lim-lib(s) limousine liberal(s)

limnol limnology

limo lemonade; limousine

LIMO Limousine Industry Manufacturers Organization

limon lime-and-lemon (hybrid citrus fruit)

Limón Puerto Limón, Costa Rica

limos limousines

limp limp cloth binding; limp cloth bound

Limpopo the Limpopo or Crocodile River of East Africa where in 1497 Vasco da Gama named it Rio do Espiritu Santo

LIMRA Life Insurance Marketing and Research Association

LIMRF Life Insurance Medical Research Fund

LIMS Logistic Inventory Management System

limvr linear-induction motor vehicle research

lin lineal; linear

lín *línea* (Spanish—line)

Lin Lincoln; Linda; Lindenberg(er); Lindley; Lindolfo; Lindon; Lindsay; Linley; Linnaeus; Linsley, Linton; Linus

L i N *Lokalhistorisk institutt Norge* (Norwegian Local History Institute)

LIN *Linjeflyg* (Swedish airline); Milan, Italy (Linate Airport)

Lina Angelina; Carolina; Caroline

linac linear accelerator

Linacre Linacre College, Oxford

linc laboratory instrument computer

Linc Lincoln; Lincoln College, Oxford

LINC Learning Institute of North Carolina; Logic and Information Network Coupler; Licensed Independent Network of CPA financial planners

Linc Coll Lincoln College Oxford

LINCO Linearly-Organized Chemical Code (for computer system)

Lincoln Abraham Lincoln, 16th President of the United States; Nebraska's capital

Lincoln's Abraham Lincoln's Birthday (February 12)

lincompex linked compressor and expander

Lincs Lincoln automobiles; Lincolnshire

LINCS Language Information Network and Clearinghouse System

Lindbergh Charles A Lindbergh; Lindbergh Field (San Diego's international airport)

LINDE Linde Air Products

Lindy Colonel Charles A Lindbergh

Line The Line—the Equator

Lines Line Islands in the equatorial mid-Pacific Ocean where they include Caroline, Christmas, Fanning, Flint, Kingman Reef, Malden, Palmyra, Starbuck, Vostock, and Washington Islands

lines/m lines per minute

lines/mm lines per millimeter

lines/s lines per second

L-Infre Loire-Inférieure

lin ft linear feet; linear foot

ling linguist(ics)

Linguis Linguistics

linim liniment

Linlithgow West Lothian, Scotland

Linn Linné; Linnaeus

lino linoleum; linotype; linotypist

linol linoleum

lino oper linotype operator

LINOSCO Libraries of North Staffordshire in Cooperation

LINS Laser Inertial Navigation System

L Inst Phys Licentiate of the Institute of Physics

LINTAS Lever's International Advertising Service

L'Intran *L'Intransigeant*

LINWR Lake Ilo National Wildlife Refuge (North Dakota)

lin yd linear yard

LIO Legislative Information Office; Lionel Corporation (stock exchange symbol); Lions International organization; Livestock Improvement Organization

LIOB Licentiate of the Institute of Building

LIOCS Logical Input/Output Control System

LION Local Integrated Optical Network

lip. lease in perpetuity; life insurance policy

lip (Latin prefix—fat)—lipectomy

LIP Local Initiatives Program

Lipari Islands Italian penal colony northeast of Sicily; islands include Stromboli and Vulcano; also called Aeolian Islands

Liparis Lipari Islands

LIPC Livestock Industry Promotion Corporation

LIPI *Lembaga Ilmu Pengetahuan Indonesia* (Indonesian Academy of Sciences)

lipl (LIPL) linear information programming language

LIPM Lister Institute of Preventive Medicine

lipo lipogram(matic)

Li Po Li T'ai-po

Lippi Lippizaner horse

Lippincott J B Lippincott Company

lips logical inferences per second

LIPS Leitner International Performance Scale

lip sync lip synchronization (in sound films)

lipup backward pupil (pupil spelled backwards)

liq liquid; liquor

liq f rkt liquid fuel rocket

liqn *liquidación* (Spanish—liquidation)

liqt liquid transient

LIR Liaison Investigation Report; Library of International Relations

L & IR Legislation and Intergovernmental Relations

lira monetary unit of Italy, San Marino, Turkey, and Vatican City

lira. loft-type infrared analysis

LIRA Linen Industry Research Association; Logging Industry Research Association; Low Income Ratepayer Assistance

lirbm liver, iron, red bone marrow

LIRES Literature Retrieval System

LIRES-MS Literature Retrieval System-Multiple Searching

LIRI Leather Industries Research Institute

lirl low-intensity runway lights

liroc last instruction readout cycle

lirod lightweight radar and optronic director

LIRR Long Island Railroad

LIRS Lutheran Immigration and Refugee Service

LIRT Library Instruction Round Table

lis laser isotope separation; lobar in situ

Lis Lisboa (Portuguese-Lisbon); Lisbon

LIS Liberian Information Service; Library and Information Science; Light Industry Services; Lisbon, Portugal (airport); Livestock Incentive Scheme; Locate in Scotland; Lockheed Information System(s); Long Island Sound

lisa (LISA) low-input sustainable agriculture

LISA Linear Systems Analysis; Long Island Schizophrenia Association

LISA *Library and Information Science Abstracts*

Lisb Lisboa (Portuguese or Spanish—Lisbon); *Lisbona* (Italian—Lisbon)

LISC Lions International Stamp Club; Local Initiative Support Corporation; London Institute for the Study of Conflict

LISD Library Information Science Division (World Information Systems Exchange)

LISM Licentiate of the Incorporated Society of Musicians

LISO Library Information Service Office(r)

lisp. list processor (computer language)

LISP List Processing (for text manipulation)

LISPA Long Island Sound Pilots Association

LISR Line Information Storage and Retrieval

LISS London Institute of strategic Studies

list. laser and isotope separation technology

LIST Library and Information Services—Tees-side

LIST *Library and Information Science Today*

'listed enlisted

LISTEN Low-Income-Schools Teacher Education

'listment enlistment

lit literacy; literate

lit. liter; literal; literally; literary; literature; litter; little

l it lire italiane (Italian lire)

lit litauisch (German—Lithuanian)

lit. litterae (Latin—letters)

Lit Litvak (Yiddish—Lithuanian)—person of Judaic origin from Lithuania or nearby regions

Lit Eng Literary English

LIT Light Intratheater Transport (aircraft); Little Rock, Arkansas (airport)

LITA Library and Information Technology Association

litcrit literary critic(ism)

litcy literacy (ability to read and write)

lite light

LITE Legal Information Through Electronics

litex searchlight illumination exercise

lith lithograph; lithography; lithology

Lith Lithuania; Lithuanian

Lith Yid Lithuanian Yiddish

litharge lead oxide (PbO)

litho lithograph

lithol lithology

LITINT Literacy International

litr lighter

litrg literage

Lits Lithuanians; Litvaks

litt litteratur or litteraer (Dano-Norwegian—literature or literary)

Litt.B. *Litterarum Baccalaureus* (Latin—Bachelor of Letters)

Litt.D. *Litterarum Doctor* (Latin—Doctor of Letters)

Little Little, Brown

Litt. M. Master of Letters

litur liturgical; liturgy

liturg liturgical; liturgistic; liturgy

Litvak (Yiddish—Lithuanian Jew)

Litvak(s) Lithuanian(s)

Litvinov Maksim Maksimovich (originally Meir Walach)

Litz Litzendraht (German—wire)

LIU Long Island University

LIUNA Laborers International Union of North America

LIUP Long Island University Press

liv. lived; living

liv liver

liv le livre (French—book); *la livre* (French—pound)

Liv Liverpool

Liv Titus Livius (Roman historian often referred to as Livy)

LIV Light Infantry Volunteers

liv. abt. lived about

Liver Liverpool; Liverpudlian(s)

Liverpool Liverpool Prison (also called LP)

livex live exercise (military)

Liv Phil Liverpool Philharmonic

livr livraison (French—issue of a journal, part of a book or serial)

liv rm living room

liv st livre sterling (French—pound sterling)

Liv St Liverpool Street (rail terminal)

lix lixiviation

liz Lizard; lizzie (as in *tin lizzie*, an old Ford Automobile)

Liz Eliza(beth); Liza; Lizette

Liza Liza Minnelli

Lizard Lizard Head, Lizard Peninsula, Lizard Point, Lizard Town—at the tip of southwest Cornwall

LIZARDS Library Information Search and Retrieval Data System

lj life jacket; long jump

LJ La Jolla; Law Judge; Libby, McNeil & Libby (stock exchange symbol); Library Journal; Lord Justice; Sierra Leone Airways (2-letter coding)

LJ *laufen Jahre* (German—current year); *Law Journal*; *Library Journal*

LJC Lackawantia Junior College; Laredo Junior College; Lincoln Junior College; London Juvenile Court

LJMCA La Jolla Museum of Contemporary Art

ljp localized juvenile periodentitis

LJR *Law Journal Reports*

LJ/SLJ *Library Journal/School Library Journal*

LJT Lear jet airplane

LJTSA Library of the Jewish Theological Seminary of America (NYC)

LJU La Jolla University

Ljub Ljubljana (capital of Slovenia)

lk link

Lk Lake; Luke

LK Lockheed Aircraft Corporation (stock exchange symbol); Sri Lanka (Internet code)

LKAB *Luossavaara-Kiirunavaara Aktiebolag* (iron-ore mines in Luossa-Kiiruna range of northern Sweden)

LKB Link-Belt Company (stock exchange symbol)

lkd locked

lked linkage editor

lkg locking

lkg & bkg leakage and breakage

lkge leakage

LKGR Lake Kyle Game Reserve (Rhodesia)

lk-n lock-in

LKPA *Landeskriminalpolizeiamt* (German—Land Criminal Office)—Prussian organization (1940s)

LK & PRR Lahaina-Kaanapali and Pacific Railroad

LKQCPI Licentiate of the King and Queen's College of Physicians of Ireland

lkr locker

Lkr Landskrona

lks links; liver, kidney, spleen

Lks Lakes

Lksde Lakeside

lkt lookout

lk up lock up

Lkw *Lastkraftwagen* (German—lorry, truck)

lkwash lockwasher

ll land lines; light lock; limited liability; lines; live load, long lead; lower left; low(er) level; lower lid; lower limit

ll (Ll) landlord

l/l library labels; line-by-line; looseleaf; lower left; lower limit

l & l leave and liberty; look and listen

'll (contraction of till and will)

ll *lectiones* (Latin—readings); *llegada* (Spanish—arrival)

LL Lake Louise; Law Latin; Language Laboratory; Language Lessons; Law List(ing); Lebanese pound; Lending Library; Linda Love; Little League (baseball); Loftleidir (Icelandic Airlines); Lord Lieutenant; Low Latin

LL (Ec) Ecclesiastic Late Latin

L/L *Lutlang* (Norwegian—limited company)

lla left lower arm; limiting lines of approach

LLA Latin Liturgy Association; Lend-Lease Administration; Louisiana Library Association; Luther League of America

llama long-life atmospheric-motoring airship

Llanfairp Llanfairpwllgwyngllgogershwymdro-bwllabtysiliogogoch (Welsh—Church of St Mary near the Raging Whirlpool and the Church of St Tysilio by the Red Cave)—the longest place-name in the world

L Lat Late Latin; Low Latin

LLat Law Latin

llb long-leg brace

LLB Little League Baseball; L L Bean

LL.B. *Legum Baccalaureus* (Latin—Bachelor of Laws)

LLBA *Language and Language Behavior Abstracts*

llbcd left lower border of cardiac dullness

LLBO Liquor License Board of Ontario

l-l brace long-leg brace

llbs low-level bombsight

llc lower left center

LLC Libertarian Law Council; Library Learning Center; limited liability company; Living Learning Center (Indiana University)

ll. cc. *locis citatis* (Latin—in the places cited)

llcca long-life cycle-cost avionics

LLCM Licentiate of the London College of Music

LLCO Licentiate of the London College of Osteopathy

Ll & Cs Lloyd's and Companies

LLCUNAE Law Library of Congress United Association of Employees

LL.D. *Legum Doctor* (Latin—Doctor of Laws)

LLDEF Lambda Legal Defense and Education Fund

LlD factor *Lactobacillus lactis* Dorner factor (vitamin B_{12})

lle left lower extremity

lle *llegada* (Spanish—arrival)

LLE Laboratory for Laser Energetics (University of Rochester)

LLEI *Lincoln Library of Essential Information*

L Lett Licentiate of Letters

Lleyn Pen Lleyn Peninsula in Wales

L-L f Laki-Lorand factor

LLF Laubach Literacy Fund

llfm land line frequency modulation

LLGMA Luggage and Leather Goods Manufacturers of America

lli latitude and longitude indicator; long lead items

LLI Laubach Literacy International; Lord Lieutenant of Ireland

LLIL Long Lead Item List(ing)

LLJ Leaf Library of Judaica

LLJJ Lords Justices

lll left lower limb; left lower lobe; light load line; looseleaf ledger; low-level logic

l/ll line-by-line libretto

LLL Lawrence Livermore Laboratories; Lutheran Laymen's League

LLL Love's Labour's Lost

lllb left long-leg brace

LLLI La Leche League International

llll left lower lung lobe

lllp leased long-lines program

llltv low-light-level television,

llm localized leucocyte mobilization; lunar landing module

LL. M. Legum Magister (Latin—Master of Laws)

LLN League for Less Noise

LLNL Lawrence Livermore National Laboratory

LLNNR Loch Leven National Nature Reserve (Scotland)

LLNWR Long Lake National Wildlife Refuge (North Dakota)

lloc land line of communications

Lloyd David Lloyd George (British politician)

Lloydbras Lloyd Brasileiro

Lloyd's Lloyd's Register of Shipping

Lloyd's Bank Lloyd's Bank International Ltd

LLP Lifetime Learning Publications; limited liability partnership

L L & P of H Life, Liberty, and the Pursuit of Happiness (original draft of the *Declaration of Independence* read: "Life, Liberty, and the Pursuit of Profit")

LLPI Linen and Lace Paper Institute

llps low-level pumping station

llq left lower quadrant

llqa limiting lines of quiet approach

llr lender of last resort; line of least resistance; load-limiting resistor; log-likelihood ratio; long-length record(ing)

llr (LLR) latent lethality of radiation

LLRS Laser Lightning-Rod System

llrv (LLRV) lunar landing research vehicle

lls low-level solids

l & l's losers and lunatics

LLS Lunar Logistics System

LLSS Low-Level Sounding System

llsv (LLSV) lunar logistics system vehicle

llt long lead time

LLT London Landed Terms

llti long lead time items

lltruw low-level transuranic waste(s)

lltv low-light-level television

llu lending library unit

LLU Loma Linda University

LLUU Laymen's League-Unitarian Universalist

llv (LLV) long-life vehicle (postal vans); lunar landing vehicle

llw lower low water (LLW); low-level waste

LLWI lower low water interval

llwl light load water line

LLWM Low-Level Waste Management

LLWSAS Low-Level Wind Shear Alert System

Lly Llanelly

llyp long-leaf yellow pine

llz localizer

lm land mine; light metal(s); liquid metal(s); long meter; longitudinal muscle; lower motor; lumen(s)

l/m lines per minute

lm livello del mare (Italian—sea level)

l.m. locus monumenti (Latin—place of the monument)

Lm middle latitude

LM Legion of Merit; Life Master (bridge); Liggett Myers Tobacco (stock exchange symbol); Lincoln Memorial; Lord Marquis; Lord Mayor; Lourenço Marques; Lunar Module

L.M. Licentiate in Midwifery

L & M Linotype and Machinery

LM Lacus Mortis (lunar area)

LM-1 Fuji Heavy Industries trainer plane

lma left mento-anterior; local marketing agreement

LMA Last Manufacturers Association; League for Mutual Aid; Lingerie Manufacturers Association; Linoleum Manufacturers' Association; London-Midlands Association

LMAA Lift Manufacturers Association of Australia

LMAC Labor-Management Advisory Committee

lmad let's make a deal

LMAF Live Missile Assembly Facility

LMAGB Locomotive and Allied Manufacturers' Association of Great Britain

lmb local message box

LMBA London Master Builders' Association

LMBC Liverpool Marine Biological Committee

l & m bond labor and material bond

LMBP Lake Manyas Bird Paradise (Turkey)

lmc liquid-metal cycle; low middling clause

lmc (LMC) large magellanic cloud

LMC Lake Michigan College; Liberia Mining Company; Lloyd's Machinery Certificate; Lutheran Medical Center

LMC (LMC) Lloyd's Machinery Certificate (temporarily suspended when enclosed in parentheses)

LMCA Lorry Mounted Crane Association

LMCC Licentiate of the Medical Council of Canada

LMCT Licensed Motor Car Trader

lmd leafmould; local medical doctor

LMD Laboratory of Meteorological Dynamics; Landscape and Maintenance District

LMDC Lawyers Military Defense Committee

lmds local multipoint distribution services (wireless technology)

lme liquid-metal embrittlement

LME Late Middle English; London Metal Exchange

LMEC Liquid Metal Engineering Center (AEC)

L Med Licentiate in Medicine

L Med Ch Licentiate in Medicine and Surgery

LMEE Light Military Electronic Equipment (department of General Electric)

lmf language media format

l/mf low and medium frequency

lmfbr liquid-metal fast-breeder reactor

lmfr liquid metal fuel reactor

lm/ft² lumen per square foot

lmg liquefied methane gas

Lmg *Leichtesmachinengewehr* (German—light machine gun)

LMG light machine gun

LMH Lady Margaret Hall, Oxford

LMHC Lady Margaret Hall College (Oxford)

lm hormone lipid mobilizing hormone

lm-hr lumen-hour

L Mi Leo Minor (constellation)

LMI Lawn Mower Institute; Logistics Management Institute

l/min liters per minute

LMIS Labor Market Information System

lml left mediolateral

LML Lankard Materials Laboratory; Lerner Marine Laboratory

LMLA Lizzadro, Museum of Lapidary Arts

LMLI Liberty Mutual Life Insurance

lm/m² lumen per square meter

lmir load memory lockout register

lm/lrv lunar module/lunar roving vehicle (LM/LRV)

lmm localiser middle marker; locator at middle marker (compass)

l/mm lines per millimeter

LMM Library Microfilms and Materials; Luis Muñoz Marin, first native governor of Puerto Rico

lmmi like mamma made it

lmn lineman; lower motor neuron

LMNP Lake Manyara National Park (Tanzania)

LMNRA Lake Mead National Recreation Area (Arizona and Nevada)

lmo lens-modulated oscillator; light machine oil; limousine

LMO Local Medical Officer; Logistics Management Office (USA); London Meteorological Office

LMOA Locomotive Maintenance Officers' Association

lmp last menstrual period; left mento-posterior; lunar module pilot

LMP Licensed Massage Practitioner; Linea Mexicana del Pacifico; London Metropolitan Police

LMP *Literary Market Place* (Directory of American Book Publishers)

LMPA Library and Museum of the Performing Arts (Lincoln Center, New York City); London Master Printers' Alliance

LMPT Logistics and Material Planning Team

L Mq Lourenço Marques

LMR Lifetime Merit Register; London Midland Region British Railways

LMRA Labor-Management Relations Act

LMRC London Medical Research Council

LMRCP Licentiate in Midwifery of the Royal College of Physicians

LMRDA Labor-Management Reporting and Disclosure Act

LMRI Living Marine Resources, Inc

lmrp lower marine riser package (oil well)

LMRU Library Management Research Unit (Cambridge)

LMRSH Licentiate Member of the Royal Society for the Promotion of Health

lms lambs; least mean square

lms (LMS) lunar mass spectrometer

lm's lunar modules (LMs)

lm/s lumen per second

LMS Lelean Memorial School; Licentiate in Medicine and Surgery; London Mathematical Society; London Medical Schools; London Missionary Society; Lotto Management Services

LMSA Labor Management Services Administration

LMSC Lockheed Missiles & Space Company

LMSD Lockheed Missile and Space Division

LMSSA Licentiate in Medicine and Surgery of the Society of Apothecaries

lmst loom state

lmt left mento-transverse; length, mass, time; licensed massage therapist; limit

LMT Local Mean Time

LMTA Language Modalities Test for Aphasis; Library Media Technical Assistant; London Master Typefounders' Association

lmtd limited; logarithmic mean temperature difference

lmtg limiting

LMU Loyola Marymount University

LMUM *Ludwig-Maximilians Universität München* (German—University of Munich)

L Mus Licentiate in Music

L Mus TCL Licentiate in Music—Trinity College of Music

LMVD Licensed Motor Vehicle Dealer

LMVUS League of Men Voters of the United States

lm/w lumen per watt

lm/W lumen(s) per watt

ln liaison; logarithm (natural, base e)

Ln Lane; Lawrencium (symbol); London; Lyttelton

LN Air Liban (Lebanese Airlines); League-of Nations; Napierian logarithm (symbol)

L & N Leeds & Northrup; Louisville & Nashville (railroad)

L of N League of Nations

LN *Liga Nacional* (Spanish—National League)

LN₂ liquid nitrogen

LN₂cou liquid-nitrogen clip-on unit

LN₂ trailer liquid-nitrogen trailer

lna low noise amplifier

LNA Liberian National Airways; Libyan News Agency

lnb (LNB) large navigational buoy

lnc loran navigation chart(s)

LNC League of Nations Covenant; Leith Nautical College; Libertarian National Committee

lnchr launcher

L-N CP Liberal-National Country Party (Australian)

lnd. land

LNDC Lesotho National Development Corporation

Lndg Landing

lndh local nationals direct hire

lndkjng *landkjenning* (Norwegian—land ho)—in sight of land

lndrs laundress

lndry laundry

lndscp landscape; landscaping

L & NE Lehigh & New England (railroad)

LNER London and North Eastern Railway

lng length (flow chart); lining; liquefied natural gas; lounge

lng (LNG) liquefied natural gas

lnge (LNGC) liquefied natural gas carrier

LNG tanker liquid-natural-gas tanker

LNHS London Natural History Society

LNI *Lega Navale Italiana* (Italian Naval League)

LNLA Lithuanian National League of America

lnlw lowest normal low water

lnmp last normal menstrual period

LNNP Lake Nakuru National Park (Kenya)

LNNR Lindisfarne National Nature Reserve (England)

LNOC Libya National Oil Company

L-note $50 bill

lnp (LNP) lunar neutron probe

LNP Lamington National Park (Queensland); Lincoln NP (South Australia); London Northern Polytechnic

lnpf lymph node permeability factor

lnr liner; low noise receiver

LNR Loteni Nature Reserve (South Africa)

Lnrk Lanark

LNS Land Navigation System; Liberation News Service; Library of Natural Sounds

LNSW Library of New South Wales

LNT Leo Nicholas Tolstoy

Lntl lintel

LNTS *League of Nations Treaty Series*

lntwta low-noise travelling-wave tube amplifier

lnu last name unknown

LNU League of Nations Union

LNWR Lacassine National Wildlife Refuge (Louisiana); Lacreek NWR (South Dakota); London and North Western Railway; Lostwood NWR (North Dakota); Loxahatchee NWR (Florida)

lo layout; light open; local; local oscillator; locked open; longitudinal optical; low; low gear; low lights; low(er) order; lubricating oil; lubrication order

lo (LO) longitudinal optic

lo' look

'lo hello

lo *loco* (Italian—place)—in music, return to original pitch

Lo low (gear)

Lo *Lordag* (Danish—Lord's Day)—Saturday

LO Land Office; Launch Operator; Liaison Office(r); Lick Observatory (Mount Hamilton, California); Livestock Office; London Office; Louisville Orchestra; Lowell Observatory (Flagstaff, Arizona); Lubrication Order; Polish Airlines (2-letter symbol)

L/O Letter of Offer

LO *Landsorganisationen* (trade union in Norway and Sweden)

LO₂ liquid oxygen

loa leave of absence; left occiput anterior; length overall; letter of acceptance

loa (LOA) light observation aircraft

LOA Letter of Agreement; Letter of Offer and Acceptance; Letter Officers Association; Letter Offices Association; Light Observation Aircraft; Lithuanian Organists Alliance

loadex loading exercise

loadg & dischg loading and discharging

loadicator computerized ship-loading indicator

loads low altitude defense system

LOAF Lesbians Over the Age of Forty

loal lowest observed adverse level

loan/A vessel(s) loaned to Army

loan/C vessel(s) loaned to Coast Guard

loan/m vessel(s) loaned to miscellaneous governmental activities (Maritime Academy)

loan/s vessel(s) loaned to states

LOANZ Life Offices Association of New Zealand

LOAP List of Applicable Publications

lob left on base; line of balance

lob. logs on board; lumber on board.

Lob *Lobgesang* (German—Hymn of Praise)—Symphony No 2 of Mendelssohn for chorus and soloist

LOB Launch Operations Building; Loyal Order of the Boar; Loyal Order of Boors; Loyal Order of Bores

lobal long base-line buoy

lobar long baseline radar

LOBI Loop Blowdown Investigation (nuclear safety)

loboto lobotomy

lob(s) lobster(s)

loc letter of credit; lines of communication; local; locate; location; locus of control; logistics other charges

l-o-c letter of credit

LoC Library of Congress

LOC Launch(ing) Operations Center; Launch(ing) Operations Complex; Letter of Certification; Louisiana Office of Conservation; Lyric Opera of Chicago

loca loss of coolant accident (nuclear reactor)

local load on call; local area network

lo-cal low calorie

locals local people; local trains

locat location; locative; low-altitude clear-air turbulence

LOCATE Library of Congress Automation Techniques Exchange

LOCC Logistical Operations Control Center

loc. cit. *loco citato* (Latin—in the place cited)

loc. dol. *loco dolenti* (Latin—to the painful spot)

loci logarithmic computing instrument

LOCK Logical Coprocessing Kernel

loc. laud. *loco laudato* (Latin—cited in the approved place)

locn location

loco locomotion; locomotive

locp launcher operation control panel

locport lines of communications ports

loc. primo cit. *loco primo citato* (Latin—in the place first cited)

loc pr pnl local control purge panel

locpuro local purchase order

LOCS Librascope Operations Control System

loc. supra cit. *loco supra citato* (Latin—in the place cited above)

locum tens. *locum tenens* (Latin—temporary position)

locuz *locuzione* (Latin—phrase)

lod limitation on dividends; line of duty

lo-d low-density

Lod Lödose

LOD Launch Operations Directorate

LOD *Little Oxford Dictionary*

lode large-optics demonstration experiment

lodestone magnetic iron oxide; Fe_3O_4; magnetite

lodg loading; lodging

lodif long-distance infrared flash (camera)

lodor loaded (vessel) awaiting orders or assignment

loe level of effort

LOE Light-Off Examination (USN)

loel lowest observed effect level

lof lecherous old fool; lowest operating frequency

LOF Lloyd's Open Form (insurance policy); Lloyd's Open-Form (contract); London and Overseas Freighter

L-O-F Libbey-Owens-Ford

lofar low frequency analyzing and recording

lo-fi low fidelity (low-quality sound reproduction)

Lofotens Lofoten Islands

loft low-frequency radio telescope

lofti low-frequency trans-ionosphere (research satellite)

Loftleidir Icelandic Airlines

log. logarithm; logic; logical; logistic(s)

Log Logroño (Spanish province including La Rioja); Longview

LOG Legion of Guardsmen

log.$_{10}$ logarithm to the base 10

logair logistics transport by air

logairnet logistics air network

logal logical algorithmic language

Logan Logan International Airport (named for WW-II hero General Edward Lawrence Logan, who gave land to the city of Boston)

logands logic commands

logan(s) loganberry; loganberries

LOGC Logistics Center (USA)

LOGCMD Logistical Command

logcom logistic communications

Log Corn Logistical Command; Logistics Command

LOGDESMAP Logistics Data Element Standardization and Management Program (DoD)

LOGDESMO Logistics Data Element Standardization and Management Office (DoD)

LOGDIV Logistics Division

log.^{e} logarithm to the base e

logel logic-generating language

logest logistics estimate

logg loggerhead; loggia; logging; log glass

logie killogie

logipac logical processor and computer

loglan logical language

logland logistics transport by land

lo glo low glow

logo logogram [initial letter, number, or symbol used as an abbreviation or as part of an abbreviation as in Q & A (question and answer), 3M (Minnesota Mining and Manufacturing Company), c (cents)]; logotype (two or more type characters cast as one piece of type)

LOGOIS Logistics Operating Information System

logol logological; logologically; logologist; logology

logophi logophilia(c)—lover(r) of words

logother logotherapeutic(al) (ly); logotherapist; logotherapy

logp logistics plans

logr logistical ration; logistics ratio

Logr Logroño

logram logical program

Log Rep Logistics Representative (USN)

logsea logistics transport by sea

logsim logic simulation

logsup logistical support; logistics support

logsvc logistics service

loh (LOH) light observation helicopter

L o H Library of Hawaii (Honolulu)

loi limit of indemnity; loss on ignition

loi (LOI) letter of intent

LOI Lunar Orbit Insertion

loib lunar orbit insertion burn

loid celluloid (strip used by burglars to unlock doors)

LOIS Library Order Information System

lo-J low inertia

loktal locked octal tube

lol laughing out loud; length of lead (actual); little old lady

LOL Lobitos Oilfields Limited; Loyal Orange Lodge

lola lollapalooza (excellent or extraordinary person or thing)

LOLA Library On-Line Acquisition

lolita language on-line investigation/transformation of abstractions; library online information and text access

lolli lollipop

lo-lo load on-load off

lolw laid off-lack of work

lom locater at outer marker (compass)

Lom Columbus

LOM League of Mothers; List of Modifications; Loyal Order of Moose

LOMA Life Office Management Association; Lutheran Outdoor Ministry Association

LOMAC Logical Machine Corporation

Lomb Lombard; Lombardian; Lombardy

lombard loads of money, but a real dickhead

lo mi low mileage

Lompoc Federal Correctional Institution at Lompoc, California also site of a Federal Prison Camp

lon *longitud* (Spanish—longitude)

Lon Alonso; London

L o N League of Nations

LON London, England international airports (London-Central Airport)

Lon Brg London Bridge (rail terminal)

Lond *Londen* (Dutch—London); London; Londonderry; Londoner(s); *Londra* (Italian—London); *Londres* (French, Portuguese, Spanish—London)

long longeron; longitude

'long along

Long Longfellow; Longford; Long Island; Longjumeau; Long Key; Longmeadow; Longview

longl longitudinal

Long Lane Girls Long Lane School for (delinquent) Girls at Middletown, Connecticut

Longshoremen's Union International Longshoremen's Association

'longside alongside

Long Straight The Long Straight—297-mile-long (478-kilometer-long) straight stretch of railway track laid across Australia's Nullarbor Plain—the world's longest straight stretch of railroad

longv longevity

long vac long vacation

Lon Phil London Philharmonic

LONSYM London Symphony

LONRHO London and Rhodesian Mining and Land Company Limited

loo British term for toilet; looker; looker-after; looker on

LOOE Loyal Order of Overtime Experts

looktr lookout tower

LOOM Loyal Order of Moose

Loop The Loop—Chicago's business section

LOOP Louisiana Offshore Oil Port

LOOS League of Older Students

lop least objectionable program

lop. launch operator's panel; left occiput posterior

l-o-p line-of-position

LOP lunar orbiting photographic (vehicle)

lopar low-power acquisition radar

l'Opera (French—the opera)—Paris Opera House

L O P & G Live Oak, Perry & Gulf (railroad)

lopkgs loose or in packages

l'Op Mont l'Opera de Montreal

lopo local post

LOPS Lloyd's Ocean Platform System

lopuro local purchase order

lo-q low iq (IQ)

loq loquitur (Latin—he speaks)

lor level of repair; lunar orbital rendezvous

Lor Lorenzo; Loreto (Peruvian department name also used by Ecuador and Paraguay, although originated in Italy and copied in Ireland); Lorong

LOR L'Osservatore Romano (Papal Roman Observer)

lorac long-range accuracy

lorad long-range active detection

loran long-range aid to navigation

LORAPHS Long-Range Passive Homing System

LORAS Low-Range Omnidirectional Airspeed System

lord long-range and detection (radar); lordosis

Lord Laurence Sir Laurence Olivier

Lords House of Lords

lo-res low-resolution

lorg large-size organization

LORIDS Long-Range Iranian Detection System

LORINE Limited Range Imagery Networks Elements

lorl (LORL) large orbital research laboratory

Lorraine Lorraine De Sola Chervin

lorv (LORV) low orbital reentry vehicle

Lorᶻᵒ Lorenzo

los length of stay; liaison operating sheet; loss of signal

l-o-s line-of-sight

Los (Mexican-American truncation—Los Angeles)

LoS Language of Sport

LOS Lagos, Nigeria (airport); Latin Old Style; Law of the Sea; Little Orchestra Society; Lockheed Ocean Systems

Losa Los Angeles

Los Alamos Los Alamos National Laboratory

losam (LOSAM) low-altitude surface-to-air missile

Los Ang Los Angeles

Los Desastres Los Desastres de la Guerra (Spanish—The Disasters of War)—Francisco Goya's etchings

Los Guilucos Los Guilucos School for (delinquent) Girls at Santa Rosa, California

lösl löslich (German—soluble)

LOSS Large Object Salvage System; Line Operation Status System

los sys landing observer's signal system

lostf line-of-sight test fixture

lot large orbiting telescope; lateral olfactory tract; left occipito-transverse; load on top

lot. lotio (Latin—lotion)

LOT Polish Air Lines (3-letter symbol)

LOTADS Long-Term Air Defense Study (USA)

LOTCIP Long-Term Communications Improvement Plan

lote lesser of two evils

lo tech low technology

lo-temp low temperature

Lot-et-Gar Lot-et-Garonne

Loth Lothian

Lothians East Lothian, Midlothian, West Lothian

lotis logic, timing, sequencing

LOTOS Language of Temporal Ordering Specification

LOTS Lotus Development Corporation

LOTUS Ladies Organized to Unfetter Sexuality

lotw loaded on trailers or wagons

Lou Lewis; Loualta; Louanna; Louanne; Loudella; Louella; Louina; Louis; Louisa; Louise; Louisetta; Louisette; Louisiana; Louisville; Loula; Loulou; Loura; Lourane; Lourene; Lourette; Louvilla; Louvina

Lou French Louisiana French

louh light observation utility helicopter

Louie Louis; Louisa; Louise; St Louis, Missouri

Louis Louisville

Louisvillain(s) native(s) of Louisville

Lou Orc Louisville Orchestra

Louv Louvain

l'Ouverture Toussaint l'Ouverture—founder and first president of Haiti

lov limit of visibility

LOVE League of Victims and Emphathizers (pro-capital-punishment group)

lovisim low-visibility landing simulation

LoW Launch on Warning (during nuclear warfare)

lo wat low water

'lowed allowed

Lowell Florida Correctional Institution at Lowell

lower 48 48 continental United States

lower 49 lower 48 plus Hawaii

Low L Low Latin

low. log. lower loge

lowpro low protein (diet)

Low Scot Lowland Scotch
Low Tatras Low Tatra Mountains
low tec(h) low technology
low-Z low-impedance
lox also the name for smoked salmon; liquid oxygen; liquid-oxygen explosive
lox-sox liquid oxygen, solid oxygen
loxygen liquid oxygen
loy loyalty
LOYA League of Young Adventurers
Loyalists Loyalist American Colonists (Tories); Loyalist Episcopalian Traditionalists; Loyalist Spanish Republicans
Loyola Saint Ignatius de Loyola (Iñigo de Oñez y Loyola)
loz liquid ozone
Loz Lozère
lp lambing percentage; landplane; large power; last paid; latent period; launch(ing) platform; learning process; light perception; linear programming; line pair (postage stamp); liquefied petroleum; liquid propellant; liquefied propane; list(ing) price; litter patient; local preacher; local procurement; long primer (type); long-play; long-playing; low pass; low point; low power; low pressure; lumbar puncture
lp (LP) long-play (record)
l-p low-pressure
l/p lactate/pyruvate ratio; launch platform; letterpress; life policy; listening post
Lp Ladyship; lipoprotein; Lordship
LP Aeralpi (2-letter symbol); Labor Party; Labour Party; Liberal Party; Libertarian Party; Liberty Press; Library of Parliament; Licensing Plan; Limited Partnership; litter patient; Liverpool Prison; long-play (record); Lower Peninsula
L-P Lionel-Pacific
LP *lunga pausa* (Italian—long pause)
lpa low-power amplifier
LPA Labor Party Association; Labor Policy Association; Little People of America
LPAA London Poster Advertising Association

L-pam L-phenylalanine mustard (anti-cancer drug)
LPAP Local Planning and Assessment Process
lpb lighted push button
LPB La Paz, Bolivia (airport)
lpc leaf protein concentrate; least-preferred co-worker; linear predictive coding; low-pressure chamber; low-pressure compressor
lpc (LPC) linear-power controller
LPC Legal Practice Course; Linear Predictive Coding; Livestock and Pastoral Company; Livestock Publications Council; Lockheed Propulsion Company; Low Price Center
LPCC Lamb Promotion Coordination Committee
LPCG Laser Planning and Co-ordination Group (ERDA)
LPCM London Police Court Mission
lpcp launcher preparation control panel
lpcw long-pulse continuous wave
lp cyl low-pressure cylinder
lpd least perceptible difference; liquid protein diet; local procurement direct; low performance drone
LPD amphibious transport dock ship (naval symbol); Local Procurement District; low performance drone
LPE London Press Exchange
LPEA Licentiate of the Physical Education Association
L Ped Licentiate in Pedagogy
lpf leukocytosis-promoting factor; low-power field
LPF Latvian Popular Front
lpg liquid propane gas
lpg (LPG) liquefied petroleum gas
LPGA Ladies Professional Golfers Association; Liquefimed Petroleum Gas Association; Louisiana Personnel and Guidance Association
lp gas liquefied petroleum gas
lph landing personnel helicopter; lines per hour
lph (LPH) landing personnel helicopter
L Ph Licentiate of Philosophy

lpi launching position indicator; lines per inch; low-power indicator; low probability of intercept
LPI Lifetime Productivity Index; Lightning Protection Institute; Louisiana Polytechnic Institute
lpicbm (LPICBM) liquid-propellant intercontinental ballistic missile
L-pills cyanide L-pills (deadly poisonous)
lpir low probability of intercept radar
LPIU Lithographers and Photoengravers International Union
LPIW Lumber, Production and Industrial Workers
LPKS Lone Pine Koala Sanctuary (Queensland)
LPKTF London Printing and Kindred Trades' Federation
lpl lightproof louver; list processing language
LPL Liverpool Public Libraries; London Public Library; Louisville Public Library; Lunar and Planetary Laboratory (University of Arizona)
LP & L Louisiana Power and Light
LPL *Lembaga Penelitian Laut* (Indonesian—Institute for Marine Research)—Jakarta
L-plane US Army liaison aircraft
LP & LC Louisiana Power & Light Company
L Plms Las Palmas
lplr lock pillar
LPLs Liverpool Public Libraries
lpm lines per millimeter; lines per minute; liters per minute
LPM Licensing Project Manager
LPMES Logistics Performance Measurement and Evaluation System
LP/MOSS Linear Programming/Mathematical Optimization Subroutine System
LPN Licensed Practical Nurse; Longview, Portland and Northern (railway)
LPN *Leembaga Padi Negara* (Malay—National Rice Paddy)
LPNA Licensed Practical Nurses Association; Lithographers and Printers National Association

LPNI Langley Porter Neuropsychiatric Institute

lpo liquid phase oxidation; loan production office; local purchase order

LPO Licensing Project Office(r); Limited Practice Officer; London Philharmonic Orchestra; London Post Office

Lpool Liverpool

LPPM Lembaga Penilitian Pertanian Maros (Indonesian—Department of Agriculture)

LPPTFS London and Provincial Printing Trades Friendly Society

lpr (LPR) liquid-propellant rocket

LPR Lauritzen; Peninsula; Reefer (steamship line)

LPRC Library Public Relations Council

lps lightproof shade; line program selector; liters per second; low primary sequence; low-pressure sodium

lps (LPS) lipopolysaccharide

lp(s) loop(s)

lp's (LPs) long-playing records

LPS Laboratory for Planetary Studies (Cornell); Lanterman-Petris-Short Act; Lebanese Press Syndicate; Light Photo Squadron; Linear Programming System; London Philharmonic Society; Lord Privy Seal; Lyceum Performing Society

LPSA Liberal Party of South Africa

LPS Act Lanterman-Petris-Short Act (commitment procedures covering mental patients in California)

LPSO Lloyd's Policy Signing Office

LPSS amphibious transport submarine (naval symbol)

lpstt low-power schottky transistor-transistor logic

lpt limited-production test

LPT Licensed Physical Therapist

LPTA Louisiana Parent-Teacher Association

LPTB London Passenger Transport Board

lptv low-power television

lptv (LPTV) large payload test vehicle

lpu limited-production urgent

Lpud Liverpudlian (native to or inhabitant of Liverpool)

lpv launching point vertical; lightproof vent

lpw low-power window; lumens per watt

l & p wood lumber and plywood wood

LPYS Labour Party Young Socialists

Lpz Leipzig

L Pz La Paz

LPZG Lincoln Park Zoological Gardens

LPZS Lincoln Park Zoological Society

lq last quarter; letter quality; linear quantifier; lowest quartile

l.q. lege quaeso (Latin—please read)

lqdr liquidator

LQR Law Quarterly Review

lqss liquid steady state

LQST Leadership Q-Sort Test

lr latency relaxation; leave rations; letter report; lire; log run; long range; long run; lower

l/r left right; lower right

l-to-r left-to-right (photo caption abbreviation)

l R laufen Rechnung (German—current account)

Lr lawrencium

LR Laboratory Report; Land Registry; Landing Report; Lee Rubber (stock exchange symbol); Letter Report; Liaison Report; Liberia (Internet code); Little Rock

LR Law Reports; Libertarian Review; Lloyd's Register

lra long-range aviation

LRA Labor Research Association; Landing Rights Airport; Libertarian Republican Alliance; Lithuanian Regeneration Association; Lord's Resistance Army

lraam (LRAAM) long-range air-to-air missile

lrac long-run average cost

LRAD Licentiate of the Royal Academy of Dancing

LRAFB Little Rock Air Force Base

LRAM Licentiate of the Royal Academy of Music

LRB Labor Research Bureau; Laboratory of Radiation Biology (University of Washing-

ton); Legislative Reference Bureau; Loyalty Review Board

LRBA Laboratoire de Recherches Balistiques et Aérodynamiques (French—Laboratory for Ballistic and Aerodynamic Research)

LRBC Lloyd's Register Building Certificate

lrbm long-range ballistic missile

lrc longitudinal redundancy check(ing); long-range communication; lower right center

lrc (LRC) longitudinal redundancy check character (data processing)

LRC Labor Representation Committee; Ladies Recreation Club; Langley Research Center (NASA); Law Reform Commission; Learning Resource Center; Lesbian Resource Center; Lewis Research Center (NASA); Library Resource Center; Linguistics Research Center; Logistics Research Center; London Rowing Club

lrca long-range combat aircraft

LRCA London Retail Credit Association

LRCE Little Rock Cotton Exchange

lrcm (LRCM) long-range cruise missile

LRCM Licentiate of the Royal College of Music

lrco limited remote communications outlet

LRCP Licentiate of the Royal College of Physicians

LRCPE Licentiate of the Royal College of Physicians of Edinburgh

LRCPI Licentiate of the Royal College of Physicians of Ireland

lrcr longitudinal redundancy check register

LRCS League of Red Cross Societies; Licentiate of the Royal College of Surgeons

LRCSE Licentiate of the Royal College of Surgeons of Edinburgh

LRCSI Licentiate of the Royal College of Surgeons of Ireland

LRCT Licentiate of the Royal Conservatory of Toronto

LRCVS Licentiate of the Royal College of Veterinary Surgeons

lrd labelled radar display; long-range data

L-rd Lord (Hebraic contraction)

LRDC Learning Research and Development Center

LRDP Library Research and Demonstration Program

lrdr last revision date routine

lre law-related education

LREA Licensed Real Estate Agent

lrecl logical record length

LRES Linear Rocket Engine System

lrew long-range early warning

lrf latex and resorcinol formaldehyde; liver residue factor

lrf (LRE) luteinizing hormone-releasing factor

LRFI League for Religious Freedom in Israel

LRFPB Louisiana Rating and Fire Prevention Bureau

LRFPS Licentiate of the Royal Faculty of Physicians and Surgeons

LRFPSG Licentiate of the Royal Faculty of Physicians and Surgeons of Glasgow

lrg large; liquefied refinery gas; long range

lrg grdn large garden

lrh (LRH) luteinizing releasing hormone

LRHL Law Reports—House of Lords

lri left-right indicator; long-range input; long-range interceptor; lower respiratory infection

LRI Library Resources Incorporated

LRIBA Licentiate of the Royal Institute of British Architects

LRIC Licentiate of the Royal Institute of Chemistry

lrim long-range input monitor

lrip language research in progress

LRIP Low-Rate Initial Production

lrir limb radiance inversion radiometer

LRIS Lloyd's Register Industrial Services

LRJC Lake Region Junior College

LRKB Law Reports—King's Bench

LRL Lawrence Radiation Laboratory; Lunar Receiving Laboratory

LRLA La Raza Legal Alliance

lrl's living-room liberals

LRLS London Regional Library System

LRLSA La Raza Law Students Association

LRLTRAN Lawrence Radiation Laboratory Translator

lrm length register mark; liquid radiation monitor

LRMC Lloyd's Refrigerating Machinery Certificate

lrmg (LRMG) lockless-rifle machine gun

lrmp long-range maritime patrol

LRMS Liquid Radwaste Management System

LRN Landslaget for Reiselivet i Norge (Norwegian—Norway Travel Association)

LRNC Long Reference Number Code

lrp launching reference point; long-range planning

LR-P La Rochelle-Pallice

LRP Law Reports—Probate

lrpa long-range patrol aircraft

lrpg long-range proving ground

LRPGR Long-Range Planning Ground Rules

LRPL Liquid Rocket Propulsion Laboratory; Little Rock Public Library

LRPS Long-Range Planning Service

LRQB Law Reports—Queen's Bench

lrr long-range radar; lower reduced rate

LRR Location Recommendation Report

lrra low-range radio altimeter

lrrd long-range reconnaissance detachment

lrrmf long-range resource and management forecast

lrrp lowest required radiated power

LRRS Long-Range Radar Station

LRRT Library Research Round Table

LRRTS Light-Rail Rapid-Transit System

lrs long-range search; long-run supply

lr's leave rations; light refreshments

l/r/s library rubber stamps (used-book trade abbreviation indicating book may belong or may have belonged to a public library)

lRs lactated Ringer's solution

Lrs Lancers

LRS Land Registry Stamp; London Research Station (British Gas)

LRS Lloyd's Register of Shipping

lrsam (LRSAM) long-range surface-to-air missile

lrsm long-range seismic measurement

LRSM Licentiate of the Royal Schools of Music

LRSS Long-Range Survey System

lrt laser ray tube; launch, recovery, and transport

lrt (LRT) light rail transit

lrtc long-run total cost

LRTgt last resort target

LRTL Light Railway Transport League

lrtnf long-range theater nuclear forces

LRTS Library Resources and Technical Services

lru least recently used; line replacement unit

lrv (LRV) light-rail vehicles (rapid transit); light recreational vehicle; lunar roving vehicle

LRWES Long-Range Weapon Experimental Station

LRWRE Long-Range Weapons Research Establishment

LRY Liberal Religious Youth

ls landing ship; left side; light vessel; lightship; limestone; liminal sensitivity; limit switch; list; local sunset; long shot; long site; long sleeves; loud-speaker; low secondary; low smoke; low speed; lump sum

ls (LS) legal seal

l-s lumbo-sacral

l's losers (gambling short form)

l/s lecithin/sphingomyelin (ratio); liters per second

l & s launch(ing) and servicing

l.s. locus sigilli (Latin—place of the seal)

Ls Lopes; Louis

LS Lamson & Sessions; Law Society; Leading Seaman; Lesotho (Internet code); Letter Service; Licensed Surveyor; Linnaean Society

L-S Lewis-Shepard

L & S Lands and Survey (department or office)

lsa left sacro-anterior; logistic support analysis; low specific activity

lsa (LSA) lichen sclerosus et atrophicus

LSA Labor Services Agency; Labour Services Association; Land Service Assistant; Land Settlement Association; Leukemia Society of America; Licentiate of the Society of Apothecaries; Lighthouse Society of America; Limbless Soldiers Association; Linguistic Society of America; Liquor Stores Association; Lithuanian Society of America; London Salvage Association; London School of Accountancy

L&SA Law and Society Association

LSA Library Science Abstracts

LSAA Linen Supply Association of America

LSAC Law School Admission Council; London Small Arms Company

lsar local storage address register

LSAS Law School Admission Service

LSAT Law School Admission Test; Law School Aptitude Test

lsb left sternal border; lower sideband

lsb (LSB) least significant bit

LSB Launch Service Building; London School Board; Louisiana School Board

LSBA Leading Sick-Bay Attendant; Louisiana School Boards Association

LSBR Large Seed-Blanket Reactor (AEC)

lsc least-significant character; linear sequential circuit; logistic support cost

l.s.c. loco supra citato (Latin—in the foregoing place cited)

LSC Laser Systems Center; Legal Services for Children; Legal Services Corporation; Lie Scale for Children; Lower School Certificate

lsca left scapulo-anterior

LSCA Library Services and Construction Act

LSCC Library of the Supreme Court of Canada

lscp left scapuloposterior

LSCRS Law School Candidate Referral Service

lscs lower segment caesarean section

lsct low-speed compound terminal

LSCT Lamar State College of Technology

lsd landing ship deck; landing ship dock; laser-support detonation; last signifiicant data; last signifiicant digit; leadless sealed device; least significant difference; least significant digit; library system(s) development; liquid scale disintegrator; logarithmic-series distribution; long, slow, distance (jogging); loss, short, and damage; lottery stress disorder

ls & d liquor store and delicatessen

l s d librae, solidi, denarii (Latin—pounds, shillings, pence)

LSD landing ship, dock (naval symbol); League for Spiritual Discovery; lysergic acid diethylamide—powerful psychedelic drug nicknamed "acid"

L.S.D. Doctor of Library Science

LSD Lyserginsaure Diathylamid (German—lysergic acid diethylamide)

LSDAS Law School Data Assembly Service

LSDI Logistics Support Departmental Instruction

lsd li leased line (telephone)

LSDS Low-Speed Digital System

lse limited signed edition; limited special edition

LSE London School of Economics; London Stock Exchange; Louisiana Sugar Exchange

LSECS Life Support and Environmental Control System

l sect longitudinal section

LSEL London School of Economics Library

LSE & PS London School of Economics and Political Science

lse skds loose (or on) skids

LSET Logistics Supportability Evaluation Team

LSEU La Salle Extension University

lsf log super feet

LSF Literary Society Foundation; Lloyd Shaw Foundation; Lock Security Force (Panama Canal)

lsfa logistic system feasibility analysis

lsg list set generator

Lsg Lösung (German—solution)

lsgd lymphocyte specific gravity distribution

L Sgt Lance Sargeant

LSH Latter-day Saints Hospital

L-shape ell-shaped

LSHTM London School of Hygiene and Tropical Medicine

lsi large-scale integration; lateral shear interferometer

LSI Labor and Socialist International; Lake Superior & Ishpeming (railroad); landing ship-infantry; Law of the Sea Institute; Law-Science Institute (University of Texas); Lear Siegler Incorporated; Logistic Shipping Instruction(s); Lunar Science Institute

LS & I Lake Superior & Ishpeming (Railroad)

LSIA Lamp and Shade Institute of America

lsic large-scale integrated circuitry

LSIO Labor Standards Inspection Officer

LSIS Laser-Scan Inspection System

lsk liver, spleen, kidney

lsl left sacrolateral; low-speed logic

LSL landing ship, logistic; Life Sciences Laboratory; Linnaean Society of London; Lucy Stone League

lslb left short-leg brace

lsm linear synchronous motor; lysergic acid morpholide

lsm (LSM) lysergic acid morpholide

l.s.m. litera scripta manet (Latin—the written word remains)

LSM Laboratory for the Structure of Matter (USN); Lancastrian School of Management; landing ship, medium; Liberation Support Movement; Logistic Support Manager; Louisiana State Museum

LS/mft Leopold Stokowski/ means fine tone; Lucky Strike/means fine tobacco

LSMI Lake Superior Mining Institute

LSMP Logistic Support and Mobilization Plan

LSMSC Lake Superior Mines Safety Council

LSN Legal Services Network

LSNR League of Struggle for Negro Rights

LSNSW Linnaean Society of New South Wales

LSNY Linnaean Society of New York

LSO Landing Signal Officer; Leningrad Symphony Orchestra; London Symphony Orchestra

lsp left sacro-posterior; logical signal processor

LSP Last Slave Plantation; Launch Pad; Logistic Support Plant

LSPA Louisiana Sugar Planters' Association

LSPOJC La Salle-Peru-Oglesby Junior College

LSPR Library Society of Puerto Rico

L-square capital-L-shaped square; carpenter's square

lsr launch signal responder; lens shutter reflex

Lsr Luftschutzraum (German— air raid shelter)

LSR landing ship, rocket; landing ship, support

Lsr Ant Lesser Antilles (Leeward and Windward Islands)

lss liquid scintillation spectroscopy

LSS Life Saving Service; Life Saving Station; Life Support System; Lockheed Space Systems; Logistic Support Squadron

L.S.S. Licentiate of Sacred Scripture; Leopold-Sedar Senghor

lssc logistic support system characteristics; lower sideband suppressed carrier

LS Sc Licentiate in Sacred Scriptures; Licentiate in Sanitary Science

LSSC Logistic System Support Center (USA)

LSSF Land Special Security Force (USA)

LSSG Logistics Studies Steering Group; Logistics Studies Support Group

lssm local scientific surface module

LSSR Latvian Soviet Socialist Republic; Lithuanian Soviet Socialist Republic

LSSS London School of Slavonic Studies

lsst large space systems technology

L S St L Louis Stephen St Laurent (Canada's sixteenth Prime Minister)

lst large space telescope; laser spot tracker; left sacro-traverse; liquid storage tank; liquid-oxygen start tank; living structures tank

Lst Launceston

LST landing ship, tank; Local Sidereal Time; Local Standard Time

LST London Sunday Times (newspapers)

lstc low-speed trim compensation

LSTM Liverpool School of Tropical Medicine

lst wk last week

lsu launcher selector unit; livestock unit

LSU landing ship, utility; Louisiana State University

LSU-IES Louisiana State University Institute of Environmental Studies

LSUNO Louisiana State University (New Orleans)

LSUP Louisiana State University Press

lsuv lunar surface ultraviolet (camera)

LSV landing ship, vehicle

LSVP landing ship, vehicle, and personnel

lsw least significant word; limit switch(ing)

LSW Licensed Shorthand Writer

lsw lt landing signal wand light

LSWR London and South Western Railway

LSWY League of Socialist Working Youth

LSZ Limited Speed Zone; Local Slow Zone

lt landed terms; language translation; latch trip; laundry tray; less than; lid tank; light; light trap; line terminator; local time; long ton; loop test; low temperature; low tension; low torque

lt (LT) lymphotoxin

l/t loop test

lt laut (German—according to)

l.t. locurn tenens (Latin— substitute)

Lt Lieutenant

LT Land Transfer; landing team; large tug; Lithuania (Internet code); Lloyd Triestino; local time; London Transport

lta launch-through-attack; lighter-than-air

LTA Lawn Tennis Association; Library Technical Assistant; lighter-than-air; Listening Transit Analysis; Logistic Task Authorization

LTA Life of Timon of Athens

LTAA Lawn Tennis Association of Australia

ltadl launcher tube azimuth datum line

LTAS Lighter-Than-Air Society

ltb laryngo-trachael bronchitis; line terminating battery; low- tension battery

Lt. B. Bachelor of Literature

LTB Lepers Trust Board; London Tourist Board; London Transport Board

LTBP London Tanker Broker Panel

LTBT Limited Test Ban Treaty (prohibiting nuclear testing in certain environments)

ltc long-term care

ltc (LTC) locking torque converter

LTC Land Transport(ation) Commission; Lawn Tennis Club; Le Tourneau College; Library of Trinity College; Loop Test(ing) Conference

LTCB Long-Term Credit Bank (Japan)

Lt Cdr Lieutenant Commander

LTCIAC Long-Term Care Insurance Advisory Council

LTCL Licentiate of Trinity College of Music (London)
Lt Cmdr Lieutenant Commander
Lt Col Lieutenant Colonel
ltc's long-term contracts
ltd long-term depression; long-term disability
lt/d long tons per day; lower tween deck
Ltd Limited
Ltda *Limitada* (Spanish—limited)
ltd ed limited edition
LTDP Long-Term Defense Program
lte large table electroplotter; linear threshold element
Lte *Limité* (French—limited)
LTE London Transport Executive
lted letter to the editor
LTER Long-Term Ecological Research (network)
LTEU Liquor Trades Employees Union
ltf (LTF) lipotrophic factor
LTF Lithographic Technical Foundation; Logistic Task Force; tropical fresh water load line (Plimsoll mark)
ltfrd lot tolerance fraction reliability deviation
ltg lighting
LTG Leadership Training Graduate
ltgc lithographic
ltge lighterage
Lt Gen Lieutenant General
ltgh lightening hole
Lt Gov Lieutenant Governor
lth lath; lathing; less than honest (crooked, dishonest); luteotrophic hormone (LTH)
Lth Leith
L Th Licentiate in Theology
lthr leather
lti land training installation(s)
lti (LTI) light transmission index
Lti Laotian
LTI Ladder Towers Incorporated; London Taxis International; Louisiana Training Institute; Lowell Technological Institute
LTIB Lead Technical Information Bureau
Lt Inf Light Infantry
ltip long-term incentive plan
Lt JG Lieutenant Junior Grade
ltl listing time limit

ltl (LTL) less than truckload
Ltl Little (postal abbreviation)
ltla launcher tube longitudinal axis
ltm laser target marker; long-term mortgage; low thermal mass
ltm (LTM) long-term memory
LTM Licentiate of Tropical Medicine; Little Theatre Movement
ltmr laser target marker ranger
ltng lightning
ltng arr lightning arrester
lto landing takeoff
Lto *lento* (Italian—slowly)
LTO Land Transfer Office; Leading Torpedo Operator
ltof low-temperature optical facility
Lton long ton
ltp limit on tax preferences; long-term potentiation; low-temperature passivation
LTP Library Technology Program
ltpd lot tolerance percent defective
ltpp lipothiamide-pyrophosphate
ltr letter; lighter; liter; local twitch response; long-term relationship; long-term reserve
LTR Long Terminal Repeat; Long Term Reserve
LTR *Library Technology Reports*
Lt RN Lieutenant—Royal Navy
LtrO letter order
ltrom linear-transformer read-only memory
LTRP Long-Term Requirement Plan
ltrs (LTRS) letters shift (data processing)
LTRS Laser Target Recognition System
lts lights; long-term store
l'ts let's
LTs *Legal Times*
LTS Landfall Technique School; London Transport System; London Typographical Society
LTSB London Trustee Savings Bank
LT & SR London, Tilbury and Southend Railway
ltt liquid toning transfer
LTT Lymphocyte Transformation Test

ltta long-tank thrust augmented
LTTC Lowry Technical Training Center
LTTE Liberation Tigers of Thamil Eelam
lttr latter
L-T Trade Agreement Liao-Takasaki Trade Agreement
ltu line terminating unit
LTU La Trobe University
ltv loan to value; long tube vertical
lt/v light vessel
L-T-V Long-Temco-Vought (corporation)
ltvc launcher tube vertical centerline
ltwt lightweight
lu lock up; logic unit; logical unit; logistical unit; lumen
lu. *lues* (Latin—contagious disease)—plague or syphilis
Lu Lorentz unit; Lugano; Lugo; lutetium
LU Langston University; Lawrence University; Laurentian University; Laval University; Lefevre-Utile (French bakers); Lehigh University; Lethbridge University; *Ligue Universelle* (French—Universal Esperantist League); Lincoln University; Liverpool University; London University; Loyola University; Luxembourg (Internet code)
lu. I *lues I*—primary syphilis
lu. II *lues II*—secondary syphilis
lu. III *lues III*—tertiary syphilis
lua left upper arm
LuA Launch under Attack (during nuclear warfare)
LUA Life Underwriters Association; London Underwriters Association
LUAA Life Underwriters Association of Australia
LUAC Land Use Advisory Council; Life Underwriters Association of Canada
LUANZ Life Underwriters Association of New Zealand
lub lubricant; lubricate; lubrication
lub (LUB) logical unit block
lube lubricate; lubrication
lubed lubricated (intoxicated)
lub oil lubricating oil
lubs large undisturbed bottom sampler
Luc Lucan; Lucifer; Lucretius; Lucullus

LUC Land Use Commission; Land Use Committee; Louisiana University Center

LUCB Library of the University of California at Berkeley

LUCHIP Lutheran Church and Indian People

luchtv luchtvaart (Dutch—aviation)

Luci Lucifer

Lucia St Lucia

Lucia di Lucia di Lammermoor (Donizetti opera)

lucid. language used to communicate information system design

Luck Lucknow

lucom lunar communication

luc, prim luce primo (Latin—at daybreak)

Lucr The Rape of Lucrece

Lueretius Roman poet-philosopher Titus Lucretius

Lucrezia Bori Lucrecia Borja y Gonzalez de Riancho

lud liftup door

Lud Ludlow; Ludo; Ludolf; Ludolph; Ludovic; Ludovica; Ludovick; Ludovico; Ludovicus; Ludvig; Ludwell; Ludwig; Ludwig van Beethoven (German composer), Ludwik

luda land use data

Luddy Ludlow (*see* Lud)

lude quaalude (a depressant drug)

Ludendorff General Erich Friederich Wilhelm Ludendorff

Lud(s) Luddite(s)

Ludwig van Ludwig van Beethoven

lue left upper entrance; left upper extremity

LUER Land Use and Environmental Regulation

lues I primary syphilis

lues II secondary syphilis

lues III tertiary syphilis

luf lowest usable frequency

LUFTHANSA *Deutsche Lufthansa* (German Airline)

lug luggage; lugger; lugging; lugsail; lugworm

lug luglio (Italian—July)

Lugd. Bat. Lugdunum Batavorum (Latin—Leiden)—Leyden

lu h lumen hour(s)

luhf lowest usable high frequency

LUI Labor Union Insurances

Luigi Cherubini Marla Luigi Carlo Zenobio Salvatore Cherubini

LUIP London University Institute of Psychiatry

Luis Buñuel Luis Buñuel Portolés

lujb left umbilical junction box

Luke The Gospel according to St Luke

lul left upper limb; left upper lobe

LUL London University Library

LULA Loyola University of Los Angeles

LULAC League of United Latin-American Citizens

LULAC La Liga de Ciudadanos Latinoamericanos Unidos (Spanish—The League of United Latin-American Citizens)

LULOP London Union List of Periodicals

Lulu Louise

lum lumbago; lumbar; lumber; lumen; luminosity; luminous

lum (LUM) lunar excursion module

Lum Columbus

LUMAS Lunar Mapping System

lumb lumber; lumbering

LUMC Laval University Medical Center

Lumiére Auguste and Louis Lumiére

LUMIS Land Use Management Information System

lumpec lumpectomy, surgical removal of a tumor with a limited amount of associated tissue

Lumpen Lumpenproletariat (German—ragged bums, ragged street people)

lumps large urban medical practices

lun lunar; lunette

lun lundi (French—Monday); *lunedi* (Italian—Monday); *lunes* (Spanish—Monday)

Lunar Lunar Society (Birmingham, England)

lunar caustic silver nitrate ($AgNO_3$)

lunch luncheon

LUNCO Lloyds Underwriters Nonmarine Claims Office

lun int lunitidal interval

Lunnon London

luo luogo (Italian—place)—in music, return to original pitch

Lup Lupus (Latin—Wolf constellation)

LUP Liverpool University Press; Loyola University Press

lupa lupanar (Latin—brothel)

Lupe Guadalupe

Lupe Vélez Maria Guadalupe Vélez de Villalobos

lupinh lupus inhibitor assay

luq left upper quadrant (abdomen)

Luqa Malta's main airport

lurd living unrelated donor

LUS Land Utilization Survey

LUSB Land Utilization Survey of Britain

lus fin luster finish

Lu-shun (Chinese—Port Arthur)

lusi lunar surface inspection

lusing lusingando (Italian—coaxing)

Luso Lusotania (Portuguese—Lusitania)

lust. lustrous

lut launcher umbilical tower (LUT); local user terminal

lut. luteum (Latin—yellow)

LUT Launcher Umbilical Tower; Loughborough University of Technology; Ludwig Universe Tankships

LUTA Library of the University of Texas at Austin

LUTC Life Underwriter Training Council

Lutetia (Latin—Paris)—more fully Lutetia Parisiorum

LUTFCSUSTC Librarians United to Fight Costly, Silly, Unnecessary Serial Title Changes

Luth Luther(an)

Lutia. St Lucia, West Indies

Lutz Lucien

luv let us vote (popular teenage plea); lightweight utility vehicle (pickup truck)

LUV Love Uniting Volunteers

lux luxurious; luxury

Lux Luxembourg; Luxembourger; Luzon

lux aet lux aeterna (Latin—everlasting light)

LUXAIR Luxembourg Airlines

Luxem Luxembourg

Luxembourg Grand Duchy of Luxembourg (European lowland), *Grand-Duché de Luxembourg*

Lux Fr Luxembourger franc

Luz Luzon

lv land valuation; largest vessel; launch vehicle (LV); leave; left ventricle; light and variable (wind); low viscosity; low voltage; lumbar vertebra; luncheon voucher; lyric vocalist

l-v lacto-vegetarian

l/v light vessel (lightship)

lv livre (French—book)

Lv Latvia; Latvian; lev (Bulgarian currency unit)

Lv. Leviticus (Book of)

LV large-capitalization value (stocks); Las Vegas; Laser Vision; Latvia (Internet code); launch vehicle; Lehigh Valley (railroad); Licensed Victualler; light vessel (light ship); Linda Vista; Lindholmens Varv (Lindholmens Shipyard)

LV-3 Atlas launch vehicle (Convair)

lva landing vehicle airoll; landing vehicle amphibious; left visual acuity

lva (LVA) landing vehicle—assault

LVA Licensed Victuallers Association; Literary Volunteers of America

lvad left ventricular assist device

L v B Ludwig van Beethoven

L v Bthvn Ludwig van Beethoven

lvcd least voltage coincidence detection

lvd louvered door

LVD laser video-disc player

lvda launch vehicle data adapter

lvdc launch vehicle digital computer

lvdt linear variable-differential transformer; linear variable-displacement transducer

lved left ventricular end diastolic

lvet left ventricular ejection time

lvf left ventricular failure; left visual field

Lvfa low-voltage fast activity

lvgo light vacuum gas oil

lvh left ventricular hypertrophy

lvh (LVH) landing vehicle hydrofoil

lvhv low volume high velocity

lvi low viscosity index

LVI Local Veterinary Inspector

lvl laminated veneer lumber; level

LVL La Verendrye Line (Hall Corporation); Linda Vista Library

LVLO Local Vehicle Licensing Office

lvn light virgin naphtha

LVN Licensed Visiting Nurse; Licensed Vocational Nurse

LVNM Lava Beds National Monument (California)

LVNP Lassen Volcanic National Park (California); Luangwa Valley National Park (Zambia)

lvo (LVO) leveraged buyout

lvp low-voltage protection

lvp (LVP) left ventricular pressure

LVP Launch Vehicle Program(s)

lvp dr leverpak drum

lvr line voltage regulator; longitudinal video recording; low-voltage release

LVRB Launch Vehicle Reliability Board

lvrj low-volume ramjet

Lvrpl Liverpool

LVRS Longitudinal Video Recording System

lvs leaves

lv's lunch(eon) vouchers

LVs launch vehicles

LVS Licentiate in Veterinary Science

LVT landing vehicle, tracked

LVTC landing vehicle, tracked command

lvupk leave and upkeep

LVUSA Legion of Valor of the USA

lvw linked vertical wall

lw left wing (field hockey); light warning; lightweight; live weight; long wave; low water

l/w in lieu of weighing; lumens per watt

l & w living and well

l W lichte Weite (German—internal diameter)

Lw lawrencium (element 103)

LW light warning; lower berth

L-W Lee-White (method)

lwar lightweight attack and reconnaissance

L-wave long wave (usually the third major earthquake shock wave)

l'way leeway

lwb long wheelbase

lwbr light-water breeder reactor

lwc lightweight concrete

LWCA London Wholesale Confectioners Association

LWCF Land and Water Conservation Fund

lwcp lightweight coated paper

lwd larger word; leeward; left wing down; lewd; lowered

LWD Laser Welder/Driller; Liquid Waste Disposal

lwest low water equinoctial spring tide

lwf lightweight fighter

LWF Lutheran World Federation

LWFB Lake Washington Floating Bridge

lwf & c low water full and change

lwg last we've got; live-weight gain

LWG Logistic Work Group (NATO)

lwgr (LWGR) light-water-cooled graphite-moderated reactor

L-w-H Lewis-with-Harris (Outer Hebrides)

lwic lightweight insulating concrete

lwir long-wave infrared

LWJ Lowell Weicker, Jr

lwl length at waterline; load waterline; low-water line (tidal marking)

LWL Limited War Laboratory (US Army)

lwld light-weight laser designator

lwm low-water mark

LWM Leonard Wood Memorial (American Leprosy Foundation)

LWMAT Locke-Wallace Marital Adjustment Test

LWMEL Leonard Wood Memorial for the Eradication of Leprosy

LWNWR Lake Woodruff National Wildlife Refuge (Florida)

lwont low water ordinary neap tide

lwop leave without pay

lwos low-water ordinary spring

lwost low-water ordinary spring tide

lwp leave with pay; load water plane

lwpf long-wave pass filter

lwr lightweight radar; lower

lwr (LWR) light water reactor

Lwr Lower (postal abbreviation)

LWR Light Water Reactor

l'wrd leeward

lwrm (LWRM) lightweight radar missile

lwrs light-warning radar set

lwru lightweight radar unit

LWS Late West Saxon; Letter Writing System

LWSI *Lloyd's Weekly Shipping Index*

lwst low-water spring tides

lwt lightweight

Lwt Lowestoft

LWT amphibious warping tug (naval symbol); London Weekend Television

lwta laser window test apparatus

LWTMA London Wool Terminal Market Association

lwtvp laser window technology validation program

LWU Leather Workers Union

LWUI Longshoremen's and Warehousemen's Union International

LWV Lackawanna & Wyoming Valley (railroad); League of Women Voters

LWVEF League of Women Voters Education Fund

LWVUS League of Women Voters of the United States

lww launch window width

lwyr lawyer

lx lux; lymphatic invasion can't be assessed

lx. *lux* (Latin—light)

LX Los Angeles Airways (2-letter coding)

Lx^a *Lisboa* (Portuguese—Lisbon)

Lxmbrg Luxembourg

lXr limb X-ray

LXX Septuagint (70)

lxxx love and kisses

ly langley (solar heat unit); last year; last year's model

ly (LY) lethal yellowing (coconut-palm disease)

Ly Lyceum Theatre, English Opera House; Lyman; Lyon

LY Libya; Light Yeomanry; Love Year

LYC Larchmont Yacht Club

Lyd Lydia; Lydian

Lýd Isl *Lýdhveldidh* (Icelandic—Republic of Iceland)

lye potassium hydroxide (KOH) or sodium hydroxide (NaOH)

LYK Lykes Brothers Steamship company (stock exchange symbol)

LYKU Lykes Lines (container) Unit

Lylis Lilian; Lilly

ly & lt lastex yarn and lactron thread

lym last year's model(s); lymph; lymphatic(s); lymph node

lymphoks lymphokines

lympho(s) lymphocyte(s)

lymphs lymphocytes

lyn lynch (named for Captain William Lynch, also called Judge Lynch, who advocated hanging on the basis of mob action rather than legal procedure; this type of violence is also called lynch law)

Lyn Lynch; Lynde; Lyndon; Lyne; Lynn; Lynx (constellation)

Lynwood Lynwood (delinquent) Girls Center at Anchorage, Alaska

Lyo Lyons (British maritime contraction)

Lyo Isl *Lýoveldio Island* (Icelandic—Republic of Iceland)

LYONS Liquid Yield Option Notes

lyr lyric; lyrical; lyricism; lyricist; lyrics

Lyr *Lyra* (Latin—Lyre constellation)

L & YR Lancashire and Yorkshire Railway

lyric. language for your remote instruction by computer

lys lysine

LYs Light Years

lysis (Greek—dissolution or loosening)—analysis, catalytic, electrolyte, hydrolysis, paralysis; (Latin suffix— dissolve or solution)— hemolysis

lysog lysogen(ic)(al)(ly); lysogenization; lysogenize(d); lysogenizing; lysogeny

Lyt Lyttelton, New Zealand

Lytt Lyttelton

Lyttelton Lyttelton Harbour (Christchurch, New Zealand's port)

Lz Lopez

LZ Landing Zone

LZEXE Fabrice Bellard, Grabels, France trademark

lzm lysozyme

LZOA Labor Zionist Organization of America

lzp left zero point

LZSU Leningrad AA Zhdanov State University (University of Leningrad, former name for University of St Petersburg)

lzt lead zirconate-titanate

LZT Local Zone Time

L-Zug *Luxus-Zug* (German— luxury railroad train)

LZW Lempel-Ziv-Welch

lzy lazy

M

m difference of meriodional parts (symbol); magnetic dipole moment (symbol); main; maintainability; male; malignant; manual; margin: marriage: married; marsh; masculine; mass; mature; mean (arithmetical); measure; mediator (chemical); mega; mega-ohm; member; memory; mentum; meridian; mesh; metabolite; meter; mid; migrant; mile; mill; milli-(thousandth); minim; minor; minute; minutes; modulation coefficient (symbol); modulus; molal (concentration); molar; molecular weight; molecule; monkey; month; monthly: moon; morning; morphine; mother; motile; mucoid; murmur (heart); muscle; myopia

m- Meta

(m) multiple tumors

m (M) Marijuana; morphine

m/ merged into

m/1 married first

μ micron (symbol); micro

'm (contraction—am)—as in *I'm here*

m mass (symbol); *Mazda* (Japanese auto with German Wankel rotary engine); *metro* (Portuguese or Spanish—meter); *mijnheer* (Dutch—mister); *murió* (Spanish—died)

m. *macerare* (Latin—macerate)

m/ *med* (Norwegian—with)— as in *varm aplepai m/is* (hot apple pie with ice cream)

M bending moment (symbol); distant metastasis; Mach (Austrian physicist); mach number; mach speed; magnaflux; magnetic inspection; maintainability; Majesty; Malay; Malaya; Malaysia; Maldives; March; mark; Martin; materiel; Matson Navigation Company; matured bonds (in newspaper bond listings); median; medium; mega-(million); megabyte; megacycle; Member; metal; Metro(politan); metropolitan; Mike—code for letter M; Min; missile; mixture; mobile; Mohammedan; Mohammedanism; molal (concentration); molecular weight (symbol); moment; Monday; Montour (railroad); Moore-McCormack (steamship lines); morgan; mortality rate; Moslem; muscle; mutual inductance (symbol); pitching moment (symbol); refractive modulus (symbol); thousand (symbol)

M *(M)* money supply (macroeconomics symbol)

M' *Mac* (Gaelic—son of)

M *der Mörder* (German—the murderer)—Fritz Lang film; *Marcus* (Latin); *Missa* (Latin—Mass); *Monsieur* (French—Mister); *mujeres* (Spanish—women)

m_1 mitral first sound

M1 money supply including currency and trading bank demand deposits; primary motor cortex

M-1 basic money aggregate (funds for spending); mutual inductance symbol; U.S. semi-automatic service rifle used in Vietnam

M-1/M-1A1/M-1A2 U.S. main battle tank, also known as "Abrams"

m/1 married first

m1s matte one side

m/2 married second

m^2 square meter(s)

μ^2 square micron

M2 Ml plus savings and small-denomination time deposits under $100,000; money supply (Ml) plus demand deposits of savings & loan banks as well as stock firms; supplementary motor cortex

M-2/M-3 White half-track armored- personnel carrier

m2s matte two sides

m/3 middle third (long bones)

m^3 cubic meter(s)

μ^3 cubic micron

M3 M2 plus large-denomination time deposits, term repurchase agreements, and investments of institutions in money-market mutual funds; secondary motor cortex

M-3 sterling deposits of British residents plus coins and notes

M-3A1/M-5 Stuart tank armed with a 37mm gun

m³/m cubic meters per minute

m³/s cubic meters per second

M-4 Sherman medium tank with 76mm gun

M-6 American-made armored car

m8 medium octavo

M-8 Greyhound 6-wheeled armored car carrying a 37mm gun and made in the U.S.A.

M-14 U.S. fully-automatic or semi-automatic service rifle used in Vietnam

M15 gasoline extender (15% methanol plus 85% gasoline)

M-15 British secret service charged with counterespionage and security operations at home and overseas

M-16 British Foreign Service Military Intelligence (secret intelligence service); U.S. fully or semi-automatic lightweight small-bore service rifle used in Vietnam

M-18 Hellcat 76 mm gun mounted on a tracked chassis made in the U.S.A.

M-20 Mystère 20 aircraft; unarmed Greyhound 6-wheeled armored car produced in the U.S.A. during World War II

M-36 U.S.-made Slugger tank destroyer

M-44 U.S.-made self-propelled 155mm howitzer

M-47 U.S.-made Patton tank carrying a 90mm gun

M-48 later version of the M-47 medium tank

M-56 U.S. self-propelled 90mm antitank gun called Scorpion

M-60 Patton main battle tank carrying 105mm gun

M-61 20mm Vulcan aerial machine gun firing 6000 rounds per minute

M100 100% methanol

M-107 U.S.-made self-propelled 175mm gun

M-113 U.S-made 13-man amphibious armored personnel carrier

M-198 155mm howitzer (USA)

M-551 U.S.-built Sheridan assault vehicle armed with a 152mm gun

M$ Microsoft

ma machine account; machine accountant; magnetic amplifier; maleic anhydride; man-

power allotment; manufacturing assembly; map analysis; mauve; mechanical advantage; medium amphibian; menstrual age; mental age; microscopic agglutination; mill annealed; mill-anneal(ing); milliampere; mixed ages; mixed-age (ewes); monthly account(ing)

ma (MA) maleic anhydride

m/a mechanical airframe; my account

m & a maintenance and assembly; merger and acquisition

μa micro+ampere

mA milliampere(s)

ṁ milliangstrom(s)

μA microampere(s)

Ma Malayalan; Mama; Manchuria, Manchurian; Maria; masurium (symbol)

Mᵃ María

Ma Mandag (Danish—Monday)

MA Magma Arizona (railroad); Magnesium Association; Mahogany Association; Maintainability Analysis; Manpower Administration; Manpower Authorization; Maritime Administration; Marshaling Area; Marshalling Area; Massachusetts; Material Authorization; Mediterranean Area; Menorah Association; Merchandising Assessment; Metric Association; Military Academy; Military Attaché; Mountaineering Association; Morocco (Internet code)

M-A Miller-Abbott (tube)

M.A. Magister Artium (Latin—Master of Arts)

M & A Missouri & Arkansas (railroad)

MA Maison d'Arrêt (French—jail, lockup, prison); Modern Age; Musical America

maa maximum authorized altitude

maa (MAA) macroaggregated albumin

Maa Madras

Maa Maandag (Dutch—Monday)

MAA Manufacturers Aircraft Association; Master Army Aviator; Master-at-Arms; Master-of-Arms; Mathematical Association of America;

Medical Assistance for the Aged; Medieval Academy of America; Message Authenticator Algorithm; Microfilm Association of Australia; Museum of African Art (Smithsonian); Museums Association of Australia; Music Associates of America, Mutual Aid Association; Mutual Assurance Association

MA of A Motel Association of America

MAAA Metropolitan Area Apparel Association

ma ac machine accessory

MAAC Major Additions Adjustment Clause; Medical Assistance Advisory Council; Metropolitan Area Advisory Committee; Mutual Assistance Advisory Committee

MAAE Minnesota Alliance for the Arts in Education

MAAEE Ministero degli Affari Esteri (Italian—Ministry of Foreign Affairs)

MAAF Mediterranean Allied Air Force; Mediterranean Army Air Force

MAAG Military Assistance Advisory Group

MAAGB Medical Artists' Association of Great Britain

MAAH Museum of African-American History

MAAI Modeling Association of America, International

ma'am madam

ma'amselle mademoiselle

MAAN Mutual Advertising Agency Network

MAAN Much Ado About Nothing

maap maintenance and administration panel

MAAP Minority Association for Animal Protection

M.A.Arch. Master of Arts in Architecture

maarm (MAARM) memory-aided anti-radiation missile

MAARS Multi-Access Airline Reservation System

MAAS Member of the American Academy of Arts and Sciences

MAAV Multiple Abstract Analysis of Variance

MAATC Mobile Antiaircraft Training Center

mab multibase arithmetic blocks

Mab Mabel

MAB Magazine Advertising Bureau; Malfunction(ing) Analysis Board; Man and the Biosphere (UNESCO); Marine Air Base; Marine Amphibious Brigade; Medical Advisory Board; Member Advisory Board; Metropolitan Asylum Board; Missile Assembly Building; Monetary Affairs Branch; Munitions Assignment Board

MAB Manufacture d'Armes Automatiques Bayonne (French—Bayonne Automatic Arms Factory)

M.A.B.E. Master of Agricultural Business and Economics

MABF Major Adjustment Billing Factor

mabflex marine amphibious brigade field exercise

mablex marine amphibious brigade landing exercise

MABO Marianas-Bonin (islands)

mabp mean arterial blood pressure

MABRON Marine Air Base Squadron

MABs Marine Amphibious Brigades

MABS Marine Air Base Squadron; Marine Automation Bridge System

MABSC Management and Behavioral Science Center

MABYS Metropolitan Association for Befriending Young Servants

mac macadam(ize)(d); macerate; machine-aided cognition; mackintosh; maximum allowable concentration(s); maximum allowable cost; mean aerodynamic chord; message authentication code; message authorization code; minimal alveolar concentration; monitored anesthesia care; motion analysis camera; multiple-access computer

mac (MAC) mycobacterium avium-intracellulare

mac macizo (Spanish—strong, solid); *maquereau* (French—mackerel)—pimp

mac. macerare (Latin—macerate)

Mac Douglas MacArthur; Freddie Mac, Federal Home Loan Mortgage Corp; Macao, Macintosh (computer); Portuguese China; Macedonia; nickname of anyone whose surname begins with Mac

M.Ac. Master of Accountancy

MAC Maintenance Advisory Committee; Maintenance Analysis Center; Major Air Command; Management Aggregation Code; Mandatory Access Control; Marine Amphibious Corps; Maritime Advisory Committee; Material Availability Commitment; McDonnell Aircraft Corporation; Media Access Control; Mediterranean Air Command; Message Authentication Code; Miami Aviation Corporation; Middle Atlantic Conference; Military Airlift Command; Mineralogical Association of Canada; Monitored Anesthesia Care; Multidimensional Actuarial Classification; Municipal Assistance Corporation; Musical Arts Center

MACA Maritime Air Control Authority; Mental After-Care Association

MACE Mid-America Commodity Exchange

MACAE Minnesota Association of Continuing Adult Education

MACAIR Macao Air Transport

Macanese of Chinese and Portuguese descent

MACAP Major Appliance Consumer Action Panel

MACAS Magnetic Capability and Safety System

Macb Macbeth

MACBETH Memphians Against Culture Buffs Exposing Themselves Heedlessly

Macc Maccabees

MACC Mexican-American Cultural Center; Military Aid to the Civilian Community

MACCS Manufacturing Cost Collection System; Manufacturing and Cost-Control System; Marine Air Command and Control System

m. accur. misce accuratissme (Latin—mix very accurately)

MACD Member of the Australian College of Dentistry

MACDC Military Assistance Command Director of Construction

Macdonnells short form for the Macdonnell Ranges of Australia's Northern Territory

MACE Machine-Aided Composition and Editing; Massachusetts Advisory Council on Education; Mid-America Commodity Exchange; Military Aircraft Capability Estimator(s); Minnesota Association for Childhood Education; Missile and Control Equipment (North American Aviation); trade name for tear gas used by policemen and postmen

M.A.C.E. Master of Air-Conditioning Education; Master of Air-Conditioning Engineering

Maced Macedonia; Macedonian

Maced. Macedonia (Republic of Macedonia, formerly Yugoslavia)

MACG Marine Air Control Group

Macgillicuddy's Macgillicuddy's Reeks (Ireland's highest mountain range)

mach machine; machinery; machinist

mach machete (Spanish—all-purpose long knife)

Mach The Tragedy of Macbeth

machdiprt machine discharge print(ed)

machflshd machine fleshed (skins)

Machinists Union International Association of Machinists and Aerospace Workers

macho massive compact halo object

MACHO Memphians Against Chest Hair at Operas

machovprt machine over-printed

machprt machine printed

machsltd machine salted (skins)

Machu Machu Picchu (ancient Incan sanctuary and stronghold in the high Andes near Cuzco, Peru)

machwsh machine washed (skins)

maci military adaptation of commercial items

Macintosh Apple Corporation trademark

MACJC Minnesota Association of Community and Junior Colleges; Missouri Association of Community and Junior Colleges

mack mackinaw; mackintosh; maststack (marine superstructure containing mast and smokestack)

Mackenzies Mackenzie Mountains of the Canadian Northwest

mack(es) mackintosh(es)

Mackinac or Mackinaw formerly Michilimackinac

Macmillan Macmillan Publishing Co.

MACNYC Men's Apparel Club of New York City

MACOM Major Army Command; Mayor's Committee of Welcome

MACR Missing Air Crew Report

macrf macroforma(tion)

macro (Greek *makro*—great or large)—macrocyte, macromolecule, macromutation, macroscopic, macrophage

macrobio macrobiologic(al); macrobiology; macrobiotic(s)

macrobop macrobopper (underground slang—older teenager in sympathy with the modern scene)

macrocephs macrocephalics (large-headed people)

macroeco macroeconomics

macrol macrologic(al)(ly); macrologist(s); macrology

macros macroinstructions

MACRS Modified Accelerated Cost-Recovery System

MACS Maintenance Assistance Capability Software; Marine Air Control Squadron; Military Airlift Command Service; Mobile Air Conditioning Society

macship merchant aircraft ship (merchant vessel fitted with a flight deck)

MACSS Medium-Altitude Communication Satellite System; Montana Association of County School Superintendents

Mactan air field near LapuLapu, Cebú Island, Philippines

MACTU Mines and Countermeasures Tactical Unit (USN)

MAC/V Military Assistance Command, Vitetnam

Macy's RH Macy's

mad magnetic airborne detector; magnetic anomaly detector; maintenance, assembly, and disassembly; mathematical analysis of downtown (computer); mean absolute deviation; midpoint air dose; mind-altering drug

mad (MAD) music and dance (festival); mutual(ly) assured destruction (via nuclear warfare)

Mad Madam(e); Madeira; Madison; Madras; Madrid

Mad. Madagascar (has a northern port named Hell-Ville)

M. Ad. Master of Administration

MAD Madrid, Spain (airport); Maintainability Analysis Data; Maintenance Assessment District; Manufacturing Assembly Drawing; Marine Air Detachment; Marine Aviation Detachment; Memphians Against Degeneracy; Michigan algorithmetic decoder; Mine Assembly Depot; Mississippians Against Disposal (of nuclear wastes); Mongolian Asiatic Development (plan)

MAD Militarischer Abschirmdienst (German—Military Screening Service)—German counterintelligence corps

Mada Madagascar

MADA Muda Agricultural Development Authority

MADAEC Military Application Division of the Atomic Energy Commission

Madag Madagascar

Madagascar Democratic Republic of Madagascar (Indian Ocean island country long a French colony)

MADAM Manchester Automatic Digital Machine

madar malfunction analysis detection and recording

MADARS Maintenance Analysis, Detection, and Reporting System

Mad Av advertising and communications enterprises (located on Madison Avenue, New York City)

madc mini air data computer

MADD Manufactured Artificial Dog Dung; Mothers Against Drunk Drivers; Mothers Against Drunk Driving

maddam macromodule and digital differential analyzer; multiplexed analog-to-digital digital-to-analog multiplexed

MADDER Memphians Against Damsels Doing Ecdysiast Routines

maddida magnetic-drum digital-differential analysis

MADE Multichannel Analog-Digital Data Encoder

MADECO Manufacturas de Cobre (Spanish—Copper Manufacturers)

Madeiras Madeira Islands in the North Atlantic off Morocco

madevac medical evacuation

madex magnetic anomaly detection exercise

madge microwave aircraft dlgital guidance equipment

Mad I Madeira Islands

MADIAR Societé Nationale Malgache des Transports Aériens (French—Madagascar Air Transport)

MADIS Manual Aircraft Data Input System; Millivolt Analog-Digital Instrumentation System

Mad Isl Madeira Islands

Madison James Madison, author of the *Bill of Rights* and fourth President of the United States; capital of Wisconsin

madm medium atomic demolition munition

M. Admin.S. Master of Administrative Studies

MAD Policy Mutually-Assured Destruction Policy (nuclear warfare)

madr minimum adult daily requirement

Madr Madrid; Madrileño

madre magnetic-drum receiving equipment

madrec malfunction detection and recorder

mads mind-altering drugs

MADs Mothers Against Drugs

MADS Maintainability Analysis Data System; Modular Army Demonstration System

madt microalloy diffused-base transistor

mae mean absolute error; motion aftereffect

Mae Fannie Mae, Federal National Mortgage Association; Mary

Ma.E. Master of Engineering

MAE Medical Air Evacuation; Museum of Atomic Energy

MAEA Maine Art Education Association

M.A.E. Master of Aeronautical Engineering; Master of Art Education; Master of Arts in Education; Master of Arts in Elocution

M.A.Econ. Master of Arts in Economics; Master of Arts in Economic and Social Studies

MAECON Mid-America Electronics Convention

M.A.Ed. Master of Arts in Education

MAEE Marine Aircraft Experimental Establishment

MAEF Master Asphalt Employer's Federation

MAELU Mutual Atomic Energy Liability Underwriters

M.Aero.E. Master of Aeronautical Engineering

MAES Mexican American Engineering Society

maesto *maestoso* (Italian—majestically)

ma ewes mixed-age ewes

maf macrophage activation factor; major academic field; manpower authorization file; million-acre feet; minimal audible field; minimum audible field; multiplanar angular forces

MAF MacArthur Foundation; Marine Air Facility; Marine Amphibious Force; Middle Atlantic Fisheries; Midland, Texas (airport); Minister of Armed Forces; Ministry of Agriculture and Fisheries; Ministry of Agriculture and Forestry; Missile Assembly Facility; Mission Aviation Fellowship; Mobile Air Force; Mutual Adjustment Fund(ing); Mutual Asset Fund(ing)

MA & F Ministry of Agriculture and Fisheries

MAFA Manchester Academy of Fine Arts; Museum of American Folk Art

MAFAC Marine Fisheries Advisory Council

MAFB Mitchell Air Force Base

MAFC Major Army Field Command

MAFCA Model-A Ford Club of America

MAFDAL *Miflaga Datit Le'umit* (Hebrew—National Religious Party)

mafe magnesium + iron (Ma + Fe)

MAFF Minister of Agriculture, Fisheries and Food; Ministry of Agriculture, Foresty, and Fisheries

MAFFS Modular Airborne Fire Fighting System (USAF)

MAFI Medic-Alert Foundation International; Ministry of Agriculture, Forestry, and Irrigation

MAFIA *Morte Alla Francia Italia Anela* (Italian—Death to France Is Italy's Cry), acronym devised when the secret society was organized in the 1860s, to combat French forces of intervention

mafr merged accountability and fund reporting

MAFS Mobilization Air Force Specialty

MAFSI Marketing Agents for Food Service Industry

MAFVA Miniature Armored Fighting Vehicles Association

maf/yr million-acre feet per year

mag magazine; magnesia; magnesium; magnet; magnetic; magnetism; magneto; magnetron; magnum; maximum available gain

mag. *magnus* (Latin—great)

Mag Magallanes (Punta Arenas); Magallanic; Magyar; Margaret

Mag. Magistrate

Mag. Magnificat (Latin—it magnifies)—song of the Virgin Mary

MAG magnesium (machine shop style); Marine Aircraft Group; Marine Aviation Group; Metropolitan Area Government; Military Advisory Group

MAGAF Magna International Incorporated

maga magazine

Mag.Agg. *Magister Aggregatus* (Latin—Master of Aggregation)—Head Master

mag ampl magnetic amplifier

MAGB Microfilm Association of Great Britain; Mining Association of Great Britain

Mag Bay Magdalena Bay, Baja California

mag cap magazine capacity

mag card magnetic card

magcheck magneto check

mag ci magnetic cast iron

magcon magnetic concentration

mag cs magnetic cast steel

Magd Magdalene (pronounced *Modlin*) College (Cambridge or Oxford)

Magda Magdalen(a)

Magdalens Magdalen Islands in the Gulf of Saint Lawrence

Mag Dav *Magen David* (Hebrew—Shield of David; Star of David)—six-pointed star, symbol of Judaism

Magd Coll Magdalen College—Oxford

M.Ag.Ec. Master of Agricultural Economics

M.Ag.Ed. Master of Agricultural Education

MAGERT Map and Geography Round Table

magg *maggio* (Italian—May); *maggiore* (Italian—major)

magic modern analytical generator of improved circuitry; modern analytical generator of improved circuits

MAGIC Madison Avenue General Ideas Committee; Men's Apparel Guild In California; Midac Automatic General Integrated Computation; Mothers Against Gangs In our Communities; Mozambique, Angola, and Guinea Information Center

magid magnetic intrusion detector

maglev magnetically levitated (linear motor-propelled railroad trains); magnetic levitation

maglevs magnetically-levitated superfast trains; magnetically levitated vehicles

magloc magnetic logic computer

mag mod magnetic modulator

magn magnetism

magn. *magnus* (Latin—large)

magna material and geomettically nonlinear analysis

magnalium magnesium + aluminum (alloy)

magneform magnetic forming (process)

Mag Nép *Magyar Népköztárság* (Hungarian People's Republic)

magnesia magnesium oxide (MgO)

magnet. magnetics

MAG^{ni} *Magazini* (Italian—warehouse)

magno manganese-nickel alloy

magnox magnesium oxide

magnum high-powered cartridge or weapon for firing magnum ammunition; $^2/_5$-gallon champagne bottle

mag. op. *magnum opus* (Latin—major work)

M.Agr. Master of Agriculture

mags magazines; magnesium wheels

MAGSAT Magnetic Field Satellite

MAGTF Marine Air-Ground Task Forces

mag tape magnetic tape

magtig allemagtig (Afrikaans—almighty)—Almighty God

MAH Museum of American History

mah mahogany

µAh microampere hour

MAHA Malaysian Agri-Horticultural Association

MAHE Michigan Association for Higher Education

mahog mahogany

mai machine-aided index(ing); marriage adjustment inventory; mean annual increment; minimum annual income

MAI Member Appraisal Institute; Military Assistance Initiative (Philippines); Military Assistance Institute; Multilateral Agreement on Investment; Multilateral Assistance Initiative; Museum of the American Indian

MAI *Moskovskiy Aviatsionny Institut* (Russian—Moscow Aviation Institute)

MA.I. *Magister in Arte Ingeniaria* (Latin—Master of Engineering)

MAIBL Midland and International Banks Limited

maid maintenance automatic integration detector

M-Aid Marshall-Plan Aid (given European countries by the United States after World War II)

MAIG Matsushita Atomic Industrial Group

Mail Safe RSA Data Security, Redwood City, California

Maimon Maimonides

MAIN Medical Automation Intelligence System

Mainbocher Main Rousseau Bocher

Maine St Mus Maine State Museum

Maine Turn Maine Turnpike

Maine Yankee Maine Yankee Atomic Power Company

Mainichi *Mainichi Shimbun* (Japanese—Everyday Newspaper)—modern Japan's oldest periodical

MAINS Marine International Navigation System

maint maintenance

maintnce maintenance

maip memory access and interrupt processor

MAIS Maine Association of Independent Schools; Maintenance Information System; Minnesota Adaptive Instructional System

MAIT Maintenance Assistance and Instruction Team; Multidiscipline Accident Investigation Team

maître d' *maître d'hôtel* (French—head waiter)

Maître d'Hôtel of Philosophy Baron Paul-Henri-Dietrich d'Holbach (1723–1789)

maj major; majority

Maj Major

MAJ Majuro (Marshall Islands airport); Muhammad Ali Jinnah

majac maintenance antijam console

Maj Com Major Command

maj dem(s) major demon(s)—Asmodeus (lechery), Beelzebub (gluttony), Belphegor (sloth), Leviathan (envy), Lucifer (pride), Mammon (avarice), Satan (anger)

MAJECA Malaysia-Japan Economic Association

Maj Gen Major General

Major John Major (British prime minister)

mak. making

Mak Makdougall; Makoto; Maksim; Maksimovich

makbsctr make best counteroffer

Mák-i-no Mackinac (island, river, or strait as pronounced locally)

Makh. *Makhshirin*

MAKN *Mongol Ardyn Khuv'sgalt Nam* (Kalkha Mongol—Mongolian People's Revolutionary Party)

maksutsub make suitable substitutions

mal (Latin prefix—abnormal, bad, disorder)—malignant; *malayisch* (German—Malayan)

Mal The Book of Malachi; Malaga; Malagueña(o); Malawi; Malay; Malayan; Malaysia; Malaysia Airlines; Malta; Maltese

Mal Malay (the basic language of the Indo-Malayan islands and nations including Indonesia); *Maréchal* (French—Marshal)

MAL Malaysian Airways Limited; Material Allowance List

Mala Malaya; Malayan; Malaysia; Malaysian

malac malacology

malachite hydrated copper carbonate

malaprop *mal à propos* (French—out of place, unappropriate)

Malaspina Malaspina Glacier on Yukutat Bay, Alaska

Malawi Republic of Malawi (formerly Nyasaland, East African nation)

Malaya Malay Peninsula also called Malaysia

Malay Pen. Malay Peninsula, containing Burma, Malaysia, Singapore and Thailand

Malaysia Malay Peninsula (countries formerly comprising British Malaya plus Saba and Sarawak but minus Singapore)

Malaysian Federation Johore, Kedah, Kelantan, Malacca, Negri Sambilan, Pahang, Penang, Perak, Perlis, Sabah, Sarawak, Selangor, Trengganu; up to 1965 included Singapore

Mald Maldive Islands; Maldives

Mal$ Malaysian dollar

Mal $ Malaya dollar

M.A.L.D Master of Arts in Law and Diplomacy

MALDEF Mexican-American Legal Defense and Educational Fund

Maldives Republic of Maldives (Indian Ocean island nation)

MALEV (Hungarian Airline)

MALEV Magyar Legikole-kedesi Vallat (Hungarian Airlines)

malfunc malfunction

Malg Rep Malagasy Republic

Mali Republic of Mali (landlocked West African country), *République du Mali*

MALI Air Mali

malig malignant

Malindo Malaysia-Indonesia

Mal Isl Maldive Islands

mall malleable

Mall Mallorca

Mallows (British slang shortcut—St Malo)

MALODES Modern Army Logistics Data Exchange System

malor mortar-and-artillery-locating radar

maloti monetary unit of Lesotho

malpais (Spanish—badlands)—basaltic-lava wastelands

Malpartidas (Spanish—Badly Divided Lands)—short form for Malpartida de Cáceres, Malpartida de la Serena, and Malpartida de Plasencia—all in western Spain near the Portuguese borderlands

Malpaso Alto de Mal Paso (highest point on Hierro in the Canary Islands)

Mal-Port Malay-Portuguese (East African patois)

malprac(s) malpractice(s); malpractitioner(s)

MALRA Malaysian Leprosy Relief Association

MALS Medium-intensity Approach Light(ing) System

M.A.L.S. Master of Arts in Liberal Studies; Master of Arts in Library Science; Master of Arts in Library Service

Mal Sam Sis Malotuto'atasi o Samoa i Sisisfo (Samoan—Independent State of Western Samoa)

MALSCE Massachusetts Association of Land Surveyors and Civil Engineers

Mal St Malay States

malt. malted milkshake

Malt(s) Maltese sailor(s)

Malukus Maluku or Moluccas Islands of Indonesia also called the Spice Islands

Malvinas Malvinas Islands (Falklands)

mam medium automotive maintenance; milliampere; minute(s)

ma'm madam

m + am (compound) myopic astigmatism

mam mot à mot (French—word for word)

MAM Military Assistance Manual; Missile Acceptance Meeting; Montclair Art Museum

MAMA Mobile Air Materiel Area; Middletown Air Materiel Area

Mam Arab Saud al-Mamlaka al-'Arabiya as Sa'audiya (Arabic—Kingdom of Saudi Arabia)

MAMB Military Advisory Mission—Brazil

MAMBO Mediterranean Association of Marine Biology and Oceanography

MAMC Madigan Army Medical Center

MAME Michigan Association for Media in Education

MAMENIC Marina Mercante Nicaraguense (Nicaraguan Merchant Marine—Mamenic Line)

mami machine-aided manufacturing information

mamie minimum automatic machine for interpolation and extrapolation

Mamie Margaret

Mam Mag al-Mamlaka al-Maghrebia (Arabic—Kingdom of Morocco)

mammal. mammalogist; mammalogy

mammax machine-made and machine-aided index

mammog mammogram; mammograph; mammographer; mammographic(al)(ly); mammography

Mammoth any mammoth caves in Australia, California, and Kentucky; Mammoth Hot Springs in Wyoming; Mammoth Lakes in California; Mammoth Onyx Cave in Kentucky; Mammoth Spring in Arkansas; Mammoth Village in Arizona

Mammoth Cave national park in Kentucky

mamos marine automatic meteorological observing station

MAMS Missile Assembly and Maintenance Shop

Mam Urd Hash al Mamlaka al Urduniya al Hashemiyah (Arabic-Hashemite Kingdom of Jordan)

M.A.Mus. Master of Arts in Music

Mamzel Mademoiselle

man. manhold; manifest; manifold; manual; manufacture; manure

man. manipulus (Latin—handful)

m A n meiner Ansicht nach (German—in my opinion)

Man management; Manager; Isle of Man in the Irish Sea or the Man canal and river in Burma or La Mancha in Spain, or Manchester, Mangalore, Manhattan, Manila, and Manitoba

MAN Managua, Nicaragua (airport); Metropolitan Area Network; Middletown (Conn.) Mansfields (National Association); Motorcyclists Against Noise

M-A-N Maschinenfabrik-Augsburg-Nurnberg

MAN Movemento Antillese Nuevo (Papiamento—New Antillean Movement)—based in Curaçao

Man¹ Manuel (Spanish—Emanuel)

MANA Manufacturers' Agents National Association; Mexican-American National Women's Association

M.Anaes. Master of Anaesthesiology

Man Brdg Manhattan Bridge (New York City)

manc mancando (Italian—gradually softer)

Manc Machester; Mancunian—inhabitant of Manchester

Manch Manchuria

Manch Guard *Manchester Guardian* (newspaper)

mand mandamus; mandate; mandatory; mandible; mandibular; mandolin

mand mandamus (Latin—we command) writ issued by a superior court commanding the performance of a specified official act or duty

Mand Mandarin

MANDFHAB Male and Female Homosexual Association of Great Britain

Man Dir Managing Director; Managing Directress

mando mancando (Italian—gradually softer)

mandy man day

Man Ed Managing Editor

manf manifold; manufacture; manufacturer; manufacturing

MANFED Manufacturers Federation

MANFEP Manitoba Finite Element Program

MANFORCE Manpower for a Clean Environment

mang management

manganim manganese-copper-nickel alloy

mang b manganese bronze

manglish mangled English

manhr manhour

MANI Minister of Agriculture for Northern Ireland

maniac (MANIAC) mechanical and numerical integrator and computer

manif manifest

MANIFILE Manitoba File (of worldwide nonferrous metallic deposits)

Manil Marcus Manilius (Roman poet)

manip. manipulus (Latin—handful)

manit man minute

Manit Manitoba

Manitoulins Islands in Lake Huron

Manley Norman Manley International Airport serving Jamaica and named for its first native-born chief minister—an Irish-Negro lawyer

manmam manufacturing management

Man Med Dept Manual of the Medical Department (USN)

manmo man month

Manny Emanuel; Manuel

Manny Hanny Manufacturers-Hanover Corporation

mano monograph; manometer

MANO Mexican-American Neighborhood Association

MANO Movimiento Argentina Nacional Organizado (Spanish—National Organized Movement of Argentina)—terrorist group

Mano Blanca (Spanish—White Hand)—Cuban exile group

man. one first-degree manslaughter

manop manually operated; manual operation

Man Op Manitoba Opera Association; Manual of Operation(s)

Manor The Manor (British underworld slang for London)

manova multivariate analysis of variance

man. p. mane primo (Latin—early in the morning, first thing in the morning)

MANP Masai Amboseli National Park (Kenya); Mount Apo NP (Mindanao, Philippines); Mount Arayat NP (Luzon, Philippines)

Man Ray Emmanuel Radnitsky

Man Rtg Manual Rating (code)

mans. mansions

Mans Mansfield College, Oxford; Mansion

MANS Map Analysis System

mansat manned satellite

mansec man second

Mansf Coll Mansfield College-Oxford

man(s) rep(s) manufacturer(s) representative(s)

Man Sym Manila Symphony

manta manta ray (devil ray)

mant (MANT) mantissa (calculator)

MANTECH Manufacturing Technology (USN)

MANTIS Manchester Technical Information Service

MANTRAP Machine and Network Transients Program

manuf manufacture(r); manufacturing

Manutius Aldus Manutius—Latinized version of Aldo Manuzio—inventor-of italic type

manuv maneuvering

MANWEB Merseyside and North Wales Electricity Board

manwich man-sized sandwich

manwk man week

manx manx cat (almost tailless breed of cat originatiing on the Isle of Man); manx shearwater (small black-and-white oceanic bird of the eastern North Atlantic); month after next

manyr man year

MANZ Medical Association of New Zealand; Montreal-Australia New Zealand (Line); Motel Association of New Zealand

mao (MAO) monoamine oxidase

mao med andra ord (Swedish—in other words); *med andre ord* (Dano-Norwegian—in other words)

Mao Mao Zedong

MAO Master of the Art of Obstetrics; Musica Aeterna Orchestra

MAO Magyar Allami Operhaz (Hungarian State Opera)

MAOA monoamine oxidase A

MAOF Mexican-American Opportunity Foundation

maoi (MAOI) monamine oxidase inhibitor

Maoriland New Zealand

MAOs monoamine oxidase inhibitors

maot medium-aperture optical telescope

MAOT Member of the association of Occupational Therapists; Military Assistance Observer Team

Mao Zedong (Pinyin Chinese—Mao Tse-tung)—Chinese leader 1949–1976

map manifold absolute pressure; manifold air pressure; mapping; market access program; maximum average price; micro-assembly program; minimum advertised price; minimum association price; minimum audible pressure; missed approach point; missed approach procedure; mitogen-activated protein; multiaccess portal; multiple-aim point

map (MAP) machine automated protocol (system); mean arterial pressure

MAP Maintenance Analysis Program; Managing Accelerated Productivity; Manufacturing Automation Protocol; Material Analysis Plan(ning); Media Access Project; Medi-Cal Aid Post; Medical Assistance Program; Melanesian Alliance Party; Microprocessor Application Project; Middle Atmosphere Program; Military Aid Program; Military Assistance Program; Military Association of Podiatrists; Mini-Activity Plan; Ministry of Aircraft Production; Ministry of Aircraft Production; Multiple Application Procedure; Mutual African Press (agency); Mutual Agreement Processing (for parolees); Mutual Assistance Program

M-A-P Modified American Plan (breakfast and dinner included)

MAP Maghreb-Arabe Presse (Maghreh Arab Press Agency)

MAPA Malayan Agricultural Producers Association; Malaysian Airlines Pilots Association; Mexican-American Political Association

MAPI Manufacturers Alliance for Productivity and Innovation

MAPAD Map Address Directory

MAPAG Military Assistance Program Advisory Group

MAPAI Miflaget Poaley Israel (Hebrew—Israel Labor Party)

MAPAM Miflaget HaPaolim HaMe'uchedet (Hebrew—United Workers Party)

MAPC Minnesota Association of Private Colleges

mapce mobile automatic programmed checkout equipment

MAPCO Mid-America Pipeline Company

mapd maximum allowable percent defective

MAPDA Mid-American Periodical Distributors Association

MAPDFA Media-Advertising Partnership for a Drug-Free America

maped machine-aided program for the preparation of electrical power

MAP-ga Military Assistance Program-grant aid

maph manned ambient-pressure habitat

MAPHILINDO Malaysia, Philippines, Indonesia (proposed unification of these Malayan countries)

MAPI Machinery and Allied Products Institute; Manufacturers Alliance for Productivity and Innovation; Millon Adolescent Personality Inventory; Mitsubishi Atomic Power Industries

mapid machine-aided program for the preparation of instruction(al) data

mapk mitogen-activated protein kinase

MAPL Manufacturing Assembly Parts List .

Maple Lane Maple Lane School (for juvenile delinquents) in Centralia, Washington

MAPNY Maritime Association of the Port of New York

MAPOM MAP-owned materiel

mapp methylacetylenepropadiene

mapple macro-associative processor-programming language

M.App.Sc. Master of Applied Science

MAPR Manufacturing Aids Program Requirements

mapros maintain production schedule(s)

maps monopropellant accessory power supply; Mothers of AIDS Patients

MAPS Major Assembly Performance System; Management Analysis and Planning System; Memory-Archives-Programs; Middle Atlantic Planetarium Society; Military Products and Systems (RCA); Miniature Air Pilot System;

Monetary and Payments System; Monitoring Avian Productivity and Survivorship; Multiple Address Processing System; Multiple Aiming Point System; Multivariate Analysis and Prediction of Schedules (USA)

MAPU Movimiento de Acción Popular Unitaria (Spanish—Movement for United Popular Action)—Chile

MAPW Medical Association for the Prevention of War

maq monetary allowance in lieu of quarters

maq maquereau (French—mackerel)—a pimp (also abbreviated *mac)*

MAQ Measures for Air Quality (NBS)

MAQC Metropolitan Air Quality Council

maquils maquiladoras (Spanish—corn grinders)—manufacturing plants on the Mexican Border

mar maintenance action rate; (bone) marrow; marimba; marine; maritime; married; marry; memory address register; minimal angle resolution; minimum acceptable rate (of return); multiarray radar; multifunction array radar

mar marokkanisch (German—Moroccan)

mar. mardi (French-Tuesday); *martedi* (Italian-Tuesday); *martes* (Spanish—Tuesday)

Mar Marathi; March; Marseilles; Marshall Islands

Mar Marseillaise (national anthem of France)

M.Ar. Master of Architecture

MAR Baltimore Marylands (National Association); Manistee and Repton (railroad); Maracaibo, Venezuela (airport); Maritime Central Airways; Mars Excursion Module; Material Availability Report; Material Availability Request

MARA Mexican-American Research Association

MARA Majilis Amanah Raayat (Malay—Handicraft Center for the Development of Malaysian People)

MARAD Maritime Administration (US Department of Commerce)

MARAIRMED Maritime Air Mediterranean

marb marbling

Marb Marblehead; Marbleheart; Marbury

Mar Ber Mar Bermejo (Spanish—Vermillion Sea)

marbi machine-readable bibliographic information

marble calcium carbonate ($CaCO_3$)

marc monitoring and results computer

marc marcato (Italian—marked)

Marc Marcus

MARC Machine-Readable Cataloging (Library of Congress magnetic-tape catalog system); Manpower Authorization Request for Change; Matador Automatic Radar Command; Metropolitan Applied Research Center; Micronesia Area Research Center (Guam); Model-A Restorers Club (Model-A Ford autos)

MARCA Mid-Continent Area Reliability Coordination Agreement

MarCad Marine Cadet

Mar Cad Marine Cadet

MARCEP Maintainability and Cost-Effectiveness Program

March. Marchioness

March Marchese (Italian—Marquis)

M.Arch. Master of Architecture

MARCHA Methodists Associated Representing the Cause of Hispanic Americans

Marcha Real (Spanish—Royal March)—anthem of Spain

Marchbanks (British contraction—Marjoribanks)

March^{sa} Marchesa (Italian—Marchioness)

MARC(LC) Machine-Readable Catalog(ing) (Library of Congress)

MARCO Marine Construction and Design Company

MARCOM Maritime Command (Canadian)

MARCONFOR Maritime Contingency Force

MARCONFORLANT Maritime Contingency Forces—Atlantic

MARCOR US Marine Corps

Mar Cort Mar de Cortés (Spanish—Sea of Cortez)—Gulf of Mexico

MARCS Marine Computer System

MARC(S) Machine-Readable Catalog(ing) for Serials

MARC(UK) Machine-Readable Catalog(ing) in the British Library of the United Kingdom

Marcus Mark

mardan marine digital analyzer

MARDEC Malaysian Rubber Development Corporation

MARDI Malaysian Agricultural Research and Development

MARECS Marine Communications Satellites

mar eng marine engineer(ing)

MARFOR Marine Forces

marg margarine; margin; marginal; marginalia

Marg Margrave; Margravine

marge margarine (oleomargarine); margin

margen management report generator

Mar Gils Area Marshalls-Gilberts (island) Area

Marg^{ta} Margarita (Spanish—Margaret)

marg trans marginal translation

marbelilex marine helicopter landing exercise

MARI Middle America Research Institute

María Félix Maria de los Angeles Félix Guereña

Marianas Mariana Islands; once called Ladrones (thieves)

Maribo Paramaribo, Surinam

Marichu (Spanish-American nickname—María de Jesús)

maricult mariculture; mariculturist

mariculture marine culture (growing food in the sea)

marifarm maritime farm

marifex marine firing exercise

mariholic marijuanaholic (addict)

MARIN Marine Industry (application of broadband communications)

Marinsky Marinsky Theater in St Petersburg (Leningrad), renamed Kirov

Mário Mário Cláudio (Portuguese author)

Mario Mario Cuomo (orator and politician)

Mariol Mariolatry; Mariology

Marion Federal Prison Camp at Marion, Illinois; Mary; Maryjane; U.S. Penitentiary at Marion, Illinois

MARIS Maritime Research Information Service

marisat maritime industry satellite

marit maritime

marita maritime airfield

Marit Admin Maritime Administration

Marit Com Maritime Commission

Maritime Alps *Alpes Maritimes* (French)—AM

Maritimes Canada's Maritime Provinces; the Maritime Alps between France and Italy; the Soviet Union's Maritime Territory along the Sea of Japan

maritrain(s) maritime train(s)—articulated sea-going barges

Marj Marja; Marjan; Marjorie; Marjory

mark monetary unit of Germany

mark. market; marketing

Mark The Gospel according to St Mark

Marka Markarian (galaxy thirteen times larger than the Milky Way)

markkaa monetary unit of Finland

mark twain leadline sounding of two fathoms (12 feet or 3.66 meters); leadsmen announcing *mark twain* meant there was enough water to keep the average shallowdraft paddlewheel river steamer afloat; Sam(uel) L(anghorne) Clemens; *mark three* is three fathoms and *quarter twain* is two-and-a-half fathoms

Marlag Marinenlager (German—sailor's camp for prisoners of war)

Marlene Mary + Helena

marlex marine reserve landing exercise

MARLF Middle Atlantic Regional Library Federation

mar lic marriage license

MARLIS Multi-Aspect Relevance Linkage System

Marm Marmaduke

marmap marine resources monitoring, assessment, and prediction

Marmara Sea of Marmara (connecting the Black Sea with the Mediterranean via the Dardanelles Strait and separating Asiatic Turkey from European Turkey)—called Marmara Denizi by the Turks

mar merc *marina mercantile* (Italian—merchant marine)

mar mil *marina militare* (Italian—navy)

MARNR *Ministerio del Ambiente y los Recursos Naturales Renovables* (Spanish—Ministry of the Environment and Renewable Natural Resources)—Venezuela

Maro *Marocco* (Italian—Morocco)

MARO Maritime Air Radio Organization

marops maritime operations

MAROPT Marine Optical (recording system)

marots (MAROTS) marine orbital technical satellite

MARPEX Management of Repair Parts Expenditure (USA)

Marpril March and April

Marq Marquesas Islands

Marquesas Marquesas Islands of the South Pacific or the Marquesas Keys west of Key West, Florida in the Gulf of Mexico

Marquis Marquis Who's Who Books

marr marriage; minimum acceptable rate of return

Marr Marranic; Marranism; Marranoism; Marrano(s)

Marr *Marruecos* (Spanish—Morocco)

MARRES Manual Radar Reconnaissance Exploitation System

marr lic marriage license

Marro *Marrocos* (Portuguese—Morocco)

marr sett marriage settlement

Marru *Marruecos* (Spanish—Morocco)

mars master attitude reference system; mathematics anxiety rating scale; mid-air retrieval system; military affiliated radio system

Mars Marseilles

Mars' Marshals' Offices

Mars *Ares* (Latin—god of war); *Marselha* (Portuguese—Marseilles); *Marsella* (Spanish—Marseilles); *Marsiglia* (Italian—Marseilles)

MARs Middle American Radicals

MARS Magazine Affinity & Renewal Services; MagneticElectronic Automatic Reservation System; Maintenance Action Reporting System; Manned Astronautical Research Station; Master Altitude Reference System; Military Affiliate Radio System; Miniature Accurate Ranging System; Mobile Atlantic Range Station; Modern Architectural Research Society; Monitored Atherosclerosis Regression Institute

MARSAP Mutual Assistance Rescue and Salvage Plan

MARSAS Marine Search and Attack System (USMC)

marsat maritime satellite

MARSATE Maintenance and Repair of Scientific and Technical Equipment

MARSATS Maritime Satellite System

M.Ar.Sci. Master of Arts and Sciences

mar settl marriage settlement

MARSH Matching Aid to Restore States' Habitats (for waterfowl)

Marshalls Marshall Islands in the western Pacific

marsh gas methane (CH_4)

MARSIS Marine Remote Sensing Information System

MA/RSO Mobilization Augmentee/Reserve Supplement Officer (USAF)

mart maintenance analysis and review technique; mean active repair time

mart *martes* (Spanish—Tuesday)

Mart Martinique

Mart. Martyrology

Mart Marcus Valerius Martialis (Roman poet)

MART Metropolitan Area Rapid Transit

MARTA Metropolitan Atlanta Rapid Transit Authority

Marth Martha

Martí *Aeropuerto José Martí* (Havana, Cuba's airport named for the founder of the Cuban Revolutionary Party who did much to organize resistance to Spanish rule but was killed in 1895, three years before his island was liberated by American and Cuban forces)

Martin St Martin or Sint Maarten in the Leeward Islands

M.Art RCA Master of Art of the Royal College of Art

mart(s) market(s)

MARTS Master Radar Tracking Station

Mart(y) Martin

marv maneuvering reentry vehicle (MaRV in Salt Talk reports, also MARV); marvel; marvelous

Marv Marvin

Marylebone St Marylebone

mas masculine; masonry; mathematics anxiety scale; metal angle slots; military assistance sales; milliampere second; moved aboard ship

mas Malaysian Airline System's trademark

Mas Massachusetts; Massachusettsan

MAs Mothers Anonymous

MAS Malaysian Airline System; Marine Acoustical Services; Maryland Academy of Sciences; Master Activation Schedule; Medical Administration Service; Military Agency for Standardization; Ministry of Aviation Supply; Missile Assembly Site; Monetary Authority of Singapore; Municipal Art Society; Mutual Assured Security

M.A.S. Master of Applied Science

M & AS Music and Art School

MAS *Motoscafi Anti Sommergibli* (Italian—antisubmarine motor torpedo boat); *Movimiento al Socialismo* (Spanish—Movement toward Socialism)—Venezuela's leftist party

MASA Mail Advertising Service Association; Malaysian Shipowners Association; Member of the Acoustical Society of America; Michigan Association of School Administrators; Military Automotive Supply Agency; Minnesota Association of School Administrators; Mississippi Association of School Administrators; Modular Avionic Systems Architecture; Montana Association of School Administrators

MASANYC Mail Advertising Service Association of New York City

masar microwave-accurate surface antenna reflector

MASB Michigan Association of School Boards

MASBO Minnesota Association of School Business Officials

masc masculine

masc. masculus (Latin—male)

M.A. Sc. Master of Applied Science

MASC Massachusetts Association of School Committees

MASCA Middle Atlantic States Correctional Association

Mascarenes Mascarene Islands

mascon massive concentration

MASCOT Meteorological Auxiliary Sea Current Observation Transmitter

MASCS Marriage Adjustment Sentence Completion Survey

MASEA Midwest Association of Student Employment Administrators

MaSEA Massachusetts Society of Enrolled Agents

maser microwave amplification by stimulated emission of radiation

mash. mashed potatoes

MASH Medical Aid for Sick Hippies; Memphians Against Social Harassment; Mobile Army Surgical Hospital; Multiple Accelerated Summary Hearing (for alien deportation); Mutual Aid Self Help

MASHAE Member of the American Society of Heating and Air Conditioning Engineers

mash(ed) mashed potatoes

MASHVE Member of the Australian Society of Heating and Ventilating Engineers

masint measure and signature intelligence

MASIS Management and Scientific Information Service

mask maskulinum (Dano-Norwegian—masculine)

MASL Military Assistance Articles and Services List

MASME Member of the American Society of Mechanical Engineers; Member of the Australian Society of Mechanical Engineers

MASO Munition-Accountable Supply Office(r) USAF

M.A. Soc. Stud. Master of Arts in Social Studies

mas. pil. massa piluarum (Latin—pill mass)

MASQUES Medical Application Software Quality Enhancement by Standards

mass. masseter; multiple-access sequential selection

Mass Massachusetts; Massachusettsan(s)

MASS Marine Air Support Squadron; Massachusetts Association of School Superintendents; Michigan Automatic Scanning System

MASS MoCA Massachusetts Museum of Contemporary Art

M.A.S.S. Master of Arts in Social Science

Massachusetts General Massachusetts General Hospital

Massanutteas Massanuttea Mountains (on the Appalachian Trail in northern Virginia between Charlottesville and Culpepper Court House)

masscult mass culture (culture for the masses)

massdar modular analysis, speedup, sampling, and data reduction

Massey Massey-Harris; Massey University (Palmerston North, New Zealand)

MASSP Michigan Association of Secondary School Principals; Minnesota Association of Secondary School Principals; Missouri Association of Secondary School Principals; Montana Association of Secondary School Principals

MASSPIRG Massachusetts Public Interest Research Group

MASSR Mari Autonomous Soviet Socialist Republic; Mordovian Autonomous Soviet Socialist Republic

Mass Turn Massachusetts Turnpike

mast masthead; missile automatic supply technique

mast. metastasize

MAST Marine Science and Technology; Metro Arson Strike Team (San Diego): Metropolitan Arson Strike Team; Michigan Alcoholism Screening Test; Military Assistance to Safety and Traffic

MAST Minimum Abbreviations of Serial Titles

MASTARS Mechanical and Structural Testing and Referral Service (NBS)

Ma State New South Wales, Australia

master matching available student time to educational resources; multiple-access shared-time executive routine

MASTIF Multiple Axes Space Test Inertia Facility

mastir microfilmed abstract system for technical information referral

MASUA Mid-America State Universities Association

mat machine-aided translation; material; materiel; maternal; maternity; matins; maturity; microalloy transistor; molankothane (molybdenum disulfide urethane); mutual aid team

mat matemática (Spanish—mathematics); *matematik* (Dano-Norwegian—mathematics)

Mat Matadi; Matanzas; Matthew

MAT Manual Arts Therapy; Mechanical Aptitude Test; Metropolitan Achievement Tests; Military Air Transport; Miller Analogies Test

M.A.T. Master of Arts in Teaching

mata multiple-answering teaching aid

MATA Motorcycle and Allied Trades Association; Museums Association of Tropical Africa

Matagorda Pen Matagorda Peninsula south of Bay City, Texas

Mata Soc Mattachine Society

MA & TB Missile Assembly and Test Building

MATCALS Marine Air Traffic Control and Landing System (USN)

match medium-range antisubmarine torpedo-carrying helicopter

MATCH Manpower and Talent Clearinghouse

MATCOM Materiel Command (USA)

MATCOMEUR Materiel Command, Europe

MATCOMTELNET MATS Command Teletype Network

matcon microwave aerospace terminal control

mate modular avionics test equipment

maté yerba maté

Mate the Mate (Chief Officer)

Mat.E. Materials Engineer

Ma Tec Maintenance Technician

MATELO Maritime Air Telecommunications Organization

matern maternal; maternity

Mater Res Bull Materials Research Bulletin(s)

MATFA Meat and Allied Trades Federation of Australia

math mathematical(ly); mathematician; mathematics

Math Mathematics; Matthew; Matthews; Mathewson; Mathias; Mathieu; Mathilde; Mathurin; Mathys; Mattias

Math.D. Doctor of Mathematics

M.A. Theol. Master of Arts in Theology

mathn mathematician

maths mathematicians, mathematics; mathematics majors

math soc science mathematical social science

MATI Moskovskiy Aviatsionnyy Teknologicheskiy Institut (Russian—Moscow Aviation Technology Institute)

MATIC Multiple and Technical Information Center; Multi-Strategy Authoring Toolkit for Intelligent Courseware

Mat Lab Material Laboratory

mat.med. materia medica

matnav mathematics for navigators

matp masking template

MATP Military Assistance Training Program

matr. matrimonium (Latin—marriage)

MATRESS Money, Advancement, Training, Recreation, Security, Satisfaction (USAF recruiting acronym)

matric matriculate; matriculation

mats maintenance analysis test set

MATS Military Air Transport Service

Mat Soc Mattachine Society

Matt Matthew; Matthewtown, Great Inagua; The Gospel according to St Matthew

Matt Matthew, Book of

mattergy matter + energy

MATTS Multiple Airborne Target Trajectory System

Mattw Matthew

Matty Matthew

matut. matutinus (Latin—in the morning)

matv master antenna television

MATVS Master Antenna Television System

matw metal awning-type window

MATZ Military Air Traffic Zone

mau maintenance analysis unit; marine amphibious unit; multistation access unit

Mau Mauritius

MAU Marine Amphibious Unit

Maud Mathilda

Maude Morse automatic decoder

M.Au.F. Master of Automotive Engineering

maulex marine amphibious unit landing exercise

MAUM Movement Against Uranium Mining

Maur Mauritius

M.A. Urb. Plan. Master of Arts in Urban Planning

Maurit Mauritania (Islamic Republic of)

Mauritania Islamic Republic of Mauritania (West African nation)

MAUS Master of Arts in Urban Studies; Metric Association of the United States

mauw maximum all-up weight

mav manpower authorization voucher; maverick

mav (MAV) maleic anhydride value

mav maven (Yiddish—expert)

maverick. manufacturers assistance in verifying identification in cataloging

MAVIA Mothers Against Violence in America

mavica magnetic video camera

MAVIS Medical Audio-Visual Information Service

MAVOGA Malaysian Vocational Guidance Association

maw medium assault weapon

maw met andere woorden (Dutch—in other words)

Maw Mama

MAW Marine Aircraft Wing; Mission Adaptive Wing

mawb master air waybill

mawec maritime exercise weather code

MAWLOGS Models of the Army Worldwide Logistics System (USA)

MAWS Marine Air Warning Squadron

max maximal; maximum

m'ax (American contraction—my ax)

Max Maxene; Maxie; Maxim; Maxime; Maximilian; Maximiliano; Maxine; Maxwell; Maxy; Soviet Yakovlev trainer aircraft designated Yak-18

max cap. maximum capacity

maxi maximum

maxibop maxibopper (underground slang—fatter or older woman wearing miniskirts)

maxid maximize indefinite delivery (contracts)

maxill maxilla; maxillary

maximin maximum + minimum

maxis maximum-length garments (coats, skirts, etc.)

maxnet modular application executive for computer networks

max q maximum aerodynamic pressure per square foot

maxr maximum room rate desired

max trq maximum torque

May Maybelle; NATO nickname for Soviet Ilyushin transport designated Il-38

MAYA Maya Airways (British Honduras); Mexican-American Youth Association

mayday international distress call (from the French *m'aidez*—help me)

May Day May 1 (Morris Dancers in England, international worker's day in communist and socialist lands)

Mayjun May and June

May^mo Mayordomo (Spanish—butler, estate manager, steward)

mayn't may not

mayo mayonnaise

MAYO Mexican American Youth Organization

mayoralection mayoral election

maz mazda

Maz Mazatlan

mazh missile azimuth heading

mb machine blended; magnetic bearing; main battery; master bundles; may be; medium bomber; megabyte(s)—two million bytes; megabyte; meter band; methyl bromide; methylene blue; midband; midbody; millibar(s); minimum bid; motor barge; motorboat; myoglobin

mb (MB) macrobiotic; megabyte; memory buffer

m/b make-break; master batch

m & b matched and beaded; metes and bounds

m.b. misce bene (Latin—mix well)

mB male Black

m/B male Black

Mb megabar; million bytes; myoglobin

MB magnetic bearing; Manitoba; March-Bender (factor); Marine Barracks; Marine Base; Maritime Board; Marketing Board; Mechanized Battalion; Medical Board; Meridian & Bigbee (railroad); Miami Beach; middle of bow; municipal borough; Munititons Board; Music for the Blind; Myrtle Beach; Sir Mackenzie Bowell (Canada's sixth Prime Minister)

M-B Mercedes-Benz

M.B. Medicinae Baccalaureus (Latin—Bachelor of Medicine)

M/B Master Barber

M & B metes and bounds

mba multiple-beam antenna

Mba Mombasa

MBA Make or Buy Analysis; Make or Buy Authorization; Marine Biological Associa-

tion; Master Builders Association; Men's Basketball Association; Military Bases Agreement; Military Benefit Association; Monterey Bay Aquarium; Monument Builders of America; Mortgage Bankers of America; Mortgage Bankers Association; Mortgage Brokers Association

M.B.A. Master of Business Administration

MBAA Master Brewers Association of America; Mortgage Bankers Association of America

MBAC Member of the British Association of Chemists

M-Bahn Magnetbahn (German—magnetic levitation transit system)—train cars travel on a thin cushion of air providing high-speed transportation quieter than wheel-on-rail systems

MBAL Master Bookbinders' Alliance of London

m bale 1000 bales

mbar millibar

MBAR Multiple Beam Acquisition Radar

MBARI Monterey Bay Aquarium Research Institute

MBAUK Marine Biological Association of the United Kingdom

MBAWS Marine Base Warning System

mbb make before break; mortgage-backed bonds

MBB Messerschmitt-Büllkow-Blom; Museum of the Borough of Brooklyn

m bbl 1000 barrels

mbc maximum breathing capacity; multiple burst correction

MBC Malawi Broadcasting Corporation; Malaysian Building Construction; Mauritius Broadcasting Corporation; Mercantile Bank of Canada; Metropolitan Borough Council; Metropolitan Business College(s); Miname Nipon Broadcasting (Japan); Mitsubishi Bank of California

MBCA Motor Boat Club of Amenca

MBCC Massachusetts Bay Community College; Migratory Bird Conservation Commission

MBCMC Milk Bottle Crate Manufacturers Council

mbd macro-block design; management by decision; million barrels per day; minimal brain dysfunction; minimum brain damage

MBDA Minority Business Development Agency

m bd ft 1000 board feet

mbe minority business enterprise; missile-borne equipment; my own bloody efforts

MBE Mail Boxes Etc.

M.B.E. Member of the Order of the British Empire

M. B. Ed. Master of Business Education

MBELDEF Minority Business Enterprise Legal Defense and Education Fund

Mbf thousand board feet

MBF Master Builders Federation; Medical Benefits Fund(ing); Military Banking Facility; Milk Bottlers Federation

M-B factor Marsh-Bender factor

MBF et H Magna Britannia, Francia et Hibernia (Latin—Great Britain, France, and Ireland)

MBFR Mutual and Balanced Forced Reduction

MBG Midland Bank Group; Missouri Botanical Garden

mbge missileborne guidance equipment

mbh manual bomb hoist

mbH mit beschränkter Haftung (German—limited liability)

mbi may be issued

mbi (MBI) magnetic resonance infarction

Mbi Mbini (formerly Rio Muni)

MBI Medical Biology Institute; Molecular Biosystems Inc

MBIA Malting Barley Improvement Association; Municipal Bond Investors Assurance

M. Bi. Chem. Master of Biological Chemistry

M. Bi. Eng. Master of Biological Engineering

MBII Minority Business Information Institute

M. Bi. Phy. Master of Biological Physics

M. Bi. S. Master of Biological Science

MBJ Montego Bay, Jamaica (airport)

mbk missing, believed killed

mbl missile baseline; mobile; mobile branch library; model breakdown list(ing); model breastline

m & bl meat and bonemeal

Mbl Monatsblatt (German—monthly report)

MBL Marine Biological Laboratory (Woods Hole, Massachusetts); Mobile, Alabama (airport)

MBLIC Mutual Benefit Life Insurance Company

mbm thousand feet board measure

mbm (MBM) magnetic bubble memory

MBM Mac Bride Museum

M.B.M. Master of Business Management

MBMA Master Boiler Makers' Association; Metal Building Manufacturers Association

MBMRC Malcolm Bliss Mental Health Center (St Louis)

MBMS Magnetic Bubble Memory System

m.bn. marriage banns

MBNA Monument Builders of North America

MBNBR Mount Bruce Native Bird Reserve (North Island, New Zealand)

mbo management by objectives; management buyout; monostable blocking oscillator

Mbo Maracaibo (inhabitants called Maracaiberos or Maracuchos)

MBO Mutual Benefit Organization

MbO₂ oxymyoglobin

MBOC Minority Business Opportunity Committee(s)

MBOU Member British Ornithologists Union

mbp mean blood pressure

MBP Marine Biotelemetry Project; Minnesota Business Partnership

MBPA Metropolitan Bicycle Polo Association; Military Blood Program Agency

mbp antigen melitensis bovin porcine antigen

mbpfo make best possible firm offer

MBPO Military Blood Program Office(r)

MB & PR MacMillan, Bloedel & Powell River

mbps megabits per second; million bits per second

MBps megabytes per second

MBPXL Missouri Beef Packers Express Line

mbr member; memory buffer register

MBR Minerações Brasileiras Reunidas (Brazil—Brazilian Mining Reunited)

MBRBA Motor Body Repairers and Builders Association

mbr/e memory buffer register—even

M Bret Middle Breton

MBRF Mission Bay Research Foundation

mb/l million barrels

mbr/o memory buffer register—odd

Mbro Middlesbrough

mbrt methylene-blue reduction time

mbruu may be retained until unserviceable

mbrv maneuverable ballistic reentry vehicle (MBRV)

mbs magnetron beam switching; main bang suppressor; megabits per second; mortgage-backed security

mb's milk brothers (two or more males who have had sexual relations with the same female)—*see* ms's

MBS Macquarie Broadcasting Service; Mainichi Broadcasting System; Merry Birthday Syndrome; Miami Beach Symphony; Motor Bus Society; Music Broadcasting Society; Mutual Broadcasting System

MBSA Modular Building Standards Association; Munitions Board Standards Agency

MBSC Modular Building Systems Council

M. B. Sc. Master of Business Science

mbsd multi-barrel smoke discharger

mbsi missile battery status indicator

MBSI Musical Box Society International

MBSJC Metropolitan Boroughs Standing Joint Committee (of librarians)

MBSM Mexican Border Service Medal

MBSSM Maxfield-Buchholz Scale of Social Maturity

mbt main ballast tank; main boundary thrust (geology); mean body temperature; mechanical bathythermograph; metal-base transistor; murder before treason

MBT Minimum Blood Test; Modified Boiling Test

MBT-70 Main Battle Tank

MBTA Massachusetts Bay Transportation Authority; Metropolitan Boston Transit Authority; Midwest Book Travelers Association; Migratory Bird Treaty Act

MBTI Manpower Business Training Institute; Myers-Briggs Type Indicator

MBTs main battle tanks; Municipal Bond Trusts

MBTS Meteorological Balloon Tracking System

MBtu thousand British thermal units

mbu mobile tracking unit; thousand bushels

MBUCV Museo de Biología de la Universidad Central de Venezuela (Spanish—Biology Museum of the Central University of Venezuela)

M Build Master of Building

M. Bus. Ed. Master of Business Education

MBV Mexican Border Veterans

MBW Metropolitan Board of Works

MBYC Manhasset Bay Yacht Club; Mission Bay Yacht Club

mbz must be zero

mc main color; marginal check megacycle(s); marked capacity; marker card; market(ing) capacity; marriage case; married couple; material control; message composer; metal case; metal clad; meter candle; metric carat; miles on course; military characteristic(s); millicurie(s); mission control; moisture content; momentary contact; monkey cells; motorcycle; motor(ized) contact; mourning covers (envelopes); moving coil(s); multiple contact

mc (MC) magnetic center; magnetic course; Maltese cross; marginal cost

m-c medico-chirugical (surgical); mineralo-corticoid (hormones)

m/c machine(ry); marginal credit; metalling clause; middle center

m & c manufacturers and contractors; morphine and cocaine

mc mois courant (French—current month)

m/c mi cargo (Spanish—my debt, my responsibility); *mi casa* (Spanish—my home, my house); *mi cuenta* (Spanish—my account)

μC microcoulomb

Mc Mac (Gaelic—son of)

M/c metallic currency

MC Macalester College; Machinery Certificate; Madison College; Madonna College; Magistrates Court; magnetic course; Mailet College; Maine Central (railroad); Malin College; Malone College; Manatee College; Manchester College; Manhattan College; Manhattanville College; Manpower Commission; Maria College; Marian College; Marietta College; Marine Corps; Marion College; Marist College; Maritime Carrier(s); Maritime Commission; Marlboro College; Marriage Certificate; Martin College; Mary College; Marycrest College; Maryglade College; Marygrove College; Maryhurst College; Marymount College; Maryville College; Marywood College; Master of Ceremonies; Master Commandant; Material Center; Materiel Center; Materiel Command; Maunaolu College; Medical Center; Medical Certificate; Medical College; Medical Corporation; Medical Corps; Member of Congress; Member of Council; Memorial Commission; Memphis College; Menlo College; Mesa College; Mess Committee; Michigan Central (railroad); Microfilm Corporation; Microstat Corporation; mid-capitalization core;

Middlebury College; Midland College; Miles College; Military Committee; Military Cross; Milligan College; Mills College; Milsaps College; Milton College; Misericordia College; Mission Computer; Mitchell College; Mitsubishi Corporation; Monaco (Internet code); Monmouth College; Monticello College; Moravian College; Morehouse College; Morris College; Morse Code; Morse College; Muhlenberg College; Multnomah College; Mundelein College; Munitions Command; Muskingum College; Muskogee College

M-C Magovern-Cromie (prosthesis)

M.C. Military Cross

M/C Machinery Certificate; Master Card

MC Maison Centrale (French—Central Prison); *Mercado Común* (Spanish—Common Market)

M.C. *Magister Chirurgiae* (Master of Surgery)

mca minimum control airspeed; minimum crossing altitude; monetary compensation amounts; monetary compensatory account(ing); money compensatory account(ing); mud cleaning agent

mca (MCA) maximum credible accident

Mca Macassar

MCA Malacca Consumers Association; Malayan Chinese Association; Malayan Chinese Association; Manufacturing Chemists Association; Maritime Central Airways; Maritime Control Area; Massachusetts Correctional Association; Material Control Area; Material Coordinating Agency; Maternity Center Association; Mechanical Contractors Association; Mechanization Control Area; Mecro-Channel Architecture; Media Credit Association; Medical Correctional Association; Medical Council of Australia; Metal Construction Association; Michigan Council for the Arts; Micro Channel Architecture; Millinery

Credit Association; Minnesota Correctional Authority; Minnesota Corrections Association; Missouri Corrections Association; Movers Conferences of America; Multiple Classification Analysis; Muscat Control Agency; Music Corporation of America; Music Critics Association; Musicians Club of America

MCA Metric Conversion Act

MCAA Mason Contractors Association of America; Mechanical Contractors Association of America; Medical Consumers Association of Australia; Military Civil Affairs Administration

MCAAA Midland Counties Amateur Athletic Association

MCAB Marine Corps Air Base

MCACE Measurement Characterization and Control of Ambulatory Care in Europe

MCAD Military Contracts Administration Department

MCADO Micronesian Community Action Development Organization

MCAF Marine Corps Air Facility; Marine Corps Air Field; Military Construction, Air Force

MCAIR McDonnell Aircraft Company

m car 1,000 carats

MCAR Material Corrective Action Report(s)

mcarquals marine-carrier qualifications

mca's money compensatory amounts

MCAS Marine Corps Air Station

MCAT Medical College Admission Test; Midwest Council on Airborne Television

M-CATS Municipal Certificates of Accrual on Tax-Exempt Securities

m. cau. misce caute (Latin—mix cautiously)

MCAUSA Military Chaplain's Association of the U.S.A.

MCAUTO McDonnell-Douglas Automation

mcb membranes cytoplasmic bodies; miniature circuit breaker

McB McBurney's (point)

MCB Marine Corps Base; Metric Conversion Board; Metric Conversion Bureau; Mobile Construction Battalion

M.C.B. Master of Clinical Biochemistry

MCBA Master Car Builders' Association

mcbf mean cycles between failures

mcc maintenance of close contact; midcourse correction; modified close control; multilayer ceramic chip

mcc (MCC) main communication(s) center; multi-component circuit(s)

MCC Maintenance Control Center; Manual Combat Center; Marine Corps Commandant; Marine Corps Commander; Marylebone Cricket Club; Massachusetts Council of Churches; Materials Characterization Center (Battelle, Richland); Mennonite Central Committee; Mesta Machine Company (stock exchange symbol); Meteorological Communications Center; Metropolitan Community Church; Metropolitan Correctional Center; Microelectronics and Computer Technology Corporation, Microfilm Card Catalog; Mira Costa College; Missile Control Center; Mission Control Center; Monroe Community College; Mortgage Credit Certificate; Munitions Carriers Conference; Music Critics Circle

MCCA Michigan Community College Association; Minor Counties Cricket Association; Mortgage Capital Company of America

MCCA *Mercado Común Centro Americano* (Spanish—Central American Common Market)

MCCB Multinational Configuration Control Board

MCCC Metropolitan Correctional Center (Chicago); Muskegon County Community College

MCCCA Marine Corps Combat Correspondents Association

MCC-H Mission Control Center—Houston (NASA)

MCCISWG Military Command, Control, and Information Systems Working Group

McCL McCabe Library (Swarthmore)

mccp maintenance console control panel

mccs missile critical circuit simulator

MCCs Metropolitan Correctional Centers (of the Bureau of Prisons in Chicago, New York, and San Diego); Military Committee in Chiefs-of-Staff Session(s)

MCCS Medco Containment Services Incorporated; Modular Communications Control Systems

mccu multiple communications control unit

MCCW Miami Citizens Crime Watch; Military Council of Catholic Women

MCAWA McCaw Cellular Communications Incorporated

mcd magnetic crack detector; mean corpuscular diameter; median control death; metal-covered door; mine clearance dive; mine clearance diving; minimum cash deposit; miscellaneous cash deposit

mcd *mininio comune denomiatore* (Italian—least common denominator)

Mc D Mc Donald; Mc Donald's

McDD McDonnell Douglas

MCD Melbourne College of Divinity; Municipal Civil District

M.C.D. Doctor of Comparative Medicine; Master of Civic Design

McDA McDonnell Aircraft

MCDA Manpower and Career Development Agency; Motor Car Dealers Association

McDAC McDonnell Aircraft Corporation

MCDC Montgomery County Detention Center

McD Obs McDonald Observatory

MCDS Management Control Data System

mcd/slv minimum-cost-design/space launch vehicle (MCD/ SLV)

mcdt mean corrective down time

mce mean chance expectation; military characteristics equipment

MCE Memphis Cotton Exchange; Montgomery Cotton Exchange

M.C.E. Malaysian Certificate of Education; Master of Civil Engineering

MCE *Mercado Común Europeo* (Spanish—European Common Market); *Mercato Comune Europeo* (Italian—European Common Market)

MCEB Military Communications Electronics Board

M cell magnocellular (system)

M.C.Eng Master of Civil Engineering

MCEO Malayan Council of Employers Organization; Malaysian Council of Employers Organization

M. Cer. E. Master of Ceramic Engincermng

MCET Mississippi Center for Educational Television

MCEWG Multinational Communication-Electronics Working Group (NATO)

mcf magnetic card file; medium corpuscular fragility; military computer family; thousand cubic feet

MCF Master Code File;, Michigan Colleges Foundation; Missouri Colleges Fund

mcfd 1000 cubic feet of gas per day

mcfh 1000 cubic feet of gas per hour

MCFI Malaysian Chamber of Film Industries

mcfim microfilm(ing)

mcflm microfilm; microfilming

mcfm 1000 cubic feet of gas per month

MCFP Medical Center for Federal Prisoners (Springfield, Missouri); Member of the College of Family Physicians

mcfshe microfische

mcg magnetocardiogram; microgram

mc & g mapping, charting, and geodesy

MCG Mandalay Coral Gardens (Queensland)

MCGA Multi Color Graphics Array

McG-H McGraw-Hill

McGill-Queens U Pr McGill-Queens University Press

McGraw McGraw-Hill

MCGS Microwave Command Guidance System

McG U McGill University

McGUL McGill University Library

mch mail chute; mean corpuscular hemoglobin (MCH)

Mch Manchester; March

M. Ch. *Magister Chirurgiae* (Latin—Master of Surgery)

MCH Maternal and Child Health

mcha merchandise

mchan multichannel

mchc mean corpuscular hemoglobin concentration

M.Ch.D. *Magister Chirurgiae Dentalis* (Latin—Master of Dental Surgery)

M.Ch.E. Master of Chemical Engineering

M. Chem. E. Master of Chemical Engineering

M.Chir. *Magister Chirugiae* (Latin—Master of Surgery)

M.Ch. Orth. *Magister Chirurgiae Orthopaedicae* (Latin—Master of Orthopedic Surgery)

M.Ch.Otol. Master of Otorhinolaryngological Surgery

MCHP Maternal and Child Health Program

mc hr millicurie hour(s)

MCHR Medical Committee for Human Rights

MCHRD Mayor's Committee for Human Resources Development

M.Chrom. Master of Chromatics

M Ch S Member of the Society of Chiropodists

MCHS Maternal and Child Health Service

mcht merchant

Mchter Manchester

M. Chur. *Magister Chirurgiae* (Latin—Master of Surgery)

mchy machinery

mci malleable cast iron; megacurie; mild cognitive impairment; mottled cast iron; multichip integration

mCi millicurie(s)

MCI Kansas City Airport (symbol); Marine Corps Institute; Maximum Credible Incident; Massachusetts Correctional Institution (Framingham);

Metropolitan Communications Incorporated; Mexican Coffee Institute; Microwave Communications, Inc; Milk Can Institute; Motor Coach Industries; Motor Coach Institute

MCIC MCI Communications Corporation: Metals and Ceramics Information Center (DoD)

mcid multipurpose concealed intrusion detection (device)

MCIE Midland Counties Institution of Engineers

McINP McIwaine National Park (Rhodesia)

MCIS Management Control and Information System; Multi-Currency Intervention System

MCIT Member of the Chartered Institute of Transport

M.C.J. Master of Comparative Jurisprudence

MCJC Mason City Junior College

MCJCC Mayor's Criminal Justice Coordinating Council (New York City)

McKay David McKay

McKinley Mount McKinley or Mount McKinley National Park in Alaska between Anchorage and Fairbanks, containing North America's highest mountain named for President William McKinley

McKnight McKnight Publishing Co

McKS (Sir Colin) McKenzie Sanctuary (Victoria, Australia)

McKVHS McKee Vocational High School

mcl macro-creation language; maximum contaminant level; midclavicular line; midcostal line; most comfortable level

Mc L Marshall McLuhan

MCL Manchester Central Library; Marine Corps League; Master Configuration List(ing); Master Control Log; Metal Control Laboratories; Metropolitan Central Library; Mid-Canada Line (radar warning fenceline); Moore-McCormack Lines; Mushroom Canners League

M.C.L. Master of Civil Law

MCLA Marine Corps League Auxiliary

McLaughlin Youth McLaughlin Youth Center (for delinquents) at Anchorage, Alaska

McLg McLaughlin Group

mclg maximum contaminant level goal

MCLI Meiklejohn Civil Liberties Institute

M.Clin.Psychol. Master of Clinical Psychology

mcll missile compartment, lower level

MCLO Medical Construction Liaison Office

M.Cl.Sc. Master of Clinical Science

mcm military characteristics motor vehicles; mine countermeasures; minimum commitment method(ology); missile-carrying missile; multiple chip modules; thousand circular mils

mcm (MCM) missile-control module

mcm *minimo comune multiple* (Italian—least common multiple, lowest common multiple)

McM McMahon; McManus; McMaster; McMillan; Mcmurry

MCM Manual for Courts-Martial; Marine Corps Manual; Mine Countermeasure (minesweeper); Monte Carlo Method

MCMA Machine Chain Manufacturers Association; Marine Corps Memorial Commission; Metal Cookware Manufacturers Association

MCMC Marine Corps Memorial Commission

MCMF Marie Curie Memorial Foundation

MCMI Millon Clinical Multiaxial Inventory

mcml missile compartment, middle level

mcmops mine countermeasures operations

MCMS Marin County Medical Society

MCMSP Material Command Maintenance Service Publication

MCM & T Michigan College of Mining & Technology

McM U McMaster University

McMUL McMaster University Library

McMUMC McMaster University Medical Centre (Hamilton)

MCN Management Control Number; Manual Control Number; Master Control Number(ing); Museum Computer Network

MCN *Maternal Child Nursing* (Journal)

McNally McNally & Loftin

McNeil Island U.S. Penitentiary at McNeil Island, Washington

mcng meaconing

MCNLP *Museo de Ciencias Naturales de La Plata* (Spanish—La Plata Natural Sciences Museum)—La Plata, Argentina

MCNP Mammoth Cave National Park (Kentucky); Mount Cook NP (South Island, New Zealand)

MCNY Museum of the City of New York

MCNZ Medical Council of New Zealand

mco main civilian occupation; managed care organization; mills culls out; miscellaneous charges order

mco *março* (Portuguese—March)

Mco Morocco

MCO Materials Characterization Organization; Michigan Corrections Organization (prison wardens); Movement Control Office(r)

MCOAG Marine Corps Operations Analysis Group

MCODA Motor Cab Owner Drivers' Association

MCOGA Mid-Continent Oil and Gas Association

mcol musicological; musicologist; musicology

M.Com. Master of Commerce; Minister of Commerce

MCOM Mobility Command (US Army)

M. Com. Adm. Master of Commercial Administration

M.Comm.H. Master of Community Health

M. Comp. Law Master of Comparative Law

M. Com. Sc. Master of Commercial Science

MCON Military Construction—Navy

MCOO Monte Carlo Opera Orchestra

MCOP Marine Corps Ordnance Publication

mcos *marcos* (Spanish—marks), German coins

MCOW Medical College of Wisconsin

mcp main control panel; male chauvinist pig; manual control panel; mode control panel; multi-component plasma; multiple chip package

mcp (MCP) master control program; metacarpophalangeal (joint)

mCp my Cadillac payment

MCP Maine Coastal Program; Malaysian Communist Party; Management Control Plan; Maritime Company of Philadelphia; Maritime Company of the Philippines; Massachusetts College of Pharmacy; Master Control Program; Military Construction Program; Minerals and Chemicals Philipp; Model Cities Program

M.C.P. Master of City Planning

MCPA Member of the College of Pathologists of Australia

MC Path Member of the College of Pathologists

MCPC Multi-Cultural Primary Care

mcph metacarpal-phalangeal

MCPO Master Chief Petty Officer

mcps megacycles per second

MCPS Mechanical Copyright Protection Society; Member of the College of Physicians and Surgeons

MCPT Maritime Central Planning Team

McQ-E McQuaid-Ehn (grain size)

mcr master control routine; metabolic clearance rate; micrographic computer retrieval; military compact reactor; modular circuit reliability; monitor control routine; mother-child relationship

MCR Manufacturing Change Request; Marine Corps Reserve; Mass Communications Research; Master Change

Record; Master Charge Record; Mobile Control Room

M.C.R. Master of Comparative Religion

MCRA Member of the College of Radiologists of Australasia

MCRC Mass Communications Research Center (University of Wisconsin)

MCRD Marine Corps Recruit Depot

MCRE Mother-Child Relationship Evaluation

MCREL Mid-Continent Regional Educational Laboratory

mcrfsch microfische

MCRHS Mid-Continent Railway Historical Society

MCRL Master Cross-Reference List(ing)

MCRML Midcontinental Regional Medical Library (University of Nebraska)

MCROA Marine Corps Reserve Officers Association

MCRS Micrographic Computer Retrieval System; Military Command Research System

mcrt multichannel rotary transformer

mcrwv microwave

mcs meridian control signal; meter-candle second; missile checkout set; motor circuit switch; multiple chemical sensitivity; multiple column selector

mc/s megacycles per second

MCs Military Characteristics

MCS coastal minesweeper (naval symbol); Maintenance Control Section; Major Component Schedule; Management Computing Services; Management Control System; Marine Cooks and Stewards (union); Marine Corps School; Marine Corps Station; Master of Computer Science; Message Control System; Military College of Science; mine counter-measures support ship (naval symbol); Missile Commit Sequence; Mobile Checkout Station; Mobile Coastal Service; Multiple Chemical Sensitivity

M.C.S. Master of Commercial Science

MCSA Marble Collectors Society of America; Medical Computer Services Administration; MultiChannel Spectrum Analyzer

MCSB Motor Carriers Service Bureau

MCSC Medical College of South Carolina; Military College of South Carolina (The Citadel); Monell Chemical Senses Center

mc/sec megacycles per second

MCSH Manhattan College of the Sacred Heart

MCSL Marine Corps Stock List; Marine Corps Supply List

MCSP Member of the Chartered Society of Physiotherapy

mc spec motorcycle specialist(s); motorcycle specifiication(s)

mcss multiple chemical sensitivity syndrome

MCSs Memorial Cremation Societies

mcst magnetic card selectric typewriter

MCST Member of the College of Speech Therapists

MC S & T Manchester College of Science and Technology

MCSTB Motor Carriers Service Tariff Bureau

MCSW Mining Club of the Southwest

MCSWG Multinational Command Systems Working Group

mct maximum continuous thrust; mean corrective time; medium-chain triglycerides; multiple-compressed tablet

mct (MCT) modular computing typewriter

MCT Maritime Crew Trainer; Master Cycle Trader; Mechanical Comprehension Test; Minimum-Competency Test(ing)

m/cta mi cuenta (Spanish—my account)

MCTA Metropolitan Commuter Transportation Authority; Motor Carriers Traffic Association

MCTC Maritime Cargo Transportation Conference; Microelectronics and Computer Technology Corporatiion (Austin, Texas)

mctd mixed connective tissue disease

MCTE Michigan Council of Teachers of English

MCTI Metal Cutting Tool Institute

mctow maximum certificated takeoff weight

mctp missile control test panel

MC & TS Monotype Casters' and Typefounders' Society

MCTSSA Marine Corps Tactical System Support Activity

mcu mechanical condition unknown; median control unit; medium closeup; microprocessor control unit; monitor control unit

m & cu monitor and control unit

MCU Modern Churchmen's Union

MCUG Military Computer Users Group

mcul missile compartment, upper level

mcv mean corpuscular volume; model control volume

McV myelocytomatosis virus

MCV Medical College of Virginia

mcvf multichannel voice frequency

mcw metal casement window; modulated continuous wave

m & cw maternity and child welfare

MCW Mallinkrodt Chemical Works

m cwt 1000 hundredweight

mcx maximum-cost expediting

mcy machinery

MCZ Military Control Zone (Korea); Museum of Comparative Zoology

md main deck; malicious damage, manufacturing day; map distance; mathematics disabled; maximum demand; maximum design; mean deviation; memorandum of deposit; mental(ly) defective; mentally deficient; mentally disordered; message dropping; milliard; minor defect(s); minute difference(s);

mitral disease; month's date; motorized damper; movement directive; muscular dystrophy

m-d manic-depressive

m/d market day; memorandum of deposit(s); messages per day; missile driver; modulator-demodulator; month(s) after date

m & d medicine and duty

md. married

md main droite (French—right hand); *mano derecha* (Spanish—right hand); *mano destra* (Italian—right hand); *marchand* (French—good value, marketable); *milliard* (French—1000 million)

m d mano destra (Italian—right hand)

Md Maid; Maryland; Marylander; mendelevium; Muhammed

M$ Malaysia dollar (Singapore dollar)

MD Management Directive; Managing Director; Marine Detachment; Maryland; Material Division; Medical Department; Medical Discharge; Mess Deck; Meteorological Department; Middle District (court); Middle Dutch; Military District; Mine Depot; Moldova (Internet code); Music Director; Musical Director

M.D. *Medicinae Doctor* (Latin—Doctor of Medicine)

M D mano destra (Italian—right hand)

mda maintenance depot assistance; minimum detectable activity; monochrome display adaptor; multiple-docking adapter

mda (MDA) methyldiamphetamine (stimulant); minimum descent altitude

Mda Mérida (inhabitants—Meridanos)

MDA Marking Device Association; Master Dyes Association; Material Department Amendments; Material Disposal Authority; Michigan Dance Association; Middle Depot Activity; Minor Deviation Authorization; Multiple-Docking Adapter; Mural Decorators Association; Muscular Dystrophy Association; Music Distributors Association; Mutual Defense Agency; Mutual Defense Assistance

MDAA Muscular Dystrophy Association of America; Mutual Defense Assistance Act

MDAC McDonnell Douglas Astronautics Company; Mutual Defense Assistance—China area

MDAGT Mutual Defense Assistance, Greece and Turkey

MDAIKP Mutual Defense Assistance, Iran, Republic of Korea, and the Philippines

MDAN Material Department Administrative Notices

MDANAA Mutual Defense Assistance, North Atlantic Area

MDAP Mutual Defense Assistance Program

MDAPT Machover Draw-a-Person Test

MDAs Medical Doctor Anesthesiologists

MDAS Medical Data Acquisition System; Multispectral Data Analysis System

Mda Vle Maida Vale

M-day manufacturing day; mobilization day; moratorium day

mdb miniature dwarf bearded (iris); multilateral development bank(ing)

Mdb Middlesbrough

MdB Mitglied des Bundestages (German—member of the Bundestag)

MDB Movimiento Democrático Brasileiro (Portuguese—Brazilian—Democratic Movement)—political party

MDBVHS Mabel D Bacon Vocational High School

mdc maintenance data collection; more developed country

M d C Maestro di Cappella (Italian—Chapel Master); *Maître de Chapelle* (French—Chapel Master)—titles often meaning conductor or musical director

MDC Manhattan Drug Corporation; Manufacturing Development Council; McDonnell Douglas Corporation; Metropolitan District Commission; Minnesota Department of Corrections; Moncure Daniel Conway; Movement for Democratic Change

M-D C Main-Danube Canal

MDCA Master Diamond Cutters Association

MDC-W McDonnell Douglas Corporation—West

mdd major depressive disorder; mechanical dough development; medical devices division; milligrams per square decimeter per day

MDDC Military Dependents Dental Clinic

mddpm magnetic-drum dataprocessing machine(ry)

Mddx Middlesex

mde matrix difference equation

M.D.E. Master of Domestic Economy

MDE Modern Drug Encyclopedia

m'dear my dear

M de C Maître de Chapelle (French—conductor)

M.Dent.Sci. Master of Dental Science

M.Des. Master of Design

mdf mild detonating fuze; minimum diversion fuel; myocardial depressant factor

mdf (MDF) main distributing frame (data processing); manual direction finder

MDF Manitoba Development Fund; Modderfontein Dynamite Factory

MDFC McDonnell Douglas Finance Corporation

mdfd modified

mdfg modifying

mdfn modification

mdfy modify

MDG Medical Director-General

mdh minimum descent height; multidirectional harassment

mdh (MDR) malate dehydrogenase

MDHB Mersey Docks and Harbour Board (Liverpool)

Md Hist Maryland Historical Society

Mdhv Marek's disease herpes virus

mdi magnetic direction indicator; major depressive illness; manic depressive illness

MDI Material Department(al) Instruction; Material Development(al) Instruction

MDIA Material Department(al) Instruction Amendment(s)

MDIB Material Departmental Instruction Bulletin; Minimum Distribution Incidental Benefit

m. dict. more dictu (Latin—in the manner directed)

M.Did. Master of Didactics

M.Di.Eng. Master of Diesel Engineering

MDIG Multipurpose Display Indicator Group

M.Dip. Master of Diplomacy

M dis Marek's disease

mdise merchandise

M.Div. Master of Divinity

MDJC Miami-Dade Junior College; Mississippi Delta Junior College

m dk main deck

mdl master design layout; middle; model; modular design language

Mdl Middle (postal abbreviation)

M d L Mitglied des Landtages (German—member of the Landtag)

MDL Method Detection Limits; Mine Defense Laboratory

MDL Master Drug List

Mdlle Mademoiselle (French—Miss)

mdm medium-depth mine; multiplexer/demultiplexer

mdm (MDM) middiastolic murmur

Mdm Madam

MDM Mass Democratic Movement (South African); Material Division Manual; Movement (for a) Democratic Military

mdma (MDMA) methylenedioxy-methamphetamine (also known as Adam or Ecstasy)

Mdme Madame (French—Missus)

mdn median

m$n moneda (pesos) nacional [Spanish—national monetary unit(s)—Argentinian peso(s)]

mDNA mitochondrial deoxyribonucleic acid

MDNA Machinery Dealers National Association

mdnb metadinitrobenzene

mdngt midnight

MDNS Modified Decimal Numbering System

mdnt midnight

mdo medium-density overload(ing); monthly debit ordinary

MDO MARC Development Office (Library of Congress)

M-dog mine dog (trained to find buried mines)

mdp minimum-distance principle; multi-disc player

m.d.p. *mento-dextra posterior* (Latin—right mento-posterior)

MDP Mainland Data Processing; Manufacturing Development Program

MDPR Manufacturing Development and Process Request

mdr magnetic disk recorder; master-clock generator; memory data register; minimum daily requirement; multichannel data record(er); multi-drug resistance

Mdr Madras MdR Marina del Rey

MdR Marina del Rey

MDR Master Discrepancy Report; Material Deficiency Report(ing)

mdrc maximum distance for radiological consequences

MDRC Manpower Demonstration Research Corporation

MDRSF Multi-Dimensional Random Sea Facility

MDRT Million Dollar Round Table

MDR-TB multi-drug-resistant tuberculosis

mds minimum discernible signal; mission design and series

Mds *Mesdames* (French—Ladies)

M$S peso *(moneda nacional—* Argentine letter symbol)

MDS mail distribution schedule; mail distribution scheme; Main Dressing Station; Manufacturing Data Series; Medical-Dental Service; meteoroid detection satellite; Model Designator Series; Multipoint Distribution Systems (microwave)

M.D.S. Master of Dental Surgery

mdsa multiple disc-sampling apparatus

M.D.Sc. Master of Dental Science

mdse merchandise

MdSEA/DCSEA Maryland/ District of Columbia Society of Enrolled Agents

MDSF Mission to Deep Sea Fishermen

mdsg merchandising

MDSG Mood Disorders Support Group (Incorporated)

MDSI Manufacturing Data Systems Inc

MDSOs mentally disordered sex offenders

MDST Mountain Daylight Saving Time

mdt maggot debridement therapy; mean down time; moderate

m.d.t. *mento-dextra transversa* (Latin—right mento-transverse)

MDT Mobile Data Terminal; Multidisciplinary Team; Mutual Defense Treaty

MDTA Manpower Development and Training Act

mdtm medium duty target mechanism

MDTS Modular Data Transmission System

M Du Middle Dutch

MDU Medical Defence Union; Mine Disposal Unit; Mobile Development Unit

M du N *Magasin du Nord* (Copenhagen's leading department store)

MDUS Medium Data Utilization Station

Mdv Marek's disease virus

M.D.V. Doctor of Veterinary Medicine

mdw mass-destruction weapons; meadow; measured day work

MDW Chicago, Illinois (Midway Airport); Military Defense Works; Military District of Washington; Minnesota, Dakota & Western (railroad)

MDWC Mississippi Department of Wildlife Conservation

Mdws Meadows

Mdx Middlesex

mdy magnetic deflection yoke

MDY Midland Oil (stock exchange symbol)

me male employee; marbled edges; marbled edging; mathematics education; maximum effect; maximum effort; mechanical equipment; medical essay; metabolizable energy; methyl; mill edge; milligram equivalent; miter end; most

excellent; multi-engine; multiple exposure; muzzle energy; myelincephalic encephalitis

me (ME) measles encephalitis

m/e mechanical/electrical; mobility equipment

m & e mechanical and electrical; music and (sound) effects; mortality and expense

m E *meines Erachtens* (German—in my opinion)

Me Maine; Mainers; Mexican(s); Mexico

Me *Maître* (French—Master)— advocate; attorney

ME Mail Early; Maine; Managing Editor; Marine Engineer; Mechanical Engineer(ing); Medical Examiner; Methodist Episcopal; Middle East(ern)(er); Middle English; Military Engineer; Military Expenditures; Mining Engineer; Montreal Exchange; Morristown and Erie (railroad); myalgic encephalomyelitis

ME *Mouvement Européen* (French—European Movement)

M.E. Master of Education; Mechanical Engineer

mea measure(s); measuring; minimum enroute altitude; monoethanolamine (MEA)

MEA Maintenance Engineering Analysis; Malaysian Economics Association; Maritime Employers Association; Medical Exhibitors Association; Member of the European Assembly; Michigan Education Association; Middle East Airlines; Minnesota Education Association; monoethanolamine; Mobile Electronics Association; Montana Education Association; Municipal Employees Association; Music Educators Association; Musical Educators Association

M.E.A. Master of Engineering Administration

MEA *Municipalidades En Acción* (Spanish—Municipalities in Action)

MEAF Middle East Air Force

meal master equipment allowance

MEAL Master Equipment Authorization List(ing)

mean max mean maxmmum; mean maximum temperature

MEAR Maintenance Engineering Analysis Record

meas measure; measurement

Meas for M *Measure for Measure*

M-East Middle-East

M.E. Auto. Master of Automobile Engineering; Master of Automotive Engineering

meb military early bird

MEB Marine Expeditionary Brigade; Master Electronics Board; Medical Board; Melbourne, Australia (airport); Midlands Electricity Board (UK)

MEBA Marine Engineers' Beneficial Association

mec main engine cutoff, marine extension clause; measuring equipment control

méc *mécanique* (French—mechanical)

M. Ec. Master of Economics

MEC Maine Central (railroad); Maintenance Engineering Change(s); Marine Expeditionary Corps; Master Executive Council; Member of the Executive Council; Methodist Episcopal Church; Monetary and Economic Council

M.E.C. Master of Engineering Chemistry; Member of the Executive Council

meca maintainable electronics component assembly; malfunctioned equipment corrective action; mercury evaporation and condensation; multielement component array

MECA Manufacturers of Emission Controls Association; Molecular Emission Cavity Analysis

mecano mechanotherapy

MECAS Middle Eastern College for Arabic Studies (Beirut, Lebanon)

mecc *meccanica* (Italian—mechanic)

mecca master electrical common connector assembly

MECCA Minnesota Environmental Control Citizens Association

mech mechanic; mechanical; mechanism

Mech Mechanics

ME Ch Methodist Episcopal Church

MECHA *Movimiento Estudiantil Chicano de Aztlán* (Mexican-Spanish—Chicano Student Movement of Aztlán)

Mechai (Thai—condom)— named for Thailand's Mr Contraception, Mechai Viravaidya

mechanochem mechanochemical; mechanochemistry

M.E.Chem. Master of Chemical Engineering

Mech Eng *Mechanical Engineering*

Mech Illus *Mechanix Illustrated*

MEC/IES Menimac Education Center's Institute for Educational Services

Meck-Vorp Mecklenburg-Vorpommernb (German—Mecklenburg Western Pomerania)

mecl monolithic emitter-coupled logic (computer)

meco main engine cutoff

MECO Metropolitan Edison Company

mecom marine engine condition monitor(ing)

M.Econ. Master of Economics

MECON Metallurgical and Engineering Consultants

MEC/PA Maintenance Engineering Change/Problem Analysis

MECR Maintenance Engineering Change Request

MECS Middle East Container Service

mecu main engine control unit

MECU Municipal Employees Credit Union

mecz mechanized

med medal; medalist; medallion; median; median erythrocyte diameter; medic; medical; medication; medicinal; medicine; medieval; medievalism; medievalist; medium; minimal effective dose; minimal erythema dose; mobile engine diagnosis

Med Medicine; medieval; Mediterranean

Med. Mediterranean Sea, between north Africa and southern Europe

Med *Médico* (Italian, Portuguese, Spanish—doctor); *Méditerranée* (French— Mediterranean); *Mediterra-*

neo (Italian—Mediterranean); *Mediternâneo* (Portuguese—Mediterranean); *Mediterráneo* (Spanish— Mediterranean)

Méd *médicine* (French—medicine)

M.Ed. Master of Education

MED Maintenance Engineering Data; Manhattan Engineer District (cover name used by developers of the first atomic bomb); Metal-working Equipment Division (US Department of Commerce); Military Electronics Division (Motorola); Municipal Electricity Department

M.E.D. Master of Elementary Didactics

med-50 minimal effective dose in 50 percent of cases

medac medical accounting

medal micromechanized engineering data for automated logistics

Med C Medical Corps

Med CAP Medical Civil Action Program

medcat medium clear-air turbulence

MEDCO Meat Export Development Company

med col medium color

MEDCOM Mediterranean Communications System

medcrit medical critic(ism)

medda mechanized defense decision anticipation

MED-DENT Medical-Dental Division (USAF)

medevac medical evacuation

medex medical expert

médex *médecin extension* (French—doctor's aides, medics)

Medfly (Med fly) Mediterranean fruit fly

Med Gr Medieval Greek

medi (Latin prefix—middle)— median

Medi (British seamen's short form—Mediterranean)

Media Magnavox electronic data-image apparatus

MEDIA Manufacturers Educational Drug Information Association; Missile Era Data Integration Analysis; Move to End Deception in Advertising

mediacrit media critic(ism)

mediaese cultivated English spoken by many entertainment, radio, and television personalities

mediator media time-orienting and reporting

medic medical corpsman; medical doctor; medical student

medicaid medicinal aid (free medicine for the needy)

Medicaid Medical Aid (federal and state health insurance for people unable to afford medical care)

MEDI-CAL Medical Aid of California

Medical Exam Medical Examination Publishing Company

Medicare medical care

Medicare Medical Care (federal health insurance for aged and disabled persons)

MEDICO Medical International Corporation

medifraud medical fraud

Medigap uninsured medical costs often requiring supplemental coverage

mediog mediograph(ic)(al); mediography

Med Isr *Medinat Israel* (Hebrew—State of Israel)

Medit Mediterranean

Mediterranean Mediterranean Sea

Mediterranean Countries Albania, Algeria, Cyprus, Egypt, France, Greece, Israel, Italy, Lebanon, Libya, Malta, Monaco, Morocco, Spain, Syria, Tunisia, Turkey, Yugoslavia

MEDIUM Missile Era Data Integration Ultimate Method

medivac medical evacuation

medix medical students

med juris medical jurisprudence

med lab(s) medical laboratories; medical laboratory

MEDLARS Medical Literature Analysis and Retrieval System

Med. L. Medieval Latin

Med Lat Medieval Latin

MEDLINE Medical On-Line (computer retrieval system)

M. Ed. L. Sc. Master of Education in Library Science

med nec medically necessary (abortion)

Med Phys Medical Physics

med ray medullary ray

med ray par medullary ray parenchyma

med ray trac medullary ray tracheids

MEDRC Medical Reserve Corps

MEDRECO Mediterranean Refining Company

MEDRESCO Medical Research Council

meds medicaments; medicines

MEDS Maintenance Engineering Data Sheets

MEDSAC Medical Service Activity (USA)

Med. Sc. D. Doctor of Medical Science

Med Sch Medical School

Med Sea Mediterranean Sea

med show medicine show (carnival slang)

MEDSPA Mediterranean Special Program of Action

Med Supt Medical Superintendent

med tech medical technologist; medical technology

Med Tech Medical Technician; Medical Technologist

med trans medical transcriptionist

Mee methylethyl ether

M.E.E. Master of Electrical Engineering

MEECN Minimum Essential Emergency Communications Network

M.E. Eng. Master of Electrical Engineering

meerschaum hydrated magnesium silicate

MEES Middle East Economic Survey

mef maximal expiratory flow

Mef Mefisto

MEF Marine Expeditionary Force; Meal Export Federation; Mesopotamian Expeditionary Force; Middle East Forces; Musicians Emergency Fund

Mefisto Mefistofele

mefp main-engine fuel pump

mef's morality enhancing factors

mefv maximum expiratory flow volume

meg magnetoencephalography; megacycle; megaton; megawatt; megohm

Meg. Megillah

Meg Margaret; Megan

MEG Management Evaluation Group

mega 10^6

mega *megas* (Greek—great, large, powerful)—acromegaly, megacycle, megaspore, megaton

megabuck one million bucks (dollars)

megacorpses one million corpses (atomic bomb unit)

megacurie one million curies

megacycle one million cycles

megadeaths million deaths

megajoule one million joules

megal megalopolis (industrial urban area)

megameter one million meters

megamouse one million mice (statistical unit—experimental biology)

megaton one million tons

megawatt one million watts

megger megohmmeter

Meggie Margaret

ME/GNP Military Expenditures/Gross National Product

Meglin Megliola

mego megaphonal megohm(s)

MEGO my eyes glaze over

megohm one million ohms

megs megacycles

megv million volts

megw megawatt

megwh megawatt-hour

mei mathematics in education and industry

mei (MEI) marginal efficiency of investment

MEI Maintenance and Engineering Inspection; Manual of Engineering Instructions; Marine Ecological Institute; Metals Engineering Institute; Middle East Institute

MEIC Member of The Engineering Institute of Canada

Me'il. Me'ilah

MEIS Military Entomology Information Service

MEIU Management Education Information Unit

mej mejikanisch (German—Mexican); *mejuffrouw* (Dutch—Miss)

Mej Mejicana(o), Mejico (Spanish—Mexican, Mexico); *Mejuffrouw* (Dutch—Miss)

mek methyl ethyl ketone

mel *melanesisch* (German—Melanesian)

Mel Melanesia; Melanesian; Melanesian Pidjin English (Bêche de Mer); Melanie; Melba; Melbourne; Melissa; Melvil; Melville; Melvin; Melvina; Melvyn

MEL Master Equipment List(ing); Minimum Equipment List (FAA); Music Education League

M.E.L. Master of English Literature

MELA Middle East Librarians Association

Melan Melanesia; Melanesian

Melanesia islands in the western Pacific whose natives inhabit the Bismarck Archipelago, Fiji, New Caledonia, New Hebrides, and the Solomon Islands

Melb Melbourne

Melba Nellie Armstrong (of Melbourne)

MELCO Mitsubishi Electric Corporation

Meld melt + weld

MELF Middle East Land Forces

melg most European languages

mellif mellifluous

melo melodrama, melody

M. Elo. Master of Elocution

melos melodic lines

melt. pt melting point

mem member; membership; memoirs; memorial

mem (MEM) memory map (calculator)

mem. *memoria* (Latin—memory)

Mem Member (of Congress, Parliament, etc.); Memorial (postal abbreviation)

MEM Mars Excursion Module; Member; memorial; Memphis, Tennessee (airport)

mema middle ear muscle activity

MEMA Marine Engine Manufacturers' Association; Motor and Equipment Manufacturers Association

MEMAC Machinery and Equipment Manufacturers Association of Canada

memb membrane

'members remembers

MEMC Marathon Electric Manufacturing Corporation

memci model ewe for microclimate integration

MEML Master Equipment Management List

memo memoranda; memorandum

MEMO Medical Equipment Management Office

Memorial Memorial Stadium, Baltimore

Memp Memphis

Mem Pkwy Memorial Parkway

Mem Roy Astron Soc *Memoirs of the Royal Astronomical Society*

mems micro-electromechanical systems

Mem Soc Assn Memorial Society Association

Mem SP Memorial State Park

MEMSPO Michigan Elementary and Middle School Principals Organization

men. menses; menstruation; mensuration

men *meno* (Italian—less)

Men. *Menahot*

Men Mensa (constellation)

M.En. Master of English

MEN Manasco (stock-exchange symbol); Medium Energy Neutrino experiment

MEN *Middle East News*

MENA Middle East News Agency

MENC Music Educators National Conference

mend macro end

Mend Mendoza (Argentine province)

MEND Medical Education for National Defense; Mothers Embracing Nuclear Disarmament

Mendelssohn Felix Mendelssohn (Jakob Ludwig Felix Mendelssohn-Bartholdy)

Mendl Lib Mendelssohn Library

Mendy Mendelssohn

M.Eng. Master of Engineering; Mining Engineer

M. Eng. P.A. Master of Engineering and Public Administration

Menn Menninger

Mennon Mennonite

meno menopausal; menopause; menorrhoea

Menotti Gian Carlo Menotti

MENP Mount Elgon National Park (Kenya)

MENS Mission Element Needs Statement

Mensa (Latin—Table constellation)

mens san. *mens sana in corpore sano* (Latin—a sound mind in a sound body)

menst menstrual; menstruation

mensur mensuration

ment mental; mentalis

ment mentioned

MENT Mentor Graphics Corporation

M. Ent Master of Entomology

mentd mentioned

mentholyptus menthol + eucalyptus

meo (MEO) manned earth observatory

Meo Bartolomeo

MEO Maintenance Engineering Order; Marine Engineer(ing) Office(r)

MEOA Malaysian Estate Owners Association

meoh methanol; methol

meos microsomal ethanol oxidizing system

MEOW Moral Equivalent of War (President Carter's energy program)

mep mean effective pressure

MEP Main European Port; Management Engineering Program; Management Evaluation Program; Member of European Parliament; Member of the European Parliament; Micro-Electronic Education Program; Middle East Perspective

M.E.P. Master of Engineering Physics

MEP *Movemiento Electoral Popular* (Papiamento—Popular Electoral Movement)—Aruba's pro-independence party; *Movimiento Electoral del Pueblo* (Spanish—People's Electoral Movement)—Venezuelan political party

M.E.P.A. Master of Engineering and Public Administration

MEPC Maritime Environment Protection Committee; Medical Examination Publishing Company (imprint of Elsevier Science Publishing Company); Member of the Educational Publishers' Council; Metropolitan Estate and Property Coporation

MEPCOM Military Enlisting Processing Command

M.E.P.H. Master of Public Health Engineering

Mephisto Mephistopheles (The Devil)

MEPs Members of the European Parliament

MEPS Means-Ends Problem-Solving Test; Multilanguage Electronic Phototypesetting System

MEPSI Mexican-Elmhurst Philatelic Society International

mepu (MEPU) monopropellant emergency power unit

meq/l millequivalents per liter

mer meridian; minimum energy requirement(s)

mer (MER) methanol extraction residue

Mér Mérida (capital of Extremadura, Spain)

m & er mechanical and electrical room

mer mercoledi (Italian—Wednesday); *mercredi* (French—Wednesday); *meros* (Greek—part)—blastomere, centromere, isomer, polymer

Mer Mercury; Merino

MER Metropolitan Elevated Railroad; Ministry of Energy Resources

mera minimum enroute altitude

MERA Michigan Educational Research Association

MERADO Mechanical Engineering Research and Development Organization

MERAG Middle-East Research and Action Group

MERB Mechanical Engineering Research Board

merc mercury

Merc Mercantile Exchange; Mercator; Mercedes; Mercedes-Benz; Mercury

MERC Music Education Research Council

MERCEDO Merseyside Economic Development Office

merch merchantable

MERCHANT Methods in Electronic Retail Cash Handling

Merch V Merchant of Venice

'mercial commercial

mercs mercenaries

Mercurys Mercury Islands off New Zealand's North Island

MERDL Medical Equipment Research and Development Laboratory (USA)

Meredith Meredith Press

meres matrix of environmental residuals for energy systems

merfin mercerized finish

Merguis Mergui Islands off Lower Burma

Meri Merionethshire

'Merica(n) [Cockney contraction—America(n)]

merid meridian

Merigrove the *New Grove Dictionary of American Music*

MERIP Middle East Research and Information Project

MERIT Medical Relief International

meritoc meritocracy; meritocrat(ic)(al)(ly)

MERL Mechanical Engineering Research Laboratory

MERLIN Machine Readable Library Information; Multi Element Radio Linked Interferometer Network

MERMAID Marine Environment Remote-Controlled Measuring and Integrated Detection; Metrication and Resource Modeling Aid

MERMAIDS Mediterranean Eddy Resolving Modeling and Interdisc Studies

Merriam G & C Merriam

Merritt Pkwy Merritt Parkway

Merry W Merry Wives of Windsor

mer's multiple ejection racks

Mers Merseyside

mersar merchant ship search and rescue

mersex merchant shipping exchange

Mert Merton College, Oxford

MERT Maintenance Engineering Review Team; Milwaukee Electric Railway and Transit

Mert Coll Merton College-Oxford

MERU Mechanical Engineering Research Unit

Merv Mervin

merzd meridian zenith distance

MERZONE Merchant Shipping Control Zone

mes main engine start; main equipment supplier; missile engineering station; motor end support; mud estuary slick; mutual energy support

mes meseta (Spanish—plateau; tableland); *mesos* (Greek—middle)—mesentery, mesoderm(ic), Mesopotamia, Mesozoic

Mes Mesozoic; Messina

Mes Mesdames (French—ladies)

MES Michigan Engineering Society; Midwest Electronic Society

mesa modularized equipment storage assembly

mesa (MESA) mathematics, engineering, and scientific achievement;

MESA Malarial Eradication Special Account; Marine Ecosystems Analysis; Mechanics Educational Society of America; Mining Enforcement and Safety Administration

Mesabi Mesabi Range of iron ore in Minnesota

Mesa Verde Mesa Verde National Park in Colorado

mesbic (MESBIC) minority enterprise small business investment companies

mesc mescal; mescaline

M. E. Sc. Master of Engineering Science

MESC Middle Eastern Solidarity Council

MESCO Middle East Science Cooperative Office (UNESCO)

Mesd Mesdames (French—Ladies)

MESDA Museum of Early Decorative Arts (Winston-Salem, North Carolina)

MESF Mobile Earth Station Facility

mesfet metallized semiconductor field-effect transistor

mesh medical headings

MeSEA Maine Society of Enrolled Agents

MeSH Medical Staff Hospital; Medical Subject Heading (National Library of Medicine's thesaurus)

Meslier Jean Meslier—obscure parish priest whose name was used by Voltaire to escape persecution (see Jean Meslier); even the names of the parishes he served—Entrepigny and But—are not to be found in most atlases and gazetteers

meso (Latin prefix—middle, moderate)—mesoderm

Meso Mesopotamia; Mesozoic (Jurassic; Mississippean—Lower Carboniferous; Pennsylvania—Upper Carboniferous; Permian; Triassic)

Mesoamer Mesoamerica(n), Middle America(n) (Central America, Mexico, and the West Indies)

Mesop Mesopotamia (Iraq)

MESP More Effective Schools Program

MESPA Minnesota Elementary School Principals Association

Mespot Mesopotamia (Iraq)

mess maximum effective sonar speed

Mess Messidor (French—Harvest Month)—beginning June 19th—tenth month of the French Revolutionary Calendar

Messien Olivier Eugéne Prosper-Charles Messien (composer, organist, teacher)

Messner Julian Messner

messplex multiplex emission sensors

Messrs Messieurs (French—Gentlemen)

mest mestizo

mestranol methyl + estrogen + pregnane (synthetic oral contraceptive)

MESUR Mars Environmental Survey (metabolism)

met metal; metallic; metallize; metaphor; metaphysics; meteorology; methionine (amino acid) (MET); metronome; metropolitan

met metropolitana (Spanish—metropolitan)

Met Metro; Metropolitan Correction Center; Metropolitan Museum of Art; Metropolitan Opera

MET Mobile Examining Team; Multi-Environment Trainer

meta (Greek—after or beyond)—metabolism, metacarpal, metastasis, metatarsal

Meta Margarita

META Megachannel Extraterrestrial Assay; Metropolitan Educational Television Association (Canadian); Model Engineering Trade Association

metab metabolism

metadex metal abstracts index

metall metallurgy

metallog metallography

METALMA Metalúrgica Matarazzo (Brazilian company)

metaph metaphor(ical)(ly); metaphysical(ly); metaphysician; metaphysics

metaphys metaphysics

metaplan methods of abstracting text automatically programming language

metas metastasis; metastasize

metath metathesis

metb metal base

met bor metropolitan borough

metc metal curb; mouse embryo tissue culture

Met Cen Lib Metropolitan Central Library

METCO Metropolitan Council for Educational Opportunity

metd metal door

mete multiple engagement test environment

Met. E. Metallurgical Engineer

metec meteoroid technology

METEI Medical Expedition to Easter Island

meteor. meteorology

Meteorol Meteorology

meteorolo meteorology

meteosat meteorological satellite

metf metal flashing

metg metal grille

meth methadone; methamphetamine; methane; methedrine; methyl; methylated; methyprylon

Meth Methodist

methan methanol

methanol methyl alcohol or wood alcohol (CH_3OH)

Meth Epis Methodist Episcopal

meth freak methedrine freak (underground slang—habitual user of methedrine)

meth head methedrine head (underground slang—methedrine addict)

metho methodology; methyl alcohol

meths methylated spirits (denatured alcohol)

methu methuselah (8-bottle capacity)

meti major engineering test item; metal jalousie

METKIT Metrics Education Toolkit

M-et-L Maine-et-Loire

meth lab methamphetainine laboratory

metical monetary unit of Mozambique

Met Lith Assn Metropolitan Lithographers Association

metm metal mold

M-et-M Meurthe-et-Moselle

Met Man Metro Manila

Met Mus Metropolitan Museum

m. et n. mane et nocte (Latin—morning and evening); mane et nocte (Latin—morning and night)

meto maximum except takeoff

Met O Meteorological Office(r)

METO Middle East Treaty Organization

metob meteorological observation

metol methyl-p-aminophenol (photographic developer)

meton metonomy

metp metal partition

Met Pol Metropolitan Police

metr metal roof

Met R Metropolitan Railway

METR Manufacturing Engineering Test Report

metro metropolitan

métro chemin de fer métropolitain (Paris subway system)

Metro Metromedia; Metropolitan Life Insurance Company

Metro Metropolitan (Paris and Madrid subway systems—originally stood for Metropolitan District Railway—the London Underground)

METRO New York Metropolitan Reference and Research Library Agency

metroc meteorological rocket

metrocenter metropolitan center

metrocomplex metropolitan complex

Metrodome Metrodome Stadium, Minneapolis

metrocore metropolitan core

metroframe metropolitan framework

metrol metrology

METROMEX Metropolitan Meteorogical Experiment

metrop metropolis; metropolitan

metroplex metropolitan complex

metropol melropolis; metropolitan

METRRA Metropolitan Toronto Residents' and Rate Payers' Association

mets metal strip; metastasize(d)

METS Mayne's Exchange Transfer Systems; Mechanized Export Traffic System (USA)

metsats meteorological satellites

m. et sig. *misce et signa* (Latin—mix and write a label)

metso sodium metasilicate

mett (METT) manned evasive target tank (USA)

Met tec Meteorologist Technician

METU Middle East Technical University (Ankara)

Met-Vic Metropolitan-Vickers (electrical company)

MEU Marine Expeditionary Unit; Mental Evaluation Unit; Municipal and Shire Employees Union

MEU *Modern English Usage*

mev million electron volts

meV mega-electron volt(s); milli-electron volt(s); million-electron volt(s)

Mev *Mevrouw* (Dutch—Missus)

MeV megaelectronvolt; million electronvolt

mevr *mevrouw* (Dutch—Missus)

Mevr *Mevrouw* (Dutch—Missus)

MEW Microwave Early Warning; Ministry of Economic Warfare

MEWA Motor and Equipment Wholesalers Association

MEWS Maintenance Engineering Work Sheet; Missile Early Warning Station

MEWTA Missile Electronic Warfare Technical Area

mewttos main engine working to telegraph orders

mex military exchange

Mex Mexican; Mexico

MEX Mexican League (TripleA baseball); Mexico City, Mexico (airport)

Méx *México* (Spanish—Mexico)

Mex Air Mexicana Airlines

Mex C Mexico City

Mex Cy Mexican currency

Mex$ Mexican peso

Mex hair Mexican hairless dog (Spanish—*perro pelado*)

Mexi Mexicali

MEXICANA *Compañia Mexicana de Aviación* (Spanish—Mexican Aviation Company)

Mexicanos *Mexicanos al grito de guerra* (Spanish—Mexicans, the war cry) national anthem of Mexico

Mexico United Mexican States (Middle America's largest and most populated nation), *Estados Unidos Mexicanos*

mexit macro exit

Mex. S.C. Mexican Survivors' Certificates

MEXSM Mexican Service Medal

Mex. S.O. Mexican Survivor's Originals

Mex Sp Mexican Spanish

Mex. W.C. Mexican Widows' Certificate

Mexsur Mexican (automobile) insurance

mez mezcal(ine)

mez *mezzo* (Italian—half)

MEZ *mitteleuropäische Zeit* (German—Central European Time)

mezz mezzanine; mezzotint

Mezzogiórno (Italian—south)—southern Italy including Naples, Palermo, Salerno

mezzo(s) mezzosoprano(s); mezzotint(s)

mf machine finish; main feed; main force; maintenance factor; male-to-female (ratio); manufacture(d); manufacturing; mastic floor; medium frequency (300–3,000 kc); melamine formaldehyde; microfarad(s); microfiche; microfilm; milk fat; mill finish; millifarad(s); mother fucker; motor field; motor freight; multiplying factor

m/f maintenance-to-flight (ratio); male or female; manifest; marked for; marked for; milk fat

m & f male and female

μf micro + farad

mf *mezzo-forte* (Italian—half loud, moderately loud)

m/f *mi favor* (Spanish—my favor); *motorfaerge* (Dano-Norwegian—motor ferry)

μF microfarad(s)

M^f *Massif* (French—mountain mass)

MF Magazines for Friendship; Marshall Field (stock exchange symbol); Medal of Freedom; Middle Fork (railroad); Millard Fillmore (13th President U.S.); Mount Fuji

M-F Massey-Ferguson; Monday through Friday

M.F. Master of Forestry

MF *medlem af Folketinget* (Danish—Member of Parliament)

mfa malicious false alarm; multi-fiber arrangement

MFA Master Fencers Association; Metal Fixing Association for Building Insulation; Military Flying Area; Ministry of Foreign Affairs; Motor Factors' Association; Museum(s) of Fine Arts

M.F.A. Master of Fine Arts; Museum of Fine Arts

MFA *Movimento das Forças Armadas* (Portuguese—Armed Forces Movement)—military dictatorship

M Fac Hom Member of the Faculty of Homeopathy

M-factor mobility, movement, migration (automotive Americans on the move)

MFAH Museum of Fine Arts of Houston

M.F.A. Mus. Master of Fine Arts in Music

MFAR Michigan Foundation for Advanced Research

mfb message from base; metallic foreign object; moisturefree basis

mfb (M) median forebrain bundle

MFB Metropolitan Fire Brigade; MFB Mutual Insurance (Manufacturers, Firemen's and Blackstone combined)

MFBB Mexican Food and Beverage Board

MFBE Minority/Female Business Enterprise

mfbm thousand foot board measure(ment)

mfc magnetic-tape field scan(ning); measure for coffin; medicated face conditioner; membrane fecal coliform, microfilm frame card; microfunction circuit

mfc (MFC) marginal factor cost

MFC Master Facility Census

MFCA Master Fruit Carriers Association

MFCC Marriage, Family, Child Counsellor

MFCCA Master Floorcovering Contractors Association

m/fcha meses fecha (Spanish— months dated)

mfcm multifunction card machine

MFCM Member of the Faculty of Community Medicine

MFCMA Magnuson Fishery Conservation and Management Act

mfco manual fuel cutoff

mfcs mathematical foundations of computer science

MFCS Manual Flight-Control System (NASA)

mfcu multifunction card unit

mfd magnetofluid dynamics; manufactured; microfarad; minimum fatal dose (MFD); multi-function device; multi-function display

mfdf medium-frequency direction fiinder

mfdp maintenance float-distribution point

MFDP Mississippi Freedom Democratic Party

MFDT Memory for Designs Test

mfe multiflow evaluator

MFE Mouvement Fédéraliste Européen (French—European Federalist Movement); *Movimento Federalista Europeo* (Italian—European Federalist Movement)

MFECS Mediterranean Far East Container Service

M Fed Miners' Federation

MFED Manned Flight Engineering Division (NASA)

M. F. Eng. Master of Forest Engineenng

mff mighty fine fuckin'

MFF Master Freight File

mfg manufacturing; molded fiber glass

MFGT Multiple Family Group Therapy

mfh military family housing

MFH Master of Fox Hounds; Mobile Field Hospital

M/F/H Minorities/Females/Handicapped

mfhfs multi-function high-frequency sonar

mfi machine feature index; melt-flow index

MFI Master Facility Inventory; Multi-port Fuel-Injected (engine); Musicians Foundation Incorporated

MFIA Mutual Fund Investors Association

MFIANE Mutual Fire Insurance Association of New England

MFIBNE Mutual Fire Inspection Bureau of New England

MFIC Military Flight Information Center

MFIT Manual Fault Isolation Test

mfj married filing jointly

MFJ Modification of Final Judgment

MFJSA Mass Finishirug Job Shops Association

mfkp multifrequency key pulsing

m fl med flere (Dutch—and others)

Mfl Monfalcone

MFL Master Facility List; Missile Firing Laboratory; Mobile Field Laboratory; Mutual Funds Limited

MFL My Fair Lady (musical adaptation of George Bernard Shaw's *Pygmalion*)

MFLA Midwest Federation of Library Associations

MFLDA Malaysian Federal Land Development Authority

M Flem Middle Flemish

mflops million floating-point operations per second

MFM Miracle Food Mart; Modified Frequency Modulation

MFM Measure For Measure

MFMA Maple Flooring Manufacturers Association; Master Fish Merchants Association

mf method membrane or millipore filter method

mfn (MFN) most-favored nation

mf(n) microfiche (negative)

MFNO Midland Federation of Newspaper Owners

MFNP Mount Field National Park (Tasmania); Murchison Falls NP (Uganda)

MFNZ Music Federation of New Zealand

mfo missile firing order

MFOA Municipal Finance Officers Association

mfopp missile firing order patch panel

M.For. Master of Forestry

MFOWW Marine Firemen, Oilers, Watertenders, and Wipers

mfp main fuel pump; minor forest produce; multi-function peripheral

mfp (MFP) monoflurophosphate

mf(p) microfiche (positive)

mfpa monolithic focal-plane array

MFPB Mineral Fiber Products Bureau

MFPS Mobile Field Photographic Section

mfr manufacture; manufactured; manufacturer; missile firing range (MFR)

M Fr Mali franc(s); Middle French; Moroccan franc(s)

MFR Military Force Reduction(s)

mfrd manufactured

mfrg manufacturing

mfrn manufacturer's number

MFRP Midwest Fuel Recovery Plant (AEC)

mfrr manufacturer(s)

mfs magnetic-tape field search; maximum file size; missile firing simulator; mouse fatigue syndrome

mf & s magazine flooding and sprinkling

Mf's Moslem fanatics

MFS Malleable Founders' Society; Manned Flying System; Massachusetts Financial Services; Medal Field Service; Military Flight Service; Missile Firing Station; Mountain Fuel Supply; steel-hulled fleet minesweeper (3-letter naval symbol)

M.F.S. Master of Food Science; Master of Foreign Service; Master of Foreign Study

MFSA Master Floor Sanders Association; Metal Finishing Suppliers' Association

mfsk multiple-frequency shift keying

mfso main fuel shutoff

mfsov main fuel shutoff valve

MFSR Message Frame Size Register

MFSS Missile Flight Safety System(s)

mfst manifest

mft main frontal thrust (geology); major fraction thereof, mechanized flamethrower; motor freight tariff, multiprogramming with a fixed number of tasks

m. ft. *mistura fiat* (Latin—make a mixture)

MFT Microflocculation Test; Muscle Function Test; Musical Fundamentals Test(ing)

M.F.T. Master of Foreign Trade

MPTA Managed Futures Trade Association

MFTB Motor Freight Tariff Bureau

mftbf mean flight time between failure(s)

MFTD Mobile Field Training Detachment

mftl millifoot lamberts

m.ft.m. *misce fiat mistura* (Latin—mix to make a mixture)

mftv mechanical fmt test vehicle

mfu military fuckup

MFURB Maryland Fire Underwriters Rating Bureau

MFUSYS Microfiche File Update System

mfv magnetic field vector; microfilm viewer; motor fleet vessel

MFV Mars Flyby Vehicle

MfVB Museum für Volkerkunde, Berlin

MFW Maritime Federation of the World

mfy manufactory

mfz *mezzo forzando* (Italian—with moderate force)

mg machine gun; mahogany gene; marginal; middle guard (football); milligram; motor generator; multigauge

mg (MG) myasthenia gravis

mg % milligrams percent

m-g machine glazed

m/g motor generator

m & g mapping and geodesy

μg microgram

mg *main gauche* (French—left hand)

m/g *mi giro* (Spanish—my check, my draft)

mG *méridien de Greenwich* (French—Greenwich meridian)

Mg magnesium; Margrave; Margraviate; megagram (metric ton)

Mg *Molekulargewicht* (German—molecular weight)

MG Madagascar (Internet code); Maintainability Group; major general; Marine Gunner; Master Gardener; mid-capitalization growth; Middle German; Military Government; Minas Gerais; Minister General; Minister of the Gospel; Morris Garage (M-G); Murray Grey (cattle)

M-G Morris-Garage (British sports car)

M & G Mobile & Gulf

MG *Manchester Guardian* (newspaper); *Maschinegewehr* (German—machine gun)

mga melengestrol acetate

Mga Malaga; Mongolia

Mga *Mongolia* (Spanish—Mongolia)

MGA Managua, Nicaragua (Las Mercedes airport); Member of the General Assembly; Military Government Association; Monongahela (railroad); Mushroom Growers Association

mgal milligal

M-gauge meter gauge (39.37-inch) railroad track

mgawd make good all works disturbed

mgb (MGB) missile gunboat

Mg of B Margrave of Breslau; Margraviate of Breslau

MGB motor gunboat (British naval symbol); Soviet Ministry of State Security (see *VOT*)

MGB *Ministerstvo Gosurdastvennoi Bezopasnosti* (Russian—Ministry of State Security)—Soviet secret police

mgc manual gain control

MGC Machine Gun Corps; Machinery of Government (committee); Marriage Guidance Council; Marriage Guidance Counsellor

mg/cig milligrams (of nicotine tar) per cigarette

mgcir master ground-controller-interception radar

mgcr maritime gas-cooled reactor

mg/cu m milligrams (dust, fume, or mist) per cubic meter of air

mgd magnetogasdynamics; million gallons per day

mg/d million gallons per day

Mgd Magdeburg

MGD Military Geographic Documentation

mg/dl milligrams of glucose per deciliter of blood

mge (MGE) maintenance ground equipment

MGE Minneapolis Grain Exchange

M.G.E. Master of Geological Engineering

M.Geol.Eng. Master of Geological Engineering

MGES Maintenance Ground Equipment Specification

mgf macrophage growth factor

MGF Myasthenia Gravis Foundation

mgg mouse gamma globulin

mgh milligram hour(s)

MGH Massachusetts General Hospital

mgi military geographic(al) intelligence

MGI Managed Growth Initiative (urban planning); Media Group Inc; Mining and Geological Institute of India

MGIC Mortgage Guarantee Insurance Corporation

MGICA Mortgage Guaranteed Insurance Corporation of Australia

MGID Military Geographic Information and Documentation

MGk Medieval Greek

mg/l milligrams per liter

MGL Morris Geneological Library

mgm (MGM) mobile guided missile

Mg of M Margrave of Moravia; Margraviate of Moravia

MGM Metro-Goldwyn-Mayer

MGM-18A Lacrosse surface-to-surface missile

MGM-29A Sperry Sergeant surface-to-surface missile

MGM-31A Pershing surface-to-surface missile made by Martin

MGMI Mining, Geological, and Metallurgical Institute

MGMS Manchester Geological and Mining Society

mgmt management

mgn micrograin

MGN Mirror Group Newspapers

Mgna Montagna (Italian—mountain)

Mgne Montagne (French—mountain)

Mgo Mormugao

MGP Marcus Garvey Park (formerly Mount Morris Park); Mountain Gorilla Project

mgq maximum guarantee quality (a statistic used in tobacco production)

mgr mined geological repository

mgr (MGR) mobile guided rocket

Mgr Manager; Monseigneur (French—Monsignor); Monsignore (Italian—Monsignor)

M Gr Middle Greek

MGR Matusadona Game Reserve (Rhodesia); Micro-Graphic Recording

mgress manageress

mgs milligrams; missile guidance set (system); money-grubbing scum

mg('s) machine gun(s)

m-g-s meter-gram-second

Mg's Malay gavials, also called Malay gharials

MGS Minnesota Geological Survey

MGSA Military General Supply Agency

MGSMTC Mid-Gulf Seaports Marine Terminal Conference

mgt management

MGTB Mexican Government Tourist Bureau

MGTC Morgan Guaranty Trust Company

MGTD Mexican Government Tourist Delegation

mgtrn magnetron

MGU Moskovskiy Gosudarstvenny Universitet (Russian—Moscow State University)

M Gun Sgt Master Gunnery Sergeant

mgus monoclonal gammopathy of undetermined significance

mgw maximum gross weight

MGW Manchester Guardian Weekly

m'gwd my gawd (my god)

mh magentic heading; main hatch; manhole; man-hour; marital history; materials handling; meeting house; menstrual history; mental health; millihenries; millihenry; murine hepatitis

μh microhenry

mH millihenry

Mh Monatsheft (German—monthly magazine)

MH magnetic heading; Marshall Islands; Master Hosts; Medal of Honor; Military Hospital; Ministry of Health; Mission Hills; Most Honorable; Most Honourable

M-H Minneapolis-Honeywell (stock exchange symbol and trademark)

M & H Mason and Hamlin

MH Mo'etzet Hapo'alot (Hebrew—Woman Workers Council)

MH² Mary Hartman, Mary Hartman (tv show)

mha manhour accounting; maximum holding altitude

MHA auxiliary minehunter (naval symbol); Marine Historical Association; Medal for Humane Action; Member of the House of Assembly; Mental Health Administration; Mental Health Association; Mental Health Authority; Mining Houses of Australia; Multiple Handicapped Association

M.H.A. Master of Hospital Administration

MHANY Mutual Housing Association of New York

MHATA Mental Health Assistant Therapy Aide

MHb Mueller-Hinton broth

MHB Material Handling Bureau; Mental Health Branch

M-H B Mid-Hudson Bridge

mhc major histocompatibility complex

MHC coastal minehunter (naval symbol); Massachusetts Historical Commission

MHCO Mine-Hunting Control Office(r)

MHCOA Motor Hearse and Car Owners Association

mh cp mean horizontal candlepower

MHCP Multiple Habitat Conservation Plan

mhcv (MHCV) manned hypersonic cruise vehicle

mbd magnetohydrodynamics

MHD Mental Health Department; Military History Detachment

mhdg magnetohydrodynamic generator

mhdl magnetohydrodynamic laser

mhd lt masthead light

mhe materials handling equipment

MHE Mechanical Handling Engineering

M.H.E. Master of Home Economics

MHEA Mechanical Handling Engineers Association

M Heb Middle Hebrew

MHEDA Material Handling Equipment Distributors Association

M.H.E.E. Master of Home Economics Education

M. H. E. Ed. Master of Home Economics Education

mhf medium high frequency

M-H-F Massey-Harris-Ferguson

MHFNZ Mental Health Foundation of New Zealand

mhg message-header generator

μHg microns of mercury

MHG Middle High German

MHHS Mercy Home Health Services

mhhw mean higher high water

MHI Manufactured Housing Institute; Material Handling Institute; Metal Hydrides Incorporated; Mitsubishi Heavy Industy

mhic microwave hybrid integrated circuit

M.Hi.E. Master of Highway Engineering

M.Hi.Eng. Master of Highway Engineering

MHII Material Handling Institute Incorporated

MHJC Mary Holmes Junior College

MHK Member of the House of Keys (Isle of Man)

mhl metal halide lamps

MHL Manaus Harbour Limited; Mission Hills Library

M.H.L. Master of Hebrew Literature

MHLF Mutual Home Loan Funds

MHLG Ministry of Housing and Local Government

mhls metabolic heat-load simulator

mhlw mean higher low water

Mhm Mannheim

MHM Mill Hill Missionary

MHMA Mobile Homes Manufacturers Association

MHMC Mercy Hospital and Medical Center; Montefiore Hospital and Medical Center

mho unit of conductance or reciprocal ohm

M. Hor. Master of Horticulture

M. Ho. Sc. Master of Household Science

mhp millihorsepower

MHP Missouri Highway Patrol

mhpg (MHPG) 3-methoxy-4-hydroxy phenylethylene

MHQ Maritime Headquarters; Mediterranean Headquarters

mhr manhour(s); maximum heart rate; microwave hologram radar

mhr (MHU) mental health unit

MHR Member of the House of Representatives

MHRA Modern Humanities Research Association

MHRF Mental Health Research Fund

MHRI Mental Health Research Institute (University of Michigan)

MHRT Mental Health Review Tribunal

mhs medical history sheet

MHS Massachusetts Historical Society; Measurement Handicap System; Message-Handling Service; Musical Heritage Society

MHSA Military Historical Society of Australia

MHSc Master (Mistress) of Household Science

MH strain Mill Hill (viral) strain

mht mean high tide; mild heat treatment; military hospital trainee

mht (MHT) missile-handling trailer

MHT Museum of History and Technology (Smithsonian Institution)

MHTA Mental Hygiene Therapy Aide

MHTC Manufacturers Hanover Trust Company

MHTF Manhattan Homicide Task Force (NYPD)

MHTG Marine Helicopter Training Group

MHTGR Modular High Temperature Gas Reactor

mhtl mean high tide line

Mhtn Manhattan

M. Hu. Master of Humanities

mhv mean horizontal velocity; miniature homing device; murine hepatitis virus

mhw mean high water

MHW Mental Health Worker; Ministry of Health and Welfare

mhwli mean high water lunitidal interval

mhwlr mobile hostile-weapon-locating radar

mhwn mean high water neaps

mhws mean high water springs

Mhy Mahayana

M. Hy. Master of Hygiene

MH y C Miguel Hidalgo y Costilla

M. Hyg. Master of Hygiene

mhz (MHz) megahertz(es), formerly megacycle(s) per second

mi malleable iron; manual input; marginal inscription; mentally ill; metabolic index; middle initial; mildew; mile(s); mill; minor; minute(s); mitral; mitral insufficiency; monument inscription; multiple intelligences; mutual inductance

mi. mile; miles

mi (MI) myocardial infarction

m & i modernization and improvement; municipal and industrial

m of i moment of inertia

mi (Italian—third tone in diatonic scale, *E* in fixed-*do* system)

Mi Mach indicated; Mach speed indicated; Miami; Minor; Mitte

Mi *Michel* (stamp catalog)

Mi. Michah

MI Maintenance Instruction; Mare Island, Marshall Islands; Match Institute; Maturation Index; Mauritius Institute; Meat Inspection (US Department of Agriculture); Mellon Institute; Member of the Institute; Member of the Institution; Metal Industries; Michigan; Military Intelligence; Ministry of Information; Missouri-Illinois (railroad); Mounted Infantry

M-I Missouri-Illinois (railroad)

M & I Manpower and Immigration (Canada)

Mi-1 Soviet utility helicopter nicknamed Hare

mi² square mile(s)

mi³ cubic mile(s)

MI 5 (British) Military Intelligence Security Service (somewhat equivalent to American FBI)

MI-6 Military Intelligence 6 (British external intelligence organization)

Mi-8 Soviet transport helicopter nicknamed Hip

Mi-10 Soviet heavy-transport helicopter nicknamed Harke

Mi-12 Soviet heavy helicopter nicknamed Homer by NATO and in the mid-1970s allegedly the world's heaviest and largest aircraft of its kind

Mi-16 40-ton Soviet helicopter

mia missing in action (military personnel)

mia (MIA) missing in action

MIA Malaysian Institute of Art; Malleable Ironfounders' Association; Manila International Airport; Manitoba Institute of Agrologists; Marble Institute of America; Masonry Institute of America; Media Information Australia; Miami, Florida (airport); Mica Industry Association; Millinery Institute of America; missing in action; Montgomery Improvement Association; Murrumbidgee Irrigation Area (Australia); Mythmaking in America

M.I.A. Master of International Affairs

MIAA Miniatures Industry Association of America; Mortgage Insurers Association of Australia

MIAC Manufacturing Industries Advisory Council

MIA–CHI Miami—Chicago

MIA–LAX Miami—Los Angeles

MIA–NY Miami—New York

MIAP Member of the Institution of Analysts & Programmers

MIAPD Mid-Central Air Procurement District

MIARS Maintenance Information Automated Retrieval System (USN)

MIAS Major-Item Automated System (USA)

MIA–SAN Miami—San Diego

MIA–SFO Miami—San Francisco

MIASI Moore Institute of Art, Science, and Industry

MIA–TOR Miami—Toronto

MIB Management Improvement Board; Maritime Index Bureau; Meat Inspection Branch; Medical Information Bank; Medical Information Bureau; Mental Information Bureau; Metal Information Bureau; Michigan Inspection Bureau; Military Intelligence Branch; Military Intelligence Bureau; Millinery Information Bureau; Missouri Inspection Bureau; Multi-layer Interconnect Board; Sir Marc Isambard Brunel

mibg metaiodobenzylguanidine

mibk methyl isobutyl ketone

mic machine index card; micrometer; microphone; microwave integrated circuit; military-industrial complex; minimum ignition current; monolithic integrated circuit

mic (MIC) methyl isocyanate (poison gas)

Mic The Book of Micah; Microscopium (constellation)

MIC Malayan Indian Congress; Management Information Center; Marshall Islands Congress; Medical Information Center; Mitsubishi International Corporation; Monaco

Information Centre; Motorcycle Industry Council; Motors Insurance Corporation; Music Industry Conference

mica macro instruction compiler assembler; medically ill chemical abuser

mica (Spanglish—migration card)—issued by the U.S. Immigration and Naturalization Service

MICA Maternity and Infant Care Association; Meat Importers Council of America; Medical Imaging Centers of America; Medicare Insurance Counseling and Advocacy; Mobile Industrial Caterers Association; Moscow Institute for Complex Automation

micbm (MICBM) mobile intercontinental ballistic missile

micc mineral-insulated copper-covered (cable); miniature integrated circuit computer

MICC Malaysian International Chambers of Commerce

MICE Member of the Institution of Civil Engineers

Mich Michael; Michelle; Michigan; Michiganite; Michoacan; Mitchell

MI Chem E Member of the Institution of Chemical Engineers

Michig Michigander

MiCIS Microbial Culture Information Service

Michl Michael

mick manufacturer's item correlation key

Mick Michael

Mickey Mickey Mouse

Mickey D's McDonalds

MICLASS Metal Institute Classification

MICMA National Ice Cream Mix Association

MICMD Milwaukee Contract Management District

mic-min micro-mini (automobiles)

MICOM Missile Command(er)

micpac molecular integrated circuit package

mic. pan. mica panis (Latin—bread crumb)

MICPS Microfiche Interface Controller-Processor System

micr magnetic ink character recognition; microscope; microscopic; microscopy

micr (MICR) magnetic-ink character recognition

Micr Microscopium

MICR Magnetic Ink Character Recognition

micro 10⁻⁶

micro (Greek *mikros*—small) microbe, microbiology, microcephalic, micrometer, microscope

Micro Micronesia (Trust Territory of the Pacific); Micronesian

microbiol microbiology

microbop microbopper (underground slang—very young person attuned to the modern scene)—see macrobop

Microcard Microcard Editions

microcephs microcephalics (small-headed people)

microcom microcomputer (pocket calculator)

microdoc microphotography and document (reproduction)

microeco microeconomics

microelectro microelectronic(s)

Microg Microgramma

micro-id microscopic identification disk

micro-in. micro-inch

micromation microfilm + automation

micromoms micromomentaries (split-second facial expressions)

micron millionth of a meter

Micron Micronesia; Micronesian

Micronesia U.S. Trust Territory of the Pacific including the Caroline, Mariana, and Marshall islands

micropaleo micropaleontology

micropros microprocess(ing); microprocessor

micros microscopy

micro(s) microcomputer(s)

MICROSIFT Microcomputer Software Information for Teachers

microt microtome

microwav microwave(able)

micr's magnetic ink characters

MICRS Magnetic Ink Character Recognition System

mics metal-insulated copper-sheathed (cable)

mic's military-industrial complex executives; military-industrial complex salesmen

MICS Museum of the International College of Surgeons

MICTI Ministerio de Industria, Comercio, Turismo, e Integración (Spanish—Ministry of Industry, Commerce, Tourism, and Integration)—Peru

micu (MICU) medical intensive care unit; mobile intensive care unit

micv (MICV) mechanized infantry combat vehicle(s)

mid mentioned in dispatches; middle; midnight

mid (MID) minimal inhibiting dose; minimum infective dose; multi-infarct dementia; multiple-infarct dementia

Mid Midshipman

Mid. Middot

MID Merida, Yucatan (airport); Midway Islands (in mid-Pacific); Military Information Division; Military Intelligence Division; Multi-Infarct Dementia

M.I.D. Master of Industrial Design

midac management information for design and control

Midac Michigan digital automatic computer

midas modified-integration digital-analog simulator (USAF)

MIDAS Maintenance Integrated Data Access System; Management Information Dissemination Administrative System; Materials for Industry Data and Applications Service; Media Investment Decisions Analysis Systems; Meteorological Information and Dose Acquisition System; Mine Detection and Avoidance System; Missile Defense Alarm System; Missile Detection Anti-Surveillance

midcab minimally invasive direct coronary artery bypass

midcult middle-class culture

midd merchandise information desk director

Middlx Middlesex

Middx Middlesex

Middy Midshipman

Middys Midshipmen

Mideast Middle East

MIDEASTFOR Middle East Air Force (USN)

MIDEC Middle East Industrial Development Corporation

MIDELEC Midlands Electricity Board

MIDF Major Item Data File

MIDFL Malayan Industrial Development Finance Limited

Mid-Glam Mid-Glamorgan

midi (MIDI) musical instrument digital interface

midis mid-length (below-the-knee) skirts

Midl Midlands; Midlothian

Midland Midland Bank Limited

Mid Lat Middle Latin

MIDLNET Midwest Region Library Network

Mid Loth Midlothian

midmo middle of the month

Midn Midshipman

MIDP Major-Item Distribution Plan

midr mandatory incident and defect report(ing)

mids middies (middieblouses, midshipmen); missile ignition and destruct simulator

Mids Midlands—in Great Britain

mid. sag. midsagittal

Mids ND A Midsummer-Night's Dream

midssn mid-season

'mid(st) amid(st)

Midsummer Midsummer Day (Saturday nearest June 21 or 22); Midsummer Eve or Midsummer Night

MIDU Mineral Investigation Drilling Unit

midw midwestern

midwk midweek

mie military-industrial establishment

mie miércoles (Spanish—Wednesday)

M.I.E. Master of Industrial Engineering

MIEA Music Industry Educators Association

MIECO Marshall Islands Import-Export Company

MIEE Member of the Institution of Electrical Engineers

MIEL Malaysian Industrial Estates Limited

mierc miércoles (Spanish—Wednesday)

Mies Ludwig Mies van der Rohe (1886–1969)

mif merthiolate-iodine-formaldehyde (fecal examination technique); modulusirregularity factor

mif (MIF) migratory inhibitory factor

MIF Market Intervention Fund; Milk Industry Foundation; Miners International Federation

MIFCT Moscow Institute of Fine Chemical Technology

MIFI Moskovskiy Inzhenerno Fizicheskiy Institut (Russian—Moscow Engineering Physics Institute)

mifil microwave filter

MIFT Manchester International Freight Terminal

mig magnesium-inert gas; metal inert gas (welding); migrant

MIG Marine Industry Group; Mikhail Ivanovich Glinka; Soviet jet fighter aircraft named for designers Mikoyan and Gurevich

MIG-1 Moody's Investment Grade

MIGA Make It Go Away; Multilateral Investment Guarantee Agency

MIGB Millinery Institute of Great Britain

mightn't might not

Mig¹ Miguel (Spanish—Michael)

migra migración (Mexican-American slang—immigration or migration)—*la migra* means the U.S. Border Patrol or the Immigration and Naturalization Service

mi/h mile(s) per hour

M.I.H. Master of Industrial Health

mihn-baau (Cantonese Chinese—bread)

mihped microwave-induced helium-plasma emission detection

MIHS Marshall Islands High School

MIIA Medical Information and Intelligence Agency

MIIDS Maine's Integrated In-service Delivery System

MIIF Master Item Intelligence File

MI Inf Sci Member of the Institute of Information Scientists

MIIS Marshall Islands Intermediate School

mij maatschappij (Dutch—company, society)

M i J Made in Japan

MIJ Muhammad Ali Jinnah

miji meaconing, interference, jamming, intrusion

Mik Mikhail; Mikhail Bakunin (Russian anarchist)

mike micrometer; microphone

Mike letter M radio code; Michael

Mikhail Mikhaylovich

Mikimotos Mikimoto cultured pearls

mikos mindervärdighets komplex (Swedish—inferiority complex)

Mikrop Mikropunkt (German—microdot)—microfilm marvel of World War II when a page of top-secret information could be reduced to a dot no larger than the dot over a letter i and then could be enlarged when needed

mil mileage; milennium; military; militia; milieme; million; 1/1000 inch; 1/10 cent; 1/1000 Palestinian pound (currency formerly used in Israel)

m-i-l mother-in-law

Mil Milan; Milford Haven (British maritime abbreviation); Military; Milwaukee, Wisconsin

MIL Malaya Indonesia Line; Medical Information Line; Member of the Institute of Linguists; Microsystems International Limited; Milan, Italy (Malpensa Airport)

MILA Merritt Island Launch Area

MilAdGru Military Advisory Group

Mil Att Military Attaché

milc military characteristics

MILC Midwest Inter-Library Center

MILCAP Military Civic Action Program; Military Civil Action Plan(ning); Military Contract Administration Procedure(s)

MILCASE Military Career Awareness Course for Educators

milcomsat military communication satellite

MILCON Military Construction

milcrit military critic(ism)

MILDAT Military Damage Assessment Team

mildec(s) military decision(s)

Mil Dist Military District

Mil Dist 1 Virginia from 1869 to 1870

Mil Dist 2 North Carolina from 1868 to 1870; South Carolina from 1868 to 1876

Mil Dist 3 Alabama from 1868 to 1874; Florida from 1868 to 1877; Georgia from 1870 to 1871

Mil Dist 4 Arkansas from 1868 to 1874; Mississippi from 1870 to 1876

Mil Dist 5 Louisiana from 1868 to 1877; Texas from 1870 to 1873

mile mille passuum (Latin—1000 paces), a pace being a double step

Mile-High Mile-High Stadium (Denver)

MILES Multiple Integrated Laser Engagement System

Mil-Hndbok Military Handbook

MILIMETS Military Meteorological System

milit military; militia

Mil Jrn Milwaukee Journal

milk of magnesia magnesium hydroxide—$Mg(OH)_2$

mill. millinery; milling; million(s)

Mill Million(en) [German—million(s)]

milli 10^{-3}

Millo Escamillo

mil m/t million metric tons

MILNET Military (computer) Network

milob military observer

milpac military personnel accounting activity

MILPERCEN Military Personnel Center

mil pers military personnel

MILPO Military Personnel Office

M.I.L.R. Master of Industrial and Labor Relations

milrep military representative

mils missile impact locator system

mil-s milling specification(s)

MILSATCOM Military Satellite Communications

MILSIMS Military Standard Inventory Management System

milspec military specification

Mil-Spec Military Specification(s)

milstac military staff communication

milstam military staff memorandum

MILSTAMP Military Standard Transportation and Movement Procedures

MILSTAR Military Strategic, Tactical, and Relay (satellite)

Mil-Std (MIL-STD) Military Standard

MILSTRAP Military Standard Requisitioning and Accounting Procedures

MILSTRIP Military Standard Requisitioning and Issue Procedures

Mil Sym Milwaukee Symphony

Milt Milton

Milw Milwaukee

Milw German Milwaukee German

MILW Milwaukee Route (Chicago, Milwaukee, St Paul & Pacific Railroad)

Milwaukee Milwaukee County Stadium

mim micro-impulse mosaic, mimeograph(ing; y)

mim (MIM) mobile interceptor missile

M i M Morality in Media

MIM Maintenance Instruction Manual (DoD); Material Inventory Master; Mature Insurance Marketing Services Incorporated; Minorities in Media; Mount Isa Mines (Queensland)

MIM-3A Douglas Nike-Ajax surface-to-air missile

MIM-10 military designation of the Boeing Bomarc missile

MIM-14A Douglas surface-to-air missile called Nike-Hercules and armed with a heavy-explosive or nuclear warhead

MI Mar E Member of the Institute of Marine Engineers

MIMB Malaysia International Merchant Bankers

MIMC Member Institute of Management Consultants

MIMD Multiple Instruction, Multiple Data

MIME Midland Institute of Mining Engineers; Multipurpose Internet Mail Extension

MI Mech E Member of the Institution of Mechanical Engineers

mimeo mimeograph(ed)

mimic microfilm-information master-image converter

MIMIC Microwave and Millimeter Wave Monolithic Integrated Circuit

mi/min miles per minute

mimo man in-machine out; many-input many-output (computer)

MIMR May Institute of Medical Research

mims mineral-insulated metal sheathed (cable)

MIMS Major Item Management System; Multiple Independently Maneuvering Submunitions

MIMS Monthly Index of Medical Specialties

mimsy miserable and flimsy

min minim; minimum; minor; minority; minute

min minore (Italian—minor); *minuto* (Portuguese or Spanish—minute)

MIN minimum fuel requirement (aircraft code)

Min Minister; Ministry; Minoan

Min Ministerio (Portuguese or Spanish—Ministry); *Ministro* (Portuguese or Spanish—Minister)

MIN Media Industry Newsletter

Min Agric Ministry of Agriculture

min b/l minimum bill of lading

M-in-C Matron-in-Chief

MINCEX Ministerio de Comercio Exterior (Spanish—Ministry of Foreign Trade)

Min Counc Mining Councillor

mind magnetic integrator neutron duplicator

Mind Mindanao

mindac miniature inertial navigation digital automatic computer

mindd minimum due date

Min Def Ministry of Defence

MINDUR Ministerio de Desarallo Urbano (Spanish—Ministry of Urban Development)

Min. E. Mining Engineer

MINE Microbial Information Network in Europe; Minnesota Information Network for Educators

Mineap Minneapolis

minec military necessity

minelco miniature electronic component

mineola orange + tangerine (hybrid citrus fruit)

mineral. mineralogy

Mineral Soc Mineralogical Society

minex minelaying, minesweeping, and minehunting exercise

MINFAR Ministerio de las Fuerzas Armadas Revolucionarias (Spanish—Ministry of the Revolutionary Armed Forces)—Cuba

minfin minicare finish(ing)

Min Fuel Ministry of Fuel and Power

mingy mean and stingy

Minho Entre Douro e Minho (Porgutuese—Between the Douro and the Minho)—province of Portugal

Min Hous Ministry of Housing

MINI Minicomputer Industry National Interchange

minibop minibopper (underground slang—older child attuned to the modern scene)—*see* macrobop

minibra(s) miniature brassiere(s)

minibus miniature autobus

minicam lightweight miniature camera; miniature camera

minicane miniature hurricane

minidoc miniature documentary (radio or tv)

miniDOS small disk-operating system

MiniDV Mini Digital Video

minimax minimize maximum possible losses

MININT Ministerio del Interior (Spanish—Ministry of the Interior)

mininuke(s) miniature nuclear-explosive device(s)

minipal miniature polamar accelerometer

miniskirt(s) short skirt(s)

minisym miniature symphony

minium red lead (lead oxide)

Mink Minkus (stamp catalog)

Minkies Minquier Islands (Rocks)

m-in-l mother-in-law

min/mc minimum material condition

Minn Minnesota; Minnesotan

Minne Minnesota

Minn Geol Surv Minnesota Geological Survey

Minn Hist Soc Minnesota Historical Society

Minn Orch Minnesota Orchestra

Minn Trib Minneapolis Tribune

Min° Ministro (Spanish—Minister, Ministry)

Min P Minister Plenipotentiary

MINP Mallacoota Inlet National Park (Victoria, Australia); minpac; Mine Warfare Forces, Pacific (USN)

MINPAC Mine Warfare Forces, Pacific (USN)

Min PBW Ministry of Public Building and Works

Min Plenip Minister Plenipotentiary

min prm minimum premium

Min PW Ministry of Public Works

Minquiers Minquier Islands (Rocks)—also called the Minkies

Minquiers (French—The Minkies)—semi-submerged reefs and rocks in Gulf of St Malo between Jersey and port of St Malo on the English Channel

minr minimum room rate desired

Min^r Minister

Min Res Minister Residentiary

MINREX Ministerio de Relaciones Exteriores (Spanish—Ministry of Foreign Relations)

min rnfl minimum rainfall

MINRON Mine Squadron

mins minutes

mins (MINS) minor(s) in need of supervision

M Inst BE Member of the Institute of British Engineers

M Inst Met Member of the Institute of Metals

M Inst SP Member of the Institution of Sewage Purification

MINT Managing the Integration of New Technology

MINTACTS Mobile Integrated Telemetry and Tracking System

Min Tech Ministry of Technology

mintie minimum test instrumentation equipment

min till. minimum tillage

M. Int. Med. Master of Internal Medicine

min trq minimum torque

MINWR Merritt Island National Wildlife Refuge (Florida)

min wt minimum weight

mio meteoritic impact origin; minimum identifiable odor

MIO Marine Inspection Office; Metric Information Office; Mobile Issuing Office; Movements Identification Order

Mioc Miocene

MIOCA Monolithic Integrated Optics for Customer Access Applications

MIOUDO Museo del Instituto Oceanografico de la Universidad de Oriente (Spanish—Museum of the Oceanographic Institute of the University of Oriente)

mip magnetic-induced polarization; malleable iron pipe; marine insurance policy; mean indicated pressure; minor in possession; missile impact predictor; modulated interference plan; monthly investment plan; mortgage insurance premium(s)

MIP Manufacturers of Illumination Products; Marine Interdiction Program; Material Improvement Program; Metals Investigation Proprietary; Methods Improvement Program; Military Improvement Program

mipe modular information-processing equipment

mipir missile precision instrumentation radar

MIPL Mauritius Institute Public Library (Port Louis)

mip/ma missile in place/missile away

MIPO Multiple Item Purchase Order

Miporn Miami pornography (FBI investigation's code name covering billion-dollar pornographic racket)

MIPR Member of the Institute of Public Relations; Military Interdepartmental Purchase Request

MIPRO Manufactured Imports Promotion Organization (of Japan)

mips male iron-pipe size; million instructions per second

MIPS Modular Integrated Pallet System

MIPTC Men's International Professional Tennis Council

Mipu Mikropunkt (German—microdot)—espionage technique assuring transmission of microscopic messages no bigger than a dot

Miq Maiquetia (Venezuela's principal airport)

Miq. Miqva'ot

mir memory information register; minimum implementation requirement; mirror; multiple isomorphous replacement; music information retrieval

mir mirador (Spanish—window with a wonderful view)

M Ir Middle Irish

MIR Manufacturing Inspection Record; Medical Inspection Room; Missile Intelligence Report; Movement for International Reconciliation

MIR Movimiento de Izquierda Revolucionaria (Spanish—Movement of the Revolutionary Left)—active in Bolivia, Chile, Ecuador, Peru, and Venezuela

MIRA Motor Industry's Research Association

MIRA Monthly Index of Russian Accessions

mirac microfilmed reports and accounts

MIRAC Management Information Research Assistance Center

miracl mid-infrared advanced chemical laser

mirad monostatic infrared intrusion detector

MIRADOR Multinational Investment Review Agency and Department of Research

MIRADS Marshall Information Retrieval and Display System

MIRAGE Migration of Radioisotopes in the Geosphere

MIran. Middle Iranian

Miranda Rule reading of criminal suspect's rights under law before questioning

miras mortgage interest relief at source

MIRC Member of the Idle Rich Class

mird medium internal radiation dose

MIRDAB Microbiological Resource Databank

MIRE Media Information Research Exchange; Member of the Institution of Radio Engineers

mirfac mathematics in recognizable form automatically compiled

MIran Middle Iranian

Miranda Rule reading of criminal suspect's rights under law before questioning

Miri Miranda; Miriam(ne)(ah)

MIRIAM Model Scheme for Information on Rural Development Initiatives and Agree Markets

MIRINZ Meat Industry Research Institute of New Zealand

mirl medium-intensity runway lights

MIRPL Major Item Repair Parts List

MIRR Materials Inspection and Receiving Report

MIRROS Modulation-Inducing Reactive-Retrodirective Optical System

mirv multiple independent reentry vehicle

mirv (MIRV) multiple independently-targeted reentry vehicle (warhead)

mirving fitting missiles with multiple warheads

mis metal insulator semiconductor; miscarriage; missing; mistake(n)

Mis. Miserere (Latin—have mercy); *Misisipi* (Spanish—Mississippi)

MIS Management Information Science; Management Information Services; Management Information System;

Master Integrated Schedule; Material Inspection Service; Medical Improvement Standard; Met Path Information System; Migrant Information Service; Military Intelligence Service; mine issuing ship (naval symbol); Mining Institute of Scotland; Minstrel Instruction Society; Modified Initial System; Multilingual Information System

M.I.S. Master of International Service

misa (MISA) medium-income sustainable agriculture

MISAA Middle-Income Student Assistance Act

mis. accur. misce accuratissme (Latin—mix very intimately)

misad misadventure

misc miscarriage; miscellaneous; miscible

MISC Malaysian International Shipping Corporation

mis. caute misce caute (Latin —mix cautiously)

miscend. miscendus (Latin—to be mixed)

miscg miscarriage

miscld miscalculated

miscln miscalculation

miscon misconduct

miscy miscellaneously

MISD Multiple Instruction, Single Data

mis doc miscellaneous documents

MISE Member of the Institution of Sanitary Engineers

MiSEA Michigan Society of Enrolled Agents

MISEP Mutual Information System on Employment Policies

miser microwave space relay

MISER Management Information System for Expenditure Reporting; Methodology of Industrial System Energy Requirements; Moorfields Information System Exception Reporting

mis. et seg. misce et signa (Latin—mix and write a label)

misg missing

Misha (Russian—Mikhail)— Michael

MISHAP Missile High-Speed Assembly Program

MISI Member of the Iron and Steel Institute

MISL Major Indoor Soccer League

MISLIC Mid-Staffordshire Libraries in Cooperation

mismed Mismedication

mis. mei miserere mei (Latin— have mercy on me)

misn misnumbered; mistaken

Misn Bch Mission Beach

MISO Military Intelligence Service Organization

misog misogynic(al)(ly); misogynist; misogynistic(al) (ly); misogyny

misol misologist; misology

MISP Member of the Institution of Sewage Purification

mispo mission summary printout

mispronc mispronunciation

Misr (Arabic—Egypt)

M I Sr Muy Illustre Señor (Spanish—Very Illustrious Sir)

MISR Macauley Institute for Soil Research; Major Item Status Report; Material Item Status Report(ing); Multiangle Imaging SpectroRadiometer

miss. mission; missionary

Miss. Mississippi; Mississippian

MISS Management and Information System Staff, Man In Space Soonest; Medical Information Science Section; Mississippi

missilese engineering jargon of guided-missile experts

missilex missile firing exercise

Missini Mussolini's neo-fascist followers

MISSIS Mississippi Student Information System

missy missionary

mist. mistura (Latin—mixture)

Mist Mistress

MIST Manchester Institute of Science and Technology; Medical Information Service (via) Telephone

mistr management of items subsequent to repair

MISTRAM Missile Trajectory Measurement System

mistrans mistranslation

misudstd misunderstand

misudstdg misunderstanding

misudstod misunderstood

mit market if touched; master instruction tape; milled in transit; minimum individual training; monoiodotyrosine

mit. mitte (Latin—send)

Mit Mittwoch (German— Wednesday)

M i T Made in Taiwan

M It Middle Italian

MIT Mara Institute of Technology (Kuala Lumpur); Maritime Institute of Technology; Massachusetts Institute of Technology (M.I.T. preferred as periods set it apart from all other MITs); Massachusetts Investors Trust; Materials Interaction Test(s); Military Intelligence Translator; Milwaukee Institute of Technology; Miracidial Immobilization Test; Municipal Investment Trust

M.I.T. Massachusetts Institute of Technology

MITAGS Marine Institute of Technology and Graduate Studies

MITC Magdalen Island Transportation Company

mite. master instrumentation timing equipment

MITERS Minor Traffic Engineering and Road Safety Improvements

MITGS Marine Institute of Technology and Graduate Studies

MITI Ministry of International Trade and Industry; Ministry of International Trade and Industry (Japan); Ministry of International Trade and Investment (Japan); Multilingual Intelligence Interface

mit insuf mitral insufficiency

mito minimum interval takeoff; miscellaneous tool

mito (Latin prefix—thread)— mitosis

mi tp miniature template

MITRE Massachusetts Institute of Technology Research Establishment

Mitrop Dimitri Mitropoulos

Mitropa Mitteleuropäische Schlafund Speisewagen Aktiengesellschaft (German— Middle-European Sleeping Car and Dining Car Company)

MITS Missouri-Illinois Traffic Service; Museum Institute for Teaching Science

mit. sang. mitte sanguinem (Latin—bleed)

Mitsubishi Mitsubishi Bank; Mitsubishi Corporation; Mitsubishi International Corporation

mitt minutes of telephone traffic

mitt mittente (Italian—sender)

Mitt Mitteilungen (German—communications)

MITT Management Implications of Team Teaching

mit. tal. mitte tales (Latin—send such)

mitt(s) mitten(s)

mitz mitzvah (Yiddish from Hebrew *miswah*—a good deed)

MIU Maharishi International University; Micronesian Insurance Underwriters

MIV Moody's Investor Service (stock exchange symbol)

MIWE Member of the Institution of Water Engineers

MIWMA Member of the Institute of Weights and Measures Administration

mix. mixture

mixt mixture

Mizrachi Merkaz Ruchani (Hebrew—Spiritual Center) —orthodox organization

mizzle mist + drizzle

mj marijuana; megajoule

mJ millijoule

MJ Mary Jane; (underground slang—marijuana); megajoule; Michael Jordan; Ministry of Justice

M.J. Master of Journalism

M & J sexologists William Masters and Virginia Johnson

MJ Military Justice Reporter

MJA Manuel José Arce; Mortimer J Adler

MJC Manatee Junior College; Masters and Johnson Center; Metropolitan Junior College; Moberly Junior College

MJCA Mississippi Junior College Association

mjd management job description

mjg management job guide

MJI Member of the Journalists Institute

MJ/kg megajoules per kilogram

MJme megajoules metabolizable energy

MJQ Modern Jazz Quartet

MJS Member of the Japan Society

MJSA Manufacturing Jewelers and Silversmiths of America

MJV Mojud Hosiery (stock exchange symbol)

mk mark (British equivalent of type)

mk (MK) master clock

mK millikelvin(s)

Mk Mark; markka (Finnish monetary unit)

Mk Manualkoppler (German—manual coupler)—organ

MK Mackey Airlines; Member of Knesset; Mishima-Kisumi (steel company)

M-K Morrison-Knudsen

M/K Member of the Knesset

M.K. Multi-Kontact (electrical accessories)

Mk$_{3nbc}$ Mark 3 nuclear; biological, chemical (coverall suit)

MKC Kansas City, Missouri (airport)

mkd marked

MKE Milwaukee, Wisconsin (airport)

mkg meter kilogram

MKGM Milli Kütüphane Genel Müdürlügü (Turkish—National Library General Directorate)—Ankara

MKH Mackintosh-Hemphill (stock-exchange symbol)

MKK Mitsubishi Kakoki Kaishi

mkm marksman

Mkm Mohammedan-killed meat

MKM Manawatu Knitting Mills

MKNP Malawi Kasungu National Park (Malawi); Mount Kenya National Park (Kenya)

MKO Mauna Kea Observatory; Muskogee Company (stock exchange symbol)

MKPL Modification Kit Parts List

mkr marker

mkr mikroskopisch (German—microscopic)

Mkr million Swedish kroner

MKR Mkuzi Game Reserve (South Africa)

mks meter, kilogram, solar second system of fundamental standards

mks^a meter, kilogram, second, ampere system

mkt market

Mkt Market

MKT Missouri-Kansas-Texas (railroad)

mktg marketing

mktl marketable

Mkt Mgr Marketing Manager

mk tp mark template

MKU Mary Kathleen Uranium (Australian firm)

MKW Military Knight of Windsor

MKY McKee and Company (stock-exchange symbol)

ml machine language; mean level; millilambert(s); milliliter(s); mine layer; mixed lengths; mold line; molder; money list; mother language; motor launch; muzzle-loading

ml (ML) maximum load

m/l middle left; missile lift

m:l monocyte-lymphocyte (ratio)

m or l more or less

m & l matched and lost

μl microliter

ml moneda legal (Spanish—legal tender)

m/l mi letra (Spanish—my letter)

mL millilambert(s)

Ml Malay; Malaya; Malayan; Malaysia; Manuel; marl

ML Mali (Internet code); Manuel; Maori Land; Martin-Marietta (stock exchange symbol); Micro Log; Middle Latin; Military Liaison; minelayer; Mineral Lease; Missile Launcher; Mitchell Library; Morgan Library; mother-in-law; motor launch; small minesweeper (naval symbol)

M/L MacNeil-Lehrer, PBS newscasters Robert MacNeil and Jim Lehrer; Maersk Line

M.L. Medicinae Licentiatus (Latin—Licentiate in Medicine)

mla magnetic lens assembly; manpack loop antenna; microwave linear accelerator

MLA Maine Library Association; Maintenance Level Analysis; Manitoba Library Association; Marine Librarians Association; Maryland Library Association; Massachusetts Library Association; Master Locksmiths Association; Medical Library Association; Member of the Landlord's Association; Member of the Legislative Assembly; Michigan Library Association; Minnesota Library Association; Mississippi Library Association; Missouri Library Association; Modern Language Association; Montana Library Association; Music Library Association

M-LA Mont-Laurier Aviation

M.L.A. Master of Landscape Architecture

MLAA Modern Language Association of America

MLAC Multi-Level Access Control

m'lady my lady

mlaf missile-loading alignment fixture

MLAP Migrant Legal Action Program

ml ar mill arbor

M. L. Arch. Master of Landscape Architecture

mlases molasses

MLAT Modern Language Aptitude Test

mlb main load bus; multilinear board

Mlb thousand pounds

MLB Major League Baseball; Marginal Lands Board; Maritime Labor Board; Multiple Listing Board; Multiple Listing Bureau

ML & BC Montana Logging and Ballet Company

mlbf mean life before failure

mlbm (MLBM) modern large ballistic missile

Mlbo Malabo (formerly Santa Isabel)

MLBPA Major League Baseball Players Association

mlc machine level control; machine location card; main lobe clutter; mesh level control; microelectric logic circuit; missile launch computer (MLC); mixed leucocyte culture; motor load control; multilayer circuit; multilens camera; multiplanar chain link

MLC Maori Land Court; Meat and Livestock Commission; Member of the Legislative

Council; Military Liaison Committee; Mutual Life and Citizens (insurance company)

mlca machine level control address

MLCAEC Military Liaison Committee to the Atomic Energy Commission

MLCRA Model Litter Control and Recycling Act

ml cu mill cutter

mld mailed; manual lymphatic drainage; middle landing; minimum lethal dose; minimum line of detection; molded; mixed layer depths

mld (MLD) metachromatic leukodystrophy

MLD Missile Launch(ing) Detection

mld₅₀ minimum lethal (radioactive) dose

M. L. Des. Master of Landscape Design

mldg moulding

mldr molder

mle maximum likelihood estimate; maximum loss expectancy; microprocessor language editor

mle modèle (French—model, pattern)

Mle Mile

ML (Ec) Ecclesiastic Middle Latin

M. L. Eng. Master of Landscape Engineering

MLES Multiple-Line Encryption System

MLEU Mouvement Libéral pour l'Europe Unie (French—Liberal Movement for a United Europe)

mlf media language and format

m/lf medium/low frequency

Mlf thousand linear feet

MLF Mobile Land Force(s); motor launch, fast (naval symbol); Multi-Lateral Force

MLF Mouvement de Libération de la Femme (French—Women's Liberation Movement)

ml fx mill fixture

mlg mailing; main landing gear; most languages

MLG Middle Low German

MLG Ministry of Labour Gazette

mlge mileage

mlg(s) mailing(s)

mlgt marine light

Ml'H Musée de l'Homme (French—Museum of Man)—Paris

MLHA Master Ladies Hairdressers Association

mlhw mean lower high water

mli minimum line of interception

M-Li Muller-Lyer (illusion)

M. Lib. Master of Librarianship

M. Lib. Sci. Master of Library Science

MLIRB Multi-Line Insurance Rating Bureau

M. Lit. Master of Letters; Master of Literature

MLK Jr Martin Luther King, Junior

mlk mag milk of magnesia

MLL Manchester Lines Ltd; Manned Lunar Landing; Music Lovers League

Mlle Mademoiselle (French—Miss)

Mlles Mesdemoiselles (French—Misses)

mllw mean lower low water

mllws mean lower low water springs

mlm million locomotive miles

MLM Multi-Level Marketing

MLMA Metal Lath Manufacturers Association; Miners' Lamp Manufacturers' Association

MLMTT Marxism-Leninism-Mao Tse-tung Thought

mln million

Mln Milan

MLN Movimiento Liberación Nacional (Spanish—National Liberation Movement)—Guatemalan political party; *Movimiento de Liberación Nacional* (Spanish—National Liberation Movement)—Uruguay

MLNP Malawi Lengwe National Park

mlnr milliner

MLNR Ministry of Land and Natural Resources

mlns mucocutaneous lymphnode syndrome

MLNS Ministry of Labour and National Service

MLNWR Medicine Lake National Wildlife Refuge (Montana)

MLO Midland Light Orchestra; Military Liaison Office(r)

mlocr multiline optical character reader

m'lord my lord

Mloth Midlothian

M Low G Middle Low German

mlp master limited partnership; metal lath and plaster; multiple-line printing

m.l.p. *mento-laeva posterior* (Latin—left mento-posterior)

MLP Master Logistics Plan

mlpwb multilayered printed wiring board

MLP USA Marxist-Leninist Party, USA

MLQ *Modern Language Quarterly*

mlr main line of resistance; minimum lending rate; mortar-locating radar; multiple linear regression; multiple-line reading; multiple-location risk; muzzle-loading rifle

mlr (MLR) minimum lending rate; missile launch(ing) response, mixed lymphocyte response

m-l r muzzle-loading rifle

MLR Main Line Rail(way); Marine Life Resources (program)

MLR *Modern Labor Review; Modern Law Review*

MLRB Master Logistics Review Board; Mutual Loss Research Bureau

mlrc multi-level railway car

m-l rg muzzle-loading rifled gun

MLRP Marine Life Research Program

MLRS Multiple Launch(ing) Rocket System

MLRS-TGW Multiple Launch Rocket System—Terminally Guided Weapon

mls machine literature search(ing); median longitudinal section; medium life span; medium long shot; milliliters

ml's magnetically levitated railroad trains

ml's (MLs) mine layers

Mls Mills

MLS Microwave Landing System; Mixed Language System; Moon Landing Site; Multi-Language System; Multi-Level Security; Multiple Listing Service

M.L.S. Master of Library Science

M & LS Manistique & Lake Superior (railroad)

MLSA Ministry of Labour Staff Association

mlsc measured logistic support cast

MLSU Moscow MV Lomonosov State University (University of Moscow)

mlt mean low tide; median lethal (radioactive) time (MLT)

mlt (MLT) master library tape; median lethal (radioactive) time

Mlt Malta

mltl mean low tide line

mltn 1000 long tons

mltu missile loop test unit

mlty military

mlu mean length of utterance; mid-life upgrade

m'lud my lord

mlv membrane light valve; murine leukemia virus

mlv(M) murine leukemia virus (Moloney)

mlv(R) murine leukemia virus (Rauscher)

ml vs mill vise

mlw maximum landing weight; mean low water; medium level waste; military land warrant

MLW Monrovia, Liberia (airport)

M.L.W. Master of Labour Welfare

mlwli mean low water lunitidal interval

mlwn mean low water neaps

mlws mean low water springs; minimum level water stand

mix millilux

mly multiply

MLYC Moosehead Lake Yacht Club

mm made merchantable; mass market (book); megameter(s); merchant marine; metronome marking; middle marker; millimeter(s); millimicron; mismated; modified mercalli (scale); money maket; monthly meeting; mucous membrane

mμ millimicron(s)

m'm madam

m/m millimeter(s)—small-arms ammunition term meaning the diameter of a weapon's bore expressed in millimeters

m & m make and mend; morbidity and mortality

μm micrometer; micron(s)

mm *med mera* (Swedish—and so forth, etc.)

m.m. *mutatis mutandis* (Latin—with the necessary changes)

mm² square millimeter

mm³ cubic millimeter

mM millimole; millimore

m/M male Mexican

μM micromole(s)

Mm *Martyres* (Greek—witnesses, martyrs)

MM Master Mason; Machinist's Mate; Maintenance Manual; Majesties; Manufacturing Manual; Marilyn Monroe; Marine Midland (stock exchange symbol); Mariner Mars (NASA project); Martin Marietta; Martyres (martyrs); Maryknoll Missionary; Material Management; maximum misfit; Medal of Merit; mercantile marine; merchant marine; Methadone Maintenance; Messageries Maritimes; Metropolitan Museum; Mickey Mouse; Military Medal; Minister of Munitions; Minuteman (missile); Mira Mesa; Moral Majority; Myanmar (formerly Burma—Internet code)

M-M Marshall-Marchetti

M.M. Master of Music

M/M Mr and Mr; Mr and Mrs; Mr and Ms; Mrs and Mrs; Mrs and Ms

M & M Marxism and Market economy; Medicare and Medicaid; Merton and Morden

M for M *Measure for Measure*

M of M Ministry of Munitions; Museum of Man

MM *Marine Marchande* (French—Merchant Marine); *Messieurs* (French—gentlemen); *Modern Medicine*

M.M *Maelzel's Metronome*

M & M *Morbidity and Mortality* (Center for Disease Control's weekly report)

mma major maladjustment; money market account; multiple module access

MMA Maine Maritime Academy; Malaysia Medical Association; Manitoba Medical Association; Maritime Museum of the Atlantic (Halifax); Massachusetts Maritime Academy; Material Manufacturing Authorization; Merchandise Marks Act; Meter Manufacturers' Association; Metropolitan Museum of Art; Missile Maintenance Area; Monorail Manufacturers Association; Museum of Modern Art; Music Masters Association

MM of A Minute Men of America

mmac multiple model adaptive control

MMAC Material Management Aggregate Code

m-machine marijuana machine

MMAJ Metal Mining Agency of Japan

MMAL Mitsubishi Motors Australia, Ltd

MMANA Monarch Migration Association of North America

MMAS Manufacturing Management Accounting System; Massachusetts Male Aging Study

M. Math. Master of Mathematics

MMB Marine Midland Bank; Milk Marketing Board; Mitsui Manufacturers Bank

mm bat main missile battery

MMBC Maryland Motor Boat Club

mmbd million barrels per day

mmBtu/hr million Btu's per hour

mmc marine moisture control; maximum metal condition; metal matrix composite; money market certificate

MMC Malaysian Marketing Corporation; Malaysian Mining Corporation; Marine Moisture Control; Materiel Management Code; Meharry Medical College; Midwest Medical Clinic; Mitsubishi Motors Corporation; Monopolies + Mergers Commission

MMCB Midwest Motor Carriers Bureau

mmcfpd million cubic feet (of gas) per day

MMcKNP Mount McKinley National Park (Alaska)

MMCL Major Missile Component List(ing)

MMCNY Marine Museum of the City of New York

mmc's money-market certificates

MMCT maritime mobile coastal telegraphy

mmd mass median diameter; master monitor display; moving map display

MMD minelayer, fast (naval ship symbol); Movement for Multiparty Democracy

mmda money market deposit account

MMDS Multichannel Multipoint Distribution Service

mmdu multi-mode display unit

m mde *marine marchande* (French—merchant marine)

mme maximum maintenance effort

Mme *Madame* (French—Missus)

MME Manned Mars Expedition; Midland Mathematics Experiment

M.M.E. Master of Mechanical Engineering; Master of Music Education

mmect multiple-monitored electroconvulsive therapy

M. Mech. Eng. Master of Mechanical Engineering

M. Med. Master of Medicine

MMEG Meter Manufacturers' Export Group

Mmes *Mesdames* (French—ladies)

M. Met Master of Metallurgy

M. Met. E. Master of Metallurgical Engineering

mmf magnetomotive force; micromicrofarad; money market fund

μμf micromicrofarad(s)

MMF fleet mine layer (naval symbol); *Maggio Musicale Fiorentino* (Italian—Florence May Festival); Milbank Memorial Fund

MMFA Montreal Museum of Fine Arts

mmfds microfarads

MMFI Moravian Music Foundation, Incorporated

MMFPI Man-Made Fiber Producers Institute

mmg medium machine gun

MMGR Masai Mara Game Reserve (Kenya)

MMGS Mount Muhavura Gorilla Sanctuary (Uganda)

M.Mgt.Eng. Master of Management Engineering

mmh manual materials handling

mmh/fh maintenance man-hours per flight hour

mmHg millimeter of mercury

mmi management and maintenance inspection; microphage migration inhibition; modified mercalli intensity

Mmi Miami

MMI Malaysian Marine Industries; Manufacturers Mutual Insurance; Micro-Magnetic Industries; Moslem Mosque Incorporated (formerly American Mohammedan Society)

M. Mic. Master of Microbiology

M. Mi. Eng. Master of Mining Engineering

MMIJ Mining and Metallurgical Institute of Japan

MMIS Master of Management Information Systems; Medicaid Management Information System; Modified Mercalli Intensity Scale

MMJC Meridian Municipal Junior College

m mk material mark

mml multimaterial laminate

mm/l millimols per liter

MMLES Map-Match Location Estimation System

MMLME Mediterranean, Mediterranean Littoral, and/or Middle East (sector of conflict)

mmm merchandising, marketing, management; military medical mobilization; millimicron(s)

mMm mobile Minuteman missile

MMM Mauritian Militant Movement; Merseyside Maritime Museum (Liverpool); Minerals, Mining, and Metallurgy; Modern Music Masters

MMM *Membre de l'Ordre du Mérite Militaire* (French—Member of the Order of Military Merit)

MMMA Maine Merchant Marine Academy; Metalforming Machinery Makers Association

MMMC Medical Materiel Management Center; Milking Machine Manufacturers Council

mmmf money-market mutual fund

MMMF Multinational-Mixed Manned Force(s)

mmm/fhr maintenance man minutes per flight hour

mmmrpv (MMMRPV) modular multi-mission remotely piloted vehicle

MMMS Modern Music Masters Society

MMM & SA Master Monumental Masons and Sculptors Association

MMN Museum of Man and Nature (Winnipeg)

MM & N Museum of Man and Nature (Winnipeg, Manitoba)

MMNA Monthly Maintenance Needs Allowance

MMNP Mount McKinley National Park (Alaska)

mmo medium machine oil

Mmo Malmö

MMO Maine Meteorological Office; Music Minus One

MMOA Mobile Modular Office Association

MMOB Military Money Order Branch

MMOS Multi-Modal Organ Modeling System

MMOW Machinist's Mate of the Watch (USN)

mmp matrix metalloprotease

mmp (MMP) maritime mobile phone

MMP Major Medical Plan(ning); Masters, Mates and Pilots (union); Military Mounted Police

MM & P Masters, Mates and Pilots

MMPA Marine Mammal Protection Act; Midland Master Printers' Alliance

MMPC maritime mobile phone coastal

MMPDC maritime mobile phone distress and calling

MMPI Minnesota Multiphase Personality Inventory

MMPNC Medical Materiel Program for Nuclear Casualties

mmpp millimeters partial pressure

MMPP Moose Mountain Provincial Park (Saskatchewan)

mmq minimum manufacturing quality

mmr mass miniature radiography; measles, mumps, rubella (vaccine); minimum maintenance requirement; minimum management requirement

mm & r maintenance modification(s) and repair(s)

MMR Main Machinery Room (USN); Mass Media Research; Method of Mixed Ranges

MMRA Maritime Marshland Rehabilitation Administration (Canada)

MMRB Master Material Review Board

mmrbm (MMRBM) mobile medium-range ballistic missile

MMS Manpower Management System; Mass Memory System; Metabolic Monitoring System; Microfiche Management System; Minerals Management Service; Mobile Monitoring System; Modulation Measuring System; motor minesweeper; multimission ship (naval symbol); Multiplex Modulation System

M.M.S. Master of Management Studies; Master of Medical Science

MMSA Materials and Methods Standards Association; Mercantile Marine Service Association; Mining and Metallurgical Society of America

M.M.S.A. Master (Mistress) of Midwifery of the Society of Apothecaries

MMSC Mediterranean Marine Sorting Center

mmscfd million standard cubic feet per day

m & m session morbidity and mortality session

MMSR Master Material Source Record

MMSS Missile Motion Subsystem

MMSW Mine, Mill and Smelter Workers (union)

mmt main mantle thrust (geology); manual muscle test(ing); maritime mobile telegraphy; memory test(er); missile mate test(ing); multicomponent mass transport; multimodal therapy; multiple-mirror telescope

MMT Manual Muscle Test; maritime mobile telegraphy

MMTC maritime mobile telegraphy calling

MMTDC maritime mobile telegraphy distress and calling

MMTP Methadone Maintenance Treatment Program

mmtr maintenance man hours to repair

mmtv mouse mammary tumor virus; murine mammary tumor virus

mmu millimass unit(s)

MMU Manned Maneuvering Unit; McMaster University; Memory Management Unit

M.Mus. Master of Music

mmw millimeter wave

MM & W McKim, Mead & Wright (American architects)

MMWD Marin Municipal Water District

mmx memory multiplexer

mmy military man years

MMY *Mental Measurements Yearbook*

mn manual; million

m(n) microfilm negative

mn *maison* (French—house)

m.n. *mutato nomine* (Latin—the name being changed)

m/n *moneda nacional* (Spanish—national currency)

Mn Main; manganese

MN Magnetic North; meganewtons; Merchant Navy; Minnesota; Mongolia (Internet code)

M.N. Master of Nursing

MN *Magyar Nepkoztarsasag* (Hungarian People's Republic); *Musée Nationale* (French—National Museum)

mna (MNA) multi-network area (tv)

MNA Multi-National Account(s)

MNA *Matematikmaskinnämnden* (Swedish—Swedish Computing Machinery Board)

M.N.A. Master of Nursing Administration

MNAG Museo Nacional de Antropología (Spanish—National Anthropology Museum), Guatemala

MNAM Museo Nacional de Antropología (Spanish—National Anthropology Museum), Mexico

MNAOA Merchant Navy and Airline Officers' Association

M. N. Arch. Master of Naval Architecture

MNAs Members of the National Assembly (Québec)

MNAS Member of the National Academy of Sciences; Military Navigational Aids System

Mnasi Mnasidika

MNB Macias Nguema Biyogo (formerly Fernando Po); Moscow Narodny Bank

M-N BA Multi-National Business Association

MNC Major NATO Commanders; Media News Corporation; Multinational Corporation

mncpef meaning not clear; please explain fully

MNCR Mouvement National Contre le Racisme (French – National Movement Against Racism)

MNCRR Metro-North Commuter Railroad

mnc's multinational corporations

MNCS Multipoint Network Control System

mnd minimum necrosing dose

Mnd Mound

MND Ministry of National Defence

M-N D A Midsummer Night's Dream (Shakespeare)

mndth mean depth

MNDTS Member of the Non-Destructive Testing Society

M.N.E. Master of Nuclear Engineering

MNEA Merchant Navy Establishment Administration

mnem mnemonic

Mnemo Mnemosyne (goddess of memory and mother of the nine muses)

mnemon minimum unit of information; mnemoneutic(al)(ly); mnemonic(al)(ist);

mnemonician(s); mnemonicon; mnemonic(s); mnemonist(s); mnemonization(al)(ly); mnemonize(r); mnemotechnic(al)(ly); mnemoteechny

M. N. Eng. Master of Naval Engineering

MNF Menagasha National Forest (Ethiopia); Multilateral Nuclear Force (NATO navy)

mnfe missile not fully equipped

mnfg manufacturing

MNFP Multi-National Fighter Program

mnfrs manufacturers

mng managing; meaning

Mng Mongolia(n)

mnging managing

mngmt management

Mngr manager

mngt midnight

mnh mint never hinged

MNH Museum of Natural History (Smithsonian)

MNHN Museo Nacional de Historia Natural (Spanish—National Natural History Museum)—Uruguay; *Musée National d'Histoire Naturelle* (French—National Natural History Museum)—Paris

MNI Malaysian National Insurance; Member of the Nautical Institute; Ministry of National Insurance

MNIMH Member of the National Institute of Medical Herbalists

mnl marine navigating light

Mnl Manila; Manuel

MNL Main North Line; Manila, Philippines (airport)

MNLF Malayan National Liberation Front; Moro National Liberation Front

MNLL Malaysian National Liberation League

MNLO Merchant Navy Liaison Officer

MNLOA Merchant Navy and Air Line Officers Association

mnls modified new least squares

MNLS Marine Navigating Light System

mnm minimum; mnemonic (*see* mnemon)

MNM Museum of New Mexico

MNN Medical News Network

MNNP Malawi Nyika National Park

mnos metallic nitrogen-oxide semiconductor

M-note $ 1000 bill

MNP Malay National Party; Marsabit National Park (Kenya); Meru National Park (equatorial Kenya); Mikumi National Park (Tanzania); Mushandike National Park (Rhodesia)

MNPL Machinist Non-Partisan Political League

mnpo main port

MNPS Minimum Navigational Performance Specification

mnpz monopolize

mnpzd monopolized

mnpzg monopolizing

mnpzn monopolization

mnr massive nuclear retaliation; mean neap rise

Mnr Manor

Mnr Mijnherr (Dutch—Mr, Sir)

MNR Mozambique National Resistance (Renamo)

MNR Movimiento Nacionalista Revolucionario (Spanish—National Revolutionary Movement)

MNRJ Museo Nacional de Rio de Janeiro (Portuguese—National Museum of Rio de Janeiro)

MNRS Mobile Neutron Radiography System

MNRU Medical Neuropsychiatric Research Unit

mns metal-nitride-semiconductor (transistor)

Mns Manaus; Mines

M.N.S. Market News Service; Master of Nutritional Science

M. N. Sc. Master of Nursing Science

MnSEA Minnesota Society of Enrolled Agents

m'ns'l mainsail

Mnstr Munster

mnt mean neap tide

MNT Minnesota and Ontario Paper (stock exchange symbol)

mntmp minimum temperature

mntn maintain; maintenance

mntnc maintenance

mntnd maintained

mntng maintaining

MNTO Moroccan National Tourist Office

mntr monitor

MNU Maniti Sugar (stock exchange symbol)

M.Nurs. Master of Nursing

mnv mine-neutralization vehicle

MNWEB Merseyside and North Wales Electricity Board

MNWR Malheur National Wildlife Refuge (Oregon); Mattamuskeet NWR (North Carolina); Merced NWR (California); Mingo NWR (Missouri); Minidoka NWR (Idaho); Mississiquoi NWR (Vermont); Modoc NWR (California); Montezuma NWR (New York); Moosehom NWR (Maine)

mnx (short-order slang contraction—ham and eggs)

Mnx Manx (Manx Gaelic)

Mnzlo Manzanillo

mo mail order; manual operation; manually operated; masonry opening; mass observation; master oscillator; medical-only; method of operation; moment; money; money order(s); monthlies; monthly; month(s); moth eaten; mother; motor operated; mustered out

mo (MO) molecular orbital

mo' more; morning

m-o months old

m/o maintenance-to-operation (ratio)

m & o maintenance and overhaul(ing); management and organization

m.o. *modus operandi* (Latin—manner, method, or mode of operating, way of working)

m/o *mi orden* (Spanish—my order)

m/O male Oriental

Mo Missouri; Missourian; molybdenum; Monday; Morris; Moselle; Moses; Mozelle

Mo' Moses

Mo *Maestro* (Italian—master, title given any great artist, composer, conductor, or teacher)

MO Macau (Internet code); Mail Order; Marketing Organization; Mass Observation; Medical Officer; Meteorological Office; Missouri; Mobile Station; Mohawk Airlines (2-letter coding); Money Order; Monthly Order; Morale

Branch (of Secret Service); Movement Order(s); Municipal Office(r)

M-O Morris-Oxford

M & O Muscat and Oran

moa medium observation aircraft; minute of angle; missile optical alignment; mud on airstrip

MoA Ministry of Agriculture

M o A Memorandum of Agreement

MOA Marine Office of America; Metropolitan Oakland Area; Metropolitan Opera Association; Metropolitan Opera Auditions; Military Operatitons Area; Ministry of Aviation; Minnesota Orchestral Association; Municipal Officers Association; Music Operators of America

MOADS Montgomery Air Defense Sector

MOAMA Mobile Air Materiel Area

MOARS Mobilization Assignment Reserve Section

moat missile-on-aircraft test(ing)

moAt mainstream of American thought

mob make or buy; mobile; mobilization; mobilize(d)

mob. *mobile vulgus* (Latin—disorderly group of people)

Mob Mobile, Alabama (maritime abbreviation)

MOB Main Operating Base; Mobile, Alabama (airport); Montreux-Oberland-Bernois (railway)

Mo'Bay Mobile Bay, Alabama; Montego Bay, Jamaica

mobcom mobile communications

MOBCOM Mobile Command (Canadian)

mobeu mobile emergency unit

MOBIDACS Mobile Data Acquisition System

mobidic mobile digital computer

MOBIDICK Multivariable On-Line Bilingual Dictionary Kit

mobil mobility

mobilarian mobile branch librarian

mobilary mobile library

mobiles motion sculptures (plastic forms in motion)

Mobil Wl Mobil World

mobl macro-oriented business language

mobl *möbliert* (German—furnished)

moblas mobile laser satellite tracking station

mob lib mobile librarian; mobile library

mob lt man overboard and breakdown light

mobot(s) mobile robot(s)

MOBS Mobile Ocean Basing System; Multiple Orbit Bombardment System

MOB spread price differential between future municipal bonds and Treasury bond futures

MOBTA Mobilization Table of Distribution and Allowances

mobula model-building language

moc manufacturing other charges; master operation(al) control(ling); memory operating characteristics; mission operations computer; mocassin

MOC Maintenance Operation Center; Makapuu Oceanic Center (Hawaii); Mauna Olu College (Maui)

moca minimum obstruction clearance altitude

MOCA Museum of Contemporary Art

mocamp motor camp; motorists camp

MOCCC Massachusetts Organized Crime Control Council

MOCI Ministry of Commerce and Industry

Mo City Motor City (Detroit)

mocktail(s) mock cocktail(s)—free from alcohol

MoCom Mobile Command

MOCOM Mobile Command (US Army)

mocp missile out of commission for parts

mocr mission operation control room

mocs mocassins

mod magneto-optical disc; manned orbital development (MOD); mesial-occlusaldistal (dental cavities); model; moderate; modern; modernize(d); modification; modify; modular; module

m-o-d mesial-occlusal-distal (inlay)

Mod Modern
M o D Ministry of Defense (British)
MOD Mail Order Department; Medical Officer of the Day; Ministry of Defense; Ministry of Overseas Development; Miscellaneous Obligation Document
modasm modular air-to-surface missile
m-o-d-b mesial-occlusal-distalbuccal (inlay)
modcom modernity commercialized
mod cons modern conveniences
mod-cons modern-construction houses
mod/demod modulate-demodulate; modulating-demodulating (units)
ModE Modern English
MODE Mid-Ocean Dynamic Experiment
modem modulating-demodulating; modulation-demodulation; modulator-demodulator
MODEM Modeling of Emission (and consumption in urban areas)
Modern Lib Modern Library
Modern Nihilist Jean Genet
modf modification, modify
ModGr Modern Greek
ModHeb Modern Hebrew
mod/iran modification, inspection, and repair as necessary
MODIS MODerate-resolution Imaging Spectroradiometer
ModL Modern Latin
modo. moderato (Italian—moderately)
Mod Photo Modern Photography
mod. pres. modo prescripto (Latin—in the manner prescribed)
modr moderate room rate desired
mods mesial-occlusal-distal (dental cavities); models; moderates; moderators; moderns; modification; modifiers; modulators; modules
MODS Manned Orbital Development Station (or System); Manned Orbiting Development Station (or System); Medically Oriented Data System

modto moderato (Italian—moderately)
moe measure of effectiveness
Moe Moses
MoE Ministry of Education
M o E Ministry of Energy
MOE Major Organizational Entity
MOEA Ministry of Economic Affairs
Mo'ed Q. Mo'ed qatan
mof maximum observed frequency; member of (the police) force; metal oxide fillm
M o F Ministry of Finance
MOF Ministry of Food
moff multiple options funding facility
mo' fo mother fucker
Mog Margaret
MOG Metropolitan Opera Guild
M.O.G. Master of Obstetrics and Gynaecology
mogas motor gasoline
moggie mongrel
moggy mongrel
moh material overhead; maximum operating hours
M o H Ministry of Health
MOH Medical Officer of Health; Ministry of Health; Mohawk Airlines
Moham Mohammedan
MOHATS Mobile Overland Hauling and Transport System (USAF)
MOHLG Ministry of Housing and Local Government
μohm microhm
mohms milliohms
moho Mohorovicic discontinuity
Mohole a hole to the Mohorovicic discontinuity, the boundary between the earth's crust and mantle
mohs mud, oil, hooks, slings (oil well insurance)
moi maximum obtainable irradiance; military occupational information; multiplicity of infection
MoI Ministry of the Interior
MOI Military Operations and Intelligence; Ministry of Information
MOIC Medical Officer in Command
M.O.I.G. Master of Occupational Information and Guidance

moip missile on internal power
MOIS Minnesota Occupational Information System
Moish Moishe
Moish Moishe (Yiddish—Moses)
moiv mechanically operated inlet valve
Mok Mokpo
MOK Mohawk Carpet Mills (stock exchange symbol)
mol machine-oriented language; maximum output level; mole; molecular; molecule
mol. mollis (Latin—soft)
Mol Moldova (whose capital is Kishinev); Mollendo
M o L Minister of Labour; Ministry of Labour
MOL Manned Orbiting Laboratory; Mitsui-OSK Lines
M.O.L. Master (Mistress) of Oriental Languages
molab mobile laboratory
MOLAB Mobile Lunar Laboratory
mol cock molotov cocktail
Mol Crys Liq Crys Molecular Crystals and Liquid Crystals
Mold Moldavia(n)
MOLDS Management On-Line Data System
mole. molecular; molecule
molecom molecularized computer
Mo Life Missouri Life
Molink Moscow link (teletype cable circuit linking Moscow's Kremlin with Washington, D.C.'s White House), The Hot Line
moll metallo-organic liquid laser
mol/l molecules per liter
mollie mollienisia (tropical fish)
mollie(s) mare mule(s)—*see* hinny
Mollus Mollusca
MOLLUSA Military Order of the Loyal Legion of the U.S.A.
MOLNS Ministry of Labour and National Service
MOLOC Ministry of Labour Occupational Classification
Mol Phys Molecular Physics
MOLS Mirror Optical Landing System
molt. molten
Moluccas Maluku or Spice Islands of Indonesia

mol wt molecular weight

moly molybdenum

mom military ordinary mail; milk of magnesia

mom (MOM) micromation on-line microfilmer

m-o-m member of the month; middle of month; milk of magnesia

m/o m/ más o menos (Spanish—more or less)

Mom Momma

MOM Meals On the Move

MOM Musée Océanographique Monaco (French—Monaco Oceanographic Museum)

MoMA Museum of Modern Art

MOMA Methods of Moderation and Abstinence

m-o-m in a.m. if no bm by p.m. milk-of-magnesia in the morning if no bowel movement by evening

momar modern mobile army

momau mobile mine assembly unit

Moml Moslem meal

Mo-Mo MF Grant—pseudonym

MOMR Mayor's Office of Manpower Resources

moms missile operate mode simulator

moms mervaerdiomsaetningsskat (Danish—value-added tax); *mervardesomsattningsskatt* (Swedish—value-added tax)

MOMS Mothers for Moral Stability; Mothers of Murdered Sons

MOM/WOW Men Our Masters/Women Our Wonders (anti-feminist acronym reading the same upside down)

mon monetary; monsoon; monument; motor octane number

mon maison (French—house)

Mon Monaco; Monday; Monegasque; Mongol(ia)(n); Monitor; Monmouthshire; Monoceros (constellation); Monongahela; Monsieur (French—Mister); monument

Mon Mónaco (Spanish—Monaco); *Montag* (German—Monday)

MON Ministrstwo Obrony Naradowej (Polish—Ministry of National Defense)

Mona Madonna (Italian—Lady, Our Lady); (Manx—Isle of Man)

Monaco Principality of Monaco (tiny Mediterranean country famed for its gambling casino)

Monag Monaghan

Monas Monastic(ism); Monastery

Monashees Monashee Mountains of British Columbia

monbas monobasic

MONC Metropolitan Opera National Council

Mondale Walter E Mondale, 42nd Vice President of the United States

mon/dir monitoring direction

MONEVAL Monthly Evaluation Report (USA)

monex monsoon experiment

mong mongolisch (German—Mongolian)

Mong Mongol; Mongolia(n)

Mongolia Mongolian People's Republic (landlocked Asiatic nation of great antiquity), *Bügd Nayramdakh Mongol Ard Uls*

Mongoose Mongoose Gang (secret police in Grenada)

'mongst amongst

mon-H monohydrogen

Monh. Monhegan (outermost offshore island off coast of Maine)

Moni Monica; Monika

monic monocular

monik moniker

Monitor Christian Science Monitor

Mon Not Roy Astron Soc. Monthly Notices of the Royal Astronomical Society

mono mononucleosis; monophonic; monopoly; monopropellant; monorail(road); monotype; monotyper

mono (Latin prefix—alone, one, single)—monograph, monorail

Mono Monocerus (constellation)

monob (MONOB) mobile noise barge

monocl monoclinic

Monocer Monoceros (Latin—Unicorn constellation)

monocot(s) monocotyledon(s)

Monod Monon Railroad

monog monogram; monograph

monokini one-piece topless bikini (swimsuit)

monos monitor out of service

monot monotonous; monotony; monotype; monotypic

monpl monopoly

Mon River Monongahela River

mons (Latin prefix—mountain)—monstrosity

Mons Monsieur (French—Mister)

Mons Cur Monsoon Current

Monsig Monseigneur (French—My Lord)

monsoons seasonal storms of southern Asia

monstro(s) monstrosity; monstrosities

mont montane

Mont Montana; Montanan; Monterrey; Montevideo; Montgomery; Montgomeryshire; Montpelier; Montreal

Montagne Ce qu'on entend sur la Montagne (French—What one hears on the mountain)—Liszt's Symphonic Poem No 1

Montalbán Manuel Vázquez Motalbán (Spanish novelist)

Monte Montague; Monte Carlo; Montebianco (Mont Blanc); Montefiore; Montevideo; Montgomery

Montesquieu Baron de La Brède et de Montesquieu, Charles-Louis de Secondat (1689–1755)

Montezuma Castle Montezuma Castle National Monument in central Arizona

Montgom Montgomeryshire

Montie Montgomery

Montparno Montparnasse

Montr Montreal

montrg monitoring

Mont S Montreal Star

monu monument

Monty Montagu; Montague; Montana; Montgomery; Montmorency

Mony monastery

MONY Music Operators of New York; Mutual Life Insurance Company of New York

MOO Money Order Office

Moody's Moody's Investors Service

Moody and Sankey Dwight Lyman Moody and Ira David Sankey—an evangelist preacher and his organist partner

moop mechlorethamine, vincristine, procarbazine, prednisone (Hodgkin's disease treatment)

MOOP Ministerstvo Okhranenia Obshehestvennogo Poriadka (Russian—All-Union Ministry for the Preservation of Public Order)—secret police agency

Moor Dartmoor Prison, Devon, England

Moore's Adj Moore's International Adjudications

Moore's Arb Moore's International Arbitrations

Moore's Dig Moore's Digest (of international law)

MOOSE Move Out of Saigon Expeditiously (USA)

moot move(d) out of town; moving out of town

mop mother-of-pearl; mustering-out pay

mop medical outpatient; mother-of-pearl; muster-ingout pay

M o P Member of Parliament; Minister of Pensions; Ministry of Pensions; Minister of Power; Ministry of Power; Minister of Production; Ministry of Production

MOP Migrant Opportunity Program

MOP Ministerio de Obras Publicas (Spanish—Ministry of Public Works)

mopa master oscillator power amplifier

MOPA Museum of Photographic Arts (San Diego)

MoPac Missouri Pacific—Texas & Pacific (railroad)

mopar master oscillator-power amplifier radar

mopb manually operated plotting board

mopeds motorized pedals (bicycles containing auxiliary motors)

mopf missile onloading prism fixture

MOPH Military Order of the Purple Heart

MOPITT Measurements of Pollution in the Troposphere

mopms modular pack mine system

mopr manner of performance rating; mop rack

mops millions of operations per second

MOPS Merchandise Order(ing) Processing System; Missile Operations System

MOPSS Multispectral Opium Poppy Sensor System

M. Opt. Master of Optometry

mor middle of the road; morocco; mortar

mor (MOR) middle-of-the-road (tv program)

mor morendo (Italian—dying away, gradual softening of tone and slowing of tempo)

Mor Morelia; Morelos; Morisco; Moroccan; Morocco

M o R Ministry of Reconstruction

MOR Mandatory Occurence Report(ing); Military Operations Research

Morav Moravia; Moravian

Morb Morbihan

MORC Medical Officers Reserve Corps; Midget Ocean Racing Club (smallest racing cruisers)

Mord Mordecai, Mordehai

Mordhy Mordehai

mor. dict. more dicto (Latin—as directed)

Mordy Mordechai

MORE Mission for Outreach, Renewal, and Evangelism

Moreau Louis Moreau Gottschalk (1829–1869)

moreps monitor station reports

morf hemaphrodite

mor fib moral fiber

morf(ie) morphine

MORG Museo Oceanografico de Rio Grande (Portuguese—Oceanographic Museum of Rio Grande)—Brazil

morg mar morganatic marriage

MORI Market Opinion and Research International

moritzer mortar howitzer

MORL Manned (or Medium) Orbital Research Laboratory

Mor Lib Morgan Library

Morm Mormon

Morm Mormon, Book of

Mor Maj Moral Majority

morn morning

Moro. Book of Moroni

Moroc Moroccan; Morocco

Morocco Carolina Varga Dinicu; Kingdom of Morocco (North African Arab nation), *al-Mamlaka al-Maghrebia*

morph morphine; morphology

morph (Latin prefix—form, shape)—morphological

morpha hermaphrodite

morpheme smallest sound unit (linguistics)

morpho morphine

morphophysio morphophysiologic(al)(1y); morphophysiologist; morphophysiology

Morrow William Morrow

MORS Midland Operational Research Society

mor. sol. more solito (Latin—in the unusual manner)

mort mortal; mortality; mortar; mortgage; mortician, mortuary

mort t Morse taper

moRt mainstream of Republican thought

Mort Mortemart; Mortimer; Morton

mortal. mortality

mos metal-oxide semiconductor; metal-oxide-silicon (compound); missile on stand; mit-out sound (silent film); months; mosaic

mos (MOS) military occupational specialty

Mos Moscow

Mos. Book of Mosiah

Mos Mosca (Italian—Moscow); *Moscou* (French or Portuguese—Moscow); *Moscu* (Spanish—Moscow); *Moskau* (German—Moscow); *Moskou* (Dutch—Moscow); *Moslem* (Arabic—True Believer)

MOs Military Observers (UN)

MOS Magneto-Optical System; Management Operating System; Manned Orbital Station; Marine Observation Satellite; Military Occupational Specialty; Ministry of Supply

MOS Ministerstwo Opieki Spotecznes (Polish—Ministry of Social Welfare)

MOSA Medical Officers of Schools Association

Mosbas Moscow Basin

Mosby C V Mosby

mosc manned orbital systems concepts

MOSC Midland-Odessa Symphony and Chorale

Mose Moisés; Mosè; Moseley; Mosen; Moses; Moshe

MoSEA Missouri Society of Enrolled Agents

MOSES Manned Open Sea Experimentation Station

mosfet metal-oxide semiconductor field-effect transistor

mosic metal-oxide-semiconductor integrated circuit(s)

MOSID Ministry of Supply Inspection Department

mosis monolithic silicon sensor

Mosk Moscovici; Moscowitz; Moskowitz

mosm milliosmol(s)

MOSOP Missouri Sexual Offender Program

mosrom metal-oxide-silicon read-only memory

moss maintenance-operations support set

MOSS Manned Orbital Space Station; Market Opening Sector Specific

MOSST Ministry of State for Science and Technology (Canadian)

most metal-oxide semiconductor transistor

'most almost

MOST Michigan Opportunities and Skills Training

mostl metal-oxide semiconductor transistor logic

mot mean operating time; mechanical operability test; member of our tribe; middle of target; motor; motorized

M o T Minister of Transport; Ministry of Transport

MOT Military Ocean Terminal; Ministry of Transportation

mot aph motor aphasia

MOTAT Museum of Transport and Technology

MOTC Ministry of Transit and Communications (Philippines)

M o TCP Ministry of Town and Country Planning

motel hotel for motorists

moth mother

moth-in-law mother-in-law

Moth Jones Mother Jones

MOTIS Message-Oriented Text Interchange System

motiv modular propulsion orbital transfer vehicle

MOTNE Meteorological Operational Telecommunications Network, Europe (NATO)

motoboard(s) motorized skateboard(s)

motocross cross-country motorcycle race

mot op motor operated

motorcade motorized-vehicle parade

motorcross motorcyle cross (country race)

MOTOREDE Movement To Restore Decency

Mo' Town Motor Town (Detroit, Michigan)

mots minitrack optical tracking system

MOTS Military Off The Shelf

MOTU Mobile Technical Unit

mou memorandum of understanding

MoU Memorandum of Understanding

Mountbatten of Burma Admiral of the Fleet and last Viceroy of India known until 1917 as Prince Louis Francis Albert Victor Nicholas of Battenberg (1900—1979)

mounties mounted policemen (especially Royal Canadian Mounted Police)

MOUSE minimum orbital unmanned satellite

mout sick mountain sickness (headaches, nausea, overall weakness)

mov metal-oxide varistor; movable; moved; movement; moving; multiple-orifice valve

mov movimento (Italian—movement)

movem movement overseas verification of enlisted members (of the USA)

moverep movement report

Move Short Soc Movement Shorthand Society

movi movie; moving pictures

movies moving pictures

MOVIMS Motor Vehicle Information Management System

movord movement order

M o W Minister of Works; Ministry of Works

MOW Moscow, USSR (Vnukovo Airport); Movement for the Ordination of Women

mowasp mechanization of warehousing and shipment processing

MoWD Ministry of Works and Development

MOWOS Meteorological Office Weather Observing System

M o WT Minister of War Transport; Ministry of War Transport

MOWW Military Order of the World Wars

mox mixed oxides (platinum and uranium); oxidized metal explosive

moy money

Moz Mozambique

MOZ Mezhdunarodnaya Organizacia Zhurnlistov (Russian—International Organization of Journalists)

Mozam Mozambique

Mozambique People's Republic of Mozambique (formerly Portuguese East Africa)

Mozart Wolfgang Amadeus Mozart

Moz Cur Mozambique Current (Natal)

mozza mozzarella

mp mail payment; maintenance part(s); manifold pressure; medium pressure; meeting point; melting point; milepost; motion picture; multipole; multipurpose

mp mezzopiano (Italian—half soft, moderately soft)

mp (MP) marginal product

m(p) microfilm positive

m-p metal-point (bullet)

m/p milk powder

m & p materials and processes

m.p. mille pasuum (Latin—thousand paces)—the Roman mile of 1000 paces

mP polar maritime air

MP Malaita Province (Solomon Islands); Melphalan/Prednisone; Member of Parliament; Mercator's Projection; Metropolitan Police; Military Police; Mining Permit; Minister Plenipotentiary; Minister Provincial; Miscellaneous Proposal; Missouri Pacific (railroad); Mitsubishi Plastics; Mounted Police

MP Modern Philology

M/P Memorandum of Partnership

M & P Maryland & Pennsylvania (railroad)

MP Maschinenpistole (German—submachine gun, tommy gun)

mp$_l$ marginal product of labor

MP3 Motion Picture Coding Experts Group-1, audio layer-3

mpa maritime patrol aircraft; megapascal; multiple product (television) announcement

mpa (MPa) megapascal

mpa (MPA) maritime patrol aircraft

mpa Maryland Port Authority's italicized logotype

MPA Magazine Publishers of America; Magazine Publishers Association; Main Propulsion Assistant (USN); Maryland & Pennsylvania (railroad); Master Photographers Association; Master Printers of America; Mechanical Packing Association; Medical Procurement Agency; medroxyprogesterone; Metal Powder Association; Metropolitan Pensions Associations; Midwestern Psychological Association; Military Police Association, Mobile Press Association; Modern Poetry Association; Motion Picture Alliance; Music Publishers Association

M.P.A. Marine Physician Assistant; Master of Professional Accounting; Master of Public Administration; Master of Public Affairs

MPAA Motion Picture Association of America; Musical Performing Arts Association

MPAC Master Plan for Academic Computing

MPACS Management Planning and Control System

mpad maximum permissible annual dose

MPAGB Modern Pentathlon Association of Great Britain

mpai maximum permissible annual intake

mpam maritime polar air mass

m part movable partition

mpas millipascal second

MPAS Maryland Parent Attitude Survey

MPAUS Music Publishers Association of the United States

m payl maximum payload

mpb male pattern baldness

MPB Maintenance Parts Breakdown (spare parts); Miniature Precision Bearings; Missing Persons Bureau; Montpelier & Barre (railroad)

mpbb maximum permissible body burden (of radiation)

MPBC Memphis Power Boat Club

mp br multipunch bar

mpbs megabits per second

MPBS Mutual Permanent Building Society

MPBW Ministry of Public Buildings and Works

mpc marine protein concentrate; material program code; mathematics; physics, chemistry; maximum permissible concentration; military payment certificate; minimal planning chart; multipurpose carrier

mpc (MPC) marginal propensity to consume

MPC Manpower and Personnel Council; Manpower Priorities Committee; Manufacturing Plan Change; Marine Policy Center (Woods Hole, Mass.); Member of Parliament of Canada; Metropolitan Police College; Metropolitan Police Commissioner; Military Payment Certificate; Military Pioneer Corps; Model Penal Code; Montana Power Company

MPCA Magnetic Powder Core Associatioum; Marine and Ports Council of Australia; Master Pastry Cooks Association

MPCAG Military Parts Control Advisory Groups

MPCB Manufacturing Plan Control Board

mpc black medium-processing channel black

MPCC Minnesota Private College Council

MPCL Movimiento Patriótico Cuba Libre (Spanish—Free Cuba Patriotic Movement)

mpcp missile power control panel

MPCS Master Plan for Computing Services

mpcur maximum permissible concentration of unidentified radionuclides

mpd magnetoplasmadynamics; maximum permissible dose; missile purchase description; multiple personality disorder

M.Pd. Master of Pedagogy

MPD Metropolitan Park District; Metropolitan Police Department; Military Pay Division

MPDA Motion Picture Distributors Association

MPDFA Master Photo Dealers' and Finishers' Association

mp di multipunch die

MPDPIS Master Plan for Data Processing and Information Systems

MPDS Message Processing Distribution System

MPDSA Master Painters, Decorators, and Signwriters Association

MPDT Minnesota Perception Diagnostic Test

mpe maximum permissible exposure (to radiation)

M.P.E. Master of Physical Education

MPEA Motion Picture Exhibitors Association

MPEAUS Master Printers and Engravers Association of the United States

M. Pe. Eng. Master of Petroleum Engineering

MPEG2 Moving Picture Experts Group 2

M Pen Minister of Pensions; Ministry of Pensions

MPers Middle Persian

MPES Mathematical, Physical, and Engineering Science (NSF)

mpf motion-picture film; multipurpose food

MPF Malaysian Peasants Front; Metallurgical Plantmakers Federation; Metropolitan Police Force (London)

mpfg 1000 proof gallons

mpg miles per gallon

MPG Magazine Promotion Group; Max Planck Gesellschaft; Manhattan Publishing Group; Multimedia Publishers Group

MPGA Maine Personnel and Guidance Association; Maryland Personnel and Guidance Association; Metropolitan Public Gardens Association; Michigan Personnel and Guidance Association; Minnesota Personnel and Guidance Association; Missouri Personnel and Guidance Association

mpgn membrano proliferative glomerulonephritis

MPGR Mana Pools Game Reserve (Rhodesia)

MPGS Mobile-Protected Gun System

mph miles per hour

M.Ph. Master of Philosophy

MPH Meat Packing House; Methodist Publishing House

M.P.H. Master of Public Health

MPH Maintenance Parts Handbook

M. Phar. Master of Pharmacy

M. Pharm. Master of Pharmacy

MPHEC Maritime Provinces Higher Education Commission

M. Ph. Ed. Master of Public Health Education

M.P.H. Eng. Master of Public Health Engineering

M.Phil. Master of Philosophy

M. Pho. Master of Photography

mphps miles per hour per second

M. Ph. Sc. Master of Physical Science

M.P.H.T.M. Master of Public Health and Tropical Medicine

M. Phy. Master of Physics

M Phys A Member of the Physiotherapists Association

mpi magnetic particle inspection; maximum point of impulse; mean point of impact; mug photo interface; multiphasic personality inventory; multiphoton ionization

mpi (MPI) marginal propensity to invest

MPI Material Process Instruction; Maudsley Personality Inventory; Max Planck Institute; Medicine in the Public Interest; Meeting Planners International; Micro-Pigment Implantation; Mitsui Petrochemical Industries; Museum of the Plains Indians

MPI Movimiento Pro-Independencia (Spanish—Pro-Independence Movement)—Puerto Rico

MPIF Metal Powder Industries Federation

M-pill menstruation pill

MPIRO Multiple Peril Insurance Rating Organization

MP & IS Material Process and Inspection Specification

mPk polar maritime air colder than underlying surface

mpl mathematical programming language; maximum payload; maximum permissible language; maximum permissible level; maximum permissible limit; maximum practices limitations; message processing language; multiple-position lock

MPL Maintenance Parts List; Memphis Public Library; Metropolitan Police Laboratory; Miami Public Library; Milwaukee Public Library; Minnesota Power and Light; Missouri Pacific Lines; Montreal Public Library

M.P.L. Master (Mistress) of Patent Law

MPLA Mountain Plains Library Association

MPLA Movimento Popular Liberação Angola (Portuguese—Popular Movement for the Liberation of Angola)

mplm multi-purpose lightweight missile

MPLP Marxist Progressive Labor Party

Mpls Minneapolis

mpm meters per minute; missile power monitor; mole-percent metal; multipurpose meal

MPM Milwaukee Public Museum; Modest Petrovich Mussorgsky (1839—1881)

MP-M Museum Plantin-Moretus (Antwerp's museum devoted to book production and typography of Plantin and Moretus)

MPMI Magazine and Paperback Marketing Institute

mpn most probable number

MPNA Midwest Professional Needlework Association

MPNI Ministry of Pensions and National Insurance

mpo memory printout

MPO Memorandum Purchase Order; Metropolitan Police Office (Scotland Yard); Miami Philharmonic Orchestra; Military Pay Order; Military Planning Office(r); Military Post Office; Mobile Post Office; Mobile Printing Office

MPOIS Military Police Operating Information System

M.Pol.Econ Master (Mistress) of Political Economy

MPOLL Military Post Office Location List(ing)

mpp marginal physical product; massively parallel processor; most probable position

MPP Mailer's Postmark Permit; Maintainability Program Plan(ning); Massively Parallel Processing; Member Provincial Parliament (Canada); Mothers in Prison Projects

M.P.P. Master (Mistress) of Physical Planning

M & PP Manitou & Pikes Peak (Railroad)

MPPA Music Publishers Protective Association

MPPCA Maryland Probation, Patrol and Corrections Association

mppcf millions of particles per cubic foot of air

mp pl multipunch plate

mpps million pulses per second

MPPS Massive Parallel Processing System

MPPWCOM Military Police Prisoner of War Command

mpq manpower-planning quota(s)

mpr 1000 pair; medium-power radar

MPR Maintainability Program Requirements; Military Pay Record; Mongolian People's Republic

MPR Madjelis Permusjawaratan Rakat (Indonesian—People's Deliberative Assembly); *Maritime Provinces Reports*

MPRC Military Personnel Records Center

mpress medium pressure; medium pressurization

MPRL Master Parts Reference List

M. Prof Acc. Master of Professional Accountancy

MPRP Mongolian Peoples Revolutionary Party; Muslim Peoples Republican Party

mp & rs motive power and rolling stock

MPRSA Marine Protection, Research and Sanctuaries Act

mps marbled paper sides; maritime pre-positioning ships; megacycles per second; meters per second; motor parts stock

mps (MPS) marginal propensity to save; mucopolysaccharidosis

Mp's Minneapolis pimps

MPs Members of Parliament; plural of military police or mounted police

M.Ps. Master of Psychology

MPS Mail Preference Service; Manufacturing Process Specification; Marriage Prediction Schedule; Master Project Summary; Mathematical Programming System; Microprocessor System; Military Postal Service; Milwaukee Public Museum; Minimum Property Standards; Minister of Public Security; Mont Pelerin Society; Motor Products Corporation; Multiple Protective Shelters; Multiprogramming System

M.P.S. Member of the Pharmaceutical Society

MPSA Military Petroleum Supply Agency

MPSC Military Provost Staff Corps

mpsh mean pressure suction head

MPSM Master Problem Status Manual

M.Ps.O. Master (Mistress) of Psychology Orientation

MPSP Mathematical Problem-Solving Project; Military Personnel Security Program

MPSS Multiple Protective Structure System

M.P.S.W. Master (Mistress) of Psychiatric Social Work

M.Psych. Master (Mistress) of Psychology

M. Psy. Med. Master of Psychological Medicine

mpt male pipe thread; mean preventive time; melting point; microprocessing pro-

grammable terminal; midpoint; multiple pure tone; multipower transmission

mpt (MPT) miles per tankful

Mpt Maryport

MPT Marquis Public Theater; Maryland Public Television; Minister of Posts and Telecommunications

mpta main propulsion test article

MPTA Machine Power Transmission Association; Municipal Passenger Transport Association

MPTCA Motion Picture and Television Credit Association

MPTP Music Preference Test of Personality

mpu microprocessor unit (MPU); missile power unit; monitor printing unit

MPU Medical Practitioners Union; Mental Parents Union; Missing Persons Unit (of a police department)

M. Pub. Adm. Master of Public Administration

mpv (MPV) multipurpose vehicle

M-P v Mason-Pfizer virus

mPw polar maritime air warmer than underlying surface

MPW Minneapolis-Moline (stock exchange symbol)

MPWBS Master Plan Works Breakdown Structure

mpws mobile protected weapon system; multi-purpose weapon system

mpx multiplex

MPX Multimedia Pre-press Expo

mpxr multiplexor (flow chart)

mpy multiply

Mpy Maatschappij (Dutch—company)

MPZ Mid-Continent Petroleum (stock exchange symbol)

mq multiple quotient (register); multiplier quotient

mq (MQ) memory quotient; metol-quinol; metol-quinone

Mq mosque

MQ Martinique (Internet code); merit quotient

MQ Mayflower Quarterly

M & Q Mines and Quarries

MQA Manufacturing Quality Assurance; Medical Quality Assurance

MQAB Medical Quality Assurance Board

Mqe Martinique

mqf mobile quarantine facility

MQI Maiquetía (Venezuelan airport)

mqil miniature quartz incandescent lamp

mql miniature quartz lamp

MQM Master of the Queen's Music

MQO Marksmanship Qualification Order

MQS Mobile Quality Services

MQT Model Qualification Test

M Quad Charles Bertrand Lewis

MQV Ministère de la Qualité de la Vie (French—Ministry of the Quality of Life)

mqyco minimum quantity yards per color

mqyds minimum quantity yards per design

mr machine record(s); machine rifle; map reference; medium range; mental retardation; mentally retarded; metabolic rate; methyl red; mill run; mineral rubber; milliroentgen; mine run

mr (MR) marginal revenue; motivational research

m/r map reading; middle right

m & r maintainability and reliability; maintainability and repairs; maintenance and repair

mr meester (Dutch—master)—attorney-at-law; *mi remesa* (Spanish—my remittance)

mR milliroentgen

Mr Master; Mister; Mother

MR Machinery Repairman; Magnetic Resonating; Marketing Research (division, US Department of Agriculture); Mark Russell; Master of the Rolls; Mauritania (Internet code); Medical Record; Memorandum for Record; Memorandum Report; Michigan Reformatory; Military Railroad; Military Requirement; Minister Residentiary; Ministry of Reconstruction; Miscellaneous Report; Mobilization Regulation; Monon Railroad; Monthly Report; Morning Report; Multifamily Residential zone; Municipal Reform

M/R map reading; Mates Receipt(s)

M & R maintenance and repairs

M of R Minister of Reconstruction; Ministry of Reconstruction

MR Marca Registrada (Spanish—Registered Trademark); *Mobilización Republicana* (Spanish—Republican Mobilization)—Castro-controlled political party in the Nicaraguan underground; *Motormannes Riksforbund* (Swedish—Motorists' Association)

MR-13 Movimiento Revolucionario de 13 de Noviembre (Spanish—Revolutionary Movement of 13 November)—Guatemala

mra magnetic resonance angiography; medium-powered radio range (Adcock); minimum reception altitude

mra (MRA) metro rating area (tv)

MRA Manufacturers Representatives of America; Marketing Research Association; Maritime Royal Artillery; Master Retailers Association; Materials Review Area; Menswear Retailers of America; Moral Rearmament

MRAA Marine Retailers Association of America

mraam (MRAAM) medium-range air-to-air missile

mrac manifold-regulator accumulator charging

MRACP Member of the Royal Australasian College of Physicians

mrad megarad; millirad

M. Rad. Master of Radiology

M. Ra. Eng. Master of Radio Engineering

MRAF Marshal of the Royal Air Force

Mr Air Brake George Westinghouse

mra&l manpower, reserve affairs and logistics

MRAM Multimission Redeye Air-launched Missile

MRAP Management Review and Analysis Program

MRAS Manpower Resources Accounting System (USAF)

mrasm (MRASM) medium-range air-to-surface missile

mrat medium-range applied technology

MRAUSCAN Masonic Relief Association of the United States and Canada

mrb marble base; multi-role bomber

MRB Material Review Board; Mileage Rationing Board; Modification Review Board; Mutual Reinsurance Bureau

MRBA Mississippi River Bridge Authority

mrbm medium-range ballistic missile (MRBM)

MRBP Missouri River Basin Project

mrc magnetic rectifier control

Mrc Mauricio (Spanish—Mauritius)

MRC Maintenance Requirements Cards; Marine Research Committee; Market Research Council; Marlin-Rockwell Corporation; Material Redistribution Center; Material Review Crib; Materials Research Corporation; Measurement Research Center; Medical Research Center (Council); Media Research Center; Medical Reserve Corps; Men's Republican Club; Metals Reserve Company; Methods Research Corporation; Minnesota Restitution Center; Mississippi River Commission; Model Railway Club; Modern Railroad Club; Motor Racing Club; Movement Report Center

mrca multirole combat aircraft

MRCA Market Research Corporation of America

MRCC Medical Research Council of Canada

MRCF Module Repair Calibration Facility

MRCGP Member of the Royal College of General Practitioners

MRCI Medical Registration Council of Ireland; Medical Research Council of Ireland

MRCIU Men's Residence Center (Indiana University)

MRCo Malaysian Refrigerator Company

MRCO Member of the Royal College of Organists

MRCOG Member of the Royal College of Obstetricians and Gynaecologists

MRCP Maoist Revolutionary Communist Party; Member of the Royal College of Physicians

MRC Path Member of the Royal College of Pathologists

MRCPE Member of the Royal College of Physicians of Edinburgh

MRCPI Member of the Royal College of Physicians of Ireland

MRC Psych Member of the Royal College of Psychiatrists

MRCPUK Member of the Royal College of Physicians of the United Kingdom

MRCS Member of the Royal College of Surgeons

MRCSE Member of the Royal College of Surgeons of Edinburgh

MRCSI Member of the Royal College of Surgeons of Ireland

MRCVS Member of the Royal College of Veterinary Surgeons

MRCWA Midland Railway Company of Western Australia

mrd metal rolling door; metal(lic) roof(ing) deck(ing); minimum reacting dose (MRD)

MRD Main Roads Department; Medical Records Department; Medical Reference Department; Microbiological Research Department; Motorized Rifle Division

MRDA Media Research Directors Association

MRDC Military Research and Development Center

MR & DC Medical Research and Development Command (US Army)

mrdf machine-readable data files

MRDF maritime radio direction finding

MR & DF Malleable Research and Development Foundation

mrdhd maximum recommended daily human dose

MRDN Material Receipt Discrepancy Notice

MRDs Motorized Rifle Divisions

MRDTI Metal Roof Deck Technical Institute

mre mean radial error

mre (MRE) meal ready to eat (freeze-dried field ration)

MRE Microbiological Research Establishment (UK)

M.R.E. Master of Religious Education

M.Ref.Eng. Master of Refrigeration Engineering

MREI Marriage Role Expectation Inventory

MRELB Malaysian Rubber Exchange and Licensing Board

mrem milliroentgen equivalent man

mrep milliroentgen equivalent physical

mrf magnetic resonance bloodflow scanning; maintenance replacement factor; marble floor

MRF Mayo Research Foundation; Meteorological Rocket Facility; Music Research Foundation

MRFAC Manufacturers Radio Frequency Advisory Committee

MRFB Malayan Rubber Fund Board

MRFIT Multiple Risk-Factor Intervention Trial

MRFL Master Radio Frequency List

mr flight meteorological research flight

mrg magnetic radiation generator; margin; marginal; marginalia, methane-rich gas

MRG Maintainability Requirements Group; Material Review Group; Minorities Research Group; Minorities Rights Group

MRGO Mississippi River Gulf Outflow

MRGS Member of the Royal Geographical Society

MRH Member of the Royal Household

mrhm milliroentgens per hour at one meter

MRHMC Michael Reese Hospital and Medical Center

mr/hr milliroentgens per hour

MRHS Midwest Railway Historical Society

mri magnetic rubber inspection; mean rise interval; mediumrange interceptor; milstrip routing identifier; monopulse resolution improvement

mri (MRI) magnetic-resonance imaging

MRI Magazine Research Incorporated; Magnetic Resonance Imaging; Marine Research Institute; Marital Roles Inventory; Meat Research Institute; Medical Records Index(ing); Mental Research Institute; Meteorological Research Institute; MeuseRhine-Issel (cattle breed); Midwest Research Institute; Military Reform Institute; Missile Range Index; Motor Repair Insurance

MRIA Member of the Royal Irish Academy; Model Railroad Industry Association

MRINA Member of the Royal Institution of Naval Architects

MRINZ Meat Research Institute of New Zealand

MRIPHH Member of the Royal Institute of Public Health and Hygiene

mrir medium resolution infrared

MRIS Maritime Research Information System; Market Research Information System; Material Readiness Index System; Medical Research Information System; Mobile Range Instrumentation System

MRIW Medical Research Institute of Worcester

mrkd marked

mrkg marking

mrkr marker

Mrkt-Deli Market-Delicatessen

Mrkts Markets

mrl medium-powered radio range (loop radiators); motor refrigerator lighter; multiple rocket launcher (MRL)

MRL Materiel Requirements List; Medical Records Librarian; Medical Records Library; Mineral Research Laboratories

MRLA Malayan Races Liberation Army (Chinese-communist guerillas)

mrm mail readership measurement; mechanically recovered meat; miles of relative movement

MRM Maintenance Reporting and Management

MRMVA Master Retail Milk Vendors Association

Mrn Martin

MRN Material Recorder Notice; Meteorological Rocket Network

mRNA messenger ribonucleic acid

MRNET Minnesota Regional Network

mrng mooring; morning

MRNP Mount Rainier National Park (Washington); Mount Revelstoke National Park (British Columbia)

Mrnz Martinez

mro maintenance, repair, and operating

Mro Maestro

MRO Maintenance, Repair, and Operation(s); Materiel Release Order; Medical Review Officer

MROAR Modification and Repair Order and Acceptance Record

M-roof M-shaped roof

mrov moreover

mrp machine-readable passport; manned reusable payload; manned reusable product; manufacturing resource planning; marginal revenue product; maximum resolving power; maximum retail price

mrp (MRP) marginal revenue product

MRP Mobile Repair Party

M.R.P. Master in Regional Planning

MRPA Metropolitan Region Planning Authority

MRPP Maoist Reorganization Movement of the Party of the Proletariat

MRPRA Malaysian Rubber Producers Research Association

M rps Mauritius rupee(s)

Mr Q Marquardt Corporation

MRQ Marquardt Corporation (stock exchange symbol)

mrr medical research reactor

MRR Material Rejection Report; Mechanical Reliability Report(ing)

MRRAS Murder Release Risk Assessment Scale

MRRC Mechanical Reliability Research Center

MRRDB Malaysian Rubber Research and Development Board

mrs magnetic resonance blood-flow scanning

mrs (MRS) marginal rate of substitution

MRs Maintenance Reports

MRS Markét Research Society; Marseilles, France (airport); Master Repair(ing) Schedule; Material Request Summary; Material Requirement Summary; Military Railway Service; Ministry of Recreation and Sport; Mistress; Monitored Retrievable Storage; Mountain Rescue Service

MR & S Materials Research and Standards

mrsa (MRSA) medium-range surveillance aircraft

MRSA methicillin-resistant *Staphylococcus aureus* (hospital-acquired infection)

MR San Asn Member of the Royal Sanitary Association

M.R.Sc. Master (Mistress) of Rural Science

MRSH Member of the Royal Society of Health

MRSL Member of the Royal Society of Literature

MRSM Member of the Royal Society of Medicine

MRSMGB Member of the Royal Society of Musicians of Great Britain

MRSP Myakka River State Park (Florida)

mrsss manned revolving space systems simulator (MRSSS)

MRST Member of the Royal Society of Teachers

mrt mean radiant temperature; mid-range trajectory; mildew-resistant thread; military-rated thrust(ing); minimum resolvable temperature; mission readiness tester; music 'riter typewriter

Mrt Martinique

Mrt Maart (Dutch—March)

MRT Maintainability Review Team; Mass Rapid Transit; Mass Rapid Transport; Metropolitan Readiness Test; Military Review Team; Modulus of Rupture Test(ing)

MRTA Maintenance Requirements Task Analysis

mrtm maritime

Mrtnz Martinez

mrto miscellaneous reference tool(ing)

MRTPI Member of the Royal Town Planning Institute

mrts (MRTS) marginal rate of technical substitution

Mrts Mauritius

MRTS Mass Rapid Transit System; Master Radar Tracking Station

mru minimal reproductive units

mru (MRU) mass radiography unit; mobile radio unit

MRU Medical Rehabilitation Unit; mobile radio unit; mobile repair unit

MRUA Mobile Radio Users' Association

mrv material receipt voucher; missile re-entry vehicle (MRV); mixed respiratory vaccine; multiple re-entry vehicle (MRV)

MRV missile recovery vessel

mrV-P methyl red Voges-Proskauer

mrw morale, recreation, and welfare

mr/w multiple read/write

MRWA Midland Railway of Western Australia

mrwc multiple reading, writing, compiling; multiple read, write, compute

Mrylb Marylebone (railway terminal)

mrytm must have reply here by tomorrow morning

mrz marzo (Spanish—March)

MRZ Mineral Resource Zone

ms machine screw; machine steel; main switch; maintenance and service; major subject; manuscript; margin of safety; mass spectrometric; master switch; matched set; maximum stress; mean square; medium shot; medium steel; meters per second; metric system; microseismic; mild steel; millisecond; minimum stress; mint state; mission satellite; mitral stenosis; months after sight; multiple sclerosis; multiple starters; muscle strength

ms (MS) morphine sulfate; moving and storage; multiple sclerosis

m/s marking and stenciling; metal shank; meters per second; milestone; miniature sheet of stamps; month after sight

m & s maintenance and supply; model and series; mud and snow

μs microsecond(s)

ms. manuscript

m s mano sinistra (Italian—left hand)

m/s motorskib (Norwegian—motorship)

mS millisiemens (millimho)

Ms mature motion pictures (for adults); Mendes; mesothorium; (pronounced *Miz*) feminine title replacing Miss or Mrs; Seattle Mariners baseball team

MS Machinery Survey; magnetic south; Mail Steamer; major subject; Manuscript Society; Master Sergeant; Material Specifications; Medical Survey; Metallurgical Society; Meteoritical Society; Michigan State University of Agriculture and Applied Science; Military Service;, Military Standard; Ministry of Shipping; Ministry of Supply; Misair (Egyptian Airline); Mississippi; Montserrat (Internet code); Motorship

M-S Material Service (division of General Dynamics); Monday through Saturday

M.S. Master of Science; Master of Surgery

M/S Mannficher-Schoenauer; motorship

M & S Maintenance and Supply; Marks and Sparks; Marks and Spencer; Maternity and Surgical; Medical and Surgical; Medicine and Surgery

MS Material Standard (usually followed by a number); *Mittelsatz* (German—middle clause) central theme in a sonata

M-S Minshu-Shakaito (Japanese—Democratic Socialist Party)

msa master settlement agreement; medical savings account; method of steepest ascent; minimum safe altitude; mission system avionics

m.s.a. *misce secundum arten* (Latin—mix skillfully)

MSA Major Systems Acquisition; Malaysia Singapore Airlines; Management Selection Australia; Marine Safety Agency; Maritime Safety Agency; Medical Statistics Agency (US Army); Metropolitan Statistical Area; Middle States Association (colleges and schools); Mine Safety Appliances (company); Mineralogical Society of America; Motor Schools Association, Museum Store Association; Mutual Security Agency; Mutual Society of Arts

M-S-A Mine Safety Appliances

MSA *Marine Sanctuaries Act; Merchant Shipping Act*

MSAA Mower Specialists Association of Australia

MSAAB Military Services Ammunition Allocation Board

msac most seriously affected countries

MSAC Moore School Automatic Computer

MSA/CHE Middle States Association of Colleges and Schools Commission on Higher Education

M.S.Agr.Eng. Master of Science in Agricultural Engineering

MSA Inst MM Member of the South African Institute of Mining and Metallurgy

MSAIT Member of the South African Institute of Translators

m^s a^s muchos años (Spanish—many years)

MSAS Mandel Social Adjustment Scale; Modal Suppression Augmentation System

MSAT Marine Services Association of Texas

MSAUS Masonic Service Association of the United States

msaw (MSAW) minimum safe altitude warning

msb main switchboard; marginal social benefits; most significant bit

msb (MSB) minority small business; missile storage building

MSB Mackinac Straits Bridge (Michigan); Marine Safety Board; minesweeping boat (naval symbol)

MSBA Maine School Boards Association; Minnesota School Boards Association; Missouri School Boards Association; Montana School Boards Association

M.S.B.A. Master of Science in Business Administration

MSB-COD Minority Small Business-Capital Ownership Development

MSBLS Microwave Scanning-Beam Landing System

MSBO Michigan School Business Officials; Mooring and Salvage Office(r)

msbr molten-salt breeder reactor

M.S.Bus. Master (Mistress) of Science in Business

msc marginal social costs; millisecond; miscellaneous; moved, seconded, and carried

m.s.c. *mandatum sine clausula* (Latin—authority without restriction)

M. Sc. Master of Science

MSC coastal minesweeper (3-letter naval symbol); Maine Sardine Council; Manchester Ship Canal; Manned Spacecraft Center (NASA); Manpower Services Commission; Maple Syrup Council; Marine Safety Council; Marine Science Center (Lehigh University); Medical Service Corps; Medical Specialist Corps; Mediterranean SubCommission; Melbourne Steamship Company; Meteorological Service of Canada; Metropolitan Special Constabulary; Military Sealift Command; Missile and Space Council; Missile System Checkout; Mississippi Central (railroad); Mountain Safety Council; Morgan State College

M & SC Missile and Space Council

MScA Make or Subcontract Authorization

MSCA McCarthy Scales of Children's Abilities; Moore School of Automatic Computers; Mount Saint Agnes College; Murray State Agricultural College

M Scand Middle Scandinavian

mscc magnetic-strip credit card

M.S.C.E. Master of Science in Civil Engineering

M.S.Ch.E. Master of Science in Chemical Engineering

Mschr Monatsschrift (German—monthly magazine)

MSCIC Maryland State Colleges Information Center

MSCKC Measurement of Self Concept in Kindergarten Children

M. Sc. L. Master of the Science of Law

mscn misconnection

mscnd misconnected

MSCNY Marine Society of the City of New York

MSC(O) old coastal minesweeper (naval symbol)

M.S. Conv. Master of Science in Conservation

M. Sc. Ost. Master of Science in Osteopathy

M Scot Middle Scottish

mscp mean spherical candlepower; multiple species conservation plan

MSCP Master Shielding Computer Program

MSCRB Margaret Sanger Clinical Research Bureau

mscrbl manuscribble (handscribbled manuscript)

mscrg miscarriage

MSCT Member of the Society of Cardiological Technicians

MSCW Mississippi State College for Women

msd missile system development; most significant digit; multiple spark discharge; musculoskeletal disorder

MSD Management Services Department; Marine Sanitation Devices; Merck, Sharp & Dohme; Minesweeping drone (naval symbol)

M.S.D. Master (Mistress) of Scientific Didactics; Master (Mistress) Surgeon Dentist; Medical Science Doctor

M & SD Missile and Space Division (General Electric)

MSDA Marconi Space and Defence Systems

MSDC Mass Spectrometry Data Center; Molten Salts Data Center

M.S. Dent. Master of Science in Dentistry

M.S. Derm. Master of Science in Dermatology

MSDF Maritime Self-defense Force (Japanese Navy)

M & SDI Mayonnaise and Salad Dressing Institute

MSDN Microbial Strain Data Network

MSDO Major Systems Development Organization

MS-DOS Microsoft Disk-Operating System—trademark

MSDS Material Safety Data Sheet; Multi-Spectral-Scanner Data System

mse manufacturing support equipment; mean square error; military stressful era(s)

MSE Malaysia Shipyard and Engineering; Mental Status Examination; Midwest Stock Exchange; Mississippi Export Railroad (stock exchange symbol); Montreal Stock Exchange

M.S.E. Master of Sanitary Engineering; Master of Science in Education; Master of Science in Engineering

m sec millisecond

μsec microsecond

M.S.Ed. Master (Mistress) of Science in Education

MSED Mobile Source Enforcement Division (EPA)

M.S.E.E. Master of Science in Electrical Engineering

M.S.E.M. Master of Science in Engineering Mechanics

M.S. Eng. Master of Science in Engineering

MSEO Marine Services Engineering Office(r)

m/seq master sequencer

MSER Manufacturing Support Equipment Request; Mental Status Examination Report

mses *marchandises* (French—goods)

MSET Maintenance Supportability Evaluation Team

MSEUE *Mouvement Socialiste pour les États Unis d'Europe* (French—Socialist Movement for the United States of Europe)

msf minimum sector fuel; muscle shock factor

Msf thousand square feet

MSF fleet minesweeper (naval symbol); Maintenance Support Flight; Minesweeping Flotilla; mobile striking force; Motorcycle Safety Foundation; Multiple Shops Federation

M.S.F. Master of Science in Forestry

MSF *Médecins Sans Frontières* (French—doctors without borders)—international group of volunteer physicians

MSFC Marshall Space Flight Center

ms fm master form

M & SFM Maintenance and Supply Facility Manager

msfn manned space flight network

MSFT Microsoft Corporation

ms fx master fixture

msg machine stress grading; message; monosodium glutamate

msg (MSG) monosodium glutamate

MSG Madison Square Garden; Marine Systems Group (General Dynamics)

ms ga master gauge

M.S.G.E. Master of Science in Geological Engineering

msgfm messageform

MSGp Mobile Support Group

msgr messenger

Msgr Monsigneur

msgs messages

M Sgt Master Sergeant

msg/wtg message waiting

msh melanocyte-stimulating hormone (MSH)

Msh *Islas Marshall* (Spanish—Marshall Islands)

MSH Music Society for the Handicapped

M.S.H. Master of Science in Horticulture; Master of Science in Hygiene

MSHA Mine Safety and Health Administration

M.S.H.A. Master of Science in Hospital Administration

M.S.H.E. Master of Science in Home Economics

MSHFA Multiservice Health Facility Association

MSHI Mitsubishi Singapore Heavy Industries

Mshl Marshal

M. S. Hort. Master of Science in Horticulture

M. S. Hyg. Master of Science in Hygiene

msi maintenance supply item(ization); management system indicator; medium-scale integration;. military standard item; missile status indicator

MSI minesweeper, inshore (naval symbol); Motor Specialties Industries; Museum of Science and Industry

MSI *Movimento Sociale Italiano* (Italian Social Movement)—neo-fascist militants known as Missini

Msia Malaysia

MSIA Movement of Spiritual Inner Awareness

Msian Malaysian

MSIB Mountain States Inspection Bureau

m'sieur *monsieur* (French—mister, sir)

M.S.Ind.Eng. Master of Science in Industrial Engineering

msip multi-stage improvement program

MSIRI Mauritius Sugar Industry Research Institute

MSIS Multi-State Information System

M.S.J. Master of Science in Journalism

msk mission support kit

MSK Mitsubishi Shoji Kaisha

MS-K Memorial Sloan-Kettering (cancer center)

MS-KCC Memorial Sloan-Kettering Cancer Center

MSKK Mitsui Sempaku Kabushiki Kaisha (Mitsui Line)

Mskr *Manuskript* (German—manuscript)

msl mean sea level; midsternal line; missile

msl *mesela* (Turkish—for example)

Msl Marseilles

MSL Marine Science Laboratories; minesweeping launch (naval symbol); Mulla Sadra Library (Shiraz, Iran); Munitions Supply Laboratories

M.S.L. Master of Science in Linguistics

MSLC Manufacturing Specification Liaison Change

ms lo master layout

mslp mean sea level pressure

MSLS Military Standard Logistics Systems

M.S.L.S. Master (Mistress) of Science in Library Science

msm maximum safety margin; mechanically separated meat; methylsulfonylmethane; modern school mathematics

MSM Manhattan School of Music; Meritorious Service Medal; minesweeper, river (naval symbol); Montana School of Mines; Mystic Seaport Museum (Conn)

M.S.M. Master of Science in Music

MSMA Mail Systems Management Association; Maine School Management Association; Master Sign Makers' Association

MSMC Marie Stopes Memorial Centre

M.S.M.E. Master of Science in Mechanical Engineering

M.S.Med. Master (Mistress) of Medical Science

MSMM Missouri School of Mines and Metallurgy

msmq mild steel—merchant quality

MSMS Mutual Security Military Sales

msmt measurement

M.S. Mus. Master of Science in Music

M.S. Mus. Ed. Master of Science in Music Education

msn mission

Msn Mission

MSN Madison, Wisconsin (airport); Master Serial Number(ing); Material Supply Notice(s); Medicare Summary Notice

M.S.N. Master of Science in Nursing

MSNB Machine Screw Nut Bureau

M.S.N. Ed. Master of Science in Nursing Education

M.S.Nucl.Eng. Master of Science in Nuclear Engineering

MSNY Mattachine Society of New York

mso management services organization

mso (MSO) multiple-systems operator (tv)

MSO Manila Symphony Orchestra; Marine Safety Office(r); Melbourne Symphony Orchestra; Memphis Symphony Orchestra; Milwaukee Symphony Orchestra; Minneapolis Symphony Orchestra; Monetary Statistics Ordinance; Montreal Symphony Orchestra; Morale Support Office(r); ocean minesweeper (naval symbol)

M.Soc.Sci. Master (Mistress) of Social Science

M. Soc. Wk. Master of Social Work

MSOD Master of Science in Organization Development

m-sop mezzo-soprano

M.S. Ophthal Master (Mistress) of Ophthalmological Surgery

m sopr mezzo-soprano

MSORS Mechanical Solvent Oil-Spill Recovery System

M.S.Ortho Master (Mistress) of Orthopedic Surgery

msp metal splash pan

msp (MSP) missile support plane

MSP Material Support Plan(ning); Maximum Security Prison; Medical Services Plan(ning); Memorial State Park; Minneapolis, Minnesota (airport); Mutual Security Program

M.S.P. Master of Science in Pharmacy

MSPA Marin Small Publishers Association

MSPB Merit Systems Protection Board

MSpC Medical Specialist Corps

MSPE Master of Science in Physical Education

M.S. Pet. Eng. Master of Science in Petroleum Engineering

M.S.P.H. Master of Science in Public Health

M.S.Pharm. Master of Science in Pharmacy

M.S.P.H.E. Master of Science in Public Health Engineering

M.S.P.H.Ed. Master of Science in Public Health Education

ms pl master plate

mspr master spares positioning resolver

MSPRB Meteorological Satellite Program Review Board

MSPU Massachusetts State Prostitutes Union

msr main supply route; mean spring rise (tides); mechanical strain recorder; mineral-surface roof, missile site radar

ms & r merchant shipbuilding and repairs

m & sr missile and surface radar

MSR Manufacturing Specification Request; Material Stores Requisition; mean spring tide; minesweeper, patrol (naval symbol)

MSRA Multiple Shoe Retailers' Association

M.S. Rad. Master of Science in Radiology

MSRB Mississippi State Rating Bureau; Municipal Securities Rulemaking Board

M.S. Rec. Master of Science in Recreation

M.S. Ret. Master of Science in Retailing

MSRG Member of the Society for Remedial Gymnasts

MSRN Manufacturing Specification Revision Notice

msrp manufacturer's suggested retail price; massive selective retaliatory power

MSRP Manufacturer's Suggested Retail Price

msrpp multidimensional scale for rating psychiatric patients

MSRs Marketing Service Representatives

MSRS Missile Strike Reporting System

Msrte Misroute

MSRTS Migrant Student Record Transfer System

msry masonry

mss magnetic storm satellite; manual safety switch; manuscripts; message switching station; missile select(ion) switch; missing sea stores; mode selection switch(ing); multispectral scanning system

mss (MSS) magnetic storm satellite

ms's milk sisters (two or more females who have had sexual relations with the same male)—*see* mb's

ms's (MSs) mine sweepers

mss. manuscripts

Mss Misses; Mizzes (plural of Miz written Ms)

MSS Manufacturers Standardization Society of the Valve and Fittings Industry; Mass Storage Systems; Master Supporting Schedule; Maximum Security System; Medical Service School; Medical Service School (USAF); Medical Superintendents Society; Metropolitan Security Service(s); Movement Shorthand Society; Multiple Sclerosis Society; Multispectral Scanner Subsystem

M.S.S. Master of Social Science

MSS Museo Storico degli Spaghetti (Italian—Historical Museum of Spaghetti)—close to the Italian Riviera in Pontedassio

MSSA Maine School Superintendents Association; Maintenance Supply Services Agency; Manchester Scales of Social Adaptation

MSSAtl Military Sealift Service Atlantic

M.S.Sc. Master of Sanitary Science; Master of Social Science

MSSC medium seal support craft (naval symbol); Metropolitan School Study Council

msscc multicolor spin-scan cloudcover camera

MSSCS Manned Space Station Communications System

MSSD Model Secondary School for the Deaf

M & SSD Missile & Space System Division (Douglas Aircraft)

M.S.S.E. Master of Science in Sanitary Engineering

M.S. S. Eng. Master of Science in Sanitary Engineering

MsSEA Mississippi Society of Enrolled Agents

MSSGB Motion Study Society of Great Britain

MSSH Massachusetts Society for Social Hygiene

MSSInd Military Sealift Service Indian

MSSMS Munition Section Strategic Missile Squadron

MSSNY Medical Society of the State of New York

MSSP Market Segment Specialization Program; Multiple Sequential Screen Panel

MSSPac Military Sealift Service Pacific

MSSR Moldavian Soviet Socialist Republic

MSSRC Mediterranean Social Science Research Council

MSSS Maintenance Supply Services System

M.S.S.S. Master (Mistress) of Science in Social Service

MSSST Meeting Street School Screening Test

M.S. St.Eng. Master of Science in Structural Engineering

MSSTM Military Space Systems Technology Model

mssu midstream specimen of urine

MSSVD Medical Society for the Study of Venereal Diseases

MSSVFI Manufacturers Standardization Society of the Valve and Fittings Industry

mssw magnetostatic surface wave

M.S.S.W. Master (Mistress) of Science in Social Work

mst mean solar time; mean spring tide(s); mean survival time; measurement

m(st) metal-stabilized (runway)

M'st' Mister

MST Marconi Telecommunications Systems; Maximum Service Telecasters; Military Science Training; Mountain Standard Time

M.S.T. Master of Science in Teaching

MSTA Maryland State Teachers Association; Michigan State Teachers Association; Missouri State Teachers Association

M.Stat. Master (Mistress) of Statistics

mstb 1000 stock tank barrels

mstc mastic

MSTC Maryland State Teachers College; Massachusetts State Teachers College

MSTD Member of the Society of Typographic Designers

M.S. T.Ed. Master of Science in Teacher Education

msth mesothorium

M & ST L Minneapolis & St Louis (railroad)

MSTM Master of Science in Technology Management

mstn 1000 short tons

ms tp master template

M ST P & SSM Minneapolis, St Paul & Sault Ste Marie Railroad (Soo Line)

mstr master

Mstr Master

M.S.Trans Master of Science in Transportation

M.S. in Trans.E. Master of Science in Transportation Engineering

Mstr Mech Master Mechanic

MSTRP Military, Strategic, Tactical, and Relay Program

msts (MSTS) missile static test site

MSTS Military Sea Transport Service; Missile Static Test Site; Mobile System Test(ing) Site

msty mostly

msu main storage unit; mass storage unit; maximum space use; maximum space utilization; mode selector unit

msu (MSU) maximum security unit

MSU Mature Students Union; Memphis State University; Michigan State University; Mississippi State University; Montana State University

MSUC Middle South Utilities Company

msud maple-syrup urine disease

MSUL Medical Schools of the University of London; Memphis State University Library; Michigan State University Library; Mississippi State University Library; Montana State University Library

MSU Lond Medical Schools of the University of London

M. Surgery Master of Surgery

M.Surv Master of Surveying

msus midstream urine specimen

msv (MSV) magnetically-supported vehicle; Martian surface vehicle; mean square velocity; miniature solenoid valve; molecular solution volume; murine sarcoma virus

MSV Medical Society of Victoria

MSVC Mount Saint Vincent College

MSVD Missile and Space-Vehicle Department (General Electric)

msv(M) murine sarcoma virus (Moloney)

Msw Massawa

M Sw Middle Swedish

MSW Master of Social Work; Medical Social Worker; Mikheyev-Smirnov-Wolfenstein (effect)

M.S.W. Master of Social Welfare; Master of Social Work

mswa main storage work area (address)

MSX Midcourse Space Experiment (satellite); Seaboard Oil (stock exchange symbol)

msy (MSY) maximum sustainable yield

MSY New Orleans, Louisiana (airport)

msyd 1000 square yards

M-system magnocellular system

mt empty; machine translation; mail transfer; maximum torque; mean tide; mean time; measurement ton; mechanical translation; mechanical transport; medical technology; megaton (MT); membrana tympani; metal(lic) tape; metatarsal; metric ton; miniature tube; missile test; modified term; motor terminal; motor transport; mount torque; mount(ed); mounting

mt (MT) motor tanker

m/t mail transfer; manual transmission; measurement tons

m & t maintenance and test; movements and transports

mT tropical maritime air

Mt Mount; Mountain; tympanic membrane

Mt. *Matthew* (*Book of*)

MT Machine Translation; Mail Transfer; Malta (Internet code); Mandated Territory; Manning Table; Mark Twain (Samuel Clemens); Masoretic Text; Mechanical Translation; Medical Technologist; Meteorological Aids; Military

Training; Military Transport; Mining Tenement; Ministry of Transport; Montana; Moscow Time; Motor Transport; Mountain Time; Muscat Transport

MT-6 mercaptomerin (diuretic)

mta maximum time aloft; microwave transistor amplifier

mt/a million tons per annum

m^{ta} muita (Portuguese—much)—feminine form

MTA Maine Teachers Association; Manpower Training Association; Market Technicians Association; Massachusetts Teachers Association; Master Tilers Association; Message Transfer Agent; Metropolitan Transit Authority; Mica Trade Association; Mississippi Teachers Association; Mississippi Test Area; Motor Trade Association; Music Teachers Association

mtac mathematical tables and other aids to computation

MTACCS Marine Tactical Command and Control System

MTAG Manufacturing Technology Advisory Group

MTAI Minnesota Teacher Attitude Inventory

MTAK *Magyar Tudományos Akadémia Könyvtára* (Hungarian—Library of the Hungarian Academy of Sciences)—Budapest

mtam maritime tropical air mass

MTAMR Metropolitan Toronto Association for the Mentally Retarded

MT/AMT Mail Transfer—Airmail Transfer (funding)

MTASCP Medical Technologist of the American Society of Clinical Pathologists

mtb maintenance of true bearing; miniature tall bearded (iris)

MTB Major Trading Bank; Malayan Tin Bureau; Malaysian Tin Bureau; Materials Transportation Bureau; Medium Tank Battalion; Miyagi Television Broadcasting; motor torpedo boat

MTBA Machine Tool Builders' Association

MTBC Mitsubishi Trust and Banking Corporation

mtbd mean time between demand

mtbe (MTBE) methyl tertiary butyl ether (octane-ooster additive)

mtbf mean time before failure; mean time between failures

mtbfa mean time between false alarms

mtbff mean time between first failure

mtbfl mean time between function loss

mtbjr meant time between justified removals

mtbm mean time between maintenance

mtbma mean time between maintenance actions

mtbo mean time between overhaul

mtbr mean time between removals

MTBRON Motor Torpedo Boat Squadron

mtbsf mean time between system failure

mtbuma mean time between unscheduled maintenance actions

mtbur mean time between unscheduled removals

mtc memory test computer; more to come

m & tc mission and traffic control

MTC Malayan Tobacco Company; Marcus Tullius Cicero (106–43 B.C.); Marine Technology Center (Electric Boat); Maritime Transport Committee; Massachusetts Treatment Center; Materiel Testing Command; Mechanical Transport Corps; Medical Training Center; Men's Tennis Council; Metropolitan Transportation Committee; Military Training Cadets; Missile Test Center; Monsanto Chemicals (stock exchange symbol); Montreal Trust Company; Morse Telegraph Club; Motor Transport Corps; Mystic Terminal (railroad)

M.T.C. Master of Textile Chemistry

MTC *Ministerio de Transporte y Comunicaciones* (Spanish—Ministry of Transportation and Communication)

MTCA Ministry of Transport and Civil Aviation

MTCB Metropolitan Taxicab Board

mtce maintenance; million tons of coal equivalent

mt & ce missile test and checkout equipment

MTCL Metropolitan Toronto Central Library

MTCP Minister of Town and Country Planning; Ministry of Town and Country Planning

MTCR Missile Technology Control Regime

mtcu magnetic tape control unit

mtd manufactured technological demonstrator; mean temperature difference; midpoint tissue dose; mounted

mtd (MTD) maximum tolerated dose

m.t.d. mitte tales doses (Latin—send such doses)

Mtd Marstrand

MT$ Maria Theresa dollar (Yemeni currency unit)

MTD Mobile Training Detachment

M.T.D. Master of Transport Design; Midwife Teachers' Diploma

MTDB Metropolitan Transit Development Board

MTDDA Minnesota Test for Differential Diagnosis of Aphasia

mtde maritime tactical data exchange; modern technology demonstrator engine

MTDE Maintenance Technique Development Establishment

mtDNA mitochondrial deoxyribonucleic acid

MTDS Marine Tactical Data System

mte manufacturing test(ing) equipment; maximum temperature engine; maximum thermal energy; multiple-track error

*M*te *Monte* (Italian, Portuguese, Spanish—mountain)

MTE Marine Technical Education

MTEA Metal Trades Employers Association

M. Tech. Master (Mistress) in Technology

M.Tel.Eng. Master (Mistress) of Telecommunication Engineering

MTER Manufacturing Test Equipment Request

*M*tes *Montes* (Italian, Portuguese, Spanish—mountains)

M.Text. Master (Mistress) of Textiles

mtf mechanical time fuze; modulation transfer function; multiple technical force

MTF Medical Treatment Facility; Metal Trades Federation; Mississippi Test Facility; Multiracial Training Facility

mtfex mountain field exercise

mtg main turbogenerator(s); meeting; methanol to gasoline; mortgage; mounting

Mtg Meeting (postal abbreviation)

mtgc mounting center

mtgd mortgaged

mtge mortgage

mtgee mortgagee

mtgor mortgagor

mth microptic theodolite; month

M. Th. Master of Theology

MTH Master of Trinity House

mthly monthly

mthm metric tons of heavy metal

mths months

mthv must have

mthw medium-temperature hot water

mti moving target identification; moving target indicator(s); moving target information

*M*ti *Munti* (Romanian—mountain)

MTI Maglev Transit Incorporated; Metal Treating Institute; Motorola Teleprograms Incorporated

MTI Magyar Távirati Iroda (Hungarian Press Agency)

MTIA Metal Trades Industry Association

MTIB Malaysian Timber Industry Board

mtik missile test installation kit

mtime meantime

MTIRA Machine Tool Industry Research Association

mTk tropical maritime air colder than underlying surface

mtks many thanks

mtl material; materiel; mean tide level; medial temporal lobe; merged transistor logic; metal(lic); mixed thermoluminescence

mtl monatlich (German—monthly)

Mtl Montreal; Motel

MTL mean tide level; Modern Terminals Limited; Motor Traders Limited

MTLA Micropublishers Trade List Annual

mtlp metabolic toxemia of late pregnancy

mtl(s) material(s)

mtlz materialize

mtlzd materialized

mtm method, time, and motion; methods time measurement(s)

MTM Mary Tyler Moore

MTMA Modern Teaching Methods Association

MTMASR Method Time Measurement Association for Standards and Research

MTMC Military Traffic Management Command (USA); Mother Teresa's Missionaries of Charity

Mt McK NP Mount McKinley National Park

MTMCTEA Military Traffic Management Command Transportation Engineering Agency

MTMS Multi-Terminal Modular System

MTMTS Military Traffic Management and Terminal Service

mtn motion; mountain; mutton

Mtn Mountain

MTN Medical Television Network; Multilateral Trade Negotiations

MTNA Music Teachers National Association

mtnt must not

mtn vu mountain view

MTNWR Mark Twain National Wildlife Refuge (Illinois)

mto modification task outline

*m*to *muito* (Portuguese—much)—masculine form

MTO Mississippi Test Operations

mtoe million tons oil equivalent

Mton Moncton

mtons metric tons

mTorr millitorr(s)

mtp mandatory treatment program; minimum tour price

Mt P Mount Palomar (observatory)

MTP Management Training Program; Mobilization Training Program; Modification Task Proposal; Mount Tom Price

M.T.P. Master of Town Planning

mtpa million tons per annum

MTPCNA Metal Tube Packaging Council of North America

Mt P O Mount Palomar Observatory

mtpp missile-to-target patch panel

mtpy millions of tons per year

mtr magnetic tape recorder; materials testing reactor; mater, mean time to restore; meter; minimum time rate; missile-tracking radar; motor; moving target reactor; multiple track radar

mtr (MTR) marginal tax rate

Mtr Meinicke turbidity reaction; Montrose

MTr meridian transit

MTR Mass Transit Railroad; Mass Transit Railway; Mass Transportation Railroad; Materials Testing Report; Montour (railroad)

MTRB Motor Truck Rate Bureau

MTRC Mass Transit Railway Corporation

mtrcl motorcycle

mtre missile test and readiness equipment

Mt Rev Most Reverend

MTRF Mark Twain Research Foundation

mtrg metering

mtri missile test range instrumentation

mtrl material

Mt R NP Mount Rainier National Park

Mtro Maestro (Spanish—Master)

M.T.R.P. Master of Town and Regional Planning

mtr rdr meter reader

MTRS Magnetic Tape Recording System

mtr vlu meter(ing) value

mts mobile training set; modular television system; motorship twin screw; mountains

mt's empties

Mts Mountains

Mts Montes (Spanish—mountains)

MTS Machine Tractor Station; Marine Technology Society; Melanesian Tourist Services; Member(s) of the Technical Staff; Metropolitan Transit System; Middlebare Technical School; Missile Test Stand; Missile Test Station; Money Transfer Service

MTS Mashinno-Traktornye Stantsii (Russian—Machine Tractor Stations)

mt/sc magnetic-tape selectric composer

MTSC Middle Tennessee State College

MtSEA Montana Society of Enrolled Agents

MTSO Mobile Telephone Switching Office

mtst maximum treadmill stress time

mt/st magnetic-tape selectric typewriter

MTSU Middle Tennessee State University

mtt magnetic tape terminal; mean transit time; moving target tracking

MTT Maintenance Training Team; Metropolitan Transport Trust; Mobile Training Team (USN); Municipal Tramways Trust

MTTA Machine Tools Trades' Association

MTTAGB Machine Tool Trades Association of Great Britain

mtte magnetic tape terminal equipment

mttf mean time to failure

mttff mean time to first failure

mttm magnetic tape transmissions

mttr mean time to repair

mtu metric tons of uranium; mobile tracking unit; mobile training unit

mtu (MTU) multiplexer and terminal unit

MTU Maintenance Training Unit; Michigan Technological University; Michigan Training Unit (reformatory); Missile Training Unit; Mobile Training Unit

MTU Motoren und Turbinen Union (German—Motors and Turbines United)—corporation

M. tuberc. Mycobacterium tuberculosis

MTUOP mobile training unit out for parts

mtv mammary tumor virus; motor test(ing) vehicle; multichannel tv sound

mtv (MTV) music television channel

MtV Mount Vernon

M.Tv. Master of Television

MTV Motor Test Vehicle; motor torpedoboat (British naval symbol)

MTVs Motor Torpedo Vessels

mtw main trawl winch

mTw tropical maritime air warmer than underlying surface

mt we must we

Mt W O Mount Wilson Observatory

mtx methotrexate

MTX Morrell Tank Line (railway symbol)

mtxs military traffic expediting service

MTXW Maximum Taxi Weight (aircraft code)

Mty Monterrey (inhabitants—Regiomontanos)

MTY Monterrey, Mexico (airport)

mtz motorize

MTZS Metropolitan Toronto Zoological Society

mu machine unit; mail unit; marijuana user; monetary unit; mouse unit; multiple unit; museum

mu (MU) marginal utility (microeconomics symbol); mark-up (calculator)

m/u makeup; mockup,

MU Macquarie University; Maintenance Unit; Marquette University; Marshall University; Massey University; Mauritius (Internet code); Mercer University; Mercy University; Mercyhurst University; Meredith University; Merrimack University; Mesa University; Messiah University; Methodist University; Miami University; Midwestern University; Milliken University; Monash University; Mothers' Union; Murdoch University; Musicians Union

MUA Machinery Users' Association; Malayan Union Association; Monotype Users' Association; Musicians Union of Australia

MUAC Metropolitan Universities Admissions Center

muap motor unit action potential(s)

muat mobile underwater acoustic unit

muc mucilage

muc. *mucilago* (Latin—mucilage)

MUC Magee University College; Meritorious Unit Citation; Muchea, Australia (tracking station); Munich, Germany (Riem airport)

mu car multiple-unit (railroad) car

MUCC Michigan United Conservation Clubs

Much Ado *Much Ado About Nothing*

MUCIA Midwest Universities Consortium for International Activities

MUCM Medical University College of Medicine

MUCO Material Utilization Control Office

MUD Multi-User Dimension; Multi-User Dungeon (game); Municipal Utility District

'muda Bermuda

'muda grass bermuda grass

MUDPAC Melbourne University Dual-Package Analog Computer

MUDPIE Museum and University Data, Programs, and Information Exchange

muf material unaccounted for; maximum usable frequency

mufa monounsaturated fatty acids

muff muffler

MUFON Mutual UFO Network

muga multigated acquisition

Muh *Muharram* (Arabic—first month of the Mohammedan year)

Mühlhausen Mühlhausen in Thüringen (Mühlhausen in Thuringia)

Mühlheim Mühlheim am Main (Mühlheim on the Main River) or Mühlheim an der Donau (Mühlheim on the Danube)—Germany

Mujib Mujibur Rahman

Muk Mukden

mul multiply

mul *mulig(vis)* (Dano-Norwegian—eventual, probable, or possible, perhaps)

MUL Makerere University Library (Kampala, Uganda)

mulat mulatto

Mulatas Mulatas Islands

mule modular universal laser equipment

MULES Missouri Uniform Law Enforcement System

Mülheim Mülheim an der Ruhr (Mülheim on the Ruhr River) or Mülheim am Rhein (Mülheim on the Rhine)—Germany

MULS Minnesota (University) Union List of Serials

mult multiplication

MULTEWS Multiple-Target Electronic-Warfare System

multi von Willebrand factor multimer analysis

multi (Latin prefix—many or much)—multitude

multics multiplexed information and computing service

multitran multiple translation (translating one language into several target languages)

multr multimeter

mulv murine leukemia virus

mum mumble(d); mumbling; mummed; mummer(s); mummery

mum (MUM) (robot) methodology for unmanned manufacturers

Mum Mumford

MUMC McMaster University Medical Center

MUMMS Marine Corps Unified Management System

MUMPS Multi-Programming System (Massachusetts General Hospital)

mums chrysanthemums

mun munition

Mun Müngo; Munro; Munroe; Munster

Mün *München* (German—Munich)

MUN Memorial University of Newfoundland; Model United Nations

Mund Edmund

muni municipal; municipality

muni bond(s) municipal bond(s)

munic municipal; municipality

munis municipal bonds

munit munitions

Muñoz Marin Luis Muñoz Marin—democratic leader and first governor of Puerto Rico

muo myocardiopathy of unknown origin .

MUO Municipal University of Omaha

muon *mu meson* (Siamese—town)

MUP Makira/Ulawa Province (Solomon Islands); Manchester University Press; Melbourne University Press

M.U.P. Master of Urban Planning

mupo maximum undistorted power output

Mur Murcia

MUR *Mouvements Unis de la Résistance* (French—United Movements of the Resistance)

MURA Midwestern Universities Research Association

Murasaki Baroness Murasaki Shikibu *(The Tale of the Genji)*

MURFAAMCE Mutual Reduction of Forces and Armaments in Central Europe

Murgie Murgatroyd

muriatic acid hydrochloric acid (HCI)

MURIM Multi-Dimensional Reconstruction and Imaging in Medicine

Murph Murphy

mus multiunit school(s); musculoskeletal; museum; music; musical; musician

Mus Musca (constellation); Muscat; museum; music; Muslim

MUS Magnetic Unloading System; Manned Underwater Station

musa multiple-unit steerable antenna

MUSA Medical University of Southern Africa

Mus Anthro Mo University of Missouri Museum of Anthropology

Mus Art RI Museum of Art—Rhode Island

Mus. Bac. Bachelor of Music

Mus Bks Museum Books

musc muscle; muscular

MUSC Medical University of South Carolina

Musca (Latin—Fly constellation)

Muscov Muscovite(s)

muscrit music critic(ism)

mus dir music(al) director

Mus. Doc. Doctor of Music

MUSE Music Search Electronics

Mus.Ed.B. Bachelor of Music Education

Mus.Ed.D. Doctor of Music Education

Mus.Ed.M. Master of Music Education

museo museography; museological; museologist; museololgy

Music Academy of Music; High School of Music

MUSIC Maryland University Sectored Isochronous Cyclotron

musicol musicological; musicologist; musicology

mus id's musical identifications (of radio or tv feature programs)

MusiMus Musikkhistorisk Museum (Norwegian—Music History Museum)

MUSIP Multisensor Image Processor

muskie muskellunge

Mus.M. Master of Music

Mus Northern Ariz Museum of Northern Arizona

Musso Mussolini

Mussolini Benito Mussolini (Amilcare Andrea)

Mus Sys Museum Systems

must manned undersea station

MUST Medical Unit, Self-contained, Transportable

mustargen mustard-nitrogen (poison compound)

mustn't must not

MUSTRACS Multiple Simultaneous-Target Steerable Telemetry-Tracking System

mut mutation; mutilate(d); mutton; mutual

MUT New York Mutuals (National Association)

mutil mutilate; mutilated; mutilation

Mutiny Mutiny on the Bounty

mut. mut. mutatis mutandis (Latin—necessary changes were made)

mut. nom. mutatis nomine (Latin—name was changed)

mutt muttonhead

mutt (MUTT) military utility tactical truck

Mutton Birds Mutton Bird Islands off the southwest coast of New Zealand's Stewart Island where they are also called the Titis

mutu mutual; mutualism

Mutual Mutual Association for Professional Services; Mutual Benefit Life Insurance Company of Newark, New Jersey; Mutual Broadcasting System; Mutual Life Insurance Company of New York; Mutual Nurses Registry; Mutual of Omaha; Mutual Protection Trust; Mutual Security Life Insurance; Mutual Trust Life Insurance

muu mouse uterine units

muw music wire

MUWS Manned Underwater Station

mux multiplex; multiplexer

mux/aro multiplex-automatic error correction

muz muziek (Dutch—music)

Muz Muzio

Muz Muzlim (Arabic—True Believer)

Muzel (Cornish-English—Mousehole)—fishing port resort in Cornwall

muzh muzzle hatch

mv main verb; mean variation; medium voltage; mercury vapor; millivolt; miniature vehicle; monochromatic vision; motor vessel; moving magnet; multivibrator; muzzle velocity

m & v meat and vegetable

μv microvolt

m V megavolt(s); millivolt(s)

mv meervoud (Dutch—plural)

m v mezzo voce (Italian—middle voice)

Mv megavolt; mendelevium

M/V motor vessel

MV Maldives (Internet code); mid-capitalization value; Mild Violence

MV Maria Vergine (Italian—Virgin Mary); *Merchant of Venice*

M. V. Medicus Veterinarius (Veterinary Physician)

M of V Merchant of Venice

MV-678 Agricultural Research Service chemical for fighting fire ants by mimicking hormones to create drones

mva mean vertical acceleration; megavolt ampere; motor vehicle accident

MVA Machinists Vise Association; Mississippi Valley Association; Missouri Valley Authority

MVAS Milwaukee Vocational and Adult School

MVB Martin Van Buren (8th President U.S.)

MVBA Mercado de Valores de Buenos Aires (Buenos Aires Stock Exchange)

mvbd multiple V-belt drive

MVBL Mississippi Valley Barge Line

mvc manual volume control; manufacturing variation control

MVC Military and Veterans Code; Missouri Valley Conference

MVCC Mount Vernon Community College

mvd. moved

MVD Montevideo, Uruguay (Carrasco Airport); Motor Vehicles Department

MVD Ministerstvo Vnutrenniy Delo (Russian—Ministry of Internal Affairs)—(*q.v.—VOT*)

MVDA Motor Vehicle Dismantlers' Association

MVe Murray Valley encephalitis

MVE Metropolitan Vickers Electrical

M.V.E. Master of Vocational Education

MVEMJSUNP *Men Very Easily Make Jugs Serve Useful Nocturnal Purposes* (acrostic mnemonic for remembering the order of planets from the sun—Mercury, Venus, Earth, Mars, Jupiter, Saturn, Uranus, Neptune, Pluto); *My Very En-*

ergetic Mother Just Served Us Nine Pickels-phrase used to remember names of planets in order of their distance from the sun (every 216 years the order of Neptune and Pluto changes)

mvet motor-vehicle excise tax

M.Vet.Med. Master (Mistress) of Veterinary Medicine

M.Vet.Sci. Master (Mistress) of Veterinary Science

mvg most valuable girl

MVG Medal for Victory over Germany

MVHS Mergenthaler Vocational High School

mvi multi-vitamin infusion

MVI Mediovisual Incorporated

M/video Montevideo

MVJC Mount Vernon Junior College

mvlf (MVLF) motor-vehicle license fee(s)

mV/m millivolts per meter

MVM Motor Vehicle Mechanic

MVMA million vehicle miles; Motor Vehicle Manufacturers Association

MVMFB Mississippi Valley Motor Freight Bureau

mvmt movement

MVNP Mesa Verde National Park

MVNWR Monte Vista National Wildlife Refuge (Colorado)

Mvo Montevideo

MVO Member of the Victorian Order

mvp maximum-value package; millivolt potentiometer; mitral valve prolapse; most valuable performance; most valuable player

MVP Manpower Validation Program

MVPBA Mississippi Valley Power Boat Association

MVPCB Motor Vehicle Pollution Control Board

mvps (MVPS) mitral valve prolapse syndrome

MVPT Motor-Free Visual Perception Test(ing)

mvri mixed vaccine—respiratory infections

mvs multiple virtual storage

Mvs Maldivas (Spanish—Maldive Islands)

MVS Mennonite Voluntary Service; Multi-Vest Securities

M.V.Sc. Master of Veterinary Science

MVSS Motor Vehicle Safety Standard

mvt moisture-vapor transmission; multiprogramming with variable number of tasks

MVT Motor Vehicle Technician

MV & THS Manhattan Vocational and Technical High School

MVTI Mohawk Valley Technical Institute

MVUS Merchant Vessels of the United States

mvv maximum vacation value; maximum voluntary ventilation

mvw most valuable wife

mw male white; megawatt; milliwatt; molecular weight

mw. microwave

m/w manufacturing week

m/w (M/W) midwife

mw mevrouw (Dutch—Mrs)

mW megawatt(s)

m/W male White

mW meines Wissens (German—as far as I know)

Mw megawatt

MW Malawi (Internet code); Middle Welsh; Montgomery Ward

M-W Merriam-Webster

M of W Ministry of Works

MWA Metal Window Association; Modern Woodmen of America; Mystery Writers of America; Mystery Writers Association

MWAA Movers and Warehousemen Association of America

MWAI Mystery Writers of America, Incorporated

M-way Motorway (super-highway)

mwb motor whale boat

MWB Metropolitan Water Board; Minister of Works and Buildings; Ministry of Works and Buildings

MWC Ministry of War Communications; Missile Warning Center; Motorola Western Center

MWCG Metropolitan Washington Council of Governments

MWCOG Metropolitan Washington Council of Government

mwd measurement while drilling; megawatt day

MWD Metropolitan Water District; Military Works Department; Ministry of Works and Development; Mutual Weapons Development

mwdm multiwavelength distance measuring (instrument)

mwd/mtu megawatt day per metric ton of uranium

MWDP Mutual Weapons Development Program

mwe megawatts of electricity; meters of water equivalent

MWF Medical Women's Federation

mwg music wire gauge

MWGCP Most Worthy Grand Chief Patriarch

MWGM Most Worshipful Grand Master; Most Worthy Grand Master

mWh milliwatt hour

m & whm missile and warhead magazines

MWHS Martha Washington High School

mwi message-waiting indicator

MWIA Medical Women's International Association

mwir medium-wave infrared

MWJC Marjorie Webster Junior College

mwk millwork

Mwl Malawi (Spanish abbreviation)

MWL Minimum Wage Law(s); Mutual Welfare League

MWLP Meadowview Wild Life Preserve

MWMCA Michigan Women for Medical Control of Abortion

MWMFB Midwest Motor Freight Bureau

MWN Medical World News

MWNM Muir Woods National Monument

mwnt mean water neap tide

MWO Marshallese Women's Organization; Midwest Oil; Modification Work Order; Mount Wilson Observatory

mwp maximum working pressure; membrane waterproofing

MWP Most Worthy Patriarch

MWPA Married Women's Property Act

mwr mean width ratio

MWR Morton Wildlife Refuge (New York)

mWrtl milliwatt resistor-transistor logic

mws magnetic weapon sensor

MWS Manas Wildlife Sanctuary (India); Member of the Wernerian Society; Mudamalai Wildlife Sanctuary (India)

MWSC Midwestern Simulation Council

MWSG Marine Wing Support Group

M.W.T. Master of Wood Technology

mwth megawatts thermal

mwv maximum working voltage

MWV Mineralöl Wirtschafts Verband (German—Petroleum Industry Association)

mww manual wire wrap; municipal waste water

MWW Merry Wives of Windsor

MWZ Manischewitz (stock exchange symbol)

mx maxwell; motocross (rough terrain motorcycle race); multiplex

mx (MX) missile experimental

Mx maxwell; Middlesex

Mx (MX) experimental intercontinental ballistic missile (designed for in-the-air, on the-ground or under-the-sea launching)

MX Mexico (Internet code)

MX Mexicana de Aviación (2-letter code)

mxa mobile exercise area

MXC Minnesota Experimental City

mxd mixed

mxd cl mixed carload

mxdth maximum depth

Mxl Mexicali (inhabitants—Cachanias)

mxm maximum

MXP Milan, Italy (Malpensa Airport)

mxpst maximum possible storm

mxr mask index register

mXr mass X-ray

mx rnfl maximum rainfall

MXS Missile Experimental System

mxtmp maximum temperature

mxwnd maximum wind

my. million years; myopia; myopic

m/y man-year

My Malayalam; Milo; Mylan

My Malaysia (Internet code)—Medinat Yisrael (State of Israel); motor yacht

mya millions of years ago

Mya Myanmar (formerly Burma—capital Yangon); Myasishchev

Mya-4 Soviet heavy bomber named Bison by NATO

Myan Myanmar (Burmese—land of the Burmese)

Myamnar Burma

mybp million years before present

Myc Mycenaean

MYC Manchester Yacht Club; Middletown Yacht Club; Milwaukee Yacht Club; Minnetonka Yacht Club; Mobile Yacht Club

myco mycobacterium

mycol mycology

my/d my daughter

myel(s) myelocyte(s)

Myf's *Mayflower* descendants

myg myriagram

myl myrialiter

mylo mylohyoid

mym myriameter

myn million

myo (Greek *mys*—mouse or muscle)—myocardial infarction, myocardium, myoma; *mayo* (Spanish—May)

myob mind your own business

myobb mind your own bloody business

myodyn myodynamics

myoelectric myoelectrical(ly)

myo inf myocardial infarction

myol myology

myop myopia

mypo multiyear procurement objective

Myr million years; Myriopeda

Myrt Myrtle

Mys Mysore

Mys Sea Mystic Seaport

myst mystagogue; mystagogy; mysteries; mysterious; mystery; mystic; mystical; mystitics

Mystery Queen Agatha Christie

myth. mythological; mythologist; mythology

Myth Mythology

mz monozygote; monozygotic

mz Mangelszahlung (German—for nonpayment)

Mz Méndez

MZ Mail Zone; Mozambique (Internet code); Museum of Zoology; RH Macy and Company (stock exchange symbol)

M & Z Mombasa and Zanzibar

MZ Moskauer Zeit (German—Moscow Time)

mza monozygotic twins (reared) apart

MZA Madrid, Zaragoza, Alicante

mzm multiple-zone monitor

MZMA Moskva Zavod Maloitrazhkaya Automobili (Russian—Moscow Small-Engine Car Factory) producing the Moskvich auto

MZn magnetic azimuth

MZNP Mountain Zebra National Park (South Africa)

mzo marzo (Spanish—March)

M-zone manufacturing zone

MZP Marwell Zoological Park

mzs mezzo-soprano

M.Z.Sc. Master of Zoological Science

mzt monozygotic twins (reared) together

MA Mazatlán (inhabitants Mazatlecos)

mz twins monozygotic twins (genetically identical)

MZUSP Museo de Zoologia de Universidade de São Paulo (Portuguese—Zoology Museum of the University of Sao Paulo)

N

n name; nano; nasal; nasal-strip wearer (racehorse); national; nautical; naval; neap; negative; nephew; nerve; nesting record exists (symbol); neuter; neutral; neutron; new; newly issued; next; night; night game (baseball); night/weekend (rate); nominative; none; noon; norm; normal; noun; nuclear; number; refractive index (symbol); shear modulus of elasticity (symbol); transport number (code)

n' and

n/ and; number

'n' and (as in fish'n'chips, rock 'n' roll)

n index of refraction (symbol); load factor (symbol); *nació* (Spanish—born); number of samples; size of sample; revolutions per second (symbol); rotative speed (symbol)

n. haploid generation; *numerus* (Latin—number)

n/ *nuestro* (Spanish—our); *número* (Spanish—number)

nø regional lymph nodes abnormal

N cranial nerve; International Nickel (stock exchange sysmbol); naira (Nigerian cimrrency); national; nautical; naval; Navy; Negro; neon; neuroticism index; neutral; neutron(s); new issue (in newspaper stock listings); newton(s); New York Stock Exchange (newspaper stock listings); night; nimbus; Nippon; nitrogen; noon; normal; Norse; north; northern; Northwest Airlines; Norway (auto plaque); November—code for letter N; nuclear-propelled vessel (naval symbol); nucleus; Nudity; regional lymph nodes

N (*N*) employment of labor (microeconomics symbol)

(N) nuclear-powered ship (naval symbol, as in CL[N]—nuclear-powered cruiser)

N avogadro constant or number (symbol) *natus* (Latin—born); *Nebenstimme* (German—secondary line or motif)—12-tone term; *neer* (Dutch—down); *noord* (Dutch—north); *nord* (Danish, French, Italian, Norwegian, Swedish—north); *Nord* (German—north); *norre* (Danish—north); *norte* (Portuguese or Spanish—north); north; number of turns (symbol); rate of propeller rotation (symbol); revolutions per minute (symbol); yawing moment (symbol)

N- nuclear

N1, N2 etc. North One, North Two, etc. (London postal zones)

n.₂ diploid generation

N₂ nitrogen

N₂₀ nitric oxide

N¹⁴ radioactive nitrogen

n/30 net (payment) in 30 days

na naturalized; naturally aspirated; negative attitude; next assembly; nicotinic acid; no account; not absolutely; not applicable; not appropriated; not authorized; not available; nucleic acid (NA); numerical aperture; nursing assistant

n.a. *non allocatur* (Latin—not allowed)

n/a navigation and attack; next assembly; no account; no advise; not applicable

na *nestre år* (Norwegian—next year)

nA nanoampere

Na nadir; Napier; natrium (sodium); sodium (symbol)

Na. Nahum

Nᵃ *Nuestra* (Spanish—our)

NA Namibia (Internet code); Narcotics Anonymous; National Academician; National Academy; National Airlines; National Archives; National Association; Native American; Nautical Almanac; Naval Academy; Naval Architect; Naval Attaché; Naval Auxiliary; Naval Aviator; Netherlands Antilles (Aruba, Bonaire, Curaçao, Saba, Sint Eustatius, Sint Maarten); Neurotics Anonymous; North America; North Amcrican; Northrup Aircraft; Nurse's Aide

NA *Nautical Almanac*; *Nederlandse Antillen* (Dutch—Netherlands Antilles)—Aruba, Bonaire, Curaçao,

Saba, Sint Eustatius, Sint Maarten—the Dutch West Indies; *Nomina Anatomica* (Latin—Anatomical Names) —official nomenclature adopted by the International Congresses of Anatomists

Na₂CO₃ sodium carbonate (sal soda)

Na²⁴ radioactive sodium

naa neutron activation analysis; not always afloat

NAA National Academy of Arbitrators; National Aeronautic Association; National Aerosol Association; National Alumni Association; National Apartment Association; National Apple Association; National Arborist Association; National Archery Associaiion; National Association of Accountants; National Auctioneers Association; National Automobile Association; Naval Association of Australia; Naval Attache for Air; Neckwear Association of America; Newspaper Association of America; North American Aviation; North Atlantic Assembly; Nuclear Accidents Agreement; Nurses Auxiliaries Association

NAAA National Alliance of Athletic Associations; National Association of American Academicians; National Association of Arab Americans; National Auto Auction Association

NAAB National Architectural Accrediting Board

NAABC National Association of American Business Clubs

NAABI National Association of Alcoholic Beverage Importers

naabsa not always afloat but safe aground

NAAC National Agricultural Commission; National Association of Agricultural Contractors

NAACC National Association for American Composers and Conductors

NAACO North American Arms Corporation of Canada

NAACOG Nurses Association of the American College of Obstetrics and Gynecology

NAACP National Association for the Advancement of Colored People

NAAD National Association of Aluminum Distributors

NAADAA National Antique and Art Dealers Association of America

NAADC North American Area Defense Command

NAADS New Army Automatic Data System

NAAF North African Air Force (World War II)

NAAFA National Association to Aid Fat Americans

NAAFI Navy, Arrny, and Air Force Institutes

NAAFP National Association Against Fraud in Psychotherapy

NAAG National Association of Attorneys General; NATO Army Advisory Group

NAAIS North American Association of Inventory Services

NAAJ National Association of Agricultural Journalists

NAAMM National Association of Architectural Metal Manufacturers

NAAN National Advertising Agency Network

NAANACM National Association for the Advancement of Native American Composers and Musicians

NAAO National Association of Amateur Oarsmen; Navy Area Audit Office

NAAP National Association of Advertising Publishers

NAAPPA North American Association for the Protection of Predatory Animals

NAAQS National Ambient Air Quality Standards

NAARI National Aero- and Astronautical Research Institute

NAARPR National Alliance Against Racist Political Repression

NAAS National Agricultural Advisory Service; National Apprentice Assistance Scheme; Naval Area Audit Service; Naval Auxiliary Air Station

NAASC North American Aviation Science Center

NAASFEP National Association of Administrators of State and Federal Education Programs

NAAS & ID North American Aviation Space and Information Division

NAASRA National Association of Australian State Road Authorities

NAASS North American Association of Summer Sessions

NAATI National Accreditation Authority for Translators and Interpreters

NAATP National Association of Alcoholism Treatment Programs

NAATS National Association of Air Traffic Specialists; National Association of Auto Trim Shops

NAAUC National Association of Australian University Colleges

NAAUS National Archery Association of the United States

NAAV National Association of Atomic Veterans; North American Association of Ventriloquists

NAAW National Association of Accordion Wholesalers

NAAWP National Association for the Advancement of White People

NAAWS North American Association of Wardens and Superintendents

NAB National Aborigine Conference; National Alliance of Businessmen; National Assistance Board; National Association of Broadcasters; National Association of Businessmen; Naval Advanced Base; Naval Air Base; Naval Amphibious Base; News Agency of Burma; Newspaper Advertising Bureau

NAB New American Bible (Roman Catholic)

NABA National Association of Black Accountants; North American Benefit Association; North American Butterfly Association

NABACO National Association for Bank Audit, Control, and Operation

NABAE National Association of Black Adult Educators

NABB National Association for Better Broadcasting

NABBC National Association of Brass Band Conductors

NABC North American Bridge Championships; National Association of Boy's Clubs

NABD National Association of Brick Distributors; North American Band Directors

NABDC National Association of Blueprint and Diazotype Coaters

NABE National Association of Bilingual Education; National Association of Bilingual Education; National Association of Book Editors; National Association of Business Econommcs

NABEO National Association of Black Elected Officials

NABET National Association of Broadcast Employees and Technicians

NABEWD North American Board for East-West Dialogue

NABHP National Association of Black Hospitality Professionals

NABI National Association of Beverage Importers

NABIM National Association of Band Instrument Manufacturers

NABISCO National Biscuit Company

NABJ National Association of Black Journalists

NABLT National Association of Business Law Teachers

NABM National Association of Blouse Manufacturers; National Association of Boat Manufacturers

NABMA National Association of British Market Authorities

NABMO NATO Bullpup Management Office(r)

NABOB National Association of Black-Owned Broadcasters

Nabokov Vladimirovich Nabokov

nabor neighbor

NABP National Association of Basketball Players; National Association of Boards of Pharmacy; National Association of Book Publishers

NABPO NATO Bullpup Production Office(r); NATO Bullpup Production Organization

NABREP National Association of Black Real Estate Professionals

Nabrico Nashville Bridge Company

NABRT National Association for Better Radio and Television

NABS National Association of Bank Servicers; National Association of Barber Schools; National Association of Black Students; nuclear armed bombardment satellite

NABSE National Alliance of Black School Educators

NABSP National Association of Blue Shield Plans

NABSW National Association of Black Social Workers

NABT National Association of Biology Teachers; National Association of Blind Teachers

NABTE National Association for Business Teacher Education

nabu non-adjusting ballup (unsolvable confusion)

NABUG National Association of Broadcast Unions and Guilds

NABW National Association of Bank Women

NABWE National Association of Black Women Entrepreneurs

nac nacelle; negative air cushion; nozzle area control

Nac Nacional (Spanish—national)

NAC National Achievement Clubs; National Advisory Council; National Agency Check; National Agriculture Centre; National Airways Corporation (New Zealand); National Americanism Commission (American Legion); National Amusements Council; National Archives Council; National Arts Club; National Association of Cemeteries; National Association of Chiropodists; National Association of Choirs; National Association of Concessionaires; National Association of Coroners; National Association of Counties; National Aviation Club; National Aviation Corporation; Native American Church; Naval Academy; Naval Air Center; Naval Aircraftman; Non-Airline Carrier; North Atlantic Council; Northeast Air Command; Norwegian American Cruises; Norwegian-American Council; Nuclear Assurance Corporation; *(see* PNAC)

NACA National Advisory Committee for National Aeronautics; National Advisory Council for Aeronautics; National Agricultural Chemicals Association; National Air Carrier Association; National Armored Car Association; National Association for Campus Activities; National Association of Children of Alcoholics; National Association of Cost Accountants; National Association of County Administrators; New Australian Cultural Association

NACAC National Ad Hoc Committee Against Censorship; National Association of College Admissions Counselors

NACADA National Academic Advising Association

NACAE National Advisory Council for Art Education

NACAM National Association of Corn and Agricultural Merchants

NACAP National Association of Claims Assistance Professionals

NACASBVH National Accreditation Council for Agencies Serving the Blind and Visually Handicapped

NACATTS North American Clear-Air Turbulence Tracking System

NACB National Association of Convention Bureaus

NACC National Association for Core Curriculum; National Association of Catholic Chaplains

NACCA National Association for Creative Children and Adults

NACCAM National Coordinating Committee for Aviation Meteorology

NACCB National Association of Computer Consultant Businesses

NACCC National Association of Citizens Crime Commissions

NACCD National Advisory Commission on Civil Disorders

NACCG National Association of Crankshaft and Cylinder Grinders

N-accident(s) nuclear-power accident(s)

NACCT Nebraska Association of Community College Trustees

NACD National Association for Community Development; National Association of Chemical Distributors; National Association of Conservation Districts; National Association of Container Distributors; National Association of Corporate Directors

NACDE National Association for Child Development and Education

NACDI National Association of the Coat and Dress Industry

NACDL National Association of Criminal Defense Lawyers

NACDR National Association of College Deans and Registrars

NACDS National Association of Chain Drug Stores

NACE National Association for Career Education; National Association of Catering Executives; National Association of Corrosion Engineers

NACEC National Advisory Council on Extension and Continuing Education; National Association of Charitable Estate Counselors

NACEEO National Advisory Council on Equality of Educational Opportunity

NACEL Naval Air Crew Equipment Laboratory

NACEP National Alliance of Concurrent Enrollment Partnerships

NACF National Agricultural Cooperative Federation; National Art Collections Fund; Navy Air Combat Fighter (plane)

NACFE National Association of Certified Fraud Examiners

NACFI North American Council on Fishery Investigations

NACFR National Association of Casual Furniture Retailers

nach (nach) need for achievement

Nach Nachman

NACH National Advisory Council for the Handicapped

NACHA National Automated Clearinghouse Association

NACHEPO National Advisory Commission on Higher Education for Police Officers

Nachf Nachfolger (German—successor)

nachm nachmittags (German—afternoon, p.m.)

NACHM National Advisory Committee on Health Manpower

Nachr Nachrichten (German—bulletin)

NACHRI National Association of Children's Hospitals and Related Institutions

Nachtr Nachtrag (German—appendix, supplement)

NACI National Association for the Cottage Industry

NACIAD National Council on Integrated Area Development

NACIDA National Cottage Industries Development Authority (Philippines)

NACILA National Council of Indian Library Associations

NACIEMFP National Advisory Council on International Monetary and Financial Problems

Nacional El Nacional (Venezuela periodical published in Caracas)

NACISA NATO Communications and Information Systems Agency

NACISO NATO Communications and Information Systems Organization

NaCl sodium chloride (salt)

NACL National Advisory Commission on Libraries; Navy-Arpa Chemical Laser; Nippon Aviotronics Company Limited

NACLA North American Congress on Latin America

NACLIS National Commission on Libraries and Information Science

NACM National Association of Chain Manufacturers; National Association of Charcoal Manufacturers; National Association of Colliery Managers; National Association of Credit Management

naco night-alarm cutoff

NACO Name Authorities Cooperative Project (Library of Congress); National Arts Centre Orchestra (Ottawa); National Association of Charterboat operators; National Association of Counties; Noise Abatement and Control Office, San Diego

NACOA National Advisory Committee on Oceans and Atmosphere

NACOC National Arts Centre Orchestra of Canada

NACODS National Association of Colliery Overmen, Deputies, and Shotfirers

NACOM National Communications

NACOMPTU National Conference of Motion Picture and Television Unions

NACOR National Advisory Committee on Radiation

NACP National Academy of Cable Programming; National Association of Chiefs of Police

nacro night-alarm cutoff

NACRO National Association for the Care and Resettlement of Offenders

NACS National Association for Check Safekeeping; National Association of College Stores; National Association of Convenience Stores; National Association of Cosmetology Schools

NACSA North American Computer Service Association

NACSE National Association of Civil Service Employees

NACSIM NATO Communications Security Information

NACSM National Association of Catalog Showroom Merchandisers

NACSW National Action Committee on the Status of Women (Canadian)

NACT National Association of Careers Teachers; National Association of Corporate Treasurers; National Association of Craftsman Tailors; National Association of Cycle Traders; National Association of Cycle Trades

NACTA National Association of Colleges and Teachers of Agriculture

NACTST National Advisory Council on the Training and Supply of Teachers

NACUA National Association of College and University Administrators; National Association of College and University Attorneys

NACUBO National Association of College and University Business Office Associations

NACUFS National Association of College and University Food Services

NACUSS National Association of College and University Summer Sessions

NACUSA National Association of Composers—USA

NACV National Association of Concerned Veterans

NACVE National Advisory Council on Vocational Education

NACW National Advisory Committee on Women

NACWC National Association of Colored Women's Clubs

NACWPI National Association of College Wind and Percussion Instruments

nad nadir (lowest point); network addressing device; nicotene-adenine dinucleotide; no apparent defect; no appreciable difference; no appreciable disease; not on active duty; nothing abnormal detected; nothing abnormal discovered

Nad Nadine; Nedezhda

NAD National Academy of Design; National Advertising Division; National Alliance for Democracy; National Association of the Deaf; Navajo Army Depot; Naval Air Depot; Naval Air Division; Naval Ammunition Depot; nicotene-adenine dinucleotide; North Atlantic Division

NADA National Associatiomm of Dealers in Antiques; National Association of Drug Addiction; National Automobile Dealers Association

NADAA National Association of Dance and Affiliated Artists

NADABB National Alzheimer's Disease Autopsy and Brain Bank

NADAC National Anti-Drug Abuse Campaign

NADAP National Association of Drug Abuse Problems

NADAR North American Data Airborn Recorder

NADB National Aerometric Data Bank

NADC National Anti-Dumping Committee; National Association of Demolition Contractors; Naval Aide-de-Camp; Naval Air Development Center; Northern Agricultural Development Corporation

NADCA North American Die Casting Association

NADDIS Narcotics and Dangerous Drugs Intelligence File (computerized criminal file)

NADEE National Association of Divisional Executives for Education

NaDefCol NATO Defense College

NADEM National Association of Dairy Equipment Manufacturers

NaDevCen Naval Air Development Center

NADF National Alzheimer's Disease Foundation

NADFAS National Association of Design and Fine Art Societies

NADFS National Association of Drop Forgers and Stampers

NADGE NATO Air Defense Ground Environment Organization

NADGEMO NATO Air Defense Ground Environment Management Office

nadh (NADH) dihydronicotinarnide adenine dinucleotide; (same as dpnh or DPNH)

NADI National Association of Display Industries; National Association of Drama Therapy

NADL National Associatiion of Dental Laboratories; Navy Authorized Data List

NAD/NADH₂ nicotinamide adenine dinucleotide (coenzyme system affecting hydrogen transfer in biological oxidation-reduction reactions)

NADO Navy Accounts Disbursing Office

NADOA National Association of Division Order Analysts

NADOP North American Defense Operational Plan

NADOT North Atlantic Deepwater Oil Terminal

NADOW National Association for Training the Disabled in Office Work

nadp (NADP+) nicotinamide adenine dinucleotide phosphate; (same as tpn or TPN)

NADPAS National Association of Discharged Prisoners' Aid Societies

nadph (NADPH) dihydronicotinamide adenine dinuclectide phosphate

NADS National Association of Diaper Services

NADSA National Association of Dramatic and Speech Arts

NADUS National Association of Doctors in the United States

NADWARN National Disaster Warning System

nae national administrative expenses; not always excused

nAe no American equivalent

Naₑ exchangeable body sodium

NAE National Academy of Education; National Academy of Engineering; National Association of Evangelicals; Naval Aeronautical Establishment (Canadian); Naval Aircraft Establishment

NAEA National Art Education Association; National Association of Enrolled Agents; National Association of Estate Agents

NAEB National Association of Educational Buyers; National Association of Educational Broadcasters

NAEBM National Association of Engine and Boat Manufacturers

NAEC National Aerospace Education Council; National Association Executives Club; National Aviation Education Council

NAECA National Appliance Energy Conservation Act

NAEd National Academy of Education

NAED National Association of Electrical Distributors

NAEDA North American Equipment Dealers Association

NAEDS National Association of Engravers and Die Stampers

NAEE National Association for Environmental Education

NAEF Naval Air Engineering Facility

NAEFTA National Association of Enrolled Federal Tax Accountants

NAEH National Alliance to End Homelessness

NAEIR National Association for the Exchange of Industrial Resources

NAELA National Academy of Elder Law Attorneys

NAEMT National Association of Emergency Medical Technicians

NAEN National Association of Educational Negotiators

NAEP National Assessment of Educational Progress; National Association of Educational Programs (Carnegie Foundation)

NAEPC National Association of Estate Planning Councils

NAER National Association of Executive Recruiters

NAES National Association of Educational Secretaries; National Association of Episcopal Schools; National Associ-

ation of Executive Secretaries; Native American Educational Services

NAESP National Association of Elementary School Principals

NAEST National Archive of Electrical Science and Technology

NAESU Naval Aviation Engineering Service Unit

NAET Nambudripad Allergy Elimination Technique

NAEW NATO Airborne Early Warning

NAEYC National Association for the Education of Young Children

naf nonappropriated funds

NAF National Abortion Foundation; National Amputation Foundation; National Arts Foundation; Naval Aircraft Factory; Naval Air Facility; Netherland-America Foundation; Northern Attack Force

NAF Norges Automobil Forbund (Norwegian—Norway Automobile Association)

NAFA National Academic Funding Advisory; National Academy of Foreign Affairs; National Aerobic Fitness Award; National Association of Fleet Administrators

NAFAC National Association for Ambulatory Care

NAFAG NATO Air Force Advisory Group; NATO Air Force Armaments Group

NAFAS National Association of Flower Arrangement Societies

NAFB National Association of Farm Broadcasters; National Association of Franchised Businessmen

NAF-BRAT National Association for Better Radio and Television

NAFC National Association of Food Chains

NAFCA North American Family Campers Association

NAFCD National Association of Floor Covering Distributors

NAFCU National Association of Federal Credit Unions

NAFD National Air Forwarding Division (Institute of Freight Forwarders); National

Association of Flour Distributors; National Association of Funeral Directors

NAFE National Association for Female Executives

NAFEC National Association of Free-standing Emergency Centers; National Aviation Facilities Experimental Center

NAFEM National Association of Food Equipment Manufacturers

NAFEO National Association for Equal Opportunity (in higher education)

naff (naff) need for affiliation

NAFF National Association For Freedom

NAFFBIA National Association of Former FBI Agents

NAFFP National Association of Frozen Food Producers

NAFFS National Association of Fruit, Flavors and Syrups

NAFI National Association of Fire Investigators; National Association of Flight Instructors; Naval Avionics Facility

NAFINSA Nacional Financiera (Spanish—National Finance Corporation)

NAFLFD National Association of Federally Licensed Firearms Dealers

NAFM National Armed Forces Museum

NAFMB National Association of FM Broadcasters

NAFO National Association of Fire Officers; Northwest Atlantic Fisheries Organization

NAFP New Armed Forces of the Philippines

NAFPC National Academy for Fire Prevention and Control

NAFPP National Association of Fresh Produce Processors

N Afr North Africa

NAFRC National Association of Fiscally Responsible Cities

NAFRD National Association of Fleet Resale Dealers

NAFRLG National Alliance of Financially Responsible Local Governments

NAFS National Association of Foot Specialists; National Association of Forensic Sciences

NAFSA National Association of Foreign Student Advisers; National Association of Foreign Student Affairs

NAFT National Alternative Fuel Test(ing)

NAFTA New Zealand-Australia Free Trade Agreement; North American Fly Tackle Association; North American Free Trade Agreement (Canada, United States, and Mexico); North Atlantic Free Trade Area (Canada, United Kingdom, United States)

NAFWR National Association of Furniture Warehousemen and Removers

nag net annual gain

Nag Nagasaki; Nagoya

NAG National Action Group; National Association of Gag Writers; National Association of Gardeners; National Association of Goldsmiths; National Association of Groundsmen; Naval Advisory Group; Naval Applications Group (USN); Negro Actors Guild; Neighborhood Action Group

NA & G Norgulf Lines (North Atlantic & Gulf)

NAGA North American Gamebird Association

NAGARD NATO Advisory Group for Aeronautical Research and Development

Nagas Naga Hills (mountains on the Burmese border of India); Nagasaki, Japan

NAGC National Association for Gifted Children

NAGCP National Association of Greetitng Card Publishers

NAGDM National Association of Garage Door Manufacturers

N-age nuclear age

NAGE National Association of Government Employees

NAGLDCs National Association of Gay and Lesbian Democratic Clubs

NAGM National Association of Glove Manufacturers; National Association of Glue Manufacturers

NAGMR National Association of General Merchandise Representatives

Nagp Nagpur

NAGPM National Association of Grained Plate Makers

NAGPRA Native American Graves Protection and Repatriation Act

NAGRA Nationalen Genossenschaft für die Lagerung Radioaktiver Abfaelle (German—National Cooperative Society for the Storage of Radioactive Wastes)

NAGS National Allotments and Gardens Society

NAGT National Association of Geology Teachers

NAGWS National Association for Girls and Women in Sport

Nah The Book of Nahum

Nah Nahum

NAH National Association of Homebuilders

NAHA National Association of Handwriting Analysts; National Association of Health Authorities

NADAD National Association of Hose and Accessories Distributors

Nahal Na'or Halutsi Lohem (Hebrew—Fighting Pioneer Youth)—youngest section of the Israeli army

NAHB National Association of Home Builders

NAHC National Advisory Health Council; National Anti-Hunger Coalition

NAHCAC National Ad Hoc Committee Against Censorship (sometimes abbreviated NACAC)

NAHE National Association for Humanities Education

NAHEE National Association for Humane and Environmental Education

NAHFO National Association of Hospital Fire Officers

NAHJ National Association of Hispanic Journalists

NAHM National Association of Hosiery Manufacturers

NAHSA National Association for Hearing and Speech Action; National Association of Hearing and Speech Agencies

NAIISTA National Hiking and Ski Touring Association

NAHT National Association of Head Teachers

NAHU National Association of Health Underwriters

NAHWW National Association of Home and Workshop Writers

nai no action indicated; no airfare included; no address instruction

NAI National Agricultural Institute

NAI New Acronyms and Initialisms

NAIA National Association of Insurance Agents; National Association of Intercollegiate Athletics

NAIB National Association of Insurance Brokers

NAIC National Association of Insurance Commissioners; National Association of Investment Clubs; National Association of Investment Companies; National Association of Investors Corporation; Naval Aircraft Investigation Center

NAICU National Association of Independent Colleges and Universities

NAIDS North Atlantic Institute for Defense Studies (NATO)

NAIEC National Association for Industry-Education Cooperation

NAIES National Association of Interdisciplinary Ethnic Studies

NAIFE National Association of Independent Fee Appraisers

NAIG Nippon Atomic Industry Group

NAII National Association of Independent Insurers

NAIL National Association of Independent Lubes; National Association of Independent Lumbermen; Naval Aircraft Inventory Log; Neurotics Anonymous International Liaison

NAILS National Automated Immigration Lookout System

NAILSC Naval Air Integrated Logistics Support Center

NAIMA North American Insulation Manufacturers Association

naiop navigational aids inoperative for parts

NAIP National Association of Independent Publishers

NAIR National Arrangements for Incidents Involving Radioactivity

NAIRD National Association of Independent Record Distributors & Manufacturers

NAIRE National Association of Internal Revenue Employees

Nairns Nairnshire

NAIRS National Athletic Injury/Illness Reporting System

nairu nonaccelerating inflation rate of unemployment

NAIS National Association of Independent Schools; National Association of Investigative Specialists

NAISC National American Indian Safety Council

NAISS National Association of Iron and Steel Stockholders

NAIT Northern Alberta Institute of Technology

NAITPD National Association of Independent Television Producers and Distributors

NAITTE National Association of Industrial and Technical Teacher Educators

naivnik naive person or politician

NAIW National Association of Insurance Women

NAIWA North American Indian Women's Association

NAJ National Association for Justice

NAJA National Association of Jewelry Appraisers

NAJC National Assessment of Juvenile Correction (University of Michigan); Northern Australia Jockey Club; Northwest Alabama Junior College

NAJCA National Association of Juvenile Correctional Agencies

NAJE National Association of Jazz Educators; National Association of Jazz Education

nak negative knowledge; nothing adverse known

nak (NAK) negative acknowledge character (data processing)

nakl *naklad* (Polish—edition, publisher); *nakladatel* (Czech—edition, publisher)

NAKN National Anti-Klan Network

NAL National Acoustics Laboratory; National Aerospace Laboratory; National Agriculture Library (US Department of Agriculture); National Airlines; National Association of Laity (Catholic); National Astronomical League; Nigeria America Line; Norwegian America Line

NAL *New American Library*

NALA National Association of Legal Assistants

NALAA National Assembly of Local Arts Agencies

NALC National Association of Letter Carriers; National Association of Litho Clubs

NALCC National Automatic Laundry and Cleaning Council

NALCO Newfoundland and Labrador Corporation

NALCON Navy Laboratory Computer Network

NALDEF Native American Legal Defense and Education Foundation

NALEAO National Association of Latino Elected and Appointed Officials

NALED National Association of Limited Edition Dealers

NALEO National Association of Latino Elected and appointed Officials

NALGG National Association for Lesbian and Gay Gerontology

NALGO National and Local Government Officers Association

NALLA National Long-Lines Agency

NALM National Association of Lift Makers

NALP National Association for Law Placement

NALS National Association of Legal Secretaries

NALSA North American Land Sailing Association

NALSAT National Association of Land Settlement Association Tenants

NALU National Association of Life Underwriters

nam named; network access machine; non-aligned movement; number assignment module

Nam Vietnam; Namibia (SouthWest Africa)

N Am North America(n)

NAM National Aero Manufacturing; National Air Museum (Smithsonian Institution); National Association of Manufacturers; Naval Aircraft Modification; Newspaper Association Managers; Non-Alignment Movement; North America(n); North American Movement; Number Assignment Module

NAM *Nederlandsche Aluminium Maatschappij* (Dutch—Netherlands Aluminum Company)

NAMA National Account Marketing Association; National Association of Master Appraisers; National Automatic Merchandising Association; Native American Music Association; New Amsterdam Musical Association; North American Maritime Agencies

NAMAC National Alliance for Media Arts and Culture; National Association of Merger and Acquisition Consultants

NAMAD National Association of Minority Automobile Dealers

NAMB National Association of Master Bakers; National Association of Media Brokers; National Association of Mortgage Brokers

NAMBLA North American Man-Boy Love Association (of child molesters)

NAMBO National Association of Motor Bus Operators

NAMC National Association of Management Consultants; National Association of Minority Contractors; Naval Air Materiel Center; Naval Air Materiel Command; Nihon Aeroplane Manufacturing Company

NAMCC National Association of Mutual Casualty Companies

NAMCO Naval and Mechanical Company

NAMD National Association of Market Developers

NAMDI National Marine Data Inventory

NAMDT National Association of Milliners, Dressmakers and Tailors

NAME National Association of Marine Engineers; National Association of Media Educators; National Association of Medical Examiners; National Association of Metal Name Plate Manufacturers

NAMEB National Association of Marine Engine Builders

N Amer North America(n)

NAMES National Association of Medical Equipment Suppliers

NAMESU National Association of Music Executives in State Universities

NAMF National Association of Metal Finishers

NAMFI NATO Missile Firming Installation

NAMFREL National Citizens Movement for Free Elections

NAMH National Association for Mental Health; Norwegian-American Historical Museum

NAMI National Alliance for the Mentally Ill

NAMIA National Association of Mutual Insurance Agents

Namib Namibia or Namib Desert of Namibia

NAMIC National Association of Mutual Insurance Companies

NAMilCom North Atlantic Military Committee

naml namligen (Swedish—namely)—viz.

NAMM National Association of Music Merchandisers; National Association of Music Merchants

NAMMA NATO Multi-Role Combat Aircraft Development and Production Management Agency

NAMMC Natural Asphalt Mineowners' and Manufacturers' Council

NAMMO NATO Multi-Role Combat Aircraft Development and Production Management Organization

NAMMW National Association of Musical Merchandise Wholesalers

NAMOA National Association of Miscellaneous Ornamental and Architectural Products Contractors

NAMOS National Art Museum of Sport

NAMP National Association of Magazine Publishers; National Association of Married Priests; National Association of Meat Purveyors; National Association for Mature People; Naval Aviation Maintenammce Program

NAMPA NATO Maritime Patrol Aircraft Agency

nampg nautical air miles per gallon

NAMPOW Vietnam Prisoners of War (organization)

namppf nautical air miles per pound of fuel

NAMPS National Association of Marine Products and Services

NAMS National Air Monitoring System; National Association of Marine Surveyors

NAMSA NATO Maintenance and Supply Agency

NAMSB National Association of Men's Sportswear Buyers; National Association of Mutual Savings Banks

NAMSO NATO Maintenance and Supply Organization

NAMT National Association for Music Therapy

NAMTA National Art Materials Trade Association

NAMTC Naval Air Missile Test Center

NAMTRADET Naval Air Maintenance Detachment

NAMTRAGRU Naval Air Maintenance Training Group

NAMW National Association of Media Women

NAMWIB National Association of Minority Women In Business

n.a.n. nisi aliter nometur (Latin—unless it is otherwise noted)

Nan Anna; Nancy; Nanette; Nanking

NAN Nandi, Fiji Islands (airport)

nana (NANA) N-acetyl-neuraminic acid

NANA National Advertising News Association; North American Newspaper Alliance; Northwest Alaska Natives Association

NANAC National Aviation Noise Abatement Council

NANBH Non-A-Non-B Hepatitis

Nancy Anne Frances "Nancy" Robbins Davis Reagan, wife of President Ronald Reagan

nand not and

NAND NOT AND (data-processing logic operator)

Nando Fernando

NANE National Association for Nursery Education

Nanga Nanga Parbat, India

NANM National Association of Negro Musicians

nano 10^{-9}

nanova non-ordmogonal analysis of variance

NANPA North American Nature Photography Association

NANTIS Nottingham and Nottinghamshire Technical Information Service

NANVH & SWO National Assembly of National Voluntary Health and Social Welfare Organizations

NANWEP Navy Numerical Weather Prediction; Navy Numerical Weather Problems (USN)

NAO National Accordion Organization; National Association of Outfitters; Noise Abatement Office

NAOA Navy Officers Accounts Office

NAOC Nigerian Agip Oil Company

NaOH sodium hydroxide (caustic soda)

NAOHSM National Association of Oil Heating Service Managers

NAOP National Association of Operative Plasterers

NAORPG North Atlantic Ocean Regional Planning Group

NAOT National Association of Organ Teachers

NAOTC National Association of Over-the-Counter Companies

NAOTS Naval Aviation Ordnance Test Station

nap. knapsack; napalm (naphthalene and coconut oil—jellied gasoline incendiary mixture); naphtha; naval aviation pilot (NAP); nonagency purchase; not at present; nuclear auxiliary power

Nap Naples; Napoleon; Napoleonic

NAP Naples, Italy (airport); Narragansett Pier (railroad); National Aerospace Plane (hypersonic aircraft capable of flying up to 25 times the speed of sound); National Association for the Paralysed; National Association of Parliamentarians; National Association of Postmasters; National Association of Publishers; Naval Auxiliary Patrol; Naval Aviation Pilot; Non-insured Assistance Program

N.A.P. Neighborhood Awareness Program

NAP Nomina Anatomica, Paris; Nuclei Armati Proletari (Italian—Armed Proletarian Nucleus)—terrorists

NAPA National Asphalt Pavement Association; National Association of Performing Artists; National Association of Purchasing Agents; National Automotive Parts Association

NAPAAW National Association of Professional Asian-American Women

NAPAC National Program for Acquisitions and Cataloging

napalm naphthene palmitate (napththalene plus coconut oil—jellied gasoline used in flame-throwers)

NAPAMA National Association of Performing Arts Managers and Agents

NAPAN National Association for the Prevention of Addiction to Narcotics

NAPARE National Association for Perinatal Addiction Research and Education

NAPATMO NATO Patriot Management Office

NAPBIRT National Association of Professional Band Instrument Repair Technicians

NAPBL National Association of Professional Baseball Leagues

napc non-adherent peritoneal cells

NAPC National Association of Precancel Collectors

NAPCA National Air Pollution Control Administration; National Asian Pacific Center on Aging; National Association of Pipe Coating Applicators

NAPCAE National Association for Public Continuing and Adult Education

NAPCRO National Association of Police Community Relations Officers

NAPD National Association of Police Driving

NAPDEA North American Professional Driver Education Association

NAPE National Alliance of Postal Employees; National Association for Professional Educators; National Association of Port Employees; National Association of Power Engineers; National Association of Professional Engravers

NAPECW National Association for Physical Education of College Women

NAPET National Association of Photo Equipment Technicians

NAPF National Association of Pension Funds

NAPFA National Association of Personal Financial Advisors

NAPFE National Alliance of Postal and Federal Employees

naph naphtha; naphthyl

NAPH National Association of Professors of Hebrew

NAPHCC National Association of Plumbing/Heating/Cooling Contractors

NAPIA National Affiliate of Printing Industries of America; National Association of Professional Insurance Agents

NAPIM National Association of Printing Ink Manufacturers

NAPL National Association of Photo Lithographers; National Association of Printers and Lithographers

NAPLP National Association of Para-Legal Personnel

NAPLPS North American Presentation Level Protocol Syntax

NAPM National Association of Paper Merchants; National Association of Pastoral Musicians; National Association of Pharmaceutical Manufacturers; National Association of Photographic Manufacturers; National Association of Punch Manufacturers; National Association of Purchasing Management

NAPMO NATO Airborne Early Warning and Control Programme Management Organisation

NAPN National Association of Physician Nurses

NAPNAP National Association of Pediatric Nurse Associates and Practitioners

NAPNES National Association for Practical Nurse Education and Service

NAPO National Association of Performing Artists; National Association of Pizza Operators; National Association of Probation Officers; National Association of Property Owners; National Association of Purchasing Agents

Napoléon I Napoléon Bonaparte (1769–1821)

Napoléon II *l'Aiglon* (French —the Eaglet) François-Charles Joseph Bonaparte (1811–1832)

Napoléon III Napoléon le Petit (French—Little Napoleon) Louis-Napoléon Bonaparte (1808–1873)

NAPPH National Association of Private Psychiatric Hospitals

NAPPS National Association of Professional Pet Sitters

nap(py) napkin

na pr na priklad (Czech—for example)

NAPR National Association for Pastoral Renewal; National Association of Publishers' Representatives

NAPRA National Association of Progressive Radio Announcers

NAPRC National Association for the Prevention of Rape by Castration

NAPS National Alliance of Postal Supervisors; National Association of Personnel Services; National Association of Pet Sitters; Nissan Air Pollution System

NAPSA National Appliance Parts Suppliers Association; National Association of Pretrial Service Agencies

NAPSAE National Association for Public School Adult Education

Nap's bones Napier's bones (first slide rule)

NAPSS Numerical Analysis Problem Solving System

NAPT National Association of Physical Therapists; National Association for Poetry Therapy; National Association for the Prevention of Tuberculosis

NAPTC Naval Air Propulsion Test Center

NAPTIC National Air Pollution Technical Information Center

napu nuclear auxiliary power unit

NAIPUS National Association of Postmasters of the United States; Nuclear Auxiliary Power Unit System

NAPV National Association of Prison Visitors

NAPVD National Association for the Prevention of Venereal Disease

NAPVO National Association of Passenger Vessel Owners

NAQI National Air Quality Index

NAQP National Association of Quick Printers

nar narrow; net assimilation rate; no apparent reason

Nar Narragansett

NAR National Association of Realtors; National Association of Rocketry; Nelson Aldrich Rockefeller; North American Rockwell; North American Royalties; Northern Alberta Railway

NARA Narcotic Addict Rehabilitation Act; National Archives and Records Administration; Nippon Australian Relations Agreement

NARAA National Association of Recruitment Advertising Agencies

NARAD Navy Research and Development

NARAL National Abortion Rights Action League; National Association for the Repeal of Abortion Laws

NARAS National Academy of Recording Arts and Sciences

NARA/MU National Association of Review Appraisers and Mortgage Underwriters

NARB National Advertising Review Board

NARBA North Americaum Rare Bird Alert

NARBW National Association of Railway Business Women

narc narcotic; narcotics agent; narcotics; narcotics officer

narc (Latin prefix—numbness or stupor)—narcotic

NARC National Agricultural Research Center; National Archives and Records Service; National Association for Retarded Children; National Association of Retired Catholics

narco narcotic; narcotics hospital; narcotics officer; narcotics treatment center

NARCO United Nations Narcotics Commission

narcocard narcotic-addict registration card

narcodollars narcotic (traffic) dollars

Narconon Narcotics Anonymous

narcos narcotics; narcotics police officers

narcot narcotic, narcotize(d), narcotiz(ing)

narco-terr narcotics terrorism

narcotest narcotics test

narco-traf narcotics-traffick(er); narcotics trafficking

narcs narcotics; narcotics agents; narcotics hospitals; narcotics officers; narcotics treatment centers

nard spikenard

NARD National Association of Regimental Drummers; National Association of Retail Druggists

NARDA National Association of Retail Dealers of America

NARDIC Naval Research and Development Information Center

Nar Div *Narodni Divadlo* (Czech—National Theater)— Prague opera house

NAREA National Association of Real Estate Appraisers

NAREB National Association of Real Estate Boards; National Association of Real Estate Brokers

NAREBB National Association of Real Estate Buyer Brokers

narec naval research electronic computer

NAREC National Association of Real Estate Companies

NAREE National Association of Real Estate Editors

NAREIF National Association of Real Estate Investment Funds

NAREIT National Association of Real Estate Investment Trusts

NARELLO National Association for Real Estate License Law Officials

narf natural axial-resonant frequency

NARF National Association of Retail Furnishers; Native American Rights Fund; Naval Air Rework Facility; Nuclear Aircraft Research Facility

NARFE National Association of Retired Federal Employees

NARGA National Association of Retail Grocers of Australia

NARI National Association of Recycling Industries; National Association of the Remodeling Industry; National Atmospheric Research Institute; Native American Research Institute

Nar Inv Narcotics Investigation

narist. *naristillae* (Latin—nasal drops)—nosedrops

nark narcotics agent or law-enforcement officer

NARK Nikolai Andreyvich Rimsky-Korsakov

Narkomvneshtorg *Narodny Komissariat Vneshney Torgovli* (Russian—People's Commissariate of Foreign Trade)

NARL National Aero Research Laboratory; Naval Arctic Research Laboratory

NARM National Association of Recording Merchandisers; National Association of Relay Manufacturers; National Association of Retail Merchants; National Association of Restaurant Managers

NARMCO National Research and Manufacturing Company

N-armed nuclear-armed (aircraft, bomb, missile, submarine, etc.)

N-arm(s) nuclear armament(s); nuclear arms

N-arms control nuclear arms control

N-arms race nuclear arms race

NARO National Association of Royalty Owners; North American Regional Office

NAROCTESTSTA Naval Air Rocket Test Station

NARP National Association of Railroad Passengers; Nuclear Weapons Accident Report Procedures

NARPA National Air Rifle and Pistol Association

narr narratives

Nar Rep Bul *Narodna Republika Bulgaria* (Bulgarian People's Republic)

Narrows waterway between Brooklyn and Staten Island, New York—spanned by 4260-foot-long Verrazano Bridge, once the world's longest suspension bridge; strait in the Dardanelles near the Aegean; strait between American and British Virgin Islands

Narrow Seas the Channel between England and France as well as the southern end of the North Sea between England, Belgium, and the Netherlands

NARS National Archives and Records Service; National Association of Radiation Survivors; Non-Affiliated Reserve Section

NARSA National Automotive Radiator Service Association

NARSAD National Alliance for Research on Schizophrenia and Depression

NARSIS National Association for Road Safety Instruction in Schools

NARST National Association for Research in Science Teaching

NARTB National Association of Radio and Television Broadcasters

NARTC North America Region Test Center

NARTEL North Atlantic Radio Telephone Committee

NARTM National Association of Rope and Twine Merchants

NARTS National Association of Reporter Training Schools; National Association of Retail and Thrift Shops; Naval Air Rocket Test Station

NARTU Naval Air Reserve Training Unit

NARU North Australian Research Unit

NARUC National Association of Regulatory Utility Commissioners

NARVRE National Association of Retired and Veteran Railroad Employees

NARWACL North American Regional World Anti-Communist League

nas nasal; nasalis; nasology

n-a-s no added salt

NAS Nassau, Bahamas (airport); National Academy of Sciences; National Academy of Songwriters; National Adoption Society; National Advocates Society; National Aerospace Standard(s); National Agricultural Society; National Aircraft Standard(s); National Airspace System; National Association of Sanitarians; National Association of Scholars; National Association of Schoolmasters; National Association of Stevedores; National Association of Supervisors; National Audubon Society; Native American Studies; Naval Air Station; Nursing Auxiliary Service

N A S Noise Abatement Society

NᵃSᵃ *Nuestra Señora* (Spanish—Our Lady)

NASA National Acoustical Suppliers Association; National Advertising Sales Association; National Aeronautics and Space Administration; National Appliance Service Association; National Association of Schools of Art; National Association of Securities Administrators; National Automobile Salesmen's Association; North American Sailing Association

NASAA National Aeronautics and Space Administration Act; National Assembly of State Arts Agencies; North American Securities Administrators Association

NASABCA National Aeronautics and Space Administration Board of Contract Appeals

NASAC National Association of Scientific Angling Clubs

NASA-CF National Aeronautics and Space Administration—Cocoa Beach, Florida

NASA-CO National Aeronautics and Space Administration—Cleveland, Ohio

NASA-EC National Aeronautics and Space Administration—Edwards, California

NASAEN National Association for State-Enrolled Assistant Nurses

NASA-GM National Acronautics and Space Administration—Greenbelt, Maryland

NASA-HA National Aeronautics and Space Administration—Huntsville, Alabama

NASA-HT National Aeronautics and Space Administration—Houston, Texas

Nasakom Nationalist-Communist

NASA LST Natiomial Aeronautics and Space Administration Large Space Telescope

NASA-LV National Aeronautics and Space Administration—Langley Field, Virginia

NASA-MC National Aeronautics and Space Administration—Moffett Field, California

NASAO National Association of State Aviation Officials

NASAP Nonproliferation Alternative System Assessment Program; Nuclear Alternative Systems Assessment Program

NASAPR National Aeronautics and Space Administration Procurement Regulations

NASAR National Association of Search and Rescue

NASARR North American Searching and Ranging Radar

NASA-SC National Aeronautics and Space Administration-Santa Monica, California

NASA STAR National Aeronautics and Space Administration Scientific and Technical Aerospace Reports

NASA/STIF National Aeronautics and Space Administration/Scientific and Technical Information Facility

NASBE National Association of State Boards of Education

NASBO National Association of State Budget Officers

NASC National Aeronautics and Space Council; National Aeronautics Standards Committee; National Aircraft Standards Committee; National Alliance of Senior Citizens; National Association of Student Councils; NATO Supply Center; Naval Air Systems Command; North American Supply Council; Northwest Association of Schools and Colleges

NASCAR National Association of Sports Car Racing; National Association for Stock Car Advancement and Research

NASCO National Academy of Sciences Committee on Oceanography; National Automotive Service Company; North American Students of Cooperation

NASCom Naval Air Systems Command

NASCOM National Aeronautics and Space Administration tracking network, also performing command and control functions

NASCP North American Society for Corporate Planning

NASCUS National Association of State Credit Union Supervisors

NASD National Amalgamated Stevedores and Dockers; National Association of Securities Dealers; National Association of Service Dealers; Naval Aviation Supply Depot; Nippon Advanced Ship Design(ing)

NASDA National Association of State Development Agencies; National Space Development Agency (Japan)

NASDAQ National Association of Security Dealers Automated Quotation (system)

NASDCD National Association of State Directors of Child Development

NASDS National Association of Scuba Diving Schools

nase neutral atom space engine (sputtering engine)

NASE National Academy of School Executives; National Academy of Stationary Engineers; National Association of Stationary Engineers; National Association of Steel Exporters

NASEES National Association for Soviet and East European Studies

NASEN National Association for State-Enrolled Nurses

NASF National Aboriginal Sports Foundation; National Association of State Foresters

NASFAA National Association of Student Financial Aid Administrators

NASFCB National Association of Specialty Food and Confection Brokers

NAS & FCA National Automatic Sprinkler and Fire Control Association

NASFM National Association of Store Fixture Manufacturers

NASFO National Asset Seizure and Forfeiture Office

NASFT National Association for the Specialty Food Trade

NAS-GB Noise Abatement Society of Great Britain

Nash Nashville

NASH National Association of Specimen Hunters

NASHA North American Survival and Homesteading Association

NASIS National Association for State Information Systems

NASJA North American Ski Journalists Association

NASL North American Soccer League

NASM National Air and Space Museum (Smithsonian); National Association of School Magazines; National Association of Schools of Music; Naval Aviation School of Medicine

NASM *Nederlandsche-Amerikaansche Stoomvaart Maatschappij* (Dutch—Holland-American Line)

NASMD National Association of School Music Dealers

NASML National Air and Space Museum Library (Smithsonian Institution)

NASMV National Association for a Standard Medical Vocabulary

NASN National Air Sampling Network; National Association of School Nurses

NASNI Naval Air Station, North Island (Halsey Field, San Diego, California)

NAS-NRC National Academy of Science–National Research Council

NASOH North America Society for Oceanic History

NASP National Aero-Space Plane; National Airport Systems Plan; National Association of School Psychologists; Negro Anglo-Saxon Protestant

NASPA National Society of Public Accountants

NASPD National Association of Steel Pipe Distributors

Nas Pers *Nasionale Pers* (Afrikaans—National Press)—publisher of apartheid books and periodicals

NASPM National Association of Seed Potato Merchants

NASQAN National Stream-Quality Accounting Network

N^{a}Srª *Nossa Senhora* (Portuguese—Our Lady); *Nuestra Señora* (Spanish—Our Lady)

NASRC National Association of State Racing Commissioners

NASRP National Association of Special and Reserve Police

Nass Nassau

NASS National Agricultural Statistics Service; National Association of School Superintendents; National Association of Secretarial Services; National Association of Summer Sessions; Naval Air Signals School

NASSAM National Association for the Self-Supporting Active Ministry

NASSC National Alliance on Shaping Safer Cities

NASSCO National Steel and Shipbuilding Company

NASSD National Association of School Security Directors

NASSL National Association of Spanish-Speaking Librarians

NASSO National Association of Socialist Students' Organizations

NASSP National Association of Secondary-School Principals

NASSR Nahichevan Autonomous Soviet Socialist Republic

NAST National Association of Schools of Theatre; Nuclear Accident Support Team

NASTBD National Association of State Text Book Directors

NASTI Naval Air Station, Terminal Island

NASTL National Anti-Steel-Trap League

NASU National Adult School Union

NASULGC National Association of State Universities and Land-Grant Colleges

NASW National Association of Science Writers; National Association of Social Workers

NASWM National Association of Scottish Woolen Manufacturers

nat nation; national; nationalist; native; natural; naturalist; naturalization; naturalize(d); nature; normal allowed time

nat natuurkunde (Dutch—natural science)

Nat Natalia; Natalie; Nathalie; Natban; Nathanael; Nathaniel, Natasha; Nation; National; Nationalist; naturalized

Nat Hof und Nationaltheater (German—Court and National Theater, Munich); *Naturkunde* (German—natural science)

NAT National Air Transport; National Arbitration Tribunal; Washington, D.C., Nationals (National Association team)

NATA National Air Transportation Association; National Association of Tax Accountants; National Association of Tax Administrators; National Association of Testing Authorities; National Association of Transportation Advertisers; National Athletic Trainers Association; National Automated Transportation Association; National Aviation Trades Association; North American Telephone Association; North Atlantic Treaty Alliance

Nat Absten National Abstentionalist

NATAPROUBU National Association of Professional Bureaucrats

Nat Arc National Archives

NATAS National Academy of Television Arts and Sciences

Nat Assn National Association

natat natation

NATB National Automobile Theft Bureau; Naval Air Training Base

Nat Bur Econ Res National Bureau of Economic Research (Columbia and Princeton)

Nat Bur Stand Circ National Bureau of Standards Circular(s)

Nat Bur Stand Misc Pub National Bureau of Standards Miscellaneous Publication(s)

Nat Bur Stand Spec Pub National Bureau of Standards Special Publication(s)

NATC National Air Traffic Controllers; National Air Transportation Conferences; National Alcohol Tax Coalition; Naval Air Training Command

NATCA National Air Traffic Controllers Association

NATCG National Association of Training Corps for Girls

natch naturally

Natch Natchez

Natchez Trace national parkway serving Alabama, Mississippi, and Tennessee

NATCO National Automatic Tool Company, National Tank Company

natcol natural color(ing)

natcom national communications

NATCOM NATO communication

Nat Con Nature Conservancy

NATCS National Air Traffic Control Service; National Air Traffic Control System

NATD National Association of Teachers of Dancing; National Association of Telecommunications Dealers

Nat Dem National Democrats

Nate Nathan(iel)

NATE National Association for Teachers of Electronics; National Association for the Teaching of English; Native American Teacher Education

Nat Fed National Federation

NATFHE National Association of Teachers in Further and Higher Education

Nat Gal National Gallery

Nat Geog Mag National Geographic Magazine

Nath Nathan(iel)

Nath B Nathaniel Bowditch

nat hist natural history

Nathl Nathaniel

NATIDC Netherlands-Australia Trade and Industrial Development Council

NATIE National Association of Trade and Industrial Education

NATINFORM Nationalist Information Service

nation. nationality

National National Gallery in London or the National Gallery of Art in Washington, D.C.

Natl Cath Rep National Catholic Reporter

NATIONAL National Cash Register

Nations Bus Nations Business

NATIS National Information System(s); North Atlantic Treaty Information Service

Nativ Nativity

NATKE National Association of Theatrical and Kine Employees

natl national

N Atl North Atlantic

N Atl Cur North Atlantic Current

Nat Lib National Liberal; National Library of Canada (Ottawa)

NATLIBCAN National Library of Canada

NATLIBNZ National Library of New Zealand

NATMAP National Mapping

Nat Mon National Monument

Nat Mus Natal Museum; National Museum

nato no action—talk only

NATO National Association of Taxicab Owners; National Association of Theater Owners; National Association of Trailer Owners; National Association of Travel Organizations; North Atlantic Treaty Organization (Belgium, Canada, Czech Republic, Denmark, Germany, Greece, Hungary, Iceland, Italy, Luxembourg, Netherlands, Norway, Poland, Portugal, Spain, Turkey, United Kingdom, United States)

NATO-AGARD North Atlantic Treaty Organization-Advisory Group for Aeronautical Research and Develop-ment

Nat Obs National Observer

NATODC NATO Defense College

NATO-ELLA North Atlantic Treaty Organization—European Long Lines Agency

NATO-LRSS North Atlantic Treaty Organization—Long-Range Scientific Studies

NATOMILOCGRP North Atlantic Treaty Organization Military Oceanography Group

NATOPS Naval Air Training and Operating Procedures Standardization

Nat Ord Natural Order

NATO-RDPP North Atlantic Treaty Organization-Research and Development Production Program

NATOs National Association of Theatre Owners

NATPE National Association of Television Program Executives

nat phil natural philosophy

Nat Pk National Park

natr. natrium (Latin—sodium)

NATRA National Association of Television Recording Artists

Nat Rev National Review

Nats National Party members, Nationalists; naturalized citizens

Nats Natsionalnyii (Russian—national)

NATS National Association of Teachers of Singing; National Association of Temporary Services; Naval Air Test Station; Naval Air Transport Service

Nat. Sc.D. Doctor of Natural Science

Nat Sci Natural Science(s)

Nat Sci Fdn National Science Foundation

Nat Sec Soc National Secular Society (founded in 1866 by Charles Bradlaugh)

NATSEMI National Semiconductor Incorporated

NATSF Naval Air Technical Service Facility

NATSJA National Association of Training Schools and Juvenile Agencies

NATSO National Association of Truck Stop Operators

NATSOPA National Society of Operative Printers and Assistants

NATSPG North Atlantic Systems Planning Group

Nat Sup National Superannuation

N Att Naval Attaché

N-attack nuclear attack

NATTC National Tank Truck Carriers; Naval Air Technical Training Center

NATTKE National Association of Theatrical, Television, and Kine Employees

NATTS National Association of Trade and Technical Schools; Naval Air Turbine Test Station

Nat U Nations Unies (French—United Nations)

Nat Uni National University

natur naturalist

Natural Bridges Natural Bridges National Monument in Southeastern Utah

NATUSA North African Theater of Operations

Nat West National Westminster (British bank)

NATWJ National Alliance of Third World Journalists

naty naturally

nau network access unit

náu náutica (Spanish—nautical)

Nau Nauruan(s); Nauru Island

NAU Naval Administrative Unit; Northern Arizona University

NAUA National Aircraft Underwriters' Association; National Auto Underwriters Association

NAUB National Association of Urban Bankers

nauga naugahide (plastic upholstery)

Naughty Island Pulau Sajahat (resort offshore Singapore)

NAUI National Association of Underwater Instructors

NAUMD National Association of Uniform Manufacturers and Distributors

NAUPA National Association of Unclaimed Property Administrators

Nauru Republic of Nauru (western Pacific Ocean island nation), Pleasant Island

NAUS National Association for Uniformed Services

naut nautical

naut jar nautical jargon

NAUW National Association of University Women

n aux b new auxiliary boiler

nav naval; navigable; navigate; navigation; navigational; navigator

n/a/v net asset value

Nav Navaho; Navajo; naval; Navarra; Navarre; Navassa Island between Haiti and Jamaica

NAV Net Asset Value

NAVA National Audio-Visual Association; North American Vexillological Association

navaco navigation action cutout (switchboard)

NAVAE National Association for Vietnamese-American Education

NAVAERORECOVF Naval Aerospace Recovery Facility

navaid(s) navigation aid(s)

NAVAIR Naval Air (Systems Command)

NAVAIRILANT Naval Air Forces, Atlantic

NAVAIRPAC Naval Air Forces, Pacific

NAVAIRREWORKF Naval Air Rework Facility

NAVAIRSYSCOM Naval Air Systems Command

Nav. Arch. Naval Architect

Navarraise La Navarraise (French—The Girl of Navarre)—Massenet opera

NavAus navigation in Australian waters

Navar Navigation and ranging

NAVBALTAP Naval Forces, Baltic Approaches (NATO)

NAVBASE Naval Base

navbm (NAVBM) naval ballistic missile

nav brz naval bronze

Nav Bs Naval Base

NavCad Naval Cadet

NAVCAMS Naval Communication Area Master Station

NAVCENT Allied Naval Forces, Central Europe

NAVCJ National Association of Volunteers in Criminal Justice

NavCm navigation countermeasures and deception

navcom navigation communication

Nav.Const. Naval Constructor

NAVCOSSACT Naval Command Systems Support Activity

NAVD National Association of Video Distributors

navdac navigation data assimilation computer

NAVDAC Navigation Data Assimilation Center

Nav Dep Naval Deputy (NATO)

Nav.E. Naval Engineer

NavEarns navigation in the eastern Atlantic and the Mediterranean

NavEast navigation along the east coast of Asia

NAVEDTRASUPPCEN Naval Education and Training Support Center

NAVELEX Naval Electronic (Systems Command)

NAVEOFAC Naval Explosive Ordnance Disposal Facility

Navesink Highlands of the Navesink also called Atlantic Highlands on the New Jersey coast around Sandy Hook

navex navigation exercise

NAVFE Naval Forces Far East

NAVFEC Naval Facilities

NAVFECENGCOM Naval Facilities Engineering Command

NAVFOR Naval Forces

NAVFORJAP Naval Air Forces, Japan

NAVFORKOR Naval Air Forces, Korea

NAVH National Aid to Visually Handicapped; National Association for the Visually Handicapped; National Association of Voluntary Hostels

Nav I Navassa Island (uninhabited American islet in north Caribbean close to Windward Passage between Cuba and Hispaniola, navigatioimal light maintained by U.S; although recently Haiti claimed the islet)

NAVIC Navy Information Center

navicert naval inspection certificate (allowing neutral vessels to proceed through blockades)

navicert(s) navigation certificate(s)

Navidad Natividad (Spanish—Nativity)—Christmas

navig ravigation

Navigators Navigator Islands (American Samoa)

NavInd navigation in the Indian Ocean

NAVINFO Navy Information Offices

NAVINTCOM Naval Intelligence Command

NAVINTCOMINST Naval Intelligence Command Instructions

NAVISTAR formerly International Harvester

NAVLIS Navy Logistics Information System

NAVMACS Naval Modular Automated Communications Systems

NAVMAR Naval Forces, Marianas

NAVMAT Naval Materiel Command (USN)

NAVMED Naval Medicine

NAVMEDIS Naval Medical Information System

NavMisCen Naval Missile Center

NAVNON Naval Forces, Northern Norway (NATO)

NavNoPac navigation in the North Pacific

NavNorlant navigation in the North Atlantic

NAVNORTH Allied Naval Forces, Northern Europe

NavOceanO Naval Oceanographic Office (USN)

NAVOCFORMED Naval On-Call Force, Mediterranean (NATO)

NAVOCS Naval Officer Candidate School

NAVORD Naval Ordnance

NAVORDSYSCOM Naval Ordnance Systems Command

NAVPACEN Navy Public Affairs Center

NAVPERS Naval Personnel

NAVPERSRANDLAB Naval Personnel Research and Development Laboratory

NAVPHIBSCOL Naval Amphibious School

NAVPHIL Naval Forces—Philippines

NAVIPORCO Naval Port Control Officer

NAVPRO Naval Plant Representative Office(r)

NAVPUB Naval Publications

NAVREGMEDCEN Naval Regional Medical Center

NAVROM Romanian merchant marine

NAVS National Anti-Vivisection Society; North American Vegetarian Society

navsat navigational satellite

NavSat navigation in the South Atlantic

NAVSCAP Naval Forces, Scandinavian Approaches (NATO)

NAVSCOLCOM NORVA Naval Schools Command, Norfolk, Virginia

NAVSEA Naval Sea Systems Command (USN)

NAVSEACENTLANT Naval Sea Support Center—Atlantic

NAVSEACENTPAC Naval Sea Support Center—Pacific

NAVSEC Naval Ship Engineering Center

NAVSHIPCOM Naval Ship Systems Command

NavShipyd Naval Shipyard

NAVSMO Navigation Satellite Management Office

NavSoPac navigation in the South Pacifiic

NAVSOUTH Naval Forces, Southern Europe

NAVSPASUR Naval Space Surveillance (USN)

NAVSPECWARGRU Naval Special Warfare Group

NAVSTA Naval Station

NAVSTAR Navigation System using Time and Ranging

NAVSUPGRU Naval Support Group

NAVSUPORANT Naval Support Forces, Antarctica

navtac (NAVTAC) navigation tactical (aircraft)

NAVTELCOM Naval Telecommunications Command

NAVTIS National Vessel Traffic Information System

NAVTRACEN Naval Training Center

NAVTRACOM Naval Training Command

NAVTRADEVCEN Naval Training Device Center

NAVUWSEC Naval Underwater Weapons Systems Engineering Center

navvies navigators (unskilled canal builders, unskilled laborers)

NAVWAG Naval Warfare Analysis Group

NAVWEASERV Naval Weather Service

NAVWUIS Naval Work Unit Information Service

NAW National Association of Wholesalers; National Association for Women; North African Waters

NAWA National Association of Women Artists

NAWAC National Weather Analysis Center

NAWAPA North American Water and Power Alliance

NAWAS National Air Warning Service

NAWB National Association of Workshops for the Blind

NAWBO National Association of Women Business Owners

NAWCC National Association of Watch and Clock Collectors

NAWCH National Association for the Welfare of Children in Hospitals

NAWDAC National Association for Women Deans, Administrators, and Counselors

NAWDC National Association of Women Deans and Counselors

NAWESA Naval Weapons Engineering Support Activity

nawf nodes above white flower

NAWF National Aborigine Welfare Fund; North American Wildlife Foundation

NAWGA National American Wholesale Grocers Association

NAWIC National Association of Women in Construction

NAWK National Association of Warehouse Keepers

NAWLA North American Wholesale Lumber Association

NAWM National Association of Wool Manufacturers

NAWMP North American Waterfowl Management Plan

NAWND National Association of Wholesale Newspaper Distributors

NAWPA North American Water and Power Alliance

NAWS National Association of Women Students; National Aviation Weather System

NAWSA National American Woman's Suffrage Association

Naxas Naxalites (Maoist extrenmists active in India)

Nay Nayarit

NAYBC North American Youth Bridge Championship

NAYC National Association of Youth Clubs

NAYE National Association of Young Entrepreneurs

NAYRU North American-Yacht Racing Union

naz nazionale (Italian—national)

Naz Nazaire

Naze (Old Norse—Nose)— southern tip of Norway at Lindesnes

Nazi adherent of the former National Socialist German Workers' Party *(Nationalsozialistische Partei)*

nb nanobar(n); narrow band; new boiler(s); newborn; no ball (cricket); no bias (relay)

n/b narrow beam; no balls (lacking nerve); no brands; northbound

n.b. *nota bene* (Latin—note well); *nulla bona* (Latin—no good)

Nb nimbus; niobium (formerly columbium)

Nb *Noordbrabent* (Dutch—North Brabant)

NB National Bank; National Battlefield; Naval Base; Navy Band; New Brunswick; Newport Beach; Niagara Frontier Tariff Bureau; North Borneo

NB *Naviera Boliviana* (Spanish—Bolivian Shipping); *Norsk Bibliotekforening* (Norwegian Library Association)

Nb[94] radioactive niobiurn

NBA National Backpain Association; National Band Association; National Bank of Australia; National Bankers Association; National Banking Association; National Bar Association; National Basketball Association; National Boat Association; National Bowling Association; National Boxing Association; National Button Association

NBAA National Business Aircraft Association

NBAC National Black Alcoholism Council

NBAD National Bank of Abu Dhabi

N balance nitrogen balance

NBBA National Bed and Breakfast Association

NBBB National Better Business Bureau

NBBC National Brass Band Club

NBBS New British Broadcasting Station

NBBU New Brunswick Board of Underwriters

nbc non-battle casualty; nothing but chaos

n-b-c (NBC) nuclear-biological-chemical (warfare)

NBC Nagasaki Broadcasting Company; National Ballet of Canada; National Baseball Congress; National Beagle Club; National Beef Council; National Biscuit Company; National Book Committee; National Book Council; National Bowling Council; National Braille Club; National Broadcasting Commission; National Broadcasting Company; National Broadcasting Corporation; National Bulk Carriers; National Bus Company; Navy Beach Commando; Nigerian Broadcasting Corporation

NB & C Norfolk, Baltimore and Carolina Line

NBCA National Baseball Congress of America; National Beagle Club of America; National Business Circulation Association

NBCC National Book Critics Circle; National Breast Cancer Coalition

NBCCA National Business Council for Consumer Affairs

nbccw nuclear, biological, chemical, conventional warfare

nbcd nuclear, biological, and chemical defense

NBCDA *National Black Child Development Act*

nbcdx nuclear, biological, and chemical defense exercise

NBCL National Beauty Culturists' League

NBCSO National Broadcasting Company Symphony Orchestra

NBCU National Bureau of Casualty Underwriters

NBCUSA National Baptist Convention U.S.A.

nbd negative binomial distribution

NBD National Bank of Detr6it

NBDA National Bicycle Dealers Association

NB & DA National Barrel & Drum Association

NBDC National Bomb Data Cermter (Washington, D.C.); National Book Development Council

NBDEA National Beverage Dispensing Equipment Association

NBDL Naval Biodynamics Laboratory

NBE National Bank Examiner(s)

NBEA National Ballroom and Entertainment Association; National Broadcast Editorial Association; National Business Education Association

NBER National Bureau of Economic Research; National Bureau of Engineering Registration

NBET National Business Entrance Test(s)

NBF National Bank of Fiji; National Boating Federation

NBFA National Baseball Fan Association; National Bricklayers Foundation of Australia; National Business Forms Association

NB & FAA National Burglar and Fire Alarm Association

nbfi's non-bank(ing) financial intermediaries

NBF Life National Ben Franklin Life Insurance

nbfm narrow-band frequency modulation

nbfr neutral balance force reductions

NBFU National Board of Fire Underwriters; Newfoundland Board of Fire Underwriters

nbg no bloody good

NBG National Bank of Georgia; Naval Beach Group

NBGC National Ballet Guild of Canada

NBH National Bellas Hess

NBHA National Builders Hardware Association

NBHC New Broken Hill Consolidated

NBHS National Bureau for Handicapped Students

nbi no bone(y) injury; nothing but initials

NBI Nathaniel Branden Institute; National Benevolent Institution

NBI *Norges Byggforskningin-stitutt* (Norwegian—Norwegian Building Research Institute)

NBIA National Business Incubation Association

NBIPP National Black Independent Political Party

NBIS Narcotics Border Interdiction System

NBIT New Bedford Institute of Technology

nbl not bloody likely

NBL National Basketball League; National Book League; National Business League

NBLC *Nederlands Bibliotheek en Lektuur Centrum* (Dutch—Netherlands Center for Public Libraries and Literature)

NBL & P National Bureau for Lathing and Plastering

nbm no bowel movement; nonbook material(s); normal bowel movement; nothing by mouth

nbm (NBM) nuclear ballistic missile

NBM New Brunswick Museum

NBMC National Black Media Coalition

NBME National Board of Medical Examiners

NBMG Navigation Bombing and Missile Guidance System

NBMGS Navigation Bombing and Missile Guidance System

NBMV & NSL New Bedford, Martha's Vimneyard, and Nantucket Steamship Line

nbn new bad news

nbn (NBN) national book number

NBN Nagoya Broadcasting Network; National Black Network

NBNA National Black Nurses Association

NBNZ National Bank of New Zealand

NBO Nairobi, Kenya (airport); Navy Bureau of Ordnance

N-bomb neutron bomb; nuclear bomb

nbp normal boiling point

NBP National Battlefield Park; National Business Publications; Neighborhood Beautification Program; New Brooklyn Philharmonic

NBPA National Basketball Players Association; National Beverage Packaging Association; National Black Police Association

NBPC National Border Patrol Council

NBPI National Board for Prices and Income

NBPRP National Board for the Promotion of Rifle Practice

nbp's nude beach pests (prurient snoopers and voyeurs)

NBPW National Business Professional Women's (week)

nbq no broken quantities

nbr nitrile-based rubbers; nitrilebutadiene rubber

n br naval brass; naval bronze

n Br *nördliche Breité* (German—north latitude)

NBR National Bison Range (Montana); Nightly Business Report (tv)

NBR *National Business Review*

nbre *noviembre* (Spanish—November)

NBRF National Biomedical Research Foundation

NBRI National Building Research Institute

NBRMP National Board of Review of Motion Pictures

NBRPC New Brunswick Research and Productivity Council

NBRS National Beef Recording Scheme; National Beef Recording Service

NBRT National Board for Respiratory Therapy

nbs normal burro serum

NBS Nagano Broadcasting System; National Battlefield Site; National Broadcasting Service (NZ); National Bureau of Staumdards; New British Standard

NBSA National Bank of South Africa; Netherlands Bank of South Africa

NBSBL National Bureau of Standards Boulder Laboratory

NBSCCST National Bureau of Standards Center for Computer Sciences and Technology

NBSI National Bank of the Solomon Islands

NBSPA National Bark and Soil Producers Association

NBSRS *Narodna Biblioteka Socijalisticke Republike Srbije* (Serbo-Croatian—National Library of the Socialist Republic of Serbia)—Belgrade

NBSS National Bank Surveillance System

NBS-SIS National Bureau of Standards—Standard Information Services

nb st nimbo-stratus

NBST National Board for Science and Technology

NBT National Book Trust (India)

NBTA National Baton Twirlers Association; National Business Teachers Association

NBTC New Brunswick Teachers College

nbtdr narrow band time domain reflectometry

NBTL Naval Boiler Test Laboratory

NBTS National Blood Transfusion Service

nbuf not buffed (leather)

n butt national buttress (thread)

nbv net book value

nbw noise bandwidth

NBW National Book Week

NBWA National Beer Wholesalers Association

Nby Newbury

NBYWCAUSA National Board of the Young Women's Christian Association of the U.S.A.

nc national coarse (thread); natural convector; network computer; nitrocellulose; no change; no charge; no connection; noise criteria; normally closed; nose cone; not cataloged; not complete; nuclear capability; numerical control(s)

n/c new charter; new crop; no charge; numerical control (automation)

nc *non chiffre* (French—unnumbered)

nC *na Christus* (Dutch—after Christ)

NC Napa College; Nashville, Chatanooga & St. Louis (railroad); Nasson College; Natchez College; National Cash Register (stock exchange symbol); National Center; National Certificate; National Coarse (screw threads); National Congress; National Council; National Fire Waste Council; Nature Conservancy; New Caledonia; New College; Newark College; Newberry College; Newcomb College; Newnham College; Nicholls College; Nichols College; Norfolk College; Norman College; North Carolina; North Carolinian; Northern Cascades; Northern Counties; Northern County; Northland College; Northwestern College; Nuclear Congress; Nuffield College (Oxford); Nurse Corps

N.C. NC Wyeth

NC *Norske Creditbank* (Norwegian Credit Bank)

NC-17 no children under 17 admitted

nca neurocirculatory asthenia; no copies available

NCA Narcotics Control Act; National Camping Association; National Canners Association; National Capital Award; National Cashmere Association; National Cattlemens Association; National Caves Association; National Charcoal Association; National Cheerleaders Association; National Chiropractic Association; National Civic Association; National Club Association; National Coal Association; National Coffee Association; National Command Authority; National Commission on Accrediting; National Commission on AIDS; National Committee for Adoption; National Composition Association; National Confectioners Association; National Constructors Association; National Contesters Association; National Cosmetology Association; National Costumers Association; National Council for the Arts; National Council on the Aging; National Council on Alcoholism; National Coursing Association; National Cranberry Association; National Creameries Association; National Credit Association; National Cricket Association; Naval Communications Annex; Navy Contract Administrator; Nebraska Correctional Association; Nevada Correctional Association; Ngorongoro Conservation Area (Tanzania); North Central Airlines; North Central Association (of colleges and schools); Northern Consolidated Airlines

N C A National Cricket Association

ncaa no coupon at all (bond trader parlance)

NCAA National Children Adoption Association; National Collegiate Athletic Association

NCAAA National Center of Afro-American Artists

NCAB National Cancer Advisory Board

NCAB National Cyclopedia of American Biography

NCAC National Copyright Advisory Committee (Library of Congress); National Council of Acoustical Consultants

NCACME National Center for Adult, Continuing, and Manpower Education

ncad net cash against documents

NCAE National Center for Audio Experimentation; National College of Agricultural Engineering; National Council of Adult Education; National Council of Agricultural Employers; North Carolina Association of Educators

NCAI National Clearinghouse for Alcohol Information; National Congress of American Indians; National Council on Alcoholism Inc

NCAICU North Carolina Association of Independent Colleges and Universities

NCAIR National Center for Automated Information Retrieval

NCALI National Clearinghouse for Alcohol Information (USPHS)

NCAM National Center for Advanced Materials

NCAMP National Coalition Against the Misuse of Pesticides

NCAN National Coalition of American Nuns

NCANH National Council for the Accreditation of Nursing Homes

ncap nematic curvilinear aligned phase

N-CAP Nurses Coalition for Action in Politics

NCAP Natilonal Community AIDS Partnership; New Car Assessment Program

NCAPC National Center for Air Pollution Control

NCAR National Center for Atmospheric Research; National Committee for Antarctic Research

N Car North Carolina

NCARB National Council of Architectural Registration Boards

NCARL National Committee Against Repressive Legislation

NCARMD National Commission on Arthritis and Related Musculoskeletal Disease

NCARRV North Carolinians Against Racist and Religious Violence

N-carrier(s) nuclear-powered aircraft carrier(s)

NCas North Cascades

NCAS National Collegiate Association for Secretaries

NCASF National Council of American-Soviet Friendship

NCAT National Center for Alternative Technology; National Center for Audiotape; Northampton College of Advanced Technology

NCATE National Council for the Accreditation of Teacher Education

NCAVAE National Council for Audio-Visual Aids in Education

NCAW National Council for Animal Welfare

NCAWE National Council for Administrative Women in Education

ncb narcotic-centered behavior; new crime buffer; nickel-cadmium battery; no claim bonus

NCB National Cargo Bureau; National Coal Board; National Conservation Bureau; Nippon Credit Bank; Nippon Cultural Broadcasting

NCBA National Candy Brokers Association; National Catholic Bandmasters Association; National Cattle Breeders' Association; National Clydesdale Breeders Association; National Coop-

erative Business Associatiton; Northern California Booksellers Association

NCBD National Council for Balanced Development

NCBE National Clearinghouse for Bilingual Education; National Conference of Bar Examiners

NCBFAA National Customs Brokers and Forwarders Association of America

NCBH National Coalition to Ban Handguns

NCBI National Cotton Batting Institute

NCBIAE National Council of Bureau of Indian Affairs Educators

NCBL National Conference of Black Lawyers

NCBM National Conference of Black Mayors

NCBMP National Coalition of Black Meeting Planners; National Council of Building Material Producers

NCBR National Council of Black Republicans

NCBS National Cattle Breeding Station (Australian)

NCBVA National Concrete Burial Vault Association

NCBWA National Collegiate Baseball Writers Association

nec numerical control code

NCC Namhae Chemical Corporation (Korean); Nassau Community College; National Cadet Corps; National Carloading Corporation; National Castings Council; National Certified Counselors; National Civic Council; National Climatic Center; National Coaches Council; National Commission on Children; National Computer Center; National Computer Council; National Conference on Citizenship; National Consumer Council; National Container Committee; National Cotton Council; National Council of Churches; National Council of Churches of Christ in the USA; National Cultural Center; Nature Conservation Council; Navajo Community College; Newhouse Communications Center (University of Syracuse); Newspaper Comics Council; Noise Control Committee; Non-Combatant Corps; NORAD Control Center; Northwest Community College

NCC Nederlands Cultureel Contact (Dutch—Netherlands Cultural Contact)

NCCA National Chemical Credit Association; National Coil Coaters Association; National Commission for the Certification of Acupuncturists; North Carolina Correctional Association

NCCAM National Center for Complementary and Alternative Medicine

NCCAN National Center on Child Abuse and Neglect

NCCAS National Center of Communication Arts and Sciences

NCCAT National Committee for Clear Air Turbulence

NCCB National Citizens Committee for Broadcasting; National Consumer Cooperative Bank; National Council of Catholic Bishops

NCCC National Catholic Cemetery Conference; Niagara County Community College

NCCCA National Coordinating Council for Constructive Action

NCCCC Navy Command, Control, and Communications Center

NCCCD National Center for Computer Crime Data

NCCCLC Naval Command Control Communications Laboratory Center (formerly NEL-Navy Electronics Laboratory)

NCCCUS National Council of the Churches of Christ in the United States

NCCCUSA National Council of the Churches of Christ in the U.S.A.

NCCD National Council on Crime and Delinquency

NCCDS National Cooperative Crohn's Disease Study

NCCE National Commission for Cooperative Education; National Council for Catholic Evangelization

NCCEOA National Coordinating Council of Educational Opportunities Associations

NCCF National Committee to Combat Fascism (Black Panther front); National Commission on Consumer Finance

NCCG National Council on Compulsive Gambling

NCCH National Council to Control Handguns

NCCI National Committee for Commonwealth Immigrants; National Council on Compensation Insurance

NCCIHE North Carolina Center for Independent Higher Education

NCCIS NATO Command, Control, and Information System

NCCJ National Coalition for Children's Justice; National Conference of Christians and Jews

NCCJPA National Clearinghouse for Criminal Justice Planning and Architecture

NCCL National Council for Civil Liberties; National Council of Canadian Labor

NCCLS National Committee for Clinical Laboratory Standards; National Consumer Center for Legal Services

NCCMP National Coordinating Committee for Multiemployer Plans

NCCNHR National Citizens' Coalition for Nursing Home Reform

NCCOP National Corporation for the Care of Old People

nccp navigation control console panel

NCCP National Center for Children in Poverty (Columbia University); National Communities Conservation Planning Group

NCCPA National Council of College Publications Advisers

NCCPG National Council for the Conservation of Plants and Gardens

NCCPL National Community Crime Prevention League

NCCPV National Commission on the Causes and Prevention of Violence

NCCR National Council for Civic Responsibility

NCCS National Command and Control System; National Council for Constitutional Studies

NCCU National Conference of Canadian Universities; North Carolina Central University

NCCVD National Council for Combating Venereal Diseases

NCCW National Council of Catholic Women

NCCY National Council of Catholic Youth

ncd no can do; not considered disabling

n.c.d. nemine contra dicente (Latin—no one dissenting)

NCD National Commission on Diabetes; Naval Construction Department; Naval Construction Depot; New Community Development

NCD New Collegiate Dictionary

NCDA National Center for Drug Analysis; National Council on Drug Abuse

NCDAD National Council for Diplomas in Art and Design

NCDAI National Clearinghouse for Drug Abuse Information

NCDC National Capital Development Commission; National Center for Disease Control; National Communicable Disease Center; National Community Development Corporation; National Council on Crime and Delinquency; National Curriculum Development Center; New Community Development Corporation

NCDL National Canine Defence League

NCDs Negotiable Certificates of Deposit

NCDS National Center for Dispute Settlement (American Arbitration Association)

ncdu navigation control and display unit

nce normal curve equivalent

NCE National Council of Exchangors; Newark College of Engineering; Nice, France (Côte d'Azur airport)

NCE New Catholic Encyclopedia

NCEA National Catholic Educational Association; National Catholic Evangelization Association; National Center for Economic Alternatives; National Community Education Association; North Carolina Education Association

NCEB National Center for Educational Brokering; NATO Communications Electronic Board

NCEC National Committee for an Effective Congress; National Community Education Clearinghouse

NCECA National Council on Education for the Ceramic Arts

NCECS North Carolina Educational Computing Service

NCED National Center for time Employment of the Deaf

NCEDT National Council to Eliminate Death Taxes

NCEE National Catholic Educational Exhibitors; National Commission on Excellence in Education; National Congress for Educational Excellence; National Council of Engineering Examiners

NCEER National Center for Earthquake Engineering Research

ncef national calling and emergency frequencies

NCEFT National Commission on Electronic Fund Transfers

NCEI National Commission on Emerging Institutions

NCEL Naval Civil Engineering Laboratory

NCEMP National Center for Energy Management and Power

NCEN National Commission on Egg Nutrition

NCEO National Center for Employee Ownership

NCEP National Cholesterol Education Program

NCER National Center for Earthquake Research; National Council on Educational Research

NCERT National Council for Educational Research and Training

nces necessary; normal curve equivalent scores

NCES National Center for Educational Statistics (now CS Center for Statistics)

NCET National Council for Educational Technology

NCEW National Conference of Editorial Writers

ncf nerve cell food

NCF National Consumer Federation

NCFA National Cat Fanciers Association; National Commission of Fine Arts; National Consumer Finance Association; Navy Campus for Achievement

NCFC National Council of Farmer Cooperatives

NCFDA National Council on Federal Disaster Assistance

NCFILP National Coalition for Fair Immigration Laws and Practices

NCFIRB North Carolina Fire Insurance Rating Bureau

NCEM National Commission on Food Marketing

NCFP National Conference on Fluid Power

NCFPC National Center for Fish Protein Concentrate

NCFR National Council on Family Relations

NCFSU Naval Construction Force Support Unit

NCFT National College of Food Technology

NCG National Council for the Gifted; National Cylinder Gas (division of Chemotron)

NCGA National Church Goods Association; National Computer Graphics Association; National Cotton Ginners' Association; National Council on Governmental Accounting

NCGE National Council for Geographic Education

NCGG National Council for Geodesy and Geophysics

NCGME National Council on Graduate Medical Education

NCGRC National Church Growth Research Center

nch number changed (telephone)

NCH National Children's Home; National Clearing House; National Coalition for the Homeless

NCHA National Campers and Hikers Association; National Capital Housing Authority; National Culling Horse Association

NCHCS National Council for Health Care Services

NCHEC National Center for Home Equity Conversion

NCHEE National Council for Home Economics Education

NCHELP National Council of Higher Education Loan Programs

N Chem L National Chemical Laboratory

NCHEMS National Center for Higher Education Management Systems

nchg no charge

n chg normal charge

NCHI National Council of the Housing Industry

NCHMT National Capitol Historical Museum of Transportation

NCHP *Nouvelle Compagnie Havraise Peninsulaire (de Navigation)* (French—Havre Peninsula Navigation Line)

n Chr *nach Christus* (German—after Christ, A.D.)

NCHS National Center for Health Statistics

NCHSR & D National Center for Health Services Research and Development (HEW)

NCHVRFE National College for Heating, Ventilating, Refrigeration, and Fan Engineering

NCI New Community Instrument

nci naphthalene creosote iodiform (lice-control powder); no common interest; no-cost item

NCI National Cancer Institute; National Casing Institute; National Cello Institute; National Cheese Institute; Naval Cost Inspection; Naval Cost Inspector; Naval Court of Inquiry; New Community Instrument

NCIA National Council of Instructional Administrators; National Council for Islamic Affairs

NCIAC National Consumer Information and Advisory Center

NCIB National Charities Information Bureau

NCIC National Cancer Institute of Canada; National Career Information Center; National Crime Information Center

NCIES National Center for the Improvement of Educational Systems

NCIJC National Council of Independent Junior Colleges

NCILT National Centre for Industrial Language Training

NCIO National Council on Indian Opportunity

NCIP Northeast Corridor Improvement Program

nci powder naphthalene creosote iodoform powder (for killing lice)

NCIS National Chemical Information System; National Council of Independent Schools; Naval Criminal Investigative Services

NCISC Naval Counterintelligence Support Center

NCIT National Council on Inland Transport

NCJA National Criminal Justice Association; National Collegiate Judo Association

NCJAVM National Council on Jewish Audio-Visual Materials

NCJISS National Criminal Justice Information and Statistics Service

NCJMS National Center for Job Market Studies

NCJR National Coalition for Jail Reform

NCJRS National Criminal Justice Reference Service

NCJSC National Criminal Justice Statistics Center

NCJW National Council of Jewish Women

Nck Neck

N Cl New Caledonia(n)

NCL National Carriers Limited; National Central Library; National Chemical Laboratory; National Consolidated Limited; National Consumers League; National Culture League; Norwegian Caribbean Line; Norwegian Cruise Lines

NCLA National Council of Local Administrators (of vocational education and practical arts); North Carolina Library Association

NCLAN National Crop Loss Assessment Network

N-class a Soviet class of nuclear-powered attack submarines

NCLC National Caucus of Labor Committee; National Consumer Law Center; National Council of Labour Colleges; National Council of Local Administrators

NCLE National Center for Labored Enterprises

NCLIS National Commission on Libraries and Information Science

NCLR National Council of La Raza

NCLS National Clearinghouse for Legal Services

nem non-corrosive metal; non-crew member

n.c.m. *non compos mentis* (Latin—of unsound mind)—insane

NCM National Congress for Men; Nippon Calculating Machine

NCMA National Catalog Managers Association; National Concrete Masonry Association; National Contract Management Association; National Council of Music Associations; North Carolina Museum of Art

NCMC National Center on Missing Children; NORAD Cheyenne Mountain Complex

NCMDA National Commission on Marijuana and Drug Abuse

NCME National Council on Measurements in Education; Network for Continuing Medical Education

NCMEA National Catholic Music Educators Association

NCMEC National Center for Missing and Exploited Children

NCMH National Committee on Maternal Health; National Committee for Mental Hygiene

NCMHE National Clearinghouse for Mental Health Education

NCMIE National Council of Music Importers and Exporters

NCMLB National Council of Mailing List Brokers

NCMP National Commission for Manpower Policy

NCMS National Classification Management Society

nemt numerically controlled machine tool

NCMU National Commission on Marijuana Use

ncn no common name

NCN National Council of Nurses; New Caledonian Nickel; New Century Network

NCNA National Council on Noise Abatement; New China News Agency (mainland China)

NCNC National Captive Nations Committee; National Council of Nigeria and the Cameroons

NCNE National Campaign for Nursery Education

NCNP National Conference for New Politics; North Cascades National Park (Washington)

NCNW National Council of Negro Women

NCNY Newswomen's Club of New York

nco no-cost option; no crossing over (genes)

NCO Noncommissioned Officer

NCOA National Campground Owners Association; National Council on the Aging; Noncommissioned Officer Academy

NCOAUSA Non Commissioned Officers Association of the U.S.A.

NCOC National Commission on Organized Crime; National Council on Organized Crime

ncod net cash on delivery

NCOES Noncommissioned Officer Education System

NCOIC Noncommissioned Officer in Charge

NCOIL National Conference of Insurance Legislators

NCOLS Noncommissioned Officers Leadership School

N/COM Navy/Chief of Naval Operations

NCOMP National Catholic Office for Motion Pictures

NCOPM National Conference of Personal Managers

NCOR National Committee on Oceanographic Research

ncos non-commissioned officers

NCOSE National Council on Systems Engineering

ncp national cycling proficiency; network control program; nitrogen charge panel; normal circular pitch; not copy-protected; number of channel programs

NCP National Capital Parks; National Country Party; National Customs Police (Philippines); Navy Capabilities Plan; Noise Control Plan; Nutritiion Center of the Philippines

NCP Na viera Chilena del Pacifico (Spanish—Chilean Pacific Line)

NCPA National Cottonseed Products Association; National Crime Prevention Association

NCPAC National Conservative Political Action Committee

NCPC National Capital Planning Commission; National Consumer Protection Council; National Crime Prevention Coalition; National Crime Prevention Council; Northern Canada Power Commission

NCPCA National Center for the Prosecution of Child Abuse

NCPDP National Council for Prescription Drug Programs

NCPE National Committee on Pay Equity

NCPERL National Coalition for Public Education and Religious Liberty

NCPGA North Carolina Personnel and Guidance Association

NCPI National Clay Pipe Institute; National Crime Prevention Institute; Navy Civilian Personnel Instructions

NCPL National Center for Programmed Learning

NCPM National Clay Pot Manufacturers

NCPPL National Committee on Prisons and Prison Labor

NCPPR National Center for Public Policy Research

NCPRV National Council of Puerto Rican Volunteers

NCPS National Cat Protection Society; National Commission on the Public Service; National Commission on Product Safety; Non-Contributory Pension Scheme

NCPSM National Center of Preventive and Stress Medicine

NCPSSM National Commission to Preserve Social Security and Medicare

NCPT National Congress of Parents and Teachers

NCPTSD National Center for Post-Traumatic Stress Disorder

NCPTWA National Clearinghouse for Periodical Title Word Abbreviations

NCPV National Commission on the Prevention of Violence

NCPWB National Certified Pipe Welding Bureau

NCQA National Committee on Quality Assurance

NCQR National Council for Quality and Reliability

ncr natural circulation reactor; no calibration required; no carbon required; not combat ready

n Cr novo Cruzeiro (Portuguese—new cruzeiro)—Brazilian monetary unit

NCR National Capital Region; National Cash Register; National Consumer Research; National Council of Reconciliation (in Vietnam); New Christian Right; Non-Communist Resistance

NCR National Catholic Record; National Catholic Reporter

NCRA National Correctional Recreation Association

NCRC National Condor Research Center

NCRCL National Civil Rights Clearinghouse Library

NCRD National Council for Research and Development; National Council for Resource Development

NCRE Naval Construction Research Establishment

NCRFCL National Commission on Reform of Federal Criminal Laws

NCRFP National Council for a Responsible Firearms Policy

NCRI National Red Cherry Institute

NCRL National Chemical Research Laboratory

NCRLC National Committee on Regional Library Cooperation

NCROPA National Campaign for the Repeal of the Obscene Publications Act (British)

ncrp narrow cold-rolled products; nonreinforced concrete pipe

NCRP National Committee on Radiation Protection; National Council on Radiation Protection; National Council for Research and Planning

ncr paper no-carbon-required paper

NCRPM National Committee on Radiation Protection and Measurements

NCRR National Center for Resource Recovery

NCRS National Committee for Rural Schools

NCRT National College of Rubber Technology

NCRVE National Center for Research in Vocational Education

NCRY National Commission on Resources for Youth

ncs naval control of shipping; navigation control simulator; net control station

NCS National Cartoonist Society, National Cemetery System; National Chrysanthemum Society; National Communications System; Naval Communication Station; National Computer Systems; Net Control Station; Numerical Control Society

NCSA National Carl Schurz Association; National Center for Supercomputer Applications; National Computer Security Association; National Construction Software Association; National Council of Seamen's Agencies; National Crushed Stone Association; National Customs Service

Association; North Carolina School of the Arts; North Coast of South America

NCSAW National Catholic Society for Animal Welfare

NCSBA North Carolina School Boards Association

NCSBCS National Conference of States on Building Codes and Standards

NCSBEE National Council of State Boards of Engineering Examiners

NCSC National Cargo Security Council; National Center for State Courts; National Companies and Securities Commission (Australia); National Computer Security Center; National Council for Senior Citizens; National Council of Senior Citizens

NCSCEE National Council of State Consultants in Elementary Education

NCSCI National Center for Standards and Certification Information

NCSCT National Center for School and College Television

NCSDCJC National Council of State Directors of Community and Junior Colleges

NCSE National Commission on Safety Education

NCSEA North Carolina Society of Enrolled Agents

NCSEA National Community School Education Association; National Council of State Education Associations

NCSF National College Student Foundation

NCSG National Chimney Sweep Guild

NCSGC National Council of State Garden Clubs

NCSH National Clearinghouse for Smoking and Health

NCSI National Chimney Sweepers Institute; National Council for Stream Improvement; National Council of Savings Institutions; National Council of Self-Insurers

NCSJ National Conference on Soviet Jews

NCSL National Center for Service Learning; National Civil Service League; National Conference of Standards Lab-

oratories; National Conference of State Legislators; Naval Code and Signal Laboratory

NCSMC National Council for the Single Mother and Her Child

NCSNE Naval Control of Shipping in the Northern European Command Area of NATO

NCSNP National Council for a Sane Nuclear Policy

NCSO Naval Control of Shipping Office(r); North Carolina Symphony Orchestra

NCSP National Conference on State Parks

NCSPA National Corrugated Steel Pipe Association; North Carolina State Ports Authority

NCSPS National Committee for the Support of Public Schools

NCSR National Center for Systems Reliability; National Council for Scientific Research

NCSRC National Centre for Social Research and Criminology (Cairo)

ncsry necessary

NCSS National Center for Social Statistics; National Council for Social Studies; National Council of Social Service

NCSSA National Community Service Sentencing Association; Nature Conservation Society of South Australia; Naval Command Systems Support Activity

NCSSC Naval Command Systems Support Center

NCSSFL National Council of State Supervisors of Foreign Languages

NCSTAS National Council of Scientific and Technical Art Societies

NC & ST L Nashville, Chattanooga & St Louis (railroad)

NCSTRC North Carolina Science and Technology Research Center

NCSW National Conference on Social Welfare

NCSWCL National Commission on State Workmen's Compensation Laws

NCSWD National Center for Solid Waste Disposal

NCSWR National Conference on Solid Waste Research

nct natural contour theory; no charge for terms; no civil twilight

NCT National Chamber of Trade; National Childbirth Trust; National Culture Trust; Noise Cancellation Technologies

n/cta nuestra cuenta (Spanish—our account)

NCTA National Cable Television Association; National Capital Transport Agency; National Community Television Association; National Committee for Technological Awards; National Council for Technological Awards

NCTAEP National Committee on Technology, Automation, and Economic Progress

NCTC National Collection of Type Cultures

NCTE National Council of Teachers of English

NCTEC Northern Counties Technical Examinations Council

NCTEPS National Commission on Teacher Education and Professional Standards

NCTI National Cable Television Institute; Nationwide Consumer Testing Institute

NCTJ National Council for the Training of Journalists

NCTM National Council of Teachers of Mathematics

nctr non-cooperative target recognition

NCTR National Center for Toxicological Research; National Council on Teacher Retirement

NCTS National Council of Technical Schools

ncu navigation(al) computer unit; nitrogen control unit

NCU National Communications Union; National Cyclists' Union

NCUA National Credit Union Administration; National Credit Union Association

NCUC National Commission on Unemployment Compensation

NCUF National Computer Users Forum

NCUMA National Credit Union Management Association

NCUMC National Council for the Unmarried Mother and her Child

ncup no commission until paid

NCUPUFUB National Clean-up, Paint-Up, Fix-Up Bureau

NCURA National Council of University Research Administrators

NCUSA Navy Club of the U.S.A.

NCUSIF National Credit Union Share Insurance Fund

NCUTLO National Committee on Uniform Traffic Laws and Ordinances

ncv no commercial value

NCVA National Center(s) for Volunteer Action

NCVAE National Council for Audio-Visual Aids in Education

NCVO National Council for Voluntary Organizations

NCVOTE National Center for Vocational, Occupational, and Technical Education

NCVT National Crime and Violence Test

ncw nosecone warhead

NCW National Council of Women; North Central Washington; North City West

NCWA National Center for Wilderness Activities; National Council of Women of Australia

NCWC National Catholic Welfare Conference

NCWF Northern California Women's Facility

NCWSA National Council of Women of South Africa

NCWSB National Council of Wool Selling Brokers

NCWUS National Council of Women of the U.S.

NCY National Cylinder Gas (stock-exchange symbol)

NCYC National Council of Yacht Clubs

NCYMCA National Council of Young Men's Christian Associations

NCYRE National Council on Year-Round Education

nd indicates ordinal number as in 2nd Avenue or 2nd Street; national debt; natural draught; neutral density; new deck(ing); new drugs; next day; next day delivery (in newspaper stock listings); no date; no decision; no deed; no delay; no discount(ing); no drawing; no drinking; no drugs; non-delivery; non-directional; not dated; not deeded; not detected; not determined; not drawn; nothing doing; nuclear detonation

n-d non-drying

n/d neutral density

nd *niederdruck* (German—low pressure); *no hay datos* (Spanish—no data)

Nd neodymium; refractive index (symbol)

ND Environment Near Death; Narcotics Division (NYPD); National Dairy Products (stock exchange symbol); National Debt; Naval District; Navy Department; Newcastle Disease; New Drugs; North Dakota; Notre Dame

N.D. Doctor of Naturopathy; Naturopathic Doctor; Northern District (court)

ND *New Drugs; Notre Dame* (French—Our Lady)

nda new drug application; nondestructive analysis; nondestructive assay; non-disclosure agreement

nda (NDA) new drug applications

N d A *Nota dell'Autore* (Italian—Author's Note)

NDA National Dairy Association; National Dairymens' Association; National Dental Association; National Diploma in Agriculture

ndaa not dated at all

NDAA National District Attorneys Association

NDAB Numerical Data Advisory Board

NDAC National Defense Advisory Committee; National Defense Advisory Commission; Nuclear Defense Affairs Committee (NATO)

ND Agr Eng National Diploma in Agricultural Engineering

N Dak North Dakota; North Dakotan

nd AV nondefective avian leucosis virus

NDANZ National Dairy Association of New Zealand

n da r *nota da redação* (Portuguese—author's note)

NDAS *New Dictionary of American Slang*

nd ASV nondefective avian sarcoma virus

NDASSP North Dakota Association of Secondary School Principals

NDATUS National Drug and Alcoholism Treatment Utilization Survey

ndb national development bond(ing); new domestic boiler; new donkey boiler; nondirectional beacon

NDB National Development Bank; Navy Department Bulletin

NDBC National Data Buoy Center; National Duckpin Bowling Congress

NDBI National Dairymen's Benevolent Institution

NDBO NOAA Data Buoy Office

NDBS National Data Buoy System

NDC National Dairy Council; National Defense Contribution; National Defense Corps; National Democratic Club; National Development Company; National Development Corporation; National Development Council; NATO Defence College; Naval Dental Clinic; Nippon Decimal Classification; Nuclear Development Corporation

NDCA National Day Care Association; National Dry Cleaners Association

NDCC National Defense Cadet Corps; National Democratic Congressional Committee; National Drug Control Center

NDCD *National Drug Code Directory*

NDCP National Drug Control Policy

NDCS National Deaf Children's Society

N d D *Nota della Direzione* (Italian—Director's Note)

NDD National Diploma in Dairying

nddad net demand draft against documents

ndd(s) narcotic-detection dog(s)

NDDT National Diploma in Dairy Technology

nde near-death experience; nondestructive evaluation; nonlinear differential equation(s)

NDEA National Defense Education Act

N-defense nuclear defense

NDEI National Defense Education Institute

n del a *nota del autor* (Spanish—author's note)

n del e *nota del editor* (Spanish—editor's note)

n del t *nota del traductor* (Spanish—translator's note)

N de M *Nacional de México* (railroad)

N de M *Ferrocarriles Nacionales de México* (Spanish National Railways of Mexico)

NDER National Defense Executive Reserve

ndf nacelle drag efficiency factor

NDF National Democratic Front; National Diploma in Forestry

ndg *nedenfor* (Dano-Norwegian—beneath)

NDG National Dance Guild

NDGAA National Dog Groomers Association of America

NDGMH National Development Group for the Mentally Handicapped

NDGS National Defense General Staff; National Duncan Glass Society

NDH Delhi, India (airport); National Diploma in Health; National Diploma in Horticulture

NDHS New Drop High School

ndi numerical designation index

NDI National Dance Institute; National Death Index; National Democratic Institute; The National Directory of Internship

NDIB National Drug Intelligence Bureau

NDIC National Drug Intelligence Center

NDICF North Dakota Independent College Fund

NDIRS North Dakota Institute for Regional Studies

NDIS National Drug Information Service

ndl network definition language

ndl *niederländisch* (German—Netherlandic)

Ndl *Nederland* (Dutch—The Netherlands)

NDL National Development Loan; National Diet Library (Tokyo); Nuclear Defense Laboratory

NDL *Norddeutscher Lloyd* (German—North German Lloyd)

NDLA North Dakota Library Association

NDLB National Dock Labour Board

NDMA Nonprescription Drug Manufacturers Association

NDMB National Defense Mediation Board

ndml never during my lifetime

NDN National Diffusion Network; New Democrat Network

ndo negotiable delivery order

NDO National Debt Office (and Office for the Payment of Government Life Annuities); Natural Disasters Organization; Northern Dance Orchestra

ndp net domestic product; normal diametric pitch

NDP National Dairy Products; National Democratic Party; National Detective Police; National Drug Policy; New Democratic Party (Canada)

NDP *Nationaldemokratische Partei Deutschlands* (German—German National-Democratic Party)—neo-Nazi oriented

NDPA National Decorating Products Association; National Democratic Party of Alabama

NDPBC National Duck Pin Bowling Congress

NDPD *National-Demokratische Partei Deutschlands* (National Democratic Party of former East Germany)

NDPGA North Dakota Personnel and Guidance Association

NDPH National Diploma in Poultry Husbandry

NDPP National Drug Prevention Program

NDPR NATO Defense Planning Review

NDPs Narcotic Detention Pens (NYC)

NDPS National Data Processing Service

ndr net discount(ed) revenue

N^{dr} *Neder* (Dutch or Swedish—*lower); Nieder* (German—lower)

N d R *Nota della Redazione* (Italian—Editor's Note)

NDR *Norddeutscher Rundfunk* (German—North German Radio)

NDRC National Defense Research Committee

NDRG NATO Defense Research Group

NDRI Naval Dental Research Institute

ndro nondestructive readout

NDRSWG NATO Data Requirements and Standards Working Group

nds national development strategy

nds (NDS) nuclear detection satellite

NDs Northern Districts

NDS National Directory Service

NDSB Narcotic Drugs Supervisory Body

NDSBA North Dakota School Boards Association

NDSEA North Dakota Society of Enrolled Agents

NDSF North Dakota School of Forestry

NDSK *Nippon Dendo Sharyo Kyokai* (Japanese—Japan Electric-Powered Vehicle Association)

NDSL National Direct Student Loan

NDSM National Defense Security Medal

NDSs Nuclear Delivery Systems

NDSSS North Dakota State School of Science

ndt nondestructive testing

ndt *nota del traductor* (Spanish—translator's note); *nota del traduttore* (Italian—translator's note); *note du traduc-teur* (French—translator's note)

NDT National Driver's Test; Newfoundland Daylight Time; Nichigeki Dancing Team

NDT Ferrocarril Nacional de Tehuantepec (Spanish—National Railroad of Tehuantepec—symbol)

NDTA National Defense Transportation Association; Non-Destructive Testing Association

NDTAA Non-Destructive Testing Association of Australia

NDTC Nottingham and District Technical College

NDTI National Disease and Therapeutic Index

NDTS Non-Destructive Test(ing) Standard(s)

ndu navigation display unit; nuclear data unit

NDU National Defense University; Notre Dame University

N-dump(ing) nuclear-waste dump(ing)

N-dump(s) nuclear (waste-disposal) dump(s)

ndup nonduplication; nonduplicate

Ndv Newcastle disease virus

ndw net deadweight

NDW Naval District Washington (D.C.)

NdYAG neodymium, yttrium, aluminum, garnet (laser components)

ne new edition; new engine(s); nital etch(ing); not enlarged; not entitled; not equal to; not essential; not exceeding

ne (NE) norepinephrine

n/e no effects

ne non ébarbe (French—untrimmed)

Ne neon; Nepal; Nepalese; Netherlander; Netherlands

NE National Emergency; National Estate(s); Naval Engineer(ing); Nebraska (postal code); new edition; New England(er); News Editor; Niger (internet code); northeast; Northeast Airlines (2-letter coding); Nuclear Engineer(ing)

N.E. Nuclear Engineer

NE Navio Escola (Portuguese—Schoolship); *Noreste* (Spanish—northeast)

ne/4 mos new edition expected in four months

ne/6m new edition in preparation, expected in 6 months (for example)

ne/6 mos new edition expected in six months

nea net energy analysis

NEA National Education Association; National Electrification Administration; National Endowment for the Arts; National Erectors Association; Net Energy Analysis; New England Aquarium (Boston); Newspaper Enterprise Association; Northeast Airlines; Northern Electric Authority; Nuclear Energy Agency (UN)

NEAC National Energy Advisory Committee; New English Art Club

NEACAP National Emergency Airborne Command Post

NEACH New England Automated Clearing House

NEACSS New England Association of Colleges and Secondary Schools

NEAF Near East Air Force; New Era Aboriginal Fellowship

NEAFC Northeast Atlantic Fisheries Commission

NEAG New English Art Gallery

NEAHI Near East Animal Health Institute

NEAL National Electron Accelerator Laboratory

NEAP National Assessment of Educational Progress

NEA-PAC National Education Association Political Action Committee

NEAR National Emergency Aid Radio; National Emergency Alarm Repeater; Near Earth Asteroid Rendezvous (spacecraft)

NEARA New England Antiquities Research Association; New England Archeological Research Association

Near East Libya, Egypt, Sudan, Ethiopia, Jordan, Israel, Lebanon, Syria, Saudi Arabia, the United Arab Emiiates, Oman, Yemen, Iraq, Iran, Turkey, Afghanistan, Pakistan

Near North Australian equivalent of the Far East

Nears the Near Islands of the outermost Aleutians in southwestern Alaska, including Agattu and Attu

NEAS National Engineering Aptitude Search

NEASC New England Association of Schools and Colleges

neat non-exercise activity thermogenesis

NEAT National (Cash Register) Electronic Autocoding Technique; National Employment and Training

NEATE New England Association of Teachers of English

'neath beneath; underneath

NEATO Northeast Asian Treaty Organization

neb nembutal; noise-equivalent bandwidth

neb nebbisch (Yiddish—colorless, plain, retiring, socially ill at ease)

neb. nebula (Latin—spray)

NEB National Electricity Board; National Energy Board (Canada); National Enterprise Board (United Kingdom); North Equatorial Belt

NEB New English Bible

NEBAC National Ethnic Broadcasting Advisory Council

nebacs neutral engine bleed air control system

NEBB National Environmental Balancing Bureau

nEbC no-European-before-Columbus school of historic discovery despite Irish and Viking claims to the contrary

NEBHE New England Board of Higher Education

NEBI National Employee Benefits Institute

Nebr Nebraska; Nebraskan

NEBSS National Examinations Board for Supervisory Studies

nebuchad nebuchadnessar (20-quart-capacity champagne bottle)

nebul. nebula (Latin—spray)—nebulizer

nec necessary; no error check(ing); not elsewhere classified

Nec (NEC) Navy enlisted classification

NEC National Economic Commission; National Economic Council; National Egg Council; National Electrical Code; National Equestrian Centre; National Equity Corporation; National Exchange Club; National Exhibition Centre (Birmingham, England); Negro Ensemble Company; Netherlands Electrochemical Committee; Network of Employment Coordinators; New England Conservatory of Music; New England Council; Nippon Electric Company; Nippon Electric Corporation

NECA National Electrical Contractors' Association; Near East College Association; Numismatic Error Collectors of America

NECAA National Entertainment and Campus Activities Association

NECAP NASA Energy-Cost Analysis Program

NECC National Education Computer Center

NECCC New England Correctional Coordinating Council

NECCO New England Confectionary Company

NECEL New England Coalition of Educational Leaders

NECLC National Emergency Civil Liberties Committee

NECM New England Conservatory of Music

NECMD Newark Contract Management District

NECO Nuclear Engineering Company

NECOS Northern Europe Chiefs of Staff (NATO)

NECP New England College of Pharmacy

NECPA National Energy Conservation Policy Act

necr necrosis

necro (Latin prefix—corpse or dead)—necrophilia, necropholia, necrosis

necrol necrology

necr opo necropolis; necropolitan(ic)

NECS National Electrical Code Standards; New England Collectors Society

necy necessary

ned normal equivalent deviation

Ned Edmund; Edward; Edwin

Ned Nedarim; Nederland (Dutch—the Netherlands); *Nederlands* (Dano-Norwegian—the Netherlands)

NED National Endowment for Democracy; Nuclear Energy Division (GE)

NED New English Dictionary (Oxford English Dictionary)

NEDA National Economic and Development Authority; National Economic Development Association; National Electronic Distributors Association; National Electronics Development Association

Ned Ant Nederlandse Antillen (Dutch—Netherlands Antilles)—Aruba, Bonaire, Curaçao Saba, Sint Eustatius, half of Sint Maarten

Ned Buntline Edward Zane Carroll Judson

NEDC National Economic Development Council (of Great Britain where it is nicknamed Neddy); Near East Development Council

nedela network definition language

Nederl Nederland (Dutch—Netherlands)

NEDICO Netherlands Engineering Consultants

NEDL New England Deposit Library

Nedloyd Netherlands Line

NEDO National Economic Development Office; New Energy Development Organization

Ned Opera Nederlandse Opera-stichting (Dutch—Netherlands Opera Foundation)

nedt noise equivalent difference temperature

NEDT National Educational Development Tests

Ned Th Ts Nederlands Theologisch Tijdschrift (Dutch—Netherlands Theological Periodical)

NEDU Navy Experimental Diving Unit

NEEB North Eastern Electricity Board (UK)

NEEC National Export Expansion Council

need. needlework

NEED National Environmental Education Development

Needle Park open-air hangout of addicts, pushers, pimps, and prostitutes

needn't (contraction—need not)

NEEDS New England Electronic Data System

ne'er never (contraction)

Néerl Néerlandais (French—Dutch)

NEES Naval Engineering Experiment Station; New England Electric Service

NEETF National Environmental Education & Training Foundation

neev (NEEV) natural energy electric vehicle

NEEWSSOP NATO-Europe Early-Warning-System Standard Operating Procedures

nef national extra fine (screw thread); net energy for fattening; noise exposure forecast; nuclear energy factor(s)

NEF Naval Emergency Fund; Near East Foundation; New Education Fellowship

nefa nonesterified fatty acid

NEFA Northeast Frontier Agency

NEFC Near East Forestry Commission

NEFEN Near and Far East News

NEFIRA New England Fire Insurance Rating Association

NEFMO NATO European Fighter Aircraft Development, Production, and Logistics Management Organization

NEFO National Electronic Facilities Organization

Nefos New Emerging Forces

NEFP National Educational Finance Project

NEFSA National Education Field Service Association

neg negation; negative; negligent; negotiable; negotiate; negritude

nég négation (French—negation)

Neg Negro; Negroid

neg ad negative advertisement; negative advertising

neg am negatively amortized (loan)

negatron negative electron

negistor negative resistor

Negley Farson James Scott Negley Farson

nego negotiate

negobl negotiable

negod negotiated

negoin negotiation

negotn negotiating

negotng negotiating

Negrasian(s) person(s) of African and Asian parents such as Afro-Chinese, Afro-Indian, Afro-Japanese, etc.

NEGRO National Economic Growth and Reconstruction Organization

négt négociant (French—merchant)—wholesaler

negtax negative (income) tax

Neh Time Book of Nehemiah

Neh Nehemiah

NEH National Endowment for the Humanities

NEHA National Environmental Health Association; National Executive Housekeepers Association

NEHC National Extension Homemakers Council

NEHGR New England Historic Genealogical Register

NEHGS New England Historic Genealogical Society

nehi knee-high

Nehm Nehemiah

Nehru Jawaharlal Nehru

nei noise-equivalent input; not elsewhere included; not elsewhere indicated

n.e.i. non est inventus (Latin—it is not found)

NEI National Eye Institute; Netherlands East Indies; New England Institute; Nuclear Energy Institute

NEIC National Earthquake Information Center; National Energy Information Center

NEIDP National Electronic Industries Procurement

NEISS National Electronic Injury Surveillance System

NEISSS National Electronics Injury Surveillance Safety System

NEJA National Equal Justice Association

NEJM New England Journal of Medicine

nek nekton

NEK Norsk Electrotecnisk Komite (Norwegian—Norwegian Electrotechnical Committee)

NEKASA New England Knitwear and Sportswear Association

NEKDA New England Kiln Drying Association

nekolim neocolonialist-colonialist-imperialist (Indonesian acronym)

nel noise-exposure level

NEL National Electronics Laboratory; National Engineering Laboratory; Navy Electronics Laboratory (USN)

NEL New English Library

NELA National Electric Light Association; New England Library Association; Northeastern Loggers Association

NELC Naval Electronics Laboratory Center (formerly NEL)

NELDIC Nippon (Electric Company) Electric Layout Design (System) for Integrated Circuits

NELH National Energy Laboratory of Hawaii

NELIA Nuclear Energy Liability Insurance Association

NELIAC Navy Electronics Laboratory International Algol Compiler

NELINT New England Library Information Network

Nell Nellie; Nelson

NELL North East Lancashire Libraries

Nellie Mae New England Educational Loan Marketing Corporation

NELMA Northeastern Lumber Manufacturers Association

Nel-Mar Nelson-Marlborough (NZ)

NELP North East London Polytechnic

NELPIA Nuclear Energy Liability Property Insurance Association

Nels Nelson

NELS National Environmental Laboratories

Nelson Horatio Nelson; Knute Nelson; Nelson Olsen Nelson; Thomas Nelson

NELSON New Editing and Layout System of Newspaper

NELTAS North East Lancashire Technical Advisory Services

NEly north-easterly

nem net energy for milk; not elsewhere mentioned

NEM New Economic Mechanism

NEMA National Eclectic Medical Association; National Electrical Manufacturers Association; National Electrical Motors Association

nemat nematology

Nemat Nemathelminthes

NEMC New England Medical Center

NEMCA NATO Electromagnetic Compatability Agency

nem. con. nemine contradicente (Latin—no one contradicting)

nem, dis. nemine dissentiunto (Latin—no one dissenting)

NEMI National Elevator Manufacturing Industry

NEMLA New England Modern Language Association

NEMO Naval Edreobenthic Manned Observatory (for sedentary sea bottom research); Naval Experimental Manned Observatory

NEMPA North-Eastern Master Printers' Alliance

NEMPS National Environmental Monitoring and Prediction System

NEMRA National Electrical Manufacturers Representatives Association

NEMRB New England Motor Rate Bureau

nems (NEMS) near-earth magnetospheric satellite

NEMS National Exchange Market System

nen noise and exposure number

NEN New England Nuclear (corporation)

NENA National Emergency Number Association

nencl nonenclosed; nonenclosure

ne/nd new edition in preparation—no date can be given

N-energy nuclear energy

N Eng Naval Engineer(ing); New England; North England

N-engine(s) nuclear engine(s)

N Engl J Med *New England Journal of Medicine*

nenmld not enameled

NENP New England National Park (New South Wales)

neo near earth orbit

neo (Latin prefix—new or young)—neonatal

NEOA National Entertainers and Operators Association

NEOB New Executive Office Building (D.C.)

neobych neobychny (Russian—incomparable)

NEOC National Emergency Operations Center

Neo-Cath Neo-Catholic(ism)

Neo-Christ Neo-Christian(ity)

neoclas neoclassical; neoclassicism

neocol neocolonial(ism)

neocolim neocolonial-colonial-imperialist

Neo-Conf Neo-Confucian(ist)

neo-con(s) neo-conservative(s)

Neo-Dar Neo-Darwinian; Neo-Darwinist(ic)

neo-dhc neohesperidin dihydrochalcone (sweetener)

NEODTC Naval Explosive Ordnance Disposal Technical Center

neo-fasc neo-fascist

Neogaea landmass including Central and South America

Neo-Goth Neo-Gothic

Neo-Heg Neo-Hegelian

neo-imp neo-impressionism; neo-impressionistic

Neo-Kant Neo-Kantian(ism)

neol neologism(s); neologistic(al)(ly), neologize(r)(s)

Neo-Lam Neo-Lamarckian; Neo-Lamarckism; Neo-Lamarckist

Neo-Lat Neo-Latin(ism)

Neo-Luth Neo-Lutheran(ism)

Neo-Mel Neo-Melanesian (pidgin English of Melanesia, New Guinea, and North-East Australian islanders)

Neo-Nor Neo-Norwegian

Neo-Plas Neo-Plastic(ism)

Neo-Plat Neo-Platonic; Neo-Platonism

Neo-Pyth Neo-Pythagorean(ism)

Neo-Real Neo-Realism; Neo-Realistic

Neorican(s) New York American(s)

Neo-Ricans repatriated Puerto Ricans

Neo-Rom Neo-Romantic(ism)

Neo-Schol Neo-Scholastic(ism)

neotrop neotropical

neotwy (last-letter mnemonic—when, where, who, what, how, why)

nep new edition pending; noise equivalent power; not elsewhere provided; nuclear electric propulsion; nude-encounter parlor (brothel)

Nep Nepal; Nepomucene; Nepomuceno; Nepomuk; Neptune

Nep Cornelius Nepos (Roman biographer)

NEP National Education Program; National Energy Plan; New Ecological Paradigm; New Economic Policy; New England Pathology; New England Power (company); Nixon Economic Policy

nepa (NEPA) nuclear energy for the propulsion of aircraft

NEPA National Electric Power Authority; National Endowment Policy Act; National Environmental Policy Act

Nepal Kingdom of Nepal (Himalayan mountain nation)

NEPAL National Egg Packers' Association, Ltd

NEPC National Employers Policy Committee

NEPCO New England Provision Company

nepd noise-equivalent power density

NEPE National Emergency Planning Establishment (Canada)

neph nephew

neph-i-l nephew-in-law

nepho nephograph; nephological; nephologist; nephology

nephro (Latin prefix—kidney)—nephritis

NEPIA Nuclear Energy Property Insurance Association

NEPLEX New England Power Exchange

NEPNU National Engine Parts Manufacturers Association

NEPMU Navy Environmental and Preventive Medicine Unit

NEPOOL New England Power Pool

NEPR NATO Electronic Parts Recommendation

Nep Rs Nepalese rupees

nep's nude-encounter parlors

NEPSC National Employee Participation Steering Committee

Nep Soc Neptune Society

NEPSS Naval Environmental Protection Support Service (USN)

Nept Neptune

N Equ Cur North Equatorial Current

ner nervous system; noise equivalent radiance

NER National Educational Radio; National Elk Refuge (Wyoming); New Employee Registry; North Eastern Railway (England)

NERA National Economic Research Associates; National Emergency Relief Administration

NERAIC Northern European Region Air Information Center

NERBC New England River Basins Commission

NERC National Electronic Reliability Council; National Environmental Research Center; Natural Environment Research Council

NERDDC National Energy Research Development and Demonstration Council

ne. rep. *ne repetatur* (Latin—do not repeat)

NERO Near East Regional Office (FAO); Nutrition Education Research Organization

NERPG Northern European Regional Planning Group (NATO)

nerv nervous; nuclear emulsion recovery vehicle (NERV)

nerva nuclear engine for rocket vehicle application

NE-Rx Northeast Regional Exchange

nes not elsewhere specified

nEs non-English speaking

Nes Nesta; Nestor

NES National Emergency Services; National Energy Strategy; National Extension Service; Naval Education Service; New England School of Acupuncture; News Election Service; Nordic Ergonomics Society; Nucleus Estate and Smallholders

NESA National Electric Sign Association; National Environmental Study Area; Near East and South Asia; New England School of Acupuncture; New England School of Art

NESBIC Netherlands Student's Bureau for International Cooperation

NESC National Electric Safety Code; National English Syllabus Committee; National Environmental Satellite Center; National Executive Service Comps

NESCO National Energy Supply Corporation

NESDA National Electronics Service Dealers Association; National Equipment Servicing Dealers Association

NESDB National Economic and Social Development Board

NESDEC New England School Development Council

NESNE New England Society of Newspaper Editors

NeSEA Nebraska Society of Enrolled Agents

NES & L Nuclear Engineering, Safety & Licensing (Department)

NESO Naval Electronics Supply Office

Ness Agnes

NESS National Environmental Satellite Service

nest node execution selection table

NEST Naval Experimental Satellite Terminal; Nuclear Emergency Search Team

nestor neutron source thermal reactor

net network; noise-equivalent temperature; not earlier than; nuclear electronic transitor

Net Antoinette; Nettie; Netty

NET National Educational Television; Nippon Educational Television; Noise Enforcement Team (police antinoise team); Nutrition Education Trainitmg

NETA Northwest Electronic Technical Association

netanal network analysis

NETE Naval Engineering Test Establishment (Canadian)

NETF National Environmental Trust Fund; Nuclear Engineering Test Facility

Neth Netherlands (whose capital is Amsterdam)

Neth Ant Netherlands Antilles

Netherlands Kingdom of the Netherlands (North Sea nation created and enlarged by reclamation of salt marshes and lowland waters), *Koninkrijk der Nederlanden*

netlc nonretentive nonshocksensitive (alloy made for high-level attenuation)

netiquette network etiquette

netizens Internet citizens

n. et m. *nocte et mane* (Latin—night and early morning)

netma nobody ever tells me anything

NETRANZ National Endurance and Trail Riding Association of New Zealand

NETRB New England Territory Railroad Bureau

NETRC National Educational Television and Radio Center

nets network techniques

NETSO Northern European Transshipment Organization (NATO)

NETT Network for Environmental Technology Transfer

Netza *Netzahualcoyotl* (Aztec—Hungry Coyote)

neu neuter; neutral; neutrality

neubarb *neubearbeitet* (German—revised)

Neuk Neuköln

neur neuralgia; neurasthenia; neuritis; neurology

neuro neurotic

neuro. *neuron* (Greek—nerve, sinew, tendon)—neurasthenia, neuroanatomy; neurosis

neurobio neurobiological; neurobiologist; neurobiology

neurol neurological; neurologist; neurology

neuropath neuropathology

neurophys neurophysiological

neuropsychiat neuropsychiatry

neuropsycho neuropsychological

neurosci neuroscientific(al) (ly); neuroscientist

neurosurg neurosurgeon; neurosurgery; neurosurgical

neurs neurosis

NEUS Northeastern United States

neut neuter; neutral; neutralize; neutralizer; neutron bomb (mini-hydrogen bomb releasing neutrons and producing the minimum radioactive blast, fallout, and heat)

neutron neutral ion

nev neighborhood electric vehicle

Nev Nevada; Nevadan; Neville

Nevado del Ruiz (Spanish—Snowpeak of Ruiz)—snowcapped volcano, central Colombia

Never *Never on Sunday*

Never Never Never Land Cape York Peninsula, Australia

Nev Mag *Nevada Magazine*

nevrls nevertheless

new. net economic welfare; newton

New New College, Oxford

New Am Lib New American Library

Newark Newark-upon-Trent near Nottingham, England

Newberry Newberry Library (Chicago)

Newc Newcastle-upon-Tyne

New Cal New Caldeonia

New Castile (see *Castilla la Nueva*)

Newcastle Newcastle Emlyn, Newcastleton, Newcastle-under-Lyme, Newcastle-upon-Tyne, Newcastle Waters, and Newcastle West

New Cath World *New Catholic World*

New Col New Columbia (proposed name for Washington, D.C., if it became a state)

new cruzado monetary unit of Brazil

New Eng New England

New England Maine, New Hampshire, Vermont, Massachusetts, Rhode Island, and Connecticut

New England Colonies Massachusetts, New Hampshire, Rhode Island, Connecticut

Newf Newfoundland

Newfie(s) Newfoundlander(s)

New H New Hall College, Oxford

New Hamp Profiles *New Hampshire Profiles*

New Haven New York, New Haven, and Hartford Railroad

New Heb New Hebrides (Anglo-French island condominium in the South Pacific)

New Heb Con New Hebrides Condominium

New Hebrides New Hebrides Islands, *Nouvelles Hébrides*

new kip monetary unit of Laos

New Lib Newberry Library

New Lon New London, Connecticut

New London U.S. Coast Guard Academy at New London, Connecticut

New Mex New Mexico

New Mex Magazine *New Mexico Magazine*

Newn Newnham College, Oxford

NEWO National Energy Waste Office

New Orl New Orleans

new par new paragraph

New Phil Orch New Philharmonic Orchestra

NEWRADS Nuclear Explosion Warning and Radiological Data System

NEWRIT Northeast Water Resources Information Terminal

news naval electronic warfare simulator; news agency; news agent; new standards

NEWS New England Wildflower Society

newsl newsletter

New Sarum Salisbury, capital of Wiltshire, England northwest of Southhampton

newscast(er) news broadcast(er)

newscomp newspaper composition

New Sib New Siberian Islands

New Siberians New Siberian Islands in the Arctic (Novosibirskiye Ostrova)

newsp newspaper

Newt Newton

new Taiwan dollar monetary unit of Taiwan

New Test. New Testament

NEWWA New England Water Works Association

New Year's New Year's Day (January 1)

New Yorican New York Puerto Rican

New Zealand Dominion of New Zealand (western Pacific Ocean nation); New Zealand flax and New Zealand wineberry

nex not exceeding

nexis (NEXIS) news data base on-line to head

N-explosion(s) nuclear explosion(s)

N-exports nuclear exports

nexrad next-generation (weather doppler) radar

next near-end crosstalk

NEXT NATO Experimental Tactics

nez (NEZ) northern economic zone

NEZs New Economic Zones (Vietnamese forced-labor camps)

nf nanofarad; national fine; near face; near fiield; neurofibromatosis; no fool; no funds; noise factor; noise figure; non-ferrous; non-fiction; nonfundable; nose fuze; not fordable

n-f nonfordable

n/f neutrons per fission; no funds

n & f near and far

nf *nouveau franc* (French—new franc)—issued in 1960

n.f. *nyfolid* (Swedish-new series)

n/f *nuestro favor* (Spanish—our favor)

n.F. *neue Folge* (German—new series)

NF National Fine (threads); National Formulary; National Foundation; National Front; Newfoundland; Nieman Foundation; Norfolk Island (Internet code); Norfolk, Virginia (airport); Norman French; Nutrition Foundation

N-F Norman-French

NF *Neue Folge* (German—new series); *Nuestra Familia* (Spanish—Our Family)—prison racketeers also called *La Nuestra Familia*

nfa no further action

nfa (NFA) net financial assets

NFA National Faculty Association; National Farmers Association; National Federation of Anglers; National Florist Association; National Flute Association; National Food Administration; National Food Authority; National Foundry Association; National Futures Association; Nature Friends of America; Naval Fuel Annex; New Farmers of America; Night Fighters Association; Northwest Festivals Association; Northwest Fisheries Association; Northwest Forestry Association

NFAA National Fashion Accessories Association; National Field Archery Association; National Foundation for Advancement in the Arts; Navy Fighter Attack Aircraft

NFAC National Food and Agriculture Council; Native Forests Action Council

NFAH National Foundation for the Arts and the Humanities

NFAIS National Federation of Abstracting and Indexing Services

NFAL National Foundation of Arts and Letters

N-fallout nuclear fallout (radioactive fallout)

nfb nacelle fuselage base; narrow flange beam; no feedback

NFB National Federation of the Blind; National Film Board (Canada)

NFB *Nippon Fudosan Bank* (Japan Real Property Bank)

NFBA National Family Business Association; National Food Brokers Association

NFBC National Family Business Council; National Film Board of Canada; Newfoundland Base Command

NFBF National Farm Bureau Federation

NFBIC Netherlands Flower Bulb Information Center

NFBPM National Federation of Builders' and Plumbers' Merchants

NFBPWC National Federation of Business and Professional Women's Clubs

NFBTE National Federation of Building Trades' Employers

NFBTO National Federation of Building Trades Operatives

nfc not favorably considered

NFC National Fitness Council; National Football Conference; National Foundry College; National Freight Corporation; Navy Finance Center; Newspaper Features Council

NFCA National Federation of Community Associations

NFCB National Federation of Community Broadcasters

NFCC National Foundation for Consumer Credit

NFCDCU National Federation of Community Development Credit Unions

NFCG National Federation of Consumer Groups

nfcs night fire-control sight

NFCSA National Finance Corporation of South Africa

NFCTA National Federation of Corn Trade Associations; National Fibre Can and Tube Association

NFCU Navy Federal Credit Union

NFCUS National Federation of Canadian University Students (now NUS)

nfd no further description

nfd (NFD) neurofibrillary degeneration

Nfd Newfoundland

NF-D National Federation of Doctors; National Fisheries Development; Naval Fuel Depot

NFD National Faculty Directory

NFDA National Fastener Distributors Association; National Food Distributors Association; National Funeral Directors Association

nfdm non-fat dry milk

nfd(m) non-fat dry (milk)

NFDMA National Funeral Directors and Morticians Association

NFDRS National Fire Danger Rating System

NFDS National Financial Data Services; National Fire Data Center

nfe net funds employed; nose-fairing exit; not fully equipped

NFE National Front of England (racists advocating immediate deportation of all nonwhites to wherever they originated)

NFEA National Federated Electrical Association; Newspaper Farm Editors of America

n fem feminine form of a noun

NFEMC National Federation of Export Management Companies

NFER National Foundation for Education Research

nfet n-channel junction field-effect transistor

NFEWA Newspaper Food Editors and Writers Association

NFF National Farmers Federation; National Froebel Foundation; Naval Fuel Facility

NFFA National Farmers Federation of Australia; National Freight Forwarders Association; National Frozen Food Association

NFFC National Film Finance Corporation

NFFE National Federation of Federal Employees

NFFF National Federation of Fish Friers; National Firearms Freedom Fund

NFFPC National Foundation to Fight Political Corruption

NFFPT National Federation of Fruit and Potato Trades

NFFS National Foundation for Funeral Services; Non-Ferrous Founders' Society

NFFTR National Federation of Fishing Tackle Retailers

Nfg Nachfolger (German—successor)

NFGCA National Federation of Grandmother Clubs of America

NFHN National Federation of Hispanic-owned Newspapers

NFHS National Federation of Housing Societies

nfi no further information; non-bank financial intermediaries

NFI National Federation of Ironmongers; National Fisheries Institute; National Flood Insurance; Nature Friends of Israel

NFIB National Federation of Independent Business; National Foreign Intelligence Board

NFIC National Foundation for Ileitis and Colitis

NFIE National Foundation for the Improvement of Education

NFIP National Flood Insurance Program; National Foundation for Infantile Paralysis

NFIU National Federation of Independent Unions

NFJC National Foundation for Jewish Culture

nfk nothing further known

NFK Norfolk Island

Nfl Newfoundland

Nfl Nachfolger (German—successor)

NFL National Film Library; National Football League; National Forensic League; National Foresters League

NFLCC National Fishing Lure Collectors Club

Nfld Newfoundland

NFLPA National Football League Players Association; National Free Lance Photographers Association

NFLPN National Federation of Licensed Practical Nurses

NFLS Niagara Falls

NFLSV National Front for the Liberation of South Vietnam

NFLTA National Federation of Language Teachers Associations

nfm narrow-band frequency modulation; next full moon

NFMA National Forest Management Act

NFMC National Federation of Music Clubs; National Food Marketing Commission

NFMD National Foundation for the March of Dimes

NFME National Fund for Medical Education

NFMLTA National Federation of Modern Language Teachers Association

NFMPS National Federation of Master Printers in Scotland

NFMTA National Federation of Meat Traders' Associations

NFND National Foundation for Neuromuscular Diseases

nfnshd not finished

NFO National Farmers Organization; National Freight Organization; Naval Flight Officer

NFOIO Naval Field Operational Intelligence Office(r)

NFOO Naval Forward Observing Officer

nfou number of fourier coefficients

nfp not fiile protect(ed)

NFP National Federation of Parents (for drug-free youth); National Federation Party; Natural Family Planning

NFPA National Fire Protection Association; National Flaxseed Processors Association; National Flexible Packaging Association; National Fluid Power Association; National Food Processors Association; National Forest Products Association; Niagara Frontier Port Authority

NFPC National Federation of Priests Councils; Niagara Falls Power Company

NFPCA National Fire Prevention and Control Administration

NFPDB NATO Force Planning Data Base

NFPEX NATO Force Planning Exercise

NFPI National Frozen Pizza Institute

NFPW National Federation of Press Women; National Federation of Professional Workers

nfq night frequency

nfr no further requirement

NFRA National Forest Recreation Association

NFRC National Fenestration Rating Council; National Forest Reservation Commission

NFRN National Federation of Retail Newsagents, Booksellers, and Stationers

NFRW National Federation of Republican Women

nfs not for sale

NFS National Fire Service; National Forest Service; National Forest System; Network File System; Nuclear Fuel Services; Number Field Sieve

NFSA National Fertilizer Solutions Associations; National Food Service Association

NFSA & IS National Federation of Science Abstracting and Indexing Services

NFSE National Federation of the Self Employed

NFSG National Federation of Students of German

NFSHSA National Federation of State High School Associations

NFSID National Foundation for Sudden Infant Death

NFSM National Fraternity of Student Musicians

NFSNC National Federation of Settlements and Neighborhood Centers

NFSNO National Federation for Specialty Nursing Organizations

NFSO Navy Fuel Supply Office

nft no fixed time; no forwarding time; nutrient film technique

NFT National Film Theatre

NFTA National Film Theatre of Australia; Niagara Frontier Transportation Authority

NFTB Nuclear Flight Test Base

NFTC National Foreign Trade Council

NFTS National Federation of Temple Sisterhoods

nfu not for us

NFU National Farmers Union; National Film Unit; National Froebel Union

N-fuel nuclear fuel

NFUW National Farmers' Union of Wales

nfv no further visits

nfva net free ventilation area

nfw new field wildcat (oil well)

NFWA National Farm Workers Association; National Furniture Warehousemen's Association

NFWBO National Foundation for Women Business Owners

NFWI National Federation of Women's Institutes

NFYFC National Federation of Young Farrners' Clubs

nfyg notifying

nfz no fire zone; nuclear-free zone

ng narrow gauge; nasogastric; negative glow; new genus; next generation; nitroglycerine; no go; no good; no gum (on back of stamps); not given; not good; not ground; nut grounds

ng (NG) natural gas

n-g nitro-glycerine

n/g *nuestro giro* (Spanish—our draft)

Ng Norwegian

NG National Gallery; National Guard; National Gypsum; New Guinea; Nigeria (Internet code)

nga (NGA) non-gonococcal urethritis

Nga Nagoya

NGA National Gallery of Art; National Glass Association; National Glider Association; National Governors Association; National Grains Authority; National Graphical Association; National Grocers Association; National Guard Association Needlework Guild of America; Never Go Away (travel club dedicated to seeing America first)

NGAA National Gift and Art Association; Natural Gasoline Association of America

NGADA National Graphic Arts Dealers Association

NGAC National Guard Air Corps

Ngaragba Ngaragba Prison in Bangui (capital of the Central African Republic)

N-gauge narrow gauge (railroad track less than standard gauge, gauge 4 feet 8-1/2 inches)

NGAUS National Guard Association of the United States

NGAW *National Geographic Atlas of the World*

ngb negative guard board

NGB National Garden Bureau; National Guard Bureau

NGC National Gallery of Canada; National Gambling Commission; National Gypsum Company; Natural Gas Corporation

NGC *New Galactic Catalog; New General Catalog* (astronomical)

ngcil nice guys come in last

NGCM Navy Good Conduct Medal

NGCMS National Guild of Community Music Schools

NGCR Next Generation Computer Resources

NGCSA National Guild of Community Schools of the Arts

NGD New Geographical Dictionary (Webster's New Geographical Dictionary)

NGDC National Geophysical Data Center

NGDM & M New Grove Dictionary of Music and Musicians

NGE New York State Electric & Gas (stock exchange symbol)

NGEC National Gypsy Education Council

n gen new genus

ngf naval gunfire

ngf (NGF) nerve growth factor

NGF National Genetics Foundation; National Golf Foundation; Naval Gun Factory; Nordic Gunners Federation

NGFA National Grain and Feed Association

NGFLO Naval Gunfire Liaison Officer

NGFLT Naval Gunfire Liaison Team

NGFSFHP National Guaranty Fund for Self-Funded Health Plans

NGI National Garden Institute; Norwegian Geotechnical Institute

NGI Navigazione Generale Italiana (Italian—Italian General Navigation Line)

NGJA National Gymnastics Judges Association

NGJC North Greenville Junior College

N Gk New Greek

NGK Nihon Gakujutsu Kaigi (Japanese—Japan Research Council)

ngl natural gas liquids

NGL North German Lloyd Line

NGLTF National Gay and Lesbian Task Force

nglzd not glazed

NGMA National Greenhouse Manufacturers Association

N Gmc North Germanic

NGMEX Northern Gulf of Mexico

NGMP New Guinea Marine Products

ngo national gas outlet (thread); nongovernmental organization

Ngo Nagoya

NGOs Nongovernmental Organizations (UN)

NGPA Natural Gas Processors Association

NGPP National Guild of Professional Paperhangers

NGPT National Guild of Piano Teachers

ngr narrow gauge roll; non-grain rating

ngr neugriechisch (German—modern Greek)

Ngr Niger (Spanish abbreviation)

NGr New Greek

NGR Ndumu. Game Reserve (Zululand); Newbold General Refractories

NGRA National Gay Rights Activists

ngri not guilty by reason of insanity

NGRI National Geophysical Institute

NGRS Narrow Gauge Railway Society; National Greyhound Racing Society

ngs national gas straight (threading); net gas sand (oil well)

NGS National Genealogical Society; National Geodetic Survey; National Geographic Society; Nuclear Generating Site

NGSA National Gallery of South Africa; Natural Gas Supply Association

NGSD New Guinea Singing Dog

NGSDC National Geophysical and Solar-Terrestrial Data Center (NOAA)

NGSIC National Geodetic Survey Information Center (NOAA)

NGSL National Geographic Society Library

NGSR Nizam's Guaranteed State Railway

ngt national gas taper (threading); next-generation trainer; noise generator tube

ngt négociant (French—merchant)—wholesaler

NGT National Guild of Telephonists; North German Traders

ngta next-generation trainer aircraft

NGTE National Gas Turbine Establishment

NGTF National Gay Task Force

ng tube nasogastric tube

ngu nongonococcal urethritis

ngultrum monetary unit of Bhutan

NGUS National Guard of the United States

NGUT National Group of Unit Trusts

ngv nongonococcal vulvovaginitis

NGV Natural Gas Vehicles

NGV Nederlands Genootschap van Vertalers (Dutch—Netherlands' Translators Society)

nh never hinged; no hurry (hospitalese); non-hygroscopic; not held

nH nanohenry

Nh Noordholland (Dutch—North Holland)

NH Naval Home; Naval Hospital; New Hampshire; New Hampshirite; New Haven, Connecticut; New Haven Elm City (National Association); New Hebrides; New York, New Haven & Hartford (railroad); Nippon Airways (2-letter code); North Holland(er); Nursing Home

N & H Nedlloyd & Hoegh (steamship lines)

NH Norges Hjemmenfrontmuseum (Norwegian—Norwegian Home-Front Museum)—Oslo exhibit recalling anti-German resistance from 1940 to 1945; *Nueva Hampshire* (Spanish—New Hampshire)

N-H Noord-Holland (Dutch—North Holland)

NH₃ ammonia

NH₄ ammonium radical

NH₄CL ammonium chloride; sal ammoniac

NH₄OH ammonium hydroxide (ammonia)

nha never has anything; next higher assembly; next higher authority

NHA National Hay Association; National Health Association; National Hide Association; National Hockey Association; National Housing Act; National Housing Administration; National Housing Agency; National Housing Association; Negro History Association; Neighborhood House Association; New Homemakers of America; Nigerian Housing Administration; Nursing Home Administration

NHAC National Health Advisory Committees

NHAGB National Horse Association of Great Britain

NHAIAC National Highway Accident and Injury Analysis Center

NHAL National Hellenic American Line

NHANES National Health and Nutrition Examination Survey

NHAS National Hearing Aid Society

NHB National Harbours Board (Canada); National Health Board; Northland Harbour Board (New Zealand)

NHBRC National House Builders' Registration Council

NHBU New Hampshire Board of Underwriters

nhc nutritional hair complex

NHC National Health Council; National Hurricane Center; Naval Historical Center; New Hall College

NHCA National Hairdressers and Cosmetologists Association; National Hearing Conservation Association

NHCBS New Hampshire Council for Better Schools

NHCC National Hispanic Corporate Council

NHCIC National Hazardous Chemicals Information Center

NHDAC National Health Data Advisory Council

N.H.D. Doctor of Natural History

NHDC Naval Historical Display Center

nh di notch die

nhe nitrogen heat exchange

NHE National Housing Endowment

NHEA National Higher Education Association; New Hampshire Education Association

N-head(s) nuclear warhead(s)

N Heb New Hebrew

NHEF National Health Education Foundation

NHESA National Higher Education Staff Association

NHF National Hairdressers' Federation; National Headache Foundation; National Health Federation; National Health Foundation; National Heart Foundation; National Heart Fund; National Hemophilia Foundation; National Horse Festival; National Humanities Faculty; Naval Historical Foundation

NHFA National Heart Foundation of Australia; National Home Furnishings Association

NHF/Bull *National Health Federation Bulletin*

NHFNZ National Heart Foundation of New Zealand

NHFPL New Haven Free Public Library

NHG New High German

NHGA National Hang Gliding Association

NHHS New Hampshire Historical Society

nhi (NHI) no humans involved

NHI National Health Institute; National Health Insurance; National Heart Institutes

NHIC National Health Information Clearinghouse; National Health Insurance Commission; National Home Improvement Council

NHK *Nippon Hoso Kyokai* (Japanese—Japan Broadcasting Corporation)

NHKTV *Nippon Hoso Kyokai* (Japanese—Japanese Television Broadcasting)

NHL National Hockey League

NHLA National Hardwood Lumber Association; National Home Library Association

NHLBAC National Heart Lung, and Blood Advisory Council (NIH)

NHLBI National Heart, Lung, and Blood Institute

NHLI National Heart and Lung Institute

NHMA National Housewares Manufacturers Association

NHMRCA National Health and Medical Research Council of Australia

NHMS New Hampshire Medical Society

nhn neither help nor hinder

NHO National Hospice Organization; Navy Hydrographic Office

NHOS National Hellenic Oceanographic Society

nhp nominal horsepower

NHP National Corporation for Housing Partnerships; National Historic(al) Park; Natural History Park (Calgary, Alberta); Natural History Press; New Haven Police; New Hebrides Protectorate; Nursing Home Placement

NHPA National Horseshoe Pitchers Association

NHPC National Historical Publications Commission

NHPDA National Honey Packers and Dealers Association

NHPGA New Hampshire Personnel and Guidance Association

NHPL New Haven Public Library

NHPLO NATO Hawk Production and Logistics Organization

NHPMA Northern Hardwood and Pine Manufacturers Association

NHPRC National Historical Publications and Records Commission

NHQ National Headquarters

NHR National Housewives Register; National Hunt Rules; National Hurricane Research

nhra next higher repairable assembly

NHRA National Hot Rod Association

NHRC Naval Health Research Center

NHRE National Hail Research Experiment

NHRL National Hurricane Research Laboratory

NHRP National Hurricane Research Project

NHRR New Haven Railroad

NHRU National Home Reading Union

nhs net hydrocarbon sand (oil well); normal human sera

NHS National Health Service; National Historic(al) Site; National Historical Society; National Honor Society; Newport Historical Society

NHSA National Head-Start Association; National Heart Savers Association; Negro Historical Society of America

NHSAA New Hampshire School Administrators Association

NHSB National Highway Safety Bureau

NHSBA New Hampshire School Boards Association

NHSC National Health Service Corps; National Health Statistics Center; National Highway Safety Council; National Home Study Council

NHSEA New Hampshire Society of Enrolled Agents

NHSF National Hispanic Scholarly Fund

NHSO New Haven Symphony Orchestra

NHSR National Hospital Service Reserve

NHTI New Hampshire Technical Institute

NHTPC National Housing and Town Planning Council

NHTSA National Highway Traffic Safety Administration

NH Turn New Hampshire Turnpike

NHUC National Highway Users Conference

Nhv Newhaven

NHV New Haven Clock and Watch (stock exchange symbol)

NHYC New Haven Yacht Club; Newport Harbor Yacht Club

ni new impression; niece; night

ni (NI) inversion of the note series (12-tone); national income; net income

Ni Nica; Nicaragua; Nicaraguan; Nicaragüense; Nicas; nickel; Nigeria (whose capital is Lagos)

NI National Insurance; Native Infantry; Nautical Institute; Naval Instructor; Naval Intelligence; Negotiation Institute; Netherlands Indies; Neutralization Index; News International; Nicaragua (Internet code); Nicaraguan Airways (2-letter code) LANICA; North Island (New Zealand); North Island, San Diego, California; Northern Ireland;

Northern Island (New Zealand); Numerical Index; other North Islands

NI ampere turns (symbol)

nia nearest international airport

nia (NIA) noise-impact area

NIA National Institute on Aging; National Intelligence Authority; National Irrigation Administration; Neighborhood Improvement Area; Neighborhood Improvement Association

NIAA National Industrial Advertising Association; National Institute of Animal Agriculture

NIAAA National Institute on Alcohol Abuse and Acoholism

NIAB National Institute of Agricultural Botany

NIABC Northern Ireland Association of Boys' Clubs

NIAC National Insulation and Abatement Contractors; Nissho-Iwai American Corporation; Nuclear Insurance Association of Canada; Nutritional Information and Analysis Center

NIADA National Independent Automobile Dealers Association

NIAE National Institute of Agricultural Engineering; National Institute for Architectural Education

NIAG NATO Industrial Advisory Group

Niagara Fort Niagara; Niagara Falls; Niagara-on-the-Lake; Niagara River; Niagara University

NIAID National Institute of Allergies and Infectious Diseases

NIAL National Institute of Arts and Letters

NIAMD National Institute of Arthritis and Metabolic Diseases

NIAMDD National Institute of Arthritis, Metabolism, and Digestive Diseases (formerly NIAMD)

NIAOM Northwest Institute of Acupuncture & Oriental Medicine

NIASA National Insurance Actuarial and Statistical Association

NIASE National Institute for Automotive Service Excellence

nib noninterference basis

NIB National Information Bureau; Nebraska Inspection Bureau

NIBA National Industrial Belting Association; National Insurance Buyers Association

NIBESA National Independent Bank Equipment and Systems Association

NIBM National Institute of Business Management

nibo nibonitvchjo (ni boga ni tschiorta) (Russian—neither in god nor the devil)—materialist skeptics

NIBS National Institute of Building Sciences

NIBSC National Institute for Biological Standards and Control

nic negative impedance converter; network interface card; newly industrializing country; not in contact

Nic Nicaragua; Nicolayev; Nicosia

N i C Nurse in Charge

Nic Nicola (Italian—Nicholas)

NIC Natick Industrial Centre; National Incomes Commission; National Indications Center; National Industrial Council; National Information Center; National Institute of Corrections; National Institute of Creativity; National Institute of Credit; National Insurance Certificate; National Insurance Contributions; National Interfraternity Conference; National Inventors Council; National Investors Council; Navigation Information Center; Neighborhood Info(rmation) Center(s); Network Information Center; Newsprint Information Committee; Niagara International Centre; Nicosia, Cyprus (airport); Nineteen-hundred Indexing and Cataloging; Nippon International Containers

Nica Nicaragua(n)

nicad nickel cadmium

NiCad battery nickel-cadmium (rechargeable) battery

NICAP National Investigations Committee on Aerial Phenomena

Nicar Nicaragua(n)

Nicaragua Republic of Nicaragua (Spanish-speaking Central American country), *Républica de Nicaragua*

Nicas Nicaraguans

NICB National Industrial Conference Board; National Insurance Crime Bureau

Ni-Cd nickel-cadmium (rechargeable storage battey)

nice normal input/output control executive

Nice Eunice

NICE National Institute of Ceramic Engineers

nice cuppa nice cup of tea

NICEIC National Inspection Council for Electrical Installation Contracting

NICEM National Information Center for Educational Media

NICF Nebraska Independent College Foundation; Northern Ireland Cycling Federation

Nich Nicholas

NICHA Northern Ireland Chest and Heart Association

NICHHD National Institute of Child Health and Human Development

nichrome nickel-chromium alloy

NICIA Northern Ireland Coal Importers' Association

NICJ National Institute of Consumer Justice

nick name information correlation key

Nick Nicholas; Nichols; Nicodemus; Nikos

Nick-Pack *(see NICPAC)*

Nicky Nicholas; Nicole; Nikos

NICM Nuffield Institute of Comparative Medicine

NICMA National Ice Cream Mix Association

Nico Nicobar Islands

NICO National Insurance Consumer Organization; Navy Inventory Control Office(r)

Nicobars Nicobar Islands in the Indian Ocean

Nicolass Sint Nicolaas, Aruba

NICOP Navy Industry Cooperation Plan

Nicos Nicosia, Cyprus

NICP National Inventory Control Point

NICPAC National Independent Conservative Political Action Committee

NICRA Northern Ireland Civil Rights Association

NICRAD Navy-Industry Cooperative Research and Development

nic's newly industrializing countries

NICs National Institute of Corrections

NICS NATO Integrated Communications System

NICS COA NICS Control Operating Authority

NICSEM/NIMIS National Information Center for Special Education Material/National Instructional Material Information System

NICSO NATO Integrated Communications System Organization

NICSS Northern Ireland Council of Social Science

NICSSE National Information Center for Social Science Education

nicu (NICU) neonatal intensive-care unit

NICU Nippon International Container Unit

NICUFO National Investigations Committee on Unidentified Flying Objects

NICYRA National Ice Cream and Yogurt Retailers Association

nid network in dial

Nid. Niddah

NID National Institute for the Deaf; National Institute of Drycleaning; Naval Intelligence Department; Northern Ireland District

NID National Intelligence Daily; New International Dictionary (Webster's Third New International Dictionary of the English Language Unabridged)

nida numerically integrated differential analyzer

NIDA National Institute on Dramatic Art; National Institute on Drug Abuse; National

Investment and Development Authority; Northern Ireland Development Agency

NIDC National Institute of Dry Cleaning; National Investment Development Corporation; Northern Ireland Development Council

NIDCD National Institute on Deafness and other Communication Disorders

NIDDM Non-Insulin-Dependent Diabetes Mellitus

NIDER Nederlands Instituut voor Documentatie en Registratuur (Dutch—Netherlands Institute of Documentation and Filing)

NIDFA National Independent Drama Festivals Association

NIDH National Institute of Dental Health

NIDM National Institute for Disaster Mobilization

NIDR National Institute of Dental Research

nie not included elsewhere

NiE Newspaper in Education

NIE National Institute of Education; National Intelligence Estimate; Newspapers In Education

NIEA National Indian Education Association

NIECC National Industrial Energy Conservation Council

Nieder Niederlande (Dutch—Low Lands)—the Netherlands

niedr niedrig (German—low)

Niedsach Niedersachsen (German—Lower Saxony)

NIEHS National Institute of Environmental Health Sciences

Niels Niels Bohr (Danish physicist)

NIEM National Industrial Energy Management

NIEO New International Economic Order; Non-Incorporated Engineering Order

NIER National Industrial Equipment Reserve

NIEs Newly Industrializing Economies

NIESR National Institute for Economic and Social Research

NIEU Negro Industrial Economic Union

nif nickel-iron film; noise improvement factor; note issuance facility

NIF National Ignition Facility; Navy Industrial Fund

NIFA National Islamic Front of Afghanistan

NIFC National Income Forecasting Committee

nife nickel + iron (Ni + Fe)

NIFES National Industrial Fuel Efficiency Service

NIFI National Inland Fisheries Institute

nifo next in, first out

nifti near-isotropic flux-turbulence instrument

nig. *niger* (Latin—black)

Nig Nigeria

Nig *Niger* (Spanish—Niger)

niga nuclear-induced ground radioactivity

NIGC National Iranian Gas Company

NIGDA National Industrial Glove Distributors Association

Niger Republic of Niger (landlocked North African nation)

Nigeria Federal Republic of Nigeria (West African country)

nightie(s) nightdress(es); nightgown(s)

nightsoap(s) nighttime (tv) soap opera(s)

NIGMS National Institute of General Medical Sciences

NIGP National Institute of Governmental Purchasing

NIGRO Northern Ireland General Register Office

nig(s) nigger(s); renege(s); revoke(s)

nigyysob now I've got you, you SOB

nih not invented here

NIH National Institute of Hardware; National Institutes of Health

NIH 204 antimalarial drug

NIHB National Indian Health Board

NIHBC Northern Ireland House Building Council

NIHE Northern Ireland Housing Executive

nihil *nihil obstat quominus imprimatur* (Latin—nothing hinders it from being printed)

nihil obs. *nihil obstat* (Latin—nothing stands in the way)—official Catholic publications must obtain this before their publication

NIHR National Institute of Handicapped Research

NIHT Northern Ireland Housing Trust

nii national information infrastructure

NII Netherlands Industrial Institute

NIIC National Injury Information Clearinghouse

NIIG NATO Item Identification Guide

NIIN National Item Identification Number

NIIP National Institute of Industrial Psychology

NIIS Niagara Institute for International Studies

NIJ National Institute of Justice

NIJC North Idaho Junior College

NIJFCM National Institute of Jig and Fixture Component Manufacturers

NIJRs *National Institute of Justice Reports*

nik narcotic identification kit

Nik Nikolayev

Niki Nicholas

Nik Nik *Nicholas Nickleby*

Niko (Russian nickname—Nikolai)—Nicholas; Nick; Nicky

Nikola Nikola Tesla (1856–1943)

Niky Nicholas; Nicole; Nickerson; Nikerson

nil not in labor

NIL National Instrument Laboratories; National Investment Library

NI Lab Northern Ireland Labour (party)

NILECJ National Institute of Law Enforcement and Criminal Justice

NILI *Netzach Israel Lo Ishakare* (Hebrew—The eternity of Israel will not die)—acronymic password of the Nili spies who aided Britain by facilitating Turkish defeat in an effort to establish a homeland for Jews in Palestine

'nilla vanilla

N Ill U Pr Northern Illinois University Press

NILOJ National Institute for Law/Order/Justice

NILP National Institute for Labor Policy; Northern Ireland Labour Party

NILQ *Northern Ireland Legal Quarterly*

nil sig nothing significant

NILT National Institute for Lay Training

nim newspaper(s) in microfilm; newspaper(s) in microform

NIM Neurological Impress Method; North Irish Militia

NIMA National Insulation Manufacturers Association

NIMAC National Interscholastic Music Activities Commission

nimby not in my backyard

nimby's not in my backyarders

nimd not in my district

NIMFR National Institutes of Marriage and Family Relations

NIMH National Institute of Mental Health

NIMLO National Institute of Municipal Law Officers

nimm nuclear-induced missile malfunction

nimn not in my neighborhood

n imp new impression

NIMP National Intern Matching Program

nimphe nuclear isotope monopropellant hydrazine engine

NIMR National Institute for Medical Research; National Institute for the Mentally Retarded

NIMT National Institute for Music Theater

nimto not in my term of office

NIMU North Island Mutual Insurance (New Zealand)

NIN Narcotics Intelligence Network; National Information Network; Neighbors In Need

NINA No Irish Need Apply

NINB National Institute of Neurology and Blindness

NINCD National Institute of Neurological and Communicative Disorders

NINCDS National Institute of Neurological and Communicative Disorders and Stroke

NINCH National Initiative for a Networked Cultural Heritage

NINDB National Institute of Neurological Diseases and Blindness

NINDS National Institute of Neurological Diseases and Stroke

NIO National Institute of Oceanography; National Institute of Oceanology; National Intelligence Office(r); National Intelligence Organization; National Iranian Oil; Naval Institute of Oceanology; Northern Ireland Office

NIOC National Iranian Oil Company

niod network in-out dial

NIOP National Institute of Oilseed Products

NIOSH National Institute for Occupational Safety Hazards; National Institute of Occupational Safety and Health

nip normal investment practice; not in possession

nip. nipper; nipple

Nip Nippon (Japan); Nipponese (Japanese)

NIP National Industrial Policy; Neighborhood Improvement Program; Northern Ireland Parliament

NIP Norges Kommunistiske Parti (Norwegian—Norwegian Communist Party)

NIPA National Institute of Pension Administrators; National Institute of Public Affairs

NIPA National Income and Product Accounts

NIPC National Infrastructure Protection Center

NIPCC National Industrial Pollution Control Council

NIPDOK Nippon Documentesyon Kyokai (Japanese—Japanese Documentation Society)

NIPE National Intelligence Programs Evaluation

NIPFDA National Independent Poultry and Food Distributors Association

NIPG Nederlands Instituut voor Praeventieve Gneeskunde (Dutch—Netherlands Institute for Preventive Medicine)

NIPH National Institutes of Public Health

niphl noise-induced permanent hearing loss

nip nip(s) nipple nipper(s)

nipo negative input—positive output

NIPO Nederlands Instituut voor Publick Opinie (Dutch—Netherlands Institute for Public Opinion)

NIPPORO Nihon Hoso Rodo Kumiai (Japan Broadcasting Workers Union)

NIPR National Institute for Personnel Research

ni. pri. nisis prius (Latin—unless before)

nips nippers; non-impact printers

Nip(s) Nippon(ese)

NIPs Not In Profile students

NIPS National Information Processing System; National Institute of Police Science (Japanese)

NIPSSA Naval Intelligence Processing Systems Support Activity

nipts noise-induced permanent threshold shifts

niq no income qualifier

nir near infrared

N Ir Northern Ireland

NIR Northern Ireland Railways

NIRA National Industrial Recovery Act; National Industrial Recovery Administration; Newspaper Industries Research Association

NIRC National Industrial Relations Court

NIRD National Institute of Research in Dairying

N Ire Northern Ireland (whose capital is Belfast)

NIRI National Investor Relations Institute

NIRMP National Intern and Resident Matching Program

NIRNS National Institute for Research in Nuclear Science

NIROP Naval Industrial Reserve Ordnance Plant (USN)

NIRR National Institute for Road Research

NIRRA Northern Ireland Radio Retailers' Association

NIRs Norfolk International (container) Terminals

NIRS National Institute of Radiological Science; Nuclear Information and Resource Service

ni & rt numerical index and requirement table(s)

NIRT National Iranian Radio and Television

nis not in stock

Ni s nickel steel

NIS National Information System; National Institute of Science; National Insurance Scheme; National Intelligence Service; National Intelligence Survey; National Investment Strategy; NATO Identification System; Naval Intelligence Service; Naval Investigative Service; News and Information Service (NBC)

NISA National Impacted Schools Association; National Intelligence Security Authority (Philippines)

NISBS National Institute of Social and Behavioral Science

NISC National Independent Study Center; National Industrial Safety Committee; Naval Intelligence Support Center

NISD National Institute of Steel Detailing

NISGAZ National Intelligence Survey Gazetteer

NISIR National Institute of Scientific Industrial Research

NISM National Iron and Steel Mills

NISO National Industrial Safety Organization; National Information Standards Organization; Naval Investigative Service Offlce(r)

NISP National Information System for Psychology

NISRA Naval Investigative Service Resident Agent

NISS National Institute of Social Sciences

NISSPO NATO Identification System Special Project Office

NIST National Institute of Science and Technology (Philippines); National Institute of Standards and Technology (U.S.)

NISUCO Nigerian Sugar Company

nit negative income tax; none in town

nit (NIT) nautical industrial technology

nit unit of luminance (symbol)

NIT National Institute of Tourism; National Instructional Television; National Intelligence Test; National Invitation Tournament; New Information Technologies; Northrop Institute of Technology; Northrup International Terminals

Nita Juanita

NITA National Industrial Television Association

NITC National Information Transfer Center; National Iranian Tanker Company

nite night

NiteDevRon Night Development Squadron

NITEP Native Indian Teacher Education Programme (Canadian)

niter potassium nitrate

NITHC Northern Ireland Transport Holding Company

NITL National Industrial Traffic League

ni tp nibbling template

NITR National Institute for Telecommunications Research

nitre potassium nitrate (KNO$_3$)

nitric acid HNO$_3$

nitro nitrocellulose; nitroglycerine

nitros nitrostarch

nitts noise-induced temporary threshold shift

NITV National Iranian Television

NIU Northern Illinois University; Northern Interparliamentary Union

Niv Nivose (French—Snowy Month)—beginning December 21st—fourth month of the French Revolutionary Calendar

NIV New International Version (Zondervan Bible); *New Internet Version* (Bible)

NIVE Nederland Instituut voor Efficiency (Dutch—Netherlands Institute for Efficiency)

NIW National Industrial Worken Union

NIWAAA Northern Ireland Women's Amateur Athletic Association

NIWL National Institute for Work and Learning

NIWR National Institute for Water Research

NIWW National Institute for Working Women (prostitutes)

nix (NIX) nuclear inclusion X (clam parasite)

NIYC National Indian Youth Council

Nizh Nizhen (Bulgarian—lower); *Nizhni* (Russian—lower)

Nizim Nizmennost (Russian—lowland)

n J nächstes Jahr (German—next year)

NJ New Jersey; New Jerseyite

NJ National Journal

NJA National Jail Association; National Jewellers Association; National Jogging Association

NJAC National Joint Advisory Council

NJACU New Jersey Association of Colleges and Universities

NJAIS New Jersey Association of Independent Schools

NJASBO New Jersey Association of School Business Officials

NJASSPS New Jersey Association of Secondary School Principals and Supervisors

nJb nice Jewish boy

NJC Natchez Junior College; National Joint Council; National Judicial College; National Junior College; Navarro Junior College; Newton Junior College; Norfolk Junior College

NJCAA National Junior College Athletic Association

NJCC National Joint Computer Conference; Northeastern Junior College of Colorado

NJCCC New Jersey Casino Control Commission

NJCF New Jersey Conservation Foundation

NJCIRM National Jewish Center for Immunology and Respiratory Medicine

NJCMS New Jersey Chamber Music Society

NJDA National Juvenile Detention Association

NJDL New Jewish Defense League

NJE Network Job Entry; New Jersey Experiment

NJEA New Jersey Education Association

NJF Nordiske Jordburgsforskeres Forening (Nordic Agricultural Research Workers' Association)

NJFR National Joint Fiction Reserve

nJg nice Jewish girl

NJH National Jewish Hospital

NJHA National Junior Horticultural Association

NJ Hist Soc New Jersey Historical Society

NJHS National Junior Honor Society; New Jersey Historical Society

NJIT New Jersey Institute of Technology

njk not just kidding

NJLA New Jersey Library Association

NJLC National Juvenile Law Center

NJLJ New Jersey Law Journal

NJMA National Jail Managers Association

NJ Monthly New Jersey Monthly

NJMP New Jersey Marine Police

NJPA National Juice Products Association

NJPAC New Jersey Performing Arts Center

NJPBA New Jersey Public Broadcasting Authority

NJPC National Joint Practices Commission

NJPGA New Jersey Personnel and Guidance Association

NJPS National Jewish Population Study

NJROTC Naval Junior Reserve Officers Training Corps

NJRW New Jersey Reformatory for Women (Clinton)

NJSA New Jersey Student Association

NJSBA New Jersey School Boards Association

NJSD National Joint Service Delegate; National Joint Service Delegation

NJSEA New Jersey Society of Enrolled Agents

NJSO New Jersey Symphony Orchestra

NJSP New Jersey State Police

NJ Turn New Jersey Turnpike

NJWB National Jewish Welfare Board

NJZ New Jersey Zinc

nk neck; not known; not ours (publishing)

NK *Nihon Kyosanto* (Japanese Communist Party); *Nippon Gakushiin* (the Japanese Academy); *Nippon Kokan Steel* (Japanese Steel Exchange) *Nomenklatur Kommission* (Anatomical Nomenclature Commission); *Nordiska Kompaniet* (the Norse Company, Stockholm's leading department store); North Korea(n)

NKA National Kindergarten Association

NKBA National Kitchen and Bath Association

nkc natural killer cell

NKCA National Kitchen Cabinet Association

NKDR National Key Deer Refuge (Florida)

NKF National Kidney Foundation

NKG *Nordiska Kommissionen for Geodesi* (Nordic Commission for Geodesy)

NKGB People's Commissariat for State Security (*q.v.* VOT)

NKK Nippon Kokan Steel (Japan)

NKK *Nippon Kaiji Kyokai* (Japanese—Japanese Marine Classification Society)

NKL *Norges Kooperative Landsforening* (Norwegian—Norwegian Consumer Cooperative)

nklc nickel copper

NKM New Park Mining (stock exchange symbol)

NKMA National Knitwear Manufacturers Association

N.K. Naomi code for chemical and biological warfare

NKOA National Knitted Outerwear Association

NKP Nickel Plate Railroad (stock exchange symbol for New York, Chicago & St Louis Railroad)—locomotives on this line gleamed with nickel-plated ornaments

NKP *Norges Kommunistiske Partii* (Norwegian Communist Party)

NKPA National Kraut Packers Association

NKr Norwegian krone(r)

nks necks (woolen)

NKS *Norge Kjemisk Selskap* (Norwegian—Norwegian Chemical Society)

NKSA National Knitwear and Sportswear Association

NKSO *Narodniy Kommissariat Sotsialnogo Obespecheniya* (Russian—People's Commissariat of Social Security)

NKT Nihon Kai Telecasting

Nkv Nakskov

NKVD *Narodnyi Kommissariat Vnutrennikh Del* (Russian—People's Commissariat for Internal Affairs, Soviet secret police, q.v. VOT)

NKZ *Narodnyi Kommissariat Zdravokhranenia* (Russian—People's Commissariat of Health)

nl new line; no liability; no load; nonlubricant; not listed

nl (NL) new line character (data processing); not licensed (to sell liquor)

nl *nemlig* (Dano-Norwegian—namely); *nicht löslich* (German—not soluble); *non longue* (French—not so far)

n.l. *non licet* (Latin—not permitted); *non liquet* (Latin—it is not clear)

n/l *nuestra letra* (Spanish—our letter)

Nl National

NL National Lakeshore; National League (of Professional Baseball Clubs); National Liberal; National Library; naval lighter (naval symbol); Navy (US department) Library; Navy League; Navy List(ing); Netherlands (auto plaque and Internet code); New Latin; New London, Connecticut; Night Letter; North Latitude; Nuevo León

NL *Nederland* (Dutch—Netherlands); *Norddeutscher Lloyd* (North German Lloyd Line)

N.L. *non liquet* (Latin—unclear)

nla net lettable area

NLA National Landscape Association; National Leukemia Association; National Liberation Army; National Librarians Association; National Libraries Authority; National Library of Australia (Canberra); National Lime Association; National Lumbermen's Association; Nevada Library Association; Nigerian Library Association; Northwestern Lumber Association

NL-A *Nationaal Luchtvaartlaboratorium-Amsterdam* (Dutch—National Airline Laboratory)

NLAA National Legal Aid Association

NLA & DA National Legal Aid and Defender Association

NLAE National Laboratory for the Advancement of Education

NLAPW National League of American Pen Women

N Lat north latitude

NLB National Labor Board; National Library for the Blind; Northern Lighthouse Board

NLBA National Licensed Beverage Association

NLC National Lead Chemicals; National League for Cities; National Leathersellers College; National Legislative Conference; National Legislative Council; National Liberal Club; National Library of Canada; New Liberal Club; New Location Code; New Orleans & Lower Coast (railroad); Northern Land Council

NLCA Norwegian Lutheran Church of America

NLCIF National Light Castings Ironfounders' Federation

NLCMDD National Legal Center for the Medically Dependent and Disabled

NLCS National League Championship Series (baseball)

NLD National League for Democracy (Myanmar, or Burma); National Legion of Decency

NLDC Native Land Development Corporation

NLEA Nutritional Labeling and Education Act

NLEC National Lutheran Educational Council

NLETS National Law Enforcement Telecommunications System

nlf nearest landing field

NLF National Labour Federation; National League of Families (of men missing in action); National Liberal Federation; National Liberation Front; nearest landing field

nlg nose landing gear

NLG National Lawyers Guild; National Library of Greece (Athens); Netherlands Guilder(s); Numismatic Literary Guild

NLGDA National Lawn and Garden Distributors Association

NLGHF National Lesbian and Gay Health Foundation

NLGI National Lubricating Grease Institute

NLHE National Laboratory for Higher Education

NLHS *Neo Lao Hak Sat* (Lao Patriotic Front)

NLI National Lead Inc; National Library of India (Calcutta); National Library of Ireland (Dublin); National Lifeboat Institution

NLJ *National Law Journal*

NLL National Lending Library; Nature Lovers League; Nedlloyd Lines

NLL cards *National Lucht-en-ruimtevaart Laboratorium* (international card catalog devised in Amsterdam)

NLLST National Lending Library for Science and Technology (UK)

nl lt net-laying light

NLM National Liberation Movement; National Library of Medicine

NLMA National Lumber Manufacturers Association

NLMC National Labor Management Council

nln no longer needed

NLN National League for Nursing

NLNE National League of Nursing Education

NLNP Naujan Lake National Park (Philippines)

NLNZ National Library of New Zealand

nlo nonlinear optical

NLO Naval Liaison Office(r); Neighborhood Law Office(r)

NLOGF National Lubricating Oil and Grease Federation

nlp (NLP) neuro-linguistic programmers

NLP National League of Postmasters; Neighborhood Loan Program; Neuro-Linguistic Programming

NLPAD Natural Language Processing of Patient Discharge

nlpc (NLPC) n-laurylpyridinium chloride (detergent compound)

NLPI National Loss Prevention Institute

NL POW/MIA F National League of Prisoner-of-War/Missing-in-Action Families

nlq near letter quality

nlr noise load ratio; non-linear resistance

NLR *Nationaal Lucht-en Ruimtevaartlaboratorium* (Dutch—National Aeronautical and Astronautical Research Institute), Amsterdam; *Newfoundland Law Reports; Nigeria Law Reports*

NLRA National Labor Relations Act

NLRB National Labor Relations Board

nls new least squares; no-load speed; no-load start; non-linear system(s)

NLs New Leftists

NLS National Library of Scotland (Edinburgh); National Library Service (New Zealand and elsewhere); Non-Linear Systems

NLSA National Locksmith Suppliers Association

NLSB National League Service Bureau

NLSCS National League for Separation of Church and State

NLSI National Library of Science and Invention

NLSLS National Library of Scotland Lending Services

nlt new logic technology; not later than; not less than

NLT National Library of Thailand (Bangkok)

NLT *Navigazione Libera Triestina* (Italian Line)

NLTA National Lawn Tennis Association; National League of Teachers Associations

NLTB Native Land Trust Board

NLTU New London Training Unit (USN)

NLUCS National Land Use Classification System

NLUS Navy League of the United States

NLW National Lawyers Wives; National Library of Wales; National Library Week

NLWP National Library Week Program

Nly northerly

NLYL National League of Young Liberals

nm nanometer; nautical mile(s); neuromuscular; new moon; nitrogen mustards; nomenclature; nonmetallic; nonmotile (bacteria); not meaningful; nuclear magneton; nutmeg

n/m no mark

nm *nachmittags* (German—afternoon, P.M.); *namiddag* (Dutch—a.m.)

nm. name

n.m. *nocte et mane* (Latin—night and morning)

n M *nachsten Monats* (German—next month)

Nm Newtonmeter

Nm *Numbers (Book of)*

NM Natal Museum (South Africa); National Monument; National Mutual (life insurance); Neiman-Marcus; New Mexico; Nigeria Museum

n/m² newton per square meter

n/M³ normal cubic meter

nma negative mental attitude

NMA National Management Association; National Market Authority; National Medical Association; National Microfilm Association; National Micrographics Association; National Mortgage Association; Navy Mutual Aid (Association); Needle Makers' Association; Nonprofit Management Association; Northwest Mining Association

NMAA National Machine Accountants Association; National Museum of American Art; Navy Mutual Aid Association

NMAB National Market Advisory Board; National Materials Advisory Board

nmac near mid-air collision

NMAC National Medical Audiovisual Center

NMACT Nuclear Materials Accounting Control Team

NMAF National Medical Association Foundation

NMAH National Museum of American History (Washington, D.C.)

n mar *nivel del mar* (Spanish—sea level)

n masc masculine form of a noun

N-materials nuclear materials

NMB National Maritime Board; National Mediation Board; Nippon Miniature Bearing

NMBA National Marine Bankers Association

NMBC National Minority Business Council

nmbr number

nmc no more credit

NMC National Manufacturers Code; National Mapping Council; National Maritime Council; National Marketing Council; National Meteorological Center; National Museum of Canada; National Museums of Ceylon; National Music Council; Naval Material Command; Naval Medical Center; Naval Missile Center; Network Management Center; Northern Mining Corporation; Nuclear Material Convention; Nursing Mothers Counsel

NMCA National Music Camp Association; Navy Mother's Clubs of America

NMCB National Metric Conversion Board

NMCC National Military Command Center

NMCCIS NATO Military Command, Control, and Information System

NMCDA National Model Cities Directors Association

NMCO Naval Material Catalog Office

NMCP National Memorial Cemetery of the Pacific

NMCS National Military Command System

NMCSSC National Military Command System Support Center

NMD National Missile Defense

NMDA National Metal Decorators Association; National Motorcycle Dealers Association; N-methyl-D-aspartate

NMDC National Materials Development Center

NMDL Navy Mine Defense Laboratory

NMDP National Marrow Donor Program

NMDS New Music Distribution Service

NMDZ NATO Maritime Defense Zone

nme noise-measuring equipment

NME National Medical Enterprises; National Military Establishment; National Mortgage Exchange

Nmea Noumea

NMEA National Marine Education Association; National Marine Electronics Association

nmed named

N-medicine nuclear medicine

N-med tech nuclear-medicine technician

N Mem National Memorial

nmembler mnemonic assembler

NMERI National Mechanical Engineering Research Institute

N Mex New Mexico; New Mexican

N Méx *Nuevo México* (Spanish—New Mexico)

NMF National Marine Fisheries; Nonprofit Mailers Federation

NMFMA National Mutual Fund Managers Association

NMFO Navy Maintenance Field Office

NMFRL Naval Medical Field Research Laboratory

NMFS National Marine Fisheries Service

NMFSL National Marine Fisheries Service Laboratories

NMG National Management Game

NMGC National Marriage Guidance Council

nmh nautical miles per hour

NMH Northwestern Memorial Hospital

NMHA National Mental Health Association

NMHB National Materials Handling Bureau

NMHC National Material Handling Center; National Multi-Housing Council

NMHS National Maritime Historical Society

NMHSA National Mine Health and Safety Academy

nmi new (automobile) model introduction; no middle initial

n mi nautical miles

NMI National Maglev Initiative; National Mutual Insurance; New Mexico Military Institute; Northern Mariana Islands (Rota, Saipan, Tinian, Pagan, Guguan, Agrihan, Aguijan)

NMIA National Meteorological Institute of Athens

NMICA New Mexico Independent College Association

n mi/lb nautical miles per pound (of fuel)

NMIM & T New Mexico Institute of Mining and Technology

N-mishap nuclear mishap

N-missile(s) nuclear missile(s)

NMJ Northern Masonic Jurisdiction

NML National Measurement Laboratory; National Municipal League; National Museum Library; National Music League; National Mutual Life (insurance); Northwestern Mutual Life (insurance)

NMLA National Mutual Life Association; New Mexico Library Association

NMLRA National Muzzle-Loading Rifle Association

NMM National Maritime Museum (Greenwich, England)

NMMA National Macaroni Manufacturers Association; National Marine Manufacturers Association

nmn no middle name

NMN⁺ nicotinamide mononucleotide

NMNA National Male Nurse Association

nmnc nonmercuric noncorrosive

NMNH National Museum of Natural History (DC)

NMO National Mapping Office; Navy Management Office

nmoc new man on campus

nmos nitride-metal-oxide semiconductor

NMOS N-channel metal oxide semiconductor

nmp navigational microfilm projector; normal menstrual period

NMP National Military Park; Nuclear Matrix Protein

NMPA National Marine Paint Association; National Motorsports Press Association; National Music Publishers Association

NMPC National Maintenance Publications Center (USA); National Moratorium on Prison Construction

NMPGA New Mexico Personnel and Guidance Association

nmph nautical miles per hour

nmpm nautical miles per minute

nmps nautical miles per second

n. mque. *nocte maneque* (Latin—night and morning)

nmr normal mode rejection; nuclear magnetic resonance imaging

NMR Natal Mounted Rifles; National Military Representative

NMRA National Marine Representatives Association; National Model Railroad Association

NMRI Naval Medical Research Institute

NMRL Naval Medical Research Laboratory

NMRP New Mexico Research Park

NMRTC New Mexico Research and Treatment Center

nms neuroleptic malignant syndrome; nuclear materials safeguards

NMS National Market System; National Medal of Science; National Meteorological Service; Nobles of the Mystic Shrine

NMSA National Middle School Association

NMSC National Merit Scholarship Corporation; National Mountain and Safety Committee

NMSDC National Minority Supplier Development Council

NMSE Naval Material Support Establishment

NMSEA New Mexico Society of Enrolled Agents

NMSM New Mexico School of Mines

NMSO Naval Manpower Survey Office(r)

NMSQT National Merit Scholarships Qualifying Test

NMSRC National Middle School Resource Center

NMSS National Multiple Sclerosis Society; Nuclear Materials Safety and Safeguards

NMSSA NATO Maintenance Supply Service Agency

NMSST Naval Manpower Shore Survey Team

NMSU Naval Motion Study Unit; New Mexico State University

NMSWF National Manufacturers of Soda Water Flavors

nmt not more than

NMT National Museum of Transport

NMTA National Metal Trades Association; Northwest Marine Trade Association

NMTBA National Machine Tool Builders' Association

NMTF National Market Traders' Federation

NMTFA National Master Tile Fixers' Association

NMTLM Nuclear Materials Transportation Logistics Model

NMTS National Milk Testing Service

NMU National Maritime Union; National Miners Union

NMW National Museum of Wales

NMWA National Mineral Wool Association

NMWP National Migrant Workers Program

nn neutralization number; no name; nouns

n/n no number; not to be noted

nn *non numerato* (Italian—unnumbered)

n.n. *nemini notus* (Latin—known to no one); *nescio nomen* (Latin—I do not know the name)

NN Newport News; Northwestern National

N/N Northrop/Nortronics

NNA National Neckwear Association; National Needlework Association; National Newspaper Association; National Notary Association

nnad network non-addressing device

NNAG NATO Naval Advisory Group; NATO Naval Armaments Group

NNBA National Nurses in Business Association

NNBIS National Narcotics Border Interdiction System

NNBPWC National Negro Business and Professional Women's Clubs

NNC National Negro Conference; National News Council; Naval Nuclear Club; Navy Nurse Corps

NNCCVTE National Network for Curriculum Coordination in Vocational and Technical Education

NNCR North Norfolk Coast Reserves (England)

nnd neonatal death

NND New and Non-Official Drugs

NNE Net National Expenditure; north northeast

NNEB National Nursery Examination Board

NN & EB National Newark & Essex Bank

NNECH National Nutrition Education Clearinghouse

NNEU Naval Nuclear Evaluation Unit

NNF Northern Nurses Federation

NNFA National Nutritional Foods Association

NNFC National Negro Finance Corporation

NNG Netherlands New Guinea; Northern Natural Gas (company)

NNGA Northern Nut Growers Association

NNHT Nuffield Nursing Homes Trust

nni noise and number index (sound pollution)

NNI Norwegian Nobel Institute

NNI Nederlands Normalisatie Instituut (Dutch—Netherlands Standards Institute)

nnk (NNK) notify next of kin

NNL Nigerian National Line

NNLC National Negro Labor Council

nnm next new moon

NNMC National Naval Medical Center

nnn no national name; no native named

NNN Nihon News Network; Novy-Nicolle-McNeal (bacteriological culture)

NNNR Noss National Nature Reserve (Shetlands)

NNO National Night Out

NNO noord noordoost (Dutch—north northeast)

NNOC Nigerian National Oil Company

n. nov. nomen novum (Latin—new name)

nnp (NNP) net national product

NNP Nairobi National Park (Kenya); Ngezi National Park (Zimbabwe), Nimule National Park (Sudan)

NNPA National Negro Press Association; National Newspaper Promotion Association; National Newspaper Publishers Association; Nuclear Non-Proliferation Act (1978)

n-n p-i-f never-never pay-in-full (installment plan)

NNPP Naval Nuclear Propulsion Program (USN)

NNPT Nuclear Non-Proliferation Treaty

NNR New and Nonofficial Remedies

NNRC Neutral Nations Repatriation Commission

NNRI National Nutrition Research Institute

NNRO Norske Nasjonalkomite for Rasjonell Organisasjon (Norwegian National Committee for Scientific Management)

nns (NNS) Navy navigation satellite

nn's nubile nymphs

n N's nice Nellyisms, euphemisms

n-N's neo-Nazis

N Ns Newport News

NNS National Newspaper Syndicate

NNSC Neutral Nations Supervisory Commission

NNS & DDC Newport News Shipbuilding and Dry Dock Company

NNSL Nigerian National Shipping Line

nnsn no national stock number

NNSS Navy Navigational Satellite System

NNTO Netherlands National Tourist Office; Norwegian National Travel Office

NNTT National New Technology Telescope

NNW north northwest

NNW noord noorwest (Dutch—north northwest)

NNWR Necedah National Wildlife Refuge (Wisconsin); Noxubee National Wildlife Refuge (Mississippi)

nnws non-nuclear weapon states

NNWSI Nevada Nuclear Waste Storage Investigation

no natural order; normally open; north; number

no (NO) neuromyelitis optica

n-o not or

n/o no orders

no norsk (Dano-Norwegian—Norwegian); *nummer* (Dutch—number)

nº número (Spanish—number)

no. numero (Latin—number)

No nobelium; Norskie (Norwegian-American); Norway; Norwegian; number

NO Naval Observatory; Naval Officer; New Order; New Orleans; nitrous oxide; North Central Airlines; Norway (Internet code); Nuffield Observatory (Jordrell Bank, England); Nursing Officer

NO noordoost (Dutch—northeast); *Nordosten* (German—northeast); *noroeste* (Spanish—northwest)

No 1 first; first quality; first rate; first person; most important; most important person; number one

No 2 next in line; next-in rank; number two; second; second person; second quality; second rate

NO₂ nitrogen dioxide

No 10 Number 10 Downing Street (London residence of the British prime minister)

noa net operating assets; new obligational authority (NOA); not operationally assigned; not otherwise authorized

n-o-a not-or-and

NOA National Onion Association; National Opera Association; National Optical Association; National Orchestral Association

NOAA National Oceanic and Atmospheric Administration

NOAB National Outdoor Advertising Bureau

NO-AB New Orleans-Algiers Bridge

NOADS Newspapers Opposed to Advertising Death by Smoking

NOAH New Opportunities for Animal Health scientists

NOAL National Order of Arts and Letters

noala noise-operated automatic level adjustment

NOAOs National Optical Astronomy Observatories

NOASSR North Ossetian Autonomous Soviet Socialist Republic

nob no open burning; nobility; noble; not on board

nob nabob (Urdu—viceroy)

nob. nobis (Latin—to us)

NOB National Oil Board; Naval Operating Base; Naval Order of Battle

NOB Nationaal Orkest van Belgie (Flemish—National Orchestra of Belgium)

Nobelst Nobelstiftelsen (The Nobel Foundation)

no biz no business

NOBLEE National Organization of Black Law Enforcement Executives

Noble Patria Noble Patria' tu hermosa bandera (Spanish—Noble country, your lovely flag)—Costa Rican anthem

NOB spread price differential between Treasury notes and Treasury bonds

NOBE Nordstrom Incorporated

noc not otherwise classified; notation of content(s)

NOC National Oceanographic Council; National Olympic Committee; Network Operations Center

nocc navigation operator's control console

NOCC New Orleans Crime Commission

NOCHA National Off-Campus Housing Association

NOCI National Organization of Circumcision Information

NOCIL National Organic Chemical Industries

No-Clo Z No-Clone Zone

NOCM Nuclear Ordnance Commodity Manager

no cn no connection

No Co Northern Countries

NOCO Nuclear Ordnance Catalog Office

no code do not resuscitate; no cardiopulmonary resuscitation

no cpr no cardiopulmonary resuscitation

noct. nocte (Latin—by night, nocturnal)

NOCTI National Occupational Competency Testing Institute

noct. maneq. nocte maneque (Latin—night and morning)

nod network out dial; new offshore discharge; night observation device

NOD National Organization on Disability; Naval Ordnance Depot; Navigation and Ocean Development

NODA Night Operatic and Dramatic Association

NODAC Naval Ordnance Data Automation Center

Nodaks North Dakotans

NODC National Oceanographic Data Center

NODCP National Office of Drug Control Policy

NODE National Organization of Downsized Employees

NODECA Norwegian Defense Communications Agency

nodex new offshore dischargement exercise

NODL National Organization for Decent Literature (Catholic)

no do a nota do autor (Portuguese—author's note)

no do e nota do editor (Portuguese—editor's note)

no do t nota do tradutor (Portuguese—translator's note)

noe notice of exception; not otherwise enumerated

NOEB NATO Oil Executive Board(s)

noel no observed effect level

NOEL National Organization of Episcopalians for Life

NOESS National Operational Environmental Satellite System

NOF National Oceanographic Foundation; National Optical Font; National Osteopathic Foundation; Naval Ordnance Facility

NOFI National Oil Fuel Institute

noforn no foreign nationals; special handling—not to be released to foreign nationals

noft notification of foreign travel

nog noggin

Nogal Nogales, Sonora, Mexico

NOGC Nationaal Overleg voor Gewestelijke Cultuur (Dutch—National Council for Regional Culture)

Noguchi Hideyo Noguchi, orig. Noguchi Seisaku (1876–1928)

NoHo North of Houston Street (New York City)

nohp not otherwise herein provided

noi net operating income; not otherwise identified

noibn not otherwise identified by name; not otherwise indexed by name

NOIC National Oceanographic Instrumentation Center; National Organization of Internet Commerce; Naval Officer in Charge; Navy Opportunity Information Center

NOIM Nuclear Ordnance Inventory Manager

noise not only inserted in (modern) symphonic epics

NOISE National Organization to Insure Support Enforcement; National Organization to Insure Sound-controlled Environment

noisic noisy music

NOJC National Oil Jobbers Council

NOJTP National On-the-Job Training Program

nok next of kin

NOK Norsk Aero Klub

nol net operating loss; normal overload(ing)

NOL Naval Ordnance Laboratory; Neptune Orient Line; Norse Oriental Line

NOLA New Orleans, Louisiana

NOLAC National Organization of Liaison for Allocation of Circuits

NOLC Naval Ordnance Laboratory, Corona

nol. con. nolo contendere (Latin—I do not wish to contend)

nolo nolo contendere (Latin—I do not wish to contend)

NOLPE National Organization on Local Problems in Education

nol. pros. nolle prosequi (Latin—to be unwilling to prosecute)

NOLS National Oceanographic Laboratory System

nol. vol. nolens volens (Latin—unwilling or willing); willynilly

NOLWO Naval Ordnance Laboratory, White Oak (Maryland)

nom nominal; nominate; nominated; nomination; nominative

no'm no madam

NOM National Organization for Men

NOMA National Office Management Association

NOMAD Navy Oceanographic and Meteorological Device (world's fmrst nuclear-powered weather station)

nombos nonmine bottom objects

nom cap nominal capital

nom com nom commercial (French—business name, trade name)

nom. con. nomen conservandum (Latin—generic or specific name to be preserved by special sanction)

nom dam nominal damages

nom de fam nom de famille (French—family name, surname)

nom de g nom de guerre (French—assumed name)—stage-name

nom *de jf* *nom de jeune fille* (French—maiden name)

nom *de p* *nom de plume* (French—pen name, pseudonym)

nom *de t* *nom de théâtre* (French—stage name)

nom. dub. *nomen dubium* (Latin—doubtful name)

nomen nomenclature

nomin nominative

NOMMA National Ornamental and Miscellaneous Metals Association

nom. nov. *nomen novum* (Latin—new name)

nom. nud. *nomen nudem* (Latin—naked name); name for an animal or plant lacking further description

NOMSA National Office Machine Service Association

NOMSS National Operational Meteorological Satellite System

nom std nominal standard

NOMTF Naval Ordnance Missile Test Facilities

Non Nonoc

NON National Organization of Non-Parenthood

non acpc non-acceptance

non arri non-arrival

non-bels nonbelievers

non-can non-cancellable

nonce-wd nonce-word

non-coll non-collegiate

noncom(s) nonconformist(s)

non-com(s) non-commissioned officer(s)

noncon(s) nonconformist(s)

non-contigs non-contiguous states, Alaska and Hawaii

non cul. *non culpabilis* (Latin—not culpable, not guilty)

non-cum non-cumulative

nondely non-delivery

none no one; not one

None Nonesuch

non est *non est inventus* (Latin—he was not found; it is wanting)

non flam non-flammable

non-flam non-flammable film (slow-burning acetate-base film)

N/ONI Navy/Office of Naval Intelligence

non negl non-negotiable

non obs. *non obstante* (Latin—notwithstanding)

non op non-operational

n-on-p negative on positive

non-par non-participating

non perf non-perforated

nonporno not pornographic

non pos. *non possumus* (Latin—we cannot)

non pros. *non prosequitur* (Latin—does not prosecute)

non pyt non-payment

N/ONR Navy/Office of Naval Research

non-rem non-rapid eye movements

non repetat. *non repetatur* (Latin—do not repeat)

non res non-resident

non rtnl non-returnable

NONSAP Nonlinear Structural Analysis Program

non seq. *non sequitor* (Latin—it does not follow)

non-sked non-scheduled

nonstand nonstandard

non std nonstandard

nontax nontaxable

non-U not upper class

nonum national number

NOO Navy Oceanographic Office (formerly Hydrographic Office, USN)

NOOA New Orleans Opera Association

no op no opinion

no op (NO OP) no operation (data processing)

Noor-Brab Noord-Brabant (Dutch province)

Noor-Hol *Noordholland* (Dutch—North Holland)—province including Amsterdam, Haarlem

nop navigating operating procedure; normal operating procedure; not on production; not open (to the) public; not otherwise provided; not our publication

NOP National Oceanographic Program; National Opinion Poll; Naval Oceanographic Program; North Oscura Peak

NOPA National Office Products Association; National Organization of Police Associations

no par no paragraph (matter runs on)

NOPE New Orleans Port of Embarkation

NOPEC non-members of OPEC (Angola, China, Egypt, Malaysia, Mexico, Oman)

N O Phil New Orleans Philharmonic

NOPHN National Organization for Public Health Nursing

NOPL New Orleans Public Library

nopn normally open

NOPO New Orleans Philharmonic Orchestra

NOPS New Orleans Public Service

NOPWC National Old People's Welfare Council

NOQUIS Nucleonic Oil Quantity Indication System

nor normal; not or; not otherwise rated

nor' norther (Middle English contraction); north

nør *nørre* (Danish—north)

Nor Norma (constellation); Norway; Norwegian

Nor *Norge* (Norwegian—Norway); *Norr* (Swedish—north); *Noruega* (Spanish—Norway)

NoR Notice of Readiness

NOR North Central Airlines; NOT OR (data-processing logic-operator equivalent)

NORA National Osteoporosis Risk Assessment

NORAD North American Air Defense

NORAID Northern Aid (to IRA and other groups in Northern Ireland); Norwegian Agency for International Development

Nor Ant Norwegian Antarctica (Bouvet Island, Peter I Island, Queen Maud Land)

Nor Arc Norwegian Arctic (Bear, Edge, and Hope islands in Barents Sea, Jan Mayen Island in Norwegian Sea, Svalbard or Spitsbergen in Arctic Ocean)

nor'ard northward

NORASDEFLANT North American Antisubmarine Defense Force, Atlantic

Nor Atl North Atlantic

norc national ordnance research computer

NORC National Opinion Research Center (University of Chicago); Nippon Ocean Racing Club

Nor-Cor Northland-Coromandel (NZ)

Nor Cur Norwegian Current

nor'd northward

NORD Naval Ordnance

NORDEK Nordic Economic Community (Denmark, Finland, Norway, Sweden)

NORDEL Nordic Electricity Union

Norden (Scandinavian—the North)—Denmark Finland, Iceland, Norway, Sweden

Nordica Lillian Nordica, Lillian Norton (1857–1914)

Nordic Council Scandinavian union including Denmark, Finland, Iceland, Norway, and Sweden

Nordic Countries Denmark, Finland, Iceland, Norway, Sweden

NORDITA Nordic Institute for Theoretical Atomic Physics

NORDSFORSK Nordiska Samarbetsorganisationen för Teknisk-Naturventenskaplig Forening (Nordic Council for Applied Research)

Nord-West Nordrhein-Westfalen (German—North Rhine-Westphalia)

nor'easter northeaster (storm from the northeast)

noref no reference

Norelco North American Philips Company

Norex Nordic Exchanges

Norf Norfolk

Norf S Norfolk Southern, Norfolk & Western, Southern Railway (merger)

NORGRAIN North American Grain Charter

NORI National Office for the Rights of the Indigent

Norics Noric Alps in southern Austria

NORK New Orleans Rhythm Kings

N'Orleans New Orleans

Norlina North Carolina

norm normal; normalize; normalizing; not operationally ready (because of) maintenance; nuclear operational readiness maneuvers

Norm Norman

NORM National Optimism Revival Movement

Normands Norman Islands (Channel Islands)

NORML National Organization for the Reform of Marijuana Laws; National Organization for the Reinforcement of Marijuana Laws; National Organization for the Repeal of Marijuana Laws

Noroil Norwegian Oil

NORONTAIR Northern Ontario Airways

Nor Pac Northern Pacific

NORPAX North Pacific Experiment

Nor Pol Norsk Polarinstitutt (Norwegian—Norwegian Polar Institute)

norrd no reply received

nors not operationally ready, supplies (supply)

Norsker(s) Norwegian sailor(s)

Norskie Norwegian-American

NORSPA North Sea Special Program of Action

NORTEP Northern Teacher Education Program

north. northerly; northern

North Africa Africa north of the Tropic of Cancer; Algeria, Egypt, Libya, Morocco, Tunisia

NORTHAG North European Army Group

North America islands and lands extending from Canada to Colombia (Canada, Central America, Greenland, Mexico, the United States, the West Indies)

Northants Northamptonshire

North BH North Broken Hill

North Borneo Sabah and Sarawak

North Cascades North Cascades National Park in Washington

Northcliffe Viscount Northcliffe (Alfred Charles William Harmsworth)

Northeast Middle Atlantic and New England states

Northeast Corridor megalopolis extending from Boston to Washington, including Providence, New Haven, New York, Newark, Trenton, Philadelphia, Wilmington, and Baltimore

Northeast Region Middle Atlantic and New England states

Northern Bear political cartoonist's symbol for Russia or the former Soviet Union

Northern Institute Northern Region Correction Institute at Fairbanks, Alaska

Northern Ireland Ulster (six northern counties of Ireland)

northern lights aurora borealis

Northerns Burlington, Great Northern, and Northern Pacific railroads

Northern States northern United States in the Federal Union during the Civil War—The North

Northern Territories Kuril Islands seized from Japan by the USSR during World War II

North Jersey Coast Atlantic City to the Atlantic Highlands

Northld Northumberland

North Pole 90 degrees North latitude; zero degrees longitude; northernmost point on the globe; discovered by American explorers Frederick A Cook and Robert E Peary in 1909

Northum Northumberland

Northumb Northumberland; Northumbrian

Northwest northwestern United States (Washington, Oregon, Idaho, and Montana)

NORTLANT North Atlantic

Norton WW Norton & Co

Nortown WW Norton

Nortraship Norwegian Trade and Shipping Mission

Norvic. Norvicensis (Latin—of Norwich)

norw norwegisch (German—Norwegian)

Norw Norwegian

Norway Kingdom of Norway (northernmost Scandinavian country), *Kongeriket Norge*

NORWEB North Western Electricity Board

Norwegian Arctic Norway north of the Arctic Circle; Svalbard (Spitsbergen) and surrounding islands

NORWESTLANT Northwest Atlantic (project)

nos net oil sand; night operation sight; not on shelf; not otherwise specified; numbers

NOs New Orleans (British maritime abbreviation)

NOS National Ocean Survey; NATO Office of Security; Network Operating System; New Orleans; Night Observation Sight; Night Operation System; Nodal Operating System

N OS New Orleans

NOS Nederlandse Omroep Stichting (Dutch—Netherlands Broadcasting Foundation)

NOSA National Occupational Safety Association; National Office Systems Association; National Outerwear and Sportswear Association

NOSB National Organic Standards Board

NOSC Naval Ocean Systems Center (USN); Naval Ordnance Systems Command (USN)

NOSCAF New Orleans Sickle Cell Anemia Foundation

NOSE Neighbors Opposing Smelly Emissions

NOSG Naval Operations Support Group

nosh no show

NOSIE Nurses Observation Scale for Inpatient Evaluation

no sig no signature

nosigchng no significant change

nosmo no smoking

no smoke/drugs no smoking or drugs (allowed)

Nosodak North Dakota + South Dakota—the Dakotas

NOSOPEX Northern Sumatra Offshore Petroleum Exploration

NOSS National Oceanic Satellite System

NOSSOLANT Naval Ordinance System Support Atlantic

NOSSOPAC Naval Ordinance System Support Pacific

NOSTA National Ocean Science and Technology Agency

nosub not subject to load

NOS/VE Network Operating System/Virtual Environment

not noted; nucleus opticus tegmenti

Not Notary

nota none of the above (candidates)

notal not to, nor needed by, all addressees

NOTAM Notice to Airmen

NOTB National Ophthalmic Treatment Board

notg nothing

notif notification

no-till no-tillage

noto numbering tool

noto (Latin prefix—back)—notochord

Notogaea landmass including Australasia

notox non toxic; not to exceed

NOTP New Orleans Times-Picayune

notr no traffic rights

Notre-Dame Notre-Dame de Paris (French—Our Lady of Paris)—most famous of Gothic cathedrals of the Middle Ages

Notre Dames Notre Dame Mountains of Québec

not's non-classical organizational theories

NOTS Naval Ordnance Test Station

Not(t) Nottingham

Nottm Nottingham

Notts Nottinghamshire

notwg notwithstanding

NOU Noumea, New Caledonia (airport)

'nough enough

Nou Héb Nouvelles Hébrides (French—New Hebrides)

nouv nouvelle (French—new)

nov novels; novelist; novels

nov. novum (Latin—new)

n.o.v. non obstante veredicto (Latin—notwithstanding the verdict)

Nov November

Nov Nova (Bulgarian, Italian, Portuguese, Serbo-Croatian—new); *Novaya* (Russian—new); *Novo* (Portuguese or Russian—new); *Navy* (Czechoslovakian—new)

NOVA National Organization of Victim Assistance; Network of Volunteer Assistance

Nova Roma (Latin—New Rome)—Constantinople, Istanbul

Nova Scotia Pen Nova Scotia Peninsula

Novdec November and December

*nov*ᵉ noviembre (Spanish—November)

November letter N radio code

noviembre *noviembre* (Spanish—November)

NOVL Novell Incorporated

nov. n. novum nomen (Latin—new name)

Novo Novosibirsk

NOVS National Office of Vital Statistics

nov. sp. novum species (Latin—new species)

Nov. T. Novum Testamentum (Latin—New Testament)

Novy(s) Nova Scotian(s)

now (NOW) negotiable order of withdrawal (banking accounts)

NoW News of the World

NOW National Organization for Women; New Opportunities for Women

NOWAPA North American Water and Power Alliance

NOWC National Association of Women's Clubs

nox noxious

NOx nitrous oxide (smog component)

noy not out yet; (unit of noisiness)

Noy Noybr (Russian—November)

noydb none of your damn business

noz nozzle

np napalm (incendiary gasoline mixture); national pipe; neap; neap range; near point; negative pressure; net proceeds; neuropsychiatric; neuropsychiatry; new paragraph; new pattern; new police; nickelplated; nitroproof; nonprocurable; no paging; no payment; no place; no place of publication; no protest; nonparental; nonparticipating; nonpropelled; normal pressure; nose plug; not paginated; noun phrase; nucleotide pair; number of pitches; nursing prcocedure

np (NP) no parking; note payable

n/p net proceeds; new pence

np nedsat pris (Dano-Norwegian—reduced price)

Np neap; neap range; neap tide; neper; neptunium (symbol)

Np neper

NP Narragansett Pier; National Park; National Pipe; National(ist) Party; Naval Prison; Neighborhood Professional; Nepal (Internet code); New Providence, Bahama Islands; Newport, Rhode Island; no parking; North Pacific; North Park; Northern Pacific (railroad); Northern Province; not published; Notary Public; Nurse Practioner

N-P Non-Partisan

N/P nitrogen phosphorus ratio

NP Nasionale Partij (Afrikaans—National Party)

NPA National Packaging Association; National Paperboard Association; National Paperbox Association; National Parenthood Association; National Parking Association; National Parks Association; National Parks Authority (NZ); National Particleboard Association; National Pasta Association; National Pediculosis Association; National Personnel Associates; National Personnel Authority; National Pet Association; National Pharmaceutical Association; National Pigeon Association; National Pilots Association; National Pipeline Authority; National Planning Association; National Police Agency (Japan); National Preserves Association; National Proctologic Association; National Production Authority; Naval Procurement Account; Navy Postal Affairs; New People's Army (Filipino insurgency); Newsletter Publishers Association; Newspaper Publishers Association; Nigerian Ports Authority; Notice of Proposed Assessment (IRS penalty); Nurse Practice Act

NPABC National Public Affairs Broadcast Center

N Pac North Pacific

NPAC National Program for Acquisitions and Cataloging (Library of Congress)

N Pac Cur North Pacific Current

NPACI National Production Advisory Council on Industry

NPACT National Public Affairs Center for Television

NPAFC North Pacific Fisheries Commission

NPAG New Pest Advisory Group

N-panel panel of nuclear experts

NPAP National Psychological Association for Psychoanalysis

npat net profit after tax(es)

NPB National Park Board; National Parole Board (Canada); National Productivity Board (U.S.); North Pacific Bank

NPBA National Paper Box Association; National Pig Breeders' Association; Natural Product Broker Association

npbc newer predominantly black college

NPBC National Programming Black Consortium

NPBI National Pretzel Bakers Institute

NPBOA National Party Boat Owners Alliance

npc near point of convergence; New Process Company's trademark; notification of parts change

NPC National Paddling Committee; National Patent Council; National Peach Council; National Peanut Council; National Peoples Congress; National Periodicals Center; National Personnel Consultants; National Petroleum Council; National Pharmaceutical Council; National Philatelic-Collection; National Police Computer; National Ports Council; National Potato Council; National Power Company; National Press Club; National Productivity Center; Nauruan Phosphate Commission; Naval Photographic Center; New Peoples Center; Nigerian Population Commission; Nippon Petro-Chemicals

NPCA National Paint and Coatings Association; National Parks and Conservation Association; National Pest Control Association; National Precast Concrete Association

NPCC National Projects Construction Corporation; Nebraska Penal and Correctional Complex

NPCFB North Pacific Coast Freight Bureau

NPCI National Potato Chip Institute

NPCP National Press Club of the Philippines

npcr no periodic calibration required

np-ct naval personnel conversion tables

npd nonparental ditype; no payroll division; north polar distance

np or d no place or date (of publication)

N-P d Neimann-Pick's disease

NPD North Polar Distance

NPD Nationaldemokratische Partei Deutschlands (German—National Democratic Party of Germany)

NPDC National Patent Development Corporation

NPDEA National Professional Driver Education Association

NPDES National Pollution Discharge Elimination Scheme

NPDN Nordic Public Data Network (Denmark, Finland, Iceland, Norway, and Sweden)

NPDO Non-Profit Distributing Organization

np or dp no place or date of publication

npe nonylphenol ethoxylate

NIPE Navy Preliminary Evaluation

N-peace nuclear peace

npef new product evaluation form

NP en G Nederlandse Postcheque en Girondienst (Dutch—Netherlands Postal Check and Transfer Service)

NPEP National Public Expenditure Plan(ning)

npf newsprint pulp flat; no private facilities; not provided for

NPF National Park Foundation; National Parkinson Foundation; National Piano Foundation; National Poetry Foundation; National Press Foundation; National Provident Fund(ing); National Psoriasis Foundation; National Pugilistic Federation; Newspaper Press Fund

NPFA National Playing Fields Association

NPFC Naval Publications and Forms Center; North Pacific Fisheries Commission; Northwest Pacific Fisheries Center

NPFFA National Police and Fire Fighters Association; National Prepared Frozen Food Association

NPFI National Plant Food Institute

npfid nitrogen-phosphorus flame-ionization detector

NPFMC North Pacific Fishery Management Council

NPFSC North Pacific Fur Seal Commission

NPFT Neurotic Personality Factor Test

npg negative population growth

NPG National Portrait Gallery; NATO Planning Group; Nuclear Planning Group

NPGA National Propane Gas Association; Nebraska Personnel and Guidance Association; Nevada Personnel and Guidance Association

NPGPA Non-Powder Gun Products Association

NPGS Naval Postgraduate School; Net Profit Generator System

nph no profit here

n ph nuclear physics

npH neutral protamine Hegedorn (isoophane insulin)

NPHIS Nested Phrase Indexing System

NPHQ National Park (Ranger) Headquarters

npi no previous information

NPI National Paralegal Institute; National Penitentiary Institute; National Population Inquiry; National Productivity Institute; National Purchasing Institute; Neuro-Psychiatric Institute; Nippon Pulp Industry

NPIA Norfolk Port and Industrial Authority

NPIC Naval Photographic Interpretation Center

NPICPS National Policy Institute of the Center for Political Studies

NPIPF Newspaper and Printing Industries Printing Fund

NPIS National Physics Information System; New Product Information Service

N-P-K Nitrogen-Phosphate-Potash (fertilizer)

npl new processor line; new program language; nipple; no personal liability; noise-pollution level; nuclear pumped laser

n pl plural form of a noun

NPL Nashville Public Library; National Physical Laboratory; Newark Public Library; Norfolk Public Library; Not for Profit Leadership (degree)

N-plant(s) nuclear plant(s); nuclear-power plant(s)

NPLEX Naturopathic Physicians Licensing Exam

NPLGS Night Plane Guard Station

NPLO NATO Production and Logistics Organization

nplu not people like us

npm number of points in the point-matching method

NPMA National Property Management Association; Newspaper Purchasing Management Association

NPMAA National Piano Manufacturers Association of America

npn (NPN) nonprotein nitrogen

n-p-n (NPN) negative-positive-negative

npna no protest for nonacceptance

np/nd not published/no date (given)

N & PNWR Ninepipe and Pablo National Wildlife Refuge (Montana)

npo negative-positive-zero

n.p.o. *nil per os* (Latin—nothing by mouth)—sometimes written *ne per oris*

NPO National Philharmonic Orchestra (Manila); National Program Office (for nuclear waste terminal storage), Navy Post Office; Navy Purchasing Office(r); New Philharmonia Orchestra (London); Non-Profit Organization

NPOAA National Police Officers Association of America

NPOEV Nuclear-Powered Ocean Engineering Vehicle (miniature submarine)

N-pollution nuclear pollution

NPO-NIA Nonprofit Organizations National Insurance Alliance

NP & OSR Naval Petroleum and Oil Shale Reserve

N-power nuclear power

N-power plant(s) nuclear-power plant(s)

npp no passed proof

NPP National Potato Panel; National Prison Project (ACLU); Naval Propellant Plant; National Privacy Panel; New Progressive Party (Puerto Rico); Nuclear Power Plant

NPP (ACLU) National Prison Project (American Civil Liberties Union)

NPPA National Press Photographers Association

NPPAJ National Probation and Parole Association Journal

nppd nitrophenylpentadiene aldehyde (spy dust)

NPPF National Planned Parenthood Federation; National Press Photographers Foundation

NPPO Navy Publications and Printing Office

NPPR Nationalist Party of Puerto Rico

NPPS National Plants Preservation Society; Navy Publication Printing Service

N-P Pubns National Press Publications

NPQ Naviera de Productos Químicos (Spanish—Chemical Products Shipping Line)

npr night press rate

n/p/r noise/power/ratio

Npr Napier, NZ

NPR National Public Radio; Naval Petroleum Reserves; Navieras de Puerto Rico; Nickel Plate Road (railroad)

NPRA National Parks and Recreation Association; National Parks and Reserves Authority; National Petroleum Refiners Association; Naval Personnel Research Activity; Newspaper Personnel Relations Association

NPRAC National Public Radio Association of California

NPRC National Personnel Records Center; Newspaper Production and Research Center

nprd nuclear plant reliability data

NPRDS Nuclear Plant Reliability Data System

N Pres National Preserve

NPRL National Physical Research Laboratory

NPRN National Public Radio News

NPRO Navy Plant Representative Office(r)

NPROA National Police Reserve Officers Association

N-project nuclear-power project

N-proliferation nuclear proliferation

N-propulsion nuclear propulsion

NPR & OSR Naval Petroleum Reserves and Oil Shale Reserves

npr's nuclear-power reactors

nps non-professional staff; normal pipe size; no prior service

nps (NPS) nuclear-powered ship(ping)

NPs Notaries Public; Nurse Practitioners

NPS Narcotics Preventive Service; National Park Service; National Portrait Society; Naval Postgraduate School; Nuclear-Powered Ship(ping)

NPSB *National Prisoner Statistics Bulletin*

NPSC National Public Service Commission

npsh net positive suction head

npt normal pressure and temperature

npt (NPT) nocturnal penile tumescence

Npt Navy pointer tracker; Newport

NPT national (taper) pipe thread; Non-Proliferation Treaty

NPTA National Paper Trade Association; National Passenger Traffic Association; National Piano Travelers Association; Nevada Parent–Teacher Association

NPTC National Postal and Travelers Censorship

NPTRL Naval Personnel Training Research Laboratory

npu net protein utilization; not-passed urine

n.p.u. *ne plus ultra* [Latin—nothing beyond (it); the summit; the ultimate]

NPU National People's Union; National Pharmaceutical Union; National Police Union; National Postal Union

npv net present value; no par value

npv (NPV) nuclear polyhedrosis virus

NPVLA National Paint, Varnish, and Lacquer Association

npw new-pool wildcat (oil well)

NPW *Naturpark Pfalszer Wald* (German—Falls Forest Nature Park)—in western Germany near France

NPWC Navy Public Works Center

NPWS National Parks and Wildlife Service (Australia)

NPWU National Production Workers of America

NPX National Phoenix Industries (stock-exchange symbol)

npy neuropeptide Y (substance in brain)

NPY National Productivity Year

n-p-z negative-positive-zero

nq notes and queries

Nq National Association of Security Dealers Automated

NQ Neurological Quotient; North Queensland

N & Q *Notes & Queries*

nqa net quick assets

NQA Nuclear Quality Assurance

NQAPO Nuclear Quality Assurance Program Office

nqb no qualified bidders

NQB National Quotation Bureau

NQD Notice of Quality Discrepancy

NQMP National Quality Management Program

nqokd not quite our kind, dear

nqos not quite our sort

nqot not quite our type

nqr nuclear quadruple resonance

NQX North Queensland Express

nr narrow resonance; naturalized; natural rubber; near; net register; new reports; no risk; noise reduction; non-reactive (relay); non-resident; no response; norm-referenced; not rated; not recorded; number

n-r no(n) return; non resident

n/r no record; non-recoverable; not reported; not required; not responsible (for)

nr *non rogne* (French—untrimmed); *nummer* (Polish—issue, number); *nummer* (Dano-Norwegian or Swedish—number)

n.r. *non repetatur* (Latin—not to be repeated)

nR *neue Reihe* (German—new series)

Nr *Nummer* (German—number)

NR National Register; National River; National Riverway; Nauru (Internet code); North Riding

N/R Notice of Readiness

NR *National Review; Norsk Rikskringkasting* (Norwegian Broadcasting)

nra never refuse anything; no repair action; non-representational artist

nra *nuestra* (Spanish—our)

NRA National Racing Authority; National Reclamation Association; National Recovery Act; National Recovery Administration; National Recreation Association; National Recreation(al) Area; National Reform Association; National Rehabilitation Association; National Renderers Association; National Research Associates; National Resistance Army (Uganda); National Restaurant Association; National Rifle Association (of America); National Rivers Authority; Naval Reserve Association

NRAA National Rifle Association of America

NRAC National Research Advisory Council; National Resources Analysis Center; National Rural Advisory Council

NRACCO Navy Regional Air Cargo Control Office(r)

nrad no risk after discharge

NRAF Navy Recruiting Aids Facility

nral no risk after landing

NRAO National Radio Astronomy Observatory

nras no risk after shipment

NRAS Navy Readiness Analysis Section; Navy Readiness Analysis System

Nra Sra Nuestra Señora (Spanish—Our Lady)

Nrb Nordby

NRB National Religious Broadcasters; National Research Bureau; National Roads Board; National Rubber Bureau

NRB Narodna Republika Blgariya (Bulgarian—Bulgarian Peoples' Republic)

Nrbi Nairobi

NRBs National Religious Broadcasters

nrc noise-reduction circuitry; not recommended for children

NRC Nacorazi Railroad Company; National Racquetball Club; National Rainbow Coalition; National Realty Committee; National Referral Center (Library of Congress); National Reporting Center; National Republican Club; National Research Corporation; National Resources Committee; National Resources Council; National Roofing Contractors; National Rural Center; Naval Retraining Command; Neighborhood Recovery Center (for alcoholism); Neighborhood Reinvestment Corporation; Netherlands Red Cross; Newport Research Corporation; Nuclear Regulatory Commission; Nuclear Research Council

NRC Nieuwe Rotterdamse Courant (Dutch—New Rotterdam Courant)

NRCA National Recovery and Collection Association; National Resources Council of America; National Retail Credit Association; National Roofing Contractors Association

NRCC National Republican Campaign Committee; National Republican Congressional Committee; National Research Council of Canada

NRCD National Reprographic Center for Documentation

nrcf not reconfirmed

NRCL National Research Council Library; Natural Resources Conservation League

NRC-NAS National Research Council–National Academy of Sciences

nrcp non-reinforced concrete pipe; non-residential conditional purchase

NRCPC National Rural Crime Prevention Center (Ohio State University)

NRCR Northern Railway of Costa Rica *(Ferrocarril del Norte de Costa Rica)*

NR Crit Nuclear Rocket—Critical

NRCS Natural Resource Conservation Service

nrcy not received yet

nrd negative-resistance diode

NRD National Range Division; National Register of Designers; Navy Recruiting Depot; Navy Recruiting District

NRDA National Research and Development Authority (Israel); Nevada Research and Development Area

NRDB Natural Rubber Development Board

NRDC National Research Development Corporation; National Resources Defense Agency; National Resources Development Council; National Running Data Center; Natural Resources Defense Council

NRDCA National Roof Deck Contractors Association

NRDL Naval Radiological-Defense Laboratory

nrdo naval radio; Navy radio

NRDO National Research and Development Organization

NRDS Nuclear Rocket Development Station

N-reactor(s) nuclear reactor(s)

NREB Navy Reserve Evaluation Board

NREC National Resource Evaluation Center

NRECA National Rural Electric Cooperative Association

nrem (NREM) non-rapid eye movement

nrems (NREMS) non-rapid eye-movement sleep

nrem sleep non-rapid eye-movement (spindle) sleep

NREN National Research and Education Network

NRF National Relief Fund; National Retail Federation; Naval Reactor Facility; Naval Repair Facility

NRF Nouvelle Revue Française

NRFC National Railroad Freight Committee; Navy Regional Finance Center

NRFL National Rugby Football League

nrg energy

NRG National Research Group Incorporated; National Resurrection Group (Athenian rightist terrorists); Naval Research Group

NRGA National Rice Growers Association

NRh Northern Rhodesia

NRHA National Retail Hardware Association; National Roller Hockey Association

NRHC National Rural Housing Coalition

NRHS National Railway Historical Society

NRI National Radio Institute; Nomura Research Institute (Japan)

NRIAD National Register of Industrial Art Designers

NRIC National Registration and Identity Card

NRIMS National Research Institute for Mathematical Sciences

NRIS National Resource Information System

Nrk Newark

NRK Nikolai Rimsky-Korsakov

NRK Norsky Rikskringkasting (Norwegian—Royal Norwegian Broadcasting)

nrl normal rated load

NRL National Radiation Laboratory; National Reference Library (of Science and Invention); National Registry for Librarians; National Research Library; Naval Research Laboratory; Nelson Research Library

NRLA Northeastern Retail Lumber Association

NRLC National Railway Labor Conference; National Right to Life Committee

NRLCA National Rural Letter Carriers' Association

NRLDA National Retail Lumber Dealers Association

NRLM National Research Laboratory of Meteorology

NRLSI National Reference Library of Science and Invention

nrm natural remanent magnetism; next to reading matter; non-routine maintenance; normal rabbit serum

NRM Naval Reserve Medal; Northern Rocky Mountains; Northern Roller Mills

NRMA National Reloading Manufacturers Association; National Retail Merchants Association; National Roads and Motorists Association

NRMC National Records Management Council; Naval Records Management Center; Naval Regional Medical Center

NRMCA National Ready-Mixed Concrete Association

NRMG Nederlands Rekenmachine Genootschap (Dutch—Netherlands Computer Society)

nrml normal

NRMM National Register of Microform Masters

NRN National Radio Network

nro nuestro (Spanish—our)

NRO Narcotic Rehabilitation Office(r); National Reconnaissance Office; National Registration Office(r); Naval Research Objectives

NROO Naval Reactors Operations Office

NROTC Naval Reserve Officers Training Corps

nrp net rating points; no replacement part; normal rated power

NRP National Religious Party (Israel); National Renaissance Party; National Republican Party; National Research Poll

NRPA (NR & PA) National Recreation and Park Association

NRPB National Radiological Protection Board; National Research Planning Board

NRPC National Railroad Passenger Corporation

NRPL NATO Recommended Products List

NRPRA Natural Rubber Producers' Research Association

NRR Northern Rhodesia Regiment; Nuclear Reactor Research

NRRC National Restitution Resource Center

NRRE Netherlands Radar Research Establishment

NRRL Norsk Radio Relae Liga (Norwegian—Norwegian Radio Relay League)

nrs normal rabbit serum; numbers

N rs Nepalese rupee(s)

NRS National Rose Society; National Runaway Switchboard; Naval Recruiting Service; Navy Records Society; Navy Relief Society; New Reading System; Noise-Reduction System

NRS Naturhistoriska Riksmuseet Stockholm (Swedish—Royal Natural History Museum)

NRSA National Rural Studies Association; Natural Rubber Shippers Association

NRSCC National Registry System for Chemical Compounds

NRSFPS National Reporting System for Family Planning Services

nrt net register(ed) tonnage (tons); normal rated thrust; norm-referenced testing

NRT Tokyo-Narita (airport)

NRTA National Retired Teachers Association

NRTC Naval Recruit Training Center; Naval Reserve Training Center; Northrup Research and Technology Center

NRTI National Rehabilitation Training Institute; National Resources Training Institute

N R L C National Right-to Life Committee

nrtor no risk till on rail

nrts not reparable this station

NRTS National Reactor Testing Station

nrtwb no risk until waterborne

nru nuclear reactor—universal

Nru Nauru

NR-U Nederlandsche Radio-Unie (Dutch—Netherlands Union of Radio Broadcasters)

NRUCFC National Rural Utilities Cooperative Finance Corporation

nrv net realisable value; nonreturn value

NRVC National Railway Utilization Corporation

Nrvkg Nervenkrieg (German—nerve warfare)

NRVN Navy of the Republic of Viet Nam

Nrw Norwegian

NRWC National Right to Work Committee

NRWLDF National Right to Work Legal Defense Foundation

NRWP National Roundtable for Women in Prisons

nrwt non-resident withholding tax

nrx nuclear reactor, experimental

Nry Newry

NRYC New Rochelle Yacht Club

NR Yorks North Riding, Yorkshire

NRZ National Railways of Zimbabwe

nrz c (NRZ C) non-return-to-zero change (data processing)

nrzi non-return-to-zero IBM

nrz m (NRZ M) non-return-to-zero mark recording (data processing)

ns nanosecond; near side; neuropsychiatric; new series; nickel steel; no sparring; noise suppressor; nonsmoker; nonstandard; nonstop; no standard; not specified

ns (NS) neurosurgery; note series (synonymous with original or prime)

n/s neutrons per second; no service; not scheduled; not stocked; not sufficient

ns nostro (Italian—our or ours); *nouvelle serie* (French—new series)

nS neue Serie (German—new series)

Ns nimbostratus; Nunes; Nuñez

NS National Seashore; National Socialist; National Society; National Special (screw threads); National Superannuation; Naval Shipyard; Naval Station; New Style; Nippon(ese) Standard; Norfolk Southern (Norfolk & Western and Southern railroads merged); North Sea; Nova Scotia; Nuclear Ship; Nuclear Submarine; Numismatic Society

N.S. New Style; Norfolk Southern (railroad)

NS Nachshrift (German— postscript); *Nasjonal Samling* (Norwegian—National Unification)—fascist collaborationists headed by Vidkun Quisling during World War II *(see quis); Nederlandsche Spoorwegen* (Dutch—Netherlands Railway); *Notre Seigneur* (French—Our Lord); *Nuestro Señor* (Spanish— Our Lord)

nsa not seasonally adjusted

nsa (NSA) nonenyl succinic acid

NSA National Sawmilling Association; National Safety Association; National Secretaries Association; National Security Adviser; National Security Agency; National Service Acts; National Shellfisheries Association; National Sheriff's Association; National Shipping Authority; National Showmen's Association; National Silo Association; National Skating Association; National Ski Association; National Slag Association; National Slate Association; National Society of Auctioneers; National Speakers Association; National Standards Association; National Stereoscopic Association; National Stroke Association; National Students Association; Naval Stock Account; Naval Supply Account; Neurological Society of America; Norwegian Seamen's Association; Nuclear

Science Association; Nursery School Association; Nurses Supply Association

NSA Nuclear Science Abstracts

NSAA National Society of Architectural Administrators; Norwegian Singers' Association of America

NSABP National Surgical Adjutant Breast / Bowel Project

NSAC National Society for Autistic Children; National Society of Accountants for Cooperatives; Nova Scotia Agricultural College

NSACG Nuclear Strike Alternate Control Group

NSACS National Society for the Abolition of Cruel Sports

NSA/CSS National Security Agency/Central Security Service

NSAD National Society of Art Directors

NSADD National Society of Alcoholism and Drug Dependence

NSAE National Society of Art Education

NSAFA National Service Armed Forces Act

nsai (NSAI) non-steroidal anti-inflammatory

nsaids non-steroidal anti-inflammatory drugs

NSAM National Security Agency Memorandum; Naval School of Aviation Medicine

NSAS National Smoke Abatement Society

NSASAB National Security Agency Scientific Advisory Board

NSB National Science Board; Nippon Short-wave Broadcasting

NSB Norges Statsbaner (Norwegian—Norwegian State Railway)

NSBA National School Boards Association; National Sheep Breeders Association; National Small Business Association; National Sugar Brokers Association

NSBC National Student Book Club

NSBF National Scientific Balloon Facility

NSBISS NATO Security Bureau/Industrial Security Section

NSBIU Nova Scotia Board of Insurance Underwriters

NSBMA National Small Business Men's Association

nsc non-service connected

NSC National Safety Council; National Science Council; National Security Council; National Shippers Council; National Standards Commission; National Steel Corporation; NATO Steering Committee; NATO Supply Center; NATO Supply Classification; Naval School Command; Naval Supply Center; New Sessions Cases; New Solidarity Club(s); Newark State College; Nuclear Safety Concern (Program); Nutrition Society of Canada

NSCA National Safety Council of Australia; National Society for Clean Air; National Sound and Communications Association; National Strength Conditioning Association; Nevada State Council on the Arts; Nova Scotia College of Art

NSCAR National Society of Children of the American Revolution

NSCBS National Society for the Conservation of Bighorn Sheep

NSCC National Securities Clearing Corporation; National Society for Crippled Children

NSCCA National Society for Crippled Children and Adults

nscd nonservice-connected disability

NSCD National Society of Colonial Dames

NSCDA National Society of the Colonial Dames of America

NSCDRF National Sickle Cell Disease Research Foundation

NSCIA National Supervisory Council for Intruder Alarms

NSCID National Security Council Intelligence Directive

NSCLC National Senior Citizens Law Center

NSCM National Stamp Collecting Month

NSCO Network Systems Corporation

NSCP National Society of Compliance Professionals

NSCPA National Society of Certified Public Accountants

NSCR National Society for Cancer Relief

NSCs Network Switching Centers

NSCSP National Site Characterization and Selection Plan

NSCT North Staffordshire College of Technology

nsd no significant defect; no significant deviation; no significant difference; noise-suppression device; nonsoapy detergent; normal spontaneous delivery

NSD Naval Stores Department; Naval Supply Depot; Naval Support Data

NSDA National Soft Drink Association; National Supply Distributors Association

NSDAP Nationalsozialistische Deutsche Arbeiterpartei (German National Socialist [Nazi] Workers Party)

NSDAR National Society, Daughters of the American Revolution

NSDB National Science and Development Board

NSDC National School Development Council; National Serials Data Center; National Space Development Center

NSDD National Security Decision Directive

nsdf naval standard distillate fuel

NSDF National Sex and Drug Forum

NSDJA National Sash and Door Jobbers Association

NSDMs National Security Decision Memorandums

NSDO National Seed Development Organisation

NSDP National Society of Dental Prosthetists

NSDS National Shut-in Day Society

nse neuron-specific enolase

NSE Nigerian Society of Engineers

NSEA Nebraska State Education Association

nsec nanosecond

n/sec neutrons per second

NSEC National Service Entertainments Council

NSEE National Society for Experimental Education

NSEF National Student Educational Fund

NSEI Norwegian Society for Electronic Information

NSERC Natural Science and Engineering Research Council of Canada

NSERI National Solar Energy Research Institute

NSES National Society of Electrotypers and Stereotypers

NSESG North Sea Environmental Study Group

nsf not sufficient funds

NSF National Sanitation Foundation; National Science Foundation; National Sex Forum; National Sleep Foundation; Naval Stock Fund; Navy Strike Fighter

NSF Norges Standardiserings Forbund (Norwegian—Norwegian Standards Institute)

NSFA National Science Faculty Association; Naval Support Force Antarctica (USN); New Settlers Federation of Australia

NSFD National Surface Freight Division (Institute of Freight Forwarders)

NSFGB National Ski Federation of Great Britain

NSFNET National Science Foundation Network

NSFPA National Suppliers to Food Processors Association

nsftd normal spontaneous full-term delivery

nsg neurosecretory granules; not so good

NSG National Supply Group; NATO Standardization Group; Naval Security Group; Newspaper Systems Group; Nuclear Suppliers Group

NSGA National Sporting Goods Association

NSGC Naval Security Group Command

NSGD Near-Surface Geological Disposal

nsgn noise generator

NSGT Non-Self-Governing Territories; Non-Self-Governing Territory

nsh no stock on hand; not so hot

NSHA National Steeplechase and Hunt Association

NSHC North Sea Hydrographic Commission

NSHEB North of Scotland Hydro-Electric Board

N-ship(s) nuclear-powered ship(s)

NSHMBA National Society of Hispanic MBAs

nsi next sequential instruction; nonstandard item; nonstocked item; nuclear safety inspection; numeric signal insignia

NSI National Stock Exchange; Nielsen Station Index; Nuclear Safety Inspection

NSI Norsk Senter for Informatikk (Norwegian—Norwegian Information Center)

NSIA National Sailing Industry Association; National Security Industrial Association

NSIBU Nova Scotia Board of Insurance Underwriters

NSIC National Small Industries Corporation; National Solar Information Center

NSID National Society of Interior Designers

n sing singular form of a noun

NSIO Nova Scotia Information Office

NSIPA National Society of Insurance Premium Adjustors

NSJC Nuestro Señor Jesucristo (Spanish—Our Lord Jesus Christ)

nsk not specified by kind

NSK Nihon Shimbun Kyokai (Japanese—Japan Newspapers and Publishers Association); *Nippon Seiko KK* (bearings)

NSKK Nito Shosen Kabushiki Kaisha (Japanese—Japanese steamship line)

nsl non-standard label; not stock listed

NSL National Science Library; National Socialist League (American Nazi Party); National Standards Laboratory; Navy Stock List; Northrop Space Laboratory; Numidian Support League

NSLA National Society of Literature and the Arts; National Staff Leasing Association

NSLF National Socialist Liberation Front (American-Nazi student organization)

NSLI National Service Life Insurance

NSLL National Savings and Loan League

NSLS National Science Library System

nsm new smoking material (wood-substitute tobacco); noise source meter; number of similar (negative) matches

NSM National Savings Movement; National Security Medal; National Selected Morticians; National Socialist Movement; Naval School of Music; Nevada State Museum; Nova Scotia Museum (Halifax)

ns/m² newton second per square meter

NSMA National Scale Men's Association; National Seasoning Manufacturers Association; National Second Mortgage Association

NSMC Naval Submarine Medical Center

NSMHC National Society for Mentally Handicapped Children

NSMI National Special Media Institutes

NSMM National Society of Metal Mechanics

NSMP National Society of Master Patternmakers; National Society of Mural Painters; Navy Support and Mobilization Plan

NSMPA National Screw Machine Products Association

NSMR National Society for Medical Research

NSMS National Sheet Music Society

NSMSES Naval Ship Missile Systems Engineering Station

NSN National Stock Number; NATO Stock Number

NSNA National Student Nurses' Association

NSNC Nova Scotia Normal College

nso non-statutory stock options

NSO Nashville Symphony Orchestra; National Symphony Orchestra; Naval Staff Officer; Navy Subsistence Office(r); Norfolk Symphony Orchestra; Northern Sinfonia Orchestra

NSOA National School Orchestra Association

NSOC Navy Satellite Operations Center

NSOEA National Stationery and Office Equipment Association

NSOSG North Sea Oceanographic Study Group

nsp non-standard part

n sp new species

NSP National Siting Plan; National Society of Professors; National Stuttering Project; Navy Standard Part, Nebraska State Patrol; Nebraska State Police; North Solomons Province; Northern States Power

NSPA National Scholastic Press Association; National Socialist Party of America; National Society of Public Accountants; National Soybean Processors Association; National Split Pea Association; National Standard Part Association; Naval Shore Patrol Administration

NSPAR Non-Standard Parts Approval

NSPB National Society for the Prevention of Blindness

NSPC National Security Planning Commission; National Society of Painters in Casein; Northern States Power Company

NSPCA National Society for the Prevention of Cruelty to Animals

NSPCC National Society for the Prevention of Cruelty to Children

NSPD Naval Shore Patrol Detachment

NSPE National Society of Professional Engineers

nspf not specifically provided for

NSPF National Swimming Pool Foundation

NSPG National Security Planning Group

NSPI National Society for Performance and Instruction; National Society for Programmed Instruction; National Spa and Pool Institute

NSPLO NATO Sidewinder Production and Logistics Organization

NSPO Navy Special Projects Office; Nuclear Systems Project Office

NSPRA National School of Public Relations Association

NSPRM National Society of Professional Resident Managers

NSPS National Sweet Pea Society; New-Source Performance Standards

NSPSE National Society of Painters, Sculptors, and Engravers

NSPWA National Society of Patriotic Women of America

nsq neuroticism scale questionnaire

nsr natural sinus rhythm; natural stain remover; naval staff requirement (UK); normal sinus rhythm

NSR National Scenic River(way); National Scientific Register; National Security Regulation(s); Norfolk Southern Railway

NSRA National Service Robot Association; National Shoe Retailers Association; National Shorthand Reporters Association; National Smallbore Rifle Association; National Street Rod Association; National Swim and Recreation Association; North-South Reconstruction advisors; Nuclear Safety Research Association

NSRB National Security Resources Board

NSRC Natural Science Research Council

NSRD National Standards Reference Data

NSRDC National Standards Reference Data System

NSRDF Naval Supply Research and Development Facility

NSRDL Naval Ship Research and Development Laboratory

NSRDS National Standard Reference Data System

NSREF National Society for Real Estate Finance

NSRF Nova Scotia Research Foundation

NSRMA National Star Route Mail Contractors Association

nsrp non-technical support real property

NSRP National States Rights Party

nsrpie non-technical support real property installed equipment

nsrt near-surface reference temperature

nss (NSS) normal saline solution

NSS National Sample Survey(or)(s); National Sculpture Society; National Secular Society (British); National Serigraph Society; National Slovak Society; National Speleological Society; National Stockpile Site; New Shakespeare Society; Newburgh and South Shore (railroad); Nitrogen Supply System

NSSA National Sanitary Supply Association; National School Sailing Association; National Science Supervisors Association; National Skeet Shooting Association; National Sportscasters and Sportswriters Association

NSSAR National Society of the Sons of the American Revolution

NSSC National School Safety Center; National Society for the Study of Communication

NSSCC National Space Surveillance Control Center

NSS Co Northern Steam Ship Company (New Zealand)

NSSE National Society for the Study of Education; National Study of School Evaluation

NSSEA National School Supply and Equipment Association

NSSF National Shooting Sports Foundation; Navy Submarine Support Facility

NSSFA National Single Service Food Association

NSSFC National Severe Storm Forecast Center; National Society of Student Film Critics

NSSFFA National Soft Serve and Fast Food Association

NSSFNS National Scholarship Service and Fund for Negro Students

NSSGA *Nicherin Shoshu Soka-Gakkai Academy* (international peace scociety)

NSSL National Severe Storms Laboratory

NSSNU National Spanish-Speaking Management Association

NSSN National Standard Shipping Note

NSSOPAC Naval Ordnance System Support Pacific

NSSP National Severe Storms Project

NSSR New School for Social Research

nsss nuclear steam system supply

NSST Northwestern Syntax Screening Test

nsst(s) no-smoking seat(s)

NSSU National Sunday School Union

NSSWC National Severe Storm Warning Center

nst nonslip thread; no sales tax

NST National Security Team (U.S. National Security Affairs Adviser, Secretary of Defense, Secretary of State); Newfoundland Standard Time; Nigata Sogo Television

NST New Straits Times

NSTA National School Transportation Association; National Science Teachers Association; National Shoe Traveler's Association

NSTAP National Strategic Targeting and Attack Policy

NSTC Nebraska State Teachers College; Nova Scotia Technical College

nstd nested

NSTF National Science Teachers Foundation; Near-Surface Test Facility

NSTI Norwalk State Technical Institute

NSTIC Naval Scientific and Technical Information

NSTL National Software Testing Laboratory; National Strategic Target Line

NSTP National Society of Tax Professionals; Nuffield Science Teaching Project

NS Tripos Natural Science Tripos

NSTS National Sea Training Schools; National Securities Trading System

N-study nuclear study

nsu nitrogen supply unit; non-specific urethritis

NSU Neckarsulmer Fahrzeugwerke (German—NSU Motorworks)

N-sub(s) nuclear-powered submarine(s)

NSUC North Staffordshire University College

N-super nuclear-powered supercarrier (naval vessel)

nsurg neurosurgeon; neurosurgery; neurosurgical

nsv negative sequence voltage; nuclear service vessel

n/sv nonautomatic self-verification

NSV National Socialist Vanguard (neo-Nazi group, Oregon and Idaho)

NSVG National Survey of the Vietnam Generation

NSVP National School Volunteer Program; National Student Volunteer Program

NSW New South Wales

NSWAP National Socialist White America Party

NSWC Naval Surface Weapons Center (USN); New South Wales Centre

NSWG Naval Special Warfare Group; New South Wales Government

NSWGR New South Wales Government Railways

NSWGTB New South Wales Government Tourist Bureau

NSWHC New South Wales Health Commission

NSWIER New South Wales Institute for Educational Research

NSWIT New South Wales Institute of Technology

NSWMSB New South Wales Maritime Services Board

NSWNA New South Wales Nurses Association

NSWO National Socialist Women's Organization

NSWP New South Wales Police; Non-Soviet Warsaw Pact

NSWPAG New South Wales Prisoners Action Group

NSWPP National Socialist White People's Party (formerly American Nazi Party)

NSWPTC New South Wales Public Transport Commission

NSWR New South Wales Reports

NSWRL New South Wales Rugby League

NSWRU New South Wales Rugby Union

NSWTA New South Wales Transport Association

NSWTF New South Wales Teachers Federation

NSY New Scotland Yard

NSYF Natural Science for Youth Foundation

nt narrower term; narrow topic; net terms; neurotransmitter; nit (unit of luminous intensity); no trace; nontight; normal temperature; nose tackle; not tested; not titled; number of teams

n't not

n/t net tonnage; new terms

n & t nose and throat

nt Northern Telecom

nt nel testo (Italian—in the text)

Nt nitron

NT National Theater; National Trust; Neutral Zone (Internet code); New Territories (Hong Kong); New Testament; Northern Territory; Northwest Territories

NT National Times (of Australia); *Ny Testamente* (Dano—Norwegian—New Testament)

N.T. Novum Testamentum (Latin—New Testament)

nta net tangible assets; nitrilotriacetic (phosphate substitute for detergents); nuclear test aircraft (NTA)

NTA Narcotics Treatment Administration; National Tattoo Association; National Tax Association; National Technical Association; National Tourist Association; National Travel Association; National Trust of Australia; National Tuberculosis Association; New Territories Administration; Northern Textile Association; Northern Trade Association

NTAA National Travelers Aid Association

NTAC Nederlandse Touring en Auto Club (Dutch—Netherlands Touring Auto Club)

NTAMS Northern Territory Aerial Medical Service

NTAs Nielsen Television Areas

NTATB Northwestern Truck Association and Tariff Bureau

ntavl not available

ntb non-tariff barrier(s); not to be

NTB National Theatre Board; National Tobacco Board

NTB Norsk Telegrambyra (Norwegian—Norwegian News Service)

ntba name(s) to be advised

NTBL Nuffield Talking Book Library (for the blind)

NtBuStnds National Bureau of Standards

ntc negative temperature coefficient

NTC National Teacher Corps; National Theatre Conference; National Training Council; National Travel Club; Naval Training Center; Nigerian Tobacco Company

NTCA National Tile Contractors Association; National Training Council of Australia; National Tribal Chairmen's Association; National Tuberculosis and Chest Association

NTCC Nimbus Technical Control Center

NTCCL Northern Territory Council for Civil Liberties

neural tube defect; non-tight door; noted

NT$ New Taiwan dollar

NTD National Theater of the Deaf; Nuclear Training Division

NTDA National Tire Distributors Association; National Trade Development Association; National Tyre Distributors Association

NTDC Naval Training Device Center

NTDS Naval Tactical Data System; Naval Technical Data System

NTDSC Nondestructive Testing Data Support Center

nte not to exceed

nte norte (Spanish—north)

NTE National Teacher Examination

NTEA National Truck Equipment Association

N-tec nuclear technology

NTEC Naval Training Electronics Center(s); Naval Training Equipment Center

ntep not to exceed price

ntepq not to exceed price quoted

N-terror(ism)(ist) nuclear terrorism; nuclear terrorist

N-test nuclear test(ing)

NTETA National Traction Engine and Traction Association

NTEU National Treasury Employees Union

NTF Narcotics Task Force; National Turkey Federation; Navy Technological Forecast

NTFA National Track and Field Association

NTFP National Task Force on Prostitution

ntfy notify

ntg nontoxic goiter

NTGB North Thames Gas Board

NT Gk New Testament Greek

Nth Netherlands

NTH Norges Tekniske Hogskole (Norwegian—Norwegian Technical University, Trondheim)

Nthb Northumberland

Nth BHH North Broken Hill Holdings

Nth country next country of a series acquiring nuclear power

nthn northern

NTHP National Trust for Historic Preservation

N-threat nuclear threat

nti noise-transmission impairment

NTI National Theatre Institute; Nielsen Television Index (tv rating)

NTIA National Telecommunications and Information Administration

NTIAC Nondestructive Testing Information Analysis Center

NTIATA National Tax Institute of America Tax Association

NTIC National Training and Information Center (Chicago based housing group); Nondestructive Testing Information Center (Battelle)

NTID National Technical Institute for the Deaf

NTIRA National Trucking Industrial Relations Association

NTIS National Technical Information Service; National Technical Information Service (U.S. Department of Commerce); Nippon Technical Information Service

NTISBDF National Technical Information Service Bibliographic Data File

NTISearch National Technical Information (on-line computer) Search Service

NTK Nippon Toshokan Kyokai (Japanese—Japan Library Association)

ntl no time lost

NTL National Tennis League; National Training Laboratories

NTLC National Tax Limitation Committee

NTLF National Taxpayers Legal Fund

NTLS National Truck Leasing System

ntm net ton mile; non-tariff measure(s)

Ntm Nottingham

NTMA National Terrazzo and Mosaic Association; National Tooling and Machining Association

NTMs National Technical Means (for verifying compliance with arms-control treaties)

NTMS National Topographic Map Series; Northern Territory Medical Service

NTMV National Technical Means of Verification

NTNP Natchez Trace National Parkway

nto not taken out; not tried on

nto neto (Spanish—net)

NTO National Tenants Organization; National Theatre Organisation (South Africa); Naval Transport Office(r)

NTON National Transparent Optical Network

N-town Norristown

ntp normal temperature and pressure; no title page

NTP National Toxicology Program; National Transportation Policy; Network Time Protocol

NTPC National Technical Processing Center; Navy Training Publications Center

NTPI National Tax Practice Institute

ntpl nut plate

ntr noise temperature ratio; non-typing reperforator

NTR National Tape Repository; Northern Test Range

Ntra Sra Nuestra Señora (Spanish—Our Lady)

NTRB Northern Territory Reserve Board (Australia)

NTRDA National Tuberculosis and Respiratory Disease Association

NTRL Naval Training Research Laboratory

NTRMA National Tile Roofing Manufacturers Association

NTRS National Therapeutic Recreation Society

N Tr Z North Tropical Zone

nts not to scale

nts (NTS) navigation(al) technology satellite

Nts Nantes

NTS National Technical School(s); National Traffic System; National Trust for Scotland; Naval Telecommunications System; Naval Torpedo Station; Naval Transport Service; Naval Transportation System; Nevada Test Site; Nitroglycerin Transdermal System

NTS Narodnyi Trudovoy Soyuz (Russian—National Labor Union)—anti-communist Russian exiles; *Nederlandse Televisie Stichting* (Dutch—Netherlands Television Foundation)

NTSA National Traffic Safety Agency

NT & SA National Trust and Savings Association

NTSB National Transportation Safety Board

NTSC National Television Standards Code; National Television Standards Committee; National Television System Committee; North Texas State College

NTSK Nordiska Tele-Satelit Kommitten (Nordic Committee for Satellite Telecommunications)

NTSSO Nevada Test Site Safety Office

NTSWG National Training School for Women and Girls

NTT New Technology Telescope; Nippon Telegraph and Telephone

NTTC National Tank Truck Carriers

NTTS Northern Territory Teaching Service

NTT & TTI National Truck Tank and Trailer Tank Institute

ntu nephelometric turbidity unit; normal trading unit; nuts to you

NTU National Taiwan University; National Taxpayers Union; National Teachers Union; Navy Toxicology Unit

NTUC National Trades Union Congress

NTUH National Taiwan University Hospital

ntv nerve tissue vaccine

NTV Nippon Television

NTVLRO National TV License Records Office (Bristol)

NTWA National Turf Writers Association

ntwistdg notwithstanding

nt wt net weight

NTWU National Textile Workers Union

NTX Navy Teletype Exchange

nty not this year

N-type Jungian intuitive type

nt yt not yet

NTZ Neutral Zone; North Temperate Zone

Ntzrm Nutzraum (German—cubic capacity)

nu name unknown; new; nose up; nuclear; number unobtainable; nurse

Nu Nusselt number

NU National Union; Naval Unit; Niagara University; Niue Island (Internet code); Northeastern University; Northern Union; Northwestern University; Norwich University

NU Naciones Unidas (Spanish—United Nations), *Nahdatul Ulama* (Indonesian—Muslim Scholars Party); *Nations Unies* (French—United Nations)

NUA Network User Address

NUAAW National Union of Agricultural and Allied Workers

NUABA National United Affiliated Beverage Association

NUAK Nordisk Union for Alkoholfri Trafik (Scandinavian Union for Alcohol-free Traffic)

NUAUS National Union of Australian University Students

NUAW National Union of Agricultural Workers

NUB National Union of Blastfurnacemen; National Unity Board

NUBE National Union of Bank Employees

nube(s) nubile(s)

NUBSO National Union of Boot and Shoe Operatives

nuc not under command; nuclear; nucleated; nucleus

NUC National University Consortium; National Urban Coalition; Naval Undersea Center

NUC National Union Catalog

NUCA National Utility Contractors' Association

nu-car prep new-car preparation

Nuc.E. Nuclear Engineer

NUCEA National University Continuing Education Association

nucex nuclear exercise

nuc(l) nuclear; nucleus

Nucl Data Nuclear Data

nuclex nuclear loadout exercise

Nucl Fusion Nuclear Fusion

Nucl Instrum Nuclear Instruments

Nucl Instrum Methods Nuclear Instruments and Methods

Nucl Phys Nuclear Physics

Nucl Sci Eng Nuclear Science and Engineering

NUCMC National Union Catalog of Manuscript Collections

nuco numerical code; numerical coding

NUCO National Union of Cooperative Officials

NUCOM National Union Catalog of Monographs

nuc phy nuclear physics

nucpwrd nuclear powered

Nuc Reg Com Nuclear Regulatory Commission

NUCS National Union of Christian Schools

NUCSTAT Nuclear Operational Status Report

NUCUS National Union of Conservative and Unionist Associations

NUCW National Union of Commercial Workers

nud nudism; nudist

nud nudnick (Yiddish—nuisance, pest)

NUDBTW National Union of Dyers, Bleachers, and Textile Workers

NUDE National Union of Domestic Employees

NUDET Nuclear Detonation Report

NUDETS Nuclear Detonation, Detection, and Reporting System

nudies nude films; nude magazines; nude shows

NUE Nuremberg, Germany (airport)

NUEA National University Extension Association

nuevo lek monetary unit of Albania

Nue Zel Nueva Zelandia (Spanish—New Zealand)

'nuf enough

NUF National Urban Fellows

NUFCOR Nuclear Fuels Corporation

NUFCW National Union of Funeral and Cemetery Workers

Nuff Nuffield College, Oxford

NUFLAT National Union of Footwear, Leather, and Allied Trades

nufp not used for production

'nuf said enough said

NUFTIC Nuclear Fuels Technology Information Center

NUFTO National Union of Furniture Trade Operatives

nug nuggar (cargo boat used on the Nile)

NUGMW National Union of General and Municipal Workers

NUHS New Utrecht High School

NUHW National Union of Hosiery Workers

NUI National University of Ireland (*Ollscoil na h-Eireann*); Norwegian Underwater Institute

NUIC National Urban Indian Council

NUIW National Union of Insurance Workers

NUJ National Union of Journalists

nuke nuclear (slang)

nuke leak nuclear radioactive leak

nukes nuclear explosives; nuclear power plants

nul no upper limit

nul (NUL) null character (data processing)

NUL National Union for Liberation; National Urban League; Northwestern University Library

NULBA National United Licensees Beverage Association

null null idle

Nulla Nullarbor Plain of southern South Australia and Western Australia

nullies nullifiers

NULWAT National Union of Leather Workers and Allied Trades

num number; numbered; numbering; numeracy (see numcy); numeral(s); numeration(s); numerical; numerologist; numerology

num numero(s) [Portuguese or Spanish—number(s)]

Num The Fourth Book of Moses, called *Numbers*

NUM National Union of Mineworkers; New Ulster Movement

NUMA National Underwater and Marine Agency

NUMAS National Multifactor Assessment System

numb. numbered

Number 10 Number 10 Downing Street, official residence of the British Prime Minister

numcy numeracy (ability to count)

NUMEC Nuclear Materials and Equipment Corporation

numer numeral; numerative; numerical

numer order numerical order

numis numismatics

numism numismatic(s); numismatist

NUMMI New United Motor Manufacturing Inc

num order numerical order

NUMW National Union of Mine Workers

nuna not used on next assembly

nunc nuncupative

NUOS Naval Underwater Ordnance Station

NUP Negro Universities Press

NUPAC Nuclear Packaging Inc

NUPBPW National Union of Printing, Bookbinding, and Paper Workers

NUPE National Union of Public Employees

NUPGE National Union of Provincial Government Employees

NUPI Norsk Utenrikspolitisk Institutt (Norwegian—Norwegian Foreign Policy Institute)

nuplex nuclear-powered complex (of manufacturers)

NUPSA National Union of Pharmaceutical Students of Australia

NUPT National Union of Press Telegraphists

NUPW National Union of Planning Workers; National Union of Plantation Workers

NUR National Union of Railwaymen

NURA National Union of Ratepayers' Associations

NURC National Union of Retail Confectioners

NURDA National Urban and Regional Development Authority

NURE National Uranium Resource Evaluation (ERDA program)

NURT National Union of Retail Tobacconists

nus nuclear upper stage

NUS National Union of Seamen; National Union of Students; National University of Singapore; Nuclear Utility Service(s)

nusar nuclear sweep and radar

NUSAS National Union of South African Students

NUSC Naval Underwater Systems Center

NUSEC National Union of Societies for Equal Citizenship

NUSL Navy Underwater Sound Laboratory

NUSMWCHDE National Union of Sheet Metal Workers, Coppersmiths, Heating and Domestic Engineers

NUSRL Navy Underwater Sound Reference Laboratory

NUSS National Union of School Students; National Union of Small Shopkeepers

nusum numerical summary

Nu T Newcastle-upon-Tyne

N-u-T Newcastle-upon-Tyne

NUT National Union of Teachers (Great Britain)

NUTAT Nordisk Union for Alkoholfri Trafik (Nordic Union for Alcohol-free Traffic)

NUTAW National Union of Textile and Allied Workers

nu-tec nuclear detection (radiation monitoring device)

NUTGW National Union of Tailors and Garment Workers

NUTI Northwestern University Traffic Institute

NUTIS Numerical and Textual Information System

NUTN National Union of Trained Nurses

nutr nutrition

nuts (NUTS) nuclear-utilization theories

NUU New University of Ulster

nuv nuvaerende (Dano-Norwegian—present)

NUVB National Union of Vehicle Builders

NUWA National Unemployed Workers' Association

NUWAX Nuclear Weapon Accident Exercise

NUWC Naval Undersea Warfare Center

NUWT National Union of Women Teachers

NUWW National Union of Women Workers

nux vom nux vomica

Nuyorican(s) New York Puerto Rican(s)

nv naked vision; needle valve; new version; number of variables

nv (NV) nuclear vitrification

n-v non-vaccinated; non-veteran; non-voting

n/v nuclear vessel

n & v nausea and vomiting

nv. novicius (Latin—new, recent)

nV nanovolt

NV Nevada; Nevada Operations Office; Nord-Viscount

NV Naamloze Vernootschap (Dutch—corporation); *Naviera Vascongada* (Basque Navigation Company); *Norske Veritas* (Norwegian—Norwegian Register of Shipping)

nva near visual acuity

nva nueva (Spanish—new)

NVA National Viatical Association; North Vietnamese Army

NVAiO Norske Videnskaps Akademi i Oslo (Norwegian—Norwegian Academy of Science and Letters in Oslo)

NVATA National Vocational Agricultural Teachers Association

NVB National Volunteer Brigade

NVB Nederlandse Vereniging van Bedriftsarchivarissen (Dutch—Netherlands Association of Business Archivists); *Nederlandse Vereniging van Bibliothekarissen* (Dutch—Netherlands Library Association)

NVBF Nordisk Viedenskabeligt Bibliotekarieforbund (Nordic Federation of Research Librarians)

nvc non-verbal communication

NVC National Victim Center; National Video Clearinghouse; National Violence Commission

NVCA National Vehicle Conversion Association; National Venture Capital Association

nvCJD new variant Creutzfeldt-Jakob Disease

nvd night-viewing device; night-vision device

NVDA National Volunteer Defense Army

nvebw non-vacuum electron beam welding

NVF National Volunteer Force

NVFC National Vulcanized Fibre Company

nvg null voltage generator

NVGA National Vocabulary Guidance Association; National Vocational Guidance Association

nvh noise, vibration, hardness (problems)

NVI Nordic Volcanological Institute

NVICP National Vaccine Injury Compensation Program

Nvk Narvik

NVL Night Vision Laboratory

NVLA National Vehicle Leasing Association

nvm non-volatile matter

NVMA National Veterinary Medical Association

NVNS Naamloze Vernootschap Nederlandsche Spoorwagen (Dutch—Netherlands Railway Corporation)

nvo non-vessel operating

NVO Nevada Operations Office; Northern Variety Orchestra

nvocc (NVOCC) non-vessel operating common carrier

NVOCC New Version Ocean Container Control; New Version Overseas Container Control; Non-Volatile Ocean Container Control

NVOILA National Voluntary Organization for Independent Living for the Aging

NVOO Nevada Operations Office

nvp natural vegetable powder (powdered psyllium seed and dextrose laxative)

NVPA National Visual Presentation Association

NVPO Nuclear Vehicle Projects Office (NASA)

nvr no voltage release

NVRC National Victims Resource Center

NVRS National Vegetable Research Station

nvs neutron velocity selector

NVS Night Vision System

NvSEA Nevada Society of Enrolled Agents

NVT National Veld Trust

NVTS National Vocational Training Service

NVV Nederlands Verbond van Vakverenigingen (Dutch—Netherlands Trade Union Federation)

NV/VC North Vietnamese/Vietcong

nw nanowatt; net worth; no wind; number of weeks

n/w net weight

Nw New (sometimes confused with NW—Northwest)

NW Chicago & North Western Railway; Noah Webster; Norfolk & Western (railroad); Northern Wings Ltd; North Wales; Northwest; Northwest Airlines

N & W Norfolk & Western (railroad)

NW noordwest (Dutch—northwest); *Nordwesten* (German—northwest)

NWI, NW2, etc. Northwest One, Northwest 2, etc; (London postal zones)

NWA National Wrestling Alliance; Northwest Airlines; Northwest(ern) Australia

NWAA National Wheelchair Athletic Association

nwab necks with anybody

NWAC National Womens Advisory Council

NWAF New World Archeological Foundation

NWAH & ACA National Warm Air Heating and Air Conditioning Association

N-war nuclear war(fare)

NWASCO Nation Water and Soil Conservation Organization

N-waste nuclear (radioactive) waste

nwb non-weight bearing

nWb nano Weber

NWB National Westminster Bank

NWBA National Wheelchair Basketball Association

NWBW National Women's Bowling Writers

nwc nuclear war capability

Nwc Newcastle-upon-Tyne

NWC National Wages Council; National War College; National Water Commission; National Writers Club; Naval War College; Naval Weapons Center; Navy Widow's Certificate

NWCC Northern Wyoming Community College

NWCCL Naval Weapons Center—Corona Laboratories

NWCF Naval War College Foundation; New Waste Calcining Facility

NWCS NATO-wide Communications System

NWCTU National Woman's Christian Temperance Union

NWD New World Dictionary

NWDA National Wholesale Druggists' Association; National Wine Distribution Association

nwdc navigation weapon-delivery computer

NWDR Nordwestdeutscher Rundfunk (German—Northwest German Broadcasting System)

NWEAF National Women's Economic Alliance Foundation

N-weapon(ry) nuclear weapon(ry)

N-weapon(s) nuclear weapon(s)

NWEB Northwestern Electricity Board (UK)

NWEF National Women's Education Fund; Naval Weapons Evaluation Facility; Naval Weapons Evaluation Force

NWES New World Exploration Society

NWF National Welfare Fund; National Wildlife Federation

NWF National War Formulary

NWFA National Wood Flooring Association

NWLA Northern Woods Logging Association

Nwfld Newfoundland

NWFP North-West Frontier Province

nwg national wire gauge

NWG National Welfare Growth; Neighborhood Watch Group

NWGA National Wheat Growers Association; National Wool Growers Association

nwh normal working hours

NWI Netherlands West Indies

NWIDA North West Industrial Development Association

NWIP Naval Warfare Instruction Publication

NWIRP Naval Weapons Industrial Reserve Plant

NWJA National Wholesale Jewelers Association

nwl natural wavelength

NWL Naval Weapons Laboratory

NWLB National War Labor Board

NWLC National Women's Law Center

NWLEE Northwest Law Enforcement Equipment

NWLF New World Liberation Front (terrorists)

NWly northwesterly

nwm nuclear waste materials

nw/m net words per minute

NWM Nuclear Waste Management

NWMA Northwest Mining Association

NWMC Northwest Michigan College

NWMCC Nuclear Waste Materials Characterization Center

N/Wmn Night Watchman

NWMPA North Wales Master Printers' Alliance

NWMS Northwest Medical Service

Nw Ned Nieuw Nederland (Dutch—New Netherlands)

NWNSA National Women's Neckwear and Scarf Association

NWNT North Wales Naturalists' Trust

NWO New World Order; Nuclear Weapons Office(r)

nwoc new woman on campus

NWOO NATO Wartime Oil Organization

n-word nonce word (word coined for the occasion)

NWORG North Western Operational Research Group

NWP Naval Weapons Plant; North West Provinces

NWPAG NATO Wartime Preliminary Analysis Group

NWPC National Women's Political Caucus

NWPCA National Wooden Pallet and Container Association

NWPF New Waste Processing Facility

NWPFC Northwest Pacific Fisheries Commission

NWRA National Waterbed Retailers Association; National Wheel and Rim Association

NWRC National Weather Records Center; National Wildhorse Research Center; National Wildlife Research Center; Naval War Research Center (USN)

Nwprt News Newport News

NWPSC Northwestern Public Service Company

NWQAO Naval Weapons Quality Assurance Office

NWQI National Water Quality Inventory (EPA)

NWQSS National Water Quality Surveillance System (EPA)

nwr next word request

NWR National Waste Repository; National Welfare Rights; National Wildlife Refuge; National Wildlife Reserve; Nuclear Weapon Report

NWRB National Waste Repository Basalt

NWREL Northwest Regional Educational Laboratory

NWRF Naval Weather Research Facility

NWRLF New World Radical Liberation Front

NWRO National Welfare Rights Organization

NWRS National Wildlife Refuge System

nws normal water surface; nosewheel steering

NWS National Weather Service; National Weather Station; Naval Weapons Station; New World Symphony (Miami, Florida); Nimbus Weather Satellite; Norfolk & Western Southern (railways)

NWSA National Welding Supply Association; National Woman Suffrage Association

NWSC National Weather Satellite Center; Naval Weather Service Command

NWSCA National Water and Soil Conservation Authority

NWSCO National Water and Soil Conservation Organization

NWSF Nuclear Weapons Storage Facility (USA)

NWSO Naval Weapons Services Office

NWSS Nuclear Weapons Support Section (USA)

NWSY Naval Weapons Station—Yorktown, Va

nwt net weight; nonwatertight

NWT Northwest Territories

nwtb new water-tube boiler(s)

NWTB Northwestern Tariff Bureau

nwtd nonwatertight door

nwtdb new water-tube donkey boiler(s)

NWTEC National Wool Textile Export Corporation

N.W. Terr. North West Territory

NWTS National Waste Terminal Storage; Naval Weapons Test Station

NWTSR-1 NWTS Repository No 1 (high-level waste in a dome)

NWTSR-2 NWTS Repository No 2 (spent fuel in bedded salt)

NWTS-RSP NWTS Repository Sealing Program

nwu nosewheel up

NWU National Workers Union; National Writers Union; Nebraska Wesleyan University

NWUS Northwestern United States

NWVP Nuclear Waste Vitrification Program

NWWA National Water Well Association

NWWDA National Wood Window and Door Association

nwy newly

nx nonexpendable; regional lymph nodes can't adequately be assessed (symbol)

NX Notice to Marines

NXD Non-Executive Director

NXDO Nike-X Development Office (USA)

NXMIS Nike-X Management Information Office

nx mo next month

nxn no Christian name

NXPM Nike-X Project Manager

NXPO Nike-X Project Office

nxr non-crossing rule

NXSO Nike-X Support Office

nxt next

nxt ssn next season

nx wk next week

nx yr next year

ny new year; no year; nylon

Ny Niles; Nylan

NY New York; New York Airways (2-letter code); New Yorker; North Yorkshire

NY Neuyork (German—New York); *New Yorker* (magazine); *Nieuw York* (Dutch—New York); *Nova Iorque* or *Nova York* (Portuguese—New York); *Nueva York* (Spanish—New York)

Nya Nyasaland

NYA National Youth Administration; Neighborhood Youth Association; New York Aquarium

NYAB National Youth Advisory Board

NYAC New York Athletic Club

NYADS New York Air Defense Sector

NYAM New York Academy of Medicine

NYANA New York Association for New Americans

NYAO New York Assay Office

NYAP New York Assembly Program

Nyas Nyasaland

NYAS New York Academy of Science; New York Asian Society

NYATI New York Agricultural and Technical Institute

NYBFU New York Board of Fire Underwriters

NYBG New York Botanical Garden

NYBSBC New York Bureau of State Building Codes

NYC National Yacht Club; Neighborhood Youth Corps; Newburgh Yacht Club; New York Central (railroad); New York City; New York Coliseum

NYCA National Youth Council of Australia; New York City Affiliate (of the National Council on Alcoholism)

NYCB New York City Ballet

NYCC New York Corset Club; New York Cultural Center

NYCCC New York City Community College

NYCCCC New York City's Citizens Crime Commission

NYCCIW New York City Correctional Institution for Women

NYCDC New York City Department of Correction

NYCE New York Cash Exchange; New York Cocoa Exchange; New York College of Education; New York Commodity Exchange; New York Cotton Exchange; New York Curb Exchange

NYCERS New York City Employees Retirement System

NYCHA New York City Housing Authority; New York Clearing House Association

NY-CHI New York–Chicago

NYCJG Nikka Yuko Centennial Japanese Garden (Lethbridge, Alberta)

NYCMA New York City Metropolitan Area; New York Clothing Manufacturers Association

NYCMD New York Contract Management District

NYCMEO New York City Medical Examiner's Office

NYCMSL New York County Medical Society Library

NYCNHA New York City Nursing Home Association

NYCO New York City Opera

NYCOC New York City Opera Company

NY Col New York Coliseum

NYCPB New York Consumer Protection Board

NYCPD New York City Police Department

NYCPM New York City Police Museum

NYCS New York Chamber Symphony; New York Choral Society

NYCSA New York Coat and Suit Association

NYCSCE New York Coffee, Sugar, and Cocoa Exchange

NYCSE New York Coffee and Sugar Exchange

NYC & ST L New York, Chicago & St Louis (Nickel Plate Line)

NYCT New York Community Trust

NYCTA New York City Transit Authority

NYCTN New York Cotton Exchange

NYCWRU New York Cooperative Wildlife Research Unit

nyd not yet dead; not yet diagnosed; not yet dressed (poultry)

NYDCC New York Drama Critics Circle

NYDMC New York Downstate Medical Center

NYDR New York Dock Railway

NYF New York Foundation

NYFCC New York Film Critics Circle

NYFDM New York Fire Department Museum

NYFE New York Futures Exchange

NYFH New York Foundling Hospital

NYFIRO New York Fire Insurance Rating Organization

NYFQ *New York Folklore Quarterly*

NYFUO New York Federation of Urban Organizations

NYFWA New York Financial Writers Association

NYGASP New York Gilbert and Sullivan Players

NYGC New York Governor's Conference

NYGS New York Graphic Society

NYHA New York Heart Association (classification)

NYH-CMC New York Hospital—Cornell Medical Center

NYHD New York House of Detention

NY Hist Soc New York Historical Society

NYHS New York Herpetological Society; New York Historical Society

NYI New York Institute (of Photography)

NYIAS New York Institute of the Aerospace Sciences

NYIBS New York International Bible Society

NYIE New York Insurance Exchange

NYIH New York Institute for the Humanities

NYIT New York Institute of Technology

N Yk New York

NYK Nippon Yusen Kaisha Line

NYKU Nippon Yusen Kaisha (container) Unit

nyl nylon

NYLA New York Library Association

NY-LAX New York–Los Angeles

NY & LB New York & Long Branch (railroad)

nylfin nylon finish

NYLS National Yacht Listing Service; New York Law School

NYLTI National Youth Leadership Training Institute

nym *nymon* (Greek—name)— as in antonym, homonym, pseudonym, synonym, etc.

NYMC New York Maritime College

NYME New York Mercantile Exchange

NY Met New York City Metropolitan Correctional Center

NY-MIA New York–Miami

nympho nymphomania—nymphomaniac; nymphomaniacal

NYNEX New York/New England Exchange (telephone company)

N Y N H & H New York, New Haven and Hartford (railroad)

NY-NO New York–New Orleans

NYNP Northern Yukon National Park

NYNR New York National Review

nyo not yet out

NYO National Youth Orchestra

NYOC New York Opera Company

NYOGB National Youth Orchestra of Great Britain

NYOL New York Opera Library

N Yorks North Yorkshire

NYORT New York Opera Repertory Theatre

NYOSL New York Oceans Science Laboratory

NYOTBC New York Off-Track Betting Corporation

NYOW National Youth Orchestra of Wales

NYO & W New York, Ontario and Western (railroad)

nyp not yet published

NYP Neighborhood Youth Program; New York Philharmonic (orchestra)

NYPA New York Port Authority

NYPAA National Yellow Pages Agency Association

NYPC New York Pigment Club

NYPD New York Police Department

NYPDis New York Procurement District (U.S. Army)

NYPE New York Port of Embarkation; New York Produce Exchange

NYPFO New York Procurement Field Office (USAF)

NYPHR New York Physicians for Human Rights

NYPIRG New York Public Interest Research Group

NYPL New York Public Library

NYPLA New York Patent Law Association

NYPM New York Pro Musica

NYPO New York Philharmonic Orchestra

NYPs Neighborhood Youth Programs

NYPS New York Paleontological Society; New York Psychiatric Society; New York Publishing Society

NYPSS New York Philharmonic-Symphony Society

NYPUM National Youth Program Using Minibikes

Nyq Nyquist (data-processing time or rate)

nyr not yet returned; nuclear yield requirement

NYR National Young Republicans

NYRA National Yacht Racing Association; New York Racing Association

NYRB New York Review of Books

NYRF National Young Republican Federation

NYRG New York Rubber Group

NYRM New York Reformatory for Men

NYRMA New York Raincoat Manufacturers Association

NYRPG New York Rights and Permission Group

NYRs National Young Republicans

NYRW New York Reformatory for Women (Westfield Farm)

NYS New York Shavians; New York State

NYSA New York Shipping Association; New York State Assembly

NYSAA New York State Aviation Association

NYSAC New York State Athletic Commission

NYSAIS New York State Association of Independent Schools

NYSAJC New York State Association of Junior Colleges

NY-SAN New York–San Diego

NYSASBO New York State Association of School Business Officials

NYSASDA New York State Atomic and Space Development Authority

NYSASSRS New York State Association of Service Stations and Repair Shops

NYSAVC New York State Audio-Visual Council

NYSBA New York State Bar Association

NYSBB New York State Banking Board

NYSBC New York State Barge Canal (modern extension of Erie Canal)

NYSC New York Shipbuilding Corporation

NYSCC New York State Crime Commission

NYSCCJ New York State Coalition for Criminal Justice

NY Sch Indus Rel New York State School of Industrial Relations (Cornell University)

NYSCSDA New York State Council of School District Administrators

NYSDCS New York State Department of Correctional Services

NYSE New York Stock Exchange

NYSEA New York Society of Enrolled Agents

NYSERDA New York State Energy Research and Development Authority

NYSES New York State Employment Service

NYSF New York Shakespeare Festival

NY-SFO New York–San Francisco

NYSILL New York State Inter Library Loan (network)

NYSL New York Society Library

NYSM New York State Museum

NYSMM New York State Maritime Museum (New York City)

NYSNA New York State Nurses' Association

NYSNACC New York State Narcotic Addiction Control Commission

NYSNC New York State Narcotics Commission

NYSNI New York State Nutrition Institute

NYSO New York String Orchestra

NYSP New York School of Printing; New York State Police

NYSPA New York State Power Authority

NYSPGA New York State Personnel and Guidance Association

NYSPI New York State Psychiatric Institute

NYSSA New York Skirt and Sportswear Association

NYSSILR New York State School of Industrial and Labor Relations

NYSSMA New York State School Music Association

NYSTA New York State Teachers Association; New York State Thruway Authority

NY Sup New York Supreme Court Reports

NYSUT New York State United Teachers

NYS & W New York, Susquehanna and Western (railroad)

NYT National Youth Theatre

NYT The New York Times

NY Thru New York Thruway

NY Times Bk R New York Times Book Review

NYTNS New York Times News Service

NY-TOR New York–Toronto

NYT/TS New York Turtle/Tortoise Society

NYTU New York Theological Union

NYU New York underworld (used in law-enforcement circles); New York University

NYUL New York University Library

NYUMC New York University Medical Center; New York Upstate Medical Center

NYUP New York University Press

NYUSM New York University School of Medicine

NYWASH Navy Yard, Washington

NYY New York Yankees

NYYC New York Yacht Club

NYZP New York Zoological Park

NYZS New York Zoological Society

Nz Nuñez

NZ New Zealand; New Zealand dollar; New Zealand National Airways (2-letter coding); Novaya Zemlya

N-Z Nike-Zeus

NZ Nueva Zelandia (Spanish—New Zealand); *Nouvelle-Zélande* (French—New Zealand)

NZAA New Zealand Amateur Athletic Association; New Zealand Antique Arms Association; New Zealand Auto Association

NZAB New Zealand Association of Bacteriologists

NZABC New Zealand Audit Bureau of Circulation

NZABM New Zealand Anglican Board of Missions

NZAC New Zealand Accommodation Council; New Zealand Alpine Club

NZACA New Zealand Amateur Cycling Association

NZACAU New Zealand Athletics, Cycling, and Axemens Union

NZACE New Zealand Association for Community Education

NZACU New Zealand Auto Cycle Union

NZADS New Zealand Association for Disabled Skiers

NZAEC New Zealand Atomic Energy Committee

NZAEI New Zealand Agricultural Engineering Institute

NZAF New Zealand Authors Fund; New Zealand Aviation Federation

NZAHBS New Zealand Arab Horse Breeders Society

NZAHPER New Zealand Association of Health, Physical Education, and Recreation

NZALT New Zealand Association of Language Teachers

NZAPA New Zealand Airline Pilots Association

NZARA New Zealand Amateur Rowing Association

NZARE New Zealand Association for Research on Education

NZARP New Zealand Antarctic Research Programme

NZART New Zealand Amateur Radio Transmitters Association

NZAS New Zealand Aluminum Smelters; New Zealand Antarctic Society; New Zealand Arthritis Society; New Zealand Association of Scientists

NZASA New Zealand Amateur Swimming Association; New Zealand Asian Studies Association

NZASC New Zealand Administrative Staff College; New Zealand Army Service Corps; New Zealand Association of Soil Conservators

NZASF New Zealand Association of Small Farmers

NZASW New Zealand Association of Social Workers

NZATD New Zealand Association of Training and Development

NZAWA New Zealand Air Women's Association

NZb New Zealand black (mice hybrids)

NZB New Zealand Ballet

NZBA New Zealand Bankers Association; New Zealand Bowling Association

NZBC New Zealand Ballet Company; New Zealand Book Council; New Zealand Broadcasting Corporation

NZBCSO New Zealand Broadcasting Corporation Symphony Orchestra

NZBF New Zealand Basketball Federation

NZBIE New Zealand Bureau of Importers and Exporters

NZBS New Zealand Broadcasting Service

NZBTO New Zealand Book Trade Organisation

NZC New Zealand Certificate

NZCAR New Zealand Civil Aviation Regulations

NZCAS New Zealand Clean Air Society

NZCC New Zealand Chamber of Commerce; New Zealand Cricket Council

NZCD New Zealand Certificate in Draughting

NZCDC New Zealand Cooperative Dairy Company

NZCE New Zealand Certificate in Engineering

NZCEA New Zealand Combined Educational Associations

NZCER New Zealand Council for Educational Research

NZCF New Zealand Cadet Forces; New Zealand Cat Fancy Association; New Zealand Cycling Federation

NZCG New Zealand Chemists Guild

NZCGF New Zealand Coast Guard Federation

NZCGP New Zealand College of General Practitioners

NZCGS New Zealand standard Classification of all Goods and Services

NZCH New Zealand Cement Holdings

NZCLA New Zealand Childrens Literature Association

NZCLS New Zealand Certificate of Land Surveying

NZCMA New Zealand Cable Makers Association; New Zealand Concrete Masonry Association

NZCMF New Zealand Coal Merchants Federation

NZCO New Zealand Concert Orchestra

NZCOSS New Zealand Council of Social Services

NZCRA New Zealand Coal Research Association; New Zealand Concrete Research Association

NZCRS New Zealand Council for Recreation and Sports

NZCS New Zealand Certificate in Science; New Zealand Certificate in Statistics; New Zealand Computer Society

NZCSS New Zealand Council of Social Services

NZCTF New Zealand Cycle Traders Federation

NZCTOA New Zealand Container Terrninal Operators Association

NZCUL New Zealand Credit Union League

NZCWI New Zealand Country Women's Institutes

NZd New Zealand dollar

NZ$ New Zealand dollar(s)

NZD New Zealand Division; New Zealand Dollar; New Zoo Developments

NZDA New Zealand Dairy Association; New Zealand Dental Association; New Zealand Department of Agriculture; New Zealand Dietetic Association

NZDB New Zealand Dairy Board

NZDC New Zealand Dental Corps

NZDCMBA New Zealand Dairy Confectionary and Mixed Biscuits Association

NZDCS New Zealand Department of Census and Statistics

NZDE New Zealand Department of Education

NZDFA New Zealand Deer Farmers Association

NZDLS New Zealand Department of Lands and Survey

NZDRI New Zealand Dairy Research Institute

NZDS New Zealand Drama School

NZDSIR New Zealand Department of Scientific and Industrial Research

NZDT New Zealand Daylight Time

NZDVA New Zealand Dunkirk Veterans Association

NZDXRA New Zealand DX Radio Association

NZE New Zealand Engineers; New Zealand English

NZEA New Zealand Esperanto Association

N Zeal New Zealand(er)

NZEAS New Zealand East Asia Service; New Zealand Educational Administration Society

NZEB New Zealand Electricity Board

NZECF New Zealand Electrical Contractors Federation

NZED New Zealand Electricity Department

NZEF New Zealand Employees Federation; New Zealand Expeditionary Force

NZEI New Zealand Educational Institute; New Zealand Electronics Institute

NZer New Zealander

NZERF New Zealand Engine Research Foundation; New Zealand Equine Research Foundation

NZES New Zealand Ecological Society

NZESA New Zealand Education Standards Association; New Zealand European Shipping Association

nzf near zero field

NZFA New Zealand Football Association

NZFB New Zealand Foundation for the Blind

NZFCA New Zealand Farmers Cooperative Association; New Zealand Freezing Companies Association

NZFCDC New Zealand Farmers Cooperative Distributing Company

NZFCMA New Zealand Ferro-Cement Marine Association

NZFF New Zealand Farmers Fertiliser (company); New Zealand Federated Farmers; New Zealand Fruitgrowers Federation

NZFHA New Zealand Finance Houses Association

NZFKTA New Zealand Free Kindergarten Teachers Association

NZFKU New Zealand Free Kindergarten Union

NZFL New Zealand Federation of Labor

NZFMA New Zealand Ferrocement Marine Association

NZFMC New Zealand Federation of Master Cleaners

NZFMRA New Zealand Fertiliser Manufacturers Research Association

NZFP New Zealand Forest Products

NZFPA New Zealand Family Planning Association

NZFRI New Zealand Forest Research Institute

NZFS New Zealand Film Service; New Zealand Forest Service

NZFUW New Zealand Federation of University Women

NZFWA New Zealand Farm Workers Association

nzg near zero gravity

NZG New Zealand Government

NZG New Zealand Gazette

NZGA New Zealand Gliding Association; New Zealand Golf Association; New Zealand Grasslands Association

NZGenS New Zealand Genetical Society

NZGR New Zealand Government Railways

NZGS New Zealand Geographical Society; New Zealand Geological Society; New Zealand Geological Survey

NZGTB New Zealand Government Tourist Bureau

NZGTC New Zealand Government Travel Commissioner

NZGTO New Zealand Government Tourist Office

NZH New Zealand Helicopters

NZH New Zealand Herald

NZHA New Zealand Hockey Association

NZHC New Zealand High Commission

NZHF New Zealand Heart Foundation

NZHGA New Zealand Hang-Gliding Association

NZHI New Zealand Horological Institute

NZHPT New Zealand Historic Places Trust

NZHS New Zealand Horse Society

NZI New Zealand Insulators; New Zealand Insurance

NZIA New Zealand Institute of Architects; New Zealand Irrigation Association

NZIAS New Zealand Institute of Agricultural Science

NZIC New Zealand Institute of Chemistry; New Zealand Intelligence Council

NZICFM New Zealand Institute of Credit and Financial Management

NZICM New Zealand Institute of Credit Management

NZID New Zealand Institute of Draughtsmen

NZIDA New Zealand Invention Development Authority

NZIDC New Zealand Industrial Design Council

NZIE New Zealand Institute of Engineers

NZIELEC New Zealand Institute of Electricians

NZIEPC New Zealand Indonesia Economic Promotion Council

NZIER New Zealand Institute of Economic Research

NZIET New Zealand Institute of Engineering Technicians

NZIF New Zealand Institute of Foresters

NZIFST New Zealand Institute of Food Science and Technology

NZIG New Zealand Institute of Gases

NZIH New Zealand Institute of Horticulture

NZIHVE New Zealand Institute of Heating and Ventilation Engineers

NZIIA New Zealand Institute of International Affairs

NZIIS New Zealand Institute of Industrial Safety

NZILA New Zealand Institute of Landscape Architects

NZIM New Zealand Institute of Management; New Zealand Institute of Mining

NZIME New Zealand Institute of Mechanical Engineers

NZIMP New Zealand Institute of Medical Photography

NZIP New Zealand Institute of Printing

NZIPA New Zealand Institute of Public Administration

NZIPM New Zealand Institute of Personnel Management

NZIPRA New Zealand Institute of Parks and Recreation Administration

NZIPS New Zealand Institute of Purchasing and Supply

NZIRE New Zealand Institute of Refrigeration Engineers

NZIS New Zealand Information Service; New Zealand Institute of Surveyors

NZISM New Zealand Institute of Safety Management

NZIT New Zealand Institute of Travel

NZIUW New Zealand Industrial Union of Workers

NZIW New Zealand Institute of Welding

NZJCB New Zealand Joint Communications Board

NZJPA New Zealand Japan Parliamentary Association

NZJU New Zealand Journalists Union

NZK Noord Zee Kanaal (Dutch—North Sea Canal—linking the Atlantic with Amsterdam

NZKC New Zealand Kennel Club

NZKVA New Zealand Korean Veterans Association

NZL New Zealand Line

NZLA New Zealand Legal Association; New Zealand Library Association; New Zealand Loggers Association

NZLCC New Zealand Litter Control Council

NZLF New Zealand Literary Fund

NZLIRA New Zealand Logging Industry Research Association

NZLL New Zealand Light Leathers

NZLP New Zealand Labour Party

NZLR New Zealand Law Reports

NZLS New Zealand Securities; New Zealand Law Society; New Zealand Library School; New Zealand Library Service

NZLTA New Zealand Lawn Tennis Association

NZMA New Zealand Medical Association; New Zealand Modelling Association; New Zealand Motel Association; New Zealand Motorcycle Association

NZMAF New Zealand Ministry of Agriculture and Fisheries

NZMB New Zealand Meat Board

NZMBF New Zealand Master Builders Federation

NZMC New Zealand Maori Council; New Zealand Medical Corps

NZMCA New Zealand Motor Caravan Association

NZMF New Zealand Manufacturers Federation; New Zealand Military Forces; New Zealand Motel Federation; New Zealand Music Federation

NZMFA New Zealand Master Floorcovering Association

NZMGA New Zealand Mountain Guides Association

NZMGC New Zealand Marriage Guidance Council

NZMJ New Zealand Medical Journal

NZMOT New Zealand Ministry of Transport

NZMPH New Zealand Meat Packing House

NZMRC New Zealand Medical Research Council

NZMS New Zealand Mapping Service; New Zealand Meteorological Service

NZMSC New Zealand Mountain Safety Council

NZMSS New Zealand Marine Sciences Society

NZMTCB New Zealand Motor Trade Certification Board

NZMTMA New Zealand Methods Time Measurement Association

NZMWA New Zealand Maori Wardens Association

NZM & WB New Zealand Meat and Wool Board

NZMWU New Zealand Meat Workers Union

NZNA New Zealand Nurserymens Association; New Zealand Nurses Association

NZNAC New Zealand National Airways Corporation

NZNCC New Zealand Nature Conservation Council

NZNCOR New Zealand National Committee on Oceanic Research

NZNF New Zealand Neurological Foundation

NZNFU New Zealand National Film Unit

NZNPA New Zealand Newspaper Proprietors Association

NZNTA New Zealand National Travel Association

NZOA New Zealand Optometrical Association

NZOC New Zealand Opera Company

NZOCGA New Zealand Olympic and Commonwealth Games Association

NZOI New Zealand Oceanographic Institute

NZ£ New Zealand pound

NZP National Zoological Park; New Zealand Pacific; New Zealand Players; New Zealand Police

NZPA New Zealand Police Association; New Zealand Ports Authority; New Zealand Press Association

NZPARS New Zealand Prisoners Aid and Rehabilitation Society

NZPB New Zealand Pony Breeders; New Zealand Potato Board

NZPBA New Zealand Power Boat Association; New Zealand Publishers' Association

NZPBR New Zealand Pony Breeders Register

NZPBS New Zealand Pony Breeders Society

NZPC New Zealand Peace Council; New Zealand Planning Council; New Zealand Pony Club; New Zealand Press Council; New Zealand Print Council

NZPCA New Zealand Pony Club Association; New Zealand Portland Cement Association

NZPCI New Zealand Prestressed Concrete Institute

NZPEA New Zealand Port Employers Association

NZPGMF New Zealand Post Graduate Medical Federation

NZPM New Zealand Paper Mills

NZPMS New Zealand Plumbers Merchants Society

NZPO New Zealand Post Office

NZPOA New Zealand Purchasing Officers Association

NZPPA New Zealand Professional Photographers Association

NZPPTA New Zealand Post Primary Teachers Association

NZPS New Zealand Park Service; New Zealand Police Service

NZPSA New Zealand Political Studies Association; New Zealand Public Service Association

NZPsS New Zealand Psychological Society

NZPTA New Zealand Parent-Teachers Association

NZPTO New Zealand Public Trust Office

NZQHA New Zealand Quarter-Horse Association

NZR New Zealand Railways

NZRC New Zealand Red Cross

NZRDXL New Zealand Radio DX League

NZRFU New Zealand Rugby Football Union

NZRL New Zealand Rugby League

NZRLS New Zealand Railway and Locomotive Society

NZRMA New Zealand Ready-Mix Concrete Association

NZRMTA New Zealand Retail Motor Trade Association

NZRN New Zealand Registered Nurse

NZRNC New Zealand Radio Navigation Chart

NZRRS New Zealand Railways Road Services

NZRTA New Zealand Road Transport Association

NZS New Zealand Standards Institute

NZSA New Zealand Statistical Association

NZSB New Zealand Speech Board; New Zealand Survey Board

NZSBG New Zealand South British Group

NZSC New Zealand Securities Commission; New Zealand Settlers Club; New Zealand Squid Company; New Zealand Staff Corps; New Zealand Standards Council

NZSCA New Zealand Sheep and Cattlemens Association; New Zealand Society of Customs Agents; New Zealand Soil Conservation Association

NZSCC New Zealand Standard Country Code

NZSCES New Zealand Society of Certified Executive Secretaries

NZSCHA New Zealand Society of Custom House Agents

NZSCI New Zealand Standard Classification of Imports

NZS Co New Zealand Shipping Company

NZSCO New Zealand Standard Classification of Occupations

NZSCS New Zealand Senior Citizens Service

NZSDA New Zealand Sign and Display Association; New Zealand Stamp Dealers Association

NZSDST New Zealand Society of Dairy Science and Technology

NZSE New Zealand Stock Exchange

NZ Sea Fron New Zealand Sea Frontier (NZSEAFRON)

nzsg non-zero-sum game

NZSI New Zealand Seismological Institute

NZSIA New Zealand Security Industry Association

NZSIC New Zealand Standard Industrial Classification

NZSID New Zealand Society of Industrial Designers

NZSL New Zealand Steel Limited; New Zealand Shipping Line

NZSLO New Zealand Scientific Liaison Office

NZSNA New Zealand Society of National Accounts

NZSO New Zealand Symphony Orchestra

NZSRA New Zealand Surf Riders Association

NZSS New Zealand Social Security; New Zealand Speleological Society; New Zealand Standard Specification(s)

NZSSS New Zealand Society of Soil Science

NZST New Zealand Standard Time

NZSWWS New Zealand Spinning, Weaving, and Woolcrafts Society

nzt non-zero test(ing)

NZTC New Zealand Trade Commission

NZTCA New Zealand Teachers College Association

NZTCB New Zealand Trade Certification Board

NZTCI New Zealand Technical College Institute; New Zealand Technical Correspondence Institute

NZTF New Zealand Territorial Force(s); New Zealand Theatre Federation

NZUA New Zealand Underwater Association; New Zealand Underwriters Association

NZUE New Zealand Unit Express

NZV New Zealand Victoria (insurance)

NZVA New Zealand Veterinary Association

NZw New Zealand white (mice hybrids)

NZWA New Zealand Woolbuyers Association

NZWB New Zealand Wool Board

NZWEA New Zealand Workers Educational Association

NZW & PCS New Zealand Weed and Pest Control Society

NZWRAC New Zealand Womens Royal Army Corps

NZWS New Zealand Wildlife Service

NZWSA New Zealand Water Ski Association

NZWSC New Zealand Water Safety Council

NZWTA New Zealand Wool Testing Authority

NZWWC New Zealand Working Womens Council

NZWWF New Zealand Waterside Workers Federation

NZYF New Zealand Yachting Federation

NZYHA New Zealand Youth Hostels Association

NZZ *Neue Zürcher Zeitung* (New Zurich Newspaper)

O

o oath; observer; occasional; occidental; octavo; offshore; ohm; oil; oiliness; Olivetti; open; opium; orange; organism; oriental; origin; output; overcast

o' (Gaelic contraction—of, on)

'o (Gaelic contraction—also)

o (Japanese—big, great, large); *omkring* (Dano-Norwegian—about or around)

o. oculus (Latin—eye); *oeste* (Portuguese or Spanish—west); *oost* (Dutch—east); *op* (Dano-Norwegian or Dutch—up); *os* (Latin—bone); *ouest* (French—west); *ovest* (Italian—west)

o/ order (Spanish—order)

ö (Dano-Norwegian or Swedish—island); *öster* (Swedish—east)

ø *øst* (Dano-Norwegian—east)

O absence of perception of sound (symbol); New Orleans Mint (coin symbol); observation; Observer; ocean; Oceanic Steamship Company; October; office; officer; Ohio; old (newspaper options listings); Olsen Line; Omaha; Ontario; order; Oregon; ortho; Oscar—code for letter O; oxygen; prefix to many Irish names and meaning, for example, descendant of (O'Boyle, O'Connor, O'Leary)

O' (Gaelic prefix meaning of)

Ø shortage (symbol)

O center of the earth (symbol); observer (symbol); *oeste* (Portuguese or Spanish—west); *oost* (Dutch—east); *optimus* (Latin—best possible); *organo* (Italian—organ); *Ost* (German—east); *ouest* (French—west); *ovest* (Italian—west)

Ö *Österreich* (German—Eastern Empire)—Austria; *Östre* (Swedish—East); *Öy* (Swedish—island)

Ø *Øst* (Dano-Norwegian—East); *Øy* (Dano-Norwegian—island)

O1 organized seagoing naval reserve

O-1 Cessna Bird Dog liaison aircraft

O2 organized naval reserve aviation

O-2 Cessna liaison—utility aircraft

O_2 oxygen

O_2 cap oxygen capacity

O_2 sat oxygen saturation

O^3 ozone

oa occiput anterior; old age; on account; on or about; osteoarthritis; overall

o/a on account; on or about

oa och andra (Swedish—and others)

o/A oro Americano (Spanish—American gold, American money)

OA Obligation Authority; Office of Applications; Office Automation; Olympic Airways; Operations Analysis;

Osborne Association; overall noise level (symbol); Overeaters Anonymous; Overtime Authorization

O/A Office of Administration (EPA)

O & A October and April

O of A Office of Administration

OA Océan Atlantique (French—Atlantic Ocean)

oaa (OAA) oxalo-acetic acid

OaA Office of Aging

OAA Office of Air Accidents; Old Age Assistance; Older Americans Act; Organization of Athletic Administrators; Orient Airlines Association

OAA Organisation des Nations Unies pour l'Alimentation et l'Agriculture (French—United Nations Organization for Food and Agriculture); *Organización de las Naciones Unidas para la Agricultura y la Alimentación* (Spanish—United Nations Organization for Food and Agriculture)

OAAA Outdoor Advertising Association of America

OAAB Objective-Analytic Anxiety Battery

oaad ovarian ascorbic acid depletion

OAAI Office of Air Accidents Investigation

OAAU Organization of Afro-American Unity

OAB Old Age Benefits

OABA Outdoor Amusement Business Association

OABETA Office Appliance and Business Equipment Trades Association

oac on approved credit; optical aberrations compensation; outer approach channel

OAC Oceanic Affairs Committee; Operating Agency Code; Ordnance Ammunition Command; Oregon Agriculture College; Overseas Automotive Club

OACA Ontario Arms Collectors Association (of Beamsville near Toronto)

OACI Organisation de l'Aviation Civile Internationale (French—International Civil Aviation Organization); *Organización de Aviación Civil Internacional* (Spanish—International Civil Aviation Organization)

OACJC Oklahoma Association of Community and Junior Colleges

OACLD Ontario Association for Children with Learning Disabilities

OACT Ohio Association of Classroom Teachers

oad original air date; overall depth

OAD ordered, adjudged, and decreed

OADAP Office of Alcoholism and Drug Abuse Prevention

oadc oleic acid, albumin, dextrose, catalase

OAE Orzeck Aphasia Evaluation

OAEC Organization for Asian Economic Cooperation

OAESA Ohio Association of Elementary School Administrators

oaf open-air factor; overhaul attrition factor

OAFB Orfutt Air Force Base (Nebraska)

OAFIE Office of Armed Forces Information and Education

OAG Office of the Adjutant General; Office of the Attorney General

OAG Official Airline Guide

OAGB Osteopathic Association of Great Britain

oah overall height

OAH Organization of American Historians

OAHE Ohio Association for Higher Education

OAI Office of Aeronautical Intelligence; Opera America, Incorporated; Osborne Association, Incorporated

OAIA Organisation des Agences d'Information d'Asie (French—Organization of Asian News Agencies)

OAICU Oklahoma Association of Independent Colleges and Universities

oaide operational assistance and instructive data equipment

oais opinion, attitude, and interest survey

oak. oakum

Oak Oakland, Oak Park, Oak Ridge, etc.

OAK Oakland, California (Metropolitan International Airport)

OAKE Organization of American Kodaly Educators

Oakland Oakland Coliseum, Oakland, California

Oak Sym Oakland Symphony

oal overall length

OAL Office of Administrative Law; Ordnance Aerophysics Laboratory

OALJ Office of Administrative Law Judges

OALMA Orthopedic Appliance and Limb Manufacturers Association

o. alt. hor. omnibus alternis horis (Latin—every other hour)

OAM Office of Alternative Medicine; Office of Aviation Medicine; Order of Australia Medal

OAMA Ogden Air Material Area; Oil Appliance Manufacturers Association

oamce optical alignment, monitoring, and calibration equipment

oame orbital attitude and maneuvering electronics

OAMS Orbital Attitude and Maneuvering System

ÖAMTC Österreichischer Automobil-Motorrad und Touring Club (German—Austrian Automobile Motoring and Touring Club)

OANA Organization of Asian News Agencies

o-and-o one-and-only

oao off and on

OAO Orbiting Astronomical Observatory

oap ophthalmic artery pressure

OAP Office of Aircraft Production; Old-Age Pension

OAPC Office of the Alien Property Custodian

OAPEC Organization of Arab Petroleum Exporting Countries

OAPEP Organisation Arabe des Pays Exportateurs de Petrole (French—Arab Organization of Petroleum Exporting Nations)

OAPs Old-Age Pensioners

OAPU Old Age Pension Union

oapwl overall power watt level

O Ar Old Arabic

OAR Offender Aid and Restoration; Office of Aerospace Research; Open Architecture for Reasoning; Order of Augustinian Recollects; Organized Air Reserve

OARAC Office of Aerospace Research Automatic Computer

OARP Old Age Revolving Pensions (Townsend Plan)

OARS Offender's Aid Rehabilitation Services

OART Office of Advanced Research and Technology (NASA)

oas offensive avionics system; old-age security; on active service; option-adjusted spread

OAs older adults

OAS Office Automation System; Office of Advanced Studies; Office of Appalachian Studies; Old Age Security; Ordinary Ammunition Storage; Organization of American States

OAS Organization de l'Armée Secrete (French—Organization of the Secret Army)—counter-revolutionary group attempting to crush Algerian independence

OASBO Ohio Association of School Business Officials; Oregon Association of School Business Officials

OASD Office of the Assistant Secretary of Defense

OASD-AE Office of the Assistant Secretary of Defense, Application in Engineering

OASDHI Old-Age, Survivors, Disability, and Health Insurance

OASDI Old-Age, Survivors, and Disability Health Insurance (Social Security)

OASD-R & D Office, Assistant Secretary of Defense, Research and Development

OASD-S & L Office, Assistant Secretary of Defense, Supply and Logistics

OASD-T Office of the Assistant Secretary of Defense—Telecommunications

OASHDS Office of the Assistant Secretary for Human Development Services

OASI Office Automation Society International; Old-Age and Survivor's Insurance

OASIS Office of Academic Support Instructional Services; Office for Academic Support in Service; Ohio (chapters) of the American Society for Information Science; Older Adult Service and Information System; Overseas Access Service for Information Systems

oasp organic acid-soluble phosphorus

oaspl overall sound pressure level

OASSO Operational Applications of Satellite Snowcover Observations (NASA)

OAST Office of Aeronautical and Space Technology

oat operating ambient temperature; outside air temperature

OAT Office of Advanced Technology (USAF)

OATC Oceanic Air Traffic Control

OATS Office of Air Transportation Security; Old-Age Theatre Society (Great Britain); Older American Transportation System

oau (OAU) optical alignment unit

OAU Organization for African Unity

oav oculoauriculovertebral

oavg opponents' batting average (baseball)

OAVTME Office of Adult, Vocational, Technical, and Manpower Education

oaw old abandoned well; overall width

OAWM Office of Air and Water Measurement (NBS)

Oax Oaxaca

OAYR Outstanding Airman of the Year Ribbon

ob obligation; oboe; obsolete; obstetric; obstetrical; obstetrician; old boy; on board; operational base (OB); or better; order book; ordered back; out of bounds; outboard buffer; outbound; output buffer; outward bound; over bought; overboard (vent line)

ob (OB) outside broadcast (TV from a remote location)

o/b opening of books; outboard (engine)

ob (Latin prefix—against, in front of, toward)—obstruction

ob. *obit* (Latin—died)

o B off Broadway

o-B off-Broadway; off-Broadway theater

o B *ohne Befund* (German—without findings)

Ob object art (art accented with real objects, *e.g.*, a real watch chain dangling between two pockets of a man's vest in a painting); 3500-mile Siberian river entering Arctic Ocean at Gulf of Ob

Ob Obadiah; Ober (Germany—higher, upper)

OB Ocean Beach; Old Bailey; Operating Base; Operational Base; Order of Battle; Ordnance Battalion; Ordnance Board; Ormond Beach; Ox Box (corporation)

O.B. obstetrical; obstetrician; obstetrics

O'B O'Brien; O'Bryan

OB *Oranjeboom* (Dutch—orange tree)—Amsterdam-brewed beer

oba on base average (baseball); optical bleaching agent

OBAA Oil-Burning Apparatus Association

Obad The Book of Obadiah

OBAN *Operação Bandeirantes* (Portuguese—Operation Bandeirantes)—Brazilian Intelligence Service

OBAP Organization of Black Airline Pilots

OBAR Ohio Bar Automated Research

OBAWS On-Board Aircraft Weighing System

obb *obbligato*

OBB Battleship, old (3-letter naval symbol)

ÖBB *Österreichische Bundesbahnen* (Austrian Federal Railways)

obbl *obbligato*

obc old brutal con(vict); on-board checkout; outer back cover

OBC Old Boys Club; Osaka Broadcasting Corporation; Outboard Boating Club

obce on-board checkout equipment

obd omnibearing distance

ob d'am *oboe d'amore*

OBDC Otago Business Development Centre

ob dk observation deck

obds on-board diagnostic system

obdt obedient

obe open both ends; operating basis earthquake; other bugger's efforts; out-of-body experience

OBE Office of Business Economics; Officer of the British Empire; Order of the British Empire

O.B.E. Officer of the Order of the British Empire

obel obelisk

OBEMLA Office of Bilingual Education and Minority Languages Affairs

Oberst *Oberstimme* (German—soprano, treble, descant)

Oberw *Oberwerk* (German—highest organ bank; upperwork)

OBES Office of Basic Engineering Sciences

OBEV *Oxford Book of English Verse*

obf operating basis flood

obfusc obfuscated

obg oldie but goodie (musical hits)

Ob-G Obstetrician-Gynecologist

OBGA Office of Block Grant Assistance

obgn obligation

ob^{go} *obrigado* (Portuguese—thank you)

ob-gyn obstetrical-gynecological; obstetrician-gynecologist

obi omnibearing indicator

Obie off-Broadway; off-Broadway theater; Off-Broadway Theater Award

OBIPS Optical Band Imager and Photometer System

obit obituary

obits obituaries

obj object; objective

object. objective(ly)

objn objection

obl obligation; oblique; oblong; obloquy

ob/l ocean bill of lading

OBL Ohio Barge Line; Order of the Brave Librarian

oblg obligate; obligation

OBLI Oxford and Birmingham Light Infantry

oblig obligation(s); obligatory

obln obligation

obm oil-base mud (oil well)

obo oil/bulk freight/ore (multipurpose seagoing carrier); or best offer

oboe offshore buoy-observing equipment

ob ph oblique photograph(y)

OBRA Omnibus Budget Reconciliation Act (of 1987); Overseas Broadcasting Representatives' Association

OBRA '93 The Omnibus Budget Reconciliation Act of 1993

obre *octubre* (Spanish—October)

Obre *octobre* (French—October)

obro *outubro* (Portuguese—October)

obs observation; observe; observed; observer; obsolete; obstacle; obstetrical; obstetrician; obstetrics; ocean bottom suspension (oil well); omnibearing selector

obs (OBS) organic brain syndrome

obs *oboes*

Obs *The Observer*

OBS Oita Broadcasting Service; Organization Breakdown Structure

obs alt observed altitude

obsc obscure(d)

obscen obscenity

obsd observed

observ observation; observatory

obsn observation

obsol obsolescent

ob & sol objection and solution

ob. s.p. *obüt sine prole* (Latin—died without issue)

OBSP *Old Bailey Sessions Papers*

obss ocean bottom scanning sonar

obs spot observation spot

obst obstacle; obstruction

obstet obstetrical; obstetrician; obstetrics

obstl obstruction light(s)

obstr obstruction

obsv observation; observatory; observer

ob syn organic brain syndrome

o b syn organic brain syndrome

obt obedient

obt *obiit* (Latin—he died)

OBT Overseas Branch Transfer

OBTA Oak Bark Tanners' Association

obtd obtained

obts offender-based transaction statistics

obu offshore banking unit

OBU One Big Union; Operative Bootmakers Union

ÖBUB *Öffentliche Bibliothek der Universität Basel* (German—Public Library of the Basel University)—founded in 1460

O Bul Old Bulgarian

obv obverse; obvious; ocean boarding vessel; octane blending value; on-balance volume

obvy obviously

obw observation window

Obw *Oberwerk* (German—highest organ bank; upperwork)

obwo O-type backward-wave oscillator

oc ocean; odor control; on camera; on center; only child; open charter; oral contraceptive; oval cut

oc (OC) obstetrical conjugate; on camera; open cup; overdraft charge

o-c open-circuit

o'c o'clock (of the clock)

o/c open charter; open cover; organized crime; overcharge

o & c onset and course (disease)

oc (Latin prefix—against occlusion)

o.c. *opere citato* (Latin—in the work cited)

Oc Ocean

OC Air California (airline code); Oakland City; Oakwood College; Oberlin College; Oblate College; Occidental College; Odessa College; Office of Censorship; Office of the Commissioner; Office Consultation; Officer Candidate; Officer in Charge; Officer Commanding; Ohio College; Okolona College; Olcoresin Capsicum (pepper spray); Olivet College; Olympic College; Opera Company; Optometric Corporation; Order in Council; Oriel College; Orlando College; Otero College; Overseas Chinese; Overseas Commands

O.C. Officer Commanding

O of C Order of the Coif

OC *Opéra-Comique* (French—Comic Opera)—Paris

O.C. *Organo Corale* (Latin—choir organ)

OC-5 Organizing Committee for a Fifth Estate

oca ocarina (flutelike clay instrument nicknamed "sweet potato")

OCA Oceanic Control Area; Office of the City Attorney; Office of Computing Activities (NASA); Office of Consumer Advisor (of United States Department of Agriculture); Office of Consumer Affairs (ombudsman function of the U.S. Postal Service); Ohio College Association; Oil Company of Australia; Old Comrades Association; Ontario College of Art; Open Communications Architecture; Oregon Corrections Association; Owner Controlled Area

OCA *Organización de las Cooperativas de América* (Spanish—Organization of American Cooperatives)

OCAA Oklahoma City-Ada-Atoka (railroad); Organization of Central American Armies

OCAC Office of Chief of Air Corps

OCADS Oklahoma City Air Defense Sector

OCAFF Office Chief of Army Field Forces

ocal on-line cryptanalytic aid language

OCAL Overseas Containers of Australia Limited

OCALC Oklahoma City Air Logistics Command

OCAM Organisation Commune Africaine et Malgache [French—Organization of the African and Malagasy Community (of former French colonies)]

OCAMA Oklahoma City Air Materiel Area

O Canada O O Canada! terre de nos aïeux (French—O Canada! Land of our forefathers)—national anthem sung in English and in French

OCAS Office of Civil Aviation Security; Organization of Central American States

O Cat Old Catalan

OCAT Optometric College Aptitude Test; Optometry College Admissions Test

OCAW Oil, Chemical and Atomic Workers (union)

ocb oil circuit breaker

OCB Officer Career Brief (DoD resumé)

OCBC Overseas Chinese Banking Corporation

oc b/l ocean bill of lading

occ occasionally; occupation

o & cc order and change control

Occ occulting (light)

OCC Office of the Comptroller of the Currency; Oklahoma Crime Commission; Olney Community College; Onondaga Community College; Options Clearing Corporation; Orange Coast College

OCCA Oil and Colour Chemists Association

occas occasional(ly)

OCCC Oil Control Coordination Committee; Orange County Community College; Organized Crime-Control Commission (California)

Oc C Cm O Office of the Chief Chemical Officer

occd occupied

OCCDC Oregon Coastal Conservation and Development Commission

OCC-E Office of the Chief of Communications—Electronics (USA)

OCCF Oklahoma City Community Foundation

occip occipital; occiput

OCCIS Operational Command and Control Intelligence System (USA)

occl occlude(d); occluded front; occluding; occlusal; occlusion; occlusive(ness)

OCCL Ontario Community College Librarians

OCCM Office of Commercial Communications Management

OCCO Office of the Chief Chemical Officer

OCCP Outside Communications Cable Plant

OCCS Office of Computer and Communications Systems (U.S. National Library of Medicine)

OCCSA Ohio Correctional and Court Services Association

occ th occupational therapy

occup occupation(al)

ocd obsessive compulsive disorder; on-line communications driver; operational capability date; optical character definition; ovarian cholesterol depletion

oc/d other cargo damage

OCD Office of Child Development; Office of Civil Defense; Office of Collection and Dissemination (CIA)

OCDA Ordnance Corps Detroit Arsenal

OCDE Organización Común Africana, Malgache y Mauriciana (Spanish—African Common Organization including Madagascar and Mauritius); *Organización de Cooperacion y Desarrollo Económico* (Spanish—Organization of Cooperation and Economic Development)

OCDM Office of Civil and Defense Mobilization

OCDQ Organizational Climate Description Questionnaire

OCDR Office of Collateral Development Responsibility

OCDS Overseas College of Defense Studies (UK)

O/Cdt Officer-Cadet

oce operational control equipment

OCE Office of Career Education; Office of the Chief of Engineers; Ontario College of Education

OC & E Oregon, California, and Eastern (railroad)

Ocean Ocean Transport and Trading Limited; The Ocean (Antarctic, Arctic, Atlantic, Indian, Pacific)

OCEAN Oceanographic Coordination Evaluation Analysis Network

OCEANAV Oceanographer of the U.S. Navy

oceaneer(ing) ocean engineer(ing)

Ocean Inst Oceanografiska Institute (Swedish—Oceanographic Institute)—Göteborg, Sweden

oceano oceanologic(al)(1y); oceanologist; oceanology

oceanog oceanography

OCED Office of Comprehensive Employment Development

OCEE Organisation de Coopération Économique Européene (French—European Economic Cooperation Organization)

OCEL Optical Coating Evaluation Laboratory

OCEL Oxford Companion to English Literature

O Celt Old Celtic

ocf originally cultured formulation

OCF Officiating Chaplain to the Forces; Ossining Correctional Facility (Sing Sing); Owens-Corning Fiberglass

OC of F Office of the Chief of Finance

OCFR Oxford Committee for Family Relief

OCFT Office of Curriculum Frameworks and Textbooks

ocg omnicardiogram

ÖCG Österreichische Computer Gesellschaft (German—Austrian Computer Society)

och ochre

OCHAMPUS Office for the Civilian Health and Medical Program of the Uniformed Services

OCHS Old Colony Historical Society

oci organization conflict of interest

OCI Office of Computer Information (U.S. Department of Commerce); Office of the Coordinator of Information; Office of Current Intelligence (CIA); Operational Checkout Instruction

OCIA Organic Crop Improvement Association

OCIB Organized Crime Intelligence Bureau

OCID Organized Crime Intelligence Division (LAPD)

OCIMF Oil Companies International Marine Forum

OCIS Organized Crime Information System (FBI)

O. Cist *Ordo Cisterciencium* (Order of Citeaux)

OCJA Oklahoma Criminal Justice Association

OCJP Office of Criminal Justice Planning

ocl operator control language; optical communications link(age)

OCL Ocean Cargo Line; Overseas Container Line; Overseas Container Limited

OCL/ACT Overseas Container Lines and Associated Container Transport

OCLAE *Organización Continental Latino-Americana de Estudiantes* (Spanish—Continental Organization of Latin American Students)

OCLC Ohio College Library Center; On-Line Computer Library Center

OCLI Optical Coating Laboratory, Inc.

o'clock of the clock

OCLU Overseas Container Line (container) Unit

ocm oil content monitor; on-camera meteorologist

OCM *Oxford Companion to Music*

OCMA Oil Companies' Material Association

OCMH Office of the Chief of Military History

OCMMINST Office of Civilian Manpower Management Instruction (USN)

OCMS Optional Calling Measured Service (telephone)

OCN Operation Completion Notice

Ocn Bch Ocean Beach

ocnl occasional(ly)

OCNM Oregon Caves National Monument (limestone caverns near Medford, Oregon)

OCNUAD *Oficina del Coordinador de las Naciones Unidas para la Ayuda en los Desastres* (Spanish—Office of the Coordinator of the United Nations for Help in Disasters)

oco open-close-open

OCO Office of the Chief of Ordnance; Ontario College of Ophthalmology; San José, Costa Rica (El Coco Airport)

OCOA *Organismo Coordinador de Operaciones Antisubversivas* (Spanish—Coordinating Organism of Anti-subversive Operations)—Uruguay's secret service

o'coat overcoat

OCOM *Oficina Central de Organización y Metodos* (Spanish—Central Office of Organization and Methods)

OComS Office of Community Services

OConUS outside continental limits of the United States

OCORA *Office de Coopération Radiophonique* (French—Office of Radiophonic Cooperation)—French overseas radio help for former colonies

O Corn Old Cornish

ocp obsessive-compulsive personality; output control pulses; overland common points

OCP Office of the Chief of Protocol (U.S. Department of State); Office of Consumer Protection; Office of Cultural Presentations

OCP *Oficina Central de Personal* (Spanish—Central Personnel Office)

OCPCJR Office of Crime Prevention and Criminal Justice Research

OCPD Officer-in-Charge Police District

OCPL Oklahoma City Public Library

ocr optical character reader; optical character recognition

ocr (OCR) optical character reader

OCR Office of Civil Rights; Office of Civilian Requirements; Office of Coal Research; Office of Collateral Responsibility; Office of Coordinating Responsibility; Office of the County Recorder; Organization Change Request; Organization for the Collaboration of Railways

OCRA *Organisation Clandestine de la Révolution Algerienne* (French—Secret Organization of the Algerian Revolution)

OCRD Office of the Chief of Research and Development

ocre optical character recognition equipment

OCRE Organizations Concerned about Rural Education

ocrit optical character-recognizing intelligent terminal

OCRS Organized Crime and Racketeering Section (Dept. of Justice)

OCRSF Organized Crime and Racketeering Strike Force (U.S. Department of Justice)

OCRU Office of Communication and Research Utilization

OCRWM Office of Civilian Radioactive Waste Management

ocs obstacle clearance surface; on company service; outer continental shelf

oc's obscene (telephone) callers, obscene (telephone) calls

OCS Office of Civilian Supply; Office of Commercial Services; Office of Contact Settlement; Officer Candidate School; Officers' Chief Steward; Old Church Slavonic; Outer Continental Shelf; Overseas Civil Servants; Overseas Courier Service

OCS' Overseas Civil Servants (members of the British Overseas Civil Service)

OCS *Organe de Controle des Stupéfiants* (French—Narcotic Drug Control Organization)

ocsa oil-cooling system assembly

OC of SA Office, Chief of Staff, Army

OCSE Office of Child Support Enforcement

ocsf office contents special form (insurance)

OCSIGO Office of the Chief Signal Officer

ocsn occasion

ocsnl occasional

ocsnly occasionally

OCSPC Outer Continental Shelf Policy Committee (California)

ocst overcast

oct octagon; octal; octane; octave; octavo; octet; oxytoxic challenge test

Oct Octans (constellation); Octavius; October

OCT Office of the Chief of Transportation; Overseas Countries and Territories

octe optical component testing and evaluation

octᵉ octubre (Spanish—October)

Octans (Latin—Rule and Square constellation)

Octember October and November

OCTI Office Central des Transports Internationaux par Chemins de Fer (French—Central Office for International Railway Transport)

octl open-circuited transmission line

Octn Octanus (constellation)

October October Railway (Leningrad-Moscow); October Revolution (Bolshevik insurrection of October 1917)

octo(s) octoroon(s)

oct. pars octava pars (Latin—eighth part)

octr prot octrooi protectie (Dutch—patent protected)

octs occupational carpal-tunnel syndrome; open-cycle thermal systems

OCTU Officer-Cadet Training Unit

octupl octuplicate

octup. octuplus (Latin— eightfold)

octv open-circuit television

ocu operational conversion unit

OCUA Ontario Council on University Affairs

OCUC OCLC (Online Library Center Inc.) Online Union Catalog; Oxford and Cambridge Universities' Club

OCUFA Ontario Confederation of University Faculty Associations

ocul. oculis (Latin—to the eyes)

oculent. oculentum (Latin— eye ointment)

ocv open-circuit voltage

OCVs Overseas Cooperation Volunteers

oc vu ocean view

OCZ Ocean Container (terminal) Zebrugge

OCZM Office of Coastal Zone Management (NOAA)

od olive-drab; on demand; optical density; optic(al) disc; organization(al) development; original design; outside diameter; outside dimension; oven dried; overdose; overdrive

od (OD) overdrawn

o/d on demand; overdraft

o & d origin and destination

od och dylika (Swedish—and the like); *odur* (German—or)

o.d. oculus dexter (Latin— right eye)

Od Odyssey

OD Aerocondor (Aerovias Condor de Colombia); external grinding; officer of the day; Office of the Director; olive drab; Operational Directive; Optical Density; Ordnance Department; original design; outside dimension

O.D. Doctor of Optometry

ØD Økonomisk Demokrati (Danish—Economic Democracy)

oda occipito-dextra anterior

Oda Odessa

ODa Old Danish

ODA Office Document Architecture; Office of Debt Analysis; Office of the District Administrator; Office of the District Attorney; Office of Drug Abuse; Overseas Development Administration; Overseas Development Assistance

ODALE Office of Drug Abuse Law Enforcement

ODAS Ocean Data Acquisition System

odat one day at a time

odb opiate-directed behavior; output to display buffer

odc other departure cities; other direct costs; outer dead center

ODC Old Dominion College; Overseas Development Corporation; Overseas Development Council

ODCSRDA Office of the Deputy Chief of Staff for Research, Development, and Acquisition (USA)

ODCTI Old Dominion College Technical Institute

odd oculodentodigital

odd (ODD) operator distance dialing

od'd overdosed

ODDRE Office of the Director of Defense Research and Engineering

ode one-day event

Oᵈᵉ Oude (Afrikaans, Dutch, Flemish—old)

ODE Oil Drilling and Exploration

ODEC Ocean Design Engineering Corporation

ODECA Organización de Estados Centroamericanos (Spanish—Organization of Central American States)

ODECO Ocean Drilling and Exploration Company

od'ed overdosed

ODEE Oxford Dictionary of English Etymology

ODEPLAN Oficina de Planificación Nacional (Spanish—Office of National Planning)

ODESSA Ocean Data Environmental Sciences Services Acquisition

ODESSA Organisation Der Ehemaligen SS Angehörigen (German—Organization of Former Members of the SS)

ODESY On-Line Data Entry System

ODF Old Dominion Foundation; Operational Deployment Force

odfc outside diameter of female coupling

ODFI Open Die Forging Institute

O d G Ordine del Giorno (Italian—Order of the Day)

ODGSO Office of Domestic Gold and Silver Operations

ODH Ontario Department of Health

ODI Office of Defense Investigation (U.S. Department of Justice); Open-Door International (championing economic emancipation of women workers); Organization Development Institute

ODIL Overseas Development Institute Limited

ODIN Origin-Destination Information

Odin Scandinavian equivalent of Wotan, the supreme god of the Norse gods

od'ing overdosing

o-d-ing overdosing

O Div Ontario Division (RCMP)

ODJB Original Dixieland Jazz Band

o dk orlop deck

ODL Office of Defense Lending

odm ophthalmodynamometry

ODM Office of Defense Mobilization; Order of De Molay; Overseas Development Ministry

ODMA Optical Distributors and Manufacturers Association

odmc outside diameter of male coupling

ODMC Office for Dependents Medical Care

odn own doppler nullifier

Odn Odense; Odin; Odinist (member of Nordic-supremacy sect)

ODN Organization Development Network

ODO Outdoor Office(r)

ODOE Oregon Department of Energy

odom odometer

odont odontology

odop offset doppler

odoram. odoramentum (Latin—perfume)

odorat. odoratus (Latin—odorous, perfuming)

odorl odorless

ODOTS One-Day One-Trial System (for jurors)

odp occipito-dextra posterior; order dispatched

ODP Office of Disaster Preparedness; Operational Deploymept Plan(ning); Orbit Determination Program; Orderly Departure Program

odr order

ODR Office of Defense Resources

ODRC Office of Disaster Relief Coordinator (UN)

o'drive overdrive

ods oxide dispersion strengthened

o d's other denominations

ODS Ocean Data Station; Office of Defender Services; The Open Door Society (Canada); Orton Dyslexia Society

odsd overseas duty selection date

ODSE Open-Door Student Exchange

ODSI Ocean Data Systems Inc

ODSR Office of the Director of Scientific Research

odt occipito-dextra transverse; octal debugging technique; odor detection threshold; one-day trials; on-line debugging technique

ODT Otago Daily Times

ODTF Operational Development Test(ing) Facility

ODTS Operational Development Test Site

ODU Old Dominion University

od units optical-density units

ODWIN Opening Doors Wider in Nursing

ODWSA Office of the Directorate of Weapon Systems Analysis (USA)

oe oersted; offensive end (football); omissions expected; open end(ed); outdoor education; overseas experience

oe (OE) organizational effectiveriess

o/e on examination; otitis externa

o & e operations and engineering

oe organo espressivo (Italian—swell organ)

öe öesterreichisch (German—Austrian)

Oe oersted

OE Office of Education; Old England; Old English; Old Etonian; Oregon Electric (railroad)

OEA Oahu Education Association; Office of Economic Adjustment (USA); Office Education Association; Office of Environmental Affairs; Office Executives Association;

Office of Export Administration; Ohio Education Association; Optometric Editors Association; Oregon Education Association; Outdoor Education Association; Overseas Education Association

OEA Organización de los Estados Americanos (Spanish—Organization of American States)

OEAA Oil Engineering Apprentices Association

OEB Oregon Educational Broadcasting

oec organizational entity code

OEC Office of Energy Conservation; Ohio Edison Company; Oil Exporting Countries; Oribital Engine Corporation

ÖEC Österreichischer Aero Club (German—Austrian Aero Club)

OECC Office for Educational Credit and Credentials

OECCNU Organización para la Educación, la Ciencia, y la Cultura (Organization for Education, Science, and Culture)—UN

OECD Organization for Economic Cooperation and Development

OECE Organisation Européenne de Cooperation Économique (French—Organization for European Economic Cooperation)

OECF Overseas Economic Cooperation Fund

oeco outboard engine cutoff

OECON Offshore Exploration Conference

OECQ Organisation Européene pour la Contrôle de la Qualité (French—European Quality Control Organization)

OECS Organization of East Caribbean States

oecu outboard engine cutoff

OED Oxford English Dictionary

OEDA Office of Energy Data and Analysis

OEDP Office of Employment Development Programs

oee outer enamel epithelium; overtraining extinction effect

OEEC Organization for European Economic Cooperation

OEEO Office of Equal Educational Opportunities

OEF Osteopathic Educational Foundation

OEG Operations Evaluation Group

oegt observable evidence of good teaching

OEGT Office of Education for the Gifted and Talented

oei organizational entity identity

OEI Offshore Ecology Investigation

OEI Oficina de Educación Ibero-americana (Spanish—Office of Ibero-American Education)

OEIPS Office of Engineering and Information Processing (NBS)

OEIU Office Employees International Union

o-e-l owner's risk of leakage

OEL Organization Equipment List

OEL/MA Ohio Educational Library/Media Association

oem oil-emulsion mud (oil well); original equipment manufacturer

oem (OEM) optical electron microscope

OEM Office of Environmental Mediation; Office of Executive Management

OEMA Office Equipment Manufacturers Association

oemcp (OEMCP) optical effects module electronic controller and processor

OEMs Original Equipment Manufacturers

oen oenanthic; oenanthyl; oenolyn; oenology; oenological; oenologist; oenomancy; oenomel (wine and honey); oenometer; oenophilist; oenophobist; oenopoetic

oeo officer's eyes only

OEO Office of Economic Opportunity; Ordnance Engineer Overseer

OEOB Old Executive Office Building (D.C.)

OEP Office of Emergency Planning; Office of Emergency Preparedness; Optional Educational Programs; Organization, Education, and Personnel

OEPP Organisation Européenne et Méditerranéenne pour la Protection des Plants

(French—European and Mediterranean Organization for the Protection of Plants)

OEPS Office of Educational Programs and Services

OEQ Order of Engineers of Québec

OEQC Office of Environmental Quality Control

oer oersted (unit of magnetic force); original equipment replacement; owner's equivalent rent

o'er over

OER Office of Aerospace Research (USAF); Office of Energy Research; Officer Effectiveness Report; Officer Efficiency Report; Officer Engineering Reserve; Officers Emergency Reserve; Organization for European Research

oerc optimum earth-reentry corridor

OERPA Office of Exploratory Research and Problem Assessment (National Science Foundation)

OERS Organisation Européenne de Recherches Spatiales (French—European Space Research Organization)

OES Office of Economic Stabilization; Office of Emergency Services; Official Experimental Station; Order of the Eastern Star; Organization of European States

OES Organización de Estados Americanos (Spanish—Organization of American States)

oesbr oil-extended styrene-butadiene rubber

OESL Occanographic and Environmental Service Laboratory (Raytheon)

oesoph oesophagus

OESP O Estado de São Paulo (State of Sao Paulo)—Brazil newspaper

OESS Office of Engineering Standards Services

OET Office of Education and Training; Office of Emergency Transportation; Overseas Exchange Transactions

OETA Occupied Enemy Territory Administration

OETB Offshore Energy Technology Board

OEVE Office of Earthquakes, Volcanoes, and Engineering (U.S. Geological Survey)

OEW Office of Economic Warfare

OEWG Open-Ended Working Group; Operation and Evaluation Wartime Group

OEX Office of Educational Exchange

OEZ osteuropäische Zeit (German—East European Time)

of oil filter; orange fannings

of. old face (type); optional form; outside face; oxidizing flame

o/f oxidation/fermentation; oxidizer to fuel ratio

Of Ovenstone factor

OF Oceanographic Facility; Odd Fellows; Old French; Operating Forces; Ophthalmological Foundation; Osteopathic Foundation; Oxbow Falls; Oxenstierna Foundation; Oxford Foundation

OFA Office of Financial Analysis; Old Folks Association; Orthopedic Foundation for Animals

OFAC Owens Fine Arts Center (Dallas)

O-factor oscillation factor

ofb oval fat body

OFB output feedback

ofc office

OFC Overseas Food Corporation

OFCA Ontario Federation of Construction Associations

OFCC Office of Federal Contract Compliance

OFCCP Office of Federal Contract Compliance Programs

ofcl official

ofd one-function diagram; optical fire detector; orofacial dyskinesia

OFDA Office of Foreign Disaster Assistance (U.S.)

OFDI Office of Foreign Direct Investments

OFE Office of Fuels and Energy

OFEMA Office Français d'Exportation de Matériel Aéronautique (French—French Office for the Exportation of Aeronautical Material)

O'Fest October Fest (Munich)

off. office(r); official

Off Officer
OFF Office for Families
OFFAR Office of Fuel and Fuel Additive Registration (EPA)
offen offensive (ammunition)
offeq office equipment
Offeq-1 Horizon-1, Israel's first satellite
offer. offertories; offertory
offg offering
offic official(ly)
Office Pubns Office Publications
off-st pkg off-street parking
OFHA Oil Field Haulers Association
ofhc oxygen-free high conductivity; oxygen-free high-carbon (copper)
OFI Office of the Federal Inspector; Orangutan Foundation International
ofic *oficial* (Spanish—official)
OFIC Ohio Foundation of Independent Colleges
ofl official
Oflag *Offizierlager* (German—officer's prison camp)
OFlem Old Flemish
Ofly Offaly
OFM Office of Flight Missions (NASA); Office of Foreign Missions (State Dept.); Order of Franciscan Minors
OFNS Observer Foreign News Service
OFPA Order of the Founders and Patriots of America
OFPM Office of Fiscal Plans and Management
OFPP Office of Federal Procurement Policy
ofr off frequency rejection
OfR Office for Research
O Fr Old French
OFR Office of the Federal Register; Open File Report
OFR-ALA Office of Recruitment—American Library Association
OFris Old Frisian
O Frk Old Frankish
ofrs oxygen free radicals
ofs one-function sketch; outlook for space
OFS Ontario Federation of Students; Orange Free State
OFSPS Office of Federal Statistical Policy and Standards
OFST Office of the Secretary of the Air Force

oft. often
OFT Office of Fair Trade; Office of Fair Trading; Ohio Federation of Teachers; Optimal Foraging Theory
OFTC Overseas Finance and Trade Corporation
OFTS Office of Technical Services; Office of Transport(ation) Security; Officers Training School; Overseas Fixed Telecommunications System
OFY Opportunities for Youth (Canada)
og offensive guard (football); oh gee; oil gland; old girl; on ground; on guard; original gum
o-g orange-green
o/g opto-graphic; outgoing
OG Officer of the Guard; Old Gaelic; Olympic Games
O/G Opto/Graphic
ÖG Österreichische Galerie (Austrian Gallery)
OG *O Globo* (Rio de Janeiro's Globe)
O Gael Old Gaelic
OGAMA Ogden Air Materiel Area
Ogasawaras Ogasawara Islands (Bonins)
O-gauge $1-1/_4$-inch track gauge (model railroads)
OGB *Österreichischer Gewerkschaftsbund* (German—Austrian Trade Union Federation)
OGC Office of General Counsel
OGCMD Ogden Contract Management District
Ogd Ogdensburg
OGDC Oil and Gas Development Corporation
oge (OGE) operational ground equipment
OGE Office of Government Ethics
OGES Operating Ground Equipment Specification
ogf option growth fund
ogg *oggetto* (Italian—object)
OGI Opera Guilds International
ÖGI *Österreichische Gesselschaf für Informatik* (German—Austrian Society for Information Processing)
OGJ *Oil and Gas Journal*
ogl obscure glass; outgoing line
OGL Open General License

OGMC Ordnance Guided Missile Center
OGNR Oribi Gorge Nature Reserve (South Africa)
ogo orbiting geophysical observatory
OGO Orbiting Geophysical Observatory
OGPS Office of Grants and Program Systems (of United States Department of Agriculture)
OGPU *Obiedinennoye Gosudarstvennoye Politicheskoye Upravlenie* (Russian—United State Political Administration)—*q.v.m.* VOT
OGR Ontario Government Railway (Ontario Northland)
OGR *Official Guide of the Railways*
ogse operational ground-support equipment
OGSEL Operational Ground-Support Equipment List
OGSM Office of General Sales Manager
OGSR Office of Graduate Studies and Research
o-g stain orange-green stain
ogt on-going thing; outlet gas temperature
OGTT Oral Glucose Tolerance Test(ing)
OGU Occupational Guidance Unit
ogv outlet guide vane
oh office hours, on hand; open hearth; out home; oval head; overhead; over-the-horizon (communication); out home (lacrosse); outpatient hospital
oh (OH) ocular herpes
o/h overhaul
o.h. omni hora (Latin—hourly)
o-H on-Hudson
OH hydroxyl radical (symbol); Ohio; Olduvai Hominid; Omega House; opera house; San Francisco and Oakland Helicopter Airlines (2-letter code)
O/H *Overzuche Handelsmaatschappij* (Dutch—Overseas Trading Company)
O'H O'Higgins (General Bernardo O'Higgins—liberator of Chile)
OH-6 Hughes observation helicopter called Cayuse

OH-13 Bell Sioux helicopter
OH-23 Hiller Raven utility helicopter
OH-58 Bell Kiowa turbine-powered helicopter
oha outside helix angle
OHA Occupational Health Administration; Office of Hearings and Appeals; Oriental Herb Association
Ohal. Ohalot
O'Hare O'Hare International Airport (Chicago)—named for navy pilot Edward H. (Butch) O'Hare killed during World War II
OH-B Ocean Hill-Brownsville
OHBMS On Her (His) Britannic Majesty's Service
ohc outer hair cells; overhead cam
OHC Office of Humanities Communication; Ottumwa Heights College; Overseas Hotel Corporation
OHCS Office of Home Care Services
ohd organic hearing disease; organic heart disease; overhead drive
OHDETS Over-Horizon Detection System
OHDS Office of Human Development Services (HEW)
OHD & W Outer Harbor Dock and Wharf
OHE Office of Health Economics
oheat overheat
ohf overhaul factor
Ohf Omsk hemorrhagic fever
OHG Old High German
OHG Offene Handelsgesellschaft (German—ordinary partnership)
Oh Gloria Oh Gloria inmarcesible (Spanish—Oh Unfading Glory)—Colombian anthem
ohi ocular hypertension indicator
OHI Oil Heat Institute
OHI Organisation Hydrographique Internationale (French—International Hydrographic Organization)
OHIA Oil Heat Institute of America
Ohio Turn Ohio Turnpike
Ohio U Pr Ohio University-Press

OHIP Ontario Hospital Insurance Plan
OHJ Old House Journal
OHK Okayama Hoso KK
OHL Oberste Herresleitung (German—Supreme Headquarters)
ohm overhaul manual
ohm. ohmmeter
ohm-cm ohm-centimeter
OHMO Office of Hazardous Materials Operations
OHMR Office of Hazardous Materials Regulation
OHMS On Her (His) Majesty's Service; Onboard Health Monitoring System
OHN Occupational Health Nurse
OHNC Occupational Health Nursing Certificate
OHNO Occupational Health Nursing Officer
OHNS Occupational Health Nursing Sister
oho out-of-house operation
ohp overhead projection; oxygen at high pressure
oh Ped ohne Pedale (German—without pedals)
ohrf overhaul replacement factor
OHRG Official Hotel and Resort Guide
ohs open-hearth steel
ohs (OHS) hydroxy-steroids
OHS Office of Highway Safety; Ontario Humane Society; Oral Hygiene Service; Oregon Historical Society; Organization of Historical Studies; Overland Highway Society
OhSEA Ohio Society of Enrolled Agents
OHSGT Office of High-Speed Ground Transportation
OHSIP Ontario Health Services Insurance Plan
OHSPAC Occupational Health-Safety-Programs Accreditation Commission
oht ovarian hormone therapy; overheating temperature
OHTE Ohmic Heating Toroidal Experiment
ohv overhead valve; overhead vent
ohv's off-highway vehicles
oi oil-immersed; oil-immersion
o-i orgasmic impairment

o/i opsonic index
o & i organizational and intermediate
OI Office of Information; Office Instruction; Operating Instruction; Optimist International; Oriental Institute
O-I Owens-Illinois
OIA Ocean Industries Association; Office of Impact Analysis; Office of Industrial Associates; Office of International Administration; Oil Import Administration; Oil Insurance Association; Outboard Industry Associations
OIA Organización Internacional de Azucar (Spanish—International Sugar Organization)
OIAA Office of Inter-American Affairs; Office of International Aviation Affairs
OIAB Oil Import Appeals Board
OIAC Organización Internacional de la Aviación Civil (Spanish—International Civil Aviation Organization)
OIAJ Office for Improvements in the Administration of Justice
OIAS Occupational Information Access System
OIB Ohio Inspection Bureau; Oklahoma Inspection Bureau
oic oil cooler
O-i-C Officer-in-Charge
OIC Oceanographic Instrumentation Center; Offer-In-Compromise; Office of the Independent Counsel; Office of the Insurance Commissioner; Officer in Charge; Ohio Improved Chester (white swine); Oil Industry Commission; Opportunities Industrialization Centers; Overseas Investment Commission
OIC Organisation Interafricaine du Café (French—Inter-African Coffee Organization); *Organisation Internationale du Commerce* (French—International Trade Organization)
OICA Ontario Institute of Chartered Accountants; Oregon Independent Colleges Association

OICD Office of International Cooperation and Development

OIcel Old Icelandic

OICF Oklahoma Independent College Foundation; Oregon Independent College Foundation

OICI *Organización Interamericana de Cooperación* (Spanish—Inter-American Cooperation Organization)

OICJ Office of International Criminal Justice

oie's one-issue callers (call-in programs)

OICS Office of Interoceanic Canal Studies

oid (Latin suffix—resembling)—sigmoid; original issue discount

Oid mortales *Oid, mortales, el grito sagrado* (Spanish—Hear, mortals, the sacred cry)—Argentine anthem

OIE Office of Indian Education; Office of International Epizootics

OIE *Organisation Internationale des Employeurs* (French—International Organization of Employers)

OIEA *Organismo Internacional de Energía Atómica* (Spanish—International Atomic Energy Agency)—IAEA

OIER Office of International Economic Research

OIF Office for Intellectual Freedom (ALA)

OIG Office of the Inspector General

OIG *Organisation Intergouvernementale* (French—Intergovernmental Organization)

OIGS On Indian Government Service

oih (OIH) ovulation-producing hormone

OIHP *Office International d'Hygiene Publique* (French—International Office of Public Health)—UN

OII Office of Invention and Innovation; order instituting investigation

OIJ *Organisation Internationale des Journalistes* (French—International Organization of Journalists)

OIL Operation Inspection Log; Orbiting International Laboratory

OIL *Organizzazione Internazionale del Lavoro* (Italian—International Labor Organization)

oiloff oil ripoff

OILSR Office of Interstate Land Sales Registration

OIM Oriental Institute Museum (University of Chicago)

OINA Oyster Institute of North America

O-in-C Officer-in-Charge

OINC Officer in Charge

OING *Organisation Internationale Non-Gouvernementale* (French—Non-Govemmental Organization)

oint ointment

OIO Oklahomans for Indian Opportunity

oip oil in place; oxford india paper

OIP Office for Information Programs (NBS); Office of International Programs; Operations Improvement Program

OIPC *Organisation Internationale de Police Criminelle* (French—International Criminal Police Organization)—also known as Interpol; *Organisation Internationale de Protection Civile* (French—International Civil Defense Organization)

OIPH Office of International Public Health

OIr Old Irish

OIR Office of Inter-American Radio

OIRA Office of Information & Regulatory Affairs

OIran. Old Iranian

OIRB Oregon Insurance Rating Bureau

OIRM Office and Industrial Records Management

OIRSA *Organism Internacional Regional de Sanidad Agropecuaria* (Spanish—International Regional Association for Healthy Land and Cattle)

OIRT *Organisation Internationale de Radiodiffuslon et Télévision* (French—International Radio and Television Organization)

OIS Office Information System; Overseas Investors Services

OISA Office of International Scientific Affairs

OISE Ontario Institute for Studies in Education

OISS Online information Search Service

OISS *Organisation Ibéro-Américaine de Securite Sociale* (French—Iberian-American Social Security Organization)

OISTV *Organisation Internationale pour la Science et la Technique du Vide* (French—International Organization for Vacuum Science and Technology)

O i T Officer in Training (rookie police officer)

O It Old Italian

OIT Organic Integrity Test

OIT *Organisation Internationale du Travail* (French); *Organización Internacional del Trabajo* (Spanish)—International Labor Organization also known as ILO

OITF Office of International Trade Fairs

OIUC Oriental Institute of the University of Chicago

OIVV *Office International de la Vigne et du Vin* (French—International Office of Vines and Wines)

OIW Oceanographic Institute, Wellington (New Zealand)

OIWP Oil Industry Working Party

OIWR Office of Indian Water Rights

oj open-joint; open-joist(ed) orange juice

oJ *ohne Jahr* (German—without year)—no date

OJAJ October, January, April, July

OJARS Office of Justice Assistance, Research, and Statistics

OJC *Organisation Juive de Combat* (French—Jewish Combat Organization)

OJD Office de Justification de la Diffusion

OJDYD Office of Juvenile Delinquency and Youth Development

OJEC *Official Journal of the European Communities*

oji on—the-job injuries

OJJ Office of Juvenile Justice

OJJDP Office of Juvenile Justice and Delinquency Prevention

oJr old Jamaica rum

ojt on-the-job training

OJT (National) On-the-Job Training (Program)

ok all correct; okay; optical klystron; outer keel

ok ohne kosten (German—without cost); *ola kala* (Greek—all is fine, all is good)— believed to be the original okay used by Greek sailors of antiquity

OK all correct; okay; Oklahoma; Old Kinderhook (birthplace and home of President Martin Van Buren), Democratic OK Club believed to have started practice of putting "OK" on deals and documents they approved; Old Kingdom (Egypt); Oskar Kokoschka (1866–1980); Otto Klemperer (German conductor); Our King

OK *Our King*

O & K Orenstein & Koppel

Ø K Østasiatiske Kompagni (Danish—East Asiatic Company)

oka otherwise known as

OKA Okinawa, Ryukyu Islands (airport)

OKC Oklahoma City, Oklahoma (airport)

OKd okayed

Okefenokee Okefenokee National Wildlife Refuge and the Okefenokee Swamp between northern Florida and southern Georgia

OKH Oberkommando der Heeres (German—Army High Command)

Okhotsk Sea of Okhotsk between Kamchatka Peninsula, Sakhalin Island, and eastern Siberia

Okin Okinawa(n)

OKL Oberkommando der Luftwaffe (German—Air Force High Command)

Okla Oklahoma; Oklahoman

OklaC Oklahoma City

Okla Ob *Oklahoma Observer*

OKM Oberkommando der Marine (German—Naval High Command)

OkSEA Oklahoma Society ofEnrolled Agents

Okt Oktober (German—October); *Oktyabr* (Russian—October)

OKT Oslo Kommune Tunnelbanekontoret (Oslo subway system)

OKTc Ortho-Kung T-cell

Oktronics Oklahoma Electronics (corporation)

OKW Oberkommando der Wehrmacht (German—Armed Forces High Command)

ol oil level; operating license; or less; other locations

ol' old

o/l operations/logistics; outlook

ol. *oleum* (Latin—oil)

o.l. *oculus laevus* (Latin—left eye)

ö L östlich Längengrad (German—east longitude)

Ol olive; olympiad, four-year period between the Olympic Games

OL October League (communist group active in U.S.); Old Latin; Olsen Line; Oranje Line (Orange Line)

O.L. Office Lady

ola occipito-laeva anterior

OLA Office of Legislative Affairs; Ohio Library Association, Oklahoma Library Association; Ontario Library Association; Optical Laboratories Association; Osteopathic Libraries Association

OLADE Organización Latinoamericana de Energía (Spanish—Latin American Energy Organization)

Olan Olancho (Honduran department)

o'land overland

OLAPEC Organization of Latin American Petroleum Exporting Countries

OLAS Office of Arid Land Studies (University of Arizona); Organization of Latin American Solidarity; Organization of Latin American Students

OLAS Organización Latinoamericana de Solidaridad (Spanish—Latin American Solidarity Organization)

Olav Tryg Olav Trygvason

olbm (OLBM) orbital launched ballistic missile

OlBr olive brown

OLBS On-Line Banking Sheet

olc on-line computer

OLC Oak Leaf Cluster; Office of Legal Counsel

olcc optimum life-cycle costing

O L Cr Ordnance Lieutenant-Commander

OLCS On-Line Computer System

OLD Office of Legislative Development

Old Bailey London's Central Criminal Court

old-fash old-fashioned

Oldfos Old Established Forces

Old Maid's Old Maid's Day (June 4)

Old Point Old Point Comfort, Virginia

old pro(s) old professional(s)

old rep old repertory; old rep-robate

Olds Oldsmobile

OLDS On-Line Display System

Old Territorial Old Territorial Penitentiary (Santa Fe, New Mexico)

Old Test. Old Testament

OLE Office of Library Education (American Library Association); Organizational Learning in Enterprises

OLEA Office of Law Enforcement Assistance

oleo oleomargarine; oleoresins; oleum

OLEP Office of Law Enforcement and Planning

olericult olericulture

'oleum petroleum

O-levels ordinary levels (of educational tests)

OLEW Open Learning Experimental Workshop

olf olfactory; on-line filing

OLF Ohio Library Foundation; Orbital Launch Facility; Organ Literature Foundation

Olg Olga

OIG olive green

OLG Old Low German

Olgas The Olgas—mountain range west of Ayers Rock in Australia's Northern Territory

OLHMIS On-Line Hospital Management Information System

Oli Oliver

OLI Ocean Living Institute

O-license operator's license

Olig Oligocene

oligo (Latin prefix—few or small)—oligarchy

Olive Olivera; Olivia

OLIVER On-Line Instrumentation Via Energetic Radioisotopes

Oliver P Oliver (Cromwell) Protector

OLL Office of Legislative Liaison

OLMAT Otis-Lennon-Mental Ability Test

olmr (OLMR) organic liquid-moderated reactor

OLMR Office of Labor-Management Relations

OLMS Office of Labor-Management Standards

ol'n olden

OLOGS Open-Loop Oxygen Generating System

ol ol olive oil

olos out of line of sight

OLOS Office for Library Outreach Services

olow orbiter liftoff weight

olp occipito-laeva posterior; original list price

OLP Organización para la Liberación Palestina (Spanish—Palestinian Liberation Organization)—the PLO terrorists

olpar other large phased-array radar

OLPR Office for Library Personnel Resources (ALA)

OLPS On-Line Programming System

olq officer-like qualities

olr omnivorous leaf roller; overload relay

OLRB Ontario Labor Relations Board

ol res oleoresin

olrt on-line real time

ols ordinary least squares

ol's office ladies (divorcees and spinsters); old girls

OLS Optical Landing System

olsc on-line scientific computer

OLSD Office for Library Service to the Disadvantaged (ALA)

olt occipito-laeva transverse

ol & t owners, landlords, and tenants

Olt Old Italian

oltp on-line transaction processing

oltt on-line teller terminal

olv olivaceous; olive; on-line validation

OLV Onze Lieve Vrouw (Dutch—Our Lady)

o-l v's ovo-lacto vegetarians

Oly Olympia; Olympic

Olym Olympia

Olympic Olympic National Park, Washington; Olympic Stadium, Montreal

Olympics Olympic Games; Olympic Mountains, Washington

OLY Washington, D.C., Olympics (National Association)

om old man; old measurement; old men; operational monitor; optical microscope; organic matter; organized militia; our memo; outer marker

o & m (O & M) operation and maintenance

o.m. omni mane (Latin—every morning)

Om Omaha; Oman

Om. Book of Omni

OM Occupational Medicine; Old Man (colloquial); Oman (Internet code); Ordnance Map

O.M. Order of Merit

O & M Organization and Methods

OM Obermanual (German—upper manual keyboard); *Ostmark* (East German mark)

O. M. Optimus Maxum (Latin—best and greatest)—title given Jupiter by the Romans who worshipped him

oma orderly marketing arrangement

oma (Greek—swelling or tumor)—carcinoma, glaucoma, hematoma, lipoma, sarcoma

Oma Omaha, Nebraska

OMA Ocean Mining Administration (USDI); Office of Maritime Affairs; Oklahoma Military Academy; Omaha, Nebraska (airport); Ontario Medical Association; Overall Manufacturers' Association

OMAI Organisation Mondiale Agudas Israel (French—Agudas Israel International Organization)

Oman Sultanate of Oman (Arab oil-producing nation on Arabia's southeast coast), *Saltanat Oman*

omarb omarbetad (Swedish—revised)

OMARS Outstanding Media Advertising by Restaurants

OMAT Office of Manpower, Automation, and Training

Omb Ombudsman

OMB Office of Management and Budget; Ontario Municipal Board

OMBAC Old Mission Beach Athletic Club

OMBE Office of Minority Business Enterprise

om. bid. omnibus bidendis (Latin—every two days)

OMC Office of Munitions Control; Outboard Marine Corporation

omd off-market date

OMD Organic Mental Disorder

omdr off-market date received

'ome (Cockney contraction—home)

OME Office of Manpower Economics; Office of Minerals Exploration; Ordnance Mechanical Engineer(ing)

OMEF Office Machines and Equipment Federation

OMEGA Optimal Missile Engagement Guidance Algorithm (worldwide navigational system)

OMEL Orient Mid-East Lines

OMEP Organisation Mondiale pour l'éducation Préscolaire (French—World Organization for Pre-School Education)

O-Mess Officer's Mess

OMF Office of Management and Finance; Overseas Missionary Fellowship

omfp obtaining money by false pretenses

OMG Ophthalmology Medical Group

OMGE Organisation Mondiale de Gastro-Entérologie (French—World Gastro-Enterological Organization)

OMGUS Office of Military Government, United States

OMH Office of Mental Health

OMI Olympic Media Information; Operation Move-In

O.M.I. Oblate of Mary Immaculate

OMI Organización Maritima Internacional (Spanish—International Maritime Organization)

OMII Oxy Metal Industries International

omiom original meaning is the only meaning

omit orinthine-decarboxylase, motility, indole, trytophan-deaminase

omkr omdring (Norwegian—about)

oml outside mold line

OML Ontario Motor League; Orbiting Military Laboratory

omm ophthalmomandibulo-melic

OMM Office of Minerals Mobilization

OMM Organisation Météorologique Mondiale (French), *Organización Meteorologica Mundial* (Spanish—World Meteorological Organization)—WMO

OMMA Outboard Motor Manufacturers Association

OMMS Office of Merchant Marine Safety (USCG)

omn. bih. omni bihora (Latin—every two hours)

omn. hor. omni hora (Latin—every hour)

omni omnidirectional; omnirange; omnivisual

omn. man. omni mane (Latin—every morning)

omn. noct. omni nocte (Latin—every night)

omn. quad. hor. omni quadrante hora (Latin—every quarter of an hour)

omor one man, one responsibility

omp organo-metallic polymer(s)

ompa one-man pension arrangement

OMPD Office of Mineral Policy Development

OMPER Office of Manpower Policy Evaluation and Research

ompf omphaloskepsis

OMPI Organización Mundial de la Propiedad Intelectual (Spanish—World Intellectual Property Organization)

OMPO Oahu Metropolitan Planning Organization

ompr optical mark page reader

OMPRA Office of Minerals Policy and Research Analysis

OMPSA Organisation Mondiale pour la Protection Sociale des Aveugles (French—World Organization for the Welfare of the Blind)

OMPU Oficina Municipal de Planeamiento Urbano (Spanish—Municipal Office of Urban Planning)

omr office methods research; optical mark reader; optical mark recognition

OMR Officer Master Record

OMRD Overseas Mineral Resource Development

OMRs Optical Mark Readers

oms output per man shift

OMS Office of Management Studies; Orbital Maneuvering System

OMS Organisation Mondiale de la Santé (French), *Organización Mundial de la Salud* (Spanish—World Health Organization—WHO; *Otdel Mezdunarodnyk Svyazey* (Russian—International Relations Section)—network of overseas agents

OMSA Offshore Marine Service Association; Orders and Medals Society of America

OMSF Office of Manned Space Flight (NASA)

OMSIP Ontario Medical Surgical Insurance Plan

omt orthomode transducer; osteopathic manipulation therapy

OMT Old Merchant Taylors

OMtns Olympic Mountains

OMTS Organizational Maintenance Test Station

OMV Orbital Maneuvering Vehicle

on. octane number

o/n own name

on onomastikon (Greek—lexicon)

o.n. omni nocte (Latin—every night)

On Onorevole (Italian—Honorable); *Onsdag* (Danish—Wednesday)

ON Official Number; Ogden Nash; Old Norse; Ontario Northland (railway); Operation Notice

O.N. Orthopedic Nurse

O/N Order Number

O & N Oregon & Northeastern (railroad)

ÖN Österreichische Nationalbibliotek (Austrian National Library)

ona optical navigation attachment

ONA Office of National Assessment; Office of Noise Abatement; Open Network Architecture; Overseas National Airways; Overseas News Agency

on a/c on account

ONAC Office of Noise Abatement and Control

ONAP Orbit Navigation Analysis Program

on approv on approval

onbep onbepaald (Dutch—indefinite)

O-N Border Oder-Neisse Border separating Germany and Poland

onc operational navigational chart(s)

ONC Office of New Careers; *Oficina Nacional del Café* (Spanish—National Coffee Administration—Honduras); Oregon-Nevada-California (fast freight truck line)

oncol oncologic(al)(1y); oncologist(ic)(al)(1y); oncology; oncolysis; oncolytic(al)(1y)

OND Ophthalmic Nursing Diploma

ONDC Office of National Drug Control

ONDCP Office of National Drug Control Policy

ONE Office of National Estimates (CIA)

Oneg Onegin

Onega Lake Onega northeast of Leningrad, called Ozero Onezhskoye by the Russians

Oneida Oneida Community noted for the silverware and steel traps produced while practicing complex marriage and common care of their offspring in Oneida, New York; in 1881 the commune was incorporated

O'Neill Eugene O'Neill

ONEO Office of Navajo Economic Opportunity

ONERA *Office National des Etudes et des Recherches Aérospatiales* (French— space research agency)

one-spot $1 bill

ONF Offensive Nuclear Forces; Old Norman—French

onfm on nearest full moon

ong *ongaku* (Japanese—music); *ongeveer* (Dutch— about, approximately, roughly)

ONG Old North German

ONG *Organisation Non-Gouvernementale* (French—Non-Governmental Organization)

on hol(s) on holiday(s)

ONI Office of Naval Intelligence; Office of NWTS Integration

ÖNJ *Österreichische Nationalbibliothek Josefsplatz* (German—Josefsplatz Austrian National Library)

ONM Ocmulgee National Monument; Office of Naval Material

ONMSS Office of Nuclear Material Safety and Safeguards

ONNI Office of National Narcotics Intelligence

onnm on nearest new moon

ono or near offer

o-'n'-o one and only

ONO Organization of News Ombudsmen

ONO *Oesnoroeste* (Spanish— west northwest); *oost noord oost* (Dutch—east northeast)

onomast onomastic(al)(ly); onomastics; onomatologist; onomatology

onomat onomatologic(al)(ly); onomatologist(ic)(al)(1y); onomatology; onomatopoeia

O Norm F Old Norman French

O North Old Northumbrian

O Norw Old Norwegian

o noz oil nozzle

onp operating nursing procedure

ONP Office of National Programs; Olympic National Park (Washington); Open Network Provision

ONR Office of Naval Research; Official Naval Reporter

ONRL Office of Naval Records and Library

ONRRR Office of Naval Research Resident Representative

ON Rwy Ontario Northland Railway

ONSR Ozark National Scenic Riverways (Missouri)

On Sta On Station

ont ontology; ordinary neap tide

Ont Ontario

ONT Our New Thread (Clark's trademark)

ONTC Ontario Northland Transportation Commission

Ont Pen Ontario Penitentiary

Ont Sci Cen Ontario Science Center

ONU *Organisation Nations Unies* (French—United Nations Organization); *Organización de las Naciones Unidas* (Spanish—United Nations Organization)— UNO; *Organizzazione Nazioni Unite* (Italian—United Nations Organization)

ONUC *Operation des Nations Unies, Congo* (French— United Nations Operation in the Congo)

ONUDI *Organización de las Naciones Unidas para el Desarolla Industrial* (Spanish— United Nations Organization for Industrial Development)

ONUESC *Organisation des Nations Unies pour l'Education, la Science et la Culture Intellectuelle* (UNESCO)

ONULP Ontario New Universities Library Project

on w *onovrgankelijk werkwoord* (Dutch—intransitive verb)

ONW Oregon and Northwestern (railroad)

ONWI Office of Nuclear Waste Isolation

ONWM Office of Nuclear Waste Management

ONWR Okefinokee National Wildlife Refuge (Florida and Georgia); Ottawa National Wildlife Refuge (Ohio); Ouray National Wildlife Refuge (Utah)

ony onymous (opposite of anonymous)

oo (OO) office of origin

o/o oil/ore (carrier); on order

o & o owned and operated

o-to-o out-to-out

oo (Latin prefix—egg)—oocyte, oology

o(O) original

O/o Order of

OO Observation Officer; Oceanic Operators; Oceanographic Office

O/O Office of Oceanography (UNESCO)

O of O Order of Owls

ooa on or about

OOA Office of Ocean Affairs

OOAA Olive Oil Association of America

OOAMA Ogden Air Materiel Area

oob opening of business; out of bed; outs on base; out of business

o-o B off-off Broadway; off-off Broadway theater(s)

OoB Order of Battle

OOB Old Orchard Beach

oobe out of body experience; out-of-the-box experience

oobp opponents' on-base percentage

OoC Office of Censorship

OOCH Orient Overseas Container Holdings

OOCL Orient Overseas Container Line

OOD Officer of the Day; Officer of the Deck

oodb object-oriented database

oodep owners, officers, directors, and executive personnel

oodms object-oriented database management system

Oody Eunice

OO/Eng out of stock but on order from England (for example)

OoF Office of Facilitation

OOG Office of Oil and Gas; Officer of the Guard

OOH *Occupational Outlook Handbook*

OOHA Operation Oil Heat Associates

ooj obstruction of justice

ool oology; operator-oriented language

OOL Odessa Ocean Line: Orient Overseas Line

oolhmd optimized optical-link helmet-mounted display

oolr ophthalmology, otology, laryngology, rhinology

OOM Officers Open Mess

OO McIntyre Oscar Odd McIntyre (newspaper columnist: *New York By Day*)

o/o/o out of order

O o O One on One (tv program)

OOO-gauge $^3/_4$-inch track gauge (model railroads)

oop object-oriented programming; out of pocket (expenses); out of print (book)

OOP Oceanographic Observations of the Pacific

ooparts out-of-place artifacts

OOPEC Office for Official Publications of the European Communities

oops object-oriented programming; off-line operating simulator; offshore oil-pollution sleeve

OOPS Organization of Oil Producing States

OOQ Officer of the Quarters

OOR Office of Ordnance Research

oos occupational overuse syndrome; orbit-to-orbit shuttle; orbit-to-orbit stage; out of stock

o & o's owned and operated (tv broadcast) stations (controlled by a network)

OOSC Olfactronics and Odor Sciences Center (IITRI)

oot out of tolerance; out of town

OOT Office of Operational Testing

ooté Out-of-town executive

ootg one of the greats

Ooty Ootacamund, Tamil Nadu

OOW Officer On Watch

op oil pressure; old prices; open phase; open policy; opera; operating point; operation; operation plan(s); operational; operational priority; operator; operetta; opium; opposite prompt (stage left); optical probe; opus; ordinary pay; other people's (possessions); out of print; outer panel; outside production; overproof; overprune; overpuff

op (OP) outpatient

o/p off peak; optional; output; overpriced

o & p ova and parasites

Op optical art (art accented with or based on optical illusions); Oregon pine

Op. *Opus* (Latin—composition, literary or musical work)

OP Observation Post; Office of Preparedness; Office of Protocol (U.S. Department of State); Open Policy (floating cargo insurance); Oregon pine; organophosphate

O-P Oppenheimer-Phillips (process)

O.P. *Optimus Maximus* (Latin—supreme and best)— Jupiter's title as he was believed to be the king of the gods and the ruler of all rulers; *Ordo Praedicatorum* (Order of Preachers—Dominicans)

opo2 submarine warfare (Chief of Naval Operations numbers)

opo3 surface warfare (Chief of Naval Operations numbers)

op94 command and control (Chief of Naval Operations numbers)

op95 naval warfare (Chief of Naval Operations numbers)

op96 systems analysis (Chief of Naval Operations numbers)

opa optical plotting attachment; optoelectric pulse amplifier

OPA Office of Population Affairs; Office of Price Administration; Office of Public Affairs; Overall Payments Agreement

OPA *Oficina Postal Ambulante* (Spanish—Mobile Post Office)

OPAC Online Public Access Catalog

opal hydrous silica ($SiO_2 \cdot nH_2O$)

opal. optical platform alignment linkage

op amp operational amplifier

OPANAL *Organismo para la Proscripción de las Armas Nucleares en la América Latina* (Spanish—Organization for the Prohibition of Nuclear Weapons in Latin America)

op art optical art (art involving optical illusion)

OPB Occupational Pensions Board; Oregon Public Broadcasting

OPBE Office of Planning, Budgeting, and Evaluation (NIE)

OPBMA Ocean Pearl Button Manufacturers Association

opc office percentage; ordinary portland cement; other parks corrector

OPC Office of Price Control; Office of Public Communication; Ohio Power Company; Operating Center; Out-Patient Clinic; Overseas Press Club

OPCA Overseas Press Club of America

OPCC: Optical Product Code Council

op. cit. *opere citato* (Latin—in the work cited); *opus citato* (Latin—in the work cited)

OPCNM Organ Pipe Cactus National Monument

opco operating company

op code operation code (data processing)

op com *opéra-comique* (French—comic opera; operetta)

opcon(s) operation control(s)

OPCS Office of Population Censuses and Surveys

OPCW Organizations to Prohibit Chemical Weapons

opd opening delayed; optical path difference

o-p-d oto-palato-digital (syndrome)

OPD Office of Policy Development (White House); Officer Personnel Directorate; Out-Patient Department

opdar optical direction and ranging

OPDD Operational Plan Data Document

op dent operative dentistry

OPDR Oldenburg–Portugiesische–Dampfsehifs–Reiderei (steamship company)

ope open-point expanding; opium; oxidation pond effluents

OPE Office of Planning and Evaluation (FBI); Operations Project Engineer

O P & E Oregon, Pacific & Eastern (railroad)

OPEAA Outdoor Power Equipment Aftermarket Association

OPEC Oil Producer's Economic Cartel; Organization of Petroleum Exporting Countries

op ed opposite the editorials (newspaper page usually reserved for readers' letters and syndicated columns)

OPEDA Outdoor Power Equipment Distributors Association

opef overall plume-enhancement factor

OPEI Outdoor Power Equipment Institute

OPEIU Office and Professional Employees International Union

open. open circuit; opening

OPen Olympic Peninsula

opens. open circuits (electrical parlance); openings

opep (OPEP) orbital plane experiment package

OPEP Organisation des Pays Exportateurs de Pétrole (French—Organization of Petroleum Exporting Countries)

OPEP/OPEC Organización de Paises Exportadores de Petróleo (Spanish—Organization of Petroleum Exporting Countries)

oper operational

O Per Old Persian

OPER Office of Policy, Evaluation, and Research

Opera Paris Opera

Opera-Com Opéra-Comique (Paris)

operg operating

OPers Old Persian

OPers. Old Persian (rug)

OPET Organizations for the Promotion of Energy Technology

opex operational (and) executive (personnel)

OPEX Operational, Executive (and Administrative Personnel Program of the United Nations)

opfor opposition force

opg opening

O Pg Old Portuguese

OPG osteoprotegerin; Overseas Project Group

OPGA Ohio Personnel and Guidance Association; Oregon Personnel and Guidance Association

oph office phone; ophicleide; ophthalmologist; ophthalmology; ophthalmoscope; ophthalmoscopic

Oph Ophiucus (constellation)

Oph.D. Doctor of Ophthalmology

ophidiol ophidiologic(al)(1y); ophidiologist; ophidiology

OPHS Operational Propellant Handling System

ophth ophthalmologist; ophthalmology

ophthal ophthalmic; ophthalmologist; ophthalmology

Ophthalmias Ophthalmia Range of mountains in Westen Australia near Jiggalong and Mundiwindi

ophthalmol ophthalmologic-(al)(1y); ophthalmologist; ophthalmology

OPI Office of Primary Interest; Office of Programs Integration (ERDA), Office of Protective Intelligence (U.S. Secret Service); Office of Public Information; Office of Public Inquiry; Offsite Production (Purchase) Inspection; Omnibus Personality Inventory; Ordnance Procedure Instrumentation; Outside Production (Purchase) Inspection

OPIC Overseas Private Investment Corporation

opim order processing and inventory monitoring

opis opisometer

OPIS Operational Priority Indicating System

opl operational; other parts line

opl oplag (Danish—edition)

OPL Omaha Public Library; Orlando Public Library; Ottawa Public Library

OPLA Offshore Pollution Liability Agreement

OPLP Office of Program and Legislative Planning

opm operations per minute; operator programming method; optically-projected map; options pricing model; orthophoto map; other people's money

OPM Office of Personnel Management; Office of Production Management

OPMA Office Products Manufacturers Association; Open Pit Mining Association

OPMAC Operation for Military Aid to the Community

OPMCS Otto Pre-Marital Counselling Schedules

opn open (flow chart); operation

o.p.n. ora pro nobis (Latin—pray for us)

OpNav Office of the Chief of Naval Operations

OPNAVINST Office of the Chief of Naval Operations Instruction

opnd opened (flow chart)

opng opening

OPNL Osaka Prefectural Nakanoshima Library (Japan)

opnn opinion

Op. no. opus number

opo one-person operation; one price only; other programmed operations

Opo Oporto

OPO Office of Personnel Operations (U.S. Army)

O Pol Old Polish

OPOR Office of Public Opinion Research

oport operation(s) order

O por O Ojo por Ojo (Spanish—Eye for an Eye)—Guatemalan terrorists

O Port Old Portuguese

OPOs one-person-operated buses

opp opportunity; opposed; opposite; opposition, out of print at present

OPP Office of Pesticide Programs; Ontario Provincial Police; Otago Press and Produce

OPPE Office of Programming, Planning, and Evaluation; Operational Propulsion Plant Examination (USN)

OPPI Organization of Pharmaceutical Producers of India

opplan operating plan

oppor opportunity

oppo's opposite numbers

oppy opportunity

opq opaque

opr operate; operator; optical pattern recognition

OPr Old Provençal

OPR Office of Planning and Research; Office of Population Research (Princeton); Office of Primary Responsibility; Office of Professional Responsibility (FBI); Office of Professional Responsibility (INS)

OPRA Office Products Representatives Association

oprad operations research and development

oprex operational exercise

O Prov Old Provençal

opr's old prices riots

OPruss Old Prussian

ops open profiling specification (World Wide Web); operations; operations per second; opposite prompter's side (of stage);

op's other people's

OPs organophosphates

OPS Office of Pipeline Safety; Office of Price Stabilization; Office of Product Standards; Oxygen Purge System

OPS *Organization Panaméricaine de la Santé* (French—Pan-American Health Organization); *Organización Panamericana de la Salud* (Spanish—Pan-American Health Organization)

ops analysis operations analysis

Ops Atts Gen Opinions of the Attorneys-General of the United States

opscan optical scanning

Ops Comms Opinions of the Commissioners

Op Sec Operational Security

OPSM Optical Prescriptions Spectacle Makers

OPSP Office of Product Standards Policy

OPSR Office of Pipeline Safety Regulations

opstat operational status

opt optic; optical; optician; optics; optimal; optimum; option; optional

OPT Office of Promotion and Tourism

OPTA Organ and Piano Teachers Association

optacon optical-to-tactile converter

Opt Acta *Optica Acta* (Latin—Optics Gazette)

opt Commum *Optics Communications*

Opt.D. Doctor of Optometry

OPTEVFOR Operational Test and Evaluation Force

OPTEVG Operational Test and Evaluation Group

opti optimist(ic); optimize; optimum

optic. optical(ly); optician; opticociliary; opticopupillary

opticon optical tactical converter

optim optimization(al)(ly), optimize(d), optimum

Opt Lett *Optics Letters*

opt-Mekh Prom *Optika-Mekhanicheskaya Promyshlennost* (Russian—Journal of Optical Technology)

optmrst optometrist

optn optician

Opt News *Optics News*

optoel optoelectronics

optom optometer; optometric-(al)(ly); optometrist; optometry; optomyometer

optr *optryk* (Dano-Norwegian—reprint)

optrak optical tracking

Opt Spektrosk *Optika i Spektroskopiya* (Russian—Optics and Spectroscopy)

optul optical pulse transmitter using laser

OPU Unemployed Peoples Union

opur objective program utility routines

OPUS Older People United for Service; Open University System; Operating Utility System; Organization for Promoting the Understanding of Society

opv oral polio vaccine; oral polio virus

OPW Office of Public Works

oq oil quench; overmation quotient

OQ Officers Quarters

oqe objective quality evidence

oql on-line query language

OQMG Office of the Quartermaster General

OQR Officer's Qualification Record

or objective response; operationally ready; operations research; other ranks; out of range; outside radius; outside right; overseas replacement; owner's risk; oxidation-reduction

or (OR) orienting reflex; released on one's own recognizance

o/r on request; other ranks

o & r ocean and rail; overhaul and repair

or (Latin prefix—mouth)—oral

or. *oratio* (Latin—speech, discourse)

Or Oregon; Orient(al)

Ór *Óri* (Modern Greek—mountains); *Óros* (Modern Greek—mountain)

OR Oak Ridge; Officer Records; Official Records; Olympic Range; Olympic Record; omnidirectional radio range (symbol); Operating Room; Operational Requirement; Operation Rescue; Operations Requirement; Operations Research; Operations Room; Ordinance Report; Oregon; Owasco River (railroad); Oyster River

O.R. Operating Room

O/R Owner's Risk

O of R Office for Research (ALA)

ÖR *Österreichischer Rundfunk* (Austrian Radio and Television)

OR *Ontario Reports; Operations Research*

Ora Orabel(le)

ORA Oil Refiners Association; Operations Research Analyst

oracle optical reception of announcements of coded-line electronics

ORACLE Optimum Record Automation for Courts and Law Enforcement (Los Angeles, CA); Oracle Corporation trademark

ORAD Office of Rural Areas Development

ORAM Office for Research in Academic Methods

orang orangutan

orang *orangutan* (Malay—forest person)—one of the great anthropoid apes found in Borneo and Sumatra

orange light change approaching; potential danger

Oranges New Jersey's East Orange, Orange, South Orange, and West Orange; may also refer to the Orange Mountains also called the Watchungs

ORASS Offender Risk Assessment Scoring System

orat oration; orator; oratorio; oratory

ORAU Oak Ridge Associated Universities

orb omnidirectional radio beacon; owner's risk of breakage

orb (ORB) oceanographic research buoy

orbatrep order of battle report

ORBC Organic Rankine Bottoming Cycle

orbic orbicular; orbicularis

Orbis Polish Travel Office

ORBIT On-line Retrieval of Bibliographic Information Timeshared

Orbiter half-plane half-satellite space shuttle

ORBIS Orbiting Radio Beacon Ionospheric Satellite

orbs off-reservation boarding school

ORBS Orbital Rendezvous Base System

orc owner's risk of chafing

Orc Orcadian (inhabitant of or pertaining to Orkney Islands)

ORC Occan Racing Club; Officers Reserve Corps; Offshore Racing Council; Opinion Research Corporation; Organic Rankine Cycle; Overseas Research Council; Ozarks Regional Commission

ORCA Ocean Resources Conservation Association

Orcades Orkney Islands

ORCAP *Oficina Regional para Centroamérica y Panamá* (Spanish—Regional Office for Central America and Panama)

ORCB Order of Railway Conductors and Brakemen

ORCC Orangutan Research and Conservation Center, Tanjung Puting National Park, Borneo

orch orchestra; orchestral; orchestration

Orch Orchard

orch circ orchestra circle

Orch Consv *Orchestre de la Société des Concerts du Conservatoire de Paris* (French—Concert Society Orchestra of the Paris Conservatory)

Orch de l'Opera de Paris *Orchestre du Théatre National de l'Opera de Paris* (French—National Theater Orchestra of the Paris Opera)

orches orchestration

Orch H Orchestra Hall

orchi (Latin prefix—testicles)—orchid, orchiectomy

ORCHIS Oak Ridge Computerized Hierarchical Information System

orchl orchestral

Orch Nat *Orchestre National de la Radiodiffusion Française* (French—National Orchestra of French Broadcasting)

Orch Suisse Rom *Orchestre de la Suisse Romande* (French—Swiss Canton Orchestra)

Orch Symp de Mont *Orchestre Symphonique de Montreal* (French—Montreal Symphony Orchestra)

ORCL Oracle Systems Corporation

ORCMD Orlando Contract Management District

orcon organic control

ORCS Organic Rankine Cycle System

ORCUP Ontario Region Canadian University Press

ord operational ready date; order(s); ordinal; ordnance

ord. ordained; ordinary

o-r-d owner's risk or damage

Ord Order; Orderly; Ordinary Seaman

ORD Chicago, Illinois (O'Hare Airport); Office of Research and Development

ORDA Oceanographic Research for Defense Application

ORD-ALA Office of Research and Development—American Library Association

Ord Bd Ordnance Board

OrdC Ordnance Corps

Ord Dept Ordnance Department

ordfin ordinary finish

ordinst ordnance instruction

Ord Man Ordnance Manual

ordn ordnance

Ordn Surv Ordnance Survey

Ordo Ordovician

ORDP Office of Rural Development Policy

ords ordinary shares

Ord Sgt Ordnance Sergeant

ordvac ordnance variable automatic computer

ore overtraining reversal effect

Ore Oregon(ian)

ORE Ocean Research Equipment; Operational Research Establishment

OR & E Office of Research and Engineering

ORE *Office de Recherches et d'Essais* (French—Office of Research and Testing)

OREAM *Organisation d'études d'Aires Métropolitaines* (French—Organization for the Studies of Metropolitan Areas)

Oreg Oregon; Oregonian

Oregon Caves Oregon Caves National Monument

Ore-Ida pots Oregon-Idaho potatoes

o/r enema oil-retention enema

OREO Other Real Estate Owned

ORES Office of Research and Engineering Services

ORESCO Overseas Research Council

orf open reading frame; orifice; overhaul replacement factor

o-r-f owner's risk of fire

ORF Norfolk, Virgina (airport); Oceanic Research Foundation

ÖRF *Österreichischer Rundfunk* (Austrian radio and TV network)

Or F S Orange Free State

Org organ; organic; organization; organize; organizer

ORG Operations Research Group

organ. organic; organization

Organ Pipe Cactus Organ Pipe Cactus National Monument in Arizona

org art organic art(ist)

Orgburo Organizational Bureau of the Central Committee (of the Communist Party)

org exp *organo espressivo* (Italian—expressive organ part)

Org Gard *Organic Gardening*

orgl organizational

org-man organization man

orgn organization

ORGS Operational Research Group of Scotland

orgst organist

ori orientation inventory

Ori Orient(al)(ism); Oriente; Orion (constellation)

ORI Ocean Research Institute; Ocean Resources Institute; Office of Research Integrity; Office of Roads Inquiry; Office Research Institute; Operation Readiness Inspection

ORIA Oriental Rug Importers Association

ORIC Oak Ridge Isochronous Cyclotron

oride override

ORIEL Oriel College; Oxford

orient. oriental; orientation

ORIENT Orient Airways

Orient(al) Asia(tic)

Orientales Orientales—la patria o la tumba (Spanish—Eastern landsmen, our country or the tomb)—anthem of Uruguay

Orient Express (*see* Ori Exp)

Ori Exp Orient Express (formerly between Paris and Istanbul via Vienna but now called Central Kingdom Express running from London to Hong Kong via Paris, Berlin, Warsaw, Moscow, Irkutsk, Peking, Nanking, and Canton)

orif open reduction with internal fixation

orig origin; original; originally; originator

O-ring O-shaped ring

ORINS Oak Ridge Institute of Nuclear Studies

orion on-line retrieval of information over a network

oris orismological; orismologist; orismology

ORIT Operational Readiness Inspection Test

ORIT Organización Regional Interamericana de Trabajadores (Spanish—Interamerican Regional Labor Organization)

or j orange juice

Ork Orkney Islands

Orkneys Orkney Islands

orl orlon (synthetic fiber); owner's risk of leakage

'OrL '*Orlah*

ORL Orbital Research Laboratory; Ordnance Research Laboratory; Orlando, Florida (Harndon Airport)

ORL Outlook on Research Libraries

ORLA Optimum Repair Level Analysis

Orleans New Orleans

Órm Órmos (Modern Greek—bay)

ORM Ohio Reformatory for Men

ORMA Office of Refugee and Migration Affairs

ORMAK Oak Ridge Tokamak

orml oriental meal

orm('s) off-road motorcycle(s)

orn orange; ornament

orn orne (French—decorated, ornamented)

Orn Oran (British maritime contraction)

ORN Operating Room Nurse

ornith ornithology

ornithol ornithologic(al)(ly); ornithologist; ornithology

ORNL Oak Ridge National Laboratory

ORNLL Oak Ridge National Laboratory Library

ORO Oak Ridge Operations Office; Operations Research Office (Johns Hopkins University)

or. obliq. oratio obliqua (Latin—indirect speech, oblique speech)

orog orographer; orographic; orographical; orography

orp ordinary, reasonable, and prudent

ORP Okret Rzecypospolitej Polskiej (Polish—Ship of the Polish Republic)

ORPA Office of Regional and Political Affairs (CIA)

ORPC Office of Rail Public Counsel

orph orphan; orphanage; orphaned; orphans

orpiment arsenic sulfide

o-r pot. oxidation-reduction potential

orr operations research research (ORR)

o-r-r owner's risk rates

ORRA Oriental Rug Retailers of America

o-r release own-recognizance release

ORRRC Outdoor Recreation Resources Review Commission

ORRT Operational Readiness Reliability Test

ors omnidirectional range station; owner's risk of shifting

ors (ORS) orbiting research satellite; orthopaedic surgery

or's onion rings; orienting responses

ors. orationes (Latin—speeches)

ORS Office of Refugee Settlement; Office of Research and Statistics; Official Rate Standard; Old Red Sandstone; Operational Research Society

ORSA Operations Research Society of America

ORSANCO Ohio River Valley Water Sanitation Commission

ORSE Operational Reactor Safeguard Examination

OrSEA Oregon Society of Enrolled Agents

ORSIP Office of Research, Statistics and International Policy

ORSJ Operations Research Society of Japan

ORSTOM Office de la Recherche Scientifique et Technique d'Outre Mer (French—Overseas Office of Scientific and Technical Research)

ort odor recognition threshold; operational readiness training

ORT Operating Room Technician; Operational Readiness Test; Oral Rehydration Test; Order of Railroad Telegraphers; Organization for Rehabilitation through Training; Overage Retirement Training (program)

ORTF Office de Radiodiffusion Télévision Française (French—French Office of Television Broadcasting)

ortho orthochromatic; orthographic; orthography; orthopedic(s)

ortho (Latin prefix—normal or straight)— orthopedic

Ortho Greek Orthodox

orthog orthography

ortho-k orthokeratological(ly); orthokeratologist; orthokeratology

orthokera orthokeratologist; orthokeratology

orthomol orthomolecular; orthomolecularologist; orthomolecularology

orthop orthopedics

orthor orthorhombic

ORTO Occupational Rehabilitation Training for Overseas

ORTPA Oven-Ready Turkey Producers' Association

ORTS Optional Residence Telephone Service

ORTU Other Ranks Training Unit

ORU Oral Roberts University
ORuss Old Russian
ORV Ocean Range Vessel (naval symbol)
Orvidius Orvidius Naso, Roman poet
orvr on-board refueling vapor recovery
orv('s) off-road vehicle(s)
orw owner's risk of wetting
ORW Ohio Reformatory for Women
Orwell George Orwell (Eric Arthur Blair)
ory (Latin suffix—pertaining to)—sensory
ORY Paris, France (Orly Airport)
os oil solvent; oil switch; old series; old style; on station; out of stock; output secondary; outside; outsize; overseas; oversize
os (OS) operating system (data recording)
o/s out of service; out of stock
o & s operating and support (costs); operation and service; over and short
o.s. *oculus sinister* (Latin—left eye)
o-S on-Sea
Os Osmium
OS Ocean Station; Office Surgery; Old Saxon; Old Series; Operating System; Operation Sandstone; Operation Snapper; Optical Society; Ordinary Seaman; Ordnance Specifications; Ordnance Survey; Overseas Service
O.S. Old Style
OS₂ Operating System 2
osa oil-soluble acid; order for simple alert
Osa Osaka
Osa (Russian—Bee)—a Soviet class of guided-missile patrol boats
OSA Office of the Secretary of the Army; Official Secrets Act; Omnibus Society of America; Optical Society of America; Order of Saint Augustine; Osaka, Japan (airport); Overseas Sterling Area; Overseas Supply Agency; Oyster Shell Association
OSA *Ocean Shipping Act; Official Secrets Act*
osac orifice spark advance control

OSAF Office of the Secretary of the Air Force
OSAHRC Occupational Safety and Health Review Commission
OSAP Office of Substance Abuse Prevention in the Alcohol, Drug Abuse, and Mental Health Administration; Ontario Student Awards Program
Osa Pen Osa Peninsula in southern Costa Rica
OSAS Overseas Service Aid Scheme
O Sax Old Saxon
osb oriented strand board
OSB Order of Saint Benedict; Otago Savings Bank; Overseas Service Bureau
O.S.B. Order of St. Benedict
OSB *Occupational Safety Bulletin*
OSBA Ohio School Boards Association; Oregon School Boards Association
OSBM Office of Space Biology and Medicine
osc oscillator
Osc Oscan
OSC Office of Special Counsel; On-Scene Commander; Ontario Science Centre; Ontario Securities Commission; Order of St. Clare; Ordnance Systems Command (formerly Bureau of Weapons); Overseas Shipping Company
O.S.C. Oblate of Saint Charles
O of SC Order of Scottish Clans
OSCA Office of Senior Citizens Affairs
OSCA *Officine Specializzate Costruzione Automobili* (Italian—Special Office of Automobile Construction)
OSCAA Oil-Spill Control Association of America
O Scan Old Scandinavian
oscar orbital-satellite-carrying amateur radio (OSCAR); oxygen steelmaking computer and recorder
Oscar letter O radio code
OSCAR Ocean Surface Current Simulator; On-Line System for Controlling Activities and Resources; Optical Switching Systems, Components and Architecture Research; Optimum System for the Control of Aircraft Retardation

Oscar(s) Motion Picture Academy Award(s)
OSCE Organization for Security and Cooperation in Europe
ÖSCG *Österreichische Studiengesellschaft für Kybernetik* (German—Austrian Society for Cybernetic Studies)
OSCO Oil Service Company of Iran; Oil Shipment Corporation
oscope oscilloscope
oscp oscilloscope
OSCP Ocean Sediment Coring Program (NSF)
OSCT Office of Scholarly Communication and Technology
osd on-line systems driver; open shelter deck; optical scanning device; out-of-station designation
o s & d over, short, and damaged
OSD Office of the Secretary of Defense; Operational Support Directive; Ordnance Supply Depot; Original Sponsoring Distributor
OSDBMC Office of the Secretary of Defense, Ballistic Missile Committee
OSDBU Office of Small and Disadvantaged Business Utilization (U.S. Department of State)
OSDNRL Ocean Science Division—Naval Research Laboratory
osdocs over-the-shore discharge of container ships
osdp on-site data processing
OSDP Operational System Development Program
OSDSA Office of the Secretary of Defense, Systems Analysis
OSDSAC Office of the Secretary of Defense, Scientific Advisory Committee
ose operational support equipment
ose (Latin suffix—full of)—adipose
OSE Ocean Shipping and Enterprises; Office of Science Education; Office of Sex Equity (HEW); Office of Systems Engineering

OS & E Ocean Science and Engineering

OSEAP Oil Shale Environmental Advisory Panel

o'seas overseas

OSEB Orissa State Electricity Board

OSerb Old Serbian

osf operational service fee; ordinary shareholders funds

OSF Open Software Foundation; Order of St. Francis

OSFI Open Steel Flooring Institute

O.S.F.S. Oblate of Saint Francis of Sales

osg outstanding

OSG Office of Sea Grant (NOAA); Office of the Secretary General (UN)

OSG Official Steamship Guide

OSGP Office of Sea Grant Programs

OSGS On Sudan Government Service

o.s.h. omni singula hora (Latin—every hour)

Osh Ossian

OSH Office on Smoking and Health

OSHA Occupational Safety and Health Act; Occupational Safety and Health Administration

o/sheep odd sheep

o/ship ownership

OSHPD Office of Statewide Health Planning and Development

OSHRC Occupational Safety and Health Review Commission

OSHS Occupational Safety and Health Scheme

osi out of stock indefinitely

OSI Office of Samoan Information; Office of Scientific Integrity; Office of Special Investigation (USAF); Office of Special Investigations (Dept. of Justice and U.S. Army); Off-Site Instruction; Ohio Scientific Incorporated; Open Systems Interconnect; Open System Interconnection; Other Service Investigation; Owner Satisfaction Index

OSIA On-Site Inspection Agency; Order of the Sons of Italy in America

osie operational support integration engineering

OSIP Operational and Safety Improvement Program

OSIR Oil-Spill Intelligence Report

osis (Greek—condition or state of being)—arteriosclerosis, cirrhosis, halitosis, tuberculosis

OSIS Office of Science Information Service; On-Site Inspection System; Open Shops for Information Systems

Osk Oskarshamm

OSK Osaka Syosen Kaisha (Osaka Mercantile Steamship Company)

OSK Országos Széchényi Könyvtár (Hungarian—National Széchényi Library)—Budapest

osl orbiting space laboratory

Osl Oslo

OSl Old Slavonic

OSL Office of the Secretary of Labor; Orbiting Solar Laboratory; Oslo, Norway (airport)

OSLat Old-Style Latin

OSlav Old Slavic

osm osmosis; osmotic

Osm osmol(s)

O s M Orchestre Symphonique de Montréal (French—Montreal Symphony Orchestra)

OSM Office Service Manual; Office of Surface Mining; One of the Swinish Multitude (Philip Freneau, poet of the American Revolution, used this three-letter device after his name, thereby deriding similar-looking British titles); *Overzees Scheepvaart Maatschappij* (Overseas Shipping Company)

OSMA Otago-Southland Manufacturers Association

OSMM Office of Safeguards and Materials Management (AEC)

osmol osmosis + mol (standard unit of osmotic pressure)

osmos own ship's motion simulator

OSMRE Office of Surface Mining Reclamation Enforcement

OSN Office of the Secretary of the Navy

OSN Orquesta Sinfónica Nacional (Spanish—National Symphonic Orchestra)

OSNC Orient Steam Navigation Company

OSNY Oratorio Society of New York

oso (OSO) orbiting solar observatory

OSO Offshore Supplies Office; Offshore Supply Office; Omaha Symphony Orchestra; Ordnance Supply Office(r); Oregon Symphony Orchestra

OSO Oessudoeste (Spanish—west southwest); Orbiting Solar Observatory; Ordnance Supply Office

OSODS Office of Strategic Offensive and Defensive Systems (USN)

osp optimum sustainable population; outside purchased

o-sp off-street parking

o.s.p. obiit sine prole (Latin—died without issue)

Osp Old Spanish

OSP Open-Space Program; Order of St. Paul

OSP Oficina Sanitaria Panamericana (Spanish—Pan-American Sanitation Office)

OSPA Overseas Pensioners' Association

OSPA Organisation de la Santé Panaméricaine (French—Pan-American Health Organization)

OSPAAL Organización de Solidaridad de los Pueblos de Asia, Africa, y Latinoamérica (Spanish—Organization of Solidarity of the Peoples of Asia, Africa, and Latin America)

OSPIC Overseas Private Investment Corporation

OSPJ Offshore Procurement, Japan

osprd(s) oblate spheroid(s)

OSQ Office of Safety Quality Assurance and Safeguards

OSQ Orchestre Symphonique de Québec (French—Québec Symphony Orchestra)

OSQAS Office of Safety Quality Assurance and Safeguards

osr own ship's roll

OSR Office of Scientific Research; Office of Security Review; Office of Strategic Research; Oil Shale Reserves; Operational Support Requirement(s); Oversea Returnee

OSR Orchestre de la Suisse Romande (French—Orchestra of French Switzerland)

OSRA Ocean Shipping Reform Act; Overseas Shipping Representatives Association

OSRB Overseas Service Resettlement Bureau

OSRD Office of Scientific Research and Development; Office of Standard Reference Data

OSRO Office of Scientific Research and Development

OSROK Office of Supply Republic of Korea

OSRS Organization of Senegal River States (Guinea, Mali, Mauritania, Senegal)

OSRTN Office of the Special Representative for Trade Negotiations

oss order short shipped; osseus

oss (OSS) open-source software; orbiting space station

oSS operates Saturday and Sunday

OSS Object-Sorting Scales (psychological test); Office of Space Science; Office of Strategic Services; Office of Support Services; old submarine (3-letter code); Operating Supply Specification; Operation Safe Streets; Operational Storage Site; Optical Surveillance System; Orbital Space Station; Orient Shipping Services; Overseas Shipping Services

OSSA Office of Space Sciences and Applications (NASA)

OSSAD Office Support System Analysis and Design

OSSBA Oklahoma State School Boards Association

OSSC Oregon School Study Council

OSSNSS Ordnance Supply Segment of the Navy Supply System (USN)

OSSP Oregon Small Schools Program

OSSS Orbital Space Station Studies

OSSTF Ontario Secondary School Teachers' Federation

ost objectives, strategies, tactics, oldest; on same terms; optical star tracker; ordinary spring tides

öst österreichisch (German—Austrian)

Ost Ostend

Ost Ostrów (Polish—island)

Öst Österreich (German—Austria)

Øst Øterrike (Norwegian—Austria)

OST Office of Science and Technology; Old Spanish Trail (U.S. 90); Operational Suitability Test

OS & T Office of Science and Technology

osteo osteopath(ic)

osteo osteon (Greek—bone)—ossification; ossified; osteomyelitis; osteopath(ic)

osteoart osteoarthritic; osteoarthritis

osteol osteology

osteomy osteomyelitis

osteop osteopath(ic); osteopathy

otsteoporo osteoporosus

Österreich (German—Eastern Empire)—Austria (modern remnant of the once-great Austro-Hungarian Empire); *Österreichische Bundeshymne* (German—Austrian Confederation)—national anthem

OSTF Operational System Test Facility

OSTI Office for Scientific and Technical Information

OSTIV Organisation Scientifique et Technique Internationale du Vol à Voile (French—International Scientific and Technical Organization for Soaring Flight)

O.St.J. Officer of the Order of Saint John of Jerusalem

Østland (Norwegian—Eastland)—eastern and southeastern Norway

OSTP Office of Science and Technology Policy

Ostpr Ostpreussen (German—East Prussia)

OSTS Office of State Technical Services; Official Seed Testing Station

OSU Ohio State University; Oklahoma State University; Oregon State University

OSUAS Ohio State University (College of) Administrative Science

OSUK Ophthalmological Society of the United Kingdom

OSUL Ohio State University Library; Oklahoma State University Library; Oregon State University Library

OSUP Ohio State University Press

osv och sa vida (Swedish—and so forth); *og sa videre* (Dano-Norwegian—and so forth)—etc.

Osv Osvald; Osvaldo

OSV Ocean Station Vessel

OSV Orquesta Sinfónica Venezuela (Spanish—Venezuela Symphony Orchestra); *Our Sunday Visitor*

Osv Rom Osservatore Romano (Italian—Vatican newspaper)

osw operational switching

Osw Oswald

OSw Old Swedish

OSW Office of Saline Water

OSWA Off-Shift Work Authorization

osy (OSY) optimum sustainable yield

os & y outside screw and yolk

ot object technology; observer target; offensive tackle (football); oiltight; old terms; old tuberculin; on time; on track; ordinary tide(s); otitis; otology; our telegram; overtime; ovum transfer

ot (OT) occupational therapy; otolaryngology, overtime; original transposed (in a 12-tone row)

ot. oxytocin

o't (Gaelic contraction—of it)

o/t on truck; overtime

'ot hot

o-T on-Thames

O/t old term (grain market)

OT Occupational Therapist; Occupational Therapy; Ocean Transportation; Office of Territories; Office of Transportation; Old Testament; Operational Training; Oregon Trunk (railroad); Organization Table; Otis Elevator (stock exchange symbol); Overseas Tankship (Caltex Line); Overseas Trade

O of T Office of Telecommunications (OT)

OT Organisation Todt (German—Death Organization)—Hitler's extermination corps

ota operational transconductance amplifier

OTA Occupational Therapists Association; Office of Technology Assessment; Office of Territorial Affairs; Outer Transport Area

OTA Organisation Mondiale du Tourisme et de l'Automobile (French—World Tourism and Automobile Organization)

OTAC Ordnance Tank and Automotive Command

otadl outer target azimuth datum line

OTAF Office of Technology Assessment and Forecast

OTAG Office of the Adjutant General (USA)

Otago Otago Harbour (Dunedin, New Zealand's port); Otago Peninsula (southeast of the port)

OTAN Organisation du Traite de l'Atlantique Nord (French—NATO); *Organización del Tratado del Atlántico Norte* (Spanish—NAT0)— North Atlantic Treaty Organization

OTAR Overseas Tariffs and Regulations

OTAS Organización del Atlántico Septentrional (Spanish-North Atlantic Treaty Organization)—NATO

OTASE Organisation du Traite de l'Asie du Sud-Est (French—Southeast Asia Treaty Organization)-SEATO

OTAT Office of Technical Assistance and Training; Ortho toluidine Arsenite Test

OTATO One-Trip Air Travel Orders

otb off-track betting

OTB Overseas Trust Bank

OTBA Owners, Traders, Breeders Association

otbd outboard

otc objective, time, and cost; ocean transshipment cargo; one-stop charter; outer tube centerline; over the counter

otc (OTC) over the counter

OTC Officer in Tactical Command; officers Training Corps; Organization for Trade Cooperation; Ottawa Transit Commission; Overseas Telecommunications Commission

OTC Office de Tourisme du Canada (French—Canadian Government Office of Tourism)

otcbb over-the-counter bulletin board

otch obedience trial champion

otd organ tolerance dose

OTD Ocean Technology Division

otda other-than-defined adult

otdc optical target designation computer

OTDC Observational Test and Development Center (NWS)

OTD & SP Office of Technical Data and Standardization Policy

ote operational test and evaluation; overtaken by events

ote oriente (Spanish—east)

otec (OTEC) ocean thermal energy conversion

OTEC Ontario Teacher Education Colleges

OTECS Ocean Thermal Energy Conversion System

otel our telegram

OTeut Old Teutonic

otf optical transfer function

otf (OTF) off-the-film (camera metering system)

o-t-f off-the-film (light measurement)

OTF Ontario Teachers Federation

oth over the horizon

Oth Othello, The Moor of Venice

othb over-the-horizon backscatter

OTH-B Over-the-Horizon Backscatter

othf over-the-horizon forward scatter

oth-t over-the-horizon targeting

oti official test insecticide

OTI Oregon Technical Institute

OTI Organización de Televisión Iberoamericana (Spanish—Ibero-American Television Organization)

OTIA Ordnance Technical Intelligence Agency

OTID Office of Talented Identification and Development (Johns Hopkins)

OTIG Office of the Inspector General

OTIS Occupational Training Information System; Oregon Total Information System

OTIU Overseas Technical Information Unit

otj on the job

otK old tuberculin Koch

otl out to lunch; output transformerless; over the line (softball)

OTL Operating Time Log

otlx our telex

otm other than Mexican; other track material

OTM Office of Telecommunications Management; Old Turkey Mill

Otml oatmeal

otno our telegram number

oto one time only (tv); otorhinolaryngologist

oto (Latin prefix—ear)—otology

OTO Operational Testing Office

otol otology

otolaryngol otolaryngology

OTO/Neth only to order from Netherlands (for example)

otorhinol otorhinolaryngology

otp obstacle to progress; of this parish; order to plan; oxygen tanking panel

OTP Office of Telecommunications Policy

otr on the rag (underground slang—on the menstrual cycle)

OTR Ovarian Tumor Registry; Owning-the-Realty; Registered Occupational Therapist

otrac oscillogram trace

OTRACO Office de l'Exploitation des Transports Coloniaux (Congolese railway and river transportation administration)

OTRAG Orbital Transport and Rocket AG (German rocket company)

otran ocean test range and instrumentation

otrt operating time record tag

ots (OTS) orbital technical satellite

OTS Office of Technical Services; Office of Traffic Safety; Office of Thrift Supervision; Officers Training School; Operational Test Site; Organization of Tropical Studies

otsdg outstanding

OTSG Office of the Surgeon General

otsr optimum track ship routing

OTSS Operational Test Support System

ott one-time tape; otter; outgoing teletype

ott *ottava* (Italian—octave); *ottobre* (Italian—October)

Ott Ottawa

OTT Ocean Transport and Trading; Office of Traffic and Transportation

ottb owner to take back

Otto Otto von Bismarck-Schoenhausen (Prussian statesman)

otu operational taxonomic unit

otu (OTU) operational training unit

OTU Office of Technology Utilization (NASA)

O Turk Old Turkish

OTUS Office of the Treasurer of the United States

otv orbital transfer vehicle; outer television

otvct outer tube vertical center-line target

otw over the wing

oty over to you

ou oat unit; official use

o & u over and under

'ou thou

o.u. *oculus uterque* (Latin—either eye)

OU Oglethorpe University; Ohio University; Oklahoma University; Open University; Otago University; Ottawa University; Otterbein University; Owen University; Owosso University; Oxford University

OUA Order of United Americans

OUA *Organisation de l'Unité Africaine* (French—OAU); *Organización de Unidad Africana* (Spanish—OAU)—Organization of African Unity

OUAC Oxford University Appointments Committee; Oxford University Athletic Club

OUAFC Oxford University Association Football Club

OUAM Order of United American Mechanics

OUAS Oxford University Air Squadron

OUBC Oxford University Boat Club

OUCC Oxford University Cricket Club

OUDP Officer Undergraduate Degree Program (USA)

OUDS Oxford University Dramatic Society

Ouga Ougadougou, Upper Volta

OUGC Oxford University Golf Club

oughta (American slang—ought to)

oughtn't ought not

ouguiya monetary unit of Mauritania

OUHC Oxford University Hockey Club

OUHS Oxford University Historical Society

oui operating (motor vehicle while) under the influence

OULC Oxford University Lacrosse Club

OULCS Ontario Universities Library Cooperative System

OULTC Oxford University Lawn Tennis Club

OUM Oxford University Mission

OUN *Organizatsia Ukrainiskikh Nationalistiv* (Russian—Ukrainian Nationalist Organization)—anti-communist

OUP Oxford University Press

oupt output

OUR Office of University Research

OURC Oxford University Rifle Club

OURFC Oxford University Rugby Football Club

o/us over-under shotgun

o/US *oro US* (Spanish—American gold, American money)

OUS Oxford Union Society

OUSA Open University Students Association

OUSC Oxford University Swimming Club

OUSDRE Office of Undersecretary of Defense for Research and Engineering

'ouse douse; house; kouse; louse; mouse; rouse; souse; touse

OUSF Oxford University School of Forestry

OUSL Office of the Undersecretary of Labor

out. outlet; output

outbd outboard

outran output translator

out of sync out of synchronization

ouv *ouvrage* (French—work)

Ouviram *Ouviram do Ypiranga* (Portuguese—Listen to the Heart of Brazil)—Brazil's national anthem

ov observed velocity; office visit; optimum value; orbiting vehicle (OV); ovary; over; overture

ov *oi vey* (Yiddish—alas)

ov. *ovum* (Latin—egg)

Ov Ovid; Oviedo

Ov *Over* (Dano-Norwegian or Dutch—upper); *Overijssel* (Dutch—province above the Ijssel River)

Öv *Över* (Swedish—upper)

OV Office Visit

OV *Oranje Vrystaat* (Afrikaans-Orange Free State), Orbital Vehicle

ÖV *Österreichische Volkspartei* (German—Austrian People's Party)

OV-10 North American-Rockwell Bronco counterinsurgency aircraft

o^{va} *ottava* (Italian—octave)

O^{va} *Ostrova* (Bulgarian, Czechoslovakian, Russian—island)

OVA Office of Veterans' Affairs

OVAC Overseas Visual Aids Center

ovbd overboard

ovc other valuable consideration(s); overcast

ovcst overcast

ove on-vehicle equipment

ÖVE *Österreichischer Verband für Elektrotechnik* (German—Austrian Society for Electrotechnology)

over. overture

Over Overjssel (Dutch province); overmation

overs overshoes

ovf *ovenfor* (Dano-Norwegian—over)

ovfl overflow

ovflow overflow

ovh overhead; overheat

ovhd oval head; overhead

ovhdld overhandled

ovhl overhaul

ovh p overhead projector

ovht overheat

OVIS Ohio Vocational Interest Survey

ovk overkill

OVKOT *On Various Kinds of Thinking* (essay by James Harvey Robinson)

ovld overload

ovly overlay

ovm on-vehicle material

ovm *oi vayz mir* (Yiddish—woe unto me)

ovo (Latin prefix—egg)—ovovegetarian

ovolactos ovolactovegetarians (confining their diet to eggs, milk, and milk products, as well as vegetables)

ovos ovovegetarians (confining their diet to eggs and vegetables)

ÖVP *Österreichische Volkspartei* (German—Austrian People's Party)

ovpd overpaid

ovprt overprinted

ovpt overprint

OVPUS Office of the Vice President of the United States

OVR Office of Vocational Rehabilitation

OVRA *Opera Voluntaria per la Repressione dell' Antifascismo* (Italian—Voluntary Work for the Repression of Anti-Fascism)

ovrd override

ovsl overslow

ovsp overspeed

ovstfd overstuffed

ovstk overstock(ed)

OVSVA *Oranje Vrystaatse Veld Artillerie* (Afrikaans—Orange Free State Field Artillery)

ovtr operational videotape recorder

ov w *overgankelijk werkwoord* (Dutch—transitive verb)

ow off white; offer wanted; old woman (slang for wife); one way; ordinary warfare (OW); out of wedlock (born of unmarried parents); outer wing; over water

o-w oil-in-water

o/w oil/water ratio (oil well)

o:w oil-water ratio

oW *ohne Wert* (German—without value)

öW *Österreichische Währung* (German—Austrian currency)

OW Observation Ward; Old Welsh; Overseas Writers

O & W Oldest and Wisest (newspaper reporters' nickname for Ronald Reagan)

OW *Oberwerk* (German—swell organ)

OWAA Outdoor Writers Association of America

OWAEC Organization for West African Economic Cooperation

OWBA Older Workers Benefit Protection Act

owc old-world charm; owner will carry

OWC Ordnance Weapons Command; Outline of World Cultures

O-WC Oil-Water Contract (oil well)

OWCP Office of Workers' Compensation Programs

owe operating weight empty

Owen Stanleys Owen Stanley Mountains of New Guinea

OWERP Open Window Early Retirement Plans

owf optimum working frequency

owgl obscure wire glass

OWH Office of the War on Hunger

OWHA Oliver Wendell Holmes Association

OWI Office of War Information; Office of Waste Isolation

owise otherwise

owl. outlined-white letter (on tires)

OWL Ocotillo Water League; Older Women's League; Older Women's Liberation; Order of Women Legislators; Other Woman, Limited; Overland Western Limited

owm over without marks

OWM Office of Weights and Measures

OWMA Oscar Wells Museum of Art (Birmingham, Alabama)

OWO OWI Washington Office

owp outer wing panel

OWPP Office of Welfare and Pension Plans

owpr ocean wave profile recorder

OWPS Offshore Windpower System

OWPT Overpaid Windfall Profits Tax

OWR Ouse Washes Reserve (England)

OWRR Office of Water Resources Research

OWRT Office of Water Research and Technology

ows (OWS) operational weapon satellite

OWS Ocean Weather Station

OWSS Ocean Weather Ship Service

OWU Office Workers Union; Ohio Wesleyan University

OWWS Office of World Weather Systems

ow/ym older woman/younger man

ox. our telex; oxalic; oxide; oxygen

Ox. Oxford

OX oxygen (commercial symbol)

oxa oxalic acid

oxalic acid $(COOH)^2$

Oxbridge Oxford and Cambridge universities (the ultimate in British formal education)

oxd oxidation; oxidize(d)

Oxf Oxfordshire

OXFAM Oxford Committee for Famine Relief

Oxf & Bucks Oxfordshire and Buckinghamshire (light infantry)

Oxford *Oxford English Dictionary* published by Oxford University Press

Oxford UP Oxford University Press

oxim oxide-isolated monolithic technology

Oxm Oxmantown

Ox M OUP Oxford Medical (division) Oxford University Press

OXOCO Offshore Exploration Oil Company

Oxon Oxfordshire

Oxon. *Oxonia* (Latin—Oxford); *Oxoniensis* (Latin—Oxonian)

oxr oxidizer

oxwid oxyacetylene weld

oxy oxygen; oxymoron (figure of speech wherein opposites are coupled, as in *eloquent silence, final beginning, little but large*)

Oxy Occidental College; Occidental Petroleum Corporation; Oxy Metal Industries International

oxycephs oxycephalics (pointed-skulled people)

oxym oxymel (honey-water-vinegar solution)

oxymo. oxymoron (self-contradictory figure of speech, such as *make haste slowly*)

oy (OY) optimum yield

OY orange yellow

O/Y *Osakeytiö* (Finnish—limited company)

OYA *Oy Yleisradio Ab* (Finnish Broadcasting Company)

Oya Cur Oyashio Current (Kurile or Okhotsk or Oyasiwo)

OYD Office of Youth Development

oyo own your own (apartment, house, yacht)

oys oysters

OYS Outstanding Young Singaporeans

oystersan oyster sandwich

oysterwich oyster sandwich

oz ounce

oz *onza* (Spanish—ounce)

Oz Aussie(s); Australia(n); ooze; Osborn(e)

OZ Ozark Airlines (two-letter designation)

OZ *Ozean* (German—ocean); *Ozero* (Russian—lake)

OZA Ozark Airlines

oz ap apothecaries' ounce(s)

ozare ozone-atmosphere rocket

Ozarks Ozark Mountains of Arkansas, Missouri, and Oklahoma

oz avd avoirdupois ounce(s)

ozd observed zenith distance

ozf ounce-force

oz-ft ounce-foot

oz-in. ounce-inch

OZNa *Odelejenje Zastite Narodna* (Serbo-Croat—Department for National Protection)—Yugoslav

OZO *oost zuidoost* (Dutch—east southeast)

ozone O_3

ozs ounces

oz t ounce troy

ozws otherwise

Ozy Ozzie

P

p fluid density (symbol); page; paid this year (newspaper stock listings); paise (Indian coin currency); pamphlet; paperback; paragraph; parentage; parental; parents; park; parking; part; participle; pass(ed); past; paste; patient; pawn; pay phone call; payment; pebbles; pectoral; pekoe; pence; *pengii* (Hungarian monetary unit); penny; pepper; *per* (Latin—by); percentile; perceptual (speed); percussion; perforate; perforated; perforation; perimeter; period; perishable; permanent; personnel; peseta; peso; peta (P)—10^{15} (one quadrillion); peyote; piaster; piastre; picot; pic; pilaster; pimp; pink; pint; pipe; pitch; pitcher; plane; plasma; plaster; plate; plus; point; polar; pole; pond; poor; populace; population; porcelain; port, or left side of an airplane or vessel when looking forward (P or L); position; positive; post; postage; posterior; postpartum; pound; power; predicate; predict(ion); premolar; preserve; presbyopia; present; preserved; pressure; previous; primary; primitive; principal; principle; probability (ratio); product; progressive; prompter; proprionate; proton; publication; pulse; pupil; put (newspaper stock listings),

p- Para

p (P) prime; program

£ pound sterling

p. pagina (Italian, Latin, Portuguese, Spanish—page); *parte* (Latin—part); *pater* (Latin—father); *per* (Latin—by); *piano* (Italian—softly); *pondere* (Latin—by weight); *proximum* (Latin—near); *pugillus* (Latin—fistful)—handful

p % por ciento (Spanish—per hundred, percent)

P first class premium (airline code); gas pressure in blood (symbol); Pacific; pamphlet; Panama Line; Papa—code letter for P; Paris; Parisian; passenger vessel (symbol); patrol; Pennzoil; permeance (symbol); Philadelphia Mint (symbol); phosphorus; Piasecki; plate; Players' League; Pleyel; poise; polar; polarization; pole; Police; poor; Pope; port; Portugal (auto plaque); power; present value; President; Priestly (source of the Pentateuch); Prince Line; principal; priority; project; propulsion; Protestant; protozoa; pulse

P. Professor; protein(s) (dietary symbol)

P (Latin—Publius); pilot (white P on a blue flag flown on a pilot boat); *Pilot* (German); *pilota* (Italian); *pilote* (French); *piloto* or *practico* (Spanish); *Policia* (Spanish—police)

p00 program zero-zero

P_1 first parental generation

P 1/C Private First Class

P 1/C M Private First Class Marine

P2 Lockheed Neptune antisubmarine and reconnaisance naval aircraft

P_2 pulmonic second sound

P2 Panzer (German—armor, armor plated, tank)

P-2J Kawasaki version of the Lockheed Neptune antisubmarine and reconnaissance aircraft

P-3 Lockheed Orion antisubmarine and patrol aircraft

P-4 Soviet-era Komsomolets motor torpedo boats

P-5 Marlin twin-engine all-weather seaplane for long-range antisubmarine patrol and electronic reconnaissance

P-5M Martin Marlin flying boat

P-6 Soviet-era motor torpedo boats

P.08 German marking denoting the so-called Luger service pistol

P^{33} radioactive phosphorus

P-38 U.S. pursuit aircraft

P.38 German 9mm service pistol (World War II)

P_{55} partial pressure of O_2 wherein hemoglobin is half saturated with O_2

P-60 60-minute parking

P-149 Piaggo trainer aircraft built in Italy

P-166M Piaggo Albatross coastal patrol aircraft

P-333C Lockheed antisubmarine patrol plane

P3P Platform for Privacy Preferences

pa intensity of atmospheric pressure (symbol); paper; paper advance; paralysis agitans; participial adjective; particular average; patient; pattern analysis; pending availability; performance analysis; performance appraisal; permanent appointment; pernicious anemia; personal appearance; personnel assistant; personnel association; piaster; piastre; pilotless aircraft; point of aim; points against, position approximate; power amplifier; power approach; power of attorney; preliminary award; press agent; pressure altitude; private account; professional association; protected area; provisional allowance; psychoanalyst; public address (system); public accountant; public assistance; publication announcement; pulse amplifier; purchasing agent

pa (Pa) pascal

pa (PA) polyamide; posteroanterior

p-a psychogenic aspermia

p/a paid annually; payment authority; per annum; power of attorney

p & a percussion and auscultation; plugged and abandoned (oil well); price and availability

pa. partner

p in the a pain in the ass

p.a. per abdomen (Latin—by the abdomen); *per annum* (Latin—by the year)

p/a per adres (Dutch—care of)

pA picoampere

p A por autorización (Spanish—in care of)

Pa Panama; Panamanian; *Panameño;* Papa; Para; *Paré (Belem do Pará);* Pascal; Pennsylvania; Pennsylvanian; protactinium

PA Pan American; Panama (Internet code); Panic Alarm; Parents Anonymous; Passenger Agent; Pennsylvania; Pennsylvania Railroad (stock exchange symbol); Philippine

Army; Philippine Association; Physician's Assistant; Piedmont Airlines; Polled Angus (cattle); Port Agency; Post Adjutant; Prefect Apostolic; Press Agent; Press Association; Prince Albert (coal); Procurement Authorization; Production Assistant; Proprietary Association; Prosecuting Attorney; Prothonotary Apostolic; psychological age; Public Act; Public Affairs; Publishers Association; Puppeteers of America; Purchasing Agent

P-A Pacific-Atlantic Line; Pan-Atlantic Line

P/A Picatinny Arsenal

P & A Professional and Administrative

P of A Port of Anchorage

PA Priok Administration (Malay—Port Administration); *Psychological Abstracts*

P A Partij van de Arbeid (Dutch—Party of Labor)

PA$_{o2}$ alveolar oxygen pressure

p.a.a. parti affectae applicetur (Latin—apply to the affected parts or region)

PAA Pacific Alaska Airways; Pan American World Airways; System (3-letter designation); Paper Agents Association; Pharmaceutical Association of Australia; Phonetic Alphabet Association; Photographers Association of America; Plywood Association of Australia; Potato Association of America; Pre-arrangement Association of America; Prisoners Aid Association; Professional Archers Association; Purchasing Agents Association

PAAA Premium Advertising Association of America

PAAB Public Arts Advisory Board

PAAC Product Assurance Action Center; Program Analysis Adaptable Control; Public Arts Advisory Council

PAADC Principal Air Aide-de-Camp

PAAE Pennsylvania Association for Adult Education

PAAF Professional Actors Association of Florida

PAAM Prisoners Aid Association of Maryland

pa'anga monetary unit of Tonga

PAAO Pan-American Association of Ophthalmology

pab per acre bonus

pab (PAB) p-aminobenzoic acid

PAB Panair do Brasil (airline); Petroleum Administrative Board; Price Adjustment Board

PAB (CIA) Problems Analysis Branch of the CIA

paba para-amino benzoic acid

pabla problem analysis by logical approach

pabst primary adhesively bonded structure

pabx private automatic branch telephone exchange

pac packaged assembly circuit; personal analog computer; person in addition to the crew; phenacetin-aspirin-caffeine (all-purpose capsule); prearrival confirmation; production acceleration capacity; project analysis and control; pursuant to authority contained (in); put and call (stock exchange jargon)

pac (PAC) premature atrial contraction

p-a-c parent-adult-child (ego states)

p A c pure Argentinian cocaine

Pac Pacific

Pac Pacifico (Italian—Pacific); *Pacífico* (Portuguese or Spanish—Pacific); *Pacifíque* (French—Pacific)

PAC Pacific Air Command; Pacific Automotive Corporation; Pacific Telephone & Telegraph (stock exchange symbol); Palo Alto Clinic; Pan-Africanist Congress; Pan-American Congress; Performing Arts Center; Pharmaceutical Advertising Council; Philbrook Art Center; Planned Amortization Class; Political Action Committees; Public Access Catalog(ue); Public Affairs Committee; Public Assistance Cooperative

PACA Picture Agency Council of America

PACAF Pacific Air Force

Pacaraimas Pacaraima Mountains forming the Brazil-Guyana and Brazil-Venezuela borders

PACAS Patient Care System; Psychological Abstracts Current Awareness Service

PACB Pan-American Coffee Bureau

Pac Bch Pacific Beach

Pac Bell Pacific Bell (telephone)

PACC Product Administration and Contract Control; Professional Association of Custom Clothiers; Project Administration Contact Control

PACCS Post Attack Command and Control System

PacD Pacific Division

PACDA Personnel and Administration Combat Development Activity (USA)

pace (PACE) package-crammed executive; performance and cost evaluation; precision analog computing equipment; pre-launch automatic check-out equipment; program to advance creativity in education; programmed automatic communications equipment; projects to advance creativity in education

pace. pacemaker

PACE Pacific America Container Express; Police and Criminal Evidence Act; Professional and Administrative Career Examination; Professional Association of Consulting Engineers; Professional Athletes Career Enterprises; Program for Afloat College Education (USN); Program of Adult College Education; Program of Advanced Continuing Education; Public Access Cabletelevision by and for the Elders; Public Awareness Communication Exchange

PACECO Pacifc Coast Engineering Company

PACED Program for Advanced Concepts in Electronic Design

pacer. planning automation and control for evaluating requirements

PACFACS Programmed Appropriation Commitments—Fixed-Asset Control System

PacFIN Pacific Fishery Network

PACFLT Pacific Fleet

PACFORNET Pacific Coast Forest Research Information Network

Pac Gas & El Pacific Gas and Electric

'pache Apache

Pacif Pacific

Pacific Pacific Ocean

Pacific Basin Countries Australia, China, Hong Kong, Indonesia, Japan, Malaysia, New Zealand, Philippines, Singapore, South Korea, Taiwan, Thailand

PAC IOs Planned Amortization Class Interest-Only (bonds)

pack. packing

pacm pulse-amplitude code modulation

PACMD Philadelphia Contract Management District

PacO Pacific Ocean

PACO Polaris Accelerated Change Operation

Pa$_{CO2}$ arterial carbon dioxide pressure

Pac Ocean Terr Pacific Ocean Territories (Pitcairn Island, Ducie, Henderson, and Oeno)

PACOM Pacific Command

pacor passive correlation and ranging

PACOS Package Operating System

PACR Performance and Compatability Requirements

PACRA Pottery and Ceramic Research Association

Pac Rail Missouri Pacific, Union Pacific, Western Pacific (railroads merged)

PACRNB President's Advisory Commission on Recreation and Natural Beauty

PACs Political Action Committees

PACS Pacific Area Communications System

Pac Ship Pacific Shipper

pacsim performance achievement computer model for waste package

Pac Sym Pacific Symphony

pact production analysis control technique; programmed automatic circuit tester

PACT Prisoner and Community Together (for offenders and victims); Production Analysis Control Technique; Project for the Advancement of Coding Techniques; Professional Association of Canadian Talent

Pac Tel Pacific Telephone (company)

Pac-Tex Pacific-Texas (pipeline)

Pac T & T Pacific Telephone and Telegraph

PACU Pennsylvania Association of Colleges and Universities

pacv (PACV) personnel air-cushion vehicle

PACV Patrol Air-Cushioned Vehicle (naval)

PACW President's Advisory Committee on Women

PACX Private Automatic Computer Exchange

pad packet assembler/disassembler; padding; padlock; para-aminobenzoic acid (PAD); payable after death; pitch axis definition; provisional assembly date

pad (PAD) peripheral arterial disease; primary affective disorder

Pad Padre; Padstow

P Ad Port Adelaide

PAD Pacific Australia Direct (steamship line); Passive Air Defence; Patient Accounts Department; People Against Displacement (caused by urban redevelopment); Performance Analysis Department (ONWI); Pontoon Assembly Depot; Port of Aerial Debarkation; Provisional Air Division; Public Administration Division; Public Affairs Department

padal pattern for analysis, decision, action and learning

PADAP Philippine-Australian Development Assistance Program

padar passive detection and ranging

PADAT Psychological Abstracts Direct Action Terminal

PADC Pennsylvania Avenue Development Corporation

PADD Pedestrians Against Dangerous Drivers; Pedestrians Against Drunken Drivers; Political Art Documentation and Distribution

Paddo Paddington, Australia

PADDS Petroleum Administration for Defense Districts

PADF Pan-American Development Foundation

PADI Professional Association of Diving Instructors

p adj participial adjective

PADL Pilotless Aircraft Development Laboratoy

padloc passive detection and location of countermeasures

PADMIS Patient Administration Information System

PADPAO Philippine Agency Detective Protective Association

p Adr *per Adresse* (German—in care of)

padre portable automatic data-recording equipment

Padre Island National Seashore, Texas

pads passive advanced sonobuoy

PADS Precision Azimuth Determination System

Pad Sta Paddington Station (rail terminal)

padt post-alloy-diffused transistor

pae public affairs event

pa&e program analysis and evaluation

p. ae. *partes aequales* (Latin—equal parts)

PAE Peoria and Eastern (railroad); Port of Aerial Embarkation

PAEC Pakistan Atomic Energy Commission; Philippine Atomic Energy Commission

paect pollution abatement and environmental control technology

paed paediatric

paei perisocope azimuth error indicator

PAESP Pennsylvania Association of Elementary School Principals

paf peripheral airfield; pulmonary arteriovenous fistula; punishment and fine

paf (PAF) personal article floater (baggage insurance policy); Polaris accelerated flight

pa & f percussion, auscultation, and fermitus

paf *puissance au frein* (French—brake horsepower)

PAF Pacific Air Force(s); Pakistan Air Force; Palestine Arab Fund (for terrorists); Pet Assistance Foundation; Philippine Air Force; Ports Authority of Fiji

PAFA Pennsylvania Academy of Fine Arts

PAFB Patrick Air Force Base

PAFCO Pacific Fishing Company

PAFMECA Pan-African Freedom Movement of East and Central Africa

PAFS Primary Air Force Specialty

PAFSC Primary Air Force Specialty Code

PAFTA Pacific Area Free Trade Association

pag periaqueductal gray matter

pag *pagaré* (Spanish—I will pay); *pagina* (Italian—page)

pág *página* (Spanish—page)

Pag pagoda

PaG Pennsylvania-German

PAG Planning Advisory Group; Primary Analysis Group; Prince Albert's Guard

PAGASA Philippine Atmospheric Geophysical and Astronomical Services Administration

PAGB Proprietary Association of Great Britain

PAGEL Priced Aerospace Ground Equipment List

pageos (PAGEOS) passive geodetic satellite

Pa Ger Soc Pennsylvania German Society

pagg segg *pagine seguenti* (Italian—following pages)

pAgmk primary African green monkey kidney

pág(s) *página(s)* [Spanish—page(s)]

PAGT Port Authority Grain Terminal

pah polynuclear aromatic hydrocarbon(s)—(photochemical smog ingredient)

pah (PAH) para-aminohippuric acid

Pah Pahlavi

PAH Pan-American Highway (also called Inter-American Highway)

PAHC Pan American Highway Congress

Pahl. Pahlavi

PAHO Pan-American Health Organization

PAHOCENDES Pan-American Health Organization Center for Development Studies

pahs polycyclic aromatic hydrocarbons

pai parts application information; personal accident insurance; personal adjustment inventory; please airmail immediately; prearrival inspection

PAI Panama Airways Incorporated; Personnel Accreditation Institute; Photographic Administrators Incorporated; Piedmont Airlines (3-letter coding)

Pai Baj *Paises Bajos* (Spanish—Low Lands) — Netherlands

PAIGCV *Partido Africano da Independencia da Guine e Cabo Verde* (Portuguese—African Party for an Independent Guinea and Cape Verde)

PAIGH Pan-American Institute of Geography and History

PAILS Projectile Airburst and Impact Location System

PAIN Pan-American Institute of Neurology

paint. painter; painting

Painters Union International Brotherhood of Painters and Allied Trades of the United States and Canada

pair performance and integration retrofit

PAIR Personnel Administration and Industrial Relations; Psychological Audit for Interpersonal Relations

PAIRC Pacific Air Command

PAIRS Private Aircraft Inspection Reporting System

PAIS Pennsylvania Association of Independent Schools; Project Analysis Information System (AID); Public Affairs Information Service

PAIT Program for the Advancement of Industrial Technology

PAJU Pan-African Journalists Union

Pak Pakistan

PAK Pëtr Alekseevich Kropotkin

PAK2 p21-activated kinase-2

PakE Pakistani English

Paki(s) Pakistani(s)

Pakistan Islamic Republic of Pakistan (Moslem country between Afghanistan and India); *Pakistan* in Urdu means Land of the Pure; *Pak* (Persian—holy) plus *tan* (Urdu—land)—Holy Land

PAKISTAN Punjab, Afghan Border states, Kashmir, Sind, and *tan* from Baluchistan

pal paired associate learning; paleontology; passive activity loss; peripheral availability list(ing); permissive action link; phase-alteration line (color tv system); prescribed action link; professional adjustable ladder; program assembly language; programmed application library

pal. (PAL) phase alternate line; products and area locator

p-a-l prisoner-at-large

Pal Palace; Palencia; Paleozoic; Palermo; Palestine

Pal *Palacio* (Spanish—palace); *Palácio* (Portuguese); *Palais* (French—palace); *Palazzo* (Italian—palace)

PAL Pacific Aeronautical Library; Pacific Aluminium; Pakistan Airlines; Pan Asia Line; Paradox Application Language; Pensioners Advancement League; phase-alternating (television) line; Philippine Air Lines; Podiatry Arts Laboratory; Police Athletic League; Polynesian Airlines Limited; Prison Atheist League; prisoner-at-large; Programmable Array Logic; Public Archives Library

PALA Polish-American Librarians Association

Palat Palatinate

P Alb Port Alberni

PALC Point Arguello Launch Complex

paleo paleography

Paleo Paleolithic; Paleozoic (Cambrian, Ordovician, Silurian)

paleob paleobotany

paleon paleontology

Palestine southern Syria, according to many Arabs; Turkish province containing what is now Israel plus adjacent Arab countries in the Jerusalem area often called the Holy Land

PALI Pacific and Asian Linguistics Institute (University of Hawaii)

palimony alimony awarded a former common-law partner

palin palindrome; palindromic

palind palindrome (name or sentence the same forward or reverse—*Otto* or *Madam I'm Adam*)

PALINET Pennsylvania Area Library Network

PALIS Property and Liability Information Systems

Palisades Palisades Interstate Park along the west bank of the Hudson River washing the shores of New Jersey and New York; Palisades (amusement) Park near Englewood, New Jersey; Palisades Peaks in Kings Canyon National Park, California

pall. pallet

palm. palmist(ry); precision attitude and landing monitor

Palma *Palma de Mallorca* (capital of the Balearic Islands and the island of Mallorca)

Palmach *Plugot Machatz* (Hebrew—Spearhead Units)—commando units active in the establishment of Israel when still called Palestine

Palmas *Las Palmas de Gran Canaria* (capital and main seaport of the Canary islands)

Palmn Palmerston

PALMS Propulsion Alarm and Monitoring System

Pal Obs Palomar Observatory

Palomar Mt. Palomar Mountain Observatory near San Diego, California

Palos *Palos de la Frontera* (port of departure of Columbus in 1492)

palp palpable; palpitation

palpi palpitation

PALs Parcel Air Lifts (U.S. Post Office parcel-post service for military personnel)

PALS Pacific Association of Labor Support; Permissive Action Link Systems

PALSG Personnel and Logistics Systems Group

PALTC Pacific Asian and Latino Training Center

pam pamphlet; payload assist module; pledged account mortgage; precategorical acoustic memory; primary amoebic meningoencephalitis; procurement aircraft and missiles; pulse amplified modulation; pulse amplitude modulation

Pam Lord Palmerston; Pamela

PAM Palestine Archeological Museum; Pasadena Art Museum; Portland Art Museum

PAMA Pan-American Medical Association; Professional Aviation Maintenance Association; Publishers' Advertising and Marketing Association

pamac parts and materials accountability control

PAMBU Pacific Manuscripts Bureau

PAMC Pakistan Army Medical Corps

PAMCO Pacific Annuity Marketing Company

PAMELA Pricing and Monitoring Electronically of Automobiles

PAMETRADA Parsons Marine Experimental Turbine Research and Development Association

pamf programmable analog-matched filter

pam file pamphlet file

PAMI Performing Arts Management Institute

PAMIPAC Personnel Accounting Machine Installation Pacific Fleet

pamirasat (PAMIRASAT) passive microwave radiometer satellite

Pamirs Pamir Mountains of Central Asia

PAML Pan American Mail Line

PAMO Port Air Materiel Office

pamp pampas

PAMP Public Art Master Plan

pampa *pampas* (Spanish—plains)—of Argentina, southern Brazil, Paraguay, Uruguay)

PAMPA Pacific Area Movement Priority Agency (DoD)
pamph pamphlet
pams pamphlets
PAMS Plan Analysis and Modeling System
PAMT Port Authority Marine Terminal
PA-RISC Precision Architecture-Reduced Instruction Set Computing
pan personal area network; primary account number
pan. panchromatic; panorama; panoramic; pantomime; pantry
Pan Panama; Panamanian; Panameño
PAN Pan American Navigation; Parents Against Narcotics; peroxyacetylnitrate (air-pollutant poison)
PAN Partido Acción Nacional (Spanish—National Action Party)-Mexican; *Polska Akademia Nauk* (Polish—Academy of Sciences)
PANA Pan-African News Agency; Pan-Asia Newspaper Alliance
PANAFTEL Pan-African Telecommunications (network)
PANAGRA Pan American-Grace Airways
PANAIR Panair do Brasil (Brazilian airline)
Pan-Am Pan-American World Airways
Panama Republic of Panama (Spanish-speaking Central American country bisected by the Panama Canal), *República de Panamá*
Panamax maximum size Panama Canal locks can accommodate
Panamints Panamint Mountains of eastern California along the Death Valley border of Nevada
PANANEWS Pan-Asia Newspaper Alliance (Hong Kong)
pan b panic bolt
panc pancreas
Pan Can Panama Canal
Pan Canal Panama Canal
pand panderazo (Spanish—blow struck with a tambourine); *panderetero* (Spanish—tambourine maker or player); *pandero* (Spanish—tambourine)

PANDA Prestel Advanced Network Design Architecture; Prevent Abuse and Neglect (through) Dental Awareness; Professional Association of Numismatic Dealers of Australasia
pandex *pan* (Greek—all) + *dex* (from index)—all-inclusive index
pandg people are no damn good
p-and-p struggle prude-and-prurient struggle
PANEES Professional Association of Naval Electronics Engineers and Scientists
P Ang Port Angeles
PANGLOSS Parallel Architecture for Networking Gateways Linking Open Systems Interconnections
Pango (naval argot—Pago Pago, American Samoa)
panjan panjandrum
Pank Pankow
panol panology
panorams panoramas
PANPA Pacific Area Newspaper Production Association
pan (PAN) peroxyacetyl nitrate (smog ingredient)
Pan pan Pan paniscus (pygmy chimpanzee found south of the Congo River)
pans. peroxyacetylnitrates
PANS Procedures for Air Navigation Services
PANSDOC Pakistan National Scientific and Technical Documentation Center
Pan Sea Fron (PANSEAFRON) Panama Sea Frontier
PANSY Programme Analysis System
P Ant Port Antonio
panth pantheism; pantheist; pantheistic(al)(1y)
panto pantograph(ic); pantomime; pantomimic
Pan trog Pan troglodytes (common chimpanzee found north of the Congo River)
pants pantaloons
PANY Power Authority of the State of New York
pao product assurance operations
PAO Public Affairs Officer
Pa$_{o2}$ arterial oxygen pressure
PAOA Pan-American Odontological Association

PAODAP President's Action Office for Drug Abuse Prevention
Paola Paola Capriolo (Italian novelist)
pap (PAP) pension administration plan
pap. papa; papacy; papal; paper; papyrus
pap prét à porter (French—ready to wear)
p-a-p poco a poco (Italian—little by little)
Pap Papa; Papeete; Papist; Pappie; Papua; Papuan
P-a-P Port-au-Prince (capital of Haiti)
PAP Pacific Automation Products; People's Action Party; Performance Assessment Plan; *Polska Agencja Prasowa* (Polish News Agency); Port-au-Prince, Haiti (airport); Prices and Agricultural Products; Progressive Australia Party
papa parallax aircraft parking aid
Papa letter P radio code
PAPA Parents As Partners Associated; Pesticide Applicators Professional Association
PAPAs Parents as Partners Associates
PAPAS Pennsylvania Association of Private Academic Schools
PAPC Philological Association of the Pacific Coast
Pap diag Papanicolaou diagnosis
Papermac paperback book published by Macmillan
PAPF Pan-American Philatelic Federation
papi precision path indicator
papi (PAPI) polymethylene polyphenyl isocyanate
PAPI Pacific Automation Products Incorporated; Professional Association of Pet Industries
papil papilla; papillae
Pap Inf Papal Infallability
Pap Lib Paperback Library
Pap NG Papua New Guinea
p app puissance apparente (French—apparent power)
Pap(s) [Irish-Protestant English-Papist(s)—*see* Prod(s)]
PAPS Periodic Armaments Planning System

Pap smear Papanicolaou smear

Papsom Papaver somniferum (opium poppy)

PAPSS Procurement and Production Status System

Pap St Papal States

PAPTE President's Advisory Panel on Timber and the Environment

Pap Ter Papua Territory

Pap Test Papanicolaou Test (for cervical cancer)

Papua Indonesian island called Papua New Guinea in the eastern sector and West Irian on the western sector

paq position-analysis questionnaire

par parent; parents; parish

par (PAR) perimeter acquisition radar

par. parabolic aluminized reflector; paragraph; parallax; parallel; parent; parents; parenthesis; parish; per acre rental; planed all around (timber); precision approach radar; pulse acquisition radar

par (Latin prefix—bear or give birth to)—parturition

Par Paris, Parish

Par Parigi (Italian—Paris); *Parijs* (Dutch-Paris)

Par. Parah

PAr Punta Arenas

PAR Parental Awareness and Responsibility; Paris, France (Orly airport); Program Appraisal and Review; Protect Abortion Rights

PAR Partido Acción Revolucionario (Spanish—Revolutionary Action Party)

para parachute; paragraph; parallel; perceiving and recognition automation

para (Greek—alongside, beside, beyond)—paralysis, paramedic, parameter, parasite, parathyroid

Para Paraguay(an)

Pará Belém do Pará, Brazil

PARA Professional Audiovideo Retailers Association; Program for At-Risk Addicts

para I; para II; para III; etc. unipara; bipara; tripara; etc.—having given birth to one child, to two children, to three children, etc.

parab parabola

parabat parachute battalion

Paracels Paracel Islands in the South China Sea east of Vietnam

paracent paracentesis

parad paradichlorobenzene; paradigm(atic)(al)(ly); paradisiac(al)(ly); paradisal; paradise; paradisiacal(ly); paradox(ical)(ly); paradoxicalness

paradrop parachute airdrop

par. aff. pars affecta (Latin—to the part affected)

Paraguay Republic of Paraguay (Spanish-speaking South American country), *República del Paraguay*

Paraguayos Paraguayos, república o muerte (Spanish—Paraguayans, republic or death) Paraguay national anthem

paral parallax; paralysis

param. parameter(s); parametric

Parami Parsons active ring around miss indicator

paramp parametric amplifier

parapsy parapsycholic(al)(ly); parapsychologist; parapsychology

parapsych parapsychologist; parapsychology

paraquat paraquat—tainted marijuana

paras parasite(s); parasitic; parasitism; paratroopers

parasail parachute sail (steerable parachute)

parasitol parasitologic(al)(ly); parasitologist; parasitology

parasym div parasympathetic division

parasyn parametric synthesis

Parbo Paramaribo

parc progressive aircraft repair cycle

PARC Performing Arts Research Center (New York); Princeton Applied Research Corporation; Public Affairs Research Council; Public Archives Records Centre

PARCA Pan American Railway Congress Association

parch. parchment

PARCS Parking and Revenue Control System (for autos); Perimeter Acquisition Radar Characterization System

pard partner; periodic and random deviation

PARD Personnel Actions and Records Directorate

pardac parallel digital-to-analog converter

pardop passive-ranging doppler

PARDS Precision-Annotated Retrieval Display System

paregoric compound tincture of opium

paren parenthesis

parens parentheses

parent. parental(ly)

Parents Parents Magazine

parex programmed accounts-receivable extra (service)

par for par for the course (average, typical, usual)

PARFR Program for Applied Research on Fertility Regulation (Northwestern University)

Parg Paraguay; Paraguayan

pari parietal

Pariñas Pariñas Point (westernmost point of South America)

Paris O Paris Opera

Paris Pen Paris Peninsula in eastern Venezuela

PARKA Pacific Acoustic Research (Kaneoche, Alaska)

parkade parking arcade

parl parallel

Parl Parliament

PARL Palo Alto Research Laboratory (Lockheed)

Parl Agt Parliamentary Agent

Parl Const Parliamentary Constituency

par light parabolic aluminized reflector lamp

Parlor Irish well-to-do people

Parl Sec Parliamentary Secretary

parm (PARM) precision antiradiation missile

PARM Partido Autentico de la Revolución Mexicano (Spanish—Authentic Party of the Mexican Revolution)

PARMA Public Agency Risk Managers Association

parm(s) parameter(s)

parochiaid parochial-school aid (provided by tax monies)

paros passive ranging on submarines

parot parotid

parox paroxysm(al)

PARPRO Peacetime Aerial Reconnaissance Program

parq parquet

Parra Parramatta estuary, Sydney, Australia

pars paragraphs

PARS Passenger Airlines Reservation System; Prisoners Aid and Rehabilitation Society; Private Aircraft Reporting System; Programmed Airlines Reservation System

parsec parallax second (3.26 lightyears or 19.2 trillion miles)

Parsee (Arabic—Iranian, Persian)—Indian Zoroastrian descended from refugees who came to India to escape Muslim persecution

parsq pararescue

pairsyn parametric synthesis

part. parterre; partial; participate; participle; particle; partition; partner; partnership

part. partim (Latin—part)

PART Part Allocation Requirements Technic

part. aeq. partes aequales (Latin—equal parts)

partan parallel tangents

Partas Partagas cigars

part. dolent. partes dolentes (Latin—painful parts)

PARTEI Purchasing Agents of Radio, TV, and Electronics Industries

parth parthenogenesis

Parthia (Latin—parts of Assyria and Persia in northeastern Iran)

parti participle

partic participle; particular

partic exh particulate exhaust (soot)

Partisan Partisan Review

partit partitive

partn partition

partner. proof of analog results through numerical equivalent routines

Partridge Eric Partridge (compiler of *Dictionary of Slang and Unconventional English*)

Partrys Parity Mountains of western Ireland

part. vic. partibus vicibus (Latin—in divided doses)

paru postanesthetic recovery unit

par uni party unity (political utopia)

parv paravane

parv parvus (Latin—small)

PARVO Professional and Academic Regional Visits Organization

pas passive; photoacoustic spectrometer; portfolio advisory services; power-assisted steering; precategorical acoustic store; public-address system

pas (PAS) para-aminosalicylic acid; periodic acid Schiff, photo-acoustic spectroscopy

pa's public appearances

paS periodic acid Schiff

Pas Pasadena; Pascagoula; Pashto; Passage; Passaic; Passau; Pastaza (Ecuadorean province)

Pa s Pascal second

Pas. Paschae (Latin—Easter)

PAs Parents Anonymous; Police Agents

PA's purchasing agents

PAS Passive Alcohol Sensor; Percussive Arts Society; Pioneer Air System; Pontifical Academy of Science; Pregnancy Advisory Service; Primary Alerting System; Probation and Aftercare Service; Professor of Air Science; Public Address System

pasa (PASA) para-aminosalicylic acid

PASA Pennsylvania Association of School and Administrators; Pipelines Authority of South Australia

pasar psychological abstracts search and retrieval

PASAR Philippine Associated Smelting and Refining

PASB Pan-American Sanitary Bureau

PASBO Pennsylvania Association of School Business Officials

PASC Pacific Area Standards Congress; Palestine Armed Struggle Command (controlled by El Fatah); Pan-American Standards Committee

Pasca Pascagoula

PASCAL Philips Automatic Sequence Calculator

PASCAL Program Appliqué à la Selection et la Compilation Automatique de la Literature (French—Program Applied to the Selection and the Automatic Compilation of Literature)

PASCO Pan American Sulfur Corporation

p'ase alkaline phosphatase

PaSEA Pennsylvania Society of Enrolled Agents

PASF Photographic Art and Science Foundation

PASG Programs Activities and Services Guide

PASGT Personal Armor System—Ground Troops (new helmet of the US Army)

pasim pasimological, pasimologically; pasimologist; pasimology (study of gestures as means of communication)

PASL Pakistan Association of Special Libraries

PASLIB Pakistan Association of Special Libraries

PASO Pan-American Sanitary Organization; Pan-American Sports Organization

PASOK Pan-Hellenic Socialist Commune

pass. passage; passenger; passitive; passivate; passive; passport

pass. passim (Latin—far and wide, here and there, up and down)

Pass Passover

PASS Parents Against Secondhand Smoke; Passengers Automatic Selection System; Pass-through entity Automated Screening and Support System; Personal Alert Safety System (locates firefighter in trouble); Plan for Achieving Self-Support; Procurement Automated Source System; Procurement Automated Source System; Prototype Artillery Subsystem

PASSIM President's Advisory Staff on Scientific Information Management

Pass Kristyen Pass Christian, Louisiana

PASSP Pennsylvania Association of Secondary School Principals

Past Pasteurella

PASTIC Pakistan Scientific and Technological Information Center

pastram passenger traffic management

pastramasan pastrami sandwich (pickled corned-beef sandwich)

pastramwich pastrami sandwich (pickled corned-beef sandwich)

PASWEPS Passive Antisubmarine Warfare Environmental Protection System

p-a system public-address system

pat. patent(s); patented; paternal; patrol(s); pattern; points after touchdown

pat. (PAT) paroxysmal atrial tachycardia

p-à-t *pied-à-terre* (French—small occasional lodging)

Pat Patricia; Patrick

Pat Patrone (German—cartridge, round of ammunition)

PaT Parents as Teachers

PAT Pacific Air Transport; Pacific Automobile Train; Philippine Aerial Taxi; Post-availability Trials; Prescription Athletic Turf, Production Assessment Test; Progressive Achievement Tests

PATA Pacifilc Area Travel Association; Professional Aeromedical Transport Association

Patag Patagonia(n)

PATAS Publications Automated Task Analysis System

PATCA Panama Air Traffic Control Area; Professional and Technical Consultants Association

PATCO Port Authority Transit Corporation; Professional Air Traffic Controllers Association

PATCRA Papua New Guinea —Australia Trade and Commercial Relations Agreement

patd patented

pat&e production acceptance test and evaluation

PATE Philippine Association of Technological Education

path. pathological; pathologist; pathology; pituitary adrenotrophic hormone (PATH)

path (Latin prefix—disease) — pathologist, pathology

PATH Performing Arts Theater of the Handicapped; Port Authority Trans-Hudson (Hudson Tubes)

Pathet Pathétique

patho pathological

patho (Greek—suffering)—osteopath(ic), pathological, pathologist, pathology

pathogen pathogenic

pathol pathologic(al)(ly), pathologist; pathology

pathomorph pathomorpho logic(al)(ly); pathomorphologist; pathomorphology

pathy (Latin suffix—abnormality or disease)—neuropathy, psychopathy

Patk Patrick

pat. med patent medicine

PATMO Patent and Trademark Office

patn pattern

PATO Pacific-Asian Treaty Organization

Pat Off Patent Office

PATOLIS Patent On-Line Information System

pat pend patent pending

PATRA Printing, Packaging, and Allied Trades Research Association (also appears as PPATRA)

PATRIC Pattern Recognition and Information Correlation (police computer)

PATRICIA Practical Algorithm to Receive Information Coded in Alphanumeric

patron. patronym(ic)(al)(ly)

Patronat (French equivalent of National Association of Manufacturers in United States)

pats. patents

PATs Pre-Authorized (bank deposit) Transfers

PATS Phillipine Aeronautics Training School; Portable Acoustic Tracking System; Proof and Transit System

patt pattern

PATTERN Planning Assistance Through Technical Evaluation of Relevance Numbers

Patti Adelina Patti (1843—1919)

PATWAS Pilot's Automatic Telephone Weather Answering Service

PATX Private Automatic Telex Exchange

PATY Private Annuity for Term of Years

pau pattern articulation unit; programmer's analysis unit

Pau Pablo

Pau Casals (Catalan—Pablo Casals)

PAU Pan American Union; Pan American University; Police Airborne Unit

P-au-P Port-au-Prince

pav paving; phase-angle voltmeter

p/av particular average

Pav Luciano Pavarotti (Italian tenor); pavilion; Pavo (constellation)

PAV Personnel Allotment Voucher

PAV Poste Avion (French—airmail)

PAVAA Polish Army Veterans Association of America

pave. position and velocity extraction

PAVE Professional Audiovisual Education (study)

PAVEPAWS Precision Acquisition of Vehicle-Entry Phased-Array Warning System

PAVE-PAWS Precision Acquisition of Vehicle Entry— Phased-Array Warning System (early-warning radar system against submarine-launched missiles)

PAVM Potential Acquisition Valuation Method

PAVN Peoples Army of Viet Nam

pav. noc. pavor nocturnus (Latin—nightmares, night terrors)

PAVPAWS Precision Acquisition of Vehicle-Entry Phased-Array Warning System

pavt power adjusted variable track(ing)

paw. portable auxiliary workroom; powered all the way

Paw Papa

PAW People for the American Way; Pets and Wildlife; Poetic Allusion Watch

PAWA Pan American World Airways; Pan-American Womens Association

PAWO Pam-African Women's Organization

pawob passengers arriving without baggage

PAWs Prodigious Accumulators of Wealth

PAWS Phased Array Warning System; Programmed Automatic Welding System; Progressive Animal Welfare Society

pax. passenger(s); private automatic exchange

Pax Paxon; Paxton

Pax Am Pax Americana (Latin—American Peace)

Pax Amer Pax Americana

Pax Brit Pax Britannica (Latin—British Peace)—a long period of peaceful stability imposed throughout the British Empire and many adjacent parts of the world

Pax Por Pax Porfiriana (Latin—Porfirian Peace)—imposed on Mexico by its dictator-general-president—Don Porfirio Diaz from 1876 to 1910 when ousted by Madero

Pax River Patuxent River Naval Air Station, Maryland

Pax Rom Pax Romana (Latin—Roman Peace)—imposed throughout the Roman Empire

pax vob. pax vobiscum (Latin—peace be with you)

Pay Paymaster; Paymistress

Pay Cmdr Paymaster Commander

paye (PAYE) pay as you earn, pay as you enter

PAYGO Pay As You Go

payld payload

Paymr Paymaster; Paymistress

PAYS Patriotic American Youth Society

payt payment

Paz Ladislao Pazmany

pb painted base; paper base; passed ball; passed balls (baseball); patrol bombing; permanent ballast; permanent bunker(s); petrol bomb(ing); plate block; plotting board; plugged back (oil well); polybutylene; poor bastard; ports and beaches; power brake; power breaker; pull box; pulse beacon; push button

p/b paperback; pass book; poor bastard; pushbutton

pB purplish blue

Pb plumbum (Latin—lead)

PB Pacific Beach; Packard Bell; Palm Beach; patrol boat; patrol bomber; patrol bombing;

Permian Basis; Personnel Board; Planning Board; Product Bulletin; Pocket Book; police boat; Pompano Beach; Presiding Bishop; Public Bath; Publication Bulletin

P-B Pitney-Bowes

PB Paises Bajos (Spanish—Netherlands); Planta Baja (Spanish—ground floor), elevator pushbutton designation; Prayer Book

P.B. Pharmacopeia Britannica

P & B Porgy and Bess (George Gershwin's opera)

pba poor bloody assistant; pressure-breathing assister; published by arrangement

PBA Patrolmen's Benevolent Association; Philadelphia Bar Association; Plastic Bag Association; Port of Brisbane Authority; Port of Bristol Authority; Printing Brokerage Association; Professional Billiards Association; Professional Bookmen of America; Professional Bowlers Association; Provincetown-Boston Airline; Public Buildings Administration

PBAA Periodical and Book Association of America; Public Broadcasting Association of Australia

p'back paperback

pbai proyectil balístico de alcance intermedio (PBAI)—(Spanish—intermediate range ballistic missile)

P-band 225–390 mc

PBAS Post Block Aerial Survey(ing)

pbb push-button banking

PBBH Peter Bent Brigham Hospital (Boston)

pbb's (PBBs) polybrominated biphenyls

pbc peripheral bus computer; point of basal convergence; pregnancy and birth complication(s); primary biliary cirrhosis

pBc pure Bolivian cocaine

PBC Palisade Boat Club; Pen and Brush Club; Philadelphia Blood Clinic; Philadelphia Book Clinic; Power Boat Club; Provincial Bank of Canada

pbcb plugboard circuit breaker

PBCC Palm Beach Community College

pbd particle board

PBD Public Buildings Department

pbdndb perceived barking dog noise decibels

pbe present-barrel equivalent

Pbe Perlsucht bacillen emulsion

PBEA Paint, Body amd Equipment Association

PBEC Pacific Basin Economic Council; Public Broadcasting Environment Center

P.B.Ed. Bachelor of Philosophy in Education

PBEIST Planning Board for European Inland Surface Transport (NATO)

pbf permalloy-bar file

PBF fast patrol boat (naval symbol)

PBF Prins Bernhard Fonds (Prince Bernhard Fund)

PBFG guided missile fast patrol boat (naval symbol)

PBFL Planning for Better Family Living (UN)

Pbg Pittsburgh

PBGC Pension Benefit Guaranty Corporation

pbh partial bulkhead; primary borehole

pbhp pounds per brake horsepower

pbi please book immediately; polybenzimidazole (space-age fabric); poor bloody infantry; protein-bound iodine

pbi (PBI) polybenzimidazole

pbi proyectil balístico intercontinental (PBI) (Spanish—intercontinental ballistic missile)

PBI Paper Bag Institute; Paving Block Institute; Pitney-Bowes Incorporated; Plant Breeding Institute; Plastic Bottle Institute; Plumbing Brass Institute; Projected Books Incorporated; West Palm Beach, Florida (airport)

PBiP Paperback Books in Print

p-bills personal-computer counterfeit currency

pbip pulse beacon impact predictor

PBJC Palm Beach Junior College

pbk paperback

PBK Phi Beta Kappa

PBKTOA Printing, Bookbinding, and Kindred Trades Overseers' Association

pbl planetary boundary layer; probable

PBL Pacific Beach Library; Public Broadcast Laboratory

pb list phonetically balanced (word) list

P Blr Port Blair

pbm performance-based management

pbm (PBM) permanent bench mark

PBM Mariner twin-engine Navy bomber built by Martin; Pay By Mail; *Paramaribo (Surinam airport)*; Principal Beach Master; Production Bill of Material

PBMA Peanut Butter Manufacturers Association

PBMR Provisional Basic Military Requirements

PBN Provisional Buy Notice

PBN *Producto Bruto Nacional* (Spanish—National Bulk Products)

PBNPA Peanut Butter and Nut Processors Association

pbo packed by owner; paperback original; polite brushoff

P-boat Patrol Boat

P. Bor. *Pharmacopoeia Borussica* (Latin—Prussian Pharmacopoeia)

PBOS Planning Board for Ocean Shipping (USA)

pbp pay by phone; pushbutton panel

pbpGinfwmy please be patient; God is not finished with me yet

pb/ps power brakes/power steering

PBPS Program Budgeting and Planning System

pbr payment by results

pbr (PBR) power breeder reactor; precision bombing range

PBR Project Budget Report(ing); river patrol boat (naval symbol)

pbs paginated by sections; polarizing beamsplitter; production base support; program breakdown structure

pb's paperback books; petrol bombs; poor bastards

p-bs phosphate-buffered saline (solution)

PBS Pacific Biological Station (Canada); Panama Bureau of Shipping; Path Between the Seas (Panama Canal); Permanent Building Societies; Pharmaceutical Benefits Scheme; Philippine Broadcasting Service; Prevent Blindness Society; Public Broadcasting Service; Public Broadcast(ing) Station; Public Buildings Service

PBSA Partially Blinded Soldiers Association; Permanent Building Societies Association

PB & SC Power Boat and Ski Club

PBSCMA Peanut Butter Sandwich and Cookie Manufacturers Association

pbsct peripheral blood stem cell transplant

PBSE Philadelphia-Baltimore Stock Exchange

pbsp prognostically bad signs during pregnancy

pbt passenger boarding total (aircraft code); performance-based teaching; profit(s) before tax(ation)

pbt (PBT) polybutylene terephthalate

PBT President of the Board of Trade

PBTB Paper Bag Trade Board; Paper Box Trade Board

pbte performance-based teacher education

Pburg Pittsburgh

pbv predicted blood volume; pulmonary blood volume

pbw parts by weight; posterior bite wing

pbw (PBW) particle-beam weapon

PBWA Professional Basketball Writers' Association

PBWSE Philadelphia-Baltimore-Washington Stock Exchange

pbx private branch exchange

pbx's (PBXs) personal business exchanges (computerized telephones)

PBY Consolidated-Vultee PBY flying boat; vacation island near Long Beach, California—Santa Catalina

pbz phosphor bronze

pbz (PBZ) pyribenzamine (anti-histamine)

pc paper copy; parent cells; parliamentary cases; partly cloudy; parsec; patent cases; pay clerk; paycheck; percent; percentage; percentile; personal care; personal correction; petty cash; photocell; photoconductive; pica(s); picocoulomb; picocurie; piece(s); pitch circle; point-contact; point of curve; political code; politically correct; polycarbonate; polycomb; port of call; positive column; postcard; practice cases; prices current; principal component; printed circuit; privileged character; professional corporation; program counter; pubococcygeus (muscle); pull chain; pulsating current; punched card; purchasing and contracting purified concentrate

pc (PC) personal computer; pitch class; programme counter

p-c phophlogistic-corticoid; printed circuit

p/c percent; percentage; processor controller; programmer-comparator; pulse counter

p & c put and call

pC *point de congélation* (French—freezing point)

p.c. *per centum* (Latin—by the hundred; percent); *post cibum* (Latin—after a meal, after meals)

Pc Phillips curve (macroeconomics)

PC Pace College; Pacific Airlines; Pacific Coast (railroad); Pacific College; Paine College; Palmer College; Palomar College; Panama Canal; Panola College; Paris College; Park College; Parsons College; Pasadena College; Peace Corps; Pembroke College; Pepperdine College; personnel carrier; Pfeiffer College; Pharmacy Corps; Philadelphia College; Philippine Constabulary; Phoenix College; Piedmont College; Pikeville College; Pilotage Chart(s); Pineland College; Pittsburgh Corning; Plane Commander; Police College; Police Commissioner;

Pomona College; Population Council; Porterville College; Presbyterian College; Principal Celebrant (at the Mass); Principia College; Privy Council; Privy Councillor(s); Procurement Command; Producers Council; Professional Corporation; Providence College; submarine chaser patrol vessel (naval symbol)

PC (pc) Personal Computer

PC Pleas of the Crown; Preliminary Commendation

P-C Penn-Central (railroad); Production-Consumption

P.C. Penal Code; Plaid Cymru (party)

P&C Parents and Citizens Association; Pickpocket and Confidence (police department squad)

PC *Parti Communiste* (French—Communist Party); *Partido Colorado* (Spanish—Colorado Party)—the reds; *Partido Comunista* (Spanish—Communist Party); *Partido Conservador* (Spanish—Conservative Party); *Patres Conscripti* (Latin—Senators); *Penal Code; Plaid Cymru* (Welsh—Party of Wales); *Poder Chicano* (Spanish—Chicano Power); *Première Classe* (French—First Class); *Privy Council* (British Law Reports); *Publishers' Circular*

pca patient-controlled analgesia; permanent change of assignment; pest-control advisor; physical; configuration audit(ing); Porsche Club of America (uses lowercase initials); principal component analysis

pca (PCA) p-chloraphnyl-alanine

Pca Pensacola

PCA Parachute Club of America; Permanent Court of Arbitration (The Hague); Pest Control Association; Pennsylvania Coal Association; Pennsylvania Council on the Arts; Photogrammetric Consultants Association; Plaster Contractors Association; Police Complaints Authority; Pollution Control Agency; Pony Club Association; Portland Cement Association; Positive Control Area; Primary Coverage Area; Printers' Costing Association; Private Care Association; Production Code Administration; Production Credit Association; Proprietary Cremation Association; Pulp Chemicals Association

PCA *Partido Comunista Argentina* (Spanish—Argentine Communist Party)

PCAA Pacific Coast Athletic Association

PCAC Professional Classes Aid Council

PCAFV Physicians' Campaign Against Family Violence

pcam punchcard accounting machine; punchcard accounting method

PCAO Presidential Complaints and Action Office(r); President's Commission on Americans Outdoors

PCAPA Pacific Coast Association of Port Authorities

PCAPK President's Commission on the Assassination of President Kennedy

PCARS Point Credit Accounting and Reporting System

PCART Preparatory Committee for the Autonomous Region of Tibet

PCAs Progressive Citizens of America

PCAST Presidential Commission on Aviation Security and Terrorism

PCAT Philippine College of Arts and Trades

pcb petty cash book; power circuit breaker; printed circuit board

pcb (PCB) polychlorinated biphenyls

PCB Pest Control Bureau; Program Control Board

PCB *Partido Comunista Boliviano* (Spanish—Bolivian Communist Party); *Partido Comunista Brasileiro* (Portuguese—Brazilian Communist Party)

pcbb primary commercial blanket bond(ing)

PCB-BKB *Parti Communiste Belge* (French—Belgian Communist Party); *Kommunistischem Partij* (Flemish—Communist Party)

PCBL Pacific Commercial Bank Limited

PCBMA Pacific Coast Paper Box Manufacturers Association

pcb's (PCBs) polychlorinated biphenyls (industrial pollutants)

p-c b's printed-circuit boards

pcc phosphate carrier compound; pitch of cone to cone; portland concrete cement; program-controlled computer

pCc pure Colombian cocaine

pçc *plus ça change, plus c'est la même chose* (French—the more it changes, the more it stays the same)

PCC Pacific Coast Conference; Pacific Conference of Churches; Palmer Community College; Panama Canal Commission; Panama Canal Company; Pennsylvania Crime Commission; Philippine Cotton Corporation; Poison Control Center; Polynesian Cultural Center; Port of Corpus Christi; Portland Community College; Postal Customer Council; Presidents' Conference Committee; Press Complaints Commission; Price Control Council; Program Control Center; Program for Cooperative Cataloging

PCC *Partido Comunista Cubano* (Spanish—Cuban Communist Party)

PCCA Power and Communication Contractors Association

PCCC Pakistan Central Cotton Committee

PCCD PACE Center for Career Development

PCCEMRSP Permanent Commission for the Conservation and Exploitation of the Maritime Resources of the South Pacific

PCCI President's Committee on Consumer Interests

pcclot protein C activity

PCCNY *Penal Code of the City of New York*

PCCR Publishing Center for Cultural Resources

PCCT Percept and Concept Cognition Test

pccu (PCCU) progressive coronary care unit

PCCU President's Commission on Campus Unrest

pcd pitch circle diameter; pounds per capita per day

PCD Planned Community Development; Principal Criteria Document

PCDA Post Card Distributors Association

PCDG Prestressed Concrete Development Group

pc di pierce die

PC-DOS Personal Computer Disk Operating System

P Cdr Paymaster Commander

PCDS Program Control Display System (NATO)

pce pyrometric cone equivalent

pce (PCE) pseudocholinesterase

PCE patrol craft escort (3-letter coding); Personal Consumption Expenditure

PCE Partido Comunista Española (Spanish—Spanish Communist Party)

PCEA Pacific Coast Electrical Association; President's Council of Economic Advisors; Professional Construction Estimators Association

PCEH The President's Committee on Employment of the Handicapped

p cell parvocellular system

PCEM Parliamentary Council of the European Movement

PCEQ President's Council on Environmental Quality

PCER rescue escort (naval symbol)

pcf pistol center fire (greater than a .22); pounds per cubic foot; power per cubic foot

Pcf Pacifico (Italian—Pacifilc); *Pacifico* (Portuguese or Spanish—Pacific); *Pacifique* (French—Pacific)

PCF patrol craft fast (naval symbol); Personnel Control Facility; Program Checkout Facility

PCF Parti Communiste Français (French—French Communist Party)

PCFAP The President's Committee on the Foreign Aid Program

PCFLIS President's Commission on Foreign Language and International Studies

pcg phonocardiogram

PCG guided-missile coastal-escort vessel (naval symbol)

PCGG Presidential Commission on Good Government (Philippines)

PCGN Permanent Committee on Geographical Names

pch paroxysmal cold hemoglobinuria

P Ch Parish Church

PCH hydrofoil submarine chaser (3-letter coding)

pchbd patchboard

p Ch c pure Chilean cocaine

pchd. purchased

pci pattern correspondence index; pellet-cladding interaction; peripheral command indicator; perpetual cost index; photochromic micro-image; picocurie; potential criminal informant; programmed-controlled interruption

PCi Pacific Coast iris

PCI Packer Collegiate Institute; Peripheral Component Interconnect; Pilot Club International; Planning Card Index; Powder Coating Institute; Precase/Prestressed Concrete Institute; Program of Correctional Institutions (Puerto Rico)

PCI Partito Comunista Italiano (Italian—Italian Communist Party)

PCIB Pacific Cargo Inspection Bureau

PCIC Polaris Control and Information Center

PCIE President's Council on Integrity and Efficiency

PCIFC Permanent Commission of the International Fisheries Convention

PCII Potato Chip Institute International

PCIJ Permanent Court of International Justice

pCi/l picoCuries per liter

PCIM Presidential Commission on Income Maintenance

PCjr IBM's small personal computer

pck polycystic kidney

Pck conditional probability of kill (armament)

pckt printed circuit

pcl parcel; precancel; printed-circuit lamp

PCL Pacific Coast League (Triple-A baseball); Pacific Coast Line; Peoples College of Law; Police Crime Laboratory

PCLA Project Coordination and Liaison Administration

PCLEAJ President's Commission on Law Enforcement and the Administration of Justice

p-c lens perspective-correction lens

pclk pay clerk

PCLO Police Community Liaison Office(r)

PCLTT Permanent Committee on Land Transportation and Telecommunications (ASEAN)

pcm phase-change material(s); pilot control module; plug-compatible manufacturer(s); protein-calorie malnutrition; pulse-code modulation; pulse-count modulation; punchcard machine(s)

PCM Peabody Conservatory of Music; Power Control Module; President's Certificate of Merit

PCM Partido Comunista Mexicano (Spanish—Mexican Communist Party)

PCMA Post Card Manufacturers Association; Professional Convention Management Association

pcmb (PCMB) parachloromercuric benzoic (acid)

pcmi photographic microimage(s)—microdot photos

PCMIA Pittsburgh Coal Mining Institute of America; Plasterers and Cement Masons International Association (U.S. and Canada)

PCMO Principal Colonial Medical Officer

PCMP Progressive Car Manufacturing Program

pcm/pl pulse-code modulated/polarized light

pcmr patient computer medical record; photochromic microreproduction

PCMR President's Committee on Mental Retardation

PCMSER President's Commission on Marine Science, Engineering, and Resources

pcmx (PCMX) parachlorometaxylenol (antiseptic)

pcn parent-country national(s); personal-communications network; printed control number; processing control number

PCN Part Control Number; Pharmaceutical Case Network; Primary Care Network; Procurement Control Number

PCN *Partido de Conciliación Nacional* (Spanish—National Conciliation Party)

PCNB Permanent Control Narcotics Board

PCNG President's Commission on National Goals

PCNR Part Control Number Request

PCN's Planning Change Notices

PCNV Provisional Committee on Nomenclature of Viruses

PCNY Proofreaders Club of New York

pco pest control operator; post checkout operation(s)

pc/o *por ciento* (Spanish—percent)

PCO Pacific Chamber Opera; Pest Control Officer; Printing Control Office(r); Procuring Contracting Office(r); Public Carriage Office(r)

P/CO Purser/Catering Offcer

P$_{co2}$ carbon dioxide pressure (or tension)

PCOB Permanent Central Opium Board (UN)

PCOOS Pacific Coast OtoOphthalmological Society

PCOP President's Commission on Obscenity and Pornography

PCOS Primary Communication Operating System

pcp passenger control point; phospher-coated paper; pimp-controlled prostitution; polychloroprene (rubber); primary care physician; production change point

PCP Peking Central Philharmonic; Postgraduate Center of Psychotherapy; Program Change Proposal; Progressive Conservative Party

PCP *Partido Comunista Panameño* (Spanish—Panamanian Communist Party); *Partido Comunista Paraguayo* (Spanish—Paraguayan Communist Party); *Partido*

Comunista Peruviano (Spanish—Peruvian Communist Party); *Partido Communista Portugues* (Portuguese Communist Party)

pcp (PCP) pentachlorophenol; phencyclidine (sometimes called Pure California Poison); pneumocystis carnii pneumonia

pcpa parachlorophenylalanine

PCPA Panama Canal Pilots Association; parachlorophenylalanine; Philadelphia College of the Performing Arts; Protestant Church-owned Publishers Association

PC(PBC)R Pedestrian Crossings (Push-Button Control) Regulations

PCPD Portland Commission of Public Docks

PCPFS President's Council on Physical Fitness and Sports

PCPI Parent Cooperative Preschools International

PCPJ People's Coalition for Peace and Justice

PCPM Program Control Procedures Manual

PCPP President's Commission on Pension Policy

PCPS Philadelphia College of Pharmacy and Science

PC & PS Professional Credentials and Personnel Service (nursing)

pcpt perception; prostate cancer prevention trial

pcpv prestressed concrete pressure vessel

pcq production-control quantometer

PCQ Personal Control Questionnaire

pcr photoconductive relay; pollution control revenue; polymerase chain reaction

pcr (PCR) program control register

P Cr Paymaster Commander

PCR Program Change Request; Publication Contract Requirement

PCR *Partido Comunista Revolucionario* (Spanish—Revolutionary Communist Party) —Chile; *Partidul Comunist Roman* (Roman Communist Party)

PCRB Pollution Control Revenue Bond

PCRC Paraffined Carton Research Council; Primary Communications Research Center (University of Leicester)

pcrca pickled, cold rolled, and closely annealed

PC R & D C Pomona Colleges Research and Development Center

PCRI Papanicolaou Cancer Research Institute

PCRs Pedestrian Crossings Regulations; Planning and Compensation Reports

PCRS Poor Clergy Relief Society

PCRU Pacific Coast Rugby Union

pcrv prestressed concrete reactor vessel

pcs permanent change of station; personal clerk of session; personal communications service(s); personal communication system; phonocardioscan; picas; pieces; planning control sheet; power conversion system; program counter storage; program counter store

pc's protective clothes

pc's (PCs) personal computers

PCs Police Constables; Progressive Conservatives

PCS 136-foot submarine chaser (3-letter coding); Parent's Confidential Statement; Permanent Committee on Shipping (ASEAN); Petrochemical Corporation of Singapore; Pharmaceutical Card System; Polytechnic Certificate in shipping; Program Control System; Provincial Civil Service; Punch(ed) Card System; Punjab Cooperative Society

PCSA Polish Cultural Society of America

PCSCA Permanent Committee on Socio-Cultural Affairs (ASEAN)

PCSE Pacific Coast Stock Exchange; President's Council on Scientists and Engineers

PCSFA Potato Chip/Snack Food Association

pc sh pierce shell

PCSIR Pakistan Council of Scientific and Industrial Research

PCSP Permanent Commission for the South Pacific; Polar Continental Shelf Project; Princeton Cooperative School Program

PCSS Platform Check Subsystem

PCST Permanent Commission on Science and Technology (ASEAN)

PCSW President's Commission on the Status of Women

pct percent; porhyria cutanea tarda; prothrombin consumption test (blood test)

pct (PCT) portable camera transmitter

pct procent (Dano-Norwegian—percent)

Pct Precinct

PCT Pacific Crest Trail; Patent Corporation Treaty; Portsmouth College of Technology; Potash Core Test(ing)

PCT Partido Conservador Tradicional (Spanish—Traditional Conservative Party)—Nicaragua; *Programa de Cooperación Tecnica* (Spanish—Technical Cooperation Program)

PCTB Pacific Coast Tariff Bureau

pctfe polychlorotrifluoroethylene

pc tp pierce template

PCTs Panama Canal Treaties

PCTS President's Committee for Traffic Safety

PCT&S Philadelphia College of Textiles & Science

pcu photocopy unit; power-control unit; power conversion unit; pressurization-control unit

pcu (PCU) palliative care unit (for terminal patients); portable checkout unit; protective custody unit

pcur pulsating current

PCUS Propeller Club of the United States

PCUS Partido Comunista de la Unión Sovietica (Spanish—Communist Party of the Soviet Union)

PCUSA Presbyterian Church in the U.S.A.

PCU-USA Portuguese Continental Union of the U.S.A.

pcv packed-cell volume; passenger-controlled vehicle; physical control volume; pollution-control valve; positive crank-case ventilation

PCV Peace Corps Volunteer(s); Pestalozzi Children's Village; President's Commission on Violence

PCV Partido Comunista Venezolana (Spanish—Venezuelan Communist Party)

PCVC Public Citizen Visitor's Center

PC virus Port Chalmers (New Zealand) type of influenza virus

PCVs Peace Corps Volunteers

pcv valve positive crankcase ventilation valve

PCWPC Permanent Council of the World Petroleum Congress

pcx periscope convex

PCX Pacific Exchange (stocks)

PCY coastal yacht (3-letter Daval symbol); Pittsburgh, Chartiers & Youghiogheny (railroad)

PCYC Port Credit Yacht Club

PCZ Panama Canal Zone

PCZST Panama Canal Zone Standard Time

pd interpupillary distance; paid; panic disorder; paralysing dose; parental ditype; passed; peak district; period; permanent dunnage; physical distribution; pitch diameter; pitcher defense; plate dissipation; point detonating; poop deck; port dues; position doubtful; post date; post dated; postage due; potential difference; pound; pour depressant; preliminary design; pressure demand; preventive detention; prism diopter; prisoner's dilemma; procurement directive; property damage; proximity detector; public domain; pulse duration; purchase description

p-d prism diopter

p/d post dated

p&d pickup and delivery; price and delivery

pd prima donna (Italian—first lady)

p.d. per diem (Latin-by the day)

Pd palladium; Parade; Parkinson's disease; Pick's disease

PD Parish District; Parliamentary Debates; Pharmacopoeia Dublin; Phelps-Dodge; Physics Department; Police Department; Port of Debarkation; Port Director; Port Dues; position doubtful (navigation chart marking); Preliminary Design; Presidential Directive; Production Department; Program Director; Public Defender

P-D Parke-Davis

P&D Probate and Divorce; Promotion and Development (program)

P of D Port of Duluth

PD Partido Democrático (Spanish—Democratic Party); *Cleveland Plain Dealer; Probate Division* (British Law Reports)

P-D St. Louis Post-Dispatch

P.D. Pharmacopoeia Dublinensis (Latin—Dublin Pharmacopoeia)

pda patient distress alarm; personal death awareness; predicted drift angle; prescription drug abuse; public display of affection

pda (PDA) probability distribution analyzer

pda pour dire adieu (French—to say goodbye)

Pda *Prima donna assoluta* (Italian—absolute first lady)—a principal female singer in an opera or concert organization

PDA Parenteral Drug Association; Personal Digital Assistant (Apple Corporation trademark); Photographic Dealers' Association; Plywood Distributors Association; Port Development Authority; Pregnancy Discrimination Act

PDAC Prospectors and Developers Association of Canada

PDAD Probate, Divorce, and Admiralty Division

P Dal Port Dalhousie

P Dar Port Darwin

PDARS Pulsed Doppler Acoustic Radar System

pdas programmable data acquisition system

PDAS Police Department American Samoa

P-day day when rate of production of an item for military consumption equals rate required by armed forces

pdb paradichlorobenzine

Pd.B. *Pedagogiae Baccalaureus* (Latin—Bachelor of Pedagogy)

PDB President's Daily Brief

pdc preliminary diagnostic clinic; private diagnostic clinic

p&d c premium and dispersion credit(s)

Pdc probability of detection and conversion

PDC Pacific Development Corporation; Penang Development Corporation; Periodical Distributors of Canada; Petroleum Development Corporation; Pregnancy Distress Center; Prevention of Deterioration Center (National Academy of Sciences); Primary Distribution Course; Project Development Corporation; Proposal Development Center; Public Disclosure Commission

PDC *Partido Democrático Cristiano* (Spanish—Christian Democratic Party)

PDCA Painting and Decorating Contractors of America

PDCL Provisioning Data Check List

pdd pre-dental discomfort

pdd (PDD) primary degenerative dementia

Pd.D. *Pedagogiae Doctor* (Latin—Doctor of Pedagogy)

PDD Petty Delinquency Detention; Public Documents Department (GPO)

pdda power-driven decontaminating apparatus

PDDS Parasitic Disease Drug Service

pde paroxysmal dyspnea on exertion; partial differential equations

Pde Parade

PDE Post-test Disassembly Examination; Projectile Development Establishment

P de C *Pas de Calais* (French—Strait of Calais)—Dover Strait

P-de-D Puy-de-Dome

PDEIS Preliminary Draft Environmental Impact Statement

P del R Pinar del Rio (Cuban province)

P del E *Penitenciario del Estado* (Spanish—State Penitentiary)

P de M *Principauté de Monaco* (French—Principality of Monaco)

pdes pulse-doppler elevation scan

PDES Product Data Exchange Specification

P des L *Parc des Laurentides* (French—Laurentian Mountains Park)—Québec

pdf point detonating fuse; probability distribution function

PDF Panamanian Defense Force; Parkinsons' Disease Foundation

PDFA Partnership for a Drug-Free America

PDFLP Popular Democratic Front for the Liberation of Palestine

PDG Paymaster Director-General

pdga (PDGA) pteroyldiglutamic acid

pdgf platelet-derived growth factor

PDGW Principal Director of Guided Weapons

pdh past dental history

pdi point-diffraction interferometer; powered-descent initiation; pre-delivery inspection

pdi (PDI) primary depressive illness

PDI Printing Developments Incorporated

pdic periodic

PDID Public Disorder Intelligence Department (LAPD); Public Disorder Intelligence Division

PDIN *Putsat Dokumetutasi Ilmiah Nasional* (Bahasa Indonesian—National Scientific and Technical Documentation Center)

PDIS *Pusat Dokumentasi Ilmulimu Sosial* (Bahasa Indonesian—Social Sciences Documentation Center)

p dk poop deck

p.d. locks power door locks

pdl page description language; poundal; poverty datum line; power door locks

PDL Patent Depository Library; Program Design Language

PDLP Patent Depository Library Program

pdm pulse-delta modulation; pulse-duration modulation

Pd.M. Master of Pedagogy

PDMA Product Development and Management Association

PDMS Point Defense Missile System

pdn production

pdnes pulse-doppler non-elevation scan(ning)

pdo *pasado* (Spanish—past)

Pdo Pacific decadal oscillation

Pdo *Partido* (Spanish—Party)—political party

PDO Petroleum Development Oman; Property Disposal Office(r); Publication Distribution Office(r)

p/doz per dozen

pdp parallel distributed processing; plasma display panel; power distribution panel; project definition phase

Pdp Paradip

PDP Popular Democratic Party (Puerto Rico); Prescription Drug Program; Program Definition Phase; Program Development Plans

PDPA People's Democratic Party of Afghanistan

PDPS Parts Data Processing System

pd pt production pattern

pdq physician's data query

pdq (PDQ) programmed data quantisizer

p d q pretty damn (or darn) quick

PDQB P.D.Q. Bach

pdr pounder; powder; precision depth recorder (PDR)

pdr *polder* (Dutch—dike-protected lowland reclaimed from the sea or other body of water

PDR People's Democratic Republic; Philippine Defense Ribbon; Preliminary Design Review

PD & R Policy Development and Research

PDR *Physicians' Desk Reference*

pdrd procurement data requirements descriptions

PDRK People's Democratic Republic of Korea (North Korea)

pdrl procurement data requirements list

PDRL Permanent Disability Retirement List

pdrm payload distribution and retrieval mechanism

PDRP Power Distribution Reactor Program

PDRY People's Democratic Republic of Yemen

pds point detonating self-destroying; power drive system; programming documentation standards

pd's public defenders

PDs Police Departments; Program Directors

PDS Pacific Data Systems; Passive Defense System; Passive Detection System; Personnel Data System; Philadelphia Divinity School; Priority Distribution System; Project Data Sheet; Proposed Delivery Schedule

PDSA People's Dispensary for Sick Animals

PDSC Performers and Teachers Diploma—Sydney Conservatorium; Public Disaster Service Committee

PDSOC Police Department Superior Officers' Council

pdsq point detonating superquick fuse

PDSR Principal Director of Scientific Research

pdss (PDSS) post deployment software system

PDST Pacific Daylight Saving Time

pdt power distribution trailer; practice delivery torpedo

pdt (PDT) potentially dangerous taxpayer

PDT Pacific Daylight Saving Time

PDT-1 Picatinny Arsenal Detonation Trap l

PDTC Plymouth and Devonport Technical College

Pdte Presidente (Spanish—President)

PDTLO Pierre Dominique Toussaint l'Ouverture

PDTS Police Detective Training School; Program Development Tracking System

pdu pilot display unit; power distribution unit; power drive unit

pdv pure dried vacuum (salt)

pdv (PDV) pyrotechnic development vehicle

PDV Petroleos de Venezuela (Spanish—Venezuelan Petroleum)—state oil company

pdvm printing digital voltmeter

PDVSA Petroleos de Venezuela Sociedad Anónima (Spanish—Petroleum of Venezuela Corporation)

pdw physical damage waiver

pd work public domain work (of art, history, literature, publication, etc.)

PDX Portland, Oregon (airport)

PDZ Parachute Dropping Zone

pe period entry; personnel equipment; potential energy; presiding elder; private eye (private investigator); probable error; professional equipment; program element; printer's error

pe (PE) physical education; physical examination; polyethylene

p-e precipitation-environment (index)

p/e porcelain enamel; price-to-earnings (ratio)

p & e planning and estimating

pe par exemple (French—for example); *per esempio* (Italian—for example); *por ejemplo* (Spanish—for example); *por exemplo* (Portuguese—for example)

Pe Pecltet number; Pernambuco

Pᵉ Padre (Spanish—father)

PE Pacific Electric (railroad); patrol vessel (naval symbol); Peru (Internet code); Petroleum Engineer(ing); Philadelphia Electric; Pistol Expert; Plain End; Plant Engineer(ing); Port Elizabeth; Port of Embarkation; Port Everglades, Florida; Post Exchange; Prince Edward Island; probable error; Production Engineer(ing); Professional Engineer; Protestant Episcopal

P-E Perkin-Elmer

P & E Peoria & Eastern (railroad)

P of E Port of Entry

P.E. Pharmacopoeia Edinburgensis (Latin—Edinburgh Pharmacopoeia)

pea (PEA) primary expense account

PEA People Express Airlines; Physical Education Association; Plastics Engineers Association; Policewomen's Endowment Association; Potash Export Association; Public Education Association; Publication Effectiveness Audit

PEAB Professional Engineer's Appointments Bureau

PEACE People Emerging Against Corrupt Establishments; Project Evaluation and Assistance in Civil Engineering (USAF)

PEACESAT Pan-Pacific Education and Communications Experiments using Satellites

PEAL Professional Engineers Association Limited

Pea Mus Peabody Museum

PEAP Personal Egress Air Pack

PEAQ Personal Experience and Attitude Questionnaire

pearl. pearl white

Pearl Pearl Harbor—Oahu, Hawaii

PEARL (Committee for) Public Education and Religious Liberty

Pearls Pearl Islands (Las Perlas)

PEAS Production Engineering Advisory Service

PEAT Programmer Exercised Autopilot Test(ing)

PEAT Programme Elargi d'Assistance Technique (French—Enlarged Technical Assistance Program)—UN

peb phototype environmental buoy

Pe. B. Pediatriae Baccalaureus (Latin—Bachelor of Pediatrics)

PEB Physical Evaluation Board; Propulsion Examining Board (USN); Public Examination Board

pebb public employees blanket bond(ing)

pebd pay entry base date

pec photoelectric cell; position error correction; program element code

p E c pure Ecuadoran cocaine

PEBES Personal Earnings and Benefit Estimate Statement

PEC Philadelphia Electric Company; Plain English Campaign; Presidential Ethics Commission; Production Equipment Code; Protestant Episcopal Church; Psychology Examining Commission; Psychology Examining Committee

pecan pulse envelope correlation air navigator

PECE President's Emergency Committee for Employment

PECI Projects and Equipment Corporation of India

'pecker woodpecker

PECM Preliminary Engineering Change Memorandum (USAF)

PECO Philadelphia Electric Company

PECP Preliminary Engineering Change Proposal

PECS Plant Engineering Check Sheet

pecto pectoral

pecul peculated; peculating; peculation; peculator

Peculiar Institution Mount Holyoke College founded by Mary Lyon and described by her as a peculiar institution as it was for women

PECUSA Protestant Episcopal Church of the U.S.A.

ped pedagogue; pedagogy; pedal; pedestal; pedestrian; personnel equipment data

ped (Latin prefix—children)—pediatrics

Ped pedal (music); Pediatrics

P Ed Physical Education

pedag pedagogue; pedaguese (patois of pedants)

pedageese pedagogue jargon

Ped.B. Bachelor of Pedagogy

Ped.D. Doctor of Pedagogy

pediat pediatric(al)(ly); pediatrician; pediatrics

PE Dir Physical Education Director

Ped.M. Master of Pedagogy

pedo pedologic(al)(ly); pedologist(ic)(al)(ly)

pedobap pedobaptism; pedobaptist

pedog pedograph(ic); pedography

pedogen pedogenesis

pedol pedologic(al)(ly); pedologist(ic)(al)(ly); pedology

pedom pedometer; pedometric(al)(ly)

pedont pedodontic(al)(ly); pedodontist(ry)

pedop pedophile; pedophilia(c)

Pedralvez Pedro Alvarez

Pedrarias Pedro Arias

Pedro San Pedro, California

peds pediatrics

pedstl pedestal(s)

PED XING pedestrian crossing

pee photoelectric emission; pressure environment equipment; urine

PEE Proof and Experimental Establishment (British Ministry of Defence)

P & EE Proving and Experimental Establishment

Peeb Peebles

Peebl Peebleshire

peed off pissed off

peep positive end expiratory pressure

peep. (PEEP) pilot's electronic eye-level presentation

Peer Peer Gynt

PEER Planned Environment and Education Research Institute

pees former South Vietnam piasters

pef Peak expiatory flow; personal effects floater (policy)

PEF Palestine Exploration Fund; Personality Evaluation Form; Plastics Education Foundation; Presidential Election Fund; Psychiatric Evaluation Form

PEFCO Private Export Funding Corporation

pefr peak expiratory flow rate

peg price-to-earnings-to-growth (ratio)

peg (PEG) pneumoencephalogram

peg. polyethylene glycol

Peg Peggy

Peg Pegasus (Latin—Peacock constellation); *Pegunungan* (Malay—mountain range)

PEG Petrochemical Energy Group; Pittsburgh Elderly Gay

PEGE Program for Evaluation of Ground Environment

Peggy nickname for Margaret

Pegs Pegasus (constellation)

pei pointless electronic ignition; precipitation-efficiency index

PEI Petroleum Equipment Institute; Porcelain Enamel Institute; Preliminary Engineering Inspection; Prince Edward Island

PEINP Prince Edward Island National Park

PEIP Presidential Executive Interchange Program

PEIS Preliminary Environmental Impact Statement

pej premolded expansion joint

p ej por ejemplo (Spanish—for example)

PEJO Plant Engineering Job Order(s)

pejor pejorative(ly)

pek pig embryo kidney

Pek Peking; Pekinese

peke pekinese dog

pel pelagic; pellet; pelvis; permissible exposure limit; picture element

P El Port Elizabeth

PEL Petroleum Exploration License; Physics and Engineering Laboratory

P EL Port Elizabeth

Pelagies Pelagian Islands in the Mediterranean between Sicily and Tunisia

P Eliz Port Elizabeth

Pellews Pellew Islands in Australia's Gulf of Carpentaria

Pellys Pelly Mountains of the Yukon

PELNI Pelojaran Nasional Indonesia (National Shipping Company of Indonesia)

pem photoetectromagnet(ic); program element monitor; protein-energy malnutrition

Pem Pembrokeshire

PEM Production Engineering Measures; Project Engineering Memo

PEMA Process Equipment Manufacturers Association

Pemb Pembroke College, Oxford; Pembrokeshire

Pemb Coll Pembroke College—Cambridge

Pemex Petróleos Mexicanos (Spanish—Mexican Petroleum)

PEMR Petroleum Engineering Monthly Report

PEMS Portable Environmental Measuring System

PE Mus Port Elizabeth Museum

Pem Yeo Pembroke Yeomanry

pen. penal; penetrate; penology; peninsula; penitentiary; penmanship

pen (Latin prefix—lack or need)—penicillin

Pen Penang; Penarth; Peninsula; Penitentiary

Pen Peninsula (Portuguese or Spanish—peninsula); *Peninsule* (French—peninsula); *Penisola* (Italian-peninsula)

PEN Poets, Playwrights, Editors, Essayists, and Novelists (international organization often referred to as the P.E.N. Club)

PEN Presse Etudiante Nationale (French—Student National Press)—Québec's student news cooperative

pen. aids penetration aids

Pen Ala Peninsula de Alaska (Spanish—Alaska)

Pen BC Peninsula de Baja California (Spanish—Lower California Peninsula)

pencil. pictorial encoding language

PEN Club (*see* PEN)

Pen de Yuc Peninsula de Yucatan (Spanish—Yucatan Peninsula)

Pene Penelope

P/E News Petroleum/Energy News

P Eng Professional Engineer(ing)

Pen Fla Peninsula de la Florida (Spanish—Florida Peninsula)

PENGEM Penetrating the Gray Electronic Market (FBI undercover operation)

Penguin Norwegian surface-to-surface missile; Penguin Books

peni penicillin

Pen Ib Peninsula Ibérica (Spanish—Iberian Peninsula)—Portugal and Spain

penic penicillin

penic. penicillum (Latin—brush)

penic. cam. penicillum camelinum (Latin—camel's-hair brush)

Penit Penitentiary

Pen Lab Peninsula de Labrador (Spanish—Labrador Peninsula)

Pen Germ Pennsylvania German (dialect spoken by German settlers brought to rural Pennsylvania by William Penn)

Penn Pennsylvania; Pennsylvanian

Penna Pennsylvania

Pennamite(s) Pennsylvanian(s)

Penn Central Pennsylvania New York Central Transportation Company (merger of Pennsylvania, New York Central, New Haven, and Lehigh Valley railroads)

Penn Deutsch (German—Pennsylvania German)—not Pennsylvania Dutch, as many imagine; Dutch did not settle in Pennsylvania

PennDOT Pennsylvania Department of Transportation

Penn Dutch Pennsylvania Dutch (popular name for dialect spoken by German settlers brought to Pennsylvania by William Penn and many of his Quakers)

Penney JC Penney Company

Penn German Pennsylvania German (dialect spoken by German settlers brought to rural Pennsylvania by William Penn)

Pennines Pennine Alps between Italy and Switzerland; Pennine Hills ranging from southern Scotland to central England—the Pennine Chain

Penn State Pennsylvania State University

Pennsy Pennsylvania; Pennsylvania Railroad

Penn Turn Pennsylvania Turnpike

penol penological; penologist; penology

Peñon de Veléz Peñon de Valez de la Gomera (rocky islet belonging to Spain in the western Mediterranean)

penrad penetration radar

pens pensioneret or *pensionist* (Dano-Norwegian—retired or pensioner)

pensad pension administration

PENSADS Pension Administration System

Pension-expanding President Benjamin Harrison

Pensy Pensacola, Florida

pent. penetrate; penetration; pentode

Pent Pentagon; Pentecost

Pent Pentateuch

PENT Project for the Education of Native Teachers

Pentlands Pentland Hills southwest of Edinburgh or the Pentland Skerries comprising the southernmost Orkneys

pento (sodium) pentothal

penval penetration evaluation

peo professional employer organization

peo. people

PeO President ex-Officio

PEO Philanthropic Educational Organization; Plant Engineering Order; Protect Each Other (secret women's organization)

PEOC Publishing Employees Organizing Committee

pep peak envelope power; pepper; pep pill; peptide; planar epitaxial passivated

pep. pepper, peppermint; peppy; personal effects protection

pep. (PEP) phosphoenolypyruvate; polyestradiol phosphate; Public Employment Program

P e P Partija e Punes (Albanian—Workers Party)

PEP Parent Effectiveness Program; P.E.P. *Deraniyagala*; Pepsi-Cola (stock-exchange symbol); Performance Evaluation Process; Personalized Engineering Program; Personalized Exercise System; Petroleum Electric Power; Political and Economic Planning; Positron-Electron Project; Preventive Enforcement Patrol; Professional Enhancement Program; Proficiency in English Program; Program Evaluation Procedure; Public Employment Program

PEPA Petroleum Electric Power Association

Pep-Bis Pepto-Bismol

Pepco Potomac Electric Power Company

pepg piezo-electric power generator

PEPG Port Emergency Planning Group (NATO)

PEPIC Public Education Project on the Intelligence Community

PEPLAN Polaris Executive Plan (UK)

pep materials propellants, explosives, pyrotechnics

PEPP Professional Engineers in Private Practice

Peppe Giuseppe (Italian—Joe)

pepr precision encoder and pattern recognizer

peps pep pills; peptides; plant-expressed protectants

peps. pepsin

PEPs Public Employment Programs

Pepsi Pepsi Cola

PEPSI Proton Echo Planar Spectroscopic Imagery

PEPSU Patiala and East Punjab States Union

PEQC President's Environmental Quality Council

per period; periodic; peritoneum; perodicity; person; personal; personate

per perito (Italian—expert)

Per Perseus (constellation); Persia; Persian; Peru (whose capital is Lima)

Per Perciles, Prince of Tyre; Pereval (Russian—mountain pass); *Perevoz* (Russian—crossing, ferry); Persian

PER Perth, Australia (airport)

PE&R Policy, Evaluation, and Research

PERA Planning and Engineering for Repair and Alterations; Production Engine Remanufacturers Association; Production Engineering Research Association

per agrim perito agrimensore (Italian—surveyor)

per an. per annum (Latin—by the year); *per anum* (Latin—by the anus)

per art perito artistico (Italian—art expert)

p/e ratio price-earning ratio

PERB Personnel Evaluation Research Bureau; Planning and Environmental Review Board; Professional Engi-

neers Registration Board; Public Employment Relations Board

perc perchloroethylene; percolate; percussion

PERC Peace on Earth Research Center

per call perito calligrafo (Italian—handwriting expert)

per cent. per centum (Latin—by the hundred)—percent

Percept Psychophys Perception and Psychophysics

perco percobarg (barbiturate synthetic morphine derivative); percodan (synthetic morphine derivative)

per con. per contra (Italian—on the other side)

PERCOS Performance Coding System

perd perdenosi (Italian—dying away)

PERDDIMS Personal Development and Distribution Management System

perden. perdendosi (Italian—dying away)

perdi per diem

Peregil Pedro Gil

perestroi perestroika (Russian—economic restructuring)

perf perfect; perfection; perforate; perforation; perform; performance; performer; perfume(d)

PERF Planetary Entry Radiation Facility (NASA); Police Executive Research Forum; Police Executive Resource Form

perfect calc perfect calculation(s)

perfs perforations; performances; performers; perfumers

perg pergamino (Spanish—parchment)

Pergamon Pergamon Press

Per Gul Persian Gulf

perh perhaps

peri perigee; perimeter

peri (Greek—around)—pericardial, pericarp, perimeter, periosteum, peripheral, peritoneum

PERI Platemakers Educational and Research Institute

periap periapical

Peric Periclean

Perico Pedro

peridot yellow-green tourmaline

perig perigee

perih perihelion

PERINTREP Periodic Intelligence Report

perio periodontry

period menstrual period; period of rotation; period of revolution

period. periodical

periodontol periodontology

peris periscope

perjy perjury

perk percolate

perk. payroll earnings record keeping

PERK Prospective Evaluation of Radial Kerotomy

perk(s) perquisite(s)

perl principal exchange rate-linked (security); pupils equal and reactive to light

perla pupils equal—react to light and accommodation

perm permanent; permanent wave; permission

Perm Permian

permaflowers permanent (plastic) flowers

permafrost permanent frost

permafruit permanent (plastic) fruit

permatemps permanent temporary workers

Perm Ct of Arb Permanent Court of Arbitration

permed permanently waved

PERMIS Public Employees Retirement Management Information System

PERMREP Permanent Representation to the North Atlantic Council (NATO)

perms permanents; permanent waves (hair)

per nav per navale (Italian—ship expert)

PERO President's Emergency Relief Organization

per. op. emet. peracta operatione emetici (Latin—when the emetic action is over)

peroxide hydrogen peroxide (H_2O_2)

perp perpendicular; perpetrator

Perp Perpignan

per pro. per procurationem (Latin—by proxy)

perq(s) perquisite(s)

per rec per rectum (Latin—through the rectum)

perrla pupils equal, round, react to light and accommodation

pers person; personal; personality; personnel; persons

Pers Aulus Persius Flaccus (Roman satiric poet); Perseus (constellation); Persia(n)(s)

PERS Public Employees' Retirement System

Per.Sac.Lit. Peritus in Sacred Liturgy

persian white fantanyl (more powerful than morphine, also known as china white)

Persimfans *Pervi Simfonichesky Anseabl* (Russian—conductorless symphonic ensemble)

PERSIS Personnel Information System

pers n personal noun

'personation impersonation

personi personification; personified; personifier; personifying

persp perspective

pers pron personal pronoun

Persymfans *Pervyi Symfonitchesky Ansamble* (Russian—First Symphonic Ensemble)—conductorless orchestra organized in 1922 in Moscow

pert performance evaluation recall technique

pert. pertaining

pert. *pertussis* (Latin—whooping pertossis cough)

PERT Program Evaluation and Review Technique/Critical Path Technique

PERTCO Program Evaluation and Review Technics (plus) Cost Analysis

per tecn comm *perito tecnicocommerciale* (Italian—estimator)

pertest percolation test(ing)

Perths Perthshire

PERTVS Perimeter Television System

Peru Republic of Peru (Andean nation containing monumental structures left by the Incas), *República del Perú*

Peru Cur Peruvian Current

Peruv Peruvian

perv perversion; pervert; perverted

perv show pervert show

pes photoelectric scanner

pes (PES) polyether sulfone

pe's printer's errors

pe & s parts engineering and standardization

P es per esempio (Italian—for example)—e.g.

Pes. *Pesahim*

PEs Professional Engineers

PES Philosophy of Education Society

PESA Petroleum Equipment Suppliers Association

PESC Public Expenditure Survey Commission

pescado *pez pasado* (Spanish—past fish)—dead fish or fish out of water

Pescadores (Portuguese or Spanish—Fishermen) islands off Taiwan and part of the Republic of China

peseta monetary unit of Spain

Pesh Peshawar

peso monetary unit of Bolivia, Chile, Colombia, Cuba, the Dominican Republic, Guinea-Bissau, Mexico, and the Philippines

PEST Pressure for Economic and Social Toryism

pet paper equilibrium tester; personal electronic translator; petroleum; petrological; petrologist; petrology; point of equal time

pet. (PET) polyethylene terephthalate; positive-emission tomography; positron emission tomography (scan)

Pet Petén (Guatemalan department); Peter; Peterhead; Peterhouse College, Oxford, Peterkin; Petronius

Pet *Peters' United States Supreme Court Reports*

PET Parent Effectiveness Training; Pet Milk Company (stock-exchange symbol); Pierre Elliott Trudeau (former Canadian Prime Minister); Positron Emission Computed Tomography; Production Environmental Test(ing,s); Production Evaluation Test(ing,s); Prostitution Enforcement/Team (police versus pimps and prostitutes)

PETA People for the Ethical Treatment of Animals

PETANS Petroleum Training Association—North Sea

Pete Peter; St Petersburg

Petersburg Saint Petersburg (later called Petrograd and Leningrad); nickname for the Federal Reformatory at Petersburg, Virginia

peth petroleum ether

petitn. petition

petn petition

petr petrifaction; petrified

petr. petitioner

Petr Petronius Arbiter (Roman satirist)

PETR Preliminary Flight Test Report

petri petroleum

Petriburg. *Petriburgensis* (Latin—Peterborough)

Petrified Forest Petrified Forest National Park in Arizona's Painted Desert

Pëtr Makadonski (Russian—Peter the Great)—Pyotr Alekseyevich Romanov (1672–1725)

petro petrochemical; petroleum; petrology

Petro Petrograd (Russian City of Peter)—Leningrad in the Russian Revolution

PETROBAS *Petróleo Brasileiro* (Portuguese-Brazilian Petroleum Corporation)

petro-chem petroleum-chemical

petrodollars petroleum-controlled dollars

Petrofina Compagnie Financiere Belges des Pétroles (Belgian Financed Petroleum Company)

petrog petrography

Petrograd (Russian—Peter's City)—Saint Petersburg renamed Petrograd in 1914 and Leningrad in 1924

petrol. petroleum; petrological; petrologist; petrology

Petronas Petroliam Nasional (Malay—National Petroleum)

Petropolis (Greek—City of Peter)—St Petersburg, Petrograd, Leningrad

petros petrochemicals

PETROVEN *Petróleos de Venezuela* (Spanish—Venezuelan Petroleum Corporation)

pets. prior to expiration of term of service

PETS Posting and Enquiry Terminal System

pet/spect (PET/SPECT) positron-emission tomography/single photon-emission-computed tomography

pett (PETT) positron emission transaxial tomography

peua pelvic examination under anesthesia

pev propeller-excited vibration

PEVE Prensa Venezolana (Venezuelan press service)

pewter lead-tin alloy containing some antimony

p ex par exemple (French—for example)

p ext por extensão (Portuguese—by extension)

pf page footing; park factor; pekoe fannings; perfect; performance factor; pfennig; phenol-formaldehyde; pianoforte; picofarad; pneumatic float; points for; power factor; preferred; preflight; profile; profiled; proximity fuse; public funding; public funds; pulse frequency; pyrolysis fluorescence

pf (PF) page footing; page formatter; personal finance; punch-off character

p/f portfolio

pf pro forma (Latin-for the sake of the form), an advance declaration for a financial statement or overseas invoice

pf. piano e forte (Italian—soft and then loud)

p f piu forte (Italian—more loudly)

pF picofarad(s)

PF Pfennig (German-penny)

PF French Polynesia (Internet code); patrol escort vessel (naval symbol); Packaging Facility; Physician's Forum; Pioneer & Fayette (railroad); Police Foundation; Procurator Fiscal

P/F Peace and Freedom (political party)

pfa psychologic-flight avoidance; pulverized fuel ash

pfa (PFA) polyfluoro-alkoxy

PFA Polyurethane Foam Association; Prescription Footwear Association; Press Foundation of Asia; Private Fliers Association

PFA Policia Federal Argentina (Spanish—Argentine Federal Police)

P factor hypothetical pain-producing substance produced in ischemic muscle; preservation factor

PFAPC Psychological Factor Affecting Physical Condition

PFAS President of the Faculty of Architects and Surveyors

pfb prefabricate(d); preformed beam(s); pseudofillicutitis barbae

PFBMF Polaris Fleet Ballistic Missile Force

PFBrg pneumatic float bridge

pfc passed flying college; passed (with) flying colors; passenger facility charge; plaque-forming cell(s); privately financed consumption

Pfc Private first class

PFC Partnerships For Change; Pusan Fisheries College

PFCCT Pennsylvania Federation of Community College Trustees

pfce performance

pfc's perfluorocarbons

PFCS Primary Flight Control System

pfd personal flotation device; preferred, present for duty; primary flash distillate

Pfd Pfund (German—pound)

PFDF Petroleum Fuel Development Facility

pf di progressive die

pfd s preferred spelling

PFDI Preferred Funeral Directors International

PFEFES Pacific and Far East Federation of Engineering Societies

PFEL Pacific Far East Line

PFET P-channel junction Field-Effect Transistor

pff pie-fed farmer; plaque-forming factor

PFF Police Field Force

pffb pie-fed farm boy

PFFBI Pacific Fire Fighters Burn Institute (Sacramento)

PFFF Plutonium Fuel Fabrication Facility

pffg pie-fed farm girl

PFF Inc Police-FBI Fencing Incognito (traffickers in stolen goods)

pf fx profiling fixture

PFG Principal Financial Group

PFGM guided missile patrol escort vessel (naval symbol)

PFGX Pacific Fruit Growers Express

pfi physical fitness index (PFI)

PFI Pacific Forest Industries; Pet Food Institute; Photo Finishing Institute; Picture and Frame Institute; Pie Filling Institute; Pipe Fabrication Institute; Police Foundation Institute

PFIAB President's Foreign Intelligence Advisory Board

pfic passive foreign investment company

PFJM Policía Federal Judicial Mexicana (Spanish—Mexican Federal Judicial Police)

pfk (PFK) phosphofructokinase

pfl pressed-for-life (dress materials)

PFL Pacific Freight Lines

PFLAG Parents, Families and Friends of Lesbians and Gays

PFLO Popular Front for the Liberation of Oman

PFLP Popular Front for the Liberation of Palestine (Marxist)

pflr peak flow rate; peak flow reading; programmable film reader; prototype fast reactor (PFR)

pfm power factor meter; pulse frequency modulation

PFMA Plumbing Fixture Manufacturers Association; Pressed Felt Manufacturers' Association

pfn prefinish(ed)

PFN Partido Frente Nacional (Spanish—National Front Party)—Costa Rica

PFNM Petrified Forest National Monument

PFNP Petrified Forest National Park

pfo patent foramen ovale

PFOBA Paso Fino Owners and Breeders Association

PFOC Prairie Fire Organizing Committee (communist)

pfp pay for performance

PFP Personal Financial Planner; Progressive Federal Party (South African)

PFP Progresief Federaal Partij (Afrikaans—Progressive Federal Party)

PFPC Parents of Fluoride-Poisoned Children

pfr peak flow rate; peak flow reading; programmable film reader; prototype fast reactor (PFR)

PFRB Pacific Fire Rating Bureau

PFRS Programmed Film Reader System

PFRT Performance Flight-Rating Test; Preliminary Flight-Rating Test

pfs porous friction surface(d)

pfsa pour faire ses adieux (French—to say goodbye)

PFS Professional Software

PFSO Postal Finance and Supply Office(r)

pfsp private fee-for-service plan

pfst pianofortist (pianist)

P-F Study Picture-Frustration Study (Rozensweig)

pft portable flame thrower

PFT People For Trees; Pet-Facilitated Therapy

pft acct pianoforte accompaniment

PFTAW People For The American Way

PFTC Pestalozzi Froebel Teachers College

pfte pianoforte (piano)

pfu pock-forming units; preparation for use

P Fu Port Fuad

pfv physiological full value

pfv pour faire visite (French—to make a call)

PFV Pestalozzi-Froebel Verband (Pestalozzi-Froebel Association)

PFVEA Professional Film and Video Equipment Association

PFW Project Feederwatch

pfx prefix

PFX Pacific Fruit Express

pg page; paregoric; paris granite; parks, gardens; pay group; pass gas; paying guest; permanent grade; pistol grip; postgraduate; power gain; pregnant program guidance; proving ground; public gaol; pure gin

pg (PG) parental guidance; prostaglandin

pg pago (Portuguese—paid)

p.g. persona grata (Latin—an acceptable person)

Pg Paraguay; Paraguayan; petagram; Portugal; Portuguese

PG gunboat patrol vessel (naval symbol); Pan American—Grace Airways; Papua New Guinea (Internet code); Pennsylvania-German; Post Graduate; Proctor & Gamble; Project Group (NATO); Provincial Government

P.G. Preacher General

P & G Proctor & Gamble

P of G Port of Galveston

PG Prisonnier de Guerre (French—prisoner of war)

P.G. Pharmacopoeia Germanica (Latin—German pharmacopoeia)

PG-13 parental guidance suggested (movie rating)

pga pressure garment assembly

pga (PGA) pteroylglutamic acid (folic acid)

PGA Pharmacy Guild of Australia; pin grid array; Producers Guild of America; Professional Golfers Association

PG-AC Professional Group—Automatic Control

PGAH Pineapple Growers Association of Hawaii

p-gal(s) proof gallon(s)

PGA-NOC Permanent General Assembly—National Olympic Committees

PGB patrol gunboat (naval symbol)

pgbd pegboard(s)

PG-BTS Professional Group—Broadcast Transmission System

pgc per gyro compass

PGC Peoples Gas Company; Punxsutawney Groundhog Club

PGCE Post-Graduate Certificate of Education

PGCOA Pennsylvania Grade Crude Oil Association

PG-CS Professional Group—Communication System

PG-CT Professional Group—Circuit Theory

pgd paged; paradigm

PGD Past Grand Deacon

PGDF Pilot Guide Dog Foundation

pgdo pagado (Spanish—paid)

pge phenyl glycidyl ether

PGE Pacific Great Eastern (railroad); Portland Grain Exchange

PG-E Professional Group—Education

PG & E Pacific Gas and Electric

PG-EC Professional Group—Electronic Computers

PG-ED Professional Group—Electronic Devices

PG-EM Professional Group—Engineering Management

PGER Pacific Great Eastern Railway

pgh (PGH) pituitary growth hormone

Pgh Pittsburgh, Pennsylvania

PGH patrol gunboat—hydrofoil (naval symbol); Philadelphia General Hospital; Philippine General Hospital

PG-HFE Professional Group—Human Factors in Electronics

PGI Pyrotechnics Guild International

PG-I Professional Group—Instrumentation

PG-IE Professional Group—Industrial Electronics

PGIM Professional Group on Instrumentation and Measurement (NBS)

pgio Poggio (Italian—hill, hillock, hilltop)

P-girls pub girls (waitresses in British barrooms)

PGIS Project Grant Information System

PGIT Professional Group on Information Theory (IEEE)

PGJD Past Grand Junior Deacon

pgk phosphoglycerate kinase

pgl (pronounced *pee-gul*) puppy beagle

PGL Provincial Grand Lodge

P GL Port Glasgow

P Glg Port Glasgow

pglin page and line (flow chart)

pgm porous glass matrix (method of immobilizing nuclear waste); precision guided missile; precision guided munition; program

pgm (PGM) phosphoglucomutase

PGM motor gunboat (3-letter naval symbol); Past Grand Master

PGMA Private Grocers' Merchandising Association

PGmc Proto-Germanic
PG-ME Professional Group—Medical Electronics
PG-MITT Professional Group—Microwave Theory and Technics
pgm's precision-guided munitions
pgn pigeon
pgn (PGN) proliferative glomerulonephritis
PGNP Pagsanjan Gorge National Park (Philippines)
PGNS Primary Guidance and Navigation System
pgo pyrolysis gas oil; pontine, geniculate, occipital
PGOC Philadelphia Grand Opera Company
P of GP Pearl of Great Price
PGPR Provincial Guild of Printers' Readers
pgr population growth rate; psychogalvanic reaction; psychogalvanic response
pgr (PGR) precision graph record(er)
pg rating parental-guidance rating (of a motion picture or television program)
PGRO Pea Growing Research Association
pgrv (PGRV) precision-guided reentry vehicle
pgs predicted ground speed
pg's (PGs) prostaglandins
PGS Pennsylvania-German Society; Pidaung Game Sanctuary (Burma); Power Generation System; Primary Guidance System
PGSC Panel on Geological Site Criteria
PGSD Past Grand Senior Deacon
PGSW Past Grand Senior Warden
pgt per gross ton
PGT Pacific Gas Transmission (company); Program Global Table
PGTB Pierre Gustave Toutant Beauregard
Pgu Pagalu (formerly Annobon)
PGU Pontifical Gregorian University
pgut (PGUT) phosphogalactose uridyl transferase
PGW Persian Gulf War
PGWA Pottery and Glass Wholesalers' Association

ph page heading; pharmacopoeia; phase; phone; phosphor; phot; photon; physically handicapped; pinch hitter; power house; praying hands; precipitation hardening; previous hardening
ph (PH) past history
ph. parish; physician
p/h per hour
p & h postage and handling
pH hydrogen-ion concentration
Ph Pahari; phenyl; Philosophy
PH Parachute Handler; Paradise Hills; Pearl Harbor; Philharmonic Hall; Philippines (Internet code); Plane Handler; Power House; Public Health; Purple Heart
P-H Prentice-Hall
PH3 phosphine
pha (PHA) phytohemagglutinin
PHA Preferred Hotels Association; Public Housing Administration
PHADS Phoenix Air Defense Sector
Phaedr Phaedrus (Roman fabulist-poet)
phag (Latin prefix—to eat) phagocytes
phage(s) bacteriophage(s)
phal phalange; phalanx
'phant elephant
phar pharmacy
P Har Port Harcourt
Phar. B. Bachelor of Pharmacy
Phar. C. Pharmaceutical Chemist
Phar. D. Doctor of Pharmacy
pharm pharmaceutical; pharmacist; pharmacology; pharmacopoeia(s); pharmacy
Phar. M. Pharmaciae Magister (Latin—Master of Pharmacy)
Pharmaceutical Pharmaceutical Press
pharmacol pharmacology
pharm chem pharmaceutical chemistry
Pharm.D. Pharmaciae Doctor (Latin—Doctor of Pharmacy)
PHAs polyclitic aromatic hydrocarbons; Public Housing Agencies
'phasia aphasia

Ph.B. Philosophiae Baccalaureus (Latin—Bachelor of Philosophy)
Ph. B.J. Bachelor of Philosophy in Journalism
ph brz phosphor bronze
Ph. B. Sp. Bachelor of Philosophy in Speech
Ph. C. Pharmaceutical Chemist
PHC Patrick Henry College
PHC Prairie Home Companion (radio program)
PHCC Plumbing, Heating, Cooling Contracters
PHCIB Plumbing-Heating-Cooling Information Bureau
ph const phase constant
phd panty-hose distributor; piled higher and deeper; poor, hungry, driven
Ph. D. Philosophiae Doctor (Latin—Doctor of Philosophy)
PHD Port Huron and Detroit (railroad)
P.H.D. Public Health Doctor
PHDA Presently Hasn't Done Anything
PhD cameras Push-here-dummy cameras designed so anyone can take a picture
Ph. D. Ed. Doctor of Philosophy in Education
Phe Phoenix (constellation)
PHE phenylalanine (amino acid)
P.H.E. Public Health Engineer
PHEAA Pennsylvania Higher Education Assistance Agency
P-head pinhead, small-minded person; user of amphetamine
phency phencyclidine (angel dust)
pheno phenobarbital; user of phenobarbital
phenold phenomenological death
phenolp phenolphthlein
phenom phenomena; phenomenal; phenomenon
phf plug handling feature
phf's (PHFs) paired helical filaments
PHF Potomac Horse Fever
Ph. G. Graduate in Pharmacy
Ph. G. Pharmacopoeia Germanica (Latin—German Pharmacopoeia)
PHG Postman Higher Grade
phgt package height
PHHS Patrick Henry High School

phi philosophy; preharvest interval

Phi Philips

Ph I   *Pharmacopoeia Internationalis*

PHI Philadelphia White Stockings (National Association); Public Health Inspector

phial.   *phiala* (Latin—bottle)

PHIBLANT Amphibious Forces—Atlantic (USN)

PHIBPAC Amphibious Forces—Pacific (USN)

PHIGS Programmer's Hierarchical Interactive Graphics Standard

phil philosophy

phil (Latin suffix—having an affinity for)—neutrophiliac

Phil Philadelphia; Philadelphian; Philbert; Philharmonia; Philharmonic; Philip; Philippa; Philippine; Philippine Airlines; Philippines; Phillip; Phillipa; The Epistle of Paul to the Philippians

Phil Philippians

Phila Philadelphia; Philadelphian

Philada Philadelphia

Phila Free Lib Philadelphia Free Library

PHILAG Public Health Institute—London Action Group

philat philately

PHILDis Philadelphia Procurement District (US Army)

Philem The Epistle of Paul to Philemon

Philem Philemon

PHILEX Philadelphia Exchange

Phil Hung Philharmonica Hungarica

Philippines Republic of the Philippines, *Republika ñg Pilipinas* (Pilipino) or *República de Filipinas* (Spanish)

Philips' Philips' Gloeilampenfabrieken (Dutch—Philips' Electric Lamp Factory)

Philips Res Rept Philips Research Reports

PHILIRAN Philips Petroleum Iran

Phil Is Philippine Islands

Phillies Philadelphians

Phil Lib Philosophical Library

PHILLIPS Phillips Petroleum Company

Philly Philadelphia

Phil Mag *Philosophical Magazine*

philocrit philosopher critic; philosophical criticism

philol philology

Phil Orch Philadelphia Orchestra

philos philosophy

Philos Philosophy

philos educ philosophy of education

Philos Lib Philosophical Library

Philos Mag *Philosophical Magazine*

philosoph philosopher; philosophical; philosophy

Philos Pub Philosophical Publishing Co

Philos Res Philosophical Research Society

Philos Trans R Soc London *Philosophical Transactions of the Royal Society of London*

PHILSA Philippine Standards Association

Phil Soc Philharmonic Society

PHILSOM Periodical Holdings in the Library of the School of Medicine

Phil Sp Philippine Spanish

PHILSUCOM Philippine Sugar Commission

PHILSUGIN Philippine Sugar Institute

Phil Trans *Philosophical Transactions* (Royal Society of London)

Phin Phineas T. Barnum (American showman)

PHIND Pharmaceutical and Healthcare Industries News Database

phiz physiognomy

phk cells postmortem human kidney cells

Phl (Port of) Philadelphia

Ph. L. Licentiate in Philosophy

PHL Philadelphia, Pennsylvania (airport)

PHLAGS Phillips Petroleuun Load-and-Go System

phlebo (Latin prefix—vein)—phlebitis

phleg phlegm(atic)(al)(ly); phlegmaticalness; phlegmaticness; phlegmatize(d); phlegmier; phlegmiest

phl h phillips head

Phlm. Philemon

PHLS Public Health Laboratory Service

PHLX Philadelphia Stock Exchange

phm phase meter

phn (PHM) patrol hydrofoil missile

Ph. M. *Philosophiae Magister* (Latin—Master of Philosophy)

PHM patrol-combat missile (hydrofoil craft)

Phm. B. Bachelor of Pharmacy

PHMC Pennsylvania Historical and Museum Commission

Phm. G. Graduate in Pharmacy

PHMS Patrol Hydrofoil Missile Ship(s)

PHN Public Health Nurse; Public Health Nursing

pho physician-hospital organization; preferred health organization

PHO Public Hazards Office

phobe (Latin suffix—abnormal fear or dread)—felinophobic, hydrophobia

phocis photogrammetric circulatory surveys

phocl photo-initiated chemical laser

PhOD Philadelphia Ordnance Depot

Phoen Phoenix

Phoen Phoenicians

Phoenix Arizona's capital; Phoenix Islands in the equatorial mid-Pacific Ocean where they are claimed by the UK and the U.S.A.—included are Birnie, Canton, Enderbury, Gardner, Hull, McKean, Phoenix, and Sydney islands

PHOENIX Plasma Heating Obtained by Energetic Neutral Injection Experiment

phofl photoflash

phon phonetics; phonology

phone telephone

phone book telephone book

phoneme smallest sound unit (linguistics)

phones participating hybrid option note exchangeable securities

phonet phonetic(s)

Phonet Phonetics

phono phonograph

phonorecord(s) phonograph record(s)

phonos phonoscopy (voiceprint analysis and identification)

phonovision telephone television

Phons Alphonse

Phor Phoronida

phos phosphate; phosphorescent

phot. photograph; photographer; photographic; photography; photon; photostat; photostatic

phot photographie (French—photography)—plus all derivatives such as *photocopie* (photostat), *photographe* (photographer), *photogravure, phototype*, etc.

Phot Photographie (German—photography)—plus all derivatives

photac photographic typesetting and composing (AT & T)

photex photographic exercise

photint photographic intelligence

photo photograph; photographer; photography

photocomp photocomposed; photocomposition

photog photograph; photographer; photographic; photography

photogeog photogeography

photogeol photographic geology

photograv photogravure

photog(s) photographer(s)

photom photometry

photo op photo opportunity

photosyn photosynthesis

phot r photographic reconnaissance

p'house steak porterhouse steak

php pounds per horsepower; propeller horsepower

ph&p peace, heath, and prosperity

PHP Psychologists Helping Psychologists (self-help group); Public Health Plan

phr phrase; pounds per hour; preheater

Ph R Photographic Reconrmaissance

PHR Professional in Human Resources

PHRA Poverty and Human Resources Abstracts

phraseo phraseogram; phraseograph; phraseological(ly); phraseologist; phraseology

phren phrenic; phrenology

PHRI Public Health Research Institute

PhRMA Pharmaceutical Research and Manufacturers of America

ph & ru pubic hair and revealing underwear

phs peripheral nerve stimulation

Ph S Philosophical Society of England

PHS Pennsylvania Historical Society; Printing House Square; Prison Health Services; Pubic Hair Society; Public Health Service

PHSO Postal History Society of Ontario

phsp phase splitter

pht phototube; pitch, hit, and throw

Ph T putting husband through (college or university)

PHt Port Harcourt

PHT Passive Hemagglutination Test(ing)

PHTF Pearl Harbor Training Facility

PHTP Priority Health Training Programs

PHTS Psychiatric Home Treatment Service

Phu Port Hueneme

P Hur Port Huron

phv phase velocity

phv pro hoc vice (Latin—for this purpose)

phw pressurized heavy water

PHWA Professional Hockey Writers' Association; Protestant Health and Welfare Assembly

phwr (PHWR) pressurized heavy-water-moderated reactor

PHX Phoenix, Arizona (airport)

phy physical; physics

phyce photocopy-control electronics unit

phylo phylogeny

phys physic; physical; physician; physics

phy s physiological saline

Phys Chem Solids Physics and Chemistry of Solids

phys dis physical disability

phys ed physical education

Phys Ed Physical Education

physexam physical examination

Phys Fluids Physics of Fluids

PHYSH Physicians and Surgeons Hospital

physiat physiatric(s); physiatrical; physiatrist

physiog physiognomy

physiogr physiography

physiol physiology

Physiol Physiology

Phys Konden Mater Physik der Kondensierten Materie (German—Physics of Condensed Materials)

physl physiological

Phys Lett Physics Letters

phys med physical medicine

physocean physical oceanography

physog physiognomy

Phys Rev Physical Review

Phys Rev Lett Physical Review Letters

Phys S Physical Society

phys sci physical science; physical sciences

Phys Status Solidi Physica Status Solidi (Latin—Solid-State Physics)

Phys Teach Physics Teacher

phys ther physical therapy

Phys Today Physics Today

Phys Z Physikalische Zeitschrift (German—Physics Journal)

Phys Z Sowjetunion Physikalische Zeitschrift der Sowjetunion (German—Physics Journal of the Soviet Union)

phytopath phytopathologic(al)(ly); phytopathologist; phytopathology

pi paradoxical intention; parental investment, personal income; photo interpreter; photo interpretation; pigeon trainer; pig iron; pilotless interceptor; pimp; point initiating; point insulating; point of interception; point interception; poison ivy; position indicating; position indicator; preliminary inventories; present illness; private investigator; proactive inhibitor; proactive interference; production interval; programmed instruction; protamine insulin; protocol international (international protocol); public investigation

pi (PI) point of inversion; polyimide

pi. pierced; pink

p & i principal and interest; protection and indemnity

pi Greek letter symbol π indicating ratio of circumference of a circle to its diameter; the ratio itself; expressed as a number, pi is approximately 3.14159

Pi piaster

P$_i$ inorganic orthophosphate

PI Packaging Institute; Paducah and Illinois (railroad); Party Islam; Pasteur Institute; Paul Isnard (Mana River settlement, French Guiana); Periodicals Institute; Perlite Institute; Philippine Islands; physical instruction; Piedmont Airlines; Plastics Institute; Popcorn Institute; Pratt Institute; Principal Investigator; Productivity Index; Public Information

PI Printer's Ink

P-I Seattle Post-Intelligencer

P.I. Pharmacopoeia Internationalis

pia pain in the ass; peripheral interface adaptor

pia (PIA) primary insurance amount

PIA Packaged Ice Association; Pakistan International Airlines; Photographic Importers' Association; Pilots International Association; Plastics Institute of America; Printing Industries of America; Prison Industry Authority; Professional Insurance Agents; Public Information and Awareness

PIAA Pacific Index of Abbreviations and Acronyms

PIAI Printing Industry of America, Incorporated

PIANC Permanent International Association of Navigation Congresses

piang piangendo (Italian—mournful, plaintive)

pianiss pianissimo (Italian—very softly)

pianocorder piano recorder and reproduction system

PIARC Permanent International Association of Road Congresses

pias piaster

PIASA Polish Institute of Arts and Sciences in America

piat projector infantry antitank (weapon)

PIAT Peabody Individual Achievement Test(ing)

pib power ionosphere beacon

pib producto interno bruto (Spanish—gross internal product)

PIB Petroleum Information Bureau; Polytechnic Institute of Brooklyn; Prices and Incomes Board; Prison Industry Board; Publishers Information Bureau

PIBA Primary Industry Bank of Australia; Public Investors Arbitration Bar Association

PIBAC Permanent International Bureau of Analytical Chemistry of Human and Animal Food

pibal pilot balloon

pic (French—peak); piccolo; picture; polymer-impregnated concrete; positive-impedance converter; primary interexchange carrier; production inventory control; pulse-indicating cartridge; pulse-induced collapse

Pic (PIC) program-interrupt control(ler)

Pic Pictor (Latin—Painter constellation)

PIC Physics International Company; Piedmont Interfaith Council; Pocket Ionization Chamber; Poison Information Center; Poisons Information Centre (Australia); Private Industry Council

PICA Palestine Israel Colonization Association; Police Insignia Collector's Association; Printing Industry Computer Associates; Printing Industry Craftsmen of Australia; Professional Insurance Communicators of America

picar picaresque

Picasso Pablo Picasso

picc (PICC) presubscribed interexchange carrier charge

PICC Peoples Insurance Company of China; Philippine International Convention Center

Piccy Piccadilly

PICGC Permanent International Committee on Genetic Congresses

PICIC Pakistan Industrial Credit and Investment Corporation

pick possession of an instrument of a crime

pick. part information correlation key

Pick Pickens Railroad

PICL President's Intelligence Checklist

PICM Permanent International Committee of Mothers

Picnic City Mobile, Alabama

Pico 10^{-12}

PICO Person In Column One (census-taker euphemism for head of household)

PICOE Programmed Initiations, Commitments, Obligations, and Expenditures

PICOP Philippine Industries Corporation of the Philippines

pics pictures; publishers information cards

PICS Pacific Islands Central School; Personnel Information Communication System; Pharmaceutical Information Control System

pict pictorial; picture

Pictorial Satirist Supreme William Hogarth

Pictured Rocks Michigan's national lakeshore

PICUTP Permanent and International Committee of Underground Town Planning

pid pelvic inflammatory disease; prolapsed intervertebral disk

p-i-d poverty-ignorance-disease syndrome of society

PID Police Intelligence Detail; Procurement Information Digest

PID Partido Institucional Democrático (Spanish—Institutional Democratic Party)

pida payload installation and deployment aid

PIDA Pet Industry Distributors Association

PIDAS Portable Instantaneous Display and Analysis Spectrometer

PIDC Pakistan Industrial Development Corporation

PIDE *Policia Internacional e de Defesa do Estado* (Portuguese—International Police and Defense of the State)—security police

Pid Eng Pidgin English (hybrid dialect)

PIDO Primitive Indian Development Organization

pidp pilot information display panel

PIDS Parameter Inventory Display System

pie pulmonary infiltration (with) eosinophilia

pie (PIE) plug-in electronics

PIE Pacific Intercultural Exchange; Pacific Intermountain Express (fast freight); Partners In Education; St. Petersburg, Florida (airport)

PIEA Petroleum Industry Electrical Association

PIEC Public Interest Economics Center

Piedmont Piedmont Plateau or Piedmont Triad (Greensboro, High Point, and Winston-Salem, North Carolina) or place name found in Alabama, California, South Carolina, or West Virginia

PIERS Port Import/Export Recording Service

PIES Premium Income Equity Securities

pif (PIF) prolactin inhibiting factor

PIF Paper Industry Federation; Partners in Flight; Pilot Information File

pifr peak inspiratory flow rate

pig. pigment; pigmentation

pigs passive investment generators

PIG Pride, Integrity, Guts (acronym adopted by the Chicago police)

pigmi positron-indicating general measuring instrument

pigmt pigment(ation)

PIGS Poles, Italians, Greeks, Slavs

pigu pendulous integrating gyroscope unit

pih pregnancy-induced hypertension

PIJAC Pet Industry Joint Advisory Council

pik payment in kind

pil payment in lieu; percentage increase in loss

pil (PIL) procedure implementation language

pil. *pilula* (Latin—pill)

Pil Pitt interpretive language

PIL Pacific International Lines; Pakistan International Airlines; Pest Infestation Laboratory

pilc paper-insulated lead covered

PILCOP Public Interest Law Center of Philadelphia

pill the pill (birth-control pill)

pills particulate instrumentation by laser light scattering

pilnav piloting navigation

PILO Public Information Liaison Officer

pilot printing industry language for operations of typesetting

PILOT Piloted Low-speed Test; Programmed Inquiry, Learning, or Teaching

pilot-on-board flag signal flag consisting of a white and a red vertical band; letter H or Hotel in the international code

pilot-wanted flag yellow-and-blue vertically striped signal flag flown to indicate a pilot is wanted; letter G or Golf in the international code

pilp parametric integer linear program

pils pilsner

Pil Sta Pilot Station

pim penalties in minutes; personal information manager; pulse-interval modulation

PIM *Pacific Islands Monthly*

PIMA Paper Industry Management Association

PIMI Preinactivation Material Inspection

pimola pimento olive (pimento-stuffed olive)

pimpmobile pimp's automobile

PIMPS Program for Interactive Multiple Process Simulation

pims profit impact of marketing strategies

PIMS Project on Integrated Management Systems

pin page and item number; piece identification number; plan identification number; position indicator

pin (PIN) personal identification number (for computer protection)

pin. *pinguis* (Latin—fat, grease)

PIN Police Information Network

p/in.² parts per square inch

p/in.³ parts per cubic inch

PINA Pacific Islands News Association; Permaculture Institute of North America

PINAC Permanent International Association of Navigation Congresses

Pind Pindar

pines. pineapples

pino positive input—negative output

pins person(s) in need of supervision

pins. person in need of supervision

PINS Padre Island National Seashore (Texas); Palletized Inertial Navigation System

PNWR Pungo National Wildlife Refuge (North Carolina)

pinx. *pinxit* (Latin—he painted it)

PINY Polytechnic Institute of New York

Pinyin (Chinese—phonetic sound)—official spelling system adopted in 1979 for Chinese words written in Roman letters

PINZ Plastics Institute of New Zealand

pio parallel input-output; precision-interpret operation

PIO Photographic Interpretation Office(r); Public Information Office(r)

PIOA *Pacific Index of Abbreviations and Acronyms in Common Use in the Pacific Basin Area*

PIOB President's Intelligence Oversight Board

PIOCS Physical Input-Output Control System

pi-on pi-meson; pioneer

pion. pioneer

PIOSA Pan Indian Ocean Science Association

pip peripheral interchange package; personal injury protection; precise installation position; predicted intercept(ion) point; project initiation period; proximal interphalangeal; public and institutional property

pip. (PIP) picture in picture (tv)
Pip. Philip
PiP Proceedings in Print
PIP Peripheral Interchange Program; Permatite Instant Plastic; Personal Identification Program; Personnel Identification Project; Priority Information Program; Product Improvement Plan, Product Improvement Program; Product Information Package; Psychotic Inpatient Profile
PIP Policia de Investigación del Peru (Spanish—Peruvian Investigation Police)
PIPA Pacific Industrial Property Association; Pacific Islands Producers Association
PIPEF Pacific Islands Polynesian Education Foundation
piper (PIPER) pulsed intense plasma for exploratory research
pipe(s). pipe bomb(s)
pipi pipizintzintli
pipit. peripheral-interface and programme-interrupt translator
Pipo Filippo
PIPR Polytechnic Institute of Puerto Rico
pips pulsed integrating pendulums
piq program idea quotient; property in question
PIQ Performance IQ
Pir Piraeus
PIR Philippine Independence Ribbon; Phillip Island Reserve (Victoria, Australia); Preliminary Information Report
PIRA Paper Industries Research Association; Printing Industry Research Association; Provisional Irish Republican Army
pirb position-indicating radio beacon
pirf perimeter-insulated raised floor
PIRF Petroleum Industry Research Foundation
PIRG Public Interest Research Group
PIRGs Public Interest Groups
pirid passive infrared intrusion detector

PIRL PRISM Information Retrieval Language
pi rm pilot reamer
PIRS Personal Information Retrieval System; Poseidon Information Retrieval System
pis pistol
Pis Pisces
PIS Postal Inspection Service; Public Insurance Service
P Isb Port Isabel
PISC Philippine International Shipping Corporation; Phoenix International Science Center
Pisces (Latin—Fish constellation)
PISCES Production Information Stocks and Cost Enquiry System
pise pneumatically impacted stabilized earth
Pish Parish
PISO Philippines Investment Systems Organization
piss pissoir (French—urinal); pissotiére (French—public urinal)
pissoirs pissotiéres (French—public urinals for men)
pistaz piss-tinted topaz
pisw process-interrupt status word(ing)
pit pitot static; progressive inspection tag
pit (PIT) principal, interest, and taxes
Pit Pitanga; Pitcairn; Pitkin; Pitman; Piton; Pittsboro; Pittsburg; Pittsburgh; Pittsfield; Pittsford; Pittston; Pittsylvania
PIT Pasadena Institute of Technology; Personal Income Tax; Petr Ilich Tchaikovsky; Pittsburgh, Pennsylvania (airport)
PITA Petroleum Industry Training Association; Provincial Intermediate Teachers Association (Canadian)
PITAC Pakistan Industrial Technical Assistance Center
PITAS Petroleum Industry Training Association—Scotland
PITB Pacific Inland Tariff Bureau; Pacific Island Teachers Board
PITC Pacific International Trust Company

pitchblende uraninite ore (chief source of radium and uranium)
PITDC Pacific Islands Tourism Development Council
piti principal, interest, taxes, insurance
PIT FE Personal Income Tax Filing Enforcement (system)
PITL Pacific Islands Transport Line
pit. log pitot-static log
PITO Portuguese Information and Tourist Office
Pitons Piton Mountains (St Lucia)
pitr plasma iron turnover rate
pits. payload integration test set
PITS Pacific Islands Training School
PITT Polaris Integrated Test Team
Pitt Pittsburgh (baseball); University of Pittsburgh
Pitts Pittsburgh, Pennsylvania
pitu piping or tubing
PIU Public Inspection Unit (vice squad)
PIUS Process-Inherent Ultimately Safe (nuclear reactor)
piv peak inverse voltage; post indicator valve
PIV Positive Infinity Variable
PIVADS Product Improved Vulcan Air Defense System
pivs particle-induced visual sensations
PIW Petroleum Intelligence Weekly
pix photographs; pictures
pixel picture element
pix/sec pictures per second
PIYA Pacific International Yachting Association
pizz. pizzicato (Italian—plucked)
pj prune juice
PJ Police Judge; Presiding Judge; Probate Judge
P of J Port of Jacksonville
PJ Police Judiciare (French—criminal investigators, detective division)
PJA Pipe Jacking Association
P Jac Port Jackson
PJB Patrick J. Buchanan
PJBD Permanent Joint Board on Defense (Canada-U.S.)
PJC Paducah Junior College; Paris Junior College; Polydox Jewish Federation

pjex parachute jumping exercise

pjm postjunctional membrane

pjp probate judge of the peace

pj's pajamas; physical jerks

Pjs Pasajes

pk pack; park(ing); peak; peck; psychokinesis; pyruvate kinase

pk (PK) packed tight, kept right; probability of kill; sugar-coated chewing gum (symbol)

pK negative logarithm of the dissociation constant (symbol)

Pk Park; Peak; pink

Pk *Pedalkoppel* (German—pedal coupler); *Pauken* (German—kettledrums)

PK Pakistan (Internet code); Principal Keeper; probability of kill (symbol); Public Key

PK *Panama Kanaal* (Dutch—Panama Canal); *Posta Kutusu* (Turkish—post office box)

P Ka Port Kembla

P-K antibodies Prausnitz-Küstner antibodies

pkb photoelectric keyboard

Pkbanken *Post-och Kreditbanken* (Swedish—Post and Credit Bank)

pKC protein kinase C

pkd packed (flow chart); partially knocked down

PKD Parker Drilling Company (stock-exchange symbol)

pkdom pack(ed) for domestic use

p/KE Presidential/Key Executive Master of Business Administration

pkg package; packing

Pkg Port Kelang (also written Port Klang and formerly Port Swettenham)

pkge package

pki public key infrastructure

PKI *Partai Komunis Indonesia* (Communist Party of Indonesia)

Pkl Port Kelang (Port Klang formerly Port Swettenham)

PKL Possum Kingdom Lake

pkm perigee kick motor

pkmr packmaster

PKN *Polski Kometet Normalizacyny* (Polish—Polish Standards Committee)

pknghse packinghouse

PKNP Pu Kradeung National Park (Thailand)

pkp pre-knock pulse

pKp purple K powder (purple potassium-bicarbonate powder)

PKP *Partido Komunista Pilipinas* (Philipino—Communist Party of the Philippines)

PKPA Parental Kidnapping Prevention Act

pkr packer

PKR Parker Pen (stock exchange symbol)

Pk Rdg Park Ridge

P-K reaction Prausnitz-Küstner reaction

pkrg parking

pks packs; pecks

PKS Photo-Kit System (criminal identification)

pksea pack(ed) for overseas use

PKSRP Possum Kingdom State Recreation Park (Texas)

pkt packet

P-K test Prausnitz-Küstner test

PKTF Printing and Kindred Trades Federation (UK)

pkts packets

pku phenylketonuria

pkv killed poliomyelitis vaccine

Pkw *Personenkraftwagen* (German—automobile, passenger vehicle)

Pkwy Parkway

pky pecky

Pky Parkway

pl pamphlet law; parting line; party line; path length; perception of light; phase line; pipeline; place; plastic; plate; plural price list

pl (PL) party line; phone line; private telephone line; product liability

p/l partial loss; payload; pipeline; plain language

p & l profit and loss

pl. *plenarius* (Latin—complete, fully attended)

£L pound Lebanese

Pl Place

Pl *Place* (French—place, plaza); *plantage* (Dutch—plantation); *plass* (Scandinavian—place, plaza); *Platz* (German—place, plaza); *plaza* (Spanish—place,

Plaza); *plein* (Dutch—place, plaza); Titus Maccius Plautus (Roman Writer of comedies)

PL perception of light (symbol); petite large; Place; Players League; Pluto; Point Loma; Poland (auto plaque and Internet code); Port Line; Public Law; Public Library

P.L. Poet Laureate

PL *Paradise Lost; Partido Liberal* (Spanish—Liberal Party); *Pharmacopoeia Londinensis* (Pharmacopoeia of London)

PL 1 Programming Language 1

PL/1 Programming Language/version 1

pla plasma resin activity; probation and rehabilitation of airmen

Pla Plaza; Pula (Pola)

Pla *Playa* (Spanish—beach, strand)

PLA Palestine Liberation Army; Pedestrian's League of America; Pet Lovers Association; People's Liberation Army (Chinese communist); Philadelphia Library Association; Philatelic Literature Association; Port of London Authority; Port of Los Angeles; Private Libraries Association; Programmable Logic Array; Public Library Association; Pulverized Limestone Association

P of LA Port of Los Angeles

place. programming language for automatic checkout equipment

Place Pig Place Pigalle in Paris

PLADs Price Level Adjusted Deposits

PLADS Parachute Low-Altitude Delivery System

plam plastic laminate

plam (PLAM) price-level adjusted mortgage

plame (PLAME) propulsive lift aerodynamic maneuvering entry

plan. planet; planetarium

Plan *Planina* (Bulgarian or Serbo-Croatian—mountain, mountain range)

PLAN Paterson Looks Ahead Now; Planned Lifetime Assistance Network; Prevent

Los Angelization Now; Program for Learning in Accordance with Needs

Plan A North Atlantic Treaty Regional Planning Group

plane(s) airplane(s)

PLANES Programmed Language-based Enquiry System

planet. planetary

Planets The Planets, Gustav Holst's tone poem for large orchestra

Planet Space Sci Planetary and Space Science

planex planning exercise

plank plankton

PLANNET Planning Network

PLANS Programming Language for Allocation and Network Scheduling (NASA)

Plan Soc Planetry Society

plantflex plantar flexion

plantk plantkunde (Dutch—botany)

PLAP Port of London Authority Police

PLAR Partido Liberal Auten tica Radical (Spanish—Authentic Radical Liberal Party)—Paraguay

plarbage plane-floor garbage

PLARS Position-Locating-and-Reporting System

plas plaster

plasm (Greek—something formed or molded—chromoplast, dermoplasty, plasma, plasmasol, protoplast

Plasma Phys Plasma Physics

plastique (French-plastic) plastic bomb(s)

plasty (Latin suffix—reconstruction of)—rhinoplasty

plat. plateau; platinum; platoon

platf platform

PLATO Port Lincoln Advancement Trust Organization; Programmed Logic for Automatic Teaching Operations

platy Platypoecilus (genus of tropical fishes); *platysma*

platy (Latin prefix—flat or side)—platypus

Platy Platyhelminthes

Plaut Plautus

PLAV Polish Legion of American Veterans

plb personal locator beacon; plumber; plumbing; publisher's library binding(s); pull button

plb (PLB) publisher's library binding

PLB Poor Law Board

plbd plugboard

plc personal line of credit; power lever control; power-line carrier; prelaunch computer; power-line communication

PLC Pacific Lighting Corporation; Pacific Logging Congress; Point Loma College; Probe Launch Complex; Preferred Line of Credit; Products List Circular; Public Limited Company

P of L C Port of Lake Charles

P.L.C. Poeta Laureatus Caesareus (Latin—Imperial Poet Laureate)

PLCA Pipe Line Contractor's Association

PLCAA Professional Lawn Care Association of America

plco prostate, lung, colorectal and ovarian (cancer)

plcs propellant-loading control system

plcu propellant-level control unit

plcy policy

pld payload; programmable logic device

Pld Portland, Oregon

PLD Paul Lawrence Dunbar

PLDG Portuguese Language Development Group

PLDTC Philippine Long Distance Telephone Company

pldx polydox; polydoxy

ple preliminary logistics evaluation; pleura; primary loss expectancy; prudent limit of endurance; puerile light entertainment

P & LE Pittsburgh & Lake Erie (railroad)

plea. prototype language for economic analysis

PLEA Pacific Lumber Exporters Association; Poverty Lawyers for Effective Advocacy

plebe plebeian

plebs plebeians

pled pleaded

plegia (Latin suffix—paralysis or stroke)—paraplegic

PLEI Public Law Education Institute

Pleis Pleistocene

plem pipeline end manifold

Plen Plenary; Plenipotentiary

plenipo plenipotentiary

Plenum Plenum Publishing Corp

pleon pleonastical(ly)

plex plant experiment(ation)

plf polyforming

PLF Pacific Legal Foundation; Palestine Liberation Front

P-L F Pro-Life Federation

pl x fe plastic to female

plff plaintiff

plftr please furnish transportation requests

plfor please furnish

plg piling

Plg Porto Alegre

PLG Poor Law Guardian

PLGC Pension Loan Guarantee Corporation

plgl plateglass

p-lgv psittacosis-lymphogranuloma venereum

plh (PLH) palaemontes-lightening hormone

PLHS Public Library of the High Seas (American Merchant Marine Library Association)

pli preload indicating

PLI Pacific Law Institute; Plant Location International; Photo Library Inc

PLI Partido Liberal Independiente(Spanish—Independent Liberal Party); *Partito Liberale Italiano* (Italian—Italian Liberal Party); *Photo-Lab-Index*

PLIB Pacific Lumber Inspection Bureau

plic. plicata

p'lice police

PLIDCO Pipe Line Development Company

Plim l Plimsoll line

P Lin Port Lincoln

Plin C Gaius Plinius Secundus major (Roman naturalist often referred to as Pliny the Elder)

Plin L Plinius Caecilius Secundus minor (Roman writer often referred to as Pliny the Younger)

Plioc Pliocene

plis propellant-level indicating system

plk plank

PLK Phi Lambda Kappa; Poincare-Lighthill-Kuo (mathematical method)

p lkr peacoat locker

pll phase-locked loop

PLL Prince Line Limited

PLLS Portable Landing Light System

pllt pallet

plltn pollution

plm pulse-length modulation

Plm Palembang

P l M Pépé le Moko

P-L-M Paris-Lyon Méditerranée (famous French railway)

PLMA Private Label Manufacturers Association

plmb plumber; plumbing

pl mo plastic mould

plms periodic limb movement during sleep

Plms Palms (postal abbreviation)

pln posterior lymph node

pl-n place-name

Pln Plain (postal abbreviation)

PLN (aviation flight) Plan

PLN Partido Liberación Nacional (Spanish—National Liberation Party); *Partido Liberal Nacionalista* (Spanish—National Liberal Party)

PLNC Point Loma Nazarene College

plng planning

PLNP Port Lincoln National Park (South Australia)

Plns Plains

plo phase-locked oscillator

PLO Palestine Liberation Organization; Passenger Liaison Office(r); Peoples Liberation Organization; Plans Office(r); Presidential Libraries Office (Library of Congress); Provident Life Insurance

PLO Pairti Lucht Oibre (Irish—Labour Party); *Polskie Linie Oceaniezne* (Polish—Polish Ocean Lines)

plom prescribed loan optimization model

Plosk Ploskogorye (Russian—plateau)

plot. plotting

plp plastic-lined pipe

plp (PLP) pyridoxal phosphate

PLP Parliamentary Labour Party; Partners for Liveable Places; Presentation Level Protocol; Progressive Labor Party

pl & pd personal loss and personal damage

PLPG Publishers' Library Promotion Group

plpgrndg pulp grinding(s)

pl x pl plastic to plastic

PLPP Pennsylvania League for Planned Parenthood

PLP-PVV Parti pour la Liberté et le Progrès (French—Party of Liberty and Progress); *Partij voom Vrijheid en Vooruitgang* (Flemish—Party of Freedom and Progress)—Belgium

PLQ Public Library Quarterly

plr pillar; primary loss retention

Plr Pillar (postal abbreviation)

PLR Philippine Liberation Ribbon; Public Lending Right

P L & R Postal Laws & Regulations

PLR Partido Liberal Radical (Spanish—Radical Liberal Party)

PLRA Photo Litho Reproducers' Association

PLRA Partida Liberal Reforma Autentica (Spanish—Authentic Liberal Reform Party)—Paraguay

PLRE Partido Liberal Radical Ecuatoriano (Spanish—Ecuadorean Liberal-Radical Party)

PLRS Position-Location Reporting System

plry poultry

pls plates; please

PLS Purnell Library Service

plsd promotion list service date

plsfc part load specific fuel consumption

Pl Sgt Platoon Sergeant

plshd polished

plshr polisher

plss (PLSS) position location strike system

PLSS Portable Life-Support System

plstc plastic

plstr plasterer

plt past-life therapy; personal leave time; pilot; primed lymphocyte typing; psittacosis-lymphogranuloma trachoma

plt. platelet count; plaintiff

pltc political

PLTC Power-Limited Tray Cable

pltf plaintiff

pltry poultry

PLTS Pacific Lutheran Theological Seminary; Point Loma Test Site (Convair)

plu people like us; plural; plurality; price look-up; product line unit

P Lu Port Luis

PLU Pacific Lutheran University; Patrice Lumumba University (Moscow)

PLUG Public Law Utilities Group

plumb. plumber; plumbing

plumb. plumbum (Latin—lead)

plumcot plum plus apricot (hybrid)

plumr plumber

PLUNA Primeras Líneas Uruguayas de Navegación Agréa (Spanish—First Uruguayan Aerial Navigation Lines)

Plunket Plunket Society (Royal New Zealand Society for the Health of Women and Children)

pluperf pluperfect

plur plural

PLUS Parent Loans to Undergraduate Students; Physically-Limited United Students; Project Literacy United States; Professional Learning Unit System

plute(s) plutocrat(s)

pluto (PLUTO) pipeline under the ocean

Pluv Pluviose (French—Rainy Month)—beginning January 20th—fifth month of the French Revolutionary Calendar

plwd plywood

plx plexus; propellant-loading transfer

Ply Plymouth

PLYMCHAN Plymouth Subarea Channel (NATO)

Plz Plaza

pm paramilitary, particulate matter; phase modulation; post mortem; powdery mildew; premium; premolar; pre-systolic murmur; preventive maintenance (PM); program manager; project manager; publicity man; pulse modulation; pumice

pm (PM) primary memory

p-m permanent magnet; phase modulation

p.m. *post meridiem* (Latin—after noon, night)

p/m pounds per minute

p&m probate and matrimonial

pm *poids moléculaire* (French—molecular weight)

Pm maximum power (symbol); promethium

PM Pacific Mail; Past Master; Pattern Maker; Pay Master; Peabody Museum; Pére Marquette (railroad); petite medium; Physical Medicine; Police Magistrate; Pontifex Maximus; Postmaster; Prime Minister; Production Manager; Provost Marshal; publicity man; Saint Pierre and Miquelon (Internet code)

P.M. *post meridiem* (Latin—after noon); Prime Minister

P/M Pacific Molasses; Physical Medicine

PM *Pistol Makarov* (Russian—Makarov pistol); *Policia Metropolitana* (Spanish—Metropolitan Police)

P.M. *Piae Memoriae* (Latin—of pious memory); *Pontifex Maximus* (Latin—Supreme Pope)

pma positive mental attitude; primary mental abilities

pma (PMA) paramethoxyamphetamine

PMA Pacific Maritime Association; Pacific Missionary Aviation; Parts Manufacturing Associates; Peat Moss Association; Pencil Makers Association; Performance Management Association; Pharmaceutical Manufacturers Association; Philadelphia Museum of Art; Philippine Mahogany Association; Phonograph Manufacturers Association; Photo Marketing Association; Police Management Association; Politico-Military Affairs; Polyurethane Manufacturers Association; Precision Measurements Association; Precision Metal-forming Association; Primary Mental Abilities (test); Production and Marketing Administration; Property Management Association; Publishers Marketing Association

PMA *Programa Mundial de Alimentos* (Spanish—World Food Program)

PMAA Petroleum Marketers Association of America; Promotion Marketing Association of America

PMAC Provisional Military Administrative Council; Purchasing Management Association of Canada

PMAD Public Morals Administrative Division (New York City Police Department)

PMAE Peabody Museum of Archeology and Ethnology

PMAF Pharmaceutical Manufacturers' Association Foundation

PMAs Power-Marketing Administrations

PMAS Purdue Master Attitude Scales

PMATA Paint Manufacturers' and Allied Trades Association

pmb post-menopausal bleeding; private mailbox

PMB Potato Marketing Board

PMBC Pacific Motor Boat Club; Portland Motor Boat Club (Oregon)

pmbo participative management by objectives

pmbx private manual branch exchange

pmc postage and mailing center; precision mirror calorimeter; preventive maintenance contract(or); pseudomembranous colitis

pMc pure Mexican cocaine

PMC Pacific Medical Center; Pennsylvania Military Academy; Princeton Microfilm Corporation; Project Management Committee

PMCA Pennsylvania Manufacturing Confectioners Association

PMCC Post Mark Collector Club

pmcs process monitoring and control systems

pmd post-mortem dumps; projected map display

Pmd Portmadoc

PMD Pale Morning Dun

PMDA Photographic Manufacturers and Distributors Association

PMD/BMI Project Management Division/Battelle Memorial Institute

PMDC Pakistan Minerals Development Corporation

pmdd premenstrual dysphoric disorder

PMDD Personnel Management Development Directorate

pmds projected map display set

PMDS Property Management and Disposal Service

pme performance-measuring equipment; photomagnetoelectric; planning, management, evaluation; protective multiple earthing

P Me Portland, Maine

PMEA Powder Metallurgy Equipment Association

PMEF Petroleum Marketing Education Foundation

PMEL Pacifiic Marine Environmental Laboratory; Precision Measuring Equipment Laboratory

pmest personality, matter, energy, space, time (Raganathan's fundamental categories)

pmet painted metal

pmf probable maximum flood(ing); progressive massive fibrosis

PMF Pennsylvania Mutual Fund; Presidential Medal of Freedom

pmg permanent magnet generator

PmG Paymaster General; Postmaster General

PMG Provost Marshal General

PMG *Pall Mall Gazette*

pmh past medical history; probable maximum hurricane

PMHP Primary Mental Health Project

pmi photographic micro-image; point of maximum impulse; private mortgage insurance

PMI Palma de Mallorca, Balearic Islands, Spain (airport); Plumbing Manufacturers Institute; Pre-Marital Inventory; Project Management Institute

PMI *Partai Muslimin Indonesia* (Indonesian Muslim Party)

PMIA Presidential Management Improvement Award

PMIC President's Management Improvement Council

PMIG Political-Military Interdepartmental Group

PMIS Personnel Management Information System; Planning Management Information System; Product Management Information System

PMJC Pine Manor Junior College

pmk pitch mark; postmark(ed)

pml probable maximum loss; programmed magnesium line

PML Pacific Micronesian Line; Pierpont Morgan Library

Pmla Parmelia

PMLA Publications of the Modern Language Association of America

PMLO Principal Military Landing Officer

pmm permanent magnet magnetizer; pulse mode multiplex

pmma (PMMA) polymethylmethacrylate

PMMI Packaging Machinery Manufacturing Institute

pmmu paged memory-management unit

pmn polymorphonuclear neutrophil

pmn *producto material neto* (Spanish—net material product)

PMNA Pacific Mountain Network Association; Parkers Marsh Natural Area (Virginia)

PMNH Peabody Museum of Natural History

pmnl polymorphonuclear leukocyte

pmnr periadenitis mucosa necrotica recurrens

pmo printed matter only

pmo *pianissimo* (Italian—very softly)

PMO Palomar Mountain Observatory; Polaris Material Office; Principal Medical Officer; Provost Marshal's Office

PM & OA Printers' Managers and Overseers Association

PMOLANT Polaris Material Office, Atlantic

PMOPAC Polaris Material Office, Pacific

P Mor Port Moresby

pmos P-channel metal oxide semiconductor

PMOSC Primary Military Occupational Code

pmp per-member payment; precious metal plating; previous menstrual period; probable maximum precipitation

PMP Preliminary Management Plan; Procurement Methods and Practices (manual)

pmpm per member per month

PMPMA Plastic and Metal Products Manufacturers Association

pmr polymyalgia rheumatica; performance maintenance recorder; pressure-modulated radiometer

pm & r physical medicine and rehabilitation

Pmr Paymaster

PMR Pacific Missile Range

PMRAFNS Princess Mary's Royal Air Force Nursing Service

PMRC Parents' Music Resource Center

PMRL Pulp Manufacturer's Research League

PMRM Periodic Maintenance Requirements Manual

PMRS Physical Medicine and Rehabilitation Service

PMRY Presidio of Monterey

pms poor miserable soul; post-menopausal syndrome; pregnant mare's serum

pms (PMS) phenazine methosulphate; pollution-monitoring satellite; pre-menstrual syndrome; pre-millennial syndrome

pm's push monies

p-m-s processors-memories-switches

PMS Pantone Matching System; Peabody Museum of Salem; Performance Management System; Permanent Manual System; Planned Missile System; Preventive Maintenance System; Project Management System; Project Manager, Ships; Public Management Sources; Public Message Service

PMSA Pacific Merchant Shipping Association; Primary Metropolitan Statistical Area

pmsg pregnant mare's serum gonadotrophin

PMSP Plant Modelling System Program

pm specialists paramilitary specialists

PMSSMS Planned Maintenance System for Surface Missile Ships

PMST Professor of Military Science and Tactics

pmt payment; photomultiplier tubes; positive matte technique; premenstrual tension; programs, materials, techniques

PMT Perceptual Maze Test; photo mechanical transfer

PMTB Pacific Motor Tariff Bureau

PMTS Predetermined Motion Time System

pmu performance monitor(ing) unit; physical mockup; portable memory unit; productive man work unit

PMU Pattern Makers Union

PMUSAOAS Permanent Mission of the United States of America to the Organization of American States

PMVB Pocono Mountain Vacation Bureau

pmvi periodic motor vehicle inspection

pmvp *precio maximo de venta al publico* (Spanish—maximum price charged the public)

p mvr prime mover

pmv's parcel mail vans (British railways)

pmx private manual exchange (telephone)

pmyob please mind your own business

pn partition; part number; percussion note; percussive note; planetary nebulae; please note; position; presumably nests (birds); project note; promissory note; psychiatry-neurology; psychoneurotic

pn (PN) punch-on (computer character)

p-n positive-negative

p/n part number; promissory note

p & n psychiatry and neurology

Pn North Pole; North Celestial Pole; perigean range

PN Pacific Northern (airline); Pan-American World Airways (stock exchange symbol); part number; plasticity number; Pitcairn Island (Internet code); point of no return; Practical Nurse

P/N Part Number

P & N Piedmont and Northern (railroad)

PN Partido Nacional (Spanish—National Party); *Partido Nacionalista* (Spanish—Nationalist Party)

pna (PNA) pentosenucleic acid

Pna Panama

PNA Pacific Northern Airlines; Pakistan National Alliance; People's News Agency; Philippines News Agency; Project Network Analysis

PNAC President's National Advisory Committee

PNAI Provincial Newspapers Association of Ireland

pnavq positive-negative ambivalent quotient

pnb particle/neutron beam

pnb producto nacional bruto (Spanish—gross national product)

PNB Philippine National Bank

PNB Produto Nacional Bruto (Portuguese—Gross National Product)

PNBA Pacific Northwest Booksellers Association

PNBB Parc National de la Boucle du Baoule (French—Baoule River Bend National Park)—in the highlands of Mali

PNBC Pacific Northwest Bibliographic Center (American and Canadian libraries)

PNBP Parc National de la Boucle de la Pendjari (French—Penjari River Bend National Park)—in northwestern Dahomey

pnbt paranitroblue tetrazoleum

pnc penicillin; plate number coil; premature nodal contraction

P'n C Picnic 'n Chicken

PNC Palestine National Council; People's National Congress; Prohibition National Committee

PNC Parque Nacional Canaima (Spanish—Canaima National Park)—encloses Venezuela's Angel Falls—world's tallest waterfall

PNCC President's National Crime Commission

pnch punch (flow chart)

Pncla Pensacola

pnd paroxysmal noctural dyspnoea; postnasal drip

Pnd Pandjang

pndb perceived noise decibels

pndg pending

P-N-D-L-R park-neutral-drive-low-reverse (positions on automatic transmission gauge)

Pndo Pinedo

pne practical nurse's education

pne (PNE) peaceful nuclear explosion

PNe Pointe Noire

PNE Pacific National Exchange (Vancouver); Pacific National Exhibition (Vancouver)

PNE-A Parque Nacional El Avila (Spanish—El Avila National Park)—between Caracas and the Caribbean, encloses the Humboldt National Monument

P Ned Pharmacopee Nederlandsche (Dutch—Netherlands' Pharmacopeia)

PNERL Pacific Northwest Environmental Research Laboratory

Pnes Pines (postal abbreviation)

PNET Peaceful Nuclear Explosion Treaty

pneu pneumatic(s)

PNEU Parents' National Education Union

pneumato (Latin prefix—breathing)—pneumonia

pneumoccon pneumocconiosis (lung fibrosis due to dust-particle inhalation)

pneumog pneumograph; pneumographer; pneumographic(al)(ly); pneumography

pneumonoultra pneumonoultra-microscopicsilicovolcanoconiosis (miner's lung disease)

pnf proprioceptive neuromuscular facilitation

pnfd present not for duty

p.n.g. persona non grata (Latin—an unacceptable person)

Png Penang

PNG Papua New Guinea; Professional Numismatists Guild

PNG Papua Nueva Guinea (Spanish—Papua New Guinea); *Parque Nacional Guatopo* (Spanish—Guatopo National Park—near Caracas, Venezuela

PNGL Papua New Guinea Line

pnh (PNH) paroxysmal nocturnal hemoglobinuria

PNH Phnom-Penh, Cambodia (airport)

PNHA Physicians National Housestaff Association

PNHP Parque Nacional Henri Pittier (Spanish—Henri Pittier National Park)—near Maracay, Venezuela

pni positive noninterfering (alarm); psychoneuroimmunology; pulsed neutron interrogation

PNI Pharmaceutical News Index

PNI Parque Nacional Iguazu (Spanish—Iguazu National Park)—surrounding the Iguazu Falls shared by Argentina, Brazil, and Paraguay

P Nic Port Nicholson

PNITC Pacific Northwest International Trade Council

pnl panel

PNL Pacific Naval Laboratories; Pacific Northwest Laboratories; Philippine National Line

PNLA Pacific Northwest Library Association; Pacific Northwest Loggers Association

pnm pulse numbers modulation

PNM Pinnacles National Monument (California)

pno piano

pno pergamino (Spanish—parchment)

p^{no} Pantano (Spanish—bog, marsh)

P' n' O P and O (Peninsular and Occidental Steamship Company, Peninsular and Oriental Line)—P & 0

PNO Port of New Orleans; Principal Nursing Officer

PNO Parque Nacional Ordesa (Spanish—Ordesa National Park—near Spain's French frontier

PNOC Philippine National Oil Company; Proposed Notice of Change

pnp positive negative positive

p 'n' p pimping and pandering

PNP Pediatric Nurse Practitioner; People's National Party; Platt National Park (Oklahoma)

PNP Partido Nuevo Progresista (Spanish—New Progressive Party)—Puerto Rico

pnpn positive-negative positive-negative

pnpr positive-negative pressure respiration

pnr point of no return; prior notice required

Pnr Pioneer

PNR Passenger Name Record (s); (airline. Philippine National Railways; Pittsburgh Naval Reactor; Pulletop Nature Reserve (New South Wales)

PNRP Philadelphia Pulmonary Neoplasm Research Project

pns parasympathetic nervous system; peripheral nerve stimulation; peripheral nervous system

PNS Pacific Navigation Systems; Pakistan Naval Ship; Philadelphia Naval Shipyard; Philippine News Service; Professor of Naval Science

PNSN Parque Nacional Sierra Nevada (Spanish—Sierra Nevada National Park)—encloses Venezuela's Mount Bolivar

PNSTDC Pakistan National Scientific and Technical Documentation Center

PNSY Portsmouth Naval Shipyard

pnt paint(ed)

Pnt Pentagon

PNT Parque Nacional Tijuca (Portuguese—Tijuca National Park)—near Rio de Janeiro, Brazil

Pnt Anx Pentagon Annex

PNTBT Partial Nuclear Test Ban Treaty

pntd painted

P^{ntd} pointe (French—point)

PNTO Principal Naval Transport Officer

pntr painter; permanent normal trade relations

PNU Pneumatic Scale Corporation (stock-exchange symbol)

pnutbutsan peanut-butter sandwich

p-nut butter peanut butter

pnutbutwich peanut-butter sandwich

p-nut(s) peanut(s)

PNVS Pilot's Night-Vision System

PNW Parc National du W (W-shaped park on the borders of Dahomey, Niger, and Upper Volta)

PNWA Pacific Northwest Waterways Association

PNWD/BMI Pacific Northwest Division/Battelle Memorial Institute

PNWL Pacific Northwest Laboratory (AEC)

PNWR Piedmont National Wildlife Refuge (Georgia); Presquile National Wildlife Refuge (Virginia); Pungo National Wildlife Refuge (North Carolina)

pnx pneumothorax

pnxt. (Latin—he or she painted it)

PNYA Port of New York Authority

PNYCTC Pennsylvania New York Central Transportation Company (merger of Pennsylvania and New York Central railroads)

Pnz Penzance

po piss off; poetry; polarity; power oscillator; power-operated; power output; previous orders; principals only; public officer; putouts

po' poor

p-o postoperative

p/o part of

p & o paints and oil; pickled and oiled

p.o. per os (Latin-by mouth)

Po polonium; Portugal; Portuguese

P^{o} Pedro

PO Parole Officer; Passport Office; Patent Office; Personnel Office(r); Petty Officer; Philadelphia Orchestra; Police Officer; Port Office(r); Post office; Probation Officer; Project Office; Province of Ontario; purchase order

P-O Pyrénées-Orientales

P/O Parole Officer; Pilot Officer; Probation Officer

P & O Peninsular & Occidental Steamship Company; Peninsular & Oriental Line

PO Portland Oregonian

PO 1/C Petty Officer First Class

pO_2 oxygen pressure

PO—2 Soviet mine-sweeping launch; Soviet trainer aircraft nicknamed Mule by NATO

PO 2/C Petty Officer Second Class

PO 3/C Petty Officer Third Class

poa place of acceptance; primary optical area; primary optic atrophy

P o A Power of Attorney

POA Police Officers Association; Portland Opera Association; Prison Officers Association

POAC Peace Officers Association of California; Post Office Advisory Council

POADS Portland Air Defense Sector

POAG Peace Officers Association of Georgia

POAU Protestants and Other Americans United for Separation of Church and State

pob persons on board; pilot on board; point of beginning; prevention of blindness

pob población (Spanish—population)

PoB Port of Baltimore

POB post office box

Pobeda Pobeda Peak (highest mountain between China and Russia in the Tien Shan range at 24,406 feet)

po'-boy poor-boy (sandwich)

pobra pony + zebra (hybrid)

PO BX Post Office Box

poc performance operating characteristic; point of contact; principal operating component; privately owned conveyance

poc (POC) process operator console

POC Pittsburgh Opera Company; port of call; Prison Officer's Club; Public Oil Company

Pocahontas (Algonquin—Tomboy)—nickname of Matoka the daughter of Chief Powhatan; her married name was Rebecca Rolfe

Poca(loo) Pocatello, Idaho
po'ch porch
pocill. pocillum (Latin—small cup)
pock pocket
Pocket Bks Pocket Books
Pocket State Luxembourg (pocketed between Belgium, France, and Germany)
Poconos Pocono Mountains of eastern Pennsylvania
poc's ports of call
POCS Patent Office Classification System
pocul. poculum (Latin—cup)
pod payable on (or upon) death; point of deployment; point-of-origin device; port of debarkation; port of departure; probability of detection
pod (POD) power of deduction; process-oriented design; proof of delivery; proof of deposit
p.o.d. paid on delivery
pod (Greek—foot)—anthropod, cephalopod, gastropod, podiatrist, podiatry, pseudopod
POD Port of Debarkation; Post Office Department; Professional and Organizational Development (higher education network)
POD Pocket Oxford Dictionary
PODA Piloting of Office Document Architecture
PODAPS Portable Data Processing System
Pod D Doctor of Podiatry
podex photographic exercise
podia podiatrist(ic)(al)(1y); podiatry
POD Mods Power On Demand Modules
poe (POE) polyoxyethylene
POE Pacific Orient Express; port of embarkation; port of entry
poe buoy plank-on-edge buoy
poecrit poetry critic(ism)
p o'ed put out
POED Post Office Engineering Department
poet. poetical(ly); poetry
Poet Poetry
POETS Phooey On Everything—Tomorrow's Saturday
POEU Post Office Engineering Union
pof please omit flowers

pof (POF) pyruvate oxidation factor
POF Philharmonic Orchestra of Florida
POFI Pacific Oceanographic Fisheries Investigation
POG Pacific Oceanographic Group (British Columbia)
POGO Pennzoil Offshore Gas Operators; Polar Orbiting Geophysical Observatory
poh pull out of hole (oil well)
pOH alkalinity factor
Poh Pohang
POHMA Project for the Oral History of Music in America
poi poison; poisonous (spelled out and symbolized with skull and crossbones on labels)
POI Personal Orientation Inventory; Program of Instruction
poidrt poison dart
Point Point of Air, Alcock, Arena, Arguello, Ayre, Baker, Barber, Cairndoon, Chevalier, Chicot, Conception, Fortin, George, Harbor, Hueneme, Judith, Lay, Leamington, Lobos, Loma, Lookout, Pedro, Pleasant, Reyes, Sal, San Luis, San Pedro, Sur; The Point—West Point, U.S. Military Academy at West Point, New York
POINTER Particle Orientation Interferometer
Point Loma Pen Point Loma Peninsula in southwesternmost California
Point Reyes national seashore in California
pois poison
POIT Power-of-Influence Test
pol petroleum-oil-and-lubricants (POL); polar polarize(d); police; political; politician; problem-oriented language
Pol Poland; Polish
Pol Polen (Norwegian—Poland); Polish (Slavic language); *Polonia* (Italian, Latin, Portuguese, Spanish—Poland)
POL Pacific Oceanography Laboratories; Patent Office Library; petroleum-oil-and-lubricants; Polish Ocean Lines; Politics On-Line
p-ola payola (kickback, bribe)

POLA Prostitutes of Los Angeles (protective association)
Pol Ad Political Adviser
polad(s) political adviser(a)
Poland Polish People's Republic (North-European country between Germany and Russia), *Polska Rzeczpospolita Ludowa*
polang polarization angle
polar. polarity; polarization; polarize(d)
Polar BEAR Polar Beacon Experiments and Auroral Research (satellite)
POLARS Pathology On-Line Logging and Reporting System
Pol Col Police College
pol com political committee
Pol Com Police Commissaire (Interpol); Police Commissioner
polcrit political critic(ism)
poldamr petroleum, oil, and lubrication installations damage report
pol econ political economy
polem polemic; polemicist; polemical(ly); polemicize
POLEX Polar Experiment (weather)
polf parents of large families
Pol Fed Police Federation (London)
POLFER Polizia Ferroviaria (Italian—Railroad Police)
Pol Found Police Foundation (Washington, D.C.)
poli politician
poi ind pollen index
polio poliomyelitis
POLIS Parliamentary On-Line Information System
poli sci political science
polish polish sausage *(kielbasa)*
polit political; politician; politics
Politburo Politicheskoe Byuro (Russian—Political Bureau of the Central Committee)
polka. petroleum, oil, and lubricants out-of-kilter algorithm
poll. pollution
pollie(s) politician(s)
POLLS Parliamentary On-Line Library Study
pol in the pen politician in the penitentiary
poln polnisch (German—Polish)

polon Polonais (French—Polish)

Pol Rze Lud Polska Rzeczpospolita Ludowa (Polish People's Republic)

pols political prisoners; politicians

pol(s) political prisoner(s); political(s); politician(s); poll parrot(s)

POLs Problem-Oriented Languages (computer)

pol sci political science; political scientist(s)

POLSTRADA Polizia Stradale (Italian—Highway Police)

polwar political warfare

poly polyethylene; polymer; polytechnic; polytechnical; polyvinyl

po'ly poorly

poly (Greek—many)—polydactyly, polygenic, polymer, polymorphism, polypeptide

Poly Polynesia; Polynesian; Polytechnic (institute or school)

Polyb Polybius

poly bot polyethylene bottle

polyg polygraph(er); polygraphic(al)(ty); polygraphy (lie detection)

Pol Yid Polish Yiddish

polymorph polymorphous

polys polymorphonuclear leukocytes

poly sci political science

polysex polysexual(ity)

polytech polytechnic(al)

polywater polymerized water

pom polycyclic organic matter; pomeranian; pomological; pomology; pom-pom; preparation for overseas movememt; program objectives memorandum

pom (POM) polyoxymethylene

pom pomeridiano (Italian—afternoon, p.m.)

PoM Port of Miami

POM Port Moresby, New Guinea (airport)

pomato potato-tomato hybrid vegetable

pomcus (POMCUS) prepositioned material configured in unit sets

POME Prisoners of Mother England—Pommies; early convict immigrants (Australian slang)

Pomeranian anything from around the Baltic, including domestic dogs

POMFLANT Polaris Missile Facility, Atlantic

pomol pomologic(al)(1y); pomologist(ic)(al)(ly); pomology

Pomp Pompey

POMPAC Polaris Missile Facility, Pacific

Pompadour Jeanne-Antoinette Poisson—Marquise de Pompadour (1721–1764)

pom-pom antiaircraft gun

POMR Problem-Oriented Medical Record

POMS Panel on Operational Meteorological Satellites; Production and Operations Management Society

pomsee preparation, operation, maintenance, shipboard electronics equipment

POMSIP Post Office Management Service Improvement Program

pon pontoon

'pon upon

Pon Ponce

PON Program Opportunity Notice; Program Opportunity Notification

pona paraffin, olefin, naphthene, aromatic (test for petroleum octane rating)

PonBrg pontoon bridge

pond. pondere (Latin—by weight)

Pondo Pondoland

PON-Q Q paraoxonase

p-on-n positive on negative

pons profile of nonverbal sensitivity (body language)

Pont Pontevedra

pont b pontoon bridge

Ponti Pontiac

Pontiac Pontiac Silverdome, Pontiac, Michigan

Pontines Pontine Islands off Anzio, Italy or the Pontine Marshes of Italy

Pont. Max. Pontifex Maximus (Latin—Supreme Pontiff—the Pope)

PONY Prostitutes of New York (protective association)

p & oo pianistic and orchestral orgasm (as in the finale of Rachmaninoff's Concerto No 3 in D minor for piano and orchestra)

Poo Poole

POO Post Office Order

pood poodle dog (Russian—36-lb. weight)

POOD Provisioning Order Obligation Document

poof peripheral on-line-oriented function

Pool The Pool (the Thames just below London Bridge around Billingsgate Market)

poop. nincompoop

Poor's Poor's Register of Corporations, Directors, and Executives

POOS Priority Order Output System

poosslq person of opposite sex sharing living quarters

POoW Petty Officer on Watch

pop carbonated beverage; perpendicular ocean platform (POP); persistent occipitoposterior; personal ozone protection; plasma osmotic pressure; plaster of paris; popliteal; poppet; popular; population

pop. (POP) public offering price

p-op post-operative

p-o-p plaster of paris; printing-out-paper

Pop Poppa

POP Palletizing Optimization Potential; Panoramic Office Planning; Portuguese Overseas Province (Macao, China); Post Office Plan(ning)

Popa Popayan, Colombia

POPA Property Owners Protection Association

pop. advertising point-of-purchase advertising

POPAI Point of Purchase Advertising Institute

pop art popular art (advertising displays, comic strips, posters)

popb proposed operating plan and budget

POPE Product Oriented Procedures Evaluation

popex population explosion

popf prepared-on-premises flavor

popi post office position indicator (navigation system developed by British post office)

poplit popliteal

pop music popular music

Popo *Popocatepetl* (Aztec— Smoking Mountain)

pop psych popular psychiatry

popr pilot overhaul provisioning review

pops popular concerts; popular tunes

POPS People Opposed to Pornography in Schools

Pop Sci *Popular Science*

POPSER Polaris Operational Performance Surveillance Engineering Report

poq periodic order quantity

POQ *Public Opinion Quarterly*

por pay on return; point of regulation; porosity; porous; public opinion research

p-o-r pay-on-receipt; payable-on-receipt

Por Porifera; Portland; Portugal; Portuguese

Por. Porter

Por *Porogi* (Russian—rapids, waterfall)

POR *Partido Obrero Revolucionario* (Spanish— Revolutionary Workers' Party); *Policy, Organisation, and Rules* (of the Girl Guides and Scouts)

PORA Police Officers Research Association

PORAC Peace Officers Research Association of California

porc porcelain

PORC Peralta Oaks Research Center

Porcupines Porcupine Islands east of Bar Harbor, Maine

'pore Singapore

PORIS Post Office Radio Interference Station

porksan pork sandwich

porkwich pork sandwich

porm plus or minus

porn pornographic; pornography (*see* porno)

pornette(s) pornographic cassette(s)

pornfilm pornographic motion picture film

porno pornofilm; pornographer; pornographic; pornographically; pornographic bookshop; pornography

pornobio pornographic biography

pornofilm pornographic motion picture

porno mag pornographic magazine

pornos pornographic books, moving pictures, photographs, recordings, etc.

pornovel pornographic novel

pornovelist pornographic novelist

porn pub(s) pornographic publication(s); pornographic publisher(s)

Porn Squad Pornographic (Publication) Squad

porny pornographic

pornzines pornographic magazines

porp (PORP) printed on recycled paper

porp(s) porpoise(s)

PORS Post Office Research Station

port. portable; portrait; portraiture

port. (PORT) photo-optical recorder tracker

port *portugiesisch* (German— Portuguese)

Port Portland; Portugal; Portuguese

Port Portuguese

Port Ade Port Adelaide, South Australia

Port Alb Port Alberni on Vancouver Island, British Columbia

portalet portable toilet

Port Alex Port Alexander, Alaska

Portanol Portuguese-Spanish

Port Ant Port Antonio, Jamaica

Port Art Port Arthur (Manchuria, Ontario, Tasmania, or Texas)

portashed portable shed(ding)

Port Chi Port Chicago; Portuguese China (Macao)

Port Dal Port Dalhousie, Ontario

porteños (Spanish—port people)—in Argentina means the people of Buenos Aires and in Chile those of Valparaiso

Port Ind Portuguese India

Port Jack Port Jackson (seaport of Sydney, New South Wales, Australia)

Port Jeff Long Island, New York; Port Jefferson

Port Liz Port Elizabeth, New Jersey; Port Elizabeth, South Africa

Port Nick Port Nicholson (Wellington, New Zealand's harbor)

Portolá Gaspar de Portolá

Port Phil Port Phillip, Melbourne, Victoria, Australia

Port Rich Port Richmond, Staten Island, New York

PORTs Patient Outcome Research Teams

port side left side of an airplane, ship, or other craft when looking forward, symbolized by a fixed *red* light

Portsmouth U.S. Naval Disciplinary Command at Portsmouth, New Hampshire—the U.S. Naval Prison

Port Sud Port Sudan (Sudanese harbor on the Red Sea)

Port Swett Port Swettenham, Malaysia

Port Talb Port Talbot, Wales

Port Tew Port Tewfik (Egypt's Port Taufiq at the southern end of the Suez Canal)

Port Tim Portuguese Timor

Portug *Portugais* (French— Portuguese)

Portugal Republic of Portugal (Iberian country once ruling a vast colonial empire), *República Portuguesa*

Port Wash Port Washington, Long Island, New York

Port Wel Port Weller, Ontario

Port Yid Portuguese Yiddish (Sephardim)

pos point of sale; point of service; position; positive; possibility; possible; product of sums

PoS Point of Sale; Port of Service; Port of Spain

POs Police Officers; Postal Orders

POS Patent Office Society; Port-of-Spain, Trinidad (airport); Primary Operating System; Problem-Oriented System

posa payment outstanding suspense accounts

POSB Post Office Savings Bank

POSC Problem-Oriented System of Charting

POSD Post Office Savings Department

posdcorb planning-organization-staffing-directing-coordinating-reporting-budgeting

(mnemonic device for remembering the functions of management)

posdsplt positive displacement

posh permuted on subject headings; port side out, starboard side home (British slang)

posistor positive resistor

posit position; positive; positron

positron positive electron

POSIX Portable Operating System Interface for Computer Environments

posm patient-operated selected mechanisms

posn position

pos/nav positioning/navigation

POSNY People of the State of New York

pos pron possessive pronoun

poss possession; possessive

P o S S Point-of-Sale System

POSS Passive Optical Satellite Surveillance (System)

P-O-S S Point-of-Sale System; Point-of-Service System

POSSE Parents Of Students in Special Education

posses possessive

posslq person of the opposite sex (in) same living quarters

posslq's persons of the opposite sex sharing living quarters

'possum(s) opossum(s)

post. postage; postal; posterior; post mortern

post (Latin prefix—after or behind)—postwar; *posterior* (Spanish abbreviation)

POST Frederick Post Drafting Equipment; Peace Officers Standards and Training; Police Officer Student Training; Processes of Science Test

post-Aug post-Augustan

post aur. post aurem (Latin—behind the ear)

post.d posterior diameter

poster. posterior

pos terminal point-of-sale terminal

postgangl postganglionic

Postgrad Med Inst Postgraduate Medical Institute

postgrad(s) postgraduate(s)

posth posthumous

postl postlude

post-mort post mortem (autopsy)

post-op post-operative

post part. *postpartum* (Latin—afterbirth)

post-sync post-synchronization of a sound track made after a motion-picture film has been shot

POSWG Poseidon Software Working Group

pot point of tangency; portable outdoor toilet; potash; potassa (potassium hydroxide); potassium; potential; potentiometer; marijuana

Pot Potosi (Bolivian province)

pot. *potaguaya* (Mexican Indian—marijuana); *potio* (Latin—dose, draft, potion)

'potamus(es) hippopotmus(es)

potash potassium carbonate (K_2CO_3)

potash alum potassium aluminum sulfate

potass potassium

POTASWG Poseidon Test Analysis Software Working Group

potats potatoes

POTC PERT (*q.v.*) Orientation and Training Program

P o TD Port of The Dalles

POTIB Poseidon Technical Information Bulletin

potosslq persons of the opposite sex sharing living quarters (sometimes appears as posslq)

potp paralysis of the pen

potr *potrero* (Spanish—cattle ranch, pasture)

pots lobster pots; plain old telephone service

pots. potentiometers

pott pottery

Potteries The Potteries (Stoke-on-Trent)

PotUS Lyndon Johnson's acronym meaning President of the United States

POTUS President of the United States (address name used by Churchill when communicating with Roosevelt, later used by President Johnson—PotUS)

pot w potable water

pou piss on you

poul poultry

POUM *Partido Obrero de Unificación Marxista* (Spanish—Workers Party of Marxist Unification)

POUNC Post Office Users' National Council

pound monetary unit of Cyprus, Egypt, Ireland, Lebanon, Malta, Sudan, Syria, the United Kingdom, British colonies and dominions

POUR President's Organization for Unemployment Relief

pov privately owned vehicle

p-o-v point-of-view

P_{ov} *Poluostrov* (Russian—peninsula)

POV Pend Oreille Valley (railroad)

pov's privately owned vehicles

pow power; prisoner of war (POW); prohibited offensive weapon

P o W Prince of Wales; Prisoner(s) of Watergate

pow. brks power brakes

POW Country Potash, Oil, and Wheat Country around Saskatoon, Saskatchewan

powd powder; powdered; powered

power programmed operational warshot evaluation and review

POWER Professionals Organized for Women's Equal Rights

po'white poor white person

POW/MIA Prisoner Of War/Missing In Action

pows (POWS) prisoners of war

POWS Pyrotechnic Outside Warning System

pow. str power steering

pow. wind power windows

pox police (journalists' abbreviation)

poy pre-oriented yarn

Poz Poznan

pozn *poznamka* (Czech-footnote)

pp baby-talk for urinate(d); pages; painful pissing; panel point; parcel post; part paid; partial pay; partially paid; passive participle; past participle; peak power; pellagra preventive (factor); per person; perceptual performance; peripheral processor; permanent party; petticoat peeping; physical profile; physical properties; pickpocket; polypropylene; positive pressure; postage paid; postpaid;

prepositional phrase; present position; pressure-proof; private property; privately printed; professional paper; purchased part(s); push-pull; urination; urine

pp (PP) planning permission; planning permit

p-p peak-to-peak; pee-pee (urine); push-pull; pussy-power (feminine wiles)

p/p peepee (urinate, urine)

p&p payments and progress

p & p parsimonious and penurious (miserly and stingy)

p-to-p peak-to-peak; point-to-point

pp per procuration (Latin—by proxy); *pianissimo* (Italian—very softly); *propria persona* (Latin—in his own person)

p.p. piena pelle (Italian—full leather); *post partum* (Latin—afterbirth)

Pp plate power (symbol)

Pp. Papa (Latin—father or Pope)

PP Pablo Picasso; Pacific Petroleum; Parcel Post; Parish Priest; Past President; Planned Parenthood; Power Plant; Proletarian party (Communist)

P-P pellagra-preventive factor

PP The Passionate Pilgrim; Patres (Latin—Fathers); *Polizei Pistole* (German—police pistol)

P.P. Pater Patriae (Latin—Father of his Country)

PP¹ inorganic pyrohosphate

ppa palpitation, percussion, auscultation; per power of attorney; photo-peak analysis; program, project activity; preferred provider arrangement

ppa (PPA) phenylpropanolamine

pp & a palpitation, percussion, and auscultation

p. pa. per procura (Latin—by proxy)

p.p.a. phiala prius agitate (Latin—bottle having first been shaken)—shake well before using

PPA Pakistan Press Association; Paper Pail Association; Paper Plate Association; Parcel Post Association; People for Prison Alternatives; Periodical Publishers Associa-

tion; Personnel Pool of America; Popcorn Processors Association; Poultry Publishers Association; President's Professional Association; Produce Packaging Association; Professional Photographers of America; Proletarian Party of America; Public Personnel Association; Publishers' Publicity Association; Purple Plum Association

PPAB Program and Policy Advisory Board (UN)

PPAC Pesticide Policy Advisory Board (EPA)

PPATRA Printing, Packaging, and Allied Trades Research Association (also appears as PATRA)

ppb parts per billion

ppb (PPB) polybrominated biphenyl (cattle poison)

pp&b paper, printing, and binding; planning, programming, and budgeting

Ppb Pappband (German—boards, hard cover)

PPBAS Planning-Programming-Budgeting-Accounting System

PPBC Portland Problem Behavior Checklist

PPBES Planning- Programming-Budgeting-Evaluation System

PPBMIS Planning, Programming, and Budgeting Management Information System

ppbs postprandial blood sugar

PPBS Planning-Programming-Budgeting System

ppc palm-size personal computer; picture postcard; plain-paper copier; power plant change; progressive patient care

p p c pour prendre congé (French—to take leave)

pPc pure Peruvian cocaine

PPC Penang Port Commission(er)(s); Personal Productivity Center; Pet Population Control; Policy Planning Council (U.S. Department of State); Positive Peer Culture; Purchase Price Control

ppca plasma prothrombin conversion accelerator; proserum prothrombin conversion accelerator

PPCAA Parole and Probation Compact Administrators Association

PPCD Plant Pest Control Division

ppcf plasma prothrombin conversion factor

PPCLI Princess Patricia's Canadian Light Infantry

PPCS Personnel Protection and Communication Services (British anti-terrorist organization); Primary Producers' Cooperative Society

ppd permanent partial disability; postpaid; prepaid; purified protein derivative (tuberculin)

PPD Party for Peace and Democracy; Paranoid Personality Disorder; Petroleum Production Division; Portland Public Docks; Propulsion and Power Division

PPD Partido Popular Democraticó (Spanish—Popular Democratic Party)

PPDA Produce Packaging Development Association

PPDC Polymer Products Development Center

ppdi pilot's projected-display indicator

ppdo per person, double occupancy

p p꜀ₒ próximo pasado (Spanish—last month)

PPDP Preprogram Definition Phase

PPDS Publishers' Parcels Delivery Service

PPDSE Plate Printers, Die Stampers, and Engravers (union)

ppe personal protective equipment; philosophy, politics, and economics

PP & E Program Planning and Evaluation

PPES Pilot Performance Evaluation System

ppf personal property floater (policy); pitchers' park factor

PPF Panamanian Public Force (police); Plumbers and Pipefitters (union)

PPFA Planned Parenthood Federation of America; Plastic Pipe and Fittings Association

p-p factor pellagra-preventive factor

ppg planning and programming guidance

PPG Pago Pago, Samoan Islands (airport); Pittsburgh Plate Glass

ppga post-pill galactorrhea-amenorrhea

PPGA Pennsylvania Personnel and Guidance Association

pph pamphlet; parts per hour (or parts per person/hour); postpartum hemorrhage; pounds per hour; pulses per hour

P Php Port Phillip

pphpm parts per hundred parts of mix; pints per hundred parts of mix

pphr parts per hundred parts of rubber

ppi pages per inch; parcel post insured; plan position indicator; policy proof of interest

PPI Pickle Packers International; Plastic Pipe Institute; Producer Price Index; Progressive Policy Institute; Project Public Information; Protective Packaging Inc; Pulp and Paper International; Producer Price Index

PPIC Plumbing and Piping Industry Council

PPIE Pan-Pacific International Exposition

ppif photo-processing interpretation facility

p-pille praeventivpille (Dano-Norwegian—preventive pill—contraceptive

pp/in. pages per inch

P Ping Pulau Pinang (Malay—Penang Ferry)

PPIQ Personality and Personal Illness Questionnaire(s)

pPk purplish pink

ppl pipeline

ppl (PPL) polypropylene

PPL Philadelphia Public Libraty; Phoenix Public Library; Pittsburgh Public Library; Planned Parenthood League; Police Protective League; Portland Public Library; Private Pilot's License; Providence Public Library; Provisioning Parts List

PP&L Pennsylvania Power and Light (company)

PP & L Pacific Power and Light

P-plane pilotless airplane (explosive carrying and reaction propelled)

PPLC Patients Protection Law Commission

pple past participle

p-p letters poison-pen letters

pplo pleuropneumonia-like organism(s)

ppm pages per minute; parts per million; pounds per minute; pulse position modulation; pulses per minute

ppm (PPM) peak program meter

PPM Peter, Paul, and Mary (singing group)

PPM Partido Proletario de Mexico (Spanish—Proletariam Party of Mexico)—Chinese-trained guerrillas active in Mexico and from California to Texas in the Chicano community; Persutuan Perpustakaan *Malaysia* (Malay—Library Association of the Federation of Malaya)

ppma post-polio muscular atrophy

PPMS Plastic Pipe Manufacturers' Society

ppn proportion(al)

PPNA Pupil-Perceived-Needs Assessment

ppng (PPNG) penicillinase-producing Neisseria gonorrhoeac; penicillin-resistant gonorrhea-producing enzyme that inactivates most penicillins

PPNP Point Pelee National Park (Ontario)

PPNW Physicians for the Prevention of Nuclear War

PPNYC Planned Parenthood New York City

ppo polyphenylene oxide; prior permission only

PPO Preferred-Provider Organization

p-p-ola political plugola (media plugging or touting of a candidate or an ideological issue)—propaganda device in disrepute

ppom particulate polycyclic organic matter

ppo's (PPOs) preferred provider organizations

ppp petty political pismire; point-to-point protocol

p & pp pull and push plate

ppp piu pianissimo (Italian—very very softly)

PPP Peoples Party of Pakistan; Peoples Progressive Party (Guyana); Petroleum Production Pioneers; Pickford Projective Pictures; Point to Point Protocol; Population Policy Panel (Hugh Moore Fund); Private Patients Plan

pppp piu piu pia pianissimo (Italian—very, very, very softly)

pp & p's perverts, pimps, and prostitutes

ppq (PPQ) polyphenylquinoxaline

ppr present participle; present protected rights; printed paper rate; prior permission required

PPr Port Pirie

PPR Permanent Pay Record; Permanent Personal Registration; Procurement Problem Report

PPRA Past President of the Royal Academy

pprbd paperboard

PPRICA Pulp and Paper Research Institute of Canada

pps pictures per second; pounds per second; prior preferred stock; private parliamentary secretary; prospective payment system; pulses per second

pp's payless paydays

p-ps post-polio syndrome

PPs Prairie Provinces (Alberta, Manitoba, Saskatchewan)

PPS Pacific Passenger Services; Paper Publications Society; Pennsylvania Prison Society; Petroleum Press Service; Program Policy Staff (UN)

PPS Partido Popular Salvadoreño (Spanish—Salvadoran Popular Party); *Partido Popular Socialista* (Spanish—Popular Socialist Party); *Persatuan Perpustakaan Singapura* (Malay—Library Association of Singapore)

P.P.S. post postscriptum (Latin—additional postscript)

PPSA Pan-Pacific Surgical Association

PPSAWA Pan Pacific and Southeast Asia Women's Association

PPSB Periodical Publishers' Service Bureau

PPSEAWA Pan-Pacific and South-East Asia Women's Association

ppsn present position

ppso per person, single occupancy

PP Society (*see* PPTPP)

ppt parts per thousand; precipitate

PPT Papeete, Society Islands (airport); Pre-Production Test(ing)

PPT *Pericles, Prince of Tyre*

pptd precipitated

PPT MRNA preprotachhykinin messenger ribonucleic acid

pptn precipitation

PPTPP Promulgators of Public Toilets in Public Parks (also known as the PP Society)

PPTS Palestine Pilgrims Text Society; Precision Pointing and Tracking System

ppty property

ppu platform position unit

PPU Peace Pledge Union; Primary Producers Union

P & PU Peoria and Pekin Union (railroad)

ppv pay-per-view (cable tv channel); people-powered vehicle(s)

PPV plum pox virus

PPVT Peabody Picture Vocabulary Test

PPWC Pines to Palms Wildlife Committee; Pulp, Paper, and Woodcutters of Canada

PPWP Planned Parenthood-World Population

pq peculiar; permeability quotient; personality quotient (PQ); previous question; punishment quarters

p-q phenol-hydroquinone (photographic developer)

p & q peace and quiet (solitary confinement); proposal and quotation

PQ personality quotient; Province of Quebec; South Pacific Airlines of New Zealand (2-letter code)

PQ *Parti Quebecois* (French—Quebec Party); *Philosophical Quarterly*

pqa procurement quality assurance

PQAP Procurement Quality Assurance Program

PQC Production Quality Control

PQD Plant Quarantine Division

PQD *Partido Quisqueyano Demócrata* (Spanish—Democratic Quisqueyan Party)—Dominican Republic

pqe post-qualification education

pqi professional qualification index

PQIH Plant Quarantine Inspection House

PQLI Physical Quality of Life Index

PQR Personnel Qualification Roster; Program Quality Review

pqrs productivity increases, quality control, robotization, and savings (Japanese formula for economic success)

p&qs proposal and quotation schedule

PQS Percentage Quota System; Personnel Qualification Standard(s)

pr pair; parcel receipt; parish record; partial reinforcement; partial remission; partial response; partial reward; payroll; percentile rank; peripheral resistance; preferred (stock); progesterone receptor; proportional representation; public relations; purple

pr (PR) proctosigmoidoscopy

pr. prisoner; probated; proved

p/r per rectum

p & r parallax and refraction

pr *protestants* (Dutch—Protestants)

p.r. *per rectum* (Latin—by the rectum); *punctum remotum* (Latin—remote point)—far point of vision

pR purplish red

Pr Panama-red marijuana; Parana; Prairie; prandtl number; prascodymium; presbyopia; Press; Prince; Proctoscopy; propyl

Pr *Praca* (Portuguese—plaza, square); *Presbyter* (Latin—elder or priest)

PR Parachute Rigger; Park Ranger; Performance Rating; Performance Report; Photoreconnaissance; Pinar del

Rio; Plant Report; Presidential Range; Problem Report; Progress Report; Psychiatric Record; Public Relations; Puerto Rican(s); Puerto Rico; river gunboat (2-letter naval symbol)

P-R Pennsylvania-Reading (Seashore Lines)

P/R payroll

P & R Parks and Recreation

PR *Paradise Regained* by John Milton (1671); *Partido Republicano* (Spanish—Republican Party); *Partisan Review; Peking Review; Pipe Rolls; Polish Register* (of shipping); *Polskie Radio* (Polish Radio); *Puerto Rico* (Puerto Rico)

P.R. *Populus Romanus* (Latin—Roman People)

pra payroll audit(or); personal retirement account; plasma renin activity; probation and rehabilitation of airmen; progressive retinal atrophy

pra (PRA) print alphanumerically

Pra Pará (British maritime abbreviation)

Pra *Prachtausgabe* (German—de luxe edition)

PRA Pay Readjustment Act; Paymaster Rear Admiral; Personnel Research Activity; Popular Rotocraft Association; Postal Reorganization Act; Psoriasis Research Association; Psychological Research Association; Public Roads Administration; Puerto Rico Association

P.R.A. President of the Royal Academy

prac practice; practitioner

pracl page-replacement algorithm and control logic

pract practical; practice; practitioner

Prado El Prado (Madrid museum)

Praeger Frederick A Praeger

praen praenomen

prag pragmatic; pragmatism

Prag Prague (capital of Czech Republic)

pragma processing routines aided by graphics for manipulation of arrays

PRAI Pre-Reading Assessment Inventory

PRAICO Puerto Rican American Insurance Company

Prair *Prairial* (French—Meadowy Month)—beginning May 20th—ninth month of the French Revolutionary Calendar

prais passive-ranging interferometer sensor

PRAISE Prevention of Alzheimer's in Society's Elderly

pral *principal* (Spanish—principal)

pram perambulator; parameter random access memory; productivity, reliability, availability, and maintainability

Pram Poseidon random-access memory

Pr of An Principality of Ansbach

prund. *prandium* (Latin—dinner)

PRANG Puerto Rico Air National Guard

PRAT Prattsburgh (railroad)

p. rat. aet. *pro ratione aetatis* (Latin—in proportion to age)

PRATRA Philippines Relief and Trade Rebilitation Administration

PRAY Paul Revere Associated Yeoman

prb principal borehole

pRB retinoplastoma protein

PRB People's Republic of Benin; Personnel Review Board; Population Reference Bureau; Pre-Raphaelite Brotherhood

PRB *Partido de la Revolución Boliviana* (Spanish—Bolivian Revolutionary Party)

prc packed red cells; procedure

prc (PRC) polysulphide rubber compound

PRC Pain Rehabilitation Center; Palestine Red Crescent; Pay-Raise Commission; Pension Research Council; People's Republic of China; Picatinny Research Center (Picatinny Arsenal); Planning Research Corporation; Postal Rate Commission; Public Relations Club

P.R.C. *Post Roman Conditam* (Latin—after the founding of Rome)—753 Before the Christian Era

PRCA Professional Rodeo Cowboys Association; Puerto Rico Communications Authority

PRCB Program Requirement Control Board (NASA)

prcd priced

Pr Ch Parish Church

prchst parachutist

prcht parachute

prcp precipitation

PRCP President of the Royal College of Physicians

prcs process; processing

PRCS President of the Royal College of Surgeons

prcst precast

prcu power regulation and control unit

prd partial reaction of degeneration; pro-rata distribution

prd (PRD) printer dump(ing)

PRD Pesticides Regulation Division (USDA); Planned Residential Development (permit); Program Requirement Document

PRD *Partido Revolucionario Democrático* (Spanish—Revolutionary Democratic Party); *Partido Revolucionario Dominicano* (Spanish—Dominican Revolutionary Party)

PRDA Program Research and Development Announcement

PRDC Personnel Research and Development Center (USN); Power Reactor Development Corporation

PRDL Personnel Research and Development Laboratory (USN)

PRDS Processed Radar Display System

prdx paradox

pre partial reinforcement effect; photoreactivation enzyme; prefix (computer character); progressive resistance exercise

pre (Latin prefix—before)—prenatal, presuppose

PRE Psychophysiological Reeducation

PREA Pension Real Estate Association

prealateen program for children below teen age who are affected by an alcoholic family *(see* alateen)

preamp(s) preamplifier(s)

preb prebend

PREBS Pennsylvania Real Estate Brokers and Salesmen's (licensing examinations)

prec precedence; preceding; precision

Prec Precentor

preced. preceding

precip precipitate; precipitation

PRECIS Preserved Context Index System

precomdet pre-commissioning detail

pred predicate; prednisolone

PREDA Puerto Rico Economic Development Administration

pre-design prelimiminary design

predic predicate; predicative; prediction

pre-em preeminence; preeminent; preempt; preemptible; preemption; preemptive; preemptor; preemptory

preemies premature babies

preemy premature baby

pref preface; prefatory; prefecture; preference; prefix

Pref Prefect

prefab prefabricated

Pref-Ap Prefect-Apostolic

prefaz *prefazione* (Italian—foreword)

prefd preferred

preframo prepare fleet rehabilitation and modernization overhaul (USN)

preg pregnancy; pregnant

pregang preganglionic

prehis prehistoric

prej prejudice

prel prelude

prelim preliminary

prelim diag preliminary diagnosis

prelims preliminaries; preliminary pages (frontmatter)

prem premature; premium

pre-med premedical

premie premature baby

premies premature babies

Prensa *La Prensa* (Buenos Aires' Press)

'prentice apprentice

Prenzl Bg Prenzlauer Berg

pr enzyme prosthetic-group removing enzyme

pre-op preoperation; preoperational

prep preparation; preparatory; prepare; preposition

PREP Personal Radio-Equipped Police; Predischarge Education(al) Program; Preparation Rehabilitation Education Program; Pupil Record of Educational Progress

P.R.E.P. Personal Responsibility Education Program

prepd prepared

prep'ed prepared

prepn preparation

prepr *prepracovane* (Czech—rewritten)

pre-pub pre-publication

prere. prerefunded

pres present

Pres President

PRES Puerto Rico Employment Service

presby presbyopia; presbyopic

Presby Presbyterian

presc prescription

Presc Prescott

Presd$_e$ Presidente (Spanish—President)

preserv preservation

presilection presidential election

press. pressure

PRESS Pacific Range Electromagnetic Signature Studies

Presse *Die Presse* (German—Neue Freie Presse)—Vienna's Press

presstitute poison-pen prostitute of the press (columnist skilled in writing defamatory articles)

Pres Tense *Present Tense*

prestmo. *prestissimo* (Italian—very quickly)

PRESTO Program Reporting and Evaluation System for Total Operations

presv preservation; preserve

pret preterit

Pret Pretoria

pre-Teut pre-Teutonic

PRETTYBLUEBATCH Philadelphia Regular Exchange Tea Total Young Belles Lettres Universal Experimental Bibliographical Association To Civilize Humanity (initialism contrived by Edgar Allan Poe to satirize all such pseudo-intellectual devices)

pretz pretzel

prev prevalalence; previous

prevan precompiler for vector analysis

preven preventive

prevoc prevocational

prex(y) president (usually college or university)

prez president

prf proof; pulse recurrence frequency; pulse repetition frequency

prf (PRF) priority-reserved flight (air cargo); prolactin-releasing factor

prf. *praefatio* (Latin-introduction, preface)

PRF Personality Research Form; Petroleum Research Fund; Plywood Research Foundation; Porpoise Rescue Foundation; Public Relations Foundation; Puerto Rican Forum

PRF *Publications Reference File* (GPO)

prfe polar-reflection faraday effect

prfg proofing

prfnl professional

prfr proofreader

PRFT Portable Rod-and-Frame Test

PRG Prague, Czechoslovakia (airport); Provisional Revolutionary Government (of former South Vietnam)

PRHS Port Richmond High School

pri photographic reconnaissance and interpretation; primary; primer; primitive; priority; priority repair induction; private; pulse recurrence interval

PRI Paleontological Research Institute; Philosophical Research Institute; Plastics and Rubber Institute

PRI *Partido Revolucionario Institucional* (Spanish—Institutional Revolutionary Party); *Partito Repubblicano Italiano* (Italian—Italian Republican Party)

PRIA *Proceedings of the Royal Irish Academy*

Pribilovs Pribilov Islands in the Bering Sea off Alaska

Price Stern Price, Stern, Sloan

P Rich Port Richmond

PRIDCO Puerto Rico Industrial Development Company

PRIDE Parents Resource Institute for Drug Education; Personal Responsibility in Defect Elimination; Professional Recruiting (with) Integrity, Determination, and Enthusiasm; Protection of Reefs and Islands from Degradation and Exploitation

PRIDES Premium Redeemable Increased Dividend Equity Securities

Prieta *Agua Prieta* (Spanish—Dark Water)—Mexican border town

prim. primary

prim (Latin prefix—first)—primitive, primordial

primaries primary colors—blue, red, yellow

prime. precision recovery including maneuvering entry

PRIME Philadelphia Regional Introduction for Minorities to Engineering; Prescribed Right to Income and Maximum Equity; Program Independence, Modularity, Economy; Program Research in Integrated Multi-ethnic Education; Programmed Instruction for Management Education

PRIMES Pennsylvania Retrieval of Information in Mathematics Education System; Productivity Integrated Measurement System (U.S.)

primip primipara, woman bearing or who has borne her first child

primo *primero or supremo* (French, Italian, Portuguese, or Spanish—first, first place, top quality, supreme)

primogen primogeniture, exclusive inheritance belonging to the eldest son or the eldest daughter if there is no son

primo temp *primo tempo* (Italian—first tempo)—return to original tempo

prin principal

Prin Principal; Principality; Principat d'Andorra (Catalan—Andorra)

PRIN *Partido Revolucianario de Izquierda* (Spanish—Revolutionary Party of the Left)—Bolivian

PRINAIR Puerto Rico International Airlines

Prince Prince Rogers Nelson

PRINCE Parts, Reliability, and Information Center (NASA)

Prin d'And Principat d'Andorra (Catalan—Principality of Andorra)

Prin Monaco Principauté d'Monaco (French—Principality of Monaco)

prin pts principal parts

print. printed, printing

print.(PRINT) preedited interpreter (computer language)

PRINUL Puerto Rico International Undersea Laboratory

Prinz Prinzregententheater (German—Prince Regent Theater) Munich

PRINZ Public Relations Institute of New Zealand

prio priority

PRIO Peace Research Institute, Oslo (Norway)

prions proteinaceous infectious particles (believed by some to cause Alzheimer's disease)

prior. priority

PRIP Park Restoration Improvement Program; Puerto Rican Independence Party

prir parts reliability improvement route; parts reliability improvement routing

PRI & RB Puerto Rico Inspection and Rating Bureau

pris prison(er)

Prisca Priscilla

prise program for integrated shipboard electronics

PRISE Pennsylvania's Regional Instruction System for Education (intercollegiate network)

pris g prisonnier de guerre (French—prisoner of war)

prism. prismatic

PRISM Personnel Record Information System; Program Reliability Information System for Management

Prisoner The Prisoner; Prisoner No 1; The Prisoner of Shark Island; The Prisoner of Zenda; The Prisoners; La Prisonniere

Pris(sy) Priscilla

pritac primary tactical radio circuit

Pritch Pritchard

prithee I pray thee

priv privacy; private; privateer(ing); privation; privative; privet; privilege(d); privily; privy

priv pr privately printed

priv pub privately published

prix de fque prix de fabrique (French—manufacturer's price)

PRIZM Potential Rating Index for ZIP Markets

PRJC Puerto Rico Junior College

prk photorefractive keractectomy

PRK People's Republic of Kampuchea (Cambodia)

pr kassa per kassa (Norwegian—for cash)

prkg parking

prkng parking

prl periodical; pick-resistant lock

Pr of L Prince of Liechtenstein; Principality of Liechtenstein

PRL Personnel Research Laboratory; *Polska Rzeczpospolita Ludowa* (Polish Republic); Prairie Research Laboratory (Canada); Precision Reduction Laboratory; Price Reduction League; Project Records List

Prl Cmm Parole Commission

prld pick-resistant locking device

prls prepaid rental-listing service

prm parameter, portable radiation monitor; prime

Prm Promenade

PRMA Puerto Rican Maritime Authority

p-r man public-relations man

prmld premolded

prm's presidential review memorandums

prn print numerically

prn. printing (book edition)

p.r.n. pro re nata (Latin—as needed, for an emergency)

PRN Physicians Radio Network; Private Registered Nurse(s)

PRNC Potomac River Naval Command

PRNL Pictured Rocks National Lakeshore (Michigan)

PRNS Point Reyes National Seashore

prntr printer

PRNWR Parker River National Wildlife Refuge (Massachusetts)

pro procedure; proceed; procure; procurement; production; profession; professional; professionally; prophylactic

pro (PRO) print octal; proline (amino acid)

pro. probable; proved

pro (Latin prefix—before or in favor of)—prosection, protribal

Pro Provost

PRO Peer Review Organization; Personnel Relations Office(r); Plant Representative's Office; Problem Resolution Officer; Professional Review Organization; Public Record Office; Public Relations Office(r)

PROA Public Record Office Archives

pro-am professional-amateur

prob probability; probable; probably; problem; problematic; problematical

Prob Probate

probcost probabilistic budgeting and costing; probable cost

PROBES Processes and Resources of the Bering Sea Shelf

probie(s) probationer(s)

Prob Off Probation Officer

probs problems

proc procedure; proceeding(s); procure; procurement

proc (Latin prefix—anus)—proctologist

Proc Procedure; Proceedings; Proctor

Pro Cambridge Philos Soc Proceedings of the Cambridge Philosophical Society

Procd procedure

pro-celeb professional celebrity

Procellares Procellariiformes (albatrosses, fulmars, petrels)

Proc-Gam Proctor-Gamble

Proc IEEE Proceedings of the IEEE

Proc IRE Proceedings of the IRE

proclib procedure library

Proc Nat Acad Sci U.S.A. Proceedings of the National Academy of Sciences of the United States of America

proco programmed combustion (auto engine)

procomm program communication

Procop Procopius

Proc Phys Soc, London *Proceedings of the Physical Society, London*

procrast(s) procrastinator(s)

Proc Roy Soc *Proceedings of the Royal Society*

Proc R Soc London Proceedings of the Royal Society of London

procsim processor simulation language

procstep procedure step

procto proctocolitis; proctocolonoscopy; proctologist; proctology; proctosigmoidoscopy; proctosigmoidectomy; proctoplegia

PROCTOR Priority Routine, Computer Transfers, and Register Operations

prod product; production

prodac programmed digital automatic control

PRODAC Production Advisers Consortium

PRODFINA *Protection et Defense de la Nature* (French—Protection and Defence of Nature)

prodn production

Prod(s) Irish-Catholic English—Protestant-[*see* Pap(s)]; Protestant(s)

prof procession, professional; professor

prof (PROF) pupil registering and operational filling

Prof Professor

PROF Peace Research Organization Fund

profac propulsive fluid accumulator

Prof D *Profesor Don* (Spanish—Sir Professor)

Prof Dna *Profesora Doña* (Spanish—Madam Professor)

Prof Eng Professional Engineer

Proff *Professori* (Italian—Professors)

Profintern Red International of Trade Unions

profit. program for financed insurance technic; programmed reviewing, ordering, and forecasting

Prof Lib Pr Professional Library Press

profs professionals; professors

PROFS Professional Office System

prog progenitor; progeny; prognose; prognosis; prognostic; prognostication; prognosticator; program; programmer; progress

Prog Gro *Progressive Grocer*

proglang(s) progressive language(s)—usually euphemistically slanted

progr program(mer); programme; progressive

Prog(s) Progressive(s)

Prog Theor Phys *Progress of Theoretical Physics*

prohib prohibit(ion)

proj project; project return on investment; projectile; projection; projector

PROJACS Project Analysis and Control System

Prol prologue

prolan processed language

prole(s) proletarian(s)

proletcult proletarian culture

pro-lifer person against abortion

PROLLAP Professional Library Literature Acquisition Program

prolog programming in logic

prolong. *prolongatus* (Latin—prolonged)

ProLt procurement lead time

prom programmable read-only memory; promenade (concert or dance); promment; promontory; promote; promoter; promotion; promotional; prompter

prom *promedio* (Spanish—average)

Prom The Prom—Wilson's Promontory—national park at the southernmost tip of Australia

promex productivity measurement experiment

PROMIS Problem-Oriented Medical Information System; Prosecution Management Information System (U.S. Attorney's Office—Washington, D.C.)

proml promulgate

promo promotional

promo(s) promotional announcement(s)

PROMPT Project Management and Production Team

PROMS Projectile Measurement System (USA)

PROMSTRA Production Methods and Stress Research Association

Pro Mus Orc Pro Musica Orchestra

Promy Promontory

pron pronoun; pronounced; pronunciation; pronunciator(y)

PRON Procurement Request and Order Number (USA)

pronc pronounced; pronunciation

prond pronounced

prong(s) pronghorn(s)—pronghorn antelope(s)

pronom pronominal

pro note promissory note

PRONTO Program for Numeric Tool Operation

PRONTOS Programmable Network Telecommunications Operating System

pronun pronunciate; pronunciation

pronunc pronunciation

PROOF Parole Resource Office and Orientation Facility (Jersey City, New Jersey)

prop propaganda; propeller; property; proportion(al); proposed; proprietary; proprietor

prop. proper(ly)

Prop Sextus Propertius (Roman poet)

PROP Panel Review of Products; Planetary Rocket Ocean Platform; Portland Regional Opportunities Program; Preservation of the Rights of Prisoners

ProPAC Prospective Payment Assessment Commission

propaed propaedutic(al)(ly); propaedutics

Propaedia outline of knowledge in *The New Encyclopaedia Britannica*

prop art propaganda art

propay proficiency pay

pro.per. *in propria persona* (Latin—acting as one's own attorney)

proph prophetic; prophylactic; prophylaxis

Prophéte *Le Prophète* (French—The Prophet)—Meyerbeer opera

propjet propeller turned by jet engine (same as turboprop)
propl proportional
propn proportion(al)
propr. proprietor(s)
props (theatrical) properties
prop wash propeller wash
pro rat.aet. pro ratione aetatis (Latin—according to age)
pro rect pro recto (Latin—by rectum)
PRORM Pay and Records Office-Royal Marines
pros professionals; prosody; prostitute(s)
Pros Atty Prosecuting Attorney
prosc proscenium
PROSE Personal Record of School Experiences
prosig procedure signal
prosine procedure sign
prosp prospecting
pross(ie) prostitute
prost prostate; prosthetics; prostitution
prosth prosthesis
prostie(s) prostitute(s)
prot protective; protectorate; protein; protestant; protozoa; protractor
prot (PROT) protein anion
Prot Protectorate; Protestant; Protozoa
protag protagonist
Prot-Ap Protonotary-Apostolic
Protec Protectorate
pro tem. pro tempore (Latin—for the time being)
PROTEUS Propulsion Research and Open-Water Testing of Experimental Underwater Systems
prothrom prothrombin
pro time prothrombin time
Protoch Protochorda
Protocols Protocols of the Learned Elders of Zion (fraudulent document created and distributed in 1905 by the czarist secret police to incite pogroms against Russia's Jews; since used by antisemitic bigots in defense of their cause)
protozool protozoologic(al)(ly); protozoologist; protozoology
protr protractor
pro us.ext. pro uso externo (Latin—for external use)
prov proverb(ial)(ly); provide; provision; provisional; proviso

prov provincia (Spanish—province)
Prov Provençal; Provence; Proverbs, The (book of the Bible); Providence; Province
Prov Provençal (Romance language); *Proverbs; Provinz* (German—province)
Prov Eng Provincial English
prover procurement-value-economy-reliability
Prov GM Provincial Grand Master
Providence Plantation Rhode Island, full name—Rhode Island and Providence Plantation
Providence Plantations Rhode Island and Providence Plantations
provin provincial
Provincias Vascas Provincias Vascongadas (Spanish—Basque Provinces)—Alava, Guipúzcoa, and Vizcaya
provis. provision
provn provision
Provo city in Utah; Providenciales island and town in the Turks and Caicos Islands; Provisional (member of the IRA)
provos provokers (Dutch—street people engaged in militant tactics to provoke the police)
Provos Provisionals (Provisional Sinn Fein party members of Northern Ireland)
PROVOST Priority Research and Development Objectives for Vietnam Operations Support
proword procedure word
prox proximal; proximity
prox. proximo (Latin—next, *adv.*)
proxi protection by reflection optics of xerographic images
prox. luc. proxima luce (Latin—the day before)
prp peak radiated power; performance related pay; pickup (zone) release point; present participle; pseudo random pulse; psychological refractory period; pulse recurrence period; pulse repetition period
prp (PRP) platelet-rich plasma; polyribophosphate
Prp Principality

PRp Puerto Rican pimp
PRP People's Revolutionary Party (Tanzania); Production Requirements Plan; Production Reserve Policy; Public Relations Personnel
PRPA Puerto Rico Ports Authority
PRPC Public Relations Policy Committee (NATO)
PRPG Political Resident Persian Gulf (British)
PRPGA Puerto Rico Personnel and Guidance Association
prpln propulsion
prpp (PRPP) 5-phosphoribosyl 1-pyrophosphate
pr.pr. praeter propter (Latin—about, nearly)
PRp('s) Puerto Rican pimp(s)
PRPUC Philippine Republic Presidential Unit Citation
prr premature removal rate; production readiness review; pulse repetition rate
PRR Pennsylvania Railroad
PRRI Puerto Rico Rum Institute
p&rr's patriotic and religious racketeers
PRRWO Puerto Rican Revolutionary Workers Organization (communist)
prs pairs; printers
Prs Preston
PRs Pakistani rupees; Problem Reports; Puerto Ricans
PRS Park and Ride Scheme; Pattern-Recognition System; Pennsylvania-Reading Seashore (railroad); Performing Rights Society; Precision Ranging System; Property Recovery Squad (of a police department); Protective Research Section (U.S. Secret Service); Protestant Reformation Society; Public Radio Stations; Public Rehabilitation Scheme; Pupil Rating Scale
prsa personal retirement savings account
PRSA Public Relations Society of America
prsd pressed
prsd met pressed metal
prsfdr pressfeeder
prsmn pressman
PRSO Puerto Rico Symphony Orchestra

PRSP Puerto Rican Socialist Party (communist)

Prspct Prospect

PRSS Pennsylvania-Reading Seashore Lines

PRSSA Public Relations Student Society of America

PRST Puerto Rican Standard Time

Pr strain Prague (viral) strain

prsvn preservation

PRSY People's Republic of Southern Yemen

prt parachute radio transmitter; personal rapid transport; personnel research test; publication requirement table(s); pulse repetition time

prt (PRT) personal rapid transit; printer (flow chart); program reference table

p & rt physical and recreational training

Prt Port

PrT Prinzregentheater (Munich)

PRT Personnel Research Test; Philadelphia Rapid Transit; Prison Reform Trust; Production Re-evaluation Testing

PRT Partido Revolucionario de los Trabajadores (Spanish—Revolutionary Party of the Workers)—Mexican socialists; *Prinzrgententheater* (German—Prince Regent Theater)—Munich

prtd printed

prtg pretty good; printing

prtlsp printer line spacing

prtot prototype real-time optical tracker

prtov printer overflow

PRTS Personal Rapid Transit System

prty priority

pru peripheral resistance unit; prude; prudence; prudent

Pru Prudence; Prudential Life Insurance Company

PRU Polish-Russian Union

Prue Prudence

pru pru(s) prurient prude(s)

Prus Prussia; Prussian

prv peak reverse voltage; pressure-reducing valve; pressure-reduction valve

prv pour rendre visite (French—to return a call)

Prv Pravda (Russian—truth)—daily newspaper published in Moscow by Central Committee of the Communist Party

Prv. Proverbs

prw percent rated wattage

PRWAD Professional Rehabilitation Workers with the Adult Deaf

prx pressure regulator exhaust

PRY Pittsburgh Railways Corporation (stock exchange symbol)

PRZ People's Republic of Zanzibar

ps parlor snake; partly sunny; parts shipped; parts shipper; passenger service; passing scuttle; patient's serum; penal servitude; pending sale; phosphatidylserine; picosecond; pieces; piper syndrome; plastic surgery; point of switch; point of symmetry; polystyrene; power steering; power supply; proof shot; pseudo; pseudonym(s); public statute; pull switch; pulmonary stenosis

p-s pressure-sensitive

p's pennies

p/s paddle steamer; point of shipment; port or starboard

p & s paracentesis and suction; piss and shit; port and starboard

p.s. post scriptum (Latin—PostScript)

Ps Psalms, The (book of the Bible); South Pole; South Celestial Pole; static pressure

Ps Posaunen (German—trombones); *Psalms*

PS Pacific Southwest Airlines; Paleontological Society; Palm Society; Palm Springs (California); Parker Society; Paymaster Sergeant; Pennsylvania State University; petite small; Pharmaceutical Society; Philippine Scouts; Photo(graphic) Service; picket ship(s); Pilgrim Society; Pistol Sharpshooter; Pittsburg & Shawmut (railroad); Planetary Society; Plastic Surgery; Privy Seal; Public Safety; Public School; Puget Sound

PS (ps) Police Station

P-S Pullman-Standard

P.S. paddle steamer; public school

P & S Physicians and Surgeons; Pittsburg & Shawmut (railroad)

P of S Port of Spain

PS Parti Socialiste (French—Socialist Party); *Pferdestärke* (German—horsepower); *Publica Sicurédzza* (Italian—public security) police

P.S. post scriptum (Latin—written after)

PS 166 Public School 166 (for example)

ps3 plate strip of 3

ps5 plate strip of 5

psa parametric sound amplifier; passed staff college; personal savings account; personal security account; pressure-sensitive adhesive; proportion of survivors affected; psychoanalytic(al); public service advertising

psa (PSA) prostate specific antigen; public service announcement (radio or television)

PsA Pisces Austrinus (constellation)

PSA Pacific Science Association; Pacific Southwest Airlines; Packers and Stockyards Administration; Parcel Shippers Association; Phobia Society of America, Photographic Society of America; Photography Society of America; Play Schools Association; Poetry Society of America; Port of Singapore Authority; Poultry Science Association; Pretrial Service Agency; Program Study Authorization; Property Services Agency; Public Securities Association; Public Service Administration; Public Service Announcement; Public Service Association

P & SA Program and Systems Analysis

PSA Proceedings of the Society of Antiquaries

psaa post-stimulatory auditory adaptation

PSAB Public Schools Appointments Bureau

p sac pericardial cavity

PSAC President's Science Advisory Committee; Public Service Alliance of Canada

PSACPOO President's Scientific Advisory Committee Panel On Oceanography

psad prediction-simulation-adaptation-decision (data processing)

psaf private sector adjustment factor

PSAI Play Schools Association, Inc

PSAL Public School Athletic League

Psalt. Psalterium (Latin—Book of Psalms)

PSAMPP Philadelphia Society for Alleviating the Miseries of Public Prisons (founded by Benjamin Franklin, William Rush, and others)

ps an psychoanalysis; psychoanalyst; psychoanalytic-(al)(ly); psychoanalyze

p's and q's pints and quarts (in British pubs)

PSAODAP Presidential Special Action Office for Drug Abuse Prevention

PSAR Preliminary Safety Analysis Report

PSAT Palm Springs Aerial Tramway; Preliminary Scholastic Aptitude Test(ing)

psb please send a boat; public service band (radio)

PSB Paradox Salt Basin; Psychological Strategy Board; Public Service Board

P & SB Portland & South Bend (railroad)

PSBA Pennsylvania School Boards Association; Public Schools Bursars' Association

PSBLS Permanent Space-Based Logistics System

PSBO Public Savings Bond Office

PSBR Public-Sector Borrowing Requirement

psc passed staff college; per standard compass; port service charge; prestressed concrete

ps & c program scheduling and control

Psc Pisces (constellation)

P-S c Porter-Silber chromogen

PSC Pacific Sea Council; Pakistan Shipping Corporation; Peralta Shipping Corpora-

tion; Pittsburgh Steel Company; Point Shipping Company; Police Staff College; Porcelain-on-Steel Council; Potomac State College; Procurement Strategy Corporation; Product Safety Commission(er); Professional Services Corporation; Program Structure Code; Public Service Careers; Public Service Commission

PSC Partido Social Cristiano (Spanish—Social Christian Party)—Catholic actionists

PSCA Profit Sharing Council of America

pscb padded sample collection bag

PSCC Public Service Commission of Canada

PSCD Patrol Service Central Depot

PSCFB Pacific Southcoast Freight Bureau

pscg power supply and control gear

psclot protein S activity

PSCNI Public Service Company of Northern Illinois

PSCO Personnel Survey Control Office(r)

PSCP Public Service Careers Program

PSCPT Preschool Self-Concept Picture Test(ing)

PSCS Pacific Scatter Communications System

Psc's calculator Pascal's calculator (first adding machine)

PSCT Plastic Shipping Container Institute

pscu power-supply control unit

psd phase-sensitive demodulator; phase-sensitive detector; power spectral density; preshipment document; prevention of significant deterioration; promotion service date

P Sd Port Said

PSD Pittsburgh Steamship Division (United States Steel); Port of San Diego; Prevention of Significant Deterioration (of air quality); Public Safety Division (Texas)

PSD Partido Social Democrático (Portuguese and Spanish—Social Democratic Party)

PSDA Patient Self-Determination Act

ps detn particle size determination

PSDI Partito Socialista Democratico Italiano (Italian—Italian Social Democratic Party)

ps distn particle size distribution

psdo pseudo; pseudonym

psdp phrase structure and dependency parser

PSDS Primary Solar Duct(ing) System

psdu power-switching distribution unit

PSDUPD Port of San Diego Unified Port District

pse please; point of subjective equality; present state examinmation; program support environment

pse (PSE) psychological stress evaluator (voice-analysis lie detector)

PSE Pacific Stock Exchange; Public Service Employment

PSEA Pennsylvania State Education Association; Physical Security Equipment Agency; Pleaters, Stitchers and Embroiderers Association

psec picosecond

PSEG Public Service Enterprise Group

PSE & G Public Service Electric and Gas

PSE & GC Public Service Electric and Gas Company

psen pupils with special educational needs

Pseo Paseo (Spanish—boulevard)

pser production support and equipment replacement

pset permanent service on earth tides

pseud pseudandry (women using male names as pseudonyms); pseudepigraphy (attributing false names to artists, authors, or composers); pseudograph (falsely attributing a work to an artist, author, or composer); pseudonym (false name, nom de plume, pen name); pseudonyma (pseudonymous works)

pseudo (Latin prefix—false)—pseudonym

psf payload-structure-fuel (ratio); point-spread function; pounds per square foot

PSF Phelps-Stokes Fund; Presidio of San Francisco

P & SF Panhandle and Santa Fe (railroad)

P of SF Port of San Francisco

PSFC Pacific Salmon Fisheries Commission

PSFL Puget Sound Freight Lines

PSFS Philadelphia Savings Fund Society

psg production system generator; psychogalvanometer; psychogalvanometric(al)(1y)

PSGBI Pathological Society of Great Britain and Ireland

psgi permanent service on geomagnetic indices

psgr passenger

psgr lng passenger lounge

psgr(s) passenger(s)

P-Shaw George Bernard Shaw (also GBS)

PSHFA Public Servants Housing and Finance Association

pshr pusher

psi pounds per square inch; public school(s) investigation

PSI Pacific Semiconductors Incorporated; Performance Systems International (commercial Internet provider); Personalized System of Instruction; Personnel Security Investigation; Pharmaceutical Society of Ireland; Physician's Services Incorporated; Pollutants Standards Index; Population Services Incorporated; Population Services International; Private Sector Initiatives; Professional Secretaries International; Public Services International

PSI *Partito Socialista Italiano* (Italian—Socialist Party); *Pollution Standards Index*

psia pounds per square inch absolute

PSIC Pacific Scientific Information Center (Bernice Pauahi Bishop Museum, Honolulu)

psid pounds per square inch differential

PSIDC Punjab State Industrial Development Corporation

psig pounds per square inch gage

psil preferred-frequency speech interference level

PSIP Poultry Stock Improvement Plan; Private Sector Initiative Program

PSIUP *Partito Socialista Italiano di Unita Proletaria* (Italian—Italian Socialist Party of Proletarian Unity)

psk phase shift keying

p sl pipe sleeve

psl personal seat license

PSL Pacific Star Line; Peruvian State Line; Philharmonic Society of London; Pretoria State Library

PSL *Patterson Strategy Letter*

p-slips old-fashioned postcard size (3- x 5-inch) slips of paper used for filing

psl sol potassium, sodium chloride, sodium lactate solution

ps lt port side light

PSLT Picture Story Language Test

psm passed school of music; power supply module; production surveillance monitor; pulse-spacing modulation

psm (PSM) presystolic murmur

PSM People for Self Management; Product Sales Manager

psma progressive spinal muscular atrophy

PSMA Power Saw Manufacturers Association; Pressure-Sensitive Manufacturers Association; Professional Services Management Association

psmf protein-sparing modified fast

PSMFC Pacific States Marine Fisheries Commission

psmr parts specification management for reliability

psmsl permanent service for mean sea level

psn position; processor serial number; pulse-shaping network(ing)

PSn Port Sudan

PSN *Partido Socialista de Nicaragua* (Spanish—Socialist Party of Nicaragua)

PSNA Phytochemical Society of North America

PSNC Pacific Steam Navigation Company

PSNH Public Service New Hampshire

PSNS Puget Sound Naval Shipyard

P^so *Passo* (Italian—pass)

pso (PSO) polysulfone

PSO Pad Safety Officer; Pasadena Symphony Orchestra; Phoenix Symphony Orchestra; Pilot Systems Operator; Pittsburgh Symphony Orchestra; Portland Symphony Orchestra; Prague Symphony Orchestra

PSOE *Partido Socialista Obrero Español* (Spanish—Socialist Workers' Party)

p sol partially soluble; partly soluble

pson person

psp paralytic seafood poisoning; phenolsulfonaphthalein (test); pierced-steel plank; positive screened print; postsynaptic potential

psp (PSP) progressive supernuclear palsy

pS pekoe Souchong

PSP Pacific Security Pact; Pocahontas State Park (Virginia); Price-Subsidy Program; Price-Support Program; Programs Support Plan; Public School Pronunciation (British)

PSP *Pacifistisch Socialistische Partij* (Dutch—Pacifist-Socialist Party) *Policia de Segurança Pública* (Portuguese—police force)

PSPA Professional School Photographers of America; Professional Sports Photographers Association

PSPCD Puget Sound Pollution Control District

PSP & L Puget Sound Power and Light (company)

PSPMW Pulp, Sulphite and Paper Mill Workers

PSPP Proposed System Package Plan

ps & ps pimps and prostitutes

PSPS Paddle Steamer Preservation Society; Primary Solar Piping System

PSQC Philippine Society for Quality Control

psql process-screening quality level

p's & q's minding your p's & q's originated when printers instructed apprentices about similarity of lowercase p's and q's when handsetting type; also used in saloons to keep count of the number of pints and quarts of beer consumed

psr pain-sensitivity range; passenger space ratio; plow-steel rope; pulsars

PSR Pacific School of Religion; Physicians for Social Responsibility

PSRC Public Service Research Council

PSRF Profit Sharing Research Foundation

PSRI Public Systems Research Institute (UCLA)

PSRM Pacific Southwest Railway Museum

PSRMA Pacific Southwest Railway Museum Association

psro passenger standing route order

PSRO Professional Services (Standards) Review Organization; Professional Standards Review Organization

pss packet-switching service; physiological saline solution; progressive systematic sclerosis

Pss Princess

PSS Pad Safety Supervisor; Parents for Student Safety; Personal Security System; Personal Signalling System; PreSchool Screening (program); Printing and Stationery Service; Professional Services Section; Public Service System

P.S.S. Professor of Sacred Scripture

P.S.S. *postcripta* (Latin—postscripts)

pssbb public schools system blanket bond(ing)

PSSC Personal Social Services Council; Physical Science Study Committee (NSF); Pious Society of Saint Charles; Public Service Satellite Consortium

PS & SC Public Service and Safety Committee

P.S.S.C. Pious Society of Saint Charles

PSSMA Paper Shipping Sack Manufacturers Association

PSSNY Philharmonic Symphony Society of New York

psso passed slip stitch over (knitting)

Ps. Sol *Psalms of Solomon*

PSSS Philosphic Society for the Study of Sport

PSST Prairie State Standard Time; Public Sector Standardization Team

pst polished surface technique

pst (PST) prefrontal sonic treatment

PST Pacific Standard Time

PSTA Public Safety and Training Association

PSTB Picture Story Test Blank

PSTBC Puget Sound Tug and Boat Company

PSTC Pressure Sensitive Tape Council

PSTD Prison Service Training Depot (Pretoria)

£ sterling pound sterling

p stg c per steering compass

psth peristimulus time histogram

PSTIAC Pavements and Soil Trafficability Information Analysis Center (USA)

pstl postal

PSTMA Paper Stationery and Tablet Manufacturers Association

PSTN Public Switched Telephone Network

PSTO Principal Sea Transport Officer

P-strip P-shaped strip

P-stuff pcp (PCP)

pstv potato spindle tuber virus

pstz pasteurize

pstzd pasteurized

pstzg pasteurizing

psu package size unspecified; power supply unit; primary sampling unit

PSU Pennsylvania State University; Portland State University; Public Security Unit (Ugandan secret police)

PSU *Parti Socialiste Unifé* (French—Unified Socialist Party); *Partito Socialista Unitario*, (Italian—Unitary Socialist Party)

p-substance protein substance

PSUC Pennsylvania State University Center(s)

PSUC *Partido Socialista Unificado de Cataluña* (Spanish—Unified Socialist Party of Catalonia)

P Sud Port Sudan

PSU-MRL Pennsylvania State University—Materials Research Laboratory

PSUP Pennsylvania State University Press

p surg plastic surgeon; plastic surgery

psv polished-stone value; public service vehicle

PSV Petit St Vincent (Grenadines in the West Indies); Project Salt Vault

psvm pulse-spacing volt meter

PSW Psychiatric Social Worker

PSWB Plateau State Water Board

pswbd power switchboard

PSWC Pacific Tsunami Warning Center (Honolulu)

P Swet Port Swettenham (now Port Kelang or Port Klang)

PSWFA Prestige Saltwater Fly Anglers

PSWO Picture and Sound World Organization

pswr power standing wave ratio

psy psychological

Psy Paisley

psych psychiatry; psychology; psychopathology

psych/d psychological death

psychedeli psychedelicatessen (store selling the paraphernalia of drug addicts)

psychiat psychiatric; psychiatry

psycho dangerous lunatic; a psychiatric hospital or ward; a psychoneurotic personality; a psychotic individual (pseudo-scientific slang)

psycho Latin prefix—mental; psychologist

psychoan psychoanalytic; psychoanalysis; psychoanalyst

psychobab psychobabble(r)— psychological patter(er)

psychobio psychobiological; psychobiologist; psychobiology

psychobiog psychobiographer; psychobiographic(al)(ly); psychobiography

psychochron psychochronic (al)(ly); psychochronicle(r); psychochronologer; psychochronologic(al)(ly); psychochronology

psychochronicle psychiatric chronicle; psychological chronicle

psychodelics hallucinogenic drugs

psychodels psychodelics (hallucinogens)

psychogeog psychogeographer; psychogeographic(al); psychogeography

psychohist psychohistorian; psychohistorical; psychohistory

psychol psychological; psychologist; psychology

Psychol Psychology

psychomet psychometric

psychopathol psychopathological; psychopathologist; psychopathology

psychophys psychophysical; psychophysics; psychophysicist

psychophysiol psychophysiology (and derivatives)

psychoprison psychiatric hospital prison (USSR)

psychosurg psychosurgeon; psychosurgery; psychosurgical(ly)

psychot psychotic

psychother psychotherapist; psychotherapeutic(als); psychotherapy

psycho ward psychopathic ward

Psych Qtly Psychoanalytic Quarterly

psych test. psychological testing

psydoc psychiatrist doctor

psyk psykologi or *psykologist* (Dano-Norwegian—psychology or psychologist)

psyop psychological operation

psypath psychopath(ic)

psysom psychosomatic

p system parvocellular system

psywar psychological warfare

psz (PSZ) partly stabilized zirconia

pt part; part time; personal trade; personal trainer; physical therapy; physical therapist; physical training; pint(s); plenty tough; plenty trouble; pneumatic tube;

point; point of tangency; point of turn; point of turning; primary target; private terms; prothrombin time

pt. petition; pint; port

p & t personnel and training; posts and timbers

pt partie (French—part)

pt. perstetur (Latin—let it be continued)

p.t. protempore (Latin—temporarily)

£T pound Turkish

Pt part; platinum; Point; Port; Porto; Puerto

P^t Petit (French—little, small); *Pont* (French—bridge)

PT motor torpedo boat (naval symbol); Pacific Time; Peninsula Terminal (railroad); Philadelphia Transportation; Physical Therapist; physical therapy; physical training; Portugal (Internet code); Postal Telegraph; primary trainer; Provincetown-Boston Airline (2-letter coding)

P & T Pope & Talbot (steamship line)

P & T The Phoenix and the Turtle

PT-76 Soviet Amphibious tank

pta plasma thromboplastin antecedent; posttraumatic amnesia; primary target area; prior to admission; proposed technical approach; peseta (Spanish monetary unit, diminutive of peso); programmed test area

pta peseta (Spanish—monetary unit)

Pta Pretoria

Pta Punta (Spanish—Point)

P^{ta} Ponta (Portuguese—point); *Puerta* (Spanish—gate, gateway, mountain pass); *Punta* (Spanish—point)

Pt A Port Arthur, Ontario

PTA Paper and Twine Association; Parent-Teacher Association; Pet Traders Association; Philippine Travel Authority; Physical Therapist's Assistant; Pope and Talbot; Postal Transportation Association; Prevention of Terrorism Act; Protestant Teachers Association

PTA Prevention of Terrorism Act (British)—provides for seven days detention of suspects

PTAC Premium Target Advisory Commission

ptacv (PTACV) prototype air-cushioned vehicle

P Tal Port Talbot

Pt Alb Port Alberni

Pt Ant Port Antonio

PTAR Prime Time Access Rule

Pt Art Port Arthur

ptas pesetas

pta's part-time alcoholics

PTAs Passenger Transport Authorities

PTAS Productivity and Technical Assistance Secretariat

P Tau Port Taufiq (formerly Port Tewfik)

ptb patellar-tendon bearing; peach-tree borer

PTB Partido Trabalhista Brasileiro (Portuguese—Brazilian Workers Party); *Physikalisch-Technische Bundesanstalt* (German—Physical Technical Institute)

ptbl portable

PTBM PT Barnum Museum (Bridgeport, Connecticut)

PT boat patrol torpedo boat

ptbr punched-tape block reader

PTBT Partial Test Ban Treaty

ptc personnel transfer capsule; positive temperature coefficient

ptc (PTC) phenylthiocarbamide; plasma thromboplastin component (clotting factor IX)

PTC Pacific Theological College; Pacific Tin Consolidated; Paisley Technical College; patrol vessel (naval symbol); Peoria Terminal (railroad); Philadelphia Transportation Company; Pine Tree Camp; Pipe and Tobacco Council; Power Transmission Council; Press Trust of Ceylon; Private Truck Council

ptca percutaneous transluminal coronary angioplasty

PTCA Private Track Council of America

ptcldy partly cloudy

ptd painted

PTDA Power Transmission Distributors Association; Professional Tournament Directors Association

PTDC Pakistan Tourism Development Corporation

ptdl programmable tapped-delay line

PTDP Preliminary Technical Development Plan

PTDR Post-Test Disassembly Report

PTDS Photo Target Detection System

pte parathyroid extract; *poriente* (Spanish—west)

pte (PTE) pulmonary thromboembolism

p^{te} parte (Spanish—part)

Pte Pointe (French—Point); *Presidente* (Portuguese or Spanish—President)

PTE Passenger Transport Executive

pt ed patient education

pt ex. part exchange

ptf plasma thromboplastin factor

ptf. plaintiff

PTF fast patrol boat (naval symbol); Propulsion Test Facilities

ptfe polytetrafluoretyhylene

PTFMA Public Telecommunications Financial Management Association

ptfp prime-time family programming

ptg printing

Ptg Portugal; Portuguese

PTG Piano Technician's Guild; Polaris Task Group

ptg&l playing the game and loving it

ptgt primary target

pth parathormone

Pth. Parthian

Pth Perth

PTH hydrofoil motor torpedo boat (naval symbol); parathyroid hormone

pths post-traumatic hyperirritability syndrome

pti persistent tolerant infection; physical training instructor (PTI); pre-trial intervention; previously taxed income

PTI Philips Telecommunicatie Industrie; Pictorial Test of Intelligence; Press Trust of India; Protect the Innocent (anti-crime lobby)

PTIDG Presentation of Technical Information Discussion Group

Pt-Ir platiniridium

PTIS Piano Teachers Information Service

PTj (Cuerpo) Técnico de Policía Judicial (Spanish—Technical Corps of the Judicial Police)—Venezuela

Pt K Port Kiang (also written Kelang and formerly Port Swettenham)

ptl partial total loss; pass the loot; pintle; primary target line

Pt L Point Loma

P t L Praise the Lord

PTL People That Live; Photographic Technology Laboratory; Praise the Lord; People That Love (the Lord)

PTLA Publishers' Trade List Annual

PTLL Pittsburgh Toy Lending Library

ptm proof test model; pulse-time modulation

Ptm Pietermaai

Ptm (PTM) Polaris tactical missile

ptma phosphotungstomolybdic acid

PTMTCS Power-Tape-to-Magnetic-Tape Conversion System

ptmt (PTMT) polytetramethylene terephthalate (thermoplastic polyester)

ptn partition

PTNA Professional Travel Nurses Association

Ptnr Partner

pto permeability-tuned oscillator; please turn over; power takeoff

Pto Porto; Puerto; Punto

P^{to} Ponto (Italian—sea)—poetic term; *Porto* (Italian, Portuguese, Spanish—port); *Puerto* (Spanish—port); *Punto* (Italian—point)

PTO Parent Teachers Organization; Patent and Trademark Office; Public Trustee Office(r); Purdue Teacher Opinionaire

Pto Bivr Puerto Bolívar

Pto Cab Puerto Cabello

Pto Cast Puerto Castilla, Honduras

ptol peacetime operating level

Ptol Ptolemaic; Ptolemy

Ptolemy Alexandrian astronomer Claudius Ptolemaeus

Pto Rico Puerto Rico

P-town Provincetown

P Town Port Townsend

ptp paper-tape printer; paper tape punch; part-time pimp; post-tetanic potentiation

p-t-p point-to-point

PTP Pointe a Pitre, Guadeloupe (airport); Productive Thinking Program

ptpg participating

PTPs Public Traded Partnerships

pt/pt point-to-point

ptr printer; pupil-teacher ratio

ptr (PTR) photoelectric tape reader

PTR Preliminary Technical Review; pool test reactor

PTRA Power Transmission Representatives Association

ptrf peacetime rate factor(s)

ptry pantry; poetry; pottery

pts phantom taste syndrome; posttraumatic syndrome

pts. points

pts pesetas (Spanish—plural of peseta); pints; *puntos* (Spanish—degrees; periods)

Pts Portsmouth

PTS Philatelic Traders Society; Postal Transportation Service; Potential Tax Savings; Princeton Theological Seminary; Professional Travelogue Sponsors; Public Television Station(s)

PT & S Pacific Towboat and Salvage (tugs)

ptsd (PTSD) post-traumatic stress disorder

pts/hr parts per hour; pieces per hour

Ptsmth Portsmouth

Pt Sp Port of Spain

PTSS Princeton Time-Sharing System

PTSTV Prime Time School Television

ptt push to talk

ptt (PTT) partial thromboplastin time

PTT Postal, Telegraph and Telephone

PTT *Posta, Telgraf ve Telefon* (Turkish—Post, Telegraph, and Telephone); *Postes, Télégraphes, Téléphone* (French—postal, telegraph, and telephone system)

PTTA Philippine Tourist and Travel Association; Postal Telegraph and Telephone Authority

ptti precise time and time interval

PTTI Postal, Telegraph, and Telephone International

pt-tm part-time

pttnmkr patternmaker

ptu propylthiouracil

PTU Plumbers' Trade Union; Plumbing Trade Union; Psychiatric Treatment Unit

PTUC Philippine Trade Unions Council

ptv passenger transfer vehicle; public television

ptv (PTV) propulsion test vehicle

ptv's personal transportation vehicles (three-wheeled vehicles for city driving)

PTVs Public Television Stations

ptw per thousand words

Pt W Port Weller

PTWC Pacific Tsunami Warning Center

Pty Party; Proprietary

P-type Jungian perceptive type

pu passed urine; peptic ulcer; physical unit; pickup; pick up; plant unit; power unit; pregnancy urine; propellant utilization; propulsion unit; pump(ing) unit; pump unit

pu (PU) polyurethane

pu. pupil; purple

p-u phew (what a stench)

p.u. *plus ultra* (Latin—beyond the pinnacle, beyond the ultimate)

Pu plutonium

PU Pacific University; Phillips University; Princeton University; Prisoner's Union; Purdue University

P U Peter Ustinov

PUA *Punta de la Unidad Africana* (Spanish—Point of African Unity)—formerly Fernanda Point

PUAS Postal Union of the Americas and Spain

pub public; publican; publication; public house; publicity; publish; published; publisher; publishing

pub. publication

púb públicas (Spanish—publications)

Pub Publican; Public House; Publisher's Announcement

PUB Public Utilities Board

pub. aff public affairs

pub aide publication aide

pubbl *pubblicitá* (Italian—advertising, publicity)

Pub Cit public citizen

pub con public convenience (British term for a public toilet)

Pub Doc Public Document

pub ed publication editor

pubinfo public information

publ publication; publicity; publisher; publishing

Publ Astron Soc Pac *Publications of the Astronomical Society of the Pacific*

publg publishing

pubn(s) publication(s)

pub(s) public house(s) (British short form)

Pub Sect Lab Rel Public Sector Labor Relations Conference Board

pubtronics publication electronics that convert electronically produced manuscripts into type

Pub W *Publishers Weekly*

Pub Wks Public Works

puc papers under consideration; pickup car

PUC Peoples University of China; Presidential Unit Citation; Public Utilities Code; Public Utilities Commission; Public Utilities and Corporations

PUC *Post Urbem Conditam* (Latin—after the foundation of the city)—city usually means Rome

PUCC Port Users Consultative Committee

pucf polyurethane-coated fabric; polyurethane-coated fibers

pud puddle; pudding; public utilities district

pud (PUD) planned unit development

pu & d pickup and delivery

PUD Planned Unit Development

Pue Puebla

Puebla Puebla de Zaragoza (in central Mexico)

Puerto Rico islands of Puerto Rico, Culebra, Vieques

puf prime underwriting facility

PUF Presses Universitaires de France (University Presses of France)

pufa polyunsaturated fatty acid

PUFF People United to Fight Frustrations

PUFTT Purdue University Fast Fortran Translator

pug. print under glaze; puggy; pugilism; pugilist

PUHCA Public Utilities Holding Company Act

PUHS Phoenix Union High School

PUK Patriotic Union of Kurdistan; Pechiney Ugine Kuhlmann

puka *pukalolo* (Hawaiian Polynesian—crazy tobacco)- marijuana also known locally as Kauai electric, Kona gold, Maui wowie, Puna butter

pul pulley

PUL Princeton University Library (New Jersey); Punjab University Library (Lahore, Pakistan)

pula monetary unit of Botswana

pulchris pulchritudinous

'Pulco Acapulco

pulg *pulgadas* (Spanish—inches)

pulheems physical capacity, upper and lower limbs, hearing, eyesight, emotional capacity, mental stability

pul ins pulmonary insufficiency

pulm pulmonary

pulm. *pulmentum:* (Latin—gruel)

pulm a pulmonary artery

pulm emb pulmonary embolism

pulmo pulmoaortic(al)(ly); pulmology; pulmometer; pulmometric(al)(ly); pulmometry; pulmonary, pulmonectomy; pulmonic(al)(ly); pulmonitis; pulmonologist(ic)(al)(1y); pulmotor

pulmotor pulmonary motor

pulsar pulse star (pulsed radio-wave-emitting star); pulsing astronomical signal (received from outer space)

PULSAR Parking Urban Loading Unloading Standards and Rules

pul sten pulmonary stenosis

pulv pulverize(r)

pulv. pulvis (Latin—powder)

pulv. gros. pulvis grossus (Latin—coarse powder)

pulv. subtil. pulvis subtilis (Latin—smooth powder)

pulv. tenu pulvis tenuis (Latin—very fine)

pum pop-up mechanism

PUM Postal Union Mail

puma catamount, cougar, mountain lion, panther

puma. (PUMA) programmable universal mechanical assembly (robot)

PUMA Prostitutes Union of Massachusetts

Pumfret Pontefract

pumi pumice

pump. pumping

PUMP Protesting Unfair Marketing Practices

pums permanently unfit for military service

pun. puncheon

puN plasma-urea Nitrogen

PUN Partido Union Nacional (Spanish—National Union Party)

punc punctuation

pundonor punta de honor (Spanish—point of honor)

Punj Punjabi

punk eek punctuated equilibrium

Punta Gall Punta Gallinas (northernmost point of Colombia and all South America)

Puntaren Puntarenas (Costa Rican province)

puo pneumonia; pyrexia of unknown origin

pup puppy

pup. (PUP) peripheral unit processor

Pup Puppis (constellation)

PUP People's United Party (Belize); Polytechnic University of the Philippines; Princeton University Press

Pupp Puppis (constellation)

puppie(s) pregnant urban professional(s)

pups puppies

pup(s) pregnant urban professional(s)

p'up(s) pickup(s)

pur purchase; purchaser; purchasing; purifier; purification, purify; purple; purplish; pursuant; pursuit

Pur Purim

purch purchasing

Purdue Purdue University Press

pure mat. pure machine-aided translation

pure mt pure machine translation

pureq purchase requisition

PUREX Plutonium Uranium Extraction Plant

purg. purgativus (Latin—purgative)

PURL Persistant Universal Resource Locator

purp purple

purv powered underwater research vehicle

pus. permanently unfit for service

Pus Pusan

PUs Public Utilities

PUS Parliamentary Under-Secretary; Permanent Under-Secretary

PUS Pharmacopoeia of the United States

Push Aleksandr Sergeyevich Pushkin—Russian dramatist, novelist, short-story writer; Pushtu

PUSH People United to Save Humanity; People United to Serve Humanity

puss pussy; pussycat

puta(s) prostituta(s) [Portuguese or Spanish—prostitute(s)]

Putnam GP Putnam; Putnaham

PUTS Patented Underbody Testing System

putty linseed oil and powdered chalk mixture

puva psoralen (drug) + ultraviolet-A (fight)

PUVAS Plutonium Value Analysis System

puvep propellant-utilization vehicle-borne electronic package

Puy-de-D Puy-de-Dôme

pv par value; paravane; pave(d); paving; photovoltaic; pico-volt; plasma value; pole vault; position value; prime vertical; public voucher

p/v peak-to-valley; per vagina; pressure vacuum; pressure valve; profit volume (ratio)

p & v pressure and velocity

pv por vida (Spanish—for life); *prossimo venturo* (Italian—next month)

p v petite vitesse (French—slow train); *piccola velocity* (Italian-slow train)

p.v. per vaginam (Latin—by vagina)

Pv Peru; Peruvian

PV Eastern Provincial Airways (2-letter coding); patrol vessel; Positive Vetting; Post Village; post-Virgil; Priest Vicar; Puerto Vallarta

P. V. Pais Vasco (Spanish—Basque Country); *Procès verbaux* (French—official report); *Processi verbali* (Italian—official report)

PV-2 Lockheed maritime reconnaissance bomber

pva polyvinyl acetate; polyvinyl alcohol

PVA Paralyzed Veterans of America; Prison Visitor's Association

p. vag per vaginam (Latin-by the vagina)

pval polyvinyl alcohol

P-value probability value

pvb potentiometer voltmeter bridge

PVB Prison Visitors' Board

pvc polyvinyl chloride (thermoplastic)

pvc (PVC) premature ventricular contractions

PVC Philippine Volconology Commission; Precision Valve Corporation

PVCC Piedmont Virginia Community College

Pvccf polyvinyl-chloride-coated fabric; polyvinylchloride-coated fibers

pvd peripheral vascular disease; pulmonary vascular disease

PVD Providence, Rhode Island (airport)

PvdA Partij van de Arbeid (Dutch—Labor Party)

pvdc polyvinyl dichloride

pvdc (PVDC) polyvinylidene chloride

pvem pulse-vector emittance meter

pvf polyvinyl fluoride

pvH propane-vacuum hydrogen

pvi point of vertical intersection; postal validation imprinter

PVI Personal Values Inventory

pvis pneumatic vertical-indicating scale

pvm polyvinyl methyl

PVM Process Evaluation Module

PVMNM Perry's Victory Memorial National Monument

Pvmnt Pavement

pvn paraventricular nucleus

pVNs post-Vietnam syndrome

pvnt prevent; preventive

pvo private voluntary organizations

PVO Principal Veterinary Officer

pvp photovoltaic power; polyvinylpyrrolidone (plasma extender)

pvp precio máximo de venta al publico (Spanish—maximum price charged the public)

PVP President's Veterans Program

PVPMPC Perpetual Vice President and Member of the Pickwick Club

pvpp polyvinyl-polypyrrolidone

pvq personal-value questionnaire

pvr personal video recorder; portable volume-controlled respirator; precision voltage reference

p/vr profit/volume ratio

PVR Police Volunteer Reserves; Premature Voluntary Requirement

PVRC Pressure Vessel Research Committee (NBS)

pvs periventricular system; persistent vegetative state

PVS Pecos Valley Southern (railroad); Periventricular (fiber) System; Personal Value System

pvt page view terminal; pressure volume temperature; private

pvt par voie télégraphique (French—by telegraph)

Pvt Private

Pvt 1/C Private First Class

PVU Prairie View University

pvw pure virgin wool

pw packed weight; passing window; path width; patient waiting; picowatt; pivoted window; postwar; power windows; prisoner of war; private wire; projected window; psychological warfare; public works; pulse width

p/w parallel with

p & w pension and welfare (retirement benefits)

PW Pacific Western Airlines; Palau; Philadelphia & Western (railroad); Pittsburgh & West Virginia (railroad); prisoner of war; Public Works

P-W Prader-Willi (syndrome)

P & W Pratt and Whitney Aircraft Division, United Aircraft Corporation

PW Petroleum Week; Publishers Weekly

PWA Pacific Western Airlines; Pacific Windsurfers Association; People With Aids; Prison Wardens Association; Professional Writers of America; Psychic Workers Association; Public Works Administration

PWA Papierwerke Waldhof-Ashaffenburg (German—Waldhof-Ashaffenburg Paper Works)

PWAA Professional Women's Appraisal Association

pwafrr present worth of all future revenue requirements

PW AIDS person(s) with AIDS

P Wash Port Washington

p-wave pressure wave

p waves primary (earthquake) waves

pwb printed wiring board

PWBA Pension and Welfare Benefits Administration

pwc physical working capacity

pwc (PWC) pulse-width coded; pulse-width coding

PWC Parents Who Care; Prisoner of War Convention; Public Works Canada; Public Works Center (USN)

pwd plywood; powered

pwd (PWD) pulse-width discriminating; pulse-width discriminator

PWD Psychological Warfare Division; Public Works Department

PWDA Personal Watercraft Dealers Association

pwdrd powdered

PWDS Protected Wireline Distribution System

pwe (PWE) pulse-width encoder; pulse-width encoding

PWE Political Warfare Executive; Prisoner of War Enclosure

P Wel Port Weller

pwf pregnancy without fear (pillow-simulated pregnancy); present-worth factor

PWFP Prince William Forest Park (Virginia)

PWG Permanent Working Group (NATO); Province Working Group; Philip W Goetz (Editor-in-Chief of *The New Encyclopaedia Britannica*)

PWHS Public Works Historical Society

PWI Physiological Workload Index

P & W I Poets and Writers Incorporated

PWIC Professional Women in Construction

PWIF Plantation Workers' International Federation

PWJC Piney Woods Junior College

pwl power watt(age) level

PWLB Public Works Loan Board

pwm plated-wire memory; pokeweed mitogen (PWM), pulse width modulation

pwm (PWM) pulse-width modulating; pulse-width modulator

PWM Partnership for World Mission

PWMS Public Works Management System (USN)

pwmsp people with multiple social problems; person(s) with multiple social problems

PWNDA Provincial Wholesale Newspaper Distributors' Association

PWNP Parra Wirra National Park (South Australia)

PWO Principal Weapons Officer; Public Welfare Office(r); Public Works Office(r)

pwp picowatt power

PWP Parents Without Partners; Professional Women Photographers

PWPP Post-War Psychological Problems

pwr power; pressurized water reactor (PWR)

PWR Police War Reserve

PWRS Pacific War Research Society

pwr sup power supply

pws paddle wheel steamer

pw's prisoners of war

PWs Professional Warriors

PWS Periyar Wildlife Sanctuary (India); Private Wire System

pwt pennyweight; propulsion wind tunnel

PWT Picture World Test

pwtn power train

pwtr pewter

pwv pulse-width valve

P & WV Pittsburgh & West Virginia (railroad)

pww planar wing weapon

Pwy Poway

px past history; physical examination; please exchange; pneumothorax; press; prognosis

PX Aspen Airways (2-letter code); Post Exchange

PXCMD Phoenix Contract Management District

pxe (PXE) pseudoxanthoma elasticum

px in time of arrival

pxl (PXL) patrol experimental land-based aircraft

pxlst passenger list(ing)

px me report my arrival and departure

pxo próximo (Spanish—next)

px out takeoff time

PX-S Japanese reconnaissance flying boat

pxt. pinxit (Latin—he painted it)

py pitch and yaw

p/y pitch or yaw

PY commissioned and armed yacht (2-letter naval symbol); Paraguay (Internet code); program year; Surinam Airways (2-letter symbol); yacht (naval symbol)

Pya Pyatnitsa (Russian—Friday)

PYA plan, year, age (insurance)

pyc proteose-yeast castione

PYC Philadelphia Yacht Club; Portland Yacht Club (Maine); Poughkeepsie Yacht Club

PYE Protect Your Environment

Pyg Pygmalion (George Bernard Shaw play)

pyg broth proteose-yeast-glucose broth

PYHA Pakistan Youth Hostels Association

Pyj Soc Tha Mya Nai Pyidaungso Socialist Thammada Myanma Naingngandaw (Burmese—Socialist Republic of the Union of Burma)

pyo pick your own (flowers, fruit, vegetables)

pyo (Latin prefix—pus)—pyorrhea

pyof pick your own fruit

pyph polyphase

pyr pyridine

p-y-r pitch-yaw-roll

pyramid pyramid investment scheme

Pyrenees Pyrenees Mountains between France and Spain

pyrite fool's gold; iron disulfide; iron pyrites

pyrites copper, iron, tin pyrite; also known as fool's gold

pyrmd pyramid(ed)

pyro pyromaniac; pyrotechnic(s); pyroxylin

pyroglu pyroglutamic acid

pyrolag pyrolagnia(c)

pyrom pyrometer; pyrometry

Pyr-Or Pyrénées-Orientales

pyrot pyrotechnics

pyt pretty young thing

PYTA Prior Year Tax Assistance

pyx. pyxis (Latin—box, vessel)

Pyx Pyxis (constellation)

pz pancreozymin

PZ Paolei Zion(ist); Pickup Zone; Police Zone

PZ-61 Swiss medium tank armed with a 105mm gun

pza pyrazamide

pza pieza (Spanish—piece)

Pza Plaza (Italian or Spanish—Plaza)

pza Piazza (Italian—Square)

pzc point of zero charge

PZC Partido Zapatista Comunista (Spanish—Zapatist Communist Party) group active along the Mexican Border

pz-cck pancreozymin-cholecystokinin

pzi protamine zinc insulin

pzm pressure-zone microphone

PZM Polska Zegluga Morska (Polish—Polish Merchant Marine)

P^{zo} Pizzo (Italian—peak, summit)

PZPR Polska Zjednoczona Partia Robotnicza (Polish—Polish United Workers Party)

PZS President of the Zoological Society

pzt photographic zenith tube

P^{za} Piazza (Italian—Square)

Q

q coefficient of association (statistical symbol); cue; data-questionnaire data; dynamic pressure (symbol); electric charge (symbol); quality factor; quart; quarter; quarterly; quartile; quartile deviation; quarto; queer; quench; quenching; queries; query; question(s); quick; quintal, quire; quit; semi-interquartile range (symbol); stagnation pressure (symbol)

q *quaque* (Latin—each, every)

q. questionnaire

Q bankruptcy or receivership (stock exchange symbol); electric quadruple moment of atomic nucleus (symbol); Fairchild (symbol); Polaris correction (symbol); prison at San Quentin, California; quadrillion; Quaker Line; quality factor; quantity; quarantine; Quartermaster; quartile variation (symbol); Québec—code for letter Q; Queen; Queensland; question(s); *quetzal* (Guatemalan monetary unit); quotient; radio inductive reactance to resistance (symbol); semi-interquartile range (symbol); target or drone (symbol); thermoelectric power (symbol); volume of blood per unit of time

Q(*Q*) quantity (microeconomics)

Q (Latin—Quintus); pseudonym for Sir Arthur Quiller-Couch

Q. *Quintus* (Latin—fifth time)

Q1 *quintal* (Spanish—hundredweight)

Q₁, Q₂, Q₃, Q₄ first quartile, second quartile, third quartile, fourth quartile

q²h *quaque secunda hora* (Latin—every two hours)

q³h *quaque tertia hora* (Latin—every three hours)

q⁴h *quaque quarta hora* (Latin—every four hours)

qa quality assurance; quick-acting; quick assembly; quiescent aerial

q & a question and answer

QA Qualification Approval; Qualified Acceptance; Quality Assurance; Quatar (Internet code); Quarters Allowance

QA *Quai* (French—embankment or quay); *quetzal* (Guatemalan monetary unit); torque (symbol)

Q-A Quint-A

Q & A question and answer

QAA Quality Assurance Assistant

QAB Quality Assurance Board; Quality Assurance Bulletin; Queen Anne's Bounty (for indigent clergymen)

qac quaternary ammonium compound

QAC Quality Assurance Check(ing); Quality Assurance Code; Quality Assurance Coding; Queensland Arts Council

QACA Queensland Amateur Cyclists Association

QACAD Quality-Assurance Corrective-Action Document

QACC Queensland Automobile Chamber of Commerce

qad quick-attach-detach

QAD Quality Assurance Data; Quality Assurance Department; Quality Assurance Directive; Quality Assurance Division

QADC Queen's Aide-de-Camp

QADI Quality Assurance Department Instruction

qadk quick-attach-detach kit

QADS Quality Assurance Data Summary; Quality Assurance Data System

QAE Quality Assurance Engineering)

qaf quality-assurance firing

QAFCO Quatar Fertilizer Company

QAFL Queensland Australian Football League

qafo quality-assurance field operation(s)

QAG Quaker Action Group

QAGA Queensland Amateur Gymnastic Association

qage quiet automatic gain control

Qahira *El Qahira* (Egyptian Arabic—Cairo)

QAI Qualified Acquisition Indebtedness; Quality Assurance Instruction; Queen's Award to Industry

QAICG Quality Assurance Interface Coordination Group

QAIMNS Queen Alexandra's Imperial Military Nursing Service

QAIP Quality Assurance Inspection Procedure

qak quick-attach kit

qal quartz aircraft lamp; quaternary alluvium

qal quintal (French—hundredweight)

QAL Quality Assurance Laboratory; Quarterly Accession List; Québec Airways Limited; Queensland Alumina Limited

QALAS Qualified Associate of the Land Agents' Society

QALD Quality-Assurance Liasion Division (DNA)

qall quartz aircraft landing lamp

QALTR Quality Assurance Laboratory Test Request

qaly quality-adjusted life year

qam quadrature amplitude modulation; queued access method

QAM Quality Assurance Manager; Quality Assurance Manual; Quality Assurance Monitor

QAM Quality Assurance Manual

QAMIS Quality Assurance Monitoring Information System

QAMS Quad-Phase Amplitude Modulation System; Quality Assurance of Medical Standards

Qan Qantas Airways

QANTAS Queensland And Northern Territories Aerial Services

qao quality assurance operation

QAO Quality Assurance Office (USN)

QAOC Quality Assurance Overview Contractor

QAOP Quality Assurance Operating Procedure

qap quinine, atebrin, plasmoquine (malaria treatment)

QAP Quality Assurance Planning; Quality Assurance Procedure(s); Quality Assurance Program

QA & P Quanah, Acme & Pacific (railroad)

QAPL Queensland Airlines Proprietary Limited

QAPP Quality Assurance Program Plan(ning)

QAPS Queensland Association of Personnel Services

qar quick-access recorder; quick-access recording

QAR Quality Assurance Representative

QAR Quality Assurance Report; Queen Anne's Revenge

QARAFNS Queen Alexandra's Royal Air Force Nursing Service

QARANC Queen Alexandra's Royal Army Nursing Service

QARNNS Queen Alexandra's Royal Naval Nursing Service

qas quick-acting scuttle

QA's Queen Alexandra's

QAS Quality Answering System; Quality Assurance Service; Quality Assurance System; Question Answering System

QASA Queensland Amateur Swimming Association

QASAR Quality Assurance Systems Analysis Review

QASP Quality Assurance Standard Practice

QAST Quality Assurance Service Test(s)

Qat Qatar

QAT Qualification Approval Test; Quantitative Assessment and Training Center

Qatar State of Qatar (oil-producing Persian Gulf country)

QATB Queensland Ambulance Transport Brigade

QATP Quality Assurance Technical Publication(s); Quality Assurance Test Procedure(s)

Qattara Qattara Depression in northern Egypt's Libyan Desert

qavc quiet automatic volume control

QAVT Qualification Acceptance Vibration Test

QAWA Queensland Amateur Wrestling Association

qax quacks

qb qualified bidders; qualified buyer; quarterback (football); quick break

QB Quantitative (electrophysiological) Battery; Queensboro Bridge (New York City); Quiet Birdmen (glider enthusiasts)

Q.B. Queen's Bench

QbA Qualitätswein eines bestimmten Anbaugelbietes (German—quality wine from a specific region and made with added sugar)

QBA Quebecair; Queensland Bowling Association

QBAA Quality Brands Associates of America

QBAC Quality Bakers of America Cooperative

Q-band 36,000–46,000 mc

Q-bar second output of a flip-flop

QBB Queensland Butter Board

Qbc Quebec

QBD Queen's Bench Division; Queensland Book Depot

qbe query by example

qbi quite bloody impossible

QBI Queen's Bureau of Investigation

QBL Qualified Bidder's List

Q-boats mystery ships used in antisubmarine warfare by the British in World War I

qbop quality basic-oxygen process

QBRs Queen's Bench Reports

qb's quarterbacks

QBSM que besa su mano (Spanish—who kisses your hand)—used in closing personal letters

QBSP que besa sus pies (Spanish—who kisses your feet)—used in closing personal letters

qc qualification course; quality control; quantitative command; quantum counter; quartz crystal; quick connect; quit claim

q/c quick change

qc qualcosa (Italian—something)

Qc impact pressure (symbol)

qc (QC) quorum call

QC Quadrantal Correction(s); Quality Control; Quartermaster Corps; Québec Central (railroad); Québec City; Queens College; Queen's College; Quezon City; Quincy College; Quinnisiac College; Quit Claim

Q.C. Queen's Counsel

QCA Queen Charlotte Airlines; Queensland Coal Associates; Queensland Cricket Association; Queensland Croquet Association

Q-cab quiet (tractor) cab

Q-card qualification card

qcb (QCB) queue control block (data processing)

QCB Quality Control Bulletin

QCBC Queen's Commendation for Brave Conduct

qcbm quick-connects bulkhead mounting

qcc qualification correlation certification; quick-connect coupling(s)

QCC Queensborough Community College; Queensland Conservation Council; Quinsigamond Community College

QCCA Queensland Cleaning Contractors Association

QCCARS Quality Control Collection Analysis and Reporting System

qcd quality-control data-quanturn chromodynamics; quit claim deed

QCD Quality Control Directive; quit claim deed

QCDI Quality Control Departmental Instruction

QCDR Quality Control Deficiency Report

QCE Quality Control Engineering

QCEU Queensland Colliery Employees Union

qcf quartz-crystal filter

QCF Quality Control Form

qcfo quartz-crystal frequency oscillator

QCGC Queensland Cane-Growers Council

qch quick-connect handle

QCH Queen Charlotte's Hospital

qci quality-control information

QCI Quality Conformance Inspection; Queensland Confederation of Industry; Quota Club International

Q Cic Quintus Tullius Cicero (the brother of the Roman orator Marcus Tullius Cicero)

QCIM Quarterly Cumulative Index Medicus

QC Isl Queen Charlotte Islands

Q City Quezon City, Philippines

qck quick-connect kit

qcl quality-control level

Q-class Soviet Québec-type submarines

Q-clearance Department of Energy's highest security classification; highest security clearance from the FBI

QCM Quality Control Manager; Queensland Coal Mining

QCM Quality Control Manual

QCMA Queensland Cooperative Milling Association

QCMP Queens' Council Member of Parliament

QCNIC Quad-Cities Nuclear Information Center

qco quartz-crystal oscillator

Q Co Queens County

QCO Quality Completion Order; Quality Control Officer

QCOP Quality Control Operating Procedure

Q country Qatar

QCP Quality Control Procedure; Queens College Press

QCPE Quantum Chemistry Program Exchange

Qc/Ps impact/static pressure ratio (symbol)

QCPSA Quaker Center for Prisoner Support Activities

qcr quick-change response

qcr (QCR) quality control/reliability

QCR Quality Control Representative

QC/R Quality Control/Reliability

QC & R Quality Control and Reliability

QCRC Québec Central Railway Company

QC Rep Quality-Control Representative

QC Rept Quality-Control Report

qcrt quick-change real time

QC Ry Québec Central Railway

QCS Quality Control Standard; Quality Control System; Quality Cost System

QCSM Qualified Confined Space Monitor

QCSO Quality Control Stop Order

QCSR Quaker Committee on Social Rehabilitation

QCSSO Queensland Council of State School Organizations

QC Stand Quality-Control Standard

qct quality control technician; quiescent carrier telephony; questionable corrective task

QCT Quality Control Technology; Quantitative Computed Tomography

QC & T Quality Control and Test

QCTR Quality Control Test Report

qcu quartz crystal unit; quick-change unit

qcus quartz crystal unit set

qcvc quick-connect valve coupler

qcw quadrant continuous wave

QCWA Quarter-Century Wireless Association; Queensland Country Women's Association

Qcy Quincy

QCYC Queen City Yacht Club; Queensland Cruising Yacht Club

qd quarterdeck; quartile deviation; questionnaire data; questioned document; quick delivery; quick detachable (weapon)

q-d quick-disconnect

q.d. Quasi dicat (Latin—as if he should say)

q & d quick and dirty

q.d. quater in die (Latin—four times a day)

QD Sadios Transportes Aéreos

qda quantity discount agreement

qdc quick dependable communication(s); quick-disconnect cap; quick-disconnect connector; quick-disconnect coupling

qdcc quick-disconnect circular connection

qdc's quick, dependable, communications

qdd qualified for deep diving; quantized decision detection

QDG Queen's Dragoon Guards

QD/GD Quincy Division/General Dynamics

qdh quick-disconnect handle

qdk quick-disconnect kit

qdn quick-disconnect nipple

qdo quadripartite development objective

q^{do} quando (Portuguese or Spanish—when)

Qd'O Quai d'Orsay

QDO Queensland Dairymens Organisation

qdp quick-disconnect pivot

QDR Quality Deficiency Report

QDRI Qualitative Development Requirements Information (program)

qdrnt quadrant
QDRO Qualified Domestic Relations Order
qds quick-disconnect series; quick-disconnect swivel
qd's questioned documents
QDS Quality Data System; Quantitative Decision System
qdta quantitative differential thermal analysis
qdv quick disconnect valve
qe quadrant elevation; quick estimate
qe (QE) quantum electronics
q.e. quod est (Latin—which is)
QE Quality Engineer(ing); Quality Evaluation; Quebec
QE2 Queen Elizabeth 2 (passenger vessel)
QEA Qantas Empire Airways
qeav quick-exhaust air valve
QEB Quantitative Electrophyiological Battery
qec quick engine change
QEC Queen Elizabeth College
QECC Queen Elizabeth Chemical Center
qecu quick engine-change unit
qed quantitative evaluative device; quantum electrodynamics; quick-reaction dome
q.e.d. quod erat demonstrandum (Latin—that which was to be proved)
QED Quality, Efficiency, Dependability (reliability program)
qee quadruple expansion engine
qeev quantum electrodynamics electron volts
q.e.f. quod erat faciendum (Latin—that which was to be done)
QEF Queensland Employers Federation
QEFD Queen Elizabeth's Foundation for the Disabled
QEH Queen Elizabeth Hall
q.e.i. quod erat inveniendum (Latin—that which was to be discovered)
qel quiet extended life
QEL Quality Evaluation Laboratory
qem quadrant electrometer
QEM Quality Education for Minorities; Qualified Export Manager
q.e.n. quare executionem non (Latin—wherefore execution should not be ordered)

QENP Queen Elizabeth National Park (Uganda)
qeo quality engineering operations
QEONS Queen Elizabeth's Overseas Nursing Service
QEOP Quartermaster Emergency Operation Plan
QEP Quality Evaluation Program; Quality Examination Program; Queen Elizabeth Park; Queen Elizabeth Planetarium; Queensland Environmental Program
qer qualitative equipment requirements
QER Quarterly Economic Review
qescp quality engineering significant control points
QES Queen's English Society
QESP Queen Emma Summer Palace
QEST Quality Evaluation System Test(s)
QESTS Query, Update Entry, Search, Time Sharing
QET Queen Elizabeth Theatre (Vancouver)
qev quick exhaust valve
QEW Queen Elizabeth Way (highway linking Buffalo with Toronto)
qf quality factor; quench frequency; quick freeze; quick frozen
QF quick-firing
qfa quality per final article
Q-factor quality rating
q-fastener(s) quick-fastener(s)
qfc quantitative flight characteristics
qfcc quantitative flight characteristics criteria
QFD Quality Function Development
qfe quartz fiber electrometer
Q-fellows quartermaster fellows; queer fellows
Q fever query fever (of uncertain cause); Balkan grippe or nine-mile fever (viral disease with pneumonial symptoms caused by rickettsia)
qff quadruple flip-flop
QFGA Queensland Farmers and Graziers Association
QFI Qualified Flight Instructor
qrirc quick-fix interference-reduction capability
qfi quasi-fermi level

qfm quantized frequency modulation
qfo quartz frequency oscillator
qfp quartz fiber product
QFP Quick-Fix Program
QFR Quarterly Force Revision (USN)
Q-fract quick fraction (membrane potentials)
QFRI Queensland Fisheries Research Institute
QFS Queensland Fire Service; Queensland Fisheries Service
QFSM Queen's Fire Services Medal
qft quantized field theory
qg quadrature grid
QG Quartermaster General
QG Quartier Général (French —Headquarters; *Quartier Generale* (Italian—Headquarters)
qgb searchlight sonar (symbol)
QGGA Queensland Grain Growers Association
qgm quarter-girth measure
QGM Queen's Gallantry Medal
qgp quark gluon plasma
QGPO Qatar General Petroleum Organization
QGTB Queensland Government Tourist Bureau
qgv quantized gate video
qh quartz helix
q-h quartz-halogen (lights)
Q-horse Quarter Horse
q.h. quaque hora (Latin—every hour)
QH Queen's Hall
QHC Queen's Honorary Chaplain; Queensland Housing Commission
QHDS Queen's Honorary Dental Surgeon
QHM Queen's Harbour Master
QHNS Queen's Honorary Nursing Sister
QHO Queen's Hall Orchestra
QHP Queen's Honorary Physician
QHS Queen's Honorary Surgeon; Queensland Historical Society
QHV Queen's Honorary Veterinarian
qi quality improvement; quality indices; quarterly index
QI Queensland Insurance; Quota International

QI Quality Index; Quarterly Index
QIA Queensland Institute of Architects
qiam queued indexed access memory
qic quality inspection criteria; quartz-iodine crystal
QIC Quality Information Center
q.i.d. quater in die (Latin—four times a day)
Qid. Qiddushin
QIDN Queen's Institute of District Nursing
qie quantitative immuno-electrophoresis
QIE Qualified International Executive
QIER Queensland Institute for Educational Research
QIH Quality International Hotels
qil quartz incandescent lamp; quartz iodine lamp
QIMR Queensland Institute of Medical Research
Qin. Qinnim
qip quartz insulation part
QIP Quality Inspection Point
Q.I.P. Quiescat in Pace (Latin—Rest in Peace)
QIPA Queensland Institute of Public Affairs
QIPS Qualitative Incentive Procurement Service
QIR Quechan Indian Reservation (originally Fort Yuma)
qisam queued-indexed sequential-access method
qit qualification information and test (system)
QIT Queensland Institute of Technology
QITS Quality Information and Test System
QJC Quincy Junior College
QJSA Qualified Joint and Survivor Annuity
QJSA Quarterly Journal of Studies in Alcohol
qjump queue(d) jump(ing)
qk quick
Qk Fl quick flashing (light)
qkly quickly
qkm Quadratkilometer (German—square kilometers)
ql quarrel; query language; quick look; quintal
ql (QL) quantum leap
ql quilate (Portuguese—carat)
q.l. quantum libet (Latin—as much as you like)

QL Queen's Lancers; Queensland (airline code)
Q/L Quarantine Launch
Q'land Queensland
QLAP Quick Look Analysis Program
QLCS Quick Look and Checkout System
Qld Queensland
qlfy qualify
qlfyg qualifying
qlfyn qualification
QLGA Queensland Local Government Association
qli quality of life index
qlii quasi-laser-intensity interferometer
qlit quick-look intermediate tape
qll quartz landing lamp
qlm quasi-laser machine
QLOC Queensland Light Opera Company
QLPC Queensland Library Promotion Council
QLR Queen's Lancashire Regiment
QLR Québec Law Reports
QLS Queensland Law Society; Queensland Littoral Society; Quick Law Systems; Quick Loading System
qlsm quasi-laser sequential machine
qlt quantitative leak test
QLTA Queensland Lawn Tennis Association
qlty quality
qm quadrature modulation; quality management
qm (QM) quantum mechanics; query message
qm Quadratmeter (German—square meter); *quintal métrico* (Spanish—metric quintal, 220 pounds)
q.m. quaque mane (Latin—every morning); *quo modo* (Latin—in what manner)
QM Decca navigation system; Quartermaster; Queen's Messenger; Queens Museum
qma qualified military available; quality material approach
QMA Quartermasters Association; Qatar Monetary Agency
QMAAC Queen Mary's Army Auxiliary Corps
QMAC Quadripartite Material and Agreements Committee
qmao qualified for mobilization ashore only

Q-max quarantine maximum
qmb quick make-and-break
QMB Qualified Medicare Beneficiary
QMBA Queensland Master Builders Association
QMC Quartermaster Corps; Queen Mary's College (London)
QMC & SO Quartermaster Cataloging and Standardization Office
QMDEP Quartermaster Depot
qmdk quick mechanical disconnect kit
QMDO Qualitative Matériel Development Objective
QMDPC Quartermaster Data Processing Center
qme queueing matrix evaluation
QME Quantock Marine Enterprises
Q Met Quaestiones Metaphysica (Latin—Metaphysical Questions)
QMEPCC Quartermaster Equipment and Parts Commodity Center
QMFCI Quartermaster Food and Container Institute
QMFCIAF Quartermaster Food and Container Institute for the Armed Forces
QMG Quartermaster General
QMGF Quartermaster-General to the Forces
QMGMC Quartermaster General—Marine Corps
QMH Queen Mary Hospital
QMI Qualification Maintainability Inspection
QMIA Queensland Motor Industry Association
QMIMSO Quartermaster Industrial Mobilization Services Offices
qmo qualitative material objective
QMORC Quartermaster Officers Reserve Corps
QMP Qualitätswein mit Prädikat (German—quality wine with special attributes and no added sugar)
QMP Quezon Memorial Park (Philippines)
QMPA Quartermaster Purchasing Agency; Queensland Master Painters Association
QMPCUSA Quartermaster Petroleum Center U.S. Army

qmqb quick-make quick-break (connection)

qmr qualitative materiel requirement

Qmr Quartermaster

QMRC Quartermaster Reserve Corps

QMR & E Quartermaster Research and Engineering

QMRL Quartermaster Radiation Laboratory

QMs Quarterly Meetings (Quakers); quartermasters

QMS Quartermaster School (U.S. Army)

Qm Sgt Quartermaster Sergeant

QMSO Quartermaster Supply Office(r)

qmsw quartz metal sealed window

QMT Queens-Midtown Tunnel

QMTOE Quartermaster Table of Organization and Equipment

qmw quartz metal window

qn question; quotation

q.n. quaque nocte (Latin—every night); *quid nune* (Latin—what now?)—person eternally interested in getting the latest news

Qn Queen

qna quality per next assembly

QNI Queen's Nursing Institute

QNP Quezon National Park (Philippines)

qns quantity not sufficient

Qns Queens

QNS Queen's Nursing Sister

Qns Coll Queen's College

Qnsd Queensland

Qnsk Quensk (language of the Quains)

QNS & L Québec North Shore and Labrador Railway

Qnsld Queensland

Qns Pk Queens Park

qnt quantisizer; quintet

QNTM Quantum Corporation

qnty quantity

QNWR Quivira National Wildlife Refuge (Kansas)

qo quick opening; quick outlet

QO Quaker Oats; Qualified in Ordnance; Quartermaster Operation; Queen's Own (regiment)

Q & O Québec and Ontario (transportation company)

qO₂ oxygen quotient

QO₂ oxygen consumption (or quota)

QOA Quasi-Official Agencies

QOCH Queen's Own Cameron Highlanders

qod quick-opening device

QOD Québec Order of Dentists

QOF Quaker Oats Foundation

QOH Queen's Own Hussars

QOIC Quarantine Officer in Charge

QOMY Queen's Own Mercian Yeomanry

qon quarter ocean net

qopri qualitative operational requirement(s)

qor qualitative operational requirement

Qor Qoran (Koran)

QOR Queen's Own Royal (regiment)

QORC Queen's Own Rifles of Canada

QOS Quick On System

qot quote

qotn quotation

qp quality paperback; queen post; quick process(ing)

q.p. quantum placet (Latin—at discretion)

q-P quanti-Pirquet (reaction)

QP Qualification Proposal; Quarterly Position (electronic form); Queen's Printer

qpa qualitative point average; quantity per article, quantity per assembly

QPA Queensland Police Academy; Queensland Polynesian Association

QPB Quality Paperback (book club)

QPC Qatar Petroleum Company

qpd quid pro quo (Latin—something for something)

qpei quality per end item

qpf quantitative precipitation forecast

QPF Québec Police Force

QPFC Queen's Park Football Club

QPFL Queensland Professional Fishermens League

qpi quadratic performance index

QPIS Quality Performance Instruction Sheet

QPL Qualified Parts List; Qualified product(s) List(ing); Queens Public Library

qplt quiet propulsion lift technology

QPM Queen's Polar Medal; Queen's Police Medal

QPMA Quality and Productivity Management Association

Q-pon(s) coupon(s)

QPP Québec Provincial Police; Quetico Provincial Park (Ontario)

QPR Quality Progress Report; Quantity Progress Report; Quarterly Progress Report; Queen's Park Rangers

QPRI Qualitative Personnel Requirements Information

QPRT Qualified Personal Residence Trust

qps quantitative physical science

QPS Quasi-Permanent Storage; Quick Program Search

Q P & S Quaker Peace and Service

QPSA Qualified Preretirement Survivor Annuity

qpsk quad-phase shift key

qq quartos; questionable questionnaires; questions

qq quelques (French—some); *quintales* (Spanish—quintals)

qq. quaque (Latin—each); *quoque* (Latin—every)

QQ Celestial Equator; Qara Qash in Sinkiang province of China; Qara Qum, also in Sinkiang province of China, but sometimes spelled Kara Kum; Que Que, Rhodesia

q.q.d. quantum quatra die (Latin—every fourth day)

qqf quelquefois (French—sometimes)

q.q.h. quantum quatra hora (Latin—every four hours)

qq. hor. quaque hora (Latin—every hour)

qqma quality qualified military availability

qqpr quantitative and qualitative personnel requirements

q.q.v. quae vide (Latin—which see)

q/qy question/query

qr qualifications record, quarter(ly); questionnaire; quick reaction; quick receipt; quire

qr (QR) quick response

qr. quadrans (Latin—farthing)

q.r. quantum rectus (Latin—quantity is correct)

QR Queensland Railways; Quintana Roo; Quotation Request

Q & R Quality and Reliability

QR Quarterly Review

qra quality reliability assurance; quick reaction alert

QRA Queensland Rifle Association

qrbm quasi-random band model

qrc quick reaction capability

QRC Queensland Rubber Company

qrcg quasi-random code generator

QRCUP Québec Region Canathan University Press (now CUPBEQ)

QRDC Quartermaster Research and Development Command

QRDEA Quartermaster Research and Development Evaluation Agency

QRDS Quarterly Review of Drilling Statistics

qrg quick response graphic

qrga quadrupole residual gas analyzer

qri qualitative requirements information

qric quick reaction installation capability

QRICC Quick Reaction Inventory Control Center

QRIH Queen's Royal Irish Hussars

QRL Quadripartite Research List; Queensland Research League

qrly quarterly

QRMF Quick-Reacting Mobile Force

Qrmr Quartermaster

qro quick reaction operation

Qro Queretaro

QRO Quality Review Organization; Quick Reaction Operation; Quick Reaction Organization

Q Roo Quintana Roo

Q-room cue room (billiard room)

qrp qualified retirement plan

QRPA Quartermaster Radiation Planning Agency

QRPBI Qualified Real Property Business Indebtedness

QRPS Quick Reaction Procurement System

QRR Queen's Royal Rifles

QRRR extreme emergency amateur radio call signal

QRRs Qualitative Research Requirements (for nuclear weapons effects information)

qrs quarters

QR's Quality Reports

Q.R.S. trademark of Q.R.S. Music Rolls

qrt quarter

QRT Quick Reaction Team

qrtg quartering

qrtly quarterly

qrtmstr quartermaster

qrv quick-release valve

QRV Qualified Real-estate Valuer

QRX Queensland Railfast Express

qry quality and reliability year

QRZ Quaddel Reaktion Zeit (German—lump reaction time, rash reaction time, wheal reaction time)

qs quarter section; quarter sessions

qs (QS) quadraphonic stereo; quiet sleep

q.s. quantum satis (Latin—as much as is sufficient); *quantum sufficit* (Latin—as much as suffices)

Qs Conquistadores; Conquistadors; quartzes; questions

QS Quarantine Station; Quarter Section; Quarter Sessions; Quartermaster Sergeant; Quarternote Society; Queen's Scarf; Queen's Scholar; Queensland Society; Queueing System

QS Quecksilbersäule (German—mercury column)

QSA Queensland Shopkeepers Association

QSAL Quadripartite Standardization Agreements List

qsam queued sequential access method

qsbg quasi-stellar blue galaxies

qsbo quasi-stellar blue objects

QSC Québec Securities Commission

QSD Quality Surveillance Division (USN); Quincy Shipbuilding Division—General Dynamics

qse qualified scientists and engineers

qsf quasi-static field; quasi-stationary front

QSF Queensland Soccer Federation

qsg quasi-stellar galaxy

Q-ship disguised man-of-war used to decoy enemy vessels

qsi quality salary increase

qsic quality standard inspection criteria

Q-size Queen-size

QSJM Queen's Silver Jubilee Medal

qs & l quarters, subsistence, and laundry

QSL Queensland State Library

Q & SL Qualifications and Standards Laboratory

qsm quadruple-screw motorship; quarter-square multipliers

qsm (QSM) Queen's Service Medal

QSMO Quaker State Motor Oils

qso quasibiennial stratospheric oscillation; quasistellar object

QSO Québec Symphony Orchestra; Queen's Service Order; Queensland Symphony Orchestra

QSOP Quadripartite Standing Operating Procedure(s)

qsp quality search procedure

QSPP Québec Society for the Protection of Plants

qsr quick-strike reconnaissance

QSR Quarterly Status Report; Quarterly Summary Report

QSR Quartier de Securité Reinforcée (French—Maximum Security Prison)

qsra quiet short-haul research aircraft

QSRIG Quantity Surveyors Research and Information Group

qsrs quasi-stellar radio sources

qss quasi-stellar source

QSS quadruple-screw ship; Quota Sample Survey

qssa quasi-stationary-state approximation

QSSCT Queensland Society of Sugar Cane Technologists

qssp quasi-solid-state panel

QSSR Quarterly Stock Status Report

QST Québec Standard Test

QSTAG Quadripartite Standardization Agreement

Q-star quiet observation aircraft

qstn question

qstnr questionnaire

qstol quiet-and-short takeoff and landing

qstp qualified state tuition plan; qualified state tuition program

qsts quadruple-screw turbine steamship

q. suff. quantum sufficit (Latin—as much as needed, as much as will suffice)

Q-switch quantum switch

qsy quiet sun year

qt quality test(ing); quantity; quarry tile; quart; quarter; questioned trade; quick test; quiet (*see* q.t.)

qt (QT) quality test(ing); queuing theory

q.t. quiet (as "on the q.t.")

q & t quenched and tempered

QT Qualification Test(ing); Quick's Test (pregnancy or prothrombin)

qta quadrant transformer assembly

q^*ta* *quanta* (Portuguese or Spanish—how much)

QTAC Queensland Tertiary Admissions Centre

qtam queued telecommunication access method

qtaux *quintaux* (French—quintals)

qtb quarry-tile base

QTB Queensland Timber Board; Queensland Trotting Board

QTC Québec Teaching Congress; Queensland Turf Club

qtd quartered

QTDGs Quaker Theological Discussion Groups

qte quote

qted quick text editor; quoted

Q-Test(ing) Quality Test(ing)

qtf quarry-tile floor

QTF Québec Teachers' Federation

qtg quoting

QTIB Québec Tourist Information Bureau

QTIP Qualified Terminable Interest Property (trust)

QTLC Queensland Trades and Labor Council

qtly quarterly

qtm quality team movement

QTM Quechon Tribal Museum (Yuma, Arizona)

qtn quotation

qto quarto

q^*to* *quanto* (Portuguese or Spanish—how much)—masculine form

qtol quiet takeoff and landing

Q'town Queenstown

qtp quantum theory of paramagnetism

QTP Qualification Test Procedure

qtr quarry-tile roof; quarter; quarterly

QTR Quality Technical Report; Quality Technical Requirement; Quarterly Technical Report

qtrs quarters

qts quarts; quick turn stock

QTTC Queensland Tourist and Travel Corporation

qtte quartette

QTTP Q-Tags Test of Personality

qtt(s) quartette(s)

QTU Queensland Teachers Union

qty quantity

qtydesreq quantity desired or requested

qtz quartz

qtze quartzose

qtzic quartzitic

qtzt quartzite

qu quart; quarter; quarterly; query; question

qu. quasi (Latin—as it were, like)

Qu Queen

QU Queen's College (Cambridge, Oxford); Queen's University

qua quadrate; quadratus

quaal quaalude

quaalude trade name of methaqualone (hypnotic and sedative drug)

quack quacksalver (person pretending to be a doctor)

quacks quacksalvers (sixteenth-century doctors who used quicksilver or mercury in treating syphilis

Quacks CWACs [City-Wide Anti-Crime (units of the New York City Police Department)]

quack-u-p's quack acupuncturists

quackupunc quackupuncture; quackupuncturist

quad quaalude; quadrangle; quadragular; quadrant; quadraphonic; quadrat; quadruplets; quadruplex; quadruplicate(s); quadruplication

quad (Latin prefix—fourfold)—as in quadrille or quadriplegic

Qu-AD Quality-Assurance Department; Quality-Assurance Division

quad .50's quadruple .50-caliber machine guns

quad c quadrip cane

Quad Cities adjacent and across-the-river cities of Rock Island, East Moline, and Moline, Illinois, plus Davenport, Iowa

quadplex quadriplex

quadradar four-way radar (surveillance)

Quadrangle Quadrangle/The New York Times Book Company

quadrap quadraphonic(al)(1y)

quadrip quadriplegia

quadrivium the four liberal arts—arithmetic, astronomy, geography, and music

quadro quadroon

quadrup quadruped(s); quadruple

quadrupl. *quadruplicato* (Latin—four times as much)

quads quadraphonic records; quadruplets

QUADS Quality-Assurance Data System

quag quagmire

Quaker Quaker Oats; Quaker Press

Quaker MMs Quaker Monthly Meetings

Quakers Society of Friends

quake(s) earthquake(s)

'quake(s) earthquake(s)

qual qualification; qualify; quality

qual anal. qualitative analysis

quals qualifying examinations; qualifying tests

qual(s) qualification(s)

quam quadrature-amplitude modulation

Quandary Quandary Peak in central Colorado

quango (QUANGO) quasi-autonomous non-governmental organization

quant quantitative analyst; quantity; quantum

quant anal. quantitative analysis

Quantico Quantico, Virginia's FBI Academy and U.S. Marine Base

quantras question analysis transformation and search (data processing technique)

quant. suff. quantum sufficit (Latin—sufficient quantity)

quaops quarantine operations

QUAP Questionnaire Analysis Program

QUAPs Quality Assurance Publications

Quaq *Quaquero* (Spanish—Quaker)

quar quarantine; quarter

quar. pars quarta pars (Latin—one-fourth part)

quarpel quartermaster water-repellent (cloth or clothing)

quarr quarries; quarry; quarrying

quart quarter gallon; quarterly

quart. quartet; quartette; quartile

Quart Quarterly

QUART Quality Assurance and Reliability Team

Quart Ital Quartetto Italiano (Italian quartet)

quartz crystalline silica (SiO_2)

quartzite granular quartz rock

quasar quasi-stellar radio (object)

quaser quantum-amplification-by-stimulated-emission-of-radiation (acronym covering irasers, lasers, and masers varying only in operational frequency)

Quash Quashey; Quashley

quat quaternary; quaternary era

quat. quattuor (Latin—four)

Quat Quaternary

Quathlamba Quathlamba Mountains of Lesotho and South Africa where it is called Drakensberg

Quattrocento (Italian—four hundred) 15th century artistic and cultural development in Italy

Quayle Dan Quayle, 44th Vice President of the United States

QUB Queen's University of Belfast

QUD Queen's University of Dublin

Que Québec (inhabitants—Québeçois); Quechua; Quechuan

Que Quênia (Portuguese—Kenya)

QUE Quebecair

Québec letter Q radio code

Queen Charlottes Queen Charlotte Islands off British Columbia

Queen Elizabeths Queen Elizabeth Islands in the Canadian Arctic

Queensboro' Queensborough

Queen's Eng Queen's English

Queensl Queensland

Quen Quentin

Quent San Quentin (California State Prison)

Quer Querétaro

ques question

quest. quality electrical system test; questioned

QUEST Quality Electrical Systems Test; Queens Educational and Social Team

questal quiet, experimental, short-takeoff-and-landing (program of NASA)

questar quantitative utility evaluation suggesting targets for the allocations of resources

quester quick and efficient system to enhance retrieval

questn questionnaire

quetzal monetary unit of Guatemala

qufyd qualified

QUGA Queensland United Graziers Association

QUI Queen's University of Ireland; Quincy (railroad)

QUIC Question and Information Connection (telephone reference department of the St Louis Public Library)

Quich Quichua

quicha quantitative inhalation challenge apparatus

QUICK Queens University Interpretative Code

Quicklock Compaq Computer Corporation trademark

QUICKTRAN Quick Fortran (programming language)

quico quality improvement through cost optimization

QUIDS Quick Interactive Documentation System

quiktran quick fortran (programming language)

Quilmas San Quilmas (Mexican-Americanism—San Antonio, Texas)

quim *química* (Portuguese or Spanish—chemistry)

Quimigal Química de Portugal

quin quintet; quintette; quintuplet; quintuplicate; quintuplication

Quin Quincy, Quinten; Quintilianus; Quintilius; Quintillian; Quintin; Quintino; Quintius; Quintus

quinq quinque (Latin—five)

quins quintuplets

quint quintuplicate

quint. quintus (Latin—fifth)

Quint. Quintilian—Roman critic and rhetorician Marcus Fabius Quintilianus

Quintilian Marcus Fabius Quintilianus

quint(s) quintet(s); quintuplet(s); quintuplicate(s)

quintupl quintuplicate

quip. query interactive processor; questionnaire interpreter program

QUIPS Quarterly Income Preferred Securities

Qui Rev Quiet Revolution (Korea)

quis quisling (term for traitor derived from Vidkun Quisling who during World War II headed Norway's puppet government set up by the German invaders)

Quitmans Quitman Mountains of west Texas

quix quixote; quixotic(al)(1y); quixotism; quixotry

Quix Quixote

QUJ true course to station

QUL Queen's University Library

qume cue me

Q-unit one quintillion (1 × 10^{18})—equal to 38.46 billion tons of coal or 172.4 billion tons of oil or 968.9 trillion cubic feet of natural gas

QUNO Quaker United Nations Office

quo' quoth

quod. quodlibet (Latin—as you please)

Quoddy Passamaquoddy Bay between Maine and New Brunswick; Passamaquoddy Indians

Quoins Gunners Quoin and Quoin Channel north of Mauritius in the Indian Ocean; other Quoins in Australia, Burma, and South Africa

quok(s) quokka(s)

Quon Pt Quonset Point, Rhode Island

quonset quonset hut (originally built during World War II at Quonset, Rhode Island)

quor quorum

quor. quorum (Latin—of which)

quot quotation

quot. quotidie (Latin—daily)

quotes quotation marks

quote(s) quotations(s)

quote-unquote quotation marks (slang—phrase or word in quotation marks)

quotid. *quotidie* (Latin—every day)

qup quantity per unit pack

Qur *Quran* (Malay—Koran)

QUSA "Q" Airways

qv quality verification

q-v q-value

q.v. quantum vis (Latin—as much as is desired); *quod vide* (Latin—which see)

QVM Queen Victoria Museum (Launceston, Tasmania)

QVM *Que Viva Mexico!* (Spanish—Long Live Mexico!)

QVR Queen Victoria's Rifles

QVS Quality Verification Surveillance

qvt quality verification test

qw quarter wave

qwa quarter-wave antenna

qwd quarterly world day

q-wedge quartz wedge

qwerty standard typewriter keyboard

QWG Quadripartite Working Group

QWGCD Quadripartite Working Group for Combat Development (American, Australian, British, and Canadian armies)

qwl quality of working life; quick weight loss

QWMP Quadruped Walking Machine Program (US Army)

qwot quarter-wave optical thickness

qwp quarter-wave plate

qx *quintaux* (French—hundred-weights)

qy quantum yield, query

Qy Quay

QYC Quincy Yacht Club

QYO Queensland Youth Orchestra

qz quartz

Qz quartz

QZ Queen Zulu; Zambia Airways (2-letter coding)

qzabs qualified zone academy bonds

QZS Québec Zoological Society

R

r angle of reflection (symbol); declared or paid in preceding 12 months plus stock dividend (newspaper stock listings); position vector (symbol); racemic; racket(eer); radius; rain; range; rare; rate of interest; received; recipe; reconnaissance; recto; red; redemption charge may apply (newspaper stock listings); redetermination; refraction; registered; rejected; relative; relative humidity; reluctance; rental; reopen; repeat; reply; report; reprint; research; reserve; resident; resistance; response; restricted; retard; retarded; rex; right-handed pitcher (baseball); right or starboard side; ring; ringer; riparian; riser; road; robotics; rocky; rod; rook; rough; rubbed; rule; rules; runs; rupee (Indian monetary unit); solubilizing agent (symbol); symbol for reluctance, resistance, or resistor; thermal resistance

r (R) restricted (movie rating—children under 17 must be accompanied by parent or guardian); retrograde

r angular yaw velocity (symbol); front of the sheet (recto); *remotum* (Latin—far, remote)

R acoustic resistance (symbol); annual rent; electrical resistance; gas constant; ohmic resistance; product moment coefficient of statis-tical correlation; supersonic aircraft (symbol); Rabbi; radioactive range; radiolocation; Rankine; rare; ratio; Réaumur; received solid; reconnaissance; Rector; Regina (Queen); registered; Reiz; report(s); Representative; reprint; Republic; Republican, research; reserve; resistance; respiration; restricted; Rex (King); rial (Iranian monetary unit); Richfield Oil; right; ring; river; road; Road; Robin Line; rocket; Rocketdyne Division of North American Aviation; Roentgen; Roger—radio slang meaning all right or okay; Roma; Roman; Rome; Romeo—code for letter R; Rookie (baseball); Rotary International; Royal; ruble (Russian monetary unit); rupee (Indian monetary unit); Rwanda; Rydberg; supersonic aircraft (symbol); US Rubber Company

R (R) economic rent (microeconomics)

R. rand (South African monetary unit)

Rø no residual tumor (symbol)

R′ radius of circle in minutes of arc

R″ radius of circle in seconds of arc

R+ Rookie (Advanced) (baseball)

-R Rinne's hearing test negative

+R Rinne's hearing test positive

R *rechts* (German—right); *Reka* (Bulgarian, Czechoslovakian, Russian, Serbo-Croatian—river); resultant force (symbol); *rett* (Danish—right); *Ría* (Portuguese, Spanish—river mouth); *Rio* (Portuguese—river); *Rió* (Spanish—river); *Rivière* (French—river); *rogue*, branded on British convicts transported overseas in the early 1800s); *Romanus* or *Rufus* (on Latin inscriptions); *rua* (Portuguese—street); *rubeus* (Latin—red); *Rud* (Persian—river); *rue* (French—street); *Rzeka* (Polish—river); on the flag of Rwanda it stands for Rwanda, a Republic born of Revolution and confirmed by Referendum; The Book of Ruth

R₁ primary roots

R₂ secondary roots

R1 microscopic residual tumor (symbol)

R2 macroscopic residual tumor (symbol)

R-4 Recovery and Reuse of Refuse Resources (USN)

ra radio; radioactive; radioactivity; reduced area; right angle; right angulation; right ascension; right atrium; right auricle; robbery committed while armed (RA); *robustus archistratalis;* rubber-activated; ruling action

ra (RA) retrograde amnesia; rheumatoid arthritis

r/a radioactive; return to author
r & a right and above
RA radium; Range; Argentina (auto plaque); Coast Radar Station (symbol); high-powered radio range (Adcock symbol); Rabbinical Assembly; Rdeca Armada (Yugoslav—Red Army); Rear Admiral; Reduction of Area; Regular Army; Rehabilitation Act; Remington Arms; Rental Agreement; Republic Aviation; Resettlement Administration; Resident Adviser; Resident Auditor; Right Arch; right ascension; Rotogravure Association; Royal Academician; Royal Academy; Royal Arcanum; Royal Artillery; Royal Nepal (2-letter airline code)
RA(A) Rear Admiral (Aircraft Carriers)
RA(D) Rear Admiral (Destroyers)
R.A. right ascension
R/A Redstone Arsenal
R & A Research and Analysis; Royal and Ancient Golf Club at St Andrews, Fife, Scotland
RA República Argentina (Spanish—Argentine Republic)
r.a.a. reductio ad absurdum (Latin—reduction to an absurdity)—in mathematics sometimes appears as raa or RAA
RAA Rabbinical Alliance of America; Regional Airline Association; Registered Accountants Act; Rockport (Mass.) Art Association; Royal Academic Association; Royal Academy of Arts; Royal Australian Artillery
RAAA Red Angus Association of America; Relocation Assistance Association of America
RAAC Regional Affirmative Action Clearinghouse; Royal Australian Armoured Corps; Risk Adjustment Advisory Committee
RAADC Royal Australian Army Dental Corps
RAAEC Royal Australian Army Educational Corps
RAAF Royal Afghan Air Force; Royal Australian Air Force

RAAFMS Royal Australian Air Force Medical Service
RAAFNS Royal Australian Air Force Nursing Service
RAAFPO Royal Australian Air Force Post Office
RAAM Race Across America (bicycle)
RAAMC Royal Australian Army Medical Corps
RAAMS Remote Anti-Armor Mine System
RAANC Royal Australian Army Nursing Corps
RAANS Royal Australian Army Nursing Service
RAAOC Royal Australian Army Ordnance Corps
raap residue arithmetic-association processor
RAAPS Resource Allocation and Planning System
RAAS Royal Amateur Art Society
RAASC Royal Australian Army Service Corps
rab rabbet(ing)
Rab François Rabelais (c. 1483–1553); Rabat, Morocco; Rabaul, New Britain; Rabbi; Rabbinic Hebrew
RAB Radio Advertising Bureau
RAB Republik Arab Bersatu (Malay—United Arab Republic)
rabar Raytheon advanced battery acquisition radar
rabb rabbinate; rabbinic; rabbinical
rabbi rapid-access blood-blank information
Rabbit Ears short form for Rabbit Ears Mountain or Rabbit Ears Pass in northwestern Colorado
RABDF Royal Association of British Dairy Farmers
RABFM Research Association of British Flour Millers
RABI Royal Agricultural Benevolent Institution
RABPCVM Research Association of British Paint Colour and Varnish Manufacturers
rac racemic; radiometric area correlator; relative address coding; rhomboidal air controller; rectified alternating current
racr accomadage(s) [French—repair(s)]

Rac alternating current resistance (symbol)
RAC Railway Association of Canada; Rear Admiral Commanding; Recombinant (DNA) Advisory Committee; Reliability Action Center; Reliability Analysis Center; Rent-Adjustment Commission; Republic Aviation Corporation; Research Advisory Council; Research Analysis Corporation; Royal Air Cambodge; Royal Arch Chapter; Royal Armoured Corps; Royal Automobile Club; Rubber Allocation Committee; Rubber Association of Canada
RACA Recovered Alcoholic Clergy Association; Royal Automobile Club of Australia
RACAN Rubber Association of Canada (also RAC)
RACB Royal Automobile Club of Belgium
racc radiation and contamination control
racc raccomandata (Italian—registered letter)
RACCA Refrigeration and Air Conditioning Contractors Association
race random-access computer equipment; rapid automatic checkout equipment
RACE Railways of Australia Container Express; Research on Automatic Computation Electronics; Rescue, Activate, Contain, Extinguish
RACE Real Automóvil Club de España (Spanish—Royal Automobile Club of Spain)
racep random access and correlation for extended performance
races (RACES) radio amateur civil emergency service
racfire tactical fire-direction (system)
RACGP Royal Australian College of General Practitioners
Rachl Rocketdyne advance chemical laser
RACI Royal Australian Chemical Institute
RACIC Remote Area Conflict Information Center
racon radar beacon
RACP Royal Australasian College of Physicians

RACS Remote Access Computing System; Royal Australasian College of Surgeons

RACT Reasonably Available Control Technology; Royal Australian Corps of Transport

RACUK Royal Aero Club of the United Kingdom

RACV Royal Automobile Club of Victoria

rad radar; radian; radiation; radiation-absorbed data; radiation-absorbed dose; radiator; radical; radicalism; radio; radioactive; radius; radix; rapid-access disc; rapid application development; reflective array device; released from active duty; return to active duty; roentgen-administered dosage; roentgen-administered dose

rad (RAD) rapid access disc

rad. radix (Latin—root)

Rad Radnor; Radnorshire

RAD Relax And Discuss; Royal Academy of Dancing; Royal Albert Docks; Rural Area Development

RA(D) Rear Admiral (Destroyers)

rada radioactive; random-access discrete address

RADA Royal Academy of Dramatic Arts

radac rapid digital automatic computing

radal radio detection and location (system)

radalt radio altimeter

radan radar doppler automatic navigator

radant radome antenna

radar radio detection and ranging

RADAR Royal Association for Disability and Rehabilitation

RADARS Receivable Accounts Data-Entry and Retrieval System

RADAS Random Access Discrete Address System (battlefield communications system)

radat radar data transmission and ranging; radiosonde observation data

radata radar automatic data transmission assembly

RADATS Radar Data-Transmission System

RADC Rome Air Development Center; Royal Army Dental Corps

RADCC Rear Area Damage Control Center

rad-ch radical-changing

RADCM radar countermeasures and deception

RADCOLS Rome Air Development Center On-Line Simulator

radcon radar data converter

RADD Royal Association in Aid of the Deaf and Dumb

raddef radiological defense

raddol raddolcendo (Italian—growing, calmer)

radem (RADEM) random access data modulation

rad encl radiator enclosure

radep radar departure

radex radiation exclusion plot (actual or predicted fallout)

radfac radiating facility

radf(s) rapid-access data file(s)

radhaz radiation hazard(s)

radi radiological inspection

radiac radioactivity-detection-indication-and-computation

radial-ply radial-ply tire

Radiat Eff *Radiation Effects*

radic radical; radicle; radicotomy; radiculalgia; radicular; radiculectomy; radiculitis; radiculomeningomyelitis; radiculomyelopathy; radiculoneuritis; radiculopathy

RADIC Research and Development Information Center

radic-lib(s) radical-liberal(s); radical-liberationist(s)

radint radar intelligence

Radio 1 British disc jockey commentary and teenage pop music station

Radio 2 British family phone-in and pop programs station

Radio 3 British classical music programs station

Radio 4 British station featuring educational and informational programs on cooking, farming, and the theater

Radio City Radio City Book Store, Radio City Music Hall

radiog radiography

radiol radiology

Radio Sci *Radio Science*

radir random access document indexing and retrieval

radist radar distance indicator

RA Dks Royal Albert Docks

radl radiological

rad lab radiation laboratory

radlfo radiological fallout

Rad Lib Radio Liberty

radlib(s) radical liberal(s)

radlic radio link

RADLO Radiological Defense Officer

radlop radiological operations

radlsafe radiological safety

Rad Lux Radio Luxembourg

radlwar radiological warfare

R Adm Rear Admiral

RADMAPS Radiological Monitoring Assessment Prediction System

radmon radiological monitor(ing)

radn radiation

radnote ratio note

RADOC Regional Air Defense Operations Center

radome radar dome

radon daughter deadly microscopic radioactive uranium particles

radop radar operator

rad op radio operator

radose radiation dosimeter satellite

radot real-time automatic digital-optical tracker

RadPropCast radio propagation forecast

RADR Royal Association for Disability and Rehabilitation

rad rec radiator recess

RADRON Radar Squadron (USAF)

Radru. rapid-access data-retrieval unit

rad/s radians per second

Rad(s) Radical(s)

RADS Ryukyu Air Defense System

radsab radiator sabotage

radscat radiometer-scatterometer sensor

radsick radiation sickness

RadSo Radiological Survey Officer

radss radar alphanumeric-display subsystem

radsta radio station

radtel radar telescope

radtt radio teletypewriter

radu radar analysis and detection unit

radvs radar altimeter and doppler velocity sensor

radwar radiological warfare

rae (RAE) radio astronomy explorer

Rae Rachel; Raquelle

RAE Royal Aircraft Establishment; Royal Australian Engineers

RAE *Real Academia Español*a (Royal Spanish Academy)

R Ae C Royal Aero Club

RAEC Royal Army Educational Corps

RAEL *Real Academia Español*a *de la Lengu*a (Royal Spanish Academy of Language)

RAEME Royal Australian Electrical and Mechanical Engineers

RAeS Royal Aeronautical Society

raet range-azimuth-elevation-time

Raf Rafael; Rafe; Rafelz; Raffaele; Raffaello

RAF Red Army Faction (Baader–Meinhof terrorists); Regular Air Force; Royal Aircraft Factory; Royal Air Force

RAF *Rote Armee Fraktion* (German—Red Army Faction) terrorist group

RAFA Royal Air Force Association; Royal Australian Field Artillery

rafar radar-automated facsimile reproduction; radio-automated facsimile and reproduction

rafax radar facsimile transmission

RAFB Randolph Air Force Base

RAFBF Royal Air Force Benevolent Fund

RAFC Royal Air Force College

Rafe (Ralph)

RAFES Royal Air Force Educational Service

raff *raffiné* (French—exquisite, polished, refined)

Raffles Raffles Hotel; Raffles Institution (Singapore Institution and Library); Raffles Place; Sir Thomas Stamford Raffles (founder of Singapore)

RAFGSA Royal Air Force Gliding and Soaring Association

Raf₁ Rafael

RAFMS Royal Air Force Medical Services

RAFO Reserve of Air Force Officers

rafos long-range navigation system (sofar reversed)

RAFR Royal Air Force Regiment

RAFRO Royal Air Force Reserve of Officers

RAFS Royal Air Force Station

RAFSAA Royal Air Force Small Arms Association

RAFSC Royal Air Force Staff College

RAFSE Royal Air Force School of Education

raft. recom algebraic formula translation; recom algebraic formula translator

RAFT Regional Accounting and Finance Test

RAFTC Royal Air Force Technical College; Royal Air Force Transport Command

RAFVR Royal Air Force Volunteer Reserve

rag ragtime; ring airfoil grenade; runaway arresting gear

rag *ragioniere* (Italian—accountant)

RAG Red Army Group *(see B M B)*; River Assault Group; Royal and Ancient Game (of golf)

RAGA Royal Australian Garrison Artillery

RAGB Refractories Association of Great Britain

RAGC Royal and Ancient Golf Club (St Andrews, Scotland)

RAGE Radio Amplification of Gamma Emissions

rah hurrah

RAH Royal Albert Hall

RAHS Royal Australian Historical Society

rai radioactive interference; radioactive iodine; random access and inquiry

RAI Reading Association of Ireland (actually the International Reading Association); Royal Australian Infantry

RAI *Radiotelevision Italiana* (Italian—Italian Radio Television)—broadcasting system; *Réseau Aérien Interinsulaire* (Tahiti); Royal Albert Institution; Royal Anthropological Institute

RAIA Royal Australian Institute of Architects

RAIAD *Reverse Acronyms, Initialisms, and Abbreviations Dictionary*

RAIC Royal Architectural Institute of Canada

raidex raiders exercise

rail. railroad; railway

RAIL Religion In American Life

rails runway alignment indicator lights

railwayac railway + maniac (railway fan)

Railway Employees Union Brotherhood of Railway, Airline, and Steamship Clerks, Freight Handlers, Express, and Station Employees

Rainbow Bridge Rainbow Bridge National Monument (world's largest natural bridge located in southern Utah on the Colorado River close to the Arizona border)

RAIOMA Resource Assessment Investigation of the Mariana Archipelago

rair remote access/immediate response

rair (RAIR) ram-augmented interstellar rocket

RAIRS Recordak Automated Information Retrieval System

RAISE Rigorous Approaches to Industrial Software Engineering

RAI-TV *Radio Audizioni Italiane TV* (Italian—Italian Radio Audition TV)

raiu radioactive iodine uptake

Raj Rajasthan

Raj *Rajah* (Arabic—seventh month of the Mohammedan year); *Rajah* (Hindi—king, prince, ruler); Rajasthani (culture, language, or people); the period of British rule in India

RAJ Royal Association of Justices

ra k raised keel

RAK *Rikets Allmanna Katverk* (Swedish—Geographical Survey Office)

ral resorcyclic acid lactone

Ral Raleigh

RAL Refund Anticipation Loan; Resort Airlines; Royal Air Laos

Ralegh Sir Walter Raleigh (who spelled his name Ralegh)

RALES Randomized Aldactone Evaluation Study

Raliks Ralik Chain of Islands in the west-central Pacific, including Bikini, Eniwetok, Jaluit, Kwajalein, Rongerik

RALIP Resource and Land Information Program; Resources and Land Investigations Program

RALLA Regional Allied Long Lines Agency

rallo. rallentando (Italian—slower by degrees)

ralph reduction and acquisition of lunar pulse heights

RALPH Royal Association for the Longevity and Preservation of the Honeymooners

ralu register and arithmetic logic unit

ralv rat leukemia virus

ram radio attenuation measurement; random access memory; rapid area maintenance; right ascension of the meridian

ram (RAM) research and applications module; reverse annuity mortgage; rolling airframe missile

Ram Raman effect in spectrum analysis; Ramona; Ramsgate

RAM Reliability, Availability, Maintainability (program); Reverse Amortization Mortgage; Reverse Annuity Mortgage; Revolutionary Action Movement; Rodrigo A Muñoz, Royal Academy of Music; Royal Air Maroc; Royal Arch Masons; Royal Australian Mint

R.A.M. Richar Allen Morris (American artist)

RAMA Remote Access to Museum Archives; Retail Advertising and Marketing Association; Rome Air Materiel Area

ramac random access memory accounting

ramac (RAMAC) random access memory and accounting and control

Ramapos Ramapo Mountains of New Jersey and New York

Rama's Bridge also called Adam's Bridge; 18-mile chain of shoals between Coromandel Coast of India and Mannar Island off Ceylon; Hindus relate Rama built causeway across these shoals so his Indian army could invade Ceylon and rescue his wife Sita from the demon king Ravana; Moslems insist building this bridge was Adam's first task after his expulsion from paradise

ramb(s) rambler(s)

ramc rob all my comrades

RAMC Royal Army Medical College; Royal Army Medical Corps

ramd reliability, availability, maintainability, durability

RAMIS Rapid-Access Management Information System; Rapid-Automatic Malfunction-Isolation System

ramit rate-aided manually implemented tracking

ramont radiological monitoring

ramp rate-acceleration measuring pendulum

RAMP Radar Mapping of Panama; Radiation Airborne Measurement Program; Rating Maintenance Phase; Resource Allocation and Management Program; Reverse Annuity Mortgage Program

RAMPAC Realty and Mortgage Investors of the Pacific

rampallion ramp + rapscallion

RAMPC Raritan Arsenal Maintenance Publication Center

RAMPI Raw Material Price Index

ramps resources allocation and multiproject scheduling

RAMPS Resources Allocation and Multiproject Scheduling

rams. right ascension of mean sun

Rams Ramsgate

RAMS right ascension mean sun

RAMSA *Radio Aeronáutica Mexicana S.A.*

RAMSS Royal Alfred Merchant Seamen's Society

ramt rudder-angle master transmitter

ran (RAN) revenue anticipation note; reverse anticipation note

ran reconnaissance-attack navigator; request for authority to negotiate

Ran Rangoon

RAN Rainforest Action Network; Royal Australian Navy

rana rheumatoid arthritis nuclear antigen

Ranally Rand McNally

RANAS Royal Australian Naval Air Squadron

ranc radar attenuation, noise, and clutter

RANC Royal Australian Naval College

Rance Ransom(e)

RANCHO Rural Area Nonprofit Community Housing Organization incorporated

rancom random communication satellite

rand monetary unit of South Africa

Rand Rand McNally; Witwatersrand (Johannesburg)

randam random-access nondestructive advanced memory

RAND Corporation Research and Development Corporation

randid rapid alphanumeric digital indicating device

Random Random House

RANF Royal Australian Nursing Federation

'rang(s) boomerang(s)

Ranier Ranier Bancorporation (National Bank of Commerce of Seattle)

RANN Research Applied to National Needs

RANR Royal Australian Naval Reserve

RANRL Royal Australian Navy Research Laboratory

ran's revenue anticipation notes

RANSA Royal Australian Naval Sailing Association

RANSA *Rutas Aéreas Nacionales* (Spanish—National Airlines)

RANT Reentry Antenna Test(ing)

RANVR Royal Australian Naval Volunteer Reserve

RANZCP Royal Australian and New Zealand College of Psychiatrists

rao radio astronomical observatory

RAO Regional Administrative Office(r); Regional Airways Office(r); Rudolf A Oetker (steamship line)

RaOb radiosonde observation
RAOC Royal Army Ordnance Corps
raomp report of accrued obligations—military pay
raot rocker-arms oiling time
RAOU Royal Australasian Ornithologists' Union
rap. talking frankly about any topic; rapid; rapport; reactive atmosphere processing; rear area protection; relative accident probability; rupees, annas, pies (Indian currency)
rap regulatory accounting principles
rap (RAP) random access programming; random access projector
rap rapido (Spanish—rapid)—fast train
Rap H Rap Brown; Rapids
RAP Radical Alternatives to Prison; Radiological Assistance Plan (AEC); Rapid Assessment Program; Regimental Aid Post; Release Aid Plan; Royal Army Post
RAPC Royal Army Pay Corps
RAPCAP Radar Picket Combat Air Patrol
rapcoe random access programming and checkout equipment
rapcon radar approach control
RAPCs Regional Action Planning Commission
rapec rocket-assisted personnel ejection catapult
rape rep rape report
Raph Raphael
Raphael Raffaello Sanzio
RAPI Royal Australian Planning Institute
rapid random-access person-ruel information device; relative address programming implementation device; retrieval through automated publication and information digest(ing)
RAPID Register for the Ascertainment and Prevention of Inherited Diseases; Rocketdyne Automatic Processing of Integrated Data
RAPIDS Random-Access Personnel Information System
RAPM Russian Association of Proletarian Musicians
rapp rapport; rapporteur; rapprochement

RAPP Radical Alternatives to Prison Plan; Radiologists, Anesthesiologists, Pathologists, and Psychiatrists
rappelling rapidly lowering
rappi random-access plan-position indicator
RAPPORT Rapid-Alert Programmed-Power-Management of Radar Targets
rapr radar processor
RAPRA Rubber and Plastics Research Association
rap's rocket-assisted projectiles
RAPS Radar Automatic Plotting System; Risk Appraisal of Programs System
rap. & sup. rapport and support
raptap random access parallel tape
raptus rapid thorium-uranium-sodium (reactor)
rar radio acoustic ranging; rapid-access recording; right arm reclining
RAR Reliability Action Report; Revenue Agent's Report; Rhodesian African Rifles; Royal Australian Regiment(s)
rarad radar advisory
RARDE Royal Armament Research and Development Establishment
rare ram air rocket engine
RARE Rare Animal Relief Effort; Rehabilitation of Addicts by Relatives and Employers
rarep radar report
RARG Regulatory Analysis Review Group
RARO Regular Army Reserve of Officers
ras radome antenna structure; radula sinus; rapid audit summary; rectified air speed; requirements allocation sheet; rheumatoid arthritis serum
ras (RAS) reticular activating system
ras. rasurae (Latin—shavings)
RAs Resident Agencies; Resident Agents
RAS Report Audit Summary; Reticular Activating System; Royal Aeronautical Society; Royal Agricultural Society; Royal Asiatic Society; Royal Astronomical Society; Rubber Association of Singapore

RASA Railway and Airline Supervisors Association
RASAR Resource Allocation System for Agricultural Research
RASB Royal Asiatic Society of Bengal
RASC Royal Army Service Corps; Royal Astronomical Society of Canada
RASC/DC Rear Area Security and Damage Control
RASD Reference and Adult Services Division (American Library Association)
rase rapid automatic-sweep equipment
RASE Royal Agricultural Society of England
raser range and sensitivity extending resonator
rash. rain shower(s)
Rash Rashomon
RASK Royal Agricultural Society of Kenya
rasn rain and snow
RASNZ Royal Agricultural Society of New Zealand
RASP Reliability and Aging Surveillance Program (USAF)
RASPB Royal and Ancient Society of Polar Bears (Hammerfest, Norway's town-hall club)
RASS Radio Acoustic Sounding System; Rock Analysis Storage System; Royal-Alfred Seafarers' Society
RAST radio-allergo-sorbent test
rastac random access storage and control
rastad random access storage and display
Rastafians Rastafarians
RASTAS Radiating Site Target Acquisition System
rat ram air turbine; ratchet; rate; rated; rattan; rotational automatic tester; runs (scored) away by team
rat rating; ration(s); rocket-assisted torpedo (RAT)
rat (RAT) repeat-action tablet
RAT Remote Associates Test
ratac radar analog target acquisition computer
ratan radio television aid to navigation

RATAS Research and Technical Advisory Services (Lloyd's Register of Shipping)

ratc radar-aided tracking computer

RATCC Radar Air Traffic Control Center

RATCF Regional Air Traffic Control Facility

ratcon radar terminal control

rate remote automatic telemetry equipment

ratel radiotelephone

ratelo radio telephone operator

ratepayer(s) [Canadian English—taxpayer(s)]

rat/epr ram air temperature/engine pressure ratio

RATER Raytheon Acoustic Test and Evaluation Range

ratfor rational fortran

ratg radiotelegraph

ratn recursive augmented transition network

RATNET Rural Alaska Television Network

rato rocket-assisted takeoff

Ratons Raton Mountains of Colorado and New Mexico

RATP *Régie Autonome des Transports Parisiens* (Le métro—Paris subway system)

RATR *Reliability Abstracts and Technical Reviews*

rats. repeat-action tablets

Rats Rat Islands (Amchitka, Kiska, Rat, etc.)

RATS Ram Air Turbine Systems

ratscat radar target scatter site

RATSEC Robert A Taft Sanitary Engineering Center

ratt radioteletypewriter

RAU Rand Afrikaans University; River Assault Unit (USN)

RAU *Repubblica Araba Unita* (Italian—United Arab Republic)—Egypt

RAUS Retired Association for the Uniformed Services

'raus mit i'm *heraus mit ihm* (German—out with him)

R Aux AF Royal Auxiliary Air Force

Rav Roux-associated virus

RAVA Rochester Audiovisual Association

RAVC Royal Army Veterinary Corps

rave radar acquisition vocal-tracking equipment

rave (RAVE) research aircraft for visual environment (USA)

RAVE Register And Vote Easily

RAVEC Regional Adult and Vocational Education Council

raven. ranging and velocity navigation

RAVES Rapid Aerospace Vehicle Evaluation System

ravir radar video recorder; radar video recording

raw right attack wing (lacrosse)

RAW Reconnaissance Attack Wing (USN)

RAWA Renaissance Artists and Writers Association

Rawal Rawalpindi

rawarc radar and warning coordination

RAWI Radio American West Indies (Virgin Islands)

rawin radar wind sounding

raws radar altimeter warning set

RAWs Replenishment Agricultural Workers

rawx returned account of weather (aviation)

rax random access (computing system)

'ray hurray

Ray Rachel; Raymond

RAYCI Raytheon Controlled Inventory

raze range, azimuth, elevation

razon range and azimuth only

razz razzberry (slang for raspberry)

rb read backward; read buffer; relative heating; retinoblastoma; return to bias; right back (field hockey); rigid boat; road bend; rubberbase(d); running back; run rating for batters

r/b reentry body

r & b rhythm and blues; right and below; room and board

Rb base resistance (symbol); rubidium

RB Rancho Bernardo; reconnaissance bomber; Regiment Botha; Renegotiation Board; Republica Boliviana (Bolivian Republic); Republic of Burma; Rifle Brigade; *Ritzaus Bureau* (Danish news agency); Robert Burchfield, editor of the *Oxford English Dictionary* supplement; *Royaume de Belgique* (Kingdom of Belgium)

RB *Revue Biblique* (French—Biblical Review)

R.B. Robert Browning

R$_B$ Rockwell hardness (B-scale)

Rb-08 Saab surface-to-surface missile

rba relative batting average

RBA Rabat, Morocco (airport); Religious Booksellers Association; Reserve Bank of Australia; Retail Bakers of America; River Boards Association; Road Bitumen Association; Roads Beautifying Association; Roadside Business Association; Royal Brunei Airlines

RBAC Rule-Base Access Control

RBAF Royal Belgian Air Force

rbb room, board, and beverages

RBB Richard Bedford Bennett (Canada's fourteenth Prime Minister)

RBB *Reference Books Bulletin*

rbbb right bundle branch block

rbbsb right-bundle-branch system block

rbc red blood cell; red blood cell (count); red blood corpuscle

RBC Rhodesian Broadcasting Corporation; Richard Bland College; Roller Bearing Company; Royal Bank of Canada

RBCA Russian Book Chamber Abroad

rbcd right border of cardiac dullness

RBCM Royal British Columbia Museum (Victoria)

rbcs remote barcoding system

rbd rapid beam deflector; required beginning date; right border of dullness (heart response to percussion)

RBD Rittenhouse Book Distributors

rbde radar bright-display equipment

RBDS Radio Broadcast Data System

rbe relative biological effectiveness

RBEC Roller Bearing Engineering Committee

rbelet relative biological effectiveness linear energy transfer

R Bern Rancho Bernardo

rbf renal blood flow

RBF Rockefeller Brothers Fund

RBFC Rural Banking and Finance Corporation

RBG Royal Botanic Gardens (Kew Gardens)

RBGS Radio Beacon Guidance System

RBH Rutherford Birchard Hayes (19th President U.S.)

rbi reply by indorsement; request better information; runs batted in

rbi recibí (Spanish—I received)

RBI Reserve Bank of India; Rochester Business Institute

rb imp rubber-base impression

RBK Royal Borough of Kenington

rbl ruble

RBL Royal British Legion

RBLC Royal British Legion Club

R Bn radio beacon

RBN Registry of Business Names

RRNA Royal British Nurses' Association

RBNM Rainbow Bridge National Monument (Utah)

RBNSW Rural Bank of New South Wales

RBNZ Reserve Bank of New Zealand

rbo right back outside

RBO Russian Brotherhood Organization

rboc rapid bloom off-board chaff

RBOT Rotating Bomb Oxidation Test

rbox rail box car (rolling-stock pool)

rbp ration breakdown point

RBP Registered Business Processor

RBP *Raffinerie Belge de Petroles* (French—Belgian Petroleum Refinery)

RBPP Rotor-Burst Protection Program (NASA)

rbr risk-to-benefit ratio; rubber

rBr reddish brown

RBR Renegotiation Board Regulation

RBR *Reference Book Review*

RBRF Reproductive Biological Research Foundation

rbrvs resource-based relative value scale

rbs radar bomb score; radar bomb scoring; request blocks

Rbs Rutherford back-scatter(ing)

RBS Ranganthittoo Bird Sanctuary (India); Research for Better Schools; Royal Bank of Scotland; Royal Botanical Society

RBSA Royal Birmingham Society of Artists

rbsn (RBSN) reaction-bonded silicon nitride

rbt rabbet; rabbit; resistance bulb thermometer; roundabout

RBT Rational Behavior Therapy; Rose Bengal Test(ing)

rbtwt radial-beam travelling-wave tube

RBU Rabindra Bharati University

rbv return-beam videcon

RBZ Red Badge Zone

rc radio code; radio coding; radio controlled; rate of change; ready calendar; red cell; red corpuscle; regional controller; reinforced concrete; remote controlled; resin coat(ed); resin coating; resistance capacitance; resistor–capacitor; respiratory center; reverse course; right center; rigid center; rock-crushed; rubber-cushioned; runs created

r/c reconsign(ed); recredit(ed)

r & c rail and canal

r/c *rés-do-chão* (Portuguese—ground floor)

R_c Rockwell hardness (C-scale)

Rc conditioned response

RC Radcliffe College; Radio City; Radio Code; Reception Center; Reconstruction Commission; Red China; Red Cross; Regina College; Regional Commissioner; Regis College; Reinhardt College; Renison College; *República de Chile*; *República de Colombia*; *República de Cuba*; Ricker College; Ricks College; Rider College; *Río Colorado*; Ripon College; Rivier College; Roanoke College; Rockefeller Center; Rockford College; Rockhurst College; Rockmount College; Rollins College; Roman Catholic;

Rosary College; Rosemount College; Rosenwal College; Rust College

R, C Cauchy constant

R of C Republic of China

RC *República Centroafricana* (Spanish—Central African Republic)

R.C. *Rendiconti* (Italian—proceedings or reports)

rca replacement cost accounting

R^{ca} *Rocca* (Italian—rock; tower)

RCA Rabbinical Council of America; Radio Club of America; Radio Corporation of America; Radio Council of America; Radiologically Controlled Area; Reformed Church in America; Revenue Conciliation Act (of 1993); Rocket Cruising Association; Rodeo Cowboys Association; Roofing Contractors Association; Royal Canadian Academician; Royal Canadian Academy; Royal Canadian Artillery; Royal College of Art; Rug Corporation of America; Rural Crafts Association

RCAA Royal Cambrian Academy of Art; Royal Canadian Academy of Arts

RCAC Radio Corporation of America Communications

RCACS Readiness Command and Control System

RCAF Royal Canadian Air Force

RCAM Royal Canadian Artillery Museum

R Cam A Royal Cambrian Academy of Art

RCAMC Royal Canadian Army Medical Corps

R Can Rio Canario

RCAR Religious Coalition for Abortion Rights

RCA Rev *RCA Review*

RCAS Royal Central Asian Society,; Rutgers Center of Alcohol Studies

RCA Satcom RCA Domestic Communications Satellite

RCASC Royal Canadian Army Service Corps

rcat remote-controlled aerial target

RCAT Royal College of Arts and Technology

RCA Vic RCA Victor

RCB Ready-Crew Building; Regiment Christiaan Beyers; Retail(ers) Credit Bureau

RCBB Royal Commission on Bilingualism and Biculturalism (Canada)

rcbf (RCBF) regional cerebral blood flow

rcc read(er) channel continue(d); reader common contact; reception and care center; remote communications complex; rough combustion cut-off

r & cc riot and civil commotion

RCC Radio-Chemical Center; Radiological Control Center; Rag Chewers Club; Rape Crisis Center; Reply Coupon Collector(s); Rescue Control Center; Rescue Coordination Center; Rockland Community College; Roman Catholic Church; Royal Crown Cola

R & CC Ross and Cromarty Constabulary

RCCA Rickenbacker Car Club of America

rccb remote-controlled circuit breaker

RCCC Regular Common Carrier Conference; Republican County Central Committee

RCCE Regional Congress of Construction Employers

rC Ch Roman Catholic Church

RCCL Royal Caribbean Cruise Line

RCCLS Resource Center for Consumers of Legal Services

RCCP *Royal Commission on Criminal Procedure*

rccs revenue consequences of capital schemes; riots, civil commotions, and strikes

rcd received; relative cardiac dullness

rcd (RCD) record(ing)

RCD Regional Cooperation for Development (Pakistan, Iran, Turkey)

RCDA Retail Coin Dealers Association

RCDC Royal Canadian Dental Corps

RCDEP Rural Civil Defense Education Program

RCDI Reliability Control Departmental Instruction

RCDMS Reliability Central Data Management System

rcdr. recorder

RCDs Royal Canadian Dragoons

RCDS Royal College of Defense Studies (UK)

rce rapid circuit etch(ing); remote-controlled equipment; right center entrance

RCE Reliability Control Engineering

RCEEA Radio Communications and Electronic Engineers Association

RCEME Royal Canadian Electrical and Mechanical Engineers

RCEP Royal Commission on Environmental Pollution

RCET Royal College of Engineering Technology; Rugby College of Engineering Technology

rcf recall finder; recall finding; relative centrifugal force

RCF Remote Call Forwarding (telephonic); Residential Care Facility

RCFA Reliability Control Failure Analysis; Royal Canadian Field Artillery

RCFCA Royal Canadian Flying Clubs Association

rcfm radiocommunication failure message

RCG Reception Guidance Center

RCGA Royal Canadian Golf Association

RCGP Royal College of General Practitioners

RCGS Royal Canadian Geographical Society

Rch Rochester

RCH Railway Clearing House; Resource Center for the Handicapped

RCHM Royal Commission on Historical Monuments (England)

rci radar coverage indicator; read channel initial(ize)

RCI Radio Canada International; Range Communications Instructions; Reichold Chemicals Incorporated; Research Council of Israel; Resident Cost Inspection; Resident Cost Inspector; Residential Communities Initiative (U.S. Army); Retail Confectioners International; Roof Consultants Institute; Royal Canadian Institute

RCIA Retail Clerks International Association; Retail Credit Institute of America

RCIC Rumor Control and Information Center

R-C IP Roosevelt-Campobello International Park near Eastport, Maine, in southern New Brunswick

rcirc recirculate

RCIs *Recontres Culturelles Internationales* (French—International Cultural Meetings)

RCIU Retail Clerks International Union

rcj reaction-control jet

RCJ Royal Courts of Justice

RCJCLDS Reorganized Church of Jesus Christ of Latter Day Saints

RCK *Research Centrum Kalkzandsteen Industrie* (Dutch—Research Center for the Calcium Silicate Industry)

rcl runway center line

RCL ramped cargo lighter (naval designation); Royal Canadian Legion

R-class Soviet submarines named Romeo by NATO

rclm reclaim; reclamation

rcm radar countermeasure(s); radio-controlled mine; radio countermeasure(s); right costal margin

RCM Reliability Control Manual; Royal College of Midwives; Royal College of Music

RCMA Religious Conference Management Association

RCMA Roof Coatings Manufacturers Association

RCMF Royal Commonwealth Military Forces

RCMM Registered Competitive Market Maker

RCMP Royal Canadian Mounted Police

RCMPM Royal Canadian Mounted Police Museum, Regina, Saskatchewan

rcn reticulum cell neoplasms

RCN Reactor Centrum Nederland; Record Control Number; Republic of China Navy; Royal Canadian Navy; Royal College of Nursing

RCN *Radio Cadena Nacional* (Spanish—National Radio Chain)—Mexican broadcasting system

RCNC Royal Corps of Naval Constructors

RCNM Russell Cave National Monument

RCNR Royal Canadian Naval Reserve

RCNT Registered Clinical Nurse Teacher

RCNVR Royal Canadian Naval Volunteer Reserve

rco rendezvous compatible orbit

rco (RCO) remote-control oscillator; representative calculating operation

RCO Radio Control Office; Royal College of Organists

RCOA Radio Club of America; Record Club of America; Royal Concertgebouw Orchestra of Amsterdam

RCOC Royal Canadian Ordnance Corps

RCOG Royal College of Obstetricians and Gynecologists

R-complex reptilian complex (evolutionarily most recent part of the forebrain)

rcp recording control panel; reinforced concrete pipe; remote communications processor; reserved circuits program

RCP Regional Community Physician; Returns Compliance Program; Revolutionary Communist Party; Royal College of Pathologists; Royal College of Physicians; Royal College of Psychiatrists

RCPA Royal College of Pathologists of Australia; Royal College of Physicians of Australia

rcpc regional check processing center

RCPI Royal College of Physicians—Ireland

RCPL Realtors Co-op Photo Listing

RCPS Royal College of Physicians and Surgeons

rcpt receipt

rcr reader control relay; reverse contactor

RCR República de Costa Rica

RCRA Resort and Commercial Recreation Association; Resource Conservation and Recovery Act

RCRBSJ Research Council on Riveted and Bolted Structural Joints

rcrd record

rcs radar cross-section; reloadable control storage; remote computer service

RCs Roman Catholics

RCS Reaction Control System; Reactor Coolant System; Rearward Communications System; Reentry Control System; Reliability Control Standard; Report Control Symbol; Residential Conservation Service; Royal College of Science; Royal College of Surgeons; Royal Commonwealth Society (formerly Royal Empire Society)

RCSB Royal Commonwealth Society for the Blind

RCSD Regional Council for Social Development

RCSE Royal College of Surgeons—Edinburgh

RCSI Royal College of Surgeons—Ireland

RCSS Random Communication Satellite System

RCST Royal College of Science and Technology

rct reversible counter

Rct Recruit

RCT Regimental Combat Team(s); Registered Clinical Teacher; Rorschach Content Test; Royal Corps of Transport

rctl rectal; resistor capacitor transistor logic

RCTT Regional Center for Technology Transfer (UN)

rcu remote control unit; research coordination unit

RCU Road Construction Unit

RCUEP Research Center for Urban and Environmental Planning (Princeton U)

rcv receive

rcv (RCV) radar control van; remote-controlled vehicle

rcvr receiver

RCVS Royal College of Veterinary Surgeons

RCWP Rural Clean Water Program

RCYB Revolutionary Communist Youth Brigade (Trotskyite)

RCYC Royal Canadian Yacht Club; Royal Corinthian Yacht Club; Royal Cork Yacht Club

R Cy N Royal Ceylon Navy

RCYP Revolutionary Communist Youth Brigades

RCZ Radiation Control Zone; Rear Combat Zone

rd indicates ordinal number as in 3rd Avenue or in 3rd Street; reaction of degeneration; readiness date; red dust (tea); release of dower rights; renal disease; required date; research and development (R & D); respiratory distress; restricted data; retinal detachment; roof drain; round; rutherford.

rd (RD) red devil (seconal tablet)

r & d reamed and drifted; research and development

Rd Road

RD Air Lift International; Radio Denmark; Registered Dietician; Restricted Data; Royal Dragoons; Royal Dutch Petroleum (stock exchange symbol); Rural Dean; Rural Delivery

RD República Dominicana

R.D. Royal (Naval Reserve) Decoration

R/D Research/Development

R & D research and development

R of D Report of Debate

rda recommended daily allowance; recommended dietary allowance; right dorso-anterior

rd a (Rd A) reading age

RDA Railway Development Association, Reliability Design Analysis; Respiratory Diseases Association; Royal Docks Association

R & D A Research and Development Association

RDA Reader's Digest Almanac; República Democrática Alemana (Spanish—German Democratic Republic)—East Germany

RDAF Royal Danish Air Force

Rdam Rotterdam

RDAR Reliability Design Analysis Report

rdb research and development bond

rdb (RDB) radar decoy balloon

RDB Ramped Dump Barge; Research and Development Board; Royal Danish Ballet

rdbl readable

RDBMS Relational Database Management System

rd bot rubber diaphragm (stoppered) bottle

rdc rail diesel car; repository design condition; research diagnostic criteria; running down clause

RDC Rand Development Corporation; Research Diagnostic Criteria; Rural District Council

RDCA Rural District Councils' Association

rd/chk read/check

RDCO Reliability Data Control Office

rdd required delivery date

rd & d (RD & D) research, development, and demonstration

RD$ República Dominicana peso (Dominican currency)

rde receptor-destroying enzyme

r d & e research, development, and engineering (usually R D & E)

RDE Research and Development Establishment

R de C Radiodiffusion du Cameroun (French—Radio Network of Cameroon)

R de F Republica de Filipinas

R de J République de Djibouti (formerly French Somaliland or the Territory of Afars and Issas); Rio de Janeiro

R de O Rio de Oro (Spanish Sahara)

R de P República de Panamá; República del Paraguay; República Portuguesa

rdf radio direction finder

RDF Rapid Deployment Force; Royal Dublin Fusiliers; Remote Database Facility

Rdg Reading; Ridge

RDG Reading Railroad

R d'H République d'Haiti

rd hd round head

rdi recommended daily intake; reference daily intake

RDI Royal Designer for Industry

RDJTF Rapid Deployment Joint Task Force

RDL Radiocarbon Dating Laboratory (Florida State University); Ritter Dental Laboratories

RDLI Royal Durban Light Infantry

rdline read a line

RdlR Regiment de la Rey

rdm root drum

RDM Rand Daily Mail (Johannesburg)

Rdm3c Radarman, third class

rdmu range-drift measuring unit

rdn resource decision network

RDN Royal Danish Navy

rdo research and development objectives

RDO Radiological Defense Office(r)

rdo('s) regular day(s) off; research and development objective(s)

rdp radar detector processor; right dorso-posterior

RDP Regional Development Program(s); Repository Development Plan(ning)

RDPC Research Data Publication Center

rdpe radar data-processing equipment

RDP Lao República Democrática Popular Lao (Spanish—Lao Popular Democratic Republic)

RDPP Repository Development Program Plan(ning)

rd/q reading quotient

rdr radar

rdr (RDR) receiver data register

r dr rive droite (French—right bank)

RDR Reliability Diagnostic Report; Research and Development Report

rdr rel radar relay

rdrsmtr radar transmitter

rds respiratory distress syndrome

Rds Rixdllar; Roads; Roadstead

RDs Revolutionary Development teams; Royal Dockyards

RDS Research Defence Society; Royal Dublin Society; Rural Development Service; Rural Development Society

RD/S Royal Dutch/Shell

RD & S Research, Development, and Studies (USMC)

RD/SG Royal Dutch/Shell Group

rdt reserve duty training

rdt (RDT) remote data transmitter

RDT Regiment Danie Theron; Reliability Demonstration Test

R.D.T. Registered Dental Technician

RDT Repubblica Democratica Tedesca (Italian—German Democratic Republic)—East Germany

rdt & e (RDT & E) research, development, test, and evaluation

RDTF Rapid Deployment Task Force (US Marines)

rdu research and development utilization

RDU Royal Development Unit

RDUP Research and Development Utilization Project

R du Z République du Zaïre (French—Republic of Zaire)

rdvu rendezvous

rdw right defense wing (lacrosse)

RDW Regiment De Wet

Rdwy Roadway

rdx cyclonite (research department explosive)

RDX Research and Development Exchange

rdy ready

RDY Royal Dock Yard

RDZ Radiation Danger Zone

RDZ République Democratique du Zaïre (French—Democratic Republic of Zaire)—formerly the Belgian Congo

rdz(s) (RDZ or RDZs) radiation danger zone(s)

re radium emanation; real estate; red; reinforce(d); reinforcing; research and engineering (R & E); reticuloendothelium; retinol equivalent; right eye

re in re (Latin—in the matter of; in reference to)

re. report

re (RE) revised edition

r/e rate of exchange

reap recklessly endangering another person

re B in diatonic scale, D in fixed-do; (Italian—second tone) (Latin prefix—again or back)—reflect, repair, restate

Re real part (symbol); Reno; Reynold's Number; rhenium; rupee (Ceylon, India, Pakistan currency)

R_e *récipe* (Spanish—recipe; prescription)

RE Radio Eireann (Radio Ireland); Reformed Episcopal (church); Reliability Engineering; Religious Education; Réunion (Internet code); Rifle Expert; Right Excellent; Royal Engineers; Royal Exchange

RE *República de Ecuador*

rea right ear advantage

REA Railway Express Agency; Request for Engineering Authorization; Resources Exchange Association; Retirement Equity Act (of 1984); Rice Export Association; Rubber Export Association; Rural Education Association; Rural Electrification Administration (US Department of Agriculture)

reac reactor

REAC Real Estate Aviation Chapter; Reeves electronic analog computer; Reliability Engineering Action Center

REACH Rape Emergency Aid and Counseling for Her; Retired Executives Action Clearinghouse

reack receipt acknowledged

react reactance; reaction; reactor; register-enforced automated-control technique

REACT Radio Emergency Associated Citizens Team; Register-Enforced Automated Control Technique; Resource Allocation and Control Techniques

READ Real-Time Electronic Access and Display

Read Dig *Reader's Digest*

readi rocket-engine-analyzer-and-decision-instrumentation

readm readmission

READS Reno Air Defense Sector

Reagan Ronald Reagan, 40th President of the United States

Reaganomics economic policy of the administration of President Reagan

REAL Rape Emergency Assistance League; *Real-Aerovias do Brasil* (Portuguese—Brazil Air Lines); Residential Experience in Adult Living

realcom real-time communication(s)

real est real estate

realgar arsenic sulfide

Realm of the Chinese Alligator lower Yangtze River valley

ream rapid excavation and mining

REAMS Ramond Electronically Applied Maintenance Standards

reap recklessly endangering another person

REAP Revenue Enforcement and Protection Program; Rural Environmental Assistance Program

reapt reappoint; reappointment

REAR Reliability Engineering Analysis Report

Rear Adm Rear Admiral

reas reasonable

reasm reassemble

REAT Radiological Emergency Assistance Team

Réau(m) Réaumur

reb rebel; rebellion

Reb Reba; Rebecca; Rebekah

REB Regional Examining Body

REB *Revised English Bible*

Reba Rebecca

rebar reinforcing (steel) bar

Rebilds Denmark's Rebild Hills including the Rebild National Park

reblt rebuilt

reb(s) rebel(s)

rec radio electronic combat; receipt; receive; recessed; record; recorded; recreation

rec. *recens* (Latin—fresh)

Rec Recife

REC Recife, Brazil (airport); Rural Electrification Corporation

R & EC Research and Engineering Council

reca repetitive-element column analysis

recap recapitulate; recapitulation; recycling of automobile plastics

RECAP Reliability Evaluation Continuous Analysis Program

RECC Rhine Evacuation and Control Command (NATO)

rec chg record change(r)

recco reconnaissance

recd received

recep reception

recg radioelectrocardiograph

R & ECGAI Research and Engineering Council of the Graphic Arts Industry

rech *recherche* (French—research)

rec hall recreational hall

reci recitation

recid recidivism; recidivist(ic); recidivous

recids recidivists

recip reciprocating

recipe. recomp computer interpretive program expeditor

recip & lp turb reciprocating steam engine and low-pressure turbine

recirc recirculate; recirculation

recit. *recitativo* (Italian—recitative)

reclam reclamation

recm recommend

recmark record mark(ing)

RECMF Radio and Electronic Component Manufacturers Federation

recncln reconciliation

recog recognition; recognize

recol retrieval command language

recom recommendation; recommend(ed)

recomp recomplement(ary); repairs completed; retrieval composition

recompen recognized company pension

recon reconcentration; reconciliation; recondite; recondition; reconduction; reconnaissance; reconnoiter; reconsign; reconsigned; reconsignment; reconstruct; reconstructed; reconstruction; reconversion; reconvert; reconverted; reconvey; reconveyance; reconveyed

RECON Regional Communication Outreach Network; Retrospective Conversion of Bibliographic Records (Library of Congress)

recond recondition

R Econ S Royal Economic Society

RECONS Reliability and Configurational Accountability System

reconst reconstruct

recop remarketed certificates of participation

recov recover; recovery
recp receptacle; reciprocal; reciprocating
RECP Rural Environmental Conservation Program
recpt receptionist
recr receiver
rec rm recreation room
rec room receiving room; reception room; record room; recreation room
recryst recrystallize
Rec S Record of Survey
RECSAM Regional Center for Education in Science and Mathematics
Rec Sec Recording Secretary
RECSTA Receiving Station
recsys recreational systems analysis
rect (Latin prefix—straight)— rectified; rectifier; rectify; rectitude
rect. rectificatus (Latin—rectified)
Rect Rector(y)
recto obverse; right-hand page
rectr recommends transfer
recur. recurrence; recurrent; recurring
rec vehicle(s) recreation vehicle(s) campers, dune buggies, snowmobiles, trailers, vans, etc.
red. reduce; reduction
red redaktör (Swedish—editor); *redigé* (French—compiled; edited)
Red Rederi (Scandinavian— shipowners); Red Sea, between Arabian Peninsula and Egypt
RED Real Estate Department
REDAR R E Darling (Company)
red burgee red signal flag flown when explosives or flammable fuel is being loaded aboard a vessel; letter B or Bravo in the international code
redcape readiness capability
redcat readiness requirement
redcon readiness condition
Redcraft Red aircraft (communist-controlled aircraft)
redec redecorate
redig redigerat (Swedish—edited)
redig. in pulv. redigatur in pulverem (Latin—reduce to Powder)

REDIS Redistricting System
redisc rediscount
redist redistilled
REDLARS Reading Literature Analysis and Retrieval Service
red light danger signal; port side; stop signal; warning signal
red ochre reddle (hematite red)
redox reduction oxidation
red. in pulv. reductus in pulverem (Latin—reduced to a powder)
red ru red kangaroo
redsg redesign; redesigned; redesigning
redsh reddish
redund redundant
redup(l) reduplicate; reduplication
redux reduction
Redwood Redwood City, Redwood Empire, Redwood National Park
ree rare-earth elements
REE Regional Economic Expansion (Canada)
REEA Real Estate Educators Association
REECO Reynolds Electrical and Engineering Company
Reed Reederei (German— shipowners)
reef The Reef—Australia's Great Barrier Reef off the coast of Queensland
Reefer(s) inhabitant(s) of the Great Barrier Reef
reeg radioelectroencephalograph
REEGT Registered Electroencephalographic Technicians
Reen Irene
reenl reenlist
reep range estimating and evaluation procedure
ref refer; referee; reference; reformatory; refraction; refresher
ref (REF) renal erythropoietic factor
ref refondue (French—reorganized)
Ref low-frequency resistance (symbol); reference
Ref Referate (German—abstract, compendium)
REF Railway Engineers Forum; Reject Errors in Football; rat embryo fibroblast; Romanian Engineers Forum

refash refashion(ed)
Ref Ch Reformed Church
refcom refuse conversion to methane
REFCORP Resolution Funding Corporation
refd refund
refd conc reinforced concrete
ref dent referring dentist
refd met reinforced metal
ref doct referring doctor
refd ply reinforced plywood
ref eso reflux esophagitis
reffo refugee from Europe
refg refrigerating; refrigeration
refi refinance
refl reflection; reflective; reflector; reflex; reflexive
ref l reference line
reflecs retrieval from literature on electronics and computer science
Ref Libr Reference Librarian
refl pron reflexive pronoun
Reform Reformatory
Reforma National Association of Spanish-Speaking Librarians in the United States
reforst reforestation
refphocon reference to telephone conversation
ref phys referring physician
ref press reference pressure
refr refraction; refractive; refractory; refrigerate; refrigerator
refrg refrigerate; refrigeration; refrigerator
refrig refrigeration; refrigerator
Refrig Eng Refrigerating Engineering
refs references
ref temp reference temperature
reftra refresher training
refurb refurbish(ed)
refy refinery
Ref Zhu Referativnyi Zhurnal (Russian—Abstract Journal)
reg region; regular; regulate; regulation
reg (REG) register (flow chart)
Reg Registered
Reg Regina (Latin—queen)
RegAF Regular Air Force
regal range and elevation guidance for approach and landing; remote generalized appli-

cation language; remotely guided autonomous lightweight torpedo

Reg Arch Registered Architect

Reg Bez Regierungsbezirk (German—administrative district)

reg bot regular bottle (3/4-liter of wine)

regd registered

regen regenerate; regeneration

Regent's Regent's Zoo in London's Regent's Park

Regg Reggimento (Italian—Regiment)

Reg. Gen Registrar General

Reg. Gen. Regula Generalis (Latin—general rule of the court)

Reggie Regina(ld)

Reg(gie)(y) Reginald

Reggio Reggiu di Calabria; Reggio nel'Emilia

regis register; registered; registration; registry

Reg. Jud. Registrum Judiciale (Latin—register of judicial writs)

reg'lar regular

Regnery Henry Regnery

Reg P Regent's Park College, Oxford

Reg Pl, Regula Placitandi (Latin—rule of pleading)

Reg Prof Regius Professor

Regr Registrar

regs regions; regulars; regulations

regt regiment

Reg TM Registered Trade Mark

regu regulable; regular; regularize; regularly; regulate; regulation; regulator

regurg regurgitant; regurgitate; regurgitation

REGY Regional Employment Growth (program for) Youth

reh rehearsal

rehab rehabilitate

Rehab Department of Rehabilitation

Rehab Dept Rehabilitation Department

rehob rehoboam (6-bottle capacity)

rei re-entry interval

REI Recreational Equipment Incorporated

REI Régie Aérienne Interinsulaire

R & EI Religion and Ethics Institute

REIC Radiation Effects Information Center; Rare Earth Information Center (Atomic Energy Commission, Ames Laboratory, Iowa State University)

Reichenhall Bad Reichenhall

reig rare-earth iron garnets

reils runway end identification lights

reimb reimburse; reimbursement

reincorp reincorporate(d)

reinf reinforce(d); reinforcing

reinfmt reinforcement

reins. radio-equipped inertial navigation system

REINS Radio-Equipped Inertial Navigation System

Rein Unid Reino Unidos (Spanish—United Kingdom)

reit reiteration

REIT Real Estate Investment Trust

REIWA Real Estate Institute of Western Australia

rej reject; rejected; rejection

rejase re-using junk as something else

rejn rejoin

REK Reykjavik, Iceland (airport)

rekenk rekenkunde (Dutch—arithmetic)

rel rate of energy loss; relation; relative; relay; release; relief; relieve; religion; religionist

rel relie; reliure (French—bound, binding)

REL Radio Engineering Laboratories; Robert Edward Lee (1807–1870)

RELACS Radar Emission Location Attack Control System

rel adv relative adverb

RELC Reformation Evangelical Lutheran Church; Regional Educational Laboratory for the Carolinas

RELCV Regional Educational Laboratory for the Carolinas and Virginia

reld. relieved

RELHS Robert E Lee High School

rel hum relative humidity

reliab reliability

relig religion; religious

rel-i-l relative-in-law

reliq. reliquus (Latin—remainder)

reloc relocate; relocated; relocation

rel pron relative pronoun

Rel R Reliability Report

RELS Rapidly Extensible Language System

rem rapid eye movements; remain(ing); remission; remit; remittance; removable; remove; removed; roentgen equivalent, man

Rem Remington; roentgen equivalent, man

REM Radioactive Environmental Monitoring; Registered Equipment Management

REMA Refrigeration Equipment Manufacturers Association

remab radiation equivalent manikin absorption

remad remote magnetic anomaly detection

Remarkables Remarkable Range of mountains in New Zealand's South Island

remc resin-encapsulated mica capacitor

REMC Radio and Electronics Measurements Committee; Regional Educational Media Center

remcal radiation equivalent manikin absorption

remd rapid eye movement (sleep) deprivation

REME Royal Electrical and Mechanical Engineers

REMIC(s) Real Estate Mortgage Investment Conduit(s)

REML Radiation Effects Mobile Laboratory

REMP Research Group for European Migration Problems

rem(s) rémora(s)

rems (REMS) rapid-eye-movement sleep

REMS Registered Equipment Management System

REMSA Railway Engineering Maintenance Suppliers Association

rem sleep rapid-eye-movement (paradoxical) sleep

remstar remote electronic microfilm storage transmission and retrieval

REMT Radiological Emergency Medical Teams

Rem-UMC Remington-Union Metallic Cartridge (company)

remus routine for executive multi-unit simulation

REMUS Reference Models for Usability Specifications

ren. *renovetur* (Latin—renew); renunciation

Ren Renaissance

RENAMO *Resisténcia Nacional Moçambicana* (Portuguese—Mozambique National Resistance)

rene rocket-engine nozzle ejector

Renf Renfrew

RENFE *Red Nacional de los Ferrocarriles Españoles* (Spanish—National Network of Spanish Railroads)

RENS Reconnaissance Electronic Warfare and Naval Intelligence System

ren. sem. *renovetum semel* (Latin—renew only once)

rent. leuntry nore tip

renv renovate; renovation

reo rare-earth oxide; regenerated electrical output

Reo early American automobile named after initials of its maker, Ransom E Olds of Oldsmobile fame

REO Ransom Eli Olds, automobile inventor and manufacturer (1864–1950); Regional Education Officer

reoc report when established on course (aviation)

reopt reorder point

REORG reorganization; reorganize; reorganized

reorg. reorganization

REOs Real-Estate-Owned banking departments

REOS Reflective Electron Optical System

reo viruses respiratory-entericorphan viruses

rep repair; repeat; repertory; represent; representative; reprint; reprinted; reputation

rep. reparation; report; representative

r-ep rational-emotive psychotherapy

rep *reparto* (Italian—department)

rep. *repetatur* (Latin—let it be repeated)

Rep Representative; Republic; Republican; Republican Party; roentgen equivalent, physical

REP Radiation Exposure Permit; Radical Education Project; Recovery and Evacuation Program; Republic Corporation (stock exchange symbol); Research Expenditure Proposal; Reserve Enlisted Program; River Engineering Program

Rep Arabe Yem *República Arabe del Yemen* (Spanish—Arabic Republic of Yemen)—formerly British Crown colony of Aden

Rep Arg *República Argentina* (Spanish—Argentine Republic)

Rep de Bol *República de Bolivia* (Spanish—Republic of Bolivia)

Rep Bot Republic of Botswana

Rep Cabo Verde *República de Cabo Verde* (Portuguese—Cape Verde Republic)

Rép Cent *République Centrafricaine* (French—Central African Republic)

Rep Chile *República de Chile* (Spanish—Republic of Chile)

Rep Col *República de Colombia* (Spanish—Republic of Colombia)

Rép Côte d'Ivoire *Répulique de la Côte d'Ivoire* (French—Republic of the Ivory Coast)

Rep CR *República de Costa Rica* (Spanish—Republic of Costa Rica

Rep Day *Republik Dayti* (Haitian Creole—Republic of Haiti)

Rep de Cuba *República de Cuba* (Spanish—Republic of Cuba)

Rep Dem Mal *Repoblika Demokratika Malagasy* (Malagasy—Democratic Republic of Madagascar)

Rep Dem Pop Yem *República Democratica Popular del Yemen* (Spanish—Popular Democratic Republic of Yemen)

Rep Dem Sao Tome Prin *República Democrática de Sao Tome e Principe* (Portuguese—Democratic Republic of Sao Tome and Principe)

Rép d'Haiti *República d'Haiti* (French—Republic of Haiti)

Rep Dom *República Dominicana* (Spanish—Dominican Republic)

Rep Ecu *República del Ecuador* (Spanish—Republic of Ecuador)

Rep El S *República de El Salvador* (Spanish—Republic of El Salvador)

Rep Fed Bra *República Federativa do Brasil* (Portuguese—Federative Republic of Brazil)

Rép Fran *République Francaise* (French Republic)

Rép Gab *République Gabonaise* (French—Gabonese Republic)

Rep Ghana Republic of Ghana

Rep Gua *República de Guatemala* (Spanish—Republic of Guatemala)

Rép Gui *République de Guinée* (French—Republic of Guinea)

Rep Gui-Bis *República de Guiné-Bissau* (Portuguese—Republic of Guinea-Bisau)

Rep Gui Ecu *República de Guinea Ecuatorial* (Spanish—Republic of Ecuatorial Guinea)

Rep Hond *República de Honduras* (Spanish—Republic of Honduras)

Rep Ind *Republik Indonesia* (Malay—Republic of Indonesia)

Rep Isl Maur *Républiquu Islamique de Mauritanie* (French—Islamic Republic of Mauritania)

Rep Ital *Repubblica Italiana* (Italian Republic)

Rep Kiri Republic of Kiribati

Rep Lib Republic of Liberia

Rep Mal Republic of Malawi

Rép Mali *République du Mali* (French—Republic of Mali)

Rep Malta *Repubblika ta'Malta* (Maltese—Republic of Malta)

Rep Nauru Republic of Nauru

Rep Nic *República de Nicaragua* (Spanish—Republic of Nicaragua)

Rép Nig *République du Niger* (French—Republic of Niger)

Rep Ori Uru *República Oriental del Uruguay* (Spanish—Oriental Republic of Uruguay)

Rep Öst *Republik Österreich* (German—Austrian Republic)

Rep de Pan *República de Panama* (Spanish—Republic of Panama)

Rep Para *República del Paraguay* (Spanish—Republic of Paraguay)

Rep Peru *República del Peru* (Spanish—Republic of Peru)

Rep Phil Republic of the Philippines

Rep Pop de Ang *República Popular de Angola* (Portuguese—People's Republic of Angola)

Rep Pop Ben *République Populaire du Benin* (French—People's Republic of Benin)

Rép Pop Con *République Populaire du Congo* (French—People's Republic of the Congo)

Rep Pop Moç *República Popular de Maçambique* (Portuguese—People's Republic of Mozambique)

Rep Pop Soc e Shq *Republika Popullore Socialiste e Shqipërissë* (Albanian Popular Socialist Republic)

Rep Port *República Portuguesa* (Portuguese—Republic of Portugal)

Rep Rwa *Republika y'u Rwanda* (Swahili—Republic of Rwanda)

Rép Sénég *République du Sénégal* (French—Republic of Sénégal)

Rep Sey Republic of the Seychelles

Rep Sierra Leone Republic of Sierra Leone

Rep Singa Republic of Singapore

Rep Soc Rom *Republica Socialista România* (Romanian Socialist Republic of Romania)

Rep Suid-Afrik *Republiek van Suid-Afrika* (Afrikaans—Republic of South Africa)

Rep Sur *Republiek Suriname* (Dutch—Suriname Republic)

Rep Suri Suriname Republic

Rép Tch *République du Tchad* (French—Republic of Chad)

Rep The Gam Republic of The Gambia

Rép Togo *République Togolaise* (French—Republic of Togo)

Rep T & T Republic of Trinidad and Tobago

Repub Republic; Republican

repud. repudiated

Rep Ugan Republic of Uganda

Rep V Repair Locker 5 (Engineering)—USN

Rep Ven *República de Venezuela* (Spanish—Republic of Venezuela)

Rep y'Ub *Republika y'Uburundi* (Rundi—Republic of Burundi)

Rep Zambia Republic of Zambia

REPA Research and Engineers Professional Employees Association

REPC Racial Ethnic Parent Councils; Regional Economic Planning Council

repcon rain repellant and surface conditioner

REPE Radio Engineering Europe

reperf reperforator

repet repetition; repetitive

repl replace(d); replacement; replacing

repltr report (by) letter

repm repairman; repairmen

REPM Representatives of Electronic Products Manufacturers

repo repossess; repossessed; repossession

repo men repossession men

repop repetitive operation(s)

repo(s) repurchase agreement(s)

reppac repetitively-pulsed plasma accelerator

Rep Prog Phys *Reports on Progress in Physics*

repr repairman; representative; reprint; reprinted; reprinting

repro reproduce; reproducing; reproduction

reprosex reproductive sex

repro typ reproduction typist; reproduction typing

reps repetitive electromagnetic pulse sirnulator; representatives

Rep(s) Republican(s)

REPS Rail(way) Express Parcel Service

rep. sem. *repetatur semel* (Latin—let it be repeated once)

rept report; reprint; reptile; reptilia(n)

rept (Rept) report

rept. *repetatur* (Latin—let it be repeated)

Rept Reptilia

repub republication; republish(ed)

REPUBLIC Republic Aviation Corporation

Republocrat Republican Democrat

repud repudiated

Repubs Republicans

req request; require

reqafa request advise as to further action

reqd required

reqdi request disposition instructions

reqfolinfo request following information

reqid request if desired

reqmad request mailing address

reqmt requirement

reqn requisition

reqrec request(ed) recommendation

reqs requires

reqssd request supply status (and expected delivery) date

reqsupstafol request supply status of following

reqt requirement

reqtat requested that

REQUEST Reliability and Quality of European Software

requint request interim (reply)

rer (RER) radar effects reactor

RER Railway Equipment Register

RERC Real Estate Research Corporation

REREI Redwood Empire Research and Education Institute

rereq reference requisition

RERF Radiation Effects Research Foundation

rerl residual equivalent return loss

RERO Royal Engineers Reserve of Officers

res rescue; research; researcher; reservation; reserve; reservoir; residence; resilient; resistant; respiratory; reticuloendothelial system (RES)

res. resides

res (RES) restore (computer character)

Res Reservation; Reservoir

RES Elizabeth (N.J.) Resolutes (National Association); Royal Economic Society; Royal Entomological Society

RES República de El Salvador

RESA Regional Educational Service Agencies; Regional Educational Service Areas; Research Society of America

ResAF Reserve of the Air Force

Res Aud Resident Auditor

resc rescue

RESC Regional Educational Service Centers

RESCAM Regional Center for Education in Science and Mathematics

rescan reflecting satellite communication antenna

RESCO Refuse Energy Systems Company

rescu rocket-ejection seat catapult upward

RESCU Radio Emergency Search Communications Unit; Remote Emergency Satellite Cellular Unit

rescue remote emergency salvage and cleanup equipment

Res & Educ Research and Education Association

reser reentry system evaluation radar

resgnd resigned

resid residual; residual oil

resig resignation

RESIG Research and Engineering System Integration Group

resil resilient

resist. resistance; resistor

resistojet resistance-connective jet engine

resojet resonant pulse jet

resp respective; respelling; respiration(s); respirator; respire; responder; responsibility; responsible; responsive

RESPA Real Estate Settlement Procedures Act

respectiv. respectively

Res Phys Resident Physician

respir respiration; respiratory

respirol respirologic(al)(ly); respirologist; respirology

respirom respirometer; respirometric(al)(ly); respirometrist; respirometry

RESPO Responsible Property Officer

RESPONSA Retrieval of Special Portions from Nuclear Science Abstracts

respub responsible Republican(ism)

Resrt Resort

RESS Radar Echo-Simulation Study; Radar Echo-Simulation System

Res Sec Resident Secretary

RESSI Real Estate Securities and Syndication Institute

rest restrict; restricted; restriction

rest (REST) regressive electric shock therapy

REST Radar Electronic-Scan Technique; Reentry Environment and Systems Technology; Reentry System Test Program; Routine Execution Selection Table

resta reconnaissance, surveillance, and target acquisition

restr restaurant

ResTraCen Reserve Training Center

RESTTA Restitution Education, Specialized Training, and Technical Assistance program, funded by the Office of Juvenile Justice and Delinquency Prevention, U.S. Department of Justice

resub resublimed

resup resupply

resvr reservoir

RE system reticuloendothelial system

ret rapid eye therapy; rational emotive therapy; retainer; retaining; retire; retirement

ret. retired

ret (RET) rational-emotional therapy; return (flow chart)

r-et rational-emotive psychotherapy

Ret Reticulum (constellation); retired

RET R Emmett Tyrrell, Jr

RET Rotterdamse Elektrische Tram (Dutch—Rotterdam Electric Tramway)—electric surface car and subway system

reta retrieval of enriched textual abstracts

RETA Refrigerating Engineers and Technicians Association

Retail Clerks Union Retail Clerks International Association

retain. remote technical assistance and information network

retard. retardation; retarded

retc railroad equipment trust certificate

RETC Regional Employment Training Center; Regional Employment and Training Consortium

retd retired

rete (Latin prefix—network)—retinal

R. et I. Regina et Imperatrix (Latin—Queen and Empress)—title of Vitctoria—Queen of England and Empress of India—The Queen

retic reticulate(d); reticulation; reticule

retic count reticulocyte count

retics reticulocytes

red retail

RETL Rocket Engine Test Laboratory

RETMA Radio-Electronics Television Manufacturers Association

retng retraining

retnr retainer

retort. (RETORT) reason and equity in tort

ret p retired pay

RETP Reliability Evaluation Test Procedure

retpd retention period

retr retractable

RETRA Radio, Electrical, and Television Retailers Association

Ret Res Retirement Research

retro retroactive; retrofit; retrograde; retrorocket

retro (Latin prefix—backward or behind)—retroactive, retrograde

retros retrogrades; retrorockets

RETS Renaissance English Text Society

Reun Reunion Island

Reuter's Reuter's international news agency

rev reverse; reversed; review; revise; revised; revision; revolute; revolution

rev (REV) reentry vehicle

rev revisado (Spanish—revised)

Rev Reverend; Review, Revised, Revolutionary War; Revue; The Revelation of St John the Divine

Rev Revelation

reva recommended vehicle adjustment

rev a/c revenue account

Revd Reverend

rev'd reversed

Rev d'Opt Revue d'Optique (French—Optics Review)

rev ed revised edition

revel. reverberation elimination

Revell Fleming H Revell

revid reviderad (Swedish—revised)

Revilla Gigedos Revilla Gigedo Islands off Mexico's West coast, not to be confused with Revilia Gigedo Island off Alaska

rev/min revolutions per minute

Rev Mod Phys Reviews of Modern Physics

revocon remote volume control

revol revolution(ary); revolver

Rev. Proc. Revenue Procedure

revr reviewer

Rev Rul Revenue Ruling

revs revolutions

rev(s) revolution(s)

rev/s revolutions per second

REVS Rotor-Entry Vehicle System

Rev Sci Instrum Review of Scientific Instruments

revisec revolutions per second

Rev Stat Revised Statutes

rev of sym review of symptoms

Rev Ver Revised Version of the Bible

Rev War Revolutionary War

rew reward; rewind(ing)

REWARD Recycling of Waste Research and Development

rewdac retrieval by title words, descriptors, and classifications

rewk rework

rewrc report when established well to right of course

REWSON Reconnaissance Electronic Warfare Special Operations and Naval Intelligence Processing System(s)

rex real-time executive routine; reduced exoatmospheric cross-section

Rex Reginald

REX Rexall Drug and Chemical (stock exchange symbol)

rexs (REXS) radio-exploration satellite

Rex trem Rex tremendae (Latin—King of Tremendous Majesty)

Reykjvk Reykjavik

Reynall Reynal & Co

rf radio frequency; range finder; rapid fire; rat fink; reception fair; reflight; relative flow; replacement factor; replicative form; representative fraction; reticular formation; rheumatic fever, rheumatoid factor; right field; right fullback; rim fire; rubber-free; run factor

r-f radiofrequency

r/f right front

r_f rate of flow

rf rinforzando (Italian—reinforcing)

Rf Reef, Rufiyaa (Maldives currency); rutherfordium, also known as unnilquadium

RF Reserve Force; Rockefeller Foundation; Rocky Flats; Rodeo Foundation; Royal Fusiliers

RF République Française

R-F Reitland-Franklin (unit)

rfa radiofrequency attenuator; radiofrequency authorization(s); request for application(s); request further airways; right fronto-anterior

RFA Royal Field Artillery; Royal Fleet Auxiliary

RFA República Federal de Alemania (Spanish—Federal Republic of Germany); *République Fédérale Allemande* (French—Federal Republic of Germany)

RFAC Royal Federation of Aero Clubs; Royal Fine Arts Commission

R factor resistance factor

rfad release for active duty

rfa's return(ed) for alterations (tailoring)

rfb request for bid

RFB Recording for the Blind

RFB República Federativa do Brasil (Portuguese—Federal Republic of Brazil)

rf black reinforcing furnace black

rfc radiofrequency choke

RFC Rare Fruit Council; Reconstruction Finance Corporation; Resolution Funding Corporation; River Forecast Center; Royal Flying Corps; Rugby Football Club

RFCL Referral Form Checklist

rfcs radio-frequency carrier shift

RFCs Request for Comments

RFCWA Regional Fisheries Commission for Western Africa

rfd raised foredeck; reentry flight demonstration; reference dose; refund; reinforced; reporting for duty

RFD Radio Frequency Devices; Rural Free Delivery

rfd con reinforced concrete

rfd met reinforced metal

rfd ply reinforced plywood

rfdr rangefinder

RFDS Royal Flying Doctor Service

rfe request for estimate

RFE Radio Free Europe

RFED Research Facilities and Equipment Division (NASA)

rff remote-fiber fluorimetry

R f F Rat für Formgebung (German—Fashion Council)

RFF Rede Ferroviária Federal (Portuguese—Federal Railway System)—Brazil

RFFS River and Flood Forecasting Service

RFFSA Rede Ferrocarril Federal Sedada Anonima (Portuguese—Federal Railway Route Company)—Brazil

rfg reformulated gasoline; roofing

RFH Royal Festival Hall

rfi radiofrequency interference; ready for issue; request for information

rfing royal fucking

rfi/pf radio frequency interference/pulse frequency

rf/ir radiofrequency/infrared

R Fix running fix

rfl refuel(ing); right fronto-lateral

RFL Refrigerated Freight Lines; Rugby Football League

Rflmn Rifleman

RFLP restriction fragment length polymorphism

RFLPs Restriction-Fragment Length Polymorphisms

rfls rheumatoid factor-like substance

rfm radio frequency management

r-f m ripple-flow mill (grain)

RFMA Reliability Figure of Merit Analysis

RFMF Royal Fiji Military Forces

Rfn Rifleman

RFN Registered Fever Nurse

rfna red-fuming nitric acid

rfnip reduced-flow nominal-inlet pressure

rfnop reduced-flow nominal output pressure

RFNZJ Royal Federation of New Zealand Justices

rfo radiofrequency oscillator; request for factory order

RFO Regional Fisheries Office(r)

rfp right frontoposterior

RFP Request for Proposal

RF & P Richmond, Fredericksburg and Potomac (railroad)

RFPs Requests for Proposals

RFPS(G) Royal Faculty of Physicians and Surgeons of Glasgow

RFQ Request For Qualifications; Request for Quotation

rfr refraction; reject failure rate; required freight rate

R fr Rwanda franc(s)

RFR Royal Fleet Reserve

rfrd referred

rfs radio-frequency surveillance; ready for sea; regardless of future size

Rfs Reefs

RFS Registry of Friendly Societies; Royal Forestry Service

rf scale representative fraction scale

rfs/ecm radio-frequency surveillance/electronic countermeasures

RFSU *Riksføbundet før Sexuall Upplysning* (Norwegian—National League for Sexual Education); Rugby Football Schools' Union

rft right frontotransverse

RFT Rod and Frame Test

RFT *Repubblica Federale Tedesca* (Italian—German Federal Republic)

rfts radiofrequency test set

rfu ready for use

RFU Rugby Football Union

R-F unit Reitland-Franklin unit

rfw rapid-filling wave

RFW Radio Free Women

rfwe ring-finished with engines

Rfy Refinery

RFYC Royal Forth Yacht Club

rfz restrictive fire zone

rfz *rinforzando* (Italian—with extra emphasis)

rg real girl (not a birl); registered genealogist; regummed; repetitive group(ing)

r g *rive gauche* (French—left bank)

Rg *Ruckgang* (German—retrogression)—in sonatas

RG Reserve Grade

RG *Reader's Guide to Periodical Literature*; Rive Gauche (French—Left Bank) *Regula Generalis* (Latin—general rule of the court); *República de Guatemala*

rga rate gyro assembly

Rga Riga

RGA Republican Governors Association; Royal Garrison Artillery; Rubber Growers' Association

RGAHS Royal Guernsey Agricultural and Horticultural Society

R-gauge Russian gauge (5-foot) railroad track

rgb red-green-blue color separation; red-orange, green, blue-violet (television's triad of primary colors)

RGC Reception and Guidance Center

rgd reigned

R Gd Rio Grande

RGDATA Retail Grocery, Dairy, and Allied Trades Association

RG do S Rio Grande do Sul

rge relative gas expansion

Rge Range, Ridge

RGE *Rat der Gemeinden Europas* (German—Council of European Municipalities); *República de Guinea Ecuatorial* (Spanish—Republic of Equatorial Guinea)

RGEB Rockefeller General Education Board

R Gen Registrar General

RGEPS Rucker-Gable Educational Programming Scale

rgf range-gated filter

RGF Red Guerrilla Family (black terrorists)

RGG Royal Grenadier Guards

RGH Royal Gloucestershire Hussars

RGI Robert G Ingersoll; Royal Glasgow Institute of Fine Arts

RGJ Royal Green Jackets

rgi regulate; regulation; regulatory

rgm residential growth management

RGM Revenue Generation and Management

rgn region

Rgn (Port of) Rangoon

RGN Rangoon, Burma (airport); Registered General Nurse

RGNR Rugged Glen Nature Reserve (South Africa)

RGO Royal Greenwich Observatory

RGP Riegel Paper Company (stock-exchange symbol)

RGPL *Readers' Guide to Periodical Literature*

RGPM Regional Geological Project Manager

rgr reference geological regime

rgs radar ground stabilization

r/gs run support per game started (baseball)

RGS Rio Grande do Sul; Royal Geographical Society

RGSA Royal Geographical Society of Australasia

rgstr. registrar

rgt resonant gate transistor

Rgt Regiment

RGTC Robert Gordon's Technical College

RGTF Royal General Theatrical Fund

Rgtl Regimental

rg tp rough template

RGV Rio Grande Valley Gas Company (stock exchange symbol)

rgz recommended ground zero

RGZ Rio Grande Zoo (Albuquerque)

rh rheumatic; rheumatism; rheumatoid; rhyming; right-hand (RH); right halfback (field hockey); right hind; roundhead

r/h relative humidity; roentgens per hour

rh. *rhonchi* (Latin—rales)

Rh Rhesus factor (symbol); rhodium

Rh+ Rhesus positive

Rh- Rh- Rhesus negative

Rh *Rhein* (German—Rhine)

RH Air Rhodesia; Random House; Round House; Royal Highlanders; Royal Highness; Ryan Herco

RH *Rechte Hand* (German—right hand); *República de Honduras;* Research Highlights

RH$_{106}$ radioactive rhodium

RHA Regional Health Authority; Road Haulage Association; Royal Hibernian Academy; Royal Humane Association; Rural Housing Alliance

RHAF Royal Hellenic Air Force

RHAMM Receptor for Hyaluronan Mediated Motility

R Hamps Royal Hampshire (regiment)

rhap rhapsody

R. ha-Sh. *Ro'sh ha-shanah*

RHAWS Radar Homing and Warning System

rhb. rehabilitation

RHB Regional Hospital Board; Robin Hood's Bay

rhbdr rhombohedral

rhc respirations have ceased; rubber hydrocarbon

RHC Road Haulage Cases; Rosary Hill College

RHC *Radio Habana Cuba* (Spanish—Havana, Cuba Radio)

RHCSA Regional Hospitals Consultants' and Specialists' Association

rhd radioactive health data; relative hepatic dullness; rheumatic heart disease

RHD Robin Hood Dell (Philadelphia)

RHD *Random House Dictionary*

RHDEL *Random House Dictionary of the English Language*

RHDO Robin Hood Dell Orchestra

rhe reversible hydrogen electrode

RHE Reliability Human Engineering

RHEL Rutherford High-Energy Laboratory

Rhenish Symphony No 3 by Schumann

rheo rheostat

rheol rheological; rheology

rhet rhetoric; rhetorical; rhetorician

rheu rheumatic; rheumatism; rheumatoid

rheu fev rheumatic fever

rheu ht dis rheumatic heart disease

rheum rheumatic; rheumatism

rhf right heart failure

R-hf high-frequency resistance (symbol)

RHF Royal Highland Fusiliers

Rh factor Rhesus group of red cell agglutinogens

rhg reactive hypoglycemia

RHG Royal Horse Guards

RHGPS Rhodesian Hunters and Game Preservation Society

RHHI Royal Hospital and Home for Incurables

rhi range height indicator

RHIB Rain and Hail Insurance Board; Rain and Hail Insurance Bureau

rhic relativistic heavy ion collider

rhin (Latin prefix—nose)—rhinitis

rhino range height indicator not operating

rhinol rhinologic(al)(ly); rhinologist; rhinology

rhino(s) rhinoceros(es)

rhip rank has its privileges

rhir rank has its responsibilities

R Hist S Royal Historical Society

RHIT Rose-Hulman Institute of Technology

RHK Radio Hong Kong

RHKAAF Royal Hong Kong Auxiliary Air Force

RHKP Royal Hong Kong Police

RHKPF Royal Hong Kong Police Force

RHKR Royal Hong Kong Regiment

RHKTV Royal Hong Kong Television

RHKYC Royal Hong Kong Yacht Club

rhl rectangular hysteresis loop

RHL Radiological Health Laboratory; Rape Help Line

rhm roentgen per hour per meter

RHMG Rogers House Museum Gallery

RHMS Royal Hibernian Military School

RHN Royal Hellenic Navy

Rho Rhoda

RHO Regional Hospital Office(r); Rickwell Hanford Operations; Rural Health Office(r)

RHOB Rayburn House Office Building

Rhod Rhodesia

Rhoda Rhodacella; Rhodacelle

Rhodes Cecil John Rhodes (1853–1902)

rhodies rhododendrons

rhodo(s) rhododendron(s)

Rhodos (Greek—Rhodes)—island in the Aegean

RHOFLIGHT Rhodesian Air Services

rhoin rhombic; rhomboid; rhombus

rhp rated horsepower; right-handed pitcher

RHQ Regimental Headquarters

rhr roughness height reading

r/hr roentgens per hour

RHR Race and Human Relations; Royal Highland Regiment (Black Watch)

rhs righthand side; round-headed screw

RIUS Radio Ham Shack (amateur radio operator's station); Royal Historical Society; Royal Horticultural Society

RHSI Royal Horticultural Society of Ireland

RHSNZ Royal Humane Society of New Zealand

RHSV Royal Historical Society of Victoria

rht runs (scored) at home by team

Rhumba (stock exchange form for Royal McBee Company whose symbol is RMB)

Rhus tox Rhus toxicodendron

RHV *République de Haute-Volta* (French—Republic of Upper Volta)

rh & w radar homing and warning

RHYP Runaway and Homeless Program

ri radio interference; random interval; reflective insulation; refractive index; reliability index; require identification; respiratory illness; retroactive inhibition; right inner (field hockey); rubber-insulated; rubber insulation

ri (RI) retrograde inversion

Ri input resistance (symbol)

RI Recruit Instruction; Refractories Institute; Religious Instruction; Republic of India; Rhode Island (R.I.); Rhode Islanders; Rice Institute; Rock Island (Chicago, Rock Island & Pacific Railroad); Rotary International; Royal Institute

R & I Rural and Industries (bank)

RI Readers International, Registro Italiano (Italian—Italian Register)—of shipping; *Repubblica Italiana* (Italian—Italian Republic); *Républicains Independants* (French—Independent Republicans); *Republik Indonesia; Ring Index*

ria radioimmunoassay

RIA Railroad Insurance Association; Recording Industry Association; Research Institute of America; Robot Institute of America; Rock Island Arsenal; Royal Irish Academy

RIAA Record Industry Association of America; Recording Industry Association of America

RIAAE Rhode Island Alliance for Arts in Education

RIAC Research Information Analysis Corporation

RIAEC Rhode Island Atomic Energy Commission

RIAF Royal Indian Air Force; Royal Iranian Air Force; Royal Iraqui Air Force

riah monetary unit of Iran

riah (RIAH) radioimmunoassay of hair (for drug detection)

RIAI Royal Institute of Architects of Ireland

rial monetary unit of Yemen

rial (RIAL) revised individual allowance list

RIAL Religion In American Life; Rock Island Arsenal Laboratory

rial omani monetary unit of Oman

RIAM Royal Irish Academy of Music

RIANZ Record Industry Association of New Zealand

RIAS Research Initiation and Support (National Science Foundation)

RIAS Rundfunk im amerikanischen Sektor (German—Radio in the American Sector), Berlin

RIASBO Rhode Island Association of School Business Officials

RIASC Rhode Island Association of School Committees

RIASLP Rattlesnake Island Air Service Local Post

RIASSP Rhode Island Association of Secondary School Principals

rib range in a box; ribbon

RIB Railway Information Bureau; Referee in Bankruptcy; Roanoke Iron & Bridge; Rural Industries Bureau

RIB Rijksinkoopbureau (Dutch—Government Purchasing Office)

Rib^a Ribeira (Portuguese—brook; creek; riverside; river valley, stream); *Ribera* (Spanish—bank, beach, riverside, shore)

RIBA Royal Institute of British Architects

RIBNY Republic International Bank of New York

RIBS Restructured Infantry Battalion System

ric radar intercept calculator; ritual infant circumcision; routine infant circumcision

ric ricevuta (Italian—receipt)

Ric Ricardo; Richard; Richmond

RIC Republic Industrial Corporation; Republic of the Ivory Coast; Richmond, Virginia (airport); Royal Institute of Chemistry; Royal Irish Constabulary

RICA Research Institute on Communist Affairs (Columbia University)

RICASIP Research Information Center and Advisory Service on Information Processing

RICE Rhode Island College of Education

Rich Richard; Richards; Richardson; Richford; Richmal; Richmond

Rich Rich Stadium, Buffalo, New York

Rich II King Richard II

Rich III King Richard III

Richd Richard; Richmond

Rich-Pete Turn Richmond-Petersburg Turnpike (Virginia)

Rick Richard

ricksha(w) *jinrikisha* (Japanese—man-drawn two-wheeled carriage)

Ricky Richard

ricm right intercostal margin

RICM Registre International des Citoyens du Monde (French—International Registry of World Citizens)

RICMD Richmond Contract Management District

RICMO Radar Input Countermeasures Officer

'Rico Enrico; Puerto Rico; Ricardo

RICO Racketeer-Influenced Corrupt Organization; Racketeer-Influenced and Corrupt Organizations

RICs Regulated Investment Companies

RICS Royal Institute of Chartered Surveyors

RICU Russian Institute, Columbia University

rid review item disposition

RID Regimented Inmate Discipline (program for educating felons); Registry of Interpreters for the Deaf; Remove Intoxicated Drivers; Riddle Aviation

RIDA Rural and Industrial Development Authority

ridac range interference directing and control

RIDE Research Institute for Diagnostic Engineering

Riders Riders of the Purple Sage

Riding Mountain Riding Mountain National Park in southwestern Manitoba

Ridley Henry Nicholson Ridley (1855–1956), established rubber industry in Malaya, developed Botanic Gardens in Singapore, and for whom a Pacific Ocean sea turtle is named

ridp radar-iff (if friend or foe) data processor

rie range of incentive effectiveness; resources in education

RIE Royal Institute of Engineers

RIEC Royal Indian Engineering College

RIEI Republic Industrial Education Institute (Republic Steel); Roofing Industry Educational Institute

riel monetary unit of Cambodia

RIEM Research Institute for Environmental Medicine

rif reading is fundamental; reduction in force; rifle; right iliac fossa

rif (RIF) resistance-inducing factor

rif rifatto (Italian—restored; repaired)

RIF Reading Is Fundamental; Royal Irish Fusiliers

RIFA Royal Institute of Foreign Affairs

Rif Brig Rifle Brigade

rifc rat intrinsic factor concentrate

Riff mountainous region of northern Morocco opposite Straits of Gibraltar

riffed reduced in force (dismissed or fired)

rifi radio interference field intensity

rifl random item file locater

RIFM Research Institute for Fragrance Materials

rifma roentgen-isotope-fluorescent method of analysis

rift (RIFT) reactor-in-flight test

RIFT Rhode Island Federation of Teachers

rig radioisotope generator

Rig Riga

RIG Restricted Interagency Group

RIGB Royal Institution of Great Britain

RIGHT Rhodesian Independence Gung-Ho Troops

right on right on the nose (exactly)

rih repetition-induced hypnosis; right inguinal hernia

RIH Royal Institute of Horticulture

RIHS Rhode Island Historical Society

rihsa radioactive iodinated human serum albumin

RIIA Royal Institute of International Affairs

RIIC Research Institute on International Change

RIISOM Research Institute for Iron, Steel, and Other Metals

rikisha jinrikisha

ril record input length

RIL Royal Interocean Lines

RILSS Rapid Integrated Logistic Support System

rim radar input mapper; receiving, inspection, and maintenance; rubber insulation material

RIM Relevant Instructional Material; Resident Industrial Manager

RIMAC Research Institute for Medicine and Chemistry

RIMB Roche Institute of Molecular Biology

RIMES Road Information and Management Eco-System

RIMR Rockefeller Institute for Medical Research

RIMS Risk and Insurance Management Society

RIMV Registrar and Inspector of Motor Vehicles

Rin Rintintin

RIN Royal Institute of Navigation

RIN Registro Italiano Navale (Italian—Italian Naval Register)—bureau of shipping

rina reinitiation

RINA Royal Institution of Naval Architects

RINA Registro Italiano Navale e Aeronautico (Italian—Italian Air and Shipping Registry)

RIND Research Institute of National Defense

rinf rinforzando (Italian—with additional emphasis)

Ring Ring Lardner

Ring Ringstrasse (German—Ring Street)—tree-lined boulevard encircling inner Vienna

ringgit monetary unit of Malaysia

ringkasan (Malay—abbreviation)—also called *kependekan* or *singkatan*

RINM Resident Inspector of Naval Material

rinq. relinquished

rin (RIN) report identification number

RINS Research Institute for the Natural Sciences

RINSMAT Resident Inspector of Naval Stores and Materiel

rint rap in the nuts (kick in the scrotum)

Rio Rio de Janeiro, Brazil

RIO Reporting In and Out; Rhodesian Information Office; Rio de Janeiro (Galeao Airport)

Rioj La Rioja

riometer relative ionospheric opacity meter

Rio Neg Rio Negro (Argentine province)

RIOP Royal Institute of Oil Painters

RIOPR Rhode Island Open-Pool Reactor

Riós originally Rodríguez

riot real-time input-output transducer (translator); retrieval of information by online terminal (data processing)

rip raster image processor; radar identification point; radioisotope precipitation

rip ripieno (Italian—filling up)

RIP Reduction in Implementation Panel; Reduction in Personnel (layoffs); Reliability Improvement Program; Reserve Intelligence Program; Riker's Island Penitentiary; Rockefeller Institute Press

R.I.P. requiesca[n]t in pace [Latin—may one (they) rest in peace]

RIPA Royal Institute of Public Administration

Rip Blong Van Ripablik Blong Vanuatu (Bislama—Republic of Vanuatu)—formerly New Hebrides

RIPGA Rhode Island Personnel and Guidance Association

RIPH Royal Institute of Public Health

RIPHH Royal Institute of Public Health and Hygiene

R I Phil Rhode Island Philharmonic

RIPO Rhode Island Philharmonic Orchestra

ripple radioactive isotope-powered pulsed-light equipment (RIPPLE)

RIPPR Reliability Improvement Program Progress Report

RI & Prov Plant Rhode Island and Providence Plantation (Rhode Island's official name)

ripr viet riproduzione vietata (Italian—reproduction forbidden)

RIPS Radar-Impact Prediction System; Range-Instrumentation Planning Study; Range Instrumentation Planning System

rip viet riproduzione vietata (Italian—reproduction forbidden)

RIPWC Royal Institute of Painters in Water Colours

RIQS Remote Information Query System

rir recordable incident rate; reduction in requirement

rir (RIR) receiver input register

RIR Riverside International Raceway

R Ir AM Royal Irish Academy of Music

rirb radio-iodinated rose bengal

ririg reduced-excitation inertial reference-integrating gyroscope

ris (RIS) racially isolated school)(s)

RIS Radio Information Service; Range Instrumentation Ship; Redwood Inspection Service; Regulatory Information System; Research Information Service; Royal Imperial Society; Royal Infantry Society

risa radioiodinated serum albumin

RISB Rotter Incomplete-Sentence Blank

RISC Reduced Instruction Set Computer; Rockwell International Science Center

RISCA Rhode Island State Council on the Arts

RISCO Rhodesian Iron and Steel Company

RISCOM Rhodesian Iron and Steel Commission

RISD Rhode Island School of Design

rise reliability improvement selected equipment; reusable inflatable salvage equipment

RISE Research Information Services for Education; Responsible Industry for a Sound Environment; Rio Institute for Senior Education

RISEA Rhode Island Society of Enrolled Agents

RISM Research Institute for the Study of Man (USA)

RISOS Research in Secured Operations Systems; Research in Secured Operating Systems

risp rispettivamente (Italian—respectively)

RISP Ross Ice Shelf Project

RISS Range Instrumentation and Support System

RISSA Rhode Island School Superintendents Association

RISTA Reconnaissance, Intelligence, Surveillance & Target Acquisition

RISW Royal Institution of South Wales

rit ritard; ritardando; ritornello; ritual; ritualism; ritualistic; ritualization; ritualize

rit (RIT) retrograde inversion transposed (12-tone)

rit ritardando (Italian—holding back, retarding)

RIT Radio Information Test; Radio Network for Inter-American Telecommunication; Rochester Institute of Technology; Rorschach Inkblot Test; Royal Institute of Technology

RIT Red Interamericana de Telecommunicaciones (Spanish—Inter-American Telecommunication Network); *Roget's International Thesaurus*

Rita Margaret; Margarita

RITA Rand Intelligent Terminal Agent; Rural Industrial Technical Assistance

ritard ritardando (Italian—holding back, retarding)

RITC Rehabilitation Investment Tax Credit

Ritchie Ward Ritchie Press

RITE Rapid Information Technique for Evaluation

riten ritenuto (Italian—retaining time tempo)

RITES Rail India Technical and Economics Services

RITR Rework Inspection Team Report

RITS Rapid Information Transmission System; Reconnaissance Intelligence Technical Squadron

RITU (Profintern) Red International of Trade Unions

ritz ritzier; ritziest; ritziness; ritzy

Ritz Ritz-Carlton

riv radio influence voltage; river; rivet(ed)

riv riveduto (Italian—revised)

Riv River; Riviera; Rivington; Rivke

Riv Rivke (Yiddish—Rebecca)

Rivadavia Comodoro Rivadavia, Argentina

Riverfront Riverfront Stadium, Cincinnati, Ohio

Riverside Riverside County Jail (California)

rivu river view

RIW Reliability Improvement Warranty

riyal monetary unit of Qatar and Saudi Arabia

RIZ Radio Industry Zagreb

rj (RJ) ramjet

RJ Rio de Janeiro; Royal Jordanian (airlines)

R & J Romeo and Juliet

RJA Reform Jewish Appeal; Retail Jewelers of America; Royal Jersey Artillery

RJAF Royal Jordanian Air Force

RJAS Royal Jersey Agricultural Society

RJC Rochester Junior College; Rosenwald Junior College; Roswell Junior College

rje remote job entry

RJIS Regional Justice Information System

Rjk Reykjavik

RJM Royal Jersey Militia

rjp realistic job preview

RJR RJ Reynolds

RJR Nab R J Reynolds Nabisco

rk radial keratotomy; rock; run of kiln

r/k (R/K) radial keratotomy

rk rooms-katholiek (Dutch—Roman Catholic)

r-k rooms-katholiek (Dutch—Roman Catholic)

RK Air Afrique (2-letter coding); Radio Kabul; Rock

RK Rdeci Kriz (Yugoslavian—Red Cross)

Rka Rijeka

rkg radiocardiogram

RKN Republic of Korea Navy

RKO Radio-Keith-Orpheum (theater circuit)

rkp record key position

rkt rocket

Rkt Sta Rocket Station

RKU Ruprecht-Karl-Universität (Heidelberg)

RKV Rose Knot (tracking station vessel)

rkva reactive volt-ampere

rky rocky; roentgen kymography

rl coarse rales; radiation length; rail(ing); reduction level; relay logic; revised laws; rocket launcher

r/l radio location; random length(s)

r & l rail and lake

r-to-l right-to-left (photo caption abbreviation)

Rl Raphael

RL high-powered radio range loop radiator(s); Radiation Laboratory; Radio Liberty; Reading List; Record Librarian; Record Library; Regent's Line; Republic of Liberia; Research Laboratory; River Lines (railroad); Roland Line; Roman Law; Rupert Line; Rutland Line

RL *Rape of Lucrece; Rijksuniversiteit Limburg* (Dutch—State University of Limburg)

rl₁ few line rales

rl₂ moderate number of rales

rl₃ many coarse rales

rla restricted landing area; right lower arm

RLA Religious Liberty Association ,

RLAA Red Light Abatement Act

rladd radar low-angle drogue delivery

RLAF Royal Laotian Air Force

RLB Sir Robert Laird Borden (Canada's ninth Prime Minister)

rlbcd right lower border of cardiac dullness

rlbm (RLBM) rearward-launched ballistic missile(s)

R.LC Radio Liberty Committee

RLCA Rural Letter Carriers' Association

RLCS Radio-Launch Control System

rld radar laydown delivery; rolled

rld (RLD) relocation list dictionary

RLD Raymond L Ditmars

RLDPAS Royal London Discharged Prisoners' Aid Society

rld's retail liquor dealers

rle relative luminous efficiency; right lower extremity

Rle Ramble

rl est real estate

rletfl report leaving each thousand-foot level

rlf relief; retrolental fibroplasia

RLF Royal Literary Fund

rlg railing; ring laser gyro

rlg *rilegato* (Italian—bound)

RLG Research Library Group; Royal Laos Government

rlgn realign; religion

rlgn dfld religion defiled

RLHS Railway and Locomotive Historical Society

RLHTE Research Laboratory of Heat Transfer in Electronics (MIT)

RLI Realtors Land Institute; Rhodes-Livingstone Institute

RLIN Research Libraries Information Network

rll right lower limb; right lower lobe (lung); run length limited

rllb right long-leg brace

RLM Regional Library of Medicine (PAHO)

rlmd rat-liver mitochondria

RLNWR Rice Lake National Wildlife Refuge (Minnesota); Ruby Lake National Wildlife Refuge (Nevada)

RLO Regional Liaison Office(r); Returned-Letter Office

rlp rail loading point

RLPAS Royal London Prisoners' Aid Society

RLPO Royal Liverpool Philharmonic Orchestra

rlq right lower quadrant (abdomen)

rlr right lateral rectus (eye muscle)

rls reels (flow chart)

Rls rial (Iranian currency unit)

RLS Robert Louis Stevenson; Royal Lancastrian Society

rlse release

RLSS Royal Life Saving Society

RLT Revokable Living Trust

rltr realtor

RLTS Radio-Linked Telemetry System

rltv relative

rlty realty

rlv relieve

Rlv Rauscher leukemia virus

rly relay

Rly Railway

rm range mark(s); raw material; ream; receiving memorandum; recovered memory; respiratory movement; ring micrometer; room; rubber marker(s)

rm (RM) record mark (flow chart); remaindered mark (books)

r/m revolutions per minute

r & m redistribution and marketing; reliability and maintainability; reports and memoranda

Rm meter resistance (symbol); Romania (Rumania); Romanian (Rumanian)

RM Radioman; Raybestos-Manhattan; Registered Magistrate; Registered Mail; Registered Midwife; Reichsmark (German currency); Research Memorandum; Ringling Museum; Rocky Mountains; Royal Mail; Royal Malta; Royal Marine; Royal Marines

R/M Raybestos/Manhattan

R & M Robbins & Myers

rma right mento-anterior

RMA Radio Manufacturers Association; Regional Manpower Administration; Retread Manufacturers Association; Rice Millers Association; Ringling Museum of Art; Robert Morris Associates (Bank Loan Officers and Credit Men's Association); Royal Marine Artillery; Royal Military Academy; Royal Musical Association; Rubber Manufacturers Association

RMADB Reactor Maintenance and Disassembly Building

RMAF Royal Malaysian Air Force; Royal Moroccan Air Force

RMAG Rocky Mountain Association of Geologists

RMAI Radio Manufacturers' Association of India

rm ar reaming arbor

RMAS Rochester Museum of Arts and Sciences

r mast radio mast

RMB Royal McBee

RMB Renminbi (Chinese—people's money)

RMBAA Rocky Mountain Business Aircraft Association

RMBN Rocky Mountain Broadcasting Network

rmc rod memory computer

RMC Radio Monte Carlo; Republic Movement of Crimea; Revolutionary Military Council; Reynolds Metal Company; Rochester Manufacturing Company; Royal Military College

RMCC Royal Military College of Canada

RMCM Royal Manchester College of Music

RMCMI Rocky Mountain Coal Mining Institute

RMCPA Rocky Mountain College Placement Association

RMCS Royal Military College of Science

rmct rat mass cell technique

RMCU Royal Mail Container Unit

rmd ready money down; required minimum distribution; retromanubrial dullness

RMD Reaction Motors Division (Thiokol Chemical Corporation); Research Management Division (D of E)

RME Relay Marine Experiment

RMEA Rubber Manufacturing Employers' Association

R-meter radiation meter

R Met S Royal Meteorological Society

RMFVR Royal Marine Forces Volunteer Reserves

rmi radio magnetic indicator; reliability maturity index(ing); repetitive-motion injury

RMI Rack Manufacturers Institute; Reaction Motors Incorporated; Reactive Metals Incorporated; Republic of the Marshall Islands; Roll Manufacturers Institute

RMIC Republican Majority Issues Committee

rmicbm (RMICBM) road-mobile intercontinental ballistic missile

r/min revolutions per minute

RMIS Resource Management Information System

RMIT Royal Melbourne Institute of Management

RMJC Robert Morris Junior College

rmks remarks

rml right mediolateral; right middle lobe

RML Rand Mines Limited; Royal Mail Lines; Royal Malta Library (Valetta)

RMLF Robert M La Folette

RMLI Royal Marine Light Infantry

RMM & EA Rolling Mill Machinery and Equipment Association

RMMNH Regar Memorial Museum of Natural History (Anniston, Alabama)

RMN Registered Maternity Nurse; Registered Mental Nurse; Richard Milhouse Nixon (37th President of the United States and first to resign the presidential office); Royal Malaysian Navy

RMNP Rhodes Matopos National Park (Rhodesia); Riding Mountain National Park (Manitoba); Rocky Mountain National Park (Colorado)

RMNS Royal Merchant Navy School

RMO Regimental Medical Officer; Regional Medical Officer; Resident Medical Officer; Royal Marine Office

RMOGA Rocky Mountain Oil and Gas Association

R'mond Richmond

rmp right mento-posterior

RMP Radio Motor Patrol; Reentry Measurement Program; Regional Medical Program; Research Management Plan; Research and Microfilm Publications; Royal Marine Police; Royal Mounted Police

RMPA Royal Medico-Psychological Association

rmpc rubber-mold plaster casting

RMQ Records Management Quarterly

rmr resting metabolic rate

RMR Royal Marines Reserve

RMRA Royal Marines Rifle Association

Rmrs Ramirez

RMRS Rocky Mountain Radiological Society

rms root mean square

RMS Radiation Monitoring System; Railway Mail Service; Records Management System; Rate Monotonic Scheduling; Remote Manipulator System; Resources Management System; Royal Mail Service; Royal Mail Ship; Royal Microscopical Society

RMSA Rural Music Schools Association

RMSC Royal Marines Sailing Club

RM Sch Mus Royal Marines School of Music

rmsd root-mean-square deviation

rmse root mean square error

RMsf Rocky Mountain spotted fever

RMSM Royal Marines School of Music; Royal Military School of Music

RMSP Royal Mail Steam Packet (company)

rmt right mento-transverse

rmte remote

rmu remote maneuvering unit

rmv respiratory minute volume

RMWAA Roadmasters and Maintenance of Way Association of America

RMWC Randolph-Macon Woman's College

RM-W/MB Rijksmuseum Meermanno Westreenianum/ Museum van het Boek (Dutch—Merrmanno-Westreenianum Royal Museum and the Museum of the Book)—The Hague

rn reception nil; removal note; research note; round-nose (bullet); running noose; running nose

r of n range of neap (tides)

Rn negative resistance (symbol); radon; Rangoon

RN radionavigation; Registered Nurse; República de Nicaragua; Reynold's number; Royal Navy

RN *Registered Nurse* (periodical)

rna (RNA) ribonucleic acid

RNA Registered Nurse Anesthetist; Religion Newswriters Association; Research Natural Area; Romantic Novelists' Association; Royal Naval Association

R/NAA Rocketdyne/North American Aviation

RNAC Royal Nepal Airline Corporation

RNADC Royal Netherlands Air Defense Command

RNAF Royal Naval Air Force

RNAFF Royal Netherlands Aircraft Factories Fokker

RNAO Registered Nurses Association of Ontario

RNAS Royal Naval Air Station

rnase ribonuclease

RNAV Royal Naval Artillery Volunteers

RNAW Royal Naval Aircraft Workshop

RNAY Royal Naval Aircraft Yard

rnb received—not billed

RNB Royal Naval Barracks

RNBT Royal Naval Benevolent Trust

RNC Republican National Committee; Royal Naval College (Greenwich)

Rnch Ranch

Rnchs Ranches

RNCM Royal Northern College of Music

RN & CR Ryde, Newport, and Cowes Railway

RNCS Royal Netherlands Chemical Society

RNCSRL Ralph Nader Center for the Study of Responsive Law

rnd round

RND Royal Naval Division

RND *Rijksnijverheidstdienst* (Dutch—Government Industrial Advisory Service)

rnd(s) round(s)

RNE *Radio Nacional de España* (Spanish—National Radio Broadcasting System)

RNEC Royal Naval Engineering College

RNES *Radiodifusora Nacional de El Salvador* (Spanish—National Radio Network of El Salvador)

rnf receiver noise figure

Rnf Renfrew

RNF Royal Northumberland Fusiliers

rnfp radar not functioning properly

RNFU Rhodesia National Farmers' Union

rng range

R ng P *Republika ng Pilipinas* (Pilipino—Republic of the Philippines)

rngt renegotiate

rni recommended nutrient intake

RNIB Royal National Institute for the Blind

RNID Royal National Institute for the Deaf

rnit radio noise interference test

RNL Raffles National Library (Singapore); Royal Netherlands Line

RNLAF Royal Netherlands Air Force

RNLI Royal National Lifeboat Institution

RNLO Royal Naval Liaison Office(r)

rnm radionuclide migration

rnm (RNM) radionavigation mobile

RNMC Royal Naval Medical Corps

RNMD Registered Nurse for Mental Defectives

RNMDSF Royal National Mission to Deep-Sea Fishermen

RNMI Realtors National Marketing Institute

RNMS Registered Nurse for the Mentally Subnormal; Royal Naval Medical School

RNMWS Royal Naval Minewatching Service

RNN Royal Nigerian Navy

RNNP Royal Natal National Park (South Africa)

RNO Resident Naval Officer

RNoAF Royal Norwegian Air Force

RNOC Royal Naval Officers Club

R No N Royal Norwegian Navy

rnp required navigational performance; ribonucleic protein

RNP Redwood National Park (California); Rondane National Park (Norway); Ruaha National Park (Tanzania); Ruhana National Park (Ceylon)

R.N.P. Registered Nurse Practitioner

RNP *Radio Nacional de Peru* (Spanish—National Radio of Peru)

RNPFN Royal National Pension Fund for Nurses

RNPL Royal Naval Physiological Laboratory

RNPS Royal Naval Patrol Service; Royal Navy Polaris School

rnr runner

r-'n'-r rock-and-roll

RNR Royal Naval Reserves

RNRA Royal Naval Rifle Association

RNRRA Royal Naval Reserve Rifle Association

RNRS Royal National Rose Society

rns radar netting station

RNS Religious News Service; Royal Naval School; Royal Numismatic Society

RNSA Royal Naval Sailing Association

RNSC Royal Naval Staff College; Royal Netherlands Steamship Company

RNSR Royal Naval Special Reserve

RNSS Royal Naval Scientific Service

RNSYS Royal Nova Scotia Yacht Squadron

rnt residual nitrogern time; roentgenologist; roentgenology

RNT Registered Nurse Tutor

RNT *Revised New Testament* (Roman Catholic)

RNTE Royal Naval Training Establishment

rnth raised non-tight hatch

RNTU Royal Naval Training Unit

rnu radar netting unit; radio noise voltage

rnvc reference number variation code

RNVR Royal Naval Volunteer Reserve

RNW Radio Navigational Warning

RNWMP Royal Northwest Mounted Police

RNWR Ravalli National Wildlife Refuge (Montana)

rnwy runway

RNYC Royal Northern Yacht Club; Royal Norwegian Yacht Club

RNZ Radio New Zealand

RNZAC Royal New Zealand Aero Club; Royal New Zealand Armoured Corps

RNZAEC Royal New Zealand Army Education Corps

RNZAF Royal New Zealand Air Force

RNZAMC Royal New Zealand Army Medical Corps

RNZAOC Royal New Zealand Army Ordnance Corps

RNZAS Royal New Zealand Astronomical Society

RNZASC Royal New Zealand Army Service Corps

RNZCD Royal New Zealand Chaplains Department

RNZC Sigs Royal New Zealand Corps of Signallers

RNZDC Royal New Zealand Dental Corps

RNZE Royal New Zealand Engineers

RNZEME Royal New Zealand Electrical and Mechanical Engineers

RNZIH Royal New Zealand Institute of Horticulture

RNZ Inf Royal New Zealand Infantry Corps

RNZIR Royal New Zealand Infantry Regiment

RNZN Royal New Zealand Navy

RNZNC Royal New Zealand Nursing Corps

RNZNR Royal New Zealand Naval Reserve

RNZNVR Royal New Zealand Naval Volunteer Reserve

RNZPC Royal New Zealand Provost Corps

RNZSHWC Royal New Zealand and Society for the Health of Women and Children (Plunket Society)

RNZYS Royal New Zealand Yacht Squadron

ro rancho; receive only; recto (frontside of page); reddish orange; reverse osmosis; right opening; right orifice; road oil; rough opening; runover

ro (RO) readout (flow chart)

r/o roll out (final turn of an interceptor); routing order; rule out

r & o rail and ocean

ro. *recto* (Latin—front of the page, right-hand page)

rᵒ *recto* (Portuguese—face of page; right-hand page; this side)

Ro output resistance (symbol)

RO Radar Observer; Radar Operator; Radio Observer; Radio Operator; Recorder's Office; Recruiting Officer; Reserve Order; Romania (Internet code)

R-O Reporting Officer; Ritter Oleson (technique)

RO *Republik Osterreich* (Republic of Austria); *Resedentie Orkester* (Dutch—Resident Orchestra)—The Hague

R-O *Residentie-Orkest* (Dutch—Residency Orchestra)—at The Hague

roa received on account; return on assets; right occiput anterior

roa (ROA) rights of accumulation

RoA Record of Acquisition

ROA Reserve Officers Association; Retired Officers Association; Royal Order of Altruists

ROA *Russkaya Osvoboditelnaya Armiya* (Russian—Russian Liberation Army)

roaa return on average assets

ROAD Reorganization Objective Army Division; Re-Organize Army Division

roads. roadstead

Roads ports of Hampton Roads (Portsmouth, Newport News, Norfolk, Sewells Point)

ROA/I Received On Account/ Insurance

roam. return of assets managed (banking)

ROAMA Rome Air Materiel Area

ROA/P Received On Account/ Private

roar. right of admission reserved

ROAR Royal Optimizing Assembly Routine

ROARE Reeducation of Attitudes and Repressed Emotions

roast-beefsan roast-beef sandwich

roast-beefwich roast-beef sandwich

ROAUS Reserve Officers Association of the United States

rob remaining on board (aircraft or ship cargo)

Rob Robert; Robinson College, Oxford

ROB Regional Office Building

Robby Robert(a)

robc readiness objective code

robe. wardrobe

robeps radar operating below prescribed standards

Robert F Kennedy Robert F Kennedy Stadium, Washington, D.C.

Roberts Roberts International Airport serving Monrovia, Liberia

robin. (ROBIN) rocket-balloon instrument

robo rocket orbital bomber

robrep robbery report

Robt Robert

roc rate of climb; receiver operating characteristic (curve); required operational capabilities; return on capital; rotatable optical cube; run on crap (fuel of the future)

RoC Register of Copyrights

RoC (ROC) Republic of China (offshore China); Republic of the Congo (formerly the French Congo)

R o C Republic of Congo

ROC Reach Out for Christ; Regional Occupation Center; Rochester, New York (airport); Royal Observer Corps; Russian Orthodox Church

R o Cam Republic of Cameroon

ROCAPPI Research on Computer Applications for the Printing and Publishing Industries

roc coc rock cocaine

roce return on capital employed

Roch Rochester

R o Ch Republic of Chad

Rochambeau Cayenne, French Guiana's airport; Count Jean Baptiste Donatien de Vimeur de Rochambeau

Roch Phil Rochester Philharmonic

rocid reorganization of combat infantry divisions

rock rock-'n'-roll music; a form of cocaine

Rock Rock of Gibraltar

rock-a-billy rock-'n'-roll + hillbilly (music)

Rockaways Long Island, New York's south shore beaches— Far Rockaway, Rockaway Beach, Rockaway Park, Rockaway Point—plus other Rockaways in California, New Jersey, and Oregon

Rockefeller Nelson A Rockefeller, 41st Vice President of the United States

rockex rocket exercise

rockrest rock music festival

Rockies Rocky Mountains— major mountain system of western North America extending from Alaska and Canada to central New Mexico

rockoon(s) balloon-supported rocket(s)

ROCMD Rochester Contract Management District

Rocosas Rocallosas (Spanish—Rockies)—Rocky Mountains

rocp radar (or radio) out of commission for parts

rod required operational data; required operational date; rusting, oxidation, discoloration

Rod Roderick; Rodney; Rodrigo; Rodrigues; Rodriguez

RoD Record of Decision

ROD Rosskoye Osvoboditelnoye Dvizheniye (Russian— Russian Liberation Movement)

Rodale Rodale Books

rodar rotor-blade radar

Roddy Roderick; Rodney

rodeocade rodeo parade

Rodg Roger

rodiac rotary dual input for analog computation

roe return on employees

roe (ROE) reflector orbital equipment; return on equity

RoE Rules of Engagement

ROE Royal Observatory Edinburgh

roentgen roentgenology

ROEP Refugee Orientation and Employment Program

rof reporting organizational-file

ROF Royal Ordnance Factory

ROFA Radio of Free Asia

rofor route forecast

roft radar off target

ROFT Record Of Federal Tax (liabilities)

rog reactive organic compound; rise-off-ground

R o G Republic of Guinea

Rog Roget's Thesaurus

roger your message received and understood

r o/h regular overhaul(ing)

ROH Royal Opera House (Covent Garden)

roi return on investment

ROI Range Operating Instructions

Rois Rodrigues

Roiz Rodriguez

roi range on jamming

roid rage steroid rage

Rok a South Korean

RoK Republic of Kiribati in the western Pacific between Micronesia and the Solomon Islands

ROK Republic of Korea; Rockford (Illinois) Forest City (National Association)

roka return on knowledge assets

ROKA Republic of Korea Army

ROKAF Republic of Korea Air Force

ROKAMS Republic of Korea Army Map Service

ROKN Republic of Korea Navy

ROKPUC Republic of Korea Presidential Unit Citation

roksonde rocket sounding

rol record output length; right occipitolateral

rolet reference our letter

Rolf Rudolf, Rudolph

rol k rolling keel

Rolls Rolls-Royce

ROLS Recoverable Orbital Launch System

rom radar operator mechanic; range of motion; range of movement; roman (type); rough order of magnitude

rom (ROM) read-on memory; read-only memory; roll-over mortgage

Rom The Letter of Paul to the Romans; Roman; Romance language; Romania(n)

Rom Book of Romans (New Testament); (German— Rome)—capital of Italy; Romanian (Romance language)

RoM Republic of Macedonia; Republic of the Marshall Islands

ROM Rome, Italy (Fiumicino airport); Royal Ontario Museum

R O M Republic of Malagasy

roman remotely operated mobile manipulator (acronym); roman candle (firework display); roman number (I, II, III, IV, V, etc;); roman type

Roman Empire in A.D. 117 comprised Hispania (Spain and Portugal); Gallia (France and the Lowlands); Germania (much of Gerunany and Austria); Illyricum (the Balkans); Italia (Italy); Turkey, the Middle East, Egypt, North Africa; and Britannia (some of England, Wales, and southern Scotland)

Romania Republic of Romania, formerly Socialist Republic of Romania (Balkan state) also spelled Rhumania or Rumania

Rom Ant Roman Antiquities

Romamtisch (German—Romantic)—Bruckner's Symphony No 4

ROMBI Results of Marine Biological Investigations

rom bios read-only memory basic input-output system

Rom Cath Roman Catholic

romemo refer to our memorandum

Romeo Romeo and Juliet (Shakespearean tragedy inspiring a dramatic symphony by Berlioz, a five-act opera by Gounod, a ballet by Prokofiev, an overture-fantasia by Tchaikovsky)

Rom Hist Roman History

Rom & Jul Romeo and Juliet by William Shakespeare

romon receiving-only monitor

ROMT Range-of-Motion Test

romv return on market value

Rom Yid Romanian Yiddish

ron remain overnight; research octane number

Ron Ronald

Ronald Ronald Press

rond rondeau; rondeaux; rondel; rondels

RONDA Royal Oriental Nut Date Association

Röntgen Wilhelm Conrad Röntgen

roo kangaroo

Roo Roosevelt

ROO Range Operations Office(r)

rooi return on original investment

roor released on own recognizance

roo(s) kangaroo(s)

'roo(s) kangaroo(s)

root. relaxation oscillator optically tuned

rop record of production; registered options principal; retinopathy of prematurity; right occiput posterior; run of press

RoP Republic of Palau, located between Papua, New Guinea and the Philippines

ROP Regional Occupational Program

ropeval readiness-operational evaluation

ropp receive-only page printer

Roques Los Roques Islands

ror rocket-on-rotor (device for assisting helicopter takeoffs)

ror (ROR) release on own recognizance

Ror Rorschach (inkblot test)

RORA Reserve Officer Recording Activity

RORC Royal Ocean Racing Club

rord return on receipt of document

roreq reference our requisition

ro/ro roll on/roll off

RoRoRo *Rowolt Rotations Romäne* (German—Rowalt's Rotary Romances)

ros reduced operational status; return on sales

ros (ROS) run of schedule (radio or television)

Ros Rosamund; Roscommon; Rosemary; Rostock

R o S Republic of Senegal

ROS Range Operating Station; Range Operation Station; Royal Order of Scotland

rosa recording optical-spectrum analyzer

ROSAMES Road Safety Management Expert System

RÖSAT *Röntgensatellite* (German—Röntgen satellite)

Rosc Roscommon

ROSCOE Remote Operating System Conversational Operating Environment

ROSCOP Report on Observations/Samples Collected by Oceanographic Programs

rose residuum-oil supercritical extraction; rising observational sounding equipment; rose cut(ting); rose engine;

rose fever; rose gum; rose hips; rose lathe; rose leaf; rose leaves; rose mill; rose oil; rose quartz; rose reamer; rose window; rose wine; rose worm; rose wort; roseate; rosebud(s); rose-cake; rose-colored; rosemary; rosette; rosewood

ROSE Research Open Systems Europe

Rosh Hash *Rosh Hashanah* (Hebrew—New Year)

Rosh Hod *Rosh Hodesh* (Hebrew—beginning of the new month beginning at the new moon)

rosie (ROSIE) reconnaissance by orbiting ship-identification equipment

Rosie Rosa; Rosamund; Rose; Rosemarie; Rosemary

rosla raising of school-leaving age

ROSPA Royal Society for the Prevention of Accidents

ross remote ocean sensing system

Ross Ross and Cromarty

Rostov Rostov-on-Don

ROSY Recession Over Six Years

rot remedial occupational therapy; right occipito-transverse; rolary; rotate; rotation; rotor

rot. (ROT) rate of return(ing)

Rot Rotterdam

Rotary Flight Pioneer Juan de la Cierva

ROTC Reserve Officers Training Corps

rotcc receiver-off-hook-tone connecting circuit

roti recording optical tracking instrument

rotis rotisserie

rotmh raised oil-tight manhole

rotn rotation

roto rotary press; rotogravure

Rot Phil Rotterdam Philharmonic

rotr (ROTR) receive-only typing reperforator (data processing)

ROTS Reusable Orbital Transport System

rotsal rotate and scale

Ron Rouen

ROU *República Oriental del Uruguay*

roul roulette

Rouin Roumanian

'round around

rout routine

rov remotely-operated vehicle

Rov Rover(s)

Rover(s) Coloradan(s)

rov's (ROVs) remotely-operated vehicles

row. reverse-osmosis water; risk of war

RoW (ROW) Right of Way

R-O-W disease Rendu-Osler-Weber disease

Rox Roxburgh; Roxburghshire; Roxbury

Roxy Roxana; SL Rothafel

Roy Royal

Roy Alb Hall Royal Albert Hall

Royals Royals Stadium, Kansas City, Missoun

Royal Society The Royal Society of London for Improving Natural Knowledge (incorporated 1662)

Roy Bel *Royaume de Belgique* (French—Kingdom of Belgium)

Roy Bot Gards Royal Botanic Gardens

Roy Com Soc Royal Commonwealth Society (formerly Royal Empire Society; formerly Royal Colonial Institute)

Roy Dan Royal Dansk (Royal Danish)

Roy Fest Hall Royal Festival Hall

ROY G BIV (acronymic mnemonic for recalling spectral colors —red, orange, yellow, green, blue, indigo, violet)— see vibgyor

Roy Liv Phil Orch Royal Liverpool Philharmonic Orchestra

Roy Opera Royal Opera House Orchestra (Covent Garden)

Roy Phil Royal Philharmonic Orchestra

Roy Soc Royal Society

Rozh Rozhdestvensky (the admiral or the conductor)

rp plate resistance (symbol); raid plotter; rally point; rear-screen projection; received pronunciation (RP); reception poor; redundancy payment; relay paid; release point; relief pitcher; relief pitching; reply paid; reporting post; reprint; repurchase agreement; response pattern; responsible party; retained personnel; retinitis pigmentosa, return of post; return premium; rhodium plating; rhodium-plated; rocket projectile (RP); rocket propellant; role playing; run rating for pitchers; rust preventive

r-p reprint; reprinting

rP reddish purple

Rp Rappen (Swiss—centime); *rupiah* (Indonesian currency unit)

RP remote pickup (broadcast); rocket projectile; Rules of Procedure

R-P Rhône-Poulenc

R/P Registered Plumber; Reporting Person; Royal Provincial (Tory American troops)

RP Radiotelevisão Portugesa (Portuguese—Radio-Television); *República de Panama; República del Paraguay; República del Peru; República Portuguesa* (Portugal); *Révérend Pére* (French—Reverend Father)

RP-1 rocket-propellant type-l fuel (kerosene)

rpa radar performance analyzer; random phase approximation

RPA Rationalist Press Association; Record of Personal Achievement; Regional Planning Association; Registered Plumbers' Association; Respiratory Protection Administrator; Rubber Proofers' Association

R & PA Rifle and Pistol Association

RPAA Radiata Pine Association of Australia

rpar rebuttable presumption against registration (dangerous substance examination)

RPB Regional Preparedness Board; Research to Prevent Blindness (fund)

rpc radar planning chart; remote position control; remote procedure call; reply postcard; request (the) pleasure (of your) company; reversed phase column

RPC Regional Processing Center; Reliability Policy Committee; Republican Party Conference; Respiratory Protection Coordinator; Royal Pay Corps; Royal Pioneer Corps

RPC République Populaire du Congo (French—Popular Republic of the Congo) formerly the French Congo

RPCC Reactor Physics Constants Center

RPCFT Reiter Protein Complement Fixation Test

RP China República Popular China (Spanish—People's Republic of China)

RPCV Returned Peace Corps Volunteer

rpd radar planning device; rapid parcel delivery

RPD Regional Port Director; Regius Professor of Divinity; Rocket Propulsion Department; Rocket Propulsion Division

R.P.D. Rerum Politicarum Doctor (Latin—Doctor of Political Science)

RPD Cor República Popular Democratica de Corea (Spanish—People's Democratic Republic of Korea)—North Korea

RPDL Radioisotope Process Development Laboratory

Rpds Rapids

rpe range probable error; related payroll expense

RPE Radio Propagation Engineering; Rocket Propulsion Establishment

RPEA Regional Planning and Evaluation Agency

rpedl repetitively pulsed electric-discharge laser

rpf radiometer performance factor; relaxed pelvic floor; renal plasma flow

RPF Rassemblement du Peuple Français (French—Rally of the French People)—de Gaulle's party; Gaullists

RPFMA Rubber and Plastics Footwear Manufacturers' Association

rpfod reported for duty

RP-FS Rozensweig Picture-Frustration Study

rpg radiation protection guide; report program generator; rifle-propelled grenade; rocket-propelled grenade; rounds per gun

RPG Regional Planning Group; Report Program Generator

rph revolutions per hour

rph (RPH) remotely piloted helicopter

RPh Registered Pharmacist

RPH Royal Perth Hospital

rpha reversed passive hemmagglutination

R Phil S Royal Philharmonic Society

RPHST Research Participation for High School Teachers

rpi radar precipitation integrator; random procedure information; rated position identifier; real progress Index(ing)

RPI Railway Progress Institute; Rensselaer Polytechnic Institute; Retail Price Index; Rose Polytechnic Institute; Royal Pakistan Institute; Ryerson Polytechnical Institute

RPIA Rocket Propellant Information Agency

RPIC Rock Properties Information Center (Purdue)

rpie (RPIE) real property installed equipment

rp index respiratory rate index; respiratory pulse index

RPK Regiment President Kruger

rpl running program language

RPL Radiation Physics Laboratory (NBS); Regina Public Library; Repair Parts List; Richmond Public Library; Roanoke Public Library; Rochester Public Library; Rocket Propulsion Laboratory; Rockhampton Public Library; Rockport Public Library

rplca replica

rpm radiation polarization measurement; reliability performance measure(ment); remote performance monitoring; repairman; respiration, pulse, mental status; revenue passenger miles; revolutions per minute; rotations per minute

RPM Raven's Progressive Matrices (test); Regional Plant Manager; Regulatory Project Manager; Rustenburg Platinum Mines

R & PM Research and Program Management (NASA)

rpmb (RPMB) remotely piloted miniature blimp

RPMF Radiation Pattern Measurement Facility

rpmi revolutions-per-minute indicator

RPMI Roswell Park Memorial Institute

rpn reverse polish notation

RPN Registered Psychiatric Nurse

rpo revolutions per orbit

RPO Railway Post Office, Repository Program Office(r); Rochester Philharmonic Orchestra; Rotterdam Philharmonic Orchestra; Royal Philharmonic Orchestra

RPO *Rotterdams Philharmonisch Orkest* (Dutch—Rotterdam Philhamonic Orchestra)

rpoa recognized private operating agencies

rpoc report proceeding on course

R point restriction point

rpp radar power programmer; reply paid postcard; request present position; return paid postal

RPP Radio Propagation Physics; Registered Polarity Practitioner

rppe research, program, planning, evaluation

rppi repeater plan-position indicator

RPPI Rubber and Plastics Processing Industry

RPPMP Repair Parts Program Management Plan

RPQ Request for Price Quotation

rpr read printer

rpr (RPR) rapid plasma reagin

RPR *Rassemblement pour la République* (French—Assembling for the Republic); *Republica Populara Romana* (Romania)

RPRA Railroad Public Relations Association

RPRAGB Rubber and Plastics Research Association of Great Britain

rprt report

RPRT Rapid Plasma Reagin Test

rps revolutions per second; rotational position scanning; rotational position sensing

rp's rice planters; rubber planters

RPs repurchase agreements at commercial banks

RPS Radiological Protection Service; Railway Progress Society; Rapid Processing System; Registered Publication Section; Reliability Problem Summary; Retired Persons Services; Revenue Protection Strategy; Roadway Package System; Royal Philatelic Society; Royal Philharmonic Society; Royal Photographic Society

RPS *Republika e Shqipërisë* (Albania)

rp shortly reprinting shortly

RPSM Resources Planning and Scheduling Method

RPSs Reliability Problem Summary Cards; Republic of the Philippines Ships

rpt repeat

Rpt Report

RPT Registered Physical Therapist

RPTA Recycled Paperboard Technical Association

RPTI Rewards Program for Terrorism Information

rpt's (RPTs) rapid-phase transformations

RPU Radio Propagation Unit (USA)

rpv remotely-piloted vehicle; remote pilotless vehicle

rpw ranked positional weight

RPYC Royal Perth Yacht Club

rq remoulded-regraded quality (tires); rose quartz

rq (RQ) respiratory quotient

RQ romance quotient

R/Q Request for Quotation

R & QA Reliability and Quality Assurance

RQAS Royal Queensland Art Society

RQBA Royal Queensland Bowls Association

rqd rock-quality designation; rock-quality determination

r qd raised quarterdeck(ing)

rqdcz request clearance to depart control zone

rqecz request clearance to enter control zone

rqiac requires immediate action

rql reference quality level

RQMS Regimental Quartermaster Sergeant

rqmt requirement

rqr require; requirement

rqs ready qualified for standby

RQS Rate Quoting System

rqtao request time and altitude over

rqto request travel order

RQYS Royal Queensland Yacht Squadron

rr radiation response; radio range; radio ranging; railroad; rapid rectilinear; rear; rearward; relief runs; respiratory rate; rifle range; rural route; rush release; rush and run

r/r right rear

r & r rape and robbery; rate and rhythm (pulse); rest and recreation; rest and recuperation; rest and rotation (of military personnel); rock and roll; rock and rye (whiskey); rush and run

r & r (R & R) rape and ruin; rest and recovery; rest and recreation

RR Railroad; Raritan River (railroad); Recommendation Report; Recovery Room; Recruit Roll; Reliability Requirements; Remington Rand; Renegotiation Regulations; Research Report; Rifle Range; Right Reverend; Rolls-Royce; Ronald Reagan—40th President of the United States whose full initials are RWR (Ronald Wilson Reagan); Rural Route

R-R Rolls-Royce

rra (RRA) radio relay aircraft; ready reserve advances

rRA specific acoustic resistance

RRA Radiation Research Associates; Revenue Reconciliation Act

R/RA Repair/Rework Analysis

RRAF Royal Rhodesian Air Force

R-rated moving picture restricted to adults

RRB Race Relations Board; Railroad Retirement Board; RR Bowker

RRBC RR Bowker Company

RRBS Rapid-Response Bibliographic Service

rrc radar return code; reference repository conditions; reports of rating cases

rr & c records, reports, and control

RRC Race Relations Commission; Race Relations Conciliator; Recruit Reception Center; Regional Resource Center; Requirements Review Commmittee; Rocket Research Corporatiion; Royal Red Cross; Rubber Reserve Committee; Rubber Reserve Company; Rubber Reserve Corporation; Russian Research Center (Harvard)

R.R.C. Lady of the Royal Red Cross

rrcc reduced-rate contribution clause

RRCC Redwood Region Conservation Council

rr cells radiation reaction cells

rrd receive, record, display

rr & d reparations, removal, and demolition

RRD Reliability Requirements Directive

rrda rendezvous retrieval, docking, and assembly (of orbital station or space vehicle)

RRDS Rough Rock Demonstration School

rr & e round, regular, and equal (eye pupils)

RRE Railroad Enthusiasts; Royal Radar Establishment

RREA Rural/Regional Education Association

rr/eo race relations/equal opportunity

R Rep Records Repository (USAF)

RRF Reading Reform Foundation; Refrigeration Research Foundation

rrhage (Latin suffix—excessive flow)—hemorrhage

rrhea (Greek—derived suffix—to flow)—diarrhea, gonorrhea

rri range rate indicator

RRI Radio Republik Indonesia; Rocket Research Institute; Rubber Research Institute

RRIC Rubber Research Institute of Ceylon

rrid reverse radial immunodiffusion

RRIM Rubber Research Institute of Malaya; Rubber Research Institute of Malaysia

RR-IM Research and Reports-Intelligence Memo

RRIS Remote Radar Integration Station

rrl reference—repository location

RRL Regimental Reserve Line; Registered Record Librarian; Reserve Retired List; Road Reserve Laboratory; Roodeplaat Research Laboratory (South Africa)

R.R.L. Registered Record Librarian (hospital)

RRLNWR Red Rock Lakes National Wildlife Refuge (Montana)

RRLs Registered Record Librarians

rrm('s) renegotiable-rate mortgage(s)

rrm(s) [RRM(s)] renegotiable rate mortgage(s)

rrna (RRNA) ribosomal ribonucleic acid

rRNA ribosomal RNA (ribonucleic acid)

rrp reader and reader-printer; recommended retail price; reverse repurchase agreement

RRP Reduced Repayment Program; Riot Reinsurance Program; Rotterdam-Rhine Pipeline

RRPC Reserve Reinforcement Processing Center (USA)

RRPS Ready Reinforcement Personnel Section (USAF)

rrr risk-reward ratio

r & rr range and range rate

RRRA Regional Rail Reorganization Act

rrr's rapid runway repairs

RRS Radiation Research Society; Rational Recovery Systems (for alcoholics); Reaction Research Society; Resource and Referral Service; Retired Reserve Section; River and Rainfall Station (NWS); Royal Research Ship

RRSP Registered Retirement Savings Plan (Canadian)

rrt rendezvous radar transponder; rotating room test

RRTA Railroad Retirement Tax

RRU Radio Research Unit (USA); Road Research Unit

rrv rate of rise of voltage

RRW Royal Regiment of Wales

rs radio station; reading of standard; ready service; rear spar; receiver station; receiving ship; receiving station; re-

ception station; record separator; regulating station; reinforcing stimulus; residual sugar (wine), response stimulus; right side; road space; rocky shore; rubble stone

rs (RS) report separator character (data processing)

r/s range safety; rejection slip; revolutions per second

r & s rapport and support; re-enlistment and separation; research and study

Rs restricted motion pictures (adults only); rupees; source resistance (symbol)

RS Radio Station; Receiving Ship; Receiving Station; Reception Station; Reconnaissance Squadron; Reconnaissance Strike; Recording Secretary; Recruiting Station; Regular Station; Regulating Station; Regulation Station; Republic Steel; Research Summary; Revised Statutes; Ringer's Solution; Rio Grande do Sul; Roberval & Saguenay (railroad); Royal Scots; Royal Society; St. Louis Red Stockings (National Association)

RS *Rengo Sekigun* (Japanese—United Red Army)—urban guerrilla group active in the Middle East

RS-70 reconnaissance-strike bomber (formerly B-70)

rsa radar signature analysis; remote station alarm; right sacro-anterior

'r SA around South America

RSA Railway Supervisors Association; Railway Supply Association; Redstone Arsenal; Regional Science Association; Rehabilitation Services Administration; Renaissance Society of America; Rental Service Association; Returned Services Association; Road Safety Act; Royal Scottish Academy; Royal Society of Arts; Royal Society of Australia

RSA *Republiek van Suid-Afrika* (Republic of South Africa)

RSA (AFL-CIO) Railway and Airline Supervisors Association

RSAA Remote-Sensing Association of Australia

rsac radar significance analysis code

RSAC Reactor Safety Advisory Committee (Canada)

RSAF Royal Saudi Air Force; Royal Small Arms Factory; Royal Swedish Air Force

RSA/HEW Rehabilitation Services Administration—HEW

RSAI Royal Society of Antiquaries of Ireland

rsalt running, signal, and anchor lights

RSAM Royal Scottish Academy of Music

RSAS Royal Sanitary Association of Scotland; Royal Surgical Aid Society

RSASA Royal South Australian Society of Arts

rsb range safety beacon; rudder speed brake

RSB Regimental Stretcher Bearer; Revolutionary Student Brigade

RSBA Rail Steel Bar Association; Royal Society of British Artists

rsbe ring-standby engines

RSBS Radar Safety Beacon System

rsbt rhythmic sensory bombardment therapy (RSBT)

rsc range-safety command; range-safety control; rational self-counseling; referee stopped contest; regular slotted container; rigid-steel conduit

RSC Range Safety Command; Records Service Center; respiratory syncytial virus; Richard Strauss Conservatory (Munich); Royal Shakespeare Company; Royal Society of Canada; The Royal Society of Chemistry

rsca right scapuloanterior

rscd request to start contract definition

RSCDS Royal Scottish Country Dance Society

rsch research

R.S.C.J. Societas Sacratissimi Cordis Jesu (Religious Society of the Sacred Heart)

RSCM Royal School of Church Music

RSCN Registered Sick Children's Nurse

rscp right scapuloposterior

RSCS Rate Stabilization and Control System

RSCT Rhode Sentence Completion Test

rsd robustness semantic differential; rolling steel door

rs & d receipt, storage, and delivery

RSD Riverside Drive; Royal Society of Dublin

RSD-ALA Reference Services Division—American Library Association

RSDLP Russian Social-Democratic Labor Party

rsdp remote-site data processor

rsds reflex sympathetic dystrophy syndrome

RSDS Range Safety Destruct System

RSE Royal Society of Edinburgh

rsea reference sensing-element amplifier

RSEC Regional Science Experience Center

RSES Refrigeration Service Engineers Society

rseu remote scanner-encoder unit

RSF Religious Society of Friends; Royal Scots Fusiliers; Russell Sage Foundation; Russian Socialist Forces, Russian Soviet Forces

RSFA Roller Skating Foundation of America

RSFPP Retired Serviceman's Family Protection Plan

RSFS Royal Scottish Forestry Society

RSFSR Rossiskaya Sovietskaya Federatvnaya Sotsialisticheskaya Respublika (Russian—Russian Soviet Federal Socialist Republic)

rsg reassign; receiver of stolen goods; receiving stolen goods; regional seat of government

RSG Royal Scots Greys

RSGB Radio Society of Great Britain

RSGS Royal Scottish Geographical Society

rsh radar status history

Rsh Rosyth

RSH Recreational Services for the Handicapped; Royal Society for the Promotion of Health

RSHA Reichssicherheitshauptampt (German Secret Police headed by Heinrich Himmler); (German—Reich Central Security Department)—combined Gestapo, Kripo, and SD secret police

RSHWC Royal Society for the Health of Women and Children (New Zealand's Plunkett Society)

rsi radarscope interpretation; radial-shear interferometer; reflected signal indication; repetitive-strain injury; repetitive-stress injury; replacement stream input; reusable surface insulation

rs & i rules, standards, and instructions

RSI Research Studies Institute; Royal Sanitary Institute

rsia reference site initial assessment

RSIC Radiation Shielding Information Center; Radiation Standards Information Center; Redstone Scientific Information Center

RSID Recruiting Station Identification

R Sigs Royal Signals

RSIS Reference, Special, and Information Section (Library Association)

rsivp rapid sequence intravenous pyelogram

rsj rolled-steel joist

rsl right sacrolateral

RSL Radio Standards Laboratory; Red Star Line; Revolutionary Socialist League; Royal Society of Literature; Royal Society of London

rsla range safety launch approval

rslb right short-leg brace

rslt result

rsm (RSM) reconnaissance strategic missile

RSM Regimental Sergeant Major; Royal Scottish Museum; Royal Society of Medicine; Royal Society of Musicians

RSM Repubblica di San Marino (Italian—Republic of San Marino)

RSMA Railway Systems and Management Association; Royal School of Mines; Royal Society of Medicine;

RSMA *Republica di San Marino* (San Marino—world's smallest republic)

rsn reason

RSN Radiation Surveillance Network (USPHS)

RSNA Radiological Society of North America

RSNC Royal Society for Nature Conservation

RSNP Rancho Seco Nuclear Plant; Registered Student Nurse Program

RSNZ Royal Society of New Zealand

rso railway sorting office; railway suboffice; research ship of opportunity

RSO Radiation Safety Office(r); Range Safety Officer; Rehabilitation Service Office(r); Research Ships of Opportunity; Richmond Symphony Orchestra

RSO *Radio-Symphonie-Orkester* (German—Radio Symphony orchestra)—Berlin

rso's regional sharing organizations

RSOs Resident Surgical Officers

rsp rain stops play; rear-screen projection; raspberry; respirable particulate matter; right sacro-posterior

RSP Repository Sealing Program; Resources Specialist

RSPA Research and Special Programs Administration; Royal Society for the Prevention of Accidents

RSPB Royal Society for the Protection of Birds

RSPCA Royal Society for the Prevention of Cruelty to Animals

RSPCC Royal Society for the Prevention of Cruelty to Children

RSPE Royal Society of Painter-Etchers and Engravers

RSPH Royal Society for the Promotion of Health

rspk recurrent spontaneous psychokinesis

rspl radar significant power line

RSPN Royal Society for the Protection of Nature

rspp radio simulation patch panel

RSPP Royal Society of Portrait Painters

RSPR Royal Society for the Protection of Rats (mythical society created by Hans Werner Henze for his opera *The English Cat*)

RSPWC Royal Society of Painters in Water Colours

rsq rescue

r-sq r-squared

rsr regular sinus rhythm; required supply rate

RSR Range Safety Report; Request for Scientific Research; Research Study Requests

r-s ratio response-stimulus ratio

R-SR B Richmond-San Rafael Bridge

RSRC Remote Sensing Research Center (UCB)

RSRE Radar and Signals Research Establishment

RSROA Roller Skating Rink Operators Association

RSRS Radio and Space Research Station

rsrv (RSRV) rotor systems research vehicle

rss ready service spares; remote safing switch; root-sum square; rotary stepping switch

R s-s Russian spring-summer (encephalitis)

RSS Range Safety System; Reactant Service System; Regional Support System; Rehabilitation Support Schedule; Remote Sensing Society; Remote Sensing System; Resource Security System; Royal Security Service; Royal Statistical Society; Rural Sociological Society

RSSA Royal Society of South Africa; Royal Society of South Australia

RSSAILA Returned Sailors, Soldiers, and Airmen's Imperial League of Australia

RSSC Rand School of Social Sciences

R s-s e Russian spring-summer encephalitis

RSSF Retrievable Surface Storage Facility

Rssl Raytheon Scientific simulation language

RSSL Recruitment Subsidy for School Leavers

RSSPCC Royal Scottish Society for the Prevention of Cruelty to Children

RSSRT Russell Sage Social Relations Test

RSSS *Regiae Societatis Socius Sodalis* (Latin—Fellow of the Royal Society)

RSST Recruiter-Salesman Selection Test

rst radius of safety trace; reinforcing steel; right sacrotransverse

r-s-t readability—signal strength-tone (amateur radio signal)

Rst Rest (postal abbreviation)

RST Royal Society of Teachers

RST *Republica Socialista Romania* (Romanian Socialist Republic)

R Sta radio station

RSTMH Royal Society of Tropical Medicine and Hygiene

rstr restricted

rstrt restart

RSTS Registry of Scientific and Technical Services

rsu remote services unit; road safety unit

RSU Radical Student Union; Regional Service Unit; Road Safety Unit

rsv respiratory syncytial virus

rsv (RSV) research safety vehicle

Rsv Rous sarcoma virus

RSV Revised Standard Version (Bible)

rs virus respiratory syncytial virus

rsvp rapid serial visual presentation; research-selected vote profile; restartable solid variable pulse

RSVP Response System with Variable Prescription; Retired Senior Volunteer Persons; Retired Senior Volunteer Program

R.S.V.P. répondez s'il vous plaît (French—please reply)

rsvr reservoir

rswc (RSWC) right side up with care

R Sw N Royal Swedish Navy

RSWS Royal Scottish Water-Colour Society

rt radio telephone; radio telephony; rate; reaction time; real time; receive-transmit;

reduction table(s); related term; related topic; remote terminal; right; rocket target; room temperature; round table; round trip; runup & taxi

rt (RT) recreational therapy; respiratory therapy; transposed retrograde (of a 12-tone row)

r/t radar trigger; radiotelephone

R/T radiotelephone

Rt thermal resistance (symbol); total resistance (symbol)

RT Radio Technician; Ranger Tab; Reading Test; Recreational Therapy; Registered Technician; Registered X-ray Technician; River Terminal (railroad); Rubber Technician

R/T Record of Trial

RT *radio en televisie* (Dutch—radio and television); *République Togolaise* (French—Togo Republic)—Togo

rta reliability test(ing) assembly; road traffic accident; rumor told about

RTA Rail Travel Authorization, Railway Tie Association; Reciprocal Trade Agreements; Refrigeration Trade Association; Road Traffic Act; Royal Thai Army; Rubber Trade Association

RTA *Radiodiffusion et Télévision Algérienne* (French—Algerian Radio and Television Network)

RTAC Regional Technical Aids Center

rt ad router adapter

RTAF Royal Thai Air Force

RTAM Resident Terminal Access Method

rtb return to base

RTB Rural Telephone Bank(ing)

RTB *Radio diffusion-Télévésion Belge* (French—Belgian Radio-Television Network)

RTB/BRT *Radiodiffusion-Télévision Belge/Belgische Radio den Televisie* (French and Dutch—Belgian Radio and Television Network)

RTBL Richard Thomas and Baldwins Limited

rtc ratchet; reader tape contact(ing)

RTC Rail Travel Card; Real Time Command; Replacement Training Center; Reserve Training Corps; Resolution Trust Corporation; Revenue and Taxation Code; Rochester Telephone Corporation; Royal Trust Company

R & TC Regulation & Tax Court

RTCA Radio Technical Commission for Aeronautics; Radio-Television Correspondents Association

rtcc real-time computer complex

RTCEG Rubber and Thermoplastic Cables Export Group

rtcp radio transmission control panel

RTCs Return-to-Custody facilities

rtcu real-time control unit

rt cu router cutter

rtd remote temperature detector; resistance temperature detector(s); returned; righted

RTD Rapid Transit District; Research and Technology Division

RTDA Retail Tobacco Dealers of America

RTD/CCS Resources and Technical Services Division/Cataloging and Classification Section (American Library Association)

rtdd real-time data distribution

RTDHS Real-Time Data Handling System

rtd ht retired hurt

rt dr returnable-trip drum

RTDS Real-Time Data System

rte route

r-t-e ready-to-eat (foods)

Rte Route

RTE Research Training and Evaluation

RTE *Radio Telefis Eireann* (Irish Radio Television)

RTEB Radio Trades Examination Board

RTECS Registry of Toxic Effects of Chemical Substances

R te G *Rijksuniversiteit te Groningen* (Dutch—State University at Groningen)—Netherlands

rtel radiotelemetry; radio telephone; radiotelephony

R te L *Rijksuniversiteit te Leiden* (Dutch—State University of Leiden)—Netherlands

rtem radar tracking error measurement

RTES Radio and Television Executives Society

RTESO *Radio Telefis Eireann Symphony Orchestra* (Irish Radio Television Symphony Orchestra)

R test reductase test

R te U *Rijksuniversiteit te Utrecht* (Dutch—State University of Utrecht)—Netherlands

rtf radiotelephone; resistance transfer factor; rubber-tile floor(ing); rubber-tile foundation

RTF *Radiodiffusion-Télévision Française* (French—French Television Broadcasting)

rt fm router form

RTFR Reliability Trouble and Failure Report

rtfv radar target folder viewer

rtg radioactive thermal generator; radioisotope thermoelectric generator; rare tube gas; reusable training grenade

RTG Royal Thai Government

RTG *Radiodiffusion Télévision Gabonaise* (French—Gabonese Television Broadcasting)

rtgd room temperature gamma detector

rt gu router guide

rtgv real time generation of video

RTHK Radio Television Hong Kong

Rt Hon Right Honourable

RTHPL Radio Times Hulton Picture Library

rti respiratory tract infection; rise time indicator; rotor temperature indicator

RTI Reliability Trend Indicator; Research Triangle Institute; Roanoke Technical Institute

RT-1 *Radiodiffusion Télévision Ivoirienne* (French—Ivorian Television Broadcasting)—Ivory Coast

rtip radar target identification point

RTIR Reliability Trend Indicator Report

RTITB Road Transport Industry Training Board
rtk right to know
RTK Ras Tafari Makonnea
(Haile Selassie)
R Tks Royal Tank Regiment;
Royal Tanks
rtl reinforced tile lintel; resistor transistor logic
rtl (RTL) register-transfer language; resistor-transistor
logic
RTL Right to Life (party)
RTLA Road Transport Lighting Act
RTLO Regional Training Liaison Office(r)
rtls return to launch site
rtm running time meter
RTM Rotterdam, Netherlands
(airport)
*RTM Radiodiffusion Télévi-
sion Marocaine* (French—
Moroccan Television Broadcasting)
RTMA Radio and Television
Manufacturers Association
rtms repeated transcranial magnetic stimulation
RTMS Radar Target Measuring System
rtmso real-time multiprogramming support operation
rtn retain; return
rtn (RTN) routine (flow chart)
RTN registered trade name;
Royal Thai Navy
RTNA Radio and Television
News Association
RTNDA Radio-Television
News Directors Association
Rtnst Rottnest
rto radio-telephone operator
RTO Railway Transport Office
r-to-d forms right-to-die forms
rtol restricted takeoff and
landing
rtor right turn on red (traffic
light)
rtp records turnover package;
reinforced thermoplastic
R Tp radio telephone
RTP Rehabilitation Through
Photography; Request for
Technical Proposal (DoD)
*RTP Radiotelevisão Portu-
guesa* (Portuguese Radio
Television)
r-tpa recombinant tissue-type
plasminogen reactivator
RTPI Royal Town Planning Institute

rtpr (RTPR) reference theta-
pitch reactor
rtqc real-time quality control
rtr returning to ramp
R Tr radio tower
RTR Reliability Test Re-
quirement(s); Royal Tank
Regiment
*RTR Radiodifuziunea Televisi-
unea Romana* (Romanian Radio-Television Network)
RTRA Radio and Television
Retailers' Association; Road
Traffic Regulation Act
rtrc radio telemetry and remote control
RTRC Regional Technical Report Centers
Rt. Rev. Right Reverend
r/t room radio/telegraph room
(radio shack)
RTRSY Reuters Holdings
PLC
rtrsw rotary switch
rts radar target simulation; radar tracking station; request
to send; run time system
rt's rubber tappers
rt&s raise taxes and spend
RTS Repair Technical Service
(tractor stations—USSR era);
Repair Tracking Service;
Royal Television Society;
Rubber Traders Society
RTSA Retail Trading Standards Association
RTSD Resources and Technical Services Division (American Library Association)
RTSRS Real-Time Simulation
Research System
rtt radiation tracking transducer
*RTT Radiodiffusion Télévision
Tunisienne* (French—Tuni-
sian Broadcasting)
RTTC Road-Time Trials
Council
RTTDS Real-Time Telemetry
Data System
rt tp router template
RTTPS Real-Time Telemetry
Processing System
rttv research target and test
vehicle
r-ttv real-time television
rtty (RTTY) radio-teletype-
writer communication(s)
rtu remote terminal unit; returned to unit

RTU Rahway Treatment Unit;
Railroad Telegraphers Union;
Reinforcement Training Unit;
Reserve Training Unit
rtv reentry test vehicle (RTV);
room-temperature vulcanizing
rtv (RTV) radio television
*RTVE Radio Televisión Es-
pañola* (Spanish—Spanish
Radio Television)
RTVHK Radio-Television
Hong Kong
RTVS Royal Television Society
rtw ready to wear; right to
work; round-the-world
rtx rapid-transit experimental
(bus); report time crossing
rty rarity; realty
RTYC Royal Thames Yacht
Club
rtz return to zero
RTZ Rio Tinto Zinc
ru are you?; radium unit; rat
unit; roentgen unit; rusted
ru (RU) railroad underwriter;
railway urderwriter
Ru Rumania (Romania); Ru-
manian (Romanian); Russia;
Russian; ruthenium
Ru Ruth (Book of)
RU Radford University;
Readers Union; Revolutionary Union (communists ac-
tive in Puerto Rico and the
United States); Rice Univer-
sity; Rhodes University;
Roosevelt University;
Rugby Union; Rumanian
Union; Russia (Internet
code); Rutgers University
RU Regno Unito (Italian—
United Kingdom); *Reino
Unido* (Spanish—United
Kingdom)
RU 486 Roussel-Uclaf 486
(French abortion pill)
rua right upper arm
RUA Royal Ulster Academy
RUAS Royal Ulster Agricul-
tural Society
rub. rubber
rub rubato (Italian—with
varying tempo); *ruber*
(Latin—red)
Rub Rubbestadneset
RUB Radio Ulan Bator
rubbers rubber bullets
rubd rubberized
Rube Reuben
rubel rubella (german measles)
Rubg Rummelsburg

rubisco ribulose-1,5-bisphosphate carboxylase-oxygenase (plant protein)

ruble monetary unit of the Russian Federation

RUBN Russian, Ukrainian, and Belorussian Newspapers

RuBP ribulose-1-5-biphosphate carboxylase

rub. rm rubber room (padded cell)

ruby spinel red spinel gemstone

RUC Royal Ulster Constabulary

RUC République Unie du Caméroun (French—United Republic of Cameroon)

RUCA Rijksuniversitair Centrum Antwerpen (Flemish—Antwerp State University Center)

Ruch Ruchel

Ruch Ruchel (Yiddish—Rachel)

Rucos Russian Communists

RUCR Royal Ulster Constabulary Reserve

rud rudder

Rud Rudd; Rúdiger; Rudolf; Rudolph; Rudulph; Rudyard

Rud(dy) Rudyard

rudis reference your dispatch

Rud Kip Rudyard Kipling

Rudy Rudolf; Rudolph

rue right upper entrance; right upper extremity

RUE Regional Urban Environment

ruf revolving underwriting facility

RUF Revolutionary United Front (Sierra Leone)

RUFAS Remote Underwater Fisheries Assessment System

Rufe Rufus

rufiyaa monetary unit of the Maldives

rug red under gold

rugger rugby football

RUI Royal University of Ireland

RUKBA Royal United Kingdom Beneficent Association

rul right upper limb; right upper lobe (lung); rule

RUL Rutgers University Library

rulet reference your letter

rum (RUM remote underwater manipulator

rum rumänisch (German—Romanian)

Rum Rumania (Romania); Rumanian (Romanian)

RUM Ranger Uranium Mines; Royal University of Malta

rumem reference your memo

rumnog rum-flavored eggnog

run. rewind(ing) and unload(ing)

RUN Revolutionary United Nations

runcible revised unified new computer with its basic language extended

R und J Romeo und Julia (German—Romeo and Juliet)

R unit millimeter of mercury divided by milliliters per second; unit of resistance in the cardiovascular system

ruok 4 y2k Are you okay for year 2000 (computer problem)

RUP Rice University Press; Rockefeller University Press; Rutgers University Press

rupee monetary unit of India, Mauritius, Nepal, Pakistan, Seychelles, Sri Lanka

rupho reference your telephone (call)

rupiah monetary unit of Indonesia

rupp road used as public path

rupt rupture(d)

ruq right upper quadrant (abdomen)

rur reliably unreliable

RUR Rossum's Universal Robots (acronym—titled play by Karel Capek)

Rural Educ Rural Education Association

rureq reference your requisition

rur's rural and urban reformers

rurti recurrent upper-respiratory-tract infection

Rus Russ; Russia; Russian

Rus Russian (Slavic language)

Rus Dem Russian Democracy

Rusdic Russian dictionary

Rus Fed Russian Federation

rush remote use of shared hardware

Rush Rushdi; Rushmore; Rushton; Rushworth

RuSHA Rasse und Siedlungshauptamt (German—Race and Resettlement Department)

RUSI Royal United Service Institution

RUSM Royal United Service Museum

Rus Rep Russian Republic

russ russet; russian (leather)

russ russisch (German—Russian); *russisk* (Dano-Norwegian—Russian)

Russ Russia(n)

Russ Russland (Norwegian—Russia)

Russell Cave Russell Cave National Monument in northeastern Alabama

Rus Soc Russian Socialism

rúst rústico, a la (Spanish—paperback, paperbound)

rust of iron iron oxide

Rus Yid Russian Yiddish

rut. are you there

Rut Rutland Railroad; Rutlandshire

rutile titanium dioxide

RUU Ryksuniversiteit Utrecht (Dutch—Utrecht State University)

RUWS Remote Underwater Work System

rv rear view; recoil velocity; recreation vehicle; reentry vehicle; relief valve; residual volume; retroversion; right ventricle

rv (RV) recreational vehicle; reentry vehicle

r/v reentry vehicle

Rv Rendezvous

Rv. Revelation (Book of)

RV Rahway Valley (railroad); Reading and Vocabulary Test; *República de Venezuela;* Revised Version; Rifle Volunteer(s)

R/V rendezvous; research vessel

RV Radkikale Venstre (Danish—Radical Left)—Radical Liberal Party; *Revised Version*

rva reactive volt-ampere (meter); right ventricular apex; right visual acuity

RvA Rouva (Finnish—Madam)

RVA Regular Veterans' Association

R & VA Rating and Valuation Association

rvb radar video buffer; red venous blood

rvbr riveting bar

rvc random vibration control; relative velocity computer

RVC Rifle Volunteer Corps; Royal Veterinary College

RVCI Royal Veterinary College of Ireland

rvd radar video digitizer; residual vapor detector; right vertebral density

RVDA Recreational Vehicle Dealers of America

rvdo right ventricular diastolic overload

rvdp radar video data provessor

rvdt rotary variable displacement transducer

rve radar video extractor

rvedp right ventricular end diastolic pressure

rvedv right ventricular end diastolic volume

Rv. Ezr. Revelation of Ezra

rvf rate variance formula; right visual field

RVFN Report of Visit of Foreign Nationals

rv fx riveting fixture

rvh right ventricular hypertrophy

RVH Royal Victoria Hospital (Belfast)

RV(H)R Road Vehicles (Headlamps) Regulations

RVI Recreational Vehicle Institute

RVIA Recreation Vehicle Industry Association; Recreational Vehicle Institute of America; Royal Victoria Institute of Architects

Rvik Reykjavik

RVL Royal Viking Line

RVLP Rift Valley Lakes Park (Ethiopia)

RVLR Road Vehicles Lighting Regulations

rvm reactive voltmeter

Rvn Ravenna

RVN Republic of Vietnam

RVNAF Republic of Vietnam Air Force; Republic of Vietnam Armed Forces

RVNF Republic of Vietnam Forces

rvo relaxed vaginal outlet; runway visibility observer

R v O Rijksinstituut voor Oorlogsdocumentatie (Dutch—Netherlands State Institute for War Documentation)

RVO Regional Veterinary Officer; Royal Victorian Order

rvp radar video preprocessor

Rvp Reid vapor pressure

rvpa rivet pattern

RVPA Rape Victims Privacy Act

rvr runway visual range

R v R Rembrandt van Rijn

R & VR Rating and Valuation Reports

RVRA Recreation Vehicle Rental Association

rvrse reverse

rvs reported visual sensation

rv's recreation vehicles

Rvs Riverside

RVS Relative Value Scale; Relative Value Study

rvsc reverse self check

RVSN Raketny Voiska Strategicheskovo Naznacheniya (Russian—Strategic Rocket Forces)

rvsr revenue per voting share ratio

R.VS.VR répondez vite, s'il vous plaît (French—please reply at once)

rvsz riveting squeezer

rvtd riveted

Rvtn Riverton

rvtol rolling vertical takeoff and landing

rvu relief valve unit

rvx reentry vehicle—experimental

RVYC Royal Vancouver Yacht Club; Royal Victoria Yacht Club

rw radiological warfare; railwater (transport); random widths; raw water; recreation and welfare; recruiting warrant; right wing (field hockey); riparian woodland; rotary wing; runway

r/w read/write; right-of-way

r & w rail and water

Rw Rwanda

RW radiological war; radiological warfare; rain shower (aircraft code); Recruiting Warrant; redwood; Richard Wagner; Right Worshipful; Right Worthy; Royal Welsh; Rwanda (Internet code)

rwa (RWA) rotary-wing aircraft

Rwa Rwanda

RWA Railway Wheel Association; Regional Water Authority

RWAFF Royal West African Frontier Force

Rwanda Republic of Rwanda (landlocked East African country)

R War R Royal Warwick Regiment

RWAS Royal Welsh Agricultural Society

rwb rear wheel brake

RWB Rand Water Board; Royal Winnipeg Ballet

rwbh records will be hand-carried

rwc rainwater conductor; read, write, compute; read, write, continue; receive with code

RWC Roberts Wesleyan College

rwc's round-wire cables

RWCS Royal Water Colour Society

rwd rearward; rear wheel drive; rewind(ing); right wing down; right word(ing)

RWDGM Right Worshipful Deputy Grand Master

RWDSU Retail, Wholesale, and Department Store Union

RWE Ralph Waldo Emerson

RWEMA Ralph Waldo Emerson Memorial Association

RWF Royal Wholesalers' Federation; Royal Welch Fusiliers

rwg rigid waveguide

RWG Radio Writers' Guild; Reliability Working Group; Roebling Wire Gage

rwgl rough wire glass

RWGM Right Worshipful Grand Master

RWGR Right Worthy Grand Representative

RWGT Right Worthy Grand Templar; Right Worthy Grand Treasurer

RWGW Right Worthy Grand Warden

rwh radar warning and homing

rwi read, write, initial; real world interval; remote weight indicator

rwi (RWI) radar warning installation

R Wilts Yeo Royal Wiltshire Yeomanry

RWJC Roger Williams Junior College

RWJF Robert Wood Johnson Foundation

RWJGW Right Worthy Junior Grand Warden

rwk rework

RWK Royal West Kent (regiment)

rwl recommended weight limit; relative water level

r/w/l random widths and lengths

rwlr relative water-level recorder

rwm rectangular wave modulation; resistance welding machine; roll wrapping machine

RWMA Resistance Welding Manufacturers' Association

r/w memory read/write memory

rwms radioactive-waste management site

rwp radio wave propagation

RWQCB Regional Water Quality Control Board

rwr radar-warning receiver

r-w-r rail-water-rail

RWR rail-water-rail; Rockefeller Wildlife Refuge; Ronald Wilson Reagan—40th President of the United States

rwrc remain well to right of course

rws range while search; reaction wheel scanner; reaction wheel system; release with services

rws (RWS) release with services

RWS Regional Weather Service; Royal Water Colour Society

RWSGW Right Worshipful Senior Grand Warden

r/w storage read/write storage

rwt read-write-tape

R-W Test Rideal-Walker Test

rwth raised watertight hatch

rwv read-write-versify

rwy railway; runway

RWY Royal Wiltshire Yeomanry; runway number (aircraft code)

rv reverse; rix dollar; tens of rupees

r/x receiver

Rx recipe; prescription; unknown resistance (symbol)

RX residual tumor at primary site can't be assessed

rxb roxburgh (binding)

rxp radix point

rxs radar cross-section

ry railway; relay; rydberg

Ry railway; rydberg(s); Ryukyu (islands)

RY Royal Air Lao (coding); Royal Yeomanry

RYA Railroad Yardmasters of America; Royal Yachting Association

Ry Age Railway Age

ryal relay alarm

Ryan Ryan Acronautical Company (coding)

RYAN Ryan's Family Steak Houses

Rybinsk Andropov

RYC Richmond Yacht Club; Rochester Yacht Club; Royal Yacht Club

Ry I Ryukyu Islands

rym refer to your message

RYM Revolutionary Youth Movement

RYM-I Revolutionary Youth Movement (Weathermen)

RYM-II Revolutionary Youth Movement (Marxist–Leninist)

Ryojun Japanese equivalent of Port Arthur

ryrqd reply requested

Rys Railways

RyS Royal Yacht Squadron

ryt reference your telegram; reference your telex

Ryu Ryukyu; Ryukyuan

Ryukyus Ryukyu Islands between Japan and Taiwan

R y'u R Republika y'u Rwanda (Kinyarwanda—Rwanda)

rz return to zero

Rz Rodriguez

RZ Pacific Seaboard Airlines (2-letter symbol) doing business as Bay Area Helicopter Airlines and Los Angeles Helicopter Airlines; *République du Zaire* (formerly Belgian Congo)

R-Z Royal Australian New Zealand

R of Z Republic of Zambia

RZ *Referativnyi Zhurnal* (Russian—Reference Journal)—printed in Russian with abstracts also in their original language

RZA Religious Zionists of America

rzl return to zero level

rzm return to zero mark

RZMA Rolled Zinc Manufacturers Association

RZn relative azimuth

RZS Royal Zoological Society

RZ S Royal Zoological Society of Scotland

RZSI Royal Zoological Society of Ireland

RZSNSW Royal Zoological Society of New South Wales

RZSS Royal Zoological Society of Ireland

RZSSA Royal Zoological Society of South Australia

S

s displacement (symbol); sacral; saline; salt; sand; save; saves; schilling (Austrian currency); scuttle; sea-air temperature difference correction (symbol); second; secret; section; sections; sedimentation (coefficient); self timer; semiannually; sen (Japanese currency unit); sensation; sensitive; separate; separation; share(s); shilling (British monetary unit); ship; shunted blood (symbol); sign; signed; silicate; silver; simultaneous transmission of range signals and voice (symbol); single; sinistral; slope; slow; small; smooth; snow; soft; sol (Peruvian monetary unit); sold; soldier; solo; soluble; son; sonar; sou (French monetary unit); space; spar; specific; specific factor; speed; spherical; spherical lens; spinster; split or stock dividend (newspaper stock list); start; stay-over; steel; stere; stimulus; stock; string; subject; submitted for closure; substrate; succeeded; successor; sucre (Ecuadorian monetary unit); sum, summary; summer; sun; sunny; supravergence; surface; surgeon; survivor; symbol; symbol surface; synthesis; syphilis (sometimes indicated in reports by a Greek sigma)

s+ positive stimulus

s- negative stimulus

's (contraction—does, has, is)

s *signa* (Latin—write); *signeture* (Latin—label; let it be written); *sinister* (Latin—left)

s. *sinister* (Latin—left)

S antisubmarine (symbol); giving rise to Smooth-looking clones; no option offered (newspaper option listing); percentage saturation of hemoglobin with CO or C02 (symbol); sailing vessel (symbol); San; San Francisco mint (coin symbol); Santa; Santo; satisfactory; Saturday; Saturn; Saxon; Schilling (Austrian currency); school; Schweitzer; Schweizer Aircraft; Scotland; Scout (military); Seaman; seaplane; search and rescue; Sears, Roebuck (stock exchange symbol); Seatrain Lines; secondary and source electrode (symbol); secondary winding (symbol); secret; Section; See; semiannually; sen (Japanese currency); Senate; Senate Bill; Senator; Shinto; Shintoism; Shintoist; ship; siemens (mho); Sierra—code for letter S; Sigma; sign; Signaller (military); Signor (Italian—mister); Sikorsky; silver; Silver Lines; Sinclair; Sister; Sniper; snow (aircraft code); Socialist; sol (Peruvian monetary unit); solo; solubility; son; soprano; Souchong; south; southern; spar buoy; specific factor; specifica-

tion(s); Sperry; Staff, Statesman's Party; Statute; steamer; steamship; Steinway; stokes; stop; subject; sucre (Ecuadorian monetary unit); summer; sun; Sunday; sune; sunur; Surgery; Surveyor (military); Svedberg (unit to measure rate of sedimentation); Sweden (auto plaque); Sylvania; total entropy (symbol); wing plan area (symbol)

S(S) supply (microeconomics)

S/ sol (Peru); sucre (Ecuador)

:/S/ sign (music)

S general area (symbol); *Sábado* (Spanish—Saturday); *Sacrum* (Latin); *San* or *Santo* (Italian, Spanish—saint, *m*); *Santa* (Italian, Portuguese, Spanish—saint, *f*); *São* (Portuguese—saint, *m*); *semis* (Latin—half); *sinister* (Latin—left); *sisälle* (Finnish—in); *söder* (Swedish—south); *sor* (Norwegian—south); south; *strada* (Italian—street); *subir* (Spanish—to go up, mount); *sud* (French or Italian—south); *Süd* (German—south); *sul* (Portuguese—south); *sur* (Spanish—south); *syd* (Danish—south)

s_1 first heart sound

S-1 military personnel; personnel officer

Slc Seaman, first class

s 1 s 1 e smooth 1 side 1 edge; surfaced on one side and one edge (lumber)

S1, S2, S3, etc. first sacral nerve, second sacral nerve, third sacral nerve, etc.

s₂ second heart sound

S-2 Grumman Tracker antisubmarine search-and-attack aircraft; intelligence officer; military intelligence

S2F Tracker twin-engine antisubmarine aircraft flown from carriers

s2s surfaced two sides

S-21 Khmer Rouge name for Tuol Sieng near Phnom Penh, Cambodia where only 7 of the 20,000 prisoners escaped the mass executions

S-3 military operations and training; military operations and training officer

S₃ Systems, Science, and Software

S-4 military logistics; military logistics officer

s 4 s smooth 4 sides; surfaced on four sides (lumber)

S-35 Saab double-delta-wing supersonic fighter or fighter bomber built in Sweden and named Draken (Dragon)

S₃₅ radioactive sulfur

S-51 Sikorsky four-seat helicopter

S-60 Soviet antiaircraft system consisting of one 57mm cannon mounted on a towed carriage

S-61 Sikorsky civilian or military helicopter

sa sack(s); sail area; sailor; seasonally adjusted; second attack (lacrosse); see also; semi-annual(ly); semiautomatic; sex appeal; shaft alley; slugging average (baseball); sinoatrial; small arms; software applications; soluble in alkaline; special activities; spectrum analyzer; stone arch; subject to approval; subsistence allowance; sun-affected; superabnormal; supraabdominal; sustained action

s-a single-action (handgun); sinoatrial

s/a safe arrival; storage area; subject to approval

s & a safety and arming (mechanism)

sa siehe auch (German—see also)

s.a. secundum artem (Latin—according to the art); *sine anno* (Latin—undated)

sª Señora (Spanish—Madam)

Sa samarium; Sara; Sarah; Sarita; Saturday; Serra; Sierra

Sa Summa (German—total)

Sª Serra (Portuguese or Spanish—mountain range); *Sierra* (Spanish—mountain range)

SA Safeway Stores (stock exchange symbol); Salvation Army; Saudi Arabia; Saudia Arabian; Savage Arms; Savannah & Atlanta (railroad); Seaman Apprentice; search amphibian; second attack (lacrosse); Secretary of the Army; sex appeal; Security Administrator; Shipping Authority; Smithsonian Archives; Society of Actuaries; Society of Authors; South Africa; South African; South African Airways (2-letter coding); South America; South American; South Australia; South Australian; (Spanish—National Air Routes Corporation); Special Agent; Special Artificer; Springfield Armory; State's Attorney; Stores Accountant; Studio Address (public address system); Sugar Association; Supplemental Agreement; Supplementary Agreement; Supply Accountant; Surgery Assist; Sweep Accountant

S-A Stokes-Adams (disease)

S/A Special Agent; State Agent

S of A Society of Actuaries

SA Société Anonyme (French—limited company); *Sudáfrica* (Spanish—South Africa)

S.A. Sociedad Anónima (Spanish—corporation); *Sturmabteilung* (German—Storm troopers); *Sucursales Asociados* (Spanish associated branches)

S/A Societa Anonima (Italian—limited company)

SA-2 Soviet surface-to-air missile called Guideline by NATO

SA-3 Soviet air-defense missile system nicknamed Goa by NATO

SA4 Soviet missile system nicknamed Ganef by NATO

SA-5 Soviet surface-to-air missile called Griffon by NATO

SA-6 Soviet air-defense missile system nicknamed Gainful by NATO

SA-7 Soviet shoulder-fired surface-to-air missile called Grail by NATO

SA-8 Soviet missile system nicknamed Guideline by NATO

SA-9 Soviet air-defense missile system nicknamed Gaskin by NATO

SA-315 Aerospatiale helicopter made in France and called Lama

SA-341 Aerospatiale observation helicopter built in Brazil by Embraer

saa small arms ammunition; special arbitrage account

SAA Saudi Arabian Airlines; Shakespeare Association of America; Shelter Advertising Association; Signal Appliance Association; Singapore Aftercare Association (for ex-convicts); Society for Academic Achievement; Society for American Archeology; Society of American Archivists; Society for Applied Anthropology; Society for Asian Art; South African Airways; South Arabian Army; Southern Ash Association; South Australian Artillery; Speech Association of America; State Archeological Area; Sunglass Association of America; Surety Association of America; Suzuki Association of the Americas; Swedish-American Association; Systems Application Architecture

SAA Single-Article Announcement (American Chemical Society),

SAAA Salvation Army Association of America

SAAARNG Senior Army Advisor, Army National Guard

SAAAS South African Association for the Advancement of Science

SAAASE South African Association for the Administration and Settlement of Estates

SAAB *Svenska Aeroplan Aktiebolaget* (Swedish—Swedish Airplane Company)

saac simulator for air-to-air combat

SAAC Sciences and Arts Camps; Seismic Array Analysis Center (IBM); Special Assistant for Arms-Control (DoD)

SAAD Sacramento Army Depot; Small-Arms Ammunition Depot; Society for the Advancement of Anesthesia in Dentistry

SAAEB South African Atomic Energy Board

SAAF Saudi Arabian Air Force; South African Air Force

SAAI Specialty Advertising Association International

SAAL Syrian Arab Airlines

SAALIC Swindon Area Association of Libraries for Industy and Commerce

saam simulation, analysis, and modelling

SAAM Seattle Asian Art Museum

SAAMA San Antonio Air Materiel Area

SAAMI Sporting Arms and Ammunition Manufacturers Institute

SAAN South African Associated Newspapers

SAANYS School Administrators Association of New York State

SAAP Saturn-Apollo Applications Program; South Atlantic Anomaly Probe (NASA)

SAAPCC South African Administration, Pay and Clerical Corps

sa ar saw arbor

Saar *Saarland*

SAARC South Asia Association of Regional Cooperation—Bangladesh, Bhutan, Maidives, Nepal, India, Pakistan, Sri Lanka

SAARF Special Allied Airborne Reconnaissance Force

SAAS Science Achievement Awards for Students; Society of African and Afro-American Students; Southern Association of Agricultural Scientists; Standard Army Ammunition System

SAASC South African Army Service Corps

SAAST Self-Administered Alcoholism Screening Test

SAAT Society of Architects and Allied Technicians

SAAU South African Agricultural Union

SAAVS Submarine Acceleration and Velocity System

SAAWK *Suid Afrikaanse Aka demie vir Wetnenskap en Kuns* (Afrikaans—South African Academy for Science and Art)

sab sabbath; sabbatical; soprano, alto, baritone (SAB); special assessment bond

s-a b steel-arch bridge

sáb *sábado* (Portuguese or Spanish—Saturday); *sabato* (Italian—Saturday)

Sab Sabah; Sabbatarian; Sabbatarianism; Sabbath; Sabelian; Sabine; Sabra(s)

Sab *Sabkhat* (Arabic—salt flats)—also appears as *Sebkhat*

S-A b South-American blastomycosis

SAB Sabena; School of American Ballet; Scientific Advisory Board; Society of American Bacteriologists

SAB *Sveriges Allmänna Biblioteksforening* (Swedish Library Association)

s-aba science—a basic approach

Saba Sheba

SABA Scottish Amateur Boxing Association; South African Black Alliance

sabbat sabbatical

SABC Scottish Association of Boys Clubs; South African Broadcasting Corporation

SABCO Society for the Area of Biological and Chemical Overlap

SABCOA Screw and Bolt Corporation of America

SABE Society for Automation in Business Education

SABENA *Société Anonyme Belge d'Exploitation de la Navigation Aérienne* (Belgian World Airlines)

saber (SABER) semiautomatic business environment research

SABEW Society of American Business Editors and Writers

sabh simultaneous automaticbroadcast homer

SABHATA Sand and Ballast Haulers and Allied Trades Alliance

SABIC Saudi Basic Industries Cooperation

sabin square-foot unit of absorption

sabir semi-automatic bibliographic information retrieval

Sable Cape Sable; Sable Island

SABMIS Seaborne Anti-Ballistic Missile Intercept System (USN)

SABMS Safeguard Anti-Ballistic Missile System

sabo sabotage

sabot sabotage; saboteur

SABR Society for American Baseball Research

SABRA South African Bureau of Racial Affairs

sabre self-aligning boost and reentry

SABS South African Bureau of Standards

SABTS Shared-Aperture Breadboard Test System

Sabu Sabu Dastagir

SABW Society of American Business Writers

sac sacral; sacrament; sacramental; sacred; surface air consumption

Sac Sacramento, California; S-allyl-cysteine; stimulus as coded

SAC Sacramento, California (airport), San Angelo College; San Antonio College; School of Army Cooperation; Science Applications Corporation; Seaboard Athletic Conference; Senior Aircraftman; Sexual Assault Center; Society of Analytical Chemistry; Society of Arts and Crafts; South African Constabulary; Southwest Athletic Conference; Southwest Automotive Company; Special Agent in Charge (FBI); Statistical Analysis Center; Strategic Air Command; Suburban Authorization Committee; Swedish-American Cooperative

SAC *Service Action Civique* (French—Civil Action Service); *Sociedad de Albizu Campos* (Puerto Rican terrorists); *Sveriges Arbetares Centralorganisation* (Swedish—Swedish Workers Central Organization)

saca store and clear accumulator

SACA Steam Automobile Club of America; Supreme Allied Commander Atlantic

SACA Servicio Aereo Colombiano (Spanish—Colombian Airline Service)

sacad stress analysis and computer-aided design

SACANGO Southern Africa Committee on Air Navigation and Ground Operation

SACARTS Semi-Automated Cartographic System

Sacate Sacatepéquez, Guatemala

SACB Subversive Activities Control Board

S Acc Societá in Accomandita (Italian—limited partnership)

SACC South African Council of Churches; Supplemental Air Carrier Conference; Supporting Arms Coordination Center

saccm slow-access charge-coupled memory

SACCR Southeastern Association of Community College Researchers

SACCS Strategic Air Commammd Control System

sace serum angiotensin-converting enzyme; systems acceptance check-out equipment

SACEM Société des Auteurs, Compositeurs et Éditeurs de la Musique (French—Society of Authors, Composers, and Editors of Music)

SACEUR Supreme Allied Command, Europe

sach solid ankle cushion heel (prosthetic foot)

SACH Small Animal Care Hospital

Sach-Anh Sachsen-Anhalt (German—Saxony-Anhalt)

sach foot solid-ankle-and-cushion-heel foot

Sachs Sachsen (German—Saxony)

saci secondary address code indicator

SACI Sales Association of the Chemical Industry; South Atlantic Cooperative Investigations

Sackpig *See* SACPG

SA & CL South Atlantic & Caribbean Line

SACLant Supreme Allied Commander, Atlantic

SACLANTCEN SACLANT Anti-Submarine Warfare Research Centre (NATO)

sacm simulated aerial combat maneuver

SACM South African College of Music; South African Corps of Marines; South Arabian Common Market

SACMA Suppliers of Advanced Composite Materials Association

SACMA Société Anonyme de Construction de Moteurs Aéronautiques (French—Aeronautical Engine Construction Corporation)

SACMED Supreme Allied Command Mediterranean

SACMP South African Corps of Military Police

SACNAS Society for the Advancement of Chicano and Native American Scientists

saco select address and contract operate

SACO Sino-American Cooperative Organization

SACO Sveriges Akademikers Centralorganisation (Swedish—Swedish Central Professional Organization)

SACP South African Communist Party

SACPG Senior Arms Control Policy Group (pronounced "Sackpig")

Sacr Sacramento

Sacramentos Sacramento Mountains of New Mexico and Texas

SACRO Scottish Association for the Care and Resettlement of Offenders

SACs Solar Appliance Centers

SACS South African College System; South African Corps of Signals; Southern Association of Colleges and Schools

SACSA Special Assistant for Counter Insurgency and Special Activities

Sac-San Sacramento—San Diego

SACSEA Supreme Allied Command South-East Asia

Sac-Sfo Sacramento—San Francisco

SACSIR South African Council for Scientific and Industrial Research

Sacto Sacramento

SACTU South African Congress of Trade Unions

SACU Service for Admission to College and University

SACUBO Southern Association of College and University Business Officers

SACVT Society of Air Cushion Vehicle Technicians

SACW Senior Aircraftwoman

sad. safety analysis document; safety, arming, destruct; safety and arming device; single administrative document; situation attention display; somatosensory affectional deprivation

SAd (SAD) St Augustine decline (grass virus)

SAD Seasonal Adjustment Deficiency; Seasonal Affective Disorder; simple, average, or difficult; Social Affairs Department (Communist China's espionage agency) Standard American Diet

S & AD Science and Applications Directorate (NASA)

SAD South African Digest

sadap simplified automatic data plotter

sadarm search and destroy armor (projectile)

SADC Sector Air Defense Commander; Singapore Air Defense Command

SADCC South African Development Coordination Conference

SADD Students Against Drunk Drivers, Students Against Drunk Driving

sade sensitive acoustic-detection equipment

SADE Specialised Armour Development Establishment

SADE Sociedad Argentina de Escritores (Spanish—Argentine Writers' Society)

SA de CV Sociedad Anónima de Capital Variable (Spanish—Variable Capital Society)

SADF South African Defense Forces

sadic solid-state analog-to-digital computer

sadie scanning analog-to-digital input equipment; semi-automatic decentralized intercept environment Sadism for the libertine and novelist the Marquis de Sade (French—1740–1814)

sadm special atomic demolition munition

sado-maso sado-masochism; sado- masochist

sado-sex sado-sexual(ity)

SADS Schedule for Affective Disorders and Schizophrenia; Swiss Air Defense System

sadsac sampled data simulator and computer

sadsact self-aligned descriptors from self and cited titles (automatic index)

sad sam (SAD SAM) sentence appraiser and diagrammer—semantic analyzer machine

sadt surface alloy diffused (base) transistor

SADTC Shape Air Defense Technology Center

s-a-d test sugar-acetone-diacetic acid test

sae San Diego Aircraft Engineering (corporate symbol); self-addressed envelope; shaft angle encoder; signal-averaged electrocardiogram; stamped, addressed envelope; standard average European

SAE Sigma Alpha Epsilon (fraternity); Society for the Advancement of Education; Society of American Etchers; Society of Automotive Engineers; Solar Atmospher(ic) Explorer; Specialised Armour Establishment; Standard American English; Suntanning Association for Education

S.A.E. *Société Anonyme Egyptienne* (French—Egyptian limited compary)

SAEA Southeastern Adult Education Association

saeb self-adjusting electric brake

SAEB Spacecraft Assembly and Encapsulation Building (NASA); Special Army Evaluation Board

saec. saeculum (Latin—century)

SAEC South African Engineer Corps; Sumitomo Atomic Energy Commission (Japan)

SAEH Society for Automation in English and the Humanities

SAEI Sumitomo Atomic Energy Industries (Japan)

SAEL South African Emergency League

SAemc South African endomyocardiopathy

SAEMR Small Arms Expert Marksmanship Ribbon

SAEST Society for the Advancement of Electrochemical Science and Technology

SAET Spiral Aftereffect Test

saew ship's advanced electronic warfare

saf safety

SAF Secretary of the Air Force; See America First; Singapore Air Force; Social Affairs Federation; Society of American Florists; Society of American Foresters; Strategic Air Force

SAF Svenska Arbetsgivareforeningen (Swedish—Swedish Employers' Confederation)

safa solar-array failure analysis; soluble-antigen fluorescent antibody

SAFA School Assistance in Federally Affected Areas; Society for Automation in the Fine Arts

SAFAA South African Fine Arts Association

SAFB Scott Air Force Base; Shaw Air Force Base

saf black super-abrasion furnace black

SAFC South African Flying Corps

SAFCA Safeguard Communications Agency

SAFCB Secretary of the Air Force Correction Board

SAFCMD Safeguard Command (USA)

SAFCO Standing Advisory Committee on Fisheries in the Caribbean Organization

safe satellite alert force employment; stamped, addressed foolscap envelope; system, area, function, equipment

SAFE Braathens South American & Far East Air Transport; Security Against Fatal Encounter; Security Assured For Each; Sexual Assault Special Enforcement; Solvent Abuse Foundation for Education; South African Friends of England; Standard Authoring Facility Environment; Survival and Flight Equipment Association; System for Automated Flight Efficiency

S.A.F.E. Society of Aeronautic Flight Engineers

SAFENET Survivable, Adaptable Fiber Network

SAFEORD Safety of Explosive ordnance Databank (USN)

SAFE TRIP Students Against Faulty Tires Ripping in Pieces

SAFF South African Frontier Force

SAFI Senior Air Force Instructor

SAFISY Space Agency Forum International Space Year

SAFMARINE South African Marine (corporation)

SAFN Semi-Automatic FN (rifle)

SAFO Senior Air Force Officer (present)

SAFOH Society of American Florists and Ornamental Horticulturists

SAFN Semi-Automatic FN (rifle)

SAF£ South African pound

S Afr South Africa(n)

SAFR Senior Air Force Representative

SAFRAS Self-Adaptive Flexible-Format Retrieval And Storage System

S-Afr Du South-African Dutch (Afrikaans)

SAFS Secondary Air Force Specialty; selective automatic feed stripe (knitting machine)

SAFSL Secretary of Air Force Space Liaison

SAFSO Safeguard System Office(r)

SAFSR Society for the Advancement of Food Service Research

SAFTI Singapore Armed Forces Training Institute

SAFU Scottish Amateur Fencing Union

SAFUS Secretary of the Air Force, United States

sa fx saw fixture

Sag Sagittarius

SAG Scientific Advisory Group; Screen Actors Guild; Society of Arthritic Gardeners; Surface Action Group (USN); Systems Analysis Group

SAGA Sand and Gravel Association; Scout and Guide Activity; Society of American Graphic Artists

SAGB Spiritualist Association of Great Britain

sag. d saggital diameter

sage semi-automatic ground environment (for defense against air attack); solar- assisted gas energy (for heating)

SAGE Senior Action in a Gay Environment; Senior Actualization and Growth Exploration; Skylab Advisory Group for Experiments (NASA); South African General Electric; Soviet-American Gallium Experiment; Stratospheric Aerosol and Gas Experiment

SAGE/BUIC Semi-Automatic Ground Environment and Back-Up Interceptor Control (systems)

SAGGA Scout and Guide Graduate Association

SAGP Society for Ancient Greek Philosophy

SAGS Semiactive Gravity Gradient System (NASA)

SAGSET Society for Academic Gaming and Simulation in Education and Training

sagt systematic approach to group technology

SAG & U San Antonio, Gulf& Uvalde (railroad)

sagw surface-to-air guided weapon

sah subarachnoid hemorrhage

SAH Society of American Historians; Society of Automotive Historians; Supreme Allied Headquarters

S&AH Sutherland and Argyll Highlanders

SAHAND Society Against Have A Nice Day

Sah Esp Sahara Español (Spanish Sahara)

sahf semiautomatic height finder

SAHQ Supreme Allied Headquarters

SAHR Society for Army Historical Research

SAHSA Servicio Aéreo de Honduras SA (Spanish—Air Service of Honduras Inc)

sahyb simulation of analog and hybrid computers

sai self-appraisal instrument; self-appraisal inventory; sell (sold) as is; statement of additional information

sai (SAI) standard advertising invoice

Sai Saigon

SAI Schizophrenics Anonymous International; Science Applications Inc; Science Applications International; Self-Analysis Inventory; Social Adequacy Index; South African Irish (regiment); Stern Activities Index

SAI Societá Anonima Italiana (Italian—Italian Incorporated Company); *Son Altesse-Impériale* (French—Her or His Imperial Highness); *Su Alteza Imperial* (Spanish—Your Imperial Highness)

SAIA South Australian Institute of Architects

SAIC Science Applications International Corporation; South African Intelligence Corps; Special Agent in Charge (Secret Service)

said speech auto-instructional device

Said Port Said, Egypt

SAIDET Single-Axis Inertial Drift Erection Test

SAIF Savings Association Insurance Fund; South African Industrial Federation; South African Institute of Foundrymen

sail structural analysis input language

SAIL Sea-Air Interaction Laboratory

Sails ship's sailmaker

SAILS Simplified Accelerated Intelligence Learning System; Software-Adaptable Integrated-Logic System

SAIM South African Institute of Management

SAIMC South African Institute for Measurement and Control

SAIMENA South African Institute of Marine Engineers and Naval Architects

SAIMR South African Institute for Medical Research

SAIMS Selected Acquisition Information and Management System

SAINT Systems Analysis of an Integrated Network of Tasks (USAF)

Saint-Barth Saint-Barthélemy (French—Saint Bartholomew)—a Caribbean island; a massacre of 3,000 Protestants instigated by Catherine de Médicis on August 23, 1572

Saint-Ex Antoine de Saint-Exupéry

Saint Joe Saint Joseph, Missouri

Saint John St John, New Brunswick

Saint Johns St Johns, Antigua

Saint John's St John's, Newfoundland; St John's University, New York

Saint Kitts and Nevis Saint Christopher and Nevis (West Indies islands)

Saint-Laurent (French—Saint Lawrence)—Canadian American river

Saint Lawrence St Lawrence River

Saint Lawrence Islands Saint Lawrence Islands National Park on the Canadian islands and nearby shore of the Saint Lawrence River

Saint Loo Saint Louis, Missouri

Saint Lucy St Lucia, West Indies

Saint P St Pancras (London railway station); St Paul, Minnesota

Saint Patrick's Saint Patrick's Day (March 17)

Saint Pete St Petersburg, Florida

Saint-Pierre (French—Saint Peter)—island; Rome's basilica

Saint Stephen's Saint Stephen's Day (December 26)

Saint Vince St Vincent (West Indies)

SAIR South African Irish Regiment

SAIRR South African Institute of Race Relations

SAIS School of Advanced International Studies (Johns Hopkins University)

SAIT Southern Alberta Institute of Technology
SAIT Service D'Analyse de l'Information Technologique (French—Technological Information Analysis Service)
SAJ Shipbuilders Association of Japan; Society for the Advancement of Judaism
SAJ Suome Ammattijärjestö (Finnish—Finnish Federation of Trade Unions)
SAJAC South African Jewish Association of Canada
SAJC Southern Association of Junior Colleges
sa ji saw jig
SAK Serge Alexandrovich Koussevitsky
SAK Suomen Ammattilittojen Keskulitto (Finnish—Finnish Trade Union Confederation)
Sakura-3A Japanese CS-3A communications satellite
sal salary; saleslady; salt; salicylate; saloon
sal (SAL) surface and airlift
sal salida (Spanish—departure)
s.a.l. secundum artis leges (Latin—according to the rules of art)
Sal Salamanca; Salaverry; Salem; Sallie; Sally; Salomon
Sal Islas Salomón (Spanish—Solomon Islands); *salida* (Spanish—departure; exit); *Salmonella*
SAL San Salvador, El Salvador (airport); Seaboard Airline Railroad; Society of Antiquaries of London; South African Library (Cape Town); Symbolic Assembly Language
SAL Svenska-Amerika Linien (Swedish—American Line)
SALA Scientific Assistant Land Agent; South African Library Association; Southwest Alliance for Latin American(s)
SALALM Seminars on the Acquisition of Latin American Library Materials
salam salamanzar (12-bottle capacity)
sal ammoniac ammonium chloride (NH_4Cl)
SALB South African Library for the Blind (Grahamstown)
sale simple algebraic language for engineers

sal gal saloon girl
SALH South Alberta Light Horse
salic salicional (French—soft string-toned organ stop)
salicyl salicylate
SALINET Satellite Library Information Network
Salisbury Harare, Zimbabwe; plain in southern England on which Stonehenge is located
SALJ South African Law Journal
Sall Gains Sallustius Crispus (Roman historian often referred to as Sallust)
Sallie Mae Student Loan Marketing Association
Sallust Roman historian Gaius Sallustius
Sally Army Salvation Army
salm single-anchor leg mooring
Salm Salamon
Salm Salmonella
SALM Society of Airline Meteorologists
salmiak sal ammoniac (ammonium chloride)
salmonsan salmon sandwich
salmonwich salmon sandwich
Salomons Salomon Islands in the Chagos Archipelago in the Indian Ocean
SALP South Afiican Labour Party
salpingect salpingectomic (sterilization); salpingectomy (removal of the fallopian tubes)
salr saturated adiabatic lapse rate
SALR South African Law Reports
SALRC Society for the Assistance of Ladies in Reduced Circumstances
SALSAC Structural Analysis of the Language of School-Age Children
salt sodium chloride (NaCI)
salt suggestive-accelerative learning and teaching
Salt Salta (Argentine province)
SALT Society for Applied Learning Technology; Strategic Arms Limitation Talks
Saltees Saltee Islands in St George's Channel off Wexford, Ireland
Salt 'Uman Saltanat 'Uman (Arabic—Sultanate of Oman)

salts of lemon oxalic acid
Saludemos Saludemos la patria (Spanish—We salute the country)—anthem of El Salvador
salut salutation; sea-air-land-and-underwater targets (SALUT)
salv salvage
Salv Salvador
Salvador Brazilian port of Bahia or São Salvador de Todos os Santos; Central American republic of El Salvador
Salv Army Salvation Army
Salve Salve Oh Patria (Spanish—Hail, oh country)—anthem of Ecuador
Salve a tí Salve a tí Nicaragua (Spanish—Hail Nicaragua)—national anthem of Nicaragua
Salz Salzburg, Austria
sam scanting auger microscope; scuba ascent method; self-advising materials; serial access memory; served available market; small (secondary) annular mirror; space-available mail (SAM); spacecraft anomaly manager; student accountability model; surface-to-air missile (SAM); synchronous amplitude modulation
sam (SAM) shared-appreciation mortgage; standard academic monograph
sam (SaM (SAM) sales and marketing
sam samedi (French—Saturday)
Sam Samoa; Samoan; Samson; Samoyed; Samuel; Samuelito
S-a-m S-adenosyl-methionine
Sam Samstag (German—Saturday); *Samuel*
SAM School Administrators of Montana; School of Aerospace Medicine; Seattle Art Museum; Society for the Advancement of Management; Society of American Magicians; Special Air Mission; Student Accountability Model; Surface-to-Air Missile(s)
SAM Societa Aerea Mediterranea (Italian—Mediterranean Airline)
SAMA Sacramento Air Materiel Area; Saudi Arabian Monetary Agency; Scientific Apparatus Makers Association; Student American Medical Association

SAMANTHA System for the Automated Management of Text from a Hierarchical Arrangement

Samarians Samarian Mountains

SAMB School of Aviation Medicine—Brooks AFB

SAMBA Special Agents Mutual Benefit Association (FBI); Systems Approach to Managing Bureau of Ships Acquisitions (USN)

SAMC South African Marine Corporation; South African Medical Corps; South African Military College

SAM/CAR South America/ Caribbean

SAM-D surface-to-air missile for field air defense

SAM-e S-adenosyl-methionine

same: specific area message encoding

SAME Society of American Military Engineers

SAMECS Structural Analysis Method for Evaluation of Complex Structures

Sam'el Samuel

S Am(er) South America(n)

samex surface-to-air missile exercise

SAMF Seaborne Army Maintenance Facilities; Seaborne Army Materiel Facilities

samfu self-adjusting military fuckup

SAMH Scottish Association for Mental Health

sami socially-acceptable monitoring instrument

SAMI Scanner-Augmented Market Intelligence; System Acquisition Management Inspection

SAMIL South African Military

samizdat samizdatel'stvo (Russian—self-published and self-distributed)—literature that was suppressed by the Soviet government

Sam J Dr Samuel Johnson

SAMJ South African Medical Journal

Saml Samiel; Samuel

SAML Standard Army Management Language

SAMLA South Atlantic Modern Language Association

samm semi-automatic measuring machine

Sammie Bee Small Business Administration guaranteed loans securities

Samml Sammlung (German—collection)

SAMNS South African Military Nursing Service

Samoa national park on American Samoa, features a paleotropical rain forest

Samoas Samoa Islands

samos (SAMOS) satellite and missile observation system

SAMP Stuntmen's Association of Motion Pictures; Stuntwomen's Association of Motion Pictures

SAMPAM System for Automation of Materiel Plans for Army Materiel

SAMPE Society of Aerospace Material and Process Engineers

SAM & PE Society for the Advancement of Material and Process Engineering

sample simulation and modeling of profiles in lithography and etching

SAMR Special Assistant for Materiel Readiness (USA); South African Mounted Rifles; South Australian Mounted Rifles

samrt shared-aperture medium-range tracker

sams stratospheric and mesopheric sounder

Sams Howard W Sams and Company

SAMS Sample Method Survey; Satellite Automation System; Satellite Auto-Monitor System; South American Missionary Society; Standard Army Maintenance System

SAMSA Silica and Moulding Sands Association

SAM-SAC Special Aircraft Modification for Strategic Air Command

SAM/SAT South America/ South Atlantic

SAMSO Space and Missile System Organization (USAF)

SAMSON Strategic Automatic Message-Switching Operational Network

SAM/SPAC South America/ South Pacific

sam(s) [SAM(s)] shared-appreciation mortgage(s)

SAMTC South Atlantic Marine Terminal Conference

SAMTEC Space and Missile Test Center

SAMU Service Aide Médicale Urgente (French—Urgent Medicaid Service)

SA Mus South African Museum (Cape Town)

san sandwich; sanitary; styreneacrylonitrile copolymer

San Santos (British maritime abbreviation)

San. Sanhedrin

SAN San Diego, California (Lindbergh Field); South African Navy

SAN Standard Address Number

SAN Space Age News

SANA Scientists Against Nuclear Arms; Scottish Association of Nurse Administration; State (Department), Army, Navy, Air (Force)

San Andreas San Andreas Fault of western California

sanat sanatoria; sanatorium

San Augustins San Augustin Mountains of southern New Mexico

SANB South African National Bibliography

Sanc. Sanctus (Latin—holy)

SANCAD Scottish Association for National Certificates and Diplomas

SANCAR South African National Council for Antarctic Research

San Carlo Teatro di San Carlo—Naples' opera house

San Carlo Teatro San Carlo (Italian—San Carlo Theater)—Naples opera-house theater

Sanche St Charles

San-Chi San Diego—Chicago

San Clem San Clemente

SANCOB South African Foundation for the Conservation of Birds

SANCOG San Diego Council of Governments

SANCOR South African National Committee for Oceanographic Research

SANCOT South African National Commission on Tunnelling

sand silicon dioxide-SiO_2

sand. sandalwood; sandstone

Sand Sandford's—New York Reports

San. D. Doctor of Sanitation

SAND Sampling Aerospace Nuclear Debris

SANDA Supplies and Accounts

SANDAG San Diego Association of Governments

Sand Eng Sandalwood English (Polynesian Pidgin English)

SANDER San Diego Energy Recovery

SANDERP San Diego Energy Recovery Project (garbage and trash converted to energy)

San Di San Diego

SANDIA Sandia National Laboratories

San Domingo Santo Domingo (Dominican Republic)

sand(s) sandwich(es)—invented by the Earl of Sandwich, who disliked leaving the gaming table to eat, and had thin slices of cheese or meat brought to him between two pieces of bread; his culinary invention was devised around 1776 when he was First Lord of the Admiralty

Sands the Sands, the Godwin Sands off England's Channel coast of Kent

SANDT School of Applied Non-Destructive Testing

sane severe acoustic noise environment

SANE National Committee for a Sane Nuclear Policy; Security Against Nuclear Extinction; South African National Antarctic Expedition

SANER San Diego Energy Recovery

Sa Nev Sierra Nevada(s)

SANF South African Naval Forces

San Fran San Francisco

San Francisco Pen San Francisco Peninsula

SANFREE San Diegans for Fiscally Responsible Elected Employees

San Gabriels San Gabriel Mountains of southern California

Sangre de Cristos Sangre de Cristo Mountains extending from Colorado to New Mexico

sanguin (Latin prefix—blood)—sanguine

San Insp Sanitation Inspection; Sanitary Inspector(ate)

sanit sanitar; sanitation; sanitize

San J San Juan (Dominican Republic province)

San Jac San Jacinto

San Jo (Mexican-American—San Jose, California)

San Juans San Juan Islands (Washington); San Juan Mountains (Colorado and New Mexico)

sanka sans kaffeine (coffee without caffeine)

San-Lax San Diego-Los Angeles

SANLC South African Native Labor Corps

San Le San Leandro, California

San Lucas Cabo San Lucas, Baja California

s-a-n man stop-at-nothing man (dangerous criminal)

San Mar San Marino (whose capital is San Marino)

San Marco D atmospheric research satellite carrying experiments from Italy, Germany, and the United States

San Marino Most Serene Republic of San Marino (tiny country surrounded by Italy, on the slopes of Mount Titano near the Adriatic), *La Serenissima Repubblica di San Marino*

San Martin José de San Martin—patriot-soldier who fought to liberate Argentina, Chile, and Peru from the Spanish rule

San Met San Diego Metropolitan Correctional Center

S Ann St Anne's College, Oxford

San-NO San Diego—New Orleans

San-NY San Diego—New York

s-a node sino-atrial node

Sanpaolo Instituto Bancario San Paolo di Torino (Italian—San Paolo Banking Institute of Turin)

SANPAT San Diego Plans for Air Transportation

sanr subject to approval—no risks

sans sans serif

Sans Sanskrit

SANS South African Naval Service; System Administration Networking and Security

San-Sac San Diego—Sacramento

San Sal San Salvador (capital of El Salvador and department of El Salvador)

Sansan San Diego to San Francisco (city complex)

San-Sfo San Diego—San Francisco

Sansk Sanskrit

san sou *sans souci* (French—without a care)—also name of the palace built by Frederick the Great of Prussia

Sant Santander; Santiago

S Ant St Antony's College, Oxford

SANTA South African National Tuberculosis Association; Souvenir and Novelty Trade Association

Santa Barbaras Santa Barbara Islands off Santa Barbara, California

Santa Fe Atchison, Topeka & Santa Fe (Railway)

Santa Monicas Santa Monica Mountains of southern California

Santa Ritas Santa Rita Mountains of southeastern Arizona

SANTAS Send A Note To A Serviceman

Santa See (Spanish—Holy See)—Vatican City, Rome

Santayana George Santayana, Jorge Augustin Nicolás Ruiz de Santayana (1863–1952)

Santiago (Portuguese or Spanish—Saint James)—Santiago do Boqueirão, Brazil; Santiago de Calatrava and Santiago de Compostela, Spain; Santiago de Chile and Santiago de Cuba, Santiago de los Cabelleros, Dominican Republic; Santiago Ixcuintla, Mexico; Santiago Sacatepéquez, Guatemala; Santiago Zamora, Ecuador

Santiagos Santiago Mountains in the Big Bend National Park in Texas

Santo Dom *Santo Domingo* (Spanish—Santo Domingan)

Sant *Santuario* (Spanish—sanctuary)

San-Tor San Diego—Toronto

SANU Sudanese African National Union

San-Van San Diego—Vancouver

SANWR Santa Ana National Wildlife Refuge (Texas)

San Ysidros San Ysidro Mountains of southern California

SANZ Standards Association of New Zealand

SAO São Paulo, Brazil (airport); Secret Army Organization; Senior Administrative Officer; Senior American Officer (POW camps); Smithsonian Astrophysical Observatory

Sa$_{o2}$ arterial oxygen saturation

SAOC South African Ordnance Corps

SAODAP Special Action Office for Drug Abuse Prevention

SAORC Supreme Assembly of the Order of the Rainbow for Girls

SAOS Scottish Agricultural Organization Society

São Tom and Prin São Tomé and Principe (São Tomé the capital)

São Tomé and Principe Democratic Republic of São Tomé and Principe (West African coastal islands)

sap saphead; semi-armor piercing; scruple, apothecaries; simplified astro pattern; soon as possible; specific action potential; strong anthropic principle

SA£ South African pound

SAP Safety Assessment Plan; San Pedro Sula, Honduras (airport); Scottish Academic Press; Share Assembly Program, Society for Applied Spectroscopy; South African Police; Symbolic Assembly Program; Systems Assurance Program

s-apa sciences—a project approach

SAPA South African Press Association; South African Publishers' Association; Substance Abuse Program Administration

SAPARLI Saudi Arabian Parsons Limited

SAPAT South African Picture Analysis Test

SAPE Society for Automation in Professional Education

SAPF South African Permanent Force; South African Police Force

sapfu surpassing all previous foul ups

sapi semi-armor-piercing incendiary

SAPI Sales Association of the Paper Industry

SAPL San Antonio Public Library; Society for Animal Protective Legislation; South African Public Library

SAPM Scottish Association of Paint Manufacturers; Society for the Aid of Psychological Minorities

sap. no. saponification number

sapon saponification; saponify

saponite soapstone (hydrous magnesium aluminum silicate)

sapp sapphic; sapphist(ic)(al)(1y)

SAPRI South African Plain Research Institute

SAPS South African Price Schedule

sar seasonal allergic rhinitis; search and rescue; semiautomatic rifle; short-term acquisition and retrieval; specific absorption rate; stock appreciation right; submarine advanced reactor; suspicious activity report

Sar Saracen; Saracenic; Sardinia; Sardinian

SAR Safety Analysis Report; Society of Authors' Representatives; Solar Aircraft (company); Sons of the American Revolution; South African Railways; South African Republic; South Alberta Regiment; South Australian Railways; Summary Annual Report (benefits); Synthetic Aperture Radar

S-AR Sud-Africaine Républic (French—South African Republic)

Sara Sarah; Saratoga

Sara Sarajevo

SARA Superfund Amendments and Reauthorization Act

sarac steerable array for radar and communications

SARAH Search and Rescue and Homing (radio lifesaving beacon)

Saraj Sarajevo (capital of Bosnia and Herzegovina)

Sarasate Pablo de Sarasate (Pablo Martin Melitón Sarasate y Navacuez)

Saraw Sarawak

SARB South African Reserve Bank

SARBE Search and Rescue Beacon Equipment

SARBICA Southeast Asian Regional Branch of the International Council on Archives

sarc (Latin prefix—flesh)—sarcoma

SARC Sexual Assault Referral Centre (Australia); South Asian Regional Cooperation

SARCCUS South African Regional Committee for the Conservation and Utilisation of the Soil

sarcol sarcological; sarcologist; sarcology

SARD Special Airlift Requirement Directive

SARD Statistical Analysis and Reports Division (of Administrative Office of the United States Courts)

SARDA State and Regional Disaster Airlift

sardonyx chalcedony consisting of alternate layers of onyx and sard

sardsan sardine sandwich

sardwich sardine sandwich

sare self-addressed return envelope

SARE Sustainable Agriculture Research and Education (program)

sarge sergeant

Sargent Porter Sargent, Inc

sarh semi-active radar homing

SAR & H South African Railways and Harbours

SARHA South African Railways, Harbours, and Airways

SARHWU South African Railway and Harbor Workers Union

sarie selective automatic-radar-identification equipment

SARL Sociedade Anónima de Responsabilidadem Limitada (Portuguese—Limited Liability Corporation)

SARLANT Search-and-Rescue, Atlantic

Sarmiento Domingo Faustino Sarmiento—Argentinian educator and early president

SARMS Self-Adapting Account Receivable Management System

SARPAC Search-and-Rescue, Pacific

sarps standards and recommended practices

sarra short-arc reduction of radar altimetry

SARs Selected Acquisition Reports

SARS Ship Attitude Record System

sarsat search-and-rescue astronomical satellite system

SARSAT Search-and-Rescue Satellite-Aided Tracking; Search-and-Rescue Satellite Aided Tracking

SART St Alban's Repertory Theater; Strategic Arms Reduction Talks

sartac search radar device

sartel search and rescue telephone

SARTS Switched-Access Remote Test System

SARU Systems Analysis Research Unit

sas so and so; sodium aluminum sulfate (baking powder)

sas (SAS) small astronomy satellite; supersonic attack seaplane; surface-air-surface (second-class international mail service)

Sas Sasebo

SAs Special Agents (FBI)

SAS Scandinavian Airlines System; Science Attitude Scale; Seattle Audubon Society; Sherwood Anderson Society; Sklar Asphasia Scale; Small Area Statistics (U.K.); South African Submarine; Special Air Service; Special Armed Service(s); Statistical Analysis System; Studio Address System; Systems Assessment Survey

SAS *Societa in Accomandita Semplice* (Italian—Limited Partnership Company)

SASA South African Sugar Association

SASBO Southeastern Association of School Business Officials

SASC Senate Armed Services Committee; Small Arms School Corps (UK); South African Service Corps; South African Staff Corps; Steamship Association of Southern California

SASCOM Special Ammunition Support Command (USA)

SASD School Adnministrators of South Dakota

sase self-addressed stamped envelope

SASI Society of Air Safety Investigators

SASIDS Stochastic Adaptive Sequential Information Dissemination System

SASIS Semi-Automatic Speaker-Identification System

SASJ South African Society of Journalists

Sask Saskatchewan

SASL South American Saint Line

SASLO South African Scientific Liaison Office

SASM Smithsonian Air and Space Museum

SASMIRA Silk and Artificial Silk Mills Research Association

SASO San Antonio Symphony Orchestra; Saudi Arabia Standards Organization; Senior Air Staff Officer; South African Students Organization; South Australia Symphony Orchestra; Superintending Armament Supply Officer

sasol South African (coal-based synthetic) oil

SASOL South African Coal, Oil, and Gas Corporation

SASP State Agencies for Surplus Property

SASR Special Air Service Regiment

Sass *Sassenach* (Gaelic—English, Saxon)

SASS San Antonio Symphony Society; Shanghai Academy of Social Sciences; Society for the Advancement of Scandinavian Study; Swarthmore Afto-American Students Society

SASSO Senior Air Staff Officer

SASSY Supported Activity Supply System

sast single asphalt-surface treatment

SAST Society for the Advancement of Space Travel; Strategic Analysis in the Field of Science and Technology

sat sampler address translator; satellite; satisfactory; saturate; saturation; service acceptance trials; system alignment tool; systems approach to training

sat. (SAT) satellite; systematic assertive therapy

Sat Satan; Satanic; Saturday; Saturn

S At South Atlantic

SAT San Antonio, Texas (airport); Scholastic Aptitude Test; Scholastic Assessment Team; School of applied Tactics; Security Air Transport; Sound-Apperception Test; Southern Air Transport; Specific Aptitude Test; Spiral Aftereffect Test; Stanford Achievement Test; Support Analysis Test

SAT Scholastic Assessment Test

SATA *Sociedade Acoriana de Transportes Aéreos* (Portuguese—Azores Air Transport Line)

SATAF Second Allied Tactical Air Force; Site Activation Task Force

satan satellite automatic tracking antenna; sensor for airborne terrain analysis

satanas semi-automatic analog setting

Satanic names Abaddon, Amon, Apollyn, Asmodeus, Azazel, Balaam, Beelzebub, Behomoth, Coyote, Dagon, Diabolus, Dracula, Fenriz, Gorgo, Hecate, Ishtar, Kali, Lilith, Loki, Mammon, Marduk, Mephisto (Mephistopheles), Moloch, Nija, Pluto, Prosepine, Samiel, Shiva, Tezcalipica, Typhon, Yaotzin

satar (SATAR) satellite for aerospace research

SATASM Soviet Advanced Tactical Air-to-Surface Missile

satb (SATB) soprano, alto, tenor, bass

SATC South African Tank Corps; South African Tourist Corporation

satco signal automatic air traffic control

SATCO Senior Air Traffic Control Officer

satcom satellite communication

SATCOM Satellite Communications Agency (US Army)

satd saturated

satel satellite

SATENA *Servicio Aeronavegación a Territorios Nacionales* (Spanish—National Air Service)—Bogotá, Colombia

SatEvePost *Saturday Evening Post ·*

satex semi-automatic- telegraphic exchange

sat. fix. (SAT FIX) satellite (aircraft or ship position) fix

satfy satisfactory

SATGA *Société Aérienne des Transports Guyane Antilles* (French—Guinea Air Transport Society)

satgci satellite ground-controlled interception

SAT-HI Stanford Achievement Test for the Hearing Impaired

SATIF Scientific and Technical Information Facility (NASA)

SATIN Sage Air Traffic Integration

SATIRE Semi-Automatic Technical Information Retrieval

SAtk strike attack

S Atl South Atlantic (Falkland Islands and Dependencies)—British

S Atl Cur South Atlantic Current

SAT-M Scholastic Aptitude Test-Mathematical

satn saturation

satnav satellite navigation; satellite navigator

SATO South American Travel Organization; Southern Africa Treaty Organization

SATOUR South African Tourist Corporation

Sat Pax Pax Lao *Sathalanalat Paxathipatai Paxax-on Lao* (Lao People's Democratic Republic)

satpic satellite picture

SATRA Shoe and Allied Trade's Research Associatiion

Sat Rev *Saturday Review*

sats (SATS) short airfiield for tactical support

SATs Scholastic Aptitude Tests

SATS Satellite Antenna Test System (NASA); Small Arms Target System; South African Training Ship; South African Transport Services

SATSC South African Technical Service Corps

satsim saturation countermeasures simulator

sat sol saturated solution

sattr satisfactory to transfer

SATU Singapore Air Transport Union; South African Typographical Union

SAT-V Scholastic Aptitude Test-Verbal

SATW Society of American Travel Writers

saty satyagraha; satyriasis; satyr(ic)(al)(ly); satyrid

sau surface attack unit

sau (SAU) standard advertising unit

Sau Saudi Arabia

Sau Arab Saudi Arabia(n)

SAUCERS Saucer and Unexplained Celestial Events Research Society

Saudi Saudi Arabian(s)

Sandia Saudi Arabia

Saudi Arabia Kingdom of Saudi Arabia (largest Middle Eastern country), *al-Mamlaka al'—Arabiya as-Saudiya*

Saudis Saudi Arabians

SAUL South African Unattached List

'sault assault

'sault & assault and battery

Saunders W.B. Saunders Co

SAUS *Statistical Abstract of the United States*

S Austral South Australia(n)

sav savings; stock at valuation

sa/v surface area/volume

Sav Savannah

SAV Savannah, Georgia (airport)

SAVAK *Sazemane Etelaat va Aminate Kechvar* (Persian—Iranian Security and Intelligence organization)

SAVC Society for the Anthropology of Visual Communication; South African Veterinary Corps

SAVE Service Activities of Volunteer Engineers; Society of American Value Engineers; Special Action Program for Vigorous Energy efficiency; Stop Addiction through Voluntary Effort; Student Action Voters for Ecology; Systematic Alien Verification for Entitlements

savi science activities for the visually impaired

SAVICOM Society for the Anthropology of Visual Communications

savor signal-actuated voice recorder

SAVS Scottish Anti-Vivisection Society

Savus Savu Islands of Indonesia

saw sample assignment word; satellite attack warning; space at will; squad automatic weapon; surface acoustic wave

SAW Society of Architects in Wales; Society of Australian Writers; Special Agricultural Workers; Special Air Warfare

SAWA Screen Advertising World Association; Soil and Water Management Association

SAWANS South African Women's Auxiliary Naval Service

SAWARA Southern Arizona Water Resources Association

SAWAS South African Women's Auxiliary Services

Sawatches Sawatch Mountains of central Colorado

SAWC Special Air Warfare Center

sawd surface acoustic(al) wave device

SAWE Society of Aeronautical Weight Engineers

sawes small-arms weapons effects simulator

SAWF Special Air Warfare Force

SAWG Special Advisory Working Group; Special Air Warfare Group

SAWI Society Against World Imperialism (Beirut-based Arabic terrorists)

sawo surface acoustic-wave oscillator

s-a-w q seeking-asking-and-written questionnaire

saw(s) (SAWS) special agriculture worker(s)

SAWS Satellite Attack Warning System; Small Arms Weapons Study; Squad Automatic Weapon System; Submarine Acoustic Warfare System

Sawtooths Sawtooth Mountains of south-central Idaho

SAWTRI South African Wool Textile Research Institute

Sawy *Sawyer's United States Circuit Court Reports*

sax saxophone; strong anion exchange

Sax Saxon

Sax Duc Saxon Duchies; Saxon Dukes

saxist saxophonist

SAY Salisbury, Rhodesia (airport)

SAYCO South African Youth Conference

saye save as you earn

SA y P *San Andrés y Providencia* (Spanish—San Andres and Providence)—Caribbean island possessions of Colombia

SAZF South African Zionist Federation

sb savings bond; sideband; simple blessing; simultaneous broadcast(ing); single-bayonet (lamp base); single-breasted (coat or jacket); small bore (weapon); small business; smooth bore; solid body; southbound; special bibliography; stepbrother; stolen bases; stove bolt; stretcher bearer; subbituminous; submarine (fog) bell; switchboard

s/b should be; surface based

sb *styrbord* (Swedish—starboard; right side of a vessel looking forward, from Viking steering oar on right side of their long boats)

Sb *stibium* (Latin—antimony)

SB Air Caledonia International; Savannah Beach; Savings Bank; scouting-bombing (aircraft); Seaboard World Airlines (2-letter coding); Secondary Battery; Section Base; Selection Board; Senate Bill; Service Bulletin; Short Bill (payable on demand); shipbuilding; Signal Battalion; Signal Boatswain; Small Borough; Solomon Is-

lands (Internet code); South Bronx; South Buffalo (railroad); Soviet Bloc; Soviet Branch; Special Branch; Standard Brands (stock exchange symbol); Stanford-Binet (intelligence test); Submarine Base

S-B Stanford-Binet (intelligence test)

S & B sterilization and bath

SB *San Bartolomeo* (Italian—Saint Bartholomew)—Naples opera house; *Schweizerischer Bankverein* (German—Swiss Bank); *Sitzungbericht* (German—report of a proceeding)

S.B. *Scientiae Baccalaureus* (Latin—Bachelor of Science)

sba stolen-base average

Sba Surabaya

SBA School Bookshop Association; School of Business Administration; Show Business Association; Sick Bay Attendant; Sick Berth Attendant; Small Business Administration; Small Businesses Association; Sovereign Base Area; Systems Builders Association

SBAC Society of British Aerospace Companies

sbae stabilized bombing approach equipment

S-bahn *Stadt-Schnellbahn* (German—State Rapid Transit)—Berlin's electric railway system

SBAMA San Bernardino Air Materiel Area

S-band 1550–5200 megahertz radio-frequency band

S Bapt Southern Baptist

SBAs Sick Bay Attendants

SBAS Statewide Budgeting and Accounting System

SBAW Santa Barbara Academy of the West

Sbb. *Sabbatum* (Latin—Sunday)

SBB Sudan Black B (stain)

SBB *Schweizerische Bundesbahnen* (German—Swiss Federal Railways)

SBBNF Ship and Boat Builders' National Federation

sbc silicon blue cell; single board computer; small bomb crater; small business computer

SBC Sam Browne's Cavalry; School Based Clinics; Senate Budget Committee; Service Bureau Corporation; Small Business Council; Southern Baptist Convention; Southern Building Code; Sumitomo Bank of California; Supplementary Benefits Commission; Surinam Bauxite Company; Sweet Briar College; Swiss Bank Corporation

SBC *Société de Banque Suisse* (French—Swiss Bank Corporation)

SBCC Santa Barbara City College

SBCCI Southern Building Code Conference International

SBCCOE State Board for Community College and Occupational Education

SBCPO Sick-Bay Chief Petty Officer

SBCR State Board of Charities and Reform (Wyoming); Stock Balance Consumption Report

sbd seriously behaviorally disabled; standard bibliographic description

sbdp salary and bonus deferral plan

sbdt surface-barrier diffused transistor

sbe soft-boiled egg(s); standby engine(s); subacute bacterial endocarditis

s-b-e standby engine(s)

SBE Society of Broadcast Engineers; State Board of Equalization

SBEA Southern Business Education Association

sbec single board engine controller

sbei starch-branching enzyme

S-bend S-shaped bend

sbf surface burst fuze

SBFA Small Business Foundation of America

sbfc standby for further clearance

sbg selenite brilliant green; small box girder (bridging)

Sbg Solvesborg

SBGI Society of British Gas Industries

sbh supermassive black hole

SBH Scottish Board of Health; State Board of Health

sbi space-based interceptor

SBI Security Bureau Incorporated; Southern Burn Institute (Baton Rouge); State Bank of India

SBIC Small Business Investment Corporation

sbic's small business investment companies

SBII *Serikat Buruh Islam Indonesia* (Central Islamic Labor Union of Indonesia)

SBIR Small Business Innovation Research

sbis (SBIS) satellite-based interceptor systems

SBIW Sybil Brand Institute for Women (Los Angeles correctional facility)

sbl space-based laser

Sbl Setubal

SBL Stephen B(utler) Leacock

SBLI Savings Bank Life Insurance

sblo strong black liquor oxidation

sbm submission; submit

SBM school-based model

SBM *Société Anonymes des Bains de Mer et du Cercle des Etrangers à Monaco* (company managing gambling casino of Monte Carlo)

SBMA Santa Barbara Museum of Art

SBME Society of Business Magazine Editors; State Board of Medical Examiners

SBMF Santa Barbara Mariculture Foundation

SBMI School Bus Manufacturers Institute

sbml small bore muzzle loading

SBMNH Santa Barbara Museum of Natural History

sbn standard book number(ing)

Sbⁿ Sebastián (Spanish—Sebastian)

SBN South Bend, Indiana (airport); Standard Book Number

SBNA Standard Book Numbering Agency

SBNB Subic Bay Naval Base

S Bno San Bernardino

SBNO Senior British Naval Officer

SBNS Society of British Neurological Surgeons

sbo secure base of operations; specific behavioral objectives

Sbo Sasebo

SBO Senior British Officer

SBOA Specialty Bakery Owners of America

s'board starboard

sbom soy bean oil meal

sbp shoulder belt plate; slotted-blade propeller; sugar-beet pulp; systolic blood pressure

SBP Society of Biological Psychiatry

SBPIM Society of British Printing Ink Manufacturers

sbr small box respirator; space-based radar; stolen-base runs; styrene-butadiene rubber

s Br *südliche Breite* (German—south latitude)

SBR Society of Biological Rhythm

SBRC Santa Barbara Research Center

sbre *septiembre* (Spanish—September)

SBRI Simon Baruch Research Institute

sbs simulated borehold specimen; special boat section; surveyed before shipment

sbs (SBS) small business satellite

sb's sonic booms; space brothers

s-b-s side-by-side (double-barrel shotgun)

SBS Satellite Business System(s); Singapore Bus Service; Special Boat Squadron; Strategic Balkans Service; Swiss Broadcasting Society

SBS *Société de Banque Suisse* (French—Swiss Bank)

SBSA Standard Bank of South Africa

Sbsc Schottky-barrier solar cell

sbsp subspecies

sbss (SBSS) space-based surveillance system

SBSUSA Sport Balloon Society of the United States

sbt screening breath tester (for drunken drivers); segregated ballast; submarine bubble target; surface-barrier transistor

SBT Screening Breath Test

sbtg sabotage

sbti soy bean trypsin inhibitor

sbtow standby tow(ing) ship

sbt's segregated ballast tanks

sbv sea-bed vehicle

SBV Space Biospheres Ventures

sbw stolen base wins

SBW Seaboard & Western (Airlines); single-engine scout bomber (3-letter naval symbol)

SBWR Seal Beach Wildlife Refuge (near Long Beach, California); South Bay Wildlife Refuge (south end of San Francisco Bay)

sbx S-band transponder

SBX Student Book Exchange

sby standby

sb% stolen base percentage (baseball)

sc sad case; same case; scratched; separate cover; service certificate; shaped charge; sine cosine; single circuit; single contact; single crochet; single crystal; sized and calendered; slow cool; small caps (small capital letters); smooth contour; statistical control; step cut; supercycle; superimposed current; suppressed carrier; synaptonemal complex

sc (SC) service charge; site contractor; spinal cord; systolic click

s/c short circuit (electrical); single-column (bookkeeping); suspicious circumstances

s & c search and clear; shipper and carrier; sized and calendered

sc. *scilicet* (Latin—mainly)

s/c *su cuenta* (Spanish—your account)

Sc conditioned stimulus; scandium; Scotch (whiskey); Scot(s); Scotland; Scottish; stratocumulus

Sc *La Scala (Teatro alla Scala)*—Milan's opera house; *Scoglio* (Italian—reef, rocky reef)

SC Sacra Congregatio (Sacred Congregation); Sacramento City; Salem College; San Carlos; Sandia Corporation; Sanitary Corps; Santa Claus; Scripps College; Sea Cadets; Seamen's Center; Security Council (United Nations); Selwyn College; Service Club; Service Command; Seychelles (Internet code); Shasta College; Shaw College; Shell Transport; Shelton College; Shenandoah College; Shepherd College;

Sheridan College; Shimer College; Ship's Cook; Shorter College; Siena College; Sierra Club; Sierra College; Signal Corps; Simmons College; Simpson College; Sinclair College; Sister(s) of Charity; Skidmore college; small-capitalization core (stocks); Smith College; Somerville College; South Carolina; South Carolinian; Southern Cascades; Southern California; Southern Californian; Southern Command; Southern Conference; Southwestern College; Special Constable; Spellman College; Springfield College; Staff Captain; Staff College; Staff Corps; Star of Courage (Australia); Stephens College; Sterling College; Stockton College; Stonehill College; Stratford College; Strike Command; submarine chaser; Submarine Coxswain; Suffolk & Cambridgeshire Regiment; Sullins College; Summary Court; Sumter & Choctaw (railroad); Superintending Clerk; Supply Corps; Supreme Councillor; Suomi College; Supply Corps; Support Command; Supreme Court; Surgical Corporation; Swarthmore College; Systems Command

S-C Serbian-Croatian (people); Serbo-Croat (language); Stromberg-Carlson

S/C Star & Crescent (excursion steamer, ferry, towing, water-taxi service)

S & C search and clear;

SC Scott Catalog; Statistics Canada

sc & save situation conversion percentage (baseball)

sca sequencer control assembly; sickle-cell anemia; small-caliber ammunition; subchannel adapter; subsidiary communications authorization; sudden cardiac arrest

sca (SCA) supersonic cruising aircraft

SCA Schipperke Club of America; School and College Ability (test); Science Clubs

of America; Screen Composers Association; Senior Citizens of America; Shipbuilders Council of America; Society of California Accountants; Society of Consumer Affairs; Soldiers Christian Association; Southern Cotton Association; Soybean Council of America; Speech Communication Association; Standard Consolidated Area; State Commemorative Area; Stock Company Association; Student Conservation Association; Sub-Contract Authorization; Suez Canal Authority; Survey of College Achievement; Svenska Cellulose AB; Switzerland Cheese Association; Synagogue Council of America

SCAA Specialty Coffee Association of America, Spill Control Association of America; State Communities Aid Association

SCAAP Special Commonwealth African Assistance Plan

SCAC School and College Advisory Center; Sunrise Cultural and Art Center (Charleston, West Virginia)

SCACOP Southern California Area Construction Opportunity Program

scad schedule, capability, availability, dependability; subsonic cruise armed decoy

SCAD State Commission Against Discrimination (New York)

scadar scatter detection and ranging

SCADS Sioux City Air Defense Sector

SCADTA Sociedad de Colombo-Alemana de Transportes Aereos (Spanish—Colombia-German Air Transport Society)

SCAEF Supreme Commander Allied Expeditionary Force

SCAF Supreme Commander of Allied Forces

SCAG Sandoz Clinical Assessment-Geriatric; Southern California Association of Governments; Supplier Corrective Action Group

SCAGL Société Cinématographique des Auteurs et Gens de Lettres (French—Cinematic Society of Authors and Writers)

sc al steel-cored aluminum

SCALA Society of Chief Architects of Local Authorities

scaler statistical calculation and analysis of engine removal (USN)

scam strike camera

scama (SCAMA) switching; conferencing, and monitoring arrangement

SCAMF Standing Committee on Army Manpower Forecasts

scams scanning microwave spectrometer

scan self-correcting automatic navigation; suspected child abuse and neglect; switched-circuit automatic network

Scan Scandinavia; Scandinavian

SCAN Scheduling and Control by Automated Network; Selected Current Aerospace Notices (NASA-computerized dissemination of information); Self-Correcting Automatic Navigator; Service Center for Aging Information; Southern California Answering Network; Switched-Circuit Automatic Network

SCANCAP System for Comparative Analysis of Community Action Programs

Scand Scandinavia; Scandinavian

Scand Balts Scandinavian Baltics, including Denmark, Finland, Norway and Sweden

Scandinavian Pen Scandinavian Peninsula, containing Norway and Sweden plus a bit of Finland

ScanDoc Scandinavian Documentation Center

scanit scan-only intelligent terminal

scan. mag. scandalum magnatum (Latin—defamation of high-placed persons)

SCANNET Scandinavian (computer) Network

SCANPED System for Comparative Analysis of Programs of Educational Development

SCANs Southern California Answering Networks (cooperative library information—retrieval system)

SCANS Scheduling and Control Automation by Network Systems; Stockmarket Computer Answering Service

scantie submersible-craft acoustic-navigation and track-indication equipment

SCAO Senior Civil Affairs Office(r); Standing Conference on Atlantic Organizations

scap scapula; scapular; scapuloid

SCAP Supreme Commander, Allied Powers

Scapa Scapa Flow naval anchorage in the Orkney Islands off Scotland's north coast between Hoy, Orkney, and South Ronaldsay

SCAPA Society for Checking the Abuses of Public Advertising

'scape escápe(ment); land scape; seascape; skyscape

scaphocephs scaphocephalics (narrow-skulled people)

s caps small capital letters

SCAQMD South Coast Air Quality Management District (California)

scar subcaliber aircraft rocket; submarine celestial altitude recorder

S Car South Carolina

SCAR Scandinavian Council for Applied Research; Scientific Committee for Antarctic Research; Supersonic Cruise Airplane Research (NASA)

scarab (SCARAB) submersible craft assisting repair and burial (of underwater telephone cables)

Searboro' Scarborough

scard signal conditioning and recording device

scare sensor-control anti-antiradiation-missile radar evaluation

SCARF Special Committee on the Adequacy of Range Facilities

scarp escarpment

S-car(s) Swedish car(s)

SCARWAF Special Category Army With Air Force

sca's subsidiary communications authorizations

SCAS Senior Citizen Audiological Service

scat share compiler assembler and translator

scat (SCAT) speed-control attitude range; supersonic commercial air transport

scat. scatula (Latin—box)

SCAT School and College Ability Test; Science College Ability Test(ing); Service Command Air Transportation (USN); Sport Competition Anxiety Test

scata survival sited casualty treatment assemblage

SCATANA Security Control of Air Traffic and Air Navigational Aids

SCATE Stromberg-Carlson automatic test equipment

scatha spacecraft charging at high(er) altitude(s)

scato (Greek—excrement)—scatologic(al); scatology

scat. orig. scatula originalia (Latin—original box or package)

scats (SCATS) sequentially controlled automatic transmitter start (data processing)

scat's supersonic commercial air transports

SCATs Southern California Acrobatic Teams

SCATS Simulation, Checkout, and Training System

scatt. scattered; scattering

SCAULWA Standing Conference of African University Libraries—Western Area (Ghana)

scav scavenge

Scaw Fells Scaw Fell (or Scafell) Mountains of the Cumbrians in England's Lake District

scb state-capacity building; strictly confined to bed (q.v. fob)

sc b screw base (lamp)

Sc.B. Scientia Baccalaureus (Latin—Bachelor of Science)

SCB Sawyer College of Business; Sierra Club Books; Southern California Bookbuilders

SCB Sociedad Bolivariana de Venezuela (Spanish—Bolivarian Society of Venezuela)

SCBA Savings and Community Bankers of America; Self Contained Breathing Apparatus; Southern California Booksellers Association

SCBC Somerset Cattle Breeding Centre

SCBCA Small Claims Board of Contract Appeals

scbf spinal-cord blood flow

SFCBQ Science Classroom Behavior Q-sort

scbu (SCBU) special-care baby unit

SCBWI Society of Children's Book Writers and Illustrators

scc single-channel controller, specific clauses and conditions; stress corrosion cracking

Sc C Scottish Command

SCC Sea Cadet Corps; Second-Class Certificate; Security Coordination Committee; Select Cases in Chancery; Senior Command Course; Ship Control Center; Shoreline Community College; Sierra Conservation Center; Sitka Community College; Society of Cosmetic Chemists; Spokane Community College; Stafford Cadet Corps; Standard Commodity Classification; Standing Consultative Commission (U.S.–former USSR group created to monitor disputes); Stromberg-Carlson Corporation; Student Coordinating Council; Surveillance Coordination Center

S & CC Suicide and Crisis Counseling

SCCA Society of Company and Commercial Accountants; South Carolina Correctional Association; Southeastern Cottonseed Crushers Association; Sports Car Club of America

SCCAPE Scottish Council for Commercial, Administrative, and Professional Education

SCCC Singapore Chinese Chamber of Commerce; Suffolk County Community College; Sullivan County Community College

SCCCI Singapore Chinese Chamber of Commerce and Industry

SCCE Society of Certified Credit Executives

SCCF Security Clearance Case Files

SCCG Southern California Culinary Guild

SCCOP State Consulting Company for Oil Projects

SCCPG Satellite Communications Contingency Planning Group

SCCPT Subcommittee on Computer Program Terminology (Association for Computing Machinery)

sccrt sub-zero cooled, cold-rolled, and tempered

scd screen door; screwed; service computation date; standard change dispenser

scd (SCD) security coding device

Sc.D. *Scientiae Doctor* (Latin—Doctor of Science)

SCD Specification Control Drawing

SCD *Standard College Dictionary*

scda scapula-dextra anterior

SCDA Scottish Community Drama Association

SCDC Senior Citizen's Dental Clinic; South Carolina Department of Correction

scde's schools, colleges, and departments of education

SCDL Scientifiic Crime Detection Laboratory

scdp scapula-dextra posterior

SCDS Shipboard Chaff-Decoy System

sce sister chromatid exchange; situationally caused error; standard calomel electrode

SCE Schedule Compliance Evaluation; Society for Clinical Ecology; Southern California Edison

S.C.E. Scottish Certificate of Education

SCEA Service Children's Education Authority; Society of Cost Estimating and Analysis; South Carolina Education Association

SCEI Safe Car Educational Institute; Special Libraries Committee on Environmental Information

SCEL Signal Corps Engineering Laboratories

scen scenario(s); scenarist(s); scenographic(al)(ly)

SCENT System Customs Enforcement Network

SCEPC Senior Civil Emergency Planning Committee (NATO)

sceps stored chemical energy propulsion system

SCES State Cooperative Extension Service

SCET Scottish Council for Educational Technology

SCF Saba Conservation Foundation; Save the Children Federation; Sectional Center Facility (USAF); Senior Chaplain to the Forces; Station Code File; Stephen Collins Foster

sc f & a screw forward and aft

SCFA Southern California Fishermen's Association

scfd standard cubic feet per day

scfh standard cubic feet per hour

SCFIC South Carolina Foundation of Independent Colleges

scfm standard cubic feet per minute

scfs standard cubic feet per second

scg scoring

SCG Screen Cartoonists Guild; Social Credit Group; Society of the Classic Guitar; Special Consultative Group

SCGA Southern Cotton Ginners Association

Sc Gael Scottish Gaelic

SCGB Ski Club of Great Britain

SCGC Southern California Gas Company; Southern Counties Gas Company

SCGM Senior Cook General Mess

SCGR Sale Common Game Refuge (Victoria, Australia)

SCGRL Signal Corps General Research Laboratory

SCGSA Signal Corps Ground Signal Agency

SCGSS Signal Corps Ground Signal Service

sch school

sch (SCH) schedule

Sch Schiedarn; School (postal abbreviation)

Sch *Schauspielhaus* (German—playhouse, theater)

Sch Arts *School Arts*

SCHAVMED Schcol of Aviation Medicine (USN)

Schbg Schönberg

schd scheduled; scheduling

sched schedule

scheepv *scheepvaart* (Dutch—navigation, shipping)

scheik *scheikunde* (Dutch—chemistry)

schem schematic

Schen Schenectady

scherz *scherzando* (Italian—jesting, in a sportive manner)

schilling monetary unit of Austria

S China South China Sea, between Hong Kong and the Philippines

Schipol Amsterdam airport

Schirmer EC Schirmer (Boston); G Schirmer (New York)

Sc Hist Scottish History

schizo schizoid; schizophasia; schizophrenia; schizophrenic

schizzy schizoid; schizophrenia; schizophrenic

schl school

SCHLA School of Latin America

Schlags *Schlagobers* (Austrian German—whipped cream)

schlem *schlemiel* (Yiddish—person afflicted with bad luck)

Schles-Hols Schleswig-Holstein (German—Schleswig Holstein)

Sch Lib J *School Library Journal*

Sch Lib Sci School of Library Science

schim schematic

Sch M School Master

Schmarg Schmargendorf

Sch Mist School Mistress

schmoo space cargo handier and manipulator for orbital operations

schol *schola cantorum*; scholar(ly); scholarship; scholastic(ally); scholasticate; scholasticism; scholiast(ic); scholium

SCHOLAR Schering-Oriented Literature Analysis and Retrieval System

Schotl *Schotland* (Dutch—Scotland)

schott *schottisch* (German—Scottish)

schr schooner

Schr *Schriften* (German—publication; script, text, writing)

SCHS Senior Citizen Hospital Service; South Carolina Historical Society

Schupo Schutzpolizei (German—defense police used as a paramilitary force by Hitler)

Schwann Schwann-1 Record & Tape Guide

Schweitzerpsalm (German—Swiss Psalm)—national anthem of Switzerland also sung in French, Italian, and Romansh (Swiss dialect)

schweiz schweizerisch (German—Swiss)

Schwyz Schwyzer(tütsch)

Schwyzd Schwyzerdütsch (Swiss—German language)

sci science; scientific; scientist; ship-controlled interception; smoke curtain installation

sci (SCI) secret confidential informant; sensitive compartmented information

sci scientifique (French—scientific)

SCI Scalable, Coherent Interface; School of Counter-Insurgency; Science Citation Index; Seamen's Church Institute; Service Civil International; Service Corporation International; Shipping Container Institute; Shipping Corporation of India; Simulation Councils Incorporated; Society of Chemical Industries; Society of the Chemical Industry; Sponge and Chamois Institute; Staff Corps Indian; State Commission of Investigation; Supervisory Cost Inspector

SCI Science Citation Index; Servicio Central de Inteligencia (Spanish—Central Intelligence Service)

SCIA Signal Corps Intelligence Agency; Smart Card Industry Association

SCIAC Southern California Intercollegiate Athletic Conference

Sci Am Scientific American

SCI/ARC Southern California Institute of Architecture

scicrit scientific criticism)

scics semiconductor integrated circuits

scid severe combined immune deficiency; severe combined immunodeficiency

Sci D Doctor of Science

Sci D Com Doctor of Science in Commerce

Scidgie Sicilian-Italian (dialect)

Sci D Met Doctor of Science in Metallurgy

Science Academy of Science; High School of Science

scient scientific; scientist

sci-fi science-fiction

sci-fic science-fiction

SCII Strong-Campbell Interest Inventory

scil. scilicet (Latin—namely)

SCIL Support Center International Logistics (USA)

Scillies Scilly Islands, Isles of Scilly, the Sorlings

scim standard cubic inches per minute

Sci M Science Master

SCIM Selected Categories in Microfiche

Sci Mist Science Mistress

scimp. self-contained-imaging microprofiler

scinti scintillate; scintillation

SCIO Staff Counterintelligence Officer

scioneer scientist + engineer

SCIOP Social Competence Inventory for Older Persons

SCIP School Campus Interaction Programme

SCIPA Servicio Cooperativo Interamericano de Producción de Alimentos (Spanish—Interamerican Cooperative Service for the Production of Food)

sci-phi science-philosophy

SCIPIO Sales Catalog Index Project Input Online (Research Libraries Information Network)

scipp sacrococcygeal-to-inferior pubic point

SCIPP Santa Cruz Institute for Particle Physics

SCI & RB South Carolina Inspection and Rating Bureau

Sci Res Assoc Science Research Associates

SCIRP Select Commission on Immigration and Refugee Policy

SCI(s) Success Motivation Institutes

SCIS Science Curriculum Improvement Study

SCISP Servicio Cooperativo Interamericano de Salud Pública (Spanish—Interamerican Cooperative Public Health Service)

Sci-Tec Science-Technology Division (American Libraries Association)

SCITEC Association of the Scientific, Engineering, and Technological Community of Canada

SCI-TECH-SLA Science-Technology Division of the Special Libraries Association

SCIXF Scitex Corporation Limited

sc & j signal collection and jamming

SCJ School of Criminal Justice

SCKD Society of Certified Kitchen Designers

scl scleroderma; space charge limited; stepchild

Scl Sculptor (constellation)

SCL Santiago, Chile (airpprt); Scottish Central Library; Seaboard Coast Line; Society of County Librarians; Southeastern Composers' League; Springfield City Library

scla scapula-laeva anterior

sclc space charge limited current

SCLC Southern Christian Leadership Conference

SCLED South Carolina Law Enforcement Division

SCLERA Santa Catalina Laboratory for Experimental Relativity by Astrometry

SCLH Standing Committee for Local History; Standing Conference for Local History

SCLI Seaboard Coast Line Industries; Somerset and Cornwall Light Infantry

sclp scapulo-laeva posterior

SCLS Serra Cooperative Library System

scm samarium cobalt magnet; small-core memory; soluble cytotoxic mediator; steam-cure mortar; sternocleidomastoid

scm (SCM) specification change memo(randum); strategic cruising missile

Sc.M. Scientiae Magister (Latin—Master of Science)

SCM Section Communication Manager; Smith-Corona-Marchant; Society of Connoisseurs in Murder, Software Configuration Management; Southampton City Museum; Special Court-Martial; Squadron Corporal-Major; Summary Court-Martial

S.C.M. State Certified Midwife

SCM *Su Católica Majestad* (Spanish—Your Catholic Majesty)

SCMA Southern California Marine Association; Southern Cypress Manufacturers Association

SCMAI Staff Committee on Meditation, Arbitration, and Inquiry (ALA)

SCMC Senior Citizen's Medical Clinic

SCMES Society of Consulting Marine Engineers and Ship Surveyors

SCMP Society of Corporate Meeting Professionals

SCMP *South China Morning Post*

SCMR South Canterbury Mounted Rifles

scn scan (flow chart); suprachiasmatic nuclei

Scn Scunthorpe

SCN System Control Number

SCNAWAF Special Category Navy with Air Force

SCNM Sunset Crater National Monument (Arizona)

SCNO Senior Canadian Naval Officer

SCNR Scientific Committee of National Representatives (NATO)

scns self-contained navigation system

scn/sin sensitive command network/sensitive information network

SCNUL Standing Conference of National and University Libraries (UK)

SCNVYO Standing Conference of National Voluntary Youth Organisations (UK)

SCNWR Squaw Creek National Wildlife Refuge (Missouri)

sco single crossover (genes); subcarrier oscillator; sustainer cutoff

Sco Scorpius (constellation)

ScO Scientific Officer

SCO Sales Contracting Office(r); Statistical Control Office(r)

SCOC Senior Citizen Otolaryngological Clinic; Support Command Operations Center

scoda scan coherent doppler attachment

SCODS Standing Committee on Ocean Data Stations

ScoE Scottish English

SCOFF Society for the Conquest of Flight Fear

SCOGS Select Committee on (Generally-Regarded-As-Safe Substances)

SCOH Staff Corporal of Horse

SCOLCAP Scottish Libraries Cooperative Automation Project

SCOLE Standing Committee on Library Education

S Coll Staff College

SCOLLUL Standing Conference of Librarians of Libraries of the University of London

SCOLMA Standing Conference On Library Materials on Africa

SCOM Scientific Committee (NATO)

SCOMP Secure Communications Processor

scon self-contained

scond semiconductor

SCONMEDLIB Standing Conference of Mediterranean Libraries

'Sconsin Wisconsin

SCONUL Standing Conference of National and University Libraries

scoop scientific computation of optimum procurement

scop (SCOP) single copy order plan

scope microscope; oscilloscope; periscope; telescope; telescopic gunsight

SCOPE Scholarly Communication—Online Publishing and Education; School-to-College Opportunity for Post high-school Education; Scientific Committee on Problems of the Environment; Selected Contents of Periodicals for Educators; Simple Checkout-Oriented

Program Language; Special Committee on Problems of the Environment (ICSU); Software Certification of Program Spelling in Europe; Student Council on Pollution and Environment

SCOPES Squad Combat Operations Exercise Simulation (USA)

scor skin-conductance orienting response

Scor Scorpio

SCOR Scientific Committee on Oceanographic Research; Standing Conference on Refugees

score signal communications by orbiting relay equipment; special claim on residual equity; spectral combinations by reconnaissance exploitation

SCORE Service Corps of Retired Executives; Special Covert Operations for Resale; System Capability over Requirement Evaluation

SCORES Scenario-Oriented Recurring-Evaluation System (USA)

scorpio subject-content-oriented retrieval for processing information on-line

SCOS Scottish Certificate in Office Studies; Senior Citizen Optometrical Service

scot steel car of tomorrow

Scot Scotch; Scotland; Scots; Scotsman; Scotswoman; Scottie(s); Scottish; Scotty

SCOTAPLL Standing Conference of Theological and Philosophical Libraries in London

SCOTBEC Scottish Business Education Council

ScotGael Scots Gaelic

Scotiabank Bank of Nova Scotia

Scot Jop Scott Joplin

ScotNats Scottish Nationalists

Scots wha ha'e *Scots* *wha ha'e wi' Wallace bled* (Scottish—Scots who have with Wallace bled)—national anthem

Scott Scott, Foresman; Scott Publications; William R Scott

Scotts Bluff Scotts Bluff National Monument in western Nebraska on the Oregon Trail

SCOTUS Supreme Court of the United States

Scot virus Scottish type of influenza virus sometimes called Scotland virus

'scouse lobscouse (sailor's stew)

scp secondary control point; single-cell protein; spherical candlepower; supervisor's control panel

SCP Sea Containers Pacific; Senior Companion Program; Site Characterization Plan; Social Credit Party; Survey Control Point

SCP (AFL-CIO) Sleeping Car Porters

SCP/2 Secure Communications Processor/2

SCPA Scottish Chick Producers' Association; South Carolina Ports Authority

SCPAs State Criminal-Justice Planning Agencies

scpc single channel per carrier

SCPCU Society of Chartered Property and Casualty Underwriters

SCPD Service-Connected Physical Disability; Staff Civilian Personnel Division (USA)

SCPE State Committee on Public Education

SCPEA Southern California Professional Engineering Association

SCP-EGG Standard Communications Protocol for Computerized Electrocardiography

SCPGA South Carolina Personnel and Guidance Association

SCPI Structural Clay Products Institute

SCPL Social Credit Political League (New Zealand Party)

SCPN Society of Certified Professional Numismatists

SCPO Senior Chief Petty Officer

SCPR Scottish Council of Physical Recreation

SCPS Senior Citizen Podiatric Service; Society of Civil and Public Servants (British)

SCPt security cortrol point

SCPU Sea Containers Pacifiic Unit

scpv (SCPV) silkworm cytoplasmic polyhedrosis virus

SCQ Coastal Sentry (tracking station vessel—naval symbol)

scr screw; scruple; silicon-controlled rectifier; skin-conductance response

s-c r short-circuit radio

SCR Signal Corps Radio; Site Characterization Report; Standardized Casualty Rate; Suffolk & Cambridgeshire Regiment

S Cr Staff Commander

SCRA Southern California Restaurant Association; Stanford Center for Radar Astronomy

scram self-contained radiation monitor; supersonic combustion ramjet (engine)

SCRAM Special Criteria for Retrograde Army Materiel; Synanon Committee for Responsible American Media

scrap simple-complex reaction-time apparatus

SCRAP Society for Completely Removing All Parking (Meters); Students Challenging Regulatory Agency Proceedings

SCRATA Steel Castings Research and Trade Association

scr bh screen bulkhead

SCR brick Structural Clay Research brick

SCRC Southern California Renewal Communities; Southern California Research Council; Study Circles Resource Center

SCRCC Soil Conservation and Rivers Control Council

SCRDE Stores and Clothing Research and Development Establishment

SCRDT Stanford Center for Research and Development in Teaching

SCRE Scottish Council for Research in Education

SCREAM Society for the Control and Registration of Estate Agents and Mortgage Brokers

SCREAMS Society to Create Rapprochement among Electrical, Aeronautical, and Mechanical Engineers

screenex screening exercise

SCRIF Scripps Clinic and Research Foundation; Small Craft Repair Facility (USN)

Scribner Charles Scribner's Sons

SCRID Southern California Registry of Interpreters for the Deaf

scrim scrimmage

scrip scriptural; scripture

script manuscript; prescription

Script Scriptural; Scripture

SCRIPT Stanford Computerized-Researcher Information-Profile Technique; Support for Creative Independent Production Talent

SCRIS Southern California Regional Information Study (Bureau of the Census)

SCRL Signal Corps Radar Laboratory

SCRLC South Central Research Library Council

scrn screen; screening; screens

scr's silicon-controlled rectifiers

SCRS Society of Collision Repair Specialists

Scrt Sanskrit

SCRTD Southern California Rapid Transit District

Scrtrt the Secretariat (UN)

Scrubs Wormwood Scrubs

scrum scrummage

S Cruz Santa Cruz (Bolivian province)

scs satellite control system; secret cover sheet; silicon controlled switch; space command station; stabilization control system

scs (SCS) sea-control ship

sc & s strapped, corded, and sealed

SCS Scientific Control System(s); Screening and Costing Staff (NATO); Secondary Control Ship (USN); Serving Christian Scientists; Society for Cinema Studies; Society of Civil Servants; Society of Clinical Surgery; Society for Computer Simulation; Soil Conservation Service; Southern California Skeptics; Student Counseling Service

SCSA Soil Conservation Society of America; Southern California Symphony Association

SCSBM Society for Computer Science in Biology and Medicine

SCSC South Carolina State College

sc-se smooth curve-smooth earth

SCSE Society of Casualty Safety Engineers

ScSEA South Carolina Society of Enrolled Agents

SCSEA Southern California Solar Energy Association

SCSEP Senior Community Service Employment Program

SCSF Surface Cask Storage Facility

SCSI Small Computer System Interface

Sc.Soc.D. Doctor of Social Science

SCSP Site Characterization and Selection Plan; State Center Service Program; System Calibration Support Plan (USAF)

SCSPA South Carolina State Ports Authority

SCSS Scottish Council of Social Service

SCSU Southern Connecticut State University

sct staff continuation training; structural clay tile; sub-zero cooled and tempered

sct (SCT) subroutine call table; surface charge transistor

Sct Scutum (constellation)

Sc & T Science and Technology

S Ct *Supreme Court Reporter*

SCT scattered clouds (aircraft code); Society of Cleaning Technicians; Society of Commercial Teachers

s/cta *su cuenta* (Spanish—your account)

SCTA Steel Carriers Tariff Association

SCTE Society of Cable Television Engineers

sctl short-circuited transmission line

Sctl Schottky coupled-transistor logic

sctr sector (flow chart)

SCTR Standing Conference on Telecommunications Research

sctrd scattered

sct's sugar-coated tablets

SCTS Sycamore Canyon Test Site (Convair)

Sctsmn *The Scotsman* (Edinburgh)

sctt submarine-command team trainer

SCTTF Small-Core Triaxial Test Facility

scty security

SCU Santa Clara University; Selector Checkout Unit, Special Care Unit; Sharecroppers' Union

SCUA Suez Canal Users' Association

scuba self-contained underwater breathing apparatus

scubasub scuba-diver's submarine; scuba-diver's submersible

S-cubed serial-signalling scheme; serial-signalling system

SCUK South Coast of the United Kingdom

sculp sculptor; sculpture

sculp *sculpsit* (Latin—he carved or engraved it)

SCUM Society (for) Cutting Up Men

scup scupper

SCUP Society for College and University Planning

S-curve S-shaped curve

SCUS Supreme Court of the United States

'scutcheon escutcheon

scv single concave

s-c-v single-capsulated-virulent (bacteria)

s & cv stop and check valve

SCV Sons of Confederate Veterans

SCV *Santa Città Vaticana* (Italian—Holy Vatican City)

S.C. V. *Stato della Città del Vaticano* (Italian—Vatican City State)

SCVANYO Standing Conference of Voluntary Youth Organizations

SCVIR Society of Cardiovascular and Interventional Radiology

scvtr scan-converting video tape recorder

SCW State College of Washington

SCWC Special Commission on Weather Modification

SCWPH Students Concerned With Public Health

scwr (SCWR) supercritical water reactor

SCWS Scottish Co-operative Wholesale Society

scx single convex

SCXU Sea Containers Atlantic Unit

SCYC South Coast Yacht Club

S Cz Salina Cruz

sd sagebrush desert; second defense (lacrosse); self-destroying; semidiameter; septal defect; serum defect; service drawing; service dress; shell-destroying; shit disturber (troublemaker); sight draft; silver dollar; single deck; skin dose; sound; special duty; spontaneous delivery; stage door; standard deduction; staff duties; standard deviation; storm detection; storm drain(age); streptodornase; submarine detector; sudden death; system demonstration; systolic to diastolic; systolic discharge

s-d slow-drying

s'd said

s/d sea-damaged; systolic-to-diastolic

s & d search and destroy; song and dance

sd *siehe dies* (German—see this)

s.d. *sine die* (Latin—without date)

sD *samme Dato* (Danish—same date)

Sd discriminative stimulus; drive stimulus; Sound

S$ Singapore dollar; Solomon Islands dollar

S^d Sound

SD Salt Domes; San Diegan; San Diego; Secretary of Defense; Senior-Deacon; Sight Draft; Signals Division; snare drum; Southern District; Specification for Design; Spectacle Dispenser (oculist); Standard Oil Company of California (stock exchange symbol); State Department; Stores Depot; Sudan (Internet code); Superintendent of Documents; Supply Depot; Supply Detachment

SD *Sicherheitsdienst* (German—Intelligence Service); Social(ist) Democrat(ic) (party); *Stofarts Directoratet* (Norwegian—Directorate of Shipping); *Stronnictwo Demokratyczne* (Polish—Democratic Party)

sda sacro-dextra anterior; source data acquisition; source data automation; specific dynamic action; succinic dehydrogenase activity

SDA Scottish Development Agency; Scottish Dinghy Association; Scottish Diploma in Agriculture; Seventh Day Adventist; Ship Destination Authority; Soap and Detergent Association; Social Democratic Alliance; Source Data Automation; Students for Democratic Action

SDAC San Diego Art Center

Sdad Sociedad (Spanish—Society)

SD & AE RR San Diego & Arizona Eastern Railroad

SDAF San Diego Architectural Foundation

SDAG San Diego Association of Governments

S Dak South Dakota; South Dakotan

SDAM San Diego Aerospace Museum

sdaml send by airmail

SDAP Systems Development Analysis Program; System Development and Performance

SDASBO South Dakota Association for School Business Officials

sdAt (SDAT) senile dementia of the Alzheimer's type

SDAT Senile Dementia of the Alzheimer Type

S-day submarine-deployment day (NATO)

sdb seaward defense boat; special district bond; standard dwarf bearded (iris)

SDB Salesian of Don Bosco; Society for Developmental Biology

SDB Sluzba Drzavne Bezbednosti (Serbo-Croat—State Security Service)

sdbl sight draft bill of lading

sd bl sandblast

SDBL Sight Draft with Bill of Lading

SDBRI San Diego Biomedical Research Institute

sdby standby

sdc shipment detail card; single drift connection; space defense center; submersible decompression chamber

sdc (SDC) signal data converter

SDC Southern Defense Command; Space Development Corporation; Special Devices Center; State Defense Council; State Department of Corrections (Alabama, Colorado, Virginia); Strategic Defense Command; Support Design Change; Systems Development Corporation

SDCA Society of Dyers and Colourists of Australia

SDCB State Dissemination Capacity Building

SDCC San Diego City College

SD/CC Security Designation/Custody Classification

SDCL System Distress Check List

SD Class Superintendent of Documents Classification

SDCMS San Diego County Medical Society

SDCS San Diego City Schools

SDCSO San Diego County Symphony Orchestra

s-d curve strength-duration curve

sdd store-door delivery

SDD Scottish Diploma in Dairying; System Definition Directive; System Design Description

sddl saddle(d); sorted data-definition language

sde self-disinfecting elastomer; simple designational expression

SDE Society of Data Educators; State Department of Education

SDEA South Dakota Education Association

's'death god's death

S de B Simone de Beauvoir

S de C *Société des Cuisiniers* (French—Society of Cooks)

SDEC San Diego Ecology Center; San Diego Engineering Council; San Diego Evening College

SDECE Service de la Documentation Extérieure et du Contre-Espionage (French equivalent of American CIA)

SDEE Société de la Diffusion d'Equipements Electroniques (French—Electonic Broadcasting Society)

SDEI San Diego Eye Institute

S del E Santiago del Estero (Argentine province)

S de M Salvador de Madariaga

SDEO Salt Domes Exploration Office

sdf single-degree-of-freedom (gyroscope); special duty flight; standard data format; static direction finder

sdf sans domicile fixe (French—without address; without a fixed living place)

SDF Louisville, Kentucky (airport); Self-Defense Forces (Japan); Sudan Defense Force

SDFD San Diego Fire Department

SDFMC San Diego Foundation for Medical Care

sdg siding

Sdg Siding

SDG Sacred Dance Guild; Self Development Group; Stormont, Dundas and Glengarry (Highlanders)

S.D.G. Solo Deo Gloria (Latin—Glory to God Alone)

SDG & E San Diego Gas & Electric

SDGP State Dissemination Grant Program

sdh (SDH) sorbitol dehydrogenase

SDH Scottish Diploma in Horticulture

SDHA San Diego Hospital Association

SDHC San Diego Housing Commission

sdhe spacecraft data-handling equipment

SDHRC San Diego Human Relations Commission

sdi selective dissemination of information

SDI Saudi Arabian Airlines; Secret Diplomatic Initiative; Selective Dissemination of Information; Senior Drill Instructor (USMC); Standard Data Interface; State Disability Insurance; Strategic Defense Initiative (Star Wars);

SDIBM San Diego Institute for Burn Medicine

SDIC San Diego Improvement Association; South Dakota Intercollegiate Conference

S Diego San Diego

sdiline selective dissemination of information on-line

SDIO Strategic Defense Initiative Organization

SD & IV San Diego & Imperial Valley Railroad

SDJC San Diego Junior Colleges

sdk shelter deck

Sdk (SDK) San Diego (container symbol)

sdl saddle

sdl (SDL) state-dependent learning

SDL Special Duties List(ing); Systems Dimensions Limited

SDLA South Dakota Library Association

sdlc synchronous data-link communication(s)

SDLP Social Democratic and Labour Party

sdm selective discrimination on microfiche

SDM Su Divina Majestad (Spanish—Your Divine Majesty)

SDMA San Diego Museum of Art; Surgical Dressing Manufacturers' Association

SDMC San Diego Mesa College

SDMI Secure Digital Music Initiative

SDMICC State Defense Military Information Control Committee

SDMJ September, December, March, June

sdml seaward defense motor launch

SDMM San Diego Museum of Man

SDMNH San Diego Museum of Natural History

SDMS San Diego Memorial Society

sdn sedan

SDN System Designation Number

SDN Société des Nations (French—League of Nations)

Sdn Bhd Sendirian Berhad (Malay—Private Limited)—limited corporation

SDNCO Staff Duty Non-Commissioned Officer

SDNHM San Diego Natural History Museum

SDNS Scottish Daily Newspaper Society

SDO San Diego Opera; Santo Domingo (Dominican Republic); Signal Distribution

Officer; Squadron Duty Office(r); Staff Duty Officer; System Design Objectives

S Doc Senate Document

SDOC Space Defense Operations Center (Cheyenne, Wyoming)

sdof single degree of freedom

SDOG San Diego Opera Guild

Sdom Sodom

SDOP San Diego Organizing Project

sdp sacro-dextra posterior; social, domestic, and pleasure

Sd £ Sudanese pound (currency unit)

SDP Social(ist) Democratic Party; Subseabed Disposal Program

SDP Sozialdemokratische Partei Deutschlands (German—German Social-Democratic Party)

SDPCC San Diego Poison Control Center

SDPD San Diego Police Department; Schizoid Personality Disorder

SDPGA South Dakota Personnel and Guidance Association

SDPL San Diego Public Library

S Dpo Station Depot

SDPO Site Defense Project Office(r)

SDPOA San Diego Police Officers Association

SDPT Structured Doll Play Test

SDQ Santo Domingo, Dominican Republic (airport)

sdr scientific data recorder; self-decoding readout; simple detection response; size-to-diameter ratio; sodium deuterium reactor; sonar data recorder; splash-detection radar; strip domain resonance; successive discrimination reversal

SDR Service Difficulty Report, Special Despatch Rider; Special Dispatch Rider; Special Drawing Right; Special Drilling Rights; Strategic Defense Response; System Design Review

SDRAM Synchronous Dynamic Random-Access Memory

sdrl supplier data requirement list

SdRng sound ranging

SDRs Special Drawing Rights; Special Drilling Rights

sds self-directed search; sodium dodecyl sulfate; speech discrimination score; sudden death syndrome

s-d s single-day surgery

SDS San Diego Symphony; Scientific Data Systems; sodium dodecyl sulfate; Sons and Daughters of the Soddies; Spatial Data System(s); Special District Services; Students for a Democratic Society

SDSC San Diego State College; San Diego Steamship Company; San Diego Supercomputer Center

SdSEA South Dakota Society of Enrolled Agents

sd sms clsd side seams closed

SDSMT South Dakota School of Mines and Technology

SDSNH San Diego Society of Natural History

SDSO San Diego Symphony Orchestra

SDSRU Soil Data Storage and Retrieval Unit

SDSS Self-Deploying Space Station

SDSU San Diego State University

sdt sacro-dextra transversa; scientific distribution technique; sea depth transducer; serial data transmission; signal-detection theory; source distribution technique, surveillance data transmission

SDT San Diego Transit; Society of Dairy Technology

SDTC San Diego Transit Corporation

SDTD San Diego Transit District

sdtdl saturating drift transistor diode logic

sdti selective dissemination of technical information

SDTI San Diego Technical Institute

SDTS Satellite Data Transmission System

SDTTS San Diego Turtle and Tortoise Society

SDTU Sign and Display Trades Union

sdtv standard-definition television

sdu shelter decontamination unit; signal display unit; spectrum display unit; starter drive unit; sub-carrier display unit

SDU Rio de Janeiro, Brazil (Santos Dumont Airport); Social Democratic Union

SDUK Society for the Diffusion of Useful Knowledge; Spoiled Duck (according to Edgar Allan Poe in his essay on *How to Write a Blackwood Article*)

SDUPD San Diego Unified Port District

SDUSD San Diego Unified School District

SDV slowed-down video; swimmer delivery vehicle

sdw swept delta wing

SDWA Safe Drinking Water Act

SDWAP San Diego Wild Animal Park

SDX Stromberg DatagraphiX; Sunray Mid-Continent Oil Company

SDX *Sigma Delta Chi,* society of professional journalists

SDYC San Diego Yacht Club

SDYS San Diego Youth Symphony

SDZ San Diego Zoo

SDZS San Diego Zoological Society

se safety equipment; second entrance; semiannual; servant; service entrance; single end; single-ended; single engine; single entry; special equipment; spherical equivalent; standard error; straight edge

se (sem) standard error of the mean

s/e standardization/evaluation

s & e scour and ebb; services and equipment

sE standard English

Se selenium

SE Sanford & Eastern (railroad); Sanitary Engineer(ing); Scouting Experimental; Select Edition; Self Employment; Serotonin; Servel (stock exchange symbol); Site Exploration; Southeast; Sports Edition; Staff Engineer; Stock Exchange; Student Engineer; Sweden (Internet code)

S-E Starr-Edwards (prosthesis)

SE *Son Eminence* (French— His Eminence); *Sureste* (Spanish—southeast)

SE1, SE2, etc. Southeast One, Southeast Two, etc. (London postal zones)

s-e 22 silencer-equipped .22— caliber revolver

sea sheep erythrocyte agglutination; spontaneous electrical activity

Sea (Port of) Seattle; Sea of Arabia, Galilee, Islands, Japan, Marmora, Okhotsk, Rybinsk, the Plain, Straw; The Sea (Andaman, Baltic, Bering, Black, Caribbean, Japan, Mediterranean, North, Okhotsk, South China)

SEA Safety Equipment Association; Science and Education Administration; Sea Containers Inc; Sea Education Association; Seattle, Washington (Seattle-Tacoma Airport); Senior Executives Association; Service Employers Association; Ships Editorial Association; Single European Act; Society for Education through Art; Society of Evangelical Agnostics; Southeast Airlines; Southeast Asia; Southern Economic Association; Special Equipment Authorization; State Education Agencies; State Education Agency; Students for Ecological Action; Subterranean Exploration Agency

SEA *Sociedad Española de Automoviles* (Spanish—Automobile Society of Spain)

SEAAC South-East Asia Air Command

SEAAF South East Asia Air Forces

seac standards electronic automatic computer

SEAC Save Europe's Asiatic Colonies; Southeast Asia Command

seacel silver-chloride/magnesium cell (battery)

SEACOM South East Asia Commonwealth Cable

seacon seafloor construction

sead suppression of enemy air defenses

SEADAC Seakeeping Data Analysis Center

SEADAG Southeast Asia Development Advisory Group

seadex seaward defense exercises

SEADS Seattle Air Defense Sector

SEAES South-East Asian Ergonomics Society

Sea Ill Seaforth Highlanders

seal. sea-air-land

SEAL South-East Area Libraries

sealab sea laboratory (underwater research vessel)

SEALF South-East Asia Land Forces

SEALs Sea, Air, Land commandos; Sea, Air, and Land Teams (US Navy frogmen engaged in covert infiltration and surveillance)

SEALS Sea-Air-Land Forces (counterinsurgents)

SEAM Seattle Environmental Arts Museum

SEAM *Servicios de Equipos Agricolas Mecanizados* (Spanish—Mechanized Agricultural Equipment Service)

SEAMEC Southeast Asian Ministers of Education Council

SEAMEO South East Asian Ministers of Education Organisation

seamount sea mountain

SEAP Scientific Event Alert Program; South-East Asia Peninsula

SEAQ Stock Exchange Automated Quotations

searam semi-active radar missile

SEARCC South-East Asia Regional Computer Conference

SEARCH System for Electronic Analysis and Retrieval of Criminal Histories; Systematized Excerpts, Abstracts, and Reviews of Chemical Headlines

searchex sea/air search exercise

Sears Sears Tower building of Chicago

SEARS Sears, Roebuck; Socioeconomic Assessment for Repository Siting

SEAS Senior Emergency Alert System; Strategic Environmental Assessment System

seasat sea satellite

seascarp undersea escarpment

S-E Asia Southeast Asia [Burma (Myanmar), Cambodia, Hong Kong, Indonesia, Laos, Malaysia, Philippines, Singapore, Thailand, Vietnam]

Sea Sym Seattle Symphony

SEAT *Sociedad Español de Automoviles de Turismo* (Spanish Society of Touring Automobiles)—manufacturer's name

SeaTac Seattle-Tacoma International Aiport

seatainer(s) seagoing container(s)—theftproof steel containers for overseas cargo

Seatl Seattle

SEATO Southeast Asia Treaty Organization

SEAU Sea Containers Incorporated Unit

seb special engagement bonus (military); static error band

seb (SEB) surface-effect boat

Seb Sebastian(o)

Seb *Sebjet* or *Sebkhat* or *Sebkra* (Arabic—salt flats)—also appears as *Sabkhat*

SEB Society for Experimental Biology; South Equatorial Belt; Southern Electricity Board

SEB *Skandinaviska Enskilda Banken* (Swedish—Scandinavian Loan Bank)

S & EBC Ship and Engine Building Company

SEBM Society of Experimental Biology and Medicine

SEBT South-Eastern Brick and Tile (federation)

sec secant; second; secondary; secret; section; sector; security

sec. *secundum* (Latin—according to)

Sec Secretary; section

SEC Section Emergency Coordinator; Securities and Exchange Commission, State Electricity Commission; State Energy Commission; Strategic Economic Council; Supreme Economic Council (former USSR)

S.E.C. Springfield Equipment Company

SecA Secretary of the Army

SECA Southern Educational Communications Association

SecAgi Secretary of Agriculture

Sec Air Secretary of the Air Force

SECAIR Secretary of the Air Force

secam *séquential couleur à mémoire* (French—sequential color memory)—Franco-Russian television color transmission standard, sometimes translated as the system contrary to the American method (SECAM)

SECAM *Séquential Couleur à Mémoire* (French—sequence and memory color television system)

secar secondary radar

sec. art. *secundum artem* (Latin—according to the art)

SecCom Secretary of Commerce

secd second

SECDA Southeastern Community Development Association

SecDef Secretary of Defense

SECDEF Secretary of Defense

SecEdu Secretary of Education

SecEne Secretary of Energy

secesh secessionist

SECFLT Second Fleet (Atlantic)

SECFO Systems Engineering and Consensus Formation Office

Sec-Gen Secretary-General

sech hyperbolic secant

SecHHS Secretary of Health and Human Services

SecHou Secretary of Housing and Urban Development

Sec Hum Secular Humanism (doctrine stressing the achievements of man and ignoring the practice of religion or its teaching or the role of a deity)

secinsp security inspection

SecInt Secretary of the Interior

SecLab Secretary of Labor

sec. leg. *secundum legem* (Latin—according to law)

Sec Leg Secretary of the Legation

SECMA Stock Exchange Computer Managers Association

sec. nat. *secundum naturam* (Latin—according to nature)

SecNav Secretary of the Navy

SECNAV Secretary of the Navy

seco second-stage engine cutoff; sustainer engine cutoff

seco (SECO) self-regulating error-correct coder-decoder

secondaries secondary colors—green, orange, violet

Second International Second International Workingmen's Association (of socialists convening in Paris in 1889)

secor (SECOR) sequential collation of range

secr secret

SE & CR Southeastern and Chatham Railway

sec. reg. *secundum regulam* (Latin—according to regulations, according to rule)

secret[a] *secretaria* (Spanish—secretariat)

secs secants; seconds; sections

sec's soft elastic capsules

Secs sections

Sec Soc Foun Second Society Foundation

SecSta Secretary of State

sect section; sector

sect (Latin suffix—cut)—dissect

sectemp temporary secretary

Sectra Secretary of Transportation

SecTre Secretary of the Treasury

Secty Secretary

SECUS Sex Education Council of the United States

SecWar Secretary of War

SecWel Secretary of Welfare

Sec'y Secretary

sed sedative; sediment; sedimentation; severely emotionally disturbed; skin erythema dose

sed. *sedes* (Latin—a chair; a stool)

SeD Socioeconomic Democracy

SED Scientific Equipment Division (Westinghouse); Scottish Education Department; Special Enforcement Detail (law enforcement team)

SED *Sozialistische Einheitspartei Deutschlands* (Germany's Socialist Unity Party)—Soviet-oriented East German Party

sedar submerged electrode detection and ranging

SEDEIS *Société d'Etudes et de Documentation Economiques, Industrielles et Sociales* (French—Society of Economic, Industrial and Social Studies and Documentation)—Paris

sedi sediment(ation)

SEDIS Surface-Emitter-Detection Identification System

sedi time sedimentation time

SEDOS Software Environment for the Design of Open Distributed Systems

sed rate sedimentation rate

sed('s) seeing-eye dog(s)

sedtn sedimentation

seduc seduction

see secondary electron emission; stop-everything environmentalists; survival, evasion, and escape; systems efficiency expert(ise)

SEE Signals Experimental Establishment; Society of Environmental Engineers; Society of Explosives Engineers

SEE Société des Eléctriciens des Electroniciens et des Radioélectriciens (French—Society of Electricians, Electronicians, and Radio Electricians)—electric, electronic, and radio technicians

SEEA Société Européenne d'Energie Atomique (French—European Atomic Energy Society)

SEEB Southeastern Electricity Board (UK)

Seec Saburo exhaust-emission control

SEECA State Environmental Education Coordinators Association

SEECB Solar Energy and Energy Conservation Bank

SEECC Standards for Educators of Exceptional Children in Canada

seecom sensible, economical, electrical commuter (electric automobile)

SEECTS Subaru Exhaust Emission-Control Thermal System

seed summer of experience, exploration, and discovery

SEED Scientists and Engineers in Economic Development (National Science Foundation); Skills Escalation and

Employment Development; Special Elementary Education (for the underdeveloped)

SEEJ Slavic and East European Journal

SEEK Search for Elevation and Educational Knowledge (NY State dropout program); Sooner Exchange for Educational Knowledge; Systems Evaluation and Exchange of Knowledge

seeo sauf erreur et omission (French—excepting errors and omissions)

s.e.e.o. salvis erroribus et omissis (Latin—excepting errors and omissions)

seep seagoing jeep

seer seasonal energy efficiency ratio; submarine explosive echo ranging

SEER Surveillance, Epidemiology and End Results; System for Electronic Evaluation and Retrieval

seex systems evaluation experiment

sef small end first

SEF Shipbuilding Employers' Federation; Southern Education Foundation; Space Education Foundation

SEFA Scottish Educational Film Association

SEFT Society for Education in Film and Television

seg segment; segmentation; segmented; segments; segregate; segrated; segregation; segregationist; segue (blending without interruption); special-effects generator

seg (SEG) sonoencephalogram

seg segno (Italian—sign); *seque* (Italian—comes after; follows)

Seg Segovia

SEG Screen Extras Guild; Society of Economic Geologists; Society of Exploration Geophysicists; Systems Engineering Group

SEGB South Eastern Gas Board

SEGBA Servicios Eléctricos del Gran Buenos Aires (Spanish—Electrical Services of Greater Buenos Aires)

seggy secobarbital sedative, also nicknamed seccy)

segm segment; segmented

Segovia Andrés Segovia (guitarist)

Segr Segretario (Italian—Secretary)

Segr^{to} Segretariato (Italian—Secretariat)

segs segmented neutrophils; segments

SEH St Elizabeth's Hospital

SEH Société Européenne d'Hématologie (French—European Society of Haematology)

seha specific emotional hazards of adulthood

sehc specific emotional hazards of childhood

SEHMF South of England Hat Manufacturers' Federation

SEI Safety Equipment Institute; Scientific Engineering Institute; Self-Employment Income; Socio-Economic Index

SEIA Security Equipment Industry Association; Solar Energy Industries Association; Solar Energy Institute of America

SEIC Solar Energy Information Center; System Effectiveness Information Center

SEIE Submarine Escape Immersion Equipment

SEIF Secretaria de Estado da Informação e Turismo (Portuguese—Secretariat of Information and Tourism)

SEIFSA Steel and Engineering Industries' Federation of South Africa

Seiji Seiji Ozawa

seis seismograph; seismography; seismology; submarine emergency identification signal (SEIS); submarine-escape immersion suit

SEISA South Eastern Intercollegiate Sailing Association

Seiscor Seismograph service Corporation

seismo seismograph(er); seismographic(al)(ly); seismologist; seismology

seismol seismology

SEIT Search for Extra Terrestrial Intelligence

SEW Service Employees International Union

sel select(ed); selectee; selectivity; selector, semi-effective list; socioeconomic level; sound exposure level (SEL)

sel (SEL) socio-economic level

Sel Selby

SEL Seoul, Korea (airport); Signal Engineering Laboratories; Southeastern Education Laboratory; Stanford Electronics Laboratories; Systems Engineering Laboratories

SELA Southeastern Library Association

SELA Sistema Económica Latino Americana (Spanish—Latin American Economic System)

SELC South Eastern Louisiana College

selcall selective calling

sel-cl self-closing

SELDAMS Selective Data Management System

seleac standard elementary abstract computer

selen selenography, selenology

SELEX Systematic Evolution of Ligands by Exponential Enrichment

SELF Société d'Ergonomie de Langue Francaise (French—Language Ergonomy Society)

self-prop self-propelled

Selk Selkirk

Selkirks Selkirk Mountains of British Columbia

SELMA SEL Maduro

SELNEC South-East Lancashire North-East Cheshire

S/ELPS Spanish/English Language Performance Screening

sels selsyn

selsyn self-synchronous

Selvagens Selvagen Islands between the Canaries and Madeira

Selw Selwyn College, Oxford; Selwyn College—Cambridge

Sely southeasterly

SEly south-easterly

sem scanning electron micrograph; scanning electron microscope; semi; semicolon; seminal; slow eye movements; standard error of mean, systolic ejection murmur

sem (SEM) scanning electron microscope; systolic ejection murmur

sem. semen (Latin—seed); *semper* (Latin—always, ever)

Sem Semarang; Seminary, Semitic

SEM Society for Ethno-Musicology; Standard Electronic Modules

SEMA Specialty Equipment Market Association; Spray Equipment Manufacturers' Association; Storage Equipment Manufacturers Association

seman semantic(s)

semcor semantic correlation

SEMDA Surveying Equipment Manufacturers and Dealers Association

SEME School of Electrical and Mechanical Engineering (military)

SEMFA Scottish Electrical Manufacturers' and Factors' Association

semi semicolon

semi- semi-detached house (town house)

semi (Latin prefix—halo)—semilunar

SEMI Semiconductor Equipment and Materials International

semicol semicolon

semidr. semidrachma (Latin—half drachma)

semidur semiduration

semih. semihora (Latin—half hour)

Seminex Seminary in Exile

semiot semiotic(al)(ly); semiotician(s); semiotics (study of signs and symbols)

semipro semiprofessional(ly)

semis semifinished; semitrailers

SEMKO Svenska Elektriska Materielkontrollanstalten (Swedish—Swedish Institute for Testing and Approval of Electrical Equipment)

Semmelweis Ignaz Philipp Semmelweis (brought antisepsis into medical practice)

semp self-erecting marine platform

semp sempre (Italian—always)

SEMP Superconducting Electromagnetic Propulsion; Systems Engineering Master Plan

Semper Fi Semper Fidelis (Latin—Always Faithful)—U.S. Marine Corps motto

Semper Paratus (Latin—Always Ready)—U.S. Coast Guard motto

sems screw and washer assemblies

SEMS Systems Engineering Master Schedule

SEMT Société d'Etudes des Machines Thermiques (French—Society for the Study of Thermal Machines)

SEMTA Southeastern Michigan Transportation Authority

sem ves seminal vesicle

sen sense (flow chart)

sen seno (Italian—sine); *senza* (Italian—without)

Sen Senate; Senator; Senegal (whose capital is Dakar)

Sen Marcus (or Lucius) Seneca (Roman rhetorician) or his second son Lucius Annaeus Seneca (Roman author); *Senatore* (Italian—senator)

SEN State-Enrolled Nurse

Sen Aph Sensory Aphasia

S en C Sociedad en Comandita (Spanish—limited partnership)—silent partnership; *Société en Commandite* (French—limited partnership)

Sen Clk Senior Clerk

Sen Doc Senate Document

Seneg Senegal; Senegalese

Senegal Republic of Senegal (West African nation), *République du Sénégal*

Senegambia Senegal + Gambia

senel single-event noise-exposure level

S Eng O Senior Engineering Office(r)

SENGO Senior Engineer Officer

SENI Society for the Encouragement of National Industry

senior dent senior-citizen dental care

senior(s) senior citizen(s)

Sen M Senior Master

Sen Mist Senior Mistress

S en NC Société et Nom Collectif (French—joint stock company)

senr senior

Sen Rept Senate Report

Senr Tech Weld I Senior Technician of the Welding Institute

sens sensitivities (test)

SENSEX Stock Exchange Sensitivity Index

sensistor semiconductor resistor

sent. sentence

Sent Sentyabr (Russian—September)

SENTAC Society for ear, Nose, and Throat Advances in Children

sentimiento trágico *sentimiento trágico de la vida* (Spanish—tragic perception of life)—conflict between faith and reason

Sen Wt O Senior Warrant Officer

seo (SEO) satellite for earth observation

seo *salvo errori e omissioni* (Italian—excepting errors and omissions)

Seo Seoul

SEO Senior Executive Officer; Senior Experimental Officer; Snake Ender's Organization

SEODSE Special Explosive Ordnance Disposal Supplies and Equipment (USA)

SEOG Supplemental Educational Opportunity Grant

seoo *sauf erreurs ou omissions* (French—excepting errors and omissions)

SEOOs State Economic Opportunity Offices

seos (SEOS) synchronous earth observation satellite

seou *salve error u omisión* (Spanish—except for error or omission)

sep separate; separation solar electric propulsion; surrendered enemy prisoners

sep (SEP) solar electric power, somatosensory-evoked potential

Sep September

SEP Selective Employment Payments (UK); Self-Employment Plan (retirement plan); Simplified Employee Pension; Society of Engineering Psychologists; Society of Experimental Psychologists; Source Evaluation Panel; Student Expense Program

SEP *Saturday Evening Post; Secretaía de Educación Publica* (Spanish—Secretary of Public Education)

SEPA Southeastern Power Association; Spanish Evangelical Publishers Association; State Elementary Principals Association; State Environmental Protection Agency

separ. *separatum* (Latin-separately)

SEPB Southern Europe Ports and Beaches

SEPD Scottish Economic Planning Department

SEPE Seattle Port of Embarkation

SEPEL Southeastern Plant Environment Laboratories

Seph *Sephardim* (Hebrew—Jews from Portugal and Spain)

Sephard *Sephardim* (Hebrew—Jews from Portugal and Spain who were forced to emigrate during the Inquisition)

SEPO Space Electric Power Office (AEC)

SEPP Simplified Employee Pension Plan

SEPP *Société d'Étude de la Prévision et de la Planification* (French—Society for the Study of and Planning for the Future)

SEPR *Société pour l'Etude de la Propulsion par Réaction* (French—Society for the Study of Jet Propulsion)

SepRos separation processing

Seps (SEPS) Smithsonian earth physics satellite

SEPs Simplified Employee's Pensions

SEPSA Society of Educational Programmers and Systems Analysts

sept. *septem* (Latin—seven)

Sept September

SEPTA Southeastern Pennsylvania Transportation Authority

sept^e *septiembre* (Spanish—September)

septel separate telegram

septen septentrionale (northern)

sepfi *septicos* (Greek—infected or rotten)—antiseptic, aseptic, septic, septicemia

September September and October

seq sequence

seq. *sequens* (Latin—the following); *sequente* (Latin—what follows); *sequitur* (Latin—it follows)

seq. *luce* *sequenti luce* (Latin the following day)

Seq NP Sequoia National Park

S Equ Cur South Equatorial Current

Sequoia Sequoia National Park in east-central California

ser serial; series; somatosensory-evoked response

set (SER) serine (amino acid)

ser *série* (French—series)

Ser series; Serpens (constellation)

Ser *Serranía* (Spanish—mountains)

SER Safety Exploration Report; Service, Employment, Redevelopment; Society for Educational Reconstruction; Soil Erosion Service; Student Eligibility Report

SER *Sociaal Economische Raad* (Dutch—Social Economic Council); *Sociedad Española Radiodifusión* (Spanish—Spanish Broadcasting Society)

Sera Seraphim

SERA Services, Education, Rehabilitation for Addiction

Serb Serbia; Serbian

Serb-Croat Serbo-Croatian (slavic language)

SERC Science and Engineering Research Council

SERCH State Education Research Clearinghouse

serd support equipment recommendation data; support equipment requirements data

SERE Survival, Evasion, Resistance, and Escape (U.S. Naval Training Base)

SEREB *Société pour l'Etude et la Réalisation d'Engins Balistiques* (French—Society for the Study and Development of Ballistic Missiles)

serendip serendipitous(ly); serendipity

Serengeti Serengeti Plains of Tanzania

Serg *Sergente* (Italian—Sergeant)

Sergey Sergeyevich

serm selective estrogen receptor modulator

Serg Magg *Sergente Maggiore* (Italian—Sergeant Major)

Serg(t) Sergeant

SERI Solar Energy Research Institute; Solar Energy Research Institute (ERDA)

Series 88 IBM trademark

serj space electric ramjet

SERL Services Electronics Research Laboratory

SERLANT Service Forces, Atlantic (USN)

serline serials on-line

serm sermon

SERM Society of Early Recorded Music

serol serology

serp simulated ejector-ready panel

SERPAC Service Forces, Pacific (USN)

SERPLANT Service Forces, Atlantic (USN)

serr serrate

Ser Rep San Mar Serenissima Repubblica di San Marino (Italian—Most Serene Republic of San Marino)

SE-RRT Southern Europe—Railroad Transport (NATO)

ser sect serial sections

sert space electronic rocket test

SE-RT Southern Europe-Road Transport (NATO)

SERTOMA Service To Mankind

serv servant; service

serv. serva (Latin—keep; preserve)

Serv Servia(n)

serv chge service charge

serv clg service ceiling

SERVE Serve and Enrich Retirement by Volunteer Experience

servo device using a servomechanism; servoamplifier, servocontrol, servodyne, servomotor, servosystem

serv⁰ servicio (Spanish—service)

Servomation Service America Corporation

Serv⁰ʳ servidor (Spanish—servant)

servos servomechanisms

ses secondary engine start; single-ended scotch (boilers); socio-economic status; socio-economic strata; solar environment stimulator; surface-effect ship

ses (SES) surface-effect ship

SES Seafarers' Education Service, Seagrass Ecosystem Study; Self-Esteem Score(s); Senior Executive Service; Society of Engineering Science; Solar Energy Society; Standards Engineers Society; State Employment Service; Steam Engine Systems; Suitability Evaluation Scale

SES Service des Études Scientifique (French—Scientific Studies Service)

SESA Social and Economic Statistics Administration; Society for Experimental Stress Analysis; Solar Energy Society of America

SESAC Society of European Stage Authors and Composers

sesame. service, sort, and merge

SESAME Search for Excellence in Science and Mathematics Education; Standardization in Europe on Semantical Aspects in Medicine

sesco secure submarine communications

SESDA Small Engine Servicing Dealers Association

SESL Space Environment Simulation Laboratory

SESO Senior Equipment Staff Office(r); Ship Environmental Support Office(r)

sesoc surface-effects ship for ocean commerce

SESPO Space Environmental Support Project Office(r)

sesquih sesquihora (Latin—an hour and a half)

sesquilin sesquilingual (ability to use one-and-a-half languages)

sess session

SESS Society of Ethnic and Special Studies; Space Environmental Support System; Summer Employment for Science Students

sest short effective-service time

SEST Systems Engineering Support Team

set settlement

set septiembre (Spanish—September); *setembro* (Portuguese—September)

SET Scientists, Engineers, Technicians; Security Escort Team; Selective Employment Tax(ation); Self-Instructional Training; Senior Electronic Technician; Senior Evaluation Treatment; Simplified Engineering Technique; Synchro Error Tester

S.E.T. Selective Employment Tax

seta set arithmetic (value)

SETAF Southern European Task Force

setb set binary (value)

setc set character (value)

SETC Submarine Escape Training Center

SETCO Summit and Elizabeth Trust Company

se/td system engineering/technical direction (SE/TD)

set⁰ septiembre (Spanish—September)

SETEP Science and Engineering Technician Education Program

SETI Search for Extraterrestrial Intelligence

SETIL Société de l'Equipement de Tahiti et des Iles (French—Equipment Company of Tahiti and the Islands)

S-et-L Saône-et-Loire

S-et-M Seine-et-Marne

Set-Nai Seto-Naikai (Japanese—Inland Sea)

S-et-O Seine-et-Oise

SETP Society of Experimental Test Pilots

SETS Solar Energy Thermionic Conversion System

sett settler; settlers; settling

sett settembre (Italian—September)

seu smallest executable unit; subjective expected utility

SEU Southeastern University; Special Engineering Unit

SEUA South Eastern Underwriters Association

seuo salvo error u omisíon (Spanish—errors and omissions excepted)

SEUS Southeastern United States

sev seven; sevenfold; seventeen(th); seventy; sever; several; severally; severance; severe; severity; surface-effect vehicle

sev sever (Russian—north)

Sev Sevilla; Seville

Sev Sever or Severnaya (Russian—north, northern)

SEV Soviet Ekonomischeskoy Vzaimopomoschchi (Russian—Soviet Council for Mutual Economic Aid)—the COMECON

Seven Continents Africa, Antarctica, Asia, Australia, Europe, North America, South America

Seven Sisters Pleiades

Seventh Seventh Seal

Severnaya Zemlya (Russian—North Land—*Zemlya Imperatora Nikolaya II* (Russian—Emperor Nicholas II Land)
SEVFLT Seventh Fleet, Pacific (USN)
sevocom secure voice communications
sew safety equipment workers; sewage; sewer; sewerage
sewido surface electromagnetic-wave-integrated optics
SEWT Simulator for Electronic Warfare Training
sex. sextet; sexual
Sex Sextans (constellation)
Sexag Sexagesima
sexational sexually sensational
sexcite excite sexually
sexcitement sexual excitement
sexclusive sexually exclusive
sex ed sex(ual) education
sexercises sexual exercises
sexgregation sexual segregation
sexhibit sex exhibit
sexhibitors sex exhibitors
SExO Senior Experimental Officer
sexones sex odors
sexorgies sexual orgies
sexpensive sexually expensive
sexperience sexual experience
sexpert sex expert; sexual expert; sexpertise
sexpionage sexual exploitation in espionage
sexplanatory sexually explanatory
sexplicit sexually explicit
sexploitation sex(ual) exploitation
sexploiter sex exploiter
sexploit(s) sexual exploit(s)
sexplosion sexual explosion
SEXPOL Sexual Equality and Politics (German communist movement originated by Wilhelm Reich)
s. expr. *sine expressione* (Latin—without expressing; without pressing)
sex psycho sexual psychopath(ic)
sexquisite sexually exquisite
sexsation sexual sensation
sexslanguage sexual slang
sext sextant
Sext Sextans (constellation)
Sey Seychelles (whose capital is Victoria)
Seychelles Seychelle Islands

sez (SEZ) southern economic zone
SEZ Special Economic Zone (China)
sf sacrifice fly; safety factor; salt free; science fiction; scoring factor; semifinished; servicing flight; single feeder; single-feed; single frequency; sinking fund; snow flurries; sound and flash; special facilities; spent fuel; spinal fluid; spotface; square foot; standard form; stepfather; stress formula; sulphation factor; sunkface; sustained fire
s/f shift forward; store and forward
s & f stock and fixtures
sf *sans frais* (French—without expense); *sforzando* (Italian—accented strongly; forced; reinforced)
s.f. *sub finem* (Latin—near the end)
Sf Svedberg flotation (units)
SF San Franciscan; San Francisco; Santa Fe (Atchison, Topeka & Santa Fe Railway.); Santa Fe, New Mexico; Scots Fusiliers; Scouting Force; Security Force(s); Sherwood Foresters; Shipfitter; Soumi Finland; Special Facilities' Special Forces; Standard Frequency; State Facilitator; Swedenborg Foundation; Swiss Federation (auto plate); Syrian Forces
SF Slovenska Filharmonica (Serbo-Croat—Slovene Philharmonic); *Socialistisk Folkeparti* (Dano-Norwegian—Socialist People's Party); *Système français* (French—French system)—screw threads
S/F Sinn Fein (Irish Gaelic—Ourselves Alone)
SF-5 Spanish version of the F-5 Northrup Freedom Fighter
sfa simulated flight automatic; single frequency amplifier; slow flying aircraft; spatial frequency analyzer
sfa (SFA) serum folate; suppressive factor of allergy
s & fa shipping and forwarding agent
SFA Saks Fifth Avenue; Scandinavian Fraternity of America; Scientific Film Associa-

tion; Scottish Football Association; Show Folks of America; Slide Fastener Association, Snack Food Association; Solid Fuels Administration; Soroptimist Federation of the Americas; Southeastern Fisheries Association; Speech Foundation of America; Stuttering Foundation of America (formerly Speech Foundation of America); Symphony Foundation of America
SFA Société Française d'Astronautique (French—French Astronautical Society)
SFAAW Stove, Furnace, and Allied Appliance Workers (International Union of North America)
S-fac sensitivity factor
SFAC Société des Forges et Ateliers du Creusot (French—Schneider-Creusot Forges and Factories)
SFAD Society of Federal Artists and Designers
SFAI San Francisco Art Institute; Steel Furnace Association of India
SFAO San Francisco Assay Office
sfar sound fixing and ranging
SFAR System Failure Analysis Report
SFB San Francisco Ballet
SFB Sender Freies Berlin (German—Free Berlin Broadcasting Station); Spencer Fullerton Baird
SFBARTD San Francisco Bay Area Rapid Transit District
sf bh surface broach
SFBMS Small Farm Business Management Scheme
SFBNS San Francisco Bay Naval Shipyard
sfc S-bank frequency converter; sight fire control; specific fuel consumption; supercritical fluid chromatography; switching filter connector; synchronized framing camera
sfe (SFC) spinal fluid count
Sfc Sergeant First Class
SFC Saint Francis College; Sioux Falls College; Securities & Futures Commission; Small Faith Communities; Space Flight Center
SFC San Francisco Chronicle

SFCA Southwest Flight Crew Association

SFCC San Francisco City College

SFCI State Farms Corporation of India

SFCJ San Francisco City Jail

SFCM San Francisco Conservatory of Music

SFCMD San Francisco Contract Management District

SFCP Shore Fire Control Party

SFCS Survivable Flight Control System

SFCTA San Francisco Classroom Teachers Association

sfcw search for critical weakness

SFCW San Francisco College for Women

sfd super-fine dust

Sfd San Fernando

sfd/algol system function description/algol (language)

SFDS Spent-Fuel Disposal System

sfe safety function earthquake; seller-furnished equipment; stacking fault energy; surface-energy

SFE Society of Fire Engineers

SFE Société Française des Electriciens (French—French Society of Electricians)

SFEA Survival and Flight Equipment Association

SFEL Standard Facility Equipment List

SFEM Southern Farm Equipment Manufacturers

SFEN Société Française d'Energie Nucléaire (French—French Nuclear Energy Society)

SFET Schottky Field-Effect Transistor

sff se faz favor (Portuguese—please)

SFF Solar Forecast Facility; Special Frontier Force

sfff salt-free fat-free (diet)

sffw self-forming fragment warhead (mine)

SFG Studien und Förderungsgesellschaft (German—Studies and Advancement Society)

sfga single floating-gate amplifier

sfgd safeguard

SFGGB San Francisco Golden Gate Bridge

SFGH San Francisco General Hospital

SFHP Spent-Fuel Handling and Packaging; Spent-Fuel Handling Project

SFHR San Francisco Historic Records

SFHS Stephen Foster High School

sfi sequential fuel injection

SFI Sport Fishing Institute

SFI Société Financiére Internationale (French—International Finance Corporation)

SFIAE San Francisco Institute of Automotive Ecology

SFIB Southern Freight Inspection Bureatm

SFINX Software Factory Integration and Experimentation

SFIO Section Française de l'Internationale Ouvriere (French—French section of the Worker's International)—former name of the French Socialist Party

SFIS Small Firms Information Service

SFIT Standard Family Interaction Test

sfl sequenced flashing lights (airport runways)

s fl Surinam florin

SFL Scottish Football League; Sexual Freedom League; Society of Federal Linguists

sfm surface feed per minute; surface feet per minute

SFM Sinai Field Mission; Society for Foodservice Management

SFMA San Francisco Museum of Art; School Furniture Manufacturers' Association

SFMC San Francisco Medical Center (University of California)

SFMR San Francisco Municipal Railway (operates the cable cars)

SFMS Shipwrecked Fishermen and Mariners (Royal Benevolent Society)

SF & NV San Francisco & Napa Valley (railroad)

sfo simulated flame out; single frequency oscillator; submarine fog oscillator

S Fo (Port of) San Francisco

SFO San Francisco, California (airport); San Francisco Opera; San Francisco

Operations (office); San Francisco-Oakland Airlines; Santa Fe Opera; Senior Flag Officer; Service Fuel Oil; Space Flight Operations

SFOB Special Forces Operating Base

SF-OBB San Francisco-Oakland Bay Bridge (Transbay Bridge)

SFOD San Francisco Ordnance District; Special Forces Operational Detachment

SFOF Space Flight Operations Facility

SFOLDS Ship-Form On-Line Design System

sfp spent fuel pool

SFP Sherbrooke Forest Park (Victoria, Australia)

SFP Société Française de Photogrammétrie (French—French Society of Photogrammetry)

SFPD San Francisco Police Department

SFPDis San Francisco Procurement District (US Army)

sf pe surface plate

SFPE San Francisco Port of Embarkation; Society of Fire Protection Engineers

SFPF Spent-Fuel Packaging Facility

SFPL San Francisco Public Library

sfpm surface feet per minute

SFPO Spent-Fuel Project Office (Savannah River Operations Office)

SFPR Society of Friends of Puerto Rico

sfprf semifireproof

sfps single failure point summary

SFPs Sinn Fein Provisionals (Provos)

sfga (SFQA) structurally fixed question-answering system

sfr single frequency receiver; sinking fund rate (of return); star-formation rate

sfr (SFR) submarine fleet reactor

SFR Safety of Flight Requirement; Substitute-For-Return (unfiled taxes)

SFRA Science Fiction Research Association

S Fran San Francisco

SFRJ *Socijalisticka Federativna Republika Jugostavija* (Socialist Federated Republic of Yugoslavia)

sfrr sinking fund rate of return

sfr(s) schweizerfranc(s) [Dano-Norwegian—Swiss franc(s)]

SFRS Sea Fisheries Research Station (Haifa)

sfs strictly for suckers; surfaced four sides

sfs sine fraude sua (Latin— without fraud on his part)

SFs Special Forces (Green Berets); State Facilitators

SFS San Francisco State; San Francisco Symphony; Senior Foreign Service; Society of Fleet Supervisors

SFSA Scottish Field Studies Association; Steel Founders' Society of America

SFSAFBI Society of Former Special Agents of the Federal Bureau of Investigation

SFSC San Francisco State College

SF & SC Standard Fruit & Steamship Company

SFSE San Francisco Stock Exchange

SFSO San Francisco Symphony Orchestra

SFSP Spent-Fuel Storage Program

SF/SP Santa Fe/Southern Pacific (merged railroads)

SFSS Satellite Field Services Stations (National Oceanographic and Atmospheric Administration)

SFSSP Society of the Friendly Sons of St Patrick

sft soft; specifed financial transactions; stop for tea; superfast train

SFT Society of Forensic Toxicologists; Spent-Fuel Test(ing); System Fault Tolerant

SFTA Scientific Film Television Award; Society of Film and Television Arts

SFTAA Short-Form Test of Academic Aptitude

SFTB Southern Freight Tariff Bureau

sftgfop special finest tippy golden fancy orange pekoe (tea)

SFTI San Fernando Technical Institute (Trinidad)

SFTP Science For The People

SFPS San Francisco Theological Seminary

SFTS Service Flying Training School

SFTW Stamps For The Wounded

sftwd softwood

sftwr software

SFU Signals Flying Unit; Simon Fraser University

SFUG Security Features User's Guide

S$_f$ units Svedberg flotation units

sfv sight feed valve

SFv Semliki Forest virus

SFVAH San Francisco Veterans Administration Hospital

SFVSC San Fernando Valley State College

SFWA Science Fiction Writers of America

sfwd slow forward

SFWR Stewardesses for Women's Rights

sfx sound effects

sfxd semifixed

sfxr superflash X-ray

sfy standard facility years(s)

SFYC San Francisco Yacht Club

sfz sforzando (Italian—emphasized chord or note)

sfz p sforzato piano (Italian— emphasis followed by a soft note or chord)

sg screen grid; single groove; singular; smoke generator; soluble gelatin; specific gravity; steam generator; steel girder; structural glass; swamp glider

s-g sub-generic; sub-genus

sg selon grandeur (French— according to size); on menus, sg or SG indicates an item is priced according to the size of the serving

SG Song of Songs

s.g. salutis gratia (Latin—for safety's sake)

s/G sur Garonne (French—on the Garonne)

Sg goal stimulus; spring range of tide; Surgeon

SG Scots Greys; Scots Guards; Seaman Gunner; Singapore (Internet code); small-capitalization growth; Solicitor Gen-

eral; South Georgia (railroad); Standing Group; sub-group; Sudan Government; Sunset Gun; Surgeon General

S-G Sachs-Georgi (test); Saint Gobain; Space-General (Corporation)

SG Stanley Gibbons (stamp catalog)

sga substantial gainful activity

sg & a selling, general and administrative (expenses)

SGA Saskatchewan Government Airways; Society of the Graphic Arts; Southern Gas Association; Special Grant Application; Standards of Grade Authorization; Stock Growers Association; Student Government Association

SGAA Stained Glass Association of America

SGAC State Governmental Affairs Council

SGAE Sociedad General de Autores de España (Spanish—General Society of Authors of Spain)

SGAT Seagate Technology Incorporated

S-gauge standard gauge (4-foot 8 1/2-inch) railroad track

SGB Societe Générale de Banque (Belgian Bank); *Société Générale de Belgique* (French—General Society of Belgium)

SGBIP Subject Guide to Books in Print

sgc screen grid current; simulated generation control; spartan guidance computer (SGC); spherical gear coupling; stabilizer gyro circuit

Sg C Surgeon Captain

SGC Saint Gregory College; South Georgia College

S-G C Space-General Corporation

SGCA Secrétariat Général à l'Aviation Civil (French— Secretary General of Civil Aviation)

SGCC Safety Glazing Certification Council

SGCD Society of Glass and Ceramic Decorators

Sg Cr Surgeon Commander

sgcs silicon gate controlled switch

sgd signed

SGD Senior Grand Deacon

sgdg sans garantie du gouvernement (French—patent issued without government guarantee)

sg di swaging die

Sge Sagitta

S Ge South Georgia

sgemp system-generated electromagnetic pulse

SGF Scottish Grocers' Federation; State Guaranty Funds

SGF Sveriges Gummitekniska Forening (Swedish—Swedish Rubber Industry Association)

sgg sustainer gas generator

sghwr steam-generating heavy—water reactor

SGI Small Group Instruction; Spring Garden Institute

SGINDEX System Generation Cross-Reference Index (NASA)

SGIO State Government Insurance Office

sgl signal; single

SGL Society of Gas Lighting

S Glam South Glamorgan

Sg L Cr Surgeon Lieutenant Commander

SGLI Servicemen's Group Life Insurance

SGLS Space-Ground Link Subsystem

sgm spark gap modulator

SGM Sea Gallantry Medal; Society of General Microbiology

sg md swaging mandrel

SGMEX Southern Gulf of Mexico

SGML Standard Generalized Markup Language

SGMP Society of Government Meeting Planners

SGMT Société Générale des Transports Maritimes (French—General Society of Maritime Transport)

sgn scan gate number; signum function

Sgn (Port of) Saigon

SGN Saigon, Vietnam (airport); Surgeon General of the Navy

Sgno Stagno (Italian—pond; pool)

sgnr signature

sgo spark gap oscillator; surgery, gynecology, and obstetrics

SGO Squadron Gunnery Officer; Surgeon General's Office

sgot serum glutamic oxaloacetic transaminase

sgp starch graft polymers

SGP Shell Gasification Process; Society of General Physiologists

SGP Secretario General del Partido (Spanish—Secretary General of the Party); *Staatkundig Gereformeerde Partij* (Dutch—Political Reformed Party)

SGPA Scottish General Publishers Association

SGPH Signal/Graphics Processor Hybrid

sgpt serum glutamic pyruvic transaminase

sgr steam gas recirculation (oil-from-shale removal process)

Sgr Sagittarius (constellation)

SGR Sumbu Game Reserve (Zambia)

Sg RA Surgeon Rear Admiral

SGRAM Synchronous Graphic Ram

SGRS Stockton Geriatric Rating Scale

SGS Society of General Surgeons; Sunderbans Game Sanctuary (Bangladesh)

SGSB Stanford Graduate School of Business

SGSR Society for General Systems Research

sgt special gas taper (threading)

Sgt Sergeant

SGT Society of Glass Technology

Sgt 1/C Sergeant First Class

S-G Test Sachs-Georgi Test

SGTIA Standing Group Technical Intelligence Agency (NATO)

Sgt Maj Sergeant Major

sgu single gun unit

SGU Scottish Gliding Union; Scottish Golf Union; Singapore Golfers Union

SGU Sveriges Geologiska Undersokning (Swedish—Swedish Geological Survey)

SGUs Special Guerrilla Units

Sg VA Surgeon Vice Admiral

SGVHS Samuel Gompers Vocational High School

SGW Senior Grand Warden

SGX Seeger Refrigerator Express (stock exchange symbol)

sh sacrifice hits (bunts); scleroscope hardness; serum hepatitis; share; shelf, shelving; ship; ship's heading; shop; shopping; short; showers; sick in hospital; social history; somatotrophic hormone speech handicapped; surgical hernia

sh (SM sexual harassment

s/h shorthand

s & h son and heir

Sh shells; shilling (British East Africa)

Sh Sh'aib (Arabic—ravine; road); *Shatt* (Arabic—river; riverbank); *Shima* (Japanese—island); *Suid Holland* (Dutch—South Holland)

SH Saint Helena (Internet code); Schenley Industries (stock exchange symbol); Scinde Horse; Scottish Horse; Sherbrooke Hussars; Soldier's Home; Southland Hussars; Station Hospital; Symphony Hall

S-H Schleswig-Holstein, Scripps-Howard

S & H Sperry & Hutchinson; Sundays and Holidays

SH Sa Hautesse (French—Her or His Highness)

sha (SHA) sidereal hour angle

Sha Shanghai

SHA Safety and Health Administration; Secular Humanist Association; Society for Humane Abortion; Southern Historical Association; State Historic Area

SHAA Society of Hearing Aid Audiologists

shab soft and hard acids and bases

Shab Shabbat (Sabbath)

sh abs shock absorber

SHAC Seale-Hayne Agricultural College; Shelter Housing Aid Center

shaco shorthand coding

SHAD Sharpe Army Depot

shade (SHADE) shielded hot-air-drum evaporator

SHAEF Supreme Headquarters, Allied Expeditionary Forces

shag simplified high-accuracy guidance

shags shaggy carpets or rugs

Shah Shahanshah (Persian—King of Kings)

Shak(e) Shakespeare

Shakes Shakespeare

shale standoff high-altitude long endurance

Sham Shamrock

shamateur(s) sham amateur(s)

shamburger hamburger containing more additives and adulterants than meat

SHAME Save, Help Animals Man Exploits; Society to Humiliate, Aggravate, Mortify, and Embarrass Smokers

shandy shandygaff (beer-and-ginger-ale mixture)

Shang Shanghai

shan't shall not (colloquial)

Shanty Irish poor people

SHAPE Scanning Hartmann Aperture Plate Experiment; Supreme Headquarters, Allied Powers, Europe

SHAR Sea Harrier

SHARE Scottish Health Authorities Revenue Equalisation; Self-Help and Resources Exchange

SHARP Senior-High Assessment of Reading Performance; Ships Analysis and Retrieval Project

SHARPS Ship/Helicopter Acoustic Range-Prediction System (USN)

SHAS Shared Hospital Accounting System

Shav Shavuot

Shaw George Bernard Shaw (dramatist and music critic)

SHAWCO Students Health and Welfare Centers Organization

SHB *Svenska Handelsbanken* (Swedish—Swedish Bank of Commerce)

shbd serum X-hydroxy-butyrate dehydrogenase

shbg sex-hormone-binding globulin

shc spontaneous human combustion

SHC Sacred Heart College; Seton Hall College; Siena Heights College; Spring Hill College; Streets and Highways Code; Surveillance Helicopter Company

SHCC Statewide Health Coordinating Council

SHCJ Society of the Holy Child of Jesus

shco sulfonated hydrogenated castor oil

sh con shore connection

SHCS School of Health Care Sciences (USAF)

shd should

Sh.D. Doctor of Showbizology (Frank Sinatra)

SHD Scottish Home Department; State Hydroelectric Department

she signal handling equipment; standard hydrogen electrode

she (SHE) sodium heat engine

SHE Shelter for Help and Emergency

Shea Shea Stadium, New York city

Sheba Saba

SHEBA Surface Heat Budget of the Arctic Ocean

she'd she had; she would

Shedd Shedd Aquarium (Chicago)

Sheed Sheed & Ward

SHEEO State Higher Education Executive Officers

Sheet Metal Workers Union Sheet Metal Workers International Association

Sheff Sheffield; Sheffield Scientific School (Yale)

shekel monetary unit of Israel

shelf super-hardened extremely low frequency

she'll she will

SHELL Royal Dutch Shell Oil Company; Shell Oil Company

shellrep shelling report

SHELREP Shelling Report

Shenandoahs Shenandoah Mountains of Virginia and West Virginia

Shen NP Shenandoah National Park

Shep Shep(p)ard; Shepton

Sheq. *Sheqalim*

Sher Sherbrooke

Shere Shirley

Sherm Sherman

sherm(s) sherman(s)—pcp-soaked marijuana cigarette(s)

she's she has; she is

SHES School Health Education Study

Shet Shetland

Shetland Shetland Island, Shetland Islands called the Zetlands

Shetlands Shetland Islands off northern Scotland

Shets Shetland Islands, Scotland

Shex Sundays and holidays excepted

shf super high-frequency—300–30,000 mc

Shf Sheffield

SHF Soil and Health Foundation

SHFF Scottish House Furnishers' Federation

shftg shafting

S-H-G diet Sauerbruch-Herrmannsdorfer-Gerson (tubercular) diet

SHH Security Hardware Handler

SHH *Sociedad Honoraria Hispánica* (Spanish—Honorary Hispanic Society)

SHHV Society for Health and Human Values

Shi Shanghai

SHI Strategic Homeporting Initiative (USN)

SHIBA State Health Insurance Benefits Advisors

Shickshocks Shickshock Mountains of the Gaspé Peninsula of New Brunswick

SHIELD Sylvania High-Intelligence Electronic Defense

Shig *Shigella*

shil (SHIL) shillelagh (surface-to-surface missile of the U.S. Army)

shilling monetary unit of Kenya, Somalia, Tanzania, Uganda

Shim Shimonoseki

Shin Bet Israel's domestic security agency

Shin Bet *Sherut Habitachon* (Hebrew—Security Department)—Israel

S & h inc Sundays and holidays included

shinerium shoe-shine stand

ship shipment; shipping

SHIP Self-Help Improvement Program; State Health Insurance Program

shipcon shipping control; shipping convoy

ShipDTO ship on depot transfer order

shipmt shipment

SHIR Shah of Iran (Chieftan tank)

SHJ Society for Humanistic Judaism

SRJC Sacred Heart Junior College

shk shank

Shl Shields; shoal
Sh L Shipwright Lieutenant
SHL Society for Humane Legislation
shld shoulder
shl dk shelter deck
SHLM Society of Hospital Laundry Managers
shlp shiplap
Shls Shoals (postal abbreviation)
shm simple harmonic motion
Shm Shimizu; Shoreham
SHM *Service Hydrographique de la Marine* (French—Naval Hydrographic Service)
SHMO Senior Hospital Medical officer; Social Health Maintenance Organization
shmt shock mount
SHNC Scottish Higher National Certificate
SHND Scottish Higher National Diploma
SHNHP Signal Hill National Historical Park, St John's, Newfoundland
SHNHs Sagamore Hill National Historic Site
SHNNR Studland Heath National Nature Reserve (England)
shnoz shnozzle; shnozzola; nose
ShNP Shenandoah National Park
sho shore(d); shoring
sho shutout(s)
SHO Senior House Officer; Student Health Organization
SHOC Self-Help Opportunity Center
SHOCK Students Hot on Conserving Kilowatts
shocks shock absorbers
Shol Asch Sholem Asch (celebrated or modern Yiddish literature)
S Holmes, Esq Sherlock Holmes
shootin shooting
SHOPA School and Home Office Products Association
shop-op shopping opportunity
SHORAD Short Range Air Defense
SHORADs Short-Range All-Weather Air-Defense System
shoran short-range navigation
SHORS School/Home Observational Referral System

shorted short circuited
shortg shortage
short(s) short circuit(s)
SHOT Society for the History of Technology
shouldn't should not
show biz show business
show exhibs show exhibitions
shp shaft horsepower
Shp Sharpness
SHP Sandy Hook Pilots; Society of Hospital Pharmacists; State Historic Park
SHPBG Small Horticultural Production Business Grant
SHPC Scenic Hudson Preservation Conference
SHPDA State Health Planning and Development Agency
shpmt shipment
shpng shipping
shpng/hndlg shipping & handling (charges)
SHPO State Historic Preservation Office
shps seahead pressure simulator
shpt shipment
SHP Test Strongin-Hinsie-Peck (salivary secretion) Test
SHQ squadron headquarters; Station Headquarters
shr share(s)
Shr Shore (postal abbreviation)
shram (SHRAM) short-range air-to-surface missile
shrap shrapnel
shrd shredded
shrimpsan shrimp sandwich
shrimpwich shrimp sandwich
SHRM Society for Human Resource Management
SHRMA South Hampton Roads Metropolitan Area (Norfolk, Portsmouth, Chesapeake and Virginia Beach)
Shrops Shropshire
Shrs Shores
shrtg shortage
shs ship's heading servo
SHs Secured Horizons (Division of Pacificare)
SHS Sacred Heart Seminary; Scottish History Society; Senior High School; *Srba, Hrvata, i Slovenaca* (Serbo-Croatian—Serbs, Croats, and Slovenes)—Yugoslavia; State Historic(al) Site; Stuyvesant High School

SHSA Steamship Historical Society of America
SHSL Sherlock Holmes Society of London
SHSLB Street and Highway Safety Lighting Bureau
SHSN Sod House Society of Nebraska
SHSP Sam Houston State Park (Louisiana)
SHSS Sanford Hypnotic Susceptibility Scale
SHSSI Steamship Historical Society of Staten Island
SHSW State Historical Society of Wisconsin
shswc sample-and-hold square-wave converter
sht sheet(ing)
SIIT Society for the History of Technology
shtg shortage
shtgn shotgun
shth sheatlming
sht irn sheet iron
sht mtl sheet metal
sh tn short ton
SHU Security Housing Unit; Seton Hall University
Shula Shulamite; Shulamith
SHUR System of Hospital Uniform Reporting
shv solenoid hydraulic valve
s.h.v. *sub hoc voce* (Latin—under this work)
SHVA Satellite Home Viewer Act
shvg shaving(s)
shw safety, health, and welfare
SHW Sherwin-Williams (stock exchange symbol)
shwrs showers
S & H x Sundays and Holidays excepted
SHYC Sachem's Head Yacht Club
si salinity indicator; shift in; short interest; silicone; simple interest; sister; slight imperfection; small inclusion; spark ignition; straight-in (aircraft landing approach); subicteric; subindex; subinguinal; surface interval
Si (SI) shift-in character (data processing)
S-i semiconductor-integrated (circuits)
s/i signal/intermodulation; subject issue
s & i stocked and issued

Si Silas; silicon (symbol); Simon; Simone

Sⁱ Sidi (Arabic—My Lord)—title of honor also written *Saiyidi*

SI Sandwich Islands; Saturday Inspection; Secret Information; Secret Intelligence; Sergeant Instructor; Serra International; Sertoma International; Service Instruction; Shipping Instruction(s); Silver Institute; Slovenia (Internet code); Smithsonian Institution; Society of Illustrators; Solomon Islands; South Island (New Zealand); Spokane International (railroad); Staff Inspector; Star of India (order); Staten Island; Stevens Institute; Sulfur Institute; Summer Institute; Survey Instruction(s)

S-I Spokane International (railroad)

SI *Scheepvaart Inspectie* (Dutch—Shipping Inspection); *Sports Illustrated; Système International des Unités* (French—International System of Units)

sia subminiature integrated antenna

sia (SIA) storage instantaneous audimeter

SIA Sanitary Institute of America; Scaffold Industy Association; School of International Affairs (Columbia University); Securities Industries Association; Self-Insurers Institute; Semiconductor Industry Association; Singapore Airlines; Ski Industries of America; Society of Insurance Accountants; Soroptimist International Association; Special Interest Area; Spinal Injuries Association; Sprinkler Irrigation Association; Standard Instrument Approach; Strategic Industries Association; Structural Insulation Association

SIA *Schweizerischer Ingenieur und Architekten Verein* (German—Swiss Institute of Engineers and Architects); *Société Internationale d'Acupuncture* (French—International Acupuncture Society)

SIAC Securities Industry Automation Corporation

SIAD Society of Industrial Artists and Designers

SIAE *Società Italiano degli Autori ed Editori* (Italian—Italian Society of Authors and Editors)

sial silicon + aluminum (Si + Al)

siam self-initiated anti-air missile; signal information and monitoring

SIAM Society for Industrial and Applied Mathematics

SIAO Smithsonian Institution Astrophysical Observatory

SIAP *Sociedad Interamericcana de Planificación* (Spanish—Interamerican Planning Society)

sib satellite ionospheric beacon(s), sibilant; sibling; sibship

Sib Siberia; Siberian; Sibyl; Sybil

SIB Securities and Investment Board; Shipbuilding Industry Board; Society of Insurance Brokers; Soviet Information Bureau; Special Investigation Branch (Police)

SIB *Sveriges Investeringsbank* (Swedish—Swedish Investment Bank)

SIBC Solomon Islands Broadcasting Corporation

SIBC *Société Internationale de Biologie Clinique* (French—International Society of Clinical Biology)

SIBIA Salk Institute Biotechnology/Industrial Associates

SIBIS Smithsonian Institution Bibliographic Information Service

Sib Or Sibylline Oracles

Sibr Siberia

sibs siblings

SIBS Salk Institute for Biological Studies

sic semiconductor integrated circuits; specific inductance capacity; standard industrial classification

sic. siccus (Latin—dry)

Sic Sicilian; Siciliana; Siciliano; Sicily

SiC silicon carbide

SIC Scientific Information Center; Secret Intelligence Command; Security Intelligence Corps; Senate Intelligence Committee; Sisters in Crime (mystery writers); *Société Intercontinental des Containers* (French—Intercontinental Container Society); *Société International de Cardiologie* (French—International Cardiology Society); *Société Internationale de Chirurgie* (French—International Surgery Society); Standard Industrial Classification; State Insurance Commission; Survey Information Center

SIC *Société Internationale de Cardiologie* (French International Cardiology Society); *Société Internationale de Chirurgie* (French—International Surgery Society); *Société Internationale de Criminologie* (French—International Criminology Society)

SICA Society of Industrial and Cost Accountants

sicbm (SICBM) super-intercontinental ballistic missile

SICC Staten Island Community College

Sic Chan Sicilian Channel between Sicily and Tunisia

sicklemia sickle-cell anemia

SICOT *Société Internationale de Chirurgie Orthopédique et de Traumatologie* (French—International Society of Orthopedic Surgery and Traumatology)

SICR Specific Intelligence Collection Requirement

sicsva sequential-impaction cascade-sieve volumetric air (sampler)

sic transit *sic transit gloria mundi* (Latin—so passes away the glory of the world)

sicu (SICU) surgical intensive care unit

Sic Vesp Sicilian Vespers (massacre of the French in 1282)

sid sidereal; standard instrument departure; status-income disequilibrium; sudden infant death; sudden ionospheric disturbance

s & id surveillance and identification

Sid Sidney; Sidney Sussex College, Oxford; Sydney

S.i.D. *Spiritus in Deo* (Latin— His Spirit is with God)—he's dead

SID Security and Intelligence Department; Society for Information Display; Society for International Development; Society for Investigative Dermatology; Standard Instrument Departure; Sudden Ionospheric Disturbance Division

SIDA Swedish International Development Agency

sidar selective information dissemination and retrieval

sidase significant data selection

SIDEC Stanford International Development Education Center

Siding Spring Siding Spring Observatory, Australia

SIDINSA *Siderurgia Integrada SA* (Spanish—Integrated Iron-and-Steel Industry Corporation)

SIDQR *Siderúrgica del Oriente* (Spanish—Oriente Iron and Steel Industry)—Venezuela

Sidro San Ysidro, Califorrmia

sids sudden infant-death syndrome

SIDs Sports Information Directors

SIDS Ships Integrated Defense System; Shrike Improved Display System; Space Identification Device System; Space Investigations Documentation System

SIDS *Société Internationale de Défense Sociale* (French—International Society of Social Defense)

sie single instruction execute

SIE Science Information Exchange (Smithsonian); Scientific Information Exchange; Society of Industrial Engineers; Southwestern Industrial Electronics

SIEC Scottish Industrial Estates Corporation

SIECUS Sex Information and Educational Council of the United States

SIEE Student of the Institution of Electrical Engineers

Siern Siemensstadt

Sier Leo Sierra Leone (whose capital is Freetown)

Sierra letter S radio code

Sierra Leone Republic of Sierra Leone (West African nation established by the British as a native home for freed slaves)

Sierra Madre high mountains of western Mexico

Sierra Nevadas Sierra Nevada Mountains in California, Nevada, Spain, and Venezuela

Sierras Sierra Nevada Mountains; Sierra Mountains

SIES Soils and Irrigation Extension Service

SIETAR Society for Intercultural Education, Training, and Research

Sieur de Cadillac (French— Mr Cadillac—Antoine de la Mothe Cadillac (1658–1730)

SIEX *Superintendencia de Inverciones Extranjeras* (Spanish—Superintendence of Foreign Investments)

SI Exy Staten Island Expressway

sif selective identification feature

SIF Senate Inactive File; Society for Individual Freedom

SiF₄ silicon tetrafluoride

SIFA *Seguridad e Inteligencia de las Fuerzas Armadas* (Spanish—Security and Intelligence of the Armed Forces)—Venezuela

SIFE Society of Industrial Furnace Engineers

SIFF *Soumen Illmailuliitto Finlands Flygforbund* (Finnish—Finnish Aeronautical Association)

sif/iff selective identification feature/identification friend or foe

sifl standard industry fare level

SIFO *Statens Institut för Opinionsundersökning* (Swedish—State Institute for Opinion Research)

SIFs Stock Index Futures

SIFS Special Instructors Flying School

sift share interval fortran translator; simplified input for toss

sig sigmoidoscopy; signal; signaling; signature; special interest group

sig. *signetur* (Latin—mark with directions)

Sig Siegfried; Sieglinde; Sigdrifa; Sigmund; Sigmund Freud (Austrian founder of psychoanalysis); Sigmunt; Sigsbee; Sigurd; Sigyn

Sig *Signor* (Italian—Mister; Sir); *Signore* (Italian—Gentlemen, Our Lord, Sir); *Signori* (Italian—Gentlemen, Lords)

SIG Secret Intelligence Group; Senior Interagency Group (Space); Snowy Irrigation Scheme (Snowy Mountains Authority—Australia); Special Interest Group; Special Interrogation Group

SIG *Schweizerische Industrie Gesellschaft* (German— Swiss Industry Society)

siga sigatoka (banana leaf spot disease)

Sigᵃ *Signora* (Italian—Missus) —Mrs

SIGACT Special Interest Group on Automata and Computability Theory

SIGARCH Special Interest Group on Architecture of Computer Systems

SIGART Special Interest Group on Artificial Intelligence

SIG/BDP Special Interest Group for Business Data Processing

SIGBIO Special Interest Group on Biomedical Computing

SIG/BIOM Special Interest Group on Biomedical Information Processing

SigC Signal Corps

SIGCAPH Special Interest Group on Computers and the Physically Handicapped

SIGCAS Special Interest Group on Computers and Society

SIGCOMM Special Interest Group on Data Communication

SIGCOSIM Special Interest Group on Computer Systems Installation Management

SIGCPR Special Interest Group on Computer Personnel Research

SIGCSE Special Interest Group on Computer Science Education

SIGCUE Special Interest Group on Computer Uses in Education

SIGDA Special Interest Group on Design Automation

Sig Div Signal Division

sigex signal exercise

Sigg Signori (Italian—Messrs)

SIGGRAPH Special Interest Group on Computer Graphics

SIGI System of Interactive Guidance and Education

sigill. sigillum (Latin—seal)

sigint signals intelligence

sigint (SIGINT) signals intelligence (facsimile communications)

SIGIR Special Interest Group on Information Retrieval

Sig L Signal Lieutenant

SIGLASH Special Interest Group on Language Analysis and Studies in the Humanities

SIGLE System for Information on Grey Literature in Europe

sigligun signal-light gun

Sigm Sigmund

SIGMA Science in General Management; Sealed Insulating Glass Manufacturers Association; Society of Independent Gasoline Marketers of America

SIGMAP Special Interest Group on Mathematical Programming

SIGMETRIC Special Interest Group on Metrication

SIGMICRO Special Interest Group on Microprogramming

SIGMINI Special Interest Group on Minicomputers

Sigmn Signalman

SIGMOD Special Interest Group on Management of Data

sigmoido sigmoidoscopy

sign signature

Sig^na Signorina (Italian—Miss)

signif signifiable; signifiably; significance; significancy; significant(ly); signification; significative(ly); signifier; signify

sig. nom. pro. signa nomine proprio (Latin—label with the proper name)

SIGNUM Special interest Group on Numerical Mathematics

Sig O Signal Officer

SIGOIS Special Interest Group on Office Information Systems

SIGOPS Special Interest Group on Operating Systems

SIGPLAN Special Interest Group on Programming Languages

SIG/REAL Special Interest Group on Real-Time Processing

SIGs Special Interest Groups

SIGS Sandia Interactive Graphics System

SIGSAM Special Interest Group on Symbolic and Algebraic Manipulation

Sig Sam Lib Sigmund Samuel Library (Toronto)

SIGs-ASIS Special Interest Groups of the American Society for Information Science—AH: Arts and Humanities; ALP: Automated Language Processing; BSS: Behavioral and Social Sciences; BC: Biological and Chemical; CB: Costs, Budgeting, Economics; CR: Classification Research; ED: Education for Information Science, FS; Foundations of Information Science; IAC: Information Analysis Centers; IP: Information Publishing; ISE: Information Services to Education; LAN: Library Information and Networks; LAW: Law and Information Technology; MGT: Management Information Activities; MR: Medical Records; NDB: Numerical Data Bases; NPM: Non-Print Media; PPI: Public-Private Interface; RT: Reprographic Technology; SDI: Selective Dissemination of Information; TIS: Technology, Information, Society; UOI: User On-line Interaction

Sig Saus Sig Sauer (pistols)

SIGSDI Special Interest Group on Selective Dissemination of Information

SIGSIM Special Interest Group on Simulation

SIGSOC Special Interest Group on Social and Behavioral Science Computing

SIGSPAC Special Interest Group on Urban Data Systems, Planning, Architecture, and Civil Engineering

Sig Sta signal station

SIG/TIME Special Interest Group on Time Sharing

SIGUCC Special Interest Group on University Computing Centers

Sig Und Sigrid Undset

SIG/UPACE Special Interest Group on Urban Planning, Architecture, and Civil Engineering

SIH Samuel Ichiye Hayakawa; South Irish Horse

SIH Société Internationale d'Hématologie (French—International Hematology Society)

SIHS Society for Italian Historical Studies

SII School Interest Inventory; Security-Insecurity Inventory; Self-Interview Inventory; Standards Institution of Israel; Staten Island Institute; Structural Impediments Initiative (Japan and the United States)

SIIA Self-Insurance Institute of America; Software & Information Industry Association; Stevenson Institute of International Affairs

SIIAS Staten Island Institute of Arts and Sciences

SIIP Systems Integration Implementation Plan

SIIRS Smithsonian Institution Information Retrieval Service

SIIS Shanghai Institute of International Studies

SIJD Subcommittee to Investigate Juvenile Delinquency (U.S. Senate)

Sik Sikkim

Siks single income, kids

sil silver; speech interference level; squamous intraepithelial lesion

s-i-l sister-in-law; son-in-law

Sil Silesia; Silesian; Silurian

SIL Society for Individual Liberty; Society for International Law; Summer Institute of Linguistics; System Implementation Language

SIL Société International de la Lèpre (French—International Leprosy Society)

Silas Silvanus

silcads silver-cadmium batteries

Sile Cecilia

SILI Standard Item Location Index

SILIA South Island Livestock Improvement Association

silic silicate; siliceous

silica silicon dioxide (SiO_2)

silicos silicosis (sickness caused by stone-dust inhalation)

silkool silk + wool (Japanese synthetic textile combining qualities of silk and wool)

silos side-looking sonar

sils silver solder

sil(s) speech interference level(s)

SILs Smithsonian Institution Libraries

silv silver; silvery

silvercel silver-zinc cell (battery)

silvicult silviculture

sim self-inflicted mutilation; similar; simile; simple; simulate; simulated approach

Sim Simm(s); Simmy; Simon(d); Sims; Syme(s); Symme; Syms; etc.

SIM Sergeant Instructor of Musketry; Society for Industrial Microbiology; Society for Information Management; Sudan Interior Mission

SIM *Servicio Inteligencia Militar* (Spanish); *Servizio Informazioni Militari* (Italian—Military Intelligence Service); *Société Internationale de Musicologie* (French—International Musicological Society)

SIMA Scientific Instrument Manufacturers' Association; Steel Industry Management Association; Suburban Insurance Managers' Association

SIMAGB Scientific Instrument Manufacturers Association of Great Britain

SIMAJ Scientific Instrument Manufacturers Association of Japan

SIMBAD Set of Identifications, Measurements, and Bibliography for Astronomical Data

SIMBIOSE Scientific Improvement of Biofilters and Sensors

SIMC *Société Internationale pour la Musique Contemporaine* (French—International Society for Contemporary Music)

SIMCA *Société Industrielle de Mécanique et Carosserie Automobile* (French—Industrial Society of Automobile Manufacturers)

simcas simulated casualty

simch single mach change

simcon simplified control; simulated control

simd single-instruction multiple-data stream

SIME Security Intelligence Middle East (British)

sim excu simulated execution

SIMG *Societas Internationalis Medicinae Generalis* (Latin—International General Medicine Society)

SIMHA *Société Internationale de Mycologie Humaine et Animale* (French—International Society for Human and Animal Mycology)

SIMILE Simulator of Immediate Memory in Learning Experiments

simm single inline memory module

simmo simulated ammunition

'simmon(s) persimmon(s)

SIMNET Simulation Network

Simón Simón Bolívar (liberator of Colombia, Ecuador and Venezuela from Spanish rule plus Peru and founding of Bolivia)

Simons Simonstown; Simonstown naval base near the Cape of Good Hope in South Africa

simp simpleton

simp. simplex (Latin—simple)

simpac simulated package

SIMPER Structured Information Management Processing and Effective Retrieval

SIMPL Scientific, Industrial, and Medical Photographic Laboratories

Simplon Simplon Pass in the Swiss Alps

Simpson Simpson Desert in the southeast sector of Australia's Northern Territory

Sims secondary ion mass spectroscopy

SIMS Surface-to-Air Intercept Missile System

simstrat simulation strategy

sim sui simulated suicide

simul simulation

simula simulation language

simulcast simultaneous broadcast (am & fm)

simulcast(ing) simultaneous broadcast(ing) of the same program on radio and television

Simyens Simyen Mountains of Ethiopia

sin sewer in; sine; single

sin. sinfonia (Italian—symphony); *sinister* (Latin—left)

sin' sino (Italian—as far as; until)

Sin Sinaloa (inhabitants—Sinaloens); Singapore; Sinhalese

SIN Singapore (airport); Single Incident Number; Society for International Numismatics; Stop Inflation Now

SIN Scientific Information Notes (National Science Foundation); *Société Industrielle et Navale* (French—Industrial and Naval Society); Spanish International Network (tv)

SINB Southern Interstate Nuclear Board

SI-NAA National Anthropological Archives—Smithsonian Institution

Sinai Pen Sinai Peninsula

S-in-C Surgeon-in-Chief

SINCGARS Single-Channel Ground-Air Radio System

Sind Sindhi

S Ind Cur South Indian Current

sin Dich sinfonische Dichtung (German symphonic poem)

sinf sinfonia (Italian—symphony)

S Infre Seine-Inférieure (Lower Seine River)

sing singer; single; singing; singular

sing. singulorum (Latin—of each)

Sing Singapore; Singapore Airlines

singan singularity analyzer

Singapore Republic of Singapore (island nation at the southernmost tip of the Malay Peninsula)

Singer Isaac Bashevis Singer

Singlish Singapore English

Sing U Singapore University

sinh hyperbolic sine

Sinh Sinhalese

Sinjent St John

Sink Sinkiang

SINK Single Income, Numerous Kids

sins ship-inertial-navigation systems

SINS Ship's Inertial Navigation System; Situational Inertial Navigation System (USN)

s int senza interruzione (Italian—without interruption)

Sint Eust Sint Eustatius

Sint Maart Sint Maarten

SINTO Sheffield Interchange Organisation

si n. val. si non valet (Latin—if of no value)

sio satellite in orbit; serial input-output; staged in orbit

si/o star input/output

SIO Scripps Institution of Oceanography; Senior Intelligence Officer; Ship's Information Office(r); Special Intelligence Office(r)

sioh supervision, inspection, and overhead

SIOH standard IO Hybrid

SIOP Single Integrated Operations Plan

si op. sit si opus sit (Latin—if necessary)

SIOR Society of Industrial and Office Realtors

Sioux Falls Pen South Dakota Penitentiary at Sioux Falls

sip schedule-induced polydipsia; single inline package; standard inspection procedure; step in place; structural insulated panel

SIP Sectoral Import Program; Share of Impact Panel (measuring alcohol consumption); Smithsonian Institution Press; Special Internal Predisposition; Society of Integral Psychoanalysis; Standard Inspection Procedure; State Improvement Plan(ing); Street Improvement Program

SIP Sociedad Interamericana de la Prensa (Spanish—Interamerican Press Association); *Société Interaméricaine de Psychologie* (French—Interamerican Society of Psychology)

SIP/AG Sri lanka, India, Pakistan/Arabian Gulf (freighter route)

SIPC Securities Investor Protection Corporation

SIPE System Internal Performance Evaluation

SIPG Société Internationale de Pathologie Geographique (French—International Society of Geographical Pathology)

SIPI Southwestern Indian Polytechnic Institute

sipl scientific information processing language

SIPL Solomon Islands Plantations Limited

Sipo security police (Nazi)

Sipo Sicherheitspolizei (German—State Security Police)

SIPRC Society of Independent Public Relations Consultants

SIPRE Snow, Ice, and Permafrost Research Establishment

SIPRI Stockholm International Peace Research Institute

SIPROS Simultaneous Processing Operation System

SIPS State Implementation Plan System

siq superior internal quality

sir selective information retrieval

sir (SIR) submarine intermediate reactor; supplemental inflatable restraint (automobile air bag)

Sir Siria (Italian, Latin, Spanish—Syria); *Siria* (Portuguese—Syria)

SIR Shuttle Imaging Radar; Society for Individual Responsibility; Society for Individual Rights; Society of Industrial Realtors; Society of Insurance Research; Staten Island Rapid Transit (railroad code); Suspect Information Report

SIR Société Italiana Resine (Italian—Italian Resin Association)

SIRA Scientific Instrument Research Association

Sir Adrian Sir Adrian Boult

Sir Ambrose Sir John Ambrose Fleming (1849–1945)

Sir Charles Sir Charles Hallé who made music in Manchester and overseas

Sir Colin Sir Colin Davis

Sir Georg Sir Georg Solti

SIRC Spares Integrated Reporting and Control System

SIRCS Shipboard Intermediate-Range Combat System

Sir David Sir David Lean

sire space infrared experiment

SIRE Small Investors Real Estate (plan); Society for the Investigation of Recurring Events

Sir Edmund Sir Edmund Hillary

Sir Ernest Sir Ernest Henry Shackleton

Sir Francesco Sir Francesco Paolo Tosti

Sir George Sir George Grove (1820–1900)

Sir Henry Sir William Henry Hadow (1859–1937)

Sir Isaac Sir Isaac Newton (1642–1727)

Sirin Vladimir Nabokov (1899–1977)

SIRIUS Sociopolitical Implications of Road Transport Information Implementation and Use Strategies

Sir Jehudi Sir Jehudi Menuhin

Sir John Sir John Barbirolli the cellist and conductor; Sir John Pritchard

SIRMA Small Independent Record Manufacturers Association

Sir Malcolm Sir Malcolm Sargent

Sir Neville Sir Neville Marriner (1924-)

Sir Noël Sir Noël Coward (1899–1973)

SIRP Salon International de la Recherche Photographique (French—International Photographic Research Show)

Sir Peter Sir Peter Pears (1910–1986); Sir Peter Ustinov

SIRR Spokane International Railroad

Sir Rex Sir Rex Harrison (1908–1990)

Sir Richard Sir Richard Attenborough

SIRS School Information and Research Service; Security Information Retrieval System; Ship-Installed Radiac System; Sorption Information-Retrieval System; Student Information Record System

Sir Stephen Sir Stephen Spender

Sir Steven Sir Steven Runciman

SIRT Staten Island Rapid Transit

SIRTF Spacelab Infrared Telescope Facility

Sir Victor Sir Victor Sawdon Pritchett

Sir William Sir William Schwenck Gilbert

Sir Yehudi Sir Yehudi Menuhin

sis sensory information store; sister; skin immune system; shock insulation support; sterile injectable suspension

sis (Latin suffix—action or process)—dialysis

Sis Cecilia; sister

SIs Sandwich Islands; Service Instructions; Shipping Instructions; Solomon Islands; Survey Instructions

SIS School of information Studies; Secret Intelligence Service(s); Service Improvement Section (Veterans Administration), Shut-In Society; Signal Intelligence Service; Special Industrial Services (UN), Special Isotope Separation; Social Insurance Substitute; Standard Indexing System (DoD); Standards Information Service; Stockholm Information Service; Strategic Intelligence School; Strategic Intelligence Service; Strategic Intelligence Summary; Student Information System; Submarine-Integrated Sonar (system); Swedish Information Service

S & IS Space and Information System(s)

SISAL Società Italiana Sistemi a Lotto (Italian— Italian Lotteries)

SISD Single Instruction, Single Data

SISGAP Scottish Industrial Safety Group Advisory Council

sisi short-increment sensitivity index

Sisister (British contraction— Cirencester)

sis-in-law sister-in-law

sisp sudden increase of solar particles

siss single-item single-source

SISS Semiconductor-Insulation Semiconductor System; Senate Internal Security Subcommittee; Submarine Improved Sonar System; System Integration Support Service

SISS Société Internationale de la Science du Sol (French— International Society of Soil Science)

SISTER Special Institution for Scientific and Technological Education and Research

Sistine pertaining to Pope Sixtus

SISUSA Scotch-Irish Society of the United States of America

sit silicon intensifier target; situation; spontaneous ignition temperature; statement of inventory transaction; stopping in transit

SIT Senate Intelligence Committee; Singapore Improvement Trust; Slosson Intelligence Test; Society of Industrial Technology; Stevens Institute of Technology; Street Interface Transmission; Sugar Industry Technicians

SITA Solomon Islands Tourist Authority; Students International Travel Association

SITA Société Internationale de Telecommunications Aeronautiques (French—International Society of Aeronautic Telecommunications)

SITC Standard International Trade Classification

SITCEN Situation Center (NATO)

sitcom situation comedy (tv)

site shipboard information, training, entertainment

SITE Satellite Instructional Television Experiment; Society of Incentive Travel Executives; Superfund Innovated Technology Evaluation

SITES Smithsonian Institution Traveling Exhibition Service

SITF Shuttle Infrared Telescope Facility

sitol sitological; sitologist; sitology

sitp scheduled into production

SITP Shipyard Installation Test Procedure

sitpro simplification of international trade procedures

sitr silent treatment

SITRA South India Textile Research Association

sitrag situation tragedy

sitrep situation report

SITS Securities Instruction Transmission System

SITS Société Internationale de Transfusion Sanguine (French—International Organization for Blood Transfusion)

SITSUM Situation Summary (NATO Intelligence)

sitt sitting room

SITT System Integration of Triad Technology

SITU Society for the Investigation of the Unexplained

sitv (SITV) system-integration test vehicle

SIU Seafarers International Union; Southern Illinois University; Special Investigating Unit (NY Police Bureau of Narcotics)

SIU Société Internationale d'Urologie (French—International Urological Society)

SIUE Southern Illinois University at Edwardsville

SIUL Southern Illinois University Library

SIUM Southern Illinois University Museum

SI unit Système International unit (French—International System of Units)

SIUP Southern Illinois University Press

siv survey of interpersonal values

siv (SIV) simian inmune virus

si vir. perm. si vires permitant (Latin—if the strength will permit)

siw (SIN) self-inflicted wounds

Six-Day Six-Day War (between Israel, Egypt and Syria)—June 5 to 10, 1967

SIXFLT Sixth Fleet (USN)

six-pac six-pack
SIXPAC System for Inertial Experiment Pointing to Attitude Control
SIY Shropshire Imperial Yeomanry; Staffordshire Imperial Yeomanry; Sussex Imperial Yeomanry
SIYC Shelter Island Yacht Club; Staten Island Yacht Club
SIZ Security Identification Zone
SIZS Staten Island Zoological Society
sj slip joint; subject(s)
s.j. subjudice (Latin—under judicial consideration)
SJ San José; San Juan; Society of Jesus (S.J.—Jesuits); *Statens Järnvägar* (Swedish State Railways); Svalbard and Jan Mayen islands (Internet code)
S-J Stevens-Johnson (syndrome)
SJ Solicitor's Journal
SJA Staff Judge Advocate
SJAA St John Ambulance Association
SJAC Society of Japanese Aircraft Constructors
SJB Sluzba Javne Bezbednosti (Serbo-Croat—State Security Service)—Yugoslav
SJC San Jose, California (airport); San Juan Carriers; Snead Junior College; Spartanburg Junior College
SJCC San Jose City College
S.J.D. Scientiae Juridicae Doctor (Latin—Doctor of Juridical Science)
sjd silicon junction diode
sje swivelling jet engine
sJf single Jewish female
Sjf Sandefjord
Sjf Sjofartsverket (Swedish—Shipping Inspection Bureau)
SJI Steel Joist Institute
SJIs San Juan Islands
SJJC Sheldon Jackson Junior College
SJLA Studies in Judaism in Late Antiquity
sJm single Jewish male
SJMO Smithsonian Jazz Masterworks Orchestra
S Jn San Juan
SJO San José, Costa Rica (La Sabana Airport)
SJPC South Jersey Port Cornmmssmon

SJPL San Jose Public Library
S-J-R Shinawora-Jones-Reinhart (units)
SJSC San Jose State College
SJSO San Jose Symphony Orchestra
SJT Scottish Journal of Theology
SJU San Juan, Puerto Rico (airport); St John's University
sk sick; skein; sketch; skip (knitting instruction); skip (punched card)
sk (SK) streptokinase
Sk Skizze (German—sketch)
SK end of transmission (telegraphic symbol); Saskatchewan; Sccret Key; South Korea(n)
S-K Sloan-Kettering
SK Stuttgarter Kammerorchester (German—Stuttgart Chamber Orchestra); *Suomen Kansallisoopera* (Finnish—Finnish National Opera)
SK-37 Saab Thunderbolt or Vigen multimission combat aircraft also known as AJ-37, JA-37, and S-37
SK-60 Saab attack-type jet aircraft design based on the A—60
s-ka spolka (Polish—association, company)
SKA Switchblade Knife Act
skachet skinning knife, hammer, hatchet and hunting knife all-purpose utility tool
Skag Cape Skagen or The Skaw; Skagway, Alaska
Skager Skagerrak (North Sea between Denmark and Norway)
skamp station keeping and mobile platform
Skaw Cape Skagen or The Skaw-northernmost Denmark
skb skindbind (Dano-Norwegian—leatherbound)
SKBF Svensk Kärnbränsleförsörjning (Swedish Nuclear Fuel Supply Company)
skc sky clear
SKC Scottish Kennel Club
SKCC Sloan-Kettering Cancer Center
SkCsr Státní knikhovna Ceské socialistické republiky (Czechoslovakian—State Library of the Czech Socialist Republic)—communist-era Prague

skd skilled; station-keeping distance
skdn shakedown
sked schedule
skedcon schedule conference
skel skeletal; skeleton
S Ken South Kensington
skep skeptic(al)(ly); skepticism
SKF Svenska Kullagerfabriken (Swedish—ball-bearing factory)
SK & F Smith Kline & French
ski skin
SKI Sloan-Kettering Institute
skil science keyboard input language
skill. satellite kill; skin, kidneys, intestines, liver, lungs
skinmag magazine featuring nudes of both sexes
Skins Skinheads (neo-Nazi organization)
SKIP Skimmer Investigation Platform
skiv skiver
SKJ Savez Komunista Jugoslavije (Yugoslavian Communist League)—political party
SKKCA Supreme Knight of the Knights of Columbus of America
skl skylight(ing); spleen, kidney, liver
Skm Stockholm
SKM Süleymaniye Kütüphahesi (Turkish—Suleiman Mosque Library)—Istanbul
skmr (SKMR) hydroskimmer
S^knoll Seaknoll
Skop Skopje (capital of Montenegro)
skort short skirt
skp station-keeping position
skp (SKP) skip
SKP Suomen Kommunistinen Puolue (Finnish—Communist Party)
skpo slip one, knit one, pass slipped-stitch over (knitting)
SKQ Sexual Knowledge Questionnaire
skr standardized kill rate; station-keeping radar
skr sanskrit (German—Sanskrit)
Skr Sanskrit; Saturn kilometric radiation; Skipper; Skire (Thursday).
Skr Skrifter (Swedish—publication)
SKr Swedish krona (kronor)
SKR South Korea Republic

sks sacks
SKS Soren Kierkegaard Society; station-keeping ship
SKS Savvezna Komisija za Standardizaca (Serbo-Croatian—Federal Commission for Standardization)
Skt Sanskrit
Skt Sankt (German—saint)
SKU(s) stock-keeping unit(s)
Sky Skywest Airlines
SKY Skyways Limited (aviation symbol)
Skydome Skydome Stadium, Toronto
skyjack skyjacked; skyjacker; skyjacking (aircraft hijacking)
Skynet 4B British military communications satellite
skys'l skysail
Sky & Tel Sky & Telescope
sl liability; safety lighting; sales letter; sand-loaded; sea level; searchlight; security light; shipowner's; slang; slightly; slip (knitting instruction); sold; son-in-law; sound locator; stock length; support line
sl (SL) sprinkler leakage; standard label; straight line
s-l short-long (signals); sound-locator sublease
s/l self-loading
s & l sale and leaseback; savings and loan; signed & limited; supply and logistics
s.l. secundum legem (Latin—according to law); *sensu lato* (Latin—in the broad sense); *sine loco* (Latin—no place of publication)
s/l sobreloja (Portuguese—mezzanine floor); *su letra* (Spanish—your letter)
s/L sur Loire (French—on the Loire)
Sl Slovak; Slovakian; small diurnal range
SL San Luis Obispo; Sandia National Laboratories; Savings and Loan; Sea-Land (America's seagoing motor carrier); Session Law; Sierra Leone; Solicitor-at-Law; Squadron Leader; Statute Law; Sub-Lieutenant; Support Line; Sydney & Louisburg (railroad)
S-L Sea-Land (Line); shortlong
S & L Savings and Loan; Supply and Logistics

SL Schweizerische Landesbibliothek (German—Swiss State Library)—in Bern; *Sierra Leone* (Spanish—Sierre Leone)
sla sacro-laeva anterior; single-line approach
SLA Sandia Laboratories—Albuquerque; School Library Association; Scottish Library Association; Showmen's League of America, Sleep-Learning Association; Southeastern Library Association, South Lebanon Army; Southwestern Library Association; Special Libraries Association; Standard Life Association; State Liquor Authority; Supply Loading Airfield; Supply Loading Airport; Symbionese Liberation Army
SLAA Surf Lifesaving Association of Australia
SLAB Students for Labelling Alcoholic Beverages
SLAC Stanford Linear Acceleration Center; Student Labor Action Coalition
SLAD Society of London Art Dealers
SLADE Society of Lithographic Artists, Designers, Engravers, and Process Workers
slado system library activity dynamic optimiser
SLAET Society of Licensed Aircraft Engineers and Technologists
slam ship-launched anti-aircraft missile; surface-launched air missile
slam (SLAM) scanning-laser acoustic microscope; supersonic low-altitude (nuclear-powered) missile
s.l.a.m. sine loco, anno, nomine (Latin—without place, year, or name) ,
SLAM Society's League Against Molestation (of children)
SLAMC Sri Lanka Army Medical Corps
SLANG Systems Language
slanguage slang language (according to Carl Sandberg language which takes off its coat, spits on its hands-and goes to work); slum language

S Lan R South Lancashire Regiment
SLANT Student League Against Narcotic Traffic
SLAP Student Loan Abuse Prevention
SLAPS Serious Literary, Artistic, Political or Scientific (value); Spatial Variability of Land Surface Processes
slar side-looking airborne radar
slarg slargando (Italian—broadening of the tempo)
SLASC Sri Lanka Army Service Corps
slat surface-launched aerial target
S Lat south latitude
slate small lightweight altitude-transmission equipment
SLATE Structured Learning and Teaching Environment; Systems for Learning by Applications of Technology to Education
SLATS Safe, Loft, and Truck Squad (of a police department)
slay slavonisch (German—Slavic
Slav Slavic; Slavonic
slax slacks
slb short-leg brace
slbm sea-launched ballistic missile
slbm (SLBM) submarine-launched ballistic missile
slc searchlight control; shift left and count (instructions); straight-line capacity; straight line capacitance
sl & c shipper's load and count
SLC Salt Lake City, Utah (airport); Scout Launch Complex; Space Launch Complex; Stanford Linear Collider (pronounced slick)
SLCL Sierra Leone Council of Labour
SLCMP Sri Lanka Corps of Military Police
SLCMD St Louis Contract Management District
SLCPL Salt Lake City Public Library
SLCR Scottish Land Court Reports
SLCS Sea Level Canal Study
slc-sat (SLC-SAT) submarine laser communication satellite

sld sailed; sea-landing division; solid; specific learning disability

sld (SLD) serum lactate dehydrogenase

Sld Sunderland

SLD Special Low Dispersion

sldf solidification

sl di slot die

S Ldr Squadron Leader

sld's specific learning disabilities

SLDVS Scanning Laser Doppler Vortex System

sle (SLE) systemic lupus erythematosus

S le Sierra Leone leone(s)—monetary unit(s)

SLe St Louis encephalitis

SLE Society of Logistics Engineers; Sri Lanka Engineers

SLEAT Society of Laundry Engineers and Allied Trades

SLED State Law Enforcement Division (South Carolina)

Sleepers Sleeper Islands in Hudson Bay just north of the Belchers

Sleeping Bear Sleeping Bear Dunes national lakeshore in Michigan

SLEME Sri Lanka Electrical and Mechanical Engineers

slent slentando (Italian—slackening of time tempo)

SLEP Service Life Extension Program (USN)

s.l. et a. sine loco et anno (Latin—without place and year)

S level scholarship level

slew static load error washout

slf straight-line frequency; symmetric filter

SLF Scottish Landowners' Federation; Silcock and Lever Feeds

SLF Skandinaviska Lacktknickers Förbund (Federation of Scandinavian Paint and Varnish Technicians)

SLFCS Survivable Low Frequency Communications System

S-L Fl short-long flashing (light)

SLFP Sri Lanka Freedom Party

slg satellite landing ground; slugging average (baseball); state or local government

SLGB Society of Local Government Barristers

SLGLW St Lawrence and Great Lakes Waterway

slh severe legislative hypocrisy

SLHC St Luke's Hospital Center

sli specific language impairment; suppressed-length indication

Sli Sligo

SLI Shropshire Light Infantry; Slick Airways

slic selective listing in combination

SLIC Sober Live-In Center (for drug-troubled teenagers); Supreme Life Insurance Company

SLICE Southwestern Library Interstate Cooperative Endeavor; Surrey Library Interactive Circulation Experiment

slickums sea-launched cruise missiles (SLCMs)

slid scanning light-intensity device

SLID Student League for Industrial Democracy

Slide Slide Mountain (highest in the Catskills)

slim surface-launched interceptor missile

slim (SLIM) submarine-launched inertial missile

SLiM Selected Library in Microfiche

SLIM South London Industrial Mission; Spanish Language Immersion (teaching program)

slip symmetric(al) list processor

SLIP Serial Line Internet Protocol; Skills Level Improvement Plan

slithy lithe and slimy (Lewis Carroll's portmanteau word from *Through the Looking Glass*)

SLJ School Library Journal

SLKP Supreme Lodge of the Knights of Pythias

SLL Socialist Labour League

SLLA Scottish Ladies Lacrosse Association; Sri Lanka Library Association

slld specific language and learning disability

slm single-level masking

slm (SLM) ship-launched missile

slm sul livello del mare (Italian—at sea level)

SLMA Southeastern Lumber Manufacturers Association; Student Loan Marketing Association (Sallie Mae)

SLMC Scottish Ladies' Mountaineering Club

SLMD(RA) Searchlight Militia Depot (Royal Artillery)

slmm submarine launched mobile mine

slo-mo slow-motion playback (television)

SLMR Speed Limit on Motorways Regulations

slms selective level measuring set

SLMSC South London Medical Staff Corps

SLMSU Scientific Library of Moscow State University

SLMTA St Louis Municipal Theatre Association

sln standard library number

slnd sans lieu ne date (French—without place or date of publication)

s.l.n.d. sine loco nec data (Latin—without indication of date)

SLNM Statue of Liberty National Monument

SLNSW State Library of New South Wales (Sydney)

SLNWR Sand Lake National Wildlife Refuge (South Dakota); San Luis NWR (California); Swan Lake NWR (Missouri)

slo searchlight operator; stop-limit order; stop-loss order

Slo Saltillo (inhabitants—Saltilleños or Saltilleros); Slovak; Slovakia; Slovenia (whose capital is Ljubljana); Slovene(s)

SLO San Luis Obispo; Senior Liaison Officer

SLOA Steam Locomotive Operators Association

slob satellite low-orbit bombardment

Slob Sloboda (Russian—big village, suburb)

SLOBB Stop Littering Our Bays and Beaches

sloc sea lanes of communication

SLOC Source Lines of Code

SLOE Special List of Equipment

slomar space logistics, maintenance, and rescue

slo-mo slow-motion playback (television)

s'long so long (from the Arabic *salaam* or the Hebrew *shalom,* both meaning *peace be with you*)

s'loon saloon

SLORC State Law and Order Restoration Council (Myanmar)

SLOS Stabilized Long-Range Observation System

s/loss salvage loss

Slot The Slot—San Francisco's downtown Mission Street off Market Street

slotperson head copy reader on a newspaper

Slov Slovakia (inaugurated January 1, 1993); Slovene; Slovenian

Slov Phil Slovenian Philharmonic

SLOWPOKE Safe Low-Power Critical Experiment (AEC)

slp sacro-laeva posterior; sea level pressure; super long play

s.l.p. *sine legitima prole* (Latin—without legitimate issue)

SLP San Luís Potosí; Scottish Labour Party; Socialist Labor Party

Slphr Sulphur

SLPL St Louis Public Library

slr self-loading rifle; side-looking radar; single-lens reflex (camera)

slr (SLR) storage limits register

s-l r sea-level resident(s)

SLR Sierra Leone Regiment; State Liaison Representative

S & LR Sydney and Louisburg Railway

SLR *Scottish Land Reports*

SLRA Sierra Leone Royal Artillery

SLRB State Labor Relations Board

SLRC San Luis Rey College

sl rd searchlight radar

SL Rev *Scottish Law Review*

slrp strategic long-range planning

SLRP Society for Long-Range Planning

SLRP *St Lawrence River Pilot*

slrr sideways-looking reconnaissance radar

SLRs *Scottish Land Reports*

SLRV South London Regiment of Volunteers

sls sequential light switch; side lobe suppression; slide set

S & Ls Savings and Loan banks

SLS School of Library Science; School of Library Service; School of Library Studies; Sea-Land Service; Stephenson Locomotive Society; St Lawrence Seaway; St Louis Symphony; Supplemental Loans for Students

sl sa slotting saw

SLSA Saint Lawrence Seaway Authority; Surf Life Saving Association

SLSC Sri Lanka Signal Corps; Surf Life Saving Club; Swedish Lloyd Steamship Company

SLSDC Saint Lawrence Seaway Development Corporation

sls2e surface(d) long side and two edges

SLSENY School Librarians of Southeastern New York

SLSF St Louis-San Francisco (railroad)

SLSFC Severe Local Storm Forecast Center

slsi super-large-scale integration

slsmgr salesmanager

slsmn salesman; salesmen

sl st slip stitch

SLST Sierra Leone Selection Trust

s-l stil spring-loaded stiletto

sl stoplt sell stop limit

sl st(s) slip stitcher(es)—knitting

SLSU Sea Land Service (container) Unit

SLS-UBC School of Library Science—University of British Columbia

slt sacro-laeva transversa; searchlight

sl & t shipper's load and tally

SLT Solid-Logic Technology; Stress Limit Test(ing)

SLT *Scots Law Times*

SLTA Scottish Licensed Trade Association; Soltau-Lüneburg Training Area

SLTAN *Società Lloyd Triestino per Azioni di Navigazione* (Lloyd Triestino)

SLTC Society of Leather Trades Chemists

slto sea-level takeoff

sl tr silent treatment

slu special liaison unit

Slu slough

SLU Saint Lawrence University; Saint Louis University, Southeastern Louisiana University; Southern Labor Union

slug superconducting low-inductance undulatory galvanometer

slumlord slum landlord

slumpfla slumpflation (high inflation coupled with high unemployment)

slumpflation slump + inflation (economic decline coincident with rising inflation)

slurb slum suburb; suburban slum

slurp. self-levelling unit to remove pollution

SLUs Special Liaison Units

SLUSSR State Library of the USSR (Lenin Library, communist-era Moscow)

slutt surface-launched underwater transponder target

slv satellite launching vehicle; space launch vehicle; standard launch vehicle (SLV)

SLV-3 Atlas standard launch vehicle (Convair)

slw straight-line wavelength

slwt side loading warping tug (naval symbol)

sly. safety, liquidity, yield; slowly

Sly southerly

slyp short-leaf yellow pine

SLZG St Louis Zoological Gardens

sm service member; service module; servomechanism; sheet metal; small; standard missile; state militia; statute mile; strategic missile (SM); stepmother; streptomycin; sustained medication; systolic murmur; syzygy mathematical

sm (SM) secondary memory

s-m sadist-masochist; sado-masochism

s/m sensory-to-motor (ratio)

s & m sadism and masochism; sausages and mashed potatoes; surface and matched stock and machinery; supply and maintenance

s/M *sur mer* (French—by the sea); *sur Marne* (on the Marne River, France); *sur Maroni* (on the Maroni River, French Guiana); *sur Meurthe* (on the Meurthe River, France); *sur Moselle* (on the Moselle River, France)

Sm anti-Smith (lupus erythematosus test); samarium

Sm Seemeil (German—nautical mile)

SM mine-laying submarine; Salvage Mechanic; San Marino; Santa Monica; Scientific Memorandum; Senior Magistrate; Sergeant-Major; Service Module; Shipment Memorandum; Signalman; Silver Medal; Society of Mary; Society of Medalists; Soldier's Medal; Special Memorandum; Spiritual Mobilization; Staff Memorandum: State Militia; State Monument; States Marine (steamship lines); Structures Memorandum, submarine; Submarine Miners; Summary Memorandum; *Suomi Merivorma* (Finnish Seapower); Supply Manual; Surgeon-Major; Svenska Metall-verken (Swedish Metal Works)

S-M Seine-Maritime (formerly Seine-Inférieure)

S.M. *Scientiae Magister* (Latin—Master of Science)

S. M. *Sanctae Memoriae* (Latin—of sacred memory); *Su Majestad* (Spanish—Her/His Majesty)

SM-4 Polish three-place helicopter

SM-65 Atlas intercontinental ballistic missile (Convair)

SM-68 Titan intercontinental ballistic missile (Martin)

SM-75 Thor intermediate-range ballistic missile (Douglas)

SM-78 Jupiter intermediate-range ballistic missile (Chrysler)

SM-80 Minuteman intercontinental ballistic missile (Boeing)

sma small-motion accelerometer; special miscellaneous account; subject matter area; supplementary motor area

SMA Safe Manufacturers Association; San Miguel Arizona (railroad); Santa María, Azores (airport); Scale Manufacturers Association; Screen Manufacturers Association; Senior Military Attaché; Service Merchandisers of America; Sheffield Metallurgical Association; Socialist Medical Association; Society of Makeup Artists; Society of Management Accountants; Society of Motor Auctions; Solder Makers Association; Squadron Maintenance Area; Steatite Manufacturers Association; Steel Manufacturers Association; Stoker Manufacturers Association; Stucco Manufacturers Association

SMAA Submarine Movement Advisory Authority

SMAB Solid Motor Assembly Building

SMAC Scientific Machine Automation Corporation

SM & ACCNA Sheet Metal and Air Conditioning Contractors National Association

s mach sounding machine

SMACNA Sheet Metal and Air Conditioning Contractors National Association

SMAE Society of Model Aeronautical Engineers

SMAJ Sugar Manufacturers' Association of Jamaica

smalgol small computer algorithmic language

SMAMA Sacramento Air Materiel Area

smap surprised middle-aged person

S Mar San Marino

SMARA Surface Mining and Reclamation Act of 1975

smarea (SMAREA) Squadron maintenance area

smart special methods for attacking the right targets

SMART Scheduled Maintenance At Regular Times; School Management Appraisal and Rating Technique; Silent Majority Against Revolutionary Tactics; Software for Market Analysis and Restriction on Trade; Stop Marketing Alcohol on Radio and Television; Supersonic Military Air Research Track; Su-

personic Missile and Rocket Track; System for the Mechanical Analysis and Retrieval of Text

smartie simple-minded artificial intelligence

SMASH Students Mobilizing on Auto Safety Hazards

smashex search for simulated submarine casualty exercise

s-m-a showing suggested-for-mature-adult showing - (motion picture producers code)

smat see me about this

smatv (SMATV) satellite master antenna television

smaw shoulder-launched multi-purpose automatic weapon

smaze smoke + haze *(see smog)*

smb server message block

SMB Service Maintenance Bureau; Straits of Mackinac Bridge

SMB *Sa Majesté Britanique* (French—Her/His Britannic Majesty)

SMBA Scottish Marine Biological Association

smbl semimobile

SMBO Small and Minority Business Office

SMBW Society of Mineral and Battery Works

smc sheet-molding compound; small magellanic cloud; sperm (spore) mother cell; standard mean chord; standard military course; sub-machine carbine; super multi-coated

Smc Samic (Lapp)

SMC Saugus Marine Corporation; Scientific Manpower Commission; Smiety of Marine Consultants; State Medical Society; Strategic Military Council

S & MC Supply and Maintenance Command (US Army)

smca suckling-mouse cataract agent

sm caps small capital letters

SMCC Saint Mary's College of California; Santa Monica City College

SMCCL Society of Municipal and County Chief Librarians

SMCL Southeastern Massachusetts Cooperating Libraries

smcln semicolon

SMCRC Southern Motor Carriers Rate Conference

smd storage module device; submanubrial dullness; surface-mounted device

smd (SMD) senile macular degeneration

SMD Submarine Mine Depot

SMDA Sewing Machine Dealers' Association

SMDC Saint Mary's Dominican College

SME School of Military Engineering; Small- and Medium-Sized Enterprise; Society of Manufacturing Engineers; Society of Mining Engineers; Special Minister of the Eucharist; Standard Medical Examination

S.M.E. Sancta Mater Ecclesiae (Latin—Holy Mother Church)

SMEAR Span/Mission Evaluation Action Request

SMEC Snowy Mountains Engineering Corporation; Strategic Missile Evaluation Committee

SMEG Spring Makers' Export Group

smel single and multiengine license

smelerience smell(ing) experience

smelt. srnelter; smelting

SMEMA Surface Mount Equipment Manufacturers Association

sm-er (SM-ER) surface missile—extended range

s/mer sur mer (French—by the sea)

SMERC San Mateo Educational Resource Center

SMERSH Smert Shpionam (Russian—Death to Spies)—Soviet organization for murdering political enemies

smes superconducting magnetic energy storage

S Met O Senior Meteorological Officer

SMF Sacramento, California (airport); Shaker Museum Foundation; Snell Memorial Foundation; South Moluccan Force; System Management Facility

SMfVL Stuttgart Museum för Volker and Landerkunde (German-Stuttgart Museum for National and Regional Studies)

smg speed made good; submachine gun

Smg Samarang

SMG Sterling Machine-Gun

SMG Stato Maggior Generale (Italian—General Staff)

SMGB State Mining and Geology Board

SMGO Senior Military Government Officer

SMH Sydney Morning Herald

SMHEA Snowy Mountains Hydro-Electric Authority

smi standard measuring instrument

s mi statute mile(s)

SmI Solidaritet med Israel (Dano-Norwegian—Solidarity with Israel)

SMI Scale Manufacturers Institute; School Management Institute; Secondary Metal Institute; Sergeant-Major Instructor; Shippers Management International; Spring Manufacturers Institute; Success Motivation Institute; Super Market Institute

SMI Sa Majesté Imperiale (French—Her/His Imperial Majesty)

SMIA Sheet Metal Industries Association

SMIAC Soil Mechanics Information Analysis Center (Corps of Engineers)

SMIC Study of Man's Impact on Climate

smicbm (SMICBM) semi-mobile intercontinental ballistic missile

smice smoke + ice (ice-crystal-laden fog)

SMIG Sergeant-Major Instructor of Gunnery

SMILE Something Meaningful in Local Effort (predelinquency file kept in Orange County, California); Space Migration, Intelligence, and Life Extension (achieved by settling on other planets)

S-mine shrapnel-filled mine

SMIPP Sheet Metal Industry Promotion Plan

SMIS Society for Management Information Systems

smist smoke + mist

smit spin-motor interruption technique

SMIT Sherman Mental Impairment Test

SMITES State-Municipal Income-Tax Evaluation System

Smith Coll Smith College

Smith Coll Lib Smith College Library

Smith Inst Smithsonian Institution

Smithsonian Smithsonian Institution (United States National Museum)

Smitty Smith

SMJ Southern Masonic Jurisdiction

SMJAB State Medical Journal Advertising Bureau

SMJC Saint Mary's Junior College

smk smoke

Smk Shimonoseki

smk gen smoke generator

smkls smokeless

smkstk(s) smokestack(s)

sml simulate; simulation; simulator; small; symbolic machine language

sml sammenlign (Danish—compare)

Sml Samuel

SML Science Museum Library; Security Market Line; States Marine Lines

SMLA Samoa Muamua Le Atua (Samoan—In Samoa God Is First)

SMLC Save Mono Lake Committee

SMLE short-model Lee Enfield (British service rifle used in both world wars)

sml grdn small garden

smlm simple-minded learning machine

SMLM Soviet Military Liaison Mission

smls seamless

SMLS Saint Mary of the Lake Seminary, SciMed Life Systems Incorporated; Seborne Mobile Logistic System

smm standard method of measurement

smm (SMM) solar maximum mission

SMM Science Museum of Minnesota; Security Management Module; Solar Maximum Mission

S.M.M. *Sancta Mater Maria* (Latin—Holy Mother Mary)

SMMA Small Motor Manufacturers Association

s:m::m:b: soybean is to milk as margarine is to butter

SMMB Scottish Milk Marketing Board

smmc system maintenance monitor console

smmp screw machine metal part

smmr (SMMR) surface missile—medium range

SMMT Society of Motor Manufacturers and Traders

smmw submillimeter wave

SMN Société Maritime Nationale (French—National Maritime Society)

SMNA Safe Manufacturers National Association

SMNH Saskatchewan Museum of Natural History

SMNO Singapore Malays National Organization

SMNP Simien Mountains National Park (Ethiopia)

SMNRA Shadow Mountain National Recreation Area (Colorado)

Smnry Seminary

SMNWR Saint Marks National Wildlife Refuge (Florida)

SMO Senior Medical Officer; Squadron Medical Officer

SMO Servicio Militar Obligatorio (Spanish—Compulsory Military Service)

SMOA Ships Material Office—Atlantic

smog smoke + fog (*see* smaze); smoky air (with or without fog)

smogway smog-polluted automobile freeway

SMOH Society of Medical Officers of Health

smoker smoking car

smoketaz smoke-tinted topaz

Smokies Smoky Mountains between North Carolina and Tennessee

smokin' smoking

SMOM Sovereign Military Order of Malta (claiming to be the world's smallest country, founded in 1048 before the first crusade and located at 68 Via Conditti in downtown Rome, where in 1981 it reported a population of 80)

smon subacute myelo-optic neuropathy

SMOOSA Save Maine's Only Official State Animal (the moose)

smoothies smooth ones

SMOP Ships Material Office-Pacific

SMOPS School of Maritime Operations

smor standard mean ocean water

S'more Swarthmore

smörgas smörgasbord (Swedish appetizers or delicatessen-style meal)

smorz smorzando (Italian—dying away)

smotherlove smothering mother love

smow standard mean ocean water

smp scanning measuring projector; social marginal productivity; sound motion picture(s)

smp (SMP) special multi-peril (insurance) policy

s.m.p. sine mascula prole (Latin—without male issue)

SMP Science Manpower Project; School Mathematics Program; Society of Mural Painters; St Martin's Press

SMP Soviet Military Power (published by the Pentagon); *Suomen Masseudun Puloue* (Finnish Rural Party)

SMPC Saint Mary of the Plains College

SMPR Supply and Maintenance Plan and Report

smps switched-mode power supply

SMPS Society for Marketing Professional Services; Society of Master Printers of Scotland; Special Mobile Provost Section

SMPSD Systematic Management Plan for School Discipline

SMPTE Society of Motion Picture and Television Engineers

smpx smallpox

smr senior maintenance rating; somnolent metabolic rate; specialized mobile radio; standard mortality rate; submucous resection

sMr (SMR) standard Malaysian rubber

SMR Student Master Record; Source, Maintenance and Recoverability; South Manchurian Railway

SMR Sa Majesté Royale (French—Her/His Royal Majesty)

SMRA Spring Manufacturers' Research Association

SMRC South Manchurian Railway Company

smrd spin-motor rotation-detector

SMRE Safety in Mines Research Establishment

SMRI Solution Mining Councils of America; Sugar Milling Research Institute

SMRL Submarine Medical Research Laboratory

SMRMIS Supply, Maintenance, and Readiness Management information System

SMRV South Middlesex Rifle Volunteers

sms silico-manganese steel; subject matter specialist; synchronous meteorological satellite (SMS)

sm's (SMs) submarines

SMs subway musicians

SMS Sacramento Medical Society; School of Military Survey; Sequence Milestone System; Software Monitoring System

SMS Seine Majistäts Schiffe (German—His Majesty's Ship)

smsa standard metropolitan statistical area

SMSA Standard Metropolitan Statistical Area (any city and surrounding suburbs with a population of 50,000 or more)

SMSB Strategic Missile Support Base

SMSG School Mathematics Study Group

SMSgt Senior Master Sergeant

SMSO Senior Maintenance Staff Officer

SMSP Spring Mill State Park (Indiana)

SMSSS Sheet Metal Screw Statistical Society

smstrs seamstress

smt ship's mean time; surface mount technology

Smt Summit

S^{mt} Seamount

SMT Scottish Motor Traction; Shipboard Marriage Test; Stabilized March Technique; System Maintenance Test

SMTA Scottish Motor Trade Association; Surface Mount Technology Association

Smte Subcommittee

SMTF Scottish Milk Trade Federation

smti selective moving target indicator

SMTO Senior Mechanical Transport officer

SMTP Simple Mail Transfer Protocol

SMTRB Ship and Marine Technology Requirements Board

SMTS Scottish Machinery Testing Station

SMTWTFS Sunday, Monday, Tuesday, Wednesday, Thursday, Friday, Saturday

SMU Southern Methodist University

SMUD Sacramento Municipal Utility Department; Sacramento Municipal Utility District

Smu Gul Smuggler's Gulch (Monument Road, San Diego, California—the last road in the southwestern United States)

SMUN Soviet Mission to the United Nations

SMUP Southern Methodist University Press

SMUSE Socialist Movement for the United States of Europe

smust smoke + dust

S. mutans Streptococcus mutans

smutcom smut communication

smw standard metal window

SMW Society of Magazine Writers; Society of Military Widows

SMWIA Sheet Metal Workers International Association

SMWIU Steel and Metal Workers Industrial Union

smx serial microxerography; submultiplexer unit

smx (SMX) sulphamethoxazole

sn sanitation; sanitary; semiconductor network; service number; solid neutral; stock number; supernovae

sn (Sn) snow

s/n serial number; service number; signal-to-noise ratio

s-n *sin número* (Spanish—unnumbered, without number)

s.n. *secundum naturam* (Latin—according to nature); *sine nomine* (Latin—without name)

Sn San; Santa, Santo; *stannum* (Latin—tin)

S_n labor supply (macroeconomics)

S^n San (Spanish—saint)

SN Sacramento Northern (railroad); Scientific Note; Scope Note; Secretary of the Navy; Senegal (Internet code); Sergeant-Navigator; Serial Number; Service Number; Sierra Nevada; Standard Oil (stock exchange symbol)

S/N Serial Number; Service Number; Shipping Note; stress versus number of cycles (to failure); successes versus total number of trials

S of N Sons of Norway

SN *Sûreté Nationale* (French —National Security)— law-enforcement agency

S-N stress versus number of cycles

sna systems network architecture

SNA Society of Naval Architects; Suburban Newspapers of America; System of National Accounts (UN); Systems Network Architecture

SNAC *Syndicat National des Auteurs et Compositeurs* (French—National Union of Authors and Composers)

SNACS Share News on Automatic Coding Systems; Society for the North American Cultural Survey

snafu situation normal, all fouled up; situation normal— all fucked up

snag sensitive new age guy

SNAI Standard Nomenclature of Athletic Injuries

SNAM *Societá Nazionale Metanodotti* (Italian—National Natural Gas Company)

SNAME Society of Naval Architects and Marine Engineers

snap simplified numerical automatic processor; simplified numerical automatic

programmer; subroutine(s) for natural actuarial processing

SNAP Shelter Neighborhood Action Project; Society of National Association Publications; Society of National Publications; Space Nuclear Auxiliary Power; Student Naval Aviation Pilot; Suffolk Network on Adolescent Pregnancy; Systems for Nuclear Auxiliary Power

Snapp *Serviços de Navegação da Amazonia e de Administração do Porto do Pará* (Portuguese-Amazon Navigation and Administration of the Port of Pará)

SNAPP Social Nationalist Aryan People's Party

snapper(s) snapping turtle(s)

snappies snappy stories

snap(s) snapshot(s)

snark snake and shark (Lewis Carroll)

snc severe noise environment; standard navigation computer

SNC *Sistema Nervioso Central* (Spanish—Central Nervous System); *Société Navale Caennaise (L'amy et Cie)* (French—Caen Naval Society)

SNCASCO *Société Nationale de Constructions Aéronautique de l'Ouest* (French— National Society of Western Aeronautical Construction)

SNCC Student Nonviolent Coordinating Committee (also called SNIC)

SNCF *Société Nationale de Chemins de Fer* (French— National Railways)

SNCFB *Société Nationale des Chemins de Fer Belges* (Belgian State Railways)

SNCFF *Société Nationale des Chemins de Fer Français* (French—State Railways)

SNCO Senior Non-Commissioned Officer

snd sound

SND Society of Newspaper Design

SNDA Sunday Newspaper Distributing Association

SNDO Standard Nomenclature of Diseases and Operations

sndp   *sin nota de precio* (Spanish—without indication of price)

sn dr   snare drum

sndv (SNDV)   strategic nuclear delivery vehicle

SNE   Society for Nutrition Education

SNEA   Student National Education Association

sneaks   sneakers (tennis shoes)

SNEC   Sub-Group on Nuclear Export Coordination

SNECMA   *Société Nationale d'Etude et de Construction de Moteurs d' Aviation* (French—National Air Motors Research and Construction Company)

SNEMSA   Southern New England Marine Sciences Association

SNEP   Saudi Naval Expansion Program

SNev   Sierra Nevada

snf   skilled nursing facility; solids-non-fat

SNF   Serbian National Federation; Short-range Nuclear Forces; Skilled Nursing Facility; Southern National Front

SNFA   Standing Naval Force, Atlantic

SNFCC   Shippers National Freight Claim Council

SNFU   Scottish National Farmers' Union

sng   synthetic natural gas

sng   *sans notre garantie* (French—without our guarantee)

Sng   Singapore

sngl   single (flow chart)

SNHM   Stanford Natural History Museum

sni   sequence-number indicator

SNI   San Nicolas Island; Selective Notation of Information; Selective Notification of Information; Sports Network Incorporated

SNI   *Secretariado Nacional da Informação* (Portuguese-State Tourist Bureau); *Syndicat National des Instituteurs* (French—National Union of Teachers)

SNIC   Student Non-Violent Coordinating Committee (SNCC)

SNICER   State of Nebraska Information Center for Educational Resources

SNIE   Special National Intelligence Estimate

SNIEs   Special National Intelligence Estimates

snif   standby note issuance facility

sniffex   sniffer exercise

snirt   snort of laughter

SNL   Sandia National Laboratories; Singapore National Library; Standard Nomenclature List

SNL   *Saturday Night Live; Science News Letter*

snlr   services no longer required

SNLS   Society for New Language Study

snlv (SNLV)   strategic nuclear-launch vehicle

snm   shielded nonmetallic (sheathed cable); signal-to-noise merit; special nuclear material(s)

snm   *sobre el nivel del mar* (Spanish—above sea level)

SNM   Saguaro National Monument (Arizona); Senior Naval Member; Sitka National Monument (Alaska); Society of Nuclear Medicine

SNMP   Simple Network Management Protocol

SNMT   Society of Nuclear Medical Technologists

SNN   Shannon, Eire (airport)

sno   snow

s no   serial number

SNO   Scottish National Orchestra; Senior Naval Officer; Senior Navigating Officer; Senior Nursing Officer; Singapore National Orchestra

snob   *sine nobilitate* (Latin—without nobility)—anyone trying to outdo the manners and style of the nobility

SNOB   Senior Naval Officer on Board

snobol   string-oriented symbolic language

snoe   smart noise equipment

snok   secondary next of kin

Snooks   surname contracted from Seven Oaks

SNOOP   Students Naturally Opposed to Outrageous Prying

snoopervise   snoop and supervise

SNPO   Society for Nonprofit Organizations

SNOP   Standard Nomenclature of Pathology

snorkex   snorkel exercise

SNORT   Supersonic Naval Ordnance Research Track

Snowys   Snowy Mountains of New South Wales

Snowy Scheme   Snowy Mountains Scheme (Australian hydroelectric and irrigation system)

snp   single nucleotide polymorphism; soluble nucleoprotein

SNP   Salorp National Park (Thailand); Scottish Nationalist Party; Sebakwe NP (Rhodesia); Sequoia NP (California); Serengeti NP (Tanzania); Shenandoah NP (Virginia); Sivpuri NP (India); Sitka NP (Alaska); Snowdonia NP (Wales); Swiss NP (Switzerland)

SNPA   Scottish Newspaper Proprietors' Association; Southern Newspaper Publishers Association

SNPO   Space Nuclear Propulsion Office

SNPX   SynOptics Communications Incorporated

snr   signal-to-noise ratio

Snr   *Senhor* (Portuguese—Mister)

Sñr   *Señor* (Spanish—Mister)

SNR   Society for Nautical Research

Snra   *Senhora* (Portuguese—Missus)

Sñra   *Señora* (Spanish—Missus)

SNRA   Sanford National Recreation Area (Texas)

snrf   sunroof

snRNAs   small nuclear ribonucleic acids

Snro   *Senhoro* (Portuguese—Mister)

Snrta   *Senhorita* (Portuguese—Miss)

Sñrta   *Senhorita* (Spanish—Miss)

Sñrto   *Senhorito* (Spanish—Master)

sns   sympathetic nervous system

SNS   Senior Nursing Sister

SNSC   Scottish National Ski Council

S'n Simons Saint Simons Island off the coast of Brunswick, Georgia

SNSN Standard Navy Stock Number

SNSO Superintending Naval Stores Officer

SNSP Syrian National Social Party

SNSPS Scottish National Sweet Pea Society

snt sealant

snt *so nota* (Japanese—and so forth)—etc.

Snt Santander

SNT Society for Nondestructive Testing

snto spinning tool

SNTO Spanish National Tourist office; Swedish National Tourist Office; Swiss National Tourist Office

SNTPC Scottish National Town Planning Council

SNUPPS Standardized Nuclear Unit Power Plant System

SNURPs Small Nuclear Ribonucleoproteins

SNVBA Scottish National Vehicle Builders Association

SNVDO Standard Nomenclature of Veterinary Diseases and Operations

SNW Symphony of the New World

SNWMA Stillwater National Wildlife Management Area (Nevada)

SNWR Sabine National Wildlife Refuge (Louisiana); Sacramento NWR (California); Santee NWR (South Carolina); Savannah NWR (South Carolina); Seedskadee NWR (Wyoming); Seney NWR (Michigan); Sherburne NWR (Minnesota); Shiawasse NWR (Michigan); Slade NWR (North Dakota)

snwt steel non-watertight

So Ysdr San Ysidro

so (SO) shift-out character (data processing)

so second opinion; seller's option; senior officer; sex offender; shift out; shipping order; ship's option; shop order; show off; soiled; south(ern); special operations; special order; staff officer; standing order; strikeout; suboffice; supply office(r); survivors' original

s/o standoff; shipping order; solvent-to oil (ratio); son of

s-o shutoff

so. *siehe oben* (German—see above)

s/o *su orden* (Spanish—your order)

So Somali(a); Sunday only

So *Sondag* (Danish—Sunday)

SO Section Officer; Scottish Office; Scottish Opera; Scouting-Observation (naval aircraft); Seattle Orchestra; Secretary's Office; Senior Officer; Signals Officer; Shipment Order; Shipping Order; Shop Order; Somalia (Internet code); somalo (Somalian currency unit); Southern Airways (letter coding); Southern Company (stock exchange symbol); Special Order(s); Sphincter of Oddi; Staff Officer; Standard Oil; Standing Order(s); Stationery Office; Stores Officer; Supply Office(r); Steward Observatory

S(O) Seaman Operator

SO₂ oxygen saturation

SO (I) Staff Officer (Intelligence)

SO (O) Staff Officer (Operations)

S/O Station Officer

SO *Staatsoper* (German—State Opera); *sudoeste* (Spanish—southwest); *Südösten* (German—southeast); *Suroeste* (Spanish—southwest)

SO₂ sulfur dioxide

SO₄ sulfate

soa speed of advance; speed of approach; state of the art; stimulus-onset asynchrony

SOA Society of Actuaries; Seattle Opera Association; Shoe Corporation of America (stock exchange symbol); Staff Officer, Administration

soaa state-of-the-art advancement

SOAA Solus Outdoor Advertising Association

SOAD Staff Officer—Air Defence

So Afr South Africa(n)

soap symbolic optimum assembly programming

SOAP Society of Airway Pioneers; Student Opportunity and Access Program

SOAPD Southern Air Procurement District

soaps suction, oxygen, apparatus, pharmaceuticals, saline (anesthetist's mnemonic for checking equipment)

soap(s) soap opera(s)

SOAR Save Our American Resources; Society of Authors' Representatives

SOAs Separate Operating Agencies

SOAS School of Oriental and African Studies (University of London)

SOASIS Southern Ohio (chapter of) ASIS

sob see order blank; shipped on board; shortness of breath; souls on board (passengers and crew aboard an aircraft); still on board; suboccipito-bregmatic

s-o-b son of a bitch

SOB Senate Office Building; State Office Building; Society of Bookmen; son of a bitch

sobe sober; sobriety

SoBe South Beach (Miami, Florida)

SOBHD Scottish Official Board of Highland Dancing

soblin self-organizing binary-logic network

sob's silly old buggers; sons of bitches; souls on board

SOBs Sons of Bosses

SOBS Society for Office-Based Surgery

soc sector operations center; social; society; sociology; socket; state of consciousness (SoC)

Soc Socialist; Social Security number; Society; Socrates

Soc *Sociedad* (Spanish—society); *Sociedade* (Portuguese—society); *Società* (Italian—society); *Société* (French—society)

S o C Society of Cyprus

SOC Save Our Children; Scottish Ornithologists Club; Servicemen's Opportunity College; Southwestern Oregon College; Space Operations Center (Colorado Springs,

Colorado); Special Operations Command (counterinsurgency forces); Stamp Out Crime

So Ca South Carolina's old abbreviation

SOCAD School of Crafts and Design

SOCAL Standard Oil of California (Chevron)

Soc An Société Anonyme (French—corporation)

SOCAP Society of Consumer Affairs Professionals (in business)

SOCARE Seniors Only Comprehensive Assessment and Retirement Evaluation

Soc. Chr. Societas Christi (Latin—Christian Society)

soc/d social death; sociological death

Soc-Dem Social-Democrat(ic) (Party)

SOCEM Save Our City from Environment Mess; Society of Objectors to Compulsory Egg Marketing

SOCGPA Seed, Oil Cake, and General Produce Association

Soc I Society Islands

Societies Society Islands of Polynesia in the South Pacific

Socinus Laelius Socinus (Lelio Sozzini, Italian theologian and anti-Trinitarian whose nephew Faustus Socinus developed Socinianism, the forerunner of Unitarian-Universalism)

sociobio sociobiologic(al) (1y); sociobiologist; sociobiology

socioecol socioecologic(al) (1y); socioecologist; socioecology

sociol sociological; sociologist; sociology

Soc Isl Society Islands

socks soccer teams

SOCMA Synthetic Organic Chemical Manufacturers Association

SOCMEU Special Operations Capable Marine Expeditionary Unit

Soc Mining Eng Society of Mining Engineers

Soc NC sociedad en nombre colectivo (Spanish—general partnership under a collective name)

So Co Southern Counties

SOCO Standard Oil Company of California

socom solar communication

SOCONY Standard Oil Corporation of New York

soc psych social psychology

Soc Qua Society of Quakers

SOCRATES System for Organizing Content to Review and Teach Educational Subjects

Socred Social Credit (party of Canada)

socrit social critic(ism)

socs survey of clerical skills

soc sci social science; social scientist

Soc Sec Social Security

sod sexual orientation disturbance; sodium; sodomite; sodomy; super oxide dismutase

Sod West Virginia town named after its first postmaster—Samuel Odell Dunlap—SOD

S o D Society of Dilettanti

SOD Some of the Digits; Special Operations Division (CIA)

soda (SODA) source-oriented data acquisition

SODAC Society of Dyers and Colourists

sodar sound-detecting and ranging

soda water water charged with carbon dioxide (CO_2)

SO-DIMM Small Outline Dual Inline Memory Module

SODOMEI Nihon Rodo Kumiai Sodomei (Japanese—Japanese Trade Union Federation)

SODRE Servicio Oficial de Difusión Radio Eléctrica (Spanish—Uruguayan radio and tv network)

soe special operations executive

SoE Secretary of Energy

SOE Senior Officer Escort; Special Operations Executive (World War II British intelligence operation for rescuing scientists from Hitler)

SOED Shorter Oxford English Dictionary

SOEEA Saskatchewan Outdoor and Environmental Education Association

SOE/F SOE in France

SOEKOR Southern Oil Exploration (South Africa)

soep (SOEP) solar-oriented experiment package

SOES Small Order Execution System

sof sound on film; supervisor of flying

sof (SOF) succinic oxidase factor

Sof Sofia

SoF Soldier of Fortune

SOF Special Operations Forces

S o F Society of Friends

S of A School of Artillery

SOFA Socially Oriented For Action; Strongly Oriented For Action; Student Overseas Flights for Americans

SOFAA Society of Fine Art Auctioneers

sofar sound fixing and ranging

sofas (SOFAS) survivable optical forward acquisition sensor

SOFCS Self-Organizing Flight Control System

S of G School of Gunnery

S of I School of Infantry

SOFINA Société Financière de Transports et d'Entreprises Industrielles (French—Belgian investment syndicate)

sofnet solar observing and forecasting network

SOFRATOME Société Francaise d'Études et de Réalisation Nucléaires (French—French Society for Nuclear Study and Realization)

soft signature of fragmented tanks; software

SOFT Status of Forces Treaty; Swedish Orienteering Federation

softech software technology

softlenses soft contact lenses

soft porn soft-core pornography

sog small operations group; speed over (the) ground

sog sogenannt (German—so called)

SOG Seat of Government (Washington, D.C.); Special Operations Group

SOGAT Society of Graphical and Allied Trades

SOGC Society of Gynecologists and Obstetricians of Canada

SO & GC Signal Oil and Gas Company

Sogd. *Sogdian* (extinct Iranian people and language)

sogg *soggettivo* (Italian—subjective); *soggetto* (Italian—subject)

soh (SOH) start of heading character (data processing)

soha soft hard

sohf sense of humor failure

SOHIO Standard Oil of Ohio

soho small office, home office

SoHo South of Houston Street (New York City artist's colony in lower Manhattan)

SOHO Save Our Heritage Organization; Solar and Heliospheric Observatory (Stanford University)

SOHYO *Nihon Rodo Kumiai Sohygikai* (Japanese—Japanese General Council of Trade Unions)

soi silicon on insulation; space object identification

SOI Signal Operation Instruction(s); Southern Indiana (railroad); Specific Operating Instruction(s); Staff Officer—Intelligence

SOI *Statistics of Income* (Internal Revenue Service bulletin)

SoJ Sea of Japan

sojs standoff jammer system

sok *sokak* (Turkish—lane; street)

sol solar; soldier; solenoid; soluble; solubility; solution; solvent(s)

sol (SOL) shipowners liability; simulation-oriented language

s-o-l short of luck

sol. *solutio* (Latin—solution)

Sol Solicitor; Solomon; Solomon Islands

SoL Secretary of Labor; Solicitor of Labor

SOL Slightly Older Lesbians; Systems Optimization Laboratory

SOL *Svenska Orient Line* (Swedish—Swedish Orient Line)

SOLACE Sales Order and Ledger Accounting (using) Computerline Environment

SOLAR Semantically Oriented Lexical Archive; Shop Operations Load Analysis Report(ing)

SoLaS Safety of Life at Sea (international conference)

solb start of line block

sold solder; soldering

Soldier Soldier Field Stadium, Chicago

Sol Gen Solicitor General

sol hgt solid height

sol htg solar heating

solidif solidification

Solid-State Commun *Solid State Communications*

Solid-State Electron *Solid-State Electronics*

Solid-State Phys *Solid-State Physics*

SOLINET Southeastern Library Network

solion solution of ions

Sol Isl Solomon Islands

SOLIT Society of Library and Information Technicians

Sol J *Solicitors' Journal*

SOLL Selma Ottiliana Louisa Lagerlöf

Sol(ly) Solomon

soln solution

solo status of logistics offensive

SOLO System for Ordinary Life Operations

SOLog standardization of certain aspects of operations and logistics

sologs standardization of operations and logistics

solomon simultaneous-operation-linked ordinal modular network

Solomons Solomon Islanders; Solomon Islands (nation in the western Pacific)

Solovetskis Solovetski Islands (penal colonies in the Archangelsk Region of the former USSR—part of the Gulag Archipelago)

Solovki Solovetski Islands

Sol Phys *Solar Physics*

solr solicitor

solrad solar radiation

solut solution

solv solvent

solv. *solve* (Latin—dissolve)

Solv Solveig

soly solubility

som serous otitis media; somatology; start of message

som (SOM) standoff missile

Som Somali(a); Somaliland(er); Somerset; Somerville College, Oxford

SoM School of Musketry

SOM Society of Occupational Medicine; Standing Group on Oil Markets

SOMA Sharing of Ministries Abroad; Society of Mental Awareness

Somal Somali(a)(n)—Somalia formerly British and Italian Somaliland

Somalia Somali Democratic Republic (East African nation)

somat somatic

somat or some (Greek—body) —centrosome; chromosome, somatic

SOME Senior Ordnance Mechanical Engineer

Somerset Somersetshire

SOMEX *Sociedad Mexicana de Credito Industrial* (Spanish—Mexican Industrial Credit Society)

som-h start of message—high precedence

SOMIW Secure Open Multimedia Integrated Workstation

som-l start of message—low precedence

Som Ll Somerset Light Infantry

somm (SOMM) standoff modular missile

somnam somnambulant, somnambulating, somnambulation, somnambulism, somnambulistic(al)(ly), somnambulist(s)

SOMOS Senior Officer Minesweepers; Society of Military Orthopedic Surgeons

Somos libres (Spanish—We are free)—Peruvian national anthem

SOMPA System of Multicultural Pluralistic Assessment

SOMS Senior Officer Minesweeper; Standing-Order Microfiche Service

Som sh Somali shilling

son statement of need

son. sonata

Son Sonora

Son *Sonntag* (German—Sunday)

S o N Spear of the Nation (armed force of the African National Congress)

SON Snijders-Oomen Nonverbal (intelligence scale); Sultan of Oman's Army

sona suggested optimal daily nutritional allowance

sonac sonacelle (sonar nacelle)

SONAP Sociedade Nacional de Petroleos (Portuguese—National Petroleum Company)

sonar sound navigation and ranging

Sonbrit Simfonischen orkestur na bulgarskoto radio i televiziya (Bulgarian Radio and Television Symphony Orchestra)

SONDE Society of Non-Destructive Examination

sonet synchronous optical network

SONGS San Onofre Nuclear Generating Station

Song Sol The Song of Solomon

Sonia Sophia

sonmc sonar countermeasures and deception

Sonn Sonnets of Shakespeare

Sonny George Bernard Shaw (1856–1950)

sono sonobuoy; sonography

sonoan sonic noise analyzer

SONPP San O'nofre Nuclear Power Plant

son(s) sonata(s)

Sons Sonnets

SOO Staff Officer Operations

SO(O) Staff Officer (Operations)

soob sitting out of bed

SOOP Submarine Oceanographic Observation Program

soot solar optical observing telescope

sop sleeping-out pass; soprano; sum of products; surgical outpatient

s-o-p standard operating procedure

SOP Senior Officer Present; Standard Operating Procedure; Study Organization Plan

SOPA Senior Officer Present Afloat

SOPAC Southern Pacific; South Pacific

sop glock soprano glockenspiel

Soph Sophocles

SOPHE Society of Public Health Educators

soph(s) sophomore(s)

SOPL Save Our Public Libraries

SOPLASCO Southern Plastics Company

sop met soprano metallophone

Soppnata Sociedade Portuguese de Navios Tanques (Portuguese Tankers)

sop xyl soprano xylophone

sor sale or return; sequential occupancy rate; sorority; specific operating requirement(s)

s-o-r stimulus-organism-response

Sor Soerabaya; Sorong

Sor Señor (Spanish—Mister)

Sor Sênior (Portuguese—Mister)

SOR Sandia Optical Range; Special Order Request; Specific Operational Requirement

SORB Subsistence Operations Review Board

sord submerged object recovery device

SORD Southeastern Order Retrieval and Distribution Center

SORDID Summary of Reported Defects, Incidents, and Delays

SORE Staff Officer Royal Engineers

Sores Señores (Spanish—gentlemen)

SORG Southern Operations Research Group

SORI Southern Research Institute

Sor Juana Sor Juana Inés de la Cruz (1651-1695)

Sorlings Sorling Islands (Isles of Scilly)

SORO Special Operations Research Office

Sorolla Spanish painter Joaquin Sorolla y Bastida (1863–1923)

SORT Ship's Operational Readiness Test(ing); Slosson Oral Reading Test; Staff Organizations Round Table; Structured-Objective Rorschach Test

sorti satellite orbital track and intercept

sos same old stew; same old stuff, same only softer (musical direction); save our savings; save our seamen; save our streets; save our submarines; save our suburbs; save our supplies; scouting, observation and sniping; shit on a shingle; silicon-on-sapphire; slag on a shingle (military de-

scription of creamed chicken or beef on a slice of toast); struck off strength

s.o.s. si opus sit (Latin—if necessary)

sos sostenuto (Italian—sustained)

SoS Secretary of State; Senior Officer's School; Service of Supply; Source(s) of Supply

S-o-S Southend-on-Sea

S o S Society of Separationists (of church from state)

SOS Song of Solomon

SOs Sheriff's Offices

SOS Safety Observation Station; Save Our School(s); Save Our Shore; Senior Officer's School, Senior Opportunities and Services; Share Our Spectacle(s); Ships Ordnance Summary; Squadron Officer School(ing); Stamp Out Smog; Student Oriented Studies (National Science Foundation); Supervisor of Shipbuilding; Supplementary Ophthalmic Service(s); Support Our Schools

SOS international distress signal—three dots, three dashes, three dots; popularly translated as meaning Save Our Souls

sosc safety observation station display console

SOSC Smithsonian Oceanographic Sorting Center, Source of Supply Code

So sh somali shilling(s)

SOSIAC Singapore-Soviet Shipping Agency

SOSS Shipboard Oceanographic Survey System

sost sostenuto (Italian—sustained)

Sost Sostavitel (Russian—compiler)

SOSTAC Scottish Industrial Safety Training Advisory Council

SOSUS Sound Surveillance Underwater System; Sound and Surveillance System

sot shower over tub; sound on tape

sot (SOT) solar optical telescope (NASA)

Sot Sotah

SoT Secretary of Transport(ation); Secretary of the Treasury; Sons of Temperance

SOT Solar Optical Telescope; Special Operations Team

sota state of the art

SOTA Statewide Organization of Third-world Artists

SOTAA State-of-the-Art Association

SOTAS Stand-Off Target-Acquisition System

sotd stabilized optical tracking device

SOTDAT Source Test Data System (EPA)

SOTFE Special Operations Task Force, Europe

sotim sonic observation of the trajectory and impact of missiles

SOTO Society of Transporter Owners

Soton Southampton

SOTP Ship(yard) Overhaul Test Program; System Overhaul Test(ing) Program

SoTT School of Technical Training

sotus (SOTUS) sequentially-operated teletypewriter universal selector (data processing)

Sou Southampton

SOU Southern Airways

Sou Afr South Africa(n)

Sou Amer South America(n)

Sou Aus South Australia(n)

Sound The Sound (Arctic straits in the Canadian sector such as Lancaster Sound, Smith Sound, Viscount Melville Sound; Long Island Sound; Block Island, Rhode Island, Nantucket, and Vineyard Sounds; North Carolina's Albemarle, Bogue, Currituck, and Pamlico Sounds; Sundet—also called Öresund between Denmark and Sweden)

soundamp sound amplification; sound amplifier

Sound River old name for New York City's East River—an extension of Long Island Sound linking the Sound with New York Bay, the Harlem River, and the Hudson

SOUP Students Opposed to Unfair Practices

Sou Pac South Pacific; Southern Pacific

SOUR Stamp Out Urban Renewal

source the source (nickname for a baton with a rechargeable flashlight on one end and shock terminals on the other, used for jolting criminals with a 10,000-volt charge)

Source of the Sun Japan (called Nihon by the Japanese as it means Source of the Sun and is emblazoned on their flag)

soussa steady, oscillatory, and unsteady, subsonic, and supersonic aerodynamics

s/out sleep out (porch)

South southern American states from Virginia to Texas

South Africa Republic of South Africa (*Republiek van Suid—Afrika* in Afrikaans), formerly Union of South Africa

South African inhabitant of South Africa; Afrikaans language or people; South African Republic now the Transvaal

South African Commonwealth South Africa and its territories

South African Dominion South Africa and its territories

South-African Dutch Afrikaans

South African Ports (large, medium, and small from west to south to east) Walvis Bay, Luderitz, Cape Town, Simontown, Mosselbaai, Port Elizabeth, East London, Port St Johns, Durban

South Africa's Principal Port Cape Town

South Africa's Spine Drakensburg Mountains

South America islands and lands extending from Cape Horn to Colombia (Argentina, Bolivia, Brazil, Chile, Colombia, Ecuador, French Guiana, Guyana, Paraguay, Suriname, Uruguay, Venezuela)

South America's Largest Country Brazil

South Arabia Southern Yemen

South Atlantic ocean between South America and Africa

South Atlantic States Delaware, Florida, Georgia, Maryland, North Carolina, South Carolina, Virginia, and West Virginia

South Britain England and Wales

South Carolina Port Charleston

South Carolina's Capital City Columbia

South Central States Arkansas, Louisiana, Oklahoma, Texas

South China Sea between Indochina, Indonesia, and Philippines

SouthCom Southern Command (U.S. Air Force, Army, and Navy bases straddling the Panama Canal for its protection when under U.S. jurisdiction)

Southeast southeastern United States (North Carolina to Florida, Atlantic Coast to Mississippi River)

South Eastern Region South Eastern Region Correctional Institute at Juneau, Alaska

Southeast Sun Belt Alabama, Arkansas, Florida, Georgia, Louisiana, Mississippi, North Carolina, South Carolina, Tennessee, and Virginia

South End Boston, Massachusetts slum

souther storm from the south

Southern Southern Railway

Southern Alps mountain range on South Island of New Zealand

Southern California California south of the Tehachapis

Southern Colonies Virginia, Maryland, North Carolina, South Carolina, Georgia

Southern Cone Argentina, Chile, Paraguay, Uruguay

Southern Cross outstanding constellation of the Southern Hemisphere where it is emblazoned on the flags of Australia, Brazil, New Zealand, Papua New Guinea, the Solomon Islands, and Western Samoa as well as the state of Victoria in southern Australia

Southerns Southern Alps of New Zealand's South Island

South Ken South Kensington Imperial Institute (London's museum of science and industry)

South Orkneys South Orkney Islands in British Antarctica

South Sandwiches South Sandwich Islands

South Shetlands South Shetland Islands off British Antarctica

sou'wester southwester (waterproof oilskin hat and/or coat); southwestern wind

sov shutoff valve; special orientation visit

Sov Soviet; Sovietic; Soviets

SOV Single-Occupancy Vehicle

SOVA Savant of Virginia

s-o vlv shutoff valve

sow sent on (their) way

SoW Statement of Work

SOW Sunflower Ordnance Works

SOWC Senior Officers War Course (UK)

SOWETO Southwestern Townships (South Africa)

SoWMs Southern White Males

SOWSD Statement of Work, Specifications, and Design

soyd sum of years digits method

sox socks; solid oxygen; stockings

SoX School of Xerography (Xerox)

SOXAL Singapore Oxygen Air Liquids

SOZ Soviet Occupied Zone

sp same point; same principle; self-propelled; selling price; shear plate; single-phase, single-pole; signal processor; signal publication; single-purpose; small paper; smokeless powder; smokeless propellant; solid-propellant; space; spare; spare part; special; special paper; special performance (airliners); special proficiency; special propellant(s); special-purpose; specie; species; specific; speed; standard play; starting pitcher; starting pitching (baseball); starting point; starting price; static pressure; stop payment; summary plotter; summary programmed; supervisory personnel

sp (SP) space character (data processing); spelling; superperformance

s-p sequential-phase

s/p soft-point (bullet with lead core exposed to increase expansion)

s & p salt and pepper; systems and procedures

sp *sans prix* (French—without price); *spanisch* (German—Spanish)

sp. *species* (Latin—species)

s.p. *sine prole* (Latin—without issue)

s p *senza pedale* (Italian—without pedal)

Sp Space (trailer-court address); Spain; Spanish; Spring(s)

Sp *Spalten* (German—column; division); Spanish; *Spanji* (Dutch—Spain); *Spitz* (German—point)—pointed high-velocity bullet

SP San Pedro, California; São Paulo, Brazil; Scientific Paper; Section Control; Security Publication; Service Patrol; Shore Party; Shore Patrol; Shore Police; Socialist Party; Society of Protozoologists; Sociolinguistics Program; Southern Pacific (railroad); Specialist; Special Publication; Staff Paymaster; Standard Practice(s); State Park; State Police; Station Police; Strategic Plan(ning); subliminal perception; Submarine Patrol; subprofessional (civil service rating)

S-P Studebaker-Packard

S & P Standard & Poor's Corporation

S of P Society of Philaticians

SP *Senterpartiet* (Norwegian—Centrist party); *Social-demokratiet Parti* (Danish—Social Democratic Party); *Sozialistische Partei* (German—Socialist Party)

S.P. *Sanctissimus Pater* (Latin—Most Holy Father); *Summus Pontifex* (Latin—Supreme Pontiff, the Pope)

S/P Spouse or Partner (living as though married)

Sp/1 Specialist, 1st class

Sp3c Specialist, third class

spa (SPA) stimulation-produced analgesia

spa. subject to particular average; sudden phase anomaly

S p A *Società per Azioni* (Italian—joint stock company)

SPA Salt Producers Association; School of Performing Arts; *Società per Azioni* (Italian—joint stock company); Seaplane Pilots Association;

Singles Press Association; Society of Participating Artists; Society for Personnel Administration; Society of Philatelic Americans; Songwriters Protective Association; Software Publisher's Association, South Pacific Area; Southern Pine Association; Southwestern Power Administration; Standard Practice Amendment(s); State Principals Association

SPA *Société Protectrice des Animaux* (French—Society for the Protection of Animals)

SPAA Systems and Procedures Association of America

SPAAMFAA Society for the Preservation and Appreciation of Antique Motor Fire Apparatus in America

SPAB Society for the Protection of Ancient Buildings

spae spatial computer

SPAC Saratoga Performing Arts Center

S Pac South Pacific

S Pac Cur South Pacific Current

Spacenet III communications satellite network

SPACES Scheduling Package and Computer

spad (SPAD) space patrol air defense

SPAD Seafarers Political Activity Donation; Space Patrol Air Defense; Support Planning and Design

SPADETS Space Detection and Tracking System

Spag Spagnuolo [*see* Lad(s)]

SPAG Society for the Preservation of American Grandchildren

SPAI Screen Printing Association International

Spain Spanish State (Iberian nation once the center of an almost global colonial empire), *Estado Español*

spal stabilized platform airborne laser

spam nickname for electronic junk mail

Spam trademark for a processed meat product

SPAM Society for the Publication of American Music

spams spiced hams

SPAMS Ship Position and Altitude Measurement System

span. space navigation

Span Spanish

Span *Spania* (Norwegian—Spain)

SPAN Science Physics Analysis Network; Solar Particle Alert Network; South Pacific Action Network; System for Procurement and Analysis

SPANA Society for the Protection of Animals in North Africa

SpanAm Spanish American

SPANC Society for St Peter the Apostle for Native Clergy

Spandan Spanish dance (castanets clipping, feet stomping, fingers snapping, guitars thumbing, hands clapping) originated in Andalucia

spandar space-and-range radar

Spanglish Spanish + English (Latin American mixture of the two tongues) ·

Spanish Africa Ceuta, Peñón de Vélez de la Gomera, Peñón de Alhucemas, Villa Sanjurjo, Mellila, Islas Chafarinas, Isla Alborán; formerly Spanish Guinea, Spanish Morocco, and Spanish Sahara

Spanish Creole localized Spanish spoken in many former Spanish colonies; person of Spanish ancestry

Span Neth Spanish Netherlands

span(s) spaniel(s)

SPANS Sealift Procurement and National Security

Spansule span + capsule (prepared so different drugs encapsulated are released at various times)

Spantran Spanish translation (programming language)

Span Yid Spanish Yiddish (Ladino)—Sephardim

spar (SPAR) space processing applications rocket; space reactor; store port allocations register; submersible pipe-alignment rig; super-precision-approach radar

SPAR Seagoing Platform for Acoustics Research; Selection Program for ADMIRAL

Runs (*see* ADMIRAL); Society of Photographer and Artists Representatives; Standard Parts Approval

sparc steam power automation and results computer

SPARC Space Program Analysis and Review Council

Sparks ship's radio operator

sparm (SPARM) sparrow antiradiation missile

sparr steerable paraboloid altazimuth radio reflector (Jordrell Bank Radio-Telescope, Cheshire, England)

SPARS Women's Coast Guard Reserve (from the Coast Guard motto, *Semper Paratus*—Always Ready); Society of Professional Audio Recording Services

SPARTAN Special Proficiency at Rugged Training and National Building (Green Beret training program); System for Personnel Automated Reports, Transactions, and Notices (NASA)

SPAS *Societatis Philosophicae Americanae Socius* (Latin—Fellow of the American Philosophical Society)

SPASM Society for the Prevention of Asinine Student Movements

spasur space surveillance

spat self-protective antitank (weapon); silicon precision alloy transistor

spat (SPAT) self-propelled anti-tank gun

SPAT Submarine Processing Action Team

SPATC South Pacific Air Transport Council

spats spatterdashes

spau signal processing arithmetic unit

S Pau Saõ Paulo

Spauld Turn Spaulding Turnpike

spb special boiling point

SPB Space Science Board; Special Branch Policeman (British English—detective); State Personnel Board

SPBA Society of Professional Benefit Administrators

spbd springboard

SPBF Scientific Peace Builders Foundation

spc salicylamide-phenacetincaffeine; special fuel consumption; suspended plaster ceiling

Spc Specialist (United States Army)

SPC Service Processing Center (Immigration and Naturalization Detention Center); Sheriff's Posse Comitatus (ablebodied men deputized by a sheriff to assist in pursuing a gang or quelling a riot); Society of Photographers in Communications; Society for the Prevention of Crime; Software Productivity Consortium; Solar Power Corporation (Exxon); South Pacifilc Commission; Space Projects Center; Standard Products Committee; State Planning Council; Station Program Cooperative; Subcontract Plans Committee

spca serum prothrombin conversion accelerator

SPCA Society for the Prevention of Cruelty to Animals

spcat special category

SPCB Structural Pest Control Board

SPCC Ships Parts Control Center; Society for the Prevention of Cruelty to Children; Standardization, Policy, and Coordination Committee (NATO)

sp cd spinal cord

SPCH Society for the Prevention of Cruelty to Homosexuals

SPCK Society for Promoting Christian Knowledge

spcl special

SPCM Special Court-Martial

SPCMO Special Court-Martial Order

SPCO St Paul Chamber Orchestra; St Paul Civic Opera

spcr spacer

SPCs Suicide Prevention Centers; Suicide Prevention Clinics

Sp Cttee 24 Special Committee of 24 (United Nations' 24-member Special Committee concerning Granting Independence to Colonial Countries and Peoples)

SPCW Society for the Prevention of Cruelty to Women

spd separation program designator; ship pays dues; silicon photo diode; silver plated; sober and properly dressed (military guard); speed; subject to permission to deal; surface potential difference

Spd Spandau

SPD Sales Promotion Department; Society of Publication Designers; South Polar Distance; Summary Plan Description (benefits); System Program Director

SPD *Sozialdemokratische Partei Deutschlands* (German—Social Democratic Party of Germany)

spda single-premium deferred annuity

SPDC Spare Parts Distributing Center

sp del special delivery

spdl spindle

spdltr speedletter

spds seller's property disclosure statement

sp dt single pole, double throw

spdtdb single-pole double-throw double-break (switch)

spdtncdb single-pole double-throw normally closed double-break (switch)

spdtno single-pole double-throw normally open (switch)

spdtnodb single-pole double-throw normally open double-break (switch)

spdtsw single-pole double-throw switch

SPDV Site Preliminary Design Validation (program)

spe serum protein electrophoresis; solid polymer electrolyte; special purpose equipment

spe (SPE) sucrose polyester

Spe San Pedro

SPE Society of Petroleum Engineers; Society for Photographic Education; Society of Plastics Engineers; Society for Pure English

SPEA Southeastern Poultry and Egg Association

SPEAK Society for Preserving and Encouraging Arts and Knowledge

SPEARS Satellite Photo-Electronic Analog Rectification System

SPEBSQSA Society for the Preservation and Encouragement of Barber Shop Quartet Singing in America

spec special(ly); specialty; specie; species; specific(ally); specification; specimen; spectacle; speculation; speech-predictive encoded communication(s)

's'pec' suspect

Spec Speculative Society (of debaters)

SPEC Society for Pollution and Environmental Control; South Pacific Bureau for Economic Cooperation; Standard Performance Evaluation Corporation; Support Program for Employment Creation; Systems and Procedures Exchange Center

spec appt special appointment

specat special category

spec emp specially employed

special. specialization; specialized

special ops special operations (assassinations and sabotage)

specif specific; specifically

specl specialist; specialize

specs specifications; spectacles

SPECS School Planning, Evaluation, and Communication Service; Specification Environment for Communications Software

spect (SPECT) single photonemission-computed tomography

SPECTRE Single-Pulse CO_2 Transient Experiment; Special Executive for Counterintelligence, Terrorism, Revenge, and Extortion (fictional organization created by Ian Fleming for his James Bond books)

spectrog spectrography

SPECTROL Scheduling, Planning, Evaluation, and Cost Control (USAF)

spectrophotom spectrophotometry

spectros spectroscopy

SPECWAR Special Warfare

SPEDE System for Processing Educational Data Electronically

S Pedro San Pedro

SPEEA Society of Professional Engineering Employees in Aerospace

Speech Comm Assn Speech Communication Association

speed methamphetamine

speed. (SPEED) simplified profile enlargement from engineering drawing(s)

SPEED Signal Processing in Evacuated Electronic Devices; Systematic Plotting and Evaluation of Enumerated Data

speedalyzer automatic radar-controlled automotive-vehicle speed analyzer (for detecting speeders on byways and highways)

speedo speedometer

spef single-program-element fund(ing)

speleo speleology

SPELL Society for the Preservation of the English Language and Literature

Spelman Spelman College

spelpat spelling pattern(s)

Spel Soc Am Speleological Society of America

Spen Spencer, Spencerian

Spence Spencer

Sperrins Sperrin Mountains of Northern Ireland

Sperry Sperry Rand Coporation

SPERT simplified program evaluation and review task (technique)

SPES Stimulation Plan for Economic Science

S Pete St Petersburg

Spett *Spettabile* (Italian—Dear Sir)

Spett ditta *Spettabile ditta* (Italian—Messrs)

Spezia La Spezia naval station near Genoa in northern Italy

spf security processing facility; special procedures function; sun-protection factor; synthesis-phase faction

s-p-f spruce-pine-fir

SPF Science Policy Foundation; Society for the Propagation of the Faith; South Pacific Forum

SPFA Steel Plate Fabricators Association

spf/db superplastic forming/diffusion bonding

sp fl spinal fluid

spfw single phase full wave

spg self-propelled gun; specific gravity; sponge; spring; sprung

spg (SPG) sex-hormone-binding globulin

Spg Spring

SPG Society for the Propagation of the Gospel; Special Patrol Group

SPGA Scottish Professional Golfers' Association

SPGB Socialist Party of Great Britain

Spgfld Springfield

spgg solid-propellant gas generator

sp gr specific gravity

Spgs Springs

SPGS Spare Guidance System

sph sphenoidal

SPH Special Psychiatric Hospital

sphd special pay for hostile duty

sp hdlg special handling

SPHE Society of Packaging and Handling Engineers

SP & HE Society of Packaging & Handling Engineers

sphen sphenodon (tuatara lizard); sphenoid; sphenoidal

spher spherical; spheroid

sp—hl sun present—horizon lost

SPHR Senior Professional in Human Resources

SPHS Seward Park High School; Swedish Pioneer Historical Society

sp ht specific heat

sphw single phase half wave

sphyg sphygmic(al)(1y); sphygmogram(atic)(al)(ly); sphygmograph(ic)(1y); sphymography; sphygmoid; sphygmology; sphygmomanometer; sphygmomanometic(al)(ly); sphygmometer; sphygmometric(al)(ly); sphygmophone; sphygmoscope; sphygmus

spi scientific performance index; ships plan index; solid propellant information; specific polarization index

spi (SPI) serum precipitable iodine

SPI Sisters of Perpetual Indulgence; Smoking Policy Institute; Society of Photographic Illustrators; Society of the Plastics Industry; Society of Professional Investigators; Southern Police Institute; Spanish Paprika Institute; Strategic Planning Institute; Superintendent of Public Instruction

SPI Secretariats Professionnel Internationaux (French—International Professional Secretariats); *Service Pedagogique Interafricain* (French—Inter-African Teaching Service)

spia single-premium insurance annuity

SPIA South Pacific Island Airways

SPIB Society of Power Industry Biologists

spic ship position-interpolation computer

SPIC Society of the Plastics Industry of Canada; Society for the Promotion of Identity on Campus

spicbm (SPICBM) solid-propellant intercontinental ballistic missile

SPICE Scientific Personal Interactive Computing Environment; Spacelab Payload Integration and Coordination in Europe

spid submerged portable inflatable dwelling

spidac specimen input to digital automatic computer

SPIDER Spontaneous Psychophysical Incident Data Electronic Recorder

spiders Standard & Poor's depositary receipts

SPIDR Society of Professionals in Dispute Resolution

spids sensor personnel intrusion devices

spie self-programmed individualized education

SPIE Society of Plmotographic Instrumentation Engineers

SPIL Society for the Promotion and Improvement of Libraries

SPIN Searchable Physics Information Notes; Searchable Physics Information Notices; Speech Interface; Submarine Program Information Notebook

sp. indet. *species indeterminata* (Latin—specics indeterminate)

spindex selective permutation index(ing)

SPIndex Subject Profile Index (ABC-Clio's innovative indexing system)

spinel magnesium aluminum oxide

sp. inquir. *species inquirendae* (Latin—species of doubtful status)

spins special inquiries (FBI)

SPINSTRES Spencer Information Storage and Retrieval System

spintcomm special intelligence communication(s)

spip special position identification pulse

s'pipe standpipe

spir spiral

spir. *spirituoso* (Italian—spirited)—animated; *spiritus* (Latin—spirits)

Spirals Spiral Tunnels of the Canadian Pacific in Yoho National Park

spire space inertial reference equipment

SPIRES Standard Personnel Information Retrieval System

SPIRGs Student Public Interest Groups

Spirid Spiridione

spirit sales processing interactive real-time inventory technic

spirit *spiritoso* (Italian—spirited)

Spirit Spiritualism

spirt solar-powered isolated radio transceiver

spis service packaging instruction sheet

spis. *spissus* (Latin—dried)

spit selective printing of items from tape

Spit Spithead Channel joining The Solent and Southampton Water between the Isle of Wight and Portsmouth

spital (Early English contraction—hospital)

Spits Spitalsfields, England; Spitsbergen Islands in the Norwegian Arctic

spiu ship position-interpolation unit

spiw special-purpose infantry weapon

SPJ Society of Professional Journalists (Sigma Delta Chi)
SPJC Saint Petersburg Junior College
spk speckled
Spk Spokane
S^{pk} Seapeak
SpK Staatsbibliothek Prevssicher Kulturbesitz (German—Prussian Culture Treasure State Library)—Berlin
spkir(s) sprinkler(s)
spkr speaker
spl self-propelled launcher; simplex; small parts line; sound pressure level; special; spelling
s.p.l. sine prole legitima (Latin—without legitimate offspring)
Spl Sevastopol
SPL Sacramento Public Library; Saskatoon Public Library; Scan Pacific Line; Seattle Public Library; Space Programming Language; Spokane Public Library; Springfield Public Library; Syracuse Public Library
SPLA Sudan People's Liberation Party
splad (SPLAD) self-propelled light air-defense gun
SPLAN School Organization Budget-Planning System
SPLASH Special Program to List Amplitudes of Surges for Hurricanes
SPLC Southern Poverty Law Center; Standard Point Location Code
splcf sustained-peak low-cycle fatigue
splf simplification
SPLIT Sundstrand Processing Languages Internally Translated
splsm single-position letter-sorting machine
spm self-propelled mount(ing); sequential processing machine; set program mask; single-point mooring; source program maintenance; strokes per minute
s.p.m. sine prole mascula (Latin—without male issue)
Spm suppressor-mutator
SPM Saint-Pierre et Miquelon
SPM Scuola Professionale Marittima (Italian—Professional Maritime School)

SPMA Sewage Plant Manufacturers' Association
SPMMS Software Production and Maintenance Management System
Sp Mor Spanish Morocco
SPMRL Sulfite Pulp Manufacturers' Research League
SPMS System Program Management Surveys
SPMU Society of Professional Musicians in Ulster
sp/mva status post motor vehicle accident
spn sponsor; spoon; stop-press news
sp.n. species nova (Latin—new species)
Spn Spain; Spaniard; Spanish
SPN Saipan, Trust Territory of the Pacific (airport); Satellite Program Network; Separation Program Number; Student Practical Nurse
SPNB Security Pacific National Bank
SPNC Society for the Promotion of Nature Conservation
SPNEA Society for the Preservation of New England Antiquities
SPNET Science Parks Network
SPNI Society for the Protection of Nature in Israel
SPNM Society for the Promotion of New Music
sp. nov. species novum (Latin—new species)
SPNR Society for the Promotion of Nature Reserves
Spn Riv Spoon River in central Illinois
SPNS Standard Product Numbering System
SPNWR Salt Plains National Wildlife Refuge (Oklahoma)
spo sausages, potatoes, and onions
S Po São Paulo
SPO Sea Post Office; Senior Press Officer; Site Program Office(r); Special Project(s) Office; Staff Planning Office(r); Stoker Petty Officer; System Program Office(r)
SPO Socialistische Partei Osterreichs (German—Austrian Socialist Party)
spoc single-point orbit calculator
SPOE Society of Post Office Engineers

SPOIE Society of Photo-Optical Instrumentation Engineers
Spoke Spokane, Washington
spoke(s). spokesperson(s)
Spol Fest USA Spoleto Festival USA (Charleston, SC)
spont spontaneous
SPOOK Supervisory Program Over Other Kinds
spool simultaneous peripheral operation on-line
Spoon River Spoon River Anthology by Edgar Lee Masters
spoorw spoorwegen (Dutch—railway car)
S por A Sociedad por Acciones (Spanish –limited liability company)
Sporades Sporades Islands
Spore Singapore
spork spoon + fork (combination utensil)
spork(s) spoon-shaped fork(s)
sport. sporting; sportsman; sportsmanship; sportswoman
sportscast(er) sports broadcast(er)
s'pose suppose
SPOSS Society for the Promotion of Science and Scholarship
spot satellite position and tracking; spotlight
SPOT Satellite Pour Observation de la Terre (French—Satellite for Earth Observation); Signal Processing for Optical and Cordless Transmission
SPOTPABA Society for the Propagation of Tax Preparers and Bookkeepers of America
spots spotlights
spp special proficiency pay; species; surplus personal property
spp species (Latin—two or more species) singular is *sp.*
SPP Society of Professional Pilots; Southern Pacific Properties; Suicide Prevention Program; System Package Program
SPPA Society for the Preservation of Poultry Antiquities
SPPL St Paul Public Library; St Petersburg Public Library
sppo scheduled program printout

SPPPQCT Society for the Protection, Preservation, and Propagation of the Queensland Cane Toad

spps stable plasma protein solution (SPPS)

Sp Pt Sparrows Point

spqr small profits and quick returns

S.P.Q.R. Senatus Populusque Romanus (Latin—the Senate and People of Rome)

spr solid-propellant rocket (SPR); sponsor; spring; strategic petroleum reserve

Spr Spring; Springfield; Spruce

SPR Simplified Practice Recommendation(s); Society for Pediatric Research; Society for Psychical Research; solid propellant rocket; Society for Psychophysical Research; Special Project Report; Strategic Petroleum Reserve; Supplementary Progress Report

sprat small portable radar torch

spr Bog springender Bogen (German—bouncing bow)

SPRC Society for the Prevention and Relief of Cancer

SPRD Science Policy Research Division (Library of Congress)

sprdng spreading

SPRDO Service Parts Repairable Disposition Order

spre siempre (Spanish—always)

SPRE Society of Park and Recreation Educators

spread spring evaluation analysis and design

SPREd Society of Picture Researchers and Editors

SPRI Scott Polar Research Institute; Single Ply Roofing Institute

Spring Symphony No 1 by Schumann

Spring Bank Spring Bank Holiday (last Monday in May in Great Britain)

Springs The Springs (Palm Springs, California)

SPRINT Strategic Program for Innovation and Technology Transfer

sprint (SPRINT) solid-propellant rocket-intercept missile

SPRITE Sequential Polling and Review of Interacting Teams of Experts

sprklg sparkling; sprinkling

SPRL Société de Personnes à Responsibilité Limitée (French—limited company)

SPRO Services Public Relations Officer

sprr self-propelled recoil-less rifle

spr's small parcels and rolls

Sprs Springs

SPRs Strategic Petroleum Reserves

SPRS Sate Police Radio System (South Dakota)

sps secondary power system; ship program schedule; standby power system; student-paced statistics; super proton synchrotron (for smashing atoms)

sps (SPS) service propulsion system

s.p.s. sine prole supersite (Latin—without surviving issue)

SpS Special Services

SPs Shore Patrol vans; single-premiums (life insurance)

SPS Scottish Painters' Society; Sensorimotor Peripheral Speediness; Service Propulsion System; Society of Pelvic Surgeons; Society of Plastic Surgeons; Society of Saint Patrick; Southwestern Public Service; Special Public School; Spokane, Portland & Seattle (railroad); Standard Pressed Steel; Steam Power Systems; String Process System, Submerged Production System; Symbolic Programming System; System of Procedure Specifications

SP & S Spokane, Portland & Seattle (railroad)

SPSA Senate Press Secretaries Association

SPSC Scottish Prison Service College

SPSE Society of Photographic Scientists and Engineers

SPSHS Stanford Profile Scales of Hypnotic Susceptibility

SPSI Senate Permanent Subcommittee on Investigations

SPSL Society for the Protection of Science and Learning

SPSO Senior Principal Scientific Officer; Senior Personnel Selection Officer

SPSS Statistical Package for the Social Sciences

spst single-pole single-throw (switch)

SPST Symonds Picture-Story Test

spstnc single-pole single-throw normally closed (switch)

spstno single-pole single-throw normally open (switch)

spstsw single-pole single-throw switch

spt seaport; soldered piezo-electric transducer, strength-probability-time; support

spt. spiritus (Latin—alcohol; spirits)

Spt Split (Yugoslavia)

SPT Society of Photo Technologists; Sportsmen's Battalion

SPTA Salisbury Plain Training Area; Southern Pressure Treaters Association

sptc specified period of time contract

sptg sporting

spti senior physical training instructor

sptl (SPTL) superconducting power transmission line

SptL support line

SPTL Society of Public Teachers of Law

sptr spectrum

sptt single-pole triple-throw (switch)

spu swimmer propulsion unit

SPU Seattle Pacific University

SPUC Society for the Protection of Unborn Children

spud speech perception under distraction/distortion

spud. solar power unit demonstrator

SPUD St Paul Union Depot

SPUK Special Projects—United Kingdom

spur. spuria

SPUR Space Power Unit Reactor; Special People United to Ride

SPURT Short Public Responsibility Theory

spurv self-propelled underwater research vehicle

sputnik *iskustvennyi sputnik zemli* (Russian—artificial fellow-traveler around the earth)—Soviet satellite launched October 4, 1957

SPV Society for the Prevention of Vice

SPVA Self-Propelled Vehicles Association

SPVD Society for the Prevention of Venereal Disease

Sp Vly Spring Valley

sp vol specific volume

spw self-protection weapon

SPW Sillonian Plant Watchers; Society for the Protection of Whitey; Society of Protestant Wardens

SpWAfr Spanish West Africa

spwl single-premium whole life (insurance)

SPWLA Society of Professional Well Log Analysts

spx simplex(ed); stepped piston crossover

spx cimuit simplex circuit (data processing)

Spz Spezia

sq squadron; square; staff qualified; stereo-quadraphonic; superquick

sq. sequens, sequentia (Latin—what follows; result; sequel)

Sq Square

SQ Sick Quarters; Specialist Qualifications; stereo-quadraphonic (discs and recordings)

SQ Secondo Quantità (Italian—according to the quantity consumed)—menu abbreviation

sq3r survey, question, read, recite, review (psychological sequence)

sqa stereo-quadraphonic amplifier

SQA Software Quality Assurance

sqc self-quenching control; statistical quality control

sq cell ca squamous cell carcinoma

sq cm square centimeter(s)

SQCP Statistical Quality Control Procedure

sqd squad

SQD Squared Difference

sqdc special quick-disconnect coupling

Sqdn Ldr Squadron Leader

sq ft square foot (feet)

sq hd square head

sq in square inch (inches)

sq km square kilometer

SQL Standard Query Language; Structured Query Language

SQMC Squadron Quartermaster Corporal

SQO Senior Quarters Officer

sq m square meter, square mile

SQMS Staff Quartermaster Sergeant

sqn squadron

Sqn Ldr Squadron Leader

SqNP Sequoia National Park

SQNT Sequent Computer Systems Incorporated

Sq O Squadron Office(r)

SQP San Quentin Prison (California)

sqr square; square root; supplier quality rating

SQR Site Qualification Report

sq rd square rod

sq rt square root

sq's stereo-quadraphonic recordings; stereo-quadraphonic records

SQS Stochastic Queuing System; Supplier Quality Services; Supplemental Qualifications Statement

Sqs SM Squadron Sergeant-Major

sqt square rooter

SQT Ship Qualification Test (USN)

squa squamoid; squamous

squak squall and squeal

square symbol of four corners of the earth; four points of the compass; male symbol; quadrature; slang term for someone with unsophisticated tastes

squares'l square-sail

s quark strange quark

squarson squire + parson

SQUID Superconducting Quantum Interference Device

squidsan squid-cutlet sandwich

squidwich squid-cutlet sandwich

SQUIRE System for Quick Ultra-fiche-based Information Retrieval

'squitoes mosquitoes

sq yd square yard

sr scaling ratio; scientific research; sedimentation rate; selective ringing, semi regular; sensitization response;

separate rations; service rifle; sex ratio; shift register; shipment request; short range; sigma reaction; silicon rectifier; silicon rubber; single-reduction (geared turbine); sinus rhythm; slow release; socially responsible; sound ranging; spares requirement; split ring; standard range (aviation landing); starter runs; steradian; Stimulus response

sr srovnej (Czech—compare)

sr (SR) saturable reactor; special reserve; surveillance radar

s-r send-receive; learning-stimulus-response learning

s/r (S/R) safety representative

s/R sur Rhone (French—on the Rhone)

Sr Saudi Arabia; Saudi Arabian; Senior; strontium

Sr Señor (Spanish—mister, sir); *Sredniy* (Russian—mid; middle)

Sr Sønder (Danish—southern); *Søndre* (Swedish—southern)

SR saturable reactor; Savannah River (Operations Office); Scientific Report; Scottish Rifles; Seaman Recruit; seaplane reconnaissance (naval aircraft); Section Report; Senate Resolution; Senior Registrar; Service Record; Service Report; Shipping Receipt; Simulation Report; Signed Road; Society of Radiologists; Society of Rheology; Sons of the Revolution; Sound Report; Southern Railway; Special Regulation(s); Special Report; Specification Requirement(s); Staff Report; Standardization Report; Star Route (rural postal delivery); Status Report; Study Requirement; Summary Report; Supporting Research; Suriname (Internet code); surveillance radar

SR Statsjanstemannens Riksforbund (Swedish—National Association of Salaried Government Employees, Sweden); *Sveriges Radio* (Swedish radio broadcast network); Swissair

S-R Saunders-Roe; stimulus-response; Schopper-Riegler (paper-pulp scale)

SR Saudi Arabian riyal (currency unit)

SR-71 Lockheed Blackbird jet reconnaissance aircraft

Sr⁸⁵ radioactive strontium

sra sulforicinucleic acid

sra (Sra) sierra

Sra Senhora (Portuguese—Missus); *Señora* (Spanish—Missus; Mistress)

SRA Science Research Associates; Screw Research Association; Society of Residential Appraisers; Special Refractories Association; Spelling Reform Association; Station Representatives Association; Surgeon Rear-Admiral

SRAA Senior Army Advisor

sraam short-range air-to-air missile

SRAB Sveriges Radio AB (Swedish Broadcasting Corporation)

srac short-run average-cost curve

srac (SRAC) short-run average cost

SRAC Social Research Applications Corporation

Sra Dⁿᵃ Señora Doña (Spanish—Lady Madam)

SRAFO Senior Royal Air Force Officer

s'raight straight

sram static random access memory

sram (SRAM) short-range attack missile

sran short-range aids to navigation

Sranangtong Sranangtongo Suriname dialect spoken by former slaves

Sras Señoras (Spanish—ladies)

SRAs Senior Resident Agents

srats (SRATS) solar radiation and thermospheric structure (satellite)

sr auth senior author

srb selective reenlistment bonus; stabilized rice bran

srb (SRB) short-range booster; solid-rocket booster

srbc sheep red-blood cell

SRBC Susquehanna River Basin Compact

Srb-Crt Serbo-Croat (Yugoslavian)

srbm (SRMB) short-range ballistic missile

srbp synthetic resin-bonded paper

src sample return capsule; sample return container; solvent-refined coal; submarine rescue chamber

SRC Science Research Council; Signal Reserve Corps,; Southern Regional Council; Southwest Research Corporation; Space Research Corporation; Standard Requirements Code; Strict Regime Camp (for Soviet-era prisoners); Sul Ross State College; Swiss Red Cross

SRC Santa Romana Chiesa (Italian—Holy Roman Church)

srcc strikes, riots, and civil commotions

SRCD Society for Research in Child Development

s-r cells sensitization-response cells

srch search (computer)

srcr sonar control room

SRCs Strict-Regime Camps (Soviet-era imprisonment centers)

SRCS Special Reverse Charge Service

srd single radial diffusion; super red dust (tea)

Sr D Señor Don (Spanish—Sir, Mister)

SRD Secret Restricted Data; Society for the Right to Die; State Registered Dietician; Systems Requirements Definition

SRD Standard Rate and Data

SRDA Scottish Retail Drapers Association

SRDC Standard Reference Data Center

SRDE Signals Research and Development Establishment

Sr Dr Señor Doctor (Spanish—Mister Doctor)

SRDS Standard Reference Data Service

SRDT Single Radial Diffusion Test

sre single-round effectiveness; single-round effectivity

Sre Sreda (Russian—Wednesday)

SRE Society of Recreation Executives; Society of Reproduction Engineers

S.R.E. Sancta Romana Ecclesia (Latin—Holy Roman Church)

SR EB Southern Regional Education Board

Sr Ed Senior Editor

SRED Scientific Research and Experiments Department

SREEC Southern Regional Environmental Education Council

SREL Savannah River Ecology Laboratory

srem sleep with rapid eye movements

Sres Señores (Spanish—Messrs)

SRES Seniors Real Estate Specialist

srev slow reverse

srf self-resonant frequency; semi-reinforced furnace; solar radiation flux; stable radio frequency; state revolving fund; submarine range finder; supported ring frame; system recovery factor

SRF Self-Realization Foundation; Ship Repair Facility (USN)

srf black semireinforcing furnace black

srg sound ranging

SRG System Review Group

SRGM Solomon R Guggenheim Museum

srh single radial hemolysis

SRHE Society for Research into Higher Education

SRHL Southwestern Radiological Health Laboratory

Sr HS Senior High.School

sri servo repeater indicator; silicone rubber insulation; socially responsible investing; spectrum resolver integrator; surface roughness indicator

Sri Sri Lanka (Ceylon); Srinagar, India

SRI Scientific Research Institute; Southern Research Institute; Southwestern Research Institute; Space Research Institute; Stanford Research Institute

SRI Sacro Romano Impero (Italian—Holy Roman Empire)

Sria Secretaria (Spanish—secretariat)

srif somatotropin release-inhibiting factor

Sri Lan *Sri Lanka* (Singhalese—Resplendent Land)—Ceylon

Sri Lanka Republic of Sri Lanka (Asian island off India's southern tip)

SH Lan Pra Sam Jan *Sri Lanka Prajathanthrika Samajavadi Janarajaya* (Sinhala—Democratic Socialist Republic of Sri Lanka)—formerly Ceylon

SRILTA Stanford Research Institute Lead Time Analysis

SRIM Selected Research in Microfiche

Sri Nep Sar *Sri Nepala Sarkar* (Nepali—Kingdom of Nepal)

SRINF Shorter Range Intermediate-Range Nuclear Force

Srio *Secretario* (Spanish—Secretary)

SRIS Safety Research Information Service; School Research Information Service

sri self-restraint, joint; static round jet

SRJC Santa Rosa Junior College

srl (SRL) systems reference library

Srl Sorel

SRL Savannah River Laboratory; Save-the-Redwoods League; Science Reference Library (Chancery Lane, London); Scientific Research Laboratory; Study Reference List

SRL *Saturday Review of Literature; sociedad de responsabilidad limitada* (Spanish—limited liability company)

Srls Saudi Arabian riyal(s)

srm solid rocket motor; speed of relative movement; spontaneous rapture of membrane; survey radiation monitor

srm (SRM) short-range missile

SRM Society for Range Management; Standard Reference Material; Southern Rocky Mountains

SR & M Safety, Reliability, and Maintenance

SRM *Su Real Majestad* (Spanish—Your Royal Majesty)

SRME Society for Research in Music Education

Sr M Sgt Senior Master Sergeant

SRMU Space Research Management Unit

SRN State Registered Nurse; Student Registered Nurse

SRN-6 British Hovercraft hovercraft designation

srna (SRNA) soluble ribonucleic acid

sRNA soluble or transfer RNA (same as tRNA)

SRNA Shipbuilders and Repairers National Association

SR NC Severn River Naval Command

srnms. surnames

SRNP Stirling Range National Park (Western Australia)

sro self-regulatory organization; sex-ratio organism

sro (SRO) single-room occupancy; standing room only

SRO Savannah River Operations Office(r); Senior Ranking Officer; Superintendent of Range Operations; Supplementary Reserve of Officers

srob short-range omnidirectional beacon

s rod stove rod

sro's single-room-occupancy hotels

SROTC Senior Reserve Officers Training Corps

srp salary reduction plan; supply refuelling point

SRP Salt River Project; Saturday Review Press; Savannah River Plant; Scholarship and Recognition Program; Scientific Research Proposal; Stratospheric Research Program

SRPA Southwestern Regional Planning Association

srpi server requester programming interface

s-r psychology stimulus-response psychology

srr short range radar; survival, recovery, and reconstitution

srr (SRR) skin resistance response

SRR Site Recommendation Report; Supplementary Reserve Regulations; System Requirements Review

SRRA Scottish Radio Retailers' Association

SRRC Sperry Rand Research Center

srrcs surface raid reporting control ship

Srrnto Sorrento

SRRS Social Readjustment Rating Scale

SRRT Social Responsibilities Round Table

srs safety restraint system; slow reacting substance

srs (SRS) short-run supply

SRs Socialist Revolutionaries (moderates in czarist Russia)

SRS Scoliosis Research Society; Seat Reservation System; Sight Restoration Society; Social and Rehabilitation Service; Software Requirements Specification; Songwriters Resources and Services; Special Revenue Sharing; Sperry Rail Service; Sperry Rand Service; Statistical Reporting Service; Structural Research Series; Structural Research Service; Supplemental Restraint System

S.R.S. *Societatis Regiae Sodalis* (Latin—Fellow of the Royal Society)

SRSA Scientific Research Society of America

SRSC Sul Ross State College

sir/sep salary-reduction simplified employee pension

SRSM *Serenissima Repubblica di San Marino* (Italian—Most Serene Republic of San Marino)—official name of San Marino

SRSNY Sons of the Revolution in the State of New York

S-R strain Schmidt-Ruppin (viral) strain

srt simple reaction time; speech reception threshold; spousal remainder trust

SRT Short Range Transport (aircraft); Social Relations Test(ing); Speech Reception Test(ing); Strategic Rocket Troops; Stroke Rehabilitation Technician; System Reliability Test(ing)

SRT *Standard Radio och Telephon* (Swedish—Standard Radio and Telephone)

Srta *Señorita* (Spanish—Miss)

SRTC Salford Royal Technical College; Southern Rhodesia Transport Corps

SRTN Solar Radio Telescope Network

Srto *Señorito* (Spanish—master; young gentleman)

SRTOS Special Real-Time Operating System

SRTS Science Research Temperament Scale

sru sea reconnaissance unit; servo(mechanism) repeat unit; shop-replaceable unit

SRU Scottish Rugby Union

SRUBLUK Society for the Reinvigoration of Unremunerative Branch Lines in the United Kingdom

srv space rescue vehicle

srv (SRV) submarine research vehicle

SRV Socialist Republic of Vietnam

srvlv servovalve

SRW Sherwin-Williams Company of Canada (stock exchange symbol); State Reformatory for Women

SRY Sherwood Rangers Yeomanry

SR y C Santiago Ramón y Cajal

srypu (SRYPU) sour puss

ss saline soak; same size; sample size; semi-steel; sandy shore; sea service; setscrew; short-stop; simplified spelling; single seater; single signal; single source; single strength; single-seated; software systems; sole source; sparingly soluble; spin-stabilized; stainless steel; standard score; stepson; sterile solution; straight shank; superspeed; supersport (automobile model); sword stick; sworn statement

ss (SS) space shuttle; suspended sentence

ss (s/s)(SS)(S/S) steamship

s-s learning-stimulus-stimulus-learning; surface-to-surface missiles

s-s solid-state

s/s same size; souvenir sheet; steamship; suspended sentence

s & s signs and symptoms; slings and springs

s of s source of sex (also appears as sos)

s to s ship-to-shore; station-to-station

ss. senza sordini (Italian—without mutes); *siglos* (Spanish—centuries)

ss scilicet (Latin—namely); *semis* (Latin—one-half); *supra scriptum* (Latin—written

above)—usually printed to left of signature line in sworn statements)

s.s. sensu stricto (Latin—in the strict sense)

sS siehe Seite (German—see page)

s/S sur Seine (French—on the Seine)

Ss students; subjects

SS diesel-powered attack submaritime (naval symbol); Saints; Science Service; Secret Service; Secretary for Scotland; Secretary of State; Selective Service; Sharpshooter; Ship Service; Ship's Stores; Silver Star; Social Security; Somaliland Scouts; Special Service; Special Staff, Specification(s) for Structure; Staff Surgeon; Standard Score; steamship; Straits Settlements; Submarine Scout; Submarine Studies; Sunday School; Superintending Sister; supersonic; Support Services; Support System; Surveillance Station; sworn statement

S-S Camille Saint-Saëns (1835–1921); Sans-Serif

S & S Simon & Schuster; Steen & Strom

S of S Society of Separationists

SS Saints; Schutzstaffel (German—Nazi blackshirt elite corps); *Santa Sede* (Spanish—Holy See); *Seitensatz* (German—second theme in a sonata or symphony); *Statens Skipstilsyn* (Danish—State Shipping Inspection)

SS. Sanctissimus (Latin—most holy)

SS-4 Soviet medium-range ballistic missile called Sandal by NATO

SS-5 Soviet intermediate-range ballistic missile called Skean by NATO

SS-6 Soviet intercontinental ballistic missile nicknamed Sapwood by NATO

SS-7 Soviet intercontinental ballistic missile called Saddler by NATO

SS-8 Soviet two-stage intercontinental ballistic missile named Sasin by NATO

SS-9 Soviet imtercontinental ballistic missile called Scarp by NATO and capable of releasing warheads below early-warning radar range

SS-10 Soviet three-stage intercontinental ballistic missile named Scrag by NATO

SS-11 Nord antitank missile built in France where its air-launched version is called AS-11; Soviet liquid-fuel intercontinental ballistic missile; U.S. antitank missile called AGM-22A

SS-12 Nord antitank missile with greater range than the SS-11

SS-13 Soviet three-stage intercontinental ballistic missile code-named Savage by NATO

SS-14 Soviet two-stage intercontinental ballistic missile code-named Scapegoat by NATO

SS-18 Soviet intercontinental ballistic missile

SS-20 intermediate-range nuclear missile developed by the USSR

SS-21 tactical nuclear missile developed by the USSR

ssa smoke-suppressant additive; solid-state amplifiier

ssa (SSA) skin-sensitizing antibodies

Ssa Sjogren's syndrome antibody

SSA Sauna Society of America; Scottish Schoolmasters' Association; Secretary of State for Air, Seismological Society of America; Semiotic Society of America; Smallest Space Analysis; Soaring Society of America; Social Security Administration; Social Services Administration; Society of Scottish Artists; Society for the Study of Addiction; Sommelier Society of America; Southern Surgical Association; Sportswear Salesman's Association; Star of South Africa (medal); Subscriber Savings Account; Suspension Specialists Association

S-S A Self-Sufficiency Association

SSA *Secretaria de Salubridad y Asistencia* (Spanish—Secretariat of Health and Assistance); *Section Sanitaire Anglaise* (French—English Sanitary Section)

ssaa screened-shift-and-add

SSAA Shoe Suppliers Association of America

SSAB SONGS Site Access Badge

SSAC Soldier's, Sailor's, and Airmen's Club

SSADM Structured Systems Analysis and Design Methodology

SSAFA Soldiers', Sailors', and Airmen's Families Association

SSAG Social Security Action Group

SS agar *Shigella* and *Salmonella agar*

SSAGO Student Scout and Guide Organisation

SSAP Statement of Standard Accounting Practice(s)

ss ar spotface arbor

SSAR Society for the Study of Amphibians and Reptiles

SSARR Streamflow Synthesis and Research Regulation; Streamflow Synthesis and Reservoir Regulation

SSAS Special Signal Analysis System; Static Stability Augmentation System

SSASA Social Services Association of South Africa

SSAT Secondary School Admission Test(s)

SSATB Secondary School Admission Test Board

ssb single side band; single-strength B (quality glass); subseabed

S Sb San Sebastian

SSB fleet ballistic missile submarine (3-letter naval symbol); Security Screening Board; Selective Service Board; single-strand binding (protein); Society for the Study of Blood; Source Selection Board; Space Science Board; Special Service Battalion; Subseabed (project)

S-S B Sino-Soviet Bloc

SSBIC Special Small Business Investment Companies

SSBN nuclear-powered fleet ballistic missile submarine (4-letter naval symbol)

ssbsc single sideband suppressed carrier

SSBS S-2 French intermediate-range ballistic missile launched from an underground silo

ssc small saver certificate; safeshielded cask; sealed storage cask; shape-selective cracking; short-service commission; station-selection code (data processing)

s & sc sized and supercalendered

SSC Sacramento State College; Sarawak Shipping Company; Sculptors' Society of Canada; Ships Systems Command (formerly Bureau of Ships); Sidney Sussex College (Cambridge); Solicitor before the Supreme Courts; Southern States Conference; Straits Steamship Company; Strong Sexual Context; Super-conducting Supercollider (atom smasher); Supply Systems Command (formerly Bureau of Supplies and Accounts); Surveillance Support Center

S.S.C. *Societas Sanctae Crucis* (Latin—Society of the Holy Cross)

SSCA Southern Speech Communication Association; Southern States Correctional Association

sscc spin-scan cloud camera

SSCC Space Surveillance Control Center

S.Sc.D. Doctor of Social Science

SSCD Society of Small Craft Designers

SSCDS Small Ship Combat Data System

SSCI Steel Service Center Institute

SSCI Steel Shipping Container Institute

SSCI *Social Sciences Citation Index*

SSCNS Ship's Self-Contained Navigation System

sscp system services control point

SSCQT Selective Service College Qualification Test

ss cr stainless-steel crown

sscrn silkscreen

sscrng silkscreening

sscs strain-sensitive cable sensor

SSCS Sea Shepherd Conservation Society; Shipboard Satellite Communications System

ssd solid-state disk; source skin distance

ssd (SSD) sentence-structure determination

SSD Schizophrenic Spectrum Disorders; Science Services Department; Scientific Services Department; Social Security Disability; Social Services Department; Space Systems Division (USAF); System for System Development

SSD *Staatssicherheitsdienst* (German—State Security Service)—former East German political police

SS.D. *Sanctissimus Dominus* (Latin—Most Holy Lord)—the Pope

S.S.D. *Sacrae Scripturae Doctor* (Latin—Doctor of Sacred Scripture)

SSDA Self-Service Development Association; Service Station Dealers of America

SSDB Social Security Disability Benefits

SSDC Social Science Documentation Center (UNESCO)

ss/dd single-sided/double density (computer disks)

SSDF Single Soldiers Dependents Fund

SSDHPER Society of State Directors of Health, Physical Education, and Recreation

SSDI Social Service Disability Insurance

SSDL Society for the Study of Dictionaries and Lexicography

SS DNA single-stranded (viral) deoxyribonucleic acid

SSDO Social Security District Office

ssdr subsystem development requirement

SSDS Ship Structural Design System

sse safe-shutdown earthquake; signal security element; surface support equipment; switching single element

SSE Scale of Socio-Egocentrism; south southeast; Support System Evaluation; Seed Savers Exchange; Submarine Scout Experimental

S.S.E. Society of Saint Edmund

SSEB South of Scotland Electricity Board

ssec selective-sequence electronic calculator; spatially selective enhancement effect

SSEC Secondary School Examination Council; Social Science Education Consortium; Solar System Exploration Committee; South Seas Evangelical Church

SSEES School of Slavonic and East European Studies

ssef solid-state electro-optic(al) filter

SSEL Space Science and Engineering Laboratory

SSEM South Seas Evangelical Mission

SSEN Senior State Enrolled Nurse

ss enema soap-suds enema

SSET Steady-State Emission Test(ing)

ssf saybolt seconds furol; single-seated fighter; standard saybolt furol (viscosity)

SSF Service Storage Facility; Seven-Step Foundation; Ships Service Force; Social Science Foundation (University of Denver); Society of Saint Francis; Special Service Force

SSFA Scottish Schools' Football Association; Scottish Steel Founders' Association

SSFC Severe Storms Forecast Center (Kansas City, Missouri)

SSFF Solid Smokeless Fuels Federation

ss fx spotface fixture

ssg second-stage graphitization

SSG guided missile submarine (3-letter naval symbol)

SSGN nuclear-powered guided—missile submarine (4-letter naval symbol)

SSgt Staff Sergeant

ssgw (SSGW) surface-to-surface guided weapon

SSH Sailor's Snug Harbor

S Sh A Soyedinennye Shtaty Ameriki (Russian—United States of America)

SSHA Scottish Special Housing Association

SSHRC Social Sciences and Humanities Research Council (Canadian)

SSHSA Steamship Historical Society of America

ssi sites of scientific importance; small-scale integration

Ssi Surekasi (Turkish—company)

SSI Saint Simon's Island; Scuba Schools International; Social Security Income; Social Security Insurance; Society of Scribes ard Illuminators; Space Studies Institute; Supplemental Security Income

SSI Service Social International (French—International Social Service); *Social Sciences Index*

SSIA Shoe Service Institute of America

SSIB Seaway Skyway International Bridge

ssic small-scale integrated circuit

SSIC Southern States Industrial Council; Standard Subject Identification Code

SSIDC Small-Scale Industries Development Corporation (Indian)

SSIE Smithsonian Science Information Exchange

SSIG State Student Incentive Grant(s)

SSIH Société Suisse pour l'Industrie Horlogère (French—Swiss Society of the Horological Industry)

SSI/ITL SSI Container Corp/ITEL

ssip system setup indicator panel

SSIS Squibb Science Information System

SSISI Statistical and Social Inquiry Society of Ireland

SSI/SSP Social Security Income/State Supplemental Program

ssit (SSIT) semi-submarine icebreaking tanker

SSJ Sisters of St Joseph

ssk set storage key; soil stack; solid-state keyboard

ssl secure socket layer; solid-state lamp; spent sulfite liquor

SSL Saguenay Shipping Limited; Sapphire Steamship Lines; Seven Stars Line; Space Science Laboratory (Convair); Space Sciences Laboratory (GE)

S.S.L Sacrae Scripturae Licentiatus (Latin—Licentiate of Sacred Scripture)

sslc short-service limited commission

ssld single-seat laser designator

s sleep synchronized sleep

S-sleep slow-wave sleep

SS loran sky-wave synchronized loran

SSLS Solid-State Laser System

ss lt starboard side light

sslv (SSLV) standard space-launched vehicle

ssm sea-skimmer missile; set system mask; solid-state material(s); spread spectrum modulation

ssm (SSM) surface-to-surface missile

SSM Saturday(s), Sunday(s), Monday(s); Security Service Module; Singer Sewing Machine; Squadron Sergeant-Major; Staff Sergeant-Major; System Support Management; System Support Manager

ssma solid-state microwave amplifier

SSMA School Science and Mathematics Association; Stainless Steel Manufacturers' Association

ssme space-shuttle main engine

ssmm space station mathematical model

SS MM Sus Majestades (Spanish—Their Majesties; Your Majesties)

SSMS Submarine Safety Monitoring System

ssmt supersonic magnetic (railroad) train

SSN Space Surveillance Network; Social Security Number; Standard Serial Number; Station Serial Number

SS(N) nuclear-powered submarine (3-letter naval symbol)

SSNC Scindia Steam Navigation Company

ssnd solid-state neutral dosimeter

ssnf source spot noise figure

SSno escribano (Spanish—court clerk, notary, scribe)

SSNP Syrian Social Nationalist Party

SSNS Standard Study Numbering System

SSO Sacramento Symphony Orchestra; Savannah Symphony Orchestra; Seattle Symphony Orchestra; Senior Supply Officer; Shanghai Symphonic Orchestra; Short Service Officer; Shreveport Symphony Orchestra; Source Selection Official; Special Service Officer; Spokane Symphony Orchestra; Springfield Symphony Orchestra; Squadron Signals Officer; Staff Signals Officer; Station Staff Officer; Sydney Symphony Orchestra; Syracuse Symphony Orchestra; System Staff Office(r)

SSO *Seguro Social Obligatorio* (Spanish—Obligatory Social Security); *sudsudoeste* (Spanish—south southwest)

SSOA Subsurface Ocean Area

SSOFS Smiling Sons of the Friendly Shillelaghs

S of Sol *Song of Solomon*

s sord *senza sordini* (Italian—without mutes)

SSORM Standard Ship's Organization and Regulations Manual (USN)

ssorts ship's systems operational requirements

ssos (SSOS) severe-storm-observing satellite

SSOs Student Services Organization members

ssp seismic section profiler; ship's stores proft; single-shot probability; standby-status panel; steam service pressure; subspecies; sustained superior performance; system-support program

ssp (SSP) statutory sick pay

S-S p Sanarelli-Schwartzman phenomenon

SSP scouting seaplane (3-letter naval symbol); Seashore State Park (Virginia); Site Selection Report; Society for Scholarly Publishing; Society of St Paul; Source Selection Panel; Species Survival Plan; Stockpile Stewardship Program; S.S. Pierce; Sunshine State Parkway

S.S.P. Society of Saint Paul

sspc solid-state power controller

SSPC Steel Structures Painting Council

SSPCA Scottish Society for the Prevention of Cruelty to Animals

sspe subacute sclerosing panencephalitis

sspe (SSPE) subacute sclerosing panencephalitis

SSPFC Stainless Steel Plumbing Fixture Council

SSPHS Society for Spanish and Portuguese Historical Studies

SSPMA Sump and Sewage Pump Manufacturers Association

SSPN Satellite System for Precise Navigation

SSPP Society for the Study of Process Philosophies

S-spring S-shaped spring

SSPU Self-Service Postal Unit

SSPV Scottish Society for the Prevention of Vivisection

ssq simple sinusoidal quantity

SSQ Station Sick Quarters

SSQT Selective Service Qualification Test

ssr secondary surveillance radar

SSR Site Safety Report; Soviet Socialist Republic(s); Software Specification Review

SSR *Sovétskaya Sotsialísticheskaya Respúblika* (Russian—Soviet Socialist Republic)

SSRA Scottish Squash Rackets Association

SSRB Soil Survey Research Board

SSRC Social Science Research Council

SSRCAS Secondary-Surveillance-Radar Collision—Avoidance System

SSRCC Social Science-Research Council of Canada

ssri selective serotonin-reuptake inhibitor

SSRI Social Science Research Institute

SSRL Systems Simulation Research Laboratory

SSRP Stanford Synchrotron Radiation Project

SSRs Safe Secure Railcars

SSRS Society for Social Responsibility in Science; Submarine-Sand Recovery System

sss single-screw ship; specific soluble substance; sterile saline soak

s/ss sector/subsector

sss (SSS) strategic satellite system

sss (SSS) *su seguro servidor* (Spanish—your sure servant; yours truly)

s.s.s. *stratum super stratum* (Latin—layer upon layer)

SSS Secretary of State for Scotland; Selective Service System; Simplified Spelling Society; Special Social Services; System Safety Society

S-S-S *Schweiz-Suisse-Svizzera* (Switzerland in the three languages of the country)

S.S.S. *Societas Sanctissimi Sacramenti* (Latin—Congregation of the Most Blessed Sacrament)

SS-20s Soviet supersonic medium-range nuclear missiles

SSSA Simplified Spelling Society of America; Soil Science Society of America

S-S SA Singapore-Soviet Shipping Agency

SSSB System Source Selection Board

sssc soft-sized super-calendered (paper)

sss & c sin, syph(ilis), sulfa, and cystoscopes

SSSC Space Science Steering Committee (NASA)

sssd second-stage separation device; solid-state solenoid driver

sssi sites of special scientific importance

SSSI Science Supervisory Style Inventory

SSSJ Student Struggle for Soviet Jewry

SSSL Solid State Sciences Laboratory (USAF)

sssm site space surveillance monitor

sssm (SSSN) standard surface-to-surface missile

SSSM South Street Seaport Museum (New York City)

SSSP Space Shuttle Synthesis Program

SSSR Society for the Scientific Study of Religion

SSSR Soyuz Sovétskikh Sotsialiísticheskikh Respúblik (Russian—Union of Soviet Socialist Republics)

SSSRU School Safety and Security Resource Unit

SSSS Society for the Scientific Study of the Sea; Society for the Scientific Study of Sex

ssst symbol for the sound of an aerosol spray

SSSU Seaspeed Sea Services Unit

SSSWP Seismology Society of the South-West Pacific

sst safe-secure trailer(s); sea surface temperature; social skills training; solid state transmitter; solid-state triangulation (automatic focusing system); stainless steel; stimulus-sampling theory; supersonic transport (airplane)

SST Samoan Standard Time; Society of Silver Collectors; Source Selection Team; Space Systems Center (Douglas); Submarine Supply Center; supersonic transport (airplane); target and training submarine (naval symbol)

SSTA Scottish Secondary Teachers' Association; Secondary School Theatre Association; Special Services Transportation Agency

SSTAR Society for Sex Therapy and Research

SSTC Specialized System Test Contractor

SSTEP System Support Test Evaluation Program

ssto single-stage to orbit

SSTO Superintending Sea Transport Office(r)

SSTP Student Science Training Program

sst's safe (and) secure trailers used to haul nuclear weapons on American highways

sstu seamless steel tubing

ssu saybolt seconds universal; self-serving unit

ssus spring-solid upper stage

ssv (SSV) semi-submersible support vessel; ship-to-surface vessel; submarine support vessel

s.s.v. sub signa veneni (Latin— under a poison label)

SSV ship-to-surface vessel

SSvd Selective Service

SSV/GC & N Space Shuttle Vehicle/Guidance, Control and Navigation

ssvs slow-scan video simulator

ssw safety switch

SSW south southwest; S.S. White

SSWA Scottish Society of Women Artists

SSWS Seismic Sea Wave Warning System

SSX South Coast Corporation (stock exchange symbol)

SSXTF Solar Soft X-ray Telescope Facility

ssz specified strike zone

ssz (SSZ) pocket submarine; specified strike zone

SSZ Society of Systematic Zoology

st indicates ordinal number as in 1st Avenue or 1st Street; sedimentation time; service test; short ton; shrubby thicket; single tire; single-throw; slight trace; sound track; sounding tube; space telescope; special text; special translation; stained; standard stimulus; stanza; statement(s); steel; steel truss; stock transfer; stone; strata; surface tension; survival time; syncopated time

s & t science and technology; sink and laundry tray; supply and transport

st. stet (Latin—let it stand)— usually referring to what has been mistakenly crossed out

St Saint; Sainte; Stanton number; State; status; Street; strontium; stratus

Sᵗ Sint (Afrikaans, Dutch, Flemish—saint); *Staryy* (Russian—old)

ST São Tomé and Principe (Internet code); Seaman Torpedoman; Service Test(ing); Shipping Ticket; Software Technology; Sons of Temperance; Speech Therapist; speech therapy; Standardized Test; Summer Time; *Suomen Tsavalta* (Finnish—Finland); Syrian Territory

S.T. sidereal time

S & T Supply and Transport

S of T Sons of Temperance

sta static; station; stationary; stationery; stator; submarine tender availability; status

Sta Santa (Italian, Portuguese, Spanish—Saint)—feminine; *Señorita* (Spanish—Miss)

STA Scottish Typographical Association; Society of Typographic Arts; Southern Textile Association; Supersonic Tunnel Association

STAA Survey Test of Algebraic Aptitude

STAAS Surveillance and Target Acquisition Aircraft System

stab stabilizer

STAB Svenska Tandsticks Aktiebolaget [Swedish—Match (stick) Company]

Sta'b'd starboard

STABEX System of Stabilization of Export Earnings

stabiles static abstract sculptures

stac sensor transmitter automatic choke

stac staccato (Italian—separately and with great distinction)

St AC Saint Anne's College; Saint Anthony's College

STAC Science Teacher's Adaptable Curriculum; Science and Technology Advisory Committee (NASA)

STACO Society of Telecommunications Administrative and Controlling Officers

STACS Satellite Telemetry and Computer System

STADIUM Statistical Data Interchange Universal Monitor

sta eng stationary engineer

stafex staff exercise(s)

STAFF Smart Target Activated Fire & Forget; Stellar Acquisition Flight Feasibility (guidance system)

Staffs Staffordshire

staflo stable-flow (free-boundary electrophoresis apparatus)

stag stagger; staggered

STAG Special Task Air Group; Standards Technical Advisory Group; Strategy and Tactics Analysis Group

stagfla stagflation (high inflation coupled with high unemployment)

stagflation stagnant economy marked by rising unemployment and spiralling inflation;

stagnation and inflation

stagmag magazine featuring nude women

STAGS Sterling Transferable Accruing Government Securities

STAI State-Trait Anxiety inventory

STAIFA St Anselm's International Friendship Association

Stairs Storage and Information Retrieval Systems

Sta L *Santa Luciá* (Spanish— St Lucia)

stalac stalactite

stalag stalagmite

Stalag *Stammlager* (German— base camp, for military prisoners)

stam sequential thermal anhysteric magnetization; stammer(er); stammering

sta mi statute miles

STAMM Standards for In-Vehicle Man Machine interface

stamp. small tactical aerial-mobility platform

STAMP Systems Tape Addition and Maintenance Program

Stampa *La Stampa* (Turin's Press—one of Italy's leading newspapers)

STAMPS Structural Thermal and Meteorite Protection System

stan stanchion; standard; standing

Stan Standard; Stanford; Stanley; Stanleyville; Stanton

STANAG Standardization Agreement (NATO)

stanal statistical analysis

STANAVFORCHAN Standing Naval Force Channel (NATO)

STANAVFORLANT Standing Naval Force Atlantic (NATO)

St And St Andrews

standard standardization

Standard and Poor's *Standard and Poor's Corporation Records*

Stand Engl Standard English

STANDINAIR Standing Instructions for Air Attachés

stanine score standard-nine score (USAF standard psychological score)

STANORM Statistical Normalization

Stan Psychiat Nomen Standard Psychiatric Nomenclature

STANVAC Standard Vacuum (oil company)

STAO Science Teachers Association of Ontario

STAPFUS Stable Axis Platform Follow-Up System

staph staphylococcus

staq security-traders automatic quotations

star symbol of perfection

star (STAR) special tactics against robbery (police program)

STAR Scan to Automate Receipt; Selective Training and Retention (program); Serial Titles Automated Record (National Agricultural Library); Ship-Tended Acoustic Relay; Space Thermionic Auxiliary Reactor; Special Tactics and Response; Special Telecommunication Action for Regional development; Study of Tamoxifen and Raloxifene; submersible test and research (Electric Boat)

STAR *Scientific and Technical Aerospace Reports*

STARFIRE System to Accumulate or Retrieve Financial Information Random Extract

STARLAB Space Technology Applications and Research Laboratory (NASA)

starquake star + earthquake

stars specialized training and reassignment students; stationary automotive road stimulator (Toyota)

STARs Scientific and Technical Aerospace Reports

STARS Satellite Telemetry Automatic Reduction System; Small Tethered Aerostat Relocatable System; Standard Terminal Automation Replacement System; Student Tuition and Repayment System

Stars and Stripes *Stars and Stripes Forever* (American march); American military newspaper

START Space Technology and Reentry Test(s); Space Transport and Reentry Test(s); Spacecraft Technology and Advance Reentry Test; Strategic Arms Reduction Talks (U.S.A.-USSR); Strategic Arms Reduction Treaty (U.S.A.—USSR)

STARTS Safety Technology Applied to Rapid Transit Systems

START-UP Supplies for Technological Advanced Requirements Through Users Protocols

Star War(s) Strategic Defense Initiative(s), SDI

stas staff-to-arm signal

Stash Stanislas; Stanislaus

STASH Student Association for the Study of Hallucinogens

Stasia Anastasia

stasis or *stat* or *stato* (Greek— stand)—colonic stasis, electrostatic, hydrostatic, metastasis, thermostat

stat electrostat; electrostatic; microstat; photostat; static; stationary; statistic(al); statuary; statue; statute

stat. *statim* (Latin—immediately, right now)

Stat Publius Papinius Statius (Roman poet)

Stat Can *Statistics Canada*

state simplified tactical approach and terminal equipment (STATE)

Statesman's *Statesman's Year Book*

Statesville Statesville Correctional Center (Joliet, Illinois)

Stat Hall Stationers' Hall

STATIC Student Taskforce Against Telecommunication Concealment

STATLIB Statistical Computing Library (Bell System)

stat mux statistical multiplexor

Stat Off Her (His) Majesty's Stationery Office

stats statistics

Stats statutes

Statsbib *Statsbiblioteket* (Dano-Norwegian—State Library)

STATUS Subscriber Traffic and Telephone Utilization System

St AU University of St Andrew

STAUK Seed Trade Association of the United Kingdom

Stav Stavanger, Norway

sta wgn station wagon

St A YC St Augustine Yacht Club

Sta Ysbl Santa Ysabel

stb special tax bond

s-t b steel-truss bridge

St B Státni Bezpečnost (Czech—State Security)—secret police

STB Suriname Tourist Bureau

STB Sandatahang Tanod ng Bayan (Filipino—People's Home Defense Guard)

S.T.B. Sacrae Theologiae Bacalaureus (Latin—Bachelor of Sacred Theology)

stba selective top-to-bottom algorithm

St Bart's Day Saint Bartholomew's Day—date in 1587 when French Huguenots were attacked by Catholics as they left their churches

stbd starboard

St Ben St Benet's Hall, Oxford

st brz statuary bronze

stbt steamboat

stc said to contain; security time control; sensitivity time control; short time constant; sound transmission class; stepchild

stc (STC) supplemental type certificate

STC Satellite Television Corporation; Satellite Test Center; Satellite Tracking Committee; Scandinavian Travel Commission; Short Title Catalog; Society for Technical Communication; Southwestern Technical College; Standard Telephone and Cables; Standard Transmission Code; Sunderland Technical College

S.T.C. Samuel Taylor Coleridge

STC Short Title Catalogue

STCA Stereo Tape Club of America

St Cat St Catherine's College, Oxford

STCC Springfield Technical Community College

STCCM Sistema de Transporte Colectivo Ciudad de México (Spanish—Mexico City Collective Transportation System)

Stckhlm Stockholm

st cl storage closet

St C & N Saint Christopher and Nevis (Leeward Islands)

STCS Society of Technical Civil Servants

std salinity, temperature, depth; sexually-transmitted disease; short-term disability; skin test dose; standard; standard test dose; state-of-the-technology design; subscriber trunk dialing

St D Stage Director

STD Science and Technology for Development; Society for Theological Discussion; Software Test Document; Subscriber Trunk Dialing

S.T.D. Sacrae Theologiae Doctor (Latin—Doctor of Sacred Theology)

std by stand by

St DC St David's College

STDC Society of Typographic Designers of Canada

Stde Stunde (German—hour)

StdE Standard English

stder social introversion, thinking introversion, depression, cycloid tendencies, rhathymia (personality traits)

st diap stopped diapason (organ)

stdn standardization

Std Oil Cal Standard Oil of California

std P stand pipe

stdr steam turbine double reduction

st dr single-trip drum

std's sexually transmitted diseases

STDSD Solar-Terrestrial Data Services Division (NOAA)

Stdy Saturday

Ste Suite

Ste Sainte (French—saint, *f*)

Sté Société (French—Society)

St E St Etienne

STE Society of Telecommunications Engineers; Society of Tractor Engineers; Support of Theological Education

steakwich steak sandwich

steamers steamed clams

STECC Scottish Technical Education Consultative Council

steelie steel ball-bearing playing marble

steeving stevedoring (loading or unloading a ship's cargo)

Stef (Joseph) Lincoln Steffens; Stefan(i)(e); Vilhjalmur Stefansson (William Stevenson)

STEFER Società della Tranvia e Ferrovia Elettrica di Roma (Italian—Rome transportation system)

STEG Supersonic Transport Evaluation Group

St E H St Elizabeth's Hospital

St EHC Saint Edmund's Hall College (Oxford)

ste/ice simplified test equipment/internal combustion engines

Steiermark Styria, Austria

Stein Steinway

Steinbeck Grosssteinbeck (original name of author John Steinbeck's family)

stel short-tern exposure limit

STEL Studenta Tutmonda Esperantista Liga (Esperanto—Worldwide Esperanto Students League)

STELA System for Tracking Export License Applications

Stell Estella; Estelle

Stella Estella; Estelle

stellar star tracker for economical-long-life attitude reference

STELO Studenta Tutmonda Esperantista Ligo, (Esperanto--World League of Esperanto Students)

stem scanning transmission electron microscope; storable tubular extendible member

STEM stay time excursion module

sten stencil

Sten (Swedish—cliff); *Stenón* (Greek—pass, strait)

Sten gun Sheppard and Turpin Bren gun (submachine gun)

steno stenographer; stenography; stenotype; stenotypy

steno. (Latin prefix—narrow)—stenosis

stent stentando (Italian—delaying)

Step Stephen

STEP Safety Test Engineering Program; Science and Technology for Environmental Protection; Scientific and Technical Exploitation Program; Secondary Teachers Education Program; Sequential Tests of Educational Progress; Short-Term Elective Program; Solutions to Employment Problems; Systematic Training for Effective Parenting; Systems to Encourage Potential

Steph Stephen

STEPS Solar Thermionic Electric Power System; Specialized Training and Employment Placement Service

ster stereoscope; stereotype; sterilization; sterilize; sterilizer; sterling

stereo stereophonic; steroprojection; stereoprojector; stereoscope; stereoscopic

STERILE System of Terminology for Retrieval of Information through Language Engineering

stet let stand what has been crossed out; stetted; setting

stet (Latin—let it stand)— proofreader's mark

STETF Solar Total Energy Test Facility (ERDA)

Stetson Stetson hat [broadbrim high-crown hat made by John B Stetson (trademark)]

stev stevedore; stevedoring

Steve Stephan, Stephen; Steven

Stew Stewart

stewbum man sexually attracted to flight attendant(s)

stew(s) steward(esses), flight attendant(s)

STEWS Shipboard Tactical Electronic Warfare System

stewzoo hotel catering to flight attendants resting between flights

St Ex Stock Exchange

stf soluble thymic factor; staff

STF Salt Test Facility; Sycamore Test Facility

STF Svenska Turisforeningen (Swedish—Swedish Tourist Information)

s-t fibers slow-twitch fibers

st fm stretcher form

stg seating; stage; staging; steering; sterling; storage

STG Study Group

STG Schiffbautechnishe Gesellschaft (German—Shipbuilding Technical Association)

stg ar staging area

stge storage; strings

stgg staging

Stgo Santiago

Stgo de C Santiago de Chile (Compostela, Cuba)

stgr stringer

STgt secondary target

STGWU Scottish Transport and General Workers' Union

sth south(ern); straight to hell

sth (STH) somatotropbic hormone

Sth Stockholm

St Hel St Helena; St Helens; St Helier

St Hil St Hilda's College, Oxford

Sthlm Stockholm

St Hug St Hugh's College, Oxford

sti service and taxes included; sure to inquire; sure to investigate; surface transfer impedance

sti (STI) scientific and technical information

s & ti scientific and technical information

St I St Ives

STI Service Tools Institute; Software Technology Incorporated; Space Technology Institute; Steel Tank Institute

STIA Scientific, Technological, and International Affairs Directorate (National Science Foundation)

STIAD Scientific, Technological, and International Affairs Directorate (NSF)

stic serum trypsin inhibitory capacity

STIC Scientific and Technical Intelligence Center

STICAP Stiff Circuit Analysis Program

stiction static friction

STID Scientific and Technical Information Division (NASA)

STIF Scientific and Technical Information Facility (NASA)

stiff. stiffener; stiffened corpse

Stikines Stikine Mountains of British Columbia

stillat. stillatim (Latin—by drops, in small amounts)

stilli stillicide; stillicidium; stilliform

stillson stillson wrench

stim stimulant

stimn stimulation

STIMS Scientific and Technical Modular System

STINA Steel Tube Institute of North America

stinfo scientific and technical information

STING Stellar Inertial Guidance (System)

STINGS Stellar Inertial Guidance System (USAF)

stip short term incentive plan; stipend(iary); stipulation

STIP Science Teaching Improvement Program; Skills Training Improvement Program

STIPIS Scientific, Technical, Intelligence, and Program Information Service (HEW)

Stir Stirling

Stirner Max Stirner whose original name was Kaspar Schmidt

STIS Scientific and Technological Information Services; Specialized Textile Information Service

STISS Scientific and Technical Information Services and Systems

St J St John (New Brunswick)

STJ Special Trial Judge

STJC South Texas Junior College; Southwest Texas Junior College

STJM St. Jude Medical Incorporated

St Joe St Joe Minerals Corporation (energy and metals plus natural gas)

St Joh St John's College (Oxford)

StJU St John's University

stjw stretcher jaws

stk sticky; stock

Stk Stockton

STK Standard Test Key

St Kitts West Indian islands of Anguilla, Nevis, and St Christopher

St K-N St Kitts officially St Christopher; St Kitts-Nevis (West Indian island nation gained independence in 1983)

St K-N-A St Kitts-Nevis-Anguilla (Caribbean island federation)

stl steel; studio transmitter link

Stl Schottky transistor logic

St L St Louis

STL Seatrain Lines; Space Technology Laboratories (Thompson - Ramo - Wooldridge); Speech Transmission Laboratory; Standard Telecommunication Laboratories;

St. Louis Brown Stockings (National Association); St Louis, Missouri (airport); studio transmitter link (FM); Swedish Transatlantic Line

StLe St Louis encephalitis

StLGR Saint Lucia Game Reserve (South Africa)

STLL Submarine Tender Load List

St Lo St Louis

STLO Scientific and Technical Liaison Office(r)

STLOs Scientific/Technical Liaison Offices

STLOUISPDis St Louis Procurement District (US Army)

St L P-D *St Louis Post-Dispatch*

stlr semi-trailer

St L SW St Louis Southwestern (railroad)

STLT studio transmitter link—TV

St LU St Lucia; St Louis University

STLU Seatrain Line (container) Unit

St Luc St. Lucia (whose capital is Castries)

St L YC St Louis Yacht Club

St L ZG St Louis Zoological Garden

stm sternocleidomastoid muscle; supersonic tactical missile

stm (STM) scanning tunneling microscope; scientific, technical, and medical; shielded tunable magnatron; short-term memory; special test missile; surface-to-target missile; synthetic timing mode

St M St Malo

STM Science Teaching Museum (Franklin Institute); System Training Mission

S.T.M. *Sacrae Theologiae Magister* (Latin—Master of Sacred Theology)

St Martin's St Martin's Press

stmev storm evasion

stmftr steamfitter

stmn stimulation

Stmn *The Statesman* (Calcutta)

stmnt statement (flow chart)

STMP Scientific, Technical, and Medical Publishers

stmrs steamers

STMSA Scottish Timber Merchants' and Sawmillers' Association

stmt statement

stn stain

Stn Station

St N St Nazaire

STN The Scientific and Technical Information Network

stnd stained

stnry stationary

stnwr stoneware

sto standard temperature and pressure; standing order; stoker; stop; stoppage

Sto *Santo* (Spanish—saint); *Señorito* (Spanish—master; young gentleman)

St° *Santo* (Portuguese or Spanish—Saint)

STO Sea Transport Office(r); Stockholm, Sweden (Arlanda Airport)

STO *Service Travail Obligatoire* (French—Obligatory Labor Service)—Vichy-instituted law giving the Germans a massive labor force during World War II

Stock Stockholm

Stokowski Leopold Antoni Stanislaw Boleslawowicz (1882—1977)

stol short takeoff and landing

stolport short-takeoff-and-landing airport

stol/ved short takeoff and landing/vertical climb and descent

stom stomach

stoma (Greek—mouth or opening)—cyclostome, prostostome, stomatic

stomat stomatology

STOMP Short-Term Offshore Measurement Program

stomy (Latin suffix—surgical opening)—tracheostomy

S'ton Southampton

STon short ton

Stonys Stony Mountains (early American name for the Rockies)

S'toon Saskatoon

stop slight touch on pedal; spin tires on pavement

STOP Single Title Order Plan; Strategic Orbit Point; Study of Protection (against nuclear warfare)

STOPP Society of Teachers Opposed to Physical Punishment

stops stabilized-terrain optical-position sensor

STOPS Self-contained Tanker Offloading System

stor storage; stored

STOR Scripps Tuna Oceanographic Research

storet storage and retrieval

Storm *Storm Over Asia*

STORM Stormscale Operation and Research Meteorology

Stormont Stormon Castle—official Belfast residence of Northern Ireland's prime minister; Northern Ireland's capital district near Belfast, contains the home and office of the governor general, the House of Commons, and the Senate

stovi short takeoff with vertical landing

stow stowage

stp service time prediction; set-top box; solar-terrestrial physics; solar—terrestrial probe; step; stop

stp (STP) seawater treatment plant; shielded-twisted pair; solar thermal power; standard temperature and pressure

St P St Paul

St & P São Tomé and Principe

STP nickname of dangerous drug—methyl methoxyamphetamine; Scientifically Treated Petroleum (gasoline additive); Software Test Plan; sodium tripolyphosphate (water softener); Space Test Program (USAF); State Testing Program(s); stop the police

STP *Santo Tomé y Principe* (Spanish—St Thomas and Prince)—islands off Africa, formerly São Tome

S.T.P. *Sacrae Theologae Professor* (Latin—Professor of Sacred Theology)

st part steel partition

St Pat Saint Patrick; Saint Patrick's Day (March 17)

STPB Singapore Tourist Promotion Bureau

stpd standard temperature and pressure—dry (0°C, 760mm Hg)

St Pet St Peter's College, Oxford

St Pete St Petersburg

St-P-et-M *Saint-Pierre et Miquelon* (French—Saint Pierre and Miquelon)—French islands off Newfoundland

STPL Space Tracking Pty Ltd

St P & M St Pierre and Miquelon Islands

STPPTP Society to Protect Professional Tax People

stpr short taper; stumper

s tpr short taper

stps specific thalamic projection system

StP Sta St Pancras Station (rail terminal)

str steamer; straight; strainer; strait; strawberry; strength; structural; structure; submarine test reactor (STR)

str (STR) synchronous transmitter receiver (data processing)

str strana(y) [Czech—page(s)]

Str Strait; Stranraer; Street

Str Strasse (German—street); *Streichinstrumente* (German—stringed instrument); *Streptococcus*

STR Science and Technical Research; section, township, range; Society for Theatre Research; Southern Test Range; Stuttgart, Germany (airport); submarine test reactor

STRA State Teacher's Retirement System

strabad strategic base air defense

STRAC Strategic Army Corps

STRACNET Strategic Rail Corridor Network

STRACS Surface Traffic Control System

strad stradivarius (violin made by Antonio Stadivari or his sons Francesco and Omobono)

strad (STRAD) signal transmitting—receiving and distributing

stradap storm radar data processor

STRADS Switching, Transmitting, Receiving, and Distribution System

STRAF Strategic Army Forces

strag straggler; strategic; strategist; strategy

StragL straggler line

Strait Strait of Bab el Mandab, Bali, Bass, Belle Isle, Bering, Bosporus, Canso, Dardanelles, Denmark, Dover, Florida, Formosa, Georgia, Gibraltar, Hainan, Juan de Fuca, Korea, Lombok, Luzon, Magellan, Makassar, Mal-

acca, Messina, Molucca, Otranto, Palk, Sunda, Tiran, Torres

Straits Straits Settlements (Malaysia and Singapore); Straits of Tiran (at entrance to the Gulf of Aqaba or Eilat)

Strangeways Strangeways Prison in Manchester, England

STRAP Stretch Assembly Program

Stras Strasbourg

STRATAD Strategic Aerospace Division (USAF)

STRATCOM Strategic Communications Command (USA); Stratospheric Composition (program)

Strath Strathclyde

stratig stratigraphy

strato stratosphere

straw strawberry

STRAYS Society To Rescue Animals You've Surrendered

strbd stuurboord (Dutch—starboard)

STRC Science and Technology Research Center; Scientific, Technical, and Research Center; Scientific, Technical, and Research Commission

STREAK Surfaces Technology Research in Energetics, Atomistics, and Kinetics

Stream the Stream (Gulf Stream)

STREAMS Science Teams in Rural Environments for Aquatic Management Studies

Street The Street—London's Fleet Street (center of periodical publishing); New York's Wall Street (financial center)

Streetcar Streetcar Named Desire

strep streptococcus

STREP Ship's Test and Readiness Evaluation Procedure

stress (STRESS) structural engineering system solver

STRESS Stop the Robberies, Enjoy Safe Streets (program of the Detroit Police Department)

stret stretto [Italian—squeezed together; more rapid (as musical notes), strait]

STRI Smithsonian Tropical Research Institute

STRICOM Strike Command (US Army)

STRIDE Science and Technology for Regional Innovation and Development in Europe

strikeops strike operations

strikex strike exercise

STRIKFLANTREPEUR Striking Fleet Atlantic Representative in Europe (NATO)

STRIKFORSOUTH Striking and Forces Support, Southern Europe (USN)

string string-processing systems, technics, languages

string stringendo (Italian—accelerate)

strings stringed instruments: balalaikas, banjos, bass viols, 'cellos, dulcimers, guitars, harps, mandolins, samisens, vinas, violins, violas, zithers

STRINGS Statistical Report Integrated Generation Service

strip standard taped routines for image processing

strip (STRIP) string processing language

Strip The Strip—main street of Las Vegas, Nevada

STRIPE Swap Transferring Risk with Participating Element

STRIPS Separate Trading of Registered Interest and Principal of Securities

strl straight line

S-t-R L Save-the-Redwoods League

str lgths straight lengths

Strm Stream

STRN Standard Technical Report Number

strobe satellite tracking of balloons and emergencies

strobed stroboscopically illuminated; stroboscopically measured

strobes shared-time repair of big electronic systems

strobo stroboscope

strobotron stroboscope + electron (tube)

str off fixt store (or) office fixtures

STRS Short Tandem Repeats; State Teachers Retirement System

struc structure

struct structural

STRUT Safe *TRU* Transit

's' truth god's truth

Strv-74 Swedish light tank armed with 75mm gun

Strv-S Bofors-built Swedish medium tank with 105mm gun

strwbrd strawboard

sts scour the shower; ship-to-shore; short-term store; special treatment steel; surfaced two sides

st's sanitary towels

Sts Streets

STS Science Talent Search; Scottish Text Society; Serological Test for Syphilis; Space Transport(ation) System; Standard Test for Syphilis; Stockpile-to-Target Sequence

STS-26 26th space-shuttle mission

STS-27 27th space-shuttle mission

STSA State Technical Services Act

STSC Southwest Texas State College

STScI Space Telescope Science Institute

STSD Society of Teachers of Speech and drama

stsg split-thickness skin graft(ing)

STSI Space Telescope Science Institute

STSO Senior Technical Staff Officer

st st stocking stitch (knitting)

stt scrub the tub

St T (Port of) St Thomas

STT Medical Stenographer (USN); St Thomas, Virgin Islands (airport); Sensitization Test

S-T T Skin-Temperature Test(ing)

STTA Scottish Table Tennis Association

STTC Sheppard Technical Training Center

sttch(es) stitch(es)

ST T NHS St Thomas National Historic Site

sttr stator

STIT Space Telescope Task Team (NASA)

stu service trials unit; skin test unit; student; submersible test unit

Stu Stewart; Stuart

STU Seatrain (container) Unit

STU Styrelsen foer Teknisk Utveckling (Swedish—Board for Technical Development)

STUs Secure Telephone Units

STUC Scottish Trades Union Congress

stucco calcium sulfate

stud student

Stud Studebaker; Studies

Stud Studii (Russian—studies)

stude(s) student(s)

stud(s) student(s)

stuff system to uncover facts fast

Stuka Sturzkampfflugzeug (German—dive bomber)

stump. submersible, transportable, utility marine pump(ing)

stuns'l studdingsail

stupidental(ly) stupidly accidental(ly)

Sturt Sturt Desert in the northwest sector of New South Wales, Australia

stuvs standard unit variance scale

stv subscription television

stv (STV) submersible transport vehicle; subscription television

St V Stavanger; St Valentine; St Vincent

STV Scottish Television; Separation Test Vehicle

STV Solidaridad de Trabajadores Vascos (Spanish—Solidarity of Basque Workers)

St Val Saint Valentine; St Valentine's Day

St Val's Day Saint Valentine's Day (February 14)

stvd r stevedore

St V & G St Vincent and the Grenadines

s tv i subliminal television intoxication

STVPS Salinity, Temperature, Sound Velocity, and Pressure-Sensing System

st w storm water

STW Society of Technical Writers

ST WAPNIACLE old abbreviation mnemonic for U.S. departments in order of their creation before new ones were added and some were consolidated: State, Treasury, War, Attorney General (Justice), Post Office, Navy, Interior, Agriculture, Commerce, Labor, Education

Stwd Steward

STWE Society of Technical Writers and Editors

STWP Society of Technical Writers and Publishers

stwy stairway

stx start of test (data processing); static test stand

STX St Croix, Virgin Islands (airport)

Sty Stymie

STYCAR Screening Tests for Young Children and Retardates

S-type Jungian sensate type

Styria Steiermark, Austria

STZ South Temperate Zone; South Tropical Zone; Sterling Drugs (stock exchange symbol)

su sensation unit(s); sensor unit; service unit(s); setup; strontium unit(s); sulfur unit(s); summer

su. sumat (Latin—let him take)

s u siehe unten (German—see below)

Su Sudan; Sudanese; Sunday; unconditioned stimulus

SU Saybolt Universal; Scripture Union; Seattle University; Shaw University; Skinner Union; Southeastern University; Southwestern University; Soviet Union; Standord University; Stanford University; Stetson University; Student Union; Suffolk University; Sydney University; Syracuse University

SU Stati Uniti (Italian—United States)

SU-7 Soviet ground-attack fighter aircraft designated Fitter by NATO

SU-9 Soviet all-weather jet fighter aircraft called Fishpot by NATO

SU-11 Soviet delta-wing fighter aircraft called Flagon-A by NATO

SU-76 Soviet 76mm assault gun used in World War II and thereafter in Korea and Vietnam

SU-85 Soviet 85mm assault gun

SU-100 Soviet 100mm assault gun

SU-122 Soviet 122mm assault-gun howitzer also designated JSU-122

SU-152 Soviet 152mm assault-gun howitzer (JSU-152)

sua shipped unassembled

sua (SUA) serum uric acid

S-u-A Stratford-upon-Avon

SUA Shan United Army (Burma revolutionary force); Silver Users Association; State Universities Association

SUA Stati Uniti d'America (Italian—United States of America)

SUAB Svenska Utvecklinasaktiebolaget (Swedish—Swedish Development Corporation)

SUADPS Shipboard Uniform Automatic Data Processing System (USN)

sub subcontract(or); submarine; submerse; subordinate; substitute; suburb; subway

sub (SUB) substitute character (data processing)

sub (Latin prefix—below, beneath, under)—subterranean, subway

Sub Subic Bay; Subway

SUB Supplemental Unemployment Benefit (fund)

SUB Subbota (Russian—Saturday)

SUBABUSE (through with) substance abuse

subac subacute

SUBACLANT Submarine Allied Command, Atlantic (NATO)

SUBACS Submarine Advanced Combat System

SUBAD Submarine Air Defense

subalp subalpine

SUBAN Scottish Union of Bakers and Allied Workers

subassy subassembly

Sub Base Submarine Base

sub-bell submarine fog bell

sub chap subchapter

subcontr subcontract(or)

subcrep subcrepitant

subcut subcutaneous(ly)

subd subdivide; subdivision

subdeb subdebutante

SUBDIV Submarine Division (naval)

SUBDIZ Submarine Defense identification Zone

sub-ed sub-editor

subex submarine exercise; submerged exercise

sub. fin. coct. *sub finem coctionis* (Latin—at the end of boiling)

subfusc subfuscous (dark and dingy)

subgen. *subgenus* (Latin)

subic (SUBIC) submarine integrated control program

subing substituting

subj subject; subjunctive

subject subjective(ly)

subl sublimes

SUBLANT Submarine Forces, Atlantic (USN)

subling sublingual

sublse sublease

Sub Lt Sub-Lieutenant

subm submission; submit

submand submandibular

SUBMED Submarines Mediterranean (NATO)

SUBMEDNOREAST Submarines—Northeast Mediterranean (NATO)

submgd submerged

submtl submittal

subn subscription; substitution

SUBNOTE Submarine Notice (USN)

subor subordinate

sub-osc submarine oscillator

subot submarine bottom

SUBPA Submarine Patrol Area (USN)

SUBPAC Submarine Forces, Pacific (USN)

sub para sub paragraph

subplane submersible seaplane

sub-pro subprofessional

subprog subprogram(ming)

sub pub(s) subsidy publisher (s) (vanity publisher(s))

SUBPZ Submarine Patrol Zone (USN)

subq subsequent

subroc (SUBROC) submarine rocket

subrog subrogation

Subron Submarine Squadron

subrqmt subrequirement

subs submarines; subscription(s); subsidiary; subsistence; substantial violations; substitutes

subsafe submarine safety (program)

subsan submarine sandwich (also called sub)

sub sec subsection

subseq subsequent(y)

subset subscriber set

subsis subsistence

subsp *subspecies* (Latin)

SUBSS Submarine Schoolship (USN)

subst substantive

substa substation

substance P polypeptide found in the brain

substand substandard

substd substandard

substr substructure

subsunk submarine sunk

subsys subsystem

SUBTACGRU Submarine Tactical Group(ing)

subtopia suburban utopia

subtr subtraction

SUBTRAFAC Submarine Training Facility

sub u substitute unit

suburb suburban; suburbanite; suburbia; suburbian

SUBWESTLANT Submarine Force—Western Atlantic (NATO)

suc succeed; success; successor

suc. *succus* (Latin—juice)

Suc Sucre (Bolivian capital along with La Paz)

SUC Society of University Cartographers; Sussex University College

Succ Successori (Italian—Successors); *Succursale* (Italian—Branch)

Sucr Sucursal (Spanish—subsidiary, branch)

sucre monetary unit of Ecuador

Sucre Antonio José de Sucre—South American liberator fighting with Bolívar for freedom of Venezuela, Colombia, Ecuador, Peru, and Bolivia from Spanish rule; Mariscal Sucre (Quito, Ecuador's airport named for Marshal Sucre)

suct suction

SUCU Society for Universal Cosmic Uncertainty

sud sudden unexpected death; sudden unexplained death

Sud Sudan; Sudanese

SUD Aerovias Sud Americanas (3-letter airline coding)

Sudaf Sudáfrica (Spanish—South Africa)

SUDAM Superintêndencia do Desenvolvimento da Amazonia (Portuguese—Superintendency for the Development of Amazonia)

Sudan Democratic Republic of Sudan (Africa's biggest country), *Jumhuryat es-Sudan Al*

Democratia—formerly the Anglo-Egyptian Sudan known as Nubia in Roman times

SUDAN Sudan Airways

SUDENE Superintêndencia do Desenvolvimento do Nordeste (Portuguese—Superintendency for the Development of North-East Brazil)

SUDS Silhouetting Underwater Detecting System; Submarine Detecting System

Sud Tas Sudmen Tasavalta (Finnish—Republic of Finland)

Sue Susan; Susannah; Suzanne

SUE Standardized Unexpected Earnings

suec suéco (Spanish—Swedish); *sueco* (Portuguese—Swedish)

Suec Suecia (Spanish—Sweden); *Suécia* (Portuguese—Sweden)

SUEL Sperry Utah Engineering Laboratory

Suet Gaius Suetonius Tranquillus (Roman biographer)

suf sufficient; suffix

Suff Suffolk

suffoc suffocating

sug suggest(ion)

SUG Southern California Gas Company (stock exchange symbol)

SUGAR Services, (to diabetics through) Understanding, Grants, Assistance, Recreation

SuH Sundays and Holidays

Sui Suiza (Spanish—Switzerland)

SUI State University of Iowa

suicidol suicidologist(ic); suicidology

suid sudden unexplained infant death (crib death)

sui rep suicide report

SUIT Scottish and Universal Investment Trust

suiv suivant (French—following)

Suk Sukkot

Suky Susan; Suzanne

sul simplified user logistics; small university libraries

Sul Suleiman (Arabic—Solomon)

SUL Stanford University Libraries

Sula Sulawesi (Celebes)

sulcl set up in less than carloads

sulf sulfate; sulfur

sulfa sulfanilamide

sulfd sulfide(s)

sulfuric acid H_2SO_4

Sulli Sullivan

Sullivan Sullivan Stadium, Foxboro, Massachusetts

Suit Sultan(a)

Sulu Jolo

Sulus Sulu Islands between Indonesia and the Philippines

sum (SUM) surface-to-underwater missile

sum summary; summer; surface-to-underwater missile (SUM)

sum. sume (Latin—take)

Sum Sumatra; Sumatran; Sumer; Sumeria; Sumerian

SUM Servicio Universitario Mundial (Spanish—World University Service)

SUMCMO Summary Court-Martial Order

Sumi Sumitomo Bank

Sumitomo Sumitomo Shoji America; Sumitomo Shoji

summ summarization; summarize; summarizing

Summer Bank Summer Bank Holiday (last Monday in August in Great Britain)

SUMOC Superintendencia da Moeda e do Crédito (Portuguese—Superintendency of Money and Credit)

sumr summer

sum res summer resident

sums. summons

SUMS Sperry Univac Material System

sum. tal. sumat talem (Latin—take one like this)

sun symbolic unit number (SUN)

Sun Sunday

Sun The Baltimore Sun

SUN Solar Usage Now; Symbols, Units, and Nomenclature Commission

SUNA Sudan News Agency

Sund Sunda Islands; Sundanese

Sundarbans Sundarban creeks, half-reclaimed islands, marshes, rivers, and swamps in the Ganges delta country between Bangladesh and India

Sundas Sunda Islands of Indonesia

Sun Devil Sun Devil Stadium, Tempe, Arizona

SUNFED Special United Nations Fund for Economic Development

Sungaria Dzungaria or Zungaria region between Mongolia and Russia

SUNOCO Sun Oil Company

SunOS Sun Operating System trademark

sunrf sunroof

SUNS Sonic Underwater Navigation System

Sunset Crater Sunset Crater National Monument in north-central Arizona

SUNW Sun Microsystems Incorporated

SUNY State University of New York

SUNYAB State University of New York at Buffalo

Sun Yat-Sen's Dr Sun Yat-Sen's Birthday (November 12)

sup superb; superfine; superior; superlative; supersede(s); supine; supplement(ary); supplies; supply; support; supposition; supreme

sup supérieure (French—higher; superior, upper)

sup. supra (Latin—above)

SUP Sailors Union of the Pacific; Socialist Unity Party; Southern University Press; Stanford University Press; Sussex University Press; Syracuse University Press

SUPCE Syracuse University Publications in Continuing Education

supchg supercharger

Sup Ct Superior Court; Supreme Court

supdel superdelicious

Sup Dpo Supply Depot

supe (slang) superintendent; supernumerary; supervisor

super superficial; superfine; superheterodyne; superimposition; superintendent; superior; supermarket; supernumerary; supersede; supersession

super (Latin prefix—above, beyond, upper)—superficial, superior, superpowers; *supermercado* (Spanish—supermarket)

SUPER Skills Upgrading Program for Educational Reinforcement

superaero superaerodynamics

superan superannuated

superconduct superconductive; superconductivity; superconductor(s)

Superdome Superdome Stadium, New Orleans

superf superficie (Italian—area; surface, surface area)

superhet superheterodyne

superjet(s) supersonic jet airplane(s)

superl superlative

super(s) supercargo(s); supercharger(s); superheater(s); superheterodyne(s); superhighway(s); superhuman(s); superintendent(s); superior(s); superior court(s); superior planet(s); superlative(s); superliner(s); supermarket(s); superorganism(s); superpatriot(s); superpower(s); superscript(s); supersonic(s); superstition(s); superstructure(s); supervisor(s)

superstr superstructure

Super^te Superintendente (Spanish—superintendent)

superv supervisor

supgon super gonorrhea (resistant to all antibiotics)

suphtr superheater

SUPIR Supplementary Photographic Interpretation Report

sup. lint. super linteun (Latin—on lint)

Sup O Supply Office(r)

SUPOPS Supply Operations (DoD)

supp supplement; suppuration

supp suppositorium (Latin—suppository)

Sup P Supply Point

suppl supplement (French—supplement)

Suppl supplement

suppos suppository

supps supplementary produres; supplements

SupPt supply point

suppy supplementary

supr superior; supreme

supra (Latin prefix—above or over)—suprarenal

supra cit supra citato (Latin—cited above)

supsd supersede(d)

Sup Ship Supervisor of Shipbuilding

supt superintend; superintendent

Supt Docs Superintendent of Documents

supv supervise; supervisor

supvr supervisor

supvry supervisory

sur surface; surfacing

Sur surgery; Suriname (Netherlands Guiana)

Sur Surabaya (Indonesian—Soerabaya)

Suralco Surinam Aluminum Company

surano surface radar and navigation operation

sur art surrealistic art

surbage subway-floor garbage

surcal surveillance calibration (satellite)

Sur Cdr Surgeon Commander

SURE Symbolic Utilities Revenue Environment

sureq submit requisition

surf spent unreprocessed fuel

Sur f Suriname florin (guilder)

surf a surface area

surfactant surface active ingredient

SURFF Spent Unreprocessed Fuel Facility

SURFPA Surface Patrol Area

SURFPZ Surface Patrol Zone

surg surgeon; surgery; surgical

Surg Cdr Surgeon Commander

surge sorting, updating, report generating

Sur Gen Surgeon General

Surg Gen Surgeon General

surgiserv surgical service(s)

Surg Lt Cdr Surgeon Lieutenant Commander

Surg Maj Surgeon Major

Suri Suriname (formerly Dutch Guiana)

suric surface ship integrated control

Suriname formerly Dutch or Netherlands Guiana

surpic surface picture

surr surrender

Surr Surrogate

surrept surreptitious(ly)

SURS Surface Export Cargo System

SURSAN Superintendência de Urbanismo e Saneamento (Portuguese—Superintendency of Urbanism and Sanitation)

SURTASS Surveillance-Towed-Array Sonar System

surv survey; surveying; surveyor

SURV Standard Underwater Research Vessel

Surv Gen Surveyor General

survll surveillance

sus supressor sensitive; suspect(ed); suspected person; suspend(ed)

Sus Saybolt universal second; Susanna, The (Apocryphal) History of Sussex

SUS Scottish Union of Students; Society of University Surgeons

SUSA Scouting USA (formerly the Boy Scouts of America—BSA)

susfu situation unchanged—still fouled up

susie surface and underwater ship-intercept equipment

susp suspect(ed); suspend; suspend(ed)

susp b suspension bridge

sus. per coll. suspensio per collum (Latin—hanging by the neck)

suspn suspension

suspnd suspending

susp(s) suspect(s) [person(s) suspected]

Susque Susquehanna River flowing from western New York through Pennsylvania and Maryland before entering Chesapeake Bay

Süss Franz Xaver Süssmayr

SUSS Society of Utah School Superintendents

sust sustainer

SUSTA Southern United States Trade Association

Susx Sussex

SUT Society for Underwater Technology

Suth Sutherland

s'uth'ard southward

SuU Staats and Universitäts-bibliothek (German—State and University Library)—Hamburg

suud sudden unexpected unexplained death

suv sport utility vehicle

SUV Saybolt Universal Viscosity; Sons of Union Veterans; Suva, Fiji Islands (Nandi Airport)

SUVCW Sons of Union Veterans of the Civil War

SUX Sioux City, Iowa (airport)

Suz Suez

sv (RCA patent); safety valve; sailing vessel (SV); save(s); scientific visualization; security violator; selectavision (SV); sheet vinyl; simian virus; single vibrations; sinus venosus; stroke volume; survey; surveyor

sv% save percentage

s/v surrender value; survivability/vulnerability

sv sotto voce (Italian—in an undertone, in a whisper); *svacek* (Czecim—volume); *svensk* (Dano-Norwegian—Swedish)

s.v. spiritus vini (Latin—alcohol); *sub verbo* or *sub voce* (Latin—under the word; under the voice)

Sv Svaty (Czechoslovakian—holy); *Sveti* (Serbo-Croatian—holy)

SV El Salvador (Internet code); sailing vessel; Selective Volunteer; small-capitalization value (stocks); Sons of Veterans

S & V Sinclair and Valentine

SV Standard Version

sv 40 simian virus 40

Sva Suva

SVA Schweizerische Vereinigung für Atomenergie (German—Swiss Association for Atomic Energy)

Sval Svalbard (Spitsbergen)

Svalbard (Norwegian—Spitsbergen)—Arctic islands

svar shareholder value at risk

Svb Svendborg

SVB Stephen Vincent Benét

svc service; superior vena cava

svc (SVC) service (flow chart); supervisor call(ing)

SVC Skagit Valley College; Society of Vaccum Coaters

svcbl serviceable

SVCP Special Virus Cancer Program

svcs superior vena cava syndrome

svd spontaneous vaginal delivery; spontaneous vertex delivery; swine vesicular disease

SVD Schweizerische Vereinigung für Dokumentation (German—Swiss Documentation Association)

sve secure voice equipment

Sve Sveits (Norwegian—Switzerland)

SVE Society for Visual Education

Sven Sven Delblanc (Swedish novelist)

Sven Akad Svenska Akademien (Swedish Academy)

Sver Sverdlovsk; Sverige (Swedish Academy); *Sverige* (Norwegian—Sweden)

svg saving

SVG San Vincente y las Granadinas (Spanish—St Vincent and the Grenadines)

s. v. gal, spiritus vini gallici (Latin—brandy)

SVHs Senior Vacation Hotels

svi stroke volume index

s.v.i. spiritus vini industrialis (Latin—industrial alcohol)

SVIA Specialty Vehicle Institute of America

svib strong vocational interest blank

SVIOC South Varanger Iron Ore Company

SVL Scripps Visibility Laboratory

s.v.m. spiritus vini methylatus (Latin—methyl alcoholic)

SVN Student Vocational Nurse

Syn Dag Svenska Dagbladet (Swedish Daily Blade)

SVnese South Vietnamese

SVNV Societa Veneziana di Navigazione a Vapore (Venetian Steamship Company)

SVO Moscow, Russia (Sheremetyevo Airport); Special Vehicle Operation

SVP Social Venture Partners; Society of Vertebrate Paleontology

S V P s'il vous plâit (French—if you please)

SVPs Senior Vice Presidents

svr super video recorder

s.v.r. spiritus vini rectificatus (Latin—rectified spirit of wine)

SVR Suomen Valtion Rautatiet (Finnish—Finnish State Railways)

SVRA Sportscar Vintage Racing Association

svs games entered in save situation (baseball)

sv's security violators

SVS Society for Vascular Surgery; Society for Visiting Scientists; Still-camera Video System

SVS Sveriges Standardiseringkommission (Swedish—Swedish Standards Commission)

s.v.t. spiritus vini tenuis (Latin—proof alcohol; proof spirit)

svstd sound velocimeter/salinity temperature depth (recorder)

SVT Self-Valuation Test

SVTL Services Valve Testing Laboratory

svtol (SVTOL) short/vertical takeoff and landing

svtp sound, velocity, temperature, pressure

svtt surface-vessel torpedo tube

s.v.v. sit venia verbo (Latin—forgive the expression)

svws synthetic voice warning system

svy survey

sw salt water; sea water; sent wrong; shipper's weights; short wave; shotgun wedding; single weight; slow wave; special weapon; spotweld; spotwelding; station wagon; steelworker; stock width; swear; swell organ; switch; switch hitter (baseball); switchband wound; sworn

s-w shortwave

s/w salt water; sea water; seaworthy; standard weight

s & w salaries and wages; surveillance and warning

SW Swarthmore; Sweden; Swedish; Sadler's Wells (London theater); Secretary of War; Security Watch; Senior Warden; Shelter Warden; Ship's Warrant; South Wales; southwest; Southwest Airways (2-letter coding); Stone & Webster (stock exchange symbol)

S-W Sherwin-Williams

S & W Seaboard & Western (airlines); Smith & Wesson; Stone & Webster

SW1, SW2, etc. Southwest One, Southwest Two, etc. (London postal zones)

swa single-wire armored; superwide angle

Swa Swahili

SWA Seaboard World Airlines; South-West Africa; Southwest Airways

SWAA Southwestern Aeronautical Association

swabk sealed with a big kiss

swac special warhead arming control

Swac Standards western automatic compiler (NBS)

SWAC South-West Africa Company

SWACS Space Warning and Control System

SWAFAC Southwest Atlantic Fisheries Advisory Commission

swag(s) scientific wild-assed guess(es)

SWAI South-West African Infantry

swak sealed with a kiss

SWALCAP South-West Academic Libraries Cooperative Automation Project

swalk sealed with a loving kiss

swami software-aided multifont input

Swans Swan Islands off Honduras

SWANU South-West Africa National Union

SWANUF South-West Africa National United Front

swap selective wide-area paging

SWAPO South-West Africa People's Organization

swash sea wash (scouring surf running up a beach after a wave breaks)

SWAT Special Weapons and Tactics (team of law-enforcement officers trained to combat guerrillas and terrorists); Students Working Against Tobacco

swath small waterplane-area twin hull

SWATH Small Waterplane—Area Twin Hull (craft designed for stability in rough seas)

swatson so what's on?

s waves secondary (earthquake) waves

S-waves shear waves

Swaz Swaziland

Swaziland Kingdom of Swaziland (landlocked South African country)

swb short wheelbase; single with bath; swing bridge

SWB South Wales Borderers

swbd switchboard

swbld switchblade (knife or stiletto)

swbm still-water bending moments

S & W bracelets Smith and Wesson handcuffs

SWBRC Southwest Border Regional Commission

swc specific water content; stall warning computer

SWC Simon Wiesenthal Center; Soil and Water Conservation (US Department of Agriculture); Special Weapons Command; Supreme War Council

SWCEL Southwestern Cooperative Educational Laboratory

Swch Switch

SWCHS Simon Wiesenthal Center for Holocaust Studies (Yeshiva University)

SWCL special warfare craft, light (naval symbol)

SWCLR Southwest Council of La Raza

SWCM special warfare craft, medium (naval symbol)

swd sawed; sewed; short-wave diathermy; switching duty (circuit breaker)

SWD South Wales Docks

SWDA Scottish Wholesale Druggists' Association; Solid-Waste Disposal Act (EPA)

swdfshtg sawed-off shotgun

Swe Swede(n); Swedes; Swedish

SWE Society of Wine Educators; Society of Women Engineers

sweat(s) sweatshirt(s)

SWEB South Wales Electricity Board; South West Electricity Board

SWEC Stone & Webster Engineering Corporation

Swed Swede; Sweden; Swedish

Swed *Swedish* (Germanic language)

Sweden Kingdom of Sweden, *Konungariket Sverige*

SWEDL Southwestern Educational Development Laboratory

Sweetwaters Sweetwater Mountains of California and Nevada

SWETM Society of West End Theatre Managers

swf single white female

SWF Stockholders for World Freedom

SWFB Southwestern Freight Bureau

Sw Fr Swiss franc

sw fx spotweld fixture

SWG Society of Women Geographers; Standard Wire Gauge

Sw-Ger Swiss-German (derived from Alemannic)

swi stroke work index

Swi Swietochlowice

SWI Spring Washer Institute

SWIE South Wales Institute of Engineers

swife sexual wife

swift selected words in full title

SWIFT Society for Worldwide Interbank Financial Telecommunication

swift lass signal word index of field and title—literature abstract specialized search

swift signal word index of field and title—scientific information retrieval

SWIM Saturation Work-Incentive Model

swinc softwire integrated numerical control

SWINE Students Wildly Indignant (about) Nearly Everything (cartoonist Al Capp's contribution to contemporary acronyms)

Swinglish Swedish-English

SWIO SACLant War Intelligence Organization

swir short-wave infrared

SWIR Special Weapons Inspection Report

SWIRL South Western Industrial Research Limited

SWIRS Solid Waste Information Retrieval System

Swiss Swissair

SWISSAIR Swiss Air Transport

Swiss Confed Swiss Confederation

switch switchblade knife

Switz Switzerland

Switzerland Swiss Confederation of Cantons (Alpine nation of great productivity and high-quality workmanship)—

Schweiz (German or Romansch), *Suisse* (French), *Svizzera* (Italian)

swives sexual wives

sw.Jf single white Jewish female

sw.Jm single white Jewish male

Sw kr Swedish krona (monetary unit)

swl short wave listener

SWL safe working load (for cargo booms and derricks; SWL 5T 15 deg means the safe working load is 5 tons at 15 degrees off the horizontal); Swedish American Line

SWLA Southwestern Library Association

SWLI Southwestern Louisiana Institute

swlolak's sealed with lots of love and kisses

SWly south-westerly

swm single white male; standards, weights, and measures; surface-water management

SWM Southwest Museum

SWMA Steel Wool Manufacturer's Association

swmbo she who must be obeyed

SWMF South Wales Miners' Federation

SWMFB Southwestern Motor Freight Board

SWMP Solid Waste Management Plan

Swn Swinoujscie

SWN Synoptic Weather Network

Swnbne Swanbourne

SWO Solid Waste Office (Environmental Protection Agency)

SWOA Scottish Woodland Owners' Association

swoc subject word out of context

swog special weapons overflight guide

SWOOPE Students Watching Over Our Planet Earth

SWOPSI Stanford Workshops on Political and Social Issues

SWORCC Southwestern Ohio Regional Computer Center

's' word god's word

SWORDS Shallow-Water Oceanographic Research Data System

SWORL Southwestern Ohio Regional Libraries

swot strengths, weaknesses, opportunities, threats

's' wounds god's wounds

swp safe working pressure; sewer(age) planned; state water project; sweep; sweeper; sweeping

SWP Saskatoon Wheat Pool; Sherwin-Williams Paints; Society for Women in Plastics; Socialist Workers Party; South Wales Ports; Southwest Pacific; Special Weapons Project

SWPA Southwest Pacific Area; Southwestern Power Administration; Submersible Wastewater Pump Association; Surplus War Property Administration

swpf short wave-pass filter

swr serum wassermann reaction; sewer(age); standing-wave ratio; steel-wire rope; switch rails

swrf sine wave response filter

SWRI Sealant Waterproofing and Restoration Institute; Southwest Research Institute (Boulder, Colorado)

S-W RI Sterling-Winthrop Research Institute

swrj split wing ramjet

SWRL Southwest Regional Laboratory

sws seam-welding system; service-wide supply; slow-wave sleep; solar-wind spectrometer; still water surface

Sws Swansea

SWS Sariska Wildlife Sanctuary (India); Space Weapons System; Special Weapons System

SWSC Schlumberger Well Surveying Corporation

S & W S C Space and Warfare Systems Command (USN)

s-w sleep slow-wave sleep

swt short-wave transmission; short-wave transmitter; single weight; spiral(ly)wrap(ped) tubing; steel watertight; switch(ing)

SWT School of Welding Technology; Scottish Wildlife Trust

SWTB Surface Wellbore Test Bank

SWTC Scottish Woolen Technical College

swtchmn switchman

SWTEA Scottish Woolen Trade Employers' Association

swtg switching

SWTMA Scottish Woolen Trade Mark Association

SWTS Seabury Western Theological Seminary

SWUS Southwestern United States

swv swivel

SWWJ Society of Women Writers and Journalists

swy slipway; stopway

swymmd see what you made me do

sx section; simplex

Sx: (medical) signs and symptoms

SX Essex; sex; Southern Pacific (stock exchange symbol)

sxa stored index to address

SXC super sex combs

SXC Saint Xavier College

sxl short-arc xenon lamp

SXM St Maarten, Netherlands Antilles (airport)

sxn section

SXO Senior Experimental Officer

sxr soft X-ray region

sxrm straight reamer

sxs stellary X-ray spectra

SXS Sigma Xi Society

sxt sextant; stable X-ray transmitter

sy shipyard; square yard; sticky; supply; sustainer yaw

Sy Shipyard; Syria; Syrian

SY San Ysidro; South Yorkshire; steam yacht (naval symbol); (U.S. State Department) Security Office; Syria (Internet code)

Syb Sybil

SYB *Statesman's Year-Book*

SYC Sandusky Yacht Club; Savannah Yacht Club; Seattle Yacht Club; Springfield Yacht Club; Stamford Yacht Club

SYCATE Symptom-Cause Test

sycom synchronous communication(s)

sy crs sundry creditors

syc&s show your card & save

syd see your doctor; sum of the year's digits

Syd Sydney

Syd *sydlig* (Danish—southerly)

SYD Scotland Yard; Sydney, Australia (airport)

S Yem South Yemen
SYEP Summer Youth Employment Program
syf syphilis
syfa system for application
SyG Secretary General
SYGN Synergen Incorporated
syh see you home
SYHA Scottish Youth Hostels Association
Sy'kat Syarikat
syl syllogism
syla-iawc see you later, alligator—in a while, crocodile
syll syllabication (syllabification)
syllo syllogism; syllogistic(al)(ly); syllogist; syllogize(d); syllogizing
SYLP Support Your Local Police
Sylv Sylva; Sylvain; Sylvan(der); Sylvanus; Sylvester; Sylvius
sym symbol; symbolic; symbolism; symmetric; symmetrical; symmetry; symphonic; symphony
sym (Latin prefix—together)—symphony; *symphonie* (French—symphony)
sym. *symbolus* (Latin-token; sign)
symb symbol; symbolic; symbolism
symbal symbolic algebra
Sym Fan *Symphonie Fantastique*
symp symposia; symposium
sympac symbolic program for automatic control
sympath sympathetic; sympathy
Symph Mont *Orchestre symphonique de Montréal* (French—Montreal symphony Orchestra)
symphon *symphonia* (Greek or Latin—symphony)
symps symptoms
sympt symptom(s)
SYMRAP Symbolic Reliability Analysis Program
SYMRO System Management Research Operation
SYMS Symmetrical System
Sym & Signs Symbols & Signs
SYNWARR System for Estimating Wartime Attrition and Replacement Requirements

syn syndicate(d); synagogue; synesthesia; synonym; synonymous; synonymy; syntax; synthetic
syn (SYN) synchronous idle character (data processing)
syn (Greek—together or with)—synapsis, syndrome
Syn Synagogue
Synanon anti-drug addiction group
sync synchronize; synchronous
synchro synchronize; synchronous
synchros synchronous devices
synco syncopate(d); syncopation; syncopative; syncopator
syncom synchronous communication (satellite)
syncon synergistic convergence
syncop syncopate(d); syncope
syncrude(s) synthetic crude oil(s)
synd syndicalism; syndicate
syndet(s) synthetic detergent(s)
syndex syndicated exclusivity
syndro syndrome
syne syntactic elements
synec synecdoche
Synfuel U.S. Synthetic Fuel Corporation
synfuel(s) synthetic fuel(s)
syn gas synthetic gas
SYNMAS Synchronous Missile Alarm System
syn oil synthetic oil
synon synomymous; synonym
synonym. synonymous
synop synopsis; synopic
synroc synthetic rock
syns synopsis
synscp synchroscope
synt syntax
syntan synthetic tanning
synth synthesis; synthetic
synth-pop synthesized popular musmc
syntol syntagmatic organization of language
syntrain synthetic training (aviation)
syntran syntax translation
S Yorks South Yorkshire
SYP Society of Young Publishers
syph syphilis; syphilitic
syphil syphilology
Sy PO Supply Petty Officer

SYPR Southern Yemen People's Republic
syr syrup
syr. *syrupus* (Latin—syrup)
Syr Syracusan; Syracuse; Syria; Syriac; Syrian
SYR Syracuse, New York (airport)
Syrac Syracusan; Syracuse
syrg syringe
Syria Syrian Arab Republic (Middle Eastern nation), *al-Jamhouriya al Arabia as-Souriya*
syrm save-your-rear memorandum
sys system; systematic; systematization; systematize; systemic; systems
SYS Sun Yat-sen
sysabend system abnormal end(ing)
sysep system card punch(ing)
sysda system direct access
sysgen systems generation
sysin system input
syslib system library
syslined system linkage editor
syslmod system load module
sysop systems operator
sysout system output
SYSP Sixth-Year Specialist Program (library science)
Sys PO Systems Program Office(r)
syssq system sequential
syst system; systematic; systemic; systems
System ABC System of Automation of Bibliography through Computerization
systo systems officer
systol systolic
systran systems analysis translator
sysut system utility (data sets)
syt sweet young thing
syz syzgetic; syzygial; syzygium; syzygy (alignment of the Earth, Moon, and Sun resulting in unusually high tides and weather disturbances)
sz schizophrenia; schizophrenic; seizure; size; stratum zonal
s Z *seinerzeit* (German—at that time)
Sz Swiss; Switzerland
SZ Swaziland (Internet code)
sza solar zenith angle
SZA Student Zionist Association

Szb Salzburg

SZG Salzburg, Austria (airport); Soviet Zone (in) Germany

Szle *Szemle* (Hungarian—journal, review)

Szn Szczecin (formerly Stettin—Stn)

SZO Student Zionist Organization

SZOG Soviet Zone of Occupation in Germany

szr (SZR) sodium-cooled zirconium-hydride moderated reactor

szvr silicon zener voltage regulator

T

t airfoil temperature thickness (symbol); hour angle (symbol); meridian angle (symbol); table; tabulated (loran); tackle; tardy; tare; target; taxpayer; teaspoon; teeth; telecommunications; telephone; temperature; temporary; tenor; tense; tensor; tentative; tentative target; territorial; term; territory; tetratype; thunder; thunderstorm; tide; tide tips; time; title; today; tonnage; tons; torn; toward; town; township; trace of precipitation; tradesman; train; transferred; transformer; transit; transitive; translation; tread; tropical; troy; true; tug; tugline; turf

t(t) units of land (microeconomics)

t (T) tea (marijuana)

t' the; to

't it

t tome (French—volume); tomo (Spanish—volume)

t. ter (Latin—three times; thrice)

't het (Dutch—the)

tø no evidence of primary tumor (SYMBOL)

T Northrup Aircraft (symbol); Pacific Transport Lines; primary tumor (symbol); propeller thrust (symbol); tablespoon; tactical; Tango—code for letter T; tanker; Taoism; Taoist; T-bar; tee; teletype; Telegraphist; temperature; temple; temporary magnitude; tension of eyeball; Tera (trillion); Tesla; Testla; Texaco; Texas; Texas Company; thunderstorm (aircraft code); Thursday; thymine; torpedo; total gas (symbol); town(ship); trainer; training; Transamerica (airline); transducers; transport number; Treasury (as in T-bill, T-bond, T-note); triangle; triple bond; true; truss; Tuesday; turboprop; Turk; Turkey; Turkish

T (Latin—Titus); Teil (German—division, part); thrust (symbol); Time (magazine); transformer (symbol); tulo (Finnish—arrival)

T+ severe thunderstorm (aircraft code)

T-1 absolute temperature or transformer symbol; Canadian income-tax return

$t^1/_2$ radioactive half life

T – 1, T – 2, T – 3, etc. decreasing stages of interocular tension

T + 1, T + 2, T + 3, etc. increasing stages of interocular tension

T_1, T_2, T_3, etc. first thoracic vertebra, second thoracic vertebra, third thoracic vertebra, etc;

T2 stabilized

T-2 North American-Rockwell Buckeye trainer aircraft; tricothecenes used in biochemical warfare

T2g Technician (second grade)

T_3 triiodothyronine

t-4 therefore

T4 heat treated

T-4 Canadian statement of employment income recorded for tax purposes

T_4 thyroxine

T6 heat treated and aged

T-6 North American-Rockwell Harvard or Texan trainer aircraft

T7 heat treated and stabilized

T-7 Beechcraft navigational-training aircraft

T-10 Soviet heavy tank armed with a 122mm gun

T-11 Beechcraft bomber-training aircraft

T-28 North American Trojan trainer aircraft

T-29 Convair military transport also called Samaritan

T-33 Lockheed Shooting Star trainer aircraft

T-34 Beechcraft Mentor trainer aircraft; Soviet medium tank armed with an 85mm gun

T-37 Cessna Dragonfly twin-engine jet trainer

T-39 North American Sabreliner transport aircraft

T-41 Cessna 172 Mescalero trainer-utility aircraft

T-42 Beech Cochise transport aircraft

T-43 Boeing navigational trainer and transport aircraft; Soviet fleet minesweeper

T51 specially aged

T-54 Soviet medium tank

T-55 Soviet medium tank armed with a 100mm gun

T-59 mainland-China-made medium tank modeled after Soviet T-54 tank

T-62 Soviet medium tank with a 115mm gun

T-64 Soviet medium tank within a 120mm gun

T-104 Tupolev 104 aircraft

T-144 Tupolev 144 (Soviet supersonic transport)

T-301 Soviet coastal minesweeper

T-1824 Evans blue

T3-RIA Triiodothyronine by radioimmunoassay (symbol)

T3-UR Triiodothyronine uptake ratio

T4-RIA Thyroxine (determination by) radioimmunoassay (symbol)

ta alkaline; axillary; talus; target area; temperature, test accessory; third attack (lacrosse); time and attendance; toxinantitoxin; trade acceptance; training allowance; transactional analysis; transverse acoustic; travel allowance; true altitude; tuberculin

ta (TA) teaching assistant; terephthalic acid; transactional analysis

t-a toxin-antitoxin

t/a trading as

t & a taken and accepted; time and attendance; tonsillectomy and adenoidectomy, tonsils and adenoids

t of a terms of agreement

ta transit authority (New York City Transit Authority)

t.a. *testantibus actis* (Latin—as the records show)

Ta tantalum; Tasmania; Tasmanian

TA Table of Allowances; tactical air (missile); Tax Amortization; Teaching Assistant; Technical Assistance; Tel Aviv, Israel; Territorial Army; Trade Agreement(s); Transactional Analysis; Trans-Air; Trans-America Corporation (stock exchange symbol); Transfer Agent; Transit Authority; Truth in Advertising; Turkish Army

T-A Tacna-Arica (on the border of Peru and Chile)

T/A Teaching Assistant; Temporary Assistant

T of A *Timon of Athens*

taa turbine-alternator assembly

TAA Technical Assistance Administration; Temporary Assistance Authority; Territorial Army Association; Trade Adjustment Assistance; Trade Agreements Act; Trans-Australia Airlines; Transit Advertising Association; Transportation Association of America

TAAA Texas Amateur Archeologists Association

TAACOM Theater Army Area Command

TAAF *Terres Australes et Antarctiques Françaises* (French—French Austral and Antarctic Territories)— Adélie Land in Antarctica plus the islands of Amsterdam and St Paul, the Crozets, and the Kerguelans in the south Indian Ocean

TAAFA Territorial Army and Air Force Association

TAAG *Transportes Aéreos de Angola* (Portuguese—Air Transports of Angola)

taalk *taalkunde* (Dutch—linguistics)

TAALODS The Army's Automated Logistic Data System

TAALS The American Association of Language Specialists

TAAN Transworld Advertising Agency Network

TAAP Total Action Against Poverty

TAARS The Army Ammunition Reporting Service

taas three-axis attitude sensor

TAAS Telfair Academy of Arts and Sciences (Savannah)

TAASA Tool and Alloy Steels Association

TAASP The Association for the Anthropological Study of Play

tab. table; tablet; tabulate(d); tabulation; tabulator; tax anticipation bill; technical assistance broker(age); therapeutic abortion

tab. *tabella* (Latin—small board; tablet)

Tab Tabascan; Tabasco

Tab *Tabelle* (German—table; index)

TAB Technical Assistance Board (UN); Tobago (airport); Totalisator Agency Board; Totalizator Board

TAB *Technical Abstract Bulletin*

TABA The American Book Award(s)

TABA *Transportes Aéreos Buenos Aires* (Spanish—Buenos Aires Air Transport)

TABAN Table Analysis

tabasco tabasco sauce

Tabby Tabitha

tabc typhoid-paratyphoid A, B, and C vaccine (TABC)

tabel *tabella* (Latin—tablet)

TABL Tropical Atlantic Biological Laboratory

tabl(s) tablet(s)

TABMM Traffic Audit Bureau for Media Measurement

tab run tabulator run

tab(s) tablet(s)

Tabs Cantabrigians or Cantabs—Cambridge University undergraduates

TABS Transatlantic Book Service

tabsim tabulating simulator

TABSO *Transport Aerien Civil Bulgare* (Bulgarian Civil Air Transport)

tabsol tabular systems-oriented language

tabt tab vaccine plus tetanus toxoid (TABT)

tabtd combined tab vaccine plus tetanus and diphtheria toxoid

TAB vaccine typhoid plus paratyphoid A and B vaccine (triple vaccine)

tabwx tactical air base weather

tac tactic; tactical; tactician; tactics; total automatic color (tv); try and collect

Tac Tacitus; Tacoma

TAC Tactical Air Command; Talent-Assistance Cooperative; Technical Advisory Committee; Technical Assistance Center; Terrain Analysis Center; Thai Airways Company; The Architects Collaborative; The Atlantic Council (of the United States); Total Allowable Catch; Trade Agreements Committee

TACA Texas and Central American Airlines

TACAMO Tactical Communication Air Mobile; Take Charge And Move Out (USN)

tacan tactical air navigation

Tac Brdg Tacoma Bridge

TACC Tacna-Arica Copper Consortium; Tactical Air Command Center; Tactical Air Control Center; Technology Assessment Consumerism Center

taccar time-averaged clutter-coherent airborne radar

tacco tactical coordinator

TACCP Tactical Command Post (USA)

TACCTA Tactical Air Commander's Terrain Analysis

tacden tactical data-entry device

TACELIS Transportable Emitter Location and Identification System

tacelron tactical electronic warfare

TACEST Tactical Test(ing)

TACG Tactical Air Control Group

tach tachometer

Tacho Anastasio

tachy tachygraphy (shorthand)

tachy (Latin prefix—rapid or swift)—tachycardia

tachycard tachycardia

tacit. tacitus (Latin—unmentioned)

tacjam tactical jammer; tactical jamming

TACIS Technical Assistance to the Commonwealth of Independent States

TACL Tactical Air Command Letter

taclan tactical landing system

TACLET Tactical Law Enforcement Team (Coast Guard)

tacmar tactical malfunction-array radar

tacnav tactical navigation

tacnuc tactical nuclear (weapon)—also written *taknuk*

TACO Tactical Coordinator

tacoda target coordinate date

tacol thinned-aperture computed lens

TACOM Tank-Automotive Command (USA)

TACOMEWS Tactical Communications Electronic Warfare Systems

Taconics Taconic Mountains ranging from New York to Vermont but called the Berkshires in Connecticut and Massachusetts

TACOS Tactical Airborne Countermeasures of Strike (USAF); Tactical Air Command Simulation

TACP Tactical Air Control Party

tacpol tactical procedure-oriented language

TACR Tactical Air Command Regulation

TACRON Tactical Air Control Squadron

TACs Technical Assistance Committees (UN)

TACS Tactical Air Control System; Total Access Communication System

t-a-c salad turkey-avocado-cheese salad

tacsat (TACSAT) tactical satellite

tacsatcom tactical satellite communications

TACSS tactical schoolship (USN)

tact. technological aids to creative thought

TACT Texas Association of College Teachers; Truth About Civil Turmoil

tactas (TACTAS) Tactical Towed Array Sonar

TACTIC Technical Advisory Committee to Influence Congress (Federation of American Scientists)

TACTICS Technical Assistance Consortium to Improve College Services

TACUS The Atlantic Council of the United States

tacv tracked air-cushion vehicle

tad tadpole; telemetry analog-to-digital (information converter); terminal area distribution (processing); traffic analysis and display; transaction application driver; throwaway detector; time available for delivery

tad (TAD) temporary additional duty

Tad Thaddeus; Theodore

TAD Theta Alpha Delta; Thrust-Augmented Delta

TAD The Anglican Digest

TADA Teletypewriter Automatic Dispatch System

TADARF Toronto Alcoholism and Drug Addiction Research Foundation (Canadian)

TADARS Tropo Automated Data Analysis Recorder System

TADC Tactical Air Direction Center, Texas Association of Developing Colleges; Training and Distribution Center

tadic telemetry analog-to-digital information computer

tad(s) tadpole(s)

TADS Tactical Air Defense System

TADS Teletypewriter Automatic Dispatch System

TADSYS Turbine Automated Design System

Tadz Tadzhik; Tadzhikistan; Tadzhikistanian

Tadzhik SSR Tadzhik Soviet Socialist Republic (Tadzhikistan)

TAE National Greek Airlines; Trans-Antarctic Expedition

TAEA Texas Art Educators Association

TAEC Turkish Atomic Energy Commission

TAEDS Texas Association for Educational Data Systems

TAEG Training Analysis and Evaluation Group (USN)

TAEHS Thomas A Edison High School

Taeko Taeko Kohno (Japanese novelist)

Tae Kin Taehan Min'guk (Korean—Republic of Korea)—South Korea

ta'en taken

TAERF Texas Atomic Energy Research Foundation

taf terminal aerodrome forecast

taf (TAF) toxoid-antitoxin floccules

Taf Tessar auto focus

Taf Bildtafel (German—list of illustrations)

TAf Tuberculin Albumose frei (German—albumose-free tuberculin)

TAF Tactical Air Force; Taxpayers Against Fraud

TAFA Territorial and Auxiliary Forces Association

tafcsd total active federal commissioned service date

Taf-d Tessar auto focus dating

tafg two-axis free gyro

TAFI Technical Association of the Fur Industry

tafmsd total active federal military service date

tafor terminal aerodrome forecast

TAFSEA Technical Applications for Southeast Asia

TAFSONOR Tactical Air Force, Southern Norway (NATO)

tafubar things are fouled up beyond all recognition

ta fx tapping fixture

tag the acronym generator (RCA device)

Tag Tagalog (the language of the Philippines)

TAG The Adjutant General; The Alzheimer Group; The Association for the Gifted; Telegraphist Air Gunner; Test Analysis Guide; Timken Art Gallery

T A & G Tennessee, Alabama & Georgia (railroad)

TAG Transport Aeriens Guyanais (French—Guiana Air Transport)

TAGA Technical Association of the Graphic Arts

Tagal Tagalog

tagawi try and get away with it

TAGCEN The Adjutant General's Center (USA)

TAGG Taxpayers Against Government Giveaways

tagl täglich (German—daily; per day)

Tagore Rabindranath Tagore

TAGP Transportes Aéreos do Guine Portuguesa (Portuguese—Air Transport of Portuguese Guinea)

TAGS Time-Automated Grid System

tagw takeoff gross weight

tah temperature, altitude, humidity; total abdominal hysterectomy

Tahiti formerly *Otaheite*

Tah Pac Tahitian Pacific (area around Tahiti)

TAHq Theater Army Headquarters

TAHRI Tobacco and Health Research Institute

tai taiga (coniferous evergreen forests of subarctic America, Asia, and Europe)

Tai Taipei; Taiwan (Formosa)

Tai Tailandia (Spanish—Thailand)—Siam

TAI Thai Airways International; Travel Agents International

TAI Transports Aériens Intercontinentaux (French—Intercontinental Air Transport)

TA & IC Texas Arts and Industries College

TAICH Technical Assistance Information Clearinghouse

taid (TAID) thrust-augmented improved delta

TAIDET Triple-Axis Inertial-Drift Erection Test

TAIDHS Tactical Air Intelligence Handling System

tail tailpiece

'taint it aint

Taipas Taipa Islands off Macao in the South China Sea

TA-ISSA Travelers Aid—International Social Service of America

TAJAG The Assistant Judge Advocate General (USA)

Taju Tajumulco

tak. taken

take 5 take a rest

take 10 take a rest

tako terms and conditions of employment

tal traffic and accident loss

tal (TAL) tetra-alkyl lead

tal. talis (Latin—such)

Tal Talara, Peru's westernmost point and westernmost point of South America; Talcahuano; Tallinn (capital of Estonia)

Tal Talmud (Hebrew canon and civil lawbook)

TAL Transair Limited

tala monetary unit of Western Samoa

TALA The American Lyceum Association (currently the International Platform Association)

Talamancas Talamanca Mountains of Costa Rica

talar tactical landing-approach radar

talbe talk and listen beacon

talc hydrous magnesium silicate (agamatolite); take a look see

TALC Tank-Automotive Logistics Command (USA); Tax Agency Liaison Committee; Texas Association for the Advancement of Local Culture

TALCM Tomahawk Air-Launched Cruise Missile

Talco Talcahuano

talff total allowable level of foreign fishing

TALIC Tyneside Association of Libraries for Industry and Commerce

talisman transfer accounting and lodgment for investors and stock management for jobbers (London Stock Exchange)

talkies talking motion pictures

Talla Tallahassee

Talladegas Talladega Mountains of Alabama

Tallahassee Institution Tallahassee Correctional Institution in Florida

TALMA Truck and Ladder Manufacturers Association

'Talo Italo

TALOA Transocean Airlines

'talpa(s) catalpa(s)

tal. qual. talis qualis (Latin—as they come, average quality)

TALUS Transportation and Land Use Study

tam tactical air missile; tambourine; tam-o'-shanter; tam-tam; total available market

tam (TAM) tactical air missile

t-a m toxoid-antitoxin mixture

Tam Tamar; Tamara; Tamil; Tamba; Tampan; Tampico (inhabitants-Tampiqueños); Tamualipas (inhabitants—Tamualipecos)

Tam. Tamid

TAM Tel Aviv Museum; Television Audience Measurement

TAM Transporte Aéreo Militar (Spanish-Paraguayan Military Air Transport)

TA & M Texas A & M University

TAMA Third Avenue Merchants' Association; Training-Aids Management Agency (USA)

TAMAR Tartaruga marinha (sea turtle)—informal name for Brazil's National Sea Turtle Conservation Program

tamb tambourine

tamb *tambor* (Spanish—drum or drummer)

TAMBF TransAmerica Migratory Bird Fund

tambo tambourine

TAMC Tripler Army Medical Center

tamco training aid for morbidic console operations

TAME Television Accessory Manufacturers Institute

tami tip air mass injection

Tamiami Tampa-Miami area

Tamiami Trail trans-Florida highway between Tampa and Miami

TAMIS Technical Meetings Information Service

Tammies Tamburitzans

Tamp Tampa, Florida; Tampico; Tampico, Mexico

Tamps Tamaulipas

TAMRC Tank-Automotive Materiel Readiness Command (USA)

TAMS Token and Medal Society

Tam Shrew *Taming of the Shrew*

TAMTU Tanzania Agricultural Machinery Testing Unit

TAMU Texas A & M University

tan. tangent; tangential; tannery; tanning; total ammonia nitrogen; twilight all night

tan (TAN) tax anticipation note

Tan Tanganyika (a former country in E Africa); Tangier; Tanzania (whose capital is Dar es Salaam)

TAN Transportes Aéreos Nacionales

Tanan Tananarive

tan. bkt tangency bracket

tandel tandem + parallel

TANESCO Tanzania Electric Supply Company

Tang Tanganyika; Tangier

Tangas Tanga Islands in the southwest Pacific near New Ireland

tangelo tangerine + pomelo (tangerine-grapefruit hybrid citrus fruit)

tanglo(s) tangelo(s)

Tango letter T radio code

tanh hyperbolic tangent

Tania Tatiana

Tanimbars Tanimbar Islands of Indonesia

Tano Cayetano

tan's tax anticipation notes (TANs)

TANS Tactical Air Navigation System; Terminal Area Navigation System; Territorial Army Nursing Service

tanstaafl there ain't no such thing as a free lunch

TANU Tanganyika African National Union

TANY Typographers Association of New York

Tanz Tanzania (Tanganyika + Zanzibar)

Tanzam Tanzania-Zambia (railway)

Tanzania United Republic of Tanzania (East African country combining Tanganyika and Zanzibar; includes the island of Pemba north of Zanzibar island

tao tactical air observation; thromboangiitis obliterans; tropical atmosphere/ocean

TAO Tactical Air Office(r); Technical Assistance Operations; Test Analysis Outline; The Athenaeum of Ohio

TAO *Taxi Aéreo Opita* (Spanish—Opita Air Taxi)—Bogotá, Colombia

TAOC Tactical Air Operations Center

TAOCC Tactical Air Operations Control Center

TAOI Tactical Area of Interest

tap tackled attempting to pass

taor tactical area of responsibility

tap technical assessment phase; telephone tap(ping); temporal adrenal profile; transient analysis program

TAP Table of Authorized Personnel; Tax Action Planning; Technical Advisory Panel; Technical Assistance Program; Telecommunication Action Program; Telephone-A-Partner; Test Analysis Program; Timesharing Assembly Program; Total Action Against Poverty; Trans-Alaska Pipeline; Trend Analysis Program; Tuition Assistance Program

TAP *Transportes Aéreos Portugueses* (Portuguese—Portuguese Air Transport)—airline; *Tunis Afrique Presse* (French—Tunis Africa Press)

tapa three-dimensional antenna-pattern analyzer

tapac tape automatic positioning and control

tape. tape automatic-preparation equipment

TAPE Target Profile Examination (USAF); Transactional Analysis of Personality and Environment; Trust for Agricultural Political Education

taphon taphonomist(ic)(al)(1y); taphonomy

TAPLAN Tax Action Planning

TAPLine Trans-Alaska Pipe Line

TAPLINE Trans-Arabian Pipeline

Taplinger Taplinger Publishing Co

TAPPI Technical Association of the Pulp and Paper Industry

taps the last bugle call, the *taptoo*, meaning *lights out* or sounding the last honors at a military funeral

taps *tapaderos* (Spanish—leather hoods covering stirrups to protect the feet while riding through thorny cactus or mezquite)

TAPS Teacher Audio Placement System; Trajectory Accuracy Prediction System (USAF); Trans-Alaska Pipeline System

TAPSC Trans-Atlantic Passenger Steamship Conference

tapvc total anomalous pulmonary venous connection

TAQ *The African Queen*

tar (TAR) tariff(s); tarpaulin(s); terminal area radar; terrain-avoidance radar

TAR Technical Action Request (USA); Territorial Army Regulations; Trans-Australian Railways

TARA Technical Assistant—Royal Artillery; Territorial Army Rifle Association

taran test and replace as necessary

TARC Tactical Air Reconnaissance Center

TARDC Tank-Automotive Research and Development Command (USA)

TARDIS Traffic and Roads Drive Integrated Systems

tare transistor analysis recording equipment

tarex target exploitation

tarfu things are really fouled up

targ target
TARGET Team to Advance Research for Gas Energy Transformation
tarmac tar plus macadam (tarred road or runway)
Tar-Man Taranaki-Manawatu (NZ)
tarn. tarnish; tarnishes; tarnishing
TARO Territorial Army Reserve Office(r)(s)
TAROM Transporturile Aeriene Romine (Romanian Air Transport)
TARP Test and Repair Processor; Transitional Aid Research Project
tarp(s) tarpaulin(s)
Tarr Tarragona
Tarryalls Tarryall Mountains of central Colorado
tars. (TARS) three-axes reference System
TARS Technical Assistance Recruitment Service
tart. tartaric
TART Test Analysis Reduction Technique (USN)
tart. a tartaric acid
tas target acquisition system; therapy abuse survivors; true airspeed
tas (TAS) torpedo anti-submarine
Tas Tasmania
TAs teaching assistants
TAS The Asia Society; Texas Academy of Science; Traveler's Aid Society
TAS The American Spectator
tasa test area support assembly
TASA Texas Association of School Administrators
TASAMS The Army Supply and Maintenance System
tasc terminal area sequence and control; treatment alternatives to street crimes
TASC The Alumni Service Cooperative; The Analytic Sciences Corporation; Telecommunications Alarm Surveillance and Control; Test Anxiety Scale for Children; Treatment Alternatives to Street Crime
tascon television automatic sequence control
TASCOM Theater Army Support Command

TASD Terminal (Railway) Alabama State Docks
TASDA The American Safe Deposit Association
TASDC Tank-Automotive Systems Development Center (USA)
tase tactical support equipment
taser taser gun (electronically activated stunning device used by law-enforcement officers)
TASES Tactical Airborne Signal Exploitation System
TASF Teachers Association of San Francisco
Tash Tashkent
TASHAL Tseva Hagana Le-Israel (Hebrew—Defense Army of Israel)
tasi time-assignment speech interpolation
TASI Torpedo and Anti-Submarine Instructor
TASK Test of Academic Skills (Stanford)
TASKFLOT task flotilla; Task Flotilla (NATO) (USN)
TASKFORNON Task Force—Northern Norway (NATO)
tasm (TASM) tactical air-to-surface missile
Tasm Tasman; Tasmania; Tasmanian
Tasmans Tasman Mountains of New Zealand's South Island
TASO Television Allocations Study Organization; Training Aids Service Office (USA)
TASP The Army Studies Program; The Army's Study Program
taspac total analysis system for production accounting and control
tasr terminal area surveillance radar
tass tactical air support squadron; technical assembly
tass (TASS) towed array surveillance system
TASS Telegrafnoie Agenstvo Sovietskavo Soyuza (Russian-Soviet News Agency)
TASSO Tactical Special Security Office(r)
TASSq Tactical Air Support Squadron (USAF)
TASSR Tartar Autonomous Soviet Socialist Republic; Tuva Autonomous Soviet Socialist Republic

TAST Tactical Assault Supply Transport
TASTE Thermal Accelerated Short Time Evaporator
tat (TAT) tetanus antitoxin; tyrosine amino transferase
t & at tank and antitank
Tat Tatar (Turkestan)
TAT Tappas Acupressure Technique; tetanus antitoxin; Thematic Apperception Test; Thrust-Augmented Thor; Touraine Air Transport; Trans-Atlantic Telephone; *Transportes Aéreos de Timor* (Timor Air Transport)
TATA Tobacco Accessories Trade Association (formerly PTA—Paraphernalia Trade Association)
Tat Aut Sov Rep Tatar Autonomous Soviet Socialist Republic
TATC Tactical Air Traffic Control; Trans-Atlantic Telephone Cable
tatca trialkoxytricarballate
tatce terminal air-traffic-control element
TATCO Tactical Automatic Telephone Central Office
'tater(s) potato(es)
TATL Trust for Appalachian Trail Lands
TATPAC Trans-Atlantic Trans-Pacific (telecommunications network linking London, Montreal, New York, Tokyo, Hong Kong, and Sydney)
Tatras Tatra Mountains of Czechoslovakia
TATSA Transportation Aircraft Test and Support Activity
Tatts Tattersalls
TATU Tanganyika African Traders Union
Tau Taurus
TAU Task Assessment Unit; Tel Aviv University
Taughannock Taughannock Falls State Park on Cayuga Lake in central New York
TAUN Technical Assistance of the United Nations
taurom tauromachia
taurom tauromaquia (Spanish—art of bullfighting)
Taurus (Latin—Bull constellation)
TAUSA Tea Association of the U.S.A.
taut. tautology

tav tavern

tav (TAV) transatmospheric vehicle (McDonnell-Douglas aircraft designed to travel at up to 20 times the speed of sound)

T-a-v *Tout-à-vous* (French— Yours truly)

TAVA The Americas Ventures Associates (international drug cartel)

Tave Octave; Octavius

Tavia Octavia

TAVINA *Trans-Colombiana de Aviación* (Spanish— TransColumbian Aviation)

Tavita Octavita

T Aviv Tel Aviv

Tavo Gustavo

T & AVR Territorial and Army Volunteer Reserve

tav(s) tavern(s)

TAVSS Toward, Away, Versus Selection System

Tavy Octavius

taw thrust-augmented wing; twice a week

T A & W Toledo, Angola & Western (railroad)

TAW *Times Atlas of the World*

TAWACS Tactical Airborne Warning and Control System

TAWC Tactical Air Warfare Center

TAWG Target Acquisition Working Group

tax. taxation; taxes; taxonomic; taxonomy

Taxco Tuxco de Alarcón

taxi taxicab; taxiing

taxid taxidermy

taxir taxonomic information retrieval

taxis or taxo (Greek—to arrange in an orderly manner)—geotaxia, phototaxis, taxonomy

taxon taxonomic(al)(ly); taxonomist(ic)(al)(ly); taxonomy

Tay Tayside

Taz Tazmania(n)

TAZ Tactical Alert Zone

taz(es) topaz(es)

tb tailback (football); tall bearded (iris); temporary buoy; terminal board; thymol blue; tile base; time bandwidth; time bank; torpedo bomber; total bases (baseball); total bouts; tractor biplane; training battalion; trial

balance; true bearing; tubercle bacillus; tuberculosis; turbine; turretbase; turretbased

t/b title block

t & b top and bottom; turned and bored

Tb terbium; Tobit

TB Tank Battalion; temporary bouy; Terabyte; Torpedo Boat; Treasury Bill; Troop Basis; Twin Branch (railroad); Tyburn (reports)

TB *Technical Bulletin*

tba to be announced; to be approved; to be assigned; to be audited; terminal board assembly; tires-batteries-accessories

TBA Tables of Basic Allowance; Television Bureau of Advertising; Torrey Botanical Association; Triborough Bridge Authority

tbab to be approved by

tbab (TBAB) tryptose blood agar base

tban to be announced

T-bar T-shaped bar

tbawrba travel by aircraft, military and/or naval water carrier, commercial rail and/or bus is authorized (USA)

tbb to be billed

TBB tenor, baritone, bass; Transnational Broadband Backbone

TBB *Television Blue Book*

tbc to be crated; to be culled

tbc (TBC) time base corrector

TBC The British Council; Trinidad Broadcasting Company

TBC Co Tropical Belt Coal Company (invented by Joseph Conrad for use in his novel *Victory*)

tbcf to be called for

tbd to be determined; to be discontinued; thousand barrels daily

TBD torpedo-boat destroyer

TBDS Test Base Dispatch Service

tbe to be edited; to be encoded; to be executed; to be expanded; to be expended; to be expired; to be expunged; time base error

tbe (TBE) tuberculin bacillen emulsion

TBE Thread Both Ends; Toronto Board of Education

T-beam T-shaped beam

tb ex tube expander

tbf to be furnished; total batters faced (baseball)

TBF single-engine torpedo bomber (naval symbol); Teachers Benevolent Fund

tbfx tube fixture

tbg to be garnished; to be gathered; testosterone-binding globulin; thyroxine-binding globulin

t & bg top and bottom grille

Tbg Tönsberg

tbh to be had (sexually available); to be held

tbhq tertiary butylhydroquinone

tbi to be invented; to be inventoried; tooth-brushing instruction; total body irradiation; traditionally black institutions

TbI Tax-based Income

TBI Tennessee Bureau of Investigation; Texas Board of Insurance; The Business Institute; The Tobacco Institute

Tbil Tbilisi, Georgia

T-bill(s) Treasury bill(s)

T-bird Thunderbird

tbj to be joined

tbk to be killed

t-bk talking-book

tbl to be labelled; table; tablet; through back of loops (knitting); through bill of lading

tb lc term birth, living child

tbm tactical ballistic missiles; temporary bench marks; to be manufactured; to be monitored; tuberculous meningitis; tunnel-boring machine

tbm (TBM) terabit memory; tired businessman

TBM Ten Broeck Mansion (Albany)

TBMA Textile Bag Manufacturers Association; Timber Building Manufacturers' Association

tb md tube mandrel

TBMD Terminal Ballistic Missile Defense (USA)

TBMEC Tientsin British Municipal Emergency Corps

TBMs Tunnel-Boring Machines

TBMS Turtle Bay Music School

tbmt transmitter buffer empty

tbn to be named; to be nominated; to be noted

TBN Trinity Broadcasting Network

tbo to be ordered; time between overhaul(s); turnaround buyout

TBO Test Base Office

tboip tentative basis of issue plan(ning)

T-bolt bolt with T-shaped square head

T-bonds Treasury bonds

tbone trombone

T-bone T-bone steak; T-shaped bone; trombone

T-bowl toilet bowl

tbp to be promoted; to be purchased; true boiling point

tbp (TBP) telephone-bill payment

tbpa thyroxine-binding prealbumin

TBPA Textile Bag and Packaging Association

Tbps Terabits per second

TBps Terabytes per second

tbq to be queried

tbr team batting rating; to-be-remembered (word); to be rented; to be restored; torpedo, bombing, reconnaissance; total baseball ranking

TBR Test of Behavioral Rigidity; Treasury Bill Rate; Treasury Bond Receipt

TBRI *Technical Book Review Index*

tbs to be sold; tablespoon; talk-between-ships (radiotelephone)

tb's tuberculosis patients; tuberculosis victims

tb & s top, bottom, and sides

TBs Torpedo Boats (World War I)

TBS Tokyo Broadcasting System; Turner Broadcasting System

tb sa tube saw

tbsd thermal-blooming slow dither

TBSI The Baker Street Irregulars

tbsn tablespoon

tbsp tablespoon

tbt to be tested; target-bearing transmitter; tolbutamide test(ing); total bottom time; trachcobronchial toilet; tributyltin

tbt (TBT) torpedo-bearing transmitter

TBT Terminal Ballistic Track

TB & TA Triborough Bridge & Tunnel Authority

tbto (TBTO) tributyl tin oxide

TBTS Tracker Breadboard Test System

tbu to be used

tbv to be vacated; to be vented; tubercle bacillus vaccine

TB & VD C Tuberculosis and Venereal Diseases Clinic

tbw thrown by wave; to be weighed; to be withheld; total body washout; total body water

tbx to be x'd (out)

tby to be young

Tbytes Terabytes

tbz to be zoned; to be zonked

tc temperature classification; temperature controlled; terra cotta; tetracycline; therapeutic community; thermocouple; thermocoupled; thermocoupling; thrust chamber; tierce(s); time check; time closing; top chord; total chances (baseball); transportation cask; trash compactor; trial color; trip coil; troop carrier; true course (TC); type certification

tc (TC) total cost

t/c tabulating card; temperature coefficient; thermocouple; transformer rectifier; trim coil; type certificate

t & c threads and couplings; turn and cough

tc *tre corde* (Italian—three strings)

Tc technetium; tropic tides

TC Air Canada (formerly TCA); The Citadel; Tabor College; Tactical Command; Taft College; Talladega College; Tank Corps; Tariff Commission; Tarkio College; Tax Court; Tea Council; Teachers College; Technical Circular; Technical Communication; Tennessee Central (railroad); Texarkana College; Texas College; Thiel College; Tift College; Time Charter; Traffic Control; Training Center; Training Circular; Training Corps; Transaction Code; Transport Command; Transportation Corps; Transylvania College; Trial Counsel; Trinity College; Tri-State College; troop carrier; Trucial Coast (Arabian sheikdoms); True Course; Trusteeship Council;

Turret Captain; Turks and Caicos Islands (Internet code); Tusculam College

T & C Turks and Caicos Islands

T of C Tournament of Champions

TC *Technical Communications; Tragedy of Coriolanus; Tre Corde* (Italian—three strings)

T & C *Troilus and Cressida*

TC 1 Traffic Conference 1—North and South America, Greenland, Bermuda, West Indies, Hawaiian Islands

TC 2 Traffic Conference 2—Europe, adjacent islands, Ascension Island, Africa, and Asia west of and including Iran

TC 3 Traffic Conference 3—Asia, adjacent islands, East Indies, Australia, New Zealand, Pacific Islands except Hawaiian

tca telemetering control assembly; to come again; track crossing angle; trichloro-acetate; trichlorolthane

tca (TCA) terminal control area; tri-cyclic anti-depressant

TCA Tanners Council of America; Technical Cooperation Administration; TeleCommunications Association; Television Consumer Audit; Television Corporation of America; Television Critics Association; Temporary Change Authorization; Tennessee Correctional Association; Terminal Control Area; Texas Corrections Association; Textile Converters Association; Theater Commander's Approval; Thoroughbred Club of America; Tile Council of America; Tiltup Concrete Association; Tissue Culture Association; Trailer Coach Association; Trans-Canada Airlines; tricyclic antidepressant

TCAA Technical Communication Association of Australia; Tile Contractors' Association of America; Truck Cap and Accessory Association

tcam telecommunications access method

TCAs Terminal Control Areas (establishing airfield-safety flight paths)

TCAS The College of Advanced Science; Traffic-alert and Collision-Avoidance System

tcb take care of business

tcb (TCB) task-control block

TCB Thames Conservancy Board; Trusted Computing Base

TCBC Ty Cobb Baseball Commission

TCBG Training Centre Brigade of Gurkhas

TCBI Television Center for Business and Industry

tcbs (TCBS) thiosulfate-citratebile salt sucrose

tcc tactical control computer; television control center; teratocarcinoma; test conductor console; topical cocaine compound

tcc (TCC) transitional cell carcinoma

Tcc Tagliabue closed up

TCC Telecommunications Coordinating Committee; Transcontinental Corps; Transport and Communications Commission; Transport Control Center; Transportation Control Committee; Troop Carrier Command

T-C C Tri-Continental Corporation

TCCA Textile Color Card Association

TCCB Test and County Cricket Board

TCCP Thirteen College Curriculum Program

TCCS Texaco Controlled-Combustion System; Tide Communication-Control Ship

tcd task completion date; ternary coded decimal; tungsten carbide depositing

TCD Trinity College, Dublin

TCDA Texas Civil Defense Agency

tcdb turning, coughing, and deep breathing

tcdd (TCDD) tetrachlorodibenzo-p-dioxin

tcdf (TCDF) tetrachlorodibenzofurans

tcd's time certificates of deposit (TCDs)

TCDU Transport Command Development Unit

tce ton-coal equivalent; total composite error

tce (TCE) trichloroethylene

Tce Terrace

TCE Tax(ation) Counseling for the Elderly

T-cell thymus-derived cell

tcet transcerebral electrotherapy

tcf trillion cubic feet (natural gas)

TCF 20th-Century Fox; Territorial Cadet Force; Transparent Computing Facility; Twentieth Century Fund

TCF Touring Club de France (French—Touring Club of France)

TCFA The Cat Fanciers Association

TCFB Transcontinental Freight Bureau

tcfy trillions of cubic feet per day

TCG Theatre Communications Group

T C & G B Tucson, Cornelia & Gila Bend (railroad)

tcgf (TCGF) T-cell growth factor

TCG Troop Carrier Group

tch travel counselor's handbook

Tch (TCH) Tacoma (container symbol)

TCH Trans-Canada Highway

tchg teaching

TcHHW tropic higher high water

TcHHWI tropic higher high water interval

TcHLW tropic higher low water

tchr teacher

Tchrs Coll Pr Teachers College Press

tci trip-cancellation insurance

TCI Takeda Chemical Industries; Technical Correspondence Institute; Tele-Communications Incorporated; The Combustion Institute; The Containerization Institute; Theoretical Chemistry Institute

T & CI Turks and Caicos Islands

TCI Touring Club Italiano (Italian—Italian Touring Club)

tcj terminal coaxial junction

TCJC Texas Criminal Justice Council

tcl transfer chemical laser; transistor-coupled logic

Tcl Tymshare conversational language

TCL Tokyo Commercial University; Transatlantic Carriers Limited; Trinity College Library; Turkish Cargo Lines

TcLHW tropic lower high water

TcLLW tropic lower low water

TcLLWI tropic lower low water interval

tcm terminal-to-computer multiplexer; thermocouple meter; trajectory-correction maneuver

TCM Texas Citrus Mutual; Traditional Chinese Medicine; Trinity College of Music; Troop Corporal-Major

TCMA Telephone Cable Makers' Association

TCMB Tomato and Cucumber Marketing Board

TCMP Taxpayer Compliance Measurement Program (IRS); Tightly Coupled Multiple Processors

TCN Transportation Control Number

TCNA Turks and Caicos National Airline

TCNCO Test Control Noncommisioned Officer

TCNM Timpanagos Cave National Monument (Utah)

tco thrust cutoff

TCO Tactical Control Officer; Termination Contracting Office(r); Test Control Office(r); Torpedo Control Officer; Train Conducting Officer; Trinity College—Oxford

TCO Tjänstemännens Centralorganisation (Swedish—Salaried Employees' Central Organization)

TCO₂ total carbon dioxide

tcoc transverse cylindrical orthomorphic chart

TCOC Tri-Cities Opera Company (Binghamton)

TCOM Tethered Communications

TCOMA Tele-Communications Incorporated

T-conn T-shaped connection

TCOS Toronto Classroom Observation Schedule

tcp timing and control panel; traffic control panel; traffic control post; training control(ler) panel

tcp (TCP) trichlorophenyliodomethylsalicylates; tricresylphosphate

TCP Task Change Proposal; Task Control Proposal; Technical Cooperation Program (between Australia, Canada, the United Kingdom, and the United States); Temporary Change Proposal; Terminal Care Project; Traffic Control Post; Transitional Community Placement; Transmission Control Protocol

TCPA Town and Country Planning Association

tcpc tab card punch control

TCPC Tennessee Council of Private Colleges

TCP/IP Transmission Control Protocol/Internet Protocol

TCPL Trans-Canada Pipe Lines

tcr temperature coefficient of resistance; total controlled return

TCR Tennessee-Central Railway

TCRB *Touring Club Royal de Belgique* (French—Royal Belgian Touring Club)—automobile club

TCRMG Tripartite Commission for the Restitution of Monetary Gold (American-British-French commission, headquartered in Brussels)

tcs temperature control system; temporary change of station; terne-coated stainless steel; tierces

TCs Tax Cases; transit cops

TCS Target Cost System; The Costeau Society; Torpedo Control System; Twin-City Secularists; Typesetting Consultation Service

T & CS Transportation and Communication Service

TCS *Touring Club Suisse* (French—Swiss Touring Club)

tcsa (TCSA) tetrachlorosalicy-lanilide

TCSEC Trusted Computer System Evaluation Criteria

tcsev (TCSEV) twin-cushion surface-effect vehicle

TCSFL Turkish Cypriot Security Farm Line (Cyprus)

TCSO Tri-City Symphony Orchestra

tct total-controlled tabulation

TCTA Texas Classroom Teachers Association

tctl tactical

TCTO Time Compliance Technical Order(s)

TCTS Trans-Canada Telephone System

tcu tape-control unit; teletypewriter control unit; test(ing) computer unit; transmission control unit; transport conversion unit; threshold control unit; training combustion unit; typewriter control unit

TCU Texas Christian University; Tokyo Commercial University

TCUS Tax Court of the United States

T-cushion T-shaped cushion

tcv temperature-control valve

TCV Terminal-Configured Vehicle (NASA)

TCVA Terminal Configured Vehicles and Avionics (NASA program)

tcvr transceiver

tcw time code work

TCW Troop Carrier Wing

TCWG Telecommunications Working Group

TCWH Teamsters, Chauffeurs, Warehousemen and Helpers (union)

TCWIB Trans-Continental Weighing and Inspection Bureau

TCWP Texas Committee for Wildlife Protection

tcxo temperature-compensated crystal oscillator

td tank destroyer; technical data; test data; tetanus-diphtheria; third defense (lacrosse); tile drain; time delay; time of departure; time deposit; time disintegration; tod (28 pounds of wool); tool design; tool disposition; touchdown (football); tracking dog; transmitter distributor; trust deed; turbine drive; 'tween deck

td (TD) tardive dyskinesia; technical director; tracking dog

t/d table of distribution; telemetry data; time deposit; transmission and distribution

t & d taps and dies

t.d. *ter die* (Latin—thrice daily)

Td townsend

TS Taiwan dollar(s)

TD Chad (Internet code); Table of Distribution; Tactical Division; tank destroyer; Teachers Diploma; Technical Director; Telegraphist Detector; Territorial Decoration; Testing and Development (USCG); Topographic Draftsman; Town District: Castletown, Douglas, Peel, and Ramsey in the Isle of Man; Training Detachment; Treasury Decision; Treasury Department; Treasury Division; Trinidad and Tobago; Tyne Division; Typographic Draftsman

TD *Teachta Dala* (Gaelic—Member of the House of Commons)

tda tax-deferred annuity; tunnel-diode amplifier

t & da tracking and data acquisition

TDA Timber Development Association; Titanium Development Association; Toa Domestic Airlines; Train Dispatchers Association

tdana time-domain automatic-network analysis; time-domain automatic-network analyzer

T-day day for time schedule testing; truce day

T-Day Transition Day

tdb total disability benefit (TDB)

TDB Toronto-Dominion Bank; Toxicology Data Bank; Trade Development Bank; Trade and Development Board

TDBG Training Depot Brigade of Gurkhas

tdc top dead center; total distributed control; total distribution costs, transverse directional control

tdc (TDC) through-deck cruiser; torpedo-tracking computer

TDC Telemetry Data Center; Texas Department of Corrections; The Discovery Channel (cable tv)

TDCs Trust Deed Counselors

td cu tinned copper

tdd telecommunication device for the deaf

TDD Diploma in Tubercular Diseases

tddl time-division data link(age)

tddlpo time division data link printout

TDDRA Telephone Disclosure and Dispute Resolution Act

TDDS Teacher Development in Desegregating Schools

TDE Technology Development and Engineering

T del F Tierra del Fuego

T de M Teléfonos de México (Spanish—Telephone System of Mexico)

T de S Teatro delta Scala (La Scala)

tdf two-degree-of-freedom (gyroscope)

TDF Testes-Determining Factor; Tonga Defense Force

TDF 1 first French direct-broadcast satellite

TDF Télédiffusion de France (French-French Television Broadcasting)

TDFS Terminal Digit Fitting System

tdg twist drill gauge

tdg (TDG) test data generator

TDG Test Documentation Group; Transport Development Group

tdh total dynamic head

Tdh Trondheim

tdi time delay and integration; toluene di-isocyanate

TDI Target Data Inventory; Taxpayers Delinquency Investigation; Telegraphist Detector Instructor; Tool and Die Institute; Transportation Displays Incorporated; Trusted Database Interpretation

tdic target data input computer

TDIS Travel Document and Issuance System (for processing passports)

tdiu target data input unit

TDK Turk Dil Kurumu (Turkish Language Association)

t dk(s) 'tween deck(s)

tdl total damn loss; translation definition language

TDL Topographic Developments Laboratory

tdlr terminal-descent-landing radar

tdm tandem; teacher-developed materials; time division multiplexing

tdma time division multiple access

TDMA Trophy Dealers and Manufacturers Association

tdmg telegraph(ic) and data message generator

tdm/pcm time-division multiplex (using) pulse-code modulation

tdn totally digestible nutrients

TDNT Theological Dictionary of the New Testament

tdo tornado

TDO Technical Development Objective

tdol (TDOL) tetradecanol

TDOP Truck Design Optimization Program

TDOT Thorndike Dimensions of Temperament

tdp target director post; technical data package; technical development plans; thermal death point

TDP Technical Development Plan; Trade and Development Program

td passes touchdown passes

tdpfo temporary duty pending further orders

tdpj truck discharge point jet

tdr time-delay relay; time domain reflectometry

tdr (TDR) transmit data register

tdr tous droits réservés (French—all rights reserved)

TDR Technical Deficiency Report; Technical Documentary Report; tender (naval symbol)

TDRL Temporary Disability Retired List

t/d rly time-delay relay

tdrs (TDRS) tracking and data relay satellite

TDRSS Tracking and Data Relay Satellite System

tds telemetering decommutation system; total dissolved solids

tds (TDS) temperature, depth, salinity

t.d.s. ter die sumendum (Latin—to be taken three times daily)

TDS Tanami Desert Sanctuary (Northern Territory, Australia); Telemetering Decommunication System; Tennessee Department of Safety; Transaction-Driven System

TDS Toronto Daily Star

tdsa telegraphic data signal analyzer

TDSCC Tidbinbilla Deep Space Communication Complex

TDSTS Tidbinbilla Deep Space Tracking Station

tdt thermal death time

TDT Transport Department Tasmania; Turret Director Trainer

tdtcu target designation transmitter and control unit

tdtl tunnel diode transistor logic

TDTS Technical Data Transfer System

tdu target detection unit

TDU Teamsters for a Democratic Union

TDUP Technical Data Usage Program

TdV Teatro dal Verme (Milan)

tdw tons deadweight (tare of a ship)

TDWR Terminal Doppler Weather Radar

tdwy treadway

tdx tracking dog excellent

tdy temporary duty; toady

TDZ Touch-Down Zone

te table of equipment; tank element, task element; teal; technical exchange; tenants; tenants by the entirety; thermal efficiency; thermoelectric; tight end; tinted edge; trailing edge; transverse electric; transverse wave (symbol); trial and error; turbine electric; turboelectric; tweezer electrolysis; twin engine

t/e time expired

t & e testing and evaluation; thorough and efficient; training and evaluation; travel and entertainment; trial and error

Te tellurium

TE Table of Equipment; Task Element; Technical Exchange; Topographical Engineer; Training Establishment

TE (te) Technical Editor

TE Telefís Eireann (Television Ireland)

T & E Toledo & Eastern (railroad)

tea tetraethylammonium; transversely excited atmospheric

tea. triethanolamine

TEA Tennessee Education Association; Theatre Equipment Association; Transportation Equity Act; Tucson Education Association

TEAA Tax Equity for American Abroad

teac turbine engine analysis check(ing)

teach. teacher; teaching

TEAL Tasman Empire Airways, Limited

TEAM Technique for Evaluation and Analysis of Maintainability; Terminology, Evaluation, and Acquisition Method; The European—Atlantic Movement; Trend Evaluation and Monitoring

Teamsters Teamsters Union (International Brotherhood of Teamsters, Chauffeurs, Warehousemen, and Helpers of America)

teas trailing edge actuation system

TEAS Texas Energy Advisory Council; Threat Evaluation and Action Selection (program)

tease tracking errors and simulation evaluation (radar)

teatr *teatrale* (Italian—theatrical)

Teatro Colón (Spanish—Columbus Theater)—South American opera house

TEA21 Transportation Equity Act for the 21st Century

teb tape error block

TEB Tax Exemption Board; Textile Economics Bureau

tec technic; technical; technician; technics; technological, technology; total environmental control; total estimated cost

'tec detective

Tec Tecate

TEC Technical Education Council; Technician Education Council; The Executive Committee

TECAUS Temporary Emergency Court of Appeals of the United States

TECE Trans-Europe Container Express (train)

tech technic; technical; technician; technics; technique(s); technological; technology

Tech.CEI Technician of the Council of Engineering Institutions

tech ed technical editing; technical editor

Tech Eng Technical English (application of good English to any technical writing task)

techie(s) technician(s); technologist(s)

tech memo technical memorandum

techn technician

technocrit technological criticism; technology critic

technol technological; technologist; technology

techno-pop technolosized popular music

tech rep technical representative

tech rept technical report

Tech Weld Inst Technician of the Welding Institute

tech writer technical writer

TEC-NACS Teachers Educational Council—National Association of Cosmetology

TECOM Test and Evaluation Command (US Army)

tecquinol hydroquinone

tecr technical reason

'tecs detectives

TECS Treasury Enforcement Communications System; Treasury Enforcement Computer File

Tec Sgt Technical Sergeant

tecspert technical expert

ted transferred electron device

ted (TED) turtle excluder device

ted *tedesco* (Italian—German)

TED Tenders Electronic Daily; Territorial Efficiency Decoration

Teddy Theodore Roosevelt

TEDS Tactical Electronic Decoy System

tee transesophageal echocardiography

TEE Telecommunications Engineering Establishment; Theological Education by Extension; Torpedo Experimental Establishment; Trans Europe Express

TEEM Trans-Europe Express Merchandise (train)

'teens thirteen through nineteen

teenybop teenybopper (slang—young child attuned to the modern scene)—*see* macrobop

TEEP Teacher Education Examination Program

tef thermal effects of food

teeto teetotaler

TEFL teaching English as a foreign language

teflon tetrafluoroethylene (polymerized synthetic plastic resin)

TEFRA Tax Equity and Fiscal Responsibility Act

teg top edge gilt

Teg Tegel

te ga taper gauge

TEGMA Terminal Elevator Grain Merchants Association

teg(s) thermoelectric generator(s)

Teh Teheran

Tehachipis Tehachipi Mountains traversing south-central California

TEI Texaco Experiment Incorporated; The Entreprencurship Institute

TEIL/ESL Teaching English as an International Language/ English as a Second Language

TEJA *Tutmonda Esperantista Jurnalista Asocio* (International Association of Esperantist Journalists)

TEJO *Tutmonda Esperantista Junulara Organizo* (International Organization of Esperantist Youth)

tekn *teknisk* (Dano-Norwegian—technical)

tel telegraph; telegraphic; telegraphy; telephone; telephonic; telephony; teletype; teletypewriter; television; tetraethyl lead

tel (TEL) transporter-erector launcher

Tel Telefunken; Telescopium (constellation); Telugu

Tel *Teluk* (Indonesian or Malay—bay, bight, riverbend)

TEL Tests for Everyday Living

TELAM *Telenoticiosa Americana* (Argentine press service)

telaut telautograph; telautography

TELBRAS *Telecommunicaões Brasileiras* (Portuguese—Brazilian Telecommunications)

Tel Can Television Canada

telco telephone company

telcos telephone companies

TELDEC Telefunken + Decca (video disc)

tele television

tele (Latin prefix—far)—telegraph

Tele Telescopium (constellation)

telec thermo-electronic laser energy converter

telecast(er) television broadcast(er)

telecom telecommunication

Telecom 1C French commercial and communications satellite

telecon telephone communication

teleconcert televised concert; television concert

telecopy telephonic copying process (developed by Xerox)

telecourse television-constructed course

teledis teletypewriter distribution

teledrama televised drama; television drama

telef telefon (Norwegian—telephone); *telefone* (Portuguese—telephone)

telefac television facsimile

telefilm television film

teleg *telegramas* (Portuguese—telegrams); telegrapher; telegraphy

telegen telegenic

telegr telegrafie (Dutch—telegraphy)

Tel Eir Teleflis Eireann (Gaelic—Irish Television)

telemark telemarketing

telemorality television morality

teleol teleology

teleopera televised opera; television opera

teleosts teleostomist fishes (bony fishes)

telep telephathic(ally); telepathy

telepak telemetering package

teleph telephony

teleplay televised play; television play

teleran televised radar aerial navigation

telesex telephone(d) sex (fantasy conversations contracted for and carried out by telephone)

telesurance television insurance

tele tape television tape

telethon television marathon

teletrial television trial

telev (TV) television

telev televisão (Portuguese—television); *televisión* (Spanish—television)

televangelist television evangelist

telex (tex) teletype exchange

Tel-Law Telephone-Law (free over-the-telephone answers to many legal questions provided by many county bar associations in the U.S.)

tellie(s) television (sets)

telly television

Telly Telegonus; Telemachus; Telemus; Telephus; Telesphorus

Tel-Med Telephone-Medical (free over-the-telephone answers to many medical questions provided by many hospitals and county medical societies in the U.S.)

tel no telephone number

TELOPS Telemetry On-Line Processing System

TELS Tokyo English Language Society

telsat telecommunications satellite

tel sec telephone secretary

telsim teletypewriter simulator

tel sur telephone survey

telw telwoord (Dutch-word count)

tem technical error message; temporal; temporary; transmission electron micrograph; transmission electron microscope; transverse electromagnetic

tem. tempus (Latin—time); *tempo* (Italian—time)

tem¹ *tempo prima* (Italian—tempo at the start of a musical composition)

Tem temple

Tem. Temurah

TEM Territorial Efficiency Medal

TEMA Telecommunications Engineering and Manufacturing Association

temadd temporary additional duty

temar thermoelectric marine application

TEMIS Targets Engineering Management Information System (USN)

temp temper; temperature; tempered; tempering; template; temporary; temporize

temp. tempo (Italian—time)— musical time; *tempore* (Latin—in the time of)

Temp Tempest, The

temp. dext *tempori dextro* (Latin—to the right temple)

temping substituting

tempistors temperature compensating resistors

Temple central London's law-court area

tempo total evaluation of management and production output

TEMPO Technical Military Planning Operation

tempos temporary buildings, houses, offices, officials, workers, et cetera

temp prim tempo primo (Italian—tempo or time in the musical sense as at the start)

temps tempests; temperatures; temporary workers; transportable electromagnetic pulse simulator

temp(s) temporary worker(s)

temp sec temporary secretary

temp. sin. tempori sinistro (Latin—to the left temple)

tempy temporary

tems dub temporary musical score run in a film

TEMPUS Trans-European Mobility Program for University Students

ten. tenant; tender; tenderize(d); tenement; tenor

ten (TEN) toxic epidermal necrolysis; trans-European night (flight)

ten. tenuto (Italian—to hold, a chord or tone)

Ten Ten Commandments; Tenente (Italian or Portuguese—Lieutenant); *Teniente* (Spanish—Lieutenant)—Lieutenant

TENCAP Tactical Exploitation of National Capabilities

T(en) Col Tenente Colonnello (Italian); *Tenente Coronel* (Portuguese); *Teniente Coronel* (Spanish)—Lieutenant Colonel

ten. com tenant(s) in common

tency tenancy

tend. tendon

ten. ent tenant(s) by the entireties

TENES Teaching English to Non-English Speaking

Teng Teng Hsiao-ping

Ten Gen *Tenente General* (Portuguese); *Tenente Generale* (Italian); *Teniente General* (Spanish)—Lieutenant General

Tenn Tennessee; Tennessean

tenna(s) antenna(s)

Tenneco Tennessee Gas Companies

Tenn-Tom Tennessee-Tombigbee Waterway

Tennyson Alfred, Lord Tennyson (1809–1892)

TENOC ten years of oceanography (1961–1970)

tenot tenotomy

TENRAC Texas Energy and Natural Resources Advisory Council

tens tensile; tension

tens (TENS) transcutaneous electrical nerve stimulation

tens (Latin prefix—stretch)—tensor

ten-spot $10 bill

tens str tensile strength

tent. tentative

Ten^{te} *Teniente* (Spanish—Lieutenant)

Ten Vasc *Tenente di Vascello* (Italian—Lieutenant of the Vessel)—Navy Lieutenant

Teol Teología (Portuguese, Spanish—Theology)

TEOO Territorial Economic Opportunity Office(r)

Teor Teoretyczna (Russian—Theoretical)

TEOSS Tactical Emitter Operational Support System (USAF)

tep transparent electrophotographic process(ing); transparent electrophotography

TEP Teacher Education Program; Tucson Electric Power

tepi training equipment planning information

TEPIAC Thermophysical and Electronic Properties Information Analysis Center

TEPIGENS Television Picture Generation System (computer-controlled)

TEPS Teacher Education and Professional Standards

ter terminal; terminate; termination; terrace; terrazzo; territory; teritary

ter. tere (Latin—rub)

Ter Terrace; Territory; Teruel

Ter. Terumot

Ter *Terence* (Publius Terentius Afer)—Roman writer of comedies

tera 10^{12}

TERA The Electrical Research Association

terat teratology

TERC Technical Education Research Centers

terco telephonic rationalization by computer

tercom terrain contour matching

t & e rec time and events recorder

TERL Transit Expressway Revenue Line (mass transportation)

term terminal; terminate; terminology

te rm taper reamer

Term Terminal

Terminal Terminal Island (Bureau of Prisons correctional facility between Long Beach and San Pedro, California)

TERMS Terminal Management System

tern terminal and enroute navigation

TERPACIS Trust Territory of the Pacific Islands

TERPES Tactical Electronic Reconnaissance Processing and Evaluation System

terps elixir of terpin hydrate and codeine—cough mixture and codeine combination

terps (TERPS) terminal instrument approach

TERPS Terminal Inquiry/Response Programming System

terr terrace; territory; terrorist

Terr Terrace

TERRA Terricide Escape by Rethinking, Research, Action; the Earth Regeneration and Reforestation Association

Terra Nova Terra Nova National Park in Newfoundland

terrs terrorists

terry terrycloth(ing)

Terr' Territory

tersab terrorist sabotage; terrorist saboteur

tersabs terrorist saboteurs

Tersch Terschelling

ter. sim. tere simul (Latin—rub together)

TERSSE Total Earth Resources System for the Shuttle Era (NASA)

Tert. Tertiary

Tertullian Quintus Septimus Florens Tertullianus

tes thermal energy storage

tes tesorero (Spanish—treasurer)

TES Telemetering Evaluation Station

TES Times Educational Supplement

TESA Television and Electronic Service Association

tesac temperature-salinity-currents

tesl (TESL) teaching English as a second language

tesla technical standards for library automation

TESM Trinity Episcopal School for the Ministry

TESO Texel's Eigen Stoomboot Onderneming (Dutch—Texel's Own Steamship Society)

TESOL Teachers of English to Speakers of Other Languages

tess tessili (Italian—textiles)

tessit tessitura (Italian—texture, tissue, weaving)—range of a musical work

test testament; test-oriented engineering symbol(ic) translator

TEST Thesaurus of Engineering and Scientific Terms

TESTCOMDNA Test Command Defense Nuclear Agency

test^{mto} testamento (Spanish—testament)

test^0 testigo (Spanish—witness)

testran test translator (data processing)

TESYS Terminal Editing System

tet test equipment tool; tetanus; tetrachloride

TET Teacher of Electrotherapy; Teacher Evaluation Testing

TETAM Tactical Effectiveness Testing of Antitank Guided Missiles (USA)

T-et-G Tarn-et-Garonne

tetmtu (TETMTU) tetramethyl thiourea

TETOC Technical Education and Training for Overseas Countries

tetr tetragonal

tetra (Latin prefix—four or four-fold)—tetrachord

tetrac tetraiodothyroacetic acid

tetrah tetrahedral

tetroon tetrahedral balloon

tet tox tetanus toxin

teu twenty-foot equivalent unit(s) (container measurement)

TEU Test of Economic Understanding; Tropical Experimental Unit

Teut Teuton; Teutonic

tev tevatron (atom smasher speeding electrons to an energy level of 1000-billion electron volts)

tev (TeV) trillion electron volts

teV tetra-electron volt(s)

tew (TEW) tactical early warning; tactical electronic warfare (aircraft)

tewa threat evaluation and weapons assignment

TEWDS Tactical Electronic Warfare Defense System

tews tactical electronic warfare suite

TEWS Tactical Electronic Warfare System

TEWT Tactical Exercise Without Troops

tex telex (teletype exchange); textile(s)

t ex *till exempel* (Swedish—for example)

Tex Texan; Texas

TEX Corpus Christi, Texas (tracking station)

TEXACO The Texas Company

Tex A & M Texas Agricultural and Mechanical University

Tex A & M Pr Texas Agricultural and Mechanical University Press

Texas Texas Stadium, Irving, Texas

TEXAS Trained Experienced Area Specialist

Texas RRC Texas Railroad Commission

Tex Chr U Texas Christian University

Tex Chr U Pr Texas Christian University Press

Texcoco Texcoco de Mora

Texhoma Texas + Oklahoma

Texican Texas-Mexican or anyone from the Texas side of the Mexican Border

Texico Texas + New Mexico

Tex Instr Texas Instruments (Corporation)

Tex-Mex Texan-Mexican; Texas-Mexico

Tex Mon *Texas Monthly*

Tex Ob *Texas Observer*

Texola Texas + Oklahoma

Texoma Lake Texoma between Texas and Oklahoma

texp time exposure

text. textile

text ed text edition

Textel Trinidad and Tobago External Telecommunications Company

Textile Mus Textile Museum

textir text indexing and retrieval

text. rec. *textus receptus* (Latin—received text)

Tex W Pr Texas Western Press

tey taxable equivalent yield

TEZ Total Exclusion Zone

tf tabulating form; tactile fremitus; temporary fix; thin film; tile floor; till forbidden (run ad until stopped by advertising client); training flight; transfer function; treasured friend; tuberculin filtrate

t/f true/false

TF French Southern Territories (Internet code); Tallulah Falls (railroad); Task Force; Tax Foundation; Territorial Force; Test Flight; Tolstoy Foundation; torpedo-fighter (airplane); trainer-fighter (airplane); training film; tropical freshwater (vessel loadline marking); Twentieth Century-Fox Films (stock exchange symbol)

TF *Travail Forcé* (French—penal servitude)

TF-1 *Télévision 1 Français* (French tv network)

tfa total fatty acids; transfer function analyzer

TFA Task Force on Alcoholism; Teaching For America; Territorial Force Association; Textile Fabrics Association; Tie Fabrics Association; Tobacco-Free America; Trout Farmers Association

TFAA Track and Field Athletes of America

TFAI *Territoire Français des Afars et des Issas* (French—French Territory of Afars and Issas)—formerly French Somaliland

TFB Thatcher Ferry Bridge (over Panama Canal)

tfc traffic

TFCF Twenty-First Century Foundation

TFCNN Task Force Commander—Northern Norway (NATO)

TFCRI Tropical Fish Culture Research Institute

tfcsd total federal commissioned service date

tfd tactical fusion division; target-to-film distance

tfe tetrafluoroethylene (halon or teflon plastic)

TFE Toronto Futures Exchange

TFEA Texas Free Enterprise Act

TFF Tropical Fish Farm

TFFW The Foundation for Wellness

tfg typefounding

TFI Table Fashion Institute; Tax Foundation Incorporated; Textile Foundation Incorporated; The Fertilizer Institute; Traditional Family Ideology scale

tfio thin film integrated optics

tfis theft from an interstate shipment

TFL Trans Freight Line

TFLA Texas Foreign Language Association

TFLC Tulane Factors of Liberalism-Conservatism

tfm transmit frame memory

TFNS Territorial Force Nursing Service

tf/p tubular fluid divided by plasma concentration (concentration of a substance in renal tubular fluid divided by its concentration in plasma)

TFP Trees for People

TFP *Tradicion, Familia, y Propiedad* (Spanish—Tradition, Family, and Property)—rightwing movement

tfr terrain-following radar; transfer

TFr Tunisian franc

TFR Territorial Force Reserve

TFR/CAR Trouble and Failure Report/Corrective Action Report

tfs time and frequency standard

TFS Transport Ferry Service

tfsg trickling-filter suspended growth

TFSK Turkish Federated State of Kibris

TFSR Tools for Self-Reliance

tft thin-film technology; thin-film transistor

TFT Transfer Factor Test(ing)

TFTA Textile Finishing Trades Association

tft lcd thin-film transistor liquid crystal display

TFTP Trivial File Transfer Protocol

tfu telecommunications flying unit

TFW Tactical Fighter Wing

TFX variable geometry supersonic fighter-bomber

tg tail gear; telegram; telegraph; thyroglobulin; tollgate; tongue and groove; transformational grammar; transformational generative; type genus

tg (TG) transformational generative; transformational grammar

t/g tracking and guidance

t & g tongue and groove

tg tangente (Italian—tangent)

Tg Tanjung (Malayan—cape)

TG Task Group; Tate Gallery; Texas Terminal Gap; Gulf Sulphur (stock exchange symbol); Thai Airways International (airline code); Theatre Guild; Togo (Internet code); Torpedo Group; Townguard; Traffic Guidance; Translators' Guild

T & G Traveres & Gulf (Florida railroad); Tremont & Gulf (Louisiana railroad)

tga thermal gravimetric analysis; thermogravimetric analysis; transient global amnesia

Tga Tonga (Spanish abbreviation)

TGA Toilet Goods Association; Turpentine Growers of America

t'gallant topgallant (sail)

t'gal'n't topgallant (sail)

t'gansail topgallant sail

TGAOTU The Great Architect of the Universe

tgarq telegraphic approval requested

tgb tongued, grooved, and beaded

TGC Travel Group Charter(s)

tgca transportable ground-control approach

tgd (TGD) thiodiglycol

tge transmissible gastroenteritis

TGF Transonic Gasdynamics Facility (USAF)

tgfb transforming growth factor beta

tgfbop tippy golden flowery broken orange pekoe

tgfd transforming growth factor alpha

TGG temporary geographic grid

TGH Toronto General Hospital

tgif thank goodness it fits; toes go in first

tGiF thank God it's Friday (TGIF)

tgl toggle

TG loran traffic guidance loran

TGM Thomas G Masaryk; Torpedo Gunner's Mate; Training Guided Missile

TGMLI Tussock Grasslands and Mountain Lands Institute

tgn tangent

Tgo Tsingtao

TGO Timber Growers' Organization

TGP Terminal Guidance Program

TGPLC Transcontinental Gas Pipe Line Corporation

TGR Tiger International; Total Gross Receipts

T-Group Training Group

tgs thermal growing season

TGS Taxiing Guidance System; Translator Generator Service; Turkish General Staff

tgt target; teams-games-tournaments; turbine gas temperature

TGT Tennessee Gas Transmission

tgtp tuned grid tuned plate

TGU Tegucigalpa, Honduras (airport)

tgurq telegraphic authority requested

tgv thoracic gas volume

TGV Train de Grande Vitesse (French—Train of Great Speed)—high-speed railroad train; *Two Gentlemen of Verona*

tgw terminally guided weapon

TGWU Transport and General Workers' Union

th indicates ordinal number as in 4th, 5th, 6th, etc; tee handle; true heading

th' the

t & h transportation and handling

Th Thai (Siamese); Thailand (Siain); Thomas; thorium; Thursday

Th Theil (German—part)

TH Thailand (Internet code); Town Hall; Toynbee Hall; Transport House; Trinity House; true heading

T-H Taft-Hartley

T & H Thames and Hudson

T H Technische Hochschule (German—technical college)

tha total hydrocarbon analyzer

tba (THA) tetrahydroaminoacridine

Th A Theological Association

THA Transvaal Horse Artillery

THAA Tourist House Association of America

THAAD Theater High-Altitude Area Defense

Thad Thaddeus

Thai Thai(land)(er)(ers)

THAI Thai Airways International

Thailand Kingdom of Thailand (*Muang-Thai* or *Prathes Thai*) formerly Siam

Thaler (German abbreviation—Joachimsthaler)—Joachim's dollar—Bohemian coin struck in 16th century at Czech town of Jachymov (Joachimsthal)—its name has become *dollar*

THANACAP Funeral Service Consumer Action Program (*Thana* is the Greek word for death)

thanat thanatology

thanatol thanatologic(al)(ly); thanatologist(ic)(al)(ly); thanatology

than ever than ever before

Thang-Pho (Vietnamese—city)—Saigon, Ho Chi Minh City

Thanksgiving Thanksgiving Day (fourth Thursday in November in the United States)

Thatcherism policy and work of Britain's former prime minister, Margaret Thatcher

that's that is

that's 30 (journalistic jargon—that's all)—the end of the article, report, or story

Th.B. Theologiae Baccalaureus (Latin—Bachelor of Theology)

TH & B Toronto, Hamilton and Buffalo (railroad)

TH & BA Toll, Highways and Bridge Authority

thc tetrahydrocannabinol (active ingredient in hashish, indian hemp, and marijuana)

THC Toledo House of Correction; Toronto Harbour Commission; Toronto Harbour Commissioners; Tourist Hotel Corporation (NZ); Trinity Hall College (Cambridge)

thccre tetrahydrocannabinol cross-reacting cannabinoids

thd thread; threaded; threads; total harmonic distortion

Th.D. *Theologiae Doctor* (Latin—Doctor of Theology)

THD *Technisch Hogeschool te Deyt* (Dutch—Technological University of Delft)

th di thread die

the. (THE) tetrahydrocortisone

The. Theodora; Theodore

THE Technical Help to Exporters

thea theater

Thea Theadora; Theodeline; Theodosia; Theresa

T-head Texas-tea head (slang—marijuana user)

theat theater; theatrical

theatcrit theatrical criticism

THEC Tennessee Higher Education Commission

the E the Equator

THEN Those Hags Encourage Neuterism

theo theoretical; theoretician

Theo Theobald; Theobold; Theocritus; Theodoor; Theodor, Theodora; Theodore; Theodorus; Theodosia; Theodosius; Theodoric; Theodric; Theodule; Theophil; Theophile; Theophilus; Theophraste; Theophrastus

THEO They Help Each Other

Theoc Theocritus

theod theodolite

Theodore Roosevelt Park national park in North Dakota

theol theologian; theological; theologist; theology

Theol Theology

Theoph neophrastus

theophilanthro theophilanthropic(al)(1y); theophilanthropist; theophilanthropy (Thomas Paine's deistic religion combining belief in a god with service to mankind)

theor theorem; theoretical; theory

Theor *Theorique* (French—theoretical)

theos theosophical; theosophist; theosophy

Theo Soc Theosophical Society

ther therapy

therap therapeutic; therapeutics; therapy

there's there is

therm thermometer; thermostat(ic)

therm (Latin prefix—heat)—thermometer

Therm *Thermidor* (French—Hot Month)—beginning July 19th—eleventh month of the French Revolutionary Calendar also called the *Fervidor*

thermistor thermal resistor

thermo thermostat

thermoc thermocouple

thermochem thermochemical; thermochemistry

thermodyn thermodynamics

thermonuc thermonuclear

thermo win thermopane window

THES *Times Higher Education Supplement*

THESIS Thematic Elementary Science Individualized Studies

The Source trademark of Source Telecomputing

thesp(s) thespian(s)

Thess *Thessalonians*

thetcrit theater critic; theatrical criticism

The Troubles Ireland's efforts to separate from the United Kingdom

they'd they had; they would

they'll they will

they're they are

they've they have

thf (THF) tetrahydrocortisol

t$_{hf}$ Trust Houses Forte (British motel chain)

THF Berlin, Germany (Tempelhof Airport)

THG *Technische Hochschule Graz* (German—Technical University of Graz)

th ga thread gauge

THHS Townsend Harris High School

THhwm Trinity House high-water mark

thi temperature-humidity index

Thi Thailand (whose capital is Bangkok)

THI Texas Heart Institute

Thief *The Thief of Bagdad* (Douglas Fairbanks' motion picture)

Thim Thimbu, Bhutan

things three-dimensional input of graphical solids

TH & IS Traveler Health & Immunization Services

THIWRP The Hoover Institution on War, Revolution, and Peace

thixo thixotropic

Th:J Thomas Jefferson (initials written by him as shown)

thk thick(ness)

THK *Turk Hava Kurumu* (Turkish Air Association)

Th. L. Theological Licentiate

THlwm Trinity House low-water mark

thm (THM) trihalomethane

Th.M. *Theologiae Magister* (Latin—Master of Theology)

THMA Trailer Hitch Manufacturers Association

thms trihalomethanes

Thn Trollhättan

tho' though

'tho' although

Tho Thomas; Thorshavn

Thomas St Thomas, American Virgin Islands

Thomas' *Thomas' Register of American Manufacturers*

THOMIS Total Hospital Operating and Medical Information System

thor thorax; thoracic

Thor medium-range ballistic missile

THOR Tandy High-Intensity Optical Recording

thorac (Latin prefix—chest)—thoracic

Thoreau Henry David Thoreau

Thoreau Foun Thoreau Foundation

thoro thorough

thoro' thorough

Thoro thoroughfare

Thorstein Thorstein Veblen (American economist)

Thos Thomas

Thos Jeff Thomas Jefferson

thou. thousand

Thousands Thousand Islands in the St Lawrence River between New York and Ontario

thp thrust horsepower; track history printout

THPD *Tragedy of Hamlet, Prince of Denmark*

THq theater headquarters

thr target heart rate; their; threonine (amino acid) (THR); threshold; through; thrust

THR Teheran, Iran (airport); Tower Hamlets Rifles

Thrac. Thracian

Three Kings Three Kings Islands bird sanctuary in the South Pacific off New Zealand's North Island

Three King's Three King's Day (January 6—Epiphany)

thrmst thermostat

thro' through

THRO Throw the Hypocritic Rascals Out

thro'b/l through bill of lading

Throgs Throgs Neck (site of New York State Maritime College)

Throg Street Throgmorton Street, London

thrombo thrombosis

thrombo (Latin prefix—clot or lump)—thrombosed

thrombol thrombolytic therapy

throt throttle

thru through

Thru Thruway

thruout throughout

THS Technical High School; Titanic Historical Society; Tiwi Hot Springs (Philippines); Tottenville High School

tht (THT) tetrahydrothiopen

THT Teacher of Hydrotherapy

th ta thread tap

thtr theater

THTRA Thorium High-Temperature Reactor Association

Thu Thursday

THU The Hebrew University (Jerusalem)

Thuc Thucydides

THUMS Texaco, Humble, Union, Mobil, Shell (oil-drilling complex dominating Long Beach, California)

Thur Thuringia(n); Thursday

Thurs Thursday

Thursday Thursday Island pearl-shell fishery in Torres Strait near Cape York, Australia

Thus (nickname—Calcutta Steam Tug); Thursday

thv thoracic vertebra

Thv Theravadin; Thorvald(sen)

THW Technische Hochschule Wien (German—Technical University of Vienna)

THwm Trinity House water mark

Thwy Thruway

THY Turk Hava Yollari (Turkish airline)

THz Terahertz

thz (tHz) tetraherz

ti target identification; target indicator; temperature indication; temperature indicator; termination instruction; thermal imager; tricuspid insufficiency

t/i target identification; target indicator

ti Texas Instruments (trademark); *tudni illik* (Hungarian—that is)

Ti titanium

Ti Tirsdag (Danish—Tuesday); (Latin—Tiberius)

TI Taxable Income; Taxpayer Information (system); Technical Inspection; Technical Institute; Technical Intelligence; Terminal Interval; Terminal Island; Termination Instruction; Terrorist International; Texas Instruments; Textile Institute; Thread Institute; Title Insurance (and Trust Company); Toastmasters International; Tobacco Institute; Tonga Islands; Training Instruction; Treasure Island; Tungsten Institute; Tuskegee Institute

TI Théâtre-Italien (French—Italian Theater)—Paris

T of I Times of India

TI-67 Israeli designation for captured built-in-the-USSR tanks (T-54 and T-55 models armed with 100mm guns)

tia transient ischemic attack

tia (TIA) trading investment area; transient ischemic attack

TIA Tax Institute of America, Tobacco Institute of Australia; Tortilla Industry Association; Trans International Airlines; Travel Industry of America; Tricot Institute of America; Trouser Institute of America; Typographers International Association

TIA Tutukuvul Isukul Association (Melanesian—United Farmers Association)—Papua New Guinea coconut planters united

TIAA Teachers Insurance and Annuity Association of America

TIAACREF Teachers Insurance and Annuity Association—College Retirement Equities Fund

TIAC Thrift Institutions Advisory Council

TIAP Total Ischemia Awareness Program

TIAS Treaties and Other International Acts Series (U.S. Department of State)

tib tibia(l); trimmed in bunkers

tib tibetisch (German—Tibetan)

Tib Isabel; Tibet; Tibetan

Tib Albius Tibullus (Roman poet)

TIB Tasmanian Imperial Bushmen; Technical Information Bulletin; Temple (Records) Index Bureau; Tennessee Inspection Bureau; Thousand Islands Bridge; Tourist Information Bureau

tibc total iron-binding capacity

Tibet. Tibetan

tic target intercept computer; total information card; true interest cost

TIC Teacher Information Center; Technical Information Center; Technical Institute Council; Technical Intelligence Center; Texas Industrial Commission; Troops in Contact; Tyne Improvement Commission

TICA Technical Information Center Administration; The Independent Cat Association; The International Cat Association

TICACE Technical Intelligence Center Allied Command Europe (NATO)

TICC Technical Intelligence Coordination Center

TICCI Technical Information Center for the Chemical Industry

ticcit time-shared interactive computer-controlled information television

TICF Tennessee Independent Colleges Fund; Transient Installation Confinement Facility

tick. tickler

tick(er) ticker tape

tictac time compression tactical communications

TICUS Tidal Current Survey System

tid task initiation date

t.i.d. tres in die (Latin—thrice a day)

TIDE Technology for the Socio-Economic Integration of the Disabled and Elderly; Toledo Institute for Development and Environment (Belize)

tideda time-dependent data analysis

tidskr tidskrift (Swedish—periodical)

TIDU Technical Information and Documents Unit

tidy teletypewriter integrated display

tie technical integration and evaluation

tie (TiE) (TIE) telephone interconnect equipment

TIE Technology Information Exchange; The Institute of Technology; Total Interlibrary Exchange (California Library Network); Traveler's Information Exchange; Truck Insurance Exchange

Tiempo EL Tiempo (Time—Bagotá newspaper)

Tien Tientsin

Tien Shan high mountain ranges north of Pamirs and Himalayas between Siberia and Turkestan

tier. tierce

tier tierce (French—third)

Tierg Tiergarten

TIES Transmission and Information Exchange System

tif telephone influence factor; telephone interference factor; tumor inducing factor

Tif Tiflis

TIF Turtle Island Foundation

TIFA Tamburitzan Institute of Folk Art

tiff (TIFF) tagged image file format

Tiff Tiffany

Tiff Tiffany's Reports

TIFI Technology Insight Foundation Incorporated

tifr total investment for return

TIFR Tata Institute of Fundamental Research

tifs technology for instructional feedback

tig time in grade; tungsten-inert gas

TIG The Inspector General

Tiger Tiger Stadium, Detroit

TIGER Topologically-Integrated Geographic-Encoding-and-Reference System (Bureau of the Census)

TIGERS Telephone Information Gathering for Evaluation and Review System

tigon offspring of tiger and lioness

TIGRs Treasury Investment Growth Receipts

tigt turbine inlet-gas temperature

TIH Their Imperial Highnesses

TII Texas Instruments Incorporated; Toastmasters International Incorporated

TIIAL The International Institute of Applied Linguistics

TIIRS Title-I Information and Reporting System

TIJ Tijuana, Mexico (airport)

'til until

TIL Taylor Institution Library (Oxford); Technology Information Library; Tube Investments Limited

tili translunar injection

TILS Technical Information and Library Service

tilt. tilt steering wheel

tim technical information on microfiche; technical information on microfilm; time is money

tim (TIM) transient intermodulation

Tim Timor; Timothy

Tim Timon of Athens

Tima Fatima

TIMA Thermal Insulation Manufacturers Association

timation time navigation

timations time navigation artificial satellite

timb timbales (French—kettledrums)

TIMC The Industrial Management Center

TIME Telecommunication Information Management Executive

time imm time immemorial (time beyond memory; time out of mind)

Time-Life Time-Life Books

Times The New York Times; The Times (leading British newspaper, published in London);

local designation for all other newspapers containing *Times* in their title

TIMES The Institute of Mining and Engineering Surveyors

Times Roman Times Roman type (sometimes abbreviated T-R)

timet titanium metal(s)

timms thermionic integrated micromodules

TIMMS Third International Math and Science Study

Timmy Timothy

timp timpani (Italian—kettledrums)

Timpanogos Timpanogos Cave National Monument in north-central Utah or Mount Timpanogos in the same area

timps timpani (kettledrums)

TIMS The Institute of Management Sciences; Thermal Infrared Multispectral Scanner

Tim-Tim (Portuguese—Timor, Timur)—former colony in the Lesser Sunda islands of Indonesia

tin pipe-tobacco tin filled with marijuana

TIN Taxpayer Identification Number; tinnitic-induced noise; Transaction Identification Number

Tina There is no alternative (nickname for former British Prime Minister Margaret Thatcher)

tinc tincture

tin can(s) submarine(s)

tinct tincture

tinct. tinctura (Latin—tincture)

TINFO Tieteellisen Informoinnin Neuvosto (Finnish—Council for Scientific and Research Libraries)

'tini Martini (cocktail)

tin in tinnitus instrument

tins thermal-imaging navigation set

TINs Temporary Instruction Notices

tint international practical temperature

Tintagel Tintagel Head, Cornwall (legendary birthplace of King Arthur)

tiny terrs tiny terrorists (children used by terrorists to run errands or spot their enemies)

tio take it off; temporary immovable obstruction; time interval optimization; time in office (TIO)

TIO Target Indication Office(r); Television Informatmon Office(r); Test Integration Office(r); Troop Information Office(r)

tip tax information plan; theory in practice; threat image projection; to insure promptness (a gratuity given to insure promptness); translation-inhibiting protein (TIP)

tip tipografia; tipografico (Italian—printing firm; typographic); truly important person (TIP)

Tip Thomas P (Tip) O'Neill, former Speaker of the U.S. House of Representatives

TIP Target Interactive Project; The Institute of Physics; Tax-based Income Policy; Technological Institute of the Philippines; Terrorist Information Project; Times of Increased Probability; Trans-Israel Pipeline; Transportation Improvement Program; Trauma Intervention Program; Tripoli, Libya (airport); Troop Information Program(s); truly important person(age)

TIP Türkiye Işçi Partisi (Turkish Labor Party)

TIPAC Texas Instruments Programming and Control

tip.bkt tipping bracket

Tipp County Tipperary, Ireland

TIPRO Texas Independent Producers and Royalty Owners

tips Tax Intercept Programs; to insure prompt service (gratuities); topical information packages; truly important persons (TIPS)

TIPs Tax-based Income Policies

TIPS Technical Information Processing System; Telemetry Impact Prediction System; Total Integrated Pneumatic System; Training for Intervention Procedures by Servers; truly important persons

tiptap target input panel (and) target assign panel

tiptop tape input—tape output

TIP & TPS Institute of Physics and The Physical Society

tir target illuminating radar; total indicator reading

TIR Transport International des Marchandises par la Route (French—International Transport of Merchandise by Road)—twenty-six nation custom agreement permitting trucks marked TIR to avoid customs until reaching their final destination; *Transport Internationale Routiers* (French—International Truck Routes)

TIRB Transportation Insurance Rating Bureau

TIRC T Tauri Infrared Companion (pronounced *turk*); Tobacco Industry Research Committee

tire burner hot pursuit of one vehicle by another

T-iron T-shaped iron or steel section

Tiros American meteorological satellite designed to observe cloud coverage and infrared heat radiation of the earth; television and infrared observation satellite

TIRR Texas Institute of Rehabilitation and Research

tirs thermal infrared scanner

tis carcinoma *in situ* (in original position); tissue(s); total integrated scattering

'tis it is

TIs Thousand Islanders; Thursday Islanders; Tonga Islanders; Turks Islanders

TIS Technical Information Service; Total Information System; Transactional Information Systems; Transdermal Infusion System

TISC Technology Information Sources Center

TISI Thai Industrial Standards Institute

ti-slash tire slash(ing)

TISPM Territorie des Iles St Pierre et Miquelon (French territory offshore Canada)

TISS Title-I Support System

tit título (Spanish—title)

tit. title; titular; titulary; transitive inference training

tit títre (French—title)

Tit Titus, The Epistle of Paul to

Tit Titus

TIT Tokyo Institute of Technology; Tustin Institute of Technology

Tit A Titus Andronicus

titanox titanium dioxide

Titis Titi Islands also called the Mutton Birds, off the southwest coast of New Zealand's Stewart Island

tit° título (Spanish—title)

TITUS Textile Information Treatment Users Service

tiu trigger inverter unit

TIU Telecommunications International Union; Tokyo Imperial University

tiv total indicator variation

Tiv Tivoli

tix ticket(s)

tixi turret-integrated xenon illuminator

TIYC Thousand Island Yacht Club

tj tomato juice; triceps jerk; turbojet (TJ)

tj (TJ) talk jockey

tj to jest (Polish—that is)

Tj Tijuana, Baja California, Mexico

TJ Tajikistan (Internet code); Thomas Jefferson—third President of the United States

TJAG The Judge Advocate General

tjc trajectory

TjC trajectory chart

TJC The Jockey Club; Trenton Junior College; Tyler Junior College

TJC Tragedy of Julius Caesar

TjD trajectory diagram

TJFF Trans-Jordan Frontier Force

TJG Travel Journalists Guild

TJHS Thomas Jefferson High School

Tji Tjirebon (Cheribon)

TJM The Jewish Museum; Thomas Jefferson Memorial

tjp (TJP) turbojet propulsion

TJPOI Twisted Jute Packing and Oakum Institute

tjs tight-jean syndrome (impotence and sterility)

TJS Tactical Jamming System

TJSUSA Thomas Jefferson Society of the United States of America

tjt tactical jamming transmitter

TJTA Taylor-Johnson Temperament Analysis

TJTC Targeted Jobs Tax Credit

tk thymidine kinase; track; truck; trunk

tk (TK) transkelotase

tk to kum (printer's expression meaning material is *to come*)

Tk Turkmenian; Turkmenistan

Tk Teluk (Malay—bay; bight; riverbend)

TK Tokelau Islands (Internet code)

tkbd tackboard(ing)

tkd tokodynamometer

tkg tanking; tokodynagraph(y)

Tki Takoradi

TKK Teikoku Kayi Kyokai (Imperial Japanese Marine Corporation, ship classifiers)

TKL Tragedy of King Lear

tko technical knockout

TKP Turkiye Komünist Partisi (Turkish Communist Party)

tkr tanker; terrestrial kilometric radiation

T KR II Tragedy of King Richard II

tks thanks

TKTF Tanker Task Force

tkt(s) ticket(s)

tl terminal limen; test link; throws left; thrust line; time length; time limit; time loss; total load; transmission level; transmission line; truckload; truck loading

t-l trade last (slang, a compliment)

t/l total loss

t.l. tukus lecker (Yiddish—ass licker)—flatterer; sycophant

Tl thallium

TL Technical Letter; Technical Library; Texas League; The Leprosarium (U.S. Public Health Service, Carville, Louisiana); Torpedo Lieutenant; Townland (UK); Turk lirasi (Turkish pound)

T/L Telegraphist/Lieutenant; Torpedo Lieutenant

T & L Thames and London

T+L Travel + Leisure

TL Teatro Lirico (Italian—Lyric Theater)—Milan; *Théâtre-Lyrique* (French—Lyric Theater)—Paris

T-L Time-Life (books, magazines, recordings)

tla translumbar aortogram

TLA The Library Association (of the United Kingdom); Texas Library Association; Theatre Library Association; Trial Lawyers Association; Trinidad Lake Asphalt

TLAM-C Tomahawk Land Attack Missile—Conventional

TLAM-N Tomahawk Land Attack Missile—Nuclear

Tlax Tlaxcala (inhabitants—Tlaxcaltecas)

tlb torque limiting brake

TLB temporary lighted buoy

tlbl tape label

TLBs Time-Life Books

tlc talcum; tender loving care; thin-layer chromatography; total lung capacity

TLC Television Licensing Center; Thin-Layer Chemistry; Total Life Care; Total Life Center; Trades and Labour Club

TLCPA Toledo-Lucas County Port Authority

TLCs Tire and Lube Centers

TLC for Seniors Total Life Care for Senior Citizens

tld thermoluminescent detector; thermoluminescent dosimeter; tooled

tl dating thermoluminescent dating

tle theoretical line of escape; thin-layer electrochemistry; thin-layer electrophoresis

tlev transistional low-emissions vehicle

tlf telefon (Norwegian—telephone)

TLFB Texas-Louisiana Freight Bureau

tlg tail landing gear; telegraph

TLG Theatrical Ladies' Guild; Tiger Leasing Group

TLH Tallahassee, Florida (airport); Tembuland Light Horse

tli transport level interface

TLI Trinidad Light Infantry

tlli tank liquid-level indicator

tlm telemeter; telemetry

TLMA Tag and Label Manufacturers Association

Tln Tallinn

tlo total loss only

TLO Tank Liaison Officer; Technical Liaison Officer

tlp tension-leg platform; term-limit pricing; threshold learning process; time-lapse photography

tlp (TLP) tension-leg petroleum (oil rig)

TLP Theraputic Learning Program

TLP Telefones de Lisboa e Porto (Portuguese—Lisbon and Oporto Telephone Company)

tlr trailer; twin-lens reflex (camera)

TLR Tool Liaison Request

tls tank laser sight; testing the limits for sex; typed letter signed

TLS Technical Library Service; Technical Library System; Terminal Landing System; Territorial Long Service (medal); The Law Society; Trinity Lighthouse Service

TLS Times Literary Supplement

TLSP Transport Layer Security Protocol

tlt transportable link terminal

TLTB Trunk Line Tariff Bureau

tltr translator

tlu table look up

tlv threshold limit value(s)

tlv (TLV) tracked levitated vehicle

TLV Tel Aviv, Israel (airport)

tlvsn television

tly tally

tlz titanium, lead, zinc; transfer on less than zero

tm standard mean temperature; tactical missile (TM); team; temperature meter; thematic mapper; therapeutic massage; time modulation; tractor monoplane (TM); trademark; transport mechanism; transverse magnetic; trench mortar; true mean; twisting moment

tm (TM) tertiary memory

t/m test and maintenance

t & m time and material(s)

tm tonelada métrica (Spanish—metric ton, 2,200 pounds)

Tm melting temperature (symbol); midpoint of denaturization (symbol); thulium

TM tactical missile; Technical Manual; Technical Memoranda; Technical Memorandum; Technical Minutes; Technical Monograph; Telemetering; Test Manual; Texas Mexican (railroad); The Maccabees; Thurgood Marshall; Toledo Museum; Town Major; tractor monoplane; trademark; Training Manual; Training Mission(s); Train-master, Transcendental

Meditation; Tropical Medicine; Turkmenistan (Internet code)

T/M (t/m) trailmobile (automobile trailer)

TM *Technical Manual; Turk Mali* (Turkish—Made in Turkey); *Théâtre de la Monnaie* (French—Theater of the Currency)—Brussels; *Tragedy of Macbeth*

tma tissue mineral analysis; total material assets; total military assets

Tma Tema

TMA Ten Million Americans (mobilizing for justice); Texas Maritime Academy; Theatrical Mutual Association; Tile Manufacturers Association; Tobacco Merchants Association; Toiletery Merchandisers Association; Toy Manufactuiers Association; Trans-Mediterranean Airways (Lebanese); Transportation Management Association; Treasury Management Association; Turnaround Management Association; Twine Manufacturers Association

TMAMA Textile Machinery and Accessory Manufacturers' Association

T-man Treasury Department special agent of the IRS

tmar trial marriage

TMAS Taylor Manifest Anxiety Scale

TMB Travelling Medical Board; Trench Mortar Battery

TMBC Toronto Motor Boat Club

tmbr timber

tmc total manufacturing cost; total market coverage

TMC Tata Memorial Center; Technical Measurement Corporation; Texas Medical Center (Houston); Trans Mar de Cortés (Mexican airline); Transportation Materiel Command

TMCA Tabulating Card Manufacturers Association; Titanium Metals Corporation of America

tmcd tetramethylcyclobutanediol

TMCI Target Medical Cost Index

tmcp trimethylenecyclopropane

tmd temporomandibular disorder

TMD Tactical Missile Defense; Theater Missile Defense

tmdl total maximum daily load

TME Teacher of Medical Electricity

T-men Treasury Department law-enforcement officers

TM-Eng Technical Manual—Engineering

t'ment tournament

tmf the mushroom factor

TMF The Menninger Foundation

tmh thermomechanical hydraulic; tons per manhour

tmi technical market index (TMI)

Tmi Tsurumi

TMI Telemeter Magnetics Incorporated; Texas Military Institute; Three-Mile Island, Pennsylvania—site of nuclear reactor meltdown March 28, 1979; The Manhattan Institute; Tool Manufacturing Instruction; Trucking Management Incorporated; Tube Methods Incorporated; Turkish Military Institute; Turkish Military Intelligence

TMI *Technical Manual Index* (USN); *Technical Market Index*

TMIC Toxic Materials Information Center

TMIF Three-Mile Island (nuclear-power) Facility

TMIS Technical Meetings Information Service

tmj temporomandibular joint

tmj (TMJ) temporomandibular joint (syndrome)

TMJ *Trade Marks Journal*

tmjd temperomandibular joint dysfunction

tmkpr timekeeper

tml (TML) three-mile limit

tm/l team loses (baseball)

TML Transport Managers License; Transmanche Link (Eurotunnel)

tmldr team leader

TMM *Transportación Maritima Mexicana* (Spanish—Mexican Maritime Transportation)

TMMC Theater Materiel Management Center

TMMG Teacher of Massage and Medical Gymnastics

tmn transmission (flow chart)

Tmn Tamano

TMNP Tamborine Mountain National Parks (Queensland)

TMNT Teenage Mutant Ninja Turtle

tmo (TMO) telegraph money order

TMO Table Mountain Observatory; telegraph money order; Traffic Management Officer

TMORN Texaco Metropolitan Opera Radio Network

tmos the man on the street

tmp temperature; temporary; thermomechanical pulp(ing); transportation motor pool; trimethoprim; trimethyl phosphate (male contraceptive)

tmp (TMP) total mind power

Tmp Tampico

TMP temperature (aircraft code)

tmpry temporary

tmp's transcedental meditation practitioners

TMPS Trans-Mississippi Philatelic Society

tmr timer; total materiel requirement; total mixed ration (cow feed); trainable mentally retarded (semi-autistic children)

TMR Transvaal Mounted Rifles

TMRB Tropical Medicine Research Board

tmrbm (TMRBM) transportable midrange ballistic missile

tms tension myositis syndrome; terms; total markets served; type, model, and series

tms *tai muuta semmoista* (Finnish—and so on)

TMs Temporomandibular Disorders

TMS Tactical Missile Squadron; Technical Museum, Tesla Memorial Society; Stockholm; Tramway Museum Society; Transmatic, Money Service

TMS *Tribunal Maritime Special* (French—Special Maritime Court)—disciplinary prison court in French Guiana

TMSA Technical Marketing Society of America

tmsd total military service date

tmt technology, media, telecoms; turbine-motored train

T^mt Tablemount

TMT transonic model tunnel

TMTB The Malayan Tin Bureau

tmtc through-mode tape converter

tmtf temporal modulation transfer function

TMU Tokyo Metropolitan University

TMUS Toy Manufacturers of the United States

tmv true mean value

tmv (TMV) tobacco-mosaic virus

TMV Transportadora Maritima Venezolana (Spanish—Venezuelan Line)

tmw thermal megawatts; tomorrow

tm/w team wins (baseball)

TMW Textile Machine Works

TMWC Trial of the Major War Criminals

tn tariff number; telephone number; thermonuclear; town; township; train; true north; twisted nematic

Tn thoron (chemical symbol); Ton; transposons

TN Task Number; Technical Note; Tennessee

T & N Turner and Newhall

TN Twelfth Night

TNA The National Archives; Thermal Neutron Analysis (system for detecting bombs in baggage)

tnad thermal-neutron activation device

TNAS Tuberculosis Nursing Advisory Service

TNB Tsentral'naya Nauchnaya Biblioteka (Russian—Central Scientific Library)—Kiev

tnc total numerical control

TNC Thai Navigation Company; The Nature Conservancy; Threaded And Coupled

tnct tincture

tnd tender

TNDC Thai National Documentation Center

tndr tendered

TNEC Temporary National Economic Committee

t^nes *tonnes* (French—tons)

tnf transfer on no overflow; tumor necrosis factor

tnf (TNF) theater nuclear forces; tumor necrosis factor (natural substance killing cancer cells)

TNF Theater Nuclear Forces (NATO); Toiyabe National Forest

tng training

Tng Tandjung (Malay—Cape)

TNG Tangier, Morocco (airport); The National Grange; The Newspaper Guild

TNG The New Grove Dictionary of Music and Musicians (20-volume 1981 edition)

tnge tonnage

TnI troponin-I

TNI Trusted Network Interpretation

TNI Tentara Nasional Indonesia (Indonesian National Army)

TNIAU Tentara Nasional Indonesia Angkatan Udara (Bahasa Indonesian—Indonesian Armed Forces—National Air Force)

Tn IOB Technician of the Institute of Building

TNIU Trusted Network Interface Unit

tnm tumor, node, metastasis

tnm (TNM) tactical nuclear missile

TNM Texas-New Mexican; Texas-New Mexico; Tokyo National Museum; Tumacacori National Monument

TNM Teléigrafos Nacionales de México (Spanish—National Telegraph of Mexico)

TNNP Taman Negara National Park (Malaysia); Terra Nova National Park (Newfoundland)

Tno Taranto

T No (TNO) Track Number

T & NO Texas and New Orleans (railroad)

t no c threads no couplings

tnp (TNP) trinitrophenol

TNP Tarangire National Park (Tanzania); Taroba National Park (India); Tonariro National Park (North Island, New Zealand); Tsavo National Park (Kenya)

TNP Théâtre National Populaire (French—Popular National Theater)

tnpg trinitrophloroglucinol

TNPG The Nuclear Power Group

Tnpk Turnpike

TNPO Terminal Navy Post Office

tnr tonic neck reflex; trainer; tumor necrosis factor

tnr toneladas de registro neto (Spanish—net registered tonnage)

TNR Tananarive, Malagasy (airport); Tucki Nature Reserve (New South Wales)

TNR The New Republic

TNRIS Transportation Noise Research Information Service

Tnry Tannery

tns transcutaneous nerve stimulator

Tns Townsville; Tunis

TNS Tennessee Nuclear Specialties; Transit Navigation System

TnSEA Tennessee Society of Enrolled Agents

tnt (TNT) trinitrotoluene

t-n t trans-national terrorism; trans-national terrorist

t'n't tequila and tonic (mixed drink)

TNT Tactical Narcotics Team (NYC)

tntc too numerous to count

TNTC Thames Nautical Training College

t-n t's trans-national terrorists

tntv tentative

tnw theater nuclear weapon

tnw (TNW) tactical nuclear warfare

TNWA The New World Alliance

tn wep(s) thermonuclear weapon(s)

TNWR Tamarac National Wildlife Refuge (Minnesota); Tewaukon NWR (North Dakota); Tishomingo NWR (Oklahoma)

tnx thanks

tnz transfer on non zero

to. telephone order (TO); time off; time opening; time out; tool order (TO); transverse optic; turn off, turn over

tø no evidence of primary tumor

t/o (TO) takeoff

t & o taken and offered; technical and office (workers)

t.o. tinctura opii (Latin—tincture of opium)

t^o tomo (Spanish—volume)

To Togo; Toronto

To Torsdag (Danish—Thursday)

TO Table of Organization; take-off, Technical Observer; Technical Officer; Technical Order(s); Theater of Operations; The Order (neo-Nazi organization); Third Order (of a religious congregation); Toledo, Ohio; Tonga (Internet code), Tool Order; Torpedo Officer; Trained Operator; Trans Arabian Air Transport (cargo); Transportation Office(r); Travel Order

T/O Table of Organization

TO Technical Order

toa total obligational authority

TOA Theater Owners of America; The Orchestral Association; Toledo Opera Association

toac tool accessory

tob tobacco

tob (TOB) tender option bond

Tob Tobago; The (Apocryphal) Book of Tobit

T o B Tour of Britain (bicycle)

TOBA Theater Owners' Booking Association

tobac tobacco; tobacconist

Tobaccos Tobacco Root Mountains of southwest Montana

Tobag Tobagonian

TOBE Test of Basic Education

Tobio Gorria anagrammatic pseudonym of Arrigo Boito (1842–1918)

TOBWE Tactical Observing Weather Element (USAF)

toc table of contents; top-blown oxygen converter; top of concrete; total organic carbon

Toc Tagliabue open cup

TOC Tactical Operations Center; Technical Order Compliance; Television Operating Center; Timber Operators' Council

TOCA Tohono O'odham Community Action

TOCCWE Tactical Operations Control Center Weather Element (USAF)

Toch Tocharian

TOCHR Terminal Operators Conference of Hampton Roads

Toco Tocopilla, Chile

TOCS Terminal Operating Control System

TOCU Tornado Operation Conversion Unit

tod technical objective document(s); time of day; time of delivery; time of dispatch; transferable on death

Tod Todhunter

TOD Technical Objective Document; The Open Door

to'ds toads; towards

toe term of enlistment; theory of everything; ton-oil equivalent; total operating expense

TOE Table of Equipment; Target of Evaluation; Thread One End

to & e table of organization and equipment

T O & E Texas, Oklahoma & Eastern (railroad)

TOEFL Test of English as a Foreign Language

TOES The Other Economic Summit; Tradeoff Evaluation System

TOET Test of Elementary Training

tof time of flight; top of form

tofc trailer on flatcar (or piggyback)

tog together; toggle; to order grog

Tog Togo (whose capital is Lomé)

TOGA Tests of General Ability; Tropical Ocean/Global Atmospheric Program

to'gal'nt topgallant (mast or sail)

Togo Admiral Togo Heihachiro (victor of the Battle of Tsushima where his forces annihilated the Russian fleet in 1905); Republic of Togo (West African coastal country) *République Togolaise*

Togoland (German—Togo)— West African colony under German domination from 1884 to 1916

togr together

togs thermal observation and gunnery sight

togw takeoff gross weight

tog/wi together with

Toh. Tohorot

tohp takeoff horsepower

toj track on jamming

Tojo Premier Tojo Hideki (Japanese general and premier during World War II)

Tok Tokyo

Tokelaus Tokelau Islands of the Pacific also called the Union Islands including Atafu, Fakaofu, and Nukunono

Token-ring IBM trademark

toke(s) token(s)

Tok Uni Tokyo University

tol tolerance; toluene

Tol Toledo; Toledan

T o L Tower of London

TOL Toledo, Ohio (airport); Trans-Ocean Leasing (corporation); Trucial Oman Levies

tol'able tolerable

TOLCCS Trends in On-Line Computer Control Systems

Tolly Tolliver

Tol Orc Toledo Orchestra

to lt towing light

TOLU Transoccan Leasing (container) Unit

tom tom cat,

t-o-m the old man (the boss; the captain, the chief, the father)

tom *tomo* (Spanish—volume)

Tom $2 bill; Thomas

TOM Territoire d'Outre-Mer (Overseas Territory)

tom(at) tomato

tomats tomatoes

tomb technical organizational memory bank

Tombigbee Tombigbee River of Alabama and Mississippi

tomcat (TOMCAT) theater-of-operations missile continuous-wave anti-tank (weapon)

tome (Greek—a cutting or a slice)—anatomy, dichotomy, lobotomy, microtome

Tommie Thomas

toms male turkeys; males of various animals; tired old movies

TOMS Total Ozone Mapping Spectrometer; Total Ozone Mapping System

TOMV Tragedy of Othello, the Moor of Venice

tomy (Latin suffix—cut)—appendectomy

ton *toneel* (Dutch—scene, set, stage); *toneladas* (Spanish—tonnage, tons); *tyurma osobogo naznacheniya* (Russian—special-purpose prison)

Ton Tonga or Friendly Islands

TON Top of the News

TONACS Technical Order Notification and Completion System

Tonga Kingdom of Tonga (South Pacific island nation)

Tongariro Tongariro National Park in New Zealand's North Island or an active volcano in the same area

Tongas Tonga Islands in the South Pacific

Tongass Tongass National Forest in southern Alaska

tonguesan tongue sandwich

tonguewich tongue sandwich

Ton Isl Tonga Islands

tonk honky tonk

Tonka Minnetonka, Minnesota

tonn tonnage

too time of origin

'toon(s) cartoon(s)

top temporarily out of print; topographer; topographic-(al)(1y); topographica (threedimensional) art; topography; torque oil pressure

t-o-p temporal-occipital-parietal (lobes of the brain)

Top Topeka; Topology

ToP Taxonomy of Programs

TOP Targeted Outreach Program; Technical Office Protocol; Third Order of Penance; Trade Opportunities Program

top 10 top 10 best sellers (books or recordings of classical, jazz, or popular music)

topa tooling pattern

topaz hydrous aluminum fluorosilicate

TOPAZ Technic for the Optimum Placement of Activities in Zones

TOPCOPS The Ottawa Police Computerized On-line Processing System (Canada)

Top End northernmost Australia

TOPICS Tables of Periodical Indices Concerning Schools; Test of Performance in Computational Skills

TOPIX Tokyo stock Price Index

to po topographic; topography

Topo Topolobampo, Sonora, Mexico

TopoCom Topographic Command (USA)

topog topography

topol topology

topon toponym

topony toponym(ic)(al); toponymist; toponymy

topo(s) toponym(s)

TOPP Terminal-Operated Production Program

tops (TOPS) take off pounds sensibly

TOPS Task-Oriented Processing System; Teen-age Opportunity Programs in Summer; Tested Overhead Projection Series; Total Operations Processing System; Training Opportunities Scheme

Top Sec Top Secret

tops'l topsail

TOPSTAR The Officer Personnel System—The Army Reserve (USA)

Toquemas Toquemas Mountains of central Nevada

tor terms of reference; time of receipt; torque; torquing; torquing up

tor (TOR) teletype on radio; transmitter output register

Tor Toronto

TOR Third Order Regular

Toray Tokyo Rayon Company (tradename)

torbie tortoise-shell tabby (cat)

TORCH Toronto Orthopaedic Recreational Center's Headquarters; Toxoplasma, Rubella, Cytomegalovirus, Herpes simplex (test for mother and newborn)

Torch of Liberty Statue of Liberty

Tor-Chi Toronto—Chicago

Tor Dep Torpedo Depot

Tor Dom Toronto Dominiom (bank)

Tor Int Air Toronto International Airport

Tor-Lax Toronto—Los Angeles

Tormentine Cape Tormentine—easternmost point in New Brunswick, Canada

Tor-Mia Toronto-Miami

torn. tornado

Torngats Torngat Mountains of Labrador

Tor-NY Toronto-New York

torp torpedo; torpedoman

Torport Toronto (container) Port

torr 1 mm of mercury

Torr toor

Tor-San Toronto-San Diego

Tor-Sea Toronto-Seattle

Tor-Sfo Toronto—San Francisco

Tor Sym Toronto Symphony Orchestra

tortie tortoise-shell (cat)

Tortugas Tortuga Islands (Dry Tortugas and Wet Tortugas)

tos tactical operations system; taken on strength; term of service; track operational surcharge; temporarily out of stock

TOS Tape Operating System; The Orion Society; Tiros Operational Satellite; Trucial Oman Scouts

Tosa Tsunetaka

TOSBAC Toshiba Scientific and Business Automatic Computer

tosc toscano (Italian—Tuscan)

TOSCA Toxic Substances Control Act

Toscanini Arturo Toscanini (outstanding orchestra conductor of the first half of the 20th century)

TOSCO The Oil Shale Corporation

Tosc synd Toscanini syndrome

tose tooling samples

Toshiba Tokyo Shibaura Electric

toss. takeoff safety speed

TOSS Tiros Operation Satellite System

tot time on (over) target; total; totalize; totalizer

t-o-t tip-of-the-tongue

TOT Tourist Occupancy Tax; Tourist Organization of Thailand Transient Occupancy Tax

totalism totalitarianism

TOTCO Technical Oil Tool Corporation

tote. totalizator

TOTE Task-Oriented Teacher Education; Test-Operate-Test-Exit (unit); Totem Ocean Trailer Express (to Alaska)

TOTES Test-Operate-Test-Exit System

t'other the other

TOTO Tongue of the Ocean (deep-water channel in Great Bahama Bank); Totable Tornado Observatory (National Severe Storms Laboratory, Norinan, Oklahoma)

totp tooling template

Tou Toulon

TOU The Open University; Tractor Oils Universal

tour. tourism, tourist

tourn tournament

TOUS Test on Understanding Science

tov ten opzichte van (Dutch—with regard to)

TOVALOP Tanker Owner's Voluntary Agreement concerning Liability for Oil Pollution

tow tug of war

tow (TOW) tube-launched optically-tracked wire-guided (anti-tank missile)

TOW Top Antitank Weapon

Tower The Tower of London (formerly a prison and now a museum by the Thames in London); Tower Publications; Tower Records

TOWER Testing, Orientation, and Work Evaluation in Rehabilitation

Towers Charters Towers

townet towing net

tox toxemia; toxic; toxicant; toxicologist; toxicology

tox (Latin prefix—poison)—toxicology

toxback toxicology information backup

toxicol toxicology

toxline toxicology hot line (public information program); toxicology on-line

TOXLINE Toxicology On-Line (computer retrieval system)

toz tidelands overlay zone

tp target practice; teaching practice; technical paper; telephone; teleprinter; thermoplastic elastomers; title page; toilet paper; total points; total protein; transport pilot; treaty port; triple plays (baseball); turning point

tp (TP) tape (computer flow chart); teleprocessing; total product; transaction processing

t/p test panel

t & p theft and pilferage

tp tempo primo (Italian—speed as at the outset)

Tp Township; Troop

TP East Timor (Internet code); Technical Pamphlet; Technical Paper; Technical Problem; Technical Publication; Technographic Publication; Temota Province (Solomon Islands); Texas & Pacific (railroad); Thomas Paine; Thompson Products; Torrey Pines (Institute); True Position

T & P Texas and Pacific (railroad)

T.P. Tempore Pachale (Latin—Easter time); *Tribunicia Potestas* (Latin—Tribune of the People)

tpa third-party adjuster; tissue polypeptide antigen; total plate appearances (baseball); travel by privately owned conveyance authorized

tpa (TPA) tissue plasminogen activator

t-pa tissue plasminogen activator (for stroke patients)

TPA Tampa, Florida (Tampa International Airport); Tampa Port Authority; Telephone Pioneers of America; Trans-Pacific Airlines (Aloha Airline); Travelers' Protective Association

TPAC Thomas Performing Arts Center (Akron)

TPAO Türkiye Petrolleri Anomin Ortakligi (Turkish Petroleum Corporation)

TPAM Technical Purchasing of Avionic Material

tpb tryptone phosphate broth

TPB Transportation Programs Bureau

TPBA Transit Patrolmen's Benevolent Association

TPBC Toledo Power Boat Club

tpc treated-paper copier

TPC The Peace Corps (US Department of State)

TPCC Trade Promotion Coordinating Committee

TPC/JCA Texas Public Community/Junior College Association

TPCNA Titanium, Palladium, Copper, Nickel, Au (gold) telephone circuit-board plating

TPCP Test Plan Change Procedure

TPCUS The Pacific Council of the United States

tpd temporary partial disability; tons per day

tp'd toilet papered (some teenagers' idea of house-and-garden decoration)

TPDC Tanjong Pagar Dock Company (Singapore)

TPE Taipei, Formosa (airport)

TPEQ Task of Public Education Questionnaire

TPF Tactical Police Force; The Planning Forum; Thomas Paine Foundation

TPFH Tasmanian Pulp and Forest Holdings

TPGA Texas Personnel and Guidance Association

tpgh tons per gang hour

tph tons per hour; total petroleum hydrocarbons

TPH Theosophical Publishing House

TPH Television Production Handbook

Tpha Treponema pallidum hemagglutination

tphasap telephone as soon as possible

tphayc telephone at your convenience

TPH & PCA Toy Pistol, Holster, and Paper Cap Association

TPHS Thomas Paine High School

tpi teeth (threads, tons, or turns) per inch; total pitcher index; tracks per inch; treponema pallidum immobilization (test)

t-p i title-page, index

t-plus 3 trade day plus three (time allotted to deliver check or shares to a broker)

Tpi Taipei; Treponema pallidum immobilization

TPI Tax-and-Price Index; Tennessee Polytechnic Institute; Torrey Pines Institute; Truss Plate Institute

Tpi test *Treponema pallidum* immobilization (for the detection of syphilis)

Tpk Turnpike

Tpke Turnpike

TPL Tallahasee Public Library; Tampa Public Library; Toledo Public Library; Toronto Public Libraries; Tucson Public Library; Tulsa Public Library

TPLA Turkish People's Liberation Army

tplab tape label

TPLF Tigré People's Liberation Front; Togrean People's Democratic Front (Ethiopia), Turkish People's Liberation Front

TPLs Trust for Public Lands

tpm tape preventive maintenance; title page mutilated; tons per minute; toxoplasmosis

TPM Technical Performance Measures; Timber Products Manufacturers

tpmark tapemark(ing)

tpn trigger price mechanism

tpn (TPN) triphosphopyridine nucleotide; (same as nadp or NADP⁺)

tpn total parental nutrition

TPN Tatrzanskiego Parku Narodowego (Polish—High Tatra National Park)—in the Tatra Mountains of Poland

tp & nd theft, pilferage, and non-delivery

tpnh (TPNH) reduced triphosphopyridine nucleotide

TPNHA Thomas Paine National Historical Association (New Rochelle, NY)

TPNHS Thomas Paine National Historical Society

tpnl test panel

tpo terminal-performance objective; transmitter (signal) power output

tpo tiempo (Spanish—time)

TPO Travelling Post Office; Tulsa Philharmonic Orchestra

tpob true point of beginning

TPOR Teacher Practices Observation Record

tpp (TPP) thiamine pyrophosphate

TPP Tax Preparers Program; Technical Program Plan(ning); Total Package Procurement

TPPC Total Package Procurement Concept; Trans-Pacific Passenger Conference

TP-PL Technical Publications Planning (USN)

TP-PU Technical Publications—Public Utilities (USN)

tpqi teacher-pupil question inventory

tpr tape programmed raw; team pitching rating; telescopic photographic recorder; temperature profile recorder; temporary price reduction; thermal plastic rubber; thermoplastic recording

tpr (TPR) temperature, pulse, respiration

Tpr Trooper

TPRC Thermophysical Properties Research Center

tpri teacher-pupil relationship inventory

TPRI Tropical Pesticides Research Institute

T & P Ry Texas and Pacific Railway

tps technical problem summary; technopolymer structures; terminals per station; text processing service; thermal protection system; throttle position sensor; transactions per second; tree-pruning system (computer language)

tp's taxpayers

TPS Technical Preservation Services, Technical Publishing Society; Telephone Preference Service; Test Pattern Set; Text Processing Service; The Physical Society

tpt tetraphenyl tetrazolium; total protein tuberculin; transport; trumpet

TPT Tactual Performance Test(ing); Toy Preference Test; Transonic Pressure Tunnel (NASA)

tptg tuned plate tuned grid

tptn toilet partition

tpto tripropyl tin oxide

tptr trumpeter

tpu tape preparation unit; thermoplastic urethane

TPU Travel Planning Unit

TPUS Transportation and Public Utilities Service

tpw title page wanting

TP & W Toledo, Peoria & Western (railroad)

t.q. tale quale (Latin—as is)

TQ Test Quotient

TQCA Textile Quality Control Association

tqcm thermoelectric quartz-crystal microbalance

TQE Technical Quality Evaluation; Total Quality Excellence

TQM Total Quality Management

TQMS Technical Quartermaster-Sergeant; Troop Quartermaster-Sergeant

TQS Total Quality Service (certificate program)

t quark top quark

tr temperature, rectal; test ran; time remaining (at depth); tons registered; throws right; tilt rack; toothed ring; trace; tracer (bullet); tracking radar; transformer/rectifier; transitive; translated; translation; transmit-receive; transmitter-receiver; transpose; troop; tuberculin R

tr (TR) total revenue

t-r transmit-receive

t/r transmit(ter)/receive(r)

tr (Latin—tincture); *trillo* (Italian—rolled or shaken, as in drumming or when shaking a tambourine); *traduit* (French—translated); *trykkeri* (Dano-Norwegian—printing office); *tryckt* (Swedish—printed); *trykt* (Dano-Norwegian—printed)

Tr Transcript; Trench; Trieste; Trough

TR Tasmanian Railway; Technical Regulation; Technical Report; Territorial Reserve; Test Report; Texas Gulf Production Company (stock-exchange symbol); Theodore (Teddy) Roosevelt (26th President U.S.); therapeutic radiology; torpedo reconnaissance (naval aircraft); Training Regulation(s); Transportation Request; Travel Request; Trieste; Trip Report; Triumph (British auto or motorcycle); Turkey (auto plaque and Internet code)

TR-1 Theater Reconnaissance-1

T-R Times-Roman

tra transformer-reactor assembly

Tr A Triangulum Australe (constellation)

TRA Tactical Response Association; Tax Reform Act (of 1986); Technical Report Authorization; Teledyne Ryan Aeronautical; Textile Refinishers Association; Theodore Roosevelt Association; Thoroughbred Racing Associations; Tire and Rim Association; Trade Relations Association; Travel Research Association

TRAA Towing and Recovery Association of America

traac transit-research and altitude-control (satellite)

trac terminal rental adjustment clause; text-reckoning and compiling (computer language); tracer; tracing; tractor

TRAC Tolerance Review Advisory Committee

TRACALS Traffic Control and Landing System

tracap transient circuit-analysis program

tracdr tractor-drawn

trace tape-controlled recording and automatic checkout equipment; task reporting and current evaluation; time-shared routines for analysis, classification, and evaluation; total-risk assessing-cost estimate(s)

TRACE Task Reporting and Current Evaluation; Trane Air Conditioning Economics

trach trachea; tracheal; tracheate; tracheation; tracheoscopy; tracheostomy; tracheotomy

TRACIS Traffic Records and Criminal Justice Information System (Iowa)

trackex tracking exercise

tracon terminal radar control

TRACON Terminal Radar Approach Control

TRACS Telemetry Receiver Acoustic Command System; Telescoping Rotor Aircraft System; Total Royalty Accounting and Copyright Systems

tract (Latin prefix—drag or draw)—traction

tractorcade tractor vehicle parade

trad tradition(al)

trad traducido (Spanish—translated); *traduzione* (Italian—translation)

TRADA Timber Research and Development Association

tradex target resolution and discrimination experiment

tradic transistor digital computer

TRADOC Training and Doctrine Command

traf traffic

Trafalgar Cape Trafalgar in southwestern Spain at the western entrance to the Strait of Gibraltar

TRAFFIC Trade Records Analysis of Fauna and Flora in Commerce (endangered species)

trafphobia traffic phobia (fear of driving in traffic)

trag tragedy

T-rail T-shaped rail

train. trainee; trainer; training

TRAIN Telerail Automated Information Network; To Restore American Independence Now; Tourist Railway Association Incorporated

TRAIS Transportation Research Activity Information Service (Department of Transportation)

trai vai (Vietnamese—lychee)—tasty fruit

TRALA Truck Renting and Leasing Association

tram. tracking radar automatic monitoring; tramcar; trammel; tramway

TRAM Test Reliability and Maintenance Program (USN); Treatment Rating Assessment Matrix; Treatment Response Assessment Method

tramp. temperature regulation and monitor panel

tramps. temperature regulator and missile power supply

tran transient

tran (TRAN) tax revenue anticipation note; transmit (data processing)

tran. transsexual

tran tranvia (Spanish—tramway)—streetcar or streetcar line

trandir translation director (computer language)

tran flap transverse rectus abdominal muscle flap

tranny transistor radio; transmission (automobile)

trans transactions; transfer; transit; translation; translator; transport; transportation; transpose; transposition

trans (Latin prefix—across or over)—transalpine, transatlantic

Trans Transactions; Transmission; Transvaal

transac transaction(s)

Trans Am Cryst Soc Transactions of the American Crystallographic Society

Trans Am Geophys Union Transactions of the American Geophysical Union

Trans Am Inst Min Metall Pet Eng Transactions of the American Institute of Mining, Metallurgical, and Petroleum Engineers

Trans Am Nucl Soc Transactions of the American Nuclear Society

Trans Am Soc Mech Eng Transactions of the American Society of Mechanical Engineers

Trans Am Soc Met Transactions of the American Society for Metals

Transan Transandean Railway

Transandine Transandean Railway connecting Argentina and Chile

transatl transatlantic

Transbai Transbaikal Railway

Trans Br Ceram Soc Transactions of the British Ceramic Society

transc transcription

Trans-Carib Trans-Caribbean Airways

Trans-Caspian Trans-Caspian Railroad linking the Caspian Sea region with the southern Urals

Transcau Transcaucasia (Armenia, Azerbaijan, Georgia); Transcaucasian Railway

transceiver transmitter-receiver

TRANSCOM Transportation Coordinated Management

transcr. transcribed

transcrit transportation critic(ism)

trans d transverse diameter

TRANSDEC Transducer Electronic Center

transec transmission security

transf transfer; transference; transformer

Trans Faraday Soc Transactions of the Faraday Society

transfax facsimile transmission

transie(s) transvestite(s)

TRANSIS Transportation Safety Information System

Transisthmian Transisthmian Highway (flanking the Panama Canal and the Panama Railroad)

transistor transfer resistor

transit. transitive

Transj Transjordan; Transjordanian

Transk Transkei

Trans-Ky Exp Trans-Kyusho Expressway

transl translation; translator

translit transliteration

translu translucent

translun translunar; translunarian; translunarite

transm transmission

Transmark Transportation Systems and Market Research (British rails)

Trans Metall Soc AIME Transactions of the Metallurgical Society of the American Institute of Mechanical Engineers

transmog transmogrification; transmogrify(ing)

Transnistria Trans-Dniestria

Transocean California-Hawaii-Orient Airline; Transoceanic

transp transparent; transportation

transpac transpacific

transpl transplant(ation); transplanted

transport. transportation

transputer transistor computer

Transron Transport Squadron

trans sect transverse section

transsexual(s) transvestite homosexual(s)

Trans-Sib Trans-Siberian Railroad linking European Russia with its North Pacific coast

Trans Soc Rheol Transactions of the Society of Rheology

TRANSUB Translation and Publishing Corporation (China)

transv transverse

Transv Transvaal

transvest transvestic(al)(ly); transvestism; transvestite

transv sect transverse section

Transylvanians Transylvanian Alps of Romania

transyt traffic network study tool

trany transparency

trap. trapdoor; trap drums; trapeze; trapezoid(al); trapezium

TRAP Tracker Analysis Program

traps. trap drums; trap drummer(s)

TRASOP Tax Reduction Act Stock Ownership Plan

tratel tracking through telemetry; trailer motel

tratt trattenuto (Italian—detained or held back)

trau traumatic

TRAUS Thoroughbred Racing Association of the U.S.

trav. travel

Trav Travancore; Travis

Trav Travessa (Portuguese—Lane)

TRAWL Tape Read-and-Write Library

trb tribunal; tribune; trombone

trb toneladas de registro bruto (Spanish—gross registered tonnage)

TRB *New Republic's* pseudommymic initials standing for columnist Richard Strout; Technical Review Board; Transportation Research Board

trbn trombone

trc total response to crisis

Tr & C Troilus and Cressida

TRC Tape Relay Center; Technical Review Committee; Technology Reports Center; Telegram Retransmission Centre; Trans-Caribbean Airways; Transportation Research Command; Truth and Reconciliation Commission

TRCA Toronto Region Coordinating Agency (Hamilton to Oshawa)

trccc tracking radar central control console

Tr Co Trust Company

tr coil tripping coil

Tr Coll Training College

TRCS Trade Relations Council of the United States

TRCUD Technical Review Committee on Underground Disposal of Radioactive Wastes

trcver transceiver

Trd Trinidad

TRD Test Requirements Document

TRD Teatro Regio Ducal (Italian—Royal Ducal Theater)—Milan

TRDA Timber Research and Development Association

TRDCOM Transportation Research and Development command

trdto tracking radar data takeoff

Tʳᵉ Torre (Italian or Portuguese—tower)

TRE Telecommunications Research Establishment

TRE Tempore Regis Edwardi (Latin—in the time of King Edward)

treas treasure; treasurer; treasury

Treas Treasurer

trec tracking radar electronic components

TRECOM Transportation Research and Engineering command

tree trustee

trees. (TREES) transient radiation effects on electronic systems

TREF Trauma Research and Education Foundation

trem tremolando (Italian—trembling)

trem card transport or truck emergency card

Tren Trenton

trend. tropical environment data

treph trephining (trepanning)

Trep. pal. Treponema pallidum—the spirochete of syphilis

Tres Hermanas Tres Hermanas Mountains of southwestern New Mexico

TREVI Terrorisme, Radicalisme, et Violence International (French—Terrorism, Radicalism, and International Violence)—EEC police network

T rex Tyrannosaurus rex

trf transfer; tuned radio frequency

trf (TRF) thyrotropin-releasing factor

TRF Task Request Form; Teacher Rating Form; Timber Relief Fund; Transportation Research Foundation; Tuna Research Foundation; Turf Research Foundation

trg training; triangle

TRG Tactial Reconnaissance Group

tr & g transmit, receive, and guard

trgt target

trh tension-reduction hypothesis; thyrotrophic-releasing hormone

trh (TRH) thyrotrophin-releasing hormone

Tr H Trinity Hall, Oxford

TRH Their Royal Highnesses

TRHS Theodore Roosevelt High School

tri total response index (TRI); triangle; triangulation; tricolor; tricycle; triode

tri (Latin prefix—three)—triangle

Tri Triangulum (constellation); Trieste

Tri Tohtori (Finnish—doctor)

TRI Technical Report Instruction; Television Reporters International; Textile Research Institute; The Rockefeller Institute; Tin Research Institute; Tire Retreading Institute; total response index; Toxics Release Inventory

triad air, sea, and land defense

TRIAL Technique for Retrieving Information from Abstracts of Literature

trian triangle; triangulation

Trias Triassic

trib tribade; tribadism; tribal; tribalism; tribalist; tribasic; tribology; tribunal; tribune; tributary

TRIB Temple Records Index Bureau

Trib Tribune

Tri B Triborough Bridge

TRIB Tire Retread(ing) Information Bureau

tribas tribasic

TRIBE Teaching and Research in Bicultural Education

trib¹ Tribunal (Spanish—tribunal; court of justice)

Tribeca Triangle Below Canal Street (New York City)

tric trachoma inclusion conjunctivitis; trichloroethylene

Tric Tricolare (French or Italian—tricolor flags of French in vertical blue, white and red bands, whereas Italian is in green, white and red)

tricaphos tricalcium phosphate

trice. transistorized real-time incremental computer expandable; trichomoniasis (protozoan vaginal infection)

trich (Latin prefix—hair)—trichosis

Trich Tiruchchirappalli or Trichinopoly (famous for its Indian cigars)

Trichi Trichinopoly, Hindustan; cigar from that area

trick (slang—trichomoniasis)

tricl triclinic

trico trichomoniasis

Tri Com Trilateral Commission (Council of Foreign Relations)

TRICON Tri-Service Container (program)

trid. triduum (Latin—three days)

Trident Trident Region (Berkeley, Charleston, and Dorchester counties comprising the Charleston, South Carolina area)

tridundant triple redundant

TRIEA Tea Research Institute of East Africa

trig trigarnist; trigamy; trigger(man); trigonal; trigonometric; trigonometry

Trig trigonometry

triga trigger reactor

trihem trihemeral; trihemirer

tri ins tricuspid insufficiency

trik trichloroethylene

trike tricycle

trilat trilateral; Trilateral Commission (Council on Foreign Relations); trilateralist(ic)(al)(ly)

trillion *American*—a million million—10¹²; *British*—a million million million—10¹⁸

tril(s) trillion(s)

trim. trimetric

trim (TRIM) test rules for inventory management

trim. trimestre (Latin—quarter; three months)

TRIM Targets, Receivers, Impacts, and Methods; Technical Requirements Identification Matrices; Tax Reform Immediately

TRIMIS Tri-Service Medical Information System

TRIMMS Total Refinement and Integration of Maintenance Management Systems (USA)

TRIMS Texas Research Institute of Mental Sciences

trimtu (TRIMTU) trimethyl thiourea

Trin Trinidad(ian); Trinitarian(ism); Trinity; Trinity College

TRIN Trader's Index

Trinco Trincomalee

Trin Col Trinity College

Trin H Trinity Hall

Trinity Trinity Christian College; Trinity Church; Trinity College; Trinity House (Pilot Service); Trinity Parish; Trinity Parish School; Trinity School; Trinity University; Trinitytide

Trin-Tob Trinidad and Tobago

triol triolism; triolist

triols triolists (also called troilists)

trip. triple; triplicate; triplication; tripos

trip. (TRIP) technical reports indexing project

TRIP The Road Improvement Program

triphib triphibian; triphibious (land, sea, air)

tripl triplication; triplicate

Triple-A Anti-Aircraft Artillery

triple-A S AAAS (American Association for the Advancement of Science)

tris tris (hydroxymethyl) aminomethane

Tris Tristán; Tristō; Tristram

trishaw tricycle rickshaw

trisk triskelion

TRISNET Transportation Research Information Services Network

TRISTAN Tri-Ring intersecting Storage Accelerators in Nippon

Tristan da Cunha Tristan da Cunha Islands (Gough, Inaccessible, Nightingale, Tristan da Cunha)

tri sten tricuspid stenosis

trisyll trisyllable

trit. tritura (Latin—triturate)

TRI-TAC Tri-Services Tactical Communications Program (DoD)

tritic tritical (trite); triticale (*Triticum* + Secale hybrid between wheat and rye); triticeous; triticeum; tritish; triticum; tritium

triv trivia(l)

TRJ *Tragedy of Romeo and Juliet*

trk track; truck; trunk

Trk Turk; Turkey; Turkic; Turkish

trkdr truck-drawn

trkg tracking

trkhd truckhead

TRK T Track Time

trl trailer; transistor-resistor logic

Trl Trail

TRLB temporarily replaced by lighted buoy

trlfsw tactical-range landing-force support weapon

trlr trailer

Trlr Trailer (postal abbreviation)

trm task response module (engineer's desk area); thermoremanent magnetism

Trm Trincomalee

trml terminal

trmn trainman

trmr trimmer

TRMS Technical Requirements Management System

trmt treatment

trn transfer

Trn Troon

TRN Technical Research Note

tRNA transfer RNA (same as sRNA)

trnbkl turnbuckle

trng training

TRNMP Theodore Roosevelt National Memorial Park

trnsp transport; transportation

TRO Technical Reviewing Office; Temporary Restraining Order; Troy Haymakers (National Association)

TROA The Retired Officers Association

troch troche

troch *trochiscus* (Latin—cough drop, lozenge, troche)

Troch Trochelminthes

troil troilism; troilist

Troj Trojan

trol tapeless rotorless on-line cryptographic equipment

trom tromba; trombone

T Rom Times Roman

trombst trombonist

tromp *trompette* (French—trumpet)

troms trombones

T-room (American slang—toilet)

trop tropic; tropical; tropics

trop *tropos* (Greek—to turn or to turn toward)—entropy, geotropism, phototropism, tropic(al), tropism

troparium tropical aquarium

Trop Can Tropical of Cancer—$23^1/_2°$N Lat

Trop Cap Tropic of Capricorn—$23^1/_2°$S Lat

tropec tropical experiment

trophe (Greek—nutrition)—atrophy, autotrophe, heterotrophe, trophic level

trophy (Latin suffix—relating to nutrition)—hypertrophy

tropic (Latin prefix—pertaining to a turn)—tropical, tropicolitan; (Latin suffix—turning toward)—gonadotropic

TROPICS Tour Operators Integrated Computer System

trop med tropical medicine

troposcatter beyond-the-horizon communication

TROSCOM Troop Support Command (USA)

Trots Trotskyite(s)

trp troop; tubular reabsorption phosphate

trp (TRP) tryptophan

Trp Tripoli

tr pl treatment plan

trr teaching and research reactor; train repetition rate

TRR Test Readiness Review; Three Rivers Regiment (Canada)

TRRA Terminal Railroad Association (of St Louis)

TRRB Test Readiness Review Board (NASA)

TRRE Training Regiment Royal Engineers

TRRG Tax Reform Research Group

trs target range servo(mechanism); text retrieval system; transfer; transparency; transpose; tropical revolving storm; trustees

trs (TRS) tetrahedral research satellite

TRs Tax(ation) Reports; Technical Reports; Temporary Reserves

TRS Ticket Reservation System; Transair Limited

TRSA Terminal Radar Service Area

TRSAA Textile Rental Services Association of America

trsb time reference scanning beam

trsd total rated service date

tr sh trim shell

TrSMS triple-screw motor ship

trsp transport

TRSP Turtle River State Park (North Dakota)

trsr taxi and runway surveillance radar

TrSS triple-screw steamer

trssgm tactical range surface-to-surface guided missile

trsv (TRSV) tobacco-ringspot virus

trt total response to trauma; treatment; turret

TRTA Trader's Road Transport Association

TRTC Tropical Radio Telegraph Company

trtch tape-recording technic

tru transformer/rectifier unit

tru (TRU) transuranic (contaminated) waste

Tru Trucial; Trucial Sheikdoms; Truman; Truman Capote (1924–1984); Truro

Tru *Truman's Railway Reports*

TRU The Rockefeller University

TRUB temporarily replaced by unlighted buoy

Tru Cst 1 Trucial Coast Number 1

Tru Cst 2 Trucial Coast Number 2

Trud time remaining until dive (of satellite into Earth's atmosphere)

Trudy Gertrude

TRUE Teachers Resources for Urban Education

Truemid Movement for True Industrial Democracy

truf transferrable underwriting facility

tru-fi tru fidelity (sound reproduction)

Truman Harry Truman Field (U.S. Virgin Islands airport near Charlotte Amalie on St Thomas)

trump. trumpet

TRUMP Target Radiation Measurement Program

trumps trumpets

trun trunnion

trunc truncate; truncated; truncation

trunch truncheon

tr unit turbidity reducing unit

Truron (Church Latin—Truro)

tru(s) trustee(s)

trust. trusteeship

TRUST Trieste United States Troops

Trusted Xenix Trusted Information Systems trademark

truthsayer(s) honest person(s)

trv torpedo recovery vessel

trveh tracked vehicle

trw trawler

trwov transit without visa

TRW SL *TRW Space Log*

trx trithorax

trxrx transmitter-receiver

try. truly

try. (TRY) tryptophan

TRY Teens for Retarded Youth (juvenile correctional program)

Tryg Trygve Lie

tryp (TRYP) tryptophan

tryp(s) trypanosome(s)

ts tan-striped; taper shank; temperature salinity; temperature switch; tenor sax; tensile strength; terminal sensation; test solution; tilt and shift; time shack; time sharing; too short; tool steel; total solids; tough situation; traffic signal; trained soldier; training ship; training squadron; transit storage; transmitter station; triple strength; tubular sound; type specification(s); typescript

ts (TS) thesis

t's twins

t/s test stand; third stage; tranship(ed)(ment)

t/s (T/S) thyroid serum

t & s toilet and shower

TS Tasmania (airline code); Tasmanian Steamers; Telophase Society; Tentative Specification; Terminal Service; Test Summary; Theosophical Society; Thoreau Society; Tidewater Southern (railroad); top secret; Topical Search; Tourette Syndrome; Training Ship; Transmittal Sheet; Transvaal Scottish; Tyneside Scottish; Type Specification

T S Taming of the Shrew; tasto solo (Italian—play without accompaniment)

T & S Transport and Supplies

tsa tax-sheltered annuity; total survey area; two-step antenna

tsa (TSA) total survey area (radio and tv)

TSA Teacher on Special Assignment; Tourist Savings Association; Track Supply Association; Transportation Service, Army; Transportation Service Authority; Transportation Standardization Agency; Transuranic Storage Area

tsac title, subtitle, and caption

TSAC Target Signature Analysis Center

tsar time scanned array radar

TSB Trustee Savings Bank(s)

TSBA Trustee Savings Banks Association

TSBD Texas School Book Depository

TSBI Texas Social Behavior Inventory Form

TSBR Thomas Stamford Bingley Raffles

tsc (TSC) transmitter start code (data processing)

TSC Technical Service Corps (South Africa); Texas Southmost College; Transamerican Steamship Corporation; Transportation System Center

TSCA Tactical Satellite Communications System; Top Secret Control Agency; Toxic Substance Control Act; Tri-State College of Acupuncture

TSCC Telemetry Standards Coordination Committee

tscf top secret cover folder

TSCO Thomas Scherman's Concert Opera; Top Secret Control Officer

TSCS Tennessee Self-Concept Scale

t-s curve temperature-salinity curve

tsd tactical and staff duties; tactical simulator display; target skin distance; treatment, storage, and disposal

Tsd Tausend (German—thousand)

TSd Tay-Sachs disease (TSD)

TSD Technical Services Division (CIA); towed submersible drydock (naval symbol)

TSD-CIA Technical Services Division—Central Intelligence Agency

tsdd temperature-salinity-density-depth

tsds two-speed destroyer sweeper

tse transmissible spongiform encephalopathig

tse (TSE) test support equipment

TSE Texas South-Eastern (railroad); T(homas); S(tearns) Eliot; Tokyo Stock Exchange; Toronto Stock Exchange

TSE Tribunal Supremo de Elecciones (Spanish—Supreme Election Tribunal)

T-sect cross-section; transverse section

TSES Thumb-Signature Endorsement System

tsf technical supply flight; tower shield facility

tsf telegrafia sem fios (Portuguese), *telegrafo senza fili* (Italian), *télégraphie sans fil* (French)—radio or wireless telegraphy

TSF Tertiary of the Society of St Francis

tsfr transfer

TSG Television and Screen Writers' Guild; Tri-Service Group

TSGAOTU The Sovereign Grand Architect of the Universe

TSGEE Tri-Service Group on Communications and Electronic Equipment

TSgt Technical Sergeant

tsh thyroid-stimulating hormone; thyrotropin

tsh telegrafía sin hilos (Spanish—wireless telegraphy)—radio

T sh Tanzanian shilling(s)

TSH Their Serene Highnesses

TSHA Texas State Historical Association

T-shirt T-shaped shirt; T-shaped undershirt

t-shower thundershower

tsi test structure input; tons per square inch

TSI Test of Social Insight; Test of Social Intelligence; The Socialist International; Theological School Inventory; Transport(ation) Safety Institute

T & SI Technical and Scientific Information (UN)

tsi agar triple sugar (glucose, lactose, sucrose) iron agar

tsiaj this scherzo is a joke (abbreviation devised and used by composer Charles Ives)

TSID Technical Service Intelligence Detachments

Tsj Tsjeko-Slovakia (Norwegian—Czechoslovakia)

TSJC Trinidad State Junior College

TSKK Tsentralnya Kontrolnaya Komissiya (Russian—Central Control Commission)

TSL Terrestrial Sciences Laboratory, Texas Short Line (railroad)

TSLNP Tung Slang Luang National Park (Thailand)

TSM Tax Systems Modernization; Troop Sergeant-Major

TSM *Treasure of the Sierra Madre*

TSMG Thompson Sub-Machine Gun

tsms twin-screw motor ship

tsmt transmit

TSMTS Tri-State Motor Tariff Service

Tsn Tientsin

TSN Tape Serial Number

TSNHS Touro Synagogue National Historic Site

tso technical specification order; time-sharing option; time since overhaul

Tso Tsingtao

TSO Taiwan Symphony Orchestra; Tasmania Symphony Orchestra; Teheran Symphony Orchestra; Toronto Symphony Orchestra; Tucson Symphony Orchestra

TSOR Tentative Specific Operational Requirements

TSOS Time-Sharing Operating System

tsp teaspoon; total suspended particles; tracking station position

tsp (TSP) tropical spastic paresis

TSP thyroid-stimulating (hormone of) prepituitary; trisodium phosphate (Na_3PO_4)

tspa tally and special precinct analysis

tspn teaspoon

T square T-shaped ruler for making right angles

tsr tactical strike reconnaissance; temperature-sensitive resistor; terminate and stay resident (computer programs); torpedo spotter reconnaissance

TSR Sir Thomas Stamford Raffles (founder of Singapore as well as the London Zoo); Terminate and Stay Resident; Toronto Scottish Regiment; Trans-Siberian Railway

TSRB Top Salaries Review Body

T & SRC Tubular and Split Rivet Council

TSRL Turbine Systems Research Laboratory

tss tangential-signal sensitivity; target-selector switch(ing); time-sharing system(s)

tss (TSS) tactical surveillance system; toxic shock syndrome

t/ss turbine steamship

TSS Time-Sharing System(s); Traffic Safety Service; Trident Submarine System; turbine steamship; twin-screw ship

TSS-1 Tethered Satellite System 1

TSS-2 Tethered Satellite System 2

tssa (TSSA) tumor specific surface antigen

TSSC Technical Supply Sub-Committee

tssm total ship simulation model

tsspar time-sharing system-performance activities record(s)

TSSR Tadzhikistan Soviet Socialist Republic; Turkmenistan Soviet Socialist Republic

tst test (computer flow chart)

tsta tumor specific transplantable antigen (TSTA)

TSTA Texas State Teachers Association

t-storm thunderstorm

TSTP Test of Selected Topics in Physics

tstr tester

tstrms thunderstorms

t's t's & t's tortoises, terrapins, and turtles [tortoises are terrestrial chelonians with domed shells and elephantine feet; terrapins are semiaquatic chelonians with depressed shells, rudder-like tails, and webbed feet; turtles are marine chelonians with streamlined shells and paddle-like flippers; the term *turtle(s)* is often applied to all the chelonians]

Ts & Ts Trinidadians and Tobagonians

tsu tape search unit; this side up

tsu (TSU) triple sugar urea (agar)

TSU Texas Southern University; Tulsa-Sapulpa Union (railway)

tsu's thermosetting urethanes

TSUS *Tariff Schedule of the United States*

Tsushima Tsushima Current flowing northeasterly between Japan and Korea; Tsushima Strait where in 1905 Admiral Togo's Japanese fleet defeated Admiral Rozhdesvenski's Russian fleet

tsvp *tournez s'il vous plaît* (French—please turn over)

TSW Tactical Supply Wing; tropical summer winter (load line mark)

TSWE Test of Standard Written English

tsx time-sharing executive

TSX Telecommunications Satellite Experiment

tt tablet triturate; target tower; target tug; technical term(inology); technical test(ing); teetotaler; telegraphic transfer; teletype; teletypewriter; testamentary trust; tetanus toxoid; thrombin time; torpedo tube(s); traditional tennis; transit time; tree top(s); tuberculin tested; turret trainer

tt (TT) telephone transfer; train time

t-t tube-in-tube

t/t time to turn

t & t time and temperature

tt. *tantum* (Latin—fixed allowance, so much)

t.t. *totus tuus* (Latin—all yours)

TT tam-tam (Chinese gong); target-towing (naval aircraft); technical test(ing); Telecommunications Technician; Toledo Terminal (railroad); Trailer Train; Trans-Texas (Airways); Trinidad and Tobago (Internet code); Troop Test; Tyne and Tees (50th Division)

TT *Tidningarnas Telegrambyra* (Swedish News Agency)

T/T twin turbine (steamship)

T & T Trinidad and Tobago

tta test target array

TTA Taiwan Telecommunication Administration; Tanzania Tea Authority; TransTexas Airways; Travel Time Authorization; Travel Trade Association

TTA *Tragedy of Titus Andronicus*

ttab Trademark Trial and Appeal Board (US Patent Office)

ttac tracking, telemetry, and command; tracking, telemetry, and control

ttad transtracheal aspiration

TTAF Technical Training Air Force

ttb (TTB) tanker transport bomber (training aircraft)

TTBT Threshold Test Ban Treaty

ttc temperature test chamber; tetrazolium chloride; tight tape contact; tin telluride crystal; tow target cable; transient temperature control; tube temperature control

TTC Tank Training Center; Tariff Trade Code; Technical Training Center; Technical Training Command; Teletypewriter Center; Texas Technological College; Tobacco Tax Council; Tokyo Tanker Company; Toronto Transit Commission; Transportation Technology Center; Tuition Tax Credits

ttce tooth-to-tooth composite error

ttci transient temperature-control instrument

TTCS Truck Transportable Communications Station

ttd transponder transmitter detector

ttdr tracking telemetry data receiver

tte temporary test equipment; trailer test equipment

Tte Teniente (Spanish—Lieutenant)

TTE Tropical Testing Establishment

Tte Cnel teniente coronel (Spanish—Lieutenant Colonel)

TTEX Trailer Train Express

ttf target-towing flight; time to failure; tone telegraph filter; transistor text fixture

ttf (TTF) tetrathiafulvalene

TTF Taiwan Textile Federation; Timber Trade Federation; Townsend Thoresen Ferry

ttfn ta-ta for now

ttg time to go

TT-gauge Tiny Tim Gauge—$^1/_4$ inch track gauge (model railroads)

ttgd time-to-go engine dial

tth thyrotropic hormone

tthy transthyretin (prealbumin)

tti time-temperature indicator; trait treatment interaction

TTI The Technological Institute; Transition Technology, Inc.

T-time takeoff time

TTIO Turkish Tourism and Information Office

TTJC Tyne Trade Joint Committee

ttk two-tone keying

ttl to take leave; transistor-transistor logic

ttl (TTL) through-the-lens (camera-flash monitor)

tt & l treasury tax & loan

TTL Tokaido Trunk Line (Japanese railroad running trains at 125 miles per hour)

ttm two-tone modulation

TTMA Trinidad-and-Tobago Medical Association; Truck-Trailer Manufacturers Association

tto this transaction only

Tto Toronto

TTO Tanzania Tourist Office

ttp time-temperature parameter; total taxable pay

ttp (TTP) thymidine triphosphate

TTPI Trust Territory of the Pacific Islands

ttr type token ratio

ttr (TTR) target-tracking radar; thermal test reactor

TTRI Telecommunication Technical Training and Research Institute

T & T RR Tijuana and Tecate Railroad

tts teletypesetter (TTS); teletypesetting; temporary threshold shift

tts (TTS) teletypesetting

TTS Terminal Transparent System; Transdermal Therapeutic System

TTS The Truth Seeker

TTSU Taxi-Truck Surveillance Unit (NYPD)

ttt telemetry time transposition; time to target; time to think; time to turn

t t & t tortoise, terrapin, and turtle (*see* t's t's & t's)

TTT Transamerica Trailer Transport; Tyne Tees Television

TT & T Texas Transport and Termninal

t't'ta triple-note trumpet flourish

TTTB Trinidad and Tobago Tourist Board

TTTC Technical Teachers Training College

T & T TS Trinidad and Tobago Television Service

ttu tape transport unit; timing terminal unit

TTU Texas Technological University

TTUT Through-Transmission Ultrasonic Test(ing)

TTV Taiwan Television (offshore China)

ttvm thermal transfer voltmeter

ttw time of tension to war; total temperature and weight

TTW Tennesee-Tombigway Waterway

ttwl twin-tandem wheel loading

ttx tritated tetrodotoxin

tty teletypewriter

T-type Jungian thinking type

tu tape unit; thermal unit; toxic unit; trade union (TU); traffic unit; transfer unit; transmission unit; tuba; turbidity unit

Tu thulium; Tudor; Tuesday; Turkey; Turkish

TU Taylor University; Temple University; Tiffin University; Trade Union; transmission unit; Trinity University; Tufts University; Tulane University; Tunis Air; Typographical Union

T.U. tuberculin unit(s)

TU Technische Universität (German—technical university); *temps universel* (French—universal time)

Tu-4 Soviet Tupolev bomber inspired by the Boeing B-29 Superfortress aircraft

Tu-16 Soviet Tupolev bomber code-named Badger by NATO

Tu-20 Soviet Tupolev heavy bomber named Bear by NATO

Tu-22 Soviet Tupolev bomber named Blinder by NATO

Tu-28 Soviet Tupolev long-range interceptor aircraft named Fiddler by NATO

Tu-104 Soviet Tupolev medium-range transport aircraft called Camel by NATO

Tu-114 Soviet Tupolev long-range transport plane named Cleat by NATO

Tu-124 Soviet Tupolev jet-transport aircraft named Cookpot by NATO

Tu-144 Tupolev supersonic transport

Tu-154 Tupolev 154 supersonic aircraft

TUAC Trade Union Advisory Committee

Tuamotus Tuamotu Islands of Polynesia in the South Pacific, once called the Dangerous Islands

tu ar turning arbor

tuav tactical unmanned aerial vehicle

tub. tubing

TUB temporary unlighted buoy

TUBA Tubists Universal Brotherhood Association

Tu bandera *Tu bandera es un lampo del cielo* (Spanish— Your flag is a lamp of the sky)—Honduran anthem

tube subway; television; tunnel

Tube The Tube (London's Underground subway system)

TUBE Terminating Unfair Broadcasting Excesses

tuberc tuberculosis

tublr tubular

Tubuais Tubuai Islands of Polynesia in the South Pacific, also called the Australs

tuc time of useful consciousness (the brief moments amid oxygen deficiency when a pilot can save a troubled plane); transportation, utilities, communications

Tuc Tucana (constellation); Tucson

TUC Trades-Union Congress (British)

tu ca turning cam

TUCASI Triangle Universities Center for Advanced Studies Incorporated (Duke University, the University of North Carolina at Chapel Hill, North Carolina State University)

TUCC Temple University Community College; Triangle Universities Computation Center

TUCGC Trades Union Congress General Council

TUCSA Trade Union Council of South Africa

Tucsons Tucson Mountains of southeastern Arizona

tudor two-door

Tue Tuesday

TUEL Trade Union Educational League

Tues Tuesday

TUF Tamil United Front; Tokyo University of Fisheries; Trade Union Federation (British)

TUFEC Thailand-Unesco Fundamental Education Center

tuff tape update of formatted files

tu fx turning fixture

tug. tape update and generator

TUG Transac Users Group

tugrik monetary unit of Mongolia

tug(s) tugboat(s)

TUH Taiwan University Hospital

TUI Trade Union International

TUIAFW Trade Unions International of Agricultural and Forestry Workers

tuifu the ultimate in foulups

Tul Tulsa

TUL Tokyo University Library; Tulane University of Louisiana; Tulsa, Oklahoma (airport)

Tularosas Tularosa Mountains of western New Mexico

TULF Tamil United Liberation Front

Tul Phil Tulsa Philharmonic

TULRA Trade Union and Labor Relations Act

tum tummy (stomach); tumor

TUM Panama City, Panama (Tocumen Airport); Trades Union Movement

Tumacacori Tumacacori National Monument south of Tucson, Arizona

Tumuc-Humacs Tumuc-Humac Mountains between Brazil and the Guianas

tun tuning

Tun Tunis; Tunisia; Tunisian; Tunnel

Tun *Túnez* (Spanish—Tunisia)

tunasan tuna sandwich

tunawich tuna sandwich

tund tundra

tung tungsten

Tunic Tunicata

TUNICS Tunnel Integrated Control System

Tunisia Republic of Tunisia (North African Arab country), *Al-Djoumhouria Attunusia*— called Carthage in Roman times

Tunl Tunnel

tuos trained under other schemes

TUP Temple University Press; Trinity University Press; Tulane University Press

TUPE Tobacco Use Prevention Education

Tupper Tupper Creek in eastern British Columbia or Tupper Lake in northern New York

Tupun Tupungato

tur transurethral resection (TUR); turbine; turret

Tur Turin; Turkish

turb transurethral resection of the bladder (TURB); turbine

TURB Trainer Update Review Board

turbid. turbidity

turboalt turboalternator

turbo-elec steam turbine connected to electric motor

turbogen turbogenerator

turbojet turbine-driven jet (airplane engine)

turboprop turbine-driven jet engine (moving the) propeller

turbosuch trubosupercharger

turbotrain turbine-driven railroad train

turbpmp turbopump

turbu turbulence; turbulent

Turch *Turchia* (Italian—Turkey)

turk turkey

türk *türkisch* (German—Turkish)

Turk. Turkey; Turkish

Turk Turkish language

Turkana Lake Turkana (formerly East Rudolf)

Turk Cum *Turkiye Cumhuriyeti* (Turkish—Republic of Turkey)

Turkey Republic of Turkey (formerly the center of the Ottoman Empire extending from Morocco to Persia), *Türkiye Cumhuriyeti*

Türk-Is *Tükiye Isçi Sendikalari Konfederasyonu* (Turkish Confederation of Trade Unions)

Turkish instruments bass drum, cymbals, kettledrums, and triangle

Turkm Turkmenistan

Turkmen Turkmenian

turks turkeys

Turks Turkish people; Turks Islands east of the Bahamas and northeast of the Windward Passage

Turks and Caicos Turks and Caicos Islands northeast of the Windward Passage between Cuba and Haiti

Turk-Sib Turkestan-Siberian (railroad)

Turk-Tat Turko-Tataric

Turku formerly Abo

Turk Yid Turkish Yiddish (Ladino)

turn. turning

Turn Turnpike

TURN Toward Utility Rate Normalization

Turner Turn Turner Turnpike

turp transurethral resection of the prostate (TURP); turpentine

turps elixir of terpin hydrate; turpentine

TURPS Terrestrial Unattended Reactor Power System

turq turquoise

Turq Turquía (Spanish—Turkey)

Turtles Turtle Islands in the Sulu Sea south of the Philippines; Turtle Islands off Africa's Sierra Leone; Turtle Mountains between northern North Dakota and southern Manitoba

TUs Tenant's Unions

TUS Tuscon, Arizona (airport)

TUSAFG The United States Air Force Group (American Mission for Aid to Turkey)

TUSC Technology Use Studies Center

Tusca Tuscaloosa

Tuscans Tuscan people; Tuscan Islands

Tushars Tushar Mountains of central Utah

Tuskegee Tuskegee University

TUSLOG The United States Logistic Group

TUSM Tufts University School of Medicine

tuss. tussis (Latin—cough)

tut tutor; tutorial

Tut Tutankhamen

TUT The University of Tokyo

Tut Books Charles E Tuttle Co. books

TUTF Technology Use Task Force

TUTI Temple University Technical Institute

Tutu Tutuila, American Samoa

TUUL Trade Union Unity League

Tuv Tuvalo (Ellice Islands); Tuvalu Islands

TUV Technischer Uberwachungs Verein (German—Testing Organization)—Berlin

tuwr turning wrench

tux tuxedo (dinner jacket)

Tuzigoot Tuzigoot National Monument in central Arizona

tv transvestite

tv (TV) television; terminal velocity; test vehicle; tetrazolium violet; total volume; transverse; trichomonas vaginalis; true view; tuberculin volution

t/v temperature to voltage ratio; thrust-to-weight

t & v terrorism and vandalism

TV television; test vehicle; Tidal Volume; Tidewater Oil (stock exchange symbol); transport vehicle; Trinidad Volunteers; Tuvalu (Internet code)

TV Totenkopfverbände (German—Death's Head formations)—concentration-camp guard units

tva thrust vector alignment

tva taxe sur la valeur ajoutée (French—tax value added)

TVA Temporary Variation Authorization; Tennessee Valley Authority

tvac time-varying adaptive correlation

tv a.m. morning television

TVAs Temporary Variation Authorizations

TVB Television (Advertising) Bureau

TVBS Television Broadcast Satellite

tvc temperature valve control; thermal voltage converter, throttle valve control; thrust vector control; time-varying coefficient; timed vital capacity; torsional vibration characteristics

tvc (TVC) total variable cost

TVC Technical Valve Committee

TVCC Treasure Valley Community College

tvcrit television critic(ism)

tvd toxic vapor damper; toxic vapor detector; tuned viscoelastic damper

tvdc test volts—direct current

TVDC Tidewater Virginia Development Council

tv'dict(s) television addict(s)

tvdp thrust-vector display (unit)

tvdy television deflection yoke

tve test vehicle engine; thermal vacuum environment; township and village enterprise

TVE Televisión Española (Spanish TV network)

tvel track velocity

TVERS Television Evaluation and Renewal Standards

tvft television flyback transformer

tvg television video generator; threshold voltage generator; triggered vacuum gap

TVG T V Guide

tvhh (TVHH) television households

TV household television-equipped home

tvi television interference

TVIC Television Interference Committee

tvid televised identification; television identification; television identity

tvig television and inertial guidance

tvist television information-storage tube

tvk terminal volume kill

T v K Theodore von Karman

tvl television listener; tenth value layer; travel

Tvl Transvall

tvm tachometer voltmeter; track via missile; trailer van mount; transistorized voltmeter

TVM Television Martí

TVN Television News

TVNZ Television New Zealand

tvop television observation post

tvor terminal visual omnirange; very high frequency terminal omnirange station

tvp television poor; textured vegetable protein; time-varying parameter

tvp (TVP) textured vegetable protein

TVPA Thames Valley Police Authority

tv p.m. evening television

tvq top visual quality

tvr temperature variation of resistance; textured vegetable protein

TVRB Tactical Vehicle Review Board (USA)

TVRI Television Rating Inventory

TV-RI TV-Republik Indonesia (Bahasa Indonesia-Republic Indonesia Television)

tv rm television room

Tvrn Tavern

tvr's television recordings

tvs tactical vocoder system; telemetry video spectrum; television viewing system; transvaginal sonography

tv's television dinners; transvestites

TVSAT Television Satellite

tvsd time-varying spectral display

tvsg television signal generator

tvsm time-varying sequential measuring (apparatus)

tvso television space observatory

TVSTI Thames Valley State Technical Institute

tvsu television sight unit

tvt television typewriter

TVTV Top Value Television

tvu total volume urine

tw tail warning; tail water; tail wheel; tail wind; tankwagon; taxiway; tempered water; terrawatt; tile wainscot; torpedo water; total weight; traveling wave; twice a week; twin(s)

tw (TW) typewriter (computer flow chart)

tw tussenwerpsel (Dutch—intedection)

Tw Twaddell

TW Taiwan (Internet code); Trans World Airlines (2-letter coding)

T & W Tyne and Wear

TW3 *That Was the Week That Was* (television program)

twa time-weighted average; trailing-wire antenna

TWA Textile Waste Association; Thames Water Authority; Tooling Work Authorization; Toy Wholesalers Association; Trans World Airlines

TWAD Twadell

'twas it was

twb twin with bath

twbp transcribed weather broadcast program

TWC Tail Waggers' Club; Texas Wesleyan College

TWC Trials of War Criminals

TWCIS Transuranic-contaminated Waste Container Information System

twcrt travelling-wave cathode ray tube

TWCS Test of Work Competency and Stability

twd tail wags dog

twds tradewinds

twe tap-water enema; technology, work and environment; trading with the enemy

TWE Textile Waste Exchange

TWEA Trading With the Enemy Act

'tween between

Twel N Twelfth Night

Twelve Tribes Twelve Tribes of Israel—Reuben, Simeon, Judah, Zebulun, Issachar, Dan, Gad, Asher, Nephtali, Benjamin, Ephraim, and Manasseh

twens twenties (store catering to people in their twenties)

'twere it were

twerl tropical wind, energy conversion, and reference level

TWF Twin Falls airport

TW & FS The Wine and Food Society

twh typically wavy hair

twhl tailwheel

twi training within industry

TWI The West Indies

'twill it will

twimc (TWIMC) to whom it may concern

'twixt betwixt

twi zn twilight zone

twk typewriter keyboard

twl top water level

twm traveling-wave maser

Twn Taiwan—Republic of China consisting of offshore islands; Town

twn hse town house

two. this week only

two-0 $20 bill

two-fer two for the price of one

Two Gent Two Gentlemen of Verona

T-word tax

two-spot $2 bill

twot travel without troops

'twould it would

Twp Township

TWP True Whig Party (Liberia)

TWPD Tactical and Weapons Policy Division

twr tower

Twr Tower

TWR Trans-World Radio

tws timed wire service; track while scan

tw/s twin-screw (ship)

TWSO Transuranic Waste Systems Office(r)

twsr track-while-scan radar

twsrs track-while-scan radar simulator

twt torpedo water tube; traveling-wave tube; travel with troops

t/wt tare weight

TWT Toy World Test(ing); Transonic Wind Tunnel

twta travelling-wave-tube amplifier

TWU Tactical Weapons Unit; Tata Workers Union; Transport Workers Union

TWUA Textile Workers Union of America; Transport Workers Union of America

TWW Theater Without Walls

T WW Thick Weather Watch (Coast Guard)

twx time-wire transmission

twx (TWX) teletypewriter exchange (message)

TWX teletypewriter exchange (message)

TWX (TWXS) Teletypewriter Exchange Service

twy taxiway; twenty

twych travel with your children

twyl taxiway link(age)

twzo trade-wind-zone oceanography (term of derision by experts or about armchair oceanographers)

tx tax(ation); telex; time; torque transmitter; traction; tumor can't be adequately assessed (symbol)

tx (TX) transmitter

Tx Texas; treatment

TX Texas

txclk (TxCLK) transmit data clock

txe telephone exchange electronic

txh transfer on index high

txi transfer on index incremented

txl transfer on index low

Txl Texel

TXM Tax Systems Modernization

TXMODA Tax Module

txn taxation

TxSEA Texas Society of Enrolled Agents

txt text; textbook; textile; textual(ly); textualism; textualist; textuary; texture(d); texturize; texturizing

ty territory; thank you; truly; type

ty (TY) tax year

ty tysk (Dano-Norwegian—German)

Ty Territory; Tybalt; Tyler, Tyndall; Tyonek; Tyrone; Tyrus Raymond Cobb

TY Territorial Yeomanry

tyc tycoon

TYC Thames Yacht Club; Thomas Y Crowell; Toledo Yacht Club

TYCOM Type Commander (USN)

tydac typical digital automatic computer

tyg (TYG) trypticase yeast glucose

tylenol acetaminophen (trade name for an analgesic)

tymp tympanic(ity); tympany

tymp memb tympanic membrane

tyng topping

tyo two-year-old (horse)

TYO Tokyo, Japan (airport)

typ typical; typing; typist; typographer; typography; typewriter

TYP Ten-Year Plan; Twenty-Year Plan; etc.

type. typewriter; typewriting

typer typewriter

typewriters Chicago-gangster (Scarface) Al Capone's nickname for submachine guns

typh typhoon

typo typographical (error)

TYPOE Ten-Year Plan for Ocean Exploration

typog typographer; typographical; typography

typol typological(ly); typologist; typology

typout typewriter output

typr typewritten

typw typewriter

tyr (TYR) tyrosine (amino acid)

Tyr Tyrol; Tyrolean; Tyrolese; Tyrone

Tyr Tyrkia (Norwegian—Turkey)

Tyrol Tyrol(ean); Tyrolese

tys tensile yield strength

TYS Knoxville, Tennessee (airport)

tysd total years service date

Tysk Tyskland (Norwegian—Germany)

Tyskl Tyskland (Danish—Germany)

tytipt tape training in port (USN)

tyurzak tyuremnoye zakyucheniye (Russian—prison confinement)

tyvm thank you very much

TYZ Toronto, Ontario airport

tz terrazzo; tidal zone; time zero

Tz tuberculin zymoplastiche (symbol)

TZ Tactical Zone; Tanzania (Internet code); Transair Limited, Canada (2-letter code)

tzd true zenith distance

tze transfer on zero

tzg thermofit zap gun

TZIK Tzentrainy Ispolnitelny Kommitet (Russian—Central Executive Committee)

tzj tubular zippered jacket

TZm true azimuth

TZM titanium-zirconium-molybdenum (alloy)

tzp time zero pulse

tzt te zijner tijd (Dutch—in due time)

tzv tetrazolium violet

U

u density of radiant energy (symbol); ugly threatening weather (symbol); umpire; uncle; uncommon; unified atomic mass (symbol); unit(s); universal set (symbol); unknown; unoccupied; unread; unsymmetrical; unwatched; upper; velocity (symbol); you

u & lc upper and lowercase

u und (German—and); viscosity (symbol)

U Chance Vought Aircraft (symbol); kilourane (1000 uranium units—symbol); overall co-efficient of heat transfer (symbol); potential energy (symbol); total internal energy (symbol); U Thant; U-boat; unclassified; Underground (London's subway system); Uniform—code for letter U; Union Association; University; up; uracil; uranium; Utah; Utahans; utility; you

U Uad (Arabic—wadi)—gulley, ravine, riverbed; *ud* (Danish—out); uit (Dutch—out); ulos (Finnish—out); *Université* (French—University); *unter* (German—down); up; *upp* (Swedish—up); *ute* (Swedish—arrival); *violaceus* (Latin—violet-color)

U-1A American version of De Havilland Otter utility aircraft

u 1 b unit 1 bedroom

u 1 r unit 1 rental

u1s uroporphyrinogen-1-synthase

u-2 you too

u 2 b unit 2 bedrooms

u 2 r unit 2 rentals

U-2 high-altitude high-performance photo-reconnaissance airplane

u/3 upper third

U-3 Cessna 6-passenger aircraft

U₃O₈ uranium oxide

U-4 Aero Commander transport aircraft

U 4 T union (coupling) 4 tons

U-6 De Havilland Beaver transport aircraft

U-8 Beech Seminole transport aircraft

U-17 Cessna Skywagon aircraft

U-17A Cessna 6-passenger Skywagon

U-22 Beech Bonanza trainer aircraft

U234 trace component of natural uranium

U235 0.7 percent of natural uranium (atomic energy source)

U238 99.3 percent of natural uranium (atomic energy source)

ua unauthorized absence; unauthorized absentee; underage; unidentified aircraft; uniform allowance; upper arm; urine aliquot; user area

ua (UA) urinalysis

u/a unit of account

ua unidad(es) astronomica(s) (Spanish—astronomical unit(s))

u a uden ar (Dano-Norwegian—without date); *und andere(s)* (German—among other things, and others, inter alia); *und ähnliche(s)* (German—and the like)

u.a. usque ad (Latin—as far as; up to)

uA und andere (German—and others)

U/a underwriting account

UA Ukraine (Internet code); Ulster Association; Underwater Association; United Aircraft; Union Association; United Air Lines (2-letter coding); United Artists; University of the Americas, University of Auckland; User Agent

U-A Universal-American

U of A University of Aberdeen; University of Adelaide; University of Akron; University of Alabama; University of Alaska; University of Alberta; University of the Americas; University of Arizona; University of Arkansas

UA Universidad de las Americas (Spanish—University of the Americas)

UAA United Arab Airlines; University of Alaska, Anchorage; University Aviation Association

UAAGM University of Alberta Art Gallery and Museum

UAASUS Ukrainian Academy of Arts and Sciences in the United States

UAB Underwriters Adjustment Board; Unemployment Assistance Board; United Asian Bank; Universities Appointments Board; University of Aston in Birmingham

UABS Union of American Biological Societies

uac underwriters adjusting company

UAC Ulster Automobile Club; United Africa Company; United Aircraft Corporation; Urban Affairs Council; Utility Aircraft Council

UACA United American Contractors Association

UACC Upper Area Control Center

UACL United Aircraft of Canada, Limited

uacte universal automatic control and test equipment

UADPS Uniform Automatic Data Processing System

UADW Universal Alliance of Diamond Workers

UAE United Arab Emirates (Trucial Sheikdoms of Trucial States)

UAEMS University Association for Emergency Medical Services

UAESP Utah Association of Elementary School Principals

uaf unit authorization file

UAF University of Alaska, Fairbanks

uafs/t universal aircraft flight simulator/trainer

UAFM University of Alaska Museum (Fairbanks)

UAFT United Agency for Fair Treatment

UAG Universidad Autónoma de Guadalajara (Spanish University of Guadalajara)

UAH University of Alabama at Huntsville

UAHC Union of American Hebrew Congregations

uai universal azimuth indicator

UAI Urban America Incorporated (Action Council for Better Cities); Urban Art International

UAI União Astronomica Internacional (Portuguese—International Astronomical Union); *Union Académique Internationale* (French International Academic Union);

Union des Associations Internationales (French—Union of International Associations); *Union Astrónomica Internacional* (Spanish—International Astronomical Union); *Union Astronomica Internazionale* (Italian—International Astronomical Union)

uaide uses of automatic information display equipment

UAISEGR University of Alaska Institute of Social, Economic, and Government Research

UAJAPPFI United Association of Journeymen and Apprentices of the Plumbing and Pipe Fitting Industry (U.S. and Canada)

UAK University of Alaska

ual upper acceptance limit

UAL United Air Lines; University of Aberdeen Library; University of Akron Library; University of Alabama; University of Alabama Library; University of Alaska Library; University of Alberta Library; University of the Americas Library; University of Arizona Library; University of Arkansas Library; University of Auckland Library

UALL University of Arizona Lunar Laboratory

U of Alla University of Allahabad

uam (UAM) underwater-to-air missile

uam und andres: mehr (German—and so forth)

UAM *Union Africaine et Malgache* (French—African and Malagasy Union); United American Mechanics

UAMC United Arab Maritime Company

UAMPT Union Africaine et Malagactie des Postes et Telecommunications (French—Union of African and Malagasy Postal Service and Telecommunication)

uan uric-acid nitrogen

UANA Unión Amateur de Natación de las Americas (Spanish—Amateur Swimming Alliance of the Americas)

UANC United African National Council

uao unexplained aerial object

UAOD United Ancient Order of Druids

UAOS Ulster Agricultural Organisation Society

uap unexplained atmospheric phenomenon

Uap Micronesian name for Yap

UAP Union of American Physicians; Union of Associated Professors; United Australia Party

UAPD Union of American Physicians and Dentists

U of A Pr University of Alabama Press; University of Alaska Press; University of Arizona Press

uar underwater acoustic resistance; underwater angle receptacle; upper air route; upper atmosphere research

UAR Uniform Airman Record; United Arab Republic; University of Arkansas

UARAEE United Arab Republic; Atomic Energy Establishment

UAREP Universities Associated for Research and Education in Pathology

UARL United Aircraft Research Laboratories

UARRSI Universal Aerial Refuelling Receptacle Slipway Installation

UARS Upper Atmosphere Research Satellite

uart universal asynchronous receiver–transmitter

UARTO United Arab Republic Tourist Office

uas unmanned aerial surveillance; upper air space; upstream activation sites; upstream activator sequence

UAS Unit Approval System; University Air Squadron; University of Alaska, Southeast

UASC United Arab Shipping Company

UASCS United States Army Signal Center and School

UASIF Union des Associations Scientifiques et Industrielles Françaises (French—Union of French Scientific and Industrial Associations)

UASM University of Arkansas School of Medicine

UASS Unmanned Aerial Surveillance System

UASSP Utah Association of Secondary School Principals

UASSR Udmurt Autonomous Soviet Socialist Republic, new part of central Russia

uat ultraviolet acquisition technique

UAT *Union Aéromaritime de Transport*

UATI *Union des Associations Techniques Internationales* (French—Union of International Technical Organizations)

UATO United Airlines Tour Order

UATP Universal Air Travel Plan

UAU Underwater Archaeology Unit; Universities Athletic Union

uav unmanned aerial vehicle; urban assault vehicle

UAW United Automobile Workers

uAwg *um Antwort wird gebieten* (German—reply requested)

uax (UAX) unit automatic exchange

UAZ University of Arizona

UAZEES University of Arizona Engineering Experiment Station

ub up(ward) bound, urine bilirubin

Ub *Universiteitsbibliotheek* (Dutch—University of Library)—Amsterdam

UB Ulan Bator (Mongolia); Union Bank; Union of Burma; United Bank (of Arizona); United Biscuit; Upper Bench; Upward Bound

U of B University of Baltimore; University of Bath; University of Birmingham; University of Bombay; University of Bradford; University of Bridgeport; University of Bristol; University of Buffalo

UB *The University Bookman; Universiffit Basel* (Basel University); *Universitat Berne* (Berne University)

uba undenatured bacterial antigen

UBA Underwater Breathing Apparatus; Union of Burmah Airways; United Business Associates

UBA *Universidad de Buenos Aires* (Spanish—University of Buenos Aires)

UBAF *Union de Banques Arabes et Francçaises* (French—Union of Arab and French Banks)

U-bahn *Untergrundbahn* (German—underground road) subway system

UBAV United Buddhist Association of Vietnam

UBB Union Bank of Bavaria

UBBA United Boys' Brigades of America

UBBC Unsaturated vitamin B12 Binding Capacity

ubc universal buffer controller

UBC Uniform Building Code; United Baltic Corporation; Universal Bibliographic Control; University of British Columbia

U of BC University of British Columbia

UBC *Uniform Building Code* (legal); *Universidad de Baja California* (Spanish University of Baja California)

UBC & J United Brotherhood of Carpenters and Joiners

UBCL University of British Columbia Library

UBCMA University of British Columbia Museum of Anthropology

UBCP Union Bag-Camp Paper; University of British Columbia Press

ubd utility binary dump

UBD *Universal Business Directories*

ubdi underwater battery director indicator

UBDMA United Better Dress Manufacturers Association

UBEA United Business Education Association

U-beam U-shaped beam

UBEM *Union Belge d'Enterprises Maritimes* (French—Belgium Union of Maritime Enterprises)

ubers *übersetzt* (German—translated)

ubf universal boss fitting

UBF Union of British Fascists

ubfc underwater battery fire control

ubi ultraviolet blood irradiation; universal battlefield identification

UBI United Business Investments

UBI *Unione Bocciofila Italiana* (Italian—Italian Bocce-Ball (Bowling) Association); *Unione Bibliografica Italiana* (Italian—Italian Bibliographical Society)

Ubib Wien *Universitätsbibliothek Wien* (German—Vienna University Library)

ubip ubiquitous immunopoietic polypeptide

ubitron undulating beam interaction electron tube

UBL Union Barge Line; United Benefit Life

UBLS University of Botswana, Lesotho, and Swaziland

ubm ultrasonic bonding machine; unit bill of material

UBM United Biscuit Manufacturing (company)

U-boat *Unterseeboot* (German—submarine)

U-bolt capital-U-shaped bolt

U-bomb uranium-cased atomic or hydrogen bomb

U Books University Books

UBP United Bermuda Party; United Business Publications

UBR University Boat Race

UBS Unclassified But Sensitive; United Bank of Switzerland; United Bible Societies; United Business Service

UBSA United Business Schools Association (formerly American Association of Commercial Colleges)

UBSO Uinta Basin Seismological Observatory

ubt universal book tester

ubu you be you

ubv ultraviolet

UBVS Ultraviolet-Blue Visual System

Ubx ultrabithorax

uc ulcerative colitis; underclass; undercover (agent); unemployment compensation; universal coarse (screw thread); upper case (capital letters); upperclass

u/c upper center

UC Ulster College (Northern Ireland); Umpqua College; Underfashion Club; Union Carbide; Union College; University of California; University of Canterbury; University of Ceylon; University of Cincinnati; University College;

University of Colorado; University of Connecticut; Upland College; Upper Canada; Upsala College; Urban Council; Urgent Care; Ursinus College; Ursuline College; Utica College

U of C University of Calcutta; University of Calgary; University of California; University of Cambridge; University of Chattanooga; University of Chicago; University of Cincinnati; University of Colorado; University of Connecticut; University of Corpus Christi

UC una corda (Italian—one string)—soft pedal

uca upper control area

UCA United Chemists' Association; United Conservatives of America; United Consumers of America; University of California; Utah Correctional Association

UCAB Universidad Católica Andrés Bello (Spanish—Andrés Bello Catholic University)

UCAE Universities Council for Adult Education

UCAF You See America First

UCAN Utility Consumers Action Network

UCAR Union of Central African Republics; University Corporation for Atmospheric Research

UCAS Uniform Cost Accounting Standards; Union of Central African States

UCATT Union of Construction, Allied Trades, and Technicians

UCAV Unmanned Combat Aerial Vehicle

ucb unless caused by

UCB Unemployment Compensation Board; United California Bank; University of California at Berkeley; University College at Buckingham

UCBHM United Church Board for Homeland Ministries

UCBILR University of California at Berkeley—Institute of Library Research

UCBR University of California Board of Regents

ucc unadjusted contractual changes; universal copyright convention

UCC Uniform Code Council; Uniform Commercial Code; Union Carbide and Carbon; Union Carbide Corporation; United Cancer Council; United Church of Christ; United Community Campaign; United Electric Coal Companies (stock exchange symbol); Universal Copyright Convention; University College (Cork)

U-CC Upper Canada College

UCC Union de la Critique Cinématographique (French—Society of Cinema Criticism)

UCCA United Citizens Concerned with America; Universities Central Committee for Admissions; Universities Central Council on Admissions

UCCC Ulster County Community College; Uniform Consumer Credit Code

UCCD United Christian Council for Democracy

UCCELLO Paolo di Dono

UCCCM University of Cincinnati College Conservatory of Music

UCCJA Uniform Child Custody Jurisdiction Act

UCC-ND Union Carbide Corporation—Nuclear Division

UCCS Universal Camera Control System

ucd usual childhood diseases

UCD University of California at Davis; University College, Dublin

UCDA University and College Designers Association

U c de L Université Catholique de Louvain

UC de L Université Catholique de Louvain (French—Catholic University of Louvain)

ucdp uncorrect data processor

uce unforeseen circumstances excepted

UCEA University College of East Africa (Makerere College); University Council for Educational Administration

UCEMT University Consortium in Education Media and Technology

U of Cey University of Ceylon

UCF United Cat Federation; United Community Funds; University of Central Florida

UCFE Unemployment Compensation for Federal Employees

UCFGB University Catholic Federation of Great Britain

UCFH University College of Fort Hare

UCG University College, Galway; University College of Ghana

UCGSM University of California Graduate School of Management

UCH University College Hospital

U-channel U-shaped channel

UCHCIS Urban Comprehensive Health Care Information System

uchd usual childhood diseases

U Chi University of Chicago

U Chi Lib University of Chicago Library

UCHS University City High School

uci unit construction index

UCI Utility Communicators International

UCI Union Cycliste Internationale (French—Cyclists International Union)

UCIDT University Consortium for Instructional Development and Technology

UCIIR University of California Institute of Industrial Relations

UCIIS University of California Institute of International Studies

UCIrv University of California at Irvine

UCIW Union of Commercial and Industrial Workers

UCIWP United Cannery and Industrial Workers of the Pacific

ucj unsatisfied claim and judgment

ucl upper control limit; upper cylinder lubricant; urea clearance test

UCL Union Castle Line; Union Central Life; Union Oil Company of California (symbol); Universal Color Language; University of California Library; University College, London

UCLA University of California at Los Angeles; University of Caucasians Lost Among Asians

U-class upperclass
uc & lc upperclass and lower-class
UCM University Christian Movement
UCM Universidad Complutense de Madrid (Spanish—Alcalá de Henares University of Madrid)
u-c man undercover narcotics agent
UCMC University of Colorado Medical Center
UCMEA Ufficio Centrale di Meteorologia e di Ecologia Agraria (Italian—Central Office of Meteorology and Agrarian Ecology)
UCMJ Uniform Code of Military Justice
UCMP University of California at Berkeley Museum of Paleontology
UCMS Unit Capability Measurement System
U-C M S Union-Castle Mail Steamship
UCN University College of Nigeria
UCNL University of California Nuclear Laboratory
UCNW University College of North Wales
uco universal code; universal coding
UCO University of Colorado
U Conn University of Connecticut
UCOR Uranium Enrichment Corporation
UCP Unified Command Plan; Uniform Customs and Practice; United Cerebral Palsy; United Country Party; Universal Citizen Plan; University of California Press
UCPA United Cerebral Palsy Associations
UCPP Urban Crime Prevention Program
U of C Pr University of California Press; University of Chicago Press
ucr unconditioned response; utilization care review
UCR Uniform Crime Reports; University of California at Riverside; Utah Coal Route (railroad)
UCRA University Centers for Rational Alternatives

UCRC Underground Construction Research Council
UCRG Uniform Contractor Reporting Guidelines
UCRI Union Carbide Research Institute
UCRL University of California Radiation Laboratory
UCRN Unique Consignment Reference Number
UCR & N University College of Rhodesia and Nyasaland
UCRS Uniform Contractor Reporting System; Uniform Crime Reporting Section (FBI); University, College, and Research Section (Library Association)—also appears as UCR
ucs unconditioned stimulus; unconscious; unit-count system; universal card scanner; universal character set
uc's uterine contractions
UCs Urban Coalitionists
UCS Union of Concerned Scientists; United Community Service(s); Universal Child Survival; Universal Classification System; Universal-Cyclops Steel; University Computer Systems (computerized real estate listings); Upper Clyde Shipbuilders
UCSB University of California at Santa Barbara
UCSC University of California at Santa Cruz; University City Science Center
UCSD University of California at San Diego
UCSF University of California at San Francisco
UCSL University College of Sierra Leone
U of C SL University of California School of Law
UCSW University College of South Wales
uct unit compatability test(ing)
UCT United Commercial Travelers; Universal Coordinated Time; University of Cape Town; University of Connecticut
UCTA United Commercial Travellers' Association
UC & U Union College and University
uc & uc underclass and upperclass

UCUC University College of the University of Cincinnati
ucv uncontrolled variable
UCV Universidad Central de Venezuela (Spanish—Central University of Venezuela)
UCVs United Confederate Veterans
UCW Union of Communication Workers; University College of Wales
UCWC University College of the Western Cape
UCWI University College of the West Indies
UCWP University College of the Western Province
ucwr upon completion will return
UCWRE Underwater Countermeasures and Weapons Research Establishment
UCX Unemployment Compensation for Ex-Servicemen
UCY United Caribbean Youth
UCZ University College of Zululand
ud unfair dismissal; update, upper berth (double occupancy); upper deck; urethral discharge; uroporphyminogen decarboxylase (UD); utility dog (UD)
u/d under deck
u.d. ut dictum (Latin—as directed)
Ud Udjung (Malay—point); *usted* (Spanish—you)
UD Underground (London's subway), Undesirable Discharge; United Dairies; Universal Declaration (of human rights); University of Denver; University of Detroit; Urban District
U of D University of Dallas; University of Dayton; University of Delaware; University of Delhi; University of Denver; University of Detroit; University of Dublin; University of Dubuque; University of Dundee; University of Durham
UD Unlisted Drugs
UDA Ulster Defence Association (Protestant counterpart of the IRA); Urban Development Authority
udaa unlawfully driving away auto

U da C Uriel da Costa (Uriel Acosta)

UDAC User-Directed Access Control

UDAG Urban Development Action Grants

UDAL *Union de Universidades de América Latina* (Spanish—Union of Latin American Universities)

udam universal digital of avionics module

udarg *udarbeidet* (Danish—prepared)

UDB *Uprava Drzavne Bezbednosti* (Serbo-Croat—Administration for State Security)—Yugoslavian Secret Service

udc universal decimal classification (UDC); upper dead center; usual diseases of childhood

U d C *Universidad de Carababo* (Spanish—Carabobo University)—Venezuela

UDC United Daughters of the Confederacy; United Dye & Chemical; University of the District of Columbia; universal decimal classification; Urban District Council

UDCA Urban District Councils' Association

UDD Ulster Diploma in Dairying

'Uddersfield (Cockney contraction—Huddersfield)

udd's undisposed diapers; undumped diapers

UDE Underwater Development Establishment; *Union Douaniére Equatoriale* (French—Equatorial Customs Union); University of Delaware

U de A *Universidad de Alcala* (Spanish—Alcala University); *Universidad de Antioquia* (Spanish—Antioch University)

UDEAO *Union Douaniére des Etats de l'Afrique de l'Ouest* (French—Customs Union of West African States) former French colonies

U de B *Universidad de Barcelona* (Spanish—University of Barcelona); *Université de Bâle* (French—University of Basel)

U de BA *Universidad de Buenos Aires* (Spanish—University of Buenos Aires

udec unitized digital electronic calculation

U de C *Universidad de Cartagena; Universidad de Cauca, Universidad de Chile; Universidad de Córdoba, Universidad de Cuzco; Universidade de Coímbra*

U de CR *Universidad de Costa Rica*

U de F *Université de Fribourg*

U de G *Universidad de Granada; Universidad de Guadalajara; Universidad de Guanajuato; Université de Genéve; Université de Grenoble*

U de H *Universidad de la Habana* (Havana)

U de L *Universidad de Lérida; Universidad de Lima; Universidade de Lisboa* (Lisbon); *Université de Lausanne*

UDEL *Union des Editeurs de Littérature* (French—Literature Editors Union)

U de LA *Universidad de Los Andes*

U de M *Université de Montreal*

U de Monc *Université de Moncton*

U de O *Universidad de Oviedo*

U de Pan *Universidad de Panamá*

U de Q *Universidad de Quito (Universidad Central)*

U de S *Universidad de Salamanca; Universidad de San Andrés* (La Paz); *Universidad de San Agustín* (Arequipa); *Universidad de San Javier* (Panama); *Universidad de San Marcos* (Lima); *Universidad de Santiago; Universidad de Santo Tomaás* (Bogotá or Santo Domingo)

U de SC de G *Universidad de San Carlos de Guatemala*

U de SD *Universidad de Santo Domingo*

U de SM *Universidad de San Marcos* (Lima, Peru)

U de SP *Universidade de São Paulo*

U de ST *Universidad de Santo Tomás* (Manila)

U de T *Universidad de Toledó; Universidad de Trujillo* (Peru)

U de V *Universidad de Valencia; Universidad de Valladolid*

U de Z *Universidad Zaragoza*

udf *und die folgende* (German—and the following)

UDF Ulster Defence Force; Ulster Defence Regiment; Union Defence Force; United Democratic Front (South Africa)

UDF *Union pour la Démocratic Françoise* (French—Union for French Democracy)

udg *udgave* (Danish—edition)

u dg1 (m) *und dergleichen (mehr)* (German—and the like)

U of D GSIS University of Denver Graduate School of International Studies

Ud'H *Université d'Haiti* (French—University of Haiti)

UDHR Universal Declaration of Human Rights

udi undivided interest

UDI Unilateral Declaration of Independence

UDI *Unione Donne Italiane* (Italian—Italian Women's Alliance)

U di A *Universitá di Arezzo*

UDIA United Dairy Industry Association

U di B *Universitá di Bologna*

U d F *Universitá di Firenze* (University of Florence)

U di G *Universitá di Genova* (Genoa)

U di N *Universita di Napoli* (Naples)

U di P *Universitá di Padova* (Padua); *Universitá di Perugia; Universitá di Piacenza; Universitá di Pisa*

U di R *Universitá di Roma*

U di S *Universitá de Siena*

U di T *Universitá di Torino*

UDITPA Uniform Division of Income for Tax Purposes Act

U di V *Universitá di Venezia* (Venice); *Universitá de Vicenza*

u dk upper deck

udk *udkom* (Dano-Norwegian—Published)

udl up-data link

udm upright drilling machine

udM *unter dem Meeresspiegel* (German—below sea level)

UDM United Merchants and Manufacturers (stock exchange symbol); Universal Drafting Machine (corporation)

Udm Aut Sov Soc Rep Udmurt Autonomous Soviet Socialist Republic, now part of central Russia

udmh (UDMH) unsymmetrical dimethyl hydrazine

Udmurt member of central Russian people who live mainly in Udmurtia

udn ulcerated dermal necrosis

UDN Underwater Doppler Navigation

UDN *União Democrática Brasileira* (Portuguese—Brazilian Democratic Union)

udo unwilling drop-out

U d O *Universidad de Oriente* (Spanish—Oriente University)—Venezuela

U do B *Universidade do Brasil* (Portuguese—University of Brazil)—Brasilia

udom udometer; udometric; udometrical

udop ultrahigh-frequency doppler (system)

U do P *Universidade do Pôrto* (University of Oporto)

UDP United Democratic Party; User Datagram Protocol

udpg uridine diphosphogalactose

udpg (UPDG) uridine diphosphoglucose

UDP-gal uridine diphosphate galactose

UDIP-glu uridine diphosphate glucose

UDPH Ulster Diploma in Poultry Husbandry

UDPS Utah Department of Public Safety

udr universal data report(er); universal digital readout; usage data report; utility data reduction

UDR Ulster Defence Regiment

UDR *Union des Democrates pour la cinquiéme Republique* (French—Union of Democrats for the Fifth Republic)

udre utility data retrieval control

UDRI University of Dayton Research Institute; University of Denver Research Institute

UDRI-A University of Dayton Research Institute—Albuquerque

udro utility data retrieval output

Uds *ustedes* (Spanish—you, *pl.*)

UDS Ultraviolet Detection System; Underwater Demolition School; United Drapery Stores

udt underdeck tonnage; underwater destruction team

UDT Underwater Demolition Team; Union for a Democratic Timor; United Dominions Trust

UDTC University of Dublin Trinity College

U of D TC University of Dublin Trinity College

UDU Underwater Demolition Unit

udw ultra-deep water

UDW United Domestic Workers

UD-W University of Durban-Westville

Udy Oodie; Uddevalla

UDY United Dye and Chemical Corporation (stock exchange symbol)

ue underexcitation; unexpired; unit equipment; unit establishment; unit exception; unit extremity; upper entrance

u E *unseres Erachtens* (German—in our opinion)

UE United Electrical Workers; University Extension

U of E University of the East (Manila), University of Edinburgh; University of Essex; University of Exeter

uea unattended equipment area

UEA Universal Esperanto Association; University of East Africa; University of East Anglia; University Entrance Examination; Utah Education Association

U of EA University of East Anglia

ueac unit equipment aircraft

ueb ultrasonic epoxy bonder

UEB *Union Économique Benelux* (French—Benelux Economic Union)

UEC United Engineering Center (NYC)

UEC *Union Européene de la Carrosserie* (French—European Union of Coach-builders)

UECC United Electric Coal Companies

UECM Union Electric Company of Missouri

UECU Union for Experimenting Colleges and Universities

uee unit essential equipment

UEE *Unione Economica Europea* (Italian—European Economic Union)

uef universal extra fine (screw thread)

UEF *Union Européennedes Féderalistes* (French—European Union of Federalists); *Union Européenne Féminine* (French—European Union of Women)

UEFA Union of European Football Associations

UEI Union of Educational Institutions

UEIC United East India Company

uel upper explosive limit

UEL Unilever Export Limited; United Empire Loyalists

u enr uranium enrichment

UEO *Union de l'Europe Occidentale* (French—Western European Union); *Universala Esperanto-Asocio* (Universal Esperanto Association)—Rotterdam

uep underwater electrical potential; uniform external pressure

UEP Union Electric Power Company; *Union Européenne des Payements* (French—European Payments Union—EPU)

UEPA Utility Electric Power Association

UEPMD *Union Européenne des Practiciences en Médécine Dentaire* (French—European Union of Practitioners of Dentistry)

UER University Entrance Requirements; Unsatisfactory Equipment Report

UER *Unione Europea di Radiodiffusione* (Italian), *Union Européenne de Radiodiffusion* (French)—European Broadcasting Union

UERD Underwater Explosives Research Division (USN)

UERMWA United Electrical, Radio, and Machine Workers of America

UES Underground Experiment Subcommittee (AECL); United Engineering Societies

uesk unit essential spares kit

uet unattended earth terminal; unit equipment table

UET United Engineering Trustees

ueta (UETA) universal engineer tractor—armored

UETP University Enterprise Training Partnerships

uetrt (UETRT) universal engineer tractor—rubber-tired

UEW United Electrical Workers

uex unexposed

u/ext upper extremity

Uey U-turn (traffic)

uf urea-formaldehyde; underground feeder; used for

u/f urea-formaldehyde resin

UF Uniformed Force (police); United Fruit

U-F Ugro-Finnic

U of F University of Florida

UF_6 uranium hexafluoride

ufa until further advised

ufa (UFA) unesterified free fatty acid

UFA Uniformed Firefighters Association; University Film Association

UFA Universum-Film-Aktiengesellschaft (German—Universe Film Company)

ufac unlawful flight to avoid custody

UFAC Upholstered Furniture Action Council

UFACCC United Faculty Associations of California Community Colleges

ufaed unit forecast authorization equipment data

ufap unlawful flight to avoid prosecution

ufat unlawful flight to avoid testimony

UFAW Universities Federation for Animal Welfare

ufc uniform freight classification; urinary free cortisol

UFC Uni-Flex Container(s); United Fruit Company

UFCA Uniform Fraudulent Conveyance Act

UFCc United Free Churches

UFCE Union Fédéraliste des Communautés Ethniques Européennes (French—Federal Union of European Nationalities)

UFCS Underwater Fire-Control System

UFCT United Federation of College Teachers

UFCU Uni-Flex Container Unit

ufdr (UFDR) universal flight data recorder

ufe (UFE) unducted fan engine

UFEL United Farmers Educational League

UFERP Union Fraternelle Entre les Races et les Peuples (French—Fraternal Union Between Races and Peoples)

ufet unipolar field-effect transistor

uff ufficiale (Italian—officer; official); ufficio (Italian—bureau, office); und folgende (German—and the following)

UFF Ulster Freedom Fighters; United Freedom Front; University Film Foundation

uffi urea-formaldehyde foam insulation

UFH University of Fort Hare

UFI University Foundation International

UFI Union des Foires Internationales (French—Union of International Fairs)

UFIPTE Union Franco-Ibérique pour la Production et le Transport de l'Électricité (French—Franco-Iberian Union for the Production and Transmission of Electricity)

UFIRS Uniform Fire-Incident Reporting System

ufl upper flammable limit

UFL United Farmers League; University of Florida

UfM University for Man

UFMCC Universal Fellowship of Metropolitan Community Churches

ufn until further notice

ufo unfiltered oil; unforeseen obstacle; unidentified flying object

UFOA Uniformed Fire Officers Association

UFOD Union Française des Organismes de Documentation (French—French Union of Documentary Organizations)

ufol ufologic(al)(ly); ufologist(ic)(al)(ly); ufology

UFON Unidentified Flying Object Network

UFORA Unidentified Flying Objects Research Association

ufo's unidentified flying objects

uf p unemployed full pay

UFP United Federal Party

UFPA University Film Producers Association

UFPC United Federation of Postal Clerks

UFPO Underground Facilities Protective Organization

U-frame U-shaped frame

UFS University Film Society

UFT United Federation of Teachers

UFTA Uniform Fraudulent Transfer Act

UFTAA Universal Federation of Travel Agents Associations

UFTM Ulster Folk and Transport Museum

UFU Ulster Farmers' Union

UFVA University Film and Video Association

UFW United Farm Workers; United Furniture Workers

UFWU United Farm Workers Union

ug undergraduate; underground; universal grammar; urogenital

Ug Uganda; Ugandan; Ugric; Ugus

Ug Udjung (Malay—point)

UG Uganda (Internet code); Underground Railroad—secret system to aid slaves seeking freedom; United Gas

U of G University of Georgia; University of Glasgow; University of Guam; University of Guelph; University of Guyana

UG Universität Graz

UG3RD Upgraded Third-Generation System (for air-traffic control)

uga unity gain amplifier

UGA University of Georgia

Ugan Uganda

Uganda Republic of Uganda (East African country)

ugb unity gain bandwidth

ugc ultrasonic grating constant; unity grain crossover

UGC United Gas Corporation; University Grants Committee

UG & CW United Glass and Ceramic Workers

UGDP University Group Diabetes Program

UGE Unified Global Enterprises

UGEQ Union Generale des Estudiants du Québec (French—General Union of Students of Quebec)

ugf unidentified growth factor

UGGI Union Géodésique et Géophysique Internationale (French—International Geodesic and Geophysical Union)

ugi upper gastrointestinal

UGI Unione Geografica Internazionale (Italian), *Unión Géografica Internacional* (Spanish), *Union Geographique Internationale* (French)—International Geographical Union

UGLE United Grand Lodge of England

UGLIAC United Gas Laboratory Internally-Programmed Automatic Computer

U of G Lib University of Georgia Libraries

ugm underwater-launched guided missile

UGM Union of Graduates in Music

UGMA Unified Gift to Minors Account

ugmit you got me into this

UGMS Utah Geological and Mineral Survey

UGPL United Gas Pipe Line

U of G Pr University of Georgia Press

ugr ultrasonic grain refinement; universal graphic recorder

UGR Umfolozi Game Reserve (South Africa)

UGRR Underground Railroad (Quaker—organized means of aiding fugitive slaves escaping from southern slave states to Canada and northern free states)

ugs uniaxial gyrostabilizer; urogenital system

Ugs Ugus

UGS United Girls' School

ugt urgent; urogenital tract

UGT Unión General de Trabajadores (Spanish—General Union of Workers)—Socialist trade union

ugtl ugentlig (Dano-Norwegian—weekly)

UGU University of Guam

UGW United Garment Workers

uh upper half

uh (UH) utility helicopter

U of H University of Hartford; University of Hawaii; University of Houston; University of Hull

UH University Heights

UH Universidad de la Habana; Universität Hamburg

UH-1 Bell 204B Iroquois military helicopter

UH-19 Sikorsky transport helicopter called H-19 or Chickasaw

UH-23 Hiller Raven utility helicopter H-23

uha upper-half assembly

UHA Union House of Assembly

UHAA United Horological Association of America

UHAB Urban Housing Assistance Board

uhc under honorable conditions

UHCBCN United Hebrew Congregations of the British Commonwealth of Nations

URCC Upper House of the Convocation of Canterbury

uhcs ultra-high-capacity storage

URCY Upper House of the Convocation of York

uhel ultra-high-efficiency lamp

uhf ultra-high frequency—300–3000 mc

UHF United Health Foundation; United Holyland Fund (for Arab terrorists); United Hospital Fund

uhfdf ultra-high-frequency direction finder

uhff ultra-high-frequency filter

uhfg ultra-high-frequency generator

uhfj ultra-high-frequency jammer

uhfo ultra-high-frequency oscillator

uhfr ultra-high-frequency receiver

UHI University of Hawaii

UHK University of Hong Kong

U of HK University of Hard Knocks

uhl user header label

uhmw ultra-high molecular weight

uhmwpe ultrahigh-molecular weight polyethylene

UHOIA University of Houston Office of International Affairs

uhp ultra-high purity

UHP University of Hawaii Press

uhr ultra-high resistance; ultra-high resolution

uhrn ultra-high radio navigation

uhs ultra-high speed

UHS International Union of the History of Science; Union High School; University for Humanistic Studies

uht ultra-high temperature; ultrasonic hardness tester; universal hand tool

uht milk ultra-high temperature milk (capable of keeping without refrigeration)

uhtv unmanned hypersonic test vehicle

UHU Unhappy Hookers United

uhv ultra-high vacuum

uhvc ultra-high vacuum chamber

UHVS Ultra-High Vacuum System

ui ultrasonic industries; unit indicator; you (and) I

u/i unit of issue

ui unidades internacionales (Spanish—international units)

u.i. ut infra (Latin—as below)

UI Ube Industries; Unemployment Insurance; Universität Innsbruck; Urban Institute

U of I University of Idaho; University of Illinois; University of Iowa; University of Israel; University of Istanbul

UIA Ultrasonic Industry Association; Union of International Associations; United Israel Appeal; University of Iowa; Urban Intervention Associates

UIA Union Internationale des Architects (French—International Alliance of Architects); *Union Internationale des Avocats* (French—International Alliance of Attorneys)

UIAA Union Internationale des Associations d'Alpinisme (French—International Union of Alpinism Associations)

UIAB Unemployment Insurance Appeals Board

UIACM Union Internationale de Automobile-Clubs Médicaux (French—International Union of Medical Auto Clubs)

UIAS Union of Independent African States

UIATF United Indians of All Tribes Foundation

U i B Universitet i Bergen

UIB Unemployment Insurance Benefits; United International Bank

UIB Union Internationale des Maîtres Boulanger (French—International Union of Master Bakers)

uibc unsaturated iron-binding capacity

uic ultraviolet image converter

UIC Unemployment Insurance Code; Union International Company; University of Illinois in Chicago; Utah Innovation Center

UIC Unio Internationalis Contra Cancrum (International Union Against Cancer)

UICA Union of Independent Colleges of Art

UICC Union Internationale Contre le Cancer (French—International Union Against Cancer); *Unione Internazionale Contro il Cancro* (Italian—International Union for the Control of Cancer)

UICF Union Internationale des Chemins de Fer (French International Union of Railways)

UICIO Unit Identification Code Information Office(r)

UICN Union Internationale pour la Conservation de la Nature (French—International Union for the Conservation of Nature)

UI Comm Unemployment Insurance Commission

UICPA Union Internationale de Chimie Pure et Appliquée (French—International Union of Pure and Applied Chemistry)

UICPS Uniform Inventory Control Points System

UICR Union Internationale des Chauffeurs Routiers (French—International Coach and Lorry Drivers' Association)

UICT Union Internationale Contre la Tuberculose (French—International Union Against Tuberculosis)

UICWA United Infants' and Children's Wear Association

UID University of Idaho; User Identification

UIE UNESCO Institute for Education

UIEIS Union Internationale pour l'Etude des Insectes Sociaux (French—International Union for the Study of Social Insects)

UIEO Union of International Engineering Organizations

UIES Union Internationale pour l'Education Sanitaire (French—International Union for Health Education)

uif ultraviolet interference filter; unfavorable information file; universal intermolecular force

UIF Unemployment Insurance Fund

UIFI Union Internationale des Fabricants d'Imperméables (French—International Union of Rainwear Manufacturers)

Uig Uighur; Uigur

U i G Universitet i Göteborg

UIHL Union Internationale de l'Humanisme Laïque (French—International Union for Ethical Humanism)

UIHPS Union Internationale d'Histoire et de Philosophie des Sciences (French—International Union of the History and Philosophy of Science)

UIIG Union Internationale de l'Industrie du Gaz (French—International Gas Industry Union)

UIII Urban Information Interpreters Incorporated

U i L Universitet i Lund

UIL University of Idaho Library; University of Illinois; University of Illinois Library; University of Indiana Library; University of Iowa Library

UIL Unione Italian del Lavoro (Italian—Italian Labor Union)—republican and social-democrat

U of Ill Lib Sci University of Illinois Graduate School of Library Science

U of Ill Pr University of Illinois Press

UIM Union Industrielle & Maritime (Société Française de l'Armement) (French—Industrial and Maritime Union, French Ordnance

Company); *Union Internationale Motonautique* (French—International Motorboating Union)

UIMC Union Internationale des Services Médicaux des Chemins de Fer (French—International Union of Railway Medical Services)

UIMNH University of Illinois Museum of Natural History

UIN United States and International Securities (stock exchange symbol); University of Indiana

UINF Union Internationale de la Navigation Fluviale (French—International Union for River Navigation)

Uintas short form for the Uinta Mountains of northeastern Utah and southwestern Wyoming

uio underlying investment option

U i O Universitet i Oslo; Universitetsbiblioteket i Oslo (Norwegian—University Library in Oslo)

UIO Union Internationale des Orientalistes (French—International Union of Orientalists)

UIOOT Union Internationale des Organismes Officiels de Tourisme (French—International Union of Official Travel Organizations)

U of Iowa Pr University of Iowa Press

UIP United Irish Party; University of Illinois Press

UIP Union Internationale de Patinage (French—International Skating Union); *Union Internationale de Physique* (French—International Union of Physics)

UIPC Utah Industrial Promotion Commission

UIPC Union Internationale de la Presse Catholique (French—Catholic Press International Union)

UIPD Ulrich's International Periodicals Directory

UIPE Union Internationale de Protection de l'Enfance (French—International Union for the Protection of Children)

UIPM *Union Internationale de la Presse Médicale* (French—International Union of the Medical Press)

UIPVT *Union Internationale contre le Péril Vénérien et les Tréponématoses* (French—International Union against the Peril of Venereal Diseases and Syphilis)

uir upper information region

UIR University Industrial Research

uis (UIS) urban industrial society

U i S *Universitet i Stockholm*

UIS Unemployment Insurance Service; Unit Identification System

UIS *Union International de Secours* (French—International Relief Union)

UISAE *Union Internationale des Sciences Anthropologiques et Ethnologiques* (French—International Union of Anthropological and Ethnological Sciences)

UISB *Union Internationale des Sciences Biologiques* (French—International Union of the Biological Sciences)

uisc unreported interstate shipment of cigarettes

UISE *Union Internationale de Secours aux Enfants* (French—International Child Welfare Union)

UISN *Union Internationale des Sciences de le Nutrition* (French—International Union of Nutritional Sciences)

UISP *Union Internationale des Syndicats de Police* (French—International Union of Police Trade Union)

uit unit impulse train; unit investment trust

uit *uitgaaf* (Dutch—publication)

UIT *Unión Internacional de Telecomunicaciones* (Spanish), *Union Internationale des Télécommunications* (French), *Unione Internazionale Telecomunicazione* (Italian)—International Telecommunications Union—ITU

UITAM *Union Internationale de Mécanique et Appliquée* (French—International Union of Theoretical and Applied Mechanics)

uitg *uitgegeven* (Dutch—published)

UITS *Unione Italiana Tiro e Segno* (Italian—Italian Rifle Association)

U i U *Universitet i Uppsala*

UIU Quito, Ecuador (airport)

UIUNA Upholsterers' International Union of North America

UIUPGWA United International Union of Plant Guard Workers of America

UJ Union Jack (United Kingdom flag incorporating crosses of St Andrew for Scotland, St George for England, and St Patrick for Northern Ireland); University of Judaism

U of J University of Judaism

UJ *Universidad Javeriana* (Bogotá and Sucre)

UJA United Jewish Appeal

UJC Union Jack Club

U.J.D. *Utriusque Juris Doctor* (Latin—Doctor of Civil and Canon Law)

ujf unsatisfied judgment fund(ing)

U-joint(s) U-shaped joint(s)

ujr unijunction rectifier

UJSCs Union Jack Services Clubs

ujt unijunction transistor

uk unknown

uk (UK) urokinase

UK United Kingdom; *Universita Karlova* (Karl University—University of Prague)

U of K University of Kansas; University of Keele (formerly University College of North Staffordshire); University of Kent; University of Kentucky

UK *Universiti Kebangsaan* (Malay—National University)

UKA Ulster King of Arms; United Kingdom Alliance; United Klans of America

UK(A) United Kingdom All-comers (athletics)

UKAC United Kingdom Automation Council

UKADGE United Kingdom Air Defence Ground Environment

UKADR United Kingdom Air Defense Region (NATO)

UKAEA United Kingdom Atomic Energy Authority

UKAPE United Kingdom Association of Professional Engineers

ukb universal keyboard

UKBC United Kingdom Bomber Command

UKBG United Kingdom Bartenders' Guild

UKC United Kennel Club; University of Kent at Canterbury

U of KC University of Kansas City; University of King's College

UKCA United Kingdom Citizens Association

UKCATR United Kingdom Civil Aviation Telecommunications Representative

UKCBDA United Kingdom Carbon Block Distributors' Association

UKCSBS United Kingdom Civil Service Benefit Society

UKCTA United Kingdom Commercial Travellers' Association

UKDA United Kingdom Dairy Association

uke ukelele

UK fo United Kingdom for orders

UKGBNI United Kingdom of Great Britain and Northern Ireland

UKGPA United Kingdom Glycerine Producers' Association

UKHH United Kingdom-Havre-Hamburg (range of ports)

UKHS United Kingdom Hovercraft Society

UKIAS United Kingdom Immigrants Advisory Service

UKIRT United Kingdom Infrared Telescope

UKISC United Kingdom Industrial Space Committee

UKITO United Kingdom Information Technology Organisation

UKJATFOR United Kingdom Joint Airborne Task Force

UKJGA United Kingdom Jute Goods Association

UKKKK United Kingdom Klu Klux Klan

UKL University of Kansas Library; University of Khartoum Library

UKLF United Kingdom Land Force

UKLFS United Kingdom Low-Flying System

UKM University of Kansas Museums

UKMC University of Kansas Medical Center

UKMF United Kingdom Mobile Force

UKML United Knitwear Manufacturers League

UK(N) United Kingdom National (athletics)

UKOP United Kingdom Oil Pipelines

U K£ United Kingdom pound

UKPA United Kingdom Pilots' Association

UKPM United Kingdom Prime Minister

Ukr Ukraine; Ukrainian

Ukr Acad Pr Ukrainian Academic Press

Ukraine whose capital is Kiev; Ukrainian

Ukrainian SSR Ukrainian Soviet Socialist Republic (former name of Ukraine)

UKRAS United Kingdom Railway Advisory Service

UKS University of Kansas

UKSATA United Kingdom South Africa Trade Association

UKSM United Kingdom Scientific Mission; University of Kansas School of Medicine

UKSMA United Kingdom Sugar Merchants' Association

UKSMT United Kingdom Sea Mist Test(ing)

UKSTC United Kingdom Strike Command

Ukulele (UK) stock exchange slang for Union Carbide

ukv underground keybox vault

UKW *Ultra-Kurzwellen* (German—ultra-short wave)

UKY University of Kentucky

ul up link; upper left; upper leg; upper level; upper lid

ul (UL) user language

u/l upper left; upper limit

u & l upper and lower

UL Urban League; Underwriters Laboratories; Universal League; University Libraries; University Library

U of L University of Lancaster; University of Laval; University of Leeds; University of Leicester; University of Leth-

bridge; University of Liverpool; University of London; University of Louisville

UL Union List

ula uncommitted logic array

ULA Ulster Launderers' Association; United Labor Agency; University of Louisiana

ULA Uniform Laws Annotated; *Universidad Los Andes* (Spanish—Andes University)—Venezuela

ULAA Ukrainian Library Association of America

Ulaan Ulaanbaatar (Khalkha Mongol—Red Hero)—capital city of Ulan Bator, Mongolia

ULAD Unilever Limited Accounts Department

u-land udviklingstand (Dano-Norwegian—development land)

ULAP University-wide Library Automation Program (University of California)

ULAS University of London Air Squadron

ULAST Unión Latino Américana de Sociedades de Tisiology (Spanish—Latin-American Union of Societies of Phthisiology)

ulb universal logic bloc

ULB Université Libre de Bruxelles (French—Free University of Brussels)

ulc unit load container; unsafe lane change (vehicular code); upper left center

u & lc upper and lower case

ULC Ulster Loyalist Council; Underwriters' Laboratories of Canada; Urban Library Council

ULCA United Lutheran Church of America

ulcc ultra-large crude carrier

ULCC Ultra Large Cargo Carrier (bulk freighter or tanker of 400,000 or more tons)—superfreighter or supertanker

ULCI Union of Lancashire and Cheshire Institutes

uld ultralow distortion

ULD Unit Load Device

uldb ultralight-displacement boat; ultralong duration balloon

uldest ultimate destination

ule ultra-low expansion

ulev ultra-low-emission vehicle

ulf ultra-low frequency; unfair labor practice

ulft ultra-low-flush toilet

uli ultra-low interstitial

ULI Urban Land Institute; Uttam Language Institute (Kathmandu, Nepal)

ULI Union pour la Langue Internationale Ido (French—Union for the International Language Ido)

ULICS University of London Institute of Computer Science

ULII Union pour la Langue Internationale Ido (French—Union for the International Language Ido)

ull ullage

'Ull (Cockney contraction—Hull)

ULL Unitarian Laymen's League; University of Liverpool Library; University of London Library; University of Lund Library

ulla ultra-low-level airdrop

ullv (ULLV) unmanned lunar logistics vehicles

ulm ultrasonic light modulator, universal logic module

ULM University Library of Manchester (includes John Rylands Library)

ULMS Underwater Long-range Missile System

ULO United Licensed Officers (union); Unmanned Launch Operations

ULOTC University of London Officers Training Corps

ULP Université Louis Pasteur; University of London Press

ulpa ultra-low penetrating air

ULPA Uniform Limited Partnership Act

ulpr ultra low-pressure rocket

ULPZ Upper Limits for the Prescriptive Zone

Ulrich Ulrich's Books

uls unsecured loan stock

Uls Ulsan; Ulster (ancient Irish province comprising 6 of the 9 original counties)

ULS Universities Libraries Section (Association of College and Research Libraries)

ULS Union list of Serials

ulsi ultra-large-scale integration

ult ultimate; ultimo

ult ultimo (Latin—at last)

ULT United Lodge of Theosophists

Ult Bod Ultra Bodoni
ulto ultimo
ult° *último* (Spanish—last)
ult. praes. *ultimum praescriptus* (Latin—last prescribed)
ult pun ultimate punishment (death by execution)
ultra (Latin prefix—beyond or in excess)—ultramontane, ultrasonic
ultracom ultraviolet communications system
ultra hi-fi ultra-high fidelity
ultralight ultralight flying machine; ultralight luggage; ultralight wearing apparel
ultrason ultrasonic(s)
ultra-x universal language for typographic reproduction applications
ULTS Universal Lifeline Telephone Service
ult ts ultimate tensile strength
U of Luck University of Lucknow
ULUCLA University Library of the University of California at Los Angeles
ULUM University Library, University of Michigan (Ann Arbor)
ulv ultra-low volume
um umpire; unmarried
u/m unit of measure
ü/M über dem Meeresspiegel (German—above sea level)
UM United States' outlying islands (Internet code); Universal Match; Universal Mill; University of Malaysia (University of Malaya –Raffles Institute); University of Manitoba; University of Melbourne; University Museum(s)
U of M University of Maine; University of Malaysia; University of Manchester; University of Manitoba; University of Maryland; University of Massachusetts; University of Miami; University of Michigan; University of Minnesota; University of Mississippi; University of Missouri; University of Montreal
UM Universiti Malaya (University of Malaya)—Raffles Institute
U Ma Ursa Major (Big Bear)

UMA Ultrasonic Manufacturers Association; *Union de Mujeres Americanas* (Spanish—United Women of the Americas); Union Maghreb Arabe; University of Massachusetts; University Museum of Anthropology (Philadelphia, Pennsylvania)
UM-A University of Mid-America
U-magnet U-shaped magnet
U of Mand University of Mandalay
UMAS United Mexican-American Students
umass unlimited machine access from scattered sites
UMass University of Massachusetts
U of Mass Pr University of Massachusetts Press
umb umber; umbilical; umbilicus
Umb Umbrian
UMB Union Mondiale de Billard (French—World Billiards Union)
UMBA United Mortgage Bankers of America
umb gun umbrella gun
UMBIR University of Michigan Bureau of Industrial Relations
umbl umbilical
UMBR Umbria(n)
UMC Uniform Mechanical Code; United Metallic Cartridge (company); United Methodist Church; Universal Match Corporation; Upstate Medical Center; Urban Mobility Corporation
UMCA United Mining Councils of America; Urabá, Medellín and Central Airways
umd unitized microwave device
UMD Unit Manning Document; University of Maryland
UMDA United Micronesian Development Association
U of Md Lib Serv University of Maryland School of Library and Information Services
U of Mdrs University of Madras
UME Uniform Manufacturers Exchange; University of Maine
umf ultramicrofiche
UMF Umbrella-Makers' Federation

UMFC United Methodist Free Churches
umgearb *umgearbeitete* (German—revised)
UMHK Union Miniére du Haut-Katanga (French—United Mines of Upper Katanga)
UMHP Union Mondiale des sociétiés d'Histoire Pharmaceutique (French—World Union of Pharmaceutical History Societies)
U Mi Ursa Minor (Little Bear)
UMI University of Michigan; University of Microfilms Incorporated; University Microfilms International; Utah Management Institute
U of Miami Pr University of Miami Press
U of Mich Bus Res University of Michigan Graduate School of Business Research
U of Mich Inst Labor University of Michigan Institute of Labor and Industrial Relations
U of Mich Pr University of Michigan Press
U of Mich Soc Res University of Michigan Institute for Social Research
umids universal mine dispensing system
U/min Umdrehungen in der Minute (German—revolutions per minute)
U of Minn Bell Mus University of Minnesota Bell Museum of Pathology
U of Minn Pr University of Minnesota Press
UMIST University of Manchester Institute of Science and Technology
UML University of Michigan Library; University of Minnesota Library; University of Missouri Library
umler universal machine language
UMLS University Microfilm Library Service
UM & M United Merchants and Manufacturers
UMMS University of Maine (or Manchester, Manitoba, Maryland, Massachusetts, Michigan, Minnesota, Mississippi, Missouri, Montana) Medical School
UMMZ University of Michigan Museum of Zoology

umn upper motor neuron

UMN University of Minnesota

UMNO United Malay National Organization

UMO University of Maine at Orono; University of Missouri

umoc ugly man on campus

U of Monc University of Moncton

U of Mo Pr University of Missouri Press

ump umpire

UMP Upper Mantle Project; Upper Merion and Plymouth (railroad); University of Massachusetts Press

'Umphrey (Cockney contraction—Humphrey)

UMPO Upper Manhattan Planning Office

umr under main roof

U MR Umvoti Mounted Rifles

UMREL Upper Midwest Regional Educational Laboratory

UMRRC Universities Mobile Radio Research Corporation (Bath, Birmingham, Bristol)

UMRWFR Upper Mississippi River Wildlife and Fish Refuge (Minnesota)

ums unmanned machinery space

UMS Undersea Medical Society; Unfederated Malay States; Universal Military Service; University of Mississippi

UMSM University of Michigan School of Music

UMSU University of Malaya Student's Union

UMT Universal Military Training; University of Montana

UMT *Union Marocaine du Travail* (French—Moroccan Labor Union)

UMTA Urban Mass Transportation Administration

umtd using mails to defraud

UMTRAP Uranium Mill Tailings Remedial Action Program

UMTS Universal Military Training and Service; Universal Mobile Telecommunications System

umw ultramicrowave

UMW United Mine Workers

UMWA United Mine Workers of America

U of Mys University of Mysore

un units

un (UN) unsatisfactory

Un Union (postal abbreviation)

UN pound unit of measurement for fuel (aircraft code); Union Twist Drill (trademark); United Nations; University of the North; unsatisfactory

U of N University of Natal; University of Nebraska; University of Nevada; University of Newcastle; University of Nottingham

UN *União Nacional* (Portuguese—National Union); *Unificación Nacional* (Spanish—National Unification)—Costa Rica

UNA Underwear-Negligee Associates; United Nations Association; United Native Americans; United Natives Association

UNAA United Nations Association of Australia

UNAAF Unified Action Armed Forces

unab unabridged

unabr unabridged

UNAC United Nations Appeal for Children

UNACC United Nations Administrative Committee on Coordination

unaccomp unaccompanied

UNACIL United Africa Commercial and Industrial Limited

UNACOMS Universal Army Communications System

UNAIDS United Nations, Acquired Immune Deficiency Syndrome

UNAIS United Nations Association International Service

unalot unallotted

UNAM *Universidad Nacional Autónoma de Mexico* (Spanish—National University of Mexico)

unamace universal automatic map compilation equipment

UNAMIC United Nations Advanced Mission Cambodia

unan unanimous

UNAPO United National Association of Post Office (Craftsmen)

UNARCO United Nations Narcotics Commission

unasgd unassigned

unatt unattached

UNAUS United Nations Association of the United States

UNAUSA United Nations Association of the United States of America

unauthd unauthorized

UNAVEM United Nations Angola Verification Mission

unb unbound; universal navigation beacon

UNB United Nations Bookshop; University National Bank; University of Nebraska

U of NB University of New Brunswick

UN Bank International Bank for Reconstruction and Development

unbd unbound

unbels unbelievers

Unbib van Amsterdam *Universiteitsbibliotheek van Amsterdam* (Dutch—Amsterdam University Library)

unblkng unblanking

UNBRO United Nations Border Relief Operation

unb's unbelievers

UNBSA United Nations Bureau of Social Affairs

unc uncertain; unconscious; undercurrent; unified coarse (thread)

unc (UNC) unconditional (computer flow chart)

unc. uncle

Unc Uncle

UNC United Nations Command; United Nuclear Corporation; University of North Carolina; University of Northern Colorado

U of NC University of North Carolina

UNC *Union Nationale Camerounaise* (French—Cameroon National Union)—party; *Universidad Nacional de Colombia* (Spanish—National University of Colombia)

UNCA United Nations Correspondents Association

UNCAST United Nations Conference on the Applications of Science and Technology

UNCC United Nations Cartographic Commission

UNCCP United Nations Commission on Crime Prevention; United Nations Conciliation Commission for Palestine

UNCED United Nations Conference on Environment and Development

uncert uncertain

UNCF United Nations Children's Fund (formerly UNICEF); United Negro College Fund

unch unchanged

U of NC Inst Gov University of North Carolina Institute of Government

UNCIO United Nations Conference on International Organization

UNCIP United Nations Commission on India and Pakistan

uncir uncirculated

UNCIRSS University of North Carolina Institute for Research in Social Science

UNCITRAL United Nations Commission on International Trade Law

UNCIWC United Nations Commission for the Investigation of War Crimes

UNCL University of North Carolina Library

unclas unclassified

U.N.C.L.E. United Network Command for Law Enforcement (fictional organization created for television)

UNCLOS United Nations Conference on the Law of the Sea

UNCMAC United Nations Command Military Armistice Commission

unco uncouth

UNCO United Nations Civilian Operations Mission (to the Congo)

UNCOK United Nations Commission on Korea

uncol universal computer-oriented language

uncom uncommon

uncomp uncompensated

uncond unconditioned

UNCOPUOS United Nations Committee on the Peaceful Uses of Outer Space

uncor uncorrected

uncov uncover; uncovered; uncovers

U of NC Pr University of North Carolina Press

un cs unconditioned stimulus

UNCs United Neighborhood Centers

unct. unctus (Latin—smeared)

UNCTAD United Nations Conference on Trade and Development

uncult uncultivated

UNCURK United Nations Commission for the Unification and Rehabilitation of Korea

und under

UND University of National Defense; University of North Dakota

U of ND University of North Dakota; University of Notre Dame

UNDAT United Nations Development Advisory Team

UN Day United Nations Day (October 24)

UNDC United Nations Disarmament Commission

UNDCC United Nations Development Cooperation Cycle

undeco underground economy (composed of persons who report less than they earn, or who file no income tax returns)

unded underdeduction

undercover narc undercover narcotics agent

undergrad undergraduate

Under Sec Nav Under Secretary of the Navy

Undex United Nations Index

UNDI United Nations Document Index

undies underthings (underwear)

undoc(s) undocumented alien(s)—illegal alien(s)

UNDct United Nations Document

UNDOF United Nations Disengagement Observer Force

UNDP United Nations Development Program

und pkng underground parking

U of ND Pr University of Notre Dame Press

undrgrnd underground

UNDRO United Nations Disaster Relief Office

undrwrld underworld

undsgd undersigned

UNDSM University of North Dakota School of Medicine

undtkr undertaker

undw underwater

undwrtr underwriter

UNE University of New England (New South Wales)

UNEAS Union of European Accountancy Students

U of Neb Pr University of Nebraska Press

UNEC United Nations Education Conference

UNECA United Nations Economic Commission for Asia

UNECLA United Nations Economic Commission for Latin America

UNECOLAIT Union Européenne du Commerce Laitier (French—European Milk Trade Union)

UNEDA United Nations Economic Development Association

unef unified national extra fine (screw thread)

UNEF United Nations Emergency Forces

UNEF Union Nationule des Étudiants Francais (French—National Union of French Students)

UNEO United Nations Emergency Operation

UNEP United Nations Environment(al) Program

UNESCO United Nations Educational, Scientific, and Cultural Organization

UNESEM Union Européenne des Sources d'Eaux Minérales du Marché Commun (French—European Union of Natural Mineral Water Sources of the Common Market)

UNETAS United Nations Emergency Technical Aid Service

U of Nev Pr University of Nevada Press

unex unexecuted

unexpec unexpected

unexpl unexplained; unexploded; unexplored

unexpur unexpurgated

UNEXSO Underwater Explorers Society

unf unfinished; unfuzed (munition); unified fin thread

UNF United National Front

U of NF University of North Florida

UNFAO United Nations Food and Agricultural Organization

unfav unfavorable

UNFB United Nations Film Board

UNFC United Nations Food Conference

unfd unfurnished

UNFDAC United Nations Fund for Drug Abuse Control

UNFICYP United Nations (Peace-Keeping) Force in Cyprus

unfin unfinished

Unfin Unfinished (Schubert's Symphony No 8)

UNFPA United Nations Fund for Population Activities

UN Fund International Monetary Fund

unfurn unfurnished

ung unguent

ung ungarische (German—Hungarian)

ung. unguentum (Latin—ointment)

Ung Ungava; Ungavan

Ung Ungarn (Norwegian—Hungary)

UNGA United Nations General Assembly

Ungar Frederick Ungar Publishing Company

unh uranyl nitrate hexahydrate

UNH University of New Hampshire; University of New Haven

U of NH University of New Hampshire

UNHCR United Nations High Commissioner for Refugees

UNHQ United Nations Headquarters (Geneva, New York, Vienna)

uni (Latin prefix—one)—unilateral

Uni University

UNI United News of India; United Nuclear Industry

UNI Unione Naturista Italiana (Italian—Italian Naturist Association); *Ente Nationale Italiano di Unificazione* (Italian Unification Council)

UNIA Universal Negro Improvement Association (Garveyites)

UNIC United Nations Information Center

UNICCAP Universal Cable Circuit Analysis Program

UNICE Union des Industries de la Communauté Européenne (French—Industrial Union of the European Community)

UNICEF United Nations International Children's Emergency Fund

unicike unicycle

UNICIS Unit Concept Indexing System; University of Calgary Information Systems

unicom underwater integration communication; universal communication

UNICOM aeronautical advisory station operating on 122.8 mc

UNIDIR United Nations Institute for Disarmament Research

UNIDO United Nations Industrial Development Organization

unif uniform; uniformity

unif coef uniformity coefficient

UNIFEM United Nations Development Fund for Women

Unif Gift Min Act Uniform Gifts to Minors Act

UNIFIL United Nations Interim Force in Lebanon

Unif L Ann Uniform Laws Annotated

Uniform letter U radio code

unihedd universal head-down display

UN I-I MOG UN Iran-Iraq Military Observer Group

UNIKOM United Nations Iraq–Kuwait Observation Mission

unilat unilateral

Unilatcorps Unilateral Corps

UNIMA Union Internationale de grands Magasins (French—International Union of Department Stores)

UNIMERC Universal Numeric Coding System

UNIMS Univac Information Management System

UNINCO Union Internationale des Corps Consulaires (French—International Consular Corps Union)

unincorp unincorporated

UNIO United Nations Information Organization

Union Coll Pr Union College Press

Unions Union Islands of the Pacific also called Tokelaus

UNIP United Independence Party

uniparse universal parser

UNIPEDE Union Internationale des Producteurs et Distributeurs d'Energie Electrique (French—International Union of Producers and Distributors of Electric Energy)

unipol universal procedure-oriented language

UNIPOM United Nations India-Pakistan Observer Mission

uni(s) unisexual(s)

unis unisoni (Italian—unison)

unis 8ᵛᵃ (Italian—in unison with the octave)

UNIS United Nations International School; Univac Industrial System

UNISCAN United Kingdom and Scandinavia

UNISIST Universal System for Information in Science and Technology

UNISOMI Universal Symphony Orchestra and Music Institute

UNISTAR User Network for Information Storage Transfer

UNISYM Unified Symbolic Standard Terminology for Mini Computer Instructions

Unit Unitarian

unit. united; uniting

UNIT Union Nationale des Ingénieurs Techniciens (French—National Union of Engineers and Technicians)

UNITA Unión Nacional para la Independencia Total de Angola (Portuguese—National Union for the Total Independence of Angola)

UNITAR United Nations Institute for Training and Research

UNITE Union of Needletrades, Industrial and Textile Employees

United United Airlines

United Auto Workers International Union, United Automobile, Aerospace, and Agricultural Implement Workers of America

United Kingdom United Kingdom of Great Britain and Northern Ireland (England, Scotland, Wales, the Isle of Man, colonies and dependencies such as Belize; Bermuda; British Antarctica; the British Indian Ocean Territory; the British West Indies; the Channel Islands; Gibraltar; Hong Kong; the Gilberts, New Hebrides; Pitcairn; Ascension, the Falklands, St Helena, Tristan da Cunha)

United Mine Workers United Mine Workers of America

United Provinces United Provinces of the Netherlands (Friesland, Gelderland, Groningen, Holland, Oberyssel, Utrecht, Zeeland)—the Seven Provinces

United Rubber Workers United Rubber, Cork, Linoleum, and Plastic Workers of America

United States United States of America (North American nation); United States of Brazil; the United States of Colombia; the United States of Indonesia; the United States of Mexico; the United States of North America (the U.S.A.); the United States of Venezuela

UNITS United Nations Information for Teachers

univ universal

Univ Universal; Universalist; University; University College, Oxford

univac universal automatic computer

univar universal valve action recorder

Univ-Buchdr *Universitats Buchdrukerei* (German—university press)

Univ C University College (Oxford)

Univ. D. Doctor of the University (degree)

universal languages mathematics and music

Univ Mus of UP University Museum of the University of Pennsylvania

UNIX Universal Inner-Active Executive (IBM)

unjc united national J-series coarse (thread)

unjef united national J-series extra fine (thread)

unjf united national J-series fine (thread)

unjs united national J-series special (thread)

unk unknown

Unk Uncle

unkn unknown

UNKRA United Nations Korean Reconstruction Agency

UNL Universal Networking Language; University of Nairobi Library; University of Nebraska at Lincoln

UNLA *Unione Nazionale per la Lotta contro l'Analfabetisma* (Italian—National Association for the Fight Against Illiteracy)

UNLC United Nations Liaison Committee

unld unload (flow chart)

unldh underloading

unlib unliberated

unliq unliquidated

unlk unlock

UNLL United Nations League of Lawyers

UNLOS United Nations Law of the Sea (conference)

UNLOSC United Nations Law of the Sea Conference

unltd unlimited

UNLV University of Nevada at Las Vegas

unlwfl unlawful(ly)

unm unmarried

UNM Ukrainian National Museum (Chicago); United Nations Medal; University of New Mexico

U of NM University of New Mexico

UNMC United Nations Mediterranean Commission; University of Nebraska Medical Center

UNMEM United Nations Middle East Mission

U of NM Gen Lib University of New Mexico General Library

UNMOGIP United Nations Military Observer Group in India and Pakistan

U of NM Pr University of New Mexico Press

UNMSC United Nations Military Staff Committee

UNMSM de L *Universidad Nacional de San Marcos de Lima* (Spanish—University of Lima)

unmtd unmounted

unnilhexium unnamed element 106

unnilseptium unnamed element 107

UNO United Nations Organization; United Neighborhood Organization; University of Nebraska at Omaha; University of New Orleans

UNO *Union Nacional Odría* (Spanish—Odria National Union)—political party; *Un-*

ión Nacional Oposición (Spanish—National Opposition Union)—Nicaraguan political party

UNOC United Nations Operations in the Congo

UNOCAL Union Oil Company of California

unodir unless otherwise directed

unof unofficial

UNOGIL United Nations Observer Group in Lebanon

UNOID United Nations Organization for Industrial Development

unoindc unless otherwise indicated

UNOLS University-National Oceanographic Laboratory System

U or non-U upperclass or not upperclass

unop unopposed

unoreq unless otherwise requested

unorg, unorganized

UNOS United Network for Organ Sharing

unp unpaged; unpaid

UNP United National Party; University of Nebraska Press; Urewara National Park (North Island, New Zealand)

UNPA United Nations Postal Administration; United Nations Protected Area

UNPC United Nations Palestine Commission

unpd unpaid

UNPF United Nations Population Fund

UNPHU *Universidad Nacional Pedro Henriguez Urena* (Spanish—Pedro Henriquez Urena National University)—Dominican Republic

unpkd unpacked (flow chart)

unpleas unpleasant

UNPOC United Nations Peace Observation Commission

UNPP United Nations Partition Plan

UNPROFOR United Nations Protection Force

unpub unpublished

unqte unquote

unqual unqualified

UNR & EC United Nuclear Research and Engineering Center

UNREF United Nations Refugee Emergency Fund

unrel unreliable

unrep unreported; unrepresented

UNRISD United Nations Research Institute for Social Development

UNRPR United Nations Relief for Palestine Refugees

UNRRA United Nations Relief and Rehabilitation Administration

UNRWA United Nations Relief and Works Agency

uns unified special (thread); unsymmetrical

UNS Unified Numbering System

UNSA United Nations Specialized Agencies; University of Nottingham School of Agriculture

unsat unsatisfactory; unsaturated

unsatfy unsatisfactory

unsatis unsatisfactory

UNSC United Nations Security Council

UNSCC United Nations Standards Coordinating Committee

UNSCCUR United Nations Scientific Conference on the Conservation and Utilization of Resources

UNSCEAR United Nations Scientific Committee on the Effects of Atomic Radiation

UNSCOB United Nations Special Commission on the Balkans

UNSCOM United Nations Special Commission

UNSCOP United Nations Special Commission on Palestine

unscv unserviceable

UNSDRI United Nations Social Defense Research Institute

UNSF United Nations Security Force; United Nations Special Fund for Economic Development

UNSG United Nations Secretary General

unsgd unsigned

unskd unskilled

UNSM United Nations Service Medal; University of Nebraska State Museum

unsol unsolicited

UNSO United Nations Sahel Office (*see* Sahel); United Sabah Organization

UNSR United Nations Space Registry

unst unstable

un stim unconditioned stimulus

unstr unstressed

unsus-look(ing) unsuspicious look(ing)

unsvc unserviceable

UNSvM United Nations Service Medal

UNSW University of New South Wales

UNSY United Nations Statistical Yearbook

unsym unsymmetrical

Unt Unter (German—lower; under)

UNTA United Nations Technical Assistance

UNTAA United Nations Technical Assistance Administration

UNTAB United Nations Technical Assistance Board

UNTAC United Nations Transitional Authority in Cambodia

UNTAG United Nations Transition Assistance Group

UNTAM United Nations Technical Assistance Mission

UNTC United Nations Trusteeship Council

UNTEA United Nations Temporary Executive Authority

unthd unthreaded

UNTS United Nations Treaty Series

UNTSO United Nations Truce Supervision Organization

UNTT United Nations Trust Territory

UNTTA United Nations Trust Territory Administration

UNU United Nations University

UNUP United Nations University Press

UNUSA United Nations Association of the United States of America

UNV United Nations Volunteers; University of Nevada

UNWCC United Nations War Crimes Commission

unwmk unwatermarked

UNYOM United Nations Yemen Observation Mission

UNYNJSHPBA United New York and New Jersey Sandy Hook Pilots Benevolent Associations

u/o used on

u & o use and occupancy

uo und öfters (German—and often)

UO Ulster Orchestra (Belfast); University of Otago (at Dunedin, New Zealand); University of Ottawa

U of O University of Ohio; University of Oklahoma; University of Omaha; University of Oregon; University of Ottawa; University of Oxford

uoa use of other automobiles

UOB United Overseas Bank

U of O B University of Oregon Books

uoc ultimate operational capability

UOC Uniform Offense Classification

UOCO Union Oil Company

uod ultimate oxygen demand

UOFS University of the Orange Free State

UOH University of Ohio

uohc under other than honorable conditions

UOJCA Union of Orthodox Jewish Congregations of America

UOK University of Oklahoma

U of Okla Pr University of Oklahoma Press

uol underwater object locator

UOL United Olympic Life

uoo undelivered orders outstanding

UOP Universal Oil Products

UOPH Unaccompanied Officer Personnel Housing

UOPWA United Office and Professional Workers of America

UOR Uniform Officer Record; University of Oregon; Unusual Occurrences Report

UORI University of Oklahoma Research Institute

uos Underwater Ordnance Station (USN)

uo's undelivered orders

uot uncontrolled overtime

UOT United Ocean Transport (Daido Line)

UOTC University Officers Training Corps

UOTS United Order of True Sisters

uov unit of variance

up uncorrected proof

up. underproof; underproofed; underproofing; unpaged; upper

u/p urine-plasma concentration

u & p uttering and publishing

Up Upper

UP Ulster Parliament; Ulster Party; Union Pacific (railroad); Union Postale (Postal Union); United Party; United Presbyterian; United Press; United Province; Universal Pictures; University of Paris; University of Pennsylvania; University of Pittsburgh; University Press; Upper Peninsula; Uttar Pradesh; Uptown Planners

U of P University of the Pacific; University of Pennsylvania; University of Pittsburgh; University of Portland; University of Pretoria; University of Puget Sound

UP *Unidad Popular* (Spanish—Popular Unity)—political party; *Unión Panamericana* (Spanish—Pan-American Union); *Unión Patriótica—* (Spanish—Patriotic Union) —old Colombian Communist party; *Union Postale* (French—Postal Union)—international mail organization

UPA Ultimate Players Association; Uniform Partnership Act; United Productions of America; University of Pennsylvania; University Photographers Association

UPA *Union Postale Arabe* (French—Arab Postal Union); *Unions Professionnelles Agricoles* (French— Professional Agricultural Unions)

UPAA University Photographers Association of America

UPAC Union of Pan-Asian Communities

UPADI *Unión Panamericana de Asociaciones de Ingenieros* (Spanish—Panamerican Union of Engineers Associations)

UPAE *Union Postal de las Américas y España* (Spanish—Postal Union of the Americas and Spain); *Union*

Postale des Amériques et de l'Espagne (French—Postal Union of the Americas and Spain)

Upan Upanishad

UPAO University Professors for Academic Order

UPASI United Planters Association of South India

upc universal product code

UPC Unesco Publications Center; Uniform Practice Code; Uniform Plumbing Code; United Power Company; United Presbyterian Church; Universal Postal Convention; Universal Product Code

upd unpaid

UPD Unified Port District; Urban Planning Directorate

UPD *Union Periodistas Democratica* (Spanish—Union of Democratic Journalists)— Mexico

UPDW United Piece Dye Works

U of PE University of Port Elizabeth

UPE *Union Parlementaire Européenne* (French—European Parliamentary Union)

UPEP Undergraduate Preparation of Educational Personnel

UPF United Plastic Fabricating

UPFDA United Products Formulators and Distributors Association

UPGA Utah Personnel and Guidance Association

uphd uphold

uphol upholsterer; upholstery

UPI United Press International (merger of United Press and International News Service)

UPICA University of Pennsylvania Institute of Contemporary Art

UPIGO *Union Professionnelle Internationale des Gynécologistes er Obstétriciens* (French—International Professional Union of Gynecologists and Obstetricians)

UPIN United Press International Newsfeatures

upk's unpopped kernels

upl unauthorized practice of law

UPL United Philippine Line; University of Pennsylvania Library; University of the Philippines Library (Quezon

City); University of Pittsburgh Library; University of Portland Library

up. log. upper loge

upm uninterruptible power module; units per mile

UPM Unit Production Manager

UPNE University Press of New England

UPNG University of Papua and New Guinea

upo undistorted power output; unidentified paleontological object

UPO United Partisans' Organization; Unit Personnel Office(r); Unit Press Officers

UPOA Ulster Police Officers Association; Ulster Public Officers Association

UP of M Upper Peninsula of Michigan

UPOV Union for the Protection of New Varieties of Plants

UPOW Union of Post Office Workers

upp *upplaga* (Swedish—edition)

UPP University of Pennsylvania Press; University of Pittsburgh Press

UPPC Union Pacific Petroleum Corporation

Upper Volta Republic of Upper Volta (landlocked West African country), *République de Haute-Volta;* name changed in 1984 to Burkina Faso

uppp uvulopalatopharyngoplasty

UPPPP Underprivileged Peoples' Public Pool

upr (most); unsaturated polyester resin; upper

Upr Upper

U Pr University Press (Washington, DC)

UPR Union Pacific Railroad; University of Puerto Rico

UPREAL Unit Property Record and Equipment Authorization List

U Presses Fla. University Presses of Florida

U Pr Hawaii University Press of Hawaii

U Pr Kan University Press of Kansas

U Pr Ky University Press of Kentucky

U Pr Miss University Press of Mississippi

U Pr NE University Press of New England

U Pr Va University Press of Virginia

U Pr Wash University Press of Washington

ups ultraviolet photoemission spectroscopy; uninterruptible power supply

UPS Underground Press Syndicate; Underground Publication Society, Underwater Production System(s); United Parcel Service; United Publishers' Services; Universal Press Syndicate; Universities and Public Schools (Battalions); University of Puget Sound

UPSA Ukrainian Political Science Association

UPSEB Upper Pradesh State Electricity Board

UPSG universal polar stereographic grid

UPSM University of Pennsylvania School of Medicine

UPSS United Postal Stationery Society

UPSTC Upper Pradesh State Textile Corporation

UPSTEP Undergraduate Pre-Service Teacher Education Program

Up Swn Upper Swan

up tor upper torso

up tr up train

UPU United Prisoners Union; Universal Postal Union

UPU Unión Postal Universal (Spanish—Universal Postal Union)

Up V Upper Volta

UPV Ulster Protestant Volunteers (paramilitary counterpart of the IRA)

upvc unplasticized polyvinyl chloride

UPW Union of Postal Service Workers

UPWA United Public Workers of America

UPWIU United Paper Workers International Union

uq upper quartile

UQ University of Queensland

U of Q University of Québec; University of Queensland

UQP University of Queensland Press

u quark up quark

ur ultrared; unconditioned response, up right (stage direction); upper right; urinal; urinary; urine; utility rectifier

ur (UR) unemployment rate

u/r upper right

ur ouron (Greek—urine)— urea, uremia, ureter, urethra, urine, urology

Ur Urania; uranium; Uranus; Urdu; Uruguay; Uruguayan

UR Uganda Railway; Uganda Rifles; Uniform Regulations, Unsatisfactory Report; Urban Renewal; Utilization Review

U of R University of Reading; University of Redlands; University of Richmond; University of Rochester

UR Universidad de la República (Spanish—University of Uruguay)

URA United Republicans of America; Universities Research Association, Urban Redevelopment Authority; Urban Renewal Administration

urad your radio (message)

u-rail U-shaped rail

Urals Ural Mountains dividing Asia from Europe

URAMEX Mexican-government's uranium company

uran (Latin—tail)—anuran, urochordate

Uran Uranus

ur anal. urine analysis

U of Rang University of Rangoon

uranog uranographer; uranographic; uranography

urb urban; urbanism; urbanist; urbanistic; urbanite; urbanization; urbanize; urbicultural; urbiculture

Urb Urbanización (Spanish— Urbanization)

Urban Inst Urban Institute

Urbank Urban Bank (National Development Bank)

urbanol urbanologic(al); urbanologist; urbanology

urb guer(s) urban guerilla(s)

urbm (URBM) ultimate-range ballistic missile

urbol urbanologist; urbanology

urb ter urban terrorism; urban terrorist(s)

urc upper right center

URC United Republic of Cameroon; Universal Resources Corporation; Urban Renewal Commission; Urban Research Center (NYU)

urclk universal receiver clock

URCLPWA United Rubber, Cork, Linoleum, and Plastic Workers of America

urd upper disease (head cold)

Urd Urdu (literary language of Pakistan)

Ur$ Uruguayan peso

URD Unión Republicana Democrática (Spanish— Democratic Republican Union)—political party in Venezuela

urdis your dispatch

ure unintentional radiation exploitation

URE Undergraduate Record Examination

URESA Uniform Reciprocal Enforcement of the Support Act (for the collection and enforcement of child support)

uret urethra(l)

urf (URF) uterine-relaxing factor

URF Union des Services Routiers des Chemins de Fer Européens (French—Union of European Railways Route Services)

urg urgent

Urga Ulaanbaatar, Mongolia

uri upper respiratory illness (head cold)

Uri not an abbreviation but a Swiss canton

URI Union Research Institute (Hong Kong); University of Rhode Island

U of RI University of Rhode Island

uria (Latin prefix—urine)— urinal, urinalysis; (Latin suffix—urine)—polyuria

URIMA University Risk and Insurance Managers Association

urinalysis urine analysis

URISA Urban and Regional Information System Association

url (URL) uniform resource locator; user requirements language

URL Underground Research Laboratory (Canadian); Unilever Research Laboratory; University of Rhodesia Library (Salisbury)

urltr your letter

urmgm your mailgram

urmsg your message

urn. ultra-high radio navigation; urnal, urnary, urned, urnement(al)(ly), urnful

Urn Urnest (Ernest)

uro urological; urology

uro (Latin prefix—urine)—urinal, urinary

URO United Restitution Organization

urodyn urodynamic(s)

urogen urogenital

urol urologic(al)(ly); urologist; urology

U-room U-boat room (petty officer's quarters)

uro or uran (Greek-derived prefix or suffix from *oura* meaning tail)—anuran or urochordate

urp unique radiolytic product

URP Unit Reporting Program; United Revolutionary Party

Urq Urquhart

urr (URR) ultra-rapid reader (computer program)

URR Union for the Resurrection of Russia

URRVS Urban Rapid-Rail-Vehicle Systems

urs unit reference sheet

URs Unsatisfactory Reports; University Rationalists

UR's Unsatisfactory Reports

URS Universal Reporting System; Universal Reference System

Ursa (Latin—Bear constellations, major and minor)

urser your serial (number or reference)

URSI *Union Radio Scientifique Internationale* (French—International Scientific Radio Union)

urspr *ursprünglich* (German—originally)

URSS *União das Repúblicas Socialistas Soviéticas* (Portuguese—Union of Socialist Soviet Republics)—the former USSR; *Unión de Repúblicas Socialistas Soviéticas* (Spanish—Union of Soviet Socialist Republics); *Union des Republiques Socialistes Soviétiques* (French—Union of Socialist Soviet Republics)- the former USSR

urt upper respiratory tract; utility radio transmitter

URT United Republic of Tanzania (Tanganyika and Zanzibar)

urtel your telegram

urti upper respiratory tract infection (common cold; influenza)

URTI *Université Radiophonique-Télévisuelle Internationale* (French—International Radio-Television University)

URTU United Road Transport Union

Uru Uruguay; Uruguayan

Uruguay Oriental Republic of Uruguay, *República Oriental del Uruguay*

urv underseas research vehicle

URWA United Rubber Workers of America

us underspeed

us. ultrasound; under seal; undersize; uniform sales; united service

us. (US) unconditioned stimulus

u-s upper-stage

u/s unserviceable

u.s. *ubi supra* (Latin—where mentioned above); *ut supra* (Latin—as above)

US Uncle Sam; United States; University of Stellenbosch

U.S. United States

U of S University of Salford; University of Saskatchewan; University of Scranton; University of Sheffield; University of Sherbrooke; University of the South (Sewanee, Tennessee); University of Southampton; University of Stirling; University of Strathclyde; University of Sudbury; University of Surrey; University of Sussex; University of Swansea; University of Sydney

U.S. Ufficio Stampa (Italian—Press Agency)

USA Underwriters Service Association; Union of South Africa; United States of America (more correctly U.S.A.); United States Army; United States Attorney; United Steelworkers of America; United Swaziland Association; United Switzerland Association; Universal Savings Account; University of South Africa

US of A United Steelworkers of America; United Synagogue of America

U.S.A. United States of America

U.S. of A. United States of America; United Secularists of America

U of SA University of South Africa

USA *Unser Shtickel Arbeit* (Yiddish—Our Bit of Work)—rifle grenade produced in Palestine

U.S.A. (title of trilogy by John Dos Passos—*42nd Parallel, 1919, The Big Money*—describing first three decades of American life in the twentieth century)

USAA United Services Automobile Association

USAAA US Army Audit Agency

USAABMDA United States Army Advance Ballistic Missile Defense Agency

USAAC United States Army Air Corps (now USAF)

USAACDA United States Army Aviation Combat Development Agency

USAAD US Army Airmobile Division

USAADC United States Army Air Defense Center

USAADEA US Army Air Defense Engineering Agency

USAAF United States Army Air Force

USAAFINO United States Army Aviation Flight Information and Navigation Aids Office

USAAFO US Army Avionics Field Office

USAAMR & DL United States Army Air Mobility Research and Development Laboratory

USAAPSA United States Army Ammunition Procurement and Supply Agency

USAASD United States Army Aeronautical Service Detachment

USAASO United States Army Aeronautical Services Office

USAAVNC United States Army Aviation Center

USAAVNS United States Army Aviation School

USAAVSCOM United States Army Aviation Systems Command

USAB United States Activities Board; United States Army Berlin

USABAAR United States Army Board for Aviation Accident Research

USABC United States Advanced Battery Consortium

USABRL US Army Ballistic Research Laboratories

USAC United States Aircraft Carriers (air cargo line); United States Auto Club; US Air Conditioning Corporation

USA/Can Inst. United States of America/Canada Institute

USACAA United States Army Concepts Analysis Agency

USA CAC United States Army Continental Army Command

USACC United States Army Communications Command; U.S.-Arab Chamber of Commerce

USACCS United States Army Chemical Center and School

USACDA United States Arms Control and Disarmament Agency; United States Army Catalog Data Agency

USACDC US Army Combat Developments Command

USACDCCA United States Army Combat Development Command Combined Arms Agency

USACDCEC United States Army Combat Development Command Experimentation Command

USACDCFAA United States Army Combat Developments Command Field Artillery Agency

USACDCNG United States Army Combat Developments Command Nuclear Group

USACDCOA United States Army Combat Developments Command Ordnance Agency

USACDCQA United States Army Combat Developments Command Quartermaster Agency

USACDCSWCAG United States Army Combat Developments Command Special Warfare and Civil Affairs Group

USACE US Army Corps of Engineers

USACENDCDSA United States Army Corps of Engineers National Civil Defense Computer Support Agency

USACIC United States Army Criminal Investigation Command

USACIDC United States Army Criminal Investigative Command

USACMA United States Army Club Management Agency

USACMR United States Army Court of Military Review

USACPEB United States Army Central Physical Evaluation Board

USACRR United States Army Crime Records Repository

USACSA US Army Combat Surveillance Agency

USACSLA United States Army Communications Security Logistics Agency

USACSSEA United States Army Computer Systems Support and Evaluation Agency

USAD US Army Dispensary

USADIP United States Army Deserter Information Point

USADSC US Army Data Services and Administrative Systems Command

USAE United States Army Engineer(s); United States Army, Europe

USAEC United States Army Engineer Command; United States Atomic Energy Commission; US Army Electronics Command

USAECA United States Army Engineer Construction Agency

USAECBDE United States Army Engineer Center Brigade

USAECLRA United States Army Electronics Command Logistics Research Agency

USAED United States Army Engineer Division

USAEDC United States Army Engineer Division—Caribbean

USAEDH United States Army Engineer Division—Huntsville, Alabama

USAEDLMV United States Army Engineer Division—Lower Mississippi Valley

USAEDM United States Army Engineer Division—Mediterranean

USAEDMR United States Army Engineer Division—Missouri River

USAEDNA United States Army Engineer Division—North Atlantic

USAEDNC United States Army Engineer Division—North Central

USAEDNE United States Army Engineer Division—New England

USAEDNP United States Army Engineer Division—North Pacific

USAEDOR United States Army Engineer Division—Ohio River

USAEDPO United States Army Engineer Division—Pacific Ocean

USAEDSA United States Army Engineer Division—South Atlantic

USAEDSP United States Army Engineer Division—South Pacific

USAEDSW United States Army Engineer Division—Southwest

USAEEA United States Army Enlistment Eligibility Activity

USAEL US Army Electronic Laboratories

USAEMA US Army Electronics Materiel Support Agency

USAEMCA United States Army Engineer Mathematical Computation Agency

USAEMSA United States Army Electronics Materiel Support Agency

USAENGCOM United States Army Engineer Command

USAENPG United States Army Engineer Power Group

USAEPG US Army Electronic Proving Ground

USAERA United States Army Electronic Command Research Agency

USAERC United States Army Enlisted Records Center

USAERDAA United States Army Electronics Research and Development Activity (Fort Huachuca, Arizona)

USAERDL US Army Electronics Research and Development Laboratory

USAERG United States Army Engineer Reactor Group

USAES United States Association of Evening Students

USAETDC U.S. Army Engineer Topographic Data Center (D.C.)

USAEUR United States Army Europe

USAEVD United States Alliance for the Eradication of Venereal Disease

U S Af Union of South Africa

USAF United States Air Force; United States Armed Forces

USAFA US Air Force Academy

USAFABD United States Army Field Artillery Board

USAFAC United States Army Finance and Accounting Center

USAFACS US Air Force Aircrew School

USAFAGOS US Air Force Air Ground Operations School

USAFAPS US Air Force Air Police School

USAFAS United States Army Field Artillery School

USAFB United States Army Field Bank

USAFBMS US Air Force Basic Military School

USAFBS US Air Force Bandsman School

USAFC United States Army Forces Command

USAFD *United States Air Force Dictionary*

USAFE US Air Forces in Europe

USAFECI United States Air Force Extension Course Institute

USAFESA United States Army Facilities Engineering Support Agency

USAFEURPCR United States Air Force European Postal and Courier Region

USAFFE United States Army Forces Far East

USAFIA United States Armed Forces in Australia

USAFS United States Army Field Station

USAFFGS US Air Force Flexible Gunnery School

USAFFSR US Air Force Flight Safety Research

USAFHRC US Air Force Historical Research Center

USAFI United States Armed Forces Institute

USAFIGED United States Armed Forces Institute Tests of General Educational Development

USAFIT US Air Force Institute of Technology

US AFLANT US Air Force, Atlantic

USAFMPCR United States Air Force Mideast Postal and Courier Region

USAFNS US Air Force Navigation School

USAFO United States Army Field Office

USAFOCS US Air Force Officer Candidate School

USAFOF United States Army Flight Operations Facility

USAFPACPCR United States Air Force Pacific Postal and Courier Region

USAFPS US Air Force Pilot School

USAFSAAS United States Air Force School of Applied Aerospace Sciences

USAFSAB US Air Force Scientific Advisory Board

USAFSACS United States Air Force School of Applied Cryptologic Sciences

USAFSAM US Air Force School of Aerospace Medicine

USAFSAWC US Air Force Special Air Warfare Center

USAFSC US Air Force Systems Command; United States Army Food Service Center

USAFSE US Air Force Supervisory Examination

USAFSG United States Air Field Support Group

USAFSO US Air Forces, Southern Command

USAFSOC United States Air Force Special Operations Center

USAFSOF United States Air Force Special Operations Force

USAFSOS United States Air Force Special Operations School

USAFSS US Air Force Security Service

USAFSTC United States Army Foreign Science and Technology Center

USAFSTDS US Army-Air Force Standards

USAFSTRIKE US Air Force Strike Command

USAFTS US Air Force Technical School

USAGETA United States Army General Equipment Test Activity

USAGMPC United States Army General Materiel and Parts Center

USAH United States Army Hospital

USAHAC United States Army Headquarters Area Command

USAHC United States Army Health Clinic

USAHSC United States Army Health Services Command

USAHSDSA United States Army Health Services Data Systems Agency

USAIA United States Army Institute of Administration

USAIC US Army Infantry Center; US Army Intelligence Corps

USAICA US Army Interagency Communications Agency

USAICS United States Army Intelligence Center and School

USAID United States Aid for International Development

USAIG United States Aircraft Insurance Group

USAIIA United States Army Imagery Interpretation Agency

USAIIG United States Army Imagery Interpretation Group

USAILG United States Army International Logistics Group

USAIMS United States Army Institute for Military Systems

USAINTA United States Army Intelligence Agency

USAINTS US Army Intelligence School

USAIPSG US Army Industrial and Personnel Security Group

USAir formerly Allegheny Airlines

USAirA United States Air Attaché

USAIRE United States of America Aerospace Industries Representatives in Europe

USAir MilComUN United States Air Force Representative, UN Military Staff Committee

USAIS United States Army Infantry School; United States Army Intelligence School

USAISC United States Army Intelligence and Security Command

USAJ United States Army, Japan

USAJPG United States Army Jefferson Proving Ground

USAK United School Administrators of Kansas

USALC United States Army Logistics Center

USALEA United States Army Logistics Evaluation Agency

USALSA United States Army Legal Services Agency

USAMAA United States Army Memorial Affairs Agency

USAMC United States Army Material Command; United States Army Medical Corps; United States Army Missile Command

USAMRIID United States Army Medical Research Institute of Infectious Diseases

USAMBRDL United States Army Medical Bioengineering Research and Development Laboratory

USAMCFG United States Army Medical Center—Fort Gordon

USAMC-ITC United States Army Materiel Command—Interim Training Center

USAMDRC United States Army Materiel Development and Readiness Command

USAMDW United States Army Military District of Washington

USAMEDCOM United States Army Medical Command

USAMEOS United States Army Medical Equipment and Optical School

USAMFSS United States Army Medical Field Service School

USAMIDA United States Army Major Item Data Agency

USAMIIA United States Army Medical Intelligence and Information Agency

USAML United States Army Medical Laboratory

USAMMA United States Army Medical Materiel Agency

USAN United States Adopted Name

USA NC United States Army Nurse Corps

USAO United States Assay Office

USAPA United States Army Procurement Agency

USAPACDA United States Army Personnel and Administration Combat Development Activity

USAPDC United States Army Petroleum Distribution Command

USAPEB United States Army Physical Evaluation Board

USAPEQUA United States Army Production Equipment Agency

USAPHC United States Army Primary Helicopter Center

USAPIA United States Army Personnel Information Activity

USAPO United States Antarctic Projects Office

USAPRO United States Army Personnel Research Office

USAR United States Army Reserve

USARA United States Army Reserve Affairs

USARADCEN United States Army Air Defense Center

USARADCOM United States Army Air Defense Center; United States Army Air Defense Command

USARAE United States Army Reserve Affairs—Europe

USARAL United States Army, Alaska

USARB United States Army Retraining Brigade

USARC United States Army Recruiting Command

USARCS United States Army Claims Service

USAREC United States Army Recruiting Command

USAREUR United States Army, Europe

USARIBSS United States Army Research Institute for the Behavioral and Social Sciences

USARIEM United States Army Research Institute of Environmental Medicine

USARJ United States Army, Japan

USARP United States Antarctic Research Program

USARPA United States Army Radio Propagation Agency

USARPAC United States Army, Pacific

USARPACINTS United States Army Pacific Intelligence School

USARSA United States Amateur Roller Skating Association

USARSC United States of America Roller Skating Confederation

USARSO United States Army, Southern Command

usart universal synchronous-asynchronous receiver-transmitter

USARV United States Army, Vietnam

USAS United States Army South; United States of America Standard

US ASA United States Army School of the Americas; United States Army Security Agency

USASACDA United States Army Security Agency Combat Development Activity

USASADEA United States Army Signal Air Defense Engineering Agency

USASAE United States Army Security Agency—Europe

USASAFO United States Army Signal Avionics Field Office

USASATCOMA United States Army Satellite Communications Agency

USASC United States Army, Southern Command—Caribbean; United States Army Support Center

USASCAF United States Army Service Center for Army Forces

USASCC United States Army Strategic Communications Command

USASCII USA Standard Code for Information Interchange (data processing)

USASCSA United States Army Signal Communications Security Agency

USASG United States Army Standardization Group

USASI United States of America Standards Institute

USA Sig C United States Army Signal Corps

USASMC United States Army Supply and Maintenance Command

USASMSA United States Army Signal Corps Material Support Agency

USASRDL United States Army Signal Research and Development Laboratory

USASSA United States Army Signal Supply Agency

USASSG United States Army Special Security Group

USAT Ultra-Small-Aperture Terminal; United States Army Transport

USATA United States Army Transportation Aviation

USATC United States Army Training Center; United States Army Traffic Command

USATDC United States Army Training and Doctrine Command

USATEA United States Army Transportation Engineering Agency

USATEC United States Army Test and Evaluation Command

USATECOM United States Army Test and Evaluation Command

USATIA United States Army Transportation Intelligence Agency

USATISU United States Army Troop Information Support Unit

USATL United States Army Technical Library

USATMACE United States Army Traffic Management Agency—Central Europe

USA TopoCom United States Army Topographic Command

USATRATCOM United States Army Strategic Communications Command

USATSC United States Army Terrestrial Sciences Center

USATTC United States Army Tropic Test Center

USATTU United States Army Transportation Terminal Unit

USAU United States Aviation Underwriters

usaw (USAW) underwater security advance warning

USAWC United States Army War College; United States Army Weapons Command

USAWES United States Army Waterways Experiment Station

USAWF United States Amateur Wrestling Society

usb unified S-band; universal serial bus; upper sideband

USB United States Borax (company)

USBA United States Billiard Association; United States Boomerang Association; United States Brewers Association

USBC United States Bureau of the Census; U.S. Bancorp; United States Bureau of Customs

USB & C United States Borax and Chemical (company)

USBCSC United Society of Believers in Christ's Second Coming (Shakers)

USBE Universal Serials and Book Exchange (formerly United States Book Exchange)

USBG United States Botanic Garden

USBGN United States Board on Geographic Names

USBH United States Bureau Highways

USBIS United States Border Inspection Station

USBLS United States Bureau of Labor Statistics

USBM United States Bureau of Mines

USBP United States Board of Paroles; United States Border Patrol; United States Bureau of Prisons

USBPA United States Bicycle Polo Association

USBPR United States Bureau of Public Roads

USBS United States Border Station; United States Bureau of Standards

USBTA United States Board of Tax Appeals

USBuStand United States Bureau of Standards

usc under separate cover; upper stage center

USC Underwater Systems Center (Groton, Conn.); United Services Club; United Shipping Company; United States Code; United States Congress; United States of Colombia; United Steamship Company; University of South Carolina; University of Southern California

USC United States Catalog; United States Code (legal)

USCA Ulster Special Constabulary Association; United States Copper Association; United States Courts of Appeals

USCA United States Code Annotated

USCAC US Continental Army Command

USCANS Unified S-band Communication and Navigation System

USCB United States Customs Bonded

USCC United States Catholic Conference; United States Chamber of Commerce; United States Circuit Court; United States Commercial Company; United States Customs Court

USCCA United States Circuit Court of Appeals

USCCAN United States Code-Congressional and Administrative News

USCCPA United States Court of Customs and Patent Appeals

USCCR United States Commission on Civil Rights

USCE US Coast Guard Reserve; US Commissioner of Education

USCEA United States Council for Energy Awareness

USCF United States Chess Federation; United States Churchill Foundation

USCG United States Coast Guard

USCGA US Coast Guard Academy

USCGAD United States Coast Guard Air Detachment

USCGAS United States Coast Guard Air Station

USCG Aux United States Coast Guard Auxiliary

USCGC United States Coast Guard Cutter

USCGI United States Coast Guard Institute

USCGMSC United States Coast Guard Marine Safety Council

USC & GS United States Coast and Geodetic Survey

USCHS United States Capitol Historical Society; United States Catholic Historical Society

USCI United Satellite Communication Inc

USCIB United States Council on International Banking; United States Council for International Business

USCIIC United States Civilian Internee Information Center (USA)

USCINCEUR United States Commander-in-Chief, Europe

USCINCSO United States Commander-in-Chief, Southern Command

USCM United States Conference of Mayors

USCMA United States Cheese Makers Association; United States Coal Mines Administration; United States Court of Military Appeals

USCMI United States Commission of Mathematical Instruction

usco underwriters salvage company

USCO Union Steel Corporation (South Africa)

US Comm UNICEF United States Committee for UNICEF

USCONARC US Continental Army Command

US Const Constitution of the United States

USCP University of South Carolina Press; University of Southern California Press; U.S. Capitol Police (DC)

USCP United States Coast Pilot

USCPFA United States China People's Friendship Association

USCPSC US Consumer Product Safety Commission

USCR United States Committee for Refugees

USCRC United States Civil Rights Commission

USCRS United States Cotton Research Station

USCS United States Civil Service; United States Claims Service; United States Conciliation Service; United States Customs Service; Universal Ship Cancellation Society

USCSC United States Civil Service Commission

USCSup United States Code Supplement

USCT United States Colored Troops (1862–1865)

USCUN United States Committee for the United Nations

USCUNICEF United States Committee for UNICEF

USCWF United States Council for World Freedom

USCWHO United States Committee for the World Health Organization

US Cy United States currency

usd ultimate strength design; uninhibited sexual desire

US $ American dollar(s); United States dollar

USD Unified School District; University of San Diego; University of South Dakota

USD United States Dispensatory

USDA United States Department of Agriculture; United States Disarmament Agency

USDA/CRIS US Department of Agriculture/Current Research Information System

USDB United States Disciplinary Barracks

USDC United States Department of Commerce; United States District of Columbia; United States District Court

USDCFO US Defense Communication Field Office

USDEA United States Drug Enforcement Administration

USDF United States Dressage Federation

USDHEW United States Department of Health, Education, and Welfare (HEW)

USDHUD United States Department of Housing and Urban Development

USDI United States Department of the Interior

USDJ United States District Judge

USDL United States Department of Labor

USDLGI United States Defense Liaison Group—Indonesia

USDOCO United States Document Officer

USDoD United States Department of Defense

USDP University of San Diego Press; University of South Dakota Press

USDR United States Divorce Reform

USDSA United States Deaf Skiers Association

USDSEA United States Dependent School European Area

USDT United States Department of Transportation

USE United States English. United States Envelope (corporation); Univac Scientific Exchange; U.S. English

usea undersea

u/Sec Under Secretary

USELMCENTO United States Element Central Treaty Organization

USEP United States Escapee Program

USERC United States Environment and Resources Council

userid user identification

USES United States Employment Service

USEUCOM United States European Command

usf und so fort (German—et cetera)—and so forth

USF United States Forces

U of SF University of South Florida

USFA United States Fire Administration; United States Food Administration (World War I); United States Forces in Austria (World War II)

USFAA United States Fronton Athletic Association

USFC United States Foil Company

USFET United States Forces—European Theater

USFF United States Flag Foundation

USF & G United States Fidelity-Guaranty (insurance underwriters)

USFGC United States Feed Grains Council

USFHA United States Field Hockey Association

USFIS United States Foundation for International Scouting

USFJ United States Forces, Japan

USFL United States Football League

USForAz United States Forces in the Azores

USFORDOMREP United States Forces, Dominican Republic

USFPL United States Forest Products Laboratory

USfs United States frequency standard

USFS United States Foreign Service; United States Forest Service

USFSA United States Figure Skating Association

USFWS United States Fish and Wildlife Service

USG Ulysses Simpson Grant (18th President U.S.); Uniform Selection Guidelines; United States Government; United States Gypsum (company)

U.S.G. United States Government (railroad)

USGA United States Golf Association

US gal United States gallon

USGC United States Gold Commission

USGF United States Gymnastics Federation

USGLI United States Government Life Insurance

USGM United States Government Manual

USGOM United States Government Organization Manual

USGPO United States Government Printing Office

USGRDR United States Government Research and Development Report(s)

USGRR United States Government Research Reports

USGRS United States Graves Registration Service

USGS United States Geological Survey

USGSMMS United States Geological Survey and Minerals Management Service

usgw underwater-to-surface guided weapon

ush usher

Ush Ugandan shilling(s)

USHA United States Handball Association

USHCC U.S. Hispanic Chamber of Commerce

USHDA United States Highland Dancing Association

U of Sherb University of Sherbrooke

USHGA United States Hang Gliding Association

USHHFA United States Housing and Home Finance Agency

USHL United States Hygienic Laboratory

USHMC U.S. Holocaust Memorial Council

USHR United States Highway Research

USHS United States Historical Society; United States Hospital Ship

USHWA United States Harness Writers Association

USI United States of Indonesia; United States Industries

USIA United States Information Agency

USIan United States Statesian

USIAS Union Syndicale des Industries Aéeronautiques et Spatiales (French—Aeronautic and Space Industry Union)

USIB United States Intelligence Board

USIBR United States Institute of Behavioral Research

usic undersea instrument chamber

USIC United States Industrial Chemicals; United States Industrial Council; United States Instrument Corporation

USICA United States International Communication Agency

USIF United States Investment Fund

USIH United States Indian Health Service

USILA United States Intercollegiate Lacrosse Association

USI & NS United States Immigration and Naturalization Service

USIOSLCC United States Inter-Oceanic Sea-Level Canal Commission

USIP University of Stockholm Institute of Physics

USIS United States Information Service

USISL United States Information Service Library

USISS United States Institute of Space Studies

usit unit share investment trust

USITA United States Independent Telephone Association

USITC United States International Trade Commission

USITT United States Institute for Theater Technology

USIU United States International University

USJ United States Jaycees

USJC United States Job Corps

USJCC United States Junior Chamber of Commerce

USJF United States Judo Federation

USJPRS United States Joint Publications Research Service

USL Union Steamships Limited; United States Legation; United States Lines; University of Singapore Library; University of Sydney Library

U-slag upperclass slang

USLant United States Atlantic Subarea

USLANTCOM United States Atlantic Command

USLICASC United Services Life Insurance Company Annuity Service Center

USLICO United Services Life Insurance Company

USLO United States Liaison Office(r); University Students for Law and Order

USLP U.S. Labor Party

USLSA United States League of Savings Associations; United States Livestock Sanitary Association

USLSI United States League of Savings Institutions; U.S. League of Savings Institutions

USLTA United States Lawn Tennis Association

USLU United States Lines (container) Line

usm (USM) underwater-to-surface missile

USM United Shoe Machinery; United States Mail (U.S.M.); United States Marines; United States Marshall; United States Mint; University of Southern Mississippi; Unlisted Securities Market

USMA United States Maritime Administration; United States Metric Association; United

States Military Academy (West Point); Utah State Medical Association

USMACTHAI United States Military Assistance Command, Thailand

USMACV United States Military Assistance Command, Vietnam

USMB United States Metric Board

USMBPHA United States-Mexico Border Public Health Association (of American and Mexican public health officials)

USMC United States Marine Corps; United States Maritime Commission; United States Microfilm Corporation (company)

USMCCCA United States Marine Corps Combat Correspondents Association

USMCR United States Marine Corps Reserves

USMD United States Medical Doctor

USMeMilComUN United States Military Members, UN Military Staff Committee

USMH United States Marine Hospital

USMICC United States Military Information Control Committee

USMilComUN United States Delegation, UN Military Staff Committee

USMilLias United States Military Liaison Office

USAMILTAG United States Military Technical Advisory Group

USML U.S. Marxist-Leninists (left-wing youth party)

USMM United States Merchant Marine

USMMA United States Merchant Marine Academy

USMMCC United States Merchant Marine Cadet Corps

USMO United States Marshal's Office

USMS United States Maritime Service; United States Marshals Service

USMSMI United States Military Supply Mission to India

USMSPB United States Merit Systems Protection Board

USMUN United States Mission to the United Nations

usn ultrasonic nebulizer

Usn Ulsan

USN United States Navy

USNA United States Naval Academy; United States Naval Archives

USNAM United States Naval Academy Museum

USNARS United States National Archives and Records Service

USNAS United States Naval Amphibious School

USNAVEUR United States Navy, Europe

USNB United States National Bank; United States Naval Base

USNC United States National Committee; United States Navigation Company (North German Lloyd—Hamburg-American Line); United States Nuclear Corporation

USNCB United States National Central Bureau (Interpol); United States Naval Construction battalion (Seabees)

USNCC United States Naval Correction Center

USNCCC United States National Council of Churches of Christ

USND United States Navy Department

USNDCS United States National Drug Control Strategy

USNDRC United States Navy Drug Rehabilitation Center

USNEC United States National Earthquake Center, Golden, Colorado

USNEL United States Naval Electronics Laboratory

U.S. News U.S. News and World Report

USNFEC United States National Fruit Export Council

USNG United States National Guard

USNH United States Naval Harbor; United States Naval Hospital; United States North of Hatteras

USNHO United States Navy Hydrographic Office

USNI United States Naval Institute

USNIAAA United States National Institute on Alcohol Abuse and Alcoholism

USNII United States National Indian Institute

USNIS United States Naval Investigative Service

USNJA United States National Jogging Association

USNL United States Navy League

USNLM United States National Library of Medicine

USNM United States National Museum (Smithsonian Institution); United States Naval Museum; United States Navy Memorial

USNMR United States National Military Representative

USNO United States Naval Observatory

USNOO United States Naval Oceanographic Office

USNPC United States Naval Photographic Center

USNPS United States Naval Postgraduate School

USNR United States Naval Reserve

USNRC United States Nuclear Regulatory Commission

USNRDL United States Naval Radiological Defense Laboratory, United States Navy Research and Development Laboratory

USNS United States Naval Ship (Military Sea Transport Service); United States Nuclear Ship

USNSA United States National Student Association; United States Naval Sailing Association

USNSMC United States Naval Submarine Medical Center

USNTAF United States Navy Training Aids Facility

USNTS United States Naval Torpedo Station

USNUSL United States Navy Underwater Sound Laboratory

USNWC United States Naval War College

USNWR Union Slough National Wildlife Refuge (Iowa); Upper Souris NWR (North Dakota)

USN & WR U.S. News & World Report

uso unmanned seismological observatory

USO United Service Organizations; Utah Symphony Orchestra

U-soc upperclass society

USOC United States Olympic Committee

USOE United States Office of Education

USOEO United States Office of Economic Opportunity

USofAF Under Secretary of the Air Force

USOICP United States Oil Import Control Program

USOID United States Oversea Internal Defense (USA)

USOM United States Operations Mission

usp unique selling proposition

USP U.S. Penitentiary (Atlanta, Georgia; Leavenworth, Kansas; Lewisburg, Pennsylvania; Marion, Illinois; McNeil Island, Washington; Terre Haute, Indiana among others); United States Plywood (company); University of the South Pacific (Fiji)

USP United States Pharmacopeia

USPA United States Philatelic Agency; United States Pilots Association; United States Polo Association

USPACAF United States Pacific Air Forces

US Pat United States Patent

USPB United States Parole Board

USPC Ulster Society for the Preservation of the Countryside; United States Parole Commission; United States Peace Corps; United States Pharmacopeial Convention

USPCA United States Police Canine Association

USPCS US Philatelic Classics Society

USPDO United States Property and Disbursing Office(r)

U-speech upperclass speech

USP & F United States Pipe and Foundry (company)

USPFO United States Property and Fiscal Officer

USPG United Society for the Propagation of the Gospel

US Phar United States Pharmacopeia

USPHS United States Public Health Service

USPHSC United States Public Health Service Clinic

USPHSH United States Public Health Service Hospital

USPIG United States Public Interest Group

USPIS United States Postal Inspection Service

USPLS United States Public Land Surveys

USPO (U.S.P.O.) United States Post Office

USPP United States Probation and Parole

USPPS US Possessions Philatelic Society

USPQ United States Patents Quarterly

U.S. Pros United States Prostitutes collective

USPs United States Penitentiaries

USPS United States Postal Service; United States Power Squadron

USPSTF United States Preventative Services Task Force

USPTO U.S. Patent and Trademark Office

USPUN United States People for the United Nations

USPWIC United States Prisoner of War Information Center

usque. usquebaugh (Old Scottish—whisky)

usr underwater search and recovery; unheated serum reagin

USR United States Reserves; United States Rubber

USR United States Supreme Court Reports

USRA United States Racquetball Association; United States Railway Association; United States Revolver Association; Universities Space Research Association

USRB United States Renegotiation Board

USRD Underwater Sound Reference Division (USN)

USRDA United States Recommended Daily Allowance

USREDCOM United States Readiness Command

USRepMilComUN United States Representative, UN Military Staff Committee

usr/grp user groups for UNIX operating systems

USRL Underwater Sound Reference Laboratory

USRS United States Rocket Society

USRS United States Revised Statutes

usrt universal synchronous receiver/transmitter

USS Under-Secretary of State; Underway Saturdays and Sundays (destroyer crews' definition of USS); Union Switch and Signal; United Scholarship Service; United States Senate; United States Ship (U.S.S.); United States Shoe (company); United States Standard; United States Steel (company); United States Swimming

US & S Union Switch and Signal

U of SS University of the Seven Seas (Chapman College)

USS Union Syndicale Suisse (French—Swiss Federation of Trade Unions); *Union Syndicate Suisse* (French—Swiss Trade Union Syndicate)

USSA United States Salvage Association; United States Ski Association; United States Student Association

USSAF United States Strategic Air Force

USSB United States Savings Bond(s); United States Shipping Board (World War I)

USSBD United States Savings Bonds Division

ussc upper sideband suppressed carrier

USSC United States Secret Service; United States Sentencing Commission; United States Space Command, Paterson Air Force Base, Colorado Springs, Colorado; United States Strike Command; United States Supreme Court

USSCC United States Senate Computer Center

USS Co Ulster Steam Ship Company; Union Steam Ship Company (New Zealand)

USSCS United States Soil Conservation Service

USSDP Uniformed Services Savings Deposit Program

USSEI United States Society of Esperanto Instructors

USSF Ulster Special Service Force; United States Science Foundation; United States Soccer Federation; United States Steel Foundation; United States Special Forces (Green Berets)

USSFA United States Soccer Football Association

USSFC United States Synthetic Fuel Corporation

USSG United States Standard Gauge

USSIC United States Sex Information Council

USS & LL United States Savings & Loan League

US Soc Fed United States Soccer Federation

USSOUTHCOM United States Southern Command

USSPA United States Student Press Association

USSR Union of Soviet Socialist Republics (the former Soviet Union)

USSRA United States Squash Rackets Association

USSS United States Secret Service; United States Steamship

USSSA United States Social Security Administration

USSSM United States Sinai Support Mission

USSST United States Salt Spray Test(ing)

USSTA United States Sail Training Association

USSTRICOM United States Strike Command

USSTS United States Student Travel Service

ust. ustus (Latin—burnt)

UST undersea technology; United States Treaties; University of Santo Tomás (Manila)

UST UnderSea Technology; The Magazine of Oceanography, Marine Sciences, and Underwater Defense

U of St A University of St Andrews

USTA United States Tennis Association; United States Trademark Association; United States Trotting Association

USTC United States Tariff Commission; United States Tax Court; United States Testing Company

USTCRDWWA United Slate, Tile, and Composition Roofers, Damp, and Waterproof Workers Association

USTC & TBA U.S. Tennis Court and Track Builders Association

ustd underspeed time delay

USTD United States Transportation Department

USTDC United States Taiwan Defense Command

USTEMC United States Territorial Expansion Memorial Commission

USTES United States Training and Employment Service

USTF United States Tuna Foundation

USTFF United States Track and Field Federation

USTIS Ubiquitous Scientific and Technical Information System

USTMA United States Trade Mark Association

USTOA United States Tour Operators Association

ustol ultra short takeoff and landing

USTR United States Trade Representative

USTS United States Travel Service

USTTA United States Table Tennis Association; United States Travel and Tourism Administration

usu usual; usually

Usu Usulután (province of El Salvador)

USU Uniformed Services University; Utah State University

usuf usufruct(uary)

USUHS Uniformed Services University of the Health Sciences

USUN United States Mission to the United Nations

usurp. usurpandus (Latin—to be used)

USV United States Volunteers

USVA United States Veterans Administration; United States Volleyball Association

USVB United States Veterans Bureau (former name of the Veterans Administration)

USVH United States Veterans Hospital

USVI United States Virgin Islands (St Croix, St John, St Thomas)

USVIDT United States Virgin Islands Division of Tourism

USVMS Urine Sample Volume Measurement System

usw ultra short wave; underwater submarine warfare

usw und so weiter (German—and so forth)

USW United Show Workers

USWA United Steel Workers of America

USWAC United States Women's Army Corps

USWACC United States Women's Army Corps Center

USWACS United States Women's Army Corps School

USWB United States Weather Bureau

USWD Undersurface Warfare Division

USWI United States West Indies (Virgin Islands—St Thomas, St John, St Croix, and smaller islands)

USWLA United States Women's Lacrosse Association

USWLS United States Wild Life Service

USWP Ultra-Short-Wave Propagation Panel

USWPA United States Works Progress Administration

USWV United Spanish War Veterans

USX United States Steel and Marathon Oil, Texas Oil and Gas, US Diversified Group

USY United Synagogue Youth

USYRU United States Yacht Racing Union

usysf United States Youth Symphony Federation

ut universal trainer; urinary tract; user test; utilitarian; utility

u/t under training; untrained

UT Union Terminal (railroad); United Territories; United Territory; United Utilities (stock exchange symbol); Universal Time (Greenwich Mean Time); Universal Tubes; University of Texas; Utah; Utilities Man

U.T. U Thant

U of T University of Tampa; University of Tasmania; University of Tennessee; University of Texas; University of Toledo; University of Toronto; University of Tulsa

U of T (Austin) University of Texas in Austin

U of T (El Paso) University of Texas in El Paso (also UTEP)

U-T Union-Tribune (newspapers)

uta upper terminal area

UTA Ulster Transport Authority; United Tribes of Alaska; United Typothetae of America; University of Texas—Austin; Urban Transportation Administration

UTA Union des Transports Aériens (French—Air Transport Union)

utacv (UTACV) urban-tracked air-cushion vehicle

UTAD Utah Army Depot

Utagawa Utagawa Toyokuni

Utah St Hist Soc Utah State Historical Society

Utah St U Pr Utah State University Press

Utamaro Kitagawa Utamaro

utarb utarb eidet (Norwegian—prepared)

UT/AT Underway Trial/Acceptance Trial (USN)

UTB United Tariff Bureau; United Technocratic Board; Universal Technological Bureau

utc unit type code; unit type coding

utc (UTC) universal time coordinated

UTC United Tank Car; United Technologies Corporation; United Technology Center (United Aircraft); United Transformer Corporation; Universe Tankships Corporation (National Bulk Carriers); University Training Corps

UT-C University of Tennessee—Chattanooga

utclk universal transmitter clock

utd united

UTD University of Texas—Dallas

UTDA Ulster Tourist Development Association

UTDC Urban Transportation Development Corporation

ut dict. ut dictum (Latin—as ordered)

utdne. mor. sol. utendus more solito (Latin—use in the usual way)

Utd Tech United Technology

ute Australian utility truck

UTE underwater tracking equipment

uten utensil(s)

utend. utendus (Latin—to be used)

U of Tenn Pr University of Tennessee Press

UTEP University of Texas—El Paso

U of Tex Pr University of Texas Press

UTF Underground Test Facility

utg utgave (Norwegian—edition)

uti (UTI) urinary tract infection

UTI Union Title Insurance; Unit Trust of India

UTIAS University of Toronto Institute for Aerospace Studies

util utility; utilization

utilit utilitarian(ism); utilities

ut inf. ut infra (Latin—as below)

util rm utility room

utl unit transmission loss; universal transpor(er) loader; user trailer label

UTL University of Tampa Library; University of Tennessee Library; University of Texas Library; University of Tokyo Library; University of Toronto Library; University of Tulsa Library

UTLAS University of Toronto Library Automation System

utm universal testing machine; universal test(ing) module; universal transverse mercator

UTMA Uniform Transfer to Minors Act

UTN University of Tennessee

UTO United Thank Offering; United Town Organisation

U of Tok University of Tokyo

utop utopia (from the Greek *utopia*—no place)—ideal place

U of Tor Pr University of Toronto Press

UTP Unified Test Plan; University of Texas Press; University of Toronto Press; Unshielded Twister Pair

UTQG Uniform Tire Quality Grading

utr (UTR) university training reactor

Utr Utrecht (Dutch province)

UTR United Tire and Rubber

UTRC United Technologies Research Center

uts ultimate tensile strength; unit training standard

UTS Underwater Telephone System; Unified Transfer System (Russian-to-English translation); Uniform Thread Standard; Union Theological Seminary; Universal Time Sharing System; University of Toronto Schools

UtSEA Utah Society of Enrolled Agents

UTSSM University of Texas-Southwestern School of Medicine

ut sup. ut supra (Latin—as above)

UTTAS Utility Tactical Transport Aircraft System

uttc universal tape-to-tape converter

UTTR Utah Test and Training Range

UTU United Transportation Union

U-tube U-shaped tube

U-turn U-shaped turn

utv (UTV) underwater television

UTV Universal Test Vehicle

utw under the wing

UTWA United Textile Workers of America

UTX 4-engine jet utility transport; University of Texas

uty utilities

uu (UU) urine urobilinogen

u U unter Umständen (German—circumstances permitting)

UU Ulster Unionist, Union University; Universal Underclass; University of Utah

U-U Unitarian–Universalist

U & U Underwood and Underwood

U of U University of Uppsala; University of Utah

UU *Uppsala Universitetsbiblioteket* (Swedish—Uppsala University Library); *ustedes* (Spanish—you, pl)

UUA Unitarian-Universalist Association; Univac Users Association

UUCM University of Utah College of Medicine

UUCP Unix to Unix Copy

uue use until exhausted

uuf micromicrofarad

UUI United Utilities Incorporated

UUIP Uppsala University Institute of Physics

uum (UUM) underwater-to-underwater missile

UUP Ulster Unionist Party

uu's universal undevelopers

UUSC Unitarian Universalist Service Committee

uut unit under test

U of Utah Pr University of Utah Press

UUUC United Ulster Unionist Coalition

uuv unter üblichen vorbehalt (German—errors and omissions excepted)

UUWF Unitarian Universalist Women's Federation

uv ultraviolet; umbilical vein; under voltage; urinary volume

u-v ultraviolet

UV Ulster Vanguard; Unadilla Valley (railroad); Upper Volta

U of V University of Vermont; University of Victoria; University of Virginia

UV Una Via (Spanish—One Way)

UVA University of Virginia

U van A Universiteit van Amsterdam

U van A Universiteit van Amsterdam (Dutch—Amsterdam University, University in Amsterdam)

uvas ultraviolet astronomical satellite (UVAS)

uvaser ultraviolet amplification by stimulated emission of radiation

UVC United Veterans Council

u-v camera ultraviolet evidence camera

UVCM University of Vermont College of Medicine

UVCT University of Vermont College of Technology

uvd undervoltage device

UVDC Urban Vehicle Design Competition

UVE Unión Velocipédica Española (Spanish—Spanish Bicycle Union)

UVF Ulster Volunteer Force

UVH University of Virginia Hospital

UVI Unione Velocipedistica Italiana (Italian—Italian Cycling Association)

uviol ultraviolet

uvl ultraviolet light

UVL University of Virginia Library

UVM University of Vermont

U-vocab upperclass vocabulary

uv prom (UV PROM) ultraviolet programmable read-only memory

uvr ultraviolet radiation

UVR Uitenhage Volunteer Rifles

uvs ultraviolet spectrometer; universal versaplot software

UVSA Unie van, Suid Afrika (Union of South Africa)

uvsc ultraviolet solar constant

UVSM University of Virginia School of Medicine

UVT University of Vermont

uw unconventional warfare; unconventional weapons; underwater; underwing; underwriter; unwound

u/w underwater; underway; underwear; underwriter; used with

U/w Underwriter

UW University of Waikato; Uppity Women

U of W University of Wales; University of Warwick; University of Washington; University of Waterloo; University of Wichita; University of Windsor; University of Winnipeg; University of Wisconsin; University of Witwatersrand; University of Wollongong; University of Wyoming

UW Universität Ween (German—University of Vienna)—see *Ubib Wien*

UWA United Way of America; United World Atheists; University of Washington; University of Western Australia

U of Wash Pr University of Washington Press

UWC University of the Western Cape

UWCC Universal Wireless Communications Consortium

uwce unconventional warfare counter-effects

UWCE Underwater Weapons and Countermeasures Establishment

U-wear underwear

U-weld U-shaped weld

UWF United World Federalists

UWFL University of Washington Fisheries Laboratory

UWGB University of Wisconsin at Green Bay

UWH University of Washington Hospital

UWI University of the West Indies (Jamaica); University of Wisconsin

UWIL University of the West Indies Library (Kingston, Jamaica)

U of Wis Pr University of Wisconsin Press

UWIST University of Wales Institute of Science and Technology

UWIUB undrinkable wine in unusable bottles

UWL University of Wales Library; University of Washington Library; University of Wichita Library; University of Wisconsin Library; University of Witwatersand Library; University of Wyoming Library

UWM United World Mission; University of Wisconsin at Milwaukee

UWMI University of Wisconsin Management Institute

UWO University of Western Ontario

uwoa unclassified without attachments

UWP University of Wales Press; University of Washington Press; Up With People

UWSM University of Washington School of Medicine

uwtr underwater

UWTU Underwater Training Unit

UWUA Utility Workers Union of America

UWV University of West Virginia

UWW University Without Walls (Antioch College)

UWY University of Wyoming
ux. uuxor (Latin—wife)
uxb (UXB) unexploded bomb
uxgb unexploded gas bomb
uxib unexploded incendiary bomb
'Uxley (Cockney contraction—Huxley)
uxo unexploded ordnance
uxor uxoricide
UY Universal Youth; Uruguay (Internet code)

U of Y University of York
UYA University Year for Action
UYL United Yugoslav Lines
Uz Uzbek; Uzbekistan; Uzbekistanian
Uz Uhrzuender (German—clockwork fuze)
UZ University of Zululand; Uzbekistan (Internet code)
UZ Universität Zürich
Uzbek Uzbekistan

Uzbek SSR Uzbek Soviet Socialist Republic (former name for Republic of Uzbekistan)
Uzi Uziel Gal
UZM Universitet Zoologiske Museum (Copenhagen)
UZRA United Zionist Revisionists of America
U zu B Universität zu Berlin
U zu G Universität zu Göttingen
Uzz Uzziah

V

v vacuum; vacuum tube; vagabond; vagrant; Valium; value; valve; van; vapor; variable; variance; variation; vector; vein; velocity; vent; ventilator; ventral; verb; verbal; verbal ability; verse; version; vertex; vertical; very; vice; vincinal; violent (motion picture); violet; violin; virus; viscosity; vise; visibility; vision; visual acuity; voice; volt; voltage; voltmeter; volume; volunteer; vowel; vox

v *van* (Dutch—of); *verso* (Latin—back of page or sheet; left-hand page); *versus* (Latin—against); vibrational quantum number; *voltare* (Italian—turn, turn the page); *von* (German—of; from; used in titles)

v/ *vostra* (Italian—your)

V coefficient of vibration (symbol); five-dollar bill; Lockheed (symbol); gas volume (symbol); potential (symbol); relative wind velocity (symbol); stalling velocity (symbol); Standard Fruit & Steamship Company (Vaccaro Line); vanadium; Venerable; Ventzke; Venus; Verdet constant; Vicar; Vice; Victor—code for letter V; Victory; Village; Village District; volt; Volta; volume (symbol)

V airspeed, forward velocity (symbol); speed (symbol); vacuum tube (symbol); *varm* (Dano-Norwegian or Swedish—hot); *väst* (Swedish—west); *Venstre* (Danish or Norwegian—Left)—Liberal Party; *vertrek* (Dutch—departure); *vest* (Dano-Norwegian—west); *Via* (Italian—highway road, way); *Villa* (Spanish—village); *vialaceus* (Latin—violet color); *viridis* (Latin—green); *vrouw* (Dutch—woman)

v1 efferent veins contain tumor (symbol)

v2 distant veins contain tumor (symbol)

v-1 vernier engine 1

Vø no tumor in veins (symbol)

V1 Voyager 1 satellite with close Titan flyby

V₁ decision speed (go-no-go) for aircraft to continue takeoff run or abort flight; valve-current voltage

V¹ *violino primo* (Italian—first violin)

v-1 p vernier engine 1 pitch

V-1, V-2 rockets launched by the Germans in World War II

v-1 y vernier engine 1 yaw

V₂ aircraft takeoff speed or position where nose is lifted so plane becomes airborne

V² *violino secondo* (Italian—second violin)

V2 Voyager 2 with Uranus option

V-4 four-cylinder engine with two cylinders in each side of V-shaped engine block

V-6 six-cylinder engine with three cylinders in each side of V-shaped engine block

V-8 eight-cylinder engine with four cylinders in each side of V-shaped engine block

V-10 Viscount 10-jet airplane

v 26 d M *von 26 dieses Monats* (German—of the 26th instant; of the 26th of this month)

va variable; variable annuity; variance; verb active; verbal adjective; very abundant; victualling allowance; viola; voltampere(s); vulnerable area

v-a volt-ampere(s)

v/a verbal auxiliary; volts per ampere; voucher attached

v/a (V/A) vulnerable area

v.a. *vixit—annas* (Latin—he lived—years)

Va Virginia; Virginian

Va Vila (Portuguese—Villa; Village); *Villa* (Italian or Spanish—Villa, Village)

Vₐ Vila (Portuguese—small town, villa); *Viuda* (Spanish—widow)

VA Vatican City (Internet code); Veterans Administration (United States); Veterans' Affairs (Canada); Vice Admiral; Virginia; Voice of America; voltaic alternative (symbol); Volunteer Army; Volunteers of America

V-A Vickers-Armstrong Limited

V.A. Order of Victoria and Albert; Vicar Apostolic

V & A Victoria and Albert (Museum); Victoria and Albert (Order)

V of A Volunteers of America

V & A *Venus and Adonis*

VAA Vaccination Assistance Act; Vietnamese-American Association

VAACR Vietnamese Association for Asian Cultural Relations

vaap vaccine-associated paralytic polio

vab voice answer back

VAb Van Allen belt (zone of high-intensity radiation surrounding the earth at altitudes of about 500 miles)

VAB Vandenberg Air Force Base; Vertical Assembly Building (world's largest all-steel structure of its type; used for assembling missiles and space exploration vehicles on Merritt Island at Cape Kennedy, Florida)

Vabd Van Allen belt dosimeter

Va Bk Virginia Book Company

VABM vertical angle bench mark

vac vacant; vacate; vacation; vacuum; vector analog computer; volts alternating current (volts AC preferable)

VAC Video Amplifier Chain; Voluntary Action Center; Volunteer Advisor Corps

VACAB Veterans Administration Contract Appeals Board

vacc vaccination; vaccine; value-added common carrier

Vaccaro Standard Fruit & Steamship Company

vacci vaccinate; vaccination; vaccine

vac-dist vacuum-distilled

Vaclav (Czech–Wenceslas)

vac pmp vacuum pump

VACRP Victorian Association for the Care and Resettlement of Prisoners

vacs vacuum cleaners

v/act. verb active

vad value-added dealer; variable abbreviated dialing; velocity azimuth display; voltmeter analog-to-digital converter

VAd Veterans Administration

VAD Voluntary Aid Detachment

vada versatile automatic data exchange

V Adm Vice Admiral

vad. mec. *vade mecum* (Latin—go with me)–companion volume; handbook; manual; ready reference for readers and reference librarians

vae vinyl-acetate ethylene

VAEA Virginia Adult Education Association

VAF Vendor Approval Form; Vincent Astor Foundation

VAFB Vandenberg Air Force Base

vag vagabond; vagina; vaginal; vaginitis; vagrant; vagrancy

vag charge vagrancy charge

vag hist vaginal hysterectomy

VAGA Visual Artists and Galleries Association

vagonzak *vagon zaklyuchennykh* (Russian–railroad prisoner car)

vags vagabonds; vagrants

VAH Veterans Administration Hospital

VAHS Victorian Aboriginal Health Service

vai video-assisted instruction; vorticity area index

va & i verb active and intransitive

VAI Video Artists International

VAIS Virginia Association of Independent Schools

vakt visual-auditory-kinesthetic and tactual (imagery applied to teaching reading)

val valance; valence; valenciennes (lace); valentine; valise; valley; valuation; value; valued; valve; valvular

val (VAL) valine (amino acid)

Val Valencia; Valentina; Valentine; Valentino; Valerie; Valerius

VAL Vehicle Authorization List; Veterans Administration Library

VALA Viewers and Listeners Association

VALB Veterans of the Abraham Lincoln brigade

valc visual approach and landing chart

Vald Valdivia

Val Fl Gaitus Valerius Flaccus (Roman epic poet)

valid. validate; validation

Valka Valentin

Vall Valladolid

VALNET Veterans Administration Library Network

Valpo Valparaiso

valsas variable-length word symbolic assembly system

valt vtol approach-and-landing technic

VALUE Visible Achievement Liberates Unemployment (Air Force program for disadvantaged youth)

val vu valley view

vam visual approach monitor

vam. volt ammeter

Vam Vogel's approximation method

VAMC Veterans Administration Medical Center; Veterans Affairs Medical Center

VAMCO Village and Marketing Corporation

VAMOS Verified Additional Military Occupational Specialty

vamp vampire; vampirism; volume, area, mass properties

VAMP Voluntary Association of Master Pumpers (mid 19th-century English firefighters)

vamps vampires; seductive enemy agents, also called swallows

vam's vision-aid magnifiers

van (VAN) value-added network

van caravan; value-added network; vanguard; vanilla; vanillin

Van Vanessa

Van (VAN) Vancouver, British Columbia

VAN Value-Added Network

VAN Vereniging van Archivarissen in Nederland (Dutch Association of Archivists in the Netherlands)

Vanc Vancouver

Vanc I Vancouver Island

Vancoram Vanadium Corporation of America

Vandy Vanessa

Vanechka (Russian nickname–Ivan)

Vang Vickers-Armstrong Vanguard (aircraft)

Vang Esp Yanguardia Española (Barcelona's Spanish Vanguard)

Vanguard Vanguard Press

Vanier Centre Vanier Centre for Women (criminals) at Brampton, Ontario

Vanka (Russian diminutive—Ivan)

Van-Lax Vancouver–Los Angeles

Van-Mia Vancouver–Miami

Van Op Vancouver Opera

van. pub. vanity publisher; vanity publishing

VANS Value-Added Network Service(s)

Van-San Vancouver–San Diego

Van-Sea Vancouver–Seattle

Van-Sfo Vancouver–San Francisco

Van Sun Vancouver Sun

Van Sym Vancouver Symphony Orchestra

Van-Tor Vancouver–Toronto

Vanu Vanuatu (island republic in the southwestern Pacific)

vap value-added process

VAP Vendor Assistance Phase; Victims Assistance Program; Victims Assistance Project

vapi visual approach path indicator

vapor. vaporization

vap prf vaporproof

var variable; variant; variation; variety; variometer; various; varying; visual-aural range; volt-ampere reactive

var (VAR) value-added reseller; vertical air rocket

var variazione (Italian—variation)

Var Varna

VAR Volunteer Air Reserve

varactor variable capacitor

varad varying radiation

VARC Visual Arts Research and Resource Center Relating to the Caribbean

var con variable condenser

var dial. various dialects

var ed & trans various editions and translations

VARES Vega-Aircraft Radar-Enhanced System

vari VariType(r)

VARIG Empresa de Viação Aérea Rio Grandense (airline in southern Brazil)

varistors variable resistors

varizistor variable resistor

var. lect. varia lectio (Latin—variant reading)

varn varnish

VARP Veterans Administration Procurement Regulations

varr variable-range reflector

Varr Marcus Terentius Varro (Roman writer on agriculture and natural history)

vars varieties

Vars Varsavia (Italian or Latin—Warsaw); *Varsovia* (Spanish—Warsaw); *Varsóvia* (Portuguese—Warsaw)

VARS Vertical and Azimuth Reference System

vas vasectomy

vas (Latin prefix–vessel)-vasoconstriction

Vas Vasteras

VAs Voluntary Aids

VAS Virginia Academy of Science; Vocational Advisory Service

VAS Vedette Anti-Sommergibile (Italian—Anti-Submarine Sentry)-naval craft; *Vereniging van Accountancy Studenten* (Dutch—Society of Accountancy Students)

VASA Virginia Association of School Administrators

vas bund vascular bundle

vasc vascular

VASC Verbal Auditory Screen for Children

VASCA (electronic) Valve and Semi-Conductor (manufacturers') Association

vascar visual average-speed computer recorder

VASCO Vanadium-Alloys Steel Company

VaSEA Virginia Society of Enrolled Agents

VASEC vasectomy

vasi visual approach slope indicator

vasim voltage and synchro-interface module

VASP Viação São Paulo (São Paulo airline)

VASSP Virginia Association of Secondary School Principals

VASSS Van Allen Simplified Scoring System

vast versatile avionics shop test; vibration and static analysis

Västtyskland (Swedish—Germany)

vas vit. vas vitrium (Latin—glass vessel)

vat value-added taxes (VAT); ventricular activation time; vinyl asbestos tile; vinyl asbestos tiling

Vat Vatican; Vatican City

VAT Value-Added Tax; Vertical Assembly Tower; Veterinary Aptitude Test, Visual Apperception Test

vate versatile automatic test equipment

VATI Vermont Agricultural and Technical Institute

Vatic Vatican

Vatican Bank Institute for Religious Works, Vatican City, which failed in 1982 when

Archbishop Paul C Marcinkus and two other officials made off with its funds

Vat Lib Vatican Library (Rome)

VATLS Visual Airborne Target Location System

vatpayer value-added taxpayer

VATS Vertical-lift Airfield for Tactical Support; Video-Augmented Tracking System

Vat Sta Vatican State

VATTR Value-Added-Tax Tribunal(s)

vaud vaudeville

v aux verb auxiliary

vav variable air volume

vavbd *vavband* (Swedish—clothing)

v/a v/e value-analyst value-engineer

vavp variable-angle variable pitch

VAWA Violence Against Women Act

vax virtual address extension

VAX trademark of Digital Equipment Corporation

vb valence band; verb; verbal; vertical bomb (VB); vibration

v/b vehicle-borne

VB Navy bomber (2-letter naval symbol); Vero Beach; very bad; Virginia Beach; Volunteer Battalion; Vulgate Bible

vba verbal adjective

VBA Veterans Benevolent Association

vbaC vaginal birth after Caesarean

V-band 46,000–56,000 mc

vbc ventrobasal complex

VBC Vancouver British Columbia

VBCO Vector Biology and Control Office (California)

vbcr vanished black community remnant

VBEC Venezuelan Basic Economy Corporation

V-belt V-shaped belt (cross-section of belt is V-shaped)

VBFNPVGFPMTF *Véndmaire, Brumaire, Frimaire, Nivôse, Pluviôse, Ventôse, Germinal, Floréal, Prairial,*

Messidor, Thermidor, Fructidor (as abbreviated on the French Revolutionary Calendar—*see Vend, Brum, Frim, Niv, Pluv Vent, Germ, Flor, Prair, Mess, Therm, Fruc*)

VBI Venetian Blind Institute; Vicente Blasco Ibánez (famous for his novel *Los Cuatro Jinetes del Apocalipsis—The Four Horsemen of the Apocalypse*)

vbl verbal

V-block V-shaped block

VBMA Vacuum Bag Manufacturers Association

vbn verbal noun

V-bomb German long-range missile-type bomb used during World War II; designated as V-1 and V-2

vbos veronal-buffered oxalated saline

V-bottom V-shaped bottom

VBP Vietnam Boat People

vbr (VBR) ventricular-brain ratio

V B R Virginia Blue Ridge (highway)

VBRA Vehicle Builder's and Repairers' Association

vbs visual basic script

VBS Vacation Bible School; Vedanthangal Bird Sanctuary (India); Vocabulary Building System

vc valuation clause; vector control; venereal case; venture capital; venture capitalist; very common; vice chancellor; violoncello; visual communication

vc (VC) variable cost; vital capacity

vc *vuelta de correo* (Spanish—by return mail)

v/c *vuelta de correo* (Spanish—return mail)

vC *voor Christus* (Dutch—Before Christ)

Vc Vietcong

VC acuity of color vision (symbol); Saint Vincent and the Grenadines (Internet code); Vassar College; Vatel Club; Vatican City; Vehicle

Code; Vennard College; Ventura College; Vermont College; Veterinary Corps; Vice Consul; Victoria College; Victoria Cross; Viterbo College; Volusia College

VC *Vehicle Code*

VC-10 British BAC long-range transport aircraft

VC-137 USAF designation of the Boeing 707

vca voltage-controlled amplifier

VCA Vermont Council on the Arts; Virginia Correctional Association; Volunteer Civic Association

VCAR Vendor Corrective Action Request

VCAS Vice-Chief of Air Staff

VCB Victim Compensation Board

vcc vasoconstrictor center; vice chancellor's court; video compact cassette; vocational career concept

vcc *vin de consommation courante* (French-table wine)

Vcc supply voltage

VCC Value Control Coordinator; Vancouver Community College; Variable-Cycle Control; Violent Crime Control; Visual Communications Congress

vc card index (or reader) visual coincidence index (or reader)

vcce variable contrast/constant exposure

VCCL Victoria Council for Civil Liberties

vcco voltage-controlled crystal oscillator

vccs voltage-controlled current source

vcd variable-capacitance diode

v-c d voluntary-closing device

VCDS Vice Chief of Defence Staff

vce (VCE) variable-cycle engine

Vce Venice

VCE Venice, Italy (airport)

vcf vaginal contraceptive film; voltage-controlled filter

vcg vectorcardiogram; vertical line through center of gravity; voltage-controlled generator

VG Vice-Consul General

vch vehicle; vinyl cyclohexane (VCH)

VCH Victoria County History

vchp variable-conductance heat pipe

v Chr vor Christis (German—before Christ)

vci visual communication instructor; volatile corrosion inhibitor

VCI Variety Clubs International; Vision Conservation Institute

VCIC Vermont Crime Information Center

VCIGS Vice-Chief of the Imperial General Staff

VCIP Veterans Cost-of-Instruction Program

VCK Verenigo Cargodoorskantoor

vcl vertical center line; visual comfort light(ing); voluntary compliance level

VCL Vancouver Public Library

vcllo violincello

VCLU Virginia Civil Liberties Union

vcm vacuum; vinyl chloride monomer

VCMA Vacuum Cleaners Manufacturers Association

VCN Vendor Contact Notice

VCNS Vice-Chief of Naval Staff

vcnty vicinity

vco voltage-controlled oscillator

VCO Viceroy's Commissioned Officer

vcod vertical-carrier on board delivery

vcoi veterans cost of instruction

VCOS Vice Chiefs of Staff

v coul volt coulomb

vcp vehicle check point; vitrified clay pipe

VCP Vendor Change Proposal; São Paulo, Brazil (Viracopas Airport)

vcpi virtual control program interface

vcr variable compression ratio; voluntary compliance resolution

vcr (VCR) videocassette recorder

Vcr Vancouver

VCR Victor Comptometer (stock exchange symbol)

vcr's video cassette recorder owners

vcr('s) video cassette recorder(s)

vcs vasoconstrictor substances; voices

vc's viejos cristianos (Spanish—old Christians—Spaniards who believe they are without Jewish or Moorish blood

VCs Viet Congs; Vigilance Committeemen; Vigilant Committeemen; Vigilante Committeemen

VCS Vernier Control System; Veterans Canteen Service; Vice Chief of Staff; Video Cassette System

V & C S Virginia & Carolina Southern (railroad)

vcsr voltage-controlled shift register

vct vinyl-composition tile

Vct Victoria

vctv view-controlled television; vocative

vcty vicinity

VCU Virginia Commonwealth University

vcxo voltage-controlled crystal oscillator

V Cz Vera Cruz

vd vapor density; various dates; venereal disease (VD); videodisc; void; voltage drop; volume deleted

v/d vandyke reproduction

Vd vanadium

Vd usted (Spanish—you)

VD Village District, Isle of Man

V.D. Volunteer Officer's Decoration

vda venereal disease awareness; video distribution amplifier, visual discriminatory acuity

Vda Venda (Spanish abbreviation) Bantu homeland, South Africa; *Viuda* (Spanish—widow)

VDA Vermont Department of Agriculture

VDA *Verband der Automobilindustrie* (German—Automobile Industry Association); *Volksbund für das Deutschtum im Ausland* (German—League for Germans Abroad)

V-day day of victory

vdB velocity decibel

VDB Venereal Disease Branch (U.S. Public Health Service); *Vêrband Deutscher Biologen* (German—Association of German Biologists)

VDBC Vertol Division, Boeing Company (helicopter design and manufacturing)

vdc volts direct current (*volts DC* preferable)

vdc (VDC) vinylidene chloride

VDC Venereal Disease Clinic; Virginia Department of Corrections; Volunteer Defence Corps

vdcm (VDCM) vinylidene chloride monomer

vdcw working direct current voltage (symbol)

VDE *Verband Deutscher Elektrotechniker* (German—Association of German Electrical Engineers)

v def verb defective

VDEH *Verein Deutscher Eisenhüttenleute* (German—German Foundry Society)

VDEL Venereal Disease Experimental Laboratory

vdem vasodepressor material

v dep verb deponent

V De S Vittorio De Sica

VdF *Vigili del Fuoco* (Italian—Fire Brigade)

vdfg variable diode function

vdg vertical display generator

vd-g venereal disease—gonorrhea

VDG Fifth Dragoon Guards (Inniskilling)

vdh very-deep hold

vdh (VDH) valvular disease of the heart

vdi vegetation draught index; vehicle deformation index; venereal disease inhibition; video display input; virtual device interface

VDI Verein Deutscher Ingenieure (German—Association of German Engineers)

V-dies V-shaped dies

vdif very difficult

V di R Virtuosi di Roma

vdisk virtual disk

VdK Verband der Kriegsbeschadigten (German—League of War Invalids)

Vdkhr Vodokhranilishche (Russian—reservoir)

vdl ventilation deadlight

VDL Van Dieman's Land (Tasmania)

vdm vector-drawn map

vdm (VD) vasodepressor material

v.d.m. verbi dei minister (Latin—preacher of the world)

Vdm Veendam

VDMA Verein Deutscher Maschinenbau Anstalten (German—Mechanical Engineering Association)

vdm('s) video disc machine(s)

VDN Varudeklariosnamnden (Swedish—Institute for Informative Labelling); *Vin Doux Naturel* (French—fortified wine, natural sweet wine)

vdo video

vdp vehicle deadlined for parts; vertical data processing; videodisc player

VDPV Verband Deutscher Prädikatswein-Versteigerer (German—German Natural Wine Association)

VDQS Vin Délimité de Qualité Supérieur (French—superior-quality wine)

vdr variable-diameter rotor; voltage-dependent resistor; voyage data recorder

VDRL Venereal Disease Research Laboratories

VDRS Verdun Depression Rating Scale

VDRT Venereal Disease Reference Test

vds variable depth sonar

vd-s venereal disease—syphilis

Vds ustedes (Spanish—you, plural)

VDSCRC Very Dirty and Small Coal Railway Company (created by Dickens for service in *The Uncommercial Traveller*)

VDSI Verein Deutscher Sicherheits Ingenieure (German—Association of Safety Engineers)

vdt variable density (wind) tunnel; video data terminal

vdt (VDT) video display terminal

VDT Visual Distortion Test

VDTA Vacuum Dealers Trade Association

vdt's video display terminals

vdu visual display unit

ve vaginal examination; value engineering; varicose eczema; vernier engine; very excellent

've have

ve veuve (French—widow)

v.e. venditioni exponas (Latin—you expose to sale)

Ve Venezuela; Venezuelan

VE Value Engineer(ing); Victory in Europe

VE Vasileion tis Ellados (Greek—Kingdom of Hellas—Greece); Venezuela (Internet code)

V-E Verzlibolovo-Eydtkuhnen (Russo-German railway frontier for passengers and freight changing from wide gauge to standard European gauge rolling stock and tracks)

ve/a value engineering/analysis (program)

VEA Valve Engineering Association; Vermont Education Association; Veterans Education Administration, Virginia Education Association; Vocational Education Act

VEA-H Vocational Education Act— Handicapped

vealsan veal sandwich

vealwich veal sandwich

VEAP Veterans' Educational Assistance Program

veb variable elevation beam

VEB Volks Eigener Betriebe (German—Peoples-Owned Companies)

VEBA Voluntary Employees' Beneficiary Association

vec vector

veco vernier engine cutoff

vecp visually evoked cortical potential

VECP Value Engineering Change Proposal

VECR Vendor's Engineering Change Request

VECS Vocation Education Curriculum Specialists

vecto vectograph; vectographic; vectographical

vecu valve electronic control unit

ved vacuum erection device

ved vedova (Italian—widow)

Ved Vedic

VED Vickers Electric Division

VEDA Victorian Eastern Development Association

V-E Day Victory in Europe Day—May 8, 1945

VEDC Vitreous Enamel Development Council

vedr vedrorende (Danish—concerning)

VEDS Vocational Education Data System

Vee Venezuelan equine encephalomyelitis

vee dee venereal disease; visiting dignitary

VEENAF (South) Vietnamese Air Force

VEEP Voluntary Ethnic Enrollment Program

veg vegetable; vegetarian; vegetarianism; vegetation

vegan vegetarian who eats no meat, fish, fowl or dairy products

Vegas Las Vegas

vegf vascular endothelial growth factor

veggies vegetables

Veg Soc Vegetarian Society

vegtan vegetable tanning

veh vehicle; vehicular

VEH Vocational Education for the Handicapped

vehic. vehiculum (Latin—vehicle)

vehic manslgtr vehicular manslaughter

veh pt(s) vehicle part(s)

VEIN Vocational Education Information Network

VEIS Vocational Education Information System

vel vellum; velocity; velvet

Vel Vela (constellation)

Vel Yelikiy (Russian—large)

Vela (Latin—The Sails constellation)

Velázquez Diego Rodriguez de Silva Velázquez (1599–1660)

Vell Gaius Velleitus Paterculus (Roman historian)

velo velodrome

veloc velocity

vem vasoexciter material

ven veneer; veneering; venerable; venereal; venery; venetian; venetian blind(s); venison; venom; venomous; ventral; ventricle

ven vendredi (French—Friday); *venerdi* (Italian—Friday); *venesianisch* (German Venetian)

Ven Venetian; Venezuela (whose capital is Caracas); Venice; Venus

vend vending; vending machine; vendor(s)

Vend Vendémaire (French—Vintage Month)—beginning September 22nd—first month of the French Revolutionary Calendar

vend. ex. venditioni exponas (Latin—you expose to sale)

vend. mach vending machine

Venez Venezuela; Venezuelan

Venezuela Republic of Venezuela (oil-producing Spanish-speaking South American nation), *República de Venezuela*

Venezuelan Islands Aves, Blanquilla, Cubagua, Hermanos, Margarita, Monjes, Orchila, Testigos, Tortuga

V-engine V-shaped engine

VENISS Visual Education National Information Service for Schools

vent. ventilate; ventilating; ventilation; ventilator; venting; ventral; ventricle; venture

Vent Venffôse (French—Windy Month)—beginning February 19th-sixth month of the French Revolutionary Calendar

vent. fib. ventricular fibrillation

ventric ventricular

vent(s). ventilators; ventriloquist(s)

vent. tachy ventricular tachycardia

Venus (Latin—Aphrodite)—goddess of beauty and love

vep visual-evoked potential

VEP Veterans Education Project; Voter Education Project

VEPCO Virginia Electric and Power Company

VEPM Value Engineering Program Manager

ver verification; verify; verse(s); versine; vertex (Ver); visual-evoked response

Ver Vera Cruz

Ver Verband; Verein (German—association)

VER Voluntary Export Restraint (agreements)

Vera Veratchke; Veronica

VERA Vision Electronic Recording Apparatus (videotape)

verand verandert (German—revised)

verb verbesserte (Dutch or German—improved)

verb. et lit. verbatim et literatim (Latin—exact copy; word for word)

verb. sap. *verbum satis sapienti* (Latin—a word to the wise is sufficient)

Verdi Giuseppe Verdi; Victor Emmanuel Re d'Italia

verdigris copper acetate

Verds Cape Verde Islands

verdt verdict

Verf Verfasser (German—author)

Verg Publius Vergilius Maro (Roman poet often referred to as Virgil)

Vergl Vergleische (German—compare)

Verh Verhandlungen (German—proceedings)

VERIC Vocational Educational Research Information Center

verisim verisimilar; verisimilitude; verisimilitudinous

Veritas Det norske Veritas (The Norwegian Bureau of Shipping)

Verkh Verkhniy (Russian—upper)

verk v verkorting van (Dutch—abbreviation, abridgement, shortening)

Verl Verlag (German—publisher)

Verlagshdlg Verlagshandlung (German—book-publishing house)

verlort very-long-range tracking (radar)

verm vermiculite

verm (Latin prefix—worm) vermiform; *vermehrte* (German—enlarged)

Verm Vermont

vermilion mercury sulfide

vern vernacular

Vern Vernay; Verne; Verney; Vernon

Vern Vernon's Law Reports

vernae vernacularism); vernacularly

vers versed sine; verses; versification; versine (versed sine)

versine versed sine

verso reverso (left-hand page; reverse side of a page) opposite of recto

Ver St Vereinigte Staaten (German—United States)

vert vertebra; vertebrate; vertical; vertigo

verticam vertical camera

Verzh Vershbolovo (railway town shared by Poland and Russia)

ves vertical electric soundings; vessel

ves. vesica (Latin—bladder)

VES Veterans Employment Service; Voluntary Euthanasia Society

VESC Vehicle Equipment Safety Commission

vesca(s) vessel(s) and cargo

vesda (VESDA) very early smoke detection apparatus

VESIAC Vela Seismic Information Analysis Center

vesic. vesicula (Latin—blister)

VESO Value Engineering Services Office

vesp. vesper (Latin—evening)

vesper. vehicles, equipment, and spares provision—economics and repairs

VESPER Voluntary Enterprises and Services and Part-time Employment for the Retired

vest vestibule

VEST Volunteer Engineers, Scientists, and Technicians (organization)

Vesters Vester Islands

ves. ur. vesica urinaria (Latin—urinary bladder)

vet veteran; veterinarian; veterinary

v. et. vide etiam (Latin—also see)

Vet Veterinary Medicine

VET Verbal Test

Vet Admin Veterans' Administration

Veterans Veterans Day (November 11)—commemorating armistice to end World War I on the 11th hour of the 11th day of the 11th month of 1918, originally called Armistice Day; Veterans Stadium, Philadelphia, Pennsylvania

Vet M. B. Bachelor of Veterinary Medicine

vet med veterinary medicine

VETMIS Vehicle Technical Management Information System (USA)

vet reg veterans' regulations

vet rep veteran's representative

vets veterans; veterinaries

VETS Veterans' Employment and Training Service

vet sci veterinary science

Vets Info Veterans Information Service

Vet Surg Veterinary Surgeon

vett vetted; vetting

'vette corvette

vetted (English contraction—veterinary inspected)—inspected and investigated

vev voice-excited vocoder

VEV Vietnam Era Veterans

V Ex^a Vossa Excelência (Portuguese—Your Excellency)

vexdex vexation index

vexil vexillogical; vexillologist; vexillology

vf ventricular fibrillation; vertical file; very fair; very fine; video frequency; visual field; voice frequency; vulcanized fiber

Vf Verfasser (German—author)

VF fixed-wing fighter airplane (2-letter naval symbol); Valley Forge

V/F Voltage Frequency converter

V.F. Vicar Forane

VF Vigili del Fuoco (Italian—Fire Brigade)

V f A Voice for America (Alistair Cooke)

VFA Victoria Football Association (Australia); Video Free America; Visiting Forces Act; Voluntary Foreign Aid

V-FA Vietnamese–France Association

V-factor verbal (comprehension) factor; violence factor

v-f band voice-frequency band

vfc video video frequency carrier; video frequency channel; visual field control; voice frequency carrier

VFC Victorian Film Commission

VFD Volunteer Fire Department

vfdr viewfinder

vfet vertical field-effect transistor

vff black very-fine furnace black (rubber filler)

VFHS Valley Forge Historical Society

vfi visual field information

VFI Vocational Foundation Incorporated

VFI Vertical File Index

VFIC Virginia Foundation for Independent Colleges

vfl variable focal-length

vfl (VFL) variable field length

VFL Victoria Football League (Australia)

VFMJC Valley Forge Military Junior College

vfn very-flowery no

VFNP Victoria Falls National Park (Rhodesia)

vfo variable-frequency oscillator

VFOAR Vandenberg Field Office of Aerospace Research (USAF)

vfp variable-factor programming

VFP Volunteers for Peace

vfr vehicle fuel refinery

VfR Verein für Raumschiffahrt (German—Space Travel Society)

VFR Visual Flight Rules

vfr's visiting friends and relatives

VFSTC Valley Forge Space Technology Center (General Electric)

vftg voice frequency telegraph

vfu vertical format unit

VFU Vancouver Free University

VFW Vereinigte Flugtechnische Werke; Veterans of Foreign Wars

vfy verify

vg variable geometry; velocity gravity; very good (VG)

v.g. verbi gratia (Latin—for example)

vg verbigracia (Spanish—for example); *virgen* (Spanish—virgin)

Vg. Virgo (Latin—virgin)

VG Virgin Islands, British (Internet code); Vocational Guidance

V.G. Vicar General

VG Vaisseau de Guerre (French—warship)

vga variable gain amplifier; videographics array

VGA Victor Gruen Associates

VGAA Vegetable Growers Association of America

VGB Vandenberg Air Force Base

vgc viscosity gravity constant

VGC Veterans Guard of Canada

vge visual gross error

VGH Vancouver General Hospital

VGHR Velocity, G-force, Height (altitude) Recorder

VGIK *Vsesoyuznyi Gosudarstvenyi Institut Kinematografl* (Russian—All-Union State Institute of Cinematography)

V-girl vice girl

Vgk Vegesack

vgl vergelijken (Dutch—compare); *vergleiche* (German—compare)

VGLI **Veterans** Group Life Insurance

Vgm Vizagapatam

vgo vacuum gas oil

Vgo Vigo (British maritime abbreviation)

VGO Vickers Gas-Operated

VGP Van Gelder Papier; Volunteer Grandparent Program

vgpi visual glide-path indicator(s); visual ground-position indicator

Vgr Voyager (robot spacecraft)

V gr verbigracia: (Spanish—for example)

V-groove V-shaped groove

VGSA Viola da Gamba Society of America

vgu vorgelesen-genehmigt-unterschrieben (German—read, confirmed, signed)

vgw voegwoord (Dutch—conjunction)

vh very high; veterinary hospital

v/h vulnerability/hardness

v/h vorheen (Dutch—formerly)

v H vom Hundert (German—percent; per hundred)

VH Veterans Hospital

VHA Vermont Headmasters Association; Voluntary Hospitals of America

vhb very heavy bombardment

vhc very highly commended

vhcl vehicle

vhclr vehicular

vhd very high density; video high density

VHDL VHSIC (Very High Speed Integrated Circuits) Hardware Descriptor Language

vhf very high frequency (30,000 kc–300 mc)

VHF very high frequency (British educational tv)

vhf/df very high frequency direction finding

vhf/fm very high frequency/frequency modulated

vhf/uhf very high and ultra high frequency

VHI Value Health Incorporated

VHIS Vaal-Hartz Irrigation Scheme

VHMCP Voluntary Home Mortgage Credit Program

vhmwpe very-high-molecular-weight polyethylene

Vhn Vickers hardness number

vho very high output

vhocm very-heavy oil-cut mud

vhp very high performance

vhs very high speed; video helical scan; video home system(s)

VHS Vocational High School

vhsbw very-high-speed black-and-white (photography)

VHSIC Very High Speed Integrated Circuit

vhtr very-high-temperature reactor

V-hut inverted V-shaped hut (sometimes called A-hut)

vi in bankruptcy or receivership (newspaper bond or stock listings); variable interval; verb intransitive; viscosity index; visual editor; volume index; volume indicator

vi. visitor

v/i verb intransitive

v.i. vide infra (Latin—see below)

Vi input voltage (symbol); Viola; Violet; Virginia; Vivian

VI Vancouver Island; Vermiculite Institute; Victoria (airline code); Vinegar Institute;

Virgin Islander(s); Virgin Islands, United States (Internet code); Virgin Islands (V.I.)

VI Veiligheids Institut (Dutch—Safety Institute)

via virus inactivating agent

Via Viaduct

VIA Vancouver, British Columbia's Vancouver International Airport; VIA Rail Canada; Vision Institute of America; Visually Impaired Association; Vocational Interests and Aptitudes

VIAC Vienna Allied Command

viad viaduct

vi antigen virulence antigen

VIAR Volcani Institute of Agricultural Research (Israel)

Via Rail Canadian National + Canadian Pacific

VIARCO Venezuelan International Airway Reservations Computerized

VIAs Vocational Information Agencies

VIAS Voice Interference Analysis System

VIASA Venezolana Internacional de Aviación SA

vib vibrate; vibration; vibratory; vocational interest bank

VIB Vertical Integration Building; Volunteer Infantry Brigade

vibes vibraphones; vibrations

vibgyor (mnemonic for remembering the spectral colors—violet, indigo, blue, green, yellow, orange, red)-*see* ROY G BIV

vibra vibraphone

vibs vocabulary-information-block-design similarities

vib/s vibrations per second

VIBS Virgin Islands Broadcasting System

vic convict; value-incentive clause; vicinal; vicinity; victim; victor; victorious; victory (V)

vic vices (Latin—times)

Vic RCA Victor; Vicar; Victor; Victoria; Victorine

VIC Virginia Intermont College; Virgin Islands Corporation

VICA Vision Council of America; Vocational Industrial Clubs of America

Vic Adm Vice Admiral

vicci voice-initiated cockpit control and integration

Vic Hist Victoria History of the Counties of England

vicoed visual communication education

vicom visual communication management

VICORP Virgin Islands Corporation

Vic Pk Victoria Park

vic(s) convict(s)

Vic Sta Victoria Station (rail terminal)

Vict Victor(ia)

Viᵗᵃ **Victoria** (Spanish)

Vicᵗᵉ **Vincente** (Spanish—Vincent)

victimol victimological(ly); victimologist; victimology

Victoria Alexandrina Victoria, Queen of the United Kingdom of Great Britain and Ireland, and Empress of India (1819–1901); La Victoria (Santo Domingo City prison of the Dominican Republic)

vid video

vid. vide (Latin—see); *Viuda* (Spanish—widow)

VID Volunteers for International Development

vidac: visual information display and control

vidat visual data acquisition

VIDC Virgin Islands Department of Commerce

VIDD Virgin Islands Development Department

videocomp videocomposition (high-speed phototypesetting controlled by programmed digital-control unit)

videot(s) video (television), idiot(s)

vidiac visual information display and control

vidisc video disc

Viditel Videotelevision viewdata system (Dutch)

vie viernes (Spanish—Friday)

VIE Vienna, Austria (airport)

VIEDS Virgin Islands Educational Dissemination System

Vien Vienna

vier viernes (Spanish—Friday)

Viet Vietnam

Viet Vietnamese (oriental language)

Viet Cong Vietnam Congsan (Vietnamese—Vietnamese Communists)

Vietminh Vietnam Doc Lap Dong Ming (League for the Independence of Vietnam)

Vietnam Socialist Republic of Vietnam (Indo-Chinese country), *Cong Hoa Xa Chu Nghia Viet Nam*

Vietnam congsam Vietnamese communist (see *congsam*)

Vietsyn Vietnam syndrome

Vietyet(s) Vietnam veteran(s)

Vieux Henri Vieuxtemps; *Vieux Carre* (French—Old Quarter)—French Quarter of New Orleans

VIEW Vital Information for Education and Work (education-on-microfilm program)

viff vectoring in forward flight

vig video image generator; vigilante; vigorish

vig (VIG) vaccine-immune globulin

VIG Video Integrating Group; Virgin Islands Government

Vlg Com Vigilance Committee (men); Vigilant(e) Committee (men)

VIGIC Virgin Islands Government Information Center

vigilant. (VIGILANT) visually guided infantry light antitank (missile)

vign vignette

VIGOPRI Virgin Islands Government Office of Public Relations and Information

vigs vigilantes

vii viscosity index improver

VIJ Vera Institute of Justice

Vik Vickers; Vikelas; Vikenti; Vikentievich; Viki; Vikie; Viking; Viktor; Viktoria; Vikramaditya; Viktorovich

Vikes Vikings

Viking Pr Viking Press

vil vertical injection logic; village

Vil Las Villas (Santa Clara); Vilnius (capital of Lithuania)

vill village

Vill Voice Village Voice

Villa Pancho Villa, orig. Doroteo Arango (1878–1923)

VIM Venture in Missions; Vertical Improved Mail; Virgin Islands Museum; Visible Impact Management; Voyager Interstellar Mission

VIMI Virgin Islands Medical Institute

v imp verb impersonal

v imper verb imperative

VIMS Vertical Improved Mail Service; Virginia Institute of Marine Science

vim/var vacuum-induction melt/vacuum-arc remelt

vin vehicle identification number; vinegar; vinyl

vin. vinum (Latin—wine)

Vin Vincent

VIN Vehicle Identification Number

Vina Viña del Mar, Chile

VINB Virgin Islands National Bank

Vince St Vincent, West Indies; Vincent

Vincent Vincent Van Gogh

vind vindicate; vindication

VINES Virtual Networking Software

VINHS Virgin Islands National Historic Site

vini viniculture

VINITI Vsesoyuznyi Institut Nauchnoi Tekhnicheskoi Informatsii (Russian-All Union Institute of Scientific and Technical Information)

VINP Virgin Islands National Park (West Indies)

Vinson Vinson Massif (Antartica's highest mountain)

vio viola

VIO Veterinary Investigation Office(r)

viol violin

viol violino (Italian—violin)

vip value improving product(s); variable information processing; variable input phototypesetting (VIP); vasoactive intestinal peptide; vehicle insurance plan; very important passenger; very important people; very important person; visual identification point; visual inspection protection

vip Virgil I Partch

VIP Value Improvement Project(s); Variable Information Processing; Very Important Person; Very Important Program; *Vías Internacionales de Panamá* (Panamanian airline); Viewers in Profile; Virgin Islands Police; Visitor Information Phone; Vocabulary Improvement Program; Volunteer in Parks; Volunteers in Probation

VIPAC Virgin Islands Public Affairs Council

VIPI Volunteers in Probation, Incorporated

vilpp variable-information processing package

VIPPS Verified Internet Pharmacy Practice Sites

vipre visual precision

VIPRE Very-Intense Pulsed Radiation Experiment

vips voice interruption priority system

VIP-VIP Value in Performance through Very Important People (motivational program)

viq verbal iq

vir vertical interval reference (automatic television color system)

vir. viridis (Latin—green)

Vir Virgil; Virgo

VIR Vendor Information Request

V.I.R. Victoria Imperatrix Regina (Latin—Victoria Empress and Queen)

vira vehicular infrared alarm

VIRB Virginia Insurance Rating Bureau

Virg Virgil; Virgin; Virginia

Virgil Roman poet Publius Virgilius Maro

Virginias Virginia and West Virginia

Virgins Park Virgin Islands National Park, Saint John Island

Virgo (Latin—The Virgin constellation)

virol virology

virr verb irregular

v/irr verb irregular

vis viscera; visible; visibility; visitor, visual; voice information service

vis. viscount; viscountess

Vis Visayan; Vista (postal abbreviation)

VIs Virgin Islands

VIS Veterinary Investigation Service; Video Information System; Visual Instrumentation Subsystem

VISAR Visual Inspection System for the Analysis of Reports

visc viscosity

Visc Viscount(ess)

viscer (Latin prefix—organ)— visceral

Viscount Bolingbroke Henry St John (1678–1751)

VISIT Visit to Innovative Schools for Interested Teachers

vismins visual minorities (Africs, Asiatics, racially mixed Hispanics)

vispa virtual storage productivity aid(s)

vissr visible infrared spin-scan radiometer

vista viewing instantly security transactions automatically

VISTA Volunteers in Service to America

vit vital; vitamin; vitreous

vit (Latin prefix—life)—vitamins

vit A carotene vitamin

VITA Volunteers for International Technical Assistance; Volunteers In Tax Assistance

vit A₁ nutritive vitamin found in egg yolk, milk, and milk products such as butter

vit A₂ freshwater fish-liver-oil vitamin

VITAL Variably Initialized Translator for Algorithmic Languages

vitamin(s) vital amine(s)

VITAP Voluntary Income Tax Assistance Program

vit B nutritive vitamin essential to digestive and nervous systems; found in breads, egg yolk, lean meats, fruits, nuts, green vegetables

vit B₁ thiamine

vit B₂ riboflavin

vit B₃ nicotinamide

vit B₆ pyridoxine

vit B₁₂ cobalmine-cyancobalmine

vit Bc folic-acid

vit B cx vitamin B complex (water-soluble vitamins B₁, B₂, etc.)

vit C ascorbic acid

vit cap. vital capacity

vit D antirachitic

vit D₁ calciferol and lumisterol

vit D₂ calciferol

vit D₃ cholecalciferol (natural vitamin D)

vit E antisterility vitamin; tocopherol

vitel. vitellus (Latin—egg yolk)

vit G riboflavin

vit H biotin

viti viticulture

vit K coagulant

vit K₁ blood-clotting vitamin

vit M folic-acid vitamin

vit. ov. sol. vitello ovi solutus (Latin—dissolved in egg yolk)

vit P permeability vitamin (bioflavonoid found in paprika)

vit PP pellagra-preventive vitamin (nicotinamide nicotinic acid)

vitr vitreous

Vitr Vitruvius Pollio (Roman writer on architecture)

vit rec vital records

vitriol concentrated sulfuric acid (oil of vitriol); copper sulfate (blue vitriol); ferrous sulfate (green vitriol); zinc sulfate (white vitriol)

vits vertical-interval test signals

vits & mins vitamins and minerals

vit stat vital statistics

vit U cabagin (anti-ulcer vitamin)

VIUS Virgin Islands of the United States

viv vivace

viv vivienda (Spanish—apartment house; dwelling)

Viv Vivian; Vivien; Vivienne; Vivyan; Vivyanni

VIV Virgin Islands View

VIVA Virgin Islands Visitors Association; Voices in Vital America (organization)

VIVB Virgin Islands Visitors Bureau

Vivette Genevieve

vivi vivisection

VIX Volatility Index

vix. vixit (Latin—he/she lived)

viz. videlicet (Latin—namely)

Viz Vizcaya (Biscay); Vizcayan (Biscayan)

Viz Bay Bahía Sebastían Vizcaíno, Mexico

Vizc Vizcaya

vj jet velocity

v J vorigen Jahres (German—last year)

V-J agar Vogel-Johnson agar

VJC Vallejo Junior College

VJ Day Victory in Japan Day—August 15, 1945

V-joint angular V-shaped masonry joint

vj's video jockeys

Vjschr Vierteljahrschrift (German—quarterly)

vk vertical keel; volume kill

V of K Voice of Kenya (radio-television network)

VKC Von Karman Center

VKI Von Karman Institute

VKIFD Von Karman Institute for Fluid Dynamics

VKM Van Kampen Merritt (U.S. Government Fund)

VKO Moscow, Russia (Vnukovo Airport)

vkr video kinescope recording(s)

VKR Vodennaya Kontr Rozvedka (Russian—Counter-Infiltration Organization)

vl viola; vision, left

v/l vapor-to-liquid

VL Ville

V/l vapor-liquid ratio

Vl Volino (Italian—violin); *Vlaanderen* (Dutch—Flanders)

VL Vaasa Line; Vaasan Laiva; Venezuelan Line; Viking Line; Volcano Line; Vulgar Latin

Vla very low altitude; very-large array (radio telescopes)

vla viola (Italian—viola)

Vla Venezuela; Vlaardingen

VLA Very Large Array (Radio Astronomy Observatory); Veterans' Land Administration (Canada); Volunteer Lawyers for the Arts

VLAA Volunteer Lawyers and Accountants for the Arts

vlad vertical line array directional (sonobuoy)

Vlad Vladimir; Vladivostok

Vlad Vladivostok (Russian—Rule the East)

vladd visual low-angle drogue delivery

Vlad(i) Vladimir

vlb very long baseline

v-l b vertical-lift bridge

vlba very long baseline array

VLBC very large bulk carrier

vlbi very-long baseline interferometry

vlbti very long-burning target indicator

VLCC very large cargo carrier (bulk freighter or tanker)

vlchv (VLCHV) very-low-cost harassment vehicle

Vlcs voltage-logic-current switching

VLCT Very Large Commercial Transport

vld visual laydown delivery

vldl (VLDL) very-low-density lipoproteins

vldz Valdez

Vle Vale

V^le Viale (Italian—Avenue; Boulevard)

vlf vertical linear foot; very low frequency (to 30 kc)

vlf (VLF) vectored lift fighter

VLF Vehicle License Fee

Vlg Village

vlh very lightly hinged

VLI Port Vila, Vanuatu (airport)

vllo violoncello (Italian—cello)

vln very low nitrogen; violin

Vln Valenciennes

vlnt van links naar rechts (Dutch—from left to right)

vlo vertical lockout

vlp video long play(er) (videodisc)

VLPC Very Large Product Carrier

vlr very long range; very low resistance

vlrc very long range commuter

vls vertical liquid spring

VLS Vertical Launching System

vlsi very-large-scale integration

vlsic very large-scale integrated circuit

vlt vehicle license tax; very large telescope; violet

vltg voltage

vlv valve; valvular

vl/vs voltage logic/voltage switching

Vly Valley

vm vertical (position) multiplier; voltmeter

vm (VM) virtual machine

v/m various marks; volts per meter

vm voormiddag (Dutch—P.M.); *vormittags* (German—forenoon; A.M.)

v M vorigen Monats (German—last month)

VM Value Management; Victory Medal; Viet Minh; Vulcan Materials

V & M Virgin and Martyr

VM Völkerkundemuseum der Universität Zürich (German—Ethnographic Museum of Zurich University)

V.M. Votre Majesté (French—Your Majesty); *Vuestra Majestad* (Spanish—Your Majesty); *Vuestra Merced* (Spanish—Your Worship)

vma vanyllilmandelic acid

VMA Valve Manufacturers Association

VMAG Vanderpoel Memorial Art Gallery

V-Mann Vertrauensmann (German—Trusted Man) intelligence agent

vmap video map equipment

V max maximum flight velocity

vmc visual meteorological conditions

VMC Viet Montagnard Cong

VMCCA Veteran Motor Car Club of America

vmd vertical magnetic dipole

V.M.D. Veterinariae Medicinae Doctor (Latin—Doctor of Veterinary Medicine)

VMDP Veterinary Medical Data Program

VME Versa Module Europa

VMG Vikers Machine Gun

vmh (VMH) ventromedial nucleus of the hypothalamus

VMH Victoria Medal of Honour

vmi visual motor integration; visual motor interaction

VMI Video Music Inc; Virginia Military Institute

VMIC Vermont Maple Industry Council

v/mil volts per mil

V min minimum flight velocity

VMLI Veterans Mortgage Life Insurance

vmm virtual machine monitor

v & mm vandalism and malicious mischief

VM Molotov Vyacheslav M Skryabin

vmos V-groove metal-oxide semiconductor

vmos (VMOS) vertical metaloxide semiconductor

VMOS Virtual Memory Operating System

vmp value of the marginal product

vm & p varnish makers and painters

VMR Victorian Mounted Rifles

vms vertical-motion simulator

VMS Veterinary Medical Society; Vibrational Medicine Specialist; Vigital Memory

System; Virtual Memory System; Voluntary Medical Services

VMSC Volunteer Medical Staff Corps

vmt vehicle miles travelled; very many thanks; video matrix terminal

vn vulnerability number; violin

v/n verb neuter

vn vellón (Spanish—copper-silver alloy)

VN Václay Neumann; Vietnam; Vietnamese; Vladimir Nabokov (1899–1977); Vocational Nurse

vna volatile nitrosamines

vna (VNA) ventral noradrenergic bundle

Vna Vienna

VNA Air Vietnam; Vietnam News Agency; Visiting Nurses Association

VNAF Vietnamese Air Force

vnav volumetric area navigation (three-dimensional)

VNB Valley National Bank

V-N B Verrazano-Narrows Bridge

Vnc (VNC) Vancouver, Washington

VN$ Vietnamese dollar

VN de B Vasco Nuñez de Balboa (first European to discover the Pacific Ocean)

V-neck V-shaped neck (line)

Vnese Vietnamese

vnf very near field

Vng Vereeniging

vni variable name initialization

Vni Violini (Italian—violins)

Vnla Venezolana (Spanish—female Venezuelan)

Vnlo Venezolano (Spanish—male Venezuelan)

VNM Victoria National Museum (Ottawa)

VNMC Vietnam Marine Corps

VNN Vietnam Navy

VNNBS Vietnamese National Broadcasting Service

Vno Violino (Italian—violin)

VNO Vital National Objective

V-note $5 bill

VNP Vietnamese piastre; Voyageurs National Park (Minnesota)

vnr variable navigation ratio

VNR Van Nostrand Reinhold

VNRC Vegetarian Nutritional Research Center

VNs Vietnamese

VNS Voter News Service

VNS Vereenigde Nederlands Scheepvaartmaatschappij (Dutch—United Netherlands Navigation Company)

vnw voornaamwoord (Dutch—pronoun)

VNWR Valentine National Wildlife Refuge (Nebraska)

vo voluntary opening

vo. verso (Latin—back of the page, lefthand page); *violino* (Italian—violin)

v/o vossa ordem (Portuguese—your order)

vº verso (Portuguese—lefthand page, other side, over, reverse)

Vo output voltage (symbol)

VO Valuation Officer(r); verbal order(s); very old; Veterinary Office(r); Victorian Order; voice over

VO Volksoper (German—People's Opera)—Vienna

voa vetoed on arrival (at the President's desk); voltohm ammeter

V o A Voice of America

VOA Vancouver Opera Association; Vasa Order of America; Virginia Opera Association; Voice of America; Volunteers of America

VOA Vereeniging Ontwikkeling Arbeldstechniek (Dutch—Work Study Association)

vo-ag vocational agriculture (educators' jargon)

vob vacuum optical bench

vobanc voice band compression

VºBº vista bueno (Spanish—approved, okay)

VºBº visto bueno (Spanish—okay)

voc vocal; vocalist; vocation; vocational; vocative; volatile organic compound

VOC Vehicle Observer Corps

VOC *Vereenigde Oostindische Compagnie* (Dutch—United East India Company)—often called the Very Old Company

VOCA Visiting Orchestras Consultative Association (London)

vocab vocabulary

VOCAL Vessel Ordnance Allowance List

vocat vocation(al); vocative

voc ed vocational education

Voc Foun Vocational Foundation

vocg verbal orders—commanding general

voco verbal order—commanding officer

vocoder voice coder

VOCOSS Voluntary Organisations Cooperating in Overseas Social Service

voc rehab. vocational rehabilitation

vocs verbal orders—chief of staff

voctl vocational

vod video on demand; vision of right eye (d standing for *dexter*—Latin for right)

v od voice-operated device; voluntary-opening device

vodacom voice data communication(s)

vodactor voice data compactor

vodaro vertical ozone distribution (from) absorption and radiation of ozone

vodat voice-operated device for automatic transmission

voder voice-operated demonstrator

VÖEST Vereinigte Österreichische Eisen and Stahlwerke (United Austrian Iron and Steel Works)

vof variable-operating frequency

vog volcano smog

Vog Vogue

VoG Voice of Germany

VOG Vanguard Operations Group

vogad voice-operated gain-adjusting device (data processing)

VOHI Vancouver Oral Health Index

VOICE Voice of Informed Community Expression

VOICES Voice-Operated Identification and Computer Entry System

VOIR Venus Orbiting Imaging Radar

VOIS Visual Observation Instrumentation Subsystem

voit voiture (French—railroad coach, truck, wagon, etc.)

vol volume; volunteer

vol % volume percent

vol. volatilis (Latin—volatile)

Vol Volans (constellation); Volcan; Volcano; volume

Vol Volcán (French—volcano); *Volcon* (Spanish—volcano); *Vulcano* (Italian—volcano)

VOLAG Voluntary Agency

VOLAR Volunteer Army

vol ash volcanic ash

volat volatile; volatizes

volc volcanic; volcano; volcanology

Vol Isl Volcano Islands (south of Japan and Bonin Islands)

Volks Volkswagen

volkst volkstaal (Dutch—slang; vernacular)

vollst vollstandige (German—complete)

Voln Volans (constellation)

vols volumes

vols. volunteers

VOLS Voluntary Overseas Libraries Service

Volta Voltaic Republic (Republic of the Upper Volta)

volts AC volts alternating current

volts DC volts direct current

volum volumetric

volvar volume variety

volvend. volvendus (Latin—to be rolled)

Volvo (Latin—I roll)—Swedish automobile

voly voluntary

vom volt milliammeter; voltolun microammeter; voltohm milliammeter; vomer; vomerine; vomit; vomitory; vomitus

vom. vomitus (Latin—vomit)

VOM Vereniging voor Opervlaktetechnieken Metalen (Dutch—Metal Finishing Association)

VOMI Volksdeutsche Mittelstelle (German—Racial Assistance Office for Germans Abroad)

vom neg vomito negro (Spanish—black vomit)—last stage of yellow fever

VON Victorian Order of Nurses (public health)

vona vehicle of the new age (computer-controlled rapid-transit shuttle)

von K Herbert von Karajan (1908–1989); Theodor von Kârman (1881–1963)

V.O.N.O. Vendor of Oysters in New Orleans (Walt Whitman's invention)

vop valued as in original policy

VOP very oldest procurable

Vo-Po *Volks Polizei* (East German Police)

VOQ Visiting Officer's Quarters

vor very high frequency omnidirectional range (VOR); vestibulo-ocular reflex; visual omnirange

VORAD Vehicle O-Board Radar

vordme very-high-frequency omnirange distance-measuring equipment

vorm *vormals* (German—formerly); *vormittags* (German-forenoon, A.M.)

Vor Mus. Voortrekker Museum (Pietermaritzburg)

VORP Victim Offender Reconciliation Program

Vors *Vorsitzender* (German—chairman)

vort vortex; vortices

vortac visual omnirange and tacan

vos vision of left eye (s standing for *sinister*—Latin for left)

vo(s) verbal order(s)

vos *vostok* (Russian—east, as in Vladivostok)

v.o.s. vitello ovi solutus (Latin—dissolved in egg yolk)

Vos *Voskresene* (Russian— Sunday)

VOS Victims of Superstition; visual observation airplane (naval symbol)

Vost *Vostochnyy* (Russian— eastern)

vot voice-onset time; voluntary overtime

vot. *votivus* (Latin—promissory or votive)

VOT Foreign Operational Center of Soviet Intelligence forces (formerly called MGB, MVD, NKGB, NKVD, OGPU, GPU, VECHEKA, and CHEKA—founded in December 1917)

votc volume table of contents

VOTE Voters Organized to Think Environment

voterm voice-operated typewriter employing morse

vou voucher

vow vowel(s)

VOW Voice of Women

VOWS Vilas-Oneida Wilderness Society

vox voice-operated transmission

vox pop. *vox populi* (Latin— voice of the people)

voy voyage

Voyageurs Voyageurs National Park on the Canadian border of Minnesota

Vozv *Vozvyshennost'* (Russian—uplands)

vp vanishing point; variable pitch; verb phrase; vertically polarized; vistaphone; vulnerable point

v/p verb passive; verb phrase

v & p vagotomy and pyloroplasty

V$_p$ valve-position voltage

VP British United Air Ferries fixed-wing fighter airplane, Valencia Park; Ville de Paris; Vice-President

VP (NSC) Verification Panel (National Security Council)

V-P Voges-Proskauer (reaction)

VP *Vigilancia de la Pesca* (Spanish—Fishery Patrol)

VPA Vancouver Public Aquarium (British Columbia); Videotape Production Association; Virginia Port Authority

v pag various paging

VP & B *Veterinary Pharmaceuticals and Biologicals*

vpc volume-packed cells

VPCP Volunteer Probation Counseling Program

vpd vapor-phase degrease; variation per day; vehicles per day

vpe vapor-phase epitaxy

vpg very pregnant guppy (NASA); voltage pressure gradient

VPGA Vermont Personnel and Guidance Association; Virginia Personnel and Guidance Association

vph variation per hour; vehicles per hour; vertical photography

vpi vapor-phase inhibitor

VPI Veterinary Pet Insurance; Virginia Polytechnic Institute; Vocational Preference Inventory

VPIRG Vermont Public Interest Research Group

vpl. visible panty line

VPL Van Pelt Library (University of Pennsylvania)

vpm vehicles per mile; versatile packaging machine; vertical panel mount; vibrations per minute; volts per meter; volts per mile

VPM Vendor Part Modification

Vpn Vickers pyramid number

V P/N vendor('s) part number

vpo vapor-phase oxidation

Vpo Valparaiso

VPO Vienna Philharmonic Orchestra

vpp viral porcine pneumonia

vpr vacuum pipette rig

VPR *Vanguarda Popular Revolucionaria* (Portuguese— Popular Revolutionary Vanguard)—Brazilian terrorist organization

V Pres Vice President

v-prez vice-president

vps vectors per second; vibrations per second; video projection system; volume pressure setting; voluntary placement services

VPS Visual Programme Systems

VPSA Vertebrate Paleontological Society of America

VPT Verbal Productive Thinking

V-P test Voges-Proskauer test

vq virtual quantum; visual quotient

vqa vendor quality assurance

vqc vendor quality certification

vqd vendor quality defect

VQMG Vice Quartermaster General

VQPRD *Vin de Qualiti Produit dans une Region Determinee* (French—quality wine produced in a defined region)

vqzd vendor quality zero defects

vr variable ratio; variable response; ventilated rib; very rare; vintage racing; vision, right; vital record(s); voltage regulator; volunteer regiment; volunteer reserve; vulcanized rubber

vir (VR) virtual reality; voluntary return (voluntary deportation of illegal aliens)

v/r verb reflexive

vr *vedi retro* (Italian—please turn over)

VR fixed-wing transport airplane; Victoria Railways (Australia)

V-R Veeder-Root

V.R. Victoria Regina

VR *Valtionrautatiet* (Finnish— State Railways)

V.R. *Victoria Regina* (Latin— Queen Victoria)

vra *vuestra* (Spanish—your, *f.*)

VRA Vocational Rehabilitation Administration; Voluntary Restraint Agreement

VRAM Video RAM (Random Access Memory)

vras *vuestras* (Spanish— your, pl.)

vrb voice rotating beacon

VRBG Viceroy's Bodyguard

vrbl variable

vrbl mnmncs verbal mnemonics (abbreviations and acronyms)

vrc vertical redundancy check(ing); visible record computer

VRC Vehicle Research Corporation; Victoria Rifles of Canada; Volunteer Rifle Corps

VRCAMS Vehicle/Road Compatibility Analysis and Modification System

VRCI Variable Restrictive Components Institute

VRCS Veterinary and Remount Conducting Section

vrd vehicle reception depot

v-r'd voluntarily returned (deported)

VRD (Royal Naval) Volunteer Reserve Decoration

VRDN Variable Rate Demand Note

vrdo variable rate demand obligation

vre vanomycin-resistant enterococci; voltage-regulator exciter

vir & e vocational rehabilitation and education

v refl verb reflexive

VR et I Victoria Regina et Imperatrix (Latin—Victoria, Queen and Empress)

V Rev Very Reverend

VRF Vehicular Research Foundation

vrg veering

Vrg Varig (Brazilian Airlines)

vri virus respiratory infection

vri (VRI) visual rule instrument landing

Vri Vrijdag (Dutch—Friday)

VRI Vehicle Research Institute; Victorian Railways Institute

V-ring V-shaped ring

VRIS Vietnam Refugee and Information Services

vrm variable-rate mortgage(s)

VRM Van Riebeeck Medal (South Africa); verbal response mode

vrml (VRML) virtual reality modeling language

v rms volt(s) root mean square

VRMS Visual Resources Management System

Vroni Veronica

vros vuestros (Spanish—your, pl.)

vrp valuable-record protector; very reliable product; voluntary reporting percentage

VRP Volta River Project

vrps voltage-regulated power supply

vrr visual radio range

VRR Veterans Reemployment Rights

vrs velocity response shape

VRS Van Riebeeck Society; Vanguard Recording Society; Veterinary Remount Service; Video Response System

V & RS Vocational and Rehabilitation Service

vrt vehicle reconnaissance, tracked; visual recognition threshold

VRT Voluntary Reserve Training; Volunteer Recruitment Team

vru voice recognition unit; voltage readout unit

vr vnw vragend voornaamwoord (Dutch—interrogative pronoun)

vrx virtual resource executive

Vry Viceroy

ys variable speed; vein shot (intravenous injection); venesection; ventricles; versus; visual signaling; vital statistics; volumetric solution

vs (VS) vital signs; voluntary simplicity

v.s. very soluble

vs. ve soire (Turkish—and so forth); *versus* (Latin—against)

v.s. vide supra (Latin—see above)

VS scouting airplane (2-letter symbol); Vancouver Symphony; Victoria Symphony

V.S. Veterinary Surgeon

V & S Valley & Siletz (railroad)

VS Vereinigte Staaten (German—United States); *Verenigde Staten van Amerika* (Dutch—United States of America); *Vostra Signoría* (Italian—Your Honor)

V S volti subito [Italian—turn (music page) swiftly]

vsa vehicle stability assist; victualling store allowance; voice stress analyzer

VSA Victorian Society of America; Volunteer Services to Animals

vsam virtual storage access method

VSAT Very Small Aperture Terminal

VSAP Vehicle Structure Analysis Program

vsb vestigial sideband

vs. b. venesectio brachii (Latin—bleeding in the arm)

VSB VME (Versa Module Europa) System Bus

VSBA Vermont School Boards Association; Virginia School Boards Association

VSBP Voluntary Sterilization Bonus Plan

vsby visibility

vsc vehicle skid control; virtual speech control

v.s.c. vidi siccam cultam (Latin—I have seen a dried cultivated specimen) botanic term

VSC Virginia State College; Vocations for Social Change; Volunteer Staff Corps

VSCC Vintage Sports Car Club

vscf variable-speed constant-frequency

VSCU Vatican Secretariat for Christian Unity

vsd vehicle supply depot; ventricular septal defect

VSD Vancouver School of Design; Vendor's Shipping Document(s); Veteran Services Division; Veterans Affairs Department

VSDA Video Software Dealers Association

VSE Vancouver Stock Exchange

vsep very superior extra pale

vsf vestigial sideband filter

vsff volte, se faz favor (Portuguese—please turn over)

VS f U Vatican Secretariat for Unbelievers

VSGLS Vehicle Space Ground Link Subsystem

V-shape V-shaped

vshps vernier solo hydraulic power supply

vsi variable-speed indicator; very seriously ill; very slight imperfection; very slight inclusion; very small inclusion

VSI Vinyl Siding Institute

V-sign victory sign (raised index and middle fingers)

v signs vital signs (blood pressure, pulse, temperature, respiration)

vs jw vise jaws

vsl variable safety level

VSL Venture Scout Leader

vsm vibrating-sample magnetometer

vsmf visual search microfilm file

VSMF Vendor Spec Microfilm File

VSMS Vermont State Medical Society; Vineland Social Maturity Scale

vsn vision

V S/N vendor('s) serial number

VSNAP Vermont State Nuclear Advisory Panel

vso very special old; very superior old

VSO Vancouver Symphony Orchestra; Victoria Symphony Orchestra; Victor Symphony Orchestra; Victualling Stores Officer; Vienna State Orchestra; Vienna Symphony Orchestra

VSOE Venice-Simplon Orient Express

vsop very superior old pale (cognac)

vsp vertical seismic profile

VSP Vespertina (Latin—Vespers)

VSP VS Pritchett

VSPA Virginia State Port Anthority

vspc virtual storage personal computing

vsq very special quality (VSQ)

vsr variable speed reversible; very short range; visual security range

vss versions

vss (VSS) v/stol support ship

v.s.s. vidi siccam spontaneam (Latin—I have seen a dried wild specimen)—botanic term

VSS Vancouver Symphony Society; Vermont State Symphony; Voluntary Social Services

VSSSN Verification Status Social Security Number

vst violinest

V St A Vereinigte Staaten von Amerika (German—United States of America)

v/stol vertical and/or short take-off and landing

V-S TV Viewer-Supported Television

vsula vaccination scar upper left arm

Vsv vesicular stomatitis virus

vsw vitrified stoneware

VSW Visual Studies Workshops

vswr voltage standing wave ratio

VSX heavier-than-air antisubmarine warfare carrier-based aircraft (naval symbol)

vt vacuum technology; vacuum tube; variable time; velocity; ventricular tachycardia; verb transitive; vinyl tile; vinyl tiling; voice tube

vt (VT) vertical tabulation character (data processing)

v-t vacuum technology; variable time (fuse); velocity-time (diagram); volume-time

v/t verb transitive

v & t volume and tension (of the pulse)

vt vaart (Dutch—canal); *viz tez* (Czech—see also)

v T vom Tausend (German—per thousand)

Vt Vermont, Vermonter; Vietnam(ese)

VT fixed-wing trainer-type airplane; *Reseau Aérien Interinsulaire* (Tahiti airline); Vermont; Virgil Thomson (1896–1989)

VT Vetus Testamentum (Latin—Old Testament)

vta ventral tegmental area

v^ta vuelta (Spanish—turn)

VTA Virginia Teachers Association

VTA Voenno-Transportnayaviatsiya (Russian—Air Transport Aviation)

v/tab vertical tabulation

VTAE Vocational Technical and Adult Education (System)

VTB Vereniging voor het Theologisch Bibliothecariaat (Dutch—Association of Theological Librarians)

vtc voting trust certificate

VTC Vermont Technical College; Volunteer Training Corps

vte vertical-tube evaporator (for producing freshwater from the sea); vicarious trial and error

Vte Vicomte

V-TECS Vocational-Technical Education Consortium of States

Vtesse Vicomtesse

V-test Voluter test

vtf vertical test fixture

vt fuse variable-time fuse

vtg voting

VTG Vehicle Technology Group

vti volume thickness index

VTI Valparaiso Technical Institute

vtl variable threshold logic; vertical turret lathe

VTLs Vehicular Traffic Laws

vtm voltage tuned magnetron

VTM Victorian Tourist Ministry (Australia); Virtual Trade Mission

VTN Video Tape Network; Voorheis, Trindle, and Nelson

vto vertical takeoff; viable terrestrial organism; voltage-tuned oscillator; voltage-turntable oscillator

v^to vuelto [Spanish—change (money)]

Vto Vtornik (Russian—Tuesday)

vtoc volume table of contents (data processing)

vtohl vertical takeoff and horizontal landing

vtol vertical takeoff and landing

vtolport vertical-takeoff-and-landing airport

V-to-V Vancouver-to-Vladivostok

vtovl vertical takeoff vertical landing

vtp voluntary termination of pregnancy (abortion)

vtp viajes todo pagado (Spanish—all trips paid)

vtpir vertical temperature profile radiometer

vtr video tape recorder; video tape recording

vtr. vitreum (Latin—glass)

VTR Vermont Railway

VTRS Video Tape Recorder System

vtr sot. videotape recorder sound on recorder tape

VtSEA Vermont Society of Enrolled Agents

VTS Viewfinder Tracking System; Virginia Theological Seminary

VTSRS Verdun Target Symptom Rating Scale

VTTA Veteran's Time Trial Association

VTU Volunteer Training Unit

vtvm vacuum-tube voltmeter

vu varicose ulcer; very uncommon; view; voice unit; volumetric unit; volume unit

vu von untem (German—from the bottom)

VU Air Ivoire; fixed-wing utility airplane; Valparaiso University; Vanderbilt University; Vanuatu (Internet code); Vice Unit (police); Victoria University; Villanova University; Vincennes University

VU Vigile Urbano (Italian—Traffic Policeman)

VUA Valorous Unit Award

VUA Vrije Universiteit, Amsterdam (Dutch—Free University—Amsterdam)

vue d'opt vue d'optique (French— optical view) multidimensional art

VUH Vanderbilt University Hospital

vu indicator volume-unit indicator (data processing)

Vul Vulgate; Vulpecula (constellation)

vulc vulcanize(d,r)

vulcan vulcanization; vulcanize; vulcanizer; vulcanizing

vulcani vulcanist

vulcanol vulcanologist

vulg vulgar; vulgar fraction; vulgarian; vulgarism; vulgarist; vulgarization

Vulg Vulgar Era (Christian Era); Vulgar Latin; Vulgate

vulp vulpine

Vulp Vulpecula (constellation)

Vulpec Vulpecula (Latin—Little Fox constellation)

v-u meter volume-unit meter

VUNC Voice of United Nations Command

v u p (VUP) very unimportant person

VU-PD Vice Unit-Police Department

VU Pr Vanderbilt University Press

VUSM Vanderbilt University School of Medicine

vuv vacuum ultraviolet

VUW Victoria University of Wellington, New Zealand

vv vagina and vulva; verbs; verses; vice versa

viv volume for volume

v & v verification and validation; vintage and veteran (automobiles)

v. v. vice versa (Latin—conversely); *violini* (Italian—violins)

Vv. Virgines (Latin—Virgins)

VV Villa Viscaya (Dade County Art Museum, Miami, Florida); Voice of Vietnam (Hanoi)

VV ustedes (Spanish—you, pl.).; *Viva Verdi!; Viva Vivaldi!* (Italian—Long live Verdi; Long live Vivaldi)

VVA Vietnam Veterans Association; Vietnam Veterans of America

VVAF Vietnam Veterans of America Foundation

VVAW Vietnam Veterans Against the War

v.v.c. vidi vivant cultam (Latin— I have seen a living cultivated specimen)—botanic term

vvcd voltage-variable capacitor diode

VVCP Victims of Violent Crimes Program

VVD Volkspartij voor Vrijheid en Dentocratie Dutch— People's Party for Freedom and Democracy)—Liberal Party

vvds video verter decision storage

Vve Veuve (French—widow)

vvhr vibration velocity per hour

Vvl Varavel

vv. ll. *variae lectiones* (Latin— variant readings)

VVLP Vietnam Veterans Leadership Program

VVMF Vietnam Veterans Memorial Fund

VVN Verein der Verfolgten des Naziregimes (German— League of Victims of Naziism)

VVND Velogenic Viscerotropic Newcastle Disease

VVO very, very old

vvr variable-voltage rectifier

vvrm vortex valve rocket motor

vvs very, very superior

vv's varicose veins

v.v.s. vidi vivam spontaneam (Latin—I have seen a living wild specimen)—botanic term

vvsi very very small inclusion

V v V Volkspartij voor Vrijheid en Dentocratie (Dutch—Peoples' Party for Freedom and Democracy)

VVS Veteran's Vigil Society (Vietnam-era veterans)

V-VS Voenno-Vozdushniye Sily (Russian—Air Forces of the USSR)

vvsf very very slightly flawed (gems)

vvsi very very slight imperfection; very very slight inclusion

vvsop very very superior old pale (cognac)

vvt variable valve technology; variable valve timing

VVT Visual-Verbal Test

vvti variable valve timing-intelligent

VV UU *Vigili Urbani* (Italian—Traffic Police)

v.v.v. *veni; vidi, vici* (Latin—I came, I saw, I conquered)

VVV Vasili Vasilievich Vereschagin

vw vessel wall; volts working

vw *voegwoord* (Dutch—conjunction)

Vw View (postal abbreviation)

VW Very Worshipful; Volkswagen (People's Car)

vWd von Willebrand's disease

VWD *Vereinigte Wirtschafte Dienst* German—German News Agency)

vWf von Willebrand factor

vWfag von Willebrand factor antigen

vwg vibrating wire gage

VWG Verification Working Group

vwl variable word length

VWOA Volkswagen of America

vwp variable width pulse

VWP Victim/Witness Project; Visualize World Peace

VWPI Vacuum Wood Preservers Institute

vws ventilated wet suit; vibrating-wire stressmeter(s)

VWWI Veterans of World War I

vx venous invasion can't be assessed (symbol); vertex

VX Experimental Squadron (symbol)

vxo variable crystal oscillator

Vxtmps Vieuxtemps

vy various years; very

VY Air Cameroun; Victualling Yard

vyd *vydani* (Czech—edition)

Vygr Voyager (robot spacecraft)

Vy Rev Very Reverend

vyt *vytah* (Czech—abstract)

vz virtual zero

v-z varicella-zoster

vzd vendor zero defect(s)

VZP Venezuelan Petroleum Company (stock exchange symbol)

vxt visual zenith tube

W

w   loading (symbol); transverse acoustical displacement (symbol); wall; war; warm; waste; water; water vapor constant; watt; weather; week; weekly; weight; wet; white; wide; widow; widowed; width; wife; will; win; wind; windy; wine; winning pitcher; wins; with; won; wood; word; work; work (symbol); worn; wrong

w %   weight percent

w +   weakly positive

w −   weakly negative; will factor

w   *vatios* (Spanish—watts)

W   Canadian Car & Foundry (naval designator symbol); College of Wooster; gross weight (symbol); irradiance (symbol); tungsten (Wolfram); very wide (symbol); Wales; Ward; Ward Line; warning; Washington; water; Waterman Steamship Line; watt(s); weather reconnaissance; Weber fraction; Wednesday; Welsh; west; western; Westinghouse; Weyerhaeuser; Whiskey—code for letter W; WillysOverland; Woolworth; Writhing number; Writer; Wu

W(W)   wage rate (microeconomics)

W   *Wadi* (Arabic—gulley, ravine, riverbed); *Wald* (German—forest, wood); *Wan* (Chinese or Japanese—bay; bight); *warm* (Afrikaans, Dutch, German—hot); west;

west (Afrikaans, Dutch, German—west); Wilhelmsen (steamship line); women

W1, W2, etc.   West One, West Two, etc; (London postal zones)

wa   warm air; wire armored; with average; work energy; writing ability

w/a   welded assembly

Wa.   Warden

Wa   *Waffenamt* (German—Ordnance Department)—Third Reich marking followed by a code number and stamped on all military equipment

WA   Wabash Railroad (stock exchange symbol); Warrant Armourer; Washington; Watchmen's Association; Weapon Armourer; Weapons Analyst; Welfare Administration; West Africa; West African; Western Airlines; Western Approaches (to British Isles); Western Australia; Wheeler Airlines; Wire Association; Woolknit Associates; Workshop Assembly

W of A   Western of Alabama (railroad)

W A   *World Almanac and Book of Facts*

waa   wartime aircraft activity; welded aluminum alloy

waa (w/a/a)   with added adhesive

WAA   War Assets Administration; Warden's Association of America; Western Amateur

Astronomers; Western Australia Artillery; Women's Auxiliary Association

WAA   *World Aluminum Abstracts*

WAAA   Women's Amateur Athletic Association

WAAAF   Women's Australian Auxiliary Air Force

WAABI   World Association of Alcohol Beverage Industries

WAAC   West African Airways Corporation; Western Association for Art Conservation; Women's Auxiliary Army Corps

WAACs   Women's Auxiliary Army Corps

WAADC   Women's Auxiliary Army Drivers Corps

WAADS   Washington Air Defense Sector

WAAE   World Association for Adult Education

WAAF   Women's Auxiliary Air Force

WAAFB   Walker Air Force Base

waaj   water-augmented air jet

WAAM   Wide Area Anti-Armor Munitions

WAAP   World Association for Animal Production

waapm   wide-area anti-personnel mine

WAAS   Women's Auxiliary Army Service; World Academy of Art and Science

WAAVP   World Association for the Advancement of Veterinary Parisitology

wab water-activated battery; when authorized by

WAB Wabash (railroad); Wage Adjustment Board; Wage Appeals Board; Western Actuarial Bureau; Westinghouse Air Brake; White American Bastion (neo-Nazi group); Wine Advisory Board; Women's Abolition Bureau (for the abolition of adultery, alcoholism, and discrimination)

WABCO Westinghouse Air Brake Company

WABF *World Almanac and Book of Facts*

wablics waterborne logistical craft (junks, sampans, wallawallas)—Hong Kong harbor craft

wac wage analysis and control; waste acceptance criteria; weapon assignment console; weighted average coupon; write address counter

WAC West Africa Command; Western Athletic Conference; Women's Action Coalition; Women's Army Corps (USA); Women's Auxiliary Corps; Worked All Continents; World Aeronautical Chart; World Affairs Council

WAC(I) Women's Army Corps (India)

WACA Western Agricultural Chemicals Association; World Airline Clubs Association

WACB Women's Army Classification Battery

WACC Washington Association of Community Colleges

WACCC Worldwide Air Cargo Commodity Classification

wack wait before sending positive acknowledgement

WACL World Anti-Communist League

WACM Western Association of Circuit Manufacturers

waco written advice of contracting officer

WACO World Air Cargo Organization

WACRI West African Cocoa Research Institute

WACSM Women's Army Corps Service Medal

WACSSO Western Australian Council of State School Organisations

WACVA Women's Army Corps Veterans Association

Wad Wadham College, Oxford

WAD World Association of Detectives; Wright Aeronautical Division (Curtiss-Wright Corporation)

WADC Western Air Defense Command; Wright Air Development Center

wadd with added (costs, freight, etc.)

WADD Westinghouse Air Arm Division; Wright Air Development Division (USAF)

Wade-Giles Sir Thomas Wade (1818–1895) and Herbert Giles (1845–1935) (English lexicographers)

wadex word and author index

WADF Western Air Defense Force

Wadh Wadham College, Oxford

WADS Wide Area Data Service; Wide Area Dialing Service

Wadsworth Wadsworth Atheneum (Hartford)

wae when actually employed

WAE West African English

WAED Westinghouse Aeorospace Electrical Division

WAEMA Western and English Manufacturers Association

WAES Workshop on Alternative Energy Strategies

waf with all faults

WAF West African Forces; Women's Auxiliary Force; Women in the Air Force

WAFB Warren Air Force Base

WAFC West African Fisheries Commission

WAFF West African Frontier Force

waffle wide-angle fixed-field locating equipment

W Afr West Africa(n)

waf(s) waffle(s)

WAFS Women's Air Force Services; Women's Auxiliary Ferrying Service

wag. wagoner

WAG Walters Art Gallery; Winnipeg Art Gallery; Wireless Air Gunner; Writers' Action Group

W A & G Wellsville, Addison & Galeton (railroad)

WAGBI Wildfowlers' Association of Great Britain and Ireland

WAGGGS World Association of Girl Guides and Girl Scouts

Wag hrn Wagner horn

wagn wagon

wagr windscale advanced gas-cooled reactor

WAGR Western Australian Government Railways

WAGRO Warsaw Ghetto Resistance Organization

wags weighted agreement scores

WAGS Wireless Air Gunner School

wai walk-round inspection; water installed

wai. waitress

WAI Work in America Institute

WAIF World Adoption International Fund

WAIR Western Australia Infantry Regiment

WAIS Wechsler Adult Intelligence Scale; Wide Area Information Server

WAIT Wechsler Adult Intelligence Test(ing); Western Australian Institute of Technology

WAITR West African Institute for Trypanosomiasis Research

waj water-augmented jet

WAJ World Association of Judges

wak water analyzer kit; wearable artificial kidney; with all knowledge

Wakefield Wakefeld Prison south of Leeds in Yorkshire, England

wal walnut; wide-angle lens

Wal Wallace; Wallach; Wallachian; Wallsend-on-Tyne

WAL Western Airlines; Westinghouse Astronuclear Laboratory; Westland Aircraft Limited

W-AL Westinghouse-Astronuclear Laboratory

WALA West African Library Association

WALDO Wichita Automatic Linear Data Output (Boeing)

Wales section of Great Britain; The Wales—The Bank of New South Wales

Wal I Wallops Island

WALIC Wiltshire Association of Libraries of Industry and Commerce

walk-in robes walk-in wardrobe closets

Wall Walloon

Wall *Wallace* (U.S. Supreme Court Reports)

Wallenstein Albrecht Wenzel Eusebius von Wallenstein (Bohemian general)

Wallis and Futuna Wallis and Futuna Islands in the southwest Pacific near Samoa

Wall-Wall prison in Walla Walla, Washington

Wally Wallace; Walter

Walnut Canyon Walnut Canyon National Monument in north-central Arizona

walopt weapons allocation optimizer

Walpurgis Walpurgis Night (April 30 in Finland and Sweden)

WALST Western Alaska Standard Time

Wal Sta Wallops Station

Walt Walter; Walton

wam walk-around money; weighted average maturity; wife and mother; words a minute; wrap-around mortgage

wAm white American male

WAM We Aint Metric; Western Australian Museum (Perth); Wolfgang Amadeus Mozart; Women Against Men; Women in Advertising and Marketing; Worcester Art Museum; Working Association of Mothers

WAMI Washington, Alaska, Montana, Idaho

WAML Watertown Arsenal Medical Laboratory

wamoscope wave-modulated oscilloscope

WAMP Wire Antenna Modelling Program

wampum wage and manpower process utilizing machines

WAMRU West African Maize Research Unit

WAMY World Assembly of Muslim Youth

wan wide area network

WAN West Africa Navigation (steamship line); Wide-Area Network

WANA We Are Not Alone

WANAP Washington National Airport

Wand Wanderers

WAND Women's Action for Nuclear Disarmament

WANDPETLS Wandsworth Public Educational and Technical Library Services

Wandsworth Wandsworth Prison, London, England

Wankie Wankie National Park in Rhodesia

WANL Westinghouse Astronuclear Laboratories

wanna (American slang) want to

wannabe (American slang) want to be

WANR Wadi Amud Nature Reserve (Israel)

WANS Women's Australian National Service; Women's Australian Nursing Society

WANYNJ Warehousemen's Association of New York and New Jersey

wao wet-air oxidation

WAO Weapons Assignment Office(r)

WAOB World Agricultural Outlook Board

WAOS Welsh Agricultural Organization Society; Wide-Angle Optical System

wap water planned; weak anthropic principle; wide-angle panorama; wireless application protocol

WA£ West African pound

WAP Women Against Pornography; Work Assignment Plan; Work Assignment Procedure; Writing Associates Program

WAPA Western Area Power Administration; White American Political Association

WAPC Women's Auxiliary Police Corps

WAPD Westinghouse Atomic Power Division

WAPET Western Australia Petroleum Pty Ltd

WAPO White American Political Organization

WAPOR World Association for Public Opinion Research

WAPPRI World Association of Pulp and Papermaking Research Institutes

WAPs Work Assignment Plans

WAP's Work Assignment Plans

WAPS World Association of Pathology Societies

WAPSD Westinghouse Electric Corporation Advanced Power Systems Division

WAPT Wild Animal Propagation Trust

WAPV gunboat (USCG symbol)

war we are ready

war. warrant; with all risks

War War Department; Warsaw; Warwickshire

WAR West African Regiment; Western Australia Rifles; White Aryan Resistance (neo-Nazi group); William A Rusher; Women Against Rape

WARC Western Air Rescue Center; World Alliance of Reformed Churches

warcat workload and resources correlation analysis technique(s)

WARDA West African Rice Development Association

WARES Workload and Resources Evaluation System

warex (WAREX) we have a warrant and will extradite

warf warfare

Warf Warfarin (rodenticide)

WARFI Wisconsin Alumni Research Foundation

WARF Western Alumni Research Foundation Institute; Wisconsin Alumni Research Foundation Institute

wargasm war + orgasm; sudden outbreak of war

warhd warhead

WARI Waite Agricultural Research Institute

Warks Warwickshire

warla wide-aperture radio location array

WARLOCE Wartime Lines of Communication—Europe

warn. warning

WARN Worker Adjustment and Retraining Notification (Act of 1988)

warr warranty

WARRS West African Rice Research Station

WARS Worldwide Ammunition Reporting System

wat. weighted average remaining term

was. wide-angle sensor; wideband antenna system

WAS Worked All States

WASA Washington Association of School Administrators; Wyoming Association of School Administrators

WASAL Wisconsin Academy of Sciences, Arts, and Letters

WASAMA Women's Auxiliary to the Student American Medical Association

Warsaw Pact Former mutual defense alliance among Bulgaria, Czechoslovakia, East Germany, Hungary, Poland, Romania, and the USSR

was wheel alignment stopper

WASB Wisconsin Association of School Boards; Women's Auxiliary Service Burma

WASBO Washington Association of School Business Officials; Wisconsin Association of School Business Officials

WASC Western Association of Schools and Colleges

wascala wide-angle scanning-array lens antenna

WASCO War Safety Council

WASDA Wisconsin Association of School District Administrators

WaSEA Washington Society of Enrolled Agents

Wash Washington; Washingtonian

WASH White Anglo-Saxon Hebrew

Wash Corr Cen Washington Correctional Center

Wash DC Washington, D.C.

Washington's George Washington's Birthday (February 22)

Washmic Washington (DC) military-industrial complex

WASHO Western Association of State Highway Officials

Wash Post *The Washington Post*

Wash St Hist Soc Washington State Historical Society

Wash St U Pr Washington State University Press

Wash U Med Lib Washington University school of Medicine Library (St Louis)

WASI Wage and Salaries Index

WASL Washington Assessment of Student Learning

wasn't was not

WASO West Australian Symphony Orchestra

wasp. weightless analysis sounding probe; window atmosphere sounding projectile

WASP Waddell Sea Project; War Air Service Program; Water and Steam Program; White Anglo-Saxon Protestant; Williams Aerial System Platform; Women Against Soaring Prices; Women's Air Force Service Pilots; Women's Auxiliary Service Pilots; Workshop Analysis and Scheduling Program; Wyoming Atomic Simulation Project

WASP(S) White Anglo-Saxon Protestant(s)

Wass Wasserman

WASS Washington Association for Scientific Security

Wassermann August von Wassermann—German bacteriologist who devised test to determine diagnosis of syphillis

WASSP Washington Association of Secondary School Principals

WAST Western Australian Standard Time

WASU West African Student's Union

wat weight, altitude, temperature; wins above team

Wat Waterford

WAT Word Association Test; World Airport Technology

WATA World Association of Travel Agencies

watashi watakushi (Japanese—I, me, myself)

Watchungs Watchumg Mountains of northern New Jersey

WATDA Western Australia Tourist Development Authority

water H_2O

WATER Working Alliance To Equalize Rates

Water Gap Delaware Water Gap between New Jersey and Pennsylvania

water res water resistant

watertec water technologist; water technology

Waterton Waterton-Glacier International Peace Park on the Alberta-Montana border or Waterton Lakes National Park in the same area

watg wave-activated turbine generator

WATPL Wartime Traffic Priority List

wats wide-area telephone service

WATS Wide Area Telecommunications Service; Women's Auxiliary Territorial Service

Wat Sta Waterloo Station (rail terminal)

watt's wide-area telephone transmission lines

W Aust Western Australia

W Aust Cur West Australian Current

WAVA World Association of Veterinary Anatomists

WAVAW Women Against Violence Against Women

WAVES Women Accepted for Volunteer Emergency Service (USN)

WAVFH World Association of Veterinary Food Hygienists

WAW Warsaw, Poland (airport)

WAwa West Africa wins again

WAWF World Association of World Federalists

wax. weapon assignment and target extermination

'way away

WAY World Assembly of Youth

WAYC Welsh Association of Youth Clubs

Wayne St U Pr Wayne State University Press

'ways always

WAZ Worked All Zones

wb warehouse book(ing); water ballast(ing); waybill; weber; wheelbase; whole blood; widebeam; will book; wingback; winner's bitch

w&b weight and balance

w/b westbound; will be

Wb weber

WB Wage Board; Warner Brothers; Weather Bureau; Weekly Bulletin; Western Blot; Women's Bureau; World Bank for Reconstruction and Development (UN)

W&B Works and Buildings (Royal Air Force)

W-B Wilkes-Barre, Pennsylvania

wba wideband amplifer

WCSRC

CNYH Waterfront Commission of New York Harbor

CO Warrant Communications Officer; Weapons Control Office(r)

COTP World Confederation of Organizations of the Teaching Profession

P welder control panel; white combination potentiometer

P Weapon Control Plan; World Council of Peace; Work Control Panel; Work Control Plan

PA Western College Placement Association; Western Crop Protection Association; World Constitution and Parliament Association

PS Women's Caucus for Political Science; World Confederation of Productivity Sciences

T World Confederation of Physical Therapy

water-cooled reactor; water-cooled rod; water cooler; wire contact relay; word-control register

Western Communication Region (USAF); Women's Council of Realtors

A Weather Control Research Association; Western College Reading Association; Women's Cycle Racing Association

World Climate Research Program; World Council of Religion for Peace

waste collection system; center section

Weapons Control Station; Weapons Control System; Wisconsin Correctional ce

West Coast of South ica

weapon control switch-

weapon control system e

World Correctional e Center

A Wisconsin Cheese Specialty Food Merchant Association

World Center for Scientific Information

Wild Canid Survival Research Center

WCA

W B T & S Waco, Beaumont, Trinity & Sabine (railroad)

wbtv weather briefing television

wbv wideband voltage

wbvco wideband voltage-controlled oscillator

WBW World Bowling Writers

W By Walvis Bay

wc wadcutter; wage change; water closet (English euphemism for lavatory); weapon carrier; wet coniferous (forest); wheelchair; will call; without charge; wood casing; working capital; working circle; workmen's compensation

w/c wave change; with corrections (correct proof before printing)

WC Wabash College; Wagner College; Waldorf College; Walker College; Walsh College; War Cabinet; War College; Wartburg College; Washington College; Waynesburg College; Weatherford College; Webber College; Weber College; Webster College; Wellesley College; Wells College; Wesley College; West African Airlines; West Coast Airlines; Western Command; Westmar College; Westminster College; Westmont College; Wheaton College; Wheeling College; Wheelock College; Whitman College; Whittier College; Whitworth College; Wiley College; Wilkes College; Williams College; Wilmington College; Wilson College; Windham College; Wings Club; Winthrop College; Wofford College; Woodbury College; Woodstock College; Workers' Compensation; World Court; Wycliffe College

W/C Weapons Controller; Wing Commander

W/Cdr Wing Commander

WC1, WC2, etc. West Central One, West Central Two, etc. (London postal zones)

wca wideband cassegrain antenna; worst case analysis

WCA Wackenhut Corrections Corporation; Washington Correctional Association; Washingtonian Center for

Addiction; Western Correctional Association; Wisconsin Correctional Association; Women's Correctional Association; Workers' Compensation Act; World Calendar Association

WCAA West Coast Athletic Association; Window Coverings Association of America

w cab wall cabinet

WCAC West Coast Athletic Conference; Women's Crusade Against Crime (St Louis)

WCAFS Wideband Cassegrain Antenna Feed System

WCAP Westinghouse Commercial Atomic Power

WCas Western Cascades (mountains)

WCAT Welsh College of Advanced Technology

WCB Workmen's Compensation Board

WCBA West Coast Bookmen's Association; Western College Bookstore Association

WCBHS William Cullen Bryant High School

wcc water-cooled copper; wilson cloud chamber

WCC War Crimes Commission; Wayne County Community College; Westchester Community College; Western Cartridge Company; Westminster Choir College; White Citizens' Council; Women's Cultural Center; World Council of Churches

wcca worst-case circuit analysis

WCCE West Coast Commodity Exchange

WCCI World Council for Curriculum and Instruction

WCCU World Council of Credit Unions

WCD Workers Compensation Department

wcdb wing control during boost

wcdo war consumable distribution objective

wce weapon control equipment

WCE Winnepeg Commodities Exchange

WCED World Commission on Environment and Development

WCEMA West Coast Electronic Manufacturers' Association

WCEU World's Christian Endeavor Union

wcf white cathode follower

WCF Waste Calcining Facility; Winchester Center Fire (rifle shell designation)

WCFPR Washington Center of Foreign Policy Research

WCFST Weigl Color-Form Sorting Test

WCFTB West Coast Freight Tariff Bureau

WCG Women's Cooking Guild

wci white cast iron; wind chill index

WCI Wildlife Conservation International

WCIA Watch and Clock Importers Association

WCIR Workers Compensation and Insurance Report(ing)

WC & IR Workmen's Compensation and Insurance Report(ing)

WCJE World Council on Jewish Education

WCK West Virginia Coal and Coke (stock exchange symbol)

wcl watercooler

WCL West Coast Line; World Confederation of Labor

W-class Soviet class of submarines named Whiskey by NATO

wcld watercooled

WCLIB West Coast Lumber Inspection Bureau

wcm welded cordwood module; wired-core matrix; wired-core memory; word combine and multiplexer

WCM Worshipful Company of Musicians

WCMA Wiping Cloth Manufacturers' Association; Wisconsin Cheese Makers' Association

WCML Women's Caucus for Modern Languages

WCMR Western Contract Management Region

WCN Women's Consumer Network

WCNA West Coast of North America

WCNM Walnut Canyon National Monument

WCNP Wind Cave National Park (South Dakota)

WC & S's S & EBC William Cramp & Son's Ship and Engine Building Company

WCST Wisconsin Card-Sorting Test

WCT World Championship Tennis; World Championship Tour

WCTA Western Coal Transportation Association

WCTB Western Carriers Tariff Bureau

WCTL Western Center Telecommunications Laboratory

WCTU Wild Cats and Tigers United; Women's Christian Temperance Union

WCU West Coast University; Western Carolina University

WCUK West Coast of United Kingdom

wcv water check valve

WCW William Carlos Williams

WCWB World Council for the Welfare of the Blind

WCY Westmoreland and Cumberland Yeomanry

wd water damage, waveform deterioration; weed; well deck(ing); when distributed (newspaper stock listings); white dwarf (star); whole depth; will dated; wind; window; winner's dog; withdrawn; wood; word; would; wound

wd. ward; widow

w/d warranted; washer and dryer; weight-displacement ratio; wind direction

Wd weeds

WD War Department; Water Department; Waterworks Department; Western Division

wda wheeldrive assembly; withdrawal of availability; writing down allowance (tax)

WDA Warranty Disclosure Act; Welsh Development Agency; Wholesale Distributors Association

WDAF Western Desert Air Force

WDALMP Warehouse Distributors Association of Leisure and Mobile Products

WDC War Damage Commission; War Department Constabulary; Western Defense Command; Women's Detention Center; World Data Center

WDC-A World Data Center-A (Washington, D.C.)

WDC-B World Data Center-B (Moscow, Russia)

wdd Western Development Division (USAF Air Research and Development Command)

wdf wood door and frame

WDF Western Desert Force

wdg winding; wording

wdi water district

WDIF Women's Democratic International Federation

wdk wives don't know

WDL Western Defense Laboratories (Philco subsidiary of Ford Motor Company)

wdl. withdrawal

WDM Western Development Museum (Saskatoon); World Development Movement

wdmf wall-defective microbial forms

WDNR Wadi Dishon Nature Reserve (Israel)

wdo willing dropout

wdp wood door panel

WDP Women in Data Processing

WDPC Western Data Processing Center

wdr white drum

Wdr Wardmaster

WDRC Women's Defense Relief Corps

Wdr L Wardmaster Lieutenant

wds wood-dye stain; word discrimination score; words; wounds

WDS Washington index of Double Stars

wd sc wood screw

wdsprd widespread

wdt width

wdtahtm (wahm, for short) why does this always happen to me?

WDTC Western Defense Tactical Command

WDTS War Dogs Training School

WDTU War Dogs Training Unit

wdu window de-icing unit

wdv written-down value (tax)

W$W *Wall Street Week* (educational tv program)

wdwn well developed, well nourished

wdwrk woodwork

wdy wordy

we. watch error; weekend; white edge; while edging

w/e weekend

w & e windage and elevation

We Wednesday; Welsh

WE War Establishment; Weapons Engineering; Western Electric; World Education; Write Enable

W E Wärmeeinheit (German—thermal unit)

wea weapon(s); weather

WEA Washington Education Association; West End Avenue; Wisconsin Education Association, Workers Educational Association; Wyoming Education Association

WEAAC Western European Airports Association Conference

WEAL Women's Equity Action League

WeAPD Western Air Procurement District

WEARCONS Weather Observation and Forecasting Control System

weat weathertight

Weaver Weaverscope (rifle telescope)

Web Webster's Third New International Dictionary of the English Language Unabridged

WEBA Women Exploited by Abortion

WEBDBC WEB Du Bois Club(s)

webelos we'll be loyal scouts

Webelos We'll be loyal scouts.

webrock weather buoy rocket

WEBS Weapons Effectiveness Buoy System; World Equity Benchmark Shares

Webster's Webster's Dictionary (published in many editions by G & C Merriam of Springfield, Massachusetts)

wec wide energy conversion

WEC Westinghouse Electric Corporation; Women of the Episcopal Church; Women's Emergency Corps; World Energy Conference

WECAF Western Central Atlantic Fishery

WECAN Walking Enforcement Campaign Against Narcotics

WeCen Weather Center (USAF)

WECEP Work Experience Career Exploration Program

WECO Western Electric Company

WECOM Weapons Command (USA)

wecpnl weighted-equivalent continuous-perceived noise level

WECR Wellington, East Coast Rifles (New Zealand)

WECS Wind Energy Conversion System

we'd we had; we would

Wed Wednesday

Wed *Weduwe* (Dutch—widow)

WED Walter Elias Disney

WEDA Wholesale Engineering Distributors' Association

wedar water-damage reduction; weather-damage reduction

WEDC World Economic Development Congress

Wedd Wedding (Berlin borough)

Wednes Wednesday

Wedy Wednesday

Wee Western equine encephalitis

WEEA *Women's Education Equity Act*

WEEAP Women's Educational Equity Act Program

WEECN Women's Educational Equity Communications Network

WEEP Work Education Evaluation Project

WEETA Women's Educational Equity Technical Assistance

wef with effect from

WEF War Emergency Formula; World Education Fellowship

WE & FA Welsh Engineers' and Founders' Association

wefax weather facsimile

WEFC West European Fisheries Conference

weft wings, engine, fuselage, tail

weg war emergency grant

weg(s) wild-eyed guess(es)

WEH William Ernest Henley

WEHS Wadleigh Evening High School

WEI World Economic Institute; World Education Incorporated

weia wife's earned income allowance (tax)

weir wife's earned income relief (tax)

Weiss Weissensee

WEIU Women's Educational and Industrial Union

Wel Welsh

WEL Weapons Effects Laboratory (USA)

Wel Adm Welfare Administration

Wel Can Welland Canal

weld welding

Wel Dept Welfare Department

we'll we shall; we will

Well Wellington

WELL Whole Earth 'Lectronic Link

Wells H(erbert) G(eorge) Wells—English author

Welly Wellington

WELS Wisconsin Evangelical Lutheran Synod

WelshE Welsh English

Welt *Die Welt* (Hamburg's World)

Welts *Weltschmerz* (German—world pain)—universal misery

WEMA Western Electronic Manufacturers Association

WEMSB Western European Military Supply Board (NATO)

WEMTA Wisconsin Emergency Technician's Association

Wen Wendel; Wendell; Wendy

WEN Western Educational Network; Wien-Alaska Airlines

Wenatchees Wenatchee Mountains of central Washington

Wend *Wendell's Reports*

WENOA *Weekly Notice to Airmen* (CAA)

WEO Weapons Engineer Officer

WEOG Western European and Other Groups

wep water-extended polyester

WEP Wisconsin Electric Power Company

WEPA Welded Electronic Packaging Association

WEPCO Weather-Proof Company

wepex weapons exercise

WEPR Women Executives in Public Relations

WERA World Energy Research Authority

WERC World Environment and Resources Council

we're we are

weren't were not

WERM *World's Encyclopaedia of Recorded Music*

Werner Egk Werner Mayer

WERPG Western European Regional Planning Group (NATO)

Wes Wesley; Weston

WES Keokuk (Iowa) Westerns (National Association); War Equipment Scale; Water Electrolysis System; Waterways Experiment Station (corps of Engineers); Weather Editing Section (FAA); Women's Engineering Society; World Economic Survey

WESCOM Weapon System Cost Model

WESCON Western Electronics Show and Convention

wesentl *wesentlich* (German—essential, main)

WESO Weapons Engineering Service Office

Wes Pac Western Pacific

WESRAC Western Research Application Center

Wes Sam Western Samoa (formerly British Samoa)

West Sahr Western Sahara

West states west of the Mississippi; western bloc countries of Europe and North America; Western States (Mountain and Pacific Divisions); Wild West

WEST Western Educational Society for Telecommunications; Western Energy Supply and Transmission (Association); Women's Enlistment Screening Test

WESTAF Western States Arts Federation; Western Transport, Air Force

WESTCOMMRGN Western Communications Region

wester storm from the west

Western Isles Hebrides off Scotland's west coast

West Horse *Western Horseman*

WESTIS Westinghouse Teleprocessing Interface System

WestLant Western Atlantic Area

Westlaw computerized legal research service offered by West Publishing Co

West LB *Westdeutsche Landesbank* (West German Land Bank)

Westm Westminister; Westmorland

Westminster British Parliament; the Palace of Westminster and Westminster Abbey

Westmld Westmorland

Westo West Countryman

West Pac Western Pacific (ocean or railroad)

WESTPAC Western Pacific

Westrain Western Australian Trains

Westralia Western Australia

Westralia(n) Western Australia(n)

West's *West's Annotated Education Code*

West Sahr Western Sahara

West Sam Western Samoa

West Symp Orch Westphalia Symphony Orchestra

Wes Univ Wesleyan University

4wetc four-wheel electronic transaction control

WET Water Engineering Trading; Weapon(s) Effectiveness Test(ing)

WETA Washington Educational Television Association

wetensch *wetenschap* (Dutch—knowledge, science)

WeTip We Turn in Pushers (of narcotics)

Wet Mary Western Maryland Railway (stock exchange slang)

wets. Tory moderates

WETS Weekend Training Site(s)

WETUC Workers' Educational Trade Union Committee

WEU Western European Union

we've we have

WEWO Wing Electronic Warfare Officer

WEWP West European Working Party (Book Development Council)

Wex Wexford

WEX Westinghouse Electric Company (stock exchange nickname)

Wexf Wexford

Wey Weymouth

wez (WEZ) western economic zone

WEZ *westeuropäische Zeit* (German—West European Time); Greenwich Mean Time

wf white female; wide flange; winner's female; write forward; wrong font

w/f white female

w/f (W/F) withdrawing and failing; withdrawn/failed

w & f water and feed; wow and flutter

WF Wake Forest; Wake Forest College; Wallis and Futuna Islands (Internet code); Welch Fusiliers; Wells Fargo & Company

W-F Weil-Felix (reaction)

W.F. White Father

W & F Wallis and Futuna Islands

wfa wilderness first aid

WFA War Food Administration (World War II); Western Fairs Association; White Fish Authority; Wire Fabricators Association; World Federalists Association; World Friendship Association

w factor will factor

WFALW *Weltbund Freiheitlicher Arbeitnehmerverbände auf Liberaler Wirtschaftsgrundlage* (German—World Union of Liberal Trade Union Organizations)

WFAOSB World Food and Agricultural Outlook and Situation Board

WFAW World Federation of Agricultural Workers

WFB Wells Fargo Bank; World Federation of Buddhists

WFBI Wood Fiber Blanket Institute

WFBMA Woven Fabric Belting Manufacturers Association

wfc wide field camera; wolf first class (woman chaser)

WFC Wake Forest College; Water Facts Consortium; Women's Forage Corps; World Food Council; World Forestry Contresses

WFCA Western Fire Chiefs Association

wfd wool forward (knitting)

WFD World Federation of the Deaf

WFDY World Federation of Democratic Youth (communist)

wfe with food element

WFEA World Federation of Educational Associations

WFEB Worcester Foundation for Experimental Biology

WFEO World Federation of Engineering Organizations

WFEX Western Fruit Express

wff well-formed formula

WFF Western Frontier Force; World Friendship Federation

WFFL World Federation of Free Latvians

WF & FSA Wholesale Florists and Florist Suppliers of America

wfg waveform generator

WFGA Women's Farm and Garden Association

WFHE Washington Friends of Higher Education

WFI Wheat Flour Institute

WFIA Western Forest Industries Association

WFIC Wisconsin Foundation of Independent Colleges

WFIP Women's Financial Information Program

WFJCC World Federation of Jewish Community Centers

wfl worshipful

WFL Women's Freedom League; World Football League

W Flem West Flemish

WFLRY World Federation of Liberal and Radical Youth

WFM Walter F Mondale; Western Federation of Miners

WFMH World Federation for Mental Health

WFMW World Federation of Methodist Women

wfn well-formed net

WFN World Federation of Neurology

wfna white-fuming nitric acid

WFNS World Federation of Neurosurgical Societies

wfnt waterfront

wfo wide-field optics

wf/o wife of

WFO Washington Field Office (FBI); Weather Forecast Office

WFOA Western Fishboat Owners of America

wfof wide-field optical filter

WFOT World Federation of Occupational Therapists

wfp warm frontal passage

WFP World Food Program (UN)

WFP *Winnipeg Free Press*

WFPA Washington Forest Protection Association; World Federation for the Protection of Animals

WFPMA World Federation of Personnel Management Associations

WFPMM World Federation of Proprietary Medicine Manufacturers

WFPN Women in the Future Priesthood Now

WFPT World Federation for Physical Therapy

W Fris West Frisian

WFS World Future Society

WFSA World Federation of Societies of Anaesthesiologists

WFSF World Future Studies Federation

WFSPL Wright Field Special Projects Laboratory

WFSW World Federation of Scientific Workers

wft wandering finger trouble; waterfront

WFT Washington Federation of Teachers

wfttngs with fittings

WFTU World Federation of Trade Unions

WFUNA World Federation of United Nations Associations

WFW Woltföderation der Wissenschaftler (German— World Federation of Scientific Workers)

WFWFTHI World Federation of Workers in Food, Tobacco, and Hotel Industries

WFY World Federalist Youth

wg water gauge; weighing; weight guaranteed; wing; wire gauge

Wg Wolfgang

WG Welsh Guards; West German; Western Gear (company); WG Grace (cricketer and physician); Working Group; Writers Guild

WG Westminster Gazette

wga wheat-germ agglutinin

WGA Waterfront Guard Association; Western Governors Association; Writers' Guild of America

WGA Webster's Guide to Abbreviations

w-gal(s) wine gallon(s)

WGAO World Guide to Abbreviations of Organizations

W-gauge wide-gauge railroad track (exceeding the standard gauge of 4 feet 8½ inches)

WGB Weltgewerkschaftsbund (German—World Federation of Trade Unions)

wgbc waveguide operating below cutoff

WGC War Graves Commission; West Georgia College; World Gas Conference

Wg-Comdr Wing-Commander

WGCTA Watson-Glaser Critical Thinking Appraisal

WGD Webster's Geographical Dictionary

WGDS Warm Gas Distribution System

W Ger West Germany

WGER Working Group on Extraterrestrial Resources

wgf waveguide filter; wound glass filter

WGGB Writers' Guild of Great Britain

WGH William Gamaliel Harding (29th President U.S.)

WGI Work Glove Institute; World Geophysical Interval

WGIPP Waterton-Glacier International Peace Park (Alberta, Canada, and Montana, U.S.A.)

wgj wormgear jack

WGJB World's Greatest Jazz Band

w gl wireglass

WGL Weapons Guidance Laboratory

W Glam West Glamorgan

WGM Worthy Grand Master

WGMA Wet Ground Mica Association

WGmc West Germanic

WGMEX Western Gulf of Mexico

wgn wagon

WGP Western Gas Processors

WGPMS Warehousing Gross Performance Measurement System

WGPORA Western Gas Processors and Oil Refiners Association

wgr wide gauze roll

WGR War Guidance Requirements

Wg & Rgn Comdr Wing and Regional Commander

W Grnld Cur West Greenland Current

wgs waveguide glide slope; web guide system

WGs Welsh Guards

WGS Western Gerontological Society; World Geodetic System

wgsj wormgear screw jack

WGSPR Working Group for Space Physics Research (NATO)

wgt weight

WGTA Wisconsin General Testing Apparatus

WGTW Won't Go To Wembleys (anti-imperialist group)

WGU Welsh Golfing Union

WGVN Willard Gibbs Van Name

wgw waveguide window

WGWC Working Group for Weather Communications (NATO)

WGWP Working Group for Weather Plans (NATO)

wh water heater; watt hour; white; withholding

wh. which

w/h withholding

Wh Whig Party

WH Welsh Horse; White House

wha wounded by hostile action

wha' what

WHA Welsh Hockey Association; Western Hardwood Association; Western History Association; Women's Hockey Associatiion; World Health Assembly; World Hockey Association

wham winning the hearts and minds (of the listeners)

whamo winning the hearts and minds of (Vietnam)

W'hampton Wolverhampton

whap when or where applicable

WHASA White House Army Signal Agency

whate'er whatever

what's what has; what is

whatso'er whatsoever

WHC White House Conference (on libraries and information services)

WHCA White House Communications Agency; White House Correspondents Association

WHCF White House Conference on Families

WHCLIS White House Conference on Library and Information Services

WHCOA White House Conference on Aging

WHCT West Ham College of Technology

whd warehead

WHD Western Hemisphere Defense; Women's House of Detention (NYC)

WH DIV Wage and Hour Division

whdm watt-hour demand meter

whe water hammer eliminator

Wheat *Wheaton's* (US Supreme Court Reports)

wheats wheatcakes

wheatstone wheatstone bridge (electrical measuring device named for its inventor—Sir Charles Wheatstone—an English physicist)

whecon wheel control

whene'er whenever

where'er wherever

wheresoe'er wheresoever

whf wharf

WHF Women in Housing and Finance

WHFAM William Hayes Fogg Art Museum

whfg wharfage

whfr wharfinger

whf(s) white homosexual female(s)

WHH William Henry Harrison (9th President U.S.); William Henry Hudson (*Green Mansions*)

WHHA White House Historical Association

Whi Whitehall

WHI Western Highway Institute; Women's Health Initiative

WHIG Wildlife Habitat Improvement Group

Whigs Whigamores (originally a group of West Scottish revolutionaries against church and king)

WHIM Western Humor and Irony Membership; World Humor and Irony

WHIMSY *Western Humor and Irony Membership Yearbook*

whis whistle (fog)

Whiskey letter W radio code; Soviet class of diesel submarines; stock exchange (slang); Western Kentucky (coal company)

Whit Whitaker; Whitbread; Whitcomb; Whitman

Whitaker's *Whitaker's Almanac*

White Carpathians White Carpathian Mountains

Whiteman Marjorie Millace Whiteman's 15-volume *Digest of International Law*

White Mts White Mountains

White Pines White Pine Mountains of eastern Nevada

white precipitate ammoniated mercury

White Sands White Sands National Monument in southeastern New Mexico

white snow cocaine nickname

Whitman Albert Whitman (Chicago); Whitman Publishing Company (Racine)

WHL Western Hockey League

wh lt white light

WHMA Women's Home Missionary Association

WHML Wellcome Historical Medical Library

whm(s) white homosexual male(s)

whmstr weighmaster

WHMV & NSA Woods Hole, Martha's Vineyard and Nantucket Steamship Authority

Whn Whitehaven

WHNEWS White House News Service

WHNPA White House News Photographers Association

WHO White House Office; World Health Organization (UN)

WHOA Wild Horse Organized Assistance

WHOA? Who Hammers Out Acronyms?

who'd who had

WHODAP White House Office of Drug Abuse Prevention

WHOI Woods Hole Oceanographic Institution

WHOIRP World Health Organization International Reference Preparation

whol wholesale(r)

who'll who shall; who will

whoretel whore hotel

who's who is

who've who have

whp water horsepower; whirlpool

W & H & PC Wage and Hour and Public Contracts

wh pl whole plate (silver)

whr watt hour

WHRA Welwyn Hall Research Association; Western Historical Research Associates; Western Housing Research Association

WHRC World Health Research Center

whrlp whirlpool

whs warehouse

WHS Walton High School; Washington Headquarters Services; White Sands, New Mexico (tracking station)

whse warehouse

whsl wholesale

whsle wholesale

whsmn warehouseman

whsng warehousing

whs rec warehouse receipt

WHSRN Western Hemisphere Shorebird Reserve Network

Wht White

WHT William Howard Taft (27th President of the U.S.)

WHTHS William Howard Taft High School

whtm(s) white heterosexual male(s)

WHTSO Welsh Health Technical Services Organization

whvs wharves

why. what have you?

why'd why did

whyinel why in hell

wi when issued (newspaper stock listings); winter; wrought iron

wi' (Gaelic contraction–with)

w & i weighing and inspection

wi. windows

WI Wake Island; Washington Institute for Values in Public Policy; West India; West Indian; West Indies; Windward Islands; Wine Institute; Wire Institute; Wisconsin; Women's Institute

W & I *Welfare and Institutions (Code)*

wia (WIA) wounded in action

WIA Western Interpreters Association

WIAB Wistar Institute of Anatomy and Biology

Wib Wibbert; Wilbert

WIB War Industries Board

WIB *Werkgroep Instrument Beoordeling* (Dutch–Working Group on Instrument Behavior); *Wissenschaftliche Internationale Bibliographie* (German–International Scientific Bibliography)

WIBC Women's International Bowling Congress

wic women, infants, children

wic (WIC) war insurance corporation

WIC Welfare and Institutions Code; West India Committee; Women in Cable; Women in Construction; Women's Interart Center

wich sandwich

WICHE Western Interstate Commission for Higher Education

Wichitas Wichita Mountains of Oklahoma and Texas

WICI Women in Communications, Incorporated

Wick Wicklow

Wicklows Wicklow Mountains in eastern Ireland

WICP Women, Infants, and Children Program

WICS Women's Institute for Continuing Study

wid widow; widower

WID Waste Isolation Division; West India Docks

WIDF Women's International Democratic Federation

Widm *Widmung* (German–dedication)

WIDU Wireless Intelligence and Development Unit

WIF West India Fruit and Steamship Company; West Indies Federation; Women in Film

WIFL World Indoor Football League

wig periwig

Wig Wigtown(shire)

wige wing-on-ground effect

wigo what is going on?

Wigorn. *Wigorniensis* (Latin–of Worcester)

wih went in hole

WIHM Wellcome Institute of the History of Medicine

WIHS Washington Irving High School

wiifm what's in it for me

Wil Wilber; Wilbert; Wilbur; Wilburn; Wiley; Wilford; Wilfred; Wylie

WIL West India Lines

Wil Blvd Wilshire Boulevard

wilco will comply

WILD What I Like to Do (psychological test)

wilde *wildebeest* (Afrikaans—gnu)

Wilder's d novelist-playwright Thornton Wilder's dictum declaring *Who can count the prayers that have ascended to gods who do not exist? Mankind has himself created sources of help where there is no help and sources of consolation where there is no consolation.*

Wiley John Wiley & Sons

Wilhelmina Wilhelmina Helena Pauline Maria (1880–1962), Queen of the Netherlands 1890–1948

Wilhelmina *Wilhelmus van Nassouwe* (Dutch—William of Nassau)—national anthem of the Netherlands

Will Willard; William; Willis

Will and Ariel William James Durant and Ariel Durant—authors

Will & Mar King William and Queen Mary

Will Rogers Turn Will Rogers Turnpike

Wilm Wilmersdorf, Wilmington

WILPF Women's International League for Peace and Freedom

WILS Wisconsin Interlibrary Loan Service

Wilson HW Wilson

Wilson Lib Bull *Wilson Library Bulletin*

Wilts Wiltshire

Wilts R Wiltshire Regiment

WIM Waste Isolation Manager(s); Women In Management

W I & M Washington, Idaho & Montana (railroad)

WIMA Western Industrial Medical Association; Writing Instrument Manufacturers Association

Wimb Wimborne

w i m c whom it may concern

WIMS Wartime Instructions for Merchant Ships

win winter

win. window(s)

Win Winchester Arms; Winterthur

WIN Whip Inflation Now; Work Incentive Program

WINA Webb Institute of Naval Architecture

win'ard windward

WINBA World International Nail and Beauty Association

WINBAN Windward Islands Banana Growers Association

WINBAN(GA) Windward Islands Banana Growers Association

Winch Winchester

WINCO Wing-Commander

wind. windlass

W Ind West Indian; West Indies

WIND wind direction and velocity (aircraft code)

Wind Cave Wind Cave National Park in southwestern South Dakota

Wind I Windward Islands

'winds woodwind instruments: bagpipes, bassoons, clarinets, English horns, fifes, flutes, harmonicas, jew's harps; kazoos, oboes, piccolos, recorders' saxophones, whistles

Windwards Windward Islands

WINE Webb Institute of Naval Engineering

Wing Cdr Wing Commander

winkle(s) periwinkle(s)

Winn Winnipeg; Winnipegger

wino alcoholic addicted to wine

win'rd windward (pronounced *win-urd* by sailors)

WINS Western Integrated Navigation System; Women's Intervention Nutrition Study

wint winter; wintry

Winterthur Winterthur Museum

Wint Gard Winter Garden

wintr winter

Wint T *The Winter's Tale*

WIO Wyoming Infrared Observatory

wip work in process; work in progress

WIP Wage Insurance Program; Wall Improvement Project; West Indian Process (for sorting ripe coffee berries); Women in Production;

Work Incentive Program; World Internationalist Party; World International Partisan

WIPAP Waste Isolation Performance Assessment Program

WIPO World Intellectual Property Organization

WIPP Waste Isolation Pilot Plant; Wool Incentive Payment Program

WIPSEP Waste Isolation Program and System Evaluation Project

WIPTC Women's International Professional Tennis Council

WIR War Information Report; West Indian Regiment

WIR Weekly Intelligence Report

WIRA Wool Industry Research Association

WIRDS Weather Information Reporting and Display System

WIRE Western Installation Requirements Evaluation (DoD); Wisconsin Information Resources for Education

WIRGA West Indian Royal Garrison Artillery

WIRO Wyoming Infrared Observatory

WIRs West Indian Reports

Wis Wisconsin; Wisconsinite

WIS Waste Isolation System; Weizmann Institute of Science; West Indies Shipping

WISA West Indian Sugar Association; West Indies Students Association

WISAP Waste Isolation Safety Assessment Program

Wisc Wisconsin

WISC Wechsler Intelligence Scale for Children

WISCo West Indies Sugar Company

Wisconsin Dells Dells of the Wisconsin

WISC-R Wechsler Intelligence Scale for Children–Revised

Wisd of Sol Wisdom of Solomon (apocryphal book of the Bible)

WISE Weapon Installation System Engineering; World Information Systems Exchange; Worldwide Information System for Engineering

WiSEA Wisconsin Society of Enrolled Agents

WISH Women in Senate and House

wisk wiskunde (Dutch—mathematics)

wisp wide-range-imaging spectrometer

WISP Waste Isolation Systems Panel (NAS); Winter and Icing Study Program; Wisconsin inventory of Science Processes; Women in Scholarly Publishing

WISPr Women in Scholarly Publishing Newsletter

Wiss Wissenschaft (German—science)

Wis Trails Wisconsin Trails

wit. witness

WIT West India Tankers; World International Tennis

WITC Women's International Tennis Council

WITCH Women's International Terrorist Conspiracy (from) Hell

WITCO What Is This Thing Called Opera? (Seattle Opera Association)

withdrl withdrawal

with(out) hype with(out) hyperbole [with(out) exaggeration]

witht without

witned witnessed

wineth witnesseth

wits witkars (Dutch—white cars)—drive-it-yourself two-seater electric vehicles facilitating clean inner-city transportation

WITS Weather Information Telemetry System; Westinghouse Interactive Time-Sharing System; West Integrated Test Stand

Wits U Witwatersrand University

wittos women in the transition of separation

WIU Western International University

WIVAB Women's InterVarsity Athletic Board

wiz wizard

WIZO Women's International Zionist Organization

WJA Women's Jewelry Association; World Jazz Association

WJB William Jennings Bryan (1860–1925)

wjc wife's judicial separation

WJC Westbrook Junior College; William Jefferson Clinton (42nd president of the United States Bill Clinton); World Jewish Congress

W & JC Washington and Jefferson College

WJCB World Jersey Cattle Bureau

WJCC Western Joint Computer Conference

wJf white Jewish female

WJFITB Wool, Jute, and Flax Industry Training Board

wJm white Jewish male

wjs wife's judicial separation

wk walk; warehouse keeper; weak; week; well-known; work; wreck

Wk Walk; wreck

WK Western Alaska Airlines

W-K-B Wentzel-Kramers-Brillouin

wkbk workbook

wkd worked

W-K disease Wilson-Kimmelstiel disease

wkds weekdays

wkg working

wking working

WKKC Who Killed Kennedy Committee

wkly weekly

wkm workman

wkn weaken

WKNR Wadi Kziv Nature Reserve (Israel)

wkr workers; wrecker

wks weeks; white-knuckle seminarian; works; workshop(s)

Wks Works; wreckage (navigational abbreviation)

WKS Wernicke-Korsakoff Syndrome

WKSC Western Kentucky State College

wkshp workshop

Wk/Site Work Site

wkt wicket

wk vb weak verb

WKY Western Kentucky (coal company); Wall Street slang for this company is *Whiskey;* West Kent Yeomanry

W Ky Pkwy Western Kentucky Parkway

wl wall lavatory; water level; waterline; waterplane coefficient; wavelength; wing length; working level

w-l wins-losses

w L *westlichst Längengrad* (German–west longitude)

WL Sir Wilfred Laurier (Canada's eighth Prime Minister); Waiting List; West Lothian; Women's Legion; Women's Liberation; Women's Lobby

W-L Westfal-Larsen Line

W & L Washington and Lee University

WL *Wagon Lits* (French–sleeping cars)

WLA Washington Library Association; Welsh Library Association; Western Literature Association; Wisconsin Library Association; Women's Land Army

wlb wallboard

WLB War Labor Board; Women's Liberation Party

WLB *Werkgroep Instrument Beoordeling* (Dutch—Working Group on Instrument Behavior); *Wilson Library Bulletin; Wissenschaftliche Internationale Bibliographie* (German—International Scientific Bibliography)

WLC World Liberty Corporation (Niarchos)

WLCJ Women's League for Conservative Judaism

WL & Co Westfal-Larsen & Company (steamship line)

wl coef waterline coefficient

wld west longitude date; would

wld ch world championship

WLDF Women's Legal Defense Fund

wldmt weldment

wldr welder

WLF Washington Legal Foundation; Women's Liberation Front; World Law Fund

WLFA Wildlife Legislative Fund of America

WLFNWR William L Finley National Wildlife Refuge (Oregon)

wl fwd wool forward

WLG Wellington, New Zealand (airport)

WLGS Women's Local Government Society

WLHB Women's League of Health and Beauty

wli workload index

WLI Women's Law Institute; Wyoming Law Institute

WLJBP William Langer Jewel Bearing Plant

Wlk Walk

W-L LL Washington-Lincoln Laurels for Leaders

wlm working level month

WLM Women's Liberation Movement

WLMI Wildlife Management Institute

WLMK William Lyon Mackenzie King (former Canadian Prime Minister)

WLMO Worldwide Logistics Management Office (USA)

Wlmsbrg Brdg Williamsburgh Bridge

Wln Wellington

WLN Washington Library Network

W Long west longitude

W'Loo Waterloo

W Loth West Lothian

WLP Wallops Island, Virginia (tracking station)

WLPB War Labor Policies Board

WLPS Wild Life Protection Society

WLPSA Wild Life Preservation Society of Australia

wlr wrong-length record(ing)

Wlr Walter

WLR *Weekly Law Reports; World Labor Report*

WLRI World Life Research Institute

WLRs *Weekly Law Reports*

Wls Wells

WLS Wild Life Sanctuary

WLSC West Liberty State College

WLSP *World List of Scientific Periodicals*

WLSR Wild Life Society of Rhodesia; World League for Sexual Reform

WLTBU Watermen, Lightermen, Tugmen, and Bargemen's Union

WLU Wilfrid Laurier University; World Liberal Union

W & LU Washington and Lee University

WLUC Women Life Underwriters Confederation

WLUS World Land Use Survey

WLW Women Library Workers

WLWH Workshop Library on World Humour

Wly westerly

wlz waltz

wm wattmeter; wavemeter; wet meadow; white male; white metal; winner's female; wire mesh; wordmark (flow chart)

w/m weight or measure; white male

w & m weight and/or measurement

Wm William

WM War Medal; Warrant Mechanician; Weapon Mechanician; Western Maryland (railroad); White Motors; Whitney Museum; William McKinley (25th President of the U.S.); Women Marines; Worshipful Master

W & M College of William and Mary; War and Marine; Washburn & Moen (wire gauge)

WM *World Monitor* (*Christian Science Monitor* magazine and television program)

WMA Waterbed Manufacturers Association; Wildlife Management Area; Women Marines Association; World Medical Association

WMAA Whitney Museum of American Art

WMAC Waste Management Advisory Council

WMARC World Maritime Administrative Radio Conference

WMATA Washington Metropolitan Area Transit Authority

WMATC Washington Metropolitan Area Transit Commission

WMB Walnut Marketing Board; War Mobilization Board

WMBL Wrightsville Marine Biomedical Laboratory

WMC Ways and Means Committee; Western Maryland College; World Meteorological Center (WMO); World Methodist Council

WMCCA Washington Metropolitan Coalition for Clean Air

WMCE Western Montana College of Education

WMCIU Working Men's Club and Institute Union

WMcK William McKinley (25th President of the U.S.)

WMCL William Mitchell College of Law

WMCP Women's Medical College of Pennsylvania

wmd weapons of mass destruction; wind measuring device

Wmd Willemstad

WMD Waste Management Department; Water Management District; Weights and Measures Division

WMEAT World Military Expenditures and Arms Transfer

WMECO Western Massachusetts Electric Company

Wmg Cal Wilmington, California

Wmg Del Wilmington, Delaware

Wmg NC Wilmington, North Carolina

WMI Webbing Manufacturers Institute; Wildlife Management Institute

W Mid West Midlands

wmk watermark

w/m°k watt per meter degree kelvin (thermal conductivity unit)

WMM World Movement of Mothers

WMMA Wood Machinery Manufacturers of America

WMMPA Wood Moulding and Millwork Producers Association

wmn women

Wmn Wilmington, North Carolina

WMNF White Mountain National Forest

WMO World Meteorological Organization

WMOAS Women's Migration and Overseas Appointments Society

W of Mormon Words of Mormon

wmp with much pleasure

WMR Wasatch Mountain Railway

WMS Waste Management System; Webster Memory Scale; Wechsler Memory Scale; Women in Medical Service; Women's Medical Specialist; Work Measurement System; World Magnetic Survey

W & MS Wisconsin & Michigan Steamship (company)

WMS *Willem Mengelberg Stichiting* (Dutch—Willem Mengelberg Foundation)

WMSC Women's Medical Specialist Corps

WMSDI Western Mood & Sleep Disorders Institute

W & M SS Co Wisconsin & Michigan Steamship Company

wmt weighing more than

WMT Wilson Marine Transit

WMTB Western Motor Tariff Bureau

WMTC Women's Mechanized Transport Corps

Wmth Westmeath

WMU Western Michigan University

WMUSE *World Markets for US Exports*

W M W & NW Weatherford, Mineral Wells & Northwestern (railroad)

WMWR Wichita Mountains Wildlife Refuge (Oklahoma)

w/n well-nourished

WN Worlds of Nature (Amarillo botanical and zoological gardens)

WN *Weekly Notes*

WNA Washington, DC, National Airport; Western North America; Winter North Atlantic (loadline marking for ships crossing the North Atlantic in winter)

WNAP Washington National Airport

wnb will not be

WNBA Women's National Book Association

WnBanc Western Bancorporation

WNCCC Women's National Cancer Control Campaign

wndml windmill

WNDO Weather Network Duty Officer

wndp with no down payment

WNE Welsh National Eisteddfod

wng warning

WNGA Wholesale Nursery Growers of America

WNID *Webster's New International Dictionary*

wnl within normal limits

WNLF Women's National Liberal Federation

WNLSC Women's National Land Service Corps

wnm white noise making

WNM Washington National Monument

WNMC Weather Network Management Center (USAF)

WNNP Walpole-Nornalup National Park (Western Australia)

WNO Welsh National Opera

WNP Wankie National Park (Zimbabwe); Warrumbungle NP (New South Wales); Welsh National Party; Westland NP (South Island, New Zealand); Wilpattu NP (Ceylon); Wyperfeld NP (Victoria, Australia)

WNRE Whiteshell Nuclear Research Establishment

WNS Washington National Symphony (District of Columbia); Women's News Service

WNSB White Nile Scheme Board (Sudanese cotton production)

WNSR West Nova Scotia Regiment

WNW west northwest

WNW *west noordwest* (Dutch–west northwest)

WNWDA Welsh National Water Development Authority

WNWR Wapanocca National Wildlife Refuge (Arkansas); Washita National Wildlife Refuge (Oklahoma); Wheeler National Wildlife Refuge (Alabama); Willapa National Wildlife Refuge (Washington)

WNY West New York, NJ

WNYNRC Western New York Nuclear Research Center

WNYNSC Western New York Nuclear Service Center

wo wait order; warning order; water-in-oil (emulsion); widow's original; wine of origin; without; work order; write out; written order

wo *wie oben* (German—as previously mentioned)

wo' war; wore

w/o white/others; wife of; without

WO War Office; Warrant Officer; Weapons Officer; Welfare Officer; Welsh Office; Wing Officer; Wireless Officer; Writer Officer

WO *World Oil*

WOA Web Offset Association;
Wharf Owners' Association
WOAR Women Organized
Against Rape
wob washed overboard
wobndr(s) without binder(s)
wobo(s) without blowout(s)
Wobs Wobblies
woc without compensation
WOC Warrant Officer Candidate
WOCCI War Office Central
Card Index
wocg weather outline contour
generator
WOCIT We Oppose Computers In Tournaments
WOCL War Office Casualty
List
WOCS Women's Officer Candidate School
wod wind over deck
W & O D Washington & Old
Dominion (railroad)
WODA World Organization of
Dredging Associations
WODECO Western Offshore
Drilling and Exploration
Company
woe without equipment
Woe Woensdag (Dutch–
Wednesday)
WOFIWU World Federation
of Industrial Workers Unions
wofttngs without fittings
wog golliwog; polliwog; water
or gas (valve); wily oriental
gentleman [a confidence man
from the Far East]; with other
goods
'wog golliwog; polliwog
WOG Wily Oriental Gentleman (offensive slang term applied to Farouk I of Egypt and
similar monarchs of the area)
WOGA Western Oil and Gas
Association
wogs workers on government
service
wogs (British slang–wily oriental gentlemen; wily oriental
peoples)
WOGS War Office General
Stores
WOGSC World Organization
of General Systems and Cybernetics
woh work on hand
WOHC Warrant Officer, Hospital Corps
WOIS Wisconsin Occupational Information System

WOJG Warrant Officer, Junior
Grade
WOK Warren O Kessler
wol wharf owners' liability
WOL War Office Letter
WOLA Washington Office on
Latin America
Wolf Wolfgang; Wolfmar;
Wolfrad; Wolfram; Wolfred
Wolfs Wolfson College, Oxford
wom wireless operator mechanic
WOM Woomera, Australia
(tracking station)
WOMAD World Music Arts &
Dance
WOMAN World Organization
of Mothers of All Nations
WOMBAT Weapons of Magnesium Battalion Anti-Tank
Women's Lib Women's
Liberation Movement
womi women on words and
images
womlib women's liberation
WOMP Western Ocean Meeting Point
womzer womanizer
won monetary unit of Korea;
wool on needle (knitting)
W-o-N Walton-on-Naze
WONAAC Women's National
Abortion Action Coalition
WONARD Women's Organization of the National Association of Retail Druggists
wong weight on nose gear
won't will not
WOO Warrant Ordnance Officer; Western Operations Office (NASA); World Oceanographic Organization
Wood Woodbine; Woodbridge; Woodburn; Woodbury; Woodfield; Woodfin;
Woodhill; Woodley; Woodrow; Woodruff; Woodson;
Woodville; Woodward;
Woodworth
Wood Buffalo Wood Buffalo
National Park in northern
Alberta
Woody Woody Allen; Woody
Guthrie
woof(s) woofer(s)
Wool Woolworth's (Circuit
Court Reports)
Woolwich Royal Arsenal at
Woolwich on the south bank
of the Thames near London
woool words out of ordinary
language

Wooster(sheer) (British contraction—Worcestershire)
wop with other property; without (immigration) papers;
without personnel
WOP Wireless Operator
WOPAG Wireless Operator Air
Gunner
wopar(s) without partition(s)
wope without personnel or
equipment
WOPN Women Officers Professional Network
wopo without purchase order
WOQT Warrant Officer Qualification Test
wor without our responsibility
Wor Worcester (baseball);
Worshipful
worbat wartime order of battle
Worc Worcester College
WORC Washington Operations Research Council
Worc Coll Worcester
College—Oxford
Worc Reg Worcester Regiment
Worcs Worcestershire
WORD Women's Opportunity
and Resource Development
word proc word processor
Words Wordsworth
WORDS Western Operational
Research Discussion Society
words/min words per minute
words/sec words per second
WORK Wanted Older Residents (with) Knowhow
Work. Comp Workmen's
Compensation
workfare welfare programs
aimed at returning people to
the workforce; working for
welfare (alternative to high-
cost-assistance welfare)
workh workhouse
workingclass English modern cockney
Workmen's Workmen's Circle; Workmen's Compensation
World World Almanac
World Bank International
Bank for Reconstruction and
Development (IBRD)
worm write once, read many
times
Wormald Wormald International Security
WORSAMS Worldwide
Organization Structure for
Army Medical Support

worse word selection

WOS Washington Opera Society; Western Orchestral Society; Wilson Ornithological Society

wosac worldwide synchronization of atomic clocks

WOSB War Operations Selection Board

WOSD Weapon Operational Systems Development

WOSL Women's Overseas Service League

wot wide-open throttle

WOTAG Women's Taxation Action Group

W-o-t-N Walton-on-the-Naze

wott wolves on the track (prowling males)

wouldn't would not

W & O V Washington & Ouachita Valley (railroad)

wow waiting on weather; worst of worst (prisoners)

w-o-w worst-on-worst (worst on top of the worst possible disaster, etc.)

WoW *Werke ohne Opuszahl* (German–works without opus number)

W o W War on Want

WOW Wider Opportunities for Women; Women's Opportunity Week; Women On Wine; Woodmen of the World

w/o wn without winch

wp waste package; waste pipe; waterproof; water repellency; water repellent; way point; weather permitting; white phosphorus; wild pitches (baseball); will probated; will proceed; will proved; working paper; working party; working point; working pressure

wp (WP) word processing; word processor

w-p waterproofed

w/p without prejudice

w/p (W/P) withdrawing and passing; withdrawn/passed

Wp Worship(ful)

WP War Plan(s); Warsaw Pact; Western Pacific (railroad); Western Province (Solomon Islands); West Point; West Virginia Pulp and Paper (stock exchange symbol); William Penn (founder of the Quaker colony of Pennsylvania); Worthington Pump; Worthy Patriarch

WP *Wiener Philharmoniker* (German—Vienna Philharmonic Orchestra); *Winkler Prins Encyclopedieen* (Dutch—Winkler Prins Encyclopedia)

wpa with particular average

WPA Western Pine Association; Western Psychological Association; William Penn Association; Women's Prison Association; Works Progress Administration; World Parliament Association; World Psychiatric Association

WPAFB Wright-Patterson Air Force Base

WPA & H Women's Prison Association and Home

wpar(s) with partition(s)

WPAS Work Package Authorization System

wpb wastepaper basket

WPB War Plan Basic; War Production Board (World War II)

WPBA Western Power Boat Association; Women's Professional Billiard Association

WPBIC Walker Problem Behavior Identification Checklist

WPBL Women's Professional Basketball League

WPBS Welsh Plant Breeding Station

wpc water pollution control; watts per candle; wood plastic combination; world planning chart

WPC War Pensions Committee; Washington Press Club; William Penn College; Women's Press Club; World Peace Council

WPCA Water Pollution Control Act

WPCC Wage and Price Control Council; Western Pharmaceutical and Chemical Corporation

WPCF Water Pollution Control Federation

WPD War Plans Division; Work Package Department (ONWI)

wpe white porcelain enamel

WPEC World Plan Executive Council

w/p equipment word-processing equipment

WPF Warsaw Pact Forces; World Peace Foundation

WPFC Western Pacific Fisheries Commission; World Press Freedom Committee

Wpfl Worshipful

wpg waterproofing

WPg West Point graduate

WPG gunboat (USCG symbol)

WPGA Wisconsin Personnel and Guidance Association; Women's Professional Golfers' Association; Wyoming Personnel and Guidance Association

WPGR Willem Pretorius Game Reserve (South Africa)

WPGT Women's Professional Golf Tour

WPHC Western Pacific High Commissioner

WPHI Western Pennsylvania Horological Institute

WPHOA Women Public Health Officers Association

wpi wholesale price index

WPI Wall Paper Institute; Wedding Photographers International; Western Psychiatric Institute (Pittsburgh); Worcester Polytechnic Institute; World Press Institute; Waxed Paper Institute

WPI *World Port Index*

Wpk Ward's (mechanical tissue) pack

WPK Workers Party of (North) Korea

wpl warning point level

WPL Weapons Propulsion Laboratory; Wichita Public Library; Winnipeg Public Library; Worcester Public Library

WPLC Wisconsin Power and Light Company

WPLO Water Port Liaison Office(r)

wpm words per minute

WPMA Wood Products Manufacturers Association

WPMSF World Professional Marathon Swimming Federation

wpn weapon

WPN West Penn Traction (stock exchange symbol)

WPN *World Press News*

wpns weapons

wpo world public opinion

wpo (WPO) without purchase order

WPO Water Programs Office (Environmental Protection Agency); Wiener Philharmonic Orchestra (Vienna Philharmonic Orchestra); World Ploughing Organization; World Presidents Organization

WPOD Water Port of Debarkation

WPOE Water Port of Embarkation

wpp waterproof paper packing

WPP West Penn Power Company; White Patriot Party; Witness Protection Program Work Package Plan(ning)

WPPC West Penn Power Company

WPPDA Welfare and Pension Plans Disclosure Act

WPPO Work Package Program Office(r)

WPPP Work Package Program Plan(ning)

WPPSI Wechsler Preschool and Primary Scale of Intelligence

WPPSS Washington Public Power Supply System

wp & r work-planning-and-review (discussions)

WPRA Wallpaper and Paint Retailers' Association; Waste Paper Recovery Association; Women's Professional Rodeo Association

WPRL Water Pollution Research Laboratory

wpr's wartime personnel requirements

WPRS Wittenborn Psychiatric Rating Scale

wps with prior service; words per second

WPs Warsaw Pact members; Warsaw Pact nations

WPS Waveform Processing System; Widowed Persons Service; Wildlife Preservation Society; Wildlife Preserve Society; World Peanut Syndicate; World Porpoise Society

WPSA World Professional Squash Association; World's Poultry Science Association

WPSL Western Primary Standard Laboratory

WPSP White People's Socialist Party

WPT Windfall Profits Tax

WPTB Wartime Prices and Trade Board

WPTF World Peace Tax Fund

wpu with power unit; write punch

WPW-af Wolff-Parkinson White atrial fibrillation

WPW-avrt Wolff-Parkinson White atrioventricular reentrant tachycardia

wpwod will proceed without delay

W-P-W syndrome Wolff-Parkinson-White syndrome

WPY World Population Year (1974)

WP & Y White Pass & Yukon (railroad)

WP & YR White Pass & Yukon Route

WPZ Woodland Park Zoo (Seattle)

wq water quench

WQA Water Quality Association

WQCB Water Quality Control Board

WQF Wider Quaker Fellowship

wr war risk; water repellent; wide receiver; write (flow chart); write out

w/r water and rail; water resistant

w & r water and rail; welfare and recreation

Wr Walter

WR Ward Room; Warehouse Receipt; War Reserve; Wassermann Reaction; Wessex Regiment; Western (railway) Region; West Riding; William Randolph Hearst; Windsor Regiment; Witwatersrand Rifles

WR *Weekly Reporter*

W.R. *Wilhelmus Rex* (Latin– King Wilhem, King William)

WRA War Relocation Authority; Ward Room Attendant; Water Research Association; Western Railroad Association; Western Railway of Alabama; Winchester Repeating Arms (company)

WRA *Water Resources Abstracts*

WRAAC Women's Royal Australian Army Corps

WRAAF Women's Royal Australian Air Force

WRAC Women's Royal Army Corps

wraceld wounds received in action combat with enemy or in line of duty

WRAF Women's Royal Air Force

WRAIN Walter Reed Army Institute of Nursing

WRAIR Walter Reed Army Institute of Research

WRAMA Warner-Robins Air Material Area

WRAMC Walter Reed Army Medical Center

Wrangell Wrangell Island *(Vrangelya Ostrov)* In the Russian Arctic, once an American possession

Wrangell-St Elias largest-area national park in the U.S.

WRANS Women's Royal Australian Naval Service

WRAP Weapons Readiness Analysis Program; Weighted Record Analysis Program; Workforce Reduction Assistance Program

WRAS Women's Reserve Ambulance Society

WRAT Wide-Range Achievement Test

WRB War Refugee Board; Water Resources Board

WRBC Weather Relay Broadcast System

wrc water-retention coefficient

WRC Water Research Center; Water Resources Commission; Weather Relay Center; Welding Research Council

WRCB Water Resources Control Board

WRCNS Women's Royal Canadian Naval Service

wrcr wife's restitution of conjugal rights

WRCUP Western Region Canadian University Press

WRDA Water Resources Development Act

WRDC Western Rural Development Center; Westinghouse Research and Development Center

WRE Weapons Research Establishment (Woomera, Australia)

WREE Women for Racial and Economic Equality

WREEC Western Regional Environmental Education Council

w ref with reference

w reg with regard (to)

WREN Women's Royal Naval Service

wresat weapons research establishment satellite

W-response whole response

WRF World Rehabilitation Fund; World Research Foundation

wrfg wharfage

WRGH Walter Reed General Hospital

WRH Walter Reed Hospital

WRHS Western Reserve Historical Society

wri war risk insurance

WRI War Resisters' International; Weatherstrip Research Institute; Wellcome Research Institute; Will Rogers Institute; Wire Reinforcement Institute; Wire Rope Institute; Women's Rural Institute; World Reserves Institute; World Resources Institute

WRI *World Research INK* (monthly publication)

Wrigley Wrigley · Field, Chicago

WRINS Women's Royal Indian Naval Service

WRIR Walter Reed Institute of Research

WRIT Waste-Rock Interactions Technology (program)

Writer's Dig *Writer's Digest*

wrk work (flow chart)

Wrk Workington

WR Knottman (abbreviated signature–we are not man and wife)

wrkshp workshop

wrl wing reference line

WRL Wantage Research Laboratory; War Readiness Material; War Resisters League; Westinghouse Research Laboratories; Willow Run Laboratories (University of Michigan)

WRLC World Role of Law Center (Duke University)

wrm war readiness materiel

WRM Wasatch Railway Museum

wrmn wireman

wrms weighted root mean square

WRMT Woodcock Reading Mastery Test

wrn western; wool round needle (knitting)

WRNGA William Rockhill Nelson Gallery of Art (Kansas City)

WRNR Women's Royal Naval Reserve

WRNS Women's Royal Naval Service

wrnt warrant

WRNVR Women's Royal Naval Volunteer Reserve

WRNWR White River National Wildlife Refuge (Arkansas)

wro war risk only

WRO Washington Regional Office; Weed Research Organization; Welfare Rights Organization

wros with right of survivorship

WRP Workers' Revolutionary Party (British communists)

WRPA Water Resources Planning Act

WRPC Weather Records Processing Center(s)

WRRA Women's Road Records Association (cycling)

WRRC Willow Run Research Center

WRRI Water Resources Research Institute

WRRR Walter Reed Research Reactor

WRRS Wire Relay Radio System

wrs war reserve stock(s)

WRS Warning and Report(ing) System; Worldwide Reference Sources

WRSA Western Regional Science Association

WRSIC Water Resources Scientific Information Center

wrsk war-readiness spares kit

WRSP Wage Reporting Simplification Project

WRSP *World Register Scientific Periodicals*

wrt wrought

wrtd warranted

WRTF Waste Retrieval and Treatment Facility

wrtr writer

wru who are you?

WRU Western Reserve University

wrv water relief valve

WRV West Riding Volunteers

WRVS Women's Royal Voluntary Service

WRVT Wide-Range Vocabulary Test

wr(w) war reserve (weapon)

WRX Western Refrigerator Express (railroad code)

WRY World Refugee Year

W Ry A Western Railway of Alabama

WR Yorks West Riding, Yorkshire

ws war substantive; water supply; weather station; white-striped; wingspan; wobbling slowly; working space; working storage; workstation

w/s weapon system; weather ship

w & s wage & salary; whiskey and soda

WS Wallops Station (NASA); Ware Shoals; Warner & Swasey; weapon system(s); Western Samoa; West Saxon(y); West Sussex; Wilderness Society; Wildlife Society; windspeed; Writer to the Signet (Scottish lawyer)

WS *Wiener Stadtbibliothek* (German—Vienna State Library)

W S *Washington Star*

wsa weapons system analysis

WSA War Shipping Administration; Western Soccer Alliance; Weed Society of America; Worker-Student Alliance; World Sign Associates

WSA *Wasser und Schiffahrtsampt* (German—Water and Ship Canal Authority)

WSAAA Western States Advertising Agencies Association

WSAC Washington State Arts Commission; West of Scotland Agricultural College

WSAD Weapon System Analysis Division (USN)

WSAG Washington Special Action Group (personnel in Situation Room in White House basement)

W Sam Western Samoa

WSAO Weapon System Analysis Office

WSAP Weighted Sensitivity Analysis Program (EPA)

WSAVA World Small Animal Veterinary Association

wsb water-soluble base; wheatsoy blend; will send boat

WSB Wharton School of Business (University of Pennsylvania); World Scout Bureau

WSBA Western Sovereign Base Area (Cyprus); Wyoming School Boards Association

WSBI Western Sciences Behavioral institute

wsc weapon system contractor

WSC Western Simulation Council; Western Society of Criminology; Winona State College; Winston Spencer Churchill; Wisconsin State College; World Series Cricket; Writing Services Center

WSCC Western State College of Colorado

WSCF World Student Christian Federation'

Wschr *Wochenschrift* (German—weekly magazine)

WSCS Woman's Society for Christian Service

wsd working stress design

wsdb world studies data bank

WSDC Women's Self-Defense Council

WSDL Weapons System Development Laboratory

wse word-superiority effect

WSEC Washington State Electronics Council

WSECL Weapon System Equipment Component List

wsed weapon system electrical diagram(s)

WSED Weapon Systems Evaluation Division

WSEG Weapons Systems Evaluation Group

WSEL Weapons System Engineering Laboratory

WSEP Waste Solidification Engineering Prototype Plant (AEC); Weapon System Evaluation Program

WSET Writers and Scholars Educational Trust

wsev (WSEV) winged surface-effect vehicle

WSF Washington State Ferries; Western Sea Frontier; Women's Strike for Peace; World Sephardic Federation

WSFI Water Softener and Filter Institute; Wood and Synthetic Flooring Institute

WSFR Worcestershire and Sherwood Foresters Regiment

wsg worthiest soldier in the group

WSG Wesleyan Service Guild; Wire Service Guild

WSGA Wine and Spirits Guild of America

WSGE Western Society of Gear Engineers

wshr/dryr washer/dryer

WSHS Wisconsin State Historical Society

WSI Writers and Scholars International

WSIN Western State Information Network

WSJ *Wall Street Journal*

WSL Warren Spring Laboratory; Washington State Library

WSLA World Savings and Loan Association

WSLCB Washington State Liquor Control Board

WSLF Western Somali Liberation Front (communist)

WSLO Weapon System Logistics Office(r)

Wsm Wesermünde

WSM Weapon System Manager; Western Society of Malacologists; William Somerset Maugham

WSM *Weapon System Manual*

WSMA Western States Meat Association

WSMAC Weapon System Maintenance Action Center

WSMC Western Space and Missile Center, Vandenberg Air Force Base, California; Western States Movers Conference

WSMO Weapon System Material Office(r)

WSMR White Sands Missile Range

WSMSA Washington Standard Metropolitan Statistical Area

WSN Washington, DC

WSNA Washington State Nurses Association

WSNM White Sands National Monument

WSO Warrant Stores Office(r); Weapon System Office(r); Western Support Office (NASA); Wichita Symphony Orchestra; Winnipeg Symphony Orchestra; World Simulation Organization

WSO *Wiener Symphonisches Orchester* (German—Vienna Symphony Orchestra)

WSOC Wider Share Ownership Council

wsp water-soluble polymers; water supply point; working steam pressure

WSP Witness Security Program; Women Strike for Peace; Work Study Program; Work Systems Package (naval salvage device); Work Systems Program; World Series of Philately; Wyoming State Parks

WSPA Western States Petroleum Association; World Society for the Protection of Animals

WSPACS Weapon System Program and Control System

WSPB Western Society of Business Publications

WS Pen Washington State Penitentiary

WSPG White Sands Proving Ground

WSPL Winston-Salem Public Library

WSPO Weapon System Project Officer

WSPOP Weapon System Phase-Out Procedure

WSPU Women's Social and Political Union

wsr (WSR) weapon system reliability

w/sr watt(s) per steradian

Wsr Wesermünde

WSR Weapons Status Report; Weather Surveillance Radar

W & S R Warren & Saline River (railroad)

WS & RB Washington Surveying and Rating Bureau

WSRI World Safety Research Institute

w/srm² watt(s) per steradian square meter

wss wing sweep system

WSS Warfare Systems School; Winston-Salem Southbound (railroad); World Ship Society

WSSA Weapon System Support Activities; Wine and Spirits Wholesalers of America; World Secret Service Association

WSSC Weapon System Support Center

WSSCA White Sands Signal Corps Agency

WSSO Winston-Salem Symphony Orchestra

WSSS Weapon System Storage Site

WSSSP Western States Small Schools Projects

WSS & YP White Sulphur Springs & Yellowstone Park (railroad)

WST Whitworth Standard Thread

WSTA Washington State Trustees Association; White Slave Traffic Act

WSTC Winston-Salem Teachers College

WSTF White Sands Test Facility (NASA)

WSTI Waterbury State Technical Institute; Welded Steel Tube Institute

WSTNRA Whiskeytown Shasta-Trinity National Recreation Area (California)

WSU Washington State University; Wayne State University; Western State University; Wichita State University

w sup water supply

W Sus West Sussex

WSUSM Wayne State University School of Medicine

WSV *Wiener Stadtwerke Verkehrsbetriebe* (Vienna transportation system)

wsw white sidewall (tires)

WSW west southwest

WSWA Wine and Spirits Wholesalers of America

WSWL Warheads and Special Weapons Laboratory

WSWMA Western States Weights and Measures Association

WSWS Wexford Slobs Wildfowl Sanctuary (Ireland)

WSY West Somerset Yeomanry

WSYC West Somerset Yeomanry Cavalry

wt warrants (newspaper stock listings); war transport; watch time; waterproofed); waterproofing; watertight; weapon training; weight; wild-type (allele); withholding tax (WT)

wt % weight percent

w/t wireless telegraph(y)

w/t (W/T) walkie/talkie

w & t wear and tear

Wt. waiter

WT Warrant Telegraphist; war time; wealth tax; Western Time; winterization test; withholding tax

WT *The Winter's Tale*

W & T Wrightsville & Tennille (railroad)

W/T Wireless Telegraphist

WTA Washington Technological Associates; Washington Trails Association; Women's Tennis Association; World Trade Association; World Transport Agency

WTTA World Trade Alliance Association

WTAU Women's Total Abstinence Union

wtawtar where there's a will there's a relation

w/tax withholding tax

Wtb Whitby

Wtb *Wörterbuch* (German—dictionary)

WTBA Washington Toll Bridge Authority; Water-Tube Boilermakers' Association

WTB & TS Watchtower Bible and Tract Society (Jehovah's Witnesses)

WTC War Transport Council; Women's Timber Corps; World Tanker Corporation (Niarchos); World Trade Center; World Trade Commission

WTCA Wood Truss Council of America

wtchmn watchman

wtd watertight door; weighted

WTD War Trade Department

WTD *World Trade Directory*

WTDAOT What to Do About Old Town

WTDR World Traders Data Reports

wte wartime extension

wte (WTE) world time equivalent

WTE World Tapes for Education

Wt Eng Warrant Engineer

wtf waterfront; will to fire

Wtf Waterford

WTFDA Worldwide TV-FMDX Association

WTFP Wolf Trap Farm Park (Vienna, Virginia)

WTG *Welt-Tierärztegesellschaft* (German—World Veterinary Association)

wthr weather

WTIC World Trade Information Center

WTIS World Trade Information Service

WTJ *Westminster Theological Journal*

WTL Wyle Test Laboratories

wtm write tape mark

WTMA West Texas Museum Association; Wood Tank Manufacturers Association

wtmh watertight manhole

wtn. witness

WTNID *Webster's Third New International Dictionary*

WTNR Wadi Tabor Nature Reserve (Israel)

WTO Warsaw Treaty Organization; Weapon Training Officer; World Tourism Organization; World Trade Organization

Wt Ofcr Warrant Officer

w/t office wireless/telegraph office (aboard ships in the 1920s became the radio room)—the radio shack

WTP Weapons Testing Program; Western Tropical Pacific

WTPO Western Tropical Pacific Ocean

wtqad watertight quick-acting door

wtr waiter; winter; writer

Wtr Water

WTR Western Test Range (formerly Pacific Missile Range)

WTRC Wool Textile Research Council

wtrz winterize

wtrzn winterization

wts word terminal synchronous

WTS Watchtower Society; William Tecumseh Sherman (1820–1891); Women's Transport Service

WTSC West Texas State College

wtspt waterspout

WTT World Tennis Team

WTTA Wholesale Tobacco Trade Association

WTU Washington Theological Union

WTUC World Trade Union Conference

WTVN Worldwide Television News

wtw   wet tissue weight
WTWA World Trade Writers Association
wu work unit
WU Washington University; Weather Underground Organization; Wesleyan University; Western Union; Wilberforce University; Wittenberg University
W/U Western Union
WUA Western Underwriters Association
wuaa wartime unit aircraft activity
WUAA Wartime Unit Aircraft Activity
wuc work unit code
WUCM Work Unit Code Manual
WUCOS Western European Union Chiefs of Staff
WUCT World Union of Catholic Teachers
WUCWO World Union of Catholic Women's Organizations
WUD Water Utility Department; Western United Dairymen
WUDO Western European Defense Organization
WUF World Underwater Federation; World Union of Free Thinkers
WUI Western Union International
WUIS Work Unit Information System
WUJS World Union of Jewish Students
Wuli Xuebao (Acta Phys Sin) *Acta Physica Sinica* (Chinese Journal of Physics)
WULTUO World Union of Liberal Trade Union Organizations
WUM Women's Universal Movement
WUMP(S) White Urban Middleclass Protestant(s)
WUNS World Union of National Socialists
WUO Weather Underground Organization
WUOSY World Union of Organizations for Safeguarding Youth
Wupatki Wupatki National Monument in northern Arizona
WUPJ World Union for Progressive Judaism

WUPO World Union of Pythagorean Organizations
WUS Western United States; World University Service
WUSL Women's United Service League
WUSM Washington University School of Medicine
wut warmup time
WUT Washburn University of Topeka
wuts work-unit time standard
WUX Western Union (teleprinter) Exchange
wv wall vent; whispered voice; wind velocity; with view (room with view)
w/v weight in volume
WV West Virginia; West Virginia Pulp and Paper Company
W Va West Virginia; West Virginian
WVA World Veterinary Association; Wyoming Vocational Association
WVAESP West Virginia Association of Elementary School Principals
WVAS Wake-Vortex Avoidance System
W Va Turn West Virginia Turnpike
WVa U Lib West Virginia University Library
WVAWRD West Virginia Water Resources Division
WVC Wenatchee Valley College
wvd waived
WVD *Werelverbond van Diamantbewerkers* (Dutch—World Alliance of Diamond Workers)
wvdc working voltage—direct current
WVEA West Virginia Educational Association
WVEE Wheeled Vehicles Experimental Establishment
wveh wheel(ed) vehicle
wvem water-vapor electrolysis module
WVF World Veterans' Federation
WVFIC West Virginia Foundation for Independent Colleges
WVIT West Virginia Institute of Technology
WVL Warfare Vision Laboratory (USA)

WVLA West Virginia Library Association
WVMA Women's Veterinary Medical Association
Wvn Wivenhoe
W V N West Virginia Northern (railroad)
WVPA World Veterinary Poultry Association
wvr within visible range
WVR Wellington Volunteer Rifles; Women's Volunteer Reserve
WVRB West Virginia Rating Bureau
WVRO World Vision Relief Organization
WVS Women's Voluntary Service
WVSBA West Virginia School Boards Association
WVSC West Virginia State College; Wisconsin Vocational Studies Center
WvSEA West Virginia Society of Enrolled Agents
WVSP West Virginia State Police
WVSSPC West Virginia Secondary School Principal's Commission
wvt water vapor transfer; water vapor transmission
WVT Watervliet Arsenal
wvtr water vapor transmission rate
w/vu with view
WVU West Virginia University
WVWC West Virginia Wesleyan College
ww walking wounded; warehouse warrant; water white; waterworks; widowed; wirewound; with warrants (newspaper stock listings); woodwind(s); wrong word
ww. widow
w/w wall-to-wall (carpet, floor covering, linoleum, tile); weight for weight
ww *werkwoord* (Dutch—verb)
Ww *Witwe* (German—widow)
WW Walter White (who spoke up for blacks for nearly half a century); Walworth (trademark); Warrant Writer; Woodmen of the World; Woodrow Wilson (28th President of the U.S.); world war; world wide
W-W Winchester-Western

W & W Waynesburg & Western (railroad); Winchester & Western (railroad)

WW *Who's Who*

WW I World War I (1914–1918)

WWIVM World War I Victory Medal

WW II World War II (1939–1945)

WWIIHSLB World War II Honorable Service Lapel Button

WWIIVM World War II Victory Medal

wwa with the will annexed

WWA Western Writers of America; World Waterpark Association

WWABNCP Worldwide Airborne Command Post (USAF)

wwap worldwide asset position

WWB Walt Whitman Bridge

WWBA Walt Whitman Birthplace Association; Western Wooden Box Association

WWBCN World-Wide Business Centers Network

wwc wall-to-wall carpeting

WWC Walla Walla College; Warren Wilson College; William Woods College; World Weather Centers (Melbourne, Moscow, Washington, D.C.)

WWCP Walking Wounded Collecting Post

WWCT Wellington, West Coast and Taranaki (regiment)

WWCTU World's Women's Christian Temperance Union

wwd weather working days; windward

WWD *Women's Wear Daily*

WWDC World War Debt Commission

W Wdr Warrant Wardmaster

wwdShex weather working days Sundays and holidays excluded

Wwe *Weduwe* (Dutch—widow); *Witwe* (German—widow)

WWEMA Water and Wastewater Equipment Manufacturers

wwf welded wire fabric

WWF Welder Wildlife Foundation; Woodrow Wilson Foundation; World Wildlife Fund

WWFN World-Wide Fund for Nature

WWG *World Wildlife Guide*

WWHS Wilbur Wright High School; Woodrow Wilson High School

wwi whirlwind computer

WWI Weight Watchers International; World Watch Institute

WWIB Western Weighing and Inspection Bureau

WWICS Woodrow Wilson International Center for Scholars

wwio worldwide inventory objective

WWJC Western Wyoming Junior College

WWJD what would Jesus do

WWM WW Morrow

WWMB Woodrow Wilson Memorial Bridge

WWMC Woodrow Wilson Memorial Commission

WWMCCS Worldwide Military Command and Control System

WWMHA World-Wide Miniature Horse Association

WWMMP Western Wood Moulding and Millwork Producers

w/wn with winch

W Wnd Drft West Wind Drift (Antarctic)

WWNFF Woodrow Wilson National Fellowship Foundation

WWNSSS World-Wide Network of Stand and Seismograph Stations

WWNT West Wales Naturalists Trust

w/wo with or without

ww/o widow of

WWO Warrent Writer Officer; Wing Warrant Officer; Wing Weapon Officer; World Weather Organization

wwp water wall peripheral; working water pressure; write without program

WWP Washington Water Power company; Workers World Party (leftwing)

WWPA Western Wood Products Association; World Wide Philatelic Agency

WWPSA Western World Pet Supply Association

wwr. widower

WWR *Washington Week in Review* (educational television)

WWRP Win-Win Rap Pack (exam)

ww's walla wallas (Hong Kong harbour launches)

WWSA Walt Whitman Society of America; Women's War Service Auxiliary

WWSC Western Washington State College

WWSN World-wide Seismology Net (NBS)

WWSPIA Woodrow Wilson School of Public and International Affairs (Princeton University)

wwss water wall side skegs

WWSSN World-Wide Standardized Seismograph Network

WWSU World Water Ski Union

wwt whitewall tires

WWTP Waste Water Treating Process

WWTVN World-Wide Television News

WWTVS World-Wide Television Service

WWU West Washington University

W W V call letters of United States Bureau of Standards worldwide radio time signal; Walla Walla Valley (railroad)

WWVH World Wide Time (US Bureau of Standards, Hawaii)

WWW World Weather Watch; World-Wide Web; World Without Walls; World Without Wars

WWW *Who Was Who*

W. W. Wash Walla Walla, Washington

WWWF Worldwide Wrestling Federation

WWWV Women World War Veterans

WWWVA Wild, Wonderful West Virginia

WWWW Women Who Want to be Women

WWWW *Worldwide What & Where*—geographic glossary and traveler's guide

wwwwwh who, what, when, where, why, how—reporters' mnemonic for encompassing elements of a news story

WWY Warwickshire and Worcestershire Yoemanry

wx watts second; waxy; weather report

Wx weather; Wilcox (formation)

wxb wax bite

WXD meteorological radar station

wxg warning

wxp wax pattern

wy wey (14 pounds of wool)

Wy Way; Wyatt; Wycliffe

WY *Wy-dit-Joli-Village* (French—Wy called Pretty Village)—near Paris

WY Warwickshire Yeomanry; West Yorkshire; Wyoming

Wya Whyalla

WYACL World Youth Anti-Communist League

wyaio will you accept (the position) if offered

Wyantskill Wyantskill Center for (delinquent) Girls at Wyantskill, New York

Wyc Wycliffe; Wycliffe College

WYC Warwickshire Yeomanry Cavalry; Washington Yacht Club; Winthrop Yacht Club

WYCF World Youth Crusade for Freedom

Wycl Wycliffe

wye Y (as in wye circuit)

WYF World Youth Forum

wyo what's your opinion

Wyo Wyoming; Wyomingite

Wyo Sem Wyoming Seminary (college preparatory school)—Kingston, Pennsylvania

W Yorks West Yorkshire

WYR West Yorkshire regiment

WYRV West Yorkshire Rifle Volunteers

WySEA Wyoming Society of Enrolled Agents

wysiwyg what you see is what you get

wysiwygmol what you see is what you get more or less

wz war zone

WZ *Welt Zeit* (German—world time)

WZO World Zionist Organization

WZOA Women's Zionist Organization of America

WZW *west zuidwest* (Dutch—west southwest)

X

x an abscissa (symbol); an unknown quantity (symbol); any point on a great circle; by (in measurements); cross; cross reactance (symbol); exchange; execute(d); extra; frost; gang territorial mark(er) or place where one gang will Fight another; historical records; no recent sightings (symbol); mole ratio; no wind distance; parallactic angle; specific acoustic reactance; hoar-frost (meteorology); kiss; mechanical defect; motion picture not suitable for viewing by minors; multiply by; the spot where a crime was committed (x marks the spot); the position of objects on a chart or map, the signature of the illiterate; by (as in 3 X 5 file card)

x (X) $10 bill; Christ; Christian; Christianity; cross; execution(er); experiment; experimental (symbol); explosive (symbol); extra; extract(ed); Kienbock unit (symbol); magnification power; movie rating–no one under 17 admitted; reactance (symbol); research aircraft (symbol); single strength; times (multiplied by); univalent negative (symbol); unknown quantity; U.S. Steel Corporation (stock exchange symbol); X ray; Xavier; X ray—code for letter X

X ex-interest (newspaper bond listings); no connection (symbol)

X longitudinal axis

X-2 counterintelligence

X-15 rocket-propelled research aircraft

X 17 mortality table

X.25 International Protocol for Packet-Switched Networks

xa chiasma; extended architecture; transmission adapter

XA Crucible Steel (stock exchange symbol); experimental (USAF symbol)

xaam experimental air-to-air missile

xact exact(ly); X (in any computer) automatic code translation

XAE merchant ammunition ship (naval symbol)

xafh X-band antenna feed horn

XAK merchant cargo ship (naval symbol)

XAKC merchant coastal cargo ship, small (naval symbol)

xal xenon arc lamp

Xalapa Jalapa

Xalisco Jalisco

Xalostoc San Cosme Xalostoc, Tlaxcala, Mexico

Xaltocan San Martin Xaltocan, Tlaxcala, Mexico

XAM merchant ship converted to minesweeper (naval symbol)

x-a mix. xylene-alcohol mixture (insect larva killer)

x-a mixture xylene-alcohol mixture

xan xanthic; xanthine; yellow

Xan Xanthe; Xanthian; Xanthippe; Xanthus

Xana Xanadu

xanth xanthoma(tosis)

Xantip Xantippe (archetype of the scolding termagant shrew as she was the peevish wife of Socrates)

XAP merchant transport (naval symbol)

XAPc merchant coastal transport, small (naval symbol)

x arm cross arm

XAS X-band Antenna System

xasm experimental air-to-surface missile (XASM)

xat X-ray analysis trial

Xav Xaver; Xavier; Xaviera

XAV auxiliary seaplane tender (naval symbol)

X-axis horizontal axis on a chart, graph, or map

xb crossbar; exploding bridgewire

XB experimental bomber

xbag excess baggage

Xbal Cristobal

X-band 5,200–10,900 mc

xbar crossbar

X bear grizzly bear

Xber December

xbr experimental breeder reactor

X-bracing cross bracing

X^{bre} *dicembre* (French—December)

xbt expendable bathythermograph

XBT Expendable Bathythermograph

xbts exhibits

xc cross country; ex coupon; X-chromosome

X-c X-chromosome

X$_c$ capacitive reactance

XC experimental cargo aircraft (naval symbol); multi-capitalization growth; Xavierian College

Xca Xcalac, Quintana Roo, México

xcar from the railroad car

XCG experimental cargo glider (naval symbol)

xch exchange

xchgr exchanger

X-chromosome female-producing gene found in male sperm

xcit excitation

X-City site of UN Headquarters along New York's East River between 42nd and 49th streets

xcl excess current liabilities

XCL armed merchant cruiser (naval symbol)

xclu exclusive; exclusivity

x-con ex-convict

xconn cross connection

xcp without coupon

XCR Extraterrestrial Research Center

X-craft midget submarines

xcs cross-country skiing

xct X-band communications transponder

xcu excuse; extra-care unit; extreme close-up

xc & uc exclusive of covering and uncovering

xcvr transceiver

x cy cross country

xd ex dividend; expiration date

x'd executed

X'd crossed out

XD Executive Development

X-day launching day

xdcr transducer

xder transducer

xdf X-band flow detection

X & DFLOT Experimental and Development Flotilla

xdh xanthine dehydrogenase

xdis ex distribution, without distribution

xdiv without dividend

X division branch of society consisting of swindlers and thieves

x'd out crossed out, deleted

xdp X-ray density probe; X-ray diffraction powder

xdpc X-ray diffraction powder camera

xdps X-band diode phase shifter

xdr expanded dynamic range; transducer

Xdr Crusader

x drs ex drawings

XDS Xerox Data Systems; X-ray Diffraction System

xdt xenon discharge tube

xe ex entitlement

Xe experimental engine; xenon

xecf x-e cold-flow engine

xeg X-ray emission gage

XEG Xerox Education Group

Xen Xenia; Xenik; Xeno-cratic; Xenophon (c.434-c.355); Xenos

xeno xenodiagnosis; xenodiagnostic; xenogenic; xenograft; xenolith; xenolithic; xenophile; xenophilia; xenophobe; xenophobia

Xeno Xenocrates; Xenophanes; Xenophon

xenobio xenobiologic(al)(ly); xenobiologist; xenobiology

xenodiag xenodiagnosis

xenop xenophile—person attracted to what is foreign

Xenop Xenophon

xenop(s) xenophobe(s); xenophobia(s); xenophobic(s)

XEP Xerox Educational Publications

xer Xerox reproduction

Xer Xerxes

xerocops xerocopies (books reproduced by xerography)

xerodups xerographic duplicates

xerog xerograph(ic)(al)(ly); xerography

xeromamo xeromammograph (also called xerox mammograph—xerographic process used in diagnosis of breast cancer)

xerorads xerographic radiographs

xes X-ray emission spectra

xf ex offer; extra fine

XF experimental fighter (naval symbol)

xfa crossed-field acceleration; X-ray fluorescence absorption

xfc X-band frequency converter

xfd crossfeed; X-ray flow detection

xfer transfer

xfh X-band feed horn

Xfher Christopher

XFL Extreme Football League

xflt expanded flight-line tester

xfm X-band ferrite modulator

xfmr transformer

xformer transformer

xfqh xenon-filled quartz helix

xfrmr transformer

xft xenon flash tube

xg crossing

XG multi-capitalization growth

xgam experimental guided air missile (XGAM)

XGP Xerox Graphic Printer

XGprt xanthine-guanine phosphoribosyl transferase (enzyme)

xh extra hard; extra heavy; extra high

Xh Xhosa

XH experimental helicopter (naval symbol)

x heavy extra heavy

X-height height of central portion of lowercase letters exclusive of ascenders and descenders

xhf extra high frequency

x-high of a height equal to a lowercase x of the same face and size

xhil xenon high-intensity light

xhm X-ray hazard meter

xhmo extended huckel molecular orbit

xhr extra-high reliability

X-hr X-hour (when shipping evacuation is ordered from major ports by NATO)

Xhs Xhosa

Xst exhaust

xhv extremely high vacuum

x hvy extra heavy

xi interest; extra innings (baseball); xi particle

xia X-band interferometer antenna

Xianggang (Chinese—Victoria)—on Hong Kong island

Xian(s) Christian(s)

xic transmission interface converter

Xico Xicoténcatl

xil xilography; xilogravure (woodcuts)

xim X-ray intensity meter

xin without interest

Xin Xingu

Xina Christina

XING crossing (highway or railroad)

xio execute input-output

Xiph xiphoid; xiphoidal.

xipho Xiphosura

Xiq-Xiq Xique-Xique, Brazil

xirs xenon infrared searchlight

xis xenon infrared searchlight

xist xistoma; xistomiasis

xistor transistor

xk X-band klystron

xl crystal; crystalline; extra large; extra long

Xl inductive reactance

xla X-band limiter anntenuator

xlam cross-laminate(d)

xlc xenon lamp collimator

xld experimental laser device

xldt xenon laser discharge tube

xlf ex lady friend

xli extra-low interstitial

X-lib ex-library copy (book)

xlnt excellent

XLO Ex-Cell-O (precision products, trade name)

xlps xenon lamp power supply

xlr experimental liquid rocket

xls extreme long shot; xenon light source

XLSS Xenon Light-Source System

xlt cross-linked polyethylene; excellent; xenon laser tube

xltn translation

xl & ul exclusive of loading and unloading

xlwb extra-long wheelbase

xm crossmatch; examine; experimental model

xm (XM) experimental missile

xmm X-ray multiple mirror

Xm Christmas

XM experimental missile

XM-1 main battle tank (USA)

XM-706 Cadillac-Gage amphibious armed car and military personnel carrier called the Commando

XM-723 cavalry of infantry fighting vehicle (USA)

Xma$ Christmas (commercialized)

Xmas Christmas

X-matching cross matching

xmfr transformer

xmit transmit

xmitter transmitter

xml extensible markup language

x mod experimental module

X-mod experience modification

xmp (XMP) experimental extraordinary multiprocessor

xms X-band microwave source

XMS Experimental Development Specification; Xavier Mission Sisters

xmsn transmission

xmt exempt; transmit; X-band microwave transmitter

xmtg transmitting

xmtl transmittal

xmtr transmitter

xmt-rec transmit-receive

xmtr-rec transmitter-receiver

xn ex new

Xn Christian

XN experimental (USN)

Xndu Xanadu

xnor gate exculsive-nor gate

X-note $10 bill

XNS Xerox Network Services

xnt excellent

Xnty Christianity

xo crystal oscillator

XO Executive Officer; Experimental Office(r); Turner's syndrome wherein one of the sex-determining pair of XX chromosomes is missing

X-O cross-out test

xob xenon optical beacon

Xochi Xochimilco

x-off transmitter off

xoloiz xoloizcuintli (pronounced *sholloizquintly*)—Mexican hairless dog both hotblooded and flealess as well as faithful

XOMA XOMA Corporation

x-on transmitter on

xon/xoff transmitter on/transmitter off

xoophorec xoophorectomic (sterilization); xoophorectomy (removal of the ovaries)

xor exclusive or (data processing)

xos extra outside clothing; extra outsize (clothing)

X-out cross out; delete; strike out

xover cross over

X-over cross over

xp express paid; xerodema pigmentosum.

Xp fire-resistive protected cabinet, safe, or vault

XP (Greek—chi rho)—first two letters of the Greek word for Christ

xpa X-band parametric amplifier; X-band passive array; Xband planar array; X-band power amplifier

xpaa X-band planar-array antenna

XPARS External Research Publication and Retrieval System

XPC inshore patrol cutter (naval symbol)

xpd cross-polarization discrimination; cross-pollination discrimination; expedite(d)

xped Christened

xper without privileges

Xper Christopher

xpert expert

XPG converted merchant ship (naval symbol)

xpl explain; explanation; explosion; explosive

xplo explosion

xplos explosive

xplt exploit

XPM Xerox Planning Model

xpn expansion

Xpo Cristo (Spanish—Christ)

xpond transponder

xpp exprés payé lettre (French—express-paid letter)

xppa X-band pseudo-passive array; X-band pulsed-power amplifier

xpr ex privileges; without privileges

X-press Express

xprs express

xprts expertise; experts

xprtz expertize

xps X-band phase shifter; X-ray photoelectric spectroscopy; X-ray photoelectron spectroscopy

xps (XPS) X-ray photomission spectroscopy

xpt except; X-band pulse transmitter

xpt exprés payé télégraphe (French—express-paid telegraph)

XPto Cristóbal (Spanish—Christopher)

X-punch punch in X row (11th row) of an 80-column punchcard

xq cross-question

XQ Experimental Target Drone

xqh xenon quartz helix

xr ex rights; index register; Xerox radiography

Xr Christian; Christopher; examiner

XR External Relations (UNESCO)

X-rated movie moving picture not recommended for minors

X-rated shops sex-oriented establishments such as massage parlors and adult bookstores

xray execution recorder analyzer

Xray letter X radio code

X-ray letter X radio code; photograph or photography made by X-rays; radiograph; radiography, roentgemmograph; roentgenography; roentgen ray

xrb X-band radar beacon

xrcd X-ray crystal density

xrd X-my diffraction

X rds crossroads

X-rea X-ray events analyzer

xref cross-reference

xrep auxiliary report

xrf X-ray fluorescence

xrfs X-ray fluorescence spectrometer

XRH Tenth Royal Hussars

xrii X-ray image intensifier

xrl extended-range lance (missile)

xrm X-ray microanalyzer

xro xeroradiography

X-roads crossroads

xrp X-ray and photofluorography

xrpm X-ray projection microscope

xrpt X-ray and photofluorography technician

XRPT X-Ray and Photofluorography Technician (USN)

xrspec X-ray spectograph

xrt ex-rights; without rights; X-ray technician

Xrx Xerox (corporation or copying process)

xs cross-section; excess; extra strength; extra strong

Xs atmospherics

XS Xerces Society

xsa X-band satellite antenna

xsal xenon short arc lamp

XSB Xavier Society for the Blind

X-scale scale of a line parallel to the horizon

x sec extra sec (*trés sec*)—dry champagne

xsect cross-section

xsf X-ray scattering facility

xsistor transistor

XSL Experimental Space Laboratory

xsm experimental strategic missile; experimental surface missile

xsoa excess speed of advance authorized

X-sonad experimental sonic azimuth detector

X-spot $10 bill

xspv experimental solid-propellant vehicle

xsr X-band scatterometer radar

XSS Experimental Space Station

xsta X-band satellite-tracking antenna

xstd X-band stripline tunnel diode

xstda X-band stripline tunnel diode amplifier

xstr transistor

x str extra strong

xstrat cross-stratified

xt crosstalk; X-ray tube

xt (XT) extra technology (computer)

Xt Christ

xta chiasmata; X-band tracking antenna

xtal crystal

XTE X-ray Timing Explorer (satellite)

Xth tenth

Xtian Christian

xtk cross track

xtlo crystal oscillator

xtnd extend

xtnd wknd extended weekend

xtntn extension

xto X-band triode oscillator

X-town X names the town unnamed

xtr extra (computer flow chart)

XTR Xtra Inc

xtra extra

xtran experimental language; experimental translation

xtrm extreme

XTRU Xtra Inc (container) Unit

xtry extraordinary

xtwa X-band traveling-wave amplifier

xtwm X-band traveling-wave masser

Xty Christianity

xu X-ray unit; x-unit

Xu fire-resistive unprotected cabinet, safe, or vault

XU Xavier University

XUL Xavier University of Louisiana

Xulla formerly the Sula Islands of Indonesia

xuloc *xuloctzcuintle* (Aztec—hairless dog)—breed imported to Mexico from China in the late 1 6th century; popular among ranchers for its loud bark and its two-degree warmer body useful in the feet of sleeping bags

XUM Xerox University Microfilm

xut crosscut

xuv extreme ultraviolet

xva X-ray videcon analysis

Xvers transverse

XVP Executive Vice President

xvtr transverter

xw experimental warhead; ex warrants; without warrants

X-wave extraordinary wave

Xway (XWAY) Expressway

X-way expressway

X-weld X-shaped weld

X/Windows X-Window System

XWS Experimental Weapon System

xx without securities or warrants

XX doublecross; double strength; female (see X chromosome); twenty (Roman numerals)

X-X Xai-Xai (Mozambique seaport)

XX *Dos Equis* (Spanish—Two X)—Mexican beer

XXer doublecrosser

xxh double extra hard; double extra heavy

xxl cancel

XX-note (double-X note) $20 bill

xxos extra-large outside (clothing)

xxs extra-extra strength

xxx international urgency signal

XXX triple strength; triple-X; Triple X condition; triple X syndrome; thirty (Roman numerals)

XXXX quadruple strength

XXXXX quintuple strength

XXY Klinefelter's syndrome wherein the sex-determining chromosomes are XXY instead of the normal XY

xy xylography

XY male

xya x-y axis

xyat x-y axis table

xyl ex young lady; xylene; xylography; xylophone

xylo xylophone

xyloc xylocain (lidocaine)

xylog xylography

xyp x-y plotter

xyr x-y recorder; xyridaceae; xyridaceous; xyridales; xyris

x yr dev ten-year device (US Army service badge)

xyt x-y table

xyv x-y vector

xyz examine your zipper (your fly is open)

XYZ XYZ Affair leading to undeclared naval war between France and the United States from 1798 to 1800

X zone adrenal cortex inner zone (of some young mammals)

Y

y   economy fare

y   income (microeconomics); (Spanish—and)

Y   admittance (symbol); Convair (symbol); ex-dividend and sales in full (newspaper stock listings); service test (symbol); yacht; Yankee–code for letter Y; Yard (The Yard—Scotland Yard); yen (Japanese money unit); Yeomanry; YMCA; YMHA; YWCA; YWHA; Young's modulus; Yorkshire (Division); yen (Chinese money unit)

Y   admittance (symbol); lateral axis (symbol); ylös (Finnish—up)

Y1C   Yeoman First Class

Y2C   Yeoman Second Class

Y2K   Year 2000

Y3C   Yeoman Third Class

Y-18   Ilyushin 18 aircraft

Y-40   Yak 40 aircraft

Y-5 molars   lower molars characteristic of apes, hominoids, and humans

Y62   Ilyushin Il-Y62 jet airplane

ya   yaw axis; young adult

YA   Yasser Arafat; Young Adults; Youth Aliyah; Youth Authority

Y/A   York-Antwerp Rules

YAA   Yachtsmen's Association of America

YAAP   Young Americans Against Pollution

YABA   Yacht Architects and Brokers Association

yac   yeast artificial chromosome

YAC   Young Adult Council; Young Alumni Club; Youth Advisory Council

YACA   Youth and Adult Correctional Agency

YACC   Young Adults Conservation Corps

YACH   Yugoslav-American Cooperative Home

yactoff   yaw-actuator offset

YAD   Youth Aid Division; Youth Authority Department

Yad Fiz   Yaderna Fizika (Russian—Journal of Nuclear Physics)

yadh   yeast alcohol dehydrogenase (YADH)

YAEC   Yankee Atomic Electric Company

YAF   Young Americans for Freedom

YAF-PAC   Young Americans for Freedom–Political Action Committee

yag   yttrium aluminum garnet

yag (YAG)   yttrium, aluminum, garnet (surgical laser)

YAG   district auxiliary miscellaneous (3-letter naval symbol)

yagl   yttrium-aluminum garnet laser

yag laser   yttrium-aluminum garnet laser

YAI   Young Adult Institute

YAIC   Young American Indian Council

yak   yakushimanum

Yak   Yakolev; Yakov; Yakovlevich

Yak   Yakarta   (Spanish—Djakarta)

YAK   Yakovlev aircraft (named for its designer)

Yak-11   Soviet Yakovlev two-place trainer aircraft named Moose by NATO

Yak-12   Soviet Yakovlev two-place trainer aircraft

Yak-18   Soviet Yakovlev two place aircraft used as a trainer and named Max by NATO

Yak-25   Soviet Yakovlev all-weather interceptor fighter aircraft named Flashlight by NATO

Yak-26   Soviet Yakovlev tactical reconnaissance aircraft named Mangrove by NATO

Yak-28   Soviet Yakovlev tactical bomber aircraft named Brewer by NATO

Yak-28P   Soviet Yakovlev all-weather interceptor aircraft named Firebar by NATO

yaku   yakushimanum

Yakutia   Yakut Autonomous Soviet Socialist Republic (Soviet-era name for eastern Siberia, in Russia)

yal   yttrium-aluminum laser

YAL   Young Australia League

YALDS   Young Australia Language Development Scheme

Yale LJ   Yale Law Journal

Yallo Ballys   short form for the Yallo Bally Mountains of northern California

Yam   Yamaha

YAM   Yates American Machine (company)

YAN Yancey (railroad); Young American Nazis

YANCON Yankee Conference (intercollegiate sports)

Yang Yangon (Burmese—Rangoon)

Yangpat Yangtze Patrol

Yank Yankee; Yankel

Yank Yankel (Yiddish—Jacob)

YANK Youth of America Needs to Know

Yankee letter Y radio code; Soviet class of nuclear-powered submarines—Yankee or Y-class—similar to U.S. Polaris-type subs

Yanks British nickname for Americans

yap. yaw and pitch

yaps yaw and pitch sensor

Yar Yarmouth

YAR Yemen Arab Republic (Sana—capital); York-Antwerp Rules (insurance)

YARA Young Americans for Responsible Action

yard prison yard

'yard shipyard

Yard Scotland Yard; Yardley

YARD Yarrow-Admiralty Research Department

Yarden yard + garden

YARDS Yard Activity Reporting and Decision System

YARN Young Adult Resource Notebook

yas yaw-attitude sensor

YAs Young Adults (young people)

YAS Yorkshire Agricultural Society

YASD Young Adult Services Division (ALA)

YASSR Yakut Autonomous Soviet Socialist Republic

Yat Yatyiopia (Amharic—Ethiopia)

yavis young, attractive, verbal, intelligent, and successful

YAWF Youth Against War and Fascism

Y-axis vertical axis on a chart, graph or map

yb yardbird (confined to a military camp)

Yb ytterbiurn

YB yearbook; Youngstown Sheet & Tube (stock market symbol)

YBA Young Buddhist Association; Youth Basketball Association

YBC Yerba Buena Center

Ybk Yearbook

Ybor Ybor City (Florida town near Tampa and named for a cigar-maker – Vivente Martinez y Bor)

YBPC Young Black Programmers Coalition

yBr yellowish brown

YBR sludge-removal barge (naval symbol)

YBRA Yellowstone-Bighorn Research Association

Y-branch Y-shaped pipe fitting

YB(RS) Year Books (Rolls Series)

VBs Young Boys Inc—drug ring using young boys to promote street sales

YBS Yale Bibliographic System (computer cataloging)

yc yaw channel; yaw coupling; yellow chrome

Y-c Y-chromosome

YC open lighter (naval symbol); Yacht Club; Yankton College; Yeomanry Cavalry; York College; Youth Club; Yuba College

YCA Yachting Club of America; Young Citizens' Army; Young Concert Artists; Youth Camping Association; Youth Correction Act

YCC Youth Conservation Corps; Youth Correctional Center; Yuba Community College

YCCA Youth Council on Civic Affairs

YCCC Yui Chui Chan Club

YCCIP Youth Community Conservation and Improvement Projects

YCD feuling barge (naval symbol); Youth Correction Division (U.S. Dept Justice)

YCF car float (naval symbol); Yankee Critical-Facility; Young Calvinist Federation

YCGJ Young Christians for Global Justice

Y-chromosome male-producing gene found in male sperm

YCI Young Communist International; Youth Correctional Institution

YCI Yacht Club Italia (Italian Yacht Club)

YCia Ybarra Compañía (steamship line)

YCK open cargo lighter (naval symbol)

YCL Yarmouth Cruise Lines; York City Library; Young Communist League

YCLA Young Circle League of America

Y-class Soviet class of nuclear-powered submarines, also called Yankee, similar to U.S. Polaris-type subs

YCM Young Christian Movement

YCNM Yucca House National Monument

ycp yaw-coupling parameter

YCP Youth Challenge Program

YCS Young Catholic Students; Young Christian Students; Youth and Community Services

YCSM Young Christian Student Movement

yct yacht

YCTF Younger Chemists Task Force

YCU aircraft transportation lighter (naval symbol)

YC & UO Young Conservative and Unionist Organization

YCV aircraft transportation lighter (naval symbol); Young Citizen Volunteers

ycw you can't win

YCW Young Christian Workers

ycz yellow caution zone (airport runway lighting)

yd yard

yd. graveyard

y/d yaw damper

YD floating derrick (naval symbol); Yorkshire Dales; Yorkshire Dragoons; Young Democrat; Yugoslav dinar

Y & D Yards and Docks (USN)

Yd$_2$ square yard(s)

yd$_3$ cubic yard(s)

yda yesterday

YDA Dawson City, Yukon Territory (airport)

ydaa yellow dinitrophenyl aspartic acid

yday yesterday

ydb yield-diffusion bonding

ydc yaw-damping computer

YDCA Youth Democratic Clubs of America

YDF floating drydock (naval symbol)

ydg yardage; yarding

YDG degaussing vessel (naval symbol)

ydi yard drain inlet

YDI Youth Development Incorporated

YDL Young Development Laboratories

ydmn yardman

ydmstr yardmaster

yds yards

Yds Yards

YDs Young Democrats

YDS Yale Divinity School

YDSD Yards and Docks Supply Depot (USN)

YDSO Yards and Docks Supply Office

YDT diving tender (naval symbol)

Y-duct Y-shaped duct

ye yellow edges; yellow edging; yellow enzyme; yellow-edged; youth entry

y^e (Early English–thou)–also written ye

YE aircraft homing system; Yemen (Internet code)

yea. yaw-error amplipfier

YEA Yale Engineering Association

yearb yearbook

YEB Yorkshire Electricity Board

YEDPA Youth Employment and Demonstration Projects Act

YEFC Youth Education for Citizenship (American Bar Association)

yeg yeast extract–glucose

YEG Edmonton, Alberta (International Airport)

yegg yeggman (burglar specializing in opening safes and vaults)

yel yellow

Yel Boris Yeltsin

Yell Yellowstone National Park

Yellowjackets Yellowjacket Mountains of eastern Idaho

Yellow Peril danger to Western civilization held to arise from the influx of Oriental laborers willing to work for low wages

yellows yellowtail (fish)

Yellowstone Yellowstone County in Montana, Yellowstone Lake in Wyoming, Yellowstone National Park (Idaho, Montana, and Wyo-

ming), Yellowstone River (Montana, North Dakota, and Wyoming)

Yel NP Yellowstone National Park

yelsh yellowish

yem yeast extract–malt

Yem Yemen; Yemenite

Yemen Republic of Yemen (*Jumhuryah al-Yemen*)

Yem RA *República Árabe del Yemen* (Spanish-Arab Republic of Yemen)

Yem RDP *República Democrática Popular del Yemen* (Spanish—Popular Democratic Republic of Yemen)

yen monetary unit of Japan

Yeo Yeoman

YEO Young Entrepreneurs Organization; Youth Employment Office(r)

Yeoman F Yeoman Female (naval rating)

yeomn yeomanry

yep your educational plans

YEp yeast-episomal plasmids

yepd yeast extract–peptone, dextrose

Yer Yerevan (capital of Annenia)

YES Youth Educational Services; Youth Education Systems; Youth Employment Service; Youth Exchange Scheme (for Europe); Youth to End Smoking; Youth Engaged in Service

yesty yesterday

YETP Youth Employment Training Program

YEWTIC Yorkshire, East and West Ridings, Technical Information Centre

YEX-ZA World's Largest Blimp

yf wife (orthographic contraction proposed by Benjamin Franklin)

yf (YF) yellow fever

YF covered lighters (naval symbol)

YF-16 air-superiority single-engine lightweight-fighter aircraft (USAF)

YFB ferryboat or launch (naval symbol)

YfC Youth for Christ

YFC car float (naval symbol); Young Farmers' Club

YFCU Young Farmers' Clubs of Ulster

YFD yard floating drydock (naval symbol)

YFFP Yarrawonga Flora and Fauna Park (Australian Northern Territory)

YFN covered lighter, non-self-propelled (naval symbol)

YFNB large covered lighter (naval symbol)

YFND drydock companion craft (naval symbol)

YFNX special-purpose lighter (naval symbol)

YFP floating power barge (naval symbol); Youth For Progress

YFR self-propelled refrigerated covered lighter (naval symbol)

YFRN refrigerated covered lighter, non-self-propelled (naval symbol)

YFRT covered lighter, range tender (naval symbol)

YFT torpedo transportation lighter (naval symbol)

yfu yard freight unit

YfU Youth for Understanding (teenage exchange program)

YFU harbor utility craft (naval symbol)

y fwd yarn forward (knitting)

yG yellowish green

Yg *Young's Literal Translation of the Holy Bible*

YG garbage lighter (naval symbol); yellow green

YGC Youth Guidance Center

yggt y-gamma-glutamyl transferase

YGH Yankee Go Home

ygl yttrium-garnet laser

ygmd yaw-gimbal command

YGN garbage lighter, non-self-propelled (naval symbol)

YGR Yankari Game Reserve (Nigeria)

YGS Young Guard Society

ygt y-glutamyl transpeptidase

Y-gun Y-shaped gun used aboard ships for firing depth charges

YH Yorkshire Hussars; Youth Hostel

YHA Youth Hostels Association

Yhama Yokohama

YHANI Youth Hostel Association of Northern Ireland

YHB houseboat (naval symbol)

YHLC salvage lift craft, heavy (naval ship symbol)

YHt Young-Helmholtz theory

YHT heating scow (naval symbol)

YHVH *see* **JHVH**

Yi Yiddish

YIC Yardney International Corporation

Yid Yiddish; Yiddish-speaking person

Yid Yiddish (German dialect spoken by many persons of Judaic origin and augmented by the languages of the countries where they have emigrated)

Yidgin-English Yiddish + English

Yid lang Yiddish language

Yie Young interference experiment

YIEP Youth Incentive Entitlement Project

yig yttrium iron garnet (ferrite)

yigib your improved group insurance benefits

YIIJS Young Israel Institute for Jewish Studies

YIJR Yivo Institute for Jewish Research

YIJS Young Israel Institute for Jewish Studies

YIKOR Yadishe kultur-organizatsye (Polish-Yiddish Culture Organization)

yil yellow indicator lamp

Ying Yinglish (English and Yiddish combined)

Yinglish Yiddish-English

yip yeast integrative plasma

YIP Detroit, Michigan (Willow Run Airport); Youth International Party

YIR Yearly Infrastructure Report

Yis Yisroel (Yiddish-Israel)

YI & S Yawata Iron and Steel

Yivo Inst Yivo Institute for Jewish Research

yj radar homing beacon (map symbol)

YJC York Junior College

Y-joint Y-shaped joint

yk radar beacon (map symbol)

Yk Yakut; York

YK Yankee Airlines

Yka Yokohama

YKAA Young Keyboard Artists Association

YKF Yiddisher Kulture Farband (Yiddish Culture Club)

YKKK Yamashita Kisen Kabushiki Kaisha (Japanese—steamship line)

Ykn Yukon

Yko Yokosuka

Yks Yorkshire

Ykt Yakut

yl yellow; yield limit; young lady

Y & L York and Lancaster

Y+L York and Lancaster

YLA open landing lighter (naval symbol)

YLC Young Life Campaign

yld yield

YLI Young Ladies Institute; Yorkshire Light Infantry

YLJ Yale Law Journal

YLL Yerkes Language Laboratory

YLLC salvage lift crane, light (naval ship symbol)

YLM Yale Literary Magazine

YLO Young Lords Organization

YLP Young Lords Party

Y & LR York and Lancaster Regiment

yl's young ladies

Ylstn Yellowstone

ym yacht measurement; yawing moment; yellow metal; your measurement; your message

YM dredge (naval symbol); Yehudi Menuhin

YMA Yarn Merchants Association; Young Menswear Association

ymb yeast malt broth

YMBA Yacht and Motor Boat Association

YMCA Young Men's Christian Association

YM Cath A Young Men's Catholic Association

YMCU Young Men's Christian Union

ymd your message date

Yme Young's modulus of elasticity

YMF Young Musicians Foundation

YMFS Young Men's Friendly Society

YMHA Young Men's Hebrew Association

YMHAL Young Men's Hebrew Association Library

YMI Young Men's Institute

YML Yang Ming Line

YMLC salvage lift craft, medium (naval ship symbol)

YMLU Yamashita Line (container) Unit

ymmv your mileage may vary

ymoyl your money or your life

YMP motor mine planter (naval symbol); Young Management Printers

YMPA Young Master Printers' Alliance

yms yield measurement system

YMs Yearly Meetings (Quakers)

YMS motor minesweepers (naval symbol)

YMT motor tug (naval symbol)

Ymu Ymuiden

YMV Montreal, Quebec airport; Yazoo and Mississippi Valley (railroad)

YM & YWHA Young Men's and Young Women's Hebrew Association

yn yen

y-n yes-no

Yn Yeoman; Yeowoman

YN net tender (naval symbol); Youngstown & Northern RR

Y network wye network

yng young

YNG gate vessel (naval symbol)

YNHA Yosemite Natural History Association

Y-NHH Yale-New Haven Hospital

ynhl why in hell

YNP Yellowstone National Park (Idaho, Montana, Wyoming); Yoho NP (British Columbia); Yosemite NP (California); Youth National Party

YNSO Yomiuri Nippon Symphony Orchestra

YNT net tender, tug (naval symbol)

Ynv Ynvar (Russian—January)

YNWR Yazoo National Wildlife Refuge (Mississippi)

yo yarn over (knitting); year old; years old

yo' yore; you; your

y/o years old

YO fuel-oil barge (naval symbol); Yerkes Observatory

YOAN Youth Of All Nations

yob year of birth

YOB Youth Opportunities Board

yoc young officers course

YOC Youth Opportunity Campaign; Youth Opportunity Center(s); Youth Opportunity Corps; Youth Ornithologists' Club

YOC-RSPB Young Ornithologists' Club—Royal Society for the Protection of Birds

yod year of death

YOG gasoline barge, self propelled (naval symbol)

YOGN gasoline barge, non-self-propelled (naval symbol)

Yok Yokohama

Yoko Yokohama; Yoko Ono

Yokuska (navalese—Yokosuka, Japan)

yom year of marriage

Yom Yomiuri (Japanese—News Crier)—Tokyo newspaper

YOM yellow oxide of mercury

Yom Kip Yom Kippur (Hebrew—Day of Atonement)

yon yonder

YON fuel-oil barge, non-self-propelled (naval symbol)

yood (slang pronunciation—iud)—intrauterine device

YOP Youth Opportunity Program

YOPB Youthful Offender Parole Board

York New York; New York State

York Yorkshire Post

Yorkie Yorkshire terrier

Yorks Yorkshire

Yos Yosu

YOS oil storage barge (naval symbol)

Yosemite Yosemite National Park in California; natural attractions such as the Yosemite Falls or the Yosemite Valley

Yoshino-kumano Yoshino-kumano National Park in southern Honshu, Japan

Yos NP Yosemite National Park

yot (YOT) youthful offender treatment

YOU Youth Opportunities Unlimited; Youth Organizations United; Youthful Offender Unit

you'd you had; you would

you'll you shall; you will

Youngs Youngstown

you're you are

youthploit youth exploitation

you've you have

YOW Ottawa, Ontario (airport)

yp yellow pine; yield limit; yield point (psi)

YP patrol craft (naval symbol); yellow peril; young people; young person(s)

ypa yaw-precession amplifier

YPA Young Pioneers of America

YPCS Young Peoples Computer Society

ypd yaw-phase detector

YPD floating pile driver (naval symbol)

YPEC Young Printing Executives Club

YPF Yacimientos Petroliferas Fiscales (Spanish—Government Oil Deposits)—Argentina

YPFB Yacimientos Petrolíferos Fiscales Bolivianos (Spanish—Bolivian Government Oil Deposits)

YPFP Yunnan Phosphate Fertilizer Plant

YPG Yuma Proving Ground

yPk yellowish pink

YPK pontoon stowage barge (naval symbol)

YPM Yale Peabody Museum

YPO Young Presidents' Organization; Youth Programs Office (Bureau of Indian Affairs)

YPPA Yellow Pages Publishers Association

Yps Ypsilanti

YPSCE Young People's Society of Christian Endeavor

YPSL Young People's Socialist League

Y-punch punch in Y row (12th row) of an 80-column punchcard

YQX Gander, Newfoundland (airport)

yr year; younger; your

y-r yaw roll

YR district patrol vessel (naval symbol); floating workshop (naval symbol); Young Republican(s)

YRA Yacht Racing Association

yrb year built

Yr B Year Book

YRB submarine repair and berthing barge (naval symbol)

yrbk yearbook

YRBM submarine repair—berthing and messing barge (naval symbol)

YRC submarine rescue chamber (naval symbol)

YRD submarine repair and berthing vessel (naval symbol)

YRDH floating drydock hull workshop (naval symbol)

YRDM floating drydock machinery workshop (naval symbol)

YRL covered repair lighter (naval symbol)

yrly yearly

YRNF Young Republican National Federation

Yr obt servt Your obedient servant

yrp yeast-replicating plasmids

YRR radiological repair barge (naval symbol)

yrs years; yours

Yrs Yours

YRs Young Republicans

YRS Yugoslav Relief Society

YRST salvage craft tender (naval ship symbol)

yrs ty yours truly

yrt yearly renewable term (insurance)

YRU Yacht Racing Union

ys yellow spot (on retina); yield strength; young soldier

Ys Yugoslavia; Yugoslavian

YS Yard Superintendent; Young Socialists

Y-S Yamashita-Shinnihon

Y & S Youngstown & Southern (railroad)

YS-11 Japanese medium-range transport plane

YSA Young Socialist Alliance; Youth Service America; Youth Services Administration (District of Columbia)

ysb yield-stress bonding

YSB Yacht Safety Bureau; Youth Service Bureau

YSC Yugoslav Seamen's Club; Youth Studies Center juvenile correctional facility in Philadelphia)

YSD seaplane wrecking derrick (naval symbol); Youngstown Steel Door (company); Youth Services Division

ysdb yield-stress diffusion bonding

yse yaw-steering error

ysh yellowish

YSI Yellow Springs Instrument (company)

Ysl Ysrael

YSL Young Socialist League; Yves Saint Laurent

Y-S Line Yamashita-Shinnihon Line (steamships)

YSM Yale School of Music; Yangtze Service Medal

yso young stellar object

YSO Youngstown Symphony Orchestra

ysp years service for severance pay purposes

YSP pontoon salvage vessel (naval symbol)

ysr you're so right

YSR sludge-removal barge (naval symbol)

YSS Young Scots Society

YSSAS Yale Summer School of Alcohol Studies

yst youngest

YST Yukon Standard Time

YS & T Youngstown Sheet & Tube

YSTO Yugoslav State Tourist Office

YSU Youngstown State University

yt yoke top

Yt yttrium

Yᵗ *Ytre* (Dano-Norwegian or Swedish—outer)

YT harbor tug (naval symbol); Yukon Territory; Yukoi Time

Y & T Tale & Towne

YTA Yiddish Theatrical Alliance

ytb yarn to back; yield to broker

YTB large-harbor tug (naval symbol)

ytc yield to call

YTCA Yorkshire Terrier Club of America

ytd year to date

YTEP Youth Training and Employment Project

ytf yarn to front

YTL small-harbor tug (naval symbol)

ytm yield to maturity

YTM medium-harbor tug (naval symbol)

YTP *Yeni Türkiye Partisi* (New Turkish Party)—socialist oriented

YTPM Yuma Territorial Prison Museum

YTS Youth Training Scheme; Youth Training School; Yuma Test Center; Yuma Test Station

YTT torpedo-testing barge (naval symbol)

Y-tube Y-shaped tube

YTV Yokohama Television

Yu Yugoslav; Yugoslavian

YU Yale University; Yeshiva University; York University; Youngstown University; Yugoslavia (auto plaque and Internet code)

YUAG Yale University Art Gallery

yuan monetary unit of China

Yuc Yucatan (natives nicknamed boxitos—Maya term meaning darks)

Yuca Yucatan, Mexico

YUCA Young Upscale Cuban-American

Yuc Pen Yucatan Peninsula in southeastern Mexico

Yud Yudel

Yud *Yudel* (Yiddish—Judah)

yuffie(s) young urban failure(s)

Yug Yugoslavia(n); Yugoslavic

Yugo Yugoslav; Yugoslavia; Yugoslavian

Yuk Yukon

YUK Youth Uncovering Krud (antipollution society)

Yukon Canadair version of the Britannia designated CC-106; Yukon River; Yukon Territory

YUL Montreal Quebec (airport); Yale University Library

YULRC Yale University Lung Research Center

yumpie(s) young upwardly mobile professional(s)

YUN *Yearbook of the United Nations*

YUO Yale University Observatory

yup you're uncommonly perceptive

YUP Yale University Press

yuppie flu debilitating chronic fatigue syndrome

yuppie(s) young urban professional(s)

yup(s) young urban professional(s)

Yur Yuri; Yurievich

Yus Yussel

Yus *Yussel* (Yiddish—Joseph)

YUSM Yale University School of Medicine

Yuzh *Yuzhnaya* (Russian—southern)

Yv Yvette; Yvonne

Yv *Yvert ef Tellier* (stamp catalog)

YV Young's Version

YVA Young Volunteers in Action

yvc yellow-varnish cambric

YVC Yakima Valley College

YVF Young Volunteer Force

YVHS Yorkville Vocational High School

YVJC Yakima Valley Junior College

YVP Youth Voter Participation

YVR Vancouver, British Columbia (airport)

YVRL Yakima Valley Regional Library

YVT Yakima Valley Transportation (railroad)

y v v *y viaje vuelta* (Spanish—and return trip)

YW water barge (naval symbol); Yeoman Warder; Yreka Western RR

YWAA Youth Welfare Association of Australia

YWAM Youth With A Mission

YWCA Young Women's Christian Association

YWCAUSA Young Women's Christian Association of the U.S.A.

YWCTU Young Women's Christian Temperance Union

ywd you would

YWF Young World Federalists

YWFD Young World Food and Development (UN)

YWG Winnipeg, Manitoba (airport)

YWHA Young Women's Hebrew Association

YWHS Young Women's Help Society

YWLL Young Workers Liberation League

YWN Non-self-propelled barge (naval symbol)

YWPG Young World Promotion Group

YWS Young Wales Society

YWU Yiddish Writers Union

YX Yannis Xenakis

y-y yaw axis

y & y yin and yang

YY pseudonymous initials of Robert Lynd noted for his *New Statesman* essays; Yevgermy Yevtumshanko

YYC Calgary, Alberta (airport)

YYZ Toronto, Ontario (airport)

Z

z cracked (moving and storage symbol); electro-chemical equivalent (symbol); ounce (truncation of oz-ounce); complex variable (symbol); z-bar; zed (British usage); zee (American usage); zero; zinc; zone

z (Z) zloty (Polish currency unit)

z *zu* (German—closed, shut)

Z atomic number (symbol); azimuth (symbol); azimuth angle (symbol); gram equivalent weight (symbol); impedance (symbol); lighter-than-air aircraft (symbol); obsolete (symbol); radius of circle of least confusion (symbol); zenitim; zenith distance; zero meridian time; Zionism; Zionist; Zoroaster; Zoroastrian; Zoroastrianism; Zulu—code for letter Z

Z normal axis (symbol); *Zeit* (German—time); *Zeitschrift* (German—periodical publication ; *zuid* (Dutch—south)

Z¹, Z², Z³ first degree of contraction, second degree of contraction, third degree of contraction

Z39 Library Work, Documentation, and Related Publishing Practices (American National Standards Institute Standards Committee)

za B-flat (Tartini's scale); zero absolute; zero and add

za *zirka* (German—about; approximately)

Za *Zéro absolu* (French—absolute zero)

ZA South Africa (Internet code); Zambia Army

Z-4 *Zuid Afrika* (Afrikaans or Dutch—South Africa)

Zaa Zeeman-effect atomic absorption (spectrometry)

zaap zero anti-aircraft potential

zab zabaglione; zinc-air battery

zab *zabaglione* (Italian—egg-yolk-and-wine dessert)

Zab Greater Zab or Lesser Zab river in Iraq; Zaboj

Zab *Zabriskie's Reports*

Zac Zacatecas

ZAC Zale Award Committee; zinc ammonium chloride

Zacatecas purple Zacatecas purple marijuana from central Mexico

Z-account Zurich account (bank deposits in Zurich, Switzerland identified only by number, not by the depositor's name)

Zach Zachary; Zachariah; Zacharias; Zachary; Zachris;

Zack Zachariah; Zacharias; Zachary

ZACR Zambia Armored Car Regiment

'zactly exactly

Zad Zadar; Zadock

ZADCA Zinc Alloy Die Casters' Association

ZADCC Zone Air Defense Control Center

ZAED *Zentralstelle für Atomkernenergie Dokumentation* (German—Atomic Energy Documentation Center)

zaf zero-alignment fixture

zaf *zaftig* (Yiddish—well-rounded girl or woman)

Zafarinas Zafarinas Islands (also spelled Chafarinas)—off the Mediterranean coast of Morocco

Z-Afrika *Zuid-Afrika* (Dutch—South Africa)

zag *zaguán* (Spanish—passageway from street door to central patio of homes in Mexico and American Southwest)

Zag Zagreb

ZAG Zagreb, Yugoslavia (airport)

Zahal *Zva Hagana Leyisrael* (Hebrew—Israel Defense Forces)

Zahlentaf *Zahlentafeln* (German—table of illustrations)

zai zero address instruction

zai *zaibatsu* (Japanese—money clique)—plutocratic oligarchy of wealthy families

Zai Zaire

zaire monetary unit of Zaire

Zaire Monetary unit of Democratic Republic of the Congo, formerly Belgian Congo, later known as Zaire

zak *zaklyuchenny* (Russian—prisoner)—pronounced *zek*

zal *zaliv* (Russian—bay)

Zal *Zalmen* (Yiddish—Solomon)

ZALIS Zinc and Lead International Service

zam Z-axis modulation; zinc, aluminum, magnesium

Zam Zambia; ZamTmboanga; Zamiel; Zamora

ZAMAL *Zva Maganah Le Israel* (Hebrew—Israel Defense Force)

Zamb Zambia

Zambia Republic of Zambia (landlocked southern African nation formerly Northern Rhodesia)

Zambo Zamboanga

Zamp Zampa

ZAMPA Zanzibar and Madagascar Peoples Airway

zams zero-age main sequence

zam(s) examination(s)

ZAMS Zambia Army Medical Service

Z Anal Chem *Zeitschrift für Analytische Chemie* (German—Analytical Chemistry Periodical)

ZANC Zambia National Congress

ZANLA Zimbabwe African National Liberation Army

ZANU Zimbabwe African National Union

Zanzi Zanzibar

zap zero and add packed; zero antiaircraft potential

zap *zapad* (Russian—west)

Zap Zapotec; Zapotecan

Zapata Emiliano Zapata—champion of Mexican farmers

zapb zinc-air primary battery

zapp zygo automatic pattern processor

ZAPU Zimbabwe Africa People's Union

zar zeus acquisition radar

Zar Zaragoza

Zara Sara(h); Zarathustra (Zoroaster)

Zara (Italian—Zadar)—Yugoslavian port city

ZARPS *Zuid-Aftikaansche Republiek Polisie* (Afrikaans—South African Republic Police)

zas zero-access storage

ZASM *Zuid Afrikaansche Spoorweg Maatschappij* (Afrikaans—South African Railway serving the Transvaal at the turn of the century)

Z Astrophys *Zeitschrift für Astrophysik* (German—Astrophysics Periodical)

zasts Zastrugas

zat zinc atomspheric tracer

ZAT *Zaterdag* (Dutch—Saturday)

Z-A test Zondek-Ascheim test (for pregnancy)

ZAW *Zuid-Afrikaansche Weehuis* (Afrikaais—South African Orphan Asylum)

Zazen Zen meditation

zb zero beat

z B *zum Beispiel* (German—for instance)

ZB Zen Buddhist

ZBA Zero Bracket Amount

Z-bar Z-shaped bar

zbb zero-base budget(ing)

ZBBS Zero-Based Budgeting System

zbe zinc battery electrode

Zbig Zbigbies Brzezinski (political science author and educator)

zbl zero-based linearity

Zbl *Zentralblatt* (German—central publication)

zbr zero-base review; zero-beat reception; zero-bend radius; zone bit recording

ZBS Zambia Broadcasting Services

zbSd zero-bias Schottky diode

zc zone capacity

z of c zones of communication

ZC Zale Corporation; Zinc Corporation; Zionist Congress; Zoning Commission; Zonta Club; Zouave Corps; Zuñian Club

ZCA Zirconium Corporation of America

Z-car police car (British slang)

zcb zinc-coated bolt

ZCBC Zambian Consumer Buying Corporation

zcc zirconia-coated crucible

zcd zero crossing detector

zcic zirconia-coated iridium crucible

ZCL *Zona di Commercio Libero* (Italian—Free Trade Zone)

Z-class Soviet class of submarines named Zulu by NATO

Z-clip Z-shaped clip

zem zero cerebral muscle

ZCMI Zion's Cooperative Mercantile Institution

zcn zinc-coated nut

ZCNP Zion Canyon National Park

Z country Zimbabwe

zcr zero-temperature coefficient resistor

zcs zinc-coated screw

ZCS Zim Container Service

ZCSU Zim Container Service Unit

zcw zinc-coated washer

ZCX Zone Center Exchange

zd zener diode; zenith description; zenith distance; zenith distance; zero defects; zipper duffel (garment bag); zonal depot; zone description

Zd Zdeimck

ZD zenitim description; zero defects (quality-control goal); zond description

ZDA Zero Defects Association; Zinc Development Association

zdc zinc die casting

ZDC Zero Defects Council

Z de T *Zutano de Tal* (Spanish—so and so)

zdg zinc-doped germanium

Zdm Zaandam

ZDP Zero Defects Program; Zero Defects Proposal

zdpa zero defects program audit

zdpg zero defects program guideline

zdpo zero defects program objective

zdpr zero defects program responsibility

zdr zeus discrimination radar

ZDR *Zentraldeutsche Rundfunk* (German—Central German Radio)

ZDS Zinc Detection System

ZDSI Zung Depression Status Inventory

zdt zero-ductility transition

ze zero effusion; zone effect

zE *zum Exempel* (German—for example)

Ze Jóse

ZE Zenith Radio (stock-exchange symbol)

Z-E Zollinger-Ellison (syndrome)

zea zero-energy assembly

Zeb Zebedee; Zebulon

zebra. zero-energy breeder reactor assembly

zebrass zebra + ass—hybrid of zebra and jenny ass or zebress and jackass

zebroid zebra + horse (hybrid)

Zebrule zebra + horse—hybrid of male zebra and domestic mare

zeb(s) zebra(s)

zec zero-energy coefficient

zecc zinc electrochemical cell

Zech. Zechariah (book of the Bible)

Zech *Zechariah*

zed z; zero

Zed New Zealand; Zedekiah

Zedong Mao Zedong

Zee Zeeland (Dutch province); Zellerbach

Zee *Zeeland* (Dutch—Sea Land)

Zeeb Zeebrugge (Belgian canal town)

zeep zero energy experimental pile

zeg zero economic growth

zei zero environmental impact

Zeichn *Zeichnung(en)* [German—drawing(s)]

zel (ZEL) zero-length launcher

Zel Zelaya (Nicaraguan department); Zelia; Zelide

Z Elektrochem *Zeitschrift für Elektrochemie* (German—Electrochemistry Periodical)

zell zero-length launching

Zem *Zemlya* (Russian—earth; land)

Zemlya Imperatora (Russian—Emperor Land)—Severnaya Zemlya

Zempo Zempoaltepetl (11,142-foot peak near Oaxaca in southern Mexico)

zen nickname for lsd (LSD); zenith (highest point)

Zen Zen Buddhism; Zen Buddhist; Zengo; Zenith; Zenobe; Zenobia; Zenobio; Zenon; Zenophon; Zentippe; Zenus

ZEN EIEN *Zenkoku Eiga Engeki Rodo Kumiai* (Japanese—National Movie and Theater Workers Union)

Zenga *Zengakuren* (Japanese leftwing students)

zenith zero-energy nitrogen-heated thermal reactor

ZENKO *Zen Nihon Kinsoku Kazan Rodo Kumiai Rengokai* (Japanese—All-Japan Federation of Metal Miners Union)

ZENRO *Zen Nihon Rodo Kumiai Kaigi* (Japanese—All-Japan Trade Union Congress)

ZENTEI *Zen Teishin Rodo Kumiai* (Japanese—Postal Workers Union)

Zentr *Zentralblatt* (German—journal)

zeony zebra + pony (hybrid)

Zep Giuseppe

Zeph. Zephanialm (book of the Bible)

Zeph *Zephaniah*

zephyr warm westerly breeze

ZEPHYR Zero-Energy Plutonium-Fueled Fast Reactor

zepp zeppelin

zep(s) zeppelin(s)

zer zero-energy reflection

ZERA Zero-Energy Critical Assemblies Reactor(s)

zerc zero-energy reflection coefficient

zero-g zero gravity (weightlessness)

zert zero-reaction tool

ZES Zenith Enterprise Service (telephonic); Zero Energy System

zet zetetic(s)

zeta. zero energy thermonuclear assembly

Zetland Zetland Island or the Zetland Islands; the Shetlands

zetr zero-energy thermal reactor

zeug zeugma; zeugmatic; zeugmatically

ZEUS Zero-Energy Uranium System

zev zero-emission vehicle

zf zero frequency

z/f zone of fire

Z-F Zermelo-Fraenkel (set time-cry)

ZF *Zagrebacka Filharinonija* (Croatian—Zagreb Philharmonic)

zfb signals fading badly

zfc zirconia fuel cell

ZFGBI Zionist Federation of Great Britain and Ireland

ZFMA Zip Fastener Manufacturers' Association

Z f N *Zeitschrift für Namenforschung* (German—Journal for the Study of Place-names)

ZFO *Zone Francaise d'Occupation* (French—French Occupation Zone)

zfp zonal flow pattern; zyglofluorescent penetrant

zfpt zyglo-fluorescent penetrant testing

zfs zero field splitting

Zf's Zionist fanatics

ZFV *Zentrale jitr Fremdenverkehr* (German—Central Tourist Association)

zg zap gun

z/g zoster-immune globulin

Zg Zug

ZG Zoological Gardens

Z-gauge super-miniature model railway scale

Z-gas Zyklon-B gas (deadly)

zge zero-gravity effect; zero-gravity environment; zero-gravity expulsion

zget zero-gravity expulsion technique

ZGF Zero Gravity Facility

zgg zero gravity generator

zgh zero-gravity harmonic

ZGM Zeitner Geological Museum

zgs zero-gravity simulator

zgs (ZGS) zero gradient synchrotron

zgt (ZGT) zero-gravity trainer (NASA)

Z-gun anti-aircraft rocket gun

zh zinc heads (freight); zonal harmonic; zone heater

zH zu *Hdnden* (German—care of, deliver to)

Zh *Zuidholland* (Dutch—South Holland)

ZH lighter-than-air search and rescue aircraft (naval symbol)

ZH *Zone d'Habitation* (French—residential area)

Zh Eksp Teor Fiz *Zhurnal Eksperimental'noi i Teoreticheskoi Fiziki* (Russian—Journal of Experimental and Theoretical Physics)

Zh Fiz Khim *Zhurnal Fizicheskoi Khimd* (Russian—Journal of Physical Chemistry)

Zhg *Zhongguo* (Chinese—China)

Zho Ren Gon Guo *Zhonghua Renmin Gonghe Guo* (Mandarin Chinese—People's Republic of China)

Zh Prik Spektrosk   *Zhurnal Prikladnoi Spektroskopii* (Russian—Journal of Applied Spectroscopy)

zhr   zenith hourly rate; zirconium hydride reactor

Z hr   zero hour

ZHRC   Zinsmaster Hol-Ry Company

zhs   zero hoop stress

ZHS   Zion Historical Society

Zh Tekh Fiz   *Zhurnal Tekhnicheskoi Fiziki* (Russian—Journal of Technical Physics)

zi   zero input; zonal index

Zi   Zollner illusion

ZI   Zim Israel (steamship line); Zinc Institute; Zone of the Inferior; Zone of the Interior; Zonta International

Z of I   Zone of the Interior

ZI   *Zone Industrielle* (French—industrial zone); *Zone Interdite* (French—prohibited zone)

ZIA   Zone of the Interior Armies

Zia   Zia Khaleda (of Bangladesh)

zic   zirconia-iridium crucible

ZID   Zionist Immigration Depot

Zier   Ziervogel process

ZIF   ZyCAD Intermediate Format

zig   zero immune globulin

zig (ZIG)   zero immune globulin; zoster-immune globulin

Zig   Ziegfield; Zigfield; Zigfrid; Zigfirids

ziggur(s)   ziggurat(s)

zig(s)   *zigaboo(s)* [British West Indian—Black(s)]

zig-zag   Zig-Zag cigarette paper; zig-zag rule(r); zig-zag sewing machine attachment for making zig-zag stitches

ZiJ   *Zeitschrift für Instrumentenbau* (German—Journal for Instrument Builders)

zil   zillion (a number beyond belief)

ZIL   (Russian—*Zavod Imieni Likhatov*)—Likhatov Auto Factory producing a Packard-like luxury car formerly named for Stalin–the ZIS (*Zavod Imieni Stalin*)

Zilw   Zilwaukee

zim   zero-interest mortgage; zimbalon; zonal interdiction missile

Zim   Zimbabwe (whose capital is Harhar); Zimmerman(n)

Zim   *Zinibel* (German—cymbal); Zi Mischari (Hebrew—merchant fleet)

ZIM   *Zeitschrift der Internationalen Muskgesellschaft* (German–Journal of the International Music Society)

ZIMA   Zimbabwe Medical Aid

Zimb   Zimbabwe

Zimb   *Zimbalon* (German—cimbalom or dulcimer)

Zimbabwe   formerly Zimbabwe-Rhodesia; Rhodesia; Southern Rhodesia

Zimco   Zambia Industrial and Mining Company

Zim-Rho   Zimbabwe-Rhodesia (formerly Rhodesia or Southern Rhodesia)

Zim Tim   *Zimbabwe Times*

zin   zinfandel (grapes or wine)

ZINC   Zim Israel Navigation Company (Zim Israel Line)

zinco   zincograph

ZINCO   Zim Israel Navigation Company

zincog   zincography

zinc white   zinc oxide (ZnO)

zineb   zinc ethylenebis (fungicide)

zine(s)   magazine(s)

Zingi Zingari   (Italian—Gypsies)

Zinj   Zinjanthropirs

Zinovyev   Grigory Yevseyevich

Zion   Zion National Park, Utah

zip   zero (slang); zinc impurity photodetector; zipper (slide fastener or similar device)

ZIP   Zone improvement Plan (US Post Office Zip Code)

ZIPA   Zimbabwe People's Army

ZIPRA   Zimbabwe People's Revolutionary Army

zir   zero internal resistance

ZIR   Zug Island Road (Delray Connecting Railroad)

ziram   zinc dimethyldithiocarbamate (fungicide)

zircaloy   zirconium alloy

ZIRCOA   Zirconium Corporation of America

zircon   zirconium silicate ($ZrSiO_4$)

Zirk Hagen   *Zirkus Hagenbeck* (German—Hagenbeck Circus)

zirox   zirconium oxide (ZrO_2)

ZISS   Zebulon Israel Seafaring Society

zith   zither

zix   zinc isopropyl xantimate

zj   zipper(ed) jacket

zj   *zonder jaartel* (Dutch—without date of publication)

ZKD   Zagreb Kajkavian Dialect (Serbo-Croatian)

zkrat   *zkratka(y)* [Czech—abbreviation(s)]

Z Kristallog Kristallgeom Krystallphys Kristallchemie   *Zeaschrift für Kristallographie, Kristallgeometrie, Kristallphysik, Kristallchemie* (German—Periodical for Crystallography, Crystallographic Geometry, Crystallographic Physics, Crystallographic Chemistry)

ZKSK   *Zentrale Kommissionfar Staatliche Kontrolle* (German—Central Commission for State Control)—communist

zl   freezing drizzle (meteorological symbol); zero lift

Zl   zloty (Polish ruble)

ZL   freezing drizzle (symbol)

ZLA   Zambia Library Association

zlc   zero lift cord

zld   zero level drift; zero lift drag; zodiacal light device

Zld   *Zeeland* (Dutch—Sea Land)—old province made of land captured from the sea

zlg   zero line gap

zll   zero length launch

zloty   monetary unit of Poland

Zlsm   Zeiss light-section microscope

zm   zoom; zoomar (variable focus lens)

ZM   Zambia (Internet code); Zubin Mehta

Z-M   Zuckerman-Moloff (sewage treatment)

ZM   *Zeevaart Miratschappij* (Dutch—navigation company); *Zona Militare* (Italian—Military Zone)—restricted area

Z-man   U.S. Army reserve

zmar   zeus malfunction array radar

Z-marker   zone marker

Zmbbw   Zimbabwe (Rhodesia; Southern Rhodesia)

ZMC   Zion Mule Corps

Zmd Zung measurement of depression

Z Metallk *Zeimschrift für Metallkunde* (German—Metallurgy Periodical)

zmkr zone marker

ZMMD Zurich, Mainz, Munich, Darmstadt (algol processor joint effort of universities in those cities)

ZMRI Zinc Metals Research Institute

zmrz *zmrzlina* (Slovak—ice cream)

ZMT Zip (Zone Improvement Plan) Mail Translator (post office sorting device)

ZMW *Zeitschrift für Musikwissenschaft* (German—Journal for Musical Knowledge)

zn zenith; zone (computer flow chart)

zn *zelfstandig naamwoord* (Dutch—substantive noun) any group of words or a pronoun serving as a noun

Zn true azimuth (symbol); zinc

ZN *Zuid-Nederlands* (Dutch—southern Netherlands)—Belgium

Znak (Polish—Sign)—Roman Catholic pro-government party

Z Naturforsch *Zeitschrift für Naturforschung* (German—Natural History Periodical)

zng zoning

ZnO zinc oxide

ZNP Zanzibar Nationalist Party; Zimbabwe National Park; Zion National Park (Utah)

Zn$_{pgc}$ azimuth per gyro compass

ZNPM Zion National Park Museum

ZNPP Zamzibar and Pemba People's Party

ZNPS Zion Nuclear Power Station

znr zmnc resistor; zirconium nitride

zns zero-negative sequence

ZNS Zodiac News Service

znsr zero-negative sequence relay

ZNZ *Zanatska Nabarnoproajna Zadruga* (Yugoslavian—Procurement Sales Cooperative)

zo characteristic impedance (symbol); zero output; zobo (yak + zebu hybrid)

Zo Zoa; Zoe(belle); Zoela; Zoeta; Zofta; Zohora; Zohra; Zoila; Zona; Zormula; Zora (bel); Zorah; Zoraida; Zorana; Zorayda; Zore; Zorica; Zoril; Zorislava; Zorna; Zoruna; Zosa; Zosia; Zosimia; Zowart

ZO Zionist Organization

ZO *Zone Occupée* (French—Occupied Zone); *zuidoost* (Dutch—southeast)

zoa zero-ohms adjustment

ZOA Zionist Organization of America

ZOB *Zentral Omnibus Bahnof* (German—Central Bus Depot); *Zydowska Organizacja* (Polish—Jewish Fighting Organization)—anti-Nazi ghetto forces in World War II

zoba bull + yak—hybrid offspring of common bull and yak cow

zobo cow + yak—hybrid of yak bull and common cow

zoc *zócalo* (Mexican Spanish—public square)

Zoc Zocalo (Mexico City's great plaza)

zod. *zodiacus* (Latin—circle of animals)—the zodiac

zoe zero energy; zinc-oxide eugenol

Zoe Zoebelle; Zoela; Zowta; Zofia

zof zone of fire

Zog Ahmed Zogu

ZOG Zionist Occupation Government

Zoh *Zohar* (The Book of Splendor)

Zolá Emile Zolá—French novelist (1840–1902)

zon zoning

Zon *Zondag* (Dutch—Sunday)

Zondervan Zondervan Publishing House

Zone Panama Canal Zone

zoo zoological (garden); zoology

zoo (Latin prefix—animal)–zoological, zoologist, zoology

zoochem zoochemistry

zoogeog zoogeography

zool zoologic; zoological; zoologist; zoology

zool *zoologi(sk)* (Dano-Norwegian—zoology or zoologist)

Zool Zoology

zoomorph zoomorphic initial letter

zoopal zoopaleontology

zoopar zooparasitology

zoopath zoopathology

zooph zoophytology

zoopharm zoopharmacology

zop zero-order predictor; zinc oxide pigment

zopi zero-order polynomial interpolator

zopp zero-order polynomial predictor

zor zinc-oxide resistor; zone of reconnaissance

Zor Zoram(ites); Zoroastrian

Zora Zorabel(la); Zorah; Zoraida; Zorana; Zorayda

Zorba *Zorba the Greek*

zos zoster; zosteriform; zosteriformal

ZOS Zapata Corporation (stock exchange symbol)

zot (slang–zero)

zounds (euphemistic contraction–god's wounds)

zox zirconium oxide

zoz *zie ommezijde* (Dutch—please turn the page)

ZP lighter-than-air patrol and escort aircraft (naval symbol); Zellerbach Paper

Z & P Zanzibar and Pemba

ZP *Zagrebian Philharmonic* (Yugoslavian—Zagreb Philharmonic Orchestra)

zpa zeus program analysis; zone of polarizing activity

ZPA Zeus Program Analysis; Zoological Parks and Aquariums

zpar zeus-phased array (radar)

zpb zinc primary battery; zore portion of the byte (leftmost four bits of an eight-bit byte)

ZPC Zellerbach Paper Company

ZPDA Zinc Pigment Development Association

zpe zero-point energy

ZPEN Zeus Project Engineer Network

zpg zero population growth

ZPG Zero Population Growth

zpgn zip gun

ZPH Zondervan Publishing House

Z Phys *Zeitschrift für Physik* (German—Physics Periodical)

Z Phys Chem *Zeitschrift für Physikalische Chemie* (German—Physical Chemistry Periodical)

zpi zone position indicator

zp & j *zonder plaws en jaar* (Dutch—without place of publication or date)

ZPKK *Zentrale Parteikontrollkommission* (German Central Control Commission of the Party)—communist

zpl *zonder plaats* (Dutch—without place of publication)

Z Plz Zellerbach Plaza

zpo zinc peroxide

ZPO Zeus Project Office

Zpp Zeiss projection planetarium

zpph zinc protporhrin heme ratio

zppr zero-power plutonium reactor

zpr zero-power reactor

zprf zero-power reactor facility

ZPRSN Zurich Provisional Relative Sunspot Number

zpt zero-power test(ing); zoxazolamine paralysis time

ZPT Zero Power Test

ZPU-4 Soviet anti-aircraft weapon combining firepower of four 14.5 mm heavy machine-guns

zr freezing rain (meteorological symbol); zone refined

Zr zirconium

ZR freezing rain (symbol); Zenith Radio; zero coupon issue (security) (newspaper bond listings)

Z-R Zimbabwe-Rhodesia (Zimbabwe; formerly Rhodesia or Southern Rhodesia)

Z/R Zone of Responsibility

Zr⁹⁵ radioactive zirconium

zrc zirconium carbide

ZRC Zenith Radio Corporation

ZRCL Zlac Rowing Club Limited

ZRH Zurich, Switzerland (airport)

zrn zirconium nitride

ZRO Zone Recruiting Officer

zrp zero radial play

zrt zero-reaction tool

ZRU *Zone de Rénevation Urbaine* (French—Urban Redevelopment Zone)

zrv zero relative velocity

zs zero shift; zero and subtract; zero suppress; zero suppression (of non-significant zeros in computer-printed numerals)

z S *zur See* (German——of the navy)

Zs *Zeitschrift* (German—periodical)

ZS Zoological Society

zsa zero-set amplifier

zsat zinc-sulfide atmospheric tracer

zsb zinc storage battery

zsc zero sub-carrier chromactivity; zinc silicate coat(ing)

ZSC Zeeland Shipping Company; Zoological Society of Cincinnati

Z-scale height determination scale

zsd zebra-stripe display; zinc sulfide detector

ZSDS Zinc Sulfide Detection System

ZSE Zagreb Soloists Ensemble *(Solisti di Zagreb)*

zsf zero skip frequency

zsg zero-speed generator

zsi zero-size image

ZSI Zoological Society of Ireland

Zsig Zsigmond

ZSL Zoological Society of London

ZSL *Zjednoczone Stronnictwo Ludowe* (Polish—United Peasant Party)

ZSM Zoar State Memorial

ZSN Zoological Station of Naples

ZSP Zoological Society of Philadelphia

zspg zero-speed pulse generator

ZSS Zero-Sum Society; Zinc Sulfide System

ZSSD Zoological Society of San Diego

Zssg(n) *Zusammensetzung(en)* [German—compound word(s)]

zst zero strength time (measurement); zinc-sulfide tracer

ZST Zone Standard Time

ZSU-23 Soviet self-propelled antiaircraft gun including quadruple 23 mm cannon

ZSU-23-4 Soviet antiaircraft system mounted on a tank and carrying four 23 mm cannons

ZSU-57 Soviet self-propelled antiaircraft gun including twin 57 mm cannon

zt zero tolerance (drugs); zipper tube; zipper tubing

z T *zum Ted* (German—partly)

Zt *Zeit* (German—time)

ZT lighter-than-air training aircraft (naval symbol); Zachary Taylor (12th US President); zero time; zone time

ZT *Zone Torride* (French—torrid zone)

ZTA Zulu Territorial Authority

Z-table mortality table

Z-test Zulliger test

Ztg *Zeitung* (German—newspaper)

Z-time zebra time or zulu time (Jargon for Greenwich Mean Time)

ztlp zero-transmission level point

ZTO Zone Transportation Office(r); *Zürich Tonhalle Orchestera* (Zurich Concert Hall Orchestra)

Ztp zero temperature plasma

Ztr *Zentner* (German—hundred-weight)

Ztschr *Zeitschrift* (German—periodical)

Z-TWIST Z-shaped open-band twist

Zu Zublena; Zudegi; Zula; Zuleika; Zulena; Zulima; Zulu; Zuma

ZU lighter-than-air utility aircraft (naval symbol)

ZU-23 Soviet antiaircraft system having a maximum fire power of 2000 rounds per minute

Zuck *Zuckung* (German—contraction)—sometimes abbreviated Z

'zuco bazuco

ZUF Zapata Urban Front (Mexican terrorist group)

Zulo Felix Zuloaga (Mexican-president and soldier); Ignacio Zuloaga (Spanish painter)

Zulu code word for Greenwich mean time (Zulu time); letter Z radio code; Soviet Z-class attack submarines; Zululand(er)

ZUM Zimbabwe Unity Movement (ZANU and ZAPU united); Zone Usage Measurement

Z und Z *Zar und Zimmermann* (German—Czar and Carpenter)—opera Albert Lortzing

ZUP *Zone a urbaniser en prioriti* (French—Priority Urbanization Zone)—slum cleanup or demolition zone

Zur Zurab; Zurich; Zuriel; Zurr

Zur *Zurigo* (Italian—Zurich)

Zür Zürich (Swiss capital and largest city in Switzerland)

Zur Col Mus *Zurich Collegium Musicum* (Latin—Zurich Music College)

Zurl Zuriel

zus *zusammen* (German—together)

Zus *Zusammenfassung* (German—summary)

Zuschr *Zuschrift(en)* [German—communication(s)]

Zut Zutphen

Zuverl *zuverlassig* (German—authentic)

zuzzur *zuzzurullón(a)* (Italian—overgrown child, just a big kid)

zv zika virus

zv *zu verfugung* (German—at disposal)

Zv *Zolverein* (German—customs union)

Z-value degrees Fahrenheit required to reduce thermal death time by 90% or one log value

ZVEI *Zentralverband der Elektrotechnischen Industrie* (German—Central union of the Electrotechnical Industry)

zvr zener voltage regulator

zvrd zener voltage regulator diode

zw zero wear

zw *Zwart* (Dutch—black); *zwischen* (German—between; within)

Zw *Zwischensatz* (German—insertion or interpolation or parenthesis)

ZW Zimbabwe (Internet code)

ZW *zuidwest* (Dutch—south West)

zwc zone wind computer

zwitt zwitterion (diplole ion)

zwl zero wave length

ZWO Zionist World Organization

ZWO *Zuiver Wetenschappelijk Onderzoek* (Dutch—Netherlands Organization for the Advancement of Pure Research)

Zwol Zwolle

zwp zone wind plotter

zwv zero wave velocity

Zy Zylota; Zyma

ZYA Zionist Youth Association

zyg zygote

Zyg Zygmunt

zygo zygomatic; zygomaticus

zygo (Latin prefix—join or union)

zym zymbalon; zymurgy

zymb *zymbalum* (Hungarian—dulcimer)—also known as the cimbalom, cymbalon, czimbalom, zimbalon

Zymb *Zymbel* (German—cymbal)

zymo zymogen(esis); zymogenic(al)(ly); zymogenous; zymologic(al)(ly); zymologist; zymology; zymolysis; zymometer; zymoplastic; zymosis; zymosthenic(al)(ly); zymotic(al)(ly); zymotic disease

zymol zymology

ZYP Zefkrome Yarn Program

zyr zyrian(s)

Zyr Zyrian (Finno-Ugric language spoken by Zyrians in Komi SSR)

zyth zythlum (ancient beer beverage)

zythep zythepsary (obsolete term for brewery)

zythia zythiaceae

zyz zyzzyva

zyzo zyzogeton; zyzomys

zz increasing degrees of contraction (symbol); zigzag

z-z longitudinal axis/roll axis

zz *zingiber* (Latin—ginger)

z Z *zur Zeit* (German—at present, for the time being)

ZZ Arian Afghan Airlines; longitudinal or roll axis (symbol); zed-zed; zz-approach

Z & Z Zulch and Zulch

ZZ *Zentralbibliothek Zürich* (German—Zurich Central Library)—combines the canton state, and university libraries

zza zamack zinc alloy

ZZB Zanzibar (tracking station)

zzc zero-zero condition

zzd zig-zag diagram

z-z fold zig-zag fold (concertina fold)

ZZO *zuidzuidoost* (Dutch—south southeast)

zzr zig-zag rectifiier

Z-z's Zionist zealots

z Zt *zur Zeit* (German—at present, for the time being)

zzv zero-zero visibility

ZZV Zanesville, Ohio (aiport)

ZZW *zuidzuidwest* (Dutch—south southwest)

ZZZ Zayda, Zorayda, Zorahayda—The Three Beautiful Princesses in Washington Irving's *Alhambra*

ZZZ-ZZZ-ZZZ sawing or snoring (cartoonist symbol)

Airlines of the World

An open space after a two-letter entry means an airline so coded has been discontinued or the code is available for new airlines. Please note airline codes change frequently.

AA American Airlines	**BQ** Eurobelgian Airlines	**DF** New ACS
AB Falcon Airlines	**BR** EVA Airways	**DG** Eastern Pacific
AC Air Canada	**BS** Gamair	**DH** Jet World Aviation
AD Lone Star Airlines	**BT** Sweden Airways	**DI** Deutsche BA
AE Mandarin Airlines	**BU** Braathens SAFE Airtrans-	**DJ** Air Nordic Sweden
AEI	port	**DK** Eastland Air
AF Air France	**BV** Sun Air	**DL** Delta Air Lines
AG Air Contractors Ltd.	**BW** BWIA International	**DM** Maersk Air
AH Air Algerie	**BX** Coast Air	**DN** Air Exel Belgique
AI Air India	**BY**	**DO** Dominicana de Aviacion
AJ Air Belgium	**BZ**	**DP** Air 2000
AK Galena Air Service	**C2** Air Caribbean Ltd.	**DQ** Coastal Air Transport
AL Alsair S.A.	**C4** Airlines of Carriaco	**DR** Hyeres Aero Services
AM Aeroméxico	**C8** Chicago Express	**DS** Air Senegal
AN Ansett Airlines of Australia	**C9** Cie Aero Europeene	**DT** TAAG-Angola Airlines
AO Aviaco	**CA** Air China International	**DU** Hemus Air
AP Aliadriatica	**CB** Suckling Airways	**DV** Nantucket Airlines
AQ Aloha Airlines	**CC** Air Atlanta Iceland	**DW** Helicopter Shuttle
AR Aerolineas Argentinas	**CD** Seaview Air	**DX** Danair
AS Alaska Airlines	**CE** Care Airlines	**DY** Alyemda-Yemen Air
AT Royal Air Maroc	**CF** Faucett Peruvian	**DZ** Air Metro North
AU Austral Lineas Aéreas	**CG** Milne Bay Air Ltd.	**E2** Everest Air
AV *Avianca—Aerovias Nacion-*	**CH** Bemidji Airlines	**E3** Domodedovo Airlines
ales de Colombia (Spanish—	**CI** China Airlines	**E4** Aero Asia International
National Airlines of Colom-	**CJ** China Northern Arlines	**E5** Branson Airlines
bia)	**CK** Gambia Airways	**E6** Air Company Elf-Air
AW Horizon Airways	**CL** Lufthansa CityLine	**E7** Express Airlines
AX Binter Mediterraneo	**CM** COPA (*Compañia Pan-*	**E8** USAfrica Airways
AY Finnair	*ameña de Aviacíon*)	**E9** Ensor Air Ltd.
AZ Alitalia	**CN** Islands Aviation	**EA** Eastern Airlines
B2 Belavia	**CO** Continental Airlines	**EC** Flight Line
B3 Bellview Airlines Ltd.	**CP** Canadian Airlines Interna-	**ED** CC Air Inc.
B4 Bhoja Airlines	tional Ltd.	**EE** Euroberlin
B7 Makung Airlines	**CQ** Air Alpha	**EF** Far Eastern Air Transport
B8 Caribbean Airlines	**CS** Continental Micronesia	**EG** Japan Asia Airways
B9 Caribair	**CT** Cave	**EH** SAETA (*Sociedad Ecuato-*
BA British Airways	**CU** Cubana Airlines	*riana de Transportes Aeros*)
BAX	**CV** Air Chathams	**EI** Aer Lingus
BB Balair Cta	**CW** Air Marshall Islands	**EJ** New England Airlines
BC Brymon Aviation	**CX** Cathay Pacific	**EK** Emirates Airlines
BCAL	**CY** Cyprus Airways Ltd.	**EL** Air Nippon
BD British Midland Airways	**CZ** China Southern Airlines	**EM** Empire Airlines
BF Markair	**D2** Damania Airways	**EN** Air Dolomiti
BG Biman Bangladesh	**D3** Daallo Airlines	**EO** Express One International
BH Augusta Airways Ltd.	**D5** Nepc Airlines	**EP** Blackhawk Airways
BI Royal Brunei Airlines	**D6** Inter Air	**EQ** Tame CA
BJ Prospair	**D7** Dinar Lineas Aereas	**ER** DHL Worldwide Express
BK Chalks International	**D8** Diamond Sakha Airline	**ES** City Air Bus Ltd.
BL Pacific Airlines	**D9** Donavia	**ET** Ethiopian Airlines
BM Aero Transporti Italiani	**DA** Air Georgia	**EU** Ecuatoriana
BN Tropical Sea Air	**DB** Brittany Air International	**EV** Atlantic Southeast
BO Bouraq Airlines	**DC** Golden Air Flying	**EW** Eurowings AG
BOAC	**DD** Conti-Flug	**EX** Dallas Express
BP Air Botswana	**DE** Prime Air Inc.	**EY** EAS Europe Airlines

EZ Sun-Air of Scandinavia
F2 Family Airlines
F3 Flying Enterprise AB
F4 International Flying Services
F5 Archana Airways
F7 Alberta Express
F9 Frontier Airlines
FA Finnaviation
FB Promair Australia
FC Berliner Spezialflug
FD Cityflyer Express
FE Crane Air
FF Tower Air Inc.
FG Ariana Afghan Airlines
FH Air Midwest
FI Icelandair
FJ Air Pacific Ltd.
FK Flamenco Airways
FL Airtran Airways
FM
FN FS Air Service
FO Western NSW
FP Aeroleasing S.A.
FQ Air Aruba
FR Ryanair Ltd.
FS Sunbird
FT Flying Tiger Line
FU Air Littoral
FV Viva Air
FW Isles of Scilly
FX FedEx
FY Metro Airlines
FZ Air Facilities
G2 Tronderfly Air Service
G4 Guizhou Airlines
G5 Island Air Ltd.
G6 Avatlantic
G7 Guinee Airlines
G8 Air Great Wall
G9 Grant Aviation
GA Garuda Indonesia
GC Lina Congo
GD TAESA (*Transportes Aeros Ejecutives*)
GE TransAsia Airways
GF Gulf Air Co.
GG Air Holland
GH Ghana Airways
GI Air Guinee
GJ Saõ Tomé e Príncipe
GK Go One Airways
GL Greenlandair Inc.
GM Germania
GN Air Gabon
GO Gambia Air Shuttle
GP Gestair Executive Jet
GQ Big Sky Airlines
GR Aurigny Air Services
GS
GT GB Airways
GU Aviateca

GV Riga Airlines
GW Avia Airlines
GX Pacificair
GY Guyana Airways
GZ Air Rarotonga
H2 Heli Inter Riviera
H3 Harbour Air Ltd.
H4 Hainan Airlines
H5 Magadan Airlines
H9 Blade Helicopters
HA Hawaiian Airlines
HB Malitas
HC Naske-Air
HD New York Helicopter
HE LGW Airlines
HF Hapag-Lloyd
HG Harbor Airlines
HH Euro Direct Airlines
HI Papillon Airways Inc.
HJ Holmstroem Air
HK Swan Airlines
HL
HM Air Seychelles
HN KLM Cityhopper BV
HO Airways International
HP America West
HQ Business Express
HR China United Airlines
HS Highland Air
HT Air Tchad
HU Antigua Paradise Air
HV Transavia Airlines
HW Havasu Airline
HX Hamburg Airlines
HY Uzbekistan Airways
HZ Euroflight Sweden
IA Iraqi Airways
IB Iberia
IC Indian Airlines
ID Air Normandie
IE Solomon Island Air
IF Great China Airlines
IG Alisarda
IH Falcon Aviation AB
II Business Air
IJ TAT European
IK Roadair Lines
IL Istanbul Airlines
IM Carib Express
IN Inversija
IO TAT-Export
IP Airlines of Tasmania
IQ Interot Airways
IR Iranian Airlines
IS Eagle Air Ltd.
IT Air Inter
IU Air Straubing
IV Genesis Airways Ltd.
IW Best Airlines
IX Flandre Air
IY Yemenia Yemen Air
IZ Arkia Israel

J2 Azerbaijan Hava
J5 Aviaprima-Sochi Air
J7 Valujet Airlines
J8 Berjaya Air
J9 Algoma Airways
JA Norway Airlines Charter
JB Helijet Airways
JC Rocky Mountain Air
JD Japan Air System
JE Manx Airlines
JF Lab Flying Services
JG Air Teranga
JH Nordeste Linhas
JI Midway Airlines
JJ TAM-Transportes Aereos
JK Spanair
JL Japan Airlines
JM Air Jamaica
JN Rich International Airways
JO Twente Airlines
JP Adria Airways
JQ Trans-Jamaican Air
JR Aero California
JS Air Koryo
JT Jaro International
JU Jat/Yugoslav Airlines
JV Bearskin Lake Air Service Ltd.
JW Arrow Air
JX Jes Air
JY Jersey European Airways
JZ Skyways AB
K1 Channel Express
K2 Kyrghyzstan Airlines
K4 Kazakhstan Airlines
K7 Air Yakutaviatrans
K8 Skyway Airlines
K9 Skyward Aviation
KA Dragonair
KB Druk Air
KC Cook Islands International
KD Kendell Airlines
KE Korean Air
KF Kelowna Flightcraft Air Charter
KG King Island Airlines
KH Kyrnair
KI Air Atlantique
KJ British Mediterranean
KK TAM
KL KLM Airlines (*Koninklijke Luchtvaart Maatschappij*)— Royal Dutch Airlines
KLM KL
KM Air Malta
KN Morris Air
KO Skynet Airways
KP Kiwi International Air
KQ Kenya Airways
KR Kitty Hawk Air Cargo
KS Penair
KT Turtle Airways

KU Kuwait Airways
KV Transkei Airways
KW Carnival Airlines
KX Cayman Airways
KY Water Wings Airways
L2 Love Air
L4 Atlant-SV
L5 AS Lufttransport
L6 Air Maldives Limited
L7 Liberty Airlines
L8 Leisure Air
L9 Air Mali SA
LA LAN Chile
LB Lloyd Aero Boliviano
LC Loganair Ltd.
LD Lauda Air Italy
LE
LF Uzu Air Pty. Ltd.
LG Luxair (Luxembourg Airlines)
LH Lufthansa
LI LIAT (Leeward Islands Air Transport)
LJ Sierra National Air
LK Goldfields Air Services
LL Bell-Air
LM *ALM (Antillianaanse Luchtvaart Maatschappij)*— Dutch—Antillean Airline Company
LN Jamahiriya Libyan
LO LOT (Polish Airlines)
LP Euro City Line
LQ Cohlmia Aviation
LR *LACSA (Lineas Aéreas Costarricenses)*—Costa Rican Airlines
LS Lliamna Air Taxi
LT LTU International Airways
LU Shasta Air
Lufthansa LH
LV *LAV (Linea Aeropostal Venezolana)*—Venezuelan Acropost Lines
LW Air Nevada
LX Crossair AG
LY El Al (Israel Airlines)
LZ Balkan-Bulgarian
M4 Interimpex-Avioimpex
M6 Air St Martin
M8 Moscow Airways
M9 Modiluft
MA Malev Hungarian
MB Western Airlines Australia
MC Military Airlift Command
MD Air Madagascar
ME Middle East Air
MF Xiamen Airlines
MG MGM Grand Air Inc.
MH Malaysia Airlines
MI Silkair

MJ LAPA (*Lineas Aéreas Privadas Argentinas*)
MK Air Mauritius
ML Aero Costa Rica
MM Sociedad Aeronautica Medellin
MN Commercial Airways
MO Calm Air International Ltd.
MP Martinair Holland
MQ Simmons Airlines
MR Air Mauritanie
MS Egyptair
MT Direct Aviation Pty.
MU China Eastern Air
MV Great American
MW Maya Airways
MX Mexicana De Aviación
MY Helifrance
MZ *Merpati Nusantara*
N2 Aero Internacionales
N4 National Airlines Chile
N5 Nations Air Express
N6 Aero Continente
N7 Nordic East Airways
N9 North Coast Aviation
NA Executive Airlines
NB Sterling Airways
NC Wideroe Norsk Air
ND Airlink Pty. Ltd.
NE Knight Air
NF Air Vanuatu (Operations)
NG Lauda Air
NH All Nippon Airways
NI Portugalia
NJ Vanguard Airlines
NK Spirit Airlines
NL Shaheen Air International
NM Mount Cook Airlines
NN Cardinal Airlines
NO Aus Air
NP Piccolo Airlines
NQ Orbi Georgian Arways
NR Norontair
NS Cape York Air Services
NT Binter Canarias
NU Japan Transocean
NV Northwest Territorial Airways
NW Northwest Airlines
NX Northsouth Aviation
NY New York Air
NZ Air New Zealand
OA Olympic
OB Monarch Air
OC Sunshine Aviation
OD
OE West Air
OF Sunstate Airlines
OG Air Guadeloupe
OH Comair Inc.
OI Paradise Airways
OJ Air St. Barthelemy

OK Czechoslovak
OL *ÖLT (Östfriesische Lufttransport)*
OM Mongolian (MIAT)
ON Air Nauru
OO Skywest Airlines
OP Chalks International Air
OQ Arizona Pacific Air
OR Air Comores
OS Austrian Airlines
OT Evergreen Alaska
OU Croatia Airlines
OV Estonian Air
OW Metavia Airlines
OX Air Hudik
OY Sunaire Express
OZ Asiana Airlines
P2 Panama Air International
P3 Promech
P5 Aerorepublica
P8 Pantanal Linhas
P9 Topair
PA Pan American
PB Air Burundi
PC Air Fiji
PD Pemair Ltd.
PE Air North West Pty.
PF Vayudoot Ltd.
PG Bangkok Airways
PH Polynesian Airlines Ltd.
PI Sunflower Airlines
PJ Air St. Pierre
PK Pakistan International
PL Aero Peru
PM Tropic Air
PN Air Martinique
PO Aeropelican Air Service
PP Pacific Airlines
PQ Pacific Coast Arlines
PR Philippine Airlines
PS Ukraine International Airlines
PT West Air Sweden
PU *PLUNA (Primeras Lineas Uruguayos de Navegación Aérea)*—First Uruguayan Aerial Navigation Lines
PV Latvian Airlines
PW Challeng' Air
PX Air Niugini
PY Surinam Airways
PZ Lapsa
Q2 North American
Q3 Sandaun Air Services
Q4 Mustique Airways
Q7 Qatar Airways
QA Aerocaribe
QB Inter Quebec
QC Air Zaire
QD Grand Airways
QE Air Moorea
QF Qantas Airways

QG Peel Valley Air
QH Qwestair
QI Cimber Air
QJ Jet Airways
QK Air Nova Inc.
QL Lesotho Airways
QM Air Malawi Ltd.
QN Royal Airlines
QO Air Maroochy
QP Air Kenya
QQ Reno Air
QR Quebec Aviation
QS Propheter Aviation
QT SAR Avions Taxis
QU Uganda Airlines
QV Lao Aviation
QW Turks and Caicos Airways
QX Horizon Air
QY Aero Virgin Islands
QZ Zambia Airways
R2 State Orenburg Aviation
R3 Armenian Airlines
R4 Russian Civil
R5 Malta Air Charter
R7 Aserca
RA Royal Nepal Airlines
RB Syrian Arab Airlines
RC Atlantic Airways
RD Avianova
RE Air Arann
RF Emirates Air Service
RG *VARIG (Viacão Aérea Rio Grandense)*—Portuguese—Rio Grande Airlines
RH SAL Luftverkehrs
RI P.T. Mandala Air
RJ Royal Jordanian (ALIA)
RK Air Afrique
RL Ultrair
RM Wings West Airlines
RN Euralair
RO TAROM (Romanian Air)
RP Macair
RQ Air Engiadina
RR Royal Air Force Transport Group
RS Intercontinental de Aviacion
RT Lincoln Airlines
RU Northern Commuter
RV Reeve Aleutian Air
RW Rheinland Air Service
RX AVIAEXPRESS Airlines
RY Air Rwanda
RZ Pan Am Express
S3 Sky Service
S5 Virgin Islands Airways
S6 Air Saint Martin
S7 Siberia Airlines
S8 Estonian Aviation
SA South African
SB Air Caledonie International
SC Shandong Airlines

SD Sudan Airways
SE Wings of Alaska
SF Shanghai Airlines
SG Sempati Air
SH SAHSA (Servicio Aéro de Honduras SA)
SI Sierra Pacific Airlines
SJ Air Swift Ltd.
SK SAS (Scandinavian Airlines)
SL Rio-Sul-Servicios Aereos Regionais
SM Taino Airways
SN Sabena (Belgian World Airlines)
SO Austrian Air Service
SP *SATA (Sociedade Acoriana de Transportes Aéreos)*—Portuguese—Azores Air Transport Line
SQ Singapore
SR Swissair
SS Shabair
ST Yanda Airlines
SU Aeroflot-Russian International Airlines
SV Saudi Arabian Airlines
SW Namibia Air
SX Aeroejecutivo
SY Sun Country Airlines
SZ China Southwest
T2 Taba
T4 Transeast Airlines
T5 Avia Turkmenistan
T6 Trinity Air Bahamas
T7 Air Trans Ireland
T8 Transportes Neuquen
T9 Triton Airlines
TA Taca International
TB Trump Shuttle Inc.
TC Air Tanzania
TD Lakeside Northwest
TE Lithuanian Airlines
TF Chieftain Airlines
TG Thai International
TH *EUROAIR (Companhia Europeia de Transportes Aéreos S.A.)*
TI Baltic International
TJ TAS Airways S.P.A.
TK Turkish Airlines
TL Airnorth
TM *LAM (Linhas Aereas de Mocambique)*
TN Australian Airlines
TO Alkan Air Ltd.
TP *TAP (Transportes Aéros Portugueses)*—Portuguese Air Transport
TQ Transwede
TR Trans Brasil S/A Linhas Aereas

TS Samoa Aviation Inc.
TT Air Lithuania
TU Tunis Air
TV Haiti Trans Air
TW Trans World Airlines (TWA)
TX Air Guadeloupe 94
TY Air Caledonie
TZ American Trans Air
U3 Uralinteravia
U6 Ural Airlines
UA United Airlines
UB Myanmar Airways International
UC *LADECO* Airlines (*Linea del Cobre*)—Spanish—Copper Line
UD Hex'Air
UE Air LA
UF *SARO (Servicios Aéreos Rutas Oriente S.A. de C.V.)*
UG Tuniter
UH Transport Air Centre
UI Norlandair
UJ Aerosanta Airlines
UK KLM (UK)
UL Air Lanka Ltd.
UM Air Zimbabwe
UN Unitair
UO Northern Star Airlines
UP Bahamasair
UQ Gambiers Airlines
UR British International Helicop
US USAir
UT Union De Transportes
UU Air Austral
UV Airtransit Pty. Ltd.
UW Perimeter Airlines
UX Air Europa
UY Cameroon Airlines
UZ Atlantic Island Airways
V2 Valdresfly A/S
V3 Vanair
V4 VLM
V5 Vnukovo Airlines
V6 Vision Airways
VA *VIASA (Venezolana Internacional de Aviación)*—Venezuelan Airlines
VB Maersk Air Ltd.
VC Servivensa
VD Air Liberte
VE *Avensa*
VF British World Airlines
VG VLM Nederland
VH Air Burkina
VI Vieques Air Link
VJ Royal Air Combodge
VK Air Tungaru
VL North Vancouver Air
VM Regional Airlines

VN Hang Khong Vietnam
VO Tyrolean Airways
VP *VASP* Airlines (*Viacão São Paulo*)—São Paulo Airline
VQ Impulse Airlines
VR *TACV* (*Transportes Aéreos de Cabo Verde*)—Cape Verde Airlines
VS Virgin Atlantic Airways
VT Air Tahiti
VU Air Ivoire
VV Flex Air
VW Aeromar Airlines
VX Aces Airlines
VY Pacificair
VZ Arizona Airways
W2 Southwest Air
W3 Swiftair SA
W9 Eastwind
WA Newair
WB *SAN* (*Servicios Aéreos Nacionales*)—Spanish—National Air Services .
WC Islena Airlines
WD
WE Rheintalflug Seewald Gesellschaft mbH
WF Wideroes Flyveselskap
WG Taiwan Airlines Ltd.
WH China Northwest Air
WI Rottnest Airlines
WJ Labrador Airways
WK Southern Air Ltd.
WL Aeroperlas
WM Windward Islands Airways International N.V.
WN Southwest Airlines
WO World Airways
WP Aloha Islandair
WQ SAF/Service Aerien Français

WR Royal Tongan Airline
WS Gray Line Air
WT Nigeria Airways
WU Wuhan Airlines
WV Air South
WW Whyalla Airlines
WX Cityjet
WY Oman Air
WZ BASE Business Airlines
X2 China Xinhua Airline
X3 Baikal Airlines
XD UATP
XE Cambodia International Air
XF Australia Air International
XG Yorkshire European
XJ Mesaba Airlines
XK CAC Mediterranee
XL Country Connection
XM Servicios Especiales
XO Xinjiang Airlines
XP Casino Express
XQ Action Airlines
XT Air Exel Netherlands
XU Link Airways
XV Air Express
XW Walker's Cay International
XY Ryan Air
XZ Flugfelag Austurland
YB Air Aquitaine
YC Flight West Airlines
YD Cologne Air Trans
YE European Aviation
YG Cocesna
YH Air Baffin Ltd.
YI Air Sunshine
YJ National Airlines
YK Cyprus Turkish Airways
YL Long Island Airlines
YM Blue Sky Carrier
YN Air Creebec
YO Heli Air Monaco

YP Aero Lloyd
YQ Euro Air Heli
YR Scenic Airlines
YS Proteus
YT Skywest Airlines
YU *Dominair—Aerolineas Dominicanas S.A.*
YV Mesa Airlines
YW Air Nostrum
YX Midwest Express
YY Airinc Test Airline
YZ Aer Guinea-Bissau
Z2 Zhong Yuan Airlines
Z9 Aero Zambia
ZA ZAS Airline of Egypt
ZB Monarch Airlines
ZC Royal Swazi National
ZD Ross Aviation
ZE Arcus Air
ZF Airborne of Sweden
ZG Sabair Airlines Pty.
ZH Turk Hava Tasimaciligi
ZI Lucas Aigle Azur
ZJ Teddy Air
ZK Great Lakes Aviation
ZL Hazelton Airlines
ZM Scibe-Airlift
ZN Peninsula Airlines
ZO Southflight Aviation
ZP Air St Thomas
ZQ Ansett New Zealand
ZR Muk Air
ZS Hispaniola Airways
ZT SATENA
ZU Freedom Air
ZV Air Midwest
ZW Air Wisconsin
ZX Air BC Ltd.
ZY ADA Air

Air-Pollution Index

psi (PSI) pollution standards index
0 to 50 good air
above 50 moderately good air
above 100 hazardous and unhealthful air
200 to 299 very hazardous and unhealthful air
300 and over hazardous enough to cause premature death to elderly and sick persons
Stage I Alert unhealthful air pollution (200 to 270 psi)
Stage II Alert unhealthful and hazardous air pollution (275 to 390 psi)
Stage III Alert unhealthful and extremely hazardous air pollution (400 to 500 psi), often deadly

Airports of the World

AAA Anaa, French Polynesia
AAB Arrabury, Queensland, Australia
AAC Al Arish, Egypt
AAE Annaba, Algeria
AAF Apalachicola, Florida
AAK Aranuka, Kiribati
AAL Aalborg, Denmark
AAM Mala Mala, South Africa
AAN Al Ain, United Arab Emirates
AAO Anaco, Venezuela
AAR Aarhus, Denmark
AAS Apalapsili, Indonesia
AAT Altay, China
AAU Asau, Samoa
AAV Alah, Philippines
AAX Araxa, Brazil
AAY Al Ghaydah, Yemen
ABA Abakan, Russia
ABD Abadan, Iran
ABE Allentown, Pennsylvania
ABF Abaiang, Kiribati
ABG Abingdon, Queensland, Australia
ABH Alpha, Queensland, Australia
ABI Abilene, Texas
ABJ Abidjan, Ivory Coast
ABK Kabri Dar, Ethiopia
ABL Ambler, Alaska
ABM Bamaga, Queensland, Australia
ABN Albina, Suriname
ABO Aboisso, Ivory Coast
ABP Atkamba, Papua New Guinea
ABQ Albuquerque, New Mexico
ABR Aberdeen, South Dakota
ABS Abu Simbel, Egypt
ABT Al-Baha, Saudi Arabia
ABU Atambua, Indonesia
ABV Abuja, Nigeria
ABW Abau, Papua New Guinea
ABX Albury, New South Wales, Australia
ABY Albany, Georgia
ABZ Aberdeen, Scotland
ACA Acapulco, Mexico
ACB Bellaire, Michigan
ACC Accra, Ghana
ACD Acandi, Colombia
ACE Lanzarote, Canary Islands
ACH Altenrhein, Switzerland
ACI Alderney, Channel Islands

ACK Nantucket, Massachusetts
ACL Aguaclara, Colombia
ACM Arica, Colombia
ACN Ciudad Acuna, Mexico
ACO Ascona, Switzerland
ACT Waco, Texas
ACU Achutupo, Panama
ACV Eureka/Arcata, California
ACY Atlantic City, New Jersey—International
ADA Adana, Turkey
ADB Izmir-Adnan, Mend, Turkey
ADD Addis Ababa, Ethiopia
ADE Aden, Yemen
ADG Adrian, Michigan
ADH Aldan, Russia
ADI Arandis, Namibia
ADK Adak Island, Alaska
ADL Adelaide, South Australia, Australia
ADM Ardmore, Oklahoma—Municipal
ADN Andes, Colombia
ADP Anuradhapura, Sri Lanka
ADQ Kodiak, Alaska
ADR Andrews, South Carolina
ADV Andover, England
ADW Camp Springs, Maryland
ADX St. Andrews, Scotland
ADY Alldays, South Africa
ADZ San Andrés Island, Colombia
AEA Abemama, Kiribati
AEP Buenos Aires, Argentina
AES Aalesund, Norway
AET Allakaket, Alaska
AEY Akureyri, Iceland
AFA San Rafael, Argentina
AFD Port Alfred, South Africa
AFI Amalfi, Colombia
AFL Alta Floresta, Brazil
AFO Afton, Wyoming
AFY Afyon, Turkey
AGA Agadir, Morocco
AGB Augsburg, Germany
AGC Pittsburgh, Pennsylvania
AGD Anggi, Indonesia
AGE Wangerooge, Germany
AGF Agen, France
AGG Angoram, Papua New Guinea
AGH Hälsingborg, Sweden—Angelholm
AGI Wageningen, Suriname
AGJ Aguni, Japan

AGK Kagua, Papua New Guinea
AGL Wanigela, Papua New Guinea
AGM Angmagssalik, Greenland
AGN Angoon, Alaska
AGP Málaga, Spain
AGQ Agrinion, Greece
AGR Agra, India
AGS Augusta, Georgia
AGT Ciudad Del Este, Paraguay
AGU Aguascalientes, Mexico
AGV Acarigua, Venezuela
AGX Agatti Island, India
AGY Argyle Downs, Western Australia, Australia
AHB Abha, Saudi Arabia
AHN Athens, Georgia
AHO Alghero, Italy
AHT Amchitka, Alaska
AIC Airok, Marshall Islands
AIE Aiome, Papua New Guinea
AIF Assis, Brazil
AIG Yalinga, Central African Republic
AII Ali Sabieh, Djibouti
AIM Ailuk, Marshall Islands
AIN Wainwright, Alaska
AIO Atlantic, Iowa
AIS Arorae Island, Kiribati
AIT Aitutaki, Cook Islands
AIU Atiu, Cook Islands
AIY Atlantic City, New Jersey
AIZ Lake of the Ozarks, Missouri
AJA Ajaccio, Corsica, France
AJF Jouf, Saudi Arabia
AJN Anjouan, Comoros
AJO Aljouf, Yemen
AJY Agades, Niger
AKB Atka Island, Aleutian Islands, Alaska
AKE Akieni, Gabon
AKI Akiak, Alaska
AKJ Asahikawa, Japan
AKK Akhiok, Alaska
AKL Auckland, New Zealand
AKM Zakouma, Chad
AKO Akron, Colorado
AKR Akure, Nigeria
AKS Auki, Solomon Islands
AKU Aksu, China
AKV Akulivik, Québec, Canada

AKX Aktyubinsk, Kazakhstan	**ANQ** Angola, Indiana	**ARW** Arad, Romania
AKY Sittwe, Myanmar	**ANR** Antwerp, Belgium	**ARX** Asbury Park, New Jersey
ALB Albany, New York	**ANU** Antigua, West Indies	**ARY** Ararat, Victoria, Australia
ALC Alicante, Spain	**ANV** Anvik, Alaska	**ASA** Assab, Ethiopia
ALD Alerta, Peru	**ANY** Anthony, Kansas	**ASB** Ashkhabad, Turkmenistan
ALE Alpine, Texas	**ANZ** Angas Downs, Northern	**ASD** Andros Town, Bahamas
ALF Alta, Norway	Territory, Australia	**ASE** Aspen, Colorado
ALG Algiers, Algeria	**AOB** Annanberg, Papua New	**ASF** Astrakhan, Russia
ALH Albany, Western Austra-	Guinea	**ASG** Ashburton, New Zealand
lia, Australia	**AOD** Abou Deia, Chad	**ASH** Nashua, New Hampshire
ALI Alice, Texas	**AOH** Lima, Ohio	**ASJ** Amami-O-Shima, Ryukyu
ALK Asela, Ethiopia	**AOI** Ancona, Italy	Islands, Japan
ALL Albenga, Italy	**AOJ** Aomori, Japan	**ASK** Yamoussoukro, Ivory
ALM Alamogordo, New Mex-	**AOK** Karpathos, Greece	Coast
ico–Holloman	**AON** Arona, Papua New	**ASM** Asmara, Ethiopia
ALN Alton, Illinois	Guinea	**ASO** Asosa, Ethiopia
ALO Waterloo, Iowa	**AOO** Altoona, Pennsylvania	**ASP** Alice Springs, Northern
ALP Aleppo, Syria	**AOR** Alor Setar, Malaysia	Territory, Australia
ALQ Alegrete, Brazil	**AOS** Amook, Alaska	**ASR** Kayseri, Turkey
ALR Alexandra, New Zealand	**AOU** Attopeu, Laos	**AST** Astoria, Oregon
ALS Alamosa, Colorado	**APA** Denver, Colorado—Arap-	**ASU** Asunción, Paraguay
ALW Walla Walla, Washington	ahoe	**ASV** Amboseli, Kenya
ALY Alexandria, Egypt	**APB** Apolo, Bolivia	**ASW** Aswan, Egypt
AMA Amarillo, Texas	**APC** Napa, California	**ASX** Ashland, Wisconsin
AMB Ambilobe, Madagascar	**APE** San Juan Aposento, Peru	**ATB** Atbara, Sudan
AMC Am Timan, Chad	**APF** Naples, Florida	**ATC** Arthur's Town, Bahamas
AMD Ahmedabad, India	**APH** Bowling Green, Virginia	**ATD** Atoifi, Solomon Islands
AME Alto Molocue, Mozam-	**API** Apiay, Colombia	**ATH** Athens, Greece
bique	**APK** Apataki, French Polyne-	**ATI** Artigas, Uruguay
AMH Arba Mintch, Ethiopia	sia	**ATJ** Antsirabe, Madagascar
AMJ Almenara, Brazil	**APW** Apia, Samoa	**ATL** Atlanta, Georgia—Harts-
AML Puerto Armuelles, Pan-	**APX** Arapongas, Brazil	field
ama	**APZ** Zapala, Argentina	**ATM** Altamira, Brazil
AMM Amman, Jordan	**AQA** Araraquara, Brazil	**ATN** Namatanai, Papua New
AMN Alma, Michigan	**AQG** Anqing, China	Guinea
AMO Mao, Chad	**AQI** Qaisumah, Saudi Arabia	**ATO** Athens, Ohio
AMP Ampanihy, Madagascar	**AQJ** Aqaba, Jordan	**ATP** Aitape, Papua New
AMQ Ambon, Indonesia	**AQM** Ariquemes, Brazil	Guinea
AMR Arno, Marshall Islands	**AQP** Arequipa, Peru	**ATQ** Amritsar, India
AMS Amsterdam, Netherlands	**AQS** Saqani, Fiji	**ATR** Atar, Mauritania
AMU Amanab, Papua New	**AQY** Alyeska, Alaska	**ATT** Atmautluak, Alaska
Guinea	**ARB** Ann Arbor, Michigan	**ATU** Attu Island, Alaska
AMW Ames, Iowa	**ARC** Arctic Village, Alaska	**ATV** Ati, Chad
AMX Ammaroo, Northern Ter-	**ARD** Alor, Indonesia	**ATW** Appleton, Wisconsin
ritory, Australia	**ARE** Arecibo, Puerto Rico	**ATY** Watertown, South Dakota
AMY Ambatomainty, Mada-	**ARF** Acaricuara, Colombia	**ATZ** Assiut, Egypt
gascar	**ARH** Arkhangelsk, Russia	**AUA** Aruba, Aruba
AMZ Ardmore, New Zealand	**ARI** Arica, Chile	**AUC** Arauca, Colombia
ANC Anchorage, Alaska	**ARL** Arly, Burkina Faso	**AUD** Augustus Downs, Queens-
AND Anderson, South Carolina	**ARM** Armidale, New South	land, Australia
ANE Angers, France	Wales, Australia	**AUE** Abu Ruedis, Egypt
ANF Antofagasta, Chile	**ARN** Stockholm, Sweden—	**AUG** Augusta, Maine
ANG Angouleme, France	Arlanda	**AUH** Abu Dhabi, United Arab
ANH Anuha Island Resort, So-	**ARO** Arboletas, Colombia	Emirates
lomon Islands	**ARP** Aragip, Papua New	**AUI** Aua, Papua New Guinea
ANI Aniak, Alaska	Guinea	**AUJ** Ambunti, Papua New
ANJ Zanaga, Congo	**ARQ** Arauquita, Colombia	Guinea
ANK Ankara, Turkey	**ARR** Alto Rio Senguerr, Ar-	**AUK** Alakanuk, Alaska
ANL Andulo, Angola	gentina	**AUL** Aur, Marshall Islands
ANM Antalaha, Madagascar	**ARS** Aragarcas, Brazil	**AUN** Auburn, California
ANO Angoche, Mozambique	**ART** Watertown, New York	**AUQ** Atuona, French Polynesia
ANP Annapolis, Maryland	**ARU** Aracatuba, Brazil	**AUR** Aurillac, France

AUS Austin, Texas
AUT Atauro, Indonesia
AUU Aurukun Mission, Queensland, Australia
AUW Wausau, Wisconsin
AUX Araguaina, Brazil
AUY Aneityum, Vanuatu
AUZ Aurora, Illinois
AVB Aviano, Italy
AVG Auvergne, Northern Territories, Australia
AVK Arvaikheer, Mongolia
AVL Asheville, North Carolina
AVN Avignon, France
AVO Avon Park, Florida
AVP Wilkes-Barre/Scranton, Pennsylvania
AVU Avu Avu, Solomon Islands
AVV Avalon, Victoria, Australia
AVW Tucson, Arizona
AVX Catalina Island, California
AWA Awassa, Ethiopia
AWB Awaba, Papua New Guinea
AWD Aniwa, Vanuatu
AWE Alowe, Gabon
AWH Awareh, Ethiopia
AWK Wake Island, Pacific Ocean
AWM West Memphis, Arkansas
AWN Alton Downs, South Australia, Australia
AWP Austral Downs, Queensland, Australia
AWR Awar, Papua New Guinea
AWZ Ahwaz, Iran
AXA Anguilla, West Indies
AXC Aramac, Queensland, Australia
AXD Alexandroupolis, Greece
AXL Alexandria, Northern Territory, Australia
AXM Armenia, Colombia
AXR Arutua, French Polynesia
AXT Akita, Japan
AYK Arkalyk, Kazakhstan
AYN Anyang, China
AYP Ayacucho, Peru
AYQ Ayers Rock, Northern Territory, Australia
AYT Antalya, Turkey
AYW Ayawasi, Indonesia
AYZ Amityville, New York
AZD Yazd, Iran
AZG Apatzingan, Mexico
AZN Andizhan, Uzbekistan
AZO Kalamazoo, Michigan
AZT Zapatoca, Colombia
AZZ Ambriz, Angola

BAA Bialla, Papua New Guinea
BAE Barcelonnette, France
BAG Baguio, Philippines
BAH Bahrain, Bahrain
BAK Baku, Azerbaijan
BAL Batman, Turkey
BAM Battle Mountain, Nevada
BAN Basongo, Democratic Republic of Congo
BAO Ban Mak Khaeng, Thailand
BAP Baibara, Papua New Guinea
BAQ Barranquilla, Colombia
BAS Balalae, Solomon Islands
BAT Barretos, Brazil
BBA Balmaceda, Chile
BBC Bay City, Texas
BBF Burlington, Massachusetts
BBK Kasane, Botswana
BBM Battambang, Cambodia
BBN Bario, Sarawak, Malaysia
BBO Berbera, Somalia
BBQ Barbuda, West Indies
BBR Basse-Terre, Guadeloupe
BBS Blackbush, England
BBU Bucharest-Baneasa, Romania
BBV Bereby, Ivory Coast
BBZ Zambezi, Zambia
BCA Baracoa, Cuba
BCE Bryce Canyon, Utah
BCG Bemichi, Guyana
BCJ Baca Grande, Colorado
BCK Bolwarra, Queensland, Australia
BCM Bacau, Romania
BCN Barcelona, Spain
BCO Jinka, Ethiopia
BCS Belle Chasse, Louisiana
BCT Boca Raton, Florida
BCU Bauchi, Nigeria
BCY Bulchi, Ethiopia
BCZ Bickerton Island, Northern Territory, Australia
BDA Bermuda, Atlantic Ocean
BDB Bundaberg, Queensland, Australia
BDF Bradford, Illinois
BDI Bird Island, Seychelles Islands
BDJ Banjarmasin, Indonesia
BDK Bondoukou, Ivory Coast
BDL Hartford, Connecticut—Bradley
BDM Bandirma, Turkey
BDN Badin, Pakistan
BDO Bandung, Indonesia
BDP Bhadrapur, Nepal
BDQ Vadodara, India
BDR Bridgeport, Connecticut

BDS Brindisi, Italy
BDT Gbadolite, Democratic Republic of Congo
BDU Bardufoss, Norway
BDV Moba, Democratic Republic of Congo
BDY Bandon, Oregon
BED Bedford, Massachusetts
BEG Belgrade, Yugoslavia
BEH Benton Harbor, Michigan
BEL Belem, Brazil
BEO Newcastle, New South Wales, Australia
BEP Bellary, India
BER Berlin, Germany
BES Brest, France
BET Bethel, Alaska
BEV Beersheba, Israel
BEW Beira, Mozambique
BEY Beirut, Lebanon
BEZ Beru, Kiribati
BFC Bloomfield, Queensland, Australia
BFI Seattle, Washington—Boeing Field
BFL Bakersfield, California
BFR Bedford, Indiana
BFS Belfast, Northern Ireland
BFT Beaufort, South Carolina
BFX Bafoussam, Cameroon
BGA Bucaramanga, Colombia
BGF Bangui, Central African Republic
BGH Boghe, Mauritania
BGI Barbados, Barbados
BGJ Borgarfjordur Eystri, Iceland
BGL Baglung, Nepal
BGM Binghamton, New York
BGO Bergen, Norway
BGR Bangor, Maine
BGW Baghdad, Iraq
BGY Milan, Italy
BHB Bar Harbor, Maine
BHC Bullhead City, Arizona
BHD Belfast, Northern Ireland
BHL Bahia De Los Angeles, Mexico
BHM Birmingham, Alabama
BHN Beihan, Yemen
BHO Bhopal, India
BHP Bhojpur, Nepal
BHX Birmingham, England
BIB Baidoa, Somalia
BID Block Island, Rhode Island
BIH Bishop, California
BII Bikini Atoll, Marshall Islands
BIK Biak, Indonesia
BIL Billings, Montana
BIM Bimini, Bahamas
BIO Bilbao, Spain

BIQ Biarritz, France
BIS Bismarck, North Dakota
BIU Bildudalur, Iceland
BIZ Bimin, Papua New Guinea
BJA Bejaia, Algeria
BJF Batsfjord, Norway
BJH Bajhang, Nepal
BJJ Wooster, Ohio
BJM Bujumbura, Burundi
BJN Bajone, Mozambique
BJR Bahar Dar, Ethiopia
BJS Beijing, China
BJW Bajawa, Indonesia
BJX León, Guanajuato, Mexico
BJZ Badajoz, Spain
BKA Moscow, Russia—Bykovo
BKJ Boke, Guinea
BKK Bangkok, Thailand
BKL Cleveland, Ohio
BKM Bakalalan, Sarawak, Malaysia
BKN Birni Nkoni, Nigeria
BKO Bamako, Mali
BKR Bokoro, Chad
BLB Balboa, Panama
BLF Bluefield, West Virginia
BLI Bellingham, Washington
BLK Blackpool, England
BLQ Bologna, Italy
BLR Bangalore, India
BLT Blackwater, Queensland, Australia
BLX Belluno, Italy
BMA Stockholm, Sweden—Bromma
BMG Bloomington, Indiana
BMI Bloomington, Illinois
BMJ Baramita, Guyana
BMK Borkum, Germany
BML Berlin, New Hampshire
BNA Nashville, Tennessee
BNC Beni, Democratic Republic of Congo
BNE Brisbane, Queensland, Australia
BNF Baranof, Alaska
BNI Benin City, Nigeria
BNJ Bonn, Germany
BNQ Baganga, Philippines
BNR Banfora, Burkina Faso
BOB Bora Bora, French Polynesia
BOC Bocas Del Toro, Panama
BOD Bordeaux, France
BOE Boundji, Congo
BOG Bogota, Colombia
BOH Bournemouth, England
BOI Boise, Idaho
BOJ Bourgas, Bulgaria
BOK Brookings, Oregon
BOM Bombay, India

BON Bonaire, Netherlands. Antilles
BOS Boston, Massachusetts—Logan
BOU Bourges, France
BPH Bislig, Philippines
BPT Beaumont/Pt. Arthur, Texas
BPU Beppu, Japan
BQN Aguadilla, Puerto Rico
BQO Bouna, Ivory Coast
BQS Blagoveschensk, Russia
BQT Brest, Belarus
BRE Bremen, Germany
BRN Berne, Switzerland
BRO Brownsville, Texas
BRS Bristol, England
BRU Brussels, Belgium
BRV Bremerhaven, Germany
BRW Barrow, Alaska
BRX Barahona, Dominican Republic
BRY Bardstown, Kentucky
BSA Bossaso, Somalia
BSB Brasilia, Brazil
BSH Brighton, England
BSK Biskra, Algeria
BSL Basel, Switzerland—Mulhouse
BSR Basra, Iraq
BTK Bratsk, Russian Federation
BTL Battle Creek, Michigan
BTM Butte, Montana
BTR Baton Rouge, Louisiana
BTS Bratislava, Slovakia
BTU Bintulu, Sarawak, Malaysia
BTV Burlington, Vermont
BTZ Bursa, Turkey
BUD Budapest, Hungary
BUE Buenos Aires, Argentina
BUF Buffalo, New York
BUG Benguela, Angola
BUH Bucharest, Romania
BUN Buenaventura, Colombia
BUR Burbank, California
BUX Bunia, Democratic Republic of Congo
BUY Bunbury, Western Australia, Australia
BVY Beverly, Massachusetts
BWB Barrow Island, Western Australia, Australia
BWG Bowling Green, Kentucky
BWI Baltimore, Maryland
BWO Balakovo, Russian Federation
BWP Bewani, Papua New Guinea

BWQ Brewarrina, New South Wales, Australia
BWS Blaine, Washington
BWY Woodbridge, England
BXA Bogalusa, Louisiana
BZE Belize City, Belize
BZG Bydgoszcz, Poland
BZI Balikesir, Turkey
BZN Bozeman, Montana
BZV Brazzaville, Congo
BZY Beltsy, Moldova
BZZ Brize Norton, England
CAA Catacamas, Honduras
CAB Cabinda, Angola
CAC Cascavel, Brazil
CAD Cadillac, Michigan
CAE Columbia, South Carolina
CAG Cagliari, Italy
CAH Ca Mau, Vietnam
CAI Cairo, Egypt
CAJ Canaima, Venezuela
CAK Akron/Canton, Ohio
CAL Campbeltown, Scotland
CAM Camiri, Bolivia
CAN Guangzhou, China
CAQ Caucasia, Colombia
CAS Casablanca, Morocco
CAT Cat Island, Bahamas
CAU Caruaru, Brazil
CAX Carlisle, England
CBE Cumberland, Maryland
CBF Council Bluffs, Iowa
CBG Cambridge, England
CBH Bechar, Algeria
CBJ Cabo Rojo, Dominican Republic
CBK Colby, Kansas
CBL Ciudad Bolivar, Venezuela
CBN Cirebon, Indonesia
CBR Canberra, Australian Capital Territory, Australia
CBV Coban, Guatemala
CCJ Calicut, India
CCK Cocos-Keeling Islands, Indian Ocean
CCP Concepción, Chile
CCR Concord, California
CCS Caracas, Venezuela
CCU Calcutta, India
CCV Craig Cove, Vanuatu
CDC Cedar City, Utah
CDG Paris, France—De Gaulle
CDK Cedar Key, Florida
CDU Camden, New South Wales, Australia
CDV Cordova, Alaska
CEB Cebu, Philippines
CEC Crescent City, California
CEN Ciudad Obregon, Mexico
CEP Concepción, Bolivia

CEQ Cannes, France—Mandelieu
CER Cherbourg, France
CFF Cafunfo, Angola
CFG Cienfuegos, Cuba
CFN Donegal, Ireland
CFU Corfu, Greece
CGE Cambridge, Maryland
CGF Cleveland, Ohio—Cuyahoga
CGH São Paulo, Brazil—Congonha
CGJ Chingola, Zambia
CGK Jakarta, Indonesia—Soekarno
CGM Camiguin, Philippines
CGN Cologne/Bonn, Germany—Koeln
CGO Zhengzhou, China
CGP Chittagong, Bangladesh
CGQ Changchun, China
CGS College Park, Maryland
CGT Chinguitti, Mauritania
CGU Ciudad Guayana, Venezuela
CGX Chicago, Illinois—Meigs
CGY Cagayan De Oro, Philippines
CGZ Casa Grande, Arizona
CHA Chattanooga, Tennessee
CHC Christchurch, New Zealand
CHD Chandler, Arizona
CHF Chinhae, South Korea
CHG Chaoyang, China
CHI Chicago, Illinois (airports)—CGX (Meigs), ORD (O'Hare), MDW (Midway)
CHJ Chipinge, Zimbabwe
CHO Charlottesville, Virginia
CHS Charleston, South Carolina
CIA Rome, Italy—Ciampino
CIB Catalina Island, California
CIC Chico, Californa
CID Cedar Rapids/Iowa City, Iowa
CIP Chipata, Zambia
CIQ Chiquimula, Guatemala
CIU Sault Ste. Marie, Michigan
CIY Comiso, Italy
CJS Ciudad Juarez, Mexico
CJU Cheju, Korea
CKB Clarksburg, West Virginia
CKG Chongqing, China
CKM Clarksdale, Mississippi
CKV Clarksville, Tennessee
CLD Carlsbad, California
CLE Cleveland, Ohio—Hopkins
CLG Coalinga, California

CLM Port Angeles, Washington
CLQ Colima, Mexico
CLS Chehalis, Washington
CLT Charlotte, North Carolina
CLU Columbus, Indiana
CMB Colombo, Sri Lanka
CMF Chambery, France
CMH Columbus, Ohio—International
CMI Champaign, Illinois
CMN Casablanca, Morocco
CMW Camaguey, Cuba
CMY Sparta, Wisconsin
CNG Cognac, France
CNM Carlsbad, New Mexico
CNO Chino, California
CNY Moab, Utah
COE Coeur D'Alene, Idaho
COK Cochin, India
COS Colorado Springs, Colorado
COU Columbia, Missouri
CPE Campeche, Mexico
CPF Cepu, Indonesia
CPH Copenhagen, Denmark
CPM Compton, California
CPR Casper, Wyoming
CPS St. Louis, Missouri
CPT Cape Town, South Africa
CQF Calais, France
CQP Cape Flattery, Queensland, Australia
CRG Jacksonville, Florida—Craig
CRL Charleroi, Belgium
CRP Corpus Christi, Texas—International
CRS Corsicana, Texas
CRW Charleston, West Virginia
CSG Columbus, Georgia
CSN Carson City, Nevada
CTG Cartagena, Colombia
CTS Sapporo, Japan—Chitose
CTT Le Castellet, France
CUE Cuenca, Ecuador
CUF Cuneo, Italy
CUG Orange, New South Wales, Australia
CUL Culiacan, Mexico
CUN Cancun, Mexico
CUO Caruru, Colombia
CUR Curacao, Netherlands, Antilles
CUS Columbus, New Mexico
CUU Chihuahua, Mexico
CVG Cincinnati, Ohio
CVL Cape Vogel, Papua New Guinea
CVM Ciudad Victoria, Mexico

CVN Clovis, New Mexico
CVO Corvallis, Oregon
CVR Culver City, California
CVT Coventry, England
CWA Wausau, Wisconsin
CWL Cardiff, Wales
CXH Vancouver, British Columbia, Canada—Harbour
CXI Christmas Island, Kiribati
CXL Calexico, California
CXY Cat Cay, Bahamas
CYS Cheyenne, Wyoming
CZA Chichen Itza, Mexico
CZH Corozal, Belize
CZM Cozumel, Mexico
CZT Carrizo Springs, Texas
CZU Corozal, Colombia
CZX Changzhou, China
CZY Cluny, Queensland, Australia
DAB Daytona Beach, Florida
DAC Dhaka, Bangladesh
DAD Da Nang, Vietnam
DAE Daparizo, India
DAL Dallas/Ft. Worth, Texas—Love Field
DAM Damascus, Syria
DAN Danville, Virginia
DAR Dar Es Salaam, Tanzania
DAY Dayton, Ohio—International
DAZ Darwaz, Afghanistan
DBQ Dubuque, Iowa
DBV Dubrovnik, Croatia
DCA Washington, D.C.—Ronald Reagan National
DCR Decatur, Indiana
DDC Dodge City, Kansas
DEC Decatur, Illinois
DEL Delhi, India
DEN Denver, Colorado—Stapleton
DEO Dearborn, Michigan
DET Detroit, Michigan—City
DFW Dallas/Ft. Worth, Texas—International
DGA Dangriga, Belize
DGL Douglas, Arizona—Municipal
DGO Durango, Mexico
DHM Dharamsala, India
DIJ Dijon, France
DIL Dili, Indonesia
DIM Dimbokro, Ivory Coast
DJA Djougou, Benin
DJB Jambi, Indonesia
DJE Djerba, Tunisia
DJG Djanet, Algeria
DJJ Jayapura, Indonesia
DJM Djambala, Congo
DKR Dakar, Senegal

DLH Duluth, Minnesota/Superior, Wisconson—International
DLI Dalat, Vietnam
DLN Dillon, Montana
DLS The Dalles, Oregon
DME Moscow, Russia
DMO Sedalia, Missouri
DMR Dhamar, Yemen
DNB Dunbar, Queensland, Australia
DND Dundee, Angus, Scotland
DNP Dang, Nepal
DNS Denison, Iowa
DNV Danville, Illinois
DOH Doha, Qatar
DOL Deauville, France
DOM Dominica, West Indies
DOR Dori, Burkina Faso
DPA West Chicago, Illinois
DPK Deer Park, New York
DPL Dipolog, Philippines
DPO Devonport, Tasmania, Australia
DPS Denpasar, Bali, Indonesia
DRB Derby, Western Australia, Australia
DRM Drama, Greece
DRO Durango, Colorado
DRS Dresden, Germany
DRT Del Rio, Texas
DRW Darwin, Northern Territory, Australia
DSM Des Moines, Iowa
DTD Datadawai, Indonesia
DTH Death Valley, California
DTL Detroit Lakes, Minnesota
DTM Dortmund, Germany
DTT Detroit, Michigan
DTW Detroit, Michigan—Wayne (metropolitan)
DUB Dublin, Ireland
DUG Douglas, Arizona
DUR Durban, South Africa
DUS Duesseldorf, Germany
DUT Dutch Harbor, Alaska
DVO Davao, Philippines
DVT Phoenix, Arizona
DXB Dubai, United Arab Emirates
DXR Danbury, Connecticut
DZA Dzaoudzi, Comoros
DZI Codazzi, Colombia
DZN Dzezkazgan, Kazakhstan
DZO Durazno, Uruguay
DZU Dazu, China
EAA Eagle, Alaska
EAB Abbse, Yemen
EAE Emae, Vanuatu
EAM Nejran, Saudi Arabia
EAS San Sebastian, Spain
EAT Wenatchee, Washington

EAU Eau Claire, Wisconsin
EBA Elba Island, Italy
EBB Entebbe/Kampala, Uganda
EBD El Obeid, Sudan
EBU St Etienne, France
EBW Ebolowa, Cameroon
EDI Edinburgh, Scotland
EDW Edwards, California
EED Needles, California
EFK Newport, Vermont
EFL Kefalonia, Greece
EGE Vail/Eagle, Colorado
EGM Sege, Solomon Islands
EGV Eagle River, Wisconsin
EJH Wedjh, Saudi Arabia
EKA Eureka/Arcata, California
EKB Ekibastuz, Kazakhstan
EKD Elkedra, Northern Territory, Australia
EKI Elkhart, Indiana
EKO Elko, Nevada
ELM Elmira, New York
ELN Ellensburg, Washington
ELP El Paso, Texas
ELU El Oued, Algeria
ELY Ely, Nevada
EMN Nema, Mauritania
EMT El Monte, California
ENA Kenai, Alaska
ENC Nancy, France
ENL Centralia, Illinois
ENO Encarnacion, Paraguay
ENT Eniwetok, Marshall Islands
ENW Kenosha, Wisconsin
EOI Eday, Orkney Island, Scotland
EPI Epi, Vanuatu
ERC Erzincan, Turkey
ERF Erfurt, Germany
ERI Erie, Pennsylvania
ERV Kerrville, Texas
ESB Ankara, Turkey
ESE Ensenada, Mexico
ESF Alexandria, Louisiana
ESO Espanola, New Mexico
ESS Essen, Germany
ETZ Metz-Nancy, France
EUG Eugene, Oregon
EVE Evenes, Norway
EVG Sveg, Sweden
EWR New York, New York—Newark International
EXT Exeter, England
EYW Key West, Florida
EZE Buenos Aires, Argentina
EZS Elazig, Turkey
FAE Faroe Islands, Denmark
FAI Fairbanks, Alaska
FAR Fargo, North Dakota
FAT Fresno, California

FAV Fakarava, French Polynesia
FAY Fayetteville, North Carolina
FBU Oslo, Norway
FCA Kalispell/Glacier National Park, Montana
FCO Rome, Italy—Da Vinci
FDU Bandundu, Democratic Republic of Congo
FEZ Fez, Morocco
FFT Frankfort, Kentucky
FGI Apia, Samoa
FHZ Fakahina, French Polynesia
FIG Fria, Guinea
FIH Kinshasa, Democratic Republic of Congo
FJR Al-Fujairah, United Arab Emirates
FKI Kisangani, Democratic Republic of Congo
FKJ Fukui, Japan
FLD Fond Du Lac, Wisconsin
FLG Flagstaff, Arizona
FLL Fort Lauderdale, Florida
FLR Florence, Italy
FMH Falmouth, Massachusetts
FMN Farmington, New Mexico
FMO Muenster, Germany
FMS Fort Madison, Iowa
FMY Fort Myers, Florida
FNA Freetown, Sierra Leone
FNC Funchal, Portugal
FNI Nimes, France
FNT Flint, Michigan
FOB Fort Bragg, California
FOD Fort Dodge, Iowa
FOE Topeka, Kansas
FOG Foggia, Italy
FOK Westhampton, New York
FPO Freeport, Bahamas
FPR Fort Pierce, Florida
FRA Frankfurt, Germany
FRD Friday Harbor, Washington
FRU Bishkek, Kyrgyzstan
FRW Francistown, Botswana
FSD Sioux Falls, South Dakota
FTA Futuna Island, Vanuatu
FTW Dallas/Ft. Worth, Texas
FTX Owando, Congo
FUK Fukuoka, Japan
FUL Fullerton, California
FUN Funafuti Atol, Tuvalu
FUT Futuna, Wallis & Futuna Island
FWA Fort Wayne, Indiana
FXE Ft. Lauderdale, Florida
FYM Fayetteville, Tennessee
FYN Fuyun, China
FYT Faya, Chad

FYV Fayetteville, Arkansas
FZO Filton, England
GAA Guamal, Colombia
GAJ Yamagata, Honshu, Japan
GAO Guantanamo, Cuba
GAQ Gao, Mali
GAX Gamba, Gabon
GAY Gaya, India
GBO Baltimore, Maryland
GBZ Great Barrier Island, New Zealand
GCM Grand Cayman, West Indies
GCN Grand Canyon, Arizona
GDL Guadalajara, Mexico
GDN Gdansk, Poland
GDT Grand Turk, Turks & Caicos Islands
GED Georgetown, Delaware
GEG Spokane, Washington
GEN Oslo, Norway
GEO Georgetown, Guyana
GGT George Town, Bahamas
GIB Gibraltar, Gibraltar
GID Gitega, Burundi
GIG Rio De Janeiro, Brazil
GJT Grand Junction, Colorado
GLA Glasgow, Scotland
GLH Greenville, Mississippi
GLO Gloucester/Cheltenham, England
GLS Galveston, Texas
GMB Gambela, Ethiopia
GMC Guerima, Colombia
GNB Grenoble, France
GND Grenada, Windward Islands
GNE Ghent, Belgium
GNV Gainesville, Florida
GOA Genoa, Italy
GOB Goba, Ethiopia
GOC Gora, Papua New Guinea
GOH Nuuk, Greenland
GOI Goa, India
GOL Gold Beach, Oregon
GOM Goma, Democratic Republic of Congo
GON New London/Groton, Connecticut
GOT Gothenburg, Sweden
GPS Galapagos Islands, Ecuador
GPT Gulfport/Biloxi, Mississippi
GPZ Grand Rapids, Minnesota
GRB Green Bay, Wisconsin
GRR Grand Rapids, Michigan
GRT Gujrat, Pakistan
GRU São Paulo, Brazil
GRX Granada, Spain
GRY Grimsey, Iceland
GRZ Graz, Austria

GSE Gothenburg, Sweden
GSP Greenville/Spartanburg, South Carolina
GTF Great Falls, Montana
GTY Gettysburg, Pennsylvania
GUA Guatemala City, Guatemala
GUB Guerrero Negro, Mexico
GUM Guam, Guam
GUP Gallup, New Mexico
GVA Geneva, Switzerland
GWY Galway, Ireland
GYE Guayaquil, Ecuador
GYL Argyle, Western Australia, Australia
GYM Guaymas, Mexico
GYY Gary, Indiana
GZM Gozo, Malta
GZO Gizo, Solomon Islands
GZT Gaziantep, Turkey
HAA Hasvik, Norway
HAC Hachijo Jima Island, Japan
HAD Halmstad, Sweden
HAG The Hague, Netherlands
HAJ Hanover, Germany
HAM Hamburg, Germany
HAN Hanoi, Vietnam
HAV Havana, Cuba
HBA Hobart, Tasmania, Australia
HDB Heidelberg, Germany
HDD Hyderabad, Pakistan
HDM Hamadan, Iran
HDN Steamboat Springs, Colorado
HEA Herat, Afghanistan
HEL Helsinki, Finland
HEN Hendon, England
HEZ Natchez, Mississippi
HFA Haifa, Israel
HFD Hartford, Connecticut
HFT Hammerfest, Norway
HHR Hawthorne, California
HIJ Hiroshima, Japan
HKG Hong Kong, China
HKS Jackson, Mississippi
HKT Phuket, Thailand
HLN Helena, Montana
HLP Jakarta, Indonesia
HLT Hamilton, Victoria, Australia
HLZ Hamilton, New Zealand
HMA Malmo, Sweden
HMO Hermosillo, Mexico
HMR Hamar, Norway
HND Tokyo, Japan
HNL Honolulu, Oahu, Hawaii
HNM Hana, Maui, Hawaii
HNN Honinabi, Papua New Guinea
HNS Haines, Alaska

HOM Homer, Alaska
HOT Hot Springs, Arkansas
HOU Houston, Texas—(airports)—IAH (Intercontinental), HOU (Hobby)
HSH Las Vegas, Nevada
HST Homestead, Florida
HSV Huntsville/Decatur, Alabama
HTV Huntsville, Texas
HUF Terre Haute, Indiana
HUG Huehuetenango, Guatemala
HUT Hutchinson, Kansas
HVN New Haven, Connecticut
HYA Hyannis, Massachusetts
HYD Hyderabad, India
HZG Hanzhong, China
HZK Husavik, Iceland
IAD Washington, D.C.—Dulles
IAG Niagara Falls, New York
IBZ Ibiza, Spain
ICT Wichita, Kansas
IDA Idaho Falls, Idaho
IEV Kiev, Ukraine—Zhulhany
IFA Iowa Falls, Iowa
IFN Isfahan, Iran
IGL Izmir, Turkey
IKB Wilkesboro, North Carolina
IKI Iki, Japan
ILE Killeen, Texas
ILY Islay, Scotland
IMI Ine, Marshall Islands
INB Independence, Belize
INC Yinchuan, China
IND Indianapolis, Indiana
INL International Falls, Minnesota
INN Innsbruck, Austria
INV Inverness, Scotland
INW Winslow, Arizona
IOM Isle Of Man, United Kingdom
IPC Easter Island, Pacific Ocean
IPE Ipil, Philippines
IPW Ipswich, England
IQT Iquitos, Peru
ISB Islamabad/Rawalpindi, Pakistan
ISC Isles of Scilly, United Kingdom—St Marys
ISP Long Island, New York—Macarthur
IST Istanbul, Turkey
ITH Ithaca, New York
ITO Hilo, Hawaii, Hawaii
IWS Houston, Texas
IXD Allahabad, India
IXQ Kamalpur, India

IXW Jamshedpur, India
IXY Kandla, India
IZM Izmir, Turkey
IZT Ixtepec, Mexico
JAA Jalalabad, Afghanistan
JAC Jackson Hole, Wyoming
JAG Jacobabad, Pakistan
JAH Aubagne, France
JAI Jaipur, India
JAL Jalapa, Mexico
JAN Jackson, Mississippi
JAO Atlanta, Georgia
JAT Jabat, Marshall Islands
JAX Jacksonville, Florida
JBC Boston, Massachusetts—
  City
JBK Berkeley, California
JBP Los Angeles, California—
  Commercial
JBT Bethel, Alaska
JCA Cannes, France—Croi-
  sette
JCD St. Croix, Virgin Islands
JCI Kansas City, Missouri
JDB Dallas/Ft. Worth, Texas—
  Downtown
JDH Jodhpur, India
JDM Miami, Florida—Down-
  town
JDP Paris, France—De Paris
JED Jeddah, Saudi Arabia
JER Jersey, Channel Islands,
  United Kingdom
JFK New York, New York /
  Newark, New Jersey—
  Kennedy
JFM Fremantle, Western Aus-
  tralia, Australia
JGC Grand Canyon, Arizona
JGQ Houston, Texas
JGX Glendale, California
JHY Cambridge, Massachu-
  setts
JIB Djibouti, Djibouti
JKT Jakarta, Indonesia
JMK Mikonos, Greece
JMM Malmo, Sweden—Har-
  bor
JMY Freetown, Sierra Leone
JNB Johannesburg, South Af-
  rica
JNP Newport Beach, California
JNU Juneau, Alaska
JOG Yogyakarta, Indonesia
JOL Jolo, Philippines
JOT Joliet, Illinois
JPD Pasadena, California
JPJ Paterson, New Jersey
JPT Houston, Texas
JPU Paris, France—La De-
  fense
JRS Jerusalem, Israel

JSZ St. Tropez, France
JWH Houston, Texas—West-
  chase
JXN Jackson, Michigan
JYV Jyvaskyla, Finland
KAB Kariba, Zimbabwe
KAC Kameshli, Syria
KAD Kaduna, Nigeria
KAE Kake, Alaska
KAF Karato, Papua New
  Guinea
KAI Kaieteur, Guyana
KAJ Kajaani, Finland
KAP Kapanga, Democratic Re-
  public of Congo
KAQ Kamulai, Papua New
  Guinea
KAT Kaitaia, New Zealand
KAU Kauhava, Finland
KAV Kavanayen, Venezuela
KAW Kawthaung, Myanmar
KBL Kabul, Afghanistan
KBP Kiev-Borispol, Ukraine
KBR Kota Bharu, Malaysia
KBS Bo, Sierra Leone
KBT Kaben, Marshall Islands
KCK Kansas City, Kansas
KCZ Kochi, Japan
KDA Kolda, Senegal
KDC Kandi, Benin
KDD Khuzdar, Pakistan
KDE Koroba, Papua New
  Guinea
KDK Kodiak, Alaska
KDN N'dende, Gabon
KEB English Bay, Alaska
KEC Kasenga, Democratic Re-
  public of Congo
KED Kaedi, Mauritania
KEE Kelle, Congo
KEF Reykjavik, Iceland
KEI Kepi, Indonesia
KEM Kemi/Tornio, Finland
KEN Kenema, Sierra Leone
KGL Kigali, Rwanda
KGN Kasongo/Lunda, Demo-
  cratic Republic of Congo
KGS Kos, Greece
KHG Kashi, China
KHH Kaohsiung, Taiwan
KHI Karachi, Pakistan
KHJ Kauhajoki, Finland
KHK Khark, Iran
KHL Khulna, Bangladesh
KHM Khamti, Myanmar
KIB Ivanoff Bay, Alaska
KID Kristianstad, Sweden
KIJ Niigata, Japan
KIN Kingston, Jamaica—Man-
  ley
KIO Kili, Marshall Islands
KIP Wichita Falls, Texas

KIX Kansai International Air-
  port, Japan
KIY Kilwa, Tanzania
KKN Kirkenes, Norway
KLO Kalibo, Philippines
KMI Miyazaki, Japan
KMJ Kumamoto, Japan
KMQ Komatsu, Japan
KND Kindu, Democratic Re-
  public of Congo
KNZ Kenieba, Mali
KOA Kona, Hawaii
KRE Kirundo, Burundi
KRK Krakow, Poland
KRS Kristiansand, Norway
KRT Khartoum, Sudan
KRU Kerau, Papua New
  Guinea
KRV Kerio Valley, Kenya
KRZ Kiri, Democratic Repub-
  lic of Congo
KSD Karlstad, Sweden
KSE Kasese, Uganda
KSF Kassel, Germany
KSJ Kasos Island, Greece
KTF Takaka, New Zealand
KTG Ketapang, Indonesia
KTM Kathmandu, Nepal
KTN Ketchikan, Alaska
KTO Kato, Guyana
KTP Kingston, Jamaica—Tin-
  son
KTU Kota, India
KTV Kamarata, Venezuela
KTW Katowice, Poland
KTX Koutiala, Mali
KUC Kuria, Kiribati
KUD Kudat, Malaysia
KUE Kukundu, Solomon Is-
  lands
KUH Kushiro, Japan
KUL Kuala Lumpur, Malaysia
KUM Yaku Shima, Japan
KUQ Kuri, Papua New Guinea
KUU Kulu, India
KUV Kunsan, Korea
KUW Kamusi, Papua New
  Guinea
KVA Kavala, Greece
KVB Skovde, Sweden
KWA Kwajalein, Marshall Is-
  lands
KWE Guiyang, China
KWI Kuwait, Kuwait
KWY Kiwayu, Kenya
KWZ Kolwezi, Democratic
  Republic of Congo
KXA Kasaan, Alaska
KXE Klerksdorp, South Africa
KXF Koro, Fiji
KXR Karoola, Papua New
  Guinea

KYA Konya, Turkey
KYD Orchid Island, Taiwan
KYE Tripoli, Lebanon
KYL Key Largo, Florida
KYP Kyaukpyu, Myanmar
KYS Kayes, Mali
KYT Kyauktaw, Myanmar
KYU Koyukuk, Alaska
KYX Yalumet, Papua New Guinea
KZC Kompong, Chnang, Cambodia
KZD Krakor, Cambodia
KZF Kaintiba, Papua New Guinea
KZH Kizhuyak, Alaska
KZI Kozani, Greece
KZK Kompong Son, Cambodia
KZN Kazan, Russian Federation
KZS Kastelorizo, Greece
LAA Lamar, Colorado
LAB Lablab, Papua New Guinea
LAD Luanda, Angola
LAE Lae, Papua New Guinea
LAF Lafayette, Indiana
LAG La Guaira, Venezuela
LAH Labuha, Indonesia
LAI Lannion, France
LAJ Lages, Brazil
LAK Aklavik, Canada
LAL Lakeland, Florida
LAM Los Alamos, New Mexico
LAN Lansing, Michigan
LAO Laoag, Philippines
LAP La Paz, Mexico
LAR Laramie, Wyoming
LAS Las Vegas, Nevada
LAU Lamu, Kenya
LAV Lalomalava, West Samoa
LAW Lawton, Oklahoma
LAX Los Angeles, California—International
LAY Ladysmith, South Africa
LBA Leeds/Bradford, England
LBB Lubbock, Texas
LBC Luebeck, Germany
LBF North Platte, Nebraska
LBG Paris, France—Le Bourget
LBI Albi, France
LBJ Labuan Bajo, Indonesia
LBK Liboi, Kenya
LBL Liberal, Kansas
LBM Luabo, Mozambique
LBN Lake Baringo, Kenya
LBO Lusambo, Democratic Republic of Congo
LBQ Lambarene, Gabon
LBV Libreville, Gabon

LBW Longbawan, Indonesia
LBX Lubang, Philippines
LBY La Baule, France
LCA Larnaca, Cyprus
LCC Lecce, Italy
LCH Lake Charles, Louisiana
LCI Laconia, New Hampshire
LCK Columbus, Ohio—Rickenbacker
LCL La Coloma, Cuba
LCS Las Canas, Costa Rica
LCV Lucca, Italy
LCY London, England—City
LDA Malda, India
LDE Lourdes/Tarbes, France
LDR Lawdar, Yemen
LDY Londonderry, N. Ireland
LED St. Petersburg, Russia
LEG Aleg, Mauritania
LEH Le Havre, France
LEI Almeria, Spain
LEJ Leipzig, Germany
LEK Labe, Guinea
LEO Leconi, Gabon
LEQ Lands End, England
LEW Lewiston, Maine
LEX Lexington, Kentucky
LEZ La Esperanza, Honduras
LFT Lafayette, Louisiana
LFW Lome, Togo
LGA New York, New York/Newark, New Jersey—La Guardia
LGB Long Beach, California
LGD La Grande, Oregon
LGG Liege, Belgium
LGI Deadman's Cay, Long Island
LGK Langkawi, Malaysia
LGO Langeoog, Germany
LGP Legaspi, Philippines
LGQ Lago Agrio, Ecuador
LGR Cochrane, Chile
LGT Las Gaviotas, Colombia
LGU Logan, Utah
LGW London, England—Gatwick
LGX Lugh Ganane, Somalia
LGY Lagunillas, Venezuela
LGZ Leguizamo, Colombia
LHE Lahore, Pakistan
LHN Lishan, Taiwan
LHP Lehu, Papua New Guinea
LHR London, England—Heathrow
LIF Lifou, Loyalty Island, Pacific Ocean
LIG Limoges, France
LIH Lihue, Kauai, Hawaii
LII Mulia, Indonesia
LIM Lima, Peru
LIN Milan, Italy—Linate

LIO Limon, Costa Rica
LIQ Lisala, Democratic Republic of Congo
LIS Lisbon, Portugal
LIT Little Rock, Arkansas
LJU Ljubljana, Slovenia
LKA Larantuka, Indonesia
LKO Lucknow, India
LLY Mount Holly, New Jersey
LMM Los Mochis, Mexico
LMN Limbang, Malaysia
LMP Lampedusa, Italy
LMS Louisville, Mississippi
LMT Klamath Falls, Oregon
LNK Lincoln, Nebraska
LNS Lancaster, Pennsylvania
LNY Lanai City, Lanai, Hawaii
LOF Loen, Marshall Islands
LOG Longview, Washington
LOL Lovelock, Nevada
LON London, England
LOO Laghouat, Algeria
LOQ Lobatse, Botswana
LOS Lagos, Nigeria
LOY Loyangalani, Kenya
LPA Gran Canaria, Canary Islands
LPB La Paz, Bolivia
LPC Lompoc, California
LPL Liverpool, England
LPS Lopez Island, Washington
LPT Lampang, Thailand
LPU Longapung, Indonesia
LRA Larisa, Greece
LRD Laredo, Texas
LRU Las Cruces, New Mexico
LSI Shetland Islands, Scotland
LTN London, England—Luton International
LTO Loreto, Mexico
LUG Lugano, Switzerland
LUK Cincinnati, Ohio
LUS Lusanga, Democratic Republic of Congo
LUX Luxembourg, Luxembourg
LVK Livermore, California
LVM Livingston, Montana
LWC Lawrence, Kansas
LXR Luxor, Egypt
LXS Lemnos, Greece
LXU Lukulu, Zambia
LYH Lynchburg, Virginia
LYS Lyon, France
LZA Luiza, Democratic Republic of Congo
LZH Liuzhou, China
LZI Luozi, Democratic Republic of Congo
LZO Luzhou, China
MAA Madras, India
MAD Madrid, Spain

MAF Midland/Odessa, Texas
MAG Madang, Papua New Guinea
MAH Minorca, Spain
MAI Mangochi, Malawi
MAJ Majuro, Marshall Islands
MAK Malakal, Sudan
MAL Mangole, Indonesia
MAM Matamoros, Mexico
MAN Manchester, England
MAR Maracaibo, Venezuela
MAS Manus Island, Papua New Guinea
MAT Matadi, Democratic Republic of Congo
MAU Maupiti, French Polynesia
MAV Maloelap, Marshall Islands
MAW Malden, Missouri
MBA Mombasa, Kenya
MBC M'bigou, Gabon
MBE Monbetsu, Japan
MBJ Montego Bay, Jamaica
MBS Saginaw, Michigan
MCE Merced, California
MCI Kansas City, Missouri—International
MCJ Maicao, Colombia
MCK McCook, Nebraska
MCL Mount McKinley, Alaska
MCM Monte Carlo, Monaco
MCN Macon, Georgia
MCO Orlando, Florida—International
MCT Muscat, Oman
MCU Montlucon, France
MDE Medellin, Colombia
MDT Harrisburg, Pennsylvania
MDW Chicago, Illinois—Midway
MDY Midway Island, Pacific Ocean
MEB Melbourne, Victoria, Australia—Essendon
MEC Manta, Ecuador
MED Madinah, Saudi Arabia
MEH Mehamn, Norway
MEI Meridian, Mississippi
MEL Melbourne, Victoria, Australia
MEM Memphis, Tennessee
MEN Mende, France
MEX Mexico City, Mexico—Juarez
MEY Meghauli, Nepal
MEZ Messina, South Africa
MFA Mafia Island, Tanzania
MFB Monfort, Colombia
MFN Milford Sound, New Zealand
MFR Medford, Oregon

MFS Miraflores, Colombia
MFU Mfuwe, Zambia
MGM Montgomery, Alabama
MGN Magangue, Colombia
MGO Manega, Gabon
MGQ Mogadishu, Somalia
MGY Dayton, Ohio—Montgomery
MHB Auckland, New Zealand
MHG Mannheim, Germany
MHI Musha, Djibouti
MHT Manchester, New Hampshire
MIA Miami, Florida—International
MID Merida, Mexico
MIE Muncie, Indiana
MIL Milan, Italy—Malpensa
MIQ Omaha, Nebraska—Millard
MKJ Makoua, Congo
MKK Molokai/Hoolehua, Hawaii
MKL Jackson, Tennessee
MLA Malta, Mediterranean Sea
MLB Melbourne, Florida
MLH Mulhouse/Basel, France
MLM Morelia, Mexico
MLN Melilla, Spain
MLO Milos, Greece
MLP Malabang, Philippines
MLQ Malalaua, Papua New Guinea
MLU Monroe, Louisiana
MLW Monrovia, Liberia—Payne
MLX Malatya, Turkey
MMA Malmo, Sweden
MMB Memanbetsu, Japan
MMC Ciudad Mante, Mexico
MMJ Matsumoto, Japan
MMN Stow, Massachusetts
MMO Maio, Cape Verde Islands
MMP Mompos, Colombia
MMQ Mbala, Zambia
MMU Morristown, New Jersey
MMW Moma, Mozambique
MMX Malmo, Sweden—Sturup
MMY Miyako Jima, Japan
MND Medina, Colombia
MNI Montserrat, Montserrat
MNL Manila, Philippines
MNN Marion, Ohio
MOD Modesto, California
MON Mount Cook, New Zealand
MOP Mt. Pleasant, Michigan
MOR Morristown, Tennessee
MOW Moscow, Russia

MOY Monterrey, Colombia
MOZ Moorea, French Polynesia
MPA Mpacha, Namibia
MPB Miami, Florida
MPC Muko-Muko, Indonesia
MPL Montpellier, France
MPM Maputo, Mozambique
MPN Mount Pleasant, Falkland Islands
MPO Mt. Pocono, Pennsylvania
MPV Montpelier, Vermont
MQI Quincy, Massachusetts
MRI Anchorage, Alaska
MRS Marseille, France
MRU Mauritius, Mauritius
MRY Monterey, California
MSL Muscle Shoals, Alabama
MSN Madison, Wisconsin
MSO Missoula, Montana
MSP Minneapolis/St. Paul, Minnesota—International
MSQ Minsk, Belarus
MST Maastricht, Netherlands
MSY New Orleans, Louisiana—International
MSZ Namibe, Angola
MTA Matamata, New Zealand
MTB Monte Libano, Colombia
MTG Mato Grosso, Brazil
MTH Marathon, Florida
MTN Baltimore, Maryland
MTP Montauk Point, New York
MTX Fairbanks, Alaska—Metro
MTY Monterrey, Mexico
MUC Munich, Germany
MUD Mueda, Mozambique
MUG Mulege, Mexico
MVD Montevideo, Uruguay
MVJ Mandeville, Jamaica
MVL Stowe, Vermont
MVO Mongo, Chad
MWA Marion, Illinois
MXL Mexicali, Mexico
MXP Milan, Italy—Malpensa
MYF San Diego, California—Montgomery Field
MYQ Mysore, India
MYR Myrtle Beach, South Carolina
MYV Marysville, California
MZD Mendez, Ecuador
MZE Manatee, Belize
MZF Mzamba, South Africa
MZG Makung, Taiwan
MZH Merzifon, Turkey
MZI Mopti, Mali
MZK Marakei, Kiribati
MZM Metz, France
MZT Mazatlan, Mexico

MZV Mulu, Malaysia
MZX Mena, Ethiopia
MZZ Marion, Indiana
NAD Macanal, Colombia
NAE Natitingou, Benin
NAF Banaina, Indonesia
NAG Nagpur, India
NAH Naha, Indonesia
NAI Annai, Guyana
NAK Nakhon Ratchasima, Thailand
NAN Nadi, Fiji
NAO Nanchong, China
NAP Naples, Italy
NAR Nare, Colombia
NAS Nassau, Bahamas—International
NBL San Blas, Panama
NBO Nairobi, Kenya—Jomo
NCE Nice, France
NCG Nueva Casas Grandes, Mexico
NCL Newcastle, England
NCP Luzon Island, Philippines
NDD Sumbe, Angola
NDE Mandera, Kenya
NDM Mendi, Ethiopia
NDR Nador, Morocco
NDU Rundu, Namibia
NEV Nevis, Leeward Islands
NEW New Orleans, Louisiana
NGN Nargana, Panama
NGO Nagoya, Japan
NGS Nagasaki, Japan
NGW Corpus Christi, Texas
NIO Nioki, Democratic Republic of Congo
NIT Niort, France
NIX Nioro, Mali
NKG Nanjing, China
NLA Ndola, Zambia
NLD Nuevo Laredo, Mexico
NLK Norfolk Island, Norfolk Island
NMB Daman, India
NMU Namu, Marshall Islands
NNT Nan, Thailand
NOC Connaught, Ireland
NOG Nogales, Mexico
NOP Mactan Island, Philippines
NOV Huambo, Angola
NPT Newport, Rhode Island
NRM Nara, Mali
NRT Tokyo, Japan—Narita
NSB Bimini, Bahamas
NTA Natadola, Fiji
NTE Nantes, France
NUE Nuremberg, Germany
NUQ Mountain View, California
NUU Nakuru, Kenya

NVK Narvik, Norway
NWH Newport, New Hampshire
NWI Norwich, England
NYC New York City; New York, New York—E-EWR (Newark); J-JFK (Kennedy); L-LGA (La Guardia)
NYE Nyeri, Kenya
NYI Sunyani, Ghana
NYK Nanyuki, Kenya
NZE Nzerekore, Guinea
NZO Nzoia, Kenya
NZW South Weymouth, Massachusetts
OAG Orange, New South Wales, Australia
OAJ Jacksonville, North Carolina
OAK Oakland, California
OAM Oamaru, New Zealand
OAN Olanchito, Honduras
OAX Oaxaca, Mexico
OBC Obock, Djibouti
OBD Obano, Indonesia
OBE Okeechobee, Florida
OBM Morobe, Papua New Guinea
OBN Oban, Scotland
OBO Obihiro, Japan
OBS Aubenas, France
OCC Coca, Ecuador
OCE Ocean City, Maryland
OCF Ocala, Florida
OCN Oceanside, California
OCV Ocana, Colombia
ODB Cordoba, Spain
ODM Oakland, Maryland
ODS Odessa, Ukraine
ODW Oak Harbor, Washington
OFJ Olafsfjordur, Iceland
OFK Norfolk, Nebraska
OFU Ofu Island, American Samoa
OGD Ogden, Utah
OGN Yonaguni-Jima, Japan
OHA Ohakea, New Zealand
OIM Oshima Island, Japan
OIR Okushiri, Japan
OIT Oita, Japan
OJC Kansas City, Missouri
OKA Okinawa, Ryukyu Islands, Japan
OKC Oklahoma City, Oklahoma—Rogers
OKD Sapporo, Japan—Okadama
OKE Okino Erabu, Japan
OKF Okaukuejo, Namibia
OKG Okoyo, Congo
OKI Oki Island, Japan
OKJ Okayama, Japan

OKQ Okaba, Indonesia
OLD Old Town, Maine
OLJ Olpoi, Vanuatu
OLM Olympia, Washington
OLS Nogales, Arizona
OLU Columbus, Nebraska
OMA Omaha, Nebraska—Eppley
OME Nome, Alaska
OMK Omak, Washington
OMN Osmanabad, India
ONH Oneonta, New York
ONI Moanamani, Indonesia
ONM Socorro, New Mexico
ONP Newport, Oregon
OPO Porto, Portugal
ORB Orebro, Sweden
ORC Orocue, Colombia
ORD Chicago, Illinois—O'Hare
ORE Orleans, France
ORF Norfolk/Virginia Beach, Virginia
ORH Worcester, Massachusetts
ORK Cork, Ireland
ORL Orlando, Florida—Herndon
ORM Northampton, England
ORN Oran, Algeria
ORO Yoro, Honduras
ORP Orapa, Botswana
ORQ Norwalk, Connecticut
ORY Paris, France—Orly
ORZ Orange Walk, Belize
OSA Osaka, Japan
OSB Osage Beach, Missouri
OSD Ostersund, Sweden
OSH Oshkosh, Wisconsin
OSK Oskarshamn, Sweden
OSL Oslo, Norway—Gardermoen
OSM Mosul, Iraq
OSP Slupsk, Poland
OST Ostend, Belgium
OSU Columbus, Ohio—University
OSX Kosciusko, Mississippi
OSY Namsos, Norway
OTA Mota, Ethiopia
OTC Bol, Chad
OTP Bucharest, Romania
OTS Anacortes, Washington
OUD Oujda, Morocco
OUN Norman, Oklahoma
OUT Bousso, Chad
OUU Ouanga, Gabon
OVE Oroville, California
OVL Ovalle, Chile
OXR Oxnard, California
OYS Yosemite National Park, California
OZA Ozona, Texas

OZC Ozamis City, Philippines
OZZ Ouarzazate, Morocco
PAB Bilaspur, India
PAC Panama City, Panama
PAD Paderborn, Germany
PAE Everett, Washington
PAF Pakuba, Uganda
PAG Pagadian, Philippines
PAH Paducah, Kentucky
PAI Pailin, Cambodia
PAO Palo Alto, California
PAP Port-Au-Prince, Haiti
PAQ Palmer, Alaska
PAR Paris, France—Orly
PAS Paros, Greece
PAT Patna, India
PBA Point Barrow, Alaska
PBC Puebla, Mexico
PBF Pine Bluff, Arkansas
PBH Paro, Bhutan
PBI West Palm Beach, Florida
PBL Puerto Cabello, Venezuela
PBR Puerto Barrios, Guatemala
PCT Princeton, New Jersey
PCU Picayune, Mississippi
PCV Punta Chivato, Mexico
PDA Puerto Inirida, Colombia
PDF Prado, Brazil
PDG Padang, Indonesia
PDK Atlanta, Georgia
PDT Pendleton, Oregon
PDX Portland, Oregon
PDZ Pedernales, Venezuela
PEG Perugia, Italy
PEK Beijing, China
PEN Penang, Malaysia
PEQ Pecos, Texas
PER Perth, Western Australia, Australia
PFB Passo Fundo, Brazil
PFN Panama City, Florida
PGA Page, Arizona
PGB Pangoa, Papua New Guinea
PGV Greenville, North Carolina
PGX Perigueux, France
PGZ Ponta Grossa, Brazil
PHF Newport News/Williamsburg, Virginia
PHL Philadelphia, Pennsylvania/Wilmington, Delaware
PHN Port Huron, Michigan
PHR Pacific Harbour, Fiji
PHT Paris, Tennessee
PHX Phoenix, Arizona
PIA Peoria, Illinois
PID Nassau, Bahamas—Paradise
PIE Tampa/St. Petersburg, Florida

PIH Pocatello, Idaho
PII Fairbanks, Alaska
PIK Glasgow, Scotland
PIR Pierre, South Dakota
PIS Poitiers, France
PIT Pittsburgh, Pennsylvania
PIU Piura, Peru
PKP Puka Puka, French Polynesia
PKR Pokhara, Nepal
PKS Paksane, Laos
PLH Plymouth, England
PLI Palm Island, Windward Islands
PLP La Palma, Panama
PLW Palu, Indonesia
PLZ Port Elizabeth, South Africa
PME Portsmouth, England
PMF Parma, Italy
PMI Palma, Mallorca Island, Spain
PMO Palermo, Italy
PMS Palmyra, Syria
PMX Palmer, Massachusetts
PNA Pamplona, Spain
PNE Philadelphia, Pennsylvania
PNQ Poona, India
PNS Pensacola, Florida
PNT Puerto Natales, Chile
PNY Pondicherry, India
POL Pemba, Mozambique
POM Port Moresby, Papua New Guinea
POR Pori, Finland
POU Poughkeepsie, New York
PPG Pago Pago, American Samoa
PPM Pompano Beach, Florida
PPT Papeete, French Polynesia
PQM Palenque, Mexico
PRB Paso Robles, California
PRC Prescott, Arizona
PRG Prague, Czech Republic
PRH Phrae, Thailand
PRI Praslin Is., Seychelles Islands
PRJ Capri, Italy
PRX Paris, Texas
PRY Pretoria, South Africa
PSA Pisa, Italy
PSD Port Said, Egypt
PSL Perth, Scotland
PSM Portsmouth, New Hampshire
PSP Palm Springs, California
PSR Pescara, Italy
PSY Port Stanley, Falkland Islands
PSZ Puerto Suarez, Bolivia
PTK Pontiac, Michigan

PTT Pratt, Kansas
PTY Panama City, Panama
PUB Pueblo, Colorado
PUF Pau, France
PUL Poulsbo, Washington
PUW Pullman, Washington
PVC Provincetown, Massachusetts
PVD Providence, Rhode Island
PVF Placerville, California
PVR Puerto Vallarta, Mexico
PVU Provo, Utah
PWA Oklahoma City, Oklahoma
PWK Chicago, Illinois
PWM Portland, Maine
PWT Bremerton, Washington
PXM Puerto Escondido, Mexico
PXU Pleiku, Vietnam
PYA Puerto Boyaca, Colombia
PYB Jeypore, India
PYM Plymouth, Massachusetts
PYN Payan, Colombia
PYO Putumayo, Ecuador
PYR Pyrgos, Greece
PYV Yaviza, Panama
PYX Pattaya, Thailand
PZA Paz De Ariporo, Colombia
PZB Pietermaritzburg, South Africa
PZE Penzance, England
PZH Zhob, Pakistan
PZO Puerto Ordaz, Venezuela
PZU Port Sudan, Sudan
QBC Bella Coola, British Columbia, Canada
QDU Duesseldorf, Germany
QFE Columbus, Georgia
QKB Breckenridge, Colorado
QKL Cologne/Bonn, Germany
QKS Keystone, Colorado
QVU Viru, Solomon Islands
QWP Winter Park, Colorado
QXE Sora, Italy
QYN Byron Bay, New South Wales, Australia
QZC Smiggin Holes, New South Wales, Australia
RAA Rakanda, Papua New Guinea
RAB Rabaul, Papua New Guinea
RAC Racine, Wisconsin
RAE Arar, Saudi Arabia
RAF Ras An Naqb, Egypt
RAG Raglan, New Zealand
RAH Rafha, Saudi Arabia
RAI Praia, Cape Verde Islands
RAJ Rajkot, India
RAK Marrakech, Morocco
RAL Riverside, California

RAP Rapid City, South Dakota
RAQ Raha, Indonesia
RAR Rarotonga, Cook Islands, South Pacific
RAU Rangpur, Bangladesh
RAW Arawa, Papua New Guinea
RAX Oram, Papua New Guinea
RAY Rothesay, Scotland
RBA Rabat, Morocco
RBF Big Bear, California
RBG Roseburg, Oregon
RBI Rabi, Fiji
RBJ Rebun, Japan
RBL Red Bluff, California
RBO Robore, Bolivia
RCE Roche Harbor, Washington
RCS Rochester, England
RCY Rum Cay, Bahamas
RDD Redding, California
RDU Raleigh/Durham, North Carolina
REP Siem Reap, Cambodia
REU Reus, Spain
REW Rewa, India
REX Reynosa, Mexico
REY Reyes, Bolivia
RFD Rockford, Illinois
RFK Anguilla, Mississippi
RGI Rangiroa, French Polynesia
RGN Yangon, Myanmar
RGT Rengat, Indonesia
RHA Reykholar, Iceland
RHE Reims, France
RHG Ruhengeri, Rwanda
RHI Rhinelander, Wisconsin
RHO Rhodes, Greece
RHP Ramechap, Nepal
RIC Richmond, Virginia
RIG Rio Grande, Brazil
RIJ Rioja, Peru
RIL Rifle, Colorado
RIO Rio De Janeiro, Brazil
RIS Rishiri, Japan
RIT Rio Tigre, Panama
RJB Rajbiraj, Nepal
RJH Rajshahi, Bangladesh
RJI Rajouri, India
RKC Yreka, California
RKD Rockland, Maine
RKE Copenhagen, Denmark
RKH Rock Hill, South Carolina
RKI Rokot, Indonesia
RKO Sipora, Indonesia
RKP Rockport, Texas
RKT Ras Al Khaimah, United Arab Emirates
RLD Richland, Washington
RMB Buraimi, Oman

RMD Ramagundam, India
RMG Rome, Georgia
RMI Rimini, Italy
RMS Ramstein, Germany
RNB Ronneby, Sweden
RNE Roanne, France
RNJ Yoron-Jima, Japan
RNO Reno, Nevada
RNS Rennes, France
RNT Renton, Washington
ROA Roanoke, Virginia
ROB Monrovia, Liberia—Roberts
ROC Rochester, New York
ROM Rome, Italy—Da Vinci
RON Rondon, Colombia
ROP Rota, Mariana Islands
ROR Koror, Palau Island
ROU Rousse, Bulgaria
ROW Roswell, New Mexico
RPA Rolpa, Nepal
RPR Raipur, India
RSE Sydney, New South Wales, Australia
RSJ Rosario, Washington
RST Rochester, Minnesota
RSW Fort Myers, Florida
RTI Roti, Indonesia
RTM Rotterdam, Netherlands
RUH Riyadh, Saudi Arabia
RUI Ruidoso, New Mexico
RUK Rukumkot, Nepal
RUM Rumjartar, Nepal
RUS Marau, Solomon Islands
RUT Rutland, Vermont
RUU Ruti, Papua New Guinea
RUV Rubelsanto, Guatemala
RUY Copan, Honduras
RVJ Reidsville, Georgia
RVK Roervik, Norway
RVN Rovaniemi, Finland
RVS Tulsa, Oklahoma—Jones
RXA Raudha, Yemen
RYK Rahim Yar Khan, Pakistan
RYN Royan, France
RYO Rio Turbio, Argentina
RZE Rzeszow, Poland
RZR Ramsar, Iran
RZZ Roanoke Rapids, North Carolina
SAA Saratoga, Wyoming
SAB Saba, Netherlands Antilles
SAC Sacramento, California
SAE Sangir, Indonesia
SAF Santa Fe, New Mexico
SAI San Marino, San Marino
SAJ Sirajganj, Bangladesh
SAK Saudarkrokur, Iceland
SAL San Salvador, El Salvador
SAM Salamo, Papua New Guinea

SAN San Diego, California—Lindbergh Field
SAO São Paulo, Brazil
SAP San Pedro Sula, Honduras
SAQ San Andros, Bahamas
SAR Sparta, Illinois
SAS Salton City, California
SAT San Antonio, Texas
SAU Sawu, Indonesia
SAV Savannah, Georgia
SAX Sambu, Panama Republic
SAY Siena, Italy
SAZ Sasstown, Liberia
SBA Santa Barbara, California
SBG Sabang, Indonesia
SBH St. Barthelemy, Guadeloupe
SBI Koundara, Guinea
SBN South Bend, Indiana
SBP San Luis Obispo, California
SBQ Sibi, Pakistan
SBS Steamboat Springs, Colorado
SBT San Bernardino, California
SBV Sabah, Papua New Guinea
SBW Sibu, Sarawak, Malaysia
SCA Santa Catalina, Colombia
SCC Prudhoe Bay/Deadhorse, Alaska
SCF Scottsdale, Arizona
SCI San Cristobal, Venezuela
SCK Stockton, California
SCL Santiago, Chile
SCS Shetland Islands, Scotland
SCT Socotra, Yemen
SCU Santiago, Cuba
SCV Suceava, Romania
SCZ Santa Cruz, Solomon Islands
SDA Baghdad, Iraq—Saddam
SDD Lubango, Angola
SDF Louisville, Kentucky
SDJ Sendai, Japan
SDL Sundsvall, Sweden
SDM San Diego, California—Brown Field
SDN Sandane, Norway
SDO Ryotsu Sado Island, Japan
SDQ Santo Domingo, Dominican Republic
SDR Santander, Spain
SDS Sado Shima, Japan
SDT Saidu Sharif, Pakistan
SDU Rio De Janeiro, Brazil
SDV Tel Aviv-Jaffa, Israel
SDW Sandwip, Bangladesh
SDX Sedona, Arizona
SEA Seattle/Tacoma, Washington—Sea-Tac
SEB Sebha, Libya

SEC Serre Chevalier, France	SKE Skien, Norway	SUN Sun Valley, Idaho
SED Sedom, Israel	SKG Thessaloniki, Greece	SUS St. Louis, Missouri
SEE San Diego, California— Gillespie Field	SKH Surkhet, Nepal	SUT Sumbawanga, Tanzania
SEF Sebring, Florida	SKI Skikda, Algeria	SUV Suva, Fiji
SEJ Seydisfjordur, Iceland	SKL Isle Of Skye, Scotland	SUX Sioux City, Iowa
SEL Seoul, South Korea	SKU Skiros, Greece	SUY Sudureyri, Iceland
SEN London, England— Southend	SLC Salt Lake City, Utah	SVD St. Vincent, Windward Islands
SES Selma, Alabama	SLE Salem, Oregon	SVG Stavanger, Norway
SFB Sanford, Florida	SLJ Chandler, Arizona	SVO Moscow, Russia—Sheremetye
SFC St. Francois, Guadeloupe	SLP San Luis Potosi, Mexico	SVP Kuito, Angola
SFF Spokane, Washington	SLU St Lucia, West Indies	SVQ Seville, Spain
SFG St. Maarten, Netherlands Antilles—Esperanto	SLV Simla, India	SVR Svay Rieng, Cambodia
SFH San Felipe, Mexico	SLW Saltillo, Mexico	SWA Shantou, China
SFI Safi, Morocco	SLX Salt Cay, Turks & Caicos Islands	SWD Seward, Alaska
SFL Sao Filipe, Cape Verde Islands	SMF Sacramento, California— Metro	SWO Stillwater, Oklahoma
SFO San Francisco, California	SMV St. Moritz, Switzerland	SXB Strasbourg, France
SFP Surfers Paradise, Queensland, Australia	SMX Santa Maria, California	SXC Catalina Island, California
SFQ Sanliurfa, Turkey	SNA Orange County, California—John Wayne International	SXF Berlin, Germany
SFT Skelleftea, Sweden		SXJ Shanshan, China
SFU Safia, Papua New Guinea	SNN Shannon, Ireland	SXM St. Maarten, Netherlands Antilles—Juliana
SFX San Felix, Venezuela	SNQ San Quintin, Mexico	SXR Srinagar, India
SGD Sonderborg, Denmark	SOC Solo City, Indonesia	SYD Sydney, New South Wales, Australia
SGF Springfield, Missouri	SOF Sofia, Bulgaria	SYO Shonai, Japan
SGM San Ignacio, Mexico	SON Espiritu Santo, Vanuatu	SYR Syracuse, New York
SGN Ho Chi Minh, Vietnam	SOU Southampton, England	SYZ Shiraz, Iran
SGR Sugar Land, Texas	SPC Santa Cruz La Palma, Canary Islands	SZD Sheffield, England
SGS Sanga Sanga, Philippines	SPG Tampa/St. Petersburg, Florida	SZG Salzburg, Austria
SGT Stuttgart, Arkansas		SZH Senipah, Indonesia
SGU St. George, Utah	SPI Springfield, Illinois	SZK Skukuza, South Africa
SGY Skagway, Alaska	SPJ Sparta, Greece	SZP Santa Paula, California
SGZ Singora, Thailand	SPK Sapporo, Japan—Chitose	SZS Stewart Island, New Zealand
SHA Shanghai, China	SPN Saipan, Mariana Islands	
SHB Nakashibetsu, Japan	SPS Wichita Falls, Texas— Sheppard	SZU Segou, Mali
SHC Indaselassie, Ethiopia		SZX Shenzhen, China
SHE Shenyang, China	SQV Sequim, Washington	SZZ Szczecin, Poland
SHF Shanhaiguan, China	SRK Sierra Leone, Sierra Leone	TAA Tarapaina, Solomon Islands
SHI Shimojishima, Japan	SRL Santa Rosalia, Mexico	TAB Tobago, Trinidad & Tobago
SHJ Sharjah, United Arab Emirates	SRT Soroti, Uganda	
SHV Shreveport, Louisiana	SSB St. Croix, Virgin Islands	TAC Tacloban, Philippines
SIG San Juan, Puerto Rico	SSM Sault Ste. Marie, Michigan—County	TAD Trinidad, Colorado
SIN Singapore, Singapore		TAI Taiz, Yemen
SIT Sitka, Alaska	STI Santiago, Dominican Republic	TAJ Tadji, Papua New Guinea
SJB San Joaquin, Bolivia		TAK Takamatsu, Japan
SJC San Jose, California	STJ St. Joseph, Missouri	TAL Tanana, Alaska
SJF St. John, Virgin Islands	STL St. Louis, Missouri	TAM Tampico, Mexico
SJI San Jose, Philippines	STN London, England—Stansted	TAP Tapachula, Mexico
SJJ Sarajevo, Yugoslavia	STO Stockholm, Sweden—Arlanda	TAR Taranto, Italy
SJO San Jose, Costa Rica		TAS Tashkent, Uzbekistan
SJS San Jose, Bolivia	STR Stuttgart, Germany	TAT Tatry/Poprad, Slovakia
SJT San Angelo, Texas	STS Santa Rosa, California	TAU Tauramena, Colombia
SJU San Juan, Puerto Rico	STT St. Thomas, Virgin Islands	TAV Ta'u Island, American Samoa
SJV San Javier, Bolivia	STU Santa Cruz, Belize	
SJW Shijiazhuang, China	STV Surat, India	TAW Tacuarembo, Uruguay
SJX Sartaneja, Belize	STX St. Croix, Virgin Islands— Hamilton	TAX Taliabu, Indonesia
SJY Seinajoki, Finland		TAZ Tashauz, Turkmenistan
SKB St. Kitts, Leeward Islands	SUB Surabaya, Indonesia	TBO Tabora, Tanzania
		TBP Tumbes, Peru

TBR Statesboro, Georgia
TBS Tbilisi, Georgia
TBZ Tabriz, Iran
TCB Treasure Cay, Bahamas
TCI Tenerife, Canary Islands
TCL Tuscaloosa, Alabama
TCN Tehuacan, Mexico
TCO Tumaco, Colombia
TDK Taldy-Kurgan, Kazakhstan
TDT Tanda Tula, South Africa
TDW Amarillo, Texas
TET Tete, Mozambique
TEX Telluride, Colorado
TFN Tenerife, Canary Islands—North Losrodeo
TFS Tenerife, Canary Islands—Reinasofia
TFT Taftan, Pakistan
TFY Tarfaya, Morocco
TGD Titograd, Yugoslavia
TGE Tuskegee, Alabama
TGF Tignes, France
TGS Chokwe, Mozambique
TGT Tanga, Tanzania
TGU Tegucigalpa, Honduras
THF Berlin, Germany
THO Thorshofn, Iceland
THR Tehran, Iran
THU Thule, Greenland
TIJ Tijuana, Mexico
TIP Tripoli, Libya
TIQ Tinian, Mariana Islands
TIR Tirupati, India
TJA Tarija, Bolivia
TJV Thanjavur, India
TKA Talkeetna, Alaska
TKB Tekadu, Papua New Guinea
TKF Truckee, California
TKG Bandar Lampung, Indonesia
TKH Takhli, Thailand
TKK Truk, Caroline Islands
TKN Tokuno Shima, Japan
TKO Tlokoeng, Lesotho
TKS Tokushima, Japan
TKT Tak, Thailand
TKU Turku, Finland
TKV Tatakoto, French Polynesia
TKW Tekin, Papua New Guinea
TLC Mexico City, Mexico—Morelos
TLH Tallahassee, Florida
TLN Toulon/Hyeres, France
TLS Toulouse, France
TLV Tel Aviv, Israel
TLW Talasea, Papua New Guinea
TLX Talca, Chile

TML Tamale, Ghana
TMX Timimoun, Algeria
TMY Tiom, Indonesia
TMZ Thames, New Zealand
TNG Tangier, Morocco
TNI Satna, India
TNN Tainan, Taiwan
TNO Tamarindo, Costa Rica
TOA Torrance, California
TOE Tozeur, Tunisia
TOL Toledo, Ohio
TOP Topeka, Kansas
TPA Tampa/St. Petersburg, Florida—Tampa
TPC Tarapoa, Ecuador
TPE Taipei, Taiwan—Shek
TPG Taiping, Malaysia
TPQ Tepic, Mexico
TPS Trapani, Italy
TPT Tapete, Liberia
TRA Taramajima, Japan
TRB Turbo, Colombia
TRC Torreon, Mexico
TRD Trondheim, Norway
TRN Turin, Italy
TRS Trieste, Italy
TRV Trivandrum, India
TRW Tarawa, Kiribati
TSA Taipei, Taiwan—Sung Shan
TSF Venice, Italy—Treviso
TSJ Tsushima, Japan
TSM Taos, New Mexico
TSN Tianjin, China
TSS New York, New York/Newark, New Jersey
TST Trang, Thailand
TTA Tan Tan, Morocco
TTB Tortoli, Italy
TTC Taltal, Chile
TTJ Tottori, Japan
TTL Turtle Island, Fiji
TTN Trenton, New Jersey
TUF Tours, France
TUG Tuguegarao, Philippines
TUI Turaif, Saudi Arabia
TUL Tulsa, Oklahoma
TUN Tunis, Tunisia
TUP Tupelo, Mississippi
TUS Tucson, Arizona
TVL Lake Tahoe, California
TVU Taveuni, Fiji
TWD Port Townsend, Washington
TWT Tawitawi, Philippines
TXK Texarkana, Arkansas
TXL Berlin, Germany—Tegel
TYO Tokyo, Japan—Narita
TYR Tyler, Texas
TYS Knoxville, Tennessee
TZA Belize City, Belize—Municipal

TZM Tizimin, Mexico
TZN South Andros, Bahamas
TZX Trabzon, Turkey
UAC San Luis Rio Colorado, Mexico
UAE Mount Aue, Papua New Guinea
UAH Ua Huka, French Polynesia
UAI Suai, Indonesia
UAK Narssarssuaq, Greenland
UAL Luau, Angola
UAP Ua Pou, French Polynesia
UAS Samburu, Kenya
UAX Uaxactun, Guatemala
UBA Uberaba, Brazil
UBJ Ube, Japan
UBS Columbus/Starkville/West Point, Mississippi
UCA Utica, New York
UCE Eunice, Louisiana
UCN Buchanan, Liberia
UDD Palm Springs, California
UDE Uden, Netherlands
UDR Udaipur, India
UEE Queenstown, Tasmania, Australia
UEL Quelimane, Mozambique
UEO Kume Jima, Japan
UES Waukesha, Wisconsin
UET Quetta, Pakistan
UGA Bulgan, Mongolia
UGN Waukegan, Illinois
UGO Uige, Angola
UIH Quinhon, Vietnam
UIO Quito, Ecuador
UKB Kobe, Japan
UKI Ukiah, California
UKK Ust-Kamenogorsk, Kazakhstan
UKU Nuku, Papua New Guinea
ULG Ulgii, Mongolia
ULL Mull, Scotland
ULQ Tulua, Colombia
ULS Mulatos, Colombia
ULU Gulu, Uganda
UMA Punta De Maisi, Cuba
UMD Uummannaq, Greenland
UME Umea, Sweden
UNK Unalakleet, Alaska
UNN Ranong, Thailand
UNR Underkhaan, Mongolia
UNU Juneau, Wisconsin
UOL Buol, Indonesia
UON Muong Sai, Laos
UOS Sewanee, Tennessee
UOX University/Oxford, Mississippi
UPA Punta Alegre, Cuba
UPC Puerto La Cruz, Canary Islands

UPN Uruapan, Mexico
URA Uralsk, Kazakhstan
URI Uribe, Colombia
URM Uriman, Venezuela
URO Rouen, France
URY Gurayat, Saudi Arabia
URZ Uruzgan, Afghanistan
USI Mabaruma, Guyana
USS Sancti Spiritus, Cuba
UST St. Augustine, Florida
UTC Utrecht, Netherlands
UTI Kouvola, Finland
UTK Utirik, Marshall Islands
UTL Torremolinos, Spain
UTN Upington, South Africa
UTU Ustupo, Panama
UUU Manumu, Papua New Guinea
UVA Uvalde, Texas
UVF St. Lucia, West Indies—Hewanorra
UVL New Valley, Egypt
UVO Uvol, Papua New Guinea
UYN Yulin, China
UZH Unayzah, Saudi Arabia
UZU Curuzu Cuatia, Argentina
VAA Vaasa, Finland
VAB Yavarate, Colombia
VAF Valence, France
VAG Varginha, Brazil
VAI Vanimo, Papua New Guinea
VAK Chevak, Alaska
VAL Valenca, Brazil
VAN Van, Turkey
VAP Valparaiso, Chile
VAR Varna, Bulgaria
VAS Sivas, Turkey
VAV Vava'u, Tonga Island
VAW Vardoe, Norway
VAZ Val D'isere, France
VBV Vanuabalavu, Fiji
VBY Visby, Sweden
VCC Victoria, Cameroon
VCD Victoria R. Downs, Northern Territory, Australia
VCE Venice, Italy—Marco Polo
VCH Vichadero, Uruguay
VCP São Paulo, Brazil
VCR Carora, Venezuela
VCT Victoria, Texas
VDA Ovda, Israel
VDB Fagernes, Norway
VDE Valverde, Canary Islands
VDI Vidalia, Georgia
VDR Villa Dolores, Argentina
VDS Vadso, Norway
VDZ Valdez, Alaska
VEE Venetie, Alaska
VEG Maikwak, Guyana
VEJ Vejle, Denmark

VEL Vernal, Utah
VER Veracruz, Mexico
VEV Barakoma, Solomon Islands
VEX Tioga, North Dakota
VFA Victoria Falls, Zimbabwe
VGA Vijayawada, India
VGO Vigo, Spain
VGT Las Vegas, Nevada
VGZ Villagarzon, Colombia
VHC Saurimo, Angola
VHM Vilhelmina, Sweden
VHN Van Horn, Texas
VHY Vichy, France
VHZ Vahitahi, French Polynesia
VIC Vicenza, Italy
VID Vidin, Bulgaria
VIE Vienna, Austria
VIG El Vigia, Venezuela
VIH Vichy, Missouri
VIJ Virgin Gorda, British Virgin Islands
VIL Dakhla, Morocco
VIQ Viqueque, Indonesia
VIS Visalia, California
VIT Vitoria, Spain
VJQ Gurue, Mozambique
VKO Moscow, Russia—Vnukovo,
VKT Vorkuta, Russia
VKW West Kavik, Alaska
VLA Vandalia, Illinois
VLC Valencia, Spain
VLD Valdosta, Georgia
VLE Valle, Arizona
VLI Port Vila, Vanuatu
VLL Valladolid, Spain
VLM Villa Montes, Bolivia
VLN Valencia, Venezuela
VLV Valera, Venezuela
VNA Saravane, Laos
VNC Venice, Florida
VNE Vannes, France
VNG Viengxay, Laos
VNO Vilnius, Lithuania
VNS Varanasi, India
VNX Vilanculos, Mozambique
VNY Los Angeles, California—Van Nuys
VOG Volgograd, Russia
VOL Volos, Greece
VPS Ft Walton Beach, Florida
VPZ Valparaiso, Indiana
VQS Vieques, Puerto Rico
VRA Varadero, Cuba
VRB Vero Beach, Florida
VRC Virac, Philippines
VRE Vredendal, South Africa
VRK Varkaus, Finland
VRL Vila Real, Portugal
VRN Verona, Italy

VRS Versailles, Missouri
VRU Vryburg, South Africa
VRY Vaeroy, Norway
VSA Villahermosa, Mexico
VSE Viseu, Portugal
VSF Springfield, Vermont
VSG Lugansk, Ukraine
VSO Phuoc Long, Vietnam
VST Vasteras, Sweden
VTA Victoria, Honduras
VTE Vientiane, Laos
VTF Vatulele, Fiji
VTL Vittel, France
VTU Las Tunas, Cuba
VTZ Vishakhapatnam, India
VUP Valledupar, Colombia
VVI Santa Cruz, Bolivia
VVK Vastervik, Sweden
VVO Vladivostok, Russia
VVZ Illizi, Algeria
VXC Lichinga, Mozambique
VXE São Vicente, Cape Verde Island
VXO Vaxjo, Sweden
VYD Vryheid, South Africa
VYS Peru, Illinois
WAA Wales, Alaska
WAB Wabag, Papua New Guinea
WAC Waca, Ethiopia
WAD Andriamena, Madagascar
WAE Wadi-Ad-Dawasir, Saudi Arabia
WAF Wana, Pakistan
WAG Wanganui, New Zealand
WAN Waverney, Queensland, Australia
WAO Wabo, Papua New Guinea
WAP Alto Palena, Chile
WAQ Antsalova, Madagascar
WAR Waris, Indonesia
WAS Washington, D.C. (airports)—DCA (National), IAD (Dulles)
WAT Waterford, Ireland
WAW Warsaw, Poland
WAZ Warwick, Queensland, Australia
WBU Boulder, Colorado
WCA Castro, Chile
WCH Chaiten, Chile
WCR Chandalar, Alaska
WEA Weatherford, Texas
WES Weasua, Liberia
WET Wagethe, Indonesia
WEX Wexford, Ireland
WFB Ketchikan, Alaska
WGO Winchester, Virginia
WGP Waingapu, Indonesia
WGY Wagny, Gabon

WHO Franz Josef, New Zealand
WHR Vail/Eagle, Colorado
WHS Whalsay, Shetland Islands, Scotland
WIC Wick, Scotland
WIL Nairobi, Kenya—Wilson
WIN Winton, Queensland, Australia
WJF Palmdale/Lancaster, California
WJR Wajir, Kenya
WKA Wanaka, New Zealand
WKN Wakunai, Papua New Guinea
WKR Walker's Cay, Bahamas
WLD Winfield, Kansas
WLG Wellington, New Zealand
WLH Walaha, Vanuatu
WLK Selawik, Alaska
WLM Waltham, Massachusetts
WLS Wallis Island, Wallis & Futuna Islands
WMC Winnemucca, Nevada
WNA Napakiak, Alaska
WNP Naga, Philippines
WNZ Wenzhou, China
WOI Wologissi, Liberia
WOK Wonken, Venezuela
WOT Wonan, Taiwan
WPB Port Berge, Madagascar
WPO Paonia, Colorado
WPR Porvenir, Chile
WPU Puerto Williams, Chile
WRA Warder, Ethiopia
WRE Whangarei, New Zealand
WRG Wrangell, Alaska
WRL Worland, Wyoming
WRO Wroclaw, Poland
WRY Westray, Scotland
WSA Wasua, Papua New Guinea
WSB Steamboat Bay, Alaska
WSH Shirley, New York
WSX Westsound, Washington
WSY Airlie Beach, Queensland, Australia
WSZ Westport, New Zealand
WTD West End, Bahamas
WTE Wotje, Marshall Islands
WTK Noatak, Alaska
WTO Wotho, Marshall Islands
WTP Woitape, Papua New Guinea
WUU Wau, Sudan
WVI Watsonville, California
WVK Manakara, Madagascar
WVL Waterville, Maine
WVN Wilhelmshaven, Germany
WWD Cape May, New Jersey

WWK Wewak, Papua New Guinea
WWR Woodward, Oklahoma
WXN Wanxian, China
WYE Yengema, Sierra Leone
WYN Wyndham, Western Australia, Australia
WYS West Yellowstone, Montana
XAL Alamos, Mexico
XAR Aribinda, Burkina Faso
XAY Xayabury, Laos
XBG Bogande, Burkina Faso
XBJ Birjand, Iran
XBL Buno Bedelle, Ethiopia
XBN Biniguni, Papua New Guinea
XBO Boulsa, Burkina Faso
XBR Brockville, Ontario, Canada
XBW Killineq, Ontario, Canada
XCH Christmas Island, Indian Ocean
XCM Chatham, Ontario, Canada
XCN Coron, Philippines
XDJ Djibo, Burkina Faso
XES Lake Geneva, Wisconsin
XFN Xiangfan, China
XGL Granville Lake, Manitoba, Canada
XGN Xangongo, Angola
XIC Xichang, China
XIE Xieng Lom, Laos
XKY Kaya, Burkina Faso
XLO Long Xuyen, Vietnam
XLS St. Louis, Senegal
XLU Leo, Burkina Faso
XMD Madison, South Dakota
XMH Manihi, French Polynesia
XMI Masasi, Tanzania
XMS Macas, Ecuador
XNG Quang Ngai, Vietnam
XNU Nouna, Burkina Faso
XPR Pine Ridge, South Dakota
XRY Jerez De La Frontera, Spain
XSC South Caicos, Turks & Caicos Islands
XSE Sebba, Burkina Faso
XSM St. Mary's, Maryland
XSO Siocon, Philippines
XSP Singapore, Singapore—Seletar
XTO Taroom, Queensland, Australia
XTR Tara, Queensland, Australia
XUZ Xuzhou, China
XVL Vinh Long, Vietnam

XYE Ye, Myanmar
XZA Zabre, Burkina Faso
YAG Fort Frances, Ontario, Canada
YAI Chillan, Chile
YAK Yakutat, Alaska
YAL Alert Bay, Canada
YAM Sault Ste Marie, Ontario, Canada
YAN Yangambi, Democratic Republic of Congo
YAO Yaounde, Cameroon
YAP Yap, Caroline Islands
YAS Yasawa Island, Fiji
YAY St. Anthony, Newfoundland, Canada
YAZ Tofino, British Columbia, Canada
YBA Banff, Alberta, Canada
YBL Campbell River, British Columbia, Canada
YBZ Toronto, Ontario, Canada—Downtown
YCD Nanaimo, British Columbia, Canada
YCJ Cape St. James, British Columbia, Canada
YCR Cross Lake, Manitoba, Canada
YCV Cartierville, Québec, Canada
YCW Chilliwack, British Columbia, Canada
YDA Dawson City, Yukon Territory, Canada
YDN Dauphin, Manitoba, Canada
YDQ Dawson Creek, British Columbia, Canada
YDS Desolation Sound, British Columbia, Canada
YDT Vancouver, British Columbia, Canada—Boundary
YEA Edmonton, Alberta, Canada
YEC Yechon, Korea
YEG Edmonton, Alberta, Canada—International
YFA Fort Albany, Ontario, Canada
YFL Fort Reliance, Northwest Territories, Canada
YFX Fox Harbour/St Lewis, Newfoundland, Canada
YGA Gagnon, Québec, Canada
YGJ Yonago, Japan
YGK Kingston, Ontario, Canada
YGL La Grande, Québec, Canada
YGY Deception, Québec, Canada

YHB Hudson Bay, Saskatchewan, Canada

YHD Dryden, Ontario, Canada

YHE Hope, British Columbia, Canada

YHF Hearst, Ontario, Canada

YHG Charlottetown, Newfoundland, Canada

YHH Campbell River, British Columbia—Harbor

YHM Hamilton, Ontario, Canada

YHU Montreal, Québec, Canada

YHZ Halifax, Nova Scotia, Canada—International

YIB Atikokan, Ontario, Canada

YIH Yichang, China

YIP Detroit, Michigan—Willow

YJN St. Jean, Québec, Canada

YKA Kamloops, British Columbia, Canada

YKF Kitchener, Ontario, Canada

YKL Schefferville, Québec, Canada

YKM Yakima, Washington

YKZ Toronto, Ontario, Canada—Buttonvie

YLA Langara, British Columbia, Canada

YLF Laforges, Québec, Canada

YLG Yalgoo, Western Australia, Australia

YLL Lloydminster, Alberta, Canada

YLT Alert, Northwest Territories, Canada

YMB Merritt, British Columbia, Canada

YMF Montague Harbor, British Columbia, Canada

YML Murray Bay, Québec, Canada

YMM Ft. McMurray, Alberta, Canada

YMP Port McNeil, British Columbia, Canada

YMS Yurimaguas, Peru

YMT Chibougamau, Québec, Canada

YMX Montreal, Québec, Canada—Mirabel

YNA Natashquan, Québec, Canada

YNB Yanbu, Saudi Arabia

YNC Wemindji, Québec, Canada

YND Gatineau/Hull, Québec, Canada

YNG Youngstown, Ohio

YNK Nootka Sound, British Columbia, Canada

YNT Yantai, China

YOK Yokohama, Japan

YOL Yola, Nigeria

YOW Ottawa, Ontario, Canada

YPA Prince Albert, Saskatchewan, Canada

YPE Peace River, Alberta, Canada

YPG Portage La Prairie, Manitoba, Canada

YPQ Peterborough, Ontario, Canada

YPR Prince Rupert, British Columbia, Canada—Digby

YPS Port Hawkesbury, Nova Scotia, Canada

YPT Pender Harbor, British Columbia, Canada

YPW Powell River, British Columbia, Canada

YPX Povungnituk, Québec, Canada

YPY Ft. Chipewyan, Alberta, Canada

YQB Quebec, Québec, Canada

YQC Quaqtaq, Québec, Canada

YQD The Pas, Manitoba, Canada

YQE Kimberley, British Columbia, Canada

YQF Red Deer, Alberta, Canada

YQG Windsor, Ontario, Canada

YQI Yarmouth, Nova Scotia, Canada

YQK Kenora, Ontario, Canada

YQL Lethbridge, Alberta, Canada

YQN Nakina, Ontario, Canada

YQR Regina, Saskatchewan, Canada

YQS St. Thomas, Ontario, Canada

YQT Thunder Bay, Ontario, Canada

YQU Grande Prairie, Alberta, Canada

YQY Sydney, Nova Scotia, Canada

YRF Cartwright, Newfoundland, Canada

YRL Red Lake, Ontario, Canada

YRQ Three Rivers, Québec, Canada

YRV Revelstoke, British Columbia, Canada

YSB Sudbury, Ontario, Canada

YSC Sherbrooke, Québec, Canada

YSD Suffield, Alberta, Canada

YSE Swan River, Manitoba, Canada

YSF Stony Rapids, Sask, Canada

YSI Sans Souci, Ontario, Canada

YSJ Saint John, New Brunswick, Canada

YSM Ft. Smith, Northwest Territories, Canada

YSN Salmon Arm, British Columbia, Canada

YSP Marathon, Ontario, Canada

YTA Pembroke, Ontario, Canada

YTE Cape Dorset, Northwest Territories, Canada

YTF Alma, Québec, Canada

YTH Thompson, Manitoba, Canada

YTJ Terrace Bay, Ontario, Canada

YTK Tulugak, Québec, Canada

YTO Toronto, Ontario, Canada—Pearson

YTP Tofino, British Columbia, Canada

YTQ Tasiujuaq, Québec, Canada

YTR Trenton, Ontario, Canada

YTS Timmins, Ontario, Canada

YTZ Toronto, Ontario, Canada—Toronto

YUA Yuanmou, China

YUE Yuendumu, Northern Territory, Australia

YUL Montreal, Québec, Canada—Dorval

YUM Yuma, Arizona

YVB Bonaventure, Québec, Canada

YVD Yeva, Papua New Guinea

YVG Vermilion, Alberta, Canada

YVO Val D'or, Québec, Canada

YVP Kuujjuaq, Québec, Canada

YVR Vancouver, British Columbia, Canada—International

YVZ Deer Lake, Ontario, Canada

YWF Halifax, Nova Scotia, Canada—Waterfront

YWG Winnipeg, Manitoba, Canada

YWH Victoria, British Columbia, Canada

YWN Winisk, Ontario, Canada
YWP Webequie, Ontario, Canada
YWR White River, Ontario, Canada
YWS Whistler, British Columbia, Canada
YXD Edmonton, Alberta, Canada
YXE Saskatoon, Saskatchewan, Canada
YXH Medicine Hat, Alberta, Canada
YXK Rimouski, Québec, Canada
YXL Sioux Lookout, Ontario, Canada
YXS Prince George, British Columbia, Canada
YXU London, Ontario, Canada
YXX Abbotsford, British Columbia, Canada
YYB North Bay, Ontario, Canada
YYC Calgary, Alberta, Canada
YYJ Victoria, British Columbia, Canada—International
YYL Lynn Lake, Manitoba, Canada
YYM Cowley, Alberta, Canada
YYQ Churchill, Manitoba, Canada
YYR Goose Bay, Newfoundland, Canada
YYT St. Johns, Newfoundland, Canada
YYU Kapuskasing, Ontario, Canada
YYW Armstrong, Ontario, Canada
YYY Mont Joli, Québec, Canada
YYZ Toronto, Ontario, Canada—Pearson
YZA Ashcroft, British Columbia, Canada
YZF Yellowknife, Northwest Territories, Canada
YZG Salluit, Québec, Canada
YZH Slave Lake, Alberta, Canada
YZR Sarnia, Ontario, Canada

YZS Coral Harbour, Northwest Territories, Canada
YZT Port Hardy, British Columbia, Canada
YZX Greenwood, Nova Scotia, Canada
ZAC York Landing, Manitoba, Canada
ZAG Zagreb, Croatia
ZAH Zahedan, Iran
ZAJ Zaranj, Afghanistan
ZAK Chiusa/Klausen, Italy
ZAL Valdivia, Chile
ZAM Zamboanga, Philippines
ZAO Cahors, France
ZAR Zaria, Nigeria
ZAT Zhaotong, China
ZAZ Zaragoza, Spain
ZBF Bathurst, New Brunswick, Canada
ZBL Biloela, Queensland, Australia
ZBN Bozen, Italy
ZBS Mesa, Arizona
ZBV Beaver Creek, Colorado
ZBY Sayaboury, Laos
ZCL Zacatecas, Mexico
ZCO Temuco, Chile
ZEM East Main, Québec, Canada
ZEN Zenag, Papua New Guinea
ZER Zero, India
ZFB Old Fort Bay, Québec, Canada
ZFD Fond Du Lac, Saskatchewan, Canada
ZFW Fairview, Alberta, Canada
ZGC Lanzhou, China
ZGF Grand Forks, British Columbia, Canada
ZGM Ngoma, Zambia
ZGS Gethsemani, Québec, Canada
ZGU Gaua, Vanuatu
ZHA Zhanjiang, China
ZHM Shamshernagar, Bangladesh
ZHP High Prairie, Alberta, Canada
ZIC Victoria, Chile

ZIG Ziguinchor, Senegal
ZIH Ixtapa/Zihuatanejo, Mexico
ZIN Interlaken Ost, Switzerland
ZJG Jenpeg, Manitoba, Canada
ZKB Kasaba Bay, Zambia
ZKE Kaschechewan, Ontario, Canada
ZKG Kegaska, Québec, Canada
ZKL Steenkool, Indonesia
ZKM Sette Cama, Gabon
ZKP Kasompe, Zambia
ZLG El Gouera, Mauritania
ZLO Manzanillo, Mexico
ZLT La Tabatiere, Québec, Canada
ZMM Zamora, Mexico
ZMR Meran, Italy
ZNA Nanaimo, British Columbia, Canada—Harbour
ZNC Nyac, Alaska
ZND Zinder, Niger
ZNG Negginan, Manitoba, Canada
ZNZ Zanzibar, Tanzania
ZOS Osorno, Chile
ZPB Sachigo Lake, Ontario, Canada
ZPC Pucon, Chile
ZPH Zephyrhills, Florida
ZQN Queenstown, New Zealand
ZRH Zurich, Switzerland
ZRI Serui, Indonesia
ZRM Sarmi, Indonesia
ZSA San Salvador, Bahamas
ZSP St Paul, Québec, Canada
ZSZ Spiez, Switzerland
ZTA Tureira, French Polynesia
ZTH Zakinthos, Greece
ZTJ Interlaken West, Switzerland
ZUL Zilfi, Saudi Arabia
ZVK Savannakhet, Laos
ZYI Zunyi, China
ZYL Sylhet, Bangladesh
ZZU Mzuzu, Malawi
ZZV Zanesville, Ohio

American States and Capitals

Ala. (Alabama) Montgomery
Alas. (Alaska) Juneau
Ariz. (Arizona) Phoenix
Ark. (Arkansas) Little Rock
Calif. (California) Sacramento
Colo. (Colorado) Denver
Conn. (Connecticut) Hartford
Del. (Delaware) Dover
Fla. (Florida) Tallahassee
Ga. (Georgia) Atlanta
Ha. (Hawaii) Honolulu
Ia. (Iowa) Des Moines
Id. (Idaho) Boise
Ill. (Illinois) Springfield
Ind. (Indiana) Indianapolis
Kans. (Kansas) Topeka
Ky. (Kentucky) Frankfort
La. (Louisiana) Baton Rouge
Mass. (Massachusetts) Boston

Md. (Maryland) Annapolis
Me. (Maine) Augusta
Mich. (Michigan) Lansing
Minn. (Minnesota) St Paul
Miss. (Mississippi) Jackson
Mo. (Missouri) Jefferson City
Mont. (Montana) Helena
N.C. (North Carolina) Raleigh
N.D. (North Dakota) Bismarck
Nebr. (Nebraska) Lincoln
Nev. (Nevada) Carson City
N.H. (New Hampshire) Concord
N.J. (New Jersey) Trenton
N.M. (New Mexico) Santa Fe
N.Y. (New York) Albany
Oh. (Ohio) Columbus

Okla. (Oklahoma) Oklahoma City
Oreg. (Oregon) Salem
Pa. (Pennsylvania) Harrisburg
R.I. (Rhode Island) Providence
S.C. (South Carolina) Columbia
S.D. (South Dakota) Pierre
Tenn. (Tennessee) Nashville
Tex. (Texas) Austin
Ut. (Utah) Salt Lake City
Va. (Virginia) Richmond
Vt. (Vermont) Montpelier
Wash. (Washington) Olympia
Wis. (Wisconsin) Madison
W.Va. (West Virginia) Charleston
Wyo. (Wyoming) Cheyenne

Foregoing fifty states plus two possible states if their citizens approve and Congress concurs:

D.C. (District of Columbia) Washington

P.R. (Puerto Rico) San Juan

Astronomical Constellations and Stars

And Andromeda (Princess Enchained), also called Mirach

Ant Antlia (Air Pump)

Aps Apus (Bird of Paradise)

Aql Aquila (Eagle); contains Altair

Aqr Aquarius (Water Carrier)

Ara (Altar)

Ari Aries (Ram); contains Hamal

Aur Auriga (Charioteer); contains Capella

Boö Boötes (Herdsman); contains Arcturus

Cae Caelum (Chisel)

Cam Camelopardalis (Giraffe)

Cap Capricornus (Horned Goat)

Car Carina (Keel of Argo); contains Canopus

Cas Cassiopeia (Queen Enthroned); contains supernova 1572

Cen Centaurus (Centaur); contains Alpha Centauri, Proxima Centauri

Cep Cepheus (Monarch)

Cet Cetus (Whale); contains Mira

Cha Chameleon

Cir Circinus (Compasses)

CMa Canis Major (Great Dog); contains Sirius

CMi Canis Minor (Little Dog); contains Procyon

Cnc Cancer (Crab); contains Praesepe

Col Columba (Dove)

Com Coma Berenices (Berenice's Hair)

CrA Corona Australis (Southern Crown)

CrB Corona Borealis (Northern Crown), also called Gemma

Crt Crater (Cup)

Cru Crux (Southern Cross); Black Magellanic Cloud nearby

Crv Corvus (Crow)

CVn Canes Venatici (Hunting Dogs); contains Cor Caroli

Cyg Cygnus (Swan); contains Deneb, Northern Cross

Del Delphinus (Dolphin)

Dor Dorado, also called Xiphies (Swordfish); Large Magellanic Cloud

Dra Draco (Dragon)

Equ Equuleus (Colt)

Eri Eridanus (Great River); contains Achernar

For Fornax (Furnace)

Gem Gemini (The Twins); contains Castor, Pollux

Gru Grus (Crane)

Her Hercules; contains Ras Algethi

Hor Horologium (Clock)

Hya Hydra (Marine Monster); contains Alphard

Hyd Hydrus (Water Snake)

Ind Indus (Indian)

Kif Aus Kiffa Australis (Southern Breadbasket); contains Zuben el Genubi

Kif Bor Kiffa Borealis (Northern Breadbasket); contains Zubeneschamali

Lac Lacerta (Lizard)

Leo (Lion) contains Regulus, Denebola

Lep Lepus (Hare)

Lib Libra (Balance or Scales)

LMi Leo Minor (Little Lion)

Lup Lupus (Wolf)

Lyn Lynx

Lyr Lyra (Lyre); contains Vega

Men Mensa (Table), also called Mons Mensae (Table Mountain)

Mic Microscopium (Microscope)

Mon Monoceros (Unicorn)

Mus Musca (Fly)

Nor Norma (Rule)

Oct Octans (Octant)

Oph Ophiuchus (Serpent Bearer); contains supernova 1604

Ori Orion (Hunter); contains Betelgeuse, Rigel

Pav Pavo (Peacock)

Peg Pegasus (Winged Horse)

Per Perseus (Rescuer or Champion); contains Algol

Phe Phoenix

Pic Pictor (Painter's Easel)

PsA Piscis Austrinus (Southern Fish); contains Formalhaut

Psc Pisces (Fishes)

Pup Puppis (Stern), in Argo

Pyx Pyxis (Mariner's Compass Chest or Binnacle), in Argo

Ret Reticulum (Net)

Scl Sculptor (Sculptor's Workshop)

Sco Scorpius (Scorpion); contains Antares

Sct Scutum (Shield)

Ser Serpens (Serpent)

Sex Sextans (Sextant)

Sge Sagitta (Arrow)

Sgr Sagittarius (Archer), Center of Galaxy

Tau Taurus (Bull); contains Hyades—Aldebaran; Pleiades

Tel Telescopium (Telescope)

TrA Triangulum Australe (Southern Triangle)

Tri Triangulum (Triangle)

Tuc Tucana (Toucan); Small Magellanic Cloud

UMa Ursa Major (Great Bear— Big Dipper); contains Dubhe, Mizar

UMi Ursa Minor (Little Bear— Little Dipper); contains Polaris (Pole Star)

Vel Vela (Sail), in Argo

Vir Virgo (Virgin)

Vol Votans (Flying Fish)

Vul Vulpecula (Little Fox); also called Vulpecula cum Ansere (Little Fox with Goose)

Astronomical Symbols

⊖☾ center

☄ comet

◑ crescent moon (first quarter)

◐ crescent moon (last quarter)

⊕ Earth (symbol shows globe bisected by meridian lines into four quarters)

○ full moon

◔ gibbous moon (first quarter)

◑ gibbous moon (last quarter)

◐ half moon (first quarter)

◑ half moon (last quarter)

♃ Jupiter (symbol said to represent a hieroglyph of the eagle, Jove's bird, or to be the initial letter of Zeus with a line drawn through it to indicate its abbreviation)

☉☾ lower limb

♂ Mars (symbol represents shield and spear of the god of war, Mars; it is also the male or masculine symbol)

☿ Mercury (symbol represents head and winged cap of Mercury, god of commerce and communication, surmounting his caduceus)

♆ Neptune (symbolized by the trident of Neptune, god of the sea)

● new moon

☽ moon (symbol depicts crescent moon in last quarter)

♇ Pluto (symbol is monogram made up of P and L; initials of astronomer Percival Lowell, who predicted its discovery) Saturn (symbol thought to represent an ancient scythe or sickle, as Saturn was the god of seed sowing and hence also of time)

☆ star

☆-P star–planet altitude correction

☉ sun (symbolized by a shield with its boss; some believe this boss represents a central sunspot)

☉☾ upper limb

♅ Uranus (symbolized by combined devices indicating the sun plus the spear of Mars, as Uranus was the personification of heaven in the Greek mythology, dominated by the light of the sun and the power of Mars)

♀ Venus (designated by the female symbol, thought to be the stylized representation of the hand mirror of this goddess of love)

Bafflegab Divulged

Bafflegab consists of the ambiguous and euphemistic fig leaves of language and literature. **Bafflegab** begins where abbreviations, acronyms, and other short forms leave off.

abandoned woman prostitute

above critical in danger of exploding; out of control

absolute tripe official gibberish

abstinence syndrome withdrawal from alcohol or other addictive substances

acceptable deception(s) half truth(s)

accessible passenger loading zone wheelchair loading

accommodation house(s) whorehouse(s)

account executive(s) pimp(s) offering prostitutes to executives on liberal expense accounts

accounting aberrations profit-making mistakes made at the expense of the customer

achieved status social position gained by ability, accomplishments, and personal effort

acid indigestion heartburn, acidosis, acute indigestion, cardialgia, colic, dyspepsia, gripes, pyrosis, stomach condition, tormina, water qualm

acoustical problem deafness; loss of hearing

active defense offense

activities incompatible with diplomatic status spying

activity booster pep pill

acute heroin-morphine intoxication death due to an overdose of the drugs

acute irregularity constipation

adjustment center(s) prison; solitary confinement cell(s)

adjustment downward price drop

administrative domain collectivist system of managing and taxing distilleries, factories, and farms

administrative error bureaucratic bungle

administrative professional secretary

administrative segregation solitary confinement

adult bookstore emporium of sexually explicit publications

adult briefs disposable diapers for incontinent adults

adult entertainment erotic dancing

adult fiction sexually explicit novels

adult institution(s) prison(s)

adult movies sexually explicit films

adult relaxation center bathroom

adult retreat bedroom

adult undergarments diapers for adults

advanced in years old

advance to a strategic position retreat

advantaged gainfully employed; rich; successful

aerialist trapeze artist; trapeze performer; trapezist

aerial mishap(s) aviation accident(s)—airplane crashes, explosions

aerial visitation nuclear attack

aesthetic surgery plastic surgery

affaire de coeur (French—affair of the heart)—love or lust adventure

affaire de voyage (French—love trip)—shipboard romance

affirmative action programs designed to admit minority people to jobs and schools

affordable merchandiser's term for anything attractive enough to make potential customers become buyers although they may not be able to pay in full

after he (she) left after he (she) died

aftershock post-release prisoner program(s)

ageism prejudice against the elderly

Agent Orange deadly defoliant used in tropical warfare believed to cause cancer and birth defects in humans

agricultural laborer(s) farmhand(s)

airsickness nausea followed by vomiting when traveling in an aircraft

aisle manager floorwalker

Alaska sable skunk fur

Albany beef Hudson River sturgeon

a little sensitive neurotic; very touchy

all about the birds and the bees sex education

all-devouring element fire

all-out strategic exchange atomic bombs in nuclear warfare

all this and more, right after this important message please stay tuned

alpha alcoholic person who drinks in an effort to drown problems

alter castrate, sterilize

alternative to birth abortion

amalgamation interracial mixing; miscegenation

amenity center public toilet

Americaid welfare

American broadtail domestic lamb's wool

American tweezers burglar's tools

amply proportioned fat

amply rewarded well paid

an association sexual contact

animal controller(s) dog catcher(s)

animal shelter(s) dog pound(s)

anointing the sick Roman Catholic euphemism for unction in extremis (service rendered at the point of death); last rites

answer the final summons die; drop dead

anthropogenically-derived acidic substances acid rain

anticipatory retaliation first strike in a conflict

antisocial offender(s) convict(s); criminal(s)

antisocial restructuring candidates convicts; felons

antsy nervous feeling, jitters

arbitrary deprivation of life extermination; murder

archivist(s) library clerk(s)

ardent spirits alcohol; alcoholic drinks

area of operations battlefield

Argtec Argentine technique that disposed of political prisoners by casting them, shackled, from aircraft flying over the La Plata estuary

armed emergency small-scale war

Armenian mink dyed cat fur

arms-reduction specialists arms-control agents

articulating on paper writing

artificial dentures false teeth

artillery injection paraphernalia

artistic success box-office failure

assignation point(s) place(s) where pimps and street-walkers accost potential customers

assisted living retirement home

athletic supporter(s) jock strap(s)

at that future juncture when

at that point in time then

at this juncture now

at this moment of history now

at this point in time now

at this present juncture now

attitude adjustment time for drinking liquor

audience augmentation device for producing a capacity audience; attained by generous distribution of complimentary tickets, and cut-rate prices

audiovisually qualified able to run a motion-picture projector

auditorially handicapped deaf

auditory problem deafness; loss of hearing

au naturel naked

Australian buck processed rabbit fur

Australian steak mutton

Australian treat kangaroo meat

authoritarian totalitarian regime denying human rights to its people and persecuting all dissidents

autocompress automobile metal compression device for reducing junk vehicles into a square mass of metal ready for the smelter

auto insurance problem revocation of driver's license and subsequent loss of auto insurance

automanipulation masturbation

automotive internist auto mechanic

avian propagation facility an incubator

avoid polarizing people don't rock the boat

awardee(s) person(s) imprisoned in a ship's brig, also known as the ccu or correctional custody unit

away from one's desk absent; not accepting calls

awful acronym AIDS

axed fired

back-alley butcher abortionist

backhouse privy; toilet

backward country poor country; third-world country

bad good

bad actors cesium, plutonium, and strontium wastes from high-level fusion in nuclear power plants

bad scene unhappy experience

bafflegab bewildering language

bagno (Italian—bath)—prison or whorehouse

Baltic leopard spotted-cat fur dyed to look like leopard

Baltimore beefsteak broiled liver

bang keen drug satisfaction

banheiro (Portuguese—bath)— men's toilet

baños de caballeros (Spanish—gentlemen's baths)— men's toilet(s)

baños de damas (Spanish—ladies' baths)—women's toilet(s)

bar hustler(s) male prostitute(s) plying trade in cocktail bars

bathroom toilet (in the United States, although in the British Isles and on the Continent a

bathroom is fitted primarily for bathing although it may contain a toilet)

bath(room) tissue toilet paper

battering domestic violence

battle fatigue shell shock; a variety of psychotic and neurotic disorders associated with the stress of combat

beautician beauty parlor owner; hairdresser; make-up artist

beauty culturist beauty parlor operator; make-up artist

beauty parlor hairdresser's shop

beaverette processed cat or rabbit far

bedded sexually connected

bedroom activities sex

bedswever unfaithful spouse

been inside served time in prison

behind stuff using heroin

behind the iron door behind bars, jailed

bernice cocaine

beyond the black stump Australia's far outback; beyond the boondocks

big time prosperous drug business

billiard academy pool hall

billiard lounge pool hall

binary weapons two chemicals kept separate until mixed for firing

biological soldiering use of biological (bacterial) agents against an enemy

biosolid sewage

birth attendant midwife

black-and-whites skunks

blade man knife-wielding criminal

blaze to leave

blemishes acne; blackheads; pimples

blister agents chemicals that burn the skin and cause blisters

blood agents chemical substances that enter the body via the respiratory system and attack the blood cells

blood box ambulance

blood disease syphilis

body moisture sweat

body odor stinking sweat

Bombay duck dried and salted lizardfish

bone box ambulance

boobs breasts
book it move
book of pseudonyms hotel register
boot camp minimum-security jail
borborygmus bowel rumblings
borne by the stork born
born into a better world died
born out of wedlock illegitimate
borrow without intent to return steal
bottle baby alcoholic
boys on the border South African euphemism for soldiers defending the borders
boy's room men's toilet
breach of the peace to agitate, to arouse, to assemble unlawfully, to awaken, to hinder, to incite to riot, to molest, to obstruct traffic, to trespass
breathe one's last die
brief encounter one-night stand
bronzing sunbathing
brown nose curry favor
buddy man drug supplier
building engineer janitor
bush telegraph (Australian—word of mouth)
burn rate corporate spending
business moratorium lockout
business slowdown depression
bus one to leave
buy the farm die
by-product of the arts of peace war
caballeros (Spanish—gentlemen)—men's toilet
cabinet (French—closet; small room; water closet)—toilet
cabinet d'aisance (French—cabinet of comfort)—public toilet
cable cruising random viewing of cable television channels
call boy(s) high-priced male prostitute(s); gigolo(s)
call girl(s) high-priced female prostitute(s)
call of nature need to go to the toilet
Cambodian Incursion Cambodian Invasion
campo santo (Spanish—sacred field)—cemetery

can going to the can; going to the toilet
candy man drug dealer
canine seclusion habitat doghouse
canned cow condensed milk
Cape Cod turkey codfish
carbon monoxided killed by carbon-monoxide poisoning
cardiac arrest cessation of heart beat
cardiovascular accident stroke
career deceleration loss of job opportunities or jobs
carnal desire sexual passion; lust
carnifex hangman
cash in your chips die
casketing putting a dead person in a coffin
cast up one's cookies vomit
cast up your accounts vomit
catastrophic reaction psychiatric euphemism for angry frustration marked by crying, pulling, pushing, yelling, and even trying to commit suicide or kill others; often includes cataleptic convulsions, loss of memory and vocabulary, coronary disorders, and death
categorical imperative absolute and binding moral law
cathouse house of prostitution
cavalheiros (Portuguese—gentlemen)—men's toilet
cavalry police reinforcements
cease to purchase boycott
cemetery worker gravedigger
cerebral insufficiency brain deficiency
cerebrovascular accident stroke
C-girl call girl; hundred-dollar girl
chalet de nicessiti (French—public toilet)
channels trails of interoffice memos
channel surfing random viewing of television channels
Chapter 7 liquidation
Chapter 11 bankruptcy
character deficiency fault
check out die
cheesy appearing cheap
chemical abuse alcohol or drug addiction
chemical agent deployment throwing tear gas

chemical, electrical, or physical duress chemical, electrical, or physical torture
chicken ranch house of prostitution
chic sale country privy; outhouse
children with latent ability students with underdeveloped intellects
children with untapped potential students with underdeveloped intellects
chinchillette processed rabbit fur
chipping injecting drugs under the skin but not into a vein
chippy prostitute
chirtonsor(s) barber(s)
chopped physically unattractive
chronic alcoholic sot
chronic irregularity constipation
chronologically challenged old
churchyard cemetery
circular protector condom
circumorbital haematoma black eye
civilian irregular mercenary
civilian irregular defense soldier mercenary
Civil War the War Between the States, the War of the Secession
clandestine connection sex in the shadows
clandestine exhumation body snatching; grave robbing
clap gonorrhea
clean off drugs
clean bombs bombs that destroy structures rather than people; nonradioactive bombs; neutron bombs
cleanup policy burial insurance
client(s) prison inmate(s); users of professional services
client(s) of the correctional system convict(s)
climb the golden stairs die
cloakroom toilet (British)
closet water closet (toilet)
cluster bomb multimedia ad presentation
cocktail lounge saloon, bar
Code Blue medical jargon frequently used to mean cardiac arrest
coffin nails cigarettes

cognitive behavioral modification using task analysis to define mental problems

cognitive services any professional service involving counseling, talking, or thinking

cold war deterrent threat

collateral damage people killed in war

collection correspondent bill collector

colonic infirmity cancer of the colon

colonic stasis constipation

combat emplacement evacuator shovel

combat fatigue (*see* battle fatigue)

comfort station toilet

command of nature feeling it's time to use the toilet

commercial statement advertisement; sales pitch

commercial travellers traveling salespersons

commission agent bookmaker

commit no nuisance do not defecate or urinate

commits a nuisance defecates in a public place

commode toilet

communication arts reading, speaking, and writing

communication problem deaf and dumb

communications engineer tv repairman

community treatment center(s) prison(s)

companionate marriage free love; trial marriage

companion in misfortune fellow prisoner

Company The Company (Washington euphemism for the Central Intelligence Agency—CIA)

complete elimination mass murder

complete fabrication a lie

complete liquidation mass murder

compromising situation caught with your pants down

conchie conscientious objector

condominium (Spanish euphemism—jail or prison)

conflict with the law breaking the law, committing a misdemeanor or felony

congal cuarto con gal (Mexican-Spanish—room with girl)—house of prostitution

conjugal relations sex within marriage

conjugal visits sexual visits to prisoners by spouses

connubial bliss happy marriage; sex within marriage

connubial rites marriage ceremony

constant companion lover

constructive engagement interchange that has positive results

contact issues sex

Contragate 1980s political scandal involving the secret sale of arms to Iran in order to fund the Contra uprising in Nicaragua after Congress decided the U.S. should no longer support the Contras

contributor taxpayer

controlled substances addictive drugs; dangerous drugs, narcotics

contusion bruise

convalescent home nursing home

convenience toilet

cools cigarettes dipped in pcp (PCP)

coordinated national intelligence spying program aimed at radicals

coordinator person whose desk is flanked by those of two expediters

cop a plea plead guilty to a lesser crime; plea bargain

cop-out convenient excuse to avoid unwanted responsibility

coronary heart attack; heart failure

corpocracy corporate bureaucracy, middle management

corpulent fat

correctional custody facility brig, jail, lockup, penitentiary, prison, reformatory, etc

correctional custody unit ship's brig

correctional facilities brigs, jails, prisons

correctional institution penitentiary; prison

correctional officer(s) prison guard(s); prison warden(s)

corrective labor camps Soviet forced-labor prison camps

correct within an order of magnitude incorrect; wrong

cosmetician beauty parlor operator

counterproductive hindrance

courier service technician messenger

Cousins the Cousins (British secret service term for American intelligence organizations such as the CIA and the FBI)

cover up conceal the truth

cover-up words euphemisms

cow chips dried cow dung

cranial catastrophe stroke

crapola cover-up phrases and words characterized by their ambiguity, insincerity, and mendacity; nonsense

crapper toilet (believed by some to be named for an English turn-of-the-century plumber named Crapper and inventor of the water-flush system)

crash and burn fail utterly

creative conflict civil rights demonstration; nonviolent confrontation

creative financing where the seller loans some of the costs to the buyer in order to facilitate a sale; questionable accounting practices

creative unresponsiveness sullen indifference

credibility gap widespread disbelief and distrust due to impossible promises and lies uttered by bureaucrats and politicians

crib home

criminal conversation British euphemism for fornication

criminally attacked raped

criticalese language and style of professional critics

crowd engineer(s) police dog(s)

cruising in the corn drinking and driving

culminating experience death by nuclear weapons

culturally deprived poor

culturally deprived environment ghetto; poor section; slum

cuspidor spittoon

custodial care janitorial services; providing services to hospital and nursing home patients

custodial engineer janitor

custodian janitor

customer drug buyer and user

customer-assistance specialist debt collector

cynical ailment cynicism, doubt, disbelief

dairy nutrients manure

damas (Italian, Portuguese, Spanish—ladies)—women's toilet

Damen (German—ladies)—women's toilet

damer (Danish or Swedish—ladies)—women's toilet

dames (Afrikaans, Dutch, French—ladies)—women's toilet

damska (Polish—ladies)—women's toilet

dancing on air hanging by the neck until dead

dang(ed) damn(ed)

darn it damn it

dating street slang term meaning prostituting

daughter of joy prostitute

d_d damned (used in polite prose)

deaccessioning selling; divesting

dead cat on the line something is not right

dealer drug salesperson

dearly departed dead

death list persons to be assassinated

death machinery Hitler's extermination practices

debris disposal technician dustman; garbage man; trash collector

debt of nature death

decapitator beheader; guillotineur; guillotine

deceased dead

decimate destroy; kill off

decollation beheading

decruit fire

deek detect; look at; observe

defenestration jumping-out-the-window suicide

defense officer(s) soldier(s)

defensive aggression hitting the enemy before you are hit

defensive aircraft bombers

deferred maintenance putting off to tomorrow what should be repaired today

deferred realization company earnings spread over several quarters

deferred schedule slowdown

dehire fire from a job

deinstitutionalization discharge of felons from prisons and patients from mental institutions into surrounding communities

delicate condition pregnant

demise death

demoto not motivated

den of iniquity house of prostitution

dental discomfort toothache

deodorant of language euphemism

departed died

department of defense war department

depart this mortal coil die

deplane exit from an airplane

depopulate(d) slaughter(ed)

dermasurgeon restorative embalming artist in a morgue

designer drugs synthetic recreational drugs

detained in federal custody jailed in a federal penitentiary

détente (French—slackening of tensions)—easing of discord and warlike threats

detention center prison

deterrence policy of scaring the enemy by maintaining a larger armed force in the hope it will deter the enemy from war

developing nations have-not countries; poor countries; underdeveloped nations

deviation from the truth lying; untruth

devil's breath poison gas

diagnostic misadventure of high magnitude a patient's death

dibs residence

dick police officer

died of lead poisoning died from bullet wounds; shot

died of target practice executed by a firing squad

diesel physically attractive

dilution reduction in earnings

diminished capacity slightly insane; impaired mental functioning

dining-room attendant busboy

dip to leave

dipsomaniac alcoholic

directory assistance telephone number information

direct reduction killing off excess

dirty dishes evidence planted to incriminate another or others

dirty tricks sinister activities

dis to show disrespect

disadvantaged poor; unemployed

disappeared into the dust died

disassemble carcasses cut meat

discontinuance of student populations school closures

disengage retreat

disinflation falling inflation, reduced rate of inflation

disinform mislead

disinterested free from bias

dislike the cut of one's jib dislike one's looks

disorderly house house of prostitution

disposable undergarments diapers

disposal area dump

disposal center junkyard

dissemble hide under a false appearance

dissolution death; end

distanced themselves ran for cover; left the scene; remained uninvolved

diversion embezzlement

diverted stolen from

D-notice death notice; death report; obituary

documented immigrant legal alien

dog behavior modification dog training program for guard or police work

doggydo dog dung

dog nuisance dog dung

dog tutor dog trainer

dog warden dogcatcher

domestic functionary servant

dones (Catalan—ladies)—women's toilet

don't polarize people don't rock the boat

do one's business move one's bowels

dope something good

do the dutch act commit suicide

doublespeak the language of contradiction, coverup distortion, misdirection, and omission

doubter an agnostic, atheist, freethinker, skeptic, unbeliever

downside bad news

downsize lay workers off

down the drain (or tube, or waterslide) down the toilet (beyond recovery); too late

downward revision decrease in prices or wages

dreamless sleep death

drinking problem alcohol addiction

dripper syringe (also called a dropper)

dropout person living outside conventional society; person who does not complete an educational program; spot in a magnetic tape indicating loss of information

dropsey term for unspecific ailments

drug problem narcotic addiction

Drugville anyplace where drugs are used

doggy fashionably dressed

dump in a tin cup send to a charity hospital

duplication of the double number 4

dustbin garbage can

dustbinman garbage collector (British)

dutch gold imitation gold leaf

dyspepsia acid indigestion; heartburn

easy money stolen goods

easy way out suicide

eat it absorb a loss

ebonies ebony phonetics (black English)

ecological receptacle garbage can; trash barrel; trash container

economic action picket line; sit down; slowdown; strike

economical with the truth untruthful; misleading

economically deprived poor

economically marginalized poor

economic decline depression; recession

economic lull depression

economic regression loss

economic return profit

economic slowdown depression

educationist(s) bureaucratic educator(s) believed to be good at communicating with students

eduflation rising cost of education

effect a separation fire (from a job); dismiss

effecting linkages coordination

efficiencies firings

effluents contaminating gaseous discharges, industrial wastes, and liquid sewage pollution

effluvium stench; stink

egress exit

Egyptian calisthenics getting into bed and staring up at the ceiling or sky until the forces of sleep prevail

electrolethe electric chair

electronic surveillance wire tapping or computer snooping

electronic technician electrician

eleemosynary organization charity fund-collecting business

elegancies words believed by some to be more genteel than the words they replace

elevated intoxicated (by drugs and liquor); high

elevated to a lower level demoted; lowered

eliminate kill; murder

ellas [Spanish—they (feminine)]—women's toilet

ellos [Spanish—they (masculine)]—men's toilet

elopement risk psychiatric euphemism for patients escaping or trying to escape from the hospitals, institutions, or wards to which they are confined

embarazada (Spanish—embarrassed)—pregnant

emerging nations backward countries; poor nations

emolument salary; wage

employment center commercial district

employment decreases firings

emporium department store

Endlösung die Endlösung (German—final solution)—death; Hitlerian euphemism masking the annihilation of the Jews

energetic disassembly explosion

engine opium pipe

engineering landfill(s) garbage dump(s)

enlargement of the lower back fat buttocks

enlisted dining facility mess deck aboard American naval vessels

ensanguined undergarment bloody shirt

entertainment consumers casino gamblers

entombment burial

entomology specialist(s) bug sprayer(s)

entrapment luring anyone into the commission of a crime

enuresis bedwetting

environmental control specialist(s) janitor(s)

environmentally handicapped people ghetto people; poor people; street people; vagrants

environmental services janitorial work

equal-opportunity ailments venereal diseases

equine fertilizer horse manure

equitable compensation living wage

erase assassinate; kill; murder

erminette dyed rabbit fur

erring brethren Confederate soldiers and states as defined by their Northern neighbors

error in overamplification an error in the cover-up of the cover-up arranged by Nixon's aides after Watergate

eruct(ation) belch(ing)

escalated interpersonal altercation murder

escort vessel(s) destroyer(s); frigate(s)

eternal rest death

ethical lackers those with no scruples

ethnic cleansing elimination of an unlike culture, through killing or terrorism ,

ethnic evacuation forcible removal of a people from their native homeland

ethnic minority in the United States usually means African-Americans, Hispanics, and native Americans

ethnophaulisms slurs aimed against various ethnic groups

Eufemia Greek goddess of fair speech or good report—Euphemia

eugeria happy old age

euphemia softening of phrases or words thought to be coarse or offensive

euphemism affected niceness; goody-goodyism; overrefinement; purism; substitution of an auspicious phrase or word for inauspicious ones

euphuisms high-flown terms in the style of the English author John Lyly

euthanasia medically induced death for persons suffering from incurable and painful diseases

euthanize kill

eutrophication freshwater polluted by chemical plant or sewage runoff

evidentiary material evidence

excellent compensation good pay

exceptional child retarded child

excrement dung

excusado (Spanish—reserved; set apart)—toilet

executive action removal of an executive by assassination

executive explanation explaining the unexplainable by bending the truth

exfiltration retreat

exotic entertainer striptease dancer

expecting pregnant

expectorate spit

experienced automobiles second-hand autos

experienced tires recapped tires; retreads

expire die

exsanguination death due to loss of blood

exterminating engineer bug and rat killer; termite remover; hired killer

extinguishment death

extrajudicial executions murders

extramarital sex fornication with a married person

extrapolation educated guess; guesstimate

extreme penalty death; execution

extreme power ordered killings of anti-government people

eye service working when the boss is looking

fabrication lie

face time exposure to public

facial blemishes pimples

facial dew sweat

faded gradually shaved (hair)

fag male homosexual

failed to reply accurately, completely, and fully lied

fall asleep die

far out crazy; wild

fashion stylist(s) dress designer(s)

featherbedding employing excess employees

federalese federal bafflegab

feisty abusive, outspoken

fell in battle killed in action

feminine protection menstrual cloth, napkin, or tampon

fenestration arrangement of doors, openings, and windows in a building

ferfak (Hungarian—men)—men's toilet

ferly wonderfully great

fig leaves expressions contrived to conceal anything deemed to be indecorous, offensive, or of questionable taste

filteration radiation

filthy good

final exit death

final injection execution by lethal drug injection

finalization conclusion

finalize end

final solution Hitler's euphemism masking his plan to murder all the Jews imprisoned in concentration camps

financial consultant broker

financially embarrassed broke; without funds

financially undernourished broke; without funds

finite period of future time later

fir (Gaelic—men)—men's toilet

first strike first attack

first use initial firing of nuclear weapons

fiscal policy government spending and tax collection

fishpole microphone boom

five-finger discount stealing

five-O police officers

flame out to wear only red

flasher man who indecently exposes himself

flex to leave

flight attendant airline stewardess or steward

flight host(ess) airline steward(ess)

floral tribute wreath

food-order expediter short-order cook

food preparation center kitchen

food science specialist(s) short order cook(s)

footballs amphetamine

forensic laboratory morgue

freedom fighter terrorist; rebel; revolutionary

freeway tax-supported toll-free expressway; tax-supported toll-free highway

French welcome smallpox or syphilis

friendly discussion argument

friendly fire when your side is shooting fellow soldiers

friendly house British euphemism for brothel

front-and-rear-end wipers babies' diapers

frontin' lying

fruits of treason convict being strung up, cut down when nearly strangulated, then disemboweled, beheaded, and the torso cut into four quarters by an ax

full-figured fat; obese

funart funerary art

functions under the auspices of takes orders from; is sponsored by

funding creativity using taxes to support the arts

funeral decorations flowers

funeral director undertaker

fusilated executed by a firing squad

future points in time the future; when

future unpleasantness anticipated war

F-word fuck

gabinetto (Italian—cabinet)—toilet

Gallic disease syphilis

garbologist garbage collector

gathered gathered unto death, died

gathered to his fathers died

gay bashing physical and verbal assaults on homosexuals

gay bit prostitute

gay house brothel (Victorian)

G_D God

gelding castrated male horse

general paresis of the insane central nervous system syphilis

generously proportioned fat

genteelism euphemism

gent's gentlemen's toilet

genuine simulated imitation

geriatric(s) old person(s)

German disease syphilis

gestation control abortion

get off your butt get off your buttocks (originally); get off your ass (currently)—stop loafing and get to work

get rid of kill

getting along in years aging

getting your girl in trouble impregnating your unmarried girlfriend

giblets edible entrails of fowl

GI bride foreign-born wife of an American soldier

gifted children intelligent and studious children

girl's room women's toilet

given the pink slip fired

given special treatment exterminated (in the vocabulary of Nazis as well as Soviet commissars)

give up the ghost die

give us the benefit of your present thinking what do you think?

giving informational numbers writing parking tickets

giving the whole picture describing entire situation fully

glow Victorian euphemism for a woman's sweat

goatish dirty-old-man or unwashed-male smell

god's acre cemetery

going out with staying in bed with; dating

going to see a man about a dog going to the toilet

golden gauntlet early retirement

golly God

gone dead

gone on vacation suspended from active duty

gone to a better world died

gone to brush their teeth gone to the toilet

gone to Davy Jones' locker drowned

gone to heaven died

gone to rest died

go(ne) to the bank to make a deposit go(ne) to the toilet

go(ne) to the bathroom go(ne) to the toilet

gone to the great beyond died

gongorisms overly ornate expressions in the style of the Spanish poet Luis de Gongora y Argote

goody-goodyism affected niceness

goo-goo eyes amorous glances

gorilla sexually-aggressive person

gosh God

go to one's reward die

got to be destroyed must be killed

go under die

governmentalese federal bafflegab

government scheme government-sponsored program

grant-in-aid diplomatic euphemism for a handout; give away

grayfish shark meat

gray propaganda mixture of half truths and truths whose source is hidden

greengoods (man) paper money; counterfeiter or passer of counterfeit money

greenhouse greenhouse effect (carbon dioxide release causing atmospheric changes)

greenies amphetamines (stimulants)

grief therapist funeral director skilled in comforting the family and friends of the dead

grip money

group sex orgy

ground elements soldiers

guest house boarding house

guest of the governor inmate in a state penitentiary

guest-service employee bellhop

gunk aerosols, glues, and solvents inhaled by would-be addicts

gustatorial distinctions bitter, salty, sour, or sweet

had an accident defecated or urinated in bed or clothes

hairologist barber; hair stylist; hairdresser

hair stylist barber

halitosis bad breath

hamburger ground meat

handicapped crippled or retarded people

handout charity patient

hane heinous

happy hour time for drinking liquor (usually at discount)

hardbody targeted missile hidden from infrared detection by huge plume of gas and heat

hardcore to the max sexually explicit to the maximum

hard drugs addictive substances causing psychological and/or physical harm

hard liquor beverage high in alcoholic content

harmonize it musician's phrase meaning fake it

harvest animals kill animals

harvesting marine mammals killing dolphins, dugongs, manatees, polar bears, porpoises, sea lions, seals, walruses, and whales

harvest trees cut trees down

harvest worker fruit-and-vegetable picker

have to sharpen my skates have to go to the toilet

haystack haycock

head the head (the toilet)—originally the head of a ship where human wastes dropped over the vessel's bow from an open or partially enclosed toilet bench

health alteration assassination

hearing problem deafness

heated argument fist fight

heli-hiking helicopter transport to remote areas where one wants to hike

hemped hanged

hempen collar hangman's noose

hempen cravat hangman's noose

hemp stretcher hangman

here (Afrikaans—gentlemen)—men's toilet

heren (Dutch—gentlemen)—men's toilet

Her (His) Majesty's carriage prison van

heroic measures life-sustaining medical procedures

herrar (Swedish—gentlernen)—men's toilet

Herren (German—gentlemen)—men's toilet

herrer (Danish—gentlemen)—men's toilet

he stopped eating he died (French)

high intoxicated on alcohol or other drug

high achiever(s) good student(s)

high coefficient of slip slippery

highly scenario-dependent success of a plan depends on whatever develops

high negatives aspects that detract from a candidate's popularity

high positives aspects that enhance a candidate's popularity

high-risk activities life-threatening practices

high-risk behavior life-threatening practice

hirondelle de nuit (French—night swallow)—prostitute

hirsute adornments hairy appendages such as beards, mustaches, long tresses, sideburns (also called louseladders); unisexual hairdos

his crown was shorn he was beheaded

Hispanic time slower than standard time; also known as mañana time or Mexican time

histrionic art acting

ho prostitute

holiday headache alcohol-induced ailment

Holy Toledo Holy Jesus or Holy Moses

hombres (Spanish—men)—men's toilet

homes (Catalan—gentlemen)—men's toilet

hommes (French—gentlemen)—men's toilet

homogenized adulterated or watered

honey dip attractive woman

honorarium fee; payment for services

hooker prostitute

hookshop brothel

hoopty car

horizontally challenged fat

hospital environmental services janitorial work

hot blooded lustful

house guest boy friend; girlfriend; live-in lover

household technician cook or other domestic servant

house of all nations brothel offering women of every color and nationality

house of assignation brothel wherein rooms were assigned to prostitutes and their clients

house of confusion brothel

house of correction criminal reformatory; reform school

house of ill repute brothel

house proud clean and tidy

housitosis household halitosis (unpleasant odors)

Hudson seal muskrat fur

human assets Soviet citizens who worked undercover for the U.S.

humane shelter dog pound; place for stray cats, dogs, and other creatures

human-factor engineering designing equipment for maximum user comfort; ergonomics

human resources people

human-rights abuse murder

human-rights violation murder

hustler male prostitute

hygienic paper toilet paper

hype hyperbole (gross exaggeration)

hypersexualist oversexed person

iatrogenic disorders ailments inadvertently induced by a physician

I'd appreciate it very much Do not expect me to pay you or to reciprocate

identity-achieved person adult

I have a little problem with what you did I dislike what you did

I hear you I understand what you are saying

illegitimate bastard; child(ren) born of unwed parents

illicit love adultery

I'll let you go now shut up and get off the telephone

imbibed too much drank too much

immolation a fiery death symbolizing extreme sacrifice (*see* self-immolation)

immurement burial

I'm not crazy about it I don't like it

impaired person person suffering from Alzheimer's, Parkinson's, Pick's, or pre-senile dementia involving loss of memory and even everyday words; drunk(s)

impecunious broke; penniless; without funds

impure extraneous matter scum

in an interesting condition pregnant

incarcerate imprison; jail

inclusionary building low-cost housing included in building development projects funded in part or wholly by federal, state, or local funds

incontinence inability to control the bladder

increase your equity increase your indebtedness in the hope of profit

incrustation build-up of crud (filth, grease, refuse of any kind)

incursion hostile entrance into a territory; invasion; raid

indisposed recovering from an alcoholic hangover; suffering menstrual-cycle pains; unavailable

indisposed at the moment on the toilet

individual confinement solitary imprisonment

individualized learning center student's classroom desk

industrial action sit-in; slow down; strike

industrial consultant(s) lobbyist(s)

industrial disharmony strike

industrial espionage stealing company formulas or secrets pertaining to manufacture or sale

industrialist(s) successful speculator(s)

industrial tourism vacationing at a nuclear-power plant to soak in water that has cooled it (for pain relief)

inebriated drunk

inflation depression in the purchasing power of your currency; cost increase

in-filling building houses

informant a snitch

information processing duplicating and typing pool

information scientist librarian

information specialist librarian

infrastructure fundamental framework of an organization or system; permanent facilities such as airports, bridges, gas and electric lines, highways and streets, railways, sewers, tv cables, water mains, etc.

inheritance tax   death duty

in long supply   plentiful

inmate(s)   prisoner(s)

inner city   economically deprived residential areas formerly called barrios, ghettos, or slums; also called the innercore city

inoperate   broken down

inoperative statement   a lie

in pocket   to hold drugs

in point of fact   in fact

in reasonable supply   available

in short supply   scarce

insinuendo   insinuated

instantaneous respiratory arrest   death

institutionalize   shut up in an insane asylum

institutional superintendent(s)   prison warden(s)

intact   virginal

intellectually underprivileged   uninformed

intelligence acquisition   spying

intensive care unit   locked unit for juvenile delinquents

inter   bury

interactive discrimination   system transmitting neutral particle beams to distinguish decoys from warheads emitting gamma rays

interfacing   talking

intergenerational communication   talk between young and old

interment   burial

intermodal interface   when you get off the plane, ship, or train a bus awaits

internal exile   Soviet term meaning exile in Siberia or some other remote area

international disposal   hired killer working for a government espionage agency

internationalization of norms   conscience

interpretive dancer(s)   striptease dancer(s)

interrogation equipment   devices and drugs used to obtain confessions

intestinal distress   cramps and/or gas and/or loose bowels

intestinal fortitude   guts; stomach

in the can   in the toilet

in the education field   teacher

in the family way   pregnant

in the time period   when

intimacy   adultery; sexual intercourse

intimate relations   sexual intercourse

intimate wear   underwear

intoxicated   drunk; under the influence of drugs

intoxication   drunkenness

in trouble   about to be arrested; pregnant

intrusion detector   burglar alarm

intrusion device   burglar's tool

investment consultants   stock brokers

investor-owned hospitals   hospitals run for profit

invisible handicap   deaf and/or dumb

invisible risk   deafness-producing noise generated by aircraft, traffic, loud music, and other sounds

involuntary audience   captive audience

involuntary unemployment   layoff

Iranamok   (*see* Contragate)

Iranscam   (*see* Contragate)

ironmongery department   Her (His) Majesty's prison

irritable colon   spastic constipation

isolation booth(s)   prison cell(s)

it girl   woman with sex appeal

it has long been known   take my word for it

it is believed   I think

it is generally believed   two other guys agree with me

it is thought   I think

I wobbly wobbly   Chinese immigrants' name for the IWW (Industrial Workers of the World) in 1905

jag   a loner

jagger   tattoo artist

jakes   intoxicants (British)

jam   party

jaundiced eyes   hateful eyes; prejudiced eyes

jelly beans   amphetamine

j.o.   job

job action   sit-down or strike

job problem   demoted or fired

john   prostitute's male client; toilet

joint   brothel; prison

joints   popular brands of sneakers

journey's end   death

Judas priest   Jesus Christ!

juice   influence

juice joints   beer bars, cocktail lounges, saloons, etc.

jump boys   young Filipino street sellers of cigarettes

jumpstart   perform cardiopulmonary resuscitation

junglish   jungle English, bafflegab

junior executive(s)   clerk(s)

kaleidoscopic career   checquered career

Katzenjammer   (German— alcoholic hangover headache; tuning-up sounds produced by a symphony orchestra)

kept in custody   jailed

kept woman   mistress

keyboard   type; input on a computer

key money   bribery (to get an apartment)

kick the bucket   die

kick over the traces   cast off all restraint

kick upstairs   promote to a higher but less sought-after position

kick up your heels   have a good time

kleptomaniac   shoplifter

Klosett   *das Klosett* (German— the closet)—the water closet or toilet

knocked up   pregnant

kvinner   (Norwegian— women)—women's toilet

labor unrest   picket lines; sabotage; sit-downs; slowdowns; wide-scale discontent resulting from layoffs, poor wages, and poor working conditions

lab rats   focus group

lacerations   cuts

lack of suppression of paradoxical sleep   unsuppressed drowsiness due to bad side effects of some pills

ladies   ladies in waiting; old women; older women; most women

ladies'   ladies' toilet

ladies' lunch lament   the man shortage (all the good men are either gay or married)

ladies' man   womanizer; gigolo

ladies of the night   prostitutes

ladies' room   women's toilet

lady dog   bitch

lady of the evening   prostitute

laid back   easy-going; relaxed

laid off fired

laid to rest buried

lampin' relaxing

la langue du Coca-Cola (French—the Coca-Cola language)—English

land developer land despoiler; builder

landfill garbage dump

landscape architect gardener

landscape conversion draining wetlands, plowing a prairie or clear-cutting a forest for construction

landscaping specialist gardener

lane bowling alley

language arts reading, speaking, and writing

lapin rabbit fur (French *lapin* rabbit)

lapsus calami (Latin—slip of the pen)—error exposed and reduced to writing

lapsus linguae (Latin—slip of the tongue)—saying what you meant to say but didn't want to say, Freudian slip

lapsus memoriae (Latin—lapse of memory)—in acute instances this is called Alzheimer's disease or pre-senile dementia

large maritime targets oil tankers

large naval targets aircraft carriers

last curtain call death

last debt death

last mile walk from the prison cell to the place of execution

last obsequies funeral

last roundup death

last sleep death

last taboo incest

last waltz condemned prisoner's march to the place of execution

late lamented remains corpse

late unpleasantness American Revolution; Civil War or War of the Secession

latrine outdoor toilet

laugh lines facial-wrinkles between the nose and mouth

launched into eternity died

lav lavatory (toilet)—common British euphemism

lavabo (Spanish—washroom)—toilet

Lavabos de Homens (Portuguese—gentlemen's lavatory)—men's toilet

Lavabos de Senhoras Portuguese—ladies' lavatory)—women's toilet

lavatorium toilet

lawmen detectives, investigators, law-enforcement officers, police, secret service, sheriffs, or other security forces

lazy colon constipation

learning facilitator teacher

left the scene died

legal problems drunken-driving arrests

legislative work lobbying

less affluent people people lacking financial resources

less than able unable

less than accurate inaccurate

less than adequate inadequate

less than appetizing repulsive; unappetizing

less than artistic inartistic; ugly

less than attractive unattractive

less than bearable unbearable

less than beautiful plain or ugly

less than believable improbable; incredible; unbelievable

less than broadminded bigoted; narrowminded

less than candid crafty; disingenuous; misleading; scheming; untruthful

less than charismatic uncharismatic; without charm or the look of leadership

less than charming disagreeable

less than convincing unconvincing

less than decisive undecisive

less than delicious burned; half baked; nauseating; sour; unwholesome

less than diligent lazy, sloppy, careless

less than effective ineffective

less than elegant inelegant; shoddy

less than enthusiastic unenthusiastic

less than exciting dull as dish water; unexciting

less than fastidious dirty; sloppy

less than gentlemanly ungentlemanly, impolite

less than honest crooked; dishonest

less than honorable (discharge) dishonorable (discharge)

less than hospitable inhospitable; standoffish; unfriendly

less than ideal awful; terrible; unbearable

less than informative cryptic; obscure; uninformative

less than intelligent dumb; stupid; unintelligent

less than interesting dull; uninteresting

less than knowledgeable uninformed; unknowledgeable

less than ladylike unladylike

less than lawful illegal

less than legal illegal

less than lovable hateful; unlovable

less than navigable impassible

less than palatable inedible; repulsive

less than perfect short of ideal

less than quenched unquenched, thirsty

less than receptive unreceptive

less than reliable unreliable

less than sensitive cool, brusque

less than sheltered unsheltered, open

less than strong weak, ineffective

less than sublime shabby

less than a success a failure

less than supportive uncooperative

less than sympathetic hostile; uncomprehending; unsympathetic

less than thrilled depressed; disappointed; uninterested

less than total enthusiasm lukewarm, cool, indifferent, halfhearted

less than totally cooperative reluctant

less than truthful false; lying; untruthful

let go fired or laid off

lethal assistants deadly weapons

liaison illicit sexual connection

libation alcoholic drink

liberal spender spendthrift

liberty cabbage sauerkraut
lieu lieux d'aisance (French—comfort room)—toilet (or in British slang, *loo*)
lift attendant elevator operator
light-fingered gentry pickpockets; thieves
likvidatsiya (Russian—liquidation)—execution
linguistic fig leaves euphemisms
liquidate kill
liquidation assassination
liquidation of undesirable elements murder
liquid refreshment alcoholic drink, beverage
literary agent author's representative
literary fig leaves euphemisms
little boy's room men's toilet
little girl's room women's toilet
live-in live-in friend, unmarried lover
live-in companion a concubine, gigolo, kept man or woman, mistress, paramour
loaded under the influence of a drug
load factor percent of available aircraft seats occupied by paying passengers
loan expert pawnbroker
locus of evaluation classroom
lonely couch companionless bed; half-empty bed; solitary bed
loo toilet (British)
lost at sea presumed drowned
lost her husband her husband died; she was widowed
lost his wife his wife died
lounge toilet
love child child born of unwed parents
loved one corpse
love handles flesh over hips
low achiever poor student
low economic background poor
lower ability group slow learners
lower the boom fire
low frustration tolerance easily irritated
low-income poor
low-income neighborhood barrio; ghetto; slum
low priced cheap

lubritorium, automotive service station
lues cholera, the plague, syphilis
luetic infection of the central nervous system central nervous system syphilis'
lung affliction tuberculosis
L-word liberal
machinery drug-injection paraphernalia
mackin' well dressed
mad good
madam the operator or owner of a brothel
madamas (Italian—ladies)—women's toilet
mad dogging staring at someone until you get a reaction
made redundant laid off
maiden lady old maid
make one's exit die
make-out artist(s) fornicator(s)
make redundant fire someone
make water urinate
maladaptive behavior unacceptable or criminal acts
maladjusted youths backward or troublesome youngsters
male organ penis
Malta dog diarrhea.
mañana time (*see* Hispanic time)
Marble City the cemetery
marginalized excluded (usually, poor)
marketing analysis sales promotion
marketing engineer sales person
marketing manager sales manager
marketing representative salesperson
marquee tent
mass vigilance widespread police-informant system
matured waste garbage
measurable end products results
meat-eater's euphemisms beef—cow or steer meat; cabrito—young goat meat; lamb or mutton—sheep meat, pork—pig meat; veal—calf meat
media access center the library

media coordinator someone who operates a cassette player-recorder, motion-picture camera or projector, tv equipment, etc.
medical center hospital
medical examiner coroner
meditation solitary confinement
meet your maker die
megacity metropolis with a million or more people
mellowspeak soft-spoken euphemisms
member of the lower socioeconomic bracket poor person
member(s) of a minority group nonwhite(s)
memorial park cemetery
mendacious tendency for lying
mendacity lie
menn (Norwegian—men)—men's toilet
men's men's toilet
men's restroom men's toilet
mental institution insane asylum; psychiatric hospital
mentally disturbed crazy; psychotic
merchandise shrinkage shoplifting
mercurial temperament easily enraged
meretricious traffic prostitution
méski (Polish—gentlemen)—men's toilet
messieurs (French-gentlemen)-men's toilet
met in an unfriendly manner collision of aircraft, automobiles, boats or other vehicles
met one's maker died
metropolitan rehabilitation center downtown penitentiary or federal prison
Mexican brown marijuana
Mexicancellation Mexican divorce
miced all crytee Christ Almighty
mild irregularity constipation
mineral extraction mining
minimal prescriptiveness with fewest number of restraining rules blocking monetary grants to interest groups or states
minor misstatement lie
misadventure shootout resulting in death
misalliance mismatch

misappropriate funds steal

miscarriage spontaneous abortion

misconduct adultery

misspeak inadvertently mislead; blunder

mixed bag assortment

mixer(s) bartender(s)

mná (Gaelic—women)—women's toilet

mobile home house trailer

moderately priced cheap

modified limited hangout temporary office or headquarters

mom duke mother

money laundering concealing source of funds

monthly pains menstrual pains, cramps

monument memorial tombstone

moonlighting getting paid for a second job

mopping up military operation involving capturing or killing of enemy personnel

moral renovation the honest way; truthfully

moral turpitude depravity; immorality

morbus gallicus (Latin—French disease)—syphilis

the more you buy the more you save spend more

mortical surgeon undertaker

mortician burier; funeral director; undertaker

most effective painkiller death

motion discomfort motion sickness; nausea; seasickness

motion discomfort container vomit bag

motivating through fear coercion

moto motivated

motor vehicle operator chauffeur; driver

mountain herring whitefish

mountain oysters bull, hog, or sheep testicles used as food and famed for their imagined aphrodisiac powers

mujeres (Spanish—women)—women's toilet

multifunction wrist instrument wristwatch

multiprisoner transportation unit black lady; black maria; paddy wagon; police patrol van

multiversities large universities

mumblespeak mumbled bafflegab

muzhay (Bulgarian—men)—men's toilet

muzi (Czechoslovakian—men)—men's toilet

my back teeth are almost floating I must urinate

my peeps parents

my statement is inoperable I lied

mythomaniac liar

nappies babies' napkins; diapers

narco-military military system protecting narcotics trafficking

nasal discharge snot

nasal problem loss of smell

natal day birthday

national assistance financial aid and surplus food for the poor and the unemployed

national razor the guillotine of France

national test bed high-tech computer and video operation simulating space-battle scenarios

natural euthanasia death by starvation

naturally fed breast-fed

nature break go to the toilet

navigational errors bombing friendly territory

Neapolitan pox syphilis (*see* French disease)

necklacing setting an automobile tire ablaze around a prisoner's neck

neck oil whiskey

necktie hangman's noose

necktie party lynching; lynch mob

necropsy autopsy; post mortem

negative inducements in terms of price fees

negative patient outcome death

negative savers people who spend more than they earn

neighborhood of color minority residents in an American urban area

neighborhood revitalization slum clearance

neither here nor there irrelevant, nowhere, unimportant

neoplasm tumor

nerve agents lethal chemical agents attacking the body's nervous system

nether garments underwear

netherweed marijuana grown in the Netherlands

netherworld underworld; world of the dead

neuter castrate; remove or tie off procreative organs

neutralize assassinate; kill

news specialist(s) reporter(s)

Nicaraguan eggs iguana lizard eggs dried, pickled, or fried

nice-nellyisms euphemisms

night house brothel

night-soil excreta; feces; human dung

No 1 urine

No 2 excrement

No Code do not resuscitate

nocturnal emissions wet dreams

nök (Hungarian—women)—women's toilet

no longer with us dead

non-ambulatory bedridden; unable to walk

nonch no chance; objectionable

non compos menfis (Latin—not of sound mind)—mentally incompetent

nondiscernible microbionoculator concealed poison-dart gun

nonpermissive environment battlefield

nonterrorists unarmed and uninvolved civilian victims of terrorism

not acceptable forbidden; not tolerated; unlawful

not a happy camper someone in great distress

not all that impressive unimpressive

not an exact statement, to put it mildly a lie, to put it bluntly

not completely accurate inaccurate, false

not completely truthful untruthful, disingenuous

not entirely accurate inaccurate

not entirely safe risky

not entirely satisfied dissatisfied

not entirely truthful untruthful

Notice of Reduction in Force pink slip inserted in the pay envelope to announce layoffs

not my favorite person someone I dislike intensely

not overly bright stupid

not precisely accurate inaccurate; untrue, false

not precisely truthful untruthful, lying

not quite kosher dishonest (in the ethical sense); unclean (in the culinary or physical sense)

not so young elderly; old

not too accurate inaccurate

not too attractive ugly

not too profitable unprofitable

not too tall from average height to short

not too truthful untruthful

nude-encounter parlor brothel; massage parlor

nudie film film featuring nudity

nuptial couch marriage bed

nuptials marriage ceremonies

nut factory insane asylum; mental hospital; psychiatric hospital or ward

nuthouse insane asylum or psychiatric hospital

nutpicker(s) psychiatrist(s)

nutria coypu fur (from a giant South American rodent)

oblique love adultery; clandestine action, secret love

obsequies funeral ceremonies

obstetric paramedic midwife

obstitute part-time prostitute; substitute for a prostitute; whore

occasional interruption acute constipation

odoriferous smelling; stinking

official gibberish bureaucratic bafflegab

offshore production overseas investments in foreign countries

old girl mother

old goat dirty old man; satyr; whorehound; woman chaser; lecher

old hare rabbit meat

olfactoristic smelly

one's spirit departed one died

on the nod sleepy sensation following use of drugs

opportunity school school for the handicapped or retarded

optical problem blindness; failing eyesight

optimize performance improve

organ male organ; penis

organic receptors sense organs

orthographical errancy misspelling

osculation kiss(ing)

osmidrosis foul-smelling sweat

other prisoner(s) prison guard(s)

outdoorsmen vagrants; homeless

outhouse backyard toilet; country privy

out in left field disoriented; out of touch with reality

outlaw party illegal nightclub

out of the ballpark out of touch; unaware of reality

out of town in prison

outplaced fired (from a job)

outright fabrication lie

outsourcing hiring cheaper labor

out to lunch daffy

overachiever embarrassingly brilliant student; successful person

overindulgence gluttony

overkill possessing more than enough weapons needed to destroy enemy targets

over the blue wall confined to a hospital for the criminally insane

over the river dead

over the side buried at sea

overwrought drunken; agitated

oxymoron(s) combination of contradictory or incongruous words; seeming contradiction(s) e.g., peacekeeper missile(s)

pacification elimination of disturbances or disturbers of the peace; peace at the point of a bayonet

package store liquor store

paid the debt of nature died

pain-compliance device nunchaku

pain in the butt pain in the ass

Panamá chicken chicken-flavored iguana lizard tail

panel house whorehouse (whose bedrooms had sliding panels close to the bedstead in easy reach of an accomplice who could snatch a man's wallet or watch while he was preoccupied)

paramnesia *déjà vu*

paraphernalia papers, pipes, scales, and spoons used by drug addicts

park under construction dump

parsimonious frugal; niggardly; pennypinching; stingy

pass away die

passed over died

pass into the unknown die

passive euthanasia letting the patient die

passive resistance response deliberate procrastination

pass on die

pass water empty the bladder; urinate

paying guest boarder; lodger

peace dividend money from defense-spending reductions

peaceful person lazy or lethargic person

Peacekeeper LGM 118 intercontinental ballistic missile

Peacemaker MX missile's

peace pill pcp (phencyclidine)

peaches amphetamine

pears amyl nitrite

peculator thief

peculiar institution slavery in the United States

peculiar members lexicographer Noah Webster's euphemism for *testicles*

peculiar service espionage

pecuniary distress poverty

pedophiliac adult molester attacking children sexually

peeping tom voyeur

pejorative demeaning or derogatory

pentagonese Pentagon bafflegab

people expressways sidewalks

people mover light-rail rapid transit system

people of color blacks, Asians, etc.; those of the non-white minority in the United States

people's republic political prefix in many totalitarian nations run by communists, not by the people

people with conspicuous access to wealth in their own right rich people

perform one's ablutions shampoo, shave, and shower; wash oneself

period menstrual period

period of reconsideration cooling-off period

permanent color hair dye

perp perpetrator (of a crime)

perpetrator criminal; offender; suspect

personal emergency must go to the toilet immediately

personal protection dog attack dog

personal relationship sexual relationship

perspiration sweat

perspire sweat

pertussis whooping cough

Peruvian perfume cocaine

pet snatcher dogcatcher

phantom proliferators African, Asian, and South American would-be atomic powers

phased out brought to a halt; cancelled; closed; concluded; discontinued; ended; finished; retired; stopped; terminated

phased withdrawal scheduled retreat

phat good

Philippine Pacification Philippine Subjugation, according to Filipino historians

physically challenged disabled

picture car vehicle that appears in motion picture

pigeon drop money scam

pimpmobile automobile converted for use by pimps and prostitutes

pink slip a dismissal notice

pixillated alcoholically befuddled

planned withdrawal defeat; retreat

plaque bacteria-laden mucus-hardening film endangering one's teeth

plastic master person climbing, mainly in indoor gyms

plastic money credit cards

platform pushers persons who push others from the platform to the tracks of a railway or subway when the train is coming into the station

pleasingly plump fat

plight one's troth pledge marriage; promise to marry

ploy clever maneuver

plump fat

plunger hypodermic syringe

plural marriage polygamy

plural relations South African euphemism for *apartheid*

podium a platform designed to elevate a conductor or a speaker above the ensemble or the audience; a lectern

point in time moment; time

police state law and order imposed by law-enforcement agencies who overlook civil liberties

political unrest assassination; bombing; martial law; mass demonstrations; riots; sabotage; terrorist attacks; wide-scale subversion often leading to revolution

polychromatic alphasite colored chalk

polygraph test(ing) lie-detection test(ing)

poorly compensated poorly paid

poor man's nuclear weapon chemical weaponry; fuel–air bomb

pop duke father

popular price cheap

population-control equipment riot-control equipment such as high-pressure water hoses, rubber bullets, and tear gas

population limitation birth control

pop-up fast launching of a missile carrying a nuclear weapon to generate an X-ray laser able to shoot from space

portable john portable toilet

posey vest straitjacket

position situation job; job task

posterior backside; behind

powder her nose a woman going to the toilet

powder room toilet

power outage power failure

precursor bursts nuclear explosions in space designed to foil defense systems by creating a background of nuclear emissions, magnetic pulses, and heat capable of fooling sensors

predacide chemical agent for killing predatory animals

preemptive strike bombing the enemy before being attacked

pregnancy termination abortion

prelighten bleach the hair

prematurely retired fired

preorgasmic females frigid females

pre-owned previously owned; secondhand; used

preparation room undertaker's embalming room

press the flesh shake hands

preventive detention jail without bail to assure defendants appear for trial and commit no crimes in the meantime

pretext calling lying to get information

preventive war sneak attack

previously owned auto secondhand automobile; used car

prices you can afford prices salesmen tell you you can afford

pricey expensive

prioritize arrange in chronological order; choose; select

prison officer gaoler (UK); jailer (US)

private dick private investigator

private enterprise business

privates private parts (sexual and urinary organs)

privy outdoor toilet

problem child(ren) juvenile delinquent(s)

problem drinker alcoholic

problem pregnancy clinic abortion center

problem skin marred by acne, eczema, pimples, psoriasis, syphilitic scars, or their combination

pro-choice in favor of a woman's right to choose when to bear children

proliferation the spread of nuclear weapons

property control officer watchman

prophylactic device for preventing venereal infection, condom; a contraceptive device

proprietary institutions institutions, such as clinics and hospitals, run for profit

props respect

prosecutorial agent attorney general; city attorney; county attorney; district attorney; federal attorney; magistrate; public prosecutor; state prosecutor

prossie prostitute

protect cover up

protective bombing bombing the enemy first

protective coating engineer house painter

protective reaction bombing raids

protective reaction strike attack

protective residence fallout shelter

provisional period probation

prurience lust

pseudologist liar

psychiatric center insane asylum

psychic reader fortune teller

public convenience outdoor or public toilet (British)

public house a bar where drinks and food are sold

public property property controlled by politicians

public relations adviser press agent

public relations assistant office receptionist

public utilities contraception and family-planning clinics (Dutch)

puffer auction booster

pulling the plug euthanasia; mercy killing; turning off life support systems

pupil station desk

purloin steal

purple passage flowery speech or writing

purveyors of meat butchers

pushing up daisies dead

put one down bring about death, as in the case of a pet

put to sleep kill

put out of misery kill

quadruplication of the quadruplicate number 16

quality-control engineer inspector

quarter 25 dollars; 25-dollar narcotics-filled balloon

quean (Middle English—prostitute)

queen-size bigger than large but smaller than king-size

quiet room solitary-confinement cell

quietus death

quite a few many, several

racial hygiene eugenics

racial unrest race riots

raffs raffish bohemians

Rainier Factor slang term for an employee's acceptance of lower pay in exchange for living near a scenic locale (refers to Mount Rainier in Washington state)

'rak attack also called rack or raki

Rangoon runs diarrhea

rap buddies easy-to-talk-to people

rap parlor place of prostitution

ratepayer taxpayer

rather elderly old

rational defense planned warfare

real estate open space

realistic fees low fees

rear end ass; buttocks

rear yard environment backyard

receipts strengthening raising taxes

recession depression

recommend the matter be studied action to move controversial issues to the back burner

reconciliation room church confessional

reconditioned secondhand

reconnaissance in force search out and destroy the enemy

rectification of the front retreat

redeployment retreat; withdrawal of armed forces

reduced circumstances poor

reduced to ashes cremated

reduction in the work force firing; layoff

redundancies layoffs

re-education camps forced-labor work camps

reemployable annuitant specialist called back from retirement

reflation ascending inflation

registered warrant government-issued i-o-u issued when money is not available to pay for goods or services

regurgitation upchucking; vomiting

rehabilitation center jail; lockup; penitentiary; prison

rehabilitation laboratory modern prison

relationship sexual affair

relocation specialist realtor

remains cadaver; corpse; dead person

remand institution jail; penitentiary; prison

remediation remedial reading

remittance person living on monies sent from home; ticket-of-leave man

remove exterminate

rent boys British teenage male prostitutes

repose beneath the sod die

repossessed taken back by creditor

residuals leftover funds, monies, or profits

resource center library

resources control bombing of dams, defoliation of forests, poisoning of sources of drinking water during the course of a war

rest home convalescent or retirement home

resting in marbletown buried in the cemetery

restroom toilet

resurrectionist grave robber

retire for the night go to bed

retirement allowance pension

retirement home institution catering to retired people

retiring room toilet

retreat toilet

retrete (Spanish—retreat) toilet

returnee defector

revenue enhancement higher tax; sales; tax increase

reverse engineering industrial espionage achieved by taking apart the products of competitors to find out how they work and how they may be imitated

revised upward increased

revolverize to shoot someone

rhumba steam sweat

riffed reduced in force (dismissed from duty; fired)

right decision decision favored by the public despite massive opposition by bureaucrats and lobbyists

right in the ballpark feasible

rightsizing laying off workers

rip off cheat; overprice; rob; steal

rodent operators rat catchers

'roid rage violence brought on by use of steroids

rolling through the rye drinking and driving

roll in the hay copulation

room attendant maid

roommates unmarried lovers

rotund fat

rubbed out assassinated

rubber mirror computerized thin-glass mirror on honey comb panel controlled by mechanical arms and microchips enabling them to compensate for the distorting effects of the earth's atmosphere on laser beams

ruddy-faced drunk

run around the corner go to the toilet

sacrament of penance the rite of reconciliation (Roman Catholic)

sailor's curse constipation

salami amputation slices of a limb are removed until an uninfected area is reached

Salisbury steak World War I euphemism for *hamburger*

saliva spit

salvage assassinate; kill, murder

sanitation engineer garbage collector

sanitized made acceptable

sauce sanctuaries beer bars; cocktail lounges, saloons

sauna massage brothel (Danish)

scam non-violent confidence game involving pilferage, robbing, stealing

search and clear search and destroy

seasickness dizziness followed by nausea and vomiting

seclusion solitary confinement

secret disease syphilis

Section 8 discharged from the armed forces because of insanity or intoxication; government-subsidized decent, safe, and sanitary housing

secular humanism agnostic or atheistic humanism

secular humanist person believing in doing good outside the bonds of religious teachings; freethinkers, skeptics

secularist agnostic; atheist; disbeliever; doubter; humanist; iconoclast; nonbeliever; rationalist; skeptic; truth seeker; unbeliever

security host security guard

security zone occupied territory

see a man about a dog go to the toilet

selected out dismissed from duty; fired

selective use of violence to neutralize an intended victim assassination plan

self-ablution masturbation

self-immolation sacrifice attained by setting oneself on fire or plunging into a fire

self-liberation suicide

senior(s) senior citizen(s)—older person(s)

señoras (Spanish—ladies)—women's toilet

señores (Spanish—gentlemen)—men's toilet

senyores (Catalan—ladies)—women's toilet

senyors (Catalan—gentlemen)—men's toilet

separate from school expel

separate from the payroll fire

sequencing quitting the workplace to return later

serious headache gunshot wound in the head

service observing electronic eavesdropping on employees by management

servicio (Spanish—service)—toilet

servicios de caballeros (Spanish—gentlemen's services)—men's toilet

servicios de damas (Spanish—ladies' services)—women's toilet

sexual expressionist pornographer; purveyor of smut

sexual gradualism masturbation

sexual variety promiscuity

shading the truth lying

sharing group sex or orgy

sharpen one's skates go to the toilet

sharps needles, scalpels, pointed objects; glass pipettes, slide pipettes

she dog bitch

sheer prevarication outright lie

shelter seeker refugee

shero female hero

sherutim legvarim (Hebrew—men's services)—men's toilet

sheradin lenashim (Hebrew—women's services)—women's toilet

she's a plain girl but she means well she's ugly but not malicious

shoe rebuilder shoe repairman

shorts in a bundle upset

short term 180 days or less; an IOU due in 6 months or less

shredding party destroying documents

shuffle off this mortal coil die

sick good

significant other lover

signore (Italian—ladies)

signori (Italian—gentlemen)—men's toilets

sin against chastity birth control

single unmarried

sink of iniquity brothel

sips beer, liquor, wine

skeezer prostitute

skidlid motorcycle helmet

skillful inveracity artful lie

skimming secretly diverting unreported profits

skin doctor dermatologist, syphilologist

skin problems acne; blackheads; pimples; syphilitic sores

skin trade peep shows, sexually explicit literature, and X rated videos

skinz attractive woman

slatternly, slovenly, sluttish, or untidy individual slob

sleep engineer(s) mattress maker(s)

sleepingly warm recently dead

sleeping partner of life death

sleep with copulate

slender skinny

slip out of life die

slippery slope difficult-to-defend argument

sloshy drunk

slumbermobile hearse

slumber robe funeral shroud

slumber room area in a funeral establishment where the embalmed may be viewed

sly language of evasion euphemisms

small untruth little lie

smart rocks small kinetic-energy projectiles designed to be hurled at missiles or war heads

smartsizing laying off workers to increase profits

smilers and defilers, reekers and leakers Ambrose Bierce's euphemistic description of dogs

smoking gun observable evidence

snuff spoon tobacco snuff spoon; cocaine sniffing spoon; also coke spoon

social ailments venereal diseases

social capital shared values

social diseases AIDS (Acquired Immune Deficiency Syndrome), chancroid, gonorrhea, granuloma inguinale, lymphogranuloma venereum, and syphilis

socially disadvantaged poor

socially underprivileged poor

social maladjustment crime

social promotion advancing poor pupils so they will not suffer the embarrassment of sitting in classrooms with brighter and younger students

social prophylaxis Soviet euphemism for execution or imprisonment of dissidents or suspects

social safety net Medicare, Social Security, and unemployment compensation

social unrest demonstrations, hunger marches, hunger strikes, race riots, terrorism

sociobiology genetics

soft money currency of third-party barter; money available for only a short time; unrestricted political contributions

soiled linen dirty clothes

somewhat advanced in years old(er)

somewhat less than credible incredible; unbelievable

somewhat less than honest dishonest

somewhat less than reliable unreliable

somewhat less than truthful untruthful

somewhat unlettered rather ignorant

sonderbehandlung (German—special treatment)—Nazi euphemism for killing by gas

son of Onan masturbator

so perceived seen

souvenir hunters vandals

souvenir hunting vandalism

spa drugstore soda fountain

space mines orbiting explosives designed to threaten satellite defenses

Spanish disease syphilis, according to the Portuguese

speaks a low level of colloquialisms talks slang

special action murder

special agent(s) FBI or insurance agent(s)

specialized vehicle army tank

special psychiatric hospitals Soviet term for prisons where dissenters were kept

special transaction gambling bargain

sperm barrier condom

Spetsnaz Soviet undercover specialists in assassination and behind-the-lines counter insurgency operations

spontaneous abortion miscarriage

sporting house brothel

state chemist executioner charged with poisoning convicts in a gas chamber

state electrician executioner charged with electrocuting convicts in an electric chair

steel-collar labor robots

stepped out for a few minutes gone to the toilet

still warm died recently

stones testicles

stonewall act or argue defensively; conceal; coverup; hide the truth

straightforward and above board blunt, honest

straitened circumstances poor; of limited resources

strategic deterrents weapons of war

strategic hamlet refugee camp

strategic misrepresentation lying; speaking with a forked tongue

strategic redeployment retreat

strategic withdrawal retreat

street hustler(s) male prostitute(s)

street orderlies garbage and trash collectors

street people the homeless

street sweepers large-magazine semi-automatic shotguns

streetwalker prostitute

stretches the truth exaggerates; lies

stretch hemp hang by the neck until dead

stretch the truth lie

submarine below street-level prison cell

subsidy publisher vanity press

substitute worker scab; strike breaker

subversive delinquent political prisoner

subway pushers (*see* platform pushers)

succulent viands appetizing food

succumbed to hypertensive cardiovascular disease died of a heart attack

suddenly retired fired without warning

suicycles all-terrain three-wheel vehicles

summer complaint diarrhea; loose bowels

sunshine units nuclear radiation units

supreme measure capital punishment; execution

supreme sacrifice giving one's life; losing one's virginity

surreptitious entry break-in

surveillance spying

swallowed his birth certificate died (French)

swinger promiscuous person

synergistic(al) advance(s) when the whole is said to be worth more than the sum of its parts

system car audio system

take a dump defecate

take a leak urinate

take a powder leave hurriedly; leave suddenly

take leave of abstinence get drunk

takeoff inject narcotics

take one's last sleep die

take the electric cure be electrocuted

take to the tall tules evade arrest; hide out

talent promoter high-class pimp

tavern beer bar; cocktail lounge; saloon

tax reform often means tax increase

tax withholdings earning sharing

technical correction downward skidding of the stock market

technicals roving armed vehicles or group in a vehicle—coined in Somalia

technology transfer assistance giving advice

techspeak technical talk

telegenic tv attractive

temporary marriage rape

temporary work cessation layoff

terminal causing death or occurring at the end of life

terminal cell prison cell designed to kill the inmate(s)

terminal communication death certificate; death notice; order of execution; death warrant

terminal illness fatal illness

terminate tire; remove from job or office

terminate pregnancy have an abortion

termination with extreme prejudice assassination

terminological inexactitude lie

terpsichorean trauma dancer's pain

territorial extermination mass murder of Jews and others in German-occupied countries during World War II

Texas turkey armadillo meat

that point in time then

that time frame then

theater of war battlefield

The Company the CIA

therapeutic accident drug-induced death due to an overdose

therapeutic correctional community prison

therapeutic misadventure operation resulting in the patient's death

therapeutic termination abortion

thermalicide killing with nuclear weapons

these points in time now

The Troubles longtime fighting in Northern Ireland

thick good

thinking the unthinkable contemplating nuclear warfare

think it ain't I agree

thinning out killing wildlife or trees

this is a serious issue a sin

this job offers great experience the pay is lousy and working conditions are the worst

this point in time now

those points in time then

threat dog dog trained to bark and snarl at intruders

thundermug chamberpot; pottie

time to retire time to go to bed

tinkle urinate

tissue toilet paper

tocador (Spanish—boudoir; dressing room; dressing table; vanity)—often means toilet

toilet tissue toilet paper

toilet water perfume-scented cologne

tossed up the cookies vomited

total and complete immobilization assassination

totalitarian state tyranny

touched insane

trade of kings war, as defined by John Dryden

traffic revision detour

trainsickness nausea followed by vomiting; motion sickness

transient street person; vagrant

transient population vagrants

traumatic amputation of the lower extremities loss of the legs in an accident

tree surgeon tree trimmer

trendy chic

trial marriage free love

tribal scarification circumcision

trick prostitute's customer

triplication of the triple number 9

trots the trots—botulism, diarrhea, dysentery

truth stretcher liar

tube steak frankfurter; hotdog

turf accountant bookmaker

turistas (Portuguese or Spanish—tourists)—traveller's diarrhea caused by toxic-producing bacteria found in contaminated food or water;

if prolonged, a protozoan invasion may occur and produce amebic dysentery

turn on sell narcotics

turn one's face to the wall die

twenty toes ten toes up and ten toes down (British euphemism for a bedroom diversion of some antiquity)

two pc two-physician certification (of insanity)

ubornaya (Russian—adornment place)—toilet

umbrage to take offense

unaccompanied officer personnel housing bachelor officers quarters

uncharismatic without ability to command or lead

uncivil rude

unclothed bare; naked

underachiever unsuccessful person

underarm damp sweat

underdeveloped countries poor countries

underdeveloped nation(s) poor nation(s)

underfunded broke; in debt; lacking money

undergoing emotional retraining under psychiatric treatment

under par feels or looks awful; looks like death warmed over

underprivileged poor; unemployed

under the auspices of sponsored by

under the influence drunk, high, inebriated

under the sod dead and buried

under the weather drunk; sick

underwater rental property valued at less than its mortgage

undesirable books banned or censored books

undeveloped property land in its natural state

undies underwear

undocumented immigrant illegal alien

undocumenteds undocumented aliens

unemancipated minor teenager

unfortunate activity invasion of a country; takeover of a country; warlike preparations

unfortunate female prostitute

unimpressive height short

unit (Texan English) one hundred million dollars

unlawful deprivation of life killing

unlearned ignorant

unlocking shareholder value dismembering a company

unmentionables underwear

unobscene song(s) outspoken songs

unpedigreed alley cat or mutt; one-hundred-percent pure cat or dog

unpeople depopulate

unpredictably nonplussed have nothing to say

unprosperous poor

unrenewed discontinued

unsavory scandal morally offensive disgrace

unshreddables electronic documents

unslim fat; heavy; overweight

unstructured conversations friendly chats; informal talks

untoward clinical event bad side effect

untruism falsehood; lie

untruth lie

unworthy servant slave

upchuck vomit

upper amphetamine

upper frontal superstructures a woman's breasts

up the river Sing Sing prison, up the Hudson River from New York City

uptick increase

upward adjustment rise, increase

upward revision increase

utter an untruth lie

uttering inexactitudes telling lies

utter peer rejection failure to get along with family, friends, neighbors, fellow workers

vacationing near Chappaqua Sing Sing prison

validated learning package approved textbook

vashzimmer (Yiddish—washroom)—toilet

velvet prison censorship

verification the process of confirming compliance with a treaty limiting nuclear arms

vertical transportation corps elevator operators

victims of target practice persons executed by a firing squad

viewing the remains visiting a funeral parlor to see the deceased

violated raped

viral marketing junk electronic mail

visceral reaction gut feeling

visiting spouse pill Chinese contraceptive

visually handicapped blind

visually impaired blind

voice informant

volume reduction unit city or town dump

volumetrics number of square feet of a house

voluntary compliance paying taxes

voluntary return voluntary deportation (of illegal aliens)

voluntary termination of pregnancy abortion

vulture capitalist venture capitalist

wagging the dog creating a political smokescreen

wake-ups amphetamine

walking papers dismissal notice

walking the red road taking the American Indian spiritual path

walla miscellaneous gibberish in motion pictures; nonsynchronized background voices

wallet biopsy examination of a patient's insurance coverage and financial status before admission to a hospital

wanaume (Swahili—men)—men's toilet

wanawake (Swahili—women)—women's toilet

war baby child born of unwed parents during wartime

warm glow hot sweat

wash one's hands go to the toilet

washroom facilities toilets

waste kill

water closet(s) toilet(s)

Watergate English terminology invented by President Nixon's aides to cover up lies and to spar for time when undergoing congressional examination; *at this point in time* meant *now, in point of time* was equivalent to *then* or

when, and *my statement is inoperable* really meant *don't believe a word I say*

water qualm acid indigestion; heartburn; sudden faintness sometimes accompanied by nausea

wc babad (Romanian—male water closet)—men's toilet

wc femei (Romanian—female water closet)—women's toilet

weather activity storm

weed of wisdom hashish; mariuana; other cannabis products

well-filled out fat

well-nigh impossible almost impossible

well placed at the present level of management will never be promoted

went away died

went to sleep died

went west died; was killed

wet affairs Soviet euphemism for *assassinations*

we will no longer detain you you're dismissed; you're fired; you may leave

when I'm gone when I'm dead

where the King goes alone bathroom

white-collar labor office workers

white propaganda identifiable and truthful dissemination of facts

white slave prostitute

wilderness health strip mining

wimp ineffectual weakling; insipid and spineless person

winding cloth burial shroud

winning the West killing the Indians

with balls between parenthesis Shakespeare's description of a bow-legged man

with child pregnant

with facilities cabin (aboard ship) or hotel room with toilet, tub and/or shower, and washstand

without benefit of clergy living together but unmarried

without female friends male homosexual

without male friends female homosexual

wolf to betray or deceive

wolf ticket bad report

womanizer philanderer

Woman's Question women's rights

women's women's toilet

wooden interdental stimulator and particle remover toothpick

word from our sponsor radio or television advertisement

word up the truth

work cessation strike; walkout; work stoppage

work cessation on the premises sit-down strike

working girl(s) prostitute(s)

writing instruments expensive pens including fountain pens and pencils

X-chaser mathematician in search of the value of X; aspiring naval officer

xd deleted

xerotic wrinkled

Xeroxpoxed copier paper marred by black blotches

yellow peril Oriental conquest of Western civilization by economic penetration, immigration, and war

yesterlingo language of the recent past

you know where to go go to hell

you know where to put it shove it up your ass

young adult teenager

your eyes are changing your sight is failing

your services are no longer required you're fired

youthful offender juvenile delinquent

yuppies young urban professionals

zap advertising jargon for alerting a customer by subliminal messages; kill quickly (as with a burst of machine gun or tommy gun bullets)

Zarzuela girl Hispanic playgirl; operetta performer

zeny (Czechoslovakian—women)—women's toilet

zero-defect system foolproof system

zhenny (Bulgarian—women)—women's toilet

zilch nothing, zero

zilches misspellings and typographical errors

zipcuffed caught in your pants by a faulty zipper

zonk stun, stupefy

zonked drugged; exhausted

Bell Code from Bridge or Pilothouse to Engine Room

The bell codes are used on ferries, launches, tugs, and other powered vessels.

1 bell ahead   **3 bells** astern
2 bells stop   **4 bells** full speed

Birthstones—Ancient and Modern

Relative Values. Diamonds, emeralds, rubies, and sapphires are termed precious stones; all the rest are semiprecious. Precious gems are minerals enhanced by the lapidary's art. The pearl, although not a stone, is classed with the gems and, depending on its beauty and size, may be as valuable as any of the precious stones.

	Ancient	*Modern*
January	garnet	garnet
February	amethyst	amethyst
March	jasper	aquamarine or bloodstone
April	sapphire	diamond
May	agate	emerald
June	emerald	alexandrite, moonstone, or pearl
July	onyx	ruby
August	carnelian	peridot or sardonyx
September	chrysolite	sapphire
October	aquamarine	opal or tourmaline
November	topaz	topaz
December	ruby	turquoise or zircon

British Counties Abbreviated

England, Northern Ireland, Scotland, and Wales

Adeen Aberdeenshire
Ang Angus
Angle Anglesey
Ant Antrim
Arg Argyll
Arm Armagh
Ayrs Ayrshire
Banffs Banffshire
Beds Bedfordshire
Ber Berwickshire
Berks Berkshire
Brecon Brecknock
Bucks Buckinghamshire
Bute County Bute
Caern Caernarvonshire
Caith Cathiness
Cambs Cambridgeshire and Isle of Ely
Card Cardiganshire
Carm Carmarthenshire
Ches Cheshire
Clack Clacmannanshire
Corn Cornwall
Cumb Cumberland
Denb Denbighshire
Derbys Derbyshire
Dev Devon
Dor Dorset
Down County Down
Dumf Dumfriesshire
Dunb Dunbarton
Dur County Durham
E Lothian East Lothian

E R Yorks East Riding, Yorkshire
Ess Essex
E Suffolk East Suffolk
E Sussex East Sussex
Ferm Fermanagh
Fife Fifeshire
Flints Flintshire
Glam Glamorgan
Glos Gloucestershire
Great Lon Greater London
Hants Hampshire
Herefs Herefordshire
Herts Hertfordshire
Hunts Huntingdon and Peterborough
Iness Inverness-shire
Kent County Kent
Kinc Kincardineshire
Kinross Kinross-shire
Kircud Kircudbrightshire
Lanarks Lanarkshire
Lancs Lancashire
Leics Leicestershire
Lincs Lincolnshire (Holland, Kesteven, Lindsey)
Lond County Londonderry
Merion Merioneth
Mloth Midlothian
Mon Monmouthshire
Mont Montgomeryshire
Moray County Moray
Nairns Nairnshire
Norf Norfolk

Northants Northamptonshire
Northld Northumberland
Notts Nottinghamshire
N R Yorks North Riding, Yorkshire
Ork Orkney Islands
Oxon Oxfordshire
Peebl Peebleshire
Pemb Pembrokeshire
Perths Perthshire
Rad Radnor
Renf Renfrewshire
Ross Ross and Cromarty
Rox Roxburghshire
Rut Rutland
Selk Selkirkshire
Shet Shetland Islands
Shrops Shropshire
Som Somerset
Staffs Staffordshire
Stir Stirlingshire
Sur Surrey
Suth Sutherland
Tyr Tyrone
Warks Warwickshire
Westrmld Westmorland
Wig Wigtownshire
Wilts Wiltshire
W Lothian West Lothian
Worcs Worcestershire
W R Yorks West Riding, Yorkshire
W Suffolk West Suffolk
W Sussex West Sussex

Canadian Provinces, Territories and Capitals

Alberta Edmonton
British Columbia Victoria
Manitoba Winnipeg
New Brunswick Fredericton
Newfoundland St John's

Northwest Territories Yel-
lowknife
Nova Scotia Halifax
Nunavut Iqaluit
Ontario Toronto

Prince Edward Island Char-
lottetown
Québec Québec
Saskatchewan Regina
Yukon Territory Whitehorse

Capitals of Nations, Provinces, Places, and States

Afghanistan Kabul
Aguascalientes Aguascalientes
Alabama Montgomery
Alaska Juneau
Albania Tirana
Alberta Edmonton
Alderney Alderney
Algeria Algiers
American Samoa Pago Pago
American Virgin Islands Charlotte Amalie
Andorra Andorra la Vella
Angola Luanda
Anguilla The Valley
Antigua and Barbuda St John's
Argentina Buenos Aires
Arizona Phoenix
Arkansas Little Rock
Armenia Yerevan
Aruba Oranjestad
Australia Canberra
Australia's Northern Territory Darwin
Austria Vienna
Azerbaijan Baku
Azores Angra do Heroísmo, Horta, and Ponta Delgada
Bahamas Nassau
Bahrain Manama
Baja California Mexicali
Baja California Sur La Paz (capital of the Southern Territory of Baja California—*Territorio Sur*—abbreviated BC Sur)
Balearic Islands Palma
Bangladesh Dhaka
Barbados Bridgetown
Belarus Minsk
Belgium Brussels
Belize Belmopan
Benin Porto-Novo (official) and Cotonou (de facto)
Bermuda Hamilton
Bhutan Thimphu
Black Forest Freiburg, Germany
Bolivia La Paz (de facto) and Sucre (legal)
Bophuthatswana Mmbatho
Bosnia and Herzegovina Sarajevo

Botswana Gaborone
Brazil Brasilia
British Columbia Victoria
British Virgin Islands Road Town
Brunei Darussalam Bandar Seri Begawan
Bulgaria Sofia
Burkina Faso Ouagadougou
Burma *See* Myanmar
Burundi Bujumbura
California Sacramento
Cambodia Phnom Penh
Cameroon Yaoundé
Campeche Campeche
Canada Ottawa
Canada's Northwest Territories Yellowknife
Canary Islands Las Palmas
Cape Verde Islands Praia
Cayman Islands Georgetown
Central African Republic Bangui
Chad N'Djamena
Chiapas Tuxtla Gutiérrez
Chihuahua Chihuahua City
Chile Santiago
China Beijing (Peking)
Ciskei Bisho
Coahuila Saltillo
Cocos Islands West Island
Colima Colima
Colombia Bogotá
Colorado Denver
Comoros Moroni
Congo Brazzaville
Congo, Democratic Republic of Kinshasha
Connecticut Hartford
Cook Islands Rarotonga
Corsica Ajaccio
Costa Rica San José
Croatia Zagreb
Cuba Havana
Cyprus Nicosia
Czech Republic Prague
Delaware Dover
Denmark Copenhagen
Distrito Federal México City
Djibouti Djibouti
Dominica Roseau
Dominican Republic Santo Domingo
Durango Durango

Ecuador Quito
Egypt Cairo
El Salvador San Salvador
England London
Equatorial Guinea Malabo
Eritrea Asmara
Estonia Tallinn
Ethiopia Addis Ababa
Faeroe Islands Tórshavn
Falkland Islands Stanley
Fiji Suva
Finland Helsinki
Florida Tallahassee
France Paris
French Guiana Cayenne
French Polynesia Papeete
Gabon Libreville
The Gambia Banjul
Georgia (state) Atlanta
Georgia (country) Tbilisi
Germany Berlin
Ghana Accra
Gibraltar Gibraltar
Greece Athens
Greenland Nuuk
Grenada St George's
Guadeloupe Basse-Terre
Guam Agaña
Guanajuato Guanajuato
Guatemala Guatemala City
Guernsey St Peter Port
Guerrero Chilpancingo
Guinea Conakry
Guinea-Bissau Bissau
Guyana Georgetown
Haiti Port-au-Prince
Hawaii Honolulu
Hidalgo Pachuca de Soto
Highlands Inverness, Scotland
Honduras Tegucigalpa
Hong Kong Victoria
Hungary Budapest
Iceland Reykjavik
Idaho Boise
Illinois Springfield
India New Delhi
Indiana Indianapolis
Indonesia Jakarta
Iowa Des Moines
Iran Tehran
Iraq Baghdad
Ireland Dublin
Isle of Man Douglas
Israel Jerusalem

Italy Rome
Ivory Coast Abidjan (de facto), Yamoussoukro (official)
Jalisco Guadalajara
Jamaica Kingston
Japan Tokyo
Jersey St Helier
Jordan Amman
Kansas Topeka
Kazakhstan Astana
Kentucky Frankfort
Kenya Nairobi
Kiribati Tarawa
Korea (North) Pyongyang
Korea (South) Seoul
Kuwait Kuwait
Kyrgyzstan Bishkek
Laos Vientiane
Latvia Riga
Lebanon Beirut
Lesotho Maseru
Liberia Monrovia
Libya Tripoli
Liechtenstein Vaduz
Lithuania Vilnius
Louisiana Baton Rouge
Lower Saxony Hanover, Germany
Luxembourg Luxembourg
Macau Macau
Macedonia Skopje
Madagascar Antananarivo
Madeira Funchal
Maine Augusta
Malawi Lilongwe
Malaysia Kuala Lumpur
Maldives Malé
Mali Bamako
Malta Valletta
Manitoba Winnipeg
Marshall Islands Majuro
Martinique Fort-de-France
Maryland Annapolis
Massachusetts Boston
Mauritania Nouakchott
Mauritius Port Louis
Mayotte Dzaoudzi
Mexico Mexico City
Michigan Lansing
Michoacán Morelia
Micronesia Palikir
Minnesota Saint Paul
Mississippi Jackson
Missouri Jefferson City
Moldova Kishinev
Monaco Monaco-Ville
Mongolia Ulan Bator
Montana Helena
Montserrat Plymouth
Morelos Cuernavaca
Morocco Rabat
Mozambique Maputo

Myanmar (Burma) Yangon
Namibia Windhoek
Nauru Yaren
Nayarit Tepic
Nebraska Lincoln
Nepal Kathmandu
Netherlands Amsterdam
Netherlands Antilles Willemstad
Nevada Carson City
New Brunswick Fredericton
New Caledonia Nouméa
Newfoundland St John's
New Hampshire Concord
New Jersey Trenton
New Mexico Santa Fé
New South Wales Sydney
New York Albany
New Zealand Wellington
Nicaragua Managua
Niger Niamey
Nigeria Abuja
Niue Alofi
Norfolk Island Kingston
North Carolina Raleigh
North Dakota Bismarck
Northern Ireland Belfast
Northern Mariana Islands Saipan
Norway Oslo
Nova Scotia Halifax
Nuevo León Monterrey
Nunavut Iqaluit
Oaxaca Oaxaca
Ohio Columbus
Oklahoma Oklahoma City
Old California Monterey
Oman Muscat
Ontario Toronto
Oregon Salem
Orkneys Kirkwall on Pomona Island
Palau Koror
Pakistan Islamabad
Panama Panama City
Panama Canal Balboa Heights
Papua New Guinea Port Moresby
Paraguay Asunción
Pennsylvania Harrisburg
Peru Lima
Philippines Manila
Pitcairn Islands Adamstown
Poland Warsaw
Portugal Lisbon
Prince Edward Island Charlottetown
Puebla Puebla
Puerto Rico San Juan
Qatar Doha
Québec Québec City

Queensland Brisbane
Querétaro Querétaro
Quintana Roo Chetumal
Réunion Saint-Denis
Rhode Island Providence
Romania Bucharest
Russia Moscow
Rwanda Kigali
Saint Helena Jamestown
Saint Kitts and Nevis Basseterre
Saint Lucia Castries
Saint Pierre and Miquelon St Pierre
Saint Vincent and the Grenadines Kingstown
Samoa Apia
San Luis Potosí San Luis Potosí
San Marino San Marino
São Tomé and Príncipe São Tomé
Sark La Collinette
Saskatchewan Regina
Saudi Arabia Riyadh
Scotland Edinburgh
Senegal Dakar
Seychelles Victoria
Sierra Leone Freetown
Sinaloa Culiacán
Singapore Singapore
Slovakia Bratislava
Slovenia Ljubljana
Solomon Islands Honiara
Somalia Mogadishu
Sonora Hermosillo
South Africa Bloemfontein (judicial), Cape Town (legislative), Pretoria (administrative)
South Australia Adelaide
South Carolina Columbia
South Dakota Pierre
Spain Madrid
Sri Lanka Colombo
State of México Toluca
Sudan Khartoum
Suriname Paramaribo
Swaziland Mbabane
Sweden Stockholm
Switzerland Bern
Syria Damascus
Tabasco Villa Hermosa
Taiwan Taipei
Tajikistan Dushanbe
Tamaulipas Ciudad Victoria
Tanzania Dar es Salaam
Tasmania Hobart
Tennessee Nashville
Texas Austin
Thailand Bangkok
Tlaxcala Tlaxcala

Togo Lomé
Tonga Nuku'alofa
Transkei Umtata
Trinidad and Tobago Port-of-
Spain
Tunisia Tunis
Turkey Ankara
Turkmenistan Ashkhabad
Turks and Caicos Islands
Cockburn Town
Tuvalu Funafuti
Uganda Kampala
Ukraine Kiev
United Arab Emirates Abu
Dhabi
United Kingdom London

Uruguay Montevideo
U.S.A. Washington, DC
Utah Salt Lake City
Uzbekistan Tashkent
Vanuatu Port Vila
Vatican City State Vatican
City
Venda Thohoyandou
Venezuela Caracas
Veracruz Jalapa
Vermont Montpelier
Victoria Melbourne
Vietnam Hanoi
Virginia Richmond
Virgin Islands Charlotte
Amalie

Wales Cardiff
Wallis and Futuna Mata-Utu
Washington Olympia
Western Australia Perth
Western Sahara El Aaiún
West Virginia Charleston
Wisconsin Madison
Wyoming Cheyenne
Yemen Sana'a
Yucatán Mérida
Yugoslavia Belgrade
Yukon Whitehorse
Zacatecas Zacatecas
Zambia Lusaka
Zimbabwe Harare

C-B (Citizen's-Band) Radio Frequency Shortwave Call Signs

10-1 receiving you poorly
10-2 receiving you well
10-3 stop transmitting, channel in use
10-4 OK, message received
10-5 relay message
10-6 busy, can't talk now, stand by
10-7 out of service; going off air
10-8 in service, subject to call, working well
10-9 repeat message
10-10 transmission completed, standing by
10-11 talking too rapidly
10-12 visitors are present
10-13 advise weather, road conditions
10-14 time by the clock
10-16 make pickup at
10-17 urgent business
10-18 anything for us?
10-19 nothing for you, return to base
10-20 my location is
10-21 contact me by phone
10-22 make personal contact with
10-23 stand by
10-24 assignment completed
10-25 contact another station by radio
10-26 disregard last transmission
10-27 I am moving to channel

10-28 identify your station
10-29 time is up for contact
10-30 violates regulations
10-31 no longer in violation of regulations
10-32 I will advise re signal readability
10-33 EMERGENCY TRAFFIC AT THIS STATION
10-34 TROUBLE AT THIS STATION, HELP NEEDED
10-35 matter of urgency but cannot discuss it by radio
10-36 transmission or event is scheduled for
10-37 send tow truck
10-38 ambulance needed at
10-39 your message was delivered
10-41 please tune to channel
10-42 traffic accident at
10-43 traffic congestion at
10-44 1 have a message for
10-45 stations on this channel please identify
10-46 assist motorist
10-50 break channel
10-55 intoxicated driver
10-60 what is next message number?
10-62 unable to copy, use phone
10-63 network directed to
10-64 network is clear
10-65 awaiting your next message

10-66 cancel message
10-67 all units comply
10-68 repeat message
10-69 message received
10-70 fire at
10-71 proceed with transmission in sequence
10-73 speed trap at
10-74 negative
10-75 you are causing interference
10-77 negative contact
10-81 reserve hotel room for
10-82 reserve room for
10-84 my telephone number is
10-85 my address is
10-88 advise phone number of
10-89 radio repairman needed at
10-90 I have tv interference
10-91 talk closer to mike
10-92 your transmitter is out of adjustment
10-93 check my frequency on this channel
10-94 please give me a long count
10-95 transmit dead carrier for 5 seconds
10-97 check test signal
10-99 mission completed, all units secure
10-100 restroom stop
10-200 police needed at

Chemical Element Symbols, Atomic Numbers, and Discovery Data

Symbol	Element	Atomic Number	Discovered
Ac	actinium	89	1899 by Debierne
Ag	silver (argentum)	47	Before the Christian Era
Al	aluminum	13	1825 by Oersted
Am	americium	95	1944 by Seborg and others
Ar or A	argon	18	1894 by Raleigh and Ramsay
As	arsenic	33	13th century by Magnus
At	astatine	85	1940 by Corson and others
Au	gold (aurum)	79	Before the Christian Era
B	boron	5	1808 by Davy
Ba	barium	56	1808 by Davy
Be	beryllium	4	1798 by Vauquelin
Bi	bismuth	83	15th century by Valentine
Bk	berkelium	97	1949 by Thompson, Ghiorso, and Seborg
Br	bromine	35	1826 by Balard
C	carbon	6	Before the Christian Era
Ca	calcium	20	1808 by Davy
Cd	cadmium	48	1817 by Stromeyer
Ce	cerium	58	1803 by Klaproth
Cf	californium	98	1950 by Thompson and others
C1	chlorine	17	1774 by Scheele
Cm	curium	96	1944 by Seborg and others
Co	cobalt	27	1735 by Brandt
Cr	chromium	24	1797 by Vauquelin
Cs	cesium	55	1861 by Bunsen and Kirchoff
Cu	copper (cuprum)	29	Before the Christian Era
Dy	dysprosium	66	1886 by Boisbaudran
Er	erbium	68	1843 by Mosander
Es	einsteinium	99	1952 by Ghiorso and others
Eu	europium	63	1901 by Demarcay
F	fluorine	9	1771 by Scheele
Fe	iron (ferrum)	26	Before the Christian Era
Fm	fermium	100	1953 by Ghiorso and others
Fr	francium	87	1939 by Perey
Ga	gallium	31	1875 by Boisbaudran
Gd	gadolinium	64	1886 by Marignac
Ge	germanium	32	1886 by Winkler
H	hydrogen	1	1766 by Cavendish

Symbol	Element	Atomic Number	Discovered
Ha	hahnium	105	1970 by Ghiorso and others
He	helium	2	1895 by Ramsay
Hf	hafnium	72	1923 by Coster and Hevesy
Hg	mercury (hydrargyrum)	80	Before the Christian Era
Ho	holmium	67	1879 by Cleve
I	iodine	53	1811 by Courtois
In	indium	49	1863 by Reich and Richter
Ir	iridium	77	1804 by Tennant
K	potassium (kalium)	19	1907 by Davy
Kr	krypton	36	1898 by Ramsay and Travers
La	lanthanum	57	1839 by Mosander
Li	lithium	3	1817 by Arfvedson
Lu	lutetium	71	1907 by Welsbach and Urbain
Lw	lawrencium	103	1961 by Ghiorso and others
Md	mendelevium	101	1955 by Ghiorso and others
Mg	magnesium	12	1830 by Bussy and Liebig
Mn	manganese	25	1774 by Gahn
Mo	molybdenum	42	1782 by Hjelm
N	nitrogen	7	1772 by Rutherford
Na	sodium	11	1807 by Davy
Nb	niobium (formerly columbium)	41	1801 by Hatchett
Nd	neodymium	60	1885 by Welsbach
Ne	neon	10	1898 by Ramsay and Travers
Ni	nickel	28	1751 by Cronstedt
No	nobelium	102	1958 by Ghiorso and others
Np	neptunium	93	1940 by Abelson and McMillan
O	oxygen	8	1774 by Priestley and Scheele
Os	osmium	76	1804 by Tennant
P	phosphorus	15	1669 by Brandt
Pa	protactinium	91	1917 by Hahn and Meitner
Pb	lead (plumbum)	82	Before the Christian Era
Pd	palladium	46	1803 by Wollaston
Pm	promethium	61	1945 by Glendenin and Marinsky
Po	polonium	84	1898 by P and M Curie
Pr	praseodymium	59	1885 by Welsbach
Pt	platinum	78	1735 by Ulloa
Pu	plutonium	94	1940 by Seborg and others
Ra	radium	88	1898 by P and M Curie
Rb	rubidium	37	1861 by Bunsen and Kirchoff
Re	rhenium	75	1925 by Noddack and Tacke
Rf	rutherfordium	104	1969 by Ghiorso and others
Rh	rhodium	45	1803 by Wollaston
Rn	radon	86	1900 by Dom
Ru	ruthenium	44	1845 by Claus
S	sulfur	16	Before the Christian Era
Sb	antimony (stibium)	51	1450 by Valentine
Sc	scandium	21	1879 by Nilson
Se	selenium	34	1817 by Berzelius

Symbol	Element	Atomic Number	Discovered
Si	silicon	14	1823 by Berzelius
Sm	samarium	62	1879 by Boisbaudran
Sn	tin (stannum)	50	Before the Christian Era
Sr	strontium	38	1790 by Crawford
Ta	tantalum	73	1802 by Eckeberg
Tb	terbium	65	1843 by Mosander
Tc	technetium	43	1937 by Perrier and Segre
Te	tellurium	52	1782 by von Reichenstein
Th	thorium	90	1828 by Berzelius
Ti	titanium	22	1789 by Gregor
Tl	thallium	81	1861 by Crookes
Tm	thulium	69	1879 by Cleve
U	uranium	92	1789 by Klaproth
V	vanadium	23	1830 by Sefström
W	tungsten (wolfram)	74	1783 by d'Elhuyar brothers
Xe	xenon	54	1898 by Ramsay and Travers
Y	yttrium	39	1794 by Gadolin
Yb	ytterbium	70	1878 by Marignac
Zn	zinc	30	Before the Christian Era
Zr	zirconium	40	1789 by Klaproth

Civil and Military Time Systems Compared

Civil	Military	Civil	Military	Civil	Military
12.01 A.M.	= 0001	6.00 A.M.	= 0600	1.15 P.M.	= 1315
12.02 A.M.	= 0002	7.00 A.M.	= 0700	1.30 P.M.	= 1330
12.03 A.M.	= 0003	8.00 A.M.	= 0800	1.45 P.M.	= 1345
12.04 A.M.	= 0004	9.00 A.M.	= 0900	2.00 P.M.	= 1400
12.05 A.M.	= 0005	10.00 A.M.	= 1000	3.00 P.M.	= 1500
12.15 A.M.	= 0015	11.00 A.M.	= 1100	4.00 P.M.,	= 1600
12.30 A.M.	= 0030	12.00 noon	= 1200	5.00 P.M.	= 1700
12.45 A.M.	= 0045	12.01 P.M.	= 1201	6.00 P.M.	= 1800
1.00 A.M.	= 0100	12.02 P.M.	= 1202	7.00 P.M.	= 1900
1.15 A.M.	= 0115	12.03 P.M.	= 1203	8.00 P.M.	= 2000
1.30 A.M.	= 0130	12.04 P.M.	= 1204	9.00 P.M.	= 2100
1.45 A.M.	= 0145	12.05 P.M.	= 1205	10.00 P.M.	= 2200
2.00 A.M.	= 0200	12.15 P.M.	= 1215	11.00 P.M.	= 2300
3.00 A.M.	= 0300	12.30 P.M.	= 1230	12.00 midnight	= 2400
4.00 A.M.	= 0400	12.45 P.M.	= 1245		
5.00 A.M.	= 0500	1.00 P.M.	= 1300		

Climatic Region Symbols

Typical Climatological Regional Divisions Worldwide

Climatic Symbols	Climatic Regions	
Af, Am	*Tropical Rainforest*	Tropical rainforests of the Amazon and Middle America from southern Mexico to Colombia and the West Indies; Congo and the Guinea Coast of Africa; jungles of Ceylon, India, Indonesia, Madagascar, Malaya, the Philippines, Southeast Asia
Aw	*Tropical Dry and Wet*	Grassy savannas of Middle America; Ilanos of eastern Colombia and southern Venezuela; campos of south-central Brazil; damp lowland savannas of Africa and its dry uplands; plains of northern Australia, Burma, India, Pakistan, Southeast Asia
Bsk, Bwk	*Midlatitude Dry*	Great plains and prairies of Canada and the United States; and plains of Patagonia; pampas of Argentina, Bolivia, Paraguay, and Uruguay; Gobi and Takla Makan desert dunes of Asia; Kirghizian steppe of Turkestan; Ukrainian steppe
Bwh	*Tropical Dry*	Afghan, Arabian, Atacaman, Australian, Kalihari, Sahara, Somali, Sonoran, and other subtropical and tropical desertlands of the world
Caf	*Humid Subtropical*	Southeastern United States; northern Argentina; southern Brazil, Paraguay, Uruguay; southeast Africa; southeastern China; southern Japan; eastern Australia
Cfb	*West Coast Marine*	Pacific Northwest of Canada and the United States; southern Chile; west coast of Norway and south coast of Sweden; British Isles and northwestern Europe including northern Spain; south coast of South Africa; southeast coast of Australia; New Zealand
Csa	*Mediterranean Subtropical*	Southern California; central Chile; Mediterranean region including Portugal and most of Spain, southern France, Italy, Yugoslavia, Albania, Greece, Turkey, parts of Morocco and Alegria, much of Israel; Cape of Good Hope area around Cape Town, South Africa
Daf	*Humid Continental*	Southern Canada and the northeastern United States plus much of the Midwest; much of Russia and the eastern section of China
Dcf	*Continental Subarctic*	Alaska and northern Canada; Siberia and northern Russia from the Arctic Ocean to the North Pacific Ocean
E	*Tundra*	Arctic coasts of Alaska, Canada, Greenland, northernmost Europe and Asia from northern Norway to easternmost Siberia
Ef	*Polar Icecap*	Interior of Greenland; Antarctica's northernmost tip
H	*Highland*	High valleys and mountainous areas of the world where climatic conditions are so variable they almost defy classification

Climatic Symbols Explained

A Hot and moist equatorial or tropical climate

B Dry climate with evaporation greater than precipitation

C Moist and warm with well-defined summer and winter seasons

D Cold and snowy subarctic with northern boundary the northern limit of forest growth—the taiga

E Ice climates of the icecaps where ice and snow are perpetual or of the tundra where the growing season above the permafrost is very short

H Highland climates in mountainous regions where weather conditions are extremely variable and difficult to classify

a Long and hot summers

b Short and wet winters

c Cool or short and moderate summers

d Very cold and dry winters

f Moist the year around

h Hot and moist most of the year

k Cold and dry most of the year

m Monsoon conditions

s Dry summers and wet winters

w Wet summers and dry winters

Corrections Facilities and Criminalistic Terms

"Am I my brother's keeper?"

—Genesis 4:9

This section covers correctional and penal institutions of every kind, ranging from custodial schools for delinquent juveniles to halfway houses; included are police-station lockups, county jails, and penitentiaries as well as penal colonies and rehabilitation centers. Parole and probation services are included, as are the slang names given by many inmates or former convicts. Many entries are toponyms—place-names used to describe the institution.

All of the more than 150 nations of the world are mentioned and many have several entries. All entries, except for numbered ones, are in alphabetical order.

A-to-z ready reference to correctional facilities, halfway houses, jails, penitentiaries, prisons, and reformatories around the world; sequel to the *Crime Dictionary* compiled by Ralph De Sola.

AACFO American Association of Correctional Facility Officers (publishes *The Correction Officer* newsletter)

AACP American Association of Correctional Psychologists (publishes *Journal of Criminal Justice and Behavior*)

AACTP American Association of Correctional Training Personnel

AAEOCJ American Association of Ex-Offenders in Criminal Justice

AAMHPC American Association of Mental Health Professionals in Corrections

AARC Association for the Advancement of Released Convicts

AAWS American Association of Wardens and Superintendents (formerly Wardens Association of America founded in 1870)—publishes *The Grapevine*

Abbotsford town southeast of Vancouver and location of the Matsqui Institution

Aber Aberdeen, Scotland and its prison

Aberdeen Aberdeen jail in Hong Kong, Scotland, South Dakota, and the state of Washington

Abidjan (*see* Ivory Coast's prisons)

absent without leave or permission awol (AWOL)

Abu Dhabi (*see* United Arab Emirates)

AC Administration of Correction (Puerto Rico)

ACA American Correctional Association (publishes books and the periodical *Corrections Today*)

Acapulco Mexican prison in a popular seaside resort

ACCA American Correctional Chaplains Association

Accra (*see* Ghana's prisons)

ACFSA American Correctional Food Service Association

ACHSA American Correctional Health Services Association

aci (ACI) adult correctional institution

ACJ Arlington County Jail (Virginia)

Acklington HM Prison at Acklington, Northumberland, England

ACRIM Association for Correctional Research and Information Management

ACTO Advisory Council on the Treatment of Offenders

Acuña jail in Ciudad Acuña, México

Adana (*see* Turkey)

Addis Ababa (*see* Ethiopia's prisons)

Adee Adelaide, South Australia or its jail

Adelaide Adelaide Gaol in South Australia's capital city

Aden (*see* Yemen)

Adirondack Camp Adirondack (minimum-security prison in the Adirondack Mountains near Lake Placid, New York)

adjustment center segregated center of any prison; used for the protection of inmates who refuse to be intimidated by prison gangs but cannot defend themselves

administrative segregation solitary confinement

Adobe (often pronounced *'Dobe*) Adobe Mountain School for juvenile delinquents in Phoenix, Arizona

Adrian Michigan city site of the Adrian Training School

Adrian Training School medium-security coeducational penal facility in Adrian, Michigan

Adult Diagnostic and Treatment Center Avenal, New Jersey

Adult Training Centre Milton, Ontario

Aeolian Islands Isole Eolie or the Lipari Islands used as Italian convict colonies in the Mediterranean

aerial surveillance (*see* ASTREA)

AFOSP Air Force Office of Security Police

Afyon Afyonkarahisar (Turkish—Black Castle of Opium)—prison in western central Turkey, where opium is grown

Agaña capital of Guam and location of the Adult Correctional Facility, the Community Correction Center, Cottage Homes, Juvenile Hall, the Juvenile Justice School, and the Agaña Lockup as well as the U.S. Navy Brig

Agaña Lockup Guam's jail

Agassiz British Columbia town east of Vancouver containing the Kent Institution as well as the Mountain Institution

Agra Agra Central Prison in Uttar Pradesh, India

Agua Prieta Agua Prieta, Sonora and its jail across the border from Douglas, Arizona

Aguascaliente prison in Aguascaliente, Mexico

AI Adult Institutions (New Hampshire); Amnesty International (London-based international organization concerned. with the release of political prisoners)

Aiea Halawa High Security Facility at Aiea on Oahu Island, Hawaii

air dancing hanging (execution also known as air jigging, air polka dancing, air rhumba dancing, etc.)

aislamiento penal (Spanish—penal isolation)—solitary confinement

AJA American Jail Association

AJCA Association of Juvenile Compact Administrators

AJIS Automated Jail Information System

Akron Akron, Ohio or its Summit County jail

Alabama jails scattered throughout the state's 67 counties from Prattville, Autauga County, to Double Springs in Winston County; each county has a jail plus those in Birmingham, Huntsville, Mobile, and Montgomery

Alabama State Training School at East Lake for female delinquents

Alameda Alameda County Prison and Rehabilitation Center near San Francisco, California

Alaska (*see* Anchorage, Eagle River, Fairbanks, Juneau, Ketchikan, Nome, Palmer)

Alaska lockups the state's 29 divisions each has a lockup plus the jails of Anchorage and Fairbanks

Albania prisons are located in Tirana and elsewhere

Albany names for prisons in the Albanys of California, Georgia, Indiana, Kentucky, Missouri, New York, Oregon, and Texas as well as in England and Australia

Albany, Georgia (*see* USMC)

Albany overseas prisons in England, the original Albany, and southwest Australia

Alberta Alberta Institution for Girls (Canadian prison facility)

Albertslund Herstedvester Detention Centre at Albertslund, Denmark

Albion Albion State institution and Western Correctional Facility at Albion, New York

Albuquerque site of the New Mexico Youth Diagnostic Center and the Re-Integration Center

Albuturkey slang for Albuquerque, New Mexico and its jail

Alcalá de Henares women's prison northeast of Madrid in central Spain

Alcatraz former maximum-security prison of the United States on Alcatraz Island in San Francisco Bay; today National Park Service guides conduct sight-seeing tours through its old cell blocks

Alcatraz replaced U.S. Penitentiary at Marion, Illinois

Aldershot British military-training post and prison holding soldiers discharged from service with ignominy

Alderson Federal Correctional Institution at Alderson, West Virginia (for women serving lengthy sentences)

Aleppo (*see* Syria)

Alex slang for Alexandria, and its jail or prison

Alexander Alexander Youth Service Center in Alexander, Arkansas with a capacity for 96 coed juvenile delinquents

Alexandria place-name of a jail, lockup, or prison in Australia, Egypt, Louisiana, or Virginia

Alexis Ravelin maximum-security section of the Peter-Paul fortress-prison of St Petersburg during czarist times

Algeria prisons dating from the years of French-colonial domination may be seen in Algiers, Oran, Constantine, Annaba, and even in smaller places such as Arzew

Algiers capital city of Algeria, its prison

Alhambra Alhambra Reception and Treatment Center for incoming adult felons, in Phoenix, Arizona

Alice The Alice—Alice Springs, Northern Territory, Australia—and its jail

Allentown Lehigh County, Pennsylvania, courthouse and jail

Allenwood Federal Prison Camp at Allenwood, Pennsylvania

alley corridor or hallway between cell rows

Almacen Tia Moreno Quito, Ecuador's penitentiary

a l'ombre (French—in the shadow)—in jail or in prison

Alston Wilkes Society organization aiding families of inmates in South Carolina

Alyce D McPherson School for coeducational juvenile delinquents in Ocala, Florida

Amache Japanese-American relocation center near Granada, Colorado, where they were interned after Pearl Harbor

Amarillo north Texas city, Potter County courthouse and jail

Amenia Amenia Center for Girls at Amenia, New York

American Association of Correctional Facility Officers (*see* AACFO)

American Association of Correctional Psychologists (*see* AACP)

American Association of Wardens and Superintendents (*see* AAWS)

American Correctional Association originally the National Prison Association and later the American Prison Association (*see* ACA)

American Journal of Correction formerly *Prison World*

American People for American Prisoners (*see* APAP)

American prisons (*see individual entries by Bureau of Prisons name; city, county, or state name; or nickname*)

America's Devil's Island post-Civil-War nickname of the military prison at Fort Jefferson on the Dry Tortugas about 65 miles (105 kilometers) west of Key West in the Gulf of Mexico

Am Jour Corr *American Journal of Correction*

Amman (*see* Jordan's prisons)

Amnesty International (*see* AI)

Amsterdam (*see* Netherlands)

'Aña Agaña, Guam's capital or its jail

Anamosa site of The Men's Reformatory in Iowa, east of Cedar Rapids

Anchorage Anchorage Correctional Center in Anchorage, Alaska, together with the Annex holding maximum-security felons

Andersonville Confederate prisoner-of-war camp in Georgia where nearly 14,000 Union prisoners lost their lives due to overcrowded conditions and lack of good food

Andrade Andrade, Baja California and its jail

Angleton Texas location of Retrieve Unit opened in 1919

Angola Southwest African country with old prison in Luanda and jails in smaller places; site of the Louisiana State Penitentiary

Angolite bimonthly publication by prisoners in Angola, Louisiana state prison

Ankara capital of Turkey and nickname of its several jails and prisons

ankles ankle shackles

Annaba (*see* Algeria)

Annadale New Jersey site of the Youth Correctional Institution opened in 1929

Anna's Hope Anna's Hope Detention Center on St Croix in the American Virgin Islands

Anniston Calhoun County seat and jail in Alabama

Antakya city or prison in southernmost Turkey

Antananarivo (*see* Madagascar's prisons)

Anteroom of Auschwitz nickname of Dutch transit camp established by the Nazis in World War II (*see* Lager Westerbork)

anti-penetration glazing built to resist blowtorches, gunfire, and sledgehammers (*see* detention glazing)

antiquity of Newgate site of London gaol as early as 1190

antisocial offender(s) convict(s); criminal(s)

Anto Antofagasta, Chile, jail

APA Adult Parole Authority; American Prison Association; Association of Paroling Authorities

Apalachee Apalachee Correctional Institution in Sneads, Florida

APAP American People for American Prisoners (in overseas jails and prisons)

APFO Association on Programs for Female Offenders

Apia (*see* Samoa)

APPA American Probation and Parole Association

Appleton Thorn HM Prison at Appleton Thorn in Lancashire, England

Ararat Ararat Prison in Victoria, Australia

Arcadia south-central Florida site of the De Soto Correctional Institution

Archambault Institution maximum-security facility at Ste Anne des Plaines, Québec

Argentina is replete with prisons and jails—Buenos Aires, Cordoba, Rosario, La Plata, etc.

Arizona (*see* Adobe, Alhambra, Catalina, Florence, Fort Grant, Phoenix, Tucson)

Arizona Girls School correctional facility at Phoenix for juvenile delinquents from 8 to 21 years of age

Arizona jails the state's 14 county lockups are augmented by jails in Phoenix and Tucson; the old Territorial Prison outside Yuma is a tourist attraction

Arkansas (*see* Alexander, Cummins, Pine Bluff, Tucker)

Arkansas lockups the state's 75 counties are served by jails for the county seat; city facilities serve Little Rock

Armagh HM Prison at Armagh in Northern Ireland and the penal facility for female offenders

Armley jail in Leeds, Yorkshire, England

Arohata Arohata Women's Borstal Institution, Wellington, New Zealand

Arrowhead Arrowhead Juvenile Detention Center in Duluth, Minnesota

Arthur Kill Arthur Kill Correctional Facility in Staten Island, New York—once the Drug Rehabilitation Center

Aryan Brotherhood California prison gang including members of the American Nazi Party involved in narcotics activity and racial confrontations

Arzew (*see* Algeria)

ASCA Association of State Correctional Administrators (publishes *Correctional Memo* quarterly)

Asheville Asheville, North Carolina, the seat of Buncombe County and the site of its jail

Ashford remand prison center in the London area

Ashland Federal Youth Center at Ashland, Kentucky

Ashwell HM Prison at Ashwell, Leicestershire, England

Asilo Toribio Durán Barcelona, Spain's reformatory

Asinara Italian penal colony, prison, and prison farm on Asinara Island off the northwest coast of Sardinia

ASJJA Association of State Juvenile Justice Administrators

Askham Grange HM Prison at Askham Grange, Yorkshire, England, for female offenders

Asmara (*see* Ethiopia's prisons)

Associated Marine Institutes Florida's federation of correctional programs for young offenders

Association of State Correctional Administrators (*see* ASCA)

ASTREA Aerial Support to Regional Enforcement Agency (helicopter surveillance)

Asunción Paraguay's capital containing jails and a prison

Atascadero institution for the criminally insane and mentally disordered sex offenders at Atascadero, California

Atchison Kansas Youth Center

Atheist Penologist Jeremy Bentham (1748–1832)—English philosophical radical remembered for his Panopticon—a prison designed so every cell and interior area would have natural light and air; he argued the function of punishment was not revenge but the prevention of crime; opposed the death penalty

Atlanta Atlanta Youth Development Center in Atlanta, Georgia holding female juvenile delinquents up to the age of 17; U.S. Penitentiary at Atlanta, Georgia—maximum-security; also location of the Staff Training Center of the Bureau of Prisons in Atlanta

Atlantic Avenue Brooklyn House of Detention for Men at 275 Atlantic Avenue in Brooklyn, New York

Atlantic City Atlantic City, New Jersey jail and police-station lockups

Atmore Atmore State Prison Farm northeast of Bay Minette, Alabama; Fountain Correctional Center at Atmore, Alabama northeast of Mobile

ATPE Association of Teachers in Penal Establishments

Attica Correctional Facility at Attica, New York

Auburn Auburn Correctional Facility (maximum-security institution formerly named Auburn Prison in Auburn, New York)

Auburn cell-block plan keep the most hardened convicts in solitary confinement in separate cells, keep less hardened criminals in solitude until they give evidence of repentance, and keep the so-called least guilty in separate cells at night but working in silence in workshops during the day; popular during much of the 19th century

Auburn system characterized by enforced silence at all times for all inmates (also called the silent system)

Auckland Auckland Prison in Auckland, North Island, New Zealand

Augie Augusta, Georgia jail

Auk Auckland, New Zealand or its jail

Aurora Staff Training Center of the Bureau of Prisons in Aurora, Illinois

Auschwitz-Birkenau (*German —Oswiecim-Brzezinka*)—Nazi concentration camp

Austin seat of Travis County jail

Australian prisons in the Australian Capital Territory containing Canberra, New South Wales—Sydney, Northern Territory—Darwin, Queensland—Brisbane, South Australia—Adelaide, Tasmania—Hobart, Victoria—Melbourne, Western Australia—Fremantle

Austrian prisons Vienna, Graz, Linz, Salzburg, and Innsbruck

Austro-Hungarian Empire (*see* Austria, Hungary, and Yugoslavia)

Avenel New Jersey site of the Adult Diagnostic and Treatment Center

Avon Park Avon Park Correctional Institution in south-central Florida

awa absent without authority

Awaiting Trial Facility in Cranston, Rhode Island (formerly the Providence County Jail)

away away from here, away from home, away from work (imprisoned)

awol (AWOL) absent without leave or permission

AWS (*see* Alston Wilkes Society)

axe and block decapitating equipment; broad-bladed axes and hardwood blocks with slightly hollowed-out neck rests

Aylesbury HM Prison at Aylesbury, Buckinghamshire, England

BA any of Buenos Aires, Argentina's jails and prisons

Babi Yar concentration camp outside Kiev where more than 70,000 Jews and several hundred thousand Russian troops were killed by German forces during World War II

back-gate exit dying in jail or prison and being carried out the back gate

Back Home convicts' nickname for the Tombs Prison in downtown New York City

back time unserved portion of a prison sentence any parole violator must serve once apprehended

baddie(s) bad guy(s)—incorrigible criminal(s)

bad rap(s) long prison sentence(s)

Baffin Correctional Centre in Frobisher Bay, Northwest Territories

Baghdad Iraq's oldest prison

bagne (French—convict prison; convict ship; penal servitude)

Bahamas Nassau—on New Providence Island contains an old gaol built in British colonial times

Bahrain its capital in Manama houses the old prison (*see* Manama)

Baird Andrew C Baird Detention Center (Wayne County Jail, Detroit, Michigan)

Baird House residence of the Quaker Committee on Social Rehabilitation at 135 Christopher Street in New York City

Baker Correctional Institution
offers inmates counseling, recreation, schooling, and work in Olustee, Florida

Bakersfield Kern County courthouse and jail in central southern California

Bakirkoy hospital for the criminally insane in Istanbul, Turkey

Balearic Islands (*see* Spain)

Ball site of the Louisiana Training Institute for female delinquents

Balti Baltimore, Maryland penitentiary and other penal facilities

Baltimore site of the Maryland Penitentiary, the Maryland Training School for Boys, the Reception Center, and local lockups managed by the police and the sheriff

Bamako (*see* Mali)

Banana City Brisbane, Queensland, Australia jail

banasto (Spanish—basket)— cell or prison

bandbox county workhouse

Bandung (*see* Indonesian prisons)

Bangkok (*see* Thailand)

Bang Kwang Bangkok's maximum-security prison and its oldest

Bangladesh its principal prison is in its capital city of Dacca

Bangui (*see* Central African Republic)

banishment exile in another country or to some far place belonging to the land of the person banished

banishment upheld some penologists argue banishment is better for prisoners and society than imprisonment even if this means exile to distant deserts or remote islands

Banning Banning Rehabilitation Center in Banning, California

Barbados its capital—Bridgetown—has an old prison dating back to the years of British control

barbecue stool electric chair

Barcelona (*see* Spain)

Barlinnie Glasgow, Scotland's prison and its Young Offenders Institution

Barna Barcelona, Spain, jail

Barquisimeto (*see* Venezuela)

barracoon(s) temporary prison(s)

Barranquilla Colombian prison on the Magdalena

Bartholomew Fair nickname of the solitary-confinement section of London's Fleet Prison in Elizabethan times

Bartons Mills medium-security prison near Perth, Australia

Basel (*see* Switzerland)

Basil Basil Health Systems

Bastille (French—small fortress)—La Bastille, the infamous royalist prison of Paris, destroyed by French revolutionaries on July 14, 1789— synonym for prison holding political prisoners

Bastille by the Bay inmates' nickname for San Quentin Prison in San Francisco Bay

Bastrop Federal Youth Center at Bastrop, Texas (cares for inmates under 21 years of age)

Batavia (*see* Indonesian prisons)

Bath Ontario site of the Millhaven (maximum-security) institution

Baton Rouge East Baton Rouge Parish jail, also known for the Louisiana Juvenile Reception and Diagnostic Center for delinquent and neglected juveniles

Bat Rou Baton Rouge, Louisiana jail or Juvenile Reception and Diagnostic Center

Baumes Law New York State statute requiring life imprisonment for anyone convicted four times of felonies

Baumettes France's cobblestone-walled prison in Marseilles

Bay Botany Bay penal settlement in New South Wales

Bayamón Puerto Rican city holding the Metropolitan Regional Institution

Bay ship British prison ship destined for Botany Bay in New South Wales, Australia

Bay State Correctional Center at Norfolk, Massachusetts for long-term minimum-security male felons

BC Baja California, México or British Columbia, Canada jails and prisons.

BCC Bureau of Charities and Corrections (South Dakota)

BCI Bureau of Correctional Institutions (Iowa)

BC Pen British Columbia Penitentiary in New Westminster

BCS Bureau of Criminal Statistics

Beaconsfield Marian Hall for delinquent English-Catholic juvenile offenders at Beaconsfield, Québec

bear den police station and its lockup

Beaune-la-Rolande site of a French concentration camp southeast of Pithiviers

Beccaria (*see* Father of Criminology)

Bedford HM Prison at Bedford, Bedfordshire, England

Bedford Hills Bedford Hills Correctional Facility at Bedford Hills, New York

Bedlam nickname of Saint Mary of Bethlehem, the celebrated lunatic asylum of old London, where many inmates were criminally insane

Beersheba Israel's largest prison located between Jerusalem and the Negev desert

behavioral control unit solitary-confinement cell or dungeon

beheading form of capital punishment; decapitation

behind the iron door behind bars; jailed

behind the iron house in jail

Beira (*see* Mozambique)

Beirut (*see* Lebanon)

Belfast Welfare Unit 14 euphemism for Her Majesty's Prison Camp outside Belfast, Northern Ireland

Belgian Congo now Republic of Congo (*see* entry) penal facilities date back to the days of colonial rule under Belgium

Belgrave (*see* Yugoslavia)

Belize HM Gaol facilities are in Belize as well as Belmopan

Bellefonte site of the Pennsylvania State Correctional Institution at Rockview

Belle Glade site of the Glades Correctional Institution between Lake Worth and Lake Okeechobee, Florida

Belle Isle Confederate prisoner-of-war camp in the James River near Richmond, Virginia

Bellevue Bellevue Hospital Prison Ward at First Avenue and 30th Street in New York City

Belmopan (*see* Belize)

Belsen Nazi concentration camp near Hannover, Germany

Belzec German extermination camp in this Polish village on the railway line running through Lublin province

Benghazi Libyan port city and site of a prison built almost a century ago

Benin prisons Cotonou and Porto-Novo on the north coast of the Gulf of Guinea

Bentham Jeremy Bentham (1748–1832)—English penologist-philosopher who wrote about the need for prison reform and devised the panopticon-type prison where all cells could be observed from a central site

Berdoo San Bernardino, California jail

Bergen Nazi concentration camp of Bergen-Belsen in Lower Saxony, Germany

Bergen-Belsen Nazi concentration camp in Lower Saxony, Germany

Berhala Island prisoner-of-war camp run by the Japanese in North Borneo during World War 11

Berlin Confinement Facility of the U.S. Anny in Germany

Bermuda (*see* Casemates)

Bern (*see* Switzerland)

Bernalillo Bernalillo County Detention Home in Albuquerque, New Mexico

Bess Bessemer, Alabama or its jail southwest of Birmingham

Beth Bethlehem, Pennsylvania or its city jail

Beto Unit (*see* Tennessee Colony)

Betty's Place St Elizabeth's Hospital in Washington, DC

Bexar Bexar County Jail in San Antonio, Texas

BHD Bronx House of Detention in New York City

Bhutan its capital, Thimphu, houses its principal prison

Bialoleka Polish prison camp southeast of Bialystok .

Bialystok Nazi concentration camp northeast of Treblinka in Poland

Big A nickname for the Federal Penitentiary in Atlanta, Georgia

Big D Dallas, Texas or its jail

big day visiting day in a prison

big gate(s) prison(s)

Big H Big House (any penitentiary or prison)

Big House up the River Sing Sing Prison at Ossining, New York

big pasture penitentiary

Big Spring Federal Prison Camp at Big Spring, Texas

Bilbao (*see* Spain)

bilbo(s) leg shackle(s) consisting of an iron bar fitted with adjustable fetters

Bilibid maximum-security Philippine prison at Muntinlupa in Rizal Province

Billeshave youth home for delinquent boys, on the westem edge of Denmark's island of Funen

Billings Billings, Montana, county jail

Biloxi seat of Harrison County and its jail

bing solitary confinement

Binghampton south-central New York jail

birdcage prison cell

Birkenau (German—Birch Grove)—concentration camp next to Auschwitz

Birmingham prison in Birmingham, Alabama or HM Prison in Birmingham, England plus prisons in smaller Birminghams in Iowa, Michigan, and Saskatchewan

Biscuit Factory nickname of old Reading Gaol in Berkshire, England, where it adjoined Huntley & Palmer's biscuit factory

Bismarck site of the North Dakota Penitentiary and the North Dakota State Farm for felons and first offenders

Bissau (*see* Guinea-Bissau)

bit time served in prison

BJC Bureau of Juvenile Correction (Delaware)

BKA *Bundeskriminalamt* (German—Federal Criminal Ministry)—contains computerized files of criminal histories maintained at its center in Wiesbaden

black-and-white stripes old-fashioned convict uniforms

black book prison register of its inmates

Blackburn Correctional Complex at Lexington, Kentucky, provides vocational training release programs

Black Flower of Society Nathaniel Hawthorne's nickname for any jail, penitentiary, prison, or other place of imprisonment

Black Guerrilla Family gang involved in drug trafficking within California prisons

Black Hole of Calcutta (*see* Indian prisons)

black lock solitary confinement

black maria prison van

black peter Australian slang-solitary-confinement prison cell

Blackwell's Island early name of Roosevelt Island (formerly Welfare Island) in New York City's East River under the Queensboro Bridge and long the site of correctional institutions

Bland Correctional Center at Bland, Virginia

Bledsoe Bledsoe County Regional Correctional Facility in Pikeville, Tennessee

Blonde Beast of Belsen wardress Irma Grese

Bluefields (*see* Nicaragua)

blue lights in front of all of London's police stations and lockups except at Bow Street, near the Royal Opera

Blue Ridge Pre-Release Work-Release Center in Greenville, South Carolina

Blundeston HM Prison at Blundeston, Suffolk, England

Blythe Blythe Branch of the Riverside County Jail in California

board(ed) blindfold(ed)

B o C Bureau of Correction (Pennsylvania); Bureau of Corrections (Virgin Islands)

body shake search down to the skin and into the body

Bogotá federal prison in Colombia's capital

boiling boiling in oil; boiling people alive; boiling was replaced by hanging

Boise site of the Idaho Security Medical Facility and the Idaho State Correctional Institution county seat, courthouse, and jail of Ada County; Idaho

Bolivar Bolivar County Jail in Cleveland, Mississippi

Bolívar Carcel Nacional de Ciudad Bolívar (Spanish— National Prison of Ciudad Bolivar, Venezuela)

Bolivia La Paz and Sucre each has an old prison and there are prisons in Cochabamba and Santa Cruz

Bolshoi Dom (Russian—Big House)—prison

bolt cutter heavy-duty hardware tool used to cut bolts, chainlink fencing, handcuffs, steel bars, etc.

Bom Bombay, India, prison

Bombay (*see* Indian prisons)

Bon Air Bon Air Learning Center southwest of Richmond, Virginia

boneyard graveyard

Boniato Cuban prison in Oriente Province

Bonneville Bonneville Community Corrections Center in Salt Lake City, Utah; a work-release facility

boob (Australian slang—jail; prison)

booking formal logging of inmates when they are received in jail or prison and are fingerprinted and photographed

Boonville Missouri location of the Training School for Boys for juvenile delinquents from 12 to 17 years of age

B o P Bureau of Prisons (United States Department of Justice)

Boquillas Boquillas del Carmen, Coahuila and its small jail

Bordentown New Jersey community holding the Youth Correctional Center, originally a prison farm and later a reformatory

Boron Federal Prison Camp at Boron, California in the Mojave Desert

borstal British name for a juvenile-delinquent reformatory

Borstal Borstal Prison in Kent, England where the first juvenile-delinquent reformatory was established in 1902 for boys from 16 to 21

borstals British correctional and detention centers such as Bullwood Hall in Essex, Deerbolt, Dover, East Sutton Park in Kent, Everthorpe, Humberside, Feltham and Finnamore Wood, Gaynes Hall, Glen Parva, Guys Marsh, Hatfield, Hewell Grange, Hindley, Hollesley Bay Colony, Huntercombe, Lowdham Grange, Onley, Portland, Rochester, Stoke Heath, Usk, Wellingborough, Wetherby

Bosnia (*see* Yugoslavia)

Boston Pre-Release Center in Dorchester, southwest of Boston, Massachusetts

hot (BOT) balance of time to be served by anyone violating parole and returned to prison

Botany Bay penal settlement of Sydney, New South Wales, Australia, where some 700 British convicts were landed in 1788 after an eight-month voyage from England

Botswana South African country, has prisons in its capital, Gaborone, and in Francistown

Bouaké (*see* Ivory Coast's prisons)

Bourgoin maximum-security prison in Bourgoin, France

Bournemouth seaside site of four newer prisons on England's coast southwest of London

Bowden Institution medium-security penal facility in Innisfail, Alberta

boxcar(s) prison cell(s)

boxed up (New Zealand slang—imprisoned; jailed; locked up)

box(es) prison cell(s)

Boydton Virginia location of the Mecklenburg Correctional Center

Boys Ranch Group Home for Boys at Agaña, Guam (juvenile correctional facility)

Boys Totem Town in Saint Paul, Minnesota where delinquent boys are held in minimum security while being given training

BP Board of Parole; Bureau of Prisons

B of P Bureau of Prisons (U.S. Department of Justice)

BPT Board of Prison Terms

B of R Bureau of Rehabilitation (Washington, DC)

bracelets handcuffs

Bradford federal correctional institution in Pennsylvania

Brampton Vanier Centre for Women at Brampton, Ontario

Brandenburg concentration-camp subcamp west of Berlin

Brandon Correctional Institution in Brandon, Manitoba west of Winnipeg

brank leather or rubber head harness fitted with a gag and used to prevent a prisoner from shouting or talking

Brasilia Brazil's new prison whose facilities are better than those in Belém, Belo Horizonte, Manaus, Recife, Rio de Janeiro, Salvador, Santos, or São Paulo

Braunschweig Nazi concentration camp

Brazil has as many prisons as it has cities from Belém to Rio Grande; prisons go by such generic names as *cadeia, cárcere, penitenciária estadual,* or *prisão*

Brazoria Texas site of the maximum-security Central Unit

Brazzaville the main prison of the People's Republic of the Congo

breadand bread-and-water (prison fare for those in solitary confinement)

Bread Street London prison on Bread Street in Elizabethan times

bread and water traditional diet fed to difficult-to-handle prisoners as a form of punishment

bread and whiskey last meal traditionally fed to prisoners held by the Republic of China before they are shot

Brevard Brevard Correctional Institution in Sharpes, Florida, opened in 1975 to help first offenders up to age 25 complete their education, learn a vocation, and overcome alcohol and drug abuse

briar hacksaw

bridewell British synonym for a house of correction such as the infamous Bridewell, its site may be found on London's New Bridge Street at what was St Bride's Well near the Thames

Bridge City site of the Louisiana Training School for delinquents under 17 years of age

Bridge House detention home for juvenile delinquents in Wilmington, Delaware

Bridgeport Connecticut Community Correctional Center in Bridgeport; the Fairfield County Jail is also here

Bridge of Sighs originally the enclosed passageway connecting the Doge's Palace with the prison dungeons of Venice (*Ponte dei Sospiri*); many such bridges connect courtrooms with prisons; in London it means Waterloo Bridge; in New York it is a high-level passage linking the Criminal Courts with the Tombs Prison

Bridgetown nickname of HM Prison in Barbados

Bridgewater site of the Massachusetts Correctional Institution as well as the Southeastern Correctional Center of Massachusetts

brig ship's prison

Brighton Brighton Prison in Brighton, England

brig rat(s) naval prisoner(s); shipboard prisoner(s)

Brigs (USN) (*see* USN Brigs)

Brisbane Brisbane Prison Complex (Queensland, Australia)

Brissie Brisbane, Queensland, Australia or its jail

Bristol HM Prison at Bristol, Somerset, England as well as prisons in Bristols in Connecticut, Colorado, Florida, Georgia, Indiana, New Brunswick, New Hampshire, Pennsylvania, Québec, Rhode Island, South Dakota, Vermont, Tennessee and Virginia

British detention centers Adlington, Blantyre House, Buckley Hall, Campsfield House, Eastwood Park, Erlestoke House, Foston Hall, Haslar, Kirklevington, Medomsley, New Hall, North Sea Camp, Send, Werrington House, Whatton

British Guiana (*see* Guyana)

British Honduras (*see* Belize)

British remand centers Ashford, Brockhill, Latchmere House, Low Newton, Pucklechurch, Risley, Thorp Arch

Brix Brixham, England, jail

Brixton one of London's largest prisons where the greatest jail break in British history took place in 1973

Brno prison within the old castle of former Czechoslovakia's second largest city

Bromberg Nazi concentration camp in East Prussia

Brooklyn Brooklyn Detention Center of the Immigration and Naturalization Service in New York City; Connecticut Community Correctional Center

Brookolino (Italian-American slang—Brooklyn, New York)

Brookwood Brookwood Center for juvenile offenders at Claverack, New York

Brookwood Center for Girls at Claverack, New York

Broome Broome Regional Prison (Western Australia)

Broward Broward Correctional Center at Pembroke Pines, Florida; emphasis is on educational and vocational programs

Brownwood Brownwood, Texas and the Brownwood State School for male and female juvenile delinquents

Brushes Wormwood Scrubs Prison, a suburban penal facility London

Brushy Mountain Brushy Mountain Penitentiary at Petros, Tennessee

BSSR Bureau of Social Science Research

bt's (prison) building tenders (porters, turnkeys, wing floor tenders, etc.)

bubbling executing by injecting air bubbles in the veins

Buchanan (*see* Liberia)

Bucharest (*see* Romania)

Buchenwald Nazi concentration camp near Weimar, Germany

Buckeye Youth Center in Columbus, Ohio was started in 1914 to diagnose juvenile delinquents

Buda Budapest, Hungary, prison

Budapest capital of Hungary and nickname of its old prison built during the Austro-Hungarian Empire after the defeat of the Turks in 1697

Buenaventura Colombian prison in the Pacific seaport city

Buena Vista Buena Vista Correctional Facility at Buena Vista, Colorado

Buenos Aires (*see* Argentina and Villa Devoto)

Buffalo Erie County Jail, Buffalo, New York

Buford site of the Georgia Training and Development Center

bughouse insane asylum or prison for the criminally insane

bug trap bed or cot in a jail or prison

Bujumbura (*see* Burundi)

Bukitduri women's prison near Djakarta, Indonesia

Bulawayo (*see* Zimbabwe)

Bulgaria capital city—Sofia—has a large prison and smaller ones are in Plovdid and Varna

bullpen(s) place(s) of temporary confinement while awaiting arraignment, trial, or imprisonment; large common cell(s)

Bullwood Hall a borstal in Essex, England

Bulu women's prison near Semarang on the island of Java in Indonesia

Buna forced-labor camp near Auschwitz erected by the Nazis to aid in the production of artificial rubber

Bunbury Bunbury Rehabilitation Centre (Bunbury, Western Australia)

Bundeskrímínalamt (German—Central Criminal Council)—German Interpol headquarters in Wiesbaden

Bureau of Prisons U.S. Department of Justice, Bureau of Prisons, administering U.S. penitentiaries, federal correctional institutions, a medical center for federal prisoners, federal prison camps, community treatment centers, metropolitan correctional centers, a federal detention center, and four staff training centers

buried serving a long sentence

Burlington Burlington County Jail at Mount Holly, New Jersey; largest city in Vermont and seat of Chittenden Correctional Facility

Burnaby Lower Mainland Regional Correctional Centre in Burnaby, British Columbia

burn(ed) electrocute(d)

Burrus Burrus Correctional Complex at Forsyth, Georgia

Buru Indonesian penal colony

Burundi Bujumbura prison was built during the years of German control

Bushnell Bushnell, Florida; site of the Sumter Correctional Institution

bush parole escape from confinement; escape from jail or prison

Butner Federal Correctional Institution at Butner, North Carolina; prisoners are hard-to rehabilitate repeat offenders compelled to work at a prison job and to attend group discussions about all phases of prison life and outside life styles

butt last period of a prisoner's sentence

Butte Butte, Montana, county jail

Butterworth site of a Japanese prisoner-of-war camp

Butyrskaya Moscow's blocklong four-story prison hidden behind an eight-story department store on Novoslobodskaya Street

Bu-Tyur Butyrskaya Tyurma (Russian—Butyrki Prison)—major prison in Moscow

BVR Bureau of Vocational Rehabilitation

b & w bread and water (diet often imposed on prisoners in solitary confinement)

C.3.3. Ocar Wilde's identification number while imprisoned and when he wrote his poem, *The Ballad of Reading Gaol*, and his prose apologia for being in jail, *De Profundis;* C.3.3. stood for gallery C, 3rd landing, 3rd cell

caballo (Spanish—horse)—in Mexican-American slang a person who carries drugs into jails and prisons

Cabbage Patch Victoria, Australia or its jail or prison

cabo general (Spanish—chief corporal)—head inmate or trusty in a prison

CAC Commission on Accreditation for Corrections

CACA Central After-Care Association (British society handling prisoners on parole)

cachot (French—underground prison cell)

cage jail; lockup; imprison

CAGE Convicts Association for a Good Environment

cage and key men jailers; prison guards

Cairo capital of Egypt and its prison facilities

cala calabozo (Spanish—cell, dungeon, jail)

calaboose prison

calabozo (Spanish—prison)— also called *cárcel, celda, mazmorra, presidio, prisión*

Calc Calcutta, India or its prison

Calcutta (*see* Indian prisons)

Caledonia and Odurn Complex North Carolina penal facility in Tillery

Caliente Nevada Girls Training Center (correctional facility) in Caliente

California (*see* Chino, Corona, Folsom, Frontera, Jamestown, San Luis Obispo, San Quentin, Soledad, Susanville, Tehachapi, Tracy, and Vacaville *entries*)

California Institute for Women California's only state prison for women called Frontera

California jails 58 counties are served by local lockups, city jails serve Fresno, Los Angeles, San Diego, San Francisco, and other big cities

California Medical Facility opened in 1955 at Vacaville

California Men's Colony state prison in San Luis Obispo

California State Prisons San Quentin and Folsom are the best known

Calle Marina (*see* Marine Street)

Camaguey (*see* Cuba)

Camarillo Ventura Reception Center and Clinic at Camarillo, California; near the Ventura School for Girls, also a correctional facility

Cambridge seat of Middlesex County, Massachusetts, across the Charles River from Boston; both cities have a county jail and many police-station lockups

Cameroon West African nation with prison cities such as Douala and Yaounde

camisole straitjacket

camp confinement or correctional facility; prison camp

Campbellford Ontario town containing the Warkworth Institution

Campbell Work Release Center in Columbia, South Carolina for felons and misdemeanants in minimum custody

Camp Boiro Conakry's jail in Guinea

Camp de Drancy (French—Drancy Camp)—concentration and transit camp maintained by French collaborationists and the Gestapo in World War II

Camp de Gurs (*see* Gurs)

Camp de la Transportation official name of the penal colony headquartered at Saint Laurent du Maroni in French Guiana

Camp Douglas in the 1860s a prisoner-of-war camp near Lake Michigan

Campeche prison in Campeche, Mexico

Camp Harmony Japanese-American assembly center at Puyallup, Washington

Camp Hill British reformatory on the Isle of Wight; Pennsylvania state correctional institution

Camp Iyar prison camp in Israel

Camp Lejeune (*see* USMC)

Camp O'Donnell American military encampment on Luzon in the Philippines

Campo Numero Uno (1) (Spanish—Camp No I)—maximum-security military prison near Mexico City

Camp Pendleton (*see* USMC)

Camp San Jose California detention camp near Mount Palomar

Camp Smedley D Butler (*see* USMC)

Camp Topaz Camp Topaz—the Jewel of the Desert (Japanese-American nickname for the relocation center where they were held at Topaz, Utah)

campus prison grounds

Camp West Fork correctional center for juveniles, near Warner Springs, California

Camp Westway detention center for juveniles, near Warner Springs, California

Canada every province provides penal facilities ranging from penitentiaries and prisons to halfway houses and locallockups

canine shamus(es) dog detective(s)—used for their keen sense of smell

cannery slang for prison

Canon City Centennial Correctional Facility in Canon City, Colorado; the Colorado Territorial Correctional Facility is also here along with the Colorado

Women's Correctional Facility, the Fremont Correctional Facility for medium-security felons; the Reception and Diagnostic Center for maximum-security convicts; the Shadow Mountain Correctional Facility

Can Pen Ser Canadian Penitentiary Service

Canterbury HM Prison at Canterbury, Kent, England

Canto Grande Peruvian high-security prison in Chorillos, south of Lima

Canton Canton, Ohio or its Stark County Jail

CANY Correctional Association of New York (City)

CAP Comité d'Action des Prisonniers (French—Prisoner's Action Committee)

Cape Town (*see* South Africa)

Cape Verde island republic with its prison in Praia, off the West African coast

Cap-Haitien (*see* Haitian prisons)

capital punishment the death penalty

Capron site of the Capron Correctional Unit as well as the Deerfield, St Brides, and Southampton Correctional Centers, Virginia

Cap-Rouge the Maison Notre-Dame de la Garde facility for juvenile delinquents at Cap-Rouge, Québec

captain warden of a road prison with its road-gang guards and inmates

capun capital punishment

Caracas Venezuela location of several jails and lockups

cárcel (Spanish—prison)—also called *calabozo, celda, mazmorra, presidio, prisión*

Cárcel de Mujeres (Spanish—Women's Prison)—also name of the Instituto Nacional de Orientacion Femenina situated in los Teques Ejido Miranda, Venezuela

Cárcel de Valparaiso Valparaiso, Chile's jail

carceleras (Spanish—prisoner songs)—a flamenco song form developed by prisoners incarcerated in Ronda

Cárcel Modélo (Spanish—Model Prison)—many Hispanic places throughout Latin America and the Iberian Peninsula have a so-called model prison

Cárcel Nacional de Ciudad Bolívar (Spanish—National Prison of Ciudad Bolivar) in eastern Venezuela

Cárcel Nacional de Maracaibo (Spanish—National Prison of Maracaibo)—in western Venezuela

Cárcel Nacional de Trujillo (Spanish—National Prison of Trujillo)—northeast of Merida, Venezuela

Carceri d'Innvenzione (Italian—Imaginary Prisons)—series of etchings crated by Giambattista Piranesi in the mid-1700s to show the oppressive frustration of confinement as well as the instruments of torture

Cardiff HM Prison at Cardiff, Wales

Carraca La Carraca (Cadiz, Spain's infamous prison)

Carranza Venustiano Carranza Penitentiary in Tepic, the capital of Nayarit, Mexico

Carson City Nevada State Prison together with the Northern Nevada Correctional Center, the Nevada Women's Correctional Center, the Northern Nevada Honor Camp, all in Carson city

Cartagena Colombian and Spanish prisons in seaport cities of the Caribbean and the Mediterranean

Casablanca Buenaventura, Colombia prison; (*see* Morocco)

casa de corrección (Spanish—house of correction—reformatory

Casa de Reeducacion y Trabajo Artesanal (Spanish—House of Reeducation and Artisan Work)—Venezuelan penal facility in Maracaibo as well as Caracas where it is nicknamed La Planta

casa di correzione (Italian—house of correction)—reformatory

Casemates Bermuda's prison island, formerly a fortress

Caserta women's prison north of Naples, Italy

Cassidy Lake Technical School at Chelsea, Michigan for felons over 21 and under 30 in need of academic and vocational training

Castieau's hotel nickname given the old jail in Melbourne, Victoria, Australia honoring the jail's governor—JB Castieau

Castle The Castle—U.S. Disciplinary Barracks at Fort Leavenworth, Kansas

Castle Huntly Scottish borstal—east of Dundee

Castle Thunder Castle Thunder Prison in Richmond, Virginia where political prisoners were held during the Civil War

Castries (*see* Saint Lucia)

Castro's Prison nickname applied to communist-controlled Cuba

Catalina Catalina Mountain School near Tucson, Arizona where it offers a program for male juvenile delinquents

Catete former royal palace used as Rio de Janeiro's immigration prison

Caves The Caves (HM Prison at Rockhampton in Queensland, Australia)

Cayahoga Hills Boys School juvenile-delinquent penal facility in Warrensville Heights, Ohio

cayenne (French slang— prison ship)

Cayenne French Guiana prison on the Rue Francois-Arago in Cayenne, the capital city

cc condemned cell

CCA California Correctional Association; Colorado Correctional Association

CCC Central Correctional Center in Macon, Georgia

CCD & C Commission on Crime, Delinquency, and Corrections (Nevada)

CCHS Computerized Criminal Histories System (FBI)

CCI Coastal Correctional Institution at Garden City, Georgia; Connecticut Correctional Institution at Niantic on Long Island Sound

CCOA California Correctional Officers Association; County Court Officers Association

CCTF California Correctional Training Facility

CD Corrections Department (New Mexico); Corrections Division (Hawaii, Oregon)

CEA Correctional Education Association

Cedar City north of Jefferson City, Missouri and site of the Renz Correctional Center

cela (Portuguese—prison cell)

celda (Spanish—prison cell)

Celery City Clink Kalamazoo, Michigan's jail

cella (Italian—prison cell)

cell smell usually a mixture of excreta, sweat, stale tobacco, unaired bedding, and vomit

cement tomb prison cell

Center City Detention Center Eastern State Penitentiary in Philadelphia, Pennsylvania

Central downtown San Diego's Central Detention Facility; Hong Kong's Central Police Station and lockup-, North Carolina's Central Prison in Raleigh

Central African Republic Bangui, the capital, has an old prison built by the French

Central Community Center halfway house for released prisoners in Los Angeles, California

Central Community Corrections Center in Salt Lake City, Utah

Central Correctional Center (*see* Macon)

Central Correctional Institution in Columbia, South Carolina

Central Detention Facility District of Columbia's maximum-security prison

Centralia Centralia Correctional Center east of East Saint Louis; Washington community and address of the Maple Lane School (*see entry*)

Central Missouri Correctional Center in Jefferson City

Central Ohio Regional Forensic Unit in Columbus established to provide psychiatric treatment for mentally ill offenders

Central Ohio Training Institution in Columbus for male juvenile delinquents in maximum custody

Central Oklahoma Juvenile Treatment Center in Tecumsah, Oklahoma, housing delinquent females

Central Prison in Raleigh, North Carolina, a reception center for male felons

Central Unit Texas maximum-security facility in Sugar Land

Centro de Reeducacion Agropecuario (Spanish—Center of Cattle and Land Reeducation)—penal farm facility in El Dorado, Venezuela

Centro Penitenciario de Occidente's Ejido Trujillo (Spanish—Central Penitentiary of the West in the Ejido Trujillo of Venezuela)

Centro Penitenciario de Oriente (Spanish—Central Penitentiary of the East)—Venezuela

Centro Penitenciario de Valencia, Ejido Carabobo (Spanish—Central Penitentiary of Valencia in the Ejido Carabobo)—Venezuela

Ceuta (*see* Spain)

ceza evi (Turkish—house of punishment)—prison

CFA Correctional Facilities Association

CGIC Comisaria General de Investigacion Criminal (Spanish—Commisariat General of Criminal Investigation)—Spain's Interpol office

Chad formerly part of the French Sudan when its prison was built in N'Djamena

chain gang prisoners chained together during periods of outdoor work or transportation

chair electric chair

chamber gas chamber

Champerico Pacific coast port of Guatemala and its jail

Changi Up Changi Road (Singapore's maximum-security prison)

Channings Wood HM Prison at Channings Wood, Devonshire, England

Charleston South Carolina's principal city and its jail; West Virginia's capital city and Kanawha County seat with its jail

Charlestown Charlestown, Massachusetts jail

Charlotte Charlotte, North Carolina, seat of Mecklenburg County with its jail

Charlottetown capital of Prince Edward Island, Canada and location of the Sleepy Hollow Correctional Centre

chaser(s) prison guard(s)

Chateau d'If island prison off the port of Marseilles in the Mediterranean

Chatham Island easternmost of the Galápagos long used as a penal settlement

Chattahoochee in northwest Florida and site of the River Junction Correctional Institution

Chattanooga Tennessee city in Hamilton County with its jail

CHC Chicago House of Correction

cheats gallows (Elizabethan English)

checas (Spanish slang—communist prisons)—term derived from *Cheka*—the communist secret police active in Spain during the Spanish Civil War

check out commit suicide

Cheesebox nickname for the Statesville, Illinois penitentiary

Chelmo Nazi concentration camp near Lublin, Poland

Chelmsford English prison in Chelmsford, northeast of London

Chelsea Michigan community and site of the Cassidy Lake Technical School

Cherry Hill nickname of the Eastern Penitentiary designed to insure the solitary confinement of each prisoner

Chesapeake Virginia location of the Chesapeake Correctional Unit

Cheshire Connecticut Correctional Institution at Cheshire with age limits from 16 to 21

Chetumal prison in Ciudad Chetumal, capital of the Mexican territory of Quintana Roo

Cheyenne seat of Laramie County with its jail

Chicago toponym for six divisions of Department of Corrections or Metropolitan Correctional Center

Chihuahua prison in Chihuahua, Mexico

Chile principal prisons are the Cárcel de Valparaiso and the Penitenciario de Santiago

Chillicothe Training School for (delinquent) Girls at Chillicothe, Missouri; Chillicothe Correctional Institute at Chillicothe, Ohio (for males with mental ailments)

Chillicothe Correctional Center Missouri penal facility for female felons

Chilpancingo prison in Guerrero, Mexico

Chi Ma Wan Chi Ma Wan Prison, Lantau Island, New Territories, Hong Kong

Chi Met Chicago Metropolitan Correctional Center

China, People's Republic of has a number of forced-labor camps and many prisons filled with political prisoners

China, Republic of maintains prisons in Taipei as well as in Kaohsiung, Taichung, and Tainan

Chinde (*see* Mozambique)

Chino California Institution for Men at Chino

Chistopol prison 500 miles east of Moscow

Chittenden Correctional Facility in South Burlington, Vermont

choke hold used to restrain the unruly by wrapping a forearm around the neck (the bar hold) or by putting pressure on the carotid artery in the neck (the carotid hold); both types of choke hold cut off the air supply and halt the flow of oxygen to the brain

chokey punishment or solitary-confinement cell

choky (English slang—jail)—term believed to be derived from the Hindustani *chauki* also meaning jail

chow hall prison mess hall

chow line prisoners lined up while waiting to be served food

Christchurch Christchurch Prison in Christchurch on South Island, New Zealand

Christianstadt Nazi subcamp in Germany

chronophobia fear of time

Chula Vista Chula Vista Staging Center of the INS at San Ysidro, California on a hill overlooking the Mexican Border; holds undocumented aliens awaiting deportation or admission to the U.S.

Chuna Soviet forced-labor camp in Siberia

ci (CI) cooperative individual (informant)

CIA Central Intelligence Agency; Correctional Industries Association

Cincinnati Cincinnati, Ohio or the Hamilton County jail

Cincy Cincinnati, Ohio or its jail

City College British euphemism for Newgate Gaol—an old London lockup; New Yorker slang for The Tombs prison

city watchhouse police station; police station lockup

Ciudad Acuña Ciudad Acuña, Coahuila and its jail

Ciudad Juñrez (*see* Juárez)

Ciudad Miguel Aleman Ciudad Miguel Aleman, Tamaulipas and its jail

Ciudad Trujillo capital of the Dominican Republic during the dictatorship of Trujillo and nickname of its many prison facilities

Ciudad Victoria prison in Ciudad Victoria, capital of the Mexican state of Tamaulipas

Civic Center San Jose, California's Civic Center Jail

CIW California Institution for Women at Frontera.

clandestine prison(s) improvised jail(s) frequently found in Africa, Latin America, *the* Middle East, and Southeast Asia

Clarkson Kings County Hospital Prison Ward at 435 Clarkson Avenue in Brooklyn, New York

classification center correctional unit where inmates are held while awaiting commitment to a prison or rehabilitation program

Claverack Brookwood Center for Girls at Claverack, New York

Claymont Delaware's site for the Women's Correctional Institution and the Woods Haven-Kruse School for Girls

Clemens Unit Texas maximum-security penal facility in Brazoria

Clementina nickname of the San Michele reformatory for boys from 14 to 18 on Rome's Piazza di Porta Portese

Clermont-Ferrand French prison at 1 rue de la Prison in Clermont-Ferrand where a Franco-fascist government had its capital during World War II.

Cleve Cleveland, Ohio or its jail

Cleveland Cleveland, Ohio, seated in Cuyahoga County with its jail and police-station lockups

client(s) person(s) on probation

client(s) of the correctional system convict(s)

Clink London prison formerly dominating the south bank of the Thames near London Bridge; generic nickname for all prisons

Clink *The Clink*—Tasmanian hotel trading on the island's convict past

Clinton New Jersey location of the Correctional Institution; Clinton Correctional Facility at Dannemora, New York

close custody (*see* maximum security)

CMCC Chicago Metropolitan Correctional Center

CMS Correctional Medical Systems

CNPB Canadian National Parole Board

CO Correctional Officer

COA Correctional Officers Association

Coastal Correctional Institution (*see* CCI)

Coeur d'Alene Kootenai County's jail in northern Idaho

Coffield Unit Texas maximum-security prison in Tennessee Colony

Coiba Coiba Island Panamanian prison near the western shoreline of Panama Bay

cold Auschwitzes of the North dissident Soviet poet Yuri Galanskov's phrase describing the Arctic death camps of the USSR

Coldingley HM Prison at Coldingley, Surrey, England

Colima prison in Colima, Mexico

college reformatory

Collins Bay Institution medium-security penal facility in Kingston, Ontario

Colombia almost every city has a so-called Cárcel Modelo and there are other prisons such as Gorgona on an island in the Pacific and La Pieota in Bogotá

Colombo (*see* Sri Lanka)

Colonia Dignidad Chilean political prison in Parral, south of Santiago

Colorado (*see* Canon City; Buena Vista)

Colorado jails 63 counties are served by jails; Denver and Colorado Springs have local detention facilities

Colorado Springs El Paso County courthouse and jail in central Colorado

Colorguard trade name for a fabric-coated pre-galvanized steel fencing system, which the Colorguard Corporation of Raritan, New Jersey claims cannot be penetrated by gun muzzles, knives, or rocks

Columbia Mississippi site of Columbia Campus controlled by the Department of Youth Services; South Carolina's site of Campbell Work Release Center, the Central Correctional Institution, the Kirkland Correctional Institution, the Manning Correctional Institution, the Maximum Security Center, and the Walden Correctional Institution

Columbia Campus (*see* Columbia)

Columbus Ohio's capital containing the Buckeye Youth Center, the Central Ohio Regional Forensic Unit, the Central Ohio Training Institution, the Columbus Correction Facility, the Training Center for Youth, the Women's Correctional Admission Center

Columbus Correctional Facility formerly the Ohio Penitentiary

Columbus Fire occurred in the Ohio State Penitentiary in Columbus in April 1930

Combinado Combinado del Este (large prison outside Havana, Cuba)

Combinado del Este (Spanish—Eastern Combination)—Cuban prison in Havana Province

Community Correctional Center (*see* Bridgeport, Brooklyn, Hartford, Litchfield, New Haven, Uncasville)

community facility adult, juvenile, or nonconfinement facility where residents are allowed to depart, unaccompanied by any official, to hold or seek employment or to go to school for treatment programs

Community Treatment Centers halfway houses for male and female offenders treated by the U.S. Bureau of Prisons in Atlanta, Georgia; Chicago, Illinois; Dallas, Texas; Detroit, Michigan; Houston, Texas; Kansas City, Missouri; Long Beach, California; New York, New York; Oakland, California; Phoenix, Arizona

Comoros the Federal and Islamic Republic of the Comoros, in the Indian Ocean, has its capital and prison in the port city of Moroni

compash compassionate probation officer, prison chaplain, social worker, or prison visitor

Compiègne French concentration camp controlled by the Gestapo during the German occupation in World War II

Complex The Complex (Brisbane Prison Complex in Queensland, Australia augmented by prisons in Rockhampton, Townsville, and Woodford)

con convict

Conakry (*see* Guinea prisons)

concerning the criminal element Eugene Victor Debs, convicted for opposing World War I, had this to say— *"While there is a lower class, I am in it; while there is a criminal element, I am of it; while there is a soul in prison, I am not free."*

Conciergerie (French—porter's lodge)—the great prison of Paris on the Ile de la Cité

Concord capital of New Hampshire and location of the Concord Community Corrections Center, the New Hampshire State Prison, and the New Hampshire State Prison Community Corrections Center

Concord Community Corrections Center Concord, New Hampshire

concrete womb(s) prison(s)

condao (Mexican-American Spanish—county jail)—corruption of the Spanish word for county *(condado)*

Condemned Rock Macquarie Island, Tasmania's nickname for Grummet Island when it was a penitentiary

conditional release parole

condominio (Spanish—condominium)—euphemism for jail or prison

confinee(s) prisoner(s)

confinement imprisonment

conjugal visit plan whereby a prisoner may enjoy a marital relationship with a spouse

conk a screw club a guard

Connecticut (*see* Bridgeport, Brooklyn, Cheshire, Enfield, Hartford, Litchfield, New Haven, Niantic, Somers, Uncasville)

Connecticut Correctional Institution (*see* Cheshire, Enfield, Niantic, Somers)

Connecticut lockups the state's eight counties provide detention facilities for Hartford, Bridgeport, New Haven, New London and Litchfield, Stamford, and Norwalk

Connor Correctional Center in Hominy, Oklahoma

con(s) convict(s)

Constantine (*see* Algeria)

Constanza (*see* Romania)

convict goods things produced by convict labor vehicle license plates, mail sacks, school benches, etc.)

convict labor work performed by prisoners as part of their program of rehabilitation (public works such as ecology conservation, farming, road building, vehicle registration, etc.)

convict(s) convicted felon(s) serving a prison term

convict ships (plying between Britain and ports in New South Wales and Van Diemen's Land in the late 18th and early 19th centuries) *Aboukir, Active, Admiral Barrington, Admiral Gambier Adrian, Albemarle, Albion, Alexander Alibi, Almorah, America, Ann, Ann and Amelia, Anson, Arab, Asia, Asiatic, Atlantic, Atlas, Atwick, Augusta Jessie, Aurora, Bardaster Baring, Barossa, Batavia, Bellona, Bengal Merchant, Blenheim, Britannia, Bussorah Merchant, Cadet, Calcutta, Canada, Castle Forbes, Chapman, Charlotte, Circassian, Commodore Hayes, Coromandel, Countess of Harcourt, Cressy, Dromedary, Earl Cornwallis, Earl Grey, Earl St Vincent, Eden, Edward, Egyptian, Eliza, Elizabeth, Elizabeth and Henry, Elphinstone, Emma Eugenia, Emily, Emperor Alexander, Eolus, Equestrian, Experiment, Fanny, Fortune, Frances Charlotte, Friendship, Ganges, General Hewart, General Stewart, Gilbert Henderson, Gilmore, Glatton, Grenada, Guildford, Harmony, Hector, Henry, Henry Porcher Hillsborough, Hindostan, Hyderabad, Indefatigable, Indian, Indispensable, Isabella, Jane, John, John Barry, John Brewer John Renwick, Kinnear Lady Harewood, Lady Juliana, Lady of the Lake, Lady Rowena, Lloyds, Lord Auckland, Lord Lyndoch, Lord William Bentinck, Margaret, Maria, Maria Soames, Marion, Marmion, Marquis of Hastings, Marquis of Huntley, Mary Anne, Medina, Mexborough, Minerva, Minorca, Moffat, Morley, Navarino, Neptune, New Grove, Norfolk, North Briton, Ocean, Orator, Oriental Queen, Pestonjee Bomanjee, Pitt, Portland, Prince Regent, Rajah, Ratcliffe, Recovery, Rodney, Royal Admiral, Royal Charlotte, Runnymede, St Vincent, Scarborough, Second Fleet, Sidmouth, Sir Godfrey Webster Sir Robert Peel, Sir Robert Seppings, Sir William Bensley, Somerset, Southworth, Speke, Stakesby, Surprize, Surrey, Susan, Sydney Cove, Tasmania, Tenasserim, Tortoise, Tottenham, Triton, Waterloo, Waverley, Westmorland, William Jardine, William Miles, Woodford, etc.,* including vessels with names such as *Nile, Perseus, Persia*

con wise convict who knows what's going on in the prison including illicit operations

Cook County covering the entire area of Chicago, Illinois has a Department of Corrections with six divisions; the Cook County Jail; the old House of Correction; the Women's Division; the Men's Correctional Center; an intake and reception facility; the Training Academy

cooler(s) brig(s), guardhouse(s), jail(s), lockup(s)

coop(s) prison(s)

Coos Coos County Correctional Facility, Coos Bay, Oregon

cop a broom leave in a hurry

cop a drill leave at a normal walking pace so as not to attract the attention of guards

cop a heel escape from law-enforcement officers or from correctional facilities

cop a moke escape from a jail, prison, or other facility

cop a sneak escape from confinement

Copenhagen Denmark's modern prison noted for its design insuring inmates the maximum of light and air

cop house(s) police station(s) or lockup(s)

Copper John Auburn Prison's nickname

copper shop police station and lockup

Cordoba (*see* Argentina)

Corinto Nicaragua's seaport city with a jail

Cork (*see* Irish prisons)

Cornton Vale Scottish prison

Corona California Rehabilitation Center at Corona

corralón Mexican-Spanish—corral)—detention camp where illegal entrants await deportation

corre correccional (Mexican-Americanism—correctional institution, penitentiary, reformatory)

correctional agency federal, state, or local criminal-justice agency charged with the investigation, intake screening, supervision, custody, confinement, or treatment of adjudicated or alleged offenders

Correctional Center for Women in Raleigh, North Carolina

Correctional Clinic U.S. Bureau of Prisons Correctional Clinic in New York City

correctional custody facility brig, jail, penitentiary, prison, reformatory

correctional custody unit U.S. naval euphemism for a ship's brig or lockup

Correctional Development Centre maximum-security facility in Laval, Québec

Correctional Education Association publishes quarterly *Journal of Correctional Education*

correctional institution long-term confinement facility

correctional officer(s) prison guard(s); prison warden(s)

Correctional Service Federation American wing of the International Prisoners Aid Association

correctional training school reformatory

Correction Camp Program for male and female felons housed at Grass Lake, Michigan

Correction Division Release Center in Salem, Oregon

Correction and Rehabilitation Squadron U.S. Air Force at Lowry Air Force Base, Colorado

corrections caseload number of clients registered with a correctional agency or agent during a specified time limit

corrective labor camp communist euphemism for forced-labor camp

Corrective Services all of the gaols and prisons in Australia's New South Wales are managed by the Department of Corrective Services in Sydney

CO(s) Correction Officer(s); Correctional Officer(s)

Costa Rica Central American country with its principal prison in the capital city of San José; smaller ones are in the ports—Limón on the Caribbean and Puntarenas on the Pacific

Coti Martinez clandestine Argentine prison's pseudonym

count prison population inventory

Counterpoint bimonthly publication of the National Juvenile Detention Association

country club minimum-security prison

county cooler(s) county-supported mental hospital(s) frequently confining the criminally insane

county hotel(s) county jail(s)

Cowansville Cowansville Institution holding young offenders in Cowansville, Québec

Coxsackie Coxsackie Correctional Facility (medium-security prison in Coxsackie, New York)

CPA Connecticut Prison Association

CPPCA California Probation, Parole, and Correctional Association

CPS Canadian Penitentiary Service

CPSM Colonial Prison Service Medal (British decoration)

Cracow (*see* Poland)

Cracow/Plaszow concentration camp holding many forced laborers during World War II

Cradle of the Penitentiary Philadelphia, Pennsylvania's Walnut Street Jail built in the late 1700s and providing congregate as well as individual cells and workhouses

cranky hatch(es) cell(s) reserved for mentally deranged prisoner(s)

Cranston Rhode Island city containing eleven penal facilities

crash escape from jail or prison

crazy alley(s) cell block(s) reserved for the insane

crazy hospital(s) psychiatric ward(s)

CRD Civil Rights Division, U.S, Department of Justice

creative conflict demonstration or riot

creative sentence sentence created to fit the crime (for example, making graffiti artists clean the walls they have defaced)

crime playwright George Bernard Shaw defines crime in his play *Man and Superman,* "Crime is only the retail department of what, in wholesale, we call penal law."

crimeless society Thomas Paine, author of *The Age of Reason* and *The Rights of Man,* wrote: *"When it shall be said of any country in the world, 'My poor are happy; neither ignorance nor distress is to be found among them; my jails are empty of prisoners, my streets of beggars; the aged are not in want; the taxes are not oppressive; the rational world is my friend, because I am a friend of its happiness—when these* *things can be said, then may that country boast of its constitution and its government."*

criminoso inveterado (Portuguese—inveterate criminal)—jailbird

CRMT Community Resources Management Team (parole and probation)

Croatia (*see* Yugoslavia)

Crockett Crockett State School for Girls at Crockett, Texas

cross-bar hotel(s) jail(s) or prison(s)

Cross City Cross City Correctional Institution offers academic and vocational training

Crown Point Indiana jail south of Gary

CRS Correction and Rehabilitation Squadron (U.S. Air Force Base at Lowry Air Force Base, Colorado)

cruel and unusual punishment being boiled in oil, being set afire, ducking, mutilating, pilloring, and whipping were common practices in a number of countries; as recently as the mid-1920s whipping was a punishment applied publicly in Dover, Delaware

Crumlin Crumlin Road Prison in Belfast, Northern Ireland

Crumlin Road 14 address and nickname of one of Her Majesty's prison camps outside Belfast, Northern Ireland

CSCA Central States Corrections Association

CSD Correctional Services Department; Corrective Services Department

CSF Correctional Service Federation

CSM Correctional Service of Minnesota (Minneapolis)

Cuba a *cárcel modelo* can be found in Havana, Santiago de Cuba, Camaguey, Matanzas; there is a large penal settlement on the Isle of Pines

Cuernavaca prison in Cuernavaca, capital of the Mexican state of Morelos

Cueva Panamanian island prison

cuff(s) handcuff(s)

Culiacan prison in Culiacan, capital city of the Mexican state of Sinaloa

Cummins Cummins Prison Farm (facility of the Arkansas State Penitentiary holding girls and women convicted of illegal drug use and prostitution); Cummins Unit in Grady, Arkansas

Custer South Dakota location of the Youth Forestry Camp

custodial officer(s) prison guard(s)

Cux Cuxhaven, Germany or its jail

Cuyahoga Cuyahoga County Juvenile Detention Home in Cleveland, Ohio

CYC Colorado Youth Center at Denver

Cyprus island republic whose capital, Nicosia, has an old Turkish prison and a nearby camp site where Jews were detained to keep them out of Palestine from 1945 to 1948

CZ Pen Canal Zone Penitentiary at Gamboa

DAC Department of Adult Corrections (Alaska)

Dacca principal prison of Bangladesh in the capital city of Dacca

Dachau (Old German—marsh)—site of a large Nazi concentration camp near Munich

daddy tank prison cell reserved for lesbians to keep them from being attacked by other prisoners

Dade Correctional Center at Homestead, Florida offers offenders counseling, education, and vocational training

Dago slang for San Diego, California or its jail, lockups, or prison (*see* Diego)

Dallas Texas county jail; State Correctional Institution at Dallas, Pennsylvania

Dalmatia (*see* Yugoslavia)

Damascus (*see* Syria)

Damon Israeli medium-security prison for adult repeat offenders and Arab juvenile delinquents from 14 to 20 years old

Da Nang (*see* Vietnam)

Danbury Federal Correctional Institution at Danbury, Connecticut

dance death by hanging

dance hall cell or hallway leading to an execution chamber where the condemned seems to dance when the current is applied to the electric chair

dance of death hanging

dance on air death by hanging

dangerous aliens Canadian government designation of Japanese-Canadians in the post-Pearl-Harbor period

Dannemora Clinton Correctional Facility at Dannemora, New York

Danville Youth Development Center southwest of Lexington, Kentucky with special units for male delinquents

darbies handcuffs

Dar-es-Salaam (*see* Tanzania)

Darrington Unit Texas penal facility holding maximum-security felons within its confines in Rosharon

Dartmoor Her (His) Majesty's prison near Princetown and the Dartmoor Forest of southwest England

Davao port city of Mindanao in the Philippines and nickname of its old prison

Day Dayton, Ohio or its jails and lockups

Daytona Beach Florida site of the Tomoka Correctional Institution

Dayton Forensic Hospital in Dayton, Ohio opened in 1980 to give psychiatric help to mentally ill felons

Daytop Daytop Lodge in Staten Island, New York where male narcotic violators are offered treatment

db dirt bag (an undocumented Hispanic alien)

DB Disciplinary Barracks

dbd death by drugs (execution by lethal injection)

dc death cell

DC District of Columbia jail in Washington, DC

D of C Department of Corrections; District of Columbia

DCI Donovan Correctional Institute, jail on Otay Mesa south of San Diego

DCJ Dade County Jail in downtown Miami, Florida

DCS Department of Correctional Services (Nebraska, New York)

DCWDC District of Columbia Women's Detention Center

DEA Drug Enforcement Administration

Dead Men's Cove San Diego Police Department headquarters and lockup

Dear John letter written communication to a prisoner from a lover or wife informing him their engagement or marriage is over

death by injection a massive overdose of sodium thiopental to stop breathing while pavulon, a muscle relaxer, and potassium chloride are added to stop heartbeat

death penalty execution

death row cell block reserved for prisoners awaiting execution

decap decapitation

Deerbolt HM Borstal at Deerbolt in England

Deerfield Correctional Center in Capron, Virginia

Deer Lodge site of the Montana State Prison

defective delinquent(s) criminally insane person(s)

DeHoCo Detroit House of Correction

Delaware (*see* Claymont, Dover, Georgetown, Smyrna, Wilmington)

Delaware Correctional Center in Smyrna north of Dover

Delaware jails three counties provide lockups for communities around Wilmington, Dover, and Georgetown

Delhi (*see* Indian prisons)

Delle Stinche Florentine prison dating from the early 1300s and famous for its advanced methods of inmate handling and segregation

Del Norte prison site in northwestern California

Democratic Republic of Congo penal facilities date back to the days of colonial rule under Belgium

Denmark (*see* Herstedvester and Ringe); Copenhagen and Arhus have central prisons

Denver largest city in Colorado and nickname of its penal institutions

De Quincy Louisiana Correctional and Industrial School at De Quincy in Calcasieu Parish

Descanso Descanso Detention Facility near Alpine, California

Des Moines Iowa's county seat, courthouse, and jail of Polk County; well known for its role in correctional institution planning

Des Moines Plan innovative alternative to building new prisons; less risky convicts are steered to programs allowing them to help themselves by working or attending school under probationary supervision

De Soto De Soto Correctional Institution in south-central Florida

detention legal confinement of a person subject to criminal or juvenile court proceedings until commitment to a correctional facility or release

detention center jail housing prisoners awaiting trail

detention facility generic term for county farm, detention center, honor farm, jail, juvenile hall, road camp, work camp, etc.

detention glazing chemically strengthened and plastically bonded glass used in place of bars or walls in modern penal facilities

detention screening stainless-steel screening allowing air, light, and sound to enter but preventing the entrance of contraband such as drugs and weapons

detention windows designed to admit air and/or light but to prevent the escape of prisoners (*see* detention glazing)

Detoxification Center (*see* Inebriate Reception Center)

Detroit Wayne County Jail in downtown Detroit, Michigan

Deuel Vocational Institute (*see* Tracy)

devil's front porch prison

Devil's Island penal colony in French Guiana and its group of offshore islands in use as late as 1950 for convicts and political prisoners serving lifetime or long-duration sentences

DHC Detroit House of Correction

DHS Department of Health Services

DIA Department of Institutions and Agencies (governing New Jersey's prison system)

Diagnostic and Evaluation Center in Lincoln, Nebraska holds male felons in maximum security while under diagnosis and evaluation

Diagnostic Unit in Huntsville, Texas where all incoming inmates are diagnosed and sent to appropriate prisons

Diego (Mexican-American truncation—San Diego)—California border city or its jail, lockups, or prison (*see* Dago)

Diego Met San Diego Metropolitan Correctional Center (Hispanic appellation)

die in the hot seat be electrocuted

die of lead poisoning killed by lead bullets

die of throat trouble hanged

Directory of Juvenile Detention Homes published by the National Juvenile Detention Association

Directory of Prisoners Aid Agencies published by the International Prisoners Aid Association

Directory of Residential Treatment Centers published by the International Halfway House Association

dirt prisoner's nickname for sugar

dirty towel prison barbershop or beauty parlor

Dismas House halfway house in St Louis, Missouri

District of Columbia site of the Central Detention Facility, and some ill-reputed prisons used during and after the Civil War but replaced by modern facilities in the District and in Lorton, Virginia

District of Columbia lockup the nation's capital lockup facilities in addition to those of federal agencies

dite (Mexican-Americanism—detention hall)—place where illegal aliens must wait while being investigated

Division No 1 Chicago's Cook County Jail

Division No 2 Chicago's House of Correction

Divisions No 1 through 6 (*see* Cook County)

Dix Dorothea Lynde Dix (1802–1887)—American reformer and pioneer in securing better treatment for the insane in asylums, poor houses, and prisons throughout New England; wrote books for children and served as superintendent of women nurses during the Civil War

Dixon Correctional Institute medium-security penal facility in northern Louisiana at Jackson

Diyarbakir prison in southeastern Turkey

DJ Department of Justice (*see* U.S. Department of Justice); Don Jail in County Donegal, Northern Ireland; Don Jail in the Don Mills section of Toronto, Ontario

Djibouti formerly known as French Somaliland and later the French territory of Afars and Issas; its old prison attracts strange characters

DLPS Department of Law and Public Safety (New Jersey)

do a bit serve a sentence in jail

do a dime serve a 10-year prison sentence

do a nickel serve a five-year prison term

do a pound serve a five-year prison term

do a quarter serve a 25-year prison sentence

Dobbs School North Carolina penal facility for male and female juvenile delinquents up to 18 years of age, in Kinston

do bird do time in prison; serve a prison sentence

D o C Department of Correction (Arkansas, Connecticut, Delaware, Indiana, Massachusetts, North Carolina, Tennessee); Department of Corrections (Arizona, California, District of Columbia, Florida, Guam, Idaho, Illinois, Kansas, Kentucky, Louisiana, Maine, Michigan, Minnesota, Mississippi, Missouri, New Jersey, Rhode Island, South Carolina, Texas, Vermont, Washington, West Virginia), Division of Corrections (Utah, Wisconsin)

DOCS Department of Correctional Services (New York)

Doftana Bucharest's prison

dog detective(s) [*see* canine shamus(es)]

Doha (*see* Qatar)

D o I Department of Institutions (Montana); Director of Institutions (North Dakota); Division of Institutions (Oklahoma)

Dominica large Windward island; its capital, Roseau, contains a small gaol built by the British

Dominican Republic prisons notorious during the dictatorship of Trujillo and before intervention by the U.S. Marines (*see* Ciudad Trujillo, Santo Domingo)

Don Don Jail in County Donegal, Northern Ireland; Don Jail in Don Mills section of Toronto, Ontario

Donovan Richard J Donovan CorTectional Facility, Otay Mesa, California

D o P Department of Prisons (Nevada)

DOR Department of Offender Rehabilitation (Georgia)

Dora nickname for the Nazi concentration camp west of Leipzig, Germany at Nordhausen

Dorandordhausen Nazi concentration camp west of Leipzig, Germany

Dorchester Boston Pre-Release Center in Dorchester, Massachusetts; Dorchester Penitentiary in New Brunswick, Canada; HM Prison at Dorchester, England

Dorpat (*see* Tartu)

dossier (French—information file)—usually about a case history or individual(s) involved in some criminal or political action

do the book serve a life sentence

do time serve a prison sentence

Douala (*see* Cameroon)

double-ceiling placing two prisoners in the same cell

Dover Delaware seat of New Castle County south of Wilmington; contains the Kent Correctional Institution and police-station lockups as well as the county jail; in Kent, England, is HM Borstal in Dover on the Channel

Down Down Home (Afro-American slang—federal penitentiary in Atlanta, Georgia)

Down Home (Afro-American slang—Manhattan House of Detention long known as the Tombs)—in downtown New York City

Downstate Downstate Correctional Facility at Fishkill, New York

Down South (Afro-American nickname—U.S. Federal Penitentiary in Atlanta, Georgia)

Dozier Arthur G Dozier School at Marianna, Florida for delinquents from 12 to 17 years of age

DPA Discharged Prisoners Association (Great Britain)

DPAS Discharged Prisoners' Aid Society (Great Britain)

DPs Detention Pens in downtown New York City, also known as the Manhattan Detention Pens

DPS Department of Public Safety (American Samoa)

DPSCS Department of Public Safety and Correctional Services (Maryland)

Drake Hall HM Prison at Drake Hall, Staffordshire, England

Drancy (*see Camp de Drancy*)

Draper Draper Correctional Center at Elmore, Alabama; Utah State Prison at Draper south of Salt Lake City

DRC Department of Rehabilitation and Correction (Ohio)

dropped in the bucket jailed

drowning capital punishment usually reserved for witches and other women accused of crime; drowning was replaced by beheading and hanging as less equipment was required

Drumheller Drumheller Institution in Drumheller, Alberta

drunk tank jail cell reserved for persons arrested while under the influence of alcohol or other drugs

Dry Tortugas U.S. Military Prison on one of the Dry Tortugas in the Gulf of Mexico, west of Key West

Dubai (*see* United Arab Emirates)

Dublin (*see* Irish prisons)

Dubrolag complex of some fifteen prison camps close to Potmu in the Moldavian Republic of the USSR

Ducato Milanese Milan, Italy's old prison

Duffy of San Quentin motion picture inspired by three semi-autobiographical books by Warden Clinton T Duffy of San Quentin Prison in California where he abolished airless and dungeon-like cells, fired guards for cruelty, introduced a cafeteria, a newspaper written and printed by the prisoners, and a night school; he insisted the death penalty never deterred murder and never will

Dumfries Dumfries Young Offenders Institution near the west coast of Scotland

dummy prisoner's nickname for bread

dump truck depressed, slow moving, or torpid prisoner

dungeon underground cell or prison

Durango prison in Durango, Mexico; jail in Durango, Colorado

Durban (*see* South Africa)

Durchgangslager (German— transit camp)—established in German-occupied countries for the transport of Jews and others to concentration camps

Durham HM Prison at Durham, England; prison in Durham, England; prison in Durham, North Carolina and in smaller Durhams

Dutchman Correctional Center in Enoree, South Carolina, holding felons in maximum and minimum security

Dwight Dwight Correctional Center west of Kankakee, Illinois details female felons in maximum, medium, and minimum security

DYA Department of the Youth Authority (California)

DYS Department of Youth Services (Alabama); Division of Youth Services (Arkansas, Florida)

Eagle River Alaska Women's Facility next door to the Eagle River Correctional Center east of Anchorage

Eagle Springs North Carolina site of the Samarkand Manor youth service facility

East Block No 7 address of Interpol India on Rama Krishna Puram in New Delhi

East County Regional Center El Cajon, California, jail

Eastern Hong Kong's Eastern Police Station (and lockup); Eastern Correctional Facility at Napanoch, New York

Eastern New York Correctional Facility at Napanoch for male juvenile delinquents

Eastern State Penitentiary in North Philadelphia, Pennsylvania for more than a century but now replaced by the State Correctional Institution in Graterford

Eastharn Unit maximum-security penal facility in Lovelady, Texas

East Lake Alabama State Training School for female juvenile delinquents

East London (*see* South Africa)

East Moline East Moline Correctional Center in upper Illinois opened in 1980 as a minimum-security penal facility

East Palatka East Palatka Road Prison southeast of St Augustine; opened in 1961 with programs in alcohol and drug abuse

East Sutton Park borstal for delinquent girls in Kent, England

Ebensee Austrian concentration-camp subcamp during World War II

e by i execution by injection (of air or poison)

Echo Glen Children's Center in Snoqualmie, Washington (*see* entry)

ECJ Erie County Jail (Buffalo, New York)

Ecole Notre-Dame de Laval at Laval-des-Rapides in the Province of Québec (this and the following Ecole-type entries are for juvenile reformatories)

Ecole Ste-Agnes Montreal, Québec

Ecole Ste Domitille Montreal, Québec

Ecole Ste Helene correctional school in Montreal, Québec

ECS Episcopal Community Services (job-entry program for ex-offenders)

Ecuador its Andean capital, Quito, and its port city, Guayaquil, have so-called model prisons; former penal colony at Villamil on Isabela Island in the Galápagos has been replaced by the Penitenciario Litoral, a maximum-security prison near Guayaquil

Eddyville site of the Kentucky State Penitentiary

Edinburgh prison in Scotland's capital city and its Young Offenders Institution

Edison medicine nickname for electroconvulsive shock treatment

Edmonton Institution maximum-security facility in Edmonton, Alberta

education of correctional personnel and jailers (*see* Staff Training Centers)

EFEC Efforts From Ex-Convicts (Washington, DC's parole program)

Eglin Federal Prison Camp at Eglin Air Force Base in Florida

Egypt the Arab Republic of Egypt has penal facilities in Cairo, as well as Alexandria, Giza, and other cities

Elazik Turkish prison in the central eastern part of the country

El Cajon El Cajon Detention Facility, El Cajon, California

El Centro El Centro Detention Center of the Immigration and Naturalization Service in El Centro, California

El Chipote (Spanish—the box)—Managua, Nicaragua's underground jail beneath a hill in the center of the capital

Eldora Training School medium-security juvenile-delinquent facility southwest of Waterloo, Iowa

electrocution method of execution in Alabama, Arizona, Arkansas, Colorado, Florida, Georgia, Illinois, Indiana, Kentucky, Louisiana, Massachusetts, Nebraska, New York, Pennsylvania, South Carolina, South Dakota, Tennessee, Vermont, Virginia

El Frontón (Spanish—coastal cliff)—barren island prison of Callao, Peru

Elko site of the Nevada Youth Training Center

Ellis Unit maximum-security unit complete with a death row in Huntsville, Texas

el Met (Spanish—the Metropolitan Correctional Center)

Elmira Elmira Correctional Facility in New York State; Elmira Reception Center for male prisoners

El Paso El Paso Detention Center of the Immigration and Naturalization Service in El Paso, Texas

El Pavon (Spanish—the peacock)—Guatemala's largest prison, near Guatemala City

El Reno Federal Correctional Institution at El Reno, Oklahoma

El Retén de Catía (Spanish—The Remand of Catía)—Venezuelan house of detention in the Catía suburb of Caracas

El Salvador has a prison in its capital city, San Salvador, and smaller ones in its seaports along the Pacific

El Sexto (Spanish—The Sixth Book of Canonical Decrees)—Lima, Peru prison noted for the Easter plays staged by its inmates

Endsville place convicts dream about when they imagine life outside prison

Enfield Connecticut State Prison at Enfield includes a prison farm

England (*see* United Kingdom and individual entries)

Englewood Federal Correctional Institution at Englewood, Colorado

English nicknames for penal colonies Andaman Islands, Botany Bay, Devil's Island, French Guiana, Lipari Islands, Moreton Bay, New Caledonia, Norfolk Island, Port Blair, Port Philip, Siberia, Solovetski Islands, Sydney Cove, Tasmania, Van Diemen's Land

Enoree South Carolina site of the Dutchman Correctional Institution

Ensenada regional jail in Ensenada, California

Ensisheim French penitentiary near Mulhouse

Equatorial Guinea Fernando Po island and Rio Muni on the nearby West African coast both have old prisons built by the Portuguese

ERC Elmira Reception Center (for male prisoners) at Elmira, New York

escape unlawful departure of a lawfully confined person

Esmeralda barkentine-rigged Chilean naval-training vessel used for a short time in 1973 as a prison holding communist activists

ESP Eastern State Penitentiary (Philadelphia)

Essen concentration-camp Subcamp operated by the Nazis during World War II

Essex Essex County Jail in Newark, New Jersey

Etcher of Prisons Giambattista Piranesi (*see* Carceri d'Invenzione)

Ethan Allen School in Wales, Wisconsin serves as a reception center and training school

Ethiopia's prisons Addis Ababa, capital of Ethiopia, as well as Asmara and Massawa have prisons

Eugene Eugene, Oregon or its Lane County Jail

euphemisms of penology prison or penitentiary became reformatory, which became correctional center, and now is called a rehabilitation facility

Eureka Humboldt County courthouse and jail in California

Evanston Wyoming site of the Johnson Hall Forensic Unit and the Wyoming Womens Center

even-handed justice democratic doctrine advocating equal justice for criminals irrespective of their former position in society, their race, their religion, or their wealth

Evin Evin Prison in Teheran, Iran

Évreux maximum-security prison in the French city of Évreux

EXCEL Ex-offender Coordinated Employment Lifeline (Indiana's parole project)

ex-con(s) ex-convict(s); former convict(s)

execution box container holding a body belt, a hangman's rope, a hood, and restraining straps for fastening the limbs of the condemned

Execution Dock formerly on the muddy foreshore of a bend in the Thames at East Wapping below the Tower Bridge and the Tower of London; at first all convicted pirates were pegged down at low water so the incoming tide would drown them slowly but in later times they were hanged from a tall gallows and left to the mercy of seagulls as they decomposed within the chains suspending them above the river's reach

execution by injection injection of air or poison; less complicated and less costly than more conventional methods; abbreviated as e by i

Execution of Maximilian Edouard Manet's painting depicting the Emperor Maximilian and his Mexican generals standing between a wall and a uniformed firing squad

execution methods by states *electrocution* in Alabama, Arizona, Arkansas, Colorado, Florida, Georgia, Illinois, Indiana, Kentucky, Louisiana, Massachusetts, Nebraska, New York, Pennsylvania, South Carolina, South Dakota, Tennessee, Vermont, Virginia; *hanging* in Delaware, Montana, New Hampshire, Utah, Washington; *lethal gas* in California, Maryland, Mississippi, Missouri, Nevada, North Carolina, Oregon, Rhode Island, Wyoming; *lethal intravenous injection* in Idaho, New Mexico, Oklahoma, Texas; in Utah the prisoner chooses the method of execution or the sentencing judge decides whether execution is by firing squad or hanging

execution pennant small black flag flown over British prisons when an execution is taking place

execution shed gallows area within a prison

exercise in a cooler climate translation of a Soviet euphemism for forced labor in Siberia

Exeter HM Prison at Exeter in Devonshire, England; orjails in smaller Exeters in California, Illinois, Maine, Missouri, Nebraska, New Hampshire, Ontario, Pennsylvania, and Rhode Island

exile (*see* banishment)

EXIT Ex-offenders in Transit

Ex-offender Coordinated Employment Lifeline (EXCEL) Indiana's parole project

ex-offender(s) former offender(s) no longer under the jurisdiction of any criminal justice agency

Ex-offenders in Transit (EXIT) Maine's parole project

expunge purge or seal arrest, criminal, or juvenile-delinquent records

expungement legal ablution of a criminal's record made in an effort to assist in rehabilitation and remove any prejudice from the mind of a potential employer

extreme penalty death by execution

Eye Opener inmate publication of the Oklahoma State Prison

Fabrica de Hombres Nuevos (Spanish—Factory of New Men)—Mexico City prison constructed to permit prisoner's wives to stay overnight

Fadiffolu Fadiffolu Atoll (Maldivian island used for the banishment of law breakers) .

Fagatoa Lockup Pago Pago jail on the island of Tutuila in American Samoa

fag factory homosexual-filled prison

Fairbanks Fairbanks Correctional Center in Alaska (a maximum-security facility for mate and female felons)

Fairfield Solano County courthouse and jail in California

Fairfield School for Boys (*see* Southeastern Ohio Training Center)

Fairton federal correctional institution near Bridgeton, New Jersey

Falklands Falkland Islands capital, Stanley, houses HM Gaol

Fängelse (Swedish—Prison)—title of Ingmar Bergman's thought-provoking motion picture whose English name is *Prison*

Fannie Bay Fannie Bay Labour Prison, Fannie Bay, Australian Northern Territory

Fargo Fargo, North Dakota or its Cass County jail

farm confinement or correctional facility in a rural area; prison farm

Farmingdale Turrell Residential Group Center at Farmingdale, New Jersey

Father of Criminology Cesare Beccaria (1738–1794), author of *Delitte e della Pene* (Italian—Crimes and Punishment)—advocated an end to capital punishment and widespread prison reform

Father of Parole Captain Alexander Maconochie, the governor of the Norfolk Island penal colony from 1840 to 1844

Father of Penitentiary Science Jean Jacques Vilain founded the Maison de Force built in Ghent in 1773

Fayetteville Fayetteville (locally called *Fitzville),* North Carolina, seat of Cumberland County with its jail

FBI Federal Bureau of Investigation

FBP Federal Bureau of Prisons (U.S. Department of Justice bureau)

FCCD Florida Council on Crime and Delinquency

FCF Federal Correctional Facility

FCI Federal Correctional Institute of the U.S. Bureau of Prisons (*see* Staff Training Centers)

FCI(s) Federal Correctional Institution(s)

FCIS Foreign Counterintelligence System (FBI)

FDH Federal Detention Headquarters in Florence, Arizona

featherbed padded cell

Featherstone HM Prison at Featherstone, north of Wolverhampton, Staffordshire, England

Federal Center for Correctional Research U.S Bureau of Prisons facility in Butner, North Carolina

Federal Correctional Institutions Alderson, West Virginia; Ashland, Kentucky; Bastrop, Texas; Butner, North Carolina; Danbury, Connecticut; El Reno, Oklahoma; Englewood, Colorado; Fort Worth, Texas; La Tuna, Texas; Lexington, Kentucky; Memphis, Tennessee; Miami, Florida; Milan, Michigan;

Morgantown, West Virginia; Otisville, New York; Oxford, Wisconsin; Petersburg, Virginia; Pleasanton, California; Ray Brook, New York; Sandstone, Michigan; Tallahassee, Florida; Talladega, Alabama; Terminal Island, California; Texarkana, Texas

Federal Detention Center in Florence, Arizona's desert country

Federal Law Enforcement Training Center (*see* Fleetsie)

Federal Penitentiaries (U.S.) Atlanta, Georgia; Leavenworth, Kansas; Lewisburg, Pennsylvania; Marion, Illinois; Terre Haute, Indiana

Federal Prison Camps Allenwood, Pennsylvania; Big Spring, Texas; Boron, California; Eglin Air Force Base, Florida; Maxwell Air Force Base, Montgomery, Alabama; Safford, Arizona; Seagoville, Texas

Federal Prinson Industries (*see* UNICOR)

Federal Probation Officers Association founded in 1955 to build and maintain enlightened public interest in parole, probation, and related services

Federal Training Centre in Laval, Québec for juvenile offenders

Fed Ref Federal Reformatory

Fellowship of First Fleeters Australian society whose members must prove they were descended from the first shipment of convicts landed in Botany Bay in 1788

felon anyone who has committed a felony

felonry prison-colony population

felon swell upper-class convict

felony any crime punishable by imprisonment for more than a year or by death

felony tank jail cell reserved for felons

Feltham HM Borstal at Feltham in Middlesex, England

female penal institutions of the U.S. Bureau of Prisons Alderson, West Virginia; Pleasanton, California

Ferguson Unit Texas penal facility holding first offenders in maximum security at Midway

Fernando de Noronha Brazilian penal settlement existing since the 18th century, on an island of the same name in the South Atlantic

Fernando Po prisons in the small island nation of Equatorial Guinea in its seaport cities, Malabo and Bata

fettered restrained by ankle or leg fetters or both

fetters steel cuffs placed on the ankles or legs of prisoners to keep them from escaping

Fez (*see* Morocco)

flebre carcelaria (Spanish—prison fever)—fear of imprisonment

Fiesta de la Merced (Spanish—Mercy Fiesta)—celebrated on September 24, also known as Prisoners' Saint's Day

Fijian prisons except for drunks in lockups, anti-social criminals are segregated and forced to live in the most remote parts of this island nation (*see* Suva)

Filicudi Italian isle of exile in the Lipari Islands

finishing school euphemism for a women's prison, especially one for young women

Finnamore Wood HM Borstal at Finnamore Wood in Buckinghamshire, England

Finnish prisons Finland's capital city, Helsinki, and cities such as Tampere and Turku have prisons, but the emphasis is on rehabilitation rather than incarceration

Firlands minimum-security facility at Firlands, Washington

First American Penitentiary Walnut Street Jail in Philadelphia, built in 1790

First Execution by Electrocution August 6, 1890 in New York State's Auburn Prison

First Halfway House Issac T Hooper Home opened in New York City in 1845 by the Society of Friends—Quakers

First Nazi Concentration Camp established in Dachau, near Munich, Germany, on March 23, 1933, less than three months after President von Hindenburg appointed Hitler as Reich Chancellor (prime minister); by April 26 the Gestapo was formed, and by May 10 all books by Jews, and all books opposing Nazism, were burned; the playing of music by composers of Judaic origin was forbidden

First Prison Newspaper *The* Summary published by the inmates of the New York State Reformatory at Elmira on November 22, 1883—Thanksgiving Day

fish(es) newly arrived inmate(s); pimp(s)

Fishkill Fishkill Correctional Facility near Beacon-onHudson, New York

Fitzville (*see* Fayetteville)

five spot five-year prison term (or a five-dollar bill)

FKL Frauen Konzentration Lager (German—Women's Concentration Camp)—Hitler-era prison

Flagstaff Coconino County jail in Arizona

flat bit prison sentence for a definite period of time (*see* split bit)

Fleet Fleet Prison—London's jail dating from 14th century was finally demolished in 1846, after years of service as a debtor's prison

Fleetsie nickname of the Federal Law Enforcement Training Center in Brunswick, Georgia

FLETC Federal Law Enforcement Training Center (*see* Fleetsie)

Fleury Mérgois Europe's largest prison and France's largest high-security facility

flex-cuf ties flexible plastic ankle cuffs or handcuffs for trussing prisoners

floating hells British prison ships bound for Australia and Tasmania during the late 18th and early 19th centuries; French prison ships bound for Algeria, French Guiana, and New Caledonia, and to the. convict colony in French Guiana as recently as 1950

Florence Arizona State Prison; Florence Detention Headquarters holding persons awaiting trial or serving short sentences; South Carolina site of the Palmer Work Release Center

Florida penal facilities are numerous and widespread from Apalachee and Avon Park to Vero Beach and Zephyrhills as well as in Arcadia, Belle Glade, Broward, Bushnell, Chattachoochee, Clermont, Cross City, Daytona Beach, East Palatka, Homestead, Immakalee, Lake Butler, Lantana, Lowell, Niceville, Olustee, Pembroke Pines, Polk City, Raiford, Riverview, Sneads, Sharpes, Starke, and Trenton

Florida Correctional Institution provides custody for female felons and a measure of rehabilitation through education and vocational training at Lowell

Florida lockups 67 counties are served from Gainesville in Alachua County to Chipley in Washington County; Miami, Jacksonville, and Tampa have their own facilities

Florida School Florida School at Okeechobee holding delinquents from 12 to 17 years of age

Florida State Prison at Starke, offers inmates academic and vocational training as well as on-the-job training

Flossenbürg German prison and town near Nürnberg

Floyd Floyd County Jail in Rome, Georgia

fly a kite smuggle a letter out of prison

fly the coop escape from jail or prison

Folsom California State Prison at Folsom

forçat (French—convict)—prisoner

Ford HM Prison at Ford in Sussex, England

Fordland Missouri site of the Ozark Correctional Center

Forest Hill Forest Hill Prison in Georgetown, District of Columbia, during the Civil War and shortly after

Fort Apache nickname of the police station and lockup in New York City's south Bronx, also nicknamed the Little House on the Prairie

Fort Benning Fort Benning Confinement Facility at Fort Benning, Georgia

Fort Campbell Fort Campbell Confinement Facility at Fort Campbell, Kentucky

Fort Carson Fort Carson Confinement Facility at Fort Carson, Colorado

Fort Christian Fort Christian Detention Center on St Thomas island in Charlotte Amalie

Fort-de-France Martinique's prison in the capital and port city of this French West Indian island

Fort Dimanche Haiti's infamous prison

Fort Gordon Fort Gordon Confinement Facility at Fort Gordon, Georgia

Fort Grant Fort Grant Training Center in Arizona, offers educational and vocation training to felons

Fort Hood Fort Hood Confinement Facility at Fort Hood, Texas

Fort Jefferson former military prison on an island in the Dry Tortugas near Key West, Florida

Fort Knox Fort Knox Area Confinement Facility at Fort Knox, Kentucky

Fort Lauderdale Broward County, Florida, jail

Fort Leavenworth U.S. Disciplinary Barracks at Fort Leavenworth, Kansas

Fort Lewis Fort Lewis Confinement Facility at Fort Lewis, Washington

Fort Liquordale Fort Lauderdale, Florida's jail

Fort Madison Iowa State Penitentiary at Fort Madison

Fort Margherita old Sarawak prison near Kuching on the island of Borneo or Kalimantan served as a Japanese detention camp for Australian and British prisoners during World War II

Fort Meade Fort Meade Confinement Facility, Fort Meade, Maryland

Fort Montluc old prison in Lyon, France used during World War II as Gestapo headquarters by occupying Germans

Fort Ord Fort Ord Confinement Facility, Fort Ord, California

Fort Pillow Fort Pillow State Prison Farm on Cold Creek, Tennessee

Fort Polk Fort Polk Confinement Facility at Fort Polk, Louisiana

Fort Richardson Fort Richardson Confinement Facility at Fort Richardson, Alaska

Fort Riley U.S. Army Correctional Training Facility at Fort Riley, Kansas

Fort Saskatchewan Correctional Institution northeast of Edmonton also runs forest camps

Fort Savage nickname of New York City's East Harlem police station and lockup

Fort Sill Fort Sill Confinement Facility at Fort Sill, Oklahoma

Fort Wayne Indiana seat, courthouse, and jail of Allen County

Fort Worth Federal Correctional Institution at Fort Worth, Texas west of Dallas

Fossoli di Carpi transit camp north of Modena, Italy, built by Mussolini's Black Shirts as a depot for Jews and others enroute to concentration camps

Fountain GK Fountain Correctional Center at Holman Station, Alabama

four-time loser(s) criminal(s) convicted four times of felonies and hence imprisoned for life under the New York State statute called the Baumes Law

Fox Hill HerMajesty's Prison at Fox Hill on New Providence Island in the Bahamas

Fox Lake Wisconsin Correctional Institution (medium-security prison)

FPC Federal Prison Camp (Allenwood, Pennsylvania; Eglin Air Force Base, Florida; Lompoc, California; Marion, Illinois; Montgomery, Alabama; Safford, Arizona)

FPI Federal Prison Industries

FPOA Federal Probation Officers Association

FR Federal Reformatory (El Reno, Oklahoma and Petersburg, Virginia)

Framingham Massachusetts Correctional Institution (for women) in Framingham

Freetown (*see* Sierra Leone)

free world outside prison walls

Fremantle Fremantle Gaol (Western Australian penal institution built by convicts)

French blade the guillotine

French leave act of slipping away quietly and secretly

French prisons metropolitan France has many prisons and overseas prison colonies once were notorious in French Guiana (popularly named Devil's Island) and in New Caledonia; (*see* Bourgoin, Cayenne, Chateau d'If, Clermont-Ferrand, Conciergerie, Devil's Island, Évreux, Fort-de-France, Grand Hotel, Lisieux, *maison,* Mende, Natzweiler, *prison,* Salpetriere, Santé, Tarbes, Tulle, Vincennes)

Frentes Abiertos (Spanish—Open Fronts)—minimum-security prisons in Cuba

fresh fish(es) new prisoner(s)

fresh and sweet just out of jail

Fresnes southern suburb of Paris and location of Fresnes penitentiary

Fresno Fresno County courthouse andjail, in central California

Friarton Friarton Young Offenders Institution at Perth, Scotland

fried badly burned; electrocuted; intoxicated

Friends of Assata and Sundiata underground prison-support movement in New York City

Frisco San Francisco, California or its jails and lockups

frisk search a body for concealed drugs, contraband, weapons, etc.

Frobisher Bay on the southeast coast of Baffin Land in the Canadian Arctic's Northwest Territories contains the Baffin Correctional Centre

frog's march nickname for a method of conveying hard-to-handle prisoners (four officers each grab an arm or leg and carry the prisoner along face downward)

Frontera California Institution for Women at Frontera

Frontón Peruvian maximum-security offshore prison on an island close to Callao

Frostbite Fairbanks, Alaska or its jail and lockup

FRW Federal Reformatory for Women (Alderson, West Virginia)

fry burn badly; electrocute

Fuchu Fuchu Prison (Japan's unheated maximum-security facility in Greater Tokyo)

Fukuoka (*see* Japan's prisons)

Funafuti (*see* Tuvalu)

Fungus Corners Bremerton, Washington and its jail as well as police-station lockup

funny farm insane asylum; psychiatric ward

Futility Hill prisoner's graveyard at San Quentin, California

FYC Federal Youth Center (Ashland, Kentucky and Englewood, Colorado)

Gabonese prisons Gabon has prisons in the capital and coastal city of Libreville as well as in Port-Gentil

Gadsden Etowah County seat and jail northeast of Birmingham, Alabama

Gainesville Texas site of Brownwood State School and the Statewide Reception Center

Galle (*see* Sri Lanka)

gallery 13 prisoner's grave (yard)

gallows metal or wooden framework used for the execution of criminals by hanging

gallows bird(s) criminal(s)

Galvy Galveston, Texas or its jail or lockup

Gambian prison Gambia's capital, Banjul, has a jail built by the British

Gamboa rehabilitation center flanking the Panama Canal's dredging division close to the Gaillard Cut

Gamle Swedish prison on a Baltic inlet near Vastervik

gaol (British spelling—jail)—term introduced into Britain during the Norman Conquest and equivalent of the French *geôle*

gaolage a gaoler's fee (bribe)

gaoler (British—jailer)

Garbage Dump nickname of the Great Meadow Correctional Facility at Comstock, New York and of California's San Quentin Prison

Gardner Massachusetts location of the North Central Correctional Institution

garlic and glue convict slang for beef stew

garnish(es) bribe(s) given prison guards by inmates

garrote capital-punishment device of Spanish origin wherein the prisoner is strangled with an ever-tightening iron collar

Gartree Gartree Prison in Leicestershire, England

gas chamber specially built room where prisoners are executed by poison gas

Gasre Tehran, Iran's great prison also called Ghasre or Qsar

gassing execution in a lethal gas chamber

Gates of Hell old nickname for the entrance to Macquarie Harbour on the Indian Ocean coast of Tasmania when it was a penal settlement in Van Diemen's Land

Gatesville Texas site of the Gatesville Unit holding females in maximum security

Gatun Gatun Prison for Women and Juveniles at Gatun in the Panama Canal Zone

Gavle Swedish experimental prison where inmates receive personal visits from members of their families

gcg gas-chamber green (nickname given a bilious green prevalent on the walls of many penal institutions)

GD Gaol Delivery (*see* jail delivery)

Geisenkirchen Nazi subcamp

Geneva Girls Training School, Geneva, Nebraska; Illinois State Training School for Girls at Geneva; (*see* Switzerland)

Georgetown Sussex Correctional Institution in Delaware; also the capital of Guyana and site of a prison built by the British

George Town (*see* Malayan prisons)

Georgia Diagnostic and Classification penal facility in Jackson, Georgia

Georgia Industrial Institute (*see* GII)

Georgia jails the state's 159 counties are served by prison facilities from Baxley in Appling County to Sylvester in Worth County; Atlanta, Columbus, Macon, and Savannah have local lockups

Georgia State Prison (*see* Reidsville)

Georgia Training and Development Center (*see* Buford)

geriatric institution nickname for an old prison

German prisons all of the concentration camps, extermination camps, and forced-labor camps listed elsewhere were erected and used during Hitler's regime; prisons still in use such as Berlin's Moabit or Spandau date back to the time of the kaisers

get the wind escape; take off

get the works be given a death sentence

Ghana's prisons during the 113 years of British rule jails were erected in Accra and Kumasi

ghost trains(s) late-night railroad train(s) used to transport prisoners from one place to another

Gib Gibraltar and its convicts formerly sentenced to hard labor on The Rock (an HM Gaol)

gibbet gallows for hanging prisoners from a projecting arm

Gibraltar British dependency off the south coast of Spain contains an HM Gaol

Gig Harbor Purdy Treatment Center for Women at Gig Harbor, Washington

Giglio Italian isle of exile in the Tuscan Islands; a law, enacted in post-World War II Italy, permits judges to exile suspected criminals to remote places such as Giglio

GII Georgia Industrial Institute at Alto

Gila Bend Indian reservation in Arizona used after Pearl Harbor as a relocation center for Japanese-Americans

Girls' Cottage School Chambly, Québec

Girls Rehab Girls Rehabilitation Facility (GRF) in San Diego, California

girl's school nickname for a reformatory for young female offenders

Girl's Town correctional facility for misdemeanants at Tecumseh, Oklahoma

Gitmo Guantanamo, Cuba (U.S. Naval Base jail or the jail in the nearby town of Guantanamo)

give a permanent wave electrocute

Giza (*see* Egypt)

GK Gaol Keeper

Glades Glades Correctional Institution at Belle Glade, Florida; offers English classes to Hispanic inmates as well as educational and vocational training

Glasgow prison in Scotland's seaport city, also called Barlinnie; includes the Barlinnie Young Offenders Institution

Glass House nickname for the Los Angeles County Jail in California

glazing (*see* detention glazing)

Glenochil Glenochil Detention Centre in Scotland

glop unappetizing prison food

Gloucester HM Prison at Gloucester in Gloucestershire, England; city jail in Gloucester, Massachusetts

Golden Jefferson County courthouse and jail in Pueblo, Colorado

Golden Grove St Croix penal facility on the island of St Croix in the American Virgin Islands (*see* Anna's Hope *and* Fort Christian)

Golden Prison of Paris the Louvre—the great art museum formerly a fortress whose underground vaults held hunting dogs and political prisoners

Goldsboro North Carolina site of the Wayne Correctional Center

golpe final (Portuguese or Spanish—final blow)—death blow, execution

Goochland site of the Virginia Correctional Center for Women

Goodman Correctional Institution in Columbia, South Carolina cares for geriatric and handicapped inmates

good time time taken off a prisoner's sentence in return for good behavior

go over the hill escape

go over the wall escape

Goree Goree Island off Dakar (westernmost tip of Africa) served as the shipping point for slaves headed to the New World from the 1500s to the mid-1800s; today tourists visit its dungeons and are shown the Doorway of No Return in the House of Slaves

Goree Unit maximum-custody unit in Huntsville, Texas

Gorgona (*see* Colombia)

Gorki transit prison camp in the Russian city of Gorki, formerly Nizhni Novgorod

Goshen secure center for juvenile delinquents in Goshen, New York

go stir bug go crazy while imprisoned

Göteburg (*see* Sweden)

government men Australian euphemism for former convicts

Governor British penal equivalent of Warden

Grafton site of the West Virginia Industrial School for Boys for delinquents from 10 to 18 years of age in minimum custody

Graham Graham Correctional Center in Hillsboro, Illinois; medium-security facility

Grand Forks Grand Forks, North Dakota county jail

Grand Hotel nickname of French Polynesia's prison in Tahiti

Grand Island Grand Island, Nebraska county jail

Grand Mount Custodial School for Girls at Grand Mount, Washington

Granite site of Oklahoma State Reformatory for maximum-security inmates

Grass Lake Michigan site of the Correction Camp Program

Graterford Pennsylvania Corrections Department prison southeast of Pottstown

graybar hotel jail, lockup, prison of any kind

Great Falls Great Falls, Montana county jail

Great Jailer of the Caribbean Fidel Castro

Great Meadow Great Meadow Correctional Facility—a maximum-security prison at Comstock, New York

Grecian prisons such structures in the Hellenic Republic date back to classical times in Athens, Piraeus, Patras, or smaller places

Green Bay Wisconsin location of the Green Bay Correctional Institution, holding first-offender juvenile delinquents and serving as a reception center for young adult males

Greencastle location of the Indiana State Farm

Green Haven Green Haven Correctional Facility—maximum-security prison near Stormville, New York

green lights in front of all New York City police stations and their lockups (*see* blue lights)

Greenock prison for female offenders, in Greenock, Scotland

Greensboro Greensboro, North Carolina jail in Guilford County

Greensburg Pennsylvania location of the Regional Correctional Facility

green triangle criminal identification badge worn in concentration camps controlled by the Nazis in World War II

Greenville South Carolina location of the Blue Ridge Pre-Release Work-Release Center

Grenada gaol in St George's where it was called HM Gaol during the years of British colonial administration

Grendon HM Prison Grendon at Grendon Underwood; the first psychiatric prison in the United Kingdom to have a full-time psychiatrist as its medical superintendent

GRF (*see* Girls Rehab)

Grimes County Texas facility in Navasota

Gross Rosen Nazi concentration camp known for the number of forced-labor slaves it furnished German industry

ground animal meat prison nickname for hamburgers, hot dogs, sausages

group home nonconfining residential facility for adjudicated adults or juveniles (*see* halfway house)

Gruenheid East Berlin prison

Guadalajara Mexican prison in Jalisco's capital city of Guadalajara

Guadalcanal (*see* Solomon Islands)

Guadalupe Guadalupe, Chihuahua and its jail

Guadeloupe an overseas department of France with lockups in Basse Terre and Guadeloupe

Guamanian prison Ordot is the penitentiary on the island of Guam

Guanajay Cuban prison southwest of Havana

Guanajuato prison in Guanajuato, Mexico

Guantanamo U.S. Naval Station at Guantanamo and its brig

guardhouse military jail or lockup

Guatemala City capital and site of a large prison

Guatemalan prisons Guatemala City has its *cárcel modelo* as well as lockups maintained by the military forces

Guaya Guayaquil, Ecuador or its jail

Guaymas prison in Guaymas, Mexican port on the Gulf of California

guest of the city prisoner confined to a city jail

guest of the governor prisoner in a state penitentiary

guest of the nation prisoner in a federal penitentiary

guest of the realm prisoner in any HM gaol

guest of the state prisoner in a state penal institution

guillotine (*see* louisette)

guillotineur (French—guillotiner)—executioner using the guillotine

Guinea-Bissau former Portuguese colony with an old, unimproved prison in the capital and port city of Bissau

Guinea prisons Conakry, Labe, N'Zerekore, and Kankan have prisons built by the French

Gulag Archipelago Soviet author Alexander Solzhenitsyn's name for the forced labor camps and prisons throughout the USSR

Gulag of China northwest province of Qinghai where many forced labor camps are located

gurney hospital bed on wheels used to convey a death-sentenced criminal from death row to the chamber where lethal gas or injection is used to carry out the sentence

Gurs French internment camp for Spanish Nationalists fleeing Franco's troops and later for Jews interned by the Petain-Laval police

Gusen forced-labor camp run by the Nazis in Austria

Guy Guyana (nickname for any jail in this country once named British Guiana)

Guyama Guyama Regional Detention Center on the south-east coast of Puerto Rico

Guyana has a prison in its capital city of Georgetown as well as lockups in the interior

Guyane française (French Guiana)—extends along the tropical Atlantic coast of northern South America between 1 and 6 degrees north of the Equator, it qualifies as a fever-infested hellhole and until the early 1950s was notorious for its extensive penal colony nicknamed Devil's Island

Guy's Marsh HM Borstal at Guy's Marsh in Dorset, England

gyves handcuffs or fetters

Habana Havana has prisons dating back to the years of Spanish domination from 1492 to 1898; Morro Castle at the entrance to Habana Harbor contains an old Spanish prison still in use

hack(s) jailer(s); prison guard(s)

Haddam site of a Connecticut jail built in 1786 now occupied by the Connecticut Justice Academy

Hagerstown site of the Maryland Correctional Institution and the Maryland Correctional Training Center

Hague the Hague (*see* Netherlands)

Haiphong (*see* Vietnam)

Haitian prisons Haiti has prisons built by the French in Port-au-Prince and Cap-Haitien between 1677 and 1804

Halawa Halawa High Security Facility at Aiea on the island of Oahu, Hawaii (*see* Honolulu Jail)

half a stretch six month's imprisonment

halfway house nonconfining residential facility for adjudicated adults or juveniles; facility providing an alternative to confinement for persons not suitable for probation or

needing a period of readjustment to the community after confinement

Hall Hall of Justice

Hallowell the Stevens School for female juvenile delinquents at Hallowell, Louisiana

Hamilton Hamilton County Jail in Cincinnati, Ohio

Hampton Road Fremantle, West Australian prison on Hampton Road

handcuffed and fettered held or restrained by handcuffs and fetters

handcuffs adjustable metal bracelets connected by a chain and used to restrain

hanging method of execution in Delaware, Montana, New Hampshire, Utah, Washington

hangman executioner

hangman's day customarily Friday

Hanoi (*see* Vietnam)

Hanoi Hilton nickname of Hanoi's Hoa Lo prison

Hanover Learning Center at Hanover, Virginia

Harbison Harbison Correctional Institution for Women at Irmo, South Carolina

hard labor sentence involving imprisonment plus useful labor such as road building or maintenance

Hardwick site of the Middle Georgia Correctional Complex, also the site of the Youthful Offender Unit

Harlem Valley Harlem Valley Secure Center for juvenile and youthful offenders in Poughkeepsie, New York

Harris Harris County Juvenile Detention Center (Houston, Texas)

Harrisburg Dauphin County seat of Pennsylvania's capital city; its jail is one of many through the state

Hartford Connecticut Community Correctional Center in Hartford, also the site of the county jail

Hatfield HM Borstal at Hatfield in Yorkshire, England

Hattiesburg Hattiesburg, Mississippi or its jail

Havana (*see* Habana *and* Cuba)

Haverigg HM Prison at Haverigg on the coast of England opposite the Isle of Man

Hawaii's lockups four counties comprise this state and each has it own lockup; one in Honolulu serves Honolulu County on Oahu; another in Hilo serves the county and island of Hawaii; a third in Wailuku serves Maui; a fourth in Lihue is for the island of Kauai

Hawaii Youth Correctional Facility in Honolulu on Oahu Island.

Hawalli (*see* Kuwait)

Hayes Hayes Prison Farm, Black Hills, Tasmania

Haynesville Louisiana site of the Wade Correctional Center

Heart of Midlothian nickname Sir Walter Scott gave the Tolbooth Prison in Edinburgh, Scotland the scene of his novel, *The Heart of Midlothian*

Heart Mountain Wyoming location of a Japanese-American relocation center set up after Pearl Harbor

Helena Montana site of the Mountain View School for female delinquents from 10 to 21 years of age

Helena State School for Boys in Helena, Oklahoma, holds delinquent boys from 15 to 17 years of age

helicoptered dropped into the ocean from a helicopter while manacled (reports from Argentina indicate political prisoners have been executed in this manner)

helicopter surveillance (*see* ASTREA)

Hellhole of the Pacific New Zealand's North Island port of Russell when it was called Kororareka

Hell of Macquarie Harbour Station nickname of an old penal colony on the coast of Tasmania

Hell's Gates Macquarie Harbour—Tasmania's first penal settlement

Helsinki (*see* Finnish prisons)

hempen four-in-hand hangman's noose

hemp stretcher euphemism for hangman

Hendry Hendry Correctional Institution at Immokalee, Florida

Hennepin refers to any of three Minnesota penal facilities: the Hennepin County Home School in Minnetonka, the Hennepin County Juvenile Center in Roseville, or the Hennepin County Workhouse in Wayzata

Hennigsdorf Nazi subcamp close to Berlin and larger concentration camps

hen pen reformatory for females

herder(s) prison guard(s)

Her (His) Majesty's Penitentiary in St John's initiated in 1859 in Newfoundland's capital

Hermes dismegitus (Latin—Thrice-great Hermes)—Egyptian god Thoth and the code of laws he left concerning crime and punishment

Hermosillo Sonora state prison in Hermosillo, Mexico

Herstedvester Danish psychiatric prison for its advanced methods resulting in reduced recidivism

Herzegovina (*see* Yugoslavia)

Hewell Grange HM Borstal at Hewell Grange in Worcestershire, England

hierros (Spanish—irons)—handcuffs

Highland Rim School for Girls in Tullahoma, Tennessee for females from 12 to 18 years of age

Highpoint HM Prison at Highpoint in West Suffolk, England

High Security Center in Cranston, Rhode Island

Hillcrest School of Oregon in Salem holds juvenile-court commitments

Hillsboro (*see* Graham)

Hillsborough Hillsborough Correctional Institution at Riverview, Florida

Hilo site of Hawaii Community Correctional Center; also, the Kulani Correctional Facility, formerly the Kulani Honor Camp

Hilton Head old Federal prison for Confederate prisoners of war in the 1860s on the South Carolina coast

Hindley HM Borstal at Hindley in Lancashire, Englimd

hit the fence escape from prison

hit the pit jailed

hit the sidewalk released from jail

HK Hong Kong or one of its several jails and prisons

HLPR Howard League for Penal Reform (London, England)

HMBI Her (His) Majesty's Borstal Institution

HMG Her (His) Majesty's Gaol—the royal jail

HM Gaol Her (His) Majesty's Gaol (in the British Commonwealth)

HMP Her (His) Majesty's Penitentiary; Her (His) Majesty's Prison

HM Prison Her (His) Majesty's Prison

Hoa Lo downtown prison in Hanoi, Vietnam, nicknamed the Hanoi Hilton

hobbling walking with leg irons attached

Hobo Hoboken, New Jersey or its jail

Ho Chi Minh City formerly Saigon (*see* Vietnam)

hoist to hang a person; to rob

hole(s) solitary-confinement cell(s)

Holland site of the Michigan Dunes Correctional Facility

Hollesley Bay Colony HM Borstal at Hollesley Bay Colony in Suffolk, England

Holloway HM Prison at Holloway in Derbyshire, England near Sheffield

Holman Holman Prison at Holman Station, Alabama

Holmesburg prison in Holmesburg, Pennsylvania

Holzminden German maximum-security prison northwest of Göttingen

Homantin Homantin Girls Home (formerly Hong Kong's Matauwei Girls Home)

Homestead (*see* Dade Correctional Institution)

Hominy medium-security prison officially called the Conner Correctional Center in Tulsa, Oklahoma

Homs (*see* Syria)

Honaira (*see* Solomon Islands)

Honduran prisons Tegucigalpa, the capital city, and the port of San Pedro Sula, have cárcel *modelo* prisons

Hong Kong British crown colony off China's south coast; centers for the treatment of narcotic addicts are available as well as lockups and prisons

Honolulu site of the Conditional Release Branch, the Laumaka Conditional Release Center, the Oahu Community Correctional Center

Honolulu Jail nucleus of the Halawa High Security Facility at Aiea on Hawaii's Oahu Island

hook 'em hook them (fasten with handcuffs)

Hooper Home Isaac T Hooper Home (first halfway house in the United States, opened by the Society of Friends—Quakers—in New York City in 1845)

hoosegow jail (term may derive from the Spanish—*juzgado*—court of justice)

hoosier(s) prison visitor(s)

Hope Halls halfway houses sustained by the Volunteers of America from 1896 through the 1920s

Horsemonger Horsemonger Lane Gaol (infamous London lockup and the scene of many hangings such as the one Dickens described in a letter to *The Times*)

Hotlana Atlanta, Georgia, its jail, or federal Penitentiary

hot seat electric chair

Hot Springs Garland County jail near Little Rock, Arkansas

hot squat electric chair

House 33 Soviet secret-police prison in Rostov-on-Don

House of C House of Correction

House of D House of Detention.

house(s) of darkness prison(s)

house(s) of detention jail(s); lockup(s)

Houston Texas and its jails and prisons

Howard John Howard—18th century English prison reformer who advocated vocational training and work as ways to make men honest; Rhode Island town containing the Admission and Orientation Unit, the Maximum Custody Facility, the Medium-Minimum Facility, the Rhode Island Training School for Girls, and the Rhode Island School for Boys

Hudson Hudson Correctional Facility at Hudson, New York; Hudson County Penitentiary in New Jersey; New York School for Girls at Hudson, New York

Hudson Street alimony jail on the lower west side of Manhattan where inmates were imprisoned for failure to pay alimony

Hue (*see* Vietnam)

hulk(s) prison ship(s)—usually old craft unfit for the high seas but adequate as jails

Hull HM Prison at Kingston-upon-Hull; other prisons in Canada's Hull opposite Ottawa and in Hull, Massachusetts

Humanitarian Penologist Eugene V Debs (1855–1926) (*see* Presidential Candidate and Prison Convict)

Hungarian prisons were built during the Austro-Hungarian Empire

Hunt Corrections Center St Gabriel, Louisiana's penal facility serving as an adult reception and diagnostic center together with maximum-and medium-security sections

Huntercombe HM Borstal at Huntercombe in Oxfordshire, England

Huntingdon Pennsylvania location of the State Correctional Institution

Huntsville Madison County-jail north of Birmingham, Alabama; penal institutions such as the Diagnostic Unit, Ellis Unit, Goree Unit, Huntsville Unit, Wynne Unit in Hunstville, Texas

Huntsville Unit correctional facility in Huntsville, Texas

Huron Valley Huron Valley Men's Facility or Huron Valley Women's Facility in Ypsilanti, Michigan

Huron Valley Men's Facility Ypsilanti's maximum-security prison

Huron Valley Women's Facility in Ypsilanti, Michigan

Hutchinson location of the Kansas State Industrial Reformatory

hut(s) prison cell(s)

Huttonsville West Virginia site of the Huttonsville Correctional Center holding male felons

Hyderabad (*see* Pakistan)

IAPL International Association of Penal Law

ICA Illinois Correctional Association; Indiana Correctional Association; Iowa Corrections Association

ICCC International Concentration Camp Committee—(Vienna, Austria)

icebox prison coroner's laboratory and office

ice(d) jail(ed)

Iceland the capital, Reykjavik, has one small prison

ICFPW International Confederation of Former Prisoners of War (Paris, France)

ICSPPR International Centre of Sociological, Penal, and Penitentiary Research (Messina, Italy)

Idaho jails facilities for incarceration are found in the 44 counties

Idaho State Idaho State Correctional Institution in Boise

if you can't do the sentence, don't do the job elderly convict's advice to anyone planning a criminal career

IHHA International Halfway House Association

Ikoyi Ikoyi Prison in Lagos, Nigeria

Iksha former Soviet corrective labor colony for juvenile offenders close to Moscow's notorious prison

Île du Diable (French—Devil's Island)—penal colony used to keep political prisoners completely segregated

Îles du Salut (French—Security Islands)—off mainland French Guiana; group consists of Île du Diable (Devil's Island), Île Royale, and Île Saint Joseph

Île Saint Louis (French—Saint Louis Island)—leper colony for convicts in French Guiana

Ilha de Flôres (Portugueses—land of Flowers)—Brazilian prison in Rio de Janeiro

Illinois Industrial School for Boys (*see* Sheridan)

Illinois lockups the state's 102 counties offer detention facilities, the biggest are in Chicago's Cook County

Immokalee Florida location of the Hendry Correctional Institution

immurement confinement within walls

impound imprison

Imrali Turkish prison farm on Imrali Island in the Sea of Marmara

Imros Turkish prison on Imros Island in the Aegean Sea

incorrigible person who will not be corrected, reformed, rehabilitated, or made to conform to social standards

Indiana Boys School for delinquents from 12 to 21 years of age; in Plainfield

Indiana Girls School in Indianapolis for delinquents from 12 to 20 years of age

Indiana jails the state's 92 counties have facilities for holding felons; cities such as Indianapolis have their own lockups

Indianapolis Marion County's jail as well as the Indiana Girls School

Indiana State Farm medium-security facility in Greencastle

Indiana State Prison in Michigan City holds maximum-and medium-security felons

Indiana State Reformatory a maximum-and minimum-security penal facility in Pendleton

Indiana Women's Prison in Indianapolis

Indiana Youth Center a medium-security facility in Plainfield

Indian prisons the Republic of India maintains prisons in all principal cities; the Black Hole of Calcutta within Calcutta's Fort William was where, during the Indian Mutiny in 1756, some 146 Europeans were imprisoned in such cramped quarters that by the next morning only 23 remained alive

Indian River Indian River Correctional Institution at Vero Beach, Florida (*see* Vero Beach)

Indian River School for juvenile-delinquent boys in Massilon, Ohio

indic indicateur (French—informant)

Indio Indio Branch of the Riverside County Jail in California

individual confinement solitary imprisonment

Indonesian prison island Java

Indonesian prisons one prison exists on each of the major islands along with jails on the small islands; in Jakarta, Surabaja, Bandung, Semarang, and Medan are penal institutions ranging from jails to prisons

industrial prison workshop-oriented penitentiary where inmates produce such items as highway signs, mail bags, furniture, and vehicle license plates

Industrial School for Women in Vega Alta, Puerto Rico

Indy Indianapolis, Indiana's jails

Inebriate Reception Center in San Diego, California where nonviolent abusers of alcohol and other drugs accept coffee and counseling in lieu of being jailed

informant person giving information to law-enforcement officers who may use it in the investigation of criminals

injection (*see* death by injection)

inmate convict; person in a confinement facility; prisoner

Innisfail Canadian site of the Bowden Institution

INS Immigration and Naturalization Service

In-Service Training (IST) within California state prisons

inside the tall walls inside prison

Institute The Institute (nickname for any penal facility having Institute in its title)

institutional capacity officially determined number of inmates a correctional facility is designed to house

institutional superintendent warden

Institution for Youthful Offenders in Santurce, Puerto Rico

Instituto Nacional de Orientacion Femenina (Spanish—National Institute of Feminine Orientation)—women's prison in los Teques of the Miranda Ejido of Venezuela

Insular Penitentiary in Rio Piedra near San Juan, Puerto Rico

Intake Service Center Cranston, Rhode Island facility for holding pre-trial detainees

intensive care unit locked unit for juvenile offenders

intermediate-term adult penal institutions of the U.S. Bureau of Prisons Danbury Connecticut; Fort Worth, Texas; La Tuna, Texas; Lexington, Kentucky; Milan, Michigan; Sandstone, Minnesota; Terminal Island, California; Texarkana, Texas

internados judiciales (Spanish—judicial boarding houses)—penal facilities for holding convicts; in Venezuela there are 18 such institutions

internal exile Soviet euphemism for imprisonment in some remote part of the former USSR

International Halfway House Association publisher of the *Directory of Residential Treatment Centers*

International Penal and Penitentiary Commission founded in 1872 and in 1950 became a part of the United Nations; originally known as the International Penitentiary Commission

International Prisoners Aid Association publishes the *Directory of Prisoners Aid Agencies*

in the grinder in jail; in prison

in the nick (Cockney English—in jail; in prison)

intimidation fear of punishment

Invercargill Invercargill Borstal Institution near Invercargill on South Island New Zealand

Inverness prison in Inverness, Scotland

Ionia site of the Michigan Reformatory, Michigan Training Unit, and the Riverside Correction Facility

Iowa jails 99 counties provide detention facilities; cities such as Des Moines have their own lockups

Iowa Security and Medical Facility at Oakdale

Iowa State Penitentiary at Fort Madison

Iowa Training School for Girls at Mitchellville

IPAA International Prisoners Aid Association (Louisville, Kentucky)

IPPC International Penal and Penitentiary Commission (*see* entry)

IPPF International Penal and Penitentiary Foundation (Neuchatel, Switzerland)

Iranian prisons exist in Tehran, Ishfahan, Mashad, Tabriz, and other metropolitan places

IRC (*see* Inebriate Reception Center)

Irish prisons Dublin's old prisons is Kilmainham; Cork prison is on Rathmore Road; Limerick has one on Rutland Street

Irma Wisconsin site of the Lincoln Hills School

Irmo Harbison Correctional Institution for Women at Irmo, South Carolina

iron houses jail; lockup; penitentiary; prison

ironmongery department Her (His) Majesty's prison

ISIS Investigative Support Information System (FBI)

Isla de la Juventud (Spanish—Isle of Youth)—Isla de Pinos (Isle of Pines)

Isla de Pinos (Spanish—Isle of Pines)—site of prisons housing political prisoners

Islamabad (*see* Pakistan)

Island Blackwell's (Welfare, now Roosevelt) Island in New York City; Parkhurst Prison on the Isle of Wight; Rikers Island in New York

Island of Hell Norfolk Island in the South Pacific Ocean, once the most dreaded of all Australian penal stations

Isle of Flowers (*see* Ilha de Flóres)

Isle of Pines (*see* Cuba, Isla de Pinos)

Isle of Wight Parkhurst Prison on the Isle of Wight in the English Channel

isolation tank solitary-confinement cell

isolator Soviet penal colony specializing in solitary confinement; many were in distant parts of the Siberian Arctic and in the White Sea east of Kern on the Solovetski Islands

Isole Eolie (Italian—Aeolian Islands)—the Lipari Islands off the north coast of Sicily used as penal colonies

ISP Idaho State Penitentiary; Institute of Social Psychiatry

IST In-Service Training (within California state prisons)

Istanbul formerly Constantinople (*see* Turkey)

Italian prisons Rome's Regina Coeli (Queen of Heaven), Genoa (Prigione d'Genova), Milan (Ducato Milanese), Naples (Caserta), and Venice (Pozzi)

Ivanovo, internal prison of the Soviet secret police in Ivanovo

Ivory Coast's prisons during French occupation prisons were built that are still in use in the capital, Abidjan, and in Bouaké as well as in smaller cities

Iwakuni U.S. Marine Corps Correctional Facility of Iwakuni, Japan

Izmir Turkish city jails and prison (formerly ruled by the Greeks who called it Smyrna)

jacket prisoner's case history or dossier

Jackson site of Georgia Diagnostic and Classification Facility; Jackson, Mississippi jail; Michigan site of what probably is the world's largest walled prison enclosing 57 acres and 23 hectares, as well as the Reception and Guidance Center and the State Prison of Southern Michigan; North Carolina location of the Odom penal facility

Jacksonville Duval County's jail in Jacksonville, Florida

Jaffna (*see* Sri Lanka)

jail penal facility usually run by local law-enforcement officers; often used as pre-trial detention centers

JAIL Justice Against Identificatioin Laws

jailage a jailer's fee, bribes

jail bait person(s) whose illegal activities lead to incarceration; female under legal age of sexual consent

jailbird ex-convict, prisoner, or recidivist

jail delivery clearing a jail of its inmates by bringing them to trial and then releasing them or remanding to prison

jail distemper jail fever (typhus); sickness brought on by incarceration or fear of incarceration

jail fever typhus fever occurring in jails and other crowded places

Jail Forum quarterly publication of the National Jail Association

jailhouse building used as a jail

jailhouse lawyer convict who is well-informed about the law

jailhouse punk any prisoner who becomes a homosexual while imprisoned

jail limits area or district surrounding a jail where debtors may be at large under a bond of security

jail plant narcotics concealed on a person condemned to imprisonment or visiting a penal institution

Jakarta (*see* Indonesian prisons)

Jalapa prison in Jalapa, the capital of the Mexican state or Veracruz

Jamaica penal facilities include Richmond Farm Prison, St Catherin's District Prison, and the Tamarind Farm Prison

Jamejala Soviet Psychiatric hospital

Jamesburg New Jersey location of the Training School for Boys and Girls

Jamestown Sierra Conservation Center at Jamestown, California

jamocha java + mocha (prison slang for coffee)

Janie Porter Barrett School for Girls at Hanover, Virginia

Japanese-American relocation centers caps set up after the attack upon Pearl Harbor at Manzanar and Tule Lake in California, Gila and Poston in Arizona, Topaz in Utah, Minidoka in Idaho, Heart Mountain in Wyoming, Amache in Colorado, Jerome and Roh-wer in Arkansas

Japanita Santa Anita Assembly Center's nickname when the racetrack stables were used for the incarceration of Japanese-Americans

Japan's prisons penal facilities are provided by all 47 of Japan's prefectures and are to be found in its cities such as Tokyo, Osaka, Yokohama, Nagoya, Kyoto, Kobe, Saporo, Kitakyushu, Fukuoka and Kawasaki

jaula (Spanish—cage)—slang for a jail or lockup

Jax Jacksonville, Florida lockups

Jay State Prison at Jay, Florida

J-bird jailbird

jd juvenile delinquent

JDC Juvenile Detention Center

Jean site of the Southern Nevada Correctional Center

Jeff City Jefferson City, Missouri jail

Jefferson Jefferson County Jail in Birmingham, Alabama; Jefferson Parish Jail near New Orleans, Louisiana

Jefferson City site of the Central Missouri Correctional Center, the Missouri Intermediate Reformatory, and the Missouri State Penitentiary for men

Jefftown Journal prisoner's periodical published at Jefferson City, Missouri's prison

Jen Penjara Singapore's Remand prison nickname for the road where it is located

Jerome Arkansas location of a Japanese-American relocation center during World War II

Jersey City New Jersey city lockups

Jess Dunn Correctional Center in Taft, Oklahoma

Jessup site of the Maryland Correctional Institution for Women, Maryland Correctional Pre-Release System, Maryland House of Correction

Jester Unit Texas pre-release facility in Richmond

Jesup federal correctional institution in Jesup, Georgia

Jewell Manor Jewell Manor Girls Center at Louisville, Kentucky

JHA (*see* John Howard Association)

JHAH John Howard Association of Hawaii

JHDF Juvenile Hall Detention Facility

jhj's jailhouse juvenile delinquents

Jidda (*see* Saudi Arabia)

JIS Jail Inspection Service

Jiuren former Soviet prison near Inner Mongolia

JJC Juvenile Justice Center (Los Angeles, California)

JLS Jail Library Service (California State Library)

Joburg Johannesburg, South Africa jail

Joe Harp Correctional Center in Lexington, Oklahoma

Joelton location of the Tennessee Youth Center

Joey man who takes a prisoner's place at home while the convict is imprisoned

Johannesburg (*see* South Africa)

John Saint John in the American Virgin Islands, Saint John in New Brunswick, Canada and their lockups

John Howard Association founded in 1901 to honor an 18th-century English prison reformer; meets in Chicago and its members offer professional consultation services

John Howard Association of Hawaii Honolulu's prisoner service agency

John's Saint John's, Newfoundaland, jail

Johnson Hall Forensic Unit Evanston, Wyoming, for prisoners who are dangerously mentally ill or in need of extensive psychiatric treatment

Johnson's Island Confederate prisoners of war were held on Johnson's Island in Lake Erie

Joliet site of the Joliet Correctional Center and the Statesville Correctional Center

Joliet Correctional Center maximum-security prison in Joliet, Illinois

Joplin Joplin, Missouri, county jail

Jordan's prison Jordan's capital, Amman, has one of the world's worst prisons

Jorge Navarro Nicaraguan prison in Tipitapa

Journal of Correctional Education quarterly publication of the CEA (Correctional Education Association)

Joycevill Institution in Kingston, Ontario

JPS Juvenile Probation Services

jrl jail-release information

Juárez Ciurdad Juárez, Chihuahua, jail

Juariles (Mexican-American—Ciudad Juárez)—or its jail

Jubilee Jubilee Lodge for Girls at Brimfield, Illinois

judas peephole in a prison-cell door constructed so the inmate(s) can be observed without knowing it

judicial execution execution in response to a court order

judicial hanging hanging performed in response to a court order

judicial murder capital punishment

judicium capitale (Latin—capital justice)—justice through execution

jug(ged) jail(ed)

Jugoslavia (*see* Yugoslavia)

jug(s) jail(s); prison(s)

jug tank(s) prison cell(s) for drug addicts

Julia Julia Tutwiler Prison for Women at Wetumpka, Alabama

Juneau Juneau Correctional Center in Alaska

Jungfernhof extermination facility near Riga where many Austrian Jews were murdered

junk tank prison cell for drug addicts

jus publicum (Latin—penal law; public law)

Justice U.S. Department of Justice

juve delinquent(s) juvenile delinquent(s)

Juvenile Hall holding facility for juvenile delinquents

juvenile-justice agency government agency concerned with the adjudication, care, confinement, investigation, and supervision of juvenile delinquents

Juvenile Justice Center Los Angeles, California agency designed to cope with juvenile delinquency

juvenile penal institutions of the U.S. Bureau of Prisons Ashland, Kentucky; Englewood, Colorado; Morgantown, West Virginia

juvenile record(s) official record(s) containing information concerning juvenile court proceedings and all applicable correctional and detention process ordered

juvei(s) juvenile delinquent(s); juvenile hall(s); juvenile lawenforcement officer(s)

K-9 Corps Canine Corps (police dogs)

Kabul capital city of Afghanistan and nickname for its prison

Kaiserwald Nazi concentration camp in Latvia during most of World War II

Kakogawa careless-driver prison serving Japan's Osaka-Kobe-Kyoto area; prisoners receive driving lessons daily; at dawn and dusk they retire to the prison's Park of Interrogation where they apologize to the memory of their victims and swear never to repeat their mistakes

Kalgoorlie Kalgoorlie Regional Prison (Kalgoorlie, Western Australia)

Kaluga Soviet prison 90 miles southwest of Moscow

Kampala (*see* Uganda)

Kandy (*see* Sri Lanka)

Kangaroo court prisoners' court imposing contributions, fines, and work tasks on convicts brought before it; nickname for any small court harsh defendants

kanga(s) kangaroo(s)—Australian prison warden(s)

Kansas City Kansas City, Kansas or Kansas City, Missouri jails

Kansas Correctional Institution in Lansing

Kansas Correctional-Vocational Training Center in Topeka Kansas jails the state's 105 counties provide detention facilities; Wichita and Topeka also have local lockups

Kansas State Industrial Reformatory in Hutchinson

Kansas State Penitentiary in Lansing

Kansas State Reception and Diagnostic Center in Topeka

Kaohsiung (*see* China, Republic of)

kapidiye (Turkish—hardened criminals)—the most feared in prisons where they bribe the guards and rule the other inmates

Karachi (*see* Pakistan)

Karaganda and Kolyma two of the world's largest forced-labor penal camps; both in the former USSR

Karnet Karnet Rehabilitation Centre (Western Australia)

Kars prison near the eastern border of Turkey

Kathmandu (*see* Nepal)

katorga (Russian—hard penal servitude)—forced labor

Kauai Kauai Community Correctional Center at Lihue on Kauai, Hawaii

Kaufering forced-labor subcamp in the south of Germany

Kawasaki (*see* Japan's prisons)

Kay-Cee Kay-Cee Honor Center in Kansas City, Missouri

KC Jackson County Jail in Kansas City, Missouri

KCCD Kentucky Council on Crime and Delinquency

KCJ King County Jail in Seattle, Washington

Kearney Nebraska location of the Youth Development Center

keester plant hollow suppository made of metal, plastic, rubber, or wood and used to conceal contraband in the rectum

keester stash anything hidden in the rectum

Kendall juvenile-delinquent rehabilitation center at Kendall, Florida

Kent Kent Correctional Institution in Dover, Delaware

Kent Institution maximum-security penal facility in Agassiz, British Columbia

Kentucky Correctional Institution for Women at Pee-wee Valley, site of the Assessment and Orientation Unit

Kentucky lockups 120 counties have jails; cities such as Louisville and Lexington have augmented facilities as well

Kentucky Manpower Development state agency charged with the task of developing a prisoner rehabilitation program

Kentucky State Penitentiary maximum-security prison at Eddyville

Kentucky State Reformatory medium-security facility at La Grange

Kenya's prisons date back to the time of British rule

Ketchikan Ketchikan Correctional Center, Alaska

Kharkov world's largest prison in the capital city of the Kharkov region of the Ukraine

Khartoum (*see* Sudan)

Kholmogori Arctic death camp established by Lenin near Archangel in 1921 for the exploitation and suppression of political prisoners in the Soviet Union

Kilby Kilby Corrections Facility at Montgomery, Alabama

Kilmainham Dublin jail

Kincheloe Michigan site of the Kinross Correctional Facility

kindergarten of vice epithet applied to many jails

King King County Jail in Seattle, Washington

King's Bench one of Southwark's seven prisons, now all gone

Kingston Canadian site of the Collins Bay Institution as well as the Joyceville Instittition; (*see* Jamaica)

Kingston Pen Kingston Penitentiary (and mental hospital) in Portsmouth, Ontario

Kingstown (*see* Saint Vincent and the Grenadines)

Kinross Correctional Facility near Kincheloe, Michigan

Kinston North Carolina site of the Dobbs School for juvenile delinquents

Kiribati Tarawa, the capital of this colony what was once HM Prison

Kirikiri Lagos, Nigeria's prison

Kirkham HM Prison at Kirkham in Lancashire, England

Kirkland Correctional Institution in Columbia, South Carolina

Kitakyushu (*see* Japan's prisons)

KL Konzentrationslager (German—concentration camp) —prisoners contracted this to *KZ*

klondike solitary prison cell

KMCI Kettle Moraine Correctional Institution (near Duluth, Minnesota)

KMD Kentucky Manpower and Development (*q.v.*)

knowledge factory prison school

Knoxville Knox County, Tennessee's jail

Kobe (*see* Japan's prisons)

kogus (Turkish—cell block)— (*see turist kogus*)

Kolyma USSR penal camp north of the Arctic Circle

Korydallos prison in Athens, Greece

Kot Lakhpat jail in Lahore, Pakistan

Kragshovhede Danish prison without high fences or walls; this type of facility is said to result in reduced recidivism; located near Skagen or Skaw

Krems site of a World War II concentration camp

Kresty Leningrad's central prison

KRIM the Danish association for penal reform

Kriminalstrafkunde *die Kriminalstrafkunde* (German—penology)

KROM the Norwegian association for penal reform

Krome Avenue address and nickname of the immigration and naturalization detention camp on Krome Avenue in Miami, Florida

KRUM the Swedish association for penal reform

Kryukovo USSR prison colony

Ktr *Katorzhane* (Russian—compound)—prison compound for people sentenced to hard labor

Kuala Lumpur (*see* Malaysian prisons)

Kuibyshev Russian transit prison

Kulani Kulani Correctional Facility in Hilo, Hawaii

Kumasi (*see* Ghana's prisons)

Kumla Kumla Prison, institution for long-term prisoners in south-central Sweden

Kunie French penal colony in the Southwest Pacific

Kuwait has prisons in Hawalli and Kuwait City

Kyoto (*see* Japan's prisons)

LA Los Angeles penal institutions

La Cabaña (Spanish—The Cabin)—old Spanish prison at the entrance to Havana harbor still used for the execution and imprisonment of political prisoners

La Ceiba prison in a seaport city of Honduras on the Caribbean

LACJ Los Angeles County Jail (California)

La Ferté Macé French detention camp

La Force top-security prison of Paris in the late 1700s and early 1800s

lag ticket-of-leave man; transported convict of the type Britain shipped to Australia and Tasmania

Lager Westerbork transit camp established in the Drenthe province of the Netherlands by German occupation forces during World War II (*see Durchgangslager*)

lagging serving a three-year sentence in a British prison; transportation by sea of convicts from overcrowded jails in the British Isles to those in Australia and Tasmania

Lagoinha jail in Belo Horizonte, Brazil

Lagos (*see* Nigeria)

La Grange site of the Kentucky State Reformatory as well as the Luther Luckett Correctional Complex

lag(s) transported convict(s)

La Guaira Venezuelan jail

Lahore (*see* Pakistan)

Lake Butler Florida site of the Reception and Medical Center

Lake Correctional Institute penal facility at Lowell, Florida

Lakehills Lakehills Community Corrections Center in Salt Lake City, Utah

La Mesa Penitenciaria (Spanish—La Mesa Penitentiary)—Baja California's major prison, in the eastern section of Tijuana

laminated glass principal component of detention glazing (*see* detention glazing)

La Modelo *La Carcel Modelo* (Spanish—The Model Prison)—principal prison of Caracas, Venezuela

lamster(s) escaped convict(s)

Lancaster HM Prison at Lancaster in Lancashire, England or the jail in Lancaster, Pennsylvania, or jails in smaller Lancasters in Canada and the United States; site of the Southeastern Ohio Training Center

Lancaster Correctional Center at Trenton, Florida

Lancaster Pre-Release penal facility in Lancaster, Massachusetts

Land of Death and Chains Maxim Gorki's nickname for Siberia

'Lando Orlando, Florida jail

Land of Political Exiles Yakutia, Siberia

Landsberg fortress prison on the Lech River in Upper Bavaria

Langholmen Swedish prison

Langi Langi Kal Kal Youth Training Centre at Trawalla, Victoria, Australia

Lansing site of the Kansas Correctional Institution and the Kansas State Penitentiary; State Industrial Farm for Women at Lansing, Michigan

'Lanta Atlanta, Georgia, jails, lockups, and federal penitentiary

Lantan Lantan Island (contains Ma Po Ping Prison serving Hong Kong)

Lantana Florida site of the Lantana Correctional Institution

La Nuestra Familia (Spanish-American jail jargon—Our Family)—prison racketeers who deal with homosexual prostitution, loan sharking, murder contracting, and narcotics; a Mexican-American Mafia

Laos has an old prison in Vientiane built during the years of French-colonial rule

La Paz Bolivia's capital high in the Andes where prisoners get cool mountain air but very little else; Mexico's La Paz, near the tip of Baja California

La Pica (Spanish—the stone-cutters hammer; the pike or the long lance of the picador)—nickname of Venezuela's central penitentiary of Oriente

La Pieota prison in Bogota, Colombia

La Planta house of reeducation and artisan work in the El Paraiso section of Caracas, Venezuela

La Plata (*see* Argentina)

La Porte Noire (French—The Black Gate)—main entrance to the penal colony and prison in Saint Laurent du Maroni, nicknamed The Gate of Hell

La Roquette Parisian prison for women

La Route Zéro (French—The Zero Route)—Road to Nowhere leading out of the Saint Laurent du Maroni headquarters built by convicts in French Guiana over 50 years

LAS Legal Aid Society

Las Colinas Las Colinas Girls' Facility in Santee, California east of San Diego

Las Palomas Las Palomas, Chihuahua jail, Mexico

last mile euphemism for the walk from the prison cell to the place of execution

last waltz condemned prisoner's march to place of execution

Las Vegas (*see* Vegas)

Las Ventas (Spanish—The Stalls)—Madrid's municipal prison

Latchmere House a remand center in Surrey, England

latrinogram latrine rumor

La Tuna Federal Correctional Institution at La Tuna, Texas

Laurel maximum-security juvenile facility near Laurel, Maryland

Laval Ville de Laval, Quebec, site of the Correctional Development Centre and the Federal Training Centre as well as the Leclerc Institution

Laval-des-Rapides Ecole Notre-Dame de Laval at Lavaldes-Rapides, Québec

Lawtey Florida site of the Lawtey Correctional Institution for older inmates with medical problems

Lawton Lawton, Oklahoma, Comanche County Jail

lazaretto hospital for contagious diseases; place of quarantine; tweendecks storeroom sometimes used as a hospital or even as a lockup if a vessel is without a brig

LCCC Lucas County Corrections Center (Ohio)

LCCJ Louisiana Council on Criminal Justice

Leavenworth U.S. Penitentiary at Leavenworth, Kansas

Lebanon Lebanon Correctional Institution in Lebanon, Ohio; Middle-East republic, has prisons in its capital, Beirut, and in its port city of Tripoli

Leclerc Institution in Laval, Québec

Lecumberri-Hilton nickname of Mexico City's Lecumberri prison

Leeds HM Prison at Leeds in Yorkshire, England

Leesburg location of the Lee Correctional Institution, Georgia, New Jersey

LEF Liberté, Égalité, Fraternité (French—Liberty, Equality, Fraternity)—slogan of the French Revolution, inscribed in bronze over the gates of prisons built by the French

Lefortovo prison in Moscow described in *The Gulag Archipelago by* Aleksander I Solzhenitsyn

Leicester HM Prison at Leicester in Leicestershire, England

León (*see* Nicaragua)

Leopoldville (*see* Kinshasa)

Lesotho its capital, Meseru, has a lockup

lethal gas method of execution in California, Maryland, Mississippi, Missouri, Nevada, North Carolina, Oregon, Rhode Island, Wyoming

lethal intravenous injection method of execution in Idaho, New Mexico, Oklahoma, Texas

Leuven Leuven Prison (also called Louvain) in Belgium

level four death-dealing injection used by executioners

Lewes small jail just inside the Delaware River breakwater

Lewisburg U.S. Penitentiary at Lewisburg, Pennsylvania

Lewisburg Plan the plan of the Lewisburg Penitentiary, also called the telephone-pole design, providing for maximum-and medium-security cells, inside and outside, respectively

Lex any Lexington or its correctional facilities (*see* Lexington)

Lexington Federal Correctional institution at Lexington, Kentucky; location of the Blackburn Correctional Complex as well as the special units of the Danville Youth Development Center; Oklahoma site of the Joe Harp Correctional Center and the Lexington Assessment and Reception Center; U.S. Public Health Service Hospital in Lexington, Kentucky (drug detoxification facility)

ley de fuga (Spanish—law of flight)—privilege of law-enforcement officers to kill anyone attempting to escape

Leyhill HM Prison at Leyhill in Gloucestershire, England

LGC Laminated Glass Corporation (*see* detention glazing)

LGD London Gaol Delivery

Libby Libby Prison—converted tobacco warehouse in Richmond, Virginia, used to house Union prisoners during the Civil War

libéré (French—liberated convict)—a prisoner free from confinement but not free to leave the country of confinement

Liberia site of Monrovia and Buchanan prisons

Liberty Center Ohio site of the Maumee Youth Camp

Liberty Street city jail on Liberty Street in Louisville, Kentucky

Libreville (*see* Gabonese prisons)

Libya Tripoli (the capital) and Benghazi contain prisons built during its Italian occupation from 1912 until the end of World War II

Liechtenstein contains an old prison in the capital, Vaduz

Lieutenant of the Tower Lieutenant of the Tower of London (its warden)

lifeboat commutation of a death sentence or a prison term; judicial order for a retrial

lifer prisoner sentenced to life imprisonment

lifer's lament *gruntin' don' git yuh nuttin'* (complaining is useless)

Lihue port city on Kauai Island, Hawaii, and location of the Kauai Community Correctional Center

likvidatsiya (Russian—liquidation)—Soviet euphemism for execution

Lilongwe (*see* Malawi)

Lima Lima, Ohio, Allen County Jail or Lima, Peru jails and prisons

Lima State Hospital in Lima, Ohio offering felons psychiatric services

Limón Costa Rican prison in Puerto Lim6n

Lincoln HM Prison at Lincoln in Lincolnshire, England; jails in fifty-one American cities named Lincoln; Nebraska site of four penal facilities—the Diagnostic and Evaluation Center, the Lincoln Correctional Center, the Nebraska State Penitentiary, and the Post Care Program

Lincoln Correctional Center for youthful offenders in Lincoln, Nebraska

Lincoln Hills School reception center and training school in Irma, Wisconsin

Lino Lakes site of the Minnesota Correctional Facility

Lipari Islands Italian islands off the north coast of Sicily and called Aeolian Islands *(Isole Eolie);* for centuries, a place of exile for hardened criminals and political prisoners; islands include Alicudi, Basiluzzo, Filicudi, Lipari, Salina, Stromboli, and Vulcano

Liparis (*see* Lipari Islands)

Lisbon (*see* Portugal)

Lisieux maximum-security prison in Lisieux, France

Litchfield Community Correctional Center in Litchfield, Connecticut

Little Rock Pulaski County jail in Arkansas

Liverpool HM Prison at Liverpool in Lancashire, England; jails in other places called Liverpool as in New South Wales, New York, and Nova Scotia

Ljubljana (*see* Yugoslavia)

local lockup usually the police station or county sheriff's prison

lockup a small jail or prison

Lock-Up Tree name of a huge hollow baobab tree in Derby, Australia; its 52-foot (16-meter) girth made it a natural lockup for prisoners

locus penitentiae (Latin—place of repentance)—penitentiary

Lodz (*see* Poland)

Logan Correctional Center in Lincoln, Illinois

Lomé (*see* Togo)

Lompoc U.S. minimum-security penitentiary and prison camp at Lompoc, California

London Ohio town containing the London Correctional Institution west of Columbus; (*see* United Kingdom)

Long Bay Sydney, New South Wales, Australian prison system, including the Central Industrial Prison, Her Majesty's Training Center, and the Parramatta Gaol

Long Beach Long Beach, California jails, lockups, or Community Treatment Center

long bid(s) long prison term(s)

Long Lane Long Lane School for (delinquent) Girls at Middletown, Connecticut

Long Lartin HM Prison at Long Lartin in Worcestershire, England

Longos (Mexican-American— Long Beach, California)— nickname of the city, its lockup, and jail, and the detention pens of the Immigration and Naturalization Service

Longriggend Scotland's Longriggend Remand Institution

long stretch(es) long prison sentence(s)

long-term adult penal institutions of the U.S. Bureau of Prisons Atlanta, Georgia; Leavenworth, Kansas; Lewisburg, Pennsylvania; Lompoc, California, Marion, Illinois; Terre Haute, Indiana

loopbelt restraining device used in handling and transporting unruly prisoners

loquera (Spanish slang— prison)—also called *banasto* or *chirona*

Lorton Virginia site of the Central Facility, the Maximum Security Facility, Youth Center I and II

Los Angeles metropolitan detention center in Los Angeles, California

Los Guilucos Los Guilucos School for (delinquent) Girls at Santa Rosa, California

Los Lunas site of the Central New Mexico Correctional Facility and the Los Lunas Correctional Center

Los Lunas Correctional Center in Los Lunas, New Mexico, opened in 1940 to teach inmates farming and livestock operations

Loudonville Ohio site of the Mohican Youth Camp

Loughan House penal facility in Blacklion, Ireland

Lough Kesh location of Maze Prison near Belfast, Northern Ireland

Louie Saint Louis, Missouri, jails, lockups, and other penal facilities

louisette beheading device perfected by Dr Antoine Louis of Paris at the suggestion of his colleague Dr Joseph Ignace Guillotin; the resulting device is called the guillotine

Louisiana Correctional and Industrial School minimum-security facility for first-time young offenders held in De Quincy

Louisiana Correctional Institute for Women at St Gabriel for female felons 17 years of age and up

Louisiana lockups 64 parishes have holding facilitim New Orleans and Baton Rouge have extra facilities for felons

Louisiana State Penitentiary maximum-security facility at Angola

Louisiana Training Facility juvenile-delinquent penal facilities at Ball (for females), Baton Rouge, Bridge City, and Monroe

Louisville Kentucky's principal city and county seat of Jefferson County and its jail

Lourenço Marques (*see* Mozambique)

Louvain Belgium's central prison; here long-term prisoners work in open cells or in special workshops; called Leuven by Flemish Belgians

Lovelady Texas location of the Eastham Unit

Lowdham Grange HM Borstal at Lowdham Grange in Nottinghamshire, England

Lowell site of the Florida Correctional Institution and the Marion Correctional Institution

Low Newton remand center in Durham, England

LP Liverpool Prison

LTI Louisiana Training Institute (branches at Baton Rouge, Monroe, and Pineville)

Luanda prison in the People's Republic of Angola

lubang buaya (Indonesian—crocodile hole)—Djakarta water hole infested with crocodiles and used as a place to dispose of people

Lubianka Moscow headquarters of the Ministry of the Interior—the former Soviet secret police, named for the founder of the Cheka; located one block from the Kremlin

Lublin concentration camp in Poland

Lucasville Southern Ohio Correctional Facility at Lucasville

Lud Ludgate Prison in London long ago

Lurigancho (*see* San Pedro)

Lusaka (*see* Zambia)

Luther Luckett Correctional Complex in LaGrange, Kentucky

Luxembourg capital city contains an old prison

Luzira Kampala, Uganda's prison

lynching executing someone by mob rule rather than by the rule of law; freeing a suspect in police custody (slang definition describing encounters between gangs or mobs and police)

Lynwood Lynwood (delinquent) Girls Center at Anchorage, Kentucky

M-2 Match-Two—program matching volunteers from a community with prison inmates; sponsors write to inmates and visit them regularly with the aim of establishing meaningful relationships and providing convicts with references and job support after they are paroled

Maastricht Dutch maximum-security prison

MAB Metropolitan Asylums Board (British group responsible for administration of all sorts of asylums)

MacDougall Youth Correction Center in Ridgeville, South Carolina

MacLaren School in Woodburn, Oregon holds juvenile-court commitments in medium custody

Macon Georgia location of the Central Correctional Center

Macquarie Harbour inlet on the west coast of Tasmania that in the early 1800s contained a penal colony on Settlement Island; the entrance to the inlet was nicknamed the Gates of Hell

Madagascar's prisons old prisons are in the capital, Antananarivo, and other places such as Toamasina and Majunga

Madison Dane County jail, Wisconsin

Madras (*see* Indian prisons)

Madrid (*see* Spain)

Magadan port on the Sea of Okhotsk where Soviet ships delivered prisoners enroute to forced labor camps in the Gu-

lag; one ship arrived with all of its crew intact but all of its prisoners frozen to death

Magdalena penal institute of the Argentine armed forces about 80 kilometers from Buenos Aires

Magdeburg concentration camp subcamp southwest of Berlin

Magilligan prison near Londonderry in Northern Ireland

Magilligan Camp HM prison camp outside Belfast, Northem Ireland

Ma Hang Ma Hang Prison, Hong Kong

Maidstone HM Prison at Maidstone in Kent, England

Maine Correctional Center penal facility in South Windham

Maine jails 16 counties maintain jails; cities such as Portland and Bangor have local lockups as well as county jails

Maine State Prison at Thomaston holds incorrigibles in maximum security

Maine Youth Center in South Portland holds delinquents from 11 to 18 years of age

maison (French—house)—also a jail or lockup (*maison d'arrêt*) or a borstal or reform school (*maison de correction*)

maison de arrêt (French—prison)

maison de correction (French—house of correction)—penitentiary, prison

Maison de Correction St Bernard Brussels, Belgium's house of correction

maison de force (French—workhouse)—prison where only those who work are fed

Maison Gomin women's correctional facility at St Cyrille, Québec

Maison Notre-Dame de la Garde jail for Catholic delinquents ranging from 14 to 18 years of age at Cap-Rouge, Québec

Maison Tanguay women's penal establishment in Montreal, Québec

Majdenek Nazi concentration camp near Lublin, Poland

Majunga (*see* Madagascar's prisons)

Makindye Uganda's military prison

making little ones out of big ones [*see* rock crusher(s)]

Malabar Malabar Complex of Prisons in Australia's New South Wales

Malaga (*see* Spain)

Malang women's prison in eastern Java, Indonesia

Malawi the capital, Lilongwe, and other cities, contain pnsons built by the British between 1881 and 1966

Malaysia the capital, Kuala Lumpur, and cities such as George Town have prisons built by the British between 1889 and 1904

Malchow North German concentration camp

Maldives the capital, Male, contains a prison built during British dominion (1887–1965)

Male (*see* Maldives)

Mali has a prison in Bamako

Malines Belgian transit camp near the French border and the North Sea built by German forces during World War II as a way station for concentration camp prisoners; also known as Mechlin

Malmö (*see* Sweden)

Malta the capital city and seaport, Valetta, containing a prison built during British rule (1814–1964)

manacle handcuff

Managua capital city of Nicaragua replete with a prison

Manama capital city of Bahrain and its most-humid prison

Manbarco Man Barrier Corporation of Seymour, Connecticut engaged in manufacturing electronic detection systems, and physical barriers made of coils of barbed wire and knife-edged wire used to keep prisoners within bounds

Manchester HM Prison at Manchester in Lancashire, England, New Hampshire location of the Manchester Community Corrections Center, the New Hampshire Youth Development Center; place of detention in other cities named Manchester in Con-

necticut, Georgia, Iowa, Kentucky, Massachusetts, New York, Ohio, Tennessee, and Vermont

Manchester Community Corrections Center in New Hampshire

Mandan site of the North Dakota Industrial School opened in 1903

Manhattan Manhattan Island, center of New York City and New York County; has several large correctional facilities and many police-station lockups

Manhattan House of Detention Tombs Prison in lower Manhattan; nicknamed Down Home

manhunt hunt organized to catch a criminal, an escapee, a fugitive from justice, or even a person who is lost

Manila (*see* Philippines)

Mannheim Confinement Facility of the U.S. Seventh Army in Mannheim, Germany

Manning Correctional Institution in Columbia, South Carolina

Manor English slang for London or its prisons

Mansfield site of the Ohio State Reformatory

Mansions The Mansions in Brisbane, Queensland—Australia's name for its Prison Department

Manzanar American relocation center for Japanese-Americans detained in California's Owens Valley

Manzanillo prison in Manzanillo, Mexico

Manzini (*see* Swaziland)

MAOF Mexican-American Opportunities Foundation (supporting a program to rehabilitate juvenile Chicano recidivists)

Maplehurst Correctional Centre in Milton, Ontario

Maple Lane School in Centralia, Washington south of Olympia

Ma Po Ping Ma Po Ping Addiction Centre on Lantau Island, New Territories, Hong Kong

Maracaibo Venezuela prison

Marble Hill Bollinger County jail at Marble Hill, Missouri

Marian Hall jail for English speaking juvenile-delinquent females at Beaconsfield, Québec

Marianna federal prison in western Florida

Marigot (*see* Saint Martin)

Marina Street San Juan, Puerto Rico's district jail

Marion U.S. Penitentiary at Marion, Illinois; Marion County Detention Home for juvenile delinquents in Indianapolis, Indiana; Marion Correctional Institution, Marion, Ohio

Marion Correctional Institution in Lowell, Florida, once the Florida Correctional Institution

Marion Correctional Treatment Center in Marion, Virginia

Marquette Michigan site of the State House of Correction and Branch Prison

Marrakech (*see* Morocco)

Marshall site of a Confederate prison camp in northeast Texas

Marshalsea one of Southwark's seven prisons, now all gone, but formerly dominating much of this London district on the south bank of the River Thames

Martinez Contra Costa County courthouse and jail northeast of San Francisco, California; location of one of many clandestine prisons in Argentina

Martinière combination cargo and prison ship whose tween decks and lower holds were fitted with removable cages for transporting convicts from France to French Guiana; the vessel, built in 1911, was last seen off Devil's Island early in 1950

Maryland Correctional Institutions started in 1931 as a penal farm in Hagerstown

Maryland Correctional Institution for Women in Jessup

Maryland Correctional PreRelease System in Jessup with supporting facilities in Baltimore, Church Hill, Hughesville, Quantico, and Sykesville

Maryland Correctional Training Center in Hagerstown

Maryland House of Correction in Jessup

Maryland jails the state's 23 counties provide lockups; Baltimore has its own lockup facilities

Maryland Penitentiary maximum-security prison in Baltimore, next to the Reception Center

Marysville Ohio Reformatory for Women in Marysville

MASCA Middle Atlantic States Correctional Association

Massachusetts Correctional Institutions Bridgewater holds alcoholics, criminally insane, and sexually dangerous felons; Cedar Junction is a maximum-security facility; Framingham contains females; others at Norfolk, South Carver, Orange, and Concord

Massachusetts lockups 14 counties provide imprisonment facilities; Boston has penal facilities of its own

Massachusetts State Prison (*see* South Walpole)

Massawa (*see* Ethiopia's prisons)

Massilon Ohio town holding the Indian River School for male juvenile delinquents

Matamoros the Tamaulipan city maintains one of the world's dirtiest jails

Matanzas (*see* Cuba)

Matrah (*see* Oman)

Matsqui Institution British Columbia's minimum-security facility for drug addicts at Abbotsford

Matteawan Matteawan State Hospital (for the criminally insane) near Beacon-on-Hudson, New York

Maui Maui Community Correctional Center at Wailuki on Maui Island, Hawaii

Maumee Youth Camp male juvenile-delinquent penal facility in Liberty Center, Ohio

Mauritania has an old prison built during French-colonial domination

Mauritius Port Louis has an old prison built by the British that has held criminals as well as political prisoners

Mauthausen Austrian Nazi concentration camp near Linz

maximum custody keeping prisoners in penal institutions built with tool-proof bars and cells surrounded by high walls; maximum-security prisons are manned by many guards and are run on a plan calling for rigid discipline

maximum security applied to inmates considered dangerous to correctional officers, to others, and to themselves; prisoners awaiting the death penalty are also kept under maximum security

Maximum Security Center in Columbia, South Carolina

Maximum Security Facility in Smyrna, Delaware; also in Cranston, Rhode Island

Maxwell Federal Prison Camp at Maxwell Air Force Base near Montgomery, Alabama

Mazatlan prison in Mazatlan, México

Maze Maze (top-security) Prison outside Belfast in Northern Ireland

Mbabane (*see* Swaziland)

MCA Massachusetts Correctional Association (Boston-based); Medical Correctional Association; Minnesota Corrections Authority; Missouri Corrections Association

McAlester site of the Oklahoma State Penitentiary

MCC Metropolitan Correctional Center (in Chicago, New York, and San Diego)

McCain North Carolina Prison Sanitorium at McCain in Hoke County

MCDC Montgomery County Detention Center (Maryland)

MCI Massachusetts Correctional Institution (Framingham)

McLaughlin Youth Center Anchorage, Alaska's diagnostic-program reception center for delinquents

McNeil Island McNeil Island Correctional Center in the state of Washington (formerly a U.S. maximum-security prison)

McNeil Island Correctional Center formerly a federal prison turned over to the state of Washington in 1981

MCO Michigan Corrections Organization (of wardens)

MDC Minnesota Department of Corrections

meat wagon(s) prison van(s)

Mecca (*see* Saudi Arabia)

Mecklenburg correctional center at Boydton, Virginia

Medical Center for Federal Prisoners at Springfield, Missouri

Medina (*see* Saudi Arabia)

meditation solitary confinement

medium custody institutions designed to give freedom of movement and greater scope for positive self-direction

Medium Security Facility in Cranston, Rhode Island

Medium Security Unit at Mount Pleasant, Iowa, usually holding first-term felons within a half-year of release

Melilla (*see* Spain)

meltout escape technique used in some modern prisons where certain types of doors and windows can be melted

Memphis Federal Correctional Institution at Memphis, Tennessee; and the Women's Correctional Center in Memphis

Menard Menard. Correctional center in Menard, Illinois; the Menard

Menard Time prisoner's publication issued by the Menard branch of the Illinois State Penitentiary

Mende maximum-security prison in southern France

menschenhandel (German—trade in people)—prisoner exchange program

Men's Reformatory in Anamosa, Iowa

Merced Merced County jail in California

Mercer Pennsylvania Corrections Department prison in Mercer County

Merida prison in Merida, capital of the Mexican state of Yucatan; or the Merida in Venezuela

meritorious good time promise of parole-induced good behavior on the part of convicts wanting to get out of prison

Meseru (*see* Lesotho)

Met Metropolitan Correctional Center in San Diego, California

metanoia change of heart and mind required for criminal rehabilitation

Metro Metro Correctional Institution in Atlanta, Georgia holds emotionally disturbed inmates in close security while providing psychiatric services

Metropolitan Correctional Centers Chicago, Illinois; New York, New York; San Diego, California

Metropolitan Regional Institution in Bayamón, Puerto Rico

Mexicali capital of Baja California, and its jail

Mexican Mafia controls much of the narcotic trafficking in California prisons such as Chino, Folsom, and San Quentin; also active in Mexican prisons

México nation with a long record of horrible prisons; Mexico City contains some of its most formidable penal facilities

Miami city lockups, Dade County Jail, or the Federal Correctional Institution, the Miami Detention Center

Mich Michigan, Michoacan, or their jail's and prisons

Michigan City Indiana State Prison at Michigan City

Michigan Dunes Correctional Facility in Holland, Michigan

Michigan jails 83 counties provide detention facilities; Detroit and other cities have additional lockups

Michigan Reformatory in Ionia

Michigan Training Unit at Ionia

Midlands Reception and Evaluation Center in Columbia, South Carolina

Mid-Orange Mid-Orange Correctional Facility at Warwick, New York

Midway Texas site of the Ferguson Unit for first offenders

Milan Federal Correctional Center at Milan, Michigan

Miles City Montana site of the Pine Hills School for juvenile delinquents

Milford Delaware's penal facility for juvenile delinquents; also known as Stevenson House

milieu therapy treatment given to aid convicts returning to society via halfway houses, pre-release guidance centers, and tranquilizing drugs

military execution execution by a military firing squad .

milk van nickname for police or sheriff's van for transporting prisoners

Millhaven Institution maximum-security facility in Bath, Ontario

Milton Ontario site of the Adult Training Centre and the Maplehurst Correctional Centre

Milwaukee Milwaukee County Jail in Milwaukee, Wisconsin

Mimico Mimico Correction Center (for males) in Toronto, Ontario

M-in-C Matron-in-Chief

Mineola Nassau County Jail in Mineola, New York

Minidoka relocation center for Japanese-Americans interned after Pearl Harbor in southern Idaho

minimum custody honor dormitories, prison camps, and prison farms offering inmates as much freedom from restraint as possible while preventing their escape

Minimum Security Facility in Cranston, Rhode Island

Minnesota Correctional Facilities at Lino Lakes, Oak Park Heights, Red Wing, Saint Cloud, Sauk Center, Shakopee, and Stillwater (*see individual entries*)

Minnesota lockups 87 counties have jails; Minneapolis, Saint Paul, and Duluth have extra facilities for felons

Minnie Minneapolis, Minnesota or its jails, lockups, or prisons

Mirikiri Lagos, Nigeria's prison

misdemeanant person convicted or guilty of misdemeanors

Missie Mississippi or its jails, lockups, and prisons

Mission Institution medium-security facility in Mission, British Columbia

Mississippi jails the state's 82 counties have prison facilities; Jackson has additional detention facilities

Mississippi State Penitentiary in Parchman; inmates are offered a supervised earned-release, work-release program of rehabilitation

Missiyahu Israeli minimum-security prison offering inmates a work-release program during their pre-release period

Missouri Eastern Correctional Center in Pacific

Missouri Intermediate Reformatory for male felons from 17 to 25 who are held in Jefferson City

Missouri jails 114 counties contain prison facilities; St Louis and other big cities have additional jails

Missouri State Penitentiary for Men in Jefferson City

Missouri Training Center for Men in Moberly

Mitchellville Iowa Training School for (delinquent) Girls at Mitchellville

Moabit Berlin's great prison in the Tiergarten section

Mob Mobile, Alabama or its penal facilities

Moberly site of the Missouri Training Center for Men

Mobile Mobile County jail in Alabama

Mobtown Baltimore, Maryland or its jails, lockups, and other penal facilities

Modelo one of Bogotá, Colombia's prisons

Modesto Stanislaus County jail east of San Francisco, California

Moengo (*see* Suriname)

Mogadishu (*see* Somalia)

Mohican Youth Camp in Loundonville, Ohio

Monaco has only a lockup

Mongolia the capital, Ulan Bator, was once known as Urga and contains a prison of ancient construction

Monowitz forced-labor subcamp close to Auschwitz, Poland

Monroe site of the Louisiana Training Institute at Monroe; site of the Washington State Penitentiary, the Washington State Reformatory, and the Washington State Special Offender Center

Monrovia Monrovia Central Prison (Liberia's largest penal facility)

Monsieur de Paris (French—Mr. Paris)—the guillotine operator

Montana jails the state's 57 counties offer detention facilities

Montana State Prison at Deer Lodge

Montenegro (*see* Yugoslavia)

Monterrey prison in Monterey, capital of the Mexican state of Nuevo León

Montevideo Uruguay's capital containing jails and a prison

Montey Allenwood Federal Prison Camp at Allenwood, Pennsylvania

Montgomery Federal Prison Camp at Montgomery, Alabama or its county jail or the Kilby Corrections Facility (*see* Kilby)

Montjuic Montjuic Castle in Spain has been used as a court to try anarchist assassins and to hold them in its deep dungeons

Montluc French greystone prison in Lyons

Montreal Maison Tanguay correctional facility for women prisoners in Montreal, Québec; location of a number of penal institutions including those in adjacent Ville de Laval

Montreal House of Detention at 800 Boulevard Gouin

Montreal Prevention Centre on Parthenia Street

Montrose School for Girls correctional facility at Reiserstown, Maryland

Monty Montgomery, Alabama or its jail and lockups

Moon Crescent Singapore's minimum-security prison

Moondyne Joe Australian bushranger who was the first man to cross the Swan River Bridge in Fremantle near Perth while escaping from jail

Moor The Moor—Dartmoor Prison near Princetown in Devonshire, England

Moor Court HM Prison at Moor Court in Staffordshire, England

Morelia prison in Morelia, capital of the Mexican state of Michoacan

Morgan Morgan County Regional Correctional Facility in Wartburg, Tennessee

Morgantown Federal Correctional Institution at Morgantown, West Virginia

Moriah Moriah Shock Correctional Facility, New York

Morocco France and Spain built prisons in cities such as Casablanca, Fez, Marrakech, Rabat, and Tangier

Moroni (*see* Comoros)

Morrison Mount View School for (delinquent) Girls at Morrison, Colorado

Morro Castle (*see* Habana)

Morton Hall HM Borstal at Morton Hall in Lincolnshire, England

Mother of Prison Reform Dorothea Dix (1802–1887), American reformer active in Massachusetts and other states

Mothers in Prison Projects organization affiliated with Women in Jails and Prisons

Motown Motor Town (Detroit, Michigan) or its penal facilities

Moundsville location of the West Virginia Penitentiary

Mountain Institution in Agassiz, British Columbia, designed for holding aged inmates

Mountain View Mountain View School for female juvenile delinquents in Helena, Montana

Mount Eden Mount Eden Prison, Auckland, New Zealand

Mountjoy Mountjoy Gaol (Dublin's jail)

Mount McGregor Mount McGregor Correctional Facility near Warwick, New York

Mount Pleasant Iowa location of the Medium Security Unit

Mount View Girls School at Morrison, Colorado

Mozambique all its coastal cities (Beira, Chinde, and its capital, Lourenço Marques) contain prisons of Portuguese origin

MPP Mothers in Prison Projects

MPPCA Maryland Probation, Parole and Corrections Association

MR Michigan Reformatory in Ionia

MRC Minnesota Restitution Center

MTU Michigan Training Unit (reform school)

mule smuggler carrying contraband such as heroin or weapons into jails and prisons

Mulegé Mexican minimum-security prison in Baja California where convicts are free to roam about during the day and many find work in local enterprises, so very few ever attempt escape

multiprisoner transportation unit paddy wagon; police patrol van; prison van

Muncy Pennsylvania site of the State Correctional Institution for female felons

Muntinlupa Philippine prison southeast of Manila (also spelled Muntinglupa)

Muscat (*see* Oman)

musical execution execution performed to the roll of a field drum or tenor drums

Muskegon Michigan site of the Muskegon Correctional Facility

Muskegon Correctional Facility Michigan penal facility

Mutual Welfare League convict self-government system introduced by Warden Thomas Mott Osborne at Sing Sing and later introduced at the U.S. Naval Prison at Portsmouth, New Hampshire

MWL (*see* Mutual Welfare League)

NAAWS North American Association of Wardens and Superintendents

NAB National Alliance of Businessmen (giving ex-convicts a chance by giving them jobs)

NACRO National Association for the Care and Resettlement of Offenders

NADPAS National Association of Discharged Prisoners' Aid Societies

Nafha Israel's top-security prison

Nagoya (*see* Japan's prisons)

Nail City Wheeling, West Virginia noted for its nail factory; nickname for its detention facilities

Nairobi Kenya's capital and nickname of its prison once called HM Prison Nairobi

NAJCA National Association of Juvenile Correctional Agencies

NAJJ National Assessment of Juvenile Justice (University of Michigan)

Nakhodka port near Vladivostok and a Gulag transit center

'Nam Vietnam or any of its many jails and prisons

Namibia Walvis Bay and Windhoek, the capital, have HM Gaol prisons

Napanoch site of the Eastern New York Correctional Facility

NAPO National Association of Probation Officers

NAPV National Association of Prison Visitors

Nashville Tennessee site of the Lois Deberry Correctional Institute, the Nashville Regional Correctional Facility,

the Spencer Youth Center, the Tennessee Prison for Women, the Tennessee State Prison

Nassau capital of the Bahamas on New Providence Island and toponym for its jail

National Association of Training Schools and Juvenile Agencies (NATSJA) created by the merger of the National Association of Training Schools and the National Conference of Juvenile Agencies

National Clearinghouse for Criminal Justice Planning and Architecture (NCCJPA) maintains a 10,000-volume library at the University of Illinois at Champaign

National Correctional Recreation Association (NCRA) sponsors prison postal-weight-lifting contests for inmates in Canada and the United States

National Jail Association presents annual award for the outstanding jailer and jail matron; publishes *Jail Forum* quarterly

National Jail Managers Association maintains historical archives plus an information clearinghouse and library in Eugene, Oregon

National Juvenile Detention Association publishes *Counterpoint* bimonthly and the *Directory of Juvenile Detention Homes*

National Prison Project American Civil Liberties Union's program to fix prison sentences and improve the lot of prisoners

national razor nickname for the guillotine

National Sheriff official publication of the National Sheriffs' Association

National Society of Penal Information provided some of the first aides of the Bureau of Prisons and later became part of the Osborne Association

NATSJA (*see* National Association of Training Schools and Juvenile Agencies)

Natzweiler forced-labor concentration camp in eastern France

Nauru the capital, Yaren, has a small lockup

Navasota Texas location of the Grimes County penal facility

NCA Nebraska Correctional Association; Nevada Correctional Association

NCCA North Carolina Correctional Association

NCCD National Council on Crime and Delinquency

NCCJPA National Clearinghouse for Criminal Justice Planning and Architecture

NCJRS National Criminal Justice Reference Service

NCPPL National Committee on Prisons and Prison Labor

NCRA National Correctional Recreation Association

NCW Nebraska Center for Women (in York, Nebraska)

N'Djamena (*see* Chad)

NDPS Narcotic Detention Pens in downtown New York City

Nebraska Center for Women in York

Nebraska lockups 93 counties contain jails; Omaha has additional detention facilities

Nebraska State Penitentiary in Lincoln

NECCC New England Correctional Coordinating Council

necktie hangman's noose

necktie hanger gallows

Nepal Himalayan nation with an ancient prison in Kathmandu

Netherlands has both ancient and modern prisons in cities such as Amsterdam, Rotterdam, the Hague, and Utrecht

Netherlands Antilles all islands have small lockups and a low crime rate

Neuengamme concentration camp near Hamburg

Nevada Girl's Training Center in Caliente

Nevada jails the state's 16 counties have lockups; Las Vegas, Reno, and Carson City have extra detention capability

Nevada Women's Correctional Center in Carson City

Neve Tirza Israeli maximum-security prison for women

New Albany minimum-security jail in New Albany, Indiana

New Braintree medium-security prison in New Braintree, Massachusetts

Newc Newcastle-upon-Tyne, England, detention facilities

New Caledonia French penal colony in the South Pacific from 1864 to 1894, when it , was moved to French Guiana

New Era prisoner's newspaper published at Leavenworth, Kansas

Newgate London prison razed by rioters more than a century ago, now the site of the Old Bailey law courts, officially called the Central Criminal Courts

Newgate's Angel Elizabeth Gurney Fry—lay visitor

Newgit nickname for London's old Newgate Prison

New Hampshire jails 10 counties maintain lockups; its cities also have extra detention facilities

New Hampshire State Prison in Concord

New Hampshire Youth Development Center in Manchester; delinquents in all stages of security between 11 and 18 years of age

New Haven Community Correctional Center in New Haven, Connecticut, also the site of the county jail

New Hebrides (*see* Vanatu)

New Jersey jails 21 counties have lockups; Newark and Jersey City have additional jails

New Jersey State Prison— Leesburg in use since 1913 for male felons

New Jersey State Prison— Rahway a prison for male felons

New Jersey State prisons in Leesburg, Rahway, and Trenton

New Jersey State Prison— Trenton built in 1798 but replaced in 1836

New Life New Life House (for prisoner rehabilitation at Tam Lung Chung, New Territories, Hong Kong)

New Lon(don) New London, Connecticut, detention facilities

New Mexico Boys School in Springer

New Mexico lockups 32 counties have jails

New Mexico Youth Diagnostic Center in Albuquerque

New Orleans parish seat has and jail plus police-station lockups

New Queens newer section of the Riker's Island Penitentiary in New York City

Newton Iowa site of the Riverview Release Center

New Westminster Canadian maximum-security facility at New Westminster, British Columbia

New York any of many penal institutions in either New York City or New York State

New York lockups the state's 62 counties contain jails; all cities have extra detention facilities

New York School for Girls at Hudson, New York

New York State Commission of Correction inspects and monitors the many correctional facilities, community residential facilities, secure centers, secure detention centers, detention institutions, sentence institutions, and county jails, municipal lockups, and state prisons

New York State Correctional Facilities Albion, Arthur Kill, Attica, Auburn, Bedford Hills, Clinton, Coxsackie, Eastern, Elmira, Fishkill, Great Meadows, Green Haven, Ossining, Taconic, Wallkill, Woodbourne, Downstate, Hudson, Mid-Orange, Mount McGregor, Otisville, Queensboro

New Zealand HM Gaols in cities such as Christchurch, Auckland, and Wellington

Nha Trang (*see* Vietnam)

Niamey (*see* Niger)

Niantic Connecticut Correctional Institution farm and prison in Niantic; the J. Bernard Gates Correctional Unit

NIC National Institute of Corrections (U.S. Department of Justice in Boulder, Colorado)

Nicaragua capital, Managua, and cities such as Bluefields and León have *cárceles modelos*

Niceville former name of the Niceville Road Prison before it was moved to Crestview, Florida where it is now known as the Okaloosa Correctional Center

Nickerie (*see* Suriname)

Nicosia capital of Cyprus as well as the popular name for its old prison

Nigeria capital, Lagos, contains an old prison built by the British

nippers chain-grip-actuated handcuffs

NJA National Jail Association

NJDA National Juvenile Detention Association

NJMA National Jail Managers Association

NJRW New Jersey Reformatory for Women at Clinton

NO New Orleans, Louisiana's jail and police-station lockups

Nogal(es) Nogales, Sonora jail in México

Nome Nome State Jail in Alaska

Noranside Noranside Borstal Institution in Scotland

Norf Norfolk (in Canada, England, the South Pacific, or the United States) or any of its detention facilities

Norfolk Massachusetts site of the Bay State Correctional Center and the Massachusetts Correctional Institution; State Prison Colony in Norfolk, Massachusetts (first community prison for male felons); Virginia seaport city or its county jail or naval brig

Northallerton HM Prison at Northallerton in North Ridging, Yorkshire, England

North Carolina Department of Correction maintains statewide coverage through its Division of Prisons and its Youth Services Division

North Carolina jails 100 counties contain lockups; cities have extra facilities

North Central Correctional Institution at Gardner, Massachusetts

North Dakota Industrial School in Mandan seeks to reform delinquents from 12 to 18 years of age

North Dakota lockups the state has 53 counties providing jails

North Dakota Penitentiary in Bismarck

North Dakota State Farm near Bismarck

Northern Ireland (*see* individual entries)

Northern Nevada Correctional Center prison farm near Carson City

Northern Nevada Honor Camp near Carson City

Northern Region Correction Institute Alaskan facility at Fairbanks

Northern Rhodesia (*see* Zambia)

Northeye HM Prison at Northeye in Sussex, England

Northside Correctional Center in Spartanburg, South Carolina

Norway modern penal facilities are in or near such port cities as Oslo, the capital, Bergen, and Trondheim

Norwich HM Prison at Norwich, England; jails in Norwich, Connecticut and smaller towns named Norwich

Not-So-Nice-Ville (*see* NiceVille)

Nottingham HM Prison at Nottingham in Nottinghamshire, England

Nou island prison of Nouvelle Calédonie in the South Pacific

Nouakchott capital of Mauretania and site of its old prison

Nouvelle Calédonie (French—New Caledonia)— penal colony from 1864 to 1894 when its prisoners were shipped to French Guiana

Nova Scotia School for Girls at Truro, Nova Scotia

NP Naval Prison

NPB National Parole Board (Canada)

NPCC Nebraska Penal and Correctional Complex

NPP (*see* National Prison Project)

NPPAJ National Probation and Parole Association Journal

NPSB *National Prisoner Statistics Bulletin*

NRTI National Rehabilitation Training Institute

NSA National Sheriffs Association

NSPI National Society of Penal Information

Nuestra Familia (Spanish— Our Family)—Mexican-American prison-based underground organization engaged in narcotic trafficking and jail breaks

Nuevo Guerrero Nuevo Guerrero, Tamaulipas, jail

Nuevo Laredo Nuevo Laredo, Nuevo León, jail

Nuku'alofa (*see* Tonga)

nullum crimen sine lege (Latin—no crime without law)—crime must be defined by law

number-one diet bread and water

Nuremberg forced-labor subcamp in the south of Germany, close to the site of the Nuremberg Trials of war criminals

Nusa Kambangan Indonesian island prison off the city of Tjilatjap (Cilacap) where a crocodile-infested marsh, between island and mainland, discourages escape

nut factory section of a prison where criminally insane convicts are held

nuthouse psychiatric ward

nutpicker psychiatrist

nvd night-viewing device(s)

n-v device night-viewing device

NYCCIW New York City Correctional Institution for Women

NYCDC New York City Department of Correction

NYHD New York House of Detention

Nykobing Danish prison

NYMCC New York Metropolitan Correctional Center

NY Met New York Metropolitan Correctional Center (in downtown Manhattan)

NYRM New York Reformatory for Men

NYRW New York Reformatory for Women (Westfield Farm)

NYSDCS New York State Department of Correctional Services

Oahu capital island of Hawaii containing Honolulu, Pearl Harbor, and the Oahu Community Correctional Center

Oakalla prison in Burnaby, a suburb of Vancouver, British Columbia

Oakdale Federal Detention Center, Oakdale, Louisiana; site of the Iowa Security and Medical Facility

Oakhill Correctional Center in Oregon, Wisconsin

Oakie City Oklahoma City, Oklahoma's correctional facilities

Oakland Oakland, California's jails or its Community Treatment Center

Oakley Campus' (*see* Raymond)

Oak Park Heights site of the Minnesota Correctional Facility

OAR (*see* Offender Aid and Restoration)

Oaxaca prison in Oaxaca, Mexico

obc old brutal con(vict)

Obispo California Men's Colony at San Luis Obispo

Oblatos jail in Guadalajara, Mexico

OCA Oregon Corrections Association

Ocala Florida location of the Alyce D. McPherson School for juvenile delinquents

Occoquan Minimum Security Facility of the District of Columbia located in Virginia

OCCSA Ohio Correctional and Court Services Association

OCF Ossining Correctional Facility (Sing Sing) at Ossining, New York

OCIS Organized Crime Information System (FBI)

OCJA Oklahoma Criminal Justice Association

Odenville state prison northeast of Birmingham, Alabama

Odessa Ukraine city on the Black Sea with typical prison facilities; its namesake in west Texas contains the Ector County jail

Odom Jackson, North Carolina penal facility

Odum Georgia site of the Wayne Correctional Institution

Offender Aid and Restoration conducts CIP (Citizens Involvement Project to educate and train civic leaders and sheriffs in the use of volunteers in jails

Ogden Utah site of the Ogden Community Corrections Center and the Parkview Community Corrections Center

Ohio jails 88 counties have prison facilities; cities, such as Cleveland, Cincinnati, and Columbus, have additional jails

Ohio Reformatory for Women in Marysville

Ohio State Ohio State Penitentiary in Columbus, also called River House

Ohio State Reformatory in Mansfield, Ohio

Ohrdruf Thuringian town in central Germany and site of a concentration camp as well as the underground headquarters of the German army during World War II

Oil City nickname of Bartlesville or Tulsa in Oklahoma jails; name of a town in western Pennsylvania and its jail

OIPC Organisation Internationale de Police Criminelle (French—International Criminal Police Organization)—Interpol

Ojinaga Ojingaga, Chihuahua, jail

Okaloosa (*see* Niceville)

Okeechobee location of the Florida School for juvenile delinquents

Okie Oklahoma(n) or any of its penal facilities

Okinawa Okinawa Prison in the Ryukyu Islands of Japan; (*see* USMC)

Oklahoma lockups 77 counties have prisons; extra facilities are found in Oklahoma City and Tulsa

Oklahoma State Penitentiary in McAlester

Oklahoma State Reformatory in Granite

Old Capitol Old Capitol Prison in Washington, DC where political prisoners were held during the Civil War

Old Dorp Schenectady, New York or its detention facilities

old hand Australian nickname for former convict

Old Horse Bridewell Prison

old lag person serving a three-year sentence in a British prison

Old Melbourne Gaol and Penal Museum Melbourne, Victoria makes the most out of a bad beginning

Old Newgate Prison penological museum on Newgate Road in East Granby, Connecticut

Old Queens older section of the Riker's Island Penitentiary in New York City

old smoky electric chair

Old Sparkey Florida's natural oak electric chair

Old Territorial Old Territorial Penitentiary in Santa Fe, New Mexico

Olustee Florida site of the Baker Correctional Institution

Oma Omaha, Nebraska's jail

Oman Muscat and Matrah have small prisons

Omdurman (*see* Sudan)

Omsk czarist prison near Kazakhstan; the punishments inflicted on prisoners here are described in Dostoevski's *Notes from the House of the Dead*

on the bricks out of jail and on the streets

on the ground out of jail

on ice imprisoned

on the lam escaping, evading, or hiding from the police or other law-enforcement agents

Onley HM Borstal at Onley in Warwickshire, England

Only Tennessee site of the Turney Center for Youthful Offenders from 18 to 25 years of age

Ontario Youth Training-School at Ontario, California

on the shelf in solitary confinement

Ont Pen Ontario Penitentiary (Canadian)

ooze out sneak out

open prison penal facility built without bars on the windows, locks on the doors, or walls surrounding the prison

Oran (*see* Algeria)

Oranienburg concentration camp near Berlin

Orchid Island Rehabilitation Center on Orchid Island off southeastern tip of Taiwan (Formosa)

Ordinary of Newgate Chaplain of Newgate Jail

ordinary transportation on-foot transportation of prisoners

Ordot Guam's penitentiary at Ordot

Oregon Wisconsin site of the Oakhill Correctional Institution and the Wisconsin Correctional Camp System

Oregon jails the state's 36 counties have prisons; Portland and Eugene have augmented facilities

Oregon State Correctional Institution in Salem

Oregon State Penitentiary in 1853 opened in Portland but transferred to Salem in 1866

Oregon Women's Correctional Center in Salem; began in 1965 as a section of the Oregon State Penitentiary

Oriente Mexico City prisoner-holding facility

Oroville Butte County jail in Oroville, California

ORW Ohio Reformatory for Women

Osaka (*see* Japan's prisons)

Oslo (*see* Norway)

Ossining Ossining Correctional Facility (formerly known as Sing Sing) at Ossining, New York

Osteraker Swedish prison in Osteraker near Stockholm

Ostrava (*see* Czechoslovakia)

Ostroy Vrangelya (Russian—Wrangel Island)—Soviet prison camp northwest of Alaska and north of northeastern Siberia in the Chukchi Sea

Oswiecim Polish name for Auschwitz, site of a World War H concentration camp operated by the Nazis

Otay Otay Mesa Prison in southernmost California

Otay Mesa prison Richard J Donovan Correctional Facility south of San Diego, California

other prisoner(s) correctional officer(s); prison guard(s), warden(s), etc.

Otisville Otisville Correctional Facility at Otisville, New York

oubliette (French—secret dungeon)—cell with a trapdoor in its roof so convicts can be lowered into its hole

Outlaw bimonthly publication of the Prisoner's Union

outside outside of prison

out of town in prison

over the blue wall confined to a hospital for the criminally insane

Oxford Federal Correctional Institution at Oxford, Wisconsin, a Staff Training Center of the Bureau of Prisons; HM Prison at Oxford in Oxfordshire, England; some twenty other Oxfords in America have jails

Ozark Ozark Correctional Center in Fordland, Missouri

PA Pardon Attorney (U.S. Department of Justice)

PAA Prisoners Aid Association

PAAM Prisoners Aid Association of Maryland (Baltimore)

Pachuca prison in Pachuca, Mexico

Pachuco *Pachucolandia* (Mexican-American—El Paso, Texas)—nickname of the jail, lockups, and detention pens of the Immigration and Naturalization Service

Pacific Missouri community containing the Missouri Eastern Correctional Center

paddy wagon police van for transporting prisoners

Pakistan prison facilities in Islamabad, its capital, were built during British rule; the same is true in cities such as Karachi, Lahore, and Hyderabad

p-a-l prisoner-at-large

palacio blanc (Spanish—white palace)—nickname of Mexico city's most modern prison; it has cement-floored steel-lined cells plus baths, a hospital, and a library

Palacio Negro (Spanish—Black Palace)—nickname for Mexico's Lecumberri prison

Palais de Justice (French—Palace of Justice)—Parisian court and prison

Palmer Palmer Correctional Center northeast of Anchorage, Alaska

Palmer Work-Release Center in Florence, South Carolina

Panamá Panamá City, Colón, David, and San Miguelito all have so-called model prisons

Panamanian prisons (*see* Cárcel Modelo, Cueva, and Gamboa)

Pango Pago Pago, American Samoa or its jail

Pankrác Prague's prison

panopticon prison where all cells are visible from a central point

panopticon pattern based on Bentham's panopticon inspection house where giant circular prison houses were manned by armed guards who could supervise the surrounding cells and their inmates; the Stateville penal establishment in Illinois is built on this pattern

Papenburg Nazi concentration camp west of Bremen, Germany

Papua New Guinea a jail is in Port Moresby, the capital seaport city

Paraguay the capital city, Asunción, has a big prison

Paramáribo (*see* Suriname)

Paranam (*see* Suriname)

Parchman site of the Mississippi State Penitentiary

Pardelup Pardelup Prison Farm in Western Australia

pardon executive-applied exemption from punishment for a crime or for a pending criminal conviction

Parkhurst top-security prison near Newport on the Isle of Wight off England's south coast

Park Row Metropolitan Correctional Center in New York City

Parkside New York State Correctional Facility in Manhattan

Parkview Parkview Community Corrections Center in Ogden, Utah

parole conditional release of an offender from a confinement facility before the expiration of his or her sentence; released offender usually put under supervision of a parole agency or officer

parole agency correctional agency supervising convicts on parole

parole authority correctional agency or officer having authority to release prisoners committed to confinement facilities or to discharge them from parole or to revoke parole

parolee person conditionally released from a correctional institution before the expiration of his or her sentence and placed under the supervision of a parole agency or officer

parole violation parolee's failure to conform to the conditions of parole; such violation usually results in return to prison and loss of parole

Parramatta Parramatta Gaol, Sydney, New South Wales, Australia

Parris Island (*see* USMC)

Pasca Pascagoula, Mississippi, jail

Paterson northern New Jersey mill town (often called Silk City) or generic name of its jail and its lockups

Patras (*see* Grecian prisons)

Patton California State Hospital (for the criminally insane) at Patton

Patuxent Patuxent Institution for the Criminally Insane (Patuxent, Maryland); Patuxent Institution for Defective Delinquents (Jessup, Maryland)

Patuxent Institution Maryland penal facility and pre-release center in Patuxent

Pavón El Pavón (Spanish—the peacock)—largest prison in Guatemala

Pawiak Warsaw, Poland's prison

P'burg Pittsburgh, Pennsylvania's jail

PCI Program of Correctional Institutions (Puerto Rico)

pcu (PCU) protective custody unit

PDA Polizeiliches Durchgangslager Amersfoort (German—Police Concentration Camp—Arnersfoort, Netherlands)—staging area for the transport of Dutch prisoners to Nazi Concentration camps

P del E Penitenciaria del Estado (Spanish—State Penitentiary)

peace and quiet maximum-security cell

Pedro San Pedro, California's jail

Peewee Valley Kentucky town associated with the Kentucky Correctional Institution for Women

Pelican Bay state prison near Crescent City, California

Pembroke Pines (*see* Broward)

pen. penitentiary

penal isolation solitary confinement

penalist(s) penologist(s)

penal servitude imprisonment combined with hard labor

penalty punishment for a particular offense

Pence Springs West Virginia site of the West Virginia State Prison for Women

Pendleton town that contains the Indiana State Reformatory

Penetang Penetanguishene Provincial Establishment for the Criminally Insane on Georgian Bay, Lake Huron, north northwest of Toronto

peni penitenciaríla (Spanish—penitentiary)

peniatrist(s) prison doctor(s); prison psychiatrist(s)

peniatry branch of medical science dealing with penal establishments and their prisoners

penitenciaría (Spanish—penitentiary)—prison

penitenciaría estadual (Portuguese—state penitentiary)—each of the 22 states of Brazil maintains such a penal institution

Penitenciaría General de Venezuela Venezuela's general penitentiary situated in San Juan de los Morros Ejido Guárico

Penitenciario de Santiago Santiago, Chile's penitentiary

Penitenciario Litoral Guayquil, Ecuador's maximum-security prison

penitentiaries federal or state maximum-security institutions

penitentiary house of correction or rehabilitation center where offenders are confined for detention, discipline, reformation, rehabilitation, or punishment; in the United States a penitentiary is a maximum-security penal facility designed to hold prisoners serving long sentences

Penitentiary of New Mexico in Santa Fé

Penninghame Penninghame Prison in southwest Scotland

Pennsylvania jails the state's 67 counties have lockups; cities such as Philadelphia and Pittsburgh have extra facilities

Pennsylvania State Correctional Institutions (*see* Bellefonte, Camp Hill, Dallas, Gratcrford, Huntingdon, Muncy, Pittsburgh)

Pennsylvania State Correctional Institutions and Correctional Diagnostic and Classification Centers: (*see* Camp Hill, Graterford, and Pittsburgh *entries*)

Pennsylvania System (*see* solitary system)

penol penological; penologist; penology

penologist social scientist concerned with penal institutions and the deterrent effect of punishment decreed by law

penology scientific study of penal institutions, the deterrent effect of punishments de-

creed by law, the consequences of crime, the means of changing lawbreakers into law-abiding citizens, and repairing the damage done to victims of crime

pensioner(s) of the crown Australian euphemism for any former convict(s)

Pensy Pensacola, Florida's nickname or that of its jail

Pent The Pent (nickname for Pentonville Prison in the outskirts of London's Islington Parish)

Pentonville HM Prison at Pentonville, a district of London

Pentridge Melbourne, Victoria's prison

Peoria county seat, courthouse, and jail of Peoria County, Illitiois

Perm Soviet labor camp region 700 miles (1127 kilometers) east of Moscow

Persian prisons (*see* Iranian prisons)

Perth Scottish prison close to Perth on the River Tay

Peru its penal facilities are typical of many others in Hispanic countries (*see* Cárcel Modelo, Lurigancho, Sepa, Sexto)

Peruvian prisons (*see* Cárcel Modelo, Lurigancho, Sepa, Sexto)

Pete St Petersburg, Florida's nickname or that of its jail

Peterhead prison near Aberdeen, Scotland

Peter-Paul Peter-Paul Fortress in St Petersburg; now a museum

Petersburg Federal Correctional Institution at Petersburg, Virginia

Petros Tennessee location of Brushy Mountain Penitentiary

Pewee Kentucky Correctional Institution for Women at Pewee Valley

Pforzheim German city and prison near the Black Forest

Philadelphia Prisons Detention Center at 8201 State Road, Holmesburg Prison on Torresdale Avenue in the 8200 block; House of Correction at 8001 State Road

Philippines has some penal institutions dating back to Spain's colonization and more modern ones reflecting the American period

Phillipsburg capital complete with lockup in the West Indian Leeward Island of Sint Maarten (the French have their own part of the island, Saint Martin, with a capital also complete with lockup and called Marigot)

Philly Philadelphia, Pennsylvania or the name given to any of its many penal institutions

Phnom Penh (*see* Cambodia)

Phoenix Arizona site of Adobe Mountain School for juvenile delinquents, the Alhambra Reception and Treatment Center for incoming male felons, the Arizona Center for Women, the Arizona Training Facility, the Arizona Correctional Training Facility

Phoenix Correctional Facility At Plymouth, Michigan

PHP Preventive Health Programs

PHS Prison Health Services (providing a complete system of health services to prisons)

picking oakum picking apart pieces of tarred rope for use in caulking wooden ships (a century ago this was still the task given many prisoners confined to jails along the coast of Britain as well as the United States)

Piedmont Work Release Center in Spartanburg, South Carolina

Piedras Negras Piedras Negras, Coahuila and its jail

Pikeville Tennessee-site of the Bledsoe County Regional Correctional Facility and the Taft Youth Center

Pinal Pinal County Jail at Florence, Arizona

Pinchgut prisoner's nickname for Fort Denison Prison in Sydney Harbour, New South Wales

Pine Bluff Pine Bluff, Arkansas with its Diagnostic Unit, the Pine Bluff Youth Service Center, and the Women's Unit

Pine Hills Pine Hills School for delinquents from 10 to 21 years of age in Miles City, Montana

Pine Street Baltimore, Maryland jail on Pine Street

Piraeus (*see* Grecian prisons)

Pithiviers concentration and transit camp in France (*see* Camp de Drancy)

Pitts nickname for Pittsburgh, Pennsylvania or its penal facilities

Pittsburgh Correctional Institution in Pittsburgh, Pennsylvania, built in 1826 as the Western Penitentiary; enlarged in 1982

PK Principal Keeper

Plainfield site of the Indiana Boys School, the Indiana Youth Center, and the Reception and Diagnostic Center

Plankinton site of the South Dakota Training School for delinquents up to their seventeenth year

Plaszow Nazi concentration camp northwest of Cracow, Poland

Pleasanton Federal Correctional Facility at Pleasanton, California

Pleasantville federal minimum-security facility at Maxwell Air Force Base in Alabama

Plummer Center Delaware's work-release center in Wilmington

Plymouth Michigan site of the Phoenix Correctional Facility

PMS Prison Management Systems (health-care plan offered in the United States); Prison Medical Services (under the Home Office in the United Kingdom)

POA Prison Officers Association

POC Prison officers Club

Pocaloo Pocatello, Idaho's nickname or that of its jail

poetic punishment matching the punishment to fit the crime

Point Lookout Union prison and stockade close to the Potomac where many Confederate prisoners died

Point Salines prison camp at Point Salines, Grenada

pokey jail

political(s) political prisoner(s)

politico political prisoner; politician

Polk Polk Correctional Institution at Polk City, Florida

Pollington HM Borstal at Pollington in Humberside, Yorkshire, England

Polmont Scottish borstal

POME Prisoner of Mother England (also called a Pommy when in early colonial days convicts were shipped to Australia)

Ponar site of a Nazi concentration camp in what is now Lithuania

Ponce city on Puerto Rico's south coast has an old jail as well as police-station lockups

Ponte dei Sospiri (Italian— Bridge of Sighs)—heavily barred, stone-covered, two-storied bridge arching a Venetian canal, the Rio di Palazzo, and connecting the Doge's Palace with the state prisons and dungeons

Pontiac Correctional Center northeast of Normal, Illinois

poogie jail

Poolsmoor prison near Cape Town, South Africa

Poona Indian prison in Poona

poorhouses of the twentieth century Ronald Goldfarb's apt description of jails

population movement entries and exits of adjudicated persons, or persons subject to judicial proceedings, into or out of correctional facilities

Pork Dump epithet applied to the Clinton Prison near Utica, New York

porridge British slang for jail

Portage La Prairie Correctional Center for Women at Portage La Prairie, Manitoba

Port Arthur Australia's principal penal colony from 1834 to 1853 in Tasmania, then known as Van Diemen's Land; today a museum and visitor's center replace the convict's quarters

Port Augusta Port Augusta Gaol (South Australia)

Port-au-Prince capital of Haiti whose prisons are the most noisome in the West Indies

Port Blair headquarters of a penal settlement dating from the Sepoy Rebellion of 1857 but discontinued in 1945

Porte d'Enfer (French—Gate of Hell)—convict's nickname for the prison gate at Saint Laurent du Maroni in French Guiana

Port Elizabeth (*see* South Africa)

Port Isabel Port Isabel Detention Center of the Immigration and Naturalization Service in Port Isabel, Texas

Portland HM Borstal at Portland in Dorset, England; other detention places in towns named Portland from Australia to the West Indies, including Portland Maine and Portland, Oregon

Port Laoise prison southwest of Dublin, Ireland

Port Lincoln Port Lincoln Prison (south Australia)

Port Louis (*see* Mauritius)

Port Macquarie Australian convict colony in New South Wales in the early 1800s

Port Moresby (*see* Papua New Guinea)

Porto also called Oporto (*see* Portugal)

Port-of-Spain (*see* Trinidad and Tobago)

Portsmouth site of the Royal Navy's prison; New Hampshire port and site of the U.S. Naval Prison

Port Sudan (*see* Sudan)

Portugal Lisbon, the capital, and Porto (Oporto) have some very old prisons

Post Care Program for adult felons in Lincoln, Omaha, and Norfolk, Nebraska

Poston former Indian reservation in Arizona; used as a detention camp for Japanese Americans

Potma reputed to have been the Soviet Union's largest forced-labor penal colony

Poughkeepsie seat of Duchess County and its jail

pow (POW) prisoner of war

Powell Ohio location of the Riverview School for Boys and the Scioto Village, formerly the Girls Industrial School

Poznan (*see* Poland)

Pozsony Czech name for Bratislava and site of an old prison built long before World War I

Pozzi old Italian prison in Venice

PPCAA Parole and Probation Compact Administrators Association

PPS Pennsylvania Prison Society (Philadelphia-based)

pq punishment quarters (isolated section of many penitentiaries and reformatories)

p & q peace and quiet (solitary confinement)

Praia (*see* Cape Verde)

Presidential Candidate and Prison Convict Eugene V. Debs, Socialist candidate imprisoned for making an antiwar speech in 1918

presidio (Spanish—military prison)—may also mean citadel, penitentiary, or prison; many were built in the American Southwest during Spanish and Mexican rule

Presidio Prisoner's publication issued bimonthly at the Iowa State Penitentiary in Fort Madison

presidio modelo (Spanish—model penitentiary or prison)

Pressburg German name for Pozsony, site of an old prison in Czechoslovakia

Pretoria (*see* South Africa)

Pretoria Central maximum-security Pretoria Central Prison in Pretoria, South Africa

Pre-Trial Annex Wilmington, Delaware's facility for those held in detention status

Preungeshim security prison in Frankfurt, Germany

prigione (Italian—prison)—also called *carcere*

Prigione d'Genova (*see* Italian prisons)

Prince Albert Canadian city containing the Saskatchewan Penitentiary

Prince George British Columbia penitentiary

Princetown Her (His) Majesty's Prison in Princetown in Devon, England

Prince William Prince William—Manassas Regional Adult Detention Center in Virginia

prisão (Portuguese—prison)—also called *cárcere*

prison confinement facility with custodial authority over adults sentenced to confinement for more than one year

Prison at the Bottom of the World Ushuaia, Argentina

Prison at the Top of the World Solovetski Island in the former Soviet Union

prison bird recidivist; prisoner who has been to prison before

prison break escape from prison accompanied by force and violence

prison bug person spending most of their time in prison

prison camp minimum-security camp designed to shelter convicts assigned to farm or forestry projects, road repair work, or other federal or state projects

Prison Camps the U.S. Bureau of Prisons maintains camps at Allenwood, Pennsylvania; Eglin Air Force Base, Florida; Montgomery, Alabama; and Safford, Arizona

prison coffin plain wooden box fitted with rope handles and perforated with holes facilitating disintegration once buried

prisoner person in custody in a confinement facility or in personal custody while being transported to or between confinement facilities

prisoner at large naval prisoner confined to the barracks or the ship

prisoner-of-war camps in the Confederacy Andersonville, Georgia; others in Georgia included Camp Davidson at Savannah, Camp Oglethorpe at Macon, and one at Millen; camps in South Carolina were at Charleston, Columbia, and Florence; in

North Carolina at Salisbury; in Virginia at Danville and at Libby Prison in Richmond

prisoners of the Crown old Australian euphemism meaning convicts

Prisoner's Union publishes *Outlaw* bimonthly; seeks an end to economic exploitation of prisoners and redress for convict's grievances

prison fever typhus (usually due to overcrowding)

prison house prison

prison hulk prison ship

prison labor work carried out by convicts such as producing furniture, road building and repairing; stamping out automobile license plates, etc.

prisonment imprisonment

prison officer British euphemism for gaoler

prison pallor bloodless-yellow paleness of many prisoners deprived of fresh air, sunshine, and vitamins

prison psychosis mental disturbance actuated by imprisonment and manifested by delusions, paranoid trends, and pseudo-hallucinations

Prisons U.S. Bureau of Prison

prison sentence commitment to the jurisdiction of a confinement facility

prison simple mentally deranged by imprisonment

prison smell usually compounded of excreta, grease, stale tobacco, sweat, unaired bedding, and vomit

prison van black maria or paddy wagon used to transport prisoners

prison within a prison solitary confinement cell

Prison World original name of the *American Journal of Correction*

prob probation(ary); probation officer

probation conditional suspension of imprisonment of a convicted offender who must stay in the community under the supervision of a probation officer

probation agency correctional agency supervising adults and juveniles placed on probation and investigating

adults and juveniles to prepare predisposition and presentencing reports to assist courts in determining sentences

probation officer employee of a probation agency or probation department

probation sentence court requirement that a person fulfill certain conditions of behavior and accept the supervision of a probation agency or department

probation violation probationer's nonconformance to the conditions of probation

PROOF Parole Resource Office and Orientation Facility (Jersey City, New Jersey)

PROP Preservation of the Rights of Prisoners

Providence Rhode Island's capital city and county jail; other correctional facilities are scattered throughout this state

Providence County Jail old name for what is now known as the Awaiting Trial Facility in Cranston, Rhode Island

Provo seat of Utah County jail

PSAMPP Philadelphia Society for Alleviating the Miseries of Public Prisons (founded by Benjamin Franklin, William Rush, and others)

PSTD Prison Service Training Depot (Pretoria, South Africa)

psychiatric prison Her Majesty's Prison Grendon Underwood in the South Midlands of England

psychoprison psychiatric hospital prison (USSR's place for dissidents)

PU Prisoner's Union

Pudu prison in Kuala Lumpur, Malaysia

Puebla prison in Puebla, México

Puerto Barrios prison in Guatemala's Caribbean seaport

Puerto Cabello Venezuelan seaport where there is a jail

Puerto Cabezas Nicaraguan prison and seaport on the Mosquito Coast

Puerto Rican District jails in Aguadilla, Arecibo, Humacao, and in Ponce

Puerto Rican Prison camps six of them were active in the mid-1980s

Puerto Rico lockups 76 municipios (municipalities) make up Puerto Rico; some have jail facilities dating back to Spain's occupation of the island

Puerto Vasco pseudonym for a clandestine prison in Argentina

'Pulco Acapulco, México, jail

Pul-i-charki prison in Kabul, Afghanistan

Punta Arenas Chilean prison in the Straits of Magellan

Puntarenas Pacific coast port and prison of Costa Rica

Punxey Punxsutawney, Pennsylvania's jail and lockup

Purdy Treatment Center for Women in Gig Harbor, Washington

put away imprison; remove from society

put in the hole place in a solitary-confinement cell

Puyallup assembly center for Japanese-Americans brought in from other parts of Washington

PVA Prison Visitor's Association

PWA Prison Wardens Association

pw('s) prisoner(s) of war

Pyongyang capital of North Korea containing a prison built between 1910 and 1945

Q San Quentin Prison in California

Qatar the capital, Doha, is on the Persian Gulf and its old prison is primitive

QC Quezon City, capital of the Philippines, or its detention facilities

QCPSA (*see* Quaker Center for Prisoner Support Activities)

QCSR Quaker Committee on Social Rehabilitation (in New York City)

Qsar Teheran, Iran's great prison

quad prison; prison quadrangle; prison yard

quail roost women's dormitory in a house of detention

Quaker Center for Prisoner Support Activities (QCPSA) conducts nonviolent training workshops for prisoners

Quantico (*see* USMC)

quarry cure forcing addicts to work in stone quarries far from sources of alcohol and narcotics

quarter stretch three-month's sentence

Queen of Heaven (*see* Italian prisons)

Queensboro Queensboro Correctional Facility in Long Island City, New York

queen's bus prison van

queer bird jailbird

queer-ken prison

Quent San Quentin (California State Prison)

Querétero prison in Querétero, Mexico

Questore Italy's security service and the National Central Bureau of Interpol in Rome

Quezon City also called Quezon (*see* Philippines)

Quilmas San Quilmas (Mexican-American—San Antonio, Texas)—nickname of its jail, lockups, and the detention pens of the Immigration and Naturalization Service

quod prison

R rogue (brand burned on the left shoulder of convicts transported to various British colonies from 1619 until 1868)

Rabat (*see* Morocco)

rabbit feet escaped prisoners

rabbit fever desire to break parole or leave an honor camp before completion of sentence

rabbit foot escaped prisoner

rack maximum-security cell

Radical Alternatives to Prison Plan (RAPP) British program

Rahway site of New Jersey State Prison

Raiford Florida site of the Union Correctional Institution, once called the Florida State Prison or the Raiford State Prison

railroad sending a person to jail or prison without benefit of trial or proof of guilt

Raleigh North Carolina site of the Central Prison, Correctional Center for Women, and the Triangle Correctional Center

Ramla Israeli maximum-security prison

Ranby Camp HM Prison Camp at Ranby in Nottinghamshire, England

ranch synonym for a correctional facility such as a prison camp or farm in a rural area

Rancho del Campo juvenile correctional facility for older boys in San Diego County near Campo, California

Rancho del Rayo minimum-security correctional facility for younger boys in San Diego County, California

RAP Release Aid Program

RAPP Radical Alternatives to Prison Plan

rasoir nationale (French—national razor)—the guillotine

rasphuys (Flemish—rasp house)—Ghent's workhouse-type prison—the French call this *maison de force* (workhouse)

Rathmore Road nickname derived from the address of Cork Prison, Ireland

rat row prison cells for informants

Ravensbrück Nazi concentration camp north of Berlin

Rawlins location of the Wyoming State Penitentiary

Ray Brook Federal Correctional Institution at Ray Brook, New York

Raymond Mississippi community containing the Oakley Campus run by the Department of Youth Services

Raymond Street Raymond Street Jail in Brooklyn, New York

razor ribbon razor-edged stainless-steel security fencing

Reading HM Prison at Reading in Berkshire, England; American jails in other Readings in Kansas, Massachusetts, Michigan, Minnesota, Ohio, Pennsylvania, and Vermont

Reception and Diagnostic Center maximum-security facility of Indiana's Department of Correction in Plainfield

Reception and Diagnostic Center for Children Bon Air, Virginia's facility for delinquents from 8 to 18 years of age

Reception and Guidance Center—Jackson in Jackson, Michigan to help juvenile delinquents

Reception and Guidance Center—Riverside in Ionia, Michigan, holds male felons under 21 years of age*

Reception and Medical Center in Lake Butler, Florida complete with a hospital and surgery

Reception Center Maryland penal facility in Baltimore next to the Maryland Penitentiary

reception centers World War II euphemism for camps set up to hold some 110,000 Americans of Japanese descent until they could be relocated away from the West Coast

recidivist habitual prisoner; person spending much of their life in prison

recid(s) recidivist(s)

reclusão (Portuguese—reclusion)—solitary confinement

reclusion (French—solitary confinement)

Réclusion de Saint-Joseph solitary-confinement prison on Saint-Joseph Isle off French Guiana, close to Devil's Island

réclusionnaire (French—prisoner in solitary confinement)

Reclusorio Norte (Spanish—Northern Place of Retirement)—official name of Mexico City's penitentiary

reconcentrados (Spanish—concentration camps)—established by the Spanish in Cuba in 1896 but abolished by 1898 after many protests made in England, Spain and the United States

record purge complete removal of arrest, criminal, or juvenile records

Redding Shasta County courthouse and jail in California

Redención (Spanish—Redemption)—Spain's official prison publication printed in prison workshops and subscribed to by prisoners throughout the country and in Africa

Red Wing site of the Minnesota Correctional Facility at Red Wing

Redwood City San Mateo County jail in California

Reeducation of Attitudes and Repressed Emotions treatment program for sex offenders

ref reformatory

reflection cell maximum-security cell

reformatory house of correction charged with making convicts alter their lifestyle and return to society as lawabiding citizens

Regina Coeli (Latin—Queen of Heaven)—Rome, Italy's prison

Regina Provincial Correctional Centre in Saskatchewan

Regional Correctional Facilities (*see* Greensburg, Mercer)

rehab rehabilitate; rehabilitation

rehabilitation changing the offender's character, intent, and motivation toward law-abiding conduct

rehabilitation laboratory euphemism for any modern prison

Reidsville site of the Georgia State Prison

Re-Integration Center in Albuquerque, New Mexico

Reiserstown Montrose School for Girls in Reiserstown, Maryland

Release Aid Program (*see* RAP)

rélegué (French—isolated; relegated)—convict condemned to banishment in a penal colony

relocation center camp where Japanese-Americans were confined shortly after the attack on Pearl Harbor

remand to send a prisoner back to court for a further hearing; a remand prison contains people awaiting return to court

Remand Remand Prison at Jin Penjara 3, Singapore

remand center British term for a borstal or juvenile jail where convicts undergo a period of rehabilitation before, being paroled or released

remand home synonym for a remand center

remand institution jail, penitentiary, or prison

Rembert South Carolina site of the Wateree River Correctional Institution

Reno El Reno, Oklahoma's federal detention reformatory; Reno, Nevada or its jail

rent-a-con plan hiring ex-convicts so they get a fresh start in society

Renz Correctional Center in Cedar City, Missouri

reprieve executive order suspending execution of a sentence

resilient cell padded cell preventing prisoners from injuring or killing themselves

Rest and Reverie nickname of Terminal Island, California's prison in Los Angeles Harbor

restitution center small house where convicted criminals must spend every night after going out every day to work off their debts to the victims of their crimes

restraints ankle cuffs, belly chains, belt restraints, handcuffs, leg irons, straitjackets, etc.

Retrieve Unit in Angleton, Texas

Réunion French island in the Indian Ocean used for the isolation of political prisoners

Reykjavik (*see* Iceland)

Reynosa Reynosa, Tamaulipas and its jail

RGC Reception Guidance Center

Rhode Island jails five counties have lockups; Providence has extra facilities

Rhode Island penal facilities all 11 are in Cranston

Rhode Island State Prison in Cranston, now known as the Maximum Security Facility

Rhode Island Training School for Girls at Howard, Rhode Island

Rhodesia (*see* Zimbabwe)

Richmond site of the Virginia State Penitentary or the local jail and police-station lockups; Richmond Penitentiary in St Croix, American Virgin Islands West Indies; Texas town holding the Jester prerelease unit

Richmond Farm Jamaican prison close to Annotto Bay on the north coast

Richmond Hill prison on the island of Grenada

Richmond Village Tasmanian town noted for its convict-built bridge and old gaol

Ridgeville South Carolina site of the MacDougall Youth Correction Center

Riga main concentration camp in Riga, Latvia during World War II

Riker's complex of prisons on Riker's Island in New York City's East River; complex includes the Adolescent Remand Shelter, the Riker's Island Penitentiary, and the Riker's Island Women's Detention Center

Ringe Danish state prison at Ringe, often nicknamed the Sex Prison, as inmates of both sexes are allowed sexual freedom; on the island of Funen south of Odense

Ring-Ring Copenhagen, Denmark's penitentiary

Rio Rio de Janeiro or its jail

Rio Consumnes Correctional Facility in Elk Grove, California

Rio de Oro former Spanish prison colony in northwest Africa

Riom location in France of an old and infamous prison

Rio Muni Spanish Guinea on the West African mainland and the nearby island of Fernando used as a convict settlement

Rio Piedras location of Puerto Rico's State Penitentiary

RIP Riker's Island Penitentiary

Risdon Hobart, Tasmania's prison and prison hospital

River Avenue Bronx House of Detention in New York City

River House Ohio State Penitentiary on the Scioto River

River Junction Correctional Institution in Chattahoochee, Florida

Riverside Riverside County courthouse and jail east of Los Angeles, California

Riverside Correctional Facility in Ionia, Michigan

Riverton site of the Wyoming Honor Farm

Riverview Canadian Interprovincial Home for Women (misdemeanants) in Riverside, New Brunswick; Florida site of the Hillsborough Correctional Institution

Riverview Release Center in Newton, Iowa

Riverview School for Boys in Powell, Ohio

Riyadh (*see* Saudi Arabia)

RLDPAS Royal London Discharged Prisoner's Aid Society

RLPAS Royal London Prisoners' Aid Society

Roaston, Toaston, and Duston nicknames given by the Japanese-Americans interned at prison camps in the roasting, toasting, and dusty desert of Arizona

Robben Island South African penitentiary for political prisoners

Rochester detention facilities in Rochester, Minnesota and Rochester, New York; HM Borstal at Rochester on the Medway River estuary in England's Kent

Rock the Rock (nickname for the 12-acre rock occupied by Alcatraz when it served as a prison in San Francisco Bay); nickname for the Riker's Island penal facilities in New York's East River

rock crusher prisoner assigned to hard manual labor

Rockland Rockland State Hospital for the Criminally Insane in New York's Rockland County

Rock Mountain Richard J Donovan Correctional Facility at Rock Mountain, California

Rock Spring Georgia site of the Walker Correctional Institution

Rockville Training Center Indiana correction facility in Rockville

Rockwell City Women's Reformatory at Rockwell City, Iowa

Rocky Butte Portland, Oregon's jail

Roebuck Roebuck Campus Birmingham, Alabama (academic-oriented rehabilitation program for juvenile delinquents)

Rohwer Arkansas relocation center for interned Japanese-Americans

Romania Bucharest and Constanza have prisons built when the Balkan nation was a kingdom

Romanian concentration camps during World War II the most notorious were Akmecetka, Bogdnovka, and Dumanovka

roomie cellmate

Roosevelt Roosevelt Island in New York City's East River, formerly called Welfare Island and Blackwell's Island; site of lunatic asylum, prison, and workhouse

Roosevelt Roads Puerto Rico's location of the U.S. Naval Station and its brig

rope hangman's rope; marijuana; vein

Rosario (*see* Argentina)

Roseau (*see* Dominica)

Rosharon Texas site of the Darrington Unit

Rostov-on-Don Soviet state police prison, also called House 33

Rota small Spanish port in Cadiz Bay and location of a brig maintained by the U.S. Navy

rotan (Malay—rattan)—lashing stick in Malaysia and Singapore where corporal punishment is still used in jails and street riots

Rotterdam (*see* Netherlands)

Rottnest Rottnest Island, former penal colony in the Indian Ocean

Round House Fremantle, Western Australia's oldest structure built in 1830 as a jail

Rove Solomon Islands' prison in Honiara

RTU Rahway Treatment Unit for sex offenders imprisoned in New Jersey's State Prison at Rahway

rubber room padded cell for self-destructive or violent prisoners

Rutland Correctional Facility, in Rutland, Vermont

Rutland Street nickname of Limerick's prison on Rutland Street in Ireland

Rwanda the capital, Kigali, contains an old prison built by the Belgians

S-21 Khmer Rouge name for Tuol Sleng near Phnom Penh, Cambodia where nearly 20,000 prisoners were executed by the Khmer Rouge

SA Salvation Army

SAA Singapore Aftercare Association (hostel for ex-convicts)

Sabanete nickname of Maracaibo, Venezuela's national prison (*Cáircel Nacional de Maracaibo*)

Sachsenhausen main Nazi concentration camp near Berlin

Sacramento Sacramento County courthouse and jail in California

SACRO Scottish Association for the Care and Resettlement of Offenders

Sacto Sacramento, California or its jail

safe house military jail; rehabilitation center for prostitutes

safekeeper felon preserved from escaping by being put in custody

safety cell padded cell for self-destructive or violent prisoners

safety vest straitjacket

Safford Federal Prison Camp at Safford, Arizona

Sagmalcilar Istanbul, Turkey's great prison

Sagmalcilar Hilton inmates' nickname for the principal prison of Istanbul

Said Port Said, Egypt, and its jail as well as police lockups

Saigon (*see* Vietnam)

Saint Anthony home of the Youth Services Center of Idaho

Saint Barts Saint Barthelemy in the French West Indies or its jail

Saint Bridget's Well original name of London's Bridewell prison, once a royal palace of King Edward VI

Saint Catherine's Jamaican district prison near Kingston

Saint Cloud Minnesota Correctional Facility at Saint Cloud; 26 rue Armengaud, Saint Cloud, Paris (headquarters of the general secretariat of Interpol—the International Criminal Police organization)

Saint Cyrille Maison Gomin correctional facility for women at Saint Cyrille, Québec

Saint Gabriel Louisiana Correctional Institute for Women at Saint Gabriel; the Hunt Correctional Center is also in Saint Gabriel

Saint George's (*see* Grenada gaol)

Saint Jean prison camp on the Maroni River of French Guiana, upstream from the penal colony headquarters at Saint Laurent

Saint Joe Saint Joseph, Missouri, its jail or lockups

Saint John jail and lockups at Saint John, New Brunswick

Saint Joseph and Saint Paul adjacent maximum-security prisons in Lyons, France

Saint-Laurent-du-Maroni French Guiana port on the Maroni River; headquarters of the penal settlement known as Devil's Island

Saint Lou Saint Louis, Missouri or its jail

Saint Louis du Maroni prison camp on the Maroni River in French Guiana upstream from Saint Laurent

Saint Lucia Castries, the capital, has an HM Gaol built during British dominion, 1814 to 1979

Saint Lucy nickname for Saint Lucia or its jail or police-station lockup

Saint Marguerite old prison close to the coast of Cannes

Saint Martin seaport capital, Marigot, has its own lockup

Saint Mary's Saint Mary's Honor Center in Saint Louis, Missouri

Saint P Saint Paul, Minnesota or its jail

Saint Paul (*see* Saint Joseph and Saint Paul)

Saint Pete Saint Petersburg, Florida or its jail

Saint Petersburg (*see* Tampa)

Saint Pierre former French penal colony on Saint Pierre Island in the Gulf of Saint Lawrence

Saint Vincent and the Grenadines the capital, Kingstown, has an old HM Gaol

Salaspils Nazi concentration camp near Riga, Latvia

Salem capital of Oregon contains the Correction Division Release Center, the Hillcrest School of Oregon, the Oregon State Correctional Institution, the Oregon State Penitentiary, and the Oregon Women's Correctional Center; West Virginia Home for Girls at Salem

Salinas Monterey County jail south of San Jose, California

Salisbury Salisbury (North Carolina) National Cemetery containing graves of Union soldiers who died in the Confederate prison here during the Civil War; (*see* Zimbabwe)

sally port first gate to a prison

Salpetriere Paris hospital for the criminally insane

Saltillo prison in Saltillo, Coahuila, México

Salt Lake City Utah location of the Community Corrections Centers—Bonneville, Central, Lakehills, Women's; and the State Diagnostic Unit

Salt Lake Women's Community Corrections Center in Salt Lake City, Utah

Samarkand Manor North Carolina youth service facility at Eagle Springs

Samoa Apia, its capital, has a lockup

Sanaa (*see* Yemen)

San Anto (Mexican-American—San Antonio, Texas)—jail nickname

San Antone nickname for San Antonio, Texas or its jail and its police-station lockups

San Berdoo San Bernardino, California or its jail

San Bernardino San Bernardino County courthouse and jail east of Los Angeles, California (also called San Berdoo)

San Bruno San Francisco County Jail at San Bruno, California

sand prisoner's nickname for sugar

San Diego San Diego County courthouse and jail in San Diego, California (also called Dago or Diego)

Sands Sands Prison outside Beirut, Lebanon

Sandstone Federal Correctional Institution at Sandstone, Minnesota

San Francisco San Francisco County courthouse and jail in San Francisco, California, also known as Frisco

San Francisco (Italian—Saint Francis)—name of the prison in Parma, Italy

San Jack San Jacinto, Texas or its jail

San Jo San José, California and its jail

San Jose Santa Clara County courthouse and jail southeast of San Francisco, California

San José capital of Costa Rica and site of its main prison

San Juan capital of Puerto Rico, includes Bayamón, Carolina, Cataño, Guaynabo, Rio Piedras, and Trujillo Alto, in addition to penal facilities dating back to Spanish rule; the general penitentiary of Venezuela situated in San Juan de los Morros Ejido Guarico

San Juan Detention Center in San Juan, Puerto Rico

San Juan de Ulúa old Spanish fortress on an islet about a mile (1.6 kilometers) off the shark-infested port of Veracruz, Mexico; since colonial times the fortress has served as a dungeon for political prisoners

San Luis Obispo California Men's Colony at San Luis Obispo includes a forestry camp; also the San Luis Obispo courthouse and jail

San Luis Potosi prison in San Luís Potosí, Mexico

San Luis RC San Luís Río Colorado, Sonora and its jail

San Marino has a very small lockup

San Pedro Lima, Peru's largest penitentiary

San Pedro Sula Caribbean port of Honduras and its prison

San Quentin California State Prison at San Quentin on a small peninsula in San Francisco Bay

San Quentin Daily award-winning newspaper published by inmates at San Quentin in California

San Quilmas San Antonio, Texas or its jail or lockups

San Rafael Marin County courthouse and jail in California

Santa Ana the central penitentiary of Occidente's EJido Trujillo in Venezuela; Orange County courthouse and jail southeast of Los Angeles, California

Santa Anita California racetrack that served as an internment camp for Japanese Americans

Santa Barbara Peruvian prison for women in Lima's port of Callao; Santa Barbara County courthouse and jail in southern California

Santa Clarita maximum-security jail near Valencia, California

Santa Cruz Santa Cruz County courthouse and jail north of Monterey, California

Santa Fé site of the Penitentiary of New Mexico

Santa María de la Cabeza monastery in the Sierra Morena northeast of Cordoba, Spain, defended by Franco's forces during the Spanish

Civil War, later was used as a prison to for Republican leaders

Santa Marta Colombian prison in the seaport of Santa Marta; federal prison outside Mexico City

Santa Marta Acatitla prison southeast of Mexico City

Santa Rita Santa Rita Rehabilitation Center in California's Alameda County

Santa Rosa Sonoma County seat and jail north of San Francisco, California

Santé Parisian prison at 42 rue de la Santé

Santiago de Cuba (*see* Cuba)

Santo Domingo capital of the Dominican Republic noted for the poor quality of its detention camps, jails, and prisons

Santurce Puerto Rican site of the Institution for Youthful Offenders

Sáo Tomé and Principe prison is a veritable antique

Sapporo (*see* Japan's prisons)

Sarah Anthony San Diego, California's school for delinquent boys and girls ages 8 to 18

Sarajevo (*see* Yugoslavia)

Sasabe Sasabe, Sonora, Mexico and its jail

Saskatchewan Saskatchewan Penitentiary in Prince Albert

sat in the hot seat died by electrocution

Saudi Arabia the capital, Riyadh, and Jidda have prisons dating back the 18th, 19th, and early 20th centuries

Saughter prison in Edinburgh, Scotland

Sauk Centre location of the Minnesota Correctional Facility

Saulsbury Tennessee site of a Civil War prisoner-of-war camp

Sav Savannah, Georgia or its jail

SBCR State Board of Charities and Reform (Wyoming)

SBIW Sybil Brand Institute for Women (correctional facility in Los Angeles, California)

scam escape from jail or prison

Scanray Scanray Corporation's X-ray system for seeing what is inside parcels brought into jails or prisons

scarce commodity prison space

SCCA South Carolina Corrections Association

Schenectady upper New York State site of county jail

Schlüsselburg Leningrad prison

school of crime epithet applied to many prisons

sci (SCI) secret confidential informant

Scioto Village Ohio penal facility for male and female delinquents; opened as the Girls Industrial School in 1869

Scotland (*see* United Kingdom and individual entries)

Scottish Association for the Care and Resettlement of Offenders SACRO

Scottish prisons Aberdeen, Castle Huntly Borstal Institution, Cornton Vale, Dumfries Young Offenders Institution, Edinburgh, Edinburgh Young Offenders Institution, Glasgow, Glasgow Young Offenders Institution, Glenochil Detention Centre, Greenock, Inverness, Longriggend Remand Institution, Low Moss, Noranside Borstal Institution, Penninghame, Perth (including the Friarton Young Offenders Institution), Peterhead, Polmont Borstal Institution

Scottish Prison Service College SPSC

scragsman British slang for a hangman

Scranton Pennsylvania city and Lackawanna County seat and jail

screening (*see* detention screening)

screw(s) prison guard(s)

Scrubs The Scrubs (Wormwood Scrubs Prison in West London's stadium area)

SDC State Department of Corrections (Alabama, Colorado, Virginia)

SDCJ San Diego County Jail in California

SDMCC San Diego Metropolitan Correctional Center (California)

SD Met San Diego Metropolitan Correctional Center in downtown San Diego, California

Seagoville Federal Correctional Institution at Seagoville, Texas

Sea-Tac Seattle-Tacoma, Washington or its jail

Seavy's Island U.S. Naval Prison near Portsmouth, New Hampshire

seclusion solitary confinement

security housing section of a prison where hardened and hard-to-handle prisoners are segregated

Security Islands Îles du Salut off the coast of French Guiana; three rocky islets—Île du Diable (Devil's Island), Île Royale, and Île Saint Joseph—surrounded by shark-infested waters; until the 1950s political prisoners were isolated here

segregation isolation of criminals from other members of society; racial segregation; solitary confinement

seis y uno (Spanish—six plus one)—(*see* Six plus One)

Send detention center in Surrey, England

Senegal the capital, Dakar, has an old prison built by the French

Seoul capital of South Korea contains a prison built during the Japanese occupation of Korea

Sepa maximum-security prison in the remote jungles of Peru

Serbia (*see* Yugoslavia)

serve a sentence spend time in jail or prison

serve time spend time in jail or prison

SETAF Southern European Task Force (confinement facility in Italy)

Settlement Island former penal colony within Macquarie Harbour on the coast of Tasmania

Sevastopol Sevastopol central prison in Crimea

Seven-Step Foundation halfway house for released convicts

Seville (*see* Spain)

Sex Prison nickname for the Danish state prison at Ringe where the experiments in penology are underway in an effort to treat felons as humans

Sexto Lima, Peru jail

Seychelles the capital, Victoria, has an old HM gaol facility

SFCJ San Francisco County Jail

SFDC Santa Fe Detention Center in New Mexico

Shaker Road Albany County Penitentiary on Shaker Road in the Colonie section of Albany, New York

Shakopee location of the Minnesota Correctional Facility for female felons

shamus Irish-Gaelic nickname for a detective or other law enforcement officer—[*see* canine shamus(es)]

Sharjah (*see* United Arab Emirates)

Sharpes (*see* Brevard)

Shata medium-security Israeli prison with a work-release program managed by kibbutz volunteers

Sha Tsui Sha Tsui Detention Centre on Lantau Island, New Territories, Hong Kong

Sheff Sheffield, England or its jail

shelter confinement facility for juveniles held pending adjudication

Shelton Washington State Corrections Center or the Women's Correctional Facility at Shelton

Shelton Abbey old Irish prison near Arklow on the east coast of Ireland

Shepton Mallet HM Prison at Shepton Mallet in Somerset, England

Sheridan federal correctional institution in Oregon; Sheridan Correctional Center in Sheridan, Illinois, functions as the Illinois Industrial School for Boys (17 years old and up); the Wyoming Girls School at Sheridan

sheriff chief officer of county law enforcement and the county jail

Shin Bet Israel's domestic security agency

shit on a raft naval and prison slang for creamed beef or creamed chicken on toast

shit on a shingle military and prison slang for creamed beef or creamed chicken on toast

shiv knife made in prison; switchblade knife

Shore Patrol Tank nickname for a lockup maintained by the Navy in many American seaports

short stretch short prison sentence

short-term adult penal institutions of the U.S. Bureau of Prisons Allenwood, Pennsylvania; Elgin Air Force Base, Florida; El Paso, Texas; Florence, Arizona; Montgomery, Alabama; Safford, Arizona

Shrewsbury HM Prison at Shrewsbury in Shropshire, England

shrouding cover feature of hard-to-pry-open locks, padlocks, and shackle locks

Siam (*see* Thailand)

Siberia Russian area given over largely to the exile and imprisonment of political prisoners

Siberia de las Américas (Spanish—Siberia of the Americas)—political prisoner's nickname for prisons on the Isle of Pines (*Isla de Pinos),* renamed *Isla de Juventud* (Isle of Youth)—in the Caribbean off Cuba

Siberian salt mines nickname for the forced-labor camps and prisons throughout Siberia and other places in the USSR

Sierra Leone the capital, Freetown, has a prison built by the British

silence bell evening bell rung to advise prison inmates they must cease all talking and noisemaking

silent system imprisonment characterized by enforced silence and by night confinement in small solitary cells;

inmates allowed to congregate with other prisoners during meals or when at work; also called the Auburn System

Silk City Paterson, New Jersey, or its jail

Simons Simonstown, South Africa, or its jail

Simsbury early American prison built within an abandoned copper mine near Hartford, Connecticut; in use from 1773 to 1827

Sin Angeles (Spanish—Without Angels)—Los Angeles, California or its jail, lockups, penitentiaries, and prisons

Singapore sentence death by hanging is the Singapore sentence for trafficking in drugs

singbird(s) informant(s)

Sing Sing New York State Penitentiary at Ossining (where the birds warble twice—sing sing), now known as the Ossining Correctional Facility

Sin Lam Sin Lam Psychiatric Centre for the criminally insane at Sin Lam in the New Territories of Hong Kong

Sint Maarten (*see* Phillipsburg)

Sioux Falls South Dakota penitentiary site

Sirkeci Istanbul, Turkey's police station and lockup

Six plus One Spanish prison sentence of six years plus one day for anyone found in possession of or using narcotics or drugs

sizzle die in the electric chair

sizzle seat electric chair

Skag Skagway, Alaska, jail

Skopje (*see* Yugoslavia)

Skowhegan Women's Correctional Center at Skowhegan, Maine

slammed jailed

slam(mer) jail or prison where steel doors slam shut on the inmates

slanguage slang language; slum language

Sleepy Hollow Trenton Prison in New Jersey

Sleepy Hollow Correctional Centre in Charlottetown, Prince Edward Island

Slovenia (*see* Yugoslavia)

slum slumgullion stew (served in many jails)

Smithfield Pennsylvania Corrections Department near Carbondale

smogged executed in a gas chamber

Smyrna the Delaware Correctional Center

Sneads Florida site of the Apalachee Correctional Institution

Snoqualmie Washington town containing the Echo Glen Children's Center for delinquents from 8 to 14 years of age

Sobibor extermination camp near Lublin in Poland during World War II

Society for Alleviating the Miseries of Public Prisons Quaker group that planned the Eastern State Penitentiary in North Philadelphia, Pennsylvania; prisoners were kept in individual cells, each with an adjacent exercise yard; prisoners were supposed to become reformed by reading the *Bible* and reflecting on their crimes although this solitary confinement drove many of them crazy

Sofia capital city of Bulgaria and nickname of its prison

sol solitary confinement

Soledad California Training Facility at Soledad

solitary system imprisonment designed to segregate criminals from each other so as to prolong reflection and assure self-reform; many so subjected went insane; also known as the Pennsylvania System

Solomon Islands Honiara, the capital, has an HM Gaol-type prison on Guadalcanal Island

Solovetskis Solovetski Islands

Somalia capital and its prison, both called Mogadishu

Somers Connecticut Correctional Institution in Somers

Somerville Tennessee site of the Wilder Youth Development Center

Sonderbehandlung (German—special treatment)—term meaning killing at Nazi concentration camps

Sonderkommando Umsiedlungslager (German—Special Unit Resettlement Camp—euphemism for a quick-extermination concentration camp such as Sobibor

Sonkom Sonderkommando (German—Special Commando)—Hitler organization working at concentration camp crematoria and gas chambers where some inmates were forced to assist their armed guards in killing weaker prisoners and removing their remains

Sonoita Sonoita, Sonora, Mexico

Soo Sault Sainte Marie (Michigan or Ontario) and their jails

Soria Madrid, Spain's great prison

Sou Southhampton, England or its jail

South Africa has prisons in all its principal cities, Cape Town, Durban, Johannesburg, Pretoria, and in smaller port cities such as East London and Port Elizabeth

Southampton Correctional Center in Capron, Virginia is used for youthful first offenders

South Bay Regional Center euphemistic name for San Diego, California's jail in Chula Vista

South Burlington Vermont site of the Chittenden Correctional Facility

South Carolina jails 46 counties ranging have lockups; Charleston, Columbia, and Greenville have extra facilities

South Carolina School for Girls at Columbia

South Dakota lockups 67 counties contain jails

South Dakota Penitentiary in Sioux Falls

South Dakota Training School serving juvenile offenders held in Plankinton

Southeastern Correctional Center in Bridgewater, Massachusetts

Southeastern Ohio Training Center in Lancaster, formerly the Fairfield School for Boys

South Eastern Region Correction Institute Juneau, Alaska

Southern Michigan Southern Michigan Prison at Jackson

Southern Nevada Correctional Center in Jean

Southern Ohio Correctional Facility in Lucasville

Southern Rhodesia (*see* Zimbabwe)

Southern Steel San Antonio, Texas firm engaged exclusively in the manufacture of detention equipment

South Lansing the South Lansing School for Girls at South Lansing, New York

South Walpole location of the Massachusetts Correctional Institution, known as Cedar Junction

Southwark London borough on the Thames infamous for its Clink Prison for heretics; the prison is gone but the expression—*in the clink*—remains

South Windham site of the Maine Correctional Center

Spain jails, penitentiaries, and prisons are in its capital, Madrid, and cities such as Barcelona, Bilbao, Malaga, Seville, Toledo, Valencia, and Zaragoza

Spandau Berlin's great prison in the Tiergarten section of the city

Spanish Guinea used as a convict settlement during Spain's rule

Spanish windlass straitjacket (became tighter and tighter when sprayed with water or soaked with the prisoner's sweat)

Spartanburg South Carolina site of the Northside Correctional Center and the Piedmont Work Release Center

Spassk Soviet prison camp in Kazakhstan

SPC Service Processing Center(s)—formerly called Immigration and Naturalization Detention Center(s)

Spectator prisoner's publication of the Michigan State Prison at Jackson

Spinhaus (German—workhouse)—old term for a house of correction

spinhuiz (Dutch—workhouse)—old term for a house of correction

split bit prison sentence providing for both a maximum and a minimum sentence (*see* flat bit)

Spoke Spokane, Washington or its jail or lockups

sponging house jail where debtors were kept for a day to allow them to settle their debts before being imprisoned or transported overseas

spring release from jail or prison

Springer the New Mexico Boys School, in Springer

Springfield Sangamon County courthouse and jail in Illinois; Medical Center for Federal Prisoners at Springfield, Missouri

Spring Hill HM Prison at Spring Hill in Londonderry, Northern Ireland

Springhill Institution in Springhill, Nova Scotia

sprung released from jail on bail

SPSC Scottish Prison Service College

SQP San Quentin Prison (California)

squat to be electrocuted

squirrel cage hospital for the criminally insane

src (SRC) strict-regime (prison) camp (*see* Strict-Regime Camp)

Sri Lanka HM Gaol-type facilities are in Colombo, the capital, and in Jaffna, Kandy, and Galle

SRW State Reformatory for Women (Dwight, Illinois)

SSCA Southern States Correctional Association

Stafford HM Prison at Stafford in Staffordshire, England

Staff Training Centers the Bureau of Prisons offers training to correctional personnel and jailers from all parts of the United States in its centers in Atlanta, Georgia; Aurora, Illinois; Dallas,

Texas; and Oxford, Wisconsin—the food service training center of the FCI (Federal Correctional Institute)

Stalag (German—prisoner-of-war camp)

St Albans Correctional Facility in St Albans, Vermont

Stammheim Stuttgart, Germany's maximum-security prison

Standford Hill HM Prison at Standford Hill

Stangebro the local prison at Linkoping, Sweden

Stanley Hong Kong's maximum-security prison

Stanton Thomas F Stanton Correctional Center at Elmore, Alabama (*see* Staunton)

Stapleton Staten Island Detention Pens in the Stapleton section of Staten Island, New York

Starke site of the Florida State Prison

star prisoner inmate believed susceptible to rehabilitation and even special treatment

START Special Treatment and Rehabilitation Training (program for criminals)

state chemist executioner who poisons convicts in a gas chamber

State Correctional Institution and Correctional Diagnostic and Classification Centers in Pennsylvania (*see* Camp Hill, Graterford, and Pittsburgh)

State Correctional Pre-Release Center in Tipton, Missouri

State Diagnostic Unit in Salt Lake City, Utah

state electrician executioner who operates an electric chair

State Farm Virginia site of the Deep Meadow, James River, and Powhatan Correctional Centers

State Farm Spur Illinois penal institution near Vandalia

State House of Correction and Branch Prison in Marquette, Michigan

stateliest building in all Venice the prison of the Old Republic, described in Edgar Allan Poe tale, *The Assignation*

State Penitentiary in Puerto Rico this name applies to the state penitentiary in Rio Piedras

State Prison of Southern Michigan at Jackson

State Road Philadelphia address of the Detention Center opened as Moyamensing Prison in 1835 but replaced in 1963 and also the House of Correction

Stateville Stateville Correctional Center at Joliet, Illinois

Staunton (pronounced *Stanton)* Correctional Center in Staunton, Virginia

St Brides Correctional Center in Capron, Virginia

Ste Anne des Plaines Québec town contains the Archambault institution, a maximum-security penal facility

Steilacoom Washington location of the McNeil Island Correctional Center

Stevenson House Milford, Delaware's facility for children from 8 to 18 years of age

Stevens School at Hallowell, Louisiana

stew builder jail or prison cook

Stillwater site of the Minnesota Correctional Facility

stir jail; prison

stir bug prisoner whose insanity seems linked to long confinement or the thought of long confinement

stir crazy confinement crazy; maddened by imprisonment

Stirville Ossining, New York (site of Sing Sing)

St John's capital of Newfoundland and location of HM Penitentiary

St Johnsbury Correctional Facility in St Johnsbury, Vermont

St Joseph prison in Lyons, France

St-Martin-de-Ré French island known for its old jail close to the Bay of Biscay

Stockton San Joaquin County courthouse and jail east of San Francisco, California

Stoke Heath HM Borstal in Stoke Heath, Shropshire, England

stone dump prison

Stony Mountain Institution in Winnipeg, Manitoba

Strafanstalt die Strafanstalt (German—the prison)—also called *das Gefängnis*

Strafrechtler der Strafrechtler (German—penologist)

straitjacket restraining device designed to control violent or unruly persons such as the mentally insane or hard-to-handle prisoners

Strangeways Manchester, England's prison

strapped strapped to the electric chair; penniless

Stratford Stratford, Ontario's jail

Street Haven Toronto, Ontario's center for the rehabilitation of prostitutes and wayward girls

stretch prison sentence

Stretch prisoner's periodical published at Lansing, Michigan

stretch hemp execute by hanging

Strict-Regime Camp one of 36 Soviet-era prison camps in the Urals

Stringtown Correctional Center in Stringtown, Oklahoma

stripes striped prison clothes reintroduced for labor gangs in some U.S. prisons

Stromboli one of the Lipari Islands off of Sicily used to exile convicts and political prisoners

Stutthof main concentration camp in the north of Poland on the Bay of Gdansk

St Vincent de Paul St Vincent de Paul Penitentiary across the Riviere des Prairies from Montreal, Québec in Laval

Styal HM Prison at Styal in Cheshire, England

Sudan Khartoum and cities such as Omdurman and Port Sudan on the Red Sea, have HM Gaol-type penal facilities

Sudbury HM Prison at Sudbury

Sugamo Tokyo prison where Japanese war criminals were executed at the end of World War II

Sugar Land Texas site of the maximum-security Central Unit

Sukhanovka czarist monastery converted into a prison near Gorki

Sumpter Sumpter Correctional Institution in Bushnell, Florida

supreme penalty death

Surabaja (*see* Indonesian prisons)

Suriname Paramáribo and Nickerie, Paranam, and Moengo have penal facilities

Susanville California Conservation Center at Susanville

Sussex Correctional Institution in Georgetown, Delaware

Suva capital city of Fiji contains a jail built during British colonization

swag prison jargon for contraband such as drugs, tools, fruit, pornography, or weapons

Swan Lake the Swan River Youth Forest Camp southeast of Kalispell, Montana

Swan River Swan River Youth Forest Camp near Swan Lake, Montana

Swansea HM Prison at Swansea in Glamorganshire, Wales

Swaziland Mbabane and Manzini have HM Gaol-type penal facilities

S & W bracelets Smith and Wesson handcuffs

Sweden some of the most advanced and enlightened penal facilities are found near Stockholm and Göteborg and Malmö

Swedish prisons 37 prisons, some, like Kumla, are within old fortresses while others, like Gavle, are modern and provide for connubial visits

Swift Trail Swift Trail Federal Prison Camp in the Pinaleno Mountains near Safford, Arizona

Swinfen Hall HM Prison at Swinfen Hall in Staffordshire, England

swing die by hanging

Switzerland modem penal facilities are in or near the capital, Zürich, and in other places such as Basel, Geneva, and Bern

Swivels swivel nonlocking handcuffs

Sybil Sybil Brand Institute (Los Angeles county jail for women)

Syd Sydney (New South Wales, Australia) or Sydney (Nova Scotia, Canada) or their jails and lockups

Syracuse jails in Syracuse, Indiana, Kansas, Missouri, Nebraska, New York, and Ohio as well as on the island of Sicily

Syria Damacus, Aleppo, and Homs have prisons built by the Turks

T-4 *Tiergarten 4* (German— Zoological Park 4)—Berlin address of Nazi medical killing facility and its director Josef Mengele

Tac Tacoma, Washington or its jail or its lockups

Taconic Taconic Correctional Facility at Bedford Hills, New York

Taft Oklahoma location of the Jess Dunn Correctional Center

Taft Youth Center in Pikeville, Tennessee

Tafuna American Samoan community containing the Territorial Correctional Facility run by the Department of Public Safety

Tai Taipei, Taiwan jail or its prison

Tai Lam Tai Lam Addiction Department Centre or Tai Lam Centre for Women, both at Lung Chung in the New Territories of Hong Kong

Taipei (*see* China, Republic of)

take the electric cure suffer electrocution

take the pipe commit suicide

take the rap go to prison for someone else

take to the tules hide out in the bamboo-like tall grass

Talco Talcohuano, Chile, jail

Talladega Federal Correctional Institution at Talladega, Alabama

Tallahassee Federal Correctional Institution at Tallahassee, Florida

Tamarind Jamaican farm prison

Tamp Tampa, Florida; and Tampico, Mexico jails

Tampa Florida city has police-station lockups as well as the Hillsborough County jail

Tampere (*see* Finnish prisons)

Tampico prison in Tampico, Mexico

Tanforan racetrack near San Francisco used as a relocation center for Japanese-Americans during World War II

Tangerang boy's prison in western Java, Indonesia

Tangier (*see* Morocco)

Tanjong Pagar Singapore lockup and police station on Tanjong Pagar Road

tank prison cell

Tanzania Dar-es Salaam has an HM Prison built early in the 20th century

tap code code used by prisoners who are not allowed to talk to one another but manage to communicate by tapping on cell bars or plumbing pipes

Tarawa (*see* Kiribati)

Tarbes the maximum-security Prison de Tarbes in France

Tartu Estonia city where Nazis killed more than 12,000 in a concentration camp originally known as Dorpat

Tasmania large island off Australia, once called Van Diemen's Land, served as a penal colony from 1803 until 1853

Taycheedah Correctional Institution receiving center for adult females in Taycheedah, Wisconsin

TCA Tennessee Correctional Association; Texas Corrections Association

TDC Texas Department of Corrections

Tecate Tecate, Baja California, jail

Tecumsah Oklahoma site of the Central Oklahoma Juvenile Treatment Center

teddy boy(s) male juvenile delinquent(s) in the British Isles

teddy girl(s) female juvenile delinquent(s) in the British Isles

Teguci Tegucigalpa prison in the capital city of Honduras

Tehachapi California Correctional Institution at Tehachapi (state prison in the Tehachapi Mountains close to the Mojave Desert)

Teheran (*see* Iranian prisons)

telephone-pole design type of prison design, first introduced at Lewisburg, Pennsylvania, resembles a telephone pole with its crossarms; cellblocks and workshops are at right angles to a central corridor; this design provides flexibility in layout and coordination of elements for control and supervision of the inmates; the long connecting corridor (the telephone pole) extends from the administrative building past dining rooms and shops, and is bisected by cellblocks (*see* Lewisburg Plan)

Tel Mond medium-security Israeli prison for juvenile offenders located in the Plain of Sharon

Tennessee Colony Texas location of the maximum-custody Beto Unit; the Coffield Unit is also here

Tennessee lockups 95 counties have jails; Memphis, Nashville, Knoxville, and Chattanooga have extra facilities

Tennessee State Prison in Nashville

Tennessee Youth Center for delinquents between 15 and 18 years of age in Joelton

Tepic prison in Tepic, capital of Nayarit, Mexico

Terminal Island Federal Correctional Institution on Terminal Island opposite San Pedro, California

Terre Haute U.S. Penitentiary at Terre Haute, Indiana

Territorial Correctional Facility at Tafuna on Tutuila Island of American Samoa

Territorial(s) Territorial Prison(s)

Texarkana Federal Correctional Institution at Texarkana, Texas

Texas Department of Correction Units scattered about this state from Angleton to Sugar Land, from Brazoria and Brownwood to Rosharon and Tennessee Colony; Huntsville has the most penal facilities

Texas jails 254 counties contain lockups; cities such as Houston, Dallas, and San Antonio have local lockups

Thailand Bangkok; prison seems almost as old as the land it occupies

thana (Anglo-Indian—police station)—lockup; (Hindustani—jail)

THC Toledo House of Correction (Ohio)

theater of terror any public execution or punishment

The Castle U.S. Disciplinary Barracks at Fort Leavenworth, Kansas

The Gambia (*see* Gambian prisons)

The Pas Correctional Institution for Women at The Pas, Manitoba

therapeutic correctional community prison

Theresienstadt Nazi concentration camp about 40 miles (60 kilometers) from Prague; the Czechs called it Terezin

The Rock Otay Mesa Prison south of San Diego, also known as Rock Mountain

The Verne HM Prison at The Verne in Dorset, England

The Walls Huntsville Unit in Texas State Prison

Thieves' Palace nickname for the Surrey Prison

Thimphu Bhutan prison in the Himalayan mountains

Third Street New Jersey State Prison in Trenton on Third Street

Thomas Saint Thomas, American Virgin Islands, jail built a century ago by the Danes

Thomaston site of the Maine State Prison

Thorn Nazi concentration camp in Poland where it was known as Torun

Thorp Arch HM Prison at Thorp Arch, Yorkshire, England

Three Cs Federal Prison System logotype standing for Care, Custody, and Correction

three deuces jammed three concurrent two-year sentences

Three Strikes third felony conviction leads to life term in prison—(also known as Three Strikes and You're Out law)

three deuces running wild three consecutive two-year sentences

three-time loser person returning to prison for the third time

throwaway juvenile delinquent or young adult criminal living in the same city as his or her parents but out of their care or control

thumbs thumbcuffs for controlling and holding unruly prisoners being transported from place to place

TI Terminal Island; Federal Correctional Institution at Terminal Island, California in Los Angeles harbor

ticket-of-leave permit allowing a convict to leave prison before the expiration of the sentence and to work under certain restrictions; parole certificate

ticket-of-leave man parolee

Tidewater Correctional Unit in Chesapeake, Virginia

Tihar central jail in New Delhi, India

Tijuana Tijuana, Baja California Norte, jail

Tijuana Hilton nickname of San Diego's multistory, Metropolitan Correctional Center

Tillberga Swedish prison in Tillberga, west of Stockholm

Tillery North Carolina site of the Caledonia and Odom Complex

time off time off for good behavior (while imprisoned)

time out time out of sight (in a solitary-confinement cell)

time served total time spent in confinement before and after sentencing

Tinseltown Hollywood, California, police-station lockup as well as the nearby Los Angeles County Jail

tin throne metallic toilet; prison-cell toilet; slop bucket

tintureiro (Portuguese—dry cleaner)—Brazilian term for a prison van

Tipitapa maximum-security prison outside Managua, Nicaragua

Tipton Missouri site of the State Correctional Pre-Release Center

Tiptonville Tennessee site of the Lake County Regional Correctional Facility

Tirana prison in the capital of Albania

Tj Tijuana, Baja California, México, jail

Tjipinang maximum-security prison near Djakarta on the Indonesian island of Java

Tjirebon major prison in western Java in Indonesia

Tlaxcala prison in Tlaxcala, Mexico

TM Therapia Magna (Latin—Great Therapy)—euphemism invented by Nazi doctors in concentration camps where they participated in mass killings

TO Toledo, Ohio, jail or its lockups

Toamasina (*see* Madagascar's prisons)

Tobago (*see* Trinidad)

Toco Tocopilla, Chile, jail

Tocuyito nickname of the central penitentiary of Valencia in the Ejido Carabobo of Venezuela

Togo Lomé is its capital and its prison is primitive

toil factory prison workshop

Tokyo (*see* Japan's prisons)

tolbooth (Scottish—prison)

Toledo (*see* Spain, TO)

Toluca prison in Toluca, Mexico

tombas (Spanish—tombs)—solitary confinement cells

Tombs old New York City Prison in downtown Manhattan adjacent to the Criminal Court Building on the Lower East Side

Tombs Prison Manhattan's House of Detention

Tomoka Tomoka Correctional Institution (*see* Daytona Beach)

'Tona Daytona Beach, Florida, jail

Tonga an HM Gaol was built in the capital of Nuko'alofa

Tong Fuk Tong Fuk Detention Centre, Lantau Island, New Territories, Hong Kong

Toodyay an old gaol in Western Australia, now historical museum

Topaz relocation center for Japanese-Americans interned in central Utah

Topeka location of the Kansas State Reception and Diagnostic Center as well as the Kansas Correctional-Vocational Training Center and the Youth Center in Topeka

Topenish Japanese-American relocation center near Yakima, Washington

Topo Topolobambo, Mexico, jail

Toponym place-name prisoners use when telling where they were imprisoned; for example, Atlanta usually means the federal penitentiary in Atlanta, Georgia just as Trenton may be the New Jersey state prison there or even the Training School for (delinquent) Girls at Trenton

Toronto the Don Jail and the Mimico Correctional Centre are well-known penal facilities

torture chamber jail or prison where drugs are not available at any price

total segregation solitary confinement

Toulon French seaport known as a depot for convicts awaiting passage to the penal colonies of Algeria, French Guiana, and New Caledonia

Touquet Le Touquet, France, jail

Tower (*see* Tower of London)

Tower of London ancient fortress on the River Thames used as a royal residence, then a jail for political prisoners who often entered by the Traitors Gate before being beheaded; today it is an arsenal museum housing the crown jewels as well as ancient armor and many weapons; also called the Bloody Tower

Townsville HM Prison at Townsville (Queensland, Australia)

Tracy medium-security prison at Tracy, California, officially known as the Deuel Vocational Institute

Training Center for Youth in Columbus, Ohio for males from 12 to 17 years of age

Training School for Boys in Boonville, Missouri (*see* Boonville)

Training School for Boys and Girls at Jamesburg, New Jersey

Training School for Girls Trenton, New Jersey's correctional facility for delinquents from 8 to 17 years of age

training school(s) juvenile delinquent institution(s)

Traitors' Gate Thames River waterside gateway to the Tower of London, where prisoners were rowed in before being executed or serving long terms

tramp college nickname for county jail

Trani Italian prison near Bari on the Adriatic

Transfrisker electronically actuated hand-held no-touch personal-weapons-search device made by Federal Laboratories of Saltsburg, Pennsylvania

Transnistria Romanian-Nazi administrative region between the Bug and the Dniester rivers in the Soviet Ukraine where during World War II thousands of exiled Romanian Jews were relocated

transportation movement of prisoners to overseas penal colonies

Trautenau forced-labor subcamp in western Czechoslovakia during World War II

Treasure Island first brig for women sailors at the U.S. Navy base at Treasure Island in San Francisco Bay

Treblinka Nazi concentration camp in Poland

trembler prisoner afraid of other prisoners

Trenggalek one of Indonesia's newer prisons in eastern Java

Trenton Florida site of the Lancaster Correctional Center; site of the New Jersey State Prison and the Training School for Girls, as well as the Jones Farm at West Trenton, the St Francis Hospital Unit, and the Vroom Readjustment Unit in the Trenton Psychiatric Hospital

Tres Marias (Spanish—Three Marys)—Mexican penal settlement in the Pacific Ocean off Nayarit; prisoners are kept on Maria Madre Island

Trinidad and Tobago in Port-of-Spain, Trinidad, an HM Prison recalls the years of British rule from 1802 to 1976

Tripoli (*see* Lebanon and Libya)

Trondheim (*see* Norway)

Tropez St Tropez on the French Riviera or its jail

Trostyanets site of a Nazi concentration camp near Minsk

Troy New York State city north of Albany; seat of Rensselaer County and its jail

Trubetskoi bastion of the Peter and Paul Fortress, used as a prison in Leningrad

Trucial Sheikdoms (*see* United Arab Emirates)

Trujillo *Cárcel Nacional de Trujillo* (Spanish—National Prison of Trujillo, Venezuela)

Truk prison camp erected by the Japanese during World War II in the Caroline Islands

Truro the Nova Scotia School for Girls at Truro

trusty trustworthy convict who is allowed special privileges

TRY Teens for Retarded Youth (juvenile correctional program)

T-town Tijuana, Baja California, Mexico, jail

tuchthuiz (Dutch—house of correction; workhouse)

Tucker Tucker Unit in Tucker, Arkansas

Tucson site of Arizona Correctional Training Facility where academic and vocational training is offered; Catalina Mountain School with a treatment program for juvenile males; Pima County seat and jail

Tule Lake waterless relocation and segregation center of
northern California where
many Japanese-Americans
were interned after Pearl Harbor

Tullahoma Tennessee site of
the Highland Rim School for
Girls

Tulle maximum-security
prison on Rue Souhan in
Tulle, France

Tulsa Tulsa, Oklahoma,
County Jail

Tulungagung Indonesian
prison in eastern Java

Tunis (see Tunisia)

Tunisia cities, including the
capital of Tunis, have French-
colonial jails and prisons

Tuol Sleng (see S-21)

turist kogus (Turkish—tourist
cell block)—prison section
for foreigners

Turkey penal facilities are
mainly from the time of the
Ottoman Empire and may be
seen in Ankara, Istanbul,
Izmir, and other cities

Turku (see Finnish prisons)

Turney Center for Youthful
Offenders at Only, Tennessee

turnkey anyone entrusted with
the keys to a prison

Turrell Turrell Residential
Group Center near Farm-
ingdale, New Jersey

Tuscaloosa Tuscaloosa
County jail southwest of Birmingham, Alabama

Tuvalu an HM Gaol-type penal facility is in the capital,
Funafuti

Tuxtla Gutiérrez prison in
Tuxtla Gutiérrez, capital city
of Chiapas, Mexico

Twin Maples Farm British
Columbia's facility for treating women inmates with alcohol problems

twister(s) key(s)

Two Dzerzhinsky Moscow
address of the KGB and the
Lubyanka prison

two-time loser person going
to prison for the second time

Tyburn long a favorite hanging place fitted with gallows
and gibbets, between Edg-
ware Road and the wall of

Hyde Park, three miles from
Newgate Jail in the City of
London

UCA Utah Correctional Association

Udine Italian prison northeast
of Venice

ufac unlawful flight to avoid
confinement

Uganda the capital city, Kampala, has an old HM Prison as
well as local lockups

ugly customer dangerously
quarrelsome person

UK United Kingdom (Great
Britain, Northern Ireland, and
overseas colonies and territories)—or their penal facilities

Ulan Bator (see Mongolia)

ultimate penalty death, or life
imprisonment without parole

UN United Nations and its organizations concerned with
crime and punishment

unauthorized departure escape

Uncasville Connecticut Com-
munity Correctional Center at
Uncasville, northeast of New
London

underground kite secret message circulated throughout a
prison or from one prisoner to
another

underground tunnel secret
systems for introducing contraband into a prison

under the gun under observation or surveillance

unhook unfasten the handcuffs
or fetters

UNICOR trade name of the
Federal Prison Industries
Corporation maintaining 89
industrial operations in 39 penal institutions

Union Correctional Institution at Raiford, Florida

Unit 731 Japanese biological
warfare complex at the Harbin Military Hospital in Manchuria during World War II

United Arab Emirates the
capital, Abu Dhabi, and some
older cities, such as Dubai
and Sharjah have old HM
Gaol-type prison facilities

United Kingdom some of the
best-known gaols, penitentiaries, and prisons are in the

British Isles and there are
many remnants in former colonies and protectorates

United Prisoners Union revolutionary underground organization of hard-core convicts
in prisons such as San Quentin or Soledad

universal staircase nickname
for the treadmill once operated by felons

Unterlüss Nazi subcamp close
to Bergen-Belsen near Hannover, Germany

Up Changi Road Singapore's
Prison Headquarters at Kilometer 17 outside the city,
nicknamed for the road on
which it is located

up the river Sing Sing prison,
up the Hudson River from
New York City in the town of
Ossining

UPU (see United Prisoners
Union)

Urga (see Mongolia)

USA United States Army

U.S.A. the United States of
America contains every type
of penal facility from hospital prisons for the criminally
insane to barless maximum-
security centers for hardened criminals; (see individual entries)

USAF United States Air Force

USARB U.S. Army Retraining
Brigade

U.S. Army confinement facili-
ties Fort Benning, Georgia;
Fort Campbell, Kentucky;
Fort Carson, Colorado; Fort
Gordon, Georgia; Fort Hood,
Texas; Fort Knox, Kentucky;
Fort Lewis, Washington; Fort
Meade, Maryland; Fort Ord,
California; Fort Polk, Louisi-
ana; Fort Richardson, Alaska;
Fort Riley, Kansas; Fort Sill,
Oklahoma

USBP United States Board of
Parole; United States Border
Patrol; United States Bureau
of Prisons

U.S. Bureau of Prisons (see
Bureau of Prisons)

U.S. Bureau of Prisons federal
penitentiaries Atlanta,
Georgia; Leavenworth, Kansas; Lewisburg, Pennsylva-

nia; Lompoc, California; Marion, Illinois; Terre Haute, Indiana

U.S. Bureau of Prisons institutions for juvenile and youth offenders Ashland, Kentucky; Englewood, Colorado; Morgantown, West Virginia

U.S. Community Treatment Centers halfway houses (*see* Community Treatment Centers)

USDB U.S. Disciplinary Barracks at Fort Leavenworth, Kansas, which holds Air Force, Army, and Marine Corps prisoners whose sentences include six months or more of confinement and/or a punitive discharge

U.S. Department of Justice administers the Board of Immigration Appeals, Bureau of Prisons, Civil Division, Civil Rights Division, Criminal Division, Drug Enforcement Administration, Federal Bureau of Investigation, Immigration and Naturalization Service, Justice Management Division, Land and Natural Resources Division, Office of Legislative Affairs, Office of Public Affairs, Pardon Attorney, Tax Division, U.S. Marshals Service, U.S. Parole Division

USDJ U.S. (United States) Department of Justice (*see* entry)

U.S. Eighth Army Confinement Facility in Korea

USEP United States Escapee Program

Usk HM Borstal at Usk in England

U.S. Marshals Service maintains custody of federal prisoners, from the time of their arrest to their commitment or release, and transports federal prisoners pursuant to lawful writs and direction from the U.S. Bureau of Prisons

USMC United States Marine Corps (correctional facilities in Albany, Georgia; Camp Smedley D. Butler in Okinawa; Camp Lejeune, North Carolina; Camp Pendelton, California; Parris Island, South Carolina; Quantico, Virginia)

USMS U.S. (United States) Marshals Service (*see* entry)

USN United States Navy

USN Brigs correctional centers and detention facilities brigs in naval parlance—are operated in or close to port cities, with the exception of a naval air station in Millington, Tennessee; seaport brigs include: Agaña, Guam; Charleston, South Carolina; Corpus Christi, Texas; Great Lakes, Illinois; Guantanamo, Cuba; Jacksonville, Florida; Long Beach, California; New London, Connecticut; Newport, Rhode Island; Norfolk, Virginia; Pearl Harbor, Hawaii; Pensacola, Florida; Philadelphia, Pennsylvania; Roosevelt Roads, Puerto Rico; Rota, Spain; San Diego, California; San Francisco, California; Seattle, Washington; Subic Bay, Philippines; Yokosuka, Japan

USNCC U.S. Naval Correction Center

USP United States Penitentiary

USPB United States Parole Board

USPC U.S. Parole Commission (formerly USPB)

USPD U.S. Parole Division of the Department of Justice

U.S. Penitentiaries Atlanta, Georgia; Leavenworth, Kansas; Lewisburg, Pennsylvania; Lompoc, California; Marion, Illinois; Terre Haute, Indiana

USPP U.S. Probation and Parole

USPS United States penitentiaries

Utah jails the state's 29 counties have lockups; Salt Lake City has additional facilities

Utah State Prison at Draper

Utica upstate New York prison

Utrecht Netherlands prison clinic providing psychological treatment and observation

VAC (*see* Voluntary Action Center)

Vacaresti Romanian prison; formerly a monastery on the outskirts of Bucharest

vacationing near Chappaqua doing time in Sing Sing

Vaca Valley Star prison newspaper published by inmates in Vacaville, California

Vacaville California Medical Facility at Vacaville, psychiatric prison opened in 1955 after moving from Terminal Island

VACRP Victorian Association for the Care and Resettlement of Prisoners

VACs Voluntary Action Centers

Vaduz (*see* Liechtenstein)

Valdosta Georgia site of the Lowndes Correctional Institution

Valencia (*see* Spain and Venezuela)

Valetta (*see* Malta)

Valhalla Westchester County Penitentiary at Valhalla, New York

Valpo Valparaiso, Chile, jail

Val Verde County Clink nickname for the Val Verde County Jail in Del Rio, Texas

Vancoo Vancouver, British Columbia (or Vancouver, Washington) or their jails

Vandalia Correctional Center in Vandalia, Illinois

Van Diemen's Land former name of Tasmania and generic name for the great Australian penal settlement

Vanier Vanier Centre for Women at Brampton, Ontario

Vanuatu the old HM Gaol or Maison reflects its colonial past

VCA Virginia Correctional Association

Vega Alta Industrial School for Women at Vega Alta, Puerto Rico

Vegas Las Vegas, Nevada, jail

Vehicle City Flint, Michigan's jail

Venezuela Caracas and cities such as Barquisimeto, La Guaira, Maracaibo, and Valencia have Hispanic-type penal facilities

Ventura Ventura County jail between Los Angeles and Santa Barbara, California

Ventura Reception Center and Clinic at Camarillo, California

Ventura School for Girls at Camarillo, California

Vergennes the Weeks School for delinquent and unmanageable convicts at Vergennes, Vermont

Vermillionville Lafayette, Louisiana's former name and that of its jail

Vermont Correctional Facilities in Burlington, Rutland, St Albans, St Johnsbury, Windsor, and Woodstock

Vermont lockups 14 counties have jails

Vernichtungslager (German—extermination camp)—concentration camp of the type built and operated by German Nazis during World War Il

Vero Beach Florida site of the Indian River Correctional Institution for first-felony offenders under 20 years of age

Vesterfangsel (Danish—Westem Jail)—Copenhagen prison

vettura cellulare (Italian—celled vehicle)—prison van

vic(s) convict(s)

victims of the metal age older prisoners complain they have silver in their hair, gold in their teeth, rust in their guts, steel around their cells, and lead in their asses

Victoria British Columbia location of the William Head Institution; La Victoria (Santo Domingo city prison, Dominican Republic); Victoria Reception Centre (Old Bailey Road, Hong Kong); (*see* Seychelles)

Victor Verster Cape Town, South Africa prison

Vienna Correctional Center in Vienna, Illinois

Vientiane (*see* Laos)

Vietnam southeast Asian country formerly held by the French from 1858 to 1954 and by the Japanese during World War II; penal facilities built by the French and the Japanese are still used in Ho Chi Minh City (formerly Saigon) as well as in Hanoi, Haiphong, Da Nang, Hue, Nha Trang, and Vinh

Vieux Carre (French—Old Square)—French Quarter of New Orleans, Louisiana or its lockup

Vila (*see* Vanuatu)

Vila dos Remédios (Portuguese—Town of Reparation)—Brazilian settlement for ex-convicts on the isolated island of Fernando de Noronha in the Atlantic Ocean

Villa Devoto Argentinian prison in the Devoto section of Buenos Aires

Villahermosa prison in Villahermosa, capital of the Mexican state of Tabasco

Villamil Ecuador's former penal colony on the south coast of Isabela Island in the Galapagos Islands, where convicts serving life sentences were segregated with their families

Vince Saint Vincent Island in the Windward Islands of the Lesser Antilles or its jail

Vincennes French prison notorious for its many famous inmates dating back to the 14th century, when it was a castle and dungeon

Vinh (*see* Vietnam)

ViP (VIP) (*see* Volunteers in Probation)

Virginia State Department of Corrections consists of the Division of Adult Correctional Services and the Division of Youth and Community Services; these, in turn, are divided into regions and within each there are correctional centers and units as well as halfway houses, learning centers, and work-release units scattered throughout the Tidewater State

Virginia Correctional Center for Women in Goochland

Virginia jails the state's 96 counties, and its many independent cities, have lockups

Virginia State Penitentiary in Richmond

Virgin Islands jail lockup on the waterfront of Charlotte Amalie on the island of St Thomas; jail occupies an old fort built in Danish times when the islands were settled

Visalia Tulare County jail in central California

Vista town near San Diego, California or the county detention facility there

viuva alegre (Portuguese—merry widow)—prison van

viuva-alegre (Brazilian slang—prison van)

Vladimir medium-security prison halfway between Gorki and Moscow

Vladivostok transit prison port on the Pacific coast of far Eastern Russia

voiture cellulaire (French—celled vehicle)—prison van

Voluntary Action Center device for using the skills of persons convicted for minor crimes instead of putting them in prisons

Volunteers in Probation publishes *VIP Examiner* quarterly and books about probation

Voyvodina (*see* Yugoslavia)

VPCP Volunteer Probation Counseling Program

Vridsloesellille Danish state prison noted for its programs for reducing recidivism

Vulcano southernmost island in the Liparis off the coast of Sicily where it has served since Roman antiquity as a place of penal exile

Wacol HM Prison Wacol (Queensland, Australia—sometimes called the Bane of Brisbane

Wade Correctional Center in Haynesville, Louisiana

Wailuki port city on Maui Island, Hawaii, where the Maui Community Correctional Center is located

Wakefield jails in Wakefield, Massachusetts, Michigan, or Rhode Island

Walden Correctional Institution in Columbia, South Carolina

Wales Wisconsin site of the Ethan Allen School; (*see* United Kingdom and individual entries)

walk to be acquitted; to walk out of prison

Walker Correctional Institution in Rock Spring, Georgia

wall firing wall (place of execution by a firing squad)

Walla Walla Washington State Penitentiary at Walla Walla

Wallkill Wallkill Correctional Facility at Wallkill, New York

Wallows Walla Walla, Washington, jail

Wall-Wall nickname for the Washington State Penitentiary at Walla Walla

Walnut Street Philadelphia's oldest jail built in 1790; inmates were subjected to solitary confinement to prevent association with other prisoners and to promote reflection and self-reform

Walton Liverpool, England's Walton Prison

Walvis Bay (*see* Namibia)

Wanchai Hong Kong's Wanchai Police Station and lockup

Wandsworth prison in Wandsworth southwest of London

warden prison administrator

Ware Correctional Institution (*see* Waycross)

Warkworth Institution in Campbellford, Ontario

Warrensville Heights Ohio site of the Cuyahoga Hills Boys School

Warsaw (*see* Poland)

Wartburg Tennessee site of the Morgan County Regional Correctional Facility

Washington Corrections Center in Shelton, Washington, which serves as a reception and diagnostic center and offers a reformatory-type educational program

Washington lockups 39 counties have jails; Seattle and Tacoma have extra facilities

Washington State Funnypark Washington State Prison near Walla Walla

Washington State Penitentiary in Monroe provides all degrees of security for male felons

Washington State Penitentiary in Walla Walla holding felons 16 years of age and up in all levels of custody

Washington State Reformatory in Monroe

Washington State Special Offender Center in Monroe

watchhouse police station; police station lockup

Wateree River Correctional Institution in Rembert, South Carolina

Watergate Hilton prisoners' nickname for the District of Columbia jail in Washington, DC

Watkins Pre-Release Center in Columbia, South Carolina

Waukegan Lake County, Illinois, jail

Waupon Wisconsin site of the Dodge Correctional Institution and the Waupon Correctional Institution

Waycross Georgia location of the Ware Correctional Institution

Waymart Pennsylvania Corrections Department prison northeast of Carbondale

Wayne Wayne County Jail in Detroit, Michigan

Wayne Correctional Center in Goldsboro, North Carolina for convicts needing psychiatric care

Wayne Correctional Institution (*see* Odum)

WCA Wackenhut Corrections Corporator (of private jails); Washington Correctional Association; Western Correctional Association; Wisconsin Correctional Association; Women's Correctional Association

WCS Wisconsin Correctional Service (in Milwaukee)

WCSC World Correctional Service Center (*q.v.*)

WDC Women's Detention Center

Weeks School at Vergennes, Vermont

Weisswasser forced-labor subcamp operated by the Nazis during their occupation of Czechoslovakia

Welfare Security Program guarantees safety of prisoners who provide useful information to prison authorities

Wellingborough HM Borstal at Wellingborough in Northamptonshire, England

Wellington Wellington Prison, Wellington, New Zealand

West Concord site of the Massachusetts Correctional Institution

Westerbork Dutch site of a Nazi concentration camp during World War II

Western Hong Kong's Western Street Police Station and lockup

Western Europe's Largest Prison Fleury Mergois on the outskirts of Paris

Western State Penitentiary in Pittsburgh, Pennsylvania, replaced by the State Correctional Institution and Correctional Diagnostic and Classification Center

Westfield Westfield Correctional Center in Westfield, Indiana

Westfield Farm New York Reformatory for Women

West Palm Beach Palm Beach County jail

West Sam Western Samoa jail

West Street Federal Detention Center on New York City's waterfront in past years

Westville Westville Correctional Center in Indiana

West Virginia Industrial School for Boys in Grafton

West Virginia Industrial School for Girls in Salem

West Virginia jails 55 counties have lockups; there is also a city jail in Charleston

West Virginia Penitentiary in Moundsville

West Virginia State Prison for Women in Pence Springs

Wetherby HM Borstal at Wetherby in Yorkshire, England

Wethersfield Connecticut State Prison at Wethersfield

wets wetbacks; undocumented aliens who may get their backs wet crossing the Rio Grande to enter the United States and avoid immigration officials

Wha Wha Wha Wha Prison near Gwelo, Zimbabwe

WHD Women's House of Detention (New York City)

Wheeling Ohio County's seat in West Virginia; its jail is referred to as Wheeling

Whitehorse Whitehorse Correctional Centre in Canada

White Street Manhattan House of Detention for Men at 125 White Street in New York City

Wichita Sedgwick County jail in Kansas

Wilder Youth Development Center in Somerville, Tennessee

Wilingili penal settlement on an Indian Ocean atoll of the Maldive Islands

Wilkes-Barre northeastern Pennsylvania city and seat of Luzerne County where inmates refer to the jail as Wilkes-Barre

William Head Institution in Victoria, British Columbia

Wilmas (Mexican-Americanism—Wilmington, California)—nickname for the penitentiary on Terminal Island in Los Angeles Harbor

Wilmington Delaware site of Bridge House for boys and girls in detention status, the county jail, the Ferris School for male delinquents, the Pre-Trial Annex with maximum and medium-security cells, the Plummer (work-release) Center

Windhoek (*see* Namibia)

Windsor Correctional Facility in Windsor, Vermont

Winnipeg Manitoban site of the Stony Mountain Institution

Winson Green prison near Birmingham, England

Winston-Salem Winston-Salem, North Carolina, the seat of Forsyth County with its jail

wired addicted; electrocuted

Wisconsin Home for Women near Fond du Lac

Wisconsin jails 72 counties provide prisons, Milwaukee has local as well as county jails

Wisconsin School for Girls in the south-central part of the state

Witbank South African prison 65 miles (105 kilometers) east of Pretoria in the Transvaal

witness protection U.S. Marshals protect witnesses whose lives and those of their families are jeopardized by their testimony

Witzwil Swiss prison farm without walls in the town of Witzwil

Women's Correctional Admission Center in Columbus, Ohio

Women's Correctional Center in Columbia, South Carolina

Women's Correctional Institution Claymont, Delaware's penal facility

Women's Division current name for the Rhode Island Training School for Girls in Cranston

Women's Prison Association New York City-based service agency for prisoners; publishes A *Study in Neglect* about the plight of women in prison

Women's Reformatory officially The Women's-Reformatory, in Rockwell City, Iowa

Women's Ward Oklahoma State Penitentiary at McAlester

Woodbourne Woodbourne Correctional Facility at Woodbourne, New York

Woodburn Oregon site of the MacLaren School for juvenile-court commitments

Woods Haven-Kruse School for Girls at Claymont, Delaware

Woodstock Correctional Facility in Woodstock, Vermont

Wood Street Counter one of the most notorious of London's prisons in Elizabethan times although the Clink, Newgate, and Fleet did not lag far behind in inhumane practices

Workers' Paradise derisive nickname applied to the communist-controlled USSR

workhouse originally a prison where inmates had to work if they wanted to eat; in Great Britain, they picked oakum used for caulking ships or

they sewed mailbags; in the United States, a slang name for any prison

Work Release Program rehabilitation plan whereby convicts work outside of prison during the last part of their term and receive the same pay as other workers

Work Training Facility Louisiana has two—one in New Orleans and the other in Pineville

Worland site of the Wyoming Industrial Institute

World Correctional Service Center Chicago-based information clearinghouse

World's Freest and Smallest Jail San Marino's hilltop lockup, near Rimini, Italy uses a converted monastery to hold its prisoners who, if sober, are allowed to work in town providing they keep out of bars, restaurants, and other public places and avoid meeting with one another

World's Largest Prison Kharkhov in the Ukraine where more than 40,000 prisoners have been incarcerated at one time, according to the *Guiness Book of World Records*

World's Largest Walled Prison Southern Michigan Prison spanning 54 acres (22 hectares)

World's Most Gigantic Prison (*see* Most Gigantic Prison in the World)

Wormwood Scrubs English prison for young male offenders in the West London Stadium area

WPA (*see* Women's Prison Association)

WPA & H Women's Prison Association and Home

Wroclaw (*see* Poland)

WRP (*see* Work Release Program)

Wyantskill Wyantskill Center for Girls at Wyantskill, New York

Wyndham Wyndham Regional Prison (Western Australia)

Wynne Unit in Huntsville, Texas holds felons in maximum custody

Wyoming Girls School in Sheridan

Wyoming Honor Farm in Riverton, maintained by its inmates in minimum security

Wyoming Industrial Institute in Worland for 10- to 21-year-old delinquents

Wyoming lockups the state has 23 counties and jails; Cheyenne has city and county detention facilities

Wyoming School for Girls in the north-central part of the state holds delinquents from 18 to 21 years of age

Wyoming State Penitentiary in Rawlins

Wyoming Womens Center in Evanston

x'd executed

Xining capital of China's Qinghai province is notorious for its forced labor camps

Xochi Xochimilco, Mexico, or its lockup

YACA (*see* Youth and Correctional Agency)

Yakima Yakima County Jail in Yakima, Washington

Yaounde (*see* Cameroon)

yard prison yard

Yard Scotland Yard, London

yardbird convict; ex-convict; jailbird; prisoner

yard bull prison guard; railroad detective; railroad-yard policeman

Yardville New Jersey town containing the Youth Reception and Correction Center

Yaren (*see* Nauru)

YCA (*see* Youth and Correctional Agency)

YCC Youth Correctional Center

YCI Youth Correctional Institution (Bordentown, New Jersey)

YDI (*see* Youth Development Incorporated)

Yellowknife Correctional Centre in the capital of the Canadian Northwest Territories

Yemen has penal facilities built when the Turks ruled all Arabia before the end of the World War I

YGC Youth Guidance Center (San Francisco)

Yoko Yokohama, Japan, jail, lockup, and prison

Yokohama (*see* Japan's prisons)

Yokosuka Japan's naval base and lockup in Tokyo Bay south of Yokohama; U.S. Navy Fleet Activities Brig in Yokosuka; Yokosuka Prison on Japan's Honshu Island

Yokuska (*see* Yokosuka)

York community west of Lincoln, Nebraska and site of the Nebraska Center for Women

YOU Youthful Offender Unit

young-adult penal institutions of the U.S. Bureau of Prisons El Reno, Oklahoma; Lompoc, California; Milan, Michigan; Oxford, Wisconsin; Petersburg, Virginia; Seagoville, Texas; Tallahassee, Florida

young horse prisoner's name for roast beef

Youngs Youngstown, Ohio, jail

young stir boy's reformatory

Youth and Correctional Agency combines California's Board of Prison Terms, California Youth Authority, Correctional Industries Commission, Department of Corrections, Institutional Review Board, Narcotic Addict Evaluation Authority, and Youthful Offender Control Board

youth and juvenile penal institutions of the U.S. Bureau of Prisons Ashland, Kentucky; Englewood, Colorado; Morgantown, West Virginia

Youth Center at Atchison for delinquent boys from 13 to 15½ years of age

Youth Center at Topeka for juvenile delinquents in Kansas

Youth Correctional Institutions in New Jersey—one at Annadale, the other at Bordentown

Youth Development Center in Kearney, Nebraska

Youth Development Centers throughout Pennsylvania at locations such as Bensalem Heights, Loysville, New Castle, Waynesburg

Youth Development Inc juvenile education and rehabilitation program

Youth Forestry Camp in Custer, South Dakota for offenders from 15 to 21 years of age; Pennsylvania places

such as Hookstown, James Creek, and Whitehaven hold delinquents from 15 to 18 years of age

Youthful Offender Unit holds offenders under age 25 in close security at Hardwick, Georgia

Youth Reception and Correction Center in Yardville, New Jersey

Youth Services Center at Saint Anthony, Idaho

Youth Studies Center Philadelphia, Pennsylvania's juvenile correctional facility

Yps Ypsilanti, Michigan, jail

Ypsilanti Michigan city containing the Huron Valley Men's Facility and the Huron Valley Women's Facility

YSA Youth Services Administration (District of Columbia)

YTPM Yuma Territorial Prison Museum

YTS Youth Training School

Yugoslavia most of the prisons recall the Austro-Hungarian Empire

Yuma Arizona site of the Territorial Prison built in 1867, now a museum

Yuma Territorial Prison State Historic Park open daily from 8:30 to 5:30 on the banks of the Colorado River in Yuma, Arizona

Zacatecas prison in Zacatecas, Mexico

Zag Zagreb, Yugoslavia, jail

Zagreb (*see* Yugoslavia)

Zaire former name of Republic of Congo (*see* entry)

zak zaklyuchenny (Russian—prisoner)

Zambia HM Gaol-type structures contain its convicts

Zambo Zamboanga, Mindanao, Philippines, jail

Zanzi Zanzibar, Tanzania, jail

Zaragoza Zaragoza, Chihuahua, jail (*see* Spain)

Zenith City of the Unsalted Seas Duluth, Minnesota, jail

zenkamono (Japanese—jailbirds; tramps)—the most despised elements of Japanese society; usually segregated into run-down sections of cities

Zephyrhills Zephyrhills Correctional Institution at Zephyrhills, Florida

Zero Route (*see La Route Zero*)

Zimb Zimbabwe (formerly Rhodesia) or its jails or prisons built by the British and South Africans

Zimbabwe capital city, Harare (formerly Salisbury), and Bulawayo have penal facilities superior to many on the African continent

zoo police station and its lockup

Zuchthaus *das Zuchthaus* (German—penitentiary; prison)

Zurich (*see* Switzerland)

Zwodau forced-labor subcamp operated by the Germans during their occupation of Czechoslovakia in World War II

Currencies of the World

Afghanistan Afghani
Albania Lek
Algeria Dinar
American Samoa United
States dollar
Andorra French franc, Spanish
peseta
Angola Kwanza
Anguilla East Caribbean dollar
Antigua and Barbuda East
Caribbean dollar
Argentina Argentine peso
Armenia Dram
Australia Australian dollar
Austria Schilling and euro
Azerbaijan Manat
Bahamas Bahamian dollar
Bahrain Bahrain dinar
Bangladesh Taka
Barbados Barbados dollar
Belarus Belarussian ruble
Belgium Belgian franc and
euro
Belize Belize dollar
Benin *Communauté Financière
Africaine* franc
Bermuda Bermuda dollar
Bhutan Ngultrum
Bolivia Boliviano
Bosnia and Herzegovina Di-
nar
Botswana Pula
Brazil Real
Brunei Brunei dollar
Bulgaria Lev
Burkina Faso *Communauté
Financière Africaine* franc
Burma (*see* Myanmar)
Burundi Burundi franc
Cambodia Riel
Cameroon franc
Canada Canadian dollar
Cape Verde Cape Verdean es-
cudo
Cayman Islands Cayman Is-
lands dollar
Central African Republic
*Communauté Financière Af-
ricaine* franc
Chad *Communauté Financière
Africaine* franc
Channel Islands Guernsey
pound, Jersey pound
Chile Chilean peso
China Yuan
Colombia Colombian peso

Comoros *Communauté Finan-
cière Africaine* franc
**Congo, Democratic Republic
of** zaire
Congo *Communauté Finan-
cière Africaine* franc
Costa Rica Colón
Croatia Kuna
Cuba Cuban peso
Cyprus Cyprus pound
Czech Republic Koruna
Denmark Danish krone
Djibouti Djibouti franc
Dominica East Caribbean dol-
lar
Dominican Republic Domini-
can peso
Ecuador United States dollar
Egypt Egyptian pound
El Salvador Colón
Equatorial Guinea *Commu-
nauté Financière Africaine*
franc
Eritrea Birr
Estonia Kroon
Ethiopia Birr
Faeroe Islands Faeroese krone
Falkland Islands Falkland Is-
land pound
Fiji Fiji dollar
Finland Markka and euro
France French franc and euro
French Guiana French franc
French Polynesia Pacific fi-
nancial community franc
Gabon *Communauté Finan-
cière Africaine* franc
The Gambia Dalasi
Georgia Lari
Germany Deutsche mark and
euro
Ghana Cedi
Gibraltar Gibraltar pound
Greece Drachma
Greenland Krone
Grenada East Caribbean dollar
Guadeloupe French franc
Guam United States dollar
Guatemala Quetzal
Guinea Guinean franc
Guinea-Bissau *Communauté
Financière Africaine* franc
Guyana Guyana dollar
Haiti Gourde
Honduras Lempira
Hungary Forint
Iceland Icelandic króna

India Rupee
Indonesia Rupiah
Iran Rial
Iraq Iraqi dinar
Ireland Irish pound and euro
Isle of Man Isle of Man pound
Israel Shekel
Italy Lira and euro
Ivory Coast *Communauté Fi-
nancière Africaine* franc
Jamaica Jamaican dollar
Japan Yen
Jordan Jordanian dinar
Kazakhstan Tenge
Kenya Kenyan shilling
Kiribati Australian dollar
North Korea North Korean
won
South Korea South Korean
won
Kuwait Kuwaiti dinar
Kyrgyzstan Som
Laos Kip
Latvia Lats
Lebanon Lebanese pound
Lesotho Loti
Liberia Liberian dollar
Libya Libyan dinar
Liechtenstein Swiss franc
Lithuania Litas
Luxembourg Luxembourg
franc and euro
Macao Patacá
Macedonia Dinar
Madagascar Malagasy franc
Malawi Kwacha
Malaysia Ringgit
Maldives Rufiyaa
Mali *Communauté Financière
Africaine* franc
Malta Maltese lira
Marshall Islands United
States dollar
Martinique French franc
Mauritania ouguyia
Mauritius Mauritian rupee
Mexico Peso
Micronesia United States dol-
lar
Moldova Lem
Monaco French franc
Mongolia Tugrik
Montserrat East Caribbean
dollar
Morocco Dirham
Mozambique Metical
Myanmar (Burma) Kyat

Namibia Namibian dollar
Nauru Australian dollar
Nepal Nepalese rupee
Netherlands Guilder and euro
New Caledonia Pacific financial community franc
New Zealand New Zealand dollar
Nicaragua Córdoba
Niger *Communauté Financière Africaine* franc
Nigeria Naira
Northern Ireland Pound sterling
Norway Norwegian krone
Oman Omani rial
Pakistan Pakistan rupee
Palau United States dollar
Panama Balboa
Papua New Guinea Kina
Paraguay Guarani
Peru New sol
Philippines Philippine peso
Poland Zloty
Portugal Escudo and euro
Puerto Rico United States dollar
Qatar Qatari riyal
Réunion French franc
Romania Leu
Russia Ruble
Rwanda Rwandan franc
Saint Helena Pound sterling
Saint Kitts and Nevis East Caribbean dollar

Saint Lucia East Caribbean dollar
Saint Vincent and the Grenadines East Caribbean dollar
Samoa Tala
San Marino Italian lira
São Tomé and Príncipe Dobra
Saudi Arabia Saudi riyal
Senegal *Communauté Financière Africaine* franc
Seychelles Seychelles rupee
Sierra Leone Leone
Singapore Singapore dollar
Slovakia Koruna
Solomon Islands Solomon Islands dollar
Somalia Somali shilling
South Africa Rand
Spain Peseta and euro
Sri Lanka Sri Lanka rupee
Sudan Sudanese pound
Suriname Suriname guilder
Swaziland Lilangeni
Sweden Krona
Switzerland Swiss franc
Syria Syrian pound
Taiwan New Taiwan dollar
Tajikistan Tajik ruble
Tanzania Tanzanian shilling
Thailand Baht
Togo *Communauté Financière Africaine* franc
Tonga Pa'anga

Trinidad and Tobago Trinidad and Tobago dollar
Tunisia Tunisian Dinar
Turkey Turkish lira
Turkmenistan Turkmen manat
Turks and Caicos Islands United States dollar
Tuvalu Tuvaluan dollar, Australian dollar
Uganda Ugandan shilling
Ukraine Hryvnia
United Arab Emirates Emirian dirham
United Kingdom Pound sterling
United States of America United States dollar
Uruguay Peso
Uzbekistan Som
Vanuatu Vatu
Vatican City Lira
Venezuela Bolívar
Vietnam Dong
British Virgin Islands United States dollar
U.S. Virgin Islands United States dollar
Western Sahara Moroccan dirham
Yemen Rial
Yugoslavia Yugoslav new dinar
Zambia Kwacha
Zimbabwe Zimbabwean dollar

Diacritical and Punctuation Marks

´ acute accent (as in Bogotá)

' apostrophe; single quotation mark

@ commercial-letter A standing for *at* or *for*

[ ] brackets

˘ breve

, cedilla (as in Curaçao)

^ circumflex (as in *rôle*)

: colon

) close parenthesis

, comma

¨ diaeresis (as in München)

... **or** .... ellipsis; leaders

! exclamation point

` grave accent (as in *funèbre*)

- hyphen

? interrogation or question mark

¯ macron (dictionary pronunciation symbol indicating long vowel, as in dame)

(open parenthesis

( ) parentheses

. period

" " quotation marks; quotes

' ' quotation marks, single

; semicolon

~ tilde (as in São Paulo)

— vinculum (mathematics: placed above two or more letters)

Dysphemistic Place Names

Accident, Maryland
Atomic City, Idaho
Bad Axe, Michigan
Blue Ball, Pennsylvania
Braggadocio, Missouri
Cactus, Texas
Clam Gulch, Alaska
Climax, Colorado
Copsa Mica (Romanian—Black Town)—ink-soaked town near chemical plants
Cranks, Kentucky
Death Valley, California
Decoy, Kentucky
Dime Box, Texas
D'Lo, Mississippi
Dogpatch, Arkansas
Due West, South Carolina
Dusty, New Mexico
Embarrass, Minnesota
Eros, Louisiana
False Pass, Alaska
Fate, Texas
Fixer, Kentucky
French Lick, Indiana
Frogmore, Louisiana or South Carolina

Frozen Creek, Kentucky
Gap, Pennsylvania
Gas, Kansas
Graves, Georgia
Hanging Rock, Ohio
Hell, Michigan
Hell Gate, New York
Hell's Half Acre, Wyoming
Hemp, Georgia
Ho-Ho-Kus, New Jersey
Hungry Horse, Montana
Hygiene, Colorado
Intercourse, Pennsylvania (near Blue Ball)
Justiceburg, Texas
Kettle, Kentucky
Loco, Oklahoma
Mousie, Kentucky
Mud Lick, Kentucky
Nutsville, Virginia
Oblong, Illinois
Ordinary, Virginia
Panther Burn, Mississippi
Peculiar, Missouri
Pickleville, Utah
Pie Town, New Mexico
Plaster City, California

Porcupine, South Dakota
Quicksand, Kentucky
Rains, South Carolina
Shady, New York
Slick, Oklahoma
Smoketown, Pennsylvania
Speculator, New York
Stab, Kentucky
Stumptown, West Virginia
Tall Timbers, Maryland
Temperance, Michigan
Tiplersville, Mississippi
Tobaccoville, North Carolina
Tombstone, Arizona
Trussville, Alabama
Truth or Consequences, New Mexico
Turtletown, Tennessee
Uncertain, Texas
Vesuvius, Virginia
Volcano, California or Hawaii
Whiskeytown, California
X-ray, Texas
Yucca, Arizona
Zigzag, Oregon
Zilwaukee, Michigan

Earthquake Data (Richter Scale)

The Richter Scale, devised in 1935 by Dr Charles Francis Richter, seismologist of the California Institute of Technology, is a standardized scale for defining the destructive energy of earthquakes whose force is measured by seismographs. The magnitude of such earthquakes is the logarithm of the largest deflection measured and registered during an earthquake when a seismograph is 100 kilometers (62 miles) from the center of maximum shock, the epicenter of the earthquake, whose exact location is pinpointed by several scattered seismographs.

Numbers of the Richter Scale advance logarithmically and not arithmetically, so earthquakes measuring 8, for example, are ten times greater than those measuring 7, and this relationship is constant throughout the scale.

Earthquakes occurring before 1935, or before the invention of the seismograph in 1841, are approximated in terms of the Richter Scale.

Earthquake Damage and Intensity Devastation Effects
Encountered Historically

0 No detectable or measurable earthquake effect although about 100,000 quakes a year can be felt and at least 1000 cause some damage

1 Very slight earthquake effects felt by sensitive persons who may experience dizziness or nausea; other creatures may appear disturbed; gentle swaying may affect bodies of water as well as buildings and trees

2 Slight earthquake effects sensed by sensitive persons as well as other creatures who display uneasiness; hanging lamps and pictures swing slightly; buildings and trees sway slightly

3 Very moderate earthquake effects sensed by a few persons as well as by the most nervous and the most sensitive; dishes on shelves may rattle as may many windows; canned goods stored on shelves may rattle and may fall off, parked vehicles may rock and this is true of shrubs and trees

4 Moderate earthquake sensed by many and sufficient to awaken light sleepers; house frames creak and houses sway slightly; shrubs and trees tremble; parked vehicles may rock and sway

5 Near medium-strength earthquake felt by everyone and frightening most persons who tend to leave buildings and run out of doors to avoid cracking ceilings and crumbling walls; in older buildings plaster falls, ceilings crack, and windows break; pictures may fall off their hangers, dishes and glasses tumble off shelves; heavy desks and tables move and many may topple; old and weak chimneys may crack off at the roofline; ornamental cornices fall from buildings; church bells toll by themselves

6 Full-strength earthquake causing general fright approaching panic; stone walls crack; steep slopes and riverbanks crack; chimneys and towers may crack apart and fall; trees shake violently and often fall as do limbs; the Los Angeles Earthquake of 1971 measured 6.6, caused considerable damage, and took the lives of some 65 people

7 More devastating and more severe type of earthquake such as occurred in Nicaragua and Guatemala where thousands were killed in 1972 and 1976, respectively; or in the Chile Quake of 1906, which preceded the San Francisco Earthquake and Fire by only two days; 20,000 lives were lost in Valparaiso and 503 in San Francisco; both seismic disturbances were calculated in later years as representing 7.8 on the Richter Scale, though this figure has since been revised upward to greater than 8

8 Still more devastating and more severe earthquake causing general panic and marked by widespread land and water disturbances; many dams and dikes break, discharging vast volumes of flooding water; underground cables and pipelines crack and tear apart; railway rails bend and twist; brick, glass, and masonry facades peel off buildings and fall to the ground endangering people; loss of life can be quite severe as in the Peruvian Quake of 1970, accounting for the loss of some 66,000 people, or the Alaska Quake of 1964, reported as 8.4 on the scale, and marked by heavy damage in downtown Anchorage where 131 lost their lives; earthquakes of even greater magnitude occurred in Lisbon, Portugal in 1755 when 60,000 lives were lost and lakes in far off Norway were disturbed violently; the Shensi Province Quake, occurring in China in 1566, cost some 830,000 lives, calculated to have been 8.9 on the Richter Scale as was Japan's great quake of 1923, destroying all of Yokohama and half of Tokyo, and killing 143,000 people; the sea bottom in Sagami Bay sank 387 meters or 1300 feet; earthquakes of this magnitude afflicted New Madrid, Missouri in 1811, Charleston, South Carolina in 1886, and are predicted as long overdue along the San Andreas Fault Zone of California extending from south of the Mexican Border to San Francisco and northward; overall damage might well equal or exceed the Shinsai or Great Quake felt around Tokyo in 1923; a Chinese earthquake of July 26, 1976 registered

8.2 with a 7.9 aftershock the following day; shocks affected an area in and around Peking and Tientsin and killed some 750,000 people

9 Most devastating and most intense earthquakes top of the Richter Scale, extending from 0 to 9; minor earthquakes and tremors provide stress-relief cracking of and easing of the great tectonic energy tension beneath us

Krakatoa volcanic Indonesian island between Java and Sumatra, scene in August 1883 of one of the world's worst earthquakes

Recent Earthquakes Philippines (1991); Armenia (1989); Chile (1985); Colombia-Ecuador (1979); India (1988); Iran (1990); Japan (1931, 1946, 1948, 1983, 1993, 1995); Mexico (1985, 1999); San Francisco (1906, 1990), Los Angeles (1994), western Turkey (1999)

Greatest Earthquake ? Measuring 9.5 on the Richter Scale, an earthquake in southern Chile killed an estimated 5,700 people in May 1960

Emoticons

Symbols indicating emotions, usually viewed left side up, used in computer messages

(<>..<>)	Alienated	=^D	Grin	(:+(	Scared
0:-)	Angel	%\	Hangover	:-@	Scream
>:-<	Angry	:-)	Happy	M-)	See no evil
|-|	Asleep	d :-o	Hats off	#:-o	Shocked
~:o	Baby	:{	Having a hard time	:-V	Shout
!-(	Black eye	:X	Hear no evil	%-)	Silly or dazed
:-|	Bored (or indifferent)	^5	High five	|(	Sleepy
		()	Hug	:-,	Smirk
%-6	Brain-dead	%*}	Inebriated	=====:}	Snake
:-~)	Caught a cold	:%-{	Ironic	:-M	Speak no evil
(::()::)	Bandage (comfort)	:x	Kiss	:-p	Sticking tongue out
:*)	Clowning	^^^	Laughter		
%-(	Confused	:-x	Lips are sealed	= O	Surprised
:~-(	Crying	@>--->--->	Long-stemmed rose	:-&	Tongue-tied
:-| :-|	Deja vu			(:-\	Very sad
]:->	Devil	:( )	Loudmouth	: @	What?
:-e	Disappointed	O->	Male	;-)	Winking
<:|	Dunce	$-)	Money in mind	#-)	Wiped out
(:|	Egghead	+<:-|	Monk or nun	8:-)	Wizard
:-6	Exhausted	{}	No comment	%-|	Worked all night
O+	Female	:/}	Not funny	8-]	Wow
}{	Face to face	8-[	Overwrought	|-O	Yawn
~:-(	Flame (inflammatory message)	`:-)	Raised eyebrow	>=^P	Yuck
		[:|]	Robot		
>>:-<<	Furious	(:+(	Sad		
|-{	Good grief!	M:-)	Salute		

Eponyms, Nicknames, and Geographical Names

Aardvark US F-111 fighter-bomber

A Bachelor of Arts Phyllis Bentley's pseudonym

abandoned woman nickname for a prostitute

Abba Dabba Jim Tobin (baseball player)

Abbé Sieyès Emmanuel Joseph Sieyès

abbott nembutal sleeping tablet (nicknamed for its producer, Abbott Laboratories)

Abby Abigail

Abdul the Damned Sultan Abdul-Hamud II of Turkey

Abe nickname for a $5-bill bearing Lincoln's portrait

Abigail Van Buren Pauline Esther (Popo) Phillips, better known as Dear Abby

Abolitionist Quaker Elias Hicks

Abominable Snowman big, hairy manlike creature believed by natives to inhabit the higher Himalayas (also known as yeti)

Ab-o'-th'-Yate Benjamin Brierley

Abu Zabi (Arabic-Abu Dhabi) capital of the United Arab Emirates (UAE)

Abyssinia Ethiopia

academy whorehouse

acapulco gold gold-tinted, high-priced Mexican marijuana

accommodation houses British nickname for brothels

ace marijuana cigarette

Ace of Spies Sidney Reilly

Acerbic American Critic American Mercury's HL Mencken in the 1920s and 1930s; *American Spectator's* R Emmett Tyrell, Jr in the 1970s and 1980s

Achmed Abdullah Alexander Nicholaievitch Romanov

acid freak habitual LSD user who exhibits bizarre behavior

acid head habitual LSD user

acid rock synthesized music associated with drug use

Aco Michel Accault (La Salle's lieutenant)

Acronym Islands Indonesia where politicians delight in creating acronyms for every occasion

Acropolis of America New York City's Morningside Heights-site of Columbia University

Acton Bell pseudonym of Anne Brontë

Ada Adelaida; Adelaide

Adam Hall Elleston Trevor's pseudonym

Adam Smith George JW Goodman's pseudonym; 18th century Scottish economist

Addie Ada; Adela; Adelaide; Adelina; Adeline

Addison's disease adrenal cortical deficiency

Adelina Patti Adela Juana Maria Patti

Admirable Doctor English author-philosopher Francis Bacon—*Doctor mirabilis*

Admiral of the Atlantic Kaiser Wilhelm II

The Admiral Doctor Roger Bacon

Admiral of the Ocean Sea Christopher Columbus

Admiral of the Pacific Czar Nicholas II

Admiralties Admiralty Islands

Adobe State New Mexico

Adonis from Congress Warren Magnuson

Adonis of Fifty nickname of George IV

Adrian Hadrian

Adrian Girls (delinquent) Girls Training School at Adrian, Michigan

Adrianople former name of Eridne

Adriatic bowels diarrhea acquired in Italy or other Adriatic countries

Advance Agent of Emancipation Lucretia Mott

Adventure Captain Kidd's pirate ship

AE George William Russell

Aegean Ethicist Aristotle

Aerospace Valley California's Antelope Valley

Aesthetic Post-Impressionist Vasili Kandinski

Affable Archangel Raphael

afghan afghan blanket; afghan hound (noted for its long silk-like coat and narrow head)

Africa in Miniature Cameroon

African-American Abolitionist-Author-Editor-Orator Frederick Douglass

African-American Astronomer-Inventor-Mathematician Benjamin Banneker

African-American Botanist Chemurgist-Educator George Washington Carver

African-American Explorer Matt(hew) A Henson who pushed Peary to the North Pole as well as helping him survey the Nicaraguan canal route

African-American Messiah Booker T Washington

African-American Moses Harriet Tubman

African-American Movie Pioneer Ralph Cooper

African Queen Mrs Ian Smith of Salisbury, Rhodesia (now Zimbabwe)

Africa's big five Cape buffalo, elephant, leopard, lion, rhinoceros

Afro-America's First Great Poet Paul Laurence Dunbar

A-funk acid (LSD) funk, drug depression

Agate Capital Prineville, Oregon

agates nembutal

Age of Extinction the 20th century when the human population growth infringed on wildlife habitat

Age of Reptiles early mesozoic era

Age of Uncertainty John Kenneth Galbraith's name for the late 20th century

Age of Voltaire the Enlightenment

Aggie Agatha; Agnes

Agnes Lee Mrs Otto Freer

Agnon Shmuel Yosef Agnon pseudonym of Samuef Josef Czaczkes

Agnostic Penologist Colonel Robert Green Ingersoll (1833–1899) who lectured and wrote about Crimes Against Criminals as well as Cruelty in the Elmira Reformatory

Agrarian Champion Emiliano Zapata of Mexico

Agricola Alexander Ackerman, composer; George Bauer, German minerologgist; Johannes Sneider or Schnitter, Protestant reformer; Martin Sohr, composer; Roelof Huysman, Dutch musician-painterscholar

Agricultural Wizard of Tuskegee George Washington Carver

Agritainment farms open to public tours

Aguecheek Charles B Fairbanks

A-head acid head; LSD user; amphetamine addict

Ai Florence Anthony

Air Capital of America Wichita, Kansas

Air Capital of the World Montreal, Quebéc—headquarters of the International Civil Aviation Organization and the International Air Transport Association

Air-Conditioned City Duluth, Minnesota

Air Mike Air Micronesia

Airtourer single-engine trainer plane built by Aero Engine Services of New Zealand

Ajax Annie Besant's pseudonym when working with Charles Bradlaugg for free thought and population control in Great Britain

Akhiar formerly Sevastopol

Akmechet formerly Simferopol

Alabama Port Mobile

alacranes (Spanish—scorpions)—persons from the Mexican state of Durango

Alain pseudonym of Emile Chartier

Alain-Fournier Henri-Alban Fournier's pseudonym

Alameda Bernardo O'Higgins Las Delicias

Alamo City San Antonio, Texas

Alan Alda Alphonso d'Abruzzo

Alan Hovhaness Alan Hovhaness Chakmakjian

Alan King Irving Kniberg

Alaric Cottin Voltaire's nickname for Frederick the Great

Alas Leopoldo Alas y Ureña

Alaskan Ports (south to north) Ketchikan, Wrangell, Petersburg, Sitka. Juneau, Cordova, Seward, Anchorage, Kodiak, Dutch Harbor, Adak Naval Station, Nome

Alaska's Scenic Capital Juneau

Alaska turkey Alaska salmon

Albany beef Hudson River sturgeon

Albaturkey Albuquerque, New Mexico

Albemarle Island, Galápagos Isabela

Albert the Good Prince Albert Francis Charles Augustus Emmanuel of Saxe-Coburg-Gotha, Prince Consort of Queen Victoria

Alberto Moravia Alberto Pincherle

Alberto Savinio Andrea de Chirico

Albert's disease inflammation of the bursae over the Achilles tendon

Albertville Kalima, Zaire

Albion Britain's ancient name

Albuquerque Girls New Mexico Girls Welfare Home at Albuquerque

Albuturkey Albuquerque, NM

Al Capp Alfred Gerald Caplin

Alcibiades Alfred Lord Tennyson

Alcofribas Nasier François Rabelais

Alec Waugh Alexander Raban Waugh

Aleksei Maksimovich Peshkov Maxim Gorki

Alexander Girls Arkansas Training School for (delinquent) Girls at Alexander

Alexander of the North Charles XII of Sweden

Alexander Serarimovich Alexander Serafimovich Popov

Alexander Technique massage and physical-manipulation therapy developed by F Mathias Alexander

Alexandretta Iskenderun

Alexandria Egyptian port city of El Iskandariya

Alexandrian Century the 4th century before the Christian era when Alexander of Macedonia conquered Egypt, Persia, and India as well as encouraging Greek philosophers and poets—the 300s

Alexes ten-dollar bills bearing the portrait of America's first Secretary of the Treasury, Alexander Hamilton

Alfalfa Bill Governor William Henry Murray of Oklahoma

Alfonso XIII León Fernando María Isídro Pascual António

Alfred, Lord Tennyson Alfred Tennyson (1st Baron Tennyson—poet laureate of England from 1850 until 1892)

Algerian onyx stalagmitic calcite

Algerian Ports Annab (Bone), Skikda, Bejaia, Alger (Algiers), Mostaganem, Arzew, Oran, Mers el Kebir

Alhambra (Arabic—Red House)—ancient Moorish castle in Granada whence the Moors ruled most of Spain from 711 to 1492

Algonquin Circle F(ranklin) P(ierce) A(dams), Robert Benchley, Heywood Broun, Irvin S Cobb, Edna Ferber, George S Kaufman, Ring Lardner, Harpo Marx, Dorothy Parker, Harold Ross, Robert E Sherwood, Alexander Woolcott, and others who met informally around the bar of the Algonquin Hotel or in the offices of *The New Yorker*

Al Hirt Alois Maxwell Hirt

The Alice Alice Springs, Northern Territory, Australia

Alice B Toklas brownie cookie with baked-in marijuana

Alice Cooper Vincent Furnier

Alice Faye Alice Leppert

Alice Markova Alice Marks

Alien-and-Sedition President John Adams

Alize Breguet carrier-based three-place antisubmarine warfare aircraft

Al Jolson Asa Yoelson

Alkyd Winsor & Newton's trade name for alkyd-base watercolors

All All Alligator Alley

All-American Boy Jack Armstrong

All-American Mirror Upton Sinclair

All Bunny Albany, New York

Allenwood Camp Federal Prison Camp at Allenwood, Pennsylvania

Alligator Alley trans-Florida highway between Fort Lauderdale on the Atlantic coast and Naples on the Gulf of Mexico

Alligators Floridians

Alligator State Alabama, Florida, Louisiana, Mississippi, and Texas

All-the-Talents Administration Prime Minister William Wyndham Grenville

Alma Gluck Reba Fierson

Almanac Trial 1858 case won by Abraham Lincoln when he held up an almanac to show that since there was no moon on the night of a murder, a witness could not have seen his client commit murder

Almirante del Mar Oceano (Spanish—Admiral of the Ocean Sea), title given Christopher Columbus by Queen Isabella la Católica who also made him viceroy and governor of all the lands he discovered

Almost-Impeached President Andrew Johnson, Richard M. Nixon

Aloha State Hawaii

Aloysha (Russian nickname—Aleksei)—Alex; Alexander

Alpine Principality Liechtenstein

Alpine Republic Switzerland

Altamont, Catawba Thomas Wolfe's fictitious name for Asheville, North Carolina

Alte Fritz (German—Old Fritz)—Frederick the Great of Prussia

Alter Steffl (German—Old Stevie)—Stephen's Cathedral in Vienna

Alto Peru (Spanish—High Peru)—Bolivia

Amazonia Brazil's Amazon River basin

Amazon of the Keyboard Teresa Carreño

Ambassador of the Air Charles A Lindbergh

Ambassador of Good Will Will Rogers

A.M. Bernard Louisa Mae Alcott

Amdahl's Law a computer system's speed is determined by its slowest components (for Gene Amdahl)

America of the East Poland, according to Lech Walesa

America of the North Greenland, according to Franklin Delano Roosevelt

America of the South Panama Canal

America South of the Equator American Samoa

America of the West Guam

American American aloe (century plant); American beauty (crimson rose); American buffalo (bison); American cheddar (also called American cheese or store cheese); American cheese (cheddar); American cloth (oil-cloth); American cotton (upland cotton); American-English (American-style English); American fingering (piano); American fries (hashed brown potatoes); American leopard (jaguar); American lobster (Canadian or New England large-clawed species); American Morse (code); American plan (fixed hotel rate including board and food), (food plus room and bath); American rig (oil rig); American sable (pine marten); American School (of artists, economists, etc.); American twist (tennis) other American categories or items

American Agnostics Clarence Darrow, Felix Frankfurter, Robert G Ingersoll, Ben B Lindsey, Henry L Mencken

American Apostle of Nonviolent Disobedience Martin Luther King, Jr

American Arctic Alaska north of the Arctic Circle; the North Pole reached by Americans Robert E Peary and Matthew Henson in 1909

American Atheist Epigrammatists Ambrose Bierce, Mark Twain, and HL Mencken

American Atheists Luther Burbank, Thomas Edison, Emanuel Haldeman-Julius, Ayn Rand, Margaret Sanger, Gordon Stein

American Ballad Composer Stephen Collins Foster

American Balladeer Paul Robeson

American Balzac William Faulkner

The American Beauty Billie Dove (Lillian Bohny)

American Beauty Rose official flower of Washington, D.C.; nickname sometimes given young women—American Beauty Roses

American Caesar General Douglas MacArthur

American Cato Samuel Adams

American Century the 20th century marked by invention and industrial activity, discovery of the North Pole, landing of men on the moon, victory in two world wars, devotion to the democratic ideal—the 1900s

American Chronicler John Dos Passos

American Comedians Abbott and Costello, Fred Allen, Amos and Andy, Lucille Ball, Jack Benny and Rochester (Eddie Anderson), Edgar Bergen, Milton Berle, Josh Billings, Victor Borge, Albert Brooks, Mel Brooks, (George) Burns and (Gracie) Allen, Sid Caesar, Eddie Cantor, George Carlin, Diahann Carroll, Johnny Carson, Charlie Chaplin, Bill Cosby, Billy Crystal, Sammy Davis,

Jr, Phyllis Diller, Jimmy Durante, WC Fields, Redd Foxx, Great Gildersleeve, Jackie Gleason, Whoopi Goldberg, George B Hicks, Bob Hope, Danny Kaye, Buster Keaton, (Stan) Laurel and (Oliver) Hardy, Jay Leno, David Letterman, Sam Levinson, Harold Lloyd, Sam Lucas, Jackie (Moms) Mabley, Steve Martin, the Marx Brothers (*see* Marx Brothers), Florence Mills, Eddie Murphy, Petroleum V Nasby, Bill Nye, Pat Paulson, Paula Poundstone, Will Rogers, Mark Russell, Bobby Short, Red Skelton, Lily Tomlin, Peter Ustinov, Bert Williams, Robin Williams

American Commoner William Jennings Bryan

American Composer-Pianist Louis Moreau Gottschalk

American Composers Samuel Barber, Leonard Bernstein, William Billings, Aaron Copland, Norman Dello Joio, Stephen Foster, George Gershwin, Louis Moreau Gottschalk, Morton Gould, Howard Hanson, Victor Herbert, Alan Hovhaness, Charles Ives, Edward MacDowell, Gian Carlo-Menotti, Walter Piston, Wallingford Riegger, William Schuman, John Philip Sousa. Virgil Thomson

American Conservationists John Muir, William T Hornday, Williard G Van Name

American-Cowboy Comedian-Humorist Commentator-Philosopher Will Rogers

American Critic H(enry) L(ouis) Mencken

American Crusader for Religious Liberty Roger Williams

American Demosthenes Robert Ingersoll

American disease narcotic addiction

American Documentary Film Pioneer Robert Flaherty

American (Bald) Eagle symbol of the United States

American-English Poet Thomas Stearns Eliot

American Etchers Joseph Pennell and James Abbott McNeill Whistler

American Expatriate Painters Benjamin West and James Abbott McNeill Whistler

American Filibuster William Walker (1824–1860) who invaded Baja California, Sonora, and Nicaragua attempting to become the head of Central America

American Film Pioneer David Wark Griffith

American Founder of Women's Suffrage Elizabeth Cady Stanton (founder and first president of the National Woman Suffrage Association)

American Frontier Romanticist James Fenimore Cooper

American Gateway to Alaska and the Orient Seattle

American Heartland Illinois, Indiana, Michigan, Ohio, Wisconsin

American Hellenist Frank Mortyn

American Historical Painter Emmanuel Leutzé

American Humanist Philosopher John Dewey

American Humorists George Ade, Steve Allen, Woody Allen, Steven L Anreder, Russell Baker, Robert Benchley, Ambrose Bierce, Erma Bombeck, Art Buchwald, Al Capp, Johnny Carson, Irwin B Corey, ee cummings, Finley Peter Dunne, TS Eliot, William Faulkner, Benjamin Franklin, Lewis Grizzard, Joel Chandler Harris, Bret Harte, O Henry, Oliver Wendell Holmes, Art Hoppe, Washington Irving, Vachel Lindsay, Don Marquis, Groucho Marx, HL Mencken, Gerald Nachman, Ogden Nash, George Jean Nathan, SJ Perelman, James Whitcomb Riley, Will Rogers, Leo Rosten, Damon Runyon, Morrie Ryskind, Mort Sahl, R Emmett Tyrell, Jr, Mark Twain, Artemus Ward, Diane White, Robert Yoakum

American Ice Master Jackson Haines

American Illustrators Anton Otto Fischer, Howard Pyle, Norman Rockwell, N C Wyeth

American Impressionist Childe Hassam

American Industrial Painter Charles Sheeler

American Infidel Colonel Robert G Ingersoll, agnostic attorney and public speaker, also known as the American Demosthenes; Luther Burbank

American Karl Marx Daniel De Leon—born in Curacao, educated in Germany and the Netherlands, founded the Socialist Labor Party (SLP) and the International Workers of the World (IWW) in New York City

American Landscape Painters Albert Bierstad, George Caleb Bingham, James Britton, Frederic Church, Thomas Cole, Arthur Dove, Asher Brown Durand, Edward Hopper, Henry Inman, George Inness, J Francis Murphy, Georgia O'Keeffe, Grant Wood, and Alexander Helwig Wyant

American Libertarian Philosopher, Natural Scientist, Printer, and Publisher Benjamin Franklin

American Libertarians Thomas Jefferson, Thomas Paine, Robert Ingersoll, Clarence Darrow

American Lighthouse Painter Edward Hopper

American Lithographer (Nathaniel) James (Merritt) Ives

American Medical Historian William Henry Welch

American Melodist Stephen Foster

American Modern Jackson Pollock

American Montaigne Ralph Waldo Emerson

American National Composer John Philip Sousa

American Neurologist Extraordinary Silas Weir Mitchell

American Nine best-known classical composers arranged chronologically: MacDowell,

Ives, Piston, Hanson, Gershwin, Copland, Barber, Hovhaness, Bernstein

American Operetta Composers Irving Berlin, George M Cohan, Victor Herbert, Jerome Kern, Frederick Loewe, Cole Porter, Richard Rodgers, Vincent Youmans

American Orator Extraordinary Robert G Ingersoll and Franklin D Roosevelt

American Paradise U.S. Virgin Islands

American Patriot Author Thomas Paine

American Pestalozzi Amos Bronson Alcott

American Photographers of Distinction Berenice Abbott, Ansel Adams, Diane Arbus, Peter Beard, Mathew B Brady, Julia Margaret Cameron, Robert Capa, Imogen Cunningham, Walker Evans, Robert Frank, Arnold Genthe, Ralph Gibson, JK Hillers, William Henry Jackson, Gertrude Käsebier, Jill Krementz, Dorothea Lange, Clarence John Laughlin, Annie Leibovitz, J Ghislain Lootens, Mary Ellen Mark, Edward Muybridge, Timothy O'Sullivan, Roy Pinney, Galen Rowell, Edward Steichen, Alfred Stieglitz, Paul Strand, Edward Weston, Clarence H White, Margaret Bourke White, James Van Der Zee, Willard Van Dyke

American Portrait Painters James Britton, John Singleton Copley, Henry Inman, Eastman Johnson, John Singer Sargent, Eugene Edward Speicher, the Peale family, Gilbert Stuart, Thomas Sully, James Abbott McNeill Whistler

American Ports (*see entries under states and territories such as* Alabama Port, American Samoan Port, California Ports, etc.)

American Practical Navigator Nathaniel Bowditch

American Pragmatist Trinity John Dewey, William James, Charles Sanders Pierce

American Primitive Painters Edward Hicks, Grandma Moses

American Propagandist Novelist Upton Sinclair

American Prose-Poetry Novelist Thomas Wolfe

American Railroad Barons Jay Gould; Edward H Harriman; James J Hill; Collis P Huntington; William H Vanderbilt

American Rebel Upton Sinclair

American Samoan Port Pago Pago

American Sappho Sarah Wentworth Apthorp Morton of Braintree and Quincy, Massachusetts

American Scott James Fenimore Cooper

American Sculptors Daniel Chester French, Borglum, Brancusi, Epstein, Lachaise, Manship, Moore, Smith, St Gaudens, Ward, and Zorach

American Skeptic Philosopher George Santayana

American Socrates Benjamin Franklin

American Spokesmen for Socialism Eugene V Debs, Daniel De Leon, and Norman Thomas

American Temple of Music Carnegie Hall

Americans United Americans United for Separation of Church and State (AUSCS)

American Virgin Islands St Thomas, St John, St Croix, and other Virgin Islands

American Virgins U.S. Virgin Islands

American West Indies American Virgin Islands, Commonwealth of Puerto Rico, islands of Culebra, Vieques, and Mona, Navassa, Swan Islands, Corn Islands, and certain coral reefs in Caribbean

American Womanist Alice Walker

American Women Reformers Jane Addams, Susan B(rownell) Anthony, Elizabeth Cady Stanton, Ida M(inerva) Tarbell, Lillian D Wald, and Frances Elizabeth Willard

American Woodsman John James Audubon

American Wordsworth William Cullen Bryant

The Americas North, Central, and South America; the Western Hemisphere

America's Bermuda Rhode Island's Block Island

America's Best Girl Gertrude Ederle, first woman to swim the English Channel

America's Burns John Greenleaf Whittier

America's Cicero Richard Henry Lee

America's Dairyland Wisconsin

America's Devil's Island military prison at Fort Jefferson in the Dry Tortugas west of Key West, Florida

America's Finest City San Diego, California's nickname

America's First College Harvard

America's First Colonizers Roger Williams of Rhode Island and William Penn of Pennsylvania

America's First Financier Robert Morris

America's First Great Writer Washington Irving, James Fenimore Cooper, or Edgar Allan Poe

America's First Poet Philip Frenau

America's First Resort Newport, Rhode Island

America's First Suffragist Abigail Smith Adams

America's First Woman Newspaper Publisher Elizabeth Timothy of the *South Carolina Gazette* published in Charleston

America's Foremost Freethinker Robert Ingersoll

America's Forgotten Photographer Timothy O'Sullivan

America's Godfather Thomas Paine

America's Great Winter Garden Imperial Valley, California

America's Ice Box Alaska

America's Inside Fun City New York (concert halls, theaters, opera houses, museums)

America's Largest State Alaska

America's Last Frontier Alaska

America's Last Great Wilderness Alaska

America's Leading Operetta Composer Victor Herbert

America's Leading Proletarian Writer John Dos Passos

America's Most Famous Naval Hero Admiral David G Farragut

America's Most Useful Citizen Jane Addams—author of *Twenty Years at Hull-House*

America's Newest Big City Miami, Florida

America's No 1 Composer-Musician Aaron Copland

America's Nonsense Poet Ogden Nash

America's Official Poets Laureate Robert Penn Warren, Richard Wilbur, Howard Nemerov, Mark Strand, Joseph Brodsky, Mona Van Duyn, Rita Dove, Robert Hass, Robert Pinsky, Stanley Kunitz

America's Oldest City St Augustine, Florida

America's Outback Montana

America South of the Equator American Samoa

America's Outside Fun City San Diego (bays, mountains, year-round outside sports)

America's Poet Laureate Henry Wadsworth Longfellow

America's Practical Navigator Nathaniel Bowditch—compiler of *The American Practical Navigator*

America's Premier Air Woman Amelia Earhart—first aviatrix to fly across the Atlantic

America's Premier City New York

America's Principal Port New York

America's Proudest Musical Possession Carnegie Hall

America's Road Rage Therapist Arnold Norenberg

America's Second City Chicago

America's Star-Spangled Satirist Mark Russell

America's Sweetheart Mary Pickford

America's Tropical Islands Hawaiian Islands and the Virgin Islands

America's Wintergarden Southern California's Imperial Valley

Amiable Atheist Paul Heinrich Dietrich von Holbach, known to friends as Baron d'Holbach

Ami des Hommes (French—Friend of Mankind)— Marquis de Mirabeau

Ami du Peuple (French—Friend of the People)—Jean Paul Marat; the revolutionary journal he edited

Amiri Baraka LeRoi Jones

Ammonia King Edward Mallinckrodt

Amos and Andy Freeman F Gosden and Charles J Correll

Anarchist Geographers Prince Peter Kropotkin, Elisée Reclus

Anarchist Protagonist Michael Bakunin

Anatole France Jacques Anatole, François Thibault

Anatolia Asia Minor

Anatomist of Humanity Jean Baptiste Poquelin Molière

Ancient Capital of England Winchester

Ancient Capital of Sweden Sigtuna

Ancient Universities of England Cambridge and Oxford

Andean America (north to south) Venezuela, Colombia, Ecuador, Peru, Bolivia, Argentina, Chile

Andean Common Market Bolivia, Colombia, Ecuador, Peru, Venezuela (Andean Pact Nations)

Andean Group Bolivia, Colombia, Ecuador, Peru, Venezuela

Andean Lands Argentina, Bolivia, Chile, Colombia, Ecuador, Peru, Venezuela

Andrea del Sarto Andrea Domenico d'Agnolo di Francesco

Andrei Sinyavsky Abram Tertz

Andre Maurois Emile Salomon Wilhelm Herzog

Andrew Furuseth Anders Andreassen

Andrew Garve Paul Winterton

Andrew York Christopher Nicole's

Andropov Stalinist name for Rybinsk

Angel of the Battlefield Clara Barton

Angel of the Bottomless Pit the Devil

angel dust phencyclidine (PCP)

Angelenos residents of Los Angeles, California

Angelic Doctor Thomas Aquinas

Angie Dickinson Angeline Brown

Angolan Ports Ambriz, Luanda, Porto Amboim, Novo Redondo, Lobito, Benguela, Mocamedes, Porto Alexandre

angora a breed of long-hair cat originally from Angora (Ankara), Turkey; long hair goats or rabbits originally from Angora; long and fluffy strands of wool

Angora Turkey's old name for Ankara; long-haired breeds of cats, goats, and rabbits

Angora Goat Capital Rocksprings, Texas

Angry Eagle of Aviation General William (Billy) Mitchell

Animist Nation Dahomey

Ankara Ancyra or Angora

Anna Akhmatova Anna Andreyevna Gorenko

Annaba Bone, Algeria

Annabella Suzanne Georgette Carpentier

Anna O Bertha Pappenheim—feminist crusader against white slavery and first person to be psychoanalyzed

Annapolis of the Air Pensacola, Florida

Anna Seghers Netty Radvanyi

Anne Bancroft Anna Maria Italiano

Anne Morrow Mrs Charles Lindbergh

Ann Harding Dorothy Gatley

Annie Oakley Phoebe Anne Oakley Mozee

Annie's Town Anniston, Alabama

Ann Landers Esther Pauline (Eppie) Lederer
Ann Miller Lucille Ann Collier
Ann Sothern Harriette Lake
Antananarivo Tananarive, the capital of the Malagasy Republic (Madagascar)
Antarctica's Claimants Argentina, Chile, Norway, United Kingdom, U.S.A., Russia
Antarctica's Only Known Active Volcano Mount Erebus on Ross Island in the New Zealand sector
Antelope State Nebraska
Anthony Abbot Charles Fulton Oursler
Anthony Armstrong George Anthony Armstrong Willis
Anthony Berkeley Anthony Berkeley Cox
Anthony Boucher William Anthony Parker White
Anthony Hope Sir Anthony Hope Hawkins
Anthony Quinn Anthony Rudolph Oaxaca
Anthracite City Scranton, Pennsylvania
Antid Oto Leon Trotsky
anti-gods marijuana cigarettes
anti-Semite anti-Jewish person
Anti-Slavery President John Quincy Adams
Anti-Trinitarian Author Faustus Socinus, originally Fausto Sozzini
Anti-Trinitarian Martyr Michael Servetus, originally Miguel Serveto
anti-Zionist often anti-Semite
Antsirane Diégo Suarez
Apache State Arizona
Apis code name of Dragutin Dimitrijevic, chief of Serbian army intelligence, who engineered the assassination of Austro-Hungarian Archduke Franz Ferdinand providing the pretext for World War I in 1914
Apollinaire Guillaume Apollinaire, Wilhelm Apollinaris Kostrowitski
Apollo of the Box Tony Mullane (baseball player)
Apollyon The Devil
Apostle of Absolute Beauty Jean Sibelius

Apostle of the Anglo-Saxons Saint Augustine (first Archbishop of Canterbury)
Apostle of Caledonia Irish-born Saint Columbia
Apostle of California Padre Junipero Serra
Apostle of Christian Realism Reinhold Niebuhr
Apostle of Common Sense Voltaire
Apostle of Culture Matthew Arnold
Apostle of Dissent William Penn
Apostle of Enlightenment Thomas Paine
Apostle of Free Trade Richard Cobden
Apostle of the French Saint Denys
Apostle of the Gentiles Saint Paul (formerly Saul of Tarsus)
Apostle of Humanity Thomas Paine
Apostle of the Hungarians Saint Anastasius
Apostle of the Indians Bartolomé de Las Casas
Apostle of the Indies Saint Francis Xavier
Apostle of Ireland Saint Patrick
Apostle of Liberty Benjamin Franklin
Apostle of Mexican Rebellion Francisco I Madero
Apostle of New Zealand Reverend Samuel Marsden
Apostle of Reason Peter Abelard
Apostle Rebel Dorothy Day
Apostle of the Rights of Man Thomas Paine
Apostle of Sanity Epicurus
Apostle of Science Roger Bacon
Apostle of the Scottish Reformation John Knox
Apostles of Freedom Sam Adams and Thomas Paine
Apostle of the Slavs St Cyril
Apostle of the Sword Mohammed
Apostle of Temperance Theobold Mathew
Appeasement Premier Neville Chamberlain
Apple Capital of Texas Medina
Apple Capital of the World Wenatchee, Washington

Apple Island Tasmania
Apple Islanders Tasmanians
Apple Isle Tasmania
Apples Frank Kudelka (basketball player)
Appletalk trademark of Apple Corporation
Apron War Korean housewives helped to defeat the Japanese in 1592 by carrying rocks in their aprons for ammunition
Aqueduct City Rochester, New York
Arab Africa northern Africa from Egypt to Morocco and from Mauritania to the Sudan
Arabia Deserta Desert Arabia in the northern sector of the Arabian Peninsula
Arabia Felix Fertile Arabia also known as Aden, the Hadhramaut, or Yemenite section
arabian arabian baboon; arabian camel (one-humped dromedary); arabian coffee; arabian horse
Arabian Arabian Desert; Arabian Peninsula of Arabian Sea; inhabitant of Saudi Arabia also called Saudi (plural or singular)
Arabian Gulf Persian Gulf
Arabia Petraea Rocky Arabia in the northwestern section of the Arabian Peninsula
arabic arabic numbers (1, 2, 3, 4, 5, etc., as distinct from roman numbers—I, II, III, IV, V, etc.)
Araucania region of central Chile south of Bío-Bío River
araucanos (Latin American nickname—Chileans or chilenos)—sobriquet recalls the liberty-loving Araucanian Indians who were never conquered by the Spaniards
Arch City St Louis, Missouri, dominated by the Jefferson National Expansion Memorial arch commemorating the Louisiana Purchase
Archeological Capital of Africa Cairo, Egypt
Archeological Capital of North America Chichen Itza, Mexico
Archeological Capital of South America Cuzco, Peru
Archetypal American Painter Jackson Pollock

Archie Moore Archibald Wright

Archipiélago de Colón (Spanish—Columbus Archipelago) Ecuador's official name for the Galápagos Islands

Architect of American Modern Dance Martha Graham

Architect of the Atomic Bomb J Robert Oppenheimer

Architect in Chief of St Peters Raphael (Raffaello Santi)

Architect of the Constitution James Madison

Architect of Mexican Independence Padre Miguel Hidalgo

Architect-Naturalist-Philosopher-Statesman-President Thomas Jefferson

Architect of the New Deal Franklin Delano Roosevelt

Architect of Non-Alignment Josip Broz Tito

Architect President Thomas Jefferson

Arctic Arctic Current flowing south from Baffin Bay and Greenland to cool the coasts of Labrador, Newfoundland and most of New England; Arctic Ocean washing the north coast of Asia, Europe, and North America including Alaska and Canada

Arctic big three muskox, polar bear, walrus

Arctic Canada Northwest Territories and the Yukon

Arctic Lands Alaska, Canada, Greenland, Iceland, Norway, Sweden, Finland, Russia

Arctic Territories Canadian Northwest Territories

Arctogaea landmass including Africa, Asia, Europe, and North America

Ardent City Liège, Belgium

Ardent Conservationist Rachel Carson

Area 51 unidentified flying object aficionados' term for secret U S Air Force base north of Las Vegas, Nevada

Argentina La Argentina (Antonia Mercé y Luque)

Argentina's Principal Port Buenos Aires

Argentinian First Argentina's best known classical composer—Alberto Ginastera

Argentinian Ports Santa Fé, Rosario, Zarate, Campana, Buenos Aires, La Plata, Mar del Plata, Puerto Belgrano, Ingeniero White, Puerto Madryn, Ushuaia

Argyrol King Dr Albert C Barnes

Ariel of the Piano Frederic Chopin

Aristocles Plato's original name

Aristocrat of Sports billiards

Aristocrats of Beans lima beans

Aristotle's Lantern sea-urchin mouth parts

The Ark HMS *Ark Royal*

Arkansawyer a native of Arkansas, also called Arkie

Arkie Arkansas (or resident; migratory worker from Arkansas)

Arkopolis Little Rock, Arkansas

Arletty Arlette-Léonie Bathiat (French actress)

Armageddon Meggido, Palestine where the British defeated the Turks in 1918

Arminius Jacobus Arminius (originally Jacob Harmensen)

Arms-control President George Bush

Arnhem Land northern end of Australia's Northern Territory

Arnold Bennett Enoch Arnold Bennett

Arrowhead Herman Melville's home in Pittsfield, Massachusetts

Arroyo del Ajo (Spanish—Garlic Gulch)—John Steinbeck's name for his home near Los Gatos. California

Arsenal of the Nation Connecticut

Art Capital New York City

Art Center of Rhode Island Wickford

Art Center of the Southwest Taos, New Mexico

Artemus Ward Charles Farrar Browne (19th-century American humorist)

Artesian State South Dakota

Arthritis Season autumn

Artichoke Capital Castroville, California; Watsonville, California

Artie Shaw Arthur Arshawsky

Artist of the French Revolution Jacques Louís Davíd

Arturo de Cordova Arturo Garcia

Aruba's Ports Sint Nicolaas (Lago Refinery), Paarden Baai (Oranjestad), Druif

ARW 493 Stig Sverker Foghammar

Ashcan School (*see* Eight)

Ashenden W Somerset Maugham

Ashland Henry Clay's home in Lexington, Kentucky

ash-Sham (Arabic-northern) well known as Damascus, capital of Syria

Asian Subcontinent Bangladesh, Bhutan, India, Nepal, Sikkim, and Sri Lanka; Indian Peninsula

Asia's big five elephant, leopard, rhinoceros, tiger, water buffalo

Asmus Rasmus Rasmus Meyer Museum, Bergen, Norway; Miller

Asparagus Capital Isleton, California

Asperger's Disorder form of early infantile autism

Asphaltic Lake the Dead Sea

Assemblyman from the Bowery Al Smith (Alfred E Smith)

Associated States Caribbean island states (Antigua-St Kitts-Nevis, Dominica, Grenada, St Lucia, St Vincent)

Assyrian Century the 7th century before the Christian era when Assyria ruled the Middle East and conquered Egypt—the 600s

astrakhan astrakhan cloth or astrakhan wool of the type originally clipped from sheep native to Astrakhan on the Caspian in the delta of the Volga

Astrodome City Houston, Texas

Astronaut President John Fitzgerald Kennedy

Astronomical Prophet Galileo Galilei

Asylum for Talent Jacques Copeau's *Theatre du Vieux Colombier*

Atacama Atacama Desert of northern Chile

Ataturk (Turkish—Chief Turk) General-President Mustafa Kemal—first president of Turkey

The Atheist Percy Bysshe Shelley

Atheist Penologist Jeremy Bentham

Atheist Poetess Ellen Prouse Mardan

Atheist Poets Heinrich Heine, Percy Bysshe Shelley, Walt Whitman

Atheist's Bible Thomas Paine's *The Age of Reason*

Athenian Century the 5th century before the Christian era when the Athenians destroyed the Persian fleet and completed the Parthenon—the 400s

Athens of Alabama Tuscaloosa

Athens of Arkansas Fayetteville

Athens of America Boston

Athens of Hispanic America Bogotá

Athens of the North Edinburgh, Scotland

Athens of the Northwest Faribault, Minnesota

Athens of the South Nashville, Tennessee

Athens of Texas Waco

Athens of the West Lexington, Kentucky

Atlanta-Fulton Atlanta-Fulton County Stadium

Atlantic Bitch Atlantic Beach

Atlantic Canada Labrador and Newfoundland, New Brunswick, Nova Scotia, Prince Edward Island, Québec

Atlantic Community NATO nations

Atlantic Highlands Highlands of the Navesink around Sandy Hook, New Jersey

Atlantic Narrows relatively restricted area uniting North and South Atlantic between bulge of Africa and bulge of Brazil, Freetown and Natal

Atlantic Ocean Edens West Indies before European settlement

Atlantic Provinces New Brunswick, Newfoundland, Nova Scotia, Prince Edward Island

Atlantic Scandinavia Denmark, Iceland, Norway

Atlantique (French—Atlantic Ocean)—also name of the Breguet maritime-patrol aircraft BR-1150

Atoll Soviet Sidewinder-type missile

Atoll Nation Nauru

Atomic Age Capital Los Alamos, New Mexico

Atomic Capital of America Richland, Washington

Atomic Cities Los Alamos, New Mexico; Oak Ridge, Tennessee; Richland, Washington

Atomic City Los Alamos, New Mexico and Oak Ridge, Tennessee

Atomic Energy City Oak Ridge, Tennessee

A-town Allentown

Atterdag (Danish—Another Day)—King Valdemar IV

Attic Muse the Athenian historian Xenophon

Attorney for the Damned Clarence Darrow

Audrey Hepburn Edda van Heemstra

Augustan Age Latin literature's golden era when Horace, Livy, Ovid, and Virgil flourished during the reign of the Emperor Augustus (27 B.C. to A.D. 14)

Augustina de Aragon Augustina Domenech Zaragoza

Au₂H₂O Au_2H_2O former Arizona Senator Barry Goldwater (1908–1998)

Auld Ane (Scottish Gaelic—Old One)—the devil

Auld Brig o' Don Dundee, Scotland's Brig o' Balgownie (brig = bridge)

Auld Clootie (Scottish Gaelic—Old Cloven)—cloven-footed devil

Auld Reekie (Scottish Gaelic—Old Smelly)—smogbound Edinburgh

Auld Sod (Scottish Gaelic—Old Land)—Scotland

auntie old homosexual

Aunt Jane Tia Juana (river separating Tijuana, Mexico and San Ysidro, California)

Aunty Vicky Queen Victoria

Aurelian Century the 100s—reign of Roman emperor-philosopher Marcus Aurelius—the 2nd century

Aussieland Australia

Australia Felix (Latin—Happy Australia)—fertile central Victoria in southeastern Australia

Australian Alps mountains of New South Wales and Victoria

Australian Atheist Phillip Adams

Australian Commonwealth Australia and its territories

Australian Desert 1,300,000-square-mile area (530,000 hectares) in central and western Australia

Australian Dominion Australia and its territories

Australian Ports Port Kennedy, Cairns Harbour, Townsville, Port of Bowen, Port of Mackay, Rockhampton, Gladstone, Maryborough, Brisbane, Clarence River, Port Waratah, Newcastle, Sydney, Port Kembla, Melbourne, Williamstown, Geelong, Portland, Port Adelaide, Port Vincent, Port Pirie, Port Augusta, Whyalla, Port Lincoln, Albany, Busselton, Bunbury, Freemantle, Perth, Geraldton, Carnarvon, Broome, Wyndham, Darwin, Gove, Hobart *(in Tasmania along with* Burnie, Devonport, Beauty Point, Launceton)

Australian States New South Wales, Queensland, South Australia, Tasmania, Victoria, Western Australia

Australian Territories Australian Antarctic; Australian Capital Territory (Canberra); Northern Territory, Christmas Island; Cocos Islands; Coral Sea Islands Territory; Heard and McDonald Islands; Norfold Island

Australia's Little England Tasmania

Austrian Emperor of Mexico Ferdinand Maximillian von Hapsburg, also known as Max of Baden

Austrian Quintet classical composers Haydn, Mozart, Schubert, Bruckner, Mahler

Austrian Waltz Kings Joseph Lanner, Johann Strauss, and his son Johann Strauss Jr

Author of the Declaration of Independence Thomas Jefferson

Author of the First Amendment James Madison

Author of the first draft of the Declaration of Independence Thomas Paine

Author of the Virginia Statute for Religious Freedom Thomas Jefferson

Authority on Mythology Joseph Campbell

Autocrat of All the Russias Czar Nicholas II

Autocrat of Austria Prince Clemens Wenzel Lothar von Metternich

Autocrat of the Breakfast Table Dr Oliver Wendell Holmes

Autocrat of Commerce Venice

Automobile Capital of the World Detroit, Michigan

Automobile City Detroit, Michigan

Automobile State Michigan

Automobile Wizard Henry Ford

Auto State Michigan

Aux Cayes former name of Les Cayes, Haiti

Ava Gardner Lucy Johnson

Avalon Somerset region of southwestern England believed to be Avalon of Arthurian legend; resort port of Catalina Island off Los Angeles, California

Avenue of the Americas New York City's Sixth Avenue

Aviation City Seattle, Washington

Aviation Historian Octave Chanute

Avicenna Arabian astronomer-mathematician-physician Abu ibn Sina (980–1037)

Aviocar Spanish transport aircraft designated C-212

Avocado County San Diego County, California

Avon Avon Books; Avonmouth (Port of Bristol); Avon Water (flowing from Ayrshire to Lanark); Avonwick (Devonshire)

The Awakened One Siddhartha Gautama (Buddha)

Awakener of Bulgaria George Venelin

Awakening Moon full moon in March

Axis Salty Mildred E Gillars, American traitor convicted of treason for broadcasting Nazi propaganda during World War II

axminster eponymic name for good grade carpets and rugs originally made in the English town of Axminster in Devonshire; modern axminsters often copy well-known oriental designs

Ayrshire Poet Robert Burns born in Alloway, Ayrshire, Scotland

Azalea Trail City Lafayette, Louisiana

Azania African nationalist name for the Republic of South Africa

Aztecan and Incan Century the 1000s—great stone structures in Mexico and Peru are mute witnesses to these indigenous American cultures—the 11th century

Aztec State Arizona

Aztec two-step loose bowels acquired in Mexico, the land of the Aztecs

Aztec type microcephalic idiocy

Azure Coast Côte d'Azur on the French Riviera

Azure Sea Lake Rudolf, Kenya

B B King Riley B King

Baal Shem-Tov (Hebrew—Kind Master of the Holy Name)—Israel Ben Eliezer

Babar Jean de Brunhoff's storybook elephant; Zahir ud-Din Muhammad (founder of India's Mogul dynasty)

Babe Ruth George Herman Ruth, the Sultan of Swat

Babe Ruth's Legs Sammy Byrd (baseball player)

Babeuf Francois Noël

Babinski Reflex curling of toes upward when foot is stroked

Babinski Sign curling of toes upward after infancy

Babs Barbara

Babushka (Russian—grandmother)—Ekaterina Breshkovskaya, revolutionary leader

Baby Bells phone companies formed after breakup of AT&T

Baby Boomers people born after World War II, from 1946 to 1964

Baby Boom Generation nickname for the seventy-six million people born from 1946 to 1964

The Baby Bull Orlando Cepeda (baseball player)

Baby Doc Haitian dictator Jean-Claude Duvalier

Baby Langdon Harry Langdon

Babylon Miami, Florida

Baby Lucille Lucille Vore

Baby Ruth Ruth Cleveland, daughter of Grover and Frances Cleveland

Baby State Arizona in 1912 when admitted to statehood

Bachelor Duke William George Spencer Cavendish, the Sixth Duke of Devonshire

Bachelor Painter Sir Joshua Reynolds

Bachelor President James Buchanan—fifteenth President of the United States

Back Bay Boston's old residential section built on mud flats reclaimed from Boston Bay

Backbone of Asia the Himalayas

Backbone of the Confederacy the Mississippi River

Backbone of England Pennine Hills; Pennine Ridge

Backbone of Europe the Alps

Backbone of North America the Rockies

Backbone of South America the Andes

Backfire Soviet strategic supersonic bomber equivalent to the North American-Rockwell B-1

Back-of-Beyond Australia's sparsely inhabited interior

Back Home Again State Indiana

Bactrian Sage Zoroaster (founder of the Magian religion and native of Bactria)

Bad Boy of Music George Antheil

Bad Boys of 57th Street the New York Philharmonic in its post-Toscanini period

bad cholesterol low-density lipoprotein

Badger NATO nickname for Soviet Tupolev medium bomber (Tu-16)

Badger(s) Wisconsinite(s)

Badger State Wisconsin

Badlands arid and eroded areas of Nebraska and South Dakota

Bad News Jim Barnes (basketball player); Odell Hale (baseball player)

Baffin Basin deeper parts of Arctic Ocean between Baffin Island and Greenland

bag drugs packaged in balloons or paper bags

Bagdad by the Bay San Francisco

Bagdad-on-Hudson New York

Bagdad on the Subway one of O Henry's nicknames for New York City; He also called it the City of Razzle Dazzle

Bag Town San Diego, California, where sailors tote seabags

Bahamian Ports Freeport (Grand Bahama), Bimini (Bimini Islands), Nassau (New Providence), Matthew Town (Great Inagua)

Bahraini Ports Al Manamah Harbour, Mina Sulman, Sitra

Baiá de Guanabara (Portuguese—Guanabara Bay)—Rio de Janeiro's inner harbor

Baird Leonard Mrs Harry S Clair Zogbaum

Baked Beans Massachusettsans

Baked Bean State Massachusetts

Bakst Leon Bakst (originally Rozenberg)

Balanchine Georgi Balanchivadze

Bald Eagle Frank Isbell (baseball player)

bale 100 to 500 pounds of marijuana compressed

Baleful Prophet Cassandra

Balkan Peninsula between Adriatic and Black seas and terminating between the Aegian and Ionian peninsulas

Balkans Albania, Bulgaria, Croatia, Greece, Romania, Slovenia and European section of Turkey

balloons drugs such as heroin packaged in toy balloons

Balthus Count Balthazar Klossowski de Rola

Baltic Baltic Sea between Denmark, Estonia, Finland, Germany, Latvia, Lithuania, Poland, Russia, Sweden

Baltic Republics Estonia, Latvia, Lithuania

Baltic Scandinavia Finland and Sweden (Denmark sometimes included although much of its coast is on the Atlantic)

Baltimore beefsteak broiled liver's military nickname in the US

Baltimore Oracle HL Mencken

Balzac Honoré de Balsa

Bambino George Herman (Babe) Ruth

Bamboo Curtain old name for the barrier between anti-communist and communist countries of Southeast Asia

Banaba Ocean Island near the Gilberts in the equatorial mid-Pacific Ocean

Banana Benders Queensland Australians

banana boat cargo vessel built to carry bananas

Banana City Brisbane

Bananaland Queensland, Australia

Bananalanders people of Queensland, Australia

Banda Oriental (Spanish—Eastern Ribbon)—Uruguay

Band City Elkhart, Indiana

Bandit Queen of the Old West Belle Starr

Bane of the Bureaucrats Parkinson's Law

banewort *Atropa belladonna's* (also called beautiful lady, deadly nightshade, or death's herb)

bangkok bangkok hat (straw hat of a type first woven in Bangkok); bangkok straw (Siamese straw used in making baskets and hats)

Bangladesh Ports Chalna Anchorage, Chittagong, Cox's Bazar

Banjo Andrew Barton Paterson of Australia

Banjo Eyes Eddie Cantor (Edward Israel Itskowitz)

Banjo Joe Gus Cannon

The Bank The Bank of England

Bankwire trademark of US Banking and Settlement Service

Banner Cloud forms on the lee side of a mountain

Banner State Texas

Baotou (Pinyin Chinese—Paotou)

Bapu (Gujerati—father)—Mahatma Gandhi's title

Bar Barbara

Barão de Rio Branco (Baron Rio Branco) José María de Silva Paranhos

Barbadian Ports Speightstown, Bridgetown

Barbara Bel Geddes Barbara Geddes Lewis

Barbara Stanwyck Ruby Stevens

Barbara Ward Lady Jackson (wife of Sir Robert Jackson)

Barbarossa nickname of red-bearded Frederick I of the Holy Roman Empire; Barbarossa I (Koruk) and Barbarossa II (Khaireddin) were Greek-born Algerian pirates

Barbary Coast North African coast; San Francisco's waterfront district a century ago

Barbary States Algeria, Libya (Tripolitania), Morocco, Tunisia

Barbecue Bob Robert Hicks

Barbed Wire Capital of the World DeKalb, Illinois

Barbellion WNP Barbellion (pseudonym of Frederick Cummings)

The Barber Sal Maglie (baseball)

Barber Poet Provencal poet Jacques Jasmin—a barber by profession, also called the Last of the Troubadors

Barbs Barbados island or people

Barca the Carthaginian Maharbal

bar code Universal Product Code

Bard of Avon William Shakespeare

Bard of Ayrshire Robert Burns

Bard of Olney William Cowper

Bard of Prose Boccaccio

Bard of Sheffield James Montgomery

Bard of the Stumblebum Nelson Algren

Bard of Twickenham Alexander Pope

Barefoot Diva Cesaria Evora

Barefoot King of Cocos John Clunies-Ross (owner of the Cocos or Keeling Islands in the South Indian Ocean)

Baritone-Conductor Dietrich Fischer-Dieskau

Barney Barnato Barnett Barnato (Barnett Isaacs)

Baron Britten Edward Benjamin Britten

Baron Burnham Edward Levy-Lawson

Baron Corvo Frederick William Rolfe

Baron Cuvier Georges Léopold Chrétien Frédéric Dagobert

Baron de Reuter Israel Beer Josaphat (founder of Reuter's news agency)

Baroness Orczy Mrs Montagu Bartstow—author of *The Scarlet Pimpernel*

Baronet Peel Robert Peel (former Prime Minister of Great Britain)

Baron Grenville William Wyndham Grenville (former Prime Minister of Great Britain)

Baron Lugard Frederick John Dealtry Lugard

Baron Munchausen Rudolf Erich Raspe

Baron Passfield Sidney Webb

Baron Stiegel ironmaster Henry William Stiegel

Baron Tweedsmuir John Buchan

Barr body inactivated X chromosome (for Murray L Barr)

Barrington Island, Galápagos Santa Fé

Barrio Chino (Spanish—Chinese Quarter)—Barcelona's brothel area close to the waterfront

Barry Cornwall Bryan Waller Procter

Barry Fitzgerald William Shields

Barrymore family containing some of America's best beloved actors (Lionel, Ethel, John—children of Maurice and Georgiana Barrymore)— actual surname was Blythe

Barry Perowne Philip Atkey

Barry Sullivan Patrick Barry

Barton Cannon Barton Danzilio

Basil Rathbone Lawrence Northrup

Basket of Eggs hills of Downs in Northern Ireland

Basque Provinces álava, Guipúizcoa, and Viscaya, Spain

Basra belly Persian Gulf diarrhea

Bastille by the Bay San Quentin prison in San Francisco Bay

Bastion of the Caribbean Puerto Rico

Basutoland former name of Lesotho

Bataan Death March 65-mile march by American and Filipino prisoners of Japanese troops in 1942 (World War II)

Batavia Dutch name for Djakarta; former name of Djakarta, Indonesia; Roman name for the Netherlands

Batavian Republic name for the Netherlands during the French Revolutionary wars (1795 to 1806); the Netherlands during rule of Louis Bonaparte (1806–1810)

Bath City Mt Clemens, Michigan

Bat Masterson William Barclay Masterson

Battenberg Mountbatten

Battery old seawall containing gun emplacements in Charleston, South Carolina, Manhattan Island in New York

Battle-Born State Nevada— admitted as territory in 1848 following the Mexican War

Battlefield City Gettysburg, Pennsylvania

Battlefield of the Nations Leipzig in Germany in 1813; Plain of Esdraelen in Israel between Megiddo and Nazareth; Waterloo in Belgium

Battleground of Freedom Kansas during ten years of debate over slavery

Battle of Seattle November 19, 1999, police clash with protesters at World Trade Organization conference

Battling Battalino Christopher Battalino (boxer)

Battling Bob Robert M La Follette, Sr

Battling Levinsky Barney Lebrowitz (boxer)

Battling Nelson Oscar Nielson (boxer)

Batumskaya formerly Batum

Bauhaus German school of modern industrial design linking art and technology, headed by architect Walter Gropius

Bay Bay of Bengal, Biscay (nicknamed Biscuits by sailors), Boston, Charleston, Fundy, Holland, Islands, Laig, Mission, Naples, New York, Panama, Pigs, River, San Diego, San Francisco, State, Street, Whales; The Bay—Algoa Bay or Port Elizabeth Bay in South Africa

The Bay Hudson's Bay Company (*see* Bay)

Bay of Biscuits (naval nickname—Bay of Biscay)

Bay Cities cities surrounding San Francisco Bay

Bay City San Francisco

Bay of Gold San Francisco Bay

Bayou City Houston, Texas

Bayou State Louisiana, Mississippi

Bay of Pigs Cuba's south coastal Bahía de Cochinos where an unsuccessful attempt was made to liberate Cubans from Castro's rule

Bay State Massachusetts

Bay Stater(s) Massachusettan(s)

Bay Street financial center of the Bahamas in Nassau on New Providence Island; financial center of Canada in Toronto, Ontario

BB King Riley B King

BBB Big Bronze Buddha (in Kamakura, Japan)

b-dacs armed agents of the Bureau of Drug Control

Beagles Virginians

Bean Town Boston, Massachusetts

beans barbiturates

bears law-enforcement officers

Bears Kentuckians

Bear State Arkansas, Kentucky

The Beast Jimmie Foxx (baseball player)

Beast of the Apocalypse Aleister Crowley

Beat Generation group of writers including Allen Ginsburg, William S Burroughs, Jack Kerouac and others; "beat" is short for beatitude —the idea that the downtrodden are saintly

The Beatles George Harrison, John Winston Lennon, James Paul McCartney, Ringo Starr (Richard Starkey)

Beatrice Arthur Bernice Frankel

Beautiful City by the Sea Portland, Maine

Beautiful Kingdom translation of China's name for the United States of America

Beauty Dave Bancroft (baseball player)

Beauvoir last residence of Jefferson Davis (in Mississippi)

Beaver State Oregon

bebop jazz-style music of the 1950s

Bechuana bill an African diarrhea

Beck Depression Inventory questionnaire used to diagnose depression and its severity

Beck(y) Rebecca

Bedlam nickname of the Bethlehem Royal Hospital founded in London in 1247

Beedle General Walter Bedell Smith

Beef Barons Armour, Cudahy, Morris, Swift

Beefeaters Her (His) Majesty's Honourable Corps of Gentlemen at Arms

Beef-heads Texans

Beef State Nebraska; Texas

Bee Gee British Guiana (now called Guyana)

Beehive of Industry Providence, Rhode Island

Beehive State Utah

Beelzebub The Devil

Beemer BMW (Bavarian Motor Works) car or motorcycle

Beer City Milwaukee

Beethoven Town Bonn, Germany—birthplace of Ludwig van Beethoven

Beetle Juice (American navalese—Betelgeuse)—variable giant red star of the first magnitude in the constellation of Orion

Bee Wee nickname for a British West Indian

Beggars of the Sea Dutch pirates and privateers

Beige Book regional economic survey conducted by the United States Federal Reserve Board

Beige Toaster Macintosh computer

Bela Lugosi Bela Ferenc Blasko

Bel Anglais (French—Handsome Englishman)—nickname of John Churchill—Duke of Marlborough

Belgeek(s) Belgian(s)—so-called by American, British, and Canadian armed forces during World War II

belgian Belgian griffon (small black or reddish-black terrier); Belgian horse (draft horse); Belgian hare (rabbit); Belgian sheepdog

Belgian Congo colonial possession of Belgium from 1884 to 1963; Zaire

Belgian East Africa former name of Rwanda

Belgian Ports Antwerpen (Anvers), Bruxelles (Brussel), Gente (Gand), Brugge (Bruges), Zeebrugge, Blankenberge, Oostende (Ostend), Nieuwpoort (Nieuport)

Belgium Film Pioneer Jacques Feyder

Belial The Devil

Believing Unbeliever Supreme Court Justice Felix Frankfurter

Belize formerly capital of British Honduras and continuing seaport of Belize, whose capital is inland Belmopan

Belle of Amherst Emily Dickinson

Belle City of the Bluegrass Region Lexington, Kentucky

Belle City of the Lakes Racine, Wisconsin

Belle Riviere (*see La Belle Riviere*)

Belle Starr Myra Bell Shirley

bellman buster rolling suitcase

Bell Rock Inchcape Rock off the Forfarshire coast of Scotland, the subject of Robert Southey's *Ballad of Inchcape Rock*

Bell Town East Hampton, Connecticut

Belly of Paris Les Halles

Beloved City of El Greco Toledo, Spain

Beloved Friend Nadejda von Meek (Tchaikovsky's patroness and lifetime friend he never met)

Beloved Infidel Colonel Robert Ingersoll

Beloved Violinist Fritz Kreisler

Below Sea-Level Cities Brawley and El Centro, California

Ben Block a British sailor

Ben Cur Benguela Current

Ben Day Process photoengraving process (for Benjamin Day, 1838–1916)

Bender-Gestalt Test measures personality and assesses functional and organic disorders

Benemérito de las Américas (Spanish— Meritorious Man of the Americas)—the Mexican Indian Benito Juárez

Ben-Gurion (Hebrew—Son of a Lion)—name adopted by David Green

Benin Ports Cotonou, Kpeme

Benjies hundred-dollar bills bearing the portrait of America's Benjamin Franklin

Ben Kingsley Krishna Banji

bennies benzedrine pills

Benny Leonard Benjamin Leiner (boxer)

Benoulli's Principle pressure of water against sides of a wooden pipe decreases as the velocity of the water increases (observation of Daniel Benoulli, Swiss mathematician in mid-1700s)

Ben and Pen Benjamin Franklin and William Penn

bent-nail syndrome Peyronie's disease wherein the penis is bent out of shape

Benzer's Raiders biologist Seymour Benzer's students

Bering Bering Sea extension of the North Pacific Ocean

Berks the Beautiful Pennsylvania's Berks County

Berkshires Berkshire Hills of western Connecticut and Massachusetts

Berlin Bach Karl Philipp Emanuel Bach—also nicknamed Hamburg Bach

Bermuda High high-pressure system in the Atlantic Ocean

Bermuda Triangle area between Bermuda, Cape Hatteras, and Key West in the western North Atlantic—also called the Devil's Triangle or the Limbo of the Lost because of the many planes and ships lost in the area; area is also said to be between Bermuda, Florida, and Puerto Rico

Bernese Oberland Bernese Alps

Bernhardt Sarah Bernhardt—stage name of Rosine Bernard

Berry City Woodburn, Oregon

Berserkeley Berkeley, California

Bertholt Brecht Eugen Berthold Friedrich Brecht

Bertie Bertrand A Russell—English philosopher

Bert Lahr Irving Lahreim

Bert Parks Bert Jacobson

bessel bessel equation, bessel function, or bessel method (named for the German 19th-century astronomer Friedrich Wilhelm Bessel)

bessemer bessemer converter or bessemer steel (named for its English inventor, Sir Henry Bessemer)

Bessie Love Juanita Horton

Betel Nut Island Penang, Malaysia

Betsytown Elizabeth, New Jersey

Betty Carter Lillie Mae Jones

Betty Grable Elizabeth Grasle

Beverly Sills Belle Silverman

Bharat Republic of India

Biandrata Giorgio Blandrata

Bible Belt rural areas of the southern United States

biblioklept book stealer

Bi-City Port Gdansk-Gdynia, Poland

Bicycle Capital of the United States Missoula, Montana

Bien Aime (French—Well Beloved)—Louis XV

Big A Anaheim Stadium, Anaheim, California; Atlanta, Georgia in the Southeast; Amarillo, Texas in the Southwest; underworld nickname of the Federal Penitentiary in Atlanta, Georgia

Big Apple New York City

Big Ben aircraft carrier USS *Franklin;* battleship USS *Franklin;* huge bell attached to clock in Parliament tower, Westminster district of London, named after Sir Benjamin Hall, commissioner of works in 1859 when bell was hung

Big Bend big bend of the Rio Grande—bounding southern section of the Big Bend National Park on the Texas border of Mexico

Big Benders Tennesseans

Big Bend State Tennessee

Big Bertha World War I howitzer capable of hurling a one-ton projectile nine miles. named for a member of the Krupp family

Big Bill Broonzy William Lee Conley

Big Bill Haywood William Dudley Haywood (founder of IWW)

Big Blue IBM Corporation

Big Board New York Stock Exchange

Big Boy Arthur Crudry; largest type of steam locomotive from World War II era

Big Brother big government's watchful eye as described in George Orwell's *1984*

Big Burg New York City

big C cancer; cocaine

The Big Cat Johnny Mize (baseball player)

Big Charlie Charles de Gaulle

Big Chief mescaline

Big Crocodile P W Botha

big D hallucinogen such as diethyltryptamine, dimethyltryptamine, dipropylphyptamine

Big-D Dallas, Texas

Big Dan Daniel Joseph Tobin

Big Dipper ladle-shaped constellation of Ursa Major

The Big Dipper Wilt Chamberlain (basketball player)

Big Ditch Panama Canal

Big Drink Atlantic or Pacific Ocean

Big-D of the West Denver, Colorado

Big-E aircraft carrier USS *Enterprise*

Big Easy New Orleans, Louisiana

Big Ed Ed Delahanty (baseball player)

The Big Eight Iowa State, Kansas State, Colorado, Kansas, Missouri, Nebraska, Oklahoma, Oklahoma State (college sports)

Big Finger Australia's Cape York Peninsula

Big Five (British banks) Barclays, Lloyds, Midland, National Provincial, Westminster

Big Five System extrovert, agreeable, conscientious, emotional stability, and openness to experience (five basic components of personality)

Big Four ABC, CBS, Fox, and NBC television networks (in late 1990s); at the Johns Hopkins Medical School—William Howard Welch, William Osler, Howard Atwood Kelly, William Stewart Halsted; California railroad builders Charles Crocker, Mark Hopkins, Collis P Huntington, and Leland Stanford; civil-rights leaders of the 1960s: Martin Luther King, Jr., James Farmer, Roy Wilkins, and Whitney Young; Cleveland, Cincinnati, Chicago, and St Louis; Great Britain, France, Italy, and the United States at the end of World War I, their representatives at the Peace Conference—Lloyd George, Georges Clemenceau, Vittorio Orlando, and Woodrow Wilson

Big Four Automakers General Motors, Ford, Chrysler, American Motors (nickname from the mid to late 1900s)

Big Four Opera Houses Covent Garden in London, La Scala in Milan, Staatsoper in Vienna, The Met in New York

Big Friend B-24 or B-17 bomber

Biggest City in South Georgia nickname for Jacksonville, Florida

Biggest Little City Reno, Nevada

Big Heart of Texas Austin

Bighouse Clarence Gaines (basketball)

Big Inch 24-inch pipeline carrying petroleum products from east Texas to the New York-Philadelphia area

Big Island Hawaii (largest of the Hawaiians); Vancouver Island, British Columbia

big J big John (slang— policeman or other law-enforcement officer)

Big-J battleship USS *New Jersey*

Big Jim brothel-keeper gangster James Colosimo (1877–1920); Postmaster General James Aloysius Farley

Big Joe Turner Joseph Vernon Turner

Big John St John the Divine cathedral in New York City, famous for its concerts

Big-M battleship USS *Missouri;* Memphis, Tennessee

Big Mac Mark McGwire (baseball player); McDonald's hamburger; New York's Municipal Assistance Corporation (MAC)

Big Mama Thornton Willie Mae

Big Mamie battleship USS *Massachusetts*

Big Minny Minnesota

Big Miss Mississippi River

Big Momma HMS *Ark Royal*

Big Muddy Missouri River

Big N Vladimir Nabokov

Big Nail translation of the Eskimo nickname for the North Pole

Big Narco Saint Jesus Malverde

Big Nasty Ronnie Lee (football tackle)

big O opium; orgasm

The Big O Oscar Robertson (basketball player)

Big-O attack aircraft carrier USS *Oriskany*

Big Orange Los Angeles

Big P tenor Luciano Pavarotti

Big Pat St. Patrick's cathedral in New York City

Big Poison Paul Waner (baseball player)

Big Red cardinal (the bird); racehorse Man-o'-War

Big Red Machine Cincinnati Cardinals

Big Seven Britain, Canada, France, Germany, Italy, Japan, United States

Big Six Eldon Auker; New York City's Typographical Union Number Six

Big Sky Country Montana

Big Smoke old nickname of London, England as well as Sydney, New South Wales

Big Steel U S Steel Corporation

Big T Tucson, Arizona

Big Three The Big Three (music publishers Robbins, Feist, Miller)—Robbins Music Corp; World War I peacemakers Georges Clemenceau, Lloyd George, and Woodrow Wilson; World War II peacemakers Winston Churchill, Franklin Roosevelt, and Joseph Stalin

Big Three Automakers General Motors, Ford, Chrysler

Big Town Chicago

The Big Train Walter Johnson (baseball player)

Big Windy Chicago, Illinois

Big X McGraw-Hill trademark

Bikini State Florida

Bilad al-Sudan (Arabic— Land of the Blacks)—Guinea, Sudan

Bill William

Bill Arp Charles Henry Smith

Billie Dove Lillian Bohny

Billie Holiday Eleanora Fagan

Billings Billingsgate (London waterfront market whose women delighted in using foul language)

Bill Nye Edgar Wilson Nye

Billtown Williamstown, Kansas

Billy the Kid William Bonney alias William Wright and alias Henry McCarty

Billy Rose William Samuel Rosenberg

Billy Sanders Joel Chandler Harris

Bill of Rights President James Madison

Bimshire Barbados

Bing Crosby Harry Crosby

Binky pacifier

Biographer of Thomas Paine Moncure D Conway

Biologist of the Mind Sigmund Freud

Bip Marcel Marceau

Bird Charlie Parker

The Bird medical respirator (for Forrest Bird)

Bird Dog Cessna L-19 liaison aircraft

Bird of Happiness Japan's redcrowned crane

Bird Man Mike Barnett (basketball player)

Bird of Paradise brilliantly plumed male perching birds in New Guinea area

Bird-of-Paradise Island Little Tobago (where birds of paradise may be seen in their wild state)

Birdofredum Sawin James Russell Lowell

Birds of Paradox flightless birds such as cassowaries, flightless cormorants, emus, kiwis, ostriches, penguins, rails, rheas

Birds of Prey flesh-eating predators such as condors, eagles, falcons, hawks, kites, ospreys, owls, vultures

Bird Woman Sacajawea

Birkenau German name for Brezinka next to Auschwitz, called Oswiecim by the Poles

Birken'ead drill maritime tradition dating from 1852 with the sinking of HMS *Birkenhead* when women and children were given first place in the lifeboats

Birmingham Sam John Lee Hooker

Birthplace of American Cajuns New Brunswick

Birthplace of the American Industrial Revolution Pennsylvania's Lehigh Valley

Birthplace of American Jazz New Orleans

Birthplace of American Liberty Faneuil Hall, Boston

Birthplace of American Presidents Massachusetts (4); New York (4); Ohio (7); Virginia (8)

Birthplace of Aphrodite or Venus Cyprus

Birthplace of Aviation Dayton, Ohio where the Wright Brothers built their flying machine

Birthplace of Bach, Beethoven, and Brahms Germany (Eisenach, Bonn, and Hamburg)

Birthplace of Baseball Cooperstown, New York

Birthplace of Berlioz Côte-Saint-André, Isère, France

Birthplace of Bolívar Caracas, Venezuela

Birthplace of Brahms and Mendelssohn Hamburg, Germany

Birthplace of the British Industrial Revolution Severn Valley in England and Wales

Birthplace of Burns Alloway, Scotland

Birthplace of California San Diego

Birthplace of Camões (Camoëns) Lisbon, Portugal

Birthplace of Canada Charlottetown, Prince Edward Island

Birthplace of Cervantes Alcalá de Heneres, Spain

Birthplace of Colombia Tunja, Boyacá

Birthplace of Dante Florence, Italy

Birthplace of Democracy ancient Greece

Birthplace of the Gods Greece

Birthplace of Handel Halle, Germany

Birthplace of Hans Christian Andersen Odense, Denmark

Birthplace of Hindemith Hanau, Germany

Birthplace of Kant Kaliningrad (formerly Königsberg, East Prussia)

Birthplace of Liszt Raiding, Hungary

Birthplace of Melodramatic Opera Italy

Birthplace of Mozart Salzburg, Austria

Birthplace of Paganini Genoa

Birthplace of Purcell London, England

Birthplace of Richard Strauss Munich, Germany

Birthplace of Saint-Saëns Paris, France

Birthplace of Schubert Vienna, Austria

Birthplace of Schumann Zwickau, Germany

Birthplace of Shakespeare Stratford-upon-Avon, England

Birthplace of Sibelius Tavastehus, Finland

Birthplace of Skiing Morgedal in the Telemark region of Norway

Birthplace of the Skyscraper Chicago

Birthplace of Spinoza Amsterdam, Netherlands

Birthplace of the Tuna Fishing Industry San Diego

Birthplace of the United States of America Philadelphia, Pennsylvania where the *Declaration of Independence* was signed July 4, 1776

Birthplace of Vaudeville Sainte-Mère-église (near Norman coast of France)

Birthplace of Villa-Lobos Rio de Janeiro, Brazil

Birthplace of Vivaldi Venice, Italy

Birthplace of Wagner Leipzig, Germany

Birthplace of Wilde, Shaw, Joyce, and Behan Dublin, Ireland

Bishop of Broadway theatrical producer David Belasco whose clerical dress earned him this title

bishopese writing in a dull ecclesiastical style

Bishop of Rome the Pope

Bison City Buffalo, New York

Bitter Bierce Ambrose Bierce

Bitter Sea Dead Sea between Israel and Jordan

Bituminous City Connellsville, Pennsylvania

Bix Leon Bismarck Beiderbecke

Bjørn Bjørn Bjørnstjerne Bjørnson

B Kovner Jacob Adler

Black Black Sea between Bulgaria, Russia, and Turkey

Black Africa equatorial Africa from Ethiopia, Somalia, and Kenya to Gabon, the Congo, and Zaire

Black Babe Ruth Josh Gibson

Black Bart poetic stagecoach robber Charles E Bolton

Blackbeard Edward Teach—privateer-pirate also known as Edward Thatch

Blackbeard's curse botulism, diarrhea, or dysentery contracted in the Caribbean

black beauties biphetamine

Black Belt black-soil growing area extending from South Carolina and Georgia to Alabama and Mississippi

Blackberry Capital McCloud, California

Blackbird Lockheed SR-71 jet reconnaissance aircraft

Black Carib Garifuna spoken in Belize

Black Castle of Opium Afyonkarahisar in western Turkey

Black Charley Sir Charles Napier

Black Country coal-mining iron-founding sections of southern Staffordshire and West Midlands of England; Midlands of England around smoke-blackened Birmingham

Black Dan Daniel Webster

Black Death bubonic plague

black diamond black or gray industrial diamond also called framesite bort; anthracite or hard coal

Black Diamond City Wilkes-Barre, Pennsylvania

black diamonds coal

black disease anthrax of sheep; braxy

Black Douglas Sir James de Douglas

Black Eagle Hubert F Julian

Black Emperor of Haiti Jean-Jacques Dessalines

Black Explorer Matthew (Matt) Henson

black fever kala-azar (leishmaniasis)

black flag symbol of death or emblem of piracy

Black Flower of Society Nathaniel Hawthorne's nickname for any jail, penitentiary, or prison

Black Forest Schwarzwald (in mountainous south-central Germany)

Black Friday September 24, 1869 (financial panic occurred when speculators tried to corner the gold market in the U.S.)

Black Gladiator Bo Diddley (Ellas McDaniel)

black gold petroleum

black gold of the Amazon and Malaya rubber

black gold of the Caspian caviar

Black Hand secret terrorist society linked with the Camorra and the Mafia

Black Heart of Montana Butte

Black Hole Muscovites' nickname for Moscow

Black Irish brunette descendants of Irish women and Spanish Armada sailors

blackjack card game, bubonic plague, the black flag of pirates, blackjack chewing gum, hand-held leather-covered flexible club; zinc blende or zinc sulfide

Black Jack General John Joseph Pershing

black lead cerrusite (lead carbonate)

black mayonnaise poisonous sludge contaminating bays, beaches, harbors, and oceans

Black Monday October 19, 1987 when the New York stock market declined more than it had on Tuesday, October 29, 1929, which triggered the Great Depression

Black Money money gained illegally

Black Monk Grigori Efimovich Rasputin

Black Movie Pioneer Ralph Cooper

Black Muslims religious-oriented group composed of black nationalists and some militant extremists

Black Nationalist Marcus M(oziah) Garvey

Black Nightingale Ma Rainey (Gertrude Rainey)

Black Panthers militant black party

Black Pope head of the Jesuit Order—the Jesuit General

Black Prince Edward—Prince of Wales son of Edward III who always wore black armor

black propaganda lies repeated until they seem believable

Black Republic Haiti and many emerging African nations

Black Rock Columbia Broadcasting System (CBS) named for its black granite building at 51 West 52nd Street in New York City

Black Russian Writer Aleksandr Sergeyvich Pushkin (1799–1837) whose maternal grandfather was African

Black Saturday Commander's Internal Management Review

Black Sea Countries Bulgaria, Georgia, Romania, Russia, Turkey, Ukraine

Black Sheep of Canadian Liquors 100-proof Yukon Jack

Black Shirts Mussolini's followers

Black Sox Eight Chicago White Sox players banned from baseball for life for allegedly throwing the 1919 World Series—Eddie Cicotte, Happy Felsch, Chick Gandil, Shoeless Joe Jackson, Swede Risberg, Buck Weaver, Lefty Williams, and Fred McMullan

black spots black settlements in white-owned South African land

Black Stream Japan Current

black tar deadly Mexican-made heroin, also called Mexican mud or tootsie roll

Black Tuesday October 29, 1929, when the U.S. stock market crashed, trigggering the Great Depressuon

Black Wall Street Durham, North Carolina

Black Watch Royal Highland Regiment whose tartans display dark colors

blackwater fever malaria

Blackwater State Nebraska (with its black soil)

Blackwell's Island former name of Welfare Island in New York City's East River

black widow poisonous spider *Lactrodectus mactans*

Blarney Stone Port Cork or Corcaigh in Ireland (Eire) close to Blarney Castle containing the Blarney Stone

Blaue Reiter Blue Rider, German expressionist art movement from 1911 to 1914

Blau's theory the larger the organization, the greater the proportion of total resources devoted to management, hence the proliferation of the bureaucracy

Blessed Isles (*see* Fortunate Isles)

Bligh's Islands Fiji Islands

Blind Bards Homer and Milton

Blind Blake Arthur Phelps

Blind Bomber George Glamack

Blind Boy Fuller Fulton Allen

Blind Gary Davis Reverend Gary Davis

Blind Lemon Jefferson Lemon Jefferson

Blind Poet John Milton

Blind Publisher Joseph Pulitzer

Blind Tom Thomas Bethune

Blind Willie McTell Willie Samuel McTell

Blizzard State South Dakota

Blitzen Joe Benz (baseball player)

Bloater(s) inhabitant(s) of Yarmouth on the North Sea coast of England where herrings are salted and smoked

Blockhousers oldest Afro-American regiment whose gallant assault on a well-defended blockhouse won them this nickname during the Spanish-American War

Blonde Beauty of the Lakes Milwaukee, Wisconsin

Blonde Bombshell Jean Harlow

Blondin Charles Emile Gravele—the tightrope walker who crossed Niagara Falls in the mid-nineteenth century

blood and guts red ensign of the British Merchant Marine

Blood and Guts General George S Patton, USA

blood-clotting vitamin Vitamin K

Bloodhound British surface-to-air missile

blood pudding blood sausage

Bloody Ground State Kentucky

Bloody Mary Mary I of England (Mary Tudor); tomato juice and vodka

Bloody Sunday March 5, 1965, in Selma, Alabama, where state troopers attacked defenseless civil-rights marchers, injuring sixty

Bloomsbury writers whose center of activities was London's Bloomsbury Square in the early 1900s; Clive Bell, EM Forster, Roger Fry, John Maynard Keynes, Lytton Strachey, V Sackville-West, Leonard and Virginia Woolf; synonym for snobbish aestheticism

Bloom syndrome genetic disease characterized by retardation of growth, abnormal immune function

blue amytal (barbiturate)

blue acid pale-blue liquid LSD-25

blue angels amytal (barbiturate)

Blue-backed Speller *The American Spelling Book* by Noah Webster

Bluebeard any wife killer such as the Chevalier Raoul whose seventh wife discovered the bodies of his six previous wives

bluebirds capsules of sodium amytal

bluebook volume containing names of socially prominent people

blue-collar labor factory workers

blue devils amobarbital capsules

Blue Grass Capital Lexington, Kentucky

Blue Grass Country Kentucky

Blue Grass State Kentucky

Blue Grotto marine cavern on shore of Capri island in Bay of Naples

Blue Hen Chickens Delawareans

Bluehen(s) Delawarean(s)

Blue Hen State Delaware

Blue Law State Connecticut

Blue Light District Cartagena, Colombia's red-light district *(Distrito Luz Azuil)*

Blue Moon second full moon in one month

Bluenose fisherman or sailor from Canada's Maritime Provinces

Bluenose Province Nova Scotia

Bluenose(s) native(s) of Canada's Maritime Provinces, especially Nova Scotia; puritan(s)

blue ointment mercurial ointment

blue peter blue signal flag with a white rectangle in its center; flown when a ship is ready to sail; letter P or Papa in the international code

Blue Rider *Blaue Reiter*; German expressionist art movement from 1911 to 1914

blue room aircraft toilet

blue sky statements unsupported by facts

Blues Man Roosevelt Sykes

Blue Steel Hawker-Siddeley air-to-surface missile

Bluetooth standard for short-range wireless connections among electronic devices

Bluff City Hannibal, Missouri; Memphis, Tennessee; and Natchez, Mississippi

Bluff King Hal Henry VIII

Boadbil Abu Abdallah—last Moorish king of Granada

boards book covers

boat people refugees

Bob Robert

Bob Dole Disease erectile dysfunction

Bob Dylan Robert Zimmermann

Bob Hope Leslie Townes Hope

Bobby Robert

Bobby Blue Bland Robert Calvin Bland

Bobby Darin Walden Robert Cassotto

bobo bourgeois bohemian

Bo Carter Armentu Chatmon

Bo Diddley Ellas McDaniel

The Body Paul Hoffman (basketball player)

Boer (Afrikaans or Dutch—farmer)—South African of Dutch descent whose language is Afrikaans

Bogie Maxwell Bodenheim; Humphrey Bogart

Bogland Ireland

Boglander Irishman

bogsaat bunch of guys sitting around a table

Bogside Catholic working class district of Derry (Londonderry)

Bohemia Czech Republic; habitat of gypsies and other unconventional people

Bohemian Reformer John Hus

Bojangles Bill (Bojangles) Luther Robinson

Bolingbroke Henry IV of England

Bolivarian Block Colombia, Ecuador, Peru, Bolivia

Bolivarian Republics Bolivia, Colombia, Ecuador, Peru, Venezuela

Bollywood Bombay, for India's film industry

Bolshevik Feminist Aleksandra Kollontai

Bomba (Italian—Bass Drum)— Ferdinand ll—King of the Two Sicilies

bombay bombay duck (Asiatic lizardfish, dried and salted lizardfish served with curry, also called bummalo)

Bombingham nickname for Birmingham, Alabama, for Ku Klux Klan terrorism during 1963 struggle for black civil rights

Bonaire's Port Kralendijk

Bonanza Beech U-22 trainer aircraft

Bonanza Land Fort Smith, Arkansas

Bonanza State Montana

Bonaparte of the Antilles Toussaint Louverture

Bond Street London's street of fashionable shops

Bone Algerian port city now called Annaba

boneblack animal charcoal

Boney Napoleon Bonaparte

Bon Homme Richard (French—Good Man Richard)—Benjamin Franklin

Bonnie Prince Charles Charles Edward Stuart—the Young Pretender

Bonny Johnny John Adams, second President of the United States

Bonus Marchers ex-U.S. servicemen who marched in Washington, D. C., during the Great Depression to get their bonus certificates cashed early and were fired upon by federal troops under President Herbert Hoover's command

boobtube television

Boogie Down New York City's borough of the Bronx

book book of regulations and rules

Booker Booker T Washington

Book Man nickname by fellow prison inmates for Stephen Blumberg of Iowa, who stole thousands of valuable books from U.S. libraries

Boom Boom Boris Becker (tennis)

boomer member of the baby-boom generation

boomerang bomber advanced technology bomber; stealth bomber

Boomer's Paradise Oklahoma

Boomer State Oklahoma

Booze Bourse Brooklyn, New York

Borax King Francis Marion Smith of Death Valley, California

Borda Count voting system of ranking candidates, for mathematician Jean Charles Borda (1733–1799)

Border Country the U.S.-Mexican border extending from Brownsville, Texas to San Diego, California

Border Eagle State Mississippi

Border Minstrel Sir Walter Scott

Border State Maine (bordering on Canada)

Border States Delaware, Maryland, Virginia, Kentucky, and Missouri; before the Civil War they divided the North from the South

Boris Karloff William Henry Pratt

Boris Pilnyak Boris Andreyevich Vogau

Boris Savinkov Vladimir Ropshin

borked what happens to nominations torpedoed by an onslaught of opposition—for Robert Bork, unsuccessful nominee to United States Supreme Court, 1987

Borneo Kalimantan

Borodin Mikhail Markovich Grusenberg

Borscht Belt Catskill Mountain area, New York

bosco South American narcotic compound

Bosphorus strait connecting the Black Sea with the Sea of Marmora leading to the Mediterranean

Boss of Bosses Lucky Luciano (Salvatore Lucania)

Boss Kett Charles F(ranklin) Kettering

Boss Tweed William Marcy Tweed

Boston Brahmin Historian William Hickling Prescott

Boston Strong Boy John L Sullivan

Botticelli Sandro di Botticell — palette name of Alessandro Filipepi

Bourbon Street New Orleans nightlife center

Bowen Therapy musculoskeletal-release therapy developed by Tom Bowen of Australia

Bowery north-south thoroughfare on Lower East Side of Manhattan, New York City; notorious for its shabby hotels and saloons

Bowie State Arkansas, where many of its first settlers carried bowie knives

Box 500 British secret service headquartered on Curzon Street, London

boxer's ear hematoma auris

boxitos people from Yucatan, Mexico where in Mayan the word means dark people

Boyhood Home of Mark Twain Hannibal, Missouri

The Boy Manager Stanley Bucky Harris

Boy Orator of the Platte William Jennings Bryan

The Boy Shirley Temple Bobby Breen

Boy's Town Omaha, Nebraska; redlight sections of many Mexican border towns

Boz Charles Dickens

Bozzy James Boswell—biographer and friend of Dr Samuel Johnson

bra burner militant feminist

bracelets slang for handcuffs

Brady Bonds debt instruments named for United States Treasury Secretary Nicholas Brady

Bragman's Bluff British pirate's name for Puerto Cabezas, Nicaragua sometimes called El Bluff

Brahmsburg Hamburg, Germany—birthplace of Johannes Brahms

Brains of the Confederacy Judah Philip Benjamin (attorney general, secretary of war, and secretary of state of the Confederate States of America)

Bram Stoker Abraham Stoker

Brandy Nan Queen Anne

Brann the Iconoclast William Cowper Brann, editor and publisher of *The Iconoclast*

Brass Butte, Montana

Brassai Gyula Halàsz

Brass City Waterbury, Connecticut

brat bananas, rice, applesauce, toast (diet to relieve diarrhea)

Brattle Island, Galápagos Tortuga

Bratwurst Capital Sheboygan, Wisconsitm

Brazilian Aviation Pioneer Alberto Santos-Dumont

Brazilian emerald green variety of tourmaline

Brazilian Film Pioneer Alberto Cavalcanti

Brazilian National Composer Heitor Villa-Lobos

Brazilian Ports Manaus, Belém do Pará, São Luis, Enseada de Mucuipe, Natal, Recife, Maceio, Salvador de Bahia, Ilheus, Vitoria, Rio de Janeiro, Niteroi, Angra dos Reis, Santos, Paranagua, São Francisco, Itajai, Florianopolis, Laguna, Rio Grande, Porto Alegre

Brazilian ruby topaz altered by heating so it turns purple-red to salmon-pink and hence passes for a ruby

Brazilian sapphire blue tourmaline

Braziller George Braziller

Brazil water coffee

Bread-and-Butter State Minnesota

Bread Basket Fargo, North Dakota

Breadbasket of Canada Saskatchewan

Breadbasket of Russia the Ukraine

Breadbasket of Sweden southernmost province of Skåne

Breadbasket of the World central North America (Canada and the U.S.)

Breakfast Food City Battle Creek, Michigan

Bret Harte Francis Brett Harte

Brett Halliday Davis Dresser

Brewery Capital Milwaukee

Brick Lane London, England's East End ghetto

bricks traditional businesses in buildings

bricks and clicks businesses in buildings and Internet businesses

Bride of the Adriatic Venice

Bride of the Sea Venice

Bridewell London's old house of correction

Bridge Bum Alan Sontag

Bridge House Wilmington, Delaware's detention home for juvenile delinquents

Bridge of Sighs 16th-century Venetian bridge connecting a prison with a palace where prisoners were tried; any similar structure connecting a courthouse with a prison

Brigitte Bardot Camille Jarval

bright lighters city people

Bright's disease kidney disease named for its diagnostician Dr Richard Bright of London; nephritis

Brilliant Belgian virtuoso-violinist Arthur Grumiaux

Brilliant Madman Charles XII of Sweden

Brill's disease epidemic typhus disease recurring years after the original infection and named for its American diagnostician Dr Nathan E Brill

brimstone sulfur

Bringer of Freedom's Light John Huss

Britain's First Woman Prime Minister Margaret Thatcher

Britain's Most Exclusive Club the House of Commons

Britain of the South New Zealand

Britain's Playground Blackpool (seaside resort on Irish Sea)

Britain's Premier Passenger Port Southampton

britannia britannia metal (alloy of antimony, copper, and zinc used as antifriction material and for dinnerware)

Britannia Bristol-built military transport aircraft; Britannia metal; Britannia prima (England); Britannia secunda (Wales)—symbol of Great Britain including Scotland with England and Wales; British Empire; Commonwealth of nations; Great Britain and Northern Ireland—the United Kingdom; Roman name for Great Britain

Britain's Queen of the Blues Beryl Bryden

British America British possessions in or adjacent to the Americas

British Anatomist Extraordinary Henry Gray

British Century the 19th century

British China Hong Kong before 1997

British Guiana Guyana

British Hardware Centre Birmingham

British Honduras Belize

British Invasion takeover of United States airwaves by British rock 'n' roll groups including The Beatles, The Rolling Stones, The Who, and The Kinks

British Isle Ports Whitstable, Port Victoria, Chatham, Tilbury Docks, Gravesend, Woolwich, Greenwich, London, Wivenhoe, Harwich, Parkeston, Ipswich, Felixstowe, Lowestoft, Great Yarmouth, King's Lynn, Boston, Grimsby, Immingham, Kingston-upon-Hull, Goole, Whitby, Middlesbrough, Hartlepool, Seaham, Sunderland, North Shields, Newcastle, Gateshead, Blyth, Leith, Granton, Rosyth Dock Yard, Boness, Grangemouth. Alloa, Burntisland, Kirkcaldy, Methil, Perth, Abroath, Montrose, Aberdeen, Peterhead,

Fraserburgh, Hopeman, Inverness, Cromarty, Invergordon, Port Mahomack, Helmsdale, Wick, Thurso, Scrabster, Stornoway, Oban, Campbeltown, Greenock, Finnart, Rothesay Dock, Glasgow, Ardrossan, Irvine, Troon, Cairnryan, Douglas, Bangor, Belfast, Larne Lough, Londonderry, Sligo, Westport, Galway, Kilrush, Limerick, Foynes, Cobh, Cork Harbour, Rosslare, Dublin, Silloth, Mayrport, Whitehaven, Barrow-in-Furness, Fleetwood, Preston, Liverpool, Manchester, Port Dinorwic, Holyhead, Caernarvon, Milford Haven, Llanelly, Swansea, Port Talbot, Barry, Cardiff, Newport, Sharpness, Gloucester, Avonmouth, Bristol, Portishead, Bideford, St Ives, Penzance, Falmouth, Fowey, Plymouth, Dartmouth, Portland, Weymouth, Poole, Cowes, Yarmouth, Southampton, Gosport, Portsmouth, Folkestone, Dover

British Isles Great Britain and Ireland

British Laugh Master Benny Hill

British Librettist Italian nickname for Shakespeare, many of his dramatic plots inspired Italian composers

British lion symbol of the British Commonwealth as well as of Great Britain

British North Borneo Sabah

British Riviera England's south coast from Land's End to Margate

British Rock Gibraltar

British West Indies island possessions or former possessions of Great Britain in or near the Caribbean: Bahamas; British Leeward, Virgin, and Windward islands; Jamaica, Tobago, Trinidad

Britons English, Scottish, and Welsh people

Brixton Brixton Prison southeast of Plymouth, England; London, England slum; one of London's largest prisons

broadband high-speed digital networks

Broad-bottomed Administration of Prime Minister Henry Pelham during the reign of George II

Broken Hill Kabwe, Zambia

Bronco North American-Rockwell OV-10 counterinsurgency aircraft

Bronze Age era of mankind when implements and weapons were forged from bronze; period marked by wars and widespread violence

Bronzino Agnolo di Cosimo

Brook Farm utopian community founded by leading American transcendentalists in 1841 but disbanded by 1847 near West Roxbury, now Boston, Massachusetts; nickname of similar ventures

Brook Farmers Albert Brisbane, Orestes Brownson, Charles A Dana, John S Dwight, Ralph Waldo Emerson, Margaret Fuller, Horace Greeley, Nathaniel Hawthorn, Isaac Hecker, George Ripley, Henry David Thoreau

Brother John John Bull—long the personification of the British Empire

Brother Jonathan British nickname for the United States and its citizens

The Brothers Rockefeller brothers—John D III, Nelson, Laurance, David

Brothers Goncourt Edmund and Jules de Goncourt—literary collaborators

Brothers Grimm Jacob and Wilhelm Grimm

The Brothers Hildebrandt Gregory and Timothy Hildebrandt (illustrators)

Brown Bomber Joe Lewis

brown coal lignite

Brownie Mary Mary Jane Rathbun

Brownie McGhee Walter Brown McGhee

brown lung byssinosis or cotton-dust disease

Brown Shirts Hitler's followers

Brownstone State Connecticut

Brow of the Universe Scandinavia

BR Town Baton Rouge, Louisiana

Bruce Graeme Graham Montague Jeffries

Brummagen Birmingham (colloquial)

Bruno Jakob Bronowski

Bruno Walter Bruno Walter Schlesinger

brussels brussels carpet (woven with a raised pattern by a method first used in Brussels); brussels griffon (toy dog bred in Brussels); brussels lace (high-quality floral-pattern lace); brussels sprouts (small cabbage-like vegetable)

Brussels system universal decimal classification

B-town Beantown; Blountstown; Boys Town

B Traven Berick Traven Torscan Croves

Bubble Boy David, who died, at age 12, of severe combined immunodeficiency disease after living in an isolated environment

bubble-boy disease severe combined immunodeficiency disease

Bubbles 18th-century English firms issuing speculative stock

bubbly champagne

Buccaneer Hawker-Siddeley jet aircraft for military applications

Bückeburg Bach Johann Christoph Friedrich Bach

Bucketfoot Al Al Simmons (baseball player)

Buckeye North American Rockwell trainer aircraft designated T-2

Buckeye(s) Ohioan(s)

Buckeye State Ohio

Buckie R(ichard) Buckminster Fuller

buckminster fullerine geodesic-shaped molecule named for Buckminster Fuller

Buckshot Glenn Wright (baseball player)

buckyballs geodesic dome-shaped carbon molecule named for American architect and engineer (Richard) Buckminster Fuller (1895–1983)

Bud Abbott William Abbott

Buddha Siddhartha Gautama

Buddy Rogers Charles Rogers

Buddy System trademark of Countermeasures of Hollywood, Maryland

Buerger's disease chronic inflammation of the blood vessels in a limb or limbs

Buffalo De Haviland military transport aircraft

Buffalo Bill Colonel William F Cody

Buffalonians people of Buffalo, New York

Buffalo Plains State Colorado

Bug-eaters Nebraskans

Bug-eating State Nebraska

Bughouse Square Pershing Square in Los Angeles, Union Square in New York, Washington Square in Chicago

Bugs bootlegger-burglar George Moran (1893–1957); Tom Burgmeier (baseball player)

Bugs Baer Arthur Baer

Buhl's disease fatty degeneration associated with hemoglobinuria

Built on Oil, Soil, and Toil Ponca City, Oklahoma

Bukavu Costermansville

Bulgarian Ports Michurin, Akhtopol, Bukhta Tsiganski, Burgas, Varna, Evksinograd, Kavarna

Bulgarian Quadrilateral fortress towns of Rustchuk, Schumla, Silistria, and Varna

Bulge The Bulge of Brazil consisting of South America's easternmost coast between João Pessoa and Recife

Bulldog Jim Bouton (baseball player); Scottish Aviation single-engine two-place trainer

The Bulldozer Richard Holbrooke (diplomat)

Bullet Joe Joe Bush (baseball player)

Bull Halsey Admiral William F Halsey Junior

Bullion State Missouri

bull market stock market short form indicating an upward trend in securities

Bull Moose Theodore Roosevelt—twenty-sixth President of the United States

Bullpup Maxson air-to-surface missile

Bullring of Basra white-slave market close to Iraq's Persian Gulf coast

Bull Run Manassas

Bullwood Hall borstal in Essex, England

Bumble Bee Slim Amos Easton

Bung Karno (Malay–Brother Karno)—Indonesian dictator Sukarno

Bunions Rollie Zeider (baseball player)

bunsen bunsen burner (named for the German chemist Robert Wilhelm Bunsen)

Buntline Ned Buntline—nom de plume of Edward ZC Judson

Buranello Baldassare Galuppi

The Burg New York City

Burke to kill by suffocation as in the case of William Burke, the Irish criminal who suffocated his victims so he could sell their bodies for dissection

Burkitt's lymphoma cancer that affects children; named for Dr Dennis Burkitt

Burlington Route Chicago, Burlington, and Quincy (railroad)

Burl Ives Icle Ivanhoe Ives

Burma former name of Myanmar

burmese burmese cat (orange-eye cat originating in Burma); burmese lacquer (grayish varnish); burmese ruby (peony)

Burmese Ports Sittwe, Kyaukpyu, Bassein, Rangoon, Moulmein

Burns and Allen comedians George Burns (Nathan Birnbaum) and Gracie Allen

Burping Bedpost bassoon

Burrhead Ferris Fain (baseball player)

Burt L Standish Gilbert Patten—creator of the American boy hero—Frank Merriwell

Busby Berkeley William Berkeley Enos

business-man's special 45-minute psychosis produced by dimethltryptamine

Busta Sir Alexander Bustamante

Bustees Calcutta slum

Buster Keaton Joseph Francis Keaton

butch lesbian who plays male role

Butch Fiorello H LaGuardia

Butcher of the Balkans Andrija Artukovic

Butcher of Berlin Adolf Hitler

butcher bird loggerhead shrike

Butcher of Budapest Soviet Prime Minister Nikita Khrushchev so named because of his brutal suppression of the Hungarian freedom fighters

Butcher of Cambodia Ta Mok

Butcher of the Caribbean Rafael Trujillo

Butcher of Lyon Klaus Barbie

Butcher of Prague Reinhard Heydrich—Hitler's Reichsprotektor of Bohemia

butter butts yellow-rumped warblers

Butter Capital Owatonna, Minnesota

Buttercup Carl Anderson (basketball player)

Buttermilk Tommy Tommy Dowd (baseball player)

Butternuts Tennesseans

Buys Ballot's law if someone in Northern Hemisphere stands with back to wind, atmospheric pressure will be lower to his left hand than to his right, and the reverse in Southern Hemisphere

Buzzards Georgians

Buzzard State Georgia

Bye Bye Steve Balboni (baseball player)

Byron pen name of George Gordon who used the family title of Lord Byron

Byron Janis Byron Yanks

Bytown original name of Ottawa, Canada

Byzantium (Latin—Istanbul)—Constantinople

CABAL rnembers of the secret cabinet of Charles II of England; by coincidence their initials spelled cabal—Clifford of Chudleigh, Ashley (Lord Shaftesbury), Buckingham (George Villiers), Arlington (Henry Bennet), Lauderdale (John Maitland)

cabbage cabg (coronary artery bypass grafting)

Cabbage Patch Victoria, Australia

Cabbage Town Toronto, Ontario slum

cabbage with a college education Mark Twain's nickname for cauliflower

Cabo Tormentoso (Portuguese—Cape of Storms)—Cape of Good Hope

Cacos (Spanish—pickpockets, poltroons)—nickname of a Guatemalan political party

Cactus code name for French Mach 1.2 surface-to-air missile

Cactus Blossom Capital Phoenix, Arizona

Cactus Jack Vice President John Nance Garner also called the Sage of Uvalde

Cactus State New Mexico

Caesar Augusta (Latin—Zaragoza)—called Saragossa by the British

cafeteria bug *Clostridium perfringens*

Caffarelli Gaetano Majorano

Cagliostro Giuseppe Balsamo

Cajetan Tommaso de Vio–Italian cardinal who failed to persuade Martin Luther to remain within the Catholic Church and carry on reforms from within

Cajun a descendant of French-speaking immigrants from Acadia; their dialect

Cajun Country Bayou Teche, Louisiana, also called Teche Country

Calamity Jane Martha Jane Burke also known as Canary Jane whose activities resulted in the deaths of eleven of her twelve husbands

Calhoun Gulch Charleston, South Carolina district

California Channels Channel Islands south of Santa Barbara (Anacapa, San Miguel, Santa Cruz, Santa Rosa)

California Ports San Diego, Long Beach, Los Angeles, San Pedro, Monterey, San Francisco, Alameda, Oakland, Port Richmond, Mare Island, Port Chicago, Stockton, Sacramento, Eureka

California prayerbook deck of cards

California Riviera oceanside resorts ranging from San Diego to Santa Barbara

California's Cornerstone San Diego

Caligula Gaius Caesar

Calle de la Ballesta (Spanish—Street of the Crossbow)—Madrid's nightclub neighborhood

Calle Florida (Spanish—Florida Street)—celebrated shopping center of Buenos Aires

Calpe (Phoenician—Rock of Gibraltar)—one of the Pillars of Hercules flanking Straits of Gibraltar

Calve Emma Calvé— operahouse name of the soprano Emma de Roquer

Calvin John Calvin (originally Jean Chauvin)

Calypso Capital Port of Spain, Trinidad

Cambodia formerly Preah Reach Ana Chak Kampuchea or simply Kampuchea, the Khmer Republic, or Democratic Kampuchea

Cambodian-Incursion President Richard Nixon

Cambodian Ports Kampong Saom, Kampot, Phumi Phsar Ream

Cambridge Group Ralph Waldo Emerson, Oliver Wendell Holmes, Henry Wadsworth Longfellow, James Russell Lowell, John Greenleaf Whittier

Camden and Amboy State New Jersey

Camel NATO name for Soviet Tu-104 transport aircraft

Camel-driver of Mecca the Prophet Mohammed

Camellia Capital Sacramento, California

Camellia City Greenville, Alabama

Camera Eye Max Bishop (baseball player)

Cameroon Ports Tiko, Douala, Victoria, Kribi

Camille Erlanger Fréderic Regnal

Camille Pissarro Jacob Pizarro

camisole(s) straitjacket(s)— institutional euphemism

Campagna di Roma (Italian—Roman Campagna)— undulating lowlands around Rome

Campanella Tommaso Campanello (originally Giovanni Somenico)

Campo Alegre (Papiamento or Spanish— Happy Country)— Curaçao's controlled brothel above the hills of Willemstad

campo santo (Italian, Portuguese, Spanish—sacred ground)—name of many burial places throughout the Latin world

camp robber gray jay (or other Corvidae)

Can King Canute of Denmark and England

Canada second largest nation; Canada anemone, balsam, barberry, birch, blueberry, bluegrass, buffalo berry, field pea, fleabane, ginger, goose, hare, hemp, jay, lily, lymegrass, lynx, mayflower, mint, moonseed, pea, pitch, plum, potato, rockrose, root, tea, turpentine, violet, warbler, wormwood, yew

Canada's Breadbasket Saskatchewan

Canada's Doorstep Nova Scotia

Canada's Heartland The Province of Manitoba

Canada's Ocean Playground Nova Scotia

Canada's Principal Ports Halifax and Montreal on the Atlantic, Vancouver on the Pacific

Canada's Storied Province Québec

Canada's Wonder City Toronto

Canadian Canadian bacon (boned pork loin strips); Canadian cheddar (usually smoother and spicier than American cheddar cheese); Canadian football (rouge); Canadian-French; Canadian humorist (Stephen Leacock); Canadianism; Canadianize; Canadian whiskey (rye); prefix given such items as Canadian bacon, football, French (inhabitant or language), goldenrod, goose, hemlock, holly, lynx, whiskey

Canadian Arctic area north of the Arctic Circle including islands, the North Magnetic Pole, and the Northwest Territories

Canadian Atheist Marshall J Gauvin

Canadian black Canadian-grown marijuana

Canadian Comedians Lou Jacobi; Rich Little

Canadian Commonwealth Canada and its territories

Canadian Composer-Conductor-Educator Sir Ernest Campbell Macmillan

Canadian Dominion Canada and its territories

Canadian Freethinker Marshall J Gauvin

Canadian Galapagos Queen Charlotte Islands

Canadian Gateway to the Pacific British Columbia

Canadian Georgia O'Keeffe painter Emily Carr

Canadian Humorist Stephen B(utler) Leacock

Canadian Kaleidoscope The Province of Ontario

Canadian Ports (east coast and Great Lakes) Churchill, Cartwright, Saint Anthony, Roddickton, Springdale, Baie Verte, Fortune Harbour, Botwood, Catalina, Clarenville, Harbour Grace, Wabana, St John's, Argentia, Burin, St Pierre, Grand Bank, St George's, Corner Brook, Humbermouth, Sept-Iles, Baie-Comeau, Rimouski, Tadoussac, Port Alfred, Chicoutimi, Riviere du Loup, Québec, Trois Rivieres, Sorel, Varennes, Montreal, Ottawa, Lower Lakes Terminal, Prescott, Belleville, Trenton, Cobourg, Port Hope, Oshawa, Port Whitby, Toronto, Hamilton, Port Weller, Welland, Port Colborne, Port Maitland, Rondeau Harbor, Amhertsburg, Windsor, Sarnia, Port Edward, Goderich, Owen Sound, Collingwood, Midland, Parry Sound, Little Current, Sault Ste Marie, Thunder Bay, Gaspé, Chandler, Paspébiac, Dalhousie, Bathurst, Caraquet, Chatham, Newcastle, Souris, Georgetown, Charlottetown, Summerside, Pictou, North Sydney, Sydney, Halifax, Lunenburg, Liverpool Shelburne Yarmouth, Digby, Parrsboro, Moncton, St John, Letang Harbor; (west coast)

New Westminster, Vancouver, Horseshoe Bay, Powell River, Comox, Nanaimo, Victoria, Esquimalt, Port Alice, Ocean Falls, Prince Rupert

Canadian potato artichoke

Canadian Twin Cities Fort William and Port Arthur

Canal Concessionaire Vicomte Ferdinand Marie de Lesseps—original promoter of the Suez and the Panama Canal

Canaletto Antonio Canale; Giovanni Antonio Canal

Canal of Fire Suez Canal

canalimony $25-million-dollar alimony United States paid Colombia in 1922 for separating its province of Panama in 1903 so it could proceed unhindered in constructing the Panama Canal

Canal Zone Capital Balboa Heights

canaries of the sea beluga whales

Canberra British twin-jet light bomber built by BAC

Candia Erakleion

Canecutters sugar-cane-cutting Queensland, Australians

Cane Sugar States Florida, Hawaii, and Louisiana

Canned Salmon Capital Ketchikan, Alaska

Cannery City Seattle, Washington

Cannery Row Monterey, California's fishery foreshore described by John Steinbeck in his novel

Cannon City Kannapolis, North Carolina where Cannon towels are made

Cannon King Alfred Krupp—German armament manufacturer (1812–1887)

Cannon Kings of Germany The Krupp family

Canoe City Old Town, Maine

canola Canada oil (rapeseed)

Cantabrian Surge Bay of Biscay

Cantinflas Mario Moreno

Canuck French-Canadian; twoplace jet interceptor built in Canada by Avro and designated CF-100

Canuckland Canada

Cao Tú Phan Van Khoai

The Cape Cape Ann, Cape Cod, Cape of Good Hope, Cape Hatteras, Cape Horn, Cape Province (Union of South Africa); Cape Town

Cape of the Californias Cabo San Lucas (at the lower tip of Baja California)

Cape Cod peninsula enclosing Cape Cod Bay

Cape Cod turkey codfish

Cape Dutch Afrikaans

Cape Horner(s) deep-sea sailing vessel(s) rounding Cape Horn in southernmost South America

Cape Horn fever imaginary malady of sailors who complain they are ill when the sea is rough or there is too much work

Cape Kennedy Cape Canaveral, Florida

Cape May Pen Cape May Peninsula in southernmost New Jersey

Cape Stiff Cape Horn

Cape of Storms Cape of Good Hope

Cape V Cape Verde (whose capital is Praia)

Cape Verde Island Ports Mindelo, Santa Maria, Preguiça, Praia

Cape York Pen Cape York Peninsula in northern Australia

Capi Muriel Wylie Blanchet

Capirucha Mexican-Americanism for Mexico City

Capital of Black America Harlem section of New York City

Capital of the Black Middle Class Durham, North Carolina

Capital of the Cotswolds Cirencester

Capital of the Highlands Inverness, Scotland

Capital of Hope Brasilia

Capital of the Incan Empire Cuzco, Peru

Capital of Memory Alexandria, Egypt, according to Lawrence Durrell

Capital of the Pirate Coast Ras Al-Khaimak (northernmost sheikdom of Trucial Oman at the Strait of Ormuz)

Capital of Polynesia Auckland, New Zealand

Capital Province Ontario

The Capital Punisher Frank Howard (baseball player)

Capital of the Rhineland Cologne (Köln), Germany

Capital of the World New York City—seat of the United Nations

Capstone of Negro Education Howard University

Captain Kidd William Kidd—privateer-pirate

Captain Late James Silas (baseball player)

Captain Planet Jacques-Yves Cousteau

Captain Sir Richard Captain Sir Richard Francis Burton

Captain Van George Vancouver, British sea captain, explorer

Captive of History Northern Ireland

Capucine Germaine Lefebvre

Caracas capital of Venezuela and named for its founder, Santiago de León de Caracas

Caran d'Ache Emmanuel Poiré

Carat City diamond-mining Kimberly, South Africa

Caravaggio, Michelangelo da Michelangelo Merisio

Caravaggio, Polidoro da Polidoro, Caldara

Caravan of Death Chilean leader Augusto Pinochet's traveling military death squad

carbolic acid phenol

Carbonaro Napoleon III believed to belong to the Carboneria political society

Cargo Port of the Pacific Vancouver, British Columbia

Carib Carib Indians; Caribbean countries; Caribbean Sea between the Antilles, Central America, northern South America

Caribou De Havilland twin-engine stol transport designated DHC-4 in Canada and C-7A in the United States where it is also called CV-2A

Carinthia Kärnten, Austria

Carioca(s) native(s) of Rio de Janeiro

Carl Brandes Edvard Cohen

Carleton Kendrake Erle Stanley Gardner

Carl Milles Vilhelm Carl Emil Anderson

Carlo Collodi Carlo Lorenzini
Carlo Maria Carlo Maria Giulini
Carlos Arruza Carlos Ruiz Carnino
Carmella Ponselle Carmella Ponzillo
Carmen Miranda Maria de Cormo Cunha
Carmen Silva Elisabeth Queen of Romania
Carnegie's Savior Violinist Isaac Stern who saved Carnegie Hall from oblivion
Carol Carnac Edith Caroline Rivett
Carole King Carole Klein
Carolina Game Cock Thomas Sumter
Carolina racehorse razorback hog
Carol Lombard Jane Peters
Carolus Magnus Charlemagne
Caronia La Coruña (northwestern Spain in Roman times)
carpenter fish sperm whale
Carpet Capital of the World Dalton, Georgia
Carpet City Amsterdam, New York
Carrion's disease Peruvian-sandfly anemia
Carter Curtain Tortilla Curtain
Carter Doctrine defend the Persian Gulf (named for Jimmy Carter, thirty-ninth president of the United States)
The Carthaginian Lion General Hannibal
Cary Grant Archibald Leach
Casablanca Miami, Florida
Casa Grande William Randolph Hearst's main house at San Simeon, California
Casanova Giacomo Girolamo
Casa Pacifica former President Nixon's Spanish-colonial seaside home at San Clemente, California
Casa Rosada (Spanish—Pink House)—Argentine president's office building in Buenos Aires
Casbah Algiers, Algeria's hillside redlight and underworld district
Casey Jones John Luther Jones

Casey Stengel Charles Dillon Stengel
Cash and Cary millionaire Barbara Hutton and actor Cary Grant, as known when they married
Casino City Monte Carlo, Monaco
castile castile soap (mild cleaning agent originally made in Castile, Spain from olive oil and sodium hydroxide)
Cast-Iron Commodore Matthew Calbraith Perry
The Cat Harry Brecheen (baseball)
Catch-Me-Who-Can Richard Trevithick's railway engine tested in 1808
cat gold mica; yellowish mica
Cathay China
Cathedral of Learning University of Pittsburgh's 52-story building
Cathedral of Music Carnegie Hall, New York City
catnip of the butterfly world buddleia
cats catfish
cat's eye chrysoberyl
cat's fur Guatemalan mist
catsie cat's-eye playing marble; polished agate resembling a cat's eye
Cattle Capital Willcox, Arizona
Caudillo de la Independencia de Uruguay (Spanish— Chief of the Independence of Uruguay)—José Gervasio Artigas
cauliflower ear hematoma auris
Cavaliers Virginians
Cavalier State Virginia
caviar of drugs cocaine
Cayenne French Guiana
Cayo Hueso (Spanish—Bone Key)—original name of Key West, Florida
Cayuse Hughes OH-6 observation helicopter
C Day Lewis Nicholas Blake
Cedar Crest executive mansion of the governor of Kansas and eponym for executive government throughout the state.
Cee Cee Claudia Cardinale
Celery Capital Kalamazoo, Michigan; Sanford, Florida; San Ysidro, California

Celery City Cedar Rapids, Iowa; Kalamazoo, Michigan
Celestial City John Bunyan's name for Heaven in *Pilgrim's Progress*; old traveller's name for Peking, China
Celestial Empire Chinese Empire
Celine Louis-Ferdinand Destouches
Celtic Fringe peoples of Cornwall, Ireland, Scotland, and Wales on the fringe of England
Celts Bretons, Cornish, Gaels, Irish Gaelics, Manx, Scots Gaelics
Cement City Allentown, Pennsylvania
Censor of the Age Thomas Carlyle
Centennials Coloradans
Centennial State Colorado
Center of Austria Salzburg
Center of the Copper Circle Tucson, Arizona
Center of the Nation Topeka, Kansas
Center of Scenic America Utah
Center of the Sunshine State Pierre, South Dakota
Central African Empire formerly the Central African Republic created from the Ubangi Shari territory of French Equatorial Africa
Central America land between Colombia and Mexico—Belize, Costa Rica, El Salvador, Guatemala, Honduras, Nicaragua, and Panamá
Central Americans Costa Ricans, Guatemalans, Hondurans, Nicaraguans, Panamanians, Salvadorans
Central Bureau Amsterdam's old section for illegal and illicit activities
Central Powers Austria-Hungary; Bulgaria; Germany, and Turkey (in World War I)
Central Prairie Province Saskatchewan
Central Provinces Ontario and Québec
Central State Kansas
Centurion British tank carrying a crew of 4 and guns up to 105mm

Century of Confusion the 9th century when the empire of Charlemagne disintegrated—the 800s

Century of the Exodus the 13th century before the Christian era when Moses led the Israelites out of Egypt and across the Red Sea—the 1200s

Ceramic City East Liverpool, Ohio

Cereal City Battle Creek, Michigan, and Cedar Rapids, Iowa

Cervantes Miguel de Cervantes Saavedra (Spain's foremost author)

Cesspool of Crime London or Paris around the turn of the century; New York today

Cesspool of Latin America Cayenne, French Guiana

Cesspool of Pirates John F Kennedy International Airport in New York where cargo thefts are the highest in the nation

Ceylon Sri Lanka

Chagas-Cruz disease South American sleeping sickness

Chained Lady Andromeda

Chairman of the Board Frank Sinatra; Whitey Ford (Ed Ford— baseball)

Chairman Mao Mao Tse-Tung

Champagne of Drugs cocaine

Champagne of Teas Darjeeling tea (from Darjeeling, India)

Champion of Copernican Cosmology Italian philosopher Giordano Bruno

Champion of Darwin Thomas Henry Huxley

Champion of Education Horace Mann

Champion of Freethought President John F Kennedy who believed in an America where the separation of church and state is absolute and who renounced the appointment of an American ambassador to the Vatican and tax support of parochial schools

Champion of Liberty English reformer Charles Bradlaugh

Champion of the Old South President John Tyler

Champions of Individualism John Stuart Mill and Herbert Spencer

Champion of States Rights John C Calhoun—U.S. Senator from South Carolina

Champion of the Underdog Clarence Darrow

Chance Personified Fortuna (Roman); Tyche (Greek)

Chandeleurs Chandeleur Islands off the coast of Louisiana

The Channel Beagle; English; St George's

Channel City Santa Barbara, California

Channel fever the sense of excitement evident aboard ships approaching their port

Channel Islands Jersey, Guernsey, Alderney, Brechau, Great Sark, Little Sark, Herm, Jethou, Lihou; also off California's coast near Santa Barbara

Chapino(s) Guatemalan(s)

Chappiequack Chappaqua, New York

Charcot-Marie-Tooth disease muscular atrophy

Charger Convair multipurpose short takeoff-and-landing airplane

Charioteer British World War II medium tank armed with an 83.4mm gun

Charles Atlas Angelo Siciliano

Charles the Bald Charles I of France

Charles B Child C Vernon Frost

Charles Blondin Jean François Gravelet (French acrobat who walked tightrope above Niagara River near Niagara Falls in 1855, 1859, 1860)

Charles Bronson Charles Buchinski

Charles Dalmorès Henry Alphonse Boin

Charles de Secondat Baron de la Brède et Montesquieu

Charles the Fat Charles II of France

Charles Island, Galápagos Floreana or Santa Maria

Charles J Kenney Erle Stanley Gardner

Charles the Simple Charles III of France

Charley slang for Vietcong

Charley Car St Charles Avenue trolleycar; one of America's oldest and the last in New Orleans

Charley South Charleston, South Carolina

Charley West Charleston, West Virginia

Charlie Charles; letter C radio code; NATO name for Soviet C-class submarines built to launch missiles underwater

Charlie Chaplin Charles Spencer Chaplin

Charlie Hustle Pete Rose (baseball player)

Charlot (Spanish—Charlie)—Charlie Chaplin

Charm Spot of the Deep South Mobile, Alabama

Charter Oak City Hartford, Connecticut, where the original charter was hidden in an oak tree to insure the liberty of the First settlers

Charter Oak State Connecticut

Chatham Island, Galápagos San Cristóbal

cheapies cheap goods; cheap merchandise; cheap stocks

Checkpoint Charlie former international frontier between East and West Berlin

Cheesebox convict's nickname for the Illinois penitentiary at Statesville

Che Guevara Ernesto Guevara de la Serna

Chelmer native of Chelm (ancient Jewish town in Poland known in folklore as the Town of Fools)

Chelmo Polish name for concentration camp called Kulmhof by the Germans

Chelsea Gang John Dos Passos, Suzanne La Follette, Sinclair Lewis, Mary McCarthy, Ben Stolberg, who were among the first to expose the totalitarian nature of the USSR; lived in the Chelsea Hotel, New York City

Chemical Capital Wilmington, Delaware

Chemnitz Karl-Marx-Stadt

Chequers British prime minister's country home

Cher Cherilyn Sarkisian

Chero(s) Salvadoran(s)—person or people of El Salvador, Central America

Cheryl Ladd Cheryl Stoppelmoor

Chester Conklin Jules Cowles

Cheyenne Mountain air-defense complex near Pike's Peak, Colorado

Chicago Group poets and writers born in the Chicago area around 1900—Sherwood Anderson, Willa Cather, Floyd Dell, John Dos Passos, Theodore Dreiser, Finley Peter Dunne, James T Farrell, Francis Hackett, Harry Hansen, Ernest Hemingway, Vachel Lindsay, Archibald MacLeish, Edgar Lee Masters, Harriett Monroe, Frank Norris, Burton Rascoe, Carl Sandburg, Kay Boyle, TS Eliot, Scott Fitzgerald, Sinclair Lewis, Carl and Mark Van Doren

The Chicago of the Everglades Clewiston, Florida

Chicago piano submachine gun

Chi-chi naval nickname for Christchurch, South Island, New Zealand

Chickasaw Sikorsky transport helicopter designated H-19 or UH-19

Chicken Noodle News Cable News Network's nickname in its infancy

Chicken War trade dispute in the 1960s between the United States and some European nations over frozen chicken

Chico Marx Leonard Marx

The Chief Herbert Hoover, train on Chicago-Los Angeles run of Santa Fe

Chieftain British main battle tank armed with a 120mm gun

Child of the Mississippi Louisiana

Children of Joseph Israelites

Children of Pharoah Egyptians

Chilean Ports San Juan Bautista, Arica, Iquique. Tocopilla, Antofagasta, Taltal, Valparaiso, San Antonio.

Talcahuano, Coronella, Lota, Valdivia, Puerto Montt, Puerto Quellon, Punta Arenas

Chile's Principal Port Valparaiso

Chimneyville Jackson, Mississippi

china clay kaolin (hydrous aluminum silicate)

China's Baby Capital Guangzhou (Canton), China

China's Main Street *Yangtze Kiang* (Chinese—Yangtze River)

China's Sorrow soil-eroding Yellow River

Chinatown Chinese quarter of any city outside China

China Valley San Gabriel Valley, California

china white synthetic heroin (often deadly)

China white fentanyl (more powerful than morphine, also nicknamed Persian white)

chinese chinese banana (dwarf banana), chinese cabbage *(petsai)*, chinese checkers, chinese gelatin (agar or isinglass), chinese glue (alcohol + shellac), chinese greens (vegetables), chinese ink (india ink), chinese puzzle, chinese red, chinese watermelon (wax gourd), chinese white (barium sulfate), chinese wood oil (tung oil)

Chinese anesthesia acupuncture

Chinese cows soybeans

Chinese Gordon British general Charles George Gordon who suppressed the Taiping rebels; later named Gordon Pasha for similar services in the Sudan

Chinese Mainland Ports Macao, Huang-Pu, Kuang-Chou, Hong Kong (former British Crown Colony), Shant-T Ou, HsiaMen, Lo-Hsing-Ta, Mao-Ti, Ning-Po, Shanghai, Chen Chiang, Nan-Ching, Wu-Hu, Chiu-Chiang, Hang-kou, Chang-Sha, Ching-Tao, Wei-Hai, Yen-Tai, Ta-Ku. Tieng-Ching, Chin-Huang-Tao, Hu-Lu-Tao, Ying-K-Ou, Lu-Shun, Luta (Dairen)

Chinese Offshore Ports on Formosa or Taiwan Chilung (Keelung), Kaohsiung, Su-Ao, Hua-Lien, Tso-Ying, An-Ping, Tan-Shui

Chinese Restaurant Syndrome symptoms of those sensitive to MSG (monosodium glutamate)—tingling, headache, numbness, flush, palpitations

Chinese tobacco opium

chinese white zinc oxide (ZnO)

Chinese white fentanyl

Chino Men California (correctional) Institution for Men at Chino

Chinook Boeing-Vertol twin-rotor helicopter designated CH-47

Chinook State Washington

Chipmunk Hawker-Siddeley trainer aircrraft

Chitlin Capital of the World Salley, South Carolina

Chi Town Chicago's nickname

chloride of lime bleaching powder

Chocolate City Hershey, Pennsylvania

Chocolate Coast Ghana

Choctaw Sikorsky troop-transport helicopter designated H-34

Cholera Capital Calcutta

Cholly Knickerbocker Igor Cassini

Chomolungma Tibetan—Mount Everest

Choo-Choo Town Chattanooga, Tennessee

chop shop place where stolen autos are dismantled and sold by the part

Chosen Apostle self-proclaimed radio preacher Herbert W Armstrong

Chotzie Samuel Chotzinoff

Christiania Oslo's medieval name

Christianna Brand Mary Christianna Milne Lewis

Christian Right American Coalition for Traditional Values (includes the Moral Majority)

Christmas Cove South Bristol, Maine

Christopher Columbus Cristóbal Colón (Spanish); Cristoforo Colombo (Italian)

Chromium Continent Africa

Chrysler Chrysler Building of New York at 42nd Street and Lexington Avenue

Chuck Norris Carlos Ray

Chuco Mexican-Americanism for El Paso, Texas

Chuey (Spanish-American nickname—Jesus)

Chuppies Chinese urban professionals

chutz *chutzpah* (Yiddish—unmitigated gall; some nerve)

Cicero Marcus Tullius

Cigar Capital Key West, Florida; Tampa, Florida

Cigar City Tampa, Florida

Cigarette Josiah Flynt Willard

Cincinnati oysters pig's feet

cinnamon stone hessonite

Cinquecento (Italian—five hundred)—Italy's 16th century artistic and cultural development

Cipango Japan, as it was called by early European explorers

Circus King John Ringling

Cisco Kid Duncan Reynaldo (Renault Renaldo Duncan)

Cissie Patterson Eleanor Medill Patterson

Cissy Loftus Mary Cecilia M'Carthy

The Cit The Citadel Military College of South Carolina

Citadel of Budapest Burgberg

Citians people of Minneapolis and St Paul also called Twin Citians

Cities of Culture Edinburgh and Glasgow

Cities of the Plain Admah, Gomorrah, Sodom, and Zeboim on the Jordan River Plain of ancient Israel near the Dead Sea

Citizen Capet Louis XVI (beheaded during French Revolution)

Citizen Composer Dmitri Shostakovich

Citizen of Geneva Jean Jacques Rousseau

Citizen King Louis Philippe of France

Citizen Louis Capet Louis XVI

Citizen of the World Oliver Goldsmith, Thomas Paine

Citlatepetl Orizaba

citric acid $C_8H_6O_7$

Citrus Metropolis Los Angeles

The City financial, governmental, historical, and commercial core of London, including newspaper publishing district, Bank of England, Lloyd's, many famous restaurants; San Francisco, California

City of 1000 Lakes Oklahoma City, Oklahoma

City of Abraham Hebron, Israel

City of Alexander the Great Alexandria, Egypt

City of Aluminum Arvida, Québec

City of the Angel San Angelo, Texas

City of Angels Bangkok, Los Angeles

City of the Apprentice Boys Londonderry, Northern Ireland

City of the Arctic Tromso, Norway

City of the Arts Minneapolis

City of Athena Athens

City of Baked Beans Boston, Massachusetts

City by the Bay San Francisco

City of Beaches Montevideo, Uruguay

City of Beautiful Spires Copenhagen

City of Bells Strasbourg, France

City of Berwald Stockholm, Sweden

City Beside the Broad Missouri Bismarck, North Dakota

City Between Bridges Stockholm

City of Bicycles Copenhagen

City of Big Shoulders Carl Sandburg's sobriquet for Chicago

City of Birches Umeå, Sweden

City of Birm Sym Orch City of Birmingham (England) Symphony Orchestra

City of Black Diamonds Scranton, Pennsylvania

City of Blazing Lights Shanghai

City of the Blues Memphis, Tennessee (home of WC Handy)

City of Brazilian Churches Rio de Janeiro's nickname

City of Brick Pullman, Illinois

City of Brotherly Love Philadelphia (derived from the Greek) *philos* (love) and *adelphos* (brother)

City of the Caliphs Cairo

City of the Camellias Pensacola, Florida

City of Canals and Bridges Amsterdam, Copenhagen, Leningrad, Stockholm, and Venice

City of the Carmel Haifa, Israel, on the slopes of Mount Carmel

City of Castles Copenhagen

City of Cats Venice (felines outnumber people)

City of Certainties Des Moines, Iowa

City of Champions Pittsburgh

City of Cheese Dutch cities of Alkmaar and Gouda

City of Cheese, Chairs, Children, and Churches Sheboygan, Wisconsin

City of Churches Brooklyn, New York

City College British euphemistic name for Newgate Gaol—the old London lockup; New Yorker nickname for the Tombs prison, in Manhattan

City of Coral coral reefs off the Virgin Islands

City of Corsairs St Malo, France

City of Cypresses Rome

City of David Jerusalem

City of Death Beirut, Lebanon; Kipling's name for Lahore, Pakistan

City of Destiny Tacoma, Washington

City of the Doges Venice

City of Dreadful Night Kipling's nickname for Calcutta

City of Dreaming Spires Oxford, England

City of Dreams Port Townsend, Washington

City of the Dunes Dunkerque, France

City Ed City Editor

City of Elms New Haven

City of Eternal Spring Caracas

City of Fair Breezes Buenos Aires, Argentina

City of Five Seasons Cedar Rapids, Iowa

City of Flour Buffalo, Minneapolis

City of Fountains Aix-en-Provence, France; Bratislava

City of Four Lakes Madison, Wisconsin

City of Fun and Frolic Atlantic City, New Jersey

City of Gardens Lahore, Pakistan; Victoria, British Columbia

City of Gardens and Beaches Adelaide, Australia

City of the Gods Teotihuacán— religious capital of Mexico in the fifth century before the Christian Era

City of Gold Dawson, Yukon Territory

City by the Golden Gate San Francisco

City of the Golden Horn Istanbul

City of Good Neighbors Arlington Heights, Illinois

City of Green Spires Copenhagen

City of Grieg Bergen, Norway

City Grown Too Big For Its Bridges San Francisco

City of Hans Christian Andersen Copenhagen, Denmark

City of Heat Thermopolis, Wyoming

City of Historical Charm Savannah

City of a Hundred Hills San Francisco

City of a Hundred Spires Prague

City of a Hundred Towers Pavia, Italy

City of Illicit Love Paphos on Cyprus

City of the Immortals Amarapura, Burma

City of Jade Oaxaca, México

City of Jazz and Mardi Gras New Orleans, Louisiana

City of Kielland and Bjelland Stavanger, Norway

City of Kings and Commoners Copán, Honduras

City of Lakes Dartmouth, Nova Scotia

City of Light Paris, France; Perth, Western Australia

City of Lillies Florence, Italy

City by the Lion's Gate Vancouver, British Columbia

City of Lost Angels Los Angeles, California

City of Louis Paris

City of Magnificent Distances Washington, D.C.

City of Manifold Advantages Augusta, Maine

City of Mankind Jerusalem

City of Masts Port of London

City of Millionaires Colorado Springs

City of Minarets Miknès, Morocco

City of Money Zurich, Switzerland

City of Monuments Baltimore, Maryland; Florence, Italy

City of Mosques Istanbul, Turkey

City in Motion San Diego, California

City of Mozart Salzburg, Austria

City of Music and Song Vienna

City of Nielsen Copenhagen, Denmark

City of Nine Dragons Kowloon, Hong Kong

City of Notions Boston, Massachusetts

City of Oaks Raleigh, North Carolina

City of One Hundred Hills San Francisco, California

City on the Neva St Petersburg's nickname

City of Palaces Rome, Italy, and Vatican City

City of Palms Acajutla, El Salvador; Fort Myers, Florida; and Maracaibo, Venezuela

City of the Pampas Buenos Aires

City of Peace Brunei

City of Penn Philadelphia, Pennsylvania founded by William Penn

City of Personality Cincinnati, Ohio

City of Peter the Great Saint Petersburg (later called Petrograd and Leningrad)

City of the Plains Christchurch, New Zealand

City of Poets Jérémie, Haiti, birthplace of the father of Alexandre Dumas (*Dumas pére*) and grandfather of Alexandre Dumas (*Dumas fils*)

City of Power Peking, People's Republic of China

City of Presidents Quincy, Massachusetts

City of the Prophet Medina, Saudi Arabia, where Mohammed was protected after fleeing from Mecca

City of Quays and Grieg Bergen, Norway

City of Razzle Dazzle one of O Henry's nicknames for New York City he also called Bagdad on the Subway

City of Receptions Washington, D.C.

City by the Rivers Kansas City where the Kansas River flows into the Missouri; Minneapolis–St Paul on the Minnesota and Mississippi; Philadelphia on the Delaware and Schuylkill; Pittsburgh on the Allegheny, Ohio and Monongahela; Portland, Oregon on the Columbia and the Willamette; St Louis where the Mississippi meets the Missouri

City of Rocks Nashville, Tennessee

City of Roses Portland, Oregon

City of Ruins and Roses Visby, Sweden

City of Rumors Washington D.C.

City of Rum and Sugar Georgetown, Guyana

City of Sails Auckland, New Zealand

City of Saints Montreal

City of Salt Salzburg, Austria; Syracuse, New York

City by the Sea Newport, Rhode Island

City of the Sea Venice

City of Seven Hills Rome, Italy, built on Aventine, Caelian, Capitoline, Esquiline, Palatine, Quirinal, and Viminal

City of Seventy Isles Venice

City of Shoes Brockton, Massachusetts

City of Sibelius Helsinki, Finland

City of Silver Taxco, México

City of the Silver Gate San Diego

City of Sinbad Basra, Iraq

City of Sinding Oslo, Norway

City in the Sky Macchu Picchu, Peru

City of Skyscrapers New York

City of the Slain Arlington National Cemetery in Arlington, Virginia

City of Smells Old Delhi, India

City of Smokestacks Everett, Washington

City of Soles Lynn, Massachusetts

City of Sorrow Buchenwald (concentration camp near Weimar, Germany)

City of Southern Charm Savannah, Georgia

City of Spies Beirut, Berlin, Copenhagen, Hong Kong, London, New York, Paris, Singapore, Stockholm, Vienna, Zurich, Washington, D.C.

City of Spires Copenhagen

City State Singapore and the Vatican City State

City of Steel Pittsburgh, Pennsylvania

City of St Mark Venice

City of St Michael Dumfries, Scotland

City of St Mungo Glasgow, Scotland

City of the Straits Detroit, Michigan, on the Straits of Belle Isle

City of Suds Milwaukee

City of the Sun Baalbec, Heliopolis, and Rhodes; Campanella's utopian republic

City of Sunshine Colorado Springs, Colorado; Los Angeles, California; Tucson, Arizona

City of Surprises Amsterdam

City Surreal Miami, Florida

City of Symphonies London

City of Tamales San Antonio, Texas

City of Temples Benares, India; Katmandu, Nepal

City of Ten Million Roosters Port-au-Prince, Haiti

City That Boeing Built Seattle

City That Care Forgot New Orleans

City That Knows How San Francisco

City That Swims on the Water Stockholm

City of the Thousand and One Nights Baghdad, Iraq

City of Three Capitols Little Rock, Arkansas

City of the Three Kings Cologne, Germany, where it is reputed the Magi or Three Kings are buried; Lima, Peru

City of Totems Ketchikan, Alaska

City of Trees Christchurch, New Zealand; Saratoga Springs, New York

City of the Tribes Galway, Ireland

City under Vesuvius Naples

City of the Violet Crown Athens

City of Washington Washington, D.C.

City of Walls Beijing, China

City on the Water Amsterdam, Copenhagen, Stockholm, and Venice

City of Witches Salem, Massachusetts

City without Clocks Las Vegas, Nevada

Ciudad Blanca (Spanish—White City)—Guayaquil, Ecuador's cemetery; (Spanish—White City)—Merida, Yucatan

Ciudad de los Reyes (Spanish—City of the Kings)—Lima, Peru

Ciudad Imperial y Coronado (Spanish—Imperial and Crowned City)—Toledo, Spain

Ciudad Loca (Spanish—Crazy City)—Barranquilla, Colombia during carnival

Civil Rights Leader Martin Luther King, Jr

Civil Service President Ulysses Simpson Grant

Civil Service Reform President Chester Alan Arthur

Civil War Photographer Matthew Brady

Claire Trevor Claire Wemlinger

Clamcatcher(s) New Jerseyite(s)

Clamgrabber(s) Washingtonian(s)

Clam States New Jersey and Washington

Clam Town Norwalk, Connecticut

clap gonorrhea

clap joint brothel

Clara Covell North Clara E Ellis

Claribel Charlotte Alington-Barnard

Clarin Leopoldo Alas y Urena

Clarita Clara Elena

Clark U.S. air base, Luzon Island, Philippines

Clark Gable William Gable

Classic City Kyoto, Honshu Island, Japan

Classifier and Compiler Extraordinaire Dr Peter Mark Roget

Claude Lorraine Claude Gellée of Lorraine

Claudette Colbert Lily Cauchoin

Claudio Lars Carmen Brannon de Samayoa

The Claw Otto Schnelbacher (basketball player)

Clay-eaters South Carolinians

Cleat Soviet Tupolev Tu-124 long-range transport (NATO)

clicks Internet businesses

Clifford Ashdown pseudonym shared by R Austin Freeman and John James Pitcairn

Clifton Webb Webb Parmalee Hollenbeck

Clinton's Big Ditch Erie Canal advocated by Governor De Witt Clinton of New York

Clinton's Folly Erie Canal (*see* Clinton's Big Ditch)

Clio Joseph Addison's pseudonym; in Greek mythology the muse of history or of lyre playing

Cloud Piercer New Zealand's Mount Cook (12,349 feet or 3,764 meters)

Clown of the Orchestra bassoon

Clown Prince of Music Danny Kaye

Clowns of the Canine World dachshunds

Clubland Pall Mall clubhouse section of London

Clyde the Glide Clyde Drexler (basketball)

Coach Soviet Ilyushin transport plane Il–12 (NATO)

Coal City Pottsville, Pennsylvania

Coaley Samuel Coleridge-Taylor

Coal Metropolis Cardiff, Wales

Coal State Pennsylvania

Coatzacoalcos formerly Puerto Mexico

Cobbler Poet Hans Sachs of Nuremberg, also known as Prince of the Meistersingers

Cobh Gaelic name for Queenstown

Cobra Bolkow wire-guided anti-tank missile made in Germany

Cocaine of the '90s computer chips, which have high street value and can be difficult to trace when stolen

Cocaine Capital Bogotá, Colombia and Jackson Heights, Queens, New York

Cocaine Capital of Colombia Medellin

Cochise Beech T-42 transport aircraft

Cockade City Petersburg, Virginia

Cockade State Maryland

Cockney Poet John Keats

Cockpit of the American Revolution New Jersey

Cockpit of Europe Belgium

Cockpit of the Middle East Syria

Cocky Eddie Collins (baseball)

Coco Chanel Gabrielle Bonheur Chanel

Codder(s) Cape Cod resident(s)

Code Hammurabi Babylonian civil and legal code established in the 20th century before the Christian era

Code Napoléon French civil code established in 1804

Codfishland Newfoundland

Coke City Uniontown, Pennsylvania

Col Lucius Iunius Moderatus Columella (Roman writer on agriculture)

Colette Sidonie Gabrielle Claudine de Jouvenal

Collar City Troy, New York

College of New Jersey Princeton University's original name

College of Rhode Island Brown University's original name

Collodi Carlo Lorenzini

cologne cologne brown (vandyke brown); cologne spirits (highly concentrated ethyl alcohol); cologneware

(mottled brown and gray stoneware); cologne water (eau de cologne, toilet water); cologne yellow (chrome-yellow and lead-sulfate pigment)

Colombian Colombian-grown marijuana; Colombian mountain-grown coffee

Colombian connection network of brokers, farmers, politicians, and smugglers connected with the export of Columbian-grown coca leaf products and marijuana to Canada and the United States

Colombian First Colombia's best-known classical composer—Guillermo Uribe Holguin

Colombian Ports Santa Marta, Barranquilla, Cartagena, Covenas, Buenaventura, and Ríohacha

Colombia's Principal Port Barranquilla

Colón formerly Aspinwall

Colonels natives of Kentucky

Colonne Edouard Judá Colonne, French conductor

Colossus of the African Continent the Sudan

Colossus of the Eurasian Continent the USSR

Colossus of Independence John Adams

Colossus of the Indian and Pacific Oceans Australia

Colossus of the North United States of America

Colossus of the North American Continent Canada

Colossus of the South American Continent Brazil

Colquhoun Calhoun

Colt Colt revolver (invented by Samuel Colt of Hartford, Connecticut)

Colter's Hell Yellowstone (for John Colter, mountain man, who traversed part of what is now national park)

Columbia America; South Carolina's capital; the United States

Columbia first space shuttle launched from Cape Canaveral, Florida in April 1981

Columbia City Vancouver, Washington

Columbia the Gem of the Ocean United States of America

Columbia Lou Lou Gehrig (baseball player)

Columbus Cristóbal Colón (Spanish); Cristoforo Colombo (Italian); name of places in some twelve states in the United States; Ohio's capital

Columbus of the Cosmos Russian cosmonaut Yuri Alexeyevich Gagarin

Columbus of the Subconscious Sigmund Freud

Comedian Pianist Victor Borge

Comedian's Comedian Bert Williams

Comenius John Amos Komensky

Comet British medium tank built during World War II; De Haviland four-engine jet transport aircraft

Comic Symbol of the Century Charlie Chaplin

Commando C-46 Curtiss-Wright 36-passenger transport built during World War II; Cadillac-Gage amphibious armed car and military personnel carrier (XM-706); Dodge-built military personnel carrier built during World War II

The Commerce Comet Mickey Mantle (baseball player)

Commercial Capital of Canada Toronto

commie communist

commode portable toilet

Commoner The Commoner—William Jennings Bryan

Communist Capitalist Friedrich Engels

Communist East countries dominated by China

Communist West occidental countries dominated by the former Soviet Union

Community of True Inspiration Amana, Iowa

Commy Charlie Comiskey (baseball player)

Comoro Island Ports Moroni, Patsy, Mutsamudu, Fomboni

Comoros Republic of the Comoros (island nation in the Indian Ocean northwest of Madagascar), *Etat Comorien*

The Company the CIA

Compton Mackenzie Edward Montagu Compton

computer virus computer program that "infects" other computers, destroying files and even hard drives

Conca Doro (Italian—Shell of Gold)—nickname of the hills encircling Palermo

Concha Maria de la Concepción

Conch(s) inhabitant(s) of Key West, Florida and nearby keys

Conch Town Key West, Florida

Concordance Cruden Alexander Cruden—compiler of the *Complete Concordance of the Holy Scriptures* published in 1737

Concorde Anglo-French supersonic airplane attaining cruising speeds of 1300 miles per hour

Concord Group Bronson Alcott, Ralph Waldo Emerson, Margaret Fuller, Nathaniel Hawthorn, Henry David Thoreau

The Concord Sage Ralph Waldo Emerson

Concrete Jungle modern metropolitan center

Condemned Rock Tasmanian nickname for Grummet Island, Macquarie Harbour where there is a penitentiary

condo circuit string of retirement communities at which entertainers perform

Condor North American-Rockwell air-to-surface missile (AGM-53A)

Condorcet Marie Jean Antoine Nicolas de Caritat, Marquis de Condorcet

Coney Island butter mustard

Confederate Raider Rear Admiral Raphael Semmes, CSN

Confederation Province Prince Edward Island

Confetti Canyon lower Broadway, from Bowling Green to City Hall Park, New York City where during parades confetti and waste paper are thrown from upper-floor windows

Confucius Kung Fu-tse

Congo Ports Loango, Pointe Noire, Malongo Oil Terminal

Connecticut Ports New London, New Haven, Bridgeport

Connection City Amsterdam, Marseilles, Miami, Singapore, and major airports where drugs are smuggled

Connemara poet Carl Sandburg's home in Flat Rock, North Carolina

Connie Mack Cornelius McGillicuddy

Connie Stevens Concetta Ingolia

Conniver from Kennebunkport George Bush

Conquering Lion of Judah and King of Kings Emperor Haile Selassie of Ethiopia

Conqueror of Disease Louis Pasteur

Conqueror of Mount Everest Sir Edmund Hillary

Conqueror of Suez Ferdinand de Lesseps

Conquerors of Yellow Fever Walter Reed and his colleagues Aristides Agramonte, James Carroll, and Jesse Lazear

Conrad Joseph (Korzeniowski) Conrad—English novelist

Conrad Veidt Konrad Weidt

Conscience of America Norman Thomas

Conscience of American Music Gunther Schuller

Conscience of the American Theater Brooks Atkinson

Conscience of Europe Voltaire

Conscience of the Left George Orwell

Constable Country East Berghott in England's Sussex where James Constable's award-winning landscapes were painted

Constable of France Charles De Gaulle

Constantia Judith Sargent Murray

Constantinople Istanbul

Constant Reader Dorothy Parker

Constellation Lockheed 63-passenger transport

Constitution State Connecticut

Constructionist President James Buchanan

The Consulate France under the First Consul—Napoleon Bonaparte—1799–1804

Contadora Group Colombia, Mexico, Panama, Venezuela

Contemporary Cassandra Dorothy Thompson

Continent The Continent (usually Europe; also Africa, Antarctica, Asia, Australia, North America, South America)

Continental Divide Rocky Mountain ridge separating rivers flowing eastward to the Atlantic Ocean and the Gulf of Mexico from those flowing westward to the Pacific

Continental Maritime World the Great Lakes (Erie, Huron, Michigan, Ontario, Superior)

Continental Nation Australia

Continent of Hope South America

Convair 600 Convair-Liner powered by Rolls-Royce turbo-prop engines

Cook Islands Danger, Manahiki, Penrhyn or Tongareva, Rakahanga and nearby islets in the South Pacific

Cookpot Soviet Tupolev Tu-124 jet-transport aircraft (NATO)

Cool Papa James Bell (baseball)

Coon Dog Capital Vienna, Illinois

Copa de Oro (Spanish—Cup of Gold)—pirate's nickname for Panama

Copeia periodical of the American Society of Ichthyologists and Herpetologists named for the naturalist Edward Drinker Cope

copenhagen surprise naval attack without warning of a fleet at anchor in the manner of Nelson's sortie in 1801

Copernicus Latinized name of Polish astronomer Nikolaus Kopernicki

Copper City Butte, Montana

Copper John Auburn Prison near Syracuse, New York

Coppernose Henry the VIII whose portrait exhibited a copper-colored nose on the coins minted during his reign

copper pyrites chalcopyrite (copper iron sulfide)

Copper State Arizona; Wisconsin

Coral Atoll Country Nauru

Coral Coast Fiji's hotel complex between Sigatoka and Yanuca

Cora Montgomery Jane McManus

Corbusier Charles-Édouard Jeanneret

Corn Belt Illinois, Indiana, Iowa, and Nebraska

Cornbread Cedric Maxwell (basketball player)

Corn City Toledo, Ohio

Corncob Capital Washington, Missouri

Corncracker(s) Kentuckian(s)

Corncracker State Kentucky

Cornell bread nutritious bread developed at Cornell University; soy flour, wheat germ, and nonfat dry milk are added to wheat flour

Corn Geneticist Barbara McClintock

Cornhusker(s) Nebraskan(s)

Cornhusker State Nebraska

Cornish Riviera English Riviera extending from Falmouth to the Isles of Scilly

Corn-Law Rhymer Ebenezer Elliott

Corno di Bassetto (Italian—basset horn)—pen name used by George Bernard Shaw when he was a music critic

Cornopolis Chicago

Corn State Illinois, Iowa

Cornubian Shore Cornwall, England

Corporal John John Churchill who became the first Duke of Marlborough; known to the Spaniards as Mambrú

Corregio Antonio Allegri

Corridor of Six Continents sobriquet given the Panama Canal, Suez Canal, and projected interoceanic sea-level canals across Mexico and Nicaragua

Corridor State New Jersey

Corrie Denison Eric Partridge

Corruption-Hating President Grover Cleveland

Corsair Chance-Vought single-engine fighter popular during World War II (F4U)

Corsican Napoleon Bonaparte who was born on Corsica

Corsican Ogre Napoleon

Corvette antisubmarine-warfare convoy escort ship

Cory Corazon Aquino

cos lettuce from Greek island of Cos

cosa nostra (Italian—our thing)—nickname for international criminal syndicate network

Cosimo palette name of Piero di Lorenzo who took the given name of his teacher Cosimo Roselli

Cosmopolis of the Heartland Kansas City

Cosmopolitan Canadian-built medium-range transport designed by General Dynamics and designated CC-109

Costa de la Muerte (Spanish—death coast) Spain's northwesterly stormbound coast

Costa Geriatrica (Latin—elderly coast) southeast coast of England frequented by many older people

Costa Rican Ports Limón; Puntarenas and Golfito

Costermansville former name of Bukavu

Cotton Al Brazle (baseball player)

Cotton Belt cotton-growing areas of the southern United States; also known as the Cotton Kingdom

Cotton Bowl Dallas, Texas

cotton-dust disease brown lung or byssinosis

Cottonopolis Manchester, England

Cotton State Alabama

Cottonwood City Leavenworth, Kansas

couch potatoes sedentary people who watch a lot of television

Cougar Grumman carrierbased transonic fighter aircraft (F9F-6)

Count Basie William Basie

Count No Count William Faulkner (nickname from his Mississippi neighbors)

The Country Israel (in Hebrew, *Ha'aretz*)

Country of the Blacks Sudan (Arabic—black)

Country of a Thousand Hills Rwanda

Courland Kurland

Court of St James British royal court

Cousin Jack a Cornishman; a Cornish miner

Cousin Jenny Cornish girl or woman

Covent Garden The Royal Opera House in London adjacent to Covent Garden marketplace

Cowboy Artist Charles M Russell

Cowboy Capital Dodge City, Kansas

Cowboy Philosopher Will Rogers

Cowboys Texans

cowboys of the sea porpoises

Cowboy State Wyoming

Cowtown Fort Worth, Texas; Kansas City, Kansas and Missouri; Omaha, Nebraska

Coyote Cowboy Pecos Bill

Coyote(s) South Dakotan(s)

Coyote State South Dakota

The Crab Jesse Burkett (baseball player)

Crab Capital Crisfield, Maryland

Crabtown Annapolis, Maryland

crack low-cost smokable cocaine

Cracker Quixote Arthur R Marshall

Crackers rural Floridians and Georgians

Cracker State Georgia

crack head crack addict

Cradle of American Culture and Music Carnegie Hall

Cradle of American Independence Independence Hall, Philadelphia

Cradle of American Liberty Fanuel Hall in Boston, according to Daniel Webster

Cradle of the American Revolution Faneuil Hall, Boston

Cradle of Aviation San Diego

Cradle of California San Diego

Cradle of Canadian Confederation Prince Edward Island

Cradle of Civilization Armenia, China, Egypt, Greece, India, Iran, Iraq, Israel, Italy, Jordan, Lebanon, Mexico, Peru, Syria, and Turkey

Cradle of Classical Civilization Greece

Cradle of the Confederacy Montgomery, Alabama; Charleston, South Carolina

Cradle of Democracy ancient Greece

Cradle of Electrical Engineering Berlin, Germany where the first electric railroad and first large-scale power station were built

Cradle of the French Revolution Marseille and Paris

Cradle of Human Civilization Iraq

Cradle of Islam Saudi Arabia

Cradle of Japanese Art Nara, Honshu Island, Japan

Cradle of Japanese Civilization Kyoto

Cradle of Liberty Carpenters' Hall, Philadelphia; Faneuil Hall, Boston; House of Burgesses, Williamsburg, Virginia; Holland during formation of the Dutch Republic; Switzerland in William Tell's time

Cradle of Nuclear Research Los Alamos, New Mexico

Cradle of Oriental Civilization China

Cradle of Polynesia Samoa Islands

Cradle of Psychoanalysis Berlin; Vienna

Cradle of the Renaissance Florence, Italy

Cradle of the Revolution Boston, Massachusetts; Paris, France; Petrograd, Russia

Cradle of the Russian Revolution Petrograd

Cradle of Secession Charleston, South Carolina

Cradle of Texas Liberty The Alamo in San Antonio

Cradle of the Union Albany, New York where in 1754 Benjamin Franklin presented his Plan of Union to the Albany Congress

Cradle of Violent Crime the United States

crane berry cranberry

crank. underworld nickname for methamphetamine, a mind-altering drug

Crash John Mengelt

Crate NATO nickname for Soviet Ilyushin transport Il-14

Crawfish Town New Orleans, Louisiana

Crawthumper(s) Marylander(s)

Crazy Alley San Quentin Prison's insane asylum

Creative Genius of American Architecture Frank Lloyd Wright

Creator of Chords Franz Liszt

Creator of the Female Language for Sexuality Anaïs Nin

Creator of French Existentialism Jean-Paul Sartre

Creator God *Vracocha* (Quechua—supreme god)—deity venerated in Incan and pre-Incan times

Creator of Modern Democracy Thomas Paine

Creator of Musical Laughter Rossini

Crébillon Prosper Jolyot

Creole Country southern counties of Alabama and Mississippi as well as coastal parishes of Louisiana where many people are of French or Spanish origin

Creoles Louisianans

Creole State Louisiana

Crescent City Appleton, Wisconsin; New Orleans, Louisiana

Crestwood Heights Toronto, Ontario's Forest Hill Village

Cretinsbury-on-Cesspool Tasmanian nickname for Launceston

cri-du-chat syndrome loss of part of a chromosome—infants with this syndrome have a cry similar to meowing of a cat; they may be severely mentally retarded and malformed

Crime Syndicate Chief Meyer Lansky

croak mixture of cocaine and crack; to die

The Crocodile W Averell Harriman

Crohn's disease chronic inflammation of the intestines, named for Burrill Bernard Crohn, U S physician

Cromwell's Curse Ireland (also called the Curse of Cromwell)

Cronian Sea Arctic Ocean

Cross of Geneva emblem of the Red Cross (red cross on a white field) used to show the neutrality of ambulances, hospitals, and hospital ships during wartime

Crossroads of Africa, Asia, and Europe Egypt

Crossroads of Africa and Europe Spain

Crossroads of Asia Afghanistan

Crossroads of Continents Bering Strait

Crossroads of Europe Belgium

Crossroads of the Pacific Oahu—the Aloha Islands, Hawaii

Crossroads of the Seven Seas Singapore

Crossroads of the South Pacific Fiji

Crossroads of the World Panama Canal; Straits of Gibraltar; Suez Canal

Crotale Thompson surface-to-air guided missile made in France

Croves Hal Croves (pseudonym of B Traven—nom de plume of Berick Traven Torsvan Croves)

Crow Eaters South Australians

Crown City Coronado, California

Crown City of the Valley Pasadena, California

Crown Jewel of the Adriatic Venice

Crown Prince of Cellists Lynn Harrell

Crown Prince of Keynesism John Kenneth Galbraith

Crown Prince of Psychoanalysis Carl Gustav Jung

Crucible of Civilization Iraq

Crystal City Corning, New York

Crystal Hills New Hampshire's White Mountains

C-stand century stand

C-town Chestertown; Cowtown

Cub NATO nickname for the Soviet Antonov 100-passenger cargo plane

cuban cuban heel (broad-based heel used on women's shoes)

cubanite copper iron sulfide

Cuban Ports Bahia Honda, Cabañas, Mariel, La Habana (Havana), Matanzas, Cardenas, La Isabela, Caibarien, Nuevitas, Puerto Padre; Santiago de Cuba, Manzanillo, Cienfuegos

Cuba's Principal Port Havana

Cultural Capital of Canada Montreal

Cultural Capitals of the World Berlin, London, Moscow, New York, Paris, Vienna

Cultured Pearl Archipelago Japan

Cultured Pearl of the Orient Hong Kong

Cumberland River City Nashville, Tennessee

Cunard Cunard Line transatlantic steamers such as the *Aquitania, Lusitania,* and the *Mauretania*

cupid's itch venereal disease

Curaçao's Ports Willemstad, Bullen Baai, Caracas Baai, New Port

Curator of Culture Lord Kenneth M(ackenzie) Clark; Peter Ustinov

Curmudgeon Philosopher Ambrose Bierce

Currer Bell pseudonym of Charlotte Brontë

curse the curse (menstruation)

Curse of Balboa Panamanian-style dysentery marked by acute nausea and retching

Curse of Cabral loose bowels contracted in Brazil

Curse of Cairo diarrhea plus summertime heat

Curse of Columbus botulism, diarrhea, or dysentery contracted in the American tropics

Curse of Cortez Mexican-acquired diarrhea or dysentery also called Montezuma's Revenge

Curse of Cromwell Ireland (also called Cromwell's Curse)

Curse of Pizarro loose bowels contracted in Peru

Curt Jurgens Curd Jurgens

The Curveless Wonder Al Orth (baseball player)

Curzio Malaparte pseudonym—Curzio Suckert

Cuspidor of Europe France's nickname given it by Alexander Herzen

custard apple cherimoya

Cuvier Georges Léopold Chrétien Frédéric Dagobert

cyberfakes fake designer merchandise sold over the Internet

cyberspace the Internet

cyberterrorists Internet terrorists

cybervandals Internet hackers

Cyclone Coast Australia's northwest coast

Cyclone State Kansas; South Dakota

Cyclops of the Kremlin Josef Stalin

Cyclorama City Atlanta, Georgia with its cyclorama painting of the Battle of Atlanta

Cyd Charisse Tula Finklea

Cyprian Ports (counterclockwise north to south coast) Kyrenia, Xeros, Paphos. Limasol, Larnaca, Famagusta

Cypriot Apostle Barnabas—Cyprus-born companion of Paul and Mark, according to the New Testament

Czar of Baseball Bancroft (Ban) Johnson

czar's caviar sterlet roe

Czech Capital Prague

Czech National Composer Anton Dvořák

Czech Reformers Jan Hus, Alexander Dubcek, Vaclav Havel

dachshund (German—badger dog)

Da Costa's syndrome soldier's heart

dagestan dagestan rug

Dahlia Dalila

Daisy Ashford Margaret Mary Ashford

Dakota Douglas DC-3 21-passenger transport also called Skytrain; lines spoken before singing a song

Dalton Highway Alaska's former North Slope Haul Road, built to move supplies to Prudhoe Bay (for arctic engineer James Dalton)

Dame Cicely Dr Cicely Saunders

Dame Clara Dame Clara Butt

Dame Joan Dame Joan Sutherland

Dame Margot Fonteyn Margot Hookham

Dame Myra Dame Myra Hess

Dame Ngaio Dame Ngaio Marsh

Damnable Place Hong Kong, according to GBS

Dam on the Amstel Amsterdam

damp Spain Atlantic coasts of Spain, especially along Bay of Biscay where rains are heaviest

Dana Andrews Carver Daniel Andrews

Dandy King Joachim Murat—King of Naples

Daniel Nikolai Arzhak

Daniel Stern pseudonym of Liszt's paramour, the Countess Marie d'Agoult

danish danish pastry (light pastry often filled with stewed fruit, crushed nuts, cup custard, or raisins)

Danish Capital of the United States Racine, Wisconsin

Danish Caribees colonial name for what are now the U.S. Virgin Islands

Danish King of the Waltz Hans Christian Lumbye

Danish Ports Ronne *(on Bornholm)*; Køge, Københavne (Copenhagen), Helsingor, Frederiksvaerk, Frederikssund, Roskilde, Holbaek, Nykqøbing, Kalundborg, Korsor, Skaelskor, Naestved. Vordingvorg, Stubbekøbing, Stege, Masnedsund, Nykøbing, Falster, Saksøbing, Naskov, Rudkøbing, Marstal, Svendborg, Nyborg, Kerteminde, Odense, Middelfart, Assens, Faborg, Grasten, Sonderborg, Augustenborg, Abenra, Haderslev, Kolding, Frederkicia, Horsens, Arhus, Grena, Randers, Alborg, Frederikshavn, Skagen, Esbjerg

Danish Waltz King Hans Christian Lumbye

Danish West Indies former name of the American Virgin Islands

d'Annunzio (Gabriel) Gaetano Rapagnetta

Danny Kaye David Daniel Kaminsky

Danny O'Neill fictionalized name of James T Farrell *(Studs Lonigan)*

Danny Thomas Amos Jacobs

Dante Dante (Durante) Alighieri

Danube Delta Land Romania

Danube Empire Austro-Hungarian Empire

Danubian Monarchy Austro-Hungarian Empire

Dark and Bloody Ground Kentucky

Dark Continent Africa

Darkest Africa Lake Victoria's shores (19th century)

darwin glass queenstownite (silica glass)

Darwin Island Galápagos Culpepper

Darwin's Archipelago Galápagos Islands

Darwin's Bulldog nickname of Professor Thomas Henry, Huxley, president of the Royal Society, whose defense of Darwin's *Origin of the Species* pulverized the arguments of the Bishop of Oxford, Samuel (Soapy Sam) Wilberforce, who had set out to demolish Darwin's theory of evolution

Dash 8 Boeing twin-engine turboprop airplane

Das Jud Das Judenthum in der Musik (German—The Jew in Music)—first draft in 1850; revised in 1869 by its author—Richard Wagner—an ardent Jew-hater beloved by Hitler

Date Capital Indio, California

Daughter of the Baltic Helsinki

Daughter of the Desert Gertrude Bell

Daughter of the Dream Emma Goldman

Dave the Rave Dave Stallworth

David Bowie David Jones

David Copperfield David Kotkin

David Frome Zenith Jones Brown

David St John E Howard Hunt

David Wayne Wayne McKeekan

Davy Jones' Locker traditional resting place of all who are buried at sea or who are drowned in the depths of the ocean

Dawn on the Mesabi Aurora, Minnesota

Day of Infamy December 7, 1941 (when Japanese aircraft carriers attacked Pearl Harbor, Hawaii while diplomatic negotiations were in progress in Washington, D.C.)

Dazai Osamu Tsushima Shuji

Deacon Walt Stankey (basketball player)

Deadeye Dick Nat Love, a black cowboy of the last century who was noted for his superior marksmanship

deadly nightshade belladonna

Deadman's Cove geographic placename and nickname of the San Diego Police Department headquarters

dead President slang for American paper money

Deadwood Dick Richard W Clarke—English-born South Dakota frontier pioneer

Deaf Composer Ludwig van Beethoven

Deaf Smith Erastus (Deaf) Smith, Texan patriot-soldier

Dean of American Astronomers Henry Norris Russell

Dean of American Choral Conductors Robert Shaw

Dean of American Psychiatry Dr Karl Menninger

Dean of Classical Guitarists Andrés Segovia

Dean of Classical Music Comedians Victor Borge

Dean of the English School Sir Edward Elgar

Dean of Financial Analysts Benjamin Graham, author of *The Intelligent Investor*

Dean of the French School Camille Saint-Saëns

Dean of Italian Music Gioacchini Rossini

Dean Martin Dino Crocetti

Dean of Modern American Composers Aaron Copland

Deano Dean

Dean of Russian Music Nikolai Rimsky-Korsakov

Dean of White House Press Correspondents Helen Thomas

Dear Abby Abigail Van Buren (Pauline Esther Phillips)

Death Devil Charles Manson

Death to Flying Things Bob Ferguson (baseball player)

Death Ride Charge of the Light Brigade at Balaclava in Crimea

death's head nickname of the deadly mushroom *Amanita muscaria*

Death Valley Scottie Walter Scott

Deb (Debby; Debbie) Deborah; Debra

Deborah Kerr Deborah Kerr-Trimmer

de Broglie waves matter waves named for Louis de Broglie

Debt-reduction President Calvin Coolidge

Decade of Disillusionment the 1970s

Decade of Greed the 1980s

Decca Jessica Mitford (author)

Dede Deirdre

Dee Cee Washington, D.C.

Deep Blue chess-playing IBM computer from Stamford, Connecticut

Deep North Queensland, Australia

Deep South South Carolina, Georgia, Florida, Alabama, Mississippi, Louisiana, and Texas; the conservative south coast of England

Deep Water Moon full moon in April

Defaced City once elegant New York City where so many buildings, buses, and subways are defaced by graffiti

Defender of the Damned Clarence Darrow

Defender of Darwin Thomas Henry Huxley

Defender of Democratic Socialism John Dewey in the United States or Harold Laski in Great Britain

Defender of Freethought Thomas Jefferson, Thomas Paine, Robert Ingersoll, and Clarence Darrow

Defender of the Indians Bartolomé de Las Casas

Defender of Religious Freedom Supreme Court Justice Hugo Lafayette Black

Dejerine's disease infants' interstitial neuritis

Del Delbert

Delaware named for Thomas West, Lord De La Warr (Virginia's first governor)

Delaware Port Wilmington

delft delft blue (characteristic of a popular china developed in the Dutch city of Delft); delft china; delftware

Delhi belly diarrhea suffered in India

Dellionaires fully vested employees of Dell Computer, Austin, Texas

Delta Dagger Convair F-102 single-engine turbojet interceptor aircraft

Delta Dart Convair F-106 supersonic-interceptor aircraft

Demetia South Wales

Demi Moore Demi Guymes

Demon of Deception Beelzebub

Demon of Disease Black Death (bubonic plague); hunger plague; murine plague (carried by rats); pneumonic plague; septicemic plague; sylvatic plague (carried by many species of rodents)

Demon Dog of Literature James Ellroy

Demon of Misfortune and Ruin the Sphinx

The Den Riverfront Stadium, home of the Newark (New Jersey) Bears

Denali old Russian name for Mt McKinley also called Bol'shaya

Den Haag (Dutch—The Hague)—capital of the Netherlands and seat of the International Court of Justice

Denmark's Principal Port Copenhagen

Dental Capital of Europe Vaduz, Liechtenstein

Dentist-Novelist Zane Grey

Den Vita Staden (Finnish— The White City)—Helsinki

Department of Construction Georgia nickname for its Department of Correction

Department-of-Labor President William Howard Taft

Depression-born Cartoonist Al Capp—creator of Li'l Abner

DeQuervain's syndrome tendinitis of the thumb, originally from wringing out wash—now commonly from computer use (*see* Washerwoman's Sprain)

Der Alte Der Alte Fritz (German—Old Fritz)—Frederick the Great

Der alte Steffl Old Saint Stephen's Cathedral in Vienna; begun in the 12th century

Der Blaue Reiter (German— The Blue Rider)—abstract expressionism movement centered in Munich

Derbyville Louisville (home of the Kentucky Derby)

Dercum's disease subcutaneous connective-tissue dystrophy

Der Führer (German—The Leader)—Adolf Hitler

Der Kanonenkönig (German— The Cannon King)— Alfred Krupp

Der Meister (German—The Master)—Johann Wolfgang von Goethe

Derrick City Oil City, Pennsylvania

DeSanctis-Cacchione syndrome skin-pigment abnormality with neurological symptoms

Descendants of Eagles the founders of Algeria, according to tradition

Deseret State Utah (from Mormon colony's name)

Desert Arabia Arabian Desert in the northern Arabian Peninsula

Desert Fox Field Marshal Erwin Rommel

Desert of Ice Antarctica

desert roses barytes or gypsum concretions whose shapes resemble roses

Desi Arnaz Desiderio Alberto Arnaz y de Acha

The Destroyer George Foster

Devil of Cultured Vice Mephistopheles

Devil's Chaplain Robert Taylor (1784–1844), English cleric imprisoned for blasphemy when he exposed the universality of all religious beliefs

Devil's Half Acre Augusta, Maine's slum

Devil's Island nickname for French Guiana penal colony in use up to 1950 and the Isle off its coast where Alfred Dreyfus was imprisoned from 1894 to 1899

Devil's Rock Garden California's Death Valley also called Devil's Bathtub; Devil's Parade Ground; Devil's Pulpit; Devil's Speedway and Golf Course

Devil's Son-In-Law William Bunch (Peetie Wheatstraw)

devil's testicle mandrake's nickname (also called mandragora or satan's apple)

Devil's Triangle (*see* Bermuda Triangle)

devil's trumpet nickname for jimson weed also called devil's apple or devil's weed

Devo Devon White (baseball)

De Witt De Witt Clinton (New York state governor)

De Witt Clinton's Ditch the Erie Canal

diamondback moth; rattlesnake; terrapin

Diamond Continent Africa

Diamond Head 760-foot-high extinct crater forming cape and marking entrance to Honolulu on Oahu, Hawaii

Diamond Jim James Buchanan (Diamond Jim) Brady; Jim Gentile (baseball player)

Diamond Lil Mae West

Diamond State Delaware

Diamond Street New York City's 47th Street between 5th and 6th avenues

Diamond Teeth Mary Mary Smith McClain

Diane Keaton Diane Hall

Diazpotism despotism of Porfirio Diaz during his forty years as president of Mexico

Dice City Las Vegas, Nevada

Dick Donavan Joyce Emmerson Preston Muddock

Dick Tiger Dick Thetu (boxer)

Dicky Sam(s) inhabitant(s) of Liverpool

Dictator of Nicaragua William Walker

Dictionary Johnson Dr Sam(uel) Johnson

Diedrich Knickerbocker Washington Irving

Dief the Chief John George Diefenbaker

Diego Rivera Diego Maria Concepción Juan Nepomuceno Estanislao de La Rivera y Barrientos y Rodriguez (Mexican painter)

diesel diesel engine, diesel fuel, diesel locomotive, diesel oil (all named for the German automotive engineer, Rudolf Diesel)

Digger Land Australia

Diggers Nevadans

digital convergence merging of telephone, television, and computer technologies

Digital Divide minorities' disproportionate lack of access to the Internet

Dinah Washington Ruth Lee Jones

Dino Dean (Crocetti) Martin

Dion Fortune Violet Mary Firth

Diplomat John Franklin Carter

Dirceu Tomaz Antonio Gonzaga

Dirk Bogarde Dirk van den Bogaerd

Dirty Thirties Dust Bowl era in the United States

Discoverer of Bacteria, Blood, and Parasite Cells Anton van Leeuwenhoek

Discoverer of Capillary Blood Vessels Marcello Malpighi

Discoverer of Malaria Parasite Charles Laveran

Discoverer of Yellow Fever Parasite Hideyo Noguchi

Discoverers of How Enzymes Change Starch into Sugar Carl Ferdinand Cori and Gerty Theresa Cori

Disease of the Century AIDS; Alzheimer's disease

dismal science Carlyle's nickname for economics

Dismal Swamp City Norfolk, Virginia

Dissident Publisher Henry Regnery

Distinguished Journalism Fellow Ralph de Toledano (1989)

District of Columbia named for Christopher Columbus

divi divide; dividend

Divine Madman Drukpa Kunley, 15th-century Buddhist saint (Bhutan)

The Divine Miss M Bette Midler

Divine Poet John Donne

divine Sarah Oscar Wilde's nickname for Sarah Bernhardt, who began life as Rosine Bernard

Division No 1 Chicago's Cook County Jail

Division No 2 Chicago's House of Correction

Divorce Capital of America Reno, Nevada

Dixie southern United States; the South

Dixiecrat Southern Democrat

Dixie Kid Aaron L Brown (boxer)

Dizzy Benjamin Disraeli—British Prime Minister

Dizzy Dean Jay Hanner (Dizzy) Dean

Dizzy Gillespie John Birks (Dizzy) Gillespie

Djajapura Kotabaru, formerly called Hollandia by the Dutch when they controlled western New Guinea

Djakarta Indonesian city once called Batavia

Django Jean (Django) Reinhardt

The Django Reinhardt of the Accordion Gus Viseur

Djibouti Ports Obock, Djibouti

Doctor Angelicus (Latin—Angelic Doctor)—Thomas Aquinas also known as the *Princeps Scholasticorum* (Prince of Scholastics)

Doctor Beach Stephen P Leatherman, who ranks the best beaches in the United States

Doctor Charlie Dr Charles Horace Mayo—co-founder of the Mayo Clinic (*see* Doctor Will)

Doctor of Death Henry Kissinger

Doctor Donne John Donne

Doctor Evangelicus (Latin—Evangelical Doctor)—religious reformer John Wickliffe

Doctor Holmes John Haynes Holmes (contemporary American Universalist minister)

Doctor of the Industrial Revolution Dr Erasmus Darwin

Doctor Irrefragabilis Alexander of Hales

Doctor Jameson Sir Leander Starr Jameson

Doctor Johnson Doctor Samuel Johnson—critic, conversationalist, lexicographer

Doctor Livingston David Livingstone

Doctor Mirabilis (Latin— Admirable Doctor)—English savant Roger Bacon

Doctor of Revolution Erasmus Darwin

Doctor Rizal José Rizal (intellectual leader of Philippine insurrection against Spanish rule)

Doctor Seuss author-cartoonist Theodore S Geisel

Doctor Singularis (Latin—Singular Doctor)—William Occam

Doctor Strangelove Dick Stuart (baseball player)

Doctor Subtilis (Latin—Subtle Doctor)—Duns Scotus

Doctor Universalis Albertus Magnus

Doctor Watson Dr John B Watson, M.D. of London; companion of Sherlock Holmes of 221-B Baker Street in the literary creations of Sir Arthur Conan-Doyle

Doctor Will Dr William James Mayo—co-founder with his brother Charles of the Mayo Foundation for Medical Education and Research at Rochester, Minnesota

Documentary Photographer Alfred Stieglitz

Dodo Islands Mascarene Islands of Mauritius, Réunion, and Rodrigues formerly inhabited by dodo birds

dog's nose Mayan salsa

Dogwood City Atlanta, Georgia

Dohnányi Ernst von Dohnányi

Dolf Adolf

Dollar Mark Mark Hanna

Dolley (Dolly) Mrs Dorothea (Dolley) Payne Madison (wife of President James Madison); Dorothea; Dorothy

Dolly a sheep, the first animal cloned from the cell of another adult mammal (named for singer Dolly Parton)

Dolores del Rio Lolita Dolores Asunsolo de Martinez

Dom Dominick

Dom Getulio President Getulio Dornelles Vargas of Brazil

Dominican Ports Pepillo Salcedo, Montecristi, Puerto Plata, Sosua, Santa Barbara de Samana, Sanchez, La Romana, San Pedro de Macoris, Andres, Santo Domingo (formerly Trujillo), Rio Jaina, Bahia de las Calderas, Azua, Barahona

Dominie Hawker-Siddeley HS-125 jet transport

The Don Don Juan (as in Mozart's opera *Don Giovanni*)

Doña Fela Felisa Rincón (female mayor of San Juan, Puerto Rico for twenty-two years)

Dotia Marina Malinche (Indian interpreter-mistress of Hernán Cortés—Spanish conqueror of Mexico)

Donatello Donato di Betto Bardi

Don Emilio General Emilio Aguinaldo (fighter for Philippine independence)

Donets River City Kharkov in Ukraine

Donetzk formerly Stalino, but originally Yuzovka

Don Francisco Francisco I Madero—Mexican president

Don Giovanni Don Juan

Don Juan Don Juan Tenorio of Seville (Mozart's *Don Giovanni*)

Don Muang Bangkok, Thailand's airport

Don Pepe José Figueres Ferrer—democratic leader of Costa Rica

Don Porfirio Don Porfirio Diaz—Mexican dictator-president

Don Quixote pseudonym Alonso Quixano gave himself in *The Adventures of Don Quixote—Man of La Mancha,* by Cervantes

Don Romulo Romulo Betancourt—democratic leader and recent president of Venezuela

Don't Give Up the Ship nickname of Captain James Lawrence, USN

Don Venus Don Venustiano Carranza—Mexican general-president

Don Venustiano Don Venustiano Carranza—former president of Mexico

Don Vitone Vito Genovesa (1897–1969)

Doornik Flemish place-name equivalent for Tournai

Doorstep to Canada Nova Scotia

Dope Capital of Canada Vancouver

Doris Day Doris Kappelhoff

Dornford Yates Cecil William Mercer's pseudonym

Dorothy Dix Elizabeth M Gilmer

Dorothy Gish Dorothy de Guiche

Dorothy Lamour Dorothy Kaumeyer

Dorothy Malone Dorothy Maloney

Dorothy Parker Dorothy Rothchild

Dosso Dossi palette name of Giovanni de Lutero

Double Platinum two million (recording) copies

Double-Vay Sir Henry Wilson's nickname among the French general staff of World War I

Double X Jimmie Foxx (baseball player)

Doubting Thomas St Thomas, one of the twelve apostles of Jesus Christ, who required proof of Jesus' resurrection

Douglas Fairbanks Julius Ullman

Dove Hawker-Siddeley twin engine light transport carrying up to 11 passengers

Dow Jones Industrial Average started by Charles Dow in 1884, originally an average of twelve industrial companies' closing stock prices divided by the number of issues

Down East Atlantic coast area extending from New York to Nova Scotia, particularly coastal New England; Maine

Down Easter person from east coast of New England or Nova Scotia

downers nickname for sedative drugs also called sleeping pills or tranquilizers

Down South nickname shared by the federal penitentiay at Atlanta, Georgia and the southern United States

Down syndrome mongolism resulting from extra chromosome-21 material

Down Under Australia and New Zealand

Down Where the South Begins Virginia

Down Yonder coastal North Carolina

Downtown Fred Brown (basketball player)

Doyen of European Diplomacy Prince Klemens Wenzel Nepomuk Lothar von Metternich

Doyen of Film Animators Walt Disney

Doyen of Professional Translators Ralph Manheim

Dragonfly Cessna T-37 jet-trainer aircraft

Dragon Nation Bhutan

Dragon's Mouth Port of Spain, Trinidad's harbor entrance

Draken (Swedish—Dragon)—Saab double-delta-wing supersonic fighter or fighter-bomber designated J-35 or S-35

Drapier Jonathan Swift

Dramatic Symphony *Romeo et Juliette* by Hector Berlioz

Dr Death Jack Kevorkian

The Dream Dean Memminger (basketball player)

Dream Capital of the Western World Hollywood, California

Dream King Ludwig II of Bavaria

Dress-Rehearsal Revolution Russian Revolution of 1905

Drisheen City Cork, Ireland

Dr Jinnah Mohammed'Ali Jinnah-president of All-India Moslem League and first governor-general of Pakistan

Dr Karl Dr Karl Augustus Menninger

Droch Robert Bridges

Droll Breughel Pieter Breughel the Elder

Dr Rocket Hideo Itokawa

Dr Salazar Antonio de Oliveira Salazar—dictator and prime minister of Portugal from 1932 to 1969

Dr Seuss Theodor Seuss Geisel

Dr Spot Marty Gurian, stain-removal expert

Dr X Alan E Nourse

druggie drug addict

drugola bribes given law-enforcement officers by narcotics dealers in exchange for protection from discovery and prosecution

Druid City oak-tree-filled Tuscaloosa, Alabama

Dry Guillotine French Guiana's nickname

dry Spain Mediterranean coast of Spain

D-town Dogtown; Downtown; Doyelstown

Dual Cities Minneapolis and Saint Paul, Minnesota

Dual Monarchy Austro-Hungarian Monarchy (1867–1918)

Dual Protectorate Andorra under the protection of France and Spain

Dubini's disease rapid and rhythmic muscular contraction

Du Bois W(illiam) E(dward) B(urghardt) Du Bois

Duca Minimo Gabriele D'Anunzio

Ducansby Head northernmost headland in the British isles

Duce (Italian-Leader)—Dictator Benito Mussolini

Duchenne de Boulogne Guillaume-Benjamin-Amand Duchenne—father of modern neurology

Duchess of Dupont Circle Alice Roosevelt Longworth

Duchess of Windsor Bessie Wallis Warfield

Ducky Joe Medwick

Dude Ranch Capital Wickenburg, Arizona

Dugout Doug General Douglas MacArthur

Duke George Deukmejian (former California governor); Michael S Dukakis (former Massachusetts governor and Presidential candidate)

The Duke John Wayne

Duke of the Abruzzi Italian alpinist and arctic explorer Prince Luigi Amadeo Giuseppe Maria Ferdinando Francesco

Duke of Alba Fernando Alvarez de Toledo

Duke of Buckingham George Villiers

Duke City Albuquerque, New Mexico (named for El Duque de Alburquerque)

Duke of Devonshire William Cavendish (former Prime Minister of Great Britain)

Duke Ellington Edward Kennedy Ellington

Duke of Grafton Augustus Henry Fitzroy

Duke of Newcastle Thomas Pelham-Holles

Duke of Portland William Henry Cavendish Bentinck

Duke of Shrewsbury Charles Talbot

Duke Turned Jesuit Saint Francis Borgia (1510–1572)

Duke of Vicenza Marquis Louis de Caulaincourt

Duke of Wellington Arthur Wellesley

Duke of Windsor (formerly King Edward VIII, formerly Prince of Wales when his father, George V, was king of England)

Duluthians people of Duluth

Dumas fils Alexandre Dumas (1824–1895) playwright-creator of Camille (Verdi's *La Traviata*), son of the novelists

Dumas père Alexandre Dumas (1802–1870), author of *The Count of Monte Cristo, The Three Musketeers*

Du Maurier George Louis Palmella Busson

Dumb Girl of Portici Esprit Auber opera *La Muette de Portici*

Dumbo Down Under the Manhattan Bridge Overpass

dumb tax contractor term for what people pay when they try to build a house without a contractor

Dunedin nickname for Edinburgh, Scotland and place-name for a South Island, New Zealand port as well as a Florida resort near St Petersburg

Dungeness Crab Capital Newport, Oregon

Dun Laoghaire (Gaelic—Dunleary)—Kingstown on Dublin Bay

Dunleary Dun Laoghaire or Kingstown

Dunnet Head northernmost point on Scotland's mainland

Dunsany Edward John Moreton Drax Plunkett, Lord Dunsany

Dupontonia Wilmington, Delaware

Dupont Town Wilmington, Delaware (home of EI du Pont de Nemours & Co)

Durable Dictator General José de la Cruz Porfirio Díaz (Mexico, 1876–1911)

Durazzo English and Italian place-name for the Albanian port of Durrës

Durban formerly Port Natal, South Africa

Dusty Springfield Mary Isabel Catherine Bernadette O'Brien

Dutch Ronald Reagan, fortieth president of the United States

dutch dutch belted (black dairy cattle with a broad body-encircling white belt of hair as originally bred in the Netherlands); dutch door (horizontally divided so either the top or bottom section may be closed or opened); dutch courage (inspired by alcohol); dutch lunch (cold cuts); dutch treat (where all pay their own way)

dutch act suicide

Dutch Caribees colonial name for the Netherlands Antilles

Dutch City Holland, Michigan

Dutch courage Holland gin

Dutch Cradle of U.S. Presidents the Netherlands— ancestral home of both Presidents named Roosevelt and President Van Buren

Dutch Delight Delft

Dutch East Indies Netherlands East Indies now known as Indonesia

Dutch Guiana Netherlands Guiana or Surinam

Dutch Islands Dutch West Indies—Aruba, Bonaire, Curaçao, Saba, Sint Eustatius (Statia), Sint Maarten

Dutch Masterpiece in the Caribbean Curaçao

Dutch Microscopists Zacharias Janssen, Anton van Leeuwenhoek, and Jan Swammerdam

Dutch New Guinea West Irian, Indonesia

Dutch Ports Delfzigl, Harlingen, Den Helder, Ijmuiden, Zaandam, Amsterdam, Scheveningen, Hoek van Holland, Europoort, Maasluis. Vandelingenplaat, Vlaardingen, Schiedam, Rotterdam, Dordrecht, Middelharnis, Willemstad, Middelburg, Vlissingen, Terneuzen, Hansweert, Haven Catzand

Dutch Queens Wilhelmina, Juliana, Beatrix

Dutch Reformed Dutch Reformed Church of North America where it has been called the Reformed Church since 1867

Dutch Republic the Netherlands, sometimes called Holland

Dutch Schultz Arthur Flegenheimer (gangster)

Dutch-speaking Places Flemish sections of Belgium; the Netherlands; Aruba, Bonaire, Curaçao, Saba, Sint Eustatius, Sint Maarten—the Netherlands Antilles; Netherlands New Guinea now part of Indonesia Surinam and the communmty around Holland, Michigan

dutch treat each person pays own portion of the bill

Dutch Ultramodernist Piet Mondrian

Dutch West Indies the Netherlands Antilles

Dutch William William III of Orange—Dutch-born British king

Dvořák Anton Dvořák

dweeb small, harmless person

Dwellers of the Field the Poles

Dyan Cannon Samile Diane Friesen

dyke lesbian

Dying Grass Moon full moon in September

dyno dynamic move

Eagle (*see* Columbia)

Eagle Eye Jake Beckley (baseball player)

Eagle Forgotten Governor Peter Altgeld of Illinois

Eagle of the North Swedish statesman Count Axel Oxenstierna

Eagle Pass formerly El Paso del Aguila

Eagle and Serpent *Aguila y Serpiente* (Mexican coat of arms contains pictorialization of Aztec legend stating their people could not settle until they found an island on a lake and on that island a cactus surmounted by an eagle grasping a serpent—the lacustrine island representing Mexico City, capital of Mexico—and the two creatures that struggle between celestial and earthly elements)

Eagle State Mississippi

Eam Eamon de Valera (Irish rebel)

Earl of Aberdeen George Hamilton Gordon (former Prime Minister of Great Britain)

Earl Baldwin of Bewdley Stanley Baldwin

Earl Balfour Arthur James Balfour

Earl of Beaconsfield Benjamin Disraeli (19th-century British prime minister who declared: *A conservative government is an organized hypocrisy*)

Earl of Bute John Stuart

Earl of Carlisle Charles Howard

Earl of Carnarvon George ESM Herbert, the Egyptologist

Earl of Chatham William Pitt

Earl of Chesterfield Philip Dormer Stanhope

Earl of Derby Edward Stanley

Earl of Godolphin Sidney Godolphin

Earl Grey Charles Grey

Earl of Guilford Frederick North

Earl of Halifax Charles Montagu

Earl of Liverpool Robert Banks Jenkinson

Earl of Lytton Edward Robert Bulwer Lytton, diplomat and poet whose pseudonym was Owen Meredith

Earl of Orford Robert Walpole

Earl of Oxford Robert Harley

Earl of Oxford and Asquith Herbert Henry Asquith

Earl the Pearl Earl Monroe (basketball player)

Earl of Ripon Frederick John Robinson

Earl of Rosebery Archibald Philip Primrose

Earl Russell Prime Minister John Russell, first earl; John Stanley Russell, second earl; Bertrand (Arthur William) Russell, third earl

Earl of Shaftesbury Anthony Ashley Cooper

Earl of Shelburne William Petty

Earl of Stanhope James Stanhope

Earl of Sunderland Charles Spencer

Earl of Wilmington Spencer Compton

Earmuff Capital of the World Farmington, Maine

Earth's Last Frontier Antarctica

East End congested and depressed eastern section of London

easter storm from the east

Easter Easter lily; Easter Monday; Easter Sunday; Easter vacation

Easter Island English place-name equivalent of Isla de Pascua whose Chilean settlers are called Pascuenses

Eastern Desert Arabian Desert

Eastern Empire Byzantine Empire

Eastern Hemisphere half of the world containing Africa, Asia, Australia, Europe, and associated islands

Eastern Malaysia Sabah and Sarawak

Eastern Samoa American Samoa

Eastern Sea East China Sea

Eastern Shore shore of Delaware, Maryland, and Virginia comprising the Del-Mar-Va Peninsula

Eastern States states east of the Mississippi

East Germany one-time Soviet-dominated eastern Germany, German Democratic Republic (1945–1990)

East Indies India, Indochina, Indonesia, Malay Peninsula—formerly Dutch East Indies

Eastinghouse international nickname for Soviet nuclear power plants built in Finland

East London formerly Port Rex, South Africa

East Lothian Haddington

East Malaysia Bandar Seri, Brunei, Sabah, and Sarawak now known as Kalimantan

East North Central States Indiana, Illinois, Michigan, Ohio, and Wisconsin

East Prussia old name for western Poland along the Baltic

East Rudolf Lake Turkana

East Siberian East Siberian Sea in the Arctic off East Siberia

East Side East Side of any city or place

East South Central States Alabama, Kentucky, Mississippi, and Tennessee

Eastview Canadian city now called Vanier

East Village modern euphemism for New York City's Lower East Side

Ebreo (Italian—Jew)—nickname of Salomone Rossi the composer-violinist of Mantua in the late 1500s and early 1600s

'Ebrides Cockney contraction—Hebrides

E-C Erckmann-Chatrian (Emile Erckmann and Louis-Alexandre Chatrian, collaborating novelists)

Eccentric Naturalist Constantine Rafinesque

Eckma Jan de Hartog's pseudonym

ECR Lorac Edith Caroline Rivett's pseudonym

Ecuadorean Ports Puerto de San Lorenzo, Esmeraldas, Bahia de Caraquez, Bahia de Manta, Puerto de Cayo, La Liberiad, Salinas, Guayaquil, Puna, Puerto Bolívar, Bahia Baquerizo Moreno (on Chatham or San Cristóbal)

Ecuadorian Archipelago Galápagos Islands

Ecuador's Principal Port Guayaquil

ED Emily Dickinson

Eddie Albert Eddie Albert Heimberger

Eddie Cantor Edward Israel Itskowitz

Eddie Dean Edgar Gloslup

Eden of the Orient Thailand

Edgar Box Gore Vidal

Edgar Thorn(e) Edward MacDowell

Edie Adams Elizabeth Edith Enke

Edin(a) Edinburgh's poetical name

Edinburghshire Midlothian

Edinglassie Edinburgh + Glasgow (early name of the Moreton Bay Settlement now called Brisbane)

The Edith Piaf of Cuba Omara Portuondo

Editor of Genius Max(well) E Perkins

Ed Lacy Len Zinberg

Ed McBain Salvatore A Lombino

Edmond Adam French author-editor Juliette Lamber

Edmund Crispin Robert Bruce Montgomery

Edo old name for Tokyo, also written Yedo

Edogawa Rampo Hirai Taro (Japan's Edgar Allan Poe)

Edoo (EDU) Lady Elgar's nickname for her husband Sir Edward Elgar (EDU is the title of the fourteenth section or finale of his *Enigma Variations on an Original Theme* scored for full orchestra)

Edouard Colonne Judas Colonne

Edu Sir Edward Elgar

Eduardo E Howard Hunt

Educator-Freethinker Horace Mann

Edward G Robinson Emmanuel Goldenberg

Edwardian Radical Hilaire Belloc

Edward I Prime Stevenson Xavier Mayne's pseudonym

Edward Longshanks Edward I of England

Edward O Wilson Frank J Baird, Jr

Edward the Peacemaker Edward VII, eldest son of Queen Victoria

Edward the Rake Edward VII

Edwards syndrome extra chromosome 18 material resulting in elongated skull, ears set low, recessed chin and premature death (named for John H Edwards)

Edwin Markham Charles Edward Anson Markham

Ed Wynn Isaiah Edwin Leopold

egabrag garbage spelled backwards, sometimes used as a euphemism for garbage boat or garbage scow

Egg Basket of California Petaluma

Egg Moon full moon in April

Egypt Illinois

Egyptian Badlands Assiut area about 175 miles (282 kilometers) south of Cairo and notorious for narcotic raids, religious clashes, and village vendettas

The Egyptian James Brown Ali Hassan Kuban

Egyptian Ports (on the Mediterranean) El Iskandariya (Alexandria); Bur Said (Port Said); (on the Red Sea) El Suweis (Suez)

Egyptians Illinoisans

Egypt's Principal Port Alexandria

Eiffel Alexander Gustave Eiffel and the tower bearing his name

Eight The Eight (Ashcan School of American Art comprising Arthur B Davies, William Glackens, Robert Henri, Ernest Lawson, George Luks, Maurice Prendergast, Everett Shinn, and John Sloan)

Eighteenth State Louisiana

Eightfold Path right speech; right conduct; right livelihood; right effort; right mindfulness; right concentration; right views; right intentions (Buddhism)

Eighth State South Carolina

Eighth Wonder of the World compound interest, according to Baron de Rothschild; the Panama Canal; Grand Coulee Dam, Grand Coulee, Washington; Shirley Temple

Eisenhower Doctrine contain the Soviet Union in Europe and the Middle East; for Dwight D Eisenhower, thirty-fourth president of the United States

Either/Or Søren Kierkegaard's nickname

Ekaterinburg czarist name for Sverdlovsk

El Alto (Spanish—The Tall One)—La Paz, Bolivia's airport serving the world's highest capital city

El Bosco Spanish nickname for Hieronymus Bosch

El Caballo (Spanish—The Horse)—nickname of Cuba's Communist dictator Fidel Castro Ruz

El Cabrón (Spanish—The Goat)—nickname of dissolute Dominican dictator Generalissimo Rafael Leonidas Trujillo Molino

El Caudillo (Spanish—The Chief)—sobriquet of General Francisco Franco-Bahamonde

El Cid El Cid Campeador (Spanish—The Lord Champion)—Rodrigo Díaz de Bivar

El Coco (Spanish—Coconut Palm)—San Jose, Costa Rica's airport

El Coronelazo (Spanish—the big colonel) David Siquieros

El Duque Orlando Hernandez (baseball player)

Elder Pitt William Pitt the Earl of Chatham also called the Great Commoner

Elder Statesman President Dwight David Eisenhower

Eldest Daughter of the Church France

El Discutido (Spanish—the discussed)—Diego Rivera

El Dorado the mythical land of gold sought by the Spaniards and others in the jungles and mountains of the Americas

El Dorado State California

electric cure electrocution

Electric Motor City Detroit

electromagnetic smog signals from mobile phones, microwave ovens, cable-television channels, et cetera that interrupt radio data to telescope antennae

Elegant Arthur nickname for Chester Alan Arthur, twenty-first president of the United States

Elephant Man disease neurofibromatosis

el esmog (Spanish—the smog) Mexico City

El Español (Spanish—The Spaniard)—Giuseppe Maria Crespi—Italian painter's nickname

El Españoleto Spanish painter José Ribera

Eleventh State New York

El Fatah disease virulent antisemitism

El Fenix de España The Phoenix of Spain—Lope de Vega

El Fondo (Spanish—The Fund)—International Monetary Fund—IMF

El Gran Libertador (Spanish— The Great Liberator)— Simón Bolívar—liberated Venezuela, Colombia, Ecuador, Peru, and Bolivia from Spanish rule

El Greco (Spanish—The Greek)—Kryiakos Theotokopoulos (Domingo Theotocopuli)

El Hombre (Spanish—The Man)—nickname of Dr Arnulfo Arias de Madrid of Panama

Elia Charles Lamb

Eli Edwards Claude McKay

Elijah Muhammad Robert Poole

El Ilustre Americano (Spanish—The Illustrious American)—self-title of Venezuelan dictator Antonio Guzman Blanco

El Inca (Spanish—The Inca)—Peruvian historian Garcilaso de la Vega

Elisabethville former name of Lubumbashi, Zaire

Elizabeth Arden Florence N Graham

El Juli Julian Lopez, who became a matador in Spain at age 16

Ellen Glasgow Ellen Anderson Gholson

Ellery Queen Frederic Dannay and Manfred B Lee

El Libertador (Spanish—The Liberator)—Simón Bolivar

El Licenciado Tomé de Burguillos (Spanish—Attorney Tomé de Burguillos)—a pseudonym of Lope de Vega

Ellis Bell pseudonym of Emily Brontë

El Maestro (Spanish—the master)—José Clemente Orozco

El Manco de Lepanto (Spanish—The One-handed Man of Lepanto)—Cervantes, whose left hand was maimed at the Battle of Lepanto

El Mar del Sur (Spanish— The South Sea)—Balboa's name for the Pacific Ocean

Elm City New Haven, Connecticut

elm death Dutch elm disease

Elmer short form for *Elmer Gantry* (novel by Sinclair Lewis about a corrupt preacher)

Elmers townspeople (circus slang)

Elmo Lincoln Otto Elmo Linkenhelt

El Mudo (Spanish—the mute)—Juan Fernández Navêrette, Renaissance painter

Elmwood James Russell Lowell's home in Cambridge, Massachusetts

El Niño (Spanish—the boy)— warming of surface water off Ecuador and Peru before extending westward over the tropical eastern Pacific Ocean

El Norte (Spanish—the north)—used by Latin Americans to mean the United States

Eloquent President Abraham Lincoln

El Pisshole trucker's nickname for El Paso, Texas

El Pisso trucker's nickname for El Paso, Texas

El Precursor (Spanish—the Precursor)—Francisco Miranda—fighter for Venezuelan freedom; Antonio Nariño—fighter for Colombian freedom

El Pueblo Nuestra Señora la Reina de los Angeles de Porciúncula (Spanish—The Village of Our Lady Queen of the Angels of Porciúncula)— early Spanish name for Los Angeles

Elroy American country-boy name derived from the French for king—*Le Roi* or the Spanish equivalent—El Rey—or their combination

Elsa Lanchester Elizabeth Sullivan

El Salvador's Ports La Unión, Puerto El Triunfo, La Libertad, Acajutla

El Silencio (Spanish—The Silence)—downtown Caracas where bus routes start and automotive traffic is at its noisiest

El Smoggo trucker's nickname for any smog-smitten city in the Southwest from El Paso to Los Angeles

El Stinko (*see* El Smoggo)

El Supremo (Spanish—The Supreme)—Juan Vicente Gómez, Venezuelan dictator from 1908 to 1935; also known as *El Gran Supremo*

Elton John Reginald Dwight

El Viejo (Spanish—the old man) when average ocean-water temperature cools by at least 1 degree centigrade (1.8 degrees Fahrenheit) below normal

Elvis Costello Declan Patrick McManus

Elysian Fields mythological place thought by the Portuguese to be the Madeira Islands or by the Spaniards to be the Canary Islands—also known as Isles of the Blest

Emancipator of the Serfs Czar Alexander II of Russia

Emancipator of the Slaves William Wilberforce

Emanuel Swedenborg Emanuel Svedberg

Embodiment of Chilean Culture Pablo Neruda

emerald beryllium chromium aluminum silicate (gemstone variety of beryl)

Emerald Capital Bogotá, Colombia

Emerald of the Caribbean Guadeloupe Island, French West Indies

Emerald City mythical capital of Oz as described in *The Wizard of Oz*; Seattle, Washington

Emerald Country Colombia

Emerald Empire Idaho's panhandle

Emerald Isle Ireland

Emerald Necklace 18,000 acres of parks surrounding Cleveland

emerald nickel zaratite (basic hydrated nickel carbonate)

Emerald of the Spanish Main Colombia

emery Aluminum oxide (Al_2O_3)

Émile Emile Zola (French novelist)

Emil Jannings Theodore Friderich Emil Janez

Emil Ludwig Emil Cohn

Eminent Humanist Julian Huxley

Emma Emma Goldman (Russian-American anarchist)

Emma Calve Rosa Calvet

Emma Lathen pseudonym of Mary J Latsis and Martha Hennissart

Emmet Street Brendan Behan's pen name

Emmy award for outstanding television performances in the United States; statuette named after tv entertainer Faye Emerson

Emmy Destinn Ema Kittl

Emperor of the Cossacks Emelyan Ivanovich Pugachev (1773–1774)

Emperor of Europe Napoleon Bonaparte's self-imposed title

Emperor of Japan Akihito (1989–), Hirohito (1936–1989)

Emperor of Manchukuo Henry Pu-Yi (1934–1945)

Emperor of Mexico native revolutionary Augustín de Iturbide (1822–1823)

Emperor Philosopher Marcus Aurelius

Emperor of Tenors Enrico Caruso

Emperor of the West Charles the Great

emperors imperial potentates; largest of the Alaskan geese, Antarctic penguins, Central American boas, European moths, Japanese fishes

Empire Empire State Building of New York City at Fifth Avenue and 34th Street

Empire Builder of British South Africa Cecil Rhodes

Empire City New York; Wellington, New Zealand

Empire State New York

Empire State of the South Georgia

Emporium of the West Indies Charlotte Amalie, St Thomas, Virgin Islands; Willemstad, Curaçao

Empress of the Blues Bessie Smith

Empress Carlota Marie Charlotte Amélie Augustine Victoire Clémentine Léopoldine—empress of Mexico under Maximilian

Empress Eugenie Eugénie Marie de Montijo de Guzman—empress of the French under Napoleon III

Empress of Hollywood Bette Davis

Empress of India Queen Victoria

Empress of Vice Mary Jeffries (who in the 1800s controlled London's most elegant brothels)

Empty Quarter Empty Quarter desert of Arabia

Enchanted Isles the Galápagos or Tortoise Tslands originally called *Las Islas Encantadas* by early Spanish explorers

end of the world Ushuala, Argentina

Energy City Houston, Texas

Engineer of the Animal World the busy beaver

Engineer of Fantasy Walt Disney

Engineer President Herbert Hoover

Engineers' Town Coulee City, Washington

England's Wooden Walls the Royal Navy during the Napoleonic wars

Englebert Humperdinck Arnold Dorsey (who took the name of Wagner's protégé)

English Agnostics Thomas Henry Huxley, Charles Darwin, Herbert Spencer

English Alexander nickname of King Henry V

English Atheists Charles Bradlaugh, Bertrand Russell

English Bach Johann Christian Bach (who lived in London from 1759 to 1852)

English Caribees colonial name for the British West Indies

English Channel La Manche

English Cradle of U.S. Presidents England—ancestral home of Presidents Adams, Carter, Cleveland, Coolidge, Fillmore, Ford, Garfield, Grant, Harding, Harrison, Johnson, Lincoln, Madison, Pierce, Taft, Taylor, Tyler, and Washington

English Nonsense Poet Edward Lear

English Operetta Composer Sir Arthur S Sullivan

English Opium Eater Thomas De Quincey

English penicillin a nice cuppa' tea

English Polynesia jocular nickname for the Seychelles Islands in the Indian Ocean

English Visionary William Blake

Eniwetok Kili Atoll

Enlightened Philanthropist Andrew Carnegie

Enlightenment (*see* The Enlightenment)

The Enlightenment Europe's 18th century when encyclopedias appeared in France and England; when Voltaire and Lavoisier were matched across the Channel by Paine and Priestley

Enlightenment Centuries the 17th and 18th centuries (1600s and 1700s) when in America and Europe human reason prevailed in educational, political. and religious doctrine

Entac Nord wire-guided anti-tank missile made in France

Entendard Dassault single-engine jet attack aircraft made in France

E O'N Eugene O'Neill

Epic Poet of Ancient Greece Homer

epsom epsom salt(s)—named for the English racetrack town of Epsom Downs

epsom salt magnesium sulfate ($MgSO_4 \cdot 7H_2O$)

Equality State Wyoming

Equatorial Guinea's Main Port Bata

Era of Good Feeling administration of James Monroe—fifth President of the United States

Era of Mediocrity post-World War II era

Erasmus Desiderius Erasmus (originally Geert Geerts or Gerard Gerardzoon)

Ercoli Palmiro Togliatti

Eric Evergood King Eric I of Denmark

Erich Erich Fromm

Erich Maria Remarque Erich Maria Kramer

Eric the Lamb King Eric III of Denmark

Eric the Memorable King Eric II of Denmark

Eric the Red Eric Thorvaldsson

Eric Rohmer Maurice Scherer

Erich von Stroheim Hans Erich Maria Stroheim von Nordenwall

Erie Canal New York State Barge Canal

Erin Ireland

Ernest Bramah Ernest Bramah Smith's pseudonym

Ernie Ernald; Ernest

Ernst von Dohnányi Ernö Dohnanyi

Eros Center sex supermarket nickname in Hamburg and Bonn, Germany

Erudite Americans Will and Ariel Durant

Escape King Harry Houdini

Eskimo Village Kotzebue, Alaska

Esperanto pseudonym of Dr LL Zemenhoff—inventor of Esperanto—an artificially contrived universal language

ESP Pioneer Dr JB Rhine who coined the term extrasensory perception, ESP

Estado Libre Asociado de Puerto Rico (Spanish—Free Associated State of Puerto Rico)—Commonwealth of Puerto Rico

e-tailers electronic retailers

Etcher of Disaster Francisco de Goya y Lucientes

Etcher of Prisons Giambattista Piranesi

Eternal City Rome

Ethel Barrymore Ethel Blythe

Ethel Leginska Ethel Liggins

Ethel Merman Ethel Zimmermann

Ethelred the Unready Ethelred II of England

Ethical Culturist Felix Adler

Ethiopian Ports Massawa, Port Smyth, Assab

E-town Ellis Town; English-town

Eudora electronic-mail program named for Mississippi short-story writer Eudora Welty (inspired by her short story, *Why I Live at the P.O.*)

Eugenie Marlitt Eugenie John's pseudonym

Eulenberg's disease congenital muscular spasms

Eumenides the Furies

European Community Belgium, Britain, Denmark, France, Greece, Ireland, Italy, Luxembourg, Germany, Netherlands, Portugal, Spain

European Twelve (arranged by population) Germany, Italy, Britain, France, Spain, Netherlands, Portugal, Greece, Belgium, Denmark, Ireland, Luxembourg

Europe's Cyber City Amsterdam, Netherlands

Eusebius (*see* Florestan and Eusebius)

Ev Everett; Evlin

Evangel of Liberty Thomas Paine

Eve Arden Eunice Quedens

Everest Sir George Everest, who conducted a trigonometrical survey of India, and the great mountain bearing his name

Everest of Oil Paintings the *Mona Lisa*

Everglades State Florida

evergreen a journalistic story that reappears predictably

Evergreen City Sheboygen, Wisconsin

Evergreen State Washington

Eve's curse menstruation

Evil Empire USSR

Evil Florist Charles Pierre Baudelaire—famous for his *Les Fleurs du Mal* (Flowers of Evil)

Excelsiors New Yorkers

Excelsior State New York

Executive City Washington, D.C.

Exocet Aerospatiale surface-to-surface missile for use aboard warships

Expounders of Utilitarianism Jeremy Bentham, James Mill, and John Stuart Mill

Exxon (formerly ESSO) Standard Oil

"Eye" I Street in Washington, D.C.

Eye of the Baltic Gotland

Eye of England London

Eye of Greece Athens

Eye into Europe St Petersburg more recently known as Petersburg, Petrograd, or Leningrad

Eye of Italy Rome

Eye of Paris Brassai (photographer Gyula Halasz)

eye in the sky helicopter serving police

Eyetie(s) British slang for Italian(s)

Fabien Sevitzky Fabien Koussevitzky

Fabulous Philadelphians the Philadelphia Orchestra

Fair City Perth, Scotland

Fair Deal administration of Harry S Truman—thirty-third President of the United States

Fairview Fannie The Seattle Times

fairy a male homosexual

Fairy of Dreams Queen Mab

Fairytale Land Denmark home of Hans Christian Andersen

Faithful City Worcester, Massachusetts

Falcon Dassault twin-engine executive transport made in France and called Mystere 20

falcons of the sea clipper ships

Falks Falklanders; Falkland Islands

Falls City Louisville, Kentucky

Falls of the Rhine Rheinfall or Schaffhausen

false topaz citrine (quartz with ferric iron)

Famous Potatoes State Idaho

Fanguito (Spanish—Little Muddy)—San Juan, Puerto Rico slum

Fannie Mae Federal National Mortgage Association (FNMA)

Fanny Brice Fanny Borach

Far-Flung Hubbell nature writer Sue Hubbell

Farinelli Carlo Broschi

Farmer President William Henry Harrison and George Washington

Farmer's Bible Sears, Roebuck catalog

Far North northernmost Alaska, Canada, Greenland, Norway, Sweden, Finland, and Russia

Farther India Indochina; Indochinese Peninsula

Farting Bedpost bassoon

Far West the Rocky Mountain States

Fashoda former name of Kodok, Sudan

Fastback Plus trademark of Fifth Generation Systems of Baton

Fate American nickname for Lafayette

Father of Abolition Samuel Hopkins

Father of the A-bomb J Robert Oppenheimer

Father Abraham Abraham Lincoln

Father of Air Conditioning WH Carrier

Father of Algebra Diophantus of Alexandria

Father of All Foods alfalfa

Father of Alligator Wrestling Ed Froelich of Rouge, Louisiana

Father of All Yankees Benjamin Franklin

Father of America Sam(uel) Adams

Father of American Anarchy Josiah Warren of Cincinnati, Ohio who in 1827 advocated government activities be transferred to private citizens

Father of American Anthropology Lewis Henry Morgan

Father of the American Ballet George Balanchine

Father of American Baptists John Clarke

Father of American Botany John Bartram

Father of American Boxing William Muldoon

Father of American Church Music Lowell Mason

Father of American Conchology Thomas Say

Father of American Conservatism John Jay

Father of American Dance Ted Shawn

Father of American Egyptology James Henry Breasted

Father of American Football Walter Camp

Father of American Freethought Thomas Paine

Father of American Geography Jedidiah Morse

Father of American Geology William Maclure

Father of American History George Bancroft, William Bradford

Father of American Horticulture Peter Henderson

Father of American Independence John Adams

Father of American Lexicography Noah Webster

Father of American Literature Washington Irving

Father of American Medical Association Dr Nathan Smith Davis

Father of American Medical Botany Jacob Bigelow

Father of American Mineralogy Parker Cleaveland

Father of American Mule Breeding George Washington

Father of American Naval Architecture William A Webb

Father of American Navigation Nathaniel Bowditch

Father of the American Nuclear Navy Admiral Hyman G Rickover

Father of American Nutrition W(ilbur)—O(lin) Atwater

Father of American Oceanography Matthew Fontaine Maury

Father of American Orchestral Music Johann Christian Gottlieb Graupner

Father of American Ornithology John James Audubon, Alexander Wilson

Father of American Paleontology O(thniel) C(harles) Marsh

Father of American Photography S(amuel) F(inley) B(reese) Morse

Father of American Photo-Journalism Matthew Brady

Father of American Poetry Philip Freneau

Father of American Poets William Cullen Bryant

Father of American Pragmatism Charles Sanders Pierce

Father of American Prison Reform George O Osborne

Father of American Psychiatry Dr Benjamin Rush

Father of American Psychobiology Adolf Meyer

Father of American Psychology William James

Father of American Railroads Peter Cooper

Father of the American Revolution Sam(uel) Adams

Father of American Rocketry Robert H Goddard

Father of American Surgery Dr William Halsted or Dr Philip Syng Physick

Father of American Technology Eli Whitney

Father of the American Turf Leonard Jerome

Father of American Universalism Hosea Ballou, John Murray

Father of American Zoology Thomas Say

Father of Anatomical Dissection Andreas Vesalius

Father of Andean Archeology Max Uhle

Father of Anesthetic Dentistry William T G Morton

Father of Angling Izaak Walton

Father of Annapolis George Bancroft

Father of Antarctic Whaling Captain Carl A Larsen

Father of Antiseptic Obstetrics Ignaz Semmelweis

Father of Antiseptic Surgery Sir Joseph Lister

Father of Argentina's School System Domingo Faustino Sarmiento

Father of Argentine Independence General José de San Martín

Father of the Atomic Age Enrico Fermi

Father of the Atomic Submarine Admiral Hyman Rickover, USN

Father of the Automobile Gottlieb Daimler

Father of Bacteriology Robert Koch

Father of Baseball Henry Chadwick, Alexander Cartwright, Abner Doubleday

Father of Basic Flying John Joseph Montgomery

Father of Basketball James Naismith

Father of Bebop Dizzy Gillespie

Father of Belgian Opera Andre Ernest Modeste Gretry

Father of Believers Mohammed

Father of the Bill of Rights James Madison

Father of Bird Behavior Studies Arthur Cleveland Bent

Father of Black Baseball Rube Foster (Andrew Foster)

Father of Black History Chester G Woodson

Father of Blood Banks and Blood Plasma Charles R Rich

Father of the Blues W(illiam) C(hristopher) Handy

Father of Botany Aristotle

Father of Brazilian Opera Antonio Carlos Gomes

Father of British Archeology Augustus Henry Lane-Fox Pitt-Rivers

Father of British Boxing Jack Broughton

Father of British Musicology Sir Charles Grove

Father of the British Navy Alfred the Great

Father of British Printing William Caxton

Father of British Unitarianism John Biddle

Father of Buffalo Joseph Ellicot

Father of Camouflage Abbott Henderson Thayer

Father of Canadian Confederation Sir John A Macdonald and George Brown of Ontario, Sir George S Etienne Cartier and Sir Alexander Galt of Quebec, Sir Charles Tupper of Nova Scotia, Sir Samuel Leonard Tilley of New Brunswick

Father of Chemistry Robert Boyle

Father of Chemurgy George Washington Carver

Father of Chicago John Kinzie

Father of Child Psychology Leo Kanner

Father of the Chinese Revolution Dr Sun Yat-sen

Father Christmas Santa Claus; Snow King

Father of Church History Eusebius

Father of Civilian Atomic Power Admiral Hyman Rickover

Father of the Civil Rights Movement Martin Luther King, Jr

Father of Civil Service Reform George Hunt Pendleton

Father of Classical Music Franz Joseph Haydn

Father of Comedy Aristophanes

Father of Confederation John A Macdonald—Canada's first prime minister

Father of the Constitution James Madison—fourth President of the United States

Father of the Continental Congress Benjamin Franklin

Father of the Copyright William Hogarth

Father of the Cotton Gin Eli Whitney

Father of Courtesy Richard Beauchamp—Earl of Warwick

Father of the Cowboys Charles Goodnight

Father of Cuban Independence José Marti

Father of Czechoslovakian Music Bedrich Smetana

Father of Damien Joseph Damien de Veuster

Father of Danish Opera Friedrich Kuhlau

Father of Dano-Norwegian Literature Ludvig Holberg

Father of the Declaration of Independence title shared by Thomas Paine, Thomas Jefferson, John Adams, and Benjamin Franklin

Father of Deep Ecology Arne Naess

Father of the Detective Story Edgar Allan Poe

Father of Dinosaur Art Charles R Knight

Father Divine Morgan J Divine born George Baker

Father of Dominican Independence Juan Pablo Duarte

Father of the Dominican Republic José Nuñez de Cáceres

Father of Dutch Poetry Jakob van Maerlant

Father of the Dutch Reformed Church in America John Henry Livingston

Father of Ecclesiastical History Eusebius Pamphili

Father of Egyptian Archeology Sir Flinders Petrie

Father of Electroencephalography Hans Berger

Father of Embryology Carl Ernst von Baer

Father of the Encyclopedia Diderot

Father of English Cathedral Music Thomas Tallis

Father of English Empiricism John Locke (1632–1704)

Father of English Lexicography Dr Samuel Johnson

Father of the English Novel Henry Fielding

Father of English Poetry Geoffrey Chaucer

Father of English Printing William Caxton

Father of English Song Caedmon

Father of English Unitarianism John Biddle

Father of Epic Poetry Homer

Father of the Erie Canal De Witt Clinton

Father of Ethical Culture Felix Adler

Father of Ethology Konrad Lorenz

Father of Euphuism John Lyly

Father of the Everglades National Park John Pennekamp

Father of Experimental Physiology Galen

Father of the Faithful Abraham

Father of Farm Workers Cesar Chavez

Father of Fascism Benito Mussolini; Gabriele d' Annunzio

Father of the Federal Reserve System George Carter Glass

Father of the Film Industry D(avid) W(ark) Griffith

Father of Fingerprinting Alphonse Bertillon

Father of the Flivver Henry Ford

Father of the Free School System Governor James Edward English of Connecticut

Father of Free Trade Adam Smith

Father of French-Canadian Poetry Octave Cremazie

Father of French Lyric Poetry Pierre de Ronsard

Father of French Opera Jean-Baptiste Lully

Father of the French School of Neurology Jean-Martin Charcot

Father of French Surgery Ambroise Paré

Father of French Tragedy Pierre Corneille

Father of Frozen Foods Clarence Birdseye

Father of Full-Spectrum lighting John Ott

Father of Genetics Gregor Mendel

Father of Geography Strabo the Greek Stoic who wrote seventeen books about Asia, Egypt, Libya, and Europe

Father of Geometry Pythagorus

Father of German Literature Gotthold Ephraim Lessing

Father of German Opera Christoph Willibald von Gluck

Father of the German Reformation Martin Luther

Father of German Unification Prince Otto von Bismarck

Father of Gods and Men Odin or Wotan, according to the Norse; Jove or Jupiter, according to the Romans; Zeus, according to the Greeks; etc.

Father of the Gramophone Emile Berliner

Father of Greater Philadelphia John Christian Bullitt

Father of Greek Didactic Poetry Hesiod whose poem *Theogony* describes the beginning of the world, its gods, and the five Ages of Mankind

Father of Greek Music Terpander of Lesbos

Father of Greek Sculpture Phidias

Father of Greek Tragedy Aeschylus

Father of Greenbacks Salmon Portland Chase

Father of the H-bomb Edward Teller

Father of High-Speed Photography Harold E Edgerton

Father of His Country Cicero and several Roman caesars; George Washington— first President of the United States

Father of History Herodotus

Father of Homeopathy in America Dr Constantine Hering

Father of the Household Heater Benjamin Franklin

Father of Humanism in America John Dietrich

Father of the Hydrogen Bomb Edward Teller

Father of Hypnotism Franz Friedrich Anton Mesmer

Father of Independent India Mohandas Karamchand Gandhi

Father of Independent Vanuatu Walter Lini

Father of Individual Psychology Alfred Adler

Father of Inductive Philosophy Francis Bacon

Father of the IQ Test Alfred Benet

Father of Israel Chaim Weizmann

Father of Italian Landscape Painting Andrea del Verrocchio (Andrea di Michele Cione)

Father of Italian Opera Claudio Monteverdi

Father of Japanese Caricature Toba Sojo

Father of Japanese Shipbuilding Thomas Glover known to the Japanese as Kuraba

Father of Jazz Scott Joplin

Father of Jests Joseph Miller

Father of the Juvenile Court Judge Benjamin Barr Lindsey

Father of the Kindergarten Friedrich Froebel

Father of the Kiwifruit James MacLoughlin

Father of Kudzu Georgia farmer Charming Cope

Father of Latin Song Ennius (239–169 BCE)

Father of the Legal Code David Dudley Field

Father of Lies Satan

Father of the Locomotive Richard Trevethick

Father of Logarithms John Napier

Father of Mainframe Computers Gene Amdahl

Father of Marine Geology
Francis Shepard

Father of Massachusetts
Governor John Winthrop

Father of Medicine Hippocrates

**Father of Mexican Indepen-
dence** Miguel Hidalgo y
Costilla

Father of Mexican Muralism
Dr Atl (born Gerardo Murillo)

Father of Microscopy Anton
van Leewenhoek

Father of Military Strategy
Hannibal

Father of Mineralogy Agricola

Father of Modern Anatomy
Andreas Vesalius (1514–
1564)

**Father of Modern Architec-
ture** Frank Lloyd Wright

Father of Modern Art
Masaccio (Tommaso Guidi)

Father of Modern Astronomy
Copernicus (Nikolaus Kopernicki)

Father of Modern Baseball
Alexander Joy Cartwright

Father of Modern Brazil
Getulio Vargas

**Father of Modern Conserva-
tive Thought** Edmund
Burke

**Father of Modern Contact
Lenses** Otto Wichterle

**Father of Modern Criminol-
ogy** Alphonse Bertillon and
Cesare Lombroso

Father of the Modern Daylily
Dr Arlow Burdette Stout

**Father of Modern Democratic
Philosophy** John Locke

Father of Modern Drama
Henrik Ibsen

**Father of Modern Economic
Forecasting** Trygve
Haavelmo

Father of Modern Economics
Adam Smith

**Father of Modern English
Poetry** Walt Whitman

**Father of Modern Existential-
ism** Søren Kierkegaard

**Father of Modern Explora-
tion** Fridtjof Nansen

**Father of Modern Finger-
printing** Sir Edward Richard Henry

**Father of Modern French
Poetry** Charles Baudelaire

Father of Modern Genetics
Gregor Mendel

Father of Modern Geology
Sir Charles Lyell

**Father of Modern German
Poetry** Heinrich Heine

**Father of Modern Guitar
Music** Francisco Tárrega

**Father of Modern Gynecol-
ogy** Dr J Marion Sims

**Father of Modern Italian
Poetry** Gabriele D' Annunzio

Father of Modern Magnetism
John Van Vleck

Father of Modern Medicine
Canadian-born Sir William
Osler

Father of Modern Music
Franz Liszt; Mozart

Father of Modern Navies
Captain Alfred T Mahan author of *The Influence of Sea
Power upon History* pub-
lished in 1890

Father of Modern Neurology
Guillaume-Ben-
jamin-Amand Duchenne

Father of the Modern Novel
Lion Feuchtwanger

**Father of the Modern Orches-
tra** Hector Berlioz

Father of Modern Painters
Giovanni Cimabue (Cenni di
Pepo)

Father of Modern Pedagogy
Heinrich Pestalozzi

Father of Modern Philosophy
René Descartes

**Father of Modern Photogra-
phy** William Henry Fox Talbot

Father of Modern Physics
Albert Einstein

Father of Modern Physiology
William Harvey

**Father of Modern Prose Fic-
tion** Daniel Defoe

Father of Modern Rocketry
Robert Hutchings Goddard

**Father of Modern Russian
Poetry** Nikolai Alekseevich
Nekrasov

**Father of Modern Spanish
Poetry** Rubén Darío

**Father of Modern Surgery of
the Brain** Paul Broca

Father of Modern Symphony
César Franck

Father of Modern Taxonomy
Carl von Linné also known as
Carolus Linnaeus

Father of the Modern Zoo
Carl Hagenbeck

**Father of Mongolian Democ-
racy** Sanjaasuren Zorig

Father of Moral Philosophy
Thomas Aquinas

Father of the Mormons Joseph Smith

Father of Muckrakers Upton
Sinclair, Lincoln Steffens,
and Joseph Flynt Williard

Father of Music Festivals Edinburgh

Father of Natural History
Aristotle; Carl Linnaeus

Father of Naval Aviation
Commander Robert C Whit-
ton

Father of Negro History
Carter G(odwin) Woodson

**Father of the Neighborhood
Settlement House** Jacob
August Riis

Father of Neurosurgery
American-born Canadian
Doctor Wilder Penfield

Father of New England John
Endicott—its first governor

**Father of New England Tran-
scendentalism** Ralph Waldo
Emerson

Father of the New Left Herbert Marcuse

Father of Niagara Power
William Birch Rankine

**Father of the Nuclear Subma-
rine** Admiral Hyman G
Rickover, USN

Father of Oceanography
Matthew Fontaine Maury

Father of Ontario Hydro Sir
Adam Beck

**Father of Organic Architec-
ture** Frank Lloyd Wright

Father of Organized Labor
Samuel Gompers

Father of Osteopathy Dr Andrew T Still

Father of Paleontology
Georges Cuvier

Father of Parole Captain Alexander Maconochie the governor of the Norfolk Island
penal colony from 1840 to
1844

Father of Pasteurization
Louis Pasteur

Father of the Patent Office
John Ruggles

Father of Pathology Rudolf
Virchow

**Father of Penitentiary Sci-
ence** Jean Jacques Vilain

Father of Pennsylvania William Penn—its founder
Father of Permissiveness Benjamin McLane Spock, pediatrician (1903–1998)
Father of Philippine Independence Emilio Aguinaldo
Father of Philosophy Thales
Father of the Phonograph Thomas A Edison
Father of Photojournalism Alfred Eisenstaedt
Father of Physiography William Morris Davis
Father of Pianoforte Playing Muzio Clementi
Father of Pittsburgh George Washington who proposed the location and the name
Father of Poetry Orpheus
Father of Polar Exploration Fridtjof Nansen
Father of the Potteries Josiah Wedgwood
Father of Psychoanalysis Sigmund Freud
Father of Public Relations Edward L Bernays, Ivy L Lee
Father of Radio Lee De Forest
Father of Radio Broadcasting Harry P(hillips) Davis
Father of Railways George Stephenson
Father of Reform John Cartwright
Father of the Reformed Church John Henry Livingston
Father of the Religion of Reason (Unitarianism) James Martineau in England, Ralph Waldo Emerson and Theodore Parker in the United States
Father of Religious Humanism John Dietrich
Father of the Republic of China Sun Yat Sen
Father of the Waters great rivers such as the Amazon, Amur, Congo, Euphrates, Huang, Irrawaddy, Lena, Mackenzie, Mekong, Mississippi, Niger, Nile, Ob, Volga, Yangtze, Yenisei
Father of the Western Story Zane Grey
Father of Western Theater Aeschylus
Father of West Point Sylvanus Thayer

Father of the World Wide Web Tim Berners-Lee
Father of Yellowstone National Park Nathaniel Langford
Father of Zionism Theodor Herzl
Father of Zoology Aristotle
Fats Thomas (Fats) Waller
Fats Domino Antoine Domino
Fats Waller Thomas Waller
Fat Tony Antonio Salerno
Fatty and Skinny Oliver Newell Hardy and Stan Laurel
Faulhaber Cardinal Michael von Faulhaber, who opposed the Nazis
Faulkner's County Yoknapatawpha (an invention of novelist William Faulkner)
Faultless Painter Andrea del Sarto
Faustus Socinus Fausto Sozzini
Favelas Rio de Janeiro slums
favism disease, caused by enzyme deficiency in some people, that destroys red blood cells when fava beans are eaten
Favorite Island of Columbus Jamaica
The Fed The Federal Reserve Board
Federal Capital Territory now called Australian Capital Territory (around and in Canberra)
Federal City Washington, D.C.
federalese the jargon of bureaucrats on the federal payroll
Federal Hill Providence, Rhode Island slum
Federal Republic of Germany Bundesrepublik Deutschland
Federated States of Micronesia islands of Kosrae, Phompei, Truk, and Yap
Federation of Malaysia Malaysia (formerly Brunei, Federation of Malaya, Sabah, Sarawak, and Singapore)
Federation of South Arabia Ittihad al Janub al 'Arabi—formerly Aden Colony and Aden Protectorate
Feebie (American slang—member of the Federal Bureau of Investigation)

Feingold diet diet said to calm hyperactive children (named for Dr Benjamin Feingold)
Feldenkrais Technique posture therapy developed by Moshe Feldenkrais
Female Seminary Mount Holyoke College
Feminist Reformer Lucy Stone
Feminist Revolutionist Mary Wollstonecraft also known as Mary Godwin
Ferihegy Budapest, Hungary's airport
Fernán Caballero Cecilia Francisca Josefa de Arrom
Fernandel Fernand Contandin
Fernando de Magallanes (Spanish—Ferdinand Magellan)—originally Fernão de Magalhães
Ferrel's law a body moving in any direction over Earth's surface will tend to be deflected to right in Northern Hemisphere and left in Southern Hemisphere—for American meteorologist William Ferrel
Ferret Daimler armored scout car made in Great Britain
Ferryville old name for Menzei-Bourguiba in Tunisia
Fertile Arabia Arabia Felix in the southern Arabian Peninsula, particularly in the Yemenite lands once called Aden or the Hadhramaut
Fertile Crescent Australia's coastal plain extending from southern New South Wales to southern Queensland; agricultural region extending from the Lerant to Iraq
Fiddler's Green traditional haven of drowned sailors supposedly filled with women, grog, fine food, and tobacco; some opine it is a suburb of Davy Jones' Locker while others insist it is a synonym for Fiddler's Grotto
Fiddler's Grotto music-fulled tropical marine cavern inhabited by young women; roast turkeys fly about slowly; fine beer, whiskey, and wine cascade down its marble walls; its location, according to Captain Ed Hassel, is exactly two miles this side of Hell

Fifteenth State Kentucky

Fifteen-Year War (*see* Pacific War)

Fifth Continent Africa

Fifth Disease parvovirus

Fifth Estate The Underworld of Organized Crime

Fifth State Connecticut

Fiftieth State Hawaii

Figaro Mariano José de Larra's pseudonym

Fighting Bob Rear Admiral Robley Evans; English prize-fighter Robert P Fitzsimmons; Senator Robert M La Follette, Sr

Fighting Lady USS *Lexington*

Fighting Quaker General Nathanael Greene

Fijian Ports Suva and Levuka

Filadelfia Paraguay's *City of Brotherly Love*

Filatov's disease scarlatina-like exanthematous affection

Filbert Center Hillsboro, Oregon

Filipino Libertarian Emilio Aguinaldo

Film Capital Hollywood, California

Filthydelphia Philadelphia

Finality John Lord John Russell

final solution Hitlerian truncation covering the extermination of the Jews

Financial Center of the Rockies Denver, Colorado

Financial Genius of the Underworld Meyer Lansky

Financier of the Revolution Robert Morris

Finkelstein Box a box that divides United States into Republican Party strongholds and Democratic Party strongholds (named for Arthur Finkelstein, political consultant)

Finland's Principal Port Helsinki

Finnish Ports Tornio, Roytta, Kemi, Koivoluoto, Oulu, Raahe, Kokkola, Ykspihlaja, Jakobstad, Nykarleby, Vaasa, Vasklot, Kasko, Kristinestad, Pori, Mantyluoto, Reposaari, Kaunissaari, Rauma, Uusikaupunki, Turku, Storby, Mariehamn, Hango, Lappvik, Ekenas, Helsinki, Kotka, Hamina

Fiona Macleod William Sharp's pseudonym

Fiorello Fiorello LaGuardia

Firebar NATO name for Soviet Yakovlev all-weather fighter interceptor aircraft Yak-28P

Firebrand of the Navy Lieutenant Stephen Decatur, USN

Firebrand of the World Tamerlane (Timur Lenk or Timur the Lame)

Fireclay Capital Mexico, Missouri

firecracker factory missile manufacturing plant

Fireman Joe Begg (baseball player)

Firestreak Hawker-Siddeley air-to-air missile

First All-American Poet Walt Whitman

First American Advertiser William Penn

First American Man of Letters Washington Irving

First American Penitentiary Walnut Street Jail in Philadelphia built in 1790

First American Poet Philip Frenau

First American Republic Iceland

First American Woman Novelist Charlotte Lennox

First American Woman Physician Elizabeth Blackwell

First Astrophysicist Johannes Kepler

First Atomic City Hanford, Washington

First Black American Conductor Dean Dixon

First Black Member of the U.S. Supreme Court Thurgood Marshall

First California Mission San Diego de Alcalá

First Church First Church of Christ, Scientist

First Citizen of Ghana Dr WEB Du Bois

First City of the First State Wilmington, Delaware

First City of the South Savannah, Georgia

First Concentration Camp in Germany Dachau (erected in March 1933 near Munich)

First Estate The Clergy

First Execution by Electrocution August 6, 1890 in New York State's Auburn Prison

First Family family of the President of the United States

First Fleeters Australians tracing their lineage to their ancestors' arrival in 1788

First Foreign Enclave Macao, Portuguese China

first-generation money cash; currency

First Gentleman of the Land President Chester A Arthur

First Gospel Gospel according to Saint Matthew

First Great Cheerful Giver George Peabody

First Great Operatic Composer of the New World Carlos Gomes

First Great Symphonist Franz Joseph Haydn

First Halfway House Isaac T Hooper Home opened in New York City in 1845 by the Society of Friends (Quakers)

First International First International Workingmen's Association (convening in Paris in 1864)

First Lady wife of any American President

First Lady of the Air Amelia Earhart

First Lady of America Pocahontas

First Lady of American Revolution Mercy Otis Warren also known as Philomela

First Lady of American Song Kate Smith

First Lady of American Theater Helen Hayes

First Lady of Computers Lady Augusta Ada Byron Lovelace (the poet Lord Byron's daughter)

First Lady of Country Music Tammy Wynette

First Lady of Crime Agatha Christie

First Lady of Liberty Abigail Adams

First Lady of the Library President Millard Fillmore's wife Abigail—founder of the first library in the White House

First Lady of the Skies Amelia Earhart

First Lady of Song Ella Fitzgerald

First Lady of the World Eleanor Roosevelt—wife of President Franklin D Roosevelt

First Lawyer of the Land U.S. Attorney General

First Man to Walk on the Moon Neil A Armstrong

First Mayor of Chicago William Butler Ogden

First Person to Announce His Discovery of Photography William Henry Fox Talbot on January 31, 1839

First Perspective Painter Paolo Uccello (Paolo di Dono)—known for his studies in foreshortening and linear perspective

First Picaresque Novel *Lazarillo de Tormes* (author unknown)

First Poet Laureate Ben Jonson

First Poet Laureate of the United States Robert Penn Warren (1986–1987)

First President of the United States George Washington

First Prison Newspaper *The Summary* published by the inmates of the New York State Reformatory at Elmira on November 22, 1883—Thanksgiving Day

First Quaker George Fox, founder of the Society of Friends

First Romantic Artist Giambattista Piranesi

First State Delaware

First Street in Europe Disraeli's nickname for London's Strand

First University Plato's Academy

First White House of the Confederacy temporary home of Jefferson Davis in Montgomery, Alabama in 1861

First Woman Physician Dr Elizabeth Blackwell

First Woman Reporter Anne Royale of Virginia (publisher of *Paul Pry),* Nellie Bly of Pennsylvania (reporter for the *New York World*)

First World highly industrialized countries such as Japan, the United States and most Western European nations

First Zen First Zen Institute of America

fiscal pirates inflation and taxes

Fish-Canning Capital of the World Stavanger, Norway

Fisher Equation interest rates are composed of the real rate of interest plus the expected rate of inflation (named for economist Irving Fisher)

Fishing's Eve Juliana Berners

Fishpot NATO name for Soviet SU-9 all-weather jet fighter aircraft

Fitter NATO name for Soviet SU-7 jet ground-attack aircraft

Fiume Italian name for Rijeka, Yugoslavia formerly belonging to Italy

Five Nations Cayugas, Oneidas, Onondagas, Mohawks, and Senecas (American Indian tribes on the English side in the French and Indian Wars)

Fjord Land Norway

Flag of Alfonso yellow-and-red emblem of Spain dating from fifteenth century when it was carried by Alfonso el Magnánimo

Flag Bay Bahía de Banderas, Mexico

Flagellum Dei (Latin—Scourge of God)—Attila the king of the Huns

Flagon-A NATO code name for Soviet SU-11 delta-wing fighter aircraft

flap and zap laser in-situ keratectomy

Flash Joe Gordon

Flash Hollett William Hollett

Flashlight NATO name for Soviet Yakovlev Yak-25 two-place interceptor fighter aircraft

Flatiron New York City's Flatiron Building at Broadway, Fifth Avenue, and 23rd Street

Flats Flatlands—in Great Britain

Flea Freddie Patek (baseball player)

fleishig pertaining to meat (Yiddish: *fleyshik*)

Flemish Colorist Peter Paul Rubens

Flemish Primitive Painter Gheeraert David

Fletcherism Horace Fletcher's theory that people should chew every forkful of food thirty-two times for good health

Flickertail(s) North Dakotan(s)

Flickertail State North Dakota

Flip Wilson Clerow Wilson, comedian (1933–1998)

Flitch John Pierpont Morgan

Flivver King Henry Ford

float a term for the period of time between when a check is written and funds are drawn from the check's account

Flood City Johnstown, Pennsylvania

Floral Watercolorist William Demuth

Floreana Island, Galápagos Santa María

Florence Austral Florence Wilson

Florestan Robert Schumann

Florestan and Eusebius pseudonyms invented by Robert Schumann to represent the two sides of his character Florestan was impetuous and passionate whereas Eusebius was the dreamer

Florida of Canada British Columbia, Canada

Florida Pen Florida Peninsula

Florida Ports Jacksonville, Port Everglades, Miami, Key West, Tampa, St Petersburg, Port St Joe, Panama City, Pensacola

Flour City Buffalo or Rochester, New York

Flower Capital Encinitas, California

Flower City Rochester, New York

Flower Garden of England The Sorlings or Isles of Scilly off Land's End

Flower King Carl von Linné (Linnaeus)

Flower of the Levant Zante in the Ionian Islands

Flower Moon full moon in June

Flower of Quakerism abolitionist Lucretia Mott

Flower Seed Capital of the West Santa Maria, California

Flower State Florida

Flower Town Brampton, Ontario

Flowertown in the Pines Summerville, South Carolina

Flower of the Transvaal Pretoria

Flowery Kingdom China

fluf fat little ugly fella (engineers' nickname for Boeing 737 jet)

Flying Boxcar C-119 Fairchild-Hiller transport

Flying Dutchman mythical character immortalized in Richard Wagner's opera *Der Fliegende Holländer;* nickname of the baseball batting champion of the early 1900s—Honus Wagner

Flying Finn Paavo Nurmi

Fly-up-the-Creek green heron (*Butorides virescens*)

Fly up the Creeks Floridians

Foam City Milwaukee, Wisconsin

Fog City San Francisco

Fogfoundland fog-bound Newfoundland

Foggy Bottom the U.S. State Department

Foley motion picture postproduction synchronization of body-movement sounds with the image track; Thomas S Foley

Folk Arts and Crafts Capital of Kentucky Berea

Football Capital of the South Birmingham or New Orleans

Foothill City Calgary, Alberta

Forbidden City Lhasa, Tibet

Forbidden Island privately owned Niihau, westernmost island in Hawaii

Forbidden Kingdom Bhutan

Fordham Flash Frank Frisch

Ford Madox Ford Ford Madox Hueffer

Fordtown Detroit, Michigan

Foreigner State New Jersey

Forensic Psychiatrist Richard von Krafft-Ebing

Forerunner of the Reformation John Huss

Forerunner of Spanish-American Independence Francisco Miranda

Forest Cantons Swiss cantons of Lucerne, Schwyz, Unterwalden, and Uri

Forest City Cleveland, Ohio or London, Ontario

Forest City of the South Savannah, Georgia

Forest of Forests forested belt stretching from northern Norway to eastern Siberia

Forget-Me-Not State Alaska

The Forgotten Man President Franklin D Roosevelt's description of the American voter

Forgotten Philosopher Giordano Bruno

The Forgotten War Korean War

Former Naval Person code name of Prime Minister Churchill formerly First Lord of the Admiralty

Formidable Maestra conductor Antonia Brico

Formosa (Portuguese—beautiful, delightful, perfect)—name given Taiwan by Portuguese explorers

Fortaleza formerly Caerá

Fort Dimanche Haiti's infamous prison close to Pétionville

Fort Hill John C Calhoun's country seat in the Pendleton district of South Carolina near Anderson

Fortieth State South Dakota

Fort Liquordale Fort Lauderdale, Florida (when college students make their Easter vacation visit)

Fortunate Island Monhegan, Maine

Fortunate Islands Canary Islands

Fortunate Isles Madeira Islands—also called Isles of the Blest in classical mythology

Fortune-seekers Idahoans

forty-and-eights railroad boxcars used by French in World War I to transport forty men or eight horses

Forty-eighth State Arizona

Forty-fifth State Utah

Forty-first State Montana

Forty-fourth State Wyoming

Forty Immortals collective nickname of the forty members of the French Academy

Forty-ninth State Alaska

Forty-second State Washington

Forty-seventh State New Mexico

Forty-sixth State Oklahoma

Forty-third State Idaho

Foster Mother of the Sciences Medicine

Founder of the AAAA Dr Charles L Andrews, founder American Association for the Advancement of Atheism

Founder of Agnosticism Thomas Henry Huxley

Founder of Agricultural Chemistry Justus von Liebig

Founder of the American Federation of Labor Sam(uel) Gompers

Founder of American Military Intelligence General Ralph H Van Deman

Founder of the American Navy John Paul Jones

Founder of Analytical Psychology Carl Gustav Jung

Founder of the Aniline Dye Industry Sir William Perkin

Founder of Antiseptic Surgery Joseph Lister

Founder of Art History and Criticism Giorgio Vasari

Founder of Bacteriology Ferdinand Cohn

Founder of Behaviorism John Watson

Founder of the Birth Control Movement Margaret Sanger

Founder of the Boy Scouts and Girl Guides Sir Robert Baden-Powell

Founder of Brazil Pedro Alvares Cabral

Founder of British Imperial India Robert Clive

Founder of Buddhism Prince Siddhartha (Gautama Buddah)

Founder of Buenos Aires Pedro de Mendoza

Founder of Cellular Pathology Rudolf Virchow

Founder of Chemistry Robert Boyle

Founder of Chicago Jean de Sable

Founder of Christian Science Mary Baker Eddy

Founder of Cleveland, Ohio Moses Cleaveland

Founder of the Columbia University School of Journalism Joseph Pulitzer

Founder of Comparative Anatomy Baron Georges Cuvier

Founder of Confucianism K'ung Futzu (Confucius)

Founder of Conservative Surgery Sir William Fergusson

Founder of Continental Rationalism René Descartes

Founder of Cubism George Braque, Pablo Picasso

Founder of Cybernetics Norbert Wiener

Founder of Czech School of Music Bedrich Smetana

Founder of Dada Tristan Tzara

Founder of Detroit Sieur de Cadillac

Founder of Electromagnetism Michael Faraday

Founder of Electrophysiology Emil du Bois Reymond

Founder of English Empiricism Sir Francis Bacon

Founder of Episcopalianism Henry VIII

Founder of Experimental Hygiene Max von Pettenkofer

Founder of the Faculty of Physicians and Surgeons of Glasgow Peter Lowe

Founder of Fauvism Henri Emile Benoit Matisse

Founder of First Free Medical Clinic in the U.S. Benjamin Rush

Founder of the Franco-Belgian School of Violin Playing Charles Auguste de Bériot

Founder of French Colonial Empire Louis Faidherbe

Founder of French Grand Opera Daniel François Esprit Auber

Founder of French Opera Jean-Baptiste Lully

Founder of French Socialism Compte Claude Henri de Rouvroy de Saint-Simon

Founder of the Friends George Fox

Founder of Functionalism Louis Sullivan

Founder of Georgia James Oglethorpe

Founder of Gestalt Therapy Fritz Perls

Founder of Hadassah Henrietta Szold

Founder of Histology Marcello Malpighi

Founder of Homeopathy Christian Friedrich Samuel Hahnemann

Founder of Humanistic Psychology Abraham Maslow

Founder of Hungary Arpad

Founder of Iconographic and Physiologic Anatomy Leonardo da Vinci

Founder of Impressionism Claude Monet

Founder of Islam Mohammed

Founder of Jainism Mahavira also known as Vardhamana

Founder of Japanese Color-Print Making Iwasa Matabei

Founder of Judaism Moses

Founder of the Kelmscott Press William Morris

Founder of the Lutheran Church Martin Luther

Founder of Manila Miguel López de Legazpi

Founder of Medical Statistics Pierre-Charles Alexander Louis

Founder of the Methodist Church John Wesley

Founder of Modern Astronomy Nicolaus Copernicus (Nikolaus Kopernicki)

Founder of Modern Chemistry Antoine Lauret Lavoisier

Founder of Modern Existentialism Søren Kierkegaard

Founder of Modern German Sculpture Johann Gottfried Schadow

Founder of Modern Military Medicine Sir John Pringle

Founder of Modern Philosophy René Descartes

Founder of Modern Russian Literature Aleksandr Sergeyvich Pushkin

Founder of Modern Sculpture Donatelio (Donato di Niccolo di Betto Bardi)

Founder of Mormanism Joseph Smith

Founder of Naturalist Movement in Literature Émile Zolá

Founder of Oklahoma Jean Pierre Chouteau

Founder of Optics Giovanni Battista della Porta

Founder of Osteopathy Andrew Taylor Still

Founder of Pakistan Dr Mohammed Ali Jinnah

Founder of Pediatric Cardiology Helen Brooke Taussig

Founder of Pennsylvania William Penn

Founder of Phenomenology Edmund Husserl

Founder of Philosophic Radicalism James Mill

Founder of Polaroid Photography Edwin Land

Founder of Positivism Auguste Compte; Immanuel Kant

Founder of Postimpressionism Paul Cézanne

Founder of Pragmatism C(harles) S(anders) Peirce

Founder President of Zambia Kenneth Kaunda

Founder of Professional Nursing Florence Nightingale

Founder of Providence, Rhode Island Roger Williams

Founder of Psychoanalysis Sigmund Freud

Founder of Psychology Wilhelm Wundt

Founder of Québec Samuel de Champlain

Founder of the Religious Society of Friends George Fox

Founder of Rhode Island Roger Williams

Founder of Rome Romulus

Founder of the Royal Navy Samuel Pepys

Founder of Russian Literature Alexander Pushkin

Founder of Salt Lake City Brigham Young

Founder of the Science of Eugenics Sir Francis Galton

Founder of Scientific History Thucydides

Founder of Scottish Presbyterianism John Knox

Founder of Secularism George Holyoake

Founder of Singapore Sir Thomas Stamford Raffles

Founder of Social Psychology Gustave Le Bon

Founder of Sociology Auguste Compte

Founder of State Socialism Louis Blanc

Founder of Swarthmore College Martha Ellicott Tyson

Founder of Taoism Lao-tzu

Founder of Transcendental Idealism Immanuel Kant

Founder of Transcendentalism Ralph Waldo Emerson

Founder of Troy Tros

Founder of Unitarianism John Biddle

Founder of the University of Pennsylvania Benjamin Franklin

Founder of the University of Virginia Thomas Jefferson

Founder of Uruguay José Gervasio Artigas

Founder of the U.S. Navy Captain John Paul Jones

Founder of the Venetian School of Painting Giovanni Bellini

Founder of Vermont Ira Allen

Founder of Victimology Hans von Hentig or Benjamin Mcndclsohn

Founder of Zoroastrianism Zoroaster also known as Zarathustra

Founders of Christianity disciples of Jesus Christ

Founders of Cubism Georges Braque and Pablo Picasso

Founders of Flemish Painting the van Eyck brothers—Hubrecht and Jan

Founders of French Romantic Painting Delacroix, Géricault, and Gros

Founders of the Hudson River School (of painting) Thomas Cole and Asher Brown Durand

Founders of Neo-Impressionism Georges Seurat and Paul Signac

Founders of Scientific Socialism Karl Marx and Friedrich Engels

Founding Father City Austin, Texas named in honor of Stephen F Austin

Founding Father of Israel David Ben-Gurion

Founding Fathers of Economics Adam Smith and David Ricardo

Founding Father of the Smithsonian Joseph Henry

Founding Philosopher of Modern Capitalism Adam Smith

Fountain of Youth St Augustine, Florida

Four-C City El Paso, Texas (famous for its cattle, climate, copper, and cotton)

Four Corners boundary-line junction of Arizona, Colorado, New Mexico, and Utah

four-dimensional science geology involving the application of biology, chemistry, mathematics, and physics

four-five .45-caliber gun

Four Forest Cantons Lucerne, Schwyz, Unterwalden, and Uri—all in Switzerland

The Four Horsemen senior backfield that led Notre Dame to national collegiate football championship in 1924—Jim Crowley, Elmer Layden, Don Miller and Harry Stuhldreher

Four Lakes City Madison, Wisconsin

The Four Musketeers Frenchmen who dominated men's world tennis in 1920s and 1930s—Jean Borotra, Jacques Brugnon, Henry Cochet and René Lacoste

Four Noble Truths suffering is an unavoidable part of existence; suffering is caused by desire; elimination of desire will end suffering; desire and suffering can be eliminated

Four Seasons Crossroad of New England Manchester, New Hampshire

Fourteen Points statement made by President Woodrow Wilson of American peace goals during World War I

Fourteenth State Vermont

Fourth Bureau Red Army bureau in charge of overseas intelligence-gathering activities of the Soviet Union

Fourth Estate The Media—press, radio, television

Fourth Gospel Gospel according to Saint John

Fourth International Trotsky-oriented organization rejecting the Second and Third Internationals in the direction of the class struggle

Fourth State Georgia

Fourth World poorest of the Third World nations

Four Tigers Hong Kong, Singapore, South Korea, Taiwan

Four Winds Boreas (north), Eurus (east), Notus (south), Zephyrus (west)

Foxardo (naval argot—Fajardo, Puerto Rico)

Foxes Mainers

Fox Populi Charles James Fox

Fra Angelico Giovannida Fiesole

Fra Bartolommeo Baccio della Porta

Fra Elbertus Elbert Hubbard

Fragment of Eden Seychelles Islands in the Indian Ocean

Fragrant Harbor Hong Kong

Framer of the Bill of Rights James Madison

Framer of the *Declaration of Independence* Thomas Jefferson who rewrote Thomas Paine's draft with the aid of John Adams and Benjamin Franklin

Framer of the French Revolution Denis Diderot

Frances Alda Frances Davis

France's Largest Port Marseille

Franche-Compté Burgundy

Francis Beeding John Leslie Palmer's pseudonym

Franciscan Wine Capital Wurzburg, Germany

Franco-Hispanic Co-Principality Andorra (between France and Spain)

Francoise Sagan pseudonym—Françoise Quoirez

François Villon François de Montcorbier

Francophone Africa Afars and Issas, Algeria, Burundi, Cameroon, Central African Republic, Chad, Congo, Benin, Gabon, Guinea, Ivory Coast, Madagascar, Mali, Mauritania, Mauritius, Niger, Reunion, Rwanda, Senegal, Seychelles, Togo, Tunisia, Burkina Faso, Zaire

Francophone America French Guiana; Guadeloupe; Haiti; coastal parishes of Louisiana; Martinique; northern New York, Vermont, New Hampshire, Maine, and New Brunswick; Québec; St Pierre and Miquelon

Francophone Europe Andorra; French-speaking Belgium, France, Luxembourg, Monaco; French-speaking cantons of Switzerland

Francophone Pacific French Polynesia, New Caldonia, New Hebrides, Wallis and Fatuna Islands

Francophone Province Québec

Frank Leslie business name of Henry Carter

Frank Richards Charles Hamilton's pen name

Franz Josef Land Arctic islands called Zemlya Frantsa Iosifa by the Russians

Fred Alan John Sullivan

Fred Astaire Frederick Austerlitz

Frederick Douglass Frederick Augustus Washington Bailey

Frederick the Great Frederick II of Prussia

Frederic March Frederich McIntyre Bickel

Fred Niblo Frederico Nobile

Freedom Defender U.S. Supreme Court Justice William O Douglas

Freedom Fighter former name of the Northrup Tiger II or F-5 tactical fighter plane

Free and Hanseatic City Hamburg

Free State Maryland

Freestone State Connecticut

Freeway City Los Angeles

French French bull (small breed of dog); French chalk (tailor's talc); French curve (drafting instrument); French heel (high curved heel); French polish (alcohol + shellac); French pox (syphilis); French roll (women's coiffure); French roof (mansard-style); French seam (completely covered seam); French system (spinning system); French tamarisk (salt cedar)

French-75 75-millimeter field gun developed by the French

French Aero Pioneers Etienne and Joseph de Montgolfier

French Antilles French West Indies

French Canada French-speaking Canada, mainly the Province of Québec

French Caribees colonial name for the French West Indies

French Century the 18th century—the 1700s

French Community metropolitan France together with its overseas departments, territories, and former territories *(Communauté française)*

French disease syphilis also known as the Italian disease or the Spanish disease as well as *morbus gallicus* (Latin—Gallic disease)—the French disease

French Equatorial Africa former colonies Benin or Dahomey, Cameroon, the Central African Empire, the French Congo, Gabon, and Guinea

French flu serious artist turning fey

French India former French possessions in India (Chandernagore, Pondicherr, etc.)

French Indo-China former name of area comprising Annam, Cambodia, Chochin China, Laos, Tonkin, and Vietnam

French Morocco eastern Morocco closest to Algeria and the Sahara

French paradox disparity between rich diet and apparent good health among the French

French Polynesia Austral, Marquesas, Society and Tuomoto islands in the South Pacific Ocean

French Ports Dunkerque, Calais, Boulogne-sur-Mer; Le Treport, Dieppe, Fecamp, Le Havre, Rouen, Cherbourg, Granville, St Malo, Brest, Douarnenez, Aupierne, Port Louis, Lorient, Le Palais, Le Croisic, Saint Nazaire, Donges, Paimboeuf, Basse-Indre, Les Asables Dolonne, La Pallice, La Rochelle, Rochefort, Tonnay-Charente, Le Verdon, Mortagne, Trompeloup, Paulillac, Blaye, Ambes, Le Marquis, Bordeaux, Arcachon, Boucau, Bayonne, Biarritz, Port Vendres, Port La Nouvelle, Sete, Port St Louis du Rho, Port de Bouc, Berre Letang, Marseille, LaCiotat, Toulon, Cannes, Nice, Villefranche, Bastia and Ajaccio (on Corsica), Menton

French Quarter Vieux Carré in New Orleans

French Revolutionary Calendar (*see Vend, Brum, Frim, Niv, Pluv, Vent, Germ, Flor, Prair, Mess, Therm, Fruc,* entries)

French Riviera resort areas along the Mediterranean from Marseilles to Menton, including Cannes, Monaco, and Nice

French Sahara former colonial areas such as Algeria, French Morocco, Mauritania, and Niger

French Shore Newfoundland's northern and western coasts

French Somaliland *Côte Française des Somalis* (French Coast of the Somalis)—now known as Djibouti

French-speaking Places (*see entries under* Francophone)

French Sudan former name of Mali

French Switzerland French-speaking areas of Switzerland

French Togoland former name of Togo

French Union France plus its overseas colonies and departments as well as all its former possessions

French West Africa former colonies of France: Algeria, Chad, French Morocco, Mali, Mauritania, Niger, Senegal, and Upper Volta

French West Indies Desirade, Guadeloupe, Les Saintes, Marie Galante, Martinique, Petite Terre, Saint Bartholomew (Barthelemy), Saint Martin (French half)

frequent fliers repeat admittees to psychiatric wards

The Freshest Man on Earth Arlie Latham (baseball player)

Friar Antonio Agapida pseudonym of Washington Irving

Frida and Diego Frida Kahlo and Diego Rivera

Fridjof Nansen Land formerly Franz Josef Land (Arctic island group in Queen Victoria Sea north of Barents Sea sector of Arctic Ocean)

Friend of the American Revolution Caron de Beaumarchais

Friend of Helpless Children Herbert Clark Hoover—thirty-first President of the United States

Friendliest Town in the West Geraldton, Western Australia

Friendly Island Molokai, Hawaii; St Maarten, Netherlands Antilles

Friendly Islands Tonga Islands in the South Pacific

Friendly Kingdom Tonga Islands

Frisian Frisian language spoken off coasts of Denmark, Germany, and the Netherlands

Frog(s) Anglo-American slang for French (people)

Frontier Fighter David (Davy) Crockett

Frontier Gandhi Abdul Gaffar Khan

Frontier Hero Daniel Boone

Frontier States Alaska and Hawaii

Frostbite nickname of Fairbanks, Alaska

Frosty Beaver Moon full moon in November

Fruit Belt land around Lake Michigan in the United States, where lake tempers climate and fruit ripens at a latitude where it usually would not

Fruit Bowl of the Nation Yakima, Washington

Fruit from Heaven mango

fuel of the future solar power

fugitive dust fine-grained soil that easily blows away

Führer (German—Leader)—Hitler's title

Full Buck Moon July 3 full moon

fullerenes geodesic dome-shaped carbon molecules named for American architect and engineer (Richard) Buckminster Fuller (1895–1983)

Fulton's Folly inventor Robert Fulton's steamship *Clermont*

Fum the Fourth nickname of George IV

Fun Capital of Scandinavia Copenhagen, Denmark

Fun City New York

Fundador de la República (Spanish—Founder of the Republic)—José Nuñez de Cáceres—founder and first president of the Dominican Republic (Spanish Haiti)

Fundador de Nueva Granada (Spanish—Founder of New Granada)—Francisco de Paula Santander—founder of Colombia (Nueva Granada)

funeral order maritime tradition of older persons standing back to give young people first chance when boarding lifeboats or using other life-saving equipment (*see* Birken'ead drill)

Fungus Corners (naval argot—Bremerton, Washington)—a rainy port

Fun-Loving Philosopher Mark Russell

Fun and Sun Cities Acuña, México across the Rio Grande from Del Rio, Texas

funny money cash that cannot be spent openly

Furry Lewis Walter Lewis

Fuss and Feathers General Winfiield Scott, USA

The Fuzz [slang—detective(s); law-enforcement officer(s); police]

fuzzies 10-day to 3-week-old rats

Gabon Ports Cocobeach, Libreville, Port Gentil, Sette Cama, Gamba Oil Terminal

Gabriel Israeli surface-to-surface missile

Gabriela Mistral Lucila Godoy de Alcayaga

Gabriel d'Annunzio Gaetano Rapagnetta

Gabriel Padecopeo Lope de Vega

Gaby Gabrielle Dupont

gaffer person responsible for all lighting setups on film

Gagarin Yuri A Gagarin, Russian cosmonaut and pioneer of first manned orbital flight around Earth

Gail Hamilton Mary Abigail Dodge

Gaillard Cut formerly called Culebra Cut (in the Panama Canal)

Galen (sometimes Galin) Vasily Konstantinovich Blücher

Gambia's Port Georgetown

Gambling Capital of the Far East Macao, Portuguese China

Gambling Capital of the Far West Las Vegas, Nevada

gandi(s) half-naked railroad-track worker(s)

Gang of Five (*see* Group of Five)

Gangland Chicago

Gang of Six Canada plus the Gang of Five

Gannet Westland three-place early-warning aircraft developed in Britain

garbage land mines scattered by aircraft

Garbage Dump California's San Quentin Prison; Green Meadow Correctional Facility at Comstock, NY

Garbage Man Sly Williams (basketball player)

Garden of the Andes Mendoza, Argentina

Garden of the Antilles St Croix, Virgin Islands

Garden of Argentina Tucuman

Garden of Canada Ontario

Garden of the Caribbean Puerto Rico

Garden City a town planned to include industry while maintaining low housing density and cultural and recreational amenities; Singapore (also called Lion City)

Garden City of Australia Melbourne

Garden City of Georgia Augusta

Garden City of India Mysore

Garden City of New Zealand Christchurch

Garden of Denmark Fyn or Funen—home of Hans Christian Andersen

Garden of the East Burma (now Myanmar), Malaysia, and Sri Lanka

Garden of England Kent and Worcester

Gardener Touched With Genius Luther Burbank

Garden of France Amboise and Touraine

Garden of God ancient eponym of Lebanon just north of the Holy Land

Garden of the Gods park near Colorado Springs, Colorado

Garden of the Gulf Prince Edward Island in the Gulf of St Lawrence

Garden of Ireland Carlow

Garden Island Kauai, Hawaii

Garden Isle Kauai, Hawaii

Garden Isle of Micronesia Ponape

Garden of Italy Sicily

Garden of Love Shalimar waterside garden on Kashmir's Dal Lake

Garden of Maine Aroostook County

Garden of the Morning Breeze Naseern Bagh on Kashmir's Dal Lake

Garden of Paradise in the Sea Madeira

Garden Province Canada's Prince Edward Island

garden room morgue

Garden of the Southwest Southern California

Garden of Spain fields of Andalucía and Valencia

Garden State Kansas; New Jersey

Garden of the Sun Indonesia

Garden of Sweden Blekinge

Garden of Switzerland Thurgau

Garden of Wales southern Glamorganshire

Garden of the West California; Kansas; Illinois

Garden of the World Mississippi River Valley

Gargantua Franç I of France (or) Henri d'Albret—King of Navarre

Garlic Capital Gilroy, California

Garry Moore Thomas Garrison Morfit

Gary Cooper Frank J Cooper

Gas House of the Nation Washington, D.C.

Gaskin Soviet SA-9 air-defense missile system contained in an amphibious armored vehicle (NATO)

Gasopolis Los Angeles (on smog-filled days); other places with similar air pollution

Gasparilla (Spanish—Little Gaspar)—nickname of José Gaspar—pirate active along west coast of Florida around 1750

Gastown waterfront area of Vancouver, BC

Gate City Keokuk, Iowa; Laredo, Texas; St Louis, Missouri

Gate City of the South Atlanta, Georgia

Gatemouth Clarence Brown

Gates of the Arctic national park in north-central Alaska

Gates of Hell old nickname for the entrance to Marquarie Harbour on the Indian Ocean coast of Tasmania when it was a penal settlement

Gate of Tears Bab-el-Mandeb Strait linking Gulf of Aden and Indian Ocean with the Red Sea; many sailors call it Gate of Hell because of its desert-heated winds

Gateway to Alaska Seattle, Washington

Gateway to the Alps Zurich

Gateway to America Ellis Island immigration station, New York; New York City

Gateway Arch City St Louis, Missouri

Gate to the Arctic Fairbanks, Alaska

Gateway to the Bahamas Bimini Island off Miami

Gateway to the Big Bend National Park Marfa, Texas

Gateway to the Caribbean Tampa, Florida

Gateway City nickname of Pittsburgh, Pennsylvania, after the Revolutionary War

Gateway to the Dakotas Sioux Falls, South Dakota

Gateway of the Day Fiji Islands near the International Date Line

Gateway to the East Port Said, Egypt

Gateway to Eastern India Calcutta

Gateway to the Golden Isles Brunswick, Georgia

Gateway to the Great Seaway Green Bay, Wisconsin

Gateway to the Gulag Magadan on the Sea of Okhotsk

Gateway to India Bombay

Gateway to Israel Haifa

Gateway to Japan Yokohama

Gateway to Lapland Rovaniemi, Finland

Gateway to Latin America Miami, Florida

Gateway to Moroland Zamboanga, Mindinao

Gateway to Mount Rainier Tacoma, Washington

Gateway to the Negev Beersheba, Israel

Gateway to the North North Bay, Ontario

Gateway to Northern Europe Göteborg (Gothenburg), Sweden

Gateway to the NY-NJ Market Bayonne, New Jersey

Gateway to Parris Island Beaufort, South Carolina

Gateway to the Rhine Valley Bonn, Germany

Gateway to the San Juan Islands Anacortes, Washington

Gateway to the Smokies Knoxville, Tennessee

Gateway to South America Colombia (with ports on the Atlantic and the Pacific)

Gateway to Southwest Japan Kobe

Gateway States California, Louisiana, New Jersey, New York

Gateway to the West Pittsburgh, Pennsylvania, and St Louis, Missouri

Gateway to Western India Bombay

Gateway to the Yukon Skagway, Alaska

GATF Graphic Arts Technical Foundation

Gator Bowl Jacksonville, Florida

'Gator State Alligator State (Florida)

Gaucho Land Uruguay

Gautama Buddha Prince Siddhartha

Gavin Ogilvy James M Barrie

Gay City San Francisco

Gay Gateway to Europe Copenhagen (København)

Gay Paree Paris, France

gay(s) homosexual(s)

Gay White Way New York City's Broadway in the 42nd Street and Times Square area

Gazelle Embracer of Brazil's version of the Aerospatiale SA-341 observation helicopter

Gem Beside the Amstel Amsterdam

Gem of the Caspian Sea Volga River

Gem of the Mountains Idaho

Gem of the Oregon Coast Yachats, Oregon

Gem of the South Pacific New Zealand

Gemsbok Gemsbok National Park of Botswana in southwest Africa

Gem State Idaho

Gene Autry Orvon Gene Autry

General Booth Salvation Army founder William Booth

General Bor Tadeusz Komorowski—cavalry leader of 63-day uprising of Polish underground against Germans occupying Warsaw in 1944

General Day After Tomorrow Nez Perce nickname for General Oliver O Howard, commander of the Nez Perce Campaign of 1877, who usually was a day and a half behind them

General Douglas pseudonym of Soviet corps commander Yakov Smuskevich while leading the Spanish Republican air force in 1936–37

General John nickname of the first Duke of Marlborough— John Churchill

General Kleber pseudonym of Soviet general Grigory Shtern while serving as chief advisor to the Spanish Republican army in 1936–37

General's Lady Martha Washington—wife of General George Washington

General Tom Thumb Charles S Stratton

General Tubman Harriet Ross Tubman of Underground Railroad fame

Generation X people born from 1965 to 1977 (about 40 million people)

Generation Y people born from 1978 to 1988 (about 60 million people)

Gene Tunney James Joseph Tunney (boxer)

Geneva Cross (*see Cross* of Geneva)

Gene Wilder Jerome Silberman

Genghiz Khan (Mongolian— Perfect Warrior)—his empire stretched from the China Sea to the Dnieper and his subjects called him Ruler of the World

Genie Douglas air-to-air rocket fitted with a nuclear warhead and designated AIR-2A

Genius Ray Charles

Genius of the Renaissance Leonardo da Vinci

Gentle Jeems Pud Galvin (baseball player)

Gentleman Boss Chester Alan Arthur

Gentleman Explorer Giovanni da Verrazano

Gentleman Jim prizefighter James John Corbett

Gentleman Johnny General John Burgoyne, also a noted British playwright

Gentle Peasant-Prince-the loving Cotter-King Robert Burns, according to Robert G Ingersoll in his poem *The Birthplace of Burns*

Gentle Rebel of Psychoanalysis Karen Horney

Geoffrey Crayon Washington Irving

Geoffrey Homes Daniel Mainwaring's pseudonym

Geographical Center of North America Rugby, North Dakota

Geordieland Newcastle-on-Tyne area of northeastern England

Geordies people from the coalmining and industrial area of Newcastle-on-Tyne

Geordie(s) Newcastle (persons)

Georg Brandes Morris Cohen

George Arliss Augustus George Andrew

George Bellairs Harold Blundell's pseudonym

George Bizet Alexandre César Léopold Bizet

George Brent George Nolan

George Burns Nathan Birnbaum

George Eliot Mary Ann Evans Cross

George Gershwin Jacobi Gershvin

George Gissing J Storer Glouston

George London George Burnstein

George Orwell Eric Blair

George Raft George Ranft

George Sand Amandine Aurore Lucie Dupin (Baroness Dudevant)

George Santayana Jorge Augustín Nicholás Ruiz de Santayana

Georges Duhamel Denis Thevenin

George Spelvin John Chapman

Georges Simenon (pen name— Georges Sim)

George Szell György Széll

George W George Walker Bush — forty-third president of the United States

Georgia named for King George II of England

Georgia Peach Ty Cobb (baseball player)

Georgia Ports Savannah, Brunswick, Fernandina Beach

Georgia's Oldest City Savannah

Georgi Vladimov George Volosevich

Gerard de Nerval (pseudonym—Gerard Labrunie)

German German camomile (tea); German ivy (South African ivy); German knot (figure-8 knot); German lapis (imitation lapis lazuli); German measles (rubella); German shepherd (police dog); German silver (copper-nickel-zinc alloy resembling silver)

German Aeronautical Engineering Genius Willi Messerschmitt

German Africa former colonies (Cameroons, German East Africa, German Southwest Africa, Togoland)

German Agnostic Immanuel Kant

German Cradle of U.S. Presidents Germany—ancestral home of Presidents Eisenhower and Hoover

German East Africa colonial possession of Germany from 1885 to 1916; included most of Tanganyika

German Hanseatic Seaport Cities Bremen, Danzig, Hamburg, Lübeck

German Ocean North Sea

Germanophone Countries
Austria, Germany, Liechtenstein, Luxembourg, German-speaking cantons of Switzerland, and places in the United States where German dialects such as Pennsylvania German (Pennsylvania Dutch) are spoken

German Ports Warnemunde, Rostock, Wismar

Germans the Germans [Former President Nixon's Chief of Staff—HR (Bob) Haldeman and Domestic Adviser John Erlichman]

german silver 50% copper, 30% nickel, 20% zinc

German South-West Africa colonial possession of Germany from 1884 to 1915 Namibia

German-speaking Places (*see* Germanophone Countries)

Germany Federal Republic of Germany *Bundesrepublik Deutschland*

Germany's Largest Port Hamburg

Geronimo Goyahkla

Gershwin of the Tango Astor Piazolla

Gettysburg Battlefield Painter Henri Emmanuel Félix Philippoteaux

Ghana Ports Takoradi and Tema

ghetto blaster portable radio/tape player

Ghirlandaio Domenico di Tomaso Bigordi

ghoul trust tax-shelter gift plan involving third party expected to die prematurely

Giacomo Meyerbeer Jakob Liebmann Beer

Giant Cactus Moon full moon in July

Giant of Danish Literature Hans Christian Andersen

Giant of Freethought Charles Bradlaugh

Giant of Mexican Music Carlos Chavez

Giant in the Nursery Jean Piaget

Gig Young Byron Barr

Gilbert Roland Luis Antonio Dámaso de Alonso

Gilded Age opulent post-Civil War period in the United States

Gilois French-built scissors bridge mounted on a tank and useful in spanning canals, ditches, and small streams

Ginger Rogers Virginia McMath

Ginnie Mae Government National Mortgage Association (GNMA)

Gippsland Victoria—Australia's best-endowed province

Glacier Bay Alaskan national park

Gladiator Pete Browning (baseball player)

Glass Capital of Massachusetts Boston

Glass Capital of New York Corning

Glass Capital of Ohio Toledo

Glass Capital of Pennsylvania Pittsburgh

Glass Center Toledo, Ohio

Glass House the glassed-in Los Angeles County Jail in California

Glass Menagerie on the East River United Nations headquarters

Gleason score calculation indicating prostate cancer's level of aggressiveness

Glenard's disease prolapse of one or more internal organs

Glen Ford Gwyllyn Ford

Glengariff (Irish Gaelic—rough valley)—Banty Bay

Glimmerglass James Fenimore Cooper's nickname for Lake Otsego at Cooperstown, New York

Glimmer Twins rock musicians Mick Jagger and Keith Richards

Glitter Gulch Reno, Nevada

Globemaster Douglas transport designated C-124 and built for cargo carrying or flying 200 troops

Gloomy Gus Gustav Mahler

Glorious Fifty the United States of America

Glorious Fourth July 4 (Independence Day in the U.S.A.)

Glorious Gloria Gloria Swanson

Glorious Patriarch of Esterhazy Franz Joseph Haydn

Glubb Pasha John Bagot Glubb

Gnomes of Zurich Swiss bankers

Gobbo il Gobbo (Italian—the Hunchback)—15th-century architect and sculptor named Christoforo Solari

God of Animals, Crops, Fertility, Prophecy, and Rural Life Faunus (Roman); Pan (Greek)

God of Battle, Inspiration, and Death Odin (Norse)

God of Blacksmithing and Forges Hephaistos (Greek); Vulcan (Roman)

God of Blame and Ridicule Momus (Roman)

God of Bloodshed and War Greek god Ares; Roman god Mars

God of Boundaries Terminus (Roman)

God of the Christians and Jews Jehovah

God of Corn and Grain Robigus (Roman)

God of Creation and Destruction Siva (Hindu)

God of Cunning Dexterity Hermes (Greek); Mercury (Roman)

God of the Dead and the Underworld Dis (Roman); Hades (Greek); Mantus (Etruscan); Pluto (Roman)

God of Death Mors (Roman); Thanatos (Greek)

Goddess of Agriculture Ceres or Vacuna (Roman); Demeter (Greek)

Goddess of Animals, Crops, Fertility, Prophecy, and Rural Life Bona Dea or Bona Mater or Fauna (Roman)

Goddess of Arts, Crafts, and Sciences Athena (Greek); Minerva (Roman)

Goddess of Athens Athena

Goddess of Avenging Justice Nemesis (Greek)

Goddess of Beauty and Love Aphrodite (Greek); Venus (Roman)

Goddess of Birth the Roman goddess Carmenta also known as Carmentis

Goddess of the Breeze Aura (Greek)

Goddess of Bridesmaids Juno Pronuba (Roman)

Goddess of Burials, Corpses, and Funerals Libitina (Roman)

Goddess of Cattle and Pastures Pales (Roman)

Goddess of Chance Fortuna (Roman)

Goddess of Chaos, Sickness, and Death Kali (Hindu)

Goddess of Childbirth and Prophecy Carmenta, Juno Lucina, and Postverta

Goddess of the Crops Auxesia and Demeter

Goddess of the Dead Mania (Roman)

Goddess of Death Hel (Norse folklore)

Goddess of Destiny or Fate Necessitas (Roman)

Goddess of Discord and Strife Discordia (Roman) or Eris (Greek)

Goddess of Domestic Animals Bubona (Roman)

Goddess of Earth Gaea or Rhea (Greek); Cybele, Tellus, or Terra (Roman)

Goddess of Fair Speech and Good Report Eufemia (Greek)

Goddess of Faith, Honesty, and Oaths Fides (Roman)

Goddess of Fame Fama (Roman); Pheme (Greek)

Goddess of Family Harmony Verplaca (Roman) also spelled Virplaca

Goddess of Famine Fames

Goddess of the Fertile Earth Opalia or Ops (Roman)

Goddess of Fertility Frigga (Queen of Asgard and wife of Odin in Nordic mythology)

Goddess of Fertility, Love, Lust, and War Ishtar (Assyrian and Babylonian)

Goddess of Fertility and Procreation Aphrodite (Greek); Isis (Egyptian); Mylitta (Assyrian); Venus (Roman)

Goddess of Fertility and Purity Bona Dea (Roman)

Goddess of Fire Hestia (Greek); Vesta (Roman)

Goddess of Freedom Libertas (Roman)

Goddess of Funerals Naenia (Roman)

Goddess of the Future Antevorta (Roman)

Goddess of Gardens and Fruit Trees Pomona (Roman)

Goddess of Good Faith Fides

Goddess of Groves, Orchards, and Woods Feronia (Roman)

Goddess of Harmony Concordia (Roman)

Goddess of Healing Iaso (Greek)

Goddess of Health Hygeia (Roman); Hygieia (Greek)

Goddess of the Hearth Hestia (Greek); Vesta (Roman)

Goddess of Heaven Hera (Greek); Juno (Roman)

Goddess of the Home Hera (Greek); Juno (Roman)

Goddess of Home Security Cardea or Carna who guarded the door hinges and locks

Goddess of Horses Epona (Gallic); Hippona (Roman)

Goddess of Hunting and the Moon Artemis (Greek); Diana (Roman)

Goddess of Imposters and Thieves Laverna (Roman)

Goddess of Law and Order Eunomia or Themis (Greek); Justitia (Roman)

Goddess of Leisure and Repose Vacuna (Roman)

Goddess of Lighting Fulgora (Roman)

Goddess of Love and Lust Aphroide (Greek); Venus (Roman)

Goddess of Magic, Sorcery, and the Underworld Hecate or Hekate (Greek— working afar); Trivia (Latin— of the three ways) and hence the Romans placed her wherever three roads met

Goddess of Married Women Juno Matronalis (Roman)

Goddess of Memory Mnemosyne (Greek)—mother of the muses; her name gives rise to mnemonic—an aid to memory

Goddess of Menstruation Mena (Roman)

Goddess of Midwives Deverra (Roman); Eileitia (Greek)

Goddess of Mirth Thalia

Goddess of the Moon Luna (Roman); Selene (Greek)

Goddess Mother of the World Mount Everest

Goddess of Nature Cybele (Roman) or Kubele (Greek)—sometimes called Mistress of the Animals

Goddess of Newborn Babes Levana (Roman)

Goddess of Night Nux (Greek) sometimes spelled Nyx

Goddess of Nursing Mothers Rumina (Roman)

Goddess of Passion Stimula (Roman)

Goddess of the Past Postvorta (Roman)

Goddess of Peace known to the Romans as Concordia, Irene, or Pax, and to the Greeks as Eirene

Goddess of Profit Laverna (Roman)

Goddess of Public Welfare Salus (Roman)

Goddess of the Rainbow Iris (Roman)

Goddess of Robbers Furina (Roman)

Goddess of Rome Roma

Goddess of the Sea and Seaports Matuta (Roman)— originally goddess of the dawn

Goddess of Sensual Pleasure Voluptas (Roman)

Goddess of Shepherds Pales (Roman)

Goddess of Silence Muta (Roman)

Goddess of Storms and Winds Tempestes (Roman)

Goddess of Suckling Infants Rumina (Roman)

Goddess of Treachery Fraus (Roman)

Goddess of Truth Alethia (Greek); Veritas (Roman)

Goddess of the Underworld Persephone (Greek); Proserpina (Roman)

Goddess of Vice Kakia (Greek)

Goddess of Victory Nike (Greek)

Goddess of Virgins Juno Virginalis (Roman)

Goddess of War Bellona (Roman); Enyo (Greek)

Goddess of Wisdom Athena (Greek); Minerva (Roman)

Goddess of Youth Hebe (Greek); Juventus (Roman)

God of Dreams Morpheus—Greek god of dreams

God of Drinking Comus (Roman)

God of Earth Tellumo (Roman)

Gödel's Theorem all branches of mathematics are based on propositions that can't be proved within that branch (named for mathematician Kurt Gödel)

God of Eloquence and Oratory Hermes (Greek); Mercury (Roman)

Godfather of American Invention Thomas Jefferson

Godfather of American Liberty Thomas Paine

Godfather of Modernism James Laughlin (New Directions publisher)

Godfather of Navigation Nathaniel Bowditch

Godfather of Organized Crime in the U.S. Meyer Lansky

God of Fertility Frey (Norse); Priapos (Greek); Priapus (Roman)

God of Fields, Pastures, Shepherds, and Woods Faunus (Roman); Pan (Greek)

God of Fire Agni

God of Fire and Forges Hephaestus (Greek); Vulcan (Roman)

God of Forests, Herds, Plants, and Trees Silvanus (Roman)

God of Gods and Ruler of Heaven and Earth Zeus (Greek); Jove or Jupiter (Roman)

God of Gold Mammon the Materialist

God of Good Harvests and Successful Undertakings *Eventus Bonus* (Latin—Good Results)—a Roman god

God of the Greeks Panhellenius or Zeus

God of Healing and Medicine Asclepius (Greek); Aesculapius (Roman)

God of Heaven Uranus (Greek); Coleus (Roman)

God of Heaven, Lightning, Rain, Storm, and Thunder Indra (Hindus)

God of Inanimate Dreams Phantastus (Greek)

God of the Infernal Regions Dis (Greek); Pluto (Roman); Yama (Hindu); etc.

God of the Internet Jon Postel, director of Internet Assigned Numbers Authority

God-Intoxicated Man Benedictus de Spinoza

God of Landmarks Terminus (Roman)

Godless Victorian Sir Leslie Stephen

God of Light Mithra (Aryan, Indian, Persian)

God of Love Cupid (Roman); Eros (Greek); Krishna (Hindu)

God of Marriage Hymen, according to Greek mythology, also leader of the nuptial chorus and personification of the wedding feast

God of the Mohammedans Allah

God of Music Johann Sebastian Bach, according to Pablo Casals

God of Music, Poetry, and the Sun Apollo (Roman) or Apollon (Greek)

God of the Nile and Vegetation Osiris (Egyptian)

God of Purification Februus (Roman)

God of Revelry and Wine Dionysus (Greek); Bacchus (Roman)

God of the Romans Jupiter—supreme god

God's Athlete Pope John Paul II

God of the Sea Neptune (Roman); Poseidon (Greek)

God of Ships and the Sea Nyörd (Norse)

God of Skill Hermes (Greek); Mercury (Roman)—the winged cap-and-shoes messenger of Jove or Jupiter (Zeus) in one hand he bore a rod entwined by two serpents (the caduceus)—symbol of the medical profession

God of Sleep Hypnos (Greek); Somnus (Roman)

God of Soil Fertilization Saturn or Stercutus (Roman)

God of Springs Forts (Roman)

godsquad paramedic ambulance and crew; evangelists

God of the Sun Adonis (Syrian); Apollo (Roman); Apollon (Greek); Baal (Chaldean); Helios Hyperion (Greek in Homer's time); Horus (Upper Egypt); Mithras (Persian); Moloch (Canaanite); Osiris (Egyptian); Ra or Re (Egypt's Old Kingdom); Sol Invictus (Latin—Sun Invincible)—Romans shortened this to Sol; Surya (Hindu)

God of Thunder Thor (Norse)

God of Time Cronus (Greek); Saturn (Roman)

God of Trade and Travelers Hermes (Greek); Mercury (Roman)

God of the Underworld Dis (Roman); Hades (Greek); Mantus (Etruscan); Pluto (Roman)

God of Vineyards and Wine Bacchus (Roman); Dionysus (Greek)

God of War Ares (Greek); Mars (Roman)

God of Wine Bacchus (Roman); Dionysus (Greek)

gold gold-tinted Acapulco marijuana

Golda Meier Goldie Mabovitch, later Golda Meyerson

Gold Coast Africa's Ghana; Australia's resort area extending from Coolagantta to Southport; Florida's resort coast extending from Key West to Palm Beach

Gold Diggers Californians

Golden Age era of mankind, according to mythology, when people lived in contentment and peace with arts and crafts superior to those of the Silver, Bronze, Iron, and Stone ages; mankind's age of innocence where there was springtime all the time and happiness, right, and truth prevailed; there were no bodily ailments and nobody had to work— (*see Siglo de Oro*)

Golden Age of Greece 5th and 4th centuries before the Christian era when Aristotle, Euripides, Plato, and Sophocles lived

Golden Age of Opera late 1800s and early 1900s

Golden Age of Rome the reign of Augustus from 27 B.C.E. to 14 A.D.

Golden Age of Television the 1950s, when live studio drama reigned

golden beryl heliodor

Golden Century Nineteenth Century

Golden City Johannesburg

Golden City of a Hundred Spires Praque

Golden Continent Africa

Golden Crescent opium-production area along borders of Afghanistan, Iran, and Pakistan

Golden Door Ellis Island

Golden Gate entrance to San Francisco Bay

Golden Gate City San Francisco

Golden Gate to South America Cartagena, Colombia

Golden Greek George Kaftan (basketball player)

Golden Horn Istanbul's harbor formed by the curved arm of the Bosporus

Golden Horseshoe Hamilton-Toronto-Oshawa industrial complex along Lake Ontario

Golden Hyphen Winston-Salem, North Carolina

Golden Isles Jekyll, Saint Simons, and Sea Island off Brunswick, Georgia

Golden Key to the Fjords Stavanger, Norway

Golden Mile London, New York, Tokyo

Golden Peninsula Malay Peninsula

Golden Prison of Paris the Louvre

Golden Province Canada's Ontario

Golden Rock of the Caribbean Sint Eustatius (Statia), Netherlands Antilles

Golden State California

Golden Triangle point of downtown Pittsburgh where the Alleghney and Monongahela form the Ohio River; industrialized northern Europe where the three points are Birmingham, Paris, and the Ruhr; opium-productive

fields where Burma, Laos, and Thailand meet near southern Yunnan, China

Golden Triangle of Texan Ports Beaumont, Orange, Port Arthur

Golden Trombone of Abolition Frederick A Douglass

Golden-voiced Tenors Enrico Caruso and Luciano Pavarotti

Goldhunter(s) Californian(s)

Goldilocks economy an economy that is not too hot, not too cold, just right

Gold Rush Town Nome, Alaska

Gold-standard President William McKinley

Goldy Oliver Goldsmith's nickname bestowed him by Dr Samuel Johnson

Golftown Pinehurst, North Carolina

Gonzalez Protocol diet, supplements and procedures that eliminate toxins and are used to treat cancer and other diseases (named for Dr Nicholas Gonzalez)

Goober-grabbers Georgians

Goober(s) peanut grower(s), natives of Alabama, Georgia, and North Carolina

Goober State Georgia

good cholesterol high-density lipoprotein

Good Gray Poet Walt Whitman

Good Law Buddhism

Good Luck City Vancouver, British Columbia

Good Queen Bess Queen Elizabeth I of England (1558-1603)

Good Richard Benjamin Franklin

Good Samaritan City of the Mississippi Memphis

Goodwill Ambassador of Mexico Henryk Szering (Polish-born violinist)

Goose Rich Gossage

Gopher(s) Minnesotan(s)

Gopher State Minnesota

Gorby Mikhail S Gorbachev

Gordon Homes pseudonym shared by Louis Tracy and MP Shiel

Gordon Pasha Charles George Gordon

The Gorgeous Miliza Korjus (Mrs Walter Schector)

Gorgeous George George Sisler (baseball player)

Gorki Soviet name for Nizhni Novgorod

Gorki (Russian—Bitter One)—pen name of Aleksei Maxsimovich Peskov

Gothab old name for Nuuk, Greenland

Gotham New York City

Gothamite(s) native(s) of New York City; nickname derived from *The Three Wise Men of Gotham* by Washington Irving

Governor Moonbeam former California Governor Edmund G "Jerry" Brown's nickname

Grace Greenwood Sara Jane Clarke Lippincott's pseudonym

Gracie Fields Grace Stansfield

Graduate of Oxford John Ruskin's pseudonym

Graffiti Capital New York City

Graffitic City New York City

Graham cracker named for Sylvester Graham, a Presbyterian minister

Grain Coast Sierra Leone and Liberia

Granary of Canada Saskatchewan

Granary of Portugal province of Alemtejo

Granary of Russia Ukraine

Granary of Spain lower valley of the Guadalquivir

Granary of Sweden Skåne

Gran Chaco lowlands of Bolivia, Paraguay, and Argentina

Gran Colombia (Spanish—Great Colombia)—post-colonial consolidation of Colombia, Ecuador, and Venezuela

Grand Admiral of the Ocean Sea Christopher Columbus

Grand Canyon of the Pacific Waimea Canyon, Kauai, Hawaii

Grand Canyon State Arizona

Grand Cham of Literature Dr Samuel Johnson

Grand Commanders Cayman Islanders

Grand Dame of the Piano Alicia de Larrocha

Grand Dame of Sex Education Dr Mary Steichen Calderone

Grand Divide Continental Divide

Grand Duke Duke of Wellington

Grandfathers of the Welfare State Winston Churchill, Lloyd George

Grand Hotel nickname of French Polynesia's prison in Tahiti

Grand Inquisitor Tomás de Torquemada appointed first Inquisitor General by Ferdinand and Isabella, made Grand Inquisitor by Pope Innocent VIII

Grandma Moses Anne Mary Moses

Grand Master of Humorous Verse Ogden Nash

Grand Master of the String Quartet Franz Joseph Haydn

Grandmother of Boston preacher-reformer-teacher Elizabeth Palmer Peabody

Grandmother of the Russian Revolution Katherine Breshkovska

Grandmother of Shopping Streets Calle Florída in Buenos Aires

Grandmother of the Women's Movement Margaret Mead

Grand Old Lady of Fifty-seventh Street Carnegie Hall

Grand Old Lady of Opera Ernestine Schumann-Heink

Grand Old Man William Ewart Gladstone—four times Prime Minister of Great Britain

Grand Old Man of American Labor Samuel Gompers

Grand Old Man of Africa Felix Houphouet-Boigny, who was president of the Ivory Coast for 33 years

Grand Old Man of Canadian Letters Robertson Davies

Grand Old Party Republican Party of the United States—the GOP

Grandsire of American Painting Benjamin West

Grand Trunk India's Grand Trunk Road from the mouth of the Ganges to Pakistan

Grand Zohra General Charles de Gaulle

Granger States Illinois, Iowa, Minnesota, and Wisconsin

Granite boy(s) New Hampshirite(s)

Granite Center Barre, Vermont

Granite City Aberdeen, Scotland

Granite Island Corsica

Granite State New Hampshire

Gran Libertador (Spanish—Great Liberator)—Simón Bolívar)

Grape State California

grape sugar glucose $(C_6H_{12}O_6)$

grass marijuana

Grasshoppers Kansans

Grasshopper State Kansas

Gravenstein Apple Center of the World Sebastopol, California

Graveyard of the Pacific Point Arguello northwest of Santa Barbara, California

Graveyards of the Atlantic Cape Hatteras, North Carolina and Sable Island off Nova Scotia

Gray Flamingo Tom Brennan (baseball player)

Gray Lady The New York Times

gray wave retirees

Great American Pastimes baseball, basketball, and football

Great Apes chimpanzees, gorillas, orangutans

Great Assassin Abdul-Hamid II (notorious for his participation in the Armenian atrocities)

Great Basin Nevada national park

Great Britain England, Scotland, and Wales—GB

Great Canal waterway between Australia and the Great Barrier Reef

Great Central State North Dakota

Great Cham of Literature Dr Samuel Johnson

Great Charter Magna Charta

Great Commoner Henry Clay, William Ewart Gladstone, William Pitt (the Elder Pitt also known as the Earl of Chatham), and Thomas Paine

The Great Commoner William Jennings Bryan

Great Compromiser Henry Clay

Great Debunker H(enry) L(ouis) Mencken—editor of *The American Mercury*

Great Destroyer syphilis

Great Disappointment October 22, 1844, when thousands of Christians expected the second coming of Jesus Christ

Great Dissenter Supreme Court Justice Oliver Wendell Holmes, Jr

Great Divide continental divide formed by Rocky Mountains; water on western slopes flow to the Pacific, on eastern slopes flow to Gulf of Mexico

Great Duke Duke of Wellington

Great Emancipator Abraham Lincoln

The Great Engineer Herbert Hoover

Greater Antilles Cuba, Hispaniola (Dominican Republic and Haiti), Jamaica, Puerto Rico

Greater Sunda Islands Borneo, Celebes, Java, Sumatra, and Indonesian islands

Greatest American Jurist John Marshall—Chief Justice of the Supreme Court from 1801 to 1835

Greatest Artist of the South Seas Paul Gauguin

Greatest Composer Haydn's name for Mozart

Greatest Heavyweight Boxer Jack Johnson

Greatest Show on Earth Barnum and Bailey–Ringling Brothers Circus

Greatest Snow on Earth Idaho

Great Fatherland War Soviet name for World War II

Great Helmsman Mao Zedong

Great Indian Great Indian Desert of India and Pakistan, also called the Thar

Great Jailer of the Caribbean Fidel Castro

Great Lakes Ontario, Erie, Huron, Michigan, Superior

Great Lakes Canada Ontario (on the northern shores of Superior, Huron, Erie, and Ontario)

Great Lakes Province Ontario

Great Lake States New York, Pennsylvania, Ohio, Michigan, Indiana, Illinois, Wisconsin, Minnesota

Great Lake State Michigan

Great Land The Great Land— Alaska

Great Liberator Simón Bolívar

The Great Lover Rudolph Valentino

Great Magrid (preacher) Dov Baer of Mezritch

Great Migration migration since 1916 of American blacks out of the South

Great Moralist Dr Samuel Johnson

The Great One Jackie Gleason

Great Outsider B Traven (expatriate American author)

Great Pacificator Henry Clay

Great Patriotic Struggle official Soviet name for Russia's participation in World War II

Great Pearl Hui-hai Ta-chu, eighth-century Chinese Zen master

Great Plains plains and prairies of Canada and the United States east of the Rockies

Great Poet of Democracy Walt Whitman

Great River Road 3,700-mile-long Mississippi-Missouri-Red Rock river system

Great Sage Ramana Maharshi

Great Sandy Great Sandy Desert of northwestern Australia

Great Sea Biblical name for the Mediterranean

Great Separationist Paul Blanshard

Great Smoke London before air-pollution control was enforced

Great Society administration of Lyndon Baines Johnson— thirty-sixth President of the United States

Great Stink Thames River before cleanup

Great Stone Face Daniel Webster; Old Man of the Mountain also known as Profile Mountain in New Hampshire's White Mountains

Great Street State Street that Great Street in Chicago

Great Thirst Land South Africa

Great Victoria Great Victoria Desert of Western Australia

Great Wet Ditch British nickname for the English Channel

Great White Father (American Indian term—the President of the United States)

Great White Fleet white-hulled flotilla of the United States Navy displayed in principal ports of the world during circumnavigation ordered by President Theodore Roosevelt; white-painted ships of the United Fruit Company— also called *La Gran Flota Blanca*

Great White Strip main street of Las Vegas, Nevada

Great White Way New York City's theatrical section of Broadway

Great White Wizard Dr Albert Schweitzer

Greece's Principal Port Piraeus

Greek Century 5th century before the Christian era—the 400s BCE

Greek Isles Cyclades, Dodecanese, Ionian, Sporades

Greek love pederasty

Greek Muses (*see* muses)

Greek Ports Argostolion, Patrai, Kalámai, Póros, Piraieus, Vólos, Thessaloniki, Mililini, Iraklion, Ródhos

Greek Row fraternity and sorority houses

Greeks Greek-letter fraternities or sororities

green bullets tungsten (not lead) bullets

Green Corn Moon full moon in August

Greene and Greene architectural style typifying the American Arts and Crafts Movement (named for brothers Charles and Henry Greene)

Green Flash rare view of green light from the rising or setting sun in clear atmosphere, lasting up to 3 seconds

Green Goddess Gro Harlem Brundtland (Norway's first female and youngest prime minister)

Green Hell Paraguayan Chaco

Green Irish Roman Catholics

Green Isle Ireland

Green Mountain boy(s) Vermonter(s)

Green Mountain City Montpelier, Vermont

Green Mountain State Vermont

green vitriol copperous, ferrous sulfate ($FeSO_4 \cdot 7H_2O$)

Greenwood Brooklyn's historic cemetery and eponym standing for similar burial places

Gregorian chants Liturgical music of the Roman Catholic Church, used to accompany text of the Mass, named for Pope Gregory (590–604)

Grenada's Port St George

Greta Garbo Greta Gustafson

Gresham's law Bad money drives out good

Grey Eagle Tris Speaker (baseball player)

Greyhound M-8 6-wheeled armored car carrying a 37mm gun and made in the U.S.A.

Greyhounds of the Pacific racing yachts

Grey Owl George S Belaney

Greytown San Juan del Norte

Gröfaz *Grösster Feldherr aller Zeiten* (German—greatest general of all time)—Hitler's acronymic nickname

Groperland Western Australia

Grotius Hugo de Groot

Groucho Marx Julius Henry Marx

Groundhog State Mississippi

Group of Eight United States, Japan, Great Britain, Germany, France, Italy, Canada and Russia

Group of Five Britain, France, Japan, the US, and Germany (concerned with stabilization of the dollar and mutually profitable trade agreements)

Group of Seven Britain, Canada, France, Italy, Japan, the US, Germany

grub ghost writers; grub writers of biographies, dictionaries, histories; grubstreet

G-town Georgetown; Germantown

Guadalupe Victoria Manuel Felíx Fernández

Guamanian Port Apra

Guanaco Central American nickname for a farmer or other rustic and Andean name for a member of the camel family resembling a llama

Guanahani (Lucayan—San Salvador or Watling Island)—first land discovered by Columbus in the New World

Guangzhou (Pinyin Chinese—southern city) Canton

Guardian Angel of Israel Michael

Guardian of the Gulf Oman

Guarnerius Giuseppi Antonio Guarneri

Guatemalan Ports Livingston, Puerto Barrios, Santo Tomas de Castillo; San José, Iztapa, Champerico

Guideline Soviet SA-2 missile system

The Guild The Newspaper Guild

Guinea-Bissau Ports Casheu, Bissau, Bolama

Guinea Pig State Arkansas

Guinea Port Conakry

Guitar Town Nashville, Tennessee

Gulag Archipelago Solzhenitsyn's title for the thousands of prisons in the USSR

gulag gas natural gas pipeline built by Soviet forced labor and extending from Siberia to Germany

Gulf City Mobile, Alabama

Gulf of Mexico's Principal Port New Orleans

Gull's disease myxedema resulting from atrophy of the thyroid gland

Gumbo Coast coastal parishes of Louisiana

Gunboat Smith Edward Smyth (boxer)

Gunflint(s) Rhode Islander(s)

Gussie Augusta; Augustina; Augustine

Guyana's Ports Bartica, Georgetown, McKenzie, New Amsterdam

Guy d'Hardelot Mrs WI Rhodes (Helen Guy)

Guys Marsh borstal in Dorset, England

Gyp Gypsy; Gyp the Blood; Marie Antoinette de Riquetti de Mirabeau, Countess de Martel de Janville's pseudonym

Gyppy (British slang— Egyptian)

Gypsum City Fort Dodge, Iowa

Gypsy Rose Lee Louise Hovick

Haakon the Good King Haakon I of Norway

Haakon the Old King Haakon IV of Norway

Haarlem Netherlands city and province

Hab(bie) Albert; Alberta; Halbert

Habeas Corpus Howe William Frederick Howe, also nicknamed Criminal Bar Howe

Habitants (French—Inhabitants)—Canadian farmers and fishermen of French descent

Hacha Falaya (Choctaw—Long River)—Atchafalaya River

hacktwists computer hackers with political or social agendas

Hafun formerly Dante when Somalia was Italian Somaliland

Hageman factor causes blood to clot when it comes into contact with a foreign surface (named for patient John Hageman)

Haggisland Scotland

Haiti's Principal Port Port-au-Prince

Hal Croves pseudonym of B Traven whose full name is reputed to be Berick Traven Torsvan

halfbacks retirees who had moved to Florida and later move partway back North

Hallé Hallé Orchestra

Halévy Jacques Fromental Elie Lévy

Halfway House on the Pacific Highway Hawaii, so named by Mark Twain

Hal Meredith Harry Blyth's pseudonym

Halstern's disease endemic syphilis

hamburg hamburg brandy (beet or potato alcohol flavored to imitate grape brandy); hamburger; hamburg steak

Hamburg Bach Carl Philipp Emanuel Bach also called the Berlin Bach

Hamlet's Town Helsingør, Denmark (called Elsinore by the English)

Hammerfestinger native of Hammerfest, Norway

Hammering Hank Henry Aaron

Hammerman John Henry

Hammer of Scotland Edward I

Hammond Innes pseudonym of Ralph Hammond-Innes

Hampton Roads Ports Newport News, Norfolk, Portsmouth

Handcuff King Harry Houdini

Hand of Fatima five-fingered heraldic symbol topping the emblem of Algeria

Handy William Christopher Handy (black American composer of the *St Louis Blues*)

Hanging Judge Judge Roy Bean of Langtry, Texas—Law West of the Pecos

Hangman's Day Friday

Hangtown El Dorado, California

Hank Henry

Han Kook Republic of Korea

Hanna-Barbera cartoonists Bill Hanna and Joe Barbera

Hanot's disease cirrhosis of the liver accompanied by jaundice

Hansa Ports Hanseatic League ports—Bremen and Hamburg on the North Sea, Danzig and Lübeck on the Baltic, Visby on Gotland Island in the Baltic

Hansen's disease leprosy

Hans Fallada pseudonym of Rudolf Ditzen

Hap Arnold General Henry Harley Arnold, USA and USAF

Happy Chandler High Commissioner of Baseball Albert Benjamin Chandler

Happy Home of the Bulldozer Los Angeles

Happy Jack Jack Chesbro (baseball)

Happy Land Burma (now Myanmar)

Happy Valley The Vale of Kashmir in the Himalayas

Happy Warrior Franklin D Roosevelt's nickname for New York State's Governor Alfred E Smith; Hubert Horatio Humphrey

Hapsburg Empire Austro-Hungarian Empire

Harbin Russian name for Pinkiang, Manchuria

Harbor City Erie, Pennsylvania

Harbor of the Sun San Diego, California

Hard-Case State Oregon

Hard Heart of Hickland Cleveland, Ohio, according to authors Jack Lait and Lee Mortimer

Hard Rock nickname of the American Broadcasting Company (ABC)

Hardware City New Britain, Connecticut

Hardy-Weinberg law predicts distribution of genotypes within a population

Harke Soviet Mi-10 heavy-transport helicopter

Harlem New York City neighborhood in northeast sector

Harmonica Slim Travis Blaylock

Harmony former place-name of Ambridge, Pennsylvania

Harold Bluetooth King Harold of Denmark

Harold Harefoot Harold I of Denmark and England

Harp/Hormone Soviet Ka-20 or Ka-25k helicopter for military or commercial use (Harp is military and Hormone is commercial version)

Harpo Marx Adolph Arthur Marx

Harpoon harpoon-type aircraft command and launch subsystem missle; Lockheed maritime reconnaissance bomber

Harrier Hawker-Siddeley fixed-wing fighter aircraft; McDonnell-Douglas AV-8B jump-jet bomber

Harrisburg Houdini Bob Davies (basketball player)

Harry Golden Herschel Goldhirsch

Harry Houdini Ehrich Weiss

Harry Morgan Harry Bratsburg

Harry Reems Herbert Streicher

Hartford of the West Lincoln, Nebraska

Hartford Wits Joel Barlow, Timothy Dwight, Jonathan Trumbull

Harvard's Heroic Historian John Lothrop Motley

Harvest of Death cold wind in higher regions of Bolivia that causes lung diseases

Harvest Moon full moon in September

hash hashish

Hashish Hasan-ibn-al-Sabbah (11th-century Persian founder of the Assassins)

Hattie Harriet

haunts ghosts

Haunt of Yachtsmen British Virgin Islands

Havana II Miami, Florida

Haven for Arthritics Jacumba, California

Havercake(s) native(s) of Lancashire

Hawaiian Island Ports Hilo, Hawaii; Kahului, Maui; Honolulu, Oahu; Port Allen, Kauai; Nawiliwili Bay, Kauai

Hawaiian Pineapple King James Drummond Dole

Hawaiians Hawaiian Islanders; Hawaiian Islands

Hawk Ralph Branca (baseball player)

Hawkeye airborne early-warning and fighter-control aircraft—the E-2

Hawkeye(s) Iowan(s)

Hawkeye State Iowa

Hay Capital of the World Meckling, South Dakota

Hay diet a diet of 20 percent acid-forming foods and 80 percent alkaline-forming foods, developed by William Howard Hay

Hay Moon full moon in July

Head of the Adriatic Trieste

Head of the Commonwealth Her (His) Most Excellent Majesty the Queen (King) of the United Kingdom of Great Britain and Northern Ireland and of Her (His) other Realms and Territories Queen (King)

Headless Man the nude mystery man photographed with the Duchess of Argyll who became focal point of a lurid divorce lawsuit

Head of the Sufis Abu'l-Fayd Thawban ibn Ibrahim Dhu'l-Nun al-Misri

Health City Battle Creek, Michigan

Hearst's Castle (see La Casa Grande)

Heart of America Kansas City

heart attack on a plate fettucine

Heart of California Sacramento

Heart of Canada Ontario

Heart of Central Alaska Fairbanks

Heart of Darkness Zaire (formerly called the Congo)

Heart of Dixie Alabama

Heart of England Warwickshire

Heart of Historic Virginia Charlottesville

Heart-of-it-All State Ohio

Heart of Kentucky Frankfort

Heartland of America the Midwest

Heartland of Catholicism Italy, the Vatican

Heartland City Kansas City

Heartland of Monarchy Grand Duchy of Luxembourg

Heart of Midlothian Tolbooth Prison in Edinburgh

Heart of Polynesia Samoa

Heart of Portugal Mondego Valley

Heart of the Roman Empire Italy

Heart of the South Atlanta, Georgia

Heart of South America Bolivia

Heart of Sweden Dalarna Province formerly called Dalecarlia

Heavenly Twins Hugy Duffy and Tommy McCarthy (baseball)

heavy metal driving and loud music produced electronically

Hebb's learning rule a memory is produced when two connected neurons are active simultaneously and the synapse is strengthened (named for Donald O Hebb)

Hedda Hopper Elda Furry

Hedy Lamarr Hedwig Kiesler

Heel of Italy Salentine Peninsula

Heide Adelaide

Heine-Medin disease muscular atrophy sometimes followed by permanent deformity

Helena Modjeska Helena Modrejewska

Helen Hayes Helen Hayes Brown

Helen Twelvetrees Helen Jurgens

Helium Heads blimp supporters

Hell Breughel Pieter Breughel the Younger who painted hellish scenes

Hellcat Grumman F6F single-seat fighter aircraft; U.S.-made 76mm gun mounted in a fully traversing turret on a tracked chasis (M- 18)

Hellenic Republic Greece

Hellfire Cities Hiroshima, Nagasaki

Hell in the Hills Pittsburgh, Pennsylvania

Hellhole of the Pacific Kororareka, New Zealand

Hell of Java Trinil (where Dr Eugene Dubois discovered *Pithecanthropus erectus*)

Hell of Macquarie Harbour Station old penal colony on the Indian Ocean coast of Tasmania

Hell and Maria Charles G Dawes

Hell's Forty Acres San Carlos, Arizona

Hell's Gates Macquarie Harbour, Tasmania's first convict settlement

Hell's Kitchen New York City's lower west side including San Juan Hill

Hell's Parlor Zanzibar

Hell's Passage Saint Helen's Passage, Oxford

Hell on Wheels Cheyenne, Wyoming

Hennie Henrietta

Henry Armstrong Henry Jackson (boxer)

Henry Bolingbroke Henry IV of England

Henry Cecil Henry Cecil Leon

Henry Green Henry Vincent Yorke

Henry the K Henry Kissinger

Henry the Navigator Dom Henrique o Navegador (Prince of Portugal and patron of explorers and voyagers)

Henry Wade Henry Lancelot Aubrey-Fletcher

Herald Handley-Page turbo-prop transport plane

herbal Valium kava-kava

Herblock Herbert Lawrence Block

Hercules Lockheed KC-130 tanker aircraft

Her Deepness Sylvia Earle, marine biologist who walked on the ocean floor 1,250 feet deep in 1979

Herkimer diamond gem-quality quartz from New York State's Herkimer County

Hermanus Vanderdonk Washington Irving

The Hermitage Andrew Jackson's home near Nashville, Tennessee; palace museum of art in Leningrad (formerly a czarist palace)

Hermit of Concord Henry David Thoreau

Hermit Kingdom Bhutan; Korea

Hermit of Slabsides John Burroughs

Hernandarias Hernando Arias de Saavedra

Hero of a Hundred Battles the Duke of Wellington

Hero of a Hundred Fights Admiral Horatio Nelson

Hero of Antiquity Heracles or Herakles (Greek); Hercules (Roman)

Hero of Appomattox General Ulysses Simpson Grant, USA

Hero of the Cities Alfred E(manuel) Smith—usually called Al Smith

Hero of Civilization Thomas Paine

Heroes of Guanajuato Mexican priests Aldama, Allende, and Hidalgo who led the War of Independence against the Spanish

Hero of Fort Sumter Confederate General Pierre Gustave Toutant Beauregard (known to his soldiers as Old Alphabet or Old Bore)

Hero of the Frontier George Rogers Clark

Heroic City Cartagena, Colombia *(Ciudad Heroica)*

Hero of Lake Erie Commodore Oliver Hazard Perry, USN

Hero of Manila Bay Commodore George Dewey, USN

Hero of Mobile Bay Admiral David Glasgow Farragut

Hero of Modern Italy Guiseppe Garibaldi

Heron Hawker-Siddeley 17-passenger transport plane

Hero of Nacozari Jesús Garcia

Hero of New England Captain Miles Standish

Hero of New Orleans General Andrew Jackson

Hero of the Nile Lord Horatio Nelson

Hero of the Plain People Andrew Jackson

Hero of San Juan Hill Lt Col Theodore Roosevelt, USV

Hero of the Spanish-American War Admiral George Dewey, USN

Hero of State-Church Separation Benito Juarez

Hero of Tampico General Antonio López de Santa Anna

Hero and Traitor Benedict Arnold

Hero of Upper Canada Sir Isaac Brock

Herring Chokers Newfoundlanders

Herring Pond Atlantic Ocean

Hesperides Canary and Madeira Islands

hessian hessian boots (knee-high and tasseled); hessian fly (insect feeding on grass and wheat stems)

Hidden Empire Ethiopia

Hieronymus Bosch palette name of Hieronymus van Aeken

High Lonesome southwestern Colorado

High Plains States Arkansas, Kansas, Missouri, New Mexico, Oklahoma, Texas

Highpockets George Kelly (baseball player)

High Priestess of Transcendentalism Margaret Fuller

High-Tide Province Canada's New Brunswick

Hikari (Japanese—Sun-beam)—train linking Tokyo with other coastal cities of Honshu

Hilaire Belloc Joseph Hilary Pierre Belloc

Hildegarde Neff Hildegard Knef

Hillbilly Country mountainous parts of the Carolinas, Georgia, Tennessee, Kentucky, and West Virginia

Hill City Portland, Maine

Hill of Spring Tel Aviv

Hindu Monarchy Nepal

hinny offspring of a jennet or female donkey sired by a stallion or male horse

Hi-no-maru (Japanese—Sun Flag)—emblem of Japan

Hippocrates of Pennsylvania Benjamin Rush

Hirschsprung's disease congenital colonic dilatation

Hispania (Latin—Iberian Peninsula)—land now divided between Portugal and Spain; poetic name for Spain

Hispanic America Portuguese- and Spanish-speaking countries of Latin America

Hispanic American City Los Angeles, California; San Antonio, Texas; San Diego, California

Hispanic Places Andorra, Argentina, Azores, Balearic Islands, Bolivia, Brazil, Canary Islands, Cape Verde Islands, Ceuta and Melilla, Chile, Colombia, Costa Rica, Cuba, Dominican Republic, Ecuador, El Salvador, Equatorial Guinea, Guam, Guatemala, Honduras, Macao, Madeira, Mexico, Morocco, Nicaragua, Panama, Paraguay, Peru, Philippines, Portugal, Puerto Rico, Spain, Spanish Sahara, Uruguay, Venezuela

Hispanics people of Portuguese or Spanish descent or a study of their culture and language

Histologist of the Brain Santiago Ramón y Cajal

Historian of the American Forest Francis Parkman

Historian of Liberty Lord Acton

Historian With a Camera Mathew B Brady

Historic Center of North Carolina New Bern

Hoagy Hoagland Howard Carmichael

Hob(bie) Albert

Hobson's choice this or nothing (after Thomas Hobson, 1544–1630, whose horses were never known to have left the stable out of turn)

Ho Chi Minh (Vietnamese—He Who Enlightens)—Nguyen Ai Quac whose patronym was Nguyen Van Coong but is also known as Nguyen That Thanh

Ho Chi Minh City formerly Saigon, Vietnam

hocus morphine's nickname

Hodara's disease hair splitting

Hodge nickname for the typical English farmer or for Roger

Hodgkin's disease progressive enlargement of the lymph nodes

hog large Harley-Davidson motorcycle

Hogarth's Act Act of Parliament passed in 1735 granting copyright protection

Hog Butcher for the World Chicago

Hog Country Arkansas

Hog and Hominy State Tennessee

Hog Lane Hoxton

Hogopolis Chicago

Holland the Netherlands; actually an old Dutch province comprising part of the Netherlands; Michigan summer resort and manufacturing center settled by the Dutch

Holland of America Louisiana

Holland in the Caribbean Netherlands Antilles (Aruba, Bonaire, Curaçao, Saba, Sint Eustatius, Sint Maarten)

Hollie Holiday; Holladay; Hollingsworth; Hollis; Hollister; Hollway

Hollywood's First Dauphin Douglas Fairbanks Jr

Holstein Capital Northfield, Minnesota

Holy Alliance Austria, Prussia, Russia (in 1815)

Holy Cities Mecca and Medina in Saudi Arabia

Holy City nickname given Charleston, South Carolina by its inhabitants

Holy City of India Banaras

Holy Devil Rasputin

Holy Horatio Horatio Alger, Jr

Holy Land Israel

Holy Land of Three Religions Israel

Holy Rabble Rouser Ayatullah Ruhollah Khomeini

Holy Three of Criminology Enrico Ferri, Raffaele Garofalo, Cesare Lombroso

Home of Abraham Lincoln Springfield, Illinois

Home of the Alamo San Antonio, Texas

Home of Baseball Cooperstown, New York

Home of the Bean and the Cod Boston, Massachusetts

Home of the Blizzard Adelie Land, Antarctica

Home of the Blues Memphis, Tennessee

Home of the Brave United States of America

Home of the Casbah Algiers

Home of the Casey Jones Jackson, Tennessee

Home of the Comstock Lode Virginia City, Nevada

Home of Contented Cows Carnation, Washington

Home of the Cotton Carnival Memphis, Tennessee

Home of Diamond Walnuts Stockton, California

Home of the Dinosaurs Glen Rose, Texas where petrified footprints of 30-foot-long dinosaurs are displayed

Home of Franklin Delano Roosevelt Hyde Park, New York

Home of George Washington Mount Vernon, Virginia

Home of the Giants Jotunheimen Mountains in Norway

Home of Goethe Heidelberg, Germany

Home of Holbein Augsburg, Germany

Home of Jesse James St Joseph, Missouri

Home of the Kentucky Derby Louisville

Homeland of the Bengalis Bangladesh

Homeland of Yogurt Bulgaria

Home of Old Miss Oxford, Mississippi—the home of The University of Mississippi

Home Run Frank Baker

Homer Wilbur pseudonym of James Russell Lowell

Home of the Snow Himalaya Mountains

Home of Storms Gulf of Alaska

Home of Theodore Roosevelt Oyster Bay, Long Island, New York

Home of Thomas Jefferson Monticello, Virginia

Hometown Hamilton, Ohio as defined by author Peter Davis

Home of the Waltz Vienna

homeys persons from the same hometown or region

Hondo Frank Howard (baseball)

Honduran Ports Puerto Cortés, Tela, Puerto Este, La Ceiba, Trujillo, Puerto Castilla, Roatán, Amapala, San Lorenzo

Honduras Republic of Honduras (Spanish-speaking Central American nation) *República de Honduras*

Honest Abe Abraham Lincoln

Honest Harold Secretary of the Interior Harold Le Claire Ickes, also called the Old Curmudgeon

Honest John solid-sustainer motor surface-to-surface ballistic missile produced by Douglas Aircraft

Honey Capital Uvalde, Texas

Honeydipper Roosevelt Sykes

Honey Fitz John F. (Honey Fitz) Fitzgerald

Honey Lulu Honolulu, Hawaii

Honey Moon full moon in June

Honeymoon City Niagara Falls, New York

Honey State Western Australia

Hongcouver Vancouver, British Columbia, many of whose residents are from Hong Kong

Honorary Citizen of the United States Winston Churchill (first and only foreigner to bear this title)

Hooker Boulevard El Cajon Boulevard

Hoosier Capital Indianapolis, Indiana

Hoosier City Indianapolis, Indiana

Hoosier Poet James Whitcomb Riley

Hoosier(s) name believed to be a frontier-era contraction of *Who's there?*—native(s) of Indiana

Hoosier State Indiana

Hoot Ward Gibson (basketball player)

Hormone Soviet armed helicopter in naval service (KA-25)

Horn of Africa Djibouti, Ethiopia, Somalia, Sudan, particularly northeasternmost Somalia terminating in Cape Guardafui which the Arabs call the Ras Asir

Hornet F-A-18 McDonnell Douglas fighter-attack aircraft

Horse Latitudes belts of calms about 30 or 35 degrees north or south of Equator; horses were cast overboard in these places when sailing vessels were becalmed and drinking water became scarce

Horseshoe Curve Altoona, Pennsylvania close to the celebrated Horseshoe Curve built by the Pennsylvania Railroad to cross the Alleghenies and traverse the valley of the Juniata River

Horses in Fiction or History *Altair* who served Ben Hur; *Bucephalus,* who served Alexander the Great; *Buttermilk,* who served Dale Evans; *Egypt,* who served General Ulysses S Grant; *Fritz,* who served William S Hart; *Jenny,* who served Robert Burns; *Ko-Ko,* who served Rex Allen; *Lightning,* who served Tim Holt; *Marengo,* who served Napoleon; *Nelson,* who served George Washington, *Pegasus-Rosinante,* who served Don Quixote; *Tony,* who served Tom Mix; *Topper,* who served Hopalong Cassidy; *Trigger,* who served Roy Rogers; *Xanthus,* who served Achilles; *Traveller,* who served General Robert E Lee; and don't forget *Black Jack,* who marched alone in presidential funerals

Horse Thief Hollow Oak Lawn, Michigan

Hosea Biglow pseudonym of James Russel Lowell

Hostess to the Nation Dolly Madison—wife of President James Madison

Hotbed of Secession Charleston, South Carolina

hot blood high blood pressure

hot-dog headache reaction to nitrates, found in hot dogs and other cured meats and cold cuts

Hotel letter H radio code; H-class Soviet missile-launching nuclear-powered submarines

Hotlanta Atlanta, Georgia

hot-metal men steelworkers

hot'ns Southern nickname for biscuits

Hot Potato Luke Hamlin

Hot Springs Arkansas national park

Hot Springs Country southwest Arkansas

Hotspur Sir Henry Percy

Hottest Town in Arizona Quartzite

Hottest Town in Texas Presidio

Hot Water City Hot Springs, Arkansas

Hot Water State Arkansas

Houdini Harry Houdini (real name Ehrich Weiss)—America's foremost escapologist-magician

Houdini of the Hardwood Johnny Townsend (basketball player)

Hound Dog North American-Rockwell air-to-surface missile

house apes other people's unhousebroken children

The House Christ College, Oxford

household coal bituminous coal; soft coal

House Ruth Built New York City's Yankee Stadium in the Bronx where Babe Ruth hit many home runs

Houston haze smog of the petrochemical variety sometimes called Los Angeles haze or metropolitan mist mixed with automotive exhausts

Howard Cosell Howard Cohen

Howie Howard; Howarth; Howe; Howell; Howland

Howlin' Wolf Chester Burnett

Hsinhua New China News Agency

Huáscar's revenge loose bowels contracted in Peru where the emperor, Huáscar, was betrayed by his brother, Atahualpa, and executed by the Spaniards

Hub The Hub—Boston, Massachusetts also called Hub of American Culture, Hub of New England, and Hub of the Universe

hubby husband

Hub of the Caribbean Jamaica

Hub of the Castilian Wheel Madrid (capital of Spain, equidistant from all her boundaries)

Hub of Christianity Jerusalem

Hub City of Texas Alice

Hub of Empire London

Hubey Hubert

Hub of the Golden Mile Kalgoorlie in Western Australia where gold and nickel are found

Hub of Hinduism Benares

Hubie Hubert

Hub of Islam Mecca

Hub of Islamic Culture Cairo

Hub of Judaism Jerusalem

Hub of New England Boston

Hub of New York City Columbus Circle

Hub of the Pacific Guam

Hub of the South Pacific Fiji

Hub of the Universe nickname given by Oliver Wendell Holmes to the statehouse in Boston and later by others to the entire city

Hudson Tubes Hudson & Manhattan Railroad

Huey Cobra AH-l gunship aircraft

Hughie Hugh

Hugo Hugo Lafayette Black (American Supreme Court justice)

Hugo Wast Gustavo Martinez Zuviria

Hugues Capet Hugh Capet

Human Eraser Marvin Webster (basketball player)

Human Eyeball Bris Lord (baseball player)

Human Flea Frank Bonner (baseball player)

Human Hairpin Adrian (Addie) Joss (baseball player)

Humanist Historian Harry Elmer Barnes

Humanitarian Scientist Louis Pasteur

Human Projectile Billy Gabor (basketball player)

Humist Philosopher David Hume

Hummer High-Mobility Multi-Purpose Wheeled Vehicle successor to the Jeep

Humorist-Pianists Steve Allen, Victor Borge, and Mark Russell

Humorists of Europe the Danes

humpday Wednesday (the middle-of-the-week day)

hungarian hungarian goulash (stew); hungarian paprika (red paprika—permeability vitamin or vitamin P)

Hungarian Ocean Lake Balaton—largest lake in central Europe

Hunger Moon full moon in February

Hunkyland Hungary

Hunter Hawker jet fighter-bomber

Hunter's Moon full moon in October

Huntington's chorea hereditary disease marked by chorcic movements and mental deterioration

Huskie Kaman H-43 utility helicopter

Husky Territory the Yukon

Hustleton on the Canal Houston, Texas

hypno cops Los Angeles Police Department's hypnosis squad

Hyrcanian Caspian

I-boats Japanese transport submarines used in World War II to carry small scouting airplanes

ice crystal methamphetamine (smokable form of speed)

ice bear polar bear

Iceberg Alley North Atlantic Ocean between Greenland and Labrador

Iceblink white glare on a horizon from the reflected light of distant ice

Ice Capital of America Hawaii

ice-cream fruit Peruvian cherimoya

ice-cream headache vessels spasm from intense cold of ice cream, interrupting blood flow

Ice King of the Arctic the polar bear

Icelandic Ports Reykjavik, Seydhisfjordhur, Heimaey, Eyrarbakki

Iceland spar calcite (calcium carbonate)

Ice Mine City Coudersport, Pennsylvania

ice-T iced tea

Iconoclast English freethought author-lecturer-publisher Charles Bradlaugh; Georg (Morris Cohen) Brandes

Iconoclast Poet Percy Bysshe Shelley

Iconoclasts dramatist-authors Becque, d'Annunzio, Duse, Gorki, Hauptmann, Herpieu, Huneker, Ibsen, Maeterlinck, Shaw, Strindberg, Sudermann

Idaho Lion Senator William E Borah

idiot fish Pacific red snapper

idiot pills barbiturates

IF Stone Isidor Feinstein

Ignatius Loyola Iñigo López de Recalde

Ignatz Ignatius Donnelly (American politician)

Ignatz von Aschendorf pseudonym used by Joseph Conrad and Ford Madox Ford when they wrote *The Nature of a Crime*

Ignazio Silone pseudonym of Secondo Tranquilli

Igor Igor Stravinsky

Ike Dwight David Eisenhower (nickname)—thirty-fourth President of the United States; Isaac

Ikey (*see* Ikie)

Ikie Isaac; Isaak; Isack; Izaak; Isaque

Il Cieco (Italian—The Blind One)—Italy's blind poet Luigi Groto who wrote in the mid-sixteenth century

Il Duca di Spoleto (Italian—The Duke of Spoleto)—composer-impresario Gian Carlo Menotti's nickname as he directs the Spoleto Festival

Il Duce (Italian—dictator)—Benito Mussolini

Il Furioso (Italian—the Furious One)—nickname of Tintoretto

Ilich Russian patronymic often used as the popular name for Lenin—the party name of Vladimir Ilich Ulyanov

Illinois Parrot Carolina parakeet (extinct)

Illinois Ports Chicago, Wilmette, Great Lakes, Waukegan

Illinois River City Peoria

Illusion Factory Hollywood

Illustrator of Early Twentieth-Century America Norman Rockwell

Illustrator of the Russian Underground Ilya Efimovich Repin

Illustrious Infidel Colonel Robert G Ingersoll

Il Moro (Italian—The Moor)—Ludovico Sforza's nickname

Ilona Massey Ilona Hajmassy

Il Perugino (Italian—The Perugian)—Pietro Santi Bartoli

Image Maker Thomas Nast

Imamu Amiri Baraka Le Roi Jones

Immortal Beloved Nadejda von Meck (benefactress of Tchaikovsky)

Immortal Dreamer John Bunyan

Immortal Four Italian poets Dante Alighieri, Ludovico Ariosto, Francesco Petrarca (Petrarch), Bernardo Tasso

Immortal Infidel Col Robert G Ingersoll

Immortals the forty members of the French Academy

Immortal Sarah Sarah Bernhardt (originally Rosine Bernard)

Immortal Tinker John Bunyan

Immortal Trio John Caldwell Calhoun, Henry Clay, Daniel Webster

Impala South African version of Aermacchi MB-326 counterinsurgency aircraft

Imperial City Rome

Imperial Impersonation of Force and Murder Napoleon

Imperialist Composer Sir Edward Elgar

Imperialist Poet Rudyard Kipling

Imperialist Poet-Writer Rudyard Kipling

Imperial President Franklin D Roosevelt

imprisoned authors famous authors who spent time in prison include Bunyan (*Pilgrims's Progress*), Cervantes (*Don Quixote*), Dostoevski (*Crime and Punishment*), Raleigh (*History of the World*); O Henry (*Cabbages and Kings*), Wilde (*Ballad of Reading Gaol* and *De Profundis*)

Improper Bostonian Dr Oliver Wendell Holmes

Inca Capital Cuzco, Peru

Incan and Aztecan Century the 1000s—the 11th century

Inca's curse diarrhea picked up in the Land of the Incas—Bolivia and Peru as well as Ecuador and Chile

Inchcape Rock (*see* Bell Rock)

Inchon formerly Chemulpo or Jinsen

Incomparable Infidel Voltaire

Incorruptible The Incorruptible—sobriquet given Robespierre by his followers

Indefatigable Polemicist Alexander Solzhenitsyn

india india chintz or india cotton (heavy figured fabric used by upholsterers); india ink (glue + lampblack) also called chinese ink

Indiana Dunes national lakeshore of Indiana

Indiana Ports Michigan City, Gary, Buffington, Indiana Harbor

Indian Film Pioneer Satyajit Ray

Indian Girl Guide Sacajawea who guided Lewis and Clark

Indian Ocean Eden Seychelles Islands

Indian OPEC CERT (Council of Energy, Resource Tribes) holding coal, geothermal, and oil-productive lands in many parts of the United States

Indian Ports Mamdvi, Kandla, Okha, Porbandar, Bhaunagar, Bombay, Mangalor, Cochin, Alleppy, Quilon, Kolachel,

Tuticorin, Negapatam, Madras, Kakinada, Vishakhaptnam, Paradip, Calcutta

Indian President Benito Juárez of Mexico

Indian Princess Pocahontas

Indian's Friend Roger Williams

Indian Summer calm, mild weather sometimes occurring in autumn in the United States and Great Britain; so called because areas where it was first observed were then occupied by Native Americans

Indian Territory old name of Oklahoma

India's Principal Ports Calcutta on the Bay of Bengal, Bombay on the Arabian Sea

Indochinese Pen Indochinese Peninsula including Burma, Cambodia, Laos, Thailand, and Vietnam

Indonesian Ports Tandjungpriok (Djakarta), Surabaya, Makassar

Indonesia's Largest Port Djakarta (Jakarta)—also called Tandjungpriok

Industrial Capital of Connecticut Bridgeport

Inflexible President Andrew Johnson

Information Superhighway nationwide communications network linking homes, offices and institutions

Inland Empire Illinois

Inland Sea Pacific Ocean inlet between Honshu and Kyusho islands, Japan

Inner City Peking's Forbidden or Tartar City containing the Palace Museum

Inner Libya the Sudan Desert extending from southern Egypt and the Sudan to Africa's west coast

Inner Mongolia northern China bordering on Mongolia

Innovators Russian composers Balakirev, Borodin, Cui, Mussorgsky, Rimsky-Korsakov

Inside Passage inland passage between southern Alaska and northern Washington; also called Inner Passage

Inspired Innovator Edgar Allan Poe

Instant Asia Singapore with its Chinese, Indian, Malay, Pakistani, and Singhalese mixtures and tongues making this a global crossroads

Instant Orient Singapore—crossroads of Asia

insultant(s) nickname for overpaid consultant(s)

Insurance Capital Hartford, Connecticut and Omaha, Nebraska

Insurance City nickname shared by Atlanta, Georgia and Hartford, Connecticut

Intellectual Emperor of Europe Voltaire

Intellectual Historian Sir Isaiah Berlin

Intellectual Seed Pod of the Nation Emerson's nickname for Concord, Massachusetts where he lived with such neighbors as the Alcotts, Hawthorne, and Thorcau

Interior Plains Canada's great plains

International Capital New York City—headquarters of the United Nations

International City Montreal, Québec

International Functionalists Walter Gropius and Mies van der Rohe

International Prizegiver Alfred Nobel

Interpreter of the Sea Winslow Homer

Interpreter of the Wild West Frederic Remington

Intruder Grumman electronic intelligence-gathering aircraft (EA-6B)

Inventor of Bifocals Benjamin Franklin

Inventor of Calculus Baron Gottfried Wilhelm von Leibniz and Sir Isaac Newton

Inventor of the Detective Story Edgar Allan Poe

Inventor of Science Fiction Mary Wollstonecraft Shelley, author of *Frankenstein*

Inventor of the Stethoscope René-Théophile-Hyacinthe Laennec

Inventor of the Telephone Philipp Reis although Alexander Graham Bell is usually given credit

The Invincible Spanish Armada defeated by English vessels commanded by Sir Francis Drake

invisible disease dyslexia

Iodine State South Carolina

Iola Ida B Wells

Iolo Morgannwg bardic name of Edward Williams

Ipiranga Ypiranga

Iran-Contra 1980s political scandal involving the sale of weapons of Iran to fund Nicaraguan insurrection

Iranian Ports Khorramshahr, Abadan, Bandar-e-Mahshahr, Bandar-e-Shapur, Kharg Island Terminal, Bandar Abbas

Iran's Principal Port Abadan

Iraq Ports Al Faw and Al Basrah

Iraq's Principal Port Basra

Ireland the Great Newfoundland's name given it by Irish explorers who found it in Viking times

Ireland's Principal Port Dublin

Iris Tennessee state flower and sobriquet

Irish Irish boat (cutter-rigged fishing vessel), Irish ford (paved ford), Irish coffee (spiked with Irish whiskey and topped with whipped cream), Irish moss (edible seaweed also called carrageen), Irish pennant (unwhipped rope end flying in the breeze), Irish potato (white), Irish setter, Irish sweater (fisherman's knit), Irish terrier (red-hair terrier originally bred in Ireland), Irish tweed, Irish whiskey (originally distilled in Ireland from barley), Irish wolfhound (large breed of dog noted for its courage and originally bred in Ireland)

Irish-American Bandmaster Patrick Sarsfield Gilmore

Irish Channel New Orleans waterfront

Irish confetti bricks; thrown bricks used in street fighting

Irish Cradle of U.S. Presidents Ireland—ancestral home of Presidents Jackson, Kennedy, Nixon, and Reagan as well as Arthur, Buchanan, McKinley, Polk, Truman, and Wilson from Northern Island or Ulster

Irish Dramatist-Poet William Butler Yeats

Irish Free State Republic of Ireland

Irish Navigator Saint Brendan (formerly spelled Brandon)

Irish Ports Bangor, Belfast, Larne Lough, Londonderry, Sligo, Westport, Galway, Kilrush, Limerick, Foynes, Cobh, Cork Harbour, Rosslare, Dublin

Irish turkey corned beef

Irish whist copulation

Irish wine whiskey

Iron Age era of mankind when implements and weapons were forged from iron; period of vast degeneracy, corruption, and toil following the Stone and Bronze ages

Iron Alley corridor littered with meteorites from Portland, Oregon to Mexico City in a lane about 250 miles or 400 kilometers wide

Iron Butterfly Imelda Romauldos Marcos

Iron Chancellor Prince Otto Eduard Leopold von Bismarck-Schönhausen—first chancellor of German Empire

Iron Charles Charlemagne (Carolus Magnus)

Iron City Bessemer, Alabama and Pittsburgh, Pennsylvania

Iron Curtain Countries Albania, Bulgaria, Czechoslovakia, Estonia, East Germany, Hungary, Latvia, Lithuania, Poland, Romania, Soviet Union, Yugoslavia

Iron Duke Arthur Wellesley the Duke of Wellington

Iron Eyes Cody Oscar DeCorti

Iron Gate narrow rapids in the Danube below Orsova in Romania

Iron Horse Lou Gehrig; a steam locomotive

Iron Lady Margaret Thatcher—Britain's first woman prime minister

Iron Mountain State Missouri

Ironquill Eugene Fitch Ware

Iron Range the Mesabi Range

Ironsides Oliver Cromwell

Iron Triangle Cologne (Köln), Siegen, Solingen (all noted for their steel products)

Iroquois Bell turbo-power helicopter designated UH- I

Irving Berlin Israel Baline

Irving Stone Irving Tannenbaum

Isabela Spanish name for Albemarle Island in the Galápagos

Isak Dinesen Baroness Karen Blixen-Finecke

Isherman Israeli version of Super Sherman tank

Iskander Alexander Herzen (Aleksandr Ivanovich Yakoviev)

Iskra (Russian—Spark)—Polish single-engine jet aircraft designated TS-11

Isku Finnish guided-missile patrol boat

Isla de Juventud (Spanish—Isle of Youth)—Cuba's Isle of Pines also called *Isla de Pinos*

Isla de Pascua (Spanish—Easter Island)—Chilean island in the eastern South Pacific noted for its huge stone monuments; Polynesians call it *Rapa Nui*

Isla Más Afuera (Spanish—further out island)—Alejandro Selkirk Island

Isla Más Tierra (Spanish—island closest to land)—Robinson Crusoe Island

Islamic Century the 600s—Mohammed dies in 632; Islam begins expanding throughout the Middle East and Africa—the 7th century

Island-and-Mainland Province Newfoundland

Island at the End of the World Madagascar as described by the Malagasy

Island of Bearded Figs Barbados

Island of Betelnut Palms Penang

Island of Birds Kusadasi

Island City Manhattan, Montreal, Singapore, and Stockholm

Island of Cloves Zanzibar

Island Continent Australia

Island of Copper Cyprus

Island of Death Kahoolawe, Hawaii (used for target practice by Air Force and Navy)

Island of Dragons Komodo (home of the dragon lizards)

Island of Dreams Capri

Island of Enchantment Puerto Rico

Island of Flowers Taboga, Panama

Island Fortress Malta

Island of the Gods Bali

Island of Hell Norfolk Island in the South Pacific, the most dreaded of all Australian prison stations

Island of Hope Statue of Liberty, Liberty Island, New York

Island of Knights Hospitaliers Malta

Island of Light New Caledonia

Island of Many Faces Singapore

Island Ministate Nauru in the Central Pacific

Island of Monks and Pirates Lantau or Tai Yue Shan—largest island off Hong Kong

Island of the Moon Madagascar

Island Nation nickname shared by Australia, the Bahamas, Bahrain, Barbados, the Cape Verde Islands, the Republic of China (Taiwan), the Comoro Islands, Cuba, Cyprus, the Dominican Republic, Fiji, Grenada, Haiti, Iceland, Indonesia, Ireland, Jamaica, Japan, Madagascar, the Maldives, Malta, Mauritius, Nauru, New Zealand, Papua New Guinea, the Philippines, São Tomé and Principe, the Seychelles, Singapore, Sri Lanka (Ceylon), Tonga, Trinidad and Tobago, the United Kingdom, Western Samoa

Island of Olives Cyprus

Island of Roses Rhodes in the Dodecanese

Island of Ruins and Roses Gotland, Sweden

Island of Sages and Saints Ireland

The Islands pet name given to favorite insular groups such as the Aleutians, the Bahamas, the Balearics, the Canaries, the Hawaiians, the West India islands and even to Coney, Long, Manhattan, and Staten when referring to the New York City area

Islands of Eternal Spring the Balearics (Ibiza, Formentera, Mallorca, Menorca)

Islands of the Maoris New Zealand

Islands of Perpetual June Turks and Caicos Islands between the Bahamas and Hispaniola

Island State Tasmania, Australia

Island in the Sun Key West, Florida

Island of Venus Tahiti

Isle of the Ancestors Madagascar

Isle of Cloves Zanzibar

Isle of the Dead Böklin painting

Isle of Destiny Ireland

Isle of Fragrant Waters Hong Kong

Isle of Pines Cuba's *Isla de Pinos*, France's *Ile des Pins*, Ibiza

Isle of Roses Rhodes

Isle of Saints Iona in the Inner Hebrides off Scotland's west coast; Ireland

Isle of Sappho Lesbos in the Aegean

Isles of the Blest the Canary Islands (*see* Fortunate Isles)

Isles of Devils Bermuda

Isle of Sleep Tasmania

Isle of Springs Jamaica

Isle of Tears Ellis Island, New York

ism of the modern world racism (according to anthropologist Ruth Benedict)

Isolator of Dysentery Kiyoshi Shiga

Isolator of Gangrene Shibasaburo Kitazato

Isolde Yseult

Israeli Ports Akko (Acre), Hefa (Haifa), Netanya, Tel-Aviv-Yafo(Jaffa) Ashdod (Azotus), Ashquelon (Ascalon), Elat, Sharm el Sheik

Israel's Largest Port Tel-Aviv-Yafo (Jaffa or Joppa)

Israfel Edgar Allan Poe

Istanbul formerly Constantinople and originally Byzantium, capital of the Byzantine Empire

Isthmian Nation Panama

Isthmian Waterway Panama Canal linking the Caribbean-Atlantic and the Pacific; Suez

Canal linking the Mediterranean-Atlantic and the Red Sea leading to the Indian Ocean

Isthmus of Panama formerly Isthmus of Darien

Italian Italian greyhound (toy dog); Italian hand (script originating in Italy in medieval times or craftiness); Italian dressing (salad dressing)

Italian boot boot-shaped Italian peninsula

Italian Century the 14th century—the 1300s

Italian Classicist Sculptor Antonio Canova

Italian East Africa Eritrea, Ethiopia, and Italian Somaliland from 1936 to 1941

Italian Engraver Giambattista Piranesi

Italian Family of Sculptors the della Robbias

Italian Film Magician Federico Fellini

Italian Goldsmith and Sculptor Benvenuto Cellini

Italian Illustrator-Painter Sandro di Botticelli

Italian Lakes Como, Garda, Isea, Lecco, Lugano, Maggiore, Orta

Italian National Composer Ottorino Respighi

Italian Naturalist Painter Michelangelo da Caravaggio

Italian North Africa Libya from 1912 to the end of World War II

Italian Ports (*on Sardinia*—La Maddalena, Olbia, Cagliari, Alghero, Porto Torres), Savona, Genova (Genoa), La Spezia, Livorno (Leghorn), Portoferraio (*on Elba*), Civitavecchia, Gaeta, Forio, Ischia, Bagnoli, Napoli (Naples), Torre Annunziata, Castellamare di Stabia, Reggio di Calabria, (*on Sicily*—Messina, Palermo, Trapani, Marsala, Licata, Siracusa, Augusta, Catania), Crotone, Taranto, Gallipoli, Brindisi, Monopoli, Bari, Molfetta, Barletta, Manfredonia, Ancona, Ravenna, Chioggia, Porto di Lido (Venice), Monfalcone, Trieste

Italian Pre-Renaissance Painter Giotto (Giotto di Bondone)

Italian Riviera resort area between La Spezia and Ventimiglia

Italian Somaliland Indian Ocean coast of what is now southern Somalia

Italy of America Arizona

Italy's Principal Port Genoa

It Girl Clara Bow

Itiopia Ethiopia (Abyssinia)

I-town Indiantown

Ivan (nickname for the typical Russian)

Ivan Ivanovich the typical Russian

Ivan-Kremlin disease endemic antisemitism

Ivan Lermolieff Giovanni Morelli (19th-century Italian art expert, patriot, and senator)

Ivan the Terrible Czar Ivan IV Vasilievich—ruler of Russia

Ivory Coast Ports Grand-Lahou, Jacqueville, Port-Bouet, Abidjan, Grand-Bassant

Ivy League college athletic conference consisting of Brown, Columbia, Cornell, Dartmouth, Harvard, Pennsylvania, Princeton, and Yale; students and graduates of the abovementioned schools as well as their "characteristic" style of dress, which was considered "quiet and neat"

ixey morphine

Izzie Isador; Isadora; Isadore; Isidro; Isodoro; Ysidro

Jabal Tariq (Arabic—Mountain of Tarik)—Moorish name for Gibraltar in honor of their chief Jabal Tariq who settled the Rock in the year 111; (Arabic—Tariq's Mountain)—the Rock of Gibraltar

Jacaranda Capital jacaranda tree-lined avenues and streets comprising South Africa's capital city—Pretoria

Jack the Back John Fitzgerald Kennedy, thirty-fifth president of the United States, who turned his back to press photographers until he offered them the pose he wanted

Jack Benny Benjamin Kubelsky

Jack the Dripper Jackson Pollock

Jack Frost frosty weather personifiied

Jack Higgins Harry Patterson's pseudonym

Jackie Jack Roosevelt Robinson; Jacqueline Kennedy Onassis

Jack London John Griffith London

Jack Lord John Joseph Ryan

Jack Palance Walter Palanuik

Jacksonopolis Jackson, Michigan

Jack Soo Jack Suzuki

Jack Teagarden Weldon Leo Teagarden

Jacky Jaqueline

Jacopo Jacopo Tatti

Jacques Halevy Jacques Francois Fromental Elias Levi

Jacques Offenbach Jakob Eberst

Jacques Tati Jacques Tatischeff

Jadotville former name for Likasi, Zaire

Jagananth Juggernaut or Puri on the Bay of Bengal

Jake Jacob; Jacobus

Jaksch's disease infantile anemia

Jamaica Jamaica ginger; Jamaica rum (heavy pungent rum originally distilled in Jamaica); Jamaica shorts (midthigh short pants)

Jamaica ganga Jamaica grown marijuana

Jamaican Ports Lucea, Montego Bay, Falmouth, Rio Bueno, Dry Harbour, St Ann's Bay, Ocho Rios, Oracabessa Bay, Port Maria, Annotto Bay, Buff Bay, Port Antonio, Manchioneal, Port Morant, Morant Bay, Port Royal, Kingston, Long's Wharf, Little Pedro Point, Black River, Bluefields, Savanna la Mar

Jambalaya Capital Gonzales, Louisiana

James Garner James Baumgarner

James Hadley Chase René Raymond's pseudonym

James Herriot author-veterinarian James Alfred Wight's pseudonym

James Island, Galápagos
Bartolomé, San Salvador,
Santiago

James O'Brien James
Bronterre

Jamil Abdullah Al-Amin H
Rap Brown

Jane Doe the average American female

Janet Frame Janet Peterson
Frame Clutha's pseudonym

Janet Gaynor Laura Gainer

janets middle-aged widows
(*see* jennifer)

Jane Seymour Joyce Frankenberg

Jane Welsh Mrs Thomas Carlyle

Jane Wyman Sarah Jane
Faulks

Janey Canuck Judge Emily
Murphy

Janie Jane

Jan Peerce Jacob Pincus Perelmuth

Jan Struther Joyce Anstruther

January effect small-capitalized stocks tend to outperform large-capitalized stocks
in January

Jan Valtin Richard J Krebs

Japan Japan lacquer or varnish; Japan wax also called
Japan tallow or sumac wax

Japanese Japanese gelatin
(agar also called Japanese
isinglass); Japanese paper
(high rag content quality);
Japanese silk (high quality
raw silk)

Japanese Ports Moruran, Hakodate, Otaru (*on Hokkaido*);
Tokyo, Yokosuka; Shimuzu,
Nagoya, Yokkaichi, Senboku,
Osaka, Kobe, Fukuyama,
Kure, Shimminato, Shimonoseki, Maizuru, Niigata
(*on Honshu*), Kita Kyushu,
Kagoshima, Nagasaki,
Sasebo, Karatsu, Fukuoka (*on
Kyushu*)

Japanese Riviera Enoshima
Island recreation area

Japan's Back Door Sasebo

Japan's Front Gate Yokohama

Japan's Largest Port Yokohama (including Kawasaki,
Tokyo, and Yokosuka)

jappy derived from Jewish-American Prince(s) or Princess(es)—young generation
living in luxury provided by
doting parents

Jap Sea Japan Sea containing
many islands comprising Japan

Jascha (Russian nickname—
Jacob)—Jake

Jastreb Yugoslav jet trainer
aircraft called Hawk

Javelin Gloster delta-wing jet
fighter aircraft

javelle water sodium hypochlorite solution (NaOCl)

Jawbone Flats Clarkston,
Washington

Jaybird Coleman Burl Coleman

Jayhawker(s) Kansan(s)

Jayhawker State Kansas

Jay Silverheels actor Harold J
Smith, known as Tonto in *The
Lone Ranger* series

Jazz Ambassador Louis
(Satchmo) Armstrong

Jazz Capital of the Americas
New Orleans

Jazz Capital of Europe
Copenhagen

The Jazz Singer Al Jolson

Jean Arthur Gladys Greene

Jean Baptiste French-Canadian's sobriquet

Jean Baptiste Lully Giovanni
Battista Lulli

Jean Crapaud (nickname for
the typical Frenchman)

Jean Gabin Jean-Alexis Moncorgé

Jean Hagen Jean Verhagen

Jean Harlow Harlean Carpenter

Jean-Jacques Jean-Jacques
Rousseau

Jean l' Oiseleur (French—
Jean the bird tamer)—pseudonym of Jean Cocteau

Jean Meslier Voltaire's pseudonym concealing his authorship of an heretical tract
whose title page reads—*Superstition In All Ages* by Jean
Meslier, a Roman Catholic
Priest, who after a pastoral
service of thirty years at Entrepigny and But, in Champagne, France, wholly abjured religious dogmas, and
left as his last will and testament the following pages en-

titled Common Sense (*Le Bon
Sens*); Voltaire was an assumed name for François-Marie Arouet

Jean Moreas pseudonym—
Jannis Papadiamantopolous

Jeannie Jean

Jean Paul Johann Paul Friedrich
Richter's pseudonym

Jean-Pierre Aumont Jean-Pierre
Salomons

Jean Stapleton Jeanne Murray

Jeb Stuard Major General
J(ames) E(well) B(rown) Stuart, CSA

Jefferson's Country Charlottesville, Virginia

Jefferson Territory Colorado

Jellybean Joe Bryant (basketball player)

Jelly Roll Ferdinand Joseph
Morton

Jenghis Khan Genghis Khan—
Mongol conqueror and grandfather of Kublai Khan

Jennie Jane; Jean; Jennifer;
Lady Randolph Churchill

jennifer young woman who
marries an older man

Jennifer Jones Phyllis Isley

jenny Jane; Jean; Jennifer

Jenny Lind Johanna Maria
Lind—the Swedish Nightingale

Jeremiah Stukeley Percy
Bysshe Shelley pseudonym

Jerez Jerez de la Frontera

Jerome Hines Jerome Heinz

Jerrie(s) British slang for German(s)

Jerry Gerald(ine); Governor
Edmund G Brown, Jr of California, President Gerald R
Ford; Jeremiah; Jeremy; Jerome

Jerry Lewis Joseph Levitch

jersey jersey justice (reputedly
efficient and speedy); jersey
lightning (applejack)

Jersey Blues New Jerseyites

Jersey Blue State New Jersey

Jersey Lily Lily Langtry—English actress born on the island of Jersey where her original name was Emily
Charlotte Le Breton

Jersey Shore coastal New Jersey; former name of Waynesburg, Pennsylvania

Jerusalem of the West Amsterdam

Jervis Street Toronto, Ontario's red-light district

Jesse Jesse Jackson (black civil-rights leader)

Jesselton former name of Kota Kinabalu, Sabah

Jessi James Cleveland (Jesse) Owens; Jess; Jessica

Jessup Maryland House of Corrections at Jessup

Jessup Girls Maryland Correctional Institution for Women at Jessup

Jet Provost British jet trainer aircraft designated BAC-145

Jet Ranger Bell turbine-powered helicopter also called Sea Ranger

Jet Star C-140 Lockheed light transport plane

Jet City Seattle

Jewel of the Adriatic Dubrovnik; Venice

Jewel of Africa Lake Kivu

The Jewel of American South Seas Islands Palmyra, 1,100 miles southwest of Honolulu

Jewel of the East Bali

Jewel of the Eastern Sea Sri Lanka

jewelers' putty stannous oxide

Jewel of German Cities Heidelberg

Jewel Island Ceylon

Jewel of the Kalahari Lake Okavango, Botswana

Jewell Manor Jewell Manor (delinquent) Girls Center at Louisville, Kentucky

Jewels of the Caribbean U.S. Virgin Islands

Jewish Alps Catskill Mountains, New York

Jewish ampicillin cream of chicken soup

Jewish champagne celery tonic (carbonated celery-flavored water)

Jewish Confederate Judah Philip Benjamin

Jewish doughnuts bagels

Jewish Dristan horseradish (celebrated for its ability to bring about nasal decongestion)

Jewish penicillin chicken soup

Jews of Asia ethnic Chinese overseas

jezebel abandoned and shameless woman; a prostitute

Jim Crow Laws Southern United States system of segregating blacks from whites

Jimi Hendrix James Marshall Hendrix

Jimmy James; James Earl Carter—thirty-ninth President of the United States

Jim Thorpe formerly Mauch Chunk, Pennsylvania

Jimtown Jamestown, North Dakota

jingo jingoism (chest-thumping patriotism)

Jinx Falkenburg Eugenia Falkenburg

J.J. Connington Alfred Walter Stewart's pseudonym

Joan Crawford Lucille le Sueur

joe coffee

Joe Joseph

João Pessoa formerly Parahiba, Brazil

Joaquin Miller Cincinnatus Heine Miller's pen name

Jochanan John

Jock John

Jocko Art Conlan

Joe Bananas Mafia chief Joseph Bonanno

Joey Bishop Joseph Gottlieb

Joe Doakes nickname for the average American man

Joe Doe name used for the average American male

Joel Grey Joe Katz

Joe Louis Joseph Louis Barrow

Joe Sixpack newsroom slang describing average newspaper reader

Joe Who? Canadian Secretary of State Joe Clark

Joe Zilch the average American formerly called Joe Blow or Joe Doakes

Johann Gutenberg Johann Ganzfleisch

John Archer Ralph Bowman

John Denver Henry John Deutschendorf Jr.

John I John the First (John Adams—second President of the United States)

John II John the Second (John Quincy Adams—sixth President of the United States)

John XXIII Angelo Giuseppe Roncalli

John Barleycorn personification of beer or malt liquor

John Barrymore John Blythe

John Bull Great Britain

John Bull's Other Island Ireland before its independence was declared in 1919

John Cabot Giovanni Caboto

John Calvin Jean Chauvin

John Company British East India Company's nickname

John Danger Hough Baillie (colorful journalist who rose from reporter to head of the United Press)

John Doe fictitious name used when real name is withheld or unknown

John and Emery Bonett pseudonym shared by the husband-and-wife team— John Hubert Arthur Coulson and Felicity Winifred Carter

John Ford Sean O'Fearna

John Garfield Julius Garfinkle

John Gilbert John Pringle's stage name

John Hancock signature (nickname memorializing most prominent autograph on Declaration of Independence)

John Houseman Jacques Haussmann

John le Carré David John Moore Cornwell

Johnny John; John M Grant, Jr; John von Neumann (Hungarian-born American mathematician)

Johnny Appleseed John (Johnny) Chapman

Johnny Crapaud nickname for a Frenchman or a New Orleans creole

Johnny Guitar Watson John Watson

Johnny Reb(s) Johnny Rebel(s)—Confederate soldier(s)

John Oxenham William Arthur Dunkerly

John Paul Charles Henry Webb

John Paul I Albino Luciani

John Paul II Karol Wojtyla

John Rhode Cecil John Charles Street's pseudonym

John's St John's

Johns of Geneva Jean Calvin and Jean-Jacques Rousseau

John Sinjohn John Galsworthy

John Wayne Marion Michael Morrison

Jolly Green Giant Jim Garrison, former New Orleans district attorney

Joltin' Joe Joe DiMaggio (baseball player)

Jonathan Fogarty Titulescu James T(homas) Farrell

Jonathan Oldstyle (pseudonym—Washington Irving)

Joni Mitchell Roberta Joan Anderson

Jordan's Port Al Aqabah

Jordie Jordan(a)

Jordy Jordan

j or j jam or jelly (cocaine)

José Ferrer José Vicente Ferrer y Cintron

José Greco Constanzo Greco

Joseph a Guarneri violin (short form of Giuseppe Guarneri)

Joseph Bentonelli Joseph Horace Benton

Joseph Conrad Teodor Josef Konrad Korzeniowski

Joseph Hansen James Colton's pseudonym

Josephine Josephine Baker

Josephine Bell Doris Bell Collier Ball's pseudonym

Josephus Flavius Josephus, pen name of Arius Caipurnius Piso

Josephus Flavius Josephus— apostate Jew and recorder of the Roman conquests

Joseph von Sternberg Josef Stern

Josh Joshua; (pseudonym— Samuel L Clemens)

Josh Billings stage name of humorist Henry Wheeler Shaw

Josiah Flynt Josiah Flynt Willard's pseudonym

Josie Josephina; Josephine

The Journal *The Wall Street Journal*

The Joy Train nickname in 1990s for Brazil's bureaucracy (*O Trem da Alegria*)

J-town Jamestown; Johnstown

Jualpa Camp Juneau, Alaska, Pennsylvania Camp (Alaska gold-mining site)

Juana la Loca (Spanish— Juana the Mad)——queen of Castile whose lisp became the royal style known as Castilian

Juan Bimba the typical Venezuelan

Juan Gris José Victoriano González

Juanita Juana (Jane, Joan)

Juan Pablo (Spanish—John Paul—the Pope

Jubilee Girls Jubilee Lodge for (delinquent) Girls at Brimfield, Illinois

Judas Priest Jesus Christ (rendered as a palatable exclamation or oath)

Judy Judith

Judy Garland motion-picture-reel name of Frances Gumm

Judy Holliday Judith Tuvim

Juggernaut Jagananth or port of Puri on the Bay of Bengal

Jules Romains Louis-Henri Jean Farigoule

Julia Marlowe Sarah Frances Frost's stage name

Julie Andrews Julia Wells

Julio Diniz Joaquim Guilherme Coelho

Jumbo Barnum's famous 6½ ton 11-foot-high trained elephant exhibited in the eighties

Jumbo Al Hirt

Jumbo Bill America's 27th President—300-pound William Howard Taft

Jumbo State Texas

Jumping Joe Joe Dugan (baseball player)

Jump Jet nickname of U.S. Marine Corps AV-8B fighter-bomber capable of vertical takeoff and landing

June Allyson Ella Geisman

Juneteenth June 19, 1865, when blacks in Texas were emancipated

Jungle Novels collective name given to B Traven's books (*see* Hal Croves)

Junipero Serra Miguel José Serra

junk heroin's nickname, also called smack

Junk Bond King Michael Milken

junkie drug addict

Jupiter of Wall Street JP Morgan

Jusepe José de Ribera

Jussi Björling Johan Jonaton Björling

Justicia justice personified (second goddess wife of Jupiter or Themis who held the same post under the Greek god Zeus); she stands blind-

folded, holding a balance in one hand and a palm frond in the other

The Just Society (nickname— Prime Minister Pierre Trudeau's administration of Canada)

Jute Port Dundee, Scotland

Jutland mainland of Denmark and Schleswig-Holstein

Juvenal Decimus Junius Juvenalis

Kabul River City Kabul, Afghanistan

Kabwe formerly Broken Hill, Zambia

Kaffir King Barney Barnato

Kahlbaum's disease dementia with muscular tension

Kahler's disease bone-marrow destruction

Kaiser Bill Wilhelm ll— Emperor of Germany

Kalahari Kalahari Gemsbok National Park in southern Africa

Kalatdlit-Nunat (Greenlandic Eskimo—Land of the People)—Greenland's name adopted in 1979

Kalima formerly Albertville

Kalimantan Indonesian segment of Borneo

Kalinin Soviet name for Tver

Kaliningrad formerly Königsberg

Kamenev Lev Borisovich Rosenfeld

Kampuchean Ports (*see* Cambodian Ports)

Kanakalanders nickname for Queenslanders who hired many South Sea Kanakas to work on their plantations

kanaka(s) South Sea islander(s)

Kanawha River City Charleston, West Virginia

Kangarooland Australia

Kanner syndrome early infantile autism

kansas cathedrals silos

Kareem Abdul Jabbar Lew Alcindor

Karl Johan Jean Baptiste Jules Bernadotte

Karl Malden Mladen Sekulovich

Karl Radek Karl Sobelsohn

Kashin a class of Soviet destroyer-leader ships

Kate Catherine; Katherine; Katharine Hepburn; Katrina

Katherine Mansfield pseudonym—Kathleen Beauchamp Murry

Kawasaki syndrome most common cause of noninherited heart disease in children; first described in 1967 by Dr Tomisaku Kawasaki

Keeling Islands old name for Cocos in the Indian Ocean

Keewanee peninsula on the south side of Lake Superior, Michigan

Kelly Country Australia's northern Victoria named after the nineteenth-century outlaw Ned Kelly

Kelp Capital San Diego, California

Kelp Capital of the World offshore San Diego, California

Kelper(s) Falkland Islander(s)

Kelt Soviet air-to-surface missile carried by Tu-16 bombers

Kelvin Kelvin Scale

Kemal Ataturk Mustafa Kemal Pasha

Kenitra formerly Port Lyautey, French Morocco

Kenny G Kenneth B Gorelick

Kentucky Colonel Earle Combs (baseball player)

Kentucky Rifle Glen Combs (basketball player)

Kenya Ports Mombasa, Takaungu, Malindi, Lamu

Kester Christopher

Ketchikan State Jail and Detention Home in Ketchikan, Alaska

keyboard condom plastic sheath over computer keyboard

Keyboard Titan Vladimir Horowitz

Key City Port Townsend, Washington; Vicksburg, Mississippi

Key to England Dover on the bay beneath the chalk cliffs of Kent flanking the English Channel

Key of the Gulf Cuba

Key of the Indian Ocean Mauritius

Key of the Mediterranean Gibraltar

Key to the New World 16th-century Havana

The Keys Florida Keys extending southward from Key Biscayne near Miami on the Atlantic Ocean to Key West between the Atlantic and the Gulf of Mexico containing the Marquesas Keys, the Dry Tortugas, and Fort Jefferson National Monument

Key State New South Wales

Key to Stockholm Aland Islands between Finland and Sweden

Keystone Province Manitoba

Keystoner(s) Pennsylvanian(s)

Keystone State Pennsylvania

Keystone of the South Atlantic Seaboard South Carolina

Khazarian Way Don Volga Portage

Khazar Sea Caspian Sea

Khmer Cambodia

Khmer Republic Cambodia

Khmer Rouge Red (Communist) Cambodia

Kick-'em-Jenny Diamond Island's nickname (West Indian island near Grenada)

Kid Chocolate Eligio Sardinias (boxer)

Kid McCoy Norman Selby (boxer)

Kid Ory Edward Ory (composer-trombonist)

Kiel Canal formerly Kaiser Wilhelm Canal

kieselguhr silica (SiO_2)

Kildin Soviet class of fleet destroyers

Killer Harmon Killebrew (baseball player)

killer disease dysentery

Killer Whale Moon July 3 full moon

Kilometer-high City Boone, North Carolina

Kim Novak Marilyn Novak

The King Clark Gable (William Gable)

King of Acids sulfuric acid

King Auks great auks

King of Austrian Opera Wolfgang Amadeus Mozart

King of the Band John Philip Sousa

King of Bath Richard (Beau) Nash

King of Beans adzuki bean

King of Beasts the lion

King of the Beats Jack Kerouac, writer

King of the Bees Samuel Linnaeus

King of Birds the eagle

King Bomba Ferdinand II

King of Cakes wedding cake

King of the Carnivorous Dinosaurs Tyrannosaurus rex

King of Cartoons Tex Avery

King of Cats the lion in Africa; tiger in Asia; puma in North America; jaguar in Central and South America

King of Chefs and Chef of Kings Auguste Escoffier

King of Coke Henry Clay Frick

King of Conductors Herbert von Karajan

King of the Congo lowland and mountain gorilla

King of the Conifers the sequoia

King Cotton personification of the cotton crop of the southern United States

King of Courts the forensic orator Quintus Hortensius of Rome

King of the Cowboys Roy Rogers

King Crab Capital Kodiak, Alaska

King of the Delta Blues Singers Robert Johnson

Kingdom of Cabbage Germany

Kingdom of Death Hel (name of the Queen of Death in Nordic mythology)

Kingdom of the Hellenes Greece

Kingdom of Perpetual Night Hell

Kingdom of Sardinia the Italian Peidmont and the island of Sardinia

Kingdoms of the North Denmark, Norway, Sweden

Kingdom of the Swedes, Goths, and Wendes Sweden

Kingdom of the Thunder Dragon Bhutan

Kingdom of the Two Sicilies Naples and Sicily

King of European Music Festivals Salzburg

King of Filibusters William Walker

Kingfish Senator Huey P Long of Louisiana

King of Fish wild salmon

King of the Fjords Sogne Fjord, Norway

King of French Opera Hector Berlioz

King of Fruits the mango

King of German Opera Richard Wagner

King of the Gods Jupiter (Roman mythology)

King of the Harmonica Sonny Boy Williamson (Aleck Miller)

King of the High Cs Luciano Pavarotti

King of the Huns and Scourge of God Attila

King of Indonesian Primates the orangutan

King of Instruments the human voice; the organ; the violin or other string instruments

King of Italian Opera Giuseppe Verdi

King of Jazz Louis (Satchmo) Armstrong and Paul Whiteman

King of the Jews Jesus, according to the *New Testament*

King of the Jungle the tiger

King of Kings Jehovah—God of the Christians and Jews; title of various presumptive rulers of African and Oriental lands

King Kong Charlie Keller (baseball player)

King of Kora Toumani Diabate

King of Laughter Bert Williams (Egbert Austin Williams)

King Leso Kingdom of Lesotho

King of Lizards komodo-dragon lizard

King of Metals gold

King of the Missions Mission San Luis Rey, east of Oceanside, California

King of Naples Marshal Joachim Murat

King of Oceanic Scavengers the albatross

King of the Octaves Claudio José Domingo Brindis de Sala

King Oliver Joseph (King) Oliver—Doctor Jazz

King of the One-Liners Henny Youngman

King of the Opera tenor-conductor Placido Domingo

King of Ornithological Painters John James Audubon

King of the Pianists Claudio Arrau

King of the Ragtime Writers Scott Joplin

King of Rivers North America's Colorado and South America's Amazon

King of Roads John Loudon Macadam

King of Rock 'n' Roll Elvis Presley

King of Russian Opera Piotr Ilyich Tchaikovsky

King of the Screen Farid Shawqi

king's cure-all evening primrose oil

King of the Seas Neptune (Roman) or Poseidon (Greek)

King of the Sea Turtles trunkback turtle

king's English correct English

king's evil scrofula (lymph-gland tuberculosis)

King of Snakes king snake

King of Spruces Sitka spruce

King Star Regulus, in the constellation Leo

King of Steel Andrew Carnegie

King of Swat George Herman Ruth

King Swazi Kingdom of Swaziland

King of Swing Benny Goodman; Elvis Presley

King of Tasmanian Rivers the Gordon

King of Terrors personification of death

King of Tools the lathe

King of Tortoises Galápagos tortoise

King of Torts Melvin Belli

King of Trains and Train of Kings Orient Express

King Tut King Tutankhamen of Egypt

King of the Twelve-String Guitar Leadbelly (Huddie Ledbetter)

King of the Underworld Osiris (Egyptian mythology)

King of the Vagabonds François Villon whose real name was François de Montcorbier

King of Vaudeville Jimmy (Schnozzola) Durante

King of Verismo Giacomo Puccini (celebrated composer of realistic operas)

King of the Vibraphone Lionell Hampton

King of Waters the Amazon

King of the West Saxons Alfred the Great

King Who Lost George III, who lost the American colonies

King Who Lost America Great Britain's George III

King of Wines champagne

King of Zydeco Clifton Chenier

Kinmen Chinese name for Quemoy Island

Kinshasa formerly Leopoldville, Belgian Congo

Kiowa Bell helicopter whose civil version is called the Jet Ranger

Kipling's Khyber mountain pass between Afghanistan and Pakistan

Kipper NATO nickname for air-to-surface missile carried by Tu-16 aircraft

Kirk Douglas Issur Danielovich Demsky

Kirsty Kristina; Kristine

Kisangani formerly Stanleyville, Belgian Congo

kissy-face investing relationship funds seeking investments

kit injection paraphernalia

Kit Catherine; Christopher; Kitty

Kit Carson Christopher Carson nicknamed Monarch of the Prairies as well as Nestor of the Rocky Mountains

Kit Carson City Carson City, Nevada named for the frontiersman

Kitchener formerly Berlin, Ontario but changed in World War I to honor Lord Kitchener

Kitchener of Khartum General Horatio Herbert Kitchener

Kitsi Kathryn

Kittle Katherine; Kitty Belairs

Kittsian(s) inhabitant(s) of St Kitts

Kitty Catherine

kiwi(s) New Zealander(s)

Klinefelter syndrome affecting genitalia and internal ducts, usually male; testes underdeveloped and do not produce sperm

Klondike Country the Yukon

Klong Toey Bangkok, Thailand's waterfront area

Knickerbocker Group William Cullen Bryant, James Fenimore Cooper, Washington Irving (Diedrich Knickerbocker)

Knickerbocker(s) New Yorker(s)

Knight of La Mancha Don Quixote

Knight of the Rueful Countenance Don Quixote

Knight of the Swan Lohengrin

Knockout Brown Valentine Braun (boxer)

Know-Nothing President Millard Fillmore

Knut Hamsun Knut Pedersen

Koba Stalin's party name prior to the Bolshevik takeover of Russia

Kobarid Yugoslavian name for Caporetto

Kodak City Rochester, New York

Kodamá (Japanese—Echo)—nickname of the high-speed express train linking Kyoto with the port of Osaka

Kodok Sudanese name for Fashoda

Ko-i-noor (Persian—Mountain of Light)—Nadir Shah's name for the celebrated mountain-shaped diamond

Komar (Russian—Mosquito)—guided-missile patrol craft used by the navies of Algeria, China, Cuba, Egypt, Indonesia, Syria, etc.

kona (Hawaiian-Polynesian—lee side)—side of an island out of or protected from prevailing winds

Kona Coast gold marijuana grown on the kona or lee side of any Hawaiian or Polynesian island

Kongens By (Danish—King's City)—Copenhagen

Königsberg former name for Kaliningrad

Korea Gate nickname of the exposé involving more than a million dollars in bribes given some American politicians in return for their voting money, military aid, and supplies for South Korea

Korean Ports North Korean—Chinnampo, Wonson, Konan, Kimchaek, Chongjin, Najin Up, Unggi; South Korean—Inchon, Kunsan, Mokpo, Yosu, Masan, Pusan

Korea Strait Tsushima Strait (scene of decisive Japanese naval victory over Russian fleet during Russo-Japanese War of 1905)

Korovograd formerly Zinovievsk or Elisavetgrad

Kotlin Soviet guided-missile destroyers

Kousse Sergey Koussevitsky

Kraguj Yugoslav counterinsurgency aircraft

Kremlin seat of government in Moscow

Kremlin Killer Josef Stalin

Kresta Soviet guided-missile destroyer-leader warships

Kringleville Racine, Wisconsin

Krishaber's disease dizzy-and-sleepy neurosis accompanied by fainting

Kriss Kringle Santa Claus

Kristallnacht (German—Night of Broken Glass)—nights of November 8th, 9th, and 10th in 1938 when Nazi mobs broke the windows and smashed the doors of German-Jewish stores and temples before looting them, burning them, and sending their inhabitants to concentration camps

Kristiania Oslo's previous name

Kronos (Greek—Saturn)—god of time

Kronstadt Soviet subchasers

Krung Kao Ayutthaya, Thailand

Krupny a Soviet class of destroyers

K-T Event Cretaceous extinction, which ended the age of dinosaurs (K for Cretaceous, T for Tertiary)

K-town Kutztown, Pennsylvania

Kubyshka (Russian nickname for Cuba)—a kubyshka is a jar wherein Russian peasants bury their money—Kubachka (Little Cuba) has a similar sound

Kulmhof Chelmo a Polish concentration camp

Kuriles Kurile Islands of northernmost Japan seized by Russia at end of World War II

Kuwait Ports Mina abd Allah, Ash Shuaiba, Mina al Ahmadi, Abu Hulafah, Al Kuwayt

Kwok's disease Chinese restaurant syndrome produced by monosodium glutamate and resulting in dizziness, headaches, and nausea

Kynda a Soviet class of heavily-armed destroyers

La Argentina Antonia Mercé y Luque

La Belle Epoque (French—The Beautiful Epoch)—1900 to 1914 (turn of the century to the start of the first world war)

LaBelle Province (French—the beautiful Province)—Québec

La Belle Riviére (French—The Beautiful River)—frontier nickname of the Ohio in the days of Audubon and Boone

La Bonne Louise Louise Michel remembered for good works among the poor of Paris

Labor Boss Samuel Gompers, John L Lewis, and George Meany

Labrador Canadian version of Boeing-Vertol CH-113 helicopter; Labrador Current; Labrador duck; Labrador jay; Labrador Peninsula; Labrador pine; Labrador retriever; Labrador Sea; Labrador spar; Labrador stone (another name for Labrador spar); Labrador tea

lace-curtain middle class

La Columna (Spanish—The Column)—Venezuela's highest mountain also called Pico Bolívar

Lacrosse USA field artillery MGM-18A surface-to-surface missile

La Cumbre (Spanish—The Summit)—nickname of the Uspallata mountain pass and tunnel in the high Andes linking Argentina and Chile

La Divina Maria Meneghini Callas

The Lady cocaine; Statue of Liberty, New York

Lady of 57th Street New York City's Carnegie Hall

Lady Beaverkill Louise Brewster Miller

Lady Bird Claudia Alta Taylor Johnson—wife of President Lyndon Johnson

Lady in the Chair Cassiopeia constellation

Lady of Faubourg Saint-Honoré Salle Pleyel, Parisian concert hall

Lady Hamilton Emma Lyon

Lady of the Harbor Statue of Liberty, New York

Lady of the Lakes Michigan

Lady of the Lamp Nurse Florence Nightingale

Lady of Laughter Erma Bombeck

Lady in Red Anna Sage, who betrayed gangster John Dillinger

Lady Snow cocaine

Lady South Charleston, South Carolina

Lady with a Lamp Santa Filomena (immortalized by Longfellow); Statue of Liberty officially named Liberty Enlightening the World

l'Affaire (French—The Affair)—the Dreyfus Case

Lafitte Country Baratraria Bay (an old pirate settlement south of New Orleans)

La Gioconda (Italian—The Cheerful Woman)—another name for Leonardo da Vinci's portrait—the Mona Lisa

l'Aiglon (French—the Eagle)—Napoleon II, also known as the Duke of Reichstadt

Laird of Auchinleck James Boswell

Laird of Skibo Castle Andrew Carnegie

Laird of Vailima Robert Louis Stevenson

Laird of Woodchuck Lodge John Burroughs

Lake City Madison, Wisconsin

Lake of the Four Forest Cantons Lake Lucerne (Switzerland)

Lake Poets Samuel Taylor Coleridge, Robert Southey, William Wordsworth

Lake State(s) Michigan bordering on lakes Superior, Michigan, Huron, and Erie; Minnesota with its 10,000 lakes

Lalia Eulalia

La Lollo Gina Lollobrigida

Lama Aerospatiale observation helicopter built in France and designated SA-315

Lamb of God Jesus Christ

Lamia P L Tyraud de Vosjoli (French underground fighter and chief of intelligence)

lamo strange person

Lamp of Heaven the Moon

Lana Turner Julia Jean Turner

Land of 10,000 Lakes Minnesota

Land of Acadie (*see* Land of Evangeline)

Land of the African Equator Zaire (the largest)

Land of Albert Schweitzer Gabon

Land of Alligators Florida

Land of a Million Elephants Laos

Land of Art and Mozart Austria

Land of the Aztecs Mexico

Land Between the Rivers Mesopotamia, better known as Iraq

Land Beyond the Mountains Tennessee

Land Beyond Sorrow Uganda

Land of the Bible Israel

Land of Birds Australia

Land of the Bogs and the Little People Ireland

Land of Bondage Egypt in the time of Moses

Land of the Boomerang Australia

Land of the Bulgars Bulgaria

Land of Cactus New Mexico

Land of Caimans swamplands of tropical America

Land of Cakes Scotland

Land of the Cedars Lebanon

LandCent Allied Land Forces, Central Europe

Land of Cheese, Trees, and Ocean Breeze Tillamook, Oregon

Land of the Cherry Blossoms Japan

Land of Chopin and Copernicus Poland

Land of Clear Light the American Southwest and the Mexican Northwest (*La Tierra de Luz Clara*)

Land of Columbus Colombia

Land of the Conquistadors Extremadura, Spain where Cortez and Pizarro were born

Land of the Cornstalk Australia

Land of Crocodiles Africa and Southeast Asia

Land of the Dakotas North Dakota

Land of Death and Chains Siberia

Land of the Delight-makers New Mexico

Land of Desolation Antarctica and Greenland

Land Down Under Australia and New Zealand

Land of Dragons Hong Kong

Land of Dvořák and Smetana Czechoslovakia

Land of the Eagle Albania

Land of Emeralds Colombia

Land of Enchantment New Mexico

Land of Eternal Spring Guatemala

Land of Evangeline Maine east of the Kennebec River, New Brunswick, Nova Scotia, and Louisiana's coastal parishes

Land of Farmers and Fishermen Denmark

Land of the Firebird Ancient Russia

Land of Five Peoples Suriname, formerly Dutch Guiana, containing black, brown, red, white, and yellow people from Africa, Indonesia, South America, Europe, and the Orient

Land of the Fjords Norway

Land of Flaming Waters Malawi

Land of Flowers Florida

Land of the Free United States of America

Land of Freedom Liberia

Land of the Gaucho Uruguay

Land of Gavials India and Malaysia

Land of Genghis Khan Mongolia

Land of Gitche Gumee Lake Superior

Land God Gave Cain Arctic Canada

Land of Gold California

Land of the Golden Lion Iran

Land of Grass Roots South Dakota

Land of Greek, Roman, and Modern Ruins Lebanon

Land of Hans Christian Andersen Denmark

Land of the Happy Medium Costa Rica

Land of Heart's Desire New Mexico

Land of the Heather Scotland

Land of Heroes Finland

Land of Hiawatha Michigan's Upper Peninsula

Land of Hope and Glory Great Britain

Land of Hospitality Somalia

Land of Hospitality and Charm Thailand

Land of the Hummingbird Trinidad

Land of Ice and Fire Iceland

Land of the Incas Peru

Land of the Individual and Other Endangered Species Alaska

Land of the Inland Sea Chad

Land of the Inland Seas Great Lakes country of Canada and the U.S.

Land of Instant Women Thailand

Land of Iron and Diamonds Sierra Leone

Land of Isolation Antarctica

Land of the Kangaroo Australia

Land of the Khmers Cambodia

Land of the Kiwi New Zealand

Land of Lakes Wisconsin

Land of Lakes and Fens Finland

Land of Lakes and Forests Sweden

Land of Lakes and Volcanos El Salvador

Land of the Lamas Tibet

Land of Lamentations Holy Land, Israel and Palestine

Land of the Largest Antilles Cuba

Land of Latte Stones Guam

Land of Leeks Wales

Land of Legend Canada's Yukon Territory

Land of the Lemurs Madagascar and Comoro Islands

Land of Leopold Belgium

Land of the Leprechauns Ireland

Land of Letzeburgesch Luxembourg (where the language is Letzeburgesch)

Land of Lincoln Illinois

Land of Lions Tanzania (formerly Tanganyika)

Land of Liszt and Bartok Hungary

Land of the Llamas Peru

Landlocked South African Country Lesotho

Landlocked South American Nations Bolivia and Paraguay

Land of the Longest East Coast of South America Brazil

Land of the Longest West Coast of South America Chile

Land of the Long White Cloud New Zealand

Land of the Lotus Blossom Sri Lanka (Ceylon), where the lotus blossom symbolizes Buddha

Land of the Magyars Hungary

Land of the Manchus Manchuria

Land of Many Composers Russia (birthplace of Arensky, Borodin, Bortniansky, Cui, Glazunov, Gliere, Glinka, Khachaturian, Liadov, Liapunov, Medtner, Mussorgsky, Prokofiev, Rachmaninoff, Rimsky-Korsakov, Scriabin, Shostakovich, Stravinsky, Tchaikovsky)

Land of Many Tribes Tanzania (formerly Tanganyika)

Land of the Maoris New Zealand

Land of the Marsupials Australia

Land of the Mayas Honduras

Land of Mecca Saudi Arabia

Land of Men Marquesas Islands in the South Pacific

Land of the Midnight Sun northern Alaska, Canada's Northwest Territories, Greenland, Iceland, Norway, Sweden, Finland, and Siberia

Land of Milk and Honey Israel's Jordan River Valley

Land of a Million Elephants Laos

Land of the Moors Algeria and Morocco

Land of the Mormons Utah

Land of the Morning Calm Korea

Land of Moses Israel

Land of Mountains the Austrian Tyrol, Norway, Sweden, Switzerland, Tibet

Land of My Fathers Wales

Land of Nod where Cain was exiled after killing Abel; the realm of sleep

Land o'Cakes Land of Oatmeal Cakes—nickname Robert Burns gave to Scotland

Land o' Lakes Wisconsin

Land of Opportunity Arkansas; New Mexico

Land of Oz imaginary land invented by writer L(yman) Frank Baum

Land of Pagodas Burma

Land of the Panama Canal Panamá

Land of the Pentagram Morocco

Land of the People Greenland

Land of the Pharoahs Egypt

Land of the Philistines Palestine

Land of the Plastic Lotus California

Land of Plenty South Dakota

Land of the Poinciana Jamaica

Land of Political Exiles Yakutia (northeastern Siberia)

Land of Precious Things El Salvador

Land of the Prince Wales

Land of the Prophets Israel

Land of the Quetzal Guatemala

Land of the Red People Oklahoma

Land of the Rich Coast Costa Rica

Land of the Rising Sun Japan

Land of the Rolling Prairie Iowa

Land of the Rose England

Land of the Sagas Iceland (where the art of storytelling dates from the 12th century)

Land of the Saints Utah

Land of Saints and Scholars Ireland

Land of the Sea The Netherlands

Land of Sea and Mountain Norway

Land's End Cornish cape in southwest England—westernmost England

Land of Shakespeare England

Land of the Shamrock Ireland

Land of Silence Lapland

Land of Silver Argentina

Land of Six Peoples Guyana (formerly British Guiana), containing Africans, Amerindians, Chinese, East Indians, Spaniards, and other Europeans

Land of Skillful Farmers Lithuania

Land of the Sky North Carolina

Landslide Lyndon Senator Lyndon B Johnson

Land of Small Islands Micronesia

Land of Smiles Thailand

Land of Song Italy

Land of the South American Equator Brazil and Ecuador

Land South of the Clouds Yunnan

Land of the Southern Cross Brazil

Land of Spring coastal southern California from San Diego to Santa Barbara

Lands Reclaimed from the Sea Netherlands

Land of the Suez Canal Egypt

Lands of Sunlit Nights Scandinavian countries during summertime (Denmark, Finland, Iceland, Norway, Sweden)

Land of Steady Habits Connecticut

Land of Sunburned Faces Ethiopia

Land of Sunshine New Mexico, South Africa, and southern California

Land of Symphonists Austria (birthplace or home of Haydn, Mozart, Bruckner, and Mahler)

Land of the Templars Malta

Land That Time Forgot Australia

Land of the Thistle Scotland

Land of the Thousand Lakes Finland

Land of Togetherness Kenya

Land of Tomorrow Brazil

Land of the Trade Winds U.S. Virgin Islands

Land of the Unexpected New Guinea

Land of the Vikings Norway particularly Vestfold province on the western shore of Oslo Fjord where Viking remains are plentiful

Land of Voltaire France

Land of Waterfalls Norway

Land of Waters Guyana

Land of the Wattle Australia

Land of the West Morocco (from Arabic *maghrib* meaning west)

Land of the Whispering Bushes California

Land of the White Ant Australia's Northern Territory

Land of the White Eagle Poland

Land of the White Elephant Thailand

Land of the White Mountain Kenya

Land of the Winds Iran or Persia

Lane's disease chronic constipation

Langtry formerly Vinegaroon, Texas; renamed in 1882 by Judge Roy Bean to honor actress Lillie Langtry whose name also adorned his combination courthouse and saloon—*The Jersey Lily*

Lansen (Swedish—Lance)—jet interceptor aircraft designated J-32

La Pasionaria Dolores Ibarruri famed for her impassioned speeches during the Spanish civil war

La Perla (Spanish—The Pearl)—San Juan

Lapland northernmost Finland, Norway, Sweden, and Russia

La Popessa Mother Pascalina (nurse and confidante of Pope Pius XII for forty-one years)

Larruping Lou Henry Louis (Lou) Gehrig

Larry Laurence; Lawrence

Larry King Larry Zeigler

La Salle Street Chicago's financial district

La Scala Milan's opera house

La Scala West nickname of Chicago's Lyric Opera

Las Crutches trucker's nickname for Las Cruces, New Mexico

Last Capital of the Confederacy Danville, Virginia

Last Chance Gulch gold miner's name for Helena, Montana

Last Cocked Hat James Monroe—fifth President of the United States and last to wear the cocked hat of the American Revolution

Last Continent Antarctica (last continent to be discovered)

Last Corner of Arabia Oman

Last Frontier Alaska's old nickname and current nickname of Canada's Northwest Territories; Antarctica; Miami, Florida

Last Great Place Montana

Last Great Race Iditarod Dog Sled Race

Last of the Incas Atahualpa

Last, Loneliest, Loveliest (city) Auckland, New Zealand, according to Kipling

Last Lovely City San Francisco

Last Outpost on the Mississippi Pilot Town (near Venice, Louisiana)

Last of the Prophets before Mohammed Jesus, according to the Moslems

Last Remaining Polynesian Kingdom Tonga christened the Friendly Isles by Captain Cook

Last of the Romans Rienzi

Last Romantic poet and political writer Max Eastman; Richard Strauss; Sergei Rachmaninov; W Somerset Maugham

Last Son of the 18th Century Gioacchino Rossini

Last Stronghold of the Moors Granada, Spain

La Stupenda Dame Joan Sutherland

last trump the sound of the last trumpet believers expect to hear on Judgment Day

Last Viceroy Earl Mountbatten of Burma (India's last viceroy)

Last Word Society American Academy of Forensic Sciences

Last Year of Confusion 46 B.C., which Julius Caesar ordered lengthened by 90 days to align the calendar

La Superba (Italian—The Superb)—Genoa's proud appellation dating back to the time of Columbus

Las Vegas East Atlantic City, New Jersey

Las Villas formerly Santa Clara province in Cuba

Las Wages nickname of Las Vegas, Nevada

Latter-Day Saint Joseph Smith—author of *The Book of Mormon*

Latter-Day Saints the Mormons

Laughing Philosopher Voltaire

Launcelot Langstaff pseudonym shared by Washington Irving, William Irving, and James K Paulding when they published the *Salmagundi* essays

Laura World War II code name for Majuro, still in use by Americans and Marshallese islanders

Laura Z Hobson Laura K Zametkin

Laurel and Hardy comedians Stan Laurel and Oliver Hardy

Lauren Bacall Betty Perske

Laurence Templeton Sir Walter Scott's pseudonym used in the publication of *Ivanhoe*

Lawgiver of Ancient Greece Solon of Athens

Law of Infinitesimals dilute substances can have very strong effects

Lawrence of Arabia Thomas Edward Lawrence

Lawrence L Lynch Emma Murdock Van Deventer's pseudonym

Law of Similars like cures like

Law West of the Pecos Judge Roy Bean of Langtry, Texas, also known as the Hanging Judge

Lazy B Boeing

lazy eye amblyopia

Leadbelly Huddie Ledbetter

Leader of the East USSR (1945–1989)

Leader of French Enlightenment Denis Diderot

Leader of the Renaissance World Florence (Firenze)

Leader of the West U.S.A. (1945–1989)

Lead State nickname shared by Colorado and Missouri

League-of-Nations President Woodrow Wilson

Leaning Tower Leaning Tower of Pisa in Italy

Leao do Mar (Portuguese—Lion of the Sea)—the stormy Cape of Good Hope

Lebanese Ports Tarabulus, Beirut, Sayda, Sur

Leber's disease congenital atrophy of the optic nerve

Le Boulevard des Princes (French nickname—Hortense Schneider—actress and intimate of European royalty)

L'Ebreo (Italian—The Hebrew)—nickname of Salomone Rossi—Renaissance composer and rabbi

Le Carré John Le Carré (pseudonym of David John Moore Cornwall)

Lecumberri-Hilton nickname of Mexico City's great prison

Le Douainier (French—The Custom House Officer)—nickname of Henri Rousseau the primitive painter

Leedsloiner(s) native(s) of Leeds

Lee Grant Lyova Rosenthal

Lee J Cobb Leo Jacoby

Leewards Leeward Islands

Left Bank of the '90s Prague, Czech Republic

Lefty Robert Grove

legacy system any outdated computer software

Leghorn English equivalent of Livorno on Italy's west coast

Le Grande Orange Rusty Staub (baseball player)

Le Grand K (French—the big kilogram)—platinum cylinder, housed in Paris, that is the official standard weight for the kilogram

Le Grand Siècle (French—The Great Century)—the 1600s

Lemnos Limnos island in the Aegean

Lemonade Lucy Mrs Lucy Ware Webb Hayes—wife of President Rutherford B Hayes—who served only non-intoxicating fruit drinks while at the White House

Lemur Paradise Madagascar

Lena River City Yukutsk, Siberia

Lenny Leonard Bernstein; Lenny Bruce (Leonard Alfred Schneider)

Lens-Grinder Philosopher Benedictus de Spinoza

Leonard Holton Leonard Patrick O'Connor Wibberley's pseudonym

Leonard Q Ross (pseudonym—Leo Rosten)

Leon Bakst Leon Nikolaevich Rosenberg

Leopard West German Krauss-Maffei medium tank armed with a 105mm gun

Leopold's galloping ghost Congo-attained dysentery

Leopoldville Kinshasa, Zaire

Lepke Louis Buchalter (1897–1944)

Le Roi Soleil (French—The Sun King)—Louis XIV

Le Sage (French—The Wise)— Charles V

Lesbian Poet Sappho, the poetess of Lesbos

Lesbos Lésvos or Mytilene in the Aegean

Lesbos Long Island Fire Island, Long Island, NY

Les Cayes modern name for Aux Cayes, Haiti also called Cayes

Leslie Charteris Leslie Charles Bowyer Yin

Leslie Ford Zenith Jones Brown's pseudonym

Leslie Howard Leslie Stainer

Lesser Antilles Leeward and Windward Islands extending from the Virgin Islands to Trinidad and the Dutch West Indies (Aruba, Bonaire, Curaçao)

lessie(s) lesbian(s)

Lethbridge originally Coalbanks, Alberta

LeVar Burton Levardis Robert Burton Junior

Leviathan of Literature Dr Samuel Johnson

Le Vigan Robert Coquillaud

Levittown Pennsylvania town, created by William Levitt; considered the first suburb

Lewis Carroll Charles Lutwidge Dodgson

Lewis Grassic Gibbon pseudonym of J(ames) L(eslie) Mitchell

Liar of Biblical Antiquity Ananias, struck dead for lying, according to *The Acts* in the *New Testament*

Libba Cotten Elizabeth Cotten

Liberace Wladziu Valentino Liberace

The Liberator Daniel O'Connell

Liberator of Argentina General José de San Martín

Liberator Czar Alexander II (1855–1881)—abolished serfdom in Russia

Liberator of Genoa Andrea Doria

Liberator of God and Man Baruch de Spinoza

Liberator of Italy Giuseppe Garibaldi

Liberian Ports Robertsport, Monrovia, Buchanan, Harper

Libertador de Chile (Spanish—Liberator of Chile)—Bernardo O'Higgins

Libertybellsville Philadelphia

Liberty Bowl Memphis, Tennessee

Liberty Enlightening the World Statue of Liberty in New York Bay

Liberty Island formerly Bedloe's Island in Upper New York Bay where it supports the Statue of Liberty

Library Builder Andrew Carnegie

Library of Last Resort the Library of Congress in Washington, D.C., where anyone can consult any book in any language

Librettist-Composer Arrigo Boito

Libyan Libyan Desert in Egypt, Libya, and the Sudan

Libyan Ports Bardiyah, Tobruk, Darnah, Marsa al Hilal, Marsa Susah, Benghazi, Az Zuwaytinah, Marsa al Burayqah, As Sidr, Surt, Misratah, Tarabulus, Marsa Sabratah, Zuwarah

licorice tincture of opium

licorice stick clarinet's nickname

lifeblood of the industrialized world oil

Liftmaster Douglas DC-6 92-passenger transport

liger cross between a lion and a tiger

Light of the Ages Moses ben Maimon of Cordoba also known as Maimonades

Light of Asia Gautama Buddha

Lighthorse Harry Major General Henry (Lighthouse Harry) Lee, USA—father of Robert E. Lee

Lighthouse of the Pacific *El Faro del Pacifico*—Izalco Volcano—whenever active, its fire can be seen from planes and ships several hundred miles away from El Salvador

Lightning British BAC all-weather supersonic jet

Lightnin' Hopkins Sam Hopkins

Ligurian Republic Republic of Genoa

Likasi formerly Jadotville in the Belgian Congo

Lilac New Hampshire state flower and nickname sometimes given New Hampshire girls

Lilac City Spokane, Tacoma, Washington

Lila Lee Augusta Appel

Lillian Gish Lillian De Guiche

Lillian Nordica Lilly Norton

Lillian Russell Helen Louise Leonard

Lillibet Elizabeth

Lillie Emily; Lillian; Lillie Langtry—the Jersey Lily—christened Emilie Charlotte Le Breton

Lilli Palmer Maria Lilli Peiser

lily of the field garlic

Lily of France symbolic *fleur de lis* or lily flower

Lily-Lilo Rosalie Texier

Limbo of the Lost the Bermuda Triangle, also called the Devil's Triangle

Limejuicer British sailor

Limerick Lear Edward Lear of Highgate, near London

Limeyland England

Limey(s) Limejuicer(s)—British sailor(s) or ship(s); nickname derived from their use of limejuice to ward off scurvy

Lincoln's Shrine Springfield, Illinois

Lincoln's State Illinois

Lindy Charles Augustus Lindbergh

Linnaeus Carl von Linné

Linoleum Capital of Scotland Kirkaldy

Lion of the Caribbean Sir William Alexander Bustamante

Lion City Singapore

Lionel Barrymore Lionel Blythe

Lion Flag Ceylon's emblem featuring a golden lion with an upraised scimitar in his right paw comes from the ancient name for this island

Lion Hill Norway's parliament, officially called Storting

Lion of Judah former Emperor Haile Selassie of Ethiopia

Lion of the North King Gustavus Adolphus of Sweden

Lion of Pune B K S Iyengar (yoga teacher)

Lion's Den State Tennessee

Lion's Gate harbor entrance of Vancouver, British Columbia

Lion Tamer Ian Smith

Lippy Leo Ernest Durocher

liquid candy soft drinks

Lisbeth Elisabeth; Elizabeta; Elizabeth

Listener to the Winds Edward MacDowell

Liszt Expert Alfred Brendel

Literary Emporium Boston, Massachusetts

Literary Queen of Expatriate Americans Gertrude Stein

Lithuania Baltic nation long ruled by Soviet Russians until 1991

Little Alfie Alfred Austin

Little America Antarctic camp at the edge of the Ross Ice Shelf and the Bay of Whales where Admiral Byrd headquartered; East Anglia, England (for World War II-era United States air bases); London's Grosvenor Square where John Adams lived at No 9 when he was America's first ambassador to Great Britain; now site of the U.S. Embassy

Little Belt Lillebaelt (strait separating island of Fyn from Danish mainland between Baltic Sea and the Kattegat)

Little Ben Benjamin Harrison

Little Boy A-bomb dropped on Japanese targets before the end of World War II

Little Britain Armorica or Brittany in northern France

Little Chicago Culiacán, Sinaloa, Mexico

Little Chocolate George Dixon (boxer)

Little Corporal five-foot-high Napoleon Bonaparte *(Le Petit Caporal)*

Little Denmark Solvang, California

Little Dixies the southern parts of Ohio, Indiana, Illinois, and parts of Missouri and Oklahoma

Little Egypt delta country of southern Illinois around Cairo and the confluence of the Ohio and Mississippi

Little England of the Caribbean Barbados

Little Flower Fiorello H La Guardia

Little Giant Johnny Griffin, saxophonist; Knute Nelson—populist governor of Minnesota; oratorically gifted Senator Stephen Douglas of Illinois

Little Goldilocks Shirley Temple

Little Havana Cuban-refugee-populated sections of Miami, Florida

Little Holland Garibaldi, Oregon

Little Ida Idaho

Little Inch 20-inch pipeline paralleling the Big Inch

Little India of the Pacific Fiji Islands

Little Italy Italian section of any American or Canadian city

Little Jazz Roy Eldridge (musician)

Little Joe Apollo spacecraft booster designed and produced by General Dynamics, Convair

Little John surface-to-surface rocket produced by Emerson Electric

Little Lad of Landau American political cartoonist Thomas (Th) Nast born in Landau, Germany

Little Lady in Pants Dr Mary Walker

Little Lhasa Dharamsala, India, to which the Dalai Lama fled in exile from Lhasa, Tibet

Little Louie Luis Aparicio (baseball player)

Little Lunnon Colorado Springs, Colorado

Little Luther Hans Kung

Little Mac General George B Mc Clellan

Little Magician Martin Van Buren—eighth President of the United States

Little Mermaid Edvard Eriksen's bronze statue of a maiden seated atop a rock and looking out to sea from Copenhagen's harbor—immortalized in Hans Christian Andersen's fairy tale

Little Neddies Economic Development Committees

Little Netherlands Curaçao

Little New York Miami Beach, Florida's South Beach

Little Old Lady of Pennsylvania Avenue Federal Trade Commission

Little Paradise Queen Victoria's nickname for the Isle of Wight

Little Phil General Philip Henry Sheridan

Little Poison Lloyd Waner (baseball player)

The Little Professor Dom DiMaggio (baseball player)

Little Red Book quotations of Mao Tse-tung

Little Rhody Rhode Island

Little Richard Richard Penniman

Little's disease congenital spastic paralysis

Little Steam Engine Pud Galvin (baseball player)

Little Sure Shot Annie Oakley (Mrs Frank Butler)

Little Three Williams College, Amherst, Wesleyan University

Little Tiger mainland China's hydrofoil patrol boat called Hu Chwan

Little Tokyo Japanese section of any American or Canadian city

Little Van President Martin Van Buren

Little Van Dyke Gonzales Cocx—Flemish portrait painter who imitated the style of Van Dyke but painted family groups on small canvases

Little Van's Lady Hannah Van Buren—wife of President Martin Van Buren

Little Venice Lake Maracaibo, Venezuela

Little White House Franklin D Roosevelt's farm home near Warm Springs, Georgia

Little Yellow Book quotations of Deng Xiaoping

Live-Entertainment Capital of the World Branson, Missouri

Live Free or Die State New Hampshire

Live Oak State Florida

Liverpool of the Pacific Vancouver, British Columbia

Living Declaration of Independence Thomas Paine

Livy Olivia; Roman historian Titus Livius

Liz(a) Eliza(beth)

Lizard Moon full moon in March

Lizard State Alabama

Lizards Alabamans' nickname

Lizbeth Elizabeth

Lizzie British sailors' nickname for the superliner *Queen Elizabeth*

Lizzy Elizabeth

Ljuba Welitsch Ljuba Velichkova

Lock City Stamford, Connecticut

Lock Town Stamford, Connecticut

Log Cabin-and-Hard-Cider President William Henry Harrison

Lola Dolores

Lola Montez stage name of Marie Dolores Eliza Rosanna Gilbert also known as the

Comtesse de Lansfeld, Mrs Heald, Mrs Hull, and Mrs James

Lon Chaney Alonso Chaney

london london broil (thinly-sliced flank steak broiled before serving); london brown (carbuncle gemstone)

London Bach Johann Christian Bach

London of the Scanians King Canute the Great's name for Lund, Sweden; Scandic capital he founded to match his London of the English

London by the Sea Brighton

London-super-Mare Brighton

London Town London, England

Lone Eagle Charles A Lindbergh

Lone Lion of Idaho Senator William E Borah

Lonely Iconoclast George S Schuyler

Lone Star State Texas

Longevity Belt zone stretching from Minnesota to Nova Scotia

Longhair Lair New York's Lincoln Center of the Performing Arts

Longhorn(s) Texan(s) named after the longhorn cattle characteristic of Texas

Long Night Moon full moon in December

Longshanks Edward I of England

Longshore Philosopher Eric Hoffer

Long Tom Thomas Jefferson—third President of the United States

Lon'on Town British nickname for London

looie lieutenant

The Loop downtown commercial, financial, hotel, shopping, and theater district of Chicago

looping taping voices after a motion picture is shot

Lord Acton 1st Baron John Emerich Edward Dalberg-Acton

Lord Baltimore George Calvert

Lord Beaconsfield Benjamin Disraeli

Lord Beaverbrook William Maxwell Aitken

Lord Berners Gerald Hugh Tyrwhitt-Wilson

Lord Brougham Henry Peter

Lord Byron George Gordon Byron

Lord Chesterfield Philip Stanhope

Lord De La Warr Thomas West—Lord Delaware

Lord Desart William Ulick O'Connor Cuffe

Lord Dufferin Frederick Temple Hamilton Blackwood (Lord Rector of St Andrews University)

Lord Dunsany Edward John Moreton Drax Plunkett

Lord of the East Vladivostok

Lord God bird pileated woodpecker

Lord Haw-Haw nickname of William Joyce

Lord of Hell Lucifer

Lord High Executioner of the Mafia Albert Anastasia (1903–1957)

Lord of Hokkaido Japanese red fox

Lord Kelvin William Thomson

Lord Kenneth Kenneth Clark—Lord Clark of Saltwood

Lord Keynes John Maynard Keynes

Lord Kinross John Patrick Douglas Balfour

Lord Kitchener Horatio Herbert (also known as the Earl of Khartoum)

Lord Macaulay Thomas Babington

Lord North Frederick North

Lord Palmerston Henry John Temple nicknamed Pam

Lord Passfield Sidney Webb

Lord Peter Death Brendon Wimsey Ian Carmichael

Lord Protector Oliver Cromwell—Lord Protector of England

Lord of Reason Bertrand A Russell

Lord of the Rings J(ohn) R(onald) R(euel) Tolkien

Lord Russell Bertrand A Russell

Lord Salisbury Robert Arthur Talbot Gascoyne-Cecil

Lord of San Simeon William Randolph Hearst

Lord Tweedsmuir John Buchan

Lord of War Wotan

Lorenzo da Ponte Mozart's librettist whose real name was Emanuele Corregliano

Lorenzo the Magnificent Lorenzo de Medici

Loretta Young Gretchen Young

Loris Hugo von Hofmannsthal

Lorrie Laura; Lorraine

Los Angeles' Sister City Eilat—Israel's leading oil port

Los Pinos (Spanish—The Pines)—official home of Mexico's presidents

Lost City of the Incas Machu Picchu, Perim (near Cuzco, Peru)

Lost Colony Roanoke Island, North Carolina (site of Sir Walter Raleigh's first settlement)

Lost Generation people who came of age just after World War I and lacked cultural, political, or social stability

Lost Wages Las Vegas, Nevada (gambling resort)

Lot's Wife Japanese volcanic islet in the North Pacific between Iwo Jima and Yokohama—resembles a pillar of salt; mountain of St Helena Island in the South Atlantic

Lotte Charlotte

Lottie Charlotte

Lotty Charlotte

Lou Ambers Louis D'Ambrosio (boxer)

Lou Costello Louis Cristillo

Lou Gehrig's disease amyotrophic lateral sclerosis

Louis Calhern Carl Vogt

Louis Capet King Louis XVI

Louise Homer Louise Dilworth Beatty

Louis-Ferdinand Céline Henri-Louis Destouches

Louis Graveure Wilfred Douthitt

Louisiana for Louis XIV of France

Louisiana Ports Baton Rouge, Lake Charles, New Orleans

Louis Jourdan Louis Gendre

Louis le Debonnaire Louis I of France

Louis Napoleon Napoleon III—Emperor of France

Love Bug widespread, damaging computer virus in May 2000

Love Goddess Rita Hayworth

Lovely Felix Felix Mendelssohn-Bartholdy

love machine bedroom-on-wheels type of recreation vehicle such as a camper, trailer, or van

Lover of Liberty Thomas Paine

Low Countries Belgium, Luxembourg, and the Netherlands

Lower 48 Alaskan name for the forty-eight contiguous United States

Lower 48ers Alaskan name for people in the 48 contiguous states

Lower Amazon Amazon River traversing northern Brazil from Manaus to the Amazon River Delta on the Atlantic near Belém do Pará

Lower Austria southern Austria bordering on Switzerland, Italy, Yugoslavia, and Hungary

Lower Bavaria eastern Bavaria

Lower Burma coastal Burma west of Thailand

Lower California the Baja California

Lower Canada French-speaking Québec and the lower St Lawrence region during the 19th century

Lower East Side New York City's most congested section south of Washington Square

Lower Egypt Egypt's delta area north of Cairo and including Alexandria and Port Said

Lower Franconia northwestern Bavaria

Lower Galilee Israel between the Mediterranean and the Sea of Galilee

Lower Lakes southernmost Great Lakes—Erie and Ontario

Lower Michigan peninsular Michigan south of Mackinac Strait

Lower Mississippi Mississippi River from St Louis to New Orleans and the Gulf of Mexico

Lower Nile Nile River flowing from Khartoum in the Sudan to Cairo in Egypt and the Nile River Delta emptying into the Mediterranean Sea

Lower Peninsula southern Michigan between Lake Michigan, Lake Huron, and Lake Erie

Lower Rhine Rhine River between Bonn, Germany and the North Sea coast of the Netherlands

Lower Saxony Neidersachsen including most of Brunswick, Hannover, Oldenburg, and Schaumburg-Lippe

Lower Silesia southern Silesia

Lowland Duchy Luxembourg

Low Newton remand center in Durham, England

Loyalist Province New Brunswick

Lozovsky Solomon Abramovich Dridzo

LT Meade Elizabeth Thomasina Meade Smith

L-town Lewistown, Illinois

Lubumbashi formerly Elisabethville, Belgian Congo

Lucan Roman poet Marcus Annaeus Lucanus

luciferin an organic molecule that helps fireflies produce light (named for Lucifer, the fallen archangel and bearer of light)

Lucil Gaius Lucilius (Roman—satiric writer)

Lucille Ball Diane Belmont

Lucky Lucky Luciano (Salvatore Lucania)—once America's foremost gangster

Lucky Black Swan Western Australia's city of Perth where black swans swim about in Perth Water

Lucky Capital Canberra, Australia

Lucky Country Australia

Lucky Lindy Charles Augustus Lindbergh

Lucy Lucia; Lucilla; Lucille; Lucille Ball (Diane Belmont); St Lucia, West Indies

Lucy Stone maiden name of Mrs Henry Brown Blackwell

Lukas Foss Lukas Fuchs

Lumber Capital Tacoma, Washington

Lumber City Bangor, Maine

Lumberjacks Mainers (inhabitants of Maine)

Lumber State Maine

Lum'n Abner Chester Lauck and Norris Goff

Lunatic of Libya Col Muammar Qaddafi

Lungansk old name of Voroshilovgrad

Lungs of the Earth boreal forest encircling North from Canada to Siberia

Lunik Soviet cosmic rocket landed on Moon September 14, 1959

Lusians Portuguese

Lusitania (Latin—Portugal)—Roman name often used as the poetic equivalent of Portugal

L Wica pseudonym of Vilhelm the Prince of Sweden and Duke of Södermanland

Lynn Brock Alister McAllister

Lynn Doyle Leslie Alexander Montgomery

Lyric Land Wales

Lyric Poet and Literary Critic Heinrich Heine

Maastricht Corridor southernmost projection of the Netherlands between Belgium and Germany

Ma Bell Bell System telephone companies linked by AT&T

Mac Macintosh computer

Ma Cable American Telephone and Telegraph

Macabre Film Robert Wiene's *Cabinet of Dr Caligari*

macadam macadam road (named for its Scottish inventor—JL McAdam); macadam stone (stone used in building macadam pavements or roads)

macadamia nuts pale, buttery nuts from Australia (named for Dr John MacAdam)

MacArthur's Birthplace Little Rock, Arkansas—birthplace of General Douglas MacArthur

MacGregor mechanically operated hatch cover

Machine Gun George R Kelly (1897–1954) and Jack McGurn (1904–1936); Travis Grant (basketball player)

mach numbers mach 1, et cetera—ratio of velocity of an object to the velocity of sound (named after Ernst Mach, Austrian physicist)

Macintoy Macintosh computer

Mac OS Macintosh Operating System—trademark

MacRobertson Land Australian Antarctica

Madagascar's Principal Port Tamatave

Madame Bertha Sarah Bernhardt

Madame Blavatsky Helena Petrovna Hahn-Hahn

Madame Deficit Marie Antoinette's nickname

Madame de Stael Baronne Anne Louise Germaine

Madame Récamier Jeanne Françoise Julie Adélaïde Bernard

Madame X Madame Pierre Gautreau

Mad Anthony Major General Anthony Wayne

Mademoiselle le Professeur Nadia Boulanger

Madge Margaret; Margarita

Mad Genius of Sex and Psychiatry Wilhelm Reich—inventor of the orgone box

Madhouse on the Potomac the Capitol, the Pentagon, the State Department, the White House, Washington, D.C.

Mad Ludwig Ludwig II of Bavaria (Wagner's patron)

Mad Meg nickname of the Mayer van der Bergh Museum in Antwerp, Belgium

Mad Monk Gregori Rasputin

Madonna Madonna Louise Ciccione

Mad Priest John Ball

madras madras cloth or madras cotton; madras kerchief

The Maestro Arturo Toscanini

Maestro of Abolition Brazilian composer-conductor Carlos Gomes who fought for the abolition of slavery

Maestro Crescendo Rossini

Maestro No No conductor-perfectionist Arturo Toscanini

Mae West American actress of stage and screen; lifejacket named after her shape

Magallanes former name of Punta Arenas, Chile

Magazine of History *American Heritage*

Magazinist Edgar Allan Poe's self-invented title

Magda Lupescu Elena Wolff

Magellan of Modern Music Harry Partch

Maggie Margaret; Margaret Thatcher, Prime Minister of the United Kingdom of Great Britain and Northern Ireland (1979–1990); stock market nickname for Magnavox; Warren Magnuson (U.S. senator)

maggotbox Macintosh computer

Maggot Mile Boca Raton, Florida

Magic City Miami, Florida

Magic Island Haiti on Hispaniola

Magic Johnson Earvin Johnson

Magister Fouga jet trainer built in France

Magnificent 13 New York City's red-bereted teams of vigilantes determined to enforce law and order in the subways

Magnolia City Houston, Texas

Magnolia Lady Former First Lady Rosalynn Carter

Magnolia(s) Mississippian(s)

Magnolia State Mississippi

Magyar(s) Hungarian(s)

Mahatma (Hindi—Great Souled)—sobriquet of India's greatest leader, Mohandas Karamchand Ghandi

The Mailman Karl Malone (basketball player)

Maid of Orleans Joan of Arc— *La Pucelle d'Orleans*

Maid of Zaragoza Augustina de Aragón (Augustina Domenech Zaragoza—fighter for freedom during Spain's invasion by Napoleonic armies)

Mail Soviet BE-12 Beriev amphibian reconnaissance aircraft

Maimonides Moses ben Maimon

Main Drag Main Street or the main street of any American city or town

Main Drag of Many Tears 125th Street in New York City's Harlem

Maine Ports Calais to Kittery including Bangor, Searsport, Boothbay Harbor, Bath, and Portland

mainland China People's Republic of China

Mainland Hawaiians' nickname for the 49 mainland states of the United States

Mainlanders Hawaiian name for people of the 49 mainland states

Main Staters inhabitants of Maine

Maisie Maria; Marie; Mary; Maryjane

Maison Gomin correctional facility for women at St Cyrille, Québec

Malson Tanguay Montreal's facility for women prisoners

Majocchi's disease ringlike empurplement of the lower limbs

Major Bob nickname of former Salvadoran rightist leader Roberto D'Aubuisson

Major Prophets of the Old Testament Isaiah, Jeremiah, Ezekiel, Daniel

Make-You-Cry Composer Puccini

Make-You-Laugh Composer Rossini

Malacañan (Filipino—Home of the Ruler)—Filipino presidential palace

Malacañan Palace official residence of the President of the Philippines

Malafon Latecoere surface-to-surface or surface-to-underwater naval missile made in France

Malagasy Republic Madagascar

Malayan Island Nation Singapore

Malaysian Ports Penang-Butterworth; Lumut, Kelang (Port Swettenham), Port Dickson, Melaka

Malbrook (Louis XIV's mispronunciation of Marlborough—John Churchill — Duke of Marlborough)

Malcolm X Malcolm Little

Maldive Islands Port Malé

Malgache Republic Madagascar (territory includes Amsterdam, Crozet, Kerguelen, and Saint Paul islands)

Malinche (*see* Doña Marina)— often a Mexican synonym for Quisling or traitor—with Malinchismo meaning treachery

Maltese Ports Valetta and Marsaxlokk

Maluku (Indonesian—Moluccas)—Spice Islands

Mambru (Spanish mispronunciation of Marlborough— John Churchill—Duke of Marlborough)

Man Against Everything Henry Louis Mencken (1880–1956), editor of the *American Mercury* (1924–33)

Man for All Seasons Sir Thomas More

Mañanaland Latin America

Manassa Mauler Jack Dempsey born in Manassa, Colorado,

Man of a Thousand Faces Lon Chaney

Mana-Zucca Augusta Zuckerman

Man of Blood and Iron Prince Otto von Bismarck

Man of Destiny Napoleon; William Walker, one-time dictator president of Nicaragua who planned for Central American unification and a Caribbean federation including Central America and Cuba

mandrake nickname for *Mandragora offucinarum* also called devil's testicle or satan's apple

Mandy Amanda; Manda

Man from Independence President Harry S Truman

Man from Maine James G Blaine

Man from Missouri President Harry S Truman

Man from Stratford William Shakspere

Man of God Grigori Rasputin (so named by Czar Nicholas II—last Romanov emperor of Russia)

Mangrove Coast Florida's southernmost coast between the Everglades and the Keys

Manhattan Manhattan clam chowder (minced clams plus herbs and tomatoes); Manhat-

tan cocktail (vermouth and whiskey mix topped with a maraschino cherry)

Manifest Destiny President James Knox Polk

Manila Manila hemp (abacá fiber used in, making fabrics, rope and paper); Manila paper (buff-color paper prized for its heavy-duty applications)

Man of Monach Country Fermanagh County in Ulster, Northern Ireland

Manning Coles Cyril Henry Coles

Manny Emanuel; Manuel

Man(ny) Han(ny) Manufacturers Hanover (Bank)

Manolete Manuel Rodriguez

Man on Horseback General Georges Boulanger

Man's Oldest Disease alcoholism

Mantovani d'Annunzio Paolo

Mantua English place-name for Mantova

Manuel Emmanuel

Man Who Broke the Bank of England George Soros

Man Who Invented Panama Philippe Bunau-Varilla

Man Who Made the Greatest Dictionary James AH Murray who devoted his life to the creation of the *Oxford English Dictionary (OED)*

Man Who Made Ragtime Scott Joplin

Man With the Hoe The Man With the Hoe—Edwin Markham

Man of Words lexicographer Eric Partridge

Manxmen Isle-of-Man persons

Many Islands (*see* Polynesia)

Man You Loved to Hate Erich von Stroheim

Maple City Ogdensburg, New York

Maple Leaf Canada's flag consisting of three vertical stripes—red, white, red— with a red maple leaf on the white center stripe

Ma Rainey Gertrude Rainey

Marble Capital Proctor, Vermont

Marble City Rutland, Vermont; Sylacauga, Alabama

Marble Halls of Oregon Oregon Caves National Monument

Marburg hemorrhagic fever deadly African fever

Marc Chagall Marc Segal

Marcella Sembrich Praxede Marcelline Kochanska

Marches border region of England and Wales

March King John Philip Sousa

Marco Page pseudonym of Harry Kurnitz

Marder German armored-personnel carrier fitted with a 20mm cannon

Mardi Gras Metropolis New Orleans, Louisiana

Marg Margaret

Marge Margaret; Margery

Margie Margaret

Margo Margaret; María Margarita Guadalupe Teresa Estela Bolado Castilla y O'Donnell

Margot Margaret

maria plains of the moon; archaic (for Latin *mare*, or sea)

Maria Callas Maria Kalogeropolos

Maria Ivogün Ilse von Günther

Maria Jeritza Mimi Jedlitzka

María Montez Maria Africa Vidal de Santo Solas

Marie Brema Minny Fehrman

Marie Corelli Eva Mary Mackay

Marie Curie Manya Sklodowska (1867–1934)

Marie Dressler Leila Von Koerber

Marie's disease chronic enlargement of the face, feet, and hands

Marilyn Monroe Norma Jean Baker

Marina Street San Juan, Puerto Rico's district jail

Mariner Venus-Mars fly-by space vehicle

Mariner Mystic Herman Melville

Mario Giovanni Matteo

Mario Lanza Alfred Arnold Cacozza

Marion Davies Marion Cecelia Douras

Mariscal de Ayacucho (Spanish—Marshal of Ayacucho)—Antonio José de Sucre—companion of Bolívar and first president of Bolivia

Maritime Capital of Canada Vancouver

Maritime Provinces New Brunswick, Nova Scotia, and Prince Edward Island

Maritzburg Pietermaritzburg

Mark Aldanov Mark Aleksandrovich Landau

Markland (Norse—Forest Land)—probably Labrador

Markov chains probability theory based on sequences of random variables (named for mathematician Andrei Markov)

Mark Rothko Marcus Rothkowitz

Marks & Sparks Marks & Spencer (British department store)

Mark Twain Samuel Langhorne Clemens

Mark Twain Town Hannibal, Missouri

Mark Twain in Sneakers George Sheehan

Marlboro Country Bulgaria, where cigarettes have been the currency of choice for government favors

Marlene Dietrich Magdalene von Losch

Marlin Martin P-5M reconnaissance flying boat

Marmalade Capital Dundee, Scotland

Marquis de Vaugenargues Luc de Clapiers

Marquis de Sade Donatien Alphonse Francois

Marquise de Pompadour Jeanne Poisson

Marquis of Halifax George Savile (1633–1695)

Marquis of Queensbury boxing rules formulated by John Graham Chambers supervised by the 8th Marquis of Queensbury—Sir John Shollto Douglas

Marquis of Rockingham Charles Watson-Wentworth (former Prime Minister of Great Britain)

Marquis of Salisbury Robert AT Gascoyne-Cecil

Marse Robert (southern American—Master Robert)— General Robert E Lee

Marshall Plan President Harry S Truman

Martel Hawker-Siddeley in Britain and Matra in France built this AS-37 missile

Martha Albrand Heidi Huberta Freybe

Martial Roman epigrammatist Marcus Valerius Martialis

Martin Sheen Ramon Estevez

Martov Yuli Osipovich Tsederbaum

Martyr Abolitionist Elijah Parish Lovejoy

Martyr of Mexican Independence Miguel Hidalgo

Martyr for Science Giordano Bruno

Marunouchi Tokoyo's financial center

Marvel of Marble the TaJ Mahal

Marx Brothers Chico (Leonard), Harpo (Arthur), Groucho (Julius), Gummo (Milton), and Zeppo (Herbert) Marx

Mary Astor Lucille Langehanke

Maryland named for Mary, queen of King Charles I

Maryland Port Baltimore

Mary Pickford Gladys Mary Smith's stage name

Mary Queen of Scots Mary Stuart

Mary Roe anonym used when the true name of the woman is unknown; nickname of the average American girl

Marysville Girls Ohio Reformatory for Women at Marysville

Masaccio Tommaso Guidi

Masaniello contracted name of Tommaso Aniello

Masha (Russian nickname—Mary)

The Masked Marvel Charley Patton

Mason and Dixon Line boundary between Maryland and Pennsylvania, also used to describe former demarcation between southern slave and northern free states

Massachusetts Ports Gloucester, Salem, Boston, Quincy, New Bedford

Master of Color Contrasts Bartolomé Esteban Murillo

Master Composer Robert Schumann

Master of Disguises Alec Guinness

Master of Guerrilla Warfare Toussaint l'Ouverture

Master Illusionist Lawrence of Arabia, Thomas Edward Lawrence

Master of Light and Shade Rembrandt van Rijn and Leonardo da Vinci

Master Mariner of the Imagination Joseph Conrad

Master of Melancholy Andres Segovia

Master Melvin Mel Ott (baseball player)

Master of Mexico Don Porfirio Diaz (Porfiriato lasted from 1876 to 1911)

Mastermind of Revolution VI Lenin

Master Musician/Comedian Victor Borge

Master of Neurological Anatomy Santiago Ramón y Cajal

Masterpiece of American Engineering the Panama Canal

Master Pilot Jacques Cartier *(Le Maître-Pilote)*

Master of Political Satire Mark Russell

Master of Psychic Polyphony Richard Strauss

Master of the Psychological Novel Franz Kafka

Master of Raphael II Perugino (Pietro Vannucci)

Master Scientist-Aero Engineer Theodore von Karmen

Master of Suspense Alfred Hitchcock

Master of Swing Count Basie [originally William (Bill) Basie]

Master of Tragic Sorrow Pyotr Ilich Tchaikovsky

Master of the Telecaster Albert Collins

Master of the Walking Bass Leroy Vinegar

Master of the Yosemite Ansel Adams

Mastodon of Literature Emmanuel Swedenborg so nicknamed by Ralph Waldo Emerson

Masurca French-built surface-to-air naval missile

Mata Hari Gertrud Margarete Zelle

Mateo Boz Miguel Angel Correa's pseudonym

The Material Girl Madonna Ciccione

Mato Tepee Devils Tower National Monument in northeastern Wyoming

Matriarch of Anthropology Margaret Mead

Matsqui British Columbia's minimum-security facility for narcotic addicts at Abbotsford

Matty the Great Christy Mathewson

Matty Van Martin Van Buren

Mauch Chunk Pennsylvania place now called Jim Thorpe

Maude Adams Maude Kiskadden

Maureen Forrester Katherine Stewart

Maureen O'Hara Maureen Fitzsimmons

Maurice Barrymore Herbert Blythe

Mauritanian Port Nouakchott

Mauritius Port Port Louis

Max of Baden Maximilian Alexander Friedrich Wilhelm von Hapsburg (1867–1929)

Max Brand Frederick Faust

Maxfield Parrish Fred Parrish

Maxim Gorki Aleksei Maxsimovich Peshkov

Maxim Litvinov Maxim Maximovich Wallach

Max Nordau Max Simon Südfeld

Max Pax Max, Prince von Baden (Maximilian Alexander Friedrich Wilhelm)— Germany's last imperial chancellor

Max Reinhardt Max Goldmann

Max Stirner Johann Kaspar Schmidt

Mayfair London's residential district

Mayor of Silicon Valley Robert N Noyce, co-inventor with Jack Kilby of the microchip

McAlester Ward Women's Ward in the Oklahoma State Penitentiary at McAlester

McCain Sanatorium North Carolina Prison Sanatorium at McCain

McCarthyism sensationalist tactics and unsubstantiated political attacks (named for United States Senator Joseph Raymond McCarthy, 1908–1957)

McClintock effect synchronized menstrual periods

McGuffin filmmaker Alfred Hitchcock's term for a device that keeps a thriller's plot in motion

McGuffey Reader 19th-century reading primer named for William McGuffey

McNamara's War the Vietnam War, named for Robert Strange McNamara, former United States Secretary of Defense

Meal Ticket Carl Hubbell (baseball)

Meanie nickname for a mean person

Meat-Packing Capital of the World Chicago

Mecca of Deaf Culture Gallaudet University

Mecca of Spain Santiago de Compostela

Mechanical Man Charlie Gehringer (baseball player)

Media Prophet Marshall McLuhan

Medical Essayist Oliver Wendell Holmes

Medina-Sidonia Alonso Pérez de Guzmán (Duke of Medina-Sidonia and admiral in command of the ill-fated Spanish Armada defeated by Sir Francis Drake)

Medinat Yisrael (Hebrew— State of Israel)

megacities very large city populations as in Tokyo, São Paulo, New York City, Mexico City, Shanghai, Bombay, Los Angeles, Buenos Aires, Seoul, Beijing, Rio de Janeiro, Calcutta, Jakarta to list the first 13

Meister Der Meister (German—the Master)—Johann Wolfgang von Goethe

Meister Raro Robert Schumann

Melanchthon Philipp Schwarzert

Mel Brooks Melvin Kaminsky

Mel Ferrer Melchor Gaston Ferry y Cintron

Melina Mercouri Maria Amalia Mercouri

Melisande Melusina

melon wolf coyote

The Melting Pot New York City

Melvil Dewey Melville Louis Kossuth Dewey

Melvin Douglas Melvyn Hesselberg

Member of the Unemployed Scottish socialist leader Keir Hardie—founder of the Independent Labour Party (ILP)

Memé Remedios

Memphis Minnie Lizzy Douglas

Memphis Slim Peter Chatman

Mencius Meng-tse

Menckonaclast Henry L Mencken

Mendelian genetics principles describing transmission of genes from parents to offspring (for Gregor Johann Mendel: 1822–1884)

Mendel's law of independent assortment of the genes each pair of different characters in a hybrid union is independent of the other characters

Mendel's law of the segregation of alleles hybrid peas containing both alleles in most of their cells form egg and pollen cells that contain either the dominant allele or the recessive allele in equal numbers

Men of the East Sherpas of northern India and Nepal

Meniere's disease sudden dizziness, ear ringing, and vomiting due to disturbance of the labyrinth

Menorca's Principal Port Port Mahón

Mentor Beechcraft T-34 trainer aircraft

Mentor to Parisian Intellectuals Théophile Gautier

Mercator Gerardus Mercator real name of this 16th-century Flemish geographer is Gerhard Kremer

Merchant of Death international arms contractor Sir Basil Zaharoff

Merchant of Menace Vincent Price

Merchants of Death epithetic nickname applied to alcohol and tobacco vendors, armament makers, drug pushers, munitions makers, narcotics traffickers, and others whose business may result in the death of their customers

Merchants' Haven Copenhagen, Denmark

meretricious traffic prostitution; white slavery

Meritorious Man of the Americas Benito Juárez

Merle Oberon Estelle Merle O'Brien Thompson

Merry Monarch Charles II of Great Britain also nicknamed Patron of Bawdy Houses

Mersenne primes prime numbers that fit the form 2 to the nth power minus 1 where n also is a prime number (named for Marin Mersenne, 17th-century French mathematician)

Mescalero Cessna T-41 trainer-utility aircraft

Mesopotamia (Greek—Between Rivers)—land between the Euphrates and the Tigris; formerly Assyria, Babylonia, and Sumeria but presently Iraq

Mesopotamia's Uncrowned Queen Gertrude Bell

Messenger of the Gods Hermes (Greek); Mercury (Roman)

Messenger of Mercy Swiss banker Jean Henri Dunant— founder of the Red Cross

The Met Metropolitan Opera House—New York City

Metallic Age when you have silver strands in your hair, copper pennies in your purse, iron rust in your guts, and lead weights in your bottom

Metaphysician of Modern Atheism Baruch Spinoza

Metastasio Pietro Antonio Domenico Bonaventura Trapassi

Meteor Gloster twin-engine jet-fighter aircraft

Metroland portion of London served by the Metro subway system

Metropolis of America New York City

Metropolis of the Magic Valley Brownsville, Texas on the Rio Grande

Metropolis of the Missouri Valley Kansas City

Metropolis of the South Mark Twain's nickname for New Orleans

Metropolis of the State of Oregon Portland

Metropolis of the United States New York

Metropolis of the World London

Metropolitan Museum of Art or Opera House in New York City

Metropolitan City of the Anglican Communion Canterbury

Metternich Prince Klemens Wenzel Nepomuk Lothar von Metternich - Winneburg — Austrian statesman convening Congress of Vienna at end of Napoleonic wars

Mexican Mexican apple (white sapote); Mexican ground cherry *(tomatillo);* Mexican hairless (dog used to herd cattle and keep ranchers company in bed as its hairlessness makes it flealess and its body temperature is higher than ours); Mexican jumping beans (inhabited by insect larvae whose movements make the beans jump about)

Mexican Agrarian Reformer Emiliano Zapata

Mexican Idealist Politician and Revolutionary Francisco I (dalecio) Madero

Mexican mud deadly Mexican-made heroin, also called black tar tootsie roll

Mexican Muralists José Clemente Orozco, Diego Rivera, Davíd Alfaro Siquieros, and Rufino Tamayo

Mexican National Composer Carlos Chávez

Mexican Ports Tampico, Veracruz, Coatzacoalcos, Frontera, Progreso; Salina Cruz, Acapulco, Manzanillo, San Blas, Mazatlan, Guaymas, Santa Rosalia, Ensenada

Mexican-Spanish Mexican-style Spanish enriched with more than 50,000 Mexicanisms

Mexico's Principal Port Vera-Cruz

Meyerbeer Giacomo Meyerbeer (adopted name of Jakob Liebmann Beer)

Meyer Lansky Maier Suchowljansky

Miami Beach East Tel Aviv, Israel's nickname

Michael Angelo Titmarsh Thackeray's pseudonym

Michael Arlen Dikran Kuyumjian's pseudonym

Michael Caine Maurice Joseph Micklewhite

Michael Curtiz Michael Kertesz; Mihály Kertész

Michael Fairless Margaret Fairless Barber

Michael Field Katherine Harris Bradley and her niece Edith Emma Cooper used this pseudonym for their joint poetic efforts

Michael Innes John Innes Mackintosh Stewart

Michael Keaton Michael Douglas

Michael Landon Eugene Orowitz

Michael Servetus Miguel Serveto

Michael Tilson Thomas Mike Thomashefsky

Michelangelo Michael Angelo Buonarroti

Michel Auclair Michal Vujovic

Michéle Morgan Simone Roussel

Michigan Ports Wyandotte, Rouge River, Detroit, Bay City, Saginaw, Alpena, Saulte Ste Marie, Frankfort, Manistee, Ludington, Muskegon, Grand Haven, Holland, St Joseph, Marquette

Mickey Mouse Walt Disney Productions (Wall Street nickname)

Mickey Mouse boots large black rubber insulated boots originally developed for the United States Army

Mickey Rooney Joe Yule, Jr

Mickey Spillane Frank Morrison

Micky Micaela; Michael; Michelle

Mickyland Ireland

Microcosm of Canadian Life London, Ontario

Micronesia small islands of the western Pacific also known as the Trust Territory including the Federated States of Micronesia, the Marshall Islands, and Palau in the Carolines as well as many very small islands covering an area as wide as the United States

Micro$loth Microsoft

Middle America Central America, Mexico, and the West Indies

Middle Atlantic States Delaware, Maryland, New Jersey, New York, Pennsylvania, West Virginia

Middle Border Hamlin Garlin's nickname for the American Middle West

Middle Colonies New York, New Jersey, Pennsylvania, Delaware

Middle East area extending from Afghanistan to Egypt and including India, Iran, Iraq, Saudi Arabia, Syria, Lebanon, Israel, Jordan, Kuwait, and the United Arab Emirates

Middle Kingdom China

Middle Passage route of the slavers across the middle of the Atlantic between West Africa and the West Indies

Middle States New York, New Jersey, Pennsylvania, Delaware, and Maryland

Middletown Muncie, Indiana

Middletown title of Robert and Helen Lynd's study of cultural conflicts in Muncie, Indiana

Middle West United States from the Great Lakes to the northern border of the Gulf States and from the western slopes of the Appalachians to the eastern slopes of the Rockies

Midget City Habaianinha, Brazil

Midland Capital Birmingham, England

Midlands central England including counties of Bedford, Buckingham, Derby, Leicester, Northhampton, Nottingham, Rutland, Warwick

midlifers 35- to 55-year-old people also called the sandwich generation with responsibilities for their children and elderly parents

Mightiest of Rivers the Amazon

Mighty Champion of Freedom Frederick Douglass

Mighty Mo battleship USS *Missouri*

Mighty Mstislav Mstislav Rostropovich

Mihaly Munkacsy Michael Lieb

Mike Michael

Mike Nichols Michael Igor Peschkowsky

Mike Wallace Myron Wallace

Mildred Masters Mildred Kapilow

Mile-High or More-Than Mile-High North American Cities Butte, Montana; Cheyenne, Wyoming; Colorado Springs and Denver, Colorado; Flagstaff, Arizona; Gallup, New Mexico; Mexico City; Santa Fe, New Mexico

Mile-Square City Hoboken, New Jersey

Miliano Emiliano Zapata

Military Alphabet Alfa, Bravo, Charlie, Delta, Echo, Foxtrot, Golf, Hotel, India, Juliet, Kilo, Lima, Mike, November, Oscar, Papa, Quebec, Romeo, Sierra, Tango, Uniform, Victor, Whiskey, X-ray, Yankee, Zulu

Military Space Capital of the Free World United States Strike Command

Milk City Carnation, Washington

Milk Moon full moon in May

millionaire city a city with a population of a million or more

Millionaire's Captain Titanic Captain Edward Smith

Millie Mildred; Millicent

millennium bug computer programming glitch relating to year 2000

Million-Acre Farm Prince Edward Island

Milton Berle Mendel Berlinger

Milward Kennedy Milward Rodon Kennedy Burge

Mim Mimi; Miriam; Miryam (niah)

Mima Jemima

Mimico Boys Mimico Correction Centre (for males) in Toronto, Ontario

Mina Wilhelmina

Mineral Storehouse of the Nation Canada's Hudson Bay area

Miners Nevadans

Ming-Ming Immingham, England

Mining Baron William A Clark

Mining State Nevada

Mini State Rhode Island

Minnesota Port Duluth

Minnie Minerva; Minneapolis; Minnesota

Minor Prophets of the Old Testament Hosea, Joel, Amos, Obadiah, Jonah, Micah, Nahum, Habakkuk, Zephaniah, Haggai, Zacharia, Malachi

Minstrel Composer James Bland

Minuteman solid-fuel intercontinental ballistic missile produced by Boeing

Minuteman III America's most advanced icbm in 1979

Minx of the Movies Betty Compson

Miracle of Fifth Avenue Guggenheim Museum

Miracle of Nature Queen Christina of Sweden

Mirage all-weather delta-wing supersonic-jet ground-support interceptor built by Dassault in France (Mirage III)

Mirage IV atomic-bomber version of the foregoing Mirage III

Mischa (Russian nickname—Michael)—Mike

Mischa Auer Mischa Ounskowsky

misfit river a river that appears to be too small to have carved its valley

Miss Dimples Shirley Temple

Miss Emma morphine

Missie Miss; Mississippi; Missus; Mrs

Missionary to the Lepers Father Damien

Mission City San Antonio, Texas

Mississippi Fred McDowell Fred McDowell

Mississippi John Hurt John Smith Hurt

Mississippi Ports Pascagoula, Biloxi, Gulfport

Mississippi River Painter George Caleb Bingham

Miss Nancy nickname for William Rufus Devane King, thirteenth United States vice president

Miss New Orleans Dorothy Lamour's title in 1931

Miss Tarbarrel Ida M Tarbell

Mista Klemps conductor Otto Klemperer

Mistress of Mystery Agatha Christie

Mistress of the North the Baltic Sea, Russia, and Sweden

Mitch Mitchell; Richard Mitchell

Mitchell B-25 Bomber named in honor of General William Mitchell, who in 1925 was courtmartialed for criticizing the mismanagement of the aviation service in the U.S. Army and Navy

Mitchellville Girls Iowa School for (delinquent) Girls at Mitchellville

Mitya (Russian diminutive— Dmitri)

Mitzi Margaret

Mizzou Missouri

mnemonic hormone vasopressin (reported to improve the memory if sniffed or sprayed into each nostril every day for three days)

Mockingbirds Arkansas citizens

Model of the Common Man Benjamin Franklin

Model Republic Orange Free State

Model-T automobile once the world's most popular vehicle despite its handcranking starter and its nickname— Tin Lizzie

Modern Antigone Maria Thérèse—daughter of Louis XVI

Modernizer of Navigation Lieutenant Matthew Fontaine Maury, USN

Modern Liberal Social Philosopher José Ortega y Gasset

Modern Mother of Presidents Ohio—birthplace of Presidents Grant, Hayes, Garfield, Benjamin Harrison, McKinley, Taft, Harding (*see* Mother of Presidents)

Mogen David six-pointed star on flag of Israel and in synagogues where it is called the Star of David

Mohair Capital Del Rio, Texas

Mohammed Ali Cassius Clay

Mojave Mojave Desert of Arizona and California; Sikorsky heavy helicopter designated H-37

Molière Jean-Baptiste Poquelin

Molière of Music André Grétry

Moll Mary (slang); Molly

Molly Maria; Marie; Mary

Molly Pitcher Mrs John Hays also known as Captain Molly because she took her husband's place as cannoneer when he fell mortally wounded at the Battle of Monmouth—June 28, 1778

Molotov Vyacheslav Mikhailovich Skriabin

Mona Ramona

Monaco's Only Port Monaco

Monarch of the Mountains the elk

Monarchs of Marshes and Swamps alligators, crocodiles

Monarchy of Mount Everest Nepal

Mondrian Piet Mondrian Cornelis

Mongol Conqueror Genghis Khan (1162–1227)

Monk Matthew Gregory (Monk) Lewis; Thelonious Monk

Monkey Trial Scopes Trial (in 1925, John Scopes, a Tennessee science teacher, was tried for having taught evolution; he was prosecuted by William Jennings Bryan and defended by Clarence Darrow)

Monkey Ward Montgomery-Ward

Monk Lewis pseudonym of Matthew Gregory Lewis

monorail single-rail railway

Monroe Doctrine oppose foreign attempts to gain influence in the Americas (named for James Monroe, fifth president of the United States)

Monroe Doctrine President James Monroe

Monsieur de Paris (French— Mr Paris)—guillotine operator

Monte Carlo of Asia Macau, formerly Portuguese, territory returned to China

Monterey Jack David Jacks, and the only native cheese of California whose production he controlled

Montezuma's Revenge diarrhea or dysentery nicknamed for the last Aztec ruler of Mexico; both ailments are also nicknamed the Curse of Cortez for the Spaniard who conquered Mexico

Montgomery Camp Federal Prison Camp at Montgomery, Alabama

Monumental Intellectual John Locke

Monumental State Maryland

Monument City Baltimore, Maryland

Monument to Slavery Berlin Wall (1961–1989)

moonchild(ren) person(s) born under the zodiacal sign of Cancer

Moondog Louis Thomas Hardin

Moon Goddess Luna (Roman) whose Latin name means moon; Selene (Greek)

moon rock mixture of crack and heroin

Moor Othello; The Moor (Dartmoor Prison)

Moor Court prison for female offenders in Staffordshire, England

Moore's law A semiconductor chip's computing power will double every 18 months (for Gordon Moore)

Moralist of Psychoanalysis Erich Fromm

Morand's disease paresis affecting the feet

Moravian Capital Brno

Moreton Bay Colony Queensland

Morgan Fairchild Patsy McClenny

morganization reorganization of bankrupt railroads (after John Pierpont Morgan; 1837–1913)

Morgan's Raiders students of biologist Thomas Hunt Morgan

Mormon City Salt Lake City

Mormon's Mecca Salt Lake City, Utah

Mormon Prophet Joseph Smith

Mormon State Utah

Moroccan pasta couscous

Moroccan Ports Tanger (Tangier), Kenitra, Casablanca, Safi, Agadir

Morrie Maurice; Morris

Morrison Girls Mount View (delinquent) Girls School at Morrison, Colorado

Morris Rosenfeld Moshe Jacob Alter

Morris worm mid-1990s computer virus that inspired United States statutes against attacks on computer systems

Morton's disease metatarsal neuralgia

Moses of Her People Harriet Tubman (c. 1820–1913)

Moskva (Russian—Moscow) —a Soviet class of antisubmarine—warfare cruiser and helicopter-carrier warship

Moslem India Bangladesh and Pakistan

Moslem Sultanate Oman formerly Muscat and Oman

Mosquito Coast Caribbean coast of Honduras and Nicaragua

Mosquito State New Jersey

mother homosexual dope pusher; madam of a brothel; motherfucker

Mother of American Kindergartens Susan Blow

Mother of the American Legion Ernestine Schumann-Heink

Mother of the American Red Cross Clara Barton

Mother Ann Shaker leader Ann Lee

Mother of the Arts architecture

Mother of Balboa Park Kate Sessions, San Diego horticulturist

Mother of Believers Ayesha— Mohammed's favorite wife

Mother Bickerdyke Mary Ann Bickerdyke

Mother of Birth Control Margaret Sanger

Mother Bloor Ella Reeve Bloor

Mother Cabrini Frances Xavier Cabrini

Mother Carey's chickens stormy petrels

Mother Carey's geese fulmars or great white petrels

Mother of Child Education Doctor Maria Montessori

Mother of Cities Bombay, according to Kipling

Mother of Civilized Cities London

Mother of the Civil Rights Movement Coretta Scott King

Mother Earth the Greek Goddess Gaea or Ge who, according to mythology, arose out of chaos and in turn produced the sea, the sky, and the mountains; the Romans called her Tellus or Terra and sometimes called her Vesta Prisca

Mother of Exiles Statue of Liberty overlooking New York's former immigration stations at Battery Park and Ellis Island

Mother of Feminine Psychology Karen Horney

Mother of Ghosts the Roman goddess of Death—Mania

Mother Goose legendary authoress of children's rhymes and stories

Mother of Gospel Music Willie Mae Ford Smith

mother grain quinoa (Incan grain)

Mother of Her Country Queen Maria Theresa of Austria

Mother of Israel Golda Meir

Mother of the Japanese Novel Baroness Murasaki Shikibu *(The Tale of the Genji)*

Mother Jones Mary Harris Jones

Mother Lake Leonora Marie Kearney Barry

Mother of Libraries Alexandria, Egypt

Mother Maid The Virgin Mary

Mother of Mathematics Hypatia of Alexandria

Mother of Modern Dance Isadora Duncan

Mother of Mountains Nepal's Mount Everest

Mother of Muckrakers Ida M Tarbell (*see* Father of Muckrakers)

Mother of Parliaments British Parliament

Mother of Pearl Cloud rare high cloud formed in stratosphere—delicate and iridescent after sunset

Mother of Presidents Virginia—birthplace of Presidents Washington, Jefferson, Madison, Monroe, William Henry Harrison, Tyler, Taylor, Wilson

Mother of Prison Reform Dorothea Lynde Dix

Mother of the Red Cross Clara Barton

Mother of Rivers Tibetan Highlands; New Hampshire

Mother of Rivers and Waves Tethys, wife of the god Oceanus, and mother of the rivers plus three thousand Oceanids—the waves

Mother of Russian Cities Kiev

Mother of the Russians Moscow

Mothers of Believers the wives of Mohammed

mother-ship smuggling off-loading drugs at sea from large ship to smaller ships, then to pleasure boats

Mother of the Snows Mount Everest

Mother State Virginia

Mother of Southwestern Statesmen Tennessee

Mother of Statesmen Virginia

Mother of Storms Antarctica, the Baffin Sea, the Bay of Biscay, the Caribbean, the Gulf of Alaska, the Gulf of Mexico, the South China Sea, and the Tasman Sea

Mother of Texas Mary Long

mother tongue music (according to many great musicians, philosophers, poets, and scholars)

Mother of Trusts Standard Oil

Motion Picture Capital of the World Hollywood, California

Motion Picture Palace Potentate Roxy (SL Rothafel)

Motor City Detroit

Motor Town Detroit

Motown Motown Records (named for Mo' Town— Motor Town: Detroit)

Motown Sound rhythm and blues, gospel and pop

Mound City St Louis, Missouri

Mountain City Chattanooga, Tennesse

Mountain Devils Tasmanians

Mountain Division States Arizona, Colorado, Idaho, Montana, Nevada, New Mexico, Utah, and Wyoming

Mountain of Fire Etna, Vesuvius, or any other active volcano

Mountain of the Lion Sierra Leone

Mountain State West Virginia

Mountain States Arizona, Colorado, Idaho, Montana, Nevada, New Mexico, Utah, and Wyoming

Mountain of Tarik The Rock of Gibraltar named for the Moorish chief Jabal Tariq

Mountain View Girls Mountain View School (for juvenile-delinquent females) at Helena, Montana

Mountbatten of Burma AF Admiral of the Fleet the Earl Mountbatten of Burma, KG (better known as Lord Mountbatten)—India's last viceroy

Mount Rainier Mount Tahoma towering over Tacoma, Washington; Washington volcano named for Rear Admiral Peter Rainier

Mount Vancouver on the Alaska-Yukon border (named for explorer George Vancouver)

Movie Capital Hollywood, California

Movieland Hollywood, California

Mozambique Ports Moçambique, Beira, Lourenço Marques

Mozart Town Salzburg, Austria—birthplace of Wolfgang Amadeus Mozart

Mr Automobile Gottlieb Daimler

Mr B. Billy Eckstine

Mr Color Television Peter Goldmark

Mr Common Sense Thomas Paine

Mr Conservative Barry M Goldwater

Mr Cub Ernie Banks

Mr Diesel Rudolf Diesel

Mr Dogpatch Al Capp (also known as the Mark Twain of cartoonists or the sardonic cartoonist)

Mr Dooley (pseudonym—Finley Peter Dunne)

Mr Electric Light Thomas Alva Edison

Mr Gyrocompass Elmer Ambrose Sperry

Mr Helicopter Igor Sikorsky

Mr Klemps Otto Klemperer

Mr Laser Charles Townes

Mr Linotype Ottmar Mergenthaler

Mr Long-Play Records Peter Goldmark

Mr Moves Campy Russell (basketball player)

Mr October Reggie Jackson (baseball player)

Mr Pilot Will Adams (nicknamed *Anjim Sama* by the Japanese because of his knowledge of navigation and shipbuilding)

Mr Pitiful Otis Redding

Mr Radio Lee De Forest

Mr Republican Robert A Taft (1889–1953)

Mr Sam U.S. Senator Sam Ervin; former Speaker of the House Sam Rayburn (1882–1961)

Mrs Fletcher Maria Jane Jewsbury's pseudonym

Mrs Grundy nickname for the imaginary self-appointed arbiter of morality and taste

Mrs Jack Isabella Stewart Gardner

Mrs Patrick Campbell stage name of Beatrice Stella Tanner

Mr Television Milton Berle (Mendel Berlinger)

Mr UN Carlos P Romulo

Mr Wireless Guglielmo Marconi

Mr X-Ray Wilhelm Roentgen

M-town Mertztown; Middletown; Morgantown; Morristown

Muckraker of France Émile Zolá whose anticlerical, antimilitary, and antimonarchial writings forced him to flee to England

Muckrakers American crusader journalists David Graham Phillips, Charles Edward Russell, Lincoln Steffens, Upton Sinclair, Ida M Tarbell

mud wet concrete

Mudcat(s) Mississippian(s)

Mudcat State Mississippi

Muddy Waters McKinley Morganfield

Mudheads Tennesseans

Mud Island San Diego's South Bay Wildlife Preserve

Mudwaddler State Mississippi

Mudwaddlers Mississippians

Muggsy Francis (Muggsy) Spanier

Muhammad (Arabic—The Praised)—Mahomet

Muhammad Ali Cassius Clay

Mule Capital of the World Columbia, Tennessee

mullet thunder jumping mullet (fish)

Multimedia Gulch South of Market area in San Francisco

Mum City, U.S.A. Bristol, Connecticut

Munich Expressionist Wassily Kandinsky

Municipal Muckraker Lincoln Steffens—author of *The Shame of the Cities*

Murihiku (Maori—End of the Tail)—New Zealand's southernmost city on South Island—Invercargill

murphy murphy bed (concealed-in-the-wall bed invented by William L Murphy); nickname for an Irish or white potato as well as for a confidence swindle

Murphy's Law If something can go wrong, it will

Murrumbidgee River City Canberra

Musandam Musandam Peninsula on the Arabian Peninsula

Muscat and Oman former name of the Sultanate of Oman

Muse of Astronomy and Celestial Music Urania

Muse of Comedy and Pastoral Poetry Thalia

Muse of Dancing and Choral Singing Terpsichore

Muse of Epic and Heroic Poetry Calliope, who according to Horace, could play any musical instrument

Muse of Erotic Poetry Erato

Muse of History Clio

Muse of Lyric Poetry and Music Euterpe

Muse of Oratory, Rhetoric, and Sacred Song Polyhymnia

Muse of Tragedy Melpemone

Museum of Architecture Leningrad (also Petrograd or Saint Petersburg)

Museum Cities Padua, Venice, Verona, and Vicenza

Museum Metropolis London, New York, and Paris

Mushroomopolis Kansas City

Music Director of Europe Herbert von Karajan

Muskrats Delawareans

Musky Armando Moscaritolo

Muslim Century the 8th century—the 700s when the Turkish crescent occupied the entire Hispanic Peninsula, the Levant, the Middle East, and even westernmost India

Mustang North American fighter aircraft F-51

muttnik second Soviet satellite launched in 1957; its astronaut was a mongrel dog

Myrna Loy Myrna Williams

Mystere IV Dassault jet fighter built in France

Mystere 20 Dassault twin-engine executive transport called the Falcon

Mysterious Billionaire Howard Hughes

Mytilene Aegean island of Lesbos

Nabby Abigail

Nadar Gaspard-Felix Tournachon (1820–1910)

Nail City Wheeling, West Virginia

Namby-Pamby 18th-century English dramatist-poet Ambrose Philips

Namibia modern name for South-West Africa

Nana Anna; Anne(tte); Mariana; Nanette; grandma

Nannerl Maria Anna

Nanny State Singapore

Nanty Anthony

Nanzig (German—Nancy)— French industrial center renamed by Hitler during World War II

Napoleon Bonaparte Napoleon I—Emperor of the French

Napoleon of Mexico General Antonio López de Santa Anna (1794–1876)

Napoleon of Peace Louis Philippe

Napoleon of Wall Street J Pierpont Morgan

Napoleon of the Women's Rights Movement Susan B Anthony

Narco nickname of the U.S. Public Health Service Hospital in Lexington, Kentucky where narcotics addicts are treated

Narco Saint Jesus Malverde, a Robin Hood-style bandit hanged in 1909 in Sinaloa, Mexico; venerated by Mexican drug lords

Narragansett Bay State Rhode Island

Narrow-Gauge Capital of the World Durango, Colorado

Narrow Land Between the Seas Panama

Nashville Agrarians 1920s Vanderbilt University literary critics, poets, economists, and historians

Nashville Girls Tennessee Prison for Women at Nashville

Natacha Rambova Winifred Hudnut

Natalie Wood Natasha Gurdin

National Anthem City Baltimore, Maryland

National Beverage of England beer

National Pastime baseball in America; cricket in Britain

National Poet of Norway Bjornstjerne Bjornson

National Tity navalese for National City, California

Nation of Gentleman Scotland so named by King George IV

Nation's Capital District of Columbia

Nation's Front Yard The Mall in Washington, D.C.

Nation of Shopkeepers England, according to Samuel Adams as well as Napoleon

Nation's Hottest Town Quartzsite, Arizona where July temperatures average 108°F (42°C)

Nation's Icebox International Falls, Minnesota

Nation's State District of Columbia

natural dog product of natural, rather than human, selection—mutt

natural language of the deaf American Sign Language

natural Prozac the herb St. John's wort

Natural State Arkansas

nature's own sleeping pill tryptophan

Nature's Wonderland Iceland

Naughty MacNaughton; McNaughton

Naughty Nineties the 1890s

Nauru Islands Ports Nauru Atoll, Saipan, Tinian, Rota

Naval Person Churchill's cover name used when addressing Roosevelt—POTUS—President of the United States

Navarino Italian name for the port of Pylos

Navel of the Nation Butte County, South Dakota (geographic center of the United States including Alaska and Hawaii); Smith County, Kansas (geographic center of the forty-eight conterminous states)

Navel of the World Easter Island according to its ancient Polynesian inhabitants

The Navigator Prince Henrique of Portugal (1394 to 1460)

Navigators' Islands Samoa Islands in the South Pacific

Navigator's Nightmare the Bermuda Islands

navy navy bean (small white bean); navy blue (dark blue); navy plug (tobacco)

Neapolitan Painter and Poet Salvator Rosa

Nearly Impeached President Andrew Johnson, Richard M Nixon

nebbies nembutal capsules

Neddy Edgar; Edmund; Edward; Edwin; Edwina; National Economic Development Council's nickname

Nederlanders Dutch men and women

Needle Frank Gates (basketball player)

negawatts ability of a utility to harness efficiency instead of hydrocarbons

Negri Negritude (literary movement involving blacks)

Nel Eleanor(a); Ellen; Helen(a); Nelly

Nell Eleanor(e)

Nellie Nellie McClung (pronounced McClue)—Canadian novelist and women's rights champion in the early 1900s

Nellie Melba Helen Porter Mitchell

Nello Emmanuel

Nelly Eleanor(a); Ellen; Helen

Nelly Bly Elizabeth Cochrane Seaman

Nelson Algren Nelson Algren Abraham

nembies nembutal (sodium pentobarbital sedative hypnotics)

nemish nembutal

nemmies nembutal capsules

Nemo Guillaume; Guillermo

Neon Deion Sanders (football)

Neopagan Eclectic Miguel de Unamuno

Neptune Lockheed P-2 antisubmarine and reconnaissance aircraft

Nequam Alexander Necham

Nerve Center of Alaska Anchorage

Nessa Agnes

Nessie Agnes

Nessie the Loch Ness monster's nickname

Nesta Agnes

Nestor of American Botany William Darlington

Nestor of American Pediatrics Abraham Jacobi

Nestor of Congregationalism Leonard Bacon

Nestor of Europe Leopold I of Belgium (1790–1865)

Nestor of the Rockies Kit Carson

The Net Internet

nethead a habitué of computer networks

Netherlands Antilles Aruba, Bonaire, Curaçao, Saba, Sint Eustatius, and half of Sint Maarten

Netherlands East Indies former name of Indonesia

Netherlands Guiana Dutch Guiana or Suriname

Netherlands Indies old name of Indonesia

Netherlands New Guinea former name of West Irian now part of Indonesia

Netherlands Ports (see Dutch Ports)

Netherlands Principal Port Rotterdam

Netherlands Timor formerly the western half of Timor now an island of Indonesia

netizens Internet citizens

Nettie Henrietta

Netty Henrietta

Net(ty) Antonia

Never-Never Never-Never Land (arid Australia)

Nevil Shute Nevil Shute Norway

New Albion Sir Francis Drake's name for what is now British Columbia, plus the states of Washington, Oregon, and California

New Alcatraz nickname of the maximum-security U.S. Penitentiary at Marion, Illinois

New Amsterdam former name of New York City (Nieuw Amsterdam)

The New Beirut Miami, Florida

New Caledonia British Columbia

New Colossus Statue of Liberty's sobriquet derived from the poem by Emma Lazarus— The New Colossus—proclaiming: "Give me your tired, your poor, your huddled masses yearning to breathe free, the wretched refuse of your teeming shore. Send these, the homeless, tempest-tossed to me. I lift my lamp beside the golden door!"

New Deal administration of Franklin Delano Roosevelt— thirty-second President of the United States

New Deal President Franklin Delano Roosevelt

New Economy economy dominated by technology and communications businesses

New Edinburgh on the Antipodes Dunedin, New Zealand

New England New England boiled dinner (corned beef or ham with vegetables); New England clam chowder (clams, potatoes, milk, and stock); New England pine (white pine)

New England Mystic Emily Dickinson

New England of the West Minnesota

New France old name for French Canada

New Frontier administration of John F Kennedy— thirty-fifth President of the United States

New Granada Colombia's original Spanish name— Nueva Granada

New Hampshire Port Portsmouth

New Holland old name for Australia discovered by Dutch navigators

New Jersey eagle big mosquito

New Jersey Ports Weehawken, Hoboken, Jersey City, Newark, Bayonne, Elizabethport, Port Socony, Grasselli, Cartaret, Chrome, Port Reading, Perth Amboy, South Amboy, Leonardo, Camden, Gloucester

New Jersey seashell tampon applicator

New Munster old name for New Zealand's South Island

New Netherlands old name for New York and parts of Connecticut and New Jersey

New Prometheus Immanuel Kant's nickname for Benjamin Franklin who drew lightning from the skies

new scarlet letter herpes virus (type 1 characterized by lip sores and type 2 by genital lesions)

New Spain New Jersey

New Sweden Sweden's short-lived colony in and around what is now Wilmington, Delaware but once was called Nya Sverige (New Sweden)

Newton of Electricity André Marie Ampère

New Ulster old name for New Zealand's North Island

New World North and South America

New York cut porterhouse steak (with the bone and fillet removed)

New York Persona New York City Mayor Edward Irving Koch

New York Ports Ogdensburg, Oswego, Rochester Harbor, Tonawanda, Buffalo, Albany, Kingston, Yonkers, Manhattan, Brooklyn, Gulfport, Port Richmond, Mariners Harbor, Stapleton, Tomkinsville

New York State Barge Canal Erie Canal

New Zealand Commonwealth New Zealand and its territories

New Zealand Dominion New Zealand and its territories

New Zealand Ports on North Island: Auckland, Gisborne, Napier, Wellington, Wanganui, New Plymouth, Dargaville; on South Island: Port Nelson, Port Lyttelton (Christchurch), Timaru, Oamaru, Port Chalmers, Bluff Harbour, Greymouth, Westport; plus smaller ports such as Dunedin and Invercargill

New Zealand's Garden City Christchurch

New Zealand's Principal Port Auckland

Nguyen (Vietnamese—Dynasty)

. **Niagara Frontier** Buffalo-Niagara Falls area

Niagara Fruit Belt Canadian fruit-growing region on the Niagara Peninsula between lakes Erie and Ontario

Nicaraguan Ports Cabo Gracias a Dios, Puerto Cabezas, Puerto Isabel, Bluefields, San Juan del Norte (Greytown); San Juan del Sur, Puerto Masachapa, Puerto Somoza, Corinto

Nicholas Blake C(ecil) Day Lewis' pseudonym

Nick Carter J Russell Coryell

nickel note $5 bill

Nickel-plated Paradise nickel-rich New Caledonia (*Nouvelle Calédonie)*

Nickel Plate Road New York, Chicago, and St Louis Railroad Company

Nicky and Alicky Czar Nicholas II and Czarina Alexandra Feodorovna of Russia—the last of the Romanov Czars

Nicolas Copernicus Nikolay Kopernik

Nicolas-Favre disease lymphogranuloma venerea involving inguinal lymph glands and characterized by an exuding lesion

Nicolas Lenau Nikolaus Niembsch von Strehlenau

Nicolino Nicoló Grimaldi

Nidaros Trondheim's former name

Nifty Fifty top 50 large-capitalization growth stocks

Nigerian Ports Lagos, Bonny, Port Harcourt, Douala

Nightclub Aristocrat Artist Henri Marie Raymond de Toulouse-Lautrec

Night of Glass *Kristallnacht,* Nov. 9, 1938, when Nazis rampaged across Austria and Germany breaking the windows of homes, stores, and temples belonging to Jews who were rounded up and deported to concentration camps

Nightingale C-9 McDonnell-Douglas jetliner used for medical evacuation named in honor of Florence Nightingale

Nightmare of Europe Napoleon in the 1800s followed by Hitler in the 1900s

Night Mayor James J (Jimmy) Walker

Nikaria English place-name for Marla island in the Aegean

Nike-Ajax Douglas surface-to-air missile (MIM-3A)

Nike-Hercules Douglas surface-to-air missile armed with a high-explosive or nuclear warhead (MIM-14A)

Nike-Zeus one in a series of American-made anti-missile missiles

Nik-Nik affectionate nickname for the Royal Shakespeare Company's production of *The Life and Adventures of Nicholas Nickleby* by Charles Dickens

Nikolaus Lenau (pseudonym—Nikolaus Franz Niembsch von Strehlenau)

Nicky Nicholas; Nicole; Nickerson; Nikerson

Nile River Cities Cairo, Egypt and Khartoum, Sudan

nimbies nembutal tablets

Nimrod Hawker-Siddeley four-engine jet transport

Nina Ann; Anna; Anne; Annette

Nine Keepers of the *Constitution* nine justices of the U.S. Supreme Court

nine old men nine justices of the United States Supreme Court (before women were appointed)

Nineteenth State Indiana

Ninon de Lenclos courtesan Anne Lenclos

Ninth State New Hampshire (*see* First State)

nissen nissen hut (designed by British military engineer P.N. Nissen for arctic use)

Niteroi formerly Nictheroy

nitrate of soda sodium nitrate $(NaNO_3)$

Nitro Lew Lew Burdette

Nixon Doctrine use Japan, Iran, and South Africa to oppose communist influence in their regions; for Richard M Nixon, thirty-sixth president of the United States

Nizhni Novgorod (ancient and modern name; called Gorki under Communist rule)

Noah Noah Webster (American lexicographer)

noble rot Botrytis cinerea (parasitic fungus or mold that attacks grapes but that in certain climates produces great sweet white wines)

Noddy Nicodemus

Noll Oliver; Olivera; Oliver Cromwell—Lord Protector of England

Nolly Oliver; Olivera

Nome State Nome State Jail at Nome, Alaska

Nora Eleanora

Noratlas Norad 45-passenger transport aircraft made in France

Nordic Danish, Finnish, Icelandic, Norwegian, Swedish people or things

Norfolk Norfolk coat or Norfolk jacket (made with fore-and-aft box pleats, big pockets, and a belt, first produced in England's Norfolk county)

Normalcy nickname for Warren G Harding who advocated "a return to normalcy"

Norman Norman Thomas (American socialist leader)

NORMY Norman Douglas

Norse God of Thunder Thor, whose Roman counterpart is Jove or Jupiter

North North Sea between northern Europe and the British Isles

North America's Largest Country Canada

North Baltic Nation Estonia

North Britain Scotland

North Carolina Ports Wilmington, Wrightsville

North Cuba Miami, Florida

Northern Ireland's Principal Port Belfast

Northernmost American Town Point Barrow, Alaska

Northern Rhodesia Zambia's former name

Northern Way Norway

Northland Riviera Sweden's summer beach on the Gulf of Bothnia and the Polar Route

North River Hudson River (Battery to 59th Street on New York waterfront)

North Sea Canal Amsterdam Ship Canal

North Side North Side of any city or place

North Slope Alaska north of the Brooks Range

North Star Minnesota

North Star City St Paul, Minnesota

North Star State Minnesota

North Western Line Chicago and North Western Railway

Norumbega historian John Fiske's name for what is now New York City (*see Norvegia*)

Norumbegaland New York to Nova Scotia including New England

Norvegia (Italian—Norway); (Latin—Norway)—appears on early maps of the east coast of North America over an area extending from the Bay of Fundy to Florida and known for its Norse viking explorations and settlements in pre-Columbian times; sometimes spelled Norvega or Norbegia as well as Norumbega

Norway's Most Popular Sculptor Adolf Gustav Vigeland

Norway's Principal Port Oslo

Norwegian Norwegian elkhound (dog originally bred for hunting elk and other game); Norwegian saltpeter (calcium nitrate)

Norwegian Expressionist Edvard Munch

Norwegian Ports Kirkenes, Vadso, Vardo, Honningsvaag, Hammerfest, Tromso, Harstad, Svolvaer, Narvik, Bodo, Mo, Mosjoen, Trondheim, Thamshamn, Kristiansund, Harosund, Molde, Ulsteinik, Alesund, Vaksdal, Bergen, Odda, Haugesund, Stavanger, Egersund, Flekkefjord, Kristiansand, Grimstad, Arendal, Tvedestrand, Langesund, Brevik, Porsgrun, Larvik, Sandefjord, Tonsberg, Horten, Drammen, Oslo, Moss, Sarpsborg, Frederikstad, Halden

nose candy cocaine for sniffing

nose paint alcohol (whose prolonged use breaks the arterioles of the skin and leaves a reddened nose)

nose powder a powdered drug such as cocaine, any of several hallucinogens, or heroin when snorted

Nostradamus Michel de Nostradame also called Michel de Notredame

Novalis Friedrich Leopold von Hardenberg

Nova Scotia Girls Nova Scotia School for (delinquent) Girls at Truro

November witches fierce November storms on the Great Lakes

Nuclear Falcon Hughes air-to-air missile also called Super Falcon

Nuclear-Power Admiral Hyman George Rickover, USN

Nuestra Familia (Spanish—Our Family)—Hispanic prison gang

Nuestra Señora de los Dolores de las Vegas (Spanish—Our Lady of the Sorrows of the Lowlands)—former and somewhat prophetic name of Las Vegas, Nevada

Nueva España (Spanish—New Spain)—Spanish colonial name for Mexico

Nueva Granada (Spanish—New Granada)—Spanish colonial province comprising Colombia, Ecuador, Panama, and Venezuela

Nuk · (Greenlandic Eskimo—Point)—formerly called Godthaab (Good Hope) by the Danes and still the capital on the pointed peninsula on the southwest coast of Greenland

Number-One Host of the Jersey Coast Atlantic City, New Jersey

Nuoli Finnish depth-charge and mine-laying patrol boat armed with 20mm and 40mm guns

Nutmegs Connecticuters

Nutmeg State Connecticut

Nuuk correct name for Greenland's capital—Gothab

Nyasaland old name for Malawi

Nye Aneurin

NZedder(s) [En-zed-der(s)] New Zealander(s)

O Oprah Winfrey

Oak City Raleigh, North Carolina

Oakhill Virginia home of James Monroe

Oakie migratory farm worker or sharecropper from Oklahoma

Oasis City Roswell, New Mexico

Obediah Skinflint (pseudonym—Joel Chandler Harris)

Oberon British class of diesel submarines

Objectivist Poet William Carlos Williams

Ob River City Novosibirsk, Siberia

Occam's razor What can be done with few assumptions is done in vain with more (formulated by William of Occam, 1285–1349)

The Ocean The Atlantic, Antarctic, Arctic, Indian, or Pacific Ocean

Oceania islands of central and southern Pacific

Ocean Personified Oceanus (Roman); Okeanos (Greek)

Ocean State Rhode Island

Oder-Neisse Line rivers forming boundaries between Germany and Poland

Odetta Odetta Holmes

Offenbach Jacques Offenbach (adopted name of Jakob Levy Eberst)

Offshore Capital of the World Aberdeen, Scotland—home port of many offshore oil exploration rigs

offshore China nationalist Republic of China headquartered on Taiwan

Ogden Ogden Nash (American humorist)

O Henry William Sydney Porter

Ohio Ports Conneaut, Ashtabula, Fairport, Cleveland, Lorain, Huron, Sandusky, Toledo

Ohio's Beautiful Capital Columbus

Ohio Valley Ohio, West Virginia, Kentucky, Indiana, and Illinois

Oil Baron John D Rockefeller

oil of ben fine lubricant extracted from seeds of Arabian tree called *Moringa oleifera*

oil of cade juniper oil

oil cake cottonseed, linseed, or soybean mass used for cattle feed after oil is extracted

Oil Capital of Canada Edmonton, Alberta

Oil Capital of the Rockies Casper, Wyoming

Oil Capital of the United States Tulsa, Oklahoma

Oil City Bartlesville or Tulsa, Oklahoma

Oil Dorado northwestern Pennsylvania in the Oil City—Titusville area

oilies oilskin coats; oilskin garments

Oil Islands Chagos Archipelago in the Indian Ocean just north of Diego Garcia

oil of mirbrane nitrobenzene

oil of palm bribe(s); palm grease

Oil Province Alberta

Oil State Pennsylvania

oil of vitriol concentrated sulfuric acid (H_2SO_4)

oil of wintergreen methyl salicylate

Okazaki fragments small discontinuous strands of deoxyribonucleic acid (DNA) pro-

duced during DNA synthesis (named for Reiji and Funeko Okazaki)

Okie Oklahoma; Oklahoman

Okie City Oklahoma City, Oklahoma

Oklahoma's Yodeling Cowboy Gene Autry

Olav Hunger King Olav I of Denmark

Olav the Stout Olav Haroldsson

Olav Tryggvesson King Olav I of Norway, Sweden, and Denmark

The Old King Grom of Denmark (860–935)

Old Abe Abraham Lincoln

Old Ace of Spades Lieutenant General Robert E Lee, CSA

Old Aches and Pains Luke Appling (baseball player)

Old Andy Andrew Jackson—seventh President of the United States

Old Bay State Massachusetts

Old Beeswax Captain Raphael Semmes, CSN

Old Billie Brigadier General William Tecumseh Sherman, USA

Old Blighty nickname for London before air-pollution control

Old Blood and Guts General George S Patton, USA

Old Blue Eyes Frank Sinatra

Old Bory General Pierre Gustave Toutant de Beauregard, CSA

Old Brown of Osawatomie abolitionist John Brown

Old Bruin Commodore Matthew C Perry—father of the steam navy of the United States Navy

Old Buck Admiral Franklin Buchanan; President James Buchanan

Old Buena Vista General Zachary Taylor who attacked Mexicans at Buena Vista in February 1847; later was twelfth President of the United States

Old Bullion Thomas Hart Benton

Old Cape Stiff Cape Horn

Old Catawba fictitious name Thomas Wolfe assigned North Carolina

Old Chapultepec General Winfield Scott whose victory at Chapultepec ended Mexican War in September 1847

Old Chief Henry Clay

Old Coat Hanger Melbourne-originated nickname for the Sydney Harbour Bridge

Old Colony Massachusetts—founded in 1620

Old Corndrinking Melliflu-ous William Faulkner, according to Ernest Hemingway

The Old Country wherever anyone or their family originated—especially if in Europe

Old Country Lawyer Senator Sam Ervin of North Carolina

Old Creepy bank robber-burglar Alvin Karpis

Old Curmudgeon Harold Le Claire Ickes

Old Denmark General Christian Febiger, USA

Old Dirigo Maine whose state motto is *Dirigo* (Latin—I direct)

The Old Doctor Andrew Taylor Still, who developed osteopathy

Old Dominion Virginia— oldest English colony in America—founded in 1607

Old Dorp nickname of Schenectady, New York

Old East East Asiatic Company

Old Economy industrial, retail and financial firms

Old Faithful geyser in Yellowstone National Park; spouts about every 67 minutes

Old Four Legs coelacanth

Old Fox Clark Griffith (baseball player)

Old French Town New Orleans

Old Fuss and Feathers General Winfield Scott, USA

Old Gib Gibraltar

Old Glory the American Flag

Old Greasy West Virginian nickname for the Kanawha River or K'naw

Old Guard conservatives; Napoleon's imperial guard who made the last charge at Waterloo; the establishment

Old Harry (the devil)—Satan

Old Hickory General Andrew Jackson—seventh President of the United States

Old Iron Pants Vyacheslav Molotov (1890–1986)

Old Ironsides USS *Constitution*

Old Jeb Major General J(ames) E(well) B(rown) Stuart, CSA

Old Jefferson Joseph Jefferson

Old Joe slang nickname for syphilis

Old Kinderhook Martin Van Buren—eighth President of the United States

Old Lady the boss; mother; wife

Old Lady of Eagle Bridge Grandma (Anna Mary Richardson) Moses of Eagle Bridge, NY

Old Lady of the Thames London

Old Lady of Threadneedle Street Bank of England

Old Lady White any powdered narcotic

Old Legal Lion Clarence Darrow

Old Line State Maryland

Old Lion Joshua Nkomo

Old Man the boss; the captain; father; the skipper

Old Man Eloquent Isocrates in the opinion of Milton; John Quincy Adams in the opinion of the Congress he served after being sixth President of the U.S.

Old Man of Ferney Voltaire who lived in Ferney, France

Old Man of the Mountain New Hampshire's Profile Mountain—the Great Stone Face

Old Man of the Rhine Konrad Adenauer

Old Man River the Mississippi

Old Manse Nathaniel Hawthorne's house in Concord, Massachusetts

Old Nick (the devil)—Niccolo Machiavelli's diabolic sobriquet; Satan

Old Noll Old Oliver Cromwell

Old North State North Carolina

Old Ossawatomie John Brown

Old Pam Lord Palmerston (Henry John Temple)

Old Party W(illiam) Somerset Maugham

Old Peg Leg Petrus Stuyvesant—director-general of New Amsterdam and the New Netherlands

Old Pretender James Francis Edward Stuart (son of King James II)

Old Professor Casey Stengel (baseball player)

Old Providence island in the Colombian West Indies off the Caribbean coast of Nicaragua

Old Pueblo Tucson, Arizona

Old Put General Israel Putnam

Old Reliable Tommy Henrich

Old Roman Charlie Comiskey (baseball player)

Old Rosey General William Starke Rosecrans

Old Rough-and-Ready General Zachary Taylor—twelfth President of the United States

Old Salamander Admiral David Glasgow Farragut

Old Sarum Salisbury, England

Old Scratch Satan

Old Sol the sun (*see* Sun God)

Old South southern United States before 1865

Old Spanish Trail Saint Augustine, Florida to San Diego, California; Gulf Coast and Mexican Border route to California

Old Sparky Florida's electric chair

Old Swamp Fox Brigadier General Francis Marion, USA

Old Tecumseh General William Tecumseh Sherman, USA

Old Three Stars General US Grant, USA

Old Thumb Buster Colt .41-caliber revolver

Old Tippecanoe General William Henry Harrison—ninth President of the United States

Old Ugly .45-caliber pistol

Old Vic Royal Victoria Hall in London

Old Viking Norwegian-American labor leader Andrew Furuseth

Old War Skule Louisiana State University, formerly Louisiana State Seminary

Old West American West before it was settled during the 19th century

Old World Africa, Asia, and Europe

Old Zach Zachary Taylor—twelfth President of the United States

Ole Olaf(sen); Olav(sen)

Oleander City Galveston, Texas

Oleander City by the Sea Galveston, Texas

Ole Bull Ole Bornemann Bull

Ole Miss Old Mississippi (The University of Mississippi)

Olive Fremstad Olivia Rundquist

Oliver Hardy Oliver Norvell Hardy

Oliver Optic pseudonym of William Taylor Adams

Ollie Olive(r)

Ol' Man River Mississippi River

Ol' Miss Old Mississippi (nickname of river, state, or university)

Ol' Mo Old Missouri (the great river)

Omani Ports Masquat (Muscat) and Matrah

Omar Sharif Omar Cherif, Omar Michel Shalhoub

one-armed bandits slot machines

One-Armed Man of Lepanto Miguel de Cervantes

Only Town in the U.S. with an Apostrophe in Its Name Coeur d'Alene, Idaho

onyx marble alabaster

Oom Paul (Afrikaans—Uncle Paul)—sobriquet of Stephanus Johannes Paulus Kruger—leader of Boer rebellion and president of Transvaal

Opener of Japan Commodore Matthew Calbraith Perry, USN

Operation Keelhaul Allied policy of returning escaping anti-communists to their homelands

Opium Eater Thomas De Quincey

Opium Kingdom any country where the opium poppy is cultivated for use in making heroin (Bolivia, Burma, Colombia, Ecuador, Laos, Mexico, etc.)

Opium Land poppy fields of the Golden Triangle or northeastern Burma, northern Laos, and northern Thailand

Opium's Golden Triangle opium-growing fields between borders of Cambodia, Laos, and Vietnam

Oppenheim's disease congenital lack of muscular development of the ankles and feet

Oppie Oppenheim(er); J(ulius) Robert Oppenheimer

Oprah Oprah Winfrey

Oracle of Omaha Warren Buffett, chairman of Berkshire Hathaway

Orange Bowl Miami, Florida

orange flag potential danger signal

Orange Irish Protestants; orig., followers of William of Orange

Orangemen Protestants of Northern Ireland; orig., followers of William of Orange

Orange States California, Florida, and Texas

Orator of the American Revolution Patrick Henry

Orator Jim Jim O'Rourke (baseball player)

Orchard City Burlington, Iowa also called Porkopolis of Iowa

Orchard of Ireland County Armagh

Orchid Capital of Hawaii Hilo

The Orchid Island Hawaii

Orchid Set in the Sea Sulawesi (Celebes)

Oregon Girls Wisconsin School for (delinquent) Girls at Oregon

Oregon Ports Empire, Coos Bay, Astoria, Portland

Organist-Medical Missionary Dr Albert Schweitzer

oriental amethyst purple corundum

oriental anesthesia acupuncture

oriental emerald green corundum

Oriental Republic Eastern Republic of Uruguay (*República Oriental del Uruguay*)

oriental topaz yellow corundum

Orient Express famed Paris-to-Istanbul train; hypersonic jet designed to fly at 25 times the speed of sound to carry passengers from Washington to Tokyo in two hours

Orient's Cleanest City Singapore

Original Glamour Girl Theda Bara (Theodosia Goodman) also called Queen of the Vampires in the early days of American motion pictures

Orinoco River City Ciudad Bolivar, Venezuela

Oriole Maryland's state bird and symbolic nickname of Marylanders—Orioles

Orion Lockheed P-3 antisubmarine and patrol aircraft

Orizaba Citialtepetl (Mexico's highest volcano)

Orlando di Lasso Roland de Lassus

Orwell's Year 1984

Oscar Wilde Oscar Fingal O'Flahertie Wills (also used the anonym: C.3.3.)

Oskar Werner Josef Bschliessmayer

Oslo modern name for Christiania or Kristiania

Osloenser native of Oslo

Oslo Fjord formerly Kristiania Fjord

Ossie Oswaldtwistle, England

Ossining-on-Hudson formerly Hunter's Landing or Sing Sing

Ossip Gabrilovich Salomonovich Gabrilovich

Ostend Manifesto President Franklin Pierce

Other Side of the Herring Pond British nickname for America

O-town *Our Town*

Ottawa Convention United Nations treaty banning antipersonnel land mines

Otter De Haviland utility aircraft (DHC-3 in Canada, U-1 A in U.S.)

Ottoman Empire the old Turkish Empire extending at its height from Iran to Morocco, including the Balkans, parts of Hungary and southern Russia as well as much of Spain; Turkish Empire

Ouida pseudonym of Marie Louise de la Ramée

Ouragan (French—Hurricane)—Dassault single-engine jet fighter plane

Our Crowd New York City's German-Jewish elite

Our Gracie Gracie Fields (created Dame Commander of the Order of the British Empire after years of entertaining many millions of Britons and others around the world)

Our Lady of the Harbour nickname of the Statue of Liberty

Our Lady of the Snows Kipling's nickname for Canada

our wayward sisters Confederate States

Outer Banks North Carolina's sand-dune islands separated from the mainland by Albemarle, Croatan, Pamlico, and Bogue sounds

Outer China Mongolia, Sinkiang, Tibet

Outer City metropolitan area surrounding Peking's Inner city

Outer Mongolia The Mongolian People's Republic formerly called Mongolia

Outer Ring English counties adjacent to London

Outpost of the British Empire nickname given to any remote British settlement

Outpost of the West the Philippines

Outside anywhere outside Alaska

Overthrust Belt Rocky Mountain gas-and-oil lands

Ovid Roman poet Publius Ovidus Naso

Owen Meredith Edward Robert Bulwer-Lytton's pseudonym

owlsville London's post-midnight nickname or any other place after midnight

Owl Without a Vowel Bill Mikvy (basketball player)

Oyashio (Japanese—Father Current)—cold Okhotsk Current

Oyster Center Apalachicola, Florida

Oyster(s) Marylander(s)

Oyster State Maryland

Ozark Ike Gus Zernial (baseball player)

Ozark State Missouri

Ozzie Aussy; Australian; Osborn(e); Oscar; Oswald(o)

Pablo Neruda Neftali Ricardo Reyes

Pablo Picasso Pablo Diego José Francisco de Paula Juan Nepomuceno Crispin Crispiano de la Santísima Trinidad Ruiz y Picasso

Paca Francesca

Pachuco (Mexican-Americanism—El Paso, Texas)—also called Pachucolandia

Pacific Bitch navalese for Pacific Beach, San Diego, California

Pacific Canada British Columbia and the Yukon Territory

Pacific Coast Province British Columbia

Pacific Coast States California, Oregon, Washington

Pacific Commonwealths Australia and New Zealand and their territories

Pacific Crest Trailways for hikers, historians, and naturalists—includes John Muir Trail—extends from Canada to Mexico through Washington, Oregon, and California

Pacific Division States Alaska, California, Hawaii, Oregon, and Washington

Pacific Dominions Australia and New Zealand and their territories

Pacific Northwest Alaska to California, includung the Yukon, British Columbia, Washington, and Oregon

Pacific Ocean Eden Galápagos Islands

Pacific Paradise Hawaii

Pacific Province British Columbia

Pacific Rim Japan, South Korea, China, Taiwan, the Philippines, Malaysia, Hong Kong, Singapore and other Pacific nations associated through trade

Pacific States Alaska, Washington, Oregon, California, Hawaii

Pacific War Japan's involvement in World War II beginning with the Manchurian Incident in 1931 when Japan invaded China

Paco Pancho (Francisco)

Pac-8 Conference Pacific-8 Conference consisting of University of California (Berkeley), Oregon, Oregon State, University of Southern California, Stanford, University of California at Los Angeles, University of Washington, Washington State University

Pac-10 Conference Pacific-10 Conference

Paddie(s) Irish person(s)

Paddy an Irishman; Patrick

Paddyland Ireland

Padre de Independencia (Spanish—Father of Independence)—José Marti—Cuban patriot, poet, and soldier

Padrone of the Pedernales Lyndon B Johnson

Paganini of the Double Bass Giovanni Battesini; Serge Koussevitzky

Paganini of the Guitar Fernando Sors, Andres Segovia

Paganini of the Plains Juan Reynoso

Paget's disease bone distortion or cancer of the nipples of women

Painted Desert petrified formations and colorful rock deposits on desert floor of northeastern Arizona

Painter of Japanese Prostitutes Kitagawa Utamaro

Painter of Light Joaquin Sorolla de Bastida

Painter of Prostitutes Henri Marie Raymond de Toulouse-Lautrec

Pakistan's Principal Port Karachi

Palatinate southwest German districts once ruled by counts palatinate of the Holy Roman Empire and referred to as Oberpfalz or Rheinpfalz

Palau Pelew (Pacific islands in Caroline area)

Paleontologist Priest Pierre Teilhard de Chardin

Palestinian Salt Sea the Dead Sea

Palestrina Giovanni Pierluigi da Palestrina

Palgrave Francis Meyer Cohen

Palladian windows semicircular windows divided into small sections, named for Andrea Palladio, 16th-century Italian architect

Palma Vecchio palette name of Jacopo Negreti

Palm Coast Florida's east coast from Daytona to Jacksonville

Palm Springs of Washington state Yakima

Palmerston Henry John Temple, Viscount of Palmerston

Palmetto City Charleston, South Carolina

Palmetto(s) South Carolinian(s)

Palmetto State South Carolina

palm oil bribe(s)

Pamphleteer for American Independence Thomas Paine

Panama Panama hat (made from finely plaited young palmlike leaves)

Panama Canal Ports Balboa (Pacific terminus) and Cristóbal (Caribbean terminus)

Panama Canal President Theodore Roosevelt

Panama red high-grade marijuana grown in Panama and the Canal Zone

Panama's Principal Ports Colón, Cristóbal, Panamá City, Balboa, and Puerto Armuelles

Panama turkey iguana tail meat

Pancake lee small, thin, circular cakes of ice that form on a sea's surface as it begins to freeze

Pancho Francisco; native nickname for Valparaiso, Chile

Pancho Villa Doroteo Arango

Pandemonium South Pacifiic nickname for New Hebrides islands British-French Condominium

Panhandle Southeast Alaska

Panhandle States Florida, Maryland, Oklahoma, Texas, West Virginia

Pantaleone patron saint of Venice; nickname for an Italian taxpayer or for a Venetian

Panther Grumman single-engine single-seat naval fighting aircraft

Papa Bach Johann Sebastian Bach

Papa Bok Jean Bedel Bokassa, feared former President of Central African Republic

Papa Doc Haiti's former dictator François Duvalier

Papa Haydn Franz Joseph Haydn

Papaji Hariwansh Lal Poonja

Pappas Papadmitropoulos

Pappies Papists

Papua New Guinea's Principal Port Port Moresby

Paquita Francisca (Frances)

Paracelsus Theophrastus Bombastus von Hohenheim

Paradise of New England Salem, Massachusetts

Paradise of the Pacific Hawaii

Paradox of South America Paraguay—an affluent military dictatorship

Paraguay River City Asunción

Paraguay's Principal Port Asunción

Paraiba old name of Joao Pessoa, Brazil

Paramilitary Paradise Paraguay

Parched Heart of Australia Alice Springs, Northern Territory—The Alice

Parchman Mississippi State Penitentiary at Parchman

Pareto's law income distribution in a given economy tends to remain constant

Paris Expressionist Henri Matisse

paris green copper acetoarsenite (poison)

Parisian Italian Luigi Carlo Zenobio Salvadore Maria Cherubini (1760–1842)

Paris of North America Montréal

Parkbench Philosopher Bernard Baruch

Park City Bridgeport, Connecticut

Parkinson's disease nervous tremors accompanied by muscular weakness and rigidness; also called palsy, paralysis agitans, or the shakes

Parkinson's law work expands so as to fill the time available for its completion

Park Maker Frederick Law Olmsted

Parlour Panther *New York Review of Books*

Parmigianino Francisco Massuoli

Parrish blue blue favored by illustrator Maxfield Parrish

Parrot's disease syphilitic infantile paralysis (named for a French physician—Jules Marie Parrot)

Parry's disease exophthalmic goiter

Pas de Calais (French—Calais Strait)—also called Dover Strait

Paso del Calais (Spanish—Calais Strait)—Dover Strait in the English Channel

Passionate Pilgrim John Bunyan

Passionate Skeptic Bertrand Russell

Pastoral God Pan

Pat Patricia; Patrick

Patau syndrome genetic syndrome resulting in malformation of organs (for Klaus Patau)

Pathetic Peter Pyotr Ilyich Tchaikovsky whose compositions are soaked in sorrow

Pathfinder Major General John C Frémont, USA

Pathfinder of the Seas Matthew Fontaine Maury

Path of Gold Market Street, San Francisco

Pathmaker of the West John C Frémont

Patience and Fortitude Mayor LaGuardia's nickname for the couchant lions flanking the steps of the New York Public Library

Patland Ireland

Pat Nixon Thelma Catherine Patricia Ryan Nixon (1912–1993)

Patriarca de la Independencia (Portuguese—Patriarch of Independence)—Brazil's José Bonifacio de Andrada e Silva

Patriarch of American Labor George Meany

Patriarch of Ferney Voltaire who lived from 1758 to 1778 in Ferney, France now called Ferney-Voltaire in his honor

Patriarch of the Gold Industry Nicholas Deak

Patriarch of the Modern Consumer Movement Ralph Nader

Patriarch of New England John Cotton

Patriarch of Philosophy Bertrand Russell

Patriarch of Puerto Rico Luis Muñoz Marfn

Patriarch of the West the Pope

Patricia Wentworth Dora Amy Elles Dillon Turnbull's pseudonym

Patrick O'Brian Richard Patrick Russ

Patriot Financier Robert Morris

Patriot of the Piano Polish patriot-pianist-premier Ignace Jan Paderewski

Patriot Printer of 1776 William Bradford

Patron of Bawdy House England's King Charles II, the Merry Monarch

Patron of Explorers Henry the Navigator (Dom Henrique o Navegador)—Prince of Portugal

Patron Saint of Abandoned Children Jerome Emiliani

Patron Saint of Accountants Matthew

Patron Saint of Actors Genesius

Patron Saint of Air Travelers Joseph of Cupertino

Patron Saint of All Who Work with a Hammer Cloud

Patron Saint of Alpinists Bernard of Menthon

Patron Saint of Altar Boys John Berchmans

Patron Saint of American Orchards John (Johnny Appleseed) Chapman

Patron Saint of the Americas Rosa de Lima

Patron Saint of Anesthetists Rene Goupil

Patron Saint of Angina Sufferers Swithbert

Patron Saint of Architects Thomas, Apostle

Patron Saint of Armenia Gregory the Illuminator

Patron Saint of Artillerymen Barbara

Patron Saint of Astronomers Dominic

Patron Saint of Athletes Sebastian

Patron Saint of Austria Severino

Patron Saint of Authors Francis de Sales

Patron Saint of Aviators, Porters, Seafarers, and Travellers Christopher

Patron Saint of Bakers Elizabeth of Hungary

Patron Saint of Bankers Matthew

Patron Saint of Barbers Cosmas, Damian

Patron Saint of Barren Women Felicity

Patron Saint of Bavaria Kilian

Patron Saint of Beggers Giles

Patron Saint of Belgium Joseph

Patron Saint of Black Missions in North America Benedict the Moor

Patron Saint of Blacksmiths Dunstan

Patron Saint of the Blind Raphael

Patron Saint of the Blind and the Near Blind Clara

Patron Saint of Bloodbanks Januarius

Patron Saint of Boatmen Julian the Hospitaler

Patron Saint of Bookbinders Peter Celestine

Patron Saint of Book Collectors Jerome (credited with compiling the Latin Bible)

Patron Saint of Bookkeepers Matthew

Patron Saint of Booksellers John of God

Patron Saint of Borneo Francis Xavier

Patron Saint of Boy Scouts George

Patron Saint of Brazil Peter of Alcantara

Patron Saint of Brewers and Pawnbrokers Nicholas of Myra—the prototype of Santa Claus

Patron Saint of Bricklayers Stephen

Patron Saint of Brides Nicholas of Myra

Patron Saint of Bridgebuilders Benedict

Patron Saint of Brushmakers Anthony

Patron Saint of Builders Barbara and Vincent Ferrer

Patron Saint of Bullfighters Virgen de la Macarena (also Patron Saint of Seville)

Patron Saint of Butchers Hadrian

Patron Saint of Cab Drivers Fiacre

Patron Saint of Cabinetmakers Anne

Patron Saint of Canada Anne

Patron Saint of Candlemakers Ambrose

Patron Saint of Canonists Raymond of Peñafort

Patron Saint of Carpenters Joseph

Patron Saint of Catechists Charles Borromeo

Patron Saint of Catholic Action Francis of Assisi

Patron Saint of the Catholic Press Francis de Sales

Patron Saint of Charitable Societies Vincent de Paul

Patron Saint of the Chase Hubert

Patron Saint of Childbirth Gerard Majella

Patron Saint of Children Nicholas of Myra, also known as Santa Claus—derived from the Dutch, Sant Nikolaas

Patron Saint of Choirboys Dominic Savio

Patron Saint of the Church Joseph

Patron Saint of Clerics Gabriel

Patron Saint of Clothiers Blaise

Patron Saint of Comedians Vitus

Patron Saint of Confessors John Nepomucene

Patron Saint of Convulsive Children Scholastica

Patron Saint of Cooks Lawrence and Martha

Patron Saint of Coppersmiths Maura

Patron Saint of Cripples Vitus

Patron Saint of the Dance Vitus, according to Washington Irving

Patron Saint of Dancers Vitus

Patron Saint of Dentists Apollonia

Patron Saint of Desperate Situations Jude

Patron Saint of Diseases of the Breast Agatha

Patron Saint of Domestic Animals Anthony

Patron Saint of Domestic Workers Zita

Patron Saint of the Dominican Republic Dominic

Patron Saint of Druggists James the Less

Patron Saint of Dyers Lydia and Maurice

Patron Saint of Dying Barbara

Patron Saint of Dysentery Sufferers Matrona

Patron Saint of Earthquakes Emygdius

Patron Saint of Ecologists Francis of Assisi

Patron Saint of Editors John Bosco

Patron Saint of Emigrants Francis Xavier Cabrini

Patron Saint of Engineers Francis III

Patron Saint of England Edward the Confessor or St George

Patron Saint of Epilepsy Vitus

Patron Saint of Eucharistic Congresses Paschal Baylon

Patron Saint of Expectant Mothers Gerard Majella

Patron Saint of Eye Trouble Lucy

Patron Saint of Falsely Accused Raymond Nonnatus

Patron Saint of Farmers Isadore

Patron Saint of Fathers of Families Joseph

Patron Saint of Finland Henry of Uppsala

Patron Saint of Firemen Florian

Patron Saint of Fishermen Andrew

Patron Saint of Flagpole Sitters Simon of Stylites

Patron Saint of Florists Therese of Lisieux

Patron Saint of Foresters John Gualbert

Patron Saint of Founders
Barbara
Patron Saint of Freethinkers
Chapman Cohen of London
Patron Saint of French Attorneys Ives
Patron Saint of Funeral Directors Joseph of Arimathea
Patron Saint of Gardeners
Dorotea
Patron Saint of Gardens
Dorotea
Patron Saint of Girls Agnes
Patron Saint of Glasgow Mungo
Patron Saint of Glassworkers
Luke
Patron Saint of Goldsmiths
Dunstan
Patron Saint of Gravediggers
Anthony
Patron Saint of Greetings
Valentine
Patron Saint of Grocers
Michael
Patron Saint of Hairdressers
Martin de Porres
Patron Saint of the Hard of Hearing Ovidius
Patron Saint of Hatters
Severus of Ravenna
Patron Saint of Haymakers
Gervase
Patron Saint of Headache Sufferers Teresa of Avila
Patron Saint of Heart Patients
John of God
Patron Saint of Hernias and Ruptures Drogo, according to believers who also call him Druon
Patron Saint of Hispanic America Our Lady of Guadalupe
Patron Saint of Hospital Administrators Francis Xavier Cabrini
Patron Saint of Hospital Dieticians Martha
Patron Saint of Hospital Pharmacists Gemma Galgani
Patron Saint of Hospital Public Relations Paul
Patron Saint of Hospitals
Camillus de Lellis
Patron Saint of Hotelkeepers
Julian the Hospitaler
Patron Saint of Houseworkers Anne
Patron Saint of Hunters
Hubert
Patron Saint of Impossible and Desperate Cases Rita of Cascia

Patron Saint of Interracial Justice Martin de Porres
Patron Saint of Invalids
Roch
Patron Saint of Italy Francis of Assisi, Catherine of Siena
Patron Saint of Jewelers
Eligius
Patron Saint of Journalists
Francis de Sales
Patron Saint of Jurists John Capistrano
Patron Saint of Laborers Isadore
Patron Saint of Lawyers and Thieves Nicholas of Myra
Patron Saint of Librarians
Jerome
Patron Saint of Lighthousekeepers Dunstan
Patron Saint of Lithuania
Casimir
Patron Saint of Locksmiths
Dunstan
Patron Saint of Lost Articles
Anthony of Padua
Patron Saint of Lovers
Valentine
Patron Saint of Maidens
Catherine of Alexandria
Patron Saint of Malta Paul
Patron Saint of Mariners
Michael
Patron Saint of Married Women Monica
Patron Saint of Medical Record Librarians Raymond of Penyafort
Patron Saint of Medical Technicians Albert the Great
Patron Saint of the Mentally Ill Dymphna
Patron Saint of Merchants
Nicholas of Myra
Patron Saint of Messengers
Gabriel
Patron Saint of Metalworkers
Eligus
Patron Saint of Mexico
Virgin of Guadalupe
Patron Saint of Midwives
Raymond Nonnatus
Patron Saint of Millers
Victor
Patron Saint of Missions
Francis Xavier
Patron Saint of Mississippi
Eudora Welty
Patron Saint of Monaco
Devota
Patron Saint of Mothers
Monica

Patron Saint of Motorists
Christopher
Patron Saint of Mountaineers
Bernard
Patron Saint of the Movies
John Bosco
Patron Saint of Music Cecilia
Patron Saint of Nailmakers
Cloud
Patron Saint of Navigators
Brendan of Ireland
Patron Saint of the Netherlands Willibrord
Patron Saint of Norway
Olav
Patron Saint of Notaries
Mark
Patron Saint of Nurses Agatha
Patron Saint of Orators John Chrysostom
Patron Saint of Orphans
Jerome Emiliani
Patron Saint of Painters
Luke
Patron Saint of Paris Genevieve
Patron Saint of Parish Priests
Jean-Marie-Baptiste Vianney
Patron Saint of Pawnbrokers
Nicholas of Myra
Patron Saint of Persons Afflicted with Coughs and Colds Judas
Patron Saint of Persons Afflicted with Hydrophobia Hubert
Patron Saint of Persons Afflicted with Insect Stings and Snake Bites Sebastian
Patron Saint of Persons Condemned to Death Dismas
Patron Saint of Peru Rosa de Lima
Patron Saint of Pharmacists
James the Greater
Patron Saint of Physicians
Luke
Patron Saint of Pilots Joseph of Cupertino
Patron Saint of the Plague
Roch
Patron Saint of Plasterers
Bartholomew
Patron Saint of Poets Cecilia
Patron Saint of Poland Cunegunda
Patron Saint of Policemen
Michael
Patron Saint of the Poor
Anthony of Padua
Patron Saint of Porters
Christopher

Patron Saint of Postal Workers
Gabriel
Patron Saint of Preachers
John Chrysostom
Patron Saint of Pregnant
Women Margareta
Patron Saint of Printers
Genesius
Patron Saint of Prisons
Joseph Cafasso
Patron Saint of Prisoners
Barbara
Patron Saint of Public Relations Bernardino of Siena
Patron Saint of Radiologists
Michael
Patron Saint of Radio Workers Gabriel
Patron Saint of Retreats Ignatius Loyola
Patron Saint of Rheumatics
Gervasius
Patron Saint of Rheumatism
Affliction James the Greater
Patron Saint of Russia Andrew whose cross adorns the
flag of Imperial Russia
Patron Saint of Saddlers
Crispin
Patron Saint of Sailors Elmo
Patron Saint of Scholastic Institutions Thomas Aquinas
Patron Saint of Scotland Andrew
Patron Saint of Sculptors
Claude
Patron Saint of Secretaries
Genesius
Patron Saint of Seminarians
Charles Borromeo
Patron Saint of Shepherds
Drogo
Patron Saint of Shoemakers
Crispin
Patron Saint of the Sick
Camillus de Lellis
Patron Saint of Silversmiths
Dunstan
Patron Saint of Singers Cecilia
Patron Saint of Skaters Lidwina
Patron Saint of Skiers Bernard
Patron Saint of Skin Diseases
Marculf
Patron Saint of Social Justice
Joseph
Patron Saint of Social Workers Louise de Marillac
Patron Saint of the Sore
Throat Blaise

Patron Saint of Spain Santiago de Compostela (St James
the Greater), Teresa of Avila
Patron Saint of Speleologists
Benedict
Patron Saint of Stenographers Cassian of Tangiers
Patron Saint of Stonecutters
Clement
Patron Saint of Students
Thomas Aquinas
Patron Saint of Tailors Homobonus
Patron Saint of. Tanners
Crispin
Patron Saint of Tax Collectors
Matthew
Patron Saint of Teachers
Gregory the Great
Patron Saint of Telecommunications Workers Gabriel
Patron Saint of Television
Clare of Assisi
Patron Saint of Theologians
Alphonsus Liguori
Patron Saint of Toothaches
Apollonia of Egypt
Patron Saint of Travellers
Christopher
Patron Saint of Troubled Parents Monica
Patron Saint of the United
States Our Lady of the Immaculate Conception, according to Roman Catholics
Patron Saint of Universities
Contardo Ferrini
Patron Saint of Uruguay Our
Lady of Lujan
Patron Saint of Vocations Alphonsus
Patron Saint of Wales David
Patron Saint of War Prisoners
Leonard
Patron Saint of Watchmen
Peter of Alcantara
Patron Saint of the West
Indies Gertrude
Patron Saint of Widows
Paula
Patron Saint of Winegrowers
Vincent
Patron Saint of Wine Merchants Amand
Patron Saint of Writers Lucy
Patron Saint of Yachtsmen
Adjutor
Patron Saint of Youth Aloysius Gonzaga
Patron Saints of Advertisers
Barnardine of Siena, John
Berchmans

Patron Saints of Ailments of
the Eyes Clare and Lucy
Patron Saints of Artists
Catherine of Bologna, Luke
Patron Saints of Bohemia
Wenceslaus, Ludmilla
Patron Saints of Canada
George and Joseph
Patron Saints of Cancer Victims Michael; Peregrine
Laziosi
Patron Saints of Chile Santiago (James), Virgen del Carmen
Patron Saints of Denmark
Ansgar and Canute
Patron Saints of Doctors
Cosmas and Damian
Patron Saints of France Denis, Our Lady of the Assumption, Joan of Arc, Therese
Patron Saints of Germany
Boniface and Michael
Patron Saints of Greece
Andrew and Nicholas
Patron Saints of Housewives
Anne; Martha
Patron Saints of Hungary
Blessed Virgin—Great Lady
of Hungary and Stephen
Patron Saints of Ireland
Patrick, Brigid, Columba
Patron Saints of Lawyers
Genesius, Ivo, Thomas More
Patron Saints of Moravia
Cyril and Methodius
Patron Saints of Morticians
Dismas and Joseph of Arimathea
Patron Saints of Philosophers
Justin, Catherine
Patron Saints of Poland Casimir, Florian, Stanislaus
Patron Saints of Portugal
Anthony Padua, Francis
Padua, George, Vincent
Patron Saints of Scholars
Brigida; Gregory
Patron Saints of Schoolteachers Gregory the Great and
John Baptist de la Salle
Patron Saints of Scotland
Andrew and Columba
Patron Saints of Stonemasons
Barbara; Stephen
Patron Saints of Surgeons
Cosmas, Damian, Luke
Patron Saints of Sweden
Eric, Bridget
Patron Saints of Theologians
Alphonsus Liguori and Augustine

Patron Saints of Weavers Anastasia; Paul the Hermit

Patron Saints of Victims of Abusive or Unfaithful Husbands Blessed Anna Maria Taigl, Blessed Margaret of d'Youville, Blessed Paola Gambera-Costa, Saint Elizabeth of Portugal, Saint Godelileve, Saint Monica, Saint Rita of Cascia

Patron Saints of Victims of Unmarried Parents Blessed Eustochium of Padua, Blessed Sibyllina Biscossi, Saint Briget of Ireland

Patron of Science Ptolemy (Claudius Ptolemaeus)

Patroon Stephen Van Rensselaer's nickname

Patry Williams pseudonym of M Patry and D Williams (English writers)

Patsy Cline Virginia Patterson Hensley

Patti Page Clara Ann Fowler

Patton U.S.-made M-47 or M-48 medium tanks armed with 90mm guns

Patty Martha; Patience; Patricia

Paul VI Giovanni Batista Montini

Paula Paulcela; Paulette; Paulina; Pauline; Paulita

Paul Bunyan's Capital Brainerd, Minnesota

Paul Celan Paul Ancel (1920–1970)

Paul Creston Joseph Guttoveggio

Paul Éluard Eugéne Grindel (1895–1952)

Paulette Goddard Marion Levy

Paul Horiuchi Chi Kamasa Horiuchi

Paul Klenovsky Sir Henry J Wood's pseudonym used when he presented his orchestral arrangement of Bach's Toccata and Fugue in D minor; pupil of Alexander Glazunov

Paul Lukas Pal Lukacs

Paul Muni Muni Weisenfreund

Paul Vesey Samuel W Allen's pseudonym

Pavel Ivanovich Jones John Paul Jones (when he served as rear admiral commanding Russia's Black Sea fleet for Catherine the Great)

p-bills personal-computer counterfeit currency

P'burg Philipsburg, Montana

Peacefield Quincy, Massachusetts home of John Adams and his son John Quincy Adams

Peace Garden State North Dakota

Peacekeeper LGM 118 intercontinental ballistic missile

Peacemaker William Penn

peace pill pcp, phencyclidine

Peach Bowl Atlanta, Georgia

Peach Capital of British Columbia Penticton

Peach State Georgia

Peacracker(s) native(s) of Lowestoft

Peanut Capital of Alabama Dothan

Peanut City Suffolk, Virginia

Peanut King Amadeo Obici who organized the Planters Peanut Company in 1906

Pear City Medford, Oregon

Pearl of the Adriatic Dubrovnik, Croatia

Pearl of the Antilles Cuba

Pearl of the Atlantic Madeira

Pearl of the Baltic Bornholm Island, Denmark

Pearl Buck Pearl Syndenstricker

Pearl of the Chilean Paciric Viña del Mar

Pearl Gulf Arabian Gulf, Persian Gulf

Pearl of Ireland Saint Brigit

Pearl Island of the Caribbean Margarita, Venezuela

Pearl King Mikimoto Kokichi (Japanese who discovered the secret of creating cultured pearls)

Pearl of the Lagoons Abidjan, Ivory Coast

Pearl of the Orient Sri Lanka (Ceylon)

Pearl of Persia Isfahan

Pearl and Petroleum Sheikdom El Qatar on the Persian Gulf

Pearl of the Planet Black Sea

Pearl S Buck Mrs Richard J Walsh

Pearl of the Sharon Netanya, Israel

Pearls of the Pacific 7,000 Philippine islands; Honolulu, Pago Pago, Papeete

Pearls of the South Seas Tahiti, Tonga, Samoa, and other South Sea islands

Peasant Bard Robert Burns

Peasant Breughel Pieter Breughel the Elder

Peasant with a Pen Eric Linklater's nickname

Peck's Bad Boy pseudonym of George W Peck

Pecos Bill General William Shafter, USA

Pecos Wilderness eastern New Mexico and West Texas (nothern New Mexico east of Santa Fe to the Rio Grande above Del Rio, Texas)

The Peerless Leader Frank Chamice (baseball)

Peetie Wheatstraw William Bunch

The Peg Winnipeg, Canada

Peggy Margaret

Peg Leg Petrus Stuyvesant. Dutch governor of Nieuw Amsterdam, who lost his right leg in the siege of Sint Maarten

Peg Leg Howell Joshua Barnes Howell

Peg-Leg Sam Arthur Jackson

Pelican Louisiana's state bird and symbolic nickname often given Louisianians—Pelicans

Pelican State Louisiana

Pen of the American Revolution Thomas Paine

Penang (Malay—Betel Nut)— formerly called George Town when British Malaya

Pence Springs West Virginia State Prison for Women at Pence Springs

Pencil of Nature photography

Penguin Ron Cey (baseball player)

Peninsula Portugal and Spain

Peninsular Malaysia States of the Federation of Malaysia (Federated Malay States also known as Malaya)

Peninsula State Florida

Penmen of the Revolution John Dickinson of Dover, Delaware, Thomas Jefferson, and Tom Paine

Pennie Penina

Pennsylvania named for William Penn's father

Pennsylvania Farmer John Dickinson's pseudonym

Pennsylvania Ports Erie, Philadelphia. Chester, Marcus Hook

Pennsylvania of the West Missouri

Penny Penelope

Penobscot River City Bangor, Maine

Peony Indiana state flower; Indiana girl's nickname

Peony Center Faribault, Minnesota

People of the Lion Singhalese of Ceylon

People's Lawyer Associate Justice Louis Dembitz Brandeis of the Supreme Court of the United States

People's Poet Paul Lawrence Dunbar

People's Princess Diana, Princess of Wales

Peory Peoria, Illinois

Pepe José (Joseph)

Pepita Josefa; Josefina

Pepper Jimmy Austin (baseball player)

Pepper Coast Liberia

Perce Persival; Percy

Percy Percival

Percy Grainger George Percy Grainger

Père-Lachaise Paris' best known cemetery and generic eponym for other burial places

Perfector of Opalescent Glass Louis Comfort Tiffany

Peripatetic Philosopher Aristotle

Peripatetic Pope John Paul II

perks nickname for percodan

permatemps permanent temporary workers (who go from one temporary job to another)

Pernambuco old name of Recife, Brazil

Pero (Russian—pen)—Trotsky nickname because of his skill in writing revolutionary tracts

Perry Como Pierino Como

Perse Percival; Percy

Pershing Martin surface-to-surface missile (MGM-31A)

Persia ancient name for Iran

Persian Persian blinds (exterior blinds); Persian carpet (handwoven oriental rug);

Persian cat (long-haired breed); Persian lamb (young lamb of karakul sheep); Persian melon (greenish muskmelon)

Persian Gulf States Bahrain, Qatar, and the United Arab Emirates

Personification of Death Thanatos (Greek)—whose brother was Hypnos or sleep

Personification of the Destroying Principle Siva

Personification of Justice (*see* Justice Personified)

Personification of the Preserving Principle Vishnu

Personification of Sleep Hypnos (Greek)—whose brother was Thanatos or death

Personification of the Soul Psyche in the Greek mythology

Persuasive Evolutionist Thomas Henry Huxley

Perugino Piero Vannucci

Peru's Principal Port Callao

Peruvian Peruvian-balsam used by chocolate makers, doctors, and perfumers; Peruvian bark *(cinchona)*

Peruvian Ports Iquitos on the Amazon's headwaters; Talara, Callao, Matarani, Mollendo, Ilo, Pisco, Chimbote, and Salaverry

Pesach Hebrew Passover

Pessimistic Composer Gustav Mahler (nicknamed Gloomy Gus by some music lovers)

Pessimistic Painter Hieronymus Bosch

Pessimistic Philosopher Arthur Schopenhauer

Pesthole of the Pacific Panama City, Panama; Buenaventura, Colombia; Guayaquil, Ecuador

Pete Herman Peter Gulotta (boxer)

Peter Arno Curtis Arnoux Peters

Peterhouse St Peter's College (Cambridge)

Peter Lorre Laszlo Loewenstein

Peter Lynch factor added value that a good manager may bring to a mutual fund

Peter Martyr Pietro Martin d'Anghierra's pseudonym

Peter McGill (American slang—Pedro Miguel)— Panama Canal Locks near Balboa

Peter Mennin Peter Mennini

Peter Mikhailov pseudonym of Peter the Great, which he used while travelling and working in Dutch shipyards

Peter Pan of Politics Winston Churchill

Peter and Paul St Peter and St Paul island fortress-prison on the Neva facing Saint Petersburg

Peter Pindar Dr John Wolcot

Peter Porcupine William Cobbett's pseudonym

Peter Principle in a hierarchy. every employee tends to rise to his level of incompetence

Peter Warlock Philip Arnold Heseltine

Petit Caporal (French—Little Corporal)—Napoleon I

Petrarch Francesco Petracco

Petr Makadonski (Russian— Peter the Great)—Peter Alekseyvich

Petrograd (Russian—City of Peter)—place-name during socialist revolution of February 1917 and communist takeover in following November; called Leningrad after death of Lenin until September 1991, when replaced by the original Saint Petersburg

Petroleum Emirate Kuwait

Petroleum V Nasby David Ross Locke's pseudonym

Petya (Russian nickname— Pyotr)—Peter

Pewee Kentucky Correctional Institution at Pewee Valley

Peyronies's disease (*see* bent-nail syndrome)

Peyton Place Gilmanton, New Hampshire

p funk synthetic heroin

Phantom F-4 fighter airplane

Pharaoh's Curse diarrhea

phd poor, hungry, driven

Philadelphia Jack O'Brien Joseph Hagen (boxer)

Philadelphia Lawyer Andrew Hamilton, Philadelphia attorney who in 1734 and 1735 successfully defended New York printer Peter Zenger whose newspaper criticized British colonial policy in America—Zenger had been unsuccessful in getting a New York lawyer to take his case; an attorney who will defend a case others are afraid to touch

Philadelphia Painter Thomas Eakins

Philander von der Linde Johann Burkhard Mencke

Philidor François Andre Danican

Philippine Ports Aparri, Port Legazpi. Cavite, Manila, Poro, Masbate, Tacloban, Cebu, Iloilo, Davao, Zamboanga, Ozamiz, Isabela, Jolo

Philippines Principal Port Manila

Philipp Melanchthon Philipp Schwarzerd

Philistine Temptress Dalila _ (Delilah)

Philomela Mercy Otis Warren, called the First Lady of the American Revolution

Philosopher of the Absolute Georg Wilhelm Friedrich Hegel

Philosopher of China Confucius

Philosopher of Democracy and Humanism Sidney Hook

Philosopher of Freedom John Locke

Philosopher Freethinker Elbert Hubbard

Philosopher Kung Kung Futzu (Confucius)

Philosopher of Malmesbury Thomas Hobbes

Philosopher Physician Averroes

Philosopher of Sans Souci Voltaire's nickname for Frederick the Great

Philosopher of Sex Havelock Ellis

Philosopher of the Superman Friedrich Wilhelm Nietzsche

Philosopher of the Third Reich Alfred Rosenberg

Phiz Hablot K Browne—illustrator of the *Pickwick Papers* of Dickens–Boz

Phoenix of Spain Lope de Vega

Photographer of the Himalayas Samuel Bourne (1834–1912)

Photographic Pioneer William Henry Fox Talbot

Photographic Purists Ansel Adams and Edward Weston

Photo Reporter Margaret Bourke-White

Phrasemaker of Versailles Woodrow Wilson—twenty-eighth President of the United States

Phronie Sophronia

Phyllis Diller Phyllis Driver

Physician to the Body Politic Émile Zolá

Physician Extraordinary Sir William Osler

Physician's Physician Jacob Mendez Da Costa

Pianist's Pianist Richard Buhlig

Piano Legs Charles Hickman (baseball player)

Pickle Works nickname of building occupied by Central Intelligence Agency in Langley, Virginia

Pickpocket Heroine of Daniel Defoe's *Moll Flanders*

Pick's disease brain disorder characterized by loss of memory and speech

Pico Bolívar (Spanish— Bolivar's Peak)—Venezuela's highest mountain also called La Columna

Picture Island Enoshima, Yokahama

Picture-Postcard-Landscape Land Switzerland

Picture Province Canada's New Brunswick

Picture Province of Canada New Brunswick

Piedmont Plateau Appalachian Mountain region extending from Alabama to New York, including Georgia, the Carolinas, Virginia, West Virginia, western Maryland, and Pennsylvania

Pier Angeli Anna Maria Pierangeli

Piero della Francesca Piero di Benedetto de Franceschi

Pierre Loti (pseudonym-Louis-Marie Julien Viaud)

Pierre Louÿs (pseudonym— Pierre Louis)

Pierre Nord André Léon Brouillard's pseudonym

Pierre-Paul Prud'hon Pierre Prudon

Pietermaritzburg South African city also called Maritzburg

Pieter Timmerman (Dutch— Peter Carpenter)—pseudonym used by Peter the Great of Russia while working as a shipwright in Dutch shipyards

Pig Alley Place Pigalle

pig iron slag iron, named for alignment of iron on ground that looked like piglets suckling a sow

Pig Islander New Zealander (Australian slang)

pig's ear (Cockney English— beer)

Pilgrim of the Most Illustrious Waves Caspian Sea sturgeon, according to Ovid

Pillars of Hercules promontories flanking the Straits of Gibraltar—Abyla in Africa facing Gibraltar in Europe

Pillow Lava lava solidified underwater assuming pillowlike shapes

Pilsner Country Czechoslovakia

Pinball Dave Twardzik (baseball player)

Pineapple Face Manuel Antonio Noriega, former leader of Panama

Pineapple Island Lanai, Hawaii

Pineapple Paradise Hawaiian Islands

Pine Trees inhabitants of Maine

Pine Tree State Maine

Piney Point Harry Lundeberg School of Seamanship at Piney Point, Maryland

Pink City of Rajputana Jaipur

pinks newborn rats

Pinky conductor-violinist Pinchas Zukerman's nickname

Pinky Lee Pincus Leff

Pinturicchio Barnardino Betti

Piombo palette name of Sebastiano Luciani

Pioneer deep-space probes designed for interplanetary investigation

Pioneer Aeronautical Experimentalist Otto Lilienthal

Pioneer Airship Designer Ferdinand von Zeppelin (1838–1917)

Pioneer American Composer William Billings

Pioneer in American Science Benjamin Franklin

Pioneer of Antisepsis Ignaz Philipp Semmelweis

Pioneer Bacteriologist Louis Pasteur; Robert Koch

Pioneer of Biblical Criticism Thomas Paine (1737–1809)

Pioneer of British Aviation, Designer, Industrialist Geoffrey De Havilland (1882–1965)

Pioneer of Child Psychoanalysis Dr Anna Freud

Pioneer Dutch Aircraft Designer Anthony Fokker (1890–1939)

Pioneer German Aircraft Designer Hugo Junkers (1859–1935)

Pioneer Heart Surgeon Daniel Hale Williams

Pioneer Inventor of Aerial and Space Photography Systems Brigadier General George Goddard—USAF

Pioneer of Law Hammurabi (1792–1750 B.C.)

Pioneer of Modern Geography Alexander von Humboldt (1769–1859)

Pioneer of Modern Orchestration Hector Berlioz (1803–1869)

Pioneer of Oceanography Sir John Murray

Pioneers Pioneer Mountains of Idaho and Montana

Pioneer of Technological Change Eli Whitney

Pioneer of Two Worlds Thomas Paine

Pioneer of University Surgery William Halsted

Pioneer of Visceral Surgery Theodor Billroth

Piotr (Russian—Peter)—nickname for Petersburg, St Petersburg, Petrograd

Piper Laurie Rosetta Jacobs

Pippa Philipa; Philippa

Pirandello Stefano Landi

Pirate City Tampa, Florida

Pirate Coast Trucial Coast of Arabia including Abu Dhabi, Ajam, Dubai, Fujairah, Ras al-Khaimah, Sharjah, and Umm al-Qaiwain

Pirate of the Gulf Jean Lafitte

Pirate Port Port Royal, Jamaica 17th century

Pisanus Fraxi Herbert Spencer Ashbee

Pistol Pete Pete Maravich (basketball player)

Pitcher Plant Province Newfoundland

Pitch Lake Trinidad's asphalt lake

Pitch Lake Island Trinidad

Pitigrilli Dino Segre

Pius XII Eugenio Pacelli

Pizza (PIE) Pacific Intermountain Express (stock exchange nickname)

placazos (Mexican border Spanish—graffiti originally placartes)—placards or posters

Place of Many Waters Walla Walla, Washington

Place of Plenty Indian name for what is now Toronto, Ontario

Place of the Seven Wells Beersheba

Place of the Winds Nouakchott in Mauritania on the coast of West Africa

Plague of the Twentieth Century AIDS

Plain Joe Canada's Prime Minister Joe Clark

plain vanilla conventional stock offering

Plains States Nebraska, Montana, Wyoming, Colorado, Oklahoma, North Dakota, South Dakota

Plank Island Aberdeen, Washington

Planner of the New York Public Library John Shaw Billings

Plantation State Rhode Island whose official title is the State of Rhode Island and Providence Plantations

Planting Corn Moon full moon in May

Planting Moon Full moon in April

Plant Wizard Luther Burbank

plaster of paris white, powdery gypsum—first prepared from calcinated gypsum of Paris, France

Plateau Continent Africa

Plateglasses ultra-modern style in universities

Plate River Ports Buenos Aires, Argentina and Montevideo, Uruguay

Platine States Argentina and Uruguay

Platinum one million (recording) copies

Plato (Greek—broad-shouldered)—the famous philosopher's real name was Aristocles

Plato's School the Grove of Academe near Athens, later referred to by the Romans as the Academia

Plattensee (German—flat sea, level lake)—Hungary's Lake Balaton—largest lake in central Europe

Playground of the Middle West Indiana

Pleasant former name of Nauru

Pleasure City of the South Seas Sydney, Australia

Plein-Air Painter Manet

Plejad Swedish class of fast patrol boats

Plight of Pizarro dysentery named for the Spanish conqueror of Peru

Plough-Share City York, Pennsylvania

Plow City Moline, Illinois

Plucky Pierre Salinger

Plum Sir Pelham Warner; Sir PG Wodehouse

Plumb-line Port to Panama Charleston, South Carolina (due north of the Panama Canal)

Plumed Knight Robert G Ingersoll's name for James G Blaine when nominating him for President

Plus Brave des Braves (French—Bravest of the Brave)—Napoleon's nickname for Marshal Ney

Plymouth Rock landing place of the Pilgrims in 1620 in what is now Plymouth, Massachusetts

Poet of Affection Marianne Moore

Poet of the American Revolution Philip Freneau

Poet of the Body—Poet of the Soul Walt Whitman's self-imposed nickname

Poet of Childhood Eugene Field

Poet of the Common People James Whitcomb Riley (1849–1916)

Poet of Democracy Walt Whitman

Poet of Despair James Thomson

Poet of the Excursion William Wordsworth

Poet of Friendship Robert Burns

Poet from Jersey William Carlos Williams of Rutherford, New Jersey

Poet of Harlem Langston Hughes (1902–1967)

Poet of Imperialism Rudyard Kipling

Poet of Individuality Walt Whitman

Poet King Ossian of Ireland

Poet Laureate of England Sir John Betjeman

Poet Laureate of New England John Greenleaf Whittier

Poet Laureate of Venezuela Irma De Sola Ricardo

Poet of Liberty Johann Christoph Friedrich von Schiller

Poet Naturalist Henry David Thoreau

Poet of Nature Jean Sibelius

Poet of Passion Ella Wheeler Wilcox

Poet of the Piano Frédéric Chopin

Poet of Poets Shelley ʹ

Poet Sire of Italy Dante Alighieri

Poets Laureate of the United States Robert Penn Warren; Richard Wilbur; Howard Nemerov; Mark Strand Joseph Brodsky; Mona Van Duyn; Rita Dove; Robert Hass, Robert Pinsky, Stanley Kunitz

Poet of the Subconscious Giovanni Pascoli

pog milk-bottle cap

Pogo Joe Caldwell (basketball player)

Point Coma Point Loma

Poison Ivy Upton Sinclair's nickname for publicist Ivy Lee

Poitiers formerly Poictiers

Pokanoket American Indian name for what was Mount Hope and is now Bristol, Rhode Island

Poke Poughkeepsie

Pola Appolonia; Policarpa Salabarrieta

Pola Negri Appolonia Chalupek

Polar Bear Land the Arctic

Polar Star State Maine

Polaris brightest star in the constellation of Ursa Minor; usually called the Pole Star or the Seaman's Star (Stella Maris) as within a degree or two it points to true north

Polaris-Poseidon Lockheed submarine-launched missiles

Pole Star (*see* Polaris)

Polish City Hamtramck, Michigan

Polish Ports Nowy Port, Stettin, Gdynia, Ustka, Swinoujscie

Polish Siberia Nowogrodek southwest of Minsk

Polish Story Teller Isaac Bashevis Singer

Polish Town Parma Maria. Texas—America's oldest Polish settlement

Polka King Frankie Yankovic

Polly Mary; Paula; Paulette; Pauline; Pollyanna

Polo Capital of the South Aiken, South Carolina

Pol Pot Saloth Sar

Polygon of Drought northeast Brazil along the Rio San Francisco and east of a line between Bahia and Fortaleza

Polynesia (Greek—Many Islands)—the South Pacific

Polynesian Kingdom Tonga (The Friendly Isles)

Polynesia's Sacred Isle Raiatea (in the South Pacific west of Tahiti)

Pommy British emigrant to Australia or New Zealand

Pompey Cneius Pompeius; nickname of Portsmouth, England

Pom(s) Prisoner(s) of Mother England [Australian nickname for person(s) newly arrived from Great Britain]

Pony Express Terminus Sacramento, California

Ponzi scheme named for pyramid scheme based in Boston in 1920, run by Charles Ponzi

Poonjaji Hariwansh Lal Poonja

poor man's cocaine methamphetamine

poor man's lobster monkfish

poor-man's meat *Glycine max*—soybeans

Poor Richard Richard Saunders (pseudonym used by Benjamin Franklin in writing *Poor Richard's Amanack*)

Poosh 'Em Up Tony Lazzeri (baseball player)

Popcorn Capital of the World Shaller, Iowa

Pop Warner Glenn Warner

Pope of Geneva Calvin's nickname

Pope John XXIII Angelo Giuseppe Roncalli

Pope John Paul I Albino Luciani

Pope John Paul II Karol Wojtyla

Pope Paul VI Giovanni Battista Montini

Pope of Peppers Dave DeWitt

Pope Pius XI Achille Ratti

Pope Pius XII Eugenio Pacelli

poppers amyl nitrate (also called amys, pearls, or snappers)

Pops Arthur Fiedler

Popular Porter Cole Porter

Pori Finnish name for what the Swedes call Björneborg; Polaris operational readiness instrumentation

Pork Chop Gang a 1940s confederation of rural Florida legislators and business leaders

Pork Dump nickname of Clinton Prison near Utica, New York

Porkopolis Cincinnati, Ohio

Pork Packer Philip D Armour

Porn Capital of America San Francisco

Portage La Prairie Girls Correctional Centre for Women at Portage La Prairie, Manitoba

Portcouver urban corridor from Portland, Oregon, to Vancouver, British Columbia

Porter of Heaven Janus the Two-Faced

Port Everglades Fort Lauderdale, Florida's port

Portia pen name of Abigail Smith Adams—wife of President John Adams and America's First Suffragist

Port Kelang formerly Port Swettenham and also called Port Mang

Port Klang formerly Port Swettenham, Malaya

Port Lyautey former name of Kenitra, Morocco

Porto di Lido (Italian—Port of the Lido)—the port of Venice

Port o' Missing Men San Francisco

Porto Rico original name of Puerto Rico

Port of the Pilgrims Provincetown, Massachusetts

Portrait Painter of Presidents Gilbert Stuart

portsides portsiders (lefthanded persons)

Port of the Southwest Galveston, Texas

Ports of Philadelphia Trenton, Camden, Gloucester City, Philadelphia, Chester, Marcus Hook, Wilmington

Port of St John of Acre Akko

Portugal's Principal Port Lisboa (Lisbon)

Portuguese America Brazil

Portuguese China Macao (near Hong Kong)

Portuguese East Africa Mozambique

Portuguese Guinea former name of Guinea Bissau on Africa's west coast

Portuguese India former name of the territories of Damão, Diu, Goa, Panjim, etc.

Portuguese Mars Affonso d'Alboquerque also called Affonso o Grande (Alphonse the Great)—Portuguese empire builder and viceroy of Portuguese India

Portuguese Overseas Province Macao

Portuguese Paradise Sintra near Lisbon

Portuguese Ports (large, medium, and small from north to south) Viana do Castelo, Porto de Leixoes, Porto (Oporto), Lisboa (Lisbon), Setubal, (*in the Azores*—Horta *and* Ponta Delgada), Funchal (Madeira)

Portuguese Republic República Portuguesa

Portuguese-Spanish Century the 16th century—the 1500s

Portuguese-speaking Places Angola, Azores Islands, Brazil, Cape Verde Islands, Guinea-Bissau, Macao, Madeira Islands, Mozambique,

Portugal, São Tomé and Principe Islands, Goa and Timor

Portuguese Timor former Portuguese outpost of empire on Timor Island in Indonesia

Portuguese West Africa Angola, Portuguese Guinea, St Thomas and Prince islands during colonial era

Port Veneris nickname for Port Vendre on the Franco-Spanish frontier

Port Wine Port Oporto, Portugal

Postage-Stamp Principalities Andorra, Liechtenstein, Luxembourg, and Monaco are so named by philatelists although Luxembourg is a grand duchy and is not ruled by a prince

postie post-baby-boom person

Potain's disease pleural and pulmonary edema

Potash City Saskatoon, Saskatchewan

Potentate of the Pit Lucifer

pot machine marijuana-manufacturing device

Potomac River City Washington, DC

Pot Smuggler's Paradise Florida whose waterways provide the best background for smugggling

Pott's disease vertebral inflammation

Poverty Bay Gisborne, New Zealand

Powder Keg of Europe the Balkans

Practical Political Philosopher Niccolò Machiavelli

Pragmatist Philosopher William James

Prairie Canada Alberta, Saskatchewan, and Manitoba

Prairie City Bloomington, Illinois

prairie ghost pronghorn antelope

Prairie Provinces Alberta, Manitoba, Saskatchewan

Prairies great plains between Appalachian and Rocky Mountains of North America

Prairie State Illinois

Prairie States North and South Dakota, Nebraska, Kansas, Minnesota, Iowa, and Illinois

Prayer-shawl Flag Israeli banner derived from talith or prayer shawl with horizontal blue stripes enclosing Shield or Star of David

Preacher of the Despairing Girolamo Savonarola

Preah Reach Ana Chak Kampuchea Cambodia

Precious Province Kueichow

Precursor of Dutch Painting Lucas van Leyden

Precursor of Expressionism Edvard Munch

Precursor of Japanese Art Kose no Kanaoka

Precursor of the Mexican Revolution Ricardo Flores Magon

Precursor of Pharmacology Paracelsus

Precursor of Pictorial Realism Mathias Grünewald (Mathis der Mahler)

Precursor of Sociology Charles de Secondat Baron de la Brède et de Montesquieu

Precursor of Spanish–American Emancipation Francisco Miranda

Precursor of Surrealism Hieronymus Bosch (Hieronymus van Aken)

Precursor of Venezuela Francisco de Miranda)

Pre-emption President John Tyler

Preiser's disease porosity of the wristbone

Premier Passenger Port of Great Britain Southampton

Premier Primitive Henri Rousseau

Premier of Russia Alexander Kerenski

Pre-Raphaelite Founders Holman Hunt, Sir John Everett Millias, Daniel Gabriel Rossetti

Presbyterian Jerusalem Edinburgh

The President the President of the United States

Presidents' conference conference of presidents of major Jewish organizations (in America)

President ships American President Line vessels named after such statesmen as *President Lincoln, President Roosevelt, President Taft*

Preston K Swinehart nickname—movie actor Alan Dinehart

Preston Sturges Edmund Preston Biden

Pretender Charles Stuart

Pretty Boy Charles A Floyd (1901–1934)

Pretzel City Lancaster and Reading, Pennsylvania

Pride of the Yankees Lou Gehrig

Priest of Nature Sir Isaac Newton

Prima Donna Assoluta Dame Joan Sutherland

Primate of Italy the Pope

Prime Meridian Place Greenwich, England

Prime Minister of Hell Satan

Prime Minister of Mirth Peter Sellers

Prime Minister of National Crime Frank Costello (1893–1973), born Francesco Seriglia

Prime Minister of the Underworld Frank Costello

Prince Prince Rogers Nelson

Prince of American Letters Washington Irving

Prince of the Apostles the Pope, according to the Roman Catholics

Prince of Artists Albrecht Dürer

Prince of Comic Opera Daniel François Esprit Auber

Prince Consort Albert of Saxe-Coburg Gotha (Queen Victoria's husband)

Prince of Cranks Ignatius Donnelly

Prince of Darkness Satan

Prince of Destruction Tamerlane (Timur the Lame)

Prince of Gossips Samuel Pepys

Prince of Humbugs P(hineas) T(aylor) Barnum

Prince of Humorists Mark Twain (Samuel Langhorne Clemens)

Prince of Israel Michael

Prince of Journalists Horace Greeley

Prince of Losers Dr Frederick A Cook who claimed he reached the North Pole nearly a year before Commander Robert E Peary, who was credited with the discovery

Princely Province Prince Edward Island

Prince of the Meistersingers Hans Sachs of Nuremberg also known as the Cobbler Poet

Prince of Men Robert Louis Stevenson's nickname for Henry James

Prince of Metals silver

Prince of Music Palestrina

Prince of Orange William I of the Netherlands and his male successors

Prince of Orators Demosthenes

Prince of the Oyster Pirates Jack London

Prince of Painters Raphael

Prince of Philosophers Plato

Prince of Physicians Avicenna (Abu ibn Sina)

Prince of the Pianoforte Louis Moreau Gottschalk

Prince of Pistoleers James Butler (Wild Bill) Hickok

Prince of Poets Alexander Pushkin, according to Russian literary critics; Edmund Spenser

Prince of Prose Writers John Bunyan

Prince of the Renaissance Michaelangelo

Prince of Scoffers Voltaire

Prince of Showmen PT Barnum

Prince of Siddhartha Gautama Buddha

Prince of Skeptics Voltaire

Prince of Spanish Poetry Garcilaso de la Vega

Princess of Fruits Linnaeus' sobriquet for the pineapple

Prince of Story Tellers Giovanni Boccaccio

Prince of Trees Linnaeus' nickname for the palm

Prince of Violin Virtuosos Itzhak Perlman

Prince of Wales Island Penang's previous name

Principality of the Grimaldi Monaco

Principal Port of the United Kingdom Liverpool

Principe de la Paz (Spanish—Prince of the Peace)—Manuel Godoy y Alvarez de Faria

Printmaker to the Mexican People José Guadalupe Posada

Printmakers to the American People (Nathaniel) Currier & (James Merritt) Ives—America's most famous lithographers

Prison at the Bottom of the World Ushuaia, Argentina on Beagle Channel close to Cape Horn

Prisoner of the Chillon François de Bonnivard

Prison of Gold The Louvre

Prison of Nations Austro-Hungarian Empire

Prison Poet Oscar Wilde (1856–1900), Irish author-playright-poet-wit whose conviction for sodomy brought a three-year prison term vividly described in his philosophical poem *The Ballad of Reading Gaol*

The Pritzker Pritzker Architecture Prize (founded by Jay A Pritzker in 1979)

Prodigy of Learning Dr Samuel Hahnemann

Professor Jim Brosnan (baseball player)

Professor of Blue Sky Mortimer Adler

Professor Bruno Pantoffel Jorge Mester

Professor of Earthquakes Sir William Hamilton

Professor Julius Caesar Hannibal (pseudonym—WH Levinson)

Professor Longhair Roy Byrd

Professor Seagull Joe Gould

Profit Center of the Southwest Phoenix, Arizona

Prolific Lexicographer Eric Partridge

Prolific Professor Isaac Asimov

Prolific Rationalist English ex-priest Joseph McCabe

Prolific Typographer Frederic William Goudy

Promised Land Israel, promised to the Israelites by Moses and to the Israelis by Balfour

Promoter of Agrarian Reform Emiliano Zapata

Prophet of Allah Mohammed

Prophet of the American Way Thomas Jefferson

Prophet of Christianity John the Baptist

Prophet of Democracy William Penn

Prophet of Doom Girolamo Savonarola

Prophet of Israel Moses

Prophet of Modernity Émile Zolá

Prophet of Mythology Teiresias

Prophet Outcast Leon Trotsky

Prophet-Preacher-Hero of New England Unitarianism William Ellery Channing

Prophets of Israel Moses. Samuel, Nathan, Elijah, Elisha

Prophet of the Strenuous Life Jack London

Prose Poet of Violence Jean Genet

Prosperous Paradise of the Pacific Hawaii

Prostitution Capital of the South old nickname for New Orleans

Protector from Fever Febris (Roman goddess)

Protector from Poison Gases Mephitis—Roman goddess venerated in volcanic lands

Protector of the Indians Rodrigo de Bastidas—Spanish navigator who founded Santa Marta; Las Casas and Eliot share the title—Protector of the Indians

Protector of Peru General José Francisco de San Martin (1778–1850)

Protector of the Protestants Marguerite de Navarre

Protector of Seafarers the Greek goddess Brizo

Protestant Hero Frederick the Great of Prussia

Protestant Rome Geneva

Provisional President of Africa Marcus Garvey

Provision State Connecticut in Revolutionary times

Provost Hunting reconnaissance-trainer aircraft built in Britain

Prune Picker(s) Californian(s)

prussic acid hydrocyanic acid

psikhushka (Russian slang—psychoprison)—for punishing and segregating dissidents

Psychedelphia San Francisco's Haight-Ashbury district

Psychoanalysis Capitals Berlin, New York, and Vienna

Psychologist of the Soul Søren Aabye Kierkegaard (1813–1855)

Ptarmigan Alaska state bird; symbolic nickname given some Alaskans

P-town Provincetown

Public Enemy Number One gangster Al Capone's nickname

Public Library Builder Andrew Carnegie

Publius allonymic name used by Alexander Hamilton, John Jay, and James Madison in writing *The Federalist*

Pueblo de los Angeles (Spanish—Village of the Angels)—early Spanish name for Los Angeles

Puerto Colombia formerly Savanilla

Puerto Limón Limón, Costa Rica

Puerto Rico Ports Ensenada de Honda, San Juan, Ponce, Guanica, Mayaguez

Pugetopolis industrialized urban areas surrounding Puget Sound

Puggy Booth Joseph Mallord William Turner's nickname

Pujo Committee for Arséne Pujo, who headed a United States House Banking and currency subcommittee in 1912

Pukes former name for Missourians

pukeweed *Lobelia inflata's* nickname

Puma Franco-British Aerospatiale-Westland transport helicopter

Punkie Town Punxsutawney, Pennsylvania

Punks' Paradise nickname given any gambling center and sometimes to the State of Nevada and the casino cities of Reno and Las Vegas

Punnett square divided-square use to show genotypes and phenotypes from recombination of gametes during fertilization (for Reginald C Punnett)

Punta Arenas (Spanish—Sand Point)—the one in Chile called Magallanes from 1927 to 1937; the one in Costa Rica is written *Puntarenas*

Punxey Punxsutawney

Puppet Emperor Henry P'u-yi (1906–1967) last emperor of China; made Emperor of Manchukuo by the Japanese

Puri port of Jagananth or Juggernaut on the Bay of Bengal

Puritan City Boston, Massachusetts

Puritan State Massachusetts

Puritans inhabitants of Massachusetts

Purple Islands the Madeiras

Purple Land WH Hudson's sobriquet for Uruguay

purpurite iron magnesium phosphate

Pushkin modern name for Tsarskoe Selo south of Leningrad

Putrid Sea Sivash Sea (mineralized marshes along Crimea's north coast)

Putzhead syndrome political name-calling

Pyrenees Principality Andorra

Pyrrhic victory named for King Pyrrhus of Epirus, who won many decisive battles but lost too many soldiers and thus his war against the Romans

Pythagoreanism religion founded by Pythagoras

Qatar Ports Ad Dawhah and Musayid

Q-town Queenstown

Quad Cities Davenport, Iowa; East Moline, Illinois; Moline, Illinois; Rock Island, Illinois

Quadra Island island in British Columbia named for Spanish explorer Juan Francisco de la Bodega y Quadra

Quai d'Orsay section of Paris occupied by French Foreign Ministry

Quail Californians are sometimes nicknamed Quail; California's state bird; McDonnell-Douglas decoy missile

Quail Haven Cedar Vale, Kansas

Quaintest City in the U.S. Santa Fé, New Mexico (founded by the Spaniards around 1609)

Quake City San Francisco, California

Quaker Abolitionist epithet shared by Elias Hicks, Lucretia Mott, John Greenleaf Whittier, and John Woolman

Quaker City Philadelphia

Quaker City of the West Richmond, Indiana

Quaker Dolley Mrs Dorothea (Dolley) Madison—wife of President James Madison

Quaker Founder George Fox—founder of the Society of Friends who were nicknamed Quakers by an English judge who persecuted them

Quaker Founder of Pennsylvania William Penn

Quaker Frontiersman Daniel Boone

Quaker Liberal Elias Hicks

Quaker Penologist Elias Hicks (1748–1830), who preached against cruelty to the imprisoned and the insane

Quaker Poet Bernard Barton in England and John Greenleaf Whittier in New England

Quaker Preacher Elias Hicks—founder of the Hicksite Friends championing the abolition of slavery and opposing creeds approved by the elders

Quaker Reformer Elizabeth Fry—noted for her campaigns to better the life of inmates in insane asylums and prisons; also worked for the betterment of education

Quakers members of the Society of Friends

Quaker State Pennsylvania

Quakertown Philadelphia

Quality City Rochester, New York

quarantine flag yellow flag flown when a vessel requests pratique; letter Q or Québec in the international signal code

quas methaqualone's nickname (also called quacks or quads)

Québec Q-class Soviet submarines

queen male homosexual

Queen of the Adriatic Venice

Queen Alice Alice Lee (Roosevelt) Longworth

Queen of the Amazons Hippolyta

Queen of the Angels the Virgin Mary

Queen of the Antilles Cuba

Queen of the Arabian Sea Cochin, India

Queen of Back Bay Isabella Stewart Gardner

Queen of Bases calcium oxide and related compounds known commercially as lime

Queen Bee Director of the Women's Air Force Auxiliary

Queen of Belgian Beaches Ostend

Queen Bess Queen Elizabeth

Queen of the Blues Dinah Washington

Queen of the Brazos Waco, Texas

Queen of the Caribbees Nevis

Queen of Cats the lioness in Africa; the tigress in Asia

Queen Cities of the Austro-Hungarian Empire Budapest, Prague, Vienna

Queen City Lahore (in the Punjab of Pakistan)

Queen City of Alabama Gadsden

Queen City of Canada Toronto

Queen City of the Carolinas Charlotte, North Carolina

Queen City of the Hanseatic League Lübeck

Queen City of the Hudson Yonkers, New York

Queen City of India Bombay

Queen City of the Lakes Buffalo, New York and Toronto, Ontario

Queen City of Lake Superior Marquette, Michigan

Queen City of the Lehigh Valley Allentown, Pennsylvania

Queen City of the Merrimack Valley Manchester, New Hampshire

Queen City of the Mississippi St Louis, Missouri

Queen City of the Mountains Knoxville, Tennessee

Queen City of New Zealand Auckland

Queen City of the North Edinburgh

Queen City of the Ohio Cincinnati, Ohio

Queen City of the Ozarks Springfield, Missouri

Queen City of the Pacific San Francisco, California and Seattle, Washington

Queen City of the Plains Denver

Queen City of the Rio Grande Del Rio, Texas

Queen City of the Sea Charleston, South Carolina

Queen City of the Sound Seattle, Washington on Puget Sound

Queen City of the South Atlanta, Georgia and Sydney, New South Wales

Queen City of the Southern Tier Elmira, New York

Queen City of the Trails Independence, Missouri

Queen City of Vermont Burlington

Queen City of the West Cincinnati (in the early 1800s)

Queen of the Comstock Lode Virginia City, Nevada

Queen of the Cowtowns Fort Dodge, Iowa

Queen of Crime Agatha Christie

Queen of Crossword Puzzledom Margaret Farrar

Queen of the Danube Budapest

Queen Emma Curaçao's floating bridge across Willemstad's harbor

Queen of Flowers the rose

Queen of Folk Joan Baez

Queen of the French Riviera Nice

Queen of Freshwater Fish carp

Queen of the Goldfields Melbourne, Victoria, Australia

Queen of Gulf Ports New Orleans

Queen of Heaven Ashtoreth (Semitic); Astarte (Phoenician); Hera (Greek); Inanna (Sumerian); Ishtar (Assyrian and Babylonian); Isis (Egyptian); Juno (Roman); Virgin Mary (Christian)

Queenie Regina

Queen of the Inland Sea Chicago

Queen of the Islands Cuba in the days of Columbus

Queen of Italian Cuisine Marcella Hazan

Queen on the James River Richmond, Virginia

Queen of the Kingdom of Death Hel (daughter of Loki in Nordic mythology whose name is used for the Kingdom of Death)

Queen of Kings Cleopatra

Queen of Lake Malaren Stockholm

Queen of Lake Michigan Chicago

Queen of the Lakes Chicago

Queen of Long-Distance Roads the Appian Way extending from Brindisi to Rome and begun in 312 B.C.

Queen of Love and Lust Aphrodite or Venus

Queen Maud Land Norwegian Antarctica

Queen of Metals platinum

Queen of the Missions Mission San José in San Antonio, Texas and Mission Santa Barbara in Santa Barbara, California

Queen of the Mississippi St Louis

Queen of the Mountains Helena, Montana

Queen of Mystery Writers Ngaio Marsh

Queen Nef Queen Nefertiti of Egypt

Queen of the North Edinburgh

Queen of the Ohio Cincinnati

Queen of Opera Dame Joan Sutherland

Queen of the Pacific San Francisco

Queen of the Plains Regina, Saskatchewan

Queen of the Prairies Canada's Province of Saskatchewan

Queen of Queens Brutus, nickname for Cleopatra

Queen Recluse Emily Dickinson

Queen Sarah Sarah, the Duchess of Marlborough

Queensberry (*see* Marquis of Queensberuy)

Queen's Birthday Queen Juliana's Birthday (April 30)— Netherlands; Queen Victoria's Birthday (May 20)— Great Britain and Commonwealth countries

Queen's College Rutgers University in colonial times

Queen's Corsair Sir Francis Drake

Queen of the Sea Islands Beaufort, South Carolina

Queen of the Seas Glasgow (reputed for its Clyde-built ships); Venice

queen's English correct English

Queen's House Buckingham Palace

Queen of Skyscrapers Empire State Building

Queen of the South New Orleans

Queen of the Spas Saratoga Springs, New York

Queen State Maryland (named for Henrietta Maria, wife of Charles I of England)

Queenstown former name of Cobh on Ireland's south coast

Queen of Summer Resorts Newport, Rhode Island

Queen of Talk Oprah Winfrey

Queen of Teas Darjeeling tea

Queen of Trains Orient Express

Queen of TV Comedy Lucille Ball

Queen of the Vampires Theda Bara (Theodosia Goodman)

Queen of Watering Places Brighton, England

Queen of the West Longfellow's nickname for Cincinnati; Dale Evans

queer homosexual

Queermacks Cuyamaca Mountains in California's San Diego County

quicklime calcium oxide—CaO

quicksilver mercury (Hg)

Quicksilver Bob Robert Fulton's nickname

Quiet Americans soft-voiced well-mannered Canadians

Quiet Epidemic medical nickname for Alzheimer's disease

Quiet River Russia's quiet-flowing Don

Quincke's disease edema of the skin; giant hives

Quinquad's disease inflammation of the scalp resulting in bald patches

Quintessence of Africa Serengeti National Park, Tanzania

Quintuplets Herbert Morrison, so nicknamed because he did the work of five

Quirinale (Italian—House of the God of War)—Ministry of War in Rome

Quisquellano(s) Santo Domingan(s)

Quisqueya Hispaniola

Rabbi of Swat Moe Solomon (baseball player)

Rab(bie) Robert

Rabble-Rouser of the Revolution Sam(uel) Adams

rabelaisian writer robust humorist noted for naturalism

race records early jazz and blues recordings by black musicians

Rachilde Marguerite Vallette

Radclyffe Hall Marguerite Radclyffe Hall

Radek communist-party pseudonym of Karl Sobelsohn

Radio Capital Camden, New Jersey

Radiumbad Brambach Brambach, Saxony

Raedwulf (Early English—Ralph)—alleged to be the imp of mischief in a printing house

Rafe Ralph; Ralph Vaughan-Williams (English composer)

Raffaello Raphael

Ragnarok end of the world in Norse mythology; equivalent to Twilight of the Gods (Götterdämmerung)

Rahway New Jersey State Prison at Rahway

Railbelt parts of interior and Southeast Alaska on the Alaska Railroad route

Railroad City nickname of cities such as Atlanta, Boston, Buffalo, Chicago, Cincinnati, Cleveland, Detroit, Edmonton, Houston, Indianapolis, Kansas City, Los Angeles, Milwaukee, Minneapolis, Montreal, New Orleans, New York, Omaha, Philadelphia, St Louis, San Antonio, San Francisco, Seattle, Toronto, Washington, D.C., Winnipeg; (*see* Railway City)

Railsplitter Abraham Lincoln

Railway Killer Angel Maturino Resendiz

Railway King George Hudson

Rajah Rogers Hornsby

Ralph Connor Charles W Gordon

Ralph Iron Olive Schreiner's pseudonym

Ralph Marlowe Ralph Manheim

Ralph Rashleigh James Tucker's pseudonym

Ramana Ramana Maharshi

Ramblin' Jack Elliott Elliott Charles Adnopoz

Ramón Navarro Ramón Samanlegos

The Ranch once-top-secret aircraft test site near Las Vegas, Nevada

Randolph Scott Randolph Crance

Randy Randolph

Ran Fiennes Sir Ranulph Twistleton-Wykeham-Fiennes

Ranger American program for investigation of the Moon and region between the Moon and the Earth; Texas state policeman

Rangers Texans

Rapier British BAC surface-to-air missile launched for low-altitude defense

Raquel Torres Paula Ostermán

Raquel Welch Raquel Tejada

Racquetball Capital San Diego, California

Ras Desiderius Erasmus

Ras Asir (Arabic—Cape Guardafui)—northeastern-most Africa on the coast of Somalia and the Gulf of Aden

Rasmus Erasmus

Rasputin (Russian—Dissolute)—nickname of the Siberian monk Gregory Efimovitch

Rasputitsa several weeks between winter and summer in Siberia marked by floods and mud from melting snow, making travel difficult

Rassmen Jamaicans

Rastus Erastus; Theophrastus

Ratipole nickname of Napoleon III

Rat Pack Joey Bishop, Sammy Davis Jr, Peter Lawford, Dean Martin, and Frank Sinatra

Raven Hiller utility helicopter designated H-23 and OH-23

Ray Charles Ray Charles Robinson

Ray Milland Reginald Truscott-Jones

Raynaud's disease circulatory disorder of the extremities

Raynaud's phenomenon white-finger disease brought on by long-term use of vibrating hand tools

Razor The Razor—General Hideki Tojo's nickname

Razorback(s) Arkansan(s)

Razor Clam Capital Cordova, Alaska

Reagan Doctrine undo Soviet gains in the Third World; named for Ronald Reagan, fortieth president of the United States

Realistic Recorder of Spanish Life Goya (Francisco José de Goya y Lucientes)

Realm of the American Alligator Okefenokee Swamp, Florida and Georgia

Realm of Exotic Flavors Thailand

Rebecca West Cicily Isabel Fairfield

Rebel City Charleston, South Carolina

Rebel Poet Nazrul Islam

Rebel of Salem Roger Williams

Rebel Unitarian Theodore Parker

Rebel of Walden Henry David Thoreau

Recafellow Andrew Carnegie's nickname for John D Rockefeller, Sr

Recife (Portuguese—Reef)—Pernambuco's new name

Recklonghausen's disease neurofibromatosis

Reclus' disease cystic growths in the breasts

Reconstruction President Rutherford Birchard Hayes

Red Sinclair Lewis

Red Baron Baron Manfred von Richthofen

Redbricks red-brick universities

Redbug of the Southeast chigger

Red Buttons Aaron Chwatt

Red Chamber Canadian Senate

Red China People's Republic of China

Red Crescent equivalent of the Red Cross in the Moslem world (symbolized by a red crescent on a white field)

Red Cross red cross on a white field; used on ambulances, hospitals, and hospital ships to denote their neutrality; also called the Cross of Geneva or the Geneva Cross, as its function in war is accepted by the Geneva Convention

Red Dean of Canterbury The Very Reverend Doctor Hewlett Johnson—Dean of Canterbury Cathedral from 1931 to 1963

Redd Foxx John Elroy Sanford

red-diaper babies children born to and following beliefs of communist parents

Red Duster British flag; Red Ensign flown from British merchant vessels

Redemptorist Founder Alfonso Maria de Liguori

Red Ensign British flag

Redeye General Dynamics portable surface-to-air missile carried and fired by one man

red flag danger; stop sign

Red Gap Lone Pine, California as described in *Ruggles of Red Gap* by Harry Leon Wilson

Red Geranium of British Labor Sir Oswald Mosley

red lead lead oxide—Pb_3O_4 (minium)

Red Lewis (Harold) Sinclair Lewis

redlight district whorehouse neighborhood; zone of prostitution

Red Lion and Sun Iran's equivalent of the Red Cross (symbolized by a red lion beneath a red sun on a white field)

redneck ultraconservative person

Red Planet Mars

Red Priest red-headed Antonio Vivaldi

Red Rosa Rosa Luxemburg—co-founder with Karl Liebnecht of the Spartacus League later to become the Communist Party of Germany

Red Skelton Richard Bernard Skelton

Red Square in the heart of Moscow between the GUM department store, the Kremlin, and Lenin's tomb; called Krasnaya Ploschad by the Russians

Reds communists; underworld term for sedative/hypnotic drugs

red star symbol of the Soviet Union

Redtop Hawker-Siddeley air-to-air missile

red, white, and blue American flag

Red Wine of Teas Bengal tea (from the Indian subcontinent)

reefer(s) marijuana cigarette(s); refrigerated compartment(s) in a ship; refrigerator(s)

Reginald Bliss Herbert George Wells (1866–1946)

Regiomontanus Johann Müller (1436–1476)

Region of Four Streams Szechwan Province, China

Reichian massage massage therapy, developed by Wilhelm Reich, designed to cure disorders by removing energy blockages

Reichmann's disease continuous and excessive gastric secretion

Reidsville Georgia State Prison Facility at Reidsville

Reistertown Girls Montrose School for (delinquent) Girls at Reistertown, Maryland

Rejectionist Front Arab countries such as Algeria, Libya, and Syria most opposed to U.S. efforts to gain Israel's acceptance by its neighbors

Religious Freedom Colony Rhode Island

Reluctant Imperialist John C Calhoun who half-heartedly supported the Mexican War

Rembrandt Rembrandt Harmenszoon van Rijn—or van Ryn—RvR

Rembrandt of the Roman Ruins Giambattista Piranesi

Rene Irene

René Adorée Jeanne de la Fonte

Renée Clair René Chomette

Renegade Irishman James Joyce

República Oriental (Spanish— Oriental Republic)— Uruguay

Republic of the Sacred Heart Ecuador

Research Center of the Classical World Library of Alexandria, Egypt

Resort City Miami, Florida

Resplendent Land Ceylon or *Sri Lanka* (Singhalese— Resplendent Land)

The Restoration France, from 1814 to 1848

Ret Marut B Traven's pen name in Bavaria after World War I

Return-to-Normalcy President Warren Gamaliel Harding

Réunion Indian Ocean island formerly called Bourbon

Reval or Revel old placenames for Tallinn, Estonia

Rex Harrison Reginald Carey

Rex Ingram Reginald Hitchcock

Rhapsodic Richard Richard Strauss

Rhineland Capital Cologne (Köln)

Rhode Island Ports Providence, Newport

Rhode Island Reds Rhode Islanders

Rhonda formerly Ystradyfodwg, Wales

Ricardo Cortez Jacob Krantz

Rice Birds South Carolinians

Rice Bowl southwest Louisiana

Rice Bowl of Malaysia Kedah

Rice Center Crowley and Lake Charles in coastal Louisiana

Rice State South Carolina

Richard Arlen Van Mattimore

Richard Avalon Sir John Woodroffe

Richard Burton Richard Jenkins

Richard Coeur de Lion French—Richard Lion-Heart (Richard I of England)

Richard the First Richard Wagner

Richard Hull Richard Henry Sampson's pseudonym

Richard Llewellyn Richard David Vivian Llewellyn Lloyd

Richard Saunders Benjamin Franklin (*see* Poor Richard)

Richard the Second Richard Strauss

Richard Tauber Ernst Seiffert

Rich Coast Costa Rica

Richelieu Armand Jean du Plessis

Rich Man's Panic 1903, when Dow Jones Industrial average dropped 24 percent

Rickie Admiral Hyman George Rickover, USN

Rideau Hall Ottawa residence of the Governor General of Canada

Rienzi Niccolo Gabrini

Rifle City Springfield, Massachusetts

Riga's disease ulceration of the tongue

Rigg's disease inflammation of the gums with pus deposits in the tooth sockets; also called alveolar pyorrhea

Ring of Fire volcanic zone extending from Alaska to Chile and from Siberia to New Zealand via Indonesia

Ring Lardner Ringgold Wilmer Lardner

Ringo Starr Richard Starkey

Rio Rio de Janeiro, Brazil

Rio Branco José Mariá de Silva Paranhos—Baron of Rio Branco—Brazil's great statesman

Rio de la Plata Province Paraguay

Rip Rip Van Winkle; Robert; Rupert

Rip Van Winkle State North Carolina

rising sun symbol of Japan and the Japanese

Rising Sam Japan's nickname for the United States, reflecting improved quality of its manufactured goods

Rita Hayworth Margarita Carmen Cansino

Rita Moreno Rosita Dolores Alverio

Ritter's disease skin scaling, sometimes fatal when it attacks infants

Rivalta's disease lumpy jaw

River of African Legend Congo; Nile; Orange; Zaire

River of Alaskan Legend the Yukon

River of American Legend Mississippi

River of American Midwestern Legend Missouri
River of American Northwestern Legend Columbia
River of Asian Legend Euphrates; Ganges; Tigris; Yangtze
River of the Bears Alaska's McNeil River
River of the Black Dragon Amur River on the Sino-Soviet frontier
River of Brazilian Legend Amazon
River of British Legend Severn
River of Californian Legend Sacramento
River of Canadian Legend the St Lawrence
River City Austin, Texas; San Antonio, Texas
River of Colombian Legend Magdalena
River of Destiny Rio Grande
River of English Legend Thames
River of European Legend Danube, Rhine
River of Florida Legend Suwanee
River of French Legend Seine
River of Georgia and South Carolina Legend Savannah
River of the Gods Ganges River
River of Grass Florida's Everglades
River of Hades or Hell the Styx, according to mythology it encircles the underworld nine times and the dead are ferried over its waters by Charon
River House Ohio State Penitentiary on the Scioto River near Columbus
River of Irish Legend Shannon
River of Italian Legend Arno; Tiber
River of Kings Chao Phraya flowing through Krungthep formerly called Bangkok
River of Maine Legend Kennebec; Penobscot
River of New England Legend Connecticut; Merrimack
River of New Jersey and Pennsylvania Legend Delaware
River of New York Legend Hudson

River of the North the Yukon
River of the Old Northwest Legend Ohio
River of Oregonian Legend Willamette
River Plate Republics Argentina, Paraguay, Uruguay (all on rivers flowing into Rio de la Plata estuary)
River of Portuguese Legend Tagus
River of Russia Legend Volga
River of Scottish Legend Tay
River of Spanish Legend Ebro
River of Venezuelan Legend Orinoco
Riverview Interprovincial Home for (misdemeanant) Women at Riverview, New Brunswick
River of Welsh Legend Towy
Riviera Mediterannean coasts of Italy, France, and Spain
Riviera of South America Uruguay
roach cockroach; marijuana butt
road hustler(s) card-and-dice hustler(s)
Roadrunner New Mexico state bird and nickname applied to many New Mexicans
Road of the Sun l'Autostrade del Sole (Italian superhighway linking Milan, Rome, and Naples)
Roaring Forties storm-tossed seas between 40 and 50 degrees south latitude
Robber Barons (see American Railroad Barons, Banker Barons, Mining Baron, Oil Baron, Pork Packer, Steel Baron)
Robber's Nest Berlin, according to an old German song
Robert Alda Alpnonso d'Abruzzo
Robert Capa Andrei Friedmann (1913–1954)
Robert Forsythe Kyle Crichton
Robert Henri Robert Henry Cozad
Robert Merrill Henry Lavan
Robert Rostand Robert Hopkins
Robert Stack Robert Modini
Robert Taylor Spangler Arlington Brugh

Robert Weede Robert Wiedefeld
Robin Hood rationing using revenues from the middle and upper classes to provide social programs for those in need
Robinson's Island Niihau, Hawaii
Rob Roy (Gaelic—Red Rob)— Robert Macgregor the Scottish freebooter
Rochedos São Paulo (Portuguese—Saint Paul's Rocks)—in the Atlantic just north of the Equator and far off Brazil
rochelle salts sodium potassium tartrate
Rocher du Diamant (French— Diamond Rock)—off Fortde-France, Martinique; commissioned in 1800 as HMS Diamond Rock because here British sailors withstood a French bombardment lasting more than eighteen months
Rochers du Calvados (French— Calvados Reef)—at the mouth of the Orne in the English Channel
Rochester actor Eddie Anderson
rock crack, cocaine
Rock Knute Kenneth Rockne; Mount Desert Island's nickname; Rockaway; Rock of Gibraltar; The Rock—Alcatraz (Federal Prison once occupying a 12-acre rock in San Francisco Bay; name now applies to Rikers Island New York City's Correctional Facility in the East River or to San Quentin on the shores of San Francisco Bay)
The Rock Alcatraz (former prison, now museum); The Rock of Gibraltar (British crown colony on a rocky peninsula extending south from the Spanish mainland); Saba Island, Netherlands Antilles
Rock of Chickamauga General George Henry Thomas
Rock City Nashville, Tennessee
Rocket City Huntsville, Alabama
Rocket Pioneer Robert H Goddard

Rock Hudson Roy Scherer (changed to Fitzgerald)
Rockie Nelson A Rockefeller
Rock Lizards Gibraltarians
Rock Music Metropolis Minneapolis
Rock of Notre Dame Knute K(enneth) Rockne
Rock-ribbed State Massachusetts; Maine
rock salt halite (sodium chloride)
Rock of Uluru Ayers Rock near Mount Olga, Australia
Rockwell Girls Women's Reformatory at Rockwell City, Iowa
Rocky Roccoforte; Rochester; Rockefeller
The Rocky *Rocky Mountain News* (Denver)
Rocky Arabia Arabia Petraea in the northwestern section of the Arabian Peninsula
Rocky Butte Portland, Oregon's jail
Rocky Marciano Rocco Marchegiano
Rocky Mountain Colorado national park
Rocky Mountain States Alaska, Idaho, Montana, Wyoming, Colorado, Utah, New Mexico, and Arizona
Rodney Dangerfield Jacob Cohen
Roger Williams City Providence, Rhode Island
Rogues' Island Rhode Island, so called by Puritans of the Massachusetts Bay Colony
Roi Citoyen (French—Citizen King)—Louis Philippe
Roi Soleil (French—Sun King)—Louis XIV
Roland Franco-German Nord-Bolkow surface-to-air missile whose name honors a medieval hero of song and story in the time of Charlemagne
Rolfe Boldrewood Thomas A Browne's pseudonym
Rolfing physical therapy developed by Ida Rolf
Rolls-Royce of recreational drugs cocaine
Roloff Van Ripper Washington Irving
Roman Century the 1st century before the Christian era
Romani Gypsies

Romanian Ports Mangalia, Constanta, Sulina, Isaccea, Braila, Galati, Tiglina
Romania's Principal Port Constanta
Romano Giulio Pippi de Granuzzi's palette name—Giulio Romano
Romeo letter R radio code; Soviet R-class submarines
Rome of the West St. Louis, Missouri
Ronnie Ronald; Ronda; Veronica
Ronny Ronald
Roof Garden of Texas Alpine
Rooftop of Africa Kilimanjaro in Tanzania
Rooftop of Antarctica Vinson Massif
Rooftop of Argentina Aconcagua on the border of Chile
Rooftop of Asia Everest in China and Nepal
Rooftop of Australia Kosciusko in New South Wales
Rooftop of Austria Grossglockner
Rooftop of Bolivia Ancohuma
Rooftop of Canada Mt Logan in the Yukon
Rooftop of Chile Ojos del Salado on the border of Argentina
Rooftop of Ecuador Chimborazo
Rooftop of Europe Mont Blanc in France
Rooftop of India Mt Godwin Austen, Jammu and Kashmir
Rooftop of Italy Monte Rosa on the border of Switzerland
Rooftop of Japan Fuji
Rooftop of México Citlaltépetl also called Orizaba
Rooftop of New Zealand Mt Cook on South Island
Rooftop of North America Mt McKinley in Alaska
Rooftop of Peru Huascarán
Rooftop of South America Aconcagua in Argentina
Rooftop of Spain Mulhacén in Granada
Rooftop of Switzerland Matterhorn
Rooftop of Turkey Ararat in Armenia
Rooftop of the World Pamir Plateau of central Asia

Roosevelt I Theodore Roosevelt—twenty-sixth President of the United States
Roosevelt II Franklin D Roosevelt—thirty-second President of the United States
Roosevelt Island current name for New York City's East River island formerly called Welfare and originally Blackwell's
Rooster Rick Burleson (baseball player)
Rory Aurora
Rosa Bonheur Rosalie Mazeltov
Rosa and Carmela Ponselle Rosa and Carniela Ponzillo
Rosa Parks of the Gay-Rights Movement William Jennings
Rosa Ponselle Rosa Melba Ponzillo
roscoe slang—handgun, rifle, shotgun
Rose Bowl Pasadena, California
Rose Capital of the World Tyler, Texas
Rose City Madison, New Jersey; Pasadena, California; Portland, Oregon
Rose Moon full moon in June
Rose-Red City Petra in Jordan across the Wadi al 'Arabah from the Negev of Israel
Rose and Shamrock England and Ireland
Roseto Effect abundant good health exhibited by residents of Roseto, Pennsylvania, the result of living in a close-knit community
Rossbach's disease gastric juice secreted excessively
Ross Macdonald Kenneth Millar's pseudonym
Rosy Rosalind; Rosen; Rosenbaum; Rosenberg; Rosenfeld; Rosenthal
Rothermere Viscount Rothermere (Harold Sidney Harmsworth)
Roth IRA Individual Retirement Account named for Senator William V Roth
Rough Rider Theodore Roosevelt—twenty-sixth President of the United States
Rough Riders First, United States Volunteer Cavalry organized by colonels Theodore

Roosevelt and Leonard Wood for action in the Spanish-American War

Roundheads Cromwell's followers in the Puritan Party noted for the close-cropped hair of its members

Roundup City Pendleton, Oregon

Rovers Coloradans

Rowan Oak William Faulkner's home near Oxford, Mississippi in Lafayette County he fictionalized as Yoknapatawpha

Roy Rogers Leonard Slye

Royal Bob President Garfield's name for Colonel Robert G Ingersoll

Royal Brute of Great Britain King George III of Hanover (in the opinion of Thomas Paine and many other Americans and Britons)

Royal Epicurean Emperor Hadrian

Royal Gorge Grand Canyon of the Arkansas in Colorado

Royal Martyr Charles I of England

Roy Rogers Leonard Slye

Roz Rodriguez; Rosalind(a); Rozhdestvensky

R-town Rivertown

Rub ai-Khali (Arabic—Empty Quarter)—area between Oman and Yemen

Rubber Capital of the U.S. Akron, Ohio

Rubber City Akron, Ohio

rubber room padded cell

Rube Goldberg Reuben Lucius Goldberg (cartoonist creator of fantastic inventions for accomplishing simple tasks—hence any overcomplicated mechanism is termed a Rube Goldberg)

Rube Lacy Rubin Lacy

Ruben Dario Félix Rubén García Sarmiento

Rube Waddell George Edward Waddell

ruby red corundum

ruby copper cuprite (cuprous oxide)

Rudolf Valentino Rodolpho d'Antonguolla

Rudy Vallee Hubert Prior Vallee

Ruissalo Finnish class of motor gunboats and minesweepers sometimes used as patrol launches

Ruler of the East Vladivostok

Rural Garden of Eden South Carolina

Russia Russia leather (dark-red leather used for book binding)

Russian Russian dressing (salad dressing spiced with chili, chopped pickles); Russian roulette (each player pull the trigger of a revolver held against the player's head); Russian wolfhound (large breed of hound originally bred in Russia and called *borzoi*)

Russian America Alaska before its purchase from Russia in 1868 for $7.2 million

Russian-American Capital Sitka, Alaska

Russian Bear symbol of Russia

Russian Physiologist Extraordinary Ivan Petrovich Pavlov

Russki(s) Russian(s)

Rust Belt Detroit's automobile manufacturing area

Rust's disease tuberculosis of the upper cervical vertebrae

Rust Territory Pohnpei, Micronesia Federated States; former United States Trust territory; Trust Territory of the Pacific

Ruth Saint Denis Ruth Dennis

Ryukyu Retto (Japanese—Ryukyu Islands)—also known as Loochoo or Nansei Islands

Saar River City Saarbrücken, Germany

Sabah formerly British North Borneo

Sabine River City Orange, Texas

Sabra main battle tank built by Israel Army Ordnance and armed with a 105mm gun; native-born Israeli; Sabrina

Sabre Australian-built Canadian version of the F-86 jet-fighter designated CF-86 or CA-27 and originally built by North American as a single-engine jet-fighter

Sabre 32 Australian-built F-86 jet fighter

Sabreliner North American T-39 transport aircraft

Sacha Alexander

Sacha Guitry pseudonym—Alexandre Pierre Georges

Sacred Untouchables TS Eliot, Marcel Proust, Rainer Maria Rilke, William Butler Yeats

Saddler Soviet SS-7 liquid-fuel intercontinental ballistic missile

Sadie Sara; Sarah; Sarita

Sad Side south side of Ellis Island, which fell into disrepair after decades of exposure to the elements

Saeta Hispano HA-200 twin-engine jet trainer built in Egypt and in Spain; also called E-14

Safford Federal Prison Camp at Safford, Arizona

Safir Saab-built training and utility aircraft also known as Saab 91-D

Saga City Stavanger, Norway

Saga Island Iceland

Sagamore Hill Theodore Roosevelt's home on Long Island at Oyster Bay, New York

Sagarmatha (Nepalese—mother of the snows)—Mount Everest

Sage of America Benjamin Franklin

Sage of Anacostia Frederick Douglass

Sage of Ashland Henry Clay

Sage of Auburn Secretary of State William H Seward

Sage of Baltimore H(enry) L(ouis) Mencken

Sagebrush Princess Sarah Winnemucca

Sagebrush State Nevada

Sage of Chappaqua Horace Greeley

Sage of Chelsea Thomas Carlyle

Sage of Concord Ralph Waldo Emerson—American philosopher-poet

Sage of East Aurora Elbert Hubbard

Sage of Ebury Street George Moore

Sage of Emporia William Allen White

Sage of Ferney Voltaire

Sage of Gramercy Park Samuel H Tilden—benefactor of the New York Public Library and governor of New York

Sage-hen(s) Nevadan(s)

Sage-Hen State Nevada

Sage of Jena Ernst Haeckel

Sage of Kinderhook Martin Van Buren—eighth President of the United States

Sage of Monticello Thomas Jefferson—editor of the *Declaration of Independence,* founder of the University of Virginia, third President of the United States

Sage of Montpelier James Madison—Father of the *Constitution,* fourth President of the United States

Sage of Mount Vernon George Washington—first President of the United States

Sage of Nininger Ignatius Donnelly

Sage of Philadelphia Benjamin Franklin

Sage of Popayán Francisco José de Caldas

Sage of Princeton Cleveland

Sage of Roanoke John Randolph

Sage of Samos Pythagorus

Sage of the Shakyamuni Siddhartha Gautama (Buddha)

Sage of Sullivan Street Edgar Varese

Sage of Walden Pond Henry David Thoreau

Sage of Wheatland James Buchanan—fifteenth President of the United States

Sage of Yasnaya Polyana Count Leo Nikolayevich Tolstoy (1828–1910)

Sage of Yoknapatawpha William Faulkner

Sagger Soviet antitank missile

Sahara North Africa's great desert; the Breguet 765 troop transport aircraft

Sahel Sahel Countries (Cape Verde Islands, Chad, Gambia, Mali, Mauritania, Niger, Senegal, Upper Volta)

Sailor City San Diego, California

Sailor Historian Samuel Eliot Morison

Sailor King William IV of England

Sailor on Horseback Jack London

Sailor's Poet Charles Dibden

Sailor's Friend Samuel Plimsoll

Sailor Town Norfolk, Virginia

Saint Anthony's fire gangrenous skin conditions such as ergotism, erysipelas, and hospital gangrene

Saint Augie Saint Augustine, Florida

Saint Barts Saint Barthélemy or Saint Bartholomew in the French West Indies

Saint Bart's Saint Bartholomew's Episcopal Church

Saint Croix Santa Cruz (inhabitants of this Virgin Island called Cruzans)

Saint Didacus San Diego de Alcald de Henares

Saint Gotthard's disease intestinal hookworms

Saint of the Gutters Nobelprizewinner Mother Teresa of Calcutta

Saint-John Perse Alexis Léger

Saint Martin's evil dipsomania

Saint Paddy Saint Patrick

Saint-Simon Claude-Henri de Rouvroy, Compte de Saint-Simon

Saint Vitus' dance chorea; involuntary muscular twitching

Saki Hector Hugh Munro

Salad Bowl of California Salinas

Saladin Alvis armored car built in Britain and armed with a 76mm gun

Salish Soviet surface-to-surface missile

Sallie Sarah

Sally Sara(h); South Atlantic (baseball) League (nickname)

Sally Ann Salvation Army

Sally Field Sally Mahoney

Sally Rand Helen Gould Beck

Salmon City Astoria, Oregon

Salonika Thessalonika, Greece

Salop(ian) Shrewsbury; Shropshire

Salt City Syracuse, New York

salt horse pickled meat served to sailors and soldiers

Salt Lake State Utah

Salton Sea formerly Salton Sink

saltpeter potassium nitrate

salt of tartar potassium carbonate

Salvatoriello Salvator Rosa

Saly Salvation Army

Salzburg Philosopher Balduin V Schwarz

Samar-behish (Persian—Fruit from Heaven)—the mango

Samaritan Convair 48-passenger military transport adapted from 240/440 series airliners

Samian Sage Pythagoras, Greek mathematician-philosopher

Samiel The Devil

Sammy American soldier (British slang); Samuel

Samoan Ports Apia (Samoa or Western Samoa), Pago Pago (American Samoa)

Sam Slick Thomas Chandler Haliburton's nickname

Samuel Edwards Noel Bertram Gerson

Samuel Falkland Heijermans Herman

Sam(uel) Goldwyn Samuel Goldfish

San Anto San Antonio, Texas

San Antone San Antonio, Texas

San Berdoo San Bernardino, California

San Carlos de Bariloche Bariloche, Argentina

Sanctimonious City Toronto, Ontario's nickname years ago

sanctuary city any community hospitable to refugees

Sandcutters Arizonans

Sanders Alexander

Sand Gropers Western Australians

Sandhill State Arizona

Sandhillers Georgians; Illinoisans; South Carolinians

Sandhurst Royal Military Academy at Sandhurst on the Blackwater River in southeast Berkshire, England

Sand-lapper State South Carolina

Sand-lappers South Carolinians

Sandra Alessandra

Sandra (Russian—Aleksandra or Alessandra or Alexandra)

Sandra Dee Alexandra Zuck

Sandro Alessandro

Sandstone Federal Correctional Institution at Sandstone, Minnesota

sandwich generation adults responsible for the support and care of both their children and their parents

Sandy Hook peninsula enclosing Sandy Hook Bay and New York Lower Bay

Sandwich Islands Hawaii; discovered by Capt. James Cook in 1778 and named for his sponsor John Montagu, the fourth Earl of Sandwich

Sandy San Diego, California; Sandra; Sandro; Saundra; a Scotsman

Sandy Kitty Kansas City

sangre y pus (Spanish—blood and pus)—separatist nickname for the vivid-red and golden-yellow flag of Spain

San Fran San Francisco

San Juan del Sur Greytown, Nicaragua

San Pedro Sucio San Pedro Sula, Honduras

San Quentin California State Prison at San Quentin

sansei (Japanese—third generation)—grandchild of Japanese immigrants to the United States; (*see ivvei, kibei, nisei*)

Sansovino Andrea Contucci

Santa Barbara Islands California's Channel Islands including Santa Catalina

Santa Claus originally Sint Nicolaas in Holland

Santa Fleet Grace Line ships of their fleet named *Santa* as in *Santa Barbara, Santa Clara, Santa Teresa*

Santa ships Grace Line vessels—all names begin with Santa

Santé nickname of the Parisian prison at 42 rue de la Santé

Santo Domingo City Ciudad Trujillo

San Ysidro, California formerly Tia Juana; name changed to end confusion with Tijuana, Mexico and Tia Juana River

São Tomé and Principe's Port São Tomé

Sap Moon full moon in March

Sapper pseudonym of Lt Col Cyril McNeile—creator of Bulldog Drummond

sapphire blue corundum gemstone

Sapwood Soviet SS-6 intercontinental ballistic missile

Saracen British Alvis armored personnel carrier

Sarah Bernhardt Rosine Bernard's stage name

Saratoga Saratoga chip (potato chip); Saratoga trunk (old-fashioned round-top trunk); Saratoga water (often laxative); Saratoga vichy (imbibed in mixed drinks or as a health potion); formerly Schuylerville, New York

Sardine Capital of Norway Stavanger

Sardine Capital of the United States Eastport, Maine

Sardonic Cartoonist Al Capp

Sarong Girl Dorothy Lamour

Sasha (Russian—Alexander or Aleksandr)

Sasin Soviet SS-8 intercontinental ballistic missile

Satanic City Devils Lake, North Dakota

Satchel Leroy Paige

Satchmo Satchel-Mouth—Louis Armstrong

Satirist of the Mexican Revolution José Clemente Orozco

Satirist-Skeptic Writer Anatole France

Saturday Night Massacre United States President Richard Nixon's order to fire Watergate special prosecutor Archibald Cox, who was investigating Nixon in the Watergate scandal; Nixon's attorney general, Elliot Richardson, resigned rather than obey the order

Saudi Arabian Ports Jiddah or Juddah, the Red Sea approach to Mecca; Ad Dammam, Ras at Tannurah, Ras at Mishab, and Ras at Khafji on the Persian Gulf

Sauk Centre Minnesota town described as Gopher Prairie by Sinclair Lewis in *Main Street*

Savage Soviet SS-13 three-stage intercontinental ballistic missile

Savannahians natives of Savannah, Georgia

Savior of Babies Nathan Straus

Savior of Church Music Giovanni Palestrina (c.1525–1594)

Savior of England Oliver Cromwell

Savior of the Nations sobriquet earned by the Duke of Wellington at Waterloo

Savior of the Sierras John Muir

Savior of Southern Agriculture George Washington Carver

Savoyards performers in the Savoy Operas of WS Gilbert and Arthur Sullivan

Sawbuck Sears-Roebuck; a ten-dollar bill

Sawdust City Oshkosh, Wisconsin

Sawney a Scotsman

Saxe Holm Helen Hunt Jackson

Saxon Shore English coastline including Norfolk, Suffolk, Essex, Kent, Sussex, and Hampshire

Sax Rohmer Arthur Sarsfield Wade's pseudonym

Saybolt viscosity number

Say Hey Kid Willie Mays

Say No Otto Moore (basketball player)

Scandia southern Scandinavian peninsula—southern Norway and Sweden

Scandinavia Denmark, Iceland, Norway, and Sweden (the Faeroe Islands, Finland, and Greenland are sometimes included)

Scandinavian Fun Capital Copenhagen, Denmark

Scandinavian Shield northernmost Finland, Norway, and Sweden

Scapegoat Soviet medium-range two-stage intercontinental ballistic missile SS-14

Scarface Mafia mobster Al Capone

scatterables land mines scattered by aircraft

Scenic Center of the South Chattanooga, Tennessee

Sceptered Isle England; Great Britain

Schlickstadt (German—Mud Town)—German naval nickname for Wilhelmshaven

Schnozzola Jimmy Durante

schoolboy nickname for codeine

School of Europe fifteenth-century Italian states such as Florence, Mantua, Milan, and Venice

Schoolmaster in Politics Woodrow Wilson—twenty-eighth President of the United States

Schoolmaster of the Republic Noah Webster

Scillonian(s) inhabitant(s) of the Isles of Scilly

Scoop Washington state Senator Henry Martin (Scoop) Jackson

Scorpion British Alvis tracked reconnaissance vehicle; NATO armored tank running on five roadwheels and mounting an octagonal turret gun; U.S. self-propelled 90mm antitank gun designated M-56

Scotland Yard New Scotland Yard (London Metropolitan Police)

Scotch Scotch blackface (sheep); Scotch broth (barley, mutton, and vegetable soup); Scotch mist (drizzle, fog, and mist); Scotch whisky (distilled from barley); Scotch woodcock (toast garnished with anchovy paste and scrambled eggs)

Scotch Bard Robert Burns

Scotland's Extremitude Dunnet Head the northernmost point of mainland Scotland

Scotland's Principal Port Glasgow

Scotland Yard of Wildlife Crime Forensics Laboratory of the US Fish and Wildlife Service

Scots Ports Leith, Granton, Rosyth Dock Yard, Boness, Grangemouth, Alloa, Burnt island, Kirkaldy, Methil, Dundee, Perth, Arbroath, Montrose, Aberdeen, Peterhead, Fraserburgh, Hopeman, Inverness, Cromarty, Invergordon, Portmahomack, Helmsdale, Wick, Thurso, Scrabster, Stornoway, Oban, Campbeltown, Greenock, Finnart, Rothesay Dock, Glasgow, Ardrossan, Irvine, Troon, Cairnryan

Scott Fredericks Carl Shapiro

Scottish Agnostic David Hume (1711–1776)

Scottish Cradle of U.S. Presidents Scotland, ancestral home of Presidents Hayes and Monroe

Scottish Skeptic David Hume (1711–1776)

Scouce(s) Liverpool (persons)

Scourge of God Attila

Scourge of Princes Pietro Aretino

Scout Westland army helicopter built in Britain

Scoville Test measure of chilies' hotness by how much sugar water is required to neutralize heat; invented by Wilbur Scoville

Scrag Soviet SS-10-intercontinental ballistic missile

Scrap Iron baseball catcher Clint Courtney

Scrapper Blackwell Francis Hillman Blackwell

Scrapple City Allentown, Pennsylvania

Scrappy Bill Bill Joyce (baseball players)

scream machine amusement park roller coaster

The Screamer General H Norman Schwarzkopf (for his communication style during the Gulf War)

Screamin' Jay Hawkins Jalacy J Hawkins

script kiddies computer hackers who use hacking scripts downloaded from the Internet

Scud Soviet mobile tactical surface-to-surface missile

Sculptor of the Colossal Frédéric Auguste Bartholdi (*Liberty Enlightening the World*)

Sculptor of Great American and French Scientists and Statesmen Jean Antoine Houdon

Sculptor-King Sculptor-King Pygmalion of Cyprus

The Sea the Baltic, Bering, Black, Caribbean, Japan, Mediterranean, North, Philippine, South China

Seabees Construction Battalion (USN)

Sea-born City Venice

Seacat Short and Harland short-range surface-to-air missile used by naval vessels

Sea of Cortés Gulf of California also called the Vermillion Sea (*El Mar Bermejo*)

Sea of Cortez Gulf of California (*Mar de Cortés*)

Sea of Darkness Atlantic Ocean between Cape Verde Islands and west coast of Africa; area often afflicted by dusty Harmattan blowing from the Sahara seaward

Sea Devil Count Felix von Luckner

Sea Dogs originally the nickname of British pirates and privateers but more recently applied to British seamen

Seafarer William Clark Russell's pseudonym

Seafood Center Biloxi, Mississippi

Sea-girt Isle Great Britain

Sea-girt Province Nova Scotia

Sea-green Incorruptible Carlyle's nickname for Robespierre

Seagull Utah's state bird and symbolic nickname sometimes given its citizens—Seagulls; Yugoslav two-place single-engine jet aircraft called Galeb

Seahawk Armstrong-Whitworth carrier-based fighter-bomber aircraft

Sea Islands of Georgia Jekyll, Saint Simmons, Sea Island (all near Brunswick)

Sea Islands of South Carolina Edisto, Folly, Hilton Head, Hunting, Ladies, Murphy, Parris, Port Royal, Saint Helena, Wadamalaw

Sea Killer British short-range surface-to-surface missile

Sea King Sikorsky transport helicopter

Sea Knight Boeing-Vertol helicopter designated CH-46

Sea of Lot Dead Sea

Seamen's Bible Nathaniel Bowditch's *New American Practical Navigator*

Sea of the Plains Dead Sea along the Jordan River Plain of Israel

Seaport City of West Glamorgan Swansea

Seaport on the Prairie Chicago

Sea-Power Philosopher Admiral Alfred Thayer Mahan, USN

Sea Ranger Bell turbine-powered helicopter also known as Jet Ranger

Sea of Reeds the Red Sea

Seashell Capital Sanibel Island, Florida

Seaside State New Jersey

Sea Stallion Sikorsky heavy-assault helicopter designated CH-53

Sea of Stars sparking headwaters of the Huango or Yellow River of China rising in Tibet

sea story teller (*see* Story Teller of the Sea)

Sea of Straw Tagus River estuauy

Sea Venom DeHavilland carrier-based fighter-jet aircraft

Sea Vixen DeHavilland carrier-based jet-fighter aircraft

sea water 96.4% water plus 2.8% sodium chloride (common salt) and smaller quantities of magnesium chloride, magnesium sulfate, calcium sulfate, and potassium chloride; in inland seas such as the Dead Sea and the Salton Sea these percentages vary

Sebastian Melmoth name assumed by Oscar Wilde after he was released from Reading Gaol and lived in Paris until his death three years later

seccy seconal (secobarbital sedative, also nicknamed seggy)

Secession City Charleston, South Carolina

Second City Chicago

Second Estate The Nobility

second-generation money checks; cheques

Second Reich German Republic (1919–1933)

Second Republic France under the presidency of Louis Napoleon from 1848 to 1852

Second State Pennsylvania (*see* First State)

Second World highly industrialized nations of the West such as Belgium, France, Germany, Italy, the Netherlands, the United Kingdom

Section 8 military discharge on physical or psychological grounds; government-subsidized housing; mental case (military code)

Securité France's security service in Paris, also serves the National Central Bureau of Interpol

See-berg jukebox

Seekers truth-seeking Quakers

Seeley Regester Metta Victoria Fuller Victor

Seldom Ever Caught Running nickname of the Southeastern and Chatham Railway— SE & CR

Selebes (Dutch—Celebes)— Sulawesi

Selma Shulamith

Selma Lagerlofland Sweden's province of Värmland where the Nobel prize-winning author was born

Seminole Beech U-8 light transport aircraft

Senator Sam former U.S. Senator Sam Ervin, Jr, of North Carolina

Senegal's Ports St Louis, Dakar, Rufisque; Karabane

Senior Service the British Navy

Señor Wences Wenceslao Moreno

Sentimental Rebel Clarence Darrow

Sentinel of the Bolshevik Counter-Revolution Cheka

separated brethren Christians of opposing religious beliefs

Separationist State Rhode Island

Sepia City New York City's Harlem

Serengeti Serengeti National Park in north-central Tanzania in East Africa

Serengeti of North America Yellowstone National Park

Serenissima Venice, Italy

Sergeant Sperry MGM-29A surface-to-surface missile

serpentine hydrous magnesium silicate

Serpentine Suicide Harriet Shelley—first wife of the poet—drowned herself in the Serpentine of London's Hyde Park

Servant of the Nation Secretary of the Treasury Albert Gallatin who financed the Louisiana Purchase and found funds for the War of 1812

Servetus Michael Servetus born of Spanish parents and baptized Miguel Serveto— burned by order of Calvin in 1553

Seryozha (Russian nickname— Sergei)—Serge

Set Svenholm Karl Viktor Svanholm

Seven (telephone) Sisters Ameritech, Bell Atlantic, Bell South, NYNEX, Pacific Telesis, Southwestern Bell, US West

Seven Deadly Sins Anger, Covetousness, Envy, Gluttony, Lust, Pride, Sloth

Seven-Hill Cities Lisbon, Prague, Rome, and Valparaiso

Seven Hills of Rome Aventine, Caelian, Capitoline, Esquiline, Palatine, Quirinal or Colline, Viminal

Seven Provinces (*see* United Provinces)

Seven Sages of Greece Bias, Chilon, Cleobulus, Periander, Pittacus, Solon, Thales

Seven Seas Antarctic, Arctic, Indian, North Atlantic, South Atlantic, North Pacific, South Pacific oceans; term also applied to the Andaman, Baltic, Bering, Caribbean, Mediterranean, South China, and Yellow seas

Seven Sisters Barnard, Bryn Mawr, Mount Holyoke, Radcliffe, Smith, Vassar, and Wellesley—all colleges for women when first organized; BP (British Petroleum), Exxon (Esso-Standard Oil), Gulf, Mobil, Shell, SOCAL (Standard Oil of California— Chevron), Texaco-world's leading oil companies

Seventeenth State Ohio

Seventh Continent Antarctica

Seventh State Maryland (*see* First State)

Seven Wonders of the Ancient World Pyramids of Egypt, Lighthouse of Pharos of Alexandria, Hanging Gardens and Walls of Babylon, Temple of Artemis or Diana at Ephesus, Statue of Zeus by Phidias at Olympia. Mausoleum at Halicarnassus, Colossus of Rhodes

Sewanee University of the South in Sewanee, Tennessee

Seward's Folly nickname given Alaska in 1867 when Secretary of State William H Seward purchased the area from Russia for $7,200,000; also called Seward's Polar Bear Garden

Seward's Icebox Alaska (*see* Seward's Folly)

Sex Francisco nickname of San Francisco

Sex Isle Mykonos in the Greek Islands close to Piraeus, the port of Athens

Sex Queen of Stage and Screen Mae West

Sextan a Soviet class of trawlers

Sexyola Sixaola, Costa Rica

Sexy Rexy Rex Harrison

Seybrew Seychelles Islands brew(ery)

Seychelles Port Victoria

Shackleton Hawker-Siddeley maritime reconnaissance aircraft

Shaddock Soviet surface-to-surface missile

Shafir Israeli air-to-air missile resembling the U.S. Sidewinder

Shahada Flag Saudi Arabian green standard bearing white letterin.g—the Moslem shahada: "There is no god but God—and Mohammed is his prophet."

Shakopee Minnesota Correctional Institution for Women at Shakopee

Shalimar Garden of Love on Dal Lake in Kashmir

Shalom Aleichem Solomon Rabinowitz, pseudonym; based on the Hebrew greeting—shalom *alekhem*—peace be with you

Shami Shamrock; Shulamith

Shamo (Chinese—Sandy Waste)—the Gobi Desert

Shanghai mainland-China-built torpedo boat; principal port of the People's Republic of China

Shank End Cape Peninsula below Cape Town, South Africa

Shaq Shaquille O'Neal

Shark Island Garden Key, Dry Tortugas (Fort Jefferson National Monument reached by boat from Key West)

sharpshooter casket pine coffin wide at the shoulders narrowing to the feet

Shaston Shaftesbury, England

Sheep Islands Faeroe Islands

Sheila Cecilia

Shellfare term for Royal Dutch/Shell Group payments to Brunei government

shellouts cranberry beans

shell shock battle fatigue (in World War I); variety of psychotic and neurotic disorders associated with the stress of combat

Shelly Winters Shirley Schrift

Sheltie Shetland sheepdog

Shelty Shetland pony

Shepherd of the Ocean Sir Walter Raleigh

Sheridan M-551 assault vehicle armed with a 152mm gun; Sheridan House

Sheridan Girls Wyoming (delinquent) Girls School at Sheridan

Sherlock police computer recording and releasing essential information about many criminal activities; nickname of any good detective; investigator created by novelist Sir Arthur Conan Doyle— Sherlock Holmes; Sherlockian(s)

Sherman M-4 tank armed with high-velocity 76mm gun

Shershen Soviet class of motor torpedo boats (PT boats)

Shetland Shetland pony (small but stocky long-hair pony); Shetland sheepdog (miniature collie); Shetland wool

Shevvie(s) native(s) of Sheffield, Yorkshire

Shillelagh anti-tank surface-to-surface guided missile produced by Aeronutronic

Shining Star of the Caribbean Puerto Rico

Ship of the Desert the camel

Shirley MacLaine Shirley Maclean Beaty

Shitport Norfolk. Virginia

Shoe City Auburn, Maine; Hanover, Pennsylvania; Lynn, Massachusetts

Shoeless Joe Joe Jackson (baseball player)

Shooting Star Lockheed T-33 jet-fighter trainer aircraft

shorlans armored cars

shotgun house house with rooms in a straight line

Shotgun Billy Gardner (baseball player)

Show-Me State Missouri

Showplace of the Orient Singapore

Shrike Texas Instrument air-to-surface antiradar missile

shroudies Shroud of Turin aficionados

Siam former name of Thailand

Siamese Siamese cat (fawn or pale-gray breed of short-hair cat); Siamese fighting fish (*Betta*); Siamese twins (congenitally connected twins)

Siberian Express nickname for a devastatingly cold polar wind afflicting Canada and much of the United States

Siberian salt mines nickname for forced-labor camps and prisons

Sick Man of Africa nineteenthcentury Ethiopia

Sick Man of the Americas fever-infested nineteenth-century Panama

Sick Man of Asia nineteenth century China

Sick Man of Europe Turkey in the last years of the Ottoman Empire and the reign of the sultans during most of the nineteenth century and up to 1922 when the sultanate was abolished

Sick Man of South America nineteenth-century Ecuador

Siddhartha Gautama Buddha

Sidewinder air-to-air missile produced by Motorola, Philco, and Raytheon

Sierra Leone's Ports Freetown, Pepel, Bonthe

Sierra Madre Sierra Madre Desert°of Mexico

Siete Leguas (Spanish—Seven Leagues)—famous warhorse of Pancho Villa, the Mexican bandit general

Signe Hasso Signe Larsson

Silence Dogwood Benjamin Franklin's pseudonym used by him at age 15 when he wrote articles for the *New England Courant*

Silent The Silent (William I—Prince of Orange)

Silent Cal taciturn President Calvin Coolidge

silent killer high-blood pressure

silent service submarine service

Silicon Alley South Broadway, home of New York City's Internet businesses

Silicon Beach San Diego, California

Silicon Glen Scotland's central lowland towns specializing in microelectronics

Silicon Prairie Dallas, Texas high-technology area

Silicon Valley high-technology-oriented area around San Jose, California in Santa Clara County; nickname for Santa Clara County, California

Silk City Paterson, New Jersey, and Soochow, China

Silk Country China

Silly Billy nickname of William IV

Silver Age era when adornments and implements were made of silver; period between the Golden Age and the Bronze Age

Silver City Broken Hills, New South Wales, Australia; Taxco, México

Silver City by the Sea Aberdeen, Scotland

Silver Fox Duke Snider (baseball player)

Silver Gate entrance to San Diego Bay, California

silver industry jobs and businesses centered around people age 65 and older

Silverines Coloradans

Silver King tarpon

Silver Knight Senator Burton K Wheeler

Silver Republic Argentina

Silversmith Patriot Paul Revere (also bellfounder and dentist)

Silver State official nickname of Nevada but also applied to silver-rich Colorado

Silver State of Malaysia Perak

Silver Streak the English Channel

Silver-Tongued Orator William Jennings Bryant

Simmond's disease premature senility caused by atrophy of the pituitary

Simone Pauline Benda's pseudonym

Simone Signoret Simone Kaminker

Simpson Simpson Desert of central Australia

Sinatch nickname for Frank Sinatra

Sin Angeles (Spanish—without angels)—nickname given Los Angeles

Sin Capital Singapore's nickname

Sin City Las Vegas, Nevada

Sin City of the West Geneva, Switzerland

Singapore's Ports Serangoon, Singapore, Pulau Bukum, Pulau Sebarok

Singer of Singers Enrico Caruso

Singing Cowboy Gene Autry

Singing Nun Sister Luc-Gabrielle (Jeanine Deckers)

Singing Preacher Booker T Washington White

Singing Satellite Red China's first satellite, launched in 1970, broadcast rhymed song about Communist Party chairman Mao Tse-tung

Singing Tree casuarina (sometimes called sighing tree or sobbing tree)

Sing Sing nickname of the New York State Penitentiary at Ossining formerly named Sing Sing

Sink of New England Rhode Island; uncomplimentary nickname given by Massachusetts Puritans appalled by the separation of church and state guaranteed in Rhode Island's royal charter

Sin Sin Singapore

Sin Strip San Francisco's North Beach area

sin taxes taxes on alcohol, gasoline, and tobacco

Sinyavsky Abram Tertz

Sioux Bell Model-47 helicopter built in Britain, Italy, Japan, and the United States

Sioux State North Dakota

Sippie Wallace Beulah (Thomas) Wallace

Sir Adrian Sir Adrian Boult (conductor)

Sir Alec Sir Alec Guinness (actor)

Sir Alexander Sir Alexander Korda (motion-picture producer)

Sir Alfred Sir Alfred Hitchcock (author and film director)

Sir Arnold Sir Arnold Bax (composer)

Sir Arthur Sir Arthur Bliss (composer-conductor); Sir Arthur Conan Doyle (writer and physician); Sir Arthur S(eymour) Sullivan (composer-conductor-organist)

Sir Aurel Sir Aurel Stein (archeological explorer)

Sir Benjamin Sir Benjamin Britten

Sir Bernard Sir Bernard Haitink (Dutch conductor)

Sir Charles Sir Charles Chaplin (better known as Charlie Chaplin); Sir Charles Groves (conductor); Sir Charles Hallé (conductor-pianist and founder of Manchester's Hallé Orchestra); Sir Charles Mackerras (conductor); Sir Charles Villiers Stanford (composer-conductor-organist); Sir Charles Wheatstone (physicist)

Sir Charles Morell James Ridley's pseudonym

Sir Clifford Sir Clifford Curzon (pianist)

Sir Colin Sir Colin Davis

Sir Dan Supreme Sir Dan Godfrey

Sir David Admiral Sir David Beatty—First Earl of the North Sea

Sir Edward composer-conductors Sir Edward Elgar and Sir Edward German (originally Edward German Jones)

Sirens three nymphs named Leucosia, Ligeia, and Parthenope; their seductive singing lured sailors to their death on rockbound coasts, when they failed to lure Odysseus

(Ulysses) they flung themselves into the waves and perished

Sir Ernest Sir Ernest Campbell Macmillan (composer-conductor-educator-organist)

Sir Francis Sir Francis Bacon (philosopher politician); Sir Francis Drake (admiral-explorer-navigator); Sir Francis Palgrave (historian)

Sir Frank Sir Frank Athelstane Swettenham, lexicographer of Malaya and its one-time colonial administrator

Sir Freddie Sir Freddie Laker (airline organizer and operator)

Sir Frederic Sir Frederic Cowen (composer-conductor)

Sir Georg Sir Georg Solti (conductor-pianist knighted for his services at Covent Garden where he conducted from 1961 to 1971)

Sir George Sir (Isador) George Henschel (baritone-composer-conductor, founder of the Scottish Symphony Orchestra)

Sir Granville Sir Granville Bantock (composer-conductor-teacher)

Sir Guatteral (Hobson-Jobson—Sir Walter Raleigh)—as known to many Spaniards in colonial times

Sir Hamilton Sir Hamilton Harty (conductor)

Sir Henry Sir Henry Bessemer (engineer-inventor-metallurgist remembered for bessemer steel); Sir Henry Wood (conductor)

Sir Hubert Sir Hubert Parry (composer and musicologist)

Sir Isaac Sir Isaac Newton (mathematician and natural philosopher)

Sir John Sir John Barbirolli (conductor-violincellist); Sir John Betjeman (poet laureate); Sir John Gielgud (actor)

Sir John A Sir John Alexander Macdonald (Canada's first and third Prime Minister)

Sir John Mandeville Jehan de Bourgogne

Sir John Retcliffe pseudonym of author Hermann Goedsche

Sir Kenneth Sir Kenneth McKenzie Clark

Sir Landon Sir Landon Ronald (composer-conductor-pianist)

Sir Laurence Sir Laurence Olivier (actor-director-producer)

Sir Lennox Sir Lennox Berkeley (composer)

Sir Malcolm Sir Malcolm Sargent (ballet-choral-orchestral conductor who was chief conductor of the BBC and the Promenade Concerts)

Sir Max Sir Max Beloff

Sir Michael Sir Michael Costa (composer-conductor); Sir Michael Tippett (composer-conductor-educator)

Sir Pelham P(elham) G(renville) Wodehouse

Sir Ralph Sir Ralph Richardson (actor)

Sir Thomas Sir Thomas Beecham (composer-conductor-founder of the London Philharmonic Orchestra and the Royal Philharmonic Orchestra)

Sir Victor Sir V S (Victor Sawdon) Pritchett (short-story writer, novelist, critic and travel writer)

Sir Wilfred Sir Wilfred Pelletier (conductor)

Sir William Sir William Herschel (astronomer-mathematician-musician); Sir William Walton (composer)

Sir Winston Sir Winston Leonard Spencer Churchill (former Prime Minister of Great Britain)

Sissy Cecilia; sister

Sitting Bull Tatanka Iyotanka also known as Sitting Buffalo Bill

Six Counties Northern Ireland or Ulster's counties of Antrim, Armagh, Derry, Down, Fermanagh, and Tyrone

Six-Day War June 5–10,1967, war between Israel and Arab forces

Six Nations Five Nations plus the Tuscaroras (*see* Five Nations)

Six-Shooter Junction old name of Harlingen, Texas

Sixteenth State Tennessee

Sixth Continent Australia

Sixth State Massachusetts (*see* First State)

Skate City Northbrook, Illinois

Skean Soviet SS-5 intermediate-range ballistic missile

Skeptic-Philosopher President Thomas Jefferson

Ski Capital Aspen, Colorado

Ski City Oslo, Norway's nickname

Ski Country Colorado

Skid Row by the Sea Santa Monica, California

skinheads often neo-Nazis who promote interracial violence

Skinny Joey Joseph Merlino

Skip James Nehemiah James

Skipper the Captain; the Commander

Skowse Liverpool seaman

skunk nickname of a potent variety of marijuana

Skunk's Misery former name of Scranton, Pennsylvania

Skunk Works nickname of Air Force Plant 42 at Palmdale, California

Skybright Axe Paul Bunyan

Sky City Pueblo Acoma near Alburquerque, New Mexico

Sky Crane Sikorsky crane helicopter designated CH-54 or S-64

Skymaster Douglas DC-4 44-passenger transport also called Dakota

Skyscraper Capital New York City

Skyscraper Port of the Orient Hong Kong

Skyservant Dornier utility aircraft also designated DO-27 and DO-28; both built in Germany

Skytrain Douglas DC-3 21-passenger transport also called Dakota

Skyvan turboprop transport built by Short in Great Britain

Skywagon Cessna 185E utility aircraft

Slabsides rustic cabin built by John Burroughs near Esopus, New York

slaked lime calcium hydroxide $(Ca[OH]_2)$

slanguage slang language

Slava Mstislav; Mstislav Rostropovich's nickname

Slave Capital of the South Charleston, South Carolina before the Civil War

Slave Coast West African coastal area of Togo, Dahomey, and Nigeria

Slave States former slave-holding states (Virginia, North and South Carolina, Georgia, Florida, Alabama, Mississippi, Louisiana, Texas, Arkansas, Tennessee, Delaware, Maryland, Kentucky, Missouri

Sledge and Hoe official symbol of Zaire

The Sleeping Prophet Edgar Cayce

Sleep Personified Hypnos (Greek—sleep) whose brother was Thanatos or death

Sleepy Hollow New Jersey's Trenton Prison

Sleepy Joe Attorney General Philander C Knox

Sleepy John Estes John Adams Estes

Slick Willie Bill Clinton, forty-second president of the United States

Slim Harpo James Moore

Slim Jannie Jan Christian Smuts

Slim Pickens Louis Lindley

Slinging Sammy Sam(uel) (Adrian) Baugh of baseball and football fame

slop-jar used at night instead of bathroom or outhouse

Slugger U.S.-made tank destroyer designated M-36

Slumbering Giant of Capitol Hill The Library of Congress

Slumberjay Schlumberger

Slut of the North Empress Elizabeth of Russia so nicknamed by Frederick the Great of Prussia who called her *la Catin du Nord*

Sly Fox of Kinderhook Martin Van Buren

smack nickname for heroin

Smack Henderson Fletcher Henderson

Small Islands (*see* Micronesia)

Smelly Place Hong Kong (also called Fragrant Harbor)

Smelter City Anaconda, Montana

Smiling Jim James A Farley

Smiling Stan Stan Hack

Smithy Ian Smith

Smog Capital of California Los Angeles

Smoke the Smoke (nickname for London)

Smokeless City Reykjavik, Iceland—heated by natural hot springs

Smokeless Coal Capital Beckley, West Virginia

Smoke that Thunders Victoria Falls (Zambia)

smokies smoked haddocks

Smoking Moses Shishaldin Volcano on South Umiak Island off southwestern Alaska

Smoky City Pittsburgh. Pennsylvania before its Renaissance Plan

Smörgåsbordland Sweden (famous for its cold-table fare)

snail mail mail sent through the postal service; non-electronic mail

Snapper Soviet antitank missile

Snapper Capital of the World Pensacola, Florida

Snapshot City Rochester, New York

snowbirds people who winter in temperate areas

Snow City Aspen, Colorado

Snow-covered Continent Antarctica

Snow King Gustavus Adolphus of Sweden

Snow Moon full moon in February and November

Snow Queen Christina— Queen of Sweden

snubbies snub-nosed handguns

Soap Box Derby Center Akron, Ohio

soapstone saponite (hydrous magnesium aluminum silicate)

Soapy G Mennen Williams

Soapy Sam Samuel Wilberforce, Bishop of Winchester, who debated Darwin's theory of evolution with Professor Thomas Henry Huxley, president of the Royal Society

Socialist Pope Daniel De Leon

Society Capital Newport, Rhode Island

Society of Friends the Quakers

Soccer War El Salvador's invasion of Honduras, 1969

soda ash sodium carbonate (Na_2CO_3)

Sodaks South Dakotans

Sodoma (Italian—Sodomite)— Giovanni Antonio Bazzi (1477–1549)

Sofia Loren Sofia Scicoloni

Sofu-gan Japanese equivalent of Lot's Wife—volcanic islet resembling a pillar of salt in the North Pacific between Iwo Jima and Yokohama

Soho London's avant-garde district in the West End just north of Trafalgar Square

Sojourner Truth Isabella Baumfree

Solar Energy Capital Los Angeles

Solar Energy State Arizona

Sol de Mayo (Spanish—Sun of May)—symbol of independence appearing on the flags and seals of Argentina and Uruguay

solder 50% lead, 50% tin (common solder)

soldier's heart Da Costa's syndrome

Soledad Correctional Training Facility of the State of California at Soledad

Solid City St Louis, Missouri

Solid South Southern United States usually voting as a solid bloc: Alabama, Florida, Georgia, Louisiana, Mississippi, South Carolina

Solina South Carolina

Solomon British pianist

Solomon Islands Port Honiara on Guadalcanal Island

Solomon seal six-pointed star consisting of two interlocking triangles; sometimes called the shield of David

Solon of French Prose Jean Louis Guez de Balzac

Somalian Port Berbera

Somers' Islands Bermuda

Somnolent City of the Sahara Timbuktu

Song of Songs The Song of Solomon

Son House Eddie James House

Son of Sam David Berkowitz

Son of Nature Henry David Thoreau

Son of the Ocean Yangtse River

Son of the Star Bar Kochba—military leader of the Jews who revolted against the Romans in the year 132 A.D.

Son of Valladolid José Zorilla

Sonny Boy Williamson Aleck Miller

Sonny Terry Saunders Terrell

Sonya Sophia

Sonya (Russian nickname—Sophia)

Soo Sault Ste Marie (canal and locks)

Soo Bridge Sault Ste Marie International Bridge

Soo Canals Sault Ste Marie Canals

Soo Line Minneapolis, St Paul & Sault Ste Marie (railroad)

Sooner State Oklahoma

Sopac Southern Pacific Railroad (stock exchange nickname)

Sophia Loren Sofia Scicolone

Sophie Tucker Sophia Abuza

Sorbonne University of Paris

Sorebacks Virginians

Sorghum Capital of the World Hawesville, Kentucky

Soul of American Law Clarence Darrow

Soul City Harlem district of New York City

Soul Man Bobby Blue Bland

Soupy Sales Milton Hines

Southern Gateway of New England Rhode Island

Southern Hemisphere the world south of the equator

Southern Ireland Republic of Ireland

southern lights *aurora australis*

Southern Ocean Antarctic sections of the Atlantic, Indian, and Pacifc oceans

Southern Part of Heaven Chapel Hill, North Carolina

Southern Poet Sidney Lanier (*The Marshes of Glynn, The Song of the Chattahoochee, Sunrise*)

Southern Rhodesia Zimbabwe

Southern States Virginia, North and South Carolina, Georgia, Florida, Alabama, Mississippi, Tennessee, Arkansas, Louisiana, and Texas

The Southern Valley of Medicinal Plants Bhutan (its original name)

South Jersey Coast Atlantic City to Cape May

southpaw left-handed baseball pitcher

South Pacific between Australia and South America containing South Sea Islands; South Pacific Ocean

South Pole 90 degrees South latitude; zero degrees longitude; southernmost point on the earth; discovered by Norwegian explorer Roald Amundsen in 1911

South River old name for the Delaware River

South Seas South Pacific Ocean

South Seymour Island, Galápagos Baltra

Southwest southern California and Nevada, Arizona, New Mexico, and western Texas

South-West Africa formerly German South-West Africa; Namibia

Southwest Border between northern Mexico (Tijuana to Matamoros) to southern United States (California, Arizona, New Mexico, Texas)

Southwest Sun Belt Arizona, California, Hawaii, Nevada, New Mexico, Ohlahoma, and Texas

Soybean King Dwayne Orville Andreas

Space Capital of the United States Colorado Springs

Space City Houston, Texas (NASA headquarters)

Space Coast Florida's Cape Canaveral area

Space Needle City Seattle, Washington

Space Scientist Premier Kraft A Ehricke (1917–1984)

Spain's Forgotten Forest Extremadura

Spain's Largest Port Barcelona

spam bulk unsolicited electronic mail

Spaniola(s) British slang for Spaniards(s)

Spanish Spanish bayonet (*Yucca*); Spanish cedar (fragrant neotropical wood); Spanish dagger (*Yucca glori-*

osa); Spanish fly (cantharides); Spanish grippe (influenza); Spanish heel (woman's high heel); Spanish influenza (highly infectious respiratory viral disease); Spanish lime (genip); Spanish mackeral (Jack mackeral); Spanish Moss (epiphytic plant growing in long festoons on the branches of live oak trees in the southern United States); Spanish omelet (made with green peppers, tomatoes, and seasoning); Spanish rice (made with cayenne pepper, chopped onions, and tomatoes); Spanish topaz (citrine); Spanish trefoil (alfalfa)

Spanish Africa cities of Ceuta and Melilla; term formerly included Spanish Guinea, Spanish Morocco, and the Spanish Sahara

Spanish America Spanish-speaking countries of Latin America

Spanish Film Pioneer Luis Buñuel

Spanish Founder of the Secular School System Francisco Ferrer

Spanish Guinea former West African colony on the Gulf of Guinea; included Fernando Po, Rio Muni, and offshore islets

Spanish Honduras Republic of Honduras in Central America

Spanish Main Spanish-speaking Latin America and northern South America bordering the Caribbean from Mexico to Venezuela, including Guatemala, Honduras, Nicaragua, Costa Rica, Panama, and Colombia

Spanish Monastic Painter Francisco de Zurbarán

Spanish Morocco formerly all of coastal and northwestern Morocco but now only Alhucemas, Ceuta, the Chafarinas islands, Melilla, and Peñon de Vélez

Spanish National Composer Manuel de Falla

Spanish Naturalist Painter Diego Rodriguez de Silva y Velásquez

Spanish Netherlands all the Lowland Countries (Belgium, Luxembourg, and the Netherlands) when they were under Spanish rule

Spanish Ports Pasajes, San Sebastian, Zumaya, Santurce, Portugalete—Bilbao, Las Arenas—Bilbao, El Desierto—Bilbao, Castro Urdiales, Santander, Gijón, Musel, Aviles, San Esteban, El Ferrol del Caudillo, La Coruña, Villegarcia, Pontevedra, Marín, Vigo, Santa Cruz de Tenerife and La Luz Gran Canaria (on the Canary Islands), Huelva, Bonanza, Coria del Río, Sevilla, Rota, Cádiz, Algeciras, Málaga, Motril, Adra, Almería, Cartagena, Alicante, Valencia, Castellon de la Plana, Tarragona, Barcelona, Palamós, (*and on the Balearic Islands*—Ibiza, Palma, Mahon)

Spanish Presidios Ceuta and Melilla on the Alboran coast

Spanish Riviera Spain's Mediterranean resorts

Spanish Sahara former colony on Africa's northwest coast where it included Río de Oro and Saguia el Hamra until 1976 when it was ceded by Spain and divided between Mauritania and Morocco

Spanish-speaking Places Andorra, Argentina, Balearic Islands, Bolivia, Canary Islands, Ceuta and Melilla, Chile, Colombia, Costa Rica, Cuba, Dominican Republic, Ecuador, El Salvador, Equatorial Guinea, Guam, Guatemala, Honduras, Mexico, Morocco, Nicaragua, Panama, Paraguay, Peru, Philippines, Puerto Rico, Spain, Spanish Sahara, United States, Uruguay, Venezuela, etc.

Spanish State New Mexico

Spanish Town Jamaican resort; Tampa, Florida

Spanish West Africa Sidi Ifni

Spanky George McFarland

Sparrow McDonnell-Douglas air-to-air missile

sparrow grass asparagus

Sparse Grey Hackle Alfred Waterbury Miller's pen name

Spartans of the Prairies Comanche Indians

Speaker's Corner northeast corner of London's Hyde Park

Special K Larry Kenon (basketball player); Kermit Washington (basketball player)

Specie-Payment President James Abram Garfield

speed nickname for methamphetamine drugs

Spencer Spiridione

Sphinx of Concord Ralph Waldo Emerson

Spica Swedish-built patrol boat carrying a 57mm gun and six torpedo tubes

Spice Island Grenada (noted for nutmeg, cloves, and mace)

Spice Islands Indonesia's Moluccas; West Indian islands of Grenada and the Windwards

Spider of Florence Machiavelli

Spike Jones Lindley Armstrong

Spillionaires those who make money from oil spills

Spinach Capital of the World Crystal City, Texas (replete with a statue of Popeye)

Spindle City Lowell, Massachusetts

Spindrift Ernest Toone

Spine of South Africa Drakensberg Mountains

Spirit of Freedom State Massachusetts

spirits of hartshorn ammonia water (NH_4OH)

spirits of salts hydrochloric acid

Spiritual Father of the French Revolution Rousseau

Splendid Splinter Ted Williams (baseball player)

Spoils-System President Andrew Jackson

Spokesman for the Negro Booker T Washington

Spokesman for the Oppressed George Meany

Sponge City Tarpon Springs, Florida

Spoon River Edgar Lee Masters, poetic appellation for Lewistown, Illinois

Spoon River Poet Edgar Lee Masters

Sport of Kings (horseracing—a sport only kings can afford)

Sportsman's Paradise Louisiana

Sports Town, U.S.A. San Diego, California

Spree River City Berlin

spring onions scallions

Spruce Goose nickname of Howard Hughes' Hercules H4 319-foot-wingspan flying boat powered by eight wing-mounted engines

Spud Island Prince Edward Island

Square Deal nickname for economic and political philosophy of Theodore Roosevelt

Square Grouper floating bales of marijuana

Square Mile London's financial center

Square Mile of Vice London's Soho and East End

Squaresville area inhabited mainly by citizens who frown on all types of criminal activity and cooperate with the police

squash box accordion

Squatter State Kansas

Squid Sidney Moncrief (baseball player)

Squire of Hyde Park Franklin D Roosevelt

Squire of Monticello Thomas Jefferson

Squire of Warm Springs Franklin D Roosevelt

Sri Lankan Ports Colombo, Galle, Trincomalee

Sri Lanka's Principal Port Colombo

S S Van Dine Willard Huntington Wright (1888–1939)

Stagecoach Town Fort Worth, Texas

Stagirite Aristotle the Stagirite—so named as he was born in Stagira, Macedonia)

Stagville nickname for West Hollywood, California's Santa Monica Boulevard

Stalin (Russian—steel)—Iosif Vissarionovich Dzhugashvili

Stalingrad former name of Tsaritsyn now called Volgograd

Standard Arm General Dynamics anti-radar missile

Standard Oil King John D(avison) Rockefeller, Sr

Stanislavski Konstantin Sergeevich Alekseev

Stan Laurel Arthur Stanley Jefferson

Stanley Sir Henry Morton Stanley whose original name was John Rowlands

Stanley Ketchell Stanislaus Kiecal (boxer)

Stanleyville former name for Kisangani, Zaire

Stan the Man Stan Musial

Stanton Forbes pseudonym of De Loris Stanton Forbes

Star City Lafayette, Indiana

Star City of the South Roanoke, Virginia

Star and Crescent Moslem symbol appearing on arms and flags of Algeria, Libya, Malaysia, Mauritania, Pakistan, Singapore, Tunisia, Turkey

Star of David Judaic symbol consisting of two superimposed equilateral triangles forming a six-pointed star; device also called the Seal of Solomon or the Shield of David

Star of the East Vladivostok

Starfighter Lockheed single-engine jet-fighter aircraft built in Belgium, Canada, Germany, Italy, and the Netherlands

Star of the Indian Ocean Mauritius

Starlifter C-141 Lockheed cargo and troop transport

Star of the North King Gustavus Adolphus of Sweden; Minnesota

Stars and Bars flag of the Confederate States of America

Star Spangled Banner anthem of the United States of America and nickname of its flag

Starspeak Star Wars language

Stars and Stripes flag of the United States of America

Star Wars Strategic Defense Initiative

State of Excitement Western Australia

States the United States of America

State of Spain New Jersey

State of the Thousand Islands Maldive Islands

Statia Saint Eustatius (Netherlands Antilles)

St Dymphna's disease insanity

Steak Center Kansas City, Missouri

stealth bomber B-2 advanced technology bomber

Steel Baron Andrew Carnegie

Steel Center of the South Birmingham, Alabama

Steel City Bethlehem, Pennsylvania; Pittsburgh, Pennsylvania

Steelmaker Joe Magarac

Steel-Master Philanthropist Andrew Carnegie

Steel State Pennsylvania

Stegners Stegner Fellows at Stanford University (for writer Wallace Stegner)

Stein and Toklas Gertrude Stein and Alice B Toklas

Steiny Charles Proteus Steinmetz

Stella Maris (Latin—Seaman's Star)—(*see* Polaris)

Stendhal pseudonym—Marie-Henri Beyle

Stepin Fetchit Lincoln Perry

step sister of religion superstition

sterling silver 92% silver, 8% copper

Stevie Wonder Stevland Morris

Stewart Granger James Stewart

stick baton's nickname

Sting Gordon Sumner

Stinkstein (German—stinkstone)—coal-black limestone or marble giving off a fetid odor when rubbed because of its bituminous or carbonaceous inclusions; also called anthraconite

Stirville Sing Sing prison (Ossining, New York)

St-Jean-des-Puces (French—St John of the Fleas)—nickname of the Franco-Spanish border town of St-Jean-de-Luz (St John of Light or St John of the Marshes)

St-John Perse Alexis Saint-Leger's pen name

St John's evil epilepsy; old nickname for epilepsy

St Luke's Summer fine weather that occurs around October 18, St Luke's Day

St Martin's evil dipsomania

St Martin's Summer fine weather that occurs around November 11, St Martin's Day

St Mathurin's disease epilepsy

Stoky Leopold Stokowski (conductor-impressario-transcriber—originally named Antoni Stanislaw Boleslawowics—later adopted name of Leo Stokes but since early 1900s appeared as Leopold Stokowski)

Stomach Steinway accordion (attributed to Mark Twain)

Stone Age prehistoric era when implements and weapons were made from stone—age divided into paleolithic, mesolithic, and neolithic periods

Stone of Heaven jade

Stonewall Jackson General Thomas Jonathan Jackson of the Confederate Army

Storm Norwegian high-speed motor gunboat

Stormalong Arthur Bulltop

Storm King American meteorologist James Pollard Espy

Story Teller of the Sea sobriquet shared by Conrad, Cooper, de Hartog, Forester, Innes, London, McFee; Marryat, Masefield, Melville, co-authors CB Nordhoff and JN Hall, and Verne

St Paddy Saint Patrick

St Paddy's Day Saint Patrick's Day (March 17)

Strabolgi Joseph Montague Kenworthy

Strangler wrestler Ed (Strangler) Lewis originally named Robert H Friedrich

Stratford Will William Shakspere

Stratofreighter Boeing 707/720 transports adapted for military service and designated C-135

Stratotanker Boeing jet tanker designated KC-135

Strawberry Bill Bill Bernhard

Strawberry Capital Hammond, Louisiana

Strawberry Moon full moon in June

Stream of Pleasure Thames River above London

The Street Wall Street

Street Haven Toronto, Ontario's center for the rehabilitation of prostitutes

Street of Ink Fleet Street, London with its many newspaper offices

Street of Sorrows New York City's Wall Street; old-fashioned nickname for any thoroughfare frequented by street-walkers

S-town Spanishtown

Strikemaster British BAC 167 ground-attack jet aircraft

Strix Peter Fleming

Stub Toe State Montana

Student of Democracy British Ambassador James Bryce—author of *The American Commonwealth*

Studioland Hollywood, California

Sturgeon Moon full moon in August

St Valentine's disease epilepsy

St Vitus' dance epilepsy

Styx Soviet surface-to-surface naval missile

Subbotnik (Russian—Little Saturday)—Red Saturday—annual holiday when people clean up factory sites and neighborhoods as well as parks and other public places

Subcontinent of Asia India

Sublime Porte nickname for the government of the Turkish Empire in the times of the sultans

Submarine Capital Groton, Connecticut

Subtropical Siberia Castro's Cuba

subversive delinquents political prisoners

Subway City New York

Successor of Saint Peter the Pope

Sucker State Illinois nickname dating from pioneer days when settlers sucked water from underground springs with long hollow tubes called suckers

Suckers Illinoisans

Sudan Sudan Desert of the north-central Libyan and Sahara deserts

Sudanese Port Bur Sudan

Sudanese Sister Cities Khartoum and Omdurman

Suds City Milwaukee, Wisconsin—famous for beer

Sue *Tyrannosaurus rex* skeleton dug from a cliff in South Dakota in 1990

Sugar Bowl New Orleans, Louisiana

Sugar Country Queensland, Australia

Sugar Islands Leeward Islands of the West Indies

Sugar King Claus Spreckels

sugar of lead lead acetate

Sugar State Louisiana

Suicide Capital of the United States San Francisco

Suicide Capital of the World Budapest

Suliman seal five-pointed pentagrammic star; symbol of Morocco and other Islamic lands

Sulphur King Herman Frasch

Sultan of Brunei Sultan Hassan Bolkiah

Sultan of Swat George Herman (Babe) Ruth

Summerless Southland southernmost New Zealand on its South Island south of Dunedin

Summit City Akron, Ohio

Sun Belt southern United States extending from Florida to Hawaii

Sun Bowl El Paso, Texas

Sun City St Petersburg, Florida; Yuma, Arizona

Sun Flag *Hi-no-maru*—sun flag of Japan—Land of the Rising Sun—red sun on a white field

Sunflake City Grand Forks, North Dakota

Sunflower(s) Kansan(s)

Sunflower State Kansas

Sun God Adonis (Syrian); Apollo (Roman); Apollon (Greek); Baal (Chaldean); Helios Hyperion (Greek in Homer's time); Horus (symbolized in Upper Egypt by a hawk); Mithras (Persian); Moloch (Canaanite); Osiris (Egyptian); Ra or Re (symbolized in Egypt's Old Kingdom by an obelisk); Sol Invictus (Latin—Sun Invincible)—Romans shortened this to Sol; Surya (Hindu)

Sunken Continent of the Atlantic Atlantis

Sunken Continent of the Pacific Lemuria

Sun King Louis XIV *(Le Roi Soleil)*

Sun of May *El Sol de Mayo*—revolutionary symbol on the great seals of Argentina, Ecuador, and Uruguay; standing for national emergence in the fight for freedom

sunnie(s) sunfish(es)

Sunny Alberta Canada's Province of Alberta

Sunnyside Washington Irving's home near Tarrytown, New York

Sunny South southern United States

Sunrise Poet Sidney Lanier

Sunset Land Arizona

Sunset State Arizona; Oregon

Sunshine Capital of the United States Yuma, Arizona

Sunshine City Saint Petersburg-Tampa, San Diego, Tucson, Yuma, and Durban, South Africa—City of Sunshine

Sunshine Coast British Columbia from Lund to Vancouver; Queensland's coast from Brisbane to Noosa

Sunshine Continent Australia

Sunshine Province Alberta, Canada

Sunshine State Florida, New Mexico, and South Dakota; Queensland, Australia; official nickname of Florida

sunshine vitamin Vitamin D

super aspirin Cox-2 inhibitors (anti-inflammatory drugs)

Super Constellation Lockheed transport carrying 99 passengers

Superdome City New Orleans, Louisiana

Super Falcon Hughes air-to-air missile also called Nuclear Falcon

Super Frelon Sud antisubmarine helicopter developed in France

Super G super giant slalom

Superman Philosopher Friedrich Wilhelm Nietzsche

Superman of the Prize Ring Joe Louis

Super Mystere Dassault fighter-bomber and jet interceptor built in France

Super Sherman U.S. M-4 Sherman tank modernized

Superstition Personified Abessa who sought sanctuary behind convent walls shielding her from truth, according to Spenser's *Faerie Queene*

Supertenor Luciano Pavarotti

Supporter of the Universe Atlas, in the Roman mythology; the ash tree Ygdrasil in Norse mythology

Supreme Buffoon Alfred Jarry (1873–1907)

Supreme Genius Leonardo da Vinci

Supreme Genius of Spanish Painting Diego Rodríguez de Silva y Velázquez

Supreme God of the Hindus Brahma

Supreme Governor of the Church of England the King or Queen

Supreme Pontiff of the Universal Church the Pope

Sure Shot Fred Dunlap (baseball player)

Surgeon of the Rusty Knife Dr José Pedro de Freitas Arigo of Congonhas do Campo, Brazil

Suriname Ports Nieuw Nickerie, Paramaribo, Paranam, Moengo, Albina

Surveyor American program for lunar surface and subsurface exploration

Survivor syndrome symptoms of insomnia, nightmares, anxiety, et cetera first described in Nazi concentration camp survivors

Susan Hayward Edythe Marrener

Susan Sarandon Susan Tomaling

Susie Susan; Susannah; Suzanne

Sussex Seaport Garden Resort Felixstowe

Suzy Susan; Susan B Anthony dollar; Susanna; Susanne

Svengali Squad Los Angeles Police Department's hypnosis squad

Svensker(s) Swedish sailor(s)

Sverdlov Soviet class of light cruisers

Sverdlovsk formerly Ekaterinburg where the Soviets murdered the last of the czars and his family; terminus of the Trans-Siberian railroad and place where Asia is said to look at Europe

Swamp Fox sobriquet shared by Revolutionary War general Francis Marion as well as by Confederate generals Nathan Bedford Forrest and Philip Dale Roddey

Swamp State South Carolina

Swan of Avon Ben Jonson's name for Shakespeare

Swan City Perth, Western Australia

Swanland southwestern Australia

Swan of Mantua Virgil

Swan of Meander Homer

Swan of Pesaro Gioacchino Antonio Rossini who was born in Pesaro, Italy

Swan River Colony Perth built around the River Swan in Western Australia

Swanside Perth, Western Australia

Swansider(s) inhabitant(s) of Perth on the Swan River estuary of Western Australia

Swatow mainland-China-built fast patrol craft

Swatter Soviet antitank missile

Sweden's Most Popular Sculptor Vilhelm Carl Emil Milles (originally surnamed Anderson)

Sweden's Principal Port Göteborg (Gothenburg)

Swedish Swedish massage (based on Swedish-type physiotherapeutic movements); Swedish mile (10 kilometers); Swedish movements (physiotherapeutic exercises); Swedish putty (spackle + spar varnish waterproofing mixture); Swedish turnip (rutabaga)

Swedish Film Pioneer Ingmar Bergman

Swedish Hanseatic Port City Visby on the island of Gotland

Swedish Nightingale Jenny Lind

Swedish Ports Lysekil, Uddevalla, Göteborg, Varberg, Falkenberg, Halmstad, Hoganas, Viken, Halsingborg, Landskrona, Malmö, Limhamn, Klagshamn, Trelleborg, Ystad, Simrishamn, Ahus, Solvesborg, Karlshamn, Ronnebyhamn, Karlskrona, Kalmar, Oskarshamn, Slite, Farosund, Visby, Vastervik, Mem, Norrkoping, Oxelosund, Nykoping, Sodertaije, Nynashamn, Stockholm, Vasteras, Oregrund, Skutskar, Kastet, Gavle, Vallvik, Ljusne, Sandarne, Soderhamn, Hudiksvall, Sundsvall, Harnosand, Gustavsvik, Ornskoldsvik, Pitea, Lulea, Haparanda

Swedish West Indies St Barts (St Barthélmy now a French colony)

sweetening adding new sound after movie soundtrack is finished

sweet leaf stevia

Sweet Singer of the Sierras Joaquin Miller

Sweyn Forkbeard King Svend of Denmark

Swing Era big-band era of 1930s and '40s

Swingfire British BAC air-launched or ground-launched antitank missile

Swiss Swiss chard (beetlike herb used in stews); Swiss cheese (Emmenthaler cheese characterized by its pale-yellow body and many holes); Swiss lapis (imitation lapis lazuli); Swiss muslin (curtain material); Swiss steak (thin slice of steak doused in flour and vegetables); Swiss watch

Swiss Cheese Capital of the U.S.A. Monroe, Wisconsin

Swiss Day Independence Day (August 1)

Swiss Family of Mathematicians and Scientists the Bernoullis

Swiss Family of Painters the Fuesslis

Swiss Riviera northern shores of Lake Lucerne

Switzerland of Africa Zimbabwe, because it was stable, quiet, and prosperous

Switzerland of America Colorado, Maine, New Hampshire, New Jersey, West Virginia; Eagle Cap Wilderness, Oregon

Sycamore Bristol four-place helicopter

Sycamore City Terre Haute, Indiana

symbol of suffering the Christian cross

Synthesizer of Adrenalin Jokichi Takamine

Syrian Ports Latakia and Baniyas

Taco Town San Diego, California

Tadpoles Mississippians

Taffie(s) Welsh person(s)

Taffy diminutive of David the tutelar saint of Wales; nickname for a Welshman

Taig Terence

Tailhook term for Navy and Marine Corps aviators' sexual-harassment scandal following 1991 Tailhook Association convention in Las Vegas, Nevada

Taiwan (Chinese—Terrace Bay)—island of Formosa—the Republic of China

Taj Mahal Henry Sainte Claire Fredricks

Takla Makan Takla Makan Desert of northern China

talking cure psychoanalysis

Tall Boy Gouverneur Morris—amanuensis of the *Constitution of the United States*

Tall City Midland, Texas

Talleyrand Charles Maurice de Talleyrand-Périgord

Tallinn (Estonian—Dane's Town)—formerly Reval

Tall Tactician Connie Mack (baseball player)

Talos Bendix long-range surface-to-air missile

Tamaulipas formerly Pánuco

Tammany Boss William M Tweed

Tammies Tamburitzans

Tam(my) Thomas; Tom(my)

Tanana River City Fairbanks, Alaska

Tanami Tanami Desert of Australia's Northern Territory

Tandjungpriok Djakarta (Jakarta)—formerly Batavia

Tanganyika the mainland of Tanzania

tangs tangerines

Tangerine Bowl Orlando, Florida

Tank Bob Brannum (basketball player)

Tanya (Russian nickname—Tatiana, Tatyana)

Tanyu Morinobu Kano (*see* Kano)

Tanzanian Ports Lindi, Dar es Salaam, Tanga, Chake Chake and Zanzibar

Tapatios Mexicans from the state of Jalisco or from Guadalajara—its capital

Target Island Kahoolawe, Hawaii

Tarheeler(s) North Carolinian(s)

Tar Heel State North Carolina

Tartar General Dynamics naval surface-to-air missile

tartar emetic potassium antimony tartrate

Tartu Dorpat

Tasha (Russian nickname—Natasha)

Tasmania modern name for Van Diemen's Land

Tassie(s) Tasmanian(s)

Tassy Tasmania

Tassyland Tasmania (Australian slang)

Tat Tatshenshini River, British Columbia, Canada

Tatertown Gleason, Tennessee—shipping point for potatoes grown in the region

Tavern of Europe Paris

Taxachusetts Massachusetts

Taxassee Tallahassee

Tax Day U.S. federal taxes due April 15

Tax-Reform President Ronald Wilson Reagan

Taycheedah Wisconsin Home for Women at Taycheedah

Tay Pay Irish journalist Thomas Power O'Connor

T-bills Treasury bills

T-bonds Treasury bonds

T-Bone Walker Aaron Thibeaux Walker

Tchad Chad

T Dorp Schenectady, New York

Teacher of Doctors Sir William Osler

Teacher of Germany Philipp Melanchthon (collaborator of Luther)

Teacher President James Abram Garfield—twentieth President of the United States

Teachers Day September 28 in many Oriental lands where it is also the birthday of Confucius

Teague (nickname for an Irishman); Terence

The Team of the '90s Atlanta Braves (baseball)

Teapot Dome U.S. Navy's petroleum reserve near Casper, Wyoming and name of a political scandal during the administration of President Harding

tear gas chloroacetophenone; irritant gas that causes temporary blindness as well as irritation of the mucous membranes and the skin

Tear-jerking Giacomo Giacomo Antonio Domenico Michele Secondo Maria Puccini (1858–1924)

teathers ringworms

Telman Malaysian name for the CL-41 Wasp attack-trainer aircraft

Teche Country Bayou Teche county, Louisiana

Technion Israeli Institute of Technology

techno-baby test-tube baby

Tecumseh Girls' Town correctional facility at Tecumseh, Oklahoma

Teddy Teddy Roosevelt (Theodore Roosevelt)

Ted(dy) Edward; Theodore; Theodosia

Ted (Kid) Lewis Gershon Mendeloff (boxer)

Ted Morgan Sanche de Gramont

teenie one-sixteenth (stocks)

Teenie Christina

Teflon President Ronald Wilson Reagan

Tegoose Tegucigalpa (Honduras)

Tegusi Tegucigalpa's nickname

Tehachapi Institution California Correctional Institution at Tehachapi

teledeserts countries with only a few telephone lines per hundred population

Telemaque Denmark Vesey

Television City Hollywood, California

Teller of Sea Tales Joseph Conrad

Teller of Tall Tales Nathaniel Hawthorne, E T A Hoffmann, Washington Irving, Baron von Munchausen, Edgar Allan Poe, Aleksander Sergeevich Pushkin, and Mark Twain

Tell Town Altdorf, Switzerland—reputed home of William Tell

Temple of Culture Thomas Jefferson's home, *Monticello*

Temple Mount Jerusalem's sobriquet

Temple of Music New York City's Carnegie Hall, also called the Cathedral of Music

Tennessee Williams Thomas Lanier Williams

Tenth Muse Sappho, according to Plato, who esteemed the lyric poetess of Mytilene on the island of Lesbos

Tenth State Virginia (*see* First State)

Te Rangi Hiroa Sir Peter Buck

Teri Teresa; Theresa; Therese

Terrace Bay Taiwan (Formosa)

Terrapin State Maryland

Terre Haute U.S. Penitentiary at Terre Haute, Indiana

Terrier General Dynamics naval surface-to-air missile

Terry Terence; Teresa; Terrell; Terrill; Theresa; Therese

Tess(ie) Theresa

Texarkana Institution Federal Correctional Institution at Texarkana, Texas

Texas Alexander Alger Alexander

Texas Babe Mildred Didrikson Zaharias

Texas Cow Town Fort Worth

Texas Guitar Slim Johnny Winter

Texas Mix sodium pentothal, Pavulon, and potassium chloride (solution for fatal injections for death-row prison inmates)

Texas Nightingale Sippie Wallace

Texas Raj Upper-class Texans with ties to East Coast elite

Texas tea crude oil and natural gas

Texas television campfire

Texas Tommy Thomas Andrew Dorsey

Thailand's Principal Port Krung Thep (Bangkok)

Thais Thailands

That Man Franklin Delano Roosevelt

Theda Bara Theodosia Goodman

Theory X managerial assumption that employees dislike work and must be coerced, controlled, or threatened to motivate them

Theory Y managerial assumption that employees do not dislike work and under proper conditions will accept and seek out responsibilities to fulfill their esteem, self-actualization needs, and social needs

Theory Z an organization with a record of long-term employment, shared decision making, slow promotions, and non-specialized but varied job assignments

the pill oral contraceptive for women

The Terrible Ivan IV—Czar of Russia 1547 to 1584

third age old age (the first age is infancy through adolescence, the second age is post-adolescence through middle age)

Thermaie Gulf Gulf of Salonika in the Aegean

Thespian Maids the Nine Muses

Thief of Bad Gags Milton Berle (Mendel Berlinger)

Thiefrow nickname for London's Heathrow Airport

thieves of time procrastinators

Thin Man Red Rocha (basketball player)

Third Estate The Commons—the legislature

third-generation money electronically controlled funds

Third International Lenin's organization of communists in Moscow in 1919

Third Reich Nazi Germany (1933–1945)

Third Republic France from 1871 to 1940

Third State New Jersey (*see* First State)

Third Viennese School of Psychotherapy theories proffered by Viktor E Frankl

Third World emerging nations, often former colonies, outside industrialized communist and Western nations; poorer nations

Thirstland waterless country north of Bechuanaland

Thirteen Colonies British North American colonies that became the original thirteen states of the United States

Thirteen States New Hampshire, Massachusetts, Rhode Island, Connecticut, New York, New Jersey, Pennsylvania, Delaware, Maryland, Virginia, North Carolina, South Carolina, Georgia

Thirteenth Apostle Constantine, the Roman emperor who built Constantinople (now Istanbul)

Thirteenth State Rhode Island (*see* First State)

Thirteenth Tribe Khazars converted to Judaism (*see* Twelve Tribes)

Thirtieth State Wisconsin

Thirty-eighth State Colorado

Thirty-fifth State West Virginia

Thirty-first State California

Thirty-fourth State Kansas

Thirty-ninth State North Dakota

Thirty Rock nickname of the National Broadcasting Company (NBC) at Thirty Rockefeller Center in New York City

Thirty-second State Minnesota

Thirty-seventh State Nebraska

Thirty-sixth State Nevada

Thirty-third State Oregon

This Is the Place Salt Lake City, Utah's sobriquet repeating the words of its founder—Brigham Young

thistle symbol of Scotland and the Scots

Thomas Jefferson and Abraham Lincoln of American Music Charles Ives, according to Leonard Bernstein

Thomas Jefferson Snodgrass pseudonym—Samuel L Clemens

Thomas Kyd Alfred Bennett Harbage's pseudonym

Thomas of London Saint Thomas á Becket

three-B's Bach, Beethoven, Brahms

Three Baltic Duchies Estonia, Latvia, Lithuania

Three Capitals and Five Ports Japanese numerical categories comprising the ancient and modern capitals—Kyoto, Osaka, and Tokyo plus the ports of Hakodate, Kobe, Nagasaki, Niigata, and Yokohama

three-C's Central Criminal Court

three D's dementia, dermatitis, diarrhea (symptoms of niacin deficiency)

Three Finger Mordecai Brown (baseball)

Three F's friends, family, fools

Three Kingdoms Denmark, Norway, Sweden

Three Little S's Saba, Sint Eustatius (Statia), Sint Maarten (Dutch Windward Islands—Netherlands Antilles)

Three Poisons of the Mind ignorance (misperception of the true nature of the self and all phenomena), craving, hatred (Buddhism)

Three Poles Mount Everest, North Pole, South Pole

three-R's reading, writing, arithmetic (colloquially: readin', 'ritin', 'rithmetic)

Three Rivers Three Rivers Stadium, Pittsburgh

Three Sisters Earth, Mars, Venus

The Three Tenors Placido Domingo, José Carreras, and Luciano Pavarotti

Three Virgins St Croix, St John, St Thomas (United States Virgin Islands)

Throne of Solomon Ethiopia

thruppence threepence

Thumper Ted Williams (baseball player)

Thunder Bay modern name for the Canadian twin cities of Fort William and Port Arthur on the northwest shore of Lake Superior

Thunderbird British BAC mobile surface-to-air missile

Thunderbolt Republic fighting aircraft F-47; Swedish Viggen jet fighter J-37

Thunderchief Republic single-engine fighter-bomber jet aircraft (F-105)

Thunderjet Republic fighter-bomber F-84

Thunder Moon full moon in July

Tib(by) Isabel(la); Ishbel(le)

Tiber River City Rome

Tico Costa Rican; Ticonderoga; *USS Ticonderoga* (attack aircraft carrier)

Ticos Costa Ricans

Tidewater States Maryland, Virginia, North Carolina, South Carolina, Georgia

Tien Percent Tien Suharto, wife of Indonesian President Suharto, so called for her percentage take of business with Indonesia

Tierra del Fuego (Spanish—Land of Fire)—originally the fires of Patagonian Indians but more recently the burning gases from oil rigs in southernmost South America

Tiger II Northrup F-5 twin-jet fighter aircraft

Tigercat Short and Harland surface-to-air missile

Tiger of France Georges Clemenceau (1841–1929)

Tigers of the Sun Sherpas of northern India and Nepal

Tight Little Island Great Britain

Tightrope Walkers Extraordinaire Charles Blondin who crossed Niagara Falls in 1855 on an 1100-foot (336-meter) tightrope suspended 160 feet (48 meters) above the falls and five years later carried his agent across piggyback; in 1974 Philippe Petit crossed between the twin towers of the World Trade Center in New York on a tightwire 1350 feet (412 meters) above the city sidewalk

Tigres River City Baghdad

Tilda Mathilda

Tillie Mathilda

Tilly Mathilda

Time Personified the aged Chronos of the Greeks and Romans—Father Time

Timesqueer New York City's Times Square

Time of the Troubles Russia between 1604 and 1613, from the reign of Boris Godunov to that of Michael Romanov

Timur the Lame Tamerlane

Tina Albertina; Christina; Clementina; nickname for former British Prime Minister Margaret Thatcher (stands for: There is no alternative); Valentina

Tina Turner Annie Mae Bullock

Tin Can Island Niuafo'ou, Tonga

Tin Islands Indonesia's Banka and Belitung

Tin King Simón Ituri Patiño

Tin Lizzie Model-T Ford's nickname

Tinseltown Hollywood, California

Tintoretto Jacopo Robusti

Tip Thomas Phillip O'Neill, Jr; Timothy

Tip of Canada Ellesmere Land

Tippecanoe William Henry Harrison

Tipper Gore Mary Elizabeth Aitcheson Gore

Tipton Center State Correctional Center at Tipton, Missouri

Tiradentes (Portuguese—Tooth Puller)—nickname of José Joaquim da Silva Xavier—first Brazilian fighter for independence from Portuguese rule—a dentist

Tire City Akron, Ohio

Tirso de Molina pseudonym of Gabriel Tellez

Tish Letitia

Titan two-stage intercontinental ballistic missile (Martin)

Titian Tiziano Vecellio

Tito Josip Broz(ovich)

Tito Schipa Raffaele Attilio Amadeo Schipa

Titta Ruffo Ruffo Cafiero Titta

Titulescu James T(homas) Farrell

Tobacco Capital Durham, North Carolina

Tobacco City Winston-Salem, North Carolina

Tobacco Road dilapidated and poverty-striken rural areas (sociological synonym); tobacco-raising areas of the southern United States (generic and economic meaning)

Tobacco State Kentucky

Tóbal Cristóbal

Toby Tobyhanuia; Tobias

toc tag airline boarding pass

Togo Port Lomé

To Hell and Back nickname of the Toronto, Buffalo, and Hamilton Railway

Toinette Antoinette

Tokaido Corridor urban strip between Kyoto and Tokyo (Kyoto, Kobe, Osaka, Nara, Nagoya, Hamamatsu, Shizuoka, Yokohama, Tokyo)

Tokyo (Japanese—Eastern)— formerly called Edo or Yedo and now capital of Japan as well as the world's largest metropolitan area

Tolliver Tagliafiero

Tombs old New York City Prison on the Lower East Side, connected to the Criminal Courts Building by a Bridge of Sighs

Tomb Town Moscow

Tomcat F- 14 fighter aircraft

Tom Cruise Thomas Mapother

Tom, Dick, and Harry the crowd; ordinary people; the mob; no one in particular

Tom Mix Eugene Blackman's motion-picture name

Tommy nickname for a British soldier; Thomas

Tommy Atkins nickname for a British Army private

Tommy the Cork Thomas Corcoran

Tommy gun Thompson submachine gun

Tom o'Bedlam incurable male lunatic

Toms two-dollar bills bearing the portrait of President Thomas Jefferson

tom thumb (Cockney—rum)

Toña Antonia

'Tona Daytona Beach, Florida

Tongan Ports Nukualofa Tongata, Pangai Haapai, Neiafu Vavau

Tongareva Penrhyn Island in the South Pacific

Tongue Troopers Canadian term for government's French-language enforcement squads

Toni Antonia

Toni Morrison Chloe Anthony Wofford

Tono Tomuelo

Toño Antonio

Tony Anthony; Antoinette Perry Awards (American Theatre Wing)

Tony Bennett Anthony Benedetto

Tony Curtis Bernie Schwartz

Tony Martin Alvin Morris

Tony Randall Leonard Rosenberg

Tony Sarg Anthony Frederick Sarg

tooies tuinal (half amobarbital and half secobarbital)

toonie Canadian two-dollar coin

Tooth City Florence, South Carolina

Toothpicks nickname given early settlers of Arkansas who were believed to pick their teeth with bowie knives

Toothpick State Arkansas

tootsie roll Mexican-made heroin, also called black tar or Mexican mud

Top of Europe northern sections of Finland, Norway, Russia, and Sweden near the Arctic Circle

Top Ten FBI's list of the 10 most wanted fugitives from justice

Top of the World Point Barrow, Alaska

Torchbearer of the Revolution Nathaniel Bacon of Virginia

Tor House home of Robinson Jeffers at Carmel, California

Tornado Alley area between Lawton, Oklahoma and Wichita Falls, Texas

Tortilla Curtain U.S.–Mexican border area between El Paso, Texas and San Diego, California; sometimes extended from Brownsville, Texas

Tortoise Islands Galápagos Islands (officially Archipelago de Colón)

Toscanini Arturo Toscanini (outstanding orchestra conductor of the first half of the 20th century)

t' other siders the other siders (nickname given east coast Australians by their west coast counterparts)

Toti dal Monte Antonietta Meneghel

Tough Guy (stock exchange nickname for Texas Gulf Sulphur company)

Tourette's disease convulsive facial tic

Tow Hughes antitank missile designated MGM-71A

Towel Town Kannapolis, North Carolina where Cannon towels are made

The Tower The Tower of London

Town of Floating Gardens Xochimilco, Mexico

Town of Fools Chelm (*see* Chelmer)

Town of Merchants Shanghai

Town on the Water Stockholm

Town of Roses Molde, Norway

Town Too Tough To Die Tombstone, Arizona

Town of the Vikings Manitoba

Toy Bulldog Mickey Walker

Toy Cannon Jimmy Wynn (baseball player)

Tracker Grumman S-2 antisubmarine search-and-attack aircraft

Tracy Theresa

Trader Horn nickname of Alfred Aloysius Smith

Tragic Patriot Thomas Paine (imprisoned by the Reign of Terror in France and reviled by the clergy in the United States he helped create)

Tragic Queen Marie Antoinette

Tragus Heironymus Bock (a founder of modern botany)

Trager Work for Milton Trager—combines soft-touch manipulation with movement exercises to release tension

trailing spouses husbands or wives who follow mates to new employment

tranks nickname for sedative drugs (tranquilizers)

Transcendental Philosopher Ralph Waldo Emerson

Transjordan(ia) Hashemite Kingdom of Jordan better known as Jordan

Transplants Japanese auto plants in the United States

Trapper's Moon full moon in February

trashfish monkfish

Traven B Traven pseudonym used by Berick Traven Torsvan

Treasure State Montana

Tree of Heaven *Ailanthus* tree

Tree Planter's State Nebraska

Trench Town West Kingston, Jamaica's ghetto

Trent's Town Trenton, New Jersey's capital named for its original settler—William Trent

Tres Marías (Spanish—Three Marys)—María Madre, María Magdalena, María Cleofás—istands off the west coast of Mexico serving as a convict colony

T. rex Tyrannosaurus rex

The Tribune Man pseudonym—Henry Ten Eyck White

Tribune of the People John Bright

Tricia Patricia

Tri-Cities Florence, Sheffield, and Tuscumbia on Tennessee River near Muscle Shoals in northwestern Alabama; Davenport, Iowa—Moline and Rock Island, Illinois

Tricky Bob Robert Fulton, who took credit for the first working steamboat

Tricky Dick President Richard M Nixon

tricolor flag divided into three horizontal or vertical stripes; the Tricolor, initially capitalized, refers to the Tricolor of France consisting of red, white, and blue vertical stripes

Tri-Lingual Land Switzerland, where people often speak French, German, and Italian as well as Romansh and English

Trinidad and Tobago Ports Trinidad—Chaguaramas Bay, Port-of-Spain, Pointe a Pierre, San Fernando, La Brea, Brighton, Point Fortin; Tobago—Canaan, Charlotteville, Scarborough

Trinidadians people of the West Indian island of Trinidad, close to Venezuela

Trinity of Science Experience, Observation, and Reason

Triple Alliance Argentina, Brazil, Uruguay; Austria, Germany, and Italy (before outbreak of World War I)

Triple Cities Binghampton, Endicott, Johnson City, New York (also called Tri-cities)

Tripsville Haight-Ashbury district of San Francisco

Trish Patricia; Tricia

Trixie Friganza Delia O'Callahan

Trix(ie)(y) Beatrice; Beatrix

Trojan Johnny Evers (baseball player); North American T-28 trainer aircraft

Troldhaugen (Norwegian—Troll's Hill)—Edvard Grieg's home near Bergen

Trondheim modern name of Nidaros

Trooper Turned Physician Thomas Sydenham

Troopship Fokker military version of the 40 to 52-passenger aircraft F-27

tropical forests of the sea coral reefs

Tropical North northern Queensland, Australia

Tropic of Cracker so-called north-to-south dividing line along Florida's interior ridge

Tropic Metropolis Miami, Florida

Tropies torrid lands and seas between Tropic of Cancer and Tropic of Capricorn

trots diarrhea

Trotsky Lev Davidovich Bronstein

Trucial States (*see* United Arab Emirates)

truck drivers amphetamines

Trudy Gertrude

True King of Our Storytellers Jack London, according to Upton Sinclair

True Patriot Pen name for Joseph Warren

Truman Capote Truman Streckfus Persons

Truman Doctrine defend Greece and Turkey, contain the Soviet Union in Europe (named for Harry S Truman, thirty-third president of the United States)

Trumpeter of the Last Judgment Gabriel

Trunk Murderess Winnie Ruth Judd, who in 1931 killed two women and shipped their body parts to Los Angeles in trunks and a suitcase

Trust Buster Theodore Roosevelt—twenty-sixth President of the United States

The Trust Buster William Howard Taft

Trust Territory Micronesian islands of the Pacific (Carolines, Marianas, Marshalls, Ponape, Truk, Yap, etc.) under American administration

truth serum sodium pentathol

Tsaritsyn czarist name of Volgograd formerly called Stalingrad

Tsarskoe Selo former name of Pushkin near Leningrad

T-town Tarrytown; Tijuana; Trexlertown

Tuba Player of the Century Roger Bobo of the Los Angeles Philharmonic

Tuckahoes Virginians

Tula Gertrude; Gertrudis

Tullahoma Vocational Tennessee State Vocational School for Girls at Tullahoma

Tully Marcus Tullius Cicero

Tum-l'um portly Albert Edward, HRH the Prince of Wales who later became King Edward the Seventh

Tung Tree Capital Picayune, Mississippi

Tunisian Ports Susa, Halq al Wadi, Tunis, Banzart, Bizerte, Sfax, and Gabes

tuppenny twopenny

Turkestan Desert includes Kara Kum south of Aral Sea, Kyzyl Kum southeast of Aral Sea, Ust Urt between Aral and Caspian seas

turista *Staphylococcus aureus*

Turkey Capital of the World Berryville, Arkansas and Worthington, Minnesota

Turkey's Principal Port Istanbul (Constantinople)

Turkish Turkish bath (steam bath); Turkish delight (fruit-flavored gelatin candy dusted with confectioner's sugar); Turkish rug; Turkish tobacco (highly aromatic); Turkish towel (water-absorbent long-nap towel)

Turkish Ports Istanbul (Constantinople), Hydarpasa, Izmir (Smyrna), Antalya (Adalia), Mersin, Iskenderun (Alexandretta)

Turkish Towel Actress Brigitte Bardot

Turner syndrome female genitalia and internal ducts exist, but there are crude ovaries; broad chest; short stature and neck

Turner's syndrome genetic abnormality in females inheriting only forty-five chromosomes, causes retarded sexual development

Turpentine State North Carolina

turquoise hydrargillite (basic hydrated copper aluminum phosphate)

Turtle Bay Bahía de San Blas, Mexico

Tusitala (Samoan—Teller of Tales)—Robert Louis Stevenson's nickname

Tutor Canadair-built jet-trainer aircraft designated CL-41

Tuvalu formerly New Hebrides in the South Pacifitc between Fiji and New Caledonia

Tver czarist name for Kalinin

Twelfth State North Carolina (*see* First State)

Twelve Apostles twelve Apostle Islands in Lake Superior off northern Wisconsin

Twelve-Tone Technician Arnold Schonberg

Twelve Tribes Twelve Tribes of Israel—Reuben, Simeon, Judah, Zebulun, Issachar, Dan, Gad, Asher, Nephtali, Benjamin, Ephraim, and Mannaseh

Twentieth State Mississippi

Twenty-eighth State Texas

Twenty-fifth State Arkansas

Twenty-first State Illinois

Twenty-fourth State Missouri

Twenty-ninth State Iowa

Twenty-second State Alabama

Twenty-seventh State Florida

Twenty-sixth State Michigan

Twenty-third State Maine

Twiggy Leslie Hornby

Twilight of the Gods Gotterdámmerung (German mythology); Ragnarok (Norse mythology)

Twilight Zone the Mexican Border

Twin Cities Bristol on the Tennessee–Virginia border; Central Falls and Pawtucket, Rhode Island; Champaign and Urbana, Illinois; Minneapolis and St Paul, Minnesota; Texarkana on the Arkansas–Texas border; Winston-Salem, North Carolina

Twin Maples Farm British Columbia facility for treating women alcoholics

Twin Otter DeHavilland light transport (DHC-6)

The Twins Minneapolis and St Paul

Twin Sisters North and South Dakota

Twin States New Hampshire and Vermont

twister dustwhirl, sandspout, tornado, or waterspout wherein ascending and rotating movement of air column is especially apparent

Two Eyes of Greece Athens and Sparta

Two-headed Eagle popular symbol of the Austro-Hungarian Empire, Imperial Russia, and the Holy Roman Empire

Tybalt Theobald

Tybee Savannah Beach, Georgia

Ty Cobb Tyrus Raymond Cobb—idol of baseball fans

type metal antimony-copper-lead-tin alloy

Typhoid Mary Mary Mallon (c. 1870–1938)

Ubangi Republic Central African Republic

U-boat Führer Gross Admiral Karl Doenitz

u-dog underdog

Ulan Ude formerly Verkhneudinsk

Ulster Northern Ireland (formerly an ancient province of Ireland; now containing the counties of Antrim, Armagh, Down, Fermanagh, Londonderry, and Tyrone)

Ulster Cradle of U.S. Presidents Northern Ireland—ancestral home of Presidents Arthur, Grant, Jackson, McKinley, Truman, Wilson

Ultima Thule Iceland; Mainland (largest of the Shetland Islands); Norway; or any remote northern place, according to ancient travellers

Ulysses fifty-dollar bills bearing the portrait of President Ulysses S Grant

Umbrian Historical Painter Pinturicchio (Bernardino di Betto)

Unabashed Atheist Charles Bradlaugh

The Unashamed Accompanist Gerald Moore

UN City Vienna, Austria's International Center (available to the UN cost free)

uncle pawnbroker

Uncle Arthur Arthur Henderson

Uncle Billie General William Tecumseh Sherman, USA

Uncle Dickie affectionate nickname of British military hero Mountbatten of Burma, Admiral of the Fleet and last Viceroy of India

Uncle Gene Eugene Ormandy

Uncle George George Geist

Uncle Ho Ho Chi Minh

Uncle Horace Horace Greeley

Uncle Joe U.S. Representative Joseph Gurney Cannon also known as the Watchdog of the Treasury

Uncle Kwesi Jonathan Kwesi Lamptey

Uncle Miltie Milton Berle (Mendel Berlinger)

Uncle Remus pseudonym—Joel Chandler Harris

Uncle Robert Robert E Lee; Robert L Sheppard

Uncle Sam cartoon symbol and nickname for an American citizen or the United States of America

Uncle Sam's Crib Treasury of the United States

Uncle Sam's Pocket Handkerchief Delaware—second smallest state in the U.S.

Uncle Sugar FBI's nickname

Uncle Tom Josiah Henson (slave immortalized in Harriet Beecher Stowe's *Uncle Tom's Cabin);* submissive African-American

Uncle Whiskers nickname for Uncle Sam

Unconditional Abolitionist William Lloyd Garrison

Unconditional Surrender Grant General Ulysses Simpson Grant, USA

Uncrowned King of Ireland Charles Stewart Parnell

Uncrowned King of the Jews Chaim Weizmann

Uncrowned Queen of Iraq Gertrude Bell

Underground Railroad Conductor Harriet Tubman, who conducted fleeing slaves from the South to the northern United States and Canada

Underworld Statesman and Patriot Meyer (Little Man) Lansky

Union Jack flag flown at forward jackstaff of American vessels—consists of dark-blue rectangular field with 50 five-pointed white stars; flag of United Kingdom symbolizing union of England, Northern Ireland, Scotland, and Wales—combines crosses of St George (England), St Patrick (Northern Ireland), St Andrew (Scotland)

Union of the Great Islands Comoro Islands

Unitarian Economist David Ricardo

Unitarian Quaker Elias Hicks

United Arab Emirate Ports Ash Shariqah, Dubai, Abu Dhabi (Abu Zaby)

United Arab Emirates Trucial Sheikdoms (Persian Gulf oil-producing country)

United Arab Republic name referring to the former union of Egypt and Syria

United Kingdom's Principal Port London

United Nations Capital New York City

United Provinces colors blue and white displayed in flags of El Salvador, Guatemala,

Honduras, and Nicaragua—formerly federated after their liberation from Spain

Universal Genius Leonardo da Vinci (anatomist, architect, cartographer, engineer, inventor, musician, painter, poet, sculptor, zoologist)

universal time Greenwich Mean Time

University City Cambridge, Massachusetts

Unreconstructed Rebel Senator George Carter Glass, nicknamed by President Franklin D Roosevelt

Unsainted Anthony San Antonio, Texas

Unser Fritz (German—Our Fritz)—Frederick William III of Prussia

Unspoiled Queen Saba

Upper Adige the Italian Tyrol also called the Southern Tyrol

Upper Alsace Haut-Rhin department of France

Upper Amazon Amazon River extending from the highlands of Peru to Manaus in northern Brazil

Upper Austria northern Austria bordering Bavaria, the Czech Republic, and Slovakia

Upper Burma inland Burma

Upper California in Spanish-colonial times all of California north of Monterey but today all of California except Baja or Lower California

Upper Canada English-speaking Ontario and the upper St Lawrence region during the 19th century

Upper Egypt Egypt from Cairo south to the Sudan

Upper Galilee Israel north of the Sea of Galilee

Upper Lakes northernmost Great Lakes—Huron, Michigan, Superior

Upper Michigan the upper-peninsula of northern Michigan between Lake Michigan and Lake Superior

Upper Mississippi Mississippi River from Lake Itaska in Minnesota near the Canadian border to Saint Louis, Missouri

Upper Nile Nile River from its headwaters in central East Africa to Khartoum in the Sudan

Upper Palatinate eastern Bavaria

Upper Peninsula northern Michigan between Lake Michigan and Lake Superior

Upper Peru former name for Bolivia

Upper Rhine Rhine River between Basel in Switzerland and Mainz in Germany

uppers nickname for stimulants

Upper Silesia northern Silesia once a Prussian provunce

Urga former name of Ulan Bator

Urista Uriel da Costa

Ursula Bloom Mrs ACG Robinson's pen name

Ursula Undress Ursula Andress

Uruguayan Ports La Paloma, Maldonado, Montevideo, Puerto Sauce, Nueva Palmira, Fray Bentos, Puerto Concepción, Paysandá, Salto

Uruguay's Largest Port Montevideo

Ushant Ile d'Ouessant—France's most westerly point at the Bay of Biscay entrance to the English Channel

Usonian things distinctly American, according to Frank Lloyd Wright

Utilitarian Philosopher Jeremy Bentham; John Stuart Mill

Utopian Author Bacon, Bellamy, Butler, Cabot, Campanella, Fourier, Huxley, More, Morris, Owen, Plato, Proudhon, Rabelais, Rousseau, Saint-Simon, Wells

U-town Uniontown; Uptown

Vacation City on Casco Bay Portland, Maine

Vacationland Maine

Vacationland of Opportunity Alaska

Vagen (Norwegian—Bay)—old Bergen and its waterfront along the bay

Valencia Conductor-Pianist José Iturbi (1895–1980)

Valentine State Arizona, admitted on St Valentine's Day—February 14, 1912

Valhalla Hall of the Slain Warriors (Norse or Scandinavian mythology)

Valhalla Girls Women's Correctional Unit at Valhalla, New York

valium diazepam

Valley Between Two Worlds Rio Grande Valley (between Mexico and the United States)

Valley of God's Pleasure Cleveland, Ohio's suburban section around Shaker Heights

The Valley Island Maui, Hawaii

Valley Isle Maui, Hawaii

Valley of Opportunity New York State's Triple Cities area including Binghampton, Endicott, and Johnson City

Valley of Rice Sikkimese name for state in India known as Denjong

Valley of Roses Persia (Iran)

Valley of the Sun Arizona's central valley

Valley of Ten Thousand Haystacks Big Hole, Montana

Valley of Valleys Gudbrandsdalen, Norway

Valley of Wonders Yellowstone National Park (in Idaho, Montana, and Wyoming)

Vampire De Havilland jet fighter-bomber aircraft

Vampire Capital of America Rhode Island

Van Allen Radiation Belts two belts of charged particles in the Earth's magnetic field—discovered in 1958 by J A Van Allen, United States

Van Cliburn Harvey Lavan Cliburn

Vancoo Vancouver, British Columbia

Vancouver Island, British Columbia named for Captain George Vancouver

Van Diemen's Land Tasmania

Vanier Canadian city formerly called Eastview

Vapor City Hot Springs, Arkansas

vaporware promoted computer software that has not been released

Varangians (Russian—Vikings)—Danes or Norsemen who sailed to America around year 1000

Vasco Vasco Pratolini (Italian novelist)

Vatican City State *Stato della Città del Vaticano*

Vautour Sud attack bomber and interceptor made in France

Veecees Vietcongs

Veenees Vietnamese

Veep Vice President

Vega Alta Industrial School for (criminal) Women at Vega Alta, Puerto Rico

Vegas East Atlantic City, New Jersey

vegetable cow of Asia tofu

vegetable pear chayote squash

veggie libel laws United States agricultural product-disparagement laws

Vehicle City Flint, Michigan

Velvet Breughel Jan Breughel the Elder

The Velvet Fog singer Mel Torme

Velvet Revolution Czechoslovakia's turn from Communism

Venerable Nestor of Massachusetts John Quincy Adams—sixth President of the United States

Venetian Venetian ball (glass paperweight containing colorful objects); venetian blind (horizontally slatted sun curtain); Venetian glass (ornamental glassware of the type made in Venice); Venetian red (dark orange red); Venetian window (palladian window)

Venetian Family of Painters the Bellinis and the Tintorettos

Venetian red ferric oxide (Fe_2O_3)

Venezuelan Poet Laureate Irma De-Sola Ricardo

Venezuelan Ports Maracaibo, Puerto Miranda, Bahía de Amuay, Puerto Cabello, La Guaira, Puerto de Hierro, Puerto Ordaz, Ciudad Bolívar

Venezuela's Principal Port La Guaira

Venice of France Strasbourg

Venom British de Havilland jet fighter aircraft

Venus's curse venereal disease

Vera Zorina Eva Brigitta Hartwig

Vercors pseudonym of Jean Bruller

Vermeer Jan van der Meer van Delft

Vermilionville Lafayette, Louisiana's old name

Vermillion Sea Gulf of California (Mar Bermejo)

Verneur Gouverneur

Vernon Castle Vernon Blythe

Vernon Duke Vladimir Dukelsky

Vernon Lee Violet Paget's pseudonym

Veronese Paolo Cagliari

Veronica Lake Constance Ockleman

Verrocchio Andrea di Michele Cione

Vertical Hyphen Johnny Horan (basketball player)

Ves Sylvester

Veteran Curmudgeon H(enry) L(ewis) Mencken

Vicar of Christ the Pope

Vicente Vicente Blasco-Ibánez (Spanish novelist)

Vichy Government France—1940–1944—which ruled by collaborationists Marshal Pétain and Pierre Laval who maintained their headquarters in Vichy within unoccupied France

Vicki Victoria

Victim of Religion and Revolt Northern Ireland also called Captive of History

Victor letter V radio code; Handley-Page jet bomber aircraft

Victor Borge Borge Rosenbaum

Victor-Charlie VC; Vietcong

Victoria de los Angeles Victoria Gomez Cima

Victoria Holt Eleanor Burford Hibbert

Victorian Librettist W(illiam) S(chwenck) Gilbert

Victor Seastrom Viktor Sjöström

Victor Serge Victor Lvovich Kibalchich

Victory Personified Nike the Greek goddess or her Roman counterpart Victoria

video (Latin—I see)—picture portion of a tv broadcast

Vienna Vienna brown (bronzetone gold); Vienna green (emerald); Vienna lake (carmine); Vienna lime (Magnesialime polish); Vienna red (vermillion); Vienna sausage (short thin frankfurter)

Vietnamese Ports Cam Pha, Hon Gai, Haiphong, Ben Thuy, Da Nang (Tourane), Cam Ranh Bay, Saigon (Ho Chi Minh City)

Vietnam's Principal Port Ho Chi Minh City (Saigon)

Vietnam syndrome reluctance of Americans to intervene militarily in a country as a result of United States' defeat in Vietnam War

Vieux Carré (French—Old Square)—French Quarter of New Orleans

Vieux Carré syndrome pleasant distraction

Viggen Swedish Thunderbolt jet fighter aircraft

Vigilante North American Rockwell A-5 bomber aircraft

Viki Victoria; Victorine

Viking Capital Oslo, Norway

Viking Genius John Ericsson

Viking Land Norway

Viking Program systematic investigation of Mars from orbit and from the surface with emphasis on the search for life on this planet

Vilhjalmur Stefansson Canadian-born William Stevenson's adopted name

Villa Acuña former name of Ciudad Acuña

The Village Carmel-by-the-Sea in California; Greenwich Village in New York City; La Jolla's shopping district near San Diego, California

Village of Fools Chelm, Poland

vinegar acetic acid (CH_3COOH)

Vinegar Joe General Joseph Warren Stilwell, USA

Vinland North American coast discovered by Leif Ericsson and Norse sailors in year 1000; collective name for area extending from Labrador to Martha's Vineyard

Vinnie Vincent

Vinny Vincent

Vinsanity Vince Carter (basketball player)

vinyl phonograph records

Violet Crown Acropolis Hill, Athens

Violet-crowned City Athens

viper's weed marijuana

Virgin Goddesses Artemis, Athena also known as Parthenia, and Hestia

Virginia named for the Virgin Queen, Elizabeth I

Virginia Ports Alexandria, Newport News, Norfolk, Portsmouth

Virgin Island Ports Charlotte Amalie on St Thomas, Cruz Bay on St John, Frederiksted on St Croix

Virgin Queen Elizabeth I

Virgins American and British Virgin Islands in the West Indies

Virgin Superior St Thomas Island, Virgin Islands

Virtuoso Vladimir pianist Vladimir Horowitz (1903–1989)

Viscount Vickers medium transport aircraft

Viscountess Beaconsfield Mary Anne Disraeli (Mrs Benjamin Disraeli)

Viscount Melbourne William Lamb (former Prime Minister of Great Britain)

Viscount Palmerston Henry John Temple

Viscount Sidmouth Henry Addington

Vissarion Belinsky Vissarion Fondaminsky (1811–1848)

Vistula River Cities Cracow and Warsaw

Vitalis Erik Sjöberg

Vivazza (Italian—Vivacity)— Gioacchino Antonio Rossini's nickname

Viveca Lindfors Elsa Viveca Torstensdotter

Vivien Leigh Vivian Mary Hartley

Vladimir Sirin Vladimir Nabokov's pseudonym

The Voice Alex Dreier

Voice of the American Revolution Patrick Henry

Voice of Australia Dame Edna Everage, also known as Barry Humphries

Voice of the Century Marian Anderson

Voice of Doom Gabriel Heatter (before and during World War II); Ann Watson (in the uncertain 1970s)

Voice from the Fo'c's'le Richard Henry Dana in *Two Years Before the Mast;* Herman Melville in *Whitejacket*

Voice of Israel Abba Eban

Voice of Liberty Thomas Paine

Voice of Northern Industrialism Daniel Webster

Voice of Polish Nationalism Adam Mickiewicz

Voice of the Revolution Patrick Henry

volatile alkali ammonia

Volcano Island Hawaii

Volcano Land Iceland

Volcanos Volcano Islands (one group in western Pacific and another off Batangas in the Philippines)

Volgograd formerly Stalingrad during Stalin's time and Tsaritsyn during czarist times

Volodya (Russian nickname— Vladimir)

Voltaire assumed name of François-Marie Arouet (*see* Jean Meslier)

Voltaire of the Unitarians Dr Joseph Priestley, according to William Hazlitt

Volunteer(s) Tennessean(a)

Volunteer State Tennessee

Von Economo's disease encephalitis lethargica

von Hofmanns Hugo von Hofmannsthal

von Reuter Israel Beer Josphat (founder of Reuters news agency)

Voodoo Canadian-built version of the F-101 jet interceptor

Voyager American spacecraft destined for landings on Mars and Venus

V Sirin Vladimir Nabokov

V-spot $5 bill

V-town Valverde (Canary Islands)

Vulcan Hawker-Siddeley jet bomber aircraft; US-built six-barrel 20mm cannon

Vulcan (Latin—Hephaistos)— the blacksmith

vulture trust tax-shelter gift plan involving third party expected to die prematurely

W George Walker Bush (to distinguish him from his father, George Herbert Walker Bush)

Waddy Walter

Wade Miller pseudonym of writers Robert Wade and Bill Miller

Wagon Tongue Bill Keister (baseball player)

Waistline of the Western Hemisphere Isthmus of Panama

Walden Henry David Thoreau's hut on the shores of Walden Pond near Concord, Massachusetts

The Wales The Bank of New South Wales

walking handbag(s) alligator(s); cayman(s); gavial(s); crocodile(s)

Walking Man Eddie Yost

Wallace's Line separates tropical Asian flora and fauna from that typical of Australia; runs between Bali and Lombok, Indonesia, to the east (named for A R Wallace)

Wall of Death drift gill netting

Walleye Martin-Hughes tv-guided glide bomb

Wallows Walla Walla, Washington

Wall Street New York City's financial center

Wall Street of Canada Bay Street in Toronto, Ontario

Walnut City McMinnville, Oregon

Walrussia nickname for Alaska in 1867 when it was purchased from Russia; believed by critics to have nothing but walruses

Walt Disney José Guizao Zamora

Walt Disney of Outdoor Advertising Douglas Leigh

Walter Hampden Walter H Dougherty

Walter Huston Walter Houghston

Walter Wanger Walter Feuchtwanger

The Waltz King Johann Strauss, Jr

Wandering Mathematician Johannes Kepler (1571–1630)

War Between the States Civil War; War of the Secession

Warden of the Honour of the North Halifax, Nova Scoria

Warehouse of the East free port of Penang, Malaysia

War Fury Bellona—Roman goddess of war whose Greek counterpart is Enyo

Warhorse of the Confederacy Lieutenant General James Longstreet, CSA

War of Independence American Revolution

Warlord Wotan

Warlord of the First Reich Prince Otto von Bismarck Schönhausen

Warlord of the Second Reich Kaiser Wilhem II

Warlord of the Third Reich Führer Adolf Hitler

War of the Pacific Chile vs. Bolivia and Peru (1879–1883)

War of Reparation American Revolution

warrior's herb devil's club

War of the Secession Southern synonym for Civil War, War of the Rebellion, War Between the States (of the United States)—1861 to 1865

The Wash North Sea inlet between Lincoln and Norwich on east coast of England

Washboard Sam Robert Brown

Washerwoman's Sprain DeQuervain's syndrome (tendinitis of the thumb—once commonly from wringing out wash—now from computer use)

washing soda sodium carbonate crystals (Na_2CO_3 + $10H_2O$)

Washington named for George Washington

Washington Ports South Bend, Raymond, Aberdeen, Hoquiam, Port Angeles, Port Townsend, Olympia, Tacoma, Seattle, Everett, Anacortes, Bellingham

Washington State Funnypark Washington State Prison near Walla Walla

Washoe early settler's name for Nevada during the Comstock Lode gold-and-silver rush of 1859

Washoe Giant Mark Twain

Wasp Westland naval helicopter built in Britain

WASP Frida Kahlo Emily Carr

Watch City Waltham, Massachusetts

Watchdog of Central Park *New York Times* publisher Adolph S Ochs

Watchdog of the Eastern Pacific Pearl Harbor

Watchdog of the Western Pacific Guam

water methamphetamine

Waterfront Philosopher Eric Hoffer

Waterfront of the West San Francisco

Watergab Watergate English (Nixon-era federalese)

Watergate Potomac River waterfront of Washington, DC; scandal, involving the Nixon administration, arising from the burglary of the Democratic Party's national headquarters in the Watergate apartment building

Watergate President Richard Milhous Nixon

waterglass sodium silicate (Na_2SiO_3)

Waterland the Netherlands

Waterloo battlefield near Brussels, Belgium, where Napoleon met his final defeat June 18, 1815

Watermelon Capital of the World Hope, Arkansas

water ouzel American dipper

Waters William Russell's pseudonym

water turkey anhinga

Water Wonderland Michigan

Wat(ty) Walter

W C Fields William Claude Dukenfield (1880–1946)

wealth effect outspending income because stock market boom makes one feel rich

Wealth Personified Ploutus (Greek); Plutus (Roman)

Weather Capital of the World a groundhog hole in Punxsutawney, Pennsylvania where groundhog leaves hibernation every February 2 to forecast winter's end (if he sees his shadow it means six more weeks of ice and snow, and if he doesn't then spring is near)

The Web The World Wide Web

Webfeet Oregonians

Webfoot State Oregon

Webster Ford Edgar Lee Master's pseudonym

weed nickname for marijuana; grass, pot

Weegee photographer Arthur Fellig's nom de voir

Wee Willie William Keeler

Weil's disease jaundice

Weimar Republic Germany between 1919 and 1933—the Second Reich

Welcher(s) person(s) of Welsh origin

wellies wellington boots

Wellington Arthur Wellesley (Duke of Wellington); British Hovercraft class of hovercraft

Welsh Welsh cob (horse), Welsh corgi (dog), Welsh dresser (cupboard), Welsh main (cockfight), Welsh mountain (pony or sheep), Welsh process (smelting), Welsh rabbit (cheese dish also called Welsh rarebit), Welsh runt (cattle), Welsh springer (spaniel)

Welsh Cradle of U.S. President Wales—ancestral home of President Thomas Jefferson

Welsh Landscape Painter Richard Wilson

Welsh Ports Port Dinorwic, Holyhead, Caernarvon, Fishguard, Milford Haven, Llanelly, Swansea, Port Talbot, Barry, Cardiff

Welsh Wizard David Lloyd George

Wessex Westland-built version of Sikorsky utility helicopter

West Britain Wales

West Country Southwestern England—Cornwall, Devonshire, Dorset, Somerset

West End fashionable London

Western movie or novel featuring the Wild West

Western Hemisphere half of the world containing North America, South America, and associated islands

Western Prairie Province Alberta

Western Samoan Port Apia

Western States United States west of the Mississippi River

Western Tip of Florida Pensacola

Western Tip of Texas El Paso

West Indies Cuba, Bahamas, Greater and Lesser Antilles, Jamaica, Puerto Rico

West Irian western half of New Guinea formerly Dutch or Netherlands New Guinea and now part of Indonesia

West Jersey southern and western New Jersey

West Lothian Linlithgow, Scotland

West Malaysia mainland Malaysia plus Singapore before it became independent

Westminster Abbey Collegiate Church of St Peter in Westminster

Westminster Palace Houses of Parliament in London

West North Central States Iowa, Kansas, Minnesota, Missouri, Nebraska, North Dakota, South Dakota

West Point U.S. Military Academy at West Point, New York

West Point of the Air San Antonio, Texas

West Point of Capitalism Harvard Business School

West Point of Law Enforcement FBI National Academy at Quantico, Virginia

West Point of the South Virginia Military Institute

Westport Landing pioneer name for Kansas City

West Side West Side of any city or place

West South Central States Arkansas, Louisiana, Oklahoma, and Texas

West Virginie West Virginia

Westway New York City's west side highway extending northward from Battery Park along the Hudson River

Westy Westermoreland; Westmoreland

Wet Tortugas Florida Keys

Whaling City New Bedford, Massachusetts

Whangpoo Hwang Pu

Wharf of North America Nova Scotia

Wheat Belt prairies in North America where wheat is cultivated: Alberta, Canada, to northern Texas

Wheat Crescent dry area of South American pampas where wheat is cultivated

Wheat Energy State North Dakota

Wheatland home of President James Buchanan in Lancaster, Pennsylvania

Wheat Provinces Alberta, Manitoba, Saskatchewan

Wheat State Minnesota; South Australia

Wheaty Lincoln wheat-seed copper penny

Whelps Tennesseans

Where America's Day Begins Guam

Where the Andes Greet the Caribbean Venezuela

Where It's Springtime All The Time San Diego, California

Where Mexico Meets Uncle Sam nickname of Brownsville, Texas

Where the Turf Meets the Surf Del Mar, California

Whirlwind Westland military helicopter built in Britain

Whispering Bill Bill Barrett (baseball player)

Whispering Kid Arthur Dove (American painter)

White Africa South Africa

Whitechapel London's East End

White City of the North Helsinki

White City of the South Sucre, Bolivia

White Elephant Thailand ensign bearing a green and red caparisoned white elephant

White Ensign flag of the Royal Navy and the Royal Yacht Club—St George cross on a white ground with the Union Jack in the upper canton corner

white flag symbol of surrender or truce

white goods bleached cotton and linen fabrics; colorless high alcohol-content drinks (aguardiente, akavit, brandy, cognac, gin, rum, schnapps, tequila, vodka); large household appliances (refrigerators, stoves, washing machines, etc.)

Whitehall London's street of government offices

white horses whitecaps on sea

White House executive office and residence of the President of the United States in Washington, D.C.

White Island Ibiza in the Balearics

White James Brown Wayne Cochran

white lead lead carbonate

white light signal indicating apparatus, craft, or vehicle has power and is illuminated

White Man's Grave equatorial West Africa

White Metropolis Helsinki

White Mountain State New Hampshire

white meat breast

white plague pulmonary tuberculosis

White Rat Whitey Herzog (baseball player)

White Russia Byelorussian district around Minsk

whites thick whitish vaginal discharge; synonym for leukorrhea

White Town of Lake Mjosa Gjovik, Norway

white vitriol zinc sulfate

whitewings white-uniformed street cleaners

Whittaker Chambers Jay Chambers

Whizzer White Byron White— former football player named to United States Supreme Court in 1962

Whoopi Goldberg Caryn Johnson

Whorez trucker's nickname for Ciudad Juarez, Mexico

Why Not Town Minot, North Dakota, nicknamed Why Not Minot?

Wickedest Man in the World Aleister Crowley

widow-and-orphan any reliable and safe stock

Widow at Windsor Queen Victoria who was a widow for the last 39 years of her life

Wigwam Tammany Hall

Wild Ass of Asia onager

Wild Bill William Joseph (Wild Bill) Donovan; James Butler (Wild Bill) Hickok

Wildcat Ferns Clarence McCubbins (boxer)

Wilderness of Judah western shores of the Dead Sea in Israel

Wilderness Park Nahanni National Park (Canadian Northwest Territories)

Wilderness Trail Blazer Daniel Boone

Wildflower State Western Australia

Wild Hoss of the Osage Pepper Martin (baseball player)

Wild Man of Borneo orangutan (friendliest of the great apes)

Wild Rose Iowa state flower; Iowa girl's nickname

Wildrose Country Alberta

Wild West western United States

Wild Wonderful State West Virginia

Wilhelm Xylander Wilhelm Holtzmann

Wilkes Land Australian Antarctica

Willa Wilhelmina

Willem De Merode Willem Eduard Keuning's pseudonym

William Ashenden W Somerset Maugham

William B Goodrich Roscoe (Fatty) Arbuckle's pseudonym

William Bolitho William Bolitho Ryall

William the Conqueror William I of Normandy and England

William Faulkner William Cuthbert Falkner

William Haggard Richard Henry Michael Clayton's pseudonym

William Holden William Franklin Beedle

William of Nassau William I—Prince of Orange and Count of Nassau—founder of the Dutch Republic; also called William the Silent

William Penn's Town Philadelphia

William Sharp Fiona Macleod's pseudonym

William the Silent William— Prince of Orange

William of Stratford William Shakspere

William Tell's Town Altdorf in Switzerland's Uri Canton

William Trevor William Trevor Cox

Willie William; W(illiam) (Willie) Somerset Maugham

Willie Mays Willie Howard Mays

Willies Good Will Industries

Willy Wilhelm; William

Willy Brandt Karl Herbert Frahm

Wilma Wilhelmina

Wilmington Women Correctional institution for Women at Wilmington, Delaware

Wilt the Stilt Wilt Chamberlain (basketball player)

windjammer accordion

Window into Europe Peter the Great's name for what became Russia's seaport city, Saint Petersburg

Windscale Seltafield, England's nuclear reprocessing plant

Windsurfing Capital of the World Hood River, Oregon

Windward Islands Dominica, Grenada, Grenadines, Saint Lucia, Saint Vincent—British West Indies

Windward Passage ocean passage between Cuba and Haiti; connects Atlantic Ocean with Caribbean Sea

Windwards Windward Islands

Windy City Chicago, Illinois and Wellington on New Zealand's North Island

Windy Wellington Wellington, North Island, New Zealand

Wine-Red Sea the Aegean, according to Homer

Winesburg Sherwood Anderson's name for Clyde, a small town southwest of Sandusky, Ohio and the title of his play—*Winesburg, Ohio*

Winney Winston; Sir Winston Churchill (1874–1965)

Winnie Sir Winston Churchill— British Prime Minister

Win(nie) Winslow; Winston

Winona Ryder Winona Horowitz

Wintergarden of the East southern Florida

Wintergarden of the Gulf lower Rio Grande Valley

Wintergarden of the West the Imperial Valley

Winterless Northland northernmost New Zealand

Winter Moon full moon in January

Winter Wonderland British Columbia

Wisconsin Ports Racine, Milwaukee, Port Washington, Sheboygan, Manitowoc, Sturgeon Bay, Green Bay, Marinette, Ashland, Superior

Wisest Man of Greece Socrates

Wish Book Sears, Roebuck catalog

Witchcraft City Salem, Massachusetts

Witch of Wall Street Hetty Green

Wizard of American Drama David Belasco

Wizard from Vienna Franz Anton Mesmer

Wizard of Kinderhook Martin Van Buren—eighth President of the United States

Wizard of Menlo Park Thomas Alva Edison whose research laboratory was in Menlo Park, New Jersey

Wizard of the Saddle Lieutenant General Nathan Bedford Foffest, CSA

Wizard of Scotland Sir Walter Scott

Wizard of Tuskegee George Washington Carver

Wizard of Wall Street J P Morgan

Wizard War electronic warfare

Wizard of Word Music Edgar Allan Poe

W N P Barbellion Bruce Frederick Cummings (1889–1919)

Wobblies International Workers of the World (so named because Chinese members pronounced IWW as I *Wobbly Wobbly)*

Wolf House Jack London's home in Glen Ellen, California

Wolf Moon full moon in January

wolfram iron manganese tungstate

Wolverine nickname for Michiganite

Wolverine State Michigan

Woman with the Whip Evita Perón

Womb of Nations Scandinavia

Wonder City of the World New York

Wonder State Arkansas

Wonderland of America Wyoming

wood alcohol methyl alcohol (CH₃OH)

Wooden Leg Governor Peter Stuyvesant of Nieuw Amsterdam

Wooden Nutmegs Connecticuters

Woodie Woodmansee; Woodrow

Woodland Capital Boise, Idaho

woodpile xylophone's nickname

Woodstein Bob Woodward and Carl Bernstein (of the *Washington Post,* known for uncovering the Watergate coverup)

Woody W C (William Claude) Fields; Woodrow; Woody Allen; Woody Guthrie

Woody Allen Allen Stewart Konigsberg

Woody Guthrie Woodrow Wilson Guthrie

Woody Herman Woodrow Wilson Herman

Wool and Mohair Capital of the West Del Rio, Texas

Woolworth Woolworth Building at 233 Broadway in New York City; chain of stores once called 5-and-10 or five and dime

The Woolworth's Touch appeal to the common man

Woo Poo cadet's nickname for West Point

Word Jeweler Lafcadio Hearn (1850–1904)

Word King New Zealand-born British lexicographer Eric Partridge

Workers' Paradise derisive nickname applied to the communist-controlled USSR

Workshop of the Orient Japan

World Citizen Thomas Paine

World Island African-Asian-European land mass

World's Biggest Bookend nickname of the Secretariat building of the United Nations

World's First Streamlined Ferry The Kalakala

World's Greatest Bullshit Factory Hollywood, according to John Dos Passos

World's Workshop productive nations such as Germany, Great Britain, Japan, and the United States

World War Photographer Edward Steichen

W R William Randolph Hearst's nickname

wrapper book dust jacket

Wreckers Coast England's Cornish coast, Florida's east coast, North Carolina's Outer Banks around Nag's Head, other coasts where wreckers profit from shipwrecks

W-town Watertown

Wurst City in the World Sheboygan, Wisconsin, where making sausage is a specialty

Wyoming Suffragette Esther Hobart Morris

Xanadu Xamdu (city where Kubla Khan lived and name given the Hearst Castle at San Simeon, California by Orson Welles)

Xanrof Léon Fourneau

Xavante Aermacchi jet-trainer ground-attack aircraft also designated AT-26

Xavier Joseph Xavier Boniface's nom de plume

Xavier Mayne Edward Irenaeus Stevenson

Xenius Eugenio d'Ors

Xenocoj Santo Domingo, Guatemala

Xeres Jerez de la Frontera

Xers members of the so-called X generation

Xibaro Jivaro

Xipangu Marco Polo's name for Japan

X-ray Discoverer Wilhelm Konrad Roentgen

Xtet (Swedish—the X)—Sven Erixson

XYY syndrome unusually aggressive male having an extra Y-sex chromosome

Y2K bug computer problem linked to outdated operating-system clocks

y and y yin and yang (complementary and contrasting cosmic figures/forces/principles)

Yaacov Agam Yaacov Gipstein

Yafo Tel Aviv's seaport also known as Jaffa or Joppa

Yakumo Koizumi Lafacadio Hearn's Japanese name

Yallerhammer State Alabama

Yankee Athens New Haven, Connecticut

Yankee Clipper Joe Di Maggio

Yankeedom New England; Northeastern United States

Yankee Doodle Dandy George M Cohan (born on July 4)

Yankee Gold ice that was cut into blocks and shipped to buyers

Yankee Land of the South Georgia

Yankee State Ohio

Yanko-Spanko Conflict Spanish-American War (so named in 1899 by historian Arthur Bird)

Yaptown Cleveland, Ohio

Yard(s) Montagnard(s)

Yasnaya Polyana (Russian— Clear Glade)—Tolstoy family home near Tula about 177 kilometers (110 miles) south of Moscow

Yaz Carl Yastrzemski (baseball player)

Yedo Tokyo's old name (also written Edo)

Yekké Israeli-born nickname for German Jews

Yellow Belly Youngstown Sheet & Tube's nickname

yellowcake uranium ore (U_3O_8)

Yellow Emperor Huang Ti

yellow flag yellow signal flown when a vessel requests pratique; letter Q or Quebéc in the international code; also called the quarantine flag

Yellowhammer Alabama state bird; symbolic nickname of an Alabaman

The Yellowhammer State Alabama

yellowjack quarantine flag; yellow fever; yellow flag

Yellow Kid Micky Dugan, a character in the comic strip *Hogan's Alley*

yellow rain biochemical poison dropped by enemy aircraft; deadly weapon consisting of chemicals and myotoxins

yellows nembutal

Yellow Thunder Country around the Dells of the Wisconsin River near Baraboo, Wisconsin

Yemeni Ports Qishn, Al Luhayyah, Kamaran, Al Hudayah, Al Mukalla, Perim Harbour, Aden

Yerba Buena former name of San Francisco

Yerevan Armenia's Erevan or Erivan

Yezo Japanese island of Hokkaido

Yggdrasill tree of the universe, according to Norse mythology

Yodelandia Switzerland

Yogi Lawrence Peter Berra, known as Yogi Berra

Yoknapatawpha William Faulkner's mythical Mississippi county

Yorkshire Queen of Song Susan Sunderland

York State New York State

Young Hickory James K Polk—eleventh President of the United States

Young Napoleon U.S. General George B McClellan (1826– 1885)

Young Pretender Bonnie Prince Charlie (Charles Edward Louis Philip Casimir Stuart)—son of the Old Pretender—James Stuart

Youth Personified Juventus (Latin—youth)

You've Got a Friend State Pennsylvania

Yo-Yo Luis Arroyo (baseball player)

Yseult Isolde

Y-town Yorktown; Youngstown

Yucatan Channel ocean passage between Cuba and Yucatan Peninsula of Mexico; connects Caribbean Sea with Gulf of Mexico

Yucca Country the southwestern United States

Yukio Mishima Kimitake Hiraoka's pseudonym

Yul Brynner Taidje Kahn, Jr

Yu-Lin Betty Yü-Lin Ho

Yuri Bilstin Youry Bildstein

Yves Montand Ivo Livi

Yvonne de Carlo Peggy Yvonne Middleton

Zagreb Agram

Zaire Ports Banana, Boma, Matadi

Zancle old name for Messina

Zane Grey Pearl Grey's pseudonym

zazou (French nickname— zootsuiter)

Zazul Vera Zazulich

Zbig Zbigniew Brzezinski

Zedland English shires of Devon, Dorset, and Somerset

Zeke Ezekiel

zeks (Soviet-Russian slang— prisoners)

Zelda Griselda

Zen Garden of the Atlantic England's Scilly Isles

Zenith imaginary Minnesota city whose leading citizen is George F Babbitt described by Sinclair Lewis in his novel, *Babbitt*

Zenith City of the Unsalted Sea Duluth, Minnesota on Lake Superior leading to the other Great Lakes

Zeppelin Graf von Zeppelin, builder of rigid airships bearing his name

Zeppo Marx Herbert Marx

Zero Mostel Sam Mostel

Z-grams Admiral Zumwalt's policy statements

Zih Zihuatenajo, Mexico

Zihua Zihuatenajo, Mexico

Zilia Clara Josephine Wieck (later Clara Schumann, the wife of the composer-critic Robert Schumann who gave her that pseudonym)

Zilli Cecilia

Zinoviev, Grigori Evseevich Hirsch Apfelbaum

Zipango Marco Polo's name for Japan (variant spelling of Xipangu)

zips zipper fruits (easy-to-peel targelos and tangerines)

Zoé (French nickname—atomic pile)

Zola of America Sir Arthur Conan Doyle's nickname for Upton Sinclair

Zonian(s) American(s) of the Panama Canal Zone

Zoological Attic of the World Australia, New Guinea, New Zealand, Tasmania

Zorro Zoilo Versalles (baseball player)

Zsa Zsa Gabor Sari Gabor

Zsuzsa Zsuzsa Heiligenberg

Zubie Zubin Mehta of the New York Philharmonic

Zuinglius Latinization of Ulrich Zwingli

Zulo Ignacio de Zuloaga

Zungaria Dzungaria or Sungaria region between Mongolia and Russia

Fishing Port Registration Symbols

(Distinguishing Letters)

England

	GE Goole	**PW** Padstow
AB Aberystwith	**GR** Gloucester	**PZ** Penzance
BD Bideford	**GY** Grimsby	**R** Ramsgate
BE Barnstaple	**HH** Harwich	**RN** Runcorn
BH Blyth	**HL** Hartlepool, West	**RR** Rochester
BK Berwick-on-Tweed	**IH** Ipswich	**RX** Rye
BL Bristol	**LA** Llanelly	**SA** Swansea
BM Brixham	**LI** Littlehampton	**SC** Scilly
BN Boston	**LL** Liverpool	**SD** Sunderland
BR Briggwater	**LN** Lynn	**SE** Salcombe
BS Beaumaris	**LO** London	**SH** Scarborough
BW Barrow	**LR** Lancaster	**SM** Shoreham
CA Cardigan	**LT** Lowestoft	**SN** Shields, North
CF Cardiff	**M** Milford	**SS** St Ives
CH Chester	**MH** Middlesbrough	**SSS** Shields, South
CK Colchester	**MN** Maldon	**ST** Stockton
CL Carlisle	**MR** Manchester	**SU** Southhampton
CO Carnarvon	**MT** Maryport	**TH** Teignmouth
CS Cowes	**NE** Newcastle	**TO** Truro
DH Dartmouth	**NN** Newhaven	**WA** Whitehaven
DR Dover	**NT** Newport, Mon.	**WH** Weymouth
E Exeter	**P** Portsmouth	**WI** Wisbech
FD Fleetwood	**PE** Poole	**WO** Workington
FE Folkstone	**PH** Plymouth	**WY** Whitby
FH Falmouth	**PN** Preston	**YH** Yarmouth (Norfolk)
FY Fowey	**PT** Port Talbot	

Northern Ireland

B Belfast	**LY** Londonderry
CE Coleraine	**N** Newry

Republic of Ireland

C Cork	**G** Galway	**T** Tralee
D Dublin	**L** Limerick	**W** Waterford
DA Drogheda	**S** Skibbereen	**WD** Wexford
DK Dundalk	**SO** Sligo	**WT** Westport

Greek Alphabet

ALPHA	A	α	IOTA	I	ι	RHO	P	ρ
BETA	B	β	KAPPA	K	κ	SIGMA	Σ	σ
GAMMA	Γ	γ	LAMBDA	Λ	λ	TAU	T	τ
DELTA	Δ	δ	MU	M	μ	UPSILON	Y	υ
EPSILON	E	ε	NU	N	ν	PHI	Φ	ϕ
ZETA	Z	ζ	XI	Ξ	ξ	CHI	X	χ
ETA	H	η	OMICRON	O	o	PSI	Ψ	ψ
THETA	Θ	θ	PI	Π	π	OMEGA	Ω	ω

International Civil Aircraft Markings

AN Nicaragua	JY Jordan	TG Guatemala
AP Pakistan	LN Norway	TI Costa Rica
B Taiwan (Formosa)	LV Argentine Republic	VH Australia
CB Bolivia	LX Luxembourg	VP; VQ; VR British Colonies
CC Chile	LZ Bulgaria	and Protectorates
CF Canada	MC Monte Carlo	VT India
CR; CS Portugal and colonies	N United States of America	XA; XB; XC Mexico
CU Cuba	OB Peru	XH Honduras
CX Uruguay	OD Lebanon	XT China (Nationalist)
CZ Principality of Monaco	OE Austria	XY; XZ Burma (Myanmar)
D Germany	OH Finland	YA Afghanistan
EC Spain	OK Czech Republic	YE Yemen
EI; EJ Ireland	OO Belgium	YI Iraq
EL Liberia	OY Denmark	YK Syria
EP Iran	PH Netherlands	YR Romania
ET Ethiopia	PI Philippine Republic	YS El Salvador
F France and French Union	PJ Curaqao (Netherlands An-	YU Yugoslavia
G United Kingdom	tilles)	YV Venezuela
HA Hungary	PK Indonesia	ZA Albania
HB Switzerland	PP; PT Brazil	ZK; ZL; ZM New Zealand
HC Ecuador	PZ Suriname	ZP Paraguay
HH Haiti	RX Republic of Panama	ZS; ZT, ZU Union of South
HI Dominican Republic	SE Sweden	Africa
HK Colombia	SN Sudan	₄R Ceylon
HL Korea	SP Poland	₄X Israel
HS Thailand	SU Egypt	₅A Libya
HZ Saudi Arabia	SX Greece	₉G Ghana
I Italy	TC Turkey	
JA Japan	TF Iceland	

International Conversions Simplified

area

a (acres)	x	0.4	= ha (hectares)
cm²(square centimeters)	x	0.16	= in.² (square inches)
ft² (square feet)	x	0.09	= m² (square meters)
ha (hectares)	x	2.5	= a (acres)
in.² (square inches)	x	6.5	= cm² (square centimeters)
km² (square kilometers)	x	0.4	= mi² (square miles)
m² (square meters)	x	1.2	= yd² (square yards)
mi² (square miles)	x	2.6	= km² (square kilometers)
yd² (square yards)	x	0.8	= m² (square meters)

length

cm (centimeters)	x	0.4	= in. (inches)
ft (feet)	x	30.0	= cm (centimeters)
in. (inches)	x	2.54*	= cm (centimeters) *exactly
km (kilometers)	x	0.6	= mi (miles)
m (meters)	x	3.3	= ft (feet)
m (meters)	x	1.1	= yd (yards)
mi (miles)	x	1.6	= km (kilometers)
mm (millimeters)	x	0.04	= in. (inches)
yd (yards)	x	0.9	= m (meters)

temperature (exact)

C (degrees Celsius or centigrade)	x	9/5 + 32	= F (degrees Fahrenheit)
F (degrees Fahrenheit)	–	32 x 5/9	= C (degrees Celsius or centigrade)

volume

cups	x	0.24	= l (liters)
fl oz (fluid ounces)	x	30.00	= ml (milliliters)
ft³ (cubic feet)	x	0.03	= m³ (cubic meters)
gal (British Imperial gallons)	x	4.6	= l (liters)
gal (U.S. gallons)	x	3.8	= l (liters)
l (liters)	x	2.1	= pt (pints)
l (liters)	x	1.06	= qt (quarts)
l (liters)	x	0.22	= gal (British Imperial gallons)
l (liters)	x	0.26	= gal (gallons)
m³ (cubic meters)	x	35.00	= ft³ (cubic feet)
m³ (cubic meters)	x	1.3	= yd³ (cubic yards)
ml (milliliters)	x	0.03	= fl oz (fluid ounces)
pt (pints)	x	0.47	= l (liters)
qt (quarts)	x	0.95	= l (liters)
tbsp (tablespoons)	x	15.00	= ml (milliliters)
tsp (teaspoons)	x	5.00	= ml (milliliters)
yd³ (cubic yards)	x	0.76	= m³ (cubic meters)

weight

g (grams)	x	0.035	= oz (ounces)
kg (kilograms)	x	2.2	= lb (pounds)
lb (pounds)	x	0.45	= kg (kilograms)
oz (ounces)	x	28.00	= g (grams)
st (short tons—2000 pounds)	x	0.9	= t (tonnes)
t (tonnes—1000 kilograms)	x	1.1	= st (short tons)

International Radio Alphabet and Code

A Alpha · –	**M** Mike – –	**Y** Yankee – · – –
B Bravo – · · ·	**N** November – ·	**Z** Zulu – – · ·
C Charlie – · – ·	**O** Oscar – – –	**0** (zee-ro) – – – – –
D Delta – · ·	**P** Papa · – – ·	**1** (wun) · – – – –
E Echo ·	**Q** Quebec (kaybeck) – – · –	**2** (too) · · – – –
F Foxtrot · · – ·	**R** Romeo · – ·	**3** (thuh-ree) · · · – –
G Golf – – ·	**S** Sierra · · ·	**4** (fo-wer) · · · · –
H Hotel · · · ·	**T** Tango –	**5** (fi-yiv) · · · · ·
I India · ·	**U** Uniform · · –	**6** (siks) – · · · ·
J Juliet · – – –	**V** Victor · · · –	**7** (sev-ven) – – · · ·
K Kilo – · –	**W** Whiskey · – –	**8** (ate) – – – · ·
L Lima (leema) · – · ·	**X** Xray – · · –	**9** (ni-yen) – – – – ·

International Symbols

Country	Symbol	Country	Symbol	Country	Symbol
Albania	BSA	India	ISI	Philippines	PS
Algeria	INAPI	Indonesia	YDNI	Poland	PKNiM
Australia	SAA	Iran	ISIRI	Portugal	IGPAI
Austria	ON	Iraq	IOS	Romania	IRS
Bangladesh	BDSI	Ireland	IIRS	Saudi Arabia	SASO
Belgium	IBN	Israel	SII	Singapore	SISIR
Brazil	ABNT	Italy	UNI	South Africa,	
Bulgaria	DKC	Jamaica	JBS	Rep. of	SABS
Canada	SCC	Japan	JISC	Spain	IRANOR
Chile	INN	Kenya	KEBS	Sri Lanka	BCS
Colombia	ICONTEC	Korea, Dem.		Sudan	SSD
Cuba	NC	P. Rep. of	CSK	Sweden	SIS
Czechoslovakia	CSN	Korea, Rep. of	KBS	Switzerland	SNV
Denmark	DS	Lebanon	LIBNOR	Thailand	TISI
Egypt, Arab		Malaysia	SIRIM	Turkey	TSE
Rep. of	EOS	Mexico	DGN	United Kingdom	BSI
Ethiopia	ESI	Morocco	SNIMA	United States	
Finland	SFS	Netherlands	NNI	of America	ANSI
France	AFNOR	New Zealand	SANZ	Venezuela	COVENIN
Germany	DIN	Nigeria	NSO	Yugoslavia	JZS
Ghana	GSB	Norway	NSF	Zambia	ZSI
Greece	NHS	Pakistan	PSI		
Hungary	MSZH	Peru	ITINTEC		

International Vehicle
License Letters

A Austria
ADN Aden (South Yemen, People's Democratic Republic of Yemen)
AFG Afghanistan
AL Albania
AND Andorra
AUS Australia
B Belgium
BD Bangladesh
BDS Barbados
BG Bulgaria
BH Belize (formerly, British Honduras)
BIH Bonsia-Herzegovina
BR Brazil
BRN Bahrain
BRU Brunei
BS Bahamas
BV Bolivia
BUR Burma (Myanmar)
C Cuba
CDN Canada
CH Switzerland
CI Côte d'Ivoire
CL Sri Lanka
CO Colombia
CR Costa Rica
CU Curaqao
CY Cyprus
CZ Czech Republic
D Germany
DK Denmark
DOM Dominican Republic
DY Benin
DZ Algeria
E Spain (España)
EAK Kenya
EAT Tanzania
EAU Uganda
EAZ Tanzania
EC Ecuador
ES El Salvador
ET Egypt
ETH Ethiopia
F France
FJI Fiji
FL Liechtenstein
FR Faeroe Islands
GB Great Britain
GBA Alderney
GBG Guernsey
GBJ Jersey
GBM Isle of Man
GBZ Gibraltar

GCA Guatemala
GH Ghana
GR Greece
GUY Guyana
H Hungary
HN Honduras
HK Hong Kong
HKJ Jordan
I Italy
IL Israel
IND India
IR Iran
IRL Ireland
IRQ Iraq
IS Iceland
J Japan
JA Jamaica
K Kampuchea
KWT Kuwait
KZ Kazakhstan
L Luxembourg
LAO Laos
LAR Libya
LB Liberia
LR Latvia
LS Lesotho
M Malta
MA Morocco
MAL Malaysia
MC Monaco
MEX Mexico
MS Mauritius
MW Malawi
N Norway
NA Netherlands Antilles
NEP Nepal
NIC Nicaragua
NL Netherlands
NZ New Zealand
P Portugal
PA Panama
PAK Pakistan
PE Peru
PL Poland
PNG Papua New Guinea
PY Paraguay
RA Argentina
RB Botswana
RC Taiwan
RCA Central African Republic
RCB Democratic Republic of Congo
RCH Chile
RH Haiti
RI Indonesia

RIM Mauritania
RL Lebanon
RM Madagascar
RMM Mali
RN Niger
RO Romania
ROK Korea
RP Philippines
RSM San Marino
RU Burundi
RWA Rwanda
S Sweden
SD Swaziland
SF Finland
SGP Singapore
SK Slovakia
SL Slovenia
SME Suriname
SN Senegal
SO Somalia
SU former USSR
SWA Namibia
SY Seychelles
SYR Syria
T Thailand
TC Cameroon
TG Togo
TJ Tajikistan
TM Turkmenistan
TN Tunisia
TR Turkey
TT Trinidad and Tobago
U Uruguay
USA United States of America
V Vatican City
VN Vietnam
WAG The Gambia
WAL Sierra Leone
WAN Nigeria
WD Dominica
WG Grenada
WL Saint Lucia
WS Western Samoa
WV Saint Vincent and the Grenadines
YMN North Yemen (Yemen Arab Republic)
YU Yugoslavia
YV Venezuela
Z Zambia
ZA South Africa
ZRE Republic of Congo (formerly Zaire)
ZW Zimbabwe

International Yacht Racing Union Nationality Codes

Every yacht of an international class recognized by the International Yacht Racing Union must carry on her mainsail, when racing in foreign waters, a letter or letters showing her nationality. In 1997, these letters were changed to conform to International Olympic Committee codes.

A Argentina
AR United Arab Republic
B Belgium
BA Bahamas
BL Brazil
BU Bulgaria
CA Cambodia
CY Ceylon
CZ Czech Republic
D Denmark
E Spain
EC Ecuador
F France
G Germany
GR Greece
H Holland
HA Netherlands Antilles
I Italy
IR Republic of Ireland
K United Kingdom
KA Australia
KB Bermuda

KC Canada
KG Guyana
KGB Gibraltar
KH Hong Kong
KI India
KJ Jamaica
KK Kenya
KR Zambia, Malawi, Zimbabwe
KS Singapore
KT West Indies
KZ New Zealand
L Finland
LE Lebanon
LX Luxembourg
M Hungary
MA Morocco
MO Monaco
MX Mexico
N Norway
NK Democratic People's Republic of Korea

OE Austria
P Portugal
PH Philippines
PR Puerto Rico
PU Peru
PZ Poland
RC Cuba
RI Indonesia
RM Romania
S Sweden
SA South Africa
SE Senegal
T Tunisia
TH Thailand
TK Turkey
U Uruguay
US United States of America
V Venezuela
X Chile
Y Yugoslavia
Z Switzerland

Inventions and Inventors

AC motor Nikola Testa

adding machine Blaise Pascal and William Burroughs

addressograph J S Duncan

air brake George Westinghouse

air conditioning Willis H Carrier

airline passenger ticket Richard Roselle

airplane Orville and Wilbur Wright

AK-17 (automatic rifle) Mikhail Kalashnikov

Alexander Technique massage and physical-manipulation therapy developed by F Mathias Alexander

analog computer Vannevar Bush

autogyro Juan de la Cierva

automatic rifle John Moses Browning

automobile Karl Benz

bakelite Leo H Baekeland

ballpoint pen John Loud

barbed wire Joseph Glidden

barometer Evangelista Torricelli

Ben Day Process photoengraving process; Benjamin Day

Bessemer steel Henry Bessemer

bicycle Karl Drais von Sauerbronn

bifocals Benjamin Franklin

Bowen Therapy musculoskeletal-release therapy developed by Tom Bowen of Australia

Braille printing Louis Braille

bubble gum Walter Deemer (for Fleers in 1928)

Bunsen burner Robert Wilhelm von Bunsen

calculus Wilhelm von Leibniz and Isaac Newton

carbon-filament electric lamp Thomas A Edison

carburetor Gottlieb Daimler

carpet sweeper M R Bissell

cash register James Ritty

cathode-ray tube Sir William Crookes

cellophane Jacques E Brandenberger

celluloid Alexander Parkes

celluloid-film photography Hannibal Goodwin

cement Joseph Aspdin

chronometer John Harrison

cigarette-rolling machine James A Bonsack

coal stove Jordan Mott

color photography Gabriel Lippmann

compound microscope Hans and Zacharias Janssen

computer chip Jack Kilby and Robert Noyce (according to US Patent Office)

computer cookies Lou Montulli

computer mouse Douglas Engelbart

condenser steam engine James Watt

cotton gin Eli Whitney

craniosacral therapy William Sutherland

cream separator Gustf de Laval

cyclotron Ernest O Lawrence

cylinder lock Linus Yale

cylinder phonograph Thomas A Edison

detective story Edgar Allan Poe

Diesel engine Rudolf Diesel

digital computer Howard Aiken

diode-tube radio John Fleming

disk brake Frederick Lanchester

disk phonograph Emile Berliner

dynamite Alfred Nobel

electric battery Alessandro Volta

electric bulb Thomas Alva Edison

electric fan Schuyler Wheeler

electric flat iron H W Seeley

electric generators Michael Faraday (disk), Hippolyte Pixii (coil)

electric locomotive Werner von Siemens

electric motors Zenobe Gramme (DC), Nikola Testa (AC)

electric power plant Thomas A Edison

electric razor Jacob Schick

electric stove W S Hadaway

electric telegraph Samuel F B Morse

electric washing machine Alva Fisher

electric welder Elihu Thompson

electrocardiograph Willem Einthoven

electroencephalograph Hans Berger

electromagnet Joseph Henry, William Sturgeon

electronic television Vladimir Zworykin

electron microscope Ernst Ruska

electroplating George and Henry Elkington

Esperanto Lazarus Ludwig Zamenhof

Feldenkrais Technique posture therapy developed by Moshe Feldenkrais

field ion microscope Erwin W Mueller

fluorescent light Edmund Germer

flush toilet Joseph Brahmah

foot-operated sewing machine Isaac Singer

fountain pen Lewis Waterman

fractal geometry Benoit Mandelbrot

Franklin stove Benjamin Franklin

frequency modulation Edwin H Armstrong

frozen food Clarence Birdseye

gas lighting William Murdock

gasoline-powered automobile Karl Benz

gas stove Robert Wilhelm von Bunsen

Geiger counter Hans Geiger

gramophone Emile Berliner

gyrocompass Elmer A Sperry

gyroscope Jean Bernard Léon Foucault

half-tone engraving Frederick Ives

heart pacemaker Wilson Greatbach

Heimlich maneuver upward thrust on abdomen, performed on choking victims (developed by Dr Henry Heimlich)

helicopter Igor Sikorsky

holography Dennis Gabor

homeopathy Samuel Hahnemann

hot-air balloon Jacques and Joseph Montgolfier

jet aircraft engine Sir Frank Whittle

kiss-proof lipstick movie director Preston Sturges

Kodak camera George Eastman

laser Gordon Gould

La-Z-Boy reclining chair Edwin Shoemaker

lever-fill fountain pen W A Shaeffer

lightning rod Benjamin Franklin

linoleum Frederick Walton

linotype typesetting Ottmar Mergenthaler

Linux computer operating system developed by Linus Torvalds

liquid-crystal display James L Fergason

liquid-fuel rockets Robert Goddard, Herman Oberth, Konstantin Tsiolkovsky

lithography Alois Senefelder

long-playing records Peter Goldmark Elihu

machine gun Richard J Gatling

magazine-insert advertising cards Lester Wunderman

magnetic tape recording J A O'Neill

manual washing machine Hamilton Smith

mechanical television John L Baird

metronome Dietrich Nikolaus Winkel

microchip Robert Noyce and Jack Kilby

microphone Emile Berliner

microwave oven Percy L Spencer

milking machine Anna Baldwin

miner's safety lamp Humphry Davy

modern oilwell Edwin Drake

monotype typesetting Tolbert Lanston

motion picture camera Thomas A Edison

motorcycle Gottlieb Daimler

motor scooter Greville Bradshaw

movable-type printing Johann Gutenberg

multiplex theaters Stan Durwood

music synthesizer Robert A Moog

neon lamp Georges Claude

newspaper vending machine George Thiemeyer Hemmeter

nylon Wallace Carothers

offset printing Ira Rubel

oil pipeline Samuel van Syckel

osteopathy Andrew Taylor Still

outboard engine Ole Evinrude

parallel computer processor Gene Amdahl

parking meter Carlton Magee

pendulum clock Christiaan Huygens

photoelectric cell G R Carey

photoengraving William Fox Talbot

photographic camera Joseph Niepce

photography on metal Joseph Niepce

photography on paper William Fox Talbot

piston steam engine Thomas Newcomen

pneumatic tires John B Dunlop

Polaroid camera Edwin Land

polymeric chain reaction Kary Banks Mullis

polyurethane W E Hanford and Donald F Holmes

power loom Edmund Cartwright

projected motion pictures Auguste and Louis Lumière

propeller John Ericsson

punched cards Herman Hollerith

radar Albert Taylor, Leo Young

radio Guglielmo Marconi

radio telescope Karl Jansky

railway car couplings Eli Jannery

railway sleeping cars George Pullman

rayon Hilaire Chardonnet

reaping machine Cyrus H McCormick

reflecting telescope Isaac Newton

refracting telescope Galileo Galilei

repeating rifle O F Winchester

respirator Forrest Bird

revolver Samuel Colt

rigid dirigible airship Ferdinand von Zeppelin

Rolfing physical therapy developed by Ida Rolf

rotary printing Richard Hoe

safety elevator Elisha G Otis

safety pin Walter Hunt

safety razor King C Gillette

science fiction Mary Wollstonecraft Shelley, author of *Frankenstein*

science of flight Sir George Cayley

Scoville Test measure of chilies' hotness by how much sugar water is required to neutralize heat; invented by Wilbur Scoville

Scrabble (board game) Alfred Butts

screw propeller John Ericsson

self-propelled torpedo Robert Whitehead

self-starter Charles F Kettering

semaphore telegraph Claude Chappe

sewing machine Elias Howe, Barthélemy Thimonnier

sextant John Hadley, Thomas Godfrey

shopping cart Sylvan Goldman

spectroscope Joseph von Fraunhofer

spinning frame Richard Arkwright

spinning jenny James Hargreaves

stainless steel Harry Brearley

steam locomotive Richard Trevithick

steam railway George Stephenson

steamship Robert Fulton

steam shovel William Otis

steam tractor Nicholas Cugnot

steam turbine Charles Parsons

steel plow John Deere

stethoscope René Théophile-Hyacinthe Laennec

submarine David Bushnell

suburbs William Levitt, who designed Levittown, Pennsylvania

supercomputer Seymour Cray

swim fins Benjamin Franklin

telegraph signals Samuel B Morse

telegraphy Alexander Graham Bell

telephone Alexander Graham Bell

telephone dial tone Almon Strowger (1891), an undertaker

telescope Hans Lippershey

thermometer Galileo Galilei

thermos bottle James Dewar

Trager Work soft-touch manipulation combined with exercises to release tension, developed by Milton Trager

transformer Otto Blathy

transistor John Bardeen, Walter Brattain, William Shockley

transparent paper-film photography George Eastman

triode vacuum tube Lee De Forest

tungsten filament Irving Langmuir

tungsten-filament electric lamp William Coolidge

typewriter Carlos Glidden, Christopher Sholes

vacuum cleaner I W McGaffey

videotape recorder Charles P Ginsburg

vocal radio Valdemer Poulsen

vulcanized rubber Charles Goodyear

waterproof raincoat Charles Macintosh

wire-cable suspension bridge John A Roebling

wireless telegraphy Guglielmo Marconi

xerography Chester Carlson

X-ray Wilhelm Roentgen

zipper Whitcomb Judson

Irish Counties Abbreviated

Car County Carlow
Cav County Cavan
Clare County Clare
Cork County Cork
Don County Donegal
Dub County Dublin
Gal County Galway
Ker County Kerry
Kild County Kildare

Kilk County Kilkenny
Leit County Leitrim
Leix County Leix
Lim County Limerick
Long County Longford
Louth County Louth
Mayo County Mayo
Meath County Meath
Monag County Monaghan

Off County Offaly
Ros County Roscommon
Sligo County Sligo
Tipp County Tipperary
Wat County Waterford
Westmeath County Westmeath
Wex County Wexford
Wick County Wicklow

Mexican State Names and Abbreviations

Ags Aguascalientes (inhabitants called Hidrocalidos)

BC Baja California (Baja Californianos)

BC Front Baja California Fronteriza (Frontier Baja California)

BC Sur Baja California Sur (Baja California South)

Cam Campeche (Campechanos)

Chih Chihuahua (Chihuahuenses)

Chis Chiapas (Chiapanecos)

Coah Coahuila (Coahuileños or Coahuilenses)

Col Colima (Colimenses)

DF Distrito Federal (Federal District around Mexico City; Capitolinos)

Dgo Durango (Durangueños or Duranguenses or Durangueses)

Gro Guerrero (Guerreros)

Gto Guanajuato (Guanajuatos)

Hgo Hidalgo (Hidalgos)

Jal Jalisco (Jalisciences)

Méx México (Mexicanos)

Mich Michoacán (Michoacanos)

Mor Morelos (Morelianos)

Nay Nayarit (Nayaritos)

NL Nuevo León (Nuevo Leones)

Oax Oaxaca (Oaxaqueños)

Pue Puebla (Poblanos)

Qro Querétero (Queretanos)

Q Roo Quintana Roo (Quintana Roenses)

Sin Sinaloa (Sinaloens[es]) ⸱

SLP San Luís Potosí (Potiseños)

Son Sonora (Sonorenses)

Tab Tabasco (Tabasqueños)

Tam Tamaulipas (Tamaulipecos)

Tlax Tlaxcala (Tlaxcaltecas)

Ver Veracruz (Veracruzanos)

Ync Yucatán (Yucatecos)

Zac Zacatecas (Zacatecos)

Musical Nicknames and Superlatives

ABCs of Opera *Aida, Boheme, Carmen*

Adam's Most Popular Ballet *Giselle*

Adieu Chopin's Polonaise in B-flat minor

Aegyptische *Die Aegyptische Helena* (German—The Egyptian Helen)—Richard Strauss's music drama

Aeolian Symphony Rachmaninoff's Symphony No 3 in A minor

Africaine *L'Africaine* (French—The African Girl)—Meyerbeer opera

African-American Arranger-Composer-Conductor Eva Jessye

Age of Anxiety Bernstein's Symphony No 2

Albeniz's Most Popular Piano Piece *Iberia*

Albinoni's Most Popular Piece *Adagio for Strings and Organ*

Al combate corred bayameses (Spanish—Swift in combat, men of Bayamo)—Cuban national anthem

Alfven's Most Popular Overture *Midsommervaka* (Swedish—Midsummer-Night Vigil)

All Men are Brothers Symphony No 11 by Alan Hovhaness

Alpine *Alpine Symphony* (symphonic poem by Richard Strauss—*Eine Alpensinfonie*)

Alpine Waterfall Saint-Saëns Piano Concerto No 3 in E flat major

Alsatian musicians conductor Charles Munch; organists Edvard Nies-Berger and Albert Schweitzer; composer Émile Waldteufel

Amahl *Amahl and the Night Visitors* (Menotti opera)

Amelia *Amelia Goes to the Ball* (Menotti one-act comic opera)

American Dvořák's Quartet in F (opus 96)—for two violins, viola, and cello

American Symphony Orchestras of Greatest Distinction Boston Symphony, New York Philharmonic, Philadelphia Orchestra, Chicago Symphony, Cleveland Orchestra, St Louis Symphony, Los Angeles Philharmonic, San Francisco Symphony, Seattle Symphony

An Donau *An der schönen blauen Donau* (German—On the beautiful blue Danube)—waltz by Johann Strauss, Jr

And This Is My Beloved Borodin's Nocturne from his *String Quartet No 2*

Anniversary Waltz Ivanovici's fanfare titled *Danube Waves*

Antar Rimsky-Korsakov's *Symphony No 2*

Antheil's Most Popular Work *Ballet Mecarique*

Apotheosis of the Dance Beethoven's Symphony No 7

Appassionata Beethoven's Piano Sonata No 23 in F minor (opus 57)—nicknamed for its impassioned mood

Archduke Beethoven's Trio in B minor (opus 97)—dedicated to Archduke Rudolph

Arctic Vassilenko's Fourth Symphony

Arensky's Most Popular Ballet *Egyptian Night*

Argentinian Composers Alberto Ginastera, Juan Guiterrez, Roberto Morillo, Alberto Williams

Ariadne *Ariadne auf Naxos* (German—Ariadne on Naxos)—one-act opera by Richard Strauss

Aristocrat of Orchestras The Boston Symphony Orchestra

Arlésienne *L'Arlésienne Suite No 1* and *Suite No 2* by Bizet

Arne's Most Popular Oratorio *Judith*

Auber's Most Popular Overture *Masaniello*

Aus der Neuen Welt (German—From the New World)—Dvořák's Ninth Symphony

Aus Ital *Aus Italien* (German—From Italy)—symphonic poem by Richard Strauss

Aus meinem *Aus meinem Leben* (German—From My Life)—Smetana's autobiographical String Quartet No 1 in E minor

Australian composers Arthur L Benjamin, Peggy Glanville-Hicks, Percy Aldridge Grainger

Australian Duo classical composers Percy Grainger and Arthur Benjamin

Austrian composers Alban Berg, Anton Bruckner, Franz Joseph Haydn, Wolfgang Amadeus Mozart, Arnold Schoenberg, Franz Schubert, the Strauss family (Johann, Johann, Jr, the two Josephs)

Austrian Waltz Kings Josef (Franz Karl)—Lanner and Johann Strauss, Jr

Babi Yar Symphony No 13 of Shostakovich inspired by poems of Yevtushenko

Bach's Most Popular Chaconne *Partita in D minor for Unaccompanied Violin*

Bach's Most Popular Orchestral Works 6 *Brandenburg Concerti*

Bach's Most Popular Piece *Air for the G string*

Bach's Most Popular Work *Toccata and Fugue in D minor*

Balfe's Most Popular Opera *The Bohemian Girl*

Ballo *Un Ballo in Maschera* (Italian—A Masked Ball)—three-act opera by Verdi

Banditenstreiche (German—Jolly Robbers)—von Suppé overture

Barber *Barber of Seville* (Rossini opera)

Barber's Most Popular Symphonic Work *Adagio for Strings*

Barbiere *Il Barbiere di Siviglia* (Italian—The Barber of Seville)—opera by Rossini

Baritone of the Century Dietrich Fischer-Dieskau (1925–)

Bartók's Most Popular Symphonic Work *Concerto for Orchestra*

Bass of the Century Feodor Chaliapin (1873–1938)—George London (1920–1985)

basset-horn tenor clarinet

Bassoonist of the Century Sol Schoenbach of the Philadelphia Orchestra

Bayang megiliw (Tagalog—Land of the Morning)—Philippine national anthem

Beethoven 1st, 2nd, 3rd, 4th, 5th, 6th, 7th, 8th, 9th First, Second, Third (*Eroica*)—Fourth, Fifth, Sixth (*Pastoral*)—Seventh, Eighth, Ninth (*Choral*)—symphonies composed by Beethoven

Beethoven's Most Popular Overture *Leonore No 3* from *Fidelio*

Beethoven's Most Popular Piano Concerto *No 5 Emperor*

Beethoven's Most Popular Piano Sonata *No 14 Moonlight*

Beethoven's Most Popular Song *Adelaide*

Beethoven's Most Popular Symphony *No 5 in C minor*

Beggar's *The Beggar's Opera* (ballad opera by John Christopher Pepusch with libretto by John Gay)

Belgian composers Charles Bériot, César Franck, André Grétry, Henri Vieuxtemps

Belgian Quartet Belgium's best-known classical composers, ranked chronologically, include Grétry, Vieuxtemps, Franck, and Ysaÿe

Bellini's Most Popular Opera *Norma*

Bells (*see The Bells*)

Bells of Zionice Dvořák's Symphony No 1

Benvenuto *Benvenuto Cellini*, Berlioz opera nicknamed *Malvenuto* by critics

Berg's Most Popular Opera *Lulu*

Berlin Bach Carl Philipp Emanuel Bach, also called the Hamburg Bach

Berlioz's Most Popular Opera *Damnation of Faust*

Berlioz's Most Popular Symphonic Work *Symphonie Fantastique*

Berlioz symphonies *Symphonie Fantastique* and *Symphonie Funèbre et Triomphale*

Bernstein's Most Popular Ballet *Dybbuk*

Bernstein's Most Popular Overture *Candide*

best known Russian art song (*see None But The Lonely Heart*)

Best of the Bachs Johann Sebastian Bach who fathered Carl Philipp Emanuel, Johann Christina, Johann Christoph Friedrich, Wilhelm Friedmann; known collectively as die Bäche—the Bachs

Best of the Bonn Boys Ludwig van Beethoven, born in Bonn

Big Seven America's leading symphony orchestras—Boston, Chicago, Cleveland, Los Angeles, New York, Philadelphia, Pittsburgh

Big Six America's leading symphony orchestras—Boston, Chicago, Cleveland, Los Angeles, New York, Philadelphia

Bird Haydn's String Quartet in C (opus 33, no 3)

Birds *The Birds* (Respighi's symphonic poem—*Gli Uccelli*)

Bizet's Most Popular Opera *Carmen*

Bizet's Most Popular Song *Agnus Dei*

Bizet's Most Popular Work *L'Arlesienne Suites 1 & 2*

Black Conductor of the Century Dean Dixon (1915–1976)

Black Contralto of the Century Marian Anderson (1902–1993)

Black Key Chopin's Piano Etude No 5 in G-flat major

Black Orchestra Leader of the Century Count Basie (1904–1984)

Black Pianist of the Century Andre Watts (1946–)

Black Soprano of the Century Leontyne Price (1927–)

Black Tenor of the Century Roland Hayes (1887–1976)

Black Trumpeter of the Century Louis (Satchmo) Armstrong (1900–1971)

Blind Composer Joaquin Rodrigo born near Valencia, Spain, remembered for his guitar *Concerto Aranjuez*

Bloch's Most Popular Rhapsody *Schelomo for cello and orchestra*

Bluebeard's *Duke Bluebeards Castle* (Bartók's one-act opera)

Boccanegra *Simone Boccanegra* (three-act Verdi opera)

Boheme *La Boheme* (Puccini opera)

Bohemian composers Antonin Dvořák, Leoš Janacek, Gustav Mahler, Bedrich Smetana, Josef Suk

Boieldieu's Most Popular Overture *Caliph of Bagdad*

Boito's Most Popular Opera *Mefistofele*

Boris *Boris Godunov* (Mussorgsky's opera)

Borodin's Most Popular Opera *Prince Igor*

Brahms' 4 Brahms's four symphonies

Brahms' Most Popular Overlure *Academic Festival*

Brahms' Most Popular Symphony *No 1 in C minor*

Brandenburg Bach's *Brandenburg Concertos* dedicated to Duke Christian Ludwig of Brandenburg, Germany

Brazilian composers Antonio Carlos Gomes, Mozart Camargo Guarnieri, Heitor Villa-Lobos

Brazilian Trio Brazil's three best-known classical composers-Antonio Carlos Gomes, Alberto Nepomuceno, and Heitor Villa-Lobos

Britten's Most Popular Opera *Peter Grimes*

Britten's Most Popular Work *Young Person's Guide to the Orchestra*

Brooklyn composers Aaron Copland, George and Ira Gershwin, Roger Sessions (*see* New York composers)

Bruch's Most Popular Violin Concerto *No 1 in G*

Bruckner Conductor Eugen Jochum

Bruckner's 10 Bruckner's ten symphonies comprising *Die Nullte* (The Zero)—and 1 through 9 including *Romantische* (Romantic, No 4)

Bruckner's Most Popular Symphony *No 4 Romantic*

bup-bup-bup-bum Beethovenian kettle-drumming

Butterfly Chopin's *Piano Etude No 9 in G flat*; Puccini's opera *Madame Butterfly*—a Japanese tragedy

Calife Le Calife de Bagdad (French—The Caliph of Bagdad)—one-act opera by Boieldieu

California composers Henry Cowell, Harry Partch

Calm Sea Calm Sea and Prosperous Voyage (Goethe's *Meeresstille und glückliche Fahrt*)—Beethoven's choral work with orchestra or Mendelssohn's concert overture

Campanello Il Campanello di Notte (Italian—The Night Bell)—Donizetti opera

Carpenter's Most Popular Ballet *Skyscrapers*

Cavalleria Cavelleria Rusticana (Mascagni opera)

Cavalleria espanola Massenet's verismo opera *La Navarraise*, also called *Calvélleria espanola* after Emma Calvé

Cav-Pag Cavalleria Rusticana and *I Pagliacci* (Italian operas frequently performed in succession)

Celestial Gate Symphony No 6 by Alan Hovhaness

Cellini Benvenuto Cellini (opera by Berlioz, nicknamed *Malvenuto* by critics)

Cellist of the Century Pablo Casals (1876–1973)

Cellist-Conductor-Composer Pablo Casals

Cellist-Conductors Barbirolli, Casals, Herbert, Kindler, Rostropovitch, Toscanini, Wallenstein

'Cello Symphony Symphony No 2 by Taaffe Zwilic

Cenerentola La Cenerentola (Italian—Cinderella)—Rossini opera

Chabrier's Most Popular Overture *España*

Chaconne Bach's *Partita No 2 in D minor* for solo violin; or its transcription for the guitar of Segovia by Marc Pincherle; or for piano by Brahms, Busoni, Mendelssohn, Raff, or Schumann; or for orchestra by Hubay, Stokowski, or Wilhelmj

Charpentier's Most Popular Opera *Louise*

Chausson's Most Popular Piece *Poème for Violin and Orchestra*

Chavez' Most Popular Symphony *Sinforria India*

Cherubini's Most Popular Overture *Anacreon*

Chilean composers Pedro Allende, Próspero Biquert, Domingo Santa Cruz

Chilean First Chile's outstanding classical composer Pedro Umberto Allende

Chopin's Most Popular Piano Concerto *No 1 in E minor*

Choral Beethoven's *Symphony No 9 in D minor* whose last movement contains Schiller's *Ode to Joy* sung by chorus and soloists with full orchestral support

Choreographic Symphony Ravel's name for his *Daphnis et Chloé* ballet

Christmas sobriquet of Corelli's *Concerto Grosso Opus 6 Number 8*, Rimsky-Korsakov's *Christmas Eve* opera, Bach's *Christmas Oratorio*, Haydn's *Christmas Symphony* in D minor–No 26 also called *Lamentatione*

Clara Robert Schumann's *Symphony No 4 in D minor*

Clarinetist of the Century Benny Goodman (1909–1986)

Classical Prokofiev's *Symphony No 1*

Clock Haydn's *Symphony No 101 in D major*

Columbia the Gem of the Ocean symphonic poem by Charles Ives

Compliment Beethoven's *String Quartet in G major Opus 18 No 2*

Composer-Bandmaster Edwin Franko Goldman; Ivan Ivanovici; John Philip Sousa

Composer-Chemist Alexander Borodin

Composer-Conductor Johann Sebastian Bach; Hector Berlioz; Carlos Chávez; Aaron Copland; Edward Elgar; Carlos Gomes; Morton Gould; George Friedrich Handel; Ferdé Grofé; Howard Hanson; Aram Khachaturian; Franz Liszt; Gustav Mahler; Felix Mendelssohn; Carl Nielson; Oscar Straus; Richard Strauss; Franz von Suppé; Peter Ilyich Tchaikovsky; Heitor Villa-Lobos; Carl Maria von Weber; Richard Wagner

Composer-Conductor-Cellist Pablo Casals; Victor Herbert

Composer-Conductor-Critic Hector Berlioz

Composer-Conductor-Educator Leonard Bernstein; Walter Damrosch; Howard Hanson; Edward MacDowell; André Previn

Composer-Conductor-Musicologist Hector Berlioz; Nicholas Slonimsky; Richard Wagner

Composer-Conductor-Organist-Pianist Camille Saint-Saëns, Sir Charles Villiers Stanford, Sir Arthur S Sullivan

Composer-Conductor-Pianist Beethoven, Bernstein, Britten, Damrosch, Dohnanyi, Foss, Gottschalk, Grainger, Liszt, Mendelssohn, Prokofiev, Rachmaninoff, Stravinsky, and VillaLobos

Composer-Conductor-Pianist-Statesman Ignacy Jan Paderewski

Composer-Conductor-Pianist-Violinist Georges Enesco; Bedrich Smetana

Composer-Conductor-Pianist-Violinist Teacher Georges Enesco

Composer-Conductor-Violinist Hans Christian Lumbye; Juventino Rosas; Johann Strauss; Johann Strauss Jr; Josef Strauss; Eugéne Ysaye

Composer-Critic Joseph McCabe; Robert Schumann; Carl Shapiro; Virgil Thomson

Composer-Orchestrators Hector Berlioz; Maurice Ravel; Nikolai Rimsky-Korsakov; Richard Strauss; Richard Wagner

Composer-Organist Johann Sebastian Bach; Anton Bruckner; Dietrich Buxtehude; César Franck; Charles Gounod; George Friedrich Handel; Camille Saint-Saëns

Composer-Pianist Ludwig van Beethoven; Johannes Brahms; Frédéric Chopin; George Gershwin; Percy Grainger; Franz Liszt; Edward Macowell; Wolfgang Amadeus Mozart; Sergei Prokofiev; Sergei Rachmaninoff; Robert Schumann

Composer-Pianist-Conductor Ludwig van Beethoven; Louis Moreau Gottschalk; Franz Liszt; Sergei Rachmaninoff, Camille Saint-Saëns

Composer-Violinist Kreisler, Paganini, Sarasate, Vieuxtemps, Vivaldi, and Wieniawski

Computer Composer Yannis Xenakis

Conductor-Cellist Barbirolli, Casals, Rostropovich, Toscanini, and Wallenstein

Conductor of the Century Arturo Toscanini (1867–1957)—Herbert von Karajan (1908–1989)

Conductor-Chorus Master Frank Damrosch, Robert Shaw, Roger Wagner

Conductor-Composer George Barati; Pierre Boulez, Stanislaw Skrowaczewski

Conductor-Composer-Pianist Leonard Bernstein; Walter Damrosch; André Previn

Conductor-Double-Bass Serge Koussevitsky, Henry Lewis, Zubin Mehta

Conductor-Educator Leonard Bernstein, Walter Damrosch, André Previn

Conductor-Organist Edouard Nies-Berger; Leopold Stokowski; Walter Teutsch

Conductor-Organist-Pianist Eduard Nies-Berger, Leopold Stokowski, and Walter Teutsch

Conductor-Pianist Ashkenazy, Barenboim, Dello Joio, Foss, Ganz, Hendl, Iturbi, Mitropoulos, Previn, Solti, Szell, von Karajan, Walter, and Zinman

Conductor-Singer Placido Domingo

Conductor-Violinist Boskovsky, Brusilow, Burgin, Giulini; Haitink, Katims; (Daniel), Lewis, Menuhin, Munch, Oisrakh (father and son), Ormandy, Paganini, Piastro, Schneider, Silverstein, Stern, and Zukerman

Connecticut Composer Charles Ives

Conservatory Concerto for the Piano Mendelssohn's *Concerto No 1 in G minor* opus 25

Conservatory Concerto for the Violin Mendelssohn's *Concerto in E* opus 64

Contralto of the Century Marian Anderson (1897–1993)

Copland's Most Popular Work *Appalachian Spring*

Coq Le Coq d'Or (French—The Golden Cock)—Rimsky-Korsakoff opera

Cornerstone of Western Music Beethoven's *Ninth Symphony*, according to conductor Seiji Ozawa

Coronation Mozart's *Mass in C* or his *Piano Concerto in D major* (K 537)

Cosi Cosi Fan Tutti (Italian—Thus Do They All)—opera by Mozart

Cuban composers Alejandro García Caturla, Amadeo Roldán, Eduardo Sanchez de Fuentes y Peláez

Czar Czar und Zimmermann (German—czar and carpenter)—Lortzing opera

Czech Duo Czechoslovakia's best-known classical composers Bedrich Smetana and Antonin Dvořák

d'Albert's Most Popular Opera *Tiefland* (German—lowland)

Dalmatian First Yugoslavia's best-known operetta composer of the last century—Franz von Suppé

Damnación de Fausto (Spanish—Damnation of Faust)—four-part dramatic legend composed by Berlioz

Damnation La Damnation de Faust (French—The Damnation of Faust)—four part legend by Berlioz

Damrosch's Most Popular Opera *Man without a Country*

Danish Composers Niels Vilhelm Gade, Friedrich Kuhlau, Hans Christian Lumbye, Carl August Nielsen

Danish Quintet Denmark's five best-known classical composers—Buxtehude, Kuhlau, Lumbye, Gade, Nielsen

Dannazione di Faust (Italian—Damnation of Faust)—four-part legend by Berlioz

Danse macabre (French—dance of death)—Saint-Saëns work

Dante tone poem by Liszt

Danube Waves Ivanovici's fanfare, *Anniversary Waltz*

Daphnis Ravel's ballet *Daphnis et Chloé*

Das Judenthum Das Judenthum in der musik (German—Jewry in Music)—Richard Wagner essay

Das Lied Das Lied von der Erde (German—The Song of the Earth)—Mahler work

Death and Death and Transfiguration (symphonic poem by Richard Strauss—*Tod und Verklärung*)

Death and Life Gounod's *Mors et Vita Mass*

Death and the Maiden Schubert's Quartet No 14 in D minor

Debussy's Most Popular Opera *Pelléas and Mélisande*

Debussy's Most Popular Piano Piece *Clair de Lune*

Debussy's Most Popular Symphonic Work *La Mer*

Delibes's Most Popular Aria *Bell Song in Lakmé*

Delibes's Most Popular Ballet *Coppélia*

Delius's Most Popular Tone Poem *Over the Hills and Far Away*

Dichter und Bauer (German—Poet and Peasant)—von Suppé overture

Die Frau Die Frau Ohne Schatten (German—The Woman without a Shadow)—opera by Richard Strauss

Die Nullte (German—The Zero)—Bruckner's *Symphony No 0 in D minor*

Die schöne Galathée (German—The Beautiful Galatea)—von Suppé overture

Dies Irae (Latin—Day of Wrath)—medieval mass for the dead theme used by romantic composers such as Berlioz, Liszt, and Rachmaninoff

d'Indy's Most Popular Work *Symphony on a French Mountain Air* for piano and orchestra

Dissonant Mozart's *String Quartet in C* (K 465)

Divine Poem Scriabin's *Symphony No 3*

Doàn Quân Vietnam national anthem

Dohnanyi's Most Popular Orchestral Work *Variations on a Nursery Song*

Domestica Symphonia Domestica by Richard Strauss

Donizetti's Most Popular Opera *Lucia di Lammermoor*

Double Bassist of the Century Gary Karr, Serge Koussevitsky

Dramatic Symphony Roméo et Juliette by Hector Berlioz

Dreigroschen Die Dreigroschenoper (German—The Threepenny Opera)—Kurt Weill's modern reworking of the Beggar's Opera

Dreigroschenoper (German-Threepenny Opera)—Kurt Weill work with a libretto by Bertolt Brecht

Drum Roll Haydn's *Symphony No 103 in E-flat major*

Du Gamla, Du Fria Du gamla, du fria, du fjällhöga Nord (Swedish—You an-

cient, you free, you mountainous North)—Sweden's national anthem

Dukas's Most Popular Orchestral Work *Sorcerer's Apprentice*

Dulce Patria (Spanish—sweet country)—Chile's national anthem

Dumb Girl Dumb Girl of Portici (Auber opera—*La Muette de Portici*)

Dumky Dvořák trio

Dutch composers Cornelis Dopper, Willem Pijper, Jan Pieterszoon Sweelinck, Johan Wagenaar, Bernard Zweers

Dutchman The Flying Dutchman (Wagner opera whose German title is *Der Fliegende Holländer*)

Dutch Quartet Wagenaar, Pijper, Badings, and Otterloo

Dvořák's 9 Dvořák's nine symphonies including *Bells of Zlonice* (No 1) and *From the New World* (No 9)

Dvořák's Most Popular Air *Songs My Mother Taught Me*

Dvořák's Most Popular Concerto *Cello Concerto in B minor*

Dvořák's Most Popular Symphony *No 9 From the New World*

Egyptian Piano Concerto No 5 by Saint-Saëns

Eine Alpensinfonie (German—An Alpine Symphony)—Richard Strauss tone poem

Eine Kleine Nachtmusik (German—A Little Night Music)—Mozart's *Serenade for String Orchestra* (K 525)

Ein Heldenleben (German—A Hero's Life)—autobiographical symphonic poem by Richard Strauss

Ein Morgen, ein Mittag, ein Abend in Wien (German—Morning, Noon, and Night in Vienna)—von Suppé overture

Ein Sommernachtstraunt (German—A Midsummer Night's Dream)—Mendelssohn overture

Elektra Richard Strauss tragedy

Elgar's Most Popular Orchestral Work *Enigma Variations*

Elgar's Most Popular Song *Salut d'amour*

Elisir L'Elisir dAmore (Italian—The Elixir of Love)—two-act opera by Donizetti

Elvira Madigan motion picture and nickname of Mozart's *Piano Concerto No 21 in C major* (K 467)

Emperor Beethoven's *Piano Concerto No 5 in E flat*; Haydn's *String Quartet in C* (opus 76, no 3)

Emperor Franz-Joseph of the Austro-Hungarian Empire rolls down the Prater in the imperial coach drum-roll and trumpet-punctuated finale of the *Emperor Waltz* by Johann Strauss Jr

Enescol's Most Popular Orchestral Work *Romanian Rhapsody No 1*

English composers Sir Granville Bantock, Sir Arnold Bax, Benjamin Britten, William Byrd, Frederick Delius, Sir Edward Elgar, Gustav Holst, Henry Purcell, Sir Arthur Sullivan, Thomas Tallis, Ralph Vaughan Williams, Sir William Walton

English Hornist Thomas Stacy of the New York Philharmonic

English Nonet England's nine best-known classical composers—Tallis, Byrd, Purcell, Sullivan, Elgar, Delius, Vaughan Williams, Walton, Britten

English Symphonist Ralph Vaughan Williams

Eroica Beethoven's *Symphony No 3 in E-flat major* (*Sinfonia eroica*)

Eskimo Opera Hakon Axel Einar Boørresen's opera about Greenland Eskimos (produced in Copenhagen in 1921 under the title of *Kaddara*)

Eugen Eugen Onegin (Tchaikovsky opera

Evgeny Onyegin (Russian—Eugene Onegin)—Tchaikovsky's most popular opera based on a poem by Pushkin

Falla's (de Falla's) Most Popular Work *Nights in the Gardens of Spain*

Fanciulla La Fanciulla del West (Italian—The Girl of the Golden West)—Puccini opera whose libretto recalls David Belasco's play

Fantastique Symphonie Fantastique (French—Fantastique Symphony)—composed by Berlioz who subtitled it *Episode de la Vie d'un Artiste* (Episode in the Life of an Artist)

Farewell Beethoven's *Piano Sonata No 32 in C minor* (opus 111)—Haydn's *Symphony No 45 in F-sharp minor*

Fate Beethoven's *Symphony No 5 in C minor* (*see Victory*)

Faurés Most Popular Work *Elégie for Cello and Orchestra*

Faust Damnation of Faust by Berlioz; *Faust* opera by Gounod, *Faust Symphony* by Liszt, *Faust Overtures* by Schumann and Wagner

Fausts Verdammnis (German—Damnation of Faust)—dramatic legend composed by Berlioz

Favorite American Anthems *America the Beautiful*; *God Bless America*; *My Country 'tis of Thee*; *Star-Spangled Banner*

Favorite Canadian Anthems *O Canada!*; *The Maple Leaf Forever*

Favorite Christmas Carol *Heilige Nacht* (German—Holy Night)

FBI in War and Peace Hollywood version of Prokofiev's suite entitled "The Love for Three Oranges"

Fiddlers Three Isaac Stern, Itzhak Perlman, Pinchas Zukerman

Fidelio Beethoven opera

Fingal's Höhle (German—Fingal's Hole)—Mendelssohn overture also called *Hebrides* or *Fingal's Cave*

Finnish composers Armas Järnefelt, Selim Palmgren, Jan Sibelius

Finnish First Finland's best-known classical composer—Jean Sibelius

Finnish National Composer Jean Sibelius

First Bohemian Romantic Composer Bedrich Smetana

First Czech Romantic Composer Antonin Dvořák

First Danish Romantic Composer Friedrich Kuhlau

First Dutch Romantic Composer Johannes Verhulst

First English Romantic Composer Sir Arthur Sullivan

First Finnish Romantic Composer Armas Järnefelt

First French Romantic Composer Hector Berlioz

First German Romantic Composer Ludwig van Beethoven

First Hungarian Romantic Composer Franz Liszt

First Irish Romantic Composer William Wallace

First Italian Romantic Composer Gioacchino Rossini

First Mexican Romantic Composer Manuel Ponce

First Norwegian Romantic Composer Christian Sinding

First Polish Romantic Composer Frédéric Chopin

First Romanian Romantic Composer Ludwig Wiest

First Russian Romantic Composer Mikhail Glinka

Fledermaus Die Fledermaus (German—The Bat)—operetta by Johann Strauss, Jr

Flotow's (von Flotow's) Most Popular Overture *Martha*

Flutist of the Century Jean-Pierre Rampal; James Galway

Forellen Quintet (*see Trout*)

Foremost Austrian Composer Wolfgang Amadeus Mozart

Foremost Belgian Composer César Franck

Foremost Belgian Operatic Conductor André Cluytens

Foremost Bohemian Composer Antonin Dvořák

Foremost Brazilian Composer Heitor Villa-Lobos

Foremost Dutch Composer Jan Pieterszoon Sweelinck

Foremost English Composers Sir Edward Elgar, Ralph Vaughan Williams

Foremost Finnish Composer Jean Sibelius

Foremost French Classical Composer Jean Baptiste Lully

Foremost French Composer Hector Berlioz

Foremost French Modern Composer Maurice Ravel

Foremost German Classical Composer Johann Sebastian Bach

Foremost German Composer Ludwig van Beethoven

Foremost German Modern Composer Richard Strauss

Foremost Hungarian Composer Franz Liszt

Foremost Irish Composer William Vincent Wallace

Foremost Italian Composer Giuseppe Verdi

Foremost Mexican Composer Carlos Chávez

Foremost Musical Romanticist Berlioz, Liszt, or Schumann

Foremost Norwegian Composer Edvard Hagerup Grieg

Foremost Polish Composer Frédéric François Chopin

Foremost Romanian Composer Georges Enesco

Foremost Romantic Composer Hector Berlioz

Foremost Russian Composer Pyotr Ilyich Tchaikovsky

Foremost Spanish Composer Manuel de Falla

Forza La Forza del Destino (Italian—The Force of Destiny)—Verdi four-act opera

Foss's Most Popular Work *Night Music for Brasses and Orchestra*

Fountains Le Fontane di Roma (Respighi—The Fountains of Rome)

Four Saints Four Saints in Three Acts, opera by Virgil Thompson with text by Gertrude Stein

Four Seasons Antonio Vivaldi's concerto *Le Quattro Staggioni*

Four Temperaments Hindemith composition for string orchestra; Nielsen's *Symphony No 2*

Fra Diavolo Michele Pezza—leading character in Auber's opera *Fra Diavolo*

Francesca Francesca da Rimini (Tchaikovsky symphonic fantasia, Zandonai four-act opera)

Franck's Most Popular Work *Symphonic Variations for Piano & Orchestra*

Franck symphony César Franck's *Symphony in D*

Freischütz Der FreischUtz (German—The Freeshooter)—von Weber opera

French Baroque Masters Jean Baptiste Lully, Jean Philippe Rameau

French composers Hector Berlioz, Georges Bizet, François Boïeldieu, Pierre Boulez, Emmanuel Chabrier, Ernest Chausson, François Couperin, Achille-Claude Debussy, Vincent d'Indy, Paul Dukas, Gabriel Fauré, Benjamin Godard, Charles Gounod, Jacnegger, Edouard Lalo, Jules Massenet, Giacomo Meyérbeer, Jacques Offenbach, Jean Philippe Rameau, Camille Saint-Saëns, Ambroise Thomas

French Dozen France's twelve best-known classical composers—Lully, Couperin, Rameau, Berlioz, Gounod, Offenbach, Saint-Saëns, Bizet, Massenet, Debussy, Ravel, Milhaud

Frog Haydn's *String Quartet in D* (opus 50, no 6)

From the Halls From the Halls of Montezuma to the Shores of Tripoli (US Marine Corps anthem)

From My Life Smetana's *String Quartet No 1 in E minor*

From the New World Dvořák's *Symphony No 9* (formerly No 5)

Full Moon and Empty Arms Rachmaninoff's *Second Piano Concerto*

Funeral March Sonata Piano Sonata in B-flat minor by Chopin

Gaelic Gaelic Symphony by Mrs HHA Beach (first symphonic work by an American woman)

Georgia composer Wallingford Riegger

German Baroque Masters Johann Sebastian Bach, George Frideric Handel, Heinrich Schütz

German composers Johann Sebastian Bach and his sons, Ludwig van Beethoven, Johannes Brahms, Christoph Willibald von Glück, George Friedrich Handel, Paul Hindemith, Felix Mendelssohn, Robert Schumann, Karlheinz Stockhausen, Richard Strauss, Richard Wagner

German Fourteen Germany's fourteen best-known classical composers: Telemann, Handel, Glück, Beethoven, von Weber, Mendelssohn, Schumann, Wagner, Brahms, Bruch, Strauss, Schoenberg, Hindemith, Weill

German-Polish Composer-Pianist-Teacher Franz Xavier Scharwenka

Gershwin's Most Popular Opera *Porgy and Bess*

Gershwin's Most Popular Work *Rhapsody in Blue*

Ghost Beethoven trio in D major

Gianni Gianni Schicchi (Puccini opera)

Gioconda La Gioconda (Ponchielli opera)

Glazunov 6 Alexander Glazunov's six symphonies

Glazunov's Most Popular Ballet *The Seasons*

Gliere's Most Popular Symphony No 3—*Ilya Mourometz*

Glinka's Most Popular Aria *Sussanin's* in *A Life for the Tsar Ivan Sussanin*

Glinka's Most Popular Work *Russlan & Ludmila Overture*

Glocken von Zlonice (German—Bells of Zlonice)—Dvořák's First Symphony

Glück's (von Glück's)—Most Popular Opera *Orfeo ed Euridice*

Goldberg Bach's *Goldberg Variations*; composed for a keyboard pupil named Johann Gottlieb Goldberg

Golden Flutist Georges Barrère

Goldman's Most Popular March *On the Mall*

Goldmark's Most Popular Symphony *Rustic Wedding*

Gomes's Most Popular Overture *Il Guarany*

Gould's Most Popular Orchestral Piece *American Salute*

Gounod's Most Popular Opera *Faust*

Goyescas Enrique Granados opera

Granados' Most Popular Opera *Goyescas*

Grand Canyon Grand Canyon Suite—symphonic work by Ferdé Grofé

Great The Great Symphony No 9 in C major by Schubert (formerly No 7)

Great C major Schubert's *Symphony No 9*

Great Organ Mass Haydn's E-flat *Grosse Orgelmesse*

Greatest Russian Romantic Composer Pyotr Ilyich Tchaikovsky

Greek Composers Nicholas Mantzaros, Dimitri Mitropoulos, Iannis Xenakis

Greek First Iannis Xenakis—best-known classical composer of modern Greece

Grieg's Most Popular Orchestral Suite *Peer Gynt*

Grieg's Most Popular Song *Ich liebe Dich* (German—I Love You)

Grieg's Most Popular Work *Concerto in A minor for Piano*

Griffes' Most Popular Orchestral Piece *The White Peacock*

Grimes Peter Grimes (opera by Britten)

Grofé's Most Popular Orchestral Suite *Grand Canyon*

Guatemala feliz (Spanish—happy Guatemala)—Guatemala national anthem

Guitarist of the Century Andrés Segovia (1893–1987)

Haffner Mozart's *Serenade Suite in D* or his *Symphony No 35 in D major*; both honor the Burgomeister of Salzburg—Sigmund Haffner

Halévy's Most Popular Opera *La Juive* (French—The Jewess)

Halka (Polish—Helen)—Moniuszko's most popular opera and the most popular Polish one

Hamlet funeral march by Berlioz; fantasy overture by Tchaikovsky; opera by Thomas

Hammerklavier (German hammer, keyboard)—pianoforte; Beethoven *Piano Sonatas 28 and 29 in B flat* (opus 106)

Ham 'n' Eggs musician's nickname for *Cavalleria Rusticana* and *I Pagliacci* as these two go well together

Handel's Most Popular Air *Ombra mai fu* (Italian—Shade of My Tree)—from Xerxes, best known as *Handel's Largo*

Handel's Most Popular Oratorio *Messiah*

Handel's Most Popular Orchestral Suite *Fireworks*

Hänsel Hänsel und Gretel (Humperdinck's Christmastime entertainment and opera about a brother, sister, parents, and an old witch in a gingerbread house)

Hanson's Most Popular Symphony *No 2—Romantic*

Harold Harold en Italie (Italian—Harold in Italy)—Berlioz symphony with viola solo

Harp Beethoven's *String Quartet in E-flat major* (opus 74)—Chopin's *Piano Etude in A flat* (opus 25, no 1)

Harpist of the Century Alfredo Casella (1883–1947)

Harpsichordist of the Century Wanda Landowska (1877–1959)

Harris's Most Popular Piece *Cimarron* (symphonic overture)

Hatikvah (Hebrew—The Hope)—Israeli anthem

Haydn's 104 Franz Joseph Haydn's 104 symphonies

Haydn's Most Popular Oratorio *The Seasons*

Haydn's Most Popular Symphony *No 104—London*

Hebrew Opera-Oratorio Samson et Delila by Saint Saëns

Hebrides Mendelssohn overture, also called *Fingal's Cave*

Hercules of Music Christoph Willibald Glück

Hindemith's Most Popular Orchestral Work *Mathis der Maler* (German—Mathis the Painter)

Historical Symphony No 6 of Ludwig Spohr

Hobo Composer Harry Partch (inventor of the forty-three microtone to the octave scale)

Holländer Die Fliegende Holländer (German—The Flying Dutchman)—opera by Wagner

Holst's Most Popular Orchestral Work *The Planets*

Holy Mass Haydn's *Heiligesse in B flat*

Honnegar's Most Popular Orchestral Work *Pacific 231*

Hornist of the Century Dennis Brain (1921–1957)

Horseman Haydn's *String Quartet in G minor* (opus 74, no 3)

Hugh Hugh the Drover—opera by Ralph Vaughan Williams)

Huguenots Les Huguenots (French—The Huguenots)—Meyerbeer

Humperdinck's Most Popular Opera Hänsel und Gretel

Hungarian-born Eminent American Conductors Antal Dorati, Eugene Ormandy, Fritz Reiner, Sir Georg Solti, George Szell

Hungarian composers Béla Bartók, Ernst Dohnanyi, Zoltan Kodaly, Franz Liszt

Hungarian Quartet Hungary's four best-known classical composers—Liszt, Dohnanyi, Bartók, Kodaly

Hunting Mozart's *String Quartet in B flat* (K 458)

Husitská Dvořák's overture honoring Bohemian patriot Jan Huss

Hymn of Praise Mendelssohn's *Symphony No 2 in B-flat major* (also known as *Lobgesang*)

Ibert's Most Popular Orchestral Suite *Escales* (French—Ports of Call)

Igor Prince Igor (opera by Borodin)

Il Distrato (Italian—The Absent-Minded)—Haydn's *Symphony No 60 in C major*

Ilia Mourometz Gliere's Symphony No.3

I'm Always Chasing Rainbows Chopin's *Fantasie Impromptu in C-sharp minor* (popularized)

Imperial Haydn's *Symphony No 99 in E flat*

Impresario Mozart opera

In an 18th-century Drawing Room Mozart's *Piano Sonata in C*

Indian MacDowell's *Suite No 2 for Orchestra* introducing American Indian themes

Inextinguishable Nielsen's *Symphony No 4*

Inno de Mameli (Italian—Hymn of Mameli)—Italy's national anthem honoring Goffredo Mameli

Ippolitov-Ivanov's Most Popular Orchestral Work *Caucasian Sketches*

Irish Irish Rhapsody by Victor Herbert; *Irish Symphony* by Sir Charles Villiers Stanford; *Irish Symphony* by Sir Hamilton Harty

Irish composers William Balfe, Sir Hamilton Harty, Sir Charles Villiers Stanford, William Wallace, Charles Wood

Irish First Ireland's best-known cellist-composer-conductor was Victor Herbert, remembered for his operettas produced in the United States, where he became a naturalized citizen

Israel Ernst Bloch symphony

Israeli National Composer Ernst Bloch

Italian Mendelssohn's *Symphony No 4 in A major*

Italiana Italiana in Algeri (Italian—The Italian Girl in Algiers)—Rossini opera

Italian Baroque Masters Atcangelo Corelli, Claudio Monteverdi, Antonio Vivaldi

Italian composers Tomaso Albinoni, Vincinzo Bellini, Arrigo Boito, Ferruccio Busoni, Alfredo Casella, Mario Caslnuovo-Tedesco,

Luigi Cherubini, Domenico Cimarosa, Muzio Clementi, Arcangelo Corelli, Luigi Dallapicocla, Andrea and Giovanni Gabrieli, Giuseppe and Tommaso Giordani, Umberto Giordano, Pietro Mascagni, Claudio Monteverdi, Giovanni Palestrina, Giovanni Pergolese, Giacomo Puccini, Ottorino Respighi, Gioacchino Rossini, Alessandro and Domenico Scarlatti, Francisco Paolo Tosti, Giuseppe Verdi, Antonio Vivaldi, Ermanno Wolf-Ferrari, Riccardo Zandonai

Italian Fourteen Italy's best-known classical composers—Palestrina, Monteverdi, Fescobaldi, Vivaldi, Scarlatti (2), Cherubini, Paganini, Rossini, Donizetti, Bellini, Verdi, Puccini, Respighi

Italian Girl Italian Girl in Algiers (Rossini opera)

Ivan the Terrible Rimsky-Korsakoff opera

Ives 4 four symphonies by Charles Ives

Ives's Most Popular Orchestral Work *The Fourth of July*

Jeremiah Bernstein's *Symphony No 1* commemorating the prophet Jeremiah and his prophecies

Jeu des cartes (French—deck of cards)—Stravinsky ballet

Joke Haydn's *String Quartet in E flat* (opus 33, no 2)

Juive La Juive (French—The Jewess)—Halévy opera

Jupiter (Latin—Zeus)—Mozart's *Symphony No 41 in C major*—his last

Kabalevsky's Most Popular Overture *Colas Breugnon*

Kaddish Bernstein's *Symphony No 3*

Kaiser Kaiser-Waltzer (German—Emperor Waltz)—Johann Strauss, Jr's opus 437

Kansas City Composer Virgil Thomson

Katerina Katerina Izmaylova (Russian title of Shostakovich's opera *Lady Macbeth of the Mtsensk District*)

Kettledrum Haydn's *Kettledrum Mass in C major* (*Paukenmesse*)

Khachaturian's Most Popular Ballet *Gayne*

Khovantchina Mussorgsky opera completed after his death by Rimsky-Korsakoff

Knyaz Knyáz Igor (Russian—Prince Igor)—Borodin's uncompleted opera partly orchestrated by Glazunov and Rimsky-Korsakov

Kodaly's Most Popular Suite *Háry János*

Komische Opera (German—comic opera)—Berlin Opera House

Kreutzer Beethoven's *Sonata in A minor* (opus 47)—for violin and piano; dedicated to his friend the violinist Rudolphe Kreutzer

Kutchka Mogutchaya Kutchka (Russian—Mighty Handful)—Balakirev, Borodin, Cui, Mussorgsky, and Rimsky-Korsakov

La Chasse Haydn's *Quartet in B flat* (opus 1, No 1)—Haydn's *Symphony No 73 in D major* (The Hunt)

La Damnation de Faust (French—The Damnation of Faust)—four-part dramatic legend composed by Berlioz

Lady Macbeth Lady Macbeth of the Mtsensk District; Lady Macbeth of Mzensk (Shostakovich opera known to Russians as *Katerina Izmaylova*)

Lakmé Delibes opera

Lalo's Most Popular Overture *Le Roi d'Ys* (French—The King of Ys)

Lalo's Most Popular Work *Symphonie espagnole for violin and orchestra*

Lamentatione Haydn's *Symphony No 26 in D minor* also called the *Christmas Symphony*

Land of Hope and Glory Elgar's *Pomp and Circumstance*, March No 1

La Reine (French—The Queen)—Haydn's *Symphony No 85 in B-flat major*

Largest Opera House Metropolitan Opera House in New York's Lincoln Center seats 3,800

Lark Haydn's *String Quartet in D* (opus 64, no 5)

L'Arlésienne Bizet's suites *No 1* and *No 2*

Last American Romantic Composer Edward MacDowell

Last Austrian Romantic Composer Anton Bruckner

Last Belgian Romantic Composer César Franck

Last Bohemian Romantic Composer Gustav Mahler

Last Brazilian Romantic Composer Heitor Villa-Lobos

Last Czech Romantic Composer Josef Suk

Last Danish Romantic Composer Carl Nielsen

Last Dutch Romantic Composer Johan Wagenaar

Last English Romantic Composer Sir Edward Elgar

Last Finnish Romantic Composer Jean Sibelius

Last French Romantic Composer Camille Saint-Saëns

Last German Romantic Composer Richard Strauss

Last Hungarian Romantic Composer Ernst von Dohnanyi

Last Irish Romantic Composer Charles Stanford

Last Italian Romantic Composer Giacomo Puccini

Last Mexican Romantic Composer Carlos Chávez

Last Norwegian Romantic Composer Edvard Grieg

Last Polish Romantic Composer Stanislaw Moniuszko

Last Romanian Romantic Composer Georges Enesco

Last Russian Romantic Composer Sergei Vasilievich Rachmaninoff

Last Spanish Romantic Composer Manuel de Falla

Last Swedish Romantic Composer Kurt Atterberg

Last of the American Romantic Composers Edward MacDowell

Last of the Australian Romantic Composers Percy Grainger

Last of the Austrian Romantic Composers Gustav Mahler

Last of the Belgian Romantic Composers César Franck

Last of the Bohemian Romantic Composers Antonín Dvořák

Last of the Brazilian Romantic Composers Heitor Villa-Lobos

Last of the Danish Romantic Composers Carl Nielsen

Last of the Dutch Romantic Composers Johan Wagenaar

Last of the English Romantic Composers Sir Edward Elgar

Last of the Finnish Romantic Composers Jean Sibelius

Last of the French *Opéra Bouffe* Composers Jacques Offenbach

Last of the German Romantic Composers Richard Strauss

Last of the Great Romantic Composers Carlos Chavez, Mexico; Claude Achille Debussy, France; Antonin Dvořák, Czechoslovakia; Georges Enesco, Romania; Manuel de Falla, Spain; Edvard Hagerup Grieg, Norway; Howard Hanson, United States; Gustav Mahler, Austria; August Nielsen, Denmark; Giacomo Puccini, Italy; Sergei Rachmaninoff, Russia; Richard Strauss, Germany; Ralph Vaughan Williams, Britain; and Heitor Villa-Lobos, Brazil

Last of the Hungarian Romantic Composers Franz Liszt

Last of the Italian Opera Buffa Composers Rossini

Last of the Italian Romantic Composers Ottorino Respighi

Last of the Norwegian Romantic Composers Christian Sinding

Last of the Polish Romantic Composers Ignace Jan Paderewski

Last of the Romantic Romanian Composers Georges Enesco

Last of the Russian Romantic Composers Sergei Rachmaninoff

Last of the Spanish Romantic Composers Manuel de Falla

Last of the Swedish Romantic Composers Kurt Atterberg

Leading Bel Canto Composer Gioacchino Rossini

Le Divin Poeme (French—The Divine Poem)—Scriabin's Symphony No 3

Lehár's Most Popular Operetta *The Merry Widow*

Leichte Kavallerie (German—Light Cavalry)—von Suppé overture

Lélio *Lélio, ou Le Retour á la vie* (French—Lélio, or the Return to Life)—Berlioz monodrama sequel to his *Symphonie fantastique*

Lenin Shostakovich's *Symphony No 12*

Leningrad Shostakovich's *Symphony No 7*

Leoncavallo's Most Popular Opera *I Pagliacci* (Italian—The Players)

Les Adieux Beethoven's *Piano Sonata No 23 in E flat* (opus 81a)—*Les Adieux, l'absence, et le retour*—the farewell, the absence, and the return

L'Heure *L'Heure Espagnole* (French—The Spanish Hour)—Ravel operatic farce

Liadov's Most Popular Symphonic Poem *Kikimora*

l'Impériale Haydn's *Symphony No 53 in D major*

Linz Mozart's Symphony No 36 in C major named for the Austrian town of Linz

Liszt's Most Popular Symphonic Poem *Les Preludes*

Liszt's Most Popular Waltz Mephisto

Little Schubert's *Symphony No 6 in C*

Little C major Schubert's *Symphony No 6*

Little Russian Tchaikovsky's *Symphony No 2 in C minor*

Lobgesang (German—Hymn of Praise)—Mendelssohn's *Symphony No 2 in B-flat major*

Lohengrin Wagner's three-act romantic opera

London Haydn's *Trios No 1 and 2* (for two flutes and cello)—Haydn's *Symphony No 104 in D major*; *Symphony No 2* by Vaughan Williams—*A London Symphony*

London Suite *London Again* or *London Every Day* (symphonic suite by Eric Coates)

Lone Ranger gallop music at the end of Rossini's *William Tell* overture

Longest Opera *Die Meistersinger von Nürnberg* by Wagner (performance time: 5 hours 15 minutes)

Longest Symphony *Symphony No 3 in D minor* by Mahler (performance time: 1 hour 40 minutes)

Loudest Opera *Damnation of Faust* by Berlioz with its *Ride to the Abyss*

Loudest Oratorio *Requiem* by Berlioz (score calls for a chorus of 210 and an orchestra of 217)

Loudest Symphony Mahler's *Tragic Symphony No 6 in A minor*

Loudest Undersea Songs sounds of humpback whales

Lucia *Lucia di Lammermoor* (three-act opera by Donizetti)

Lully's Most Popular Ballet *Alceste*

MacDowell's Most Popular Piano Concerto *No 2 in D minor*

Magic Flute Mozart opera

Mahagonny *Aufstieg und Fall der Stadt Mahagonny* (German—Rise and Fall of the City of Mahagonny)—opera by Kurt Weill with text by Bertolt Brecht

Mahler's 10 Mahler's ten symphonies including the Resurrection (No 2), the *Symphony of a Thousand* (No 8), and the *Unfinished* (No 10)

Mahler's Most Popular Song Cycle *Das Lied von der Erde* (German—The Song of the Earth)

Mahler's Most Popular Work *Symphony No 1 in D major*—*the Titan*

Maine Composer Walter Piston

Malvenuto nickname critics bestowed on the Berlioz opera *Benvenuto Cellini*

Mamelles *Les Mamelles de Tirésias* (French—The Breasts of Tiresias)—comic opera by Poulenc

Manfred *Manfred Overture* (Schumann)—*Manfred Symphony* (Tchaikovsky)

Manon Massenet opera

Manon Lescaut Puccini opera

Manzoni Mass Verdi's *Requiem*

Mascagni's Most Popular Opera *Cavalleria Rusticana* (Italian—Rustic Chivalry)

Masonic Composer Wolfgang Amadeus Mozart who was a freemason and alluded to the ethical laws of Masonry in his opera *The Magic Flute* (*Il Flauto Magico, La Flûte Enchantée, Die Zauberflöte*)—his Masonic Funeral Music (*Maurerische Trauermusik*)—also reveals his affiliation with the Masonic Order

Massachusetts Composers Leonard Bernstein, William Billings, Alan Hovhaness, Daniel Gregory Mason, Lowell Mason

Massenet's Most Popular Ballet *Le Cid*

Mathis Mathis der Mahler (German—Mathias Grünewald the Painter)—symphonic suite by Hindemith

Ma Vlast (Czechoslovakia—My Fatherland)—Smetana's symphonic poem

May Day Shostakovich's *Symphony No 3* also called *May First*

May Night overture by Rimsky-Korsakov

Meerestille Mendelssohn's *Calm Sea and Prosperous Voyage* overture more correctly translated as *Becalmed at Sea and Prosperous Voyage*

Meftstofele (Italian—Mephistopheles)—Boito's opera about the Faust legend

Meistersinger Die Meistersinger von Nürnberg (Wagner's opera about The Mastersingers of Nuremberg)

Helusine Dieschdne Melusine (German—lovely Melusina)—Mendelssohn overture

Mendelssohn's 5 Mendelssohn's five symphonies including *Lobgesang* (No 2), *Scottish* (No 3), *Italian* (No 4), and *Reformation* (No 5)

Mendelssohn's Most Popular Oratorio *Elijah*

Mendelssohn's Most Popular Overture *Midsummer Night's Dream*

Mendelssohn's Most Popular Symphony *No 4 in A major—the Italian*

Mendelssohn's Most Popular Work *Concerto in E minor for Violin*

Menotti's Most Popular Work *Amahl and the Night Visitors*

Mexican Composer-Conductor Carlos Chávez or Juventino Rosas

Mexican Composers Carlos Chávez, Manuel Ponce, Silvestre Revueltas, Juventino Rosas

Mexican Trio Mexico's three best-known classical composers—Manuel Ponce, Carlos Chávez, Silvestre Revueltas

Mezzo-Soprano of the Century Christa Ludwig (1928–)

Midsommarvaka (Swedish-Midsummer Fete)—Hugo Alvén's rhapsody for orchestra

Midsummer Night's Dream Mendelssohn's incidental music to Shakespeare's play

Mighty Five Balakirev, Borodin, Cui, Mussorgsky, and Rimsky-Korsakov

Milhaud's Most Popular Ballet *Le Boeuf sur le toit* (French—The Ox on the Roof)

Militaire Paganini's *Violin Caprice* (opus 1, no 14)

Military Haydn's *Symphony No 100 in G major*

Minute Chopin's *Waltz in D flat* (opus 64, no 1)

Miracle Haydn's *Symphony No 96 in D major*

Moïse (French—Moses)—Rossini opera

Montreal Composer Henry Dreyfus Brant

Moonlight Beethoven's Piano Sonata No 14 in C-sharp minor (opus 27, no 2)—*Sonata quasi una Fantasia*

Moonlight and Roses Tchaikovsky's Andante Cantabile movement, *Symphony No 5 in E minor* also called *Moon Love*

Mors et Vita Gounod's *Death and Life* requiem

Moses Moses in Egypt—Rossini's sacred melodrama in four acts (*Mosé in Egitto*)

Most Admired and Most Discussed Anglo-American Conductor Leopold Stokowski

Most Amazing Composer Wolfgang Arnadeus Mozart

Most Famous Operatic Intermezzi *Cavalleria Rusticana* by Pietro Mascagni

Most Mispronounced Symphony Tchaikovsky's *Symphony No 6—Pathétique*

Most Popular American Composers of Musicals Gershwin—*Porgy and Bess*, Rodgers and Hammerstein—*Oklahoma!* and *South Pacific*

Most Popular American Folk Opera Gershwin's *Porgy and Bess*

Most Popular American March Sousa's *Stars and Stripes Forever*

Most Popular American Musical Show *Chorus Line*

Most Popular Anglo-American Anthem Britain's *God Save the Queen*, the same tune as America's *My Country 'tis of Thee*

Most Popular Austrian Operas Mozart's *Don Giovanni*, Johann Strauss's *Fledermaus*

Most Popular Beethoven Piano Concerto *Concerto No 5 in E flat—Emperor*

Most Popular Belgian Symphony Franck's *Symphony in D minor*

Most Popular Brahms Piano Concerto *Concerto No 1 in D minor*

Most Popular Canadian Anthems *O Canada!* and *The Maple Leaf Forever*

Most Popular Cello Concertos Dvořák Cello Concerto in B minor; Lalo in D minor, Saint-Saëns No 1 in A minor

Most Popular Chopin Piano Concerto *Concerto No 2 in F minor*

Most Popular Christmas Carol *Stille Nacht, Heilige Nacht* (German—Silent Night, Holy Night)

Most Popular Czech Symphony Dvořák *Symphony No 9 in F minor—From the New World*

Most Popular English Conductor Sir Thomas Beecham

Most Popular English Operetta *The Mikado* by Gilbert and Sullivan

Most Popular English Orchestral Work Elgar's *Enigma Variations*

Most Popular French Classic Opera von Glück's *Iphigénie en Aulide* (Iphigenia in Aulus)

Most Popular French Opera Buffa *Orpheus in the Underworld* by Offenbach

Most Popular French Operas Bizet's *Carmen*, Gounod's *Faust*

Most Popular French Romantic Opera Bizet's *Carmen*

Most Popular French Symphony Franck's *Symphony in D minor*

Most Popular German Comic Opera Richard Strauss's *Der Rosenkavalier*

Most Popular German Operas Beethoven's *Fidelio*, Richard Strauss's *Rosenkavalier*, Wagner's *Meistersinger*

Most Popular Grand Opera Verdi's *Aida*

Most Popular Grieg Piano Concerto *Concerto in A*

Most Popular Italian Opera Buffa Rossini's *Barber of Seville*

Most Popular Liszt Piano Concerto *Concerto No 1 in E flat*

Most Popular MacDowell Piano Concerto *Concerto No 2 in D minor*

Most Popular Mass Beethoven's *Missa Solemnis*

Most Popular Modern German Opera *Der Rosenkavalier* by Richard Strauss

Most Popular Modern Symphony of 20th Century Prokofiev's *No 1 in D minor—Classical*

Most Popular Mozart Piano Concerto *Concerto No 20 in D minor*

Most Popular Norwegian Symphonic Suites Grieg's *Peer Gynt*

Most Popular Opera Bizet's *Carmen*

Most Popular Operas in French *Carmen* by Bizet, *Faust* by Gounod, *Samson and Delilah* by Saint-Saëns, *The Trojans* by Berlioz

Most Popular Operas in German *Fidelio* by Beethoven, *Fledermaus* by Johann Strauss, *Freischütz* by von Weber, *Rosenkavalier* by Richard Strauss

Most Popular Operas in Italian *Aida* by Verdi, *Barber of Seville* by Rossini, *Don Giovanni* by Mozart, *Madama Butterfly* by Puccini

Most Popular Operas in Russian *A Life for the Czar* by Glinka, *Boris Godounov* by Mussorgsky, *Eugen Onegin* by Tchaikovsky, *Prince Igor* by Borodin

Most Popular Overtures Berlioz's *Benvenuto Cellini*; Rosini's *William Tell*; Tchaikovsky's *1812* and *Romeo and Juliet*; Wagner's *Tannhäser*

Most Popular Piano Concertos Grieg's *Concerto in A minor*; Rachmaninoff's *Concerto No 2 in C* and *No 3 in D minor*, Tchaikovsky's *No 1 in B flat*

Most Popular Rachmaninoff Piano Concerto *Concerto No 3 in D minor*

Most Popular Piano Concertos of 20th Century Grieg in *A minor*, Rachmaninoff *No 2 in C minor*

Most Popular Requiem Mass Berlioz, Mozart, or Verdi

Most Popular Romantic Symphony of 20th Century Rachmaninoff's *No 2 in E minor*

Most Popular Russian Grand Opera Mussorgsky's *Boris Godonov*

Most Popular Russian Operas Borodin's *Prince Igor*, Mussorgsky's *Boris Godunov*, Tchaikovsky's *Eugen Onegin*

Most Popular Russian Overture Tchaikovsky's *Overture 1812*

Most Popular Russian Symphonic Suite Rimsky-Korsakov's *Scheherazade*

Most Popular Russian Symphony Tchaikovsky's *Symphony No 4 in F minor*

Most Popular Saint-Saëns Piano Concerto *Concerto No 2 in G minor*

Most Popular Spanish Suite for Piano and Orchestra *Rapsodia española* by Albéniz, *Nights in the Gardens of Spain* by de Falla

Most Popular Symphonic Poem Liszt's *Les Préludes*

Most Popular Symphonies Beethoven's *Symphony No 5 in C minor* (also known as the *Victory Symphony* and *Fifth*), Berlioz's *Fantastique*, Dvořák's *From the New World*, Tchaikovsky's *Fourth*

Most Popular Trio Beethoven's *Archduke* for cello, piano, and violin

Most Popular Viola Concerto *Harold in Italy* by Berlioz

Most Popular Violin Concertos Beethoven *D major*, Brahms *D major*, Mendelssohn *E minor*

Most Popular Waltz *Blue Danube* by Johann Strauss (1825–1899)

Most Prolific Composer Wolfgang Amadeus Mozart or Georg Philipp Telemann

Most Prolific Song Writer Cole Porter who wrote a song a day

Most Prolific Symphonist Franz Josef Haydn composer of 104 symphonies

Most Versatile Musician of Our Era Georges Enesco (Romanian composer, conductor, pianist, teacher, and violinist)

Mozart's 41 Mozart's forty-one symphonies including the *Haffner* (No 35), the *Linz* (No 36), the *Prague* (No 38), the *Jupiter* (No 41)

Mozart's Most Popular Aria *Finch'han dal vino* (Italian—Fetch the Wine)—from Don Giovanni

Mozart's Most Popular Opera *Don Giovanni*

Mozart's Most Popular Piano Concerto No 20 in D minor

Mozart's Most Popular String
Suite *Eine Kleine Nacht-
musik* (German—A Little
Night Music)
Mozart's Most Popular Sym-
phony No 41—Jupiter
Musical Charlotte Russe
Tchaikovsky's *Andante* cant-
abile from his *Symphony No
5 in E minor*
Musical Dictator of Dalmatia
Franz von Suppé (Francesco
Ezechiale Ermenegildo Cava-
liere Suppé Demelli)
Musical Philosopher Alfred
Brendél
Music Capital of America
Los Angeles and New York
Music Capital of Eastern
Europe Vienna
Music Capital of Western
Europe London
Music City, U.S.A. Nash-
ville, Tennessee
Music Man Meredith Willson
Mussorgsky's Most Popular
Opera *Boris Godunov*
Mussorgsky's Most Popular
Work *Pictures at an Exhibi-
tion*
Mysterious Mountain Sym-
phony No 2 by Alan Hovhan-
ess
Nabuco Nabucodonosor (op-
era by Verdi)
Napoleon of the Waltz Jo-
hann Strauss
National Composer of Nor-
way Edward Grieg
Nelson Mass Haydn's *Nelson-
messe in D minor*
Nerone (Italian—Nero)—
Boito opera
New Orleans Composer
Louis Moreau Gottschalk
New World Dvořák's *Sym-
phony No 9 in E minor* (for-
merly No 5)
New York composers Elliott
Carter, Norman Dello Joio,
Edward MacDowell, William
Schuman (*see* Brooklyn com-
posers)
Nicolai's Most Popular Over-
ture *Merry Wives of Wind-
sor*
Nielsen's 6 Carl Nielsen's six
symphonies including *Four
Temperaments* (No 2), *Sinfo-
nia Espansiva* (No 3), *Inex-
tinguishable* (No 4), *Sinfonia
Semplice* (No 6)

Nielsen's Most Popular Symphony *No 4—Inextinguish-
able*
Nigger non-pejorative nickname for Dvořák's *American
Quartet* filled with Black spiritual themes
Noisiest Opera Massenet's *La
Navarraise* replete with bells,
cannon, castanets, guns, tambourines, and trumpets plus
chorus and orchestra
None But The Lonely Heart
best known Russian art song
by Tchaikovsky
Nordic Hanson's *Symphony
No 1*
Norma Beffini opera
Norwegian composers Ed-
vard Hagerup Grieg, Christian Sinding, Johan Severin
Svendsen
Norwegian First Norway's
best-known classical com-
poser—Edvard Hagerup Gr-
ieg
Norwegian National Com-
poser Edvard Grieg
Nozze Le *Nozze di Figaro* (Ital-
ian—The Marriage of Fi-
garo)—opera by Mozart
Nuits Nuits d'éte (French—
Summer Nights)—song cycle
by Berlioz including Ab-
sence, Villanelle, Le spectre
de la rose, Sur les lagunes, Au
cimetiére, L'Île inconnue
Nursery Song Variations on a
Nursery Song by Ernst von
Dohnanyi
Oboist of the Century Bruno
Labate of the New York Phil-
harmonic Symphony; John de
Lancie of the Philadelphia
Orchestra; Leon Goosens,
Marcel Tabuteau, or Harold
Gomberg
Ocean Symphony No 2 by An-
ton Rubinstein
October Revolution Shostak-
ovich's *Symphony No 2*
Ode to Heavenly Joy
Mahler's *Symphony No 4 in
G major*
Ode to Joy Beethoven's *Sym-
phony No 9 in D minor*—
whose closing movement is
based on the text of Schiller's
Ode to Joy
Odysseus Symphony No 25 by
Alan Hovhaness

Offenbach's Most Popular
Opera *Les Contes d'Hoff-
mann* (French—The Tales of
Hoffmann)
Offenbach's Most Popular
Work *Gaité Parisienne*
(French—Parisian Gaiety)
O Guarani (Portuguese—the
Guarani)—opera by Carlos
Gomes
Oklahoma composer Roy
Harris
Oldest American Opera
House Le Petit Opéra Loui-
sianais (The Little Louisian-
ian Opera House begun in
1813, later known as The Old
French Opera House and the
St Charles Theatre)
Oldest American Symphony
Orchestra New York Phil-
harmonic founded in 1842
and merged with New York
Symphony in 1928
Oldest Austrian Symphony
Orchestra Wiener Philhar-
monische Konzerte (Ger-
man—Vienna Philharmonic
Concerts), 1842
Oldest British Symphony
Orchestra London's Royal
Philharmonic Orchestra,
1813
Oldest Canadian Symphony
Orchestra Philharmonic
Society of Montreal, 1848
Oldest Czech Symphony
Orchestra Prague's Ceská
filharmonie (Czech Philhar-
monic), 1864
Oldest Dutch Symphony
Orchestra Amsterdam's
Concertgebouw (Concert
Building), 1883
Oldest European Symphony
Orchestra Rome's Santa
Cecilia, 1566
Oldest French Symphony
Orchestra Société des Con-
certs du Conservatoire of
Paris, 1828
Oldest German Symphony
Orchestra Leipzig
Gewandhaus Konzerte
(Leipzig Cloth Hall Con-
certs), 1743
Oldest Hungarian Symphony
Orchestra Budapesti Fil-
harmónin Társaság (Budapest
Philharmonic Society), 1853

Oldest Midwestern Symphony Orchestra St Louis Symphony, 1880

Oldest Performing Opera House in America New York's Metropolitan Opera House, 1883

Oldest Performing Opera House in Argentina *El Teatro Colón* in Buenos Aires, 1908

Oldest Performing Opera House in Australia Sydney's Opera House, 1954

Oldest Performing Opera House in France Paris's *Opera*, 1875

Oldest Performing Opera House in Great Britain London's Covent Garden Theatre, 1732

Oldest Performing Opera House in Italy Milan's La Scala, 1778

Oldest Performing Opera House in Russia Saint Petersburg's *Kirov*, 1860

Oldest Performing Opera House in Spain Barcelona's *Teatro Liceo* (Lyceum Theater), 1862

Oldest Popular Song Composer Irving Berlin who died at 101

Oldest Spanish Symphony Orchestra Barcelona's *Orquesta Pau Casals* (Catalan—Pablo Casals Symphony), 1919

Oldest Swiss Symphony Orchestra Zürich's *Tonhalle* (German—one Hall), 1613

Oldest Symphony Orchestra Dresdener Staatskapelle whose performance history dates to 1548

Oldest Symphony Orchestra in Berlin *Berlin Philharmoniker*, 1882

Oldest Symphony Orchestra in California San Francisco Symphony, 1911

Oldest Symphony Orchestra in Eastern Pennsylvania Philadelphia Orchestra, 1900

Oldest Symphony Orchestra in English-speaking Canada Toronto Symphony Orchestra, 1906

Oldest Symphony Orchestra on the Great Lakes Chicago Symphony, 1891

Oldest Symphony Orchestra in Minnesota Minneapolis Symphony Orchestra, 1903

Oldest Symphony Orchestra in New England Boston Symphony Orchestra, (1881–)

Oldest Symphony Orchestra in Northern Ohio Cleveland Orchestra, 1918

Oldest Symphony Orchestra in Ohio Cincinnati Symphony Orchestra, 1895

Oldest Symphony Orchestra in Pennsylvania Pittsburgh Symphony Orchestra, 1896

Oldest Symphony Orchestra in Southern California Los Angeles Philharmonic, 1919

Old Maid Old Maid and the Thief—opera by Menotti

Onegin Evgeny Onyegin (Russian—Eugen Onegin)—Tchaikovsky opera based on a poem by Pushkin

Opera of Operas Mozart's *Don Giovanni*

orchestral horses Vortex and Giaour in the *Damnation of Faust* by Berlioz; Phaeton's four steeds in Saint-Saëns' tone poem; the nine horses in Wagner's Ride of the Valkyries in *Die Walküre*

Orchestral Orgasm nickname of the *Don Juan* tone poem by Richard Strauss

Orfeo opera by Monteverdi; *Orfeo ed Euridice* (Italian—Orpheus and Euridice)—Glück's most popular opera and orchestral suite

Orff's Most Popular Scenic-Cantata *Carmina Burana*

Organ Poulenc's *Concerto in G* for organ, strings, and timpani; Saint-Saëns *Symphony No 3* for orchestra and organ

Organist of the Century E Power Biggs (1906–1977)

Orleanskaya Orleanskaya deva (Russian—Maid of Orleans)—Tchaikovsky opera based on Schiller's tale about Joan of Arc

Ory Le Compte Ory (French—The Count Ory)—opera by Rossini

Otello (Italian—Othello)—Rossini opera; Verdi opera

Oxford Haydn's Symphony No 92 in G major

Oy Veh (Yiddish—Oh My God)—Mahler's *Resurrection Symphony in C minor*

Pag I Pagliacci (Italian—The Players)—opera by Leoncavallo

Paganini's Most Popular Work 24 Caprices for Violin

Paris Mozart's *Symphony No 31 in D major*

Parisian Composers Bizet, Boulanger, Charpentier, Chausson, Debussy, d'Indy, Dukas, Gounod, Ibert, Poulenc, Rabaud, Saint-Saëns

Paris symphonies Haydn's symphonies 82 through 87, commissioned in Paris, bearing such names as *l'Ours* (The Bear—82), La Poule (The Hen—83), La Reine (The Queen—85)

Parsifal seven-hour-long music drama by Wagner

Partch's Most Popular Work *Daphne of the Dunes*

Pastoral Beethoven's Piano Sonata No 15 in D (opus 28); Beethoven's *Symphony No 6 in F major* (opus 68); Dvořák's *Symphony No 8 in G major*; *Symphony No 3* by Vaughan Williams

Pathétique Beethoven's *Piano Sonata No 8 in C minor* (opus 13)—Tchaikovsky's *Symphony No 6 in B minor*

Patriotic Symphony No 2 in D major by Sibelius

Pêheurs de Perles (French—The Pearl Fishers)—opera by Bizet

Peer Gynt drama by Ibsen with incidental music by Grieg

Pellias Pelldas et Melisande (Debussy's opera)

Pennsylvania composers Samuel Barber, Stephen Foster, Peter Mennin

Pepusch's Most Popular Ballad Opera *The Beggar's Opera*

percussion bells, castanets, chimes, clappers, cymbals, drums, glockenspiels, gongs, marimbas, tambourines, triangles, wood blocks, xylophones

Pianist of the Century Artur Rubinstein (1889–1982)

Pictures Pictures at an Exhibition (Mussorgsky's piano suite frequently presented in the Ravel orchestration)

Pikovaya Pikovaya dama (Russian—La Pique Dame)—Tchaikovsky opera sometimes sung in English under the title Queen of Spades

Pinafore HMS Pinafore or The Lass that Loved a Sailor (Gilbert and Sullivan operetta)

Pines The Pines of Rome (Respighi's symphonic poem*Pini di Roma*)

Pinnacle of the Baroque Johann Sebastian Bach

Pique Pique Dame (French—The Queen of Spades)—opera by Tchaikovsky

Pique Dame (French—Queen of Spades)—Tchaikovsky opera; von Suppé overture

Pirates Pirates of Penzance (Gilbert and Sullivan operetta)

Piston's Most Popular Ballet *Incredible Flutist*

Poem of Ecstasy Scriabin's *Symphony No 4*

Polish Tchaikovsky's *Symphony No 3 in D major*

Polish composers Frédéric Chopin, Michal Kondracki, Emil Mlynarski, Ignacy Jan Paderewski, Krystof Penderecki, Karol Szymanowski, Alexander Tansman, Henryk Wieniawski

Polish First Poland's best-known classical composer—Frédéric Chopin

Polonia (Polish—Poland)—Wagner overture; Mlynarski symphony

Poulenc's Most Popular Ballet *Les Biches* (French—The Deer Does)

Poulenc's Most Popular Concerto *Concerto for Organ, Strings, and Timpani*

Prague Mozart's *Symphony No 38 in D major*

Préludes Les Préludes (Liszt symphonic poem)

Prince *Prince Igor* (Borodin's opera known to Russians as *Knyaz Igor*)

Printer's Symphony Mendelssohn's *Symphony No 2 in B-flat major* also known as the *Hymn of Praise* (*Lobgesang*) celebrating the 400th anniversary of the invention of printing

Prisoner's Chorus part of Beethoven's opera, *Fidelio*; the finale of Act I of *Fidelio*, Beethoven's only opera, is an appeal from political prisoners longing for the scent of open air as they know their prison is a tomb

prisoner's opera Beethoven's *Fidelio* has all three acts set in a Spanish prison run by a tyrant; memorable for the compassion the composer shows political prisoners

prisoner's work songs outstanding collection compiled and edited by Bruce Jackson in *Wake Up Dead Man—Afro-American Worksongs from Texas Prisons*, Harvard University Press, Cambridge, Mass, 1972

prison scenes set to music Beethoven's opera *Fidelio*, the *Damnation of Faust* by Berlioz, Boito's *Mefistofele*, Gounod's *Faust*, and Puccini's *Tosca* present some of the most musically memorable scenes although there are others by Verdi

Prokofiev's 7 Prokofiev's seven symphonies including the *Classical* (No 1)

Prokofiev's Most Popular Ballet *Romeo and Juliet*

Prokofiev's Most Popular Symphony *No 5*

Prokofiev's Most Popular Work *Peter and the Wolf*

Puccini's Most Popular Aria *E lucevan le stelle* (Italian—And the stars shone brightly) in *Tosca*

Puccini's Most Popular Opera *Madama Butterfly*

Purcell's Most Popular Opera *Dido and Aeneas*

Queen The Queen (La Reine)—Haydn's *Symphony No 85 in B-flat major*

Queen of Spades Tchaikovsky opera; English title of a Tchaikovsky opera called *La Pique Dame* by the French and *Pikovaya dama* by Russians

Queen Symphony Haydn's *La Reine* (No 85)

Quiet Quiet Flows the Don (Dzerzhinsky's opera known to Russians as Tikhiy Don)

Quinten Haydn's *String Quartet in D* (opus 76, no 2)—nickname refers to the fifth form or grade in Austrian schools

Quixote Don Quixote (Fantastic Variations on a Theme of Knightly Character by Cervantes as composed by Richard Strauss)

Rachmaninoff's 3 Rachmaninoff's three symphonies

Rachmaninoff's 4 Rachmaninoff's four piano concertos

Rachmaninoff's Most Popular Symphonic Poem *The Isle of the Dead*

Rachmaninoff's Most Popular Symphony *No 2 in E minor*

Rachmaninoff's Most Popular Work *Concerto No 2 for Piano*

railroad music *Pacific 231* by Arthur Honneger; *Little Train of the Caipira* from *Bachianas Brasileiras No 2* by Heitor Villa-Lobos

Rain Violin and Piano Sonata in G (opus 78)—by Brahms

Raindrop Chopin's *Piano Prelude No 15 in D-flat major*

Rakóczy traditional Hungarian march used by Berlioz in his *Damnation of Faust* and by Liszt in his *Hungarian Rhapsody No 15 in A minor*

Rameau's Most Popular Opera *Dardanus*

Rape Rape of Lucretia (Britten opera)

Rasumovsky Beethoven's Quartets in F major, E minor, and C major for two violins, viola, and cello (opus 59, nos 1, 2, 3), dedicated to Count Rasumovsky

Ravel's Most Popular Ballet *Daphnis et Chloé*

Ravel's Most Popular Choreographic Poem for Orchestra *La Valse* (French—The Waltz)

Ravel's Most Popular Opera *L'Heure Espagnole* (French—The Spanish Hour)

Ravel's Most Popular Work *Bolero*

Reformation Mendelssohn's *Symphony No 5 in D major*

Requiem Mass for the dead; most memorable composed by Berlioz, Brahms, Bruckner, Cherubini, Dvořák, Fauré, Mozart, Palestrina, and Verdi

Respighi's Most Popular Suite *Gli Uccelli* (Italian—The Birds)

Respighi's Most Popular Work *Fountains of Rome*

Resurrection Mahler's Symphony No 2 in C minor

Revolutionary Chopin's Piano Etude No 12 in C minor

Revolutionary Composer Pierre de Geyter best known for the formerly official communist anthem the *Internationale*

Reznicek's Most Popular Overture *Donna Diana*

Rheingold *Das Rheingold* (Wagner music drama)

Rhenish Schumann's *Symphony No 3 in E-flat major*

Riegger 4 four symphonies by Wallingford Riegger

Rigoletto Verdi opera

Rimsky-Korsakov's Most Popular Work *Scheherazade*

Ring Cycle The Ring of the Nibelungen (q. v.)

Ring of the Nibelungen Wagner's Ring Cycle consisting of *Das Rheingold* (Rhinegold), *Die Walküre* (Valkyries), *Siegfried*, and *Götterdämmerung* (Twilight of the Gods)

Rodrigo's Most Popular Work *Concierto de Aranjuez for Guitar and Orchestra*

rogues' march quickstep played when offenders are drummed out of the army, the marines, the navy, or other military units; at public floggings and executions it was the custom to have a drummer beat out the rhythm of the rogues' march

Romanian First Romania's best-known classical composer-conductor-pianist-violinist—Georges Enesco

Romanian National Composer Georges Enesco

Romantic Bruckner's *Symphony No 4*; Hanson's *Symphony No 2*

Roméo *Roméo et Juliette* (Berlioz symphony for chorus, orchestra, and solo voices)

Rosenkavalier *Der Rosenkavalier* (German—The Red Knight)—Richard Strauss's most popular opera

Rose of Venice Haydn's *Quartet in D for Strings* (opus 20, no 4)

Rossini's Most Popular Opera *Barber of Seville*

Rossini's Most Popular Oratorio *Stabat Mater*

Rossini's Most Popular Overture *William Tell*

Rubinstein's Most Popular Aria *Epithalamium of Wndex in Nero*

Ruslan *Ruslan and Ludmila* (Glinka's most popular opera)

Russian Haydn's six string quartets—Opus 33; Rachmaninoff's *Symphony No 3 in A minor*

Russian composers Anton Arerisky, Mili Balakirevl Alexander Borodin, Cesar Cui, Alexander Dargomijsky, Alexander Glazunov, Reinhold Gliere, Mikhail Glinka, Alexander Gretchaninov, Dmitri Kabalevsky, Aram Khachaturian, Anatoly Liadov, Modest Mussorgsky, Sergei Prokofiev, Sergei Rachmaninoff, Nikolai Rimsky-Korsakov, Anton Rubinstein, Alexander Scriabin, Dmitri Shostakovich, Igor Stravinsky, Pyotr Ilyich Tchaikovsky

Russian Easter Rimsky-Korsakov's *Russian Easter Festival*—concert overture

Russian Fourteen Russia's fourteen best-known classical composers—Glinka, Borodin, Cui, Balakirev, Mussorgsky, Tchaikovsky, Rimsky-Korsakov, Glazunov, Scriabin, Rachmaninoff, Gliere, Stravinsky, Prokofiev, Shostakovich

Russian National Composer Mikhail Ivanovich Glinka

Russian Symphonist Pyotr Ilyich Tchaikovsky

Russia's Most Russian Composer Tchaikovsky

Russo-American Composer Igor Stravinsky (1882–1971)

Rustic Wedding Karl Goldmark's *Symphony in E flat* (opus 26)

Sacre *Le Sacre du Printemps* (French—The Rite of Spring)—Stravinsky ballet for orchestra

Sacred operas *Mosé en Egitto* (Moses in Egypt) by Rossini; *Samson et Dalila* (Samson and Delilah) by Saint-Saëns

Saint-Saëns' 5 the five symphonies of Saint-Saëns including his *Symphony in A major*, the *Symphony No 1*, the *Symphony in F major* (*Urbs Roma*)—the *Symphony No 2 in A minor*, the *Symphony No 3 in C minor* (*Organ*) for organ and orchestra

Saint-Saëns' Most Popular Aria *Mon coeur s'ouvre a ta voix* (French—At your voice my heart unfolds)—in *Samson et Dalila*

Saint-Saëns' Most Popular Symphony *No 3—Organ*

Saint-Saëns' Most Popular Work *Danse macabre*

Saint Vartan *Symphony No 9* by Alan Hovhaness

Salome music drama by Richard Strauss

Salomon symphonies Haydn's symphonies 93 through 104 bearing such names as *Surprise* (94), *Miracle* (96), *Military* (100), *Clock* (101), *Drum Roll* (103), and *London* (104); series named for the impresario JP Salomon who secured concerts for Haydn in London

Samson *Samson et Dalila* (opera by Saint-Saëns based on the biblical legend of Samson and Delilah)

San Carlo of the Symphony Carlo Maria Giulini

Sarasate's Most Popular Piece *Carmen Fantasy*

Savoy operas Gilbert and Sullivan operettas

Saxophonist of the Century Paul Brodie

Schelomo (Hebrew—Solomon)—title of Bloch's composition for 'cello and orchestra

Schoenberg's Most Popular Orchestral Work *Verklärte Nacht* (German—Transfigured Night)

Schubert's 9 nine symphonies of Franz Schubert including *Tragic* (No 4), *Little* (No 6), *Unfinished* (No 8), *The Great* (No 9)

Schubert's Most Popular Symphony *No 8–Unfinished*

Schumann 1st, 2nd, 3rd, 4th First (*Spring*), Second, Third (*Rhenish*), Fourth symphonies composed by Robert Schumann

Schumann's Most Popular Piano Work *Fantasy in C*

Schumann's Most Popular Symphony *No 1—Spring*

Scotch Mendelssohn's *Symphony No 3 in A minor*

Scottish Mendelssohn's *Symphony No 3 in A minor*, often called *Scotch Symphony*

Scottish Composers Erik Chisholm, Sir Alexander Campbell Mackenzie, John Blackwood McEwen, Thea Musgrave, Ian Whyte

Scriabin 5 five symphonies by Alexander Scriabin including the *Divine Poem* (No 3), the *Poem of Ecstasy* (No 4), and the *Poem of Fire* (No 5)

Scriabin's Most Popular Orchestral Work *Poem of Ecstasy*

Sea Symphony No 1 by Vaughan Williams

Seasons Glazunov's ballet; Haydn's oratorio Die Jahreszeiten

Sea Symphony *Symphony No 1* by Ralph Vaughan Williams

Sentiramide Rossini opera

Shostakovich's 15 Shostakovich's fifteen symphonies including *Leningrad* (No 7), *Year 1905* (No 11), *Lenin* (No 12), *Babi Yar* (No 13)

Shostakovich's Most Popular Ballet *Age of Gold*

Shostakovich's Most Popular Symphony *No 5*

Sibelius' 7 the seven symphonies of Sibelius

Sibelius's Most Popular Symphonic Poem *Finlandia*

Sibelius's Most Popular Symphony *No 2 in D major*

Sibelius's Most Popular Violin Concerto in *D minor*

Siegfried Wagner's music drama

Silver Pilgrimage *Symphony No 15* by Alan Hovhaness

Sinfonia Antarctica *Symphony No 7* by Vaughan Williams

Sinfonia Concertante Mozart's two are most familiar

Sinfonia Domestica composition reflecting the daily life of Richard Strauss

Sinfonia Espansiva Nielsen's *Symphony No 3*

Sinfonia India (Spanish—Indian Symphony)—by Carlos Chávez (1899–1978)

Sinfonia Semplice Nielsen's *Symphony No 6*

Six-Four Time Mass Haydn's *Sechsviertelmesse in G*

Small Organ Mass Haydn's B-flat *Kleine Orgelmesse*

Smetana's Most Popular Opera *The Bartered Bride*

Smetana's Most Popular Piece *Moldau*

Snegurochka (Russian—The Snow Maiden)—Rimsky-Korsakoff opera

Song of the Night Karol Szymanowski's *Symphony No 3*; Mahler's *Symphony No 7 in E minor*

Sonnam *Sonnambula* (Italian—Sleepwalker)—Bellini opera

Sopranos of the Century Elisabeth Schumann (1894–1966), Maria Callas (1923–1977), Joan Sutherland 1926–)

Sousa's Most Popular March *The Stars and Stripes*

South African composers John Joubert, Priaulx Rainier

Soviet Symphonists Serge Prokofiev and Dmitri Shostakovich

Spain's Quartet Spain's four best-known classical composers—de Falla, Albéniz, Granados, Turina

Spanish Caprice Rimsky-Korsakov's *Capriccio espagñol*

Spanish composers Isaac Albéniz, Manuel de Falla, Enrique Granados, Felipe Pedrell, Joaquin Turina

Spanish Dances *Danzas españoles* composed by Granados for the piano

Spanish Hour Ravel's brief but witty opera—*L'Heure espagnole*

Spanish Nights de Falla's *Nights in the Gardens of Spain*

Spanish Overture Glinka's *Jota aragonesa*

Spanish Pieces de Falla's *Piezas españoles for piano*

Spanish Rhapsody Liszt's *Rhapsodie espagnole*; Ravel's *Rapsodie espagnole*

Spanish Song Ravel's *Chanson espagnole* for piano and voice

Spanish Songbook Hugo Wolf's *Spanisches Liederbuch*

Spanish Songs *Cantos de España* composed by Albeniz for the piano

Spanish Suite *Suite Española* by Albéniz

Spanish Symphony Lalo's *Symphonie espagnole* for violin and orchestra

Spirit Beethoven's *Spirit Trio* called *Das Geister Trio* by the Germans

Spirit of Man Prokofiev's name for his *Symphony No 5 Opus 100* completed in 1944

Spring Beethoven's *Sonata No 5 for Violin and Piano* (opus 24); Schumann's *Symphony No 1 in B-flat major*

Stabat Mater (Latin—standing mother)—liturgical mass set to music by Haydn, Liszt, Palestrina, Rossini, and Verdi

Steppes In the Steppes of Central Asia (symphonic sketch by Borodin)

Stokowski silver sizzle the sound of the Philadelphia Orchestra (developed by Leopold Stokowski)

storm-at-sea mus storm-at-sea music, the *Sea and Sinbad's Ship* section of Rimsky-Korsakoff's *Scherezade*

storm mus storm music (most memorable includes the Thunderstorm movement in Beethoven's *Symphony No 6–Pastoral*, the *Royal Hunt and Storm in Les Troyens* by Berlioz, the *Tempesta* interlude in Rossini's *Barber of Seville*, and the *Alpine Storm* in his *William Tell*; the Storm sometimes accompanying and often dominating the seduction scene in the *Samson and Delilah* of Saint-Saëns, the Storm movement in the *Alpine Symphony* of Richard Strauss, the opening incidental music composed by Sir Arthur Sullivan for *The Tempest*, the howling thunder and wind in Act III of Tchaikovsky's *Queen of Spades*)

Strange Music Hollywood adaptation of Grieg's "Wedding Day at Troldhaugen" in the *Song of Norway*

Stranger in Paradise *Kismet* theme adapted from Borodin's *Prince Igor*

Strauss's (Johann, Jr) Most Popular Operetta *Die Fledermaus* (German—The Bat)

Strauss's (Johann)—Most Popular Waltz *The Blue Danube*

Strauss's (Richard)—Most Popular Opera *Der Rosenkavalier* (German—The Knight of the Rose)

Strauss's (Richard)—Most Popular Piece for Piano and Orchestra *Burleske*

Strauss's (Richard)—Most Popular Symphony *Alpine*

Strauss's (Richard)—Most Popular Tone Poem *Also sprach Zarathustra* (German—Thus Spake Zarathustra)

Stravinsky's Most Popular Ballet *Firebird*

Suicide European nickname for Tchaikovsky's *Symphony No 6 in B major—the Pathétique*

Sullivan's Most Popular Orchestral Work *In Memoriam*

Sunrise Haydn's *String Quartet in B flat* (opus 76, no 4)

Suor Angelica (Italian—Sister Angelica)—one-act opera by Puccini

Suppé's (von Suppé's) Most Popular Overture *Light Cavalry*

Surprise Haydn's *Symphony No 94 in G major*

Swan Song Symphony Prokofiev's *Symphony No 7 in C-sharp minor*

Swedish composers Kurt Atterberg, Franz Adolf Berwald, Hilding Rosenberg, Wilhelm Stenhammar, Dag Wirén

Swedish Quartet Sweden's leading classical composers, Berwald, Rangstrom, Atterberg, and Wiren

Swiss composers Ernest Bloch, Frank Martin, Jean Jacques Rousseau

Swiss Quartet Switzerland's foremost composers of classical music, Raff, Bloch, Martin, and Honegger

Symphonia domestica (German—Domestic Symphony)—autobiographical tone poem by Richard Strauss

Symphonic-Poem composers Franz Liszt, Camille Saint-Sadns, Jean Sibelius, Bedrich Smetana, Richard Strauss

Symphonie Espagnole Edouard Lalo's most popular violin concerto

Symphonie fantastique (French—Fantastic Symphony)—major orchestral work of Berlioz

Symphony of a Thousand Mahler's *Symphony No 8 in E-flat major*

Symphony of Heavenly Length Schubert's *Symphony No 9*, according to Schumann

Symphony of Psalms Stravinsky's best-known symphony

Tabarro *Il Tabarro* (Italian—The Cloak)—opera by Puccini

Tales *Tales of Hoffmann* (Offenbach opera)

Tann *Tannhâser und der Sängerkrieg auf der Wartburg* (German—Tannhäser and the

Singing Contest of the Wartburg)—three-act Wagner opera

Tchaikovsky 1st, 2nd, 3rd, 4th, 5th, 6th First (*Winter Reveries*), Second (*Little Russian*), Third (*Polish*), Fourth, Fifth, Sixth (*Pathétique*) symphonies composed by Tchaikovsky

Tchaikovsky's Most Popular Ballet *Swan Lake*

Tchaikovsky's Most Popular Opera *Eugen Onegin*

Tchaikovsky's Most Popular Overture *1812*

Tchaikovsky's Most Popular Piano Concerto *No 1 in B-flat minor*

Tchaikovsky's Most Popular Song *None But the Lonely Heart*

Tchaikovsky's Most Popular Symphony *No 4 in F minor*

Tear-Jerker Composer Giacomo Puccini—opposite of Gioacchino Rossini

Telemann's Most Popular Concerto *Concerto for Trumpet and Strings in D*

Tell Rossini's opera *William Tell*

Tempest Beethoven's *Piano Sonata No 17 in D* (opus 3 1, no 2); Tchaikovsky's *Symphonic Fantasy—Tempest*

Tenors of the Century Enrico Caruso (1873–1921), Jussi Björling (1907–1960), Luciano Pavarotti (1935–)

Thaïs Massenet opera

The Bells Rachmaninoff's choral symphony based on Poe's poem *The Bells*

The Five (Russian composers Balakirev, Borodin, Cui, Moussorgsky, Rimsky-Korsakov)

The Great Schubert's *Symphony No 9 in C major*

The Isle *The Isle of the Dead* (orchestral work by Rachmaninoff inspired by Arnold Böcklin's painting of this title)

The Lamp Is Low Hollywood version of Ravel's *Pavane pour une infante de'funte*

Thomas's Most Popular Overture *Raymond*

Thousand *The Symphony of a Thousand* (Mahler's *Symphony No 8 in E-flat major*)

Three Classic Masses Bach's *B minor*, Beethoven's *Missa Solemnis*, Bruckner's *Grosse Messe in F minor*

Three Leading Conductors in America 1900–1950 Serge Koussevitzky, Leopold Stokowski, Arturo Toscanini

Three Leading Conductors in America 1950–1990 Leonard Bernstein, Zubin Mehta, Sir Georg Solti

Three Penny *Three Penny Opera* (composed by Kurt Weill, based on a modernized German version of John Gay's *The Beggar's Opera*)

Tikhiy *Tikhiy Don* (Russian— Quiet Flows the Don)—Dzerzhinsky's opera

Till *Till Eulenspiegels lustige Streiche* (German—Till Eulenspiegel's Merry Pranks)— symphonic poem by Richard Strauss

Timpanist of the 20th Century Saul Goodman of the New York Philharmonic

Titan Mahler's *Symphony No 1 in D major*—he preferred to call it his *Werther* symphony comparing it with Goethe's first novel

Tod und Verklärung (German—Death and Transfiguration)—symphonic poem by Richard Strauss

Tone-Poem Composer Richard Strauss

Tong-hai Moolkwa (Korean— Tong-Hai Sea)—anthem of South Korea

Tonight We Love popular name for Tchaikovsky's *Piano Concerto No 1 in B-flat minor*

top 25 top 25 concertos featured in many orchestral programs (Bach's concerto for two violins; Beethoven's five piano and one violin concertos; two piano and one violin concertos by Brahms; Bruch's violin concerto; Chopin's two piano concertos; Dvořák's cello concerto; Gershwin's piano concerto; Grieg's piano concerto; Liszt's piano concerto No 1; Mendelssohn's violin concerto; Mozart's piano concerto No 20; Paganini's con-

certo No 1 for violin; Rachmaninoff's piano concerto No 2; Schumann's piano concerto; the violin concerto of Sibelius; Tchaikovsky's concerto No 1 for piano and his violin concerto)

top 30 top 30 symphonic spectaculars favored on many orchestral programs [Bach's *Toccata and Fugue in D*; Beethoven's *Lenore Overture No 3*; Berlioz's *Symphonie fantastique*; Borodin's *Polovetsian Dances*; Brahms's *Variations on a Theme by Haydn*; Debussy's *La Mer*; Glinka's *Russlan and Ludmilla Overture*; Handel's *Water Music*; Liszt's *Les Preludes*; Mussorgsky's *Night on Bald Mountain, Pictures at an Exhibition*; Prokofiev's *Peter and the Wolf*; Rachmaninoff's *Rhapsody on a Theme by Paganini*; Ravel's *Bolero*; Rimsky-Korsakov's *Scheherazade*; Saint-Saëns' *Carnival of the Animals, Symphony No 3 (Organ)*; Sibelius's *Finlandia*; Smetana's *Moldau*; Richard Strauss's tone poems— *Don Juan, Don Quixote, Hero's Life* (Heldenleben), *Thus Spake Zarathustra* (Also sprach Zarathustra), *Till Eulenspiegel*; Stravinsky's *Sacre du Printemps*; Tchaikovsky's *Overture 1812, Romeo and Juliet*; Wagner's *Flying Dutchman* and *Tannhäuser* overtures, *Tristan* Prelude and Liebestod]

top 40 top 40 symphonies favored on many symphonic programs [Beethoven's 3rd. *(Eroica)*, 5th, 6th (Pastoral), and 9th (Choral); the four by Brahms; Bruckner's 4th (*Romantic*) and 9th; Dvořák's 6th and 9th (*New World*); Haydn's 94th (*Surprise*), 100th (*Military*), 101st (Clock), 103rd (*Drum Roll*), 104th (*London*); Mahler's 1st, 2nd (*Resurrection*), and 9th; Mendelssohn's 3rd (*Scottish*), 4th (*Italian*), and 5th (*Reformation*); Mozart's 35th (*Haffner*), and 41st (*Jupiter*); Prokofiev's 1st (*Classical*) and 5th; Rachmaninoff's 2nd;

Schubert's 8th (*Unfinished*)—and 9th; Schumann's 1st (*Spring*), 2nd, 3rd (*Rhenish*), and 4th; Shostakovich's 1st and 5th; Sibelius's 1st; Tchaikovsky's 4th, 5th, and 6th (*Pathitéque*)]

Tosca Puccini opera

Totentänz (German—Death Dance)—Liszt's paraphrase on the Dies Irae for piano and orchestra

TotentUnze (German— Dances of Death)—part of *Mahler's Symphony No 9*

Toy *Toy Symphony* usually ascribed to Haydn but now believed to be part of a larger work by Leopold Mozart

Tragic overture by Brahms; *Symphony No 6* by Mahler; *Symphony No 4* by Schubert

Traviata Verdi opera

Triangle Liszt's *Piano Concerto No 1 in E flat*

Tristan *Tristan und Isolde* (German—Tristan and Iseult)—music drama by Wagner

Trittico *Il Trittico* (The Tryptych)—Puccini's three short operas—*Gianni Schicchi, Suor Angelica*, and *Il Tabarro*

Trojans *Les Troyens* (French— The Trojans)—opera by Berlioz

Trombonist of the Century Tommy Dorsey

Trout Schubert's *Quintet in A major* for violin, viola, cello, double bass, and piano

Trov *Il Trovatore* (Italian—The Troubador)—Verdi opera

Troyens *Les Troyens* (French—The Trojans)—opera by Berlioz

Trumpeter of the Century Maurice André

Tubists of the Century William Bell of the New York Philharmonic, Roger Bobo of the Los Angeles Philharmonic

Turandot Puccini opera

Turina's Most Popular Work *Danzas fantásticas*

Turkish Mozart's *Violin Concerto in A major* (K 219)

Twentieth-Century Romantics Rachmaninoff, Sibelius, and Richard Strauss

Unfinished Schubert's *Symphony No 8 in B minor*

Urbs Orba Symphony in *F* of Saint-Saëns

Valurile Dunárii (Romanian—Danube Waves)—popular fanfare also called *Anniversary Waltz*

Variations on an Original Theme Enigma Variations of Elgar

Vaughan Williams' Most Popular Fantasia *Greensleeves*

Vaughan Williams' Most *Popular Opera* Hugh the Drover

Vaughan Williams' Most Popular Symphony *No 1—Sea Symphony*

Venezuelan composers Teresa Carreño, Reynaldo Hahn, José Angel Montero, Juan Bautista Plaza, Vicente Emilio Sojo

Venezuelan First Venezuela's composer-conductor Reynaldo Hahn who became music critic of *Le Figaro* and music director of the Paris Opera

Venezuelan Pianist Teresa Carreño

Verdi's Most Popular Mass *Manzoni Requiem*

Verdi's Most Popular Opera *Aida*

verismo (Italian—realism) applied to composers such as Leoncavallo, Mascagni, Puccini

Vespri I Vespri Siciliani (Italian—The Sicilian Vespers)— Verdi opera

Victory nickname for Beethoven's *Symphony No 5 in C minor*

Vie La Vie Parisienne (French—Parisian Life)— Offenbach opera

Vieuxtemp's Most Popular Concerto *No 5 for Violin*

Villa-Lobos' Most Popular Work *Bachianas Brasileiras No 5 for soprano and 8 celli*

Violinist of the Century Jascha Heifetz (1901–1987); William Primrose (1904–1982)

Violinist-Composer-Conductor Eugène Ysaÿe

Violinist-Conductor Willi Boskovsky; Richard Burgin; Sidney Harth, David Oistrakh; Igor Oistrakh, Joseph Silverstein; Isaac Stern

Violinist-Violist-Conductor Yehudi Menuhin; Pinchas Zukerman

Violin-Maker's Capital Cremona, Italy

Vishnu Symphony No 19 by Alan Hovhaness

Vivaldi's Most Popular Piece *Le Quattro Stagioni* (Italian—The Four Seasons)

Wagner's Most Popular Aria *In fernem Land* (German—In a Far Land) in *Lohengrin*

Wagner's Most Popular Music Drama *Tannhdäuser*

Wagner's Most Popular Prelude Act I—*Lohengrin*

Wagner's Most Popular Song *Träume* (German—Dreams)

Waldstein Beethhoven's *Piano Sonata No 21 in C* (opus 53), dedicated to Count von Waldstein

Walküre Die Walküre (German—The Valkyrie)—Wagner music drama

Wallace's Most Popular Overture *Maritana*

Walton's Most Popular Orchestral Work *Belshazzar's Feast*

Waltz King nickname shared by Lanner, Lehar, Lumbye, Kalman, Johann Strauss Sr and Jr, Josef Strauss, and Oskar Straus

Wanderer Schubert's *Piano Fantasie in C* (opus 15)

Warsaw Warsaw Concerto by Richard Addinsell

Weber's (von Weber's) Most Popular Overture *Oberon*

Welsh composers Alun Hoddinott, Arwel Hughes, Daniel Jones, William Mathias, Grace Williams, David Wynne

Werther Massenet opera

West Point Morton Gould's *Symphony No 4 for Band*

When the Lights Go On Again popularized version of Beethoven's *Minuet in G*

William Tell Rossini opera

Winter Reveries Tchaikovsky's *Symphony No 1 in G minor* (Rêverie d'Hiver)

Winter Wind Chopin's *Piano Etude No 11 in A minor*

Wolf's Most Popular Lieder *Italienisches Liederbuch* (German—Italian Lieder Book)

World's Largest Concert Hall Royal Albert Hall, London, with a seating capacity of 10,000

World's Largest Opera House Metropolitan Opera House, Lincoln Center, New York City

World's Most Musical Western Nation Germany

World's Oldest Orchestra Dresdener Staatskapelle founded in Dresden in 1548

Wozzeck Alban Berg music drama

Xerxes Handel opera

Year 1905 Shostakovich's Symphony No 11

Year 1917 Shostakovich's Symphony No 12

Youth Kabalevsky's *Concerto No 3 in D major; Youth Symphony in D minor* by Rachmaninoff

Zampa Hérold opera

Zarathustra Thus Spake Zarathustra (symphonic poem by Richard Strauss—Also sprach Zarathustra)

Zigeunerbaron Der Zigeunerbaron (German—The Gypsy Baron)—operetta by Johann Strauss, Jr

Zigeunerweisen (German— Gypsy Melodies)—Pablo de Sarasate work

Zingareska George Antheil's *Symphony No 1*

Z m Z Z mého Zivota (Czechoslovakian—From my Life)— Smetana's String Quartet No 1

National Capitals

Abidjan Ivory Coast
Abu Dhabi United Arab Emirates
Abuja Nigeria
Accra Ghana
Addis Ababa Ethiopia
Algiers Algeria
Amman Jordan
Amsterdam Netherlands
Andorra la Vella Andorra
Ankara Turkey
Antananarivo Madagascar
Apia Samoa
Ashkhabad Turkmenistan
Asmara Eritrea
Astana Kazakhstan
Asunción Paraguay
Athens Greece
Baghdad Iraq
Baku Azerbaijan
Bamako Mali
Bandar Seri Begawan Brunei Darussalam
Bangkok Thailand
Bangui Central African Republic
Banjul The Gambia
Basseterre St Kitts and Nevis
Beijing China
Beirut Lebanon
Belgrade Yugoslavia
Belmopan Belize
Berlin Germany
Bern Switzerland
Bissau Guinea-Bissau
Bloemfontein South Africa
Bogota Colombia
Brasilia Brazil
Bratislava Slovakia
Brazzaville Congo
Bridgetown Barbados
Brussels Belgium
Bucharest Romania
Budapest Hungary
Buenos Aires Argentina
Bujumbura Burundi
Cairo Egypt
Canberra Australia
Cape Town South Africa
Caracas Venezuela
Castries Saint Lucia
Colombo Sri Lanka
Conakry Guinea
Copenhagen Denmark
Dakar Senegal
Damascus Syria
Dar-es-Salaam Tanzania

Dhaka Bangladesh
Djibouti Djibouti
Doha Qatar
Dublin Ireland
Dushanbe Tajikistan
Freetown Sierra Leone
Funafuti Tuvalu
Gaborone Botswana
Georgetown Guyana
Guatemala City Guatemala
Haiti Port-au-Prince
Hanoi Vietnam
Harare Zimbabwe
Havana Cuba
Helsinki Finland
Honiara Solomon Islands
Islamabad Pakistan
Jakarta Indonesia
Jerusalem Israel
Kabul Afghanistan
Kampala Uganda
Kathmandu Nepal
Khartoum Sudan
Kiev Ukraine
Kigali Rwanda
Kingston Jamaica
Kingstown St Vincent and Grenadines
Kinshasa Democratic Republic of Congo
Kishinev Moldova
Kuala Lumpur Malaysia
Kuwait Kuwait
La Paz Bolivia's administrative capital
Libreville Gabon
Lilongwe Malawi
Lima Peru
Lisbon Portugal
Ljubljana Slovenia
Lomé Togo
London United Kingdom
Luanda Angola
Lusaka Zambia
Luxembourg Luxembourg
Madrid Spain
Majuro Marshall Islands
Malabo Equatorial Guinea
Male Maldives
Managua Nicaragua
Manama Bahrain
Manila Philippines
Maputo Mozambique
Maseru Lesotho
Mata-Utu Wallis and Futuna Islands
Mbabane Swaziland

Mexico City Mexico
Minsk Belarus
Mogadishu Somalia
Monaco Monaco
Monrovia Liberia
Montevideo Uruguay
Moroni Comoros
Moscow Russia
Muscat Oman
Nairobi Kenya
Nassau The Bahamas
N'Djamena Chad
New Delhi India
Niamey Niger
Nicosia Cyprus
Nouakchott Mauritania
Nuku'alofa Tonga
Nuuk Greenland
Oslo Norway
Ottawa Canada
Ouagadougou Burkina Faso
Palikir Micronesia
Panama Panama
Paramaribo Suriname
Paris France
Phnom Penh Cambodia
Port-au-Prince Haiti
Port Louis Mauritius
Port Moresby Papua New Guinea
Port-of-Spain Trinidad and Tobago
Porto-Novo Benin
Port Vila Vanuatu
Prague Czech Republic
Praia Cape Verde
Pretoria South Africa
Pyongyang North Korea
Quito Ecuador
Rabat Morocco
Reykjavik Iceland
Riga Latvia
Riyadh Saudi Arabia
Rome Italy
Roseau Dominica
St George's Grenada
St John's Antigua and Barbuda
Sana'a Yemen
San José Costa Rica
San Marino San Marino
San Salvador El Salvador
Santiago Chile
Santo Domingo Dominican Republic
São Tomé São Tomé and Príncipe
Sarajevo Bosnia and Herzegovina

Seoul South Korea
Singapore Singapore
Skopje Macedonia
Sofia Bulgaria
Stockholm Sweden
Sucre Bolivia's judicial capital
Suva Fiji
Taipei Taiwan
Tallinn Estonia
Tarawa Kiribati
Tashkent Uzbekistan
Tbilisi Georgia
Tegucigalpa Honduras

Tehran Iran
Thimphu Bhutan
Tirana Albania
Tokyo Japan
Tripoli Libya
Tunis Tunisia
Ulan Bator Mongolia
Vaduz Liechtenstein
Valletta Malta
Victoria Seychelles
Vienna Austria
Vientiane Laos
Vilnius Lithuania

Warsaw Poland
Washington, D.C. United States of America
Wellington New Zealand
Windhoek Namibia
Yamoussoukro Ivory Coast (official)
Yangon Myanmar (Burma)
Yaounde Cameroon
Yaren Nauru
Yerevan Armenia
Zagreb Croatia

National Holidays

Afghanistan Independence Day August 19

Albania Independence Day November 28

Algeria Independence Day July 5

Andorra National Day September 8

Angola Independence Day November 11

Antigua and Barbuda Independence Day November 1

Anzac Day April 25 (in Australia, New Zealand, and associated territories)

Arab Emirates Independence Day December 2

Argentina Independence Day July 9

Australia Day January 26

Austria National Day October 26

Bahama Independence Day July 10

Bahrain Independence Day December 16

Bangladesh Independence Day March 26

Barbados Independence Day November 30

Belarus Independence Day July 27

Belgium National Days September 9 and 10

Belize Independence Day September 21

Benin Independence Day December 6

Bhutan Independence Day January 4

Birmania Independence Day January 4

Bolivia Independence Day August 6

Botswana Independence Days September 30 and October 1

Brazil Independence Day September 7

Brunei Independence Day January 1

Bulgaria National Days September 30 and October 1

Burkina Faso Proclamation Day December 11

Burundi Independence Day July 1

Cameroon National Day May 20

Canada Day Dominion Day, July 1

Canada National Day January 1

Cape Verde Independence Day July 5

Central African Republic Independence Day August 13·

Central America Day Central American Independence Day, September 15 (in Costa Rica, El Salvador, Guatemala, Honduras, and Nicaragua)

Chile Independence Day September 18

China National Days October 1 and 2

Colombia Independence Day July 20

Commonwealth Day third Monday in May in parts of the British Commonwealth

Comoros Independence Day July 6

Congo National Day August 15

Costa Rica Independence Day September 15

Cuba Liberation Day January 1

Croatia Independencce Day May 30

Curaçao Day July 26 (celebrated throughout the Netherlands Antilles)

Cyprus Independence Day October 1

Czech Republic Independence Day October 28

Denmark Constitution Day June 5

Día de la Raza (Spanish—Day of the Race)—Columbus Day, October 12

Dieciséis (Spanish—Sixteenth)—September 16 (Mexican Independence Day)

Djibouti Independence Day June 27

Dominica National Days November 2 and 3

Dominican Republic Independence Day February 27

Ecuador Independence Day August 10

Egypt Proclamation Day June 18

El Salvador Independence Day September 15

Empire Day nearest Monday to May 24, celebrated in many parts of the British Commonwealth of Nations

Equatorial Independence Day October 12

Ethiopia Victory Day April 6

Fawkes Fawkes Day, November 5 (in Great Britain)

Fiji Independence Day October 10

Finland Independence Day December 6

Flag Day June 14 in the United States

France Bastille Day July 14

Gabon Independence Day August 17

Gambia Independence Day February 18

Georgia Independence Day May 26

Germany Unity Day June 17

Ghana Independence Day March 6

Greece Independence Day March 25

Grenada Independence Day February 7

Guatemala Independence Day September 15

Guinea Republic Day October 2

Guinea-Bissau National Day September 24

Guyana Republic Day February 23

Haiti Independence Day January 1

Honduras Independence Day September 15

Hostos Eugenio María Hostos Birthday, January 11 (celebrated in Puerto Rico)

Hungary Liberation Day April 4

Iceland National Day June 17

India Independence Day August 15

Indonesia National Day August 17

Iran Day of the Islamic Republic April 1

Iraq Day of the Republic July 14

Ireland's St Patrick's Day March 17

Israel Independence Day May 14

Italy National Day June 2

Ivory Coast Independence Day December 7

Jamaica Independence Day August 6

Japanese New Year December 28 through January 3

Japan National Foundation Day February 11

Jordan Independence Day May 25

Kenya Independence Day December 12

Kiribati Independence Day July 12

Kuwait Independence Day February 25

Labor Day first Monday in September in the U.S.A.

Laos National Day December 2

Lebanon Independence Day November 22

Lesotho Independence Day October 4

Liberia Independence Day July 26

Libya Revolution Day September 1

Liechtenstein National Day January 1

Luxembourg Day June 23—Grand Duke's Birthday

Madagascar Independence Day June 26

Malawi Republic Day July 6

Malaysia National Day August 31

Maldives Independence Day July 26

Mali Independence Day September 22

Malta Republic Day December 13

Mauritania National Day November 28

Mauritius National Day March 12

Mexico Independence Day September 16

Monaco National Day November 19

Mongolia National Day July 11

Morocco Independence Day November 18

Mozambique Independence Day June 25

Myanmar National Day January 1

Namibia Republic Day May 31

Nauru Independence Day January 31

Nepal National Day February 18

Netherlands Day April 30, Queen's Birthday

New Zealand National Day February 6

Nicaragua Independence Day September 15

Niger Independence Day August 3

Nigeria National Day October 1

North Korea Independence Day September 9

Norway Constitution Day May 17

Oman National Day November 18

Pakistan Independence Day August 14

Panama Day November 3—Separation from Colombia

Papua New Guinea Independence Day September 16

Paraguay Independence Days May 14 and 15

Peru Independence Day July 28

Philippine Independence Day June 12

Poland Liberation Day July 22

Portugal Day Independence Day, December 1

Portugal Liberty Day April 25

Puerto Rico-American Independence Day July 4

Qatar Independence Day September 3

Remembrance Canada's Remembrance Day, November 11—Armistice Day

Romania National Day August 23

Rwanda Independence Day July 1

St Christopher and Nevis Associate Day February 27

St Lucia Independence Day February 22

St Vincent and the Grenadines Independence Day October 27

San Marino Liberation Day February 5

São Tomé and Principe Independence Day July 12

Saudi Arabia National Day September 23

Senegal Association Day July 14

Seychelles Association Day June 29

Sierra Leone Anniversary Day April 19

Singapore National Day August 9

Slovakia Independence Day September 1

Slovenia Independence Day June 5

Solomon Islands Independence Day July 7

Somalia Foundation Day July 1

South Africa Republic Day May 31

South Korea Liberation Day August 15

Spain Hispanic Day October 12

Sri Lanka Independence Day February 4

Stanton's Elizabeth Cady Stanton's Day, November 12

Sudan Independence Day January 1

Suriname Independence Day November 25

Swaziland Independence Day September 6

Sweden Flag Day June 6

Switzerland Confederation Day August 1

Syria National Day November 16

Taiwan Revolution Day October 10

Tanzania Union Day April 26

Thailand King's Birthday December 5

Togo Independence Day April 27

Tonga Emancipation Day June 4

Transkei Independence Day October 26

Trinidad and Tobago Independence Day August 31

Tunisia Independence Day
June 1
Turkey Independence Day
October 29
Tuvalu Day October 1
Uganda Independence Day
October 9
United Kingdom Queen's
Birthday April 21

United States of America Inde-
pendence Day July 4
Uruguay Independence Day
August 25
Venezuela Independence Day
July 5
Victoria Day Queen Victoria's
Birthday, May 20
Vietnam National Day Sep-
tember 2

Western Samoa Independence
Day January 1
Zaire Independence Day June
30
Zambia Independence Day
October 24
Zimbabwe Independence
Day April 18

National Parks in Canada

British Columbia
Yoho National Park
Glacier National Park
Mount Revelstoke National
Park
Kootenay National Park
Alberta
Banff National Park
Elk Island National Park
Jasper National Park

Waterton Lakes National Park
Saskatchewan
Prince Albert National Park
Manitoba
Riding Mountain National
Park
Nova Scotia
Cape Breton Highlands Na-
tional Park
Fundy National Park

St Lawrence Islands National
Park
Prince Edward Island Na-
tional Park
Ontario
Georgian Bay Island National
Park
Point Pelee National Park

National Parks
in the United States

Acadia National Park
(1929)—Maine
Arches National Park
(1971)—Utah
Badlands National Park
(1978)—South Dakota
Big Bend National Park
(1944)—Texas
Biscayne National Park
(1980)—Florida
Black Canyon of the Gunnison
National Park (1999)—
Colorado
Bryce Canyon National Park
(1928)—Utah
Canyonlands National Park
(1964)—Utah
Capitol Reef National Park
(1971)—Utah
Carlsbad Caverns National
Park (1930)—New Mexico
Channel Islands National
Park (1980)—California
Crater Lake National Park
(1902)—Oregon
Death Valley National Park
(1994)—California and Ne-
vada
Denali National Park and Pre-
serve (1980)—Alaska
Dry Tortugas National Park
(1992)—Florida
Everglades National Park
(1934)—Florida
Gates of the Arctic National
Park and Preserve
(1980)—Alaska
Glacier National Park
(1910)—Montana

Glacier Bay National Park and
Preserve (1980)—Alaska
Grand Canyon National Park
(1919)—Arizona
Grand Teton National Park
(1929)—Wyoming
Great Basin National Park
(1986)—Nevada
Great Smoky Mountains
National Park (1934)—
North Carolina, Tennessee
Guadalupe Mountains
National Park (1972)—
Texas
Haleakala National Park
(1960)—Hawaii
Hawaii Volcanoes National
Park (1961)—Hawaii
Hopewell Culture National
Park (1992)—Ohio
Hot Springs National Park
(1921)—Arkansas
Isle Royale National Park
(1931)—Michigan
Joshua Tree National Park
(1994)—California
Katmai National Park and
Preserve (1980)—Alaska
Kenai Fjords National Park
(1980)—Alaska
Kings Canyon National Park
(1940)—Califorfiia
Kobuk Valley National Park
(1980)—Alaska
Lake Clark National Park and
Preserve (1980)—Alaska
Lassen Volcanic National
Park (1916)—California
Mammoth Cave National
Park (1926)—Kentucky

Mesa Verde National Park
(1906)—Colorado
Mount Rainier National Park
(1899)—Washington
North Cascades National
Park (1968)—Washington
Olympic National Park
(1938)—Washington
Petrified Forest NationalPark
(1962)—Arizona
Redwood National Park
(1968)—California
Rocky Mountain National
Park (1915)—Colorado
Saguaro National Park
(1994)—Arizona
Samoa National Park
(1988)—American Samoa
Sequoia National Park
(1890)—California
Shenandoah National Park
(1935)—Virginia
Theodore Roosevelt National
Park (1978)—North Dakota
Virgin Islands National Park
(1956)—Virgin Islands
Voyageurs National Park
(1971)—Minnesota
Wind Cave National Park
(1903)—South Dakota
Wrangell–St Elias National
Park (1980)—Alaska
Yellowstone National Park
(1872)—Idaho, Montana,
Wyoming
Yosemite National Park and
Preserve (1890)—Califor-
nia
Zion National Park (1919)—
Utah

Numbered Abbreviations

0^2 both eyes

0-0 zero-zero (no ceiling, no visibility for an aircraft)

000 emergency services (Australia)

007 James Bond (Ian Fleming's international sleuth)

0 deg lat zero degrees latitude (the Equator, encircling widest part of the Earth)

0-g zero gravity

0° zero degrees (the Equator)

0°lat zero degrees latitude (the Equator)

0°longitude Greenwich meridian with lines of longitude either east or west of this prime meridian in Greenwich, England

$1/4$ **d** farthing (fourth of an English penny); a fourthling

$1/4$ **h** quarter-hard

$1/4$ **ly** quarterly

$1/4$ **ph** quarter-phase

$1/4$ **rd** quarter-round

$1/4$ **S** quarters

$1/2$ **can** narcotics equal to a half can of pipe tobacco

$1/2$ **d** halfpenny (half of an English penny); ha'penny

$1/2$ **gr** half-gross

$1/2$ **h** half-hard

$1/2$ **rd** half-round

$1/2$ **sov** half-sovereign (10 shillings)

$1/2$ **sovereign** 10 shillings

$1/2$ **t** half title

1 in the beginning; in the year one; No 1 diet (bread and water); Number 1 (urine)

I first violin

1/ a bob (British slang for one shilling)

1^a *primeira* (Portuguese—first)—feminine gender; *primera* (Spanish—first)—feminine gender

I-A available for military service

I-A-O conscientious objector available only for noncombatant military service

1-armed one-armed bandit (gambling device also called a slot machine)

1b first base(man)

1-bagger one-bagger; single

I-BCE first century before the Christian era (Caesar's Century)—Julius Caesar conquered Britain and Egypt before he was assassinated in the Roman senate in the year 44

1¢ one cent

1/c single-conductor

1C member or former member of US armed forces with honorable discharge

I-C first century (Vesuvian Century)—destruction of Pompeii, Herculaneum, and nearby Neapolitan places by the volcano Vesuvius in the year 79 of the Christian era; member of the armed forces, Coast and Geodetic Survey, or Public Health Service

1ce once

1 cent 1 penny (10 mills)

1 Chron The First Book of the Chronicles

1 Cor The First Epistle of Paul the Apostle to the Corinthians

1 crown 5 shillings

1d an English penny

I-D member of reserve component or student taking military training

1 dime 10 cents

1 double eagle $20 (gold)

1/e first edition

1 eagle $10 (gold)

$1^{er(e)}$ *premier(e)* (French—first)

1ers *premières* (French—first violins)

1 Esd The First (Apocryphal) Book of Esdras

1 florin 2 shillings

1-fold one-fold (single undivided whole)

1 frogskin 1 bill

1G, 2G, 3G, etc. slang for one, two, three thousand dollars, etc.

1 Geigen (German-first violins)

1 guinea 21 shillings

1 half crown 2 shillings, 6 pence

1 half dime 5 cents

1 half dollar 50 cents

1 half eagle $5 (gold)

1 halfpenny 2 farthings

1 Hen IV First part of King Henry IV

1 Hen VI First part of King Henry VI

1 John The First Epistle General of John

1 Kings The First Book of the Kings

1/M First Mate

1ma *prima* (Italian feminine—first)

1 Macc The First (Apocryphal) Book of Maccabees

1mo *primo* (Italian—first)

1 Ne First Book of Nephi

1° *primeiro* (Portuguese—first)—masculine gender; *primero* (Spanish—first)—masculine gender

1° solo

1/O First Officer

I-O conscientious objector available only for civilian work contributing to national health, safety, or interest

1-1-1 one man, one vote, one election

1p one new penny

1-p single pole

1 penny 4 farthings

1 Pet The First Epistle General of Peter

1 ph single-phase

1-piece(r) one-piece bathing suit, coverall, or other one-piece garment

1 pound 20 shillings

1Q first quarter

1Q66 first quarter 1966

1 quarter dollar 25 cents

1 quarter eagle $2.50 (gold)

1s shilling, also called a *bob*

I-S student deferred from military service by statute until end of current school year

1 Sam The First Book of Samuel

1 shilling 12 pence

1 sixpence 6 pence

1 sovereign 1 pound sterling; 20 shillings

1-spot $1 bill

1st first

1st Asst Engr First Assistant Engineer

1st Asst Pur First Assistant Purser

1st cl hon first-class honors (in academic degrees)

1-step one-step (dance or music for such a dance)

1st h first home (lacrosse)

1st Lieut First Lieutenant

1st Naval District Boston, Massachusetts

1st Off First Officer

1-striper ensign (USN); third assistant engineer of third mate (merchant marine); private first class (US Army)

1st Sgt First Sergeant

1st State Delaware (first state of the original thirteen states to ratify the *Constitution of the United States)*

1st Vln(s) first violin(s)

1s & 2s mixed 1st- and 2nd-quality lumber

1-suit(er) one-suit (garment bag or suitcase) ,

1/10 net 30 1 percent discount off the face value of an invoice is allowed if invoice is paid in 10 days; otherwise the full amount is payable in 30 days

1 Thess The First Epistle of Paul the Apostle to the Thessalonians

1 threepence 3 pence

1 Tim The First Epistle of Paul the Apostle to Timothy

1-upmanship one step ahead of your adversary or even your best friend

I-W conscientious objector performing civilian work contributing to national health, safety, or interest, or who has completed such work

1-way one-way street; one-way traffic; unilateral

1-wd one-wheel drive

I-Y registrant does not meet present standards; available for military service only in event of war or national emergency

1-y-o one-year-old (child, pet, racehorse, etc.)

1-yTs one-year Treasury securities

1½ striper naval lieutenant, junior grade

2 Number 2 (excrement); soft lead pencil

II in music, second violin

2/ two shillings; also the coin ·called a florin

2^{af} segunda-feira (Portuguese—Monday)

II-A registrant deferred because of civilian occupation (except agriculture and activity in study) or an apprentice deferred by statute

II-B registrant deferred because necessary to war production

2-AP 2-Amino Purine

2b doubles; second base

2-bagger two bagger (double)

II-BCE second century before the Christian era (Roman Century)—Punic wars resulted in destruction of Carthage by the Roman legions—the 100s

2 bits 25 cents

2-B's boiled or bottled waters (essential in many countries)

2¢ two cents

2/c two-conductor

II-C registrant deferred because of agricultural occupation; second century (Aurelian Century)—reign of the Roman emperor-philosopher Marcus Aurelius—the 100s

2ce twice

2 Chron The Second Book of the Chronicles

2-cycle two-cycle

2d second

2-d two-dimensional

2da segunda (Italian, Portuguese, or Spanish—second)—feminine gender

2-decker double-decker (sandwich or ship); two decker

2d h second home (lacrosse)

2do segundo (Italian, Portuguese, or Spanish—second)—masculine gender

2^{ds} deuxièmes (French—second violins)

2 Dzerzhinsky Moscow address of the KGB; and the Lubyanka prison

2/e second edition

IIe deuxième, second, seconde (French—second)

2 Esd The Second (Apocryphal) Book of Esdras

2^{et} duet

2-F two-seater fighter aircraft (naval symbol)

2, 4-D 2, 4-dichlorophenoxyacetic acid

2-4-D dichlorophenoxy-acetic acid (weed killer)

2-4-5-T trichlorophenoxy-acetic acid (antiplant agent and defoliant)

2-fer two-fer (two for the price of one)

2-fold double; twofold

2g, 3g, 4g, etc. multiples of acceleration of gravity which at the surface of the Earth is .32.2 feet per second

2 Geigen (German—second violins)

2H hard lead pencil

2 Hen IV Second part of *King Henry IV*

2 Hen VI Second part of *King Henry VI*

2 i/c second in command

2 John The Second Epistle of John

2 Kings The Second Book of Kings

2/M Second Mate

2 Macc The Second (Apocryphal) Book of Maccabbees

2MASS Two Micron All-Sky Survey

2n diploid number

2nd Asst Engr Second Assistant Engineer

2nd Asst Pur Second Assistant Purser

2nd John John Quincy Adams

2nd Lieut Second Lieutenant

2nd Off Second Officer

2nd State Pennsylvania

2nd Stwd Second Steward

2nd Vln(s) second violin(s)

2 Ne Second Book of Nephi

2° segundo (Spanish—second)

2/O Second Officer

2p two new pence

2-p double pole

2 pc two-physician certification

2 Pet The Second Epistle General of Peter

2 ph two-phase

2-piece(r) two-piece(r)

2-ply two-ply

2Q second quarter

2s two shillings; also the coin called a florin

2's two-year-old children

II-S registrant deferred because of activity in study

2 Sam The Second Book of Samuel

2-sided two-sided

2/6 two-and-six (two shillings and sixpence); also called half a crown

2-some twosome (two persons or things)

2-spot $2 bill

2-st two-story

2-step two-step (dance)

2-striper corporal (US Army); lieutenant (USN); second assistant engineer or second mate (merchant marine)

2-suit(er) two-suit (garment bag or suitcase)

2T double throw

2/10-30 2 percent discount if paid in 10 days, net in 30 days

2 Thess The Second Epistle of Paul the Apostle to the Thessalonians

2-13 drug addict

2, 3-dpg 2, 3-diphosphoglycerate

2 Tim Second Epistle of Paul the Apostle to Timothy

2-time(r) two-time(r)

2-tone two-tone(d)

2/2 (music) *breve* time; cut time

2U to you

2U2 to you too

2-way two-way

2-wd 2-wheel drive

2WW Second Weather Wing (Air Force-New York)

2-y-o two-year-old (child, pet, racehorse, etc.)

2$^{1}/_{2}$ small glass of milk

2$^{1}/_{2}$ striper naval lieutenant commander

III-A registrant with child or children or registrant deferred by reason of extreme hardship to dependents

3^{a}f terça-feira (Portuguese—Tuesday)

3b third base; triples

3-bagger three-bagger (triple)

3-ball three-ball (golf match)

III-BCE third century before the Christian era (Carthaginian Century)—Hannibal crossed the Alps to defeat the Romans—the 200s

3-bravs *bravo* (shout meaning a male performer has done very well); *brava* (for a female performer), *bravi* (for male and female performers)

3-b's bang 'em, blow 'em, bow 'em—musical instruments played by banging them (percussion), blowing them (brasses or woodwinds), or bowing them (strings)

3-Bs Bach, Beethoven, Berlioz; Bach, Beethoven, Brahms; Bach, Beethoven, Bruckner

3/c three-conductor

3C Computer Control Company

III-C third century (the Chinese Century)—Chin dynasty ruled a reunited China—the 200s

3-card three-card (monte)

3ce thrice

3-color three-color (photo, photography, print, or printing process)

3-cord three parallel white cords adorning bibs of British sailors and their allies

3d English threepenny; thruppence; third

3-d dizzy, dopey, and dumb; three-dimensional

3de three-day event

3-decker three-decker (sandwich or ship); triple-decker

3d h third home (lacrosse)

3d m third man (lacrosse)

3-Ds discouragement, disillusionment, disappointment (including frustration and loss)—often leads to suicide, experts insist

3d 10th 40m 3 days 10 hours 40 minutes (Atlantic crossing of SS *United States* in July 1952)

3/e third edition

IIIe troisième (French—third)

3-e's of the 1990s economy, environment, ethics

3 × 5 3-inch by 5-inch filing card

3-fold threefold; triple

3-40s 40 mile-per-hour wind, 40-degree air temperature, 40-degree water temperature

3-gaited three-gaited (horse)

3-Gs Gegurgel, Gejodel, Geklapper (German—gurgling, yodelling, clattering)—Wagner's opinion of music composed by Mendelssohn and Meyerbeer

3H very hard lead pencil

3-H Hubert Horatio Humphrey

3-hand(ed) three-hand(ed) (card game)

3 Hen VI Third part of *King Henry VI*

3io trio

3-I voters Irish, Israeli, Italian

3 John The Third Epistle of John

3 K's Kinder, Küche, Kirche (German—children, kitchen, church)

3-l's latitude, lead, lookout; lead, log, lookout (dead-reckoning essentials)

3M Minnesota Mining and Manufacturing Company

3-M Maintenance and Material Management (USN)

3/M Third Mate

3-m l-l c's three-martini liquid-lunch clubbers (alcoholically befuddled expense-account experts contributing to the higher cost of so many things)

3-m lunch three-martini luncheon

3 mMs three musical Ms (Martinon, Monteux, Munch)—conductor-musicians par excellence

3Ms Macmurdo, Mackintosh, and Morris (British architects)

3-Ms Mozart, Mendelssohn, Mahler

3 Ne Third Book of Nephi

3º trio

3º tercero (Spanish—third)

3/O Third Officer

3-p triple pole

3ph three-phase

3-piece three-piece (garment)

3pl three-point linkage (tractor)

3-p's three phantoms (world depression, world unemployment, world unrest)

3-Ps prosecution, punishment, and persecution (often lifelong) faced by every felon

3pt three-point (tractor linkage)

3Q third quarter

3rd Asst Engr Third Assistant Engineer

3rd degree prolonged interrogation designed to produce a confession of guilt

3rd Naval District New York, New York

3rd Off Third Officer

3rd State New Jersey

3rd Stwd Third Steward

3-ring three-ring circus

3-RRs remedial reading, remedial writing, remedial arithmetic

3-Rs reading, writing, arithmetic (colloquially, readin', 'ritin', 'rithmetic)

3-Rs of productivity recognition, responsibility, rewards (for workers)

3-Rs of war relentless, remorseless, ruthless

3 S's Saba, Sint Eustatius (Statia), Sint Maarten (in the Netherlands Antilles within the Lesser Antilles near the Virgin Islands)

3-st three-story

3-star admiral or general of three-star rank

3-striper commander (USN); first assistant engineer or first mate (merchant marine); sergeant (U.S. Army)

3T triple throw

3-3 three (school) terms—three courses per term

3-way three-way

3-wheel atv's 3-wheel all-terrain vehicles

3WW Third Weather Wing (Air Force—Nebraska)

4 level 4 (death-dealing dose or injection of breath-stopping barbital or other drug used by executioners)

4a man 38 years or over and deferred from military service by reason of age

IV-A registrant who has completed service or a sole surviving son

4^{a}f quarta-feira (Portuguese—Wednesday)

4-As American Association for the Advancement of Atheism

4-AS American Association for the Advancement of Science

IV-B government official deferred by statute

4-bagger four-bagger (home run)

4-banger four-cylinder engine

IV-BCE fourth century before the Christian era (Alexandrian Century)—Alexander the Great of Macedonia defeated the Egyptians, the Persians, and the Indians; encouraged the Greek philosophers and poets—the 300s

4 bits 50 cents

4/c four-conductor

4C Community-Coordinated Child Care Program

IV-C alien; fourth century (Constantinian Century)—Roman emperor Constantine built the city of Constantinople on the site of ancient Byzantium and proclaimed it capital of the Eastern Empire—the 300s

4-class fourth-class (mail)

4-col p four-color page

4-d meat meat of dead, disabled, diseased, or dying animals

IV-D minister of religion or divinity student

4/e fourth edition

IVe quatrième (French—fourth)

IV-E conscientious objector available for, assigned to, or released for work of national importance

4-F find, feel, fornicate, and forget—code of conduct of certain men in search of casual sexual relationships

IV-F registrant not qualified for any military service

4-gls fourth-generation languages

4H very very hard lead pencil

4-H 4-H Club(s)—H standing for head, heart, health, and helping hands

4-hand four-hand(ed)

4 humors of the body blood, yellow bile, black bile, phlegm

4 Last Songs ballet based on the Four Last Songs of Richard Strauss

4-letter four-letter words (vulgar or offensive short words which refer to sex, excrement, or other bodily functions)

4-Ls latitude, lead, longitude, lookout

4-Ms Monteverdi, Mozart, Mendelssohn, Mahler

4 Ne Fourth Book of Nephi

4° quarto (a book about 9 × 12 inches)

4° cuarto (Spanish—fourth)

4/O Fourth Officer

4-p quadruple pole

4-p's product, place, promotion, price

4Q fourth quarter

4R Ceylon aircraft

4-R Act Railroad Revitalization and Regulatory Reform Act

4 spd four speed

4-Ss shit, shave, shampoo, and shower

4-st four-story

4-star admiral or general of four-star rank

4-striper captain (merchant marine or USN); chief engineer (merchant marine)

4tet quartet(te)

4th the 4th of July (American holiday celebrating the signing of the Declaration of Independence in Philadelphia on July 4, 1776)

4th Asst Engr Fourth Assistant Engineer

4th Naval District Philadelphia, Pennsylvania

4th Off Fourth Officer

4th State Georgia

4to quarto (sheet of book folded twice)

4U for you

4U2 for you too

4-way four-way

4-wd(s) four-wheel drive vehicle(s)

4-wheel four-wheel (drive)

4-ws (4WS) 4-wheel steering

4WW Fourth Weather Wing (Air Force—Colorado)

4X Israeli aircraft

5 large glass of milk

5A Libyan aircraft

V-A registrant over the age liability for military service

5^{a}f quinta-feira (Portuguese—Thursday)

5-and-10 variety store

5b bald man with baywindow, bifocals, bridgework, and bunions (humorous Selective Service rating)

V-BCE fifth century before the Christian era (Athenian Century)—Athenians destroyed Persian fleet at Salamis, completed the Parthenon in Athens—the 400s

5-Bs Bach, Beethoven, Berlioz, Brahms, Bruckner

5-B's Boston baked beans and brownbread

5-Bu 5-Bromouracil

5BX five basic exercises (Royal Canadian Air Force physical fitness program)

5 by 5 radio reception loud and clear (volume and clarity measured on a scale from 1 to 5)

V-C fifth century (Christian Century)—Christianity affirmed as the official faith by two Roman emperors—the 400s

5-Cs 5-Cs of cinematography (camera angles, closeups, composition, continuity, cutting)

5 de cinco de mayo (Mexican holiday celebrating victory over the French at the Battle of Puebla on May 5, 1862)

5 dimensions five dimensions of health: emotional, intellectual, physical, social, spiritual

5 don'ts don't kill, steal, commit adultery, become intoxicated, or lie, advised Buddah

5/e fifth edition

Vᵉ cinquième (French—fifth)

5-er five-dollar bill; fiver; five-pound note

5'er $5 bill; 5-pound note

5/50 five-year, fifty-thousand-mile protection (new car warranty)

5-finger five-finger(ed)

5-fold cinquefold(ed); five-fold(ed)

5 GLs five Great Lakes: Ontario, Erie, Huron, Michigan, Superior

5-HIAA 5-hydroxy indoleacetic acid

5-HTP 5-hydroxytryptophan

5-letter woman defined by a five-letter word—bitch

5-mc 5-methylcytosine

5 m's man/womanpower, materials, money, machinery, management

5° quinto (Spanish—fifth)

5 o's police

5p five new pence

5-percenter person who for 5 percent arranges introductions leading to valuable orders

5 Préludes Scriabin's last composition for the piano

5-Ps William Oxbery—British player, poet, publican, publisher, and printer; Pootsa Power Publishing Press Publications

5 spd five speed

5-spot $5 bill

5-star five-star (top quality)

5tet quintet

5th Naval District Norfolk, Virginia

5th State Connecticut

5ᵗᵗᵉ quintet

5 vd's five venereal diseases: chancroid, gonorrhea, granuloma inguinale, lymphogranuloma venereum, syphilis

5 w's the *who, what, when, where,* and *why* reporters attempt to include in writing summary paragraphs

6ᵃf sexta-feira (Portuguese—Friday)

6-banger 6-cylinder engine

VI-BCE sixth century before the Christian era (Babylonian Century)—Babylonians defeated Israelites and made them captive after destroying the temple of Solomon in Jerusalem—the 500s

6 bits 75 cents

6/c six-conductor

VI-C sixth century (Persian Century)—Khosru Nushirwan made peace with the Byzantine Empire and extended Persian rule throughout the Middle East—the 500s

6d English sixpenny; sixpence

6-dW Six-day War between Arab countries of Egypt, Jordan, Lebanon, and Syria vs Israel; June 5 to 10, 1967

6/e sixth edition

Vⁱᵉ sixième (French—sixth)

6'er leader of a pack of six scouts

6-fold sixfold

6-gun six-chambered revolver; sixshooter

6-hda 6-hydroxydopamine

6-mcd six-month certificate of deposit

6-mo six month(s),

6-mTs six-month Treasury securities

6° sesto; sexto (Spanish—sixth)

6-pack carton containing six of a kind (6 containers of beer, soda, etc.)

6-Rs remedial readin', remedial 'ritin', remedial 'rithmetic

6-shooter revolver holding six cartridges

666 symbol for the anti-Christ

6tet sextet

6th Naval District Charleston, South Carolina

6th State Massachusetts

6WW Sixth Weather-Wing (Air Force—Washington, D.C.)

7A Seven Arts Society

7 aa's 7 archangels (Gabriel, Michael, Raguel, Raphael, Remiel, Sariel, and Uriel)

VII-BCE seventh century before the Christian era (Assyrian Century)—when Assyria ruled Middle East and conquered Egypt—the 600s

7ber September

7ᵇʳᵉ Septembre (French—September); *septiembre* (Spanish—September)

7/c seven-conductor

VII-C seventh century (Islamic Century)—marked by Mohammed's flight from Mecca to Medina and his death in 632; Islam began expanding throughout the Middle East and North Africa as well as moving toward France and Spain—the 600s

7 cd's 7 chief devils (Aniguel, Anizel, Ariel, Aziel, Barfael, Marbuel, and Mephistopheles, according to the diabolarchy of hell)

7 Dec Pearl Harbor Day (1941)

7ds seven deadly sins—anger, covetousness, envy, gluttony, lechery, pride, sloth

7-d's deliriums, delusions, dementias, depressions, deviations, disorders, dreams

7/e seventh edition

7ᵉ septiembre (Spanish—September)

VIIᵉ septième (French—seventh)

7-fold sevenfold

7 lib ars seven liberal arts: grammar, logic, rhetoric, arithmetic, geometry, astronomy, and music

7 Ms seven (dance) movements: *élancer, étendre, glisser, plier, relever, sauter, tourner*

7° septimo (Spanish—seventh)

7 Octaves Louis Moreau Gottschalk

7 seas collective nickname of any seven seas; the Aegean, Baltic, Black, Caribbean,

Mediterranean, North, and Red seas; the Arafura, Banda, Celebes, Coral, Java, Tasman, and Timor seas; the Aegean, Baltic, Black, Mediterranean, North, Norwegian, and Red seas; the Bering, East and South China seas, the Sea of Okhotsk, the Sea of Japan, the Philippine Sea, and the Yellow Sea; the Aral, Azov or Putrid Sea, Bellinghausen, Caspian, Korean, Marmora, Sulu, White; many mariners define the seven seas as the Antarctic, Arctic, North and South Atlantic, North and South Pacific, and Indian oceans

767-300 Boeing 767 300-passenger airliner

7 smells camphoric (moth repellant), musky (angelica oil), floral (roses), peppermint (mint-flavored confections), ethereal (dry-cleaning fluids), pungent (vinegar), putrid (rotten eggs)

7tet septet

7th State Maryland

7/24 7 days a week, 24 hours a day

7-Up a carbonated beverage

7WW Seventh Weather Wing (Air Force—Illinois)

8 numerical symbol for heroin, as H is the eight of the alphabet

8ª Ottava (Italian—octave)

8-ball behind the eight ball (in a bad position); eight ball (black billiard or pool ball numbered 8)

VIII-BCE eighth century before the Christian era (Chou Century)—eastern Chou dynasty began ruling China for the next five centuries—the 700s

8 bits one dollar

8ᵇʳᵉ octobre (French—October); *octubre* (Spanish—October)

VIII-C eighth century (Carolingian Century)—Charlemagne or Charles the Great reigned as King of the Franks and Emperor of the West as well as chief patron of learning—the 700s

8/e eighth edition

8ᵉ octubre (Spanish—October)

VIIIᵉ huitième (French—eighth)

8-fold eightfold

8 great 8 great violin concertos (Bach's for 2 violins, Beethoven, Brahms, Bruch, Mendelssohn, Paganini, Sibelius, Tchaikovsky)

8-h NBC's concert hall studio used by Toscanini and the NBC Symphony of the Air and later by Mehta and the New York Philharmonic Symphony

8mm film 8-millimeter film

8N American National 8-thread series

8° octavo (a book about $9^3/_4$ inches high)

8° octavo (Spanish—eighth)

8½ × 11 paper size 8½ inches by 11 inches; standard letterhead size

8tet octet

8th Naval District New Orleans, Louisiana

8th State South Carolina

8 to 5 everyday 8 A.M. to 5 P.M. job

8UN Unified 8-thread series

8ᵛᵃ ottava (Italian—octave)

8ᵛᵃ alto ottava alto (Italian—octave higher)

8va bass ottava bassa (Italian—octave lower)

8ᵛᵉ octave

8vo octavo (sheet of book folded three times)

IX-BCE ninth century before the Christian era (Phoenician Century)—Carthage founded by the Phoenicians who traded in all areas of the Mediterranean—the 800s

9ᵇʳᵉ novembre (French—November); *noviembre* (Spanish—November)

IX-C ninth century (Century of Confusion)—Carolingian Empire of Charlemagne disintegrated; European unity dismembered and divided—the 800s

9/e, 10/e, 11/e, 12/e, etc. ninth edition, tenth edition, eleventh edition, twelfth edition, et cetera

9ᵉ noviembre (Spanish—November)

IXᵉ neuvième (French—ninth)

9ᵉᵗ nonet

9-fold ninefold

914 9 inches by 14 inches

9° nono; noveno (Spanish—ninth)

9-pins ninepin(s)

9 Ps nine known planets in our solar system according to their distance from the sun: Mercury, Venus, Earth, Mars, Jupiter, Saturn, Uranus, Neptune, Pluto

9th Naval District Great Lakes, Illinois

9th State New Hampshire

9 to 5 everyday 9 A.M. to 5 P.M. job

9-to-5 National Association of Working Women

10 deka (da)

'10 1810 (Bolivarian-type Spanish-American revolutions and wars of liberation, 1810–1826)

10⁻¹ deci(d)

10⁻² centi (c)

10⁻³ milli (m)

10⁻⁶ micro (μ)

10⁻⁹ nano (n)

10⁻¹² pico (p)

10⁻¹⁵ femto (f)

10⁻¹⁸ atto (a)

10² hecto (h)

10³ kilo (k)

10⁶ mega (M)

10⁹ giga (G)

10¹² tera (T)

10 Aug Ecuadorian Independence Day

X-BCE tenth century before the Christian era (Israelian Century)—King Solomon reigned and Israelites defeated all enemies and built the great temple of Jerusalem—the 900s

10ᵇʳᵉ décembre (French—December); *diciembre* (Spanish—December)

X-C tenth century (Mayan Century)—great American civilization left monumental ruins strewn from Honduras to Yucatan—the 900s

10 Dec Human Rights Day (Liberia)

10 Downing Street British prime minister's home in west central London

10ᵉ diciembre (Spanish—December)

Xᵉ dixième (French—tenth)

10ᵉᵗ dixet

10-fold tenfold

10-gage 10-gage shotgun

10-gallon hat cowboy hat

10 great 10 great violin concertos (Bach's for 2 violins, Beethoven, Brahms, Bruch, Mendelssohn, Paganini, Saint–Saëns, Sibelius, Tchaikovsky, Vieuxtemps)

10° decimo (Spanish—tenth)

10p 10 new pence

10-pin tenpin (bowling)

10-spot $10 bill

10th Naval District San Juan, Puerto Rico

10th State Virginia

10-V the lowest; the opposite of A-1; the worst

11 once (Spanish—eleven)— stands for aguardiente as this strong alcoholic drink has eleven letters

XI-BCE eleventh century before the Christian era (Century of Saul and David)— King Saul followed by King David as ruler of Israel—the 1000s

XI-C eleventh century (Aztecan and Incan Century)— vast monuments in the highlands of Mexico and Peru stand as mute witnesses to these great American civilizations—the 1000s

11 Downing Street official town residence of the British Chancellor of the Exchequer

11-11-11 eleventh hour, eleventh day, eleventh month of 1918 when Armistice ended World War I

11th-hr eleventh hour (last minute; latest possible time for assistance or a death-sentence reprieve)

11th Naval District San Diego, California

11th State New York

XII-BCE twelfth century before the Christian era (Trojan Century)—Troy fell to the Greeks after a ten-year siege celebrated in Homer's epic poem, the Iliad—the 1100s

XII-C twelfth century (Portuguese Century)—when Alfonso I Henriques reigned as king of Portugal, which emerged as a great maritime power—the 1100s

12 Downing Street office of the British Government Whips

12-gage 12-gage shotgun

12N American National 12thread series

12° twelvemo (a book about 7³/₄ inches high)

12th Naval District San Francisco, California

12th State North Carolina

12-tone twelve-tone music; twelve-tone row

12UN Unified 12-thread series

13 Gallery 13—prisoner's grave(yard); numerical symbol for marijuana as M is the thirteenth letter of the alphabet; police radio signal call 13 indicates an officer needs help; the boss is nearby; white bread

XIII-BCE thirteenth century before the Christian era (Century of the Exodus)—Moses led the Israelites out of Egypt—the 1200s

XIII-C thirteenth century (Mongol Century)—dominated by the reign of the Mongol emperor Genghiz Khan whose hordes conquered China and Russia—the 1200s

13th Naval District Seattle, Washington

13th State Rhode Island

14 numerical symbol for narcotics as N is the fourteenth letter of the alphabet; special (food) order

XIV-BCE fourteenth century before the Christian era (Century of the Pharaoh Tutankhamen)

XIV-C fourteenth century (Tamerlane's Century)—Mongol emperor Timur (Tamer the Lame) dominated Middle East and western India—the 1300s

14 f 13 Nazi code name for death by gassing

14N normal nitrogen

14th Naval District Pearl Harbor, Oahu, Hawaii

14th State Vermont

XV-BCE fifteenth century before the Christian era (Egyptian Century)—Egyptian kingdom extended from the Sahara to beyond the Euphrates—the 1400s

XV-C fifteenth century (Italian Century)—powerful families such as the Borgias and the de

Medicis brought about the renewal of art and architecture in Italy—the Italian Renaissance—the 1400s

15N heavy nitrogen

15th Naval District Balboa, Canal Zone

15th State Kentucky

XVI to XXXII BCE (*see* XXXII-BCE)

XVI-C sixteenth century (Spanish Century)—marked by discoveries and colonizations of much of the New World, circumnavigation of the globe, flowering of art and literature—the Golden Age or *Siglo de Oro*—as well as the defeat of the Spanish Armada by the British—the 1500s

16-gage 16-gage shotgun

16mm film 16-millimeter film

16mo sixteenmo (sheet of book folded four times)

16N American National 16-thread series

16° sixteenmo (a book about 6 ³/₄ inches high)

16's 16 revolutions per minute phonograph records

16th State Tennessee

16UN Unified 16-thread series

XVII-C seventeenth century (Dutch Century)—saw the discovery and settlement of what is now New York as well as South Africa and the East Indies by the Dutch who after a war at sea arranged a mutual defense pact with their British rivals—the 1600s (*see* the Elizabethan Age, *Le Grand Siecle, El Siglo de Oro*)

17-D modified yellow-fever virus

17-KGS 17-ketogenic steroids

17-KS 17-ketosteroids

17-OHCS 17-hydroxycorticosteroids

17 Tage und 4 Minuten (German—17 days and 4 minutes)—Werner Egk opera

17th Naval District Kodiak, Alaska

17th Parallel line of latitude dividing North and South Vietnam

17th State Ohio

XVIII-C eighteenth century (French Century)—of courtesans and kings, poets and playwrights, of great territories acquired and lost, of Louis XVI and Marie Antoinette beheaded by the guillotine only to be replaced by Napoleon—the turbulent 1700s (*see* The Enlightenment)

18mo eighteenmo (sheet of book folded so each leaf is one-eighteenth of original sheet)

18-19 Sept Chilean Independence Days

18th State Louisiana

19 banana split

XIX-C nineteenth century (British Century)—from Napoleon's defeat by Wellington at Waterloo to the defeat of the Boers in South Africa this century was marked by British advances in invention, in the success of its industrial revolution, in its colonization in all parts of the world, and its maritime supremacy on all the oceans—the 1800s

19°2'40" N lat x 62°17'20" W long 19 degrees, 2 minutes, 40 seconds North latitude by 62 degrees, 17 minutes, 20 seconds West longitude, imaginary location of *Treasure Island* created by Robert Louis Stevenson

19th State Indiana

20 twenty

20' twenty-foot-long shipping container

XX-C twentieth century (American Century)—characterized by industrial advances, victory in two world wars, as well as the development of inventions, the discovery of the North Pole, the placing of men on the moon, the elevation of living standards, the devotion to democratic ideals—the 1900s

20-gage 20-gage shotgun

'20s 1920–1929

20-spot $20 bill

20th Twentieth Century Limited (New York Central Railroad)

20th-century syndrome hypersensitivity to modern chemical compounds

20th State Mississippi

20 toes 10 down and 10 toes up (bedtime frolic)

21 blackjack; limeade

XXI-C twenty-first century (Japanese Century)—providing productivity, standard of living, and other growth factors are not disturbed by large-scale earthquakes and world wars—the 2000s

21st State Illinois

22 customer's check still unpaid

.22 .22-caliber ammunition, pistol, or rifle

22-cal killers' assassins using 22-caliber silencer-equipped automatic pistols

22nd State Alabama

22 s-e silencer-equipped 22-caliber pistol

23 get lost; I'm busy; leave me alone; scram

23°27' N lat 23 degrees, 27 minutes North latitude, Tropic of Cancer

23°27' S lat 23 degrees, 27 minutes South latitude, Tropic of Capricorn

23rd State Maine

23 deg N lat Tropic of Cancer

23 deg S lat Tropic of Capricorn

24 *24 Capricci (Opus I)*—Paganini's Twenty-four Caprices for cadenza-like unaccompanied violin

24° twenty-fourmo (a book about 5 3/4 inches high)

24-Ps 24 Parganas (Zamindari of Calcutta fiscal divisions in the Ganges delta)

24/7 twenty-four hours a day, seven days a week (all the time)

24th State Missouri

25 LSD as 25 is part of the chemical name-d-lysergic acid diethylamide tartrate 25

.25 .25-caliber ammunition or automatic

25th State Arkansas

26th State Michigan

27th State Florida

28-gage 28-gage shotgun

28th State Texas

29th State Iowa

30 finis symbol used by newspaper journalist at end of article or story

30 days, etc. (calendar mnemonic—30 days hath September, April, June, and November; all the rest have 31

save February; 28 are all its score, but in leap year one day more)

'30s 1930–1939

30th State Wisconsin

.30-'06 .30-caliber American cartridge introduced in 1906; used by U.S. Armed Forces in World Wars I and 11 for rifles and machine guns

31st State California

XXXII-BCE thirty-second century before the Christian era (Dynastic Century) when the first and second of many Egyptian dynasties began a rule lasting for at least seventeen centuries before the power of the pharaohs began to wane—the 3100s

32mo thirty-twomo (sheet of book folded five times)

32nd State Minnesota

32° thirty-twomo (a book about 5 inches high)

33rd State Oregon

33s 33 1/3 rpm phonograph records

34th State Kansas

35mm film 35-millimeter film

35th State West Virginia

36 postum

36th State Nevada

37th State Nebraska

.38 .38-caliber ammunition or pistol

38th Parallel line of latitude dividing North and South Korea

38th State Colorado

39th State North Dakota

40 40 acres

40' forty-foot-long shipping container

'40s 1940–1949

40th State South Dakota

40 winks a nap or short sleep

41st State Montana

42nd cousin a distant relative

42nd State Washington

42nd Street New York City's most blatant nightlife area

43rd State Idaho

.44 .44-caliber ammunition or pistol

44th State Wyoming

.45 .45-caliber ammunition, pistol, or submachine gun

45's 45 rpm phonograph records

45th State Utah

46th State Oklahoma

47th State New Mexico

47th Street New York City's diamond center between the Avenue of the Americas and Fifth Avenue on 47th Street

48 48-hour weekend liberty pass; the forty-eight preludes and fugues of Johann Sebastian Bach's collection known as the *Well-tempered Clavier*

48er emigrant who came to America in 1848, participant in German revolution of 1848

48° forty-eightmo (a book about 4 inches high)

48th State Arizona

48¹/₂ discharged; fired 49er gold-rush settler who came to California in 1849

49th State Alaska

.50 .50-caliber ammunition or machine gun

50 percenter auto mechanic causing the need for unnecessary repairs or recommending unnecessary replacements and then splitting the profits with gas-station operators

'50s 1950–1959

50-spot $50 bill

50th State Hawaii

51 one cup of hot chocolate

52 two cups of hot chocolate

54 54 minorities comprising the people of China

55¹/₂ small root beer

'60s 1960–1969

64° sixty-fourmo (a book about 3 inches high)

66 dirty dishes; Phillips Petroleum Company

66 deg 17 min N lat Arctic Circle

69 pictoral numerical symbol for oral-genital copulation

70mm film 70-millimeter film

'70s 1970–1979

73 best regards (amateur radio)

75's 75mm cannon

76 Union Oil

'76 1776

78's 78 rpm phonograph records

'80s 1980–1989

81 glass of water

82 two glasses of water

84 naval prison

84 Ave of Foch Paris headquarters of the gestapo during the German occupation of Paris in World War II; its top

floor contained cells and its lower floors housed offices and torture chambers

86 don't serve

87¹/₂ look at the lovely girl(s) out front

88 Column 88 (neoNazi organization based in London); love and kisses (amateur radio); (musician's slang) a piano

89d days (New York to San Francisco run of American clipper ship Flying *Cloud* in 1854)

89er Oklahoman who settled in 1889 when the territory was opened

90-day wonder officer commissioned after only 90 days of training

90 deg N lat North Pole (zero degrees longitude)

90 deg S lat South Pole (zero degrees longitude)

90° N lat 90 degrees North latitude, North Pole

90° S lat 90 degrees South latitude (South Pole)

'91 a glass of seltzer water

92 St Y 92nd Street Y School of Music

93-score best grade of butter (USDA grade AA)

95 the customer's leaving without paying

'96 1796 (Napoleonic Wars, 1796–1815)

98 assistant manager's nearby—look out

'98 the generation of 1898 (Spain's generation of cultured persons reacting to the illusions and incompetence that led to the loss of Cuba in 1898)—Attamira the historian, Azorín the author, Cossio the art historian and teacher, Ferrer the anarchist teacher, Machado the poet, Ortega y Gasset the essayist, Ramón y Cajal the histologist and surgeon, Unamuno the teacher and writer, etc.

99 manager's nearby—look out

100 Hydrographic Office Publication 100—*Merchant Marine House Flags and Stack Insignia-U.S. Navy Hydrographic Office;* one hundred

100-percenter(s) one-hundred percent patriot(s) or worker(s)

100-proof completely honest, the whole truth

100s hundreds

108-named god of the Hindus Vishnu

110 110 volt(s)

110v 110 volts

110v/60c (60h) 110-volt 60-cycle (60 hertz) electric current

111 emergency services (New Zealand); One-Eleven (British Aircraft Corporation short-take-off-and-landing fan-jet aircraft)

118-island city Venice (Venezia)

150 publication 150—*World Port Index—*U.S. Naval Oceanographic Office

180° longitude International Date Line where east meets west 180 degrees east or west of the Greenwich or prime meridian

200 200-mile Exclusive Economic Zone (EEZ)

220(v) 220 volt(s)

240 Convair two-engine transport airplane; trotting horse speed—1 mile in 2 minutes and 40 seconds; synonyms for high speed

240v 240 volts

280 copper alloy (Muntz metal); yellow metal

291 291 Fifth Avenue (address of An American Place, Alfred Stieglitz's art gallery where much avant-garde lithography, painting, and photography were first shown)

400 the four hundred; the socially elite (originally designated by Ward McAllister, who drew up a list containing the top 400 in New York society)

410 410-caliber shotgun

415 PC Section 415 Penal Code—disturbing the peace

486 French formula employed in abortion-including pill

500 Festival 500-mile (805-kilometer) auto racing event of the International Automobile Speed Classic at Indianapolis, Indiana

502 drunken driving (police code)

606 arsphenamine compound sold as Salvarsanl; 606th compound developed and tested by Paul Ehrlich for treatment of relapsing fevers and syphilis

707 Boeing Stratoliner jet-transport airplane

720 Boeing medium-range jettransport airplane

727 Boeing jet-transport with three empennage-mounted engines

737 Boeing short-range twin–jet airplane

747 Boeing jumbo jetliner

757 Boeing medium-range jet-liner

767 Boeing's fuel-saver twin-jet passenger plane

800-C-0-C-A-I-N-E 800-262-2463 (national cocaine hot-line connecting addicts with counselors and psychiatric hospital emergency rooms)

880 Convair 880 jet airplane

911 emergency telephone number

925 Office Workers Union

990 Convair 990 fan-engine jet airplane

999 emergency services (United Kingdom)

1000s Thousand Islands, some 1500 islets in the St Lawrence River

1011 Lockheed 1011 jumbo jet-liner

1080 sodium fluoroacetate

1098 supplies and equipment requisition form (British Army)

1400s fourteen hundreds

1600 Pennsylvania Avenue (Washington D.C. address of the White House)

1600s sixteen hundreds

1700s seventeen hundreds

1800s eighteen hundreds

1812 Tchaikovsky overture for full orchestra and cannon

1814 Leroux opera

1854–1954 century-long life of the French Guiana penal colony popularly known as Cayene or Devil's Island; world opinion about the inhumanities practiced here did much to stop its existence

1900s nineteen hundreds

"1905" Leon Trotsky's account of the dress-rehearsal Russian revolution of 1905; Dmitri Shostakovich's Symphony No 11—*Year 1905*

1905er Old Bolshevik; participant in the Russian Revolution of 1905; veteran communist

1917 Symphony No 12 by Shostakovich

"1919" novel by John Dos Passos depicting World War I era of American life in series of camera-eye closeups

1941 Prokofiev's symphonic suite containing three movements—*Battle, At Night, Brotherhood of Nations*

1944 bug start date for many computer operating-system clocks prior to 2000

1945 *Symphony No 9* (E-flat major) by Shostkovich

"1984" *Nineteen Eighty-four* (novel of George Orwell describing totalitarian terror in the year 1984); symbol for anti-libertarian trends

2141 2141 islands of Micronesia (greater in area than the continental U.S. but smaller in land mass than Rhode Island)—Trust Territory of the Pacific with its capital on Saipan atop Capitol Hill

2707 Boeing supersonic transport airplane

9653 Convict 9653 (American Socialist nominee for president—Eugene V Debs while in the U.S. Penitentiary in Atlanta, Georgia)

10,000 Lakes State Minnesota

22445 prisoners' petitions for judicial reviews of their cases

23102a V(ehicle) C(ode) driving under the influence of any intoxicating liquor or drug

338171 TE TE Lawrence (Lawrence of Arabia's number in the British Army; he used this number rather than his name as a final defense against a world he found hostile and unresponsive)

Numeration

	Power	Prefix	Abbreviation	Name
1,000,000,000,000,000,000	10^{18}	eva	e	one quintillion*
1,000,000,000,000,000	10^{15}	peta	p	one quadrillion
1,000,000,000,000	10^{12}	tera	t	one trillion
100,000,000,000	10^{11}			one-hundred billion
10,000,000,000	10^{10}			ten billion
1,000,000,000	10^{9}	giga	g	one billion
100,000,000	10^{8}			one-hundred million
10,000,000	10^{7}			ten million
1,000,000	10^{6}	mega	m	one million
100,000	10^{5}			one-hundred thousand
10,000	10^{4}			ten thousand
1000	10^{3}	kilo	k	one thousand
100	10^{2}			one hundred
10	10^{1}			ten
1	10^{0}			one
0.1	10^{-1}	deci	d	one-tenth
0.01	10^{-2}	centi	c	one-hundredth
0.001	10^{-3}	milli	m	one-thousandth
0.0001	10^{-4}			one ten-thousandth
0.00001	10^{-5}			one hundred-thousandth
0.000001	10^{-6}	micro	(μ-mu)	one millionth
0.0000001	10^{-7}			one ten-millionth
0.00000001	10^{-8}			one hundred-millionth
0.000000001	10^{-9}	nano	n	one billionth
0.0000000001	10^{-10}			one ten-billionth
0.00000000001	10^{-11}			one hundred-billionth
0.000000000001	10^{-12}	pico	p	one trillionth
0.0000000000001	10^{-13}			one ten-trillionth
0.00000000000001	10^{-14}			one hundred-trillionth
0.000000000000001	10^{-15}	femto	f	one quadrillionth
0.0000000000000001	10^{-16}			one ten-quadrillionth
0.00000000000000001	10^{-17}			one hundred-quadrillionth
0.000000000000000001	10^{-18}	atto	a	one quintillionth

*Quintillions are followed by sextillions, octillions, nonillions, decillions, undecillions, duodecillions, tredecillions, quattuordecillions, quinquedecillions, sexdecillions, septendecillions, octodecillions, novemdecillions, vigintillions (a thousand novemdecillions).

Oldest North American Settlements

Oldest American Settlement entries refer to the first permanent settlements.

Alabama by the French in the Mobile area around 1702

Alaska Three Saints Bay, Kodiak Island, founded in 1784 by Grigori Shelekhov

American Samoa Pago Pago settled 4,000 years ago on Tutuila Island

American Virgin Islands St Thomas Island's port of Charlotte Amalie built by the Danish in 1672

Arizona Tombstone, the old mining town booming in the 1880s

Arkansas Arkansas Post established by French settlers in 1686 east-southeast of Pine Bluff on Arkansas River

California San Diego de Alcalá founded in 1769 by Padre Junípero Serra

Colorado Denver, 1858

Connecticut Hartford, founded as a Dutch trading post by Adriaen Block in 1633

Delaware Wilmington, called Fort Christina by the Swedes, in 1638

District of Columbia cornerstone of the capitol's north wing laid by President Washington on September 18, 1793

Florida St Augustine, founded by the Spanish in 1565

Georgia Savannah settled by the English under James Oglethrope in 1733

Guam Agaña settled by the Spanish in 1668

Hawaii Honolulu, discovered by Captain Cook in 1778

Idaho David Thompson established first trading post in 1809

Illinois Fort Crève Coeur in 1680 near Peoria

Indiana Vincennes in 1702 on the Wabash River

Iowa Dubuque, 1834

Kansas Fort Leavenworth, 1827

Kentucky Harrodsburg, 1774, southwest of Lexington

Los Angeles *Nuestra Señora la Reina de los Angeles de Porciuncula* (Our Lady of the Angels of Porciuncula), 1781

Louisiana Natchitoches founded by the French in 1714

Maine Phippsburg, 1607, at the mouth of the Kennebec River

Maryland St Mary's 1634

Massachusetts Plymouth, 1620

Michigan Fort Ponchartrain, 1701, at the site of what is now Detroit

Minnesota Fort Snelling on the site of Minneapolis, 1820

Mississippi Biloxi Bay, 1699

Missouri Sainte Genevieve, 1735

Montana Virginia City, 1850

Nebraska Bellevue, 1823

Nevada Genoa, 1849, near what is now Carson City

New Hampshire Dover, 1624

New Jersey Communipaw, 1620

New Mexico Pueblo San Juan, 1598

New Orleans platted in 1718 by the Sieur de Bienville, made capital of Louisiana in 1722

New York New Amsterdam, 1644 on the southern tip of Manhattan Island

North America St Augustine, Florida founded by the Spanish in 1565

North Carolina Bath, 1705, after the loss of Roanoke Island's settlers

North Dakota Pembina, 1851

Ohio Marietta, 1788

Oklahoma Salina settled in the early 1800s

Oregon fur-trading post established in 1811 by John Jacob Astor at the mouth of the Columbia River and the site of Astoria

Pennsylvania Tinicurn Island, south of Philadephia, founded by the Swedes in 1643

Puerto Rico San Juan settled by Ponce de León in 1521

Rhode Island Providence settled by Roger Williams in 1636

San Francisco San Francisco de Asís, now called Mission Dolores, founded in 1776 by the Spanish

Seattle began as a lumber town in 1852

South Carolina Charleston 1680

South Dakota Fort Pierre, 1817

Tennessee Jonesboro, 1779 close to the site of Johnson City

Texas Ysleta, 1681, close to El Paso

Utah Salt Lake City, 1847

Vermont Fort Dummer, near Burlington, 1724

Virginia Jamestown, 1607

Washington Fort Okanogan, 1888

West Virginia Romney and Sheperdston, 1762

Wisconsin Green Bay, 1701

Wyoming Fort Laramie, 1834, a fur-trading post near Cheyenne

Phobias

Opposite of the suffix *phobia*—fear or hatred of—is *phil* or *phile*—admiration or love of, as in *Anglophile* meaning admiration or love of the English, the British Empire, the Commonwealth, or the United Kingdom, whereas *Anglophobia* means fear of or hatred of all things English.

acarophobia delusion one's skin is infested with mites, insects, or worms

acrophobia fear of high places

aelurophobia fear or hatred of cats *(also written* ailurophobia*)*

aerophobia fear of flying due to fatal accidents, fear of heights, or motion sickness; fear of high winds or violent storms *(also written* airphobia*)*

agoraphobia fear of busy streets, crowded churches and stores, stadiums; fear of the marketplace or other open spaces

agyiophobia fear of streets

aichmophobia fear of being touched by a finger or a sharp-pointed instrument or weapon

AIDS-phobia fear of AIDS (acquired immune deficiency syndrome)

ailurophobia fear of cats and other felines *(also written* aelurophobia *and synonym of* felinophobia*)*

airphobia *(see* aerophobia*)*

alcoholophobia fear of becoming an alcoholic

algophobia fear of pain or watching anyone in pain

amaxophobia fear of vehicles

Americanophobia fear of Americans or the United States of America

amphidiaphobia fear of amphibians—coecilians, frogs, salamanders, toads

amychophobia fear of being scratched by claws or sharp-pointed nails

androphobia fear of boys and men

anemophobia fear of wind

anginophobia fear of angina pectoris

Anglophobia fear of or hatred of the English, the British Empire, the Commonwealth, the United Kingdom

Angst (German—fear)

anthophobia fear of flowers

anthrophobia fear of people, persons, or society

anthropophobia fear of being eaten alive; fear of cannibalism

antlophobia fear of floods

antrophobia fear of people

anxiety hysteria (*see* phobia anxiety phobias)

agoraphobia (fear of being alone or in public places); simple or social phobias marked by irrational fear of objects or certain social situations believed to be embarrassing or humiliating

apartheidphobia fear of people who impose apartheid in South Africa

aphephobia fear of being touched by anyone

apiphobia fear of bees or other buzzing insects

apotemnophobia fear of persons with amputations

aquaphobia fear of water

arachnephobia fear of spiders

assassinophobia fear of assassination

asthenophobia fear of weakness

astraphobia fear of lightning and thunder

astrophobia fear of the heavens and the stars

ataxophobia fear of muscular hysteria and incoordination

aulophobia fear of flutes

automysophobia fear of being unclean

autophobia fear of being alone; fear of oneself

bacillophobia fear of bacilli; fear of microbes

bacteriophobia fear of bacteria

ballistophobia fear of bullets; fear of missiles

basiphobia fear of being unable to walk

basophobia fear of being unable to stand erect *(synonym for* stasiphobia*)*

bathophobia fear of depths or looking down from high places

batophobia fear of falling objects or high places

batrachiophobia fear of amphibians—coecilians, frogs and toads, newts and salamanders

belonophobia fear of needles, pins, and other sharp-pointed instruments or weapons such as daggers and swords

bibliophobia fear of books and other printed matter

botanophobia fear of forests, jungles, plantlife

boviphobia fear of bulls and other aggressive cattle, such as bison, buffalo, and yaks

brassophobia fear of brass and brassware

bromidrosiphobia fear of body odors

brontophobia fear of thunder

bufonophobia fear of toads

cainotophobia fear of new ideas or new things

cancerophobia fear of cancer

cardiophobia abnormal fear of heart disease or heart failure

cenophobia fear of empty houses or rooms

ceraunophobia fear of thunder

cheloniaphobia fear of terrapins, tortoises, turtles

cherophobia fear of fun and gaiety

chionophobia fear of snow

chirosphobia fear of writer's cramp

chirurphobia fear of surgery

cholerophobia fear of contracting cholera

chrematophobia fear of money

chromophobia fear of certain colors

chronophobia fear of time

cibophobia fear of food

claustrophobia fear of closed spaces

climacophobia fear of climbing; fear of stairs

clinophobia fear of going to bed

clithrophobia fear of being locked in

coitophobia fear of sexual intercourse

coprophobia fear of dung

cosophobia fear of dawn

cremnophobia fear of high or steep places

crocodiliophibia fear of crocodilians—alligators, caymans, crocodiles, gavials

crystallophobia fear of glass

cumacophobia fear of falling downstairs

cumophobia fear of beds

cyclophobia fear of bicycles

cymophobia fear of waves

cynophobia fear of dogs and all their relatives—coyotes, dingos, foxes, hyenas, wolves

cypridophobia fear of venereal disease

cypriphobia fear of venereal disease

decidophobia fear of making decisions

demonophobia fear of demons or devils

dermatopathophobia fear of skin diseases

dextrophobia fear of the right side

dipsophobia fear of drinking

domatophobia fear of being confined in a house

doraphobia fear of touching the hair or fur of any animal

dromophobia fear of crossing streets or wandering

dustophobia fear of dust

Dutchphobia fear of the Dutch

dysmorphophobia fear of deformity

ecophobia fear our planet is increasingly inhospitable

Egyptophobia fear of Egypt or the Egyptians

electrophobia fear of electricity

emetophobia fear of vomiting

entomophobia fear of insects

eosophobia fear of dawn

equinophobia fear of asses, horses, mules, ponies, zebras and other members of the horse family

eratophobia fear of sexual lust

eremophobia fear of being alone, monophobia

ereuthrophobia fear of blushing

ergasiophobia fear of assuming responsibility or doing work of any kind

ergophobia fear of working

erotic pyromania sexual pleasure derived from raging fires

erotophobia fear of sexual lust

erythrophobia fear of blushing or embarrassment; fear of, or aversion to, anything colored red

eurotophobia fear of the female genital organs

felinophobia fear and hatred of all cats—bobcats, cheetahs, jaguars, leopards, lions, lynxes, ocelots, ounces, panthers, pumas, tigers, wildcats; *synonym of* aelurophobia, *also written* ailurophobia

feminophobia fear of females

forgetophobia fear of forgetting

Francophobia fear of France and all things French

Gallophobia (*see* Francophobia)

gamophobia fear of marriage

gatophobia fear of cats and other felines

genophobia fear of sex

gephyrophobia fear of bodies of water, of crossing bridges over water or traveling on boats or aircraft

Germanophobia fear of the Germans or Germany

gerontophobia fear of old people

graphophobia fear of writing

Grecophobia fear of Greece or the Greeks

gringophobia fear of gringos

gymnophobia fear of nudity

gynophobia fear of women or being in their company

hagiophobia fear of religion and its saints

haphephobia fear of being touched by another person

harpaxophobia fear of robbers

hedonophobia fear of pleasure

heliophobia fear of the sun

helminthophobia fear of worms or delusion of being infested by them

hematophobia fear of blood or the sight of blood

hemophobia fear of bleeding or seeing blood

heresyphobia fear of committing heresy

herpetophobia fear of amphibians (frogs, newts, salamanders, toads) and reptiles (crocodiles, lizards, snakes, turtles)

hierophobia fear of religious objects and people connected with religion

Hispanophobia fear of Spain or Spanish—speaking people

hodophobia fear of travel

homophobia hatred of homosexuals

hydrophobia fear of water; common name for rabies, resulting from a bite by a rabid animal

hydrophobophobia fear of rabies

hyenophobia fear of hyenas

hygrophobia fear of dampness

hylophobia fear of forests

hypnophobia fear of falling asleep

hypsophobia fear of high places, acrophobia, aerophobia

iatrophobia fear of doctors and doctor-induced disorders

ickthyophobia fear of fish

ideophobia fear of mental images

illyngophobia fear of vertigo

incontiphobia fear of becoming incontinent

insectophobia fear of insects

iophobia fear of being poisoned; fear of touching anything rusty

Italophobia fear of the Italians and Italy

jumbophobia fear of bigness

kakorrhaphiophobia fear of failure

keraunophobia fear of thunder and lightning

kleptophobia fear of stealing

koniophobia fear of dust

kopophobia fear of fatigue

lacertiliaphobia fear of lizards

lalophobia fear of speaking for fear of making errors or stammering

leprophobia fear of lepers and leprosy

levophobia fear of objects on the left side of the body

linophobia fear of strings

logizomechanophobia fear of computers, computing machines, and all their works

lyssophobia fear of rabies

maieusophobia fear of pregnancy or childbirth

mechanophobia fear of machines and machinery in operation

metallophobia fear of metallic objects

meteorophobia fear of meteors

microphobia fear of germs; fear of microbes; dread of small objects

monophobia fear of being alone, eremophobia

mysophobia fear of contamination, dirt, disease, and microbes

mythophobia fear of lying and making false statements

necrophobia fear of dead bodies or of death

negrophobia fear of the black people

neophobia fear of anything new or novel

nephophobia fear of clouds

nictophobia fear of the dark

Nippophobia fear of Japan and the Japanese

nictiphobia fear of night and darkness

noisicophobia fear of noise or electronically amplified music

nomatophobia fear of names of people or place-names

nosophobia fear of a disease or illness

nostophobia fear of returning home

nucleomitiphobia fear of nuclear warfare

nudophobia fear of being naked

numerophobia fear of numbers

nyctophobia fear of darkness or night

ochlophobia fear of populated places and crowds

odontophobia fear of dental surgery or the sight of teeth

ombrophobia fear of black clouds, rain, and storms

ommatophobia fear of eyes

ophidiophobia fear of snakes

optophobia fear of opening your eyes

osmophobia fear of odors

pagophobia fear of eating

panphobia groundless fear of everything

pantophobia fear of everything

paralipophobia fear of forgetting or neglecting to perform some duty

parasitophobia fear of parasites

pathophobia. fear of disease

peccatophobia fear of sinning

pediculophobia fear of lice

pediophobia fear of children, dolls, puppets

peniaphobia fear of poverty

pharmacophobia fear of taking medicines

phasmophobia fear of ghosts

phengophobia fear of light

–phobia suffix meaning dread, abnormal fear, or aversion to a subject

phobic characterized by or pertaining to a phobia

phobism(s) affected by phobias

phobist person with a morbid fear of anything

phobophobia fear of acquiring a phobia

phonophobia fear of noise or sound; fear of speaking or hearing one's own voice

photophobia fear of light

phronemophobia fear of thinking

pnigophobia fear of choking

polyphobia fear of many ideas and things

ponophobia fear of exerting one's self; dread of pain

pornophobia fear of erotic behavior

Portugophobia fear of Portugal and the Portuguese

potamophobia fear of rivers and running water

proctophobia fear of rectal disease

psychoneurophobia fear of emotional and mental disorders

psychophobia fear of the mind; fear of psychological findings

psychrophobia fear of cold

pyrexiophobia fear of fever

pyrolagnia sexual pleasure aroused by raging fires

pyrophobia fear of fire

quizziphobia fear of being quizzed

ranidaphobia fear of frogs

rectophobia fear of rectal disease

reptiliaphobia fear of reptiles—crocodilians, lizards, snakes, tuataras, turtles

rhabdophobia fear of being hit or beaten with a rod

rhypophobia fear of defecation or fecal filth

Romanophobia fear of the Roman Empire or the Romans

Russophobia fear of Russia or Russians

satanophobia fear of the devil

scabiphobia fear of itching; fear of scabies

schoolphobia fear of going to school

scopophobia fear of being seen

scotophobia fear of darkness

sidereaphobia fear of stars

siderophobia fear of touching iron or steel

silverphobia fear of silver and silverware

simple phobia irrational fear of specific objects such as cats, dogs, mice, snakes

Sinophobia fear of China or the Chinese

sitiophobia fear of eating; fear of food (*also written* sitophobia)

social phobia fear of being watched by others

sophophobia fear of going to school; fear of learning

spatiophobia fear of space

spectrophobia fear of mirrors

spermatophobia fear of loss of semen

spinnanphobia fear of spiders

stagiophobia fear of hell

stasiphobia fear of being unable to stand erect, basophobia

stenophobia fear of open places

sudophobia fear of sweat

symbolophobia fear of doing or saying anything that may be interpreted as having symbolic meaning

syphiliphobia fear of syphilis

Syriaphobia fear of Syria and the Syrians

taeniophobia (*see* teniophobia)

taphephobia fear of being buried alive, tapophobia

tapophobia fear of being buried alive

technophobia fear of technology

teleophobia fear of the existence of a grand design in nature

teniophobia fear of being infested with tapeworms

teratophobia fear of deformed people

terroristophobia fear of terrorists

thallasophobia fear of the sea

thanatophobia fear of death and dying

theophobia fear of the wrath of a god

tocophobia fear of childbirth

tonitorphobia fear of thunder

topophobia fear of performing; fear of a particular place; stage fright

toxicophobia fear of being poisoned

traumatophobia fear of injury

tremophobia fear of trembling

triakaidekaphobia fear of thirteen

trichinophobia fear of developing trichinosis

tricopathophobia women's fear of unwanted hair

tricophobia fear of hair or touching hair

triskaidekaphobia fear of the number 13

tropophobia fear of moving or marine changes

tuberculophobia fear of being infected with tuberculosis

Turkophobia fear of Turkey or the Turks

urbanophobia fear of cities

urophobia fear of urine

ursiphobia fear of bears

vaccinophobia fear of vaccination

verbophobia fear of words

vestiophobia fear of clothing

vomitophobia fear of vomit

vulvaphobia fear of vulval inflammation

Walloonphobia fear of the Walloons

xenophobia fear of foreigners, strangers, or anything foreign or strange

Yankeephobia fear of Yankees

zelophobia fear of jealousy

zoophobia fear of animals

Ports of the World

Ports of the world are listed alphabetically by country or state in Eponyms, Nicknames, and Geographical Names. Many also appear among Superlatives and in the main body of the dictionary.

Presidents and Vice Presidents of the United States

Presidents

1st — George Washington (1789–1797)

2nd — John Adams (1797–1801)

3rd — Thomas Jefferson (1801–1809)

4th — James Madison (1809–1817)

5th — James Monroe (1817–1825)

6th — John Quincy Adams (1825–1829)

7th — Andrew Jackson (1829–1837)

8th — Martin Van Buren (1837–1841)

9th — William Henry Harrison (1841; caught pneumonia at inaguration and died a month later)

10th — John Tyler (1841–1845)

11th — James Knox Polk (1845–1849)

12th — Zachary Taylor (1849–1850; died in office after contracting cholera)

13th — Millard Fillmore (1850–1853)

14th — Franklin Pierce (1853–1857)

15th — James Buchanan (1857–1861)

16th — Abraham Lincoln (1861–1865; assassinated)

17th — Andrew Johnson (1865–1869)

18th — Ulysses Simpson Grant (1869–1877)

19th — Rutherford Birchard Hayes (1877–1881)

20th — James Abram Garfield (1881; died months after being shot by an office-seeker)

21st — Chester Alan Arthur (1881–1885)

22nd — Grover Cleveland (1885–1889)

23rd — Benjamin Harrison (1889–1893)

24th — Grover Cleveland (1893–1897)

25th — William McKinley (1897–1901; assassinated)

26th — Theodore Roosevelt (1901–1909)

27th — William Howard Taft (1909–1913)

28th — Woodrow Wilson (1913–1921)

29th — Warreṅ Gamaliel Harding (1921–1923; died in office during a trip after becoming ill—circumstances unclear)

30th — Calvin Coolidge (1923–1929)

31st — Herbert Clark Hoover (1929–1933)

32nd — Franklin Delano Roosevelt (1933–1945) (died during fourth term in office)

33rd — Harry S Truman (1945–1953)

34th — Dwight David Eisenhower (1953–1961)

35th — John Fitzgerald Kennedy (1961–1963; assassinated)

36th — Lyndon Baines Johnson (1963–1969)

37th — Richard Milhous Nixon (1969–1974; resigned amid scandal, under threat of impeachment)

38th — Gerald Rudolph Ford (1974–1977)

39th — Jimmy (James Earl) Carter (1977–1981)

40th — Ronald Wilson Reagan (1981–1989)

41st — George Herbert Walker Bush (1989–1993)

42nd — Bill (William Jefferson) Clinton (1993–2001)

43rd — George Walker Bush (2001–)

Vice Presidents

1st — John Adams

2nd — Thomas Jefferson

3rd — Aaron Burr

4th — George Clinton

5th — Elbridge Gerry

6th — Daniel D Tompkins

7th — John C Calhoun

8th — Martin Van Buren

9th — Richard M Johnson

10th — John Tyler

11th — George M Dallas

12th — Millard Fillmore

13th — William R King

14th — John C Breckinridge

15th — Hannibal Hamlin

16th — Andrew Johnson

17th — Schuyler Colfax

18th — Henry Wilson

19th — William A Wheeler

20th — Chester A Arthur

21st — Thomas A Hendricks

22nd — Levi P Morton

23rd — Adlai E Stevenson

24th — Garret A Hobart

25th — Theodore Roosevelt

26th — Charles W Fairbanks

27th — James S Sherman

28th — Thomas R Marshall

29th — Calvin Coolidge

30th — Charles G Dawes

31st — Charles Curtis

32nd — John Nance Garner

33rd — Henry A Wallace

34th — Harry S Truman	**39th** — Spiro T Agnew	**44th** — Dan Quayle
35th — Alben W Barkley	**40th** — Gerald R Ford	**45th** — Al Gore
36th — Richard M Nixon	**41st** — Nelson A Rockefeller	**46th** — Richard Cheney
37th — Lyndon B Johnson	**42nd** — Walter F Mondale	
38th — Hubert H Humphrey	**43rd** — George Bush	

Principles, Laws, Rules, Theories, and Doctrines

Amdahl's Law A computer system's speed is determined by its slowest components (named for Gene Amdahl)

Benoulli's Principle Pressure of water against sides of a wooden pipe decreases as the velocity of the water increases (observation of Daniel Benoulli, Swiss mathematician in mid-1700s)

Buys Ballot's Law If one stands in the Northern Hemisphere with his back to the wind, atmospheric pressure will be lower to his left hand than to his right, and the reverse in the Southern Hemisphere (for Christoph Heinrich Diedrich Buys Ballot, 1817–1890)

Carter Doctrine Defend the Persian Gulf (Jimmy Carter, thirty-ninth president of the United States)

Eisenhower Doctrine Contain the Soviet Union in Europe and the Middle East (Dwight D Eisenhower, thirty-fourth president of the United States)

Ferrel's Law A body moving in any direction over the Earth's surface will tend to be deflected to the right in the Northern Hemisphere and to the left in the Southern Hemisphere (for American meteorologist William Ferrel)

Four Noble Truths Suffering is an unavoidable part of existence; suffering is caused by desire; elimination of desire will end suffering; desire and suffering can be eliminated (Buddhism)

Gödel's Theorem All branches of mathematics are based on propositions that can't be proved within that branch (for mathematician Kurt Gödel, 1906–1978)

Hebb's Learning Rule A memory is produced when two connected neurons are active simultaneously and the synapse is strengthened (for Donald O Hebb)

Hobson's Choice This or nothing (for Thomas Hobson, 1544–1631)

January Effect Small-capitalized stocks tend to outperform large-capitalized stocks in January

Law of Infinitesimals Dilute substances can have very strong effects

Law of Similars Like cures like

Mendel's Law of Independent Assortment Each pair of different characters in a hybrid union is independent of the other characters (for Gregor Johann Mendel, 1822–1884)

Mendel's Law of Segregation As gametes are formed, the pairs of hereditary units separate, one going to each gamete, and genes are not influenced by the other (for Gregor Johann Mendel)

Monroe Doctrine Oppose foreign attempts to gain influence in the Americas (James Monroe, fifth president of the United States)

Moore's Law A semiconductor chip's computing power will double every eighteen months (named for George Moore)

Murphy's Law If something can go wrong, it will (Murphy was fictitious)

Nixon Doctrine Use Japan, Iran and South Africa to oppose communist influence in their regions (Richard M Nixon, thirty-sixth president of the United States)

Occam's Razor What can be done with few assumptions is preferable to what can be done with more (formulated by William of Occam, 1285–1349)

Reagan Doctrine Undo Soviet gains in the Third World (Ronald Reagan, fortieth president of the United States)

Truman Doctrine Defend Greece and Turkey, contain the Soviet Union in Europe (Harry S Truman, thirty-third president of the United States)

Proofreader's Marks

|| align; straighten ends of lines

⌄ apostrophe or single quotation mark

bf black face or bold face type (run waved lane under text matter)

cap capital letter

≡ capital letters (run triple line under material to be capitalized: george washington)

∧ caret; insertion mark

⌒ close up

⁘ colon

⋀ comma

℮ delete or dele; expunge; take out

⊔ depress or sink a letter or word

⌐ elevate or raise a letter or word

=/ hyphen

ital set in italics (material to be italicized is under-lined)

lc lower case (run / through letter or letters to be set in /lower /case)

lead insert lead spacing between lines

⊏ move to the left

⊐ move to the right

⁋ paragraph

⊙ period

⊥ push down space which prints as a mark

⟫ quotation marks

rom set in roman type

;/ semicolon

sc small caps (run double line under material: a.d.)

space; ## double space; etc.

(sp) spell out (material to be spelledout is encircled: (U.S.))

stet let stand that which has been deleted; restore crossed out material (indicate by running dots under the letters of the words to be restored)

tr transpose (indicate in text by ∼ or ⌣)

℮ turn letter right side up

wf wrong font

Railroad Conductor's Cord-Pull Signals Plus Engineer's Whistle Signals

Cord-Pull Signals

1 short cord pull or 1 short whistle toot: apply brakes—stop
1 long whistle toot *when standing*: apply brakes or brakes applied
  when running: approaching grade crossing, junction, or station
2 long cord pulls or 2 long whistle toots: release brakes—proceed
3 short cord pulls or 3 short whistle toots *when standing*: back up
  when running: stop at next passenger station
4 short cord pulls or 4 short whistle toots: call for signals
  succession of short cord pulls or short whistle toots: alarm or emergency such as persons or livestock
  on track; stop train until safe to proceed

Engineer's Whistle Signals

1 long toot followed by 3 short toots: flagman protect rear of train
3 short toots followed by 1 long toot: flagman protect front of train
4 long toots: flagman may return from west or south
5 long toots: flagman may return from east or north
1 short toot followed by 1 long toot: flagman inspect train for sticking brakes or leaks
2 long toots followed by a short toot and a long toot (t o o t t o o t toot t o o t): approaching curve,
  grade crossing, tunnel, or other obscure place; approaching a train standing on an adjacent track
1 long toot followed by a short toot (t o o t toot): blown when running against the current of traffic
  approaching curves, grade crossings, junctions, stations, and tunnels or obscure places
In the event of whistle failure, the bell must be rung continuously while the train is enroute.
When the train is approaching or leaving a station, the bell is rung to indicate the need for caution and
  to avoid the noise of the whistle.

Railroads of the World

This listing includes abbreviations, nicknames, and reporting marks.

AA Ann Arbor Railroad

AAR Association of American Railroads

A & B Antofagasta and Bolivia

ABB Akron and Barberton Belt Railroad

ABL Alameda Belt Line

AC Algoma Central Railway

ACR Algoma Central Railway

ACY Akron, Canton and Youngstown Railroad

AD Atlantic and Danville Railway

ADN Ashley, Drew and Northern Railway (also AD & N)

AE Algoma Eastern Railway (Manitoulin & North Shore Railway)

AEC Atlantic and East Carolina

AF Alma and Jonquieres Railway

AFE Administracion de los Ferrocarriles del Estado (Spanish—State Railways Administration)—Venezuela

AGS Alabama Great Southern (Southern Railway)

AL Almanor Railroad

ALC American Locomotive Company

ALM Arkansas and Louisiana Missouri Railway (also A & LM)

ALN Albany and Northern Railroad

ALQS Aliquippa and Southern

ALS Alton and Southern Railroad

AL & S Alton and Southern Railroad

Alton Route Gulf, Mobile and Ohio Railroad

AMC Amador Central Railroad

AMR Arcata and Mad River

Amtrak American (railroad) tracks—(government-sponsored program for reviving city-to-city passenger service)

AN Apalachicola Northern Railroad

Ann Arbor Detroit, Toledo and Ironton Railroad

Annie & Mary (nickname— Arcata and Mad River Railroad)—originally the Union Wharf and Plank Walk Company

ANR Angelina and Neches River Railroad; Australian National Railways

AP Allegheny Portage Railway

APA Apache Railway Company

APD Albany Port District

AR Aberdeen and Rockfish

ARA Arcade and Attica Railroad

ARC Alexander Railroad (Southern)

ARR Alaska Railroad

ART American Refrigerator Transit

ARW Arkansas Western Railway (Kansas City Southern)

A & S Abilene and Southern

ASAB Atlanta and Saint Andrews Bay Railway

ASDA Asbestos and Danville

ASLRA American Short Line Railroad Association

ASR Association of Southeastern Railroads

ATC Arnold Transit Company

ATN Alabama, Tennessee and Northern Railroad

ATSF Atchison, Topeka and Santa Fe Railway (also AT & SF)

ATW Atlantic and Western

AUG Augusta Railroad

AUS Augusta and Summerville

Austrail Railways of Australia

AVL Aroostook Valley Railroad

AW Ahnapee and Western Railway

AWP Atlanta and West Point Rail Road (includes Western Railway of Alabama and Georgia Railroad)—also A&WP

AWW Algers, Winslow and Western Railway

A y B Antofagasta y Bolivia (Spanish—Antofagasta and Bolivia)—Chilean Railway linking Pacific port with highlands of landlocked Bolivia

AYSS Allegheny and South Side

ba BART (Bay Area Rapid Transit)

B & A Boston and Albany (Penn Central)

B-A-M Baikal-Amur-Magistral (railroad in Pacific Siberia, USSR)

BAP Butte, Anaconda and Pacific Railway (also BA & P)

BARC Baltimore and Annapolis Railroad Company

B & ARR Boston and Albany Railroad

BART Bay Area Rapid Transit (San Francisco Bay Area mass transportation system)

Bay Line Atlanta and Saint Andrews Bay Railway

BB Birmingham Belt Railroad

BCE Route British Columbia Electric Route

BCH British Columbia Hydro and Power Authority

BCK Buffalo Creek Railroad

BCK Bas-Congo au Katanga (French—Lower Congo— Katanga)—railway of Democratic Republic of Congo

BCR British Columbia Railway

BC RAIL British Columbia Railway

BCRR Boyne City Railroad

BCYR British Columbia Yukon Railway

BDZ (Cyrillic transliteration— Bulgarian State Railways)

BE Baltimore and Eastern Railroad (Penn Central)

BEDT Brooklyn Eastern District Terminal Railroad

BEEM Beech Mountain Railroad

BEM Beaufort and Morehead Railroad

Bessemer Bessemer and Lake Erie

BFC Bellefonte Central Railroad

BH Bath and Hammondsport Railroad

BHS Bonhomie and Hattiesburg Southern Railroad

Big Four Cleveland, Cincinnati, Chicago and St Louis Railway (Penn Central)

Birmingham Southern Birmingham Southern Railroad

BLA Baltimore and Annapolis

BM Boston and Maine Corporation

BME Beaver, Meade and Englewood Railroad

BMRR Beech Mountain Railroad

B & MRR Beaufort and Morehead Railroad

BMT Brooklyn-Manhattan Transit (subway system)

BN Burlington Northern (combining Frisco—the St Louis-San Francisco with former Great Northern; Northern Pacific; Chicago, Burlington and Quincy; Spokane, Portland and Seattle; and Pacific Coast railroads)

B & N Bauxite and Northern Railway

BNSF Burlington Northern Santa Fe Railway

BNT Buffalo Niagara Transit

B & O Baltimore and Ohio Railroad

BOCT Baltimore and Ohio Chicago Terminal Railroad

BOYC Boyne City Railroad

BR Botswana Railways; British Railways; Burma Railways

BRB Bangladesh Railway

BRC Belt Railway Company of Chicago

BR & W Black River and Western

BS Birmingham Southern Railroad

B & S Bevier and Southern

BTA Boston Transportation Authority

BTC Baltimore Transit Company

BTN Belton Railroad

BU Budapest Underground (subway system)

Burlington Northern combining Great Northern; Northern Pacific; Chicago, Burlington and Quincy; Spokane, Portland and Seattle; and Pacific Coast railroads

Burlington Route Chicago, Burlington and Quincy Railroad

Burl N Burlington Northern and St Louis-San Francisco (railway merger)

BUSH Bush Terminal Railroad

BVG Berliner Verkehrs Betriebe (German—Berlin Traffic Management)—Berlin's subway system

BV & S Bevier and Southern

BWC Pennsylvania New York Central Transportation Company

BYR British Yukon Railway

CAD Cadiz Railroad

CAR Central Australia Railway

CARR Carrollton Railroad

CARW Caroline Western Railroad

CASO Canada Southern Railway (Penn Central)

CBC Carbon County Railway

CBL Conernaugh and Black Lick

CB & Q Chicago, Burlington and Quincy Railroad (to Burlington Northern)

C & C Columbia & Cowlitz

CC & C Cabedelo, Charleston & Coney Island

CCCSL Cleveland, Cincinnati, Chicago and St Louis Railway (Penn Central)

CCFPCS Cie des Chemins de Fer de la Plaine du Cul-de-Sac (French—Cul-de-Sac Plaine Railroad Company)—Tahiti

CC & O Caroline, Clinchfield and Ohio Railway

CC & ORSC Caroline, Clinchfield and Ohio Railroad of South Carolina

CCR Corinth and Counce Railroad

CCT Central California Traction

CDE Chemin de Fer Djibouti-Ethiopién (French—Djibouti-Ethiopian Railway)

C de F D-N Chemins de Fer Dakar-Niger (French—Dakar-Niger Railways)—Mali

C & EI Chicago and Eastern Illinois Railroad

Central (nickname—New York Central Railroad)—now part of the Penn Central

Central of Ga Central of Georgia

CF Cape Fear Railways

CF C-O Chemin de Fer Congo-Ocean (French—Congo-Ocean Railroad)—Congo (Brazzaville)

CFF/SFF/FFS Chemins de Fer Federaux Susses/Schweizerische Bundesbahnen/Ferrovie Dederali Svizzere (French, German, Italian—Swiss Federal Railways)

CFL Société Nationale des Chemins de Fer Luxembourgeois (French—Luxembourg National Railways)

CFM Caminho de Ferro de Moçambique (Portuguese—Mozambique Railroad); *Chemin de Fer Madagascar* (French—Madagascar Railroad)

CFR Caile Ferate Ramane (Romanian—General Direction of the Romanian Railroads)

CFR Chemins de Fer Royaux du Cambodge (French—Royal Cambodian Railways)

CG Central of Georgia Railway

C & G Columbus and Greenville

C of G Central of Georgia Railway

CGR Ceylon Government Railway; Cyrenaica Government Railway (Libya)

C & GTR Canada and Gulf Terminal Railway

CGW Chicago Great Western

C & H Cheswick and Harmer Railroad

Chessie System Chesapeake & Ohio/Baltimore & Ohio

Chicago Outer Belt Elgin, Joliet and Eastern Railway

Chihuahua-Pacific *Railway Ferrocarril del Chihuahua al Pacifico*—from the border of Texas at Presidio to the Pacific coast at Los Mochis via Chihuahua over route of the Kansas City, Mexico, and Orient

CH-P Ferrocarril Chihuahua al Pacifico (Chihuahua-Pacific Railway formerly Mexico Northwestern Railway and Kansas City, Mexico and Orient Railway)

CHR Chestnut Ridge Railway

CHTT Chicago Heights Terminal Transfer Railroad

CHV Chattahoochee Valley

CKW Chesapeake Western

C & I Cambria and Indiana Railroad

CIC Cedar Rapids and Iowa City Railway

CIE Coras Iompair Eireann (Gaelic—Irish State Railways)

CI & L Chicago, Indianapolis, and Louisville Railway (Monon Railroad)

CIM Chicago and Illinois Midland Railway (also C & IM)

CIND Central Indiana Railway

CIRR Chattahoochee Industrial Railroad

C & IRR Cambria and Indiana Railroad

CIW Chicago and Illinois Western Railroad

CIWL Compangie Internationale des Wagon-Lits (French—International Sleeping Car Company)

CKSO Condon, Kinzua and Southern Railroad

CLC Colombia and Cowlitz

CLCO Claremont and Concord

Clinchfield Chinchfield Railroad (Carolina, Clinchfield and Ohio Railway)

CLK Cadillac and Lake City Railway

CLP Clarendon and Pittsford Railroad

CLRR Camp Lejeune Railroad

CMO Chicago, St Paul, Minneapolis and Omaha (Chicago North Western)

CN Canadian National (includes Canadian National Railways; Central Vermont Railway; Duluth, Winnipeg and Pacific Railway; Grand Trunk Lines in U.S.A .)

C & N Carolina and Northwestern Railway

CNJ Central Railroad of New Jersey

CN & L Columbia, Newberry and Laurens Railroad

CNO & TPR Cincinnati New Orleans and Texas Pacific Railway

CNR Chiriqui National Railroad (Panama)

CNTP Cincinnati, New Orleans and Texas Pacific

CNW Chicago and North Western Railway (includes Chicago, St Paul, Minneapolis, and Omaha; Litchfield and Madison Railway; Minneapolis and St Louis)

C & NW Chicago and North Western Railway

C & O Chesapeake and Ohio (to CSX Corporation)

Coahuila-Zacatecas Railway *Ferrocarril Coahuila-Zacatecas*—Mexico

Cog Wheel Route Manitou and Pike's Peak Railway

Conrail Consolidated Rail Corporation (Ann Arbor, Central Railroad of New Jersey, Erie Lackawanna, Lehigh and Hudson River, Lehigh Valley, Penn Central, Reading)

COP City of Prineville Railway

COPR Copper Range Railroad

Corn Belt Route St Louis Southwestern Railway

Cotton Belt Cotton Belt Route (St Louis Southwestern Railway—SSW)

CP Canadian Pacific Railway (Dominion Atlantic Railway, Esquimalt and Nanaimo Railway, Grand River Railway, Lake Erie and Northern Railway, Quebec Central Railway, Vancouver and Lulu Island Branch)

CP Companhia des Caminhos de ferro Portuguese (Portuguese—Portuguese Railways)

CPA Coudersport and Port Allegany Railroad

CPF Cotton Plant-Fargo Railway

CP & LT Camino, Placerville and Lake Tahoe Railroad

CPPR Chinese People's Republic Railways

CPR Canadian Pacific Railroad; Canadian Pacific Railway

CP Rail Canadian Pacific Railroad

CPT Chicago Produce Terminal

CR Commonwealth Railways (Australia and Tasmania); Copper Range Railroad (Michigan, Wisconsin, Illinois)

CRANDIC Route Cedar Rapids and Iowa City Railway

CRC Cameroon Railways Corporation (West Africa); Cumberland Railway Company (Nova Scotia)

CRI Chicago River and Indiana

CR & IC Cedar Rapids and Iowa City Railway

CR & IR Chicago River and Indiana Railroad

CRN Carolina and Northwestern (Southern Railway)

CRP Central Railway of Peru

CRR Clinchfield Railroad

CRRNJ Central Railroad of New Jersey

CSAR Central South African Railways

CSP Camas Prairie Railroad

CSS Chicago South Shore and South Bend Railroad

CSS & SBR Chicago South Shore and South Bend Railroad

C St P M & O Chicago, St Paul, Minneapolis and Omaha (Chicago North Western)

CSX Chessie and Seaboard (railroads consolidated)

C & T Cumbres & Toltec

CTA Chicago Transit Authority (elevated and subway railroads)

CTC Canadian Transport Commission; Cincinnati Transit Company

CTN Canton Railroad

CTS Cleveland Transit System

CUTC Cincinnati Union Terminal Company

CUVA Cuyahoga Valley Railroad

CV Central Vermont Railway

CVRy Cuyahoga Valley Railway

C & W Colorado and Wyoming Railway

C & WC Charleston and Western Carolina Railway (Seaboard Coast Line Railroad)

CWI Chicago and Western Indiana

CWP Chicago, West Pullman and Southern Railroad (also CWP & S)

CWR California Western Railroad

DA Dominion Atlantic Railway (Canadian Pacific)

DB Deutsche Bundesbahn (German—German Railways)

DC Delray Connecting Railroad

DCI Des Moines and Central Iowa

DCR Delray Connecting Railroad (Zug Island Road)

DCT Washington, DC Transit

D & E De Queen and Eastern

Delay Long and Wait nickname for the Delaware, Lackawanna and Western Railroad (derived from the initials DL & W)

D & H Delaware and Hudson Railway (part of NW)

DHR Darjeeling Himalayan Railway

diner dining car

DKS Doniphan, Kensett and Searcy Railway

DL & W Delaware, Lackawanna and Western Railroad (Erie Lackawanna)

D & M Detroit and Mackinac

DM&E Dakota, Minnesota & Eastern Railroad

DM & IRR Duluth, Missabe and Iron Range Railway

DMM Dansville and Mount Morris

DMU Des Moines Union Railway

DMWR Des Moines Western Railway

DNE Duluth and Northeastern Railroad

DO Direct Orient (Orient Express)

DORR Delaware Otsego Railroad

DQ & ERR De Queen and Eastern Railroad

D & R Dardanelle and Russellville

D & RGW Denver and Rio Grande Western Railroad

DRI Davenport, Rock Island and North Western Railway

DRy Devco Railway

DS Durham and Southern Railway

D & S Durango & Silverton, Durham and Southern Railway

DSB Danske Statsbaner (Danish—Danish State Railways)

DSR Detroit Street Railways

DT Detroit Terminal Railroad

D of T Department of Transportation

DTC Dallas Transit Company

DTI Detroit, Toledo and Ironton Railroad (also DT & I)

D & TS Detroit and Toledo Shore Line Railroad

DVS Delta Valley and Southern Railway

DWP Duluth, Winnipeg and Pacific Railway

E Erie Lackawanna

EAR East African Railways

EARC East African Railways Corporation

EAR & H East African Railways and Harbours

EBR Emu Bay Railway (Tasmania)

EBRy Eastern Bengal Railway (East Pakistan)

EDLR Egyptian Delta Light Railways

EDW El Dorado and Wesson

EEC East Erie Commercial Railroad

EFA Empresa Ferrocarriles Argentinos (Spanish—Argentine Railways Enterprise)

EFE Empresa de los Ferrocarriles del Estado (Spanish—State Railways Enterprise)—Chile

EFEE Empresa de los Ferrocarriles del Estado Ecuatoriano (Spanish—Ecuadorian State Railways Enterprise)

EJ & ERy Elgin, Joliet and Eastern Railway

EJR East Jersey Railroad

El Elevated Railroad

EL Erie Lackawanna Railway (merger of Erie with Delaware, Lackawanna and Western, now part of Conrail)

ELS Escanaba and Lake Superior Railroad (also E & LSRR)

E & M Edgmoor and Marietta

EN Esquimalt and Nanaimo Railway (Canadian Pacific)

ENF Empresa Nocional de Ferrocarriles (Spanish—National Railways Enterprise)—Bolivia

ENFE Empresa Nacional de los Ferrocarriles del Estado (Spanish—State Railways of Ecuador)

ER Egyptian Railways

ERBR Eastern Region of British Railways

Erie Erie Railroad (Erie Lackawanna)

ESLJ East St Louis Junction Railroad

ETL Essex Terminal Railway

ET & WNC East Tennessee and Western North Carolina Railroad

Eurailpass European railroad pass (ticket system valid on almost all European railroads)

EW East Washington Railway

EYB Europa Year Book

FA Ferrocarriles Argentinos (Spanish—Argentine Railways)

F & C Frankfort and Cincinnati Railroad

FCAB Ferrocarril Antofagasta-Bolivia (Spanish—Antofagasta and Bolivia Railway)

FC del P Ferrocarril Central del Perú (Spanish—Central Railway of Peru)

FCDN Ferrocarril del Nacozari (Spanish—Nacozari Railroad)—Mexico

FCG Fernwood, Columbia and Gulf Railroad

FCIN Frankfort and Cincinnati

FCM Ferrocarriles Nocionales de México (Spanish—Mexican National Railways)—includes Nacional de México and Nacional de Tehuantepec

FCNM Ferrucarriles Nacionales de México (Spanish—National Railroads of Mexico)

FCP Ferrocarril del Pacifico (Spanish—Pacific Railroad)—links Arizona border with Mazatlan on west coast of Mexico

FCZ Ferrocarril Coahuila-Zacatecas (Spanish—Coahuila-Zacatecas Railway)—Mexico

FDDM Fort Dodge, Des Moines and Southern Railway

F de C Ferrocarriles de Cuba (Spanish—Cuban Railroads)—Unidad Habana (western Cuba) and Unidad Camaguey (eastern Cuba)

F de G a LP Ferrocarril de Guayaquil-La Paz (Spanish—Guayaquil-La Paz Railway)—Peru

F del N Ferrocarriles del Norte (Spanish—Northern Railways)—Paraguay

F del P Ferrocarril del Pacífico (Spanish—Pacific Railroad)—Mexico

Feather River Route Western Pacific Railroad

FEC Florida East Coast Railway

FEGUA Ferrocarriles de Guatemala (Spanish—Railroads of Guatemala)

FEMESA Ferrocarriles Metropolitanos SA (Spanish—Metropolitan Railway—Buenos Aires)

FENADESAL Ferrocarriles Nacionales de El Salvador (Spanish—El Salvador National Railways)

FEP Ferrocarril Electrico al Pacífico (Spanish—Pacific Electric Railway)—Costa Rican line linking Pacific port of Puntarenas with mountain capital of San José

FEPASA Federação Paulista Sedada Anomina (Portuguese—Paulist Federation Company)—Brazilian railroad

FER Franco-Ethiopian Railway

FES Ferrocarril de El Salvador (Spanish—El Salvador Railway)

FFAC Federação Ferrocarril Agricola Cotias (Portuguese—Agricultural Cooperative Railway Federation)—Brazil

FICA Ferrocarriles Internacionales de Centro America (Spanish—International Railways of Central America)

FIPC Ferrocarril Industrial del Potosí y Chihuahua (Spanish—Industrial Railroad of Potosi and Chihuahua)—Mexico

FJG Fonda, Johnstown and Gloversville Railroad

FLR Fayum Light Railways (Egypt)

FMS Fort Myers Southern Railroad

FN Ferrocarriles Nacionales (Spanish—National Railways—Argentina, Chile, Colombia, Cuba, Ecuador, Honduras, Mexico, Panama, Venezuela, etc.)

FNC Ferrocarriles Nacionales de Cuba (National Railroad of Cuba nationalized by Castro government and consisting of Consolidated Railroads of Cuba, The Cuba Railroad, Cuba Northern Railways, Guantanamo and Western Railroad, Guantanamo Railroad, Hershey Cuban Railway, etc.)

FN de H Ferrocarriles Nacionales de Honduras (Spanish—National Railways of Honduras)

FNM Ferrocarriles Nacionales de México (Spanish—National Railways of Mexico)

FOM Ferrocarril Occidental de México (Spanish—Western Railway of Mexico)

FOR Fore River Railroad

FPCAL Ferrocarriles Presidente Carlos Antonio López (Spanish—President Carlos Antonio Lopez Railways)—Paraguay

FPE Fairport, Painesville and Eastern Railroad

FP & ER Fairport, Painesville and Eastern Railway

FPN Ferrocarril del Pacífico de Nicaragua (Spanish—Pacific Railway of Nicaragua)

FR Feather River Railway

FRDN Ferdinand Railroad

Frisco St Louis-San Francisco Railway (merged with Burlington Northern)

FS Ferrovie dello Stato (Italian—State Railway)

FSBC Ferrocarril Sonora-Baja California (Sonora—Baja California Railroad)

FS del P Ferrocarril del Sur del Perú (Spanish—Southern Railway of Peru)

FSVB Fort Smith and Van Buren Railway (Kansas City Southern)

FtD DM & S Fort Dodge, Des Moines and Southern Railway

FUD Ferrocarriles Unidos Dominicanos (Spanish—United Dominican Railways)—Dominican Republic

FUS Ferrocarriles Unidos del Sureste (Spanish—United Railways of the Southeast)

FUY Ferrocarriles Unidos de Yucatan (Spanish—United Railways of Yucatan)—Mexico

FWB Fort Worth Belt Railway

FW & D Fort Worth and Denver

GA Georgia Railroad

GANO Georgia Northern Railway

GASC Georgia, Ashburn, Sylvester and Camilla Railway

GB & W Green Bay and Western Lines (includes Kewaunee, Green Bay and Western Railroad)

GC Graham County Railroad

GCW Garden City Western Railway

George Washington's Railroad Chesapeake and Ohio

Georgia Georgia Railroad

G & F Georgia and Florida Railway

GFS Grand Falls Central Railway

GH & H Galveston, Houston and Henderson Railroad

GJ Greenwich and Johnsonville Railway

G & J Greenwich and Johnsonville Railway

GKB Graz-Köflacher Eisenbahn und Bergbau-Gesellshaft mbh (Austrian—Graz-Köflacher Railway)

GM Gainesville Midland Railroad

GMRC Green Mountain Railroad Corporation

GN Great Northern Railway

GNA Graysonia, Nashville and Ashdown Railroad

GNW Genessee and Wyoming Railroad

GNWR Genessee and Wyoming Railroad

GO Transit Government of Ontario Transit

G & Q Guayaquil and Quito

Grand Trunk Grand Trunk Railway System (Canadian National) and Grand Trunk Western Railroad

Green Bay Route Green Bay and Western Railroad

GRN Greenville and Northern Railway

GRNR Grand River Railway (Canadian Pacific)

GR & PA Ghana Railway and Port Authority

GRR Georgetown Railroad

GRSS Guyana Railways and Shipping Services

GSF Georgia Southern and Florida (Southern)

GSW Great Southwest Railroad

GTW Grand Trunk Western Railroad (Canadian National)

G & U Grafton and Upton Railroad

G&W Genessee & Wyoming Railway

GWF Galveston Wharves

GWR Great Western Railway

GWWDR Great Winnipeg Water District Railway

HB Hampton and Branchville

HBLRR Harbor Belt Line Railroad

HBS Hoboken Shore Railroad

HBT Houston Belt and Terminal—owned jointly by Santa Fe, Missouri Pacific, and Fort Worth & Denver

HC Heber Creeper (Utah)

HE Hollis and Eastern Railroad

HER Hellenic Electric Railway (Athens–Piraeus subway system linking capital with its seaport)

HH Hamburger Hochbahn (German—Hamburg Elevated Railway)—includes subway system

HI Holton Inter-Urban Railway

HJR Hedjaz Jordan Railway

HLNE Hillsboro and Northeastern

H & M Hudson & Manhattan (Hudson Tubes)

HN Hutchinson and Northern Railway

HNE Harriman and Northeastern (Southern)

hovertrain railroad train supported by an air cushion instead of wheels

HPTD High Point, Thomasville and Denton Railroad

HRPSSh Hekurdhë Republika Popullore Socialiste e Shgipërisë (Albanian—Albanian Railways)

HRT Hartwell Railway

HS Hartford and Slocomb Railroad

HSW Helena Southwestern Railroad

HTW Hoosac Tunnel and Wilmington Railroad

HZP Hrvatsko Zeljeznicko Poduzece (Croatian—Croatian State Railway)

I Illinois Central Gulf Railroad

IAT Iowa Terminal Railroad

IB & TC International Bridge and Terminal Company

ICC Interstate Commerce Commission

ICG Illinois Central Gulf

IGA Indian Government Administration (Railway Board of India)

IHB Indiana Harbor Belt Railroad

IN Illinois Northern Railway

INCOFER Ferrocarriles de Costa Rica (Spanish—Costa Rican Railways)

IND Independent (New York subway system)

Indiana Harbor Belt "connects with all Chicago railroads"

Industrial Railway of Potosí and Chihuahua (*Ferrocarril Industrial del Potosí y Chihuahua*)—Mexico

INT Interstate Railroad

Interstate Interstate Railroad

IPE Indian-Pacific Express [Perth to Sydney—2461 miles (3960 kilometers) in 65 hours]

IR Israel Railways

IRCA International Railways of Central America (El Salvador, Guatemala, and Honduras)

IRN Ironton Railroad

IRRys Iraqi Republic Railways

IRS Iranian State Railway

IRT Interborough Rapid Transit (New York City subway system)

ITC Illinois Terminal Company

ITRC Iowa Transfer Railway Company

IU Indiana Union Railway

JE Jerseyville and Eastern

Jersey Central Lines Central Railroad of New Jersey and Lehigh and New England

JHSC Johnstown and Stony Creek Railroad

JNR Japanese National Railways

JRC Jamaica Railway Corporation

JTC Jacksonville Terminal Company

JWR Jane's World Railways

Katy Missouri-Kansas-Texas Railroad (MKT)

KBR Kankakee Belt Route

KCC Kansas City Connecting Railroad

KCMO Kansas City, Mexico and Orient Railway (*Ferrocarril Chihuahua al Pacifico*)

KCNW Kelley's Creek and Northwestern Railroad

KCPSFO Kansas City Public Service Freight Operation

KCR Kanawha Central Railway

K-C Ry Kowloon-Canton Railway (Hong Kong)

KCS Kansas City Southern Railway (includes Arkansas Western, Fort Smith and Van Buren, Louisiana and Arkansas railways)

KCT Kansas City Terminal Railway

KGB Kewaunee, Green Bay and Western Railroad (Green Bay and Western Lines)—also KGB & W

KIT Kentucky and Indiana Terminal Railroad

K & M Kansas and Missouri Railway and Terminal Company

KMRT Kansas and Missouri Railway and Terminal Company

KNR Klamath Northern Railway; Korean National Railways

KO & G Kansas, Oklahoma and Gulf Railway

KRI Kyle Railway Inc

K & T Kentucky and Tennessee

KTM Keretapi Tanah Malayu (Malayan Railway)

Kyle Kyle Railway

L & A Louisiana and Arkansas Railway (Kansas City Southern)—also LA

LAJ Los Angeles Junction Railway

LA & LR Livonia, Avon and Lakeville Railroad

LAMCO Liberian America Swedish Minerals Company (Liberian Railways)

Land of Evangeline Route Dominion Atlantic Railway

LART Los Angeles Rapid Transit

LAWV Lorain and West Virginia Railway (North and Western)

LBR Lowville and Beaver River Railroad

L & C Lancaster and Chester Railway

LEE Lake Erie and Eastern Railroad

LEF Lake Erie, Franklin and Clarion Railroad

LE & FW Lake Erie and Fort Wayne

LEN Lake Erie and Northern Railway (Canadian Pacific)

LHR Lehigh and Hudson River

LI Long Island Railroad (Metropolitan Transportation Anthority)—M

Lickenpurr (Hawaiian nickname—Lahaina-Kaanapal and Pacific Rail Road) nickname derived from abbreviations—LK & PRR.

LIRR Long Island Rail Road

LK & PRR Lahaina-Kaanapal and Pacific Rail Road (Maui, Hawaii)

LM Litchfield and Madison Railway (Chicago North Western)—also L&M

LM Leningrad Metro (Russian—Leningrad subway)

LMC Liberia Mining Company

LMRBR London Midland Region of British Railways

L & N Louisville and Nashville Railroad (a part of CSX)

LNAC Louisville, New Albany and Corydon Railroad

LNE Lehigh and New England Railway (Central Railroad of New Jersey)

L & NR Ludington and Northern Railway

L & NRY Laona and Northern Railway

L & NW Louisiana and North West Rail Road

LOPG Live Oak, Perry and Gulf (Southern)

LPB Louisiana and Pine Bluff Railway

LPN Longview, Portland and Northern Railway

LRB London Transport Board

lrc (LRC) light, rapid, comfortable (high-speed railroad trains)

LRI Lawndale Transportation Company

LRS Laurinburg and Southern

L & S Laurinburg and Southern

LS & BC La Salle and Bureau County Railroad

LS & I Lake Superior and Ishpeming Railroad

LSO Louisiana Southern Railway (Southern)

LSR Lebanese State Railroads

LST & TRC Lake Superior Terminal and Transfer Railway Company

LT Lake Terminal Railroad (also LTRR)

LV Lehigh Valley Railroad

LW Louisville and Wadley Railway

L & W Louisville and Wadley Railway

LWV Lackawanna and Wyoming Valley Railway

M Metropolitan Transit Authority (New York City's rapid-transit system); Metropolitan Transportation Authority (Long Island Railroad); Monon Railroad

MA Magyan Allamvasutak (Hungarian—Hungarian State Railways)

MAGR Minneapolis, Anoka and Guyana Range Railroad

Main Line of Mid-America Illinois Central Railroad

MARR Magma Arizona Railroad

M-A Ry Massawa-Agordad Railway (Ethiopia)

M & B Meridan and Bigbee Railroad

MBI Marianna and Blountstown Railroad

MBT Marianna and Blountstown

MBTA Massachusetts Bay Transportation Authority (Boston's subway system)

MC Michigan Central Railroad (Penn Central)

McR McCloud River Railroad

MCRR Main Central Rail Road; Monongahela Connecting Railroad

MCSA Moscow, Camden and San Augustine Railroad

MD Municipal Docks Railway of the Jacksonville Port Authority

M del P Méxicano del Pacifico (Spanish—Pacific Railroad formerly Southern Pacific of Mexico)

MD & W Minnesota, Dakota and Western Railway

M & E Morristown and Erie Railroad

MEC Maine Central Railroad

MER Metropolitan Elevated Railroad

METC Medesto and Empire Traction Company

Metro (French—short form *Chemin de fer Metropolitain*)—Paris subway system

Metropolitano Rome's subway system

Mexican Pacific Railroad *Ferrocarril Mexicano del Pacifico*-(Spanish—Los Mochis to Camp)

MF Middle Fork Railroad

MGA Monongahela Railway

MGU Mobile and Gulf Railroad

MHM Mount Hope Mineral Railroad

M & HMRR Marquette and Huron Mountain Railroad

MI Missouri-Illinois Railroad

MICO Midland Continental Railroad

MID Midway Railroad

MILW Chicago, Milwaukee, St Paul and Pacific Railroad (Milwaukee Road)

MINE Minneapolis Eastern Railway

MIR Minneapolis Industrial Railway

Mitropa Mitteleuropaische Schlaf und Speiswagen (German—Middle-European Sleeping Car and Dining Car)

MJ Manufacturers' Junction Railway

MKC McKeesport Connecting Railroad

MKT Missouri-Kansas-Texas Railroad (Katy)

MLD Midland Railway of Manitoba

MLS Manistique and Lake Superior Railroad

MMR Moscow Metro Railway (Moscow's radiating subway system famed for its beautiful stations)

MNCRR Metro-North Commuter Railroad

MNF Morehead North Fork Railroad

MNJ Middletown and New Jersey Railway

MNS Minneapolis, Northfield and Southern Railway

MOB Montreux-Oberland-Bernois (railway)

MON Motion Railroad

Monon Motion Railroad (formerly Chicago, Indianapolis and Louisville Railway)

Mon Rys Mongolian Railways

Montour Montour Railroad (Youngstown and Southern Railway)

MOP Missouri-Pacific Lines

Mo-Pac Missouri-Pacific Lines

MOV Moshassuck Valley Railroad

MOW Montana Western Railway

MP Missouri Pacific Railroad

MPA Maryland and Pennsylvania

MPB Montpelier and Barre Railroad

MPPR Manitou and Pike's Peak Railway

MR McCloud River Railroad (also McRRR)

M of R Ministry of Railways (mainland China)

MRA Malayan Railway Administration

MRL Malawi Railways Limited

MRR Mattagami Railroad (Ontario); Mossi Railroad (Upper Volta)

MRS Manufacturers Railway

MRy Malayan Railway

MSC Mississippi Central (Illinois Central)

MSE Mississippi Export Railroad

M St L Minneapolis and St Louis (Chicago North Western)

M & StL Minneapolis-St Louis (Chicago North Western)

MSTL Minneapolis-St Louis (Chicago North Western)

MSTR Massena Terminal Railroad

MSV Mississippi and Skuna Valley Railroad

MT Ministry of Transport (former USSR's administration of twenty-six railway lines including the deluxe Leningrad-Moscow and the transcontinental Trans-Siberian linking Moscow with Vladivostok)

MTC Milwaukee Transport Company; Montreal Transportation Commission (subway and surface railways); Mystic Terminal Company (Boston and Maine)

MTFR Minnesota Transfer Railroad

MTH Mount Hood Railway

MTR Montour Railroad

MTW Marinette, Tomahawk and Western Railroad

MTWCR Mt Washington Cog Railway

MWR Muncie and Western Railroad

NAJ Napierville Junction Railway

NAP Narragansett Pier Railroad

NAR Northern Alberta Railways; Northern Australia Railway

National Railroads of Cuba *Ferrocarriles Nacionales de Cuba* (includes nationalized lines of the Cuba Railroad, Cuba Northern Railways, Guantanamo Railroad, Guantanamo Western, Hershey Cuban Railway, etc.)

National Railways of Mexico *Ferrocarriles de México*

NB Northampton and Bath Railroad

NC & StL Nashville, Chattanooga and St Louis Railway (L&N)

N de M Nacional de México (National of Mexico)

N de T Nacional de Tehuantepec (Tehuantepec National)

New Haven New York, New Haven and Hartford Railroad

NEZP Nez Perce Railroad

NFD Norfolk, Franklin and Danville Railway

NGR Nepalese Government Railway

NH New York, New Haven and Hartford Railroad (Penn Central)

NHIR New Hope and Ivyland Railroad

Nickel Plate New York, Chicago and St Louis Railroad (merged with Norfolk and Western)

NJ Niagara Junction Railway

NJI & I New Jersey, Indiana and Illinois Railroad

NKP Nickel Plate (New York, Chicago and St Louis Railroad)—merged with Norfolk and Western

NLC New Orleans and Lower Coast Railroad

N'LG North Louisiana and Gulf Railroad

NM Nagoya Municipality (subway system)

NN Nevada Northern Railway

NNC Northern Navigation Company

NO de M Noroeste de México (Northwestern of Mexico)

NODM Ferrocarril Noroeste de México (Northwest Railway of Mexico—Ferrocarril Chihuahua al Pacífico)

NONE New Orleans and Northeastern Railroad (Southern)

NOPB New Orleans Public Belt Railroad

NOPS New Orleans Public Service

Norf S Norfolk Southern, Norfolk & Western, Southern Railway (merger)

NP Northern Pacific Railway

N & PB Norfolk and Portsmouth Belt Line Railroad

NR Newfoundland Railway (Canadian National); Northern Railway of Costa Rica (from mountain capital of San José to Caribbean seaport of Limón)

NRC National Rail Corporation Limited (Australia); Nigerian Railway Corporation

NRPC National Railroad Passenger Corporation (Amtrak)

NRRC National Railroad Company (of Haiti)

NRZ National Railways of Zimbabwe

NS Norfolk Southern Railway

NS Nederlandsche Spoorwagen (Dutch—Netherlands Railway Carriage)—Netherlands Railways

NSB Norges Statsbaner (Norwegian—Norwegian State Railways)

NSL Norwood and St Lawrence Railroad

NSS Newburgh and South Shore Railway

NSWGR New South Wales Government Railways

NUR Natchez, Urania and Ruston Railway

NW Norfolk and Western

N & W Norfolk and Western Railway

NWP Northwestern Pacific Railroad

NWRy North Western Railway (West Pakistan)

NWS Norfolk & Western Southern (merger)

NYC New York Central Railroad (Penn Central)

NYCTA New York City Transit Authority (subway systems include BMT, IRT, INDependent)

NYD New York Dock Railway

NYLB New York and Long Branch Railroad

NYNH & H New York, New Haven and Hartford Railroad

NY O & W New York, Ontario and Western

NYS Nepal Yatayat Samsthan (Nepali—Transport Corporation of Nepal)

NYSW New York, Susquehanna and Western Railroad (NYS&W)

NZGR New Zealand Government Railways

NZR New Zealand Railways

ÖBB Österreichische Bundesbahnen (German—Austrian Federal Railways)

OCBN Organisation Commune Benin-Niger des Chemins de Fer ed des Transports (French—Benin-Niger Passenger Freight Railway)

OCE Oregon, California and Eastern Railway

OCTRA Office du Chemin de Fer Transgabonas (French—Gabon State Railways)

OE Oregon Electric Railway (Spokane, Portland, and Seattle Railway)

OGR Official Guide of the Railways

OKT Oakland Terminal Railway

OL & BR Omaha, Lincoln and Beatrice Railway

OMTB Osaka Metropolitan Transportation Bureau (subway system)

ON Ontario Northland

ONCF Office National des Chemins de Fer (French—National Railways Office)—Morocco

ONRY Ogdensburg and Norwood Railway

ONT Ontario Northland Railway

ONW Oregon and Northwestern

O & NW Oregon and Northwestern

ÖOB Österreichischen Bundesbahnen (German—Austrian State Railways)

OPE Oregon, Pacific and Eastern

ORER Official Railway Equipment Register

OT Oregon Trunk Railway (Spokane, Portland, and Seattle Railway)

OUR & D Ogden Union Railway and Depot

Overland Route Union Pacific Railroad

PA Pittsburgh Authority (rapid transit)

PAA Pennsylvania and Atlantic Railroad

PACC Pacific Coast Railroad

Pacific Railroad *Ferrocarril del Pacifico* (linking American border at Nogales with Mazatlan on Pacific coast of Mexico)

Pacific Railway of Costa Rica from Pacific port of Puntarenas to San José

Pacific Railways of Nicaragua *Ferrocarril del Pacifico de Nicaragua*—from Corinto on the Pacific to Granada on Lake Nicaragua

Pac Rail Missouri Pacific, Union Pacific, Western Pacific (merged)

PA & M Pittsburgh, Allegheny and McKees Rocks Railroad

Panama Railroad division of the Panama Canal linking Cristóbal and Colón on the Atlantic with Balboa and Panama City on the Pacific and running parallel to the Panama Canal

P & AR Pacific and Arctic Railway

PATCO (transportation system linking Camden, New Jersey and Philadelphia, Pennsylvania)

PATH Port Authority Trans–Hudson Corporation (operates Hudson Tubes between New Jersey and New York)

PBNE Philadelphia, Bethlehem and New England Railroad

PBR Patapsco and Back Rivers

PC Penn Central (Pennsylvania New York Central Transportation Company; Pennsylvania Railroad; New York Central Railroad; New York, New Haven, and Hartford Railroad; Baltimore and Eastern Railroad; Canada Southern Railway; Cleveland, Cincinnati, Chicago and St Louis Railway; Michigan Central Railroad; Peoria and Eastern Railway; Waynesburg and Washington Railroad)

PCL Peruvian Corporation Limited

PCN Point Comfort and Northern

PCR Paraguayan Central Railway

PCY Pittsburgh, Chartiers and Youghiogheny Railway

PE Pacific Electric (interurban railway system serving entire Los Angeles area before replacement by smog-producing buses); Pacific Electric Railway of Costa Rica (links Pacific seaport of Puntarenas with mountain capital of San José)—also called *FEP*

P & E Peoria and Eastern Railway (Penn Central)

Pennsy (nickname—Pennsylvania Railroad)—became part of the Penn Central

Peoria Peoria and Pekin Union Railway

P & F Pioneer and Fayette Railroad

PGE Pacific Great Eastern Railway

PH & D Port Huron and Detroit Railroad

P & I Paducah and Illinois Railroad

PIC Pickens Railroad

Pick Pickens Railroad

Pickens Pickens Railroad

PKP Polskie Koleje Panstwowe (Polish—Polish State Railways)

P & LE Pittsburgh and Lake Erie Railroad

PLM Paris-Lyon-Mediterranée

P & N Piedmont and Northern Railway

PNKA Perusahaan Negara Kereta Api (Indonesian—Indonesian State Railways)

PNR Philippine National Railways

PNW Prescott and Northwestern Railroad

Port St Joe Route Apalachicola Northern Railroad

'Possum Trot Line Reader Railroad

POV Pend Oreille Valley

P & OV Pittsburgh and Ohio Valley

POVA Pend Oreille Valley (railway)

P & PU Peoria and Pekin Union

PR Panama Railroad

P-R Pennsylvania-Reading Seashore Lines

PRC Philippine Railway Company

PRCR Pacific Railway Costa Rica

PRR Pennsylvania Railroad (Penn Central)

PRS Pennsylvania-Reading Seashore Lines

PRTD Portland Railroad and Terminal Division of the Portland Traction Company

PRV Pearl River Valley Railroad

PS Pittsburgh and Shahmut Railroad

P & SR Petaluma and Santa Rosa

PTC Peoria Terminal Company; Philadelphia Transportation Company (also called PATCO; includes elevated and subway lines of Philadelphia area)

PTM Portland Terminal Company

PTR Parr Terminal Railroad

PTS Port Townsend Railroad

Pullman deluxe railroad cars providing lounging, observation, and sleeping facilities aboard first-class express trains

PVS Pecos Valley Southern

P&W Providence & Worcester Railroad

P & WV Pittsburgh and West Virginia Railway (Norfolk and Western)

P y RV Potosí y Rio Verde (Spanish—Potosi and Green River Railroad of Chihuahua)

QAP Quanah, Acme and Pacific

QC Quebec Central Railway (Canadian Pacific)

QNS & LRC Quebec North Shore and Labrador Railway Company

QR Queensland Railways

Quanah Route Quanah, Acme and Pacific Railway

QUI Quincy Railroad

RB Rail Box (American box car pool)

RC Railway Corporation (Nigeria)

RCFA-N Regie du Chemin de Fer Abidjan–Niger (French—Abidjan-Niger Railway Administration)—Ivory Coast

RD Railway Directorate (Albania)

RDG Reading Company (formerly Philadelphia and Reading Railroad)

REA Railway Express Agency; Reader Railroad

Reading Lines Reading Railway System (formerly Philadelphia and Reading Railroad)

Rebel Route Gulf, Mobile and Ohio Railroad

RENFE Red Nacional de los Ferrocarriles Españoles (Spanish—Spanish National Railway System)

RFFSA Rede Ferrovidria Federal SA (Portuguese—Federal Railway System Corporation)—Brazil

RFP Richmond, Fredericksburg and Potomac Railroad (RF& P)

RF & PRR Richmond, Fredericksburg and Potomac Railroad

RI Chicago, Rock Island and Pacific Railroad; Rail India

Rio Grande Denver and Rio Grande Western

RKG Rockingham Railroad

RM Rotterdam Metro (Dutch—Rotterdam Subway)

RNCF Reseau National des Chemins de Fer (French—National Railway System) Madagascar

Rock Island Chicago, Rock Island and Pacific Railroad

RR (abbreviation—Railroad or Rail Road); (reporting mark—Raritan River Rail Road); Rhodesian Railways

RRRR Raritan River Railroad

RRV&W Red River Valley & Western Railroad

RRys Rhodesian Railways

RS Roberval and Seguenay Railway

RSP Roscoe, Snyder and Pacific

R-S Pacific Route Roscoe, Snyder and Pacific Railway

RSS Rockdale, Sandow and Southern Railroad

RT River Terminal Railway

RTM Railway Transfer Company of Minneapolis

RV Rahway Valley Railway

Ry Railway

S & A Savannah and Atlanta Railway

SAN Sandersville Railroad

Santa Fe Atchison, Topeka and Santa Fe Railway

SAR South African Railways; South Australian Railways

SAR & H South African Railways and Harbours

SATS San Antonio Transit System

SAVE Swiss-Alberg-Vienna Express

SB South Buffalo Railway

SBA Subterraneos de Buenos Aires (Spanish—Buenos Aires Subways)

SBC Ferrocarril Sonora-Baja California (Spanish—Sonora-Baja California Railway)

SBK South Brooklyn Railway

SC Sumter and Choctaw Railway

SCBF Société des Chemin de Fer du Burkina (French—Burkina Faso Railway)

SCE Shanghai-Canton Express

SCL Seaboard Coast Line Railroad (Atlantic Coast Line Railroad, Charleston and Western Carolina Railway, Seaboard Air Line Railroad—former name of the Seaboard Coast Line Railroad [now part of CSX])

SC & MR Strouds Creek and Muddlety Railroad

SCT Sioux City Terminal Railway

SDAE San Diego and Arizona Eastern Railway

SD & AE San Diego and Arizona Eastern Railway

SD & IV San Diego & Imperial Valley

SDTS San Diego Transit System

SE Ferrocarril del Sureste (Spanish—Southeast Railroad)

Seashore Lines Pennsylvania-Reading Seashore Lines

SE & CR Southeastern and Chatham Railway (nicknamed Seldom Ever Caught Running)

SEMTA Southeastern Michigan Transportation Authority

SERA Sierra Railroad

SFERR San Francisco Belt Railroad

SFMR San Francisco Municipal Railway (operates the cable cars)

SFSP Santa Fe/Southern Pacific (merger)

SF/SP Santa Fe/Southern Pacific (railroad merger)

SG South Georgia Railway (Southern Railway)

SGR Saudi Government Railroad (Saudi Arabia); Suriname Government Railway

SH Steelton and Highspire Railroad

Shawmut The Pittsburgh and Shawmut Railroad

SHK Sidirodromi Hellinikou Kratous (Greek—Hellenic State Railways)—Greece

SI Spokane International Railroad

SIR Staten Island Rapid Transit Railway

SIRRI Southern Industrial Railroad Incorporated

SJ Statens Jarnvargar (Swedish—State Railways)

SJB St Joseph Belt Railway

SK St Johnsbury and Lamoille County Railroad

SJ & LC St Johnsbury and Lamoille County Railroad

SJTR St Joseph Terminal Railroad

SKSL Skaneateles Short Line Railroad

SLC San Luis Central Railroad

SLGW Salt Lake, Garfield and Western Railway

SLR Sierra Leone Railway

SLSF St Louis-San Francisco Railway

SM St Marys Railroad

SMA San Manuel Arizona Railroad

SMR South Manchurian Railway

SMX Santa Maria Valley Railroad

SN Sacramento Northern Railway (also SNRy)

SNCB Société Nationale des Chemins de Fer Belges (French—Belgian National Railways)

SNCF Société Nationale des Chemins de Fer Français (French—French National Railways)

SNCEA Société Nationale des Chemins de Fer Algeriens (French—Algerian National Railways)

SNTF Société Nationale des Transports Ferroviaires (French—Algerian National Railways)

SNY Southern New York Railway

SOE Simplon-Orient Express

SOI Southern Indiana Railway

Sonora-Baja California Railway *Ferrocarril Sonora-Baja California*—Mexicali to Benjamin Hill

SOO Soo Line Railroad

Soo Line Soo Line Railroad

SOT South Omaha Terminal Railway

Southern Southern Railway system (Alabama Great Southern Railroad; Carolina and Northwestern Railway; Cincinnati, New Orleans and Texas Pacific Railway; Georgia Southern and Florida Railway; Harriman and Northeastern Railroad; Live Oak, Perry and Gulf Railroad; Louisiana Southern Railway; New Orleans and Northeastern Railroad; South Georgia Railway)

Southern Pacific SP

South Shore Line Chicago South Shore and South Bend Railroad

SP Southern Pacific (includes Southern Pacific Lines, Sunset Railway, Texas and Louisiana Lines, Texas and New Orleans, etc.)—in fact many

school children once said the United States was bounded on the north by Canada and the Great Lakes, on the east by the Atlantic Ocean, and on the south and southwest by the Southern Pacific

SPGT Springfield Terminal Railway

SPS Spokane, Portland and Seattle Railway (including Oregon Electric and Oregon Trunk railways)

SR Southern Railway

SRBR Southern Region of British Railways

SRA State Railway Authority (New South Wales)

SRC Salvador Railway Company (El Salvador)

SRN Sabine River and Northern

SRRC Sierra Railroad Company; Strasburg Rail Road Company

SRRCO Sandersville Railroad Company

SRT State Railways of Thailand

SSDK Savannah State Docks Railroad

SSLVRR Southern San Luis Valley Railroad

SSRy Sand Springs Railway

SSW St Louis Southwestern Railway (Cotton Belt Route)

STCUM Société de Transport de la Communauté Urbaine de Montreal (French—Montreal Urban Community Commission)

STE Stockton Terminal and Eastern Railroad

STRT Stewartstown Railroad

STS Seattle Transit System

SU Stockholm Underground (subway system)

Sub Suburban; Subway

Sud Rys Sudan Railways

SUR Soviet Union Railways (was managed by Ministry of Communications and comprising some twenty-six lines including the Trans-Mongolian and the Trans-Siberian as well as the plush Leningrad-Moscow express)

Susquehanna New York, Susquehanna and Western Railroad

Syr Rys Syrian Railways

T symbol for Boston's subways; shortened from MBTA (Massachusetts Bay Transportation Authority)

TAAA Travelers Aid Association of America

TA & G Tennessee, Alabama and Georgia Railway

TAG Route Tennessee, Alabama and Georgia Railway

Tan-Zam Tanzania-Zambia Railroad

TAR Trans-Australian Railways

TAS Tampa Southern Railroad

TASD Terminal Railway Alabama State Docks

TA & W Toledo, Angola and Western Railway

TB Twin Branch Railroad

TBTMG Transportation Bureau of the Tokyo Metropolitan Government (subway)

TC Tennessee Central Railway

TCDD *Turkiye Cumhuriyeti Deviet Demiryollari Isletmesi* (Turkish—Turkish State Railways)

TCG Tucson, Cornelia and Gila Bend Railroad

TCT Texas City Terminal Railway

TEBRCL The Emu Bay Railway Company Limited

TEE Trans-Europe Express

TENN Tennessee Railroad

TEXC Texas Central Railroad

THB Toronto, Hamilton and Buffalo Railway

The Q CB&Q (Chicago, Burlington and Quincy)

TM Texas Mexican Railway; Transport Ministry (former USSR's administration of twenty-six railway lines)—TM sometimes used on engines

TMR Trans-Mongolian Railway

TN Texas and Northern Railway

T-NM Texas-New Mexico Railway

T & NO Texas and New Orleans (Southern Pacific)—also TNO

TOC Pennsylvania New York Central Transportation Company (Penn Central)

TOE Texas, Oklahoma and Eastern Railroad

TOV Tooele Valley Railway

T & P Texas and Pacific Railway (also TP)

TPMP Texas-Pacific-Missouri Pacific Terminal Railroad of New Orleans

TPT Trenton-Princeton Traction Company

TP & W Toledo, Peoria and Western Railroad

TR Tasmanian Railways

TRA Taiwan Railway Administration

Transperth Metropolitan Passenger Transport Trust (Perth)

Trans-Sib Trans-Siberian Railway

TRC Tela Railway Company (Honduras); Trona Railway Company (California)

TRRA Terminal Railroad Association of St Louis

TS Tidewater Southern Railway

TS-E Texas South-Eastern

TSR Trans-Siberian Railway

TSU Tulsa-Sapulpa Union Railway

TT Toledo Terminal Railroad

T & T Tijuana and Tecate Railway (freight cars marked TITE)

TTC Toronto Transit Commission (subway and surface railway systems)

Turk-Sib Turkestan-Siberian (railway)

TVG Tavares and Gulf Railroad

TVRy Tooele Valley Railway

Tweetsie (nickname—East Tennessee and Western North Carolina Railroad)—believed to be derived from high-pitched whistles of its engines

T-Z RA Tanzania-Zambia Railway Authority

U Underground (London's subway system)

UBR Ulan Bator Railway

UCR Utah Coal Route

U de Y *Unidos de Yucatan* (Spanish—United Railways of Yucatan, Mexico)

UFC United Fruit Company (railroads in Costa Rica and Panama)

UMP Upper Merion and Plymouth Railroad

UNIT Union Freight Railroad

UNI Unity Railways

UO Union Railroad–Oregon

UP Union Pacific Railroad (includes Oregon Short Line and Oregon–Washington Railroad and Navigation Company)

UR Uganda Railway

URR Union Railroad—Pittsburgh

UT Union Terminal Railway

UTA Ulster Transport Authority (railways of six counties in Northern Ireland)

UTAH Utah Railway

Utah Coal Route Utah Railway

UTR Union Transportation Company

V *Valtionrautatiet* (Finnish—State Railways)

VBR Virginia Blue Ridge Railway

VC Virginia Central Railway

VCS Virginia and Carolina Southern Railroad

VCY Ventura County Railway

VE Visalia Electric Railroad

VGN Virginian Railway (Norfolk and Western)

VIA VIA Rail Canada

Via Rail Canadian National + Canadian Pacific (passenger-carrying consolidation)

Virginian Virginian Railway (Norfolk and Western)

V & LI Vancouver and Lulu Island (branch of Canadian Pacific)

V-MNR Viet-Minh National Railways (former North Vietnam)

V-NR Viet-Nam Railways (former South Vietnam)

VR *Valtionrautatiet* (Finnish—Finish State Railways); Victorian Railways (Australia)

V Ry Verapaz Railway (Guatemala)

VSL Valley and Siletz Railroad

VSO Valdosta Southern Railroad

VSOE Venice-Simplon Orient Express

VTR Vermont Railway

W of A Western Railway of Alabama

WAB Wabash Railroad (Norfolk and Western)

Wabash Wabash Railroad (Norfolk and Western)

WAG Wellsville, Addison and Galeton Railroad

WAGR Western Australian Government Railways

WATC Washington Terminal Company

WAW Waynesburg and Washington Railroad (Penn Central)

WBCRR Wilkes-Barre Connecting Railroad

WBT & SRC Waco, Beaumont, Trinity and Sabine Railway Company

Western Railway of Mexico
Ferrocarril Occidentalde México—Culiacan to Limoncito

West Point Route Atlanta and West Point Rail Road

Westrail Western Australian Government Railways

Westrain Western Australian Trains

White Pass British Columbia Yukon Railway, British Yukon Railway, Pacific and Arctic Railway

White Pass and Yukon Route British Columbia Yukon Railway, British Yukon Navigation, British Yukon Railway, Pacific and Arctic Railway and Navigation Company

WIM Washington, Idaho and Montana Railway

WL *Wagon Lits* (French—sleeping cars)

WLO Waterloo Railroad

WM Western Maryland Railway (merged with CSX)

WMR Wasatch Mountain Railway

WMTA Washington Metropolitan Transit Authority (subway system)

WMWN Weatherford, Mineral Wells and Northwestern Railway

WNF Winfield Railroad

W & NO Wharton and Northern Railroad

WOD Washington and Old Dominion Railroad

W & OV Warren and Ouachita Valley Railway

WP Western Pacific Railroad

WPER West Pittston-Exeter Railroad

W&P Willamette & Pacific Railroad

WPPR Willamette & Pacific Railroad

WP & Y White Pass and Yukon Railway

WRA Western Railroad Association

WRBR Western Region of British Railways

WRNT Warrenton Railroad

WRWK Warwick Railway

WS Ware Shoals Railroad

WSR Warren and Saline River Railroad

WSS Winston-Salem Southbound Railway

WSYP White Sulphur Springs and Yellowstone Park Railway

WTR Wrightsville and Tennille Railroad

WVN West Virginia Northern Railroad

WW Winchester and Western Railroad

WWV Walla Walla Valley Railway

WYS Wyandotte Southern Railroad

WYT Wyandotte Terminal Railroad

X express; transport; transportation (as in many private bulk carriers' names such as GATX—General American Transportation)

Xing crossing (highway or railroad)—also XING

YAN Yancey Railroad

YN Youngstown & Northern (railroad)

Y & N Youngstown and Northern Railroad

YR Yucatan Railways *(Ferrocarriles Unidos del Sureste—*United Railways of the Southeast)—along the Gulf of Mexico from Coatzacoalcos to Merida

YS Youngtown and Southern Railway (Montour)

Y & S Yakutat and Southern Railway

YVT Yakima Valley Transportation Company

YW Yreka Western Railroad

ZJZ *Zajednica Jugoslovenskih Zalesnicca* (Yugoslavian—Community of Yugoslav Railways)

ZR Zambia Railways

Zug Island Road Delray Connecting Railroad (DC)

Roman Numerals

I	1	LIX	59	M	1000
II	2	LX	60	MD	1500
III	3	LXV	65	MDC	1600
IV	4	LXIX	69	MDCC	1700
V	5	LXX	70	MDCCC	1800
VI	6	LXXV	75	MCM or	
VII	7	LXXIX	79	MDCCCC	1900
VIII	8	LXXX	80	MCMX	1910
IX	9	LXXXV	85	MCMXX	1920
X	10	LXXXIX	89	MCMXXX	1930
XV	15	XC	90	MCMXL	1940
XIX	19	XCV	95	MCML	1950
XX	20	XCIX	99	MCMLX	1960
XXV	25	C	100	MCMLXX	1970
XXIX	29	CL	150	MCMLXXX	1980
XXX	30	CC	200	MCMXC	1990
XXXV	35	CCC	300	MM	2000
XXXIX	39	CD	400	MMM	3000
XL	40	D	500	MMMM or	
XLV	45	DC	600	M$\overline{\text{V}}$	4000
XLIX	49	DCC	700	$\overline{\text{V}}$	5000
L	50	DCCC	800	$\overline{\text{M}}$	1,000,000
LV	55	CM	900		

Rules of the Road—at Sea

Red to red and green to green all is safe to pass abeam.

or

Green to green and red to red perfect safety—go ahead.
If on your starboard red appear it is your duty to keep clear:
to act as judgment says is proper—
to port, or starboard, back, or stop her.
But when upon your port is seen
a steamer's starboard light of green—
there's not so much for you to do
for green to port keeps clear of you.
Both in safety and in doubt always keep a good lookout;
in danger with no room to turn ease her, stop her, go astern.

When Two Ships Meet Head On

When both side lights you see ahead
port your helm and show your red.
(Steer to starboard so your red light will pass the red light of the approaching vessel, and thus you'll
pass on the left as people do ashore.)

Russian Alphabet (transliterated)

Russian Capital Letters	English Capital Letters	Russian Small Letters	English Small Letters	Russian Alphabet Letter Names	Nearest English Equivalent
А	A	а	a	ah	*a* as in *a*rch
б	B	б	b	beh	*b* as in *b*it
В	V	в	v	veh	*v* as in *v*est
Г	G	г	g	geh	*g* as in *g*et
Д	D	д	d	deh	*d* as, in *d*ay
Е	Ye	е	ye	yeh	*y* as in *y*es
Ё	Yo	ё	yo	yo	*yo* as in *yo*lk
Ж	Zh	ж	zh	zheh	*zh* sound as in mea*s*ure
З	Z	з	z	zeh	*z* as in *z*ero
И	I	и	i	ee	*ee* as in p*ee*l
Й	Y	й	y	ee s krátkoi	(short i after vowels)
К	K	к	k	kah	*k* as in *k*ite
Л	L	л	l	el	*l* as in woo*l*
М	M	м	m	em	*m* as in *m*an
Н	N	н	n	en	*n* as in *n*ow
О	O	о	o	oh	*o* as in h*o*ax
П	P	п	p	peh	*p* as in *p*encil
Р	R	р	r	err	*r* as in *r*ye
С	S	с	S	ess	*s* as in *s*ay
Т	T	т	t	teh	*t* as in *t*en*t*
У	Oo	у	oo	ooh	*oo* as in l*oo*se
Ф	F	ф	f	eff	*f* as in *f*ancy
Х	Kh	х	kh	khan	*kh* as in lo*ch*
Ц	Ts	ц	ts	tseh	*ts* as in ha*ts*
Ч	Ch	ч	ch	chah	*ch* as in *ch*air

Ш	Sh	ш	sh	shah	*sh* as in *sh*ave
Щ	Shch	щ	shch	shchah	*shch* as in Iri*sh ch*uck
Ъ		ъ		tvyódy znak	(silent-hard sound)
Ы	Y	ы	y	yery	*e* as in it*e*m
Ь		ь		myakhki znak	(silent)
Э	Eh	э	eh	eh oborótnoye	*eh* sound as in d*e*bt
Ю	Yu	ю	yu	yoo	*yu* as in *you*
Я	Ya	я	ya	yah	*ya* as in *ya*cht

Ship's Bell Time Signals

1 bell—12:30	or	4:30	or	8:30 a.m. or p.m.
2 bells—1:00		5:00		9:00
3 bells—1:30		5:30		9:30
4 bells—2:00		6:00		10:00
5 bells—2:30		6:30		10:30
6 bells—3:00		7:00		11:00
7 bells—3:30		7:30		11:30
8 bells—4:00		8:00		12:00

On many vessels the ship's whistle is blown at noon. On some ships a lightly struck 1 bell announces 15 minutes before the change of watch, usually at 4, 8, and 12 o'clock.

The ship's day starts at noon. The *afternoon watch* is from noon to 4 p.m. The 4 to 8 work period is called the *dogwatch*. From 8 p.m. to midnight is the *first watch*. From midnight to 4 a.m. is the *middle watch*. From 8 a.m. to noon is the *forenoon watch*.

Signs and Symbols
Frequently Used

+ add; addition sign; north; plus

& and (ampersand)

&c etcetera (and so forth)

* asterisk

@ at, because

¢ centavo; centime; cent(s)

© copyright

° degree(s)

° ′ ″ degrees, minutes, seconds (used to measure latitude north and south of the Equator and longitude east or west of the Greenwich meridian)

÷ divide; divided by; division sign

$ dollar sign—used universally for monetary units as diverse as Nicaraguan cordobas; Brazilian cruzeiros; Australian, Bahamian, Barbadian, British Honduran, Canadian, Ethiopian, Guyanian, Hong Kongese, Levantine, Liberian, Malaysian, New Zealand, Taiwan, trade, Trinidadian-Tobagonian, U.S., Vietnamese, West Indian, yuan dollars; Portuguese escudos; Honduran lempiras; Brazilian milreis; Chilean, Colombian, Cuban, Dominican, Mexican, Philippine, Uruguayan pesos; Peruvian soles (often with a lower-case dollar sign, $); Chinese yuans

$A Australian dollar(s)

$b Bolivian peso(s)

$B Bahamian, Barbadian, British dollar(s)

$BH British Honduran dollar(s)

$C Brazilian cruzeiro(s); Canadian dollar(s)

$Col Colombian peso(s)

d pence

$E Ethiopian dollar(s)

$Eth Ethiopian dollar(s)

$G Guyanian dollar(s)

$HK Hong Kong dollar(s)

$K $1000 (e.g., $13K= $13,000)

$L Levant(ine) dollar(s)— Maria Theresa thaler(s); Liberian dollar(s)

$M Malay(sian) dollar(s)

$Mal Malay(sian) dollar(s)

$Mex Mexican peso(s)

$NT New Taiwan dollar(s)

$NZ New Zealand dollar(s)

$RD Republica Dominicana peso(s)—Dominican Republic monetary unit(s)

$S Singapore dollar(s)

$T Taiwan dollar(s); trade dollar(s); Trinidad(ian) and Tobago(nian) dollar(s)

$TT Trinidad(ian) and Tobago(nian) dollar(s)

$Ur Uruguayan peso(s)

$US United States dollar(s) [also shown as US$, as are other monetary units where national designations often precede dollar sign: C$— Canadian dollar(s), HK$— Hong Kong dollar(s)]

$VN Vietnamese dollar(s)

$WI West Indian dollar(s); West Indies dollar(s)

$Y yuan dollar(s)

= equality; equals; equal to

♀ female

G Paraguayan guarani(s)

K certified kosher

LC Cyrian pound(s)

LR Rhodesian pound(s)

♂ male

− minus; south; subtract; subtraction sign

× multiplication sign; multiplied by; multiply

≥ equal to or greater than

≤ equal to or less than

> greater than

< less than

≫ much greater than

≪ much less than

fracture(s) (medical); number(s) or pound(s) (commercial); sharp(s) (musical); space(s) (typographical); tic-tac-toe (game symbol); zinc (alchemical)

P Philippine peso(s)

% percent

+ plus; north

± plus or minus

£ pound (libra) sign—used universally for monetary units such as the Australian, British, Egyptian, Gambian, Ghanian, Irish, Israeli, Jamaican, Lebanese, Libyan, Malawi, New Zealand, Nigerian, South African, Sudanese, Syrian, Turkish, Western Samoan, Zambian pound

£A pound Australian

£E pound Egyptian (United Arab Republic)

£G pound Gambian; pound Ghanian

£I pound Irish; pound Israeli (also shown as I£)

£J pound Jamaican

£L pound Lebanese; pound Libyan

£M pound Malawi

£N pound Nigerian

£NZ pound New Zealand (also shown as NZ£)

£S pound sterling; pound Sudanese; pound Syrian

£SAf pound South African (also shown as SAf£)

£/s/d pounds, shillings, and pence

£T pound Turkish

£WS pound Western Samoan

£Z pound Zambian

R registered

R$_x$ prescription; receipt; recipe; response; reverse

/ shilling mark; slash; solidus; virgule

☯ T'ai-chi-T'u (Chinese—yin and yang)—ancient symbol for the diagram of the supreme ultimate

∴ therefore

Ⓤ Union of Orthodox Jewish Congregations of America (symbol for kosher product approved for detergent or dietary use)

XMA$ (symbol—commercialized Christmas)

Y Japanese yen

y & y yin and yang (*see T'ai-chi-T'u*)

States, Nations, and Territories

States, nations, and territories are listed alphabetically in Eponyms, Nicknames, and Geographical Names. Many also appear among Superlatives.

Steamship Lines

A Ahearn Shipping Ltd;
Alaska Steamship Company;
Alcoa Steamship Company;
American Export Isbrandtsen
Lines; American Mail Line;
American Oil Company;
American Steamships; Tide-
water Oil (capital A between
red wings); etc.

ABC Line Antwerp Bulk Carriers Line

ABRT A/B Rederi Transatlantic (Pacific Australia Direct
Line)

AC African Coasters

ACL American Canadian
Line; American Cruise Line;
Atlantic Container Line

ACS American Coal Shipping

ACSC Australian Coastal
Shipping Commission

AD *Armement Dieppe*

AE African Enterprises

AECL Anglo–European Container Line

AEL Afro Eurasian Line;
American Express Line

AFCL Africa Container Lines

AFS American Foreign Steamship

AH Alfred Holt (Blue Funnel
Line)

AHB Great Eastern Line

AHL Associated Humber
Lines

AJCL Australia-Japan Container Line

AL Admiral Line

Alcoa Alcoa Steamship Company

ALL Anchor Line Limited

All America Cables All America Cables and Radio

AML American Mail Line

AMOCO American Oil Company

AN Anglo Nordic

ANCAP *Administracion Nacional de Combustibles Alcohol
y Portland* (Spanish—Na-
tional Administration of
Flammable Alcohol and Portland Cement)—Uruguay

ANL Australian National Line

ANZECS Australia-New
Zealand-Europe Container
Service

ANZS Africa-New Zealand
Service

AP American Pioneer Lines

AP *Atlantska Plovidba* (Yugoslavian—Atlantic Line)

APL American President
Lines

APT Australian Pacific Traders

ASA Admanthos Shipping
Agency

ASC Alcoa Steamship Company

ASCL Australia Straits Container Line

ASFS Alaska State Ferry System

ASN Atlantic Steam Navigation

ASNC Atlantic Steam Navigation Company

ASOK *Angfartigas Svenska
Östasiatiske Kompaniet*
(Swedish—Swedish East
Asiatic Steamship Company)

AT American Trading

ATLANTIC Atlantic Refining
Company

Atlantic Container Line ACL

AUT American Union Transport

AWPL Australia West Pacific
Line

B Barber Lines; Booth Line;
Branch Lines; Bull Steamship
Lines; etc.

BACS Ben Asia Container Service

BAF Belgian African Line

BBS Barber Blue Sea

BCCS British Columbia
Coastal Service

BCF British Columbia Ferries

BCL Bermuda Container Line;
Bristol City Line

BCSC British Columbia
Steamship Company; British
and Continental Steamship
Company

BDS *Bergenske Dampskibsselskab* (Norwegian—Bergen
Steamship Line)—connect-
ing Norway and United Kingdom ports

Ben Ocean Ben Line, Blue
Funnel, and Glen Line

BFL Belgian Fruit Lines; Blue
Funnel Lines

BHP Broken Hill Proprietary

BISNC British India Steam
Navigation Company

B & I SPC British and Irish
Steam Packet Company

BL Bahamas Line; Bank Line;
Bergen Line; Bibby Line;
Booth Line; etc.

B & L Bums and Laird Lines

BLS Ben Line Steamers

Blue Star Blue Star Line

BM British Methane Limited

BMM Belfast, Mersey and
Manchester Steamship Com-
pany

BOC Burmah Oil Company

Bore Ro-Ro Bore Roll-on-
Roll-off Line

BOS British Oil Shipping

BP British Petroleum

BPC British Phosphate Commissioners

BP & Co Bums, Philip and
Company

BR British Railways (operates
many ferry steamers linking
England and Scotland with
Belgium, France, Ireland, and
Holland)

BSC Baltic Steamship Company

BSL Black Star Line; Blue Sea
Line; Blue Star Line; etc.

BSNC Bristol Steam Navigation Company

BSPL Blue Star Port Lines

BTC Bethlehem Transportation Corporation

B & W Brocklebank and Well
Lines

C Calmar Line (Bethlehem
Steel); Caribbean Steamships
Company; Clarke Line;
Clyde Line; etc.

"C" Costa Line

CA *Carregadores Açoreanos*
(Portuguese—Azorean Cargo
Carriers)

CAROL Caribbean Overseas
Lines

Carnival Carnival Cruise Lines

CAVN Compañía Anónima Venezolana de Navegación (Spanish—Venezuelan Navigation Company)—Venezuela Line

CCAL Christensen Canadian African Line

CC Co Commercial Cable Company

CCN Companhia Colonial de Navegacão (Portuguese—Colonial Navigation Company)

CCNI Cia Chilena de Navegación Interoceanica (Spanish—Chilean Interoceanic Navigation Company)

CEA Central Electricity Authority

Celebrity Celebrity Cruises

CF Compagnie de Navigation Fraissinet

CFL Container Fleets Limited

CFPO Compagnie Française des Phosphates de l'Oceanie (French—French Phosphate Company of Oceania)

CGL Canadian Gulf Line

CGM Compagnie Générale Maritime (French Line)

CGS Central Gulf Steamships

CGT Compagnie Générale Transatlantique (Cie Gle Trans) (French—General Transatlantic Company)—the French Line

CHEVRON Chevron Shipping (oil tankers)

Chilean Line (*see* CSAV)

CI Catalina Island Steamship Line; Christmas Island Phosphate Commission

Cie Gle Trans Compagnie Générale Transatlantique (French—General Transatlantic Company)—the French Line

Cities Service Cities Service Oil Company

CL Ceylon Lines; Coast Lines

Clipper Line Wisconsin and Michigan Steamship Company

CM Compañía Marítima (Spanish—Maritime Company)

CMB Compagnie Maritime Belge (French—Belgian Maritime Company)—Royal Belgian Lloyd

CMSNC China Merchants Steam Navigation Company

CMZ Compagnie Maritime du Zaire

CNC China Navigation Company

CNM Canadian National Marine (steamship line)

CNN Compagnie de Navigation Nationale

CNN Companhia Nacional de Navegacão (Portuguese—National Navigation Company)

CNP Compagnie Navigation Paquet (French—Paquet Navigation Company)—Paquet Line

CNS Canadian National Steamships

Coastal Express (*see* Hurtigruta)

COLDEMAR Compañía Colombiana de Navegación Maritima (Spanish—Colombian Maritime Navigation Company)

Columbus Line HSDG

COSCO China Ocean Shipping Company

CP Ships Canadian Pacific Steamships *(Empress* vessels)

CPV Corporación Peruana de Vapores (Spanish—Peruvian Steamship Corporation)

Crusader Crusader Line

CSAV Compañía Sud-Americana de Vapores (Spanish—South American Steamship Company)—Chile

CSC Clyde Shipping Company

CSL Canada Steamship Lines

CSO Cities Service Oil

CSS Caribbean Steamship

CSSCo Cunard Steamship Company

CT Cleveland Tankers; Cove Tankers

CT Compañía Transmediterranea (Spanish—Transmediterranean Company)

CTE Compañía Transatlantica Española (Spanish—Spanish Transatlantic Line)—The Spanish Line

CTL Coastal Transport Limited

Cunard Cunard Steam-Ship Company, Limited (includes White Star Line)

D Delta Line; Donaldson Line; Red 'D' Line; etc.

'D' Red 'D' Line (merged with Grace Line)

DAL *Deutsche-Afrika Linien* (German—German–Africa Line)

d'Amico d'Amico Line

Day Line Hudson River Day Line

DBK Daiichi Bussan Kaisha

DDSG Donau-Dampfschiffahrt-Gesellschaft (German—Danube Steamship Company)—Austria

D-F Dansk–Franske (Danish—French Line)

DFDS Det Forenede Dampskibs-Selskab (Danish—United Steamship Company)—famous for its ferries

DHX Dependable Hawaiian Express

DL Djakarta Line; Djakarta Lloyd

DPLC Dundee, Perth and London Shipping Company

DS Dominion Shipping

D-S Ditlev-Simonsen, Halfdan and Company

e El Paso Marine

E American Export Isbrandtsen Lines; Eastern Steamship Line; Exxon Tankers; Hellenic Lines and many Greek lines where the letter E stands for Ellas or Hellas— Greece— or for the last name of an owner as in other lands

E & A Eastern and Australian Steamship Co

EAC East Asiatic Company

E & B Ellerman and Buchnall Steamship Company

EDL Elder Demptser Lines

E & F Elders and Fyffes Ltd

ELMA Empresa Lineas Maritimas Argentinas (Spanish—Argentine Maritime Lines)— formerly *FANU* and uses *FANU* house flag

EMC Evergreen Marine Corporation

Empress liners Canadian Pacific ships

ENS Empresa Naviera Santa

ESL Eagle Shipping Ltd

Esso Esso Petroleum Company

EXXON formerly Esso

EY El Yam (bulk carriers)

F Fabre Line; Falcon Tankers; Falkland Islands Trading Company; Farrell Lines; Firmlines; etc.

FAA Finska Angfartygs Akiebolaget (Finnish—Finnish Steamship Company)—Finland Line

Falline Federal Atlantic-Lakes Line

FANF Flota Argentina de Navegación Fluvial (Spanish—Argentine River Navigation Fleet)

FANU Flota Argentina de Navegación de Ultramar (Spanish—Argentine High-Sea Navigation Fleet)

Far East Steamship Company FESCO

FB Franco Belgian Line

FBS Franco–Belgian Services

FCNCo Federal Commerce and Navigation Company

F de P *Ferrocarril de Panamá* (formerly the Panama Railroad)

Fedpac Federal Pacific Lakes Line

Fedsea Federal South East Asia Line

FESCO Far East Steamship Company

Finald Line (*see* FAA)

FL Ferdinand Laeisz Line; Fesco Pacific Line

FLL Finanglia Line Ltd

FMC Federal Maritime Commission

FMD Flota Mercante Dominicana (Spanish—Dominican Merchant Fleet)

FMG Flota Mercante Grancolombiana (Spanish—Great Colombian Merchant Fleet)

French Line (*see* CGT)

Frota Frota Oceanica Brasileira

FW Furness, Withy and Company

FWL Furness Warren Line

G Glynafon Shipping; Graig Shipping; Arthur Guiness (the brewer); etc.

GAL German Atlantic Line

GG Guinea Gulf Line

GL Greek Line

GMC Gulf Maritime Company

GO Gulf Oil

GPRL Gulf Puerto Rico Lines

GRACE Grace Line (Prudential-Grace Lines)

Gran Flota Blanca (Spanish—Great White Fleet)—United Fruit Company (fleet of white steamships)—United Brands

GS Galleon Shipping

GSA Gulf and South American Steamship Company

GULF Gulf Oil Corporation

GYSCo Great Yarmouth Shipping Company

H Hansa Line; Heering Line; Horn Line; etc.

HAL *Holland Amerika Lijn (NASM—Nederlandsch–Amerikaasche Stoomvaart Maatschappij)*—NASM appears on house flag

HANSA Hansa Line

Hanseatic-Vassa Line VL

HAPAG Hamburg-Amerika Paket Aktiengesselschaft (German—Hamburg-America Packet Company)—Hamburg-America Line

Hapag-Lloyd Hamburg–Amerika—North German Lloyd Lines

HB C Hudson's Bay Company

HCL Hamburg-Chicago Line

HFL Hawaii Freight Lines

HH H Hogarth and Sons

HHA HH Andersen Line

HKCL Hong Kong Container Line

HKEL Hong Kong Export Lines

HKIL Hong Kong Islands Line

HKX Hong Kong Express

HL Home Lines

H-L Hapag-Lloyd

HLC Hapag-Lloyd Container (line)

HMM Hyundai Merchant Marine

HMS Her (His) Majesty's Ship (as in HMS *Dreadnought*)

hovercraft marine craft supported by an air cushion instead of a conventional hull

HSAL Hamburg South American Line

HSDG Hamburg–Sudamerikanische Dampfs Gesell (Columbus Line)

HT Hudson Tankers

Hurtigruta (Norwegian—quick way)—coastal steamship lines linking ports from Bergen to Kirkenes

H & W Holm and Wonsild

HWAL Holland West-Afrika Line

I Incres Line; Interocean Steamship Lines; Isthmian Lines (U.S. Steel); Ivaran Lines; etc.

ICI Imperial Chemical Industries

ICSN Indo-China Steam Navigation Company

IFI Inter-Freight International

INSCO Intercontinental Shipping Corporation

Inter-Freight International IFI

IO Ltd Imperial Oil Ltd

IOM SPC Isle of Man Steam Packet Company

IOT Iron Ore Transport

IPL Ital Pacific Line

ISOS International Ship Operating Services

Italia Italian Line

ITI Inagua Transports Incorporated

J Japan Line; John I Jacobs and Company; Johnson Line; etc.

Jadroplov Jadramska Nobodna Plovida (Yugoslav—Great Lakes Line)

JBPS Jamaica Banana Producer's Steamship Co

JCL Japan Cruise Line

JL J Lauritzen; Jebsen Line

K Kavolines; Kawasaki Kisen Kaisha; Kerr Lines; Keystone Shipping (Chas Kurz); Kingsport Shipping; Kirkconnel; Klaveness Line; Knutsen Line; etc.

KG Koctug Line

KK Karlander Kangaroo Line

KKL Karlander Kangaroo Line

K Line Kawasaki Kisen Kaisha

KMTC Korea Marine Transport Company

KNC Kingcome Navigation Company

KNSM Koninklijke Nederlandsche Stoomboot Maatschappij (Dutch—Royal Netherlands Steamship Company)

Koctug Line KL

KS Korea Shipping

KSC Korean Shipping Corporation

KSN Karachi Steam Navigation Line

L Lauritzen Line; Luckenbach Line; Lykes Line; etc.

LASH Lighter Aboard Ship Handling

LB Lloyd Brasileiro

L + H Lamport and Holt Line

LL Lauro Line; Link Line

Lloyd's Lloyd's Register of Shipping (LRS)

LPR Lauritzen Peninsula Reefer (line)

LRS Lloyd's Register of Shipping

LT Loyd Triestino (Italian—Trieste Line)

M Maersk Line; Marine Transport Lines; Matson Line; Meyer Line; Montship Lines; Moore-McCormack Lines; Munson Line; etc.

Maersk Maersk Line

MAMENIC Marina Mercante Nicaraguense (Spanish—Nicaraguan Merchant Marine)

MANZ Montreal-Australia-New Zealand (Line)

Maritime Fruit Carriers MFC

MCP Maritime Company of the Philippines

MFC Maritime Fruit Carriers

MILl Micronesia Interocean Line Incorporated

Milwaukee Clipper Wisconsin and Michigan Steamship Company

MISC Malaysian International Shipping Corporation

Mitsui Mitsui OSK Lines

ML Manchester Liners

MLL Manchester Lines Ltd

M. M. Messageries Maritimes (French—Maritime Mail, Parcel, and Passenger Service)

MOLU Mitsui-Osaka Container Line

MOPAS Mitsui O.S.K. Passenger Line

M/S Motorship

MSC Mediterranean Shipping Company

MS Co Melbourne Steamship Company

MSTS Military Sea Transportation Service

MTL Marine Transport Lines

MV Motor Vessel

N Naess Shipping Company; Niarchos Tankers; Nigerian National Line; etc.

NA & G North Atlantic and Gulf Steamship Company

NAL Nigeria America Line

N-A-L Norwegian America Line

NASM (*see* HAL)

NAWAL North American-West African Line

NB Navibel (Belgian Maritime Navigation Company)

NB & C Norfolk, Baltimore and Carolina Line

NCL Norwegian Caribbean Line; Norwegian Cruise Lines

NCP Naviera Chilena del Pacifico (Spanish—Chilean Shipping of the Pacific)

Nedlloyd Nedlloyd and Hoegh Lines

NEE New England Express

NEPU Neptune Orient Container Line

New England Express NEE

NMB Navigation Maritime Bulgare (Bulgarian Maritime Navigation)

NNC Northern Navigation Company

NOL Neptune Orient Line; Norse Oriental Lines

Norasia Northeast Asia

NPCL North Pacific Coast Line

NPL Nauru Pacific Line

NPR Navieras de Puerto Rico

NTGB North Thames Gas Board

NYK Line Nippon Yusen Kaisha

NZL New Zealand Line

NZSCo New Zealand Shipping Company

O Ocean Carriers; Olsen Line; MJ Osorio; etc.

OCL Overseas Container Line; Overseas Containers Limited

Official Steamship Guide International OSGI

OG O Gross and Sons Ltd.

OK Oijekonsumenternas

ØK Østasiatiske Kompagni (Danish—East Asiatic Company)—EAL

Olympic Olympic Steamship Company

OO Orient Overseas Line

OOCL Orient Overseas Container Line

OOL Odessa Ocean Line; Orient Overseas Line

OS Ocean Steamship

OSGI Official Steamship Guide International

OSK Osaka Syosen Kaisha (Osaka Mercantile Steamship Company)—Mitsui Lines

OW Olof Wallenius Line

P Panama Line (Panama Canal Company); Pocahontas Steamships; Prudential-Grace Lines; Pure Oil; etc.

P-A Pan-Atlantic Steamship Corporation

PACE Pacific America Container Express

Pacific America Container Express PACE

Pacific Australia Direct Line (*see* ABRT)

PAD Pacific Australia Direct (Line)

PAL Pan Asia Line

Petrobras Petroleo Brasileiro (Portuguese—Brazilian Petroleum)

PFEL Pacific Far East Line

PFL Pacific Forum Line; Pacific Freight Line

P-G Prudential-Grace Lines

PIL Pacific International Line

PITL Pacific Islands Transport Line (Thor Dahls Hvalfangerselskap)

PL Polynesia Line; Port Line; Poseidon Lines; Prince Line

PLA Port of London Authority

PLL Prince Line Limited

PLO Polskie Linie Oceaniczne (Polish—Polish Line)

PM Petroleos Mexicanos (Spanish—Mexican Petroleum)

PNGL Papua New Guinea Line

PNL Philippine National Lines

P & O Peninsular and Occidental Steamship Company; Peninsular and Oriental Line

POE Pacific Orient Express Line

PPL Philippine President Lines

Princess Line Gothenburg-Frederikshavn Line

PSC Point Shipping Company

PSFL Puget Sound Freight Lines

PSNC Pacific Steam Navigation Company

PT Pope and Talbot

PURE Pure Oil Company

PV Pacific Venture

Q Qatar Petroleum; Quaker Line; Queensland; Quintessence Navigation; etc.

Q & O Quebec and Ontario Transportation

R Rasmussen; Richfield Oil; Ringdal; Robert; etc.

RCI Royal Caribbean International

RIL Royal Interocean Lines *[Koninklijke-Java-China-Paketvaart Lijnen—(Dutch— Royal Java-China-Packet Line)]*

RL Regent's Line (Grand Union Shipping)

RLR Royal Rotterdam Lloyd

RML Royal Mail Lines

Royal Netherlands Steamship Line *(see* KNSM)

RSSC Radisson Seven Seas Cruises

RVL Royal Viking Line

S Saguenay Terminals Ltd; Salen; Seatrain Lines; Sinclair Refining; Socony Mobil Oil; Standard Oil of California; States Marine Lines; States Line (seahorse-shaped red-letter S); Sun Oil; Svea Line; etc.

SA & CL South Atlantic and Caribbean Line

Safmarine South African Marine Corporation

SAL Svenska Anzerika Linien (Swedish—America Line)

Santa ships Prudential-Grace Line vessels

SC Submarine Cables Ltd

S & C Star and Crescent

SCC Shipping and Coal Company

SCI Sea Containers Incorporated; Shipping Corporation of India

Scindia Scindia Steam Navigation

SEGB South Eastern Gas Board

Shell Shell Tankers

Shipping Corporation of India SCI

SL Southern Lines

S-L Sea-Land (Line)

SLS Sea-Land Service

SML States Marine Lines

SN Sincere Navigation

SOPONATA Sociedade Portuguesa de Navios Tanques Limitada (Portuguese—Portuguese Tankships Limited)

Sovtorgflot Soviet Merchant Marine Fleet

Spanish Line *Compañia Transatlantica España*

SPL Scan Pacific Line

SS Steamship (as in SS Santa Clara)

SSS Sea Speed Service (container)

STANVAC Standard-Vacuum Oil Company

STL Seatrain Lines

SUNOCO Sun Oil Company

T Tankers Limited; Texaco (The Texas Company); Thai Mercantile Marine; Thompson Shipping; Thoren Line; Tirrenia; Transatlantic Line; etc.

TCL Transatlantic Carriers Limited

TCR Texas City Refining

Texaco The Texas Company

TFL Trans Freight Line

TH Thorvald Hansen

Thomson Thomson Cruises

Thor Dahls Havalfangerselskap Pacific Islands Transport Line

TMM Transportación Maritima Méxicana

TOTE Totem Ocean Trailer Express

Transamerica Trailer Transport TTT

TS Tasmanian Steamers

TSK Tokyo Senpaku Kaisha

TTT Transamerica Trailer Transport

U Union Oil; United Oriental Steamship Company; Universe Tankships; etc.

UA United Africa Company, Ltd

UBC United Baltic Corporation

UBL Union Barge Line

UCMS Union-Castle Mail Steamship

UFC United Fruit Company

UIL Ulster Imperial Line

U.O. Co. Union Oil Company of California

UPL United Philippine Lines

USC Union Steamship Company

USL United States Lines

USMSTS U.S. Military Sea Transport Service

USS United States Ship (as in USS Constitution)

USSCo Ulster Steam Ship Company; Union Steam Ship Company

UT United Transports

UYL United Yugoslav Lines

V Vaccaro Line (Standard Fruit); Valentine Chemical Carriers; Vinke Tankers; Von Sydow; Vulcan Shipping; etc.

VA Compañía de Navegación Vasco-Asturiana (Spanish— Basque-Asturian Navigation Company)

VC Victory Carriers

VL Vaasa Line (Hanseatic– Vassa Line)

VLC Valley Line Company

VNGC Van Niervelt, Goudriaan and Company (Rotterdam-South American Line)

VW Volkswagen (auto-carrier ships)

W Waterman Steamship Lines; West Line; Westriver Ore Transports; Weyerhaeuser Line; etc.

W & A Wiel and Amundsen

Wallenius Line OW (Olof Wallenius)

WHMV & NSSA Woods Hole, Martha's Vineyard and Nantucket Steamship Authority

WIL West India Lines

WIT West India Tankers

WL Westfal-Larsen Line

W & L Westcott and Laurance Line (Ellerman's)

WL & Co Westfal–Larsen and Company

W & M SS Co Wisconsin and Michigan Steamship Company (The Clipper Line)

WSFS Washington State Ferry System

WTC Western Transportation Company

X (funnel marking—Chandris America Lines; Southern Cross Steamship Line); Xenophon Navigation Company; etc.

Y Yamashita-Shinnihon Kisen Line; Ybarra Lines; Yukiteru Kaiun; Yung Yang Shipping; etc.

YML Yang Ming Line

YPF Yacimientos Petroliferos Fiscales (Spanish—Fiscal Petroleum Deposits)—Argentine tanker fleet

Y-S Line Yamashita-Shinnihon Line

Z Zacharissen; *Zante Naveg-ación;* Zillah Shipping; Zim Israel Navigation; Zurga Shipping Company; etc.

Zapata Zapata Bulk Transport
Zim Zim Israel Line
ZPL Zim Passenger Line

ZSC Zeeland Shipping Company; Zeeland Steamship Company

Superlatives

Afghanistan's Highest Peak Tirich Mir—25,230 feet, 7,690 meters—in the Hindu Kush on the Pakistan border

Afghanistan's Largest City Kabul

Africa's Easternmost City Hafun, Somalia

Africa's Easternmost Point Cape Guardafui (Ras Asir), Somalia

Africa's Highest Peak Kilimanjaro—19,340 feet, 5,861 meters—in Tanzania where it is called Kibo

Africa's Largest City Cairo

Africa's Largest Drainage System Congo

Africa's Largest Island Madagascar

Africa's Longest River Nile

Africa's Northernmost City Bizerta, Tunisia

Africa's Northernmost Point Ras el Abiadh (near Bizerta, Tunisia)

Africa's Southernmost City Cape Town, South Africa

Africa's Southernmost Point Cape Agulhas, South Africa

Africa's Westernmost City Dakar, Senegal

Africa's Westernmost Point Cape Almadies, Senegal

Alabama's Deepest Cavern 12-story-deep Cathedral Caverns containing a stalagmite 60 feet (18 meters) high

Alabama's Highest Point Cheaha Mountain—2,407 feet, 738 meters

Alabama's Largest City Birmingham

Alabama's Longest River Tombigbee

Alabama's Principal Port Mobile

Alaska's Highest Point Mount McKinley—20,320 feet, 6,187 meters

Alaska's Largest City Anchorage

Alaska's Longest River Yukon

Alaska's Principal Port Valdez Harbor

Alaska's Richest Agricultural Area the Matanuska Valley

Albania's Highest Peak Korab—9,066 feet, 2,747 meters

Albania's Largest City Tirana

Albania's Principal Port Durazzo

Alberta's Highest Point Mount Columbia—12,294 feet, 3,747 meters

Alberta's Largest City Edmonton

Algeria's Highest Peak Lella Khedidja—7,572 feet, 2,295 meters—in the Tell Atlas mountains

Algeria's Largest City Algiers

Algeria's Principal Port Algiers

American Samoa's Highest Point Matafao on Tutuila Island—2,142 feet, 649 meters—and Lata on Tau Island in the Manua Islands—3,056 feet, 926 meters

American Samoa's Largest City Pago Pago

American Samoa's Principal Port Pago Pago

American Virgin Islands' Highest Point Crown Mountain—1,556 feet, 472 meters—on Saint Thomas

American Virgin Islands' Largest Center Charlotte Amalie on Saint Thomas

American Virgin Islands' Principal Ports Charlotte Amalie on Saint Thomas, Cruz Bay on Saint John, Frederiksted on Saint Croix

America's Busiest Airport Chicago's O'Hare International

America's Easternmost City Eastport, Maine

America's Easternmost Point West Quoddy Head, Maine

America's Finest Opera George Gershwin's *Porgy and Bess*

America's First National Park Yellowstone

America's First Saint Elizabeth Ann Seaton

America's Foremost Modern Composer Aaron Copland

America's Highest Peak Mount McKinley—20,320 feet, 6,158 meters—in Alaska

America's Largest City New York City

America's Largest State Alaska

America's Longest River Mississippi-Missouri

America's Most Beloved Woman The Statue of Liberty

America's Most Used and Abused Drug alcohol

America's Northernmost City Barrow, Alaska

America's Northernmost Point Point Barrow, Alaska

America's Oldest Animal the horseshoe crab (*Limulus polyphenius*)

America's Oldest Tibetan Community in Seattle, Washington (dating to 1960)

America's Principal Port New Orleans

America's Smallest State Rhode Island

America's Southernmost City Hilo on the island of Hawaii

America's Southernmost Point Ka Lae (South Cape) on the island of Hawaii

America's Westernmost City Agaña, Guam

America's Westernmost Point Cape Wrangell on Attu Island, Alaska

Andorra's Highest Peak Puig de la Coma Pedrosa—9,665 feet, 2,929 meters—between France and Spain

Andorra's Largest Place Andorra la Vella

Angola's Largest City Luanda

Angola's Principal Port Luanda

Anguilla's Largest Town The Valley

Animal Longest Domesticated dog

Antarctica's Highest Peak
Vinson Massif—16,860 feet, 5,140 meters

Antigua and Barbuda's Principal Port St John's

Antigua's Largest Town St John's

Arctic's Largest Carnivore polar bear

Argentina's Easternmost City Posadas (on the Paraguay border)

Argentina's Highest Peak Aconcagua—22,835 feet, 6,920 meters

Argentina's Largest City Buenos Aires

Argentina's Largest Port Buenos Aires

Argentina's Largest Province Santa Cruz

Argentina's Northernmost City San Salvador (close to the Chilean border)

Argentina's Principal Port Buenos Aires

Argentina's Smallest Province Tucumán

Argentina's Southernmost City Ushuaia (on Beagle Channel)

Argentina's Westernmost City Mendoza (near the Chilean border)

Arizona's Grandest Canyon Grand Canyon of the Colorado River

Arizona's Highest Point Humphrey's Peak—12,655 feet, 3,835 feet

Arizona's Largest City Phoenix

Arizona's Longest River Colorado

Arkansas' Highest Point Magazine Mountain—2,753 feet, 839 meters

Arkansas' Largest City Little Rock

Arkansas's Largest Spring Blue Spring produces 144 million liters (38 million gallons) of water every day

Arkansas' Longest River Arkansas

Arkansas' Principal Port Little Rock

Asia's Highest Peak Everest—29,028 feet, 8,796 meters—(*see* China's Highest Peak)

Asia's Largest City Tokyo

Asia's Largest Drainage System Ob-Irtysh

Asia's Largest Island Borneo

Asia's Largest Nation China in terms of population

Asia's Longest River Yangtze

Australasia's Largest Island New Guinea

Australia's Easternmost City Brisbane, Queensland.

Australia's Easternmost Point Cape Byron, New South Wales

Australia's Highest Peak Mount Kosciusko—7,305 feet, 2,214 meters—in the Australian Alps

Australia's Largest City Sydney

Australia's Largest Port Sydney

Australia's Largest State Western Australia

Australia's Longest River Murray-Darling

Australia's Northernmost City Darwin, Northern Territory

Australia's Northernmost Point Cape York, Queensland

Australia's Principal Port Sydney

Australia's Southernmost City Hobart, Tasmania

Australia's Southernmost Point South East Cape, Tasmania

Australia's Westernmost City Camarvon, Western Australia

Australia's Westernmost Point Cape Inscription, Western Australia

Australia's Highest Peak Gross-glöckner—12,460 feet, 3,776 meters—on the border with Italy

Austria's Longest River Danube

Azores' Largest Town Ponta Delgada

Bahamas' Largest city Nassau

Bahamas' Principal Port Nassau

Bahrain's Largest City Manama

Bahrain's Principal Port Sitra

Balearic Islands' Largest City Palma de Majorca

Bangladesh's Largest City Dhaka

Bangladesh's Principal Port Chittagong

Barbados' Largest City Bridgetown

Barbados' Principal Port Bridgetown

Baseball's Most Famous Poem *Casey at the Bat* by Ernest Thayer

Belau's Largest Town Koror

Belgium's Largest City Brussels

Belgium's Largest Port Antwerp

Belgium's Longest River Scheldt

Belgium's Principal Port Antwerp

Belize's Highest Peak Victoria—3,681 feet, 1,115 meters—in the Cockscomb Mountains

Belize's Largest City Belize

Belize's Principal Port Belize City

Benin's Largest City Cotonou

Benin's Principal Port Porto Novo

Bermuda's Largest Town Hamilton

Best Sellers in Nazi Germany the *Bible* and *Mein Kampf* according to Benjamin B Ferencz

Best Title given by Queen Isabella the Catholic appointing Christopher Columbus Admiral of the Ocean Sea, Viceroy and Governor of the Lands You Discover (*Almirante del mar Océano, Virrey y Gobernador de las Tierras que Descubriese*)

Bhutan's Highest Peak Kula Kangri—28,780 feet, 8,721 meters—in the West Assam Himalayas

Bhutan's Largest City Thimphu

Biggest Domestic Dog Saint Bernard, standing up to 25 inches (65 centimeters) and weighing up to 170 pounds (77 kilograms)

Biggest Murder Trial in History Nuremberg Trial of Nazi war criminals

Bolivia's Highest Mountain Ancohuma—21,490 feet, 6,512 meters—in the Andes

Bolivia's Largest City La Paz

Bolivia's Largest Department Santa Cruz

Bolivia's Smallest Department Tarija

Botswana's Largest City Gaborone

Brazil's Easternmost City João Pessõa or Recife

Brazil's Highest Mountain Pico de Bandeira—9,462 feet, 2,867 meters

Brazil's Largest City Sáo Paulo

Brazil's Largest Port Rio de Janeiro

Brazil's Largest State Amazonas

Brazil's Northernmost City Belém do Pará

Brazil's Principal Port Rio de Janeiro

Brazil's Smallest State Sergipe

Brazil's Southernmost City Rio Grande do Sul (close to the Uruguay border)

Brazil's Westernmost City Rio Branco (close to the Bolivian border)

Britain's Largest City London

British Columbia's Highest Point Mount Fairweather—15,300 feet, 4,663 meters—on the Alaskan border

British Columbia's Largest City Vancouver

British Columbia's Longest River Fraser

British Columbia's Principal Port Vancouver

British superlatives (*see* United Kingdom, Largest British, *and* Smallest British *entries*)

British Virgin Islands' Largest Town Road Town

Brunei's Largest City Seri Begawan

Bulgaria's Highest Peak Musala—9,596 feet, 2,908 meters

Bulgaria's Largest City Sofia

Bulgaria's Largest Seaport Varna (formerly called Stalin)

Bulgaria's Longest River Danube

Bulgaria's Principal Port Varda

Bullfight Capital of the World Madrid

Burundi's Highest Peak Bugungu—5,025 feet, 1,523 meters—near the Tanzania border

Burundi's Largest City Bujumbura

California's Highest Point Mount Whitney—14,494 feet, 4,420 meters

California's Largest City Los Angeles

California's Longest River Sacramento

California's Most Spectacular Park Yosemite

California's Principal Port Long Beach in Los Angeles Harbor containing San Pedro, Terminal Island, and Wilmington

Cambodia's Highest Peak Tadet—3,667 feet, 1111 meters—south of Pailin

Cambodia's Largest City Phnom Penh

Cambodia's Principal Port Phnom Penh

Cameroon's Largest City Douala

Cameroon's Principal Port Douala

Canada's Biggest Port on the Atlantic Halifax, Nova Scotia

Canada's Biggest Port on the Great Lakes Sault Sainte Marie, Ontario on a canal connecting Lake Huron and Lake Superior

Canada's Biggest Port on Lake Ontario Toronto, Ontario

Canada's Biggest Port on the Pacific Vancouver, British Columbia

Canada's Biggest Port on the St Lawrence Montreal, Québec

Canada's Easternmost City St John's, Newfoundland

Canada's Easternmost Point Cape Spear, Newfoundland

Canada's Highest Peak Mount Logan—19,850 feet, 6,015 meters—in the southwest Yukon

Canada's Highest Town Lake Louise, Alberta (1,540 meters or 5,051 feet)

Canada's Largest City Montreal

Canada's Largest Province Québec

Canada's Longest River Maekenzie-Peace

Canada's Northernmost Deepwater Port Churchill, Manitoba

Canada's Northernmost Point Cape Columbia, Ellesmere Land

Canada's Northernmost Town Inuvik, Northwest Territory

Canada's Oldest City Québec City

Canada's Oldest National Park Banff National Park, including lovely Lake Louise

Canada's Principal Port Montreal

Canada's Smallest Province Prince Edward Island

Canada's Southernmost City Kingsville, Ontario

Canada's Southernmost Point Point Pelee, Ontario on Lake Erie

Canada's Tallest Mountain Mount Robson—12,972 feet or 3,954 meters—in the Canadian Rockies

Canada's Westernmost City Dawson, Yukon

Canada's Westernmost Point in Yukon Territory just east of Alaska's Demarcation Point

Canary Islands' Largest City Las Palmas de Gran Canaria

Cape Verde's Largest City Praia

Cape Verde's Principal Port Praia

Cayman Islands' Largest Place Georgetown

Central African Republic's Largest City Bangui

Central America's Highest Peak Tajumulco volcano—13,816 feet, 4,605 meters—in Guatemala

Central America's Largest City Guatemala City

Central America's Largest Island Cuba

Chad's Highest Peak (*see* Libya's Highest Peak)

Chad's Largest City N'Djaména

Channel Islands' Largest City Saint Helier on Jersey

Chile's Easternmost City
Santiago

Chile's Highest Peak Aconcagua—22,835 feet, 6,920 meters—shared with Argentina

Chile's Largest City Santiago

Chile's Largest Port Valparaiso

Chile's Largest Region Antofagasta

Chile's Longest River Maipo

Chile's Northernmost City Arica (near Tacna, Peru)

Chile's Principal Port Valparaiso

Chile's Southernmost City Punta Arenas (on the Strait of Magellan)

Chile's Westernmost City Valdivia

China's First Patented Dish baked pig's head with 30 herbs and spices

China's Highest Peak Everest—29,028 feet, 8,796 meters—called Quomolangma by the Chinese and Sagarmatha by the Nepalese, on the China–Nepal border

China's Largest City Shanghai

China's Largest Port Shanghai

China's Longest River Yangtze

China's Principal Port Shanghai

Cleanest City in the East Singapore

Cleanest City in the West Copenhagen; Oslo; Stockholm

Cleanest Port in the Orient Singapore

Cleanest and Most Decorative Subway System Moscow's

Coldest Spot Vostok, Antarctica

Coldest and Windiest Continent Antarctica

Colombia's Biggest Port on the Caribbean Barranquilla

Colombia's Biggest Port on the Pacific Buenaventura

Colombia's Easternmost City Cúlcuta

Colombia's Highest Peak Cristóbal Colón—18,947 feet, 5,742 meters

Colombia's Largest City Bogota

Colombia's Largest Department Caquetá

Colombia's Largest Port Barranquilla

Colombia's Longest River Magdalena

Colombia's Northernmost City Riohacha

Colombia's Smallest Department Quindio

Colombia's Southernmost City Cali

Colombia's Westernmost City Buenaventura

Colorado's Highest Point Mount Elbert—14,433 feet, 4,398 meters

Colorado's Largest City Denver

Colorado's Largest Flat-topped Mountain Grand Mesa and its National Forest atop the world's largest flat-topped mountain

Colorado's Longest River Colorado

Commonest British Bird blackbird

Comoros' Largest City Moroni on Grande Comore

Comoros' Principal Port Moroni

Congo's Largest City Kinshasa

Connecticut's Biggest Beach Park Hammonasset fronting on Long Island Sound

Connecticut's Highest Point Mount Frissell—2,380 feet, 725 meters

Connecticut's Largest City Bridgeport

Connecticut's Longest River Connecticut

Connecticut's Principal Port New Haven

Cook Islands' Largest Place Avarua on Rarotonga

Costa Rica's Atlantic Port Limón

Costa Rica's Highest Peak Chirripó—12,533 feet, 3798 meters—in the Talamanca. Cordillera

Costa Rica's Largest City San José

Costa Rica's Largest Province Puntarenas

Costa Rica's Pacific Port Puntarenas

Costa Rica's Smallest Province Heredia

Croatia's Principal Port Rijeka

Cuba's Easternmost Point Cabo Maisi

Cuba's Easternmost and Southernmost City Santiago

Cuba's Highest Peak Pico Turquino—6,560 feet, 1,988 meters—in the Sierra Maestra

Cuba's Largest City Havana

Cuba's Largest Province Camagüey

Cuba's Northernmost City and Point Havana

Cuba's Principal Port Havana

Cuba's Smallest Province Habana

Cuba's Southernmost Point Cabo Cruz

Cuba's Westernmost City Pinar del Rio

Cuba's Westernmost Point Cabo San Antonio

Cultural Capital of the World New York City

Curaçao's Largest Town Willemstad

Curaçao's Principal Port Willemstad

Cyprus' Highest Point Mount Troodos—6,404 feet, 1,941 meters—in the Olympic Mountains

Cyprus' Largest City Nicosia

Cyprus' Principal Port Nicosia

Czech Republic's Largest City Prague

Deepest Gorge Hell's Canyon of the Snake River in Idaho

Deepest Lake Baikal—5,315 feet, 1,620 meters—in southeastern Siberia

Deepest Part of the Arctic Ocean Eurasia Basin between Kornsomolets Island and the North Pole (2,980 fathoms or 17,980 feet or 5,450 meters in depth)

Deepest Part of the Atlantic Puerto Rico Trench north of Hispaniola and Puerto Rico (4,729 fathoms or 28,374 feet or 8,648 meters in depth)

Deepest Part of the Caribbean Cayman Trench between the Cayman Islands and Jamaica (3,833 fathoms or 23,000 feet or 7,010 meters deep)

Deepest Part of the Indian Ocean Diamantina Trench south of Western Australia (4,400 fathoms or 26,400 feet or 8,047 meters in depth)

Deepest Part of the Mediterranean off Cape Matapan, Greece (2,406 fathoms or 14,435 feet or 4,400 meters deep)

Deepest Part of the North Sea in the Skaggerak between Denmark and Norway (333 fathoms or 1,998 feet or 605 meters)

Deepest Part of the Ocean Mariana Trench in the Western Pacific east of Saipan (6,033 fathoms or 36,198 feet or 11,034 meters in depth)

Deepest Part of the Pacific (*see* Deepest Part of the Ocean)

Delaware's Highest Point Centerville—442 feet, 135 meters

Delaware's Largest City Wilmington

Delaware's Longest River Delaware

Delaware's Principal Port Wilmington

Denmark's Highest Peak Mollejoh—561 feet, 170 meters—in East Jutland

Denmark's Largest City Copenhagen

Denmark's Principal Port Copenhagen

District of Columbia's Highest Point Tenleytown—410 feet, 125 meters

Djibouti's Largest City Djibouti

Djibouti's Principal Port Djibouti

Dominica's Largest City Roseau

Dominica's Principal Port Roseau

Dominican Republic's Highest Peak Monte Tina—10,301 feet, 3,434 meters

Dominican Republic's Largest City Santo Domingo

Dominican Republic's Largest Province San Juan

Dominican Republic's Principal Port Santo Domingo

Dominican Republic's Smallest Province Salcedo

Driest Desert Atacama, Chile where at Calama rain has rarely been recorded

Dutch superlatives (*see* Netherlands *entries*)

Easter Islands' Largest Town Hanga Roa

Easternmost Capital of Africa Mogadishu, Somalia

Easternmost Capital of Asia Tokyo, Japan

Easternmost Capital of Australia Brisbane, Queensland

Easternmost Capital of Europe Moscow, Russia

Easternmost Capital of North America St John's, Newfoundland

Easternmost Capital of South America Brasilia, Brazil

Easternmost Hawaiian Island Hawaii

Easternmost Point of Canada Cape Spear, Newfoundland on Avalon Peninsula near St John's

Easternmost Point of the continental United States West Quoddy Head near Lubec, Maine

Easternmost Point of Mexico Isla de las Mujeres off the Caribbean coast of Quintana Roo

Easternmost Point of the Territorial United States East Point on Saint Croix in the Virgin Islands east of Puerto Rico

Ecuador's Easternmost City Quito

Ecuador's Highest Mountain Chimborazo—20,702 feet, 6,275 meters—in the Andes

Ecuador's Largest City Guayaquil

Ecuador's Largest Port Guayaquil

Ecuador's Largest Province Napo

Ecuador's Northernmost City Esmeraldas

Ecuador's Principal Port Guayaquil

Ecuador's Smallest Province Bolivar

Ecuador's Southernmost City Cuenca

Ecuador's Westernmost City Manta

Egypt's Highest Peak Gebel Katherina—8,651 feet, 2,622 meters—in the Sinai

Egypt's Largest City Cairo

Egypt's Principal Port Alexandria

El Salvador's Highest Peak Santa Ana—7,825 feet, 2,371 meters

El Salvador's Largest City San Salvador

El Salvador's Largest Department Usulután

El Salvador's Principal Port La Union

El Salvador's Smallest Department Cuscatlán

England's Extremitude Land's End—westernmost point in Cornwall and nearby Lizard Head the southernmost point

England's Highest Mountain Scafell Pike—3,210 feet, 973 meters—in Cumberland

England's Largest City London

England's Largest Lake Windermere

England's Longest Rivers Severn, Thames

Equatorial Guinea's Largest City Malabo

Equatorial Guinea's Principal Port Malabo

Ethiopia's Highest Peak Ras Dashan—15,158 feet, 4,593 meters

Ethiopia's Largest City Addis Ababa

Ethiopia's Principal Port Masewa

Eurasia's Easternmost Point Mys Dezhneva (East Cape), Siberia

Eurasia's Northernmost Point Rudolph Island off Franz Josef Land in the Arctic

Eurasia's Southernmost Point Rod Island in the Lesser Sundas of Indonesia

Eurasia's Westernmost Point Tearaght Island off Ireland's Dingle Peninsula

Europe's Easternmost City and Point Gomyatskiy, in Russia on the western slopes of the Ural Mountains

Europe's Highest Peak Mont Blanc—15,770 feet, 4,779 meters—between France and Italy

Europe's Largest Drainage Basin Rhine

Europe's Largest Island Great Britain

Europe's Largest Nation Russia

Europe's Longest River Volga

Europe's Most Sparsely Populated Country Iceland

Europe's Northernmost City Hammerfest, Norway

Europe's Northernmost Point Nordkyn, Norway (north of North Cape)

Europe's Southernmost City Nicosia, Cyprus

Europe's Southernmost Point south coast of Cyprus

Europe's Westernmost Place Tearaght Island off Ireland's Dingle Peninsula

Faeroe Islands' Largest Town Tórshavn

Falkland Islands' Largest Town Stanley

Fastest Clipper Ship *Lightning*—designed and built by Donald McKay in his East Boston shipyards—logged 436 nautical miles in 24 hours; McKay launched 16 of his fastest and finest clippers between 1850 and 1853

Fastest Railroad *TGV* Paris–Lyon express operating in one section at 320 miles per hour

Fastest Transatlantic Clipper Ship Donald McKay's *James Baines* sailed from Boston to Liverpool in 12 days and 6 hours (more than a 100 years ago)

Fifty Most Expensive Cities in the Late 1900s Tokyo, Kobe, Osaka, Brazzaville, Libreville, Dakar, Douala, Tehran, Abidjan, Geneva, Zurich, Vienna, Copenhagen, Oslo, Helsinki, Munich, Hamburg, Berlin, Paris, Duesseldorf, Frankfurt, Milan, Lyons, Rome, Dublin, Algiers, Amsterdam, Brussels, Taipei, Amman, Stockholm, Luxembourg, Cairo, New York, Tel Aviv, Los Angeles, Chicago, Washington,

D.C., San Francisco, Boston, London, Lima, Miami, Houston, Abu Dhabi, Tripoli, Barcelona, Madrid, Port Moresby, Singapore

Fiji's Largest City Suva

Fiji's Principal Port Suva

Finest Harbor in the World Sydney Cove, according to Captain Arthur Phillip when he founded Australia's first settlement in 1788

Finland's Highest Peak Haltia—4,343 feet, 1,316 meters—on the Norwegian border

Finland's Largest City Helsinki

Finland's Longest River Kemi

Finland's Principal Port Helsinki

First Afghan Amir Ahmad Shah

First Albanian King William of Wied

First American Cookbook in 1796 by Amelia Simmons

First American Poet Laureate Robert Penn Warren

First American President George Washington

First American Romantic Composer William Mason

First Animal Cloned from Another Adult Mammal Dolly, a sheep (named for singer Dolly Parton)

First Apartment House in the United States Pontalba Apartments facing Jackson Square in the French Quarter of New Orleans

First Argentinian President Justo José Urquiza

First Australian Prime Minister Edmund Barton

First of the Austrian Romantic Composers Ignaz Pleyel

First Austrian Ruler Otakar

First Austro-Hungarian Empress Maria Theresa

First Bahaman Governor-General Milo Broughton Butler

First Bangladesh President Abu Sayeed Chowdhury

First Barbadan Governor General John Montague Stow

First Belgian Romantic Composer Henri Vieuxtemps

First Belizian Prime Minister George C Price

First Black United States Supreme Court Justice Thurgood Marshall

First Bolivian President Antonio José de Sucre

First Brazilian Emperor Dom Pedro I

First Brazilian President Manuel Deodoro da Fonseca

First Brazilian Romantic Composer Carlos Gomes

First British Prime Minister Sidney Godolphin

First British Sovereign Egbert

First Bulgarian Ruler Prince Alexander of Battenberg

First Burmese President Sao Shwe Thaike

First Canadian Prime Minister John A Macdonald

First Chilean President Bernardo O'Higgins

First Chinese Communist Head of State Mao Zedong

First Chinese Nationalist President Chiang Kai-shek

First Chinese Provisional President Sun Yat-sen

First Colombian President Eustorgio Salgar

First Costa Rican President Rafael Iglesias

First Cuban President Tomás Estrada Palma

First Czechoslovakian President Tomás Garrigue Masaryk

First Danish King Gorm the Old

First Dominican Republic President Pedro Santana

First Dutch Monarch William I

First Ecuadorian President Juan José Flores

First Egyptian President Muhammad Naguib

First English Poet Laureate John Dryden

First English Ruler Egbert

First Ethiopian Emperor in Modern Times Theodore

First Finnish President Kaarlo Juho Stahlberg

First First Lady Martha Dandridge Custis Washington

First in Flight Orville and Wilbur Wright, Kitty Hawk, North Carolina

First French Fifth Republic President Charles de Gaulle

First German Ruler Charlemagne (Karl the Great)

First Greek King in Modern Times Otto I of Bavaria

First Grenadan Prime Minister Eric Matthew Gairy

First Guatemalan President Rafael Carrera

First Guyanan Prime Minister Lindon Forbes Sampson Burnham

First Haitian Emperor Jean Jacques Dessalines

First Honduran President José M Medina

First Hungarian Regent Admiral Nicholas Horthy de Nagybánya

First Indian Prime Minister Jawaharlal Nehru

First Indonesian President Achmed Sukarno

First Iranian Shah Agha Mohammed Khan

First Irish President William T Cosgrave

First Irish Prime Minister Eamon de Valera

First Ironworks in the New World Saugus Ironworks, Massachusetts

First Israeli President Chaim Wiezmann

First Israeli Prime Minister David Ben-Gurion

First Italian King of Sardinia and Italy Victor Amadeus II

First Jamaican Prime Minister Alexander Bustamante

First Japanese Prime Minister Hirobumi Ito

First Jordanian King Abdullah

First Lebanese President Bechara el-Khoury

First Liberian President Joseph J Roberts

First Living Person to Have an Element Named After Him Glenn Seaborg

First Mexican Emperor Augustín de Iturbide

First National Park in the United States Yosemite in California

First New Zealand Prime Minister Henry Sewell

First Nicaraguan President Frutos-Chamorro

First North Korean Premier Kim Il Sung

First Norwegian Ruler Haakon the Good

First Pakistan Governor-General Doctor Mohammed Ali Jinnah

First Panamanian President Manuel Amador Guerrero

First Paraguayan President Cirolo Rivarola

First Patent Issued in America to Sam Winslow, in 1641, for a salt-making method

First Person to Survive Going Over Niagara Falls in a Barrel Anna Edson Taylor, a 43-year-old widow, in 1901

First Peruvian President José de San Martin

First Philippine President Manuel Roxas

First Polish President Josef Pilsudski

First Portuguese Ruler Alfonso I

First President of Tuskegee Institute Booker T Washington

First President of the U.S. George Washington

First Puerto Rican Elected Governor Luis Muñoz Marín

First Radio Novelist Jean Shepherd

First Roman Ruler Octavianus, Emperor Augustus

First Romanian Ruler Prince Alexander John I

First Ruler of England and Great Britain Egbert

First Russian Ruler Ivan the Great

First Salvadoran President Rafael Zaldivar

First Saudi Arabian Ruler Abdul-Aziz

First Scot to Rule Picts and Scots Kenneth I MacAlpin

First South African Prime Minister Louis Botha

First South Korean President Syngman Rhee

First Soviet Chairman Nikolai Lenin

First Spanish Republican Prime Minister Niceto Alcalá Zamora y Torres

First Spanish Romantic Composer Felipe Pedrell

First Suburb Levittown, Pennsylvania, created by William Levitt

First Swedish Regent Earl Birger

First Swedish Romantic Composer Ivar Hallström

First Trinidad and Tobago Prime Minister Eric Eustace Williams

First Turkish Sultan Osman

First Uruguayan President José Fructuoso Rivera

First U.S. Attorney General Edmund Randolph

First U.S. Chief Justice John Jay

First U.S. Postmaster General Samuel Osgood

First U.S. President George Washington

First U.S. Secretary of Agriculture Norman J Colman

First U.S. Secretary of Commerce William C Redfield

First U.S. Secretary of Commerce and Labor George B Cortelyou

First U.S. Secretary of Defense James V Forrestal

First U.S. Secretary of Education Shirley Hufstedler

First U.S. Secretary of Energy James R Schlesinger

First U.S. Secretary of Health, Education, and Human Services Oveta Culp Hobby

First U.S. Secretary of Health and Human Services Patricia Roberts Harris

First U.S. Secretary of Housing and Urban Development Robert C Weaver

First U.S. Secretary of the Interior Thomas Ewing

First U.S. Secretary of Labor William B Wilson

First U.S. Secretary of the Navy Benjamin Stoddert

First U.S. Secretary of State Thomas Jefferson

First U.S. Secretary of Transportation Alan S Boyd

First U.S. Secretary of the Treasury Alexander Hamilton

First U.S. Secretary of War James Knox

First U.S. Speaker of the House of Representatives Frederick A C Muhlenberg

First U.S. Vice President John Adams

First Venezuelan President José Antonio Páez

First West German Chancellor Konrad Adenauer

First Yugoslav President Josip Broz, better known as Tito

Florida's Highest Point west boundary of Walton County—345 feet, 105 meters

Florida's Largest City Jacksonville

Florida's Largest Subtropical Wilderness Everglades National Park

Florida's Longest River Saint Johns

Florida's Principal Port Tampa

Foremost Atheist Orator of America Robert G Ingersoll

Foremost Atheist Philosopher of England Bertrand Russell

Foremost Atheist Philosopher of Germany Friedrich Wilhelm Nietzsche

Fourth First Lady Dorothea "Dolley" Payne Todd Madison

France's Easternmost Point Lauterbourg (on the Rhine opposite Karlsruhe)

France's Easternmost Port Menton (at the Italian border on the Mediterranean)

France's Highest Peak Mont Blanc—15,781 feet, 4,782 meters—on the Italian border in the Alps

France's Largest City Paris

France's Largest Metropolitan Region Midi-Pyrénées

France's Longest River Loire

France's Northernmost Point Malo les Bains (on the North Sea at the Belgian border)

France's Northernmost Port Dunkerque (on the Strait of Dover)

France's Principal Port Marseille

France's Smallest Metropolitan Region Alsace (ashore); Corsica (offshore)

France's Southernmost Point Cebère (opposite Spain's Port Bou on the Mediterranean)

France's Southernmost Port Port Vendres (on the Mediterranean close to the Spanish frontier)

France's Westernmost Point Île d'Ouessant (off the Britany peninsula)

France's Westernmost Port Brest (on the Brittany peninsula)

French Guiana's Highest Peak Bienvenue—2,723 feet, 825 meters

French Guiana's Largest City Cayenne

French Guiana's Longest River Maroni

French Guiana's Principal Port Cayenne

French Polynesia's Largest City Papeete

French Polynesia's Principal Port Papeete

Gabon's Largest City Libreville

Galápagos Islands' Largest Town Baquerizo Moreno on Chatham; San Cris bal island

Gambia's Largest City Banjul

Gambia's Principal Port Banjul

Georgia's Highest Point Brasstown Bald—4,784 feet, 1,458 meters

Georgia's Largest City Atlanta

Georgia's Longest River Savannah

Georgia's Most Gorgeous Swamp the Okefenokee it shares with Florida

Georgia's Principal Port Savannah

Germany's Easternmost City Zittau (close to the Czech and Polish borders)

Germany's Highest Peak Zugspitze- 9,719 feet, 2,945 meters—in the Bavarian Alps

Germany's Largest City Berlin

Germany's Largest Port Hamburg

Germany's Longest River Elbe

Germany's Northernmost Point List on Sylt (in the North Frisian Islands)

Germany's Principal Ports Hamburg and Rostack

Germany's Southernmost Region Algäuer Alps (near Oberstdorf on the Austrian border)

Germany's Westernmost City Aachen (Aix-la-Chapelle)—near the Belgian and Netherlands border

Ghana's Largest City Accra

Ghana's Principal Port Accra

Great Britain's Highest Peak Ben Nevis—4,406 feet, 1,335 meters—in Scotland

Great Britain's Largest City London

Great Britain's Longest River Severn

Greatest Capes of the northern Pacific coast Cabo San Lazaro, Cabo San Lucas, Cape Scott, Cape Cook, Cape Beale, Cape Flattery, Cape Disappointment, Cape Mendocino

Greatest Island Australia

Greatest Jewish City New York, according to Harry Golden who wrote *Greatest Jewish City in the World*

Greatest Man Who Made a Dictionary Dr Samuel Johnson

Greatest Poet of the 20th Century Rainer Maria Rilke (1875–1926)

Greatest Predator Man (*Homo sapiens*)

Greatest Primate Gymnasts gibbons (who leap through the air with the greatest of ease, unaided by any trapeze)

Greatest Sculptured Cape of the Northern Atlantic Coast Cape Cod

Greatest Sculptured Capes of the Southern Atlantic Coast Cape Hatteras, Cape Lookout, Cape Fear, Cape Canaveral

Greatest Shortcut the Panama Canal

Greatest Show on Earth Barnum & Bailey's three-ring circus (later merged with Ringling Brothers)

Greatest Tropical Killers dysentery, hookworm, malaria, roundworm, sleeping sickness, tapeworm

Greece's Highest Peak
Mount Olympus—9,570 feet, 2,900 meters—legendary abode of the gods

Greece's Largest City Athens

Greece's Principal Port Piraeus

Greenland's Highest Peak
Gunnbjorn—12,139 feet, 3,678 meters

Greenland's Largest Town
Nuuk

Greenland's Principal Port
Nuuk

Grenada's Highest Point
Mount Saint Catherine—2,749 feet, 833 meters

Grenada's Largest Town St George's

Grenada's Principal Port St George's

Guadeloupe's Largest City
Point-á-Pitre

Guadeloupe's Principal Port
Basse-Terre

Guam's Best Port Apra on its west coast

Guam's Highest Point
Mount Lamlam—1,329 feet, 402 meters

Guam's Largest Town Agaña

Guam's Principal Port Apra

Guatemala's Easternmost Point Puerto Barrios on the Caribbean Sea

Guatemala's Highest Peak
Tajumulco—13,816 feet, 4,187 meters—volcano in south, eastern sector

Guatemala's Largest City
Guatemala City

Guatemala's Largest Department El Petén

Guatemala's Principal Caribbean Port Puerto Barrios

Guatemala's Principal Pacific Port San Jose

Guatemala's Westernmost Point Ocó on the Pacific Ocean near Ciudad Hidalgo

Guinea-Bissau's Largest City
Bissau

Guinea-Bissau's Principal Port Bissau

Guinea's Highest Peak Tamgui—4,970 feet, 1,506 meters

Guinea's Largest City Conakry

Guinea's Principal Port
Conakry

Guyana's Highest Peak
Roraima—9,219 feet, 2,794 meters—mountain shared with Brazil and Venezuela

Guyana's Largest City
Georgetown

Guyana's Longest River Essequibo

Guyana's Principal Port
Georgetown

Haiti's Highest Peak Massif de la Selle—8,793 feet, 2,665 meters

Haiti's Largest City Port-au Prince

Haiti's Largest Department Town l'Artibonite

Haiti's Principal Port Port au-Prince

Haiti's Smallest Department
Nord-Est

Hawaii's Highest Point
Mauna Kea—13,796 feet, 4,205 meters—an extinct volcano, on the island of Hawaii

Hawaii's Largest City Honolulu

Hawaii's Principal Port
Honolulu

Hawaii's Superlative Volcanoes Kilauea and Mauna Loa in the Hawaii Volcanoes National Park on Hawaii

Heaviest Even-Toed Ungulate
hippopotamus

Heaviest Living Dog Saint Bernard

Highest American Mountain
Mount McKinley

Highest Andean Nation Bolivia

Highest Border Crossing in the World Khumjerab Pass, on the Pakistan/China border at 15,514 feet, 4,701 meters

Highest Bridge Royal Gorge Bridge in Colorado

Highest Canadian Mountain
Mount Logan

Highest Capital City of Africa
Addis Ababa, Ethiopia

Highest Capital City of the Americas La Paz, Bolivia

Highest Capital City of Asia
Kathmandu, Nepal

Highest Capital City of Australia Canberra

Highest Capital City of Europe
Madrid, Spain

Highest Capital City of Latin America Mexico City

Highest Capital City of North America Mexico City

Highest Capital City of South America La Paz, Bolivia

Highest Mountain Ranges and Mountains (in order of height) Himalayas (Everest; Andes (Aconcagua); Rockies (McKinley); Kilimanjaro (Kibo); Caucasus (El'brus); Ellsworth, Antarctica (Vinson Massif); Australian Alps (Kosciusko)

Highest Mountain in the Western Hemisphere Aconcagua—22,835 feet, 6,920 meters—shared by Argentina and Chile

Highest Murder Rate South Africa 53.5 / 100,000 in 1995

Highest Navigable Lake Titicaca

Highest Peak in the Eastern Hemisphere Everest

Highest Peak in North Africa
Djebel Toubkal—13,665 feet, 4,141 meters—in the High Atlas of Morocco

Highest Peak in the Northern Hemisphere Everest

Highest Peak in the Southern Hemisphere Aconcagua

Highest Peak in Tajikistan
Pik Kommunizma (Russian—Peak of Communism)—24,590 feet, 7,452 meters—in Central Asia, once named for Stalin

Highest Peak in the Western Hemisphere Aconcagua

Highest Peak in the West Indies Monte Tina—10,301 feet, 3,122 meters—in the Dominican Republic

Highest Point in Africa Kilimanjaro in Tanzania—19,340 feet or 5,963 meters above sea level

Highest Point in Antarctica
Vinson Massif—16,860 feet or 5,140 meters above sea level

Highest Point in Asia (*see* Highest Point in the World)

Highest Point in Australia
Mount Kosciusko in New South Wales (7,328 feet or 2,229 meters above sea level)

Highest Point in the Caucuses
Mount Elb'rus in Russia—18,567 feet or 5,659 meters above sea level

Highest Point in North America Mount McKinley in Alaska—20,320 feet or 6,187 meters above sea level

Highest Point in South America Mount Aconcagua in Argentina—22,835 feet or 6,960 meters above sea level

Highest Point in the World Mount Everest between Nepal and Tibet in Asia—29,028 feet or 8,848 meters above sea level

Highest Tides Nova Scotia's Bay of Fundy

Highest Volcanic Island Mountain Mauna Loa (Hawaiian—Long Mountain)—on the island of Hawaii and often ranked as the world's largest active volcano

Highest Waterfalls (in order by height) Angel in Venezuela; Itatinga in Brazil; Cuquenan in Guyana-Venezuela; Ormeei in Norway

Honduras' Highest Peak (*see* Nicaragua's Highest Peak)

Honduras' Largest City Tegucigalpa

Honduras' Principal Caribbean Port Puerto Cortés

Honduras' Principal Pacific Port Amapala

Hong Kong's Largest City Victoria

Hottest and Rainiest Continent Africa

Hottest Spot Al'Aziziyah, Libya

Hungary's Highest Point Kekes—3,330 feet, 1,009 meters—in the Matra hills

Hungary's Largest City Budapest

Hungary's Longest River Danube

Iceland's First Settler Ingolfur Arnarson, a Viking, in 874

Iceland's Highest Peak Hvannadailshnjukur—6,952 feet, 2,107 meters

Iceland's Largest City Reykjavik

Iceland's Principal Port Reykjavik

Idaho's Deepest Gorge Hells Canyon within the Grand Canyon of the Snake River with an average depth of 550 feet (1,676 meters) and a maximum of 7,900 feet (2,408 meters) in depth

Idaho's Highest Point Borah Peak—12,662 feet, 3,869 meters

Idaho's Largest City Boise

Illinois' Highest Point Charles Mound—1,235 feet, 377 meters

Illinois' Largest City Chicago

Illinois's Largest Forest Shawnee National Forest extending from the Ohio to the Mississippi rivers in southern Illinois

Illinois' Principal Port Chicago

Indiana's Highest Point Franklin—1,257 feet, 381 meters in Wayne County

Indiana's Largest Cave Wyandotte Caves complete with five floor levels and an underground mountain

Indiana's Largest City Indianapolis

Indiana's Principal Port Lake Michigan Facility cast of Gary

India's Highest Peak Nanda Devi—25,645 feet, 7,771 meters—in the Himalayas

India's Largest City Calcutta

India's Longest River Ganges

India's Principal Port Calcutta

India's Principal East Coast Port Calcutta

India's Principal West Coast Port Bombay

Indo-China's Longest River Mekong

Indonesia's Highest Peak Kinabalu—13,455 feet, 4,077 meters—in North Bomeo

Indonesia's Largest City Jakarta

Indonesia's Principal Port Djakarta

Iowa's Highest Point Ocheyedan Mound—1,675 feet, 511 meters—deposited by a melting glacier that contained much sand and gravel

Iowa's Largest City Des Moines

Iran's Highest Peak Damavand—18,602 feet, 5,637 meters

Iran's Largest City Tehran

Iran's Principal Port Khorramshahr

Iraq's Highest Point Kuhe Haji Ebrahim—11,880 feet, 3,600 meters

Iraq's Largest City Baghdad

Iraq's Principal Port Basra

Ireland's Easternmost City Belfast or Dublin

Ireland's Easternmost Point Burr Head or Wicklow Head

Ireland's Highest Mountain Carrauntoohil—3,414 feet, 1,035 meters—(*see* Northern Ireland's Highest Peak)

Ireland's Largest City Dublin

Ireland's Largest Lake Lough Corrib

Ireland's Longest River Shannon

Ireland's Northernmost City Londonderry

Ireland's Northernmost Point Malin Head

Ireland's Principal Port Dublin

Ireland's Southernmost City Cork

Ireland's Southernmost Point Fastnet Rock

Ireland's Westernmost City Galway

Ireland's Westernmost Point Tearaght Island

Islands of the World (ranked by area) Greenland, New Guinea, Borneo, Madagascar, Baffin Land, Sumatra, Honshu, Great Britain, Victoria, Ellesmere, Sulawesi, New Zealand's South Island, Java, New Zealand's North Island, Cuba, Newfoundland, Luzon, Iceland, Mindinao, Ireland

Isle of Man's Largest Town Douglas

Israel's Highest Peak Hare Meron—3,963 feet, 1,201 meters—called Jebel Jarmaq by the Arabs

Israel's Largest City Jerusalem

Israel's Principal Port Haifa

Italy's Easternmost Seaport City Trieste (shared with Slovenia)

Italy's Easternmost Town Otranto

Italy's Highest Point Monte Bianco (Mont Blanc)—15,781 feet, 4,782 meters—on the French border in the Alps

Italy's Largest City Rome

Italy's Largest Port Genova (Genoa)

Italy's Largest Southern Port Napoli (Naples)

Italy's Longest River Po

Italy's Most Historic Northern Port Venezia (Venice)

Italy's Northernmost Big City Milano (Milan)

Italy's Northernmost Town Brennero (at the Brenner Pass in Upper Adige on the Austrian frontier)

Italy's Principal Port Genoa

Italy's Southernmost Big City Messina (on Sicily)

Italy's Southernmost Town Portopalo (on Sicily)

Italy's Westernmost Big City Torino (Turin)

Italy's Westernmost Town Bardoecchia (on the French frontier)

Ivory Coast Largest City Abidjan

Ivory Coast Principal Port Abidjan

Jamaica's Highest Peak Blue Mountain—7,520 feet, 2,279 meters

Jamaica's Largest City Kingston

Jamaica's Principal Port Kingston

Japan's Easternmost Point Nashappu (on Hokkaido just south of Sakhalin)

Japan's Easternmost Town Habomai (on Hokkaido)

Japan's Highest Peak Fuji—12,388 feet, 3,754 meters—on the island of Honshu, also called Fujisan or Fujiyama

Japan's Largest City Tokyo

Japan's Northernmost Cape and Town Soya (on Hokkaido)

Japan's Principal Port Yokohama

Japan's Southernmost Island and Point Hateruma Shima in the Ryukyu Islands

Japan's Westernmost Island Yonaguni Jima (in the Ryukyu Islands close to Taiwan)

Japan's Westernmost Town Sonai (on Yonaguni Jima)

Jordan's Highest Peak Jebalal Lawz—8,465 feet, 2,565 meters

Jordan's Largest City Amman

Jordan's Principal Port Aqaba

Kansas' Highest Point Mount Sunflower—4,039 feet, 1,231 meters

Kansas' Largest City Wichita

Kansas' Most Famous Landmark Pawnee Rock on the Santa Fe Trail where Indians and pioneers battled

Kansas' Principal Port Kansas City

Kentucky's Highest Point Black Mountain—4,145 feet, 1,262 meters

Kentucky's Largest City Louisville

Kentucky's Longest River Ohio

Kentucky's Most Scenic Park Mammoth Cave National Park

Kentucky's Principal Port Paducah

Kenya's Highest Peak Mount Kenya—17,040 feet, 5,164 meters—a volcanic cone

Kenya's Largest City Nairobi

Kenya's Principal Port Mombasa

Kiribati's Largest City Bairiki

Kiribati's Principal Port Tarawa

Korea's Highest Peak (*see* North Korea's Highest Peak)

Korea's Largest City Seoul

Kuwait's Largest City Kuwait

Kuwait's Principal Port Mina al-Ahmadi

Laos' Highest Peak Phoi Loi—7,425 feet, 2,250 meters

Laos' Largest City Vientiane

Largest Active Volcano Mauna Loa on the island of Hawaii

Largest Afghan City Kabul

Largest African City Cairo

Largest African Nation Sudan, in landmass

Largest African Primate gorilla of equatorial Africa

Largest American Bird California condor (nearly extinct in the wild)

Largest American City New York

Largest American City on the Canadian Border Detroit, Michigan

Largest American City on the Mexican Border San Diego, California (opposite Tijuana, Baja California)

Largest American East Coast City New York

Largest American Estuary Chesapeake Bay

Largest American Great Lakes City Chicago

Largest American Gulf Coast City Houston

Largest American Port of Entry San Ysidro, California

Largest American Prison Population California, Texas, New York, Florida, and Ohio

Largest American Samoan City Pago Pago

Largest American Southern City Houston

Largest American State Alaska

Largest American Virgin Island Saint Croix

Largest American West Coast City Los Angeles

Largest Amphibian Chinese giant salamander *(Andreas davidianus)*

Largest Anteater giant anteater

Largest Anthropoid mountain gorilla

Largest Arboreal Mammal orangutan

Largest Arctic Ocean Island Greenland

Largest Asian Primate orangutan of Borneo and Sumatra

Largest Atlantic Ocean Island Greenland

Largest Atlantic Port New York (including Brooklyn, New Jersey, and Staten Island ports)

Largest Balkan City Athens

Largest Baltic Sea Island Gotland

Largest Canadian Arctic Island Baffin

Largest Canadian City on the American Border Windsor, Ontario

Largest Canadian Great Lakes City Toronto

Largest Canadian Province Québec

Largest Canadian West Coast City Vancouver

Largest Caribbean Sea Islands (by size) Cuba, Hispaniola (Dominican Republic and Haiti), Jamaica, Puerto Rico, Trinidad

Largest Case for a Museum Artifact four-story case for an American flag in the National Museum of American History

Largest Central American Republic Nicaragua

Largest Clipper Ships *Great Republic* built in Boston by Donald McKay and launched in 1853 as the world's largest ship

Largest Coffee Port Santos, Brazil

Largest Colombian City Bogotá

Largest Communist State People's Republic of China

Largest Continent Asia

Largest Eggs laid by North African ostriches

Largest European City Paris

Largest-Eyed Primate spectral tarsier of Indonesia and the Philippines

Largest, Fastest, and Most Powerful Present-Day Bird ostrich

Largest Fish Market in Tokyo

Largest Flightless Land Bird South African ostrich

Largest Florida Metropolitan Area Miami

Largest Flower rafflesia (found in Indonesia and Malaysia by Sir Stamford Raffles; often 3 feet, or 900 millimeters, in diameter)

Largest Flying Animal extinct pterodactyl (*Quetzocoatlus northropi*)

Largest Flying Bird condor

Largest French City Paris

Largest Freshwater Lake in Africa Victoria between Kenya, Tanzania, and Uganda

Largest Freshwater Lake in Central America Lake Nicaragua between Costa Rica and Nicaragua

Largest Freshwater Lake of Eurasia Baikal in Russia

Largest Freshwater Lake in Europe Balaton in Hungary

Largest Freshwater Lake in North America Superior between Canada and the U.S.

Largest Freshwater Lake in South America Titicaca between Bolivia and Peru

Largest Freshwater Lake in the World Superior in North America

Largest Galápagos Island Albemarle, officially Isabela

Largest of the Geese Canada goose

Largest Gorge Grand Canyon of the Colorado River in Arizona

Largest Hospital Ship USS *Mercy*—United States Navy seagoing hospital

Largest Indian Ocean Island Madagascar

Largest Indian Ocean Nation Australia

Largest Indian Ocean Port Singapore

Largest Industrial City in Mexico Monterrey, Nuevo León

Largest Invertebrate octopus (eight-armed cephalopod sometimes with an armspan of 9 meters or 30 feet)

Largest Island in Australasia New Guinea

Largest Island in Indonesia Borneo (Kalimantan)

Largest Island in the West Indies Cuba

Largest Island in the World Greenland

Largest Island Mountain Mauna Kea (Hawaiian—White Mountain)

Largest Islands (by size) Greenland, New Guinea, Borneo, Madagascar, Baffin, Sumatra, Honshu, Great Britain, Ellesmere, Victoria, Celebes (Sulawesi)

Largest Jail Complex in the Western democracies Rikers Island, New York

Largest Japanese Island Honshu

Largest Lakes (by size) Caspian Sea, Asia; Superior, North America; Victoria, Africa; Aral Sea, in Russia and Uzbekistan; Huron, North America; Tanganyika, Africa; Great Bear, Canada; Baikal, Russia; Nyasa, Africa

Largest Lake in the World Caspian Sea

Largest Land Mammal African bush elephant

Largest Living Amphibian Japanese giant salamander

Largest Living Antelope north-central Africa's giant eland

Largest Living Bat Kalong fruit bat of Indonesia and Malaysia

Largest Living Bird North African ostrich

Largest Living Canine the endangered timber wolf of the wild (the largest domestic dog is the Saint Bernard while the tallest is the Great Dane or the Irish Wolfhound)

Largest Living Carnivore brown bear, also called the Kodiak bear

Largest Living Clam giant clam of the South Pacific Ocean up to 500 pounds (227 kg) and 5 feet (150 cm) long

Largest Living Crocodilian salt-water crocodile (*Crocodylus porosus*)

Largest Living Crustacean Japanese spider crab

Largest Living Deer American moose (called elk in Europe)

Largest Living Dog Saint Bernard

Largest Living Elephant African elephant

Largest Living Feline long-furred Manchurian or Siberian tiger

Largest Living Fish whale shark

Largest Living Freshwater Fish Mekong River catfish (the Pla Buk of Laos)

Largest Living Freshwater Terrapin alligator snapping terrapin of the Mississippi River region

Largest Living Frog giant or goliath frog found in the Cameroons of Africa

Largest Living Game Bird peafowl

Largest Living Horse Belgian stallion

Largest Living Land Mammal African elephant

Largest Living Lizard Komodo dragon *(Varanus komodoensis)* of Indonesia

Largest Living Mammal blue or sulphur-bottom whale

Largest Living Marine Carnivore walrus

Largest Living Marsupial Australian red kangaroo

Largest Living Monotreme duck-billed platypus

Largest Living Penguin emperor penguin

Largest Living Primate African mountain gorilla

Largest Living Rabbit Flemish Giant rabbit

Largest Living Reptile saltwater crocodile

Largest Living Rodent capybara

Largest Living Salamander giant salamander of China and Japan

Largest Living Shark whale shark

Largest Living Snake regal python *(Python reticulatits)*

Largest Living Species of Shark 60-foot-long (18-meter) whale shark

Largest Living Starfish Gulf of Mexico's bristling starfish *(Midgardia xandaros)*

Largest Living Tiger Siberian tiger

Largest Living Tortoise Galapagos tortoise, sometimes 4 feet (120 cm) long, weighing 500 pounds (225 kg)

Largest Living Turtle leatherback turtle

Largest, Longest Coral Reef Australia's Great Barrier Reef extending 1,200 miles (1,931 km) along the northeast coast

Largest, Longest Live Snake Colossus, a reticulated python in the Pittsburgh Zoo; measured 29 feet and weighed 300 pounds (136 kg and 8.8 meters)

Largest of the Low Countries Netherlands

Largest Man-Made Lake Gatun Lake watering the Panama Canal

Largest Marine Mammal blue or sulphur-bottom whale

Largest Marsupial red kangaroo

Largest Mediterranean Port Marseille

Largest Mediterranean Sea Island Sicily

Largest Mexican City on the American Border Ciudad Judrez, Chihuahua

Largest Mexican City in Jalisco Guadalajara (Mexico's second largest city)

Largest Mississippi River City St Louis

Largest Nation Russia

Largest National Areas Russia, Canada, China, U.S.A., Brazil, Australia, India, Argentina, Sudan, Mongolia

Largest Nation in the South Pacific Australia

Largest North American Land Mammal bison (buffalo)

Largest North American Nation Canada

Largest North American Turtle alligator snapper

Largest North Atlantic Ocean Nation Canada

Largest North Pacific Ports Los Angeles (including Long Beach, San Pedro, and Wilmington) and Yokohama (including Kawasaki, Tokyo, and Yokosuka)

Largest Northwest Territory Town Yellowknife

Largest Norwegian Arctic Island Svalbard

Largest Oceanic Areas Pacific, Atlantic, Indian, and Arctic followed by the Mediterranean, South China, Bering, and Caribbean seas; the Gulf of Mexico and the Sea of Okhotsk; the East China and Yellow seas; Hud-

son Bay; the Sea of Japan; the North, Black, Red, and Baltic seas

Largest Oceanic Bird Pacific albatross

Largest of the Oceanic Dolphins killer whale *(Orcinus orca)*

Largest Oceanic Nation Indonesia

Largest of the Forty-Eight Contiguous States Texas

Largest and Oldest Subway System New York City

Largest Order of Living Birds the perching birds (containing some 6000 species)

Largest Owl fish owl of northernmost Japan

Largest Pacific Ocean Island Papua New Guinea

Largest Pacific Ocean Nation China

Largest Penguin Emperor

Largest Planet Jupiter

Largest Population China (followed by India, U.S.A., Indonesia, Brazil, Pakistan, Russia, Japan)

Largest Port in Southeast Asia Singapore

Largest Primate central Africa's *Gorilla gorilla*

Largest Producer of Wool Australia's sheep

Largest Rodent capybara

Largest Russian Arctic Island Novaya Zemlya

Largest Saltwater Lake in Asia Aral in Russia and Uzbekistan

Largest Saltwater Lake in Australia Torrens in South Australia

Largest Saltwater Lake in the World Caspian Sea in Eurasia

Largest Seas (by area) South China, Caribbean, Mediterranean, Bering, Gulf of Mexico, Sea of Okhotsk, Sea of Japan, Hudson Bay, Andaman, Black, Red, North, Baltic

Largest Shopping Mall Great Mall of America in Minneapolis, Minnesota

Largest South American Primate muriquis of Brazil

Largest South Atlantic Ocean Nation Brazil

Largest South China Sea City
Hong Kong

Largest Southeast-Asian City
Singapore

Largest South Pacific Ocean Nation Australia

Largest South Pacific Port Sydney

Largest State East of the Mississippi Georgia

Largest State in the United States Alaska

Largest State West of the Mississippi Alaska

Largest Toad marine toad (*Bufo Marinus*) of South America

Largest Tropical Island New Guinea

Largest Turtle leatherback or trunkback sea turtle

Largest Village in Europe The Hague

Largest West European Country France

Largest West Indian City Havana, Cuba

Largest West Indian Nation Cuba

Largest Whirlpool in the Western Hemisphere off Eastport, Maine

Largest Zoological Garden in the World San Diego Zoo

Last Continent Antarctica (last to be discovered)

Last Emperor of Austria-Hungary Charles I (1916–1918)

Last Emperor of Brazil Dom Pedro II (1831–1889)

Last Emperor of China Henry Pu-yi (K'ang Ti) 1934–1945

Last Emperor of Ethiopia Haile Selassie (1930–1974)

Last Emperor of France Napoleon I Bonaparte (1804–1815)

Last Emperor of Germany Wilhelm 11 (1888–1918)

Last Emperor of Mexico Austrian Archduke Ferdinand Maximilian (1864–1867)

Last Frontier Alaska's old sobriquet and current nickname of Canada's Northwest Territories, as well as nickname for Antarctica

Last Place on Earth South Pole, Antarctica

Last Remaining Polynesian Kingdom Tonga

Last Slave Plantation Louisiana State Penitentiary

Last Viceroy Louis Mountbatten, Admiral of the Fleet and 1st Earl Mountbatten of Burma

Last Wild Indian in North America nickname for Ishi, a Yahi Indian from Northern California

Least Populous Canadian Province Prince Edward Island

Least Populous State Alaska

Lebanon's Highest Peak Quernet es Sauda—10,131 feet, 3,070 meters

Lebanon's Largest City Beirut

Lebanon's Principal Port Beirut

Liberia's Highest Peak Yekepa—5,748 feet, 1,742 meters in Nimba Mountains

Liberia's Largest City Monrovia

Liberia's Principal Port Monrovia

Libya's Highest Peak Kegueur Terbi—10,335 feet, 3,132 meters

Libya's Largest City Tripoli

Libya's Principal Port Tripoli

Liechtenstein's Highest Peak Naafkopf—8,440 feet, 2,558 meters)

Liechtenstein's Largest City Vaduz

Loneliest Man in the World Leonardo da Vinci, according to George Santayana

Longest Bridge Span Verrazano-Narrows Bridge, New York

Longest Commercial Airport Runway in the United States John F Kennedy airport, New York

Longest Crocodile saltwater crocodile attaining more than 20 feet (6 meters)

Longest Eastern Siberian River Lena

Longest English Novel Richardson's *Clarissa*

Longest Fjord SognefJord (extending 110 miles or 175 kilometers into the heart of Norway)

Longest Highway Tunnel Mont Blanc Tunnel between France and Italy

Longest Indo-Chinese River Mekong

Longest Interior River China's Tarim River

Longest Living Vertebrates Galápagos tortoises

Longest Lizard 10-foot (3 meter)-long Komodo dragon lizard of Indonesia

Longest Natural Beach in the United States Long Beach, Washington

Longest North American River Mississippi-Missouri

Longest Northeast Asiatic River Amur

Longest and Oldest Shopping Street Copenhagen's Stroget

Longest Otter Giant South American Otter attaining 7 feet (213 cm) in length

Longest Peninsulas (by length) Scandinavia, Arabia, Kamchatka, Labrador, Baja California, Balkans, Iberia, Korea, Malay, Alaska, Italy, Florida, Yucatan, Nova Scotia, La Guaira

Longest Railroad in America Burlington Northern from Vancouver, BC to Mobile, Alabama

Longest Railroad Tunnel Seikan Tunnel in Japan

Longest Rivers Nile, Amazon, Mississippi-Missouri-Red Rock, Ob-Irtysh, Yangtze, Hwang Ho, Congo, Amur, Lena, Machenzie, Mekong, Niger, Yenisei, Paraná, Plata-Paraguay, Volga, Madeira, St Lawrence, Rio Grande, Orinoco, Yukon, Danube, Euphrates, Murray, Ganges, Irrawaddy, Dneiper, Negro, Don, Orange, Pechora, Marañon, Dneister, Rhine, Donets, Elbe, Gambia, Yellowstone, Vistula, Tagus (*Tajo*), Oder, Maas (*Meuse*), Seine, Guadalquivir, Hudson, Thames, Moldau

Longest Snake South American anaconda measuring 25 feet (7.9 meters) or more; (*see* Largest, Longest Live Snake)

Longest Train Trip in the World on the Trans-Siberian railway from Moscow in eastern Europe to Nakhodka near Vladivostok on the North Pacific coast

Longest Venemous Snake sea snake of western Australia (*Hydrophis belcheri*)

Louisiana's Highest Point Driskill Mountain—535 feet, 163 meters—near Arcadia

Louisiana's Largest City New Orleans

Louisiana's Largest Lake Pontchartrain on the Greater New Orleans Expressway

Louisiana's Longest River Mississippi

Lowest Birth Rate Country in the Industrialized World Italy

Lowest Murder Rate Sikkim—a Himalayan protectorate of India

Lowest Place in Africa Lake Assal in Djibouti (512 feet or 156 meters below sea level)

Lowest Place in Asia (*see* Lowest Place in the World)

Lowest Place in Australia Lake Eyre in South Australia (52 feet or 16 meters below sea level)

Lowest Place in Eastern Europe Caspian Sea (92 feet or 28 meters below sea level)

Lowest Place in North America Badwater in Death Valley between California and Nevada (286 feet or 87.5 meters below sea level)

Lowest Place in South America Valdes Peninsula of Argentina (131 feet or 40 meters below sea level)

Lowest Places in Western Europe coastal areas of the Netherlands (15 feet or 4.6 meters below sea level)

Lowest Place in the World Dead Sea between Israel and Jordan (1,302 feet or 397 meters below sea level)

Luxembourg's Largest City Luxembourg-Ville

Macau's Largest City Macau City

Madagascar's Highest Peak Tsaratanana Massif—9,450 feet, 2,864 meters

Madagascar's Largest City Antananarivo (Tananarive)

Madagascar's Principal Port Tamatave

Madeira's Largest City Funchal

Maine's Highest Point Mount Katahdin—5,268 feet, 1,605 meters

Maine's Largest City Portland

Maine's Longest River Penobscot

Maine's Outermost Islands Matinicus, Monhegan, Ragged, and Seal islands

Maine's Principal Port Portland

Malawi's Largest City Balantyre-Limbe

Malaysia's Highest Peak Gunong Tahan—7,224 feet, 2,189 meters—east of the Cameron Highlands

Malaysia's Largest Kuala Lumpur

Malaysia's Principal Port George Town

Maldives' Largest City Malé

Maldives' Principal Port Malé Atoll

Mali's Largest City Bamako

Malta's Largest City Valletta

Malta's Principal Port Valletta

Manitoba's Highest Point Mount Baldy—2,729 feet, 832 meters

Manitoba's Largest City Winnipeg

Marshall Islands' Largest Town Majuro

Martinique's Highest Peak Mont Pelée—4,429 feet, 1,342 meters

Martinique's Largest City Fort-de-France

Martinique's Principal Port Fort-de-France

Maryland's Highest Point Backbone Mountain—3,360 feet, 1,024 meters—between Virginia and West Virginia

Maryland's Largest City Baltimore

Maryland's Longest River Potomac

Maryland's Principal Port Baltimore

Maryland's Unspoiled Seashore Assateague Island National Seashore

Massachusetts's Best Known Lake Walden Pond, immortalized by philosopher–naturalist, Henry David Thoreau

Massachusetts' Highest Point Mount Greylock—3,491 feet, 1,064 meters—in the Berkshires

Massachusetts' Largest City Boston

Massachusetts' Longest River Connecticut

Massachusetts' Principal Port Boston

Mauritania's Largest City Nouakchott

Mauritania's Principal Port Nouakchott

Mauritius' Largest City Port-Louis

Mauritius' Principal Port Port-Louis

Mexico's Easternmost City Cheturnal, Quintana Roo

Mexico's Easternmost Point Cabo Catoche, Quintana Roo (first Mexican site discovered by the Spaniards)

Mexico's Highest Peak Citlaltepetl (Orizaba)—18,697 feet, 5,666 meters

Mexico's Largest City Mexico City

Mexico's Largest State Chihuahua

Mexico's Longest River Rio Bravo, also called Rio Grande

Mexico's Northernmost City Mexicali, Baja California

Mexico's Northernmost Point Los Algodones, Baja California

Mexico's Principal Gulf of Mexico Port Veracruz

Mexico's Principal West Coast Port Mazatlan

Mexico's Smallest State Tlaxcala

Mexico's Southernmost City Tapachula, Chiapas

Mexico's Southernmost Point Ciudad Hidalgo in the state of Chiapas

Mexico's Westernmost City Tijuana, Baja California

Mexico's Westernmost Point Playas de Tijuana, Baja California

Michigan's Highest Point Mount Curwood—1,980 feet, 604 meters—in the Upper Peninsula

Michigan's Largest City Detroit

Michigan's Principal Port
Detroit

Michigan's Unspoiled Forest
Wilderness Hiawatha Na-
tional Forest

Micronesia's Largest Place
Kolonia on Ponape

Middle America's Largest
Country Mexico

Minnesota's Highest Point
Eagle Mountain—2,301 feet,
701 meters—in Cook County

Minnesota's Largest City
Minneapolis

Minnesota's Longest River
Red

Minnesota's Northwestern-
most Angle Lake of the
Woods

Minnesota's Principal Port
Duluth

Mississippi's Highest Point
Woodall Mountain—806
feet, 246 meters—in Tishomingo County

Mississippi's Largest City
Jackson

Mississippi's Longest River
Mississippi

Mississippi's Most Scenic
Highway Natchez Trace
Parkway linking Alabama,
Mississippi, and Tennessee

Mississippi's Principal Port
Pascagoula

Missouri's Highest Point
Taum Sauk Mountain—1,772
feet, 541 meters

Missouri's Largest City St
Louis

Missouri's Longest River
Mississippi

Missouri's Principal Port St
Louis

Missouri's Superb Riverway
Ozark National Scenic Riverways in southeastern Missouri's Ozark Mountains

Monaco's Largest City
Monaco-Ville

Monaco's Principal Port
Monaco

Mongolia's Highest Peak Ta-
bun Bogdo—15,266 feet,
4,626 meters

Mongolia's Largest City
Ulaanbaatar

Montana's Highest Point
Granite Peak—12,799 feet,
3,902 meters—in Park
County

Montana's Largest City Billings

Montserrat's Largest Place
Plymouth

Morocco's Highest Peak (see
Highest Peak in North Africa)

Morocco's Largest City
Casablanca

Morocco's Principal Port
Tangier

Most Active Volcano in the
Continental United States
Mount St Helens in south-
western Washington close to
Portland, Oregon

Most Agnostic and Atheistic
Century 18th—the Age of
Voltaire, the Encyclopedists,
and Paine's Age of Reason

Most Dangerous Rattlesnake
Eastern North American Diamondback

Most Dangerous Volcano in
the United States Mount
Rainier, near Tacoma, Washington

Most Densely Populated
Nation in Mainland Latin
America El Salvador

Most Elegant Salt-Marsh Terrapin commercially culti-
vated diamondback terrapin
of the Carolinas and Georgia

Most Endangered Species of
Sea Turtle Hawksbill or
Tortoise-shell Turtle (Eretmo-
chelys imbricata)

Most Expensive Cities (see
Fifty Most Expensive Cities
in the Late 1900s)

Most Feared Marine Preda-
tor shark

Most Glorious Hero of Norwegian Viking Times Olav
Tryggvasson

Most Murderous Country
South Africa, with a murder
rate of 53.5 / 100,000 in 1995

Most Northern Southern City
Tulsa, Oklahoma

Most Populous American
Megalopolis Los Angeles

Most Populous Canadian
Province Ontario

Most Populous Nation China

Most Populous State Cali-
fornia

Most-Prized Sailing Trophy
America's Cup

Most Remote Continent
Australia

Most Serene Republic of the
Sea sobriquet for Venice,
Queen of the Adriatic

Mozambique's Highest Peak
Serra de Gorongosa—6,112
feet, 1,852 meters

Mozambique's Largest City
Maputo

Mozambique's Principal Port
Maputo

Myanmar's Highest Peak
Mount Victoria—10,016 feet,
3,035 meters

Myanmar's Largest City
Yangon

Myanmar's Principal Port
Yangon

Namibia's Largest City Windhoek

Namibia's Principal Port
Walvis Bay

Nauru's Largest Town Yaren

Nauru's Principal Port
Yaren

Nebraska's Highest Point
Johnson—5,426 feet, 1,653
meters—in Kimball County
near the Colorado and Wyoming border

Nebraska's Largest City
Omaha

Nebraska's Longest River
Missouri

Nebraska's Most Unusual Natural Feature the Sandhills

Nepal's Highest Peak Mount
Everest—29,028 feet, 8,796
meters

Nepal's Largest City Kathmandu

Netherlands Antilles' Largest
City Willemstad on Curaçao

Netherlands' Easternmost
Town Nieuwe-Schans

Netherlands' Largest City
Amsterdam

Netherlands' Largest Port
Rotterdam

Netherlands' Longest River
System Rhine-Meuse

Netherlands' Northernmost
City Groningen

Netherlands' Northernmost
Island Rottumeroog (in the
Frisians between the North
Sea and the Waddenzee)

Netherlands' Principal Port
Rotterdam

Netherlands' Southernmost City Maastricht

Netherlands' Southernmost Point Vaals

Netherlands' Westernmost Port Flushing or Vlissingen

Netherlands' Westernmost Town Sluis

Nevada's Highest Point Boundary Peak—13,140 feet, 4,005 meters

Nevada's Largest City Las Vegas

Nevada's Largest Forest Toiyabe National Forest (also the largest forest in the contiguous U.S.)

New Brunswick's Highest Point Mount Carleton—2,690 feet, 820 meters

New Brunswick's Largest City Saint John

New Brunswick's Principal Port Saint John

New Caledonia's Largest City Noumea

New Caledonia's Principal Port Noumea

New England's Highest Point Mount Washington—6,288 feet, 1,917 meters—in New Hampshire

Newfoundland's Highest Point Mount Cladonia—4,725 feet, 1,441 meters—in Labrador

Newfoundland's Largest City St John's

New Foundland's Principal Port St John's

New Hampshire's Highest Point Mount Washington—6,288 feet, 1,917 meters

New Hampshire's Largest City Manchester

New Hampshire's Largest Lake Winnipesaukee

New Hampshire's Longest River Connecticut

New Hampshire's Principal Port Portsmouth

New Jersey's Highest Point High Point—1,803 feet, 550 meters—in the Kittatinny Mountains

New Jersey's Largest City Newark

New Jersey's Longest River Delaware

New Jersey's Most Spectacular Park Palisades Interstate Park overlooking the Hudson River and much of New York City

New Jersey's Principal Port Newark

New Mexico's Highest Point Wheeler Peak—13,161 feet, 4,011 meters

New Mexico's Largest City Albuquerque

New Mexico's Largest Gypsum Desert the White Sands National Monument, world's largest gypsum deposit

New Mexico's Longest River Rio Grande

New York State's Most Spectacular Waterfall Niagara Falls

New York State's Highest Point Mount Marcy—5,344 feet, 1,630 meters—in the Adirondacks

New York State's Largest City New York City

New York State's Longest River Hudson

New York State's Principal Port New York City

New Zealand's Easternmost City Gisborne on North Island

New Zealand's Easternmost Point North Island's East Cape

New Zealand's Highest Peak Mount Cook—12,349 feet, 3,742 meters—Aorangi by the Maori, on South Island

New Zealand's Largest City Auckland on North Island

New Zealand's Largest South Island City Christchurch

New Zealand's Northernmost City Whangarei north of Auckland on North Island

New Zealand's Northernmost Point North Island's North Cape

New Zealand's Principal Port Auckland

New Zealand's Southernmost City Invercargill on South Island south of Dunedin

New Zealand's Southernmost Point Southwest Cape on Steward Island south of South Island

New Zealand's Westernmost City Milford Sound on South Island

New Zealand's Westernmost Point Resolution Island west of South Island

Nicaragua's Highest Peak Teotecacinte—above 7,000 feet, 2,121 meters—on the Honduran–Nicaraguan border

Nicaragua's Largest City Managua

Nicaragua's Largest Department Zelaya

Nicaragua's Principal Caribbean Port Bluefields

Nicaragua's Principal Pacific Port Corinto

Nigeria's Highest Peak Jos—5,843 feet, 1,771 meters

Nigeria's Largest City Lagos

Nigeria's Principal Port Port Harcourt

Niger's Highest Peak Iferouane—5,906 feet, 1,790 meters

Niger's Largest City Niamey

North America's Easternmost City St John's, Newfoundland

North America's Easternmost Point Cape Spear, Newfoundland

North America's Largest Cities Mexico City, New York City

North America's Largest Drainage Basin Mississippi-Missouri

North America's Largest Island Greenland

North America's Largest Nation Canada

North America's Most European City Quebec

North America's Northernmost City Barrow, Alaska

North America's Northernmost Point Canada's Cape Columbia

North America's Southernmost City David, Panamá

North America's Southernmost Point southwesternmost Panamá just northeast of Juradó, Colombia

North America's Westernmost City Seward, Alaska

North America's Westernmost Point Cape Wrangell on Attu Island in the Aleutians

North Atlantic's Largest Nation Canada

North Carolina's Highest Point Mount Michell— 6,684 feet, 2,037 meters

North Carolina's Largest City Charlotte

North Carolina's Longest River Catawba

North Carolina's Loveliest Seashore the Outer Banks including the Cape Hatteras National Seashore

North Carolina's Principal Port Morehead City

North Dakota's Highest Point White Butte—3,506 feet, 1,069 meters

North Dakota's Largest City Fargo

North Dakota's Longest River Missouri

Northeasternmost Beach of the United States Wilsons Beach, Campobello Island, Maine

Northeasternmost Point of the Continental United States West Quoddy Head near Lubec, Maine

Northern Ireland's Highest Peak Slieve Donard—2,796 feet, 847 meters

Northern Ireland's Largest City Belfast

Northern Ireland's Largest Lake Lough Neagh

Northernmost Metropolis St Petersburg in Russia

Northernmost Point in Europe Cape Norkkyn, Norway

North Korea's Highest Peak Kwanmo also called Kambo Ho—8,337 feet, 2,526 meters

North Korea's Largest City Pyongyang

North Korea's Principal Port Chonglin

Northwesternmost Point of the Contiguous United States Tatoosh Island off Cape Flattery, Washinton

Northwesternmost Point of the Continental United States Cape Wrangell on Attu Island in the Aleutians, Alaska

Northwest Territories' Highest Point Sir James MacBrien—9,062 feet, 2,762 meters

Norway's Highest Peak Galdhoppigen—8,097 feet, 2,454 meters—in the Jotunheim mountains

Norway's Largest City Oslo

Norway's Longest River Glomma

Norway's Principal Port Bergen

Nova Scotia's Highest Point North Barren—1,747 feet, 532 meters—on Cape Breton Island

Nova Scotia's Largest City Halifax

Nova Scotia's Principal Port Halifax

Ohio's Highest Point Campbell Hill—1,550 feet, 472 meters—near Bellefontaine

Ohio's Largest City Cleveland

Ohio's Longest River Ohio

Ohio's Principal Port Cincinnati

Oklahoma's Highest Point Black Mesa—4,973 feet, 1,515 meters

Oklahoma's Largest City Oklahoma City

Oklahoma's Largest Gypsum Cave in Alabaster Caverns State Park south of Freedom

Oklahoma's Longest River Red

Oldest Anglican Church Site in the Western Hemisphere St Peter's in St George Harbour, Bermuda

Oldest British Colony Newfoundland, discovered in 1497 by John Cabot

Oldest British Columbia Settlement Vancouver, founded by Hudson's Bay Company in 1825

Oldest Canadian Settlement Port Royal (Annapolis, Nova Scotia) founded by Champlain in 1604

Oldest Central American Democracy Costa Rica

Oldest Centralized Nation China

Oldest City in Denmark Ribe (whose church was built in the 12th century)

Oldest City in Germany Trier (founded by the Romans in 15 B.C.)

Oldest City in Malaysia Malacca (settled by Malays around 1400)

Oldest City in North America Mexico City (built by the Aztecs in 1325)

Oldest City in the U.S. St Augustine, Florida (founded by the Spaniards in 1565)

Oldest Continually Operating Streetcar Line St Charles Avenue line in New Orleans

Oldest Continuous Civilization Australia's aboriginals

Oldest European Settlement in the Far East Macao (leased from China by the Portuguese in 1557)

Oldest European Settlement in the New World Santo Domingo City (founded by the Spaniards in 1496)

Oldest German Ocean Harbor Bremen (created as an archbishopric in 845 A.D.)

Oldest Inhabited City Damascus, Syria

Oldest Inhabited Place in the United States Pueblo Acoma near Albuquerque, New Mexico

Oldest Known Canon *Sumer is incumenn in—lhude sing cuccu*—Reading Rota most likely composed between 1280 and 1310—still sung

Oldest Known Visual System the trilobite eye

Oldest Nova Scotia Settlement Port Royal established in 1605

Oldest Ontario Settlement Fort St Marie founded in 1639

Oldest President of the United States Ronald Reagan

Oldest Public School in America Boston Latin School

Oldest Publisher Elsevier, also spelled Elzevir, in business since 1581

Oldest, Quaintest City in the United States Santa Fé, New Mexico (built in 1621)

Oldest Quebec Settlement city of Québec founded by Champlain in 1608

Oldest Ship in the British Navy HMS *Victory*—launched in 1765 and served as Nelson's flagship at Trafalgar

Oldest Ship in the United States Navy USS *Constitution*—launched in 1797 and nicknamed Old Ironsides

Oldest Sports Club in the United States Schuylkill Fishing Company (Pennsylvania)

Oldest Synagogue in the New World Mikve Israel in Willemstad, Curagao (in continuous use since 1732)

Oldest University in the Americas Santo Tómas de Aquino in Santo Domingo, Dominican Republic, founded in 1538

Oman's Principal Port Muscat

Only American Marsupial opossum

Only Great Ape in Asia orangutan

Ontario's Highest Point Ogidaki Mountain—2,183 feet, 665 meters—overlooking Lake Superior's Haviland Bay

Ontario's Largest City Toronto

Ontario's Longest River St Lawrence

Ontario's Principal Port Toronto

Oregon's Highest Point Mount Hood—11,235 feet, 3,433 meters

Oregon's Largest City Portland

Oregon's Longest River Columbia

Oregon's Most Beautiful and Peaceful Park Crater Lake National Park

Oregon's Principal Port Portland

Pacific Ocean's Largest Nation Australia

Pakistan's Highest Peak Tirich Mir—25,230 feet, 7,645 meters

Pakistan's Largest City Karachi

Pakistan's Principal Port Karachi

Panama Canal's Most Ecologically Protected Island Barro Colorado

Panama's Highest Peak Chiriqui—11,410 feet, 3,458 meters—an inactive volcano near Costa Rica

Panama's Largest City Panama City

Panama's Largest Province Darién

Panama's Principal Caribbean Port Cristobal

Panama's Principal Pacific Port Balboa

Papua New Guinea's Highest Peak Carstensz—16,400 feet, 4,970 meters

Papua New Guinea's Largest City Port Moresby

Paraguay's Largest City Asunción

Paraguay's Largest Department Presidente Hayes

Paraguay's Longest River Paraguay

Pennsylvania's Highest Point Mount Davis—3,213 feet, 979 meters—in Somerset County

Pennsylvania's Largest City Philadelphia

Pennsylvania's Longest River Delaware

Pennsylvania's Most Pristine Park Allegheny National Forest, extending into New York State

Pennsylvania's Principal Port Philadelphia

Peru's Easternmost and Northernmost City Iquitos (on the Amazon)

Peru's Highest Peak Huascaran—22,205 feet, 6,729 meters—in the Andes

Peru's Largest City Lima

Peru's Largest Port Callao

Peru's Longest River Marafión

Peru's Principal Port Callao

Peru's Southernmost City Tacna (near the Chilean border)

Peru's Westernmost City Talara

Philippines' Highest Peak Apo—9,690 feet, 2,936 meters—volcano on the island of Mindinao

Philippines' Largest City Manila

Philippines' Principal Port Manila

Poland's Highest Peak Rysy—8,212 feet, 2,488 meters—in the High Tatra near Czechoslovak

Poland's Largest City Warsaw

Poland's Principal Port Gdansk

Poorest Nations in the World Congo (formerly Zaire), Ethiopia, Burundi, Mozambique

Portugal's Highest Peak Malhdo—6,532 feet, 1,979 meters—in the Serra da Estrala

Portugal's Largest city Lisbon

Portugal's Longest River *Tejo* (Portuguese—Tagus)

Portugal's Principal Port Lisbon

Prince Edward Island's Highest Point between Hartsville and Stanchel—450 feet, 136 meters

Prince Edward Island's Largest City Charlottetown

Prince Edward Island's Principal Port Charlottetown

Puerto Rico's Best Port San Juan

Puerto Rico's Biggest Port on the Atlantic San Juan

Puerto Rico's Biggest Port on the Caribbean Ponce

Puerto Rico's Easternmost City Fajardo

Puerto Rico's Easternmost Point Culebrita Island east of Culebra and Vieques

Puerto Rico's Highest Peak Cerro de Punta—4,400 feet, 1,467 meters—in the Cordillera Central

Puerto Rico's Largest City San Juan

Puerto Rico's Northernmost Cities Arecibo and San Juan

Puerto Rico's Northernmost Point Punta Jacinto

Puerto Rico's Principal Port San Juan

Puerto Rico's Southernmost City Ponce

Puerto Rico's Southernmost Point Caja de Muertos (southeast of Ponce)

Puerto Rico's Westernmost Cities Aguadilla and Mayagtiez

Puerto Rico's Westernmost Point Punta Higuero

Qatar's Largest City Doha

Qatar's Principal Port Doha

Québec's Highest Point Mont Jacques Cartier—4,160 feet, 1,269 meters

Québec's Largest City Montreal

Québec's Longest River St Lawrence

Québec's Principal Port Montreal

Rainiest Spot Mawsynram, Meghalaya State, India

Réunion's Largest City Saint-Denis

Réunion's Principal Port Saint-Denis

Rhode Island's Highest Point Jeremoth Hill—812 feet, 247 meters—near Connecticut

Rhode Island's Largest City Providence

Rhode Island's Most Notable Feature Narragansett Bay

Rhode Island's Principal Port Providence

Richest-Per-Capita Countries of the World Switzerland, Japan, Norway, Denmark, Singapore, U.S.

Rivers of the World (ranked by length) (see Longest Rivers)

Romania's Highest Peak Negos—8,361 feet, 2,534 meters—in the Transylvanian Alps

Romania's Largest City Bucharest

Romania's Longest River Danube

Romania's Principal Port Constanta

Russia's Easternmost Point Mys Deshneva, across the Bering Strait from Alaska

Russia's Greatest Poet Alexander Pushkin

Russia's Largest City Moscow

Russia's Longest Rivers Ob'Irtysh, Volga

Russia's Northernmost Point Mys Chelyuskin, Siberia

Russia's Principal Baltic Port St Petersburg

Russia's Principal Pacific Port Vladivostok

Russia's Principal Port St Petersburg

Rwanda's Highest Peak Gorna—14,787 feet, 4,481 meters—on the Uganda and Zaire borders

Rwanda's Largest City Kigali

Saba's Highest Peak Gunong Kinabalu—13,510 feet, 4,094 meters—east of Kota Kinabalu

Saba's Largest City Kota Kinabalu

Samoa's Best Port Pago Pago on Tutuila Island

San Marino's Largest Town San Marino

São Tomé and Principe's Largest City São Tomé

São Tomé and Principe's Principal Port São Tomé

Sardinia's Largest City Cagliari

Saskatchewan's Highest Point Cypress Hills—4,567 feet, 1,392 meters

Saskatchewan's Largest City Regina

Saskatchewan's Longest River Nelson-Saskatchewan

Saudi Arabia Mountain Peak Jebal al Lawz (Arabic—Lawrence Mountain)—8,465 feet, 2,580 meters—honoring Lawrence of Arabia

Saudi Arabia's Highest Peak in the Asir Highlands—10,279 feet, 3,133 meters

Saudi Arabia's Largest City Riyadh

Saudi Arabia's Principal Port Jidda

Scotland's Highest Peak Ben Nevis—4,406 feet, 1,335 meters

Scotland's Largest City Glasgow

Scotland's Largest Lake Loch Lomond

Scotland's Longest River Tay

Second First Lady Abigail Smith Adams

Senegal's Largest City Dakar

Senegal's Principal Port Dakar

Seychelles' Highest Peak Morne Seychellois—2,993 feet, 907 meters—on Mahe Island

Seychelles' Largest City Port Victoria

Seychelles' Principal Port Port Victoria

Shallowest Sea Baltic (average depth less than 190 feet)

Shortest Snake thread snake of Barbados, Martinique, and St Lucia (Leptophylops bilineata)

Siberia's Largest City Novosibirsk

Sicily's Largest City Palermo

Sierra Leone's Highest Peak Bintimane Peak—6,390 feet, 1,936 meters

Sierra Leone's Largest City Freetown

Sierra Leone's Principal Port Freetown

Singapore's Largest City Singapore

Singapore's Principal Port Singapore

Single Most Significant Scientific Instrument deemed to be a device that, since 1958, has tracked the amount of carbon dioxide emitted into the atmosphere by Hawaii's Mauna Loa volcano

Slovenia's Highest Peak Triglav—9,395 feet, 2,847 meters—in the Julian Alps

Smallest African Country The Gambia

Smallest African Nation Comoros

Smallest American Prison Populations North Dakota, Vermont, New Hampshire, Wyoming, and South Dakota

Smallest American State Rhode Island

Smallest American Virgin Island Cockroach

Smallest Asian Country Singapore

Smallest Asian Nation Maldives

Smallest Australian State Tasmania

Smallest British Bird wren

Smallest British Mammal pygmy shrew

Smallest British Mouse harvest mouse

Smallest Capital of the Eastern World Yaren, Nauru

Smallest Capital in the United States Carson City, Nevada

Smallest Capital of the Western World San Marino, San Marino

Smallest Central American Country El Salvador

Smallest Channel Island Brechau

Smallest Continent Australia

Smallest Domestic Dog Chihuahua (5 inches or 13 cm high and weighing 6 pounds or 2.7 kg)

Smallest East European Country Albania

Smallest European Country San Marino

Smallest European Nation Monaco

Smallest Galápagos Island Duncan, officially Pinzón

Smallest Indian Ocean Nation Seychelles

Smallest Indonesian Island Krakatoa

Smallest Living Amphibian tree frog (Hyla ocularis) of the southeastern United States

Smallest Living Antelope West African royal antelope

Smallest Living Bat Kitt's hog-nosed bat from Thailand

Smallest Living Cat rusty-spotted cat of India and Sri Lanka

Smallest Living Crocodilian dwarf caiman of the Amazon Basin

Smallest Living Deer Andean dwarf deer called *pudo* in Chile; Ecuadorean *pudu*

Smallest Living Fish half-inch long (1.3 cm) Philippine goby

Smallest Living Freshwater Turtle musk turtle

Smallest Living Horse Argentina's Falabella breed

Smallest Living Lizard British Virgin Island gecko (Sphaerodactylus elasmobranchus)

Smallest Living Mammal pygmy shrew or wood mouse

Smallest Living Marine Mammal sea otter

Smallest Living Marsupial Kimberly marsupial mouse of Western Australia

Smallest Living Primate Madagascar mouse lemur

Smallest Living Reptile pygmy gecko of Israel

Smallest Living Rodent Eurasian harvest mouse

Smallest Living Sea Turtle Kemp's Bastard or Ridley

Smallest Living Snake West Indian thread snake

Smallest Living Terrapin Colombian musk terrapin

Smallest Living Tortoise Egyptian or South African tortoise

Smallest Middle American Country Grenada

Smallest Nation Principality of Monaco—less. than 370 acres (150 hectares)

Smallest Nation in the South Pacific Nauru

Smallest Oceanic Country Nauru in the equatorial Pacific

Smallest Pacific Ocean Nation Nauru

Smallest Penguin Little Blue Penguin of Australia and New Zealand

Smallest Planet Mercury

Smallest Present-Day Bird hummingbird

Smallest Primate Madagascar's mouse lemur (Microrebus murinus)

Smallest Royal Capital of the Eastern World Kathmandu, Nepal

Smallest Royal Capital of the Western World Monaco-Ville, Monaco

Smallest Sea Baltic

Smallest South American Country French Guiana

Smallest South American Nation Suriname

Smallest South Atlantic Nation Uruguay

Smallest South Pacific Nation Nauru

Smallest Sovereign State Vatican City (*see* World's Smallest Sovereign State)

Smallest State in the United States Rhode Island

Smallest Warm-Blooded Creatures hummingbirds

Smallest West Indian Nation Grenada; St Kitts and Nevis

Solomon Island's Highest Peak Mount Balbi—10,170 feet, 308 meters—on Bougainville Island

Solomon Islands' Largest City Honiara

Solomon Islands' Principal Port Honiara

Somalia's Largest City Mogadishu

Somalia's Principal Port Mogadishu

South Africa's Highest Peak Thabantshon—11,425 feet, 3,462 meters—in the Drakensberg Range in Basutoland

South Africa's Biggest Black Ghetto Soweto near Johannesburg

South Africa's Largest City Johannesburg

South Africa's Principal Port Durban

South America's Easternmost Cities Brazil's Jodo Pessoa and Recife

South America's Easternmost Points Brazil's Cabo Branco near Jodo Pessoa and Punta de Pedra near Recife (both on the same longitude—34°37' W)

South America's Largest City Buenos Aires

South America's Largest Drainage Basin Amazon

South America's Largest Island Tierra del Fuego

South America's Largest Nation Brazil

South America's Longest River Amazon

South America's Northernmost City Coro, Venezuela

South America's Northernmost Point Punta Gallinas, Colombia

South America's Southernmost City Punta Arenas, Chile

South America's Southernmost Point Cabo de Homos (Cape Horn), Chile

South America's Westernmost City Talara, Peru

South America's Westernmost Point Punta Parihas, Peru

South Atlantic's Largest Nation Brazil

South Carolina's Highest Point Sassafras Mountain—3,560 feet, 1,085 meters

South Carolina's Largest City Columbia

South Carolina's River Savannah

South Carolina's Principal Port Charleston

South Dakota's Highest Point Harney Peak—7,242 feet, 2,208 meters

South Dakota's Largest City Sioux Falls

South Dakota's Longest River Missouri

South Dakota's Most Memorable Memorial Mount Rushmore with the carved heads of George Washington, Thomas Jefferson, Abraham Lincoln, and Theodore Roosevelt

Southeast Asia's Longest River Mekong

Southeasternmost Beach of the United States Miami Beach

Southeasternmost Point of the Continental United States Cape Florida at the southern tip of Key Biscayne near Miami

Southeasternmost Point of the Territorial United States Vagthus Point on Saint Croix in the Virgin Islands east of Puerto Rico

Southernmost American Town Naalehu, Island of Hawaii

Southernmost Beach in the United States South Beach in Key West, Florida

Southernmost Canadian Town Kingsville, Ontario

Southernmost Capital of Africa Cape Town, South Africa

Southernmost Capital of Asia Jakarta, Indonesia

Southernmost Capital of Australia Hobart, Tasmania

Southernmost Capital of Europe Athens, Greece

Southernmost Capital of North America Panamá City, Panamá

Southernmost City of South America Montevideo, Uruguay

Southernmost Europe Crete's south coast

Southernmost Island in the West Indies Grenada

Southernmost Metropolis Melbourne, Australia

Southernmost Point of Canada Middle Island, Ontario on Lake Erie

Southernmost Point of the Continental United States South Beach on Key West, Florida

Southernmost Point of the Insular United States Ka Lae or South Cape on Hawaii

Southernmost Point of Mexico Barra del Río Suchiate (Mouth of the Suchiate River)

Southernmost Province of Canada Ontario

Southernmost State Hawaii

Southernmost Town Puerto Williams, Chile

South Korea's Largest City Seoul

South Korea's Principal Port Pusan

South's Oldest Daily Newspaper Charleston's *The News and Herald*

Southwesternmost Beach in the Continental United States Imperial Beach near San Diego, California

Southwesternmost Point of the Continental United States Point Loma, California at the entrance to San Diego Bay

Southwesternmost Point of the Territorial United States Steps Point on Tutuila Island of American Samoa in the South Pacific

Spain's Easternmost Point Cabo de Creus (northeast of Barcelona)

Spain's Easternmost Port Barcelona

Spain's Highest Peaks Mulhacón—11,411 feet, 3,458 meters—on the mainland; Pico de Teide—12,200 feet, 3697 meters—in the Canary Islands

Spain's Largest City Madrid

Spain's Largest Province Andalucía

Spain's Longest River Tagus

Spain's Most Precious Jewel Toledo, according to Cervantes

Spain's Northernmost Point Cabo Ortegal

Spain's Northernmost Port El Ferrol del Caudillo

Spain's Principal Port Barcelona

Spain's Smallest Province Vascongadas (Basque Provinces)

Spain's Southernmost Point Punta de Tarifa

Spain's Southernmost Port Algeciras

Spain's Westernmost Point Cabo Finisterre

Spain's Westernmost Port Vigo

Sri Lanka's Highest Peak Pidurutalagala—8,291 feet, 2,512 meters

Sri Lanka's Largest City Colombo

Sri Lanka's Principal Port Colombo

St Helena's Largest Place Jamestown

St Kitts and Nevis' Largest Town Basseterre

St Kitts and Nevis' Principal Port Basseterre

St Lucia's Largest City Castries

St Lucia's Principal Port Castries

St Pierre and Miquelon's Largest Town St John's

St Pierre and Miquelon's Principal Port St Pierre

St Vincent's Largest City Kingstown

St Vincent and the Grenadines' Principal Port Kingstown

Strongest Surface Wind Mount Washington, New Hampshire

Sudan's Largest City Khartoum

Sudan's Principal Port Port Sudan

Suriname's Highest Peak Wilhelmina Gebergte—4,200 feet, 1,273 metrs

Suriname's Largest City Paramaribo

Suriname's Longest River Nickerie

Suriname's Principal Port Paramaribo

Swaziland's Largest City Mbabane

Sweden's Highest Peak Galdhoppigen—8,097 feet, 2,454 meters—in the Jotunheim range

Sweden's Largest City Stockholm

Sweden's Longest River Torne

Sweden's Principal Port Gothenburg

Swiftest Mammal cheetah

Switzerland's Highest Peak
Dufourspitze—15,203 feet,
4,607 meters—of Monte
Rosa
Switzerland's Largest City
Zurich
Switzerland's Longest River
Rhine
Syria's Highest Peak Mount
Hermon—9,232 feet, 2,798
meters
Syria's Largest City Dam-
ascus
Syria's Principal Port
Latakia
Tahiti's Highest Peak Mont
Orohena—7,618 feet, 2,308
meters
Tahiti's Principal Port Pap-
eete
Taiwan's Highest Peak
Mount Morrison—14,000
feet, 4,242 meters
Taiwan's Largest City Taipei
Taiwan's Principal Port
Kaositing
Tallest Domestic Dog Great
Dane, standing up to 32
inches (81 cm) and weighing
up to 150 pounds (68 kg)
Tallest Living Mammal gi-
raffe
Tallest North American Bird
whooping crane
Tanzania's Highest Peak
Kibo or Kilimanjaro—19,340
feet, 5,861 meters
Tanzania's Largest City Dar
es Salaam
Tanzania's Principal Port
Dar es Salaam
Tennessee's Highest Point
Clingmans Dome—6,643
feet, 2,025 meters—in the
Great Smoky Mountains
Tennessee's Largest City
Memphis
Tennessee's Longest River
Tennessee
**Tennessee's Most Famous
Mountain** Lookout over-
looking the Tennessee Valley
and seven states beyond
Tennessee's Principal Port
Memphis
Texas' Highest Point Guada-
lupe Peak—8,751 feet, 2,667
meters—in Culberson
County
Texas' Largest City Houston
Texas' Longest River Rio
Grande

**Texas' Most Spectacular
Park** Big Bend National
Park
Texas' Principal Port Houston
Thailand's Highest Peak In-
thanon—8,452 feet, 2,561
meters
Thailand's Largest City
Bangkok
Thailand's Principal Port
Bangkok
Third First Lady Martha
Wayles Skelton Jefferson
Togo's Largest City Lorné
Togo's Principal Port Lomé
Tokyo's Largest Slum Sanya
Tonga's Largest City
Nuku'alofa
Tonga's Principal Port
Nuku'alofa
Transkei's Largest City
Umtata
**Trinidad and Tobago's Largest
City** Port of Spain
Trinidad's Highest Peak
Mount Aripo—3,085 feet,
935 meters
**Trinidad and Tobago's Princi-
pal Port** Port of Spain
Tunisia's Largest City Tunis
Turkey's Highest Peak
Mount Ararat—16,946 feet,
5,135 meters—in Armenia
Turkey's Largest City Istan-
bul
Turkey's Principal Port Is-
tanbul
**Turks and Caicos Islands'
Largest Town** Cockburn
Town
Tuvalu's Largest City
Fongafale
Tuvalu's Principal Port
Funafuti
Uganda's Highest Peak Ru-
wenzori—16,795 feet, 5,089
meters
Uganda's Largest City Kam-
pala
Ukraine's Principal Port
Odessa
Ultimate Marine Predator
great white shark
**United Arab Emirates' Larg-
est City** Abu Dhabi
**United Arab Emirates' Prin-
cipal Ports** Abu Dhabi,
Dubai
**United Kingdom's East-
ernmost Point** Lowestoft
Ness on Norfolk's east coast

**United Kingdom's East-
ernmost Port** Lowestoft,
Suffolk
**United Kingdom's Northern-
most Point** Herma Ness
and Muckle Flugga Light on
Unst in the Shetland Islands
**United Kingdom's Northern-
most Port** Scapa Flow in
the Orkney Islands
**United Kingdom's Principal
Ports** London, Liverpool,
Cardiff, Glasgow, Belfast
**United Kingdom's Southern-
most Point** Lizard Point
**United Kingdom's Southern-
most Port** Penzance
**United Kingdom's West-
ernmost Point** Land's End,
Cornwall
**United Kingdom's Westernmost
Port** Penzance, Cornwall
United States superlatives
(*see* America's *entries in this
section*)
Uruguay's Largest City Mon-
tevideo
**Uruguay's Largest Depart-
ment** Tucuarembó
Uruguay's Largest Estuary
Rio de la Plata combining the
Paraná Uruguay rivers
Uruguay's Principal Port
Montevideo
**Uruguay's Smallest Depart-
ment** Montevideo
**U.S. Virgin Islands' Largest
City** Charlotte Amalie on St
Thomas
Utah's Highest Point King's
Peak—13,528 feet, 4,123 me-
ters
Utah's Largest City Salt Lake
City
Utah's Saltiest Inland Sea
the Great Salt Lake
Uttermost Port of the Earth
Patagonia at the southern tip
of South America
Uttermost South Cape Horn/
Tierra del Fuego region of
southern South America
south of Patagonia
Uttermost Tip of Africa
Cape Alguhas, South Africa
Uttermost Tip of Asia Singa-
pore at the tip of the Malay
Peninsula
Uttermost Tip of Australia
South East Cape, Tasmania
Uttermost Tip of Europe
Nordkin, Norway

Uttermost Tip of North America Attu Island, Alaska or Ounta Mariato, Panama

Uttermost Tip of South America Cabo de Hornos (Cape Horn), Chile or Punta Gallinas, Colombia

Vanuatu's Largest Town Vila on Efate Island

Vanuatu's Principal Port Vila

Venezuela's Easternmost City Tucupita (in the delta of the Orinoco)

Venezuela's Easternmost Town La Horqueta (on the Guyana border)

Venezuela's Highest Mountain Pico Bolivar

Venezuela's Highest Peak Bolívar—16,411 feet, 4,973 meters

Venezuela's Highest Town San Rafael de Mucuchies

Venezuela's Largest City Caracas

Venezuela's Largest Port La Guaira

Venezuela's Largest State Bolívar

Venezuela's Longest River Orinoco

Venezuela's Northernmost City Coro

Venezuela's Principal Port Maracaibo

Venezuela's Smallest State Nueva Esparta

Venezuela's Southernmost City San Cristóbal

Venezuela's Westernmost City Maracaibo

Vermont's Greatest Forest Green Mountain National Forest in western Vermont

Vermont's Highest Point Mount Mansfield—4,393 feet 1,339 meters

Vermont's Largest City Burlington

Vermont's Longest River Connecticut

Vietnam's Highest Peak Fan Si Pan—10,308 feet, 3,124 meters

Vietnam's Largest City Ho Chi Minh City (Saigon)

Vietnam's Principal Port Ho Chi Minh City (Saigon)

Virginia's Highest Point Mount Rogers—5,729 feet, 1,746 meters

Virginia's Largest City Norfolk

Virginia's Longest River Potomac

Virginia's Most Scenic Highway Blue Ridge Parkway through the southern Appalachian Mountains

Virginia's Principal Port Norfolk Harbor

Volacanos—Best Known Kilimanjaro in Tanzania, Africa; Popocatepétl, Mexico; Rainier, Washington; Pelée, Martinique; Krakatoa, Indonesia; Vesuvius and Vulcano, Italy

Wales' Highest Peak Snowdon—3,560 feet, 1,079 meters—in Caernavonshire

Wales' Largest City Cardiff

Wales's Largest Lake Bala Lake (Llyn Tegid)

Wales's Longest River the Towy

Wallis and Futuna Islands' Largest Town Mata-Utu

Wallis and Futuna Islands' Principal Port Mata-Utu

Washington State's Highest Point Mt Rainier, also called Mount Tahoma—14,410 feet, 4,392 meters

Washington State's Largest City Seattle

Washington State's Longest River Columbia

Washington State's Principal Port Seattle

Waterfalls of the World (see Highest Waterfalls)

Westernmost Hawaiian Island Nihau

Westernmost Ireland Tearagt Island off the Dingie Peninsula often called the westernmost peninsula of Europe

Westernmost Point of the Territorial United States Orate Point on Guam in the western Pacific

Westernmost State Alaska

Western Samoa's Best Port Apia on Upolu Island

Western Samoa's Largest City Apia

Western Samoa's Principal Port Apia

West Virginia's Highest Point Spruce Knob—4,862 feet, 1,482 meters

West Virginia's Largest City Charleston

West Virginia's Longest River Ohio

West Virginia's Most Historic Park Harper's Ferry National Historical Park originally surveyed by Thomas Jefferson in 1783

West Virginia's Principal Port Huntington

Widest Meteoric Crater in New Québec

Windiest Continent Antarctica

Wisconsin's Largest City Milwaukee

Wisconsin's Longest River Mississippi

Wisconsin's Most Scenic Lakeshore Apostle Islands National Lakeshore on southern Lake Superior

Wisconsin's Principal Port Superior

World's Biggest Bauxite Port Weipa, Queensland

World's Biggest Bay Bay of Bengal

World's Busiest Airport Chicago's O'Hare International

World's Busiest Border San Diego, California and Tijuana, Baja California

World's Busiest Border Crossing San Ysidro, California

World's Busiest Morgue New York City Medical Examiner's Office at First Avenue and Thirtieth Street in Manhattan

World's Busiest Seaport Rotterdam in the Netherlands

World's Coldest Places –158°F (–105.6°C) at Ornyakon, Siberia and –194°F (–125.6°C) near Vostok in Antarctica

World's Coldest Seas Arctic and Antarctic oceans (2°C or 28°F)

World's Coolest City Ulan-Bator, Mongolia

World's Crookedest River the Mississippi, according to Mark Twain

World's Deepest Gorge Hells Canyon in Idaho's Snake River—7,900 feet or 2,410 meters

World's Deepest Lake Baikal in Russia—1,742 meters or 5,714 feet

World's Driest City Arica, Chile

World's Driest Place Chile's Atacama Desert

World's First Detective François Eugène Vidocq

World's First National Park Yellowstone in Idaho, Montana, and Wyoming in 1872

World's First Nuclear-Powered Submarine USS *Nautilus*

World's First Parliament Iceland's Althing founded in 930

World's First Photographer Joseph Nicéphore Niepce

World's First Streamlined Ferry *The Kalakala*

World's First Woman Prime Minister Sirimavo Ratwatte Badaranaike of Ceylon (Sri Lanka)

World's Foremost Debating Society Oxford Union Society

World's Freest and Smallest Jail San Marino's hilltop lockup

World's Greatest Economic Choke Point Strait of Ormuz between Iran and Oman—on the Persian Gulf tanker route

World's Greatest Gorge Grand Canyon of the Colorado River

World's Greatest Investor Warren Buffett, as of the late 1990s

World's Greatest Job President of the United States

World's Greatest Library Library of the British Museum

World's Greatest Predator man

World's Greatest Railroad Terminal Grand Central Terminal in New York City

World's Greatest Tides Nova Scotia's Bay of Fundy (53 feet or 16 meters)

World's Greatest Volcanic Eruption in Ancient Times Mount Vesuvius in southwestern Italy killed 16,000 people in Herculaneum and Pompeii in 79

World's Greatest Volcanic Eruption in Modern Times Mount Pelée wiped out nearby city of St Pierre and its 40,000 people on Martinique in 1902

World's Greatest Volcanic Eruption in Recent Times Nevado del Ruiz, Colombia in mid-November 1985 killed more than 25,000 people

World's Highest Capital City La Paz, Bolivia (elevation 11,909 feet or 3630 meters)

World's Highest City Lhasa, Tibet (3,687 meters or 12,087 feet above sea level)

World's Highest Lake Titicaca in the Andes between Bolivia and Peru (3,800 meters of 12,500 feet)

World's Highest Large City La Paz, Bolivia (3,632 meters or 11,909 feet above sea level)

World's Highest Mountain Everest in Nepal (29,028 feet or 8,848 meters)

World's Highest Mountains (*see* Mountains of the World)

World's Highest Murder Rate (*see* Highest Murder Rate)

World's Highest Navigable Lake Titicaca (Bolivia)

World's Highest Point Mount Everest between Nepal and Tibet

World's Highest Tides at Burntcoat Read in the Bay of Fundy near Noel, Nova Scotia [53 feet (16 meters) and sometimes even 60 feet (18 meters)]

World's Highest Village Ancanquilca, Chile (17,500 feet or 5,334 meters)

World's Highest Waterfall Angel in Venezuela (3,297 feet or 1,005 meters)

World's Highest Waterfalls (*see* Waterfalls of the World)

World's Hottest Place Arizia, Libya (136°F or 58°C)

World's Hottest Sea Persian Gulf (365°C or 97°F)

World's Largest Active Volcano Mauna Loa on the island of Hawaii

World's Largest Archipelago Indonesia's more than 3,000 islands extending from the Indian Ocean to the western Pacific

World's Largest Art Gallery Hermitage and Winter Palace in St Petersburg

World's Largest Atoll Kwajalein in the Marshalls

World's Largest Atom Smasher Superconducting Supercollider

World's Largest Bell Tsar Kolokol III in Moscow's Kremlin

World's Largest Buddhist Nation Japan

World's Largest Cassowary single-wattled cassowary (*Casuaritts unappendiculatus*) of northern New Guinea

World's Largest (but unfinished) Cathedral St John the Divine in New York City

World's Largest Cave Big Room in New Mexico's Carlsbad Caverns

World's Largest Church Holy Roman Catholic Church

World's Largest City Tokyo-Yokohama

World's Largest Cold Current West Wind Drift (circling Antarctica and washing the southernmost shores of Africa, Australia, and South America)

World's Largest Continent Asia

World's Largest Coral Formation Australia's Great Barrier Reef

World's Largest Countries (ranked by size) Russia, Canada, China, U.S.A., Brazil

World's Largest Delta Ganges-Brahmaputra between India and Pakistan

World's Largest Democracy India

World's Largest Desert North Africa's Sahara

World's Largest Dictionary 13-volume *Oxford English Dictionary* plus supplements

World's Largest Exporter of Bananas Ecuador

World's Largest Exporter of Oil Saudi Arabia

World's Largest Fish Market Tokyo

World's Largest Flower *Rafflesia arnoldii* found in southwestern Sumatra

World's Largest Flower Auction Aalsmeer near Amsterdam's Schipol Airport

World's Largest Game Reserve Selous Game Reserve (in Tanzania)

World's Largest Gorge Grand Canyon carved by the Colorado River in Arizona

World's Largest Gothic Cathedral St John the Divine, New York City

World's Largest Green-Space City Oslo, Norway followed by Stockholm, Sweden

World's Largest Gulf Gulf of Mexico

World's Largest Hindu Nation India

World's Largest Island Greenland

World's Largest Islands (*see* Islands of the World)

World's Largest Lagoon Truk Island in Micronesia

World's Largest Lakes (*see* Largest Lakes)

World's Largest Library Library of Congress, Washington, D.C.

World's Largest Living Structure Australia's Great Barrier Reef

World's Largest Moth Atlas moth (Attacus atlas)

World's Largest Mountain Hawaii's Mauna Loa

World's Largest Museum New York City's American Museum of Natural History

World's Largest Nation Russia

World's Largest National Park Tsavo (in Kenya)

World's Largest Naval Base Norfolk, Virginia

World's Largest Number of Great Cities (with a million or more people) United States followed by China, Latin America, and middle-south Asiatic countries including India

World's Largest Ocean Pacific

World's Largest Open Sewer Mediterranean Sea from the Bosporus and Iskenderun to the Straits of Gibraltar

World's Largest Open-Space Countries the United States followed by Canada and Australia

World's Largest Passenger Ship *Queen Elizabeth 2* whose deadweight tonnage exceeds the longer *Norway* (formerly the *France*)

World's Largest Peninsula Arabian Peninsula

World's Largest Ports Port of New York and New Jersey in waterfront acreage; Port of Rotterdam in tonnage

World's Largest Postage Stamp 75-cent Marshall Islands postal adhesive measuring some 4 by 6 inches (105 by 150mm)

World's Largest Public Library New York Public Library at Fifth Avenue and 42nd Street plus its more than 80 branches

World's Largest Publisher U.S. Government Printing Office (GPO)

World's Largest River Basin Amazon

World's Largest Roman Catholic Nation Brazil

World's Largest Shark whale shark

World's Largest Shinto Community Japan

World's Largest Slum Area Mexico City's 500 slums

World's Largest System of Freshwater Lakes Great Lakes of Canada and the United States

World's Largest University State University of New York

World's Largest Walled Prison probably Southern Michigan Prison spanning 54 acres (22 hectares)

World's Largest Warm Current Gulf Stream

World's Least Populated Places Greenland, followed by French islands in the south Indian Ocean, Svalberg in the Arctic Ocean, the Falklands in the South Atlantic, the once Spanish Sahara claimed by Algeria and Morocco, French Guiana, Namibia, Mongolia, Botswana, and Libya

World's Least Populous Nation Tuvalu

World's Loftiest Metropolis La Paz, Bolivia

World's Loneliest Meeting Place Isla de Pascua (Easter Island) in the South Pacific

World's Longest Railroad Trans-Siberian from Moscow to Vladivostok

World's Longest Rail-and-Road Bridge connects Honshu and Shikoku islands

World's Longest Railway Tunnel 13-mile-long (21-kilometer-long) Dai-shimuzu in Japan

World's Longest Reef Australia's Great Barrier

World's Longest River Nile

World's Longest Rivers (*see* Rivers of the World)

World's Longest Suspension Bridge Verrazano-Narrows Bridge in New York Harbor

World's Lowest City Brawley, California (184 feet or 56 meters below sea level)

World's Lowest Lake Dead Sea between Israel and Jordan (394 meters or 1292 feet below sea level)

World's Lowest Point Dead Sea between Israel and Jordan

World's Lowest Settlement Ein Bobek on the shores of the Dead Sea (396 meters or 1299 feet below sea level)

World's Main Choke Point for the Flow of Oil Strait of Hormuz connecting Persian Gulf countries with the Indian Ocean

World's Most Active Volcano Kilauea on the island of Hawaii

World's Most Easterly City Gisborne, New Zealand

World's Most Famous Beach Daytona Beach, Florida

World's Most Famous Painting the *Mona Lisa*

World's Most Isolated City Perth, Western Australia

World's Most Northerly City Hammerfest, Norway

World's Most Northerly Settlement Alert, Canada

World's Most Populated Country China

World's Most Populated Islands Barbados, Haiti, Hong Kong, Jamaica, Java, Puerto Rico, Trinidad.

**World's Most Populated
Place** Macao
**World's Most Population-
Exploding Nation** India
**World's Most Populous
Nation** China
World's Most Southerly City
Punta Arenas, Chile
World's Most Westerly City
Nome, Alaska or Pago Pago,
Samoa
**World's Most Widely Preva-
lent Carnivore** the red fox
**World's Most Widely Spoken
Language** English
**World's Northernmost Point
of Land** Greenland's Cape
Morris Jesup
World's Oldest Capital City
Damascus, Syria
**World's Oldest Constitutional
Democracy** the United
States of America
**World's Oldest Cultivated
Grain** barley, many food his-
torians believe
World's Oldest Kingdom
Denmark
**World's Oldest Medical Jour-
nal** *New England Journal* of
Medicine
World's Oldest Monarchy
Denmark
World's Oldest Parliament
Iceland's *Althing* founded in
930
World's Oldest Profession
prostitution
World's Only Marine Lizard
Galdpagos Marine Iguana
**World's Only Poisonous Liz-
ards** gila monster of the
American southwest; Mexi-
can beaded lizard called *esc-
orpion;* Borneo's *Lantha-
notus*

World's Rainiest City
Monrovia, Liberia
World's Rainiest Place (*see*
Rainiest Spot)
World's Richest Countries
(*see* Richest Per-Capita
Countries in the World)
**World's Second-Oldest Pro-
fession** arms making and
trading
World's Shortest Poem *I—
why?* by Eli Siegal
World's Smallest Bird bee
hummingbird (*Mellisuga be-
lenae*) of Cuba at 2.17 inches
long
World's Smallest Continent
Australia
World's Smallest Monkey
pygmy marmoset about 6
inches (14 cm) from nose to
base of tail
World's Smallest Owl elf owl
**World's Smallest Sovereign
State** Vatican City within
Rome occupies 44 hectares
(109 acres)
World's Smoggiest City Mex-
ico City
World's Southernmost Town
Puerto Williams, Chile
World's Tallest Bird New
Zealand's *Diornis maximus,*
an extinct flightless ostrich-
like bird measuring 13 feet
(396 cm) in height
World's Tallest Trees Califor-
nia's redwoods, towering 367
feet (112 *meters) high*
World's Warmest City Tim-
buktu, Mali
World's Wettest City Monro-
via, Liberia
World's Wettest Place (*see*
Rainiest Spot)

World's Windiest Place
Adélie coast of Antarctica
World's Worst Shipwreck
unsinkable superliner *Titanic*
struck an iceberg on April 14,
1912 and sank in the North
Atlantic south of Newfound-
land with a loss of more than
1,500 lives
Wyoming's Highest Point
Gannett Peak—13,785 feet,
4201 meters
Wyoming's Largest City
Casper
**Wyoming's Most Spectacular
Park** Yellowstone National
Park
**Youngest Central American
Democracy** Belize (for-
merly British Honduras)
Yugoslavia's Largest City
Belgrade
Yugoslavia's Longest River
Danube
Yukon's Highest Point Mount
Logan—19,850 feet, 6,050
meters
Yukon's Largest City White-
horse
Zaire's Highest Peak Mount
Stanley—16,795 feet, 5,089
meters
Zaire's Largest City Kinshasa
Zaire's Principal Port Matadi
Zambia's Largest City Lusaka
Zimbabwe's Highest Peak
Inyanga—8,514 feet, 2,580
meters
Zimbabwe's Largest City
Harare

U.S. Naval Ship Symbols

AALC Amphibious Assault Landing Craft

AD Destroyer Tender

ADG Degaussing Ship

AE Ammunition Ship

AF Store Ship

AFDB Large Auxiliary Floating Dry Dock (non-self-propelled)

AFDL Small Auxiliary Floating Dry Dock (non-self-propelled)

AFDM Medium Auxiliary Floating Dry Dock (non-self-propelled)

AFS Combat Store Ship

AG Miscellaneous

AGDE Escort Research Ship

AGDS Deep Submergence Support Ship

AGEH Hydrofoil Research Ship

AGER Environmental Research Ship

AGF Miscellaneous Command Ship

AGFF Frigate Research Ship

AGM Missile Range Instrumentation Ship

AGMR Major Communications Relay Ship

AGOR Oceanographic Research Ship

AGOS Ocean Surveillance Ship

AGP Patrol Craft Tender

AGR Radar Picket Ship

AGS Surveying Ship

AGSS Auxiliary Research Submarine

AGT Transport Oiler

AGTR Technical Research Ship

AH Hospital Ship

AK Cargo Ship

AKD Cargo Ship, Dock

AKL Light Cargo Ship

AKR Vehicle Cargo Ship

AKS Stores Issue Ship

AKV Cargo Ship and Aircraft Ferry

ALS Auxiliary Lighter

ANL Net Laying Ship

AO Oiler

AOE Fast Combat Support Ship

AOG Gasoline Tanker

AOR Replenishment Oiler

AP Transport

APB Self-propelled Barracks Ship

APL Barracks Craft (non-self-propelled)

AR Repair Ship

ARB Battle Damage Repair Ship

ARC Cable Repairing Ship

ARD Auxiliary Repair Dry Dock (non-self-propelled)

ARDM Medium Auxiliary Repair Dry Dock (non-self-propelled)

ARG Internal Combustion Engine Repair Ship

ARL Landing Craft Repair Ship

ARS Salvage Ship

ARSD Salvage Lifting Ship

ARST Salvage Craft Tender

ARVA Aircraft Repair Ship (aircraft)

ARVE Aircraft Repair Ship (engine)

ARVH Aircraft Repair Ship (helicopter)

AS Submarine Tender

ASPB Assault Support Patrol Boat

ASR Submarine Rescue Ship

ATA Auxiliary Ocean Tug

ATC Mini-Armored Patrol Vehicle

ATF Fleet Ocean Tug

ATS Salvage Tug

ATSS Auxiliary Training Submarine

AV Seaplane Tender

AVM Guided Missile Ship

AVS Aviation Supply Ship

AVT Auxiliary Aircraft Transport

AW Distilling Ship

BB Battleship

CA Heavy Cruiser

CC Command Ship

CCB Command and Control Boat

CG Guided Missile Cruiser

CGN Guided Missile Cruiser (nuclear propulsion)

CL Light Cruiser

CLG Guided Missile Light Cruiser

CVA Attack Aircraft Carrier

CVAN Attack Aircraft Carrier (nuclear propulsion)

CVS Antisubmarine Warfare Support Aircraft Carrier

CVT Training Aircraft Carrier

DD Destroyer

DDG Guided Missile Destroyer

DE Escort Ship

DEG Guided Missile Escort

DER Radar Picket Escort Ship

DL Frigate

DLG Guided Missile Frigate

DLGN Guided Missile Frigate (nuclear propulsion)

DSRV Deep Submergence Rescue Vessel

DSV Deep Submergence Vehicle

E (Prefix) Experimental Ship

F (Prefix) Ship being built by US for a foreign nation

FDL Fast Deployment Logistics Ship

IX Unclassified Miscellaneous

LCA Landing Craft, Assault

LCC Amphibious Command Ship

LCM Landing Craft, Mechanized

LCPL Landing Craft, Personnel Large

LCPR Landing Craft, Personnel, Ramped

LCSR Landing Craft Swimmer Reconnaissance

LCU Landing Craft, Utility

LCVP Landing Craft, Vehicle, Personnel

LFR Inshore Fire Support Ship

LFS Amphibious Fire Support Ship

LHA Amphibious Assault Ship (general purpose)

LKA Amphibious Cargo Ship

LPA Amphibious Transport

LPD Amphibious Transport Dock

LPH Amphibious Assault Ship

LPR Amphibious Transport (small)

LPSS Amphibious Transport Submarine

LSD Dock Landing Ship

LSSC Light SEAL Support Craft

LST Tank Landing Ship

LWT Amphibious Warping Tug

MAC MIUW Attack Craft

MCS Mine Countermeasures Ship

MON Monitor

MSB Minesweeping Boat

MSC Minesweeper, Coastal (nonmagnetic)

MSD Minesweeper, Drone

MSF Minesweeper, Fleet (steel hull)

MSI Minesweeper, Inshore

MSL Minesweeping Launch

MSM Minesweeper, River (Converted LCM-6)

MSO Minesweeper, Ocean (nonmagnetic)

MSR Minesweeper, Patrol

MSS Minesweeper, Special (device)

MSSC Medium SEAL Support Craft

NR Submersible Research Vehicle (nuclear propulsion)

PBR River Patrol Boat

PCE Patrol Escort

PCER Patrol Rescue Escort

PCF Patrol Craft, Fast

PCH Patrol Craft (hydrofoil)

PG Patrol Gunboat

PGH Patrol Gunboat (hydrofoil)

PTF Fast Patrol Craft

QFB Quiet Fast Boat

RUC Riverine Utility Craft

SDV Swimmer Delivery Vehicle

SES Surface-Effect Ship

SLWT Side Loading Warping Tug

SS Submarine

SSBN Fleet Ballistic Missile Submarine (nuclear propulsion)

SSG Guided Missile Submarine

SSN Submarine (nuclear propulsion)

SST Target and Training Submarine (self-propelled)

STAB Strike Assault Boat

SWCL Special Warfare Craft, Light

SWCM Special Warfare Craft, Medium

T (Prefix) Military Sealift Command Ship

W (Prefix) U.S. Coast Guard Ship

X Submersible Craft (self-propelled)

YAG Miscellaneous Auxiliary (self-propelled)

YC Open Lighter (non-self-propelled)

YCF Car Float (non-self-propelled)

YCV Aircraft Transportation Lighter (non-self-propelled)

YD Floating Crane (non-self-propelled)

YDT Diving Tender (non-self-propelled)

YF Covered Lighter (self-propelled)

YFB Ferryboat or Launch (self-propelled)

YFD Yard Floating Dry Dock (non-self-propelled)

YFN Covered Lighter (non-self-propelled)

YFNB Large Covered Lighter (non-self-propelled)

YFND Dry Dock Companion Craft (non-self-propelled)

YFNX Lighter (special purpose) (non-self-propelled)

YFP Floating Power Barge (non-self-propelled)

YFR Refrigerated Covered Lighter (self-propelled)

YFRN Refrigerated Covered Lighter (non-self-propelled)

YFRT Covered Lighter (range-tender) (self-propelled)

YFU Harbor Utility Craft (self-propelled)

YG Garbage Lighter (self-propelled)

YGN Garbage Lighter (non-self-propelled)

YHLC Salvage Lift Craft, Heavy (non-self-propelled)

YLLC Salvage Lift Craft, Light (self-propelled)

YM Dredge (self-propelled)

YMLC Salvage Lift Craft, Medium (non-self-propelled)

YNG Gate Craft (non-self-propelled)

YO Fuel Oil Barge (self-propelled)

YOG Gasoline Barge (self-propelled)

YOGN Gasoline Barge (non-self-propelled)

YON Fuel Oil Barge (non-self-propelled)

YOS Oil Storage Barge (non-self-propelled)

YP Patrol Craft (self-propelled)

YPD Floating Pile Driver (non-self-propelled)

YR Floating Workshop (non-self-propelled)

YRB Repair and Berthing Barge (non-self-propelled)

YRBM Repair, Berthing and Messing Barge (non-self-propelled)

YRDH Floating Dry Dock Workshop (hull) (non-self-propelled)

YRDM Floating Dry Dock Workshop (machine) (non-self-propelled)

YRR Radiological Repair Barge (non-self-propelled)

YRST Salvage Craft Tender (non-self-propelled)

YSD Seaplane Wrecking Derrick (self-propelled)

YSR Sludge Removal Barge (non-self-propelled)

YTB Large Harbor Tug (self-propelled)

YTL Small Harbor Tug (self-propelled)

YTM Medium Harbor Tug (self-propelled)

YW Water Barge (self-propelled)

YWDN Water Distilling Barge (non-self-propelled)

YWN Water Barge (non-self-propelled)

Vehicle Registration Symbols (Index Markers)

British Isles Symbols

This listing includes registration marks for the Republic of Ireland. These marks were introduced before the creation of the Republic of Ireland, which was continued with the same system.

A	London	BW	Oxfordshire	DS	Peeblesshire
AA	Hampshire	BX	Carmarthenshire	DT	Doncaster
AB	Worchestershire	BY	London	DU	Coventry
AC	Warwickshire	BZ	Down	DV	Devon
AD	Gloucestershire	C	Yorkshire (WR)	DW	Newport (Mon)
AE	Bristol	CA	Denbighshire	DX	Ipswich
AF	Cornwall	CB	Blackburn	DY	Hastings
AG	Ayrshire	CC	Caernarvonshire	DZ	Antrim
AH	Norfolk	CD	Brighton	E	Staffordshire
AI	Meath	CE	Cambridgeshire	EA	West Bromwich
AJ	Yorkshire (NR)	CF	Suffolk (West)	EB	Cambridge
AK	Bradford	CG	Hampshire	EC	Westmorland
AL	Nottinghamshire	CH	Derby	ED	Warrington
AM	Wilshire	CI	Laoighis	EE	Grimsby
AN	London	CJ	Herefordshire	EF	West Hartlepool
AO	Cumberland	CK	Preston	EG	Huntingdon
AP	Sussex (East)	CL	Norwich	EH	Stone-on-Trent
AR	Hertfordshire	CM	Birkenhead	EI	Sligo
AS	Nairnshire	CN	Gateshead	EJ	Cardiganshire
AT	Kingston upon Hull	CO	Plymouth	EK	Wigan
AU	Nottingham	CP	Halifax	EL	Bournemouth
AV	Aberdeenshire	CR	Southampton	EM	Bootle
AW	Salop	CS	Ayshire	EN	Bury
AX	Monmouthshire	CT	Lincolnshire (Kesteven)	EO	Berrow-in-Furness
AY	Leicestershire	CU	South Shields	EP	Montgomeryshire
AZ	Belfast	CV	Cornwall	ER	Cambridgeshire
B	Lancashire	CW	Burnley	ES	Perthshire
BA	Salford	CX	Huddersfield	ET	Rotherham
BB	Newcastle upon Tyne	CY	Swansea	EU	Breconshire
BC	Leicester	CZ	Belfast	EV	Essex
BD	Northamptonshire	D	Kent	EW	Huntingdonshire
BE	Lincolnshire (Lindsey)	DA	Wolverhampton	EX	Great Yarmouth
BF	Staffordshire	DB	Stockport	EY	Anglesey
BG	Burkenhead	DC	Teesside	EZ	Belfast
BH	Buckinghamshire	DD	Gloucestershire	F	Essex
BI	Monaghan	DE	Pembrokeshire	FA	Burton-on-Trent
BJ	Suffolk (East)	DF	Gloucestershire	FB	Bath
BK	Portsmouth	DG	Gloucestershire	FC	Oxford
BL	Berkshire	DH	Walsall	FD	Dudley
BM	Bedfordshire	DI	Roscommon	FE	Lincoln
BN	Bolton	DJ	St Helens	FF	Merionethshire
BO	Cardiff	DK	Rochdale	FG	Fife
BP	Sussex (West)	DL	Isle of Wight	FH	Gloucester
BR	Sunderland	DM	Flintshire	FI	Tipperary (NR)
BS	Orkney	DN	York	FJ	Exeter
BT	Yorkshire (ER)	DO	Lincolnshire (Holland)	FK	Worcester
BU	Oldham	DP	Reading	FL	Huntingdon
BV	Blackburn	DR	Plymouth	FM	Chester

FN	Canterbury	HZ	Tyrone	KN	Kent
FO	Radnorshire	IA	Antrim	KO	Kent
FP	Rutland	IB	Armagh	KP	Kent
FR	Blackpool	IC	Carlow	KR	Kent
FS	Edinburgh	ID	Cavan	KS	Rosburghsh
FT	Tynemouth	IE	Clare	KT	Kent
FU	Lincolnshire (Lindsey)	IF	Cork (County)	KU	Bradford
FV	Blackpool	IH	Donegal	KV	Coventry
FW	Lincolnshire (Lindsey)	IJ	Down	KW	Bradford
FX	Dorset	IK	City and County of Dublin	KX	Buckinghamshire
FY	Southport	IL	Fermanagh	KY	Bradford
FZ	Belfast	IM	Galway	KZ	Antrim
G	Glasgow	IN	Kerry	L	Glamorgan
GA	Glasgow	IO	Kildare	LA	London
GB	Glasgow	IP	Kikenny	LB	London
GC	London	IR	Offaly	LC	London
GD	Glasgow	IT	Leitrim	LD	London
GE	Glasgow.	IU	Limerick	LE	London
GF	London	IW	Londonderry	LF	London
GG	Glasgow	IX	Longford	LG	Cheshire
GH	London	IY	Louth	LH	London
GI	London	IZ	Mayo	LI	Westmeath
GK	London	J	Durham (County)	LJ	Bournemouth
GL	Bath	JA	Stockport	LK	London
GM	Motherwell & Wishaw	JB	Berkshire	LL	London
GN	London	JC	Caenarvonshire	LM	London
GO	London	JD	London	LN	London
GP	London	JE	Cambridge	LO	London
GR	Sunderland	JF	Leicester	LP	London
GS	Perthshire	JG	Canterbury	LR	London
GT	London	JH	Hertfordshire	LS	Selkirkshire
GU	London	JI	Tyrone	LT	London
GV	Suffolk (West)	JJ	London	LU	London
GW	London	JK	Eastbourne	LV	Liverpool
GX	London	JL	Lincolnshire (Holland)	LW	London
GY	London	JM	Westmorland	LX	London
GZ	Belfast	JN	Southend	LY	London
H	London	JO	Oxford	LZ	Armagh
HA	Warley	JP	Wigan	M	Cheshire
HB	Merthyr Tydfil	JR	Northumberland	MA	Cheshire
HC	Eastbourne	JS	Ross & Cromarty	MB	Cheshire
HD	Dewsbury	JT	Dorset	MC	London
HE	Barnsley	JU	Leicestershire	MD	London
HF	Wallasey	JV	Grimsby	ME	London
HG	Burnley	JW	Wolverhampton	MF	London
HH	Carlisle	JX	Halifax	MG	London
HI	Tipperary	JY	Plymouth	MH	London
HJ	Southend	JZ	Down	MI	Wexford
HK	Essex	K	Liverpool	MJ	Bedfordshire
HL	Wakefield	KA	Liverpool	MK	London
HM	London	KB	Liverpool	ML	London
HN	Darlington	KC	Liverpool	MM	London
HO	Hampshire	KD	Liverpool	MN	Isle of Man
HP	Coventry	KE	Kent	MO	Berkshire
HR	Wiltshire	KF	Liverpool	MP	London
HS	Renfrewshire	KG	Cardiff	MR	Wittshire
HT	Bristol	KH	Kingston-upon-Hull	MS	Stirlingshire
HU	Bristol	KI	Waterford	MT	London
HV	London	KJ	Kent	MU	London
HW	Bristol	KK	Kent	MV	London
HX	London	KL	Kent	MW	Wiltshire
HY	Bristol	KM	Kent	MX	London

MY	London	PH	Surrey	SN	Dunbartonshire
MZ	Belfast	PI	Cork	SO	Moray
N	Manchester	PJ	Surrey	SP	Fife
NA	Manchester	PK	Surrey	SR	Angus
NB	Manchester	PL	Surrey	SS	East Lothian
NC	Manchester	PM	Sussex (East)	ST	Inverness-shire
ND	Manchester	PN	Sussex (East)	SU	Kincardineshire
NE	Manchester	PO	Sussex (West)	SV	Kinross-shire
NF	Manchester	PP	Buckinghamshire	SW	Kircudbrightshire
NG	Norfolk	PR	Dorset	SX	West Lothian
NH	Northampton	PS	Zetland	SY	Midlothian
NI	Wicklow	PT	Durham (County)	SZ	Down
NJ	Sussex (East)	PU	Essex	T	Devon
NK	Hertfordshire	PV	Ipswich	TA	Devon
NL	Northumberland	PW	Norfolk	TB	Lancashire
NM	Bedfordshire	PX	Sussex (West)	TC	Lancashire
NN	Nottinghamshire	PY	Yorkshire (NR)	TD	Lancashire
NO	Essex	PZ	Belfast	TE	Lancashire
NP	Worcestershire	QA QB QC QD QE QF QG		TF	Lancashire
NR	Leicestershire	QH QJ QK QL QM QN		TG	Glamorgan
NS	Sutherland	QO QP QS London for ve-		TH	Carmarthenshire
NT	Salop	hicles temporarily imported		TI	Limerick
NU	Derbyshire	from abroad		TJ	Lancashire
NV	Northamptonshire	R	Derbyshire	TK	Dorset
NW	Leeds (B)	RA	Derbyshire	TL	Lincolnshire (Kesteven)
NX	Warwickshire	RB	Derbyshire	TM	Bedfordshire
NY	Glamorgan	RC	Derby	TN	Newcastle upon Tyne
NZ	Londonderry	RD	Reading	TO	Nottingham
O	Birmingham	RE	Staffordshire	TP	Portsmouth
OA	Birmingham	RF	Staffordshire	TR	Southampton
OB	Birmingham	RG	Aberdeen	TS	Dundee
OC	Birmingham	RH	Kingston-upon-Hull	TT	Devon
OD	Devon	RI	City and County of Dublin	TU	Cheshire
OE	Birmingham	RJ	Salford	TV	Nottingham
OF	Birmingham	RK	London	TW	Essex
OG	Birmingham	RL	Cornwall	TX	Glamorgan
OH	Birmingham	RM	Cumberland	TY	Northumberland
OI	Belfast	RN	Preston	TZ	Belfast
OJ	Birmingham	RO	Hertfordshire	U	Leeds
OK	Birmingham	RP	Northamptonshire	UA	Leeds
OL.	Birmingham	RR	Nottinghamshire	UB	Leeds
OM	Birmingham	RS	Aberdeen	UC	London
ON	Birmingham	RT	Suffolk (East)	UD	Oxfordshire
OO	Essex	RU	Bournermouth	UE	Warwickshire
OP	Birmingham	RV	Portsmouth	UF	Brighton
OR	Hampshire	RW	Coventry	UG	Leeds
OS	Wigtownshire	RX	Berkshire	UH	Cardiff
OT	Hampshire	RY	Leicester	UI	Londonderry
OU	Hampshire	RZ	Antrim	UJ	Salop
OV	Birmingham	S	Edinburgh	UK	Wolverhampton
OW	Southampton	SA	Aberdeenshire	UL	London
OX	Birmingham	SB	Argyll	UM	Leeds
OY	London	SC	Edinburgh	UN	Denbighshire
OZ	Belfast	SD	Ayrshire	UO	Devon
P	Surrey	SE	Banffshire	UP	Durham (County)
PA	Surrey	SF	Edinburgh	UR	Hertfordshire
PB	Surrey	SG	Edinburgh	US	Glasgow
PC	Surrey	SH	Berwickshire	UT	Leicestershire
PD	Surrey	SJ	Bute	UU	London
PE	Surrey	SK	Caithness	UV	London
PF	Surrey	SL	Clarkmannanshire	UW	London
PG	Surrey	SM	Dumfriesshire	UX	Salop

UY	Worcestershire	WR	Yorkshire (WR)	YI	City and County of Dublin
UZ	Belfast	WS	Edinburgh	YJ	Dundee
V	Lanarkshire	WT	Yorkshire (WR)	YK	London
VA	Lanarkshire	WU	Yorkshire (WR)	YL	London
VB	London	WV	Wiltshire	YM	London
VC	Coventry	WW	Yorkshire (WR)	YN	London
VD	Lanarkshire	WX	Yorkshire (WR)	YO	London
VE	Cambridgeshire	WY	Yorkshire (WR)	YP	London
VF	Norfolk	WZ	Belfast	YR	London
VG	Norwich	X	Northumberland	YS	Glasgow
VH	Huddersfield	XA	London/Kirkcaldy	YT	London
VJ	Herefordshire	XB	London/Coatbridge	YU	London
VK	Newcastle upon Tyne	XC	London/Solihull	YV	London
VL	Lincoln	XD	London/Luton	YW	London
VM	Manchester	XE	London/Luton	YX	London
VN	Yorkshire (NR)	XF	London/Torbay	YY	London
VO	Nottinghamshire	XG	Teesside	YZ	Londonderry
VP	Birmingham	XH	London	Z	City and County of Dublin
VR	Manchester	XI	Belfast	ZA	City and County of Dublin
VS	Greenock	XJ	Manchester	ZB	Cork (County)
VT	Stoke-on-Trent	XK	London	ZC	City and County of Dublin
VU	Manchester	XL	London	ZD	City and County of Dublin
VV	Northampton	XM	London	ZE	City and County of Dublin
VW	Essex	XN	London	ZF	Cork
VX	Essex	XO	London	ZH	City and County of Dublin
VY	York	XP	London	ZI	City and County of Dublin
VZ	Tyrone	XR	London	ZJ	City and County of Dublin
W	Sheffield	XS	Paisley	ZK	Cork (County)
WA	Sheffield	XT	London	ZL	City and County of Dublin
WB	Sheffield	XU	London	ZM	Galway
WC	Essex	XV	London	ZN	Meath
WD	Warwickshire	XW	London	ZO	City and County of Dublin
WE	Sheffield	XX	London	ZP	Donegal
WF	Yorkshire (ER)	XY	London	ZR	Wexford
WG	Stirlingshire	XZ	Armagh	ZT	Cork (County)
WH	Bolton	Y	Somerset	ZU	City and County of Dublin
WI	Waterford	YA	Somerset	ZW	Kildare
WJ	Sheffield	YB	Somerset	ZX	Kerry
WK	Coventry	YC	Somerset	ZY	Louth
WL	Oxford	YD	Somerset	ZZ	Dublin for vehicles temporarily imported from abroad
WM	Southport	YE	London		
WN	Swansea	YF	London		
WO	Monmouthshire	YG	Yorkshire (WR)		
WP	Worcestershire	YH	London		

Weather Symbols (Beaufort Scales)
with corresponding Sea State Codes

Beaufort number	Wind speed				Seaman's tearm	U.S. Weather Bureau term	Effects observed at sea	Effects observed on land	Hydrographic Office		International	
	knots	mph	meters per second	km per hour					Term and height of waves, in feet	Code	Term and height of waves, in feet	Code
0	under 1	under 1	0.0–0.2	under 1	Calm		Sea like mirror.	Calm; smoke rises vertically	Calm, 0	0	Calm, glassy, 0	0
1	1–3	1–3	0.3–1.5	1–5	Light air	Light	Ripples with appearance of scales; no foam crests	Smoke drift indicates wind direction; vanes do not move	Smooth, less than 1	1	Rippled, 0–1	1
2	4–6	4–7	1.6–3.3	6–11	Light breeze	Light	Small wavelets; crests of glassy appearance, not breaking	Wind felt on face; leaves rustle; vanes begin to move	Slight, 1–3	2	Smooth, 1–2	2
3	7–10	8–12	3.4–5.4	12–19	Gentle breeze	Gentle	Large wavelets; crests begin to break; scattered whitecaps	Leaves, small twigs in constant motion; light flags extended	Moderate, 3–5	3	Slight, 2–4	3
4	11–16	13–18	5.5–7.9	20–28	Moderate breeze	Moderate	Small waves, becoming longer; numerous whitecaps	Dust, leaves, and loose paper rasied up; small branches move	Rough, 5–8	4	Moderate, 4–8	4
5	17–21	19–24	8.0–10.7	29–38	Fresh breeze	Fresh	Moderate waves, taking longer form; many whitecaps; some spray	Small trees in leaf begin to sway			Rough, 8–13	5
6	22–27	25–31	10.8–13.8	39–49	Strong breeze	Strong	Larger waves forming; whitecaps everywhere; more spray	Larger branches of trees in motion; whistling heard in wires	Very rough, 8–12	5	Very rough, 13–20	6
7	28–33	32–38	13.9–17.1	50–61	Moderate gale	Strong	Sea heaves up; white foam from breaking waves begins to be blown in streaks	Whole trees in motion; resistance felt in walking against wind				
8	34–40	39–46	17.2–20.7	62–74	Fresh gale	Gale	Moderately high waves of greater length; edges of crests begin to break into spindrift; foam is blown in well-marked streaks	Twigs and small branches broken off trees; progress generally impeded	High, 12–20	6		
9	41–47	47–54	20.8–24.4	75–88	Strong gale	Gale	High waves; sea begins to roll; dense streaks of foam; spray may reduce visibility	Slight structural damage occurs; slate blown from roofs	Very high, 20–40	7	High, 20–30	7
10	48–55	55–63	24.5–28.4	89–102	Whole gale	Whole gale	Very high waves with overhanging crests; sea takes on white appearance as foam is blown in very dense streaks; rolling is heavy and visibility reduced	Trees broken or uprooted; considerable structural damage occurs				
11	56–63	64–72	28.5–32.6	103–117	Storm	Whole gale	Exceptionally high waves; sea covered with white foam patches; visibility still more reduced		Mountainous, 40 and higher	8	Very high, 30–45	8
12	64–71	73–82	32.7–36.9	118–133	Hurricane	Hurricane	Air filled with foam; sea completely white with driving spray; visibility greatly reduced	Usually accompanied by widespread damage	Confused	9	Phenomenal, over 45	9
13	72–80	83–92	37.0–41.4	134–149								
14	81–89	93–103	41.5–46.1	150–166								
15	90–99	104–114	46.2–50.9	167–183								
16	100–108	115–125	51.0–56.0	184–201								
17	100–118	126–136	56.1–61.2	202–220								

Note: Since January 1, 1955, weather map symbols have been based upon wind speed in knots, at five-knot intervals, rather than upon Beaufort number.

Wedding Anniversary Symbols

1st — *Paper* (negotiable paper such as bonds, currency, trust certificates, as well as books, napkins, stationery, and towels)

2nd — *Cotton* (bedspreads, curtains, draperies, pillows, sheets, shirts, socks, underwear, etc.)

3rd — *Leather* (belts, handbags, leatherbound books, luggage, shoes, etc.)

4th — *Linen* (bedsheets, napkins, samplers, scarfs, shirts, tablecloths)

5th — *Wood* (furniture as well as boats and bungalows)

6th — *Iron* (hardware, wrought-iron furniture, ornamental ironwork)

7th — *Wool* (blankets, robes, rugs, socks, suits, sweaters, underwear)

8th — *Bronze* (bells, brassware, bronze objects, gongs, statuary)

9th — *Pottery* (kitchenware, planter's pots, pottery ornaments)

10th — *Aluminum* or *Tin* (Kitchenware and ornaments)

11th — *Steel* (automobiles, hardware, recreation vehicles, tools)

12th — *Silk* (casual clothes, scarfs, wraps)

13th — *Lace* (bedspreads, curtains, doilies, tablecloths)

14th — *Ivory* (carvings, desk sets, scrimshaw)

15th — *Crystal* (crystal sculpture and glassware)

20th — *China* (chinaware and porcelain figurines and tableware)

25th — *Silver* (silver coins and silverware)

30th — *Pearl* (jewelry and mother-of-pearl objects)

35th — *Coral* (jewelry and rare collector's items)

40th — *Ruby* (jewelry)

45th — *Sapphire* (jewelry)

50th — *Golden* (gold coins, gold-plated, solid-gold ornaments)

55th — *Emerald* (jewelry)

60th — *Diamond* (jewelry)

65th — *Diamond-and-gold anniversary* (jewelry)

70th — *Diamond-and-emerald anniversary* (jewelry)

75th — *Diamond-emerald-sapphire anniversary* (solid gold dipped in diamond, emerald, and sapphire chips or stones)

80th — (consult your nearest jeweler; contact the media and the police if you have accumulated all the foregoing wedding anniversary gifts; treat yourself to whatever you want)—this is the *time-flies anniversary* and may earn you a place in the *Guinness Book of World Records*

Winds and Rains of the World

The wind bloweth where it listeth.

—John 3:8

Aajej southern Moroccan whirlwind

Abroholos Easterly squall sweeping Brazil's southern coast between May and August

Afer hot southwest wind in Italy, so called because it comes from Africa; also called Africanus ventus (the African wind), Africino, Africo, Africuo

Alizé trade wind from the northeast in southern France

Anabatic an uphill wind resulting from local heating

Antitrades winds blowing above the trade winds but in opposite directions

Apheliotes (Greek—East Wind)

Arashi a storm wind in Japan

Arifi Sirocco wind in Morocco

Asifa-T circular tropical storm of the Arabian Gulf

Auger dust devil found in California

Austru dry, cold winter westerly wind of the Lower Danube

Avalaison steady west wind of western France

Aziab hot, wet, southerly wind of the Red Sea

Bad-i-sad-o-bist roz (Persian—120-day wind)—northerly dust-and-salt-laden wind blowing over Seistan province of Iran from June through September

Baguio circular tropical storm in the Philippines

Bali strong easterly wind in Java

Barat strong, blustery northwester in Sulawesi, Indonesia

Barber icy wind of midwestern Canada and United States

Barine westerly gale from the sea on Venezuela's coast

Bat Furan (Arabic—Open-Sea Season)—when northeast or winter monsoon wafts over

the Arabian Sea with light winds favoring small sailing vessels

Bat Hiddan (Arabic—Closed-Sea Season)—when southwest or summer monsoon agitates the Arabian Sea with high winds

Bayomo gusty northerly wind on flanks of Cuba's Sierra Maestra

Bergwind foehn wind of South Africa's south coast

Bise cold and dry northerly wind of southern France and Switzerland

Blaast cold, wet gusts in Scotland

Black Roller dust storm common to western United States

Blizzard cold northerly gale occurring during winter months in Canadian provinces and north United States; great Blizzard of 1888 covered much of Canada and northern United States; needlelike ice crystals and fine dry snow make up the blizzard's pattern of penetrating cold

Blood rain rain tinted red from dust

Blossom showers rains from March through May in monsoon regions of Southeast Asia

Bofu gale in Japan

Bohorok foehn wind of Sumatra

Bora cold north wind blowing over the Adriatic and originating in the Dinaric Alps

Boreas (Greek—North Wind)

Brave West Wind westerly winds of the southern hemisphere

Breath of the Sahara the Sirocco

Breva and Tivano afternoon and morning winds blowing over Lake Como—Breva

blows from north to south, Tivano blows from south to north

Brickfielder dusty hot wind originating in sandy wastes of central Australia

Brisa northeasterly wind along the west coast of South America

Brisote northeastern trade wind refreshing Cuba

Broboe dry southeastern trade of Sulawesi, Indonesia

Buran blizzard of Central Asia

Burster southerly wind of New South Wales

Cacimbo heavy mists and low clouds on the coast of Angola

Canterbury Northwester hot dry wind sometimes blowing over New Zealand

cat's fir Guatemalan mist

Caurus the Northwest Wind

Chemsin (Arabic—Sirocco)

Chorgui Moroccan name for the Sirocco

Chichili Algerian name for the Sirocco

Chili Tunisian name for the Sirocco

Chinook foehn wind blowing over the plains east of the Rockies from northern Canada to southern Colorado; warm southwesterly wind characteristic of the lower Columbia River of Oregon and Washington

Choclatero chocolate-colored dusty wind common about Yucatan

Chubasco rain-filled violent wind threatening west coast of Mexico from May to November

Colla fierce wind of gale strength in the Philippines

Collada strong wind in the Gulf of California

Cordonazo de San Francisco autumnal equinox falling close to St Francis Day—4 October—and often ushered in by a short but violent hurricane; blow struck with a knotted cord or rope like one worn by St Francis; storm felt along west coast of Central America and Mexico around St Francis Day during autumnal equinox

Coromuel southerly land breeze felt from November to May and from sunset to about 9 A.M. around La Paz and nearby entrance to Gulf of California

Criador moisture-laden wind in Spain

Cyclones counterclockwise winds of the northern hemisphere—often called tropical cyclones: hurricane in West Indies, typhoon in China Sea, willy-willy off northwestern Australia, clockwise winds of the southern hemisphere—frequently of great force and considerable duration

Datoo ocean wind refreshing Gibraltar

Doctor sea breeze refreshing inhabitants of African coasts and west coast of Australia

Dust Bowl area suffering from dust storms as in Oklahoma, West Texas, New Mexico, and Arizona

Dust Devil harmless whirls of dust ascending from the desert floor as high as 3000 feet; may be as wide as 10 feet

Easter wind on Oregon's coast

East Wind rainy wind characteristic of many places such as England, New England

Eecatl (Aztec—Wind)—derived from the wind god Quetzalcoatl

Eliseos northeast trade wind in Spain

Etesian Wind northerly summer wind found in the eastern Mediterranean

Euros (Greek—Southeast Wind)

Favonius (Latin—South Wind)—also known as Foehn or Föhn

Flakt a gentle breeze in Sweden

Foehn warm dry mountainous wind characteristic of the Alps where its downward rush melts snowdrifts rapidly

Fremantle Doctor cool southwest wind coming from the Indian Ocean to the Swan River Valley of Western Australia around Fremantle and Perth

Friagem (Portuguese—Cold Wave)—sometimes lasts for several days during Brazil's winter season

Furious Fifties storms ranging from west to east in the south fifty latitudes of the southern hemisphere

Gale wind of at least 35 miles per hour (56 kilometers per hour)—a high wind

Garmsal hot wind of Turkestan

Garua thick mist or drizzle in western Peru in winter

Ghibli Libyan name for the Sirocco

Greco the Greek wind—easterly wind encountered in the Mediterranean

Gregale northerly wind of south central Mediterranean area—the Greek gale—often a cold northeast wind blowing from eastern Mediterranean

Haboob Sudanese dust storm noted for its many colors and gritty intensity

Harmattan dry dusty desert wind blowing to Atlantic coast of Africa from the Sahara

Helm Wind cold northeasterly wind of northern England

Hokuto northeasterly wind in Japan

Hubbub Sudanese sandstorm

Huracán (Spanish—Hurricane)

Hurricane devastating rain-filled wind along Atlantic coast of the United States originating in the Caribbean and Gulf of Mexico; hurricane months recalled by these lines: June—too soon; July—stand by; August—look out you must; September—remember, October—all over

Ibe foehn-type wind blowing through Dzungarian Gate in western China near Lake Balkash

Irish Hurricane a flat calm when no wind blows; also called Paddy's hurricane

Jet Stream high-altitude wind

Kaikias (Greek—Northwest Wind)

Karaburan black-dust blizzard of the Gobi Desert

Khamsin Egyptian name for the Sirocco

Kochi easterly wind in Japan

Kohilo soft Hawaiian breeze

Kokaze small wind of Japan

Kona southeastern wind of Hawaii

Lake breeze wind blowing inland from a lake

Land breeze wind blowing seaward from the land

Landlash gale-force wind in Scotland

Leste Sirocco in Madeira and nearby North African coastal region

Levanter easterly wind characteristic of southern Spain and Straits of Gibraltar

Leveche hot dry wind found in southeastern Spain where it comes from North Africa

Libecio (Italian—Southwest Wind)—Genoese wind blowing inland from the Mediterranean

Lip (Greek—Southwest Wind)

Maestral cold north wind afflicting Genoa and Gulf of Genoa

Maestro northwesterly wind of central Mediterranean area about Italy and Yugoslavia

Maize rains heavy rain from February through May in East Africa

Mango showers rains from March through May in monsoon regions of Southeast Asia

Mausim (Arabic—Season)—the monsoon, a seasonal wind, is derived from *mausin*

Medina land breeze felt at port of Cadiz in southwestern Spain

Meltemi (Turkish—Etesian Wind)

Millet rains heavy rain, October–December in East Africa

Mistral (Latin—masterful; masterly)—the Master Wind—cold dry northerly wind blowing down Rhone Valley into Gulf of Lyons—cold north wind characteristic of Marseilles, southern France, and the Rhone Valley

Monsoon Asiatic wind blowing from northeast in winter and southwest in summer

Mud rain rain which has large dust particles in its drops

Nevados cold Andean winds found in Ecuador

Nor'easter storm blowing from the northeast characterized by high winds and rain of three days' duration

Norte cold north wind often experienced in Central America and Mexico

North Canadian icy-cold wind

Norther cold north wind characteristic of Texas; hot dry foehn-type wind of California

Nor'wester storm blowing from the northwest

Notus (Greek—South Wind)—the Sirocco

Oberwind katabatic wind of the Salzkammergut in Austria

Ora late morning to early afternoon wind blowing over Lake Garda in northern Italy—direction is south to north (*see* Sover or Vento)

Ox's Eye West African sailor's name for the hurricane of the Guinea Coast

Paddy's Hurricane (*see* Irish Hurricane)

Pampero cold south wind blowing offshore in South Atlantic and over adjacent coastal pampas or plains of Argentina and Uruguay—often carries much dust and rain

Papagayo cold north wind often causing crop damage in Costa Rica

Phyrhenerwind foehn-type wind occurring in the Austrian and Bavarian Alps

Plum rain early summer rain of Japan, when plums are ripening, when rice is transplanted; important to crops

Ponente west wind from the western Mediterranean; sea breeze refreshing west coast of Italy and sometimes penetrating as far inland as Rome

Prester waterspout or whirlwind encountered off the Greek Isles

Pruga intense and cold wind in Alaska

Purga the Siberian blizzard—extremely cold northerly wind filled with cutting needlelike ice crystals and fine dry snow

Quara Bulgarian west wind; also called Karajol

Reppu a warm, cyclonic wind in Japan

Roaring Forties roaring storms sweeping from west to east around the southern hemisphere in the south forty latitudes

Samiel hot, devilish, and dusty wind of northeast Africa

Samun (Egyptian—Sirocco)

Santa Ana foehn-type hot dry wind of California usually blowing in late spring, summer, and early fall; named for Mexican general who once charged from the north and seemed to take the path of this wind from north to south

Schneefresser (German—Snoweater)—foehn wind warming lower mountainsides and valleys of Switzerland where it melts the snow drifts almost as rapidly as it contacts them

Scotch mist mist and drizzle combination common in Scottish highlands

Sea breeze wind blowing inland from the sea

Seistan 120-day wind of Iran in eastern province of Seistan

Shamal northerly wind, like the Seistan, but found in Iraq over the Tigris-Euphrates plains country

Shrieking Sixties shrieking winds coming from the easterly and southerly sections of Antarctica and prevailing in the south sixty latitudes

Simoon name given the Sirocco when it is dirtier and hotter than usual as at this time natives believe it is a poisonous wind; a dry hot wind felt on deserts of Africa and Arabia during spring and summer

Sirocco south wind blowing from Sahara across North Africa, Mediterranean, and southern Europe; in North Africa it is dry, dusty, and hot but after crossing Mediterranean arrives in Europe moist and warm

Skiron (Greek—Northwest Wind)

Smokes heavy mists and low clouds on the Guinea Coast

Snoweater foehn-type Chinook wind blowing down eastern slopes of the Rockies in Canada and the Rocky Mountain states

Solano easterly rainy wind of southern Spain and Straits of Gibraltar—the Levanter

Sonora desert wind of Arizona

Sou'easter rain-filled southeast wind

South Wind along the Mediterranean this is the Sirocco

Sou'wester rain-filled southwest wind; oilskin hats, coats, and pants are also called sou'westers as they offer protection from rainy winds

Sover or Vento late afternoon winds blowing over Lake Garda in northern Italy—direction is north to south (*see* Ora)

Squall violent wind of short duration

Suchowej desert wind of the steppes of southern Russia

Sudestadas southeasterly pampero-type gales along coasts of Argentina, Uruguay, and Brazil

Sumatra squall characteristic of Malacca Strait where it occurs during the southwest monsoon season

Taino Haitian hurricane

Tatsumake Japanese tornado

Tebbad sand-laden hot wind of Turkestan

Tehuantepecer cold north wind often blowing with hurricane force around the Gulf of Tehuantepec and the peninsula of Yucatan

Terral land breeze felt in Valparaiso, Chile

Tokalau northeasterly wind in the Fiji Islands

Tornado violent storm best known for its twisting vertical wind responsible for causing great damage with little warning

Tower of Winds octagonal Greek structure near the Acropolis in Athens; each of its eight sides is decorated with a carved-marble allegorical figure representing the principal winds: North, Boreas; Northeast, Kaikias; East, Apheliotes; Southeast, Euros; South, Notus; Southwest, Lips; West, Zephyrus; Northwest, Skiron

Trades Trade Winds (Northeast Trades in northern hemisphere blow from northeastern subtropics to the equator; Southeast Trades in southern hemisphere blow from southeastern subtropics to the equator)

Trade Winds northeast in northern hemisphere and southeast in southern hemisphere; the Northeast Trades cool the West Indies and much of the Spanish Main

Tramontana Lake Maggiore's morning wind blowing from the south and followed by the afternoon wind blowing from the north and called the Inverna

Tronada (Spanish—thunderstorm)

Tropical cyclone a hurricane

Tsumujikaze whirlwind in Japan

Twister a vertical spiralling cyclonic wind often called a tornado

Typhoon the hurricane of the western Pacific

Uala-andhi dusty Bay of Bengal squall ushering in the southwest monsoon season (April through June)

Uracano (Spanish American—Hurricane)—originally *huracán*

Vendavales southwesterly winds blowing around eastern Spain and Straits of Gibraltar

Virazon sea breeze cooling Cadiz on southwestern Spanish coast; afternoon sea breeze—often reaching gale force at Valparaiso on central coast of Chile; sea breeze felt along coast of Chile and Peru

Waimea moisture-laden gusts in Hawaii

Westerlies westerly winds

Willie-Willie Indian Ocean hurricane

Williwaw violent squall afflicting mariners attempting passage through the Straits of Magellan

Willyway violent squall characteristic in Straits of Magellan

Willy-Willy Australian cyclone

Wind of One-Hundred-and-Twenty Days (*see Bad-i-sad-o-bist roz*)

Xaloch (Catalan—Sirocco)

Xaloque (Spanish—Sirocco)

Yalca Peruvian snowstorm occurring in northern Andean mountain passes

Yellow Wind cold dry wind of eastern Asia depositing loess dust over much of China

Youg hot summer wind of the Mediterranean

Zephyrus (Greek—West Wind)—the balmy Zephyr

Zobaa Egyptian dust whirl or whirlwind

Zonda westerly foehn wind characteristic of Argentina and southern Chile where it descends the eastern slopes of the Andes; enervating hot winds felt in Argentina and Uruguay where the zonda often precedes a cold pampero storm

Zip-Coded Automatic
Data-Processing Abbreviations

The abbreviations listed here may be used in addresses on mail. By using the city–state abbreviations, it is possible to enter city, state, and five-digit ZIP Code on the last line of address, within a maximum of 28 positions: 13 positions for city, 1 space between city and state abbreviation, 2 positions for state, 2 spaces between state and ZIP Code, and 10 positions for ZIP + 4 code.

Two-Letter State and Possession Abbreviations

Alabama	**AL**	Kansas	**KS**	Northern Mariana Islands	**MP**
Alaska	**AK**	Kentucky	**KY**	Ohio	**OH**
American Samoa	**AS**	Louisiana	**LA**	Oklahoma	**OK**
Arizona	**AZ**	Maine	**ME**	Oregon	**OR**
Arkansas	**AR**	Marshall Islands	**MH**	Palau	**PW**
California	**CA**	Maryland	**MD**	Pennsylvania	**PA**
Colorado	**CO**	Massachusetts	**MA**	Puerto Rico	**PR**
Connecticut	**CT**	Michigan	**MI**	Rhode Island	**RI**
Delaware	**DE**	Minnesota	**MN**	South Carolina	**SC**
District of Columbia	**DC**	Mississippi	**MS**	South Dakota	**SD**
Federated States		Missouri	**MO**	Tennessee	**TN**
of Micronesia	**FM**	Montana	**MT**	Texas	**TX**
Florida	**FL**	Nebraska	**NE**	Utah	**UT**
Georgia	**GA**	Nevada	**NV**	Vermont	**VT**
Guam	**GU**	New Hampshire	**NH**	Virginia	**VA**
Hawaii	**HI**	New Jersey	**NJ**	Virgin Islands	**VI**
Idaho	**ID**	New Mexico	**NM**	Washington	**WA**
Illinois	**IL**	New York	**NY**	West Virginia	**WV**
Indiana	**IN**	North Carolina	**NC**	Wisconsin	**WI**
Iowa	**IA**	North Dakota	**ND**	Wyoming	**WY**

Geographic Directional Abbreviations

North	**N**	West	**W**	Southwest	**SW**
East	**E**	Northeast	**NE**	Northwest	**NW**
South	**S**	Southeast	**SE**		

Abbreviations for Street Designators (Street Suffixes)

Alley	**ALY**	Bluff	**BLF**	Bypass	**BYP**
Annex	**ANX**	Bottom	**BTM**	Camp	**CP**
Arcade	**ARC**	Boulevard	**BLVD**	Canyon	**CYN**
Avenue	**AVE**	Branch	**BR**	Cape	**CPE**
Bayou	**BYU**	Bridge	**BRG**	Causeway	**CSWY**
Beach	**BCH**	Brook	**BRK**	Center	**CTR**
Bend	**BND**	Burg	**BG**	Circle	**CIR**

Cliffs	**CLFS**	Hills	**HLS**	Prairie	**PR**
Club	**CLB**	Hollow	**HOLW**	Radial	**RADL**
Corner	**COR**	Inlet	**INLT**	Ranch	**RNCH**
Corners	**CORS**	Island	**IS**	Rapids	**RPDS**
Course	**CRSE**	Islands	**ISS**	Rest	**RST**
Court	**CT**	Isle	**ISLE**	Ridge	**RDG**
Courts	**CTS**	Junction	**JCT**	River	**RIV**
Cove	**CV**	Key	**KY**	Road	**RD**
Creek	**CRK**	Knolls	**KNLS**	Row	**ROW**
Crescent	**CRES**	Lake	**LK**	Run	**RUN**
Crossing	**XING**	Lakes	**LKS**	Shoal	**SHL**
Dale	**DL**	Landing	**LNDG**	Shoals	**SHLS**
Dam	**DM**	Lane	**LN**	Shore	**SHR**
Divide	**DV**	Light	**LGT**	Shores	**SHRS**
Drive	**DR**	Loaf	**LF**	Spring	**SPG**
Estate	**EST**	Locks	**LCKS**	Springs	**SPGS**
Expressway	**EXPY**	Lodge	**LDG**	Spur	**SPUR**
Extension	**EXT**	Loop	**LOOP**	Square	**SQ**
Fall	**FALL**	Mall	**MALL**	Station	**STA**
Falls	**FLS**	Manor	**MNR**	Stravenue	**STRA**
Ferry	**FRY**	Meadows	**MDWS**	Stream	**STRM**
Field	**FLD**	Mill	**ML**	Street	**ST**
Fields	**FLDS**	Mills	**MLS**	Summit	**SMT**
Flats	**FLT**	Mission	**MSN**	Terrace	**TER**
Ford	**FRD**	Mount	**MT**	Trace	**TRCE**
Forest	**FRST**	Mountain	**MTN**	Track	**TRAK**
Forge	**FRG**	Neck	**NCK**	Trail	**TRL**
Fork	**FRK**	Orchard	**ORCH**	Trailer	**TRLR**
Forks	**FRKS**	Oval	**OVAL**	Tunnel	**TUNL**
Fort	**FT**	Park	**PARK**	Turnpike	**TPKE**
Freeway	**FWY**	Parkway	**PKY**	Union	**UN**
Gardens	**GDNS**	Pass	**PASS**	Valley	**VLY**
Gateway	**GRWY**	Path	**PATH**	Viaduct	**VIA**
Glen	**GLN**	Pike	**PIKE**	View	**VW**
Green	**GRN**	Pines	**PNES**	Village	**VLG**
Grove	**GRV**	Place	**PL**	Ville	**VL**
Harbor	**HBR**	Plain	**PLN**	Vista	**VIS**
Haven	**HVN**	Plains	**PLNS**	Walk	**WALK**
Heights	**HTS**	Plaza	**PLZ**	Way	**WAY**
Highway	**HWY**	Point	**PT**	Wells	**WLS**
Hill	**HL**	Port	**PRT**		

Zodiacal Signs

♈ Aries (The Ram), first sign of the zodiac, symbolized by the ram's horns; the sun enters this period on March 21, marking the spring or vernal equinox

♉ Taurus (The Bull), second sign of the zodiac, symbolized by the bull's head and horns; sun enters this period April 20

♊ Gemini (The Twins), third sign of the zodiac, symbolized by wooden statues of Castor and Pollux coupled by horizontal lintels; sun enters this period May 21

♋ Cancer (The Crab), fourth sign of the zodiac, symbolized by overlapping crab claws; sun enters this period June 22, marking the summer solstice, the longest day of the year

♌ Leo (The Lion), fifth sign of the zodiac, symbolized by stylized figure representing the lion's tufted tail; sun enters this period on July 23

♍ Virgo (The Virgin), sixth sign of the zodiac; symbol taken from *par* in *parthenos,* Greek for virgin; sun enters Virgo on August 23

♎ Libra (The Balance), seventh sign of the zodiac, symbolized by a stylized balance; sun enters this period on September 23, marking the autumnal equinox

♏ Scorpio (The Scorpion), eighth sign of the zodiac, symbolized by stylized representation of legs and stinger tail of the scorpion; sun enters this period on October 24

♐ Sagittarius (The Archer), ninth sign of the zodiac; symbolized by archer's bow and arrow; sun enters this period on November 22

♑ Capricornus (The Goat), tenth sign of the zodiac; symbol taken from *tr* of *tragos*, Greek for goat; sun enters Capricorn on December 22, marking the winter solstice, the shortest day in the year

♒ Aquarius (The Water Carrier), eleventh sign of the zodiac, symbolized by two parallel water waves; sun enters this period on January 20

♓ Pisces (The Fishes), twelfth sign of the zodiac; symbolized by two fishes tied by a thong; sun enters this period on February 19